SURFACE SCIENCE
The First Thirty Years

surface science
The First Thirty Years

Editor:

Charles B. Duke
Webster, NY

1994

North-Holland

Amsterdam-London-New York-Tokyo

ISBN 0-444-81827-8

Reprinted from:
SURFACE SCIENCE 299/300 (1994)

PRINTED IN THE NETHERLANDS

Surface Science: The First Thirty Years

This volume is a collection of short articles describing the historical development of selected topics in surface science during the formative years of this field, i.e., 1963–92. It was assembled under the guidance of the Advisory Editorial Board of *Surface Science* for four purposes: to provide a personalized historical record of the first thirty years of surface science as seen by active researchers who founded and developed the field; to serve as an indication of the realized and prospective value of the field for the betterment of the human condition; to commemorate the thirtieth anniversary of the founding of the journal *Surface Science*; and to honor Harry Gatos for his founding and nurturing of this journal during its first twenty eight years of existence.

The period 1963–92 was the scene of astonishing advances in the ability to characterize surfaces and study surface phenomena: truly the "birth" of surface science as a field of endeavor [1,2]. At the beginning of this period one could not determine any important characteristics of a surface, e.g., the chemical identity of surface species, their atomic geometries, the dynamics of their motions, or the nature and consequences of the surface electronic charge distributions. By 1992 all of these quantities were being determined routinely in surface science experiments. Moreover, dynamic measurements of atomic positions were being used routinely to study surface diffusion, film growth, and surface chemical reactions. In addition, the scientific results obtained from such studies were being utilized for the engineering of new materials and processes, as described, for example, in the articles herein in the section on Applications of Surface Science. Thus, the fruits of surface science research are impacting commercially important industrial problems, for example, the fabrication of microelectronics and the design of catalysts. Not only has a scientific revolution in the sense of Thomas Kuhn [3] occurred, but a technological revolution as well, which is inducing profound effects on such economically vital industries as computers and communications. This volume is the story of the key ingredients in these revolutions as told by the players who "made them happen".

The articles in this volume are written for a technically literate but non-specialist audience. Considerable effort has been invested in rendering them interesting and comprehendable to those familiar with science in general but not surface science in particular, e.g., historians and students. If you, gentle reader, have trouble with one, skip it and try another. Some of these articles are really gems. Others, well, each author was presented with the opportunity to capture your interest and attention. You are the judge of how well or poorly each succeeded. This is designed to be a fun book to browse just to glean an impression of what surface science is all about and to observe the foibles as well as accomplishments of its practitioners. Read accounts from their own hands. Read them at bedtime, on an airplane, at the beach, at your leisure,... enjoy.

The table of contents and authors were selected by concensus of the Advisory Editorial Board during the first few months of 1992; invitations were issued in April; and the volume was closed in May 1993. It is designed to encompass the full scope of the field of surface science: its foundations, major teachings, applications and current frontiers. Articles are grouped within these broad categories in such a fashion that the intellectual structure of the field should be self-evident from the table of contents.

Authors were invited to contribute to this volume based on the Advisory Editorial Board's assessment of their contributions to surface science over the years, their demonstrated ability to write interesting and lucid papers, and their personal involvements with the development of the topical areas to be covered in the volume. An amazing 80% of those invited actually contributed an article which appears in this volume. Therefore the volume represents a comprehensive collection of personalized technical histories by individuals whom the Advisory Editorial Board deemed to be leaders in developing the topical areas about which they write.

The styles of the individual articles vary widely, from personal reminiscences to compact technical reviews of the major events which define the scope and impact of the topic covered. The editor was tolerant of a wide variety of styles, provided the perspective of the writer was clearly specified in the introductory material of the article. Considerably less tolerance was exhibited for inadequate presentation and excessive length. All of the articles were reviewed by one or more members of the Advisory Editorial Board, most were revised in response to these reviews, and a few were recast completely in order to accommodate the style and/or length restrictions of this special volume. Authors were, however, encouraged to tell their stories from an entirely personal perspective without regard to whether or not the reviewers agreed that these stories reflected objective reality with regard to the topic covered. Thus, the reader will occasionally find side by side two accounts of a topic which are mutually exclusive in that neither article even mentions the work and events described in the other. Such is the nature of all human endeavor: each of us constructs his or her own personal interpretation of events within the context of which we make sense out of the happenings around us. The editor's goal in assembling the historical reviews in this volume was to render these personal value systems explicit, not to suppress them under the guise of objectivity.

Harry Gatos, the founder of the journal and one of the founders of the field, is the author of the first article. In it he describes his personal world view of the origins of the field and his group's contributions to it. I cannot envisage a more fitting testimonial to his contributions in a volume dedicated to commemorating the thirtieth anniversary of his founding of *Surface Science* and his twenty eight years as editor thereof. The Advisory Editorial Board and I are thrilled that Harry was willing to prepare this contribution to the anniversary volume. We acknowledge our debt to Harry, both personally and professionally, for his stewardship of the journal during its first twenty eight years. We wish him the very best on this happy occassion, and would like to express to him our collective "thank you" on behalf of the thousands of regular readers, contributors, and referees of *Surface Science*.

And so, on to *Surface Science: The First Thirty Years*: the story of a scientific and technological revolution which is changing the world as we know it, transforming a planet into a global village.

Charles B. Duke, Editor
Webster, New York, August 27, 1993

[1] C.B. Duke, J. Vac. Sci. Technol. A 2 (1984) 139.
[2] C.B. Duke, J. Vac. Sci. Technol. B 11 (1993) 1336.
[3] Thomas S. Kuhn, The Structure of Scientific Revolutions (University of Chicago Press, Chicago, 1962).

Contents

Surface Science 299/300 (1994) 1–23
North-Holland

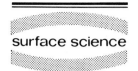

surface science

Semiconductor electronics and the birth of the modern science of surfaces

Harry C. Gatos

Massachusetts Institute of Technology, Cambridge, MA 02139, USA

Received 1 May 1993; accepted for publication 8 June 1993

Semiconductor surfaces were the launching platform for the solid state electronic revolution. The brilliant concept of surface states led directly to the discovery of the transistor in 1947. The chemical instability of the Ge surfaces, however, rendered the new devices irreproducible and unstable. This threat to the viability of semiconductor electronics, precipitated an international search, of a magnitude unparalleled in the history of science and technology, directed at the understanding and controlling of semiconductor surfaces. The roots of all of today's powerful experimental tools, techniques, procedures and fundamental concepts at the disposal of surface science and engineering can be traced to that gigantic effort on the study of semiconductor surfaces. Clearly, solid state electronics, born and nurtured on semiconductor surfaces, gave, in turn, birth to the modern science of all surfaces. In this article, I attempt to convey a general impression of the mutually constructive interplay between the evolution of solid state electronics on the one hand and surface science and engineering on the other. I take this opportunity to highlight my own experiences and research involvement with semiconductor surfaces.

1. General introductory remarks

The importance of surfaces in the applications of solids has been recognized ever since man put solids into use. However, as late as in the early years of the transistor, surface technologies and their understanding rested primarily on empiricism and phenomenological models. This situation prevailed even in the light of the fact that surface phenomena bear directly on all facets of technology and all aspects of our daily life. There were immense problems to be overcome before meaningful insight into the fundamentals of the surfaces of crystalline solids could be realized. No motives commensurate with the magnitude of such problems emerged that could amass the necessary driving force to tackle them head on. I will review how solid state electronics generated such motives.

Solid surfaces constitute an abrupt termination of the periodicity of the crystalline lattice; they are essentially giant lattice defects. Consequently,

they could not be submitted to the theoretical treatments which assume crystalline periodicity. Appropriate theoretical treatments, taking into consideration such an abrupt termination and the relevant energetics, presented complexities beyond the reaches of intellectual curiosity.

Fundamental experimental approaches simply did not exist in the 40s. Good vacuum systems reached down to 10^{-6} mm Hg and under special circumstances 10^{-8} mm Hg became possible. In the late 40s, I was very proud of my vacuum system which would consistently achieve pressures below 10^{-6} mm Hg. Thus, "clean" surfaces by today's standards could not be realized. All surface studies were just carried out on contaminated surfaces. Beyond the uncertainties in chemical composition, the actual structure of solid surfaces could not be probed with the available approaches and techniques.

These remarks should not, of course, be taken as a general condemnation of all surface studies carried out on "real" surfaces. Numerous such studies of the pre- as well as of the post-transistor eras have been very informative and have yielded very useful results. Actually, real surfaces can lead to meaningful studies and useful results, provided some key criteria are met: real surfaces need to be reproducibly prepared and sensitive to changes of the parameters pertinent to the phenomena under study.

2. The very beginning

In the middle 40s the Bell Telephone Laboratories engaged a group of scientists, under the leadership of Bill Shockley, to carry out research on semiconductors [1] with the ultimate goal the development of a solid state amplifier. Walter Brattain, a member of the group, set out to optimize the rectification characteristics of semiconductor–metal diodes (maximize the potential barrier in the semiconductor side of the contact) using Ge. According to the well accepted theory of Schottky, the magnitude of this potential barrier is equal to the difference in work function between the metal and the semiconductor. Walter Brattain, however, came to an impasse: no matter what metal he brought in contact with Ge, the potential barrier did not vary. John Bardeen, a theoretical physicist in the group, was confronted with this inconsistency. He came up with a brilliant idea, a stroke of genius, to account for the inconsistency: he reasoned that, on the surface of Ge, there are energy levels (surface states) within the forbidden energy gap. These states trap majority carriers and lead to the creation of a potential barrier (space charge region) which can extend some microns into the semiconductor (Fig. 1) [2]. This barrier is unrelated to the contacting metal and, thus, it is not affected by it. With that concept in mind, within a very short period of time, perhaps the greatest invention in the history of science was made: the invention of a solid state amplifier i.e., the transistor [3]. The original transistor was based on the appropriate bias of the newly conceived potential surface barrier by two, very closely spaced, metal point contacts (point contact or type-A transistor).

3. Surface states

With the discovery of the transistor, the surface states were catapulted to real prominence. Now, William Shockley's turn came to display his brilliant mind. He devised the field effect experiment, an approach with which not only the presence of surface states can be demonstrated but also their energy position in the energy gap and their density on the semiconductor surface can be quantitatively determined [4]. By monitoring the surface conductance of a thin Ge slab, as a function of applied field in a parallel plate capacitor configuration, it became possible to determine the number of carriers trapped in the space charge region and, thus, the density of surface states. It should be noted here that the field effect experiment, the p–n junction [5] and the concept of the field effect transistor [6] were some of the early major contributions made by William Shockley.

The origin of surface states was phenomenologically attributed to the unsaturated bonds (unpaired electrons) of the surface atoms resulting from the termination of the lattice at the surface. Actually, such an origin was corroborated by the experimentally determined surface state densities on Ge surfaces. These values (on real surfaces) were within 2 to 3 orders of magnitude from the densities of surface atoms [7]. Each surface atom

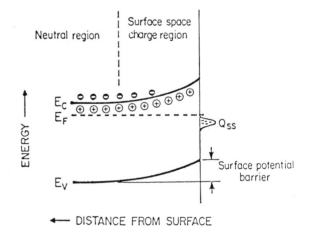

Fig. 1. Surface potential barrier resulting from the capture of electrons by the surface state (Q_{ss}). E_c and E_v is the electron energy at the bottom of the conduction band and the top of the valence band, respectively; E_F is the Fermi energy.

with an unpaired electron is a potential electron trap or a surface state. Thus, the concepts of dangling electrons and dangling bonds were formulated and, as I discuss later, they proved to be extremely useful in pursuing the study of semiconductor surfaces and semiconductor defects in general.

Intellectually at first, surface states raised numerous key questions and certainly challenges. Here are some: how are surface states affected by physical adsorption, chemisorption or oxide formation? Do adsorbed, chemisorbed species or oxide films introduce surface or interface states of their own? What is the configuration of the space charge resulting from the surface states? What are the effects of dopant impurities? What are the effects of surface stresses and/or surface deformation?

Extensive researches were undertaken to address these types of questions. By and large, however, they led to irreproducible and conflicting results. All kinds of surface states were found which exhibited various rates of charging and discharging and had various positions within the forbidden energy gap.

Three fundamental reasons were responsible for this confusing situation. First, the growth of Ge single crystals was in its embryonic stages. Thus, the defect densities were very high. For example, millions and often billions of dislocations intercepted every square centimeter of the surfaces (today's Si and Ge crystals are dislocation-free). Secondly, the crystals contained relatively large amounts of background impurities (several orders of magnitude larger than today's crystals do). Thirdly, the surfaces used in experimentation could not be characterized and assessed on an atomic scale, a process which turned out to be an essential prerequisite in some of these types of studies.

Perhaps the most critical problem was the fact that neither defect densities nor chemical composition (purity) could be reproduced from crystal to crystal within the same laboratory, not to mention the discrepancies found from one laboratory to another. To these problems we must add the uncertainties associated with the reproducibility of the surfaces under study. It is no wonder then

that, in some instances, in the early years, each major laboratory had its own family of surface states and expounded its own phenomenological theories. One can say that, in general, new results on semiconductors represented the characteristics of the individual crystal employed rather than fundamental truths. Unfortunately, this type of a situation is reencountered with the introduction of every new compound semiconductor. The reliable growth of single crystals of compound semiconductors presents immensely more complex chemical and structural problems than that of elemental semiconductors [8].

4. The transformation from the vacuum tube to solid state electronics

Before proceeding with developments in the science and engineering of surfaces, I consider it useful to comment on the nature of the vacuum tube to solid state electronics transformation, on some of its characteristics and the novel interplay among the classical and engineering disciplines precipitated by this transformation.

The transformation from the vacuum tube to solid state electronics cannot be referred to as an improvement or a change. Going from the vacuum tube to the silicon chip, which can contain many millions of vacuum-tube-equivalent devices, cannot be called an improvement. The first advanced vacuum tube computer the ENIAC, built in the 40s in Philadelphia, had 18 000 vacuum tubes, 70 000 resistors, 10 000 capacitors, 6 000 switches and a maze of connecting wires [9]. It was one hundred feet long, ten feet high and three feet deep. It weighed thirty tons. The temperature in the room soared to 120 F. Tube replacing teams searched the machine for burned out tubes. It is alleged that the lights of the city of Philadelphia dimmed when ENIAC was turned on. What ENIAC did then, a hand-held computer does much better today! A hand-held computer cannot be referred to as an improvement of ENIAC.

The transformation to solid state electronics constituted a birth of a new and different era for science and technology. The birth took place in

1947, the first infant steps were taken in the early 50s. From there on, there has been an explosive growth carrying with it motives with unprecedented powers for tackling head-on problems considered, only a few years earlier, beyond the scope and/or the capabilities of science and technology.

Let us look first at the heart of the transformation. Vacuum tube electronics is based on the generation and control of electrons in vacuum. In transistor electronics the current carriers (electrons or holes) originate in the atoms within the solid and their characteristics depend on the atomic scale structure and composition of the solid.

It is instructive to go back to about 1950. The understanding of the conduction of carriers in semiconductors and their manipulation to achieve device functions were at a respectably high level even by today's standards. In fact, the book by William Shockley, "Electrons and Holes in Semiconductors", a classic in that field, was published in 1950 [10].

What about the state of suitable materials to fabricate working semiconductor devices at that time? For all practical purposes such materials did not exist. Germanium single crystals with reproducible characteristics were necessary (Ge was then the key semiconductor). Technology for single crystal growth from the melt was hardly in existence. Starting materials were needed with background impurities less than a few parts per billion; that meant many orders of magnitude beyond the prevailing limits. No crystalline defects, planar, line or point, should be present. At that time, these requirements were just fantasies; and so was the realization of devices which were being conceived and even patented. Some of these theoretically conceived devices were fabricated many years later (as suitable materials and processes were developed) and were proven to be valid and valuable.

I quote from the book "The New Alchemist" by Dirk Hanson, a journalist, reporting on this period of the early 50s [9]. He states: "At first the financial arguments of sticking with the vacuum tube were persuasive and tube engineers could readily temper the enthusiasm of the solid state

people with the weight of experience. Maybe the transistor was not going to be such a big thing after all. The early fuss died down. For one thing, the manufacturing methods were completely ad hoc and seat-of-the-pants. Controlling electricity by rearranging the atoms was nice practice in theory but not quite so awe-inspiring when it came to the production line, where almost anything could go wrong and frequently did. It was like trying to do surgery on the head of a pin. It was wondrous that transistors worked at all, and quite often they did not. Those that did varied widely in performance, and it was some times easier to test them after production and, on that basis, find out what kind of electronic component they had turned out to be. If they failed it could have been due to any number of undesirable impurities that had sneaked into the doping process. It was as if the Ford Motor Company was running a production line so uncontrollable that it had to test the finished product to find out if it was a truck, a convertible or a sedan".

5. Surfaces on the front line

Actually, the situation was worse than that described in the above quotation. The electronic characteristics of the devices that tested well after fabrication, did not remain constant with time. A fraction of them, presumed to live forever, died within months. For a while, in the early 50s, it appeared that semiconductor electronics was heading for an early crash. The embryo of our transformation, rather than grow, was being consumed.

The cause of the instability of the p–n junction devices was found to be chemical changes of the Ge surfaces leading to changes in surface conductivity. These changes caused conduction between the p- and n-type surfaces which degraded the n–p junctions and eventually caused shorting and failure. Thus, in addition to the acute problems of the semiconductor bulk, even more acute problems emerged associated with surfaces [11].

Fortunately, it was realized by a handful of industrial organizations that solid state technology had no learned disciples and that no single existing technical discipline could accommodate it. They decided to go forward from ground zero, building on intimate interactions among the classical disciplines.

Bell Telephone Laboratories (BTL) was the world's most advanced center. The point contact, the p–n junction and the field effect transistors were discovered there. There, also, single crystals had begun to be grown from the melt [12] and purification breakthroughs (such as zone refining) were taking place [13]. In parallel, extensive studies were initiated on the preparation of clean and reproducible surfaces and on their electrical characteristics, chemical properties and structure. Following the lead of BTL, traditional disciplines were merging and joining forces in other semiconductor research centers in this country (in the 50s). Among such centers were the industrial research laboratories of General Electric, IBM, Raytheon, RCA, Texas Instruments and Westinghouse. In Europe the research laboratory of Philips followed.

Merging of the classical disciplines was not particularly smooth nor effective from the start. Communication was not at a very high level because of "technical language problems", basic training and discipline tradition. It was not uncommon to encounter physicists developing advanced measurement techniques, only to apply them on chemically very poorly prepared and assessed samples. Conversely, chemists did engage in carrying out primitive physical measurements on chemically, state of art samples.

The pressure was immense, however, to develop materials, processes and structures to bring to life phenomena and devices already theoretically conceived and even patented. Key electronic characteristics (of the bulk and of the surfaces) needed to be related quantitatively to structural and/or compositional parameters. Thus, it is easy to see that aspects of physics, chemistry, metallurgy, chemical engineering, electrical engineering and possibly other disciplines were needed to enter this undertaking and to function in an interactive mode. From such new polygamous relationships among all these disciplines, inevitably a new offspring began to take shape: materials science and engineering. One can certainly state

that surface science and engineering is a very close relative to that offspring, if not its twin.

6. The early days: an overview

In view of the critical role of the surfaces in device fabrication and device performance, the pressure was ever rising to understand and control their chemical as well as their electrical behavior. It was very clear indeed that for successful solutions the pay off was enormous. Accordingly, in the 50s and particularly in the first part of the decade, the major emphasis in nearly all semiconductor research centers was on semiconductor surfaces.

In parallel to the intensive fundamental studies, the engineering aspects of surfaces were being studied, including cutting, polishing, etching, oxidation and crystallographic orientation effects. These types of investigations were essential for the ongoing development and fabrication of devices.

Along fundamental lines and for the longer range, it was universally realized that an indispensable prerequisite for meaningfully pursuing the basic understanding of the structural, chemical and electrical properties of surfaces was the availability of atomically "clean" surfaces to begin with. The direction was clear: development of ultrahigh vacuum environments (with pressures 2 to 3 orders of magnitude lower than those then attainable) and in situ surface fabrication technology. These environments should permit removing any oxide or thin films present on the surfaces, cleaving single crystals to expose "virgin" surfaces, carrying out electrical and structural measurements and the introduction of controlled atmospheres. Only a few years earlier, such goals would have been dismissed outright (as apparently they had been) as figments of the imagination.

The above goals were essentially achieved in the middle 50s by Harry Farnsworth and his group at Brown University [14] and by the BTL group [15]. The "cleanliness" of the surfaces was assessed by low energy electron diffraction (LEED). No diffraction of the surface structure could be obained even if a fraction of a mono-layer of a contaminent was present. This technique was also used to monitor, for the first time, the truly initial stages of oxidation and other chemical reactions on Ge and Si clean surfaces by monitoring the status of the diffraction pattern. In addition, electrical measurements were carried out for the determination of the work function (as a function of orientation), the density of surface states and other physical parameters. It is very important to point out that the density of surface states on clean Ge surfaces was found to be of the same order of magnitude as the density of atoms. Thus, these results gave credence to the hypothesis that surface states were introduced by the unpaired (dangling) electrons on the surface atoms. Indeed, surface chemical reactions could be studied by monitoring the changes in the density of states. A number of very useful studies were carried out on that basis [16].

Once ultrahigh vacuum systems and relevant experimentation were demonstrated, many surface research groups were instrumented and functioning in a very short period of time (as has been invariably the case in the history of science and engineering). The development of very versatile metal vacuum systems followed which permitted the convenient introduction of a number of powerful new optoelectronic spectroscopies (now common place) to the study of surfaces. For example, Auger electron spectroscopy (AES) and X-ray photoelectron spectroscopy (XPS), introduced in the 60s, made possible the chemical characterization of surfaces (composition and bonding) on a submonolayer scale [15]. The fundamental science of surfaces, was advancing at a high rate.

Among all classes of materials, semiconductors were by far the best suited to bring about the birth of materials science and surface science. Their covalent, highly directional tetrahedral bonding is a major advantage both from the theoretical and the experimental point of view. It simplifies theoretical approaches as compared to metallic bond. It limits the solubility of other components. Thus, attaining high purity becomes much easier. (Today, single crystals of Ge and Si can readily be obtained with total impurities not

exceeding one part in 10^{12} [17].) Furthermore, prevention of defect formation is relatively simple. (Si single crystals, eight inches in diameter and a few feet long, are routinely grown dislocation-free.) As a result of their covalent bonding, semiconductors exhibit cleavage at a well defined crystallographic orientation. It was by cleavage of Ge in high vacuum that the first "virgin" surfaces were obtained and studied. In addition, tetrahedral bonding lends itself to atomistic approaches both of the bulk and of the surfaces [18]. Such approaches, in contrast to the more rigorous statistical ones, lend themselves to making predictions leading to the development of materials and structures with needed properties. For example, the discovery of the III–V semiconductor compounds was based on the simple "octet rule". The behavior of the surfaces of these compounds was, at first, accounted for on the basis of a simple atomistic model which I discuss later. Finally, it is due to their covalent bonding that their electrical properties can be readily modulated by many orders of magnitude. For example, one part per million of an impurity from either Group III or V of the periodic table increases the electrical conductivity of intrinsic Si by five orders of magnitude.

I now proceed with a discussion of my personal experiences and involvement with semiconductor surfaces. Here again, I will attempt to convey impressions rather than detailed and rigorous accounts.

7. "Real" semiconductor surfaces

7.1. Elemental semiconductors

When I joined MIT's Lincoln Laboratory in 1955 the research emphasis was still on Ge surfaces just as it was in all solid state research centers. It was generally expected or at least hoped that the problems stemming from the chemical instabilities of these surfaces would be overcome and that Ge-based solid state electronics would go forward.

Having done my doctorate research on the passivity of metal surfaces, I embarked immediately on the search for means to passivate and, thus, stabilize Ge surfaces. We found that strong oxidizing agents (such as concentrated nitric acid) which passivate metal surfaces (such as Fe and Cr) passivate also Ge surfaces and render their electrode potential more noble which is a confirmation of the state of passivity [19].

Before practical passivating processes could be developed, however, much needed to be learned about the chemical characteristics of real Ge surfaces. In view of the sp^3 tetrahedral bonding of Ge, we speculated, along with others, that the "dangling" bonds of the surface atoms must control chemical surface reactions. Since relatively pure single crystals of Ge could be readily grown by now, and, thus, samples could be prepared with surfaces of the desired orientation, we undertook to pursue some basic chemical reactions in water solutions. Our first basic finding was that Ge surfaces do not react with oxygen-free water, although the reaction is thermodynamically feasible. Furthermore, the germanium oxide, commonly encountered on real Ge surfaces, is soluble in water. Thus, in oxygen-free water we could obtain clean and reproducible Ge surfaces which we employed in our studies with liquid media. Some highlights of the results are indicated below [20].

We found that, for the three principal crystallographic faces, the order of the dissolution rates in oxygen-saturated water, in the temperature range of 30 to 40°C, is

$$\{100\} > \{110\} > \{111\}.$$

This order is the same as the order of the densities of dangling bonds on these faces (Table 1). Clearly, the unpaired surface electrons control

the reduction of oxygen to form water-soluble germanium oxide. In the light of these results it became possible to control the reactivity of Ge surfaces by the proper choice of species which upon contacting the surface atoms can share, accept or donate electrons. These findings helped put chemical etching, which was essentially an empirical process, on a rational if not on a scientific basis. Thus, chemical etchants with the desired action (polishing, defect definition and others) could be developed much more directly than before [21]. Actually, high resolution chemical etching, combined with phase contrast and interference microscopy, was for many years the only powerful tool for the identification and study of chemical and structural inhomogeneities in semiconductors on a microscale (with a resolution of about 2000 ångström units) [22].

By the late 50s it became clear that, inspite of the remarkable progress that had been made on the technology and science of Ge surfaces, their reliable and permanent stabilization, indispensable in solid state electronics, remained a moving target. Naturally, the emphasis shifted from Ge to Si. The very thin surface oxide on Si was found to be chemically refractory and, thus, assured surface chemical stability. Later on, this oxide film proved to be a godsent for achieving device isolation in integrated circuits on Si chips. I should point out, however, that it took many years of intensive efforts to overcome the electronic difficulties presented by the Si–SiO$_x$ interfaces. My group took a small part in those efforts. Methods were needed to be developed for assessing the electronic characteristics of the interfaces. Specifically, the density of interface states (within the forbidden energy gap), originally estimated to be of the order of $10^{13}/cm^2$, needed to be reduced by five orders of magnitude. The efforts were successful and extremely rewarding. Metal–oxide–semiconductor (MOS) structures became functional and today they are key structures in integrated circuitry. Here we have a truly exciting phase in the development of the science and engineering of solid interfaces [17,23].

Today, Si is used in about 95% of all electronic applications. They touch nearly all facets of our lives and culture. We can safely say that we are

Table 1
Density of free bonds on germanium surfaces and dissolution rates in oxygen-saturated water

Orientation	Free bonds (cm^{-2})	Relative free bond density	Relative dissolution rate
{100}	1.25×10^{15}	1.00	1.00
{110}	8.83×10^{14}	0.71	0.89
{111}	7.22×10^{14}	0.58	0.62

well into the Silicon Age. The reason for this outcome is not Si's semiconducting superiority but rather the superiority of the properties of its surface oxide.

7.2. Compound semiconductors

By the late 50s the solid state centers were switching from Ge to Si. At MIT's Lincoln Laboratory we decided to bypass Si and go directly to semiconductor compounds. It was for this reason that I started my long association with semiconductor compounds, although later on I too became involved with Si surfaces and Si–SiO$_x$ interfaces.

I should mention, parenthetically, that a few groups, particularly in the eastern European countries, kept on working for many years on Ge surfaces [24]. They concentrated on the formation of the tetragonal GeO$_x$ which, we, and others, found to be chemically refractory. However, the

reproducible formation of this oxide on Ge surfaces turned out to be unattainable.

The first tantalizing problem we encountered working with Group III Group V compounds (usually referred as III–V compounds, e.g. GaAs) was associated with their parallel (111) and ($\overline{1}\overline{1}\overline{1}$) surfaces. Thus, the (parallel) surfaces of an ⟨111⟩ oriented wafer exhibited totally different chemical characteristics, as seen in Fig. 2, where the two parallel (111) surfaces of a chemically etched GaAs wafer are shown. We observed such striking differences between the parallel {111} surfaces in all III–V as well as in II–VI semiconductor compounds.

7.2.1. The bonding model

We developed an atomistic bonding model [25] which not only accounted for the above differences in behavior but made it possible to predict a number of new chemical, structural and electronic phenomena (some with important techno-

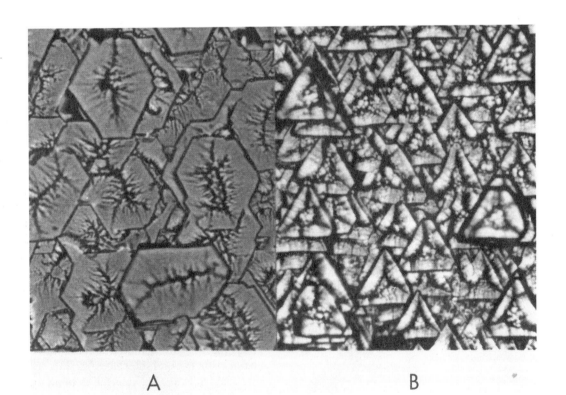

A   B

Fig. 2. Etched A and B parallel surfaces of a GaAs(111) wafer, 750 × .

logical implications), all of which were confirmed experimentally.

The outermost layer of the {111} surfaces of the III–V compounds (zinc-blende structure) must consist of either Group III (usually referred to as A atoms, e.g. Ga atoms) or Group V atoms (usualy referred to as B atoms e.g. As atoms) which are triply bonded to the lattice as seen in Fig. 3. It is readily apparent from the figure that this configuration must prevail because the {111} surfaces can be created only by cuts between planes AA and BB. The reasons are that in such cuts: (a) only one bond between atoms needs to be disrupted and (b) the surface atoms remain triply bonded to the lattice. Cuts between planes such as AA and B'B' are not possible because: (a) three bonds between atoms must be disrupted, and (b) the created surface atoms are singly bonded to the lattice. Such a configuration in a four-fold coordination must be presumed

unstable (see also the two-dimensional illustration in Fig. 4).

In the case of elemental semiconductors, when a bond is broken each atom carries one of the bond's two electrons. In the case of a III–V or A–B compounds, in addition to the above mode, it is possible that the two electrons of the bond remain with atom B as shown schematically in Fig. 5. Case II is thermodymically improbable since the ionization potential associated with the formation of B^+ (leading to charged surfaces) is much greater than the electron affinity associated with the formation of A^-. Case III is the realistic one since it takes into account the electron contributions of the A and B atoms to the tetrahedral bonding and leads to electrical neutrality of the created surfaces.

An examination of this model and its feasibility leads to some clear implications (predictions) regarding the behavior of the A and B (parallel)

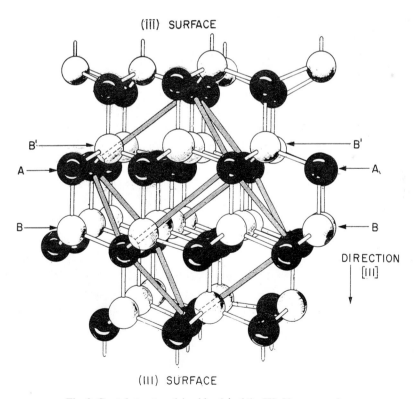

Fig. 3. Crystal structure (zinc-blende) of the III–V compounds.

B SURFACE

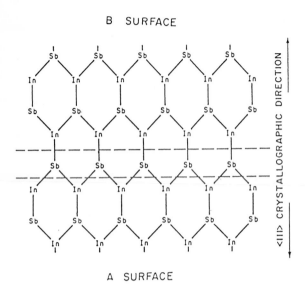

A SURFACE

Fig. 4. Two-dimensional representation of the zinc-blende structure along the polar direction.

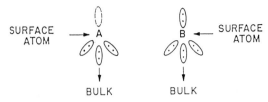

Fig. 6. Electron configuration of an A and a B atom of the parallel {111} surfaces.

reaction rates of the B surfaces in oxidizing media (etching rates) were about an order of magnitude greater than those of the A surfaces [25]. Actually, the A and B tetrahedra, although taken from the same single crystal, behaved like different materials. For example, in various etching media, the two tetrahedra of InSb exhibited an electrode potential difference of 75 mV or greater. The In tetrahedron was the more noble of the two, consistent with the above reasoning.

surfaces terminating in atoms with the electronic configuration as illustrated in Fig. 6.

7.2.1.1. Chemical implications. It is apparent from the above bonding configuration that the chemical reactivities of the A and B surfaces of a sample of a III–V (or II–VI) compound should be different. For example, the unshared pair of electrons of the B atoms should render the B surfaces much more reactive in oxidizing media than the A surfaces whose atoms have no unshared electrons. We were able to explore such differences by preparing, from a given single crystal, tetrahedral samples bounded entirely by either A or B surfaces as shown in Fig. 7. Thus, we were able to show that, at low temperatures, the

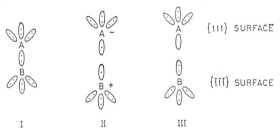

Fig. 5. The formation of an A and B surface atom by a cut of a bond between planes AA and BB of Fig. 3.

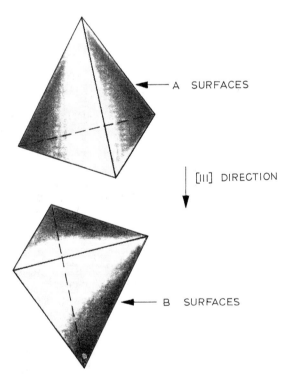

Fig. 7. Geometric relationship between tetrahedron A having exclusively group III atom surfaces, i.e., (111), ($\overline{1}1\overline{1}$), ($1\overline{1}\overline{1}$), ($\overline{1}\overline{1}1$), and tetrahedron B having exclusively Group V atom surfaces, i.e., ($\overline{1}\overline{1}\overline{1}$), ($\overline{1}11$), ($1\overline{1}1$), ($11\overline{1}$).

We were also able to account for the fact that dislocation etch pits appeared on the A but not on the B surfaces although edge dislocations intercept both types of surfaces [26]. This behavior becomes perfectly clear by extending the above bonding model to the bonding configuration of the edge dislocations terminating in A atoms (α dislocations) and those terminating in B atoms (β dislocations). In order for a pit to form, the etching rate along the axis of the dislocation must be greater than the dissolution rate of the surface. This condition can be met for the slow reacting A and not for the fast reacting B surfaces [25].

It was, thus, shown that the formation of dislocation etch pits is not due to the strain-induced increased reactivity at dislocation–surface intersection, as generally believed up to that time, but is the result of the prevailing bonding configuration. By slowing down the reactivity of the B atoms below that of the A atoms, utilizing electron sharing organic molecules, we were able to reveal dislocation etch pits on the B surfaces as well as to reveal dislocation etch pits of β edge dislocations [27].

Regarding the chemical reactivities of the principal crystallographic planes, the density of the dangling bonds is no longer, necessarily, the controlling factor, as is the case in elemental semiconductors. For example, for InSb the order of reactivities of the principles crystallographic planes was found to be:

$$B\{111\} \geq \{100\} > \{110\} > A\{111\}.$$

The $\{110\}$ surfaces have both A and B atoms triply bonded to the lattice and, therefore, are more reactive than the A surfaces but less reactive than the B. The $\{100\}$ surfaces can consist exclusively of either A atoms (with one unshared electron) or B atoms (with three unshared electrons) doubly bonded to the lattice. For this type of configuration to exist the surfaces must be atomically flat. In reality these surfaces consist of A and B atoms having an average of two unshared electrons. Accordingly, the $\{100\}$ surfaces are more reactive than the $\{110\}$ surfaces and approximately as reactive as the B$\{111\}$ surfaces.

The electronic configuration of the A and B atoms of the $\{110\}$ surfaces must be exactly the same as those of the $\{111\}$ A and B surfaces, i.e., the A atoms have a dangling empty orbital and the B atoms have a dangling pair of electrons.

Consistent with our model, we altered the relative rates of the A and B surfaces by adding in the etching media molecular species which were either electron donors or electron acceptors [27]. The latter should adsorb preferentially on the B surfaces and the former on the A. Indeed, in the presence of such electron acceptor molecules, the dissolution rate of the B surfaces was reduced by at least one order of magnitude. In this case dislocation etch pits did appear on the B surfaces consistent with our mechanism of dislocation etch pit formation.

I should point out here that the original absolute identification of the A and B surfaces was achieved by X-ray diffraction [28]. Today, for single crystals of III–V and II–VI compounds, such identification is done by chemical etching based on our results [21].

7.2.1.2. Structural implications. According to our model, the B surface atoms have an unshared pair of electrons (Fig. 6). This configuration is quite compatible with sp^3 bonding, although some distortion of the tetrahedral symmetry may occur. There is a similarity here with the well known NH$_3$ molecule.

The situation is quite different in the case of the A surface atoms. Here, there are only six electrons involved in their bonding instead of the required eight. Anomalies must then be expected. The Group III atoms (A) normally enter into sp^2 hybrid bonding which has a planar configuration. Thus, the surface A atoms, we reasoned, are caught in a predicament. They are terminal atoms of a tetrahedral configuration, but as surface atoms they have only three bonds which, normally, acquire a planar configuration. We concluded that there must be a competition (hybridization) between the two configurations leading to significant lattice strain. We proved experimentally, in several ways, the existence of such a strain which turned out to have very important consequences.

First we concluded that, if the A surfaces are

significantly strained, single crystal growth on these surfaces would be difficult, if not impossible. Indeed, we were unable to grow single crystals by immersing the A surface of an ⟨111⟩ oriented seed into the melt. The grown material was invariably twinned or polycrystalline. In contrast, by turning the seed around and immersing the B face of the seed into the melt, we invariably obtained single crystals [29]. In this way, we resolved a nagging crystal growth problem prevailing at that time: All III–V compound crystals were grown in the ⟨111⟩ direction using ⟨111⟩ oriented seeds; the single crystal yield was about 50%. Clearly, there was a 50% chance that either the A or the B face of the seed would be immersed into the melt. Ever since our discovery, all ⟨111⟩ single crystal of the non-centrosymmetric (e.g. III–V and II–VI) compounds are grown by immersing the B face of the seed into the melt.

The presence of the elastic strain associated with the A surfaces was also manifested in experiments in which both the A and B surfaces of four III–V compounds were damaged by abrasion under identical conditions. We found that the depth of damage perpendicular to the A surfaces (of the order of microns) was significantly less that that found perpendicular to the B surfaces as shown in Fig. 8 [30]. Obviously, the bonding strain renders the A surfaces more re-

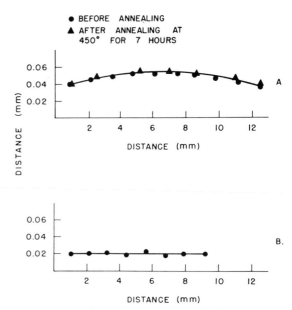

Fig. 9. (A) Curvature exhibited by an InSb(111) wafer 7.8 μm thick; annealing was carried out to rule out introduction of strain during preparation of sample. (B) Ge wafer of the same orientation and thickness as in A.

fractory. We did carry out hardness measurements on both types of surfaces. Although there are uncertainties associated with such measurements, there were clear indications that the A surfaces are actually harder than the B surfaces.

We further found that {111} wafers of InSb with a thickness of the order of 10 microns are spontaneously bent, the A surface being convex (Fig. 9) [31]. Such bending is caused by the elastic strain energy associated with the distortion of the bonding configuration of the A surfaces. The radius of curvature was found to be a function of the thickness of the wafer. On the basis of the classical theory of elasticity, we derived an equation relating the elastic strain energy with the radius of curvature, the dimensions of the wafer and the corresponding elastic constants. Some results are shown on Table 2. These energy values correspond to approximately 10^{-6} times the InSb binding energy. We pursued further the determination of the bonding elastic strain. We carried out X-ray diffraction measurements [31]; specifically we obtained the half breadth of rocking curves (using a double crystal spectrometer)

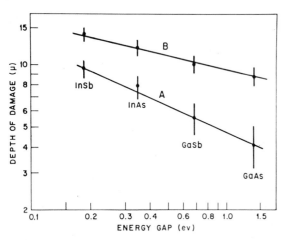

Fig. 8. Depth of damage in the A and B surfaces of III–V compounds as a function of energy gap. Cold work was introduced with 20 μm particle size abrasive.

Table 2

Radius of curvatures and the corresponding elastic strain energy (ESE) of InSb{111} wafers

Thickness of wafer (μm(± 0.3))	Measured radius (cm($\pm 10\%$))	Total ESE energy (erg$\times 10^4$)	ESE per surface gram-atom (erg/g-atom $\times 10^{-6}$)
7.8	104	3.0	0.8
7.9	81	5.2 [a]	1.4
11.1	205	2.3	0.6
13.1	194	4.1	1.1

[a] This calculated stored elastic strain corresponds to a strain produced by a load of approximately 0.2 g applied to the center of a wafer supported at both ends.

from both the A and B surfaces. By attributing all of the observed half breadth differences (5s) to elastic strain effects, we obtained strain energy values which compare very favorably with those obtained from the spontaneous bending results of Table 2.

All spontaneous bending experiments and measurements were carried out in a room atmosphere. It is not known what the chemical status of the surfaces was. It is possible that an absorbed layer of oxygen or a thin oxide layer was present. However, the presence of such layers is not likely to have affected to any great extent the energetics of the surfaces. It is of particular interest to point out that exposure of thin wafers of III–V compounds (e.g. InSb and GaAs) to NH_3 increased the radius of curvature of the wafers [32]. In the case of InSb no curvature could be detected after exposure to NH_3. In contrast, exposure to H_2S decreased the radius of curvature. The effect of these ambients is reversible, although only partially, and can be observed even after multiple exposures. On the basis of these results, it appears that NH_3 adsorbs preferentially on the A surfaces and the H_2S on the B surfaces. The striking increase (virtual disappearance) of the curvature in the presence of NH_3 indicates that these molecules provide two electrons to the dangling "empty orbitals" of the A surface atoms, leading to the relaxation of the bonding strain.

These results are perfectly consistent with our bonding model, which predicted, for the first time, the presence of surface strain due to dangling bonds, leading to surface modifications or reconstructions to accommodate such strain.

In the late 50s, when we carried out this work, neither theoretical nor experimental tools were yet developed to rigorously pursue our model on an atomic scale. Today, surface reconstruction constitutes, experimentally and theoretically, an important and dynamic field of surface science. Recent advances are, by and large, consistent with our model. In fact no finding, since we proposed our model, has contradicted it.

7.2.1.3. Electronic implications. According to our model, the atoms of clean B surfaces should be electron donor sites whereas those of the A surfaces should be acceptor sites. It was not possible, at that time, to verify directly these implications. However, we did make this verification indirectly. As pointed out above, edge dislocations in III–V compounds terminate (along the dislocation line) either in triply bonded A atoms (α dislocations) or triply bonded B atoms (β dislocations) atoms. These atoms must have the identical electron configuration as the A and B surface atoms and should, thus, be acceptors and donors, respectively.

We employed properly oriented samples of InSb in which we introduced excess of either α or β dislocations by plastic deformation [33]. We measured the carrier concentrations before and after deformation and we found that, indeed, α (or In) dislocations introduce acceptors whereas β (or Sb) dislocations introduced donors.

7.3. Surface states and electronic characteristics of compound semiconductor surfaces and interfaces

As the device applications of semiconductor compounds increased to a significant level, in particular those of GaAs, the importance of their surface electronic characteristics came into focus. GaAs became second to Si in importance regarding electronic and optical applications. In view of its large direct energy gap (1.4 eV; that of Si is indirect and equal to 1.1 eV), light emitting characteristics and high electron mobility (about 6 times larger at room temperature than that of Si), it was earmarked from the start (early 50s) as "the material of the future". Unfortunately, it has remained the material of the future for reasons I touch upon below.

We pursued the study of the electronic characteristics of GaAs surfaces and interfaces, along with the surfaces of other semiconductor compounds. I pointed out earlier that intrinsic surface states are determined by the electronic configuration of the surface atoms. In the case of elemental semiconductors, each surface atom with a dangling electron, generates a surface state within the forbidden energy gap. Indeed, the densities of surface states on clean Ge and Si surfaces were found to be comparable to the corresponding density of surface atoms. On real surfaces, where the dangling electrons partake in adsorption or film formation, the density of surface states decreases by 3 to 4 orders of magnitude. Extrinsic surface states can, of course, be introduced by foreign species on the surface. The field effect proved to be a most powerful tool in studying these surfaces [7].

In III–V (and II–VI) compounds the surface atoms have no unpaired dangling electrons (e.g. {111} or {110} surfaces). The B atoms have an unshared pair of electrons, whereas the A atoms have no unshared electrons. Indeed, in GaAs intrinsic surface states were not found within the energy gap [34]. States introduced by the As atoms are in the valence band (As sublattice) and those by Ga atoms are in the conduction band (Ga sublattice).

7.3.1. GaAs

Studies on GaAs real surfaces have been limited. Perhaps the major reason is the fact that standard field effect analysis cannot be carried out on GaAs. A minimum in surface conductance cannot be obtained, apparently because the Fermi level is pinned at the surface.

We employed pulsed field effect measurements, whereby the dynamic properties of the surface states are utilized rather than their equilibrium properties, which are utilized in the standard field effect. In an early study (1964), we

identified surface states 0.4, 0.7, and 0.9 eV below the conduction band [35]. In a later study (1971), again employing pulsed field effects, we identified and determined the characteristics of two main states 0.7 to 0.75 and 0.9 to 1.00 eV below the conduction band [36]. Their densities were found to be of the order of $10^{12}/cm^2$. It was surprising, but very interesting, to find that the presence of these states was essentially independent of the crystallographic orientation, the chemical treatment (by chemical etching) and the exposure to various gaseous ambients. Later on, we confirmed these findings employing surface photovoltage spectroscopy, a method we developed in our electronic materials group and which I discuss below.

From the above studies it is apparent, that the states on the real surfaces of GaAs are not associated with the abrupt termination of the lattice at the surface, nor are they associated with the interaction of the surface atoms with ambients. We concluded that these surface states are associated with intrinsic lattice defects in the surface space charge region. We did not speculate on their configuration because the defect structure of GaAs is far too complex. This complexity is associated with the two-component lattice, the volatile As constituent, and the high melting point (1240°C) of GaAs. Growing crystals at such a high temperature, can readily cause deviations from stoichiometry, over and above the unavoidable trapping of high densities of vacancies, which constitute the raw material for the formation of multitudes of defect structures [37].

The complex defect structure of GaAs, particularly as grown from the melt, has been a major obstacle in realizing the recognized potential of this compound in solid state electronics.

7.3.1.1. GaAs–insulator interfaces.

It was recognized some time ago, that the successful development of GaAs–insulator interfaces with high chemical and electronic stability and with low densities of interface states (of the order or $10^9/cm^2$) would lead to a new generation of, otherwise, unattainable electronic applications. Indeed we recall that central to today's solid state electronics is the Si–SiO$_2$ interface structure which exhibits such a low density of interface states.

The GaAs–native oxide system is very complex from the thermodynamic as well as from the kinetic point of view. Numerous As–O phases can exist as a function of temperature and pressure (as many as 40 such phases have been predicted). Nevertheless, many attempts have been made to achieve satisfactory oxide layers on GaAs employing thermal oxidation, anodic oxidation and plasma techniques. Actually, Ga–As oxides with satisfactory dielectric properties have been achieved, but their thermal instability and chemical reactivity render them impractical [34].

Numerous other insulating films have been studied, including Si_3N_4, SiO_2, $Si_xO_yN_z$ and Al_2O_3 with varying but limited degree of success. The studies of the interface states in such structures have relied primarily on capacitance based methods. However, in those investigations the electronic behavior of the MIS structures has exhibited a large majority carrier hysteresis and a significant frequency dispersion of the positive gate bias capacitance. Although these investigations have revealed important aspects of the GaAs-MIS structures, no consistent models have

emerged relating the anomalous behavior to the interface structures. It is difficult indeed to determine conclusively unknown parameters from anomalous characteristics particularly if the nature of the anomaly itself is not understood.

We developed a new approach to the study of interface states which eliminates the above uncertainties. It is based on the analysis of the spectra and transient responses of photostimulated currents in MIS structures [38]. Essentially, the interface states are populated with a current pulse (at low temperatures) and, subsequently, depopulated with light of the appropriate wave length. The measurements procedure is schematically illustrated in Fig. 10.

We found that the energy position and densities of the interface states were the same with those found in real GaAs surfaces; their characteristics are independent of the insulator in contact with the GaAs. We considered these results very important because we can conclude that the interface states are intrinsic to melt-grown GaAs and most likely are associated with lattice defects. This is the identical conclusion we arrived at in the case of the surface states of real surfaces.

Accordingly, we can presume that the prevail-

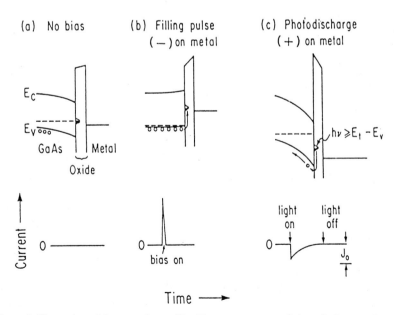

Fig. 10. Schematic illustration of the procedure utilized in measurements of photodischarge at interface states.

ing interface state density can be reduced to an acceptable level by improving the defect structure of GaAs single crystals. For example, we have grown bulk GaAs single crystals at low temperatures (400 to 500°C below the melting point) by liquid phase epitaxy. Such crystals exhibited a high level of crystalline perfection. In addition, processes need to be investigated which could lead to gettering the high densities of vacancies present in melt-grown crystals by appropriate impurities and/or by heat treatments. I believe that GaAs-MIS structures are most promising for an exciting new and better technology [34].

7.3.2. InSb

This is a very high electron mobility semiconductor with a small energy gap (0.17 eV). It is used in limited but specialized applications such as infrared detection. It has a relatively low melting point (525°C) and no volatile constituent. As a result, it can be grown in single crystal form conveniently and with a high degree of crystalline perfection. For these reasons, we have employed it as a model material in extensive studies of crystal growth, impurity segregation and defect structures [39].

We performed large signal alternating current field effect experiments on the A and B {111} real surfaces of InSb which were exposed to a variety of chemical treatments and ambients [40]. We observed discrete surface states in the energy gap near the conduction and near the valence band on both types of surfaces. The state near the conduction band edge exhibited the same density on both the A and B surfaces, while the state near the valence edge was more dense on the B surfaces. We formulated a model whereby the state near the conduction band edge is acceptor-like and characteristic of the In atom. The state near the valence band edge is donor-like and characteristic of the Sb atom. This model is consistent with our bonding model of the A and B surfaces discussed above.

Using high transverse electric fields, we determined the surface excess electron densities [41]. The observed changes of such electron densities we described with a tunnel equation, whereby tunneling currents are extracted through a sur-

face film (probably oxide) into the semiconductor bulk.

We further investigated the surface electron mobility. We utilized the field effect to produce a pronounced carrier accumulation layer at the surface, in the presence of a magnetic field, which was directed perpendicular to the surface. A decrease in surface conductance was observed in contrast to the expected field effect increase in conductance. Furthermore, the ratio of the surface to the bulk mobility decreases with increasing magnetic field beyond a given value of the magnetic field. These results were explained on the basis of cross quantization of the kinetic energy of the electrons by the narrow surface space charge potential barrier normal to the surface and the magnetic orbital quantization on a plane parallel to the surface [41].

7.4. Surface photovoltage spectroscopy and related effects: a new approach to the study of high energy gap semiconductor surfaces

The photovoltaic effect, i.e., the appearance of a voltage upon illumination of a semiconductor–metal rectifier, was observed more than one hundred years ago. The photo-excitation of band-to-band transitions associated with this effect has been studied and utilized extensively [42]. They lead to a decrease in the barrier height of any potential barrier in the semiconductor.

We discovered that surface photovoltage can be induced by illumination with energies smaller than the energy gap of the semiconductor, where no band-to-band transitions are involved [43]. We showed that this photovoltage is associated with electron transitions from the surface states to the conduction band and from the valence band to the surface states. The analysis of this photovoltage as a function of the energy of the incident light, together with the analysis of the features of the associated transitions, leads to the direct determination of the energy position and the dynamic parameters of the surface states. We called this process "surface photovoltage spectroscopy" (SPS). It has proven to be uniquely suited for the study of high energy gap semiconductor surfaces, where the standard methods (such as the field

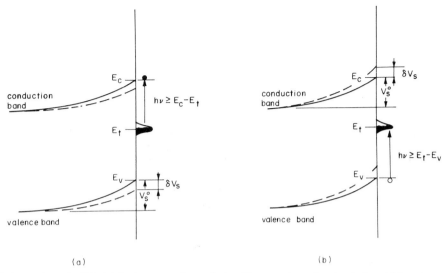

Fig 11. Origin of photovoltage in SPS. (a) Surface state depopulation (decrease in surface barrier) and (b) surface state population (increase in surface barrier, i.e., photovoltage inversion).

effect) are not applicable. We employed it successfully in the study of surfaces of semiconductor compounds such as GaAs, CdS and ZnO.

The basic processes involved in SPS are schematically illustrated in Fig. 11. Photostimulated electron transitions from or into the surface states cause a change in the electrical charge localized in the surface states. In order for charge neutrality to be maintained between the surface and the bulk, a corresponding change must take place in the net electrical charge in the space charge region. This change, in turn, leads to a change of the surface barrier height which can readily be determined by contact potential measurements. No electron transitions from the valence to the conduction band are involved in SPS.

Thus, in n-type semiconductors two types of photovoltaic effects are encountered under sub-gap illumination: (1) a decrease of the surface barrier due to photostimulated depopulation of the surface states, Fig. 11a; (2) an increase in the surface barrier due to photostimulated transitions

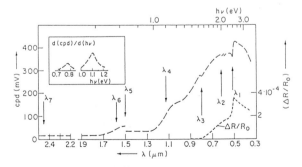

Fig. 12. Surface photovoltage spectrum and photoconductivity of the basal faces of CdS in room atmosphere. The insert shows the derivative of the surface photovoltage.

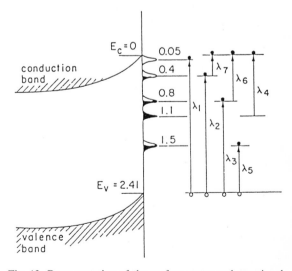

Fig. 13. Representation of the surface states and associated electron transitions in spectrum of Fig. 12 of a CdS surface.

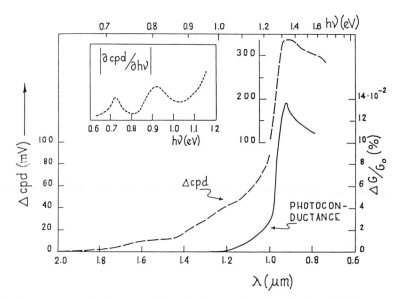

Fig. 14. Typical photovoltage spectrum (and photoconductance) of an (111) surface of n-type GaAs in room atmosphere. The insert shows the derivative of the surface photovoltage with respect to the photon energy.

from the valence band into the surface states, Fig. 11b. This latter effect we called "photovoltage inversion" since all known photovoltage effects, prior to this discovery, were associated with a decrease in the height of any potential barrier present in the semiconductor. When photovoltage inversion is present with other electron transitions we refer to it as "photovoltage quenching".

The quantitative treatment of SPS is significantly simplified by the fact that no excess free carriers are involved. The surface photovoltage spectra of CdS and GaAs are shown in Figs.

12–15. SPS is now widely used for the study of the surfaces and interfaces of high energy gap semiconductors.

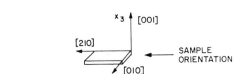

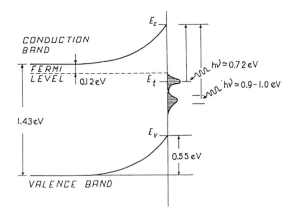

Fig. 15. Representation of the surface states and associated electron transitions in spectrum of Fig. 14 of a GaAs surface.

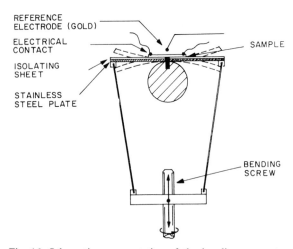

Fig. 16. Schematic representation of the bending apparatus and sample orientation.

7.4.1. Surface piezoelectric effect

We discovered a new effect whereby mechanical bending of thin semiconductor wafers, such as CdS with (001) orientation, causes pronounced changes (of the order of one volt) in the contact potential difference [44]. We called this effect "piezoelectric effect". It is due to the polarization induced in the depletion layer by the mechanical stress. Thus, a stress applied through mechanical bending of thin wafers corresponds to an applied electric field. This field causes an increase or decrease in the surface barrier height, depending on the direction of bending. A convenient experimental arrangement is shown in Fig. 16. Accordingly, it is possible, by combining controlled mechanical bending and SPS, to study the characteristics of the surface states under equilibrium conditions.

7.4.2. Photomechanical effect

We found that (001) wafers of CdS (or ZnO) about 0.015 mm thick, fastened to one end, exhibit pronounced bending upon illumination with white light [45]. This "photomechanical effect" results from the light-induced changes in the electric field at the two (001) parallel surfaces (surface barriers). Such changes induce, in turn, strains. Pronounced bending takes place because both surfaces have a depletion layer; consequently, the electric field and, thus, the strain components in the two surfaces are of opposite directions.

By modulating the surface barriers with chopped light, the wafers are excited to their fundamental vibration frequency. For subgap illumination, the amplitude of the vibration was found to be a function of the wavelength of the incident radiation i.e., a function of the changes in the surface barrier. As in SPS, changes in the surface barrier height are induced by electron transitions from the surface states to the conduction band or from the valence band to the surface states. Actually, the spectra of the amplitude of vibration versus the wavelength of the incident radiation are the same as the SPS spectra. Accordingly, the photomechanical effect provides a simpler means for determining the position of the surface states. The amplitude of the vibration is

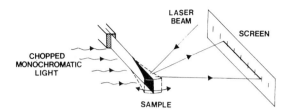

Fig. 17. Schematic representation of experimental arrangement for utilizing photomechanical vibration.

readily determined with the reflection of a laser beam from the vibrating wafer. A convenient experimental arrangement is shown in Fig. 17. The intensity of the beam is adjusted so that it does not cause any excitations in the semiconductor.

I should point out, that ambient molecules, which introduce surface states through adsorption, when present even in trace amounts, cause pronounced changes in the amplitude of the vibration. These changes occur at the energy of illumination which corresponds to the energy of the introduced surface states. Thus, the photomechanical effect can serve as the basis for extremely sensitive detectors of gaseous (or liquid) species. Employing various non-centrosymmetric semiconductors, it should be possible to detect, with high sensitivity, a number of molecular species, which, through adsorption, introduce distinct surface states. We were issued a patent on the applications of this invention [46].

7.4.3. Piezochemisorption effect

Mechanical strain (introduced by controlled bending of a thin wafer) has a striking effect on the chemisorption rate of gases on the polar (basal) surfaces of semiconductor compounds such as ZnO [47]. Depending on its sign, a strain as small as 10^{-3} can increase or decrease the chemisorption rate by more than an order of magnitude. We called this effect the "piezochemisorption effect". We found it to be consistent with a model based on strain-induced changes in the height of the surface potential barrier and the surface barrier-controlled transfer of electrons between the semiconductor and the adsorbed species. Thus, this effect can be employed for studying the adsorption processes associated with charge transfer between semiconductor and ambient, provided that a depletion layer exists (or is formed) at the surface. It is particularly suited for studying catalytic processes in which charge transfer is involved.

8. Surface recombination velocity of minority carriers and the surface denuded zone in silicon

The recombination properties of the minority carriers are basic electronic properties of semiconductors and control the performance of numerous types of devices. The minority carrier lifetime, recombination velocity and diffusion length are controlled by carrier scattering (and recombination) processes which in turn are strongly affected by lattice defects including chemical impurities. Clearly then, the recombination characteristics of the minority carriers at the surface are faster than in the bulk.

The minority carrier recombination properties at the surface (and in the bulk) are determined by the transient characteristics of excited excess minority carriers. The excitation source can be light, electron beam, electric current and others. In all instances, the linear (or spacial) resolution was limited.

In view of the ever decreasing device dimensions, I considered it important to obtain high resolution detailed profiles of the recombination properties at the surface (and in the bulk). We reached this goal by a quantitative analysis of the electron beam induced current (EBIC) in conjunction with high resolution electron beam scanning [48]. The analysis is based on the concept of the effective excitation strength of the carriers which takes into consideration all possible recombination sources. We achieved two-dimensional maps of the surface recombination velocity in Si (and three-dimensional mapping of the changes in the minority carrier lifetime in ion-implanted Si). We found pronounced variations in surface recombination. We could also trace, in some detail, mechanical damage profiles intentionally introduced with a sharp tool (Fig. 18).

In the case of GaAs we determined the surface recombination velocity from the dependence

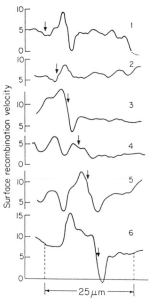

Fig. 18. Surface recombination velocity profiles on a Si surface. Arrows indicate the position of EBIC minima corresponding to positions of a mechanically introduced surface scratch.

of the minority carrier diffusion length employing scanning electron microscopy. Pronounced variations in the recombination velocity (exceeding a factor of 3) were found along real surfaces of GaAs (Fig. 19) [49].

The recombination characteristics of surfaces becomes critical in very large scale integration. Actually, for such integration suitable Si crystals cannot be obtained with the commonly used crys-

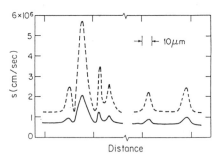

Fig. 19. Solid line represents variations of surface recombination velocity calculated from the variation of the minority carrier diffusion length as a function of the electron beam penetration depth; dashed line represents variations calculated by other means (see Ref. [49]).

tal growth techniques. The necessary surfaces or surface regions are obtained by gettering defects and impurities in the surface region (intrinsic gettering) employing high temperature heat treatments. The crystals must contain oxygen. During heat treatment, the oxygen forms precipitates (oxides) below the surface, which strain the lattice and enhance the diffusion of defects and impurities away from the surface and in the lattice strained region. In this way, the surface region is denuded from impurities and defects known to be minority carrier lifetime killers. This denuded surface region is referred to as the "denuded zone".

The need for oxygen in Si to obtain the denuded zone does exclude the use of oxygen-free (or low oxygen containing) Si which is of much higher overall quality than the oxygen containing Si. We developed a new process for surface region gettering which does not require oxygen [50]. This novel method of intrinsic gettering relies on high temperature heat treatment in an oxidizing ambient which causes the injection of Si self interstitials. These interstitials cause the formation of interstitial precipitates just below the surface which, in turn, lead to the formation of the denuded zone.

At this point I like to underline that little, if anything, is known about the chemical and/or structural changes associated with the formation of the denuded zone, either in the presence or in the absence of oxygen. We have here an area heavily populated by device oriented scientists. I believe that surface scientists can greatly benefit and at the same time make needed contributions by sharing this very fertile common ground.

9. Concluding remarks

Semiconductor electronics at a very young age gave birth to modern surface science. Soon after, the two became partners and they have remained indispensable to each other. Aging of that partnership is not in sight and should not be. Although I do not believe that this statement needs justification, here are sketches of some of the reasons for making it.

The Si chip, the driver of our civilization, is fabricated on a thin Si wafer just a few cm^2 in size. The surface of that wafer must be prepared to accommodate many millions of discrete devices. It must first be "denuded" i.e., freed from all carrier life-time killers (impurities and other defects). Then fabrication begins. The surface is submitted to several hundred steps involving surface or near-surface processes such as oxidation, diffusion, etching, metalization and others [51]. They all need to be controlled at a submicron scale. Here we have major challenges and opportunities for surface science. We also have at work surface engineering at its unprecedented best.

We must also consider the layered structures of semiconductor compounds needed for microwave, optical and laser devices. In some of these devices, the thickness and uniformity (structural and chemical) of the layers must be controlled on an atomic scale. The characterization and study of such layers, in terms of experimental approaches and disciplinary orientation fall in the domain of surface science and surface engineering. At the same time, the techniques and instrumentation developed for the preparation and monitoring of these layers (e.g., molecular beam epitaxial technology) [15] have opened new horizons for surface scientists. It has become possible to study these layers, on an atomic scale, as they form. It has also become possible to pursue the structural interactions of such layers with the substrate material and also their chemical interactions with ambients. I believe we have here a rich "goldmine" for surface scientists and engineers.

I feel gratified to have conceived, launched and, for about thirty years, sustained the growth of the journal *Surface Science*. It was in 1962 that North-Holland Publishing Company (now part of Elsevier Science Publishers) asked me to consider the publication of a new journal on semiconductor surfaces and to become its editor. My response was that time had come for a new journal devoted to the fundamentals of the surfaces and interfaces of all classes of materials and not just those of semiconductors. It was clear to me then that the experimental and theoretical developments achieved for the study of semiconductor surfaces were being rapidly transplanted to the study of the surfaces of other classes of materials. It was thus that *Surface Science* was conceived and subsequently delivered in January 1964. I am certain that this journal will continue to serve the surface science community well.

Acknowledgement

I express here my gratitude to the many members of my Electronic Materials Group family, who worked and shared with me, over the years, the excitement of meeting research challenges. I wish them the very best as they raise their own research families in so many parts of the world.

References

[1] The genesis of the transistor – memorandum for record, in: A History of Engineering and Science in the Bell System, Physical Sciences (1925–1980), Ed. S. Millman (AT&T Bell Laboratories, 1983).

[2] J. Bardeen, Surface States and Rectification at a metal semiconductor contact, Phys. Rev. 71 (1947) 717.

[3] J. Bardeen and W.H. Brattain, Transistor, a semiconductor triode, Phys. Rev. 74 (1948) 230.

[4] W. Shockley and G.L. Pearson, Modulation of conductance of thin films of semiconductors by surface charge, Phys. Rev. 74 (1948) 232.

[5] W. Shockley, The theory of p–n junctions in semiconductors and p–n junction transistors, Bell System Tech. J. 28 (1949) 435;
W. Shockley, M. Sparks and G.K. Teal, p–n junction transistors, Phys. Rev. 83 (1951) 151.

[6] W. Shockley, A unipolar field-effect transistor, Proc. IRE 40 (1952) 1365.

[7] For a review of early results, see: A. Many, Y. Goldstein and N.B. Grover, Semiconductor Surfaces (North-Holland, Amsterdam, 1965).

[8] See, for example: T. Bryskiewicz, C.F. Boucher, J. Lagowski and H.C. Gatos, Bulk GaAs crystal growth by liquid phase electroepitaxy, J. Cryst. Growth 82 (1987) 279.

[9] D. Hanson, The New Alchemists (Little Brown & Co., 1980).

[10] W. Shockley, Electrons and Holes in Semiconductors (Van Nostrand, New York, 1950).

[11] See, for example: W.H. Brattain and J. Bardeen, The surface properties of germanium, Bell System Tech. J. 32 (1953) 1.

[12] See, for example: Semiconductors, Ed. N.B. Hannay (Reinhold, New York, 1959).

[13] W.G. Pfann, Principles of zone melting, J. Met. 4; Trans. AIME 194 (1952) 747.

[14] H.E. Farnsworth, Clean surfaces, in: The Surfaces of Metals and Semiconductors, Ed. H.C. Gatos (Wiley, New York, 1959);
See also papers in: Conference on Clean Surfaces, Ann. N. Y. Acad. Sci. 101, Art. 3, p. 583–1014.

[15] Surface physics, in: A History of Engineering and Science in the Bell System, Ed. S. Millman (AT&T Bell Laboratories, 1983).

[16] See, for example: W.H. Brattain, Introduction to the physics and chemistry of surfaces, in: The Surface Chemistry of Metals and Semiconductors, Ed. H.C. Gatos (Wiley, New York, 1959).

[17] See: Papers in Semiconductor Silicon 1986, Eds. H.R. Huff, T. Abe and B. Kolbesen (The Electrochemical Society Inc., 1986).

[18] N.A. Goryunova, The Chemistry of Diamond-like Semiconductors (MIT Press, Cambridge, MA, 1965).

[19] M.C. Cretella and H.C. Gatos, The reaction of germanium with nitric acid. II. Passivity of germanium, J. Electrochem. Soc. 105 (1958) 492.

[20] W.W. Harvey and H.C. Gatos, The reaction of germanium with aqueous solutions. I. Dissolution kinetics in water containing dissolved oxygen, J. Electrochem. Soc. 105 (1958) 654.

[21] For an extensive review, see: H.C. Gatos and M.C. Lavine, Chemical Behaviour of Semiconductors: Etching Characteristics, in Progress in Semiconductors, Vol. 9, Eds. A. Gibson and R. Burgess (Temple Press Books, London, 1965).

[22] See, for example: A.F. Witt and H.C. Gatos, Impurity distribution in single crystals. II. Impurity striations in InSb as revealed by interference contrast microscopy, J. Electrochem. Soc. 113 (1966) 808.

[23] S.M. Sze, Physics of Semiconductor Devices (Wiley/Interscience, New York, 1981).

[24] Tibor Nemeth, Hungarian Academy of Sciences, private communication, 1969.

[25] H.C. Gatos and M.C. Lavine, Characteristics of the {111} surfaces of the III–IV intermetallic compounds, J. Electrochem. Soc. 107 (1960) 427.

[26] H.C. Gatos and M.C. Lavine, Dislocation etch pits on the {111} and {$\bar{1}\bar{1}\bar{1}$} surfaces of InSb, J. Appl. Phys. 31 (1960) 743.

[27] H.C. Gatos and M.C. Lavine, Etching and inhibition of the {111} surfaces of the III–V intermetallic compounds: InSb, J. Phys. Chem. Solids 14 (1960) 169.

[28] See Ref. [25], and references therein.

[29] H.C. Gatos, P.L. Moody and M.C. Lavine, Growth of InSb crystals in the ⟨111⟩ polar direction, J. Appl. Phys. 31 (1960) 212;
P.L. Moody, H.C. Gatos and M.C. Lavine, Growth of GaAs crystals in the ⟨111⟩ polar direction, J. Appl. Phys. 31 (1960) 1696.

[30] E.P. Warekois, M.C. Lavine and H.C. Gatos, Damaged layers and crystalline perfection in the {111} surfaces of III–V intermetallic compounds, J. Appl. Phys. 31 (1960) 1302.

[31] R.E. Hanneman, M.C. Finn and H.C. Gatos, Elastic strain energy associated with the A surfaces of the III–V compounds, J. Phys. Chem. Solids 23 (1962) 1553.

[32] M.C. Finn and H.C. Gatos, Spontaneous bending of thin (111) crystals of III–V compounds, Surf. Sci. 1 (1964) 361.

[33] H.C. Gatos, M.C. Finn and M.C. Lavine, Antimony edge dislocations in InSb, J. Appl. Phys. 32 (1961) 1174.

[34] H.C. Gatos, J. Lagowski and T.E. Kazior, GaAs-MIS structures – hopeless or promising?, Jpn. J. Appl. Phys 22 (1983) 11, and references therein.

[35] S. Kawaji and H.C. Gatos, Gallium arsenide surface states, Surf. Sci. 1 (1964) 407.

[36] T.M. Valahas, J.S. Sochanski and H.C. Gatos, Electrical characteristics of gallium arsenide 'real' surfaces, Surf. Sci. 26 (1971) 41.

[37] H.C. Gatos and J. Lagowski, Challenges in III–V semiconductor compounds, in III–V opto-electronics epitaxy and device related processes, Eds. V.G. Keramidas and S. Mahajan (The Electrochemical Society, Pennington, NJ, 1983), p. 1.

[38] T.E. Kazior, J. Lagowski and H.C. Gatos, The electrical behavior of GaAs insulator interfaces: a discrete energy interface state model, J. Appl. Phys. 54 (1983) 2533.

[39] A.R. Witt, H.C. Gatos, M. Lichtensteiger, M.C. Lavine and C.J. Herman, Crystal growth and steady-state segregation under zero gravity: InSb, J. Electrochem. Soc. 122 (1975) 276.

[40] H. Huff, S. Kawaji and H.C. Gatos, Field effect measurements on the A and B {111} surfaces of indium antimonide, Surf. Sci. 5 (1966) 399.

[41] S. Kawaji, H. Huff and H.C. Gatos, Field effect on magnetoresistance of n-type indium antimonide, Surf. Sci. 3 (1965) 234.

[42] See, for example: C.G.B. Garrett and W.H. Brattain, Physical theory of semiconductor surfaces, Phys. Rev. 99 (1955) 376.

[43] C.L. Balestra, J. Lagowski and H.C. Gatos, Determination of surface state energy positions by surface photovoltage spectroscopy: CdS, Surf. Sci. 26 (1971) 317.
J. Lagowski, C.L. Balestra and H.C. Gatos, Photovoltage inversion effect and its application to semiconductor surface studies: CdS, Surf. Sci. 27 (1971) 547;
J. Lagowski, C.L. Balestra and H.C. Gatos, Electronic characteristics of 'real' CdS surfaces, Surf. Sci. 29 (1972) 213.

[44] J. Lagowski, J. Baltov and H.C. Gatos, Surface photovoltage spectroscopy and surface piezoelectric effect in GaAs, Surf. Sci. 40 (1973) 216.

[45] J. Lagowski and H.C. Gatos, Photomechanical effect in non-centrosymmetric semiconductors-CdS, Appl. Phys. Lett. 20 (1972) 14;
J. Lagowski and H.C. Gatos, Photomechanical vibration of thin crystals of polar semiconductors, Surf. Sci. 45 (1974) 353.
J. Lagowski, H.C. Gatos and E.S. Sproles, Jr., Surface stress and the normal mode of vibration of thin crystals: GaAs, Appl. Phys. Lett. 26 (1975) 493.

[46] H.C. Gatos and J. Lagowski, Semiconductor Sensor, U.S. Patent No. 3 887 937 (1975).

[47] J. Lagowski, H.C. Gatos and E.S. Sproles, Jr., Piezochemisorption effect: a new method or modulating the rate of chemisorption on polar crystals, Appl. Phys. Lett. 27 (1975) 437.

[48] M. Watanabe, G. Actor and H.C. Gatos, Determination of minority-carrier lifetime and surface recombination velocity with a high spacial resolution, IEEE Trans. Electron Devices ED-24 (1977) 1172.

[49] L. Jastrzebski, H.C. Gatos and J. Lagowski, Observation of surface recombination variations in GaAs surfaces, J. Appl. Phys. 48 (1977) 1730.

[50] K. Nauka, J. Lagowski, H.C. Gatos and O. Ueda, New intrinsic gettering process in silicon based on interactions of silicon interstitials, J. Appl. Phys. 60 (1986) 615.

[51] W.R. Ranyan and K.E. Bean, Integrated circuit processing technology (Addison-Wesley, Reading, MA, 1990).

Surface Science 299/300 (1994) 24–33
North-Holland

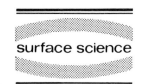
surface science

Interaction of electrons and positrons with solids: from bulk to surface in thirty years

C.B. Duke

Xerox Webster Research Center, 800 Phillips Road, 0114-38D, Webster, NY 14580, USA

Received 22 March 1993; accepted for publication 21 April 1993

In 1963 electrons and positrons were regarded as probes of the bulk properties of solids via, e.g., energy-band theory, fast electron energy loss spectroscopy, and positronium decay. During the late 1960s it was recognized that at energies in the range $10 \leq E \leq 1000$ eV, the inelastic collision mean free paths, λ, of electrons and positrons become comparable to atomic dimensions, i.e., $\lambda \approx 10$ Å, so that the elastic reflection or emission of these particles from solids probes the properties of the uppermost few atomic layers. This recognition led to the development of core level spectroscopies (e.g., Auger electron spectroscopy, X-ray photoelectron spectroscopy) for surface compositional analysis in the 1960s, to the application of low-energy electron diffraction for quantitative surface structure determination in the 1970s, and to that of low-energy positron diffraction for surface structure analysis in the 1980s. In addition, in the 1960s it was recognized that resonant field emission through adsorbed species offers a quantitative spectroscopy of the single-electron excitation spectrum of these species: a discovery which presaged the use of photoemission spectroscopy for this purpose in the 1970s and the development of scanning tunneling spectroscopy in the 1980s. This article is an exposition of how the evolution of the understanding of the nature of electron (positron) solid interactions together with advances in ultrahigh vacuum instrumentation and semiconductor microelectronics led from the dawn of the surface science revolution in the 1950s and 1960s to the quantitative electron (positron) surface spectroscopies which are routinely deployed in the 1990s.

1. Introduction

This paper is a personal account of selected aspects of the development of theories of electron and positron scattering from solids and of electron tunneling at interfaces during the period 1964–1993. In it I convey my impression of the major driving forces of the "scientific revolution", in the sense of Thomas Kuhn [1], which transpired in surface science during this period [2]. My theme is that this revolution resulted from the confluence of three factors: evolving knowledge about the details of electron [2,3] (and positron [4,5]) solid interactions, rapidly improving ultrahigh vacuum (UHV) technology [6], and the astounding increase of the performance/cost ratio of digital computing enabled by the rise of semiconductor microelectronics [7]. The focus of the paper is the first of these factors: the interaction of electrons and positrons with solids at their surfaces.

The status of surface science in 1964 may be illustrated by inspection of the book *Semiconductor Surfaces* [8] which appeared the following year. Although low-energy electron diffraction (LEED) spot patterns revealed the nearly universal reconstruction (i.e., lower than bulk symmetry) of Si, Ge, and compound semiconductor polar surfaces, neither the chemical compositions nor the atomic geometries of these surfaces were known. Surfaces were described by their accumulated charge, surface state densities and lifetimes, scattering and trapping cross sections, and surface recombination velocities: all indirect manifestations of surface physical properties as embodied in phenomenological models of space charge potential and carrier transport. UHV prepared "clean" surfaces (about which almost noth-

SSDI 0039-6028(93)E0386-9

ing quantitative was known) were distinguished from "real" surfaces characteristics of the devices in which the transport and space charge effects were measured. No quantitative assessments of surface composition, structure, or excitations (either electronic or vibrational) existed in 1964, so their effects on device performance could, at best, be surmised indirectly or speculatively.

In contrast, 1993 is characterized by surface atomic geometries on demand (for a price; of course), spatially resolved atomic composition in the optimal cases, and quantitative measurements (and predictions) of the dispersion relations of surface electronic and vibration excitation spectra. Descriptions of the achievement of this level of surface characterization are given in other contributions in this volume. They also have been summarized elsewhere, e.g., for semiconductors [9] or catalysts [10]. This remarkable transformation relative to 1964 is indeed a scientific revolution. [1,2] Our objective in this paper is to chronicle highlights of the role of our evolving knowledge of electron and positron solid interactions in achieving this revolution.

I proceed by considering the highlights in each of three consecutive decades. The 1960s were characterized by discoveries about the nature of electron–solid interactions which enabled the use of the scattering and emission of electrons in the energy range $10 \leq E \leq 1000$ eV for the description of surface (as opposed to bulk) phenomena. These discoveries are summarized briefly in Section 2. During the 1970s these discoveries were refined and consolidated to develop techniques for the quantitative determination of the composition and atomic geometry of surfaces as indicated in Section 3. An account of my personal involvement in these events may be found in the 1977 Welch Award address [11]. The 1980s were the scene of the discovery and development of new surface structure determination techniques based on electron tunneling (e.g., scanning tunneling microscopy), electron diffraction (e.g., various types of diffraction of electrons emitted from core levels by photo- or Auger-electron excitation) and positron diffraction (e.g., low-energy positron diffraction), as discussed in Section 4. The paper concludes with a brief synopsis of the

enormous changes which have occurred since 1964 in the use of electron and positron spectroscopies for the characterization of surfaces and of the major scientific and technological developments which made such changes possible.

2. Electrons in solids: determinants of surface sensitivity

The state of understanding the behavior of the electrons in solids circa 1964 is well-illustrated by the book *Quantum Theory of Solids* [12] by Kittel, which appeared the preceding year. Electrons were described as exhibiting undamped wave-like motion in a rigid, defect free crystal. Their dynamics were characterized by energy-momentum relations calculated via energy band theory from the periodic potential due to the atomic potentials. Finite electron scattering times were caused by either quantized thermal vibrations of the atoms ("phonons") or defects. For thermal electrons (excitation energies of the order of 25 meV or less) this is a valid picture. For the "low-energy electrons $10 \leq E \leq 1000$ eV characteristic of LEED or the keV electrons used in Auger electron spectroscopy (AES) [6], however, it is totally inappropriate because of the rapid creation of "plasmons", collective excitations of the valence electrons in the solid [12], by these energetic electrons. This result (known experimentally by the early 1960s [13]) had been cast into the language of conventional solid state theory by Quinn [14] in 1962, but had not been incorporated into Kittel's book. It also had been overlooked by the surface science community in this time frame, as discussed elsewhere [11]. Thus, the prevailing concept in this community circa 1964 was that fast electrons measured the energy-band structure of solids, e.g., via LEED [6,15].

In fact, the overlooked inelastic scattering of incident electrons by the valence electrons of the solid is the foundation of the surface sensitivity of most commonly used electron spectroscopies, including LEED, AES, and photoelectron spectroscopies. In a range of energies $10 \leq E \leq 1000$ eV incident electrons lose energy to plasmons with mean free paths $4 \leq \lambda_e \leq 10$ Å, as indicated in an

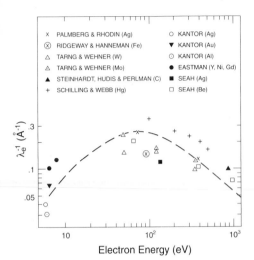

Fig. 1. Measured inelastic collision mean free paths for electrons in various materials. (After Duke, Ref. [11].)

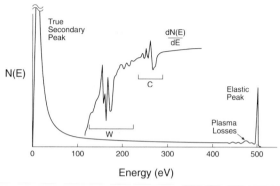

Fig. 2. Secondary electron spectrum of a tungsten surface bombarded with 500 eV electrons. The AES structure associated with the tungsten crystal and a carbon surface contaminant are shown. See Duke and Park [6] for a detailed discussion of the origin and features of this spectrum. (After Duke and Park, Ref. [6].)

early compilation [11] of measurements shown in Fig. 1, and now widely recognized [16,17]. Thus, if the elastically reflected electrons are detected with an energy resolution much less than a plasmon energy ($\hbar\omega \geq 5$ eV), only electrons which have been reflected within about λ_e of the surface are counted. The total spectrum of "secondary" electrons emitted from a solid when bombarded with energetic electrons is complicated [3,6,17], as indicated in Fig. 2, but the elastically emitted electrons could easily be discriminated from those experiencing plasmon energy losses by detectors available [3,6] in the late 1960s. Auger electron processes occur when a core hole is filled by one electron with another being emitted to conserve energy. Thus, both the elastic and Auger electron structures in this spectrum reflect the properties of surface species. The validity of this concept and its application to photoemission processes was slowly accepted by the surface science community during the late 1960s as documented, e.g., by Duke [3,18] and Powell [17].

Although most incident electrons experience inelastic losses which are associated with small-angle forward scattering, for a crystalline sample a few are reflected backwards by their diffraction from the two-dimensional planes of atoms in the vicinity of the surface. These emerge in a series

of diffracted beams characterized by momentum transfer parallel to the surface associated with the reciprocal lattice vectors of these two-dimensional periodic arrays, as indicated in Fig. 3. As depicted in Fig. 3, momentum parallel to the surface is conserved, although that normal to the

LEED Intensities From
(Clean) Surfaces

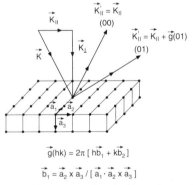

Fig. 3. Schematic illustration of an incident electron beam of wave-vector $\boldsymbol{k} = \boldsymbol{k}_\perp + \boldsymbol{k}_\parallel$, scattered elastically from a single crystal into a state characterized by the wave-vector $\boldsymbol{k}' = \boldsymbol{k}'_\parallel + \boldsymbol{k}'_\perp$. The construction of the reciprocal lattice associated with the single-crystal surface is also shown. The vectors $\boldsymbol{g}(hk)$ designate the reciprocal lattice vectors associated with the lowest-symmetry Bravais net parallel to the surface. (After Duke, Ref. [3].)

Schematic Low – Energy Electron
Diffraction Apparatus

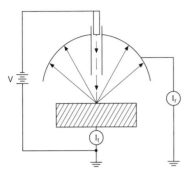

Fig. 4. Schematic diagram of a low-energy electron diffraction apparatus illustrating that most of the current drawn from the gun, $I = I_t + I_r$, passes through the target to ground (I_t) rather than being reflected back from the target (I_r). [After Duke, Ref. [3].)

surface is not because of the strong inelastic scattering. Most electrons experience multiple inelastic scatterings and eventually either pass through the crystal or emerge in the true secondary peak shown in Fig. 2. A schematic diagram of an apparatus designed to detect the backwards reflected and emitted electrons is shown in Fig. 4. The elastically diffracted beams are indicated by the lines emanating from the crystalline sample. Most of the incident electrons are inelastically transmitted through the crystal, generating "secondary" electrons in the process by exciting the valence electrons of the solid. Some of these are reemitted from the surface of the crystal as indicated by the "true secondary peak" in Fig. 2. The plasma losses occur as sidebands on the elastically diffracted beams. Descriptions of the details of the scattering processes which contribute to the secondary electron spectrum may be found in the literature [3,15] of the 1970s.

To provide the basis for quantitative surface compositional and structure analysis, it is necessary to examine the cross sections for core level excitation by the incident electron (e.g., for surface composition determination via AES and X-ray photoelectron spectroscopy (XPS)) and for elastic scattering (e.g., for surface structure deter-

mination via LEED). A detailed analysis of these cross sections was given [3,18] in the early 1970s with the result that surface sensitivity due to inelastic electron collisions in the energy range $10 \leq E \leq 1000$ eV is inevitably accompanied by multiple elastic scattering. Thus, the single-scattering linear response theory [12] used to determine bulk structure and excitations via X-ray and neutron scattering is not suitable for the determination of surface structures via electron diffraction. Rather, the development of a complete theory incorporating both multiple elastic scattering and the excitation of phonons and plasmons between the elastic scattering events was required as a precursor to quantitative surface structure determination. George Laramore and I constructed such a theory in 1970 [19,20] in which all of the elastic and inelastic scattering events were incorporated in a systematic way. We showed that because of this fact we could take limiting cases of the theory to construct useful computer programs for both elastic [21] and inelastic [22] (e.g., plasmon generating) diffraction. These programs were used to determine both the surface atomic geometries [21] and surface plasmon energy-momentum relations [23,24] of the low index faces of Al, thereby revealing several new and unexpected phenomena (e.g., the contracted top layer spacing of Al(110) [21] and negative linear term in the Al(100) surface plasmon energy-momentum relation [24]) years ahead of their time.

In the case of elastic diffraction, many of the complexities of the complete theory can be finessed by the use of a complex "optical" potential to calculate the energy-momentum relations of the incident electrons within the crystal and of effective (temperature dependent) scattering amplitudes to describe elastic scattering from thermally vibrating atoms [3,25,26]. This limiting case has gained wide acceptance and has been incorporated in many computer programs used today for surface structure determination via "dynamical" (i.e., multiple scattering) LEED intensity calculations [21,27–30]. Although multiple elastic scattering was ignored in surface compositional analyses via AES and XPS until relatively recently, it also affects these analyses as indicated

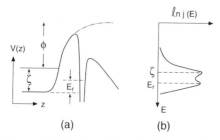

Fig. 5. Schematic indication of the potential energy (panel (a)) and isothermal field emission energy distribution (panel (b)) of an adsorbed species on a metal surface. The adsorbate exhibits a resonant energy level at E_r in the presence of the metal and the external field. Tunneling from the metal occurs preferentially at this energy because the barrier penetration probability is greatly enhanced for $E \simeq E_r$. (After Duke, Ref. [11].)

in this volume by Powell [17]. In addition, studies of various distinct types of diffraction-induced fine structure in core level excitation and emission spectra have led to additional techniques for surface structure determination [31–34], as discussed herein, e.g., by Citrin [35] and by Woodruff [36].

An alternative source of surface sensitivity is provided by resonance phenomena, e.g., the resonant tunneling of electrons through surface species [11,37]. As illustrated in Fig. 5, the discrete energy levels of atoms become resonant states upon adsorption on metal surfaces resulting in observable resonances in measured field emission energy distributions [37,38]. These states subsequently were shown to create peaks in suitably measured valence-electron photoelectron energy distributions which during the 1970s became the most utilized technique for determining their energies [39,40]. With the advent of scanning tunneling microscopy (STM) in the 1980s, the resonant tunneling phenomenon was exploited to determine the energies of both filled and empty surface states [41]. Since the tunneling current–voltage spectra (and their derivatives) can be measured with atomic precision by using the STM, the features of these spectra associated with individual surface atoms can be determined and related to the chemical behavior of each type of atom [42]. Thus, by using the combination of electron tunneling and a sharp tip, atomic spatial

resolution and resonance energy resolution are achieved simultaneously, leading to the capability of performing atomic electronic spectroscopy on individual surface species: a remarkable accomplishment for a macroscopic experimental measurement.

3. The 1970s: advent of quantitative surface structure determinations

While no surface structures were known in 1970, by 1980 quantitative surface structure determinations, mostly performed using dynamical LEED intensity analyses, had been reported for approximately 100 surfaces. The state of the art at this time is reflected in the book *Adsorbed Monolayers on Solid Surfaces* [43] by Somorjai and van Hove, which was published in 1979. These required not only dynamical LEED computer programs, as noted above [21,27–29], but also complementary assets in the form of reliable, reproducible LEED intensity data and large computers on which to calculate the intensities associated with many hypothetical surface structures for comparison with the data. Accounts of how this procedure is performed today are given in the present volume by Heinz [44] and by Tong [45].

Following our early studies of Al [21] and of clean and adsorbate-covered Ni [3,26], my personal involvement in this activity was focused on the determination of the structures of semiconductor surfaces, especially those of the (110) cleavage faces of zincblende structure compound semiconductors as shown [46] in Fig. 6. The structure of GaAs(110), initially reported in 1976 [47], was the first semiconductor surface structure ever determined and has withstood the tests of time and numerous challenges [48]. The power of quantitative surface structure determination is well illustrated by the fact that we went on to determine the atomic geometries of the (110) surfaces of all binary zincblende structure compound semiconductors, and from these extracted important general features of these geometries [49]. We found that they were universal in the sense that all materials exhibit the same geometry to within experimental error if distances are measured in units of the bulk zincblende lattice constant. We also found [46,49] that delocalized surface states, rather than local atomic coordination chemistry, were the mediators of the surface chemical bonding leading to the relaxed surface atomic geometries via an approximately bond-length-conserving tilting by $29° \pm 30°$ of the top layer chains of surface atoms, as shown in Fig. 6. Thus, quantitative surface structure determination, enabled by the theory of electron–solid scattering, its simplified elastic diffraction computational versions, and the acquisition of large reproducible data sets, has revealed the existence of new and unexpected types of chemical bonding at semiconductor surfaces.

During the 1980s the field of quantitative surface structure determination continued to grow until today when many hundreds of surface structures are known. Recent compilations of these are given by MacLaren et al. [50] and by van Hove et al. [51]. Even with the explosion in new techniques, however, most of these structures have been determined by dynamical LEED intensity analysis. The effect of the dramatic increase in the cost-effective performance of digital computers on surface structure determination by dynamical LEED intensity analysis has been profound. Calculations for a single GaAs(110) surface structure which took about an hour and cost

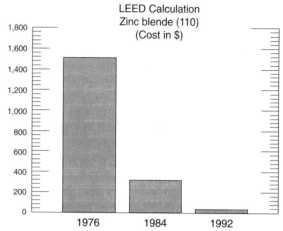

Fig. 7. The cost in current dollars of a 100-energy run for an arbitrary number of output beams of a converged calculation for a relaxed top-layer structure of GaAs(110). The 1976 data were obtained on a CDC 6600. The 1984 data were obtained on the CYBER 205 at Princeton University. the 1992 data were obtained on the Cray Y-MP at the University of Illinois. The computer program [52] is an expanded and adapted version of the one written by Laramore and Duke [21] in 1972.

$1,500 on a CDC 6600 in 1976 now takes several times longer but can be run on our Sun 470 workstation. Alternatively, this calculation takes less than two minutes on one of the CRAY-YMP supercomputers on the NSF network. The remarkable improvement in cost effectiveness of the computers used to run our dynamical LEED programs is indicated in Fig. 7. This improvement has enabled new approaches to surface structure determination via automated computer structural searches based on optimizations of goodness of fit criteria. We typically use the X-ray R factor criterion and a simplex optimization procedure [52]. Thus, although the structure of the computer programs is far more complex than the linear-response X-ray diffraction analysis of bulk structures, the actual operation of these programs is comparable to that of their X-ray analogs in the 1970s in both cost and ease of use.

4. The 1980s: new physics, techniques, and technology

In the 1980s two major new surface-science exploitations of the features of electron and

Zincblende (110)

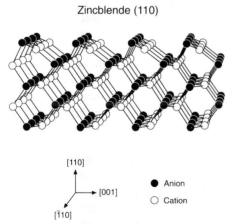

Fig. 6. Drawing of the relaxed nonpolar (110) cleavage faces of zincblende structure binary compound semiconductors. (Adapted from Duke and Wang, Ref. [46].)

positron interactions with solids were introduced: the scanning tunneling microscope (STM) [41] and low-energy positron diffraction (LEPD) [5]. A significant trend during this decade was the utilization of increasingly complex instrumentation for the fabrication of increasingly elaborate samples (e.g., superlattices) in a vacuum environment by deposition technologies like molecular beam epitaxy (MBE) and for the characterization of their growth and properties in situ, for example, by core level spectroscopies and reflection high energy electron diffraction (RHEED) [9]. Such instruments were enabled by the improved performance of both vacuum technology (i.e., large metal vacuum cans with increasingly complicated sample manipulation capabilities) and electronics which enabled digital data collection and processing as well as the operation of most, if not always all, of the instrumentation simultaneously during the preparation and characterization of a sample. Thus, a whole new generation of instrumentation became available which embodied multiple surface characterization techniques and in situ sample fabrication.

As noted in Section 2, the STM exploits the tunneling of electrons through a vacuum barrier between a tip and a sample. Since the tunneling probability depends sensitively on the spatial extent of the wave function of sample electrons (of the appropriate energy) in the vacuum gap [53,54], STM measures the electronic structure of the sample. By controlling the tip height to achieve constant current, a direct image of this structure with atomic resolution is generated. By varying the tip voltage, different full or empty states in the sample may be accessed, leading to scanning tunneling spectroscopy (STS). STM and STS have exerted a profound impact on surface science by generating images of both atomic and electronic structures and thereby enabling the in situ study of elemental growth processes like step motion. Surface states associated with local morphological features like steps or island edges can be identified. A review of the remarkable accomplishments of such instruments has been collected recently by Güntherodt and Wiesendanger [41,55]. Their impact has been assessed by Murday and Colton [56], and their current status is

reflected in articles in this volume by Webb [57] and by Williams [58].

The interactions of positrons with solids differ from those of electrons in three important respects [4,5]. First, their interactions with the valence electrons in the solid are repulsive rather than attractive and do not exhibit a contribution due to the Pauli exclusion principle (i.e., "exchange"). This fact leads to a qualitative change in the nature of the positron–atom cross sections used in LEPD calculations [5,59] and to the ability to construct more accurate model calculations of these cross sections. It also leads to positron work functions which are small and sometimes negative [4,5]. Second, the absence of Pauli principle restrictions on the positron final state in positron–electron inelastic collisions reduces the inelastic-collision mean free path of positrons relative to electrons at low ($E \leq 50$ eV) energies [5,60] thereby enhancing the surface sensitivity of their scattering in this energy range Third, the additional decay channels associated with positronium formation further increase the probability of inelastic processes near the surface, thereby augmenting the consequences of the lack of final state restrictions due to the Pauli principle. Although these three phenomena cause important quantitative differences between positron and electron solid scattering [4,5], qualitatively the same general features described in Section 2 and Figs. 2–4 pertain to positrons as well.

The implementation of low-energy positron diffraction lagged that of electron diffraction by over five decades (i.e., 1927–1979 [5]) primarily because of the lack of suitably strong positron sources. This difficulty was overcome in the mid-1980s via utilization of the concept of brightness enhancement via positron focusing, capture and reemittance [5,61] and the use of digital electronics (i.e., channel multiplier arrays plus digital data collection) to detect weak diffracted beams [5,62,63]. The resulting diffracted beams have been analyzed to recover the classic surface structure of GaAs(110) [64] as well as to determine the cleavage surface geometries of CdSe [5,62,63]. Use of the simplified complex potential model of elastic diffraction [3,21,25] yields significantly better descriptions of LEPD intensities than LEED intensities primarily because of the improved description of positron–atom scattering cross sections [59]. Since the cross sections for all atomic species are more nearly equal in the case of positrons than in the case of electrons, the use of LEPD opens new opportunities for the structure determination of surfaces with highly unlike atomic species (e.g., InP as opposed to GaAs) because it reduces the contrast between the different types of scatterers. Indeed, our recent work on this topic [65] has revealed the need for greatly improved error analysis in LEED and LEPD structure determinations in order to quantify the significance of changes in the best fit structures due to differences in the electron and positron solid interactions.

5. Synopsis

The articles collected in this volume provide more than adequate testimony to the sophistication and impact of surface science in 1993. Relative to 1964 the field has indeed experienced a "scientific revolution" in the sense of Kuhn [1,2]. Moreover, this revolution has encompassed ideas (e.g., about electron–solid interactions, atomic growth on surfaces, etc.), theoretical models (e.g., dynamical versus linear response models, ab initio three-dimensional versus empirical one-dimensional models of surface states and structures, etc.), experimental techniques (e.g, the advent of STM, LEPD, synchrotron radiation spectroscopy, quantitative atom and ion scattering spectroscopies, etc.), vacuum instrumentation (e.g., in situ growth by MBE, routine use of multiple techniques in large vacuum cans, etc.), and electronics technology (e.g., digital data acquisition and analysis, supercomputers, low-cost workstations, etc.). Surface science in 1993 is about as similar to its practice in 1964 as the modern intercontinental commercial jet aviation infrastructure is to the biplanes and pasture airports of the 1910–20 era.

New insights into electron and positron solid scattering spectroscopies have played an important role in this revolution. Most electron spectroscopies (including photoelectron spectroscopy) rely on the strong inelastic scattering of "low"

energy (or glancing incidence hence low energy normal to the surface) electrons to achieve surface sensitivity. Incorporating inelastic processes into the theory of electron–solid scattering led to today's structural spectroscopies based on elastic LEED and excitation spectroscopies (both electronic and vibrational) based on inelastic LEED. Exploitation of the unique features of tunneling into and out of surface species led to field and photoemission adsorbate spectroscopies in the 1970s and to STM and STS in the 1980s. The recognition of the unique features of positron-solid scattering is leading to new insights into particle–solid interactions and perhaps to a new level of accuracy in quantitative surface structure analysis. Therefore, the physics of electron and positron solid interactions has documentably been a leading contributor to the 1964–93 revolution in surface science.

Other complementary assets also were required, however, to effect this revolution. One of these is vacuum technology which enabled the achievement of a stable environment for the experiments. Another is the amazing improvement in semiconductor electronics which yielded ever cheaper, more powerful digital computers, real-time instrument and fabrication process control, cheap digital data acquisition and analysis, and adequate reliability of the surface characterization and sample fabrication instrumentation. A third is continual invention and innovation, leading to STM and STS, to high brightness positron sources and to MBE machines with atomic precision, among many others noted elsewhere in this volume. If this revolution is the grandchild of physics, it is the child of vacuum technology, digital electronics, and the innovative urges of its practitioners. Stimulated by a perceived need (reliable semiconductor electronics), enabled by scientific insight, and driven by inexorably advancing vacuum and electronics technology, the surface science revolution during 1964–93 has changed the meaning of "surface science", the practice of research, development, and manufacturing in electronics and certain types of chemical processing, and ultimately the nature of communications, and hence economics, in the world today.

Acknowledgements

I am indebted to my managers at Xerox Corporation, most notably M.M. Shahin, M.D. Tabak, M.B. Myers, and S.B. Bolte, for their continuing support over two decades of my participation in the research on which this article is based. Such ongoing support through numerous crises and budget cuts reflects a confidence in the ultimate value of new knowledge to Xerox customers which is in the finest traditions of industrial research worldwide and which has been validated a posteriori in the customary indirect, but traceable, ways.

References

[1] T.S. Kuhn, The Structure of Scientific Revolutions (University of Chicago, Chicago, 1962).

[2] C.B. Duke, J. Vac. Sci. Technol. A 2 (1984) 139.

[3] C.B. Duke, Adv. Chem. Phys. 27 (1974) 1.

[4] P.J. Schultz and K.G. Lynn, Rev. Mod. Phys. 60 (1988) 701.

[5] K.F. Canter, C.B. Duke and A.P. Mills, in: Chemistry and Physics of Solid Surfaces VIII, Eds. R. Vanselow and R. Howe (Springer, Berlin, 1990), p. 183.

[6] C.B. Duke and R.L. Park, Phys. Today 25 (8) (1972) 23.

[7] C.B. Duke, J. Vac. Sci. Technol. 17 (1980) 1.

[8] A. Many, Y. Goldstein and N.B. Grover, Semiconductor Surfaces (North-Holland, Amsterdam, 1965).

[9] C.B. Duke, J. Vac. Sci. Technol. B 11 (1993) 1336.

[10] G.A. Somojai, Surf. Interface Anal. 19 (1992) 493.

[11] C.B. Duke, J. Vac. Sci. Technol. 15 (1978) 157.

[12] C. Kittel, Quantum Theory of Solids (Wiley, New York, 1963).

[13] H. Raether, Ergeb. Exakt Naturwiss. 38 (1965) 84.

[14] J.J. Quinn, Phys. Rev. 126 (1962) 1453.

[15] C.B. Duke, in: LEED–Surface Structures of Solids, Vol. 2, Ed. M. Laznicka (Union of Czechoslovak Mathematicians and Physicists, Prague, 1972) p. 125.

[16] M.P. Seah and W.A. Dench, Surf. Interface Anal. 1 (1979) 2.

[17] C.J. Powell, Surf. Sci. 299/300 (1994) 34.

[18] C.B. Duke, in: Dynamic Aspects of Surface Physics, Ed. F.O. Goodman (Editrice Compositori, Bologna, 1974) p. 52.

[19] C.B. Duke and G.E. Laramore, Phys. Rev. B 2 (1970) 4765.

[20] C.B. Duke and G.E. Laramore, Phys. Rev. B 3 (1971) 3183.

[21] G.E. Laramore and C.B. Duke, Phys. Rev. B 5 (1972) 267.

[22] G.E. Laramore and C.B. Duke, Phys. Rev. B 3 (1971) 3198.

[23] C.B. Duke and U. Landman, Phys. Rev. B 8 (1973) 505.

[24] C.B. Duke, L. Pietronero, J.O. Porteus and J.F. Wendelken, Phys. Rev. B 12 (1975) 4059.

[25] C.B. Duke and C.W. Tucker, Jr., Surf. Sci. 15 (1969) 231.

[26] C.B. Duke, in: Dynamic Aspects of Surface Physics, Ed. F.O. Goodman (Editrice Compositori, Bologna, 1974) p. 99.

[27] D.W. Jepsen, P.M. Marcus and F. Jona, Phys. Rev. B 5 (1972) 3933.

[28] J.B. Pendry, Low Energy Electron Diffraction (Academic Press, New York, 1974).

[29] M.A. Van Hove and S.Y. Tong, Surface Crystallography by LEED (Springer, Berlin, 1979).

[30] M.A. Van Hove, W.H. Weinberg and C.-M. Chang, Low Energy Electron Diffraction (Springer, Berlin, 1986).

[31] P.A. Lee, P.H. Citrin, P. Eisenberger and B.M. Kincaid, Rev. Mod. Phys. 53 (1981) 769.

[32] M. De Crescenzi, Crit. Rev. Solid State Mater. Sci. 15 (1989) 279.

[33] C.S. Fadley, in: Synchrotron Radiation Research, Vol. 1, Ed. R.Z. Bachrach (Plenum, New York, 1992) p. 421.

[34] J. Stöhr, NEXAFS Spectroscopy (Springer, Berlin, 1992).

[35] P. Citrin, Surf. Sci. 299/300 (1994) 199.

[36] D.P. Woodruff, Surf. Sci. 299/300 (1994) 183.

[37] C.B. Duke and M.E. Alferieff, J. Chem. Phys. 46 (1967) 923.

[38] J.W. Gadzuk and E.W. Plummer, Rev. Mod. Phys. 45 (1973) 487.

[39] R. Gomer, Adv. Chem. Phys. 27 (1974) 211.

[40] D.E. Eastman and M.I. Nathan, Phys. Today 28 (4) (1975) 44.

[41] H.J. Güntherodt and R. Wiesendanger, Eds., Scanning Tunneling Microscopy I (Springer, Berlin, 1992).

[42] P. Avouris and I.-W. Lyo, in: Chemistry and Physics of Solid Surfaces VIII, Eds. R. Vanselow and R. Howe (Springer, Berlin 1990) p. 371.

[43] G.A. Somorjai and M.A. van Hove, Adsorbed Monolayers on Solid Surfaces (Springer, Berlin, 1979).

[44] K. Heinz, Surf. Sci. 299/300 (1994) 433.

[45] S.Y. Tong, Surf. Sci. 299/300 (1994) 358.

[46] C. B Duke and Y.R. Wang, J. Vac. Sci. Technol. B 6 (1988) 1440.

[47] A.R. Lubinsky, C.B. Duke, B.W. Lee and P. Mark, Phys. Rev. Lett. 36 (1976) 1058.

[48] C.B. Duke and A. Paton, Surf. Sci. 164 (1985) L797.

[49] C.B. Duke, J. Vac. Sci. Technol. A 10 (1992) 2032.

[50] J.M. MacLaren, P.B. Pendry, P.J. Rous, D.K. Saldrin, G.A. Somorjai, M.A. van Hove and D.D. Vvedensky, Surface Crystallographic Information Service: A Handbook of Surface Structures (Reidel, Dordrecht, 1987).

[51] M.A. van Hove, S.-W. Wang, D.F. Ogletree and G.A. Somorjai, Adv. Quantum Chem. 20 (1989).

[52] W.K. Ford, T. Guo, K.-J. Wan and C.B. Duke, Phys. Rev. B 45 (1992) 11 896.

[53] C.B. Duke, Tunneling in Solids (Academic Press, New York, 1969).

[54] J. Tersoff, in: Physics and Chemistry of Solid Surfaces VIII, Eds. R. Vanselow and R. Howe (Springer, Berlin, 1990), 335.

[55] R. Wiesendanger and H.-J. Güntherodt, Eds., Scanning Tunneling Microscopy II (Springer, Berlin, 1992).

[56] J.S. Murday and R.J. Colton, in: Chemistry and Physics of Solid Surface VIII, Eds. R. Vanselow and R. Howe (Springer, Berlin, 1990), 347.

[57] M.B. Webb, Surf. Sci. 299/300 (1994) 454.

[58] E. Williams, Surf. Sci. 299/300 (1994) 502.

[59] C.B. Duke and D.L. Lessor, Surf. Sci. 225 (1990) 81.

[60] J. Oliva, Phys. Rev. B 21 (1980) 4909.

[61] A.P. Mills, Jr., Appl. Phys. 23 (1980) 189.

[62] C.B. Duke, D.L. Lessor, T.N. Horsky, G.R. Brandes, K.F. Canter, P.H. Lippel, A.P. Mills, Jr., A. Paton and Y.R. Wang, J. Vac. Sci. Technol. A 7 (1989) 2030.

[63] T.N. Horsky, G.R. Brandes, K.F. Canter, C.B. Duke, A. Paton, D.L. Lessor, A. Kahn, S.F. Horng, K. Stevens, K. Stiles and A.P. Mills, Jr., Phys. Rev. B 46 (1992) 7011.

[64] D.L. Lessor, C.B. Duke, X.M. Chen, G.R. Brandes, K.F. Canter and W K. Ford, Vac. Sci. Technol. A 10 (1992) 2585.

[65] X.M. Chen, K.F. Canter, C.B. Duke, A. Paton, D.L. Lessor, and W.K. Ford, Phys. Rev. B 48 (1993) 2400.

Surface Science 299/300 (1994) 34–48
North-Holland

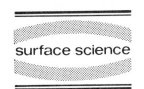

surface science

Inelastic interactions of electrons with surfaces: application to Auger-electron spectroscopy and X-ray photoelectron spectroscopy

C.J. Powell

Surface and Microanalysis Science Division, National Institute of Standards and Technology, Gaithersburg, MD 20899, USA

Received 4 March 1993; accepted for publication 18 May 1993

Electron-based probes of surface properties are used frequently in surface science since strong inelastic scattering for electron energies between about 50 and 2000 eV ensures high surface sensitivity. An overview is given of developments in the understanding of inelastic electron scattering in solids with emphasis on important surface properties. Auger-electron spectroscopy (AES) and X-ray photoelectron spectroscopy (XPS) are the two techniques most commonly used for measurements of surface composition. We give a brief description of the development of AES and XPS and then proceed to describe the role of inelastic electron scattering in AES and XPS. Attention is given to the measurement of electron attenuation lengths, to the calculation of electron inelastic mean free paths, and to intensity measurements for quantitative AES and XPS.

1. Introduction

Electron-based probes are commonly used in surface science experiments [1,2]. Such probes are used primarily for measurements (or assessments) of surface composition (by techniques such as Auger-electron spectroscopy and X-ray photoelectron spectroscopy), but are also used for measurements of surface atomic structure, surface electronic structure, surface vibrational properties, surface magnetic properties, surface topography and thin-film morphology. In addition, electron beams can stimulate surface reactions; these reactions may be desired (e.g., the processing of semiconductor devices) or undesired (e.g., the analysis of adsorbed molecules).

The phrase "electron-based probe" is intended to refer to two typical types of situations. First, an electron beam of some specified energy may be incident on a surface and measurements are made of absorbed current, of scattered electrons, or of secondary particles (often electrons, ions, neutrals or photons) emitted by the surface with certain energies or momenta. Second, the surface may be bombarded by ions, neutrals or photons, and measurements are made of emitted electrons with specific energies or momenta. Descriptions of specific probes are given in a number of articles in this volume as well as in recent books [1,2]. An overview of the features of electron–solid interactions which render the probes surface sensitive has been given by Duke [3].

The surface sensitivity of electron-based probes of surfaces most commonly arises from the very strong inelastic scattering of low-energy electrons in solids. For electrons with energy in the range 5 to 2000 eV, the inelastic mean free path (IMFP) is typically between about 2 and 100 Å (as will be discussed further below). Measurement conditions can also be varied (e.g., by adjustment of beam angles with respect to the surface or sometimes by controlled variation of an electron energy) to attain surface sensitivities of one to several atomic-layer spacings.

We present an overview in Section 2 of the development of understanding concerning inelastic scattering of electrons at surfaces. We then give a brief summary of the theory of inelastic electron scattering in solids in Section 3. Some historical remarks will be given in Section 4 on two heavily used electron-based probes, Auger-electron spectroscopy (AES) and X-ray photo-

SSDI 0039-6028(93)E0298-9

electron spectroscopy (XPS), which are the two techniques most commonly used for the measurement of surface composition (itself the most common surface measurement). In Section 5, we discuss the role of inelastic electron scattering in AES and XPS. For AES and XPS, the electron energy range of practical interest is between 50 and 2000 eV. A summary and outlook are given in Section 6.

2. Historical overview of inelastic electron scattering in solids and at surfaces

Since the discovery of secondary-electron emission by Austin and Starke in 1902, many investigators have attempted to obtain quantitative information both on this phenomenon and the processes that give rise to it [4–8]. Most electrons leaving a surface under electron bombardment are of low energy (generally around 5–10 eV).

A number of early investigators [9–11] detected electrons that had been elastically scattered by the specimen surface and others that had been inelastically scattered; the latter electrons were recognized by the fact that their energy position with respect to the peak of elastically scattered electrons did not change with incident energy [3]. The inelastic structures have been designated characteristic electron energy losses [11].

In 1936, Rudberg [12] reported new measurements with improved instrumentation of the characteristic losses observed with 50–370 eV electrons that were incident on evaporated (i.e., polycrystalline) films of Cu, Ag, Au, Ca, Ba and oxides of Ca and Ba. It was found that thin films of Ca, Ba and their oxides, with estimated thicknesses between 0.5 and 4 monolayers (ML), were sufficient to exhibit inelastic scattering characteristics of the particular material. Rudberg and Slater [13] made the first quantitative attempt to explain the origin of these losses. The losses were considered to be due to interband electronic transitions, and the copper loss spectrum was computed from its band structure. Fair agreement was found for the two smaller energy losses

observed by Rudberg, but other larger losses predicted by theory were not detected.

The first experiments of inelastic electron scattering by single-crystal surfaces were reported by Farnsworth and coworkers in 1938 and 1949 [14]. Their experiments were conducted with Ag(111) and Cu(100) surfaces and with incident electron energies between 8 and 150 eV. For both surfaces, the intensities of the energy-loss peaks depended on both the incident energy and the scattering angle, a phenonemon now known to be associated with inelastic low-energy electron diffraction [3,15].

Turning to the inelastic scattering of electrons transmitted through thin films, Ruthemann [16] performed the first experiments of this type using 2–8 keV electrons in the early 1940's. In 1944, Hillier and Baker [17] reported inner-shell excitations by 25–70 keV electrons transmitted through thin films of six materials. These authors recognized the potential of such measurements for elemental microanalysis, a technique now in common use in electron microscopy [18].

During the 1950's, a number of groups were active in the measurement of characteristic energy losses. A summary of relevant measurements and of emerging theory was published by Marton et al. in 1955 [19]. A distressing feature of this early work was the often-found large variation in energy-loss values from different laboratories for nominally the same material [19]. Experimental problems that could give rise to these variations were reviewed by Marton in 1956 [20].

In an important series of papers published in the 1950's, Bohm, Pines and Nozières [21] showed that energetic electrons incident on a solid could excite collective oscillations of the valence electrons in a valence electron–ion core plasma (analogous to the plasma oscillations known to occur in an ionized gas). For some materials, the quantum of energy loss observed experimentally agreed reasonably well with the values predicted by the plasma-excitation theory, and the losses were later termed bulk or volume plasmons. Examples of such "free-electron-like" materials included Be, Mg, Al, Si and Ge. For other materials, most notably the transition and noble metals, the measured characteristic loss spectrum consisted of a number of broad, often overlapping features which were difficult to identify as being due to the excitation of plasma oscillations. At this time, it was also thought that features in a characteristic loss spectrum could be due to single-electron excitations, as proposed by Rudberg and Slater [14]. The first direct measurements of IMFPs were reported by Blackstock et al. [22] in 1955 for the excitation of characteristic losses by 25–115 keV electrons in several materials. Their measurements for Al agreed reasonably well with values expected from the new plasma oscillation theory, but it was realized later that similar values for the IMFPS would be expected if the losses were due to single-electron excitations [23].

A more comprehensive basis for describing inelastic electron scattering in solids was proposed by Hubbard and by Frohlich and Pelzer in 1955 [24]. This theory provides a clear and unified treatment of different types of electronic excitations (i.e., characteristic energy losses due both to single-electron transitions and to plasma oscillations) and has been successfully utilized by many workers; it will be summarized in the following section. Reviews describing the use of this theory for valence-electron excitations and inner-shell excitations have been published by Raether [25], Schnatterly [26], and Sevier [27].

In 1957, Ritchie [28] reported the effects of a specimen surface on the characteristic loss spectrum of electrons transmitted through thin films. His theory showed that additional collective modes, surface plasma oscillations associated with the surfaces of the films, could be excited. The relative intensities of the bulk- and surface-plasmon losses were predicted to depend directly on specimen thickness and inversely on electron velocity. Stern and Ferrell [29] later showed that the magnitude of the surface-plasmon loss depended on the dielectric properties of the medium adjacent to the film of interest.

Starting in the late 1950's, a group at the University of Western Australia investigated inelastic electron scattering of 750–2000 eV electrons by a variety of materials [30]. They utilized a "reflection" scattering geometry and prepared their specimens by evaporation onto a substrate. Fig. 1 is an example of their work. The character-

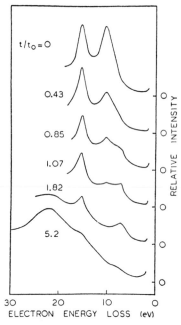

Fig. 1. Portions of the characteristic electron energy-loss spectra of aluminum in progressive stages of oxidation as reported by Powell and Swan [30]. The spectra were measured with 750 eV electrons incident on evaporated Al in a reflection scattering geometry. The number against each curve is the average value of t/t_0, where t was the elapsed time after evaporation of Al and t_0 was the time in which the measured intensity of the bulk plasmon loss decreased to half its initial value.

istic loss spectrum for an oxide-free surface of Al consists of multiples of a 15.3 eV bulk-plasmon loss and a 10.3 eV surface-plasmon loss. The magnitude of the surface-plasmon loss is close to the value expected from the theory of Ritchie [28]. As the surface was allowed to oxidize, the intensity of the original surface-plasmon loss decreased and a new surface-plasmon loss appeared at the position expected from the theory of Stern and Ferrell [29]. This result and measurements for other materials [30] were important in demonstrating that at least some of the variability in the reported values of energylosses measured in transmission through thin films [19,20] was probably associated with varying degrees of surface contamination on the specimen films. Later work [31] involving the inelastic scattering of keV electrons by the smooth surface of a drop of liquid Al showed striking variations in the intensities of the bulk-plasmon and surface-plasmon energy losses as a function of scattering angle in accord with the predictions of Stern and Ferrell [29].

Many studies of the inelastic scattering of keV electrons transmitted through thin films were reported in the 1960's and 1970's. They include measurements of differential and total inelastic scattering cross sections, measurements of dispersion relations (energy loss versus momentum transfer) for bulk and surface scattering modes, measurements of plasmon lifetime versus momentum transfer, and derivation of optical constants in the ultraviolet and soft X-ray spectral regions from characteristic loss spectra. This work has been summarized in a number of reviews [25–27,32].

3. Synopsis of the theory of inelastic electron scattering in solids

Inelastic electron scattering in solids can be described conveniently in terms of a complex dielectric constant $\epsilon(\omega, q)$ dependent on frequency ω and momentum transfer q [25–27,33] For $q = 0$, the dielectric constant is related to the familiar index of refraction n and extinction coefficient k by

$$\epsilon(\omega, 0) = (n + ik)^2 = \epsilon_1 + i\epsilon_2, \qquad (1a)$$

$$\epsilon_1 = (n^2 - k^2), \qquad (1b)$$

$$\epsilon_2 = 2nk. \qquad (1c)$$

The differential inelastic scattering cross section, per atom or molecule, for energy loss $\Delta E = \hbar\omega$ and momentum transfer q in an infinite medium is

$$\frac{d^2\sigma}{d\omega \, dq} = \frac{2e^2}{\pi N \hbar v^2} \, \mathrm{Im}\left(\frac{-1}{\epsilon(\omega, q)}\right) \frac{1}{q}, \qquad (2)$$

where N is the density of atoms or molecules and v is the velocity of the incident electrons.

The term $\mathrm{Im}(-1/\epsilon) = \epsilon_2/(\epsilon_1^2 + \epsilon_2^2)$ is known as the energy loss function. The denominator $(\epsilon_1^2 + \epsilon_2^2)$ is responsible for the differences that can occur between electron energy-loss spectra and optical absorption spectra (which are propor-

tional to ϵ_2). For $\Delta E < 100$ eV (i.e., losses predominantly due to valence-electron excitations), $(\epsilon_1^2 + \epsilon_2^2)$ is usually appreciably different from unity and there is a large difference between electron energy-loss spectra and optical-absorption spectra. For $\Delta E > 100$ eV (i.e., energy losses due predominantly to core-electron excitations), $\epsilon_1 \approx 1$ and $\epsilon_2 \ll 1$ so that $\mathrm{Im}(-1/\epsilon) \approx \epsilon_2$. Thus, electron energy-loss spectra are then similar to the corresponding X-ray absorption spectra. Maxima occur in the energy loss function when

$$(\epsilon_1^2 - \epsilon_2^2)(d\epsilon_2/d\omega) - 2\epsilon_1\epsilon_2(d\epsilon_1/d\omega) = 0$$

and $d^2[\mathrm{Im}(-1/\epsilon)]/d\omega^2$ is negative. Maxima in $\mathrm{Im}(-1/\epsilon)$ then occur near frequencies for which: (a) maxima occur in ϵ_2 (single-electron excitations of the type considered by Rudberg and Slater [14]); or (b) $\epsilon_1 = 0$ and $\epsilon_2 \ll 1$ (excitation of bulk plasmons [21]).

Model calculations have shown how small changes in the position and strength of single-electron excitations affect the position and strength of maxima in the energy loss function [34]. As a result, it can be difficult to identify a particular loss peak unambiguously. For the free-electron-like solids, comparison of the energy loss function $\mathrm{Im}[-1/\epsilon(\omega)]$ with the functions $\epsilon_1(\omega)$ and $\epsilon_2(\omega)$ often shows that $\epsilon_2(\omega) < 0.2$ for frequencies in the vicinity of that for which $\epsilon_1(\omega) = 0$; the bulk plasmon is then well defined. The general situation for non-free-electron-like solids is that the energy loss function consists of at least several broad and overlapping maxima that do not correspond obviously with the positions of maxima in $\epsilon_2(\omega)$. In such cases, it is not meaningful to identify one structure in $\mathrm{Im}(-1/\epsilon)$ as being due to bulk plasmon excitation. Fits of experimental $\mathrm{Im}(-1/\epsilon)$ data to a model dielectric function are useful in showing the relationship of structure in the energy loss function to structure in $\epsilon_2(\omega)$ [35].

In the vicinity of surfaces, the differential cross section for inelastic scattering is modified by a reduction in the term for the bulk (Eq. (2)) and by the addition of a new term to describe the excitation of surface plasmons [28]. At a solid–vacuum interface, the probability for surface-plasmon excitation is proportional to Im$[-1/(1+\epsilon)]$, the surface energy-loss function.

4. Historical remarks concerning AES and XPS

We present here some brief remarks on the history of AES and XPS. An excellent history of XPS from 1900 to 1960 has been published by Jenkin et al. [36] who describe the many contributions that led to the major instrumental developments and new discoveries (e.g., of chemical shifts [37]) at Uppsala by Siegbahn et al. [38] in the 1950's and 1960's: developments that led to the selection of Professor Siegbahn as a Nobel physics Laureate in 1981.

Siegbahn et al. [37] coined the term "Electron spectroscopy for chemical analysis" with the acronym ESCA to refer to XPS for surface analysis. This term is in widespread use, but we will use the more specific terms (AES, XPS, etc.).

In a series of papers beginning in 1923, Auger observed and interpreted the effect that now bears his name [39]. A prediction that atoms could de-excite non-radiatively also was made by Rosseland [40] in 1923, although Auger was unaware of this work until later [27]. Information on the Auger effect through the 1960s is contained in books by Burhop [41] and by Sevier [27] and in a book chapter [42].

Auger peaks in the secondary-electron spectra of electron-bombarded surfaces were first reported by Howarth [43] in 1935. Similar studies were reported in the 1950's [44–46]. Lander [44] was the first to interpret the lineshapes of core–valence–valence (CVV) Auger transitions in terms of self-convolutions of the occupied valence-band densities of states. Powell et al. [46] demonstrated that AES could be used to detect a surface contamination layer.

Steinhardt and Serfass [47] developed instrumentation in the early 1950s that was intended for surface analysis by XPS and possibly AES. While Lander foresaw the possibility of surface analysis by AES, it was not until the late 1960's that practical AES analyses were demonstrated by Harris [48] and by Weber and Peria [49]. A key development was the use of differentiation of the secondary electron signal in order to make the Auger peaks stand out from the secondary electron background [3]. Another important factor in the subsequent rapid growth of AES as a surface diagnostic tool was the wide availability at that time of instrumentation for low-energy electron diffraction [3]. Weber and Peria demonstrated that this equipment could be quickly utilized to obtain composition data as well as structural information at surfaces.

The pioneering contributions of Harris in developing AES as a useful tool for surface analysis were recognized by the American Vacuum Society through his selection as the recipient of the Medard W. Welch award in 1973. Harris [50] published some comments on the early history of AES, including the fact that extended disagreement with referees delayed publication of his early papers [48] until after the later work of Weber and Peria [49] was published.

Commercial instruments designed specifically for AES and XPS became available in the late 1960's. Hofmann [51] has identified 1969 as the effective take-off year for both AES and XPS from a plot of the number of journal articles utilizing these techniques as a function of time.

Many review articles and book chapters have been published that detail the application of AES and XPS for a wide variety of scientific and technological purposes, including two surveys in 1978 [52] and 1982 [53]. The National Institute of Standards and Technology has available a database for XPS containing data for core-electron binding energies, doublet splittings, and kinetic energies of X-ray-excited Auger electrons that appear in typical XPS measurements with excitation by Al or Mg characteristic X-rays [54]. A survey of the published literature was made for 1986–90 using the following search words and phrases for matches in published titles and abstracts: ESCA, electron spectroscopy chemical analysis; XPS; X-ray photoelectron spectroscopy; and X-ray photoemission. This search identified 4120 articles in 487 different journals. Table 1 is a listing of journal titles that contain 20 or more XPS articles from this search. While the titles of the journals in Table 1 give an idea of the likely XPS applications, the titles of the other 443 jour-

Table 1

Titles of journals containing 20 or more XPS articles for 1986–90 (see text)

Number of papers	Journal title
257	Surface and Interface Analysis
227	Physical Review B
194	Journal of Vacuum Science and Technology A
185	Surface Science
130	Applied Surface Science
129	Journal of Electron Spectroscopy and Related Phenomena
110	Journal of the Electrochemical Society
99	Journal of Applied Physics
69	Corrosion Science
65	Thin Solid Films
60	Macromolecules
60	Vacuum
55	Applied Physics Letters
51	Solid State Communications
50	Journal of Chemical Physics
47	Journal of Non-Crystalline Solids
46	Journal of Catalysis
41	Polymer
37	Journal of Materials Science
37	Journal of Vacuum Science and Technology B
33	Applied Catalysis
32	Journal of Applied Polymer Science
32	Journal of Physical Chemistry
30	Japanese Journal of Applied Physics Part 2
29	Electrochimica Acta
29	Journal of Polymer Science Part A
27	Japanese Journal of Applied Physics Part 1
27	Nuclear Instruments and Methods in Physics Research Section B
25	Physica Scripta
24	Chemical Physics Letters
24	Journal of Colloid and Interface Science
24	Physical Review Letters
23	Corrosion
23	Fresenius Zeitschrift für Analytische Chemie
22	Bulletin of the Chemical Society of Japan
22	Journal of the Chemical Society–Faraday Transactions I
22	Langmuir
21	Journal of Materials Science Letters
21	Physica Status Solidi A
21	Zeitschrift für Physik B
20	Journal of Physics–Condensed Matter
20	Journal of the American Chemical Society
20	Journal of the Physical Society of Japan
20	Synthetic Metals

nals indicate clearly the breadth of XPS applications (e.g., in coatings technology, mineral processing, biomaterials, dental research, zeolites, adhesion, wear, rubber chemistry, aerosol science, cereal science, composites, bacteriology, clays, food science, pesticides, and wood science to name but a few). There is little doubt that similar diversity would be found in a comparable AES survey.

In an article describing Siegbahn's scientific contributions on the occasion of his selection as a Nobel physics laureate, Hollander and Shirley [55] wrote in 1981 that "it is safe to estimate that (XPS) has been employed for the study of high-technology materials with annual sales volumes of several billion dollars". It is probably a safe extrapolation that the present annual volume of high-technology materials based on both AES and XPS measurements must now be in the tens of billions of dollars.

5. Role of inelastic electron scattering in AES and XPS

We will now join the two main themes of this article by examining the effects of inelastic electron scattering in AES and XPS. The remarks also are applicable to other electron-based probes of surfaces [3,56].

As noted earlier, the surface sensitivity of electron-based surface probes is due predominantly to inelastic scattering. A useful measure of the surface sensitivity in a particular experiment is the IMFP. Two other terms, the electron attenuation length (AL) and the electron escape depth (ED), to be discussed shortly, have also been utilized frequently as measures of surface sensitivity.

Beginning in the late 1960's, many experiments were performed to measure the attenuation of photoelectrons or Auger electrons from a substrate as an overlayer film was deposited. Corresponding measurements were also frequently made of the rate of increase of a photoelectron or Auger-electron signal from the overlayer film as a function of thickness. The signal intensities depended nearly exponentially on film thickness, and it was natural to refer to the exponential parameter as the attenuation length. The initial experiments were performed with XPS by Siegbahn et al. [38] and with AES by Palmberg and Rhodin [57].

Early in the development of AES and XPS, it was recognized that these techniques could be used to make measurements of composition in the near-surface regions of a specimen. Simple models were proposed with which the composition of a near-surface region could be determined by AES [58] or XPS [59]. Data on the inelastic attenuation are needed for quantitative analyses by AES or XPS [60].

5.1. Terminology

The three terms, IMFP, AL, and ED are frequently used interchangeably but each term in fact has a separate meaning [61,62]. The AL definition and the related ED definition were developed in order to describe the results of overlayer-film experiments using a model in which elastic electron scattering was neglected [60]. In the early 1980's, several groups [63–66] pointed out that elastic electron scattering could be significant under the conditions relevant to AES and XPS experiments and to the overlayer-film experiments for the measurement of ALs. As a result of the resulting longer electron paths, on average, than when elastic scattering was neglected, the exponential parameter derived from the overlayer-film experiments could be systematically *less* than the corresponding IMFP by up to about 30%.

The definitions [61,62] for AL and ED have recently been questioned [67,68] and, as a result, the definitions are undergoing revision. It has also become clear [67–69] that elastic electron scattering can cause appreciable deviations from the exponential-attenuation behavior found when elastic scattering is neglected. Monte Carlo and similar calculations [67–69] indicate that the "effective" AL, as obtained from local or average slopes of the calculated depth distribution functions for the signal electrons, can depend on the choice of the substrate-film combination (in simulations of the overlayer-film experiments to measure ALs), the experimental configuration (particularly the solid angle and direction of the analyzer) and the thickness of the overlayer film. These results indicate clearly that the "effective" AL is not simply a material-dependent parameter independent of geometrical effects. Since the revised definitions have not, as yet, been formally approved, we will use the existing terms [61] herein.

5.2. Synopsis of AL measurements

Three reviews appeared independently in 1974 that presented available AL data and in which overviews of the AL data were given from different perspectives [70–72]. By this time, AL values for different materials plotted versus electron energy showed clear trends [3]. As the electron energy increased from about 5 to about 50–100 eV, the AL decreased from about 50 to about 3–5 Å and then increased to about 30 Å at an energy of 2000 eV. In 1979, Seah and Dench [73] published a review of AL data in which they analyzed the possible dependences of AL values on different material parameters. They found least scatter about a common curve of AL versus energy when the AL, λ_{al}, was expressed as follows:

(a) for elements

$$\lambda_{al} = 538aE^{-2} + 0.41a^{3/2}E^{1/2} \text{ nm}, \tag{3a}$$

(b) for inorganic compounds

$$\lambda_{al} = 2170aE^{-2} + 0.72a^{3/2}E^{1/2} \text{ nm}, \tag{3b}$$

(c) for organic compounds

$$\lambda_{al} = (10^3/\rho)(49E^{-2} + 0.11E^{-2}) \text{ nm}, \tag{3c}$$

where a is the average monolayer thickness (nm) defined by

$$a^3 = 10^{27}A/\rho nN_a. \tag{4}$$

In Eqs. (3) and (4), E is the electron energy (eV), ρ is the bulk density (g cm^{-3}), A is the atomic or molecular weight, n is the number of atoms per molecule, and N_a is Avogadro's number. The Seah and Dench equations have been widely utilized for the prediction of AL values: as of mid-1989, their paper [73] had been cited over 800 times and was gathering about 150 citations per year [74].

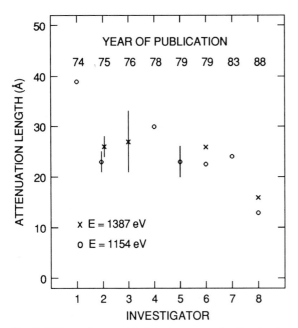

Fig. 2. Values of reported electron attenuation lengths for silicon at two electron energies E (corresponding to photoemission from the Si2p shell by Al or Mg characteristic X-rays) reported by eight investigators [77–84]. The vertical lines represent the uncertainty ranges (68% confidence limits) of the published AL values where these were given [76].

More recent reviews [62,75,76] contain comments about the difficulties in making AL measurements. Although the overlayer-film technique has an appealing conceptual simplicity, there are many potential sources of significant systematic error. As an example, Fig. 2 [76] shows the AL results of eight investigations [77–84] for Si ALs at electron energies corresponding to photoemission from the Si2p shell by aluminum and magnesium characteristic X-rays. The ratio of the highest to the lowest AL at an energy of 1154 eV is a factor of three. Even if the two extreme values were discarded (without justification), the variation among the remaining five values is a factor of 1.5.

5.3. Synopsis of IMFP calculations and measurements

Several reviews [62,72,75,85] describe IMFP calculations that have been made for solids in the energy range of interest for AES and XPS. The earliest calculations [85,86] were based on excitations of bulk plasmons. Shelton, for example, used Lundqvist's results to indicate that the IMFP depended strongly on valence-electron density [86]. More recently, a number of groups [87–90] have reported IMFP calculations that have made use of Eq. (2) and more complex formulations for $\epsilon(\omega)$ than those required for bulk plasmon excitations.

We summarize here the IMFP calculations of Tanuma et al. [91] that are based on the Penn algorithm [89]. In this approach, experimental optical data for a material are used to compute the energy loss function and thus the dependence of the inelastic scattering probability on energy loss. The theoretical Lindhard dielectric function [92] is employed to describe the dependence of the inelastic scattering probability on momentum transfer.

Tanuma et al. [91] have reported IMFP data for 50–2000 eV electrons in 27 elemental solids and 15 inorganic compounds. It was found that the calculated IMFP values for each material could be fitted satisfactorily to a modified form of the Bethe equation [93]:

$$\lambda_i = E / \left\{ E_p^2 \left[\beta \ln(\gamma E) - (C/E) + (D/E^2) \right] \right\}, \tag{5}$$

where λ_i is the IMFP (in Å), E is the electron energy (in eV), $E_p = 28.8 \ (N_v \rho/M)^{1/2}$ is the free-electron plasmon energy (in eV), ρ is the density (in g cm^{-3}), N_v is the number of valence electrons per atom (for elements) or molecule (for compounds), and M is the atomic or molecular weight. Values for the four parameters β, γ, C, and D were determined for each material. The set of values of the four parameters for each of the 27 elements was analyzed in terms of various material parameters to yield the following empirical relationships:

$$\beta = -0.0216 + 0.944 / \left(E_p^2 + E_g^2 \right)^{1/2}$$
$$+ 7.39 \times 10^{-4} \rho, \tag{6a}$$

$$\gamma = 0.191 \rho^{-0.50}, \tag{6b}$$

$$C = 1.97 - 0.91U, \tag{6c}$$

$$D = 53.4 - 20.8U, \qquad (6d)$$

$$U = N_v \rho / M = E_p^2 / 829.4, \qquad (6e)$$

and E_g is the bandgap energy (in eV) for non-conductors. Eqs. (5) and (6) (designated as TPP-2) give IMFP values for the elements that differ by 13% RMS (root-mean-square) from those calculated directly from the optical data.

More recently [94], similar IMFP calculations have been made for a group of 14 organic compounds. The IMFP values for these materials differed noticeably from the predictions of TPP-2, mainly because the organic compounds were of much lower density than the solid elements. The sets of parameter values for the fits of computed IMFPs for the elements and the organic compounds to Eq. (5) were reanalyzed to yield a modified expression for the parameter β:

$$\beta_m = -0.10 + 0.944 / \left(E_p^2 + E_g^2 \right)^{1/2} + 0.069 \rho^{0.1}. \qquad (7)$$

Eqs. (5), (6a), (6c)–(6e), and (7) constitute the modified IMFP predictive formula TPP-2M which is believed to be superior to TPP-2.

The IMFP calculations described above are for bulk solids. It would be expected from the discussion in Sections 2 and 3 that modifications to the IMFPs should occur in the vicinity of surfaces (e.g., due to the excitation of surface plasmons in free-electron-like solids). Yubero and Tougaard [95] have recently shown how the effective IMFP could change with path-length for reflection electron energy-loss spectroscopy on the basis of two models for the electron–surface interaction.

Fig. 3 shows a comparison of the AL empirical formula of Seah and Dench [73] and the IMFP calculations for 27 elements reported by Tanuma et al. [91]. The comparison can most readily be made in terms of the parameter λ_a defined as [73]:

$$\lambda_a = 0.1 \lambda a^{-3/2}, \qquad (8)$$

where λ is either the AL or IMFP (in Å) and a is given by Eq. (4). Eq. (3a) for ALs in elements then becomes:

$$\lambda_a = 1040 E^{-2} + 0.41 E^{1/2} \text{ monolayers nm}^{-1/2}, \qquad (9)$$

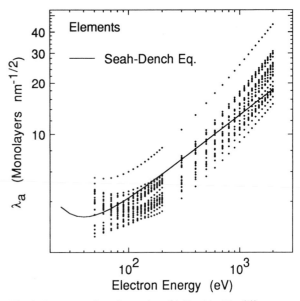

Fig. 3. Summary plot of λ_a values (defined by Eq. (8)) versus electron energy. The solid points are derived from the IMFP results of Tanuma et al. [91] for a group of 27 elemental solids and the solid line is the Seah and Dench [73] expression for ALs given by Eq. (9).

where an average value for the parameter a has been used to determine the coefficient of the first term in Eq. (9). The advantage of Eq. (9) is that the second term (which dominates the expression for $E > 150$ eV) is material-independent.

Fig. 3 indicates a general similarity of the Seah and Dench expression, Eq. (9), with the IMFP values of Tanuma et al. Nevertheless, there are some important differences. First, the energy dependence of the IMFP for individual elements is close to $E^{0.75}$ rather than to $E^{0.5}$. Second, the material-dependence of the AL suggested by Eq. (3) is not reflected by the IMFP results; the IMFP values for any energy can differ by up to a factor of two rather than following a material-independent curve such as Eq. (9). Finally, Eq. (9) has a minimum for all elements at an energy of about 40 eV while the IMFP data show minima in the range 30–125 eV; such material-dependent minima are to be expected from variations of valence-electron density for the free-electron-like solids [86] and to variations in the shapes of the electron energy-loss functions from material to material [91].

The IMFP can be determined from measurements of the elastic electron backscattering probability from surfaces [96]. The advantage of this technique for measuring IMFPs is that experiments can be performed with macroscopic specimens rather than the thin overlayer films needed for AL measurements. The relatively few published IMFP measurements agree reasonably well with calculated IMFPs [96,97] although, in some instances, the measured IMFPs vary more rapidly with electron energy than expected from the calculations.

5.4. Role of inelastic electron scattering in quantitative AES and XPS

In AES and XPS, the signal of interest lies on a background due to secondary electrons, to scattered primary electrons (AES) or photoelectrons (XPS), and to photoelectrons excited by bremsstrahlung (XPS). This background has been termed the matrix background [76] since it refers to the background for some matrix that would be observed if the signal due to a particular AES or XPS line was absent. If the AES or XPS signal is "switched on", the measured signal intensity can be considered to be of two parts, a part due to photoelectrons or Auger electrons that have not been inelastically scattered in the specimen and a part due to such electrons that have been inelastically scattered. The first part is the intensity that should be measured in a quantitative analysis. This intensity is termed the peak intensity and comprises not only the intensity of the "main" peak (i.e., the intensity of what is often a single, obvious AES or XPS feature) but also a continuous intensity and features that are each due to intrinsic excitations accompanying core-electron excitation or decay. The second part of the total signal intensity that is due to inelastic scattering (extrinsic excitations) of electrons originally at energies corresponding to the peak intensity is termed the peak background [76].

A practical problem in AES and XPS is the separation of the overlapping contributions to measured intensities for a peak and the associated peak background. A number of deconvolution schemes have been developed for this purpose [98–100] in which either an experimentally determined characteristic loss spectrum (measured on the same specimen at an incident energy corresponding to the AES or XPS feature of interest) or an energy-loss function for the material of interest (taken from literature sources) is used to obtain the peak intensity. The first approach has the advantage of ensuring that the loss spectrum for the specific material is used in the deconvolution; it has the disadvantages of additional measurement time and the fact that the loss spectrum contains information associated with electrons traversing the material–vacuum interface twice while, in the AES or XPS measurement, the signal electrons traverse this interface only once. As a result, the intensities associated with the surface loss function in the loss spectrum are enhanced with respect to the AES or XPS measurement [101]. The second approach for deconvolution has the advantage of simplicity and the disadvantages that literature data may not be available and that, until recently, it was not clear whether the deconvolution should be performed with the bulk loss function or with some (unspecified) combination of the bulk and surface loss functions.

In an important series of papers, Tougaard and coworkers [65,95,102] have developed improved algorithms for separating the peak intensity from the associated peak background intensity. This work has shown clearly the existence of a substantial intensity component in the peak intensity for XPS due to intrinsic excitations in the photoionization process. This component is (mistakenly) often ignored in quantitative XPS although it is known that the intrinsic fraction can change substantially with chemical state and be different for different photoelectron lines from the same element [103].

Tougaard et al. [95,102] have also found that a characteristic loss spectrum measured in a reflection scattering configuration cannot be regarded simply as a superposition of bulk and surface loss functions. They have developed a procedure for extracting inelastic scattering cross section information from characteristic loss spectra measured in the reflection geometry. Surprisingly, they find that a simple empirical formula for the inelastic scattering cross section is often superior to experimental inelastic cross section data for a given material in analyses to separate the peak intensity from the associated peak background intensity. This superiority is found for non-free-electron-like materials that have broad, overlapping loss spectra and, in these cases, the details of the structure are unimportant in the deconvolution process. For free-electron-like materials, however, the sharp structure due to plasmon excitation in the energy-loss function is poorly represented by the empirical formula. Yoshikawa et al. [100] have also analyzed the relation between the loss function for XPS and reflection energy-loss spectroscopy. They find that the surface component to the loss spectrum is significant and that different effective loss functions should be used for different take-off angles in XPS.

The relative intensities of a peak and its associated background will be different depending on whether the emitting atoms are homogeneously distributed through the specimen, are present only on the surface, or are present in a subsurface layer. Sickafus [104] and Tougaard et al. [102,105] have shown how the composition-depth profile for the emitting atoms can be extracted from the AES or XPS data.

As noted earlier, the effects of elastic electron scattering have been realized to be significant in quantitative AES and XPS [63–69]. As a result, the simple formulae [60] developed for quantitative applications in which elastic scattering was ignored are not necessarily valid. Jablonski [106] however, has shown that the simple expressions in common use for analyses of a homogeneous surface region by AES are not only valid but that the IMFP rather than the AL should be employed. A similar result has been obtained for XPS analyses [97] with configurations in which the angle between the directions of the incident X-rays and the detected photoelectrons is close to the "magic" angle of 54°44′ [107].

Efforts to determine peak intensities in AES and XPS have been guided by knowledge of the relevant lineshapes [108] for the observed peaks. Doniach and Sunjic [109] showed that photoelectron lines could be asymmetrical and that the degree of asymmetry was related to the density of

empty states at the Fermi level in conductors. Their expression for the lineshape has proven to be useful in analyses of photoelectron energy distributions and in the determination of "surface shifts" of core levels [110]. For AES, theories of the lineshapes for CVV Auger transitions were developed independently in 1977 by Cini [111] and Sawatzky [112]. Their theories have formed the basis of subsequent efforts [113] to describe the existence of both band-like and atomic-like effects in valence-band Auger-electron spectra [114].

6. Summary and outlook

An overview has been given of the role of inelastic electron scattering in electron-based surface probes. While the conceptual framework for inelastic scattering in solids was established prior to the advent of modern surface science, many important issues have been addressed during the past 30 years.

We have described the role of inelastic scattering in AES and XPS, the two most common techniques for the measurement of surface composition. While inelastic scattering largely determines the surface sensitivity, it is a source of complication in proper intensity measurement. During the past ten years, the complicating effects of elastic electron scattering on the transport of signal electrons in AES and XPS have been recognized. It is likely that more efficient calculational schemes will be developed [115] to assist those making AES or XPS analyses. Data on ALs and IMFPs are included in AES and XPS software [116]; this information will probably be included in expert systems under development for XPS [117] and under consideration for AES.

Perhaps one of the most urgent needs for all electron-based surface probes is the development of new or improved methods for experimental determinations of ALs and IMFPs. There are many difficulties in AL measurements [62,76], and it is now recognized that elastic electron scattering can cause significant deviations from the previously assumed exponential behavior for thin-film overlayers [69,118]. Improvements in the theory for the calculation of IMFP are also required, particularly for treating exchange and correlation effects at energies below about 200 eV. More detailed analyses are needed of the role of surface effects on the inelastic scattering of signal electrons in AES and XPS. Finally, recent experiments [119] have shown that ALs for electron energies between about 5 and 50 eV are much lower than expected from the Seah and Dench [73] empirical relation; further work is needed to explore inelastic electron scattering mechanisms and cross sections for these low energies.

AES and XPS are now rightly considered routine tools for surface science and for many applications in the development of advanced materials, devices and processes. There are, however, scientific issues concerning these techniques that have not been fully developed. For both AES and XPS, there is inadequate knowledge and documentation of the lineshapes and intensities associated with multi-electron processes in the creation and decay of core-level ionizations. There is still limited knowledge of the lineshapes of the commonly observed Auger-electron transitions and the changes that occur in different chemical environments. Although the basic lineshapes in AES and the intrinsic features in both AES and XPS are often obscured by inelastic scattering, it is hoped that recent advances in spectral analysis [102] and the development of new experimental tools [120] will lead to further advances in understanding of the significant processes.

Acknowledgement

The author thanks Dr. C.B. Duke for making many suggestions for improving this article.

References

[1] D.P. Woodruff and T.A. Delchar, Modern Techniques of Surface Science (Cambridge University Press, New York, 1986).

[2] J.C. Riviere, Surface Analytical Techniques (Clarendon, Oxford, 1990).

[3] C.B. Duke, Surf. Sci. 299/300 (1994) 24.

[4] H. Bruining, Physics and Applications of Secondary Electron Emission (Pergamon, London, 1954).

[5] R. Kollath, Handbuch der Physik 31 (1956) 232.

[6] A.J. Dekker, in: Solid State Physics, Vol. 6, Eds. F. Seitz and D. Turnbull (Academic Press, New York, 1958) p. 251.

[7] O. Hachenberg and W. Brauer, in: Advances in Electronics and Electron Physics, Vol. 11, Ed. L. Marton (Academic, New York, 1959) p. 413.

[8] H. Seiler, Scanning Microsc. 2 (1988) 1885;
H.E. Bauer and H. Seiler, Surf. Interface Anal. 12 (1988) 119.

[9] J.A. Becker, Phys. Rev. 23 (1924) 664.

[10] D. Brown and R. Whiddington, Proc. Leeds Philos. Lit. Soc. 1 (1927) 162.

[11] E. Rudberg, Proc. R. Soc. London A 127 (1930) 111.

[12] E. Rudberg, Phys. Rev. 50 (1936) 138.

[13] E. Rudberg and J.C. Slater, Phys. Rev. 50 (1936) 150.

[14] J.C. Turnbull and H.E. Farnsworth, Phys. Rev. 54 (1938) 509;
P.P. Reichertz and H.E. Farnsworth, Phys. Rev. 75 (1949) 1902.

[15] C.B. Duke and G.E. Laramore, Phys. Rev. B 3 (1971) 3183;
G.E. Laramore and C.B. Duke, Phys. Rev. B 3 (1971) 3198.

[16] G. Ruthemann, Naturwiss. 29 (1941) 648; 30 (1942) 145; Ann. Phys. 6 (1948) 113; 6 (1948) 135.

[17] J. Hillier and R.F. Baker, J. Appl. Phys. 15 (1944) 663.

[18] R.F. Egerton, Electron Energy-Loss Spectroscopy in the Electron Microscope (Plenum, New York, 1986).

[19] L. Marton, L.B. Leder and H. Mendlowitz, in: Advances in Electronics and Electron Physics, Vol. 7, Ed. L. Marton (Academic Press, New York, 1955) p. 183.

[20] L. Marton, Rev. Mod. Phys. 28 (1956) 172.

[21] See, for example: D. Bohm and D. Pines, Phys. Rev. 85 (1952) 338;
D. Pines, Rev. Mod. Phys. 28 (1956) 184;
P. Nozieres and D. Pines, Phys. Rev. 113 (1959) 1254.

[22] A.W. Blackstock, R.H. Ritchie and R.D. Birkhoff, Phys. Rev. 100 (1955) 1078.

[23] N. Swanson and C.J. Powell, Phys. Rev. 145 (1966) 195.

[24] J. Hubbard, Proc. Phys. Soc. A 68 (1955) 441;
H. Frohlich and H. Pelzer, Proc. Phys. Soc. A 68 (1955) 525.

[25] H. Raether, Springer Tracts in Modern Physics 88 (1980) 1.

[26] S.E. Schnatterly, in: Solid State Physics, Vol. 34, Eds. H. Ehrenreich, F. Seitz and D. Turnbull (Academic, New York, 1979) p. 275.

[27] K.D. Sevier, Low Energy Electron Spectrometry (Wiley–Interscience, New York, 1972).

[28] R.H. Ritchie, Phys. Rev. 106 (1957) 874.

[29] E.A. Stern and R.A. Ferrell, Phys. Rev. 120 (1960) 130.

[30] See, for example: C.J. Powell and J.B. Swan, Phys. Rev. 115 (1959) 869; 118 (1960) 640;
C.J. Powell, Proc. Phys. Soc. 76 (1960) 593;

J.L. Robins and J.B. Swan, Proc. Phys. Soc. 76 (1960) 857;
J.L. Robins, Proc. Phys. Soc. 78 (1961) 1177;
J.L. Robins and P.E. Best, Proc. Phys. Soc. 79 (1962) 110;
J.L. Robins, Proc. Phys. Soc. 79 (1962) 119;
P.E. Best, Proc. Phys. Soc. 79 (1962) 133; 80 (1962) 1308;
B.M. Hartley and J. B. Swan, Austr. J. Phys. 23 (1970) 655.

[31] C.J. Powell, Phys. Rev. 175 (1968) 972.

[32] H. Raether, Springer Tracts in Modern Physics 38 (1965) 84;
J. Daniels, C. v. Festenberg, H. Raether and K. Zeppenfeld, Springer Tracts in Modern Physics 54 (1970) 77.

[33] C.J. Powell, in: Electron Beam Interactions with Solids for Microscopy, Microanalysis and Microlithography, Eds. D.F. Kyser, H. Niedrig, D.E. Newbury and R. Shimizu (SEM Inc., Chicago, 1984) p. 19.

[34] C.J. Powell, J. Opt. Soc. Am. 59 (1969) 738.

[35] S. Tougaard and J. Kraaer, Phys. Rev. B 43 (1991) 1651.

[36] J.G. Jenkin, R.C.G. Leckey and J. Liesegang, J. Electron Spectrosc. 12 (1977) 1;
J.G. Jenkin, J.D. Riley, J. Liesegang and R.C.G. Leckey, J. Electron Spectrosc. 14 (1978) 477.

[37] E. Sokolowski, C. Nordling and K. Siegbahn, Phys. Rev. 110 (1958) 776;
C. Nordling, E. Sokolowski and K. Siegbahn, Ark. Fys. 13 (1958) 483.

[38] K. Siegbahn, C. Nordling, A. Fahlman, R. Nordberg, K. Hamrin, J. Hedman, G. Johansson, T. Berkmark, S. Karlsson, I. Lindgren and B. Lindberg, ESCA; Atomic, Molecular and Solid State Structure Studied by Means of Electron Spectroscopy (Almqvist and Wiksells, Uppsala, 1967).

[39] P. Auger, Compt. Rend. 177 (1923) 169; 180 (1925) 65; J. Phys. Rad. 6 (1925) 205; Compt. Rend. 182 (1926) 773; Compt. Rend. 182 (1926) 1215; Ann. Phys. 6 (126) 183.

[40] S. Rosseland, Z. Phys. 14 (1923) 173.

[41] E.H.S. Burhop, The Auger Effect and Other Radiationless Transitions (Cambridge University Press, Cambridge, 1952).

[42] I. Bergstrom and C. Nordling, in: Alpha-, Beta- and Gamma-Ray Spectroscopy, Ed. K. Siegbahn (North-Holland Publishing Company, Amsterdam, 1965) p. 1523.

[43] L.J. Howarth, Phys. Rev. 48 (1935) 88; 50 (1936) 216.

[44] J.J. Lander, Phys. Rev. 91 (1953) 1382.

[45] G.A. Harrower, Phys. Rev. 102 (1956) 340.

[46] C.J. Powell, J.L. Robins and J.B. Swan, Phys. Rev. 110 (1958) 657.

[47] R.G. Steinhardt and E.J. Serfass, Anal. Chem. 23 (1951) 1585; 25 (1953) 697.

[48] L.A. Harris, J. Appl. Phys. 39 (1968) 1419; 1428.

[49] R.E. Weber and W.T. Peria, J. Appl. Phys. 38 (1967) 4355.

[50] L.A. Harris, J. Vac. Sci. Technol. 11 (1974) 23.

[51] S. Hofmann, Surf. Interface Anal. 9 (1986) 3.

[52] C.J. Powell, Appl. Surf. Sci. 1 (1978) 143.

[53] M.P. Seah, Surf. Interface Anal. 2 (1980) 222.

[54] C.J. Powell, Surf. Interface Anal. 17 (1991) 308;
J.R. Rumble, D.M. Bickham and C.J. Powell, Surf. Interface Anal. 19 (1992) 241.

[55] J.M. Hollander and D.A. Shirley, Science 214 (1981) 629.

[56] D. Menzel, Surf. Sci. 299/300 (1994) 170;
D.P. Woodruff, Surf. Sci. 299/300 (1994) 183;
P.H. Citrin, Surf. Sci. 299/300 (1994) 199;
F.J. Himpsel, Surf. Sci. 299/300 (1994) 525.

[57] P.W. Palmberg and T.N. Rhodin, J. Appl. Phys. 39 (1968) 2425.

[58] H.E. Bishop and J.C. Rivière, J. Appl. Phys. 40 (1969) 1740;
P.W. Palmberg, Anal. Chem. 45 (1973) 549A.

[59] B.L. Henke, Phys. Rev. A 6 (1972) 94;
C.S. Fadley, in: Progress in Solid State Chemistry, Eds. G. Somorjai and J. McCaldin, Vol. 11 (Pergamon Press, New York, 1976) p. 265.

[60] See, for example: C.J. Powell, in: Quantitative Surface Analysis of Materials, Ed. N.S. McIntyre, ASTM Special Technical Publication 643 (American Society for Testing and Materials, Philadelphia, PA, 1978) p. 5;
M.P. Seah, in: Practical Surface Analysis, Vol. 1, Auger and X-Ray Photoelectron Spectroscopy, 2nd ed., Eds. D. Briggs and M.P. Seah (Wiley, New York, 1990) p. 201.

[61] ASTM Standard E-673 Annual Book of ASTM Standards, Vol. 3.06 (American Society for Testing and Materials, Philadelphia, PA, 1992) p. 507; Surf. Interface Anal. 17 (1991) 951.

[62] C.J. Powell, J. Electron Spectrosc. 47 (1988) 197.

[63] O.A. Baschenko and V.I. Nefedov, J. Electron Spectrosc. 21 (1980) 153; 27 (1982) 109.

[64] J. Ferron, E.C. Goldberg, L.S. De Bernadez and R.H. Buitrago, Surf. Sci. 123 (1982) 239.

[65] S. Tougaard and P. Sigmund, Phys. Rev. B 25 (1982) 4452.

[66] V.M. Dwyer and J.A.D. Matthew, Vacuum 33 (1983) 767.

[67] A. Jablonski and H. Ebel, Surf. Interface Anal. 11 (1988) 627.

[68] W.H. Gries and W. Werner, Surf. Interface Anal. 126 (1990) 149.

[69] A. Jablonski, Surf. Interface Anal. 14 (1989) 659;
A. Jablonski and S. Tougaard, J. Vac. Sci. Technol. A 8 (1990) 106;
W.S.M. Werner, W.H. Gries and H. Stori, J. Vac. Sci. Technol. A 9 (1991) 21;

Z.-J. Ding, PhD Thesis, Osaka University, 1990;
W.S.M. Werner, W.H. Gries and H. Stori, Surf. Interface Anal. 17 (1991) 693;
H.F. Chen, C.M. Kwei and C.J. Tung, J. Phys. D 25 (1992) 262;
W.S.M. Werner, Surf. Interface Anal. 18 (1992) 217;
I.S. Tilinin and W.S.M. Werner, Phys. Rev. B 46 (1992) 13739.

[70] C.R. Brundle, J. Vac. Sci. Technol. 11 (1974) 212.

[71] I. Lindau and W.E. Spicer, J. Electron Spectrosc. 3 (1974) 409.

[72] C.J. Powell, Surf. Sci. 44 (1974) 29.

[73] M.P. Seah and W.A. Dench, Surf. Interface Anal. 1 (1979) 2.

[74] J.A. Mears, Institute for Scientific Information, private communication.

[75] C.J. Powell, Scanning Electron Microsc. 4 (1984) 1649.

[76] C.J. Powell and M.P. Seah, J. Vac. Sci. Technol. A 8 (1990) 735.

[77] M. Klasson, A. Berndtsson, J. Hedman, R. Nilsson, R. Nyholm and C. Nordling, J. Electron Spectrosc. 3 (1974) 427.

[78] R. Flitsch and S.I. Raider, J. Vac. Sci. Technol. 12 (1975) 305.

[79] J.M. Hill, D.G. Royce, C.S. Fadley, L.F. Wagner and F.J. Grunthaner, Chem. Phys. Lett. 44 (1976) 225.

[80] P. Cadman, G. Gossedge and J.D. Scott, J. Electron Spectrosc. 13 (1978) 1.

[81] A. Ishizaka, S. Iwata and Y. Kamigaki, Surf. Sci. 84 (1979) 355.

[82] M.F. Ebel and W. Lieble, J. Electron Spectrosc. 16 (1979) 463.

[83] T. Hattori and T. Suzuki, Appl. Phys. Lett. 43 (1983) 470.

[84] M.F. Hochella and A.H. Carim, Surf. Sci. 197 (1988) L260.

[85] R.H. Ritchie, F.W. Garber, M.Y. Nakai and R.D. Birkhoff, in: Advances in Radiation Biology, Vol. 3, Eds. L.G. Augenstein, R. Mason and M. Zelle (Academic, New York, 1969) p. 1.

[86] J.J. Quinn, Phys. Rev. 126 (1962) 1453;
B.I. Lundqvist, Phys. Status Solidi 32 (1969) 273;
E. Bauer, J. Vac. Sci. Technol. 7 (1970) 3;
J.C. Shelton, Surf. Sci. 44 (1974) 305.

[87] J.C. Ashley, C.J. Tung and R.H. Ritchie, Surf. Sci. 81 (1979) 409;
C.J. Tung, J.C. Ashley and R.H. Ritchie, Surf. Sci. 81 (1979) 427;
J.C. Ashley, J. Electron Spectrosc. 28 (1982) 177; 46 (1988) 199.

[88] J. Szajman and R.C.G. Leckey, J. Electron Spectrosc. 23 (1981) 83.

[89] D.R. Penn, Phys. Rev. B 35 (1987) 482.

[90] Z.-J. Ding and R. Shimizu, Surf. Sci. 222 (1989) 313.

[91] S. Tanuma, C.J. Powell and D.R. Penn, Surf. Interface Anal. 17 (1991) 911; 927; Acta Phys. Pol. 81 (1992) 169; Surf. Interface Anal. 20 (1993) 77.

[92] J. Lindhard and M. Scharff, K. Dan. Vidensk. Selsk. Mat.-Fys. Medd. 27, No. 15 (1953);
J. Lindhard, M. Scharff and H.E. Shiott, K. Dan. Vidensk. Selsk. Mat.-Fys. Medd. 33, No. 14 (1963);
D. Pines, Elementary Excitations in Solids (Benjamin, New York, 1963) p. 144.

[93] H. Bethe, Ann. Phys. 5 (1930) 325.

[94] S. Tanuma, C.J. Powell and D.R. Penn, Surf. Interface Anal., to be published.

[95] F. Yubero and S. Tougaard, Phys. Rev. B 46 (1992) 2486.

[96] A. Jablonski, P. Mrozek, G. Gergely, M. Menyhand and A. Sulyok, Surf. Interface Anal. 6 (1984) 291;
P. Mrozek, A. Jablonski and A. Sulyok, Surf. Interface Anal. 11 (1988) 499;
B. Lesiak, A. Jablonski and G. Gergely, Vacuum 40 (1990) 67;
W. Dolinski, S. Mroz, J. Palczynski, B. Gruzza, P. Bondot and A. Porte, Acta Phys. Pol. A 81 (1992) 193.

[97] A. Jablonski and C.J. Powell, Surf. Interface Anal. 20 (1993) 771.

[98] H.H. Madden and J.E. Houston, J. Appl. Phys. 47 (1976) 3071.

[99] M.C. Burrell and N.R. Armstrong, Appl. Surf. Sci. 17 (1983) 53;
K.W. Nebesny and N.R. Armstrong, J. Electron. Spectrosc. 37 (1986) 355.

[100] H. Yoshikawa, R. Shimizu and Z.-J. Ding, Surf. Sci. 261 (1992) 403;
H. Yoshikawa, T. Tsukamoto, R. Shimizu and V. Crist, Surf. Interface Anal. 18 (1992) 757.

[101] J.A.D. Matthew and P.R. Underhill, J. Electron Spectrosc. 14 (1978) 371;
V.M. Dwyer and J.A.D. Matthew, Surf. Sci. 193 (1988) 549.

[102] S. Tougaard, Surf. Interface Anal. 11 (1988) 453; J. Electron Spectrosc. 52 (1990) 243;
S. Tougaard and J. Kraaer, Phys. Rev. B 43 (1990) 1651;
F. Yubero and S. Tougaard, Surf. Interface Anal. 19 (1992) 269;
C. Jansson, H.S. Hansen, F. Yubero and S. Tougaard, J. Electron Spectrosc. 60 (1992) 301.

[103] L. Ley, N. Martensson and J. Azoulay, Phys. Rev. Lett. 45 (1980) 1516;
J. Halbritter, H. Eiste, H.-J. Mathes, P. Walk and H. Winter, Solid State Commun. 68 (1988) 1061.

[104] E.N. Sickafus, Phys. Rev. B 16 (1977) 1448.

[105] S. Tougaard and A. Ignatiev, Surf. Sci. 129 (1983) 355.

[106] A. Jablonski, Surf. Interface Anal. 15 (1990) 559.

[107] R.F. Reilman, A. Msezane and S.T. Manson, J. Electron Spectrosc. 8 (1976) 389.

[108] See, for example: C.J. Powell and P.E. Larson, Appl. Surf. Science 1 (1978) 186.

[109] S. Doniach and M. Sunjic, J. Phys. C 3 (1970) 285.

[110] G.K. Wertheim and S.B. DiCenzo, J. Electron Spectrosc. 37 (1985) 57;
G.K. Wertheim, J. Electron Spectrosc. 60 (1992) 237.

[111] M. Cini, Solid State Commun. 24 (1977) 681.

[112] G. Sawatzky, Phys. Rev. Lett. 39 (1977) 504.

[113] D.E. Ramaker, Crit. Rev. Solid State Mater. Sci. 17 (1992) 211.

[114] C.J. Powell, Phys. Rev. Lett. 30 (1973) 1179.

[115] P.J. Cumpson, Surf. Interface Anal. 20 (1993) 727;
P.F.A. Alkemade, K. Werner, S. Radelaar and W.G. Sloof, Appl. Surf. Sci. 70/71 (1993) 24.

[116] See, for example: A.G. Fitzgerald, P.A. Moir and B.E. Storey, J. Electron Spectrosc. 59 (1992) 127.

[117] J.E. Castle and M.A. Baker, to be published.

[118] C.R. Brundle, private communication.

[119] E. Cartier, P. Fluger, J.-J. Pireaux and M. Rei Vilar, Appl. Phys. A 44 (1987) 43;
D.L. Abraham and H. Hopster, Phys. Rev. Lett. 58 (1987) 1352;
O. Paul, M. Taborelli and M. Landolt, Surf. Sci. 211 (1989) 724;

E. Cartier, J.A. Yarmoff and S.A. Joyce, Phys. Rev. B 42 (1990) 5191;
X. Zhang, H. Hsu, F.B. Dunning and G.K. Walters, Phys. Rev. B 44 (1991) 9133;
D.P. Pappas, K.-P. Kämper, B.P. Miller, H. Hopster, D. E. Fowler, C.R. Brundle, A.C. Luntz and Z.-X. Shen, Phys. Rev. Lett. 66 (1991) 504;
M. Donath, D. Scholl, H.C. Siegmann and E. Kay, Appl. Phys. A 52 (1991) 206;
E.E. Eklund, F.R. McFeely and E. Cartier, Phys. Rev. Lett. 69 (1992) 1407.

[120] H.W. Haak, G. Sawatzky and T.D. Thomas, Phys. Rev. Lett. 41 (1978) 1825;
D.D. Sarma, C. Carbone, P. Sen, R. Cimino and W. Gudat, Phys. Rev. Lett. 63 (1989) 656;
N. Wassdahl, J.-E. Rubensson, G. Bray, P. Glans, P. Bleckert, R. Nyholm, S. Cramm, N. Mårtensson and J. Nordgren, Phys. Rev. Lett. 64 (1990) 2807.

Surface Science 299/300 (1994) 49–63
North-Holland

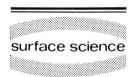

surface science

Interaction of photons with surfaces

A.M. Bradshaw

Fritz-Haber-Institut der Max-Planck-Gesellschaft, Faradayweg 4–6, 14195 Berlin (Dahlem), Germany

Received 5 March 1993; accepted for publication 23 July 1993

Many of the spectroscopic and structural techniques developed in surface science during the last thirty years rely on the absorption, reflection or scattering of photons at the first few atomic layers of a solid. The present article chronicles the introduction and development of these techniques and provides a brief, largely phenomenological description of the processes on which they are based.

1. Introduction

The various interactions of photons with clean and adsorbate-covered surfaces over a wide region of the electromagnetic spectrum not only provide us with a number of techniques aimed at probing electronic, vibrational and geometric structure, but also form the basis of surface photochemistry and surface photoprocessing. An exhaustive description of the physics associated with the absorption, reflection and scattering of photons at surfaces is, however, well beyond the scope of this short article. Furthermore, the last thirty years have shown that many of the more extensively investigated effects and processes may have produced some fascinating science, but have not necessarily furthered our understanding of surfaces per se. This article is therefore mainly concerned with some fundamental points – and related historical aspects – pertaining to those photon-based spectroscopies which have strongly influenced surface science since the "UHV-revolution" began in the 1960's.

Thirty years ago reflection spectroscopy (including ellipsometry) and transmission spectroscopy were the only commonly used optical techniques in surface science. Reflection spectroscopy in the UV and visible is not particularly surface-sensitive and was therefore mainly used under modest vacuum conditions to measure the bulk dielectric properties of solids and thin film systems [1]. Total internal reflection had been used in the infrared to investigate the vibrational spectra of thin films and, in some cases, even of adsorbed layers [2]. Thus, Becker and Gobeli had applied the method to study the adsorption of hydrogen on Si{111}, and clearly detected the Si–H stretching mode [3]. IR (external) reflection studies of adsorbates on metal surfaces were at this point not so far advanced. On the other hand, IR transmission experiments had been carried out for some years on porous materials: Eischens and Pliskin's investigations [4] of CO adsorbed on supported metal catalysts actually represent the first application of a spectroscopic technique to an adsorption problem. The high dispersion of the metal on the support, and the resulting small particle size, serve not only to maximise the effective surface area available for adsorption, but also reduce photon scattering and hence the opacity of the sample.

The electrons excited into the unoccupied states of a solid by absorption of a UV or X-ray photon may migrate towards the surface and be *photoemitted* into vacuum, if they are sufficiently energetic. An energy analysis of these electrons gives the photoemission or photoelectron spectrum. Thirty years ago it was appreciated – largely due to the work of Spicer [5] – that photoemission was essentially a bulk, or volume, effect and

SSDI 0039-6028(93)E0432-T

that it could be described in terms of a three-step model. The first step was optical excitation, the second the migration of the electron to the surface with attenuation of the electron current due to inelastic scattering and the third escape over the surface barrier into vacuum. Earlier it had been assumed that the surface played a more important role: the fundamental problem concerned momentum conservation. The momentum carried by the photon is negligible compared with the change in electron momentum during photoemission. It was therefore assumed that the source of momentum was the potential barrier at the surface, giving rise to a *surface photoeffect*. Later Fan [6] suggested that momentum can indeed be conserved in a bulk optical transition by the scattering of the photoelectron off the periodic potential of the lattice, i.e., the crystal lattice provides the required momentum in units of reciprocal lattice vector, g, via a diffraction process. In 1964 Kane [7] argued that the dispersion of energy bands $E(k)$ could perhaps be mapped directly in an angle-resolved photoemission experiment, but although the work of Gobeli et al. [8] essentially confirmed this suggestion, it was another ten years or more before "photoemission band mapping" was fully realised. What was not appreciated thirty years ago, however, is that the surface indeed contributes strongly to the measured photoemission current, but in a different way, namely due to the low mean free path for inelastic electron scattering. Thus in step (2) of the three step model the mean escape depth of the photoelectron can be as low as 5–10 Å [9,10]. This high surface sensitivity eventually became apparent in benchmark photoemission experiments on well-defined surfaces under strict ultrahigh vacuum conditions in which intrinsic surface states and adsorbate-induced levels were identified.

Before proceeding to discuss some of the more fundamental aspects of the interaction of photons with surfaces two further points should be emphasized. Firstly, many of the specifically *surface* applications of the photon-based spectroscopies developed in the last thirty years have been concerned with *adsorption*. This process affects a wide variety of phenomena that are dependent on surface properties. The most frequently encountered example in the present context is the adsorption of atoms or molecules from the gas phase on metal or semiconductor surfaces which constitutes the primary step in both corrosive oxidation and heterogeneous catalysis. Whereas the need to understand more fully the nature of clean semiconductor surfaces may well have provided the impetus for many of the more important innovations in surface science (particularly in instrumentation) [11], chemisorption studies on metals profited simultaneously. With the exception of field emission studies [12] the majority of these investigations were carried out thirty years ago on polycrystalline samples, such as films, powders or supported metals under modest vacuum conditions [13].

Secondly, fundamental studies of the interaction of radiation with surfaces and adsorbates generally require intense, wide-band tunable light sources. In the visible and in the IR out to $\lambda = 20$ μm this has not usually been a problem: lasers and/or black-body radiation have provided adequate laboratory sources for most applications. In the UV and X-ray regions some experiments can be performed with laboratory line sources, but for most applications synchrotron radiation (SR) provides the optimal solution. It is quasi-continuous, spanning the electromagnetic spectrum from the X-ray region to the IR, as well as possessing high brightness and useful polarisation properties. Tamboulion and Hartman reported the first successful use of an electron accelerator as a light source for spectroscopy in 1956 [14]. They had recorded the Be K and Al $L_{2,3}$ absorption edges in the VUV/soft X-ray region using a grazing-incidence spectrograph on the Cornell 300 MeV synchrotron. Thirty years ago SR experiments were still performed parasitically at accelerator facilities built for particle physics. In the early 1970's three low-energy electron storage rings (in Washington, Wisconsin and Orsay) were built or converted for use as dedicated SR facilities. Second generation storage rings, i.e., machines specifically designed as SR sources, went into service in the late 70's and early 80's. At the time of writing the first of the third generation sources are being commissioned. These are characterised by undulators and wigglers (periodic magnet structures inserted in the straight sections) which provide radiation of even higher brightness. Surface science experiments using SR radiation have acquired considerable importance in recent years. Indeed, they have always figured strongly in the general scientific case made for new and improved synchrotron radiation facilities.

2. Reflection and refraction

For *reflection* at a solid surface the amplitudes of the reflected and refracted waves are given by the familiar Fresnel expressions. These are derived by solving Maxwell's equations subject to the boundary conditions that the parallel component of the electric vector E and the perpendicular component of the displacement vector D ($=$ $\tilde{\epsilon}E$) are continuous across the interface. Using the Fresnel expressions the complex dielectric function $\tilde{\epsilon}(\omega)$ may be obtained from the measured reflectivity either with the help of the Kramers–Kronig relations or by making measurements at different angles of incidence. Im $\tilde{\epsilon}$ is closely related to the optical absorption coefficient and, to a first approximation, may be calculated in the visible/UV by summing over all possible direct (e.g. momentum-conserving) transitions in the electronic band structure. (Ellipsometry is somewhat more complicated but does have the advantage that two parameters are determined, thus obviating the inconvenience of measuring at more than one angle or having recourse to dispersion relations.) How can the specifically surface contribution to the reflectivity be separated out in order to determine a surface dielectric response? The simplest, but rather crude approach is to consider the substrate, surface and ambient in terms of a system of three phases separated by sharp boundaries [15]. The absorbing surface "layer" might then be composed of the outermost atomic layer(s) of the substrate or, alternatively, of an adsorbed atomic or molecular layer on the substrate. As before, the approach is to solve Maxwell's equations using the same boundary conditions and, assuming the thickness of the surface layer is much smaller

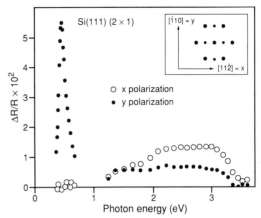

Fig. 1. $\Delta R/R$ as a function of photon energy for a single domain Si(111)(2×1) surface. The light was polarised perpendicular and parallel to the π-bonded chains. Inset: LEED pattern. After Ref. [18].

than the wavelength of the light (usually the case), to calculate the differential reflectivity (with and without surface layer) as $\Delta R/R$. Experiment shows that this works well in two situations, namely, for electronic transitions between surface state bands on semiconductors at energies below the band gap [16] and for vibrational excitations (weak absorption) in adsorbed layers [15]. In both cases, it can be easily shown that $\Delta R/R$ is simply related to the imaginary part of an $\tilde{\epsilon}$ associated with the surface layer and the spectral features thus correspond to specific elementary excitations. It is appropriate to consider two examples at this point.

Although Chiarotti, Chiarada et al. first observed the transitions across the gap formed by the surface states on Si and Ge{111}(2 × 1) using multiple total reflections in 1968 [16,17], the more recent work of this group has demonstrated that a single external reflection suffices. Fig. 1 shows differential normal incidence spectra (taken relative to an oxidised surface on which the surface states are quenched) for single domain Si{111}(2 × 1) and two orientations of the E vector [17,18]. The feature at 0.45 eV, and particularly its polarisation dependence, are evidence for the essentially one-dimensional-like nature of the surface state bands. The result confirms the π-bonded chain model of Pandey [19], as opposed

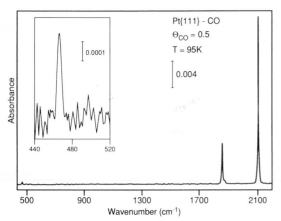

Fig. 2. Infrared-reflection absorption spectrum of CO on Pt{111} between 440 and 2150 cm⁻¹. After Ref. [23].

to the "buckling" model proposed earlier. The first IR reflection–absorption spectra [20] of adsorbed molecules were recorded by Pritchard and coworkers in the late 60's and early 70's [21]. The term absorption in this context indicates that a weakly absorbing layer on a strongly reflecting (metallic) substrate leads to a positive value of $\Delta R/R$, i.e., "positive" absorption. This may not be the case on a dielectric substrate [20]. CO adsorption on metal single crystals has been a favourite object of study in the last twenty years (partly because of the high oscillator strength associated with the C–O stretching vibration), but investigations of other molecules in a wider spectral range are now possible. The main advantage of this technique for surface vibrational spectroscopy lies in its high spectral resolution, although that of electron energy loss spectroscopy (EELS) is continuously being improved [22]. Fig. 2 shows the spectrum for CO adsorbed on a Pt{111} surface in the c(4 × 2) structure [23]. The latter consists of equal numbers of CO adsorbed in atop and bridging sites, giving rise to the two bands at 2104 and 1855 cm⁻¹ with linewidths 3.5 and 6.0 cm⁻¹, respectively. The metal–C stretch of adsorbed CO was seen for the first time in this study, but then only for the atop species (467 cm⁻¹; see also inset). From EELS it is known that the metal–C stretch of the bridging species is weaker still at 380 cm⁻¹ at a frequency where the sensitivity in the IR experiment is decreasing very

rapidly when a conventional black body source is used. Synchrotron radiation may offer some advantage here [24].

These two examples show that there are limiting cases where the simple three layer model using the Fresnel fields can give a reasonable description of the experiment. The general problem of determining a surface dielectric response remains, however. The conventional description of the dielectric function of a solid has a serious flaw: It is assumed that there is an abrupt discontinuity in $\tilde{\epsilon}$, and thus in the normal component of the electric field, E_z, at the surface which is necessarily accompanied by a sheet of induced surface charge. (Note that this problem does not exist for the E_x, E_y components, so that in the case of so-called s-polarised light the Fresnel description may be sufficient.) In fact, $\tilde{\epsilon}$ is not discontinuous at the surface, but rather, varies rapidly on a scale comparable to that of a lattice unit. This means that the usual, essentially local constitutive relation $D = \tilde{\epsilon}E$ must be replaced by a more general *non-local* expression, i.e., one which allows for the fact that the field at any one point in space is related to the field at neighbouring points in space. In the 1970's Feibelman established the basis for our understanding of surface optics by introducing a surface response function in terms of an integral equation in E_z, yielding a continuous solution for all values of z [25]. In order to explicitly calculate E_z the corresponding dielectric function ϵ_{zz} was determined within the framework of the jellium model. At and above the plasma frequency the sinusoidal plasmon field is found to dominate the behaviour of $E_z(z)$. Below the plasma frequency, however, the local field directly at the surface develops a sharp peak and Friedel oscillations become more strongly apparent. As we shall see below, the photoemission measurements of Levinson et al. [25] have confirmed the essential correctness of this picture. In principle, it would now be possible to return to the standard equations for reflectivity based on the Fresnel fields and modify them to include the surface response. Then, in a differential measurement similar to the one described above, experimental data could be compared with calculations that use a self-consistently computed $\epsilon_{zz}(\omega)$. Apart from the fact that the latter is only available for a jellium-like metal such as aluminium, the specifically surface-related contribution would probably still be obscured by the bulk signal. To the author's knowledge, there has so far been no measurement of this kind.

For reasons of space a whole range of other phenomena associated with the electromagnetic fields at surfaces has to be omitted. Most prominent are surface polaritons, non-linear effects such as second harmonic generation and, of course, surface-enhanced Raman spectroscopy (SERS) [26]. Thirty years ago it was concluded that the Raman effect was too insensitive to be useful for surface vibrational spectroscopy (except perhaps on high surface area catalysts). Yet a while ago SERS was one of the hottest topics in surface science, only to become almost totally neglected again in recent years. The largest enhancements in the Raman cross-section occur when the surface – usually silver – is roughened. Experiments on artificially roughened substrates have confirmed that plasma oscillations ("local plasmons") associated with small protrusions on the surface give rise to strong local electric field enhancements. This is often referred to as the "classical" or "electromagnetic" description. Many observations in SERS, however, are only satisfactorily explained if there is additional enhancement from some local property generally associated with the chemical bonding or the electronic states of the system involved [26c].

3. Photoemission: general remarks

The last thirty years have shown that photoemission is the most direct probe of the discrete electronic excitations which largely determine the imaginary part of the dielectric function, and also the surface dielectric response, in the visible, UV and X-ray regions. Moreover, the technique gives us to a good approximation valuable information on the ground state electronic structure. The basis of the experiment in the independent particle description is the absorption of a photon and the transfer of its total energy $h\nu$ to an electron.

The latter is excited from the state i into the state f, so that $h\nu = E_f - E_i$. If E_f is sufficiently large that the electron can surmount the work function barrier ϕ and be emitted into vacuum, its kinetic energy is given by the modified Einstein equation: $E_{kin} = h\nu - \phi - E_b$. The measurement of the photoelectron current as a function of E_{kin} (the photoemission spectrum) enables a set of photoelectron binding energies, E_b, to be determined for the various initial states in the system under investigation. In turn, the binding energies can be equated with the one electron eigenvalues of the system E_i. Since excitation takes place from an occupied electronic state to an empty one, the number of possible transitions depends on the availability of states at the energies concerned. The photoemission spectrum thus contains features which relate to the density of both the initial and final states. In practice, momentum conservation and dipole selection rules make photoemission spectra from crystalline solids easier to interpret than would seem to be the case at first sight. Further, if the momentum of the photoelectron is determined in an angle-resolved experiment, restrictions are necessarily placed on the initial and final states by defining the points at which excitation takes place in the Brillouin zone. Angle-resolved photoemission has been very successful in determining such electronic band dispersions in three-dimensional solids. If the capability also exists of determining the spin of the photoemitted electron, majority and minority spin states in ferromagnetic systems may also be distinguished. Note that the *inverse* photoemission experiment is, as the name suggests, the reverse of photoemission: Electrons incident upon a surface transiently occupy empty electronic states and then decay into other unoccupied states at lower energy via radiative transitions. Finally, at the end of this rather superficial introduction to photoemission, it should be emphasized again that this simple picture of the process is only a good approximation which has proved valuable in interpreting the photoemission spectra of solid samples in the valence region. A measured electron binding energy is in fact the difference between the total energy of the initial state and the total energy of the final, hole state. Equating E_b

with the one electron eigenvalue E_i, which is the approximation of the independent particle model, means that the final state differs only from the initial state in that one electron has been removed, while the other electrons remain in their respective configurations. This, however, neglects many-body effects, in particular the influence of electronic relaxation and the change in electron correlation accompanying photoionisation, which lead to the appearance of so-called many body effects in the spectrum, i.e., shake-up and shake-off satellites. Examples are provided by the nickel valence band satellite [27] and the intrinsic surface plasmon satellites on adsorbate core levels [28]. The reader is referred to general texts on photoemission and photoionisation (e.g. Ref. [29]).

The realisation that the photoemission process can be extremely surface-sensitive dates from the early 1970's, as noted in the Introduction. At this time emission from intrinsic as well as from adsorbate-induced *surface states* were observed for the first time. Intrinsic surface states exist in the so-called bulk band gaps of a crystal, i.e., they possess values of energy and momentum for which there are no bulk states. Surface states arise from the truncation of the crystal, are located strictly in the surface or in the intermediate sub-surface layer and decay exponentially into the bulk. If a state at the surface has the same energy, momentum and symmetry as a bulk state, hybridisation will occur. Such a *surface resonance* also has a large amplitude at the surface which decreases on going into the bulk. However, it does not die out completely, having a small oscillating amplitude like a Bloch wave. Although surface states on semiconductors had previously been detected with reflection spectroscopy [16] and photoelectron yield [30], their observation with spectroscopic photoemission simultaneously by two groups in 1972 had a profound effect on the field [31,32]. Fig. 3 shows the angle-integrated spectrum from clean and contaminated Si{111}(2 × 1) measured by Eastman and Grobman [31]. The difference curve shows the emission from the surface state (×3) centered ∼ 0.5 eV below the top of the valence band. The position of the feature was constant in the photon energy range 10 to 20 eV.

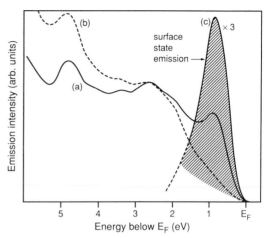

Fig. 3. Photoemission spectra of the Si{111}(2×1) surface at $h\nu = 12$ eV. (a) Clean surface, (b) after contamination from the residual vacuum. (c) Difference spectrum, ×3, down to 2 eV binding energy. After Ref. [31].

(Note that this is also one of the first examples of the use of synchrotron radiation in photoemission.) The lack of dispersion with photon energy and the quenching by adsorbed gases are now recognised criteria for identifying surface states. Surface states on metals were not discussed specifically until 1970 [33] and their first observation in photoemission appears to have been made on W{100} – also simultaneously by two groups – in 1972 [34,35].

A paper of Eastman and Cashion just a year earlier [36] was of equal significance not only for photoemission, but also for the study of chemisorption. They observed for the first time adsorbate-induced valence levels in the photoemission spectrum of CO on a nickel surface (see Fig. 4). On a surface, the molecular orbital description of chemical bonding in terms of a discrete level scheme has to be extended: levels lying in the valence band region of the substrate receive a finite width if they interact with bulk states. They are often referred to as adsorbate-induced resonances (in the sense applied above in connection with the clean surface) and can be detected with photoemission. States which are energetically lower than the valence band will be less strongly affected. Interpreting and assigning these resonances, while initially appealing, turned

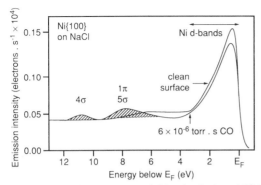

Fig. 4. Photoemission spectrum of CO adsorbed on Ni{100}. $h\nu = 40.8$ eV. After Ref. [36].

out to be more complicated due to the presence of both bonding effects and relaxation shifts. Calculations have proved useful in understanding bonding in simple systems. Thus, in a benchmark paper in 1978, Lang and Williams presented some of the first self-consistent calculations for atomic chemisorption using the jellium model [37]. Molecular chemisorption, on the other hand, has normally been studied with the help of the cluster approximation. For example, Cederbaum et al. [38] and Hermann and Bagus [39] using only a one Ni atom cluster arrived at reasonably satisfactory descriptions of the Ni–CO bond shortly after the paper by Eastman and Cashion. It is found that both the CO 5σ and 1π orbitals are significantly perturbed by the interaction with the nickel. The energy of 5σ changes most strongly, however, since charge is donated from this filled orbital into the metal. At the same time, the metal gives charge back to the antibonding C–O $2\pi^*$ orbital. This donor–acceptor model for adsorbed CO had been originally put forward by Blyholder [40] on the basis of the known bonding scheme in transition metal carbonyl compounds. In the spectrum of Fig. 4 the adsorbate-induced feature of lowest binding energy is due to the overlapping 5σ- and 1π-derived orbitals and the one at higher binding energy to the 4σ-derived orbital. Compared to the free molecule the 5σ orbital is stabilised by ~ 3 eV relative to the 1π orbital. Both orbitals are below the Ni d band; the "bonding" feature associated with the Ni d–CO $2\pi^*$ interaction has been identified in

later work as a resonance just below the Fermi level. An earlier assignment of the adsorbate-induced features in Eastman and Cashion's paper turned out to be incorrect, but this should not detract from the dramatic influence this work had at the time, particularly in the field of surface chemistry. The assignment problem in general will be taken up again below when photoemission matrix elements are discussed.

Almost at the same time it was established that core level photoelectron spectroscopy (frequently referred to as XPS = X-ray photoelectron spectroscopy or ESCA = electron spectroscopy for chemical analysis) pioneered by Siegbahn and his collaborators [41] was also sufficiently surface-sensitive to detect emission from adsorbates. The groups of Madey and Yates [42] and of Menzel [43] carried out pioneering characterisation studies on single crystal surfaces using this method. Today, it forms the basis of an important structural technique, namely, photoelectron diffraction which is discussed in the chapter by Woodruff [44]. In the scanned energy mode version of this experiment the emission intensity from an adsorbate core level is measured at a fixed angle as a function of photon energy, and thus of photoelectron energy. The directly emitted photoelectron wave interferes with that part of the wavefield which is backscattered from the substrate, leading to modulations in the photoemission current as a function of energy. Since these modulations depend on the adsorption site and the distances to neighbouring substrate atoms, surface structural information may be obtained by comparing such curves with calculated ones. Forward scattering photoelectron diffraction, which does not require the tunable photon energy provided by synchrotron radiation, has also been used to solve a number of surface structural problems, in particular, the orientation of molecules on surfaces [46] and geometric aspects of overlayer growth [46].

An interesting relationship exists between scanned energy mode photoelectron diffraction and the *photoabsorption* technique of surface EXAFS described in the chapter by Citrin [47]. In SEXAFS the photodissociation, or photoabsorption, cross-section for an adsorbate core level is given by the fraction of absorbed photon flux, but, because such a measurement is not feasible in practice, the secondary electron current resulting from the subsequent Auger decay is used. The extended fine structure above the absorption edge consists of oscillations due to interference between the emitted photoelectron wave and the waves scattered back to the emitter from the surrounding atoms. The frequency of the oscillations as a function of photon energy (wavelength) is determined by the distances to the surrounding substitute atoms; structural information can in principle be obtained directly from a subsequent Fourier analysis. SEXAFS was first applied independently by two groups in 1978 [48,49] and has since been quite successful in solving structures, particularly for high Z atomic adsorbates. In scanned energy mode photoelectron diffraction the scattered electron intensity is redistributed between the various final state manifolds (corresponding to different emission directions), as the photon energy is varied. If it were possible to integrate over the whole photoelectron current in 4π solid angle, the "diffraction" effects would be cancelled out and only the SEXAFS would remain. Using the terminology of atomic physics we measure in SEXAFS a partial cross-section and in photoelectron diffraction a differential partial cross-section. Further, we note that polarisation-dependent X-ray absorption spectra in the near edge region (as opposed to the extended region) have proved to be a useful probe of molecular orientation on surfaces. This technique, pioneered by Stöhr and coworkers [50], suffered from zealous overinterpretation in its early years – particularly in its application to larger molecules [51] – but is now used more cautiously [52].

No discussion of core level photoemission from surfaces would be complete without a brief reference to the phenomenon of the surface core level shift, i.e., the difference between the core level binding energies of atoms in the bulk and at the surface. The latter may be the clean surface or a surface covered with an adsorbed monolayer or even a thicker film. Although detected for the first time with conventional XPS photon sources [53–55], the use of synchrotron radiation has considerable advantages in such an experiment: the

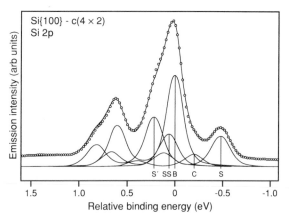

Fig. 5. Si 2p photoelectron spectrum from the Si{100}c(2×4) surface showing the decomposition into the component parts. See text for explanation of the symbols. After Ref. [57].

kinetic energy of the photoemitted electrons can be tuned to give maximum surface sensitivity. There has been a large number of publications on this topic in the last twenty years, particularly in the area of semiconductor surface physics (e.g. the pioneering experiment on GaAs by Eastman and coworkers [56]). It is therefore appropriate to show here a very recent example which demonstrates the surface atom specificity of the binding energy. Fig. 5 shows the 2p photoelectron spectrum of the c(4 × 2) reconstructed surface of Si{100} measured by Landmark et al. at $h\nu = 130$ eV as well as the decomposition into its constituent parts [57]. Five components (each showing the characteristic spin–orbit split doublet) are required for a satisfactory fit at all photon energies and emission angles. The S and SS are probably due to the up and down Si atoms, respectively, of the asymmetric dimers. S′, C and B may be assigned to second layer sub-surface and bulk atoms, respectively. Note that the combined resolution of spectrometer and monochromator was 70 meV and that the photon energy range used gives rise to photoelectron kinetic energies of 20–70 eV and thus to optimal surface sensitivity.

4. The photoemission matrix element

The differential photoionisation cross-section which describes the energy- and angle-dependent photoelectron emission characteristics, contains the square of a matrix element of the sort $\langle f \mid \boldsymbol{A} \cdot \boldsymbol{p} + \boldsymbol{p} \cdot \boldsymbol{A} \mid i \rangle$, where $\boldsymbol{p} = -i\hbar\nabla$ is the momentum operator and $\boldsymbol{A}$ the vector potential of the incident radiation [28,29]. Using the commutation relation between position and momentum this can be re-written as $\langle f \mid 2\boldsymbol{A} \cdot \boldsymbol{p} - i\hbar\nabla \cdot \boldsymbol{A} \mid i \rangle$. If the gradient of the vector potential is ignored (which is probably a good approximation except, as we have noted above, for the component A_z at the surface itself), we then have the particularly simple form of the matrix element $\langle f \mid \boldsymbol{p} \mid i \rangle \cdot \boldsymbol{A}$. The latter forms the basis for the application of dipole selection rules in photoemission, for example, in determining the allowed transition between occupied and unoccupied states in the bulk band structure. For both intrinsic and extrinsic surface states the angle-resolved photoemission matrix element acquired in the 1970's particular significance: Two groups showed that simple symmetry considerations can give valuable qualitative information on assignment and/or structure [58,59]. In the case of an adsorbed molecule, for example, $\mid i \rangle$ is a molecular orbital modified due to bonding with substrate states and $\mid f \rangle$ the photoelectron final state. The correct description of $\mid f \rangle$ must account for molecular effects (e.g. continuum resonances) as well as for the scattering of the photoelectron by the substrate atoms (photoelectron diffraction, vide supra). The calculation of the matrix element is therefore not trivial. There are, however, certain conditions which can be placed on the experiment so that $\mid \langle f \mid \boldsymbol{p} \mid i \rangle \cdot \boldsymbol{A} \mid = 0$. For an adsorbed molecule belonging to a particular point group, e.g. C_{2v}, the set of $\mid i \rangle$ will belong to the corresponding irreducible representations, e.g. A_1, A_2, B_1 and B_2 [60]. By placing symmetry constraints on $\langle f \mid$ and orienting appropriately the polarisation vector of the incident radiation, simple selection rules can be derived. For emission along the surface normal, corresponding to the C_2 symmetry axis $\mid f \rangle$ must belong to the totally symmetric representation of the point group, A_1. This follows from the requirement that the final state wave function cannot have a node in the direction of emission and must therefore be composed of $m = 0$ partial waves, i.e., s, p_z, d_{z^2} ... etc. [58,59]. Since the

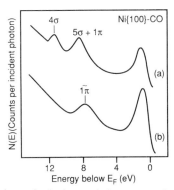

Fig. 6. Angle-resolved photoemission spectra from CO adsorbed on Ni{100}. Normal emission. (a) p-polarisation, (b) s-polarisation. After Ref. [61].

symmetry of the vector potential, and thus of the momentum operator, is also defined in the chosen coordinate frame, it is then straightforward to use direct product theory to predict which initial states are dipole-allowed in normal emission. Similar considerations apply to emission in symmetry planes. The first paper in which these effects became apparent was also concerned with the adsorption of CO on a Ni{100} surface [61]. Fig. 6 shows two normal emission photoelectron spectra for (a) p-polarised light and (b) s-polarised light. Since the C–O axis is perpendicular to the surface, the selection rules tell us that s-polarised light (containing only A_x and/or A_y) cannot excite the σ (or A_1) states. This enables the 4σ, 1π and 5σ adsorbate-induced levels to be located. These ideas were first applied more rigorously a year later [62], using as well the calculations of Davenport [63], and were then extended to a whole range of larger adsorbed molecules.

Using the commutation relation between the momentum operator and the unperturbed Hamiltonian the momentum operation $\langle f \mid \boldsymbol{p} \mid i \rangle \cdot \boldsymbol{A}$ may also be written as $\langle f \mid \nabla V \mid i \rangle \cdot 2i\boldsymbol{A}/\omega$, where ∇V is essentially the gradient of the periodically varying crystal potential but also contains a term $\delta V_{\text{surface}}/\delta z$ which gives rise to the "classical" surface photoelectric effect. In practice, this contribution to the photoemission current is small compared to the bulk emission. However, the spatial variation of A_z can be much more effec-

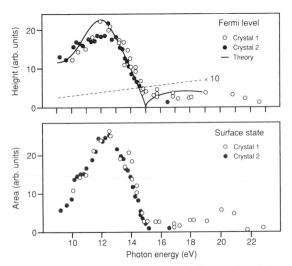

Fig. 7. Measured photoemission intensities in normal emission from Al{100} at the Fermi level, E_F, and at 2.7 eV below E_F (surface state). After Ref. [25].

tive in providing the necessary momentum for "surface" photoemission than $\delta V_{surface}/\delta z$, particularly at photon energies below the plasmon energy. That the non-local surface response function of the type introduced by Feibelman is necessary to describe the induced field was first shown in an elegant experiment on aluminium by the Plummer group [25]. Fig. 7 shows the photoemission spectra with p-polarisation and normal emission for both the surface state 2.7 eV below the Fermi energy and at the Fermi edge itself for two Al{100} crystals. For both initial states the cross section rises from threshold, peaks at $h\nu =$ 12.5 eV and then falls rapidly to a low value at the energy of the bulk plasmon. From symmetry considerations (see above) it is known that only excitations with A_z can occur in this experimental geometry. Moreover, the final state bands are such that no structure in normal emission can be attributed to direct transition from the Fermi level. The solid line is the calculated emission using the Feibelman theory; the dashed line is the calculated $\delta V_{surface}/\delta z$ term for constant $|A|$. Clearly, the "classical" surface photoelectric effect cannot account for the observed behaviour.

5. Surface band structures

As discussed briefly above, in bulk photoemission the crystal lattice acts as a source of momentum in that the photoelectron is diffracted with the addition or subtraction of a reciprocal lattice vector: $k_f = k_i \pm g$. In the angle-resolved experiment the momentum of the photoelectron can be measured from the relation $E_{kin} = h^2 k^2/2m$. For a determination of the electronic band structure in the single particle picture we are concerned with measuring the E_i (or E_b) of the system together with the corresponding initial state momentum, k_i. The photoelectron momentum is, however, changed at the surface (due to refraction at the potential barrier) so that the determination of k_i is not a straightforward matter. Fortunately, so-called specular conditions pertain, i.e., the parallel component is conserved as the electron is emitted through the surface. We thus have $k_\parallel = k_\parallel \pm g_s$, where g_s is a surface reciprocal lattice vector. Various methods are available to determine experimentally the corresponding perpendicular component of the initial state momentum, $k_\perp(i)$, in bulk band structure studies, but are outside the scope of this article. (Note, incidentally, that in this largely phenomenological description of bulk photoemission, it is immaterial whether a three-step [5] or one-step [64] process is involved.) For the case of emission from intrinsic or extrinsic surface states which may form a (pseudo-) two-dimensional band structure there is no difficulty in determining $k_\parallel(i)$, i.e., $|k_\parallel(i)| = (2mE_{kin}/h^2)^{1/2} \sin\theta$, where θ is the polar angle of emission.

This simple way of determining initial state parallel momentum was first used by Smith et al. to measure the band structure of layer compounds such as 1T-TaS$_2$ and 1T-TaSe$_2$ [65]. To the author's knowledge, however, the first measurement of surface state dispersions on a metal surface were made by Gartland and Slagsvold [66] (Cu{111}) and by Knapp and Lapeyre [67] (GaAs{110}). In keeping with the general preoccupation with silicon surfaces in this article it is perhaps, however, appropriate to show a compilation of somewhat later data for Si(111)(2 × 1) [68,69], Fig. 8 shows the dangling bond surface state dispersion along the $\bar{\Gamma}$–$\bar{J}$ and $\bar{\Gamma}$–$\bar{J}'$ directions of the surface Brillouin zone. The $\bar{\Gamma}$–$\bar{J}$ direction corresponds to [0$\bar{1}$1] in real space and thus to the direction of the π-bonded chains. The surface Brillouin zone for both the (2 × 1) and the unreconstructed, (1 × 1) surfaces is shown above. The full line is the calculated dispersion of Pandey [19]; the data points have all been moved up in energy by 0.15 eV. The hatched area corresponds to the projection of the bulk bands. As expected, the strongest dispersion is obtained along the chain direction and helps to confirm the structural model (see Section 2).

The formation of bands in adsorbed layers (CO, S, Se, Xe) was first observed by Horn, Jacobi and coworkers [70–72]. As might be expected, such effects occur only at high coverages when the orbitals on neighbouring species begin to overlap. A simple tight-binding model is usually sufficient to describe the essential features of the two-dimensional band structure [73]. A particularly instructive example is provided by ad-

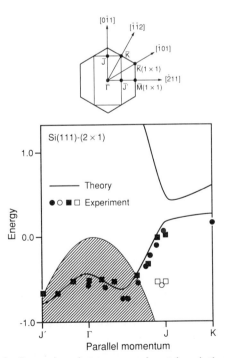

Fig. 8. Comparison between experimental and theoretical results for the surface state dispersion on Si{111}(2×1). Top: surface Brillouin zone. After Refs. [19,68,69].

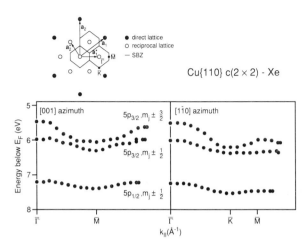

Fig. 9. Experimental Xe 5p band structure for Cu{110}c(2×2)-Xe. Top: surface Brillouin zone. After Ref. [74].

sorbed xenon [72,74]. Fig. 9 shows the experimentally determined Xe 5p band structure in the c(2 × 2) layer on Cu{110} together with the appropriate surface Brillouin zone. The $5p_{3/2}$ band is split into its $m_j = \pm 3/2$, $\pm 1/2$ components because of the lateral interactions. Note that it is possible to reach the $\bar{M}$ point either directly along $\bar{\Gamma}\bar{M}$ or along $\bar{\Gamma}\bar{K}\bar{M}$. The energy of all the bands at the $\bar{M}$ point is, however, the same (to within ~ 0.1 eV) irrespective of which azimuthal direction is chosen. If a slightly higher photon energy had been used it would also have been possible to reach $\bar{\Gamma}$ in the second Brillouin zone by continuing to measure along $\bar{\Gamma}\bar{M}$; as expected, the bands are symmetric about the $\bar{M}$ point. In recent years surface band structures have been measured for a wide variety of high-coverage adsorbate overlayers and, in some cases, compared to the results of more sophisticated calculations (e.g. Ref. [75]). Photoemission studies of adsorbate band structures provide – within the limits of the one-electron picture – a good test for such calculations, since agreement for both the energy and momentum of the electronic states is required.

6. Scattering

The 1/e penetration depth of electromagnetic radiation in solids in the UV/X-ray range is given to a first approximation by $(\sin\alpha_i')/\mu$ where μ is the absorption coefficient $(=4\pi k/\lambda)$ and $n\sin\alpha_i' = \sin\alpha_i$ (Snell's Law). The complex refractive index $\tilde{n} = n + ik$ is related to the dielectric function by $\tilde{\epsilon} = \tilde{n}^2$. This simple expression for the penetration depth assumes $n \gg k$ and may have to be modified in the UV where n and k have comparable values, an effect which serves to increase the penetration depth [76]. In the UV the penetration depth will be of the order of 100 Å or less. In the X-ray region k is much smaller and the penetration depth can be in the range 10^3–10^4 Å. This is at first sight unfortunate because X-ray scattering is after all the most important technique for determining the three-dimensional structure of bulk materials and, due to its relative simplicity (kinematic theory), would seem to offer some advantages in surface investigations. However, because $n < 1$ the refracted beam is bent away from the surface normal in the denser medium and total reflection therefore occurs at high angles of incidence. There is thus an angle α_i, termed the critical angle α_c, for which $\sin\alpha_c = n$, when $\alpha_i' = 90°$. $\alpha_i \geq \alpha_c$ is thus the condition for total reflection which is accompanied by an evanescent wave propagating along the surface and being exponentially damped into the vacuum. The effect can be used in X-ray diffraction studies in order to enhance the signal from the surface. Synchrotron radiation has proved invaluable for these studies since it provides orders of magnitude more spectral brilliance than conventional sources, thus giving substantially more flux per unit area of surface and a highly collimated beam. Both are necessary when working at very high angles of incidence, i.e., $\alpha_i > 88°$. The reader is referred to a recent review article for a more detailed account [77].

The technique of grazing-incidence X-ray scattering was introduced to surface science by Eisenberger and collaborators [78] in 1979 and has since been applied to several different structural problems, e.g. surface reconstruction of metals [79]. Only recently has the chemisorption of low Z atoms on metal surfaces become possible (Feidenhans'l et al. [80]). Fig. 10 shows the diffraction intensity along the so-called truncation rods for the (1,0) and (1,1) features together

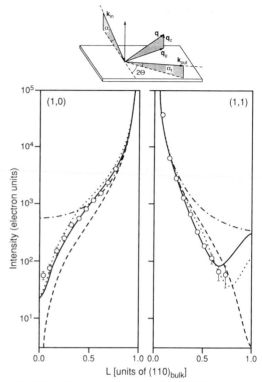

Fig. 10. Diffraction intensity along the (a) (1,0) and (b) (1,1) crystal truncation rods from the system Cu{110}(2×1)-O. Top: the experimental geometry. After Refs. [77,80].

with the results of calculations for various model structures for the Cu{110}(2 × 1)-O system. The experimental geometry is shown above. The component of momentum transfer parallel to the surface $q_{\parallel}$, corresponds to a surface reciprocal lattice vector. By measuring at the grazing incidence geometry $|q_z| \approx 0$ at various q, the integrated Bragg intensities can be determined for this plane in reciprocal space. The Patterson function, corresponding to the autocorrelation function of the electron density, is that of the surface unit cell projected into the surface plane and gives interatomic distances in a straightforward way. However, the two-dimensional periodicity associated with the surface actually gives rise to rods in reciprocal space perpendicular to the surface planes (the truncation rods). Owing to the new periodicity included by an ordered adsorbed layer some of these rods will have fractional order indices and will be well separated from the bulk Bragg peaks. By varying α_i and/or α_f, the intensity modulations along both integral and fractional order rods can be measured. This provides information on displacements perpendicular to the surface plane. Comparison of measured intensities with calculated curves for various surface models is then made. The dash-dotted curve in Fig. 10 corresponds to an ideally terminated Cu{110} surface and the dashed curve to the missing row reconstruction. The solid curve corresponds to the best fit with oxygen in the long bridge sites 0.34 ± 0.17 Å below the first Cu layer, which itself is relaxed outwards by 0.37 ± 0.05 Å. (The dotted curve corresponds to the oxygen atom in the same site but 0.34 Å above the first Cu layer.) The best fit structure gives a Cu–O nearest-neighbour distance of 1.84 ± 0.17 Å. These values are somewhat less precise than those of the corresponding LEED analyses of this system but do indicate the potential of the technique, particularly in conjunction with the next generation of synchrotron radiation sources.

Before concluding this final section we should note that the technique of X-ray standing wave analysis pioneered by Batterman [81] and first applied to adsorbates by Cowan et al. [82] has recently been performed using softer X-rays at normal incidence, thus making it interesting for the study of adsorbates on metal single crystal surfaces [83].

7. Concluding remarks

It is not possible to do justice to the large amount of material with which one is inevitably confronted when trying to write an article of this kind. As a result, certain areas have only been dealt with in passing and others have remained unmentioned (e.g. the important field of photodesorption and photofragmentation in adsorbate layers). Moreover, there is the grave danger that the judgement or, indeed, the recollection of the author becomes blurred when he attempts to establish the historical significance of a particular piece of work. So it is quite possible that in this article there are serious omissions and, at the same time, some material is included which,

viewed from somebody else's perspective, is quite superfluous. The author therefore apologises in advance to any of his colleagues who feel that their contributions to the areas covered in this necessarily superficial account have not been accorded the recognition they deserve.

References

[1] F. Bassani and M. Altarelli, in: Handbook on Synchrotron Radiation, Vol. 1a, Ed. E.E. Koch (North-Holland, Amsterdam, 1983) p. 463.
[2] N.J. Harrick, Phys. Rev. Lett. 4 (1960) 24;
J. Fahrenfort, Spectrochim. Acta 17 (1961) 698.
[3] G.E. Becker and C.W. Gobeli, J. Chem. Phys. 38 (1963) 2942.
[4] R.P. Eischens, W.A. Pliskin and S.A. Francis, J. Chem. Phys. 22 (1954) 1786.
[5] C.N. Berglund and W.E. Spicer, Phys. Rev. 136 (1964) 1030A; 136 (1964) 1044A, and references therein.
[6] H.Y. Fan, Phys. Rev. 68 (1945) 43.
[7] E.O. Kane, Phys. Rev. Lett. 12 (1964) 97.
[8] G.W. Gobeli, F.G. Asllen and E.O. Kane, Phys. Rev.Lett. 12 (1964) 94.
[9] C.B. Duke, Surf. Sci. 299/300 (1994) 24.
[10] C.J. Powell, Surf. Sci. 299/300 (1994) 34.
[11] H.C. Gatos, Surf. Sci. 299/300 (1994) 1.
[12] R. Gomer, Field Emission and Field Ionisation (Harvard University Press, Massachusetts, 1961).
[13] D.O. Hayward and B.M.W. Trapnell, Chemisorption (Butterworths, London, 1964).
[14] D.H. Tomboulian and P.L. Hartman, Phys. Rev. 102 (1954) 1423.
[15] S.A. Francis and A.H. Ellison, J. Opt. Soc. Am. 49 (1959);
R.G. Greenler, J. Chem. Phys. 44 (1966) 310;
J.D.E. McIntyre and D.E. Aspnes, Surf. Sci. 24 (1971) 417.
[16] G. Chiarotti, Surf. Sci. 299/300 (1994) 541.
[17] G. Chiarotti, G. Del Signere and S. Nannarone, Phys. Rev. Lett. 21 (1968) 1170;
G. Chiarotti, S. Nannarone, R. Pastore and P. Chiaradia, Phys. Rev. B 4 (1971) 3318.
[18] S. Nannarone, P. Chiaradia, F. Ciccacci, R. Memeo, P. Sassaroli, S. Selci and G. Chiarotti, Solid State Commun. 33 (1980) 593.
[19] K.C. Pandey, Phys. Rev. Lett. 47 (1981) 417; Physica 117 B (1983) 761.
[20] A.M. Bradshaw and E. Schweizer, in: Advances in Spectroscopy, Vol. 23, Eds. R.J.H. Clark and R.E. Hester (Wiley, Chichester, 1988) 413;
Y. Chabal, Surf. Sci. Rep. 8 (1988) 211.
[21] A.M. Bradshaw, J. Pritchard and M.L. Sims, Chem. Commun. 1519 (1968);

M.L. Sims and J. Pritchard, Trans. Faraday Soc. 66
(1970))) 427;
M.A. Chesters and J. Pritchard, Surf. Sci. 28 (1972) 460.

[22] H. Ibach, Surf. Sci. 299/300 (1994) 116;
H. Ibach and D.L. Mills, Electron Energy Loss Spectroscopy and Surface Vibrations (Academic Press, New York, 1982).

[23] D. Hoge, M. Tüshaus, E. Schweizer and A.M. Bradshaw, Chem. Phys. Lett. 151 (1988) 230.

[24] E. Schweizer, J. Nagel, W. Braun, E. Lippert and A.M. Bradshaw, Nucl. Instrum. Methods A 239 (1985);
C.J. Hirschmugl, G.P. Williams, F.M. Hoffmann and Y.J. Chabal, Phys. Rev. Lett. 65 (1990) 480.

[25] P.J. Feibelman, Progr. Surf. Sci. 12 (1982) 287, and references therein;
H. Levinson, E.W. Plummer and P.J. Feibelman, Phys. Rev. Lett. 43 (1979) 952.

[26] For a general introduction to these three topics, see: A. Zangwill, Physics at Surfaces (Cambridge University Press, Cambridge, 1988);
More detailed accounts are found in: (a) V.M. Agranovich and D.L. Mills, Surface Polaritons (North-Holland, Amsterdam, 1982);
(b) A. Liebsch and W.L. Schaich, Phys. Rev. B 40 (1989) 5401;
(c) A. Otto, in: Light Scattering in Solids IV, Eds. M. Cardona and G. Güntherodt (Springer, Berlin, 1984).

[27] C. Guillot, Y. Ballu, J. Paigne, J. Lecante, K.P. Jain, P. Thiry, R. Pinchaux and Y. Petroff, Phys. Rev. Lett. 39 (1977) 1632.

[28] A.M. Bradshaw, W. Domcke and L.S. Cederbaum, Phys. Rev. B 16 (1990) 1480.

[29] E. Plummer and W. Eberhardt, Adv. Chem. Phys. 49 (1982) 533;
S.D. Kevan, Ed., Angle-Resolved Photoemission (Elsevier, Amsterdam, 1982).

[30] F.G. Allen and G.W. Gobeli, Phys. Rev. 127 (1962) 150.

[31] D.E. Eastman and W.D. Grobman, Phys. Rev. Lett. 28 (1972) 1378.

[32] L.F. Wagner and W.E. Spicer, Phys. Rev. Lett. 28 (1972) 1381.

[33] F. Forstmann and J.B. Pendry, Phys. Rev. Lett. 24 (1970) 1419.

[34] B.J. Waclawski and E.W. Plummer, Phys. Rev. Lett. 29 (1972) 783.

[35] B. Feuerbacher and B. Fitton, Phys. Rev. Lett. 24 (1972) 786.

[36] D.E. Eastman and J.K. Cashion, Phys. Rev. Lett. 27 (1971) 1520.

[37] N.D. Lang and A.R. Williams, Phys. Rev. B 18 (1978) 616.

[38] L.S. Cederbaum, W. Domcke, W. von Niessen and W. Brenig, Z. Phys. B 21 (19 75) 381.

[39] K. Hermann and P. Bagus, Phys. Rev. B 16 (1977) 4195.

[40] G. Blyholder, J. Phys. Chem. 68 (1964) 2772.

[41] K. Siegbahn et al., ESCA – Atomic, Molecular and Solid State Structure Studied by Means of Electron Spectroscopy (Almquist and Wicksells, Uppsala, 1967).

[42] T.E. Madey, J.T. Yates and N.E. Erickson, Chem. Phys. Lett 19 (1973) 487.

[43] D. Menzel, J. Vac. Sci. Technol. 12 (1975) 313.

[44] D.P. Woodruff, Surf. Sci. 299/300 (1994) 183.

[45] H.P. Bonzel, Progr. Surf. Sci., in press.

[46] W. Egelhoff, Surf. Sci. Rep. 6 (1987) 253.

[47] P.H. Citrin, Surf. Sci. 299/300 (1994) 199.

[48] P.H. Citrin, P. Eisenberger and R.C. Hewitt, Phys. Rev. Lett. 41 (1970) 304.

[49] J. Stöhr, D. Denley and P. Perfetti, Phys. Rev. B 18 (1978) 4132.

[50] J. Stöhr, K. Baberschke, R. Jaeger, R. Treichler and S. Brennan, Phys. Rev. Lett. 47 (1981) 381.

[51] J. Somers, A.W. Robinson, Th. Lindner, D. Ricken and A.M. Bradshaw, Phys. Rev. B 40 (1989) 2053.

[52] J. Stöhr, NEXAFS Spectroscopy (Springer, Berlin, 1992).

[53] A. Barrie and A.M. Bradshaw, Phys. Lett. A 55 (1975) 306.

[54] J.C. Fuggle and D. Menzel, Surf. Sci. 53 (1975) 21.

[55] P.H. Citrin and G.K. Wertheim, Phys. Rev. Lett. 41 (1978) 1425.

[56] D.E. Eastman, T.C. Chuang, P. Heimann and F.-J. Himpsel, Phys. Rev. Lett. 45 (1980) 656.

[57] E. Landemark, C.J. Karlsson, Y.-C. Chao and R.I.G. Uhrberg, Phys. Rev. Lett. 69 (1992) 1588.

[58] J. Hermanson, Solid State Commun. 22 (1977) 9.

[59] K. Jacobi, M. Scheffler, K. Kambe and F. Forstmann, Solid State Commun. 22 (1977) 17.

[60] M. Scheffler, K. Kambe and F. Forstmann, Solid State Commun. 25 (1978) 93;
A.M. Bradshaw, Z. Phys. Chem. NF 112 (1978) 33;
K. Horn, A.M. Bradshaw and K. Jacobi, J. Vac. Sci. Technol. 15 (1978) 575;
N.V. Richardson and A.M. Bradshaw, in: Electron Spectroscopy: Theory, Techniques and Applications, Vol. 4, Eds. A. Baker and C.R. Brundle (Academic Press, London, 1981).

[61] R.J. Smith, J. Anderson and G.J. Lapeyre, Phys. Rev. Lett. 37 (1976) 1081.

[62] C.L. Allyn, T. Gustafsson and E.W. Plummer, Chem. Phys. Lett. 47 (1977) 127.

[63] J.W. Davenport, Phys. Rev. Lett. 36 (1976) 945.

[64] P.J. Feibelman and D.E. Eastman, Phys. Rev. B 10 (1974) 4932;
D.J. Spanjaard, D.W. Jepsen and P.M. Marcus, Phys. Rev. B 15 (1977) 1728.

[65] N.V. Smith, M.M. Traum and F.J. DiSalvo, Solid State Commun. 15 (1974) 211.

[66] P.O. Gartland and B.J. Slagsvold, Phys. Rev. B 12 (1975) 4047.

[67] J.A. Knapp and G.J. Lapeyre, J. Vac. Sci. Technol. 13 (1976) 757.

[68] F.J. Himpsel, P. Heimann and D.E. Eastman, Phys. Rev. B 24 (1981) 2003;
F.J. Himpsel, Appl. Phys. A 38 (1985) 205.

[69] R.I.G. Uhrberg, G.V. Hanssen, J.M. Nicholls and S.A. Flodström, Surf. Sci. 117 (1982) 1032.

[70] K. Horn, A.M. Bradshaw and K. Jacobi, Surf. Sci. 72 (1978) 719; K. Horn, A.M. Bradshaw, K. Hermann and I.P. Batra, Solid State Commun. 31 (1979) 257.

[71] K. Jacobi and C. von Muschwitz, Solid State Commun. 26 (1978) 477.

[72] K. Horn, M. Scheffler and A.M. Bradshaw, Phys. Rev. Lett. 41 (1978) 822.

[73] A.M. Bradshaw and M. Scheffler, J. Vac. Sci. Technol. 16 (1978) 447.

[74] C. Mariani, K. Horn and A.M. Bradshaw, Phys. Rev. B 25 (1982) 7798.

[75] H. Kuhlenbeck, M. Neumann and H.J. Freund, Surf. Sci. 173 (1986) 194.

[76] T. Gustafsson and E.W. Plummer, in: Photoemission and the Electronic Properties of Surfaces, Eds. B. Feuerbacher, B. Fitton and R.F. Willis (Wiley, Chichester, 1978).

[77] R. Feidenhans'l, Surf. Sci. Rep. 10 (1989) 105.

[78] W.C. Marra, P. Eisenberger and A.Y. Cho, J. Appl. Phys. 50 (1979) 6927.

[79] I.K. Robinson, Phys. Rev. Lett. 50 (1983) 1145.

[80] R. Feidenhans'l, F. Grey, R.L. Johnson, S.G.J. Mochrie, J. Bohr and M. Nielsen, Phys. Rev. B 41 (1990) 5420.

[81] B.W. Batterman, Phys. Rev. Lett. 22 (1969) 703.

[82] P.L. Cowan, J.L. Golovchenko and M.F. Robbins, Phys. Rev. Lett. 44 (1980) 1680.

[83] D.P. Woodruff, D.L. Seymour, C.F. McConville, C.E. Riley, M.D. Crapper, N.P. Prince and R.G. Jones, Phys. Rev. Lett. 58 (1987) 1450.

Surface Science 299/300 (1994) 64–76
North-Holland

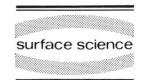

surface science

Probing surfaces with ions

E. Taglauer

Max-Planck-Insitut für Plasmaphysik, D-85748 Garching bei München, Germany

Received 16 August 1993; accepted for publication 17 September 1993

Ions which carry potential and kinetic energy to a solid surface cause interactions through electronic and collisional processes. These occur with high probabilities and can be used to probe surface properties. The various mechanisms of electronic charge exchange and stopping have been explained on the basis of quantum mechanical concepts with varying degrees of sophistication; the growing understanding of the processes involved did not result in extensive exploitation of these processes for surface characterization. Interpretation of collisional processes is based on generally simple classical concepts and these processes became most important for probing surfaces with ions. Ion induced collision cascades lead to sputtering of surface atoms, a phenomenon which became extremely important for surface cleaning, depth profiling, surface characterization by analysis of sputtered particles and surface modification, e.g. due to radiation induced segregation. Analysis of backscattered ions in the low-energy (ion scattering spectroscopy) and high-energy (Rutherford backscattering spectroscopy) regime is a unique tool for the determination of surface compositions and structures. Some significant aspects of these developments over the last thirty years are discussed.

1. Introduction

Surface science is intimately related to the interactions of ions with surfaces. For ions with kinetic energies up to several kilo electron volts, electronic charge exchange with the surface and the transfer of energy and momentum to surface atoms are very efficient processes. Therefore e.g. sputtering, i.e. the removal of surface atoms due to bombardment with energetic ions, is one of the processes which stand at the beginning of modern surface studies [1,2]. Quite naturally this effect is of importance in the first article of the first issue of Surface Science and the subject of further relevant studies in the same issue [3]. This role of ion–surface interactions, both as a powerful tool for surface preparation and analysis and as the object of fundamental studies continued over those next thirty years and maintains its importance for surface science, in the discipline as well as in the homonymous journal. Several topics of ion–surface interactions are discussed in the following sections, demonstrating the potential of ions as effective surface probes: An ion carries potential and kinetic energy to the surface and therefore charge, energy and momentum transfer can proceed through a number of processes. Electron transfer mechanims have been identified which determine the charge state of secondary particles and the energy distributions of secondary electrons. Kinetic energy transfer from impinging ions to atoms in the surface region initiates collision cascades resulting in lattice defects, relocation of atoms and in the important phenomenon of sputtering. The determination of the energy and angular distribution of backscattered ions in ion scattering experiments has become an important tool for elemental surface analysis and for investigations of surface structures.

The information obtained by probing surfaces with ions is generally based on fairly simple physical concepts. Detailed quantum mechanical treatment of charge exchange processes and energy loss have been used to describe experimental observations but the gain for understanding electronic surface properties was very limited so far. Collisional processes can be treated in the classi-

SSDI 0039-6028(93)E0529-4

cal approximation and by using purely repulsive interaction potentials. In a large number of investigations ions turned out to be very efficient probes for composition and structure of surfaces, their dynamic behaviour and surface reactions.

2. Historical remarks

The interaction of ions with surfaces was first discovered and studied in experiments with gas discharges [1]. In these early investigations (around 1910) the basic phenomena which later became the subject of extensive research were already observed: removal of material from the surface and redeposition on other parts of the experiment chamber (sputtering), emission of particles, charged and neutral, from a bombarded surface (secondary emission, scattering). Particularly the experiments using canal rays ("Kanalstrahlen") by Thomson, and Stark and Wendt can be considered as predecessors of modern surface studies with ions. The extraction of canal rays from a discharge (e.g. in hydrogen gas) represents the important step towards the use of a defined ion beam from an ion source. By applying static or dynamic mass spectrometers for the identification of secondary particles, the large field of secondary ion mass spectrometry was opened [4]. Gas discharges were also used in the pioneering systematic studies of sputtering by Wehner [5], in which yields were determined by measuring the weight losses of the target material. The detection of aniostropic particle emission from single crystalline targets ("Wehner spots") was decisive for the interpretation of sputtering as a collisional phenomenon rather than material evaporation from a "hot spot". All these experiments refer to fairly low kinetic energies of the ions, about 1 keV or less. Ion scattering experiments in the sense of single particle collisions, on the other hand, originated from nuclear physics, where ion energies in the MeV regime were used to analyse bulk target compositions [6]. Experiments with ions of several ten keV in energy then showed that atomic masses on surfaces could be identified from the energy spectra of secondary (scattered) ions on the basis of two-body collision

physics [7]. Surface analysis with ion beams then developed in two different energy regimes, both essentially based on the binary collision scheme between ions and target atoms and exploiting additional effects to obtain surface sensitivity. Using MeV ions in Rutherford backscattering, surface analysis is achieved by making use of channeling and blocking effects in single crystalline targets such that in the chosen scattering geometry only near-surface atoms can contribute to the scattered ion intensity. Exclusive surface sensitivity for low-energy (keV) ions was observed by Smith in 1967 [8] for noble gas ions, for which neutralisation at the surface is extremely important. Ions which penetrate into the target are almost completely neutralised and thus the scattered ion signal arises exclusively from the topmost atomic layer.

The experimental set-up in all cases consists of an ion beam, a manipulator which allows target positioning with several degrees of freedom to define impact and scattering angles and an ion energy analyser and detector. The large investments for MeV accelerators set of course limitations to the distribution of high-energy experiments, whereas low-energy ion sources are relatively easily incorporated into standard ultra-high vacuum chambers.

According to these developments the main efforts in the past were then directed towards the understanding of charge exchange and neutralisation processes, the collisional processes responsible for sputtering and ion scattering mechanisms. Important aspects of the development of these effects and their applications are outlined in the following sections.

3. Charge exchange processes

If an ion approaches a surface closely enough that the wave functions overlap, electronic transition processes between ion and surface occur which lead to neutralization and deexcitation of the impinging particle. The transition probabilities depend on the distance between ion and surface and on the relative position of the energy levels of the projectile and the (bulk) band struc-

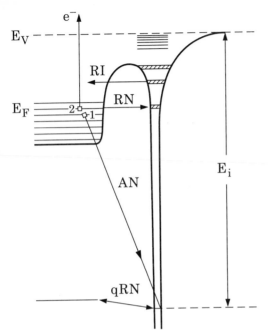

Fig. 1. Schematic energy level diagram of an ion (atom) in front of a solid, indicating the charge exchange processes. E_v, E_F, E_i: vacuum energy, Fermi energy, ionisation energy. Auger neutralization (AN) involves two electrons, resonance neutralization (RN), resonance ionisation (RI) and quasi-resonant neutralization (qRN) are discussed in the text.

ture of the solid. The situation is generally represented in energy diagrams such as shown in Fig. 1. Neutralization can proceed through resonance neutralization (RN) which involves long range delocalized valence electron states and therefore takes place at fairly large distances of the order of 5 Å. The neutralized particle can be deexcited through Auger deexcitation or reionised by resonance ionisation (RI). At closer distances around 1–2 Å Auger neutralization (AN) becomes effective. If lower levels lie energetically very close (e.g. Pb 5d and He 1s) so-called quasi-resonant charge exchange is possible and for very short distances (around 0.5 Å) reionization of previously neutralized projectiles can take place. All these processes have been investigated experimentally and described by appropriate models [9]. The first quantitative description was developed in the pioneering work of Hagstrum [10] with the aim to interpret the energy spectra of the elec-

trons emitted in the neutralization processes. The related technique was called ion neutralization spectroscopy (INS) and yields information on the density of states (DOS) of the solid surfaces. The interpretation of the data is complicated since the spectra represent convolutions of the DOS involved. It is probably for this reason (which makes it inferior to photoelectron spectroscopy) that INS has not been more widely used for probing surface electronic properties. A simpler situation is given if the incident particle is a metastable atom and can give its potential energy to a surface electron in one deexcitation step (metastable deexcitation spectroscopy, MDS [11]). It turns out, however, that for clean surfaces the deexcitation proceeds via resonance ionization and subsequent Auger neutralization. In Hagstrum's neutralization model the final state of the projectile depends on the electron transition rate R and the interaction time. R is characterized by a rate factor A (typically 10^{14} to 10^{16} s^{-1}) and the range parameter a (a^{-1} is of the order 1 to 2 Å), i.e. $R = A \exp(-as)$. This rate has to be integrated over the interaction time which results in a typical expression for the probability P to find a reflected ion in large distance from the surface

$$P = \exp\left[-\frac{A}{a}\left(\frac{1}{v_i} + \frac{1}{v_f} \right) \right], \qquad (1)$$

where v_i and v_f are the initial and final velocity components perpendicular to the surface. Expressions such as Eq. (1) have been widely used for the interpretation of data for which AN is the dominant process. In many cases it proved to be sufficient to take only the final velocity component v_f into account for a quantitative description. An example is given in Fig. 2 [12] which shows that the measured intensities of He$^+$ ions backscattered from a Ni(110) surface agree quantitatively with calculated scattering cross sections if a neutralization probability as discussed above is used. The model described is of course based on simplifying assumptions and further experimental results required various refinements. Good agreement with experimental results for situations which involve grazing incidence or exit angles could be obtained when a local process of

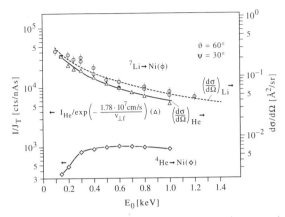

Fig. 2. Primary ion energy dependence of He$^+$ and Li$^+$ scattered from Ni(110) (left ordinate) compared with calculated differential scattering cross sections (right ordinate) including neutralization [12].

the charge exchange (at the closest impact point) and an integration over the path of the actual trajectory were taken into account [13–17].

If an ion approaches a surface the energy levels are shifted and broadened due to the interaction with the electrons of the solid (image force). This is of particular importance for RN and RI processes, since the corresponding atomic levels of the projectile are in the vicinity of the Fermi energy E_F and can be shifted from values below E_F to above the Fermi level and therefore change from RN to RI transitions. A charge equilibrium is established at the distance of closest approach and the final state only depends on the outgoing trajectory. A quantitative description requires the solution of the Schrödinger equation with a time-dependent Newns–Anderson Hamiltonian [18,19]. The trend of the dependence of the ion survival probability can be seen under simplifying assumptions. Considering free-electron-like metal states and assuming that the valence level E_i is near the Fermi level (chosen as energy zero) and temperature effects can be neglected, the probability for a final ionized state at a large distance from the surface is

$$P = 1 - (2/\pi)\, e^{-\pi E_i / \alpha v_f}, \quad \text{for } E_i > 0,$$
$$P = (2/\pi)\, e^{\pi E_i / \alpha v_f}, \quad \text{for } E_i < 0, \tag{2}$$

where α characterizes the decay of the level

broadening with increasing ion–surface distance. Eq. (2) shows that P can decrease or increase with increasing ion velocity, if E_i is above or below zero, respectively. This was also found experimentally (see Ref. [18] for a review). It is interesting to consider the role of the work function $\Phi = E_v - E_F$ in this context (E_v is the vacuum energy level, see Fig. 1). For AN or noble gas in which low-lying ground state levels are involved, the value of Φ is not very critical. But for RN (with alkali ions or noble gas ions) where P depends on the position of the ionization level relative to the Fermi level, a change of the work function can modify the ion survival probability dramatically. This has been shown for alkali ions and also for noble gas ions [18,20,21]. Although here the interaction of the ion with an extended surface is considered, it was shown that RN also has a local character and depends on the distance of closest approach between the ion and the specific scattering target atom [21].

Additional phenomena determining the ion yield occur at close encounters between projectile and target atom. Due to electron promotion reionisation of the projectile can occur, the required ionization energy being taken from its kinetic energy. These phenomena were studied experimentally [22,23] and corresponding orbital energies were calculated as a function of atomic distance of diatomic systems [24]. In these cases of close encounters the surface is essentially represented by one single target atom. This is also correct for the so-called quasi-resonant charge exchange processes. Here oscillatory ion yields are observed as a function of ion velocity [25], (see Fig. 3). These oscillations are explained by Landau–Zener type charge exchange processes which are known in atomic collision physics as Stueckelberg oscillations [26]. They arise from a quantum mechanical phase interference of the wave functions describing two energetically very close states, in analogy to the well known double slit experiment. The resulting scattered ion intensity depends on the ion velocity as was shown by comparing ^{3}He$^+$ and ^{4}He$^+$ and the similarity between scattering from an atomic Pb beam and a Pb surface has in fact also been demonstrated [27]. It has also been shown that the frequency of

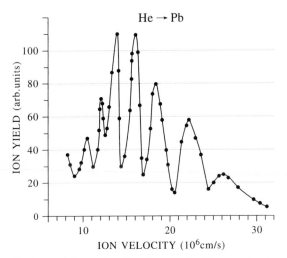

Fig. 3. Ion yield as a function of energy for He scattering from Pb. The same curve is obtained for $^4He^+$ and $^3He^+$ ions [25,27].

these oscillations depends on the chemical environment of the target atom [28]. Consequently the exploitation of this effect for obtaining chemical information for a certain mass on the surface was suggested, but it has not been extensively used.

Charge exchange effects which are based on similar processes as discussed above occur in the stepwise neutralization of highly charged ions which have become available in the recent past [29]. They are mainly used to study the multistep neutralization process itself and not for probing surface properties. This is partly true also for grazing incidence experiments in which electrons can be selectively captured into excited states of swift ions moving close to a surface. The light emitted during subsequent deexcitation shows pronounced polarization features, depending on the symmetry of the process which is governed by angular momentum effects [30]. Analysis of the spin polarization of deuterons (about 150 keV) after scattering from magnetic surfaces carries the potential of measuring the surface magnetization of ferromagnetic and antiferromagnetic material [31].

Finally, charge exchange processes were also found to be responsible for the energy loss to surface electrons which an ion suffers during

scattering from a surface (apart from the energy lost in nuclear collisions) [32]. First-principle calculations could explain the results found for low-energy He^+–Ni scattering. Yet there is not enough knowledge accumulated to use such experiments as a probe of electronic surface properties.

4. Kinetic energy transfer

In addition to charge transfer, the interaction of kinetic ions with surfaces is decisively determined by kinetic energy transfer, i.e. atomic collision processes. The explanation for many of these processes can be based on the classical collision laws that result from the conservation of energy, momentum and angular momentum for the encounter of two masses M_1 and M_2. Tremendous progress has been made over the last thirty years in detailed understanding of a variety of effects which occur at or near a surface under ion impact. This development has been accompanied by an equally remarkable increase in the applications of ion bombardment of solids. Consequently, a huge number of results were published in this time from experimental, theoretical and computer simulation studies. Some important aspects of these developments are outlined in the following and illustrated by a few characteristic examples.

4.1. Sputtering

One of the most important effects resulting from the transfer of kinetic energy to a solid surface is the release of surface atoms into the vacuum, i.e. sputtering. This phenomenon has been investigated for a long time [2] and an established theory exists [33] which was developed for the sputtering of amorphous elements. Further extensions were made with respect to monocrystals [34] and multicomponent targets [35].

The physical picture which e.g. dominates in the sputtering of metals is the development of linear collision cascades due to ion impact. Elastic nuclear collisions within the cascade volume dissipate the energy and only a small part of the atoms within this volume is in motion at any given time. If the energy transferred to a surface atom exceeds its binding energy, sputtering can occur. The description of these processes in a wide range of primary energies (from about 100 eV to MeV) considering classical scattering with purely repulsive interaction potentials gives good agreement with experimental results [36,37]. These potentials are generally screened Coulomb potentials of the form

$$V(r) = \frac{Z_1 Z_2}{r} \phi\left(\frac{r}{a}\right) \qquad (3)$$

characterized by a screening length a. The sputtering yield (atoms/ion) obtained from solving the Boltzmann transport equation for the linear cascade regime [33] depends on the energy deposited per unit depth at the surface, F and a factor Λ containing material properties:

$$Y = \Lambda F. \qquad (4)$$

F is a function of the stopping power for elastic collisions, the mass ratio M_2/M_1, the primary energy and the angle of incidence; important parameters for Λ are the surface binding energy E_B (e.g. expressed as a planar surface potential), the atomic density and the scattering cross section parameters. Y varies approximately as E_B^{-1}.

Sputtering yields can also be well reproduced by numerical Monte Carlo simulation codes for which again interatomic potentials, binding energies and electronic stopping are important input parameters [38]. Since these yields have a similar form for a large variety of ion–target combinations, namely a steep increase from a threshold energy to a broad distribution as a function of primary ion energy, empirical formulae were developed [39] which describe this behaviour quite generally, particularly at low energies (at and above the threshold energy) where the approximations used in the analytical theory [33] lose validity.

An important case in surface physics is the sputtering due to direct impact of the incoming (or reflected) projectile, the so-called single-knockon events. It is relevant for sputtering with light primary ions and has been particularly considered for the sputtering (or ion impact desorp-

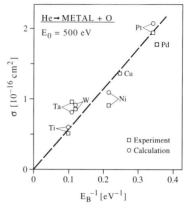

Fig. 4. Sputtering cross section as a function of the inverse metal–oxygen binding energy. Squares: experiment, circles: numerical simulation [43].

tion) of adsorbed layers [40–44]. Studying the sputtering of adlayers in the monolayer regime allows the detailed investigation of sputtering processes and also connects to the substantial part of surface science in which adsorption systems are investigated. The above-mentioned linear dependence of the sputter cross section on the inverse surface binding energy could e.g. be verified by sputtering of oxygen adsorbed on various metals as shown in Fig. 4 [43], which demonstrates agreement between the experiment, theoretical expectation and numerical simulation. Variations for cases with similar binding energies but different substrate masses are due to the differences in the contribution from reflected particles. This is just one example to show that the basic physical processes and also many detailed aspects of surface layer sputtering have been learnt and understood in the past decades.

The sputtering of surface adlayers has become one of the most widely used applications of ion bombardment in surface science, namely surface cleaning by sputtering. This effect had been first observed by Wehner [45] in glow discharges and has been introduced specifically by Farnsworth et al. [46] to study well defined surfaces. Since this first article [3], many publications in Surface Science reporting experimental work are based on sputter-cleaned surfaces. Reviews of the experimental procedures for preparing atomically clean surfaces of many elements [47] and the physical concepts for surface cleaning using sputtering [48] are given in the literature.

Surface etching by ion bombardment leads to another field of useful and widespread application of sputtering, namely sputter depth profiling. During sputter etching of a surface by ion bombardment, the surface composition is continuously monitored by one of the many surface analytical techniques. In order to obtain a concentration depth profile it is important to consider in detail to which extent the actual surface composition or the sputtered particle flux represent the initial elemental distribution [49,50]. Although this large field of applications does not directly belong to surface physics dealing with the topmost atomic layers, the understanding of the relevant physical processes has helped to remarkable achievements. E.g. an interface resolution of the order of a few nm has been demonstrated even after sputtering to a depth of 500 nm in metal layer structures [51].

Sputter etching of a surface cannot be expected to proceed through layer-by-layer removal of surface atoms, since collision cascades extend over many atomic distances. But the depth of origin for the majority of sputtered particles is only one or two atomic layers [52]. Therefore "depth profiles" in monolayer dimensions are

accessible and they have become particularly useful in the characterization of supported catalyst systems [53]. Important properties of catalysts such as efficiency, selectivity etc. are determined by the outermost atomic species on the surface. It has been demonstrated that real metal oxide catalysts as well as model systems of these so called supported catalysts can be uniquely probed by using ion beams for the analysis and at the same time for sputter etching [54,55] (see also Fig. 6 in Section 4.2). One of the results e.g. shows that the active component, i.e. Mo or W oxide, covers the support oxide (alumina or titania) under heat treatment of a physical mixture via the process of solid–solid wetting [54].

Particles emitted from a surface under ion bombardment convey information on the elemental composition and also the chemical bonding on this surface. The detection and identification of secondary ions and neutrals has developed into a large section of surface analytical techniques and is reviewed in a separate article [56].

Besides surface analysis and etching ion bombardment is important since it can produce unique surface modifications that are used to study fundamental solid state and surface properties. Ion bombardment of compound material such as alloys generally results in a change of composition within an altered layer close to the surface. This

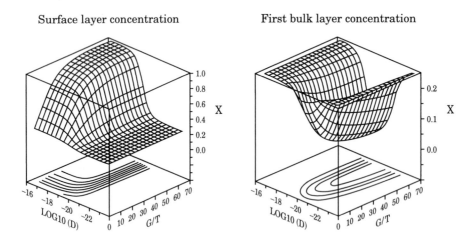

Fig. 5. Calculated surface and first bulk layer concentrations X_{Al} as a function of diffusion coefficient D and the segregation energy/temperature ratio for a Fe_3Al alloy (bulk level $X_{Al} = 0.25$ [61].

effect is caused by the difference in sputtering yields of the constituents called preferential sputtering [35,50,57]. In addition to the processes initiated by the collision cascade [58], radiation enhanced diffusion and surface segregation of one of the constituents can take place [59,60].

From such studies values for the diffusion coefficients D and the segregation energy G can be deduced. Fig. 5 shows as an example the dependence of the first and second layer composition on these quantities, for a Fe_3Al alloy, as obtained from model calculations [61] and related experi-

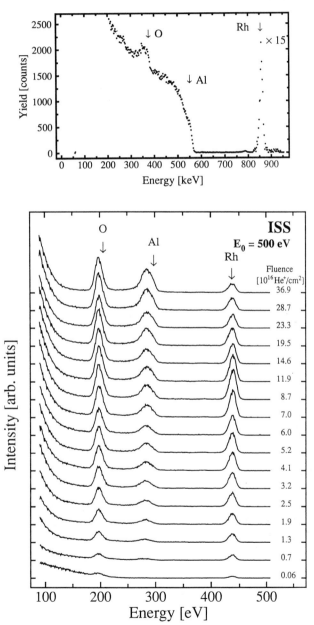

Fig. 6. Ion scattering analysis of about 1 ML Rh on Al_2O_3 as a model catalyst system. Top: RBS analysis using 1 MeV He$^+$. Bottom: ISS analysis using 500 eV He$^+$, also for sputter etching of the surface with the given fluences [55].

ments [62]. If in an experiment the steady state compositions of the first and second layer are measured, the corresponding values for D and G can be determined in a unique way.

Finally an interesting recent development in surface studies consists in the possibility to create a non-equilibrium surface state by sputtering and follow the relaxation of the system into equilibrium. E.g. in a study of a Pt(111) surface sputtered by Ar^+ ions at different temperatures and subsequent examination of the surface structure with scanning tunneling microscopy, the relative free energies of different surface steps and analogies between the equilibrium morphology of sputtered metal surfaces and crystal layer growth could be established [63]. A further example along these lines is the "smoothening" kinetics of a sputter-roughened surface by time-resolved X-ray diffraction [64]. The motivation lies in the expectation that the kinetic behaviour of such systems can be classified by universality classes in analogy to phase transitions.

4.2. Ion scattering

The analysis of the energy and angular distributions of ions backscattered from a surface represents one of the major analytical developments in surface science over the last decades. The basic concepts are simple: from the conservation laws (energy, momentum) follows a direct relation between the scattered ion energy and the target atomic mass; therefore an energy spectrum can easily be converted into a mass spectrum as long as the final ion energy is determined by one major collision. An example is given in Fig. 6 which shows for comparison RBS and ISS spectra from the same sample, a model catalyst system with a monolayer of Rh on Al_2O_3 [55]. In the high energy range of Rutherford backscattering (RBS) of typically 1 MeV additional "inelastic" energy losses to the electrons of the solid are important and therefore the scattered ion energy distribution additionally corresponds to a depth distribution of the scattering element. The depth resolution is determined by the energy resolution of the detector, for solid state detectors typical values of about 50 Å are obtained, with the

development of electrostatic analyzers a resolution of about 10 Å is achieved [65]. Therefore RBS is only a surface analytical tool if crystal effects are exploited as discussed in detail in the contribution by Feldman [66]. Some aspects are mentioned below.

Low-energy ion scattering (LEIS or ISS for ion scattering spectroscopy) on the other hand has the potential of being exclusively sensitive to the outermost atomic layer. This is due to the larger cross sections and particularly to the neutralization processes mentioned in Section 3. They cause very effective (close to 100%) neutralization of all ions which penetrate into the target below the top atomic layer and thus they are not detected with electrostatic analyzers. ISS was first introduced by Smith [67] and is discussed in detail in this volume by Rabalais [68]. For a recent review see Ref. [69]. One of the limitations with ISS lies in quantification, since the neutralization yields are not generally known well enough and also the cross sections can only be approximately calcu-

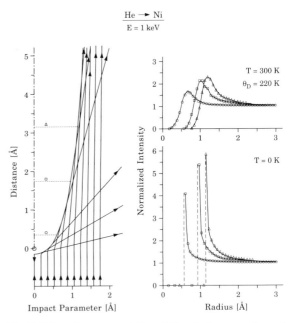

Fig. 7. Schematic representation of the shadow cone formed by ion trajectories (left) and numerically calculated intensity distributions for 1 keV Ne at different distances from the scattering center (Ni). Target temperature 0 and 300 K, Debye temperature 220 K.

lated from screened Coulomb potentials (Eq. (3)). Nevertheless substantial contributions concerning elemental distributions and surface structures could be made in the past using ion scattering techniques.

Real-space techniques such as ISS and RBS (as opposed to diffraction techniques) do not directly yield structural information since the ion beam hitting the surface is uniform. For structural information a non-uniform flux distribution is necessary and this is obtained with ion beams on an atomic scale by the effects of shadow cone formation and channeling. The radial flux distribution $f(R)$ behind a scattering atom (with R perpendicular to the initial flight direction), re-

sembles a classical rainbow-situation and for a Coulomb potential it is approximately given by

$$f(R) = 0, \quad R < R_c,$$
$$f(R) = \tfrac{1}{2}\left[(1 - R_c/R)^{-1/2} + (1 - R_c/R)^{1/2}\right],$$
$$R > R_c, \quad (5)$$

showing a flux peaking at the edge of the shadow cone. R_c is the shadow cone radius at distance d from the scatterer, $R_c = 2(Z_1 Z_2 \, e^2 \, d/E)^{1/2}$. The sharp singularity is smeared out by beam divergence and thermal vibrations of surface atoms (see Fig. 7). The peaked flux can be most effectively used for structural investigations in the so-called impact collision mode (ICISS) intro-

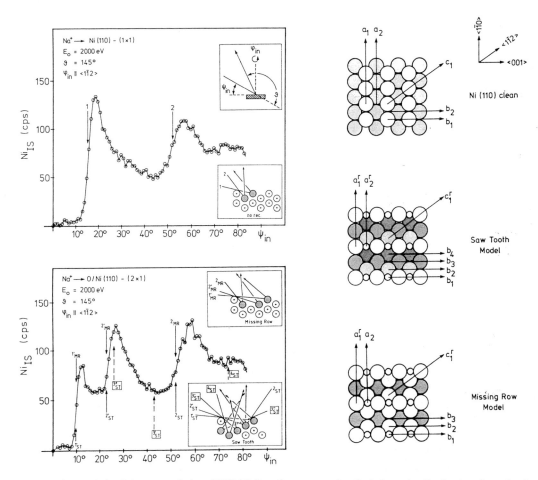

Fig. 8. Ion scattering analysis of the oxygen-induced Ni(110)-(2 × 1) reconstruction. Left: intensity distributions from the clean (top) and reconstructed (bottom) surface; right: structure models, from Ref. [71].

duced by Aono et al. [70] as discussed in Ref. [68]. An example is given in Fig. 8 [71] which nicely confirms the missing row model for the reconstructed Ni(110)-O(2 × 1) structure. The technique can also be used to locate adsorbed hydrogen which is generally hard to analyze by ion scattering unless directly recoiling particles are detected. As an example results from the measurement of the position of H on Ru(001) are shown in Fig. 9 [72]. Direct recoil spectroscopy (DRS) of H^+ ions proves that at 138 K H is adsorbed in threefold hollow sites, 1 ± 0.1 Å above the top Ru layer.

Another possibility for obtaining a non-uniform flux distribution arises from ion channeling along close-packed crystal directions [73]. The radial distribution $f(R)$ perpendicular to the channel with its center at R_0 is approximately given by

$$f(R) = \ln\left(\frac{1}{1 - (R^2/R_0^2)} \right), \qquad (6)$$

i.e. there is a strong flux peaking in the center of the channel. This can be used to advantage for the detection of adsorbates on interstitial adsorption sites as was e.g. done to determine the position of deuterium on Pd(100). In that example direct recoil detection in connection with transmission channeling of 2 MeV He^+ ions demonstrated the deuterium position in a four-fold hollow site with different vertical displacements for various adsorbate phases [74].

Ion scattering techniques are not only sensitive to thermally averaged atomic positions but can also give information about vibrational amplitudes from the width of angular distributions as indicated in Fig. 7 or from intensity distributions [75–80]. These results have been frequently expressed in values for the surface Debye temperature for specific vibrational components. Such measurements are useful in their straightforward interpretation and are not intended to compete with low-energy electron [81] or thermal atom scattering [82] by which surface phonon dispersion curves are determined.

Structure investigations using ion scattering generally are based on models which arise from

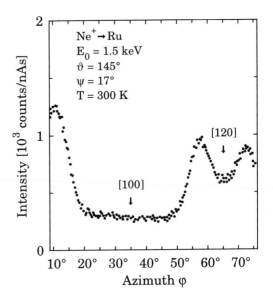

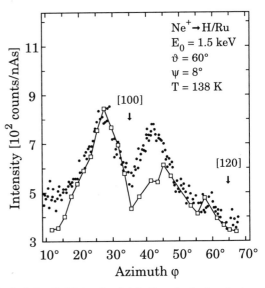

Fig. 9. Azimuthal intensity distributions for backscattering of Ne^+ from clean Ru(001) (top) and direct H^+ recoils (bottom). Experimental results are given by black dots, open squares connected by the solid line correspond to numerical simulation (MARLOWE code) with H only in fcc positions, from Ref. [72].

other techniques, mainly low-energy electron diffraction. The advantage of additional ion scattering studies is that they give a mass sensitive signal and data interpretation is straightforward, such that distinction between different structure models is generally unambiguous. Additional combination with scanning tunneling microscopy (STM) has recently been proven to be advantageous [69,83]. Ion scattering averages over a macroscopic area on the sample (beam diameter of the order of mm) and the detailed microscopic view of STM can help in interpreting ion scattering results while ion scattering contributes mass identification and quantification of atomic distances. Thus, the combination of appropriate methods again proves to be particularly successful in surface science, as has been frequently demonstrated in the past thirty years.

References

[1] W.R. Grove, Philos. Mag. 5 (1853) 203;
J.J. Thomson, Philos. Mag. 20 (1910) 752;
J. Stark and G. Wendt, Ann. Phys. (Leipzig) 38 (1912) 921, 941.

[2] R. Behrisch, Ed., Sputtering by Particle Bombardment I, Topics Appl. Phys. 47 (Springer, Berlin, 1981).

[3] J.B. Marsh and H.E. Farnsworth, Surf. Sci. 1 (1964) 3;
J.E. Boggio and H.E. Farnsworth, Surf. Sci. 1 (1964) 399;
S.P. Wolsky and E.J. Zdanuk, Surf. Sci. 1 (1964) 110.

[4] R.F.K. Herzog and F. Viehböck, Phys. Rev. 76 (1949) 855;
A. Benninghoven and E. Loebach, Rev. Sci. Instrum. 42 (1971) 49.

[5] G.K. Wehner, J. Appl. Phys. 26 (1955) 1056; Phys. Rev. 102 (1956) 690.

[6] A.B. Brown, C.W. Snyder, W.A. Fowler and C.C. Lauritsen, Phys. Rev. 82 (1951) 159.

[7] B.V. Panin, Sov. Phys. JETP 15 (1962) 215;
S. Datz and C. Snoek, Phys. Rev. 134 (1964) A347.

[8] D.P. Smith, J. Appl. Phys. 38 (1967) 340.

[9] W. Heiland and E. Taglauer, Nucl. Instrum. Methods 132 (1976) 535;
D.P. Woodruff, Nucl. Instrum. Methods 194 (1982) 639.

[10] H.D. Hagstrum, Phys. Rev. 96 (1954) 336; in: Inelastic Ion–Surface Collisions, Eds. N.H. Tolk, J.C. Tully, W. Heiland and C.W. White (Academic Press, New York, 1977) pp. 1–25.

[11] C. Boiziau, in: Inelastic Particle–Surface Collisions, Eds. E. Taglauer and W. Heiland (Springer, Berlin, 1981) pp. 48–72;
H. Conrad, G. Ertl, J. Küppers, W. Sesselmann and H. Haberland, in: Inelastic Particle-Surface Collisions, Eds. E. Taglauer and W. Heiland (Springer, Berlin, 1981) pp. 73–77.

[12] E. Taglauer, W. Englert, W. Heiland and D.P. Jackson Phys. Rev. Lett. 45 (1980) 740.

[13] D.J. Godfrey and D.P. Woodruff, Surf. Sci. 105 (1981) 438.

[14] R.J. MacDonald and D.J. O'Connor, Surf. Sci. 124 (1983) 423.

[15] G. Engelmann, E. Taglauer and D.P. Jackson, Nucl. Instrum. Methods B 13 (1986) 240.

[16] J.M. Beuken, E. Pierson and P. Bertrand, Surf. Sci. 223 (1989) 201.

[17] G. Verbist, J.T. Devreese and H.H. Brongersma, Surf. Sci. 233 (1990) 323.

[18] J. Los and J.J.C. Geerlings, Phys. Rep. 190 (1990) 133.

[19] R. Brako and D.M. Newns, Surf. Sci.108 (1981) 295; Rep. Prog. Phys. 52 (1989) 655.

[20] J.J.C. Geerlings, L.F.Tz. Kwakman and J. Los, Surf. Sci. 184 (1987) 305.

[21] M. Beckschulte and E. Taglauer, Nucl. Instrum. Methods B 78 (1993) 29.

[22] S.B. Luitjens, A.J. Algra, E.P.Th.M. Suurmeijer and A.L. Boers, Surf. Sci. 99 (1980) 652.

[23] R. Souda, M. Aono, C. Oshima, S. Otani and Y. Ishizawa, Surf. Sci. 150 (1985) L59.

[24] M. Tsukada, S. Tsuneyuki and N. Shima, Surf. Sci. 164 (1985) L811.

[25] R.L. Erickson and D.P. Smith, Phys. Rev. Lett. 34 (1975) 297.

[26] E.C.G. Stueckelberg, Helv. Phys. Acta 5 (1932) 370.

[27] A. Zartner, E. Taglauer and W. Heiland, Phys. Rev. Lett. 40 (1978) 1259.

[28] T.W. Rusch and R.L. Erickson, in: Inelastic Ion–Surface Collisions, Eds. N.H. Tolk, J.C. Tully, W. Heiland and C.W. White (Academic Press, New York, 1977) p. 73.

[29] P. Varga, Comments. At. Mol. Phys. 23 (1989) 111.

[30] H.J. Andrä, R. Fröhling and H.J. Plön, in: Inelastic Ion–Surface Collisions, Eds. N.H. Tolk, J.C. Tully, W. Heiland and C.W. White (Academic Press, New York, 1977) pp. 329–348;
J.C. Tully, N.H. Tolk, J.S. Kraus, C. Rau and R.J. Morris, in: Inelastic Particle–Surface Collisions, Eds. E. Taglauer and W. Heiland (Springer, Berlin, 1981) pp. 196–206;
H. Schröder and J. Burgdörfer, in: Inelastic Particle–Surface Collisions, Eds. E. Taglauer and W. Heiland (Springer, Berlin, 1981) pp. 207–210;
W. Graser and C. Varelas, in: Inelastic Particle–Surface Collisions, Eds. E. Taglauer and W. Heiland (Springer, Berlin, 1981) pp. 211–215;
H. Winter, in: Inelastic Particle–Surface Collisions, Eds. E. Taglauer and W. Heiland (Springer, Berlin, 1981) pp. 216–221;
H. Winter and R. Zimny, in: Coherence in Atomic Collision Physics, Eds. H.J. Beyer, K. Blum and R. Hippler (Plenum, New York, 1988) pp. 283–319.

[31] C. Rau, Appl. Surf. Sci. 13 (1982) 310.

[32] A. Närmann, W. Heiland, R. Monreal, F. Flores and P.M. Echenique, Phys. Rev. B 44 (1991) 2003.

[33] P. Sigmund, in: Sputtering by Particle Bombardment I, Ed. R. Behrisch, Topics Appl. Phys. 47 (Springer, Berlin, 1981) pp. 9–71.

[34] M.T. Robinson, in: Sputtering by Particle Bombardment I, Ed. R. Behrisch, Topics Appl. Phys. 47 (Springer, Berlin, 1981) pp. 73–144.

[35] G. Betz and G.K. (Wehner, in: Sputtering by Particle Bombardment II, Ed. R. Behrisch (Springer, Berlin 1981) pp. 11–90.

[36] H.H. Andersen and H. Bay, in: Sputtering by Particle Bombardment I, Ed. R. Behrisch, Topics Appl. Phys. 47 (Springer, Berlin, 1981) pp. 145–218.

[37] P. Sigmund, Phys. Rev. 184 (1969) 383.

[38] W. Eckstein: Computer Simulation of Ion–Solid Interactions (Springer, Berlin, 1991).

[39] J. Bohdansky, J. Roth and H.L. Bay, J. Appl. Phys. 51 (1980) 2861.

[40] H.F. Winters and P. Sigmund, J. Appl. Phys. 45, (1974) 4760.

[41] E. Taglauer, W. Heiland and J. Onsgaard, Nucl. Instrum. Methods 168 (1980) 571.

[42] H.F. Winters and E. Taglauer, Phys. Rev. B 35 (1987) 2174.

[43] E. Taglauer, W. Heiland and U. Beitat, Surf. Sci. 89 (1979) 710.

[44] Y. Yamamura and H. Kimura, Surf. Sci. 185 (1987) L475.

[45] G.K. Wehner, Phys. Rev. 108 (1957) 35.

[46] H.E. Farnsworth, R.E. Schlier, T.H. George and R.M. Burger, J. Appl. Phys. 29 (1958) 1150.

[47] R.G. Musket, W. McLean, C.A. Colmenares, D.M. Makowiecki and W.J. Siekhaus, Appl. Surf. Sci. 10 (1982) 143.

[48] E. Taglauer, Appl. Phys. A 51 (1990) 238.

[49] J.W. Coburn, J. Vac. Sci. Technol. 13 (1976) 1037.

[50] E. Taglauer, Appl. Surf. Sci. 13 (1982) 80.

[51] J. Fine and B. Navinsek, J. Vac. Sci. Technol. A 3 (1985) 1408.

[52] P. Sigmund et al. (Round Robin), Nucl. Instrum. Methods B 36 (1989) 110.

[53] H. Jeziorowski, H. Knözinger, E. Taglauer and C. Vogdt, J. Catal. 80 (1983) 286;
E. Taglauer and H. Knözinger, in: Surface Science – Principles and Applications, Eds. R.F. Howe, R.N. Lamb and K. Wandelt (Springer, Berlin, 1993) pp. 264–278.

[54] J. Leyrer, R. Margraf, E. Taglauer and H. Knözinger, Surf. Sci. 201 (1988) 603.

[55] Ch. Linsmeier, H. Knözinger and E. Taglauer, Surf. Sci. 275 (1992) 101.

[56] A. Benninghoven, Surf. Sci. 229/300 (1994) 246.

[57] H.H. Andersen, in: Ion Implantation and Beam Process- ing, Eds. S. Williams and J.M. Poate (Academic Press, Sydney, 1984) p. 127.

[58] B. Baretzky and E. Taglauer, Surf. Sci. 162 (1985) 996;
B. Baretzky, W. Möller and E. Taglauer, Vacuum 43 (1992) 1207.

[59] N.Q. Lam and H. Wiedersich, Nucl. Instrum. Methods. B 18 (1987) 471.

[60] H.H. Brongersma and T.M. Buck, Surf. Sci. 53 (1975) 649;
J. du Plessis, G.N. van Wyk and E. Taglauer, Surf. Sci. 220, (1989) 381.

[61] J. du Plessis and E. Taglauer, Nucl. Instrum. Methods B 78 (1993) 212.

[62] D. Voges, E. Taglauer, H. Dosch and J. Peisl, Surf. Sci. 269/270 (1992) 212.

[63] T. Michely and G. Comsa, Surf. Sci. 256 (1991) 217.

[64] K. Kern, I.K. Robinson and E. Vlieg, Surf. Sci. 261 (1992) 118.

[65] J.F. van der Veen, R.G. Smeenk, R.M. Tromp and F.W. Saris, Surf. Sci. 79 (1979) 212.

[66] L.C. Feldman, Surf. Sci. 299/300 (1994) 233.

[67] D.P. Smith, Surf. Sci. 25 (1971) 171.

[68] J.W. Rabalais, Surf. Sci. 299/300 (1994) 219.

[69] H. Niehus, W. Heiland and E. Taglauer, Surf. Sci. Rep. 17 (1993) 213.

[70] M. Aono, C. Oshima, S. Zaima, S. Otani and Y. Ishizawa, Jpn. J. Appl. Phys. 20 (1981) L829.

[71] H. Niehus and G. Comsa, Surf. Sci. 151 (1985) L171.

[72] J. Schulz, E. Taglauer, P. Feulner and D. Menzel, Nucl. Instrum. Methods B 64 (1992) 558.

[73] J. Lindhard, K. Dan. Vidensk. Selsk. Mat. Fys. Medd. 34 (1965) 1;
D.S. Gemmel, Rev. Mod. Phys. 46 (1974) 129.

[74] F. Besenbacher, I. Stensgard and K. Mortensen, Surf. Sci. 191 (1987) 288.

[75] E. Taglauer, Appl. Phys. A 38 (1985) 161.

[76] L.C. Feldman, Nucl. Instrum. Methods 191 (1981) 211.

[77] B. Poelsema, L.K. Verheij and A.L. Boers, Surf. Sci. 133 (1983) 344.

[78] G. Engelmann, E. Taglauer and D.P. Jackson, Surf. Sci. 162 (1985) 921.

[79] Th. Fauster, R. Schneider, H. Dürr, G. Engelmann and E. Taglauer, Surf. Sci. 189/ 190 (1987) 610.

[80] H. Dürr, R. Schneider, Th. Fauster, Phys. Rev. B 43 (1991) 12187.

[81] H. Ibach, Surf. Sci. 299/300 (1994) 116.

[82] G. Benedek and J.P. Toennies, Surf. Sci. 299/300 (1994) 587.

[83] H. Niehus, Appl. Phys. A 53 (1991) 388.

Surface Science 299/300 (1994) 77–91
North-Holland

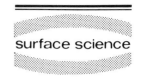

Surface scattering of thermal energy He beams: from the proof of the wave nature of atoms to a versatile and efficient surface probe

George Comsa

IGV-KFA, Forschungszentrum, D-52425 Jülich, Germany

Received 24 March 1993; accepted for publication 30 April 1993

The first conclusive molecular beam scattering experiment on surfaces has been performed in 1915 by Knudsen. This and a few other classical experiments which led to the modern period will be outlined, and the conclusions drawn from these experiments as well as their influence on the further development will be discussed in the first part. In the second part, the three main modes in which He-beams are currently used as a probe for surface investigations are commented on, with emphasis on the study of disordered surfaces.

1. Introduction

The molecular beam is known to be one of the various probes used to investigate surface properties and processes. As a dedicated supporter of this probe, I cannot accept that it is simply "one of the various probes": it is a fascinating probe. The molecular beams have features, which make up the fascination, but – as is often the case with fascinations – they are not completely free of problems. The development over decades of the molecular beam approach as a surface probe and its present capabilities outlined here will hopefully evidence both the positive features and the difficulties.

Already the procedures for the generation and the detection of the beams illustrate straightforwardly both aspects. Nothing is, in principle, simpler than to generate a well defined, narrow beam of molecules moving on straight, almost parallel trajectories toward a sample: In the simplest version one has only to keep a constant pressure in a container which has a small hole through which the molecules effuse into a vacuum; at some distance a second orifice collimates the beam to the desired shape. If the vacuum is

good enough, the molecules propagate undisturbed and no other precautions are necessary. If the pressure in the container is kept so that the Knudsen conditions are fulfilled, the distribution of the velocities and of the internal energies of the molecules is perfectly defined (Maxwell and Boltzmann, respectively) and only dependent on the container temperature. Especially for He-beams, the nature has provided us with the possibility to enhance *simultaneously* by orders of magnitude both the intensity and the monochromaticity of the beam: we have basically only to increase the pressure in the container to a couple of hundreds of bar and to give the collimator an appropriate shape. On the other hand, the weak interaction of the molecules with fields renders the manipulation of molecular beams (bending, focussing, etc.) a not very easy job. In addition, the detection of the beams of choice – the thermal He-beams – is still extremely inefficient.

A specific feature of the molecular beams is that their interaction with the surface represents a substantial part of the object to be studied. Indeed, the gas–surface dynamics (adsorption, diffusion, reaction, desorption) is one of the constituents of surface science; the molecules are at

SSDI 0039-6028(93)E0280-8

the same time probe particles *and* part of surface phenomena. In this area the molecular beams are obviously the natural probe. As a consequence, He, H_2, and H beams have been used very successfully to investigate in great detail their interaction potential with surfaces via bound state resonances. In addition, by means of a completely different procedure – the modulated beam relaxation spectroscopy – surface reactions of beams made of any kind of molecules can be in principle analyzed in unprecedented detail. On the other hand, this hybrid position of molecular beams in these types of experiments – probe and object at the same time – has led to the impression among members of the surface community that He-beams are not surface tools in the usual sense (defined as a means for supplying information on the object itself and not on the interaction probe/object). This impression lasted for a long period of time.

Another feature which is unique to molecular beams, and in this case also to scanning tunneling microscopy (STM), is the exclusive sensitivity to the outermost surface layer. This has of course consequences on the nature of the information which is obtained. Here only two remarks about consequences of this aspect on the historical development of He-scattering as a surface tool. (1) Because the systematic investigation of bulk properties has preceded that of surfaces, practically all surface methods are in one way or another a development of already existent bulk methods. Even if somewhat different, their approach could be extrapolated and thus the results gained easily credibility. This could not be the case with the molecular beam scattering and thus to reach general confidence needed time and patient effort. (2) In view also of the extreme sensitivity of He-scattering to the presence of defects and impurities on this outermost layer, He-scattering can be sensibly used only on reasonably well prepared surfaces. This has been actually the case only starting with the seventies. (The success of Stern's experiments in the thirties, discussed below, was only possible due to another gift of nature: the alkali-halide surfaces, in particular the LiF, which can be kept clean even in poor vacuum conditions.)

Writing an article for a volume like the present one is a good opportunity not only to review some new aspects of a method or of a research field – which may anyhow be found in current publications – but also to outline and comment on some classical experiments. This might be interesting (or amusing) and instructive for the reader, by learning how knowledge (and even errors) propagate in time. The experiments, which will be discussed, go far beyond "the first thirty years"; they are the fundamental preliminaries of this period. In the second part, the three approaches in which He-scattering is currently used as a tool in the usual sense will be outlined, with emphasis on the investigation of disordered surfaces.

2. History

Molecular beams have been used for the first time to investigate the gas–surface interactions in the experiments of Wood [1] and Knudsen [2], both reported in 1915. Note that this was only four years after Dunoyer [3] has reported the first realization of a molecular beam. In both experiments a Hg-beam was directed onto a glass sample, and the angular distribution of the Hg-atoms leaving the sample was measured. Likewise, in both experiments the sample was kept at a temperature high enough to cause the adsorbed Hg to desorb, while the Hg-atoms leaving the sample were collected on an appropriately cooled spherical bulb. Wood placed the sample at the center of the bulb and estimated the amount deposited onto the bulb (and thus the angular distribution) by measuring photometrically the light transmission at each point. The distribution was in agreement with a cosine one, but within a large error margin because the relation between light transmission and deposit thickness was not a priori known. In contrast, Knudsen, who started with the certainty that the distribution should be cosine, made the sample to be a part of the bulb itself. This design allows to check whether the distribution is cosine or not with high accuracy and without any additional assumption. (Because of its genial simplicity, this "zero-balanced" experiment is shown in Fig. 1.) Indeed, from simple geometry it is obvious that if and only if the

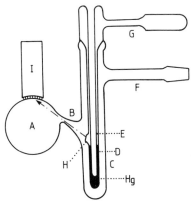

Fig. 1. The device used in the first conclusive molecular beam experiment for the investigation of the angular distribution of Hg-atoms scattered from a glass surface (from Ref. [2]). The mercury in the tube at the right was heated resistively. The Hg-vapor effusing through the hole H was collimated to a Hg-beam in B and then impinged on the upper portion of the glass sphere A. This impact region was kept by means of the heater I at about 280 K while the apparatus, and in particular A, was immersed in liquid oxygen (see text).

molecules leaving the "sample" (i.e. the warm area of the spherical bulb) have a *cosine* distribution, the deposit on the otherwise cold bulb will have a *constant* thickness.

Knudsen found eventually that the Hg-beam atoms are restituted by the sample with a cosine distribution. He was convinced – like Maxwell [4] nearly 40 years before – without any particular justification, that desorbing molecules have always a cosine distribution. Accordingly, he concluded from his experiment that *all* incident Hg-atoms were first adsorbed (i.e. sticking probability, $s = 1$) and then, due to the elevated temperature of the sample, desorbed. In this particular case the conclusion was obviously correct. However, then he generalized his conclusion by stating that the experiment invalidates Maxwell's assumption that molecules incident on a surface may be either adsorbed or reflected [4], because they are always *all* adsorbed. As we know since the late twenties from Stern's He and H_2 diffraction experiments (see, e.g., Ref. [5]), Knudsen's generalization was wrong.

Most curious has been the "knowledge propagation" which followed. It became common knowledge that desorbing molecules have always

a cosine distribution and this was named the Knudsen law. Remember that this was simply Maxwell's and Knudsen's not proved belief. Knudsen's experiment proved only that, for the system Hg/glass, $s = 1$ (this was actually Knudsen's point) or more generally that if $s = 1$, the desorption is cosine. It took nearly two decades until Clausing [6], who was aware of Stern's diffraction experiments and thus that in general $s < 1$, showed that Knudsen's law needs not to be obeyed. He demonstrated that under *equilibrium* conditions only the *sum* of the distributions of the molecules leaving the surface after having undergone various interactions (adsorption–desorption, reflection, diffraction and – as we know today – inelastic scattering) has to be cosine. Neither thermodynamics, nor another general law, impose to any of these distributions taken alone – and thus in particular to desorption – any constraint. In spite of this, the validity of Knudsen's law continued to be current knowledge of the cultivated physicist. Much later, in 1968, I have resumed Clausing's rather implicit arguments, making more explicit and provocative statements [7]. Using old He "reflectivity power" data [8] (which suggested that under equilibrium the distribution of the reflected He atoms is non-cosine) and detailed balance arguments, I concluded that desorption *has* to deviate from cosine and predicted that it should be more peaked in normal direction. The changement of the general opinion was slow. I believe that people started to accept the failure of Knudsen's law only upon the experimental evidence, which followed shortly after [9]. Now a large amount of data exists demonstrating in addition that neither the volocity distribution, nor the internal energies of desorbing molecules follow the Maxwell's and Boltzmann's "equilibrium" laws, respectively. However, the sequelae of the long standing confusion are still present in the literature: some are still calling the cosine and the non-cosine desorption distributions equilibrium and non-equilibrium desorption, respectively. This is not logical, but might be the logical fate of a beautiful but misinterpreted experiment.

A final remark: Clausing, who played a decisive positive role in this story, has emphasized in a few impressive sentences the existence of an "incomprehensible wonder machine which achieves the maintenance of the cosine law" [6]. In other words, he was astonished about the mechanism at atomic level which makes the *sum* of the angular distributions of the desorbing and of the elastically and inelastically scattered molecules to be cosine under equilibrium conditions. We have to remember in this context that for a number of systems the desorption distribution deviates dramatically from the cosine one (see, for examples, Ref. [5]), which implies that at least one of the other distributions must have a complementary distribution. To my knowledge, there is so far no theoretical attempt to explain this "wonder machine". The same holds also for the sum of the velocity distributions and for that of the internal energy distributions of the molecules leaving the surface, after having undergone the various interactions, and which – in equilibrium conditions – should follow Maxwell and Boltzmann, respectively.

The second classical molecular beam experiment, which I outline and comment on here, is "a direct measurement of the thermal molecular velocity" by Otto Stern [10]. As Stern's original schematic diagram reproduced in Fig. 2 shows and as he states in the introduction of Ref. [10], his intention was to measure *for the first time* directly the velocity (distribution) of the molecules effusing from the volume G at a temperature T. The idea of the experiment is again beautiful: the molecules, which effuse through the orifice L and are shaped to a narrow beam by the diaphragm B, form a spot on the collector plate at P; when the apparatus is rotated (see arrows) the position of the spot is displaced at P′ The lower the molecule's velocity the larger the displacement. From the position of the thickest deposit on the

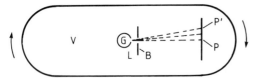

Fig. 2. Schematic drawing of the first experiment for the measurement of the velocity distribution of the molecules effusing from a Knudsen cell (from Ref. [10]); see text.

collector plate (the most probable velocity) Stern deduced the mean energy of the molecules.

There are two remarkable aspects concerning this experiment. The first is that it was again a surface experiment. Indeed, although Stern conceived initially the beam source as a Knudsen cell (G in Fig. 2), he used instead in the actual experiment a heated silver-plated Pt wire, from which the Ag-atoms were evaporating. This was thus the first measurement of the velocity distribution of atoms evaporating from a (molten [10]) surface. Stern conceded a few months later in a follow-up paper [11] that he had assumed without actual justification that the velocity distribution of Ag-atoms evaporating from molten Ag have a Maxwellian distribution. Because this was evident for him (and also probably true because of $s \approx 1$ for Ag/molten Ag), he actually intended to check his method on a "known" distribution.

The second aspect discussed here is more instructive and interesting. Stern reported in the already mentioned follow-up paper [11] that Albert Einstein has pointed out an error made by Stern in the initial paper [10]. Indeed, Stern has calculated the displacement of the maximum spot intensity by using the Maxwell velocity distribution for molecules in the *volume* of a gas in equilibrium: $v^2 \exp(-mv^2/2kT)$. This "v^2" distribution corresponds to the well known mean energy $E = \frac{3}{2}kT$ of a gas in equilibrium. Einstein pointed out that the molecules effusing through a hole from a volume in equilibrium have a higher energy, because the probability to exit the volume is proportional to v. As a consequence, Stern should have used a "v^3" distribution and correspondingly a mean energy $E = 2kT$. Einstein's statement concerning the fact that the effusing molecules have a higher mean energy than the molecules in the volume from which they originate is perfectly correct, but does not point out the exact cause of Stern's error. If Stern were not measuring the flux of the Ag-atoms (collector plate) but their number density in a small volume (e.g. ionization detector) at the same position, the use of the lower mean energy $E = \frac{3}{2}kT$ would have been correct.

This last statement needs some explanation (see for more details Refs. [5], [12] and [13]). The

"v^3" distribution and the corresponding higher mean energy $E = 2kT$ is obviously not a special property of the molecules effusing from a volume in equilibrium (i.e. of the molecules entering the beam), but a common property of the flux of molecules hitting any surface placed anywhere in a volume in equilibrium. There is thus no difference whatsoever between the velocity distributions of the molecules in a vessel in equilibrium at temperature T and in the *steady* part (see below) of the beam effusing under Knudsen conditions from this vessel: the velocity distribution (and the mean energies) of the molecules contained in an elementary volume thereof and in the flux hitting a surface are "v^2" ($\frac{3}{2}kT$) and "v^3" ($2kT$), respectively. The only aspect which has to be considered is: what is actually measured, the distribution of the molecules in a *volume* or in a *flux*.

Probably because the discussion between these two great personalities has not been conducted with accuracy up to the last consequences, the posterity concluded – as most of text books, monographies, etc. show – that beams and gases in equilibrium have to be described by different formulae, "v^3" and "v^2", respectively; the effusive beams were considered to be "warmer" than the Knudsen cell from where they originate.

Note, however, that this last sentence is correct for the beam as a *whole* but not for the *steady* part of the beam, which everybody considers when referring to non-chopped effusive beams. This apparent paradox may be the "objective" reason for the confusion. Indeed, while the beam is continuously fed with molecules leaving the orifice with a mean energy $2kT$, i.e. the *whole* beam has a mean energy $2kT$, the *steady* part of the beam has a mean energy $\frac{3}{2}kT$. Where is the energy difference? In the non-steady part. Indeed, the beam, in contrast to the equilibrium gas in the vessel, is a non-equilibrium phenomenon and thus must have a beginning: we must open the orifice at a certain time $t = 0$. Obviously, the more rapid molecules will overtake the slower ones so that the farther from the orifice we analyze the molecules the higher will be their velocity, i.e. the mean energy of the molecules increases continuously with the distance, d, from the orifice, at any given time t. Inversely, at a given distance, d, the mean energy decreases with t until it attains the "steady value" $\frac{3}{2}kT$ when the "slowest" molecule, which has left the orifice at $t = 0$, reaches the point d. It is easy to show that the "steady condition" is realized in a good approximation at any time $t > t_0$ for any distance $d < d_0$, where $d_0/t_0 \ll 10^4\sqrt{T/M}$ cm/s, with M the molecular mass, i.e., in any indoor experiment after a small fraction of a second after opening the orifice [13].

With the advent of the nozzle beams, the discussion about the equilibrium gas versus beam distributions has of course lost a part of its actuality. However, it is still important when interpreting scattering and desorption data and has in my opinion methodological and historical interest.

The two classical experiments described so far belong in fact to the prehistory of the use of molecular beams in surface research. Indeed, the time zero of this research is set to coincide with the first successful He-beam diffraction experiments by Otto Stern published in 1929 [14]. However, these experiments were by no means intended to be the first *surface* experiments but to prove "the real existence of the de Broglie waves" [15]. Indeed, after having been so sucessful with the Stern–Gerlach experiment, Stern realized the unique capabilities of the molecular beams in atomic physics research. In his first paper [15] of the famous U.z.M. (Untersuchungen zur Molekularstrahlmethode) series, after having made basic proposals for the improvement of the method (intensity increase, detection efficiency), Stern developed an eight-point program for the application of molecular beams in atomic physics. Some of these points were realized during the so-called Hamburg period which ended abruptly in 1933 by Stern's forced emigration. The most important results obtained in Hamburg within this program were the accurate measurement of the magnetic moment of the proton and the proof of the existence of the de Broglie waves for atoms and molecules. (The proof for electrons has been given by Davisson and Germer in 1927, one year after the publication of Stern's program and only two years before the proof for He [14].) Although

Stern was always very keen to emphasize exciting but also realistic new domains to which the methods he invented and the phenomena he discovered could be applied, no mention of the use of molecular beams for the investigation of surfaces (except for the study of selective adsorption which he has also discovered) can be found in his papers. This is in contrast with Thomas Johnson – who actually failed short to be the first to prove the "reality of de Broglie waves". Indeed, Johnson wrote already in 1931 [16]: "These experiments are of interest not only because of their confirmation of the predictions of quantum mechanics, but also because they introduce the possibility of applying atom diffraction to investigations of the atomic constitution of surfaces ... wave-lengths of the right magnitude, centring around 1 Å, and the complete absence of penetration of these waves will ensure that the effects observed arise entirely from the outermost atomic layer". It might be possible that Stern felt that due precisely to this extreme surface sensitivity a sound investigation of surface properties with the technical means of that time was utopic.

The experiments proving definitively the existence of the de Broglie waves of H_2 and He have been described in great detail in U.z.M. No. 15 [8] by Estermann and Stern (see also Ref. [17]). The de Broglie relation was verified within the error margins of the experiment for the two different masses used and for various molecule velocities by varying the Maxwellian beam temperature in the range 100–590 K. The grating period, giving the right wave-length, was the distance between ions of the same sign of the ionic crystals used, i.e. a first indication for the shape of the interaction potential between He (and H_2) and the surface atoms. The data contain clear evidence for Debye–Waller effects which were sensibly interpreted by the authors as an increase in surface roughness due to thermal vibrations. The discovery which – in addition to the diffraction itself – had a paramount influence on surface studies almost 40 years later, has been the observation of deep minima in the intensity of the specular beam when varying the azimuth angle – the phenomenon later called selective adsorption. Three years later Frisch and Stern (U.z.M. No. 23

[19]) reexamined this phenomenon with a monochromatized beam (see below). With his genial intuition Stern came to the conclusion (after having rejected the influence of steps or of impurities as a cause), that the sharp features should be due to the adsorption of beam molecules and thus that the corresponding adsorption probability has to depend strongly on the direction and velocity (ultimately on the momentum [20]) of the molecule. Three more years later Lennard-Jones and Devonshire [21] gave the definitive explanation. I must confess that this discovery (and the corresponding explanations) was one of the main arguments which convinced me that people in the twenties and thirties knew a great deal about phenomena we (re)discover now, and that reading their papers is very useful.

One cannot leave the Hamburg period without mentioning the ingenious and many-sided technical developments and improvements, which ultimately made the experiments so successful. They involved: source, beam collimation and finally detection. A particular emphasis deserve the two proposed and realized beam monochromatization procedures: (1) the double-crystal diffractometer and (2) the Fizeau selector. They were both described in U.z.M. No. 18 by Estermann, Frisch and Stern [22] (here also – see Ref. [17] – with the sharp remark, that Estermann did not participate at the Fizeau experiment due to a USA visit). With the advent of the monochromatic nozzle beams these two methods were used only rarely. However, it seems that they will come again in use in the frame of new projects to further improve the monochromacity of the nozzle beams.

The abrupt end of the short but amazingly productive and innovative Hamburg period represents the beginning of thirty years of stillstand in molecular beam scattering from solids. Indeed, the first paper which to my knowledge appeared after this long period has been a He and Ne scattering study on LiF [23], which basically confirmed Stern's results. A rather academic question arises: were the Hamburg period not been brutally stopped, would have been molecular beam scattering further developed to become a real surface analysis tool already in the thirties?

Probably not. Indeed, Stern and Estermann, who rebuilt a laboratory on a smaller scale at the Carnegie Institute of Technology, as well as their former coworkers as for instance J. Rabi at Columbia University, resumed essentially – often with impressive success – research on the lines of Stern's famous "Program". Remember, this program contained only one point implying (and this only implicitly) scattering with surfaces – the proof of the existence of de Broglie waves – and this point has been successfully brought to an end. An additional argument for this thesis is that people like T.H. Johnson and R.M. Zabel, who published in the same period significant contributions on molecular beam/surface scattering, but who neither had some obvious external pressure to interrupt their research nor were related to Stern's "atomic physics program", also did not continue in this direction. It seems that the obtained results were practically all what could be done with the technical means available at that time. (Estermann wrote: "...in most cases experimental techniques had to be stretched to the limit.") It appears that the monography "Molecular Rays" by Ronald G.J. Fraser [24] – another visiting scientist at Hamburg – was a kind of final report on this project, at least as far as the interactions with surfaces are concerned.

Obviously, the main technical obstacle for the further development of the molecular beam/surface research was the absence of the ultrahigh vacuum technique. Even if a clean surface could be prepared in some particular cases – e.g. by cleavage – the coverage of the surface with molecules from the gas phase was in principle a matter of seconds at the pressures which could be realized at that time. (The peculiar behavior of alkalihalide surfaces, in particular of LiF, which, when cleaned in vacuum and then heated in situ, practically do not adsorb water and other common gases at room temperature, was a lucky circumstance allowing ultimately all these beautiful experiments. However, this peculiarity has not been known at that time, so that Stern conceded that the studied surfaces might have been covered with adsorbate molecules, which should have taken a structure identical to that of LiF.) Neither the pumping speed nor the ultimate pressure

attained by the pumps available were sufficient. It seems that in sealed-off devices pressures below 10^{-8} mbar could be attained with the aid of getters. (This has been inferred indirectly from the change in time of the work function of metals.) However, sealed-off devices might have been useful for a number of experiments, but certainly not for molecular beam ones, where inherently a beam of molecules is streaming into the apparatus. It was only in 1950 that the Bayard–Alpert gauge, capable of measuring pressures below 10^{-8} mbar, was invented [25] and with it the access to UHV opened. It is this invention which allowed the improvement and the creation of new pumps as well as of the whole set of instruments (UHV-valves, feed-throughs, gauges, etc.) and procedures, which make now-a-days the generation of UHV in large and complicated machines to an almost trivial exercise. The invention of the Bayard–Alpert gauge made the birth of surface science ultimately possible.

During the sixties the needs of the aerospace development have prompted a remarkable renaissance of the molecular beam/solid surface research. The obvious goal of these activities has been the study of the friction of gases with surfaces under the conditions of space flight. Accordingly, mostly hyperthermal beam energies and technical surfaces were used. The large body of acquired information has been very useful but not particularly for the microscopic understanding of surfaces. It was only towards the end of this decade that a group involved in this type of research started to use well defined, clean monocrystalline metallic surfaces (mainly epitaxially grown on mica). One of their experiments, which probably had the most far reaching consequences, has been the measurement of the angular dependence of the adsorption *and* of the desorption of hydrogen from Ni(111) [26]. This is not only because it was the first experimental confirmation of the applicability of detailed balance in adsorption–desorption processes, but also because it demonstrated the capabilities of molecular beams to investigate adsorption–desorption processes (and surface reactions in general) by decomposing the educts and products according to their propagating direction with re-

spect to the surface. This has been followed by
Stickney and Cardillo [27] for the velocities, and
by Kubiak, Satz and Zare [28] for the internal
degrees of freedom of the molecules. (See the
article by Rendulic and Winkler in this volume,
for a detailed and up-to-date review.) If in addition the educt molecular beam is pulsed and the
shape of the resulting product-pulse is analyzed,
information on the reaction time on the surface is
obtained. This method, in which surface reactions
are investigated by supplying a pulsed and state-
selected (direction, velocity and internal energy)
molecular beam and by analyzing the pulse shape
of the likewise state selected products, has been
called modulated beam relaxation spectroscopy
(MBRS). The MBRS is probably the most exhaustive method to study surface reactions and is
of course unique to molecular beams. This has
been recognized very early [29]. In spite of this
and due especially to experimental difficulties
(analysis of the pulse shape of a manyfold differentiated molecular beam signal) the number of
experiments done so far is limited (see, for a
recent review, Ref. [30]). Very recently, Sibener
proposed and applied a method which simplifies
the procedure by the linearization of the reaction
equations (by a superposed modulation of the
educts) [31]. This may give a new momentum to
this type of research.

The other approach which is obviously also
unique to molecular beams – and which was
extremely successful in the seventies and even in
the eighties – is the investigation of the shape of
the molecule/solid surface interaction potential
via the selective adsorption discovered in 1930 [8].
As Lennard-Jones and Devonshire demonstrated
already in 1936 [21] from selective adsorption
data, the energy levels of the molecules in the
surface potential can be inferred straightforwardly. The first quantitative determination of
three energy levels of H bound on LiF has been
performed by Hoinkes, Nahe and Wilsch in 1972
[32]. They took advantage of a ZnO-detector
which was sensitive and, more important, selective for atomic hydrogen.

At the beginning of the seventies, the technique of the nozzle beams had reached sufficient
maturity to be used in surface scattering ma-

chines. As discussed further above, already Stern
felt that the very broad velocity spread of the
Maxwell distribution supplied by the Knudsen
cell was a main obstacle for quantitative studies.
He proposed *and* used two methods of
monochromatization: the diffraction on a crystal
and the Fizeau-selector. Both, however, like all
monochromatization devices for other particles
(except the laser for photons), have the essential
drawback to reduce the beam intensity by eliminating the molecules with the wrong velocity.
This is particularly unfortunate for molecular
beams generated from Knudsen-cells, where the
intensity is already severely limited by the conditions imposed by the Knudsen flow. The nozzle
beams realized the miracle to increase the intensity and to improve the monochromaticity by orders of magnitude with respect to the Knudsen
cells and this particularly with He-beams. Due to
quantum effects present essentially only with He,
high pressure nozzle beams can supply extremely
narrow distributions ($\Delta v/v < 1\%$) [33]. The use
of nozzle beams started within a few years in a
number of laboratories on both sides of the At-
lantic. This gave an unprecedented momentum to
the molecular beam/surface scattering research.
The Genova group, which built a conceptually
very modern and effective molecular beam machine (supersonic nozzle, highly sensitive bolo-
metric detection and cryogenic pumping), started
a very ambitious and successful search for bound
state resonances (selective adsorption) [34]. This
was followed by a number of other groups. As a
result, extremely accurate values of the energy
levels of He (and also of H and H_2) on alkali
halides, graphite, and rare gas coated graphite
had been obtained, and new phenomena, as for
instance band structure effects, have been uncovered (see, for a recent review, Ref. [35]). These
fascinating results have kept the general attention.

In this type of experiments (as well as in the
MBRS approach) the beam particles are at the
same time *probe* and *object* of the investigation.
This had an interesting consequence: because the
obtained information involved essentially the interaction probe/object and not properties of the
object independent of the probe nature, the sur-

face community did not start to consider He scattering as "an analytical tool in the usual sense", and the molecular-beam people continued to belong to a rather isolated club. I mean here under "analytical tool in the usual sense" a probe which supplies information primarily on the properties of the object under investigation and only occasionally on the interaction between probe and object.

3. The He-beam: an analytical surface tool (in the usual sense)

The first domain, in which the molecular beams (in particular the He-beams) gained the recognition of being a tool in the usual sense, has been the investigation of surface structures by diffraction. In spite of the fact that already in 1974 Boato et al. [34] have demonstrated the high quality of diffraction patterns obtainable by scattering He-nozzle beams from LiF [36], the recognition went on sluggishly. The main cause is probably the low lateral corrugation of the He/close-packed metal surface potential, which makes the diffraction pattern to consist mainly of a strong specular beam, the higher diffraction beams being at least 10^3 times smaller. As a consequence, during the seventies most of the diffraction studies were done on alkali-halide surfaces, which were not particularly exciting for the general surface community. After a number of unsuccessful attempts in various laboratories to see diffraction peaks – in addition to the specular one – on clean, close-packed metal surfaces, Tendulkar and Stickney [37] and Weinberg and Merill [38] succeeded in obtaining patterns with rich structure from a clean but open metal surface (W(112)) and a tungsten-carbide surface, respectively. Even after these experiments, the idea that the shallowness of corrugation of close-packed surfaces and not Debye–Waller or other effects is the main cause of the difficulties made only slowly its way. Meanwhile, however, LEED became the successful and, except for the quantitative interpretation, the straightforward method to unravel surface structures. He-diffraction had thus to find areas where its intrinsic, specific features allow to

get structure information hardly accessible to LEED or other methods.

One of these features – the scattering of the He-atoms by the electron sea outside the first atomic layer (turning point at about 3–4 Å) – had, as we have seen above, a negative influence on the development because the close-packed metal surfaces appear structureless, like an almost perfect mirror. However, this same feature appeared to be very valuable when looking at disturbances of these mirror-like surfaces, caused, for instance, by an adsorbed layer. The disturbances give a very strong effect compared to the "zero-balanced" mirror-like situation. The investigation of adsorbed layers of light atoms, in particular of hydrogen, with He-diffraction has been very rewarding. Even complicated hydrogen adlayer structures could be very effectively unraveled from the rich He-diffraction patterns obtained, while LEED approaches were difficult because of the low electron scattering cross section of H-atoms (see for a review Ref. [39] and more recent papers by Rieder's group).

The other intrinsic feature of the He-scattering which opened new analytical capabilities is the non-destructivity. The energy of the He-atoms being of only a few tens of meV and He being a rare gas, even the most unstable adsorbates and structures are not disturbed by the measuring process. This allowed, for instance, the very detailed study of various rare gas adlayers (see for a review Ref. [40]), some of them difficult to access by LEED because of the high cross section for electron-stimulated desorption of most of these adsorbates. The very high sensitivity of He-scattering to minute corrugations of the potential allowed the detailed characterization of the various incommensurate, commensurate and even high-order commensurate phases and their mutual transformation. A recent example of a very subtle phenomenon evidenced by He-diffraction is the "devil' staircase" of the Ar phases on Pt(111) [41]. Also other very sensitive layers as, for instance, the very exciting self-organizing molecular layers started to be investigated recently by He-diffraction [42].

I would not like to leave the impression that He-diffraction is competitive only in the case of sensitive or light atom overlayers. The He-scattering is highly sensitive to any deviation from the "mirror-like" behavior as, for instance, for reconstructions of the close-packed metal surfaces. Examples are the contribution of He-diffraction to the determination of the complex structures of the reconstructed Au(111) [43] and Pt(100) [44] surfaces.

There is a third feature intrinsic to He-scattering emphasized by Johnson many years ago [16]: the exclusive sensitivity to the outermost surface layer. Depending on the specific situation this might be an advantage or a disadvantage. An obvious consequence is that, for instance, the diffraction patterns are simpler, being not disturbed by scattering effects from deeper layers. For me personally this is a clear advantage because I have always a hard time to understand complicated matters.

Recently, improvements in X-ray scattering allowed to use it more and more efficiently in surface studies. With its relatively low destructivity and the very large transfer width, X-ray becomes a serious challenge for He-diffraction; likewise the recent improvements in LEED, leading to transfer widths larger than those of the best He-scattering devices. There are correspondingly recent efforts to increase the transfer width of He-machines. The most serious limitation is, at present, the nozzle beam monochromaticity ($\Delta v / v$) which, even with pressures of hundreds of bars in the nozzle, can be hardly reduced below about half a percent. It seems that we have to recourse to one of the two Stern recepies, dispersion on a crystal grating or Fizeau-selector, to monochromatize the nozzle beams; the first one seems to be more promising.

The second domain, in which He-beams gained recognition as an efficient surface tool, has been the study of surface dynamics via inelastic scattering (annihilation and creation of phonons). Like for neutrons, the energy and momentum of thermal He-atoms match well to phonons, but, in contrast to neutrons, the He-atoms are highly sensitive for the outermost surface layer. This obviously suggested that He-atoms are capable to unravel the dynamics of the surfaces, like the neutrons did for the bulk. After a few reports at

the beginning of the seventies about the observation of phonons by Williams [45] and Fisher and Bledsoe [46] on LiF, and by Subbarao and Miller [47] on Ag the progress was again slow. The breakthrough came at the beginning of the eighties when the Göttingen group, using a highly monochromatic nozzle beam, a good time-of-flight set up and efficient differential pumping, succeeded in obtaining well-resolved phonon spectra [48]. They were soon followed by other groups, so that now the dynamics of a large variety of clean and covered surfaces has been investigated. It became soon evident that besides its advantages, the He-scattering investigation of phonons had also its limitations. So for instance, energy losses higher than about 20 meV can be hardly resolved due mainly to multiphonon effects. Accordingly, a complete study of the dynamical effects has to be performed together with high high energy electron loss investigations. The many-sided aspects of the phonon inelastic He-scattering as well as the new phenomena revealed by these studies have recently been excellently reviewed by Doak [49]. (See also the paper by Benedek and Toennies in this volume [50], which illuminates in addition the theoretical aspects.)

The third domain, in which He-scattering is used as "a usual surface tool", is the investigation of *disordered* surfaces. This approach is to a large extent unique to He-scattering, in contrast to the other two ones (diffraction and inelastic scattering) which are common also to other particle beams, like electrons. The main process, which gives access to disordered surfaces, is the diffuse scattering of He-atoms at local perturbations of the surface, like impurities and defects (steps, adatoms, vacancies, etc.). The key aspect is that the individual impurities and defects have a surprisingly large cross-section for the diffuse scattering of thermal He-atoms (for most ad-molecules, adatoms, and monovacancies the cross-section is of the order of 100 $\mathring{A}^2$ and for step-lines ~ 12 $\mathring{A}$ broad), much larger than that of all other probe particles. This high sensitivity of He-scattering for the presence of individual perturbations of the ordered surface has been noticed very early. However, as long as the surface preparation procedures and the vacuum conditions were not good enough, this high sensitivity was considered as a drawback rather than an advantage, because, e.g., the diffraction features were strongly attenuated. Smith and Merrill were the first to recognize already in 1970 [51] the superiority of He-scattering in characterizing the perfection of surfaces and even in monitoring the adsorbate coverages in the range of low coverages. The question arises, why it took again more than one decade until He-scattering started to be used effectively to characterize the degree of surface perfection, to monitor coverages and many other surface properties and processes accessible through this approach (see, e.g., Ref. [52]). The probable reason is that Smith and Merrill's explanation for the unexpectedly large cross-section for He-diffuse scattering ($\Sigma = 96$ Å) from ethylene adsorbed on Pt(111), which they inferred from the attenuation of the specular He-intensity with ethylene exposure, was incorrect. Indeed, they tried to explain the value of $\Sigma = 96$ Å by assuming that it results from the area covered by the adsorbate and by the size of the He-atom, estimated from B.E.T. and Lennard-Jones sizes, respectively. In order to get the experimental value they had to assume that ethylene dissociates into an acetylenic species and two hydrogen atoms, which is now known to be not true [53]. The fact that this way to explain the high sensitivity of He-scattering for the presence of adsorbates is not appropriate became obvious after measurements have shown that non-dissociated CO molecules and even individual atoms like Xe, exhibit cross-sections for He diffuse scattering larger than 100 Å, i.e. almost one order of magnitude larger than their "geometrical" (e.g., van der Waals) size.

The experiment, which led to the correct explanation of the origin of the size of the cross-section for He (and H_2) diffuse scattering at adsorbates, has been the measurement of the He (and H_2) velocity dependence of the cross-section of CO adsorbed on Pt(111) [54]. The shape of the curves appeared to be very similar to that of their gas-phase counterparts (i.e., cross-section for He (and D_2) diffuse scattering at gas-phase CO [55]), while the absolute values were about 30% larger. Accordingly, Poelsema et al. concluded that the

cross-section for diffuse scattering from adsorbates has the same origin as that from gas-phase molecules: it is determined by small angle deflections of the beam particles caused mainly by the long range (attractive) dispersion forces. These forces being always present, the effect is completely general. Following this line of thought Jónsson et al. [56] were able to fit quantitatively the experimental data, making the reasonable assumption that the polarizability of CO (and also its electron density) is enhanced when adsorbed on the Pt(111)-surface. Note, that not only adatoms or ad-molecules exhibit very large cross-sections for He diffuse scattering but also isolated defects, like monovacancies or steps, and that they are also caused by the long range dispersion forces. This has been demonstrated both experimentally (see for various examples Ref. [52]) and theoretically [57,58].

Because of the large size of the cross-sections of the individual scatterers, the cross-sections start to overlap already at very low coverages even in the least favorable case of mutually repelling scatterers. As a consequence the effective contribution of each scatterer is obviously reduced with increasing coverage. In order to rationalize this effective contribution, it has been assumed that the overlap is geometrical and that the polarizability of the individual scatterers is not affected by the presence of the others [59]. By means of this so-called "overlap approach" [52], unequivocal relations between the attenuation of the coherent scattered He-beams and ad-molecule (or defect) coverage were deduced. In spite of the fact that this assumption is the simplest one, it has proven to be correct in the limit of experimental errors for a large number of adsorbates and defects; theoretical studies seem also to confirm that this is a satisfactory approximation (see, e.g., Ref. [58]). These relations turn out to be very useful for monitoring the coverage by the simple measurement of the height of the specular He-beam. The high sensitivity (coverages below 1%), due to the size of the cross-section, and the complete lack of influence on the processes going on, have been efficiently used, for instance, for the monitoring of the coverage in various adsorption- and desorption-kinetics experiments (see, for examples, Ref. [52]) and during modulated beam relaxation spectroscopy measurements (see, e.g., Refs. [31] and [60]).

The large cross-section for He diffuse scattering of ad-molecules and defects is not only useful for monitoring their coverages with high sensitivity, but, because the cross-section is so much larger than the van de Waals size of the same objects, new, unexpected and exciting investigation areas could be opened. This is based on the fact that by means of He-scattering, information on the lateral distribution of adsorbates, of defects, and of adsorbates and defects can be straightforwardly obtained. Indeed, let us consider for instance a surface covered with a given amount of adsorbate. The attenuation of the specular He-beam will depend on the effective cross-section for diffuse scattering which in turn depends on the amount of overlap: for a given coverage the overlap will be maximized (effective cross-section minimized) if the adsorbates, due to their mutual attraction, will condense in 2D-islands, where their mutual distance is given by the van de Waals radius; in contrast, if the adsorbates are mutually repelling, the overlap will be minimized (effective cross-section maximized). Based on such type arguments, relations between the specular He-peak height and coverage for various types of interaction between the scattering objects have been derived and checked experimentally (see, for details, Ref. [52]). As an example, the 2D-condensation of a 2D-lattice gas could be monitored in great detail even at very low coverages; this led, for instance, to the first direct determination of the latent heat of vaporization in two dimensions [61]. The cross-section overlap can be used also to determine whether the molecules are adsorbed on defect-free terraces (no overlap at low coverages → large effective cross-section) or whether they are bound at defects, like steps (substantial overlap between the cross-sections of the molecule and that of the defect → strong reduction of the effective cross-section of both partners). On this basis the activation energy for migration of ad-molecules on perfect terraces (on their way to the steps) can be straightforwardly determined at very low coverages. These and other uses of He-scattering are described in detail in Ref. [52].

The disorder of clean, nominally low indexed metallic surfaces is due to the presence of steps, which – even if mainly oriented along low-indexed directions – are distributed more or less at random along the surface. They are usually caused by the mosaic structure, always present into some degree even on the best available monocrystals, by inherent imperfections of the polishing procedure, by incomplete annealing of the damage created during sputtering–cleaning procedures, etc. A similar type of disorder is characteristic for surfaces during growth or sputtering at temperatures at which adatoms and monovacancies are mobile. The attenuation of the specular He-beam due to the diffuse scattering at steps is again, under *in-phase* scattering conditions (see below), a good measure for the step density. Moreover, by using also the diffraction capabilities of the He-beams, additional, very interesting – even if partially redundant – information is obtained. ("Redundance" is often full of exciting surprises.) It is obvious that, due to the lateral disorder, i.e. no periodicities along these surfaces, practically no diffraction features have to be expected. (Remember, the close-packed metal terraces are seen as almost perfect mirrors by the He-beam.) However, because on clean metal surfaces the steps are of monatomic height, one obtains strong interference effects between He-waves scattered from, say, adjacent terraces. (There is a well-defined periodicity in the direction *normal* to the terraces.) The He-waves from the terrace mirrors interfer constructively or destructively depending on the scattering conditions; *in phase* or *anti phase*, respectively. While the shape of the specular peak from such a disordered surface under in-phase conditions is practically identical to that from a perfectly ordered surface, the anti-phase specular peak is broadened: the smaller the average terrace width (higher step density) with respect to the transfer width of the instrument, the broader the anti-phase peak. In fact, from the detailed analysis of the shape of the anti-phase peak not only the average terrace width but even the distribution of these widths can be infered [62]. Because the peak broadening results in a reduction of the peak-height, the latter is a rough but straightforward measure for the average terrace width. The characterization of this type of disordered surfaces by measuring the specular He-peak height has been done so far in two ways: the specular peak height has been monitored as a function of incidence angle (so-called rocking curves) or alternatively as a function of time during vapor-deposition growth (or during sputtering), while the scattering geometry is maintained constant (see, for details, Ref. [52]). During the monitoring of rocking curves the scattering varies continuously between in- and anti-phase conditions, i.e. the peak-height as a function of *incidence angle* exhibits maxima and minima, respectively. The relative amplitude of the oscillations is a measure of the ratio between transfer width and average terrace width. Based on this sensitive characterization, the crystal surface preparation procedures could be improved so that crystal surfaces with average terrace widths larger than 1000 Å became currently available [63]. On the other hand, by monitoring the *time* evolution of the peak height during growth (or sputtering-removal), the growth (or removal) modes can be determined straightforwardly: an oscillatory behavior indicates layer-by-layer growth (or removal) while a monotonic decrease of the peak-height 3D-growth (or pit formation, respectively). This is similar to the monitoring of these phenomena by RHEED or MEED; however, due to the simpler nature of the He-scattering, much easier to interpret correctly. The use of the He-scattering for the investigation of growth and sputtering-removal has supplied so far a large variety of interesting information; in particular, it led to the discovery of the reentrant layer-by-layer growth of Pt on Pt(111), at low [64], and of the layer-by-layer removal at high temperatures, as well as to the development of a very precise – because interferometric – method to determine sputtering yields on almost perfect surfaces [65].

The STM is nowadays of course the ideal instrument to study disorder in microscopic detail: one is able to "see" directly the disorder and the objects of which it consists. The spreading of STM-devices in many laboratories within the last years had the important consequence that people started to realize how imperfect our "perfect" surfaces are. (This aspect was for a long time very much ignored, because the information supplied by diffraction instruments is mainly on the periodic, ordered areas, whereas that on disordered areas is mostly lost in the background.) The superiority of STM in seeing the disorder does not mean that this application area of He-scattering is out. Disorder has by definition a practicaly unlimited variety of appearances on the same surface. Thus in order to characterize the disorder on a macroscopic area the detailed STM-information has to be complemented by average information supplied by a method like He-scattering.

The investigation of surface disorder by He-scattering has been reviewed here in much more detail than the other two approaches, diffraction and inelastic scattering, in which He-scattering is also used as a usual surface tool. The reasons for this are the following: (1) diffraction and inelastic scattering are both more familiar to the potential reader, because it is common also to other probes, and both are reviewed in detail in this volume (Refs. [42,50], respectively); (2) this approach is rather unique to He-scattering; (3) I was to some extent involved in the development of the "disorder" approach and thus, because I might be not fully objective in this respect, I leave the responsability of the fourth reason to Giacinto Scoles: (4) he wrote in the Preface to Vol. 2 of "Atomic and molecular beam methods" (Oxford University Press, New York–Oxford, 1992): "chap. 16 discusses the *scattering from disordered surfaces* and is perhaps the chapter of this part that is most full of consequences and indications for the future".

4. Instead of conclusions

I feel that writing conclusions for an article like this is inappropriate. Anybody who succeeded to read it, may draw his own one. I hope that some will find in the long history of molecular beam scattering from surfaces with its exciting ideas and long lasting errors something useful for himself. The efficiency and elegance of the three ways to use He-scattering for surface investigations might convince the one or the other to try them. They may discover new phenomena, with the aid of a clear and straightforward method, and will certainly have fun, as I always had.

References

[1] R.W. Wood, Philos. Mag. 30 (1915) 300.
[2] M. Knudsen, Ann. Phys. 48 (1915) 1113.
[3] L. Dunoyer, Le Radium 8 (1911) 142.
[4] J.C. Maxwell, Philos. Trans. 170 (1879) 231.
[5] G. Comsa and R. David, Surf. Sci. Rep. 5 (1985) 145.
[6] P. Clausing, Ann. Phys. 4 (1930) 533.
[7] G. Comsa, J. Chem. Phys. 48 (1968) 3235.
[8] I. Estermann and O. Stern, Z. Phys. 61 (1930) 95.
[9] W. Van Willigen, Phys. Lett. A28 (1968) 80.
[10] O. Stern, Z. Phys. 2 (1920) 49.
[11] O. Stern, Z. Phys. 3 (1920) 417.
[12] G. Comsa, Rev. Roum. Phys. 14 (1969) 1165.
[13] G. Comsa, Vacuum 19 (1969) 277.
[14] O. Stern, Naturwissensch. 17 (1929) 391.
[15] O. Stern, Z. Phys. 39 (1926) 751.
[16] T.H. Johnson, Phys. Rev. 37 (1931) 847.
[17] A couple of statements in the introduction of this fundamental work shed light on a remarkable peculiarity of the team-work in that laboratory: the sharp description of the contribution of the individual members of the team. It is explicitly stated that although the data presented in part I are made obsolete by those in Part II, they are shown because they represent the first certain proof for the diffraction of molecular beams; and further, that although the experiments are a continuation of those produced together with Knauer [18] – which had not bring the definitive proof – the experiments in part I have been performed by Stern alone.
[18] F. Knauer and O. Stern, Z. Phys. 53 (1929) 779.
[19] R. Frisch and O. Stern, Z. Phys. 84 (1933) 430.
[20] R. Frisch, Z. Phys. 84 (1933) 443.
[21] J.E. Lennard-Jones and A.F. Devonshire, Nature 137 (1936) 1069.
[22] I. Estermann, R. Frisch and O. Stern, Z. Phys. 73 (1931) 348.
[23] J.C. Crews, J. Chem. Phys. 37 (1962) 2004.
[24] R.G.J. Fraser, Molecular Rays (MacMillan Comp., New York; University Press, Cambridge, 1931).
[25] R.T. Bayard and D. Alpert, Rev. Sci. Instrum. 21 (1950) 571.
[26] R.L. Palmer, J.N. Smith, Jr., H. Saltsburg and D.R. O'Keefe, J. Chem. Phys. 53 (1970) 1666.
[27] A.E. Dabiri, T.J. Lee and R.E. Stickney, Surf. Sci. 26 (1971) 522;
M. Balooch, M.J. Cardillo, D.R. Miller and R.E. Stickney, Surf. Sci. 46 (1974) 358.
[28] G.D. Kubiak, G.O. Sitz and R.N. Zare, J. Chem. Phys. 83 (1985) 2638.
[29] J.A. Schwarz and R.J. Madix, Surf. Sci. 46 (1974) 367.
[30] M. Ascher and G.A. Somorjai, in: Atomic and Molecular Beam Methods, Vol. 2, Ed. G. Scoles (Oxford University Press, New York–Oxford, 1992) ch. 17.
[31] K.A. Peterlinz, T.J. Curtiss and S.J. Sibener, J. Chem. Phys. 95 (1991) 697.
[32] H. Hoinkes, H. Nahr and H. Wilsch, Surf. Sci. 30 (1972) 363.
[33] D.R. Miller, in: Atomic and Molecular Beam Methods, Vol. 1, Ed. G. Scoles (Oxford University Press, New York–Oxford, 1988) ch. 2.
[34] G. Boato, P. Cantini and L. Mattera, Proc. Int. Conf. Solid Surf. 2 (1974) 553.
[35] G. Boato, in: Atomic and Molecular Beam Methods Vol. 2, Ed. G. Scoles (Oxford University Press, New York–Oxford, 1992) ch. 12.
[36] The difference between Knudsen and nozzle beams becomes evident when comparing, for instance, fig. 11.2 and 11.4 in the review paper of Valbusa, in: Atomic and Molecular Beam Methods Vol. 2, Ed. G. Scoles (Oxford University Press, New York–Oxford, 1992) ch. 11.
[37] D.V. Tendulkar and R.E. Stickney, Surf. Sci. 27 (1971) 516.
[38] W.H. Weinberg and R.P. Merrill, Phys. Rev. Lett. 25 (1970) 1198; J. Chem. Phys. 56 (1972) 2893.
[39] T. Engel and K.H. Rieder, Springer Tracts in Modern Physics, Vol. 91 (Springer, Berlin, 1982) p. 55.
[40] K. Kern and G. Comsa, in: Chemistry and Physics of Solid Surfaces VII (Eds. R. Vanselow and R.F. Howe) (Springer, Berlin, 1988) ch. 2.
[41] P. Zeppenfeld, U. Becher, K. Kern and G. Comsa, Phys. Rev. B 45 (1992) 5179.
[42] G. Scoles, to be published.
[43] U. Harten, A.M. Lahee, J.P. Toennies and Ch. Wöll, Phys. Rev. Lett. 54 (1985) 2629.
[44] K. Kuhnke, K. Kern and G. Comsa, Phys. Rev. B. 45 (1992) 14388.
[45] B.R. Williams, J. Chem. Phys. 55 (1971) 1315, 3220.
[46] W.S. Fisher and J.R. Bledsoe, J. Vac. Sci. Technol. 9 (1972) 814.
[47] R.E. Subbarao and D.R. Miller, J. Vac. Sci. Technol. 9 (1972) 808.
[48] G. Brusdeylins, R.B. Doak and J.P. Toennies, Phys. Rev. Lett. 44 (1980) 1417.
[49] R.B. Doak, in: Atomic and Molecular Beam Methods Vol. 2, Ed. G. Scoles (Oxford University Press, New York–Oxford, 1992) ch. 14.
[50] G. Benedek and J.P. Toennies, Surf. Sci. 299/300 (1994) 587.
[51] D.L. Smith and R.P. Merrill, J. Chem. Phys. 52 (1970) 5861.
[52] B. Poelsema and G. Comsa, Scattering of Thermal Energy Atoms from Disordered Surfaces, Vol. 115 of Springer Tracts in Modern Physics, Ed. G. Höhler (Springer, Berlin, 1989).
[53] Ethylene is, indeed, dissociated at room temperature on Pt(111), but into ethylidyne and only one hydrogen atom. Ethylidyne being adsorbed upright "covers" hardly more

than one unit cell of the Pt surface (6.67 $Å^2$) and even together with the hydrogen atom and the contribution of the He-atom would result in a cross-section much smaller than 96 $Å^2$.

[54] B. Poelsema, S.T. de Zwart and G. Comsa, Phys. Rev. Lett. 49 (1982) 578 and 51 (1983) 522.

[55] H.P. Butz, R. Feltgen, H. Pauly and H. Vehmeyer, Z. Phys. 247 (1971) 70.

[56] H. Jónsson, J. Weare and A.C. Levi, Phys. Rev. B 30 (1984) 2441 and Surf. Sci. 148 (1984) 126.

[57] E. Zaremba, Surf. Sci. 151 (1985) 91.

[58] A.T. Yinnon, R. Kosloff, R.B. Gerber, B. Poelsema and G. Comsa, J. Chem. Phys. 88 (1988) 3722.

[59] B. Poelsema and G. Comsa, Faraday Discuss. Chem. Soc. 80 (1985) 115.

[60] L.K. Verheij, M.B. Hugenschmidt, B. Poelsema and G. Comsa, Catal. Lett. 9 (1991) 195.

[61] B. Poelsema, L.K. Verheij and G. Comsa, Phys. Rev. Lett. 51 (1983) 2410.

[62] L.K. Verheij B. Poelsema and G. Comsa, Surf. Sci. 162 (1985) 858.

[63] U. Linke and B. Poelsema, J. Phys. E: Sci. Instrum. 18 (1985) 26.

[64] R. Kunkel, B. Poelsema, L.K. Verheij and G. Comsa, Phys. Rev. Lett. 65 (1990) 733.

[65] B. Poelsema, L.K. Verheij and G. Comsa, Phys. Rev. Lett. 53 (1984) 2500.

Surface Science 299/300 (1994) 92–101
North-Holland

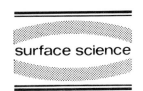

surface science

Semiconductor surface spectroscopies: the early years

Jacek Lagowski

Center for Microelectronics Research, University of South Florida, Tampa, FL 33620, USA

Received 30 April 1993; accepted for publication 15 June 1993

In the Fall of 1970, Chester Lee Balestra, a graduate student working in Harry Gatos' laboratory at the Massachusetts Institute of Technology, observed a strong photovoltaic signal on CdS surfaces illuminated with a sub-bandgap light. Part of the photovoltage spectrum disappeared for surfaces cleaved in ultra-high vacuum and reappeared at higher ambient pressure, implying the involvement of surface states. A corresponding technique, "surface photovoltage spectroscopy", was developed and used for studying a wide range of surface-related phenomena in semiconductors. Two decades later, the spin-off techniques are used on silicon IC fabrication lines for instantaneous, non-contact detection of metal contaminants with an astounding sensitivity of one part per quadrillion.

1. The fall of 1970

In October 1970, I saw my first ultra-high vacuum system. I was quite impressed. A huge stainless steel dome, covered with heating pads, contained 140 liters of 10^{-11} Torr vacuum. Inside there was a clever aluminum–stainless steel construction for cleaving the crystals. The cleaved surface was automatically positioned against a tiny (~ 1 mm) Kelvin probe at the focal point of the illuminating system. A double-prism Zeiss monochromator and a xenon lamp were mounted on the elevated platform near the viewport to the vacuum chamber.

The location was Harry C. Gatos' laboratory at MIT. I was a post-doctoral fellow, recently arrived from Poland, anxious to see new things and use them to do research. Why a Kelvin probe and not Auger, LEED or HEED, about which I had been reading so much? Harry Gatos' views were very definite and the electronic properties interested him most. Although related to structure and chemistry, they were the key to semiconductor applications, including both surfaces and bulk. My curiosity for surface analytical techniques was set aside, and I started working with Kelvin probes and surface photovoltage. The principles of the

techniques were not new to me. I had been using them extensively since 1963 in my research in Poland. What was different was the semiconductor material (cadmium sulphide and not germanium or silicon), the approach (high resolution spectral study of cleaved surfaces, not the "real" surfaces), and the experimental apparatus, of course.

I was teamed with Chester Lee Balestra, Harry Gatos' graduate student, during the final stages of his doctoral thesis devoted to the electronic properties of cadmium sulphide surfaces. Two months later, in a letter to *Surface Science* [1], we described "a method (...) for determining the energy positions of the surface states in high energy gap semiconductors (...) based on the photo-stimulated emission of carriers from the surface states into the bulk. The spectral distribution of the associated photovoltage gives directly the energy position of the surface states." The historical, very first, spectrum is shown in the upper portion of Fig. 1. It is seen that sub-bandgap illumination ($h\nu < E_g$; the energy gap, E_g, for CdS is about 2.4 eV) induces pronounced photovoltage, although virtually no generation of electron–hole pairs takes place. In Fig. 1a, the photovoltage measured in an ambient of 10^{-4} Torr

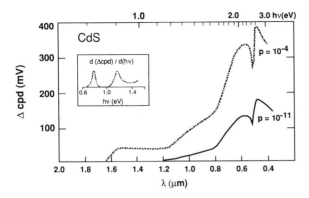

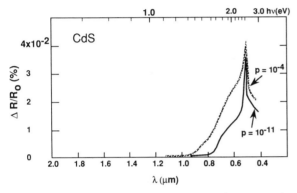

Fig. 1. The surface photovoltage spectra of CdS (upper figure) and corresponding spectra of photoconductivity. In the region $\nu > 0.8$ μm, the photovoltage is due to photo-stimulated emission of electrons from two surface states with energy positions determined by peaks of the derivative of the surface photovoltage shown in the inset.

appears at $\lambda = 1.6$ μm and reaches a plateau soon after. At $\lambda = 1.2$ μm the photovoltage rises. A plot of the derivative of the surface photovoltage with respect to the photon energy (insert in Fig. 1a) shows clearly two peaks at 0.78 and 1.15 eV, below the conduction band, which correspond to the energy positions of two discrete surface states. It is also seen that under high vacuum (10^{-11} Torr) the photovoltage vanishes in the spectral region of $\lambda \approx 1.6$ to 1.2 μm, indicating that these surface states are associated with absorbed ambient species. It should be pointed out that there is no detectable photoconductivity for λ larger than about 1 μm, as shown in Fig. 1b.

The surface photovoltage in Fig. 1 was measured as the change in the contact potential difference, CPD, with respect to a gold reference electrode going from the dark to conditions under illumination. A small 1 mm diameter gold-boss electrode was used as a vibrating Kelvin probe [2]. A low-noise MOSFET operational preamplifier (which became available in the late 1960's) and a single harmonic lock-in detection made it possible to achieve an off-null linear measurement of CPD. The changes induced by illumination as small as 0.1 mV could be readily monitored. This MIT system, set up in 1970 by Chester Balestra, had a sensitivity and stability comparable to the very best systems available today [3,4]. Needless to say, this was a very important experimental refinement critical for the development of surface photovoltage spectroscopy, SPS.

Completion of the paper for *Surface Science* coincided with an important annual event in the life of MIT's Electronic Materials Group; a Christmas chamber music concert in the Gatos' Weston residence. There I learned that playing the flute could have been another true profession of Harry Gatos.

2. The rationale behind SPS

From its beginning in 1964 to its end in 1989, Harry Gatos' Electronic Materials Group was studying semiconductors and only semiconductors. By the end of 1970, silicon has already dominated semiconductor electronics. Germanium was to a large extent a material of the past, while gallium arsenide was emerging as a potential material of the future. The cold war period generated tremendous appetite for high-resolution aerial infrared photography which established a small, but lucrative, market for mercury cadmium telluride devices. These were four established materials, while the entire semiconductor family contained many more III–V, II–V, and IV–VI compounds. Researchers believed that ideal materials for new devices are still to be found. Exploration of new possibilities was an important part of the game. With optoelectronic applications on the horizon, the high energy gap ($E_g \geq 1.8$ eV) compound semiconductors were studied as candidates for devices in the visible spectral range. The MIT group investigated CdS, ZnO, and SiC. The research was always very broad in scope. It included crystal growth; characterization of electronic properties; and phenomena and processes relevant to device applications. Cadmium sulphide and zinc oxide, with energy gaps of 2.4 and 3.3 eV, respectively, were important also for surface applications such as chemisorption and catalysis. Harry Gatos' fundamental interest in surfaces of these II–VI materials stemmed from his earlier work on $A_{III}B_V$ compounds [5,6]. Gatos' model of surface atom bonding predicted contrasting behavior of the crystallographic surfaces terminating with A atoms and those terminating with B atoms [6]. Experiments showed that these differences were indeed manifested in crystal growth, etching characteristics, and micro-hardness, in excellent agreement with the Gatos' model. Extension to II–VI compounds and to electronic surface properties was the next logical step.

Logical, yes, but not simple. Electrical techniques for surface state studies such as field effect, surface conductivity, capacitance–voltage, developed for Ge and Si, were practically useless in CdS and ZnO. Surface photovoltage, prior to our work, was used only for probing of the surface barrier height [7] or for determination of the minority carrier diffusion length [8]. It was considered only for the case of electron–hole excitation and was proclaimed insensitive to surface state parameters. The surface photovoltage in a sub-bandgap spectral range where no electron–hole pairs were generated offered a new means for surface state investigation. We felt that, for the first time, we might be able to probe for the surface states within the entire energy gap in high gap semiconductors. Why we had felt that way was related very much to the principle of SPS based on photoexcitation phenomenon and to the specific feature of the surface states in high E_g semiconductors. We termed this feature "quasi-isolation from the bulk".

As shown in Fig. 2, a localized level of the energy, E_t, below the conduction band edge, E_c,

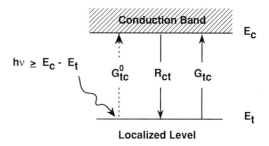

Fig. 2. Electron transitions between the localized level and the conduction band.

communicates with this band via the thermal emission, G_{tc}, and via the recombination transitions, R_{ct}. If probabilities of both transitions are large, then the localized level will quickly establish an occupational fraction, $f_n = n_t/N_t$, dictated by the Fermi statistics. If the level is shifted in energy from below (above) the Fermi energy to above (below) the Fermi energy, then its occupation, n_t, decreases (increases) to a new value after a time period depending on $R_{ct} + G_{tc}$. The larger $R_{ct} + G_{tc}$, the shorter is the equilibration time. On the other hand, if R_{ct} and G_{tc} are very small, the occupation of the level may not change appreciably even after hours or days. This latter case is a common one for the surface states in CdS and ZnO [9–12].

In our work on SPS, we described it as the "quasi-isolated surface state". For SPS it is very important that the occupation of the quasi-isolated states can be readily changed using an optical excitation of carriers from or into the surface states, G_{tc}^0.

The thermal emission, G_{tc}, decreases as an exponent, $\exp(-E_t/kT)$. This favors deep levels with large E_t/kT which are found in high energy gap semiconductors rather than in Ge or Si (respective energy gaps are only 0.66 and 1.12 eV). A decreasing of the recombination transition rate, R_{ct}, is a different matter as R_{ct} is proportional to the free carrier concentration. Above the freeze-out temperature it is not possible to achieve quasi-isolation in the bulk region because of the omnipresent free carriers.

The free carriers, however, are removed from the depleted space charge regions at the surface

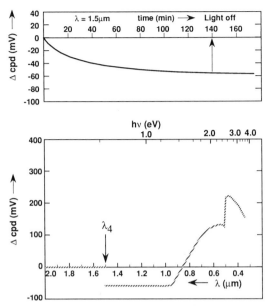

Fig. 3. The kinetics of surface photovoltage inversion, i.e., the increase of the surface barrier height (decrease of the ΔCPD) under illumination (upper figure) versus illumination time. The lower figure presents the surface photovoltage spectrum with a drop of ΔCPD at $\lambda = 1.5$ μm, corresponding to the photovoltage inversion.

or in the rectifying junction [13]. Surface depletion regions found in CdS and ZnO satisfy very well the requirement of a low recombination rate, R_{ct} (one may note that surface depletion layers are quite common in compound semiconductors).

An experimental manifestation of the quasi-isolated surface states was provided by the surface photovoltage inversion effect [9] in CdS, shown in Fig. 3. We have found that illumination of CdS with photons of the energy from about 0.8 to 1.2 eV (wavelength from 1.5 to 1.0 μm) *results in increasing of the surface barrier height*. The process was best observed on the basal Cd or S surfaces exposed to a reduced ambient pressure (about 10^{-8} Torr) for a prolonged time (days to one week). This eliminated the long wavelength photovoltage tail (see Fig. 1a) due to absorption-induced surface states at $E_c - 0.8$ eV.

The increase of the surface barrier with time was a single exponential process with a time constant inversely proportional to the incident light intensity. A mechanism of the process is illustrated in Fig. 4. The increase of the surface barrier is associated with optical pumping of electrons from the valence band into the surface states at about 0.85 eV above the valence band top. *The states are located about 30 kT below the Fermi energy and, nevertheless, are unoccupied.* Such situation was created by previous illumination with photons of high energy ($h\nu > 1.6$ eV) which emptied the levels. The levels could not be completely refilled by electrons from the conduction band due to a depletion region and high surface barrier, V_{SO}. They also could not be filled by electrons from the valence band due to a low thermal emission rate.

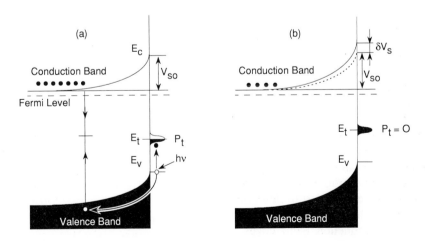

Fig. 4. Schematic representation of the transitions involved in the surface photovoltage inversion.

Prior to our observation of the surface photo-voltage inversion, it was believed that illumination can only decrease the surface barrier. We observed photovoltage inversion effect originally in CdS [1,9–11], then in ZnO [14], and later on also in other high gap semiconductors [15]. Thus, we have generally established that the surface barrier can decrease as well as increase when in n-type semiconductors electrons are optically pumped from the surface states or into the surface states, respectively.

3. Determination of the surface state parameters with SPS

Photoionization transitions from or into the surface states, E_t, cause a change, ΔQ_{ss}, of the electrical charge localized on the surface. Accordingly, the surface barrier changes by an amount, ΔV_s, which is measured as the corresponding change in the work function, $\Delta V_s \approx -(1/q)\Delta W$. The 0.1 mV sensitivity of the vibrating probe Kelvin method permits the detection of ΔQ_{ss} as low as 10^8 q/cm^2. This is extremely high sensitivity which can be visualized by considering that 10^8 q/cm^2 corresponds to one elementary charge per about 10^7 surface atoms. To our knowledge, no other technique can match this sensitivity.

As discussed above, the sign of the photovoltage depends on the type of photoionization transition. Thus, photo-ejection of electrons from the surface states decreases the negative charge on the surface and leads to a decrease of the surface barrier. Transitions from the valence band to the surface state increase the negative charge, Q_{ss}, which, in turn, increases the surface barrier.

Accordingly, the sign of the surface photovoltage permits the identification of the photoionization transition. It is also quite evident that surface state depopulation can occur for photons with energy $h\nu_1 \geq E_c - E_t$, while the surface state population transition requires $h\nu_2 \geq E_t - E_v$. Consequently, the energy position of the surface photovoltage threshold can be used to determine the energy positions of the states with respect to the valence and the conduction band edges as shown in Fig. 5.

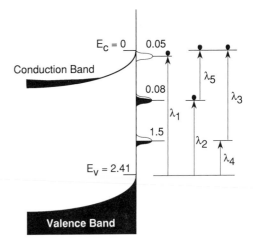

Fig. 5. Optical transitions into and from the surface states on basal CdS surfaces identified by the surface photovoltage spectroscopy. Note that surface states are observed twice and the sum of the transitions energies corresponds to E_g.

When the same surface state is involved in the above transition, $h\nu_1 + h\nu_2 = E_g$, as has been found in the case of CdS surfaces. However, $h\nu_1 + h\nu_2$ can exceed the value of the energy gap $E_g = E_c - E_v$. Localized states in compound semiconductors can exhibit strong lattice coupling, and the related photoionization transitions may involve configurational changes. In these cases not only $h\nu_1 + h\nu_2 > E_g$, but also $h\nu_1 > E_c - E_t$ and $h\nu_2 > E_t - E_v$.

The determination of the dynamic parameters of surface states [9] relies on the analysis of photovoltage transients, as shown in Fig. 6, using

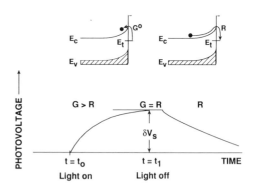

Fig. 6. Surface photovoltage transient (lower portion) used for determination of surface state parameters and corresponding electron transitions (upper portion).

an example of photodepopulation of surface states. For deep states, the thermal emission can be neglected and the electronic transitions involved in the transient consist of photoinduced transitions from the surface state to the conduction band, G^0, and the recombination transitions from the conduction band to the surface state, R. The concentration of electrons, n_t, on the state E_t varies as

$$dn_t/dt = R - G^0. \tag{1}$$

The initial increase of photovoltage for $t > t_0$ is due to G^0 which exceeds R_{ct}. Steady state value is reached when $R = G^0$. This is achieved due to decreasing of G^0 with decreasing the surface state population. At the same time, R increases due to two factors: decreasing of the surface barrier and increasing of the concentration of empty states capable of capturing electrons. The photovoltage decay in the dark $(t > t_1)$ is due entirely to recombination transitions, since in the dark $G^0 \equiv 0$. It is, thus, apparent that the analysis of the light-on, light-off transients makes possible the determination of R and G^0 and all relevant parameters such as K_n, the surface state capture cross section for electrons multiplied by their thermal velocity, and σ^0, the optical photoionization cross section, and, of course, the density of the surface states.

From a historical perspective, it is interesting to note that, in 1970, two research groups, working independently (in fact, not knowing of each other), formulated procedures for deep level defect characterization from recombination and generation transients. In the Gatos group at MIT, we have considered surface states and the surface photovoltage, while Sah and his co-workers [16] have developed procedure for bulk states based on junction capacitance and current. The work of Sah was later on refined by Lang [13] into a very practical form know as deep level transient spectroscopy (DLTS). Surface photovoltage analogy of DLTS was proposed only in 1992 in a form of a non-contact, no-preparation, wafer-scale technique [17].

SPS can also be used in a simplified mode in which AC photovoltage is generated by chopped incident light and is recorded as a function of photon energy. This mode is best suited for the study of bulk levels rather than surface states. It does permit the determination of the energy position of localized states but not the determination of their dynamic parameters.

4. Wall spectroscopy of surface states

The surface states observed in CdS and ZnO were extrinsic defects created by absorption–desorption processes. What was their relation to the surface atom bonding and how did they relate to the Gatos' model? To answer these questions, we designed an experiment for in situ monitoring of the spontaneous bending of thin CdS crystals. Carefully prepared 20 μm thick platelets with large surfaces corresponding to the basal planes of Cd and S, respectively, were attached by one end to a holder placed in an UHV chamber near the viewport window. The light spot of the He–Ne laser beam, reflected from the surface, was observed on the wall as an indicator of the shape of the platelet. We expected that a stress differential between the Cd and S surfaces would be manifested by bending of the platelet and, thus, shifting of the laser spot on the wall.

In the very first experiment, a prolonged evacuation of the chamber and subsequent addition of oxygen was used as an ambient cycling. Prior to that, we established that this cycle had a pronounced effect on surface states as manifested by surface photovoltage spectroscopy. We applied the cycle once, twice, three times with no effect at all. The laser spot was positioned exactly on the same mark on the wall. I communicated the results to Harry Gatos, but he was not convinced. Spontaneous bending was observed in several laboratories and ambient cycling should have an effect.

I decided to make sure that the laser beam was reflected by the platelet. For the visual inspection, I shined a flashlight beam on the CdS sample. Everything appeared to be correct. The tiny red spot of the He–Ne laser beam was near the free end of the platelet, as it should be. Disappointed, I looked at the laser spot on the wall. It seemed to be somewhat shifted away from the mark. I turned the flashlight off. The laser spot jumped exactly onto the mark. I turned the flashlight on and off, on and off, several times and the spot on the wall repeated the movement. An illustration of the light pressure, I thought. Then I carefully analyzed the direction of the motion. It was opposite to that expected for light pressure. The light appeared to attract the platelet instead of pushing it away. A moment later, Harry Gatos was with me in the laboratory. We replaced the flashlight with a stronger microscope lamp and now the laser beam on the wall jumped by inches. We were observing a new phenomenon in which a thin crystal of CdS was converting the light into mechanical energy. The crystal was like a magic sunflower bending toward the light.

By using a chopped light and proper tuning of the chopping frequency, we were able to achieve a resonance. Natural mode vibrations could be excited with a monochromatic light of low intensity similar to that used in surface photovoltage spectroscopy. With a simple arrangement of the "wall spectroscopy" (see Fig. 7), we could precisely monitor the amplitude of the light-induced vibrations. A comparison between the excitation spectra of the vibration, the surface photovoltage spectra, and the photoconductivity spectra measured under similar conditions is given in Fig. 8 for CdS and GaAs platelets about 20 μm thick. It was clear that the vibration spectra resembled the surface photovoltage rather than the photoconductivity spectra. We could, therefore, assign the vibration induced by illumination in a certain wavelength range to transitions involving particular surface states. Unlike SPS, however, the recombination transitions here were of importance since a chopped light was used and the depopu-

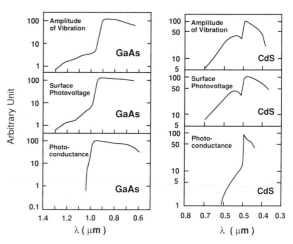

Fig. 8. Excitation spectra of the vibration of thin GaAs and CdS crystals (top figures). Note similarity to surface photovoltage spectra and differences in comparison with photoconductance spectra.

lated states had to recover its occupation during the dark period.

Understanding of the surface phenomenon leading to light-induced vibrations of thin crystals came immediately after we found no such effect on platelets with large prismatic planes. The phenomenon had piezoelectric origin. Incident light altered the surface charge and, thus, also the electric field, E_s, at the surface. Via the piezoelectric effect, this modified surface stress is shown in Fig. 9. It was clear confirmation that sub-bandgap illumination in CdS and GaAs was indeed altering the surface charge, as postulated in surface photovoltage spectroscopy. It may be appropriate to remember that it was 1971. All surface photovoltage treatments prior to ours assumed that the surface state charge remained constant under illumination.

In photoinduced vibrations, a modulation of the surface charge produced surface stress. Soon after this observation, we started using an inverse phenomenon, namely, by applying strain (mechanical bending), we modulated the surface charge [10], thereby very effectively altering the surface chemical reactions based on charge transfer (for example, the oxygen chemisorption rate on ZnO [18]).

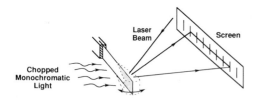

Fig. 7. Arrangement used in the wall spectroscopy of surface states.

5. Transition from the early years to the future

The research mentioned above started in the fall of 1970 and was essentially completed in the fall of 1971. The three major contributors, Chester Balestra, Harry Gatos, and myself, all left MIT. Chet joined Eastman Kodak in Rochester, New York, Harry Gatos went to Stuttgart for a sabbatical, and I returned to Poland. When I rejoined Harry Gatos' MIT group in 1976, the times were clearly different. After a very successful demonstration of the advantages of space in growing homogenous semiconductor crystals, Harry Gatos was invited to undertake a multi-year program sponsored by NASA and aimed at establishing the practical potential of a micro-gravity environment for growing-improved device-quality semiconductors.

GaAs was the material of the future and it became our choice. With Harry Gatos' passion for semiconductor materials, and his enthusiasm for imaginative projects, it was clear that the emphasis of the group would focus on the effects of micro-gravity. A shift from surfaces to the bulk was imminent. Stoichiometry, native defects,

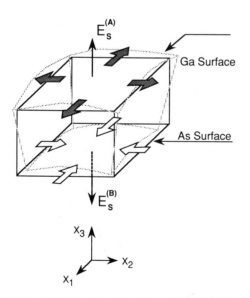

Fig. 9. Electric field on basal surfaces of GaAs and corresponding stress due to piezoelectric effect. For CdS the polarity is reversed and, thus, the Cd surface is under tension and S surface is under compression.

growth–property relationships, and the search for ideal crystal growth methods filled up about 200 of our next publications. No more than 20 of them dealt with surfaces. Defect engineering, i.e., creative manipulation of defects to benefit semiconductor properties or device manufacturing, became our new area. The results of ground-based research surpassed all expectations, however, our crystals never flew in space. The Challenger disaster occurred and took its toll on the space research program.

We used surface photovoltage spectroscopy occasionally in relation with other research projects such as charge transfer in dye sensitization, charge transfer in chemisorption, interface states in GaAs–anodic oxide, characterization of α-Si and silicon-on-sapphire. Interest in high energy gap II–VI compounds for electronic applications dropped and so did the motivation for corresponding surface state study. Outside MIT, probably the most important SPS application was done by Len Brillson who used the technique in his studies on metal-induced surface states [19]. Only recently, Baikie and his co-workers [20] have constructed a low-temperature SPS system and used it to study the initial oxidation stages on silicon. A tuneable laser version of SPS [21] has also been demonstrated this year and applied to the technologically important InAlAs ternary compound.

The MIT group was dissolved in 1989. Without Harry Gatos, without his enthusiasm and vitality, the place was not the same. My interests and contacts with the semiconductor industry directed me toward silicon. Surface photovoltage in silicon was becoming important in defect diagnostics. I realized that I could use my expertise in SPS, semiconductor characterization, and defect engineering, and combine them into a final product – a diagnostics tool for silicon wafers. Together with Lubek Jastrzebski (also a former member of Harry Gatos' group) I founded Semiconductor Diagnostics, Inc. (SDI), a small high-tech company designed to manufacture diagnostic equipment. Originally, SDI was located in the Route 128 ("High Technology Highway") area of Boston. I was about 15 miles east at MIT and Lubek Jastrzebski was about 300 miles south at the David Sarnoff Research Laboratory in Princeton,

New Jersey. As a next step, both of us moved to the newly established (and excellently equipped) Center for Microelectronics Research at the University of South Florida in Tampa and we transferred SDI from the Route 128 area to Tampa.

In the most advanced sub-micron silicon-integrated circuits, iron is a very frequent, uncontrolled contaminant with detrimental effects to manufacturing yield [22]. Iron precipitates at the

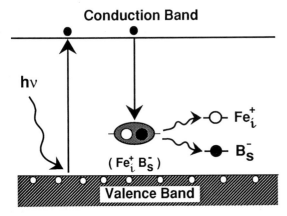

Fig. 10. An iron contamination map of a Si wafer obtained with the surface photovoltage spectrometer and optical iron activation. The Si wafer was illuminated through a mask with opening shaped as letters "Fe". Illumination with strong white light activates iron as a recombination center by splitting iron–boron pairs (see lower portion of the figure) and facilitates iron detection by SPS minority carrier diffusion length measurement.

Si–SiO$_2$ interface act as localized breakdown spots. The thinner the oxide, the more significant is the effect of Fe precipitates. For example, in 0.5 μm technology (16 Mbit IC's) based on a 10 nm thick gate oxide, the Fe concentration limit in the underlying silicon wafer is about 10^{11} atoms/cm^3 or 2 parts per trillion [23]. The thinner oxides used in 0.25 μm technology (256 Mbit IC's) and the forthcoming 0.18 μm technology (1 Gbit IC's) require a reduction in the Fe concentration below 10^{10} atoms/cm^3 or down to the ppq (part per quadrillion) range.

Surface photovoltage spectroscopy identifies the presence of iron in silicon wafers within seconds. It has been demonstrated that the method has a sensitivity for iron concentration determination approaching one ppq (5×10^7 atoms/cm^3) [24]. In the method, the minority carrier diffusion length and the surface recombination velocity are determined from the surface photovoltage spectrum. Then, a defect engineering step is used to alter the state of iron in the silicon. A strong light (10 W/cm^2) decomposes the iron–boron pairs (recombination-enhanced pair dissociation process schematically shown in Fig. 10) within seconds and creates lifetime killing iron interstitials which are detected via a corresponding decrease of the minority carrier diffusion length. In the entire process, surface photovoltage measurements and iron activation, light is the only medium which touches the wafer. The surface photovoltage is measured in a non-contact way using capacitive coupling to the wafer. It is generated on the native surface barrier and, thus, no special wafer preparation is needed. These features of SPS make the technique uniquely suited for in situ, real-time monitoring required in the IC manufacturing environment.

In 1993, about 60 surface photovoltage spectrometers are used in silicon IC fabrication lines for monitoring processing-induced contaminants in the bulk and on the surface.

6. Closing remarks

I started this article describing the early work which dealt with the problem of surface state characterization in semiconductors and the resultant surface photovoltage technique. Harry C. Gatos, the first editor of *Surface Science*, was mentioned in the article in connection with the discussion of scientific problems, however, I would like to add that Harry has also been an incredible motivator and a thorough scientific "entrepreneur".

Surface science has always been instrumental for the advancement of semiconductor microelectronics. It is, therefore, justified that the end of my article describes the beginning of surface photovoltage contamination monitoring technology; a clear product of semiconductor surface science. The idea of converting a semiconductor surface into a powerful sensor of electrically active contaminants in the bulk and on the surface was already present in our early MIT work. The need for its practical implementation, however, was only recently created with the advent of gigabit silicon-integrated circuits.

References

[1] C.L. Balestra, J. Lagowski and H.C. Gatos, Surf. Sci. 26 (1971) 317.

[2] C.L. Balestra, Electronic Properties of CdS Surfaces, Doctoral Thesis, The Massachusetts Institute of Technology, 1971.

[3] I.D. Baikie, K.O. van der Werf, H. Oerbekke, J. Broeze and A. van Silfhout, Rev. Sci. Instrum. 60 (1989) 930.

[4] I.D. Baikie, S. Mackenzie, P.J.Z. Estrup and J.A. Meyer, Rev. Sci. Instrum. 62 (1981) 1326.

[5] H.C. Gatos and M.C. Lavine, J. Electrochem. Soc. 107 (1960) 427.

[6] H.C. Gatos, J. Appl. Phys. 32 (1961) 1232.

[7] E.O. Johnson, Phys. Rev. 111 (1958) 153;
D.R. Frankl and E.A. Ulmer, Surf. Sci. 6 (1966) 115.

[8] A.M. Goodman, J. Appl. Phys. 32 (1961) 2550.

[9] J. Lagowski, C.L. Balestra and H.C. Gatos, Surf. Sci. 27 (1971) 547; 29 (1972) 203, 213.

[10] J. Lagowski and H.C. Gatos, Surf. Sci. 30 (1972) 491.

[11] J. Lagowski and H.C. Gatos, Proc. Conf. Phys. Semiconductors, Warsaw, Poland, 1972.

[12] H.C. Gatos and J. Lagowski, J. Vac. Sci. Technol. 10 (1973) 130.

[13] Note that such a condition is also commonly used in deep level transient spectroscopy (DLTS). For an analysis of the deep levels in the bulk: D.V. Lang, J. Appl. Phys. 45 (1974) 3023.

[14] J. Lagowski, E.S. Sproles, Jr. and H.C. Gatos, Surf. Sci. 30 (1972) 653.

[15] H.C. Gatos, J. Lagowski and R. Banisch, Photogr. Sci. Technol. 26 (1982) 42.

[16] C.T. Sah, L. Forbes, L.L. Rosier and A.F. Tasch, Jr., Solid State Electron. 13 (1970) 759.

[17] J. Lagowski, A. Morawski and P. Edelman, Jpn. J. Appl. Phys. Lett. 31 (1992) L1185.

[18] J. Lagowski, E.S. Sproles, Jr. and H.C. Gatos, J. Appl. Phys. 48 (1977) 3566.

[19] See: L.J. Brillson, J. Vac. Sci. Technol. 13 (1976) 325;
Also this volume.

[20] I.D. Baikie, E. Venderbosch and B. Hall, MRS Proc. 261 (1992) 149.

[21] L. Kronik, L. Bourstein, Y. Shapira and M. Oran, Appl. Phys. Lett., in press.

[22] L. Jastrzebski, W. Henley and C. Nuese, Solid State Technol. 35 (1992) 27.

[23] W. Henley, L. Jastrzebski and N. Haddad, Effects of Iron Contamination on Thin Oxide Breakdown and Reliability Characteristics, Proc. IEEE Int. Reliability Physics Symp., March 1993, Atlanta, Georgia.

[24] J. Lagowski, A.M. Kontkiewicz, W. Henley, M. Dexter, L. Jastrzebski and P. Edelman, Appl. Phys. Lett., submitted.

Surface Science 299/300 (1994) 102–115
North-Holland

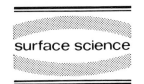

surface science

Surface electron microscopy: the first thirty years

E. Bauer

Physikalisches Institut, Technische Universität Clausthal, D-3392 Clausthal-Zellerfeld, Germany

Received 2 March 1993; accepted for publication 4 May 1993

After a brief account of the early years of surface electron microscopy the evolution of ultrahigh vacuum surface electron microscopy from the early sixties to the early nineties is described. Low energy electron microscopy is selected as a case study to illustrate the difficulties encountered in the development of a new method, but all other successful imaging methods are discussed too.

1. The prehistory

More than 30 years before the first issue of Surface Science appeared, the first images of surfaces obtained with electrons using electron lenses were published [1–6]. These images were of low magnification and resolution but demonstrated that surfaces could be imaged with thermionically emitted electrons [1,2], photoelectrons [3], secondary electrons produced by electron [4] or ion bombardment [5] and with reflected electrons [6]. Soon afterwards scanning electron microscopy (SEM) [7] and mirror electron microscopy (MEM) [8] were added to the list of direct surface imaging methods and in 1940 high energy electron reflection microscopy (REM) [9] in the mode used in todays instruments. Resolutions beyond that of the light optical microscope were, however, not obtained in direct electron optical surface imaging until the early forties [10,11], ten years after the invention of electron microscopy.

In the meantime transmission electron microscopy (TEM) had already become "Übermikroskopie" ("super microscopy"), that is microscopy with a resolution beyond that of the light microscope, and the introduction of the replica technique in 1940 [12] had produced the first "super microscopic" images of metal surfaces using TEM. Indirect surface electron microscopy by surface replication and TEM soon received a further boost by the shadow casting method [13] in which a thin heavy metal layer was deposited obliquely onto the (rough) surface before the thin electron-transparent replication layer was applied. This easy way of imaging surfaces indirectly by TEM made direct imaging techniques such as emission, mirror or reflection microscopy much less attractive. Also, SEM had already reached a high level of sophistication in 1942 [14] and world war II soon terminated further developments.

After the war widespread efforts were made to develop emission and mirror electron microscopy into viable imaging methods, in spite the rapid progress of TEM and of the replica method. The motivation was the desire to study phenomena which are inaccessible to the replica method such as electron emission, oxidation or recrystallization in situ. These developments which are very well described in several reviews (see, e.g., Refs. [15–17]) actually lead to some commercial instruments, the most sophisticated being the Balzers Metioskop KE3, an instrument which allowed several emission modes [18]. Although a large amount of interesting work was done with these instruments it soon become evident that surface contamination was a serious handicap which lim-

SSDI 0039-6028(93)E0279-4

ited their usefulness. Furthermore, in the mid-sixties the first commercial SEMs became available and soon displaced emission, mirror and reflection microscopes although having similar contamination problems.

In view of these limitations, indirect surface electron microscopy continued to be an attractive alternative. The carbon replica technique, introduced 1954 [19] and its extension, the Pt–C shadowing technique allowed replication in vacuo with a resolution not achievable by direct imaging at that time. In favorable cases, modifications of this technique gave a resolution of about 25 Å [20]. Another indirect surface imaging mode discovered in these years, however, became much more important in surface science. In my thesis [21] I had used Pd shadow casting to study the growth of thin fluoride films. On one occasion the Pd layer had become much too thin and the Pd crystals had decorated the growth steps. This was obvious to me and I did not pay much attention to it. It was Bassett who recognized the potential of the decoration technique [22] and Bethge who made it popular in surface science by his and his coworkers' beautiful studies of monatomic steps on alkali halide surfaces [23]. The decoration replica technique is still the most important technique for the study of the surface microstructure of electron-sensitive materials such as ionic crystals and of surface processes on them.

In the sixties several attempts were made to overcome the contamination problem in conventional electron microscopes. One route was to use differential pumping with well-trapped diffusion pumps augmented by sorption pumps such as Koch's photoemission electron microscope (PEEM) in Tübingen which achieved a base pressure in the 10^{-8} Torr range [24]. Another, still used method was to prepare the surface in a UHV system and to transfer it to a conventional vacuum electron microscope, hoping that the exposure to atmosphere and the subsequent contamination in the microscope would not influence the surface features of interest. None of these attempts, of course, can satisfy the pure surface scientist who wants to study surface phenomena under well-defined conditions in the system in which he prepared and characterized his surface.

This lead early, shortly after metal UHV technology had became commercially available, to the development of UHV surface electron microscopy, the main subject of this review.

2. The childhood years of UHV surface electron microscopy

As far as I can determine from the literature, my group at the Michelson Laboratory, China Lake, California, seems to have been the first one who dared to attack the task of building an UHV surface electron microscope. According to Ref. [17], this effort was soon followed by another one, also in a military laboratory, the Night Vision Laboratory, Fort Belvoir, Virginia, in which Burroughs built about 1965 the electrostatic flange-on PEEM and TEM designed by Gertrude F. Rempfer in 1963. We were not aware of this work and nothing was apparently published about it so that I can recount only the Michelson Lab Story. Our effort was a true child of surface science. Shortly after Germer et al. had revived LEED by developing the display type LEED system they reported streaked LEED patterns [25] which I subsequently attributed to linear adsorbate structures caused by preferred adsorption at surface steps [26]. It was obvious that an unambiguous interpretation of the LEED pattern required independent information on the microstructure of the surface. At this point I remembered what I had learned in my thesis: the Boersch ray path, introduced by Boersch in 1936 in TEM [27]. It allowed to take the diffraction pattern of the specimen in a TEM by imaging the back focal plane of the objective lens, in which the Fraunhofer diffraction pattern is located, onto the fluorescent screen, using an auxiliary lens. Why could not the same be done in reflection with the slow electrons used in LEED?

This idea was the birth of a new surface imaging technique, low energy electron (reflection) microscopy (LEEM) which uses elastically backscattered electrons and is based on diffraction contrast. The experience gained in writing a book on electron diffraction [28] and building an electron optical bench as well as the basic research-friendly atmosphere and the excellent workshop facilities at the Michelson Laboratory at that time gave me the necessary confidence for this endeavour. The project started in a somewhat funny way. Although I wanted to build a metal system right away our glass blowing contractor succeeded to convince management that he could build a glass system with the necessary tolerances and conductive coating (to avoid charging) much faster and much cheaper than a metal system could be built. Of course, it took much longer and probably was also much more expensive than he had promised, but nevertheless it was finished in time for presentation at a conference in 1962 [29].

Fig. 1 shows the schematic (a) and the physical appearance (b) of this curiosity. The double 90° deflection (3,4) was dictated by the limitations of precision glass blowing. The electrostatic lenses (6,11,12) were pre-aligned with precision-ground glass cylinders. Electrons from a pointed filament cathode (1) could be accelerated up to 25 keV and decelerated in the cathode lens to the desired low energy before being reflected in or before the specimen (8) for LEEM and MEM, respectively. Although I got a good direct and a distorted reflected beam on the auxiliary fluorescent screen (14) and with deflector 4 some strangely looking features on the final image screen 13, repeated accidents such as arcovers in the gun – which destroyed the emitter 1 – or in the cathode lens – which made repolishing necessary –, burnout of the filaments of the ionization getter pump 10 and of the electron bombardment heater 9 for specimen cleaning soon made the system irreparable.

This convinced management that glass was not quite the right way to go and I was allowed to build a metal system. Similar to the glass system it relied on proven lens designs: an electrostatic triode as cathode lens [30], magnetic intermediate and projective lenses [31] and a filter lens [32]. The precision possible in an all-metal construction allowed to deviate from the 90° beam deflection, but first a simple straight beam set-up was chosen in order to test the various components. One of my coworkers, George Turner, a very skilled designer and experimentalist, took over

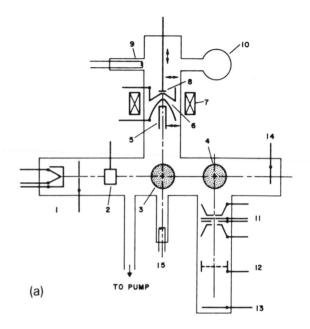

(a)

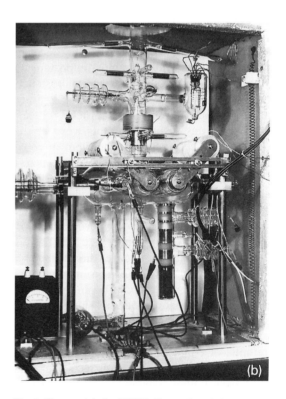

(b)

Fig. 1. First model of a LEEM. For explanation see text.

design and assembly of the instrument in early 1963. According to his somewhat sketchy notebook, assembly started in October 1964 and on Christmas we had the first beam through the system. The first mirror and secondary electron images, however, were not obtained until 8 months later after the filter lens had been removed again and an earth magnetic field compensation had been installed. They were from a polycrystalline surface on which a (BaSr)O layer was deposited through a 1000 mesh grid.

We then switched to single crystals in order to be able to obtain also LEED patterns. By that time the publication pressure had become so strong that we decided to interrupt instrument development and to study with the operation modes available at that time a specific surface science problem, the microstructural aspects of electron emission from alkaline earth oxide films on tungsten. This had become possible because (i) due to many bake-outs we had reached a base pressure in the low 10^{-10} Torr range, (ii) we could clean the W(110) crystal and deposit SrO in situ using the procedures developed in a separate LEED system and (iii) we had photoelectric and retarding field work function measurements available in addition to MEM and PEEM. The instrumental set-up used in this work is shown in Fig. 2. Some of the results were reported briefly at a conference in 1966 [33].

The instrument development work was accompanied from the very beginning by theoretical work on the optics of the cathode lens together with Cruise and on low energy electron scattering and diffraction with Browne. The major hurdle to LEEM was from the very beginning the Recknagel formula for the resolution δ of a cathode lens, $\delta \sim V_0/E$, where V_0 is the start voltage of the electrons and E the electric field strength at the cathode [34]. Applied indiscriminately to LEEM where V_0 is much larger than in PEEM or thermionic electron emission microscopy (THEEM), a much poorer resolution would be expected in LEEM than in the emission modes. I had addressed this problem early in the game by looking at the aberrations of the homogeneous field in front of the cathode but did not report the results until Cruise had confirmed my predic-

Fig. 2. Temporary straight beam set-up during the initial LEEM development phase with the then still young developers.

tions of a better resolution of LEEM than PEEM by his calculations for the Bartz lens [30] used in our instrument. Cruise's main job was computational chemistry of propellants so that he could do the calculations for me only on the side. He first developed a new computational procedure for electron optics [35], the charge density method, and applied it then to our specific problem [36] confirming that the homogeneous field calculations [37] were realistic and that the LEEM instrument design based on them [38] was promising. The homogeneous field calculations were so simple that I did not publish them until much later [39] together with some of Cruise's data in order to counter the persistent misunderstood Recknagel resolution argument from the electron microscope community.

In writing my book on electron diffraction [28] I had become not only aware that an adequate description of the diffraction of slow electrons required a dynamical theory but also that the scattering of slow electrons by the individual atoms in the crystal could not be calculated within the Born approximation. In order to understand the diffraction contrast in LEEM we embarked, therefore, early in an effort aimed at a managable multiple scattering theory for slow electrons. We started by developing effective scattering potentials V_{xc} for free atoms taking into account exchange and correlation and applied them successfully to free atoms for which measurements and more precise calculations were available [40]. For atoms in solids we used superposition of atom potentials truncated at the nearest-neighbor distances plus an energy-dependent V_{xc} from the free-electron approximation. The $I(V)$ curves obtained with these potentials from a multiple scattering calculation in a column approximation which we presented at the first LEED theory seminar in Brooklyn 1967 had much broader maxima than experiment, due to a trivial error: in the amplitude attenuation due to inelastic scattering I had used the inelastic mean free path characteristic for the intensity attenuation. This lead us to give up our theoretical efforts, in particular as in the meantime high power professional theoreticions such as Charlie Duke had entered the field. Some results of our early work have been reported much later [41,42].

Returning to the LEEM instrument development, further progress was very slow due to the many problems encountered. For example, in a period of changes on the instrument we moved it into a then low AC field region further away from the power lines which ran past our room. When we resumed microscopy again with the supposedly improved conditions, we got very poor images. There were many possible causes such as instability of the specimen holder, of power supplies, charging, etc. which we eliminated until we finally discovered that the public works department had improperly re-wired in the meantime the AC power distribution grid in the basement below us, causing AC fields not existing before. Other problems were increasing high voltage arcovers with increasing complexity of the system which frequently damaged power supplies causing considerable delays until we could reduce them with inductive damping and other tricks.

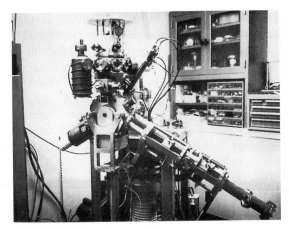

Fig. 3. Final configuration of the first LEEM.

Good THEEM was delayed by these problems until summer 1967, when we could take images of BaO layers on W(110) at magnifications up to 8000 × with about 0.3 μm resolution, while PEEM was still limited by intensity problems to magnifications below 3000 ×. Still no acceptable LEED patterns could be obtained. This lead to the unsuccessfull attempt to build a stable Zr–O–W(100) field emission gun. Although activation of a small spot source was no problem, frequent reactivation by in situ Zr deposition made operation at high voltages impractical.

We, therefore, decided to use a cored oxide cathode when we assembled the system in January 1968 in the final set-up (Fig. 3) with the 60° deflector block which separates incident and reflected beam and allows LEEM in addition to LEED, MEM and the emission imaging modes. In mid-February we had the beam to the final screen and to the specimen but it took until the middle of March before we had (poor) MEM, THEEM and what we believed to be LEEM images which later, however, turned out to be secondary electron emission (SEEM) images. Proper handling of the remanence problems of the Armco iron deflector and compensation of its astigmatism made for slow progress, but in May good THEEM images could be obtained as well as reflected or secondary electron images over an energy range from 0–40 eV. By that time management had changed and was understandably loosing patience. Nevertheless, work continued.

In autumn 1968 we finally received the channel plate image intensifier which had long been classified for military reasons and from spring to summer 1969 we made a last attempt at the Zr–O–W(100) field emission gun. Then Turner left my group. Shortly afterward I left the Michelson Lab with the permission to take all electron optical components with me which was the basis on which the work in the seventies in Clausthal was built.

I have described the first decade of LEEM in so much detail for two reasons. First of all to give recognition to George Turner who has done all the early pioneering LEEM instrument development work without receiving recognition for his achievements by the scientific community. Secondly, this example is a good illustration of how tedious and time-consuming the development of a new method is, in particular if a lot of new physics and technology is needed. Less ambitious endeavours, extending or combining known methods, obviously bring faster success. For example, by improving the vacuum using UHV technology to a large extent, Eichen et al. [43] built a MEM capable of working in the 10^{-8} Torr range. Delong and Drahos [44] used differential pumping with an Orbion pump in an emission microscope and could, thus, reach about 10^{-8} Torr which enabled them to publish the first LEED patterns in a surface electron microscope [45] with the basic set-up shown in Fig. 2. Although this mode of operation is the one which stimulated the invention of LEEM [29] we had not used it in our early work [33] which concentrated on the connection between microstructure and emission properties. By the end of the first decade of surface electron microscopy two true, bakeable emission microscopes were in operation. One, which had been built by Griffith et al. [46], partially from the components of the Fort Belvoir instrument, however, was not used in surface science, the other one by Recknagel et al. [47] apparently did not produce any additional results beyond those reported in Ref. [47]. This instrument, which incorporated a LEED and a RHEED system, was the most versatile operating surface electron microscope at that time. The impact of all these instruments on surface science, however, was small not only in the first decade but also in the one to follow.

3. The second decade: adolescence or the formative years

In the late sixties Auger electron spectroscopy (AES) had been developed as a tool for chemical surface characterization. In view of the large practical importance of the distribution of the chemical species on the surface, the seventies became the decade of scanning Auger electron microscopy (SAM). Mc Donald and Waldrop demonstrated the feasibility of SAM already in 1971 by incorporating a cylindrical mirror analyzer (CMA) into a conventional scanning electron microscope (SEM) [48]. This approach was also followed later in commercial UHV-SEMs, the JEOL-JAMP [49] and the VG HB50 which was equipped with a field emission gun [50]. There were also numerous other developments (for a review see Ref. [51]). The most successful incorporated the illumination system coaxially into the CMA and found widespread distribution over the whole world [52].

Soon most users of these instruments found out, however, that even with the best-designed illumination system the signal decreased very rapidly with decreasing spot size making it practically impossible to use the smaller spot sizes for high resolution imaging. This is an inherent problem of AES caused by the low Auger electron yield and the high background on which the Auger electron signal rides. As a result, SAM never developed into a high resolution surface imaging technique but is more or less used on the 0.1–1 μm level. Chemical analysis of smaller areas is usually accomplished by spot or line profile analysis combined with SEM imaging. Another problem is quantification in SAM. The Auger signal depends generally upon angles of incidence and emission and is, thus, very sensitive to surface roughness. Although this problem has been reduced later significantly by various correction algorithms, SAM has found no widespread application in pure surface science, as important as it has become for practical surface analysis.

The fixation in the seventies on chemical imaging had as a consequence that hardly any UHV-SEM surface imaging with true secondary electrons was done although these are much more suitable for imaging because of their usually high yield. This method is particularly well suited for fundamental in situ film growth studies at elevated temperatures at which sufficiently large three-dimensional crystals form [53]. The desire to study nucleation and growth in situ with high resolution under UHV conditions lead to the incorporation of UHV specimen chambers in conventional transmission microscopies (TEM). The most successful approach, the "cryocage" system was first used by Poppa [54] and perfected by Honjo's group [55]. A small cage surrounding the specimen and cooled with liquid nitrogen or helium was inserted in a UHV chamber between objective and condensor of a commercial TEM. High pumping speeds of Vacion, Orbion, Ti sublimation or cryopumps, combined with small apertures between the specimen chamber and the rest of the microscope allowed to maintain a pressure difference of several orders of magnitude between the conventional vacuum and UHV parts of the instrument. Thus, experiments in the 10^{-9}–10^{-10} Torr range at the specimen became possible and were done with great success in the late seventies and in the eighties. A parallel and equally important development was the introduction of UHV into REM in a very similar manner, also by Honjo's group [56].

While SAM and UHV-TEM attracted increasing attention in the seventies, interest in non-scanning emission microscopy decreased, largely because the limitations of non-UHV instruments became increasingly evident. A conference in 1979 dedicated solely to emission microscopy [57] summarized the results of the late years of non-UHV emission microscopy (see also the review [58]). Only in a few laboratories UHV instruments were developed. Griffith, Rempfer et al. started in 1975 with a second generation PEEM instrument (see Ref. [17]). Bethge et al. combined a UHV-PEEM with LEED and AES [59] and my group in Clausthal continued working on LEEM. Progress was slow because I was busy building up a surface science institute and fighting student unrest while Koch, the experienced PEEM microscopist, was so absorbed by administrative duties that the master's students working on the instrument were more or less left to themselves. Although the instrument was rebuilt by 1972 [60] it was not operating until the late seventies when a particularly good student, W. Telieps, took it over. He did a very thorough study of the electron optics of the system and by summer 1978 he obtained THEEM and PEEM images of a quality comparable to that obtained ten years before in China Lake. It took, however, two more years and many instrument improvements, in particular in the field emission illumination system, before he succeeded with MEM and LEEM, initially with a resolution of about 200 nm. By the time he finished his thesis [61], LEEM and PEEM resolution was improved to 50 and 60 nm, respectively. This leads us to the third decade of surface electron microscopy.

4. The third decade: maturity

In the early eighties the instrument development efforts of the seventies began to bear fruits and in the second half of this decade UHV surface electron microscopy was flourishing. Initially the TEM and REM studies of the Tokyo Institute of Technology group (Takayanagi, Yagi and Honjo) produced the most impressive results on the structure of metal and Si(111) surfaces and its modification by metal condensation (see the review in Ref. [62]). The scanning counterpart (SREM) of conventional REM (CREM), proposed in the seventies by Cowley (see Ref. [63] and realized in a non-UHV system [64], was adapted by Petroff's group [65] to UHV in a manner similar to that of Honjo's group. It was used successfully in scanning TEM and REM studies of reconstructed Si(111) and GaAs(100) surfaces [66]. The poor accessibility of the specimen was eliminated in a later modification of another commercial STEM which allowed AES and in situ MBE growth [67]. A quite different approach to SREM ("μ-RHEED") was taken by Ichikawa et al. [68] who started from a UHV system equipped with a CMA and other surface science tools and added a field emission (FE)-SEM to it. This allowed them to combine SREM not only with μ-RHEED but also with -AES which was very useful in the later in situ growth studies on semiconductor surfaces. A similar system, but without CMA, built by Cowley's group [69], apparently did not find much application, inspite of its better resolution. Ichikawa's approach was later extended by incorporating a FE-SEM into a large MBE system with a separate surface analysis chamber [70]. In this system compound semiconductor MBE can be studied under production conditions.

An even more versatile UHV scanning electron beam system was built by Ichinokawa et al. [71]. It also combined a field emission SEM with a CMA but in addition had also a LEED optics so that imaging could be complemented not only by AES and energy loss spectroscopy with the CMA but also by LEED and work function change measurements with the LEED optics. The system was not used in the usual SREM mode but was originally intended for low energy SEM with primary energies below 2 keV for which a high surface sensitivity is expected [72]. A natural later extension was to use the diffracted LEED beams for scanning low energy REM [73], the scanning counterpart to LEEM.

By the mid-eighties UHV-REM, both in the conventional (non-scanning) and in the scanning mode, had made important contributions to the understanding of the microstructure of clean surfaces. They are summarized in the proceedings of a 1987 NATO conference [74] and several reviews [75,76] which include also the important contributions from non-UHV-REM. UHV-TEM also was further improved [77] and continued to give information on surface microstructure and on surface processes. Its major impact, however, was in the determination of the atomic structure of surfaces, either indirectly via TED or directly, mainly via profile imaging. Thus, Takayanagi et al. [78] succeeded to analyze the TED pattern of the Si(111)-(7 $\times$ 7) structure which lead them to the famous dimer–adatom stacking-fault (DAS) model, ending a long controversy about this surface reconstruction. Atomic resolution in profile imaging had already been achieved in 1983 by

Marks and Smith [79] but this was done in conventional vacuum with the accompanying residual gas interactions. The importance of these interactions became evident in the first in situ UHV profile imaging studies [80] which showed atomic relaxations different from those found in conventional vacuum, and is further discussed in Ref. [81]. The contributions of profile imaging to surface science up to 1987 have been summarized by Smith [82]. It should be mentioned that ordinary "plan-view" TEM also continued to find important applications. One of many examples is the study of the growth of Au films on Ag surfaces in the monolayer range, using carbon films for stabilization of the Au layers [83].

Before turning to the evolution of PEEM LEEM and other low energy microscopies in the eighties one scanning imaging mode using fast primary electrons, biased secondary electron microscopy, should be mentioned [84,85]. The most adsorbate-sensitive surface imaging method with wide application range, however, remained PEEM due to the strong work function dependence of the photo yield. Bethge's instrument became productive in the early eighties [86–88]. One interesting result was the strong doping sensitivity of PEEM ranging from 10^{11} to 10^{16} Zn atoms/cm^2 in GaAs, and its enhancement by CH_4 adsorption [88]. By 1985 our LEEM instrument also produced good PEEM images [89] but LEEM was so exciting that there was little time for PEEM studies. Only later, after Telieps' untimely death in 1987, was the instrument used more frequently in the PEEM mode, one example being the study of the decoration of monatomic steps on Mo(110) by Cu and the step flow growth of the Cu monolayer on this surface [90]. Another example is the PEEM imaging of three-dimensional Cu silicide crystals on Si(111) [91]. Until the late eighties Bethge's, our and an instrument similar to ours at the Fritz Haber Institut (FHI) in Berlin were the only UHV-PEEMs in surface science and the scientific output was correspondingly low. The development of scanning-PEEM in 1989 [92,93] and its application to the study of chemical reactions on surfaces greatly stimulated the further development of PEEM in surface science. The next step was the construction of flange-on PEEMs with the specimen at ground potential which was started in Clausthal [94] and shortly thereafter in Berlin and lead to an instrument [95] which proved immediately its superiority over the scanning PEEM [96]. Ertl's group at the FHI Berlin has since obtained stunning videos of oscillatory and chaotic surface reactions.

The photoelectrons in the PEEMs discussed up to now originate from valence levels or from the conduction band. The contrast is largely determined by the work function and contains only indirect information on the chemical nature of the surface. Direct information can be obtained by imaging with photoelectrons originating from core levels which requires excitation with synchrotron radiation. The possibilities of this type of imaging have been discussed already in 1984 by Cazaux [97] but applications in surface science appeared only in the late eighties. Both modes of operation, scanning ("microspectroscopy") and non-scanning ("spectromicroscopy") were developed at the various synchrotron radiation facilities. The Wisconsin [98], Brookhaven [99] and DESY Hamburg groups [100] choose the first path, all with different photon focussing systems, Tonner's group, also at Wisconsin [101], and the Stanford group [102,103], the second route. While Tonner et al. used an immersion lens as in ordinary PEEM, the Stanford group used a commercial magnetic projection PEEM [104] based on the concepts of Turner et al. [105,106].

Returning to slow electrons, the lowest energies are encountered in MEM, in which the electrons are reflected in front of the surface. MEM is a natural by-product of LEEM but (short-lived) MEM instruments have also been developed independently [107,108]. An interesting non-UHV MEM system which allowed interferometry [109] should also be mentioned. We have used our LEEMs only occasionally in the MEM mode and only sometimes in the zero-impact energy mode for local work-function measurements because LEEM is much more powerful for surface science studies of crystalline materials. The first images of monatomic steps on Mo(110) and the measurement of the step height [89] were soon followed by a study of the Si(111)-$(7 \times 7) \leftrightarrow (1 \times 1)$ phase transition at 1100 K [110]. It revealed many more

details than the preceeding REM studies [111] which were hampered by the strong foreshortening caused by the small grazing angle of incidence. In particular the videos of the phase transition made such a strong impression on the surface science community that several groups started to build LEEMs too: at the FHI Berlin, according to our design, at IBM Yorktown Heights, at the Max Planck Institute for Plasma Physics in Munich and in several other places. The Berlin LEEM became operational with limited resolution in 1992 and soon produced interesting results on reactions in adsorbed layers [112]. The Munich and IBM instruments which contained several new design ideas were described in 1991 [113,114]. The IBM group became very rapidly productive but work on the Munich instrument was terminated unfortunately. An instrument built by Delong et al. [115] had suffered the same fate already several years before. En route to the third generation instrument which will allow energy-filtered imaging with Auger and core photoelectrons [116,117], in the meantime we had built a second generation instrument which soon after completion was used in semiconductor surface studies [118,119], similar to the IBM instrument [120,121]. In spite of the small number of LEEM instruments in operation a large amount of valuable information on surface structure and processes has already been obtained with this method (for recent reviews on instrumentation and applications see Refs. [122,123]).

The goal of the microscopies discussed up to now was the determination of the surface topography, work function, crystal structure and orientation or chemical species distribution on the surface. Only charge and energy of the electrons are used in these imaging modes. If the spin of the electron is also used in imaging, information on the magnetic microstructure of the surface can be obtained. One possibility, the spatially resolved measurement of the spin polarization of secondary electrons was already discussed in the late seventies [124]. By the mid-eighties two groups had demonstrated the power of this technique (spin-SEM or SEMPA = scanning electron microscopy with polarization analysis) [125,126].

The success of this initial work and the large interest in the magnetic microstructure of surfaces and thin films soon lead to the development of other SEMPA instruments [127,128] and to impressive results which are reviewed up to 1989 in Ref. [129]. A more recent example of the power of this technique is the demonstration of the oscillatory magnetic coupling between iron films through a Cr film as a function of its thickness [130]. A second magnetization-sensitive surface imaging mode, spin-polarized LEEM (SPLEEM), is based on the spin dependence of the elastic scattering from magnetic materials. This imaging mode is easily realized in a LEEM instrument equipped with a spin-polarized electron source. This was done recently in Clausthal in cooperation with IBM San Jose [131]. After the usual initial difficulties in situ SPLEEM studies have already produced interesting results for ultrathin epitaxial Co layers [132], for example on the evolution of the magnetic microstructure during in situ film growth or annealing [133], with much shorter data aquisition times than those needed in SEMPA. This brings us to the present state of the art.

5. Present state and future prospects

Today surface science has a large arsenal of methods for surface imaging with electrons available, with resolutions from the sub-nm to the μm range. They give information on the atomic structure, the microstructure, the chemical and the magnetic structure, all under well-defined vacuum conditions and frequently with in situ manipulation possibilities such as heating, cooling, gas exposure, vapor deposition, sputtering etc. For most surface science problems several methods are suited but one will always choose the simplest and cheapest one which can solve the problem. A good example are the oscillatory and chaotic surface reactions for which the low resolution of a simple flange-on PEEM is sufficient [96]. On the other hand, the direct measurement of atomic relaxations on a surface by profile imaging or chemical identification in the nm range by AES requires sophisticated expensive high resolution UHV instruments with field emission guns. Instruments of this kind have already been built in the second or third generation, for example for TEM [134], REM combined with PEEM

Fig. 4. Third generation, computer-operated LEEM.

[135] or SEM combined with SAM [136]. Also second and third generation PEEM [137] and LEEM instruments are now in operation. Fig. 4 shows a professionally built LEEM based on Veneklasens design [116]. All lenses are computer-controlled and a UHV preparation chamber with airlock allows easy sample exchange and preparation, e.g. outgassing or sputter-cleaning. Electron microscopists usually put a lot of emphasis on resolution. In surface science this is frequently less important than image aquisition time and interpretation. Image interpretation problems have lead to the development of multi-method imaging systems, e.g. the MULSAM [138], and of sophisticated image analysis software.

What is ahead for surface electron microscopy in surface science? One major future effort certainly will be in image processing and analysis. The amount of data produced by a well functioning microscope is so large that much information will be lost without efficient image analysis procedures which are also necessary for the extraction of quantitative data. Instrument development certainly will continue too. Examples close to my interest are new versions of LEEM/MEM/PEEM [139,140], SPLEEM [141] or scanning LEEM [142]. Further developments may be aberration-corrected LEEM systems which have been proposed recently [143]. The efforts to combine several imaging modes in the same system such as LEEM and core electron PEEM have already been mentioned and other combinations such as LEEM and STM are also obvious. In addition to these more sophisticated systems there is a definite need for simpler systems which can be incorporated in many surface science systems. This is the lesson which the experience of the past teaches us: a large number of surface imaging systems has been built at high financial and human effort cost. Many of these systems were never or only briefly used. Only few made major contributions to surface science over a period of time. The reasons for this unfortunate situation are manyfold: the complexity of many systems which causes long development times, resulting in loss of financial support and exhaustion of the instrument builder; the lack of inclination of many instrument-oriented scientists to use their instru-

ments for systematic studies; the complexity of many instruments which makes their use by measurement-oriented scientists difficult and unattractive. As human nature cannot be changed the obvious route which surface electron microscopy must go – a few sophisticated instruments with very experienced users excepted – is towards simpler affordable and dependable workhorse microscopes. Once this happens, surface science will make a major move forward again similar to the one initiated by STM.

6. Closing remarks

This historical perspective of surface electron microscopy was strongly method-oriented because the various results obtained with it are dealt with in other chapters of this volume. The review is far from comprehensive and many important aspects, in particular those involving specimen transfer from a surface science instrument to a non-UHV microscope, have only been touched upon. I have selected LEEM as a case study because I know its evolution in detail from the very beginning and because this example has many of the ingredients of unsuccessful and successful efforts of development of a new method. In conclusion I would like to dedicate this review to my late friend and coworker Wolfgang Telieps. Without him the development of LEEM may well have ended up in the category "unsuccessful efforts".

Acknowledgements

The author's work discussed here was initially supported by Navy Independent Research Funding. Later, the various aspects of the work were and still are supported by the Deutsche Forschungsgemeinschaft, the Volkswagen Foundation and by the Bundesminister für Forschung und Technologie. I wish to express my thanks also to all my coworkers in the endeavour "UHV surface electron microscopy", some but not all of whom have been mentioned in the references.

References

[1] E. Brüche and H. Johannson, Naturwiss. 20 (1932) 49, 353; Ann. Phys. 15 (1932) 145.

[2] M. Knoll, F.G. Houtermans and W. Schulze, Z. Phys. 78 (1932) 340.

[3] E. Brüche Z. Phys. 86 (1933) 448.

[4] V.K. Zworykin, J. Franklin Inst. 215 (1933) 554.

[5] M. Knoll and E. Ruska, Ann. Phys. 12 (1932) 622.

[6] E. Ruska Z. Phys. 83 (1933) 492.

[7] M. Knoll, Z. Tech. Phys. 11 (1935) 467.

[8] G. Hottenroth, Ann. Phys. 30 (1937) 689.

[9] B. von Borries, Z. Phys. 116 (1940) 370.

[10] E. Kinder, Naturwiss. 30 (1942) 591.

[11] W. Mecklenburg, Z. Phys. 120 (1943) 21.

[12] H. Mahl, Z. Tech. Phys. 21 (1940) 17; 22 (1941) 33; Metallwirtschaft 21 (1940) 488.

[13] H.O. Müller Kolloid-Z. 99 (1942) 6.

[14] V.K. Zworykin, J. Hillier and R.L. Snyder, Bull. Am. Soc. Test. Mater. 117 (1942) 15.

[15] G. Möllenstedt and F. Lenz, Adv. Electron. Electron Phys. 18 (1963) 251.

[16] A.B. Bok, J.B. le Poole, J. Roos, H. de Lang, H. Bethge, J. Heydenreich and M.E. Barnett, in: Advances in Optical and Electron Microscopy, Vol. 4, Eds. R. Barer and V.E. Cosslett (Academic Press, London, 1971) p. 161.

[17] O.H. Griffith and W. Engel, Ultramicroscopy 36 (1991) 1.

[18] L. Wegmann, J. Microsc. 96 (1972) 1.

[19] D.E. Bradley, Brit. J. Appl. Phys. 5 (1954) 65.

[20] E. Bauer, B. Fritz and E. Kinder, in: 4th Int. Conf. Electron Microscopy, Berlin, 1958, Vol. 1, Eds. W. Bargmann et al. (Springer, Berlin, 1960) p. 430.

[21] E. Bauer, PhD Thesis, University Munich, 1955.

[22] G.A. Bassett, Philos. Mag. 3 (1958) 1042.

[23] H. Bethge, Surf. Sci. 3 (1964) 33.

[24] W. Koch, Optik 25 (1967) 523.

[25] L.H. Germer and C.D. Hartmann, J. Appl. Phys. 31 (1960) 2085.

[26] E. Bauer, Phys. Rev. 123 (1961) 1206.

[27] H. Boersch, Ann. Phys. 26 (1936) 631; 27 (1936) 15.

[28] E. Bauer, Elektronenbeugung (Verlag Moderne Industrie, Munich, 1958).

[29] E. Bauer, in: Electron Microscopy, Vol. 1, Ed. S.S. Breese (Academic Press, New York, 1962) p. D-11.

[30] G. Bartz, in: 4th Int. Conf. Electron Microscopy, Berlin, 1958, Vol. 1, Eds. W. Bargmann et al. (Springer, Berlin, 1960) p. 201.

[31] G. Liebmann, Proc. Phys. Soc. B 65 (1952) 94.

[32] G. Möllenstedt and O. Rang, Z. Angew. Phys. 3 (1951) 187.

[33] G.H. Turner and E. Bauer, in: Electron Microscopy 1966, Vol. 1, Ed. R. Uyeda (Maruzen, Tokyo, 1966) p. 163.

[34] A. Recknagel, Z. Phys. 117 (1941) 689.

[35] D.R. Cruise, J. Appl. Phys. 34 (1963) 3477.

[36] D.R. Cruise and E. Bauer, J. Appl. Phys. 35 (1964) 3080.

[37] E. Bauer, J. Appl. Phys. 35 (1964) 3079.

[38] G. Turner and E. Bauer, J. Appl. Phys. 35 (1964) 3080.

[39] E. Bauer, Ultramicroscopy 17 (1985) 51.

[40] E. Bauer and H.N. Browne, in: Atomic Collision Processes, Ed. M.R.C. McDowell (North-Holland, Amsterdam, 1964) p. 16.

[41] E. Bauer, J. Vac. Sci. Technol. 7 (1970) 3.

[42] E. Bauer, in: Interaction des Electrons avec la Matière Condensee (Verbier, 1972) p. 42.

[43] E. Eichen, R.L. Forgacs and B.A. Parafin, Rev. Sci. Instrum. 37 (1966) 438.

[44] A. Delong and V. Drahos, J. Phys. E 1 (1968) 397.

[45] A. Delong and V. Drahos, Nat. Phys. Sci. 230 (1971) 196.

[46] O.H. Griffith, Proc. Natl. Acad. Sci. USA 69 (1972) 561.

[47] A. Recknagel, H. Mahnert and J. Brückner, Optik 35 (1972) 376.

[48] N.C. McDonald and J.R. Waldrop, Appl. Phys. Lett. 19 (1971) 315.

[49] A. Mogami and T. Sekine, Israel EMI (1976) (JEOL-JAMP).

[50] J.A. Venables, A.P. Janssen, C.J. Harland and B.A. Joyce, Philos. Mag. 34 (1976) 495.

[51] J.A. Venables and A.P. Janssen. 9th Int. Congr. Electron Microscopy, Toronto, 1978, Vol. III, p. 280.

[52] N.C. McDonald, C.T. Hovland and R.L. Gerlach IITRI/SEM 1977, p. 201: 7th Int. Vacuum Congr. Vienna, 1977, p. 2303 (Physical Electronics).

[53] K. Hartig, A.P. Janssen and J.A. Venables, Surf. Sci. 74 (1978) 69.

[54] H. Poppa, J. Vac. Sci. Technol. 2 (1965) 42.

[55] K. Takayanagi, K. Yagi, K. Kobayashi and G. Honjo, J. Phys. E 11 (1978) 441.

[56] N. Osakabe, Y. Tanishiro, K. Yagi and G. Honjo, Surf. Sci. 97 (1980) 393.

[57] G. Pfefferkorn and K. Schur, Eds., 1st Int. Conf. Emission Electron Microscopy, Beiträge zur Electronenmikroskopischen Direktabbildung von Oberflächen 12/2 (1979) 1.

[58] R.A. Schwarzer, Microsc. Acta 84 (1981) 51.

[59] H. Bethge, G. Gerth and D. Matern, 10th Int. Congr. Electron Microscopy, Hamburg, 1982, Vol 1, p. 69.

[60] W. Koch, B. Bischoff and E. Bauer, 5th Eur. Congr. Electron Microscopy (Institute of Physics, London, 1972) p. 58.

[61] W. Telieps, PhD Thesis, TU Clausthal, 1983.

[62] K. Yagi, K. Takayanagi and G. Honjo, in: Crystals. Growth, Properties and Applications, Ed. H.C. Freyhardt, Vol. 7 (Springer, Berlin, 1982) p. 47.

[63] J.M. Cowley, in: Microbeam Analysis 1980, Ed. D.B. Wittig (San Francisco Press, San Francisco, 1980) p. 33.

[64] J.M. Cowley, J.L. Albain, G.G. Hembree, P.E. Hojlund-Nielsen, F.A. Koch, J.D. Landry and H. Shuman, Rev. Sci. Instrum. 46 (1975) 826.

[65] R.J. Wilson and P.M. Petroff, Rev. Sci. Instrum. 54 (1983) 1534.

[66] P.M. Petroff in: Electron Microscopy of Materials (MRS Symp. Vol. 31), Eds. W. Krakow, D.A. Smith and L.W. Hobbs (North-Holland, New York, 1984) p. 117.

[67] P.M. Petroff, C.H. Chen and D.J. Werder, Ultramicroscopy 17 (1985) 185.

[68] M. Ichikawa and K. Hayakawa, Jpn. J. Appl. Phys. 21 (1982) 145, 154.

[69] C. Elibol, H.-J. Ou, G.G. Hembree and J.M. Cowley, Rev. Sci. Instrum. 56 (1985) 1215.

[70] T. Isu, A. Watanabe, M. Hata and Y. Katayama, Jpn. J. Appl. Phys. 27 (1988) L 2259; J. Vac. Sci. Technol. 7 (1989) 714; J. Cryst. Growth 100 (1990) 433.

[71] T. Ichinokawa and Y. Ishikawa, Ultramicroscopy 15 (1984) 193.

[72] T. Ichinokawa, Y. Ishikawa, N. Awaya and A. Onoguchi, in: Scanning Electron Microscopy 1981, Ed. O. Johari (SEM. AMF O'Hare) Vol. I, p. 271.

[73] J. Kirschner, T. Ichinokawa, Y. Ishikawa, M. Kemmochi N. Ikeda and Y. Hosokawa, in: Scanning Electron Microscopy 1986, Ed. O. Johari (SEM, AMF O'Hare) Vol. II, p. 331.

[74] P.K. Larson and P.J. Dobson, Eds., Reflection High Energy Electron Diffraction and Reflection Imaging of Surfaces (NATO ASI Ser. B 188) (Plenum, New York, 1988) p. 261.

[75] K. Yagi, J. Appl. Crystallogr. 20 (1987) 147.

[76] J.M. Cowley, in: Surface and Interface Characterization by Electron Optical Methods (NATO ASI Ser. B 191), Eds. A. Howie and U. Valdre (Plenum, New York, 1988) p. 127.

[77] K. Heinemann and H. Poppa, J. Vac. Sci. Technol. A 4 (1986) 127.

[78] K. Takayanagi, Y. Tanishiro, M. Takahashi and S. Takahashi, J. Vac. Sci. Technol. A 3 (1985) 1502; Surf. Sci. 164 (1985) 367.

[79] L.D. Marks and D.J. Smith, Nature 303 (1983) 316.

[80] T. Hasegawa, K. Kobayashi, N. Ikarashi, K. Takayanagi and K. Yagi, Jpn. J. Appl. Phys. 25 (1986) L366; in: Proc. 11th Int. Congr. on Electron Microscopy, Kyoto, 1986, p. 1345.

[81] L.D. Marks, R. Ai, J.E. Bonevich, M.I. Buckett, D. Dunn, J.P. Zhang, M. Jacoby and P.C. Stair, Ultramicroscopy 37 (1991) 90.

[82] D.J. Smith. in: Surface and Interface Characterization by Electron Optical Methods (NATO ASI Ser. B 191). Eds. A. Howie and U. Valdre (Plenum, New York, 1988) p. 43.

[83] M. Klaua and H. Bethge, Ultramicroscopy 17 (1985) 73.

[84] M. Hanbücken, M. Futamoto and J.A. Venables, Inst. Phys. Conf. Ser. 68 (1983) 135.

[85] M. Futamoto, M. Hanbücken, C.J. Harland, G.W. Jones and J.A. Venables, Surf. Sci. 150 (1985) 430.

[86] H. Bethge and M. Klaua, Ultramicroscopy 11 (1983) 207.

[87] H. Bethge, Th. Krajewski and O. Lichtenberger, Ultramicroscopy 17 (1985) 21.

[88] W. Driesel and H. Bethge, in: Proc. Energy Phase Modification of Semiconductors and Related Materials, Dresden, 1984, Vol. 2 (1985) 595.

[89] W. Telieps and E. Bauer, Ultramicroscopy 17 (1985) 57.

[90] M. Mundschau, E. Bauer and W. Swiech, Surf. Sci. 203 (1988) 412.

[91] M. Mundschau, E. Bauer, W. Telieps and W. Swiech, J. Appl. Phys. 65 (1989) 4747.

[92] H.H. Rotermund, G. Ertl and W. Sesselmann, Surf. Sci. 217 (1989) L383.

[93] H.H. Rotermund, S. Jakubith, A. von Oertzen and G. Ertl, J. Chem. Phys. 91 (1989) 4942.

[94] B. Weidenfeller and E. Bauer, unpublished.

[95] W. Engel, M.E. Kordesch, H.H. Rotermund, S. Kubala and A. von Oertzen, Ultramicroscopy 36 (1991) 148.

[96] H.H. Rotermund, W. Engel, S. Jakubith, A. von Oertzen and G. Ertl, Ultramicroscopy 36 (1991) 164.

[97] J. Cazaux, Ultramicroscopy 17 (1984) 43.

[98] F. Cerrina, G. Margaritondo, J.H. Underwood, M. Hettrick, M.A. Green, Jr., L.J. Brillson, A. Franciosi, H. Höchst, P.M. Deluca, Jr. and M.N. Gould, Nucl. Instrum. Methods A 266 (1988) 303.

[99] H. Ade, J. Kirz, S. Hulbert, E. Johnson, E. Anderson and D. Kern, Nucl. Instrum. Methods A 291 (1990) 126.

[100] J. Voss, H. Dadras, C. Kunz, A. Moewes, G. Roy, H. Sievers, I. Storjaham and H. Wongel, J. X-ray Sci. Technol. 3 (1992) 85.

[101] B.P. Tonner and G.R. Harp, Rev. Sci. Instrum. 59 (1988) 853; J. Vac. Sci. Technol. A 7 (1989) 1.

[102] P. Pianetta, I. Lindau, P.L. King, M. Keenlyside, G. Knapp and R. Browning, Rev. Sci. Instrum. 60 (1989) 1686.

[103] P. Pianetta, P.L. King, A. Borg, C. Kim, I. Lindau, G. Knapp, M. Keenlyside and R. Browning, J. Electron Spectroscop. Relat. Phenom. 52 (1990) 797.

[104] R.L. Chaney, Surf. Interface Anal. 10 (1987) 36.

[105] G. Beamson, H.Q. Porter and D.W. Turner, J. Phys. E 13 (1980) 64; Nature 290 (1981) 556.

[106] D.W. Turner, J.R. Turner and H.Q. Porter, Phys. Trans. Roy. Soc. A 318 (1986) 219.

[107] C. Guittard, M. Babout, S. Guittard and M. Bujor, in: Scanning Electron Microscopy 1981, Ed. O. Johari (SEM, AMF O'Hare), Vol. 1, p. 199 and references therein;
J.C. Dupuy, A. Sibai and B. Vilotitch, Surf. Sci. 147 (1984) 191.

[108] M.S. Foster, J.C. Campuzano, R.F. Willis and J.C. Dupuy, J. Microsc. 140 (1985) 395.

[109] H. Lichte, Optik 57 (1980) 35; Proc. Int. Symp. Foundations of Quantum Mechanics, Tokyo, 1983, p. 29.

[110] W. Telieps and E. Bauer, Surf. Sci. 162 (1985) 163; Ber. Bunsenges. Phys. Chem. 90 (1986) 197.

[111] N. Osakabe, Y. Tanishiro, K. Yagi and G. Honjo, Surf. Sci. 109 (1981) 353.

[112] B. Rausenberger, W. Swiech, W. Engel, A.M. Bradshaw and E. Zeitler, Surf. Sci. 287/288 (1993) 235.

[113] H. Liebl and B. Senftinger, Ultramicroscopy 36 (1991) 91.

[114] R.M. Tromp and M.C. Reuter, Ultramicroscopy 36 (1991) 99.

[115] A. Delong and V. Kolarik, Ultramicroscopy 17 (1985) 67.

[116] L.H. Veneklasen, Ultramicroscopy 36 (1991) 76.

[117] E. Bauer, Ultramicroscopy 36 (1991) 52.

[118] R.J. Phaneuf, W. Swiech, N.C. Bartelt, E.D. Williams and E. Bauer, Phys. Rev. Lett. 67 (1991) 2986.

[119] R.J. Phaneuf, N.C. Bartelt, E.D. Williams, W. Swiech and E. Bauer, Surf. Sci. 268 (1992) 227.

[120] R.M. Tromp and M.C. Reuter, Phys. Rev. Lett. 68 (1992) 820, 954.

[121] R.M. Tromp, A.W. Denier van der Gon and M.C. Reuter, Phys. Rev. Lett. 68 (1992) 2313.

[122] E. Bauer, in: Chemistry and Physics of Solid Surfaces VIII, Eds. R. Vanselow and R. Howe (Springer, Berlin, 1990) p. 267; Rep. Prog. Phys., to be published.

[123] L.H. Veneklasen, Rev. Sci. Instrum. 63 (1992) 5513.

[124] T.H. DiStefano, IBM Techn. Discl. Bull. 20 (1978) 4212.

[125] K. Koike and K. Hayakawa, Jpn. J. Appl. Phys. 23 (1984) L187; J. Appl. Phys. 57 (1985) 4244.

[126] J. Unguris, G.G. Hembree, R.J. Celotta, D.T. Pierce, J. Microsc. B 9 (1985) RP1.

[127] H.P. Oepen and J. Kirschner, J. Phys. (Paris) 49 (1988) C8-1853; in: O. Johari, Ed., Scanning Microscopy 1991, Vol. 5 (Scanning Microscopy International Chicago, 1990) Vol. 5, p. 1.

[128] R. Allenspach, M. Stampanoni and A. Bischof, Phys. Rev. Lett. 65 (1990) 3344.

[129] J. Unguris, M.R. Scheinfein, R.J. Celotta and D.T. Pierce, in: Chemistry and Physics of Solid Surfaces VIII, Eds. R. Vanselow and R. Howe (Springer, Berlin, 1990) p. 239.

[130] J. Unguris, R.J. Celotta and D.T. Pierce, Phys. Rev. Lett. 69 (1992) 1125.

[131] M.S. Altman, J. Hurst, G. Marx, H. Pinkvos, H. Poppa and E. Bauer, in: Magnetic Materials: Microstructure and Properties. Eds. T. Suzuki, Y. Sugita, B.M. Clemens, K. Onchi and D.E. Laughlin, MRS Symp. Proc. 232 (1991) 125.

[132] H. Pinkvos, H. Poppa, E. Bauer and J. Hurst, Ultramicroscopy 47 (1992) 339.

[133] H. Pinkvos, H. Poppe, E. Bauer and G.-M. Kim, in: Magnetism and Structure in Systems of Reduced Dimensions, Ed. M. Donath, NATO Workshop, Corsica, 1992, to be published.

[134] Y. Kondo, K. Ohi, Y. Ishibashi, H. Hirano, Y. Harada, K. Takayanagi, Y. Tanshiro, K. Kobayashi and K. Yagi, Ultramicroscopy 35 (1991) 111.

[135] Y. Kondo, K. Yagi, K. Kobayashi, H. Kobayashi, Y. Yanaka, K. Kise and T. Ohkawa, Ultramicroscopy 36 (1991) 142.

[136] J.A. Venables, J.M. Cowley and H.S. van Harrach, Inst. Phys. Conf. Ser. 90 (EMAG 87) 1; 119 (EMAG 91) 33.

[137] G.F. Rempfer, W.P. Skoczylas and O.H. Griffith, Ultramicroscopy 36 (1991) 196.

[138] M. Prutton, C.G.H. Walker, J.C. Greenwood, P.G. Kenny, J.C. Dee, J.R. Barkshire, R.H. Roberts and M.M. Gomati, Surf. Interface Anal. 17 (1990) 71.

[139] W.P. Skoczylas, G.F. Rempfer and O.H. Griffith, Ultramicroscopy 36 (1991) 252.

[140] O.H. Griffith, K.K. Hedberg, D. Deslodge and G.F. Rempfer, J. Microsc. 168 (1992) 249.

[141] K. Grzelakowski and E. Bauer, unpublished.

[142] I. Müllerova and M. Lenz, Ultramicroscopy 41 (1992) 399; Mikrochim. Acta 12 (1992) Suppl., 173.

[143] H. Rose and D. Preikszas, Optik 92 (1992) 31.

Surface Science 299/300 (1994) 116–128
North-Holland

Electron energy loss spectroscopy: the vibration spectroscopy of surfaces

H. Ibach

Institut für Grenzflächenforschung und Vakuumphysik, Forschungszentrum Jülich, D-52425 Jülich, Germany

Received 2 March 1993; accepted for publication 16 July 1993

Thirty years of inelastic scattering of electrons from surfaces are critically reviewed. The main emphasis is on the application of the method for the investigation of localized and collective vibrational modes.

1. Introduction

Electron energy loss spectroscopy has become one of the major tools in our arsenal of techniques for the investigation of surfaces. It is an unfortunate, but common practice to refer to experimental techniques by acronyms. This technique mostly goes under EELS, sometimes also under HREELS (high resolution...). High resolution is indeed required for the most prominent application which is in the studies of surface vibrations and vibrations of adsorbed species. In this area electron energy loss spectroscopy competes with a few other techniques. One of these techniques is Infrared Reflection Absorption Spectroscopy (IRAS), which was proposed by Greenler in 1966 [1] and realized experimentally on well defined surfaces first by A.M. Bradshaw and J. Pritchard [2]. With Fourier-transform spectrometers [3] the technique nowadays has a surface sensitivity of about 10^{-3} monolayers for strong IR-absorbers like CO. Submonolayer quantities of hydrocarbons can likewise be probed. The spectral resolution remains unsurpassed by any other surface technique. IRAS (with a single reflection) requires metallic surfaces which are probed in grazing incidence. Obviously one can observe only those vibrational modes which are associated with a dipole mo-

ment. Until recently, it was believed that the dipole moment had to be oriented perpendicularly to the metallic surface ("surface selection rule"). This, however, was a prejudice, based on an incomplete analysis of surface optics [4]. The prejudice stood for so many years because the parallel vibrational modes of strong dipole scatterers are in a spectral regime inaccessible to spectroscopy with conventional light sources. Only recent work using the synchrotron source made this spectral regime (200–500 cm^{-1}) accessible to IR-spectroscopy [5].

A second technique for studies on surface vibrations is the inelastic scattering of He atoms. The technique is described in detail in other contributions to this volume. Compared to EELS, He atom scattering has a higher resolution but is, however, limited to the lower spectral range of vibrations ($\hbar\omega < 30$ meV). Typical applications are substrate surface phonon spectroscopy and the spectroscopy of low lying vibrational modes of adsorbates, such as the so-called hindered translations. For a fair evaluation of the merits of inelastic electron and helium scattering one must include the consideration of the cross sections (section 3).

This review is organized as follows. The next section is a retrospect on the early days of electron energy loss spectroscopy with some indulgence to personal memories. Sections 3 and 4 give an account of the milestones in the development of theory and experiment, respectively.

2. The early days

Electron energy loss spectroscopy from surface vibrations was introduced by F.M. Propst and his student, T.C. Piper [6]. In their brief, but eminently important paper the authors investigated the dissociative adsorption of H_2, N_2, CO and H_2O on a clean W(100) surface. For the first time in the history of surface science, spectroscopic evidence for the dissociation of a species on a clean and well defined surface was demonstrated. The significance of this work was probably obvious to the scientific community, but so were the tremendous difficulties of the experiment. Presumably because of these difficulties, the technique was not further pursued in the United States at the time and it took over a decade before it returned to the country where the pioneering work was performed.

The idea found a more fertile ground in Germany. Through the work of H. Boersch, J. Geiger and H. Ehrhardt the technology of electron spectrometers was already better developed. In 1966 (a year before the work of Propst) H. Boersch, J. Geiger and W. Stickel [7] published a paper on the energy losses of 25 keV electrons after having passed through a thin film of LiF. The energy resolution was around 5 meV, and thus an order of magnitude higher than in the work of Propst and Piper. Later the same group achieved even resolutions below 2 meV [8], a record in energy loss spectroscopy which stood until the late 80ties. Despite the remarkable success, transmission spectroscopy was a cul-de-sac as far as applications in surface science were concerned. For one, the technique was based on fast electrons, which are less sensitive to surfaces than low energy electrons. More importantly, the transition to ultrahigh vacuum could not be made with the technology of electron beam handling of Boersch and Geiger [9]. In that regard the work of Ehrhardt et al. [10] became more instrumental to the further-

ance of surface science. Ehrhardt et al. developed a spectrometer for low energy electrons for the purpose of inelastic scattering from atoms and molecules. This spectrometer was entirely based on electrostatic deflection elements and lenses, and was compatible to the requirements of ultra high vacuum. A modified version of this spectrometer [11] proved reproducibly successful with a resolution of ~ 15 meV and achieved occasionally even resolutions better than that [12]. In retrospect, I believe I was indeed rather fortunate to pick ZnO as the first material for investigation. The results were relatively unimportant for surface science, as it turned out that the intense energy losses of low energy electrons reflected from a ZnO surface are caused by long wavelength dielectric surface phonons, which represent largely a bulk property of the material. These phonons are now known as Fuchs–Kliewer surface phonons [13]. Nevertheless the results were important for the understanding of the interaction of electrons with surface vibrations [14–16] and, even more importantly, they rendered more satisfaction and encouragement than the meager count rates from the genuine surface vibrations in the work of Propst and Piper. The reader may appreciate the excitement of a young researcher when he saw the series of intense losses in the spectrum of electrons back-reflected from a UHV-cleaved ZnO surface (Fig. 1). The next, again rather fortunate strike was to look for energy losses on the cleaved Si(111) surface [17]. Fortunate insofar, as this specific surface with its (2×1) reconstruction carries a surface phonon with a strong dipole moment. Whereas it was clear from adsorption experiments, that the surface phonon was of microscopic origin [17], details of its nature were debated until lately [18].

A systematic endeavor in the exploration of the potential of the new technique began at the Research Center Jülich after 1974. Other groups joined into this effort. Research of the early days culminated in a first International Conference of Vibrations in Adsorbed Layers, held in Jülich in September 1978. This first meeting initiated a series of international conferences which continues to serve as a forum for the presentation of papers on surface vibration spectroscopy and is-

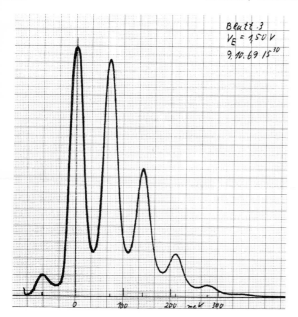

Fig. 1. Reproduction of an original recorder trace of one of the first energy loss spectra obtained from a ZnO surface on October 9, 1969, 3:30 p.m. The energy losses (and gains) are due to single and multiple excitation of Fuchs–Kliewer surface phonons.

sues related to the periodic motion of atoms at the surface [19–25].

3. Development of the theory

As mentioned before, the intense energy losses of ZnO are caused by the excitation of dielectric surface phonons. The theory of the excitation process is virtually identical to the (elementary) theory of plasmon excitations. In the limit when the electron is "fast", i.e. its energy is large compared to the quantum energy of a plasmon, the energy loss process can be calculated easily using the classical electromagnetic theory. In order to introduce individual quantum losses one has to equate the total classical loss W in Fourier representation

$$W = \frac{1}{4\pi} \int_{-\infty}^{\infty} \mathrm{d}t \, \mathrm{d}\mathbf{r} \, \mathscr{E}(\mathbf{r}, t) \dot{\mathbf{D}}(\mathbf{r}, t)$$

$$= \int \hbar\omega P(\mathbf{q}_\parallel, \hbar\omega) \, \mathrm{d}\mathbf{q}_\parallel \, \mathrm{d}\hbar\omega, \tag{1}$$

to the probability $P(\mathbf{q}_\parallel, \eta\omega)$ for an individual energy loss process [26]. In this manner the loss probability for an electron traversing a thin film was calculated by Ritchie [27] and Stern and Ferrell [28]. A systematic analysis of the classical aspects of the scattering theory including radiation effects was performed by Geiger [29]. From the classical electromagnetic theory it is easy to understand that electrons loose their energy in a different way when they are inside the solid or outside, since the screening factors for the electric field are $1/|\epsilon|$ and $1/|\epsilon + 1|$, respectively. Consequently the loss probability for an electron reflected from a surface with little penetration into the bulk is proportional to

$$P(\hbar\omega) \propto \frac{\epsilon_2(\omega)}{\left(\epsilon_1(\omega) + 1\right)^2 + \epsilon_2^2(\omega)}$$

$$= -\mathrm{Im}\left(\frac{1}{\epsilon(\omega) + 1}\right), \tag{2}$$

with $\epsilon_1(\omega)$ and $\epsilon_2(\omega)$ the real an imaginary part of the dielectric function $\epsilon(\omega)$ of the solid, which is simply considered to be a semi-infinite medium here. For a free electron gas the surface loss function $-\mathrm{Im}[1/(\epsilon(\omega) + 1)]$ has a pole at the surface plasmon frequency when $\epsilon_1(\omega) = -1$. Similarly, the loss function has a pole in the frequency range of phonons when the material is infrared active. The surface eigenmodes equivalent to the surface plasmons are the aforementioned Fuchs–Kliewer surface phonons and these are the origin of the intense losses in Fig. 1.

The high intensity of energy losses stems from the fact that many monolayers of the substrate contribute to the interaction with the electron. This can be seen from the following simple considerations. The electrostatic potential of an excitation in the bulk, which is considered to be semifinite for the moment, must obey the Laplace equation on both sides of the surface. The Fourier components of the potential must therefore be of the form

$$\phi(\mathbf{q}_\parallel, \omega) = \mathrm{e}^{\mathrm{i}\mathbf{q}_\parallel \cdot \mathbf{r}_\parallel} \, \mathrm{e}^{-q_\parallel z} \, \mathrm{e}^{\mathrm{i}\omega t}, \tag{3}$$

with $\mathbf{q}_\parallel$ the wave vector of the surface wave which also determines the decay of the potential into the solid. The electron has the strongest interaction with those surface wave components which have a parallel component of the wave vector $q_\parallel$ such that the phase velocity $\omega/q_\parallel$ matches the parallel component of the electron velocity $v_\parallel$. For typical electron energies and energy losses $\hbar\omega$ one has a decay length $\Lambda = q_\parallel^{-1}$ of the order of ~ 100 Å. Hence the number of monolayers which contribute to an energy loss for a collective excitation is of the order of 50, with the consequence that these losses are as intense as they are.

The simple classical model does not account for multiphonon excitations. In order to include those, Lucas and Sunjic [14,15] in their first treatment of the reflection case employed the venerable model of the forced harmonic oscillator, with the electron travelling on its largely unperturbed trajectory providing the force. The first completely quantum mechanical theory of dielectric losses in the reflection case was developed by Evans and Mills [16,30]. This theory included the case where only the surface itself bears dipole active modes with the bulk being infrared inactive either due to the high symmetry of the structure (e.g. the case of silicon) or because of metallic screening. Dipole activity can also result from the dipole moments associated with vibrations of adsorbates. Evans and Mills also discovered the surface selection rule for dipole active modes, by which term one alludes to the fact that modes with their dipole moment oriented parallel to the surface are screened by $1/(\epsilon + 1)^2$, with ϵ the dielectric function of the bulk. Localized modes of adsorbates with a dipole moment oriented perpendicularly to the surface belong to the totally symmetric representation of the surface point group. Hence the observation or non-observation of particular modes can serve as an instrument to determine the local point group of adsorbate complexes which constitutes an important element of the structure. Surface point groups and the methods by which one determines the number and type of modes of surface complexes are discussed in Ref. [31]. This qualitative interpretation of vibration spectroscopy does not require a detailed theoretical calculation. There are however a number of interesting applications of energy loss spectroscopy, which belong more to the realm of solid state physics and require more sophisticated theoretical methods. An example is the response of the free electron gas. In the regime of very low excitation energies these energy losses where addressed by Persson and Demuth [32]. In this work the more general surface response function $g(q_\parallel, \omega)$ [33] is used which equates the externally induced potential $\phi_{ind}(q_\parallel, \omega)$ to the external potential $\phi_{ext}(q_\parallel, \omega)$, caused by the moving electron:

$$\phi_{ind}(q_\parallel, \omega) = g(q_\parallel, \omega)\phi_{ext}(q_\parallel, \omega). \tag{4}$$

In the case of local screening the response function $g(q_\parallel, \omega)$ for a semi-infinite half-space is

$$g(q_\parallel, \omega) = \frac{\epsilon(\omega) - 1}{\epsilon(\omega) + 1}. \tag{5}$$

Based on the generalized linear response formalism Andersson et al. [34] interpreted the background intensity of Cu(100) and Ni(100) near the elastic peak. In cases where the density of electrons is much lower, as e.g. in semiconductors, the dissipative energy losses of the free carriers can lead to a substantial energy broadening of the elastically scattered peak with the consequence of a loss in resolution. This is easily seen from the probability for a single loss event $P_s(\omega)$ in the dielectric response formalism where

$$P_s(\omega) = \frac{2}{a_0 k_i \cos \vartheta_i} \frac{1}{\omega} \mathrm{Im}\left(\frac{-1}{\epsilon(\omega) + 1}\right) \tag{6}$$

for the semi-infinite half-space. Here a_0 is the Bohr radius, k_i the wave vector of the incoming electron, and ϑ_i the angle of incidence. Inserting the dielectric function for a free electron gas

$$\epsilon(\omega) = \epsilon_\infty - \frac{\omega_p^2}{\omega^2 + i\omega/\tau(\omega)} \tag{7}$$

one has

$$P_s(\omega) = \frac{2}{a_0 k_i \cos \vartheta_i} \frac{1}{(1 + \epsilon_\infty)}$$
$$\times \frac{\omega_s^2 \omega_\tau}{\left(\omega_s^2 - \omega^2\right)^2 + \omega^2\omega_\tau^2}, \tag{8}$$

where $\omega_s^2 = \omega_p^2/(1 + \epsilon_\infty)$ and $\omega_\tau = 1/\tau$, ω_p is the plasma frequency, τ the relaxation time and ϵ_∞

the dielectric constant for $\omega \gg \omega_p$. With the additional Bose factor $n(\omega)$ which enters the scattering probability, the single loss intensity diverges as $\omega \to 0$. For the total loss spectrum one therefore has to sum over all multiple loss events using [15]

$$P_{tot}(\omega) = \frac{1}{2\pi} \int_0^\infty dt\ e^{-i\omega t} \exp\Big\{ \int_0^\infty d\omega'\ P_s(\omega') \\ \times \big[(n(\omega') + 1)(e^{i\omega' t} - 1) \\ + n(\omega')(e^{-i\omega' t} - 1) \big] \Big\}. \quad (9)$$

In the case of a Gaussian profile the full width at half maximum Γ of the quasi-elastic line can be calculated directly from the single scattering probability [32]

$$\Gamma^2 = (8 \ln 2) \int_0^\infty d\omega\ P_s(\omega) \omega^2 (2n(\omega) + 1). \quad (10)$$

Fig. 2 displays the result calculated for n-type silicon with different donor concentration N_D together with experimental results for the hydrogen terminated Si(111) surface [35]. For this surface a flat band situation should prevail, as assumed in the calculation. The results do fit the theory quite well in view of the fact that no parameter was adjusted. For the sake of com-

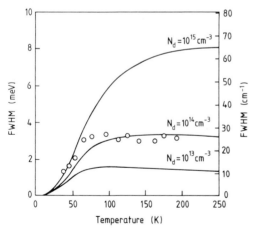

Fig. 2. Full width at half maximum of the quasi-elastic peak of electrons backscattered from a hydrogen terminated Si(111) surface versus temperature together with the theoretical curves calculated from silicon bulk transport data. The donor concentration as calculated from the conductivity was $\sim 10^{14}$ cm^{-3}.

pleteness we mention that the dielectric theory was also extended to anisotropic media [36] and to solids consisting of an arbitrary number of layers with different optical constants, such as to be found with GaAs/AlGaAs superlattices [37].

In the model presented above it is assumed that the response function $\epsilon(\omega)$ remains constant up to the surface. In most experimental situations this will not be the case. Due to the presence of surface states, semiconductors may have accumulation or depletion layers in which the carrier concentration changes with the distance from the surface. An elementary treatment of the scattering from solids covered with a surface layer of different optical constants is given in Ref. [31]. A case of special interest is a strong accumulation layer where quantum effects come to bear. The Schrödinger equation must then be solved self-consistently with the Poisson equation. In particular the quantum mechanical boundary condition, namely that the wave function must vanish at the surface, is reflected in the dependence of the scattering cross section on the momentum transfer $q_\parallel$ [38]. Boundary effects on the free electron gas are also the cause for the recently discovered multipole plasmon modes [39].

After this excursion into the area of electronic losses we now return to vibration spectroscopy. So far we have discussed only the theoretical development of dipole scattering. Hence we have focused our interest onto the inelastic scattering events to be found at scattering angles near the specularly reflected beam (and near the diffracted beams). The theoretical development of dipole scattering was stimulated by the experiments in which initially only dipole scattering was observed. The reason is, that dipole scattering gives rise to a high *differential* cross section, and consequently signals from such scattering effects are more easily picked up by instruments which for electron optical reasons necessarily accept only a very small fraction of the angular half-space. Inelastic scattering events caused by the motion of atoms and mediated by the short range part of the potential around each atom (impact scattering), on the other hand, have a small differential cross section, since the inelastic intensity is spread over the entire half-space. The total inelastic intensity due to phonon scattering divided by the elastic intensity is simply

$$I_{inel}/I_{el} = e^{2W} - 1, \qquad (11)$$

where W is the well known Debye–Waller factor. At room temperature I_{inel}/I_{el} typically exceeds one. The *total* cross section for impact scattering is therefore quite high.

The theory of phonon scattering in electron energy loss spectroscopy is an off-spring of the theory of low energy electron diffraction, since one needs to consider the combination of elastic multiple scattering with one-phonon or multi-phonon processes. The basic formalism of the theory was developed in 1980 by Tong, Li, and Mills [40]. In the meantime the theory has been applied to a number of surfaces and the calculated intensities of phonon spectra compare well with the experimental results [4145]. It is of significance for the experiment that, because of the multiple elastic scattering process involved, the cross section for the scattering from individual phonons varies quite rapidly with the scattering parameters. (See also Section 4.4 and Fig. 5.)

4. Experimental highlights

4.1. Collective dipole excitations

Collective dipole excitations give rise to intense energy loss features, as we have seen already with the example ZnO (Fig. 1). Research in this area therefore requires less advanced spectrometer technology than any other application. In Section 2 we have remarked that the results on the Fuchs–Kliewer surface phonon were, once understood, relatively boring. The physics becomes more interesting, however, with the collective excitations of the free carriers in surface space charge layers. Let us consider the case of an accumulation layer of moderate carrier density, so that the thickness of the layer is of the order of 100 nm or larger. Quantum effects as mentioned in Section 3 can then be disregarded. The dielectric response of the accumulation layer is then in its simplest form given by the free electron dielectric function. The plasmon energy (Eq. (7)) is then in the range of 10–100 meV and plasmon excitation gives rise to discrete surface plasmon losses when the relaxation time τ is large enough, e.g. at low temperatures. Since the total charge density per surface area in the accumulation layer must match the excess charge in surface states, the carrier density can be affected by changing the surface conditions. The first work capitalizing on the idea was published in 1981 by Matz and Lüth [46] (for a review see also Refs. [47,48]). For the reason of being able to affect the surface charge density most easily the authors had chosen cleaved GaAs surfaces where the density of intrinsic surface states in the band gap is small (zero for a perfect surface). For this infrared active material one has to include the phonon contribution into the dielectric response, which causes an interesting coupling phenomenon between the surface phonon and the surface plasmon. Work in this area of research continues until presently and is used to characterize space charge layers and electronic properties of new materials and layered systems [49–52].

Most infrared active materials are good insulators. Here one has the problem of charging in electron spectroscopy. The obstacle was elegantly circumvented by Liehr et al. [53] by stabilizing the potential of the sample with respect to the potential of the cathode of a flood gun at the point where the coefficient of total secondary electron emission from the sample is one. The base potential of an electron spectrometer, with its comparatively very low monochromatic current, was then set such as to obtain the desired impact energy of the probing electrons at the sample. With this technique Lambin et al. and Thiry et al. were able to obtain highly resolved spectra of crystalline Al_2O_3 [37,54]. Already in this work it was noted that the experimental spectra deviated from the spectra calculated with the bulk dielectric constant, indicating a surface modification of the dielectric properties. Such modifications become even more important when one has a thin (epitaxial) layer on a substrate. An example is CaF_2 on Si [54]. Here the loss spectrum displays substantial variations from the simple dielectric model with bulk material constants when the thickness of the layer is below 50 Å [54].

A useful application of the spectroscopy of collective excitations is also in the characterization of materials in the near surface region. Three recent examples are shown in Fig. 3. Fig. 3a shows a spectrum originating from a thin, crystallographically well ordered, Al_2O_3 overlayer on NiAl(100) [55]. This overlayer is generated by exposing NiAl to oxygen and annealing to 900°C. Similar Al_2O_3 layers, though with a different crystallographic structure, can be grown on the other faces of NiAl [56,57]. Today they serve as a model material for studies of chemisorption and catalysis on oxides. The advantage of using these oxide overlayers lies in their small thickness of a few ångströms, which is thin enough so that no charging occurs in electron spectroscopical studies [57]. Fig. 3b shows a spectrum of the Fuchs–Kliewer phonons of β-$FeSi_2$ grown on Si(111) by deposition of an Fe overlayer and annealing to 900 K [58]. The observation of the surface phonons proves [59] that this β-phase of $FeSi_2$ is nonmetallic, since otherwise the phonon modes would be screened by the metallic conductivity $(\omega_p \gg \omega_{phonon})$. The final example in Fig. 3c is a

spectrum obtained after cleaving $YBa_2Cu_3O_7$ in vacuum [60]. The spectrum is as expected from infrared data. The spectrum in Fig. 3c was however taken at 80 K, i.e. were the material is superconducting in the bulk. The spectrum proves therefore that the material cannot have metallic conductivity in the near-surface region of a few monolayers. This is presumably due to an immediate loss of oxygen in the cleavage process, so that the material converts to the insulating phase with the composition $YBa_2Cu_3O_6$ at the surface.

4.2. The CO age

The development of electron energy loss spectroscopy and its applications to surface vibration spectroscopy followed very much the path from the most intense loss features to the features giving weaker and weaker signals. Loss features next in intensity to collective losses arise from vibrational modes of dipole active vibrations of adsorbed species. Carbon monoxide has been the prototype for many years. Other strong dipole active modes are associated with the perpendicular motion of adsorbed oxygen, carbon and sulfur atoms. Early research in surface vibration spectroscopy therefore largely focused on these species adsorbed on metal surfaces. Among the early examples studied were CO, O and C on W(100) [61,62] in following trail of Propst and Piper [6]. Carbon monoxide, oxygen, and sulfur adsorption was also studied on Cu(100), Ni(100), Pd(100), Ni(111) and Pt(111) [63–73]. By using the aforementioned selection rules and by simple reasoning with regard to the observed vibration frequencies it was established that C, N, O and S tend to prefer sites of high coordination. In seeking to achieve a high coordination, the dissociative adsorption of these species frequently causes the surface to reconstruct. The less denser packed (110) surfaces of fcc metals are well-studied examples [74–76]. The oxygen induced reconstruction of fcc metal surfaces has been addressed recently also using scanning tunnelling microscopy, see e.g. Ref. [77] as one example out of many.

Carbon monoxide may or may not dissociate upon adsorption on a clean metal surface.

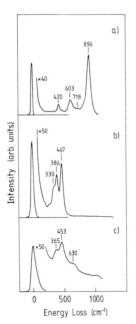

Fig. 3. Energy loss spectra of (a) θ-Al_2O_3 overlayer grown on NiAl, (b) β-$FeSi_2$ overlayer grown on Si(111) and (c) $YBa_2Cu_3O_7$ after cleaving in vacuum at 80 K.

Whether it does, or does not, largely depends on the position of the metal in the periodic table, with elements more on the left in the periodic table bearing a large propensity to dissociate CO [78]. CO dissociation also depends somewhat on the crystallographic orientation of the surface, in particular for elements on the border-line between dissociative and non-dissociative adsorption.

An issue discussed frequently for CO and also NO adsorption is that of the binding site. CO and NO may adsorb in terminal and bridging sites. As the frequency of the molecular stretching-vibration depends on the coordination [79,80], with the frequency being the highest for the lowest coordination (terminal position), one may use the observed frequency to determine the binding site. Recent studies show however, that an attempt to determine the binding site on the basis of a *single* vibrational mode is dangerous [81]. In particular the frequencies in the various higher coordination sites are as much affected by molecular interactions as they are by the coordination alone. Probably the only safe conclusion from the observation of the dipole active molecular vibrations of CO or NO alone is, that the number of binding sites occupied must be equal (or larger) than the number of observed stretching-vibrations.

Until very recently I had the impression (presumably shared by many) that the issue of binding sites for CO had been beaten to death and was about to become a bore. Our interest was renewed however in the context of the measurement of the CO-induced surface stress of Ni(100) [82]. In the course of this study it became necessary to determine the number of CO molecules adsorbed in bridging and terminal site quantitatively as a function of coverage. The work was also stimulated by a recent IR-study [83] showing CO to occupy bridging and terminal sites at nearly all coverages and temperatures, contrary to earlier work using vibration spectroscopy [73] and also at variance with earlier structure analysis on the system [84]. Very much to our surprise we have found that merely 1.4% surface coverage of carbon shifts the balance from preferential coverage of terminal sites at room temperatures significantly towards a higher coverage of bridge sites

[85]. Once the surface is clean to the limit of our detection capability, the relative occupation of bridge and terminal sites depends reversibly on the temperature. This temperature dependent shift in the site occupation is driven by the phonon entropical contribution to the free energy of the system, which is different for the two sites, since the terminal site has two low frequency eigenmodes (hindered translations), whereas the bridge site has one [85]. These results show that the issue of binding sites for adsorbed CO is more delicate than assumed previously. In particular the sensitivity to small amounts of contaminants near the limit of detectability is disastrous. One has reasons to doubt that structure analyses using EXAFS, XPD or LEED of this and related systems was ever been performed with C-contamination levels of less than 1%, in particular since CO dissociates readily when exposed to high energy electrons, which are inevitably involved in these methods for structure analysis. Here we have clearly a new case for vibration spectroscopy, though on a level of much higher sophistication with regard to sample preparation.

4.3. The hydrocarbon age

The probably most important contribution of electron energy loss spectroscopy to our present level of understanding of the surface sciences is in the area of surface chemistry of hydrocarbons. Here the unique capability of vibration spectroscopy to probe the chemical bond, including the bonding of hydrogen atoms, fully comes to bear. The basis for the use of vibration spectroscopy is that the frequencies of CH stretching and bending-vibrations e.g. depend on the number of hydrogen atoms attached to the carbon atoms and thus on the degree of saturation of the carbon bonds [86]. In addition, the frequencies associated with C–C stretching-vibrations provide unequivocal information on the bond strength. A vibrational analysis of an adsorbed ethylene molecule e.g. will therefore instruct us whether the molecule remains π-bonded at the surface or rehybridizes to engage into a sp^3-type bonding with the surface atoms. Furthermore, the vibrational analysis provides information on fragmentation and synthesis on surfaces and permits the identification of surface intermediate species. The power of vibration spectroscopy in that regard is still unparalleled by any other technique.

In probing the vibration spectra of hydrocarbons with energy loss spectroscopy one is leaving the realm of dipole scattering, since the dipole moments associated with hydrocarbon modes are typically small. Furthermore, because the molecules contain more atoms, it is necessary to probe also those vibrational modes which are non-dipole active, in order to have the full information required for an analysis. Probing non-dipole active modes via the impact scattering mechanism did require a substantial improvement of spectrometers compared to studies on adsorbed CO. Typically the intensity of a non-dipole active loss event is 10–100 times lower than the loss intensity of the CO stretching mode. The first experimental work which demonstrated a vibrational energy loss due to impact scattering was published in 1978 by Ho, Willis and Plummer [87], where it was shown that a non-totally symmetric mode of hydrogen adsorbed on W(100) peaked in intensity at an angle ∼ 15° off the specular reflected beam, which is a clear proof of the impact scattering mechanism. In order to probe the weak signals of impact scattering, the authors had to back off in the energy resolution significantly. It took several years of further development of spectrometers to improve the signal intensity and reduce the background before the impact scattering mechanism could be employed on a routine basis [88].

The first hydrocarbon studied with vibration spectroscopy on clean surfaces was acetylene on Pt(111) [89]. In this work the rehybridization of acetylene was shown and an attempt was made to determine the symmetry of the adsorption site. Shortly after this, the adsorption of acetylene, ethylene, benzene, cyclohexane and formic acid was investigated on nickel, copper, ruthenium, iron, tungsten, silver, palladium and platinum surfaces. An early synopsis of P. Thiry, published in the proceedings of the second conference on Vibrations at Surfaces held in 1982 [90], contains no less than 40 entries dealing with energy loss spectroscopy on hydrocarbons. One of the early

triumphs of the work was the discovery of ethylidyne surface species which is a stable species on several fcc (111) surfaces and forms in dehydrogenation and hydrogenation reactions. The discovery and identification of this species was rather a detective story which involved the work of many groups and the use of several techniques including LEED-structure analysis [91].

Alas, despite important discoveries on the nature of hydrogen bonding at surfaces and new surface species, it is more than likely that quite a few conclusions were premature, in particular when primarily based upon vibration spectroscopy, the reason being that the energy resolution was poor, and not adequate to the complexity of the problems until very recently. Rarely a work could claim to have detected all, or even nearly all, vibrational features of the surface species in question. I illustrate the significance of this critical statement with Fig. 4, where the spectrum of cyclohexane as obtained 1979 (which was among the best at the time, I dare say) [92,93] is compared to a spectrum obtained with the penultimate generation of spectrometers [94]

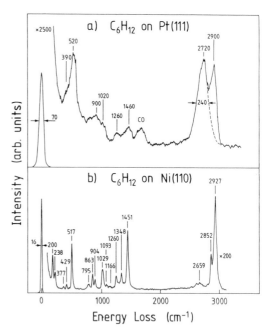

Fig. 4. The progress in resolution is demonstrated with these two energy loss spectra of cyclohexane on Pt(111) and Ni(110) obtained in 1979 and 1992, respectively.

which is commercially available now [95]. Only in the latter spectrum, the different groups of CH stretching and bending-vibrations are clearly resolved and the broad hump around 500 cm^{-1} in Fig. 4a resolves into a sequence of CC-deformation modes and modes with mixed substrate and molecular character.

Even the poorly resolved spectrum of cyclohexane in Fig. 4a has left little doubt about the nature of the adsorbed species [93], although a dehydrogenation to C_6H_9 [96] could not have been excluded. A case where an identification based on insufficiently resolved spectra was in error is the intermediate formed in the decomposition of formic and acetic acid on Ni(111) [97,98]. Here the crucial spectral features which established the nature of the intermediate as di-formic-acid-anhydride and di-acetic-acid-anhydride, respectively, are split CO- and CH-modes with the splitting ranging from 4 to 22 cm^{-1}. At the time (1988) these small splittings could be resolved only in Fourier-transform infrared spectroscopy, and even now detection of such small splittings is on the limit of electron energy loss spectroscopy. To summarize this section on hydrocarbons, I still regard only a few cases of intermediate surface species as closed issues, and I would not be surprised, if more seemingly established surface intermediates will have to be reconsidered.

4.4. The phonon age

On ordered crystal surfaces the frequencies of adsorbed species and of surface atoms depend on the wavevector parallel to the surface $q_\parallel$. The dispersion $\omega(q_\parallel)$ is a measure of the interatomic and intermolecular coupling, respectively. The dispersion is rather small for the intramolecular modes in ordered overlayers of molecules. Nevertheless the dispersion can be measured and such measurements can serve to identify symmetry elements of the surface space group and the binding site [99]. Vibrational modes where the substrate atoms are involved significantly, bear a much larger dispersion, since the vibrating substrate atoms provide the coupling to the next adsorbate [100]. The vibrational amplitude of the substrate

atoms in such a mode is the larger the closer the frequency of the adsorbate mode (typically higher than the substrate modes) is to the upper boundary of the bulk phonon band. Finally, there are surface phonons of the substrate itself. The first electron energy loss spectrum of a substrate surface phonon, the Rayleigh-mode of Ni(100), was published in 1983 [101], two years after the first dispersion of a surface phonon was determined

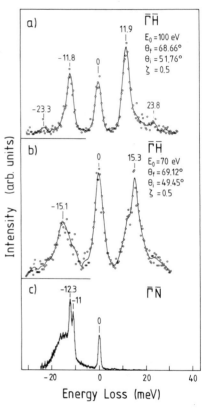

Fig. 5. (a/b) Sample spectra of surface phonons in electron energy loss spectroscopy taken with two different sets of scattering parameters at the same point of the surface Brillouin-zone on W(110). Because of the high energy of the electrons, phonon spectroscopy with electrons is a "constant $q_\parallel$-spectroscopy". All loss and gain features in (a) and (b) therefore correspond to the same wave vector $q_\parallel$ which is halfway to the zone boundary in the example here. This is different for inelastic scattering of He atoms (c) where $q_\parallel$ changes with the energy loss. The broad hump in the He spectrum is due to the excitation of multiphonons [103]. Multiphonon scattering in electron spectroscopy is weak unless the electron energy is very high (> 200 eV). In both sets of data the lower energy loss (gain) corresponds to the Rayleigh-wave, the upper to a longitudinal surface mode.

using He atom scattering [102]. Since impact scattering is employed in phonon dispersion measurements with electrons, the intensities are small and early spectra had to be taken with poor resolution (~ 7 meV). Recent progress in the technology of electron spectrometers has improved the situation and the spectra compare not too badly to those obtained with inelastic He scattering (Fig. 5). Considering the fact that one can tune selectively into particular modes with energy loss spectroscopy by changing the scattering parameters (Figs. 5a and 5b) the quality of the information is on the same footing with both techniques, except for the very soft modes, where the higher resolution makes inelastic He scattering the only technique of choice.

By measuring the dispersion of substrate surface phonons one essentially probes the interatomic forces. Thereby a close contact with theoretical total energy calculations is established [104]. In recent years interest in surface phonons was focused in particular on "anomalies" in the dispersion curves, i.e. in situations where the dispersion curves for surface phonons deviated from those calculated with bulk force constants. Such anomalies can arise from various sources such as electron–phonon coupling effects and surface stress. For overlayers of carbon, oxygen, and sulfur on Ni(100) a correlation was established between stress-induced phonon anomalies [105], adsorbate-induced reconstruction [106], and direct experimental measurements of the adsorbate-induced surface stress [107]. The largest phonon anomaly ever reported was recently found with the hydrogen covered W(110) surface, the nature of this anomaly being not understood presently [108].

5. Summary

In the past 25 years electron energy loss spectroscopy has enlarged our knowledge on surface science. Most outstanding contributions have been made to the understanding of surface reactions, the nature of interatomic interactions and the driving forces for surface reconstruction. The technology of spectrometers is now based on sci-

ence [109], and excellent, easy to operate instruments capable of resolutions down to 1 meV are commercially available. The potential of making instruments with energy resolutions even below 1 meV is being explored currently and recent results [110] seem to indicate that this endeavor may be successful.

Acknowledgement

Critial reading of the manuscript by S. Lehwald is gratefully acknowledged.

References

[1] R.G. Greenler, J. Chem. Phys. 44 (1966) 310.
[2] A.M. Bradshaw and J. Pritchard, Surf. Sci. 17 (1969) 372; 19 (1970) 198; Proc. Roy. Soc. (London) A 316 (1970) 169.
[3] Y.J. Chabal, Surf. Sci. Rep. 8 (1988) 211.
[4] B.N.J. Persson, Phys. Rev. B 44 (1991) 3277; Surf. Sci. 269/270 (1992) 103.
[5] C.J. Hirschmugl, G.P. Williams, F.M. Hoffmann and Y.J. Chabal, Phys. Rev. Lett. 65 (1990) 480.
[6] F.M. Propst and T.C. Piper, J. Vac. Sci. Technol. 4 (1967) 53.
[7] H. Boersch, J. Geiger and W. Stickel, Phys. Rev. Lett. 17 (1966) 379.
[8] J. Geiger, M. Nolting and B. Schröder, Proc. Seventh Int. Conf. on Electron Microscopy, Grenoble, 1970, p. 111.
[9] Electron energy loss spectroscopy shares this difficulty with electron microscopy where UHV conditions in the sample chamber are becoming more common only nowadays. An UHV spectrometer for fast electrons is currently operated by J. Fink, see e.g., J. Fink, Adv. Electron. Phys. 75 (1989) 121;
J. Fink, in: Topics in Applied Physics, Vol. 69, Eds. J. Fuggle and J.E. Inglesfield (Springer, Berlin, 1992).
[10] H. Ehrhardt, L. Langhans, F. Linder and H.S. Taylor, Phys. Rev. 173 (1968) 222.
[11] H. Ibach, Phys. Rev. Lett. 24 (1970) 1416.
[12] H. Ibach, in: Advances in Solid State Physics XI, Ed. O. Madelung (Pergamon, Oxford, 1971) pp. 135–174.
[13] R. Fuchs and K.L. Kliewer, Phys. Rev. 140 (1965) A2076; K.L. Kliewer and R. Fuchs, Phys. Rev. 144 (1966) 495; 150 (1966) 573.
[14] A.A. Lucas and M. Sunjic, Phys. Rev. Lett. 26 (1971) 229.
[15] A.A. Lucas and M. Sunjic, Prog. Surf. Sci. 2 (1972) 75.

[16] E. Evans and D.L. Mills, Phys. Rev. B 5 (1972) 4126; B 7 (1973) 853.

[17] H. Ibach, Phys. Rev. Lett. 5 (1971) 253.

[18] O.L. Alerhand, D.C. Allan and E.J. Mele, Phys. Rev. Lett. 55 (1985) 2700; 59 (1987) 657;
L. Miglio, P. Santini, P. Ruggerone and G. Benedek, Phys. Rev. Lett. 62 (1989) 3070;
F. Ancilotto, W. Andreoni, A. Selloni, R. Car and M. Parrinello, Phys. Rev. Lett. 65 (1990) 3148.

[19] H. Ibach and S. Lehwald, Eds., Vibrations in Adsorbed Layers, Jül-Conf-26, ISSN 0344-5798, Jülich, 1978.

[20] R. Caudano, J.M. Giles and A.A. Lucas, Eds., Vibrations at Surfaces (Plenum, New York, 1982) ISBN 0-306-40824-4.

[21] C.R. Brundle and H. Morawitz, Eds., Vibrations at Surfaces (Elsevier, Amsterdam, 1983) (ISBN 0-444-42166-1) and J. Electron Spectrosc. Relat. Phenom. Vols. 29 and 30.

[22] D.A. King and N.V. Richardson, Eds., Vibrations at Surfaces (Elsevier, Amsterdam, 1985) (ISBN 0-444-42631-0) and J. Electron Spectrosc. Relat. Phenom. Vols. 38 and 39.

[23] A.M. Bradshaw and H. Conrad, Eds., Vibrations at Surfaces (Elsevier, Amsterdam, 1987 (ISBN 0-444-42944-1) and J. Electron Spectrosc. Relat. Phenom. Vols. 44 and 45.

[24] Y.T. Chabal, F.M. Hoffmann and G.P. Williams, Eds., Vibrations at Surfaces (Elsevier, Amsterdam, 1990) and J. Electron Spectrosc. Relat. Phenom. Vols. 54 and 55.

[25] Vibrations at Surfaces, 1993, to be published.

[26] An elementary classical derivation of the loss probability for electrons reflected from a surface is given in: H. Ibach, Surface Vibration Spectroscopy in Interactions of Atoms and Molecules with Solid Surfaces, Eds. V. Bortolani, N.H. March and M.P. Tosi (Plenum, New York, 1990).

[27] R.H. Ritchie, Phys. Rev. 106 (1957) 106.

[28] E.A. Stern and R.A. Ferrell, Phys. Rev. 115 (1960) 869.

[29] J. Geiger, Elektronen und Festkörper (Vieweg, Braunschweig, 1968).

[30] For review see also, D.L. Mills, Surf. Sci. 48 (1975) 59.

[31] H. Ibach and D.L. Mills, Electron Energy Loss Spectroscopy and Surface Vibrations (Academic Press, New York, 1982) (ISBN 0-12-369350-0).

[32] B.N.J. Persson and J.E. Demuth, Phys. Rev. B 30 (1984) 5968.

[33] P.J. Feibelmann, Prog. Surf. Sci. 12 (1982) 287; Phys. Rev. B 22 (1980) 3654; B 12 (1975) 1319; B 14 (1976) 762.

[34] S. Andersson, B.N.J. Persson, M. Persson and N.D. Lang, Phys. Rev. Lett. 52 1984) 2073.

[35] Ch. Stuhlmann, Thesis, RWTH Aachen, D80 (1992).

[36] A.A. Lucas and J.P. Vigneron, Solid State Commun. 49 (1984) 327.

[37] Ph. Lambin, J.P. Vigneron, A.A. Lucas, P.A. Thiry, M. Liehr, J.J. Pireaux and R. Caudano, Phys. Rev. Lett. 56 (1986) 1842.

[38] S.R. Streight and D.L. Mills, Phys. Rev. B 40 (1989) 10488.

[39] K.D. Tsuei, E.W. Plummer, A. Liebsch, K. Kempa and P. Bakshi, Phys. Rev. Lett. 61 (1990) 44.

[40] S.Y. Tong, C.H. Li and D.L. Mills, Phys. Rev. Lett. 44 (1980) 407;
C.H. Li, S.Y. Tong and D.L. Mills, Phys. Rev. B 21 (1980) 3057;
S.Y. Tong, C.H. Li and D.L. Mills, Phys. Rev. B 24 (1981) 806.

[41] S. Lehwald, F. Wolf, H. Ibach, B.M. Hall and D.L. Mills, Surf. Sci. 192 (1987) 131.

[42] Z.Q. Wu, Y. Chen, M.L. Xu, S.Y. Tong, S. Lehwald, M. Rocca and H. Ibach, Phys. Rev. B 39 (1989) 3116.

[43] B.M. Hall, D.L. Mills, M.H. Mohamed and L.L. Kesmodel, Phys. Rev. B 38 (1988) 5856.

[44] Y. Chen, M.L. Xu, S.Y. Tong, M. Wuttig, W. Hoffmann, R. Franchy and H. Ibach, Phys. Rev. B 42 (1990) 5451.

[45] M. Balden, S. Lehwald, H. Ibach, A. Ormeci and D.L. Mills, Phys. Rev. B 46 (1992) 4172.

[46] R. Matz and H. Lüth, Phys. Rev. Lett. 46 (1981) 500.

[47] H. Lüth, Surf. Sci. 126 (1983) 126.

[48] H. Lüth, Surf. Sci. 168 (1986) 77.

[49] M.G. Betti, U. del Pennino and C. Mariani, Phys. Rev. B 39 (1989) 5887.

[50] L.H. Dubois and G.P. Schwartz, Phys. Rev. B 40 (1989) 8336.

[51] G. Annove, M.G. Betti, U. del Pennino and C. Mariani, Phys. Rev. B 41 (1990) 11978.

[52] C. Lohe, A. Leuther, A. Förster and H. Lüth, Phys. Rev. B 47 (1993), in press.

[53] M. Liehr, P.A. Thiry, J.J. Pireaux and R. Caudano, Phys. Rev. B 33 (1986) 5682.

[54] P.A. Thiry, M. Liehr, J.J. Pireaux and R. Caudano, J. Electron Spectrosc. Relat. Phenom. 39 (1986) 69.

[55] P. Gassmann and R. Franchy, to be published.

[56] R. Franchy, M. Wuttig and H. Ibach, Surf. Sci. 189/190 (1987) 438.

[57] R.M. Jaeger, H. Kuhlenbeck, H.J. Freund, M. Wuttig, W. Hoffmann, R. Franchy and H. Ibach, Surf. Sci. 259 (1991) 235.

[58] Ch. Stuhlmann, J. Schmidt and H. Ibach, J. Appl. Phys., in press.

[59] A. Rizzi, W. Moritz and H. Lüth, J. Vac. Sci. Technol. A 9 (1991) 912.

[60] R. Franchy and B. Decker, to be published.

[61] H. Froitzheim, H. Ibach and S. Lehwald, Phys. Rev. B 14 (1976) 1362.

[62] H. Froitzheim, H. Ibach and S. Lehwald, Surf. Sci. 63 (1977) 56.

[63] S. Andersson, in: Vibrations at Surfaces, Eds. R. Caudano, J.M. Gilles and A.A. Lucas (Plenum, New York, 1982) p. 169.

[64] S. Andersson, Solid State Commun. 21 (1977) 75.

[65] S. Andersson, Solid State Commun. 20 (1976) 229.

[66] S. Andersson, Solid State Commun. 24 (1977) 385.

[67] S. Andersson, Surf. Sci. 79 (1979) 385.

[68] L. Nyberg and C.G. Tengstål, Solid State Commun. 44 (1982) 251; Surf. Sci. 126 (1983) 163.

[69] H. Froitzheim, H. Hopster, H. Ibach and S. Lehwald, Appl. Phys. 13 (1977) 147.

[70] W. Erley, H. Wagner and H. Ibach, Surf. Sci. 80 (1979) 612.

[71] W. Erley, H. Ibach, S. Lehwald and H. Wagner, Surf. Sci. 83 (1979) 585.

[72] A.M. Baró and H. Ibach, J. Chem. Phys. 71 (1979) 4812.

[73] P. Uvdal, P.A. Karlsson, C. Nyberg, S. Andersson and N.V. Richardson, Surf. Sci. 202 (1988) 167.

[74] B. Voigtländer, S. Lehwald and H. Ibach, Surf. Sci. 225 (1990) 162.

[75] G. Kleinle, J. Wintterlin, G. Ertl, R.J. Behm, J. Jona and W. Moritz, Surf. Sci. 225 (1990) 171.

[76] D.J. Coulman, J. Wintterlin, R.J. Behm and G. Ertl, Phys. Rev. Lett. 64 (1990) 1761.

[77] F. Jensen, F. Besenbacher, E. Lægsgaard and I. Stensgaard, Phys. Rev. B 41 (1990) 10233.

[78] G. Brodén, T.N. Rhodin, C. Brucker, R. Bendow and Z. Hurych, Surf. Sci. 59 (1979) 593.

[79] For early work on metal carbonyls see, L.H. Little, Infrared Spectra of Adsorbed Species (Academic Press, London, 1966) p. 47;
M.L. Hair, Infrared Spectroscopy in Surf. Chemistry (Dekker, New York, 1967) p. 230.

[80] For studies of nitrosyls see discussion in, G. Pirug, H.P. Bonzel, H. Hopster and H. Ibach, J. Chem. Phys. 71 (1979) 593.

[81] S. Aminpirooz, A. Schmatz, L. Becker and J. Haase, Phys. Rev. B 45 (1992) 6337.

[82] A. Grossmann, W. Erley and H. Ibach, to be published.

[83] J. Lauterbach, M. Wittmann and J. Küppers, Surf. Sci. 279 (1992) 287.

[84] S.Y. Tong, A. Maldonado, C.H. Li and M.A. Van Hove, Surf. Sci. 94 (1980) 73.

[85] A. Grossmann, W. Erley and H. Ibach, to be published.

[86] This is probably the appropriate place to refer to the books of G. Herzberg, in particular to G. Herzberg: Molecular Spectra and Molecular Structure II, Infrared and Raman Spectra of Polyatomic Molecules (Van Nostrand-Reinhold, New York, 1945).

[87] W. Ho, R.R. Willis and E.W. Plummer, Phys. Rev. Lett. 40 (1978) 1463.

[88] H. Ibach and S. Lehwald, J. Vac. Sci. Technol. 18 (1981) 625.

[89] H. Ibach, H. Hopster and B. Sexton, Appl. Phys. 14 (1977) 21.

[90] P. Thiry, in: Vibrations at Surfaces, Eds. R. Caudano, J.M. Giles and A.A. Lucas (Plenum, New York, 1982) (ISBN 0-306-40824-4) p. 231.

[91] See Ref. [31], pp. 326 ff.

[92] S. Lehwald and H. Ibach, Surf. Sci. 89 (1979) 425.

[93] J.E. Demuth, H. Ibach and S. Lehwald, Phys. Rev. Lett. 40 (1978) 1044.

[94] H. Ibach, M. Balden, D. Bruchmann and S. Lehwald, Surf. Sci. 269/270 (1992) 94.

[95] Spectrometers are manufactured by VSW Instruments and LK-Technologies.

[96] P.D. Land, W. Erley and H. Ibach, Surf. Sci. 289 (1993) 237.

[97] W. Erley and D. Sander, J. Vac. Sci. Technol. A 7 (1989) 2238.

[98] W. Erley, J.G. Chen and D. Sander, J. Vac. Sci. Technol. A 8 (1990) 976.

[99] B. Voigtländer, D. Bruchmann, S. Lehwald and H. Ibach, Surf. Sci. 225 (1990) 151;
B. Voigtländer, S. Lehwald and H. Ibach, Surf. Sci. 225 (1990) 162.

[100] For a review see, M. Rocca, H. Ibach, S. Lehwald and T.S. Rahman, in: Topics in Current Physics, Vol. 41, Eds. W. Schommers and P.V. Blanckenhagen (Springer, Berlin, 1986) (ISBN 3-540-16252-6).

[101] S. Lehwald, J.M. Szeftel, H. Ibach, T.S. Rahman and D.L. Mills, Phys. Rev. Lett. 50 (1983) 518.

[102] G. Brusdeylins, R.B. Doak and J.P. Toennies, Phys. Rev. Lett. 46 (1981) 437.

[103] E. Hulpke, private communication.

[104] J.A. Gaspar and A.G. Equiluz, Phys. Rev. B 40 (1989) 11976;
A.A. Quong, A.A. Maradudin, R.F. Willis, J.A. Gaspar, A.G. Eguiluz and G.P. Alldredge, Phys. Rev. Lett. 66 (1991) 743;
K.P. Bohnen and K.M. Ho, Vacuum 41 (1990) 416;
Y. Chen, S.Y. Tong, K.P. Bohnen, T. Rodach and K.M. Ho, Phys. Rev. Lett. 70 (1993) 603.

[105] T.S. Rahman, M. Rocca, S. Lehwald and H. Ibach, J. Electron Spectrosc. Relat. Phenom. 38 (1986) 45.

[106] J.E. Müller, M. Wuttig and H. Ibach, Phys. Rev. Lett. 56 (1986) 1583.

[107] D. Sander, U. Linke and H. Ibach, Surf. Sci. 272 (1992) 318.

[108] E. Hulpke and J. Lüdecke, Phys. Rev. Lett. 68 (1992) 2846.

[109] H. Ibach, Electron Energy Loss Spectrometers, Vol. 63 of Springer Series in Optical Sciences (Springer, Berlin, 1991) (ISBN 3-540-52818-O).

[110] H. Ibach, J. Electron Spectrosc. Relat. Phenom., in press.

Surface Science 299/300 (1994) 129–152
North-Holland

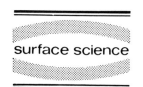

surface science

Field emission, field ionization, and field desorption

Robert Gomer

Department of Chemistry and The James Franck Institute, The University of Chicago, Chicago, IL, USA

Received 22 February 1993; accepted for publication 25 May 1993

A brief description of the essential features of the theory of field emission, field ionization, field desorption, and the field emission and field ion microscopes is presented. The principal applications of field emission including work function and surface diffusion measurements, and a brief sketch of the theory of diffusion are also given. Applications of field ion microscopy and field desorption are mentioned but not treated in detail.

1. Introduction

In the fall of 1950 I arrived at the University of Chicago as a brand new instructor, determined to change my research from gas phase free radical kinetics to "something" to do with adsorption on metals. Shortly after my arrival, Joe Mayer, the distinguished theoretical chemist, gave a chemistry colloquium. He and his wife, Maria Göppert-Mayer had been sent to Germany by the State Department to assess the state of physics and chemistry there, and Joe's colloquium reported some of the highlights. I remember that two things in this talk struck me. The first was that physics departments were making their own vacuum tubes, for lack of hard currency. The second was a short description of the work of a young physicist at the Kaiser Wilhelm (now Max Planck) Institute in Berlin, named Erwin Müller, who had invented a "needle microscope". This seemed to be just what I had been looking for, and so, initially without even checking the literature, I plunged in. My start-up funds at Chicago consisted of a room with two vacuum racks, such as organic chemists still use, a mechanical pump, a low speed Hg diffusion pump and the free services of a one-man electronics shop and also of an excellent machine shop. There was also an abundance of glass tubing and tungsten wire and

the machine shop made me a bench torch for glass blowing. It is still in occasional use. In those simpler times that was enough to get started. There was also a low temperature laboratory, ready to provide liquid He and H_2 at no cost, something that turned out to be crucial fairly soon. Most important of all, however was the close-knit intellectual and social interaction of all the physical scientists at the University. These included such people as Enrico Fermi, Gregor Wentzel, Murph Goldberger, Willy Zachariasen, Cyril Smith, Clarence Zener, the Mayers, Harold Urey, Bill Libby, Morris Karasch, Henry Taube, James Franck, Leo Szilard, as well as many others. Some like me, came as instructors, for instance Murray Gell-Mann, Morrel H. Cohen, John Kahn, and others. Ryogo Kubo and Dick Garwin were research associates. It would have been difficult not to be stimulated by this environment, but my career would most probably have taken a different turn were it not for Erwin W. Müller, who had invented the field emission microscope in 1936–1937 [1] and the field ion microscope in 1951 [2]. I am keenly aware of the debt all workers in these fields owe to him.

The field emission microscope was the first modern tool of surface science and required ultra-high vacuum for meaningful work. The field ion microscope was the first instrument providing atomic resolution of surfaces, albeit under rather special conditions. Both techniques still contribute to surface science some forty years after their initial use as this and related articles by Tsong [3] and Ehrlich [4] in this issue will try to show. Since the latter two articles deal with field ion microscopy and closely related subjects, the present one concentrates on field emission; because of my involvement with field ionization and field desorption it also treats the fundamentals of those topics.

2. Field emission

Field emission in the present context is the tunneling of electrons through a field deformed barrier at the surface of a metal, as indicated by Fig. 1. If image and correlation/exchange effects near the surface are ignored the barrier is triangular and the WKB penetration coefficient D takes the form [5]

$$D = \frac{4(\omega E_x)^{1/2}}{(\phi + \mu)} \exp\left[-\tfrac{4}{3}(2m/\hbar^2)^{1/2}\omega^{3/2}/Fe\right],$$
$$\tag{1}$$

where ω is the barrier height, F the applied field, m the electron mass, e electron charge, ϕ work function, E_x energy in the tunneling direction, and $\mu = E_F$, the Fermi energy. The emission current can then be obtained by multiplying this coefficient by the differential arrival rate $v_x N(v_x)\, dv_x$, and integrating over the energy range E_x corresponding to $v_x = 0$ to its maximum value. For $T = 0$ K, and still approximately at

moderate temperatures $E_x \leq \mu$. For this case $N(v_x)\, dv_x$ takes a very simple form, namely

$$N(v_x)\, dv_x = \frac{4\pi m^2 \Delta}{h^3}\, dv_x,$$
$$\tag{2}$$

where $\Delta = \mu - E_x$. Integration can be carried out with a few straightforward approximations [5], yielding

$$i = \frac{4}{3}\frac{16\pi m e (\mu/\phi)^{1/2}}{h^3(\phi + \mu)b^2}F^2$$
$$\times \exp\left[-\tfrac{4}{3}(2m/\hbar^2)^{1/2}\phi^{3/2}/Fe\right],$$
$$\tag{3}$$

where $b = (4/3)(2m/\hbar^2)^{1/2}/e = 6.8 \times 10^7$ eV$^{-1/2}$ cm^{-1}, or

$$i = 6.2 \times 10^6 \frac{(\mu/\phi)^{1/2}}{(\phi + \mu)}F^2$$
$$\times \exp\left[-6.8 \times 10^7 \phi^{3/2}/F\right] \text{amp/cm}^2,$$
$$\tag{4}$$

for energies in eV and F in V/cm. This is the famous Fowler–Nordheim equation without image corrections. If the potential step at the surface is assumed to be modified by a classical image potential,

$$V_{\text{im}} = -\frac{e^2}{4x} = -\frac{3.6}{x} \text{ eV},$$
$$\tag{5}$$

(for x in units of ångström) the barrier is reduced in height and the term in the exponent must be multiplied by a factor σ which is a function of the argument

$$y = 3.8 \times 10^{-4}F^{1/2}/\phi,$$
$$\tag{6}$$

for F in V/cm and ϕ in eV. A simple discussion of the magnitude of σ as well as a table of σ versus y is given in Ref. [5]. It turns out [6] that σ is nearly a linear function of y^2 so that it can be written over the F-range of most experiments as

$$\sigma(y^2) = s(y^2) - c(y^2)F,$$
$$\tag{7}$$

where $s(y^2)$ and $c(y^2)$ are slowly varying functions of y^2. Consequently only $s(y^2)$ multiplies the term in $1/F$ in the exponent of Eq. (4) which now becomes

$$i = \frac{6.2 \times 10^6 (\mu/\phi)^{1/2}}{(\phi + \mu)} \exp\left[c(y^2)b\phi^{3/2}\right](F/\sigma)^2$$
$$\times \exp\left[-6.8 \times 10^7 \phi^{3/2}s(y^2)/F\right] \text{amp/cm}^2.$$
$$\tag{8}$$

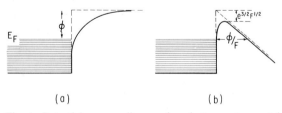

(a)   (b)

Fig. 1. Potential energy diagram for electrons at a metal surface (a) in the absence and (b) in the presence of an applied field. The Schottky saddle and corresponding reduction in barrier height are shown in (b).

For a field of 4×10^7 V/cm and $\phi = 5$ eV, $s(y^2) = 0.95$, $c(y^2)/F = 0.25$, and $\sigma(y)^2 \approx 0.58$ [6], so that the image correction, while substantial in increasing the total current, does not affect the field dependent Fowler–Nordheim exponent very much. It seems reasonable to assume that this carries over to more realistic electron–metal potentials, particularly since the main effect of barrier thinning occurs at electron–surface distances where the image potential is probably a reasonable approximation in any case. This is substantiated by the fact that Eq. (8) is obeyed quantitatively over a wide range of emission current when the classical image correction is taken into account [7]. The Fowler–Nordheim exponent has essentially the same form as the barrier penetration coefficient if $\omega = \phi$. This is not surprising since the dependence on barrier height in the exponent is so strong that most emission comes from electrons near the Fermi energy. The discussion to this point is valid for low temperatures where the fraction of electrons with energies $E_x > E_F$ is small. Since μ corresponds to some 10^4 K in most metals this is a good approximation to well above room temperature. Alternately put, field emission current remains virtually constant to absolute zero.

2.1. Energy distributions

Energy distributions of field emitted electrons are today largely of historical interest, for reasons which will become apparent presently, but must be understood in order to appreciate the potentialities and limitations of field emission. Since only total rather than distributions along the emission normal can be measured in practice, at least for readily available emission geometries [5,6], we restrict ourselves to the former. It is straightforward to show [5] that the total energy distribution $j(\epsilon)$ is given to good approximation (within the WKB barrier penetration model) by the product of an incident flux and $D(E_x)$, integrated over all E_x consistent with total E. The result, even for non-spherical energy surfaces other than extreme cases [8], is

$$j(\epsilon) = \frac{i}{d} f(\epsilon) \, e^{\epsilon/d}, \tag{9}$$

where $\epsilon = E - E_F$ (negative for $E < E_F$), $f(\epsilon)$ the Fermi-Dirac function

$$f(\epsilon) = \left(1 + e^{\epsilon/k_B T}\right)^{-1}, \tag{10}$$

$$1/d = \tfrac{3}{2}(6.8 \times 10^7 \phi^{1/2}/F), \tag{11}$$

and i the total current density.

Eq. (9) shows that the energy density of emitted electrons drops exponentially with decreasing energy below E_F. The half-width at low T can be shown to be $6.8 \times 10^{-9} F$ eV [5], which turns out to be 0.2 eV for $F = 3 \times 10^7$ V/cm.

Eq. (9) does not contain any information on the density of states in the metal because it is based on the product of a velocity, proportional to $\mathrm{grad}_p E$, and a density of states, proportional to $1/\mathrm{grad}_p E$. Density of states effects have been observed, however, in some cases and can be explained by more rigorous treatments [9], which we omit.

It was hoped at one time that energy distributions of field emitted electrons from adsorbate covered surfaces would give much needed information on local densities of state. While this is true the steep decrease in $j(\epsilon)$ with decreasing energy below E_F limits the accessible energy range to ~ 2 eV below E_F, unless extraordinarily elaborate analyzers are used, in which case one can see structure to 5–6 eV below E_F [10]. The advent of photoelectron spectroscopy has superseded this application of field emission and we therefore omit a detailed discussion of energy distributions in the presence of adsorbates [11]. It should be clear, however, that an appreciable local density of states on the adsorbate should lead to appropriate bumps in the energy distribution since the tunneling barrier for electrons in such states is reduced.

3. The field emission microscope

The exponent of the Fowler–Nordheim equation contains the factor 6.8×10^7 in the numerator; since $\phi = 4$–6 eV for most surfaces (clean or adsorbate covered) it is clear that appreciable field emission requires fields of the order of 10^7

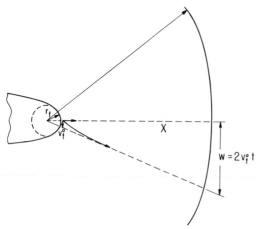

Fig. 2. Schematic diagram showing the displacement w of an electron at the screen of a field emission microscope resulting from a transverse velocity component v_t^0 at the emitter. A second ($v_t^0 = 0$) electron trajectory, illustrating the mechanism of magnification is also shown.

V/cm. The most straightforward way of obtaining such fields, at least statically, is to make the surface from which emission is desired highly curved, for instance a needle with a very small radius of curvature at its end. When surrounded by a suitable anode very high fields can then be created at the needle tip by modest voltages. If the anode consists of a fluorescent screen the device is a field emission microscope (Fig. 2). Magnification comes about because electrons just emerging from the tunneling barrier have very small kinetic energy and thus follow the lines of force (at least initially). Since the emitter surface, being metallic, is an equipotential the lines of force are orthogonal to it and, hence, diverge radially. If the emitter were truly a sphere the linear magnification would be simply x/r_t where x is the tip to screen distance and r_t the tip radius. Since actual emitter geometries correspond more nearly to hemispheres on cylindrical or conical shanks there is a compression both of the lines of force and the electron trajectories so that the actual magnification is

$$M = x/\beta r_t, \tag{12}$$

with $\beta \approx 1.5$. Closely related to the compression is also the decrease in field at the emitter apex.

For a small sphere this would be V/r_t. For actual geometries it is given by

$$F = V/kr_t, \tag{13}$$

where k ranges from 5 to 3.5, depending on the emitter–anode geometry. It is clear that the anode has relatively little effect because virtually all the rapid change in the potential occurs within a few emitter radii, when the latter are much smaller than the emitter–anode spacing. Emitters can usually be etched electrolytically from fine wires and radii of 500–2000 Å are readily obtainable. Thus, Eqs. (8) and (13) indicate that modest voltages, i.e. a few kilovolts, suffice to produce appreciable field emission and that linear magnifications in excess of 10^5 are readily obtainable. Thus a highly magnified emission map of the emitter apex region will be seen on the anode screen. Since field falls off rapidly with polar angle emission is confined to $\sim 45°$ from the emitter axis [5].

The emitting region is generally much smaller than the average grain size even for polycrystalline wires, so that it almost invariably forms part of a single crystal. For sufficiently refractory metals the surface can be cleaned simply by heating to high temperature; this also produces a smooth surface in which flats, corresponding to low surface energy planes blend smoothly into regions of varying Miller indices. Emission is governed mainly by the Fowler–Nordheim exponent, so that it varies strongly with local work function and field. If the surface is smooth (almost on the atomic scale), so that the field over the apex region varies only slightly and in any case smoothly, emission will be governed by work function, so that the pattern on the screen will be a reflection of the work functions of the various exposed crystallographic directions. Since close-packed, low energy planes have highest, and atomically rough, high surface energy regions lowest work function, the patterns will thus show crystal symmetry. Despite the compression of the pattern it is possible, from symmetry alone, to index it and thus of course also to determine the compression factor. If there are local surface asperities the field at the latter will be enhanced so that these will show extra brightness. There can of course also be additional local magnification, although the latter is not nearly as much as the effective radius of an asperity would suggest, because of very high local image compression. Figs. 3a and 3b show emission patterns from a clean smooth Ni emitter and one in which surface segregation of Si has led to local field enhancement and increased emission in the form of bright rings about various planes. The observation of such decorations gave qualitative evidence for surface segregation long before the advent of Auger spectroscopy.

3.1. Resolution

It might be thought that resolution is diffraction limited. However, the relevant wavelength is that of electrons at the screen where the image is formed and this corresponds to $12.5 \ V^{-1/2}$ Å, which for kV electrons is very small, so that in general diffraction is relatively unimportant. Much more important and in fact determining is the statistical spread in electron velocities transverse to the emission direction. For a given transverse velocity v_{yz} at the emitter surface, i.e. for an electron just emerging from the tunneling barrier, the displacement w at the screen from the original emission direction is

$$w = 2v_{yz}t, \tag{14}$$

where t is the time of flight from emitter to screen. The factor 2 arises because the transverse displacement throws the electron off a line of force so that it receives additional transverse acceleration and displacement from electric field components transverse to the original one [5]. The time of flight is given to very good approximation by

$$t = x(m/2Ve)^{1/2} \tag{15}$$

where V is the applied voltage, because almost the entire trajectory corresponds to an electron with full kinetic energy Ve. Thus, the half-width on the screen of a disk corresponding to v_{yz} will be given by combination of Eqs. (14) and (15); the corresponding radius ρ of a disk on the emitter is obtained by dividing this quantity by the magnification

$$\rho = 2v_{yz}\beta r_t(m/2Ve)^{1/2}$$
$$= 2r_t\beta(E_{yz}/Ve)^{1/2}, \tag{16}$$

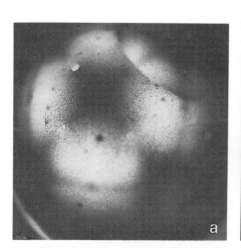
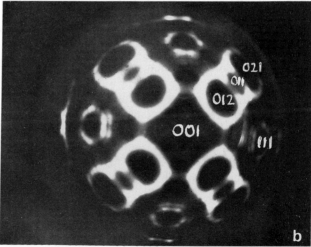

Fig. 3. (a) Clean (100)-oriented Ni emitter, showing also a grain boundary. (b) (100)-oriented Ni emitter prepared from Ni containing 0.3% Si. The pattern shown was obtained after heating to 1300 K. Note the decorations around various planes, some of which are indexed.

where the second form of Eq. (16) is written in terms of the corresponding initial kinetic energy E_{yz} transverse to the surface normal.

A good idea of the actual resolution can be obtained by calculating first the normal (not total) energy distribution of field emitted electrons $N(E_x)$ and then finding the average value of E_{yz} corresponding to the maximum in this distribution. E_{yz} and E_x are linked because at $T = 0$ K the maximum total energy $E_x + E_{yz} = E_F$, the Fermi energy. It turns out [5] that the maximum in $N(E_x)$, corresponds to a value of $\Delta = E_F - E_x$ which increases linearly with applied field F. This is not surprising since increasing the field thins the barrier and allows electrons of lower E_x to tunnel. It further turns out that for given Δ the average transverse energy $\langle E_{yz} \rangle$ is

$$\langle E_{yz} \rangle = (4/9)\Delta. \tag{17}$$

The effective resolution half-width $\delta/2$ is then found by putting the value of $\langle E_{yz} \rangle$ corresponding to the maximum in $N(E_x)$ in Eq. (16). The result is [5]

$$\delta/2 = 1.31 \times 10^{-4} \beta \left(r_t / k \sigma \phi^{1/2} \right)^{1/2} \text{ cm.} \tag{18}$$

The absence of field or voltage dependence comes about because the increase in transverse velocity, i.e. in $\langle E_{yz} \rangle^{1/2}$ with increasing F, is just cancelled by the corresponding decrease in flight time. The resolution obtainable in field emission thus depends mainly on emitter radius and is of the order of 20 Å. For very sharp emitters, and consequently low applied voltages, there will be also a diffraction contribution, but this is seldom important.

It might be thought at first that the resolution could be improved by using a relatively low field to extract electrons from the emitter thus making Δ and $\langle E_{yz} \rangle$ small and then to post-accelerate them, thereby reducing the time of flight. This scheme is practically impossible, however, for a theoretically interesting reason. Since the emitter and anode approximate concentric sphere geometry, electrons move in a central force field, and consequently their angular momentum about the emitter is a constant. Thus, calling $v_{yz} = v_t$,

$$m v_t r = m v_t^0 r_t, \tag{19}$$

so that, by the time the electron has traveled a distance of, say, $r = 10 r_t$, i.e. $\sim 10^{-4}$ cm, v_t will have been reduced by a factor of 10 from its value v_t^0 at the tip. In other words, electron trajectories very rapidly approach a radius vector, and post-acceleration can improve the resolution only if it occurs at a distance of the order of r_t in front of the emitter. The fact that electrons are thrown along a radius vector also means that any experiment designed to determine electron energy will always measure total E rather than E_x at the point of emission. To measure the latter a parallel plane geometry would be required which is virtually impossible in practice because of microscopic surface irregularities on any macroscopic specimen.

4. Applications of field emission and field emission microscopy

4.1. Summary of salient features

Any discussion of the applications of the FEM requires a brief summary of the strengths and weaknesses of the technique. Fabrication of emitters is simplest by electrolytic etching and can be carried out for many metals, with achievable tip radii of 500–2000 Å. Because of the small size of the emitting region the latter is almost invariably part of a single crystal, even if polycrystalline wires are etched. For drawn wires the emitter axis usually corresponds to a (110) direction for bcc and either to (111) or (100) for fcc materials. Other orientations require the use of single crystal wires. Refractory metals, e.g. W or Mo can be cleaned easily by heating without excessive emitter blunting. For lower melting materials this method becomes difficult if not impossible, although Pt and Ni have been cleaned thermally. The major problem is the presence of impurities which form non-volatile oxides or carbides which then segregate on the surface on heating. It is probably possible to circumvent this problem by using vapor grown whiskers [12], which can be ultrapure but the technique has not been used systematically, and is somewhat more laborious than standard tip fabrication. The problem of cleaning can be overcome in many cases by field desorption, which removes both adsorbate and substrate atoms in fairly controllable fashion from the apex region of the emitter, leaving the shank, however, untouched. Since fields of $(3-5) \times 10^8$ V/cm are required, the emitter material must withstand the high electrostatic stress s

$$s = (F/300)^2/8\pi \text{ dyne cm}^{-2} \qquad (20)$$

with F in V/cm.

For materials which can be fabricated and cleaned field emission microscopy offers some unique advantages: high linear magnification, 10^5-10^6, with resolution of 20–40 Å; extremely high sensitivity to work function changes; high sensitivity to local field enhancement, so that asperities of atomic dimensions show up; virtual temperature independence of emission; a monocrystalline substrate exposing simultaneously different crystal planes. Finally, although not directly relevant to surface science, field emitters are virtual point sources (~ 50 Å in linear dimension) of electrons with high brightness and virtual freedom from space charge spreading up to high currents, since the point source is virtual, at the center not at the surface of the emitter hemisphere. We shall now discuss applications, primarily to surface science but will briefly mention others.

4.2. Work function measurements

It is possible to place a small hole in the anode, steer the emitted beam either by electrostatic deflection or by actual rotation of the emitter about its axis and to measure current as function of voltage from small regions of single crystal planes. Application of the Fowler–Nordheim equation then allows one to obtain work functions from the slopes S of $\ln i/V^2$ versus $1/V$ plots, provided the local field–voltage proportionality $c = 1/kr_t$ from Eq. (13) is known. For this purpose the Fowler–Nordheim equation is conveniently written as

$$\ln i/V^2 = \ln B - S/V, \qquad (21)$$

where B is the field (and, hence, voltage) independent part and

$$S = 6.8 \times 10^7 \phi^{3/2}/c. \qquad (22)$$

c is generally not known a priori, since there are deviations from uniform curvature, and decreases in c with increasing apex angle, so that an average c value, easily obtainable from Eqs. (21) and (22) using total emission current and average work function, will not give accurate c values for individual planes. The problem can be overcome if energy distributions are available. The latter yield d from their slopes as indicated by Eqs. (9) and (11). If F in Eq. (11) is replaced by cV, combination with Eq. (22) then eliminates c and yields [13]

$$d \cdot S = (2/3)\phi V, \qquad (23)$$

where V refers to the voltage at which the energy distribution is taken. Once ϕ for a given region has thus been found the local c can of course be obtained as well from Eq. (22). Since photoemission has largely superseded field emission energy distribution measurements this method is largely of historical interest. The availability of work function values for clean crystal surfaces of many metals [14] allows a much simpler procedure for finding work function increments for adsorbates: local c values are first found from S of the clean surface or equivalently

$$\phi(\theta)/\phi(\theta = 0) = [S(\theta)/S(\theta = 0)]^{2/3}, \qquad (24)$$

where θ is the adsorbate coverage. This method turns out to be quite reliable on atomically open planes. On the close packed W(110) plane it seems to work for some but not all adsorbates. In particular for Xe adsorption it gives $|\Delta\phi| = |\phi(\theta) - \phi(\theta = 0)|$ much larger [15] than Kelvin probe values [16]. However, even the apparent $\Delta\phi$ values are useful as a measure of coverage.

The reasons for this discrepancy are not fully understood. The answer may lie in an effect first discussed by Duke and coworkers [17] who found that field emission through a potential well representing an adsorbate could produce resonances and antiresonances, having to do with wave function matching in this well. These resonances strongly affect transmission and could easily be very sensitive to changes in applied field thus yielding erroneous ϕ values when forced into the Fowler–Nordheim mold. It is quite possible that such effects are most pronounced on smooth, and least on atomically rough substrate surfaces, the former sharpening and the latter detuning such resonances.

In this connection another interesting phenomenon must be mentioned, namely that B in Eq. (21) is invariably dependent on adsorbate coverage and in general reduced by adsorbates. In some cases this may be connected with the resonance or antiresonance effects just discussed. However, in other cases the effect may arise from polarization of the adsorbate by the applied field, which contributes an induced dipole αF, α being the polarizability of the ad-complex. The sign of the induced dipole moment is such as to raise the work function and thus $\phi_0^{3/2}$, the zero-field work function in Eq. (8), must be replaced by

$$\phi_F^{3/2} = (\phi_0 + \gamma F)^{3/2}, \tag{25}$$

where γ contains the polarizability and adsorbate coverage. In general $\gamma F \ll \phi_0$ so that the rhs of Eq. (25) can be replaced by

$$\phi_F^{3/2} \approx \phi_0^{3/2} + \tfrac{3}{2}\gamma \phi_0^{1/2} F. \tag{26}$$

Reinsertion into Eq. (8) then shows first that S in Eqs. (21) and (22) refers to the zero-field value of ϕ, and second that the field-independent term B is reduced by $\exp\{-[(3/2)b\gamma \phi_0^{1/2}]\}$.

4.3. Adsorption–desorption kinetics

The fact that almost all adsorbates produce emission and work function changes, and that true or apparent $\Delta\phi$ versus coverage curves can be found, makes field emitters suitable for determining rates of ad- and desorption. In particular alkali and alkaline earth adsorption has been rather extensively studied in this way [18]. Since surface diffusion from plane to plane can occur at the relatively high temperatures which may be required for thermal desorption such results can be difficult to interpret quantitatively. It should be pointed out, however, that some of the first quantitative measurements of electron stimulated desorption were carried out by determining work function changes of adsorbate covered tungsten emitters after electron impact [19]. The advent of modern steel ultrahigh vacuum systems and mass analyzers equipped with electron multipliers has largely supplanted adsorption–desorption measurements by field emission.

4.4. Energy distributions

It has already been pointed out that, despite its great surface sensitivity, this method of probing electronic surface structure has largely been supplanted. However, it is worthwhile to point out some of its achievements. These include the first clear evidence for partially filled s-like LDOS on alkali and alkaline earth ad-atoms by Plummer and coworkers [20]. Also worth mentioning is the first determination of a surface resonance by Swanson et al. [21], found on the W(100) plane. Lea and Gomer [22] first noted that this resonance was quenched by adsorbates while Richter and Gomer [23] found that this resonance was not quenched by Au, but by Cu, Ir, Ta and Nb, suggesting that these metals except Au substantially perturb the potential seen by electrons at the W(100) surface. (A UPS investigation by Egelhoff and coworkers [24] showed that Hg also does not quench the W(100) resonance.) Richter and Gomer also noted that for a monolayer of Mo on W(100) the Mo(100) resonance emerged at monolayer Mo coverage and that the latter was quenched by additional disordered (but not ordered) Mo [23].

4.5. Field emitter detectors

The small physical size and high sensitivity to adsorbate coverage changes have made it possible to use field emitters as integrating detectors of desorbing or reflected gas fluxes, or to use them as an integrating pressure gauge in situations where use of a mass spectrometer or ion gauges is difficult for reasons of space or other constraints. In such applications the highly emitting, i.e. low ϕ regions of the emitter and the total emission current are utilized. It is not difficult to show from the Fowler–Nordheim equation that for coverage increments $\Delta\theta$ on the emitter a voltage increment ΔV is required to maintain a constant emission current such that

$$\Delta\theta = q(\Delta V/V_0)_i, \tag{27}$$

where $\Delta V = V - V_0$, V_0 being the voltage required to produce current i when the emitter is clean and q a constant depending on the adsorbate and the units of θ. $\Delta\theta$ on the field emitter detector is in turn proportional to the gas flux impinging on it. This flux may correspond to desorption from a surface on heating or to gas reflected from it. This method has been used by Gomer and coworkers [25] to obtain step thermal desorption spectra, as well as sticking coefficients and, in conjunction with calibrated effusion sources accurate absolute coverages. Sticking coefficients and coverages are determined by measuring $(\Delta V/V_0)_i$ for gas reflected from a macroscopic single crystal surface when gas from an effusion source impinges on the latter. The detector is placed between the crystal and the source, facing the crystal with its back to the gas source. Since its supports subtend a negligible solid angle there is no interference with the arriving flux. The integral of $\Delta V/V_0$ versus flux impinged is then compared to the signal under conditions of total reflection from the surface, i.e. when the latter is saturated. For a calibrated effusion flux (e.g. via a quartz oscillator microbalance) the absolute coverage can then be found. Sticking coefficients can be obtained as functions of both gas and surface temperature by using a heatable effusion source. Obviously the method requires very high pumping speed as well as ultrahigh vacuum. This was achieved by immersing the entire apparatus in liquid H_2 or He. As we shall see shortly in another context this provides both ultrahigh vacuum and virtually unit sticking coefficients on the cold walls, and thus infinite pumping speed. It was possible to combine a magnetically movable detector with a magnetically movable Kelvin probe in the same apparatus (Fig. 4) and thus to determine also $\Delta\phi$ as function of coverage for a number of systems.

4.6. Surface diffusion

We turn now to what is probably the most important application of field emission to surface science, namely the determination of diffusion coefficients of adsorbates, including "loose" substrate atoms. Although the focus of this review is

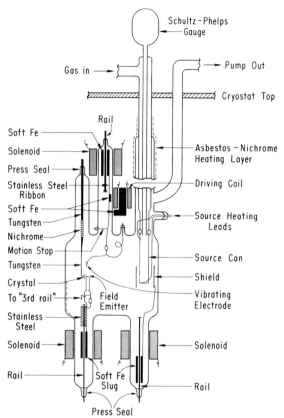

Fig. 4. Diagram of a tube containing a retractable field emitter detector, a retractable Kelvin probe, a heatable effusion source and a heatable single crystal slab. The tube, immersed in a liquid H_2 cryostat, was used for obtaining absolute coverages and sticking coefficients as function of coverage and crystal and gas temperatures, as well as work function versus coverage data. It is shown as an example of the "ship in a bottle" technique to which field emission lent itself. (From Ref. [25].)

field emission and ionization it will not be amiss to start with a very brief general discussion of diffusion. The interested reader is referred to a fairly recent review article [26] for more details. Conceptually simplest is the so called tracer diffusion coefficient D^* which refers to the mean square displacements of single ad-particles:

$$D^* = \langle (\Delta r)^2 \rangle / 4t = \frac{1}{4t} \frac{1}{N} \sum_{i=1}^{N} (\Delta r_i)^2, \qquad (28)$$

where $\langle (\Delta r)^2 \rangle$ is the mean square displacement in time t and N the number of diffusing particles. In terms of a mean jump length a and a jump frequency $\nu \exp(-E^*/k_B T)$

$$D^* = \tfrac{1}{4}a^2\nu \exp(-E^*/k_B T)$$
$$= D_0^* \exp(-E^*/k_B T), \qquad (29)$$

with $D_0^* = \tfrac{1}{4}a^2\nu$. D^* is easily measured in the computer but in real life only under special circumstances, for instance in the field ion microscope in which single metal ad-atoms can be seen and their displacement measured [4]. In virtually all other experiments the chemical, or Fick's law D is measured, i.e. the response of a system of particles to a gradient, usually in concentration. Without going into details D can be expressed by the Kubo–Green equation as

$$D = \frac{1}{2\langle(\delta N)^2\rangle}\int_0^\infty dt \left\langle \sum_{i=1}^N v_i(0) \sum_{j=1}^N v_j(t) \right\rangle, \quad (30)$$

where $v_j(t)$ is the velocity of particle j at time t. $\langle(\delta N)^2\rangle$ is the mean square number fluctuation of ad-particles in area A, containing an average $\langle N \rangle$ particles. D can also be written as

$$D = \frac{1}{4t}\left(\frac{\langle(\delta N)^2\rangle}{\langle N \rangle}\right)^{-1}\left\langle \frac{1}{N}\left(\sum_{i=1}^N \Delta r_i\right)^2 \right\rangle, \qquad (31)$$

where Δr_i is the displacement of the ith particle at time t. $\langle(\delta N)^2\rangle/\langle N \rangle$ is the inverse of the so-called thermodynamic factor and also related to the ad-layer compressibility K:

$$\left(\frac{\partial(\mu/k_B T)}{\partial \ln \theta}\right)_T^{-1} = \frac{\langle(\delta N)^2\rangle}{\langle N \rangle} = k_B T K \frac{\langle N \rangle}{A}. \tag{32}$$

Here μ is the chemical potential, θ coverage and A area. Eqs. (31) and (32) are equivalent to writing D as

$$D = \left(\frac{\partial \mu/k_B T}{\partial \ln \theta}\right)_T D_j = \left[\frac{\langle(\delta N)^2\rangle}{\langle N \rangle}\right]^{-1} D_j, \qquad (33)$$

where D_j is a jump diffusion coefficient, i.e. a complicated average over all particles of something akin to $\tfrac{1}{4}a^2\nu \exp(-E/k_B T)$. Thus D consists of a thermodynamic driving force, $(\partial\mu/k_B T)/\partial \ln \theta$, and a dynamic part, D_j. It should

also be noted that even E^* is a complicated average if ad–ad interactions exist, since these will modify diffusion barriers. For instance, if attractive nearest-neighbor interactions increase binding at a site more than at a saddle point, E^* will be raised locally. Although it can be shown that D^* and D_j can be equal only if there are no velocity–velocity correlations between different ad-particles [27] recent Monte Carlo simulations [28] for various nearest- and next-nearest-neighbor ad–ad interactions indicate that the difference between D^* and D_j is not large, certainly less than an order of magnitude. Both show similar trends with coverage and temperature. Thus a knowledge of both D and $\langle(\delta N)^2\rangle/\langle N \rangle$ is useful since it allows determination of D_j, the closest thing to a microscopic object obtainable in a multiparticle diffusion system. In view of the complicated coverage dependence of D^* and $\langle(\delta N)^2\rangle/\langle N \rangle$ it is not surprising that D is also quite coverage dependent in many instances, so that methods of measuring it at nearly constant θ are desirable.

We return now to field emission. It should be evident that the great sensitivity to work function and, hence, adsorbate coverage changes as well as the high magnification and moderately high resolution of the field emission microscope makes it suitable for the study of surface diffusion. This was in fact done very early by Drechsler [29] who evaporated Ba from the side onto an emitter and was able to follow its diffusion by the increases in emission over the area covered by Ba, which lowers work function and, hence, increases emission. In the early nineteen fifties Gomer and coworkers [5] extended this to the diffusion of O and H by immersing sealed-off field emission tubes in liquid H_2 or He baths. An electrically heatable Pt crucible filled with zirconium hydride or copper oxide, both prepared in situ before seal-off, was placed at one side of the emitter. Ultrahigh vacuum was achieved automatically, while the very high sticking coefficients of gases on the cold walls made it possible to cover only the portion of the emitter "seeing" the gas source. In later experiments sublimation sources made it possible to study all condensable gases. These sources consisted of a small, electrically heatable

Pt platform, on which gas could be condensed and subsequently evaporated in reproducible doses. In this way CO and inert gases were investigated.

This "shadowing" technique yielded qualitatively useful results and also provided the first evidence of extrinsic precursors in chemisorption since diffusion in a second, weakly bound layer could be observed as a moving boundary: for sufficient deposits in the impingement region a sharp boundary spreading at ~ 25 K could be observed for oxygen or CO, corresponding to diffusion in a second layer followed by precipitation onto the clean surface at the edge of the chemisorbed layer, thereby extending the latter and permitting further low temperature diffusion on top of it (Fig. 5). Diffusion with precipitation produces the sharp moving boundary; this is isomathematical with thermal diffusion with melting (or solidification). It was also possible from the upper temperature limit of this process to get an idea of the difference between activation energies of diffusion and desorption in the weakly bound layers [5]. In the case of inert gases it was possible in addition to diffusion to demonstrate visually the existence of multilayers, since heating led to their desorption at different rates and thus to formation of layered structures with ring-like boundaries, whose shrinking at different rates could be observed.

Although the shadowing method gives information over a single crystal, diffusion occurs on different planes and in any case over a range of coverages. We have seen that information at constant but variable θ is desirable. Obviously one would also like to look at single crystal planes. This can in fact be done. The method consists of dosing the emitter uniformly and in fact heating

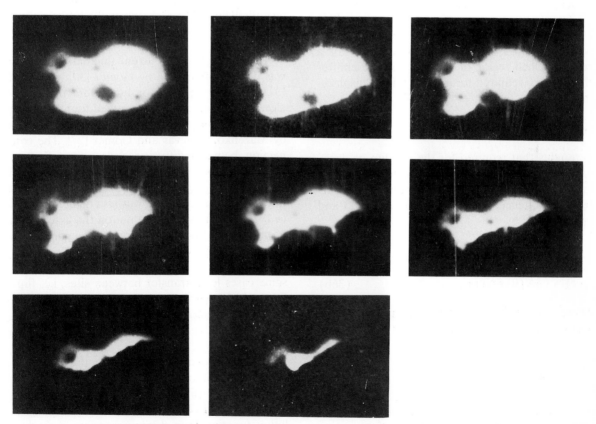

Fig. 5. Selected frames from a 16 mm movie showing second layer boundary diffusion of O_2 on a tungsten field emitter at 27 K. Note the successive spreading of the boundary so that the O covered portion (dark) increases.

it to equilibrate the adsorbate. A probehole in the screen is used so that emission from a region 50–100 Å in linear dimension on the emitter can be measured. Under conditions where the adsorbate is mobile there will be statistical fluctuations in the number of ad-atoms (or ad-molecules) in the probed region, and for a region so small the rms fluctuations will be appreciable. The number fluctuations in turn will cause emission current fluctuations of a magnitude which can easily be measured. To first approximation the number fluctuations in the probed region $\delta N = N(t) - \langle N \rangle$ are simply related to the current fluctuations $\delta i(t)$ via the Fowler–Nordheim equation:

$$\delta \ln i(t) = \frac{\delta i(t)}{\langle i \rangle}$$

$$= \left[\frac{\partial \ln B}{\partial \theta} - \frac{3}{2} \left(6.8 \times 10^7 \langle \phi \rangle^{1/2} / F \right) \frac{\partial \phi}{\partial \theta} \right] \frac{\delta N}{A},$$

$$(34)$$

where $\theta = N/A$ with A the probed area. For a round probed region of radius r_0 on the emitter the time τ_0 required for a mean square displacement r_0^2,

$$\tau_0 = r_0^2 / 4D, \qquad (35)$$

is roughly that required for the buildup or decay of a fluctuation. Thus it should be possible to measure the time autocorrelation function of current, and, hence, number fluctuations, $f_i(t)$ and $f_n(t)$ defined by

$$f_i(t) = \langle \delta i(0) \, \delta i(t) \rangle / \langle i \rangle^2, \qquad (36a)$$

$$f_n(t) = \langle \delta N(0) \, \delta N(t) \rangle, \qquad (36b)$$

where brackets denote ensemble (in practice time) averages. From Eqs. (34) and (36) $f_i(t)$ is given by

$$f_i(t) = \frac{1}{A^2} \left[\partial \ln B / \partial \theta \right.$$

$$\left. - \left(10.2 \times 10^7 \langle \phi \rangle^{1/2} / F \right) \partial \phi / \partial \theta \right]^2 f_n(t).$$

$$(37)$$

$f_n(t)$ can be shown to be expressible as a universal function of t/τ_0, namely

$$f_n(t/\tau_0) = \frac{\langle (\delta N)^2 \rangle}{\pi} \int_0^1 d^2 x \int_0^1 d^2 x'$$

$$\times \frac{\exp \left[- |x - x'|^2 / (t/\tau_0) \right]}{\pi (t/\tau_0)}, \qquad (38)$$

so that it is possible to superimpose, say, semilog-arithmic plots of the actual $f_i(t)$ and theoretical $f_n(t/\tau_0)$ curves and thus to determine τ_0 and, hence, D if r_0 is known. The latter can be found at least approximately from the emitter radius r_t and the magnification. The emitter radius can be determined reasonably well from the Fowler–Nordheim equation if the average, or better still single plane work function is known. It is clear that the method can be applied at constant coverage, to within the thermodynamic limit on the size of fluctuations and that temperature can be varied so that activation energies of diffusion can be found.

The field emission current fluctuation method was developed in 1973 by Gomer [30], with Eqs. (37) and (38) derived for non-interacting particles, and put on a firm theoretical foundation for the general case of interacting adsorbate particles by Mazenko, Banavar and Gomer [27]. The connection between diffusion and current fluctuations had already been recognized by Timm and Van der Ziel in 1966 [31]. A connection between fluctuations and diffusion was also made by Kleint in 1971 [32], but was based not on the assumption of density fluctuations in a probed region but on emission varying with adsorption site, the transfer between different sites being diffusion limited. In some cases local transfer between sites, i.e. flip–flop, which manifests itself by an exponentially decaying correlation function [30] has in fact been seen [33], but generally as a precursor to diffusion.

It is also possible to employ a thin rotatable rectangular slit in the anode, i.e. a rectangular probed region (Fig. 6). If this is lined up along the principal directions of the D tensor the correlation function then decomposes into a product of two one-dimensional ones, each with its own

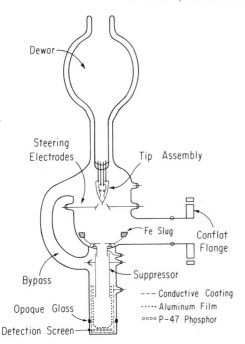

Fig. 6. Field emission tube for measuring anisotropy of surface diffusion by the FEFAC method, showing the rotatable slit in the anode screen. Electrons passing through the slit are post-accelerated onto a fast phosphor whose light output is fed into a photomultiplier (not shown) below the detection screen. The photomultiplier output is converted to a voltage, amplified, filtered and passed into a correlator. It is possible to replace the phosphor–photomultiplier–correlator chain by an electron multiplier in the tube and to obtain the output signal as pulses. The number of pulses collected in each preset time interval is then transferred in real time to a computer which calculates the correlation function digitally, thereby avoiding analogue noise. This variant was first used by T. Okano, T. Honda and Y. Tuzi (Jpn. J. Appl. Phys. 24 (1985) L764) and has recently been improved in our laboratory.

τ_0 corresponding to the long and short slit dimensions [34]. For a probe $2a \times 2b$ with a the dimension parallel to the x- and b the dimension parallel to the y-axis these are, respectively, $\tau_{0x} = a^2/D_{xx}$ and $\tau_{0y} = b^2/D_{yy}$, where the double subscripts refer to the corresponding components of the (diagonal) D tensor. For $a^2/b^2 \ll 1$ is clear that for moderate times only the decay across the a dimension, i.e. τ_{0x} and, hence, D_{xx} will be measured since the correlation function corresponding to diffusion along the y-axis will not have decayed appreciably. By rotating the slit in

the anode 90° τ_{0y} and, hence, D_{yy} can then be determined.

To this point we have ignored the problem of finite resolution. This can be taken into account by convoluting the emission over the probe area elements with the resolution. With some approximations results can be obtained in closed form [35]. It turns out that t/τ_0 must be replaced by $t/\tau_0 + 2(\delta/2r_0)^2$ for a round probe and by $t/\tau_0 + \frac{1}{2}(\delta/2a)^2$ for the one-dimensional case. It can be shown that this amounts in essence to probe dimensions increased by $0.75(\delta/2)$ [35].

The field emission fluctuation autocorrelation (FEFAC) method has the drawback that it can only be used in the presence of applied fields of the order of $(3-5) \times 10^7$ V/cm. For adsorbates with high polarizability α and/or high dipole moment P, i.e. alkali or alkaline earth atoms, there is known to be some field effect on activation energies and possibly prefactors of diffusion [36,37] since the electrostatic energy contributions to binding, $\frac{1}{2}\alpha F^2 + \boldsymbol{P}.\boldsymbol{F}$. may differ at an equilibrium binding site and a saddle point. For adsorbates like O, H, or CO the effect is probably very small, as indicated by looking at diffusion with and without applied field in the shadowing method.

Although some new methods for measuring surface diffusion coefficients have recently been developed in which concentration variations can be made small or, up to a point handled by Boltzmann–Matano analysis [38], the field emission fluctuation method is the only one developed to date which also determines $\langle(\delta N)^2\rangle/\langle N\rangle$ via Eq. (37) if absolute coverage is known at least approximately. The range of D values which can be measured by the FEFAC method is limited at the low end by excessive sampling times and, hence, low frequency cut-off and transient problems and at the high end by short sampling times to a range of 10^{-9} to 10^{-14} cm^2 s^{-1}.

A detailed discussion of the various systems examined by the FEFAC method is impossible in the confines of this article and only some highlights can be mentioned. Perhaps the most important is the discovery of tunneling diffusion of H and its isotopes on all planes so far examined, namely various W planes [39–43], and Ni(111)

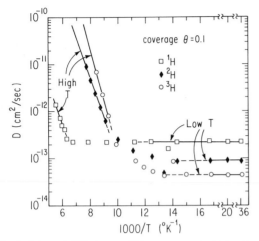

Fig. 7. Selected diffusion coefficients versus $1/T$ for H and its isotopes on W(110) at low coverage, showing both activated and tunneling diffusion. Note the small isotope effects in the tunneling region. Compiled from the results of Ref. [39].

seem able to climb and descend close-packed W steps but not open ⟨100⟩ oriented ones [47]. It was also possible to find evidence for one-dimensional diffusion (probably of kinks) on stepped surfaces, followed by roughening and two-dimensional diffusion above certain temperatures [48]. Finally the fact that on ridged (channeled) planes like Pd(110) cross channel diffusion can occur by substitution of an atom in a channel with channel wall atoms, first demonstrated elegantly by Ehrlich and coworkers [49], via the ion microscope and atom probe was shown to have an analogue on W(211) via FEFAC, using a slit probehole. It was found [50] that diffusion across channels had a lower activation energy but much lower prefactor than diffusion along channels, suggesting that the former process involved exchange with channel wall atoms thus lowering E but also D_0 because of strict phasing requirements.

and Ni(100) [44]. Fig. 7 shows typical plots of log D versus $1/T$, showing the activated and temperature independent, i.e. tunneling regimes for ^{1}H, ^{2}H, and ^{3}H on W(110). It is clear that isotope effects are small, in the tunneling regime. This was found to be the case on all other surfaces examined. The explanation is probably that substrate atom motion is involved in adsorbate tunneling because (small) lattice distortion caused by adsorption must be carried along in tunneling. If the time scales and couplings for M–M and H–M (M being a substrate metal atom) are comparable a mass renormalization occurs with the effective tunneling mass much larger than the bare one, so that the H isotope effect is small [45]. If time scales and coupling are well separated, the H and M motions can be treated separately and the small isotope effect comes about because the M–M Franck–Condon matrix elements depend on adsorbate mass because of zero-point effects. This isotope effect goes in the opposite direction from that for the "pure" H tunneling and thus can largely cancel the latter [44,46].

Diffusion of thermally liberated "loose" W atoms on some planes can be studied by the FEFAC method. Thus it was found that W atoms

5. Other applications of field emission

We list here-very briefly some applications of field emission either not or only indirectly related to surface science, or which like STM, have assumed a life of their own, but omit very specialized applications, such as the determination of Hg whisker growth kinetics [5].

5.1. The scanning tunneling microscope

Two chapters in this volume are devoted to this subject. Suffice it therefore to say that the STM is based on field emission and was in fact foreshadowed by an attempt to develop a field emission surface roughness monitor by R.D. Young.

5.2. Electron optics and array sources

The advantages of field emitter sources in electron optical applications have already been pointed out. Most modern electron microscopes do in fact use field emitter sources. Multiple sources capable of giving very high pulsed current in the form of combs were developed many years ago for flash X-ray and other applications by Dyke and coworkers at the Field Emission Corporation [51]. A more recent development made possible by MBE technology is the fabrication of square arrays of very high density, a technique developed by Spindt [52]. These can consist of diodes or triodes with microscopic tip–anode spacings, so that 10–50 V suffice for appreciable emission. These arrays have potential applications in low voltage image tubes, etc.

5.3. Field emission in liquids

It is possible to obtain field emission into various liquids. Halpern and Gomer [53] used modified emission tubes which permitted switching from ultrahigh vacuum to very pure liquids and back to ultrahigh vacuum to emit electrons into $^1H_2$, $^2H_2$, He, Ar, O_2 and C_6H_6. In all cases space charge behavior occurred at $i \geq 10^{-9}$ amperes from which negative charge carrier (in most cases free electron) mobilities could be found. For He and Ar emission onset was abrupt indicating electron avalanching and eventual gas bubble formation and tip sputtering. For O_2 electrochemical etching occurred, but for H_2, where electron energies can be dissipated by vibrational excitation of H_2 avalanching can be avoided and a true Fowler–Nordheim regime observed. The resultant effective work function suggests an energy barrier of 0.1–0.5 eV (relative to E_F) for solution of free electrons in liquid H_2, i.e. a free energy of solution of ~ 4 eV, relative to vacuum. Although not exploited to my knowledge, field emission can provide a simple method of electron injection into dielectric liquids without use of either vacuum or high energy accelerators.

6. Field ionization

6.1. Early observations

The first observations of field ionization were made by Müller [2] when he admitted hydrogen at low pressure (10^{-3} Torr) to a field-emission tube and applied a high positive voltage to the tip. A faint but highly resolved image appeared on the screen and was originally attributed by him to protons desorbed from the tip. He had been led to this experiment by some of his previous observations on the behavior of adsorbed electropositive metals under reversed fields [2], which indicated a field-induced desorption.

At the time of his early hydrogen-ion observations he believed that the improvement in resolution over electron emission resulted from the shorter de Broglie wavelength of heavy particles. Since diffraction does not limit resolution under normal circumstances, as we have seen, another explanation was advanced, based on the low zero-point energy of the bending modes of adsorbed H atoms.

These explanations were unsatisfactory, particularly because $\Delta\phi$ for H adsorbed on tungsten indicated covalent bonding. Mass-spectrometric experiments by Inghram and Gomer [54] soon justified these doubts. In order to determine the nature of the ions produced by a field emitter, a small portion of the beam was allowed to penetrate into a mass spectrometer through a small hole in the screen of an emission tube. I did the glass blowing myself and remember that it was somewhat tricky but by no means impossible to do this within an all-glass tube. Today we would undoubtedly have used machined metal inserts. In the case of hydrogen (or deuterium) the principal ions were $^1H_2^+$ (or $^2H_2^+$) with H^+ appearing only at much higher fields than those required for parent ions. Ions were also found from all other gases introduced into the tube and it was noted that organic molecules, for instance acetone, gave primarily parent rather than fragment ions. It was also noticed that ion energy distributions were fairly narrow at low fields but broadened on the low energy side as field increased. This indicated very clearly that ionization then took place farther and farther in front of the emitter, accounting for the energy deficit. This was also true for H^+ indicating that most of it came from H_2 or H_2^+ break-up. It was also found that in some cases, for instance CH_3OH, the parent peak broadened while the CH_3O^+ peak stayed sharp, indicating that the latter was formed at the same location in space, evidently on the emitter surface, thus corresponding to field desorption. We were able to confirm this by using pulsed rather than DC voltages: at high repetition rates the steady state, DC situation corresponding to gas phase ionization, adsorption, and field desorption is simulated. As the pulse repetition rate decreases the signal from the adsorbed species, i.e. CH_3O^+ increases relative to that of the gas phase product CH_3OH^+ since the density of the adsorbed phase is much greater than that of the gas phase at low pressure, and has time to become replenished between pulses.

6.2. Mechanism of field ionization

The mechanism of field ionization indicated by the result just outlined must therefore consist of field emission in reverse [5], that is of field emission from gas phase atoms or molecules into the emitter, as illustrated by Fig. 8. In the case of an atom in free space the potential deformed by the applied field can lead to tunneling from filled high energy, i.e. low ionization energy levels, as shown in Fig. 8 (left). Near a metal surface (Fig. 8 (right)) the situation is more complicated, since the tunneling electron must enter the metal at or

above the Fermi level, E_F. Consider a gas phase atom A near the metal surface. If the highest filled energy level lies below E_F in the absence of a field, i.e. if $I > \phi$, where I is the ionization potential of A, the role of the applied field is to raise the electron energy so that for an atom at a distance $x \geq x_c$ from the surface, tunneling into the metal at or above E_F can occur. Thus

$$Fex_c = I - \phi - e^2/4x_c, \qquad (39)$$

where the last term corresponds to the fact that near a surface the binding energy of any electron in A is decreased by the image potential. For given F there is a dead zone $x < x_c$ within which atoms cannot be ionized. Thus there is the possibility of atoms "field-adsorbed" on the surface and there is evidence that this is the case for He for instance [55]. It used to be believed that such atoms were held by polarization forces, but it has recently been suggested by Kreutzer [56] that the effect of the field is to raise the ionization level of He sufficiently close to E_F, so that it broadens enough to become partially emptied with He then binding to the substrate much like an alkali or alkaline earth atom.

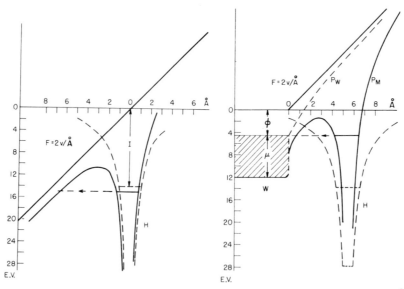

Fig. 8. Potential energy diagram for an electron in an H atom in the presence of an applied field of 2 V/Å (left) in free space (right) near a metal surface. I: ionization potential of atom, ϕ: work function of metal. Dashed lines marked H are the H-atom Coulomb potential in the absence of the field. Solid lines are the field deformed potentials. P_W is the sum of applied plus image potential, P_M the atom potential.

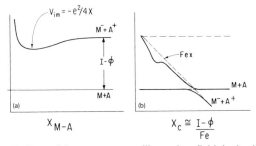

Fig. 9. Potential energy curves illustrating field ionization. Shown are the neutral and ionic curves M + A and M⁻ + A⁺, respectively, (a) in the absence and (b) in the presence of an applied field F which deforms the ionic curve so that it intersects the neutral curve near x_c.

It is also useful to consider field ionization in terms of M + A and M⁻ + A⁺ potential energy curves, as illustrated in Fig. 9. The field-distorted M⁻ + A⁺ curve "intersects" the M + A curve precisely at x_c of Eq. (39) (if polarization effects for the latter are neglected). The curves do not in fact intersect but are split near the crossing point, with the splitting in energy ΔE corresponding approximately to

$$\Delta E = \hbar/\tau,$$

where τ is the lifetime with respect to electron tunneling. The curve M⁻ + A⁺, shown in Fig. 9, corresponds to the lowest member of a quasi-continuous manifold M⁻* + A⁺, where the asterisk indicates electronic excitation in the metal: e.g. for a transition at $x > x_c$ an electron enters M above the Fermi level. For (relatively) low fields where the tunneling barrier is appreciably smaller at x_c than at $x \gg x_c$, because of the truncation of the barrier by the metal potential transitions at $x \gg x_c$ will contribute very little. However, at sufficiently high fields appreciable ion formation can occur at x appreciably larger than x_c. It should also be clear that the energy with which an ion arrives at a collector will correspond (neglecting contact potentials for the moment) to $V - Fex$, where V is the applied potential between emitter and collector (here cathode) and Fex the energy deficit resulting from ionization at x rather than at the surface, $x = 0$.

The lifetime τ with respect to field ionization can be crudely estimated by assuming an electron

frequency $\nu_e \approx 10^{15} - 10^{16}$ s⁻¹ and multiplying this by a WKB barrier penetration coefficient. The result is

$$\tau_i^{-1}(x_c) \approx \nu_e \exp\left[-6.8 \times 10^7 (I^{3/2} - \phi^{3/2})/F\right]. \tag{40}$$

The work function of the metal appears in Eq. (40) because the tunneling barrier seen by an electron in an atom at x_c is truncated by the presence of the metal, in such a way that a barrier equivalent to that of emission from the metal must be subtracted. $\tau(x)$ increases very rapidly as $x > x_c$ for two reasons. First, the field falls off with increasing distance from a spherical surface as $(r/r_t)^2$ and more importantly the term $\phi^{3/2}$ must be replaced by a decreasing term which rapidly goes to zero, so that the exponent in Eq. (40) increases rapidly with distance.

There is one interesting qualification to this statement. An examination of ion energies in a high resolution mass spectrometer by Inghram and Jason [57] revealed one main peak, at an energy corresponding to an ionization distance of x_c, but several small subsidiary peaks at lower energy, and, hence, corresponding to ionization at a greater distance from the surface. The most straightforward explanation of this phenomenon is that the sharp discontinuity in potential at the metal surface has the effect of introducing an effective reflection coefficient for tunneling electrons. This can be formally thought of as equivalent to a potential wall, so that tunneling electrons see a triangular potential well between the atom they are leaving and the metal. The tunneling probability will have abnormally high values when the atom is at just the right distance from the surface for the electron energy to coincide with one of the levels of this well (Fig. 10).

It is also possible to use perturbation theory so that

$$\tau^{-1} = (2\pi/\hbar)\rho_k|\langle k| Fex |g\rangle|^2, \tag{41}$$

where ρ_k represents the density of final states $|k\rangle$, and $|g\rangle$ represents the initial electronic state on the atom. $|k\rangle$ corresponds to free electrons for space ionization and to metal states for ionization in front of the metal surface.

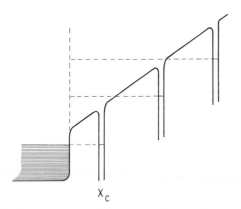

Fig. 10. Potential energy diagram for electrons in an atom near a metal surface in the presence of an applied field, illustrating the resonance effect in field ionization discussed in the text.

In connection with the visibility of adsorbates in the field ion microscope it is worthwhile to consider ionization in front of an adsorbed atom or molecule. The analogue to Eq. (41) becomes

$$\tau^{-1} = (2\pi/\hbar)\big(\rho_k\big|\langle k| Fex |g\rangle\big|^2$$
$$+\rho_a|\langle a| Fex |g\rangle|^2\big), \qquad (42)$$

where ρ_a is the local density of states on the adsorbed entity above E_F and $|a\rangle$ the electron state associated with the adsorbate. Thus, the presence of a high local density of states *at the Fermi level* could enhance the field ionization probability, provided the adsorbate orbitals stick out appreciably, normal to the surface.

It is important to realize that the field normally used in the ion microscope is ~ 2 to ~ 4 V/Å. This is sufficiently high to constitute a perturbation on the wave functions of the gas atom being ionized, of the adsorbate (if any) and of the metal. In particular, any adsorbate levels, or local densities of state will be shifted upward in energy by an amount Fex_a, i.e. 4 to 8 eV, unless screened, which is hardly a trivial shift. If there is no appreciable local density of states *at the Fermi level*, the presence of an adsorbate would generally de-enhance ionization in front of it, if only because of the fact that it may prevent the atom being ionized from approaching the

surface as closely as in the absence of adsorbate, and possibly also because of a reduction in local field. Furthermore, the adsorbate could de-enhance the total density of states near the Fermi level and, thus, contribute to a decrease in ionization probability. Alferieff and Duke [58] have also postulated the existence of resonances in field ionization, analogous to those for field emission [17]. Thus, the presence of adsorbates can be expected to increase or decrease ionization probability in their immediate vicinity, depending on the specifics of each case.

We have already seen on the basis of the WKB approximation that τ_i^{-1} decreases rapidly with distance from the surface and this is equally clear in perturbation theory, since the matrix elements will decrease very rapidly. Thus, to a good approximation the ionization probability of an atom moving past x_c, is given by

$$P_i = \Delta x/v\tau_i(x_c), \qquad (43)$$

where v is the velocity of the atom and $\Delta x \approx 0.5$ Å. The importance of the effective dwell-time t, here equal to $\Delta x/v$, can be demonstrated very clearly. When field ionization occurs from the edge of a thick adsorbed film [59], or in a liquid [60] where the dwell-time is effectively infinite, fields of the order of 1–2 V/Å are required for He, as compared to 4 V/Å for true gas phase ionization.

Ion current is limited by gas supply under most conditions of interest both for film and gas phase ionization since even in the latter case the ionization probability can be made very high. For gas phase ionization the arrival rate is the gas kinetic one, $P/(2\pi mkT)^{1/2}$, multiplied by an enhancement factor arising from the polarization attraction of gas atoms to the tip. As first shown by Southon [61], the enhancement factor f is roughly

$$f = \big(\pi\alpha F^2/2k_B T\big)^{1/2}, \qquad (44)$$

where α is the atomic polarizability.

For ionization from films the supply function $\dot{n}$ has been shown by Halpern and Gomer [59] to correspond to

$$\dot{n} = fP(2\pi mk_B T)^{-1/2}2\pi r(D_s t)^{1/2} \qquad (45)$$

for a cylindrical emitter shank of radius r. Here D_s is the surface diffusion coefficient of atoms migrating toward the high field zone at the tip and t is a time which, with D_s, effectively defines the length of the supply zone $(D_s t)^{1/2}$. If the coverage in the diffusing layer is low, i.e. if the evaporation lifetime, τ_e, is small with respect to the mean arrival rate, $t = \tau_e$ and $\dot{n} \propto P$. On the other hand, if coverage is high, i.e. if t reduces to a replenishment time $\propto (2\pi mkT)^{1/2}/P$, $\dot{n} \propto P^{1/2}$. These results can actually be demonstrated with He in the temperature range from 4–1 K [59].

In practice ion current is limited by pressure for the gas phase case, since the mean free path must be kept larger than the dimensions of the high field region in order to avoid avalanching and discharges. In general, this means pressures of ~ 1–5 mTorr, although special geometries make it possible to raise P. The currents normally attainable are of the order of 10^{-10} to 10^{-9} amp. Film ionization makes it possible to reach such values at much lower pressures, and thus may be useful for high intensity quasi-point sources of ions.

7. Applications of field ionization

7.1. Mass spectrometry

As already noted here, and first pointed out by Inghram and Gomer [62], field ionization leads to very little fragmentation and thus provides, where applicable, much simpler mass spectra than conventional impact ionization. This application has been extensively pursued by Beckey and coworkers [63].

7.2. Field ionization in liquids

Halpern and Gomer [60] found that field ionization occurred in most liquids examined. For Ar or H_2 linear $\log i/V$ versus $1/V$ regimes were seen which could be interpreted in terms of Zener breakdown, i.e. by a band model. At currents in excess of 10^{-9} amp space charge regimes were seen and yielded ionic mobilities. Field emission

and field ionization are almost certainly the mechanism of dielectric breakdown in liquids.

7.3. The field ion microscope

By far the most useful application of field ionization is the field ion microscope. We have already noted that it was invented, if not at first fully understood by E.W. Müller in 1951. In essence it consists of a field emission microscope run with the emitter positive and filled with a few mTorr of the imaging gas. At fields of 2–4 V/Å, depending on gas an image with much higher resolution than that of the FEM is then seen. In 1956 Müller and Bahadur [55] cooled the emitter to 20 K and with He as imaging gas obtained the first images with essentially atomic resolution, except on close-packed planes which appeared dark. To anticipate slightly, it is possible, however, to evaporate single or a few metal atoms onto the emitter surface; any such atoms landing on close-packed planes can then be seen in the ion image.

It is clear that the image is formed from gas atoms ionized near the surface. To a first approximation the ion trajectories will be identical with those of field emitted electrons and, hence, the magnification will be the same in the FIM as in the FEM. Contrast depends on the variation in ionization probability over the surface, which is determined by local field, and by local geometry. It is now believed that ionization is actually enhanced by tunneling through field adsorbed, e.g. He atoms. This can be understood if it is accepted that the latter present an alkali-like LDOS with appreciable amplitude at the Fermi level. Field adsorption is, for "chemical reasons" as well as because of the higher local field most probable on protruding substrate atoms. Thus, there can be contrast on the atomic scale except on close packed planes where ionization probabilities are lower and also do not vary enough over the plane. Contrast alone is not enough of course to ensure resolution. The latter is achieved not because of the smaller de Broglie wavelength of He^+ or H_2^+ ions but because of the possibility of vastly reducing the transverse momentum of image forming atoms by cooling the emitter. The actual mechanism contains a subtlety: gas atoms approaching the emitter are accelerated by polarization forces so that the incoming velocity at the tip is

$$v_{in} = (\alpha/m)^{1/2} F. \tag{46}$$

Even for He this corresponds to some 10 times thermal velocity at 300 K, and is enough to make the ionization probability of incoming gas atoms very small if the field is properly adjusted. If incoming atoms become thermally accommodated on the emitter surface their rebound velocities correspond to $\sim k_B T_t$ ($T_t = 20$ K) and consequently they will be moving slowly enough to have appreciable probability of being ionized. Thus the image will be formed mainly from slow atoms rebounding from the emitter with correspondingly small transverse momenta. The resolution is thus given by

$$\delta/2 = 2\beta r_t (k_B T_t/Ve)^{1/2} \tag{47}$$

or in terms of the field F required at the emitter by

$$\delta/2 = 2\beta r_t^{1/2}(k_B T_t/5F)^{1/2}. \tag{48}$$

Eq. (48) is of some interest in that it indicates that imaging gases with high ionization potentials, and thus requiring high F, give best resolution. The resolution for He at $T_t = 20$ K based on Eq. (48) is less than 1 Å and thus better than that actually observable, indicating that actual resolution may be be limited by spatial variations of ionization probability along the surface.

In the early days of field ion microscopy He and H_2 were the only useable imaging gases because conversion of ion kinetic energy into electronic excitation of the phosphor decreases catastrophically with increasing ion mass, most of the energy going into phonons. The advent of channel plates around 1969 made it possible to obtain images even with Ar and also obviated the need for extreme optical and photographic techniques let alone the obligatory dark adaptation period for the experimenters.

Field ion microscopy and also its application to surface diffusion studies are discussed in some detail elsewhere in this volume. For completeness I mention here only the latter, which is based on

the visibility of metal ad-atoms on close-packed planes. Motion can be frozen-in at low temperature while the viewing field is applied, with displacements occurring during heating with the field off. In this way tracer diffusion coefficients can be measured in great detail. Unfortunately this method is almost exclusively confined to metal ad-atoms since other entities are either field desorbed below viewing fields or lack the necessary empty LDOS just above the Fermi energy to become visible.

8. Field desorption

8.1. Introduction

Field ion images of thermally annealed field emitters generally show a great deal of thermal disorder except on close-packed planes, whose uniform darkness implies that close packing is maintained. If the positive field is gradually increased changes in the pattern can be observed. If the field is raised to 5 eV/Å (for tungsten), and then reduced for viewing, atomically perfect images can be seen. It is clear that the high field led to removal of substrate atoms; it is possible to watch the shrinking of close-packed plane edges, i.e. to observe collapsing atom rings and to peel back the surface essentially layer by layer [55]. Desorption of Ba from tungsten emitters had actually been reported by E.W Müller as early as 1941 and interpreted as field-induced desorption of preexisting ions [64].

We have already noted that Inghram and Gomer [54] observed field desorption of CH_3O^+ from W emitters. The theory of field desorption was established by Gomer in 1959 [65] and elaborated in 1963 in connection with observations on the field desorption of CO from W emitters [66].

8.2. Theory of field desorption

Field desorption is most easily understood in terms of metal–adsorbate potential energy curves like those shown in Fig. 11. It is clear that field desorption, at least of adsorbates, is a simple extension of field ionization and can be activated or unactivated depending on where the (avoided)

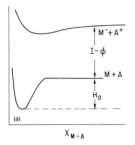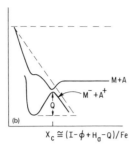

Fig. 11. Potential energy curves illustrating field desorption for the case of large $I - \phi$, where I is the ionization potential of an adsorbate A and ϕ the work function of the metal M. Neutral and ionic curves, M+A and $M^- + A^+$, respectively, (a) in the absence and (b) in the presence of an applied field F. In the latter case the field deformed curves intersect at x_c. H_a is the heat of adsorption of A, Q is the activation energy for desorption corresponding to the x_c shown in (b). Polarization effects have been ignored in this figure.

curve crossing of M + A and $M^- + A^+$ occurs. Here again A stands for adsorbate. The only difference between field ionization and field desorption is that the initial state is more complicated, because of the strong interaction of chemisorbed A with the metal, but this complicates only the calculation of matrix elements and not the conceptual scheme. There is also the possibility of some tunneling through the tip of the barrier formed by the ascending M + A curve and the descending $M^- + A^+$ curve.

The mechanism just discussed applies to cases where the ionic curve lies well above the neutral one in the absence of applied fields. For small $I - \phi$, which usually corresponds to highly polar adsorption, it may happen that the curve crossing occurs at distances from the surface smaller than that corresponding to the Schottky saddle point for the field-deformed ionic curve (Fig. 12). The Schottky saddle would then determine the activation energy. This mechanism was also postulated by E.W. Müller [55].

It is interesting to consider the energetics of field desorption. For the case of moderately large $I - \phi$ the activation energy Q (Fig. 11) is determined as already pointed out by the crossing of the ionic and neutral curves, so that Q is given by

$$Q = I - \phi + H_a - Fex_c - e^2/4x_c + \tfrac{1}{2}\alpha F^2 + P.F.,$$
$$(49)$$

assuming that the ionic curve is given by an image potential in the region of interest, where H_a is the binding energy of A. Q can also be thought of as $V(x_c)$ measured from the vibrational ground-state of the polarization deformed M + A curve. Thus a number of interesting things can be done, at least in principle. If Q is determined as function of applied field F by measuring thermal desorption rate versus T and constructing Arrhenius plots it is possible to find x_c from Eq. (49) and thus to obtain $V(x)$, i.e. a potential curve to within the polarization terms. Experiments of this kind were first carried out by Swanson and Gomer [67] for CO on W, using field emission current changes as endpoints of thermal field desorption. This work was done without a probehole and, therefore, gave only averages over the various planes of the emitter.

It is also possible to determine activation energies mass spectrometrically, again via Arrhenius plots. In this way Ernst [68] found the interesting result that the activation energy of field desorption of Rh^{1+} and Rh^{2+} from Rh emitters, i.e. "self"-field desorption was identical, indicating that the higher charge state arose from field ionization after field desorption of the singly charged ion [69]. It also could be shown that the curve crossing mechanism seems to apply.

It is also possible to measure either by suitable retardation techniques or via time-of-flight measurements in the atom probe, which will be explained shortly, the energy deficit Fex_c directly since this is equal to

$$Fex_c = -e\Delta V + (\phi_{coll} - \phi_e),$$
$$(50)$$

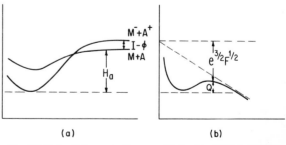

Fig. 12. Potential energy diagrams illustrating field desorption for small $I - \phi$. In the presence of a field the activation energy Q corresponds to the Schottky saddle as indicated in (b).

where ΔV is the (negative) voltage of the collector relative to the emitter at $V = 0$ just sufficient to prevent collection of ions. ϕ_{coll} is the collector and ϕ_e the emitter work function so that $\phi_{coll} - \phi_e$ represents the contact potential difference between these electrodes. (Obviously an intermediate electrode at *high* negative voltage is needed to provide the requisite field at the emitter.) If Eq. (50) is applied to field ionization, i.e. to Eq. (39) where $H_a = 0$, it is seen that

$$\left(I - e^2/4x_c\right) = -e\Delta V + \phi_{coll}, \qquad (51)$$

i.e. that the sum of positive terms on the rhs of Eq. (51) is the image corrected ionization potential of A. For this reason the quantity $-e\Delta V + \phi_{coll}$ is called the appearance potential by some authors and this terminology is retained even for cases where $H_a \neq 0$ [69].

In the case of small $I - \phi$ Q is given by

$$Q = H_a + I - \phi - e^{3/2}F^{1/2} + \tfrac{1}{2}\alpha F^2 + \boldsymbol{P.F.}, \quad (52)$$

for singly charged ions, from which H_a can also be deduced when F is such as to make $Q = 0$. In practice this mechanism is encountered only for electropositive adsorption but not for self-field desorption, which is sometimes called field evaporation. The latter is the basis of obtaining atomically perfect emitters and requires some additional discussion. For covalently bonded, somewhat electronegative, adsorbates like O or CO the curve crossing model in which the field experienced by the adsorbate is well defined seems at least qualitatively correct. For an atom forming part of the substrate itself, or in fact for a metal atom adsorbed on a metal substrate it is not clear to what extent the field is screened out by the atom to be desorbed, so that the calculation of effective polarization corrections becomes quite difficult. This point has recently been discussed by Kreutzer [56].

9. Applications of field desorption

The applications to the preparation of atomically perfect emitters and to the determination of binding energies have already been mentioned. The former requires no additional comment, except to point out that the very high electrostatic stresses, s, given by Eq. (20) amount to 10^{10}–10^{12} dyne/cm^2 and so limit the method to fairly refractory metals. Some reduction in field is possible because the introduction of H_2 significantly lowers desorption field, presumably because of hydride ion formation. The determination of binding energies via field desorption is interesting, since, in combination with the atom probe, specific sites can be picked. However, polarization contributions are very difficult to estimate accurately, as just pointed out.

9.1. The atom probe

This device, invented by E.W. Müller in 1967 [55], is discussed in great detail elsewhere in this volume, but is mentioned here for completeness since it constitutes the most spectacular application of field desorption. It consists of a field ion microscope with a small hole in the screen (these days channel plate plus screen) and a means of moving the emitter about its center so that a desired image point may be placed over the probehole. Behind the probehole is a drift tube and an ion collector, i.e. a particle multiplier. A field not quite sufficient to cause field desorption is then applied, followed by a very short small electrical pulse or a laser pulse sufficient to give desorption. The pulse starts a timing sequence so that the time interval between desorption pulse and ion arrival time at the detector can be found. Since the ion traverses the drift space with known potential difference the e/m ratio of the ion can be found.

Modern versions are quite sophisticated but are in essence elaborations on this scheme [69]. The atom probe has been used extensively in metallurgical applications and was instrumental in discovering field adsorption of He. It is also useful in metal atom diffusion studies, as the article by Ehrlich will indicate.

10. Conclusion

This paper has attempted to sketch the theory and major applications of field emission and field ion microscopy and of field desorption. Despite the proliferation of techniques and an expansion of surface science undreamed of when field emission and field ionization were first used, both retain a niche today, as this paper has tried to show. However, no paper can really describe the thrill and fun of first steps in a new land, which were the privilege of early workers in these fields.

Acknowledgement

The writing of this paper was supported in part by the Materials Research Laboratory of the National Science Foundation at the University of Chicago.

References

[1] E.W. Müller, Z. Phys. 106 (1937) 541.

[2] E.W. Müller, Ergeb. Exakten Naturwiss. 27 (1953) 290.

[3] T.T. Tsong, Surf. Sci. 299/300 (1994) 153.

[4] G. Ehrlich, Surf. Sci. 299/300 (1994) 628.

[5] R. Gomer, Field Emission and Field Ionization (Harvard University Press, 1961). Reprint edition Am. Inst. Phys. (1993).

[6] R.H. Good and E.W. Müller, Handbuch Physik 21 (1956) 176.

[7] W.P. Dyke and W.W. Dolan, Adv. Electron Phys. 8 (1956) 89.

[8] R. Stratton, Phys. Rev. 135 (1964) A794.

[9] D. Penn and E.W. Plummer, Phys. Rev. B 9 (1974) 1216.

[10] M. Isaacson and R. Gomer, Appl. Phys. 15 (1978) 253.

[11] R. Gomer, Adv. Chem. Phys. 27 (1974) 211.

[12] A.J. Melmed, J. Chem. Phys. 38 (1963) 607.

[13] R.D. Young and H.E. Clark, Phys. Rev. Lett. 17 (1966) 351.

[14] J. Hölzl and F.K. Schulte, Springer Tracts in Modern Physics 85 (1979) 1.

[15] T. Engel and R. Gomer, J. Chem. Phys. 52 (1970) 5572.

[16] C. Wang and R. Gomer, Surf. Sci. 91 (1980) 533.

[17] C.B. Duke and M.E. Alferieff, J. Chem. Phys. 46 (1967) 923;
C.B. Duke and J. Fauchier, Surf. Sci. 32 (1972) 175.

[18] R. Gomer, Solid State Phys. 30 (1975) 93.

[19] D. Menzel and R. Gomer, J. Chem. Phys. 41 (1964) 3311.

[20] E.W. Plummer and R.D. Young, Phys. Rev. B 1 (1970) 2088.

[21] L.W. Swanson and L.C. Crouser, Phys. Rev. 163 (1967) 622.

[22] C. Lea and R. Gomer, J. Chem. Phys. 54 (1971) 3349.

[23] L. Richter and R. Gomer, Surf. Sci. 83 (1979) 93.

[24] W.F. Egelhoff, Jr., D.L. Perry and J.W. Linnett, Surf. Sci. 54 (1976) 670.

[25] C. Wang and R. Gomer, Surf. Sci. 74 (1978) 389; 84 (1979) 329.

[26] R. Gomer, Rep. Prog. Phys. 53 (1990) 917.

[27] G. Mazenko, J.R. Banavar and R. Gomer, Surf. Sci. 107 (1981) 459.

[28] C. Uebing and R. Gomer, J. Chem. Phys. 95 (1991) 7626, 7636, 7641, 7648.

[29] M. Drechsler, Z. Elektrochem. 58 (1954) 340.

[30] R. Gomer, Surf. Sci. 38 (1973) 373.

[31] G.W. Timm and A. Van der Ziel, Physica 32 (1966) 1333.

[32] Ch. Kleint, Surf. Sci. 25 (1971) 394, 411.

[33] J.-R. Chen and R. Gomer, Surf. Sci. 79 (1979) 413.

[34] D.R. Bowman, R. Gomer, K. Muttalib and M. Tringides, Surf. Sci. 138 (1984) 581.

[35] R. Gomer and A. Auerbach, Surf. Sci. 167 (1986) 493.

[36] L.W. Swanson, R.W. Strayer and F.M. Charbonnier, Surf. Sci. 2 (1964) 177.

[37] H. Utsugi and R. Gomer, J. Chem. Phys. 37 (1962) 1706.

[38] M. Tringides and R. Gomer, Surf. Sci. 265 (1992) 283.

[39] S.C. Wang and R. Gomer, J. Chem. Phys. 83 (1985) 4193.

[40] C. Dharmadhikari and R. Gomer, Surf. Sci. 143 (1984) 223.

[41] E.A. Daniels, J.C. Lin and R. Gomer, Surf. Sci. 204 (1988) 129.

[42] D.-S. Choi, C. Uebing and R. Gomer, Surf. Sci. 259 (1991) 139.

[43] C. Uebing and R. Gomer, Surf. Sci. 259 (1991) 151.

[44] T.-S. Lin and R. Gomer, Surf. Sci. 255 (1991) 41.

[45] K. Muttalib and J. Sethna, Phys. Rev. B 32 (1985) 3462.

[46] A. Auerbach, K.F. Freed and R. Gomer, J. Chem. Phys. 86 (1987) 2356.

[47] D.-S. Choi, S.K. Kim and R. Gomer, Surf. Sci. 234 (1990) 262.

[48] Y.M. Gong and R. Gomer, J. Chem. Phys. 88 (1988) 1359, 1370.

[49] J.D. Wrigley and G. Ehrlich, Phys. Rev. Lett. 44 (1979) 661.

[50] M. Tringides and R. Gomer, J. Chem. Phys. 84 (1986) 4049;
D.-S. Choi and R. Gomer, Surf. Sci. 230 (1990) 277.

[51] W.P. Dyke and F.M. Charbonnier, 6th Natl. Conf. on Electron Tube Techniques (Pergamon, New York, 1963) p. 199.

[52] C. Spindt, C.E. Holland, A. Rosengreen and I. Brodie, IEEE Trans. Electron Devices 38 (1991) 2355.

[53] B. Halpern and R. Gomer, J. Chem. Phys. 51 (1969) 1031.

[54] M.G. Inghram and R. Gomer, J. Chem. Phys. 22 (1954) 1279; Z. Naturforsch. 10a (1955) 864.

[55] E.W. Müller and T.T. Tsong, Field Ion Microscopy, (Elsevier, New York, 1969).

[56] H.J. Kreutzer, in: Chemistry and Physics of Solid Surfaces VIII, Eds. R. Vanselow and R. Howe (Springer, Berlin, 1990) p. 133.

[57] A.A. Jason, Phys. Rev. 156 (1967) 266.

[58] M.E. Alferieff and C.B. Duke, J. Chem. Phys. 46 (1967) 938.

[59] B. Halpern and R. Gomer, J. Chem. Phys. 51 (1969) 5709;
A. Jason, B. Halpern, M.G. Inghram and R. Gomer, J. Chem. Phys. 52 (1970) 2227.

[60] B. Halpern and R. Gomer, J. Chem. Phys. 51 (1969) 1048.

[61] M. Southon, PhD Thesis, Cambridge University 1963.

[62] R. Gomer and M.G. Inghram, J. Am. Chem. Soc. 77 (1955) 500.

[63] H.D. Beckey, Field Ionization Mass Spectrometry (Pergamon, Oxford, 1971).

[64] E.W. Müller, Naturwissenschaften 29 (1941) 533.

[65] R. Gomer, J. Chem. Phys. 31 (1959) 341.

[66] R. Gomer and L.W. Swanson, J. Chem. Phys. 38 (1963) 1613.

[67] L.W. Swanson and R. Gomer, J. Chem. Phys. 39 (1963) 2813.

[68] M. Ernst, Surf. Sci. 87 (1979) 469.

[69] References to experimental and theoretical work relating to post-ionization are given by N. Ernst and G. Ehrlich, in: Microscopic Methods in Metals, Ed. U. Gonser (Springer, Berlin, 1986) p. 75.

Surface Science 299/300 (1993) 153–169
North-Holland

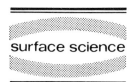

surface science

Atom-probe field ion microscopy and applications to surface science

Tien T. Tsong

Institute of Physics, Academia Sinica, Taipei, Taiwan, ROC

Received 22 March 1993; accepted for publication 13 May 1993

Atom-probe field ion microscopy is capable of imaging solid surfaces with atomic resolution, and at the same time chemically analyzing atoms selected by the observer from the atomic image. While the samples have to be restricted to those having a tip shape, this is becoming an advantage in many applications where a high electric field is needed. Its early developments and recent applications to surface science are briefly described here.

1. Introduction

The field ion microscope, invented by Müller in 1951 [1], is a very simple instrument which consists of a sample in the form of a tip and a phosphorus screen some 10 cm away. The FIM chamber is filled with an image gas such as helium or neon. It is an outgrowth of the field emission microscope [2] in an effort to improve the spatial resolution using a field desorption technique he discovered earlier [3]. Around 1957, Müller succeeded in seeing atomically resolved images of refractory metals for the first time after he introduced a cold finger to the FIM to cool the tip [4]. This represented the first time atoms could be seen directly with a microscope. It is a milestone in microscopy as well as in surface science. Ten years later, he conceived the idea of chemically analyzing atoms selected by the observer from the atomic image and developed with his assistants a prototype atom-probe FIM [5]. While the basic ideas of these instruments were known by 1967 from the work of Müller, perfection of these instruments and applying them to study scientific problems in a meaningful way turns out to be much more difficult and have to be left to others. Some of the most ardent applications of field ion microscopy are now in metallurgy and materials science. The subjects studied include defect structures, radiation damages, pre-

cipitate structures and their composition variations, grain and phase boundary structures and impurity segregations to these boundaries, oxidation, and compound formation, etc. Review of these materials related studies is not the scope of this article; fortunately they can be found in the literature [6]. Here I will describe briefly the development of various techniques in atom-probe field ion microscopy, studies of high field effects, and applications of these techniques to study a few selected problems in surface science.

2. Instrument development

The first field ion microscope used hydrogen as the image gas and there was no provision for cooling the tip sample [1]. Around 1956, the need of cooling the tip to improve the resolution was recognized [7], and atomic resolution was achieved for the first time. Up until the late 1960s, all field ion microscopes were made of Pyrex glasses, and cooling was done by filling a cold finger with liquid nitrogen or hydrogen. Fig. 1 shows a double-jacketed all glass FIM used in the late 1950s and early 1960s. Nowadays, almost all FIMs are made of a UHV compatible stainless steel chamber, and the cooling is mostly done with a Helium–Displex refrigerator. Heat conduction between the tip and the refrigerator is done by a tip

SSDI 0039-6028(93)E0300-J

mounting sapphire piece. FIM image intensity is very low without using an image intensification device. An external image intensifier was used for

a short period of time in the late 1960s and early 1970s, but with the availability of the very convenient channel plate from the Vietnam war re-

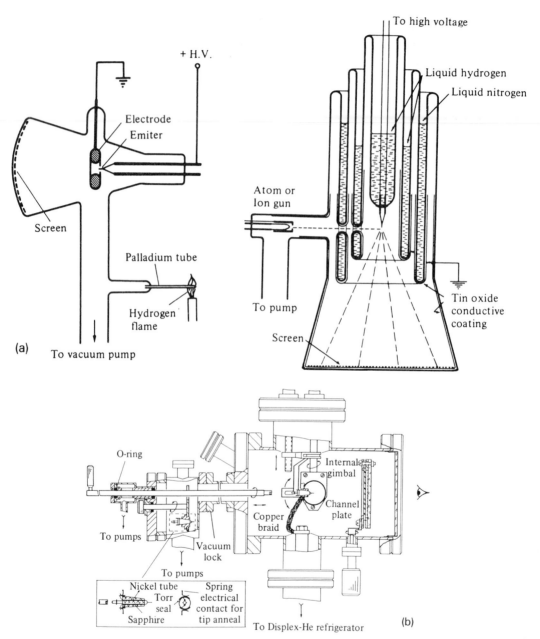

Fig. 1. (a) The first FIM built by Müller in 1951 which does not have a cold finger, and a Pyrex glass field ion microscope used by Müller and coworkers in the late 1950s and early 1960s which has a double-jacketed cold finger. (b) An UHV chamber which can be used as an FIM or the FIM chamber of an atom-probe. There is a vacuum lock for easier tip exchange. The tip is mounted on a sapphire. Cooling is done by a Displex–He refrigerator connected to the sapphire through a copper braid (from the author's laboratory).

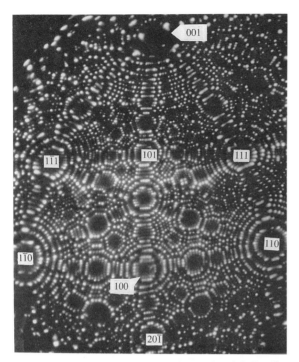

Fig. 2. A helium field ion image of an ordered PtCo alloy emitter of radius about 300 Å. This alloy has the $L1_0$ structure which is an fct lattice composed of alternating pure Pt(001) and pure Co(001) layers. A detailed analysis shows that only Pt sublattice is imaged, or Co atoms are not imaged. In other words, Pt and Co atoms can be distinguished simply from their image brightness alone [32].

search effort, almost all FIMs built now are equipped with such a device. Resolution of the FIM is, however, determined by the effective cooling of the tip, not by all these image intensification devices. Fig. 2 shows a FIM image of an ordered PtCo alloy.

The first prototype time-of-flight atom-probe FIM [5], also a glass system, while demonstrating the idea of an atom-probe, had neither sufficient mass resolution nor mechanical fineness to perform truly single atom chemical analyses. Systems with a variety of new designs were built around the late 1960s and early 1970s, but progress was slow. The reason was that in the time-of-flight atom-probe FIM, ns high voltage pulses were used for field evaporating surface atoms. These pulses produce a large ion energy spread of the field evaporated ions. Only when a flight time focusing scheme of Poschenrieder [8] was adapted to the atom-probe, was its resolution greatly improved [9]. Most atom-probes used today for metallurgical analyses use this design. Fig. 3 shows such a system. The high voltage pulse operated atom-probe, however, cannot be used for analyzing tips of electrically non-conducting materials as these pulses will not transmit across the slender tip stem to reach the tip surface. This difficulty was resolved when the idea of substituting

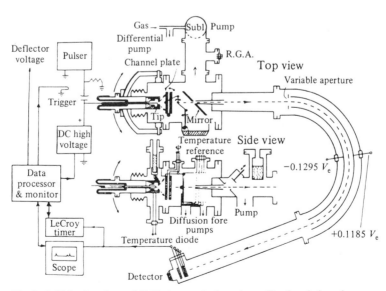

Fig. 3. A flight time focused ToF atom-probe based on a Poschenrieder scheme.

high voltage pulses with laser pulses was conceived [10]. Laser pulses can be directly aimed at the tip. Pulsed field evaporation is induced by either a heating effect or by a photo-excitation effect. For metals, only a heating effect is found when the laser intensity is not excessively high. For semiconductors, photo-excitation effects produce an excessive kinetic energy of the field evaporated ions even when the laser intensity is low [11]. Not only can semiconductors and insulators now be analyzed, but also the mass resolution is greatly improved without having to use an artificial flight-time focusing scheme. Thus, this atomprobe can also be used for ion energy analysis of field evaporated and desorbed ions as well as for measuring the site specific binding energy of surface atoms [12]. The capability of the ToF atomprobe FIM has now been expanded beyond the

original purpose of a mass analysis to include an energy analysis. Fig. 4 shows a high resolution system and a few ToF spectra.

A variation of the probe-hole-type atom-probe is the imaging atom-probe [13]. This is basically an FIM with its screen replaced by a single particle detection sensitivity Chevron channel plate which is also an imaging device. Using a flight-time gating technique of this particle detector, the spatial distribution of different chemical species on the emitter surface can be mapped. When the Chevron channel plate is replaced with a position sensitive ion detector and the pulsed field evaporation is done at a very low rate, both the location and the mass–charge ratio of each of the evaporated ions can be identified. With a suitable calibration of the depth and the use of a high speed computer for correcting the different

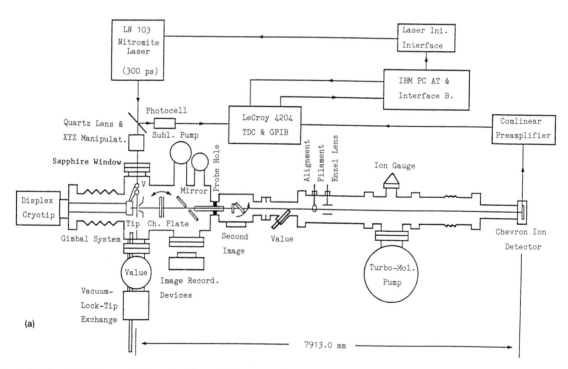

Fig. 4. (a) An ultra-high resolution pulsed-laser ToF atom-probe FIM used in many mass and energy analyses in the author's laboratory. (b) At low temperature, Si field evaporates as +2 ions. Though there are 7000 ions, no spreading of the ToF mass lines occurs. (c) At high temperature, a fraction of Mo field evaporates as Mo_2^{2+} ions. Though there are only 438 ions in this spectrum, all 15 mass lines produced by isotope mixing of the 7 Mo isotopes are clearly seen with a proper distribution of their abundances. These lines are separated by $\frac{1}{2}$ u, thus these ions are doubly charged.

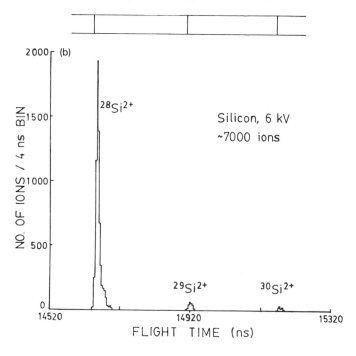

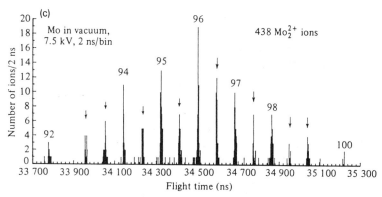

Fig. 4 (continued).

flight path lengths for ion paths at different directions, a three-dimensional distribution of chemical species in the sample can be mapped out similar to a tomography but now with a spatial resolution of a few Å [14]. As far as this author is aware, there is no other microscope capable of doing surface and bulk chemical analysis of materials with a comparable spatial resolution. Such a system will be very useful for analyzing materials, both structures and elemental distributions, with

microstructures such as micro-precipitates, and grain and phase boundaries.

3. High electric field effects at the surface and image formation

With a tip of apex radius of a few hundred Å and an applied voltage of several kV to over 10 kV, the field strength at the surface can be as high as a few V/Å. In such a high electric field, many novel high field effects can occur some of them are responsible for forming field ion images [15]. The FIM is usually filled with an image gas such as He or Ne of pressure in the 10^{-4} to 10^{-5} Torr range. Gas atoms flying around the tip will be polarized by the high field. As the field near the tip surface is not uniform, these polarized atoms will be attracted to the tip surface, hit the surface, bounce off and be attracted back to the surface again, or they will bounce around the surface. Every time an atom hits the tip surface, it will lose part of its kinetic energy, and will gradually accommodate to the tip temperature. At the apex of each of the surface atoms in the more protruding site, an image gas atom will be field adsorbed on it [16]. This adsorption may occur by a field induced dipole–dipole interaction. It may be due to a chemisorption effect when the atomic level of the image gas atom is shifted up close to the Fermi level by the electric field [17]. The adsorption energy at the image field of ~ 4.5 V/Å is about 0.15 to 0.25 eV; thus, field adsorption of He and Ne can occur below about 120 K [16]. This author believes that field adsorption is responsible for the image brightness and contrast observed in the FIM, since above 120 K field ion images start to lose their intensity and contrast but most theoretical calculations to date show that field adsorbed inert gas atoms reduce the field ionization rate [18].

Other hopping atoms, when they pass through the disk-shaped ionization regions above field adsorbed atoms, may be field ionized by the tunneling of the atomic electron into the surface. These ions are accelerated to the phosphorus screen to form a field ion image of the emitter surface. Ionization of hydrogen atoms by tunneling of the atomic electron into free space was predicted to occur in a field of ~ 2 V/Å by Oppenheimer in 1928 when quantum mechanics was still in its infancy. Experimental observation of field ionization was realized when the FIM was introduced in 1951. Field ionization in the FIM is, however, slightly different from that in free space. At low temperature, the atomic electron can tunnel into a vacant state of the tip material

only when the state is above the Fermi level. Thus the ionization rate should depend on the work function of the surface. A phenomenological theory of field ionization above a metal surface based on a one-dimensional WKB tunneling picture can be found in the article by Gomer.

Field ion images are formed by radial projection of field ions. To have a uniform image magnification, the emitter surface has to be as smooth as possible. Tips prepared by electrochemical polishing of thin wires usually have many asperities. These asperities can be removed by field desorption/field evaporation by simply raising the tip voltage. Field evaporation, desorption of surface atoms in the form of ions by high electric field, is a self-regulated process, or atoms in the more protruding positions are desorbed first thus the surface can be smoothed. It can occur for most materials in the range of 2 to 5 V/Å. Two theoretical models, an image-hump model by Müller [19] and a charge-exchange model by Gomer [20], have been very successful in explaining field evaporation data. Model calculations using these two models conclude that for most metals, ions produced are usually doubly charged [21]. Experimental results such as the charge states and critical fields of elements are in excellent agreement with calculations based on these models. New theories based on self-consistent density functional calculations have been developed but the predicting power is still limited [22]. To explain recent STM observations, the possibility of negative field evaporation of Au and Si as negative ions in the close tip-to-sample geometry of the STM has been proposed [23].

There are other interesting high field effects which are not linked directly to the FI image formation. A particularly interesting phenomenon is field dissociation of $HeRh^{2+}$ in a high electric field [24]. In this compound ion, the He atom remains neutral. It is bound to the Rh^{2+} by a polarization force. As $HeRh^{2+}$ is desorbed from a Rh tip surface, it will rotate and vibrate. In a region with a field above 4.5 V/Å, it can dissociate by atomic tunneling [25]. A careful consideration of the Schroedinger equation governing the relative motion of He and Rh^{2+} indicates that the dissociation rate will decrease when the ^{4}He is replaced with a ^{3}He. This strange behavior arises from a center of mass transformation in the equation of motion. An experiment indeed confirms this interesting behavior. From a measurement of the ion energy distribution, it is possible to calculate the rotation time by $\pi/2$ of the compound ion in the applied field, which is also the dissociation time of the compound ion. The measured time, 730 fs, represents one of the fastest reaction times ever measured, in this case dissociation by a particle tunneling effect.

In an applied high electric field, molecules tend to line up along the field. Thus polymerization of organic molecules on an emitter surface is a common phenomenon [26]. Field ionization and field desorption/evaporation can also be promoted by photo-excitation especially if the substrate is a semiconductor. When this occurs, the ensuing ions will exhibit an excess kinetic energy of a few eV to a few hundred eV depending on the laser intensity [27]. Mechanisms of these photo-excitation effects are not yet studied in detail and are not yet well understood.

4. Atomic structures of solid surfaces

The field ion microscope can give atomically resolved images of field emitter surfaces. Early field ion microscope observations, however, did not detect surface reconstructions as were concluded from studies with macroscopic techniques such as LEED and ion scattering experiments; FIM images were therefore viewed by some surface scientists to be somewhat in suspect. Only in the 1980s was the difference between a low temperature field evaporated surface and a thermally annealed surface recognized [28]. The atomic structure of a low temperature field evaporated surface reflects the details of how atoms are field evaporated from the tip surface. It is not a thermally equilibrated surface. For metals, the field penetration depth is less than 0.2 to 0.5 Å, or much less than the interlayer step heights. Atoms are removed from lattice steps one by one. The structure of the surface so prepared should have the same structure as the bulk, or they should have the (1×1) structure. Fig. 5 shows a few

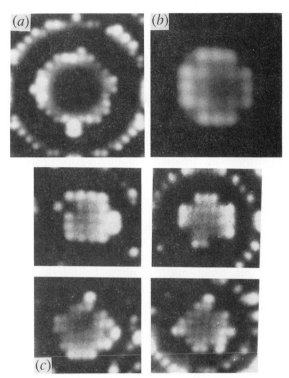

Fig. 5. (a) and (b) Field ion images of two (001) facets of a Pt emitter. These facets are prepared by low temperature field evaporation, thus both have the (1×1) structure. Atoms are fully resolved only when the facets are very small. (c) A few fully resolved non-reconstructed W(001) surfaces prepared by low temperature field evaporation.

examples. For semiconductors, the field penetration depth is very much greater than the interlayer separation. Atoms can be field evaporated almost from anywhere at the surface. The structure of the surface prepared by low temperature field evaporation should be disordered. On the other hand, if the emitter surface prepared by low temperature field evaporation is further annealed to reach the thermodynamic equilibrium structure, then the structure is found to be consistent with results obtained by other techniques. In fact, the FIM result was able to help in establishing the missing row (1×2) reconstruction of the Pt{110} and Ir{110} surfaces [29], and the quasi-hexagonal atomic arrangement of the (1×5) reconstructed Ir{001} surface [30]. For semiconductors, atomic structures of the reconstructed high index surfaces of Si can be observed, but it is still very difficult to observe the atomic structures of low index surfaces with the FIM.

5. Surface chemical analysis

Another important aspect of an atomic resolution microscope is the chemical or elemental identification. In field ion microscopy, components of a binary alloy can be identified from their evaporation fields [31] and from the image brightness if the alloy is at least partially ordered [32]. Using this preferential imaging of Pt atoms in ordered and partially ordered PtCo and Pt_3Co alloys, many defect structures such as antiphase domains, order–disorder phase boundaries, and 90° orientation domains, etc., can be easily identified. The order parameters in partially ordered alloys and the geometry of the microclusters in these alloys can be directly mapped from the chemically distinguishable atomically resolved field ion images, and the results obtained compare well with experimental data obtained with an X-ray diffraction technique [33]. The same method has been successfully applied to an STM study of partially ordered Ni_3Pt alloy [34].

The above method is useful only for identifying the elemental distribution of binary alloys where the alloy components are already known. It is not applicable to the general surface chemical analysis. In fact one of the most difficult problems in surface science is the elemental, or the chemical, analysis of the surface and near surface layers. Most macroscopic techniques lack spatial resolution, both the depth and the lateral resolution. The atom-probe FIM was originally conceived for the purpose of identifying the chemical species of individual surface atoms. As field evaporation of metals and alloys proceeds atomic layer by atomic layer, it can also be used to analyze the composition of surface layers with a true atomic layer depth resolution. Whereas the lateral resolution of the atom-probe in chemical analysis is no better than several Å because of the slight difference in the flight path of the field evaporated ions and field ionized gas ions, it can achieve a true absolute atomic layer by atomic

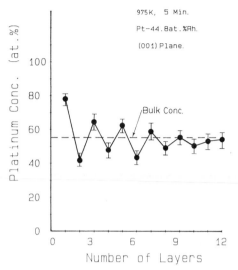

Fig. 6. When a Pt–Rh alloy is thermally annealed to 975 K to equilibrate the distribution of Pt and Rh atoms in the sample, the composition of these surface layers is found to oscillate with the depth of the layers. This oscillation persists down to the depth of the tenth surface layer.

layer chemical analysis of the near surface layers; thus the composition depth profile so obtained is an absolute profile.

In surface science, the atom-probe was successfully used to establish an atomic exchange between W adatoms and channel wall atoms of the Ir{110} surface in a surface diffusion study [35]. It has also been initiated by Tsong et al. [36] to analyze the absolute composition of the near surface layers of binary alloys in surface segregation and co-segregation of binary alloys. A recent interesting finding of this alloy study is the atomic layer by atomic layer oscillatory composition variation of the near surface layers when the Pt–Rh alloy is thermodynamically equilibrated at high temperature, as shown in fig. 6 for a Pt–44.8 at% Rh alloy equilibrated at 975 K [37]. The composition oscillation can be detected as deep as the tenth atomic layer. Other techniques cannot detect directly this kind of oscillation because of a limited spatial depth resolution. With the finite size of the facets of a field emitter, edge effects in surface segregation can be and have also been studied for this alloy; they are found to be very small. The atom-probe FIM is currently the only instrument of imaging solid surfaces with atomic

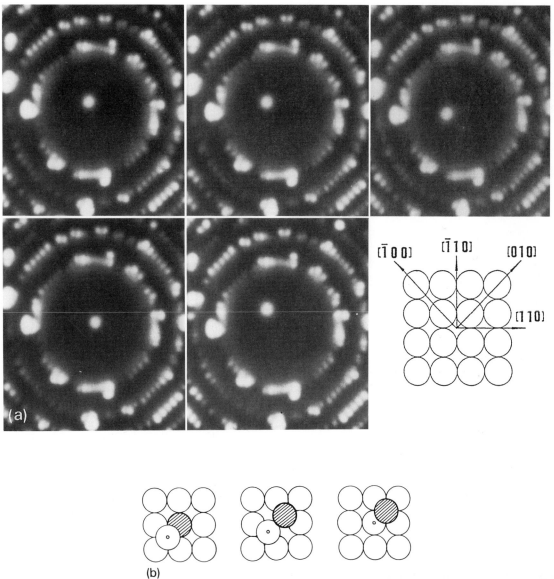

Fig. 7. (a) FI images showing random walk diffusion of one Ir adatom on an Ir(001) surface. The atomic structure of the substrate can be revealed if the facet is reduced in size. This is shown in the inset. (b) The jump direction and the displacement of the adatom can be mapped from successive images. It is found that the adatom moves in the diagonal direction of the square unit cell by concerted motion with a substrate atom as illustrate in this figure. (c) On a doubly-spaced Ir(110) atomic channel of a partially (1 × 2) reconstructed Ir(110) surface, an Ir adatom diffuses by atomic hopping, only along the (110) surface channel direction. One of the atomic rows has a vacancy which was produced by heating to induce this surface reconstruction. The adatom and the vacancy annihilate each other at the fourth picture. The annihilation process is by atomic replacement as illustrated in the line drawing. This author concludes from available FIM data that atomic replacement self-diffusion is favored for the metastable, or still reconstructible, (1 × 1) surfaces. (d) Diffusion of a W diatomic cluster is achieved through coupled but otherwise random hopping of the two atoms along the surface channels of the W(112) surface. Diffusion of a ledge atom can also be seen on the upper lattice step of the top surface layer. (Figure continued on next page.)

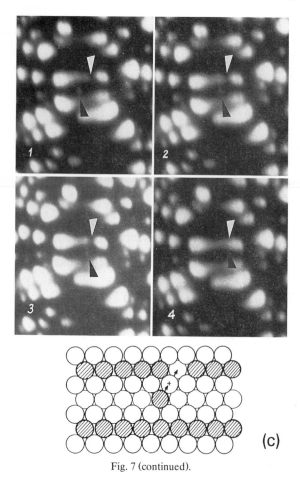

Fig. 7 (continued).

6. Surface atomic processes and dynamics of surfaces

A very successful application of FIM to surface science is the study of the dynamical behavior of metal surfaces, and the behavior of metal and semiconductor adatoms on metal surfaces. Problems studied include surface diffusion of adatoms and clusters, adatom–adatom interactions, adatom–substrate interactions, equilibrium crystal shape, and two-dimensional sublimation, etc. Beside the atomic resolution, there are several other advantages of the FIM in this type of study: (a) A nearly ideal surface can be prepared by low temperature field evaporation even though the size of the surface is very small. (b) The number of adatoms deposited on a facet can be specified by the need of the experiment and be implemented by repeated deposition and controlled field evaporation. (c) Available on an emitter surface are crystal facets of different atomic structures. Many lattice steps are also available for studying effects of the lattice steps. (d) The temperature of the sample can be easily controlled and varied. Therefore, many thermally activated processes can be studied in atomic details and quantitatively [38].

Solid surfaces are by no means static. When the temperature is raised, atoms in the near surface layers will start to move rapidly well before the melting point is reached. Mass transport at a solid surface involves many elementary atomic steps or processes. Even a simple macroscopic

resolution and at the same time of performing chemical analysis of these surfaces atom by atom and atomic layer by atomic layer.

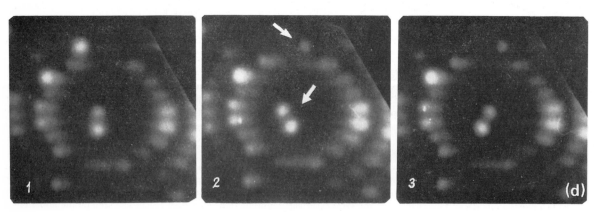

Fig. 7 (continued).

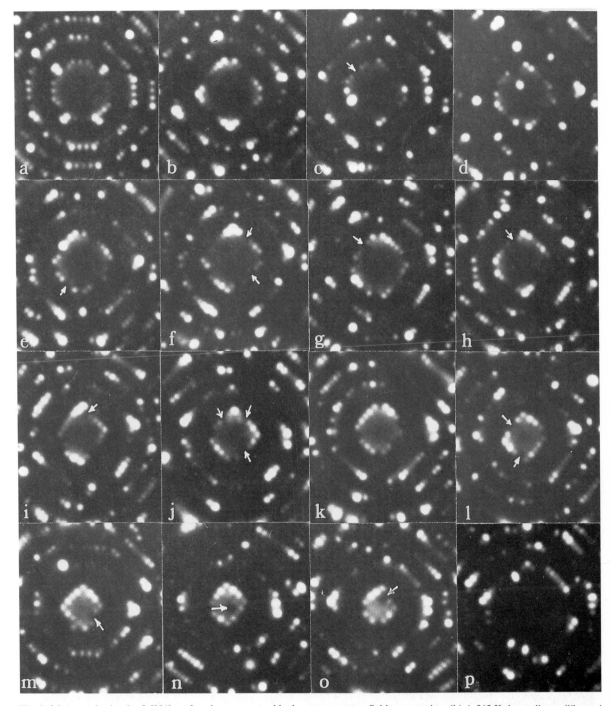

Fig. 8. (a) A nearly circular Ir(001) surface layer prepared by low temperature field evaporation. (b) A 565 K thermally equilibrated shape of the same layer, which is a perfect square with its sides parallel to the ⟨110⟩ closely-packed atomic row directions. From (c) to (p), this layer gradually reduces its size by dissociation of step atoms to the terrace. Arrows point to where atoms inside the steps have dissociated, thus producing roughened steps at ∼565 K, or the step roughening occurs at less than 1/5 the melting point of Ir!

phenomenon such as the shape change of a solid involves a series of elementary atomic steps. Atoms have to be dissociated from their rest sites first, then move on the terraces, go up or down the lattice steps, and finally settle into new sites elsewhere. In other words, even a simple macroscopic phenomenon involves a series of "elementary atomic processes or steps". Unless one studies each of these processes separately, it will be difficult to understand the dynamics of this phenomenon in terms of first principle theories. Elementary atomic processes are common to other surface phenomena involving atom transport, or they are universal in understanding the dynamical behavior of solid surfaces. It is the investigation of the mechanisms and energetics of these elementary surface atomic processes that the FIM finds its very useful applications.

Several elementary atomic processes have already been studied in detail. They are surface diffusion of single adsorbed atoms (adatoms) and small atomic clusters, adatom–adatom interac-

tions, site specific atom–substrate interactions, adatom–lattice interactions, etc. Fig. 7 shows diffusion of one Ir adatom on an Ir(001) plane, diffusion of one Ir adatom along the doubly spaced Ir(110) surface (or the surface is partially (1×2) reconstructed) and diffusion interaction of a two-atom cluster of W on the W(112) surface [39]. Fig. 8 shows how a nearly circular facet of the Ir(001) surface, prepared by low temperature field evaporation, reaches a thermally equilibrated square shape and then dissolves slowly by dissociation of atoms, one by one, from the step to the terrace. Here some of the interesting results learned from FIM studies are summarized: (a) The binding energy of an atom with the substrate is site specific. For kink site atoms, within the ± 0.2 eV accuracy of the experiment, it is found to be identical to the cohesive energy of the solid [40], as is also expected from a theoretical consideration. (b) In self-adsorption on terrace sites, the adsorption energy can be derived by measuring the dissociation energy of lattice

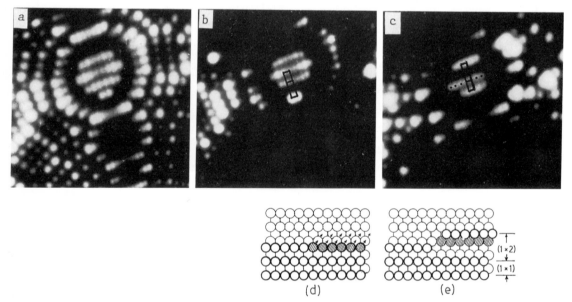

Fig. 9. (a) A small (1×1) (110) layer of a Pt(110) surface. When the sample is heated to ~ 400 K for 5 ns with a laser pulse, two atoms "jump together" to a neighbor surface channel. Another laser pulse and another small row of atoms jump to the neighbor channel again. From this experiment we conclude that in the (1×1) to (1×2) surface reconstruction of the fcc {110} surfaces, atoms in an atomic row near the lattice step of a layer can jump by atomic replacement to the next surface channel as illustrated in (d) and (e). In other words, atom transport in this reconstruction is achieved through gradual spreading of (110) atomic row fragments at the lattice step of the top (1×1) layer.

step atoms and self-diffusion energy. The adsorption energy in self-adsorption, Ir/Ir{001}, is found to be 0.56 ± 0.08 eV lower than the cohesive energy of Ir which is 6.93 eV [41]. (c) For many systems, no direct chemical bond is formed between two adatoms. However, they can interact via the substrate through an electron hopping effect. Such indirect interaction is very weak, long range, and oscillatory. It is only on the order of 0.1 eV. The oscillatory tail is only ~ 0.01 eV or less in amplitude [42]. (d) No covalent bond is formed between two Si atoms adsorbed on a metal surface [43]. (e) Surface diffusion of adatoms can occur either by atomic hopping or by atomic replacement, depending on the diffusion system [35,44]. When the atomic jump occurs by an atomic replacement, the adatom and the replaced substrate atom move in concerted motion [44]. The activation energy of adatom diffusion depends sensitively on the chemistry of the system as well as on the atomic structure of the substrate. It ranges from less than 0.2 eV to well over 1 eV. However, the frequency factor is insensitive to the diffusion system. It is always with a magnitude of $\sim kT/h$. (f) A lattice step may be absorptive or reflective to a diffusion system. When it is reflective, the extra barrier height is only about 0.2 eV or less [45]. (g) Dissociation of step atoms and step roughening of a metal surface can occur at a temperature less than 1/5 the melting point of the metal. The dissociation energies of edge atoms, ledge atoms, kink atoms, and recessed atoms are almost identical. Therefore, step roughening occurs at a temperature when dissociation of the step atoms starts to occur. The dissociation energy is about 50% larger than the self-diffusion energy. For the case of the Ir(001) surface, dissociation energy of step atoms is 1.40 ± 0.09 eV, whereas the self-diffusion energy is 0.84 ± 0.05 eV [41]. (h) In the (1 × 1) to (1 × 2) reconstruction of the Pt(110) and Ir(110) surfaces, a small ⟨110⟩ row of atoms at the [110] layer edge tend to jump together to the neighbor surface channel, presumably by atomic replacement shown in Fig. 9, thus a doubly-spaced missing row structure is formed. The richness and novelty of results obtained from this type of FIM study makes even a casual review of this subject impossible. Interested readers should consult recent papers by S.C. Wang and G. Ehrlich, G.L. Kellogg, and C.L. Chen and T.T. Tsong.

7. Applications of some high field effects

7.1. Liquid metal ion sources and scanning ion microscope

A practical application coming out of field ion emission research is the liquid metal ion source which is a high brightness point source. Ion beams of a wide variety of chemical elements from mostly low melting point metals and alloys can be produced. Basically there are two different configurations of these sources. One uses a capillary filled with a liquid metal while the other consists of a metal tip having a drop of low vapor pressure liquid metal placed at the spot-welded junction of the tip and the tip mounting loop. In the second configuration the tip surface has to be wetted with a liquid film first. Upon the application of a high positive field, the liquid surface is drawn into a conical shape protrusion, referred to as the Taylor cone, by the applied field. If the field strength is sufficiently high, ions will be emitted from the tip apex by a complicated interplay of hydrodynamic flow of a liquid layer, field evaporation, bombardment heating of the tip by electrons coming out of field ionization of liquid metal vapor, etc., depending on the current density drawn.

Liquid metal ion sources are characterized by their high brightness and point like size. The emitter tip is believed to be only about 20 to 30 Å in size when it is operated at the optimum condition although the liquid cone may whirl around slightly. They have been used in a wide variety of applications such as scanning ion microscopy, X-ray and optical mask repair, secondary ion mass spectroscopy with submicron spatial resolution for the chemical analysis of submicron material structures, ion beam lithography, and maskless implantation doping of semiconductors, etc. With a proper lens system, the ion beam can be made to be smaller than a few hundred Å in cross section, and in the scanning ion microscope, chemical maps of materials of ~ 300 Å resolution have been obtained. The advantage of this scanning ion microscope is that while the material surface is imaged, it is also being sputtered away slowly; therefore, a composition depth profile is automatically obtained. The disadvantages are the destructive nature of the microscope and its limited lateral and vertical resolutions which are no better than ~ 300 and ~ 50 Å, respectively.

7.2. Atomic and molecular manipulations by high field effects

A recent new development in science and technology is atomic and molecular manipulation using scanning tip microscopy [46]. There are many conceivable methods for this purpose among them the use of high electric field effects such as field evaporation and field gradient induced surface diffusion. To create an atomic feature on a solid surface, one will need to accomplish at least the following three tasks: (1) move an atom from one location of the surface to another; (2) deposit atoms to a surface preferably at the intended locations; and (3) remove unwanted or pre-specified atoms from the surface. For these tasks, two high field methods give a particularly good control. Consider only the first task. The first method which can be used is based on field evaporation

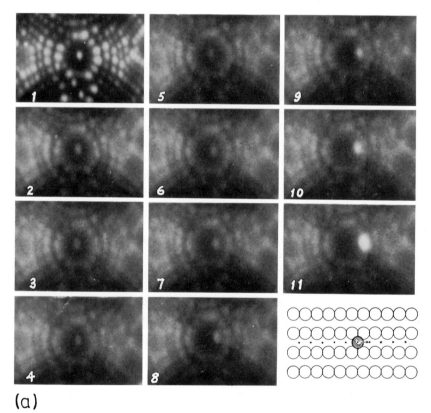

(a)

Fig. 10. (a) 270 K FI images (except the first one which is taken at ~ 80 K) showing the directional walk of one W adatom on a W(112) surface. Starting from the 5th picture, the adatom jumps toward the plane edge one step at a time. The image brightness reflects well the field variation. (b) to (d) illustrate why in the absence of an applied field, an adatom will perform a symmetric random walk, but in the presence of an applied field the adatom will drift from the central region to the plane edge because of the existence of a *field gradient*. (e) shows that for an electropositive and an electronegative adatom, the effect may be reversed. When the term $(-\mu F - \frac{1}{2}\alpha F^2)$ becomes positive, the adatom will diffuse from the plane edge toward the plane center instead. This type of diffusion has not yet been experimentally observed. As this diagram shows, the potential energy change can be very small near the turning point field. This explains why in the STM experiment, adatoms on the sample surface do not move at a certain biasing voltage. When the polarity of the biasing voltage is reversed, the adatoms start to move toward the spot right beneath the tip.

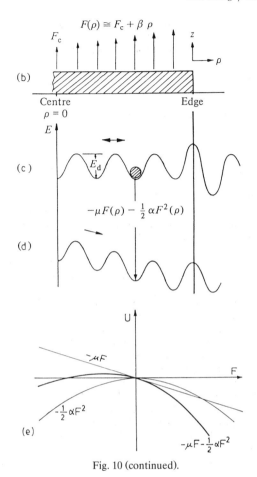

$$F(\rho) \cong F_c + \beta \rho$$

$$-\mu F(\rho) - \frac{1}{2}\alpha F^2(\rho)$$

Fig. 10 (continued).

adatoms closer to the lattice step are field evapo-rated first, one at a time as seen in Fig. 2.10 of Ref. [4b]. The critical fields and charge states of ions emitted of different materials can be calcu-lated from simple theoretical models and have been verified experimentally [4b]. Model calcula-tions conclude that soft metals like Al, Cu, Pd, Ag, Au and Pb, etc., will field evaporate as +1 ions. Most metals such as Be, Ni, Co, Fe, Ru, Rh and Pt and semiconductors such as Si and Ge field evaporate as +2 ions. Refractory metals such as Ta, W, and Re field evaporate as +3 ions. The critical fields calculated from these models agree with experimental values to within ±15%. In the close double-electrode geometry of the STM, however, the field strengths needed are only about 1/3 of those needed in the FIM. In addition, field evaporation of negative ions may occur [47].

Another method of moving adsorbed atoms using the STM is based on the observation that when a voltage bias or pulse of appropriate polar-ity is applied to the sample, Cs adatoms on a GaAs surface tend to move from the neighbor-hood of the tip to the spot directly below the tip [46]. This motion is produced by a field gradient induced surface diffusion found and studied in detail earlier in the FIM by Tsong, Kellogg and Wang [48]. In the FIM, when a random walk experiment of an adatom is carried out on a field emitter facet with the image voltage applied, the adatom will drift from the central region directly to the edge of the facet as shown in Fig. 10a. The mechanism of this directional walk is illustrated in Figs. 10b–10d, and Fig. 10e illustrates the energetics. In the absence of an applied tip volt-age, the potential energy of the adatom–surface interaction is horizontal and periodic. The adatom will perform symmetric random walk. When a voltage is applied to the tip, the field at the central region is lower than that near its edge. When the polarization energy of the adatom is included, the surface potential becomes inclined. Thus the adatom will jump preferably toward the plane edge. This method has been used to mea-sure the dipole moment and polarizability of ad-sorbed atoms [48]. While the accuracy of the experiment is still limited, those for tungsten

[15]. First, the tip is placed above the surface. An electric pulse of appropriate polarity is applied to the sample thereby the atom is emitted from the sample surface to the tip surface. The tip is moved to a desired new location and a pulse is applied to the tip thereby the atom is transferred back to the sample at this new spot. In the second method, the tip is first moved to a position near the adatom in question. A voltage bias or pulse of appropriate polarity is applied to the sample. The adatoms are found to move from the original location to a spot directly below the tip. This motion is produced by a field gradient induced directional diffusion of the adatoms.

Field evaporation can be very well controlled by the field strength. When a surface is deposited with many adatoms, as the tip voltage is raised,

adatoms on the tungsten (110) surface have been measured to be 1.0 debye and 14 $\overset{\circ}{A}^3$, respectively. With this explanation, the STM observation becomes obvious. The field gradient in this case arises from the geometrical asymmetry of the tip–sample configuration. The field is highest at the spot just below the tip. Thus when a voltage pulse is applied, the atoms in the neighborhood of the tip will move toward the spot directly below the tip. This motion is induced by an electric field gradient, not by an electric field as commonly stated. There are many other effects of the electric field which may be useful for atomic manipulations and other purposes. For an example, by using the temperature and field gradient induced surface diffusion, the tip can be processed to a single atom sharpness for use as an electron source in the field emission mode. In the field ion emission mode, it can be maintained at the same sharpness while emitting ions at a rate of 10^6 to 10^7 ions/s [49]. In the field emission mode, the development of an array of sharp tips for an image display panel is finding great interest in vacuum microelectronics [50]. Discussion of these and other applications are, however, beyond the intended scope of this article.

References

[1] E.W. Müller, Z. Physik 131 (1951) 136.
[2] E.W. Müller, Z. Physik 106 (1937) 541;
For the development of the field emission microscope and its applications, see the article by R. Gomer.
[3] E.W. Müller, Naturwissenschaften 29 (1941) 533.
[4] E.W. Müller, Phys. Rev. 102 (1956) 618;
For the historical development of the FIM, its basic principles and applications, and recent advances, see for example T.T. Tsong, Atom-probe Field Ion Microscopy (Cambridge Univ. Press, Cambridge, 1990).
[5] E.W. Müller, J.A. Panitz and S.B. McLane, Jr., Rev. Sci. Instrum. 39 (1968) 83.
[6] M.K. Miller and G.D.W. Smith, Atom-Probe Microanalysis: Principles and Applications to Materials Problems, (Mater. Res. Soc., Pittsburgh, PA, 1989);
J. Orloff, Sci. Am. 265 (1991) 96;
A. Cerezo, T.J. Godfrey and G.D.W. Smith, Rev. Sci. Instrum. 59 (1988) 862;
D. Blavette, A. Bostel, J.M. Sarrau, B. Deconihout and A. Menand, Nature 363 (1993) 432.
[7] E.W. Müller and K. Bahadur, Phys. Rev. 102 (1956) 624.

[8] W.P. Poschenrieder, Int. J. Mass Spectrom. Ion Phys. 6 (1971) 413; 9 (1972) 357.
[9] E.W. Müller and S.V. Krishnaswamy, Rev. Sci. Instrum. 45 (1974) 1053.
[10] T.T. Tsong, Surf. Sci. 70 (1978) 228;
G.L. Kellogg and T.T. Tsong, J. Appl. Phys. 51 (1980) 1184.
[11] T.T. Tsong, Surf. Sci. 177 (1986) 593.
[12] J. Liu, C.W. Wu and T.T. Tsong, Phys. Rev. B 45 (1992) 3659;
T.T. Tsong, Int. J. Mod. Phys. 5 (1991) 1871.
[13] J.A. Panitz, Rev. Sci. Instrum. 44 (1973) 1034.
[14] A. Cerezo, T.J. Godfrey and G.D.W. Smith, Rev. Sci. Instrum. 59 (1988) 862.
[15] For high electric field effects responsible for field ion image formation and the mechanisms of image formation, see R. Gomer, Field Emission and Field Ionization (Harvard Univ. Press, Cambridge, MA, 1961);
E.W. Müller and T.T. Tsong, Field Ion Microscopy, Principles and Applications (Elsevier, Amsterdam, 1969);
Also Ref. [4].
[16] Early atom-probe experiments found that in pulsed field evaporation of tungsten, helium ions could be detected. These helium ions were thought to come from the hopping gas atoms. A measurement of the temperature dependence of observing these ions by Tsong et al. agreed with the Langmuir isotherm, thus establishing the adsorption nature of these atoms which was induced by the applied field. T.T. Tsong and E.W. Müller, Phys. Rev Lett. 25 (1970) 911; J. Chem. Phys. 55 (1971) 2884.
[17] H.J. Kreutzer, in: Chemistry and Physics of Solid Surfaces VIII, Eds. R. Vanselow and R. Howe (Springer, Berlin, 1990) p. 133.
[18] D.A. Nolan and R.M. Herman, Phys. Rev. B 10 (1974) 50.
[19] E.W. Müller, Phys. Rev. 102 (1956) 618.
[20] R. Gomer and L.W. Swanson, J. Chem. Phys. 38 (1963) 1613.
[21] D.G. Brandon, Surf. Sci. 3 (1965) 1;
T.T. Tsong, Surf. Sci. 10 (1968) 102.
[22] E.R. McMullen and J.R. Perdew, Phys. Rev. B 36 (1987) 2598;
H.J. Kreuzer and K. Nath, Surf. Sci. 183 (1987) 591.
[23] N. Miskovsky and T.T. Tsong, Phys. Rev. B 46 (1992) 2640;
N. Miskovsky, C.M. Wei and T.T. Tsong, Phys. Rev. Lett. 69 (1992) 2427.
[24] See N. Miskovsky and T.T. Tsong, Phys. Rev. B 38 (1988) 11188, and references therein.
[25] A quantum mechanical theory of field dissociation was first reported for H_2^+ by R. Hiskes, Phys. Rev. 122 (1961) 1207.
A theoretical model of field dissociation of $^4HeRh^{2+}$ was reported by T.T. Tsong, Phys. Rev. Lett. 55 (1985) 2826;
T.T. Tsong and M.W. Cole, Phys. Rev. B 35 (1987) 66.

[26] H.D. Bechey, Field Ionization Mass Spectrometry (Pergamon, Oxford, 1971).

[27] Ref. [5].

[28] H.M. Liu, T.T. Tsong and Y. Liou, Phys. Rev. Lett. 58 (1987) 1535, and next reference.

[29] G.L. Kellogg, Phys. Rev. Lett. 55 (1985) 2168;
Q.J. Gao and T.T. Tsong, Phys. Rev. Lett. 57 (1986) 452.

[30] J. Witt and K. Müller first reported an FIM study of the (1×5) reconstruction of this surface (Phys. Rev. Lett. 57 (1986) 1153);
Unfortunately, it was found by Gao and Tsong that the structure they observed was probably due to contamination. Correct structures and interpretation were later given by Q.J. Gao and T.T. Tsong, Phys. Rev. B 35 (1987) 2547, 7764.

[31] D.G. Brandon, Surf. Sci. 5 (1966) 137;
H.N. Southworth and B. Ralph, Philos. Mag. 14 (1966) 383.

[32] T.T. Tsong and E.W. Müller, Appl. Phys. Lett. 9 (1966) 7;
J. Appl. Phys. 38 (1967) 545, 3531.

[33] H. Berg, Jr., T.T. Tsong and J.B. Cohen, Acta Met. 21 (1973) 1589.

[34] M. Schmidt, H. Stadler and P. Varga, Phys. Rev. Lett. 70 (1993) 1441.

[35] J.D. Wrigley and G. Ehrlich, Phys. Rev. Lett. 44 (1980) 661;
The channel wall atomic-exchange diffusion was first postulated by D.W. Bassett and P. Weber, Surf. Sci. 70 (1978) 520.

[36] T.T. Tsong, Y.S. Ng and S.V. Krishnaswamy, Appl. Phys. Lett. 32 (1978) 778;
Y.S. Ng and T.T. Tsong, Phys. Rev. Lett. 42 (1979) 588.

[37] M. Ahmad and T.T. Tsong, J. Chem. Phys. 83 (1985) 388;
D.M. Ren, C.H. Qin, J.B. Wang and T.T. Tsong, Phys. Rev. B 47 (1993) 3944, and references therein.

[38] (a) G. Ehrlich and K. Stoltz, Ann. Rev. Phys. Chem. 31 (1980) 603;
(b) T.T. Tsong, Rep. Prog. Phys. 51 (1988) 759.

[39] T.T. Tsong, Phys. Rev. B 6 (1972) 417.

[40] J. Liu, C.W. Wu and T.T. Tsong, Phys. Rev. B 45 (1992) 3659.

[41] C.L. Chen and T.T. Tsong, Phys. Rev. B 47 (1993) 15852.

[42] T.T. Tsong and R. Casanova, Phys. Rev. B 24 (1981) 3063.

[43] T.T. Tsong and R. Casanova, Phys. Rev. Lett. 47 (1981) 113.

[44] G.I. Kellogg and P.J. Feibelman, Phys. Rev. Lett. 64 (1990) 3143;
C.L. Chen and T.T. Tsong, Phys. Rev. Lett. 64 (1990) 3147;
T.T. Tsong and C.L. Chen, Nature 355 (1992) 328.

[45] S.C. Wang and T.T. Tsong, Surf. Sci. 121 (1982) 85.

[46] J.A. Stroscio and D.M. Eigler, Science 254 (1991) 1319.

[47] N.M. Miskovsky and T.T. Tsong, Phys. Rev. B 46 (1992) 2640.

[48] T.T. Tsong and G.L. Kellogg, Phys. Rev. B 12 (1975) 1343;
S.C. Wang and T.T. Tsong, Phys. Rev. B 26 (1982) 6470.

[49] Vu Thien Binh, S.T. Purcell, N. Garcia and J. Doglioni, Phys. Rev. Lett. 69 (1992) 2527.

[50] I. Brodie and C.A. Spindt, Adv. Electron. Electron Phys. 83 (1992) 1.

Surface Science 299/300 (1994) 170–182
North-Holland

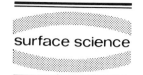

surface science

The development of core electron spectroscopies of adsorbates

Dietrich Menzel

Physik-Department E 20, Technische Universität München, D-85747 Garching bei München, Germany

Received 1 April 1993; accepted for publication 24 July 1993

A short historical survey is given of the development of applications of core electron spectroscopies (X-ray induced core photoelectron spectroscopy, XPS; and Auger electron spectroscopy, AES) in surface physics, in particular in the investigation and characterization of well-defined adsorbate layers on single crystal surfaces. A personal perspective is used to show, with this subject and partly the author's group as example, the sometimes winding ways on which ideas and insights jump between communities, and the importance of personal relations in this process. Topics mentioned include qualitative and quantitative analysis of adsorbate species, interpretation of binding energy shifts induced by adsorption and connected topics such as reference levels, local potential, relaxation/screening and charge transfer, as well as final state splitting, satellites, peak shape changes and angular effects in both photo- and Auger electron spectra. The newly accessible possibilities from very high resolution in photon source and electron analysis, coupled with selective excitation with synchrotron radiation, are briefly discussed, and emphasis is laid on the access by these means to electronic dynamics and to their coupling to the dynamics of atomic motion, as visible in fragmentation of adsorbed molecules by these electronic excitations.

1. Introduction

Core electron spectroscopies – photoelectron (XPS) and Auger (AES) – certainly belong to the most used surface techniques of today. This is due to their high surface sensitivity, their capabilities of qualitative (both in terms of atoms and of their chemical situation) and quantitative analysis which can be applied to real world samples as well as to model systems, and their relatively low destructive action on surface systems (for XPS and X-ray induced AES, XAES). Practical surface analysis by these techniques and in particular by electron induced AES has grown to a huge field with many specialties in terms of types of materials (from all types of metallic, semiconducting, and insulating solids and mixtures to polymers and even liquids) and procedures (for instance for spatial resolution by scanning probes or angular evaluations) [1]; it will not be dealt with here.

The emphasis here will be on XPS and XAES for fundamental surface physics, in particular as applied to adsorbate systems. Even here the main applications are the same as for the practical systems mentioned, namely those aimed at information about type and quantities of surface species present on a well-defined surface under certain conditions as well as their changes with heat treatment, coadsorption, reaction or other influences; therefore, these methods belong to the working horses of adsorbate physics. While such applications represent the bulk of the reported investigations up to today, I shall briefly mention only a few of these, and concentrate more on the underlying physics, trying to show the interaction between the development of the understanding of the measurement process and of the basic physics of surface systems. Since the bonding at surfaces as well as the photoelectron emission and the core hole decay processes are all many-body effects, these are important in this context. Using the recent developments in instrumentation, the dynamics of interactions come into reach in such investigations.

This short historical survey will necessarily be

SSDI 0039-6028(93)E0443-X

sketchy and subjective. There will be no attempt at a balanced presentation of all contributions to this wide field; rather, using mainly our own work (which hopefully is not too untypical), I hope not only to show the main objective trends, but by noting their genesis and the importance of persons and their specific background for it, an example may be obtained for the – haphazard, but still efficient – way such developments in new disciplines proceed. I also hope that these personal reminiscences do capture some of the flavour of the field, and of the fun connected with working in it for 25 years.

2. Some subjective and personal history

Electron spectroscopy has had a long and complicated history of over 100 years, if we include the early work of Hertz, Thomson, and Lenard [2], and then of course of Einstein [3], on the photoelectric effect. Spectroscopy proper started with the Rutherford team [4] and reached maturity with Siegbahn's Uppsala group [5,6]. This history contains periods of explosive growth as well as periods of stalling in wait for a seminal new idea, and long delay times for the acceptance and utilization of ideas developed by one community, by others. This has been very vividly described by Jenkin et al. [7], and attention is called to their papers here, as well as to the historical accounts by Spicer [8] and by Feuerbacher, Fitton and Willis [9], to name a few. The real inception of XPS can be dated from the mentioned work of Robinson and Rawlinson [4], and for AES the seminal paper by Auger [10] marks the start. However, little actual application of the effects for the understanding of systems was made until the concerted effort of Siegbahn et al. in the nineteen fifties. By these both techniques were revived and applied to the study of atoms, molecules and solids, as the two monumental monographs [5,6] of this group show. However, the application of AES and, even later, of XPS to surface and adsorbate problems started only in the late sixties and came into full bloom only in the seventies. It is an interesting question to ask why these techniques so ideally suited for surface

studies came to their own with such delay. Several influences seem to have conspired to this end.

Early surface science, i.e. that before about 1960, consisted of research of quite varied quality, ranging from the excellent work of Langmuir or Roberts which still stands today, to many papers which, for instance for the adsorbate-induced work function change, did not even get the sign right – not too surprising in view of the missing means to define the surface purity and even to measure the residual pressure in the vacuum systems used. In his interesting historic survey mentioned above, Spicer [8] attributes the fact that Langmuir never did any photoemission work to the poor understanding of photoemission existing in the thirties – even though this was the period when it was still unclear how photoemission (in the valence range) could be anything *but* a surface effect: at that time the only way to conserve momentum appeared to consist in the contribution of the surface. Probably more important was that early surface science was a low tech science utilizing all-glass vacuum systems; the lack of adequate total and partial pressure measurement, large pumps, and demountable systems, as well as of means to characterize clean surfaces and define adsorbate layers except on tungsten filaments and field emission tips, severely limited the usable techniques.

All this, however, started to change in the early sixties, when the combination of Bayard–Alpert gauges, ion pumps, and demountable stainless steel systems, and the revitalization of LEED made the still small surface community turn to bigger, more flexible machines and to single crystal work. Why, then, did it take another decade until electron spectroscopy came to surface science? For the case of UPS, Spicer [8] describes vividly the relief of people in this field when it became finally clear [11] how momentum could be conserved in photoemission from the bulk by utilization of a reciprocal lattice vector, ending the "period of misguided quantum mechanics", in Spicer's words. This may then have led to an overemphasis of the bulk aspect for quite a while. Only after vacuum techniques and sample preparation techniques became sufficiently sophisticated, the strong surface component became obvious. As to XPS and AES (or ESCA – electron spectroscopy for chemical analysis, in Siegbahn's term) surface applications may have been delayed by the conviction of the leaders in the field who, using Langmuir–Blodgett films, had concluded that the average source depth was of the order of 100 Å [12] and thus had convinced themselves that surface effects proper (i.e. the first monolayer or so) were not important at their energies. Because of this, commercial instrument makers argued even after 1970 that therefore vacuum conditions of about 10^{-7} mbar were sufficient and that carbon or gold plating of samples were acceptable – which, in a circular effect, limited the "surface" applications to catalysts and oxidation phenomena and the like, and excluded adsorption studies. Interestingly, much earlier Quinn [13] had already calculated the inelastic mean free path in aluminum quite correctly, and the sizes and energy behaviour of ionization and excitation cross sections of atoms and molecules were well known. Also, Auger emission from surfaces had been seen as early as 1953 [14], but the features were too small compared to the background to be usable. In papers around 1970 articulation of surprise can be found at the apparent difference of behaviour of photoelectrons and Auger electrons. To be sure, some groups were aware of the problems [15], but this appears not to have been general knowledge. So I believe that the delay of a decade was mainly due to crossover of know-how between communities: of vacuum techniques and sample preparation methods from the surface people to the spectroscopists, and of the understanding of the technique and its physical background and potential in the opposite direction. I will come back to this in my personal history below.

In the late 1960's, then, the barriers disappeared. It started with AES, as many surface machines by then contained LEED screens which, using the (already existing [16]) modulation technique [17], could be used for AES despite the unfavourable signal/background ratio under electron excitation. The understanding of the inelastic mean free path (see Duke [18]) supplied the basis of the surface sensitivity thus discovered, and prompted the attempts to use UPS and XPS for surface investigations on well-defined single crystal surfaces. This then led to the first demonstrations of (sub)monolayer sensitivity in UPS [19] and XPS [20,21], in the early seventies, and to explosive growth in the following years. In September 1976 a symposium held in Noordwijk reviewed this development [22], and the book to which it finally led [23] contains most of the basics still important today.

The character of this contribution and of the present volume in general may justify the addition of some personal memories which give additional points of the same questions on a lower, personal level. They may also be seen as an example of the genesis of a group working in this newly emerging field, surface science, and of the factors that conspired to further and hinder the accumulation of a particular combination of know-how and competence.

Having done work on fundamental catalysis [24] for my PhD in Darmstadt, I had concluded that for the development of a microscopic understanding of catalysis, the underlying processes such as adsorption had first to be understood in well-defined systems. Looking around for methods that allowed the definition that appeared necessary, I settled on field emission microscopy; and I had the great luck to be accepted as postdoc by Bob Gomer in Chicago in 1962. This thoroughly incorporated me into the all-fused-glass surface community believing in field emission and field ionization as the (then) only proper techniques for surface and adsorbate characterization, and heavily relying on cryotechniques, and taught me, besides many other things, that one can do what one can sensibly imagine. My work of this time [25] has contributed to the establishment of a research area today usually called DIET (desorption induced by electronic transitions; see the article of Madey on page 824 in this volume, or our recent review [26]) which still thrives today. The turning of the tide towards the much more expensive, but also much more flexible metal systems was noticeable when I left the USA in 1964. Back in Darmstadt I started, while staying partly in the glass community doing field emission and energy transfer measurements, to slowly build up the capability for development of the new types of machines. With technical and financial circumstances at that time in Germany (and in particular in a small chemistry department) being a far cry from today, but also from the USA of the time, this was not at all easy; the products were first applied to further work on stimulated desorption [27]. The basic question that drove all this work (and that still drives me today; see below) concerned the coupling, in terms of energy and charge, between an adsorbate and the substrate it sits on or collides with, and the consequences for surface reactivity. To be frank, my knowledge of electron spectroscopy in general was virtually nonexistent up to 1967. This changed drastically when I visited Uppsala in 1968 as a participant of Löwdin's summer school on quantum chemistry, where I heard about ESCA (not in connection to surface science) when talking to people there. In the same year Auger spec-

troscopy came to surface science and I learned about it, along with everyone else.

My transfer to Munich in 1969 increased the opportunities, and other methods came into reach. Having brought to surface physics the basic attitude of a chemist, who concentrates on his problem, utilizing whatever method appears best for its solution, and loves to combine many, I was very eager to acquire the capability for new methods for the solution of my general problem: the understanding of adsorption layers – their composition, geometric, electronic and dynamic properties, and their reactivity. So, when it became obvious to me as to many others in both communities that photoelectron spectroscopy (of which I now was no longer ignorant) should be able to deliver important information on this problem of mine – and the last straw to this was for me Eastman's finding of adsorbate submonolayer sensitivity in UPS [19] – I started a funding attack

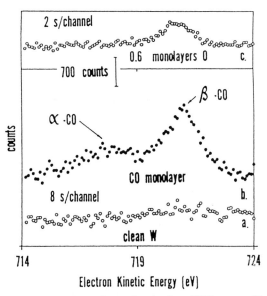

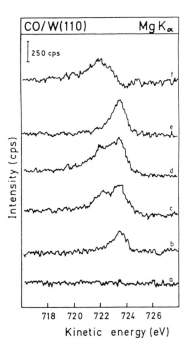

Fig. 1. Left: The first submonolayer chemisorbate XPS spectra (O and CO on polycrystalline W). From Madey et al. [21], with permission. Right: The first submonolayer chemisorbate spectra on a well-defined single crystal surface: O 1s of CO on W(110) at 300 K. (a) Clean surface; (b) 1 L CO (mostly β-CO); (c) 10 L CO (β + v-CO); (d) under 10^{-6} Torr CO (same); (e) after pump-off and heating to 600 K (β-CO); (f) (d) minus (e) to isolate α-CO. From Ref. [30], with permission. The dissociation of β-CO is not yet unquestionable.

to get a UHV electron spectrometer suitable for adsorbate research. In a competitive session of the Deutsche Forschungsgemeinschaft who had received a number of such applications but wanted to grant only one machine at first – not trusting the feasibility under the influences outlined above – we did win the funds in 1972, and started to work on a machine design. We included myself and Alex Bradshaw who was then my postdoc; he was a chemist, too, and had come from Pritchard in London, having worked with infrared spectroscopy of adsorbates. While our design [28] was not the first UHV photoelectron spectrometer [29], it was the first one fully geared to single crystal surface work. It was also realized by Vacuum Generators and formed the basis of our research in this field for more than a decade (and in fact is still in use in the lab of Eberhard Umbach). Since we set out from the start to work with well-defined single crystals, we were too late [30,31] to be first in adsorbate XPS [21] – Madey and coworkers, using a "normal" electron spectrometer, published the first chemisorbate XPS results; but we joined in soon (Fig. 1) and registered a few other firsts in this field subsequently (see below).

Then I switched departments, from chemistry to physics, while Alex remained in chemistry (and soon went to the Fritz Haber Institute in Berlin). For the next four years, the driving force of our work both in the utilization of UPS, XPS, and XAES for adsorbate characterization and in the fundamental understanding of the related physical processes, in particular of the many-body effects in production and decay of adsorbate core holes, was my next postdoc John Fuggle whose death less than two years ago was so untimely. John had come to us from the University of Strathclyde in Glasgow, with a true electron spectroscopy background (but none in surfaces), and a burning ambition and drive – a real workaholic. Ted Madey joined some of these early investigations as a guest from NBS. John was soon joined by Eberhard Umbach (and others) who later handed on the torch to Wilfried Wurth and Uli Höfer and many others; John himself went on to new horizons in materials science, first in Jülich and then in Nijmegen. Some of the work that has

come out of these early cooperations will be mentioned below.

The new possibilities offered by the use of tunable radiation from synchrotron light sources also in this field became obvious to us as to many others in the second half of the seventies. When we found it difficult to get access to the obvious place for us, DESY at Hamburg, I turned to Stanford. The collaboration, mainly with Jo Stöhr, which started in 1978, and which was strengthened by the number of excellent students who went from here to Stanford and partly back (Rolf Jaeger, Eberhard Umbach, Rolf Treichler, Wilfried Wurth) was important for our establishment in this area. This, then, contributed to the emphasis on dynamics, both of electronic evolution and of coupling to the nuclear coordinates, which has developed over the years in our thinking as a logical consequence of the emphasis on coupling and energy and charge transfer. This led to a merging of the electron spectroscopy line of our work with that on DIET which had started so much earlier and to which again many coworkers – among them Rene Franchy, Rolf Jaeger, Rolf Treichler, and in particular Peter Feulner – have contributed. While here I will only be concerned with core level spectroscopy, we did combine these with UPS valence studies from the beginning [30,31] and all along. Also, we turned to angle resolved measurements of adsorbates and their interpretation very early [32]; this was helped by the development of a particularly suited multichannel spectrometer [33] (the development of which was mainly the merit of Albert Engelhardt and the very inspired advice of Hans Liebl), which has been extremely productive in recent years in the hands of Hans-Peter Steinrück [34]. Our parallel efforts on different aspects of adsorbate physics, from geometric and electronic structure via vibrational properties to sticking and thermal desorption kinetics and dynamics, serve – in addition to their intrinsic interest – as background to our core electron studies by supplying accurate definition of the studied adsorbate systems; and vice versa. This genesis may be a not untypical example of the spreading out of a group in this field from different converging chance roots to a rather broad approach and capability.

3. Early adsorbate core spectroscopy, and what came of it

3.1. Fingerprinting

Already in the first two years of work with the new methods, we as well as other groups tried to stake out their potential, both in terms of applications to gain qualitative and quantitative knowledge about specific adsorbate systems, and of understanding of the coupling of adsorbates to a surface in general. This can be seen from the first surveys of such work in 1974 by Brundle [35], Yates et al. [36], and myself [31], as well as from my book chapter written in 1977 [37]. The clear definition of cleanliness of a single crystal, and of the composition of an adsorbate layer, which now was routinely possible for any material, widened the materials accessible to well-founded adsorbate studies and brought a relief to this artifact-ridden field which cannot be imagined by somebody who has not experienced the time of surface physics as the "Physik der Dreckeffekte", from which not even the sole use of tungsten as the substrate allowed full escape. It is not surprising, therefore, that one of the first systems to which we applied the new methods was CO on W(110), and our clear proof of dissociation of ß-CO

[30,38]), ended the belief to the contrary long held by many [39] (including myself), and proved the intuition of others right [40]. This was a typical use of XPS, UPS, XAES, and XPS satellites (see below) as fingerprints to distinguish and characterize adsorbate species qualitatively and quantitatively (Figs. 2–4). Full exploitation of the quantitative capability of XPS later [41] made the measurement of the dissociation kinetics of CO on W(110) possible, as well as, for instance, the disentanglement of the complicated states of NO on Ru(001) [42]. As a third example of the use of fingerprinting, let me mention the clear proof of the formation of N_2O on W(110) from N_{ad} and NO [43]. Many similar applications of fingerprinting, without and with quantitative information, have been carried out by many authors over the years (see the examples given in the cited reviews [35–37]). As to qualitative differentiation, vibrational spectroscopy by HREELS (high resolution electron energy loss spectroscopy, see the article by Ibach on page 116 in this volume) has now taken over, because it can usually resolve finer details (for instance in the mentioned system NO / Ru(001) [44]), except if XPS is carried out at nonstandard high resolution (see the recent differentiation of CO sites for CO/Ni(100), and derivation of a temperature-dependent redistri-

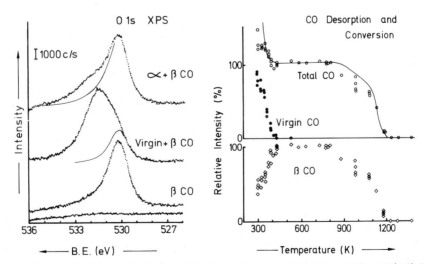

Fig. 2. Early example of the quantitative use of adsorbate XPS: Desorption and dissociation of CO on W(110). Left: O 1s peaks for conditions indicated (the improvement compared to Fig. 1 is obvious). Right: Quantitative evaluation showing the evolution with temperature of the layer produced at 300 K. From Ref. [38], with permission.

bution among them by today's Uppsala group [45]). But other than HREELS, XPS has the clear quantitative capability which makes it superior in many instances even today. As a small caveat, we only mention that under conditions of high angular resolution forward-scattering effects can be disturbing in some cases, so that for quantitative analysis a wide acceptance angle is advisable. Also, satellite intensities have to be taken into account if different molecules are compared quantitatively (see below).

3.2. Adsorbate binding energy shifts

The differentiation of adspecies by XPS rests on core level binding energy differences, and an understanding of such shifts requires physical understanding of the contributing effects. In the early years of surface photoelectron spectroscopies, such understanding was quite poor (see, e.g., Eastman's first interpretation of the two CO-induced UPS peaks seen by them, and the clarifying discussion that took place at the Faraday discussion meeting in September 1974 in Cambridge [46]; or the controversies about the "proper" reference levels for adsorbate photoemission peaks. See the references given in Ref. [37]). But the general understanding of the contributing phenomena which existed in the electron spectroscopy community [47] filtered into the surface community and was applied to surface conditions. Soon a consensus evolved, for which at least in my view Bill Gadzuk [48] was the important clarifier, and I believe that the views given by him and by me in our contributions to Refs. [37,48] can still be taken as basis today.

In short, these conceptually distinguish two initial state shifts and one final state shift. The first initial state shift is due to the changed electrostatic potential into which the adsorbate is immersed even without any interaction (potential shift, also called local work function – not a good nomenclature in my opinion), and the second is the normal "chemical shift" due to the fact that the surface bond leads to electron redistribution in the ground state. The final state adsorbate shift is the correlate to the usual relaxation or reorganisation shift seen as the difference be-

tween ΔSCF and Koopman's energies for atoms and molecules as well, and can be understood as the extra screening of the core hole by, e.g. the image charge on metal surfaces, or by polarisation of the surrounding molecules for dense adsorbates or condensates. To be sure, this disintegration is partly ambiguous, but it helps in visualisation of the physical effects; also it assumes that changes of correlation upon adsorption can be neglected. Later it became clear through the work of Schönhammer and Gunnarsson [49], building on Kotani and Toyozawa [50], that for adsorbates strongly coupled to the surface, screening happens by charge transfer from the substrate to the adsorbate; we shall come back to this below when discussing dynamic effects. Also, my view that the effects are essentially clear appears not to be shared by everybody: see the contrasting extreme views of Wandelt [51] and Jacobi [52] on the correct interpretation of BE shifts of adsorbed xenon; or the recent controversial discussion of substrate core level shifts by alkali adsorption and their bearing on the alkali charge state [53].

Which brings me to substrate core level shifts. First attempts to see those were carried out by us [31,54] and others [55,56] already in the early years – with limited success for adsorbates, due to the small values and low available resolution; but for oxidation the effects were large and allowed one to derive conclusions even then [56]. Utilization of the higher resolution offered by synchrotron sources led to success, and substrate core level shifts (in particular for the very sharp 4f-levels of the heavier transition metals) can now be followed in great detail and interpreted in terms of initial and final state effects [57]. However, as mentioned above in connection with alkali adsorption, this does not mean that all researchers agree on everything even today.

3.3. Core satellites

As clear from these few words, core level shifts are already many-particle effects. We were lucky to be able to build upon the existing extensive work on many-body effects in photoemission from atoms and molecules, in particular of the Shirley group [47]. So it was obvious from early on that

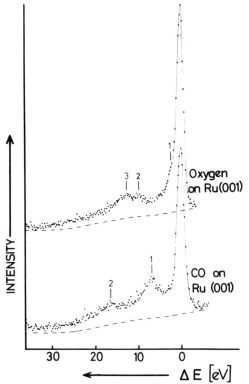

Fig. 3. First XPS satellite spectra of well-defined adsorbate layers. From Ref. [59], with permission.

partitioning between screened states and satellites. Obviously for a totally uncoupled "adsorbate", no screening will exist, and only a high energy XPS peak will result. For a strongly coupled adsorbate, a new strong "screened" peak will develop at lower BE and will carry the main intensity, while satellites will exist whose energies are governed by the electronic structure of the adsorbate complex and their intensity by the lever rule [58]; among them there will also be satellites corresponding essentially to the unscreened state. A particularly interesting case can be expected for intermediate coupling for which comparable intensities of screened and unscreened (or less well-screened) states are expected [62]. Indeed, such behaviour of "giant satellites" was found (before the theoretical prediction was known) experimentally for weakly chemisorbed N_2 on W and CO on Cu [63]. The many theoretical treatments existing by now, while all arriving at the final state splitting, disagree considerably in the detailed interpretation of the various peaks [60]. On a more down-to-earth level, the relative strength of satellites in different bonding situations has to be taken into account when quantitative comparisons of the surface concentrations are made, as mentioned above [38,64].

3.4. XAES of adsorbates

Electron induced Auger spectroscopy is, of course, the older and the more used general surface analysis technique, as indicated in the introduction; its ease of application and its easier use in an imaging fashion – two-dimensional by beam scanning and three-dimensional by sputtering – still make it the preferable technique for general surface analysis. Its application to adsorbate systems, however, is severely limited by the strong destructive beam effects observed in essentially all cases of molecular adsorbates. This is considerably improved if X-ray excitation is used. Since the Auger process is the predominant decay process of core holes in the energy range of concern for surface science, roughly one Auger electron is emitted per X-ray photoelectron for narrow-band excitation which is spread out over a considerable kinetic energy range; therefore,

the dynamical correlate of shifts [58] must exist in the form of additional satellites. Indeed we succeeded quite early to measure the characteristic satellites of adsorbates [59] (Fig. 3). The fact that there was a clear distinction between O 1s satellites in atomically adsorbed oxygen and in molecularly adsorbed species such as CO, NO or others, was used as an additional fingerprint already for the ß-CO case (see above). The later work of Plummer et al. [60] on the satellite structure of various types of carbonyls and their comparison to adsorbed CO, as well as theoretical modelling, has improved the understanding, and recently high resolution synchrotron work by the Uppsala group has continued along these lines to even more detail [61]. The coupling between the strengths and shifts of screened states and correlated shake-up states required by the sudden approximation [50,58] can then be used to examine the influence of variable coupling on the

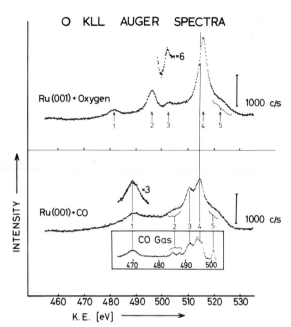

Fig. 4. First well-resolved XAES spectra of well-defined adsorbate layers, with comparison to gas phase CO. From Ref. [65], with permission.

quantitative analysis is better done by XPS in the way discussed above. Nevertheless, XAES of adsorbates is a very interesting field of research in itself. If conducted in an XPS machine, $N(E)$ spectra can directly be obtained because of the much lower background compared to electron induced AES, and good resolution of the complicated peak shapes can be obtained. This was shown for well-defined adsorbates early on [65] (Fig. 4). The first use was again as a fingerprint for a certain adsorbate situation, and the application was the differentiation of molecular and dissociated CO [38], as described above: because of the large number of possible final states, the peak shapes are much more specific than XPS, UPS, or XPS satellite spectra. The adsorbate induced shifts, which in the simplest electrostatic model should be three times those of the corresponding XPS peaks [65,66], can be used for arguments on the screening mechanisms. The peak shapes can be analysed in terms of the many possible two-hole final states and of shake states, and conclusions can be drawn from the peak shape changes upon adsorption [67]; in solids, the

limiting peak shapes ("atomic" versus "band-like") of CVV spectra are well understood in terms of Coulomb interaction of the final state holes, and of screening [68]. Finally, if the decay spectra are obtained for continuum and for bound core excitations and compared, information on the dynamics of screening and decay can be obtained, as will be discussed in Section 4.

3.5. Angle dependences in XPS and XAES

The simplest use of angular effects in adsorbate XPS is for tuning the relative contribution of the first layer, the adsorbate, to the total spectra [20]. Outside practical surface analysis, this approach has been used, for instance, to differentiate between intrinsic satellites and extrinsic losses of the outgoing photoelectron [69]. The more important direct angular effects for XPS of solids and adsorbates are dealt with by Woodruff on page 183 in this volume, or in detail in various excellent reviews [70]. It suffices here, therefore, to just mention the development of angular and normal photoelectron diffraction (PED) into practicable tools for adsorbate geometry determination.

The angular distribution of Auger electrons from adsorbate core holes has been shown by Umbach and Hussein [71] to contain very distinct information on the orientation of the admolecule, as well as on the correct assignment of the Auger peaks. The latter can then be used to determine more accurately the energy shifts compared to the gas phase molecule, i.e. the remaining Coulomb interaction of the two final state holes and their screening in the adsorbed state.

A more recent development which also involves angular measurements is that of electron holography [72], using either photoelectrons or Auger electrons. The interesting point here is that the use of (calculational) reconstruction of the atom distribution from the holograms deriving from the coherent scattering of the emitted electron from the surroundings would give a direct way of surface geometry determination, rather than the indirect diffraction methods LEED and PED which necessitate comparisons of measurement and calculation. Since the holograms arise by local interference close to the emitting centre, which will depend critically on the wave function of the outgoing electron, different distributions should arise for, e.g., photoelectrons and Auger electrons of the same kinetic energy and from the same centre. Also, chemical shifts are usable to tune the primary emitter. Difficulties arise from the need to eliminate the very strong forward scattering peaks as well as truncation effects and other problems [73], so that at present one may have doubts that this will ever be a general method which can compete in accuracy with the mentioned diffraction methods. But the use as a precursor to fullblown LEED analysis for the narrowing down of possible structures and parameters appears well possible.

4. Dynamics of core excitation and decay in adsorbates, and coupling to nuclear motion

The many-body effects mentioned above, which in a stationary-state analysis lead to peak splittings, are really evidence of dynamical properties of the total system under consideration. The hierarchy of influences of screening by polarization and charge transfer, and of satellite production in excitation and core hole decay, depend on the relative time scales of image charge formation/ charge transfer, and photoelectron removal, compared to hole life times. Another dynamical effect is concerned with the interaction of all these processes with nuclear motion. These will show up in non-Franck–Condon vibrational effects for nondissociative processes, and in relative fragmentation probabilities for dissociative events. Such dynamical processes will play a role in all electronic excitations of surface species and will contain the signatures of the couplings between adsorbate and surroundings. Their investigation in the case of core hole excitations is of particular interest, as the core life time which is roughly known can be taken as an internal time mark, and because the existence of core-to-continuum as well as of core-to-bound excitations (primary ionic and primary neutral core excitations) allows interesting conclusions from their comparison. The accessibility of neutral core resonances by

tunable synchrotron radiation therefore opened new possibilities in this field. The use of the corresponding measurements of NEXAFS (near edge X-ray absorption fine structure) or XANES (X-ray absorption near edge structure) for chemical and geometrical information on adsorbates [74] is being discussed by Citrin [75] and will not concern us here. What is of interest is the investigation of decay spectra of surface species obtained with well-defined variable primary excitation, and the extension to fragmentation patterns.

The first interesting information that was obtained here was connected with a simple conclusion from the models used above. If it is correct that core excitation of, e.g., the O 1s level of adsorbed CO on Ni or Ru or similar surfaces leads with high probability to a core-ionized CO molecule which has been neutralized by charge transfer from the substrate (see above), then this final state is not really an ionic state but a neutral state. Its decay spectrum should then really not be an Auger spectrum, but rather look more like an autoionisation spectrum obtained by decay of a neutral core-to-bound excited state. Following such reasoning, we found [76] that the decay spectra of nominal core-ions (obtained by excitation into the continuum) and of nominal neutral core states (obtained by variable core-to-bound excitation) were indeed essentially identical in the case of chemisorbed species; a small difference existed only for primary excitations into Rydberg states around the vacuum level. The simplest explanation – that the primary excited electron of the core-to-bound state and the screening electron are equivalent for the decay spectrum – cannot be correct, however. In all strongly coupled CO adsorbates investigated in detail, the lowest core-to-bound excitation is about 1–2 eV higher than the Fermi level, from where the screening charge should come. Also, while this difference vanishes for weakly bound CO (e.g., on Cu), the decay spectra are identical here as well, and no trace of a special decay of the "giant" unscreened satellite (see above) was observable. We have interpreted these findings by the conclusion that the primary excitations in the various cases (core-to-bound, ionic, satellite) are indeed different (on the time scale of excitation), but that all these differences have decayed by the time of core hole decay, i.e. during core life time. The other authors who have done similar work [77] may not fully agree with us here. There is no question, however, that the additional excitation energy in satellites essentially does not show up in decay spectra, i.e. has been dissipated during core life time. The only way some survival of such highly excited core states can be seen is in the fragment ion spectra [78]. This is one example in which surface dissociation, which is a very weak channel in well-coupled systems [25,26], selectively indicates a minority decay channel.

Other interesting effects which we have been able to find in such correlated excitation–decay–fragmentation studies [79] of adsorbed and condensed molecules and atoms include the following. Just as a primarily core-ionized chemisorbate is neutralized by charge transfer during core life time, a core-to-bound neutral resonance of a physisorbed species (e.g. a rare gas) can be ionized during core life into an image charge-screened ion with certain probability, as again shown by decay spectra [80–82]. The existence of two ionisation continua for adsorbed and condensed layers (where the "outer" continuum starts at the vacuum level, as for the isolated species, while the "inner" starts at E_F for an adsorbate on a metal and at the conduction band bottom for a condensate) becomes obvious and governs such possible evolutions. Coupling to nuclear motion not only becomes apparent from vibrational evolution during core hole life time [81] as known from free molecules [83], but also from selective dissociation of adsorbed and condensed molecules such as water, ammonia, benzene, by specific core-to-bound excitations [84]. We have explained these by "ultrafast" dissociation, i.e. dissociation on the time scale of the core hole life time [79,84]. Such investigations become particularly interesting if the time scale of the molecule–substrate interaction can be tuned, either by comparing different but similar molecules (CO–N_2) on the same substrates, or the same molecules on differently interacting substrates (CO on Cu versus on Ni; any molecule on a well-defined inert spacer layer, a mono- or multilayer of, say, Ar or Xe), or in cases where the same molecule on the same surface can interact in different ways [85]. Such investigations are in full bloom at present and promise to lead to many more interesting results [86].

5. Concluding remarks

These reminiscences about the evolution of some typical methods and concepts in a small branch of surface science, and about the interaction and evolution of people concerned with them, are obviously not intended as an objective assessment, but rather as a very subjective account on how it all happened to a specific contributing group over the years. If it became obvious that neither the flow of knowledge and ideas nor the development on a certain place are adiabatic, i.e. they do not follow the easiest path and do not even out differences of knowledge and understanding automatically, and that people are involved, not automata (which would be infallible, but without creativity), then the main purpose has been reached. The enormous importance of collaborations and cooperations, and within those of personal relations, in this process has hopefully become clear from the specific examples. It remains for me to thank all those involved in the described research over the years, some of whom have been mentioned by name, for their indispensable contributions.

Acknowledgements

Since all this happened in a real and expensive world, the financial basis was important as well. None of our work could have been carried out without the support by the Deutsche Forschungsgemeinschaft, the godfather of basic research in Germany. In the last twenty years, this happened through the Sonderforschungsbereiche 128 and 338. The synchrotron work mentioned at the end has been supported by the German Ministry of Research and Technology, BMFT. We are very grateful that, most of the time, these agencies have been very responsive, within their limits, although over the years red tape has grown more than funding.

References

[1] See, for instance: D. Briggs and M.P. Seah, Eds., Practical Surface Analysis by Auger and X-Ray Photoelectron Spectroscopy (Wiley, Chichester, 1983).

[2] H. Hertz, Ann. Phys. 31 (1887) 983;
J.J. Thomson, Phil. Mag. 48 (1899) 547;
P. Lenard, Ann. Phys. 2 (1900) 359.

[3] A. Einstein, Ann. Phys. 17 (1905) 132.

[4] H. Robinson and W.F. Rawlinson, Phil. Mag. 28 (1914) 277;
H. Robinson, Proc. Roy. Soc. Ser. A 104 (1923) 455.

[5] K. Siegbahn et al., ESCA: Atomic, Molecular and Solid State Structure by Means of Electron Spectroscopy (Almqvist and Wiksells, Stockholm, 1967).

[6] K. Siegbahn et al., ESCA Applied to Free Molecules (North-Holland, Amsterdam, 1969).

[7] J.G. Jenkin, R.C.G. Leckey and J. Liesegang, J. Electron Spectrosc. Relat. Subj. 12 (1977) 1; 14 (1978) 477;
J.G. Jenkin, J. Electron Spectrosc. Relat. Subj. 23 (1981) 187.

[8] W.E. Spicer, in: Chemistry and Physics of Solid Surfaces IV, Eds. R. Vanselow and R. Howe (Springer, Berlin, 1982) p. 1.

[9] B. Feuerbacher, B. Fitton and R.F. Willis, in: Photoemission and the Electronic Properties of Surfaces, Eds. B. Feuerbacher et al. (Wiley, Chichester, 1978) p. 1.

[10] P. Auger, C.R. Acad. Sci. 180 (1925) 65.

[11] H.Y. Fan, Phys. Rev. 68 (1945) 43.

[12] ESCA: Atomic, Molecular and Solid State Structure by means of Electron Spectroscopy (Almqvist and Wiksells, Stockholm, 1967) p. 141.

[13] J.J. Quinn, Phys. Rev. 126 (1962) 1453.

[14] J.J. Lander, Phys. Rev. 91 (1953) 1382.

[15] C.S. Fadley and D.A. Shirley, Phys. Rev. Lett. 21 (1968) 43.

[16] W.E. Spicer and C.N. Berglund, Rev. Sci. Instrum. 35 (1964) 1665.

[17] L.A. Harris, J. Appl. Phys. 39 (1968) 1419;
R.E. Weber and W.T. Peria, J. Appl. Phys. 38 (1967) 4355.

[18] C.B. Duke, Surf. Sci. 299/300 (1994) 24.

[19] D.E. Eastman and J.K. Cashion, Phys. Rev. Lett. 27 (1971) 1520.

[20] C.R. Brundle and M.W. Roberts, Chem. Phys. Lett. 18 (1973) 380;
W.A. Fraser, J.V. Florio, W.N. Delgass and W.D. Robertson, Surf. Sci. 36 (1973) 661.

[21] T.E. Madey, J.T. Yates, Jr. and N.E. Erickson, Chem. Phys. Lett. 19 (1973) 487.

[22] Photoemission, Proc. Int. Symp., Noordwijk, September 1976; R.F. Willis, B. Feuerbacher, B. Fitton and C. Backx, Eds., ESA, Paris, 1976.

[23] B. Feuerbacher, B. Fitton and. R.F. Willis, Eds., Photoemission and the Electronic Properties of Surfaces (Wiley, Chichester, 1978).

[24] D. Menzel and L. Riekert, Z. Elektrochem. Ber. Bunsenges. Phys. Chem. 66 (1962) 432;
L. Riekert, D. Menzel and M. Staib, Proc. 3rd Int.

Congr. on Catalysis (North-Holland, Amsterdam, 1965) p. 387.

[25] D. Menzel and R. Gomer, J. Chem. Phys. 41 (1964) 3311.

[26] P. Feulner and D. Menzel, in: Laser Spectroscopy and Photochemistry on Metal Surfaces, Eds. H.-L. Dai and W. Ho (World Scientific Publishers Co., Singapore, 1994), in press.

[27] D. Menzel, Ber. Bunsenges. Phys. Chem. 72 (1968) 591;
Angew. Chem. Intern. Ed. 9 (1970) 255.

[28] A.M. Bradshaw and D. Menzel, Vak.-Tech. 24 (1975) 15.

[29] C.R. Brundle, M.W. Roberts, D. Latham and K. Yates, J. Electron Spectrosc. Relat. Subj. 3 (1974) 241.

[30] A.M. Bradshaw, D. Menzel and M. Steinkilberg, Chem. Phys. Lett. 28 (1974) 516.

[31] A.M. Bradshaw, D. Menzel and M. Steinkilberg, Faraday Disc. Chem. Soc. 58 (1974) 46;
D. Menzel, J. Vac. Sci. Technol. 12 (1975) 313.

[32] J.C. Fuggle, M. Steinkilberg and D. Menzel, Chem. Phys. 11 (1975) 307.

[33] H.A. Engelhardt, W. Bäck, D. Menzel and H. Liebl, Rev. Sci. Instrum. 52 (1981) 835, 1161.

[34] See, for instance: H.-P. Steinrück, in: Modern Methods of Surface Science and Analysis, Ed. R. Smith (Pergamon, London, 1993), in press.

[35] C.R. Brundle, J. Vac. Sci. Technol. 11 (1974) 212.

[36] J.T. Yates, Jr., N.E. Erickson, S.D. Worley and T.E. Madey, in: The Physical Basis for Heterogeneous Catalysis, Eds. E. Drauglis and R.J. Jaffee (Plenum, New York, 1975) p. 75.

[37] D. Menzel, in: Photoemission and the Electronic Properties of Surfaces, Eds. B. Feuerbacher, B. Fitton and. R.F. Willis (Wiley, Chichester, 1978) p. 381.

[38] E. Umbach, J.C. Fuggle and D. Menzel, J. Electron Spectrosc. 10 (1977) 15.

[39] R. Gomer, Jpn. J. Appl. Phys. 2 (1974) 213, Suppl. 2.

[40] D.A. King, C.G. Goymour and J.T. Yates, Jr., Proc. Roy. Soc. (London) A 331 (1972) 361.

[41] E. Umbach and D. Menzel, Surf. Sci. 135 (1983) 199.

[42] E. Umbach, S. Kulkarni, P. Feulner and D. Menzel, Surf. Sci. 88 (1979) 65.

[43] R.I. Masel, E. Umbach, J.C. Fuggle and D. Menzel, Surf. Sci. 79 (1979) 26.

[44] H. Conrad, R. Scala, W. Stenzel and R. Unwin, Surf. Sci. 145 (1984) 1.

[45] A. Nilsson and N. Martensson, Solid State Commun. 70 (1989) 923; Surf. Sci. 211/212 (1989) 303.

[46] See the discussion in: Faraday Disc. Chem. Soc. 58 (1974).

[47] D.A. Shirley, J. Electron Spectrosc. Relat. Subj. 5 (1974) 135;
R.L. Martin and D.A. Shirley, in: Electron Spectroscopy: Theory, Techniques, and Applications, Vol. 1, Eds. C.R. Brundle and A.D. Baker (Academic Press, London, 1977) p. 75.

[48] J.W. Gadzuk, Phys. Rev. B 14 (1976) 2267; in: Photoemission and the Electronic Properties of Surfaces, Eds. B. Feuerbacher, B. Fitton and. R.F. Willis (Wiley, Chichester, 1978) p. 111.

[49] K. Schönhammer and O. Gunnarsson, Solid State Commun. 23 (1977) 691; 26 (1978) 147, 399.

[50] A. Kotani and Y. Toyozawa, J. Phys. Soc. Jpn. 37 (1974) 912.

[51] K. Wandelt, J. Vac. Sci. Technol. A 2 (1984) 804; in: Thin Metal Films and Gas Chemisorption, Ed. P. Wissmann (Elsevier, Amsterdam, 1987) p. 280.

[52] K. Jacobi, Surf. Sci. 192 (1987) 499.

[53] D.M. Riffe, G.K. Wertheim and P.H. Citrin, Phys. Rev. Lett. 64 (1990) 571;
G. Pacchioni and P.S. Bagus, Surf. Sci. 286 (1993) 317.

[54] J.C. Fuggle and D. Menzel, Chem. Phys. Lett. 33 (1975) 37; Surf. Sci. 53 (1975) 21.

[55] A. Barrie and A.M. Bradshaw, Phys. Lett. 55 A (1975) 306.

[56] P. Pianetta, I. Lindau, G. Garner and W.E. Spicer, Phys. Rev. Lett. 35 (1975) 1356.

[57] B. Johansson and N. Martensson, Phys. Rev. B 21 (1980) 1;
D. Spanjaard, C. Guillot, M.-C. Desjonqueres, G. Treglia and J. Lecante, Surf. Sci. Rep. 5 (1985) 1;
W.F. Egelhoff, Surf. Sci. Rep. 6 (1987) 253, and references therein.

[58] B.I. Lundqvist, Phys. Kond. Mater. 9 (1969) 236;
R. Manne and T. Aberg, Chem. Phys. Lett. 7 (1970) 282.

[59] J.C. Fuggle, T.E. Madey, M. Steinkilberg and D. Menzel, Chem. Phys. Lett. 33 (1975) 233.

[60] E.W. Plummer, W.R. Salaneck and J.S. Miller, Phys. Rev. B 18 (1978) 1673.

[61] A. Nilsson and N. Martensson, Phys. Rev. B 40 (1989) 10249.

[62] O. Gunnarsson and K. Schönhammer, Phys. Rev. Lett. 41 (1978) 1608; 42 (1979) 195; see also the survey, in: Manybody Phenomena at Surfaces, Eds. D. Langreth and H. Suhl (Academic Press, Orlando, 1984) p. 221.

[63] J.C. Fuggle and D. Menzel, Proc. 7th Int. Vac. Congr. and 3rd Int. Conf. Solid Surf., Vienna, 1977, p. 1003;
J.C. Fuggle, E. Umbach, D. Menzel, K. Wandelt and C.R. Brundle, Solid State Commun. 27 (1978) 65.

[64] See the discussion, in: E. Umbach, Surf. Sci. 117 (1982) 482.

[65] J.C. Fuggle, E. Umbach and D. Menzel, Solid State Commun. 20 (1976) 89.

[66] See, e.g.: C.D. Wagner, Faraday Disc. 60 (1975) 291.

[67] J.C. Fuggle, E. Umbach, R. Kakoschke and D. Menzel, J. Electron Spectrosc. Relat. Subj. 26 (1982) 111.

[68] J.C. Fuggle, in: Electron Spectroscopy: Theory, Techniques, and Applications, Vol. 4, Eds. C.R. Brundle and A.D. Baker (Academic Press, London, 1981) p. 85.

[69] A.M. Bradshaw, W. Domcke and L.S. Cederbaum, Phys. Rev. B 16 (1977) 1480.

[70] C.S. Fadley, Phys. Scr. T 17 (1987) 39; in: Synchrotron Radiation Research, Vol. 1, Ed. R.Z. Bachrach (Plenum, New York, 1992) p. 421;
W.F. Egelhoff, Crit. Rev. Solid State Mat. Sci. 16 (1990) 213, and references therein.

[71] E. Umbach and Z. Hussein, Phys. Rev. Lett. 52 (1984) 457;
E. Umbach, Comments At. Mol. Phys. 18 (1986) 23.

[72] A. Szöke, in: Short Wavelength Coherent Radiation: Generation and Applications, Vol. 147, Eds. D.J. Atwood and J. Boker (AIP, New York, 1986);
J.J. Barton, Phys. Rev. Lett. 61 (1988) 1356.

[73] S.A. Chambers, Surf. Sci. Rep. 16 (1992) 261.

[74] J. Stöhr, NEXAFS Spectroscopy (Springer, Berlin, 1992).

[75] P. Citrin, Surf. Sci. 299/300 (1994) 199.

[76] W. Wurth, C. Schneider, R. Treichler, E. Umbach and D. Menzel, Phys. Rev. B 35 (1987) 7741;
W. Wurth, D. Coulman, A. Puschmann, D. Menzel and E. Umbach, Phys. Rev. B 41 (1990) 12933.

[77] C.T. Chen, R.A. DiDio, W.K. Ford, E.W. Plummer and W. Eberhardt, Phys. Rev. B 32 (1985) 8434;
G. Illing, T. Porwol, I. Hemmerich, G. Dömötör, H. Kuhlenbeck, H.-J. Freund, C.-M. Liegener and W. von Niessen, J. Electron Spectrosc. Relat. Subj. 51 (1990) 149;
M. Ohno, Phys. Rev. B 45 (1992) 3865.

[78] R. Jaeger, R. Treichler and J. Stöhr, Surf. Sci. 117 (1982) 533;
R. Treichler, W. Riedl, W. Wurth, P. Feulner and D. Menzel, Phys. Rev. Lett. 54 (1985) 462;
R. Treichler, W. Wurth, W. Riedl, P. Feulner and D. Menzel, Chem. Phys. 41 (1991) 259.

[79] D. Menzel, in: Synchrotron Radiation and Dynamic Phenomena, Vol. 258, Ed. A. Beswick (AIP, New York, 1992) p. 387.

[80] D. Menzel, Appl. Phys. A 51 (1990) 163;
W. Wurth, G. Rocker, P. Feulner, R. Scheuerer, L. Zhu and D. Menzel, Phys. Rev. B 47 (1993) 6697.

[81] W. Wurth, P. Feulner and D. Menzel, Phys. Scr. T 41 (1992) 213.

[82] O. Björneholm, A. Sandell, A. Nilsson, N. Martensson and J.N. Andersen, Phys. Scr. T 41 (1992) 217.

[83] R. Murphy, In-wan Lyo and W.J. Eberhardt, J. Chem. Phys. 88 (1988) 6078.

[84] D. Coulman, A. Puschmann, U. Höfer, H.-P. Steinrück, W. Wurth, P. Feulner and D. Menzel, J. Chem. Phys. 93 (1990) 58;
D. Menzel, G. Rocker, D. Coulman, P. Feulner and W. Wurth, Phys. Scr. T 41 (1990) 588;
D. Menzel, G. Rocker, H.-P. Steinrück, D. Coulman, P.A. Heimann, W. Huber, P. Zebisch and D.R. Lloyd, J. Chem. Phys. 96 (1992) 1724.

[85] W. Wurth, J. Stöhr, P. Feulner, K.R. Bauchspiess, Y. Baba, E. Hudel, G. Rocker and D. Menzel, Phys. Rev. Lett. 65 (1990) 2426.

[86] See the articles by W. Wurth and D. Menzel, and by N. Martensson and A. Nilsson, in: High Resolution Studies of Molecules and Molecular Adsorbates on Surfaces, Ed. W. Eberhardt (Springer, Berlin, 1993), in press.

Surface Science 299/300 (1994) 183–198
North-Holland

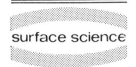

surface science

Photoelectron and Auger electron diffraction

D.P. Woodruff

Physics Department, University of Warwick, Coventry CV4 7AL, UK

Received 26 April 1993; accepted for publication 21 June 1993

The development of the techniques of photoelectron and Auger electron diffraction as means of determining aspects of the structure of surfaces are surveyed in the form of a historical review. Key features in the development are identified, and their significance described briefly. Different aspects covered include high kinetic energy forward scattering experiments aimed at determining bond angles within adsorbed molecular species and structural properties of epitaxial films, backscattering experiments aimed at quantitative adsorbate structure determination, Auger electron diffraction including recent renewals of interest in the method and its proper interpretation, and direct methods of structure determination including the idea of photoelectron holography. Some comments are included on the possible future applications of these techniques.

1. Introduction

Any attempt to survey the development of a scientific field historically by someone involved in this phase must, necessarily, be subjective. Events are seen from a particular geographical and scientific viewpoint; as such, they cannot be recounted in a disinterested fashion, and one's knowledge of events in some laboratories is inevitably far clearer than in others. This short review must therefore suffer from the disadvantages of this approach, but hopefully it also has some complementary advantages. In particular, it can be revealing to understand how the course of developments in a research area may have been influenced by constraints in instrumentation and current thinking at the time.

The basic idea of photoelectron diffraction – the phenomenon of coherent interference of a directly emitted photoelectron wavefield from a surface atom with components elastically scattered (Fig. 1) to produce variations in measured photoelectron flux as a function of emission angle and energy – can be attributed to the paper published by Leibsch in 1974 [1], who specifically proposed that such information could be used as a method of surface structure determination. Angular variations in core level photoemission [2,3] (and Auger electron emission [4]) had been observed experimentally from bulk solids prior to this time, and the idea that "diffraction" (or more properly coherent interference of scattered wavefield components) contributed to these effects, had been discussed. However, the experimental demonstration that photoelectron diffraction from adsorbed atoms (or surface localised atoms) could give rise to strong intensity modulations, and that these modulations could provide a means of obtaining structural information, was to take a further four years to reach the stage of a clear demonstration, and several more years to produce a method capable of routine application.

Each of the three research teams which provided the initial demonstrations (Fig. 2), independently within a few months of each other [5–7], took slightly different approaches to the problem. From the point of view of the basic physics, there was most similarity between the approach of the Berkeley group of Shirley and his colleagues, and the Warwick/Bell Laboratories collaboration which Neville Smith and I initiated. The common idea was that adsorbed atoms typically lie *above* the surface, so if photoelectron diffraction was to provide structural information on the adsorbate registry, backscattering was needed (see Fig. 1a), and our experience of LEED told us that this was most effective at low electron energies (say below 200 eV). Indeed, there were sceptics at the time who argued that even at these energies, the backscattering would be too weak. The argument was that LEED diffraction intensities are typically no more than a few percent of the incident beam, so the backscattering is too weak to produce measurable effects in the photoemitted signal. Such arguments neglect the fact that photoelectron diffraction relies on interfering amplitudes rather than intensities; for example, if a scattered amplitude is 20% of the incident beam, the scattered intensity is only 4%, but if this 20% signal interferes with a reference equal to the incident amplitude, the resulting variation between constructive and destructive interference in the observed intensity ($(1.0 + 0.2)^2$ and $(1.0 - 0.2)^2$) amounts to $\pm 40\%$. Nevertheless, the requirement of low kinetic energies forced our two groups to seek a solution based on synchrotron radiation, so that we could have access to variable photon energies in the range of tens to hundreds of eV. This route certainly accounted for some of the delays in realising the experiment; at that time almost no monochromators were available in this "grazing incidence" photon energy range on suitable synchrotron radiation sources. Our first experiments [5] were performed on the Tantalus facility at the University of Wisconsin–Madison

using a monochromator [10] borrowed from Daresbury in England during the period between the shutdown of one facility (NINA) and the opening of another (SRS), whilst the first results of the Berkeley group [6] were from a new monochromator on the Stanford storage ring.

A somewhat different approach was taken by Fadley's group in Honolulu [7] who realised that photoelectron diffraction could still be made to work at the higher kinetic energies (hundreds of eV to more than 1 keV) more typical of laboratory XPS instruments by using a geometry which was more reminiscent of RHEED than LEED. Specifically, if the photoemission signal is measured at near-grazing emission, scattering events can occur with substrate atoms at relatively small scattering angles which still have substantial scattering cross-sections (Fig. 1b). Although this experiment does not require special (and at the time very rare) synchrotron radiation beamlines, the grazing geometry does necessitate a high quality sample goniometer and patience to handle the low signal intensities, factors which undoubtedly left all three groups taking four years to provide the crucial first demonstrations of the viability of Liebsch's ideas.

Since that time there have been substantial developments in the methodology to turn the basic idea of photoelectron diffraction into a practical technique (or more exactly into at least two rather distinct practical techniques) for the

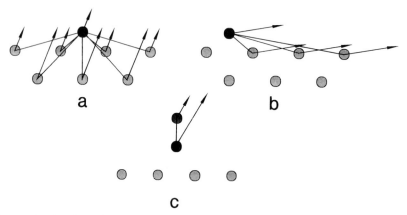

Fig. 1. Schematic diagram showing the main scattering pathways involved in producing interference in three different forms of the photoelectron diffraction experiment from adsorbate atoms. (a) True backscattering geometry, (b) near-forward scattering events which can be observed at grazing emission angles, (c) true forward scattering within an adsorbed diatomic molecule.

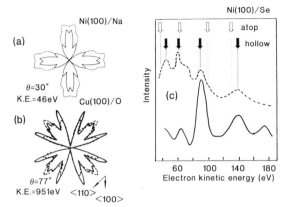

Fig. 2. Examples of the experimental data (dashed lines) from the three original demonstrations of adsorbate photoelectron diffraction all published in 1978. (a), (b) Azimuthal angle dependence of the adsorbate photoemission signal from which the minimum intensity has been removed. (a) Na 2p emission from Na adsorbed on Ni(100) recorded in a backscattering geometry; the full line is the result of a full multiple scattering calculation which was included in the original report [5]. (b) O 1s emission from O adsorbed on Cu(100) recorded in the near-forward scattering grazing emission geometry, compared with the results of a single scattering calculation (full line) also included in the original report [7]. (c) scanned energy mode photoelectron diffraction measurements at normal emission from the 3d state of Se adsorbed on Ni(100) [6]. The arrows correspond to the expected peak positions based on calculations of Li and Tong [8]. The full theory curve shown is taken from a later publication [9].

determination of various aspects of surface structure. There have also been some interesting parallel, and in some cases divergent, developments in the use of Auger electron diffraction for the same purpose. A further, relatively recent, area of work which has still not entirely reached a clear conclusion is to establish the optimum method of data analysis, and particularly the extent to which simple simulations or even direct methods may be of value in interpreting the results. In the remaining sections the developments in each of these areas are summarised.

2. Forward-scattering photoelectron diffraction

As remarked in the introduction, one virtue of the approach taken by Fadley and his colleagues

at Honolulu was that the higher electron energy forward scattering experiments could be performed using a conventional laboratory XPS source (Al or Mg Kα radiation at fixed photon energies of 1487 and 1254 eV). There are, of course, some disadvantages. For the purposes of determining the adsorbate–substrate registry via the emission from adsorbed atoms, the key problem is that the elastic scattering is quite strongly focussed in the forward direction, so large photoelectron diffraction intensity modulations can only be seen easily at grazing emission angles of less than about 20°, for which reasonably small angle scattering events involving substrate atoms can occur (see Fig. 1b). In order to build up a reasonably large data base for structure determination, the photoemission signal was therefore measured as a function of azimuthal angle at fixed kinetic energy for each of several values of the polar emission angle. This experiment places substantial demands on the precision of the sample goniometer – small "wobbles" in polar angle can lead to large intensity modulations. From the point of view of demonstrating the basic phenomenon of photoelectron diffraction, these azimuthal plots are ideal because *all* modulations must be due to diffraction if the experiment is performed well, and the quality of the experiment can be assessed from the extent to which the observed modulations reflect the known point group symmetry of the substrate. (Indeed, for precisely this reason, our own experimental data [5] using the backscattering geometry were collected in the same mode.) A further advantage of the high energy forward scattering azimuthal plots is that the data can be rather effectively simulated by simple single scattering calculations, and Fadley's group was quite successful in using this approach to determine some adsorption structures.

One disadvantage with this experiment, however, is that the strong forward scattering is only really effective if the adsorbate emitter has a small adsorbate–substrate layer spacing above the surface (the initial example of O adsorbed on Cu(100) (Fig. 2b) chosen for study by this method was, with hindsight, rather fortuitous in this respect because the O atoms lie almost coplanar with the top Cu layer); for emitters higher above the surface the data become far less sensitive to the adsorbate–substrate registry because, at least for ordered layers, the forward scattering becomes dominated by scattering within the adsorbate layer itself (in which the atoms are coplanar and, therefore, well able to scatter strongly into grazing emission angles). Thus, despite the fact that Fadley's group did perform analyses of a small number of adsorption systems using these grazing emission azimuthal measurements [11], this version of the technique has not proved generally useful for adsorbate registry determination.

A somewhat different application of the forward scattering idea, however, also developed by Fadley and coworkers [12,13], is the one which has been most extensively exploited in recent years. This is based on true (zero-angle) forward scattering, rather than the small-angle scattering exploited in the grazing emission azimuthal plots. The simplest version of this experiment is seen in intramolecular scattering in, for example, a diatomic adsorbed species such as CO, and this formed the proving ground for the method. Fig. 1c illustrates the basic scattering geometry of this experiment; in the case of adsorbed CO, for example, emission from the C atom bonded to the surface can forward scatter through the O atom. Providing the scattering phase shift for the forward scattering event is small (i.e. less than $\pi/2$), the directly emitted and scattered waves add constructively along the intramolecular (zero scattering angle) emission direction, and a zero-order diffraction peak is seen along this direction in a polar angle scan of the photoemitted signal. At reasonably high photoelectron kinetic energies (typically greater than about 500 eV), this phase condition is invariably met and the rather strong fall-off of scattering cross-section with scattering angle aids the production of a strong forward scattered peak. The rate at which this simple but rather powerful idea has been taken up for practical studies of molecular adsorbates has been surprisingly slow following the initial demonstration for CO adsorbed on Ni(100) (see Fig. 3a), but the method is now gaining favour (e.g. Refs. [14–20]). Of course, in its simplest form the information to be gained (the molecular orientation) is

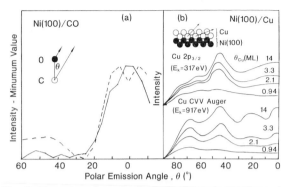

Fig. 3. Examples of results from the original true forward scattering (zero-order) diffraction experiments for the determination of molecular orientation and the form of epitaxial growth. (a) Experimental results (full line) for C1s emission from CO adsorbed on Ni(100) compared with the results of a theoretical calculation (dashed line) assuming a C–O axis perpendicular to the surface [12]. (b) Egelhoff's data [21] for Cu photoemission and Auger electron emission from Cu grown on Ni(100).

quite modest, but the equipment needed for such a measurement is simply a standard XPS machine, and the interpretation of the data can be very direct. This form of photoelectron diffraction is commonly referred to as XPD – X-ray photoelectron diffraction.

A different area of application has, however, led to wider exploitation of the same idea. This is for the characterisation of very thin epitaxial films, and the potential to use XPD seems to have been first recognised by Egelhoff in 1984 [21] (Fig. 3b). Consider the growth of a thin epitaxial layer of material A (e.g. Cu) on a single crystal substrate B (e.g. Ni); angular scans of high kinetic energy XPS (and Auger electron) emissions from the species A can be expected to show no strong diffraction-induced angular structure if the overlayer comprises only one atomic layer. Formation of a second layer, on the other hand, means that emission of atoms of A in the first layer can now be forward scattered by atoms in the second layer above, so strong forward scattering diffraction peaks can develop. Fig. 3b shows the results of some early experiments by Egelhoff using this method. The sensitivity to the formation of second and subsequent layers clearly makes the method rather valuable in attempting to establish the type and morphology of the early stages of epitaxial growth. For example, one can distinguish between layer-by-layer and island growth by such measurements. Indeed, if the measurements and interpretation can be sufficiently precise, it is possible to determine the extent of strain perpendicular to the surface in very thin epitaxial films with a (strained) mismatch at the interface, because for a known film structure the angles of the forward scattered peaks provide a direct measure of the interlayer spacing. This application of XPD has been used to study quite a number of epitaxial growth problems, as described in reviews by Egelhoff [22], Chambers [23] and Fadley [24]. One potential complication, however, in the use of these forward scattering ideas for bulk solids or thin films, is the role of multiple scattering. In the simple two-atom case, as in adsorbed molecules, one of the attractions of the XPD is the simplicity of the interpretation; much can be learnt by simply measuring the angular peak locations alone, but reasonable quantitative modelling can also be achieved using single scattering calculations. The periodic structure of solids, however, means that in emission from subsurface substrate atoms, one can have linear chains of forward scattering atoms, and each of these atoms can add further forward scattering along the chain direction. At first glance this would seem to lead to even stronger forward scattering peaks from more deeply lying emitters, although in the absence of inelastic scattering this view would ultimately suggest infinite forward scattering intensity for an infinitely deep atom, a conclusion which is evidently unphysical. The solution to this apparent dilemma is that in the multiple scattering effects which we can no longer ignore, the small forward scattering phase shifts can accumulate over several atoms to produce destructive interference. This actually means that the forward scattering peaks seen in experiments, even on bulk solids, arise mainly from emitters in the outermost two or three subsurface layers [25,26]. This is a rather attractive conclusion as it indicates that multiple scattering enhances the strong specificity to the near-surface structure, contrary to first expectations.

3. Backscattering photoelectron diffraction

Because most adsorbates lie *above* the outermost layer of atoms of the surface on which they are adsorbed, the forward scattering zero-order diffraction condition is generally of little value in determining the local adsorbate–substrate registry in such structures. Photoemission originating from substrate atoms can be forward scattered out of the crystal through adsorbed atoms, but it is difficult to distinguish such effects from the strong scattering within the substrate experienced by emission from subsurface substrate atoms described in the previous section. Grazing emission geometries can still be used to produce conventional non-zero-order diffraction interferences at high kinetic energies by using the reasonably strong scattering cross-sections on the edge of the forward scattering "cones" (Fig. 1b), as in the original Fadley group measurements of 1978 [7], but experiments designed to optimise backscattering from the substrate are clearly more generally applicable to this problem. As has already been remarked in the introduction, the experience with LEED suggested that this backscattering would be most effective if the photoelectron kinetic energies were kept low; say below 150–200 eV. The use of low energies, of course, has the disadvantage that by making the backscattering cross-sections large, the role of multiple scattering could be expected to also be substantial, greatly increasing the scale of the computational complexity. For this reason the early computational work in backscattering photoelectron diffraction was all conducted using modified LEED theory computer codes, notably developed by Holland [5,27–29], who collaborated with us at Warwick, and by Tong, at Milwaukee, who ran calculations relevant to systems studied by Shirley and coworkers [8,9,32] and us [30,31]. This theoretical complexity contrasted with the rather simple single scattering codes which proved to be rather successful for the interpretation of Fadley's high energy near-forward scattering experiments. Indeed, the computational complexity (essentially equivalent to that of LEED), combined with the short data range, whether in scanned energy or scanned angle, and the limited soft X-ray photon energy range available at most synchrotron radiation sources, conspired to make the rate of progress in backscattered photoelectron diffraction rather slow for several years.

An important step in overcoming these limitations came from the Shirley group when they first started to use much higher energy photons (around 3 keV) to probe photoemission from deeper core levels; notably the S 1s state with a nominal binding energy of 2472 eV [33,34]. The double crystal monochromators used in this energy range typically have very wide, uninterrupted scan ranges (up to 10 keV or beyond), and it was found that good scanned energy photoelectron diffraction modulations could be obtained up to kinetic energies of 400 eV or more. This conclusion was consistent with later trends in LEED; as larger data sets were found to be needed for unambiguous structure determination, there has been a tendency to extend this technique to higher energies than had hitherto been thought appropriate. The higher photon energies associated with S 1s photoelectron diffraction also overcame many of the instrumental and atomic effects which had superimposed troublesome (non-diffractive) structure on early scanned energy mode photoelectron diffraction; the photon energy dependence of the light output from the monochromator is typically less structured, and the deep 1s core level photoionisation cross-section shows only a slow monotonic decrease with increasing energy. These effects make it easier to observe the diffraction oscillations, even when they are somewhat damped by reduced scattering strength and Debye–Waller factors at the higher electron energies. By contrast, using low photon energies, as had been necessary in the early synchrotron radiation work due to lack of higher energy monochromators, we were constrained to work on relatively shallow core levels with binding energies less than 100 eV. Shallow core levels have more rapidly decaying photoionisation cross-sections, and in the case of heavier atoms one is often forced to work on high orbital angular momentum quantum state levels, with associated structure in the energy dependence of the photoionisation cross-sections, due to atomic effects such as delayed onset and Cooper minima [35,36]

One striking feature of the scanned energy mode photoelectron diffraction results obtained by Shirley and colleagues in this way was the apparent dominance of a small number of scattering pathlength differences in the observed modulations. Apart from the fact that the modulations in photoelectron diffraction are about a factor of ten larger, the spectra are reminiscent of EXAFS data (taken in the same photon energy range for the appropriate adsorbate K-edge), and this led the Shirley group to refer to the spectra as ARPEFS (angle-resolved photoemission fine structure), and to suggest that simple Fourier transforms would be valuable to analyse the data, just as they are commonly used in EXAFS and SEXAFS [33]. It is clear that there are dangers in pursuing this analogy too far, although the idea of using some form of Fourier transform is currently being discussed again, as will be described further in Section 5. Nevertheless, the apparently simple form of the wider energy range spectra (and their similarities to EXAFS) highlighted the importance of single scattering effects in these backscattering experiments [37–39]. The exact extent to which higher order multiple scattering need be included to provide a good description of such data remains a matter of debate, but it is clear that some multiple scattering can be important in refining the structure determinations based on these data, as first discussed by Barton and Shirley [40–42]. It is also now true, of course, that simulations of scanned energy mode photoelectron diffraction or ARPEFS can, like LEED, take advantage of faster and cheaper computers, as well as a range of approximate methods for calculating the scattering, which makes the difficulty of calculating multiple scattering effects less significant than it once was.

Although the relatively deep 1s core levels of the second row elements of the Periodic Table, such as S, P and Cl, do ease the instrumental problems of wide scan photoelectron diffraction, advances in instrumentation in grazing incidence grating optics monochromators [43] also allow access to the potentially more chemically important C, N and O 1s levels, and our own work in photoelectron diffraction, in a new collaboration with Alex Bradshaw's group at the Fritz Haber

Institute using the BESSY light source, has exploited this advance, coupled with the wide-range energy scan photoelectron diffraction technique, to investigate a range of atomic and molecular adsorption systems. Our first experiments were on the inevitable favourite molecule of the surface scientist, CO (on Cu(100)), but by measuring both the C 1s and O 1s photoelectron diffraction over a kinetic energy range of approximately 200 eV we were already able to find quite a good structure determination, even with a very simple single scattering calculation [44]. Notice incidentally, that the Shirley group was actually the first to demonstrate the possibility of determining the adsorption structure for CO (on Ni (100) and Ni(111)) [45], but with a somewhat shorter energy range and lower kinetic energies the analysis required more complex calculations and would now seem to have reached the incorrect solution for the CO/Ni(111) structure [46,47]. Of course, the problem of incorrect structure determinations in the early stages of development of a technique is not unique to photoelectron diffraction; similar problems were encountered in LEED [48]. In fact the origin of the problem in the two techniques is probably the same, namely insufficient data for a unique interpretation. We have recently found a rather clear example of the "multiple site coincidences", previously found in LEED [48], in a photoelectron diffraction study of PF_3 on Ni(111) [49]; the problem of uniqueness can be overcome in both techniques by appropriate enlargement of the data set. By contrast, incidentally, it now seems rather clear that the structure of chemisorbed oxygen on Cu(100) obtained in the original Fadley X-ray photoelectron diffraction experiment [7] was essentially correct, but a later (short-range) scanned energy mode experiment from the Shirley group [50] led to the wrong structural solution, although the results of experiments using other techniques appeared, for some time, to support this incorrect model. The newly accepted model in this case is based on measurements by several methods, including a more extensive scanned energy mode experiment [51] and LEED [52] analyses. It is, however, a reflection of the relative maturity of the scanned energy mode photoelectron diffraction method that much

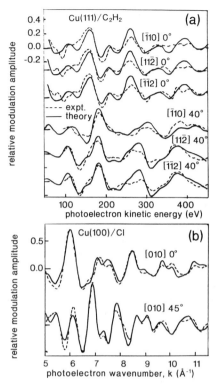

Fig. 4. Examples of the results of recent surface structure determinations using scanned energy mode photoelectron diffraction for the adsorption of acetylene on Cu(111) [54] and Cl on Cu(100) [53]. The results shown are the best fit thereoretical calculations compared with experimental data taken in several emission geometries (different azimuths and polar emission angles). In the case of the Cu(100)/Cl system, the curves are shown in terms of photoelectron wavenumber rather than energy, and the experimental curves have been Fourier filtered.

larger data sets, and more sophisticated structural optimisation methods, are now becoming integrated into the methodology. Two examples of recent results using scanned energy mode photoelectron diffraction are given in Fig. 4 which shows the best fits between experiment and theory for two rather different structural problems. Fig. 4b shows Cl 1s data taken from Cl adsorbed on Cu(100); in this case the fit has been optimised by adjusting structural parameters associated with distortion of the Cu atom positions around the adsorbed atomic Cl [53]. In Fig. 4a are shown similar experiment/theory comparisons for C 1s emission from acetylene adsorbed on Cu(111); here the principal structural parameters are those associated with the molecular adsorption site and its internal structure, and a large data set is used to minimise the possibility of non-uniqueness in the solution [54].

One rather significant new advance which has emerged from our own work is the exploitation of not only the elemental specificity intrinsic in the photoelectron diffraction method, but also its potential for *chemical state specificity*. By tuning to different chemically shifted core levels of the same element in a complex molecular adsorption system, one can obtain chemically specific structural information. Using this chemical state photoelectron diffraction we have investigated independently the site of the methyl and carboxyl carbon atoms in the acetate (CH_3COO-) and trifluoroacetate (CF_3COO-) species on Cu(110) [55], and the individual adsorption sites of coadsorbed PF_3, PF_2, and PF fragments on Ni(111) [56]. We should perhaps note that this is not the first time that differently shifted core level photoemission peaks have been studied in photoelectron diffraction; for example, surface core level shifts have been studied in this way on tungsten surfaces [57–59]. However, these surface core level shift studies were conducted in a very narrow energy range, leading to a very small data set; in addition, only one of these studies [58] appears to use a realistic calculation to interpret the data. Our recent work on chemical shift photoelectron diffraction provides a more meaningful pointer to what may well be one of the most important areas of development and exploitation of photoelectron diffraction in the future.

4. Auger electron diffraction

Insofar as Auger electron emission provides an alternative method of producing a local source, centred on an atom of a specific element, of electrons of well-defined kinetic energy, the idea that Auger electron diffraction should provide a technique of similar capability to photoelectron diffraction is an obvious one. Of course, Auger electrons occur at fixed electron energies, so a scanned energy mode experiment is not possible,

but in this regard the constraints are not significantly worse than those of photoelectron diffraction using a laboratory X-ray source, and intense electron-excited Auger electron peaks are simpler to generate in the laboratory than similar intensity XPS peaks. For low energy backscattering mode experiments, however, a significant complication in Auger electron diffraction is that of the form of the initial, unscattered, outgoing electron wavefield.

For the author, at least, this realisation came very early because our first experiments in local source diffraction were conducted using Auger electrons, and it was the difficulty of providing a good theoretical description of the unscattered wave that encouraged us to move to core level photoemission. Our original "discovery" of Auger electron diffraction [4,60,61] came about through pure curiosity. Like many surface scientists in the early 1970s, we saw the development of the technique of Auger electron spectroscopy as a major bonus because, for the cost of some relatively simple electronics, we could use our existing LEED optics as a retarding field analyser and obtain characterisation of surface composition as well as order. In our case, we had built a small movable (energy filtering) Faraday cup to measure LEED intensities, and we used this to measure angle-resolved Auger electron spectra on the low energy (63 eV) $M_{2,3}M_{4,5}M_{4,5}$ Cu Auger emissions from Cu(100) and (111) surfaces. Strong angular effects were found, but because the transition we studied involved valence electron states ($M_{4,5}\equiv 3d$), it was far from clear whether the effect was predominantly due to initial or final state (diffraction) effects.

Eventually we performed single scattering calculations [61] of the diffraction component of the problem assuming the initial unscattered wavefield could be described by a simple isotropic s-wave, and found that although detailed agreement with experiment was poor, the results certainly demonstrated that the diffraction process was capable of producing angular effects of the same scale as those seen experimentally. The paper of Liebsch referred to earlier [1] was published about this time, and almost certainly triggered our recognition of the structural potential of the measurements for adsorbate studies, although it seems that, in common with other authors at the time, we failed to fully appreciate the general significance of this paper, perhaps because it was couched in terms of rather low energy electrons [62]. In the early stages of angle-resolved photoemission, which was developing at this time, one theme being pursued was the extent to which emission and polarisation angular dependences in valence electron emission could provide local structural information on adsorbates. Rereading a number of papers written in the mid-1970s it seems that several authors thought that Liebsch's paper was concerned with this problem, although careful reading makes it clear that it is concerned with localised state emission.

Other authors pursued the idea of Auger electron diffraction much further than we had done, and showed that the angular distribution resulting from final-state scattering was sensitive to the dominant l and m quantum numbers characterising the initial unscattered wave [63–65]. More importantly, it became clear that Auger selection rules were consistent with the dominant angular momentum states in the unscattered wave being those which gave the best description of the experimental Auger angular distributions, so that a rather good quantitative description of the results of low energy Auger electron diffraction patterns from several substrate materials could be achieved [64,66]. Our own work failed to capitalise on this, however, because we had already turned our attention to core level photoemission for which this initial wave was well-understood in terms of simple dipole transitions, and which offered the advantage of kinetic energy tunability using synchrotron radiation excitation. These advantages may be the reason why little further effort has been devoted to backscattering Auger electron diffraction as a method of adsorbate structure determination.

By contrast to the complexities of interpreting low energy Auger electron angular distributions, the situation at high kinetic energies is much simpler. A number of studies concerned mainly with XPD in epitaxial layers has shown that high kinetic energy Auger electrons show a very similar angular distributions to those of high energy photoemission (e.g., Fig. 3b). The reason for this is probably that when the electron energy is high enough for forward scattering to dominate the angular distributions, the subtleties of the initial unscattered wavefield become relatively unimportant. In this area of application, work typically does not distinguish between X-ray photoelectron and Auger electron diffraction. Moreover, as the forward scattering peaks seen at high electron kinetic energies are zero order diffraction features, the angular location of which is therefore energy independent, the fact that Auger electrons are only emitted at fixed energies is not a restriction on their utility. On the other hand, all of these experiments are conducted with incident X-ray excitation; this is at least in part due to the fact that with photoionisation there are no very strong angular effects in the excitation process, so measurements can be made simply by rocking the sample crystal. By contrast there are strong incident beam diffraction effects with electron incidence [67] which demand that the experiment be conducted with a fixed incident geometry.

One further aspect of the low energy Auger electron diffraction problem, however, has been its "rediscovery" in recent years by the group of Hubbard, Frank and colleagues, at the University of Cincinnati [68], who have nevertheless chosen to ignore the diffraction effect itself and to interpret the results in terms of "atom shadowing". The heated debate which this work has produced has an almost religious fervour, and is still continuing despite published discussions [69], and even new experimental and theoretical papers to explain the diffractive nature of the process and the fact that it already rather well understood [70–72]. Indeed, it is notable in this debate that not only has the Cincinnati group "rediscovered" the phenomenon and chosen to discount all current theory, but also some of the papers correcting their interpretation have "rediscovered" ideas established in the 1970's [63,64]. In essence, the Cincinnati group attempts to interpret their results in terms of a classical shadowing picture which could, perhaps, be rationalised as a manifestation of inhomogeneities in the inelastic scattering within the solid medium. The fact that the angular dependence can be accounted for entirely in terms of a proper description of the unscattered wavefield together with a standard scattering calculation appears to have been ignored. Indeed, they make no attempt to include this known important piece of physics. One issue which has clouded the arguments is that the forward scattering effect seen at high energies has been widely understood as a *general* consequence of diffraction, so the fact that one often observes minima along interatomic vectors in low energy Auger electron diffraction has been taken as evidence that the diffraction explanation is not correct. These forward scattering minima can arise from some combination of the very different scattering phase shifts at low energies and the different form of the unscattered wave which modifies the importance of different scattered partial wave components. A full description of the controversy is not really appropriate here, as we are mainly concerned with constructive developments. On the other hand, efforts to convince the whole scientific community that Auger electron angular distributions can be readily understood in terms of simple local diffraction have led to a few new experiments and theoretical treatments which provide particularly clear demonstrations of the effects of the complex unscattered source wave in low energy core–valence–valence Auger emissions. For example, Barton and Terminello have actually measured the angular distribution of not only the $Cu\, M_{2,3}M_{4,5}M_{4,5}$ Auger emission from a Cu(100) surface (as was measured, in less detail, 20 years earlier [4,60,61]), but also photoemission from the Cu 3p level on the same crystal and at the same emitted electron kinetic energy [70]. The considerable difference of the two angular patterns shows rather clearly the effect of the source wave difference [67].

5. Direct methods and photoelectron holography

One of the limitations of any technique based on coherent interference in low energy electron scattering is that the method of structure determination is based on a trial and error process involving comparisons between the results of experiments and theoretical simulations for trial structures. This has a number of associated problems; notably, if the theoretical calculations are complex, the time and expense of calculating a sufficiently large set of trial structures can be very large, but in addition the task of optimising the structure may become unrealistic if the number of structural variables is too large. Even worse, the final result can only be as good as the imagination of the researcher, in that if the true structure is never used as a trial, the best fit will inevitably be to the wrong structure; a related problem is that of optimising on a structure which lies in the wrong part of the structural parameter space, but is locally the best fit. All of these problems are well known in LEED, but also apply to true photoelectron diffraction. Notice, of course, that the (0°) forward scattering XPD experiment does not suffer from these problems in that maxima in the measured angular distribution can be related *directly* to real-space atom pair directions. The reason that this technique does not fit the pattern is that, insofar as the angular distribution is found to be dominated by zero order diffraction peaks, it is not true diffraction at all, and because of the narrow forward scattering peak in the electron scattering cross-section, the technique would work even if electrons did not suffer wave interference!

The problems of the indirect approach to structure determination associated with LEED and photoelectron diffraction has led to searches for *direct methods*; i.e. methods in which direct processing of the experimental data leads to the structural solution. In X-ray diffraction, this is (approximately) possible through simple Fourier transforms, because all scattering phase interferences can be associated with scattering path-length differences. In electron scattering, scattering phase-shifts (and multiple scattering) prevent the use of such a simple approach. One idea which has gained very considerable attention in the last few years is to view a photoelectron diffraction angular distribution as a *photoelectron hologram*. Insofar as the angular distribution results from the interference of a reference wave (the direct outgoing photoelectron wavefield) and scattered object waves (from the surrounding atoms in the surface) this is clearly correct. The question, of course, is whether there is a viable route to the mathematical reconstruction of the image of the surface from this holographic record. The first suggestion that this view of photoelectron diffraction may offer a practical route to surface structure analysis was made by Barton [73], who himself was following up a more general idea by Szöke [74]. In effect, Barton proposed a method of Fourier transforming such angular distributions, and in his original paper he performed this transform on simulated data for backscattering photoelectron diffraction of S on Ni(100), obtained from the results of a multiple scattering calculation. The results of this inversion are shown in Fig. 5; the "images" do show features close to (within 0.2–0.3 Å of) the correct position of scatterer atoms included in the original simulation, but they also show quite a lot of structure of comparable intensity which is not located near atom sites. In addition, the resolution of the "atomic images" is only about 0.5 Å parallel to the surface, and 2 Å perpendicular to the surface. The utility of such images is debatable, but considerable effort (with some success) has been devoted to producing more effective inversions by addressing some of the problems inherent in the process. For example, atomic scattering phase shifts (which are energy, angle, and species dependent) will cause atomic displacements in a simple Fourier transform, multiple scattering will produce similar effects and add extra structure, and the limited electron momentum range of typical detection methods will produce poor resolution in images and can introduce spurious structure depending on the form of the data truncation.

At the present time it is a little difficult to tell whether these advances should be viewed with guarded optimism or healthy scepticism. This author's subjective view is slightly coloured by events of the 1970s, when the idea that simple Fourier transforms could be used to treat LEED data emerged, notably from Mason's group at Sussex [75], but was taken up in a more serious way by Landman and Adams [76,77] and for a few years was hailed as the way ahead for LEED. There

were good reasons at the time for believing that this was unlikely to be fruitful [78,79], and this is the view that prevailed with the passage of time. Holographic inversion in photoelectron diffraction therefore has a ring of familiarity, particularly as the idea has been generalised to possible applications in LEED [80]. Indeed this scientific background probably contributed to the author concluding, in a conversation at Bell Labs in 1975 with Patrick Lee, that Lee's suggestion that our Auger angular distributions were really electron holograms was probably correct but not useful! There is no doubt, however, that tests of more recent ideas for inverting photoelectron diffraction data, using simulated data, do look more promising than the early results of Fig. 5 might suggest.

To make a realistic evaluation of the prospects for routine use of photoelectron holography, we should first distinguish between the situations for forward scattering and backscattering photoelectron diffraction. In true forward (zero-order) photoelectron diffraction, it is difficult to see any substantial purpose for holographic inversion. The zero-order diffraction peaks provide direct structural information without any need for complex computational processing. Indeed, because the zero-order features carry no pathlength difference information, they must be excluded from the data if one is to perform a holographic inversion on the much weaker first and higher order diffraction features which may remain. This means removing the primary *direct* structural information in order to apply complex processing to the weak remaining information which is not (easily) directly interpretable. In principle, this procedure would allow one to determine forward scattering interatomic bondlengths as well as angles, but in general it seems that in the forward scattering geometry, successful holographic inversion methods are a technique in search of a problem. Indeed, the only obvious reason for any serious effort having been devoted to this aspect of photoelectron holography seems to be the availability of several very beautiful sets of high quality experimental data (which can be obtained on a conventional laboratory XPS instrument); attempts to perform holographic inversions of

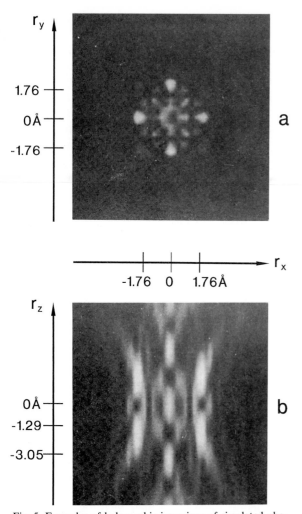

Fig. 5. Examples of holographic inversions of simulated photoelectron diffraction data for S 1s emission from S adsorbed on Ni(100) taken from Barton's original paper [73]. The S emitter is 1.29 Å above the hollow site, so the xy-plane section of the image shown in (a) 1.29 Å below the emitter passes through the top Ni layer. The four main features seen are associated with the nearest neighbour Ni atoms which should occur at coordinates (0, 1.76), (−1.76, 0) etc. (in Å). (b) Section through the emitter perpendicular to the surface. In this z-plane the emitter lies at (0, 0), the nearest-neighbour Ni atoms lie at (−1.76, −1.29) and (1.76, −1.29) and the second layer Ni atom below the emitter lies at (0, −3.05). Note the elongation of the "atomic images" perpendicular to the surface, the offset positions of atoms, and the many spurious features, in these early images.

these data have led to conflicting results (e.g., Refs. [81–83]).

In the backscattering geometry, the potential

utility of an effective direct method is beyond doubt. Moreover, this is a true diffraction regime, so although there are several basic problems to overcome as mentioned above (to which one must add the non-spherical reference (source) wave in a holography experiment), we do know that the angular pattern contains the appropriate phase information. Unfortunately, collecting the large amount of data in just one wide-angle distribution on a fine angular mesh, as has been deemed to be necessary, is a major task. This is presumably why, at least for an adsorbate system, there are still no good quality experimental data and all the developments in methodology have used simulated data. The problem of obtaining appropriate experimental data, however, is being exacerbated by the fact that the more recently proposed inversion methods which hold the promise of more meaningful "images" (with improved resolution and less spurious features) are based on the use of not just one wide-angle hologram, but perhaps 10 or more such holograms recorded at different photoelectron kinetic energies (e.g., Refs. [84–87]).

In the author's view these developments are in danger of losing sight of the objective of photoelectron hologram inversion. Certainly in the early work, holographic inversion was seen as a potential rapid and direct route to an *approximate* structure determination, which could then be refined by more conventional (indirect modelling) methods. This would overcome the major weakness with the trial and error indirect route to structure determination in which the initial search of structural parameter space is "blind". Holographic inversions were not anticipated to provide the ultimate precision currently obtainable by full multiple scattering modelling of the raw experimental data. Even the best inverted images being obtained (from simulated data) by the most recent methods do not change this expectation. In the light of this, one must call into question the practical utility of a first-order route to structure determination which requires some one to two orders of magnitude more experimental data points than is needed for the precise indirect method.

In fact there is recent evidence that simpler

methods are available to obtain a first-order structure determination, although these are not true holographic inversions. Indeed, the group of Shirley and his coworkers have, for some years, rationalised their first-order structural models in terms of the main peaks in a Fourier transform of one, or a small number of, scanned energy mode photoelectron diffraction spectra. The extent to which it really is possible to recognise the adsorption site from the main pathlength differences (uncorrected for the effect of scattering phase shifts) however, has been questioned (e.g., Ref. [37]). A slightly different approach is to try to identify the *directions* of the nearest neighbour backscattering substrate atoms. At intermediate electron energies (i.e., in the 100–400 eV range typically used in the scanned energy experiments), the modulus of the atomic electron scattering cross-section not only has a peak in the forward (0°) scattering direction, but also has a 180° backscattering peak. This means that scanned energy mode spectra collected along a nearest-neighbour backscattering direction are typically dominated by the influence of this single scatterer and show particularly strong and long energy periodicity oscillations. One way of identifying such directions is to measure angular scans at the photoelectron energies at which the 180° backscattering condition gives rise to exact constructive interference; for a known interatomic (adsorbate–substrate) spacing, it is easy to calculate the energy at which this should occur, and a polar angle scan at this energy should reveal a backscattering peak along the interatomic direction. Simple tests suggest that this method may be quite successful [88], but an alternative derivative of this idea may be more generally applicable. This [89] relies on identifying the backscatterer directions by the strong nearest-neighbour peak in a Fourier transform of a scanned energy mode spectrum taken in this direction. By recording a series of scanned energy spectra in different directions and taking Fourier transforms, one can identify the backscatterer direction as that giving the largest nearest-neighbour peak in the transforms. Because this method relies only on the nearest-neighbour peak in the transform, and is primarily concerned with identifying the direc-

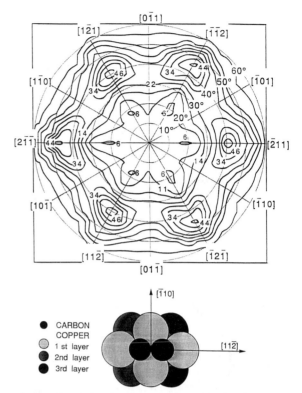

Fig. 6. Azimuthal and polar angle map of the location of the nearest-neighbour substrate scatterers for the case of C1s emission from acetylene adsorbed on Cu(111), obtained by mapping the amplitude of the shortest (4 Å) pathlength feature in Fourier transforms of scanned energy mode photoelectron diffraction spectra [54]. The local adsorption structure is shown in the schematic plan view of the surface; two different hollow adsorption sites are occupied by the C atoms, leading to the 6 nearest-neighbour peaks in the ⟨211⟩ azimuths.

tion, rather than the distance of these neighbours, it is rather insensitive to the problems of phase shift corrections and multiple scattering. An illustration of the utility of this method is provided by data from the acetylene/Cu(111) adsorption system referred to earlier [54]. Fig. 6 shows the results of mapping the amplitude of the main (4 Å effective pathlength difference) peak in the Fourier transform of a series of C1s scanned energy mode diffraction spectra recorded at 24 different emission geometries. Six main backscatterer peaks can be seen in the ⟨211⟩ azimuths at approximately 40° polar angle; these can be attributed to the Cu nearest neighbours in the two different hollow adsorption sites occupied by the C atoms, as seen in the plan view of the surface structure shown in Fig. 6. In principle, of course, this approach uses photoelectron diffraction intensities measured at many angles and energies, as do the more recently proposed holographic methods. The size of the energy and angular mesh on which the data points are required is, however, very much coarser for this approximate method. Tests of this approach on experimental data from several adsorption systems have already proved remarkably successful [54,90].

6. Conclusions

Photoelectron diffraction has developed hugely from its early beginnings almost 20 years ago. The initial basic ideas have not only been demonstrated in model experiments, but at least two different methodologies, the simple zero-order diffraction forward scattering, and the more complex but more generally applicable backscattering approach, have been developed to a point where they are being used to elucidate surface structural problems of increasing importance. High energy forward scattering Auger electron diffraction is essentially integrated with the equivalent photoelectron diffraction experiment, while we have achieved a much clearer understanding of the backscattering version of the Auger electron diffraction experiment; in this latter case, however, there has been little development of the method as a true means of surface structure determination, perhaps because the information data base available at one or two specific electron energies, would prove too small for truly quantitative studies, but also because it is not clear that very simple rules exist for the proper pre-description of the outgoing source wave for adsorbate emissions involving valence states. Novel ideas for the use of direct methods of analysis of the backscattering geometry, in particular, have been developed; it is too early to predict the full impact that these will have, and in particular, the utility of photoelectron hologram inversion remains to be established, but there are already

indications that some of the simpler direct methods may help to reduce the computation effort involved in structure determination using photoelectron diffraction.

Of course, this period of development has seen major advances in other structural methods. LEED methodology, in particular, has developed hugely, helped by approximate methods and a remarkable increase in the computing power-to-cost ratio over the last 20 years. Structures of increasing complexity in multi-dimensional parameter space are being tackled with LEED. Nevertheless, it seems likely that the insensitivity to long-range order and the elemental and chemical specificity offered by photoelectron diffraction will cause this technique to have a growing impact in the field of surface structure determination in the future. Chemical shift photoelectron diffraction, in particular, will benefit hugely from the combined improvements in both photon flux and photon energy resolution which will become available on undulator beamlines on the third generation of synchrotron radiation sources now being built. We can anticipate that the further development of chemical shift photoelectron diffraction will have a major impact on our ability to study complex problems of surface chemistry.

Acknowledgements

Twenty years of research in photoelectron and Auger electron diffraction has only been possible through collaboration with many other scientists including research students and postdoctoral researchers in my own laboratory. It is a pleasure to acknowledge the role of these many colleagues. Synchrotron radiation research also involves a lot of travelling to central facilities and collaboration with scientists outside one's own institution, and in my case has often involved fruitful interactions with, and assistance from, staff at the Tantalus facility at the University of Wisconsin-Madison, at the Synchrotron Radiation Source at Daresbury, and at the BESSY facility in Berlin. Two individuals have been particularly closely involved in my own efforts in this field: Neville Smith at Bell Labs, with whom I first became involved in trying to demonstrate the viability of the method, and Alex Bradshaw at the Fritz Haber Institute with whom I developed a collaboration which continues to produce exciting new results. I owe a special debt to these two scientific friends. Finally, I am happy to acknowledge the financial support of both the Science and Engineering Research Council and the European Community, in the form of research grants to sustain this work.

References

[1] A. Liebsch, Phys. Rev. Lett. 32 (1974) 1203.
[2] K. Siegbalin, V. Gelius, H. Siegbahn and E. Olson, Phys. Scr. 1 (1970) 272.
[3] C.S. Fadley and S.A.L. Bergström, Phys. Lett. A 35 (1971) 375.
[4] L. McDonnell and D.R. Woodruff, Vacuum 22 (1972) 477.
[5] D.P. Woodruff, D. Norman, B.W. Holland, N.V. Smith, H.H. Farrell and M.M. Traum, Phys. Rev. Lett. 41 (1978) 1130.
[6] S.D. Kevan, D.H. Rosenblatt, D. Denley, B-C. Lu and D.A. Shirley, Phys. Rev. Lett. 41 (1978) 1565.
[7] S. Kono, S.M. Goldberg, N.F.T. Hall and C.S. Fadley, Phys. Rev. Lett. 41 (1978) 1831.
[8] C.H. Li and S.Y. Tong, Phys. Rev. B 19 (1979) 1769.
[9] D.H. Rosenblatt, S.D. Kevan, J.G. Tobin, R.F. Davis, M.G. Mason, D.A. Shirley, J.C. Tang and S.Y. Tong, Phys. Rev. B 26 (1982) 3181.
[10] M.R. Howells, D. Norman, G.P. Williams and J.B. West, J. Phys. E 11 (1978) 199.
[11] See, e.g., P.J. Orders, R.E. Connelly, N.F.T. Hall and C.S. Fadley, Phys. Rev. B 24 (1981) 6163.
[12] L.-G. Petersson, S. Kono, N.F.T. Hall, C.S. Fadley and J.B. Pendry, Phys. Rev. Lett. 42 (1979) 1545.
[13] P.J. Orders, S. Kono, C.S. Fadley, R. Trehan and J.T. Lloyd, Surf. Sci. 119 (1981) 371.
[14] E. Holub-Krappe, K.C. Prince, K. Horn and D.P. Woodruff, Surf. Sci. 173 (1986) 176.
[15] D.A. Wesner, F.P. Coenen and H.P. Bonzel, Phys. Rev. Lett. 60 (1988) 1045.
[16] R.S. Saiki, G.S. Herman, M. Yamada, J. Osterwalder and C.S. Fadley, Phys. Rev. Lett. 63 (1989) 283.
[17] A. Nilsson, H. Tillborg and N. Mårtensson, Phys. Rev. Lett. 67 (1991) 1015.
[18] A.V.de Carvalho, M.C. Asensio and D.P. Woodruff, Surf. Sci. 273 (1992) 381.
[19] L.S. Caputi, R.G. Agostino, A. Amoddeo, E. Colavita and A. Santaniello, Surf. Sci. 282 (1993) 62.
[20] U. Grosche, H. Hamadeh, O. Knauff, R. David and H.P. Bonzel, Surf. Sci. Lett. 281 (1993) L341.
[21] W.F. Egelhoff, Jr., Phys. Rev. B 30 (1984) 1052.
[22] W.F. Egelhoff, Jr., Crit. Rev. Solid State Mater. Sci. 16 (1990) 213.
[23] S.A. Chambers, Adv. Phys. 40 (1991) 357.
[24] C.S. Fadley, in: Synchrotron Radiation Research: Advances in Surface and Interface Science, Vol. 1, Ed. R.Z. Bachrach (Plenum, New York, 1992).
[25] M.-L. Hu, J.J. Barton and M.A. Van Hove, Phys. Rev. B 39 (1989) 8275.
[26] A.R Kaduwela, U.S. Herman, D.J. Friedman, C.S. Fadley and J.J. Rehr, Phys. Scr. 41 (1990) 948.
[27] N.V. Smith, H.H. Farrell, M.M. Traum, D.R. Woodruff, D. Norman, M.S. Woolfson and B.W. Holland, Phys. Rev. B 21 (1980) 3119.
[28] B.W. Holland, J. Phys. C 8 (1975) 2679.
[29] R.S. Zimmer and B.W. Holland, J. Phys. C 8 (1975) 2395.
[30] H.H. Farrell, M.M. Traum, N.V. Smith, W.A. Royer, D.P. Woodruff and P.D. Johnson, Surf. Sci. 102 (1981) 527.
[31] W.M. Kang, C.H. Li and S.Y. Tong, Solid State Commun. 36 (1980) 149.
[32] D.H. Rosenblatt, S.D. Kevan, J.G. Tobin, R.E Davis, M.O. Mason, D.R. Denley, D.A. Shirley, Y Huang and S.Y. Tong, Phys. Rev. B 26 (1982) 1812.
[33] J.J. Barton, C.C. Bahr, Z. Hussain, S.W. Robey, J.G. Tobin, L.E. Klebanoff and D.A. Shirley, Phys. Rev. Lett. 51 (1983) 272.
[34] J.J. Barton, C.C. Bahr, S.W. Robey, Z. Hussain, E. Umbach and D.A. Shirley, Phys. Rev. B 34 (1986) 3807.
[35] D. Norman, D.P. Woodruff, C. Norris and G.P. Williams, J. Electron Spectrosc. Relat. Phenom. 14 (1978) 231.
[36] U. Fano and J.W. Cooper, Rev. Mod. Phys. 40 (1968) 441.
[37] P.J. Orders and C.S. Fadley, Phys. Rev. B 27 (1983) 781.
[38] D.P. Woodruff, Surf. Sci. 166 (1986) 377.
[39] M.D. Crapper, C.E. Riley, P.J.J. Sweeney, C.F. McConville, D.P. Woodruff and R.G. Jones, Surf. Sci. 182 (1987) 213.
[40] J.J. Barton and D.A. Shirley, Phys. Rev. B 32 (1985) 1892.
[41] J.J. Barton and D.A. Shirley, Phys. Rev. B 32 (1985) 1909.
[42] J.J. Barton, S.W. Robey and D.A. Shirley, Phys. Rev. B 34 (1986) 778.
[43] See, e.g., E. Dietz, W. Braun, A.M. Bradshaw and R. Johnson, Nucl. Instrum. Methods A 239 (1985) 359.
[44] C.F. McConville, D.R Woodruff, K.C. Prince, G. Paolucci, V. Chab, M. Surman and A.M. Bradshaw, Surf. Sci. 166 (1986) 221.
[45] S.D. Kevan, R.F. Davis, D.H. Rosenblatt, J.G. Tobin, M.G. Mason, D.A. Shirley, C.H. Li and S.Y. Tong, Phys. Rev. Lett. 46 (1981) 1629.
[46] L. Becker, S. Aminpirooz, B. Hillert, M. Pedio, J. Haase and D.L. Adams, Phys. Rev. B 47 (1993) 9710.
[47] K.-M. Schindler, K.-U. Weiss, P. Dippel, P. Gardner, A.M. Bradshaw, M.C. Asensio and D.P. Woodruff, to be published.

[48] S. Andersson and J.B. Pendry, Solid State Commun. 16 (1975) 563.

[49] R. Dippel, K.-U. Weiss, K.-M. Schindler, D.P. Woodruff, P. Gardner, V. Fritzsche, A.M. Bradshaw and M.C. Asensio, Surf. Sci. 287/288 (1993) 465.

[50] J.G. Tobin, L.E. Klebanoff, D.H. Rosenblatt, R.F. Davis, E. Umbach, A.G. Baca, D.A. Shirley, Y. Huang, W.M. Kang and S.Y. Tong, Phys. Rev. B 26 (1982) 7076.

[51] M.C. Asensio, M.J. Ashwin, A.L.D. Kilcoyne, D.P. Woodruff, A.W. Robinson, Th. Lindner, J.S. Somers, D.E. Ricken and A.M. Bradshaw, Surf. Sci. 236 (1990) 1.

[52] H.C. Zeng and K.A.R. Mitchell, Surf. Sci. 239 (1990) L571.

[53] L-Q. Wang, A.E. Schach von Wittenau, Z.G. Li, L.S. Wang, Z.Q. Huang and D.A. Shirley, Phys. Rev. B 44 (1991) 1292.

[54] S. Bao, K.-M. Schindler, Ph. Hoftnann, V. Fritzsche, A.M. Bradshaw and D.P. Woodruff, Surf. Sci. 291 (1993) 295.

[55] K.-U. Weiss, R. Dippel, K.-M. Schindler, P. Gardner, V. Fritzsche, A.M. Bradshaw, A.L.D. Kilcoyne and D.P. Woodruff, Phys. Rev. Lett. 69 (1992) 3196.

[56] K.-U. Weiss, R. Dippel, K.-M. Schindler, P. Gardner, V. Fritzsche, A.M. Bradshaw, D.P. Woodruff, M.C. Asensio and A.R. González-Elipe, Phys. Rev. Lett. 71 (1993) 581.

[57] D. Sébilleau, G. Tréglia, M.C. Desjonquères, D. Spanjaard, C. Guillot, D. Chauveau and J. Lecante, J. Phys. (Paris) 49 (1988) 227.

[58] Y. Jugnet, N.S. Prakash, L. Porte, T.M. Duc, T.T.A. Nguyen, R. Cinti, H.C. Poon and G. Grenet, Phys. Rev. B 37 (1988) 8066.

[59] D. Chauveau, C. Guillot, B. Villette, J. Lecante, M.C. Desjonquères, D. Sébilleau, D. Spanjaard and G. Tréglia, Solid State Commun. 69 (1989) 1015.

[60] B.W. Holland, L. McDonnell and D.P. Woodruff, Solid State Commun. 11 (1972) 991.

[61] L. McDonnell, D.P. Woodruff and B.W. Holland, Surf. Sci. 51 (1975) 249.

[62] D.P. Woodruff, Surf. Sci. 53 (1975) 538.

[63] R.N. Lindsay and C.G. Kinniburgh, Surf. Sci. 63 (1977) 162.

[64] D. Aberdam, R. Baudoing, E. Blanc and C. Gaubert, Surf. Sci. 71 (1978) 279.

[65] J.M. Plociennik, A. Barbet and L. Mathey, Surf. Sci. 102 (1981) 282.

[66] V. Fritzsche, J. Phys. Condensed Matter 2 (1990) 9735.

[67] See, e.g., A.F. Armitage, D.P. Woodruff and P.D. Jolinson, Surf. Sci. 100 (1980) L483.

[68] See, e.g., D.G. Frank, N. Batina, T. Golden, F. Lu and A.T. Hubbard, Science 247 (1990) 182.

[69] See, e.g., Letters by S.A. Chambers; W.F. Egelhoff, Jr., J.W. Gadzuk, C.J. Powell and M.A. Van Hove; X.D. Wang, Z.L. Han, B.P. Tonner, Y. Chen and S.Y. Tong; D.P. Woodruff, Science 248 (1990) 1129.

[70] L.J. Terminello and J.J. Barton, Science 251 (1991) 1281.

[71] J.J. Barton and L.J. Terminello, Phys. Rev. B 46 (1992) 13548.

[72] T. Greber, J. Osterwalder, D. Naumovic, A. Stuck, S. Hüfner and L. Schlapbach, Phys. Rev. Lett. 69 (1992) 1947.

[73] J.J. Barton, Phys. Rev. Lett. 61 (1988) 1356.

[74] A. Szöke, in: Short Wavelength Coherent Radiation: Generation and Applications, Eds. D.T. Attwood and J. Boker, AIP Conf. Proc. 147 (American Institute of Physics, New York, 1986).

[75] T.A. Clarke, R. Mason and M. Tescari, Surf. Sci. 30 (1972) 553.

[76] U. Landman and D.L. Adams, J. Vac. Sci. Technol. 11 (1974) 195.

[77] D.L. Adams and U. Landman, Phys. Rev. Lett. 33 (1974) 585.

[78] D.P. Woodruff, K.A.R. Mitchell and L. McDonnell, Surf. Sci. 42 (1974) 355.

[79] D.P. Woodruff, Disc. Faraday Soc. 60 (1975) 218.

[80] D.K. Saldin and P.L. de Andres, Phys. Rev. Lett. 64 (1990) 1270.

[81] G.R. Harp, D.K. Saldin and B.P. Tonner, Phys. Rev. B 42 (1990) 9199.

[82] S. Hardcastle, Z.-L. Han, G.R. Harp, J. Zhang, B.L. Chen, D.K. Saldin and B.P. Tonner, Surf. Sci. Lett. 245 (1991) L190.

[83] A. Stuck, D. Naumovic, H.A. Aebischer, T. Greber, J. Osterwalder and L. Schlapbach, Surf. Sci. 264 (1992) 380.

[84] J.J. Barton, Phys. Rev. Lett. 67 (1991) 3106.

[85] S.Y. Tong, H. Li and H. Huang, Phys. Rev. Lett. 67 (1991) 3102.

[86] B.P. Tonner, J. Zhang, X. Chen, Z.-L. Han, G.R. Harp and D.K. Saldin, J. Vac. Sci. Technol. B 10 (1992) 2082.

[87] L.J. Terminello, J.J. Barton and D.A. Lapiano-Smith, J. Vac. Sci. Technol. B 10 (1992) 2088.

[88] R. Dippel, D.P. Woodruff, X-M. Hu, M.C. Asensio, A.W. Robinson, K-M. Schindler, K-U. Weiss, P. Gardner and A.M. Bradshaw, Phys. Rev. Lett. 68 (1992) 1543.

[89] V. Fritzsche and D.P. Woodruff, Phys. Rev. B 46 (1992) 16128.

[90] K-M. Schindler, Ph. Hoftnann, V. Fritzsche, S. Bao, S. Kulkarni, A.M. Bradshaw and D.P. Woodruff, to be published.

Surface Science 299/300 (1994) 199–218
North-Holland

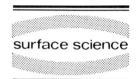

surface science

X-ray absorption spectroscopy applied to surface structure: SEXAFS and NEXAFS

P.H. Citrin

AT & T Bell Laboratories, Murray Hill, NJ 07974, USA

Received 9 June 1993; accepted for publication 19 July 1993

The application of both surface extended X-ray absorption fine structure (SEXAFS) for determining adatom–substrate distances and adsorption sites, and near-edge X-ray absorption fine structure (NEXAFS) for determining adsorbed molecular orientations and bonding changes, is simple, direct, and now routine. The events leading up to developing these methods, however, were neither simple nor direct, and by no means were they routine. Communities of X-ray spectroscopists, synchrotron radiation researchers, and surface scientists overlapped to create a particularly special and fertile environment. This article presents a brief overview of these scientific and sociological events from a personal perspective, with emphasis on some of the earlier history not familiar to many surface science practitioners. Comments about the capabilities, limitations, and accomplishments of SEXAFS and NEXAFS are also mentioned.

1. Introduction

When the Master Index of Surface Science premiered for volumes 1–10 (1964–1968) [1], the subject of "Surface Structure" was still just an idea. The topic was legitimized with its own (or at least shared) billing when it appeared in the next Master Index in 1970 [2], along with "see also, low energy electron diffraction". After fifteen years from the inception of Surface Science, the listing "X-ray absorption" appeared [3]; under that subject was a study on small clusters [4] using extended X-ray absorption fine structure (EXAFS). Immediately below, under a different heading entitled "X-ray *ad*sorption", was a paper on surface-EXAFS (SEXAFS) [5]. Such were the auspicious introductions of surface structure, X-ray absorption, and SEXAFS into this journal.

Typographical errors notwithstanding, we see that during the first half of Surface Science's thirty years, LEED was the only experimental technique for determining surface structure with atomic resolution. This was not for lack of interest. A large number of widely varying approaches have since been developed, including (i) scattering (low-/medium-/high-energy-ion-), (ii) dif-

fraction (high-energy-electron-; normal-/off-normal-/azimuthal-X-ray-photoelectron-; low-energy-He-atom-; glancing-angle-X-ray-), and (iii) microscopy (transmission-/scanning-electron-; field-ion-; scanning-tunneling-). Given this selection, it might be surprising to learn that (iv) X-ray absorption (SEXAFS, and the related technique of near-edge X-ray absorption fine structure (NEXAFS)) has been successfully applied thus far to more surface systems, both in number and variety, than any of these other methods. While this is certainly not the only criterion for evaluating the relative merits of a given approach (many others will be presented here and elsewhere), it does point out that SEXAFS and NEXAFS have had utility in a range of many different studies.

It might be tempting to ask: Since X-ray absorption is, after all, one of the oldest spectroscopies [6], is it really surprising to find it so widely applied to problems of surface structure? The answer is yes, arguably, for two reasons. The first is that, following the discovery of X-rays in 1896, it was diffraction, not absorption, which quickly found its use in solving crystal structures. The experiments were relatively simple, and the diffraction theories of von Laue and of Bragg are as correct now as they were then. By contrast, the structures near the absorption edge observed by Kossel in 1920, and those extending well beyond the edge described by Kronig in 1931, were difficult to measure and to reproduce (the measurements required weeks). Accordingly, it took more than forty years before a true theoretical understanding of these phenomena emerged. The second reason is that X-ray experiments on bulk systems using a stationary, low-powered X-ray source have historically been performed in a laboratory located somewhere near the researcher's home. This implies, in most cases, that experiments could be repeated in a timely manner, broken equipment could be fixed at leisure, and scientific problems could be discussed by the fireplace; a life of calm, and clean linens, was possible. Surface systems, on the other hand, are more dilute by factors of 10^4–10^6 and are often reactive (unstable), so a much more intense source of X-radiation is required, i.e., a synchrotron storage ring. The only such source at the time during which the SEXAFS and NEXAFS techniques were being developed was at Stanford, a source which was usually operated in a parasitic mode (unpredictable beam conditions) and was available only a few times a year in competitive allotments of a few weeks or less per experiment but – now the good news – for 24 hours a day. This implies, in most cases, all that can be imagined, including researchers who often became reactive (unstable).

This brief overview is intended to provide some insight into how and why a spectroscopy that was too laborious and unreliable to apply to structural problems, was not theoretically well described as a structural probe until quite recently, and was not convenient for or accessible to many experiments and experimenters until even more recently, evolved into one of the most accurate, straightforward, and widely used methods for determining the local structure of atoms and molecules adsorbed on surfaces. Since there are already several technical reviews on SEXAFS and NEXAFS [7–10], and since the goals of this special issue warrant a very different presentation, the emphasis here is more personal and historical rather than technical. Also, some accounts are

given about interesting scientific and sociological events and circumstances dealing with synchrotron radiation research which provided an essential background for developing SEXAFS and NEXAFS and which may not be generally familiar to the surface science community. More attention is given to the earlier history rather than later developments, because the latter are either covered elsewhere [7–10] or appear in other chapters of this issue. Finally, comments regarding capabilities, limitations, accomplishments, and directions of these two techniques are briefly summarized at the end.

2. Evolution

This historical account is presented largely from an individual perspective, so there will be an unavoidable imprinting of personal flavor, characterized by some color and a little strangeness (...still working on the charm). Accuracy is a goal, but the limitations of perspective and required brevity necessarily imply some omissions. Apologies are tendered for any unintentional slights.

2.1. Bulk EXAFS and NEXAFS: General

The evolution of SEXAFS and NEXAFS is much like the techniques themselves: similar in some respects, different in others, and often complementary. Before exploring the application of these two methods to surface structure, we first consider the general features of an absorption spectrum which serve to define them. Both types of "fine structure" are obtained in the same measurement, be it from an atom or a molecule, in the bulk or on the surface. Those structures less than 50 eV or so above the absorption edge are referred to as "near-edge", while those greater than about 50 eV are called "extended". The actual energy of this boundary is somewhat arbitrary, of course, but the reasons for introducing the division are not. In the near-edge regime, a low energy photoexcited core electron can populate Rydberg states, or unfilled molecular orbitals or bands of dipole-allowed symmetry.

Multi-electron satellites are possible, and multiple scattering is the norm. Near-edge structures are large, sometimes exceeding the absorption edge jump itself. In the extended regime, an energetic photoelectron can essentially be regarded as a plane wave which backscatters off the core-dominated potentials of neighboring atoms and interferes with the initial outgoing wave. Single scattering involving atomic potentials is the norm. The low probability for backscattering leads to weak extended structures, often $< 10\%$ of the edge jump. Understanding these very different features and their origins now, and recognizing that prior to the 1960's X-ray absorption data were obtained not only with low power sources but with photographic film for detection, it is easy to understand why Kronig (covering all bets) described those weakly oscillating intensities in the photographic film using a long-range order theory involving Bloch waves [11] *and* a short-range order theory involving nearby-atom scattering [12].

Significant advances in theoretically describing the extended and near-edge regions came many years later, at similar times, via different routes, but for the same reason: improved experimental data. In the mid-1960's, Lytle revived interest in X-ray absorption measurements [13] by implementing more modern detection and automation techniques. His higher-quality extended data, which could not be reconciled with a long-range order theory [11], prompted him to break with the term "Kronig structure" and coin the acronym "EXAFS" [14]. A collaboration with Sayers and Stern a few years later led to the intuitive and correct derivation of the basic EXAFS formula [15] in use today (see Ref. [16] for further review of EXAFS). Increased interest in the near-edge region of low-Z molecules, which are common adsorbates on surfaces (edge singularities in metals and chemical effects in compounds have a different history), came in the early 1970's with experiments by Van der Wiel and Brion [17]. These involved not X-rays, but electrons sufficiently energetic (> 1 keV) to behave like X-rays, in a technique called inner shell electron energy loss spectroscopy (ISEELS) [18]. Near-edge measurements of molecular nitrogen with this method [19] inspired an accurate theoretical treatment by Dehmer and Dill [20] (see Ref. [7] for further review).

2.2. Synchrotron radiation at SSRP: Users

At about the time when the essential understanding of the near-edge and extended absorption features was at hand in the early 1970's, a very different area of research was nearing a breakthrough, namely, synchrotron radiation. Following the 1960's, which saw synchrotron efforts emerge mainly in the vacuum-ultraviolet (VUV) and very soft X-ray region, harder X-rays were generated, briefly, in 1972 from the Cambridge Electron Accelerator in Boston. The same year, an NSF-sponsored panel recommended that a national synchrotron facility for X-radiation be built on a trial basis, operating in a "parasitic" mode to the Stanford Positron Electron Asymmetric Ring (SPEAR). By late 1973, the Stanford Synchrotron Radiation Project (SSRP), under Seb Doniach as director, was ready to accept research proposals from investigators ("Users") outside Stanford University. It was Doniach's job to prove this Project a success through substantial growth from capital input and scientific output. He had to find money and Users.

There are worse places than the Bay area in California for enticing Users. I was at (then) Bell Telephone Laboratories for more than a year and had come to know Doniach through my research in X-ray photoemission core levels and lineshapes (he and Šunjić had predicted asymmetric lineshapes in metals). He told me about some recent 8 keV X-ray photoemission results of Ingolf Lindau (then at Stanford working with Bill Spicer) using a somewhat makeshift "beamline" (it was outdoors, so he took data in the rain), and about a new beamline being built with Piero Pianetta (a graduate student of Spicer). I mentioned some research ideas to him and he asked me to propose these informally in a letter. Doing so, I unwittingly became the first official outside User of SSRP.

There were also invited "unofficial" outside User-institutions. In exchange for their valuable scientific manpower and experience – and equipment (remember that input) – they were to be rewarded with guaranteed amounts of valuable "beam time". Since there were no guarantees of success to such a User (it was very much a gamble), and since no administrative precedents existed for SSRP, there were initially few objections to this resourceful plan. Bell was one of several unofficial outside User-institutions (others included Xerox and China Lake). Representing Bell (wearing white trunks) was Peter Eisenberger, who at the time had been there about five years and was working on Compton scattering using a new 60 kW rotating anode (then the most intense laboratory X-ray source). He and I met in 1972 at an extraordinary Gordon Conference on Physics and Chemistry of Solids, where among other topics we heard Sayers talk about EXAFS. Shortly after, we interacted for a few months in an effort to improve upon that technique using his new X-ray source, but the measurements were still too time consuming and I lost interest. He might have lost interest as well, had it not been for SSRP. Initially planning to pursue Compton scattering there, Eisenberger instead became involved as part-time mentor to Brian Kincaid (a graduate student of Doniach). His thesis was going to focus on EXAFS measurements using a beamline that he and Sally Hunter (a graduate student of Artie Bienenstock) were in the process of building.

In mid-1974, I was on the mezzanine level of SSRP with Lindau and Pianetta trying to reproduce the earlier 8 keV photoemission results on their new beamline. It was frustrating enough for us then because our efforts turned out to be unsuccessful, but it became almost unbearable against the background of screaming whoops of joy coming from the EXAFS beamline immediately below us. Seemingly anything not nailed down was put into that beam line, and seemingly everything gave incredible results. The reason was clear. The tunable monochromatic beam intensity from SSRP was about 10^5 times greater than that from any laboratory source. In experimental science, rarely (if ever) has so large an improvement been realized in such a relatively short period of time. Measurements previously requiring weeks now took minutes. Synchrotron X-radiation research was about to explode [21].

For the SSRP administration, finding outside Users was now no longer a problem; explaining to Them (wearing black trunks) that substantial amounts of beam time had already been guaranteed to User-institutions was. SSRP found itself trapped between previous commitments to User-institutions and increasing pressure from other interested researchers. Compromises were eventually reached, of course, but not without some difficulties. One important outgrowth of all this was the proposal review process, where every User (in or out of Stanford) would be treated individually, not institutionally, and would have to compete for allotments of beam time based on peer-review ratings of formally written proposals. In the meantime, though, while these policies and compromises were being formulated, earlier beam time commitments were still being honored. In retrospect, I was very fortunate at the time not only to be on-site and see the beginnings of this burgeoning field of research, but equally important, to be associated with one of the User-institutions that would be receiving more of this precious beam time.

My earlier interest with Eisenberger in EXAFS and my recent overlap with him and Kincaid at SSRP led naturally to our three-way collaboration. Thinking about the most immediate scientific problems, it became clear that the accuracy of determining distances with EXAFS was limited by the accuracy of knowing the total phase shift of the photoelectron wave as it scattered off the cores of the absorbing and neighboring atoms. Earlier theoretical and empirical approaches to this problem had given only qualitatively accurate results. For EXAFS to be competitive with diffraction and provide meaningful information, distances needed to be determined to at least 0.02 Å. Fortunately, there was an easy test of whether EXAFS was in the running. If EXAFS really was described by single scattering off core potentials, then only the atomic environments would be important, i.e., the chemical nature of the atoms (their valence electrons) would not matter. Therefore, knowing the distance between absorbing and backscattering atoms in a model system allows for the empirical determination of the summed atomic phase shifts, and these, in turn, should be "chemically transferable" to any other system involving the same atom pair. Using the simplest gases for testing this concept (e.g., Br_2, Ge_2H_6, $BrGeH_3$) proved interesting, because some of the gases were pyrophoric and air-explosive. Happily, there were screams of joy (not death) from this beamline yet again as the experiments succeeded in demonstrating that EXAFS phase shifts were indeed chemically transferable.

The significance of this result was apparent in two ways. First, from a practical point of view, it meant that EXAFS could provide distances that were not only precise, i.e., reproducible, but also accurate, i.e., correct on an absolute scale (as compared with known diffraction results). Moreover, determining these distances was straightforward and purely empirical; all that was needed, apart from the requisite data from the system under study, was a model compound containing the relevant atom pair. Second, from a theoretical point of view, it meant that single scattering from atomic cores describes EXAFS extremely well, so previous discrepancies between theory and experiment must have been due to inadequacies in the calculations rather than in the physics. This point was seized upon by Patrick Lee, who had been at Bell about two years and was actively working on EXAFS theory at the time. Being in the advantageous position of learning about our results almost as soon as the experiments had been performed, he quickly developed an improved approach for calculating theoretical EXAFS phase shifts. This proved so successful, in fact, that it appeared there was little else left for him to do. Experimentally, EXAFS was rapidly finding applications in biological systems, heterogeneous catalysts, metallic alloys, and just about anything else that had previously been too difficult to study because long-range order was lacking. Were there no new theoretical or experimental areas left to which EXAFS could contribute?

2.3. Surface science: SEXAFS

We have seen that the early 1970's was an extremely active period for scientific growth and development, with the fields of X-ray absorption spectroscopy and synchrotron radiation research being just two examples. Surface science was no exception. A simple, and for this issue, appropriate way to gauge how the field of surface science was growing during that time is to look at the number of papers contributed to this journal. From the early- to late-1970's, the total number increased more than a factor of four [1–3]. In these papers it is possible to see the emergence of two major classes of study: the dynamics of adsorption/desorption of atoms or molecules on surfaces, and the preparation/characterization of clean and adsorbate-covered surfaces.

Within the category of surface characterization, three spectroscopies were dominant: Auger electron spectroscopy for atomic composition, photoelectron spectroscopy for electronic structure, and low energy electron diffraction (LEED) for geometric structure. In the early-1970's, much of the increased activity in using these spectroscopies, each already decades old by that time, can be linked to improvements in the vacuum and analyzer equipment needed to perform the measurements. (Obviously there were scientific motivations as well, discussed elsewhere in this issue.) The increased commercial availability of this equipment helped to attract many new investigators. By the mid-1970's, it was almost routine to determine which atoms were on a surface, how many of them there were, and what their general electronic properties were. Determining where they were located, however, was quite another matter. For all the relative simplicity and convenience of performing LEED measurements, it was very difficult to analyze the data. An elaborate computer simulation scheme was needed to account for all the multiple scattering pathways experienced by the low energy electrons. Structures were assigned on the basis of best agreement between simulated and actual data, and (prior to the introduction of standardized tests) much of the criteria for agreement was subjective. Several groups studying the same system were reporting different results. Recognizing the need for additional methods, researchers began to extend the application of photoemission, one of the most prevalent spectroscopies, to problems of surface structure. Prototypical systems were studied for simplicity and cross-comparison. As one familiar example then, many groups were engaged in simply determining whether or not CO adsorbed on Ni(100) with the C end down.

Sharing a common interest with Lee in looking for new applications of EXAFS, it took us little effort to realize that the structure of adsorbate-covered surfaces was an important and needed area to consider. The only question was how to make the measurement. At that time, taking the ratio of incident to transmitted photon intensities was the only method used to monitor absorption through gases or micron-thick foils or powders. This was clearly impractical for monolayer concentrations of adsorbates, even if the substrates could be appropriately thinned, because the amount of absorption by the adsorbate would be too small. A much more surface-sensitive method was needed. Lee set about calculating whether measuring the adsorbate photoelectrons would give rise to the exact EXAFS expression. He found that to cancel multiple-scattering cross terms in that expression, complete angular averaging over 4π sr was necessary [22], and so he abandoned that approach (the maximum collection in a real experiment is, of course, only 2π sr). While this was going on, another detection scheme for measuring dilute concentrations of absorbing atoms was being carried out at SSRP (from Users at Lawrence Berkeley Laboratory and at Bell), namely, fluorescence detection [23]. Here, the intensity of characteristic fluorescence X-rays from the absorbing atoms are measured, and this is just directly proportional to the absorption cross section of those atoms. Hearing this idea, Lee realized that measuring the intensity of the non-radiative by-product of absorption, namely, the characteristic Auger electrons, would also work. As a test case for this scheme, he calculated EXAFS spectra for Se on different Ni surfaces and showed how adsorption sites might be identified. It turns out that Landman and Adams [24] independently proposed using Auger electrons as well, but their focus was on using the small escape depths of these electrons to measure the structure of clean surfaces (this will not work because of unavoidable interference from Auger electrons in the bulk). They did, however, also

mention using Auger electrons from adsorbate atoms.

There were three substantial barriers to overcome before I could test this new application, then around early-1976. First, an appropriately equipped ultra-high vacuum chamber was needed at SSRP. Transporting explosive gases under dry ice seemed easy compared with this. Fortunately, much of the equipment would be graciously loaned to me by Spicer and his group and by Bob Madix (also at Stanford). Gary Schwartz (at Bell) would also provide a number of important components, as well as his own manpower in the initial experiment. Second, more beam brightness (intensity per unit area per unit solid angle) was needed to detect such low atom concentrations within the small acceptance size of the electron energy analyzer to be used. Fortunately again, Kincaid and Jerry Hastings (then on the SSRP staff) were in the process of building a new beam line incorporating a doubly-bent focusing mirror at that facility. The beamline would not be completed for several months, however, and after that they were going to use it themselves for several months more. Third, beam time was needed with X-radiation sufficiently energetic to pass through the 0.25 mm Be window isolating the ultra-high vacuum in the chamber from the atmospheric He pressure of the beamline. Here, however, I was not so fortunate. The problem was not administrative, it was scientific. The relatively low-energy psi/J particle had been co-discovered earlier at Brookhaven and at Stanford and, starting about the time the new beamline would be ready, SPEAR would be looking for more low energy particles. As a parasitic user-facility, SSRP would therefore be given much less high-energy X-ray beam. During this time, I would return to my other research interests (X-ray photoemission and X-ray edge singularities) and give some talks about EXAFS and proposed applications to surfaces.

The first time I spoke about possible surface studies using EXAFS was at a 1976 summer conference on electron spectroscopy. (Interestingly, an abstract submitted with Lee as co-author to a different meeting that summer was rejected on the grounds that it was too speculative.) I remember discussing Lee's model calculations for Se on Ni and noticing among the attendees an interested Dave Shirley (then at Berkeley) taking notes. I also remember that it was here the acronym "SEXAFS" was introduced. At that time, the mere mention of SEXAFS – a term assiduously avoided by Lee in his paper [22] but thoroughly enjoyed in the privacy of our own discussions – brought audience reactions ranging from embarrassed titters to boisterous laughter. Two months later, speaking about SEXAFS to a different cross section of attendees at a conference of the American Chemical Society, similar reactions were expressed. Name recognition would not be a problem with this technique.

By the time suitable X-ray beam time became available to me in February 1977, I had given enough talks on SEXAFS to feel as though the experiments had already been performed. Reality set in, however, and a surface system based on practical rather than theoretical considerations had to be chosen. Iodine adsorbed on Ag(111) was selected because Ag was easy to clean and the I L_3 edge at 4.5 keV was near the peak output intensity of the new focusing mirror beamline. Furthermore, that system had been studied with LEED [25], thereby providing a basis for comparison. The sample was dosed following the original study [25] to give a $\frac{1}{3}$-monolayer $(\sqrt{3} \times \sqrt{3})R30°$ structure, but no LEED apparatus was available at the time to confirm this. Also, the chamber had not been intended for use as a beamline end station, so it was not well equipped with adjustment mechanisms to move the apparatus and find the beam. Eisenberger and I had a truly memorable time with that exercise. Perhaps the most fun were our 24 hour experiences of intermittent beam (see Fig. 1). Looking back, it seems almost a miracle that anything worthwhile came of the experiment. After about a week of near futility, we were rewarded with our first SEXAFS scan of I on Ag(111). True, the scan resembled roadkill on wet pavement, but it was data nonetheless. We had to remind ourselves that this was the most dilute sample ever studied with EXAFS. With the system debugged within working limits, we began accumulating scans.

In this data collection mode, it was possible to look around at experiments of Users on other beamlines. One in particular, being carried out by Jo Stöhr (then a postdoc of Shirley) and some graduate students working on a lower-photon-energy beamline with a "grasshopper" grating monochromator, involved an experiment dealing with Se adsorbed on Ni (now I remembered Shirley's notes). It was not clear what type of

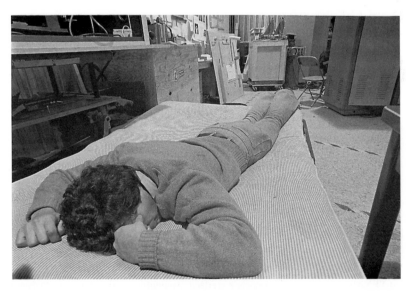

Fig. 1. Thoughtful evaluation of data analysis procedure during first SEXAFS run at SSRP, February 1977.

measurement was being attempted, but it was clear that it was not going to be completed as planned. The line feeding the poisonous dosing gas, H_2Se, into the chamber had sprung a leak, leading to a spirited mass exit from the building. For some of those experimenters, this may have been the impetus for developing an interest in oxygen as an adsorbate. Incidents and speculations like these were sometimes the only source of entertainment at SSRP.

The completion of that long-awaited SEXAFS experiment was at least well timed. Several days after the end of our beam time allotment I was to give a talk in an EXAFS symposium at the March meeting of the American Physical Society, and I had suggested in my abstract that applications to surface structure would be discussed [26]. Viewgraphs were made at the conference site from data carried on the plane from San Francisco. It was extremely gratifying to demonstrate, finally, that there would be more to remember about SEXAFS than its name.

2.4. SEXAFS development: Instrumentation

There was a great temptation to publish then, but I felt it was important to know whether we had measured the same $\frac{1}{3}$-monolayer system stud-

ied earlier [25]. A narrow slice of high-energy beam time coming up soon meant the issue could be settled but there would be no time to build an end station. We rented an equipped vacuum system for delivery at SSRP and I asked Rich Hewitt to join the effort. Hewitt, then already at Bell a number of years, had recently started working with me and would soon make invaluable contributions. The system was ready just about the time that beam was being delivered. It was now possible to see with LEED that dosing with I_2 as prescribed [25] gave a more complex pattern than the expected $(\sqrt{3} \times \sqrt{3})R30°$ structure. The SEXAFS data were so beautiful and the available beam time was so small, however, that we chose to deal with this problem off line. Back at Bell, I sought the assistance of Helen Farrell, whose expertise in LEED and willingness to contaminate her chamber with I_2 proved to be both very generous and valuable. Detailed measurements showed a variety of overlayer structures whose formation sensitively depended on sample temperature and dosing rate. The results indicated that the first sample had too little iodine while the second had too much, somewhat more than a monolayer (explaining the beautiful data). A third attempt would be needed.

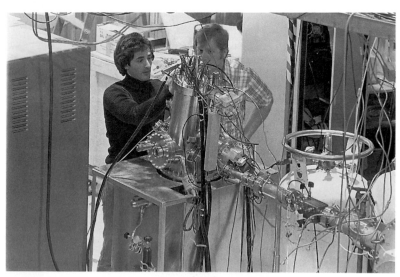

Fig. 2. Third attempt at SEXAFS, this time on a well-characterized $\frac{1}{3}$-monolayer-covered surface, December 1977. Rich Hewitt, right, guarding his custom manipulator in action.

Times like these made limited high-energy beam most exasperating. The next available slot was more than half a year away. On the positive side, this imposed vacation made time for my photoemission research (now surface-atom core level shifts) and for putting together my own SEXAFS chamber. The system would have a number of interesting features, including a near-vertically mounted analyzer to optimize surface sensitivity (see Fig. 2) and a manipulator, master-minded by Hewitt, to allow for independent off-axis sample rotations, one of which would help identify adsorption sites by varying the incident angle of the polarized beam.

In December 1977, we were finally in a position to perform the measurements on a well-characterized sample. Again, only a small amount of beam time was available, so only one polarization direction could be investigated. Nevertheless,

even with noisy data, a first-neighbor I–Ag(111) bond length was determined simply and directly to within about ± 0.03 Å. Our efforts were finally at a level for publication [27].

During this brief, pressure-filled run I learned that in September, SSRP had been promoted from "Project" to "Laboratory" status (see Fig. 3). Doniach had succeeded in his efforts, and soon Bienenstock would be succeeding him in his position as director. I also learned that Stöhr had joined the SSRL staff as of October and had become interested in applying SEXAFS to low-Z atoms, e.g., C, N, and O. With new data in hand, he wanted to hear about our theoretical under-standing, measurement techniques, and data analysis methods. I was happy to share what I knew and hear about his results. He showed me some O K-edge data from a natural oxide film of Al that he had taken on the "grasshopper" beam-

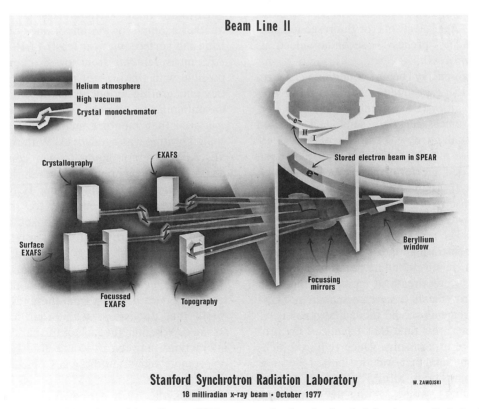

Fig. 3. Schematic view of first focused beamline at SSRL, one month after the Stanford Synchrotron Radiation Project was promoted to Laboratory status. The SEXAFS experiment was performed at the end of the beamline.

line by monitoring the O KVV Auger electrons at about 510 eV. There were peaks not understood in the data that seemed too large and sharp to be EXAFS. After some thought, we realized that with scanning photon energy the core and valence photoelectrons from O and Al were being swept through the energy analyzer window, thereby interfering with the EXAFS. (This was not a problem in the measurement of the I LMM Auger electrons at about 3300 eV because the analyzer window was inaccessible to the lower energy photoelectrons.) A way to avoid this difficulty had been demonstrated years earlier in absorption experiments < 150 eV above the edge, namely, measuring the total electron yield [28,29]. Outgoing Auger electrons inelastically scatter and create secondary electrons. Therefore, monitoring their intensity in total yield, which avoids energy discrimination and thus photopeaks, should be directly proportional to the absorption cross section. In a sense, total yield is a "tertiary" measure of EXAFS: photoelectron scattering ("primary" process) → Auger electrons (or fluorescence photons, "secondary" process) → secondary electrons ("tertiary" process).

With high-energy beam conditions not required for the experiment, Stöhr and two colleagues of Shirley (Dave Denley and Paolo Perfetti) soon remeasured their sample. They used an energy analyzer now set to accept ~ 1–3 eV secondaries, so this "partial yield" detection scheme avoided interfering photopeaks. Although the data in the published work [30] resembled Al_2O_3 because of the oxide film thickness, monolayer sensitivity was not the main point of the experiment. The real significance of those results was in demonstrating that absorption measurements from relatively dilute concentrations of low-Z atoms were already at hand, and that with further improvements in surface sensitivity SEXAFS measurements should indeed be possible.

It was about this time that I realized my present choice of experiments, which required high-energy beam time that was so limited because SSRL was parasitic to SPEAR, was only making difficult matters worse. Available low-energy beam time made low-Z SEXAFS experiments look increasingly attractive. Work of O on Ni(100)

had been recently carried out by Stöhr [31] with a coverage reported equivalent to about 1 monolayer, so it seemed that the time for trying a submonolayer low-Z SEXAFS experiment had arrived sooner than I thought. A collaboration was arranged with Shirley and some of his graduate students, who had beam time coming up on the grasshopper beamline and who were interested in learning more about SEXAFS. For the experiment, we prepared and characterized a Cu(100)c(2 × 2)-O surface (0.5 monolayer of O) in our chamber. Since this was a vacuum beamline, i.e., with no He atmosphere and no Be window needed, Hewitt and I also installed several new features for improving the adjustment of the incident beam shape and for properly normalizing its intensity. A long and frustrating week later, an O K-edge spectrum had been acquired which made our first I-on-Ag(111) roadkill scan look good. From this experience I learned: (1) The earlier O on Ni(100) data must have had much more than ~ 1 monolayer of O; this was another example of the importance of characterizing the surface with (at least) LEED. (2) SEXAFS measurements from low-Z atoms were not possible with the present parasitic beam intensities and grating monochromator, whose C- and O-contaminated, thin Au coating was giving more scattered than monochromated light. A new Pt-coated grating was needed. (3) With this showing, SEXAFS was not going to win converts. The Se-on-Ni(100) experiments that Shirley and coworkers had tried earlier using H_2Se involved photoelectron diffraction, and for their money that approach understandably looked much more favorable. (4) There was no easy way to ride out the 1977–1978 high-energy "X-ray drought" for SEXAFS. Small amounts of intense high-energy beam were better than large amounts of weak low-energy beam. Apart from these lessons, I also found that discussions with Stöhr had become strained. Whether the lack of further constructive interactions between us would ultimately impede or promote our scientific goals remains speculation.

During this 1978 period of the "drought", a number of important and more constructive developments occurred at SSRL, which significantly affected the growth of SEXAFS. Indeed, construction was the operative word. In the soft X-ray regime, two new beamlines were being built. One would have an improved "grasshopper" grating design, and the other, later to be nicknamed "JUMBO", would have a focusing mirror and a set of four interchangeable monochromator crystal pairs, all in vacuum. This would open up the previously inaccessible ~ 1–3 keV energy region that was beyond vacuum gratings and below Be-window cutoffs. In the hard X-ray regime, a new focused beamline was being built containing a device, inserted upstream into a straight section of the storage ring, called a "wiggler" (these devices, along with "undulators", are generically referred to as "insertion devices"). Comprised of a series of alternating ultra-strong magnets which accelerate the electron beam in a transverse direction, a wiggler produces radiation that is both more intense and has higher energy. Thus, even low-energy running conditions at SPEAR could produce hard X-radiation at SSRL.

The "drought" was catalytic in another respect. It focused attention not only on the need for insertion devices, but on the need for "dedicated" beam, a commodity long familiar to Users of VUV storage rings but unknown thus far to members of the X-ray synchrotron radiation community. During the summer of 1978, one week of beam time was purchased in which SPEAR was dedicated to operate solely for SSRL. The "single" beam of stored electrons, i.e., without the second beam of positrons, produced a synchrotron radiation beam that was brighter, more intense, more energetic, more stable, and longer-lived (if only all this happened to people). During this brief run, I had a rare glimpse into the future. A single 20 min scan with dedicated beam from $\frac{1}{3}$-monolayer of I was far superior to 6 h of summed data during parasitic running. I could only imagine the improvements using the new wiggler line, literally, because it would be completed after early- to mid-1979 and be available sometime after that. Until then, I would have to use the focused beamline under parasitic running conditions. The situation was reminiscent of the WWII song lyric "...How're ya gonna keep'em down on the farm, after they've seen Paris?"

During this period, anticipation and frustration were not limited to me. The new grasshopper and JUMBO beamlines were also under construction and would not be available until early- to mid-1980. This left only the original grasshopper line for experiments on low-Z atoms, and from personal experience I knew what that was like. Even freshly installed gratings were known to degrade after a few weeks of running from the ubiquitous C and O contamination that cracked in the beam. The increased C and O absorption reduced the monochromatic beam intensity, and the resulting additional structure in the transmission function made difficult normalization procedures even worse. Measurements therefore required many cross-checks to test for systematic errors and many accumulated scans to obtain even modest statistics. Overcoming these problems was very time-consuming, but finally rewarding. A year and a half after the initial study of an oxidized Al foil [30], Leif Johansson (then on the SSRL staff) and Stöhr successfully obtained SEXAFS measurements from about a monolayer of O on Al(111) using total yield detection with an electron multiplier [32]. The surface had been characterized with photoemission and LEED. These first data were not the best quality, but neither were those in the $\frac{1}{3}$-monolayer study of I on Ag(111) [27]. The emphasis of that work [32] was on the measured O–Al bond length, which distinguished between various theoretical calculations and stimulated re-analysis and correction of earlier studies. Given the circumstances, this work was truly an accomplishment.

While those measurements were underway, separated slices of high-energy parasitic beam time became available to me between late-1978 to mid-1979. I was interested in developing a systematic approach for determining adsorption sites from the polarization dependence of SEXAFS amplitudes, but I found that the quality of this beam was simply not high enough for the series of experiments needed to do the job. With little to show for my efforts, I was having difficulty explaining to myself, as well as to my family 3000 miles away, why I should continue with this ridiculous lifestyle.

I was about to throw in the towel when SSRL

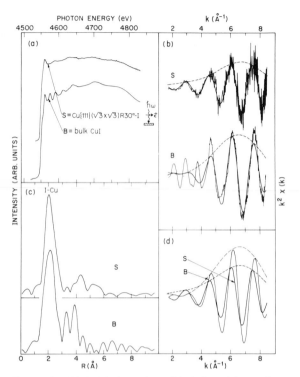

Fig. 4. First SEXAFS data taken with dedicated beam, from Ref. [33]. (a) Raw SEXAFS and EXAFS I L_3-edge absorption data, in energy space, from $\frac{1}{3}$-monolayer of I on Cu(111) surface (labelled **S**) and from bulk model compound CuI (labelled **B**). (b) Raw data from (a) (thin noisier lines), in photoelectron momentum (k) space, after truncating data above the edge, subtracting a smooth polynomial background and multiplying by k^2. The amplitudes of the SEXAFS and EXAFS have been normalized to the corresponding absorption edge jumps. (c) Fourier transforms of data from (b), in real space (R). The peak at ~2 Å (uncorrected for phase shift) is due to the first-neighbor shell of Cu atoms around I. The double peaks at ~3.3 and 3.9 Å in CuI are due to the second shell of I atoms (the second-shell Cu atoms in the surface data are not resolved). (d) Back-transformed data from (c) after filtering with a window function around the first-neighbor peak. The filtered first-shell data are also superposed on the raw data in (b), along with the amplitude functions (dotted lines). The I–Cu surface bond length is determined to be 2.66 ± 0.02 Å and the adsorption site is the three-fold hollow.

announced that starting late-1979, it would be operating in a 50% dedicated mode. The view that this was actually good news came only after the data from I on Cu(111) and Cu(100) were collected; they were stunning (see Fig. 4). My brief and earlier experience with dedicated beam,

seemingly in a former life, had returned. Back then, evaluating the signal/background efficiencies of Auger, partial, and total electron yield detection schemes for I on Ag(111) suggested that Auger electron yield was best. Now, reevaluation with a different substrate (and a more efficient electron multiplier) indicated that total yield was. From this I learned the importance of testing each surface system for the best detection scheme. Combined with improved beam conditions, the optimized signal was now more than 25 times better than that of the earlier data [27], making possible the study of the previously impractical $\frac{1}{4}$-monolayer Cu(100)p(2 × 2)-I system. This work established – finally, with our custom manipulator – that the relative polarization dependence of SEXAFS amplitudes can be used for reliably determining adsorption sites [33]. This was important because, unlike phase shifts, absolute EXAFS amplitudes (determined from bulk model compounds) depend, in general, on chemical bonding and need not be transferable [16]. Furthermore, higher-neighbor SEXAFS distances, which can in principle also be used for determining adsorption sites, are for a variety of reasons not always observable (this was the case in these systems). Therefore, it was essential to show that the first-neighbor signal could be relied upon to provide both the surface bond length – from the SEXAFS frequency, and the adsorption site – from the SEXAFS amplitude, in a direct and empirical manner.

2.5. NEXAFS: Further development

A major corner had been turned in the development of SEXAFS. The basic methodology of the technique and its applicability to high- and low-Z atoms on metal surfaces using different electron detection schemes were now established in spite of some beam time and beamline adversities. It was time to advance this development to another level. In the process, a related technique would emerge, NEXAFS, which would extend the application of X-ray absorption spectroscopy to adsorbed surface structures even further.

SSRL was now in a 50% dedicated-running mode with anticipated state-of-the-art X-ray beamlines about to become available. I could see my involvement increasing substantially there, away from home, and soon I became more aware of another phenomenon that I very much wanted to avoid. A large number of synchrotron radiation staff members and Users were becoming, or had become, part of a booming business for divorce attorneys, well above the already-high average for California. Whether the association with synchrotron radiation was a cause, a result, or a coincidence could always be debated, but in my view that was not the issue. The former period of X-ray "drought" had been catalytic for insertion devices and dedicated running, and it appeared that the impending period of X-ray "flood" would be catalytic as well. With grateful encouragement from Bell and SSRL, I wound up taking a sabbatical year with my family in Stanford starting mid-1980. Not only did we all have a great time, but as it turned out, that year-period would provide one of the largest amount of dedicated running in SSRL's history.

The SEXAFS work performed from this point forward would be focused on extending its applicability with only the highest quality data. Collaborating with Stöhr, Dave Norman (a visiting scientist from Daresbury), Rolf Jaeger (a DFG fellow), and Sean Brennan (an SSRL postdoc) used the new grasshopper beamline to determine very reliable O–Al bond lengths for O on Al(111) [34], but now in submonolayer concentrations (the former study of ~ 1 monolayer [32] had been less precise because of the coexistence of two O–Al phases). With the new JUMBO beamline operating in the formerly inaccessible 1–3 keV energy range, the prototypical Ni(100)c(2 × 2)-S system was studied [35] for comparison with previous work and found to agree very well. Now, however, the adsorption site was determined not only from absolute and relative SEXAFS amplitudes, but from a clearly observable higher-shell S–Ni distance. Collaborating with Jack Rowe (from Bell), the new and more intense wiggler beamline was used to study I and Te on Si(111)7 × 7 and Ge(111)2 × 8 [36]. In these four systems, three different adsorption sites were identified, two had not been previously reported, and none corresponded to the most common site observed for

metal substrates. That work also showed how determining the local structure of adatoms on reconstructed semiconductor surfaces, too complex to study previously, presented little difficulty with SEXAFS because the technique is inherently short-range.

Amidst these SEXAFS developments involving very different adatoms and substrates performed on newly available soft and hard X-ray beamlines, another experiment with molecules on surfaces was being carried out by Stöhr and visiting scientists Klaus Baberschke (from Freie Universität Berlin) and Rolf Treichler (from Technische Universität München) on the "old" grasshopper. Was I hearing right? This beamline? Home of the H_2Se poison-gas leak, the abortive attempt with O on Cu(100), and the marathon study of O on Al(111)? And not atoms on surfaces, but molecules?

Little time was wasted in sharing the news that CO adsorbed on Ni(100) showed very strong and very polarization-dependent features in the regions around the C and O K-edges. The features were not initially understood by these workers because they, in good company with many other surface and solid state physicists at the time, viewed CO as a complex organic molecule. Discussions with other colleagues and trips to the atomic, molecular, and chemical literature helped to explain their origin. In CO, the 1s atomic levels, called molecular 1σ (for O) and 2σ (for C) levels, are nonbonding and can still be referred to as 1s. The 2p atomic levels, on the other hand, form bonding and antibonding molecular orbitals of σ and π symmetry whose atomic 2p components lie along and perpendicular to the molecular axis, respectively (for simplicity we ignore 2s and 2p hybridization). Now, creating a 1s core hole in absorption pulls the low-energy antibonding $2\pi^*$ level below the 1s ionization potential. As a result, for absorption containing a polarization component along the direction of the $2\pi^*$ orbital, a $1s \rightarrow 2\pi^*$ transition is observed which is below, i.e., lower in energy than, the K edge itself. Following dipole selection rules, the intensity of this "π^* resonance" peak is maximized when the polarization lies along the $2\pi^*$ orbital direction and goes to zero when perpendicular to it. Similarly, for the relatively more energetic antibonding $6\sigma^*$ level, a $1s \rightarrow 6\sigma^*$ transition is observed which is above the K edge. Since the direction of the $6\sigma^*$ orbital is perpendicular to the $2\pi^*$ orbital, the intensity of its corresponding transition, referred to as a "σ^* (shape) resonance", obeys just the opposite polarization dependence. Therefore, by measuring the intensity of the π^* resonance as a function of polarization direction (this resonance is easier to use since it is below the edge and not vibrationally broadened), the orientation of the CO molecule should be easily determined. Indeed, the intensity of that resonance was found to vanish with the polarization perpendicular to the surface normal, immediately demonstrating that the C–O axis itself must lie along the normal. The experiment was repeated for NO, with similarly straightforward results [37].

The significance of these measurements was that the orientation of the adsorbed molecules could be obtained so simply and directly. Exactly the same setup was used as for detecting C, N, or O KVV Auger electrons in a SEXAFS experiment. It was well known, however, that SEXAFS could not be measured with this scheme because of interfering photopeaks, and the same problem existed for the near-edge region. To overcome this inherent difficulty, identical near-edge scans with and without the adsorbed molecules were obtained and the difference between them was taken. The results were not perfect, but they did not have to be because the near-edge structures were so large, well exceeding the size of the edge. In fact, it was because these structures were large that the old grasshopper beamline was sensitive enough for such experiments. The authors, looking for an acronym involving surfaces that was different from XANES (X-ray absorption near-edge structure, which had been around several years for bulk EXAFS systems), called their technique surface-absorption fine structure (SAFS) [37].

Shortly after these measurements, I had inadvertently become involved in some experiments that would relate to this near-edge technique applied to surfaces. The JUMBO beamline was made available to me and I chose to look at two

problems that had used theoretical explanations of photoemission data to infer surface structure. One dealt with Cl on Cu(100) [38] and the other with Cl on Si(111) and Ge(111) [39]. Rowe and I collaborated on these systems and had a great time, particularly because we found with SEXAFS that some of the earlier interpretations of the photoemission work with which he had been involved were not quite right (read: wrong). For example, Cl had been inferred to occupy the on-top adsorption site for Si(111) but not for Ge(111) (the on-top site has Cl saturating the substrate dangling bond normal to the surface). Now, we found that the polarization dependence of the Cl K-edge SEXAFS amplitude vanished for both substrates when the polarization was perpendicular to the surface normal, immediately identifying the atop site in both systems. Furthermore, the identical polarization dependence was apparent right at the edge, where transitions occurred between the Cl 1s and either the (Cl, Si) or (Cl, Ge) p-derived antibonding σ^* orbitals. Thus, although the particular orbitals and substrates here were different than those of CO and NO on Ni(100), our studying the polarization dependence of the edge structure for the purpose of identifying the adatom–substrate bond direction was very similar.

My reaction to this at the time was different than that of Rowe, who wanted to show how both the extended and near-edge regions of the data could be useful. I thought the extended region alone would suffice, and moreover, it contained adatom–substrate distances. Undaunted, he wanted to make a simple connection between the two regions and decided a new acronym was needed to do this. He did not like SAFS, because it lacked name recognition, or XANES, because it was too difficult to pronounce. (In fact, an unnamed Bell colleague of ours feigned total confusion with how to pronounce XANES, saying that rather than sounding like "z-āyn-z", to him the "X" sounded like "ex" and the "anes" rhymed with "heinous".) In an inspired flash, Rowe came up with near-edge X-ray absorption fine structure (NEXAFS). I objected, arguing that its similar sound to SEXAFS (which was exactly his point) would only lead to confusion because the near-edge region was emphatically "not" EXAFS. Hearing this, we smiled and knew something had been born. He shared this with Stöhr and N(ot-)EXAFS was quickly adopted [40,7].

By the time I returned to Bell in mid-1981, Stöhr and Eisenberger had moved, independently, to Exxon Corporate Research Science Laboratories. Fabio Comin came to join me as a postdoc soon after. Further SEXAFS work was carried out in the next few years, with studies including the study of bond length with coverage (0.25 versus 0.5 monolayers O on Ni(100) [41]), an unusual adsorption site with different crystal face (Te on Cu(111) versus Cu(100) [42]), and metal evaporation on Si (Ag and Pd [43], Ni [44]). The reactive chemisorption work with Pd and Ni leading to silicide formation [43,44] was particularly interesting because the unannealed systems had no signs of long-range order and could not be studied with any other structural technique. NEXAFS was applied to atomic adsorption (O on Ni(100) [44]) to see if, like SEXAFS, the bond length and adsorption site could be easily determined. It was found, however, that multiple scattering complicated the interpretation, thereby requiring theoretical modeling. This is just what SEXAFS avoids, so a more empirical application of NEXAFS was sought.

The success of NEXAFS from adsorbed CO and NO suggested that molecules were the area to explore. At Exxon, Stöhr could now not avoid hearing about more complex organic molecules. In collaboration with Earl Muetterties (at Berkeley) and his graduate student Allen Johnson, they studied the polarization dependence of the π^* resonance in deuterated benzene (C_6D_6) and pyridine (C_6H_5N) adsorbed on Pt(111) [45]. For benzene, the dependence was opposite to that of CO, consistent with other work suggesting the molecule lies flat on the surface. For pyridine, however, the dependence looked more like CO, indicating that the benzene ring was nearly normal to the surface. This new information clearly played into the strength of the technique. Simplicity and empiricism prevailed, and it appeared NEXAFS had found a niche for studying orientations of even complex adsorbed molecules on metals.

Applying NEXAFS for determining adsorbed molecular orientations alone without bond lengths, however, did not seem wholly satisfying, particularly because only molecules with double and triple bonds exhibit π^* resonances. All molecules exhibit σ^* resonances, but apart from qualitatively confirming the opposite polarization dependence observed for the π^* resonances, these had not been used yet in interpreting NEXAFS data. Reference to the gas phase literature again proved helpful. A theoretical description [46] of the ISEELS data from free N_2 [19] is that the σ^* resonance reflects the strong multiple scattering of the core-excited electron between the N atoms. Since the energy position of the resonance relative to the ionization potential, Δ, depends on the N–N separation, a simple correlation between it and the intramolecular bond length, r, might exist. Around the early-1980's, there were experimental [47,48] and theoretical [49,50] discussions about how Δ should depend on r. A model in which Δ varied as r^2 was proposed [47,49], and in simple low-Z gas molecules this model was qualitatively confirmed [51].

The correlation of Δ with r^{-2} for free molecules stimulated Stöhr and coworkers [52] to look for similar trends in CO, formate $(HCOO)^-$, and methoxy (CH_3O) adsorbed on Cu(100). These molecules have triple, pseudo-double, and single C–O bonds, respectively, which vary as much as 0.3 Å. The qualitative polarization dependence of the σ^* resonance intensity in the O K-edge NEXAFS data showed that all the C–O bonds lie in a plane normal to the surface. Assuming that Δ varied with r^{-2} for CO and CH_3O and that for weak adsorption their $r(C-O)$ did not change from the gas phase value, the C–O bond length in adsorbed formate was interpolated by determining the appropriate proportionality constants; the result agreed with typical values for formate in inorganic compounds. The O_2 molecule chemisorbed on Pt(111) was also studied and found to lie flat on the surface. The large change in Δ relative to its gas phase value indicated that the O–O bond length changed upon chemisorption. Assuming the same proportionality constants for O–O as found for C–O, an increase in *r* of 0.24 Å from the free molecule was determined.

Since the proportionality constants for different atom pairs should not be the same [48], Francesco Sette (then a postdoc with Stöhr) and coworkers extended the correlation of Δ with *r* to a large number of low-*Z* gas molecules by grouping them according to the summed atomic numbers of the bonded atoms. A linear, rather than inversely quadratic dependence of Δ with *r* was found [53], but these two forms were shown to be very similar [54] so long as *r* is small (as is the case for bonds between C, N, and O). Different slopes were found for each of the different groups. The linear relationship for the C–C group was then used in the interpretation of the C–C bond lengths in the hydrocarbons cyclohexane (C_6D_{12}), cyclohepatatriene (C_7H_8), ethylene (C_2H_4), and acetylene (C_2D_2) adsorbed on Pt(111) [55]. The former two spectra closely resembled their gas phase counterparts, indicating little change in *r*, whereas Δ for ethylene and acetylene showed C–C bond lengthenings of 0.15 and 0.25 Å, respectively. The polarization dependence of the σ^* resonances in the C K-edge data qualitatively showed that all four molecules lie flat on the surface. The π^* resonance was not used because it is absent in C_6D_{12} and because it exhibited a more complex polarization dependence in adsorbed C_7H_8 and C_2D_2.

These studies [52,55] of intramolecular bond lengths and orientations of adsorbed molecules on metals stimulated much activity in testing the original findings and applying NEXAFS to other systems. As a result, several limitations and corrections were uncovered, much of which is reviewed in Ref. [7] and mentioned in the next section. It is important to place this all within the proper context. At that time, around 1984, NEXAFS was still relatively young; it needed to develop further. Despite (or perhaps because of) this fact, it had already attracted a growing number of surface scientists to a part of the X-ray absorption spectrum that had been all but overlooked because it was believed to be too complicated to be useful for surface structural studies.

There were two more final developments that took place about this time, both "instrumental",

which substantially enhanced the use and utility of SEXAFS and NEXAFS. The first was the commissioning of BESSY, a dedicated synchrotron facility in Berlin. New researchers, and some already familiar with these techniques from SSRL visits, would be using a new soft X-ray monochromator (SX-700) that would open up previously impractical studies of low-*Z* systems. The second was the development of an efficient fluorescent detector capable of operating at low energies and in ultra-high vacuum [56]. This would simplify the problem of interfering photopeaks for low-*Z* atoms and significantly improve the signal/background for all absorption measurements, allowing the study of very low concentrations of absorbers ($\ll$ 1 monolayer).

After 1985–86, a number of other synchrotron facilities became operational worldwide. Higher intensities, more stable conditions, new monochromators, and of course, greater accessibility all contributed into producing a variety of new players and an expanding array of new applications.

3. Assessments

A brief summary is given of present capabilities, inherent and practical limitations, and overall accomplishments for SEXAFS and NEXAFS. The scope and format of this paper do not call for individual reference to the large body of papers exemplifying the general rules of what these techniques can do and have done (some were given above and many are covered in reviews [7–10]). However, a few selected examples of work not covered above, which illustrate either additional information or exceptions to the rule, have been cited.

In the outline below, it is useful to evaluate SEXAFS and NEXAFS not on only their accomplishments, but on how easy or difficult it is to obtain them on a scale relative to other surface structural methods. Doing this, I suggest the following maxim be considered:

Conservation of pain: If the data come easy, the analysis won't. And vice versa.

A strongly interacting probe means a strong signal, but often a weak or indirect measure of what is fundamentally sought. The SEXAFS "signal" is weak, but its single scattering formalism is straightforward. NEXAFS is stronger, but strong surface bonding can complicate its interpretation. (NEXAFS, however, looks at much more complex systems and is relatively simple compared with other approaches.) Both SEXAFS and NEXAFS measurements require intense, accessible synchrotron radiation sources, reliable beamlines and related equipment, and a supportive professional and personal environment. That's the hard part. After paying these dues, data analysis is simple and reliable.

3.1. What can and cannot be learned

SEXAFS is ideally suited for determining *local distances* and *positions* of *atoms adsorbed* on single crystal surfaces. Long-range order is not required; theoretical calculations are not needed. First-neighbor bond lengths are easily determined, usually to an accuracy (not just precision) of ± 0.02 Å, provided that (1) there is a single configuration, (2) higher-neighbor distances are separated from the first neighbor by more than ~ 1 Å, and (3) a reliable model compound for phase shifts is used. High-symmetry adsorption sites can usually be determined by polarization-dependent first-neighbor amplitudes alone, again if (1) and (2) apply. Site assignments are strengthened by also using absolute amplitudes, particularly for less anisotropic $L_{2,3}$-edge data.

If higher-neighbor distances are observable (typically < 5 Å from the absorbing atom), substrate *relaxation* and *reconstruction* can be measured and site assignments can be made with even greater assurance [57,58]. If there are distances separated by less than ~ 1 Å, existing least squares fitting routines can be used, standard in analyzing bulk EXAFS data [59,60]. If a reliable model compound containing the same absorbing and backscattering atoms is unavailable, another compound with atoms differing in Z by about ± 2 can be used after correcting (~ 0.01 Å) with theoretical phase shifts [39,61]. If molecules are studied with SEXAFS, they can be simple pseudo-atomic (containing H, which scatters negligibly), diatomic, or pseudo-diatomic [62–64].

Two nonstructural quantities obtainable from a SEXAFS measurement are relative *vibrational strengths* of adsorbate–substrate bonds and relative adsorbate *concentrations* in a given system. The former are obtained from the temperature dependence of SEXAFS amplitudes (the Debye–Waller-like term) [64,65] while the latter are obtained from the height of the absorption edge jump.

Anything not included in the above qualifiers – specifically, clean surfaces, low-symmetry sites, distances > 5 Å, unknown multiple configurations, complex molecules, and H adatoms –cannot be measured or easily determined with SEXAFS.

NEXAFS is ideally suited for determining *intramolecular* bond length *changes* and *orientations* of *molecules adsorbed* on single crystal surfaces. Long-range order is not required. Theoretical calculations are not needed for weakly- to moderately-adsorbed molecules. As with SEXAFS, NEXAFS is generally nondestructive, even for fragile molecules. Absolute intramolecular bond lengths can be determined, usually to an accuracy of about ± 0.05 Å, provided that (1) there is a single configuration, (2) gas phase data exist for identifying σ^* (and/or π^*) resonances, (3) X-ray photoemission data exist for determining 1s binding energies ($\approx$ K-edge thresholds), and (4) the molecules are diatomic or pseudo-diatomic involving only C, N, O, or F. Absolute molecular orientations can be determined to about $\pm 10°$ requiring (1) and (2), but also provided that (5) corrections are made for underlying metal-substrate structure (viz., excitations at the Fermi level), (6) the molecules are low-Z and unsaturated (i.e., contain a π^* resonance), and (7) the molecule is not very-strongly-adsorbed (which distorts the π^* resonance).

If the molecules are strongly- or very-strongly-adsorbed, theoretical (e.g., Xα-multiple-scattering) calculations can be used to interpret the data [66,67]. If the molecules are not (pseudo-) diatomic and/or not low-Z (i.e., contain atoms with $Z > 10$), qualitative *changes* (or lack thereof) in intramolecular bond lengths upon adsorption can be determined [68,69]. This makes NEXAFS less quantitative but much more generally applicable. If the molecules are not low-Z and/or saturated and/or very-strongly-adsorbed, qualitative molecular orientations can still be determined [62,66]. Again, this extends the usefulness of the technique. If NEXAFS of atomic adsorbates is studied, more detailed multiple scattering calculations are needed [44,70].

Two additional quantities obtainable from a NEXAFS measurement are *chemical identification* of adsorbed species and relative *concentrations* of different co-existing species (as opposed to the same species in a different configuration) [70,71]. The former can be determined from the edge positions and structure, similar to X-ray photoemission chemical shifts. The latter can be determined from the relative intensities of these structures using a difference spectrum, provided of course that one of the species can be measured separately.

Anything not included in the above qualifiers – specifically, quantitative intramolecular distances and molecular orientations from mid- to high-Z non-diatomics involving bond length changes upon adsorption – cannot be measured or easily determined with NEXAFS.

3.2. What has been learned

As stated in the Introduction, SEXAFS and NEXAFS have been applied to many different systems, more so than any other surface structural technique excluding LEED (because it can study clean surfaces). This may now seem even more surprising because of the apparently large number of limitations mentioned in the preceding section. One of the goals in this brief outline is to point out that there are, in fact, a large number of different systems falling within these limits.

In *number,* where a change of substrate, crystal face, adsorbate, coverage, or temperature (leading to a structural modification) constitutes a different system, more than 60 have been studied using SEXAFS and more than 150 using NEXAFS. In *adsorbate,* atoms range from C to Cs, molecules from H_2S to $C_{24}H_{12}$. In *substrate,*

metals are close- or open-packed, semiconductors are reconstructed or amorphous. In *adatom*–substrate *bonding,* interactions range from physisorption to reactive chemisorption. In *admolecule*–substrate *interactions,* intramolecular bonds vary from unchanged to dissociated. In *adsorption sites,* one-, two-, three-, four-, quasi-five, and six-fold coordinations are observed. In *molecular orientations,* they range from normal, to tilted, to flat on the surface. In *coverage,* they vary from < 0.1 to ~ 1 monolayer, and in *temperature* from < 20 to > 1100 K.

The *quality* of the information obtained is very high. In many systems SEXAFS is capable of determining surface bond lengths with an absolute accuracy unmatched by any other method. High-symmetry adsorption sites can be reliably assigned and checked using up to three different procedures. NEXAFS can identify changes in intramolecular bond length upon adsorption extremely well, typically from systems that are inaccessible with any other technique. Molecular orientations from spectra of similarly complex systems can often be established by inspection.

Because of the reliability and versatility of these techniques, a great deal of new information has been learned. Unanticipated adsorption sites and unusual substrate behavior have been discovered. Long-standing problems regarding the interpretation of various experimental results using theoretical methods or other data have been unambiguously resolved. Classes of systems exhibiting more conventional adsorption behavior have been studied, from which unifying and previously untested principles of surface bonding trends have emerged. New effects on molecular orientations and intramolecular bond lengths varying with temperature, coverage, and other co-adsorbed species have been discovered. Novel surface reaction pathways, precursors, and intermediates varying with the same parameters have been identified and structurally characterized. Finally, kinetic studies in real time and at near-atmospheric pressures have been performed.

Experimental improvements in synchrotron storage rings have played a strong role in attracting new researchers and stimulating new science. First-generation facilities were parasitic and limitedly available; second-generation machines are dedicated and primarily use conventional bending magnets. Third-generation machines, currently under construction and in some cases near completion, will primarily use insertion devices. The very much increased brightness (and, to a lesser extent, intensity) of the beam from the wigglers and undulators will open up many new areas to explore with SEXAFS and NEXAFS. Some have been touched upon already, e.g., kinetic studies and co-adsorption experiments, but the new sources will make these much easier to carry out. Smaller, more intense beams invite small samples of submillimeter to micron dimensions. Study of very dilute concentrations, or minority species found only in parts of a sample, such as at steps or other defects, could be possible. Apart from collecting data faster, higher quality statistics will make difference spectra less intimidating. The future should indeed be very "bright".

4. Concluding remarks

Reflecting on how and where and why this journey began, from early X-ray days to SSRP, to Bell, CO on Ni pointing down?, psi/J, embarrassed titters, poison gases, marathon beam runs, ...many thoughts come to mind. How might it have been different if any one of the events was altered, if one of the players there was not involved, or one of them not involved was? The melding of diverse scientific communities with people of overlapping or orthogonal interests and personalities was very special. Each failure and frustration along the way now seems as important as the successes. It is, of course, this way in all fields of science, in all phases of life. It's just that sometimes we forget.

Acknowledgements

For their experience, comradery, and support, I am indebted to my collaborators Peter Eisenberger, Rich Hewitt, Alastair MacDowell, Jack Rowe, and Francesco Sette, my postdocs Dave Adler, Fabio Comin, Tomi Hashizume, and Ro-nan McGrath, my colleagues and management at Bell, and the SSRL staff and administrative body. I also thank Ed Chaban for valuable technical help, and Don Hamann and Stefan Schuppler for useful comments. Finally, I thank my family, Karen, Lisa, and David, for their tireless understanding and encouragement.

References

[1] See Master Index, Vols. 1–10, Surf. Sci. 10 (1968) 487.

[2] See Master Index, Vols. 11–20, Surf. Sci. 20 (1970) 453.

[3] See Master Index, Vols. 81–90, Surf. Sci. 90 (1979) 691.

[4] B. Moraweck, G. Clugnet and A.J. Renouprez, Surf. Sci. 81 (1979) L631.

[5] P.H. Citrin, P. Eisenberger and R.C. Hewitt, Surf. Sci. 89 (1979) 28.

[6] For example, see A.H. Compton and S.K. Allison, X-Rays in Theory and Experiment, 2nd ed. (Van Nostrand, Princeton, 1935).

[7] J. Stöhr, NEXAFS Spectroscopy, Vol. 25 of Springer Series in Surface Science (Springer, Berlin, 1992).

[8] J. Stöhr, in: X-Ray Absorption: Principles, Applications, Techniques of EXAFS, SEXAFS, and XANES, Eds. D.C. Koningsberger and R. Prins (Wiley, New York, 1988) p. 443.

[9] D. Norman, J. Phys. C 19 (1986) 3273.

[10] P.H. Citrin, J. Phys. (Paris) 47 (1986) C8-437.

[11] R. de L. Kronig, Z. Phys. 70 (1931) 317.

[12] R. de L. Kronig, Z. Phys. 75 (1932) 468.

[13] F.W. Lytle, in: Advances in X-Ray Analysis, Vol. 9, Eds. G.R. Malett, M. Fay and M.W. Mueller (Plenum, New York, 1966) p. 398.

[14] F.W. Lytle, in: Physics of Non-Crystalline Solids, Ed. J.A. Prins (North-Holland, Amsterdam, 1965) p. 12.

[15] D.E. Sayers, E.A Stern and F.W. Lytle, Phys. Rev. Lett. 27 (1971) 1204.

[16] P.A. Lee, P.H. Citrin, P. Eisenberger and B.M. Kincaid, Rev. Mod. Phys. 53 (1981) 769.

[17] M.J. Van der Wiel, Physica 49 (1970) 411; M.J. Van der Wiel, Th.M. El-Sherbini and C.E. Brion, Chem. Phys. Lett. 7 (1970) 161.

[18] C.E. Brion, S. Daviel, R.N.S. Sodhi and A.P. Hitchcock, AIP Conf. Proc. 94 (1982) 429.

[19] G.R. Wright, C.E. Brion and M.J. Van der Wiel, J. Electron Spectrosc. Relat. Phenom. 1 (1973) 457.

[20] J.L. Dehmer and D. Dill, Phys. Rev Lett. 35 (1975) 213.

[21] For example, see Synchrotron Radiation Research, Eds. H. Winick and S. Doniach (Plenum, New York, 1980).

[22] P.A. Lee, Phys. Rev. B 13 (1976) 5261.

[23] J. Jaklevic, T.A. Kirby, M.P. Klein, A.S. Robertson, G.S. Brown and P.Eisenberger, Solid State Commun. 23 (1977) 679.

[24] U. Landman and D.L. Adams, Proc. Natl. Acad. Sci. USA 73 (1976) 2550.

[25] F. Forstmann, W. Berndt and P. Büttner, Phys. Rev. Lett. 30 (1973) 17.

[26] P.H. Citrin, Bull. Am. Phys. Soc. 22 (1977) 359.

[27] P.H. Citrin, P. Eisenberger and R.C. Hewitt, Phys. Rev. Lett. 41 (1978) 309.

[28] A.P. Lukirskii and I.A. Brytov, Sov. Phys. Solid State 6 (1964) 33.

[29] W. Gudat and C. Kunz, Phys. Rev. Lett. 29 (1972) 169.

[30] J. Stöhr, D. Denley and P. Perfetti, Phys. Rev. B 18 (1978) 4132.

[31] J. Stöhr, J. Vac. Sci. Technol. 16 (1979) 37.

[32] L.I. Johansson and J. Stöhr, Phys. Rev. Lett. 43 (1979) 1882.

[33] P.H. Citrin, P. Eisenberger and R.C. Hewitt, Phys. Rev. Lett. 45 (1980) 1948; Erratum: ibid. 47 (1981) 1567.

[34] D. Norman, S. Brennan, R. Jaeger and J. Stöhr, Surf. Sci. 105 (1981) L287.

[35] S. Brennan, J. Stöhr and R. Jaeger, Phys. Rev. B 24 (1981) 4871.

[36] P.H. Citrin, P. Eisenberger and J.E. Rowe, Phys. Rev. Lett. 48 (1982) 802.

[37] J. Stöhr, K. Baberschke, R. Jaeger, R. Treichler and S. Brennan, Phys. Rev. Lett. 47 (1981) 381.

[38] P.H. Citrin, D.R. Hamann, L.F. Mattheiss and J.E. Rowe, Phys. Rev. Lett. 49 (1982) 1712; Comment: ibid. 50 (1983) 1824.

[39] P.H. Citrin, P. Eisenberger and J.E. Rowe, Phys. Rev. B 28 (1983) 2299.

[40] J. Stöhr and R. Jaeger, Phys. Rev. B 26 (1982) 4111.

[41] J. Stöhr and R. Jaeger and T. Kendelewicz, Phys. Rev. Lett 49 (1982) 142.

[42] F. Comin, P.H. Citrin, P. Eisenberger and J.E. Rowe, Phys. Rev. B 26 (1982) 7060.

[43] F. Comin, J.E. Rowe and P.H. Citrin, Phys. Rev. Lett. 51 (1983) 2402.

[44] D. Norman, J. Stöhr, R. Jaeger, P.J. Durham and J.B. Pendry, Phys. Rev. Lett. 51 (1983) 2052.

[45] A.L. Johnson, E.L. Muetterties and J. Stöhr, J. Am. Chem. Soc. 105 (1983) 7183.

[46] J.L. Dehmer, D. Dill and S. Wallace, Phys. Rev. Lett. 43 (1979) 1005.

[47] T. Gustafsson and H.J. Levinson, Chem. Phys. Lett. 78 (1981) 28.

[48] A.P. Hitchcock and C.E. Brion, J. Phys. B 14 (1981) 4399.

[49] C.R. Natoli, in: EXAFS and Near Edge Structure, Vol 27 of Springer Series in Chemical Physics, Eds. A. Bianconi, L. Incoccia and S. Stipcich (Springer, Berlin, 1983) p. 43.

[50] A.E. Orel, T.N. Rescigno, B.V. McKoy and P.W. Langhoff, J. Chem. Phys. 72 (1980) 1265.

[51] A. Bianconi and M. Dell'Ariccia, in Ref. [49] p. 57.

[52] J. Stöhr, J.L. Gland, W. Eberhardt, D. Outka, R.J. Madix, F. Sette, R.J. Koestner and U. Döbler, Phys. Rev. Lett. 51 (1983) 2414.

[53] F. Sette, J. Stöhr and A.P. Hitchcock, J. Chem. Phys. 81 (1984) 4906; Chem. Phys. Lett. 110 (1984) 517.

[54] F. Sette and J. Stöhr, in: EXAFS and Near Edge Structure III, Eds. K.O. Hodgson, B. Hedman and J.E. Penner-Hahn, Proc. Phys. 2 (Springer, Berlin, 1984) p. 250.

[55] J. Stöhr, F. Sette and A.P. Hitchcock, Phys. Rev. Lett. 53 (1984) 1684.

[56] D.A. Fischer, J.B. Hastings, F. Zaera, J. Stöhr and F. Sette, Nucl. Instrum. Methods A 246 (1986) 561.

[57] M. Bader, A. Puschmann, C. Ocal and J. Haase, Phys. Rev. Lett. 57 (1986) 3273.

[58] F. Sette, T. Hashizume, F. Comin, A.A. MacDowell and P.H. Citrin, Phys. Rev. Lett. 61 (1988) 1384.

[59] D.J. Holmes, N. Panagiotides, R. Dus, D. Norman, G.M. Lamble, C.J. Barnes, F. Della Valle and D.A. King, J. Vac. Sci. Technol. A 5 (1987) 703.

[60] D.R. Warburton, G. Thornton, D. Norman, C.H. Richardson, R. McGrath and F. Sette, Surf. Sci. 189/190 (1987) 495.

[61] M. Richter, J.C. Woicik, J. Nogami, P. Pianetta, H.E. Miyano, A.A. Baski, T. Kendelewicz, C.E. Bouldin, W.E. Spicer, C.F. Quate and I. Lindau, Phys. Rev. Lett. 65 (1990) 3417.

[62] R. McGrath, A.A. MacDowell, T. Hashizume, F. Sette and P.H. Citrin, Phys. Rev. Lett. 64 (1990) 575.

[63] D. Arvanitis, K. Baberschke, L. Wenzel and U. Döbler, Phys. Rev. Lett. 57 (1986) 3175.

[64] P. Roubin, D. Chandesris, G. Rossi, J. Lecante, M.C. Desjonquères and G. Treglia, Phys. Rev. Lett. 56 (1986) 1272.

[65] F. Sette, C.T. Chen, J.E. Rowe and P.H. Citrin, Phys. Rev. Lett. 59 (1987) 311.

[66] J.A. Horsley, J. Stöhr and R.J. Koestner, J. Chem. Phys. 83 (1985) 3146.

[67] W. Wurth, J. Stöhr, P. Feulner, X. Pan, K.R. Bauchspiess, Y. Baba, E. Hudel, G. Rocker and D. Menzel, Phys. Rev. Lett. 65 (1990) 2426.

[68] M.D. Crapper, C.E. Riley, D.P. Woodruff, A. Puschmann and J. Haase, Surf. Sci. 171 (1986) 1.

[69] J. Stöhr, J.L. Gland, E.B. Kollin, R.J. Koestner, A.L. Johnson, E.L. Muetterties and F. Sette, Phys. Rev. Lett. 53 (1984) 2161.

[70] P.J. Durham, in Ref. [8], p. 53.

[71] R. McGrath, A.A. MacDowell, T. Hashizume, F. Sette and P.H. Citrin, Phys. Rev. 40 (1989) 9457.

Surface Science 299/300 (1994) 219–232
North-Holland

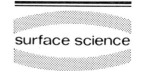

surface science

Low energy ion scattering and recoiling

J. Wayne Rabalais

Department of Chemistry, University of Houston, Houston, TX 77204-5641, USA

Received 24 January 1993; accepted for publication 2 April 1993

Ion scattering spectrometry was developed as a surface elemental analysis technique in the late 1960's. Further developments during the 1970's and 80's revealed the ability to obtain surface structural information. The recent use of time-of-flight (TOF) methods has led to a surface crystallography that is sensitive to all elements, including hydrogen, and the ability to directly detect hydrogen adsorption sites. TOF detection of both neutrals and ions provides the high sensitivity necessary for non-destructive analysis. Detection of atoms scattered and recoiled from surfaces in simple collision sequences, together with calculations of shadowing and blocking cones, can now be used to make direct measurements of interatomic spacings and adsorption sites within an accuracy of ≤ 0.1 Å. Structures are determined by monitoring the angular anisotropies in the scattered primary and recoiled target atom flux. Applications of such surface structure and adsorption site determinations are in the fields of catalysis, thin film growth, and interfaces. This article provides a short historical account of these developments along with some examples of the most recent capabilities of the technique.

1. Historical perspective

The origin of scattering experiments has its roots in the development of modern atomic theory at the beginning of this century. As a result of both the Rutherford experiment on the scattering of alpha particles (He nuclei) by thin metallic foils and the Bohr theory of atomic structure, a consistent model of the atom as a small massive nucleus surrounded by a large swarm of light electrons was confirmed. Following these developments, it was realized that the inverse process, namely analysis of the scattering pattern of ions from crystals, could provide information on composition and structure. This analysis is straightforward because the kinematics of energetic atomic collisions is accurately described by classical mechanics. Such scattering occurs as a result of the mutual Coulomb repulsion between the colliding atomic cores, that is the nucleus plus core electrons. The scattered primary atom loses some of its energy to the target atom. The latter, in turn, recoils into a forward direction. The final energies of the scattered and recoiled atoms and the directions of their trajectories are determined by the masses of the pair of atoms involved and the closeness of the collision. By analysis of these final energies and angular distributions of the scattered and recoiled atoms, the elemental composition and structure of the surface can be deciphered.

Low energy (1–10 keV) ion scattering spectrometry (ISS) had its beginning as a modern surface analysis technique with the 1967 work of Smith [1], which demonstrated both surface elemental and structural analysis for CO chemisorbed on Mo; screening of the scattering from C by the presence of O showed that CO chemisorbed on Mo with the C end down and the O end up. During the 1970's, work by Heiland and Taglauer [2], Brongersma et al. [3], Boers et al. [4], and Bronckers et al. [5] clearly demonstrated that direct structural information could be obtained from ISS.

Interest in ISS as a technique for investigating surface structure grew quickly after the 1982 work

SSDI 0039-6028(93)E0301-A

of Aono et al. [6], which showed that the use of backscattering angles near 180° greatly simplified the scattering geometry and interpretation. This allowed experimental determination of the shadow cone radii which could be applied to surfaces with unknown or reconstructed structures in order to determine surface geometry. There are two problems with this technique: (i) It analyzes only the scattered ions; these are typically only a very small fraction ($<5\%$) of the total scattered flux. Thus, high primary ion doses are required for spectral acquisition which are potentially damaging to the surface and adsorbate structures. (ii) Neutralization probabilities are a function of the ion beam incidence angle α with the surface and the azimuthal angle δ along which the ion beam is directed. This is not a simple behavior since the probabilities depend on the distances of the ion to specific atoms. As a result, it is difficult to separate scattering intensity changes due to neutralization effects from those due to structural effects. Aono and Souda [7] also proposed the use of ion scattering to probe the spatial distributions of surface electrons. This work successfully demonstrated that the phenomenon of ion neutralization could be used to probe the electron orbital angular distributions, however, ambiguity remained due to the fact that the information was derived from analysis of the scattered ions only. It was difficult to separate ion neutralization effects from shadowing and blocking effects. In 1984 Niehus [8] proposed the use of alkali primary ions which have low neutralization probabilities, leading to higher scattering intensities and pronounced focusing effects. The contamination of the sample surface by the reactive alkali ions is a potential problem with this method. Buck and coworkers [9], who had been developing time-of-flight (TOF) methods for ion scattering since the mid 1970's, used TOF methods for surface structure analysis in 1984 and demonstrated the capabilities and high sensitivity of the technique when both neutrals and ions are detected simultaneously. In 1987 van Zoest et al. [10] demonstrated that TOF analysis of both the scattered and recoiled neutrals and ions provided much more information about surface structure, however, they did not have adequate resolution in the TOF mode to separate the scattered and recoiled particles; this limited the use of the recoil data.

Having been developing TOF methods for scattering and recoiling, spectral interpretation, and surface elemental analysis in our own laboratory since the early 1980's [11], our research group began to concentrate its efforts on surface structural determinations in the late 1980's. In 1989 [12] we published structural determinations for O and H on W{211} using TOF detection of scattered and recoiled neutrals plus ions with sufficient resolution to clearly separate the recoiled and scattered particles. The first TOF spectrometer system with long flight path for separation of the scattered and recoiled particles and continuous variation of the scattering θ and recoiling ϕ angles was developed in our laboratory and published in 1990 [13]. This coupling of TOF methods with detection of both scattered and recoiled particles has led to the development of time-of-flight scattering and recoiling spectrometry (TOF-SARS) as a tool for structural analysis [14].

Several research groups [15–35] throughout the world are now engaged in surface structure determinations using some form of keV ISS. The number of research papers concerning surface structure determinations by means of some form of low energy (<10 keV) ion scattering and recoiling spectrometry published since 1975 are shown graphically in Fig. 1. It is obvious that the level of

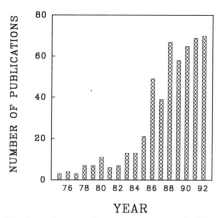

Fig. 1. Number of research papers concerned with surface structure determinations using some form of low energy (<10 keV) ion scattering or recoiling published in recent years.

research activity in this field grew tremendously during the late 1980's.

This historical overview will concentrate on time-of-flight scattering and recoiling spectrometry (TOF-SARS) [14], for it is felt that this is the technique that will be most important in future applications. Also in this overview, emphasis will be placed on surface structure determinations rather than surface elemental analysis, for it is felt that TOF-SARS is capable of making unique contributions in the area of structure determination.

2. Basic physics underlying keV ion scattering and recoiling

There are two basic physical phenomena which govern atomic collisions in the keV range. First, repulsive interatomic interactions, described by the laws of classical mechanics, control the scattering and recoiling trajectories. Second, electronic transition probabilities, described by the laws of quantum mechanics, control the ion–surface charge exchange process.

2.1. Atomic collisions in the keV range

The dynamics of keV atomic collisions are well described as binary collisions between the incident ion and surface atoms [36]. When an energetic ion makes a direct collision with a surface atom, the surface atom is recoiled into a forward direction as shown in Fig. 2. Both the scattered and recoiled atoms have high, discrete kinetic energy distributions. According to the laws of conservation of energy and momentum, the energy E_S of an incident ion of mass M_1 and energy E_0 which is scattered from a target atom of mass M_2 into an angle θ is given by

$$E_S = E_0 \left[\cos \theta + \left(A^2 \pm \sin^2 \theta \right)^{1/2} / (1 + A) \right]^2, \tag{1}$$

where $A = M_2/M_1$. For cases where $M_1 > M_2$, there exists a critical angle $\theta_c = \sin^{-1} A$ above which only multiple scattering can occur. Recoils that are ejected from single collisions of the pro-

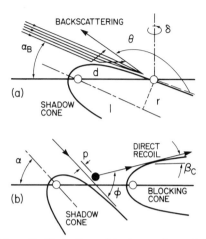

Fig. 2. Schematic illustrations of backscattering and shadowing and direct recoiling with shadowing and blocking.

jectile into an angle ϕ, that is direct recoils, have an energy E_R given by

$$E_R = E_0 \left[4A / (1 + A)^2 \right] \cos^2 \phi. \tag{2}$$

As a result of the energetic nature of the collisions, only atomic species are observed as direct recoils and their energies are independent of the chemical bonding environment.

2.2. Interatomic potentials

Although scattering in the keV range is dominated by repulsive potentials, it is not simply a hard sphere or billard ball collision where there is a clean "hit" or "miss". The partial penetration of the ion into the target atom's electron cloud results in bent trajectories even when there is not a "head-on" collision. This type of interaction is well described by a screened Coulomb potential [36] such as

$$V(r) = \left[Z_1 Z_2 e^2 / R \right] \Phi(R/CR_s), \tag{3}$$

where R is the internuclear separation and the Z_i are the atomic numbers of the collision partners. Φ is a screening function which is determined by R, the screening radius R_s, and a scaling parameter C; there are several good approximations for Φ [36]. Using such a potential, one can determine the relationship between the scattering angle θ and the impact parameter p. The p is defined as

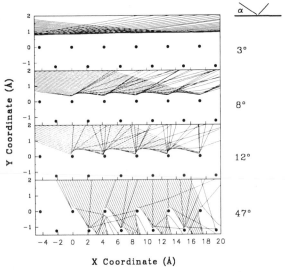

Fig. 3. Classical scattering trajectories for 4 keV Ne$^+$ imping-ing on a Ni surface at different incident angles α.

the minimum perpendicular distance from the target atom to the ion trajectory. A small value of p corresponds to a near head-on collision and backscattering and a large value corresponds to a glancing collision and forwardscattering. Simi-larly, the recoiling angle ϕ is also determined by p.

2.3. Shadowing and blocking cones

Considering a large number of ions with paral-lel trajectories impinging on a target atom, the ion trajectories are bent by the repulsive poten-tial such that there is an excluded volume, called the shadow cone, in the shape of a paraboloid formed behind the target atom as shown in Fig. 2. Ion trajectories do not penetrate into the shadow cone, but instead are concentrated at its edges much like rain pours off an umbrella. Atoms located inside the cone behind the target atom are shielded from the impinging ions as shown in Fig. 3. Similarly, if the scattered ion or recoiling atom trajectory is directed towards a neighboring atom, that trajectory will be blocked. For a large number of scattering or recoiling trajectories, a blocking cone will be formed behind the neigh-boring atom into which no particles can pene-trate. The dimensions of the shadowing and

blocking cones can be determined experimentally from scattering measurements along crystal az-imuths for which the interatomic spacings are accurately known. This provides an experimental determination of the scaling parameter C and reliable cone dimensions. The cone dimensions can also be constructed theoretically from the relationship of p with θ and ϕ [36,37]. A univer-sal shadow cone curve has been proposed [38] and cone dimensions for common ion–atom com-binations have been reported [39]. Since the radii of these cones are of the same order as inter-atomic spacings, that is 1 to 2 Å, the ions pene-trate only into the outermost surface layers.

2.4. Scattering and recoiling anisotropy caused by shadowing and blocking cones

When an isotropic ion fluence impinges on a crystal surface at a specific incident angle α, the scattered and recoiled atom flux is anisotropic. This anisotropy is a result of the incoming ion's eye view of the surface, which depends on the specific arrangement of atoms and the shadowing and blocking cones. The arrangement of atoms controls the atomic density along the azimuths and the ability of ions to channel, that is, to penetrate into empty spaces between atomic rows. The cones determine which nuclei are screened from the impinging ion flux and which exit trajec-tories are blocked as depicted in Fig. 3. By mea-suring the ion and atom flux at specific scattering and recoiling angles as a function of ion beam incident α and azimuthal δ angles to the surface, structures are observed which can be interpreted in terms of the interatomic spacings and shadow cones from the ion's eye view.

2.5. Ion–surface electronic transitions

Electron exchange [40] between ions or atoms and surfaces can occur in two regions, (i) along the incoming and outgoing trajectories where the particle is within ångströms of the surface and (ii) in the close atomic encounter where the core electron orbitals of the collision partners overlap. In region (i), the dominating processes are reso-nant and Auger electron tunneling transitions, both of which are fast ($\tau < 10^{-15}$ s). Since the work functions of most solids are lower than the ionization potentials of most gaseous atoms, keV scattered and recoiled species are predominately neutrals as a result of electron capture from the solid. In region (ii), as the interatomic distance R decreases, the atomic orbitals (AOs) of the separate atoms of atomic number Z_1 and Z_2 evolve into molecular orbitals (MOs) of a quasi-molecule and finally into the AO of the "united" atom of atomic number ($Z_1 + Z_2$). As R decreases, a critical distance is reached where electrons are promoted into higher energy MOs because of electronic repulsion and the Pauli exclusion principle. This can result in collisional reionization of neutral species. The fraction of species scattered and recoiled as ions is sensitive to atomic structure through changes in electron density along the trajectories.

3. Instrumentation

The basic requirements [13] for low energy ion scattering are an ion source, a sample mounted on a precision manipulator, an energy or velocity analyzer, and a detector. The sample is housed in an ultra-high vacuum (UHV) chamber in order to prepare and maintain well-defined clean surfaces. The UHV prerequisit necessitates the use of differentially pumped ion sources. Ion scattering is typically done in a UHV chamber which houses other surface analysis techniques such as low energy electron diffraction (LEED), X-ray photoelectron spectroscopy (XPS), and Auger electron spectroscopy (AES). The design of an instrument for ion scattering is based on the type of analyzer to be used. An electrostatic analyzer (ESA) measures the kinetic energies of ions while a time-of-flight (TOF) analyzer measures the velocities of both ions and neutrals.

Ion source and beam line – The critical requirements for the ion source are that the ions have a small energy spread, there are no fast neutrals in the beam, and the available energy is 1–5 keV. Both noble gas and alkali ion sources are common. For TOF experiments, it is necessary to pulse the ion beam by deflecting it past an aperature. A beam line for such experiments is shown

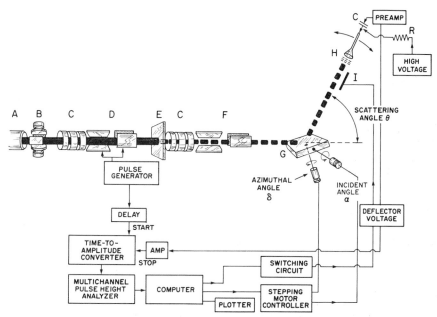

Fig. 4. Schematic drawing of TOF-SARS Spectrometer system. A = ion gun, B = Wien filter, C = Einzel lens, D = pulsing plates, E = pulsing aperture, F = deflector plates, G = sample, H = electron multiplier detector with energy prefilter grid and I = electrostatic deflector.

in Fig. 4; it is capable of producing ion pulse widths of ~ 15 ns.

Analyzers – An ESA provides energy analysis of the ions with high resolution. A TOF analyzer provides velocity analysis of both fast neutrals and ions with moderate resolution. In an ESA the energy separation is made by spatial dispersion of the charged particle trajectories in a known electrical field. ESA's were the first analyzers used for ISS; their advantage is high energy resolution and their disadvantages are that they analyze only ions and have poor collection efficiency due to the necessity for scanning the analyzer. A TOF analyzer is simply a long field-free drift region. It has the advantage of high efficiency since it collects both ions and fast neutrals simultaneously in a multichannel mode; its disadvantage is only moderate resolution.

Detectors – The most common detectors used for ISS are continuous dynode channel electron multipliers or channel plates which are capable of multiplying the signal pulses by 10^6–10^7. They are sensitive to both ions and fast neutrals. Neutrals with velocities $\geq 10^6$ cm/s are detected with the same efficiency as ions.

4. Elemental analysis from scattering and recoiling

4.1. Qualitative analysis

TOF-SARS is capable of detecting all elements by either scattering, recoiling, or both techniques. TOF peak identification is straightforward by converting Eqs. (1) and (2) to the flight times of the scattered t_S and recoiled t_R particles as

$$t_S = L(M_1 + M_2)/(2M_1E_0)^{1/2}$$
$$\times \left\{ \cos\theta + \left[(M_2/M_1)^2 - \sin^2\theta\right]^{1/2} \right\} \quad (4)$$

and

$$t_R = L(M_1 + M_2)/(8M_1E_0)^{1/2} \cos\phi, \quad (5)$$

where L is the flight distance, that is, the distance from target to detector. Collection of neutrals plus ions results in scattering and recoiling

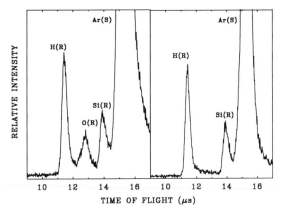

Fig. 5. TOF spectra of a Si(100) surface with chemisorbed H_2O (left) and H_2 (right). Peaks due to scattered Ar and recoiled H, O, and Si are observed. Conditions: 4 keV Ar^+, scattering angle $\theta = 28°$, incident angle $\alpha = 8°$.

intensities that are determined by elemental concentrations, shadowing and blocking effects, and classical cross sections. The main advantage of TOF-SARS for surface compositional analyses is its extreme surface sensitivity as compared to the other surface spectrometries, i.e. mainly XPS and AES. Indeed with a correct orientation and aperture of the shadow cone, the first monolayer can be probed selectively. At selected incident angles, it is possible to delineate signals from specific subsurface layers. Detection of the particles independently of their charge state eliminates ion neutralization effects. Also, the multichannel detection requires primary ion doses of only ~ 10^{11} ions/cm² or ~ 10^{-4} ions/surface atom for spectral acquisition; this ensures true static conditions during analyses.

TOF spectra – Examples of typical TOF spectra obtained from 4 keV Ar^+ impinging on a Si{100} surface with chemisorbed H_2O and H_2 are shown in Fig. 5 [41]. Peaks due to Ar scattering from Si and recoiled H, O, and Si are observed. The intensities necessary for structural analysis are obtained by integrating the areas of fixed time windows under these peaks.

4.2. Quantitative analysis

While qualitative identification of scattering and recoiling peaks is straightforward, quantitative analysis requires relating the scattered or recoiled flux to the surface atom concentration. The flux of scattered or recoiled atoms F depends upon the primary ion flux F_p and the experimental geometry $f(\theta, \phi)$ as well as the following three factors: (1) the concentration of surface atoms C; (2) the degree to which shadowing and blocking attenuates the scattered and recoiled particles, as represented by a masking factor $a(E_0, E_S, E_R, \theta, \phi))$ with range $0 \leq a \leq 1$; and (3) the differential scattering or recoil cross section $\sigma(E_0, E_S, E_R, \theta, \phi)$. This can be expressed as

$$F = F_p f C \sigma (1 - a). \qquad (6)$$

Compositional analyses by TOF-SARS and ISS have been applied in different areas of surface science, mainly in situations where the knowledge of the uppermost surface composition (first monolayer) is crucial. Some of these areas are as follows: gas adsorption, surface segregation, com-

pounds and polymer blends, surface composition of real supported catalysts, surface modifications due to preferential sputtering by ion beams, diffusion, thin film growth and adhesion.

5. Structural analysis from TOF-SARS

The atomic structure of a surface is usually not a simple termination of the bulk structure. A classification exists based on the relation of surface to bulk structure. A *bulk truncated* surface has a structure identical to that of the bulk. A *relaxed* surface has the symmetry of the bulk structure but different interatomic spacings. With respect to the first and second layers, *lateral relaxation* refers to shifts in layer registry and *vertical relaxation* refers to shifts in layer spacings. A *reconstructed* surface has a symmetry different from that of the bulk symmetry. The

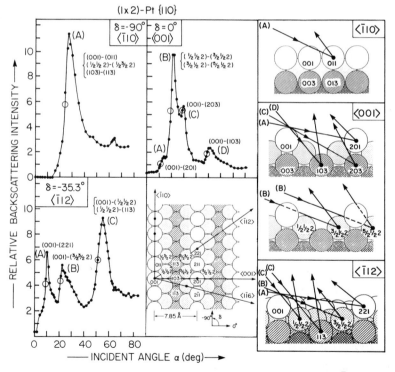

Fig. 6. Scattering intensity versus incident angle α scans for (1×2)-Pt{110} at $\theta = 149°$ along the $\langle \bar{1}10 \rangle$, $\langle 001 \rangle$ and $\langle \bar{1}12 \rangle$ azimuths. A top view of the (1×2) missing-row Pt{110} surface along with atomic labels is shown. Cross-section diagrams along the three azimuths illustrating scattering trajectories for the peaks observed in the scans are shown on the right.

methods of structural analysis will be delineated below.

5.1. Scattering versus incident angle α scans

When an ion beam is incident on an atomically flat surface at grazing angles, each surface atom is shadowed by its neighboring atom such that only forwardscattering (FS) is possible; these are large impact parameter (p) collisions. As α increases, a critical value $\alpha^i_{c,sh}$ is reached each time the ith layer of target atoms moves out of the shadow cone allowing for large angle backscattering (BS) or small p collisions as shown in Fig. 3. If the BS intensity I(BS) is monitored as a function of α, steep rises [42] with well defined maxima are observed when the focused trajectories at the edge of the shadow cone pass close to the center of neighboring atoms (Fig. 6). From the shape of the shadow cone, i.e. the radius (r) as a function of distance (l) behind the target atom (Fig. 2), the interatomic spacing (d) can be directly determined from the I(BS) versus α plots. For example, by measuring $\alpha^1_{c,sh}$ along directions for which specific crystal azimuths are aligned with the projectile direction and using $d = r/\sin \alpha^1_{c,sh}$, one can determine interatomic spac-

ings in the 1st atomic layer. The first/second-layer spacing can be obtained in a similar manner from $\alpha^2_{c,sh}$ measured along directions for which the first- and second-layer atoms are aligned, providing a measure of the vertical relaxation in the outermost layers.

5.2. Scattering versus azimuthal angle δ scans

Fixing the incident beam angle α and rotating the crystal about the surface normal while monitoring the backscattering intensity provides a scan of the crystal azimuthal angles δ [43]. Such scans reveal the periodicity of the crystal structure. For example, one can obtain the azimuthal alignment and symmetry of the outermost layer by using a low α value such that scattering occurs from only the 1st atomic layer. With higher α values, similar information can be obtained for the 2nd atomic layer. Shifts in the first/second-layer registry can be detected by carefully monitoring the $\alpha^2_{c,sh}$ values for 2nd-layer scattering along directions near those azimuths for which the 2nd-layer atoms are expected, from the bulk structure, to be directly aligned with the 1st-layer atoms. The $\alpha^2_{c,sh}$ values will be maximum for those δ values where the 1st- and 2nd-layer neighboring atoms are aligned.

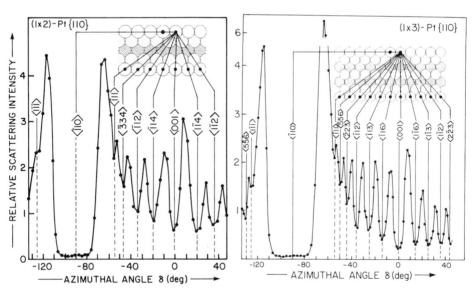

Fig. 7. Scattering intensity of 2 keV Ne$^+$ versus azimuthal angle δ scans for Pt{110} in the (1 × 2) and (1 × 3) reconstructed phases. Scattering angle $\theta = 28°$ and incident angle $\alpha = 6°$.

When the scattering angle θ is decreased to a forward angle ($< 90°$), both shadowing effects along the incoming trajectory and blocking effects along the outgoing trajectory contribute to the patterns. The blocking effects arise because the exit angle $\beta = \theta - \alpha$ is small at high α values. Surface periodicity can be read directly from these features [39] as shown in Fig. 7 for Pt{110}. Minima are observed at the δ positions corresponding to alignment of the beam along specific azimuths. These minima are a result of shadowing and blocking along the close-packed directions, thus providing a direct reading of the surface periodicity.

Azimuthal scans obtained for three surface phases of Ni{110} are shown in Fig. 8 [44]. The minima observed for the clean and hydrogen-covered surfaces are due only to Ni atoms shadowing neighboring Ni atoms, whereas for the oxygen-covered surface minima are observed due to both O and Ni atoms shadowing neighboring Ni atoms. Shadowing by H atoms is not observed because the maximum deflection in the Ne$^+$ trajectories caused by H atoms is $< 2.8°$.

5.3. Recoiling versus incident angle α scans

Adsorbates can be efficiently detected by recoiling them into forward scattering angles ϕ as shown in Fig. 2. As α increases, the adsorbate atoms move out of their neighboring atom shadow cones so that direct collisions from incident ions are possible. When the p value necessary for recoiling of the adsorbate atom into a specific ϕ becomes possible, adsorbate recoils are observed with the TOF predicted from Eq. (5). Focusing at the edge of the shadow cone produces sharp rises in the recoiling intensity as a function of α. By measuring $\alpha_{c,sh}$ corresponding to the recoil event, we can directly determine the interatomic distance of the adsorbate atom relative to its nearest neighbors from p and the shape of the shadow cone.

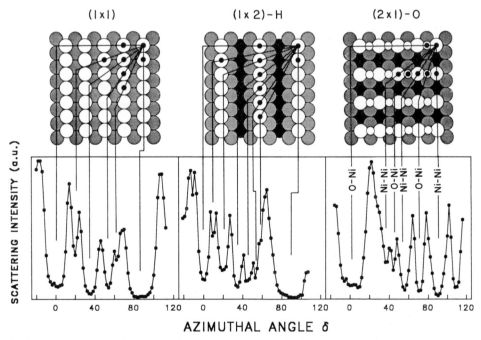

Fig. 8. Scattering intensity of 4 keV Ne$^+$ versus azimuthal angle δ for a Ni{110} surface in the clean (1×1), (1×2)-H missing row and (2×1)-O missing row phases. The hydrogen atoms are not shown. The oxygen atoms are shown as small open circles. O–Ni and Ni–Ni denote the directions along which O and Ni atoms, respectively, shadow the Ni scattering center.

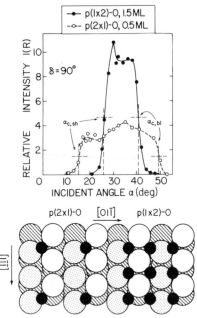

Fig. 9. Plots of oxygen recoil intensity versus incident angle α for both high (1.5 ML) and low (0.5 ML) oxygen coverages on a W{211} surface. The critical shadowing $\alpha_{c,sh}$ and blocking $\alpha_{c,bl}$ angle are indicated.

Example plots of oxygen recoil intensity versus α for two different chemisorbed O atom coverages on W{211} are shown in Fig. 9 [12]. Low dose exposure forms a $p(2 \times 1)$ structure consisting of 0.5 monolayer (ML) coverage and high dose exposure forms a $p(1 \times 2)$ structure consisting of 1.5 ML coverage. Sharp rises appear at low α and sharp decreases appear at high α. The rises are due to peaking of the ion flux at the edges of the shadow cones of neighboring atoms and the decreases are due to blocking of recoil trajectories by neighboring atoms. The critical α values for both shadowing, $\alpha_{c,sh}$, and blocking, $\alpha_{c,bl}$, can be used for determination of interatomic spacings. At high coverage, $\alpha_{c,sh} = 24°$ and $\alpha_{c,bl} = 42°$, which is considerably higher and lower, respectively, than the values $\alpha_{c,sh} = 16°$ and $\alpha_{c,bl} = 48°$ obtained at low coverage. This indicates that as coverage increases, both the shadowing and blocking effects are enhanced due to close packing of O atoms; this results from shadowing and blocking of O atoms by their neighboring O atoms. The α_c values correspond to O atoms

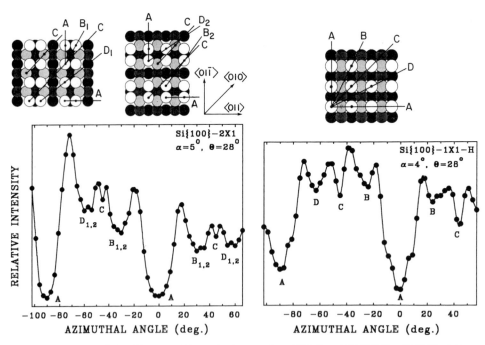

Fig. 10. Azimuthal angle δ scans of the silicon recoil intensity for the clean Si{100} and Si{100}-(1×1)-H dihydride surfaces. The minima are identified from the structural drawings above the scans. The hydrogen atoms are not shown.

separated by a distance of two W lattice constants at low coverage and one W lattice constant at high coverage.

5.4. Recoiling versus azimuthal angle δ scans

Plots of recoiling intensity versus azimuthal angle δ reveal the surface periodicity of the recoiled atoms in a manner similar to that of the scattering intensity. Azimuthal scans of Si recoils from the Si{100}-(2 × 1) and -(1 × 1)-H surfaces are shown in Fig. 10 [45]. The patterns are symmetrical about the ⟨011⟩ (δ = 0°) azimuth. The positions of the minima are consistent with the structures indicated above the figures. The repetition of the symmetry features every 90° indicates that there are two domains which are rotated by 90° with respect to each other. Azimuthal scans of H recoils from the Si{100}-(2 × 1)-H and Si{100}-(1 × 1)-H surfaces are shown in Fig. 11 [41]. The observed minima are due to Si atoms shadowing neighboring H atoms. The patterns are consistent with the structures indicated above the figures.

6. Ion–surface electron exchange from TOF-SARS

Ion–surface electron transition probabilities are determined by electron tunneling between the valence bands of the surface and the atomic orbitals of the ion. Such transition probabilities are highest for close distances of approach. Since TOF-SARS is capable of directly measuring the scattered and recoiled ion fractions, it provides an excellent method for studying ion–surface charge exchange. It has recently been shown that the charge transfer probabilities have a strong dependence on surface structure [46]. A direct method for measuring the spatial dependence of charge transfer probabilities with atomic-scale resolution has been developed using the method of direct recoil ion fractions [47]. The data demonstrate the need for an improved understanding of how atomic energy levels shift and broaden near surfaces. These types of measurements, combined with theoretical modeling, can provide a detailed microscopic map of the local reactivity of the surface. This information is of crucial importance for the understanding of vari-

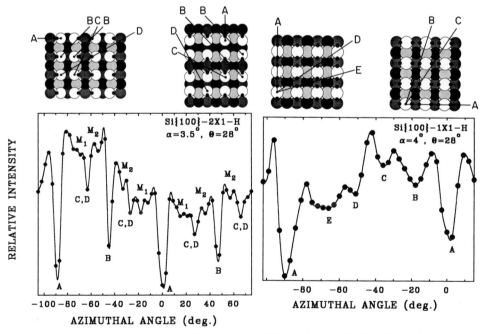

Fig. 11. Azimuthal angle δ scans of the hydrogen recoil intensity for a Si{100} surface in the (2 × 1)-H monohydride and (1 × 1)-H dihydride surfaces. The minima are identified from the structural drawings above the scans. The hydrogen atoms are indicated as small dark circles.

ous impurity-induced promotion and poisoning phenomena in catalysis and electron-density maps from scanning tunneling microscopy (STM).

7. Role of scattering and recoiling in the myriad of surface science techniques and expected new developments

Scattering and recoiling contribute to our knowledge of surface science through (i) elemental analysis, (ii) structural analysis, and (iii) analysis of electron exchange probabilities. We will consider the merits of each of these three areas.

7.1. Elemental analysis

The unique feature of scattering and recoiling spectrometry is the sensitivity to the outermost atomic layer of a surface. Using an ESA, it is possible to resolve ions scattered from all elements of mass greater than carbon. The TOF-SARS technique is sensitive to all elements, including hydrogen, although its limited resolution presents difficulties in resolving spectral peaks of high mass elements with similar masses. *The unique feature for elemental analysis is direct monitoring of surface hydrogen.* For general qualitative and quantitative surface elemental analyses, XPS and AES remain the techniques of choice.

7.2. Structural analysis

The major role of TOF-SARS is as a surface structure analysis technique which is capable of probing the positions of all elements with an accuracy of ≤ 0.1 Å. TOF-SARS is sensitive to short-range order, i.e., individual interatomic spacings along azimuths. It provides a direct measure of interatomic distances in the first and subsurface layers and a measure of surface periodicity in real space. *One of its most important applications is the direct determination of hydrogen adsorption sites by recoiling spectrometry* [12,41]. Most other structure techniques fail for H atom determinations, with the possible exception of He atom scattering, VLEED, and vibrational spectroscopy

TOF-SARS is complementary to low energy electron diffraction (LEED), which probes long-range order, minimum domain size of 100 to 200 Å, and provides a measure of surface and adsorbate symmetry in reciprocal space. Coupling TOF-SARS and LEED provides a powerful combination for surface structure investigations. The techniques of medium and high (Rutherford backscattering) energy ion scattering sample subsurface and bulk structure and are not as surface sensitive as TOF-SARS.

New developments in structural applications of ion scattering are occurring rapidly. Coaxial ion scattering, in which the ion beam goes through a hole in a channel plate detector, allowing detection of ~ 180° backscattered particles has been demonstrated [48,49]. Since this apparatus fits on a single flange, it is suitable for in situ monitoring of epitaxial film growth of atomic layers and for time-resolved analysis of dynamical surface processes. Improvements in optics will provide narrower ion pulse widths, resulting in enhanced time resolution of the spectra. Fast programs for computer simulation of scattering and recoiling trajectories have recently been developed [50]. These are extremely valuable in solving unknown structures by comparison of experimental data to simulations based on proposed structural models. Development of the simulations will make computer modeling of surface structures routine.

Future designs of scattering and recoiling instruments using large area detectors with position and time resolution of the events and fast electronics will be able to collect surface structural images on a very short time scale, allowing one to monitor dynamic processes at surfaces in real time. This will make it possible to monitor the dynamics of film deposition, chemisorption, surface reconstruction, surface diffusion, etc.

7.3. Ion–surface electron exchange probabilities

One of the unsolved problems in the interaction of low energy ions with surfaces is the mechanism of charge transfer and prediction of the charge composition of the flux of scattered, recoiled, and sputtered atoms. The ability to collect spectra of neutrals plus ions and only neutrals provides a direct measure of scattered and re-coiled ion fractions. Plots of ion fractions in inci-dent- and azimuthal-angle space provide elec-tronic transition probability contour maps which are related to surface electron density and reac-tivity along the various azimuths [47]. The appli-cation of TOF-SARS techniques to ion–surface electron exchange processes can develop into *a method* [47] *for determining local electron tunnel-ing rates within the surface unit cell.*

8. Conclusions

Emphasis in this article has been placed on the physical concepts and structural applications of TOF-SARS. TOF-SARS is now well established as a surface structural analysis technique that will have a significant impact in areas as diverse as thin film growth, catalysis, hydrogen embrittle-ment and penetration of materials, surface reac-tion dynamics, and analysis of interfaces. Surface crystallography is evolving from the classical con-cept of a static surface and the question of "where do atoms sit?" to the concept of a dynamically changing surface. The development of large area detectors with rapid acquisition of scattering and recoiling structural images, as described in Sec-tion 7, will provide a technique for capturing time-resolved snapshots of such dynamically changing surfaces.

Acknowledgements

This material is based on work supported by the National Science Foundation under grant no. CHE-8814337, the R.A. Welch Foundation under grant no. E-656, and the Texas Advanced Re-search Program under grant no. 3652022ARP.

References

[1] D.P. Smith, J. Appl. Phys. 38 (1967) 340.
[2] W. Heiland and E. Taglauer, Surf. Sci. 68 (1977) 96; 47 (1975) 234; Radiat. Eff. 19 (1973) 1; W. Heiland, F. Iberl, E. Taglauer and D. Menzel, Surf. Sci. 53 (1975) 383.
[3] H.H. Brongersma and J.B. Theeten, Surf. Sci. 54 (1976) 519; H.H. Brongersma and P. Mul, Surf. Sci. 35 (1973) 393.
[4] E.P.Th.M. Suurmijer and A.L. Boers, Surf. Sci. 43 (1973) 309.
[5] A.G.J. DeWit, R.P.N. Bronckers and J.M. Fluit, Surf. Sci. 82 (1979) 177.
[6] M. Aono, Y. Hou, R. Souda, C. Oshima, S. Otani, Y. Ishizawa, K. Matsuda and R. Shimizu, Jpn. J. Appl. Phys. 21 (1982) L670; M. Aono, Y. Hou, C. Oshima and Y. Ishizawa, Phys. Rev. Lett. 49 (1982) 567.
[7] M. Aono and R. Souda, Jpn. J. Appl. Phys. 24 (1985) 1249.
[8] H. Niehus, Surf. Sci. 145 (1984) 407; H. Niehus and G. Comsa, Surf. Sci. 140 (1984) 18.
[9] L. Marchut, T.M. Buck, G.H. Wheatley and C.J. McMa-hon, Jr., Surf. Sci. 141 (1984) 549.
[10] J.M. van Zoest, J.M. Fluit, T.J. Vink and B.A. van Hassel, Surf. Sci. 182 (1987) 179.
[11] J.W. Rabalais, J.A. Schultz, R. Kumar and P.T. Murray, J. Chem. Phys. 78 (1983) 5250.
[12] J.W. Rabalais, O. Grizzi, M. Shi and H. Bu, Phys. Rev. Lett. 63 (1989) 51; O. Grizzi, M. Shi, H. Bu, J.W. Rabalais, R.R. Rye and P. Nordlander, Phys. Rev. Lett. 63 (1989) 1408; O. Grizzi, M. Shi, H. Bu and J.W. Rabalais, Phys. Rev. B 40 (1989) 10127; H. Bu, O. Grizzi, M. Shi and J.W. Rabalais, Phys. Rev. B 40 (1989) 10147; M. Shi, O. Grizzi, H. Bu, J.W. Rabalais, R.R. Rye and P. Nordlander, Phys. Rev. B 40 (1989) 10163.
[13] O. Grizzi, M. Shi, H. Bu and J.W. Rabalais, Rev. Sci. Instrum. 61 (1990) 740.
[14] J.W. Rabalais, Science 250 (1990) 521.
[15] M. Aono, M. Katayama and E. Nomura, Nucl. Instrum. Methods B 64 (1992) 29.
[16] A.H. Al-Bayati, K.G. Orrman-Rossiter, R. Badheka and D.G. Armour, Surf. Sci. 237 (1990) 213.
[17] R. Ghrayeb, M. Purushotham, M. Hou and E. Bauer, Phys. Rev. B 36 (1987) 7364.
[18] B.J.J. Koeleman, S.T. de Zwart, A.L. Boers, B. Poelsema and L.K. Verheij, Phys. Rev. Lett. 56 (1986) 1152.
[19] Th. Fauster, Vacuum 38 (1988) 129.
[20] M. Chester and T. Gustafsson, Surf. Sci. 256 (1991) 135.
[21] W. Hetterich, C. Höfner and W. Heiland, Surf. Sci. 251/252 (1991) 731.
[22] D.J. O'Connor, B.V. King, R.J. MacDonald, Y.G. Shen and X. Chen, Aust. J. Phys. 43 (1990) 601.
[23] A.I. Dodonoy, E.S. Mashkova and V.A. Molchanov, Ra-diat. Eff. Def. Solids 110 (1989) 227.
[24] M.H. Mintz, U. Atzmony and N. Shamir, Surf. Sci. 185 (1987) 413.
[25] E. van de Riet and A. Niehus, Surf. Sci. 243 (1991) 43.
[26] H. Niehus, R. Spitzl, K. Besocke and G. Comsa, Phys. Rev. B 43 (1991) 12619.
[27] F. Shoji, K. Kashihara, K. Sumitomo and K. Oura, Surf. Sci. 242 (1991) 422.

[28] S.H. Overbury, D.R. Mullins, M.T. Paffett and B.E. Koel, Surf. Sci. 254 (1991) 45.

[29] E. Taglauer, M. Beckschulte, R. Margraf and D. Mehl, Nucl. Instrum. Methods B 35 (1988) 404.

[30] D.M. Cornelison, M.S. Worthington and I.S.T. Tsong, Phys. Rev. B 46 (1990) 4051.

[31] J. Vrijmoeth, A.G. Schins and J.F. van der Veen, Phys. Rev. B 40 (1989) 3121.

[32] J.H. Huang and R.S. Williams, Phys. Rev. B 38 (1988) 4022.

[33] M.J. Ashwin and D.P. Woodruff, Surf. Sci. 237 (1990) 108.

[34] E.G. Mcrae, T.M. Buck, R.A. Malic, W.E. Wallace and J.M. Sanchez, Surf. Sci. Lett. 238 (1990) L481.

[35] G. Bracco, M. Canepa, P. Catini, F. Fossa, L. Mattera, S. Terreni and D. Truffelli, Surf. Sci. 269/270 (1992) 61.

[36] E.S. Mashkova and V.A. Molchanov, Medium-Energy Ion Reflection From Solids (North-Holland, Amsterdam, 1985).

[37] J.F. Zeigler, J.P. Biersack and U. Littmark, The Stopping and Range of Ions in Solids (Pergamon, New York, 1985).

[38] O.S. Oen, Surf. Sci. 131 (1983) L407.

[39] C.S. Chang, U. Knipping and I.S.T. Tsong, Nucl. Instrum. Methods B 18 (1986) 11; B 35 (1988) 151.

[40] S.R. Kasi, H. Kang, C.S. Sass and J.W. Rabalais, Surf. Sci. Rep. 10 (1989) 1.

[41] M. Shi, Y. Wang and J.W. Rabalais, Phys. Rev. B 48 (1993) 1689.

[42] F. Masson and J.W. Rabalais, Surf. Sci. 253 (1991) 245; 258.

[43] F. Masson and J.W. Rabalais, Chem. Phys. Lett. 179 (1991) 63.

[44] C.D. Roux, H. Bu and J.W. Rabalais, Surf. Sci. 259 (1991) 253;
H. Bu, C.D. Roux and J. W. Rabalais, Surf. Sci. 271 (1992) 68.

[45] Y. Wang, M. Shi and J.W. Rabalais, Phys. Rev. B 48 (1993) 1678.

[46] C.C. Hsu and J.W. Rabalais, Surf. Sci. 256 (1991) 77.

[47] C.C. Hsu, H. Bu, A. Bousetta, J.W. Rabalais and P. Nordlander, Phys. Rev. Lett. 69 (1992) 188; Phys. Rev. B 47 (1993) 2369.

[48] Y. Wang, M. Shi and J.W. Rabalais, Nucl. Instrum. Methods B 62 (1992) 505.

[49] M. Aono, M. Katayama and E. Nomura, Nucl. Instrum. Methods B 64 (1992) 29.

[50] R.S. Williams, M. Kato, R.S. Daly and M. Aono, Surf. Sci. 225 (1990) 355;
S. Chaudhury and R.S. Williams, Surf. Sci. 255 (1991) 127.

Surface Science 299/300 (1994) 233–245
North-Holland

surface science

High energy ion scattering

L.C. Feldman

AT&T Bell Laboratories, 600 Mountain Avenue, Murray Hill, NJ 07974-0636, USA

Received 7 April 1993; accepted for publication 13 August 1993

High energy ion scattering has emerged as an important tool in the arsenal of surface science. This paper briefly reviews some of the underlying physics of the ion scattering/surface probe and describes the history and accomplishments of the technique.

1. Introduction

The surface structure problem – determination of the coordinates of the atoms in the first monolayer(s) of a solid – provided an exciting challenge to all physical scientists. One of the least likely communities to contribute might be the high energy ion solid scatterers, with roots in nuclear physics. The idea of MeV "cannon balls" determining eV phenomena seemed to defy simply physical sense. Nevertheless, high energy ion scattering has played a critical and prominent role in recent surface science.

"High energy" refers to positive ions with incident kinetic energies of greater than 25 keV and extends to the region of many MeV. Historically this regime is associated with nuclear physics and the basic processes are of a nuclear type. For surface analysis one uses the subset of well-understood nuclear processes to apply to material problems. The primary advantage of these ion scattering techniques is quantitative analysis, which results from the successful understanding of these processes by the nuclear physics community.

The fundamental relations that govern the particle–solid interaction and their use in surface analysis are relatively simple. Nuclear physicists were using ion scattering for the elemental analysis of their targets as early as 1951 [1]. However, it is in the last 20 years or so that these techniques have been carefully honed to a high level of sophistication as materials technology requirements became ever more stringent.

Ion scattering from a solid

The basis of MeV ion scattering for materials analysis rests on the fundamental relations that govern ion scattering in solids. Atomic and nuclear collisions can be characterized by the impact parameter of the scattering process (Table 1). The largest impact parameters possible in a solid are of the order of 1 Å. For such a distant collision, the fast ions interact primarily through excitation of valence electrons, with energy transfers, T, of the order of 10 eV/collision. These processes have cross sections of the order of atomic dimensions, $\sim 10^{-16}$ cm^2. At smaller impact parameters, the particles make "harder" collisions, inner shell excitation for example, with larger energy transfer and a smaller cross section. At the smallest impact parameter, the order of nuclear dimensions, 10^{-12} cm, the incident ion scatters from the atomic nucleus in a billiard ball type of collision with a large energy transfer, ~ 100 keV. The cross sections for these processes are $\sim 10^{-24}$ cm^2. The picture of an energetic particle traversing a solid consists of an ion losing energy gradually to electrons as it passes each monolayer, coming to rest microns below the surface. A small fraction of the particles will

SSDI 0039-6028(93)E0477-C

Table 1

Atomic and nuclear collisions characterized by the impact parameter of the scattering process

Impact parameter (b)	Scattering process	Energy transfer (T)
$\sim 1\ \text{Å}$	Inelastic excitation of valence electrons	$T_e \approx 10\ \text{eV}$
$\sim 10^{-1}\ \text{Å}$	Inelastic excitation of L-shell electrons	$T_e \approx 100\ \text{eV}$
$\sim 10^{-2}\ \text{Å}$	Inelastic excitation of K-shell electrons	$T_e \approx 1\ \text{keV}$
$\sim 10^{-4}\ \text{Å}$	Elastic scattering from nuclei	$T_{nuc} \approx 100\ \text{keV}$

undergo energetic nuclear encounters before coming to rest. It is these nuclear encounters that represent the signal in ion scattering studies. The final energy of the particle determines the depth at which the scattering event occurred, since the particle will lose energy to electrons in penetrating the solid. Electron encounters are generally called inelastic events while the nuclear backscattering event is usually denoted an elastic event.

Close collisions can initiate a number of atomic or nuclear processes which are useful in materials analysis. Small impact parameters processes may give rise to inner shell electron excitation with the subsequent emission of a characteristic X-ray or Auger electron; this analysis process is known as particle-induced X-ray emission or particle-induced Auger emission. If the impact parameter is of the order of the nuclear size, a nuclear reaction may occur; this process is part of ion beam analysis and denoted by the abbreviation NRA, nuclear reaction analysis. The most probable process at small impact parameters is elastic scattering; this is commonly known as RBS, Rutherford backscattering spectrometry. This latter process is, by far, the most used for surface characterization. Nuclear reaction analysis is also used for quantitative light atom detection (hydrogen, deuterium, carbon, oxygen) and is important for surface and thin film analysis.

The use of ion backscattering and nuclear reactions as a quantitative surface analysis tool depends on an accurate knowledge of the nuclear and atomic scattering processes. The calculation of the probability of this process assumes a uniform distribution of impact parameters. However, when the ion beam is aligned with a major symmetry direction of a single crystal the impact

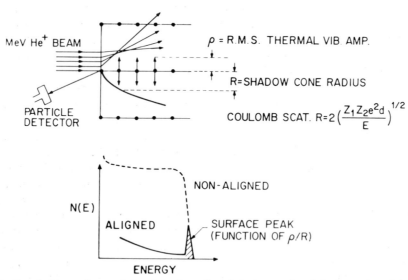

Fig. 1. Schematic showing the interactions at the surface of an aligned single crystal and the formation of the shadow cone. The energy spectra for the aligned and non-aligned case are shown below. The Coulomb shadow cone radius, R, is given by $2(Z_1 Z_2 e^2 d/E)^{1/2}$, where Z_1 and Z_2 are the atomic numbers of the beam and substrate, respectively, d is the spacing of atoms along the row and E is the incident kinetic energy.

parameter distribution *inside* the crystal is grossly modified, reducing the intensity of the small impact parameter processes. This phenomenon is known as "channeling". The impact parameter distribution remains uniform, however, at the first monolayer(s). Thus, the scattering intensity at the surface remains high, the bulk is suppressed, and the "surface peak" constitutes the surface signal. The concept of surface scattering is developed more completely in Refs. [2–5].

2. The surface peak

The discovery of the channeling process led to great activity in the "particle–solid interaction" community. The emphasis was on the phenomenon itself and its application for thin film analysis – not on surface science. Aarhus University, Denmark, became the center for channeling activity because of its excellent experimental program, tied to their strong theoretical science in the field of atomic collisions in solids.

In his early paper, entitled "Influence of Crystal Lattice on Motion of Energetic Charged Particles", Lindhard gave an elegant description of channeling [6]. Almost as an aside, he also discussed the basic effect which is used in the surface application, namely, the classical Rutherford shadow behind one atom. He considered a geometry as illustrated in Fig. 1 and showed that the intensity distribution, $f(r)$, a distance d behind the atom, is given by

$$f(r) = \begin{cases} 0, & r < R_C, \\ \frac{1}{2}\left[\left(1 - R_C^2/r^2\right)^{-1/2} \right. \\ \left. + \left(1 - R_C^2/r^2\right)^{1/2}\right], & r > R_C, \end{cases}$$

where R_C is the Coulomb shadow cone radius, and r is the perpendicular distance from the atom. The Coulomb shadow cone radius is given by

$$R_C = 2\left(Z_1 Z_2 e^2 d/E\right)^{1/2},$$

where Z_1 and Z_2 are the atomic numbers of the projectile and scattering atom, respectively, and E is the incident ion energy. The flux distribution

is illustrated in Fig. 1. There is no intensity for $r < R_C$ and then a "compensating" peak followed by an almost constant intensity. Note that the distribution is very sharp on the scale of thermal vibration amplitudes. The shadow cone radius increases as $d^{1/2}$ and the beam cannot penetrate close enough to the second atom to undergo any close encounter event. Usually one monitors this process by observing the yield of large angle Rutherford scattered ions (the close encounter process) from the surface region. This elementary picture of scattering from an aligned atomic pair may be thought of as the simplest example of the influence of atomic ordering on a close encounter probability.

This picture of scattering from an isolated string of atoms has been applied mostly to light projectiles (H or He) in the energy range of 0.1 to 2.0 MeV. In this region the shadow cone radius can be ~ 0.05 to 0.3 Å. This distance is comparable to thermal vibration amplitudes in real materials. Thus, the thermal motion of the atoms must be taken into account. Essentially one considers the probability that the second atom extends beyond the shadow cone and thus interacts with the beam. In this picture the intensity of scattering from the second atom, I_2, is given by

$$I_2 = \int f(r) P(r) \, dr$$

where $P(r)$ is the position distribution of the atom,

$$P(r) = \frac{1}{\pi \rho^2} e^{-r^2/\rho^2},$$

and ρ is the two-dimensional root mean square thermal vibration amplitude. Note that this equation assumes that the interaction being measured is for a close encounter (i.e., Rutherford backscattering) characterized by a distance which is small compared to the shadow cone or the vibration amplitude. This two-atom surface peak intensity, I_2, can be expressed as a function of one variable, ρ/R_C.

In real applications, the Coulomb scattering approximation and the two-atom approximation are not sufficient. Then the calculation of the surface peak intensity is usually carried out via numerical techniques where the governing potential is of a screened Coulomb type and many atoms along the string are considered [7,8]. One usually measures the scattering contribution from all atoms in the string so the surface peak intensity I is given by

$$I = 1 + \sum_{n=2} \int f_n(r) P_n(r) \, dr_n,$$

where $f_n(r)$ is the flux distribution at the nth atom and $P_n(r)$ the position distribution of the nth atom. A scaling of the surface peak intensity in terms of ρ / R_C was derived via the Lindhard distribution [5]. Stensgaard et al. found that the results of many computer stimulations can be parameterized in terms of ρ / R_M (Fig. 2) where R_M is the shadow cone radius using the screened Coulomb (Molière) potential [8]. All of the scattering parameters are expressed in terms of the shadow cone; the thermal motion of the solid is represented by ρ. This universal scaling of the surface peak was experimentally verified (Fig. 2) and added a concrete underpinning to the picture of the interaction of ion beams with clean surfaces.

An underlying assumption in these calculations, and our picture of the surface scattering process, is the validity of the classical approximation. This assumption was proved valid in Lindhard's discussion of the quantal shadow behind an atom. The classical picture is important. It allows for a simple conceptual description of the scattering process, helpful not only in calculation but in conceiving and interpreting experiments.

This surface peak calculation also assumes that the close encounter event is characterized by an interaction distance which is small compared to the shadow cone or vibration amplitudes. The use and power of this assumption is explicitly illustrated in the following. The two-atom interaction probability may be explicitly written as:

$$I_2 = \int f(r_b) P(r_a) I_{int}(r_a, r_b) \, dr_a \, dr_b,$$

where r_a, r_b, are the position coordinates of the crystal atom and beam, respectively. $I_{int}(r_a, r_b)$ is the interaction potential that describes the scattering. For very close encounter events, such as

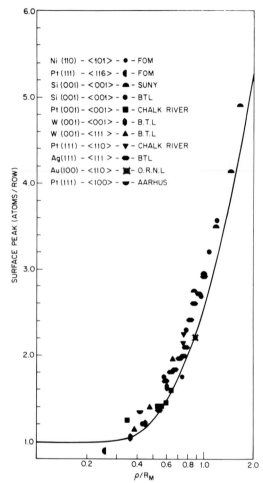

Fig. 2. Surface peak intensity (atoms/row) measurements as a function of the parameter ρ / R_M for surfaces displaying a (1×1) LEED pattern. The full line is the computer-simulation-generated "universal" curve (from L.C. Feldman, Nucl. Instrum. Methods 191 (1981) 211, and references therein).

Rutherford backscattering, it is clear that the scattering occurs as a "point" interaction and can be described by a delta function. Then I_2 reduces to the simplified form given above, $\int f(r) P(r) \, dr$. This concept was first written down by Barrett [7] and is denoted in the field as the "nuclear encounter probability approximation". It represents a great simplification in the computer simulation of all RBS/channeling phenomena, including the surface interaction. An example where the interaction potential is not "δ-like" was described by

MacDonald et al. who examined the extended interaction potential associated with Auger electron emission by H and He ion beams [9].

The use of such Monte Carlo simulations for the channeling/blocking geometry represented a further escalation in complexity, since the trajectory must be followed on the in-going and outgoing path. A discussion of such simulations for this double alignment configuration was given by Tromp and van der Veen [10].

These successful simulation methods have played an extremely important role in ion beam surface crystallography providing interpretation of data, insight into the uniqueness of physical models and the tool to design and anticipate new investigations.

2.1. Measurement of the surface peak

The surface peak intensity is measured in the channeling geometry. That is, the ion beam is aligned with a symmetry direction of the crystal, usually in an axial channeling direction. The beam scatters from the first monolayer of the solid at least. Scatterings from the next few atoms depend on ρ/R_M as described previously. Deeper in the crystal most of the ion beam acquires channeling trajectories; such channeled particles do not undergo close encounter events. The channeling effect suppresses the scattering from the bulk, thus allowing a clean definition of the surface peak. A channeling and non-channeling spectrum from W(001) is shown in Fig. 3. Note the strong suppression of scattering from the bulk of the crystal (a factor of ~ 100) and the sharp definition of the surface peak. The surface peak intensity can be easily and accurately measured with an accuracy of $\sim 3\%$.

2.2. Use of surface peak

The conceptual use of the surface peak intensity in surface structure determination is illustrated in Fig. 4. The top panel shows the surface peak intensity from an ideal, "bulk-like" single crystal. The second panel shows reconstruction, that is translation of the first atomic layer in the plane of the surface. Since the first atom no

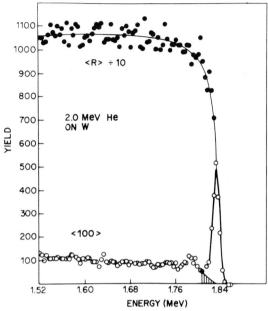

Fig. 3. Backscattering spectra for 2.0 MeV He ions incident on a clean W(100) surface along the normal [100] axis (open circles) and for the beam in a non-aligned (closed circles) direction. The method of background subtraction is illustrated by the cross-hatched area under the SP.

longer shadows the second atom in the string there is an increase in the surface peak. The third panel shows surface relaxation, a change in the spacing between the first two layers. In this case a surface peak increase is observed in an off-normal channeling direction, but not in the normal direction. Finally, the lower panel shows the case of an adsorbate, registered to the bulk-like substrate. The adsorbate shadows the substrate leading to a decrease of the substrate surface peak. All of these geometries, and many more, have been used in a large variety of surface physics studies.

2.3. The surface shadowing effect

As an example of a surface study we consider the effect shown in Fig. 4d – the shadowing effect. This is perhaps the clearest example of the classical Rutherford shadow described by Lindhard. Indeed in his 1964 paper he effectively described a "gedanken" experiment in the semiconductor ZnS. For this semiconductor, crystal directions can be found in which the surface monolayer consists of all Zn atoms, shielding the underlying S from the ion beam. As it turned out shadowing was most easily seen in metallic systems. Consider the shadowing effect associated with ~ 1 monolayer of Au epitaxially deposited onto a Ag(111) surface. This case had been reported by E. Bøgh [11] in some of the earliest ion scattering/surface studies. It was studied in detail by our group at Bell Labs [12] and represents one of the clearest examples of simple, classical epitaxy.

Fig. 5 shows a series of backscattering spectra from a Ag crystal with various Au coverages. The existence of the shadowing effect is clearly visible as a decrease in the Ag SP intensity with increasing Au coverage. The Au is sitting at fcc-like sites on bulk-like positions. Note that these measure-

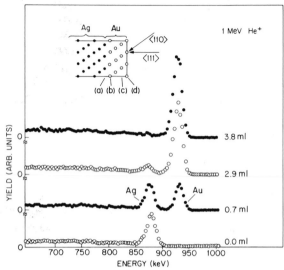

Fig. 5. Room temperature, ⟨110⟩, aligned energy spectra from 1.0 MeV He$^+$ incident on: (a) clean Ag(111); (b) Ag(111)+0.7 monolayer of Au; (c) Ag(111)+2.9 monolayer of Au; and (d) Ag(111)+3.8 monolayers of Au (from Ref. [9]).

ments are along the (110) direction to increase the surface sensitivity. In the (110) direction every atomic string begins in the surface while in the (111) direction only one in three strings begins in the surface. Careful observation of the figure will show that the Au intensity does not scale precisely with coverage. This is due to the Au–Au blocking effect in which second-layer Au atoms are blocked by surface atoms, the formation of Au epitaxial growth. Classical studies such as these built an underlying framework for the use of ion beams in surface structure determination.

3. History

As mentioned above, the earliest applications of this type of surface exploration were carried out by E. Bøgh of Aarhus University [11]. He recognized its potential and was the first to illustrate some of the advantages.

The timing was just about correct. Up to about 1975 ion scattering had its greatest influence in the thin film world. It was used for detailed studies of ion implantation, semiconductor metalization and other thin film issues, and had be-

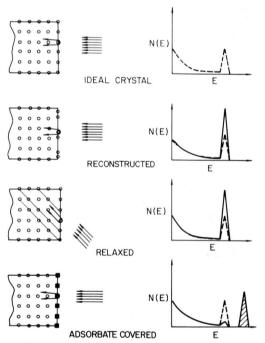

Fig. 4. Schematic of the dependence of the intensity of the SP on different crystal surface structures: (a) the ideal crystal SP from "bulk-like" surface; (b) enhanced SP observed in normal incidence for a reconstructed surface; (c) enhanced SP observed in non-normal incidence for a relaxed surface; (d) reduced SP observed in normal incidence for a registered overlayer.

come an established technique. The nuclear technology had been successfully transferred to the solid state materials community. Scientists in the field who were "technique oriented" sought another challenge. The forefront issue of solid state structure, at that time, was surfaces; thus, the ion scatterers sought to make a contribution to this challenging field.

The initial and formidable task was to make ourselves UHV compatible. For most of us this meant moving from the world of 10^{-6} Torr to 10^{-10} Torr and to accepting 24 hour bake-outs, long turn-around times and relatively slow experiments. It also meant we had to be able to talk to the surface science community – we required Auger spectroscopy, LEED apparatus, residual gas analysis, UHV compatible goniometers – all somewhat foreign to the nuclear physicists – at least at that time. I remember one colleague finally reaching state-of-the-art. As he said: "Now I can go to a surface science meeting and ask a speaker: 'What are *your* vacuum conditions?'".

In the final analysis each of these instrumental problems was solved and the ion scatterers were able to contribute to surface science.

Special recognition must be afforded to the channeling/blocking technique (double alignment) developed by Saris and co-workers at FOM [3,13]. As depicted in Fig. 6 the geometry and the information that could be obtained was qualitatively different than single alignment, the simple measure of the surface peak. The yield as a function of the exit angle could be thought of as a differential measure (relative to single alignment) with the promise of the added information accompanying differential measurements. The intricate equipment necessary to facilitate this measurement was created at the FOM lab making such measurements accessible.

In "double alignment" the surface information was primarily contained in the angular distribution of the backscattered ions. The qualitative surface structure provided by "angular shifts", was more clearly interpretable than experiments based on intensity calculations of a single alignment surface peak. One of the earliest examples of the success of this method was the determination of adsorbate position in the case of Ni(110)-

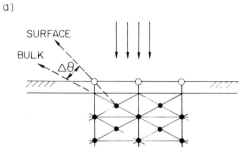

a)

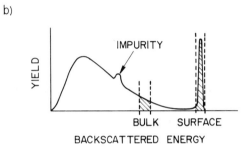

b)

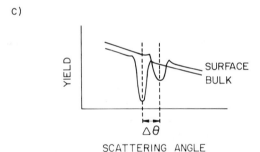

c)

Fig. 6. (a) Geometry for a double alignment experiment. (b) Backscattered energy spectrum for a light impurity on a heavier adsorbate. (c) Angular scan for scattering from the bulk and from the surface. The shift, $\Delta\theta$, indicates the relative lattice spacing between the surface layer and the bulk (redrawn from Ref. [13]).

c(2 × 2)S [14]. Such a light adsorbate surface site determination is essentially impossible by single alignment. In the long run critical quantitative information was still dependent on computer simulation with uncertainties described later in this paper.

Success and the O/A curve

Early successes were plentiful and straightforward. The published results would almost always

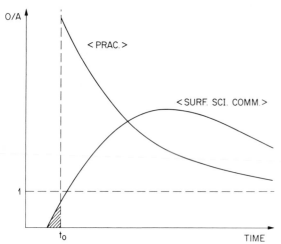

Fig. 7. The O/A (optimism/actuality) factor for surface science techniques for two communities; ⟨prac⟩, the practitioners of the technique, and ⟨surf sci com⟩, the balance of the surface scientists. The time t_0, represents the time of conception of the technique.

be classified as one of the schematics shown in Figs. 4 or 6. Led by Ivan Stensgaard, Bell Labs' first ion scattering surface science emerged from the measurement of the W(100) surface peak with and without hydrogen coverage [15]. It was a classic case of observing an adsorbate-induced reordering and was textbook in its qualitative explanation. Indeed to measure the predicted surface peak intensity was an accomplishment.

Results such as the W(001) story and other "classic" studies appeared in the literature. Since I also knew what never got published I was able to develop the O/A curve – which I used to end many a talk at that time.

Many new surface structure probes were appearing at this time and I believe the O/A curve applied to all – I know it applied to ion scattering. This curve (Fig. 7) was the optimism/actuality ratio for a surface science technique and is determined for two sets of people, the practitioners of the technique ⟨prac⟩ and the rest of the surface science community ⟨surf sci com⟩. The practitioners are the most optimistic at inception and have a high O/A. The rest of the surface science community are non-believers – they found it hard to accept that an MeV particle

beam will solve the delicate surface science problem. Difficulties in interpretation, non-reproducible surfaces, ion beam damage begin to temper the zeal of the practitioner and O/A begins to fall for them. At the same time successes are beginning to appear in the literature, engaging the rest of the surface science community and making them enthusiastic converts, high O/A. Eventually the "word" gets out, realism sets in, failures are documented and all converge. Fortunately, the final value remains finite.

Three more points need some classification in finishing this retrospective.

(i) Difficulty of thermal vibrations

Some of the simplest scenarios for the case of ion scattering depend on a straightforward measure of the surface peak. This intensity could be easily and accurately measured to 3%. However, the detailed interpretation of the intensity in terms of a surface model depends on a detailed knowledge of atomic vibrations near the surface of a solid. In general, these quantities are not well known and this uncertainty in surface vibration amplitude led to an uncertainty in surface structure analysis. The best experiments were designed around this uncertainty but this, of course, reduced the experimental space considerably.

(ii) Success in quantitative analysis

As stressed in the introduction, these nuclear-like techniques lent themselves to accurate quantitative analysis. An early example was correlation of the W surface reconstruction with deuterium coverage, measured via a ^{3}He(d,p)^{4}He nuclear reaction. This quantitative ability has been recognized and used by the community and is a lasting contribution of ion scattering. As has been said, "a technique will survive if it does at least one thing best" – and ion scattering is the best at providing coverage information. This use is still growing.

(iii) Monolayer films

Finally, the mass resolution must be noted in a different context. Combined with surface isolation via the surface peak, ion scattering provides mass dispersive, structure dependent, near monolayer resolution. Thus, many recent contributions center about the monolayer/epitaxy problem rather than the clean surface. The ion scattering probe can be ideal for this kind of issue and, indeed, the community has become more interested in this problem. The realm of clean surfaces and surface defects, is now best covered by microscopic probes such as STM and grazing-angle X-ray diffraction.

4. Accomplishments

High energy ion scattering has made numerous contributions to our understanding of clean surfaces. A few of the highlights, at least from the author's point of view, are described below. This listing is not meant to be comprehensive, rather it is meant to indicate ion scattering based investigations that have had a strong influence in surface science.

4.1. Si(100) surface

The clean Si(100) surface exhibits a (2×1) LEED pattern. Energy dependent RBS measurements by Feldman et al. [16,17] and Tromp et al. [18,19] have revealed large lateral displacements in the top layer and extensive sub-surface distortion. Fig. 8 shows the extra number of visible Si layers above that expected for a bulk-like surface as a function of the Molière shadow cone radius. Even for a radius as large as 0.4 Å about a full extra monolayer remains visible, indicating that the atoms in the top layer are, on average, displaced by ≥ 0.4 Å. As the shadow cone is tuned to smaller radii (by increasing the energy of the primary beam), the number of visible monolayers rises until the equivalent of ~ 4 monolayers is seen at a shadow cone radius of 0.06 Å. This is clear evidence of sub-surface distortions, theoretically predicted in some Si(100) models but unverified prior to these ion scattering measurements.

Upon saturation of the Si(100) surface with atomic hydrogen the LEED pattern becomes (1×1) and the effective number of displayed Si layers is seen by RBS to decrease dramatically. This means that the dimers are broken up, result-

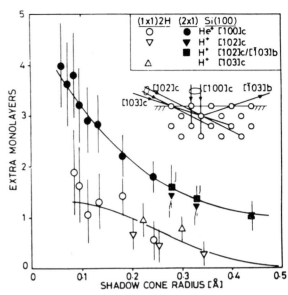

Fig. 8. The difference between the experimental and calculated SP intensity as a function of the shadow cone radius for Si(001) in the clean (full symbols) and hydrogen-covered (open symbols) state (from Ref. [3], and references therein).

ing in a more bulk-like arrangement in which (supposedly) two hydrogen atoms saturate the two dangling bonds of each surface atom. (Nuclear microanalysis of deuterium adsorption by Stengaard et al. [17] revealed a somewhat lower saturation coverage.)

4.2. Si(111)-7 × 7

The geometry of the Si(111)-7 × 7 surface has been a continuing challenge ever since the first observation of the (7×7) LEED pattern by Schlier and Farnsworth in 1959 [20]. Measurements of the surface peak intensity for normal beam incidence indicated a nearly bulk-like surface, while for non-normal incidence the equivalent of two extra monolayers was seen to be displaced from regular lattice sites by at least 0.4 Å [21,22]. This result is surprising since it is hardly conceivable that bond lengths could change by so much in a hard material such as Si. This notion led Peter Bennett [23] to propose a model based on stacking faults, i.e., stacking sequences of atom layers in the selvedge which are different from the cubic structure. When viewed along the normal, such a faulted surface would indeed appear bulk-like, but when viewed along a non-normal direction, it would exhibit atoms off cubic lattice sites, the number of atoms depending on the type and particular arrangement of stacking faults within the unit cell. However, non-cubic stacking alone cannot explain the STM tunneling microscope images. Therefore, McRae [24] proposed an arrangement of stacking faults, dividing the (7×7) unit cell in two differently stacked triangular units, with dimers of atoms lying on the sides of each unit and with deep holes at the corners, precisely where minima are found in the tunneling microscopy images. The model further assumes adatoms on positions corresponding to maxima in these images. The accepted picture of the dimer-adatom stacking fault model was put forth by Takayanagi et al. [25], including elements of all these different measurements and interpretation. The ion scattering data provided a critical role in revealing the surface stacking fault.

4.3. Surface relaxation

At first glance surface relaxation determination appears as an ideal application of ion scattering, as most of the information is contained in the angular deviation of the surface peak intensity. In many cases, however, the extraction of a surface

relaxation was not straightforward, again due to lack of knowledge of surface thermal vibrations. However, the solid theoretical basis of the technique permitted an estimate of the magnitude of this uncertainty and led to an important body of data. Ion scattering analyses of relaxation values for a number of unreconstructed (110) surfaces of face-centered cubic metals are shown in Table 2 (adapted from Ref. [3]). Comparison to low energy electron diffraction analysis is also shown. This is one of the most satisfying examples of detailed experimental agreement in surface science. Combined with LEED it provides a concrete picture of the surface relaxation process for the (110) surfaces of face-centered cubic metals.

4.4. Reconstructed metal surfaces

Among the most studied metal surfaces are the Au(110) and Pt(110) reconstructed faces. Part of the interest stems from the fact that their lower atomic number isoelectronic counterparts, (Cu(110)/Au, Ag(110)/Au and Ni(110)/Pt) are not reconstructed, thus confusing simple pictures of the driving force for surface reconstruction. Structures of these complex surfaces were extracted from channeling/blocking measurements by Gustafsson and his co-workers, showing clear evidence of a missing row reconstruction to explain the (2×1) LEED symmetry [43,44]. The

Table 2
Comparison of (multilayer) relaxation values determined by RBS and LEED for various metal surfaces (adapted from Ref. [3])

Surface	RBS			LEED		
	$\Delta d_{12}/d$ (%)	$\Delta d_{23}/d$ (%)	Ref.	$\Delta d_{12}/d$ (%)	$\Delta d_{23}/d$ (%)	Ref.
Ni(110)	-4 ± 1		[26]	-5 ± 1		[29]
	-4 ± 1		[27]	-7		[30,31]
	-4.8 ± 1.7	$+2.4 \pm 1.2$	[28]			
Cu(110)	-5.3 ± 1.6	$+3.3 \pm 1.6$	[32]	-10 ± 2.5	0 ± 2.5	[33]
				-8 ± 3		[34]
				-8.5 ± 0.6	$+2.3 \pm 0.8$	[35]
Ag(110)	-7.8 ± 2.5	$+4.3 \pm 2.5$	[36]	-7		[38]
	-9.5 ± 2.0	$+6.0 \pm 2.5$	[37]	-10		[39]
				-8		[40]
				-5.7	$+2.2$	[41]
Pb(110)	-16.6 ± 2.0		[42]			

results also show a large contraction of the first monolayer spacings, similar to the fcc non-reconstructed surfaces.

4.5. Adsorbate induced reconstruction

Adsorbates of low atomic number are known to influence surface ordering. Hydrogen, carbon monoxide, nitrogen and oxygen can radically affect the symmetry and chemical activity of a surface. The ion scattering tool is ideal for microscopic investigations of this phenomenon. The low Z adsorbates are invisible to the beam and the surface reconstruction is monitored by the surface peak. An example is the hex $\rightarrow (1 \times 1)$ transition in Pt(100) induced by CO adsorption. In a comprehensive study, using RBS, nuclear microanalysis, LEED and work function measurements, Jackman et al. [45] followed the phase transition in detail by monitoring the number of Pt atoms off bulk lattice sites and the corresponding CO coverage, while cooling the sample at fixed CO pressure (Fig. 9). The hex $\rightarrow (1 \times 1)$ transition was found to start abruptly when 0.08 $\pm$ 0.05 monolayer of CO had accumulated on the surface and to go to completion at 0.5 monolayer coverage. The hysteresis effect upon reversal of the transition (Fig. 9) was attributed to the heat of CO adsorption being larger on the (1×1) surface than on the hex surface. Other adsorbate induced transitions have been identified in W(100), Pt(111) and various Ni surfaces.

In addition to these more "classical" uses of ion surface scattering, particle beams have been employed in important studies of surface melting [46] and the lattice location of hydrogen on metallic surfaces [47].

5. Future

It is interesting to note that ion scattering studies of clean surfaces have actually diminished in number in recent years. This is probably due to a combination of three reasons:
(1) Interest in the macroscopic nature of clean surface has actually waned. The focus has drifted

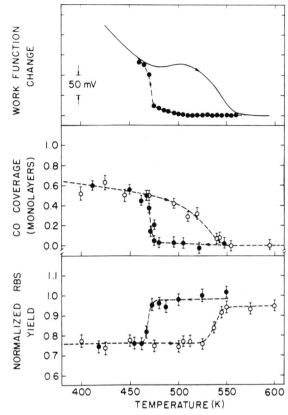

Fig. 9. The CO-induced hex $\leftrightarrow (1 \times 1)$ phase transition in Pt(100) monitored by work function change and RBS (from Ref. [45]).

to surface microscopy and surface defects, accessible to microscopic techniques.
(2) Ion scattering tends to be limited in its detail. Thus, high energy ion scatterers have probably accomplished most of what is accessible in such problems as the Si(111)-7 $\times$ 7. It is difficult to imagine a more complete ion scattering work then that represented by R. Tromp's thesis [48].
(3) In recent years ion scattering investigations were focussed more on thin film issues.

Thin film science

Ion scattering yields critical information on epitaxy, thin film reactivity and surface morphology. A clear example of the influence of epitaxy was shown in the previous section, the Au/

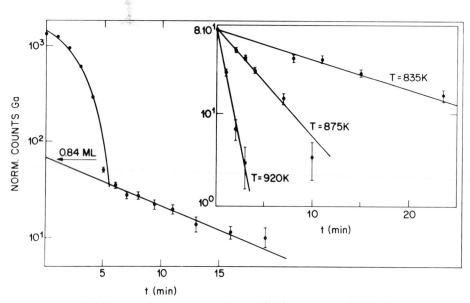

Fig. 10. Thermal desorption of 20 ML equivalent coverage of Ga on Si(111) deposited at 875 K. The upper branch represents the desorption of Ga clusters and the lower branch represents the desorption of the submonolayer Ga on Si(100) at three temperatures (from Ref. [49]).

Ag(111) case. The ability to probe an epitaxial film from the monolayer stage through to ~ 1000 Å has strong technical advantages. For strained layer epitaxy this means probing from sub-critical thickness to super critical thickness.

Thin film reactivity is critical to many technologies, particularly semiconductor devices. The outstanding example is the reaction of oxygen with silicon to form the critical Si/SiO_2 interface. Metal–semiconductor interactions and semiconductor–semiconductor interdiffusion have all been probed by ion scattering. At this point it should be noted that all such investigation are now multi-technique, involving TEM, SIMS and other materials and electrical probes.

A final example is surface morphology and surface kinetics. These kinds of analysis often involve the simplest aspect of ion scattering – quantitative coverage analysis and depth resolution. For example, in recent studies of Ga thin films on Si(111), Zinke-Allmang et al. showed large Ga cluster formation (Fig. 10) and were able to monitor the desorption rate and Ga binding energy associated with two components: the Ga within the clusters and the Ga adsorbed on the clean Si substrate [49].

Clearly, it is these thin film issues which will occupy the largest segment of ion scattering activity in the future.

Acknowledgements

One of the pleasures of doing science is the formation of personal collaborations among investigators in the same field. I have benefited from many such interactions. Some of those who have contributed to the topics in this paper are J.U. Andersen, J. Barrett, E. Bøgh, J.A. Davies, J. Lindhard, P.R. Norton, F. Saris, R.M. Tromp and J.F. van der Veen. Ion scattering collaborators at Bell Labs include N. Cheung, R.J. Culbertsen, H.-J. Gossmann, R. Haight, R.L. Headrick, Y. Kuk, T.E. Jackman, P.J. Silverman, I. Stensgaard, B.E. Weir and M. Zinke-Allmang.

References

[1] A.B. Brown, C.W. Snyder, W.A. Fowler and C.C. Lauritsen, Phys. Rev. 82 (1951) 159.

[2] L.C. Feldman and J.W. Mayer, Fundamentals of Surface and Thin Film Analysis (Elsevier, New York, 1986).

[3] J.F. van der Veen, Surf. Sci. Rep. 5 (1985) 199.

[4] L.C. Feldman, Ion Scattering from Surfaces and Interfaces, in: Ion Beams for Materials Analysis, Eds. J.R. Bird and J.S. Williams (Academic Press, Sydney, 1989) p. 413.

[5] L.C. Feldman, J.W. Mayer and S.T. Picraux, Materials Analysis by Ion Channeling (Academic Press, New York, 1982).

[6] J. Lindhard, Mat. Fys. Medd. Dan. Vidensk. Selsk. 34 (1965) 14.

[7] J.H. Barrett, Phys. Rev. B 3 (1971) 1527.

[8] I. Stensgaard, L.C. Feldman and P.J. Silverman, Surf. Sci. 77 (1978) 513.

[9] J.R. MacDonald, L.C. Feldman, P.J. Silverman, J.A. Davies, K. Griffiths, T.E. Jackman, P.R. Norton and W.N. Unertl, Nucl. Instrum. Methods 218 (1983) 765.

[10] R.M. Tromp and J.F. van der Veen, Surf. Sci. 133 (1983) 159.

[11] E. Bøgh, in: Channeling, Ed. D.V. Morgan (Wiley, London, 1975).

[12] R.J. Culbertson, L.C. Feldman, P.J. Silverman and H. Boehm, Phys. Rev. Lett. 47 (1981) 657.

[13] W.C. Turkenburg, W. Soszka, F.W. Saris, H.H. Kersten and B.G. Colenbrander, Nucl. Instrum. Methods 132 (1976) 587.

[14] J.F. van der Veen, R.M. Tromp, R.G. Smeenk and F.W. Saris, Surf. Sci. 82 (1979) 468.

[15] I. Stensgaard, L.C. Feldman and P.J. Silverman, Phys. Rev. Lett. 42 (1979) 247.

[16] L.C. Feldman, P.J. Silverman and I. Stensgaard, Nucl. Instrum. Methods 168 (1980) 589.

[17] I. Stensgaard, L.C. Feldman and P.J. Silverman, Surf. Sci. 102 (1981) 1.

[18] R.M. Tromp, R.G. Smeenk and F.W. Saris, Phys. Rev. Lett. 46 (1981) 939.

[19] R.M. Tromp, R.G. Smeenk, F.W. Saris and D.J. Chadi, Surf. Sci. 133 (1983) 137.

[20] R.E. Schlier and H.E. Farnsworth, J. Chem. Phys. 30 (1959) 917.

[21] R.J. Culbertson, L.C. Feldman and P.J. Silverman, Phys. Rev. Lett. 45 (1980) 2043.

[22] R.M. Tromp, E.J. van Loenen, M. Iwami and F.W. Saris, Solid State Commun. 44 (1982) 971.

[23] P.A. Bennett, L.C. Feldman, Y. Kuk, E.G. McRae and J.E. Rowe, Phys. Rev. B 28 (1983) 3656.

[24] E.G. McRae, Phys. Rev. B 28 (1983) 2305.

[25] K. Takayanagi, Y. Tanishiro, M. Takahashi and S. Takahashi, Surf. Sci. 164 (1985) 367.

[26] J.F. van der Veen, R.G. Smeenk, R.M. Tromp and F.W. Saris, Surf. Sci. 79 (1979) 212.

[27] E. Törnquist, E.D. Adams, M. Copel, T. Gustafsson and W.R. Graham, J. Vac. Sci. Technol. A 2 (1984) 939.

[28] R. Feidenhans'l, J.E. Sorensen and I. Stensgaard, Surf. Sci. 134 (1983) 329.

[29] J.E. Demuth, P.M. Marcus and D.W. Jepsen, Phys. Rev. B 11 (1975) 1460.

[30] Y. Gauthier, R. Baudoing, C. Gaubert and L. Clarke, J. Phys. C 15 (1982) 3223.

[31] Y. Gauthier, R. Baudoing and L. Clarke, J. Phys. C 15 (1982) 3231.

[32] I. Stensgaard, R. Feidenhans'l and J.E. Sorensen, Surf. Sci. 128 (1983) 281.

[33] H.L. Davis, J.R. Noonan and L.H. Jenkins, Surf. Sci. 83 (1979) 559.

[34] J.R. Noonan and H.L. Davis, Surf. Sci. 99 (1980) L424.

[35] D.L. Adams, H.B. Nielsen and J.N. Andersen, Surf. Sci. 128 (1983) 294.

[36] Y. Kuk and L.C. Feldman, Phys. Rev. B 30 (1984) 5811.

[37] E. Holub-Krappe, K. Horn, J.W.M. Frenken, R.L. Krans and J.F. van der Veen, Surf. Sci. 188 (1987) 335.

[38] E. Zanazzi, F. Jona, D.W. Jepsen and P.M. Marcus, J. Phys. C 10 (1977) 375.

[39] M. Maglietta, E. Zanazzi, F. Jona, D.W. Jepsen and P.M. Marcus, J. Phys. C 10 (1977) 3287.

[40] M. Alff and W. Moritz, Surf. Sci. 80 (1979) 24.

[41] H.L. Davies and J.R. Noonan, Surf. Sci. 126 (1983) 245.

[42] J.W.M. Frenken, J.F. van der Veen, R.N. Barnett, U. Landman and C.L. Cleveland, Surf. Sci. 172 (1986) 319.

[43] P. Fenter and T. Gustafsson, Phys. Rev. B 38 (1988) 10197.

[44] M. Copel and T. Gustafsson, Phys. Rev. Lett. 57 (1986) 723.

[45] T.E. Jackman, K. Griffiths, J.A. Davies and P.R. Norton, J. Chem. Phys. 79 (1983) 3529.

[46] J.W.M. Frenken and J.F. van der Veen, Phys. Rev. Lett. 54 (1985) 134.

[47] F. Besenbacher, I. Stengaard and K. Mortensen, Surf. Sci. 191 (1987) 288.

[48] R.M. Tromp and E.J. van Loenen, Surf. Sci. 155 (1985) 441.

[49] M. Zinke-Allmang, L.C. Feldman and M. Grabow, Surf. Sci. Rep. 16 (1993) 377.

Surface Science 299/300 (1994) 246–260
North-Holland

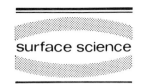

surface science

Surface analysis by secondary ion mass spectrometry (SIMS)

A. Benninghoven

Physikalisches Institut der Universität Münster, Wilhelm-Klemm-Strasse 10, 48149 Münster, Germany

Received 3 May 1993; accepted for publication 28 June 1993

Static secondary ion mass spectrometry (static SIMS) supplies detailed information on the atomic and molecular composition of the uppermost monolayer of a solid. This information is obtained by mass analysis of sputtered secondary ions emitted from this layer during keV primary ion bombardment. Damage to the uppermost monolayer is minimized by applying extremely low primary ion fluences. Progress in static SIMS over the past 30 years is described by following the most important developments in instrumentation, i.e. the introduction of single-ion counting techniques, quadrupole and time-of-flight mass analyzers, charge compensation devices, focused ion beams for imaging, UV laser postionization of sputtered neutrals and the on-line combination of SIMS with other devices for surface modification and analysis. Today, static SIMS allows high performance surface analysis of samples of any material, geometry and conductivity. In particular the atomic and molecular composition of real world samples, very often insulators and covered by complex mixtures of atomic and molecular surface components, can be obtained with high sensitivity and high lateral resolution. Present day applications of static SIMS include microelectronics, catalysis and polymer research as well as clinical analysis, environmental monitoring and all kinds of microstructure technologies. Fundamental research, in particular on the mechanism of molecular ion formation, is still in its very beginning.

1. Introduction

Secondary ion mass spectrometry (SIMS) is based on the mass analysis of positive and negative ions sputtered out of the uppermost monolayer of a solid during keV primary ion bombardment. Not only atomic species are generated but – most surprisingly and unexpectedly – even intact involatile large organic molecules survive the sputtering process, at least as large characteristic fragments [1]. SIMS is a surface analytical technique having all the advantages of mass spectrometry. Progress in SIMS over the last 30 years was strongly influenced by instrumental developments, for example, by the introduction of single-ion counting techniques, the replacement of sector fields by quadrupole and time-of-flight mass spectrometers, the introduction of laser postionization of sputtered neutrals or the use of focused ion beams for surface imaging.

My personal engagement in secondary ion mass spectrometry started in 1963, when we applied an open Geiger–Müller counter for single-ion detec-

tion in a magnetic sector field mass spectrometer [2]. With this instrument we studied the energy distribution of sputtered secondary ions and neutrals after their postionization by an electron beam. Our interest at that time concentrated on whether the sputtering process was a thermal process or the result of a direct momentum transfer. Today, we know that both can play an important role.

2. SIMS and surface analysis in the 60's

In the 60's a number of instruments for the investigation of secondary ion emission were operated. In all these instruments high primary ion current densities and correspondingly high erosion rates of many monolayers in a second were applied. This "dynamic" operation had two advantages: high secondary ion currents were generated, and surface contaminations originating from the ambient vacuum were reduced. Due to the high sputter rates applied, dynamic SIMS characterizes the bulk of an ion-bombarded sample. In order to get more information on the high energy tail of the secondary ion energy distribution, which we investigated at that time, we needed higher sensitivities. By using an open Geiger–Müller counter [2], we could increase this sensitivity by several orders of magnitude. This drastic gain in sensitivity was a most important precondition for developing "static" SIMS.

What techniques for surface analysis were available at that time? Adsorbed layers of small volatile molecules on solid surfaces could be identified by thermal desorption mass spectrometry, and the ion emission from a tip could be studied by field desorption mass spectrometry. The first real surface analytical technique, however, was Auger electron spectroscopy (AES), developing just at that time [3]. It supplied, for the first time, quantitative information on the elemental composition of the uppermost monolayer of any solid. AES, however, could not detect hydrogen and did not supply significant information on the chemical state of the detected atoms, i.e., molecular surface species could not be identified. This lack of molecular information was and is one of the most important drawbacks of AES.

3. Molecular secondary ion emission and the concept of static SIMS

Impressed by the rapid development of AES we considered the secondary ion emission at the very beginning of a sputter experiment – i.e., during sputtering of the originally uppermost monolayer – with regard to its analytical information. Applying our single-ion counting technique, we obtained details of the positive and negative spectra and recognized very soon that they contain important elemental and molecular information [4]. On the other hand, the primary ion bombardment removes surface species and in addition changes the chemical composition of the bombarded area. So we tried to reduce the primary ion current density to such a degree, that during recording a complete spectrum no significant sputtering of this uppermost monolayer occurred. This was the concept of static SIMS (Fig. 1). We realized this concept by bombarding a large surface area, in our first experiments up to 0.5 cm^2, by improving the transmission of our analyzer and, most importantly, by applying a single-ion counting technique for secondary ion detection. In addition, UHV conditions were ensured, e.g., during the investigation of surface reactions.

The static SIMS spectra which we obtained were very complex (Fig. 2a). They featured mass lines of elements, their isotopes, metal–oxygen complexes of the general composition $Me_mO_n^{\pm}$, hydrogen-containing metal–oxygen complexes, hydrocarbon ions, etc. The spectra demonstrated that static SIMS supplies information on all elements including hydrogen, their isotopic composition, and, in addition, most importantly, molecu-

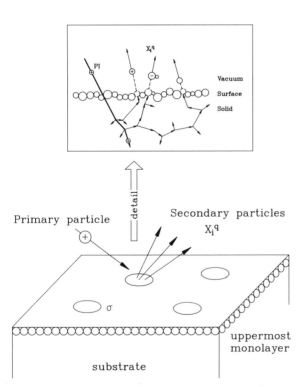

Fig. 1. Concept of static SIMS. A single primary ion impact results in the emission of charged and neutral secondary particles X_i^q (see text). The respective sputtered area of the uppermost monolayer can be characterized by a disappearance cross section σ. In a static SIMS experiment only a negligible overlap of these areas occurs.

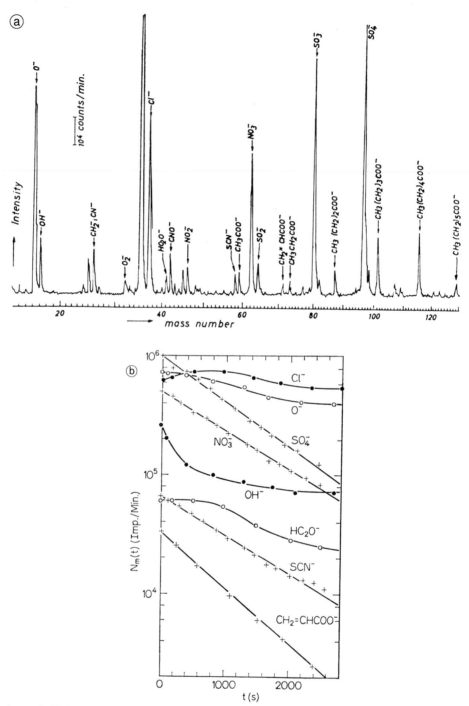

Fig. 2. (a) Negative static SIMS spectrum of a poisoned silver catalyst [4]. Besides atomic ions (O^-, Cl^-) many involatile molecular species (NO_3^-, SO_4^-, $CH_3(CH_2)_nCOO^-$) are detected. (b) Changes in secondary ion emission during sputtering of the uppermost monolayer of a silver catalyst (a) during extended primary ion fluence [4]. An exponential decrease indicates the presence of the corresponding surface species only in the uppermost monolayer.

lar information by means of the sputtered molecular ion species. The most serious shortcoming was the so-called "matrix effect", describing the fact that the characteristic secondary ion emission in SIMS does not only depend on the surface concentration of a given element but can change by orders of magnitude depending on the chemical environment of this element. A corresponding matrix effect exists for molecular surface species as well.

Progress in static SIMS over the last three decades has been the result of a continuous development in instrumentation. The application of double-focusing magnetic sector fields, quadrupoles, time-of-flight instruments and focused ion beams resulted in a corresponding increase in mass range, mass resolution, sensitivity and lateral resolution. This progress allowed the detection of more and more complex molecular surface species, e.g., synthetic polymers or large biomolecules.

4. Ion–surface interaction

In a simplified model we may assume that the impact of a primary ion on a solid surface results in the distribution of the primary ion energy in the vicinity of the point of impact (Fig. 1). This takes place within a very short period of time (<1 ps) and results in a modification of the target area around the primary ion path by primary ion and knock-on implantation, by radiation damage, etc. In addition, photon, electron and heavy particle emission occur.

If we consider a solid covered by one single monolayer of a surface species M, the sputter yield $Y(M)$ describes the average number of surface species M disappearing as a result of one single primary ion impact. $Y(M)$ corresponds to a disappearance cross section $\sigma(M) = Y(M) A_0$ (A_0 = surface area occupied by one single M). During ion bombardment with a primary ion flux density ν for a time t, an initial surface coverage $\theta(0)$ decreases to $\theta(t) = \theta(0) \exp(-\sigma\nu t)$ (Fig. 2b). The condition for a static SIMS experiment is $\sigma\nu t \ll 1$.

The disappearance of a surface species M may result in the generation of different sputtered particles X_i^q. For example, $(M + H)^+$ and $(M + H)^-$ are some of the ions X_i^q generated during sputtering of organic molecules M. The transformation probability $P(M \rightarrow X_i^q)$ describes the probability that a particle M is transformed into the secondary particle X_i^q, if it disappears from the surface during the sputtering process. For amino acid molecules on noble metals, for example, the transformation probability $P(M \rightarrow (M + H)^+)$ is on the order of 10^{-3}–10^{-4}.

As mentioned before, transformation probabilities are determined not only by the surface species M but also by its chemical environment. P can be changed by orders of magnitude only by changing this environment. Mass, energy, angle of incidence and charge of the primary ion have only a very limited influence on P. The chemical nature of the bombarding particle (e.g., Ar, Cs, O, Ga) has an influence on P only as far as the matrix is changed by primary ion implantation. In static SIMS this effect is negligible.

5. The development of static SIMS

To what extent materials and problems are accessible to static SIMS is determined and limited by the mass range, the mass resolution, the sensitivity and the lateral resolution of available instrumentation. For this reason, progress in static SIMS over the last three decades will be described by following the developments in instrumentation.

5.1. Single-ion counting

Single-ion counting instead of measuring the secondary ion current with an amplifier was the most important instrumental precondition for the development of static and in particular molecular SIMS. It resulted in an increase in sensitivity by a factor of 10^5! The first UHV instrument for static SIMS was realized by single-ion detection with an open multiplier in a single-focusing magnetic sector field instrument. With this instrument we investigated extensively the oxidation of clean metal surfaces during O_2-exposure [5]. We applied sample heating, O_{18}/O_{16}-exposure, etc., and got detailed information on the composition of surface oxides and the dynamics of their formation during O_2-exposure. During the investigation of poisoned silver catalysts in such an instrument, we found as one of the most important results that in the negative spectrum involatile surface species such as SO_4^-, NO_3^- or $CH_3(CH_2)_nCOO^-$ appeared with high yields [4]. This was a key result with regard to the development of molecular SIMS. The high sensitivity achieved by single-ion counting allowed in addition detailed investigations of thin films and interfaces (depth profiling) which we carried out in particular for evaporated metal layers.

During our work with single-focusing magnetic sector field instruments we realized very soon serious limitations originating from limits in mass resolution, mass range and transmission, as well as from magnetic stray fields, slow mass scan and

problems concerning the integration of such an instrument into a UHV system. The situation could be improved considerably by the use of quadrupole instruments for secondary ion analysis.

5.2. Quadrupole instruments

Quadrupole mass analyzers feature a fast mass scan, a relatively high mass resolution and transmission, a small size and, in addition, they produce no magnetic stray fields. Since 1971 we have applied quadrupole mass analyzers in two different types of static SIMS instruments: in UHV instruments mainly applied to the investigation of surface reactions and in instruments featuring an only moderate vacuum, equipped with a fast sample lock system and mainly used for secondary ion mass spectrometry of involatile organic compounds.

With UHV instruments we studied, for example, metal–oxygen [5] (Fig. 3) and metal–hydro-

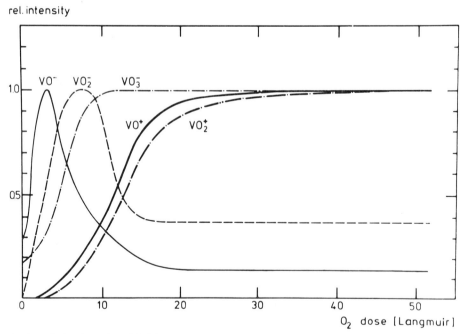

Fig. 3. Changes in the relative intensity of $VO_n^{\pm}$ emission during O_2-exposure of a sputter-cleaned vanadium surface [5]. The changes in relative intensities reflect the change in the valence state of the vanadium atoms on the surface during oxidation [7] (static SIMS condition).

gen [6] interactions, combining SIMS with thermal desorption mass spectrometry and applying $^{18}O_2/^{16}O_2$- and H_2/D_2-exposure for the investigation of the corresponding surface reactions. The SIMS results for metal–oxygen interactions could be described by a "valence" model, which allows one to calculate the average valence state of surface metal atoms Me from the $Me_nO_n^{\pm}$-emission [7].

Detailed investigations of the secondary ion emission from amino acid overlayers on atomically clean metal surfaces have been carried out by means of a UHV quadrupole instrument, additionally equipped with an electron beam for post-ionization of sputtered neutrals, and a molecular

beam device for the controlled deposition of volatile molecules on metal surfaces (Me). We found that besides the molecular secondary ions $(M \pm H)^{\pm}$ and $(M + Me)^{+}$ in addition intact neutral molecules M are sputtered from such an overlayer. By heating and H_2/D_2-exposure experiments, the origin of the proton in the protonated $(M + H)^{+}$ ion could be determined as well as different bonding states of the amino acid molecules on metal surfaces which are responsible for the $(M + H)^{+}$ and the $(M - H)^{-}$-emission, respectively [8,9] (Fig. 4).

For more detailed investigations of catalytic surface reactions, oxidation behaviour of metals and multicomponent semiconductor materials like CdHgTe, GaAs, PtSi, etc., we integrated into some of our quadrupole SIMS instruments additional surface analytical techniques like XPS, AES, ISS and TDMS [10]. These instruments became very complex and could be operated only by computer control – which was not trivial at that time. The combination of different techniques supplied very complementary information concerning molecular surface species, depth distribution of different elements, quantification, etc. We applied one of these combined instruments for the investigation of ethylene oxidation on silver catalysts. For this study we coupled the SIMS-XPS-AES-ISS-TDMS instrument with a gas-chromatograph-controlled high pressure reactor [11].

Following the detection of involatile molecular surface species on a silver catalyst, we carried out a more systematic investigation of the secondary ion emission behaviour of large organic molecules such as amino acids and some derivatives, small peptides, a variety of drugs, etc. [12], using a quadrupole instrument equipped with a fast sample lock system. Surface depositon of the sample molecules was realized by depositing a 1 $\mu\ell$ droplet containing an appropriate amount of the respective molecules on the metal target. After evaporation of the solvent, the deposited molecules remained on the substrate surface as a mono-, submono- or multilayer, depending on the deposited amount of material and the covered surface area. As a general result we found a rather high generation of $(M \pm H)^{\pm}$- and $(M +$

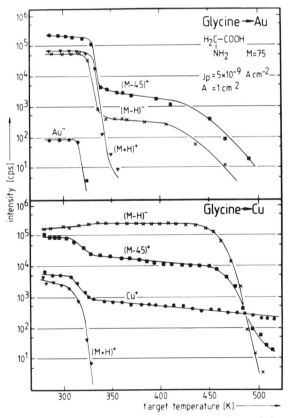

Fig. 4. Changes in the characteristic secondary ion emission from a glycine monolayer on Au and on Cu during heating [8]. The different behaviour of $(M - H)^{-}$ and $(M + H)^{+}$ for a noble (Au) and for a reactive (Cu) metal substrate reflects different bonding states of the amino acid on these metals (static SIMS condition).

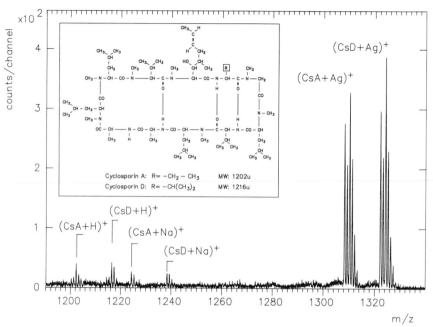

Fig. 5. Molecular ion region in the positive static TOF-SIMS spectrum of a 1 : 1 mixture of two cyclosporin derivatives, deposited from a solution on Ag [13].

Me)$^+$-ions (Fig. 5). In addition, a variety of fragments were emitted. We found that noble metals, and in particular silver, were by far the most efficient substrates promoting this molecular ion formation. Detection limits in the femtomol range and below could be achieved. By adding to the solution similar molecules as internal standards, quantification could be realized [13].

5.3. Double focusing magnetic sector field instruments

By the static SIMS results for involatile materials it became evident very soon that molecular SIMS could be a general mass spectroscopic technique for the detection and identification of involatile materials. This would mean a most significant progress in organic mass spectrometry. For this purpose, however, a higher mass range and a higher mass resolution than that available in quadrupole instruments was required. This has been achieved by two different developments: the use of liquid matrices in combination with low transmission double-focusing magnetic sector field instruments on one hand, and the development of high transmission high performance time-of-flight mass spectrometers on the other hand.

By depositing the analyte molecules with relatively high concentration in a liquid matrix of low vapor pressure, e.g., glycerol [14], a higher total number of sample molecules became available for secondary ion formation. During sputtering, a continuous regeneration of the uppermost monolayer by diffusion of analyte molecules from the bulk of the liquid matrix to the liquid surface occurs. This continuous regeneration allows higher primary ion current densities and correspondingly higher secondary ion currents over a much longer period of time – of course at the

Fig. 6. Positive static SIMS spectrum of a monolayer polydimethylsiloxane (PDMS) on silver. The upper spectrum reflects the oligomer distribution of the polymer. The lower spectrum demonstrates the high mass resolution and the accurate mass determination (ppm range) which can be obtained [17].

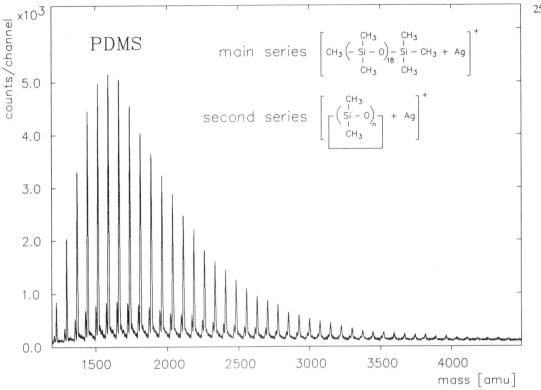

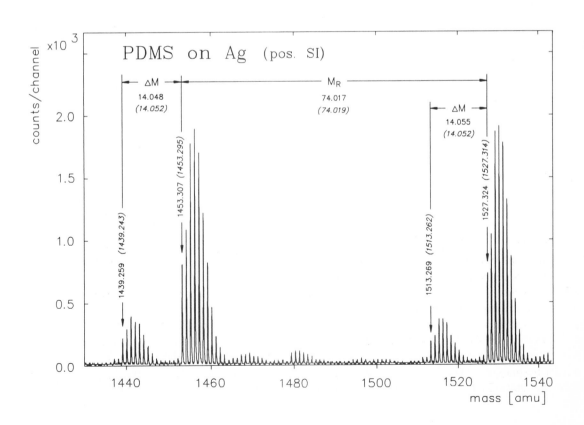

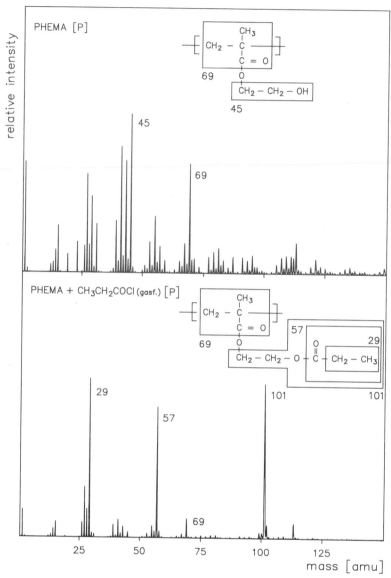

Fig. 7. Positive static SIMS spectrum of polyhydroxyethylmetacrylate in the low mass range before and after treatment of the surface with propionylchloride. The spectrum reflects directly the molecular modification of this polymer surface.

expense of a higher material consumption. By the use of liquid matrices (liquid SIMS), double-focusing magnetic sector field instruments with higher mass range and mass resolution became available for molecular SIMS.

Liquid SIMS was an overwhelming success. Most organic compounds are involatile and could hardly be mass-analyzed before. Existing double-focusing instruments could now be upgraded eas-

ily, by only adding a primary particle source. Today, liquid SIMS is a standard technique in organic mass spectrometry and is applied to all kinds of involatile, soluble materials up to a molecular weight of about 10 000 amu. Liquid SIMS was introduced into organic mass spectrometry under the misleading name "fast atom bombardment" (FAB), putting the emphasis on a relatively unimportant technical aspect (the use of neutral primary particles) and omitting the important point, the use of a liquid reservoir for the sample molecules [14].

5.4. Time-of-flight instruments

The introduction of time-of-flight (TOF) mass analyzers has to be considered as another milestone in the development of static SIMS. The quasi-parallel ion detection in these instruments overcomes a most serious limitation of both magnetic sector field instruments and quadrupoles: the "one mass at a time" principle. In these instruments, secondary ions are generated continuously by a constant primary ion beam. At a given time, the instrument is tuned to one single mass and all other ions are lost. This is completely different in a time-of-flight instrument; the target is bombarded only for a very short period of time (ns-range) by the pulsed primary ion beam. In the time between two successive

pulses (10–100 μs), the generated secondary ions are postaccelerated and mass separated in a flight tube. Different ion species arrive at different times at the detector and a quasi-parallel detection of all generated secondary ion species can be realized.

The more effective use of all secondary ion species by this quasi-parallel ion detection and the expected high transmission were the reasons why we started developing and building a time-of-flight instrument already in 1978. The design of this instrument addressed the angular and energy distribution of the emitted secondary ions by providing appropriate focusing ion optics. The instrument was equipped with a mass-separated primary ion source, mandatory for high mass resolution [15]. Later on, we developed a number of more sophisticated, reflectron-based instruments [16], equipped with laser postionization capabilities for the analysis of sputtered neutrals, a second ion detector for the investigation of

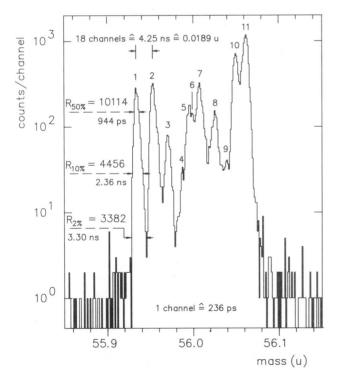

peak no.	sum formula	mass measured	mass calculated
1	^{56}Fe	55.9343	55.9349
2	$^{28}Si_2$	55.9536	55.9539
3	$^{28}SiCO$	55.9706	55.9718
4	$^{29}SiCNH$	55.9869	55.9874
	$^{23}NaO_2H$		55.9874
5	$^{28}SiCNH_2$	55.9956	55.9956
6	$^{29}SiC_2H_3$	55.9999	56.0000
7	$^{28}SiC_2H_4$	56.0079	56.0082
8	C_3OH_4	56.0256	56.0262
9	$C_2N_2H_4$	56.0385	56.0374
10	C_3NH_6	56.0500	56.0500
11	C_4H_8	56.0618	56.0626

Fig. 8. Fine structure of mass line 56 in the positive static SIMS spectrum of a Si wafer. Mass 56 is composed of 11 different components which can be separated by high mass resolution and identified by their exact masses. The Fe surface concentration was 6×10^{12} atoms/cm^2.

metastable decay of molecular ions, and fine-focused pulsed-ion sources for imaging. We integrated in this type of instruments plasma desorption mass spectrometry (PDMS), XPS and various surface preparation and modification techniques such as plasma devices, molecular beam sources and high current ion sources for more efficient depth profiling.

By a careful design of the primary ion sources, the extraction and energy focusing ion optics, as well as the detector arrangement and the corresponding electronics, we achieved mass resolutions $m/\Delta m > 20\,000$ at transmissions > 0.5, even in the high mass range $> 10\,000$ amu. The availability of these high performance time-of-flight instruments opened completely new fields of application for static SIMS, mainly due to the extended mass range, the additional increase in sensitivity by a factor up to 10^6, compared with a quadrupole, and the combination of high mass resolution and high transmission, even in the high mass range. Over the last decade, we started new fields of application that became accessible only by TOF instruments. They include the analysis of polymer materials (Fig. 6) and modified polymer surfaces (Fig. 7), trace analysis on wafer surfaces (Fig. 8), quantitative analysis of complex drugs [13] and the analysis of Langmuir–Blodgett [18] and self-assembly layers.

5.5. Postionization of sputtered neutrals

In general, most of the sputtered surface species are neutral in charge. It was always an exciting idea, to use these neutrals for analytical purposes. Electron impact postionization is not very efficient and maximum ionization probabilities $\leq 10^{-4}$ are in general too low for static SIMS [19,20]. More efficient is multiphoton ionization by UV laser pulses [21] fired into the cloud of emitted neutrals shortly after sputtering of these particles by a primary ion pulse. The generated photoions are separated from the secondary ions and mass-analyzed in a reflectron TOF instrument. The general principle is very simple. Problems, however, arise from the energy distribution of sputtered neutrals, from the influence of laser pulse length, frequency and power

density on the ionization efficiency and the fragmentation of molecular species, from the broad energy distribution of the postionized neutrals which are generated at different distances from the target, from the influence of the excitation state of the sputtered particles on the postionization and fragmentation of sputtered molecules, and from interferences with postionized residue gas components. Some of these problems have been solved by pulsed extraction, the application of high power density, ultra-short laser pulses, the use of a double reflectron for improved energy focusing, etc.

For elemental analysis, sensitivities in the range of 1 ppm of a monolayer are achieved. Compared with static SIMS, quantification is much easier, the matrix effect being greatly reduced. Postionization of sputtered neutral molecular species is more difficult because of two competing processes: ionization and fragmentation. Both increase with laser power density. The ionization and fragmentation behaviour of sputtered molecular species is determined by their spectroscopy, by their excitation state after sputtering, as well as by the length, frequency and power density of the laser pulse. This means that optimization depends strongly on the considered molecular species.

Recently, we have improved the postionization of atoms as well as of some molecular species drastically by using fs- instead of ns-laser pulses with power densities up to 10^{12} W cm^{-2}. We achieved 100% postionization at the focal point for elements and an increase in molecular ion formation by a factor of 100 for some molecular species [22,23]. It was only by this increase in ionization efficiency, that imaging with postionized atomic and molecular sputtered particles became feasable for practical applications.

5.6. Monolayer imaging

A basic problem in monolayer imaging by sputtered particles is the limitation of the amount of available material. 1 μm^2 of the uppermost monolayer contains only some million surface species. Considering typical transformation probabilities $P(M \rightarrow X_i^q)$ for medium-sized surface

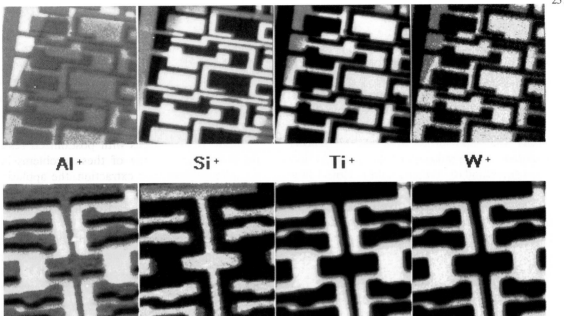

Fig. 9. Static TOF-SIMS images of the element ions Al$^+$, Si$^+$, Ti$^+$ and W$^+$, emitted from an Al–Ti–W-test structure on a Si-wafer. Imaged surface areas are 120×120 μm^2 (upper row) and 20×20 μm^2 (lower row). The corresponding lateral resolutions are 1 μm and 0.1 μm, respectively. During one experiment, several hundred images of selected masses can be stored.

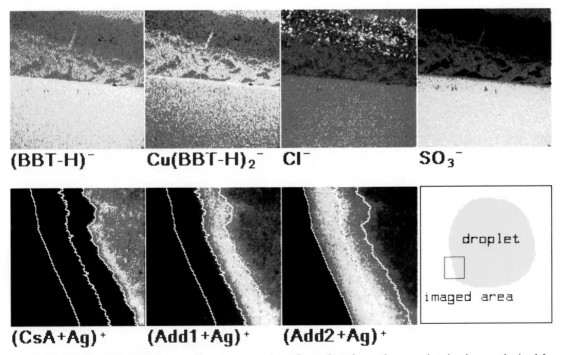

Fig. 10. Static molecular TOF-SIMS images. The upper row shows four selected negative secondary ion images obtained from a self-assembly butylbenzotriazol (BBT) layer on Cu, covering the lower part of the imaged surface. In addition to the BBT-specific secondary ions (BBT-H)$^-$ and Cu(BBT-H)$_2^-$ the contamination ions Cl$^-$ and SO$_3^-$ have been selected. The lower row shows three selected positive ion images obtained from a silver surface after depositing and drying an oily solution of the peptide cyclosporin A (CsA) and two high molecular components of the solvent (Add 1 and Add 2).

molecules M in the 10^{-4} range, it is evident that one really needs all secondary ions which can be generated from this very limited amount of sample material. This means that for imaging quasi-parallel ion detection and a high transmission are mandatory. The most efficient instrument for monolayer imaging is the time-of-flight ion microprobe. Here a small surface area ($\phi > 0.1$ μm) is bombarded by a pulsed-focused primary ion beam. The secondary ion spectrum is stored in a computer before the next pixel is analyzed. From the secondary ion spectra of all pixels of an imaged area, the lateral distribution of any secondary ion can be reconstructed (Figs. 9, 10). Lateral resolution and sensitivity of an ion microprobe are mainly determined by the primary ion source, i.e. pulse width, beam diameter and number of ions in a single pulse. If sufficiently high photoion yields can be realized, the ion microprobe can be operated successfully with sputtered neutrals, too. The development of surface imaging by sputtered charged and neutral particles opened again many new fields of applications. They include surface analysis of fiber surfaces, particle surfaces, biological samples, microstructured wafer surfaces, etc. [24].

6. The analytical capabilities and limits of static SIMS

State-of-the-art static SIMS – which means TOF-SIMS – is characterized by a unique combination of most important analytical features:

- Direct molecular information achieved by the generation of analytically useful molecular secondary ions, even from involatile and thermally labile molecular surface structures. This molecular ion emission – which occurs for virtually all kinds of materials – allows a detailed molecular surface analysis, in most cases with very high sensitivities and an excellent spectral separation and identification of different species by the combination of high mass resolution and exact mass determination.
- High sensitivity for elements as well as for molecular species, even for those with low transformation probabilities. This is achieved

by the combination of high transmission and quasi-parallel ion registration.
- High lateral resolution realized by fine-focused ion beams.
- Unrestricted applicability to insulators, achieved by pulsed charge compensation with low energy electrons.

Static SIMS can be applied to all kinds of materials, metals and semiconductors, as well as complex organic molecular surface structures. It can be applied to samples of any geometry, flat single-crystal surfaces, as well as rough polycrystalline surfaces, fibers, particles or microstructured surfaces. Present-day applications of static SIMS range from microelectronics, catalysis and polymer research to clinical analysis, environmental control and all kinds of microstructure technologies.

The main analytical limitations of static SIMS originate from the matrix effect. As long as a surface species M is in the same chemical environment, the transformation probabilities $P(M \rightarrow X_i^q)$ are constant and the secondary ion emission is proportional to the coverage of the surface by the component M. This is confirmed by many experimental results (e.g., Fig. 2b). As soon as the chemical environment changes, however, the transformation probabilities for the same surface species change too, sometimes by orders of magnitude, and quantification needs calibration, for example by the use of internal or external standards (elements as well as molecules [13]) or by additional information obtained by more quantitative techniques like AES, XPS or laser postionization of sputtered neutrals.

New instrument concepts like TOF-SIMS are not restricted to applications they have been developed for. Within certain limits TOF-SIMS instruments can also be applied for dynamic SIMS investigations. The high current densities achieved with fine-focused ion beams allow monolayer removal in a fraction of a second and a correspondingly small area depth profiling. Large area depth profiling became possible by integration of an additional high current ion source.

As a most important criterion for the usefulness of an analytical technique one may consider its degree of application by others than the inventors. High performance instrumentation for static SIMS is offered today by all major manufacturers of surface analytical equipment and is increasingly applied in industrial and academic laboratories, in all technological fields where surface phenomena play an important role. This supports and confirms our opinion that static SIMS has become a very useful surface analytical technique, in particular for "real world" problems and samples.

7. Directions of future developments

State-of-the-art modular TOF-SIMS instruments are very flexible and future developments concerning primary ion sources, secondary ion optics, detector systems, etc., can easily be integrated. Instrumental development may follow various directions; we can expect a continuous improvement of existing concepts concerning mass resolution, lateral resolution, laser postionization, ease of operation, etc. In addition, new concepts allowing, e.g., the parallel detection of positive and negative secondary ion spectra, the parallel use of backscattered primary ions, the integration of fine-focused electron beams, etc., are expected.

The unique combination of element, isotope and molecular information with high sensitivity, high lateral resolution and nearly unrestricted applicability to all kinds of samples will result in a continuously increasing analytical application of static SIMS, in particular in the areas of:
- New materials (composites, molecular overlayers such as Langmuir–Blodgett and self-assembly layers, catalysts, polymers, etc.).
- Interfaces (metal/polymer, biocompatibility, immobilized catalysts, lubrication, etc.).
- Particle and trace analysis (environmental, soil, clinical analysis, biological samples, etc.).
- Micro- and nanotechnologies (micromechanics, microelectronics, sensors, etc.).

This development will be supported by the increasing importance of molecular and in particular organic surface structures in all technological developments where adhesion, adsorption, wear and lubrication, biocompatibility, chemical reac-

tivity, etc., play an important role. For these surface structures, standard techniques like AES and XPS in most cases do not supply the required detailed molecular information.

Another challenge is surface imaging and microanalysis of biological samples. This is a most exciting field for molecular surface imaging, but it needs developments in preparation techniques, establishment of molecular SIMS libraries, development of SIMS specific "staining" techniques, improvement in and new concepts for laser postionization of sputtered molecular species, etc.

In contrast to the broad analytical application of static SIMS, there has been little activity in fundamental research. Progress in understanding the formation of secondary and in particular molecular ions needs experiments under well-defined conditions, e.g., UHV experiments with single crystals, covered by well-known molecular surface components. These experiments need complex instrumentation. In addition, theoretical models describing and predicting molecular secondary ion emission processes will become very complex. With state-of-the-art TOF-SIMS equipment, however, the required experiments can be carried out and more fundamental research in this exciting field is expected in the future.

References

[1] A. Benninghoven, F.G. Rüdenauer and H.W. Werner, Secondary Ion Mass Spectrometry (Wiley, New York, 1987).
[2] A. Benninghoven and F. Kirchner, Z. Phys. 18a (1963) 1008.
[3] R.R. Weber and W.T. Peria, J. Appl. Phys. 38 (1967) 4355.
[4] A. Benninghoven, Phys. Status Solidi 34 (1969) K169.
[5] A. Benninghoven, Surf. Sci. 28 (1971) 541.
[6] A. Benninghoven, K.H. Müller and M. Schemmer, Surf. Sci. 78 (1978) 565.
[7] C. Plog, L. Wiedmann and A. Benninghoven, Surf. Sci. 67 (1977) 565.
[8] W. Lange, M. Jirikowsky and A. Benninghoven, Surf. Sci. 136 (1984) 419.
[9] A. Benninghoven, W. Lange, M. Jirikowsky and D. Holtkamp, Surf. Sci. 123 (1982) L721.
[10] U. Kaiser, O. Ganschow, L. Wiedmann and A. Benninghoven, J. Vac. Sci. Technol. A 1 (1983) 657.

[11] O. Ganschow, R. Jede, L.D. An, E. Manske, J. Neelsen, L. Wiedmann and A. Benninghoven, J. Vac. Sci. Technol. A 1 (1983) 1491.

[12] A. Benninghoven and W. Sichtermann, Org. Mass Spectrom. 9 (1977) 595.

[13] K. Meyer, B. Hagenhoff, M. Deimel and A. Benninghoven, Org. Mass Spectrom. 27 (1992) 1148.

[14] M. Barber, R.S. Bordoli, R.D. Sedgewick and A.N. Tyler, Anal. Chem. 54 (1982) 645A.

[15] P. Steffens, E. Niehuis, T. Friese, D. Greifendorf and A. Benninghoven, J. Vac. Sci. Technol. A 3 (1985) 1322.

[16] E. Niehuis, T. Heller, H. Feld and A. Benninghoven, J. Vac. Sci. Technol. A 5 (1987) 1243.

[17] E. Niehuis, T. Heller, U. Jürgens and A. Benninghoven, J. Vac. Sci. Technol. A 7 (1989) 1823.

[18] B. Hagenhoff, M. Deimel, A. Benninghoven, H.U. Siegmund and D. Holtkamp, J. Phys. D (Appl. Phys.) 25 (1992) 818.

[19] H. Oechsner and W. Gerhard, Phys. Lett. A 40 (1972) 211.

[20] D. Lipinsky, R. Jede, O. Ganschow and A. Benninghoven, J. Vac. Sci. Technol. A 3 (1985) 2007.

[21] C.H. Becker and K.T. Gillen, Anal. Chem. 56 (1984) 1671.

[22] M. Terhorst, R. Möllers, E. Niehuis and A. Benninghoven, Surf. Interface Anal. 18 (1992) 824.

[23] R. Möllers, M. Terhorst, E. Niehuis and A. Benninghoven, Org. Mass Spectrom. 27 (1992) 1393.

[24] J. Schwieters, H.G. Cramer, T. Heller, U. Jürgens, E. Niehuis, J. Zehnpfenning and A. Benninghoven, J. Vac. Sci. Technol. A 9 (1991) 2864.

Surface Science 299/300 (1994) 261–276
North-Holland

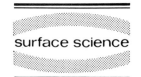

surface science

Adsorption and desorption dynamics as seen through molecular beam techniques

K.D. Rendulic* and A. Winkler

Technische Universität Graz, Institut für Festkörperphysik, Petersgasse 16, A-8010 Graz, Austria

Received 2 February 1993; accepted for publication 24 May 1993

A description of the history of knowledge about adsorption and desorption dynamics is given. The individual stations include the encounter with non-cosine, non-Maxwellian distributions of adsorbing and desorbing particles; detailed balancing in its development as a tool to relate adsorption and desorption data is described. A further section treats the concept of precursor mediated adsorption and its verification by molecular beam methods. The problem of surface defects is briefly touched. Refinements in the molecular beam techniques finally lead to the possibility to gain state resolved dynamics data for adsorption and desorption processes.

1. Introduction

In this report we want to deal with the application of molecular beam techniques both to the investigation of the dynamics of adsorption and the dynamics of desorption. In particular the dissociative chemisorption of reactive gases (H_2, O_2, N_2) will be treated. Dissociative chemisorption from a theoretical point of view is a most interesting topic because the dynamics of adsorption is profoundly influenced not only by the translational energy but also by the quantum state (vibration, rotation) of the impinging molecule. For practical purposes the attempt to elucidate the mechanism of heterogeneous catalysis involves the understanding of dissociative chemisorption.

The beginning of the insight into the dynamics of dissociative chemisorption is the legendary paper by Lennard-Jones: "Processes of Adsorption and Diffusion on Solid Surfaces" published in 1932 [1]. Everybody is familiar with the picture of the one-dimensional model of activated dissociative chemisorption, where the two diabatic potential curves for the molecular and the atomic state intersect and form the crest of an activation barrier in front of the surface. Yet there is a lot more to be found in this paper: already the variation of the barrier height across the unit cell is dealt with and the question as to the role of the vibrational and rotational states of the molecule in activated adsorption is raised for the first time.

Of course, we know now that in the context of the one-dimensional Lennard-Jones model it is impossible to describe many features (e.g. the contribution of internal energy of a molecule) of the adsorption/desorption dynamics. *The aim in understanding adsorption in today's language is to build an appropriate multi-dimensional potential energy surface which describes the dynamics of adsorption and desorption.* Such a potential energy surface has to be deduced from the experiment or, if a theoretical model is present, it has to be tested against the experiment.

How can we experimentally determine the shape of such a potential energy surface? To probe the potential energy surface *differential sticking experiments* are needed for which the particle properties can be chosen or at least analyzed at will. The sticking coefficient for example should be determined as function of the angle of incidence, the translational energy, the quantum state and perhaps even the orientation

---

* Corresponding author.

of the molecule. A measurement of this type is said to determine the *dynamics of adsorption*. Dynamics measurements involve the application of molecular beams. Ideally a well collimated beam of molecules for which some or all of the above mentioned parameters are known is made to collide with the sample surface and the fraction of adsorbed particles is determined. In turn also the desorption properties may be deduced by molecular beam techniques: a collimator selects a beam of desorbing particles under a certain desorption angle. Again analytical techniques can be applied to determine the velocity distribution and the quantum state of the desorbing particle.

This short sketch of the application of molecular beam techniques for the investigation of the adsorption/desorption dynamics looks quite simple. Nevertheless, tremendous experimental and conceptual difficulties had to be overcome until at least some of the aims could be realized. Presently the most complex experiments in surface physics are employed to elucidate the details of adsorption dynamics. The end of this development is certainly not in sight but one can state that the birth of these techniques coincides with the birth of "Surface Science".

What follows is a limited account of the progressive development in our knowledge about the dynamics of adsorption and desorption. It also is a subjective account of the development in the sense that every person has different ideas of what constitutes an important step in science; we do not attempt an encyclopedic approach but limit ourselves to several topics encompassing adsorption of reactive gases, in particular hydrogen. Nevertheless, this is not such a serious limitation after all, since basic ideas and basic experimental accomplishments need only a few characteristic examples for elucidation.

2. Activated adsorption and non-cosine distributions

Around 1960 a number of papers started to deal with the measurement of angular distributions of particles leaving a surface after having been supplied to the surface by a molecular beam [2,3]. Everybody at that time believed in the cosine distribution of desorbing molecules and the experiments seemed to verify this concept. The general theoretical arguments given in favor of a cosine distribution of desorbing particles were equilibrium considerations involving the second law of thermodynamics. Since in thermal equilibrium the particles impinging on a surface have a cosine distribution with respect to the surface normal the particles leaving the surface have to exhibit a cosine distribution too. This is generally labelled the cosine equilibrium law which has been discussed by Maxwell [4], Langmuir [5], Clausing [6] and Knudsen [7]. The same argument can be made with respect to the energy distribution of particles leaving the surface: it ought to be Maxwellian. These arguments are, of course, correct. But as Comsa [8] has pointed out, the particles leaving the surface are not only desorbing particles, there are in addition also elastically and inelastically scattered particles. Only the sum of these particles needs to have a cosine/Maxwellian distribution. Nothing can be said a priori about the angular and energy variation for each of the subsets of particles such as desorbing particles on their own. Actually, already in the 1960s there was evidence that the reflecting power of a surface was generally not isotropic, which according to Comsa would directly lead to *non-cosine* distributions for desorbing particles. A graphic description of the relation between scattered and desorbing particles is given in Fig. 1. A rather nice explicit verification of the fact that scattered particles plus desorbing particles have to add up to a cosine distribution was given quite early in the game through a molecular beam experiment by Palmer et al. [9] for the case of $D_2/Ni(111)$.

In one of the most famous experiments in surface physics van Willigen [10] in 1968 showed that molecular hydrogen desorbing from Fe, Pd and Ni exhibited strongly forward (towards the surface normal) peaked angular flux distributions. What an ingenious experiment! The hydrogen was permeated from a high pressure area through the heated sample; on the vacuum side one obtained a continuous desorption flux

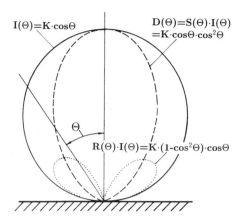

Fig. 1. Polar diagram schematically depicting the relation between total particle flux $I(\Theta) = K \cos \Theta$ impinging on a surface (solid line), the scattered flux (dotted line) and the adsorbed flux (dashed line). If for example the sticking coefficient changes like $S(\Theta) = \cos^2 \Theta$ the scattering probability $R(\Theta)$ has to vary as $(1 - \cos^2 \Theta)$ to always add up to $I(\Theta)$. In addition, detailed balancing demands the adsorbing flux $S(\Theta)I(\Theta)$ to be identical to the desorbing flux $D(\Theta)$. Data from Ref. [18].

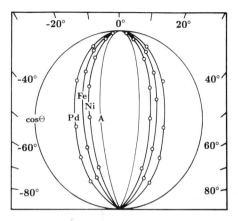

Fig. 2. Non-cosine desorption flux obtained by van Willigen for a number of permeation sources at a temperature of 900 K. Interpretation of the data was given in terms of a one-dimensional barrier desorption mechanism. An activation barrier of 1 eV height would result in a flux distribution depicted under A. Data from Ref. [10].

without the disturbance of scattered particles present in a gaseous environment. This was a true molecular beam experiment, the desorbing particles were registered in a beam detector. The results of this experiment are depicted in Fig. 2. Not only did van Willigen get the experiment right, he also presented a basically correct interpretation of the data in terms of a one-dimensional barrier model: a Maxwellian hitting a one-dimensional barrier will be attenuated in the low energy end of the spectrum. The velocity component normal to the barrier $(v \cos \Theta)$ can only be used to transgress the one-dimensional barrier of height E_A when the "normal translational energy" $E \cos^2 \Theta > E_A$. As the angle with the surface normal increases, fewer and fewer particles out of the Maxwellian are able to cross the barrier. With increasing barrier height the desorption flux $D(\Theta)$ becomes more and more concentrated about the surface normal. van Willigen explicitly formulated:

$$D(\Theta) = D(0°) \frac{E_A + kT \cos^2 \Theta}{(E_A + kT) \cos \Theta}$$

$$\times \exp[-(E_A/kT) \tan^2 \Theta]. \qquad (1)$$

Traditionally [9] the desorption flux $D(\Theta)$ has been approximated by the expression

$$D(\Theta) = D(0°) \cos^n \Theta. \qquad (2)$$

The experimentally obtained distributions are very well represented by this function; as later could be shown [11] the $\cos^n \Theta$ functions are closely related to normal energy scaling (see below).

The van Willigen paper immediately stimulated a large number of investigations involving activated adsorption, in particular the measurement of differential sticking coefficients $S(v, \Theta)$ and differential desorption probabilities $D(v, \Theta)_T$. It was, for example, immediately obvious that the energy distribution of the particles desorbing over the barrier had to exhibit a mean particle energy which is larger than the $2kT$ expected for a Maxwellian.

3. Detailed balancing

Detailed balancing, a principle of equilibrium physics frequently applied to problems of physical chemistry, was first formulated by Kirchhoff [12] and Richardson [13]. Actually, Kirchhoff's law relating absorption and emission of radiation is quite similar to the laws relating adsorption and desorption dynamics. To gain access to this important principle in its application to surface physics one best follows the historic, statistical treatment by Langmuir [14] describing the properties of adsorption isotherms. This statistical treatment very clearly shows the salient features of detailed balancing. The derivation of the adsorption isotherm is generally obtained from the dynamic equilibrium between the rate of adsorption and the rate of desorption:

$$R_{ads}(T, p, N_s) = R_{des}(T, N_s). \tag{3}$$

The variables are temperature T, pressure p and surface coverage N_s. The left side of Eq. (3) contains the *specific kinetics of adsorption* (e.g. via a precursor path), the right side the *specific kinetics of desorption*. From this formulation one would expect that the adsorption isotherm will depend on the particular kinetics involved. But as has been pointed out [14,15], from a statistical point of view an equilibrium is described by a distribution of particles over a set of available states, free or adsorbed. The physics of transition between these states, that is the particular kinetics, cannot enter in this description. This means the adsorption and desorption kinetics of Eq. (3) have to be related at all times in such a way that they cancel and do not appear in the resulting adsorption isotherm. Or as Langmuir [14] already put it in 1916: *Since evaporation and condensation are in general thermodynamically reversible phenomena, the mechanism of evaporation must be the exact reverse of that of condensation, even down to the smallest detail.*

A general definition of detailed balancing can be obtained using the concept of microscopic reversibility: for each "forward" path there exists an identical "reverse" path related to the former by time reversal. Detailed balancing demands that in thermal equilibrium each forward path occurs with the same probability as the corresponding backward path. This principle relates subsets of particles such as adsorbing and de-

sorbing particles as well as particles which are going to be scattered (elastically, inelastically) and particles which have been scattered (elastically, inelastically). Detailed balancing is particularly well suited in its application to molecular beam data. It is therefore not surprising that the widespread use of this principle started in the 1970s along with the development of molecular beam techniques [16].

An especially useful formulation of detailed balancing for the adsorbing and desorbing flux in thermal equilibrium was introduced by Stickney and Cardillo [17]:

$$S(v, \Theta, Q_i) f_M(v, Q_i) \cos \Theta$$
$$= D(v, \Theta, Q_i) f_M(v, Q_i). \tag{4}$$

Here v is the velocity of the particle, Θ the angle with the surface normal and Q_i the quantum state (which was not yet contained in the original formulation); the Maxwellian flux distribution is labelled f_M. Here, and for all other considerations in this paper, the sticking coefficient is always defined for the zero coverage limit; that means it is the initial sticking coefficient. The right hand term can be considered a definition of the desorption probability D in terms of the generated desorption flux. On the left hand side the already existing definition of the sticking coefficient as ratio of adsorbing to impinging particles introduces the $\cos \Theta$ term. One can also write:

$$S(v, \Theta, Q_i) \cos \Theta = D(v, \Theta, Q_i). \tag{5}$$

The equation formulated for thermal equilibrium is in this form of little practical use. The importance comes from the fact that one can *independently investigate adsorption dynamics and desorption dynamics* in certain situations of nonequilibrium, situations of quasi-equilibrium, and still retain the validity of Eq. (5). The term quasi-equilibrium in the context of adsorption has acquired a well-defined meaning: During the separate investigation of adsorption and desorption the thermodynamic parameters of the remaining part of the system are kept as close as possible to those of the corresponding equilibrium state. This has been discussed in detail by Stickney [17] and Comsa [18].

What is generally labelled as a "test" of detailed balancing can only involve this last point mentioned: Will the adsorbing and desorbing particles observed in a specific non-equilibrium situation (a situation of quasi-equilibrium) still retain the same adsorption and desorption paths as in thermal equilibrium and obey Eq. (5)? Experimental verification of the principle of detailed balancing for a situation of quasi-equilibrium has been performed on several levels [19]. The simplest type of experiment involves the angular variation of $S(\Theta)$ and $D(\Theta)$. If the desorption flux changes according Eq. (2) as

$$D(\Theta) = D(0°) \cos^n \Theta,$$

then Eq. (5) will lead to:

$$S(\Theta) = S(0°) \cos^{n-1} \Theta. \qquad (6)$$

Palmer et al. [9] were the first ones to apply molecular beam techniques to verify above relations for the case hydrogen/Ni(111). Similar data were obtained for H_2/Cu by Cardillo et al. in a series of papers which will be discussed in the next chapter. A systematic investigation of $S(\Theta)$ and $D(\Theta)$ was later performed in our laboratory involving H_2 adsorption and desorption for Ni [20,21], Pd [21], Cu [11], Al [22,23] and W [24]. Angular distributions $S(\Theta)$ were measured with nozzle beams or Knudsen beams while $D(\Theta)$ in all cases was determined by angle resolved flash desorption. As an example the measurements for the system H_2/Ni(111) [20] are depicted in Fig. 3.

On an other level one can check detailed balancing for the velocity distributions of adsorbing and desorbing particles. For the desorption flux a time-of-flight measurement has to be performed on molecules originating from a permeation source. These experiments were pioneered by Stickney et al. [25] and later by Comsa and David [26]. Complementary adsorption data have to be obtained by the ap-

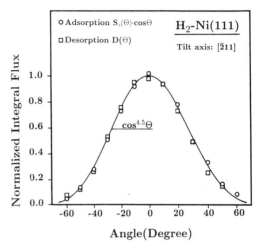

Fig. 3. Detailed balancing requires the differential adsorption flux $S(\Theta) \cos \Theta$ to be equal to the differential desorption flux $D(\Theta)$. The impinging hydrogen beam was generated by a capillary array at 300 K; angle resolved flash desorption was applied to determine the desorption flux from the Ni(111) surface. Data from Ref. [20].

plication of monoenergetic nozzle beams as described in the next chapter. Actually, there are only a few adsorption systems for which both sets of velocity resolved dynamics data have been obtained: H_2/Cu [11,27], H_2/Ni [21,28] and H_2/Pd + sulfur [21,29]. The example of hydrogen adsorbing on and desorbing from a sulfur covered Pd(100) surface is shown in Figs. 4a and 4b. The occurrence of an activation barrier at about 3 kcal/mol separates the hydrogen flux in a Maxwellian and a non-Maxwellian component, both in the adsorption and in the desorption flux.

Finally, most recently validity of detailed balancing also in terms of vibrational quantum states was found to hold. Vibrationally excited molecules have been found to be over-represented both in the desorption flux as well as in the adsorption flux for some systems of activated dissociative adsorption; this matter is treated in the last chapter.

For almost all adsorption systems investigated so far, detailed balancing seems to hold for a situation of quasi-equilibrium. Those cases showing an apparent breakdown of detailed bal-

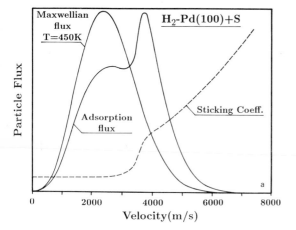

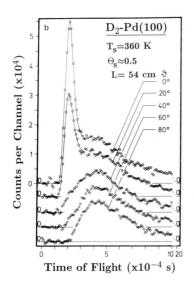

Fig. 4. Both (a) and (b) demonstrate the validity of detailed balancing in terms of the velocity distributions of the adsorbing and desorbing flux. A nozzle beam was used to determine the velocity dependent sticking coefficient of hydrogen on a sulfur covered (~ 0.5 ML) Pd(100) surface. Because of the barrier for dissociation around 3500 m/s (equivalent to ~ 3 kcal/mol) an impinging Maxwellian produces a mixture of a Maxwellian and a non-Maxwellian (high velocity) component in the adsorption flux. Data from Ref. [21]. (b) Time-of-flight measurement on deuterium originating from a sulfur covered (~ 0.5 ML) palladium (100) permeation source. The situation is roughly complementary to the adsorption experiment of (a). The desorption flux also contains a high velocity component as well as a Maxwellian fraction. Data from Ref. [29].

ancing can all be traced to some violation of the state of quasi-equilibrium in the experimental procedure: for example, it is easy to see why one can adsorb say CH_3OH on a cold surface and obtain CO and H_2 during subsequent thermal desorption. These are just non equivalent situations to be related by detailed balancing.

4. Velocity distributions of adsorbing and desorbing particles

This chapter deals with the most important developments for the understanding of adsorption and desorption dynamics. It comprises a set of seminal experiments by Stickney and Cardillo that profoundly influenced the ideas in surface science. The van Willigen experiment [10] described earlier, directly implied that adsorbing and desorbing particles had to have non-Maxwellian energy distributions. The development of nozzle beams as sources of monoenergetic molecular beams (discussed e.g. by Scoles in this volume and in Ref. [30]) opened the door for the investigation of velocity resolved sticking coefficients $S(v, \Theta)$. On the other hand time-of-flight measurements on permeation sources made it possible to check on the velocity distribution $D(v, \Theta)_T$ of desorbing particles.

The first of the Stickney papers [25] dealt with time-of-flight measurements of D_2 molecules desorbing from a hot polycrystalline nickel surface. As in the van Willigen experiment a permeation source served as a sample. The result of this crucial experiment was *the first time determination of the non-Maxwellian character* of the desorption flux generated in activated associative desorption (Fig. 5). Hydrogen molecules (D_2) desorbing from the polycrystalline nickel surface exhibited a mean energy of $3kT$ instead of the $2kT$ expected for a Maxwellian beam. In addition, the velocity distribution of the desorbing particles was narrower than expected for a Maxwellian. Both of these results are in qualitative agreement with the van Willigen formula [10] (Eq. (1)) and are also reasonably compatible with the angular distribution $D(\Theta)$ determined as $\cos^{4.5} \Theta$. Shortly af-

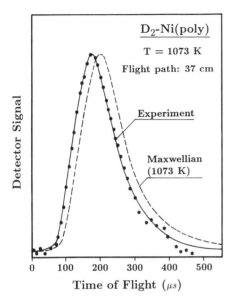

Fig. 5. Time-of-flight measurement performed by Stickney et al. on the D_2 desorption flux originating from a polycrystalline nickel permeation source. This experiment for the first time showed the non-Maxwellian character of desorbing hydrogen. The desorbing particles have a higher mean energy than expected from a Maxwellian of identical temperature. Data from Ref. [25].

terwards Cardillo and Stickney [31] made the next logical step and investigated the *adsorption* dynamics for H_2/Cu with the help of a nozzle beam. The HD exchange reaction served as a measure for the sticking coefficient; a choice that in retrospect proved to have some serious consequences. The much discussed experiment seemed to indicate an activation barrier of about 4 kcal/mol for H_2/Cu. What made this paper so fascinating was the agreement with some general principles of surface physics. Angular and energy distributions of adsorbing molecules obeyed *normal energy scaling*, the consequence of a one-dimensional barrier mechanism:

$$S(E, \Theta) = S(E \cos^2 \Theta, 0°). \qquad (7)$$

The picture of a one-dimensional barrier was still more reinforced by the measurement of the S-shaped form of $S(E)$ as predicted by the model. In a further impressive paper Cardillo and Stickney [17] discussed the above data with respect

to detailed balancing. A well rounded picture of dissociative chemisorption seemed to emerge.

What a surprise when Comsa and David [27] determined the velocity distribution of hydrogen originating from a copper permeation source. There was no compatibility in terms of detailed balancing with the Cardillo/Stickney data. What had gone wrong? Only 15 years later it was found out [11] that the activation energy of 4 kcal/mol obtained in Ref. [17] was much too small. These experiments had just been one step too far for the experimental capabilities of that time. Nevertheless, the Stickney/Cardillo papers were milestones in the investigation of adsorption/desorption dynamics and the general procedures (normal energy scaling, detailed balancing) are still applied in the way presented at that time.

During the period of 1975–1980 Comsa and David performed a series of experiments characterizing the desorption of hydrogen from permeation sources [26]. These were experiments performed under truly well defined surface and vacuum conditions. Again and again these data have proved to be some of the most accurate characterizations of desorption dynamics. The time-of-flight investigations included non-activated adsorption for H_2/Pd [29], slightly activated adsorption for $H_2/Ni(111)$ [28] and the highly activated adsorption system $H_2/copper$ [27], which exhibited mean energies for desorbing hydrogen of about $8kT$. At this point, one faced the situation that microscopic reversibility and detailed balancing were up to scrutiny again because of the H_2/Cu dilemma. A beautiful set of desorption data by Comsa and David existed for several adsorption systems but no corresponding adsorption data were available yet. In addition, there was the unanswered question if post-permeation hydrogen really originated from the same initial surface state as in the traditional desorption experiments. Was this perhaps the reason for the apparent breakdown of detailed balancing?

It therefore was not surprising that at the same time experimental efforts were directed to gain differential sticking coefficients $S(v, \Theta)$ to compare them to the permeation data. In

1985 a wealth of information on the adsorption/desorption dynamics of H_2/Ni appeared in the literature [20,32–34]. All these molecular beam results indicated that adsorption of H_2 on Ni(111) was an activated process, although exhibiting a rather small activation barrier of about 3 kcal/mol. What a relief that adsorption [20] and desorption [28] for Ni(111) agreed in terms of detailed balancing: the mean energy of adsorbing as well as desorbing hydrogen was found to be identical with $\langle E \rangle \approx 3kT$. Especially good agreement was also obtained for the angle resolved adsorption and desorption data. The $\cos^{4.5}\Theta$ obtained in angle resolved flash desorption agreed perfectly with the $\cos^{3.5}\Theta$ determined for the sticking coefficient in beam experiments [20,32]. A more refined analysis later [21] completed the picture in showing that really *all existing adsorption and desorption data* for H_2/Ni(111) could be fitted into one consistent picture: direct, activated adsorption obeying normal energy scaling and detailed balancing. At the same time these experiments proved that permeation/desorption experiments and conventional low temperature adsorption/desorption experiments yield identical results. Post-permeation hydrogen apparently originated from the same initial state as in flash desorption.

What about the now famous case of H_2/copper? It still resisted any elucidation. Nozzle beams using the HD exchange reaction and post-permeation had established non-compatible sets of data: here a picture of low activation energy [31] hardly larger than for nickel, there an immense excess energy of $8kT$ in the desorption flux [27]. In the meantime molecular beam techniques had been sufficiently refined to circumvent the HD reaction and use flash desorption to determine coverage values for adsorption of hydrogen on copper. Anger et al. [11] showed that sticking coefficients for H_2/Cu were one to two orders of magnitude smaller than previously assumed. Only above a beam energy of 0.2 eV the sticking coefficient would increase rapidly with particle energy. This rapidly increasing sticking coefficient when folded with a Maxwellian yielded a mean energy for the adsorp-

tion flux of about $7kT$, in close agreement with the permeation data. Because of normal energy scaling one could expect a rather sharp angular change of the sticking coefficient; and indeed values of about $n = 16$ were observed. Both the small value of the sticking coefficient and the high value of n implied *a rather high activation barrier* to dissociative adsorption. Needless to say that everybody was quite happy with these results, in particular the theoreticians who for long time had proclaimed that noble metals like copper should exhibit large activation barriers, perhaps 1 eV, for hydrogen adsorption [35]. The nozzle beam experiments basically confirmed these ideas.

There was one feature though in the nozzle beam results for H_2/Cu immediately pointed out [36,37]: normal energy scaling was only obeyed in an *approximate fashion*. The sticking coefficients $S(E, \Theta)$ for high energies *and high nozzle temperatures* when projected into the 0° sticking curve according to $S(E\cos^2\Theta, 0°)$ systematically gave *too high values*. To understand the implication of this result one has to realize that to produce nozzle beams of varying translational energy E one employs high pressure expansion from a nozzle of varying temperature (T_{nozzle}), whereby translational energy and nozzle temperature are related through:

$$E \approx \tfrac{5}{2}kT_{nozzle}. \qquad (8)$$

But not only translational energy, also the *internal energy* of the molecules is increased with temperature. Ostensively, the sticking coefficient for H_2/Cu as seen in Fig. 6 is not only a function of the translational energy but also of the nozzle temperature. The reason, as suggested by Harris [36], is the contribution of the vibrational energy to the dynamics of adsorption. But this is a matter to be discussed in the last chapter.

5. Precursors for dissociative chemisorption

At this point one should keep in mind that already the simple Lennard-Jones model [1]

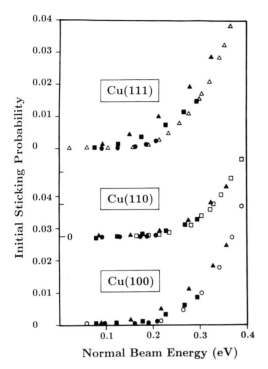

Fig. 6. Breakdown of normal energy scaling in the adsorption of H_2/Cu. Angle resolved sticking coefficients obtained at high total energies and high nozzle temperatures (solid triangles: $T = 1700$ K, $E = 0.4$ eV, solid squares: $T = 1400$ K, $E = 0.33$ eV and solid circles: $T = 1000$ K, $E = 0.23$ eV), when plotted versus normal energy $E \cos^2 \Theta$ yield different values than the sticking coefficients obtained at $\Theta = 0°$ incidence (open symbols). This indicates an explicit influence of the nozzle temperature on sticking via a vibrational contribution. Data from Ref. [11].

for dissociative chemisorption included the possibility of a molecularly adsorbed state. The molecular state might be a separate adsorption state [38] but more frequently can act as an intermediate state to dissociative chemisorption. Because of the shallow well depth the lifetime of such an intermediate state would be very short. Physisorption or trapping was initially investigated for adsorption of rare gases [39]; as a precursor mode to chemisorption it was studied only considerably later. The seminal work in this area can be traced to Becker and Hartman [40], Ehrlich [41] and Kisliuk [42]. Frequently precursors were merely considered conceptual crutches of little physical reality. One still re-

members headings: "Precursor states, myth or reality" [43] or "How real are precursors"? [44]. Although indirect evidence of precursor mediated chemisorption had been available for a long time, only molecular beam studies could get hold on the specific dynamics involved.

The physics determining trapping is the dissipation of the excess translational energy of the impinging molecule to prevent the particle from redesorption. From all we know today the most effective mechanism involved in this energy transfer is the excitation of phonons in the adsorbing surface. Actually, on an uncorrugated smooth surface it is only necessary for the impinging particle to dissipate the normal energy $E \cos^2 \Theta$ to remain trapped on the surface; parallel energy may be dissipated during a subsequent movement along the surface. This physics leads to just the opposite result as in the case of activated chemisorption. Normal energy scaling is expected, but the larger the normal energy the lower the trapping probability. The characteristic dynamics of precursor adsorption will be a sticking coefficient falling with particle energy. Normal energy scaling will then lead to sticking coefficients increasing with increasing angle and thus to values n of Eq. (6) smaller than unity.

The first application of a molecular beam technique to precursor mediated chemisorption involved the system N_2/W by King and Wells [45]. The paper by King and Wells is remarkable in several ways: from an experimental point it introduced the now generally applied "Method by King and Wells" to determine sticking coefficients. The appeal of this method is its capability to determine *absolute* values of sticking coefficients without the need of any calibration. Although only a limited range of energies was available for the effusive beam source used in the experiments, all of the pertinent features of precursor adsorption emerged from the study: the sticking coefficient clearly decreased with increasing beam energy as expected by the need of energy dissipation. Second, an observed variation of the sticking coefficient with surface temperature T_s was also in line with an intermediate precursor state. The molecule equilibrated in the precursor state can either desorb or transfer into the chemisorbed state. These competing processes will depend on surface temperature. In fact, today the condition $\partial S / \partial T_s \neq 0$ is considered a strong indication of a precursor mediated process.

By the years 1985 to 1990 a more systematic application of molecular beam techniques to the problem of precursors took place. Sticking coefficients falling with beam energy have been observed for the adsorption systems $H_2/Ni(110)$ and $H_2/Ni(997)$ [21,32,46] in our laboratory as well as in studies by Rettner and Auerbach on the adsorption of $O_2/W(110)$ [47] and $N_2/W(100)$ [48]. Further results included $O_2/Pt(111)$ [49] and propane and butane adsorption on $Ir(110)(1 \times 2)$ [50].

What about the angular variation of the sticking coefficient in chemisorption via a precursor state? As mentioned above, as a consequence of normal energy scaling the sticking coefficient should actually increase with increasing angle of incidence. Indeed, such a behavior with values of $n < 1$ (Eq. (6)) could be observed for several adsorption systems [21,51,52]. Generally though, normal energy scaling is obeyed only in a qualitative fashion and the values n are usually somewhat larger than expected from the shape of $S(E)$. This seems to imply that also the *tangential component* of the energy has to be accommodated to some extent. This will occur if the mobile precursor state is confined to a limited surface area. Finally, if complete localization of the intermediate state on the surface is encountered *total energy scaling* is the result. A rather nice example of this behavior can be seen in Fig. 7 [21]. Adsorption of H_2 on a Ni(110) surface proceeds via a mobile precursor ($n < 1$) when the beam is aimed parallel to the surface grooves. This corresponds to an orientational change of the crystal by tilting it about the [100] direction relative to the fixed beam direction. Aiming the beam perpendicular to the grooves (tilt axis [110]) suppresses any tangential movement and results in a near angle independent sticking coefficient.

One feature encountered in many dissociative adsorption processes utilizing a precursor path is the *presence of a second, direct, activated adsorption channel*. This has, for example, explicitly

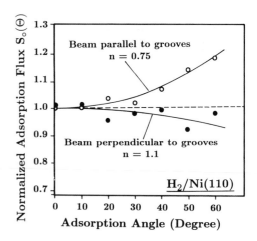

Fig. 7. Influence of a precursor state on the adsorption of hydrogen on a corrugated surface. With the beam aimed parallel to the grooves of the Ni(110) surface only normal energy has to be dissipated. This results in a sticking coefficient increasing with angle of incidence. A beam hitting the surface perpendicular to the atom rows has to dissipate the total energy, resulting in a near angle independent sticking coefficient. Data from Ref. [21].

been shown for hydrogen adsorption on all transition metals (Pd, W, Pt, Fe, Ni) as well as on all other adsorption systems already mentioned above. At low particle energy the precursor path is utilized whereas at higher beam energies the activated, direct adsorption path with a sticking coefficient increasing with increasing beam energy dominates. At the same time the angular variation of the sticking coefficient changes from values $n < 1$ to values of $n > 1$ as the particle energy is raised. As an example of a molecular beam study exhibiting most of the features expected for precursor adsorption the system N_2/W [48] is shown in Fig. 8.

As a consequence of detailed balancing one should really also see two *desorption channels* in the flux of desorbing molecules; and indeed the time-of-flight measurements by Comsa and David [26] for some cases exhibited a superposition of two sets of particles. The mean energy of desorbing particles was observed to *decrease with increasing desorption angle* for $H_2/Ni(111)$ [28,53]. The real shocker at that time was the mean energy of hydrogen desorbing at angles larger than 70°; they had an *even smaller mean*

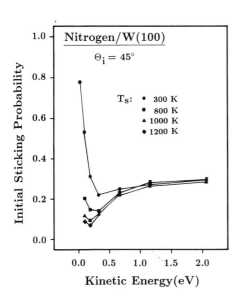

Fig. 8. The adsorption of $N_2/W(100)$ serves as an example for precursor mediated adsorption. At low beam energy adsorption is dominated by a precursor path: the sticking coefficient decreases with beam energy. A second, activated adsorption path is utilized at high beam energy (> 0.5 eV). In addition, low surface temperature will aid adsorption via the precursor path. Data from Ref. [48].

energy than the $2kT$ expected for a Maxwellian [28]. Once nozzle beam adsorption data on precursor mediated adsorption became available, also this puzzle was solved [54]. A parallel path involving a precursor will inevitably exhibit mean energies for adsorbing and desorbing particles smaller than $2kT$ because of the monotonically falling sticking coefficient. The higher desorption angles just separate the precursor path from the activated path. But where had the precursor path for $H_2/Ni(111)$ come from? For all one knew, hydrogen adsorption on $Ni(111)$ was a purely direct, activated process. The source of the precursor path proved to be surface defects, a topic that deserves a more detailed treatment in this context.

6. Dynamics of adsorption at defect sites: stepped surfaces

If we look back at the short history of surface physics, we can recognize that up to the 1960's

wires and polycrystalline foils of sometimes undefined surface composition served as samples in the adsorption experiments. Although the experiments corresponded to the situation on a real, technologically relevant surface, very little about the microscopic mechanism during adsorption could actually be learned at that time. Only after the physics of adsorption was known for a perfect crystal one could start to introduce surface defects and contaminants to study their effect on the process of adsorption [55,56].

And at that moment, one has to say, a gift from heaven arrived: the concept of stepped surfaces. Stepped surfaces or vicinal planes are located a few degrees off a smooth low index crystal plane. They consist of terraces of the low index plane several atom rows wide, separated by generally monatomic steps [57]. Field ion microscopy and the physics of crystal growth have already dealt for a long time with stepped surfaces. It was only in the 1970's that vicinal planes were introduced as models for well characterized defective, catalytically active surfaces. The true starting point of this development were the beam experiments by Somorjai [58,59]. For the first time one could see that the catalytic rate of dissociation for hydrogen became strongly dependent on the angle of incidence relative to the steps. But what particular process would yield the "promoting" action at the steps? The answer was found later during a detailed examination of the adsorption/desorption dynamics of H_2 on $Ni(111)$ vicinal planes such as the (997) plane [32,46,56]: steps introduce sites for non-activated adsorption via a precursor on a surface which normally only allows activated adsorption. The dynamics of hydrogen adsorption for a beam aimed "step up" and "step down" is completely different [46]. "Step up" the beam encounters mostly step sites and a sticking coefficient falling with beam energy is observed, whereas a beam aimed "step down" will mainly hit the terraces and exhibit most properties of the original (111) plane, e.g. activated adsorption. Again the beam energy will determine which adsorption/desorption path is taken. Molecules with low energy (< 0.1 eV) will predominantly take the precursor path via

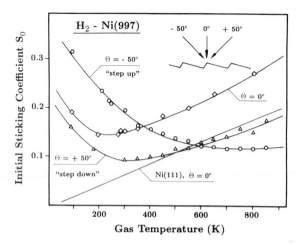

Fig. 9. Surface steps as active sites for H_2 dissociation. Adsorption dynamics for a Knudsen beam of varying temperature is different on the step sites of the Ni(997) surface and on the (111) terraces. With the beam aimed "step-up" ($\Theta = -50°$) mostly step sites are encountered and adsorption proceeds via a precursor ($\partial S/\partial T < 0$). A beam directed "step-down" ($\Theta = +50°$) hits mostly the (111) terraces exhibiting activated adsorption ($\partial S/\partial T > 0$). Data from Ref. [46].

the steps whereas high energy molecules will predominantly adsorb on terraces. This drastic change in the adsorption dynamics with temperature and angle of incidence is clearly manifested in the measurements of $S(T)_\Theta$ shown in Fig. 9. While all the evidence points to the paramount role of surface defects in adsorption dynamics, theoreticians have paid little attention to these developments. Aside from a few general ideas relating work function changes at defect sites to barrier heights, detailed work is still missing.

7. Internal energy contributions to the dynamics of adsorption and desorption

This section deals with the simple but important question already touched by Lennard-Jones [1]: what happens to the internal energy of a molecule during activated dissociative chemisorption? Can this energy perhaps be utilized to aid transgression of an activation barrier? Surface scientists owe many of their

concepts to the description of gas phase chemistry. Not only the elbow type two-dimensional potential energy surface has successfully been transposed to describe adsorption/desorption processes [60], but also the idea of vibrational contributions to the dynamics of a reaction was first explored by gas phase chemistry [61]. Several well documented cases of vibrationally assisted gas phase reactions can be found in the literature [62].

First evidence of an exchange of rotational and vibrational energy at surfaces came from molecular beam *scattering data*. The role of vibrational energy in *adsorption/desorption dynamics* was investigated much later. Again it was the famous adsorption system H_2/Cu that started it all. Kubiak et al. [63], puzzled by the discrepancies between the Cardillo/Stickney molecular beam data [31] and the Comsa/David time-of-flight experiments [27], proceeded in a now famous paper to explore the influence of rotational and vibrational quantum states on the desorption dynamics of H_2/Cu. A copper permeation source at 850 K served as source for a continuous molecular beam of desorbing hydrogen. Laser spectroscopy on this beam was employed to probe the internal state distribution of the desorbing hydrogen molecules. Whereas the rotational energy levels of hydrogen were slightly underpopulated compared to a Maxwell–Boltzmann distribution defined by the surface temperature, the vibrational state $\nu = 1$ was highly overpopulated by a factor of about 50 to 100. By way of detailed balancing this would mean that a hydrogen molecule impinging on a copper surface *would have a vastly superior sticking coefficient when in the first excited vibrational state*.

The corresponding molecular beam experiment for the case of hydrogen adsorption on copper would still have to wait another six years. How can one set up such an experiment? Obviously the influence of translational energy and of nozzle temperature have to be separated in the exploration of adsorption dynamics: one has to get away from the physics of Eq. (8). A rather nice method can be employed to change the translational energy even at a constant nozzle temperature: to slow down the beam a second heavier gas, usually an inert gas, is mixed into the primary molecular beam. These *seeded beams* were developed quite early in the short history of nozzle beams [64,65]. The original applications had of course nothing to do with the vibrational state of the molecules; the aim was strictly to produce high energy beams through the acceleration of heavy molecules seeded into a beam of light molecules. At least in a limited fashion nozzle temperature T_{nozzle} (which determines the vibrational state of the molecule) and the translational energy E can be adjusted independently. Any time the inequality:

$$\frac{\partial S(E, T_{nozzle})}{\partial T_{nozzle}} \neq 0 \qquad (9)$$

is fulfilled for the sticking coefficient, a contribution of internal energy is present.

Molecular beam experiments explicitly probing the influence of vibrational energy were first performed on the system N_2/Fe [66]. Even more complex molecules like CH_4 [67,68] and CO_2 [69] were shown to convert vibrational energy into translational energy during dissociative, activated adsorption. But again the adsorption of hydrogen on copper provided a model system for vibrationally assisted adsorption [23,37,70–73]. Actually, the adsorption of H_2/Cu had earlier already been singled out by theoreticians as a possible candidate for vibrational/translational energy conversion [74]. This conversion can come about in two different ways:

(1) A softening of the potential to which the vibrating atoms of the adsorbing molecule (e.g. hydrogen) are subjected will lead to a decreased vibrational energy at constant vibrational quantum number ν. This energy can be converted into translational energy [36,75,76].

(2) If the adsorption path in the PES is curved, a mixing of vibrational energy and translational energy can occur. Here the vibrational quantum number has to change. One can understand this mechanism even in a classical picture. For example, a desorbing particle possessing translational energy will climb up the potential

wall in the curved section of the desorption path leading to a vibrational motion. The reverse process in adsorption, although more difficult to visualize, is of course equally possible [77,78].

The translational energy gained from the conversion of vibrational energy can be used to overcome an existing activation barrier to dissociative chemisorption.

The application of seeded beam techniques to the system H_2/Cu was first performed by Hayden and Lamont [72] and demonstrated, at least qualitatively, the importance of vibrational energy in the dynamics of adsorption. In short sequence further investigations on this topic followed [70,71,73,79]. The most striking demonstration of the internal energy contribution to the dynamics of hydrogen adsorption and a clear verification of inequality Eq. (9) is obtained when the *nozzle temperature is changed while the translational energy is kept constant* [23,73,80], using the seeded beam technique (Fig. 10). The explicit variation of the sticking coefficient with nozzle temperature can only be caused by a change in occupation for the individual vibrational states of the hydrogen molecules, each state exhibiting a different sticking probability. An evaluation of the activation energy from the Arrhenius plots leads to the conclusion [23] that for adsorption of hydrogen on Cu(110) mainly the *first vibrational level* (activation energy 0.52 eV) and for adsorption of D_2 the *second vibrational level* (activation energy 0.72 eV) results in the observed temperature dependence of the sticking coefficient. Transgression of the activation barrier for dissociative chemisorption is aided by the partial conversion of vibrational energy into translational energy, as a consequence *less of the directly supplied translational energy is needed* for the particle to get over the barrier. One can say, the effective barrier height to dissociation has been lowered [75]. A lowering of the barrier will directly lead (via normal energy scaling) to a widening of the angular variation of the sticking coefficient $S(\Theta)_E$ with increased nozzle temperature and the increased occupancy of the excited vibrational states. This widening of $S(\Theta)$ has indeed been observed for all cases of vibrationally as-

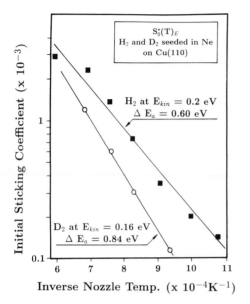

Fig. 10. An explicit manifestation of the vibrational contribution to the adsorption dynamics of hydrogen on copper. A change of nozzle temperature *alone* introduces a drastic change in the sticking probability, while the translational energy of the hydrogen molecules is kept constant by a seeding procedure. The activation energy observed in this experiment points to the predominant influence of the first vibrational state in the case of H_2 adsorption; deuterium in contrast uses mainly the second vibrational level to facilitate barrier transgression. Data from Ref. [23].

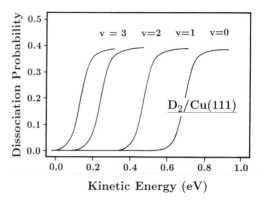

Fig. 11. State resolved sticking coefficients for the individual vibrational states in the system $D_2/Ni(111)$. To obtain the result shown above, a particular S-shaped dependence of the sticking coefficient on the kinetic energy has to be introduced in the deconvolution procedure. Data from Ref. [79].

sisted hydrogen adsorption: $H_2/Cu(111)$ [70], $H_2/Cu(110)$ [23] and $H_2/Fe(100)$ [73]. In this context it is interesting to note that normal energy scaling has been observed for the sticking coefficient $S(E,\Theta)_{\nu=0,1,2}$ in the *individual vibrational states* [73].

Seeded hydrogen beam experiments will directly lead to the information $S(E)_T$. It is not too difficult to transform this function into a plot of *state resolved* sticking coefficients $S(E)_{\nu=0,1,2}$ as has been done [73] for $H_2/Cu(110)$, $H_2/Cu(111)$ and H_2/Fe. As an example state resolved sticking coefficients for $D_2/Cu(111)$ obtained by Rettner [79] are depicted in Fig. 11.

What about an isotope effect in the sticking coefficients? Obviously we are dealing with the quantum nature of the H_2 and D_2 molecules exhibiting different vibrational levels. As expected,

the sticking data on H_2/Cu for $S(E)_T$ obtained in a seeded beam experiment clearly show an isotope effect between H_2 and D_2 [23]. The sticking coefficient for H_2 is larger than the one for D_2 at identical translational energy E and constant T_{nozzle}. This is a consequence of the smaller population of the predominantly utilized $\nu = 2$ level for D_2 compared to the $\nu = 1$ level for H_2. In addition this also reflects the smaller tunnelling probability for D_2. Interestingly, if a *conventional beam experiment* to obtain sticking coefficients in the form $S(E(T))$ is evaluated, only relatively small isotope effects are obtained. This is due to partial compensation through opposing behavior of adsorption parameters with translational energy and internal energy as shown by Brenig [81]. Isotope effects have also been observed for the adsorption systems H_2/Fe [73], CH_4/Ni [67] and CH_4/W [68].

While the vibrational contributions leave a clear mark on the *adsorption* as well as on the *desorption dynamics*, the detection of rotational contributions has been confined mostly to laser spectroscopy of *desorbing* molecules. Almost in all cases investigated, rotationally excited molecules are underrepresented in the desorption flux, implying a retarding effect of rotational energy on the process of adsorption [63,82,83]. Recently, though, indications appeared that in the adsorption system H_2/Cu the high rotational states ($j > 8$) might actually contribute to *lower* the effective height of the activation barrier [84].

It certainly seems to be a safe prediction that adsorption/desorption dynamics in the near future will be investigated along the line of state resolved quantities pointed out in this last section. Also the trend in the theoretical description of adsorption dynamics seems to point into the same direction.

8. Concluding remarks

What we have seen is that the advancement in a field generally comes from concepts, not necessarily from the first time formulation of ideas, but from the incorporation of concepts in the general line of thought. Experimental developments per se have a smaller impact than usually assumed. An example are the van Willigen and the Cardillo/Stickney papers. These people were thinking in terms of differential adsorption quantities, that means dynamics, while the rest of the world was still pondering kinetics. In a quantitative fashion their results have been replaced by more accurate measurements; but despite many experimental shortcomings these papers have charted the course to follow.

What will the future bring? Certainly the major steps will again happen unexpectedly. Nothing is more difficult to predict than the future – especially in advance. Perhaps around the next corner we might meet some unexpected but exciting surprises.

Acknowledgements

The work on this project has been supported by grants from the Austrian "Fonds zur Förderung der wissenschaftlichen Forschung" and from the "Jubiläumsfonds der Österreichischen Nationalbank".

References

[1] J.E. Lennard-Jones, Trans. Faraday Soc. 28 (1932) 333.
[2] J.N. Smith, Jr. and W.L. Fite, J. Chem. Phys. 37 (1962) 898.
[3] F.C. Hurlbut, Recent Research in Molecular Beams, Ed. I. Estermann (Academic Press, New York, 1959) p. 145.
[4] J.C. Maxwell, Philos. Trans. 170 (1879) 231.
[5] I. Langmuir, Phys. Rev. 8 (1916) 149.
[6] P. Clausing, Ann. Phys. 4 (1930) 533.
[7] M. Knudsen, Ann. Phys. 48 (1915) 1113.
[8] G. Comsa, J. Chem. Phys. 48 (1968) 3235.
[9] R.L. Palmer, J.N. Smith, Jr., H. Saltsburg and D.R. O'Keefe, J. Chem. Phys. 53 (1970) 1666.
[10] W. van Willigen, Phys. Lett. A 28 (1968) 80.
[11] G. Anger, A. Winkler and K.D. Rendulic, Surf. Sci. 220 (1989) 1.
[12] G. Kirchhoff, Pogg. Ann. 109 (1860) 148.
[13] O.W. Richardson, Philos. Mag. 27 (1914) 476.
[14] I. Langmuir, J. Am. Chem. Soc. 38 (1916) 2221.
[15] R.H. Fowler and E.A. Guggenheim, Statistical Thermodynamics (Cambridge University Press, Cambridge, 1949) p. 431.
[16] E.P. Wenaas, J. Chem. Phys. 54 (1971) 376.
[17] M.J. Cardillo, M. Balooch and R.E. Stickney, Surf. Sci. 50 (1975) 263.
[18] G. Comsa, in: Proc. 7th Int. Vacuum Congr. and 3rd Int. Conf. on Solid Surfaces, Vienna, 1977, p. 1317.
[19] K.D. Rendulic, Surf. Sci. 272 (1992) 34.
[20] H.P. Steinrück, K.D. Rendulic and A. Winkler, Surf. Sci. 154 (1985) 99.
[21] K.D. Rendulic, G. Anger and A. Winkler, Surf. Sci. 208 (1989) 404.
[22] A. Winkler, G. Požgainer and K.D. Rendulic, Surf. Sci. 251/252 (1991) 886.
[23] H.F. Berger and K.D. Rendulic, Surf. Sci. 253 (1991) 325.
[24] H.F. Berger, Ch. Resch, E. Grösslinger, G. Eilmsteiner, A. Winkler and K.D. Rendulic, Surf. Sci. Lett. 275 (1992) L627.
[25] A.E. Dabiri, T.J. Lee and R.E. Stickney, Surf. Sci. 26 (1971) 522.
[26] G. Comsa and R. David, Surf. Sci. Rep. 5 (1985) 145.
[27] G. Comsa and R. David, Surf. Sci. 117 (1982) 77.
[28] G. Comsa, R. David and B.-J. Schumacher, Surf. Sci. 85 (1979) 45.
[29] G. Comsa, R. David and B.-J. Schumacher, Surf. Sci. 95 (1980) L210.
[30] G. Scoles, Ed., Atomic and Molecular Beam Methods, Vol. 1 (Oxford University Press, New York, 1988).
[31] M. Balooch, M.J. Cardillo, D.R. Miller and R.E. Stickney, Surf. Sci. 46 (1974) 358.
[32] H.P. Steinrück, M. Luger, A. Winkler and K.D. Rendulic, Phys. Rev. B 32 (1985) 5032.
[33] H.J. Robota, W. Vielhaber, M.C. Lin, J. Segner and G. Ertl, Surf. Sci. 155 (1985) 101.

[34] A.V. Hamza and R.J. Madix, J. Phys. Chem. 89 (1985) 5381.

[35] J. Harris, Appl. Phys. A 47 (1988) 63.

[36] J. Harris, Surf. Sci. 221 (1989) 335.

[37] H.A. Michelsen and D.J. Auerbach, J. Chem. Phys. 94 (1991) 7502.

[38] S. Andersson, L. Wilzén and J. Harris, Phys. Rev. Lett. 55 (1985) 2591.

[39] F.O. Goodman, Prog. Surf. Sci. 5 (1975) 261.

[40] J.A. Becker and C.D. Hartman, J. Phys. Chem. 57 (1953) 157.

[41] (a) G. Ehrlich, J. Phys. Chem. 59 (1955) 473;
     (b) G. Ehrlich, J. Phys. Chem. Solids 1 (1956) 1.

[42] (a) P. Kisliuk, J. Phys. Chem. Solids 3 (1957) 95;
     (b) P. Kisliuk, J. Phys. Chem. Solids 3 (1957) 78.

[43] M. Grunze and H.J. Kreuzer, Eds., Kinetics of Interface Reactions, Springer Series in Surface Science, Vol. 8 (Springer, Berlin, 1986).

[44] D. Menzel, Proc. Symp. Surf. Sci., Eds. P. Braun et al., Obertraun 1983, p. 218.

[45] D.A. King and M.G. Wells, Proc. Roy. Soc. (London) A 339 (1974) 245.

[46] H. Karner, M. Luger, H.P. Steinrück, A. Winkler and K.D. Rendulic, Surf. Sci. 163 (1985) L641.

[47] C.T. Rettner, L.A. DeLouise and D.J. Auerbach, J. Chem. Phys. 85 (1986) 1131.

[48] D.J. Auerbach and J.T. Rettner, in: Kinetics of Interface Reactions, Springer Series in Surface Science, Vol. 8, Eds. M. Grunze and H.J. Kreuzer (Springer, Berlin, 1986).

[49] M.D. Williams, D.S. Bethune and A.C. Luntz, J. Chem. Phys. 88 (1988) 2843.

[50] A.V. Hamza, H.P. Steinrück and R.J. Madix, J. Chem. Phys. 86 (1987) 6506.

[51] K.D. Rendulic, A. Winkler and H. Karner, J. Vac. Sci. Technol. A 5 (1987) 488.

[52] G. Anger, H.F. Berger, M. Luger, S. Feistritzer, A. Winkler and K.D. Rendulic, Surf. Sci. 219 (1989) L583.

[53] G. Comsa, R. David and K.D. Rendulic, Phys. Rev. Lett. 38 (1977) 775.

[54] K.D. Rendulic, Appl. Phys. A 47 (1988) 55.

[55] See, for example: K. Wandelt, Surf. Sci. 251/252 (1991) 387.

[56] K.D. Rendulic and A. Winkler, Intern. J. Mod. Phys. B 3 (1989) 941.

[57] H. Wagner, Springer Tracts Mod. Phys. 85 (1979) 151.

[58] S.L. Bernasek, W.J. Siekhaus and G.A. Somorjai, Phys. Rev. Lett. 30 (1973) 1202.

[59] M. Salmeron, R.J. Gate and G.A. Somorjai, J. Chem. Phys. 67 (1977) 5324.

[60] J.K. Nørskov, A. Houmøller, P.K. Johansson and B.I. Lundqvist, Phys. Rev. Lett. 46 (1981) 257.

[61] J.C. Polanyi and W.H. Wong, J. Chem. Phys. 51 (1969) 1439.

[62] See, for example: R.N. Zare and R.B. Bernstein, Phys. Today (1980) 43.

[63] G.D. Kubiak, G.O. Sitz and R.N. Zare, J. Chem. Phys. 83 (1985) 2538.

[64] N. Abuaf, J.B. Anderson, R.P. Andres, J.B. Fenn and D.R. Miller, in: Proc. 5th Int. Symp. on Rarefied Gas Dynamics, Ed. C.L. Brudin (Academic Press, New York, 1967).

[65] S. Yamazaki, M. Taki and Y. Fujitani, Jpn. J. Appl. Phys. 18 (1979) 2191.

[66] C.T. Rettner and H. Stein, J. Chem. Phys. 87 (1987) 770.

[67] M.B. Lee, Q.Y. Yang and S.T. Ceyer, J. Chem. Phys. 87 (1987) 2724.

[68] C.T. Rettner, H.E. Pfnür and D.J. Auerbach, Phys. Rev. Lett. 54 (1985) 2716.

[69] M.P.D'Evelyn, A.V. Hamza, G.E. Gdowski and R.J. Madix, Surf. Sci. 167 (1986) 451.

[70] H.F. Berger, M. Leisch, A. Winkler and K.D. Rendulic, Chem. Phys. Lett. 175 (1990) 425.

[71] B.E. Hayden and C.L.A. Lamont, Surf. Sci. 243 (1991) 31.

[72] B.E. Hayden and C.L.A. Lamont, Phys. Rev. Lett. 63 (1989) 1823.

[73] H.F. Berger, E. Grösslinger and K.D. Rendulic, Surf. Sci. 261 (1992) 313.

[74] J. Harris, S. Holloway, T.S. Rahman and K. Yang, J. Chem. Phys. 89 (1988) 4427.

[75] M.R. Hand and S. Holloway, J. Chem. Phys. 91 (1989) 7209.

[76] J.E. Müller, Surf. Sci. 272 (1992) 45.

[77] W. Brenig, S. Küchenhoff and H. Kasai, Appl. Phys. A 51 (1990) 115.

[78] D. Halstead and S. Holloway, J. Chem. Phys. 93 (1990) 2859.

[79] C.T. Rettner, D.J. Auerbach and H.A. Michelsen, Phys. Rev. Lett. 68 (1992) 1164.

[80] A. Winkler and K.D. Rendulic, Intern. Rev. Phys. Chem. 11 (1992) 101.

[81] S. Küchenhoff, W. Brenig and Y. Chiba, Surf. Sci. 245 (1991) 389.

[82] L. Schröter, R. David and H. Zacharias, Surf. Sci. 258 (1991) 259.

[83] K.W. Kolasinski, S.F. Shane and R.N. Zare, J. Chem. Phys. 96 (1992) 3995.

[84] H.A. Michelsen, C.T. Rettner and D.J. Auerbach, Phys. Rev. Lett. 69 (1992) 2678.

Surface Science 299/300 (1994) 277–283
North-Holland

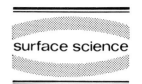

surface science

Molecular beam scattering from solid surfaces:
He diffraction, hyperthermal scattering, and surface dynamics

Mark J. Cardillo

AT&T Bell Laboratories, Murray Hill, NJ 07974, USA

Received 7 May 1993; accepted for publication 14 June 1993

I discuss some qualitative aspects of the interaction of atoms/molecules at surfaces as probed with molecular beam techniques, using as a basis selected experiments done in my laboratory. These include a brief assessment of He diffraction, a description of the phenomenology associated with hyperthermal scattering, and brief comments on activated dissociative adsorption and surface diffusion.

1. Introduction

Over a period of 25 years considerable progress has been made in our knowledge of the chemical processes which occur at surfaces. The specific area of the dynamics of atom or molecule–surface interactions has provided many detailed and often sophisticated descriptions of the elementary events which underlie broader topics, such as surface rate process (chemical kinetics and their mechanisms), adsorption/desorption, and energy transfer. Molecular beam techniques have played a crucial role in this development as they allow complex processes to be partitioned and then studied in terms of fundamental components. For example experiments can be arranged to isolate the effects of incoming particle kinetic energy, momentum, and possibly quantum state and flux. A similar and even finer partition can be made for the outgoing species. Further, an accurate clocking ($\sim 10^{-5}$ s) of the residence time on the surfaces is easily accomplished.

Molecular beam techniques have been applied to a wide range of surface studies ranging from atom diffraction [1] and single phonon scattering [2] to adsorption [3] and reaction [4]. In general, as the interaction with the surface becomes more complex, the experimentally derived information becomes more averaged, and at some point the relationship between the emitted particle and the incident particle becomes confounded by the many degrees of freedom of the solid which come into play. Thus the goals of experiments in these different regimes are quite varied. Yet molecular beam techniques have had a profound influence on our understanding in each area.

With these introductory remarks in mind, I will describe in the paragraphs below a selection of experiments, which span part of this range of complexities of interaction between atoms and molecules and surfaces. This selection is based on molecular beam experiments which have been carried out in my laboratory, and thus represent quite a personal view of the field. I note that in many laboratories around the world, in which molecular beams are employed to study dynamics at surfaces, a comparable or broader scope of experiments have been carried out. However, the nature of this issue of Surface Science is such as to allow the author the convenience of a more personal selection of topics.

2. He diffraction from surfaces

Thermal He scattering experiments from surfaces are the most direct beam studies to carry out and, in the case of diffraction, conceptually

SSDI 0039-6028(93)E0302-B

the simplest. However, as we see below the apparent simplicity is only at the surface. The earliest and most famous of such experiments were He diffraction measurements from LiF surfaces carried out by Otto Stern and coworkers around 1928 [5], which triumphantly demonstrated the wave-nature of atoms. Exploitation of this development as a probe of surface properties was slow to come about. About 25 years ago there arose several groups working to investigate He diffraction in greater depth. There were clever approaches employed to exploit less expensive components or to take advantage of special detector selectivities. These experiments restarted the field, but experimentally it turned out that supersonic jet sources (more correctly – free jet expansions) with differentially pumped quadrupoles in an ultra-high-vacuum (UHV) apparatus became the techniques of choice. Arguments citing versatility and signal-to-noise considerations are the usual explanation for this evolution, but the one overlooked factor was the good funding of the transposed gas-phase molecular beam crowd.

At AT&T Bell Laboratories in 1977 we had just completed building a (relatively) simple apparatus of this kind, to extend the studies of activated dissociation adsorption started at MIT in R. Stickney's laboratory. As we started up we were using, as a convenience, specular He scattering from a Pt(100) crystal for alignment purposes. We stumbled upon a complex diffraction pattern which it turned out was due to a trace Si surface precipitate [6]. With the enthusiastic in-house environment supporting the understanding of semiconductor surfaces, and noting the relative experimental ease of carrying out He diffraction experiments, we could not resist launching a series of He diffraction studies on silicon. Eventually we covered Si(100)c(2 × 4) [7], Si(111)7 × 7 [8], GaAs(110) [9], and finally Ge(100)c(2 × 4) [10]. The structures, and in some cases even periodicities, of these surfaces were to various degrees unresolved at the time (except GaAs(110)), and so our emphasis was on the (static) geometry of these very corrugated surfaces.

There were touted advantages of He diffraction, i.e. high sensitivity to both structural nuances and surface order, and particularly the non-penetration of the probe which was presumably going to lead to a "simple" inversion process for analysis. However, it soon became apparent that there were significant difficulties in determining actual geometries due mostly to the same physical origin as the advantages. The effective cross-section of a He atom at an incident energy of 25–100 meV is of the order of 10 Å^2, comparable to the "area" of the target surface unit cells. Although this allows diffraction to be easily observed the interpretation is complicated as the probe is rather large. To make it smaller (higher energy) costs dearly in coherent elastic scattering. Secondly, at 25–100 meV the distance of closest approach of the probe to the outermost nuclei is ~ 3–4 Å. Thus the He atom is scattering from a periodic potential which may be quite smoothed due to electronic delocalization, which at the vertical distance of closest approach includes significant contributions of many bodies of the substrate surface. The surface geometry is defined by the substrate nuclei, and the low energy He atoms never get "close" to the nuclei. In order to obtain an accurate geometry from He diffraction patterns, the scattering potential (which is nearly the electron density) at this considerable distance from the nuclei has to be calculated accurately on the scale of the He energy (25–100 meV). Once the potential is available, the diffraction patterns need be calculated [11] using strongly coupled dynamics. In summary the requisite theory for analysis of this "simple" experiment is quite substantial and difficult. It became clear that the development of He diffraction as a quantitative structural probe would be as complex in analysis as it was experimentally simple.

This point of view was dramatically demonstrated to ourselves ("Bad Day at Black Rock" – D.R. Hamann – unpublished) in a study of the Ag(001)c(2 × 2)Cl [12] structure. At the time, there were two "competing" models (a classic – AT&T vs friends of IBM!), one with Cl in the plane of the Ag surface [13] and the second with Cl sitting over the four-fold hollows [14]. A low energy electron diffraction (LEED) study had deduced the latter. Almost from inspection of the He scattering rainbow patterns, it was immediately clear the latter (F–O–IBM) model was roughly correct. We pushed further, however, and attempted to accurately determine the height of the Cl atom above the Ag surface. It was soon clear that this was nearly futile. We were surprised to find that beyond some Cl–Ag vertical distance ($\sim 1\frac{1}{2}$ Å), there was essentially no difference in the calculated He scattering potential corrugation between Ag(100)c(2 × 2)Cl and an isolated c(2 × 2)Cl layer floated into the vacuum! The consequence of this realization was first the above quote but thereafter an effort to establish the regimes of sensitivity of He diffraction to surface structure (lemons → lemonade!) [15]. In retrospect we might have done this first. We were able to derive simple forms and graphs which showed the regimes where the He diffraction technique was sensitive to geometry parameters. In our final geometry study we applied our best theoretical effort and experimental abilities to determine the structure of the reconstructed Ge(100)c(2 × 4) surface [10]. The surface was thought to be structurally similar to the Si(100). We were able to confirm this and worked hard to determine a tilt angle of $\sim 10°$ for the dimer bond. However, the overall conclusion about the optimum role of He diffraction in surface structure remains.

In conclusion it became apparent that the advantages of He diffraction as a surface structural probe were most opportune for problems for which the theory was far more tractable. For the most part these are in distinguishing between models for which there are significantly different gross corrugations (of electron density) at van der Waals distances, and/or as a sensitive probe of surface order and periodicity. For these problems He diffraction is quick and sure. As a general or precise structural probe, the level of theory and computational effort and accuracy required is too great. Thus the technique of He diffraction has a specific and useful role to play in surface geometry, but at present it is not likely to be a general geometric structure tool.

A significant and useful area of He scattering outside of this generalization should be noted. The use of the attenuation of the (0, 0) He beam, i.e. simply specular scattering, as a measure of surface order has proved to be of great usefulness

as a non-invasive and precise monitor of absorbate coverage, surface damage, and residual order. This was pioneered by Comsa and coworkers in Julich [16].

A second useful indirectly derived benefit from He diffraction is associated with the development of calculational techniques to generate the rarefied electron densities (10^{-5} au) at the thermal energy He turning point. A linear proportionality to this density of charge is the leading term on the He–surface potential [11]. Thus the qualitative visualization of potential contours is available, i.e. a set of electron density contours are close to the shape of contours of the He–surface potential. It is a simple argument that the van der Waals potential for any closed shell molecule, such as CX_4, SF_6, etc., is essentially repulsive and similar in form and magnitude to the He–surface potential. Consequently, the same surface charge densities used in He scattering provide a conceptual and perhaps semiquantitative basis for consideration of both physisorption phenomenology and activated dissociation adsorption (which arises for closed shell species). The ability to generate a reasonable picture of the shape of the potential and the distances from the surface plane at which dynamical chemical events occur is sure to be of widespread pedagogical value [17].

3. Hyperthermal scattering from surfaces

In a free jet expansion, mixed gases exit the nozzle with a common uniform velocity determined by the average mass [18]. Thus by "seeding" a small amount of a heavy gas into a light gas expansion (or vice versa) a wide kinetic energy range of the seed gas can be obtained (10^{-2}–10^{+2} eV) with the relatively narrow energy distribution characteristic of free jet expansions. If the carrier gas causes no perturbation to the experiment, then a form of "kinetic energy spectroscopy" can be carried out for atom surface dynamics over a wide range of kinetic energy.

Using these techniques a series of experiments was carried out to examine the energy dissipation at surfaces of atoms in the hyperthermal energy regime (> 1 eV), primarily at semiconductor surfaces. Over a range of energies $1 < E_i$ (eV) < 15, several remarkable effects were revealed to us. These included direct excitation of electron–hole pairs in the substrate [19], the efficiency of which far exceeds that of a general thermal effect (bolometer). In addition ions (and neutrals) were ejected into the vacuum with significant efficiency ($\sim 10^{-3}$) at energies far below what has often been cited as a sputtering threshold [20]. The energy loss of projectiles was massive and exceeded 90% for hyperthermal Xe. Finally, despite this array of excitations and energy dissipation phenomena, scattered projectiles displayed sharply featured angular distributions, similar to those of low energy elastic rainbow scattering, which is characteristic of the underlying structure of the surface [21,22].

A key aspect to understanding much of this phenomenology is the corrugation of the target, which if sufficient, exposes the target atom center of mass to the incident projectile. A hyperthermal heavy atom impact near the target atom center of mass will transfer a large fraction of its energy in a binary-type collision. This results in a recoil that is sufficiently fast that the lattice cannot at first dissipatively respond. The target atom then rebounds from its local lattice, nearly along the line of centers, rehits the still incoming heavy projectile atom and repels it (John Tully figured much of this out but I opt to write about it). Although the energy transfer is large and the projectile leaves the surface at much slower velocity, it is scattered with an angular probability resembling that of classical elastic scattering, i.e. with rainbow-like maxima, which is a consequence of this double-hit mechanism.

Returning to the target atom, we recall that the entirety of the energy transfer is accomplished within about a vibrational period. Several electron volts of lattice excitation are initially transiently contained within a few atoms. The lattice energy of the region has been temporarily raised by $> 10^3$ K and thus electronic excitations, even in the sense of a local equilibration, would seem reasonable. A simple thermodynamic type model based on this transient local excitation (local hot spot) results in an estimate for the absolute electronic excitation efficiency. This esti-

mate was found in remarkable agreement (better than an order of magnitude) with experimental derivations of the excitation yield, based on combining measured excitation current transients and surface recombination lifetimes in the same experiment [23].

The third phenomenon of note in these experiments was the observation of substantial ion emission associated with these hyperthermal surface collisions. As the angle of incidence (θ_i measured from the surface normal) of the projectile increases (approaching grazing incidence), the yield of ejected ions rises dramatically. Near $\theta_i = 70°$ and $E_i \approx 13$ eV, the ejected positive ion yield in some cases approaches 10^{-3}. The fact that the ion yield is complementary, in angle of incidence of the projectile, to the electron–hole pair excitation probability, when combined with our understanding of the origin of the rainbow-like scattering features in angle, yields the following picture. Recall the argument that the incident projectile hits the exposed target atom near the center of mass (either because of the intrinsic surface structural corrugation or the exposure of adsorbate adatoms) and drives it hard into the bulk transferring a large fraction of the incident kinetic energy. The target rebounds quickly from a lattice which has insufficient time to dissipate the energy. If the angle of incidence of the (massive) projectile is near the normal, it is slowed but remains roughly located over the impact site during the target atoms first round trip excursion. The rebounding atom rehits the projectile and the energy transfer is complete resulting in substantial local transient heating with corresponding electronic excitation. As the incident particle angle approaches grazing, after its initial collision it may traverse and clear the unit cell in a time comparable to the target atom round trip time. Some fraction of the target atoms rebounds with more energy than the binding energy and, if unencumbered by the incident atom, escapes the surface. Because of the mixing of the many substrate degrees of freedom, a proportionate fraction of low ionization potential substrate atoms will emerge as ions. This is somewhat like a knock-off process, but with mixing in of many more degrees of freedom of the lattice. The fraction of these sputtered atoms which will emerge as ions, whether considered in terms of a thermodynamic-type argument or a dynamic picture such as with the introduction of an artificial impurity type level due to the dislocated target atom, are issues which require further experiments and theoretical guidance.

4. Aspects of surface chemical dynamics

4.1. Activated dissociative adsorption

Molecular beam techniques have been broadly employed to understand certain aspects of fundamental molecular events which underlie surface chemical rate processes. They do not directly reveal what is on the surface nor how it evolves in time. Thus most experiments which first explored the capabilities of beam techniques dealt with the "sticking" or adsorption process, particularly activated dissociation adsorption, and conversely with angle, time, and sometimes quantum-state resolved desorption. The interpretation of these experiments deals directly with the multidimensional gas–surface potentials and their traversal.

In 1971 at MIT we carried out a series of molecular beam experiments designed to explore the activated dissociation of H_2 on Cu. Dave Miller actually stimulated this experiment in the Stickney Lab and Mehdi Balooch was nearly heroic in the struggle to extract useful data. Previous desorption experiments in that laboratory and elsewhere showed peaked angular distributions about the surface normal, characteristic of the postulated energy barrier to dissociation. The apparatus available was crude but use of H_2 and D_2 isotopes allowed several aspects of the dynamics to be revealed [24,25].

Although the quantitative results of those experiments have been (appropriately) challenged in recent years, many of the concepts which were broached have remained vigorous issues in the field. Some of these include the dissociation probability scaling with the square of normal momentum, the detailed balance relationship between adsorption and desorption, fly-by times, and the role of crystal structure in determining

the accessible regions of the surface potential. A variety of activated adsorption studies have been carried out since with much improved experimental techniques. The specific H_2/Cu system has been recently revisited by several groups with exciting and far more accurate results under better defined conditions (low crystal temperatures, specified vibrational and translational distributions) [26–28]. This in turn has stimulated increasing theoretical discussion and more realistic calculations of the interaction potentials and the associated dynamics [29–31]. (The authors of the original study hereby apologize to one of the above theorists for likely errors in the estimate of the magnitude of the dissociation probabilities. This it is said caused a widespread lack of appreciation for his "accurate" calculations involving two "entire" Cu "valence only" atoms and two entire H atoms) [32].

4.2. Surface diffusion

The diffusion of atoms and molecules at a crystal surface is a central dynamical step in surface chemistry including the growth of new (semiconductor) materials. There are several techniques which have historically contributed to our knowledge of transport at surfaces [33]. The more recent exploitation of current fluctuations in field emission microscopy [34], laser-induced desorption [35,36], and Fourier transform infrared [37] techniques have greatly extended the range of times and nature of materials that can be studied. One important goal of this research is to characterize the associated experimental activation energies and to relate them to the energy barriers which determine the topography of the adsorbate surface interaction potential. Of comparable importance is the characterization of surface diffusion as a rate process, which may often be in competition with other parallel rates such as reaction or desorption.

We have used the temporal resolution of molecular beam techniques to reveal two regimes of surface diffusion, by measuring the competition between diffusion to defects and desorption. The first made use of flat metal (Pt) surfaces and monitored the onset of non-linearities in desorption kinetics of NO as the coverage became *low* (comparable to the defect density) [38]. In effect as the small density of defects sites (steps) began to depopulate, the effect of lateral interactions in the one-dimensional defects was revealed in the kinetics. This effect was used as a monitor of the ability of adparticles to reach defects prior to desorption. Based on detailed kinetic modeling [39], these effects provided experimental confirmation of a previous simulation study [40] indicating that adparticle diffusion lengths for the mean residence time prior to desorption on smooth metal surfaces were very long when compared to the characteristic length between defects or steps. The conclusion is that for low Miller index (smooth) metal surfaces, all defects are always sampled at low coverage.

For strongly corrugated substrates, this conclusion is not valid. We were able to directly demonstrate this for the case of NO_2 on GaAs(110) [41]. Observing the evolution of an initial temporal delay with temperature for the desorption of NO_2, which was long (s) compared to the measured terrace residence time (ms), we were able to show that desorption became competitive with diffusion to defects at moderate temperatures (350 K). This is of course in complete contrast to smooth surfaces. Changing the defect density via sputtering yielded longer onset times for desorption and provided direct confirmation. Although this kind of experiment does not lead to the traditional parameters associated with surface diffusion such as activation energies, as it provides an interesting point of view in terms of competitive rates, particularly with regard to special site access as is required in the growth of semiconductor materials or in catalysis.

5. Conclusions

I have made use of some of the highlights (in my own mind) of research in my laboratory to illustrate the nature of progress in our understanding of elementary atom–surface interactions based on molecular beam experimentation. This selection represents a small fraction of in-depth and provocative studies making use of these techniques in the surface science community. There is a general scientific fascination associated with understanding the motion of and forces upon individual particles, as is the goal of molecular beam experiments. Indeed essentially everyone who has viewed a well-produced movie of a molecule–surface simulation, whether accurate or not, has experienced this. And so there will very likely be continual vigorous pursuit of this area of research over the next 25 years.

If one were to ask if, as a result of more than 20 years of research in this field, there is yet a molecule–surface system in chemical dynamics for which there is a relatively complete understanding of the forces and responses, the answer is probably not. But this is generically true of most active scientific areas. A more appropriate measure of accomplishment is the continually deepening level and sophistication of questions being asked and answered in the area of molecule–surface dynamics. By this measure there has been great progress since the first issue of the journal Surface Science.

Acknowledgements

The author has benefited from and often been led by many collaborators. Some who contributed specifically to those topics mentioned here are A. Amirav, C. Bahr, M. Balooch, G. Becker, E. Chaban, C. Ching, A. Gelb, D.R. Hamann, W.R. Lambert, R.G. Laughlin, D.R. Miller, A. Sakai, J. Serri, S. Sibener, R.E. Stickney, J.L. Tersoff, P. Trevor, J. Tully, A. vom Felde, and P. Weiss.

References

[1] T. Engel and K.H. Rieder, Springer Series in Modern Physics, Vol. 91 (Springer, New York, 1982).
[2] G. Brusdeylin, R.B. Doak and J.P. Toennies, Phys. Rev. Lett. 44 (1980) 1417.
[3] J.C. Tully and M.J. Cardillo, Science 223 (1984) 445.
[4] M.P. D'Evelyn and R.J. Madix, Surf. Sci. Rep. 3 (1983) 413.
[5] J. Estermann and O. Stern, Z. Phys. 61 (1930) 95.
[6] M.J. Cardillo and G.E. Becker, Surf. Sci. 99 (1980) 269.

[7] M.J. Cardillo and G.E. Becker, Phys. Rev. B 21 (1980) 1497.

[8] M.J. Cardillo and G.E. Becker, Phys. Rev. Lett. 42 (1979) 508.

[9] M.J. Cardillo, G.E. Becker, S.J. Sibener and D.R. Miller, Surf. Sci. 107 (1981) 469.

[10] W.R. Lambert, P.L. Trevor, M.J. Cardillo, A. Sakai and D. Hamann, Phys. Rev. B 35 (1987) 8055.

[11] R.B. Laughlin, Phys. Rev. B 25 (1982) 2222.

[12] M.J. Cardillo, G.E. Becker, D.R. Hamann, J.A. Serri, L. Whitman and L.F. Mattheiss, Phys. Rev. B 28 (1983) 494.

[13] H.S. Greenside and D.R. Hamann, Phys. Rev. B 23 (1981) 4879.

[14] E. Zanazzi, E. Jona, D.W. Jepsen and P.M. Marcus, Phys. Rev. B 14 (1976) 432.

[15] J. Tersoff, M.J. Cardillo and D.R. Hamann, Phys. Rev. B 32 (1985) 5044.

[16] G. Comsa and B. Poelsema, Appl. Phys. A 38 (1985) 153.

[17] M.J. Cardillo, Langmuir 1 (1985) 1.

[18] See J.B. Anderson, in: Molecular Beams and Low Density Gas Dynamics, Gas Dynamics, Vol. 4, Ed. P. Wegener (Dekker, New York, 1974) ch. I.

[19] A. Amirav and M.J. Cardillo, Phys. Rev. Lett. 57 (1986) 2299.

[20] A. Amirav and M.J. Cardillo, Surf. Sci. 198 (1988) 192.

[21] A. Amirav, P.L. Trevor, C. Lim, J.C. Tully and M.J. Cardillo, J. Chem. Phys. 87 (1987) 1796.

[22] C. Lim, J.C. Tully, A. Amirav, P.L. Trevor and M.J. Cardillo, J. Chem. Phys. 87 (1907) 1808.

[23] P. Weiss, P.L. Trevor and M.J. Cardillo, Phys. Rev. B 38 (1988) 9928.

[24] M. Balooch, M.J. Cardillo, D.R. Miller and R.E. Stickney, Surf. Sci. 46 (1974) 358.

[25] M.J. Cardillo, M. Balooch and R.E. Stickney, Surf. Sci. 50 (1975) 263.

[26] G. Anger, A. Warbler and K.D. Rendulic, Surf. Sci. 220 (1989) 1.

[27] B.E. Hayden and C.L.A. Lamont, Chem. Phys. Lett. 160 (1989) 331.

[28] C.T. Rettner, D.J. Auerbach and H.A. Michelsen, Phys. Rev. Lett. 68 (1992) 1164.

[29] J. Harris, Surf. Sci. 221 (1989) 335.

[30] M.R. Hand and S. Holloway, J. Chem. Phys. 91 (1989) 7209.

[31] W. Brenig and H. Kasai, Surf. Sci. 213 (1989) 170.

[32] (One of references [29–31] – frequent and not-so-private communication.)

[33] R. Gomer, Rep. Prog. Phys. 53 (1990) 917.

[34] R. Gomer, Surf. Sci. 38 (1973) 373.

[35] S.M. George, A. DeSantolo and R.B. Hall, Surf. Sci. 159 (1985) L425.

[36] G.A. Reider, U. Höfer and T.F. Heinz, Phys. Rev. Lett. 66 (1991) 1994.

[37] J.E. Reutt-Robey, D.S. Doren, Y.J. Chabal and S.B. Christman, Phys. Rev. Lett. 61 (1988) 2883.

[38] J.A. Serri, G. Becker and M.J. Cardillo, J. Chem. Phys. 77 (1982) 2175.

[39] J.A. Serri, J.C. Tully and M.J. Cardillo, J. Chem. Phys. 79 (1983) 1530.

[40] J.C. Tully, Surf. Sci. 111 (1981) 461.

[41] A. vom Felde, C.C. Bahr and M.J. Cardillo, Chem. Phys. Lett. 203 (1993) 104.

Surface Science 299/300 (1994) 284–297
North-Holland

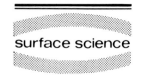

surface science

Density-functional studies of metal surfaces and metal–adsorbate systems

N.D. Lang

IBM Research Division, Thomas J. Watson Research Center, Yorktown Heights, NY 10598, USA

Received 22 March 1993; accepted for publication 14 June 1993

A historically organized review is given of theoretical work done by the author and co-workers on metal surfaces, metal–adatom systems and scanning tunneling microscopy. The calculations described are all based on the density-functional formalism for the inhomogeneous electron gas. Studies of the electron density distribution, work function and surface energy calculated using the uniform-background (jellium) model of a metal surface are described. The ground-state properties of an atom adsorbed on such a surface are analyzed, and the extension to a study of core holes in these adatoms is presented. Based on the adsorption studies, a theoretical account is given of a variety of results obtained in scanning tunneling microscopy.

1. Introduction

This article is an account of density-functional studies that I have done starting in the late 1960's, many of them with Walter Kohn or Art Williams, on metal surfaces and metal–adatom systems. The body of work I describe is characterized by a simplicity of formulation sufficient to render numerical approximation essentially unnecessary, yet rich enough to capture the physical and chemical effects of importance. Its aim is to understand the most basic ground-state properties of metal surfaces and their bonds to atomic adsorbates.

Density-functional calculations of surface electronic structure have provided much of our current understanding of the properties of crystalline surfaces. The dominance of this approach in the theoretical analysis of surface electronic properties could hardly have been anticipated at the time, nearly thirty years ago, when Hohenberg and Kohn [1], and Kohn and Sham [2] proved the fundamental theorem and introduced the key variational principle and its wave-mechanical form that underlie the method.

In performing the first wave-mechanical density-functional calculations for the properties of simple metal surfaces, Kohn and I hoped to learn whether fundamental surface properties such as work functions and surface energies obeyed regularities that could be understood using these methods. The success of this work in accounting for the most basic ground-state properties of metal surfaces provided a foundation for a vast amount of later work on the electronic structure of clean surfaces.

As interesting as the properties of such surfaces may be, one of the prime goals of surface science is to understand, and ultimately learn to control, surface chemical behavior. With this in mind, Art Williams and I embarked on density-functional calculations of the properties of an adsorbed atom on a metal surface. This work provided valuable insight into the nature of the surface chemical bond. Ultimately, it also proved to be important in the interpretation of data from one of the most exciting experimental developments in the history of surface studies, the scanning tunneling microscope.

2. Density-functional formalism (wave-mechanical treatment)

Kohn and Sham [2] have shown that the many-body problem for the ground-state density distribution of an inhomogeneous electron gas can be solved formally by obtaining the self-consistent solution to a set of one-particle equations. Starting with the total non-electrostatic energy of the system, $G[n]$, which had been shown [1] to be a universal functional of the electron distribution $n(r)$, they defined an exchange–correlation energy functional as

$$E_{xc}[n] \equiv G[n] - T_s[n], \qquad (2.1)$$

where $T_s[n]$ is the kinetic energy of a non-interacting system of electrons with the same density distribution as the interacting system. The set of one-particle equations to be solved is then (using atomic units, where $|e| = \hbar = m = 1$)

$$\left\{ -\tfrac{1}{2}\nabla^2 + v_{\text{eff}}[n; r] \right\}\psi_i(r) = \epsilon_i \psi_i(r), \qquad (2.2a)$$

$$n(r) = \sum_i |\psi_i(r)|^2 n_i \qquad (2.2b)$$

(the ψ_i in (2.2b) are orthonormal solutions of (2.2a) and n_i is 1 for the N lowest-lying solutions and zero otherwise, with N the number of electrons), where

$$v_{\text{eff}}[n; r] \equiv \phi(r) + \frac{\delta E_{xc}[n]}{\delta n(r)} \qquad (2.3)$$

is a local potential, with $\phi(r)$ the total electrostatic potential in the system including that due to the nuclei and to any external electrostatic fields.

The functional $E_{xc}[n]$ can be expanded in a series in gradients of the density. A common approximation that is made in using Eqs. (2.2)–(2.3) is to omit all of the gradient terms in this series (the "local-density" approximation); in this case, the effective potential becomes

$$v_{\text{eff}}[n; r] = \phi(r) + \mu_{xc}(n(r)), \qquad (2.4)$$

where $\mu_{xc}(n)$ is the exchange–correlation part of the chemical potential of a uniform electron gas of density n.

3. Uniform-background (jellium) model of a surface

One of the simplest models of a metal surface is that in which a semi-infinite lattice of positive ions is replaced by a semi-infinite uniform posi-

tive background which terminates abruptly along a plane (the uniform-background or "jellium" model) [3]. This model is most appropriate for simple (s–p bonded) metals. It should be noted that the sharpness of the termination of the background does not have a very important influence on the electronic wave functions, since an abrupt termination leads to a discontinuity only in the second derivative of the potential, which is of course the quantity that enters the Schrödinger equation. (Note also that the background represents a *smoothing* of the actual positive charge density in the crystal.)

It is useful to define a parameter r_s by the relation $(4/3)\pi r_s^3 \equiv \bar{n}^{-1}$, where $\bar{n}$ is the number density of the background charge (equal to the mean interior electron number density). This parameter ranges from about 2 to 6 bohr for simple metals.

3.1. Electron density, work function and surface energy

The electron density distribution that was calculated [4,5] for the uniform-background surface model using Eqs. (2.2) and (2.4) is shown in Fig. 1. The density tails off exponentially into the vacuum; in the metal, it exhibits quantum density (Friedel) oscillations, which have the characteristic wavelength of half the Fermi wavelength, and an amplitude that decreases asymptotically with

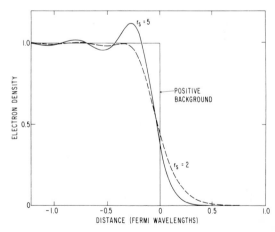

Fig. 1. Electron density in surface region of uniform-background model. (From Ref. [4].)

the square of the distance from the surface. The work function Φ can be computed easily once the electron density distribution is known; the comparison with experimentally determined values for polycrystalline simple metals is shown in Fig. 2 [6,7]. The agreement is quite reasonable, and is improved if the discrete ionic lattice of the metal is taken into account using first-order perturbation theory [6]. It should be noted that the Thomas–Fermi treatment (equivalent to using a local-density approximation for the entire non-electrostatic energy $G[n]$) applied to this surface model [7] gives a work function that is identically zero for all values of r_s, indicating that this classic method is wholly inadequate in the present context [8].

The surface energy σ found in this calculation is negative at high densities ($r_s \lesssim 2.4$), which is the proper behavior for the unmodified uniform-background model [9]. It stems simply from the fact that the electrons lower their kinetic energy when they "spill out" through the new crystal surfaces that are formed by cleavage (which is how the surface energy is defined). The reason that the kinetic-energy contribution dominates the final result is that two important electrostatic

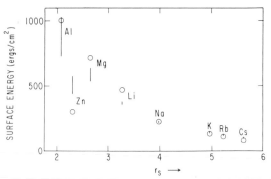

Fig. 3. Comparison between computed values of the surface energy and extrapolations to zero temperature of measured liquid-metal surface tensions. Vertical lines give computed values: the lower end-point represents the value appropriate to an fcc lattice, the upper end-point that appropriate to a bcc lattice. The surface plane in both cases is taken to be the most closely packed plane of the lattice. (For the alkali metals of lower density, the lines are contracted almost to points.) (After Ref. [5].)

contributions to the surface energy of an actual metal are treated incorrectly in the uniform-background model: the ion–ion interaction energy (a type of Madelung energy), and the electron–ion interaction energy. The former contribution can be calculated using classical electrostatics, the latter by using first-order perturbation theory (i.e. no changes in charge density involved). The results obtained [5] for σ after including discrete lattice effects in this way are shown in Fig. 3. The agreement with experiment is quite reasonable.

Since the early success of this work in accounting for the most basic ground-state properties of metal surfaces, there have been many calculations of these same properties which have taken into account the full lattice structure and potentials (or pseudopotentials) of the constituent atoms. Some of the earliest such studies on metals were those of Appelbaum and Hamann (Na(100)) [10], Alldredge and Kleinman (Li(100)) [11], and Chelikowsky et al. (Al(111)) [12]. Recently, there have been studies of whole series of metals such as those by Methfessel et al. [13], and Skriver and Rosengaard [14].

3.2. Response to electric fields

The response of a metal surface to a normally directed electric field was studied by solving Eqs.

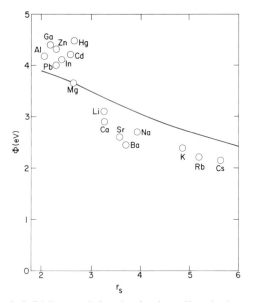

Fig. 2. Solid line: work function for the uniform-background model. Circles: measured work function for polycrystalline samples. (From Ref. [7].)

(2.2) and (2.4) with the potential due to a uniform external field included in $\phi(\mathbf{r})$ [6,15]. It will be recognized that this method is not restricted to the linear-response region.

A typical curve of the field-induced density change $\delta n(z)$, with z the coordinate normal to the surface, is shown in Fig. 4 for the weak-field region. The center of gravity of this distribution is denoted z_0. The point z_0 can be considered the effective location of the metal surface. In a parallel-plate capacitor, for example, the effective distance between the plates would be the separation of the z_0-points associated with the two plates; and it was proved as well that z_0 gives the location of the image plane for a small point charge q at a position z well outside the surface:

$$U = -\frac{q^2}{4(z - z_0)} + O\!\left(\frac{q^2}{(z - z_0)^3}\right), \qquad (3.1)$$

with U the image potential [15]. The location of z_0 relative to the actual crystal lattice is discussed in Ref. [15].

3.3. Extension to adsorption of alkali layers

The first use of the wave-mechanical density-functional formalism to study chemisorption was an analysis of a problem that had been studied experimentally since the 1920's, namely the

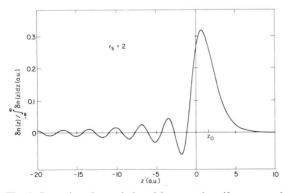

Fig. 4. Screening charge induced by a weak uniform normal electric field. The positive background occupies the $z < 0$ half-space. The location of the center of mass of the screening charge is denoted z_0. One atomic unit of length, i.e. one bohr, is equal to 0.529 Å. (After Ref. [15].)

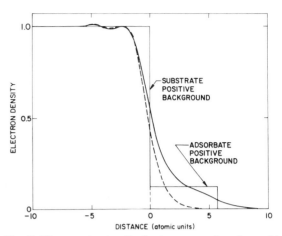

Fig. 5. Electron density distribution at metal surface with (solid line) and without (dashed line) full adsorbed Na layer, computed for the uniform-background model of the substrate–adsorbate system (substrate $r_s = 2$). (After Ref. [16].)

change in work function of a high-work-function substrate as alkali atoms are chemisorbed on its surface. It was found in these experiments that the work function drops rapidly as a function of alkali coverage; reaches a minimum, and then increases to roughly the value for the bulk alkali as the first full layer of chemisorbed atoms is completed. Now just as the uniform-background model for the substrate in this problem would consist of replacing each layer of substrate ions by a slab of positive charge (thereby building up the semi-infinite positive background), so also did this analysis replace the layer of alkali ions by such a slab [16]. This yielded the positive-background configuration shown in Fig. 5; the electron distribution, also shown, is computed self-consistently just as in the case of the bare substrate. The adsorbate (alkali) slab thickness was fixed at the spacing of the closest-packed planes in the bulk alkali [17] and the slab density was varied to simulate changes in coverage (since the alkalis are monovalent, the charge per unit area of the slab is equal to the number of adsorbed alkali atoms per unit area). Fig. 6 shows the work function as a function of coverage for adsorbed sodium and cesium. A minimum is observed, just as in the experiments; Fig. 7 compares the minimum values with those obtained experimentally. The agreement is seen to be quite reasonable. A

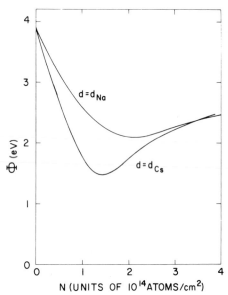

Fig. 6. Curves of work function Φ versus coverage N for Na and Cs adsorption, calculated using the uniform-background model for the substrate–adsorbate system. The adsorbate slab thickness in this model is denoted by d. (After Ref. [16].)

discussion of why in fact there is a minimum is given in Ref. [16].

4. "Atom–jellium" model for single-atom chemisorption

In the mid-1970's, Art Williams and I began an extensive study of the chemisorption of a single atom on a metal surface using the density-functional formalism. The uniform-background model was used for the surface; and the aim was to treat this simple model as carefully and completely as possible [18]. We felt that this would enable us to understand the basic physics and chemistry of adsorption, which had not hitherto been clear, either because the models had been treated too approximately, or had not been treated self-consistently, or had been extensively parameterized.

Whereas the problems discussed above were translationally invariant parallel to the surface, this problem has cylindrical symmetry about a line perpendicular to the surface through the nucleus of the atom. The solution proceeds conveniently via an analysis of a Lippmann–

Schwinger equation corresponding to each single-particle equation (2.2a), for states degenerate with the conduction band of the substrate; the details of this treatment are given in Ref. [19].

All of the core states of the adsorbed atom were computed in the full, non-spherical potential of the chemisorption system; and the calculation was done fully self-consistently.

4.1. Ground-state properties

Figs. 8 and 9 show calculated results for the chemisorption of Li, Si, and Cl, illustrating the three basic types of chemisorption bond – positive ionic, covalent, and negative ionic [19]. A high-density substrate is used, with $r_s = 2$, which is representative of a metal like aluminum ($r_s = 2.07$). Fig. 8 shows the difference in eigenstate density of Eq. (2.2a) between the bare metal and the metal with the chemisorbed atom. The atom valence states which are (or become) degenerate

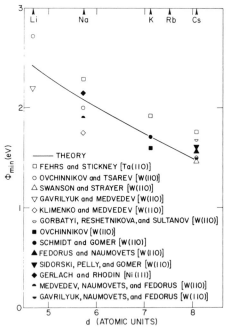

Fig. 7. Symbols: measured values of the minimum work function for alkali adsorption on close-packed metal substrates of initially high work function. Solid line: minimum value computed using the uniform-background model of alkali adsorption. The adsorbate slab thickness in this model is denoted by d. (From Ref. [16].)

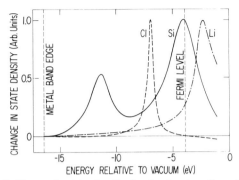

Fig. 8. Change in state density due to chemisorption of one atom on a uniform-background substrate ($r_s = 2$). Curves each correspond to metal–adatom distance which minimizes the total energy. (From Ref. [18].)

with the metal conduction band broaden into resonances. Since the center of the Li valence resonance is well above the Fermi level, this suggests that it has an occupation lower than that of the 2s state of the free atom, so long as the presence of a 2p component in the resonance does not materially alter the picture, which in fact it does not, as discussed in Ref. [20]. This would indicate that charge transfer has taken place from the Li to the metal. The states constituting the Cl 3p resonance are below the Fermi level, and are therefore occupied, which implies that charge transfer has taken place from the metal toward the Cl, as would be expected from its electronegativity. The prohibitively large energy required either to fill or empty the Si 3p level forces this resonance to straddle the Fermi level, resulting in the formation of a covalent (or metallic) rather than ionic bond.

The electron densities associated with these adsorbed atoms are exhibited for comparison in Fig. 9. In the upper row of the figure are con-

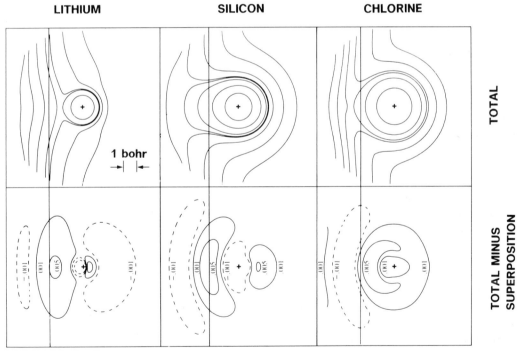

Fig. 9. Electron-density contours for chemisorption of one atom on a high-density uniform-background substrate. Metal–adatom distances shown minimize the total energy. Upper row: Contours of constant density in (any) plane normal to the metal surface containing the adatom nucleus (indicated by +). Metal is to the left-hand side; positive-background edge indicated by vertical line. Contours are not shown outside the inscribed circle of each square; contour values were selected to be visually informative. Lower row: total electron density minus the superposition of atomic and bare-metal electron densities (electrons/bohr^3). The polarization of the core region, shown for Li, was deleted for Si and Cl because of its complexity. (The bulk metal density is that corresponding to $r_s = 2$, i.e. ~ 0.03 electrons/bohr^3.) (After Ref. [18].)

tours of constant total electron density. These contours rapidly regain their bare-metal form away from the immediate region of the atom, illustrating the short range of metallic screening.

The detailed charge rearrangements associated with chemical-bond formation are displayed by the contour maps of the difference between the electron density in the chemisorption system and the superposition of bare-metal and free-atom densities shown in the lower row of Fig. 9. The solid contours indicate regions of charge accumulation; broken contours indicate regions of charge depletion.

The density-difference contours in the case of Li indicate that electrons have been displaced from the vacuum side to the metal side of the atom. The fact that the Li 2s resonance is substantially unoccupied [20] leads, as noted above, to the interpretation of this displaced charge as transferred to the metal, rather than as some sort of covalent bond charge; the bond in this low-coverage case should be regarded as basically ionic [21].

The kidney-shaped depletion contour on the vacuum side of the Li adatom, and even the reverse-dipole contours in the core region, are very similar to those found by Bader et al. [22] in difference plots for the LiH and LiF molecules. In these ionically bonded diatomics, as may be seen by consulting the reference, just as in the case of the adsorbed Li atom, the charge transferred away from the alkali does not go very far spatially; it remains in fact rather close to the Li. The sequence of contours continues into the metal in the form of Friedel oscillations induced by the perturbing atom.

The difference contours for Cl show that charge has been transferred from the metal to form a polarized negative ion. In the case of Si, there is a central region of charge depletion, and accumulations on both the bond and vacuum sides. The same general configuration of contours seen here for an adsorbed Si atom is found in difference plots for covalently bonded diatomic molecules in which p orbitals play a significant role [22].

A quantity discussed in Ref. [19] that is particularly relevant to understanding the ionicity of

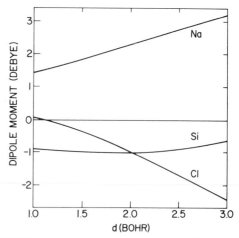

Fig. 10. Dipole moment as a function of metal–adatom distance d for Na, Si, or Cl atom chemisorbed on a uniform-background substrate ($r_s = 2$). The sign of the dipole moment is defined so that a positive moment corresponds to a decrease in substrate work function. The equilibrium distances are, in bohr, 3.1 (Na), 2.6 (Cl), and 2.3 (Si). (From Ref. [19].)

chemisorption bonds is the dipole moment as a function of distance. This is shown in Fig. 10 for Na, Si, and Cl. A "dynamic" charge on the adatom can be defined as the slope of such a curve: if the dipole moment is thought of as just the product of a fixed charge and a distance, then taking the derivative with respect to distance gives the value of this charge. For distances in the central part of the graph, the charge defined in this way is about $+0.4$ for Na, about -0.5 for Cl and ~ 0 for Si (units of the magnitude of the electron charge). The value for Na has now been verified approximately by Lindgren et al. [23] using electron-energy-loss spectroscopy (they find $+0.5$ for Na adsorbed on Cu(111)). These numbers, however, do not provide a precise measure of static charge transfer because of the importance of such contributions as the distance dependence of polarization effects. Nonetheless, considering the region around the equilibrium position for each adatom, it seems clear that a substantial slope is to be associated with ionic bonding (Na is similar to Cl in this respect, except of course for the sign), and a negligible slope is to be associated with covalent (or metallic) bonding.

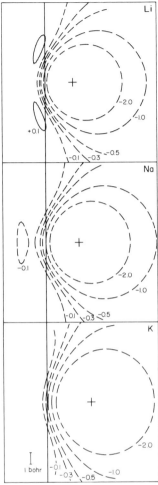

Fig. 11. Contour maps of the induced electrostatic potential for Li, Na and K at their calculated equilibrium distances outside a uniform-background ($r_s = 2$) surface. The term "potential" in the present article means the potential energy of an electron (so most of the contours here, with values in eV, are negative). The vertical line denotes the positive-background edge and the crosses the atomic positions. The vacuum is to the right. Contours are shown within a sphere centered at the atomic position. (Corrects a similar figure in Ref. [24].)

These charge displacements cause electrostatic fields in the vicinity of the adatoms, which can be important in promotion or poisoning of chemical reactions at surfaces. This is discussed in Ref. [24], and in the article in this volume by Nørskov [46]. The contours of the electrostatic potential near three different adsorbed alkali atoms are shown in Fig. 11.

4.2. Core holes

It was a natural extension of the work described above to study the problem of core holes in chemisorbed atoms. One of the interesting unknowns in this connection was the form of the charge distribution that screens the core hole. Now the density-functional formalism described above is, strictly speaking, not valid in general for excited states of the system [25]; but it has given results in good agreement with experiment when applied for example to the study of deep core holes in transition metals [26]. The justification for its use for core holes is thus only empirical; and there are particular cases, such as those in which core-hole fluctuations are important, where, in an unmodified form, it can be expected to be inappropriate.

Applying Eqs. (2.2) to this problem involves solving these equations for the case in which the n_i corresponding to a particular core level is set equal to 0 instead of 1. There will in general be a self-consistent solution for this problem just as there is for the ground state. It is of interest to consider the electron density $n(r; n_i)$ and the total energy $E(n_i)$ for the cases in which n_i is 0 and 1. These quantities are evaluated both for the metal–adatom system (superscript MA below) and the free atom (superscript A).

The experimental quantity usually discussed in this problem is the difference in core-level binding energies between the free and adsorbed atoms:

$$\Delta = I^{\mathrm{A}} - I^{\mathrm{MA}}, \tag{4.1}$$

where

$$I^{\mathrm{MA}} = E^{\mathrm{MA}}(0) - E^{\mathrm{MA}}(1), \tag{4.2a}$$

$$I^{\mathrm{A}} = E^{\mathrm{A}}(0) - E^{\mathrm{A}}(1), \tag{4.2b}$$

with the electron removed from the core taken to the vacuum level in both cases. Calculations of Δ for deep core levels were performed for several different atoms [27].

The change in binding energy Δ can be thought of as having two components: the change which takes place prior to the removal of the electron (chemical or initial-state shift) and the change

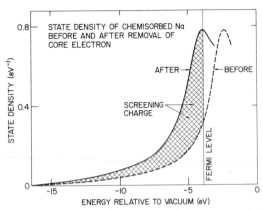

Fig. 12. State density (relative to that of the bare metal) for Na atom chemisorbed on a uniform-background substrate ($r_s = 2$), before and after removal of 2s core electron. Discrete portion of spectrum not shown. (From Ref. [27].)

associated with the screening of the core hole (relaxation or final-state shift). Fig. 12 shows the state density (relative to that of the bare metal) for Na chemisorbed on a high-density substrate, both before and after the creation of a deep core hole. It is the additional charge which occupies the resonance when it is pulled down through the Fermi level that screens the hole and thus produces the relaxation shift.

The screening of a core hole is sometimes visualized as the linear response of the bare metal to a point positive charge at the position of the adatom nucleus, i.e., the screening charge is taken to have an image-like form. The calculations show that this is usually not correct, and that instead the screening charge resembles the distribution obtained by subtracting the electron density of the free atom with a core hole from that of the excited atom (free atom with a core hole plus an extra electron in the lowest unoccupied valence orbital). Such an atomic-like result is in fact suggested by Fig. 12, and will generally be true so long as there is a partially empty valence level close to the Fermi level in the ground state. (It is true, e.g., for Si, but not for Cl, whose valence resonance is filled.)

For the cases of Na, Si and Cl ($r_s = 2$ substrate) considered in Ref. [27], the calculated relaxation shifts were found to be ~ 5 eV, about twice the magnitude of the chemical shifts obtained in that analysis.

5. Theory of the scanning tunneling microscope

5.1. Topography in the tunneling-Hamiltonian regime

The chemisorption of a single isolated atom, while theoretically elegant, was, from an experimental standpoint, largely neglected in favor of the adsorption of ordered layers of adsorbates, until the invention of the scanning tunneling microscope [28]. (The major exception to this was work in field emission and field ion microscopy.) Since it had been suggested from time to time that the study of single-atom chemisorption was a waste of time because it was experimentally irrelevant, I welcomed the opportunity provided by this new instrument to link the earlier work on chemisorption more closely to experiment.

A simple model for the scanning tunneling microscope consists of two planar metallic electrodes each of which has an adsorbed atom, with one electrode corresponding to the tip, and the other to the sample [29]. A small bias is taken to be present between the two electrodes [30].

The method developed in Ref. [31] (equivalent to a spatial decomposition of the Bardeen tunneling-Hamiltonian formalism [32]) was used to calculate the current density distribution in terms of the wave functions determined separately for each electrode in the absence of the other. These wave functions were calculated just as in the work on atomic chemisorption described above. The tunneling-Hamiltonian formalism is a perturbative treatment appropriate for the case in which tip and sample do not interact too strongly (i.e. the case in which the tip–sample separation is not too small).

I mention first results obtained when the same atom (Na) on one electrode, taken as the tip, is scanned past several *different* atoms on the other electrode: Na, S and He [29]. These chemically very different atoms produce characteristically different tunneling-current behaviors. Curves of tip displacement versus lateral separation for constant current are shown in Fig. 13.

Note first in Fig. 13 that the maximum change in tip distance Δs is much smaller for S than for Na. In part this is due to the fact that S sits closer

to the surface than Na, and in part to the fact
that the Fermi-level state density for S is appreciably smaller than that for Na. (For the case of
low bias considered here, only states near the
Fermi level participate in the tunneling process.)
Even more striking, however, is the fact that the
change in tip distance for He is slightly negative.
The He atom sits rather far from the surface, and
so would cause a large enhancement of the tunneling current if there were to be even a small
increase of the Fermi-level state density due to
broadened levels of the atom. The closed valence
shell is very far down in energy, however, and its
only effect is to polarize metal states away from
the Fermi energy, producing a decrease in
Fermi-level state density, which leads to the negative tip displacement. This was the first demonstration of this rather unusual behavior.

It is shown in Ref. [29] that the tip displacement curves for Na and S are quite close to the
contour of constant Fermi-level local density of
states due to the sample, evaluated at the tip
position, which confirmed the s-wave tip model of
Tersoff and Hamann [33] for such cases. It was

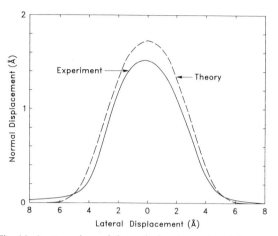

Fig. 14. A comparison of theoretical and experimental curves
of normal tip displacement (Å) versus lateral tip displacement
(Å) for Xe adsorbed on a metal surface. (From Ref. [36].)

found also in these studies that the local density
of states at the Fermi level for certain adsorbed
atoms in addition to helium, such as oxygen [34],
could be lower than that for the bare metal; that
is, the presence of these atoms can make a negative contribution to the local density of states at
this energy. This of course would also cause these
atoms to look like holes or depressions in the
surface in a topographic image; this is seen in the
studies of Kopatzki and Behm [35] for oxygen on
Ni(100).

The heavier rare gas atoms like Xe present an
interesting test of the theory, in that the low-bias
images of adsorbed Xe consist of quite distinct
protrusions [36], but yet Xe, like He, would seem
to have no states near the Fermi level. Now if a
contour of constant Fermi-level local density of
states is calculated [36] and compared with the
measured tip trajectory across the center of a Xe
atom, the two are seen to be very close (Fig. 14).
The physical origin of the Fermi-level state density which gives rise to the tip displacement is
seen in Fig. 15. This plots the additional local
density of states due to the presence of the Xe, at
a point well outside the surface. A barrier penetration factor has been divided out. It is seen that
the local state density at the Fermi level is due to
the broadened 6s resonance, the peak of which

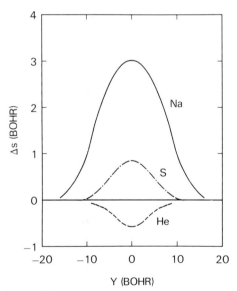

Fig. 13. Tip displacement Δs versus lateral separation Y for
constant current. Tip atom is Na; sample adatoms are Na, S,
and He. The component along the surface normal of the
distance between nuclei of tip and sample atoms is 16 bohr at
large Y. (1 bohr = 0.529 Å.) (From Ref. [29].)

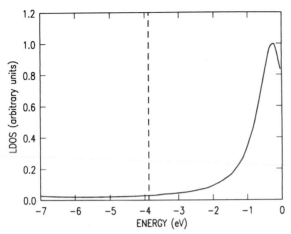

Fig. 15. Additional local density of states due to an adsorbed Xe atom, with a barrier penetration term divided out, versus energy measured from the vacuum level, calculated at a position 11 bohr directly above the center of the atom. Dashed line indicates Fermi level position. (From Ref. [36].)

lies less than 0.5 eV below the vacuum level. (The adsorbed He atom shows no such resonance.)

Despite the fact that the 6s resonance is virtually unfilled, it dominates the local density of states on the vacuum side of the adsorbed Xe atom because Xe binds at a relatively large distance from the surface and because the 6s orbital is relatively large. Thus even if the Fermi level is well out in the low-energy tail of the 6s resonance, the 6s *local* state density at this energy can be substantially greater than the bare-metal *local* state density on the vacuum side of the atom.

5.2. From tunneling to point contact

Consider what happens when the tip–sample distance is reduced into the region of transition from tunneling to point-contact. The initial contact takes place, in the experiments mentioned below, between a single tip atom and the sample surface [37–39]. This was studied by calculating the current that flows between the two flat metallic electrodes in the uniform-background model, one of which has an adsorbed atom (the tip electrode), as a function of distance between them, in the presence of a small applied bias voltage V.

Now in contrast to the studies described above, it is not possible in this case to use the tunneling-Hamiltonian formalism since the overlap of the wave functions of the two electrodes is no longer small. It is necessary therefore to treat the tip and sample together as a single system in computing the wave functions [37].

The Lippman–Schwinger equation was solved using the procedure of Ref. [19], but this time to obtain the current-carrying states at the Fermi level in the presence of the atom, starting with the current-carrying states of the bimetallic junction without the atom. The potential that goes into this equation is the difference between the self-consistently determined potentials in the bimetallic junction with and without the atom. These wave functions are then used to compute the total additional current δI due to the presence of the atom, and an additional conductance is defined as $\delta G = \delta I/V$. It is also convenient to define an associated resistance $R \equiv 1/\delta G$.

In Fig. 16, the resistance R is shown as a function of tip–sample separation s for the simple case of a Na tip atom. The atom is kept fixed at a distance d, measured from the nucleus to the positive-background edge of the tip electrode, of 3 bohr; s is measured from the nucleus to the positive-background edge of the sample electrode. At large separations, R changes exponentially with s. As s is decreased toward d, the resistance levels out at a value of 32 000 Ω. (For $s = d$, the atom is midway between the two metal

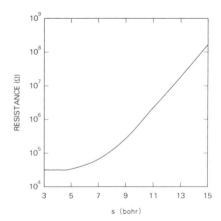

Fig. 16. Calculated resistance $R \equiv 1/\delta G$ as a function of tip–sample separation s for a Na tip atom. (From Ref. [37].)

surfaces, so this can in some sense be taken to define contact between the tip atom and the sample surface).

This leveling out, including the order of magnitude of the resistance, can be understood from the discussions of Imry and of Landauer [40]. These authors point out that there will be a "constriction" resistance $\pi\hbar/e^2 = 12\,900\ \Omega$ associated with an ideal conduction channel, sufficiently narrow to be regarded as one-dimensional, which connects two large reservoirs. The atom, in the instance in which it is midway between the two electrodes, contacting both, forms a rough approximation to this.

Just such a plateau in the resistance was found in the experiments of Gimzewski and Möller [38] using a Ag sample surface and an Ir tip (though the identity of the tip atom itself was not determined). These authors fix the voltage on the tip at some small value, and measure the current as a function of distance as the tip is moved toward the sample. They find a plateau in the current at about the same distance from the surface that it is found in the calculation described here, with the resistance at the plateau $\sim 35\ \mathrm{k}\Omega$. Analogous results were found by Kuk and Silverman [39] with a minimum resistance of $\sim 24\ \mathrm{k}\Omega$.

It should also be noted that the plateau in the resistance in the near-contact region implies that the apparent barrier height measured in scanning tunneling microscopy, defined as

$$\varphi_A \equiv \frac{\hbar^2}{8m}\left(\frac{\mathrm{d}lnI}{\mathrm{d}s}\right)^2, \tag{5.1}$$

where I is the tunneling current and s is the tip–sample separation, will be zero in this region. The apparent barrier height only approaches the sample work function (Φ) for appreciably larger tip–sample separations. The dependence of φ_A on separation has been measured by Kuk and Silverman [39], and their results are in agreement with those that can be derived from Fig. 16 using Eq. (5.1) [41].

5.3. Atom transfer

The procedure and model discussed above has been used to analyze the transfer of an atom

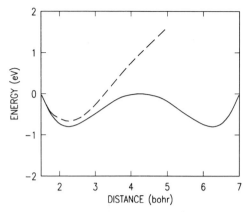

Fig. 17. Solid curve: energy for Si atom between two uniform-background electrodes ($r_s = 2$) with zero bias, as a function of distance of atom from left electrode. Electrode separation held fixed at 8.5 bohr. Zero of energy is taken to be value at atom distance of 1.5 bohr. Dashed curve: same as for solid curve except right electrode is absent. Energy value far to right is 3 eV above the minimum in the curve (i.e. the heat of adsorption is 3 eV). (From Ref. [43].)

between two electrodes under the influence of an applied bias (which in relevant experiments is usually several volts) [42]. The first application was to the case of a Si atom [43], whose transfer between tip and sample in the scanning tunneling microscope had been studied by Lyo and Avouris [44]. The calculated energy for a Si atom, as a function of atom position, between two electrodes whose spacing corresponds to that used in the experiment is shown in Fig. 17 for the zero-bias case. The curve for the case in which the right electrode is absent (i.e. the chemisorption case) is given for comparison. This figure shows that just the proximity of tip and sample reduces the activation barrier for an atom to leave the surface from 3 to 0.8 eV; in addition, this activation barrier is calculated to be a strong function of electrode separation (i.e. tip–sample spacing) [43].

When a bias in the range 3–6 V is applied, a small (of order 0.1–0.2 eV) asymmetry is introduced in the energy curve, which favors activated transfer of the atom off of the positively biased electrode (i.e. in the direction opposite to that of the tunneling electrons). The calculated results show the same direction of atom transfer relative

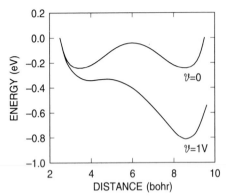

Fig. 18. Energy for Na atom between two uniform-background electrodes ($r_s = 2$), as a function of distance of atom from left electrode, for bias values V of 0 and 1 V. Electrode separation held fixed at 12 bohr. Zero of energy is taken to be value at atom distance of 2.5 bohr. (From Ref. [45].)

to bias polarity as the experiment, a similarly sharp dependence of activation barrier (or transfer rate) on distance, and a comparable magnitude for the activation barrier.

The case of a Na atom shows a much more dramatic effect of bias on the activation barrier than seen for Si [45]. Such results are shown in Fig. 18, where it is seen that a bias of only 1 V decreases the activation barrier nearly to zero. The contrast with Si can be understood both from the fact that Na is less strongly bound to the metal surface than Si, and on the basis of the rather different charge values obtained from the dipole moment curves of Fig. 10. As noted in the discussion of that figure, the charge on the adsorbed Si was very small, whereas that on Na was about $+0.4\,|e|$. Thus the applied field has little pre-existing (i.e. not field-induced) charge on which to act to cause a transfer in the Si case, while in the Na case this charge is substantial and its interaction with the applied field is a driving force behind the transfer.

6. Conclusion

A historically organized review has been given of studies by the author and coworkers of the basic ground-state properties of metal surfaces and metal–adatom systems, with a number of

applications to scanning tunneling microscopy. These studies, because of their notable success in explaining such properties, have led to a vast amount of further theoretical work, much of which takes into account the full lattice structure of the surface, including relaxation of the atoms away from their ideal positions and surface reconstruction. Many of the basic properties are independent of such additional complications, however, and the uniform-background model discussed here provides a simple, quantitative, and physically transparent picture of them. The fact that density-functional methods enable us to investigate relatively complicated inhomogeneous systems with accuracy suggests that these methods will remain at the heart of theoretical surface studies well into the future.

Acknowledgment

I am very grateful to Peter Feibelman for his comments on the manuscript.

References

[1] P. Hohenberg and W. Kohn, Phys. Rev. 136 (1964) B 864.
[2] W. Kohn and L.J. Sham, Phys. Rev. 140 (1965) A 1133.
[3] Other work done around this time using a uniform-background model, which showed the utility of the density-functional formalism in surface and interface studies, were A.J. Bennett and C.B. Duke, Phys. Rev. 162 (1967) 578 (interfaces);
J.R. Smith, PhD Thesis, Ohio State University, Columbus, 1968 (unpublished);
Phys. Rev. 181 (1969) 522 (extended Thomas–Fermi study of surfaces).
[4] N.D. Lang, Solid State Commun. 7 (1969) 1047.
[5] N.D. Lang and W. Kohn, Phys. Rev. B 1 (1970) 4555.
[6] N.D. Lang and W. Kohn, Phys. Rev. B 3 (1971) 1215.
[7] N.D. Lang, in: Solid State Physics, Vol. 28, Eds. F. Seitz, D. Turnbull and H. Ehrenreich (Academic Press, New York, 1973) p. 225.
[8] The extended Thomas–Fermi treatment of Smith [3], however, gives reasonable work function values.
[9] X.-P. Li, R.J. Needs, R.M. Martin and D.M. Ceperley, Phys. Rev. B 45 (1992) 6124.
[10] J.A. Appelbaum and D.R. Hamann, Phys. Rev. B 6 (1972) 2166.
[11] G.P. Alldredge and L. Kleinman, Phys. Rev. B 10 (1974) 559.

[12] J.R. Chelikowsky, M. Schlüter, S.G. Louie and M.L. Cohen, Solid State Commun. 17 (1975) 1103.

[13] M. Methfessel, D. Hennig and M. Scheffler, Phys. Rev. B 46 (1992) 4816.

[14] H.L. Skriver and N.M. Rosengaard, Phys. Rev. B 46 (1992) 7157.

[15] N.D. Lang and W. Kohn, Phys. Rev. B 7 (1973) 3541.

[16] N.D. Lang, Phys. Rev. B 4 (1971) 4234.

[17] The extensive experimental studies of A. Hohlfeld and K. Horn, Surf. Sci. 211/212 (1989) 844, find in fact that this is a reasonable choice for near full layer coverages, but that better agreement with experiment can be obtained if, for low and moderate coverages, the slab thickness is taken to be closer to the ionic diameter of the alkali.

[18] Preliminary accounts of this work were given in N.D. Lang and A.R. Williams, Phys. Rev. Lett. 34 (1975) 531; 37 (1976) 212.
A somewhat similar study for hydrogen chemisorption was done by O. Gunnarsson, H. Hjelmberg and B.I. Lundqvist, Phys. Rev. Lett. 37 (1976) 292; Surf. Sci. 63 (1977) 348;
H. Hjelmberg, O. Gunnarsson and B.I. Lundqvist, Surf. Sci. 68 (1977) 158.

[19] N.D. Lang and A.R. Williams, Phys. Rev. B 18 (1978) 616.

[20] N.D. Lang, in: Physics and Chemistry of Alkali Metal Adsorption, Eds. H.P. Bonzel, A.M. Bradshaw and G. Ertl (Elsevier, Amsterdam, 1989) p. 11.

[21] For additional discussion of the degree of ionization of alkali atoms adsorbed on metal surfaces see M. Scheffler, Ch. Droste, A. Fleszar, F. Máca, G. Wachutka and G. Barzel, Physica B 172 (1991)143;
G. Pacchioni and P.S. Bagus, Surf. Sci. 269/270 (1992) 669.

[22] R.F.W. Bader, W.H. Henneker and P.E. Cade, J. Chem. Phys. 46 (1967) 3341.

[23] S.-Å. Lindgren, C. Svensson and L. Walldén, Phys. Rev. B 42 (1990) 1467.

[24] J.K. Nørskov, S. Holloway and N.D. Lang, Surf. Sci. 137 (1984) 65;
N.D. Lang, S. Holloway and J.K. Nørskov, Surf. Sci. 150 (1985) 24.

[25] It is, however, valid for the energetically lowest state of each symmetry: O. Gunnarsson and B.I. Lundqvist, Phys. Rev. B 13 (1976) 4274.

[26] N.D. Lang and A.R. Williams, Phys. Rev. Lett. 40 (1978) 954.

[27] N.D. Lang and A.R. Williams, Phys. Rev. B 16 (1977) 2408.

[28] G. Binnig, H. Rohrer, Ch. Gerber and E. Weibel, Phys. Rev. Lett. 49 (1982) 57.

[29] N.D. Lang, Phys. Rev. Lett. 56 (1986) 1164.

[30] Studies for larger biases, but still using the tunneling-Hamiltonian approximation, are given by N.D. Lang, Phys. Rev. Lett. 58 (1987) 45; Phys. Rev. B 34 (1986) 5947.

[31] N.D. Lang, Phys. Rev. Lett. 55 (1985) 230; 55 (1985) 2925(E); IBM J. Res. Dev. 30 (1986) 374.

[32] J. Bardeen, Phys. Rev. Lett. 6 (1961) 57.

[33] J. Tersoff and D.R. Hamann, Phys. Rev. B 31 (1985) 805; Phys. Rev. Lett. 50 (1983) 1998.

[34] N.D. Lang, Comments Condensed Mater. Phys. 14 (1989) 253.

[35] E. Kopatzki and R.J. Behm, Surf. Sci. 245 (1991) 255.

[36] D.M. Eigler, P.S. Weiss, E.K. Schweizer and N.D. Lang, Phys. Rev. Lett. 66 (1991) 1189.

[37] N.D. Lang, Phys. Rev. B 36 (1987) 8173.

[38] J.K. Gimzewski and R. Möller, Phys. Rev. B 36 (1987) 1284; and data of Gimzewski and Möller reproduced in Ref. [37].

[39] Y. Kuk and P.J. Silverman, J. Vac. Sci. Technol. A 8 (1990) 289.

[40] Y. Imry, in: Directions in Condensed Matter Physics: Memorial Volume in Honor of Shang-keng Ma, Eds. G. Grinstein and G. Mazenko (World Scientific, Singapore, 1986) p. 101;
R. Landauer, Z. Phys. B 68 (1987) 217.

[41] N.D. Lang, Phys. Rev. B 37 (1988) 10395.

[42] Another use that has been made of this procedure and model is the analysis of the current from a field-emission tip terminated by a single atom: N.D. Lang, A. Yacoby and Y. Imry, Phys. Rev. Lett. 63 (1989) 1499.
These calculations yield a relatively focused, high-current beam with a narrow energy distribution, in agreement with experimental studies by H.-W. Fink, Phys. Scr. 38 (1988) 260.

[43] N.D. Lang, Phys. Rev. B 45 (1992) 13599.

[44] I.-W. Lyo and Ph. Avouris, Science 253 (1991) 173.

[45] N.D. Lang, in: Nanosources and Manipulation of Atoms under High Fields and Temperatures: Applications, Eds. Vu Thien Binh, N. Garcia and K. Dransfeld (Kluwer, Dordrecht, 1993) p. 177.

[46] J.K. Nørskov, Surf. Sci. 299/300 (1994) 690.

Surface Science 299/300 (1994) 298–310
North-Holland

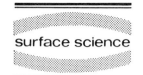

surface science

Tight-binding methods

Walter A. Harrison

Applied Physics Department, Stanford University, Stanford, CA 94305, USA

Received 22 December 1992; accepted for publication 24 February 1993

The history of tight-binding theory is traced, from the corresponding band models of Bloch, to the development of universal parameters, derivable from a combination of tight-binding theory and free-electron theory. In the end, tight-binding theory becomes an independent, and in that sense ab initio, method for studying virtually all of the properties of surfaces. It is this conceptual aspect, rather than semiempirical tight-binding theory as an approximate computational method, which is emphasized here. We outline the application first to surface energy, and its dependence upon structure, for semiconductors, ionic insulators, and simple metals, finding wide differences from the rudimentary view of one bond energy per broken bond. We turn to photothresholds, work functions and electron affinities, finding direct tight-binding predictions but necessary corrections for image potentials, which also are derivable in terms of tight-binding theory. For the particular case of electron affinities there are additional corrections to the band gap, usually associated with correlation energy, but readily and generally estimated as the intra-atomic Coulomb U divided by the dielectric constant. We turn to history of the understanding of heterojunction band line-ups, which in the end can be obtained by matching "neutrality levels" in each band structure, those levels being the average sp^3-hybrid energy in tight-binding theory and in the case of metals, the Fermi energy. Alternate views are also discussed as is the bonding of individual atoms to the surface. Finally, the successes and failures of tight-binding theory in understanding semiconductor surface reconstructions are reviewed.

1. Introduction

During the years when our understanding of solid surfaces was growing, the techniques for carrying out first-principles calculations of solid-state properties were also growing rapidly, as described in the preceding section by Lang. This went hand in hand with the developments of the large-scale computers needed for these calculations. Although these methods were nominally "from first principles" one should not forget that they required continual empirical checks to see if the very serious approximations – local-density theory, or the one-electron approximation itself – were adequate to give meaningful predictions. As it turned out they were, and it became possible to answer many questions concerning surfaces through complete careful calculations. A notable example was the determination of the two-by-one silicon (111) surface reconstruction by Pandey [1] by comparison of the total energies of a wide

range of possibilities. This was one of the cases of a genuinely new prediction rather than a theoretical confirmation of a known fact.

At the same time that these first-principles approaches were developing, simplified theoretical understandings were arising, notably those based upon pseudopotentials for metals (thinking of the electrons as free with small corrections for constituent atoms) and tight-binding theory for semiconductors and insulators (thinking of the electrons as bound to atoms with small corrections for the overlap of neighboring atoms). The basic ideas were old. Bloch [2] had formulated a rudimentary tight-binding representation of electronic states in solids in the early days of quantum mechanics. The nearly-free-electron-view of metals dates from still earlier [3] and Fermi had utilized the basic pseudopotential idea in particle scattering. These simplified views had long been considered naïve, but with careful comparison of the results of theories based upon them with

SSDI 0039-6028(93)E0371-Z

more accurate calculations and with experiment, they came to be recognized as meaningful theories of the properties in their own right.

There were in fact two rather separate developments of tight-binding theory. One was as an approximate semiempirical scheme for total-energy computations – a short-cut to the time-consuming self-consistent local-density theories. Pandey [1], for example, used these extensively before his final complete calculation for the suggested structure. There have been a number of such successful studies. Such methods for semiconductors have been thoroughly reviewed by LaFemina [4], along with the more accurate methods. The second approach was development of a universal set of parameters and approximations which allowed "back-of-the-envelope" calculations aimed more at understanding and semiquantitative prediction than at high accuracy. It is the story of this second approach, and its application to surfaces, which we wish to tell here.

2. The history of the tight-binding approach

2.1. The early days

A necessary precursor to any tight-binding theory of electronic structure was of course the discovery of the electron by Thomson in 1897 and the subsequent ideas of the structure of the atom by Bohr in 1913 and two-electron molecular bonds proposed by Lewis in 1916. Once a quantitative formulation of the quantum theory was given by Schroedinger in 1926, Bloch's study of solids [2], including a tight-binding theory, followed almost immediately. In the subsequent years, chemists continued to use the tight-binding view, or the description of molecular states as linear combinations of atomic orbitals, notably, Debye and Hückel [5], Coulson [6], and Hoffmann [7]. Physicists tended to take the free-electron view as illustrated by Wilson's text [8], particularly in the late fifties when it so successfully explained the observed Fermi surfaces of simple metals [9].

Although physicists tended not to take the tight-binding approach seriously, they did use it in describing the results of full calculation; i.e., to note the width of the "d-bands" in transition metals. To make any quantitative statements it was necessary to adjust the parameters of the tight-binding theory to fit, for example, the bands from a full calculation so that it could be regarded as little more than a "fitting scheme", and only useful for interpretation of band structures.

2.2. Ionicity theories

Somewhat independently of this, Phillips [10] in the late 1960's began discussing semiconductor properties in terms of "ionicities". Ionicity theory itself seems to be largely due to Pauling [11]. The idea was to associate an intuitive scale with compounds, such as the increasing "ionicity" in a series of isoelectronic, isostructural compounds such as Ge, GaAs, ZnSe, and CuBr. These come from elements all in the same row of the periodic table, but with increasingly different atomic numbers. Certainly a property such as an elastic constant would be expected to vary smoothly from one of these systems to the other and it is possible, without any real understanding whatsoever, to interpolate this property if it were known for some of these compounds and not the others. Ionicity is a scale invented to organize such interpolations. Similarly a "metallicity" scale can be formed to describe variations vertically in the periodic table, such as in the series C, Si, Ge, Sn, and Pb. There have long been efforts to represent compounds on both scales by a point in a plot with one scale on the abscissa and one on the ordinate. Not surprisingly, compounds of the same structure tend to be clustered in the same region of these structure plots. These are methods of organizing our knowledge, and using it to make predictions, but the conceptual trappings which went along with it were in some sense superfluous. The predictions were arising from experimental interpolation rather than basic understanding.

Pauling had taken his ionicity scale from cohesive energy of the compounds, but Phillips recognized that a more suitable scale might be based upon the dielectric constant, which was intimately connected to the basic electronic structure from which all properties derive. The loose connection was made through Penn's model of the dielectric constant [12]. This model began with the free-electron picture of a semiconductor (as if it were a metal), but with the electrons diffracted by a pseudopotential when their wavenumbers lay near Bragg reflection planes. The important planes then were assumed to be the (220) planes, which form a dodecahedron (called the Jones Zone) of exactly the volume in wavenumber space needed to accommodate the eight valence electrons per atom pair. The pseudopotential was imagined to give rise to a single gap between the valence and conduction bands at this Jones Zone. It was assumed that that gap had a homopolar contribution E_h and an ionic contribution C, as $\sqrt{E_h^2 + C^2}$. Using Penn's formula one could fit the E_h for the systems with no ionic contribution (C, Si, Ge, and Sn) to see that it varied approximately as $1/d^{2.48}$. One could then use the corresponding formula to define E_h for any compound and C for that compound could be adjusted to fit the experimental dielectric constant. Upon more careful consideration [13] it turned out that this form was quite inappropriate for Penn's concept based upon the gap at the Jones zone, but it did not matter since it was in the end only used to define an ionicity as $C^2/(E_h^2 + C^2)$. It was based upon a property representative of the electronic structure, and as long as it was systematic it would provide a good scale for interpolating properties.

2.3. Tight-binding for semiconductors

By 1971 it was also noted, in a number of studies [14], that a simple tight-binding expansion of the electronic states in semiconductors, based only upon an s-state and three p-states on each atom, and coupling only between nearest neighbors gave a very reasonable fit to the true band structure. We noted then [15] that if one constructed the traditional sp³-hybrids in terms of these orbitals and included only the coupling between two hybrids directed into each bond, one had an approximate description of the electronic structure in terms of which one could calculate directly many properties. This "bond–orbital model" was based upon the well-defined mathe-

matical approximations of representing the states in terms of this limited atomic basis, and what was later known as the "bond–orbital approximation" [16] of neglecting some of the smaller matrix elements in the Hamiltonian matrix. To proceed one needed to determine the parameters to be used in the tight-binding description. In fact the two principal parameters were the coupling between two hybrids directed into the same bond (analogous to E_h) and half the difference in energy of the two hybrids (analogous to C). Then the bonding–antibonding splitting became twice the square root of the sum of the square of the two parameters (analogous to $\sqrt{E_h^2 + C^2}$). It was natural to derive these two parameters from the experimental dielectric constants following directly the Phillips approach.

There were, however, substantive differences between the two approaches which in the end made the tight-binding approach a powerful method in the study of surfaces, among other properties. The most essential was that as a (crude) approximation to the real electronic structure, it was possible to estimate virtually all properties of the semiconductors directly in terms of the parameters, [15,16] rather than simply interpolate them between similar compounds. This meant, for example, that one could compare the energies of different surface geometries, where the Jones-zone picture of the perfect semiconducting solid had no bearing. This was done already in 1976 [17]. The fact that the two approaches were based on the quite opposite free-electron and tight-binding points of view seemed less essential.

2.4. Universal parameters

In expanding to treat more complete sets of properties, such as the energy bands [18], it became necessary to have values for a more complete set of parameters, the s- and p-state atomic energies, and the matrix elements coupling s-states and p-states on neighboring atoms, not just the single matrix element between hybrids in an individual bond and their energy difference. It was natural to look to fits to the known energy bands of semiconductors to obtain them. This also was done in 1976 [19]. Somewhat surprisingly it was found that the free-atom term values were quite consistent with the fits to the bands and that the interatomic matrix elements were quite insensitive to how "ionic" the semiconductor was (nearly the same for Ge, GaAs, ZnSe, and CuBr) and, thus, only depended upon which row of the periodic table elements were in. It was quite natural then to again follow Phillips [10] in plotting the fit values against bond length (internuclear distance) to find that they all varied as $1/d^2$, similar to his $1/d^{2.48}$. The result was somewhat prettier, though, in that each matrix element – say the coupling between neighboring s-states – could be written as

$$V_{ll'm} = \eta_{ll'm} \frac{\hbar^2}{md^2},$$ (1)

with $\eta_{ll'm}$ a universal dimensionless constant given, for example, for coupling between s-states by $\eta_{ss\sigma} = -1.40$. These "universal parameters" made it possible to address directly and simply, though quite approximately, virtually any property of a semiconductor.

It is remarkable in hindsight that it was not for another three years that it became obvious why Eq. (1) is appropriate [20]. Both the Phillips pseudopotential picture and the tight-binding picture had proven appropriate for understanding properties, and in fact it was known that the semiconductor bands at the same time were nearly-free-electron-like and describable in tight-binding theory. For this to be true, the splitting between the $k = 0$ s-like bands in silicon, for example, given by tight-binding theory $(-8V_{ss\sigma})$ must equal the free-electron value $(9\pi^2\hbar^2/8md^2)$. Thus, Eq. (1) applies and $\eta_{ss\sigma}$ should equal $-9\pi^2/64 = -1.39$, and this should be true for all semiconductors. Similar identifications could be made for the other three coefficients, $\eta_{sp\sigma}$, $\eta_{pp\sigma}$, and $\eta_{pp\pi}$ [20]. The tight-binding approximation for sp-bonded materials had become an independent approximate formulation of the electronic structure and properties, without fits to experimental data nor more complete solid-state calculations.

Interestingly enough, the occurrence which first suggested that there must be a theoretical basis

for Eq. (1) was the completion of a theory of transition-metal systems based upon transition-metal pseudopotentials. This led to matrix elements for the coupling between d-states on neighboring atoms given by [21]

$$V_{ddm} = \eta_{ddm} \frac{\hbar^2 r_d^3}{md^5}.$$ (2)

Here $\eta_{dd\sigma} = -45/\pi$, $\eta_{dd\pi} = 30/\pi$, and $\eta_{dd\delta} = -15/2\pi$, and r_d is a d-state radius characteristic of each transition-metal element. If such a formula could be derived for such a complicated system then surely there must be a derivation of Eq. (1), and once the question arose it was obvious what it must be.

Similar formulae were obtained for the coupling between d-states and s- and p-states, proportional to $r_d^{3/2}/d^{7/2}$, and for coupling involving f-states, depending upon an f-state radius for each rare earth or actinide [23]. The stand-alone tight-binding theory was available for all classes of materials.

3. Surface energies

The simplest view of surface energies is that they are the excess energy due to broken surface bonds. One takes the bond energy equal to the cohesive energy (heat of atomization) per nearest-neighbor bond and simply counts the number of broken bonds. Such a picture correctly indicates that the minimum-energy surface for silicon is the (111) surface, which is then the cleavage surface and the growth surface.

3.1. Tight-binding theory for elemental semiconductors

As appealingly simple as this view is, a slightly deeper look in terms of tight-binding theory indicated that it will get very little else right. We look first at an elemental semiconductor such as silicon which we might expect to be the best case. It is necessary to consider the cohesive energy and to use the tight-binding concepts and parameters given above; we follow Ref. [24].

One starts with the free silicon atoms and calculates the change in energy as the solid is formed as the sum of the changes in the one-electron energies. (The Coulomb energies which are double counted and the correlation energy are little changed since the charge distribution in the solid is so close to a superposition of atomic charge densities.) The first step is the promotion of electrons in each atom to sp^3-hybrids, costing an energy $(\epsilon_p - \epsilon_s)/2$ per bond, equal to 3.61 eV for silicon. We then bring the atoms together and form bonds between each pair of hybrids, coupled by $-3.22\hbar^2/md^2$ [22,23] for universal parameters, and equal to -4.44 eV for silicon. The gain in energy per bond is twice this, but approximately half of the total is canceled by the repulsion which prevents collapse of the crystal. Thus, the contribution of these two terms in the cohesion, 3.61 and -4.44 eV, very nearly cancel. However, if we split the solid, forming two surfaces, we will still have expended the promotion energy for every broken bond at the surface since each dangling orbital must also be an sp^3-hybrid to remain orthogonal to the other three hybrids. Thus the cost in energy is 4.44 eV per bond for silicon, not the cohesive energy of $4.44 - 3.61$ eV per bond. Use of the cohesive energy per bond greatly underestimates the surface energy. In fact, this cancellation is so great that the next-order corrections are required to obtain a meaningful cohesive energy [24] and it would be necessary to include such corrections at the surface to obtain a good value for surface energy, and this has not been done.

The presence of these surface energies much larger than the energy per bond has an important consequence. The formation of the ideal surface is so costly in energy that semiconductors find alternative arrangements to recover some of this lost energy; surface reconstruction occurs on essentially all surfaces of elemental semiconductors.

3.2. Ionic crystals and polar semiconductors

The simplest view of one cohesive-energy-per-bond lost for every broken bond fares as badly in other systems. In the tight-binding formulation of cohesion in ionic crystals [22], such as rock salt, the neutral atoms are brought together and at the equilibrium spacing an electron is transferred from the sodium to the chlorine for a gain in energy of $\epsilon_s(\text{Na}) - \epsilon_p(\text{Cl})$ equal to 8.83 eV for NaCl, in comparison to the experimental cohesive energy of 8.04 eV per atom pair. (This is also the tight-binding band gap, experimentally 8.5 eV for NaCl.) With the tight-binding view, there is no change in the energy due to the formation of the surface and this prediction of a negligible surface energy in comparison to the cohesion per bond is essentially true, and just the opposite of the case for elemental semiconductors. Because the expected surface energy is so small in ionic crystals, one does not expect reconstruction of the ideal surface, and it seems not to occur.

Born's approach to ionic-crystal cohesion, based upon the cation ionization energies minus the anion electron affinity and the Madelung energy gives the same qualitative result. The Coulomb surface energy of $0.0422 e^2/d$ per surface atom for a (100) surface, obtained by Madelung himself [25], is tiny compared to the full Madelung energy of $1.75 e^2/d$ per atom. The surface must be neutral (the (100) surface is, but a (111) surface would not be), or the Madelung energy would diverge [26]. Thus the (100) surface has the lowest Madelung energy and is the growth and cleavage surface.

This same exclusion of charged surfaces applies also to polar semiconductors such as gallium arsenide [26]. Thus the (111) surface which we found for silicon is not allowed for gallium arsenide, the lowest energy surface of which is a (110) surface, neutral but with a slightly higher broken-bond density than the (111) surface.

3.3. Metal surfaces

Tight-binding theory also provides a formulation of surface energy for metals [27]. We consider a tight-binding s-band with a single atom per primitive cell. From an infinite crystal we select a large number N of atomic planes parallel to the surface we wish to construct. We consider wavefunctions which vary in the z-direction, perpendicular to the plane, as $\sin(k_z sn)$, with s the spacing between planes and n an integer. Such states, as well as states of the form $\exp(ik_z sn)$, are exact eigenstates for the infinite crystal, and in both cases we take propagating waves for motion parallel to the surface planes. k_z may be chosen such that the sine is zero for $n = 0$ and for $n = N + 1$. Then removal of the atoms for $n < 1$ and $n > N$ from the infinite crystal leaves these as exact eigenstates of the now finite slab.

Having formed the states in the slab, we may return to a free-electron dispersion, $\epsilon(k) = \hbar^2 k^2 / 2m$. Then, for each k_z allowed by the condition $k_z s(N + 1)$ equal to an integral multiple of π, we have a subband occupied over an energy range $\hbar^2 (k_F^2 - k_z^2)/2m$. We may sum the energies and evaluate the result for large N. The leading term, proportional to N, is the cohesive energy obtained earlier by Wills and Harrison [28] in rather good accord with experiment. The term independent of N may be divided by two to obtain a surface energy of

$$E_{\text{surf}} = \frac{k_F^2 E_F}{16\pi}\left(1 - \frac{16 k_F s}{15\pi}\right). \qquad (3)$$

The first term dominates and the surface energy is positive. The second term arises from some decompression of the electron gas when part of the crystal is removed; we could have missed it in a pure free-electron picture. This term indicates that the surface energy is lowered when the spacing between planes is increased, corresponding to more closely-packed planes. A (111) surface is favored for an fcc lattice. Eq. (3) appears to underestimate the real surface energy of the metal and the bond-breaking picture is closer for the case of metals, the case we might least expect to succeed.

Perhaps the most important message is that the simplest view – in this case broken bonds – can be very far off the mark, but the almost-as-simple tight-binding view, based upon free-electron fits to the parameters, is generally qualitatively correct, and frequently semiquantitative It does of course not compete for accuracy with the complete local-density calculation.

4. Photothresholds, work functions, and electron affinities

The tight-binding electron states are based upon free-atom term values, which approximate the removal energy of an electron from the atom. We might expect that the eigenvalues, as modified for the solid, might give good estimates of the removal energies for the solid. For variation from one material to another this is qualitatively the case, but there is an important correction which is easy to understand [29], though it is technically a many-body effect.

4.1. Correction to the tight-binding photothreshold

We may see that the removal energy for an electron from an atom is considerably reduced if that atom is near a metal or dielectric medium. This is done in steps for an atom near, but not touching, a metal. We first carry the atom far from the metal, at no cost in energy, remove an electron, requiring an energy approximately equal to the magnitude of the free-atom term value, and then return the charged atom to the vicinity of the metal. It is this last step which gives the correction. The atom feels an attractive image force, given by $e^2/(2z)^2$ if the electron is a distance z from the metal. Integrating the effect as we bring the atom to the surface, a distance d away, we find an energy gain $e^2/4d$. Of course, removing the electron directly from the atom near the metal, without removing the atom first, must give the same result so this is the reduction in the removal energy. It seems clear that a similar reduction is appropriate to removal of an atom imbedded in the surface or a removal of an electron from the surface. Had the same calculation been performed for an atom near a dielectric, it would have been the same except that the image force is reduced by a factor $(\epsilon - 1)/(\epsilon + 1)$, which is very close to one for a semiconductor, so the effect is essentially the same. The value of this correction is about 1.5 eV for silicon, with a d of 2.35 Å.

Tight-binding theory gives directly the valence-band maximum ϵ_v in terms of the term values ϵ_p^c for cation (Ga) and ϵ_p^a for the anion (As) as

$$\epsilon_v = \frac{\epsilon_p^c + \epsilon_p^a}{2}$$

$$- \sqrt{\left(\frac{\epsilon_p^c - \epsilon_p^a}{2}\right)^2 + \left(\frac{4V_{pp\sigma}}{3} + \frac{8V_{pp\pi}}{3}\right)^2}. \quad (4)$$

A plot of these values from semiconductors, along with experimental photothresholds (Ref. [22], p. 254), indicates a relative shift of 3.8 eV, within half an eV. This is somewhat larger than our estimate of 1.5 eV. Indeed there is also an increase in all eigenvalues due to the interatomic repulsion discussed in Section 3.1. For the bonding state in silicon this was 2.22 eV for each of the two electrons in the bond; it could account for the difference. In addition, there can be corrections from surface relaxations and reconstructions which can shift the photothreshold, and which make it different at different crystalline interfaces. All of these findings are equally applicable to metals.

The work function is defined to be the energy required to remove an electron *at the Fermi energy* from the crystal, so it is equal to the photothreshold for a metal. If the Fermi energy in a semiconductor is referenced to the valence-band maximum, the work function can be obtained from the photothreshold by direct subtraction. There is an additional complication for the electron affinity in semiconductors.

4.2. Corrections to electron affinities

There turns out to be an additional problem concerning electron affinities which was long in being discovered. In 1983 it was noted by Sham and Schlüter [30] and by Perdew and Levy [31] that a proper density-function-theory band calculation would give correct electron densities, total ground-state energy, and removal energy for electrons at the Fermi energy, but that the eigenvalues corresponding to the conduction band in a semiconductor would not correspond to the electron affinity; that is, the band gap would be

incorrectly predicted. It was obvious, once pointed out, but the entire solid-state community had been fooled by a long line of ostensibly proper band calculations which gave gaps equal to the observed values.

The point is most easily seen in the free atom of, for example, silicon. A density-functional – or a Hartree–Fock – calculation will give an eigenvalue for the valence 3p-state, which will approximately equal the removal energy. However, the electron affinity is higher by an intra-atomic U of about 7.6 eV [32] because an electron added to a neutral atom sees one more electron than one added to a positive ion (the latter energy gain being equal to the removal energy for a neutral atom). We could correspondingly say that for the electron affinity there was an extra potential, equal to e^2/r_0, with r_0 the atomic radius, to be added to the eigenvalue. In the solid this potential is $e^2/(\epsilon r_0)$ with ϵ the dielectric constant, so a shift U/ϵ should be added to the conduction-band eigenvalues to obtain the real energy for putting an electron there [32]. Note that the screening comes from the surrounding bonds, not from electrons within the atom in question. The corresponding shift $7.6/12 = 0.63$ eV is approximately equal to the correction which is required to obtain the observed band gap. The universal parameters described above were obtained [23] from fits to band calculations made before this correction was understood and which, therefore, consciously or not, had the correction built in to obtain the observed band gap.

This simple correction, U/ϵ, illustrates the contrast between tight-binding theory and full local-density theory. It makes the physical origin of the effect clear and makes it clear how it should vary from material to material (largely through the dependence upon ϵ). In contrast, it can be calculated more accurately in the density-functional formalism where it appears as a discontinuity in the energy-dependent exchange and correlation potential [33]. This is a correct description of the gap enhancement, as it is of the difference in electron affinity and ionization energy in the atom, but it obscures the origin of the effect and makes it difficult to know how it will vary from system to system.

5. Heterojunction band line-ups and Schottky barriers

Ideal heterojunctions have a lattice structure which continues through the interface where the material changes, as for germanium grown on gallium arsenide. The band gap is different on the two sides, corresponding in each case to the bulk band gap, and the fundamental remaining number to characterize it is the valence-band discontinuity. Given that, the entire band structure of the system is defined. Given that discontinuity, the electrons will flow in such a way as to produce band bending, equating the Fermi energies at very much larger distances as is familiar, but this has no effect on the discontinuity which occurs just at the heterojunction.

5.1. Natural band line-ups and neutrality levels

The earliest estimate of this discontinuity seems to have been given by Anderson [34] in 1960. He suggested matching the electron affinities determined separately for the two materials. By subtracting the experimental gaps, this becomes equivalent to matching the measured photothresholds. Almost by definition this gives the correct result for the line-up if the two materials are separated by vacuum and have no net charge so that there is no field between them. However, there may be corrections to the photothreshold at the surface from surface dipoles which disappear when the two are joined. It seemed preferable to proceed in the same spirit but to use the tight-binding bands discussed in Section 4, which put all bands on the same scale, to make the match [33]. These avoided the extra complications of the vacuum surface and of surface reconstructions to be discussed in the next section, and they were called "natural band line-ups".

In that tight-binding analysis [35] we were very careful to note that there can be corrections due to interface dipoles arising from the interface bond between materials of different electronegativity. These were estimated for the silicon–germanium interface and found to be quite small. We also discussed the effect of "metal-induced gap states" which had initially been discussed by

Heine [36] and could also influence heterojunction band line-ups. They arise, for example, if the valence band on the left is higher than that on the right. Then some valence-band states from the left (with energies in the gap on the right) have exponentially decaying tails on the right and their corresponding negative charge tends to raise all the energies on the right, decreasing the initial off-set. Tersoff [37] had used this idea to define neutrality levels in the band gap of a semiconductor, the energies at which band tails at large distances are equally valence-band-like and conduction-band-like. It could be demonstrated that the effect of the corresponding gap states was to shift the bands in the direction of bringing the two neutrality levels into alignment. This provided a new "natural band line-up" on a physically different basis. We, on the other hand, were able to show [35] that these neutrality levels were dominated by the light-mass bands and had negligible effect on the band line-ups.

The disagreement was resolved in private conversations where it became clear that the distant tails of the gap states were indeed not important, but Tersoff's neutrality levels as calculated were dominated by short distances and in fact by what we had distinguished as interface dipoles. These were indeed small in germanium–silicon but could be very important in some heterojunctions. All of this became clear in the context of a tight-binding model and we could see that in tight-binding terms Tersoff's neutrality level was the energy of the dangling sp^3-hybrid [38].

Thus, though the initial natural band line-ups were often correct, when they disagreed with the neutral-point view, the latter prevailed. In tight-binding theory the neutrality level was the average hybrid energy which should be matched on the two sides. This new neutrality level did not even need to lie in the gap, and certainly did not in the case of InAs–GaSb, where the conduction band in the former was found even in the first treatment [35] to lie below the valence-band maximum in the latter – a "type-II line-up".

5.2. Other views

At any real interface there may be defects and it is certainly true that they may control the band line-up, as suggested by Spicer and coworkers [39]. The effects of such defects can readily be treated in tight-binding theory [35] if it is known what the defects are, but it remains primarily the job of the experimentalist to determine what defects are present in interfaces he has produced.

Perhaps because of such defects there has been some spread in experimental line-ups produced by different workers. There have been a wide range of semiempirical and theoretical models for determining heterojunction line-ups. From the tight-binding point of view, it is the average hybrid energy which aligns and any theory consistent with that is a correct theory. One fact of particular interest seemed not to be consistent: it was found that transition-metal levels could serve as reference levels for band line-ups [40–43] as if these levels provided some absolute reference for the energy bands. Upon careful examination with a self-consistent tight-binding approach [44] it was found that in contrast, these transition-metal levels were fixed relative to the same dangling-hybrid energies by a charge-neutrality condition arising from the large Coulomb U of the transition-metal atom. Thus, these levels *should* provide a suitable reference, but only because they were consistent with the same neutrality-level view.

5.3. Adsorbed atoms and Schottky barriers

The problem seems somewhat different when the heterojunction is replaced by metal atoms, or a bulk metal, on the semiconductor surface, but the same tight-binding approach can be used. For band line-ups with the bulk metal we may simply let the gap in the second semiconductor go to zero, with its neutrality level remaining near the gap and becoming the Fermi level. Thus, the Fermi level in the metal plays precisely the role of the dangling-hybrid energy. It had in fact been long known that the Schottky-barrier heights for different semiconductors correlated with the band line-ups between those semiconductors, so this was no major revelation.

Of perhaps more interest is the question of individual atom adsorption on a semiconductor surface. A careful tight-binding view, rather than

general bonding considerations, can be essential. If we ask whether an oxygen atom should adsorb on a gallium or on an arsenic site at a gallium arsenide surface, a general consideration might have us compare the gallium–oxygen bond energy with the arsenic–oxygen bond energy. However, if we start with the tight-binding view of this surface as a doubly occupied arsenic dangling hybrid and an empty gallium dangling hybrid, we are led to expect the oxygen to go to the arsenic. For this case it is the only place to acquire electrons, irrespective of the question of reactions of oxygen with elemental gallium and arsenic.

These can be made quantitative with universal parameters, and care with the Coulomb shifts. Not so much has been done here [45], but in one study Klepeis and Harrison [46] found that for Cs and Au on gallium arsenide (110) the arsenic site is favored at very low coverages, but at coverages of the order of one-tenth monolayer the gallium site is favored. The latter was consistent with tunneling-microscope studies by Feenstra [47], but the former has not been tested. Such a prediction requires at least a systematic tight-binding study, if not a full local-density calculation.

6. Reconstruction on semiconductor surfaces

There has been intense activity in studying reconstruction over the past thirty years. One major impact was the observation by Binnig et al. [48] of the seven-by-seven reconstruction with the scanning tunneling microscope, which demonstrated convincingly for the first time that this microscope had truly atomic resolution. The fact that this showed the adatoms which had been predicted for this reconstruction using tight-binding theory [49] was not of major consequence. This same systematic study of general reconstructions by tight-binding theory turned out not to have been so successful on all other systems. However, the use of tight-binding theory has been constructive and some interesting lessons have been learned. We recount that history briefly, noting what new came from each case. A recent authoritative account of semiconductor reconstruction is given by LaFemina in Ref. [4]. We noted already in Section 3.1 that the large surface energies for semiconductors help explain how ubiquitous these reconstructions are.

6.1. The silicon (111)2 × 1 and general (110) surfaces

The freshly cleaved (111) silicon surface has been known for thirty years to show a 2 × 1 reconstruction [50]. Tight-binding theory suggests an immediate explanation [49]. On the (111) surface sp^3-hybrids from each surface atom form bonds with the three neighboring atoms below the surface, leaving a single dangling sp^3-hybrid and one electron to occupy it. The same is true of the (110) surface. This is a prime candidate for a "negative-U center": if we were to allow alternate atoms to move slightly in and slightly out (from their relaxed equilibrium position), the energy will increase in proportion to the square of the displacement. This follows from the fact that they are displacements from the minimum energy and is consistent with an elasticity view. However, the hybrid energies shift alternately up and down *in proportion* to the displacement. This results as the hybrid energy varies continuously from the s-state energy for back-bond angles equal to 90° to the p-state energy for back-bond angles equal to 120° [49]. The reason that the total energy still varies in proportion to the square of the displacement is that a single electron occupies each and the linear shifts cancel. However, if we shift an electron from each hybrid rising in energy to each hybrid lowering in energy, the net energy decreases linearly with displacement and we are guaranteed a gain in energy, as in the Jahn–Teller effect. As long as the maximum gain in energy as a function of displacement exceeds the cost in Coulomb energy, a screened U, from doubling up the electrons in one hybrid, this is the expected surface, reconstructed in a two-by-one pattern. The electron behavior is as if the repulsive energy U were a negative energy, attracting the electrons to occupy the same hybrid.

Such a distortion had early been suggested by Haneman [51] for somewhat analogous reasons. A similar argument could be made in terms of a Peierls-like distortion opening a gap in the half-filled surface band. There seemed little doubt that this was the explanation, although calculations in Cohen's group [52] indicated that the repulsive U exceeded any gain which could be obtained. He proposed an alternative, antiferromagnetic rearrangement which did not involve this U.

At this point, Pandey suggested a totally different explanation, the π-bonded chain model [53]. This suggestion came purely from a theoretical effort, interestingly enough mostly carried out using the simple tight-binding model which Pandey had used earlier [54]. This enabled him to explore a large range of geometries, and when the most promising one came along, he carried out careful and complete local-density calculations to confirm this startlingly different suggestion. There is no doubt now that he was correct, and the π-bonding energy gained from this structure overwhelms the distortion energy it requires, just as the π-bonding energy in graphite more than makes up for the broken σ-bonds relative to diamond. For tight-binding theory there was the lesson that the Coulomb U's and Madelung shifts do not necessarily cancel at the surface, as assumed in Section 3.2.

The corresponding in–out relaxation can occur if the outmoving atom has extra positive charge, as on the (110) surface of a polar semiconductor. This is not considered a reconstruction since the translational symmetry is not modified, but the corresponding distortion does occur at a gallium arsenide (110) surface with the arsenic moving out as expected. It presumably does not occur on the (110) surfaces of silicon for the same reason it did not on the (111).

6.2. The silicon (100)2 × 1 and polar interfaces

On the (100) surfaces of silicon each surface atom has two neighbors below the surface so there are a pair of electrons to occupy the surface bands. This is quite a different situation. Nonetheless a 2 × 1 reconstruction is observed and was attributed by Lander and Morrison [55] to alternate vacancies. This seemed likely to us also, and we studied a range of vacancy-based reconstructions using tight-binding theory [49].

The central point was that making a vacancy on a (100) surface did not break extra bonds: the surface atom with two broken bonds was removed, leaving two broken bonds behind. The modified geometry allowed for other paths to lower energy.

There had been an earlier proposal by Schlier and Farnsworth [56] which we regarded as a misuse of tight-binding theory, though it turned out to be correct. They thought of the two broken bonds for the surface atom as two dangling hybrids, each with a single electron. Then two neighboring surface atoms might lower their energy by moving together, like hydrogen atoms, forming two-electron bonds and leaving a single dangling bond on each atom. In fact the situation is more analogous to two helium atoms. The remaining two orbitals on the silicon atom at a (100) surface, after the formation of the two bonds below the surface, are one even in reflection about the surface normal, an sp-hybrid, and one odd in reflection, a p-state. The sp-orbital is several volts lower in energy than the p-state and is doubly occupied. Bringing the two together will initially split the lower state, but with both occupied the shifts cancel and one expects a net upward shift (arising from the non-orthogonality which produced the repulsion discussed in Section 3). The upper state is also split, but unoccupied, so it does not affect the energy. This is as for a pair of helium atoms. Only if the distortion proceeds to the point that the bonding combination of upper states drops below the antibonding combination of lower states, can the energy drop. This is the circumstance discussed by Woodward and Hoffmann in *"Conservation of Orbital Symmetry"* [57] in which absorption of light is required to take an electron into an antibonding state before the bonding can occur, and in cases such as helium it does not go in any case. It hardly seemed a likely explanation but it could not be ruled out and it is now established theoretically and experimentally (see Ref. [4] for details). It was learned that with competing plausible alternatives, one should not be confident in predicting the outcome.

The (100) surface of a compound semiconductor was another problem since it tended to lead to an unacceptable charged surface as indicated

in Section 3.2. For gallium arsenide the last plane would be entirely arsenic or entirely gallium. If it was arsenic we might expect two electrons in each dangling sp-hybrid, since that hybrid energy is well below the tight-binding valence-band maximum. Then the internal fields would correspond to half an electronic charge per surface atom. To remove half an electron per surface atom from the valence band, which would be required to neutralize the surface, would be energetically expensive. Since it breaks no extra bonds to make a vacancy on this surface, we expected [49] that the equilibrium structure would have vacancies at half the surface sites, leading to a neutral surface with doubly occupied sp-hybrids on the arsenic atoms and empty sp-hybrids on the gallium. This seems to have been the case. The situation on the (111) surfaces is much less clear.

6.3. The silicon (111)7 × 7

The real challenge was the exotic 7 × 7 reconstruction observed on (111) surfaces after annealing at high temperature, observed first by Schlier and Farnsworth [56]. Lander and Morrison [58] early proposed a pattern of thirteen vacancies in the 7 × 7 surface unit cell, but this required the breaking of three bonds for each vacancy and seemed impossible. We took a clue from the findings of Lander and Morrison [58] that added aluminum and phosphorous atoms on silicon (111) surfaces appeared to form on bridging sites, over three dangling hybrids, to replace the three broken bonds by three bent bonds and a single broken one, forming a $\sqrt{3} \times \sqrt{3}$ pattern. We suggested [49] that silicon adatoms might similarly be energetically favorable, with a concentration determined by a compromise between the energy gained by each and a repulsion between them arising from the distortion of the substrate. The pattern could not be predicted, but the same pattern as that of the thirteen vacancies proposed by Lander and Morrison seemed plausible.

This aspect of the 7 × 7 pattern was strikingly confirmed by the scanning tunneling microscope images by Binnig et al. [48], but with one of the adatoms missing. Subsequent studies by Takayanagi et al. [59] established that there was in addition a stacking fault buried beneath one half of the 7×7 cell. This stacking fault does not occur on the germanium $(111)2 \times 8$ structure which similarly is a pattern of adatoms [60]. M.B. Webb had in fact on the basis of LEED studies suggested this stacking fault for the silicon 7×7 structure earlier to the author. It seemed difficult to believe that if such a stacking fault were there, the crystal could successfully grow as an unflawed crystal, but that appears to be the case.

Acknowledgement

This author's work and the preparation of this manuscript were supported by the ONR under Grant No. N00014-92-J-1231.

References

[1] K. Pandey, Phys. Rev. Lett. 47 (1981) 1913.

[2] F. Bloch, Z. Phys. 62 (1928) 555.

[3] P. Drude, Ann. Phys. 1 (1900) 566.

[4] J.P. LaFemina, Surf. Sci. Rep. 16 (1992) 137.

[5] E. Hückel, Z. Phys. 70 (1931) 204.

[6] For a review, see: C.A. Coulson, in: Physical Chemistry, an Advanced Treatise, Eds. H. Eyring, D. Henderson and W. Jost (Academic Press, New York, 1970).

[7] R. Hoffmann, J. Chem. Phys. 39 (1963) 1397.

[8] A.H. Wilson, Theory of Metals (Cambridge University Press, London, 1936).

[9] See, in particular, the Fermi Surface, Eds. W.A. Harrison and M.B. Webb (Wiley, New York, 1960).

[10] J.C. Phillips, Bands and Bonds in Semiconductors (Academic Press, New York, 1973).

[11] L. Pauling, The Nature of the Chemical Bond (Cornell University Press, Ithaca, NY, 1960).

[12] D.R. Penn, Phys. Rev. 128 (1962) 2093.

[13] W.A. Harrison, review of: J.C. Phillips, Bands and Bonds in Semiconductors, in: Phys. Today (Jan. 1974) p. 67.

[14] D. Weiere and M. Thorpe, Phys. Rev. B 4 (1971) 2508, for example, gave such a description of energy bands in the diamond structure.

[15] W.A. Harrison, Phys. Rev. B 8 (1973) 4487.

[16] W.A. Harrison and S. Ciraci, Phys. Rev. B 10 (1974) 1516.

[17] W.A. Harrison, Surf. Sci. 55 (1976) 2.

[18] S.T. Pantelides and W.A. Harrison, Phys. Rev. B 11 (1975) 3006.

[19] W.A. Harrison, Bull. Am. Phys. Soc. 21 (1976) 1315.

[20] S. Froyen and W.A. Harrison, Phys. Rev. B 20 (1979) 2420.

[21] W.A. Harrison and S. Froyen, Phys. Rev. B 21 (1980) 3214.

[22] For a review, see: W.A. Harrison, Electronic Structure and the Properties of Solids (Freeman, San Francisco, 1980) (reprint: Dover, New York, 1989);
For a briefer review, including new parameters from Ref. [23], and improved treatment of semiconductor bonds from Ref. [24], see: W.A. Harrison, Pure Appl. Chem. 61 (1989) 2161.

[23] W.A. Harrison, Phys. Rev. B 24 (1981) 5835.

[24] W.A. Harrison, Phys. Rev. B 27 (1983) 3592.

[25] E. Madelung, Z. Phys. 19 (1918) 524.

[26] W.A. Harrison, J. Vac. Sci. Technol. 16 (1979) 1492.

[27] W.A. Harrison, to be published.

[28] J.M. Wills and W.A. Harrison, Phys. Rev. B 29 (1984) 5486.

[29] Ref. [22], p. 252, ff.

[30] L.J. Sham and M. Schlüter, Phys. Rev. Lett. 51 (1983) 1888.

[31] J.P. Perdew and M. Levy, Phys. Rev. Lett. 51 (1983) 1884.

[32] W.A. Harrison, Phys. Rev. B 31 (1985) 2121.

[33] See, for example, C.S. Wang and W.E. Pickett, Proceedings of the International Conference on the Physics of Semiconductors, San Francisco, 1984, Eds. D.J. Chadi and W.A. Harrison (Springer, New York, 1985) p. 993, or Refs. [30,31].

[34] R.L. Anderson, Proceedings of the International Conference on Semiconductors, Prague, 1960 (Czechoslovakian Academy of Sciences, Prague, 1960) p. 563.

[35] W.A. Harrison, J. Vac. Sci. Technol. B 3 (1985) 1231.

[36] V. Heine, Phys. Rev. A 138 (1965) 1689.

[37] J. Tersoff, Phys. Rev. Lett. 52 (1984) 465.

[38] W.A. Harrison and J. Tersoff, J. Vac. Sci. Technol. B 4 (1986) 1068.

[39] W.E. Spicer, P.W. Chye, P.R. Skeath, C.Y. Su and I. Lindau, J. Vac. Sci. Technol. 16 (1979) 1422.

[40] A. Zunger, Annu. Rev. Mater. Sci. 15 (1985) 411.

[41] M.J. Caldas, A. Fazzio and A. Zunger, J. Appl. Phys. 45 (1984) 671.

[42] J.M. Langer and H. Heinrich, Phys. Rev. Lett. 55 (1985) 1414.

[43] J. Tersoff, Phys. Rev. Lett. 56 (1986) 675.

[44] J. Tersoff and W.A. Harrison, Phys. Rev. Lett. 58 (1987) 2367.

[45] For example, J.E. Klepeis and W.A. Harrison, J. Vac. Sci. Technol. B 6 (1988) 1315.

[46] J.E. Klepeis and W.A. Harrison, Phys. Rev. B 40 (1989) 5810.

[47] R.M. Feenstra, private communication.

[48] G. Binnig, H. Rohrer, Ch. Gerber and E. Weibel, Phys. Rev. Lett. 50 (1983) 120.

[49] W.A. Harrison, Surf. Sci. 55 (1976) 1.

[50] J.J. Lander, G.W. Gobeli and J. Morrison, J. Appl. Phys. 34 (1963) 2298.

[51] D. Haneman, Phys. Rev. 170 (1968) 705.

[52] M.L. Cohen, private communication.
[53] K.C. Pandey, Phys. Rev. Lett. 47 (1981) 1913.
[54] K.C. Pandey and J.C. Phillips, Phys. Rev. Lett. 32 (1974) 1433.
[55] J.J. Lander and J. Morrison, J. Appl. Phys. 33 (1962) 2089.
[56] R.E. Schlier and H.E. Farnsworth, J. Chem. Phys. 30 (1959) 917.
[57] R.B. Woodward and R. Hoffmann, Conservation of Orbital Symmetry (Verlag Chemie, Weinheim, 1971).
[58] J.J. Lander and J. Morrison, J. Appl. Phys. 34 (1963) 1403.
[59] K. Takayanagi, Y. Tanishiro, M. Takahashi and S. Takahashi, J. Vac. Sci. Technol. A 3 (1985) 1502.
[60] N. Takeuchi, A. Selloni and E. Tosatti, Phys. Rev. Lett. 69 (1992) 648.

Surface Science 299/300 (1994) 311–318
North-Holland

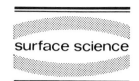

surface science

Energy-minimization approach to the atomic geometry of semiconductor surfaces

D.J. Chadi

NEC Research Institute, 4 Independence Way, Princeton, NJ 08540-6620, USA

Received 2 April 1993; accepted for publication 20 April 1993

A brief review of the main results obtained from applications of a quantum mechanical total-energy-minimization approach to surface structural determination of tetrahedral semiconductors is given. The tight-binding based method has proved quite reliable in providing information on the atomic and electronic structure of a variety of surfaces including the (110), (111), and (100) surfaces of zincblende and elemental semiconductors. For the Si(100) surface, a breaking of the mirror symmetry at the surface was predicted and this led to a natural explanation for the c-4 × 2 unit cells observed at this surface. The first quantitative estimates for the anisotropic monolayer and bilayer step formation energies on Si(100) surfaces were derived.

Questions about the nature of the surface atomic structure of semiconductors began with the first electron diffraction studies which showed that, in general, the surfaces of many metals and semiconductors were "reconstructed", i.e., they had surface unit cells larger than those expected from corresponding bulk planes. The first observations of a reconstructed semiconductor surface were made by Schlier and Farnsworth in 1959 [1]. Their low-energy-electron-diffraction (LEED) studies showed that the Si(100) surface had a 2 × 1 reconstruction, i.e., a unit cell twice as large as might have been expected. A reconstruction with a larger c-4 × 2 periodicity occurring at lower temperatures was later discovered by Lander and Morrison in 1962 [2]. In both instances the reports of these observations were accompanied by structural models. The 2 × 1 surface was suggested to result from dimerization of surface atoms and the c-4 × 2 from vacancy formation.

The development of quantitative LEED analysis for surface structural determination in the 1970's was an important step in surface structural studies. Initial results from various LEED studies on the Si(100) surface, gave differing views on the surface structure. Dimer [1,3,4] vacancy [2,5] and

"conjugated-chain" models [6] were proposed to explain the data. Additional information related to surface electronic structure that was provided by surface sensitive photoemission spectroscopy proved invaluable in distinguishing among the various models. Comparison of the energies and dispersions of surface electronic states with theoretical predictions for the various models provided a means of testing the adequacy of the models. This approach proved quite effective in deciphering the structures of many surfaces, in particular of the Si(100)-2 × 1 and Si(111)-2 × 1 surfaces.

My own work in the area of surface atomic structure started in 1977 after I joined the Surface Science group at the Xerox Palo Alto Research Center. At the time the atomic structure of the Si(100)-2 × 1 and Si(111)-2 × 1 surfaces were thought to be well understood and attention was shifting to the cleavage planes of compound semiconductors such as GaAs which were assumed to be more complex. The (110) cleavage planes of zincblende semiconductors were observed by LEED to be unreconstructed, i.e., they had the same periodicity as a (110) bulk plane. However, since the unit cell contained two atoms, one Ga and one As atom, the relative positions of the atoms in the unit cell could be very different from their "ideal" bulk terminated values.

The first quantitative estimates of atomic relaxations at this surface came from LEED studies [7,8]. In particular, three different sets of structural data became available at approximately the same time. The data all showed that the surface As atoms moved out of the surface and the Ga atoms moved towards bulk atoms but there were substantial differences in the actual atomic coordinates suggested by these data. To determine the optimal structure, I carried out electronic structure calculations [9] for all three models to find which model provided the best fit to experimental data from angle-resolved photoemission spectroscopy [10]. The most important result to emerge from this exercise was the finding that the energies of surface electronic states were very sensitive to atomic relaxation. For the "ideal" surface with no atomic relaxations, both occupied and unoccupied surface states occurred in the band gap. The energies of these states changed by nearly 1 eV when the surface atoms were allowed to relax. The inclusion of relaxation greatly improved the agreement between the calculated and experimental surface-derived energy bands. Atomic relaxation had the effect of moving occupied states below the valence-band-maximum and unoccupied states above the conduction-band-minimum. The prediction that the relaxed surface did not give rise to electrically active states was initially controversial and there were arguments as to whether this could be possible. Core level excitation spectroscopy from a Ga 3d core level into a Ga-derived surface state seemed to show that an empty Ga-derived dangling-bond surface state was in fact within the band gap. It was later demonstrated that this was caused by the excitonic binding of the electron to the core hole and the band gap was, in fact, free from any surface states. The history of events of the interplay between the early structural results and measurements of excitation spectra for the GaAs(110) surface have been recounted by Duke [11].

The sensitivity of the surface electronic states to atomic relaxation and the differences between the different sets of LEED data led me to think about a total-energy-minimization approach for surface structural determination. In an earlier work on bulk semiconductors Richard Martin and I had shown that total-energy changes resulting from atomic displacements could be determined fairly accurately from changes in electronic energy levels [12]. Initially displacements resulting in only bond-angle but no bond length changes were considered and we showed that reasonably accurate values for elastic coefficients and phonon frequencies could be obtained from a simple tight-binding scheme. I generalized this approach to the general case of both bond length and bond angle changing atomic distortions and applied it to semiconductor surfaces [13].

The first application of the tight-binding based energy-minimization approach was to the (110) surfaces of GaAs and ZnSe [13]. The resulting structure for the GaAs(110) surface was in good agreement with the primary features derived from LEED [7,8]. It showed that the As atoms moved out of the surface and Ga atoms towards the bulk. The magnitude of the atomic displacements and energy changes were found to be large at the surface (about 0.5 Å) but decayed rapidly into the bulk and it was necessary to consider the relaxation of only a few, typically 2–3, atomic layers at the surface. The displacements normal to the surface were in good agreement with the existing LEED data which were most sensitive to these types of motions. The overall structure determined for GaAs was subsequently found to be indistinguishable from the best structure obtained from a dynamical LEED fit [14]. Extension of the calculations to a variety of group IV, III–V and II–VI semiconductors showed that the (110) surfaces in all cases could be specified by a bond rotation angle of $28° \pm 2°$ [15]. This prediction has been generally well borne out by subsequent studies [16]. The effect of the relaxation is to change the bond angles around the surface Ga (As) atoms from 109.47° to 120° (94°). The energy lowering resulting from a rehybridization of the atomic orbitals on the cations (anions) from sp^3 to sp^2 (s^2p^3) is the driving force for the surface atomic relaxation and occurs on surfaces other than the (110) surface. A recent LEED study has shown that, surprisingly, these types of relaxations occur even for a I–VII semiconductor such as CuCl [17]. An important success of the tight-binding method was the prediction that the magnitude of the experimentally determined [18] surface relaxations scaled with lattice constant [19] and not with ionicity, for example.

The (110) surface structural results from the tight-binding calculations have been found to be in good agreement with those from subsequent first-principles calculations [20]. The (110) atomic structure of GaAs is now the best understood surface of any semiconductor. The most extensive studies of the cleavage planes of compound semi-conductors with zincblende and wurtzite crystal structures have been carried out by Duke [16] and coworkers, initially using LEED and more recently a combination of LEED and tight-binding energy-minimization approaches. The invention of scanning-tunneling-microscopy (STM) and its application to GaAs, both in a conventional as well as in a spectroscopic mode [21], has not resulted, as in the case of many other surfaces, in any revision of our understanding of its (110) surface atomic and electronic structure.

The total-energy-minimization was next applied to the Si(100) surface. As mentioned earlier, Schlier and Farnsworth [1] had suggested a dimer model for the 2 × 1 reconstruction of this surface. Theoretical studies by Appelbaum et al. [22] had shown that among several models, this structure gave the best agreement between the surface electronic density of states and experimental [23] photoemission data. A competing "conjugated-chain" model suggested by LEED [6] was found by Kerker et al. [24] to have an electronic structure inconsistent with experiment and was not further considered. An apparently serious flaw with the dimer model was revealed, however, by surface sensitive angle-resolved photoemission experiments [25]. The experiment showed that the surface electronic structure was nonmetallic whereas theory predicted a semimetallic electronic structure. This discrepancy led to strong doubts about the dimer model and left the question of the Si(100) surface structure wide open. I started work on the Si(100) surface by first testing whether the reliability of

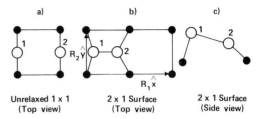

Fig. 1. The structure of the ideal (a), and dimerized (b) and (c) Si(100) surface is shown. The motions of the surface Si involved in dimer formation are not equivalent. Comparison to III–V surfaces shows that as far as the atomic relaxations are concerned atom 2 becomes "Ga-like" while atom 1 becomes "As-like".

the tight-binding model for this surface. The results of calculations for the surface electronic structure of the symmetric dimer model [26] turned out to be in very good agreement with those obtained from the previous self-consistent calculations [22]. My total-energy calculations suggested, however, that the symmetric dimer was unstable with respect to a buckling of the dimer atoms which broke mirror symmetry [26]. The buckling was found to be appreciable (about 0.65 Å for the c-4 × 2 surface) and it made the two atoms of the dimer quite inequivalent in the sense that, as shown schematically in Fig. 1, one atom became nearly sp^2 bonded (i.e., it became "Ga-like") while the other atom of the dimer became nearly s^2p^3 bonded (i.e., it became "As-like"). The tilting of the dimers could explain the nonmetallic nature of the 2 × 1 surface and it also led to a simple model for the c-4 × 2 (and 2 × 2) structures seen frequently at low temperatures on Si(100) and especially on Ge(100) surfaces. Structurally, the 2 × 1 and c-4 × 2 structures were shown to differ only in the relative phase of nearest-neighbor dimers. Energetically, each dimer behaved as a small dipole and the energy difference between the 2 × 1 and c-4 × 2 two structures resulted from their different dipole–dipole interactions. The c-4 × 2 structure was found to be 0.12 eV/dimer more stable than the 2 × 1 structure.

The suggestion that the dimers on a Si(100) surface were asymmetric led to great controversy. Over the last 14 years there have been numerous experimental and theoretical investigations of this question. The application of STM to this surface has removed any doubt that dimers are the basic building blocks of the surface [27]. The c-4 × 2 structure is clearly seen in STM to be made up of asymmetric dimers; without the asymmetry the structure would only have 2 × 1 periodicity. The transformation from the 2 × 1 to the c-4 × 2 structure at low temperatures suggests that asymmetric dimer formation does indeed lower the total-energy. STM images of the 2 × 1 surface show both buckled and unbuckled dimers and the ratio between the two appears to be sensitive to the defect density at the surface. It has been suggested that rapid oscillations of dimers between the two inequivalent buckled geometries can give an image that appears as a symmetric dimer. Theoretical results obtained from ab initio methods, particularly over the last three years have all shown that buckling lowers the total-energy [28–30]. Plane-wave convergence and Brillouin zone sampling are found to be important elements in obtaining a reliable total-energy. Initial pseudopotential calculations suggested that the tight-binding method overestimated the degree of dimer buckling by a factor of nearly 2, but results from the most recent self-consistent calculation [30] are in excellent agreement with the predictions of the tight-binding theory on the degree of dimer tilting at the c-4 × 2 surface.

Another aspect of the Si(100) surface which has come under increased interest over the last several years is step formation. Vicinal surfaces in which the surface normal deviates slightly away from the (100) surface are characterized by (100) terraces and by steps. The density of steps increases with increasing tilt of the surface normal from a (100) axis. Depending on the orientation of dimers with respect to the step edge four types of single and double layer steps (as shown in Fig. 2) are possible. The stepped surface has the interesting property that after annealing single layer steps disappear and only D_B type double layer steps shown in Fig. 2d are observed to occur [31]. Initially this was thought to be due to the lower energy of double layer steps as compared to single layer steps. Using the tight-binding total-energy method I showed, however, that a particular type of single layer step (the S_A step shown in

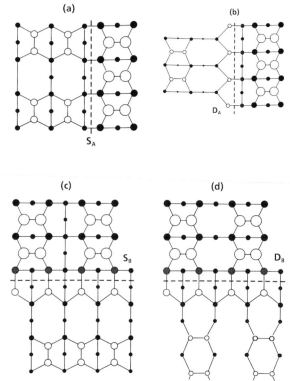

Fig. 2. Top views of the four possible single and double layer step geometries on a Si(100) surface are shown in (a)–(d). The dimers on the upper terraces are shown as larger circles. Alternating layer of atoms are shown as open and dark circles. The step energies are ordered as $E(S_A) \ll E(D_B) \ll E(S_B) \ll E(D_A)$ with $E(D_B) \ll [E(S_A) + E(S_B)]$. The latter inequality leads to a conversion of single layer steps into double layer ones with annealing.

Fig. 2a) had the lowest formation energy [31]. The effect of the tilting was to force an alternation between low energy S_A and high energy S_B single layer step configurations and this made the surface unstable against a double layer D_B step configuration. The energy of the D_A step was found to be very high, explaining why this type of step was never experimentally observed. The calculated step energies were found to range from 0.01 eV per step atom for S_A to 0.54 eV for D_A. The atomic structure obtained from the calculations were found to be in very good agreement with STM measurements [32]. The step energies could explain the roughness or smoothness of the various steps [33]. Recent experimental studies

[34,35] have yielded an S_A step energy of 0.023–0.028 eV per step atom as compared to the theoretical value [31] of 0.01 ± 0.01 eV. Considering the fact that the calculation involved taking a difference between two large numbers, the degree of agreement between the theoretical and experimental step energy is very satisfying (despite the fact that at a recent technical conference the theory was criticized for giving a result 50% smaller than experiment). Recent studies have shown the dynamics of step formation to be quite fascinating and a number of important recent contributions have deepened our understanding of stepped surfaces [36–40].

The efforts in understanding the (100) surfaces of Si and Ge were overshadowed by the more intense experimental and theoretical effort devoted in the 1980's to the (111) cleavage plane of Si. This surface has a metastable 2×1 reconstruction after cleavage which transforms to the well known 7×7 structure upon annealing. Haneman [41] originally proposed a buckling model for the 2×1 surface in which alternate rows of surface atoms were raised and lowered. My tight-binding calculations [13] for this surface showed a large charge transfer of almost one electron from the lowered surface atom to the raised one (this is nearly 3 times larger than that calculated later for the buckled dimer on the Si(100) surface). This was initially worrisome to me because the simple tight-binding method ignored the repulsive Coulombic interaction of two electrons localized on the same site. A self-consistent pseudopotential calculation of the optical spectra for the model seemed to show, however, exactly the same type of charge flow suggesting that Coulombic effects were not crucial [42]. Later on it turned out that this result was incorrect. The electronic charge on the surface was flip-flopping between the two raised and lowered atoms with each iteration toward self-consistency and the correctly converged result was that Coulombic effects in fact prevented charge transfer between the atoms and made the buckling reconstruction energetically unfavorable [43]. Another problem with the Haneman model was that the calculated dispersion for the surface band had the wrong sign as compared to experiment.

The calculated surface level in the band gap moved to lower energies as the electron wave vector increased in contrast to experiment which seemed to show an upward dispersion. This led Pandey [44] to suggest a completely new and novel type of reconstruction model for the surface which he called the π-bonded chain model. In this model the surface structure is transformed as a result of bond breaking and rebonding in such a way that the surface dangling-bonds become, in effect, nearest-neighbors instead of second nearest-neighbors. This results in a significant enhancement of the π-bonding interaction between the atoms which stabilizes the surface. Many experimental observations including the anisotropy of surface optical absorption spectra [45], angle-resolved photoemission [46], and STM [47] all strongly support this model for the surface. Hydrogen chemisorption at the surface which effectively quenches the π-bonding interaction removes the 2×1 reconstruction.

As for the more complex Si(111)-7×7 surface, the structure was determined from a combination of STM [48] and transmission electron diffraction experiments [49]. The currently accepted DAS (dimer–adatom–stacking-fault) model of Takayanagi et al. [49] has 12 adatoms, 9 dimer bonds, and a stacking fault in each 7×7 cell. The surface reconstruction reduces the number of dangling bonds from 49 on the "ideal" surface to only 19. Our calculation of the surface energies of the 2×1 π-bonded and the 7×7 DAS models using the tight-binding approach showed that the surface energies for the two surfaces were very close. The 7×7 surface energy was found to be only 0.04 eV per 1×1 cell lower than the 2×1 surface [50]. Recently it has become possible to apply ab initio methods to systems containing a large number of atoms including the 7×7 surface [51,52]. The energy difference between the two surfaces from these calculations is about 0.06 eV per 1×1 cell. The optimized 7×7 structure from the tight-binding energy-minimization calculations has also been found to be in very good agreement with data from surface X-ray diffraction studies [53,54].

The total-energy-minimization approach has also proved useful in studies of the atomic struc-

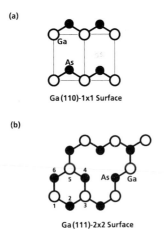

Fig. 3. Top views of the GaAs(110) and (111) surfaces are shown in (a) and (b), respectively. The vacancy structure for the (111)-2×2 surface is, from a nearest-neighbor macroscopic point of view, nearly the same as the (110) surface. This similarity in atomic structure is also reflected in the electronic structure of the two surfaces.

ture of the *polar* surfaces of compound semiconductors. The cation terminated surfaces of zincblende semiconductors exhibit a 2×2 surface reconstruction. The LEED analysis of Tong et al. [55] for the GaAs(111)-2×2 surface gave strong evidence for a vacancy induced reconstruction, i.e, each 2×2 cell had a missing Ga atom. Examination of this model showed that with a $1/4$ monolayer of vacancies, the (111) surface became topologically the same as the nonpolar (110) surface [56]. As shown in Fig. 3, at each surface, one finds threefold coordinated Ga and As atoms which are bonded to each other and to a bulk (i.e., fourfold coordinated) atom. The zigzag chains of Ga and As surface atoms on the (110) surface is transformed to hexagonal rings on the (111) surface, but from a nearest-neighbor atomic point of view the two surfaces are very similar. Optimization of the structure showed exactly the same trends as obtained previously on the (110) surface, i.e., the surface Ga atoms became essentially planar with their three nearest-neighbor As atoms and the As atoms became more s^2p^3 bonded. Vacancy creation was found to be exothermic by about 0.45 eV per 2×2 cell. The angle-integrated photoemission spectra of the two surfaces were observed to be nearly the same, as

expected from the similarities of the atomic structures [57].

The As-terminated (111)-B surface of GaAs is more complex. It exhibits a metastable 2×2 and a stable $\sqrt{19} \times \sqrt{19}$ structure. LEED [55] and ion-neutralization studies had suggested that the structure of the 2×2 surface was quite different from that of the (111)-A vacancy reconstructed surface. A possible 2×2 structure for this surface consisting of multiple vacancies was suggested from the tight-binding calculations [58]. Scanning-tunneling-microscopy studies [59] have shown however that the 2×2 periodicity arises from the chemisorption of an As trimer at the surface in agreement with a previous theoretical suggestion [60]. The structure with multiple vacancies turns out to be a building block for the $\sqrt{19} \times \sqrt{19}$ surface as seen by STM.

With the experience gained from working on the (110) and the (111) surfaces of GaAs, the total-energy-minimization approach was next applied to the GaAs(100) surface [61]. The (100) surface like the (111) surface can be either Ga or As-rich. Many different reconstructions as a function of temperature occur on this surface. Under normal growth conditions the surface periodicity for an As-stabilized surface is either 2×4 or a closely related c-2×8. X-ray and LEED studies had shown that the twofold periodicity along one axis was consistent with a dimerization of the surface atoms. The tight-binding calculations showed that the fourfold periodicity resulted from either one or two dimer vacancies per 2×4 cell. The electronic structure for the missing dimer models was found to be nonmetallic and in generally good agreement with experimental data. The occurrence of a c-2×8 instead of a 2×4 structure could be easily explained within the dimer vacancy model. Subsequent STM studies [62,63] have verified the missing dimer models for the (100)-2×4 and c-2×8 surfaces and have shown that various types of vacancy structures are quite common for other periodic structures seen on the (100) surface. The effect of the vacancies in each case is to transform the surface into an essentially nonpolar one.

Over the last two decades, the development of new experimental and theoretical tools have revolutionized the field of surface science. The atomic structure of the principal surfaces of Si and Ge and of many zincblende and wurtzite structure semiconductors are now known with high precision. A simple tight-binding energy-minimization approach has proved very effective in analyzing the structural and electronic properties of many of these surfaces. A primary result of these investigations has been the demonstration that, for semiconductors, whenever a particular surface structure gives rise to occupied electronic states with energies high in the band gap, atomic relaxations and reconstructions which lead to an elimination or lowering of the energies of these states can be anticipated with near certainty. This idea has proved quite useful in understanding rebonding effects at what may be called "internal" surfaces induced by impurities and point defects in *bulk* semiconductors. The large bond-breaking lattice distortions at: Ga-antisite [64] and at As-antisite [65] defects in GaAs (the *EL*2 defect), donor impurities in AlGaAs alloys [66] (*DX* centers), acceptor impurities [67] in ZnSe, and of self-interstitial bonding in Si and GaAs [68] may be justly considered to be examples of atomic reconstructions in the bulk induced by high energy electronic states introduced by impurities or native defects. The tight-binding total-energy-minimization approach has proved to be a powerful tool in the determination and analysis of the atomic and electronic properties of both "external" and "internal" surfaces of semiconductors.

References

[1] R.E. Schlier and H.E. Farnsworth, J. Chem. Phys. 30 (1959) 917.

[2] J.J. Lander and J. Morrison, J. Chem. Phys. 37 (1962) 729.

[3] S.J. White and D.P. Woodruff, Surf. Sci. 64 (1977) 131.

[4] S.Y. Tong and A.L. Maldondo, Surf. Sci. 78 (1978) 459.

[5] T.D. Poppendieck, T.C. Ngoc and M.B. Webb, Surf. Sci. 43 (1974) 647.

[6] F. Jona et al., J. Phys. C 10 (1977) L67;
R. Seiwatz, Surf. Sci. 2 (1964) 473.

[7] A.R. Lubinsky, C.B. Duke, B.W. Lee and P. Mark, Phys. Rev. Lett. 36 (1976) 1058;
A. Kahn, E. So, P. Mark and C.B. Duke, J. Vac. Sci. Technol. 15 (1978) 580.

[8] S.Y. Tong, A.R. Lubinsky, B.J. Mrstik and M.A. Van Hove, Phys. Rev. B 17 (1978) 3303.

[9] D.J. Chadi, Phys. Rev. B 18 (1978) 1800; J. Vac. Sci. Technol. 15 (1978) 631, 1244.

[10] A. Huijser, J. van Laar and T.L. van Rooy, Phys. Lett. 65 A (1978) 335.

[11] C.B. Duke, Appl. Surf. Sci. 11/12 (1982) 1.

[12] D.J. Chadi and R.M. Martin, Solid State Commun. 19 (1976) 643.

[13] D.J. Chadi, Phys. Rev. Lett. 41 (1978) 1062; Phys. Rev. B 29 (1984) 785.

[14] R.J. Meyer, C.B. Duke, A. Paton, A. Kahn, E. So, J.L. Yeh and P. Mark, Phys. Rev. B 19 (1979) 5194.

[15] D.J. Chadi, Phys. Rev. B 19 (1979) 2074.

[16] C.B. Duke, J. Vac. Sci. Technol. A 10 (1992) 2032;
C.B. Duke, in: Surface Properties of Electronic Materials, Eds. D.A. King and D.P. Woodruff (Elsevier, Amsterdam, 1988) pp. 69–118.

[17] A. Kahn, S. Ahsan, W. Chen, M. Dumas, C.B. Duke and A. Paton, Phys. Rev. Lett. 68 (1992) 3200.

[18] C.B. Duke, J. Vac. Sci. Technol. B 1 (1983) 732.

[19] C. Mailhiot, C.B. Duke and D.J. Chadi, Surf. Sci. 149 (1985) 366.

[20] J.P. LaFemina, Surf. Sci. Rep. 16 (1992) 133. This review article contains an extensive discussion and list of references.

[21] R.M. Feenstra, J.A. Stroscio, J. Tersoff and A.P. Fein, Phys. Rev. Lett. 58 (1987) 1192.

[22] J.A. Appelbaum, G.A. Baraff and D.R. Hamann, Phys. Rev. Lett. 35 (1975) 729; Phys. Rev. B 14 (1976) 588.

[23] J.E. Rowe, Phys. Lett. A 46 (1974) 400;
J.E. Rowe and H. Ibach, Phys. Rev. Lett. 32 (1974) 421.

[24] G.P. Kerker, S.G. Louie and M.L. Cohen, Phys. Rev. B 17 (1978) 706.

[25] F.J. Himpsel and D.E. Eastman, J. Vac. Sci. Technol. 16 (1979) 1302.

[26] D.J. Chadi, Phys. Rev. Lett. 43 (1979) 43; J. Vac. Sci. Technol. 16 (1979) 1290.

[27] R.A. Wolkow, Phys. Rev. Lett. 68 (1992) 2636;
R.J. Hamers, R.M. Tromp and J.E. Demuth, Phys. Rev. B 34 (1986) 5343;
R.M. Tromp, R.J. Hamers and J.E. Demuth, Phys. Rev. Lett. 55 (1985) 1303.

[28] N. Roberts and R.J. Needs, Surf. Sci. 236 (1990) 112;
M.T. Yin and M.L. Cohen, Phys. Rev. B 26 (1982) 5668.

[29] J. Dabrowski and M. Scheffler, Appl. Surf. Sci. 56 (1992) 15;
P. Kruger and J. Pollmann, Phys. Rev. B 47 (1993) 1898.

[30] J.E. Northrup, Phys. Rev. B 47, in press.

[31] D.J. Chadi, Phys. Rev. Lett. 59 (1987) 1691, and references therein.

[32] P.E. Wierenga, J.A. Kubby and J.E. Griffith, Phys. Rev. Lett. 59, 2169 (1987).

[33] Low energy step configurations such as an the S_A step are generally much less jagged than high energy ones like the S_B. The meandering of an S_A step would expose a high energy S_B step thereby suppressing its occurrence.

[34] B.S. Schwartzentruber, Y.-W. Mo, R. Kariotis, M.G. Lagally and M.B. Webb, Phys. Rev. Lett. 65, 1913 (1990).

[35] D.J. Eaglesham, A.E. White, L.C. Feldman, N. Moriya and D.C. Jacobson, Phys. Rev. Lett. 70 (1993) 1643.

[36] O.L. Alerhand, D. Vanderbilt, R.D. Meade and J.D. Joannopoulos, Phys. Rev. Lett. 61 (1988) 1973.

[37] O.L. Alerhand, A.N. Berker, J.D. Joannopoulos, D. Vanderbilt, R.J. Hamers and J.E. Demuth, Phys. Rev. Lett. 64 (1990) 2406.

[38] T.W. Poon, S. Yip, P.S. Ho and F.F. Abraham, Phys. Rev. Lett. 65 (1990) 2161.

[39] B.S. Schwartzentruber, Y.W. Mo and M. Lagally, Appl. Phys. Lett. 58 (1991) 822.

[40] J. Tersoff and E. Pehlke, Phys. Rev. Lett. 68 (1992) 816; R.W. Tromp and M.C. Reuter, Phys. Rev. Lett. 68 (1992) 820;
H.J.W. Zandvliet, H. Wormeester, D.J. Wentink, A. van Silfhout and H.B. Elswijk, Phys. Rev. Lett. 70 (1993) 2122.

[41] D. Haneman, Phys. Rev. 121 (1961) 1093.

[42] M. Schluter, J.R. Chelikowsky, S.G. Louie and M.L. Cohen, Phys. Rev. B 12 (1975) 4200.

[43] J.E. Northrup and M.L. Cohen, Phys. Rev. Lett. 49 (1982) 1349; 47 (1981) 1910; J. Vac. Sci. Technol. 21 (1982) 333.

[44] K.C. Pandey, Phys. Rev. Lett. 47 (1981) 1913.

[45] P. Chiaradia, A. Cricenti, S. Selci and G. Chiarotti, Phys. Rev. Lett. 52 (1984) 1145;
M.A. Olmstead and N. Amer, Phys. Rev. Lett. 52 (1984) 1148.

[46] F. Houzay, G.M. Guichar, R. Pinchaux and Y. Petroff, J. Vac. Sci. Technol. 18 (1981) 860;
R.I.G. Uhrberg, G.V. Hansson, J.M. Nicholls and S.A. Flodström, Phys. Rev. Lett. 48 (1982) 1032.

[47] R.M. Feenstra, W.A. Thompson and A.P. Fein, Phys. Rev. Lett. 56 (1986) 608.

[48] G. Binnig, H. Rohrer, Ch. Gerber and E. Weibel, Phys. Rev. Lett. 50 (1983) 120.

[49] K. Takayanagi, Y. Tanishiro, M. Takahashi and S. Takahashi, J. Vac. Sci. Technol. A 3 (1985) 1502.

[50] G.-X. Qian and D.J. Chadi, Phys. Rev. B 35 (1987) 1288; J. Vac. Sci. Technol. B 4 (1986) 1079.

[51] K.D. Brommer, M. Needels, B.E. Larson and J.D. Joannopoulos, Phys. Rev. Lett. 68 (1992) 1355.

[52] I. Stich, M.C. Payne, R.D. King-Smith and J.-S. Lin, Phys. Rev. Lett. 68 (1992) 1351.

[53] I.K. Robinson, W.K. Waskiewicz, P.H. Fuoss, J.B. Stark and P.A. Bennett, Phys. Rev. B 33 (1986) 7013.

[54] R. Feidenhans'l, J.S. Pederson, J. Bohr, M. Nielsen, F. Grey and R.L. Johnson, Phys. Rev. B 38 (1988) 9715;
J.S. Pederson et al., Phys. Rev. B 38 (1988) 13210.

[55] S.Y. Tong, G. Xu and W.N. Mei, Phys. Rev. Lett. 52 (1984) 1693.

[56] D.J. Chadi, J. Vac. Sci. Technol. A 4 (1986) 944; Phys. Rev. Lett. 52 (1984) 1911.

[57] A.D. Katnani and D.J. Chadi, Phys. Rev. B 31 (1985) 2554.

[58] D.J. Chadi, Phys. Rev. Lett. 57 (1986) 102.

[59] D.K. Biegelsen, R.D. Bringans, J.E. Northrup and L.E. Swartz, Phys. Rev. Lett. 65 (1990) 452.

[60] E. Kaxiras, Y. Bar-Yam, J.D. Joannopoulos and K.C. Pandey, Phys. Rev. B 35 (1987) 9625.

[61] D.J. Chadi, J. Vac. Sci. Technol. A 5 (1987) 834;
G.X. Qian, R.M. Martin and D.J. Chadi, Phys. Rev. B 38 (1988) 7649.

[62] M.D. Pashley, K.W. Haberern, W. Friday, J.M. Woodall and P.D. Kirchner, Phys. Rev. Lett. 60 (1988) 2176.

[63] D.K. Biegelsen, R.D. Bringans, J.E. Northrup and L.E. Swartz, Phys. Rev. B 41 (1990) 5701.

[64] S.B. Zhang and D.J. Chadi, Phys. Rev. Lett. 64 (1990) 1789.

[65] D.J. Chadi and K.J. Chang, Phys. Rev. Lett. 60 (1988) 2187;
D.J. Chadi, Phys. Rev. B 46 (1992) 15053.

[66] D.J. Chadi and K.J. Chang, Phys. Rev. Lett. 61 (1988) 873; Phys. Rev. B 39 (1989) 10063;
S.B. Zhang and D.J. Chadi, Phys. Rev. B 42 (1990) 7174; D.J. Chadi, Phys. Rev. B 46 (1992) 6777.

[67] D.J. Chadi and K.J. Chang, Appl. Phys. Lett. 55 (1989) 575.

[68] D.J. Chadi, Phys. Rev. B 46 (1992) 9400.

Surface Science 299/300 (1994) 319–331
North-Holland

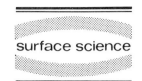

surface science

Tight-binding methods as applied to transition metal surfaces

G. Allan

Institut d'Electronique et de Microélectronique, Département ISEN, 41, Boulevard Vauban, F-59046 Lille Cedex, France

Received 20 April 1993; accepted for publication 7 June 1993

A few simple models are described from the earliest cubium to recent applications to metallic superlattices. It is shown how these simple models have allowed us to understand the properties of transition metals surfaces such as surface tension, core level shifts and surface magnetism. Simple theoretical methods as moment or recursion are also explained as they have been developed at the beginning for surface studies.

1. Introduction

In 1967, following Professor J. Friedel's advice, our laboratory, whose director was G. Leman, started to develop theoretical research on surfaces. Two main directions were chosen which have some common features: phonons and electrons. Surface experiments became available due to recent improvements to obtain ultra-high vacuum during a time sufficiently long to make measurements on a well-defined "clean" surface.

As a young student beginning research, to get what was at that time in France equivalent to the Ph.D., I was in charge of the electronic part and was rather enthusiastic about this new field whose technological applications were so numerous. Among all the reasons we had to study surfaces, catalysis was certainly the most important and the most used to get contracts and money. I am not sure that we still nowadays completely understand this phenomenon but it is sure that much progress has been made during these 25 years.

Transition metals are well known for their catalytic properties. Our laboratory had some good experience of these materials. So it was natural to start with these metals. The augmented plane wave method was the state of the art to calculate bulk electronic structure of transition metals, but it was difficult to apply in the surface

case. We wanted a much less sophisticated method to develop simple models. The tight-binding method which was "la spécialité de la maison" seemed simple enough to be applied near surfaces.

One must also remember that at that time computers were not so fast and so powerful as nowadays. In the simplest tight-binding version, a great part of the calculations was analytical and the computational effort remained reasonable. But it was difficult to publish results obtained by this empirical method as, for bulk crystals, "a-priori" APW calculations became available. This antagonism has now disappeared as everyone has realized that both methods are complementary. A simple tight-binding model is often used to explain the results obtained from a much more sophisticated calculation. It can also be quantitative if properly used. Then it is often applied to more complicated systems where a-priori methods are not yet applicable even with the recent improvements of computers.

So due to our lack of fast computers and also because no theory for surface electronic structure of transition metals had been done, we started with very simple models. A hypothetical material was created: cubium, by analogy to jellium which was developed for alkali metals. The success of cubium was not only due to its simplicity but also to its implications. It allowed us to show the existence of a surface electrostatic dipole (measured later by surface core level spectroscopy) and of the contraction of the interatomic distances close to the transition metal surfaces.

At the same time, to improve the simple electronic structure models, one was obliged to develop new computational methods like moments to calculate the local surface densities of states without using the reciprocal lattice. With these new methods, it was possible to study a real transition metal surface. The existence of a surface peak in the local (100) bcc transition metal surface density of states was first demonstrated by another french group in Grenoble with M.-C. Desjonqueres and F. Cyrot-Lackmann. Around these years, after a winter school in Les Houches in 1972, a french school in surface physics was really created. The lecture notes which have been collected by the students themselves and corrected by the lecturers unfortunately have not been published, but copies are still available from the participants.

If in theory, it is possible to get the density of states from its moments, in practice, this is more difficult. The solution was independently found at about the same time by J.-P. Gaspard in Grenoble and in Cambridge (Great Britain) by R. Haydock, M. Kelly and V. Heine. The recursion method was born and it has been applied to a lot of crystal without translational symmetry.

Some other properties directly related to the surface density of states were studied in the following years such as surface energy, adatom binding energy, etc. Notably, I showed the existence of so-called "live" magnetic layers near the (001) chromium surface in 1977. At that time, this looked more like a theoretician's speculation even if there was already some experimental evidence available for powders. A more convincing experimental proof was only obtained later. Meanwhile, a more complete and exact investigation of surface magnetism was done mainly around A.J. Freeman at Northwestern University in Evanston (USA). This has led nowadays to the explosive development of metallic superlattices which, for example, show interesting properties for magnetic recording.

2. Cubium

2.1. The tight-binding method and its application to surfaces

In the tight-binding method, the solution ψ of the Schrödinger equation is expressed as linear combination of atomic orbitals (LCAO). Derived from the well-known Hückel method used by chemists, it assumes that the basis functions $\phi_l(r - R_j)$ are orthogonal, where R_j is a lattice site and l an atomic quantum number (for s, p or d states). The core functions which are not interacting in the solid are not included in the basis set. A minimal basis of valence states is generally chosen: s and p states for semiconductors and d states for transition metals. In this latter case, the s and p orbitals are often neglected as their band width in the bulk material is large and the corresponding density of states small so they do not contribute too much to the transition metal properties [1].

In this orthogonal basis, one can distinguish among the Hamiltonian terms the "intraatomic" diagonal ones $\epsilon_l = \langle \phi_l(r - R_i) | H | \phi_l(r - R_i) \rangle$ which behave as ''atomic levels'' for an atom in the solid and "interatomic" non-diagonal terms $\beta_{ij}^{lm} = \langle \phi_l(r - R_i) | H | \phi_m(r - R_j) \rangle$ or hopping parameters which broaden the atomic levels into bands. These Hamiltonian matrix elements are treated as band parameters and nowadays fitted to "a-priori" or so-called first-principles band structures. Some general laws like the Harrison ones [2] are also used to evaluate the interatomic parameters. One must also notice that there is a good transferability of these parameters from one compound to the other if the two atoms defining the bond are identical.

The method is not only a mathematical trick to make a good fit of the known band structures, it also takes into account experimental results. In transition metals, for example, the d band width is rather small and the d electrons in the solid keep a strong atomic character. In tetrahedral semiconductors, the electrons are localized along the bonds between atoms and are well described by hybridized sp_3 orbitals. The β parameters keep a physical meaning only if they are limited to atoms nearest or next-nearest neighbors.

To perform a first calculation of the transition metal surface electronic structure, the complete five d band model was at that time a bit too difficult to handle. We also wanted to get the most important effects of the surface on the electronic structure. Complete calculations would come later on. As many properties only depend on the integral of the density of states, the d band was replaced by a non-degenerate s band and a simple cubic lattice was used: "cubium" was created.

2.2. The Green function method applied to surfaces [3–5]

The first idea to create a surface was to cleave a bulk infinite crystal into two pieces. This is still the experimental way to get clean surfaces and this can be also done in theory by a perturbation which splits an infinite perfect crystal into two pieces. Starting from a bulk infinite crystal Hamiltonian H_0, we switch on a perturbation potential V which suppresses the interactions between atoms on each side of the cleavage plane (Fig. 1). Such a perturbation can be exactly treated by the Green function method. Let us define the Green operator G for the two semi-infinite crystals as

$$G = (E - H_0 - V)^{-1} \qquad (1)$$

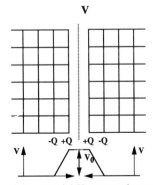

Fig. 1. The cleavage potential V creates charge transfers near the surfaces and a dipole layer V_0.

and G^0 the Green operator for the bulk infinite crystal

$$G^0 = (E - H_0)^{-1}. \qquad (2)$$

One can show that the imaginary part of the diagonal matrix elements G_{ii} is proportional to the local density of states on orbital i. If the range of V is limited (that is the case here in the tight-binding approximation as only a few interactions between atoms on both sides of the cleavage plane are suppressed), one can easily calculate the G_{ii} matrix elements for the surface atom orbitals as a function of the perfect bulk crystals G_{ij}^0 using the Dyson equation

$$G = G^0 + G^0 V G. \qquad (3)$$

At this point, one can introduce the periodicity along the surface and define Bloch waves in each plane z_i parallel to the surface [6]:

$$\phi(k_{\parallel}, z_i) = \frac{1}{N} \sum_{R_{i\parallel}} e^{ik_{\parallel}R_{i\parallel}} \phi(r - R_{i\parallel} - z_i), \qquad (4)$$

where N is the number of atoms in the surface plane and $R_{i\parallel}$ the lattice sites in this plane.

Such a procedure reduces the three-dimensional crystal to a linear chain whose Hamiltonian has to be solved for each wave vector $k_{\parallel}$ in the surface Brillouin zone [5]. Ref. [5] is often quoted as the first example of the Green function technique applied to calculate the surface electronic structure, but in fact the results for the cubium (100) surface had been published before [3–4]. However, one can find in Ref. [5] the first clear definition of the surface Brillouin zone.

For the cubium (100) surface, the cleavage potential does not depend on $k_{\parallel}$ and is simply equal to:

$$\begin{bmatrix} V_0 & +\beta \\ +\beta & V_0 \end{bmatrix}, \qquad (5)$$

where the non-diagonal elements anneal the interaction between atoms on each side of the cleavage plane. The diagonal elements are due to charge transfers between atoms induced by the surface [4,5,7]. Such an effect is well known near bulk defects as Friedel oscillations [8]. The corresponding variation of the potential must be calculated by the Coulomb law in order to make the model self-consistent.

In a transition metal or in a semiconductor, the influence of a defect is screened at very short distances as the electron density is large. This means that charge transfers only occur close to the surface. Near a surface, the self-consistent potential can be approximated by a dipole layer (Fig. 1). The Friedel sum rule [8] implies that the Fermi level is not modified by the surface. As the crystal remains neutral, the total crystal electronic charge below the Fermi level $N(E_F, V_0)$ obtained after summation over the surface Brillouin zone is constant. Its variation [4] is then equal to zero:

$$\Delta N(E_F, V_0) = 0. \tag{6}$$

Relation (6) determines in this simple model the potential step V_0 at the surface. This potential contributes to the work function anisotropy. It strongly reduces the charge transfer near the surface to values of the order of 0.01 electron [4].

When the potential has been determined, one can calculate other properties like, for example, the surface tension per atom γ_s which is the energy difference between the bulk infinite crystal and the two semi-infinite ones [3,4]:

$$2\gamma_s = \int^{E_F} E \, \delta n(E, V_0) - V_0 N(E_F)$$
$$- \tfrac{1}{2} V_0 \, \delta N_0(E_F). \tag{7}$$

The factor 2 on the left side of (7) comes from the fact that the cleavage potential creates two surfaces. $\delta n(E, V_0)$ is the change of the density of states for the whole crystal when one creates the surfaces ($= dN(E, V_0)/dE$), and in the Green function method this was obtained from the phase shift [9]. The last two terms on the right side of Eq. (7) are due to the fact that electron–electron interactions are counted twice in the one-electron energy [10].

The perturbation due to the surface can also give rise to localized surface states. Contrary to bulk states which are reflected by the surface, these localized states are confined near the surface and do not propagate into the bulk. They occur at energies outside the bulk band (at a

fixed $k_\parallel$). When $k_\parallel$ varies, they form a surface state band which can then overlap the total bulk band. One must also mention that in the cubium model, these states only occur when one takes into account the self-consistent potential V_0 and do not appear when this potential is neglected [7]. This is not the case for the complete five d band model: surface bound states exist even when the self-consistent potential is neglected.

All the results one can get with this method will not be developed here for they will be obtained below more simply as examples of the moment method applied to surfaces. The Green function method was used later on mainly by J. Pollmann in Germany and applied to semiconductor surfaces. We shall see below that this is the starting-point of new computational methods in tight-binding as recursion or decimation techniques.

2.3. The moment method applied to surfaces [11–12]

Let us define μ_i^n the nth moment of the local density of states on orbital i:

$$\mu_i^n = \int E n_i(E) \, dE. \tag{8}$$

One can show that μ_i^n is equal to a diagonal element of the nth power of the Hamiltonian H:

$$\mu_i^n = \langle i | H^n | i \rangle. \tag{9}$$

So the moments of the local density of states can be evaluated in direct space. The key point is that the Hamiltonian has not to be diagonalized to get information about the local density of states. So the method can be applied to crystals without any translation symmetry. In the cases where the translation symmetry exists, one can use it to simplify the moment computation [13].

The very first moments of the density of states can be evaluated by hand even for the five d band model. For the non-degenerate s band of the cubium and for a bulk atom, we get:

$$\mu^0 = 1,$$
$$\mu^1 = \epsilon_0,$$
$$\mu^2 = \epsilon_0^2 + N_b \beta^2, \tag{10}$$

where ϵ_0 is the atomic level, β the hopping parameter between atoms nearest neighbors and N_b the bulk coordination number. These moments are related to the band barycenter ϵ_0 (μ^1) and to the band width which is roughly proportional to $\sqrt{N_b \beta^2}$ $[\mu^2 - (\mu^1)^2]$. All the moments can be analytically evaluated for cubium but this is not the general case, and for a true d band it is difficult to calculate more than 6 moments by hand but a larger number (~ 40) is easy to get using a numerical computation.

The main problem is in fact to reconstruct the density of states from its few first moments. For the three moments given by relation (10), rectangular band or Gaussian representations of the density of states have been used. For a higher number of known moments, the only satisfactory representation of the density of states was obtained later by J.-P. Gaspard in Grenoble as a continued fraction expression [14]. This will be developed below in relation to the recursion method.

In fact, physical properties which are obtained by integration of the density of states like the cohesive energy or the Fermi level position do not depend very much on the details in the density of states. So a very small number of moments is sufficient to get accurate values or at least to determine the behavior of a property (like cohesive energy, surface tension, etc.) as a function of the band filling. Let us use a Gaussian, as it was done at the beginning. For the sake of simplicity, we take ϵ_0 equal to zero. To apply such a model to a d band, one must take into account the band degeneracy. The bulk density of states is then:

$$n_b(E) = \frac{10}{\sqrt{2\pi N_b \beta^2}} \exp\left(-\frac{E^2}{2 N_b \beta^2}\right). \qquad (11)$$

Due to N_{db} dangling bonds, the number of neighbors of a surface atom N_s is reduced ($N_b = N_s + N_{db}$). The first effect of the surface is to reduce the band width as $N_s < N_b$. The atomic level is shifted by the surface dipole layer $\mu^1 = V_0$.

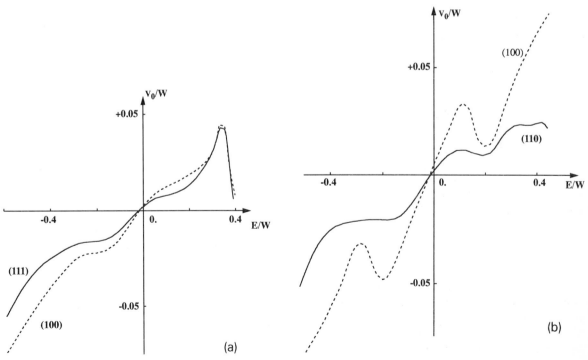

Fig. 2. Surface core level shifts: (a) bcc lattice; (b) fcc lattice.

The perturbation due to the surface is, in this approximation, localized to the surface atoms and the surface density of states is equal to:

$$n_s(E) = \frac{10}{\sqrt{2\pi N_s \beta^2}} \exp\left[-\frac{(E - V_0)^2}{2N_s \beta^2}\right].$$ (12)

If the atomic configuration of a transition atom is $d^n s^2$, the number of d electrons for the metal N_d is close to $n+1$ as there is almost one s electron in the bulk. The Fermi level which is given by:

$$N_d = \int^{E_F} n_b(E)\, dE$$ (13)

is constant in a crystal. The electronic charge in the surface plane which is equal to

$$N_s = \int^{E_F} n_s(E)\, dE$$ (14)

can be used to calculate the dipole layer V_0 using the Poisson equation. To simplify, we can assume that the charge is spread in a plane and using the parallel plate capacitance formula, one gets a simple relation between the potential V_0 and the net charge $N_s - N_d$ in the surface plane:

$$V_0 = \frac{4\pi d}{S}(N_s - N_d),$$ (15)

where S and d are respectively the area per surface atom and the distance between planes parallel to the surface. If V_0 is expressed in eV and $N_s - N_d$ in electrons, we get $V_0 \approx 50(N_s - N_d)$. If V_0 is assumed to be of the order of half the band width (a few eV), the charge transfer will be very small (< 0.01 electron per atom). So a good approximation is to assume that each atom remains neutral and $N_s = N_d$. Eqs. (11) to (15) determine the dipole layer V_0 [15]:

$$V_0 = E_F\left(1 - \sqrt{N_s/N_b}\right),$$

$$V_0 \approx E_F N_{db}/2N_b$$ (16)

which, as assumed above, is smaller than half the band width. The core levels are also shifted by this potential and this can be measured by photoemission spectroscopy [16,17]. The experimental values agree very well with this simple relation (16). The core level shift increases with surface

roughness and is roughly antisymmetric with respect to Cr, Mo or W for which $N_d = 5$ and $E_F = 0$. A more complete calculation (Fig. 2) using the true d band still improves the agreement with the measurements [18].

2.4. Surface tension [3,12] and surface relaxation [19,20]

With this simple model, one can determine the bulk cohesive energy E_C and the surface tension γ_S. The electronic contribution to the bulk cohesive energy is given by:

$$E_A^b = \int^{E_F} E n(E)\, dE - N_d \epsilon_0,$$ (17)

and as here ϵ_0 is equal to zero:

$$E_A^b = -10\sqrt{\frac{N_b \beta}{2\pi}} \exp\left(-\frac{E_F^2}{2N_b \beta^2}\right).$$ (18)

This shows that the attractive part of the cohesive energy is proportional to $\sqrt{N_b}$ and not to the number of nearest neighbors as in a pair potential model. In the bulk, all the β are equal but near a defect or a surface, the interatomic distance R can vary. It is generally assumed that β follows an exponential law:

$$\beta = \beta_0 \exp(-qR).$$ (19)

The crystal stability at short distances is ensured by Born–Mayer repulsive potentials C between nearest neighbor atoms [21]:

$$C = C_0 \exp(-pR).$$ (20)

The parameters p, q, β_0 and C_0 must satisfy the equilibrium condition and they are also fitted to the experimental values of the cohesive energy and bulk modulus. The ratio p/q is close to 3 [21]. Near a surface, both repulsive and attractive terms are reduced. The electronic E_A^s part is

$$E_A^s = E_A^b \sqrt{N_s/N_b}$$ (21)

and the repulsive term is equal $N_s C$. To first-order, one gets that the surface tension per atom is roughly proportional to the number of dangling bonds:

$$\gamma_S = |E_C| N_{db}/4N_\nu$$ (22)

as in a pair potential model. But within such approximation for the crystal energy, one would get twice the value given by Eq. (22). The difference is due to the fact that the attractive energy contribution is not proportional to the number of nearest neighbors. Expression (23) also shows that the surface tension is parabolic like the cohesive energy as a function of the d band filling. It is anisotropic and increases with the surface roughness. This value calculated with only a few exact moments is very close to the result obtained for the cubium by the Green function technique. The self-consistent and non-self-consistent results are also very close [3,4].

One can also calculate in this model the interatomic surface relaxation. A simple analytic expression is then obtained [19], where the forces due to the surface are limited between the atoms in the surface plane and their neighbors in the plane(s) below the surface. The interatomic forces are proportional to $(\sqrt{N_s/N_b} - 1)$, so roughly to the number of dangling bonds N_{db}, and increases with the surface roughness. The relative relaxation amplitudes $\Delta d/d$ vary from 1% (for a dense surface such as the fcc (111)) to about 10% for a non-dense one (such as the bcc (100) or (111)). The main point is that in this model the forces lead to a "surface contraction": the distance between the atoms lying in the surface plane and their neighbors in the plane(s) below the surface are reduced by the relaxation. The existence of such a surface contraction near the surface of transition metals was published at about the same time as the Finnis and Heine model valid for free electron metals [22]. Both models predict a surface contraction and are in disagreement with pair potential models which lead to no surface relaxation in a first-nearest neighbor approximation and to a surface dilatation when second nearest interactions are taken into account [15,23,24].

Even at the time these results were published, there were already some experimental evidences for surface contraction, but sometimes surface dilatations were also measured. This discrepancy has disappeared since and has been explained by chemisorption. The effect of oxygen adsorption was carefully studied later and measured for example with Rutherford back scattering (RBS) [25]. If the contraction of a clean surface is due to the reduction of nearest neighbors for a surface atom, the adsorption of oxygen which makes strong bonds with the transition metal substrate is equivalent for a surface atom to an increase of the number of its nearest neighbors. This is just the opposite effect of dangling bonds. One must also notice that self-consistency is important. A non-self-consistent model, we have first published, predicted a surface dilatation for nearly empty or filled d bands. Measurements have been done with RBS and low energy electron diffraction (LEED) for many transition metal surfaces. From these results, it now seems well established that almost all the transition metal surfaces are contracted. These results have also been confirmed by more detailed calculations [26].

The relaxation decreases and oscillates as one goes into the bulk. For example, the first interatomic distance near a Ni(100) surface is contracted, whereas the second is dilated. In the case of Fe(111), the periodicity is more complicated. It can be shown that the period of oscillation is related to the wave vector of bulk phonons with zero frequency and can also be obtained from a simple second moment approximation [27].

2.5. Effect of the relaxation on surface phonons

This surface contraction has two effects. First, it increases the tight-binding interactions between atoms in the surface plane and in the plane just below the surface and then reduces the band narrowing due to the reduction of nearest neighbors [28].

The surface contraction also gives rise to a shift of the surface phonons to higher frequency [25,29]. This effect is due to anharmonicity which can also be included in the expression of the energy as a function of the atomic displacements. The main contribution to the force constants between the atoms is due to the repulsive potentials between the atoms which are enhanced by the surface contraction. The energy of the surface phonons increases due to this variation of the force constants. This effect has been notably shown on Ni(100) surfaces [29]. This also reduces the surface atomic mean square displacements [25].

The effect is still more important near surface steps. If, due to dangling bonds, surface localized phonons like Rayleigh waves appear below bulk phonon frequencies then the step edge localized phonons have been measured on Pt(111) step surface above the bulk frequencies [30]. The atoms along the step edge have fewer neighbors than the surface ones. One then can expect a relaxation near the step edge to be larger than the bulk one. The effect on the force constants between atoms close to the step edge is also larger [31]. As pR_0 is close to 9, a 10% contraction increases the force constants which varies as $\exp(-pR_0 \delta R/R_0)$ by a factor 2 or 3. Below the step edge, some atoms have kept all their nearest neighbors but due to step contraction, some of the corresponding force constants are much larger than the bulk ones. Then this leads to a step localized phonon with a frequency higher than the bulk ones.

3. Transition metal surfaces

Even if these results that are obtained by a second moment approximation agree quite well with the experimental measurements, it seemed necessary to improve the band structure model, at least to get a more exact surface density of states. This had been possible with two new methods which have been developed and applied first to surfaces and later to aperiodic systems. These techniques have also been widely used since for bulk or interface properties. Both use the Green function method as a starting point.

3.1. The recursion method [32]

Recursion can be used in a three-dimensional crystal without any translation symmetry. At the origin, it is based on the properties of the finite linear chain with one s orbital per atom. When a new atom is adsorbed at the end of the chain, the "surface" properties are not modified but simply translated on the new atom. Without self-con-

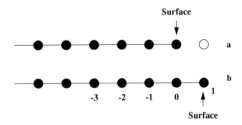

Fig. 3. When an adatom is adsorbed to the end of a linear chain, the surface is simply translated.

sistent potential, the Green function matrix element on the "adatom" (1) can be simply expressed as a function of the Green function matrix element on the surface atom (0) (Fig. 3) before adsorption:

$$G_{11} = \frac{1}{E - \epsilon_0 - \beta^2 G_{00}^0}. \tag{23}$$

But $G_{11} = G_{00}^0$ as we have simply displaced the surface from atom 0 to 1, we get without any more calculation:

$$G_{00}^0 = \frac{E - \epsilon_0 \pm \sqrt{(E - \epsilon_0)^2 - 4\beta^2}}{2\beta^2}. \tag{24}$$

All the Green function matrix elements can be obtained by repeated adsorption of atoms at the end of the linear chain. Two semi-infinite chains can be used to get the results for an infinite linear chain.

For a semi-infinite chain, the calculation is analytic and feasible only because the Hamiltonian is tridiagonal. This is not the case for a three-dimensional crystal in the usual atomic basis. However, one can use the Lanczos method to tridiagonalize the 3D Hamiltonian. The main difference with the linear chain Hamiltonian is that now the diagonal elements a_i are all as different as the non-diagonal ones b_i. Starting from one atomic orbital labelled $|0\rangle$, the new orthogonal basis is obtained by the following recurrence equation:

$$|n + 1\rangle = H |n\rangle - a_n |n\rangle - b_n |n - 1\rangle, \tag{25}$$

with $|-1\rangle = 0$. The coefficients a_n and b_n are

calculated such that the new function $|n+1\rangle$ is orthogonal to $|n\rangle$ and $|n-1\rangle$:

$$a_n = \frac{\langle n | H | n \rangle}{\langle n | n \rangle} \tag{26}$$

$$b_n = \frac{\langle n-1 | H | n \rangle}{\langle n-1 | n-1 \rangle}. \tag{27}$$

In this new basis, the Hamiltonian is tridiagonal:

$$\begin{bmatrix} a_0 & b_1 & 0 & 0 & \cdots \\ b_1 & a_1 & b_2 & 0 & \cdots \\ 0 & b_2 & a_2 & b_3 & 0 \\ 0 & 0 & b_3 & a_3 & b_4 \\ 0 & 0 & 0 & b_4 & \cdots \end{bmatrix}. \tag{28}$$

Using a repeated "adsorption" process at the end of this chain, one gets an expression for G_{00} as a continued fraction. For example, after the third adsorption step, one gets

$$G_{00} = \cfrac{1}{E - a_0 - \cfrac{b_1^2}{E - a_1 - \cfrac{b_2^2}{E - a_2 - b_3^2 g_3(E)}}} \tag{29}$$

The coefficients a_n and b_n converge (if there is no gap in the band) to constant values a_∞ and b_∞ which are related to the band limits E_B and E_T [14]:

$$a_\infty = \frac{E_B + E_T}{2}, \tag{30}$$

$$b_\infty = \frac{E_T - E_B}{4}. \tag{31}$$

So, for large n, $g_n(E)$ converges to the linear chain expression (11) where we put $\epsilon_0 = a_\infty$ and $\beta = b_\infty$. In general, the computation is stopped at a finite level ($n \approx 20\text{--}30$) as the computing times rapidly increases with n. The limit values a_∞ and b_∞ are sometimes not reached but some methods have been developed to extrapolate these coefficients [33] or to find appropriate expressions for $g_\infty(E)$ when the band limits are known [34]. The continued fraction coefficients can also be evalu-

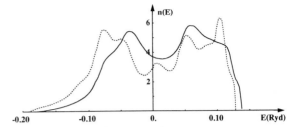

Fig. 4. Ni fcc lattice: comparison of the bulk density of states (dotted line) to the (111) surface one (full line) [35].

ated from the moments of the density of states but the computation is less stable due to the finite number of significant digits in a numerical computation and the nth stage is equivalent to $2n$ exact moments.

Let us recall again that the imaginary part of the Green function is equal to $-\pi$ times the local density of states. For the five d bands model, the (111) surface density of states of nickel [35], which is an fcc lattice, is shown in Fig. 4. The (111) fcc surface is a dense one (there are only 3 missing neighbors for a surface atom). The surface density of states is close to the bulk one. One can just notice that the surface band width is smaller than the bulk one. This is in agreement with the reduction of the number of nearest-neighbors for a surface atom. There is one more dangling bond for a (100) surface atom. One can observe (Fig. 5) a sharp and broad peak in the middle of the d band [35] at an energy close to the d level. It seems to be due to the fact that the surface atoms have fewer neighbors. This peak is far from the Fermi level (which is close to the top of the d band for fcc metals) and it has no effect on properties depending on the density of states at the Fermi level.

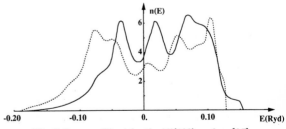

Fig. 5. Same as Fig. 4 for the Ni(100) surface [35].

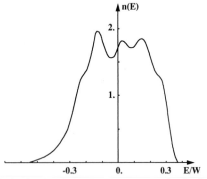

Fig. 6. fcc lattice: local density of states on a (111)×(010) step edge atom [36].

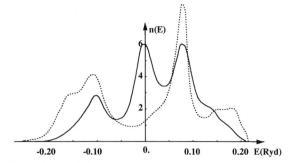

Fig. 8. bcc lattice: comparison of the bulk density of states (dotted line) to the (100) surface one (full line) [35].

This rather small difference between bulk and surface densities of states at the Fermi level for fcc materials was rather disappointing as its change for example could not explain the catalytic properties of materials like Ni or Pt. However, perfect flat surfaces are not good catalysts and the metallic clusters which are used for catalysts surely present many defects like steps or kinks. The local density of states on step edges are shown in Fig. 6 [36]. The band width is still smaller than on a surface atom. The densities for different steps (dense or not) are very close and they even show less structures than the surface ones. This is still more obvious on kink atoms (Fig. 7) where the local density of states is equal to a broad peak. All the peaks have disappeared. This is certainly due to the lack of symmetry near a kink, which suppresses all interferences and

Van Hove singularities. One can show that, for a non-degenerate s band, the local density of states on kink atoms do not depend on the kink type. This is not exactly true for a degenerate band like the d one. However, it is possible to show that the six first moments of the total density of states on a kink atom are constant and independent of the kink type. This explains why the local density of states on kink sites are so close [36].

The case of the bcc (100) surface is more interesting because a large surface peak also appears at the middle of the d band in a region close to the Fermi level where the bulk density of states is small (Fig. 8) [35]. For such a surface, a surface atom has only four neighbors instead of eight for a bulk atom. This reduction of the number of nearest-neighbors is the largest one can expect for a surface atom so the effect of the surface is quite important. These states are localized close to the surface plane (their density in the plane below the surface is quite small). In a pure d band model, they are strictly localized sates as they appear for a given $k_{\parallel}$ in a band gap. They become resonant when one takes into account the s–d interaction. For metals like vanadium or chromium whose Fermi level is localized in this energy region, this could lead to surface magnetism.

3.2. Transition metal surface magnetism

This is certainly the most striking effect of a surface on the electronic structure. Moreover, this has been the starting point of research on metallic magnetic superlattices.

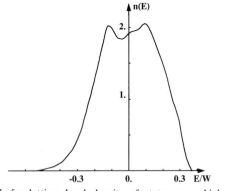

Fig. 7. fcc lattice: local density of states on a kink in a (111)×(010) step [36].

For a long time, surface magnetism was confused with a purely two-dimensional phenomenon. It had been shown that two-dimensional magnetism could not exist [37] and "dead layers" (non-magnetic) had been observed at the surface of ferromagnetic iron [38]. However, due to anisotropy, one can get two-dimensional magnetism. Moreover, the surface plane is never isolated and is always bound to a substrate.

In fact, the occurrence of a large surface peak at the Fermi level leads to what has been called "live layers" (magnetic layers at the surface of non-magnetic materials). The first calculation was done in 1977 for the (100) surface of chromium [39–40]. Bulk chromium is antiferromagnetic with an almost commensurate magnetic moment along the (100) direction. Due to the oscillating magnetic moment, the resultant moment in the bulk is equal to zero. However, the surface region is a good candidate to locally satisfy the Stoner criterion $(Un(E_F) > 1)$ for ferromagnetism. The exact condition involves bulk and surface susceptibilities [41] and a large magnetic moment (2.8 bohr magnetons instead of 0.6 in the bulk) has been calculated in the surface plane. This magnetic moment oscillates when one goes inside the crystal and the amplitude decreases to reach the bulk value after a few layers. A non-zero net magnetic moment is then expected near the surface. This has not been observed. This discrepancy with experiment could be due to steps. If there are some steps at the surface, the surface plane is no longer flat and each patch between two steps can be up or down according to the oscillation of the magnetic moment.

However, two surface peaks have been observed by spin non-resolved photoemission [42]. Each peak seems to correspond to a different spin. The energy difference between the two peaks decreases with temperature to disappear above a transition temperature close to 900 K.

Vanadium (which is a bulk paramagnetic metal) seemed to be also a good candidate for surface live layers. More sophisticated calculations made notably in Art Freeman's group seem to show that this is not the case even if vanadium is close to surface magnetic instability [43]. Such calculations have been applied to many other

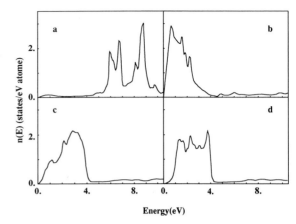

Fig. 9. Local densities of states for adsorption of a Cr layer on an Ag substrate: Cr adatom (a) and Ag atoms in the surface plane (b) and in the planes below the surface (c) and (d).

situations. They show that the surface magnetic moments of magnetic metals like iron and nickel are slightly increased near the surface [44–46]. It has also been shown that for Cr the largest magnetic moment is obtained for an isolated plane. From an experimental point of view, this could be difficult to realize. However, this result can be applied, for example, for a chromium layer adsorbed on a noble metal surface. As the charge transfer between the chromium plane and the substrate is small, the Fermi level is close to the middle of the chromium d band and also lies in the s band well above the noble metal d band. The situation is schematized in Fig. 9. Starting from an isolated chromium plane, one can switch on the coupling with the substrate. Due to the difference between the d bands, the effect of the coupling is small. Moreover, it still tends to reduce the width of the chromium d band as the interaction is larger with the states close to the bottom of the band than with the states near the top [47]. So the resulting magnetic moment in such a layer is close to the value calculated for an isolated plane. If the (100) vanadium surface is not magnetic such as bulk vanadium, its interface with Ag could be magnetic [43]. This seems to be due to the reduced lattice constant of a vanadium layer adsorbed on a silver substrate.

Metallic chromium–iron superlattices have been, for example, grown these last years. They show a large magnetoresistance [48] which can be used in magnetic recording. This is attributed to the interlayer antiferromagnetic coupling between Fe layers across the Cr layer. There are two problems to grow such layers: (i) the materials must be lattice matched to avoid strain in the deposited layers; (ii) in principle, if it is possible to grow a layer of material 1 on material 2 if its surface tension γ_{S1} is less than γ_{S2}, the inverse is not possible (one can neglect the interface tension between the two materials which is generally small). However, in certain situations, a metallic superlattice can be grown, which can be in a metastable situation. Moreover, the relation between the material surface tensions is valid only for thick layers as a thin deposited layer can have a surface tension which is slightly different from the bulk one.

4. Conclusion

This is not an exhaustive list of all the research which has been done to study the electronic structure of transition metal surfaces using the tight-binding method. But at least, one can find here some of the main milestones for these 25 years. All these results have been confirmed later on when first-principles calculations of the surface electronic structure became available. I hope that each one will realize that much progress has been done since the early years. This has also been possible thanks to the computer power which has been increased by several orders of magnitude. A lot remains to do, for example, to understand chemisorption or chemical reactions near surfaces for example. The interest in this research field has not decreased. To be convinced, one can look to the attendance of big surface science conferences. I am sure that the next 25 years will certainly be as exciting as the last ones.

References

[1] J. Friedel, in: The Physics of Metals 1. Electrons, Ed. J.M. Ziman (Cambridge University Press, Cambridge, 1969).

[2] W.A. Harrison, in: Electronic Structure and the Properties of Solids: The Physics of the Chemical Bond (Freeman, New York, 1980).

[3] G. Allan and P. Lenglart, Surf. Sci. 15 (1969) 101.

[4] G. Allan, Ann. Phys. (Paris) 5 (1970) 169.

[5] D. Kalkstein and P. Soven, Surf. Sci. 26 (1971) 85.

[6] R.A. Brown, Phys. Rev. 156 (1967) 889.

[7] G. Allan and P. Lenglart, Surf. Sci. 30 (1972) 641.

[8] J. Friedel, Nuovo Cimento, Suppl. 7 (1958) 287.

[9] G. Toulouse, Solid State Commun. 4 (1966) 593.

[10] M. Lannoo and G. Allan, J. Phys. Chem. Solids 32 (1971) 637.

[11] F. Cyrot-Lackmann, Adv. Phys. 16 (1967) 393; J. Phys. Chem. Solids 29 (1968) 1235.

[12] F. Cyrot-Lackmann, Surf. Sci. 15 (1969) 535.

[13] G. Allan, Solid State Commun. 19 (1976) 1019.

[14] J.P. Gaspard and F. Cyrot-Lackmann, Phys. Rev. B 6 (1973) 3077.

[15] G. Allan, in: Handbook of Surfaces and Interfaces, Ed. L. Dobrzynski (Garland STPM, NY, 1978).

[16] Tran Minh Duc, C. Guillot, Y. Lassailly, J. Lecante, Y. Jugnet and J.C. Vedrine, Phys. Rev. Lett. 43 (1979) 789; J.F. van der Veen, F.J. Himpsel and D.E. Eastman, Phys. Rev. Lett. 44 (1980) 189; P.H. Citrin, G.K. Wertheim and Y. Baer, Phys. Rev. Lett. 41 (1978) 1425.

[17] For a review see D. Spanjaard, C. Guillot, M.C. Desjonqueres, G. Treglia and J. Lecante, Surf. Sci. Rep. 5 (1985) 1.

[18] M.C. Desjonquères and F. Cyrot-Lackmann, J. Phys. (Paris) 36 (1975) L45.

[19] G. Allan and M. Lannoo, Surf. Sci. 40 (1973) 375.

[20] G. Allan and M. Lannoo, Phys. Status Solidi B 81 (1977) 681.

[21] F. Ducastelle, J. Phys. (Paris) 31 (1970) 1055.

[22] M.W. Finnis and V. Heine, J. Phys. F (Met. Phys.) 4 (1974) L37.

[23] A. Blandin, Nobel Symp. 24 (1974) 194.

[24] J. Friedel, Ann. Phys. (Paris) 1 (1976) 257.

[25] J.W. Frenken, F. van der Veen and G. Allan, Phys. Rev. Lett. 51 (1983) 1876.

[26] J.S. Luo and B. Legrand, Phys. Rev. B 38 (1988) 1728.

[27] G. Allan and M. Lannoo, Phys. Rev. B 37 (1988) 2678; in: The structure of Surfaces II, Eds. J.F. van der Veen and M.A. Van Hove (Springer, Berlin, 1988).

[28] G. Allan and J.N. Decarpigny, Phys. Lett. A 65 (1978) 143.

[29] S. Lehwald, J.M. Szeftel, H. Ibach, T.S. Rahman and D.L. Mills, Phys. Rev. Lett. 50 (1983) 518.

[30] H. Ibach and D. Bruchmann, Phys. Rev. Lett. 41 (1978) 958.

[31] G. Allan, Surf. Sci. 85 (1979) 37.

[32] R. Haydock, V. Heine and M.J. Kelly, J. Phys. C (Solid State Phys.) 5 (1972) 2845.

[33] G. Allan, J. Phys. C (Solid State Phys.) 17 (1984) 3945.

[34] P. Turchi, F. Ducastelle and G. Treglia, J. Phys. C (Solid State Phys.) 15 (1982) 1891.

[35] M.C. Desjonquères and F. Cyrot-Lackmann, Surf. Sci. 50 (1975) 257.

[36] G. Allan, Surf. Sci. 115 (1982) 335.

[37] R.E. Peierls, Ann. Inst. Henri Poincaré 5 (1935) 177; L.D. Landau, Phys. Z. Sowjetunion 11 (1937) 26.

[38] L.N. Liebermann, D.R. Fredkin and H.B. Shore, Phys. Rev. Lett. 22 (1969) 539; L.N. Liebermann, J. Clinton, D.M. Edwards and J. Mathon, Phys. Rev. Lett. 25 (1970) 232.

[39] G. Allan, Proc. 3rd Int. Conf. on Solid Surfaces (Vienna, 1977) Eds. R. Dobrozemsky, F. Rüdenauer, F.P. Viehböck and A. Breth.

[40] G. Allan, Surf. Sci. 74 (1978) 79.

[41] G. Allan, Phys. Rev. B 19 (1979) 4774.

[42] L.E. Klebanoff, S.W. Robey, G. Liu and D.A. Shirley, Phys. Rev. B 30 (1984) 1048.

[43] See for example A.J. Freeman and C.L. Fu, in: Magnetism Properties of Low-Dimensional Systems, Eds. L.M. Falicov and J.L. Morán-López (Springer, Berlin, 1986).

[44] U. Gradmann, J. Magn. Magn. Mater. 100 (1991) 481.

[45] A.J. Freeman and R. Wu, J. Magn. Magn. Mater. 100 (1991) 497.

[46] R. Wu and A.J. Freeman, Phys. Rev. B 47 (1993) 3904.

[47] G. Allan, Phys. Rev. B 44 (1991) 13641.

[48] M.N. Baibich, J.M. Broto, A. Fert, F. Nguyen Van Dau, F. Petroff, P. Etienne, G. Creuzet, A. Friederich and J. Chazelas, Phys. Rev. Lett. 61 (1988) 2472.

Surface Science 299/300 (1994) 332–345
North-Holland

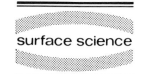

surface science

Surface Green function matching

F. García-Moliner * and V.R. Velasco

Instituto de Ciencia de Materiales, CSIC, Serrano 123, 28006 Madrid, Spain

Received 12 April 1993; accepted for publication 13 May 1993

The evolution of the surface Green function matching (SGFM) method is reviewed in relation to the experimental and computational developments which influenced its evolution or changed its scope. Starting with electronic surface/interface states, the theory is extended to arbitrary elementary excitations. The other extension of the formalism is from continuous to discrete media. Finally, the extension from single to multiple interfaces provides a way of dealing with the new layered systems, e.g., quantum wells and superlattices.

1. Introduction: early stages

On studying the modification of the spectrum of eigenstates of a bulk medium due to the surface we may be interested in any kind of "state" or elementary excitation, such as electronic states, elastic or piezoelectric or magnetoelastic waves, plasmons or electromagnetic waves, phonons, spin waves and so on. Green functions can be used in different ways and the various problems can be separately studied in a seemingly disconnected manner. But it is also possible to build up a systematic Green function approach in which there is a running theme linking up the different specific cases with a conceptual unity throughout.

The method of surface Green function matching (SGFM) provides such a possibility. This paper presents the evolution and development of the SGFM approach to surface states, the features it has in common with some other developments in Green function formulations and its relationship to some advances in experimental or theoretical techniques which influenced its evolution or changed its scope.

In the late sixties interest in electronic surface states was gathering momentum. Green functions had been used in matrix form, mostly to study general abstract issues [1] and often based on elementary aspects and simple models [2]. Wavefunction matching appeared to be in a more advanced stage in practice and had brought to the fore the importance of the analytic continuation of the bulk band structure to complex wavevectors [3]. Its usefulness had been borne out by an early application to the study of the metal/semiconductor junction [4]. The SGFM method started with the study of electronic surface states. The idea was to express the matching conditions in terms of Green functions. The first formulation [5] used an involved pseudopotential kind of argument and assumed a specularly symmetric surface. In spite of these shortcomings it soon proved a fruitful notion, as it demonstrated the fact that $G(E)$, for all energies E, contained the full information included in the analytical continuation of the band structure to complex momenta and it was thus possible to derive the key features of Shockley surface states [6] and resonances at the metal/semiconductor interface [7] in a simple straightforward manner. These early applications also demonstrated some attrac-

* Corresponding author.

tive features of the Green functions, as one can establish general considerations like conservation rules which stem directly from general key analytical features of G and are fairly model independent. Moreover, it soon became clear that the general principles involved were not restricted to the electronic states, as was demonstrated by an early application to surface plasmons [8]. The method was also soon applied to the two-surface problem [9] in a rather elementary way which was eventually superseeded by other developments to be commented later.

It soon dawned on several authors [10–13] that the original formulation of SGFM [5] was unnecessarily opaque and restrictive and the secular equation was derived in more direct and transparent ways which did not require specular symmetry. It was also stressed [11,12] that the SGFM idea is equally applicable to surfaces – or interfaces – of arbitrary geometry and this was demonstrated with an early introduction to a spherical well model of an impurity [11]. This work had an interesting follow-up in applications to alloy theory [14] which bear on significant aspects of the electronic structure from the point of view of the local environment [15] and also demonstrate how G is ideally suited to study the local density of states (LDOS).

Given two media with bulk Green functions G_1 and G_2 matched at a surface the SGFM analysis proceeds in two steps [12]: (i) The full G_s of the entire 3D matched system is written in terms of G_1, G_2 and $\mathscr{G}_s$, which is the surface projection of G_s, so this is known in full, provided just its surface projection is known. (ii) The *matching formula* for $\mathscr{G}_s$ is obtained from the matching boundary conditions. Its inverse, $\mathscr{G}_s^{-1}$ is the *secular matrix* and the roots of its determinant yield the surface or interface state eigenvalues. The results of step (i) are quite general, independent of the physical problem under study and they constitute an extension of standard scattering theory [12,16,17] in which the scattering events are classified in reflection and transmission; both amplitudes can be obtained from $\mathscr{G}_s$. The matching formula – step (ii) – has a general form [16,17] but its explicit expression differs in each specific case. From G_s we obtain all the

spectral functions of interest by applying the standard formula

$$N_s(E, r) = -\frac{1}{\pi} \operatorname{Im} \lim_{\epsilon \to \infty} G_s(E; r, r)$$

in its various forms. For planar interfaces, after 2D Fourier transform, this yields the κ-resolved LDOS $N_s(E, \kappa, z)$. Its surface projection $\mathscr{N}_s(E, \kappa)$ can be used as an alternative to the secular determinant, since a matching eigenvalue corresponds to a peak in $\mathscr{N}_s$; formally a δ-function, in practice a very sharp peak, as a small imaginary part is added to the real energy variable. This procedure in fact often proves more practical than seeking roots of the secular determinant. Integration of these spectral functions over κ yields the complete LDOS. Alternatively, integration over z yields the total DOS of a matched system, which can be κ-resolved or integrated. The interest of these formulae is that they take the form of bulk terms plus surface-induced changes. For instance, the surface matching term in the total DOS is [12]

$$\Delta\mathscr{N}_{sm} = -\frac{1}{\pi} \lim_{\epsilon \to \infty} \frac{d \arg \det \left| \mathscr{G}_s^{-1}(\kappa, E + i\epsilon) \right|}{dE},$$

which can again be κ-resolved or integrated. This is not the complete surface effect, as there also is a *hard-wall* term [12,16,17] due to the abrupt termination of the bulk media at the surface, but it is the most interesting one and it displays another interesting connection with scattering theory, showing that the phase function, $-\arg \det |\mathscr{G}_s|^{-1}$, is a generalised phase shift [16–19].

While these and other concepts related to SGFM were elaborated in the course of time, numerous applications were started since the early period, often based on the simple models then in use, which served to clarify several elementary issues [20]. Thus, studies involving the LDOS helped to establish the role of the useful concept of charge neutrality in many surface and interface problems [7,20], as well as in alloy theory [14], and in the study of surface atomic displacements and the electronic response to them [21]. Other studies related to surface atomic and electronic motions were carried out in the early period [22,23] and the relationship of $\mathscr{G}_s$ to the reflection amplitude was used to obtain basic features of LEED which depend on analytical features of the bulk G and not on details of the potential [24]. Similar considerations explained rather simply the disappearance of Si surface states upon oxidation of the surface [25]. Several surface state calculations were performed, some still using rather elementary models [26] and others based on less elementary, although still relatively simple, pseudopotential models [27]. These studies were useful in sorting out some basic features related to surface states and ionicity [28] and the limitations of the abrupt barrier model [29]. Optical absorption data due to transitions from occupied to unoccupied surface states [30] provided very timely and significant information: (i) there was evidence of a gap in the surface bands and (ii) the type of Van Hove singularities of the joint DOS was unmistakably characteristic of a spectrum of states with a 2D quantum number. These features were very suitable for the kind of SGFM calculations one could do at the time [31] as they stem from general analytical features of G and are thus to a fair extent model independent [32]. Other interesting applications to different electronic problems which took place in the early years included the study of simple molecules starting with Wigner–Seitz atomic potentials and then using SGFM to match the atoms [33]; some aspects of the surface many-body problem, studied to various degrees of approximation [34]; the study of virtual states and resonances at free surfaces or in thin films [35] and a formulation of tunneling theory avoiding the use of a tunneling Hamiltonian [36].

2. Practical Green function calculations for a Schrödinger equation

The practical evaluation of the Green functions and their normal derivatives, needed for the matching formula *when differential calculus is used* is quite straightforward for many physical problems which will be discussed later. The same holds for electronic states starting from a Schrödinger equation when the model is fairly

simple, as in many of the early applications and for a class of potentials amenable to analytical treatment [37]. However, the numerical evaluation of the Green functions and their derivatives is no trivial task when an elaborate model is used. Efforts to use fairly realistic models started very early, as for instance in a rather reliable calculation of surface states for CsBr [38]. Cu surfaces were also studied by starting from a careful description of the bulk band structure [39] and it was thus possible to analyse successfully directional photoemission data. A very complete study of interface states and LDOS at twin faults in Cu was also carried out on a similar basis [40]. Perhaps the most articulate effort to do SGFM calculations based on a very realistic description of the bulk band structure took place a few years later [41] but, while SGFM went through many other developments, its practical use for electronic state calculations involving differential calculus tended to dwindle for a while.

Two interesting theoretical developments took place some time later. One was the formulation of the *embedding method*, first used for a rather careful study of the Al(001) surface [19]. The method of embedding is conceptually related to SGFM but uses a hybrid approach in which a plane is defined beyond which one has as a substrate a perfect semi-infinite crystal. On the other side of the matching plane is the "surface region", which includes the smooth surface potential profile. This region is numerically studied separately and selfconsistently and then it is joined to the substrate which is essentially represented by the surface projection $\mathscr{G}$ of the bulk crystal G. This in turn is calculated, via the connection between $\mathscr{G}$ and the reflectance, from the LEED scattering amplitude. This method has been successfully applied to study the electronic structure of metal surfaces. A review of most of the these applications can be found in Ref. [41].

Another interesting development was the combination of Green functions and the *full transfer matrix M* [42], which transfers amplitudes *and* derivatives. This is suitable and useful to study second-order differential equations and can be evaluated from standard algorithms for the integration of differential systems. The first successful step was the study of electronic states in parabolic quantum well [43]. By relating the elements of G intervening in the secular equation to the corresponding elements of M across the parabolic potential it was possible to calculate the eigenstates of the well with great ease. The full connection between G and M came soon thereafter [44] and thus a new way to do SGFM calculations was started. This has been used for various applications, reviewed in Ref. [45], as well as for selfconsistent calculations [46] and continues to be applied. This is a practical way of using the SGFM analysis for the theoretical formulation combined with the calculation of the intervening Green functions from the numerical evaluation of the corresponding transfer matrices. It has the considerable advantage that the numerical evaluation of the normal derivatives of G is altogether avoided and exactly transformed into matrix algebra, done with submatrices of M [44,45]. The matched wavefunctions can also be obtained from the same calculation [44,17]. The practical usefulness of this approach rests on the efficiency of the numerical algorithms one uses for the integration of the differential system. In the first practical applications [47] Runge–Kutta methods were employed, with a *predictor–corrector* algorithm. Lately it has been established (current research) that multistep techniques which allow for backward differentiation methods are more practical. We want to remark that the numerical integration of differential systems is a problem to which a vast amount of effort has been devoted and which still presents inherent numerical difficulties, constituting in fact a substantial subject of active research [48]. The future of the combination of SGFM and full transfer matrix M depends in practice on further "instrumental" developments in this field of computer science. A useful auxiliary device which helps to simplify the calculations is based on the notion of *extended pseudomedia*, which was initially stressed in Ref. [12], implicitly used in Refs. [19,41], and explicitly discussed and demonstrated in Refs. [44,17]. The practical usefulness of this concept goes beyond the simple notion that one has arbitrary disposable normal derivatives of the constituent Green functions before matching, as is demonstrated by studies of surface electrodynamics [16] in which Green functions are used in a way which is very similar to SGFM in spirit but different in technique.

3. A wide range of surface problems

It was mentioned above that an early study of surface plasmons [8] demonstrated that the formal framework of the SGFM analysis is quite generally applicable. This was further expounded in Refs. [49,50] and demonstrated by a study of surface polaritons [51]. Not only surface elastic and electromagnetic waves, but also viscoelastic surface waves at fluid surfaces and interfaces [50,52] can be readily studied by SGFM. An interesting development had been meanwhile taking place independently [53] in which the Green functions are derived via the fluctuation–dissipation theorem and a method to calculate surface Green functions is thus built up. These appear then related to physical concepts directly relevant to surface spectroscopy, such as the *power function*, essentially a measure of the surface scattering power which, on the other hand, and by virtue of the same fluctuation–dissipation theorem, is related to the imaginary part of the corresponding response function, i.e., the surface projection $\mathscr{G}_s$. But its imaginary part, as was stressed in Section 1, is just the surface LDOS, which is the spectral function of interest for the study of the scattering of external probes, as in the theory of Brillouin scattering [54,55]. Thus there is an interesting conceptual link between SGFM and the method initiated in Ref. [53], which has been applied to many studies, mostly of liquid surfaces, reviewed in Ref. [54]. Indeed basic formulae for spectral functions were independently derived by both methods [50,52,54] and of course with full agreement. SGFM has also been applied to the study of capillary waves in electrically charged liquids [56] and to the clarification of an existing controversy concerning spectral functions for liquid surfaces [57]. It is interesting that the *form* of the matching formula is the same for the various physical problems and this admits a general interpretation in terms of surface impedances [17] which gives it a formal appeal.

One of the fields where SGFM has been mostly applied is that of elastic, piezoelectric and magnetoelastic waves at surfaces, interfaces and superlattices, to be discussed later. An early application to surface elastic waves [58] provided a simple surface Debye model of surface thermodynamics. Surface effects in the theory of surface waves can be attributed to special surface rheological coefficients and conveniently described as a change in the matching boundary conditions which can be very simply introduced as an appropriate modification of the SGFM matching equation. This study was made for solids and fluids [59]. For a time the most meaningful experiments concerning surface vibrational modes gave long-wave information – e.g., ultrasonic diffraction [60] and Brillouin scattering [61]. Although some interesting instrumental developments had achieved sources capable of producing sufficiently intense and monochromatic He atom beams [62], information on surface phonons was still to come and the practical applications of SGFM to surface wave problems looked more at long waves. The study of Brillouin scattering off polycrystalline Al surfaces demonstrates the usefulness of the spectral function $\mathrm{Im}\ \mathscr{G}_s(\kappa, \omega)$ to analyse the full surface spectrum, including bulk threshold effects [55]. This turned out to be coincidental with a very similar study based on the fluctuation–dissipation approach [63]. The detailed study of all the spectral information contained in $\mathrm{Im}\ \mathscr{G}_s(\kappa, \omega)$ proved very useful in the analysis of anisotropic crystal surfaces. Experimental data on the main surface peak presented at the time some apparently puzzling aspects. An SGFM study [64] provided an explanation in terms of a satellite structure in which the spectral strength is differently distributed, depending on the propagation direction – vector κ –, between the generalised Rayleigh mode and a resonance appearing between the two transverse bulk thresholds, often termed a pseudo-surface-wave. This satellite structure has been repeatedly corroborated by Brillouin scattering experiments for a variety of anisotropic crystal surfaces including Ge, GaAs, InSb, PbS and some high-T_c superconductors [65].

Piezoelectric surface and interface waves have

also been studied by SGFM [66]. The coupling of the vibrational amplitude u and the electrostatic potential φ adds another coupled differential equation and with it some extra complication. However, it proved rather useful to define a tetrafield (u, φ) with corresponding total Green function. This is then the resolvent of a 4×4 differential matrix and, hence, also a 4×4 matrix. However, it can still be analytically evaluated and a compact SGFM analysis can be easily carried out [67]. Grounded piezoelectric surfaces can then be studied by a specification in the matching boundary condition resulting in a simpler matching formula for $\mathscr{G}_s$. The tetrafield formulation has been used to study surface thermodynamics of piezoelectrics [68]. Other applications of SGFM to similar problems, mostly for superlattices, are reviewed in Refs. [17,45]. Magnetoelastic surface and interface waves can be similarly studied. With this it was possible to give an exact account of finite frequency effects in rare earth compounds [69]. The coupling of u and φ, taking place in piezoelectrics, has a parallel [70] in the long-wave phenomenological model of polar optical modes in ionic semiconductors, where u is the relative displacement of anions and cations in the same unit cell. Within the long-wave model the simultaneous satisfaction of all matching boundary conditions has originated some confusion in fairly recent literature but all these matching conditions can be handled without conflict by means of a compact SGFM analysis [71] which follows essentially the same methodology as that employed in the study of piezoelectrics [67], and works rather well in practice, giving a correct account of the observed symmetry pattern of the confined and interface normal modes in GaAs-based quantum wells [72]. A simpler 2×2 differential matrix appears on using Bologioubov's model of elementary excitations in superconductors. The SGFM method has been used to study tunneling and proximity effects for normal metal–superconductor interfaces based on this model [73]. From $\mathrm{Im}\ G_s(E; \kappa; z, z' = z)$, for instance, one can have a direct picture of the distribution in real space of the strength of the elementary excitations induced on the normal side by the proximity effect. This demonstrates the usefulness of the spectral functions one obtains directly from the Green functions.

Alternative Green function approaches were developed roundabout the same period and often used to study elastic surface and interface elastic waves. In one of these G_s is obtained by direct integration of the differential system with the surface boundary conditions [74]. In another one [75] the differential system is written down for z-dependent stiffness coefficients. Consider the stress tensor of elements

$$\tau_{ij} = C_{ijkl} \frac{\partial u_l}{\partial x_k}$$

and let x_3 be z. Then $C_{ijkl} \equiv \theta(x_3)C_{ijkl}$ describes a semiinfinite medium terminated at $x_3 = 0$. The components of the force per unit volume, given by the divergence of the stress tensor, then contain a surface term with $\delta(x_3)$ and the Green function describes the semiinfinite medium. There have been many applications of this method, reviewed in Ref. [76].

4. SGFM for discrete systems

We define as discrete systems problems like lattice dynamics or electronic structure in a tight binding model with a discrete *Hamiltonian matrix*, where one uses finite differences instead of differential calculus. The choice between a tight binding model and a Schrödinger equation to study electronic states is optional, but the experimental developments based on inelastic atom scattering and electron energy loss spectroscopy [77] provided new and interesting direct information on surface phonons which can only be studied by lattice dynamics. Although other discrete Green function methods, which will be commented later, had been developed for surface problems, it seemed desirable to extend SGFM to discrete systems. To be able to obtain the phonon G_s in terms of the bulk G_1 and G_2 seemed an attractive prospect, but how to do it was not quite so obvious. SGFM focuses, by definition, on matching but, how does one write down a "*matching formula*" for $\mathcal{G}_s$ in terms of finite

difference calculus? It took some time to realise that all that is needed for discrete systems is to express the interactions in the surface or interface region and this is achieved mathematically by projecting the equations of motion onto this region. The use of identities resulting from the general form of the SGFM analysis [17] then yields the formula for $\mathcal{G}_s$. It is important to note that in discrete systems one starts from matrices and matrix equations and then defines G as the inverse of a matrix which is, say, $E - H$ for electrons or $M\omega^2 - \Phi$ for phonons, Φ being a dynamical matrix of force constants. This is a very important observation which will be commented later. The first version of SGFM for discrete systems was cast in a language specifically designed for lattice dynamics [78] as the attention was mainly drawn by further experimental information on surface phonons, mostly from inelastic atom scattering, which was by then becoming more abundant [79]. Surface and interface phonons had also been studied by means of the *invariant Green function* (IGF) method [80], where the surface is treated as a perturbation of the bulk; a perturbation theoretic argument is given in which the corresponding t matrix is used, and from it the surface projection – there denoted $\tilde{g}$ – is obtained, whence the surface eigenvalues and LDOS. The analysis is interestingly related to invariance arguments and rests on the observation, particularly useful for crystals with long-range forces, that the 2D Fourier components of the surface perturbation matrix – and, hence, also of the corresponding t matrix – decay with distance from the surface much more rapidly than the interatomic force constants do with atom–atom distance. It is obvious that both methods are identical in spirit and rest on two key concepts, namely: (i) the focus is on the surface projection of the Green function and (ii) the surface perturbation is treated exactly in the scattering theoretic formulation. In fact the $\tilde{g}$ of the IGF method is identical to the $\mathcal{G}_s$ of the SGFM formalism, as was later explicitly proved in applications to study inelastic atom scattering data for ionic crystal surfaces [81]. The difference between both methods is in the technical way in which $\tilde{g}$, or $\mathcal{G}_s$, is obtained and this opens an interesting possibility to study surface and interface lattice dynamics in practice. For instance, for ionic crystals one could use the IGF method to calculate $\tilde{g}$: one has $\mathscr{G}_s$ and can then switch to the SGFM analysis where the rest of the physical information contained in the full G_s is obtained from $\mathscr{G}_s$.

A more generally suitable formulation of SGFM was later developed [82] and successfully applied to electronic surface structure calculations for semiconductors and metals [83] with empirical tight binding (ETB) Hamiltonians. This method proves quite flexible and useful in practice. This is demonstrated significantly by the ease with which one can study disordered interfaces. Theoretical arguments and experimental evidence [84] indicate that one should expect stoichiometric disorder at the Ge–GaAs interface. This can be modelled by a virtual crystal description of the disordered material in the interface domain which amounts simply to a perturbation of the ETB Hamiltonian in this region. This in turn results in a simple modification of the matching formula from which a straightforward SGFM calculation shifts the local position of the band edges relative to the Fermi level in agreement with experimental evidence [85]. Another example is the explanation of significant differences in the local magnetism of the Cr(001) surface and the Cr–GaAs(001) interface, which has also been studied by means of a SGFM-ETB calculation [86].

As in the continuous case, the ultimate practical question is the evaluation of the G elements intervening in the SGFM formulae. With discrete matrices this turns out to be very easy due to the development of extremely useful computational techniques based on a concept of transfer matrix T which transfers *amplitudes only*, and is defined by $G_{n+1,n} = T \cdot G_{n,n}$, where n is the layer index playing the role of the discrete position variable. This definition has a built-in directionality. It only holds when the first index *increases* and is never smaller than the second one, that is to say, it is associated to propagation in the positive direction – to the right. For propagation to the left one defines $G_{n-1,n} = \bar{T} \cdot G_{n,n}$. Similar transfer matrices, S and $\bar{S}$ are defined to act on the

second index and there are relationships involving the four transfer matrices, so it suffices to know how to calculate T. The relationship between discrete and continuous systems, and the corresponding appropriate transfer matrices, is discussed in detail in Ref. [17]. For a proper perspective of the various alternative methods existing in the literature it is convenient to note that in discrete as well as in continuous media the corresponding transfer matrices need not only be auxiliary devices for the calculation of G elements – and derivatives, in the continuous case. In the SGFM method one formulates the surface or interface problem entirely in terms of (bulk) G's of the constituent media and then uses T, or M, as a very convenient auxiliary device for the computation of the intervening G's. However, a surface problem can also be entirely studied in transfer matrix language. It is a matter of optional strategy, in which different combinations of the two extremes are often chosen according to preference. What is very powerful and has greatly facilitated practical progress is the combination of G and T. In its beginnings T was used first to study bulk and surface electronic states [87] and then surface phonons [88]. Studies of disordered linear chains showed that the convergence of the algorithms to calculate T improved significantly with the introduction of renormalisation group ideas [89]. This was later taken up for surface electronic states in full 3D semiinfinite crystals and developed into very fastly convergent algorithms [90]. In the many applications of SGFM to discrete systems – Refs. [17,32,83,85,86] and others to be discussed later – this is the current state of the art. As a computational method it proves very efficient in practice for all discrete systems [91].

There are several alternative Green function approaches to study surfaces and interfaces for discrete systems, mostly – though not only – for electronic structure. The notion of *atom removal* as a formal way of creating a surface has been used in various applications [92]. The corresponding perturbation is used in a surface projected Dyson equation to obtain from it the surface projection of the Green function of the semiinfinite crystal. This idea has been carried further in a formulation which stresses more explicitly the scattering theoretic viewpoint and which has been used for many rather more elaborate calculations of the electronic structure of semiconductor surfaces and interfaces [93]. The bulk G's appear to be calculated by direct evaluation of the spectral formula. A deeper formal insight into the meaning of this method was given in a general analysis [94], where it was shown that the basic formula for G_s in terms of the bulk G in the atom removal method of Ref. [93] for the ideal surface follows from a purely formal manipulation from the entire space of the bulk crystal to the reduced space of the half crystal. This looks very different from the SGFM formula for G_s and caused at first some confusion [95] which was clarified in Ref. [96] by stressing the basic difference, originally emphasised in Ref. [94], between the Green function defined as the resolvent of a differential operator, which may be later represented in some basis, and the Green function defined as the inverse of the matrix representing $E - H$ from the start. In fact the formal construction – ideal Green function – of Ref. [94], which is the same as the one obtained from the atom removal argument, can be proved to be identically equal to the SGFM result for the ideal surface [17]. This holds only for tight binding Hamiltonian matrices. Although the general analysis holds formally for all discrete systems the case of phonons has some specific features which invalidate the literal application of the atom removal formula to this case [95,17]. The notion of *cleavage perturbation* is of course widely used for all problems and, with due account of the particular details of each case, always involves the notion of reduced space, which has been widely discussed [94,97,98]. In this respect the SGFM analysis differs from all perturbative Green function methods, in which the unperturbed system is the infinite crystal and then the cleavage or atom removal perturbation cuts out the unwanted part and leads to the reduced space of the semiinfinite crystal. In SGFM one only deals with the actual system under study from the start and the entire analysis is in its own space. This concerns only the formal understanding, with an obvious extension to the case of heterostructures. In practice, the formulae for

making calculations are sometimes even identically equal to those derived from other formalisms, as is the case with the invariant Green function method for phonons [80,81] or the ideal atom removal formula for electrons [92–94,17]. The latter can also be conceptually related [99] to the ideas of the embedding potential method [19,41]. Indeed, considering a semiinfinite crystal starting at the layer, or principal layer, $n = 1$ and contained in the half space $n > 0$, in the SGFM version one finds

$$\mathscr{G}_s^{-1} = (E - H)_{11} - H_{12} \cdot T.$$

$(E - H)_{11}$ would be the secular matrix if the layer $n = 1$ stood alone. But this is coupled to the rest: H_{12} measures the strength of this coupling and T transfers this information to the right. Thus $H_{12} \cdot T$ acts as an embedding potential for the surface layer.

Other discrete Green function approaches have been developed for electrons [100] or phonons [74,97] in which the idea is always essentially the same, that is to introduce some cleavage perturbation and then to use a Dyson equation approach to obtain the truncated Green function. Another formalism which is similar in spirit to SGFM can be found in Ref. [101].

In all the discrete Green function methods so far discussed one seeks closed formulae for the desired Green functions. An alternative proposal is to set up an algorithm designed to yield the desired Green function by means of path counting techniques based on a random walk approach [102]. The "other half" of the crystal, that is missing the corresponding paths linking the surface atoms to the missing one, is not counted, thus, the Green function of the semiinfinite crystal results. Sometimes moment techniques and recursion algorithms are designed to yield not the Green function but directly the LDOS, i.e., essentially Im G [103].

A common feature of all these Green function calculations is that one starts from an empirical tight binding (ETB) Hamiltonian, essentially of the Slater–Koster type, although the formal theory is not necessarily restricted to an orthogonal basis [94,104]. The point is that the matrix elements of the ETB Hamiltonian have the nature of empirical parameters normally obtained by some fitting procedure. The question of self-consistency has been faced in this context. The idea as initially proposed [105] focuses on the fact that a charge redistribution near a surface or an interface should modify the Hamiltonian matrix elements in this region. More articulate proposals how to carry on this programme have been made in Refs. [106], which also reviews many applications of this method, and [107].

A very different approach was initiated in Ref. [108]. Although the Green function formalism as such is very similar to that of Ref. [93] the intervening Green functions are calculated from first principles by means of advanced techniques. An essential ingredient of this method – the so called linear muffin tin orbital (LMTO) method – is the use of "tight binding muffin tin orbitals" obtained from the original MTO's by means of an exact transformation [109]. The calculation starts from a first principles band structure, so no fitting procedure is involved. Further elaboration of this method can be found in Ref. [110]. The real significance of this approach is that it constitutes a very powerful way to calculate ab initio the matrix elements of a tight binding Hamiltonian, reaching in this respect an advanced state of the art. Thus having obtained a reliable Hamiltonian matrix one could then use this as the starting point for any chosen discrete Green function method.

5. Matching at more than one interface

A free surface or a heterojunction constitutes essentially the same problem, as the free surface is the interface of the semiinfinite medium with the vacuum. The obvious next step beyond the one-interface problem is that of two interfaces and subsequently multiple interfaces. Extensions of the SGFM analysis to the two-interfaces problem took place at a very early stage [9,12,13,111]. Various approaches were used, but in general the formulations were rather elementary and mostly concentrated on the secular equation. They proved useful in the study of different problems like the disappearance of Si surface states upon oxidation [25]; virtual surface states when an elec-

tric field is applied to a thin semiconductor film [35]; surface effects in elastic surface waves represented by a thin layer on the surface [57]; or piezoelectric overlayers on piezoelectric or non-piezoelectric substrates [65]. A very different approach, presently discussed, laid still in the future. Attempts to extend SGFM to the N-interfaces problem – e.g. a stratified medium or a layered structure – started somewhat later [112]. In one way or another these formulations involved repeated matching, often combined with a Dyson equation approach. For some time the most intensely studied layered structure has been the superlattice, where the Bloch (super) periodicity condition is of great help. For a Kronig–Penney superlattice one can easily write down the eigenfunctions in this way and obtain from them the superlattice Green functions [113], but it is not so easy to carry out this programme for a less elementary model. The issue is best understood by keeping in mind that the matching problem is related to reflection and transmission. Consider, for instance, a structure of the type A–B–C, which includes the overlayer problem (A = vacuum), the quantum well (A = C), or any sandwich-type structure. For propagation starting at some point of B we seem to encounter off-hand the problem of multiple reflection which requires to sum an infinite series. This is much more so for a layered structure and, thus, the direct study of the propagator for a superlattice is apparently beset by the problem of the infinite series for multiple reflection and transmission, as with the multiple images in classical electrostatics.

The SGFM approach to sandwich structures and superlattices took a completely different turn in 1986 both for continuous [114] and for discrete [115] systems, which entirely avoids the summation of series while keeping exact account of reflection and transmission. This approach yields a formalism which can be used to do practical calculations for any model, provided one knows how to calculate the intervening G's. Consider, for instance, a bulk medium perturbed by the creation of a surface. The form of G_s inside the crystal is of the Dyson type: $G_s = G + GRG$, in terms of the unperturbed bulk G and of an object R related to transmission, which can be in turn related to $\mathscr{G}_s$ [12,16,17]. The details do not matter, the point is that G is the unperturbed bulk propagator everywhere on the rhs. Therefore, the effect of the perturbation due to the surface has been summed to infinite order and R is a (reflection) total scattering matrix. A full discussion of this, for reflection and transmission, even when G near the surface or interface differs from the perfect bulk G, is given in Ref. [17]. When this idea is carried over to sandwich structures and superlattices then one obtains the full SGFM analysis, in which matching is effected simultaneously, and in a compact form at two consecutive interfaces, in a way which takes exact account of all reflections and transmissions to infinite order without any series summation. Combined with transfer matrix algorithms when the intervening G's are difficult to calculate otherwise, this provides a very useful computational scheme for practical calculations for quantum wells and superlattices. This method has been applied to many different problems, e.g., Acoustic and Bleustein–Gulyaev modes in piezoelectric superlattices [116]; phonons in periodic intercalation compounds in graphite [117] and in transition metal superlattices [118]; effects of biaxial strains on phonons in thin layer Cu/Ni superlattices [119]; electronic structure of transition metal superlattices [120] and of strained [121] and unstrained [122] semiconductor superlattices. It has also been used to study quantum well problems like long-wave polar optical modes [65] and electro-optical properties of semiconductor quantum wells in an external electric field [123] – see Ref. [124] for a general discussion of the use of Green functions for heterostructures. Many other applications of this technique are currently going on.

6. Back to the single interface

We started with the free surface and ended up with a superlattice. But this is grown on a substrate and further experimental work [125] is focusing interest on the buried substrate/superlattice interface. This is a new single-interface problem which, from the point of view of the

SGFM analysis, involves a significant subtlety related to the concept of the *structural Green function*. This concept was first introduced in the KKR method of the electronic band structure of bulk crystals [126] which, technically, is the first elaborate matching calculation, only that the matching is effected at the atomic muffin tin spheres. A rigorous analysis of the general aspects of the application of the KKR method to surfaces and thin films was given in Ref. [127].

The affinities between the KKR argument and the SGFM analysis were early recognised [128,16], but this only provides a new derivation of the same secular equation. However, the idea proves rather fruitful when it comes to the study of a SL. In fact, the superlattice eigenstates constructed by explicitly using the Bloch superperiodicity condition are described in the *reduced-zone scheme*. The well-known picture of the folded modes is literally a 1D reduced-zone representation of the band structure for fixed momentum parallel to the interfaces (κ) and variable momentum in the perpendicular – growth – direction (q). This implies [17] that the G one obtains from the SGFM analysis is precisely the (1D) structural Green function $G_{SL}(\kappa, q; n, n')$ where the dependence on the eigenvalue is understood and the discrete layer indices (n, n') can be replaced by the continuous position variables (z, z') for the continuous case. Suppose one wants to study the free surface of the superlattice as a bulk medium, or its interface with another medium. The point is that this cannot be done with G_{SL} as it stands. We cannot match at $n = n' = 0$ or ($z = z' = 0$) and still have a q-dependence in the matching formula. We must first eliminate q by "de-structuring" G_{SL}, i.e., by unfolding the picture of the superlattice band structure from the reduced- to the extended-zone scheme, which is achieved mathematically by going over to another Green function, namely [17]

$$G_{\overline{SL}}(\kappa, q; n, n') = \int_{1DBZ} \frac{dq}{2\pi} G_{SL}(\kappa, q; n, n').$$

With this we can now match at $n = n' = 0$ and obtain κ-dependent matching formulae. This programme has just been initiated with an ETB study of the surface electronic structure of AlAs/GaAs superlattices [129]. One can then study with great ease the evolution of the surface properties, as the superlattice terminates at different atomic layers of one or the other material. This work is in progress at the moment of writing. The study of the free surface, besides having a possible interest of its own, constitutes a first step to tune the technique to practical working conditions. With this one can proceed in a straightforward manner to the study of the substrate/superlattice interface. Thus, from the bulk of the superlattice we are back at its surface.

Acknowledgement

A significant fraction of the work here reviewed was done with partial support of the Spanish CICYT through Grant No. MAT91-0738.

References

[1] B.W. Holland, Philos. Mag. 8 (1963) 87.
[2] J. Koutecký, Adv. Chem. Phys. 9 (1965) 85;
M. Tomášek, in: The structure and chemistry of solid surfaces, Ed. G.A. Somorjai (Wiley, New York, 1969) p. 15-1.
[3] V. Heine, Proc. Phys. Soc. 81 (1963) 300.
[4] V. Heine, Phys. Rev. A 138 (1965) 1689.
[5] F. García-Moliner and J. Rubio, J. Phys. C: Solid State Phys. 2 (1969) 1789.
[6] F. García-Moliner, V. Heine and J. Rubio, J. Phys. C: Solid State Phys. 2 (1969) 1797.
[7] F. Yndurain, J. Phys. C: Solid State Phys. 4 (1971) 2849.
[8] F. Flores, F. García-Moliner and J. Rubio, Solid State Commun. 8 (1970) 1065.
[9] F. Flores, F. García-Moliner and J. Rubio, Solid State Commun. 8 (1970) 1069.
[10] B. Velický and I. Bartoš, J. Phys. C: Solid State Phys. 4 (1971) L104.
[11] J.E. Inglesfield, J. Phys. C: Solid State Phys. 4 (1971) L14.
[12] F. García-Moliner and J. Rubio, Proc. R. Soc. London A 324 (1971) 257.
[13] W. Klose, P. Entel and P. Hertel, Z. Phys. 245 (1971) 347.
[14] J.E. Inglesfield, J. Phys. F: Metal Phys. 2 (1972) 63; 878; 1291; 9 (1979) 1551.
[15] V. Heine, Solid State Phys. 35 (1980) 1.
[16] F. García-Moliner and F. Flores, Introduction to the Theory of Solid Surfaces (Cambridge University Press, Cambridge, 1979).
[17] F. García-Moliner and V.R. Velasco, Theory of Single and Multiple Interfaces. The method of Surface Green Function Matching (World Scientific, Singapore, 1992).
[18] L. Dobrzynski, V.R. Velasco and F. García-Moliner, Phys. Rev. B 35 (1987) 5872.
[19] J.E. Inglesfield and G.A. Benesh, Surf. Sci. 200 (1988) 135.
[20] I. Bartoš, M. Hietschold and G. Paasch, Phys. Status Solidi (b) 87 (1978) 111;
V. Marigliano Ramaglia, B. Preziosi and A. Tagliacozzo, Nuovo Cim. 53B (1979) 411.
[21] F. Flores, E. Louis and F. Yndurain, J. Phys. C: Solid State Phys. 6 (1973) L465;
C. Tejedor, E. Louis and F. Flores, Solid State Commun. 15 (1974) 587;
E. Louis, F. Yndurain and F. Flores, Phys. Rev. B 13 (1976) 4408;
C. Tejedor and F. Flores, J. Phys. C: Solid State Phys. 11 (1978) L19;
F. Flores and C. Tejedor, J. Phys. C: Solid State Phys. 12 (1979) 731;
I. Bartoš and J. Koukal, Czech. J. Phys. B 26 (1976) 677.
[22] J.E. Inglesfield, J. Phys. C: Solid State Phys. 11 (1978) L69;
J.E. Inglesfield and A. Tagliacozzo, J. Phys. C: Solid State Phys. 17 (1984) 5227.
[23] G. Iadonisi, Nuovo Cim. 58B (1979) 233.
[24] I. Bartoš and B. Velický, Surf. Sci. 47 (1975) 495.
[25] F. Yndurain and J. Rubio, Phys. Rev. Lett. 26 (1971) 138.
[26] G. Iadonisi and B. Preziosi, Phys. Rev. B 9 (1974) 4178;
G. Iadonisi and B. Preziosi, Nuovo Cim. 27B (1975) 193;
G. Iadonisi, V. Marigliano Ramaglia and G.P. Zuchelli, Surf. Sci. 64 (1977) 141;
I. Bartoš and F. Máca, Phys. Status Solidi (b) 99 (1980) 755.
[27] M. Elices and F. Yndurain, J. Phys. C: Solid State Phys. 5 (1972) L146;
F. Flores and J. Rubio, J. Phys. C: Solid State Phys. 6 (1973) L258;
E. Louis and M. Elices, Phys. Rev. B 12 (1975) 618.
[28] E. Louis and F. Yndurain, Phys. Status Solidi (b) 57 (1973) 175.
[29] F. Flores, E. Louis and J. Rubio, J. Phys. C: Solid State Phys. 5 (1972) 3469;
M. Elices, F. Flores, E. Louis and J. Rubio, J. Phys. C: Solid State Phys. 7 (1974) 3020;
F. Flores, F. García-Moliner, E. Louis and C. Tejedor, J. Phys. C: Solid State Phys. 9 (1976) L429.
[30] G. Chiarotti, S. Nannarone, R. Pastore and G. Chiaradia, Phys. Rev. B 4 (1971) 3398.
[31] F. Yndurain and M. Elices, Surf. Sci. 29 (1972) 540;
V. Bortolani, C. Calandra and A. Sghedoni, Phys. Lett. A 34 (1971) 193.

[32] F. Flores and J. Rubio, Phys. Status Solidi (b) 68 (1975) K17.

[33] J.E. Inglesfield, J. Chem. Phys. 67 (1977) 505; Mol. Phys. 37 (1979) 873; 889.

[34] J.E. Inglesfield and J. Wikborg, J. Phys. C: Solid State Phys. 6 (1972) L158;
J. Wikborg and J.E. Inglesfield, Phys. Scr. 15 (1977) 37;
J.P. Muscat and G. Allan, J. Phys. F: Metal Phys. 7 (1977) 999;
I Bartoš, M. Kollar and G. Paasch, Phys. Status Solidi (b) 115 (1983) 437.

[35] S.G. Davison, H. Ueba and R.J. Jerrard, J. Phys. C: Solid State Phys. 13 (1980) 1351;
R.J. Jerrard, H. Ueba and S.G. Davison, Phys. Status Solidi (b) 103 (1981) 353;
I. Bartoš and S.G. Davison, Solid State Commun. 56 (1985) 69.

[36] T.E. Feuchtwang, Phys. Rev. B 13 (1976) 517.

[37] M.L. Glasser, Surf. Sci. 64 (1977) 141.

[38] C.M. Bertoni and C. Calandra, Phys. Status Solidi (b) 50 (1972) 527.

[39] J.A. Vergés and E. Louis, Solid State Commun. 22 (1977) 663.

[40] E. Louis and J.A. Vergés, J. Phys. F: Metal Phys. 10 (1980) 207.

[41] J.E. Inglesfield, Prog. Surf. Sci. 25 (1987) 57.

[42] M. Mora, R. Pérez-Alvarez and Ch.B. Sommers, J. Phys. (Paris) 46 (1985) 1021.

[43] R. Pérez-Alvarez, H. Rodriguez-Coppola, V.R. Velasco and F. García-Moliner, J. Phys. C: Solid State Phys. 21 (1988) 1789.

[44] V.R. Velasco, F. García-Moliner, H. Rodriguez-Coppola and R. Pérez-Alvarez, Phys. Scr. 41 (1989) 375;
F. García-Moliner, R. Pérez-Alvarez, H. Rodriguez-Coppola and V.R. Velasco, J. Phys. A: Math. Gen. 23 (1990) 1405.

[45] F. García-Moliner and V.R. Velasco, Phys. Rep. 200 (1991) 83.

[46] L. Chico, W. Jaskólski and F. García-Moliner, Phys. Scr. 47 (1993) 284.

[47] W. Jaskólski, V.R. Velasco and F. García-Moliner, Phys. Scr. 42 (1990) 495; 43 (1991) 337.

[48] L.F. Shampine and M.K. Gordon, Computer Solution of Ordinary Differential Equations. The Initial Value Problem (Freeman, San Francisco, 1975);
D. Zwillinger, Handbook of Differential Equations, 2nd ed. (Academic Press, San Diego, 1992).

[49] F. Flores, Nuovo Cim. 14B (1973) 1.

[50] F. García-Moliner, Ann. Phys. (Paris) 2 (1977) 179.

[51] P.R. Rimbey, J. Chem. Phys. 67 (1977) 698.

[52] G. Platero, V.R. Velasco and F. García-Moliner, Phys. Scr. 23 (1981) 1108.

[53] R. Loudon, J. Raman Spectrosc. 7 (1978) 10.

[54] R. Loudon, in: Surface Excitations, Vol. 9, Modern Problems in Condensed Matter Sciences, Eds. V.M. Agranovich and A.A. Maradudin (North-Holland, Amsterdam, 1984) p. 589.

[55] V.R. Velasco and F. García-Moliner, Solid State Commun. 33 (1980) 1.

[56] V.R. Velasco and G. Navascués, Phys. Scr. 34 (1986) 435.

[57] V.R. Velasco and F. García-Moliner, J. Phys. Lett. 46 (1985) L733.

[58] V.R. Velasco and F. García-Moliner, Surf. Sci. 67 (1977) 555.

[59] V.R. Velasco and F. García-Moliner, Phys. Scr. 20 (1979) 111.

[60] R.G. Pratt and T.C. Lim, Appl. Phys. Lett. 15 (1969) 403.

[61] J.R. Sandercock, Solid State Commun. 26 (1978) 547.

[62] J.P. Toennies and K. Winkelmann, J. Chem. Phys. 66 (1977) 3965.

[63] R. Loudon and J.R. Sandercock, J. Phys. C: Solid State Phys. 13 (1980) 2609.

[64] V.R. Velasco and F. García-Moliner, J. Phys. C: Solid State Phys. 13 (1980) 2237.

[65] V.V. Aleksandrov, T.S. Velichkina, V.G. Mozaev and I.A. Yakovlev, Solid State Commun. 77 (1991) 559;
V.V. Aleksandrov, T.S. Velichkina, V.G. Mozhaev, Yu.B. Potapova, A.K. Khmelev and I.A. Yakovlev, Phys. Lett. A 162 (1992) 418;
V.V. Aleksandrov, T. S. Velichkina, Yu.B. Potapova and I.A. Yakovlev, Phys. Lett. A 170 (1992) 165;
V.V. Aleksandrov and Yu.B. Potapova, Solid State Commun. 84 (1992) 401;
V.V. Aleksandrov, T.S. Velichkina, Yu.B. Potapova, V.I. Voronkova, I.A. Yakovlev and V.K. Yanorski, Solid State Commun. 84 (1992) 517.

[66] V.R. Velasco, Surf. Sci. 128 (1983) 117.

[67] V.R. Velasco and F. García-Moliner, Surf. Sci. 143 (1984) 93.

[68] L. Fernández, V.R. Velasco and F. García-Moliner, Surf. Sci. 172 (1986) 525.

[69] V.R. Velasco, J. Phys. C: Solid State Phys. 18 (1985) 4923;
V.R. Velasco and F. García-Moliner, Surf. Sci. 161 (1985) 342.

[70] M. Born and K. Huang, Dynamical Theory of Crystal Lattices (Oxford University Press, Oxford, 1968).

[71] F. García-Moliner, in: Proceedings Nato Advanced Research Workshop on Phonons in Nanostructures (Kluwer, Dordrecht), to be published;
R. Perez-Alvarez, F. García-Moliner, C. Trallero-Giner and V.R. Velasco, to be published.

[72] A.K. Sood, J. Menéndez, M. Cardona and K. Ploog, Phys. Rev. Lett. 54 (1985) 2111.

[73] E. Louis, J.A. Vergés and F. Guinea, J. Phys. Condensed Matter 2 (1990) 4143;
H. Rodriguez-Coppola, V.R. Velasco, F. García-Moliner and R. Pérez-Alvarez, Phys. Scr. 42 (1990) 115.

[74] Ya.A. Iosilevskii, Phys. Status Solidi (b) 44 (1971) 513; 60 (1973) 39; 64 (1974) 431.

[75] A.A. Maradudin and D.L. Mills, Ann. Phys. (New York) 100 (1976) 262;

L. Dobrzynski and A.A. Maradudin, Phys. Rev. B 14 (1976) 2200.

[76] A.A. Maradudin, R.F. Wallis and L. Dobrzynski, Eds., Handbook of Surfaces and Interfaces, Vol. 3 (Garland, New York, 1980);
M.G. Cottam and A.A. Maradudin, in: Surface Excitations, Vol. 9, Modern Problems in Condensed Matter Sciences, Eds. V.M. Agranovich and A.A. Maradudin (North-Holland, Amsterdam, 1984) p. 1.

[77] G. Brusdeylins, R.B. Doak and J.P. Toennies, Phys. Rev. Lett. 44 (1980) 1417;
H. Ibach and D.L. Mills, Electron Energy-loss Spectroscopy and Surface Vibrations (Academic Press, New York, 1982);
H. Ibach, in: Dynamical Phenomena at Surfaces, Interfaces and Superlattices, Eds. F. Nizzoli, K.-H. Rieder and R.F. Willis (Springer, Berlin, 1985) p. 109.

[78] F. García-Moliner, G. Platero and V.R. Velasco, Surf. Sci. 136 (1984) 601;
G. Platero, V.R. Velasco and F. García-Moliner, Surf. Sci. 152/153 (1985) 819.

[79] J.P. Toennies, J. Vac. Sci. Technol. A 2 (1984) 1055.

[80] G. Benedek, Surf. Sci. 61 (1976) 603;
G. Benedek and V.R. Velasco, Phys. Rev. B 23 (1986) 6691.

[81] G. Platero, V.R. Velasco, F. García-Moliner, G. Benedek and L. Miglio, Surf. Sci. 143 (1984) 243;
A.C. Levi, G. Benedek, L. Miglio, G. Platero, V.R. Velasco and F. García-Moliner, Surf. Sci. 143 (1984) 253.

[82] F. García-Moliner and V.R. Velasco, Phys. Scr. 34 (1986) 257.

[83] M.C. Muñoz, V.R. Velasco and F. García-Moliner, Phys. Scr. 35 (1987) 504;
R. Baquero, V.R. Velasco and F. García-Moliner, in: Lectures on Surface Physics, Eds. M. Cardona and R.G. Castro (Springer, Heidelberg, 1986) p. 32; Phys. Scr. 38 (1988) 742.

[84] W.A. Harrison, E.A. Kraut, J.R. Waldrop and R.W. Grant, Phys. Rev. B 18 (1978) 4402;
R.S. Bauer and J.C. Hikkelsen, Jr., J. Vac. Sci. Technol. 21 (1982) 491;
A.D. Katani, P. Chiaradia, H.W. Sang, Jr., P. Zurcher and R.S. Bauer, Phys. Rev. B 31 (1985) 2146.

[85] M.C. Muñoz, V.R. Velasco and F. García-Moliner, Progr. Surf. Sci. 26 (1987) 117.

[86] M.C. Muñoz and M.P. López Sancho, Phys. Rev. B 41 (1990) 8412.

[87] L.M. Falicov and F. Yndurain, J. Phys. C: Solid State Phys. 8 (1975) 147;
F. Yndurain and L.M. Falicov, Phys. Rev. Lett. 37 (1976) 928;
V.T. Rajan and F. Yndurain, Solid State Commun. 20 (1976) 309.

[88] V.R. Velasco and F. Yndurain, Surf. Sci. 85 (1979) 107.

[89] C.E.T. Gonçalves da Silva and B. Koiller, Solid State Commun. 40 (1981) 215.

[90] F. Guinea, C. Tejedor, F. Flores and E. Louis, Phys. Rev. B 28 (1983) 4397;
M.P. López Sancho, J.M. López Sancho and J. Rubio, J. Phys. F: Metal Phys. 14 (1984) 1205; 15 (1985) 851.

[91] R.A. Brito-Orta, V.R. Velasco and F. García-Moliner, Surf. Sci. 209 (1989) 492.

[92] M. Lannoo, Ann. Phys. (Paris) 3 (1968) 391;
M. Lannoo and P. Lenglart, J. Phys. Chem. Solids 30 (1968) 2409;
G. Allan, Ann. Phys. (Paris) 5 (1970) 169;
B. Velický and J. Kudrnovský, Surf. Sci. 64 (1977) 411.

[93] J. Pollmann and S.T. Pantelides, Phys. Rev. B 18 (1978) 5524;
J. Pollmann, Festkörperprobleme (1980) 117;
M. Schmeits, A. Mazur and J. Pollmann, Phys. Rev. B 27 (1983) 5012;
P. Krüger and J. Pollmann, Phys. Rev. B 30 (1984); 38 (1988) 10578.

[94] A.R. Williams, P. Feibelman and N.D. Lang, Phys. Rev. B 26 (1982) 5433.

[95] F. García-Moliner and V.R. Velasco, Prog. Surf. Sci. 21 (1986) 1.

[96] W.R.L. Lambrecht, Phys. Scr. 35 (1987) 724.

[97] A.A. Maradudin, E.W. Montroll, G.H. Weiss and I.P. Ipatova, Theory of Lattice Dynamics in the Harmonic Approximation. Solid State Physics Suppl. 3, Eds. H. Ehrenreich, F. Seitz and D. Turnbull (Academic Press, New York, 1975).

[98] G. Leibfried and N. Breuer, Point Defects in Metals. I. Introduction to the Theory, Springer Tracts in Modern Physics (Springer, Berlin, 1978);
R.E. Allen and M. Menon, Phys. Rev. B 33 (1986) 5611.

[99] G.A. Baraff and M. Schlüter, J. Phys. C: Solid State Phys. 19 (1986) 4383;
M. Fisher, Solid State Phys. 21 (1988) 3229.

[100] L. Dobrzynski and D.L. Mills, Phys. Rev. B 7 (1973) 2367.

[101] L. Fernández, V.R. Velasco and F. García-Moliner, Surf. Sci. 188 (1987) 140.

[102] A. Robledo and C. Varea, Phys. Rev. B 21 (1980) 1469.

[103] F. Cyrot-Lackmann, J. Phys. (Paris), Suppl. C1 (1970) 67;
R. Haydock, V. Heine and M. Kelly, Surf. Sci. 38 (1973) 139;
R. Haydock, Solid State Phys. 35 (1980) 215.

[104] D. Lohez and M. Lannoo, Phys. Rev. B 27 (1983) 5007.

[105] C. Tejedor, F. Flores and E. Louis, J. Phys. C: Solid State Phys. 10 (1977) 2163;
C. Tejedor and F. Flores, J. Phys. C: Solid State Phys. 11 (1978) L19;
F. Flores and C. Tejedor, J. Phys. C: Solid State Phys. 12 (1978) 731;
F. Guinea, J. Sánchez-Dehesa and F. Flores, J. Phys. C: Solid State Phys. 16 (1983) 6499;
J. Sánchez-Dehesa, F. Flores and F. Guinea, J. Phys. C: Solid State Phys. 17 (1984) 2039.

[106] F. Flores, J.C. Durán and A. Muñoz, Phys. Scr. T19 (1987) 102.

[107] R. Haussy, C. Priester, G. Allan and M. Lannoo, Phys. Rev. B 36 (1987) 1105.

[108] W.R.L. Lambrecht and O.K. Andersen, Surf. Sci. 178 (1986) 256.

[109] O.K. Andersen and O. Jepsen, Phys. Rev. Lett. 53 (1984) 2571;
O.K. Andersen, O. Jepsen and D. Glötzel, in: Highlights of Condensed Matter Theory, Eds. F. Bassani, F. Fumi and M.P. Tosi (North-Holland, Amsterdam, 1985) p. 59.

[110] F. Máca and M. Scheffler, Comput. Phys. Commun. 38 (1985) 403;
B. Wenzien, J. Kudrnovský, V. Drchal and M. Šob, J. Phys. Condensed Matter 1 (1989) 9893.

[111] I. Bartoš and B. Velický, Czech. J. Phys. B 24 (1974) 981.

[112] H. Ueba and S.G. Davison, J. Phys. C: Solid State Phys. 13 (1980) 1175;
I. Bartoš, Phys. Status Solidi (b) 85 (1978) K127;
S.G. Davison and K.W. Sulston, Phys. Status Solidi (b) 120 (1983) 415.

[113] I. Bartoš and S.G. Davison, Phys. Rev. B 32 (1985) 7863.

[114] F. García-Moliner and V.R. Velasco, Surf. Sci. 175 (1986) 9.

[115] F. García-Moliner and V.R. Velasco, Phys. Scr. 34 (1986) 252.

[116] L. Fernández, V.R. Velasco and F. García-Moliner, Surf. Sci. 185 (1987) 175; 188 (1987) 140.

[117] V.R. Velasco, F. García-Moliner, L. Miglio and L. Colombo, Phys. Rev. B 38 (1988) 3172.

[118] R.A. Brito-Orta, V.R. Velasco and F. García-Moliner, Phys. Rev. B 38 (1988) 9631;
L. Fernández, V.R. Velasco and F. García-Moliner, Surf. Sci. 251/252 (1991) 685.

[119] L. Colombo, L. Miglio, R.A. Brito-Orta, V.R. Velasco and F. García-Moliner, J. Appl. Phys. 70 (1991) 2079.

[120] V.R. Velasco, R. Baquero, R.A. Brito-Orta and F. García-Moliner, J. Phys. Condensed Matter 1 (1989) 6413.

[121] J. Arriaga, G. Armelles, M.C. Muñoz, J.M. Rodriguez, P. Castrillo, M. Recio, V.R. Velasco, F. Briones and F. García-Moliner, Phys. Rev. B 43 (1991) 2050;
J. Arriaga, M.C. Muñoz, V.R. Velasco and F. García-Moliner, Phys. Rev. B 43 (1991) 9626; Phys. Scr. 46 (1992) 466.

[122] M.C. Muñoz, V.R. Velasco and F. García-Moliner, Phys. Rev. B 39 (1989) 1786;
G. Armelles, M.C. Muñoz, V.R. Velasco and F. García-Moliner, Superlattices and Microstructures 7 (1990) 23;
D.A. Contreras-Solorio, V.R. Velasco and F. García-Moliner, Phys. Rev. B 47 (1993) 4651.

[123] D.P. Barrio, M.L. Glasser, F. García-Moliner and V.R. Velasco, J. Phys. Condensed Matter 1 (1989) 4339.

[124] M.L. Glasser, F. García-Moliner and V.R. Velasco, J. Appl. Phys. 68 (1990) 4319.

[125] H. Ohno, E.E. Mendez, J.A. Brum, J.M. Hong, F. Agulló-Rueda, L.L. Chang and L. Esaki, Phys. Rev. Lett. 64 (1990) 2555;
H. Ohno, E.E. Mendez, A. Alexandrou and J.M. Hong, Surf. Sci. 267 (1992) 161.

[126] W. Kohn and N. Rostoker, Phys. Rev. 94 (1954) 1111.

[127] W. Kohn, Phys. Rev. B 11 (1975) 3756.

[128] J.E. Inglesfield, J. Phys. C: Solid State Phys. 10 (1977) 3141.

[129] J. Arriaga, F. García-Moliner and V.R. Velasco, Prog. Surf. Sci. 42 (1993) 271.

Surface Science 299/300 (1994) 346–357
North-Holland

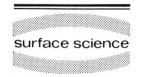

surface science

Quasiparticle theory of surface electronic excitation energies

Steven G. Louie

Department of Physics, University of California at Berkeley, and Materials Sciences Division, Lawrence Berkeley Laboratory, Berkeley, CA 94720, USA

Received 6 May 1993; accepted for publication 7 June 1993

An important factor contributing to the phenomenal advances in surface science in the past three decades has been the development and applications of numerous high resolution spectroscopies to the electronic properties of surfaces and interfaces. Over the same period, equally impressive progress has been made in the theoretical front in understanding and predicting surface electronic structure. In this paper, the development of a first-principles quasiparticle approach to the electronic excitation energies in crystals and at surfaces is reviewed. The approach allows an accurate computation of the particle-like excitations (quasiparticle states) of the interacting electrons in real materials. It is the quasiparticle energies which essentially determine the spectral features measured in photoemission, optical, scanning tunneling, and various other spectroscopic experiments. Applications of this approach to the analysis of selected bulk and surface systems are presented. Significant self-energy corrections arising from many-electron effects to the excitation energies are found. Employing atomic positions from total energy minimization, the calculated excitation energies have been used to explain and predict the experimental spectra of a variety of systems.

1. Introduction

Since the founding of *Surface Science* in 1963, there has been tremendous progress in the theoretical understanding and the development of methods for calculating the properties of surfaces. It is now possible to compute basically from first principles the structure, bonding properties, vibrational spectra, electron excitation energies, and even thermodynamical quantities and structural phase transitions of many surface systems. None of these ab initio calculations were possible in the early 1960's or even the 1970's. These theoretical advances were made possible by the efforts of many research groups and by the availability of increasingly powerful computers.

As for other condensed matter systems, the physical and chemical properties of surfaces are ultimately governed by their electronic structure. Hence, the ability of theory to predict electronic excitation spectra is essential for the understanding and application of surface systems. This review focuses on a brief account of the development of a first-principles quasiparticle method [1]

for calculating the electron excitation energies in crystals and at surfaces. (Other articles in this volume undoubtedly will account for the developments of other theoretical methods.) Owing to many-electron effects [1,2], the spectroscopic properties of electrons can be significantly different from those of the independent-particle or self-consistent-field picture. This first-principles quasiparticle method, which was developed in the mid 1980's, has shown to be capable of yielding very accurate excitation energies and, thus, provides an important theoretical link between structural studies and spectroscopic probes of surfaces.

The approach is based on an expansion of the electron self-energy operator to first order in the dressed electron Green's function and the dynamically screened Coulomb interaction with local field effects included in the so-called GW approximation [2]. Ab initio calculations using this approach have successfully explained and predicted results from optical, direct and inverse photoemission, scanning tunneling, and other spectroscopic measurements for a variety of systems including bulk crystals, surfaces, interfaces, and small clusters. Here we briefly discuss the theoretical formulation of the approach and present results for several prototype systems to illustrate some of the accomplishments of the theory in surface studies.

2. Historical perspective of self-consistent electronic structure calculations

2.1. Early surface and interface studies

The electronic and structural properties of surfaces and interfaces are dominated by electronic states that are localized near the surface or interface. Many of the early conceptual developments on surface states were done in the 1930's and 1940's by Tamm, Shockley, Davydov, Mott, Schottky, Bardeen, and others [3]. Subsequent to this pioneering work, there were a number of specific calculations on simple model systems. However, it was not until the 1970's that realistic models for real systems were analyzed in detail. Advances in ultra-high vacuum technology made well-characterized and reproducible surface studies possible in the 1970's. The theoretical studies, together with new experimental techniques such as low energy electron diffraction (LEED) to study surface geometry and photoemission to determine electronic structure, created an exciting era for research in surface science.

In the 1970's, the theoretical efforts were focused on extending standard bulk band structure techniques to surfaces using either empirical or self-consistent-field methods. Two major constraints at the time were (1) the lack of translation symmetry because of the surface and (2) the necessity to perform the calculations self-consistently. The first constraint has been overcome either by matching the electron wavefunctions across the surface region allowing both extended and decaying states [4] or by using a thin slab of the material to simulate the surfaces in a supercell geometry [5] with the slabs repeated indefinitely with certain separation. The supercell method allows the use of standard band structure techniques since periodicity is mathematically re-

stored. It has since become a standard tool in surface calculations. The second constraint is a statement that changes in electronic interactions resulting from the rearrangement of the electrons at the surface must be fed back and added to the total potential seen by the electrons. The self-consistent requirement has been dealt with by performing self-consistent-field calculations.

Many of the surface and interface calculations in the 1970's, including the first self-consistent calculation for a semiconductor [4] and a transition metal [6] surface, were performed using the self-consistent pseudopotential method. In these studies, the electronic structure was determined self-consistently using semi-empirical ionic pseudopotentials and a local density functional approximation (LDA) for exchange and correlation effects. The geometric structures were either inferred from experiments or taken to be those of idealized models. Even with these limitations, a great deal has been learned from these calculations, and the results often provided insights in interpreting experimental observations [7]. Studies were carried out for semiconductor and metal surfaces, metal–semiconductor interfaces (Schottky barriers), semiconductor–semiconductor interfaces (heterojunctions), and internal interfaces such as stacking faults. The calculations yielded information on the surface-state bands, charge densities, and local densities of states (LDOSs). For example, Fig. 1 illustrates how the calculated changes in the surface density of states of the Pd(111) surface [8] explained the observed adsorbate-induced changes in photoemission intensities [9].

Another prototypical calculation of this era is given in Fig. 2, which shows the calculated LDOS for an Al/Si(111) interface with the aluminum metal modeled by jellium [10]. Each region in Fig. 2 corresponds to a thickness of two Si layers with Region IV being the Si region closest to the interface. An important concept emerged from this study [10] is that of the metal-induced gap states (MIGSs). The MIGSs at a metal–semiconductor interface correspond to semiconductor surface states in the band gap hybridized with the continuum states of the metals. These states have characters which are in between those of

Bardeen's [3] surface-state picture and those of Heine's [11] decaying-metal-state picture. Based on the existence and nature of the MIGS, a theory [12] for the Schottky barrier heights was constructed which, for example, explains the behavior of Schottky barrier heights with metal electronegativity for semiconductors with different ionicity. The MIGS have played a central role on many subsequent theoretical work in this area.

The advent of ab initio total energy methods in the early 1980's propelled theoretical studies to yet another level. In the total energy calculations, the exact geometry is no longer a required input. The structure is determined by minimizing the total energy with respect to the atomic positions for a given topology. With these calculations, it became possible to determine with accuracy crystal structures, lattice constants, bulk and shear moduli, phonon frequencies, structural phase transformations, electron–phonon and phonon–phonon couplings, and a host of other properties related to the electronic ground state of a solid or surface [13]. Several important factors contributed to this development. Among these were the refinements in band structure calculation techniques, development of approximations to the density functional formalism, invention of the ab initio norm-conserving pseudopotentials, and de-

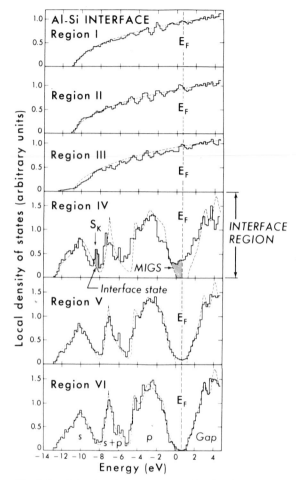

Fig. 2. Local density of states near an Al/Si(111) interface.

velopment of techniques for calculating total energies and forces. Yin and Cohen [14] first demonstrated in 1982 that one can obtain highly accurate structural energies from ab initio pseudopotential LDA total energy calculations. Calculations of this kind have been successfully employed to determine the structure and properties of a variety of surface systems. Fig. 3 compares the calculated structural parameters for the Si(111)2 × 1 surface [15] with experimental data from LEED and medium energy inelastic scattering (MEIS) measurements [16,17]. (In Fig. 3, z_i is the coordinate of the ith atom along the surface normal.) More recently, calculations have been performed to determine the structure of surfaces as complex as the Si(111)7 × 7 reconstruction with

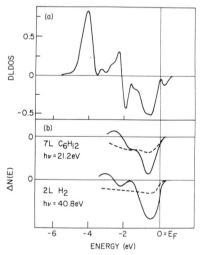

Fig. 1. (a) Calculated difference in LDOS between layer 4 and the surface layer for the Pd(111) surface. (b) Adsorbate-induced differences in photoemission intensities.

Si (111) 2 x 1 Surface

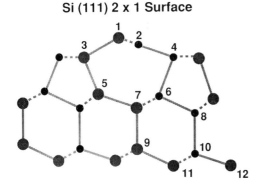

Structural Parameters (in Å)

	Experiment		Theory
	LEED	MEIS	
$Z_1 - Z_2$	0.38	0.30	0.47
$Z_3 - Z_4$	-0.07	-0.15	-0.05
$Z_5 - Z_6$	-0.07	-0.10	-0.07
$Z_7 - Z_8$	0.20	0.27	0.28
$Z_9 - Z_{10}$	0.13	0.14	0.15

Fig. 3. Calculated surface geometry of the Si(111)2×1 reconstructed surface (see text). Experimental data are from Ref. [16] (LEED) and Ref. [17] (MEIS).

supercells containing more than 700 effective atoms [18,19] using the Car–Parrinello method [20].

2.2. A crisis in ab initio calculation of electron excitation energies

Despite the dramatic progress in electronic structure theory, it became clear in the early 1980's that there was a major problem with theory in describing excited-state properties of solids. Although excitation spectra may be fitted empirically, the empirical methods, in general, lack the predictive capability for studying new materials or systems such as surfaces. On the other hand, the existing ab initio methods at that time which had given highly accurate structural energies and other ground-state properties failed even the simple task of predicting the band gap of a semiconduc-

tor. The source of the problem was traced to an inadequacy of these methods in treating many-electron effects for excited state properties [1]. These ab initio methods, being ground-state theories mostly based on the LDA, do not directly give electron excitation energies.

The common practice of interpreting spectroscopic data with the LDA Kohn–Sham eigenvalues had often led to rather severe discrepancies. As shown in Table 1, the LDA calculations incorrectly predicted Ge to be a metal and gave a band gap of 0.5 eV for Si instead of the experimental value of 1.17 eV. In general, the band gaps of semiconductors and insulators are underestimated in the LDA by 50–100%. Other self-consistent-field methods such as the Hartree–Fock (HF) approach yielded even worse results. This was known as the "band gap" problem. This problem existed not only for the minimum gap of semiconductors but also for any excitation spectra of solids which involve promoting an electron above the ground state. In the case of surfaces, the LDA surface-state bands are, in general, significantly misplaced in energy and the surface-state gaps are often off by a factor of two or more as compared to measurements. As we shall see below, in some cases, the LDA calculations even give incorrect band dispersions for the surface states.

This situation where ab initio methods could not even tell a metal from a semiconductor led to a real crisis in electronic structure theory of materials in the early 1980's. It was recognized in the late 1950's and early 1960's that a quantitative description of spectroscopic data of solids requires the concept of quasiparticles, the parti-

Table 1
Comparison of calculated band gaps E_g (in eV) with experiment

	HF	LDA	Quasiparticle theory [a]	Experiment [a]
Diamond	13.6	3.9	5.6	5.48
Si	6.4	0.5	1.29	1.17
Ge	4.9	< 0	0.75	0.74
LiCl	16.9	6.0	9.1	9.4

[a] Ref. [1].

cle-like excitations in an interacting many-electron system [1,2], but because of the complexity of the strong electron–electron interaction, it remained a major challenge to calculate the quasiparticle energies of real materials from first principles. Considerable theoretical efforts [21] had been devoted to this problem. A successful solution was obtained with the development of the first-principles quasiparticle approach in 1985. In this approach [1], the only inputs to the calculations are the atomic numbers of the constituent elements and the crystal or surface symmetry. The method has been applied quite successfully to the study of semiconductors and metals as well as surfaces, interfaces, and small metal clusters. In the remaining part of this article, we will present the basic ideas behind the method and discuss some selected examples from these studies.

3. First-principles self-energy approach to quasiparticle energies

Because of electron–electron interactions, the energy for creating a particle-like excitation with a well-defined wavevector k in a solid can be quite different from that of an independent-particle picture. The excited electron, or quasiparticle, is dressed with an electron polarization cloud giving rise, in general, to a renormalized energy, an effective mass, and a finite lifetime. An accurate treatment of the exchange and dynamical correlation effects is thus crucial in calculating the excitation energies. The Green's function formalism provides a rigorous and systematic way of including these many-electron effects. In this formalism, the quasiparticle energies and wavefunctions are obtained by solving a Dyson equation [2]:

$$[T + V_{ext}(r) + V_H(r)]\psi(r)$$

$$+ \int dr' \, \Sigma(r, r'; E^{qp})\psi(r') = E^{qp}\psi(r), \quad (1)$$

where T is the kinetic energy operator, V_{ext} the external potential due to the ions, V_H the average electrostatic Hartree potential, and Σ the elec-

tron self-energy operator. The self-energy operator Σ contains the many-electron effects. In general, Σ is nonlocal, energy-dependent, and non-Hermitian with the imaginary part giving the lifetime of the quasiparticles.

The approach of Ref. [1] involves taking the self-energy operator Σ as the first-order term in a series expansion of the screened Coulomb interaction W and the dressed Green function G of the electron:

$$\Sigma(r, r'; E) = \frac{i}{2\pi} \int d\omega \, e^{-i\delta\omega} G(r, r'; E - \omega)$$

$$\times W(r, r'; \omega), \quad (2)$$

where δ is a positive infinitesimal. This is the so-called GW approximation. The basic idea [1] is to make the best possible approximations for G and W, calculate Σ, and obtain the quasiparticle energies without any adjustable parameters. The screened Coulomb interaction $W = \epsilon^{-1}V_c$ (where ϵ is the dielectric response function and V_c the bare Coulomb interaction) incorporates the dynamical many-body effects of the electrons. Hence, the dielectric response function $\epsilon(r, r', \omega)$ is a key ingredient in determining the electron self-energy. For a crystal, the dielectric function is a matrix $\epsilon_{GG'}(q, \omega)$ in the reciprocal lattice vectors. The off-diagonal elements of this matrix give the so-called local field effects which distinguish the variations in screening properties in different parts of the crystal. These local fields are physically very important [22] and are an essential component in the quantitative evaluation of the self-energy operator for a real material. A major factor that made possible our first ab initio calculation of quasiparticle energies in semiconductors in the fall of 1984 was the coming together of having developed techniques for calculating the static dielectric matrices with an idea for extending them to finite frequencies.

Since the quasiparticle energies E_{qp} and wavefunction ψ enter the electron Green's function G [1], they need to be calculated self-consistently with Σ and G. In practice, G is usually constructed initially using the LDA Kohn–Sham eigenvalues and iteratively updated with the quasiparticle spectrum from Eq. (1). The quasiparticle wavefunctions, on the other hand, have been shown to be virtually identical to the LDA eigenfunctions [1] in all the cases considered. The dynamical dielectric matrix is typically obtained in a two-step process. First, the static dielectric matrix is computed as a ground-state property within the LDA [22]. Second, each element of the dielectric matrix is extended to finite frequencies using a generalized plasmon pole model employing exact dispersion and sum rule relations. There are no adjustable parameters in this procedure. Comparison with results from subsequent calculations using alternative methods [23] to compute the ω dependence of ϵ showed that the generalized plasmon pole scheme is highly accurate.

4. Selected applications

4.1. Band gaps and excitation spectra of semiconductors and insulators

The first-principles quasiparticle approach was first employed to the calculation of the band gaps and excitation spectra of bulk semiconductors and insulators. Table 1 presents the calculated band gaps [1] of several representative crystals together with the experimental values. As mentioned earlier, the LDA gaps in general significantly underestimate the experimental gaps. The Hartree–Fock results show the opposite trend of typically too large by several folds because of the neglect of electron correlations. On the other hand, with the excitation energies properly interpreted as transition energies between quasiparticle states, the calculated quasiparticle energy gaps in Table 1 are in excellent agreement with experiment. The results in Table 1 were from the 1985 work [1] and were obtained with the atomic numbers and crystal lattice parameters as the only input. These calculations demonstrated for the first time that quasiparticle excitation energies may be calculated from first principles to the level of accuracy of 0.1 eV. Subsequent calculations have obtained similarly accurate band gaps for a range of other semiconductors and insulators [22,24].

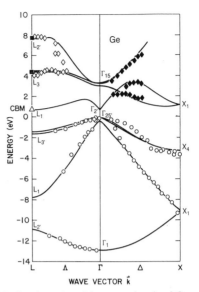

Fig. 4. Calculated quasiparticle energy bands of Ge compared to data from direct (open circles, Ref. [26]) and inverse (open and closed diamonds, Refs. [27,28]) photoemission experiments.

The theory also yields excellent optical transition energies and band dispersions. The calculated optical transition energies (neglecting excitonic effects) are generally within 0.1–0.2 eV of the observed features in the optical spectra. This kind of agreement is, in fact, comparable to that of empirical methods [25] in which typically several parameters are used to fit to the experimental data. Fig. 4 shows the calculated quasiparticle energy bands of Ge as compared to results from angle-resolved direct [26] and inverse [27,28] photoemission measurements. We see that the agreement between theory and experiment for the occupied states is well within the experimental and theoretical error bars. Moreover, the 1985 predicted values for the unoccupied conduction bands have been well reproduced by the subsequent angle-resolved inverse photoemission data. Ortega and Himpsel [28] have recently also performed band mapping using direct and inverse photoemission for several other semiconductors and obtained similarly good agreement with our calculated quasiparticle band structures.

4.2. Semiconductor surfaces

Shortly after the calculations for bulk crystals, the quasiparticle method was extended to the more complex systems including surfaces and interfaces. In this section, we give several samples of surface studies with emphasis on the Si surfaces. The calculations were carried out using the supercell method. Typically, about a dozen layers of the crystal were used with a vacuum region equivalent to several layers separating the slabs. The geometric structure was determined by minimizing the total energy in a LDA ab initio pseudopotential calculation as discussed in Section 2. Quasiparticle energies and wavefunctions were then computed for both the bulk and surface states.

Si(111):As and Ge(111):As. The first surface application of the quasiparticle approach was to the study of the As capped Si(111) and Ge(111) surfaces [29,30]. At saturation coverage, the surface is unreconstructed with the As atoms substituting for the last layer of the host atoms forming a very stable and geometrically simple surface [31]. The stability of this surface arises from the fully occupied lone pair orbital of the As adatoms. The calculated quasiparticle surface bands are shown in Fig. 5 for Si(111):As together with the corresponding LDA surface bands. Two surface-state bands are found near the gap region: an occupied As lone pair surface-state band and a well-defined empty surface-state band split off from the continuum near the zone center.

Compared to the LDA results, the quasiparticle surface-state band for the occupied lone pair states has similar dispersion but is slightly broader and lower in energy. Both are required for a better agreement with experiment. (Although the self-energy corrections to the occupied surface states are not too large for this particular surface, this is, however, not generally true. As shown below, the position and dispersion of occupied surface states can be considerably changed due to self-energy corrections as for the case of the H/Si(111) surface.) The effects of self-energy corrections to the empty surface states are strikingly larger in Fig. 5. These states are substantially shifted upward in energy, opening up the energy gap between the occupied and empty surface states by nearly an extra 1 eV at some *k*-points. Very similar results are obtained for the Ge(111):As case.

Fig. 6 gives a comparison of the occupied quasiparticle surface-state energies with angle-resolved photoemission experiment [31]. For both Si(111):As and Ge(111):As, excellent agreement between theory and experiment is achieved. For the empty surface states, scanning tunneling [30] and inverse photoemission [32] experiments have further given confirmation to the predicted values. Specifically, the scanning tunneling spectroscopy data gave measured surface-state gaps of 1.9 and 0.8 eV in reasonable agreement with the calculated gaps of 2.2 and 0.8 eV for Si(111):As and Ge(111):As, respectively. A very important conclusion from these results is that the self-energy corrections to the surface-state band gaps are not the same as those for the bulk band gaps. We will return to the significance of this point later.

Si(111)2 × 1. This surface is one of the most studied semiconductor surfaces. In addition to having an unusual π-bonded chain reconstruction [33], its surface electronic structure is also quite intriguing. The surface-state band gap from optical measurements [34–36] for this surface appears to be very different from that measured in

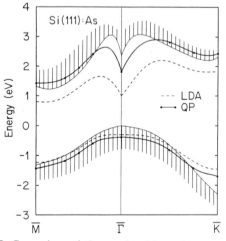

Fig. 5. Comparison of the quasiparticle surface-state band energies of Si(111):As to those derived from the LDA. The quasiparticle bulk projected band structure is also shown.

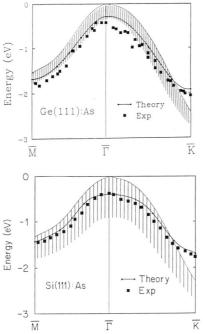

Fig. 6. The calculated occupied surface band is compared to photoemission results (Ref. [31]) for the Ge(111):As and the Si(111):As surfaces.

photoemission [37,38] and scanning tunneling [39] experiments. This difference indicates that electron–hole (excitonic) interactions may be considerably stronger in the surface optical processes than those in bulk processes.

The quasiparticle calculation was performed using the minimal energy geometry depicted in Fig. 3. The π-bonded chain reconstruction corresponds to rearrangement of atomic positions extending five layers into the crystal. Fig. 7 depicts the calculated surface-state bands together with the surface-state energies from direct [37] and inverse [38] photoemission experiments. The surface electronic structure is dominated by a pair of π (occupied) and π^* (empty) surface bands. These surface states arise from the π bonding of the dangling orbitals on the three-fold coordinated surface atoms and are dispersive only along the chain direction resulting in one-dimensional-like bands for the surface states. As evident in Fig. 7, both the position and the dispersion of the surface states are given very accurately by theory.

The surface-state band gap from photoemission in Fig. 7 is larger than the theoretical value. However, a subsequent analysis of the effects of the limited resolution in the angle-resolved inverse photoemission [38] indicated that the true energy of the π^* state at the k-point $\bar{\text{J}}$ could be lower than that given in Fig. 7 by as much as 0.15 eV yielding a surface-state gap in good agreement with the calculated value of 0.62 eV. (The LDA surface-state gap is only 0.27 eV for this surface.) A direct gap of 0.6 eV has also been observed in scanning tunneling spectroscopy [39].

The measured onset energy [34–36] for electron–hole creation on the Si(111)2 × 1 surface is, however, significantly smaller than the calculated surface-state gap. The optical gap measured at low temperature [35] is only 0.47 eV which disagrees with theory and with direct/inverse photoemission and tunneling results. Another problem was that the observed optical absorption spectrum, which has only one asymmetric peak, does not agree with a two-peak structure predicted by the joint quasiparticle density of states of the calculated π and π^* bands. These discrepancies led to the proposal that there may be enhanced excitonic effects [15,40] of order 0.1 eV in the surface optical data. Physically, the reduced screening and the quasi-one-dimensional dispersion of the π-bonded surface states of this

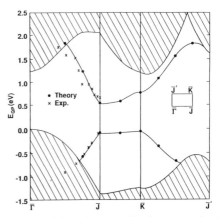

Fig. 7. Quasiparticle surface-state band structure for Si(111)-2×1 compared to photoemission (Ref. [37]) and inverse photoemission (Ref. [38]) experiments.

particular surface are expected to enhance the electron–hole interactions.

Fig. 8 shows the theoretical surface-state absorption spectra $\alpha(\omega)$ [15] for the Si(111)2 × 1 surface for light polarized along the surface atomic chain. The spectra were calculated using exciton energies and wavefunctions obtained with a model which employs the calculated quasiparticle surface-state bands and a statically screened electron–hole interaction. Curve (b) in Fig. 8 corresponds to a complete neglect of the electron–hole interaction and exhibits a two-peak structure which arises from the critical points in the joint density of states of the π and π^* surface-state bands. Curve (a) is $\alpha(\omega)$ calculated from an excitonic spectrum for which the lowest energy exciton has a binding energy of 0.13 eV obtained with an electron–hole Coulomb interaction screened by an effective dielectric constant of 6.5. With excitonic effects included, the absorption spectrum is dominated by the ground-state exciton, and the spectrum exhibits a single asymmetric peak in agreement with experimental results from differential reflectivity measurements [35].

The lesson from this study is that the calculated quasiparticle energies correspond to the energy needed to create an isolated electron or hole in the solid. Hence, the calculated quasiparticle energies describe well the photoemission and scanning tunneling data since these experiments probe single-particle excitations. However, these quasiparticle energies need not be the same as the optical transition energies if electron–hole

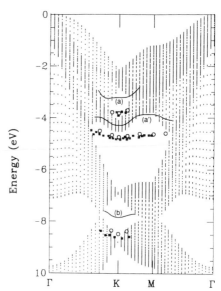

Fig. 9. Calculated surface-state bands of H/Si(111): LDA (full lines) and quasiparticle (open circles) results. The photoemission data (black dots) are from Ref. [41].

interaction effects are not negligible as in the present case of the Si(111)2 × 1 surface.

H / Si(111)1 × 1. In this study [41], it was found that many-electron effects can very significantly affect the position and dispersion of the occupied surface states even for a system as simple as the H-chemisorbed Si(111) surface. There is presently much renewed interest in this system because of the recent development of a wet chemical method [42] for preparing ideally hydrogen-terminated Si(111) surfaces which are highly stable, easily transportable, and structurally perfect over large areas. Fig. 9 depicts the calculated quasiparticle surface-state bands as compared to those from a LDA calculation as well as the measured surface-state energies from photoemission [41]. Two striking features are revealed. First, the shifts in the surface-state energies due to self-energy corrections are very large. The shifts are, in fact, larger by a factor of 2–3 as compared to typical self-energy corrections to occupied surface states seen in previous studies [15,29,30,43,44]. This is related to the very localized hydrogen 1s orbital forming the surface states in this system. Second, the self-energy corrections to the LDA results give rise to an unexpectedly large change in the

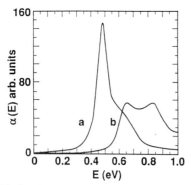

Fig. 8. Calculated surface-state optical absorption spectra obtained with (a) and without (b) excitonic correlations.

band dispersion of the surface bands. In particular, the LDA surface band (a') in Fig. 9 shows a 0.42 eV dispersion going from K to M which disagrees with the near dispersionless data from experiment. This discrepancy is completely removed in the quasiparticle calculation. Analysis of the theoretical results showed that this large change in dispersion may be understood from the sensitivity of the nonlocal self-energy operator to the localization of the electron wavefunction. The wavefunction of the surface state (a') at M is much more localized than that at K leading to a larger self-energy correction for states near M.

From the above and other surface studies, several conclusions may be drawn. These calculations have demonstrated that very accurate quasiparticle energies may be obtained for semiconductor surfaces and chemisorption systems. With the calculated electronic structure, one can then go on to study other phenomena such as electron–hole interactions in optical transitions. Generally, many-electron effects can significantly change the position and band dispersion of both the occupied and empty surface states. As a result, the quasiparticle band gaps between occupied and empty surface states are typically much larger than the LDA values. Some of these changes, such as the surface-state band gap enhancement, are qualitatively similar to the self-energy corrections to the bulk-state energies. However, the self-energy corrections to the surface states are, in general, quite different from those to the bulk states [29]. The differences arise both from changes in screening at the surface which lead to a change in the self-energy operator Σ and from changes in the characters of the surface-state wavefunction which can be substantially different from those of the bulk states. This means that, in general, an accurate theoretical determination of the surface excitation energies will require a full quasiparticle calculation.

4.3. Heterojunctions and superlattices

The quasiparticle approach has also been applied to the study of the electronic structure of several common semiconductor heterojunctions and superlattices [45–47]. A prototypical study is

the calculation of the band offsets or band discontinuities at the GaAs–AlAs(001) interface [46]. The band offsets given in transport or spectroscopic measurements are just the differences in the quasiparticle (or quasihole) energies across the interface for the band edge states. The self-energy corrections to the LDA band offsets may be expressed in terms of the differences in the self-energy corrections to the bulk states of the two materials forming the junction. That is, in the case of the valence band offset, $\Delta E_v = \Delta E_v(\text{LDA}) + \delta_{\text{vbm}}$, where the δ_{vbm} is the difference in the self-energy corrections to the valence band maximum states between the two materials. For the GaAs–AlAs(001) interface, δ_{vbm} was found to be 0.12 eV, which is about 30% of the LDA value for the valence band offset. Thus, the self-energy correction is relatively large even for this junction of nearly perfectly lattice matched materials. This correction can be understood in terms of a more localized valence band wavefunction for AlAs which leads to a more negative self-energy correction for AlAs as compared to GaAs and, hence, a positive δ_{vbm} between GaAs and AlAs. For the conduction band offsets, the self-energy corrections are, in general, much larger. The self-energy corrections to band edge states for a variety of semiconductors have been tabulated by Zhu and Louie [24] for use in analyses of band offsets and related properties.

Quasiparticle calculations have also been performed for short period semiconductor superlattices including the GaAs/AlAs superlattices and the Si/Ge strained layer superlattices [45,47]. As in the case of surface studies, the calculations have detailed the electronic structure of these systems and contributed to the interpretation of experimental excitation spectra, in particular the various band orderings at different superlattice periodicity and the optical response functions. The effects of disorders and strains have also been analyzed.

5. Summary and future prospects

In the past three decades, theory has played an important role in our understanding of the electronic and structural properties of surfaces. Empirical and self-consistent electronic structure calculations in the 1970's, in conjunction with experiments, elucidated the nature of many surface and interface systems. Ab initio total-energy methods developed in the early 1980's made possible the ab initio determination of structure and other ground-state properties. The quasiparticle approach described here provided the means for computing and predicting spectroscopic properties from first principles.

The first-principles quasiparticle approach, which involves a first-order expansion of the electron self-energy operator in the screened Coulomb interaction, is shown to be well-founded theoretically as well as generally applicable in practical computations for real materials. The approach has been applied to semiconductors, ionic insulators, and simple metals as well as to various surfaces, interfaces, and superlattices. Although not discussed here, applications to the study of clusters [48] and of pressure-induced insulator–metal transitions [49] have also been made. In the cases considered, the method has been successful and of predictive power. Results in excellent agreement with experiments have been obtained for band gaps, optical transition energies, direct and inverse photoemission spectra, and scanning tunneling spectra. In some cases, the calculations also have shed light on the possible importance of processes going beyond the quasiparticle picture such as electron–hole interactions in the optical response of the Si(111)2 × 1 surface.

This ability of theory to compute accurate excitation energies from first principles is a development available only since 1985 and is a major advance over previous ab initio methods. Its successes so far have been quite impressive and encouraging. There remain, however, several issues for exploration and improvement. Most of the studies have been on semiconductor systems. The application of the present quasiparticle method to transition metals and other more highly correlated electron systems remains to be made. The question of quasiparticle lifetimes due to many-electron interactions has yet to be addressed seriously in quantitative studies. Al-

though quasiparticle calculations have been done for surfaces and interfaces with supercells containing tens of atoms and for the fullerites which have 60 carbon atoms per molecular unit [50], it would be desirable to have further algorithmic developments to allow computation of even more complex systems with perhaps hundreds or thousands of atoms. If the rate of past progress is any indication, these issues will soon be addressed. In any event, the quasiparticle approach for excitation energies together with total energy methods for structural properties has provided a powerful theoretical framework for the study of surfaces and interfaces.

Acknowledgements

This work was supported by National Science Foundation Grant DMR91-20269 and by the Director, Office of Energy Research, Office of Basic Energy Sciences, Materials Sciences Division of the U.S. Department of Energy under Contract No. DE-AC03-76SF00098.

References

[1] M.S. Hybertsen and S.G. Louie, Phys. Rev. Lett. 55 (1985) 1418; Phys. Rev. B 34 (1986) 5390.
[2] L. Hedin and S. Lundqvist, Solid State Phys. 23 (1969) 1. This is an excellent review of pre-1969 work on excitation energies in solids.
[3] I. Tamm, Phys. Z. Sowjetunion 1 (1932) 733;
    W. Shockley, Phys. Rev. 56 (1939) 317;
    B. Davydov, J. Phys. (Moscow) 1 (1939) 167;
    N.F. Mott, Proc. R. Soc. London A 171 (1939);
    W. Schottky, Z. Phys. 113 (1939) 367;
    J. Bardeen, Phys. Rev. 71 (1947) 717.
[4] J.A. Appelbaum and D.R. Hamann, Rev. Mod. Phys. 48 (1974) 3.
[5] M. Schlüter, J.R. Chelikowsky and S.G. Louie and M.L. Cohen, Phys. Rev. B 12 (1975) 4200.
[6] S.G. Louie, K.M. Ho, J.R. Chelikowsky and M.L. Cohen, Phys. Rev. Lett. 37 (1976) 1289; Phys. Rev. B 15 (1977) 5627.
[7] For a review, see M.L. Cohen, in: Advances in Electronics and Electron Physics (Academic Press, New York, 1980) p. 1.
[8] S.G. Louie, Phys. Rev. Lett. 41 (1979) 476.
[9] H. Conrad, G. Ertl, J. Küppers and E.E. Latta, Surf. Sci. 58 (1976) 578.

[10] S.G. Louie and M.L. Cohen, Phys. Rev. Lett. 35 (1975) 866; Phys. Rev. B 13 (1976) 2461.

[11] V. Heine, Phys. Rev. 138 (1965) A1689.

[12] S.G. Louie, J.R. Chelikowsky and M.L. Cohen, Phys. Rev. B 15 (1977) 2154.

[13] For a review, see S.G. Louie, in: Electron Structure, Dynamics and Quantum Structural Properties of Condensed Matter, Eds. J. Devreese and P. van Camp (Plenum, New York, 1985) p. 335.

[14] M.T. Yin and M.L. Cohen, Phys. Rev. B 26 (1982) 5668.

[15] J.E. Northrup, M.S. Hybertsen and S.G. Louie, Phys. Rev. Lett. 66 (1991) 500.

[16] F.J. Himpsel, P.M. Marcus, R.M. Tromp, I.P. Batra, M. Cook, F. Jona and H. Liu, Phys. Rev. B 30 (1984) 2257.

[17] R.M. Tromp, L. Smit and J.F. van der Veen, Phys. Rev. Lett. 51 (1983) 1672.

[18] I. Stich, M.C. Payne, R.D. Kingsmith, J.S. Lin and L.J. Clarke, Phys. Rev. Lett. 68 (1992) 1351.

[19] K.D. Brommer, M. Needels, B.E. Larson and J.D. Joannopoulos, Phys. Rev. Lett. 68 (1992) 1355.

[20] R. Car and M. Parrinello, Phys. Rev. Lett. 55 (1985) 2471.

[21] C.S. Wang and W.E. Pickett, Phys. Rev. Lett. 51 (1983) 597;
C. Strinati, H.J. Mahausch and W. Hanke, Solid State Commun. 51 (1984) 23;
S. Horsch, P. Horsch and P. Fulde, Phys. Rev. B 29 (1984) 1870;
M.S. Hybertsen and S.G. Louie, in: Proceedings of the 17th International Conference on the Physics of Semiconductors, Eds. D.J. Chadi and W.A. Harrison (Springer, New York, 1985) p. 1001.

[22] M.S. Hybertsen and S.G. Louie, Phys. Rev. B 35 (1987) 5585; 5602.

[23] R.W. Godby, M. Schlüter and L.J. Sham, Phys. Rev. B 35 (1987) 4170.

[24] S.B. Zhang, D. Tomanek, M.L. Cohen, S.G. Louie and M.S. Hybertsen, Phys. Rev. B 40 (1989) 3162;
M.P. Surh, S.G. Louie and M.L. Cohen, Phys. Rev. B 43 (1991) 9126;
X. Zhu and S.G. Louie, Phys. Rev. B 43 (1991) 14142;
E.L. Shirley, X. Zhu and S.G. Louie, Phys. Rev. Lett. 69 (1992) 2955.

[25] M.L. Cohen and J.R. Chelikowsky, in: Electronic Structure and Optical Properties of Semiconductors, 2nd ed. (Springer, Berlin, 1989).

[26] A.L. Wachs et al., Phys. Rev. B 32 (1985) 2326.

[27] D. Straub, L. Ley and F.J. Himpsel, Phys. Rev. B 33 (1986) 2607.

[28] J.E. Ortega and F.J. Himpsel, Phys. Rev. B 47 (1993) 2130.

[29] M.S. Hybertsen and S.G. Louie, Phys. Rev. Lett, 58 (1987) 1551; Phys. Rev. B 38 (1988) 4033.

[30] R.S. Becker, B.S. Swartzentuber, J.S. Vickers, M.S. Hybertsen and S.G. Louie, Phys. Rev. Lett. 60 (1988) 116; unpublished.

[31] R.D. Bringans, R.I.G. Uhrberg, R.Z. Bachrach and J.E. Northrup, Phys. Rev. Lett. 55 (1985) 533;
R.I.G. Uhrberg, R.D. Bringans, M.A. Olmstead, R.Z. Bachrach and J.E. Northrup, Phys. Rev. B 35 (1987) 3945.

[32] W. Drube, R. Ludeke and F.J. Himpsel, in: Proceedings of the 19th International Conference on the Physics of Semiconductors, Ed. W. Zawadski (Inst. of Phys., Polish Acad. of Sci., 1988) p. 637.

[33] K.C. Pandey, Phys. Rev. Lett. 47 (1981) 1913; 49 (1982) 223.

[34] M.A. Olmstead and N.M. Amer, Phys. Rev. Lett. 52 (1984) 1148.

[35] F. Ciccacci, S. Selci, G. Chiarotti and P. Chiaradia, Phys. Rev. Lett. 56 (1986) 2411.

[36] J. Bokor, R. Storz, R.R. Freeman and P.H. Bucksbaum, Phys. Rev. Lett. 57 (1986) 881.

[37] R.I.G. Uhrberg, G.V. Hansson, J.M. Nicholls and S.A. Flodstrom, Phys. Rev. Lett. 48 (1982) 1031;
F.J. Himpsel, P. Heimann and D. Eastman, Phys. Rev. B 24 (1981) 2003;
F. Houzay, G.M. Guichar, R. Pinchaux and Y. Petroff, J. Vac. Sci. Technol. 18 (1981) 860.

[38] P. Perfetti, J.M. Nicholls and B. Reihl, Phys. Rev. B 36 (1987) 6160;
A. Cricenti, S. Selci, K.O. Magnusson and B. Reihl, Phys. Rev. B 41 (1990) 12908.

[39] R.M. Feenstra, Surf. Sci. 251, (1992) 401.

[40] R. Del Sole and A. Selloni, Phys. Rev. B 30 (1984) 883.

[41] K. Hricovini, R. Gunther, P. Thiry, A. Taleb-Ibrahimi, G. Indlekofer, J.E. Bonnet, P. Dumas, Y. Petroff, X. Blase, X.-J. Zhu, S.G. Louie, Y.J. Chabal and P.A. Thiry, Phys. Rev. Lett. 70 (1993) 1992.

[42] G.S. Higashi, Y.J. Chabal, G.W. Trucks and K. Raghavachari, Appl. Phys. Lett. 56 (1990) 656;
G.S. Higashi, R.S. Becker, Y.J. Chabal and A.J. Becker, Appl. Phys. Lett. 58 (1991) 1656.

[43] X.-J. Zhu, S.B. Zhang, S.G. Louie and M.L. Cohen, Phys. Rev. Lett. 63 (1989) 2112.

[44] X.-J. Zhu and S.G. Louie, Phys. Rev. B 43 (1991) 12146.

[45] M.S. Hybertsen and M. Schlüter, Phys. Rev. B 36 (1987) 9683.

[46] S.B. Zhang, D. Tomanek, S.G. Louie, M.L. Cohen and M.S. Hybertsen, Solid State Commun. 66 (1988) 585.

[47] S.B. Zhang, M.S. Hybertsen, M.L. Cohen, S.G. Louie and D. Tomanek, Phys. Rev. Lett. 63 (1989) 1495;
S.B. Zhang, M.L. Cohen, S.G. Louie, D. Tomanek and M.S. Hybertsen, Phys. Rev. B 41 (1990) 10058.

[48] S. Saito, S.B. Zhang, S.G. Louie and M.L. Cohen, Phys. Rev. B 40 (1989) 3643.

[49] H. Chacham, X. Zhu and S.G. Louie, Europhys. Lett. 14 (1991) 65; Phys. Rev. Lett. 66 (1991) 64; Phys. Rev. B 46 (1992) 6688.

[50] E.L. Shirley and S.G. Louie, Phys. Rev. Lett. 71 (1993) 133.

Surface Science 299/300 (1994) 358–374
North-Holland

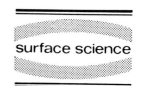

surface science

Electron-diffraction for surface studies – the first 30 years

S.Y. Tong

Laboratory for Surface Studies and Department of Physics, University of Wisconsin-Milwaukee, Milwaukee, WI 53201, USA

Received 3 May 1993; accepted for publication 20 May 1993

This article traces the historical development of the dynamical theory of electron diffraction as it relates to surface studies. A personal account will be given of the evolution of ideas which led to the first successful comparison between calculated low-energy electron diffraction (LEED) spectra and measured spectra. The broad impact of dynamical methods developed in the 1970's for LEED on other areas of surface studies also will be discussed. Personal accounts will be given on two specific areas. In addition, new ideas on how the inversion of diffraction spectra to produce real space information will be covered.

1. Introduction

Surface science, circa the late 1960's, was a budding branch of intellectual pursuit encompassing main-stream areas of condensed matter physics, physical chemistry and materials science. Of a handful of surface analytical techniques available at the time, low-energy electron diffraction (LEED), a phenomenon discovered by Davisson and Germer in 1927 [1], was enjoying a resurgence of interest propelled largely by advances in ultra-high vacuum technology [2] and the invention of post-acceleration display-type apparatus [3–5]. By 1968, Weber and Peria [6], and Palmberg and Rhodin [7], had just introduced Auger spectroscopy into the LEED chamber. Combining Auger spectroscopy with LEED was a particularly useful improvement because the cleanliness of a sample now could be quantitatively monitored in situ. While quantitative surface information on an atomic level was essentially lacking at the time, some notions about surface structure were known: e.g. clean semiconductor surfaces were observed to reconstruct and exhibit a different two-dimensional period, the most famous being the Si(111)7 × 7 structure (1959, Schlier and Farnsworth) [8]. Surface spacings were thought to be different from bulk spacings, although the prevailing idea was based on

pairwise potential results which predicted that surface layers would most likely displace outwards. Foreign atoms were known to induce surface reconstruction and surface relaxation. (A very readable account of the prevailing ideas on LEED theory and experiment, circa 1968, can be found in an article written by Jona [9].) My interest in LEED started in late 1969. Part A of this article is a historical account of the development of dynamical LEED theory, drawn from a personal perspective developed from my involvement in the field during this period. In this account, I shall identify key individuals and key works that most influenced my thoughts on the subject. As much as possible, I shall reference original papers rather than cite later reviews. Part B of this article will cover broader areas of surface science that involve electron diffraction in solids. I shall discuss the influence of dynamical LEED theory on these other areas of surface science.

2. Part A: Dynamical LEED theory

2.1. The beginning

My interest in LEED started in September, 1969 when I began a three-year postdoctoral fellowship with Thor Rhodin at Cornell University.

I had just finished my PhD thesis with Alexei Maradudin at the University of California, Irvine. My thesis work was a theoretical determination of surface and localized vibrational modes of a 25 layer thick slab of NaCl [10]. The thesis involved extending the Wallis–Mills–Maradudin method [11] of lattices dynamics for short-range forces to include long-range Coulomb interactions. I derived a set of interlayer and intralayer dynamical matrix sums over infinitely extended two-dimensional atomic planes in the slab and solved the resulting dynamical matrix to obtain eigenvectors and eigenvalues for the vibrational modes of the slab. In the thesis, optical and acoustic modes were mapped out along high-symmetry directions in the Brillouin zone. The theoretical thesis was ahead of its time because when we submitted a paper based on the thesis to Physical Review, a referee commented that *"surface phonons have never been observed experimentally and the prospect of ever measuring such excitations seems dim. ..."*. The paper was published anyway [10]. I did not return to the area of my thesis until the early 1980's, when advances in inelastic helium atom scattering and high-resolution electron energy loss spectroscopy allowed the mapping of dispersion curves of surface and localized phonons. It was serendipitous that the dynamical LEED methods developed in the 1970's were adapted in the 1980's to calculate the inelastic electron-phonon loss cross-sections. This latter development will be covered in part B of this article.

When I arrived at Ithaca, Rhodin was out of town for a meeting. I was his house guest until I could find a place for myself in Ithaca. Rhodin returned that evening and we sat in the kitchen of his rustic Tudor home discussing physics until late into the night. I gathered from that conversation that his interests at the time were in chemisorption and LEED, and that I was to choose one of the two areas to work on.

The following two weeks involved meeting people and reading the literature on LEED and chemisorption. Cornell in 1969 was a hotbed for surface science and a remarkable group of talented people were associated with Cornell at the time. Paul Palmberg had left Cornell after finishing his postdoctoral appointment with Rhodin.

Ward Plummer had finished his PhD thesis (with Rhodin) and was at his new job at the National Bureau of Standards (now NIST). I was introduced to Rhodin's two new graduate students: Joe Demuth and Alex Ignatiev. They were making arrangements to measure LEED *IV* spectra. Demuth's (with Rhodin) subsequent data of chalcogens chemisorption on Ni(001), i.e., the c(2 × 2) and p(2 × 2) phases of O, S, Se and Te on Ni(001) [12], were probably the most important sets of *IV* spectra – in terms of defining the capabilities of the LEED technique in the early days – ever collected for chemisorption systems [13–16], much like the significance of Jona's Al(001), (011) and (111) data [17] for defining the capabilities of LEED on clean surfaces. Ignatiev would measure the first detailed *IV* spectra for the study of temperature effects of noble gas crystals [18–21]. Other experimentalists at Cornell interested in the LEED problem included David Adams (a postdoctoral fellow), and Jack Blakely (on the faculty of Cornell), and Lester Germer, the co-discoverer of electron diffraction. Theorists at Cornell interested in surface problems included Neil Ashcroft, John Wilkins, John Strozier and Robert Jones (both postdoctoral fellows), and Roald Hoffman, who was testing the application of the extended Hückel method to surface problems.

I decided quickly to work on the LEED problem over chemisorption. This decision was not difficult because, while some very interesting papers on chemisorption had appeared in the literature (notably a 1967 paper by Bill Gadzuk [22] on a quantum mechanical model of charge transfer), it seemed to me that without detailed information of the adsorption site, any theory on chemisorption must be qualitative. Thus, I decided to work on LEED theory – to see if a dynamical theory accurate enough to produce calculated spectra in good agreement with experiment could be developed.

In the fall of 1969, thoughts on LEED theory were at a crossroad. The largest group of theorists working on LEED had a background in either bulk band theory or pseudopotential theory [23–32]. This was not surprising because reflected peaks in LEED *IV* spectra were identified early on as being caused by gaps in the band structure of a solid, therefore, LEED theory was viewed as an extension of bulk band structure to energies above vacuum. It was argued that the strong ion-core scattering required construction of new pseudopotentials $\langle k | V | k' \rangle$ applicable in the 0–300 eV range. A typical dynamical theory at the time involved selecting an ion-core pseudopotential and solving for Bloch waves and eigenvalues inside a semi-infinite solid, then matching the solutions to an incoming plane wave and outgoing scattered waves at the vacuum–solid interface.

I decided against pursuing these lines. My first objection was that a 100 eV electron backscattered by an atom had to probe deeply the strong core region of the potential. I was doubtful that any energy-independent pseudopotential would work in the energy range covered by LEED. My second objection was of a practical nature: if LEED *IV* spectra were to provide structural information, any dynamical theory would have to deal with multilayer relaxations and reconstructions of surface layers. A Bloch-wave eigenvalue solution for a semi-infinite crystal seemed to me to be the wrong approach for the problem at hand. Perhaps I was influenced by my thesis work: that of a 25-layer slab with dynamical matrix elements linking together interlayer and intralayer atomic forces and the energy of the slab was minimized to produce multilayer relaxations. Therefore, for the LEED problem, I was looking for a finite-slab scattering approach.

Such a scattering approach did exist. Beeby, in 1968, published a multiple scattering theory using a Green's function propagator approach [33]. He started with the *t*-matrix of a single ion core:

$$t_l(k) = -\frac{\hbar^2}{2m}\left\{\frac{e^{2i\delta_l} - 1}{2ik}\right\}, \qquad (1)$$

where δ_l was the lth partial wave phase shift for the real-space potential. Multiple scattering within a plane of atoms was given by:

$$\tau_{LL'}(k_i) = t_l(k)\{1 - G^{sp}(k_i)t_l(k)\}^{-1}_{LL'}, \qquad (2)$$

where $G^{sp}_{LL'}(k_i)$ was an intraplanar structural

propagator which linked together atoms in the plane. The contributions from different planes were solved self-consistently by a matrix equation:

$$T^{\nu}_{LL'}(k_i) = \tau^{\nu}_{LL'}(k_i)$$

$$+ \sum_{\substack{L_1 L_2 \\ \nu' \neq \nu}} \tau^{\nu}_{LL_1}(k_i) G^{\nu\nu'}_{L_1 L_2}(k_i) T^{\nu'}_{L_2 L'}(k_i),$$

$$(3)$$

where $G^{\nu\nu'}_{L_1 L_2}(k_i)$ were structural propagators linking together different planes in the slab. Pseudopotentials were not used and spherical-wave basis functions were used in t, τ, G^{sp} and $G^{\nu\nu'}$.

Beeby presented his method without consideration of inelastic damping. Without absorption, the sum over ν' in Eq. (3) would not converge because T^{ν} had to run over all layers of a three-dimensional crystal. Beeby published the paper without numerical demonstration [33] nor discussions on how the various quantities were to be evaluated.

Inelastic damping was important to LEED – a fact known to many experimentalists at the time. In 1966–68, two particularly illuminating papers demonstrating the effects of inelastic damping were published. The first paper, by Jones, McKinney and Webb (1966) [34], measured LEED intensity dependence on temperature of Ag(111) as a function of incident energy. These authors showed that the effective Debye temperature of Ag(111) was 15% higher at 180 eV than at 120 eV, indicating a shorter mean-free-path for electrons at the lower energy. From such data, Jones et al. [34] derived a formula for the absorption coefficient: $\mu(=k_2) \propto E^{-1/2}$. The second paper, by Palmberg and Rhodin (1968) [35], measured the attenuation of Auger peaks as a function of deposition of a metallic film on a different metal substrate. From such measurements, Palmberg and Rhodin directly determined that the mean escape depth for Auger electrons in Ag varied between 4 to 8 Å for energies of 72 and 362 eV, respectively.

In November, 1968, Duke and Tucker introduced their s-wave inelastic collision model of

LEED [36]. They replaced the electron Green's function propagator in Beeby's method by

$$G_0(\boldsymbol{r}_1 \boldsymbol{r}_2)$$
$$= \frac{1}{(2\pi)^3} \int \frac{e^{i\boldsymbol{k}\cdot(\boldsymbol{r}_1 - \boldsymbol{r}_2)}}{E - \frac{\hbar^2}{2m}k^2 - \Sigma(\boldsymbol{k}, E)} \, d^3k, \quad (4)$$

where $\Sigma(\boldsymbol{k}, E)$ was the electron self-energy containing single-particle and plasmon excitations. They approximated Beeby's method by taking the $l = 0$ partial wave only and treating the scattering phase shift $\delta_0(E)$ as an energy-independent adjustable parameter [36–39]. A detailed account of earlier theoretical treatments of inelastic damping can be found in the Introduction of a 1970 paper by Tucker and Duke [39].

The s-wave inelastic collision model of Duke and coworkers [36–39] represented a major shift in the approach from the dominant thought of LEED at the time. In a series of four papers published between November 1968 and May 1970 [36–39], Duke and coworkers presented the case that because of strong attenuation of incident electrons, LEED IV spectra were less sensitive to details of the ion-core potential. Therefore, they introduced models which contained surface as well as bulk scattering factors, δ_0^{B}, δ_0^{S}, respectively, both treated as energy-independent parameters.

By March 1970, I concluded that it might be possible to obtain quantitative agreement between theory and experiment if Beeby's theory was used with the inclusion of realistic phase shifts from a band-structure potential. The use of adjustable parameter for the phase shifts was too crude. Inelastic damping and phonon vibrational effects would be included by the methods of Duke and coworkers [36–41]. The intralayer and interlayer structural propagators of Beeby, given by

$$G_{LL'}^{\mathrm{sp}}(\boldsymbol{k}_i) = -4\pi i \left(\frac{2m}{\hbar^2} \right) k \sum_{L_1} \sum_{\boldsymbol{P} \neq \boldsymbol{0}} i^{l_1} a(LL'L_1)$$
$$\times h_{l_1}^{(1)} \left(k | \boldsymbol{P} + \boldsymbol{d}_\gamma - \boldsymbol{d}_\beta | \right)$$
$$\times Y_{L_1}(\boldsymbol{P} + \boldsymbol{d}_\gamma - \boldsymbol{d}_\beta) \, e^{-i\boldsymbol{k}_i \cdot (\boldsymbol{P} + \boldsymbol{d}_\gamma - \boldsymbol{d}_\beta)}$$
$$(5)$$

and

$$G_{LL'}^{\gamma\beta}(\boldsymbol{k}_i) = -4\pi i \left(\frac{2m}{\hbar^2} \right) k \sum_{L_1} \sum_{\boldsymbol{P}} i^{l_1} a(LL'L_1)$$
$$\times h_{l_1}^{(1)} \left(k | \boldsymbol{P} + \boldsymbol{d}_\gamma - \boldsymbol{d}_\beta | \right)$$
$$\times Y_{L_1}(\boldsymbol{P} + \boldsymbol{d}_\gamma - \boldsymbol{d}_\beta) \, e^{-i\boldsymbol{k}_i \cdot (\boldsymbol{P} + \boldsymbol{d}_\gamma - \boldsymbol{d}_\beta)},$$
$$(6)$$

where $a(LL'L_1)$ were the Clebsch–Gordon coefficients, $h_l^{(1)}(r)$ the spherical Hankel functions of the first kind and $Y_L(\boldsymbol{k})$ the spherical harmonics, needed to be numerically evaluated. From these, Eqs. (1)–(3) could be evaluated. The reflected IV intensity was given by

$$R_{\boldsymbol{k}_g^-}(\boldsymbol{k}_i) = \left| \frac{8\pi^2 i}{a} \left(\frac{2m}{\hbar^2} \right) \sum_{LL'} \frac{Y_L(\boldsymbol{k}_g^-) Y_{L'}^*(\boldsymbol{k}_i)}{k_{\perp g}^-} \right.$$
$$\left. \times \sum_\alpha e^{i(\boldsymbol{k}_i - \boldsymbol{k}_g^-) \cdot \boldsymbol{d}_\alpha} T_\alpha^{LL'}(\boldsymbol{k}_i) \right|^2, \quad (7)$$

where a is the unit cell area.

Of course, there was no guarantee that such an approach would work. Indeed, many papers in 1969–70 indicated that the dynamical ingredients chosen here might not work at all. For example, there were theoretical considerations which suggested that the broad widths of LEED peaks were not due to inelastic damping, but to phonon scattering [32] or variations of interlayer spacings in the surface region [42]. The papers of Duke and coworkers [36–39] showed that the IV spectra were very sensitive to changes in the surface phase shifts δ_0^{S}. This suggested that my model, which included only phase shifts from a bulk potential, might not work at all. In the end, however, I decided that the only sensible course of action was to carry through with a realistic calculation.

There was, however, a practical problem. Eq. (3) required solving a set of N inhomogeneous equations with complex matrices at each energy. Using 8 atomic planes and 4 partial waves for the ion-core potential (the absolute minimum for a realistic calculation), the size of the complex matrices were 128×128 [43]. To solve Eq. (3) at each energy was beyond the computer budget of

Rhodin's group. Therefore, I decided to use a perturbation method which included exact intralayer multiple scattering, and up to second-order in interlayer scattering, while the number of layers were summed to convergence. The perturbation method, of course, required additional programming – it was in fact much simpler to calculate the exact $T_{LL'}^{\nu}(\boldsymbol{k}_i)$ by inverting Eq. (3). Tucker and Duke [39] had shown that within the s-wave collision model, the perturbation method failed to converge properly for strong scatterers. It was thus necessary to select a material with a weak ion-core potential. We had at hand two sets of data: Jona's measurements on Al(001), (011) and (111) [17]; and Demuth's data on Ni(001), (011) and (111) [12]. Because this was the first application of a realistic calculation, we felt that in order to assess the accuracy of the dynamical factors used in the model, it was important that the test case be done on a material for which the perturbation approach would converge well. I therefore chose Jona's Al(001) data [17], even though Demuth's Ni(001) data [12] were measured in Rhodin's own group.

Because I had no practical background in band-theory, particularly in KKR theory, every special function, including the Clebsch–Gordon coefficients, had to be programmed from scratch. These included the analytical continuation into the complex plane for the spherical harmonics, Hankel functions, and evaluating the layer propagator matrices $G_{LL'}^{\mathrm{sp}}(\boldsymbol{k}_i)$, $G_{LL'}^{\nu\nu'}(\boldsymbol{k}_i)$, etc. By August 1970, the program was ready for execution. At the time, Beng Lundqvist was spending a year at Cornell after finishing his PhD thesis with Stig Lundqvist at Chalmers University, Sweden. For his thesis work, Beng evaluated the self-energy of a quasiparticle in an interacting electron gas [44–47]. I used his results of the self-energy [44] for Al in my calculation. For the atomic phase shifts, Rhodin obtained a set of Al phase shifts generated by Pendry and Capart [31]. Pendry was visiting Cornell after he finished his PhD degree at Cambridge with V. Heine. The Al phase shifts were generated from a non-local Hartree–Fock calculation constructed by Pendry and Capart for their pseudopotential LEED theory [31,48]. We also included temperature corrections using a

vertex renormalization approach of Duke and Laramore [40,41]. Looking back at those days, it was rather odd that although Pendry and I spent the summer together at Cornell – our offices were on the same floor of Clark Hall – we had hardly ever discussed the LEED problem. I did not even know if John was aware of my approach to the problem. I was, however, too involved in my work to seek out John's thoughts on the matter.

My paper with Rhodin on Al(001) was written during the Christmas of 1970. It was submitted to Physical Review Letters on January 18, 1971 and appeared March 22, 1971 [49]. In the paper, we wrote that:

"In this Letter, we wish to report on the first complete calculation of LEED spectra for a free-electron metal in terms of the multiple-scattering approach involving strong inelastic damping with no adjustable parameters. The significance of this work is that within the constraints inherent to the approach it produces spectral curves in meaningful agreement with specific details of experimental results obtained by Jona for a relatively simple metal, aluminum, when we compare to the detailed conditions of his measurement...".

In the abstract, we summarized the essential ingredients in a dynamical LEED theory that would lead to meaningful agreement with experiment [49]:

"Inclusion of energy-dependent higher order phase shifts obtained from a realistic potential, of energy-dependent strong inelastic damping, and of temperature effects contributes significantly to the agreement of the calculation with experiment...".

Of the perturbation approach, we wrote:

"The use of this approach greatly simplifies the computation and increases its speed, and its accuracy is within the limits imposed by the interpretation of the experimental data....".

In Fig. 1, I reproduce a figure from the 1971 Tong–Rhodin paper [49], which compared calculated IV spectra of the (00) beam for Al(001) at $\theta = 0°$, $6°$ and $10°$ and $T = 0$ and 293 K with the room-temperature data at $\theta = 6°$ and $10°$ of Jona [17]. We ourselves were quite surprised by the

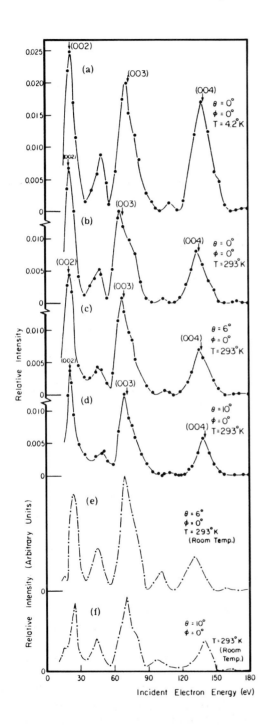

Fig. 1. (a)–(d) are calculated LEED spectra for Al(001). The curves are drawn by joining the calculated values (dots) with a smooth line. (e), (f) are experimental spectra of Jona, Ref. [17]. This figure is reproduced from Ref. [49].

degree of agreement obtained, considering the crudeness of the dynamical inputs for the surface region. An important message from the comparisons shown in this figure was that the LEED *IV* spectra were much more sensitive to the phase of the structural factors, i.e., $e^{i(k_i - k_f) \cdot d_s}$, where d_s were the surface-region interlayer vectors, then changes of the ion-core potential in the surface region. This was an important point because the trial-and-error approach of LEED structural determination, as we presently use the method, involves searching through thousands of structural variations *while keeping the ion-core potential fixed*.

Rapid progress was made on the Bloch-wave approach in 1971. In 1970, Jennings and McRae [50] abandoned the pseudopotential representation of ion-core scattering. They used Kambe's method [27] to treat the strong monolayer scattering. Kambe's method of monolayer scattering was equivalent to Beeby's expression of $\tau_{LL'}(k_i)$ (Eq. (2)), – the difference was that without damping, Beeby's formula was nonconvergent while Kambe's method converged because he partitioned the sum over atomic sites into real-space and reciprocal-space components [27]. A plane-wave representation was used by Jennings and McRae [50] in the weak scattering regions between layers and Bloch functions for a semiinfinite solid were solved and matched at the vacuum–solid interface. Phase shifts for W(001) were calculated from a real-space potential and used directly without converting to a pseudopotential. Inelastic damping, however, was not included. Again, the calculations of Jennings and McRae [50] produced peak intensities too high, close to 100% at band gap energies, and peak widths too narrow, of the order 2–3 eV, to correspond with the experiment.

Jepsen, Marcus and Jona, in a paper submitted to Physical Review Letters on April 26, 1971 and appeared on May 31, 1971 [51], showed that by incorporating inelastic damping in a method similar to that of Jennings and McRae [50], good agreement with Jona's data [17] on Al(001) was obtained. I met Don Jepsen, Paul Marcus and Franco Jona for the first time in March 1971 at the Fifth LEED Meeting held at the National

Bureau of Standards. At this LEED meeting, my paper (with Rhodin) and the Jepsen–Marcus–Jona paper were read. Pseudopotential applications in LEED became history: their use were not to be found in LEED papers published after the fall of 1971 [24–26,28–31,52–54].

By today's standards, the Tong–Rhodin [49] and Jepsen–Marcus–Jona [51] results lacked accuracy. For example, the Tong–Rhodin calculation used only 4 phase shifts and the second-order perturbation method [49] worked well only for free-electron metals and some semiconductor surfaces. The Jepsen–Marcus–Jona calculation neglected temperature and inner potential corrections. They produced calculated peaks which were split [51] while the corresponding peaks in the data were not [17]. Nevertheless, the two papers were important because they clarified the role of the LEED technique. Before these papers were published, an ongoing debate was whether LEED would be useful for studying bulk band gaps in solids [9]. Another debate was whether LEED would even be sensitive to surface spacings at all: Bloch-wave calculations [29] without absorption had shown that calculated LEED peaks were insensitive to 5% displacements in the surface interlayer spacing. On the opposite end, a debate was whether LEED would be so sensitive to surface dynamical effects [36–39] that one would always have to know surface scattering and damping factors exactly before meaningful agreement with experiment could be obtained – a Catch 22 situation for LEED, if true. The two 1971 papers [49,51] demonstrated that with reasonably accurate inputs of dynamical factors such phase shifts from a bulk potential, a uniform inelastic damping and temperature corrections, good agreement with experiment could be obtained, and these dynamical factors could be held fixed while surface structural parameters were varied [55–57]. Quoting again the Tong–Rhodin paper [49]:

> "It is indicated that LEED spectra for which the above considerations are applicable present the distinct possibility of providing meaningful information on the atomic geometry of the surface layer. This is probably one of the most important single objectives of current

studies in the application of the LEED approach to the surface physics of crystals".

These words perhaps marked the philosophical underpinnings of modern surface crystallography by LEED. Indeed, the first structural results were soon reported on three faces of Al by Laramore and Duke (January 1972) [58]. This was shortly followed by a study also on Al(001), (011) and (111) in November 1972 by Jepsen, Marcus and Jona [59] who used the layer KKR method, i.e., the Kambe–Jennings–McRae Bloch-wave method with absorption. Extension of the Beeby matrix-inversion method to include layer-dependent temperature corrections was made by Tong and Kesmodel [60] the following year for Ni(001) and Ni(011) – the calculations were compared to the data of Demuth and Rhodin [12].

I characterize 1971 as a watershed year signalling the beginning of modern LEED crystallography. By the end of 1971, the role of LEED was clearly defined: we knew how the technique should be used and what LEED would be useful for. The next phase in the development of theory was the introduction of revolutionary methods for assembling layer-by-layer diffraction processes and for handling the strong multiple scattering within closely-spaced composite layers.

2.2. Efficient intralayer and interlayer methods

In a paper published in Physical Review Letters in September 1971 [61], Pendry introduced a new perturbation method for handling interlayer multiple scatterings. The method was a clever and elegant way to systematically sum all forward-scattering events until the wavefield inside a slab was sufficiently attenuated, and back-scattering events were summed in a perturbation series. This method, known as the renormalized forward-scattering (RFS) perturbation theory, and a more convergent but less speedy method known as the layer-doubling method, also introduced by Pendry (in 1974) [62], revolutionized the way layer-by-layer scatterings were handled in LEED theory. RFS and layer-doubling became standard methods for treating interlayer multiple scatterings in a finite, but usually rather thick (20–25 layers) slab.

It was March 1974 when I went to the Milwaukee airport to meet my new postdoctoral associate who was arriving from England. The previous fall, I had moved to a faculty position at the University of Wisconsin-Milwaukee. His neat but minuscule handwriting did not prepare me for the sight of a tall, young man with a broad smile who greeted me at the arrival gate. Michel Van Hove, my new postdoc, had just finished his PhD thesis with Pendry at Cambridge University and came to work with me for the next two years.

Van Hove and I began the task of making the sums over intralayer multiple scatterings more flexible and efficient. In treating scatterings between atoms within a unit cell, because of the close proximity of these atoms, a plane-wave representation did not always converge. Both RFS and layer-doubling methods were based on the plane-wave representation, which worked well in the weak scattering regions between well-separated layers. For scatterings within a composite layer, Van Hove and I introduced the reverse-scattering perturbation (RSP) method of Zimmer and Holland [63] into the Beeby method. The spherical-wave representation was used throughout. We then transformed the resulting layer scattering matrices of composite layers into the plane-wave representation so to join to either the RFS or layer-doubling methods. To accomplish this, the matrices $T^{\nu}(k_i)$, $G^{\nu\nu'}(k_i)$, etc., needed to be evaluated not only at k_i but also at $(k_i + g)$ directions. Details of the combined-space method can be found in a 1977 paper by Tong and Van Hove [64], and books written by Van Hove and coworkers [56–57].

Van Hove proceeded to build an elaborate suite of programs based on the combined-space method which allowed flexible and efficient handling of intralayer and interlayer multiple scatterings, either by perturbation methods, exact methods or a mixture of such methods- the choices were controlled by input codes. Since their creation in 1978, these programs, known as the Van Hove–Tong tape [56], have been used by researchers in over 21 countries. Since January 1991, Van Hove upgraded the suite of programs to include Tensor LEED [65] and automated search routines. Besides Van Hove and Pendry,

other researchers who helped in creating this new version included Adrian Wander, Pedro de Andres, Phillip Ross and Angelo Barbieri. In a separate effort in 1987, Hong Huang at Milwaukee introduced full symmetry [66–68] into the Van Hove–Tong programs [69,70]. From the beginnings in 1971 of modest LEED codes [49,51], to the present day sophisticated Van Hove–Tong/ Automated Tensor LEED Tapes, and efficient LEED codes of others (Moritz and coworkers, Tear and coworkes, etc.) [71–74], LEED theory and applications have come a long way.

3. Part B: Slab methods for other analytical techniques of surface science

Because the short mean-free path of low-energy electrons provides surface sensitivity, many UHV surface-oriented techniques involve the propagation of low energy electrons in solids. It is not surprising that the slab methods developed in the 1970's for the LEED problem have found ready applications in many other areas of surface science. In this article, I shall cover two such areas, again drawn largely from my own involvement in these areas.

3.1. Photoelectron diffraction: EDPD spectra from core levels

In 1974–76, Ansgar Leibsch published two papers on photoelectron diffraction [75,76]. In these papers, he suggested that diffraction patterns of photoelectrons emitted from localized sources (e.g. an atomic core level) contained structural information of the near-neighbor atoms within the mean-free path of the emitted electrons. To demonstrate such effects, Leibsch carried out a number of model calculations. The methods he used were first- and second-order perturbation schemes.

I was interested in the ideas presented in Liebsch's papers. I also knew that the photoelectron diffraction calculations should be done more accurately in order to obtain meaningful agreement with experiment. Particularly, I wished to introduce quantitative evaluation of the photon-electron excitation matrix elements in the layer-by-layer photo-electron diffraction calculations. In 1977, I (with postdocs C.H. Li and A.R. Lubinsky) developed a dynamical method of photoelectron diffraction [77] in which the source wave from an atomic layer containing an emitting atom was written as [78]:

$$A^\alpha_{k^\pm_g}(k) = \left(\frac{2m}{\hbar^2}\right)\frac{2\pi i}{a}\sum_{LL'}\frac{Y_L\left(k^\pm_g\right)}{k^+_{\perp g}}\left(1 - t(k)\right.$$

$$\left. \times G^{sp}(k_i)\right)^{-1}_{LL'} e^{i k^\pm_g \cdot d_\alpha} F_{L'L_i}(k), \qquad (8)$$

where d_α is the vector from the master origin to the emitting atom in layer α. The photon-electron excitation matrix elements were evaluated as:

$$F_{LL_i}(k) = (-1)^l i^{l+1}\left(\frac{eh}{mc}\right)\frac{1}{(E_f - E_i)}$$

$$\times \int e^{i\delta_l} R^f_l(r) Y^*_L(r) A$$

$$\cdot \nabla V(r) R_{l_i}(r) Y_{L_i}(r)\, d^3r. \qquad (9)$$

In Eq. (9), $R^f_l(r)$ and E_f referred to the final state electron while $R_{l_i}(r)$ and E_i referred to the initial state core electron. Also, in Eq.(8), multiple scattering of the final-state electron within the layer of the emitting atom was handled in the spherical wave basis, via the term $(1 - t(k)G^{sp}(k_i))^{-1}$, but the final expression was cast in the plane-wave representation. The source terms $A^\alpha_{k^\pm_g}(k)$ could then be conveniently joined to the layer-by-layer methods of either RFS or layer doubling. If the source atom was embedded in a composite layer, the expressions for the source terms were more complicated. These expressions could be found in the thesis work of my former student Dr. W.M. Kang [79].

This dynamical theory was first applied to analyze the low-energy data of Weeks and Plummer [80]. From 1976 to 1979 [81–89] most photoelectron diffraction measurements were taken by fixing the photon's energy and mapping out the photoelectron intensity as the function of either polar or azimuthal variations. The data from the latter mode of measurement were known as "flower patterns" – to describe the highly sym-

metrical diffraction patterns which reflected the point group symmetry of the emitting site. For example, Fadley et al. measured flower patterns at XPS energies [87–89] while Woodruff et al. measured such patterns at very low photoelectron energies ($E \approx 40$ eV) [83]. From my experience in LEED, I knew that the diffraction fringes most directly connected to atomic spacings were those in which the emission angles were held fixed while the outgoing electron's wavelength was varied. I called such data energy-dependent photoelectron-diffraction (EDPD) spectra. In the spring of 1978, I had calculated a set of EDPD spectra for c(2 × 2) and p(2 × 2) Se–Ni(001), the emission core level was Se(3d).

In May of 1978, I taught a course on surface science at the University of California, Berkeley. I discussed the theories of LEED and photoelectron diffraction and the EDPD calculations I made on c(2 × 2) and p(2 × 2) Se-Ni(001). In my class were Steven Kevan and Denny Rosenblatt, both then students of David Shirley. Kevan et al. were in a position to measure EDPD spectra at the Stanford Synchrotron Radiation Center. In August and September 1978, a theoretical [90] and experimental [91] paper were submitted respectively for publication. In Fig. 2, I reproduce a figure from a 1979 paper [92] which compared the measured spectra of Kevan et al. [91] with the calculated EDPD spectra corresponding to different adsorption heights of the Se overlayer on Ni(001). The high sensitivity of EDPD spectra to the adatom distance from the substrate was because every scattering path originated from the adatom and was a measure of its distance from neighboring atoms which scattered the photoelectron. EDPD spectra were soon also measured by Gerry Lapeyre's group for c(2 × 2) Na-Ni(001) at the Wisconsin Synchrotron Radiation Center, Stoughton [93] (Lapeyre et al. had a different name for EDPD spectra: constant initial state (CIS) spectra) [94,95].

The use of EDPD spectra to study molecular adsorbate was demonstrated in 1981 [96]. Recently, analysis of EDPD spectra were extended to the study of clean surfaces, by measuring photoelectrons emitted from surface-shifted core levels [97]. There is currently increased interest in

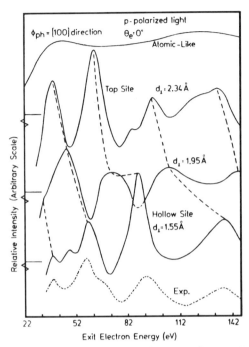

Fig. 2. Calculated EDPD spectra (solid lines) for Ni(001)-p(2×2) Se, with Se at the hollow site, $d_\perp = 1.55$ and 1.95 Å, and at the top site, $d_\perp = 2.34$ Å, respectively (middle three curves). Atomic-like curve corresponds to infinite distance between Se layer and Ni substrate. Chained line is experimental spectra of Kevan et al., Ref. [91]. The emission core state is Se(3d), $\theta_{ph} = 60°$. This figure is reproduced from Ref. [92].

the use of EDPD spectra for structural analysis. Leading these efforts in the US are research groups headed by James Tobin at Lawrence Livermore Laboratory [98,99], Gerry Lapeyre at the Wisconsin Synchrotron Radiation Center, Chuck Fadley at the Advanced Light Source and in Europe, the research groups headed by Alex Brashaw and Phillip Woodruff [100]. Advances in photoelectron diffraction as a structural tool were greatly aided by dynamical methods developed in the 1970's for the LEED problem.

3.2. Dynamical theory of electron-phonon loss cross-sections

Surface phonons in slab geometries were actively studied theoretically in the 1960's and 70's [10,11,101–103]. However, it was not until 1978–80 that it became possible for surface phonons to

be measured by inelastic helium atom scattering [104] and high-resolution electron energy loss spectroscopy (HREELS) [105]. I was particularly excited by the HREELS measurements because I knew that it would not be difficult to develop a dynamical theory to calculate the electron-phonon loss cross-sections in a slab geometry.

In February 1980, C.H. Li, D.L. Mills and I published the first paper on electron-phonon loss cross-section calculations [106]. We showed that the outgoing EELS wave-function corresponding to the displacement in the αth coordinate of an atom at $\boldsymbol{R}_A$ was represented by:

$$|\psi_{\mathrm{EELS}}^{(\alpha)}\rangle = g_{\mathrm{PE}} \left(\frac{\partial V}{\partial R_A^{(i)}} \right)_0 |\psi_{\mathrm{LEED}}^{\mathrm{I}}\rangle, \qquad (10)$$

where $|\psi_{\mathrm{LEED}}^{\mathrm{I}}\rangle$ was the wavefunction from dynamical LEED theory which described the incoming electron's propagation and g_{PE} was the photoemission propagator that described the outgoing part of the wavefunction. We then cast the layer matrices in which a single phonon-loss event occurred in layer A to the form:

$$
\begin{aligned}
Q_{k_g k_{g'}}^{(A,\alpha)\pm\pm} = {} & \frac{16\pi^2 i m}{\alpha \hbar^2} \sum_{L_1 L_2} \sum_{L_3 L_4} \frac{Y_{L_1}(\boldsymbol{k}_g^{(F)\pm})}{k_{\perp g}^{(F)}} \\
& \times \left(1 - t_A G^{AA}(\boldsymbol{k}_F)\right)_{L_1 L_2}^{-1} (F_A^\alpha)_{L_2 L_3} \\
& \times \left(1 - G^{AA}(\boldsymbol{k}_I) t_A\right)_{L_3 L_4}^{-1} Y_{L_4}^*(\boldsymbol{k}_{g'}^{(I)\pm}),
\end{aligned}
\qquad (11)
$$

where the loss matrix elements were given by:

$$
\begin{aligned}
(F_A^\alpha)_{LL'} = {} & (\mathrm{i})^{l'-l} \exp\left[\mathrm{i}(\delta_l^A + \delta_{l'}^A)\right] \\
& \times I(L, \alpha, L') \int_0^{R_{\mathrm{MT}}} \mathrm{d}\rho\, \rho^2 \\
& \times R_l^{(F)}(\rho) \frac{\mathrm{d}v_A(\rho)}{\mathrm{d}\rho} R_{l'}^{(I)}(\rho),
\end{aligned}
\qquad (12a)
$$

and

$$I(L, \alpha, L') = \int \mathrm{d}\Omega\, Y_L^*(\hat{\boldsymbol{\rho}}) (\hat{x}_\alpha \cdot \hat{\boldsymbol{\rho}}) Y_{L'}(\hat{\boldsymbol{\rho}}). \qquad (12b)$$

In Eq. (12a), R_{MT} was the muffin-tin radius. Because the phonon-loss layer-matrices in Eq.

(11) were expressed in the plane-wave representation, they joined conveniently to the layer-by-layer diffraction methods of LEED. One needed to keep track of two sets of diffraction processes: inelastic processes with one-phonon losses and elastic processes with regular LEED diffractions. Details of the method were described in the papers of Li et al. [107,108].

Using this method, the analysis of off-specular scattering, where the wavevector of the phonon modes excited may extend out to the boundary of the surface Brillouin zone, revealed the conditions required for such experiments to be realized in the laboratory. Most particularly, we found the off-specular excitation cross-sections *to be strong in the LEED energy range (50–300 eV)* [106]. Furthermore, we found the energy and angular variations of the off-specular cross-sections to be sensitive to surface structure, and that modes forbidden in the near specular geometry by the dipole selection rule can be excited with substantial cross sections in an off-specular measurement [106]. The first HREELS study of the dispersion relation of a surface phonon throughout the Brillouin zone was carried out under conditions similar to those predicted by the theory [109,110]. The dynamical theory also aided the observation of a surface phonon that had never been measured before. For a fcc (001) surface, lattice dynamics yielded a Rayleigh surface mode (S_4) and a longitudinal gap mode (S_6), among other localized modes at the zone boundary $\overline{X}$. Until 1985, only the Rayleigh (S_4) mode was measured. The gap mode (S_6) had not been detected, but, according to the Born approximation, the S_6 mode should have a cross-section well below the background noise level and hence should not have been seen (see, for example, broken lines, Fig. 3). Although for a long time the gap mode was hidden from measurements, there was no real effort to look for it.

According to results of dynamical calculations [111], it was possible to enhance the S_6 cross-section by multiple scattering, to a magnitude that would easily be measured under current experimental conditions. This was done by either sending an electron in at a small angle from the surface or measuring at a final direction with a

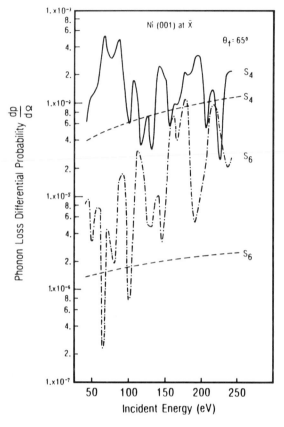

Fig. 3. Calculated electron-phonon loss differential probability for the Rayleigh (S_4) and gap (S_6) modes versus incident beam energy at $\overline{X}$ of Ni(001). The Born approximation results are shown by dashed lines. (Reproduced from Ref. [111].)

small angle from the surface. In such scattering geometries, the electron scattered effectively into directions parallel to the surface. Through subsequent Umklapp processes, the electron could couple strongly to the parallel displacements of the S_6 mode. According to calculation, the S_6 mode for Ni(001) could be seen within such scattering geometries at three energy intervals in the 50–250 eV range. These were from 110 to 115 eV, 155 to 180 eV and 210 to 230 eV (see Fig. 3, chained line). Subsequent measurements uncovered the S_6 mode at exactly the energy intervals predicted by theory [111].

In Fig. 4, we show the measurement of the S_4 and S_6 modes at $\overline{X}$ for Ni(001) at 155 eV and $\theta_f = 65°$, exactly according to the scattering geom-

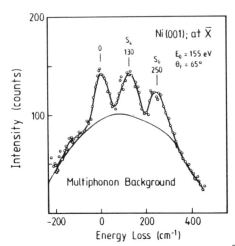

Fig. 4. Experimental electron-phonon loss spectrum at $\overline{\mathrm{X}}$ and the incident $E = 155$ eV, $\theta_{\mathrm{f}} = 65°$, showing both S_4 and S_6 modes. The multiphonon background is calculated in the minima between the peaks by fitting Gaussian functions to the peaks with a half-width taken from the resolution. The full curve is generated by the assumption that the multiphonon back-ground is a smooth function of energy. Above 295 cm^{-1} the entire intensity is due to multiprocesses. (Reproduced from Ref. [111].)

etry predicted by theory. The measurement of the S_6 mode and the determination of its frequency provided valuable input to the surface force constants. A relaxation of the intralayer surface force constants, for example, would lower the S_6 mode, moving it from the gap at $\overline{\mathrm{X}}$ into the bulk band. The gap mode (S_6) has since been experimentally studied in a number of surfaces, such as Cu(001), Cu(111), Ag(001), etc. [112–116].

Again, development of the HREELS technique was greatly aided by quantitative theory, which from Eq. (10), had its roots in LEED and photoemission diffraction methods.

4. Inversion of LEED *IV* spectra

The use of LEED as a structural tool requires choosing trial-and-error geometries for the surface atoms until a satisfactory fit between theory and experiment is obtained. The process is much like that used in X-ray crystallography. However, it has long been a subject of great interest and

debate to find methods which successfully invert LEED *IV* spectra to provide directly atomic arrangements in real space. The strong multiple scattering present in LEED *IV* spectra long has been recognized as a severe obstacle for direct methods. For example, in the historic 1968 paper [33] by Beeby, he wrote in the Introduction:

"It is unlikely that such data will prove capable of inversion to give directly details of the crystal surfaces involved...".

An attempt was made in the 1970's to average intensities over angles [117]. The purpose was to wash out multiple scattering features in integral-order *IV* spectra. This was not a direct method because even after averaging, the end product was a composite *IV* curve which still had to be fitted by trial-and-error model calculations. Another attempt at data inversion was to scalar Fourier transform [118] a specularly reflected *IV* spectrum, which of course contained multiple scattering features. This attempt failed because even in the kinematical limit, a scalar Fourier transformation of a (00) beam *IV* curve with respect to $s_\perp = (\mathbf{k}_i - \mathbf{k}_f)_\perp$ would produce peaks corresponding to interlayer distances $d_{1\perp}$, $d_{2\perp}$, etc. as well as differences of these distances ($\mathbf{d}_2 - \mathbf{d}_1)_\perp$, $(\mathbf{d}_3 - \mathbf{d}_2)_\perp$, $(\mathbf{d}_3 - \mathbf{d}_2)_\perp$, etc. The situation is like using a meter stick to measure layer distances $d_{i\perp}$, but the origin of the meter stick is at unknown and varying positions $d_{j\perp}$.

The modern method of LEED spectra inversion is based on the principle of electron holography [119]. The analogy between photoelectron diffraction and electron holography was pointed out in 1986 by Szoeke [120]. Barton used an optical formula, the Helmholtz–Kirchhoff (HK) integral, to Fourier transform single-energy angular spectra to real space [121]. The single-energy HK integral was also used by Saldin and de Andres to invert diffuse LEED spectra [122]. However, because of multiple scattering and strong angular anisotropies in the atomic scattering factors and/or photoelectron matrix elements, the optical formula (i.e. HK integral) did not generally produce good quality images. To eliminate artifacts due to multiple scattering, Barton suggested phase-summing HK transforms over energies [123].

A more accurate and efficient method, particularly in eliminating scattering factor anisotropies, consisted of (vector) Fourier integrals over wavenumbers of energy-dependent diffraction spectra [124]. This is because in the scan-energy mode of data reduction, the reference wave $I_0(E, \Omega)$ is more simply determined at each exit direction. The vector integral was applied to calculated photoelectron diffraction (EDPD) spectra [124], scan-energy DLEED spectra [125] and fractional-order LEED IV spectra [126] with very encouraging results.

An example of LEED spectra inversion is shown in Fig. 5. The images show individual Cu atoms in the first substrate layer of Cu(001) in the Cu(001)-c(2 × 2) Se system. The images were reconstructed by inverting calculated fractional-order IV spectra of LEED (upper panels) and low-energy positron-diffraction (lower panels), respectively [126]. The images shown, from left to right, are reconstructed by using increasingly more

energies. For electrons (top panels), strong artifacts appear at nonatomic positions with two energies. As more energies are added, the images at the atomic positions grow brighter while the intensities of artifacts become dimmer. With five energies the images, at least for the four nearest-neighbor atoms, are sharply formed, although one can still detect (weak) artifacts approximately midway between the farther atoms. By contrast, when using the same energies with positrons, the atom images are already formed at approximately the correct positions with only two energies. With increasing number of energies, the atomic positions marked by the images become more accurate while the intensities of the artifacts become weaker. Using only five energies in the range 114–166 eV, the images for the first twelve neighbors are sharply formed for the positron case, and the view is essentially free from serious artifacts. These images provide a direct view of the surface structure. Each circle in Fig. 5 has a

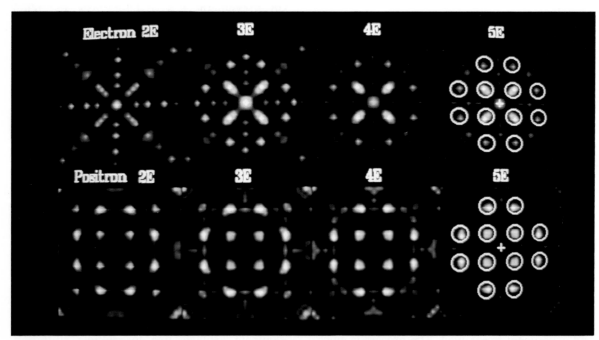

Fig. 5. Images of copper atoms from holographic reconstruction of fractional-order low-energy electron diffraction spectra (upper panels) and low-energy positron diffraction spectra (lower panels) as more energies are added, from left to right. (Reproduced from Ref. [126].)

diameter of 1.1 Å and its center marks the correct atomic position. The images from positron holography are shifted by less than 0.1 Å for the nearest-neighbor atoms and by ≤ 0.3 Å for the next shell of atoms. The "beam splitter", in the above two examples, is a selenium atom at 1.8 Å above the copper surface.

It is apparent that the complementary nature between imaging and diffraction methods makes their combination a promising approach. Holographic imaging offers a three-dimensional view of the surface geometry while quantitative diffraction methods can then be used to determine more accurately the bond lengths and angles. The obstacle of having to deal with massive amounts of trial geometries is eliminated with this combination. Positron holography has an additional advantage due to the weak scattering of positrons in solids which leads to substantial savings in data acquisition. Details of the scan-energy inversion method are found in a number of articles [124,125]. A comprehensive review of the early approaches of internal-source electron holography is covered in an article by Scott Chambers [127].

5. Epilogue and outlook

Before 1972, we knew very little about surface structure. There was practically no information on the distances between surface atoms. Calculations of dynamical properties were based on bulk terminated structures. The first structural results on clean metal surfaces by LEED were obtained in 1972–73 [58,59] and showed that the spacing between layers was usually *less* than it was in the bulk. Later calculations that allowed the participation of several layers in the relaxation process indicated that multilayer relaxation was a common phenomenon on many metal surfaces [128–130]. The earliest structural determinations of chemisorbed atoms on metal surfaces were done in 1973–75 [13–16,131]; the first structural determination of a semiconductor surface, GaAs(110), was done in 1976–78 [132–135]. In each case, LEED was the technique that first found the correct surface spacings. As with other techniques, some mistakes were made along the way. But the technique was robust enough that incorrect determinations were quickly discovered and correct answers found. Current versions of the LEED method are accurate and efficient enough to determine, for example, multilayer structural parameters of large unit cell surfaces such as Si(111)7 × 7 [69,70] and Ge(111) c(2 × 8) [136].

In the twenty years since the beginning of LEED crystallography, researchers have amassed a large body of information on surface structure. We now have a much more realistic and quantitative view of surface atomic arrangement, adsorbate binding sites and even the distribution of steps on surfaces. We are at the threshold of actually being able to look at a surface at the atomic level through new imaging techniques. Looking back at the development of the field, we see achievements beyond the most optimistic expectation of many researchers.

When I researched much of the literature of the past 30 years for this article, I was impressed by the diversity and richness of ideas contained in these papers. As is usual in the development of any science, there were false starts and blind alleys; but, out of the confusion, there emerged enduring ideas that left a valuable scientific legacy. This legacy will be passed on. Future generations will benefit from this knowledge, which is based on sound and fundamental principles of multiple scattering of electrons that are used to study surface and interface problems.

Acknowledgements

I dedicate this article to the memory of Lester Germer, the man who co-discovered electron diffraction in 1927. Shortly after I arrived at Cornell in the fall of 1969, Germer came to my office to introduce himself to me simply because he heard that I was planning to work on the LEED problem. That was the first of many visits. Germer patiently explained the results of his pioneering experiments to me. As I write this article nearly 24 years after the first meeting, the fond memories of these pleasant afternoon visits are still vivid in my mind. This work is supported in part by NSF Grant no. DMR 8805938, DOE Grant no. DE-FG02-84ER45076 and ONR Grant no. N00014-90-J-1749.

References

[1] C.J. Davisson and L.H. Germer, Phys. Rev. 30 (1927) 705.

[2] H.E. Farnsworth et al., J. Appl. Phys. 26 (1955) 252; 29 (1958) 1150; 30 (1959) 917.

[3] L.H. Germer, Sci. Am. 212 (1965) 32.

[4] J.J. Lander, Progr. Solid State Chem. 2 (1965) 26.

[5] W. Ehrenberg, Philos. Mag. 18 (1934) 878.

[6] R.E. Weber and W.T. Peria, J. Appl. Phys. 38 (1967) 4355.

[7] P.W. Palmberg and T.N. Rhodin, J. Appl. Phys 39 (1968) 2425.

[8] R.E Schlier and H.E. Farnsworth, J. Chem. Phys. 30 (1959) 917.

[9] F. Jona, Helv. Phys. Acta 41 (1968) 960.

[10] S.Y. Tong and A.A. Maradudin, Phys. Rev. 181 (1969) 1318.

[11] R.F. Wallis, D.L. Mills and A.A. Maradudin, in: Localized Excitations in Solids, Ed. R.F. Wallis (Plenum, New York, 1968) p. 403.

[12] J.E. Demuth and T.N. Rhodin, Surf. Sci. 42 (1974) 261; 45 (1974) 249.

[13] S. Andersson, J.B. Pendry, B. Kasemo and M. Van Hove, Phys. Rev. Lett 31 (1973) 595.

[14] J.E. Demuth, D.W. Jepsen and P.M. Marcus, Phys. Rev. Lett. 31 (1973) 540.

[15] C.B. Duke, N.O. Lipari, G.E. Laramore and J.B. Theeten, Solid State Commun. 13 (1973) 579.

[16] M. Van Hove and S.Y. Tong, J. Vac. Sci. Technol. 12 (1975) 230.

[17] F. Jona, IBM J. Res. Devel. 14 (1970) 444.

[18] A. Ignatiev, J.B. Pendry and T.N. Rhodin, Phys. Rev. Lett. 26 (1971) 189.

[19] A. Ignatiev and T.N. Rhodin, Phys. Rev. B 8 (1973) 808.

[20] S.Y. Tong, T.N. Rhodin and A. Ignatiev, Phys. Rev. B 8 (1973) 906.

[21] A. Ignatiev, T.N. Rhodin and S.Y. Tong, Surf. Sci. 42 (1974) 37.

[22] J.W. Gadzuk, Surf. Sci. 6 (1967) 133.

[23] E.G. McRae, J. Chem. Phys 45 (1966) 3258; Surf. Sci. 7 (1967) 41; 8 (1967) 14.

[24] D.S. Boudbreaux and V. Heine, Surf. Sci. 8 (1967) 426.

[25] V. Heine, Surf. Sci. 2 (1964) 1; Proc. Phys. Soc. 81 (1963) 300.

[26] P.M. Marcus and D.W. Jepsen, Phys. Rev. Lett 20 (1968) 925.

[27] K. Kambe, Z. Naturforsch. 22a (1967) 322, 422.

[28] J.B. Pendry, J. Phys. C 2 (1969) 1215.

[29] J.B. Pendry, J. Phys. C 2 (1969) 2273.

[30] J.B. Pendry, J. Phys. C 2 (1969) 2283.

[31] J.B. Pendry and G. Capart, J. Phys. C 2 (1969) 841.

[32] V. Heine and J.B. Pendry, Phys. Rev. Lett. 22 (1969) 1003.

[33] J.L. Beeby, J. Phys C 1 (1968) 82.

[34] E.R. Jones, J.T. McKinney and M.B. Webb, Phys. Rev. 151 (1966) 467.

[35] P. Palmberg and T.N. Rhodin, J. Appl. Phys. 39 (1968) 2425.

[36] C.B. Duke and C.W. Tucker, Jr., Surf. Sci. 15 (1969) 231.

[37] C.B. Duke and C.W. Tucker, Jr., Phys. Rev. Lett 23 (1969) 1163.

[38] C.B. Duke, J.R. Anderson and C.W. Tucker, Jr., Surf. Sci. 19 (1970) 117.

[39] C.W. Tucker, Jr. and C.B. Duke, Surf. Sci. 24 (1971) 31.

[40] C.B. Duke and G.E. Laramore, Phys. Rev. B 2 (1970) 4765.

[41] C.B. Duke and G.E. Laramore, Phys. Rev. B 2 (1970) 4783.

[42] P.M. Marcus, D.W. Jepsen and F. Jona, Surf. Sci. 17 (1969) 442.

[43] S.Y. Tong, in: Progress in Surface Science, Vol. 7 (Ed. S.G. Davison) (Pergamon, New York, 1975).

[44] B.I. Lundqvist, Phys. Status Solidi 32 (1969) 273.

[45] J.J. Quinn, Phys. Rev. 126 (1962) 1453.

[46] D.E. Beck, Phys. Rev B 4 (1971) 1555.

[47] P.J. Feibelman, Phys. Rev. 176 (1968) 551.

[48] G. Capart, Surf. Sci. 26 (1971) 429.

[49] S.Y. Tong and T.N. Rhodin, Phys. Rev. Lett. 26 (1971) 711.

[50] P.J. Jennings and E.G. McRae, Surf. Sci. 23 (1970) 63.

[51] D.W. Jepsen, P.M. Marcus and F. Jona, Phys. Rev. Lett 26 (1971) 1365.

[52] J.A. Strozier, Jr. and R.O. Jones, Phys. Rev. Lett. 25 (1970) 516.

[53] J.A. Strozier, Jr. and R.O. Jones, Phys. Rev. B 3 (1971) 3228.

[54] R.O. Jones and J.A. Strozier, Jr., Phys. Rev. Lett 22 (1969) 1186.

[55] J.B. Pendry, Low-Energy Electron Diffraction (Academic Press, London, 1974).

[56] M.A. Van Hove and S.Y. Tong, Surface Crystallography by Low Energy Electron Diffraction: Theory, Computation and Structural Results (Springer, Berlin, 1979).

[57] M.A. Van Hove, W.H. Weinberg and C.-M. Chan, Low-Energy Electron Diffraction: Experiment, Theory and Structural Determination, Vol. 6 of Springer Series in Surf. Sciences (Springer, Berlin, 1986).

[58] G.E. Laramore and C.B. Duke, Phys. Rev. B 5 (1972) 267.

[59] D.W. Jepsen, P.M. Marcus and F. Jona, Phys. Rev. B 5 (1972) 3933.

[60] S.Y. Tong and L.L. Kesmodel, Phys. Rev. B 8 (1973) 3753.

[61] J.B. Pendry, Phys. Rev. Lett. 27 (1971) 856.

[62] J.B. Pendry, Low-Energy Electron Diffraction Theory (Academic Press, London, 1974).

[63] R.S. Zimmer and B.W. Holland, J. Phys. C 8 (1975) 2395.

[64] S.Y. Tong and M.A. Van Hove, Phys. Rev. B 16 (1977) 1459.

[65] P.J. Rous, J.B. Pendry, D.K. Saldin, K. Heinz, K. Müller and N. Bickel, Phys. Rev. Lett 57 (1986) 2951.

[66] J. Rundgren and A. Salwen, J. Phys. C 7 (1974) 4247; 9 (1976) 3701.

[67] M.A. Van Hove and J.B. Pendry, J. Phys. C 8 (1975) 1362.

[68] W. Moritz, J. Phys. C 17 (1983) 353.

[69] H. Huang, S.Y. Tong, W.E. Packard and M.B. Webb, Phys. Lett. A 130 (1988) 166.

[70] S.Y. Tong, H. Huang, C.M. Wei, W.F. Packard, F.K. Men, G. Glander and M.B. Webb, J. Vac. Sci. Technol. A 6 (1988) 615.

[71] G. Kleinle, W. Moritz and G. Ertl, Surf. Sci. 238 (1990) 119.

[72] W. Moritz, H. Over, G. Kleinle and G. Ertl, in: The Structure of Surfaces III, Eds. S.Y. Tong, M.A. Van Hove, X. Xide and K. Takayanagi (Springer, Berlin, 1991) p. 128.

[73] H. Over, U. Ketterl, W. Moritz and G. Ertl, Phys. Rev. B 46 (1992) 15438.

[74] The CAVLEED program suite is used; see, for example, H. Zhao and S.P. Tear, in: Structure of Surfaces-IV (Eds. X. Xie and S.Y. Tong) (World Scientific, New Jersey, 1994).

[75] A. Liebsch, Phys. Rev. Lett. 32 (1974) 1202.

[76] A. Liebsch, Phys. Rev. B 13 (1976) 544.

[77] S.Y. Tong, C.H. Li and A.R. Lubinsky, Phys. Rev. Lett. 39 (1977) 498.

[78] C.H. Li, A.R. Lubinsky and S.Y. Tong, Phys. Rev. B 17 (1978) 3128.

[79] W.M. Kang, University of Wisconsin-Milwaukee PhD Thesis (1982).

[80] S.P. Weeks and E.W. Plummer, Solid State Commun. 21 (1977) 695.

[81] D.P. Woodruff, Surf. Sci. 53 (1975) 538.

[82] N.V. Smith, P.K. Larsen and S. Chiang, Phys. Rev. B 16 (1977) 2699.

[83] D.P. Woodruff, D. Norman, B.W. Holland, N.V. Smith, H.H. Farrell and M.M. Traum, Phys. Rev. Lett. 41 (1978) 1130.

[84] N.V. Smith and M.M. Traum, Phys. Rev. B 11 (1975) 2087.

[85] D. Norman, D.P. Woodruff, N.V. Smith, M.M. Traum and H.H. Farrell, Phys. Rev. B 18 (1978) 6789.

[86] M.M. Traum, J.E. Rowe and N.V. Smith, Vac. Sci. Technol. 12 (1975) 298.

[87] S. Kono, C.S. Fadley, N.F.T. Hall and Z. Hussain, Phys. Rev. Lett. 41 (1978) 117.

[88] S. Kono, S.M. Goldberg, N.F.T. Hall and C.S. Fadley, Phys. Rev. Lett. 41 (1978) 1831.

[89] L.G. Petersson, S. Kono, N.F.T. Hall, C.S. Fadley and J.B. Pendry, Phys. Rev. Lett. 42 (1979) 1545.

[90] C.H. Li and S.Y. Tong, Phys. Rev. B 19 (1979) 1769.

[91] S.D. Kevan, D.H. Rosenblatt, D.R. Denley, B.C. Lu and D.A. Shirley, Phys. Rev. Lett. 41 (1978) 1565.

[92] C.H. Li and S.Y. Tong, Phys. Rev. Lett. 42 (1979) 901.

[93] G.P. Williams, I.T. McGovern, F. Cerrina and G.J. Lapeyre, Solid State Commun. 31 (1979) 15.

[94] G.J. Lapeyre, A.D. Baer, J. Anderson, J.C. Hermanson, J.A. Knapp and P.L. Gobby, Solid State Commun. 15 (1974) 1601.

[95] G.J. Lapeyre, R.J. Smith, J.A. Knapp and J. Anderson, J. Phys. (Paris) 39 (1978) C4-134.

[96] S.D. Kevan, R.F. Davis, J.G. Tobin, D.A. Shirley, C.H. Li and S.Y. Tong, Phys. Rev. Lett. 46 (1981) 1629.

[97] R.A. Bartynski, D. Heskett, K. Garrison, G. Watson, D.M. Zehner, W.N. Mei, S.Y. Tong and X. Pan, Phys. Rev. B 40 (1989) 5340.

[98] J.G. Tobin, J.C. Hansen and M.K. Wagner, J. Vac. Sci. Technol. A 8 (1990) 2494.

[99] J.G. Tobin, G.D. Waddill, H. Li and S.Y. Tong, Symp. Proc. Mat. Res. Soc. (1992).

[100] R. Dippel, D.P. Woodruff, X.-M. Hu, M.C. Asensio, A.W. Robinson, K.M. Schindler, K.-U. Weiss, P. Gardner and A.M. Bradshaw, Phys. Rev. Lett. 68 (1992) 1543.

[101] R.E. Allen, G.P. Alldredge and F.W. de Wette, Phys. Rev. B 4 (1971) 1661.

[102] R.E. Allen, G.P. Alldredge and F.W. de Wette, Phys. Rev. Lett. 23 (1969) 1285.

[103] R.F. Wallis, Prog. Surf. Sci. 4 (1973) 233.

[104] G. Brusdeylins, R.B. Doak and J.P. Toennies, Phys. Rev. Lett. 44 (1980) 1417.

[105] W. Ho, R.F. Willis and E.W. Plummer, Phys. Rev. Lett. 40 (1978) 1463.

[106] S.Y. Tong, C.H. Li and D.L. Mills, Phys. Rev. Lett. 44 (1980) 407.

[107] C.H. Li, S.Y. Tong and D.L. Mills, Phys. Rev. B 21 (1980) 3057.

[108] S.Y. Tong, C.H. Li and D.L. Mills, Phys. Rev. B 24 (1981) 80.

[109] S. Lehwald, J.M. Szeftel, H. Ibach, T.S. Rahman and D.L. Mills, Phys. Rev. Lett. 50 (1983) 518.

[110] J.M. Szeftel, S. Lehwald, H. Ibach, T.S. Rahman, J.E. Black and D.L. Mills, Phys. Rev. Lett. 51 (1983) 268.

[111] M.L. Xu, B.M. Hall, S.Y. Tong, M. Rocca, H. Ibach, S. Lehwald and J.E. Black, Phys. Rev. Lett. 54 (1985) 1171.

[112] Y. Chen, S.Y. Tong, M. Rocca, P. Moretto, U. Valbusa, K.P. Bohnen and K.M. Ho, Surf. Sci. Lett. 250 (1991) L389.

[113] M.H. Mohamed, L.L. Kesmodel, B.M. Hall and D.L. Mills, Phys. Rev. B 37 (1988) 2763.

[114] B.M. Hall, D.L. Mills, M.H. Mohamed and L.L. Kesmodel, Phys. Rev. B 38 (1988) 5856.

[115] Y. Chen, S.Y. Tong, K.P. Bohnen, T. Rodach and K.M. Ho, Phys. Rev. Lett. 70 (1993) 603.

[116] Y. Chen, S.Y. Tong, J.S. Kim, L.L. Kesmodel, T. Rodach, K.P. Bohnen and K.M. Ho, Phys. Rev. B 44 (1991) 11394.

[117] M.B. Webb and M.G. Lagally, Solid State Phys. 28 (1973) 301.

[118] U. Landman and D. Adams, Phys. Rev. Lett. 33 (1974) 585.

[119] D. Gabor, Proc. Roy. Soc. (London) A 197 (1949) 454.

[120] A. Szoeke, in: Short Wavelength Coherent Radiation: Generation and Applications, Eds. D.T. Attwood and J. Boker, AIP Conf. Proc. No. 147 (AIP, New York, 1986).

[121] J.J. Barton, Phys. Rev. Lett. 61 (1988) 1356.

[122] D.K. Saldin and P.L. de Andres, Phys. Rev. Lett. 64 (1990) 1270.

[123] J.J. Barton, Phys. Rev. Lett. 67 (1991) 3106.

[124] S.Y. Tong, H. Huang and C.M. Wei, Phys. Rev. B 46 (1992) 2452.

[125] S.A. Chambers, V.A. Loebs, H. Li and S.Y. Tong, J. Vac. Sci. Technol. B 10 (1992) 2092.

[126] S.Y. Tong, H. Huang and X.Q. Guo, Phys. Rev. Lett. 69 (1992) 3654.

[127] S.A. Chambers, Surf. Sci. Rep. 16 (1992) 261.

[128] D.L. Adams, H.B. Nielsen, J.N. Andersen, I. Stensgaard, R. Feidenhans'l and J.E. Sorensen, Phys. Rev. Lett. 49 (1982) 669.

[129] H.L. Davis and J.R. Noonan, Surf. Sci. 126 (1983) 245.

[130] J. Sokolov, H.D. Shih, U. Bardi, F. Jona and P.M. Marcus, Solid State Commun. 48 (1983) 739.

[131] For a discussion of these early results, see: T.N. Rhodin and S.Y. Tong, Phys. Today 28 (1975) 10.

[132] A.R. Lubinsky, C.B. Duke, B.W. Lee and P. Mark, Phys. Rev. Lett. 36 (1976) 1058.

[133] S.Y. Tong, A.R. Lubinsky, B.J. Mrstik and M.A. Van Hove, Phys. Rev. B 17 (1978) 3303.

[134] A. Kahn, G. Cisneros, M. Bonn, P. Mark and C.B. Duke, Surf. Sci. 71 (1978) 387.

[135] M.W. Puga, G. Xu and S.Y. Tong, Surf. Sci. 164 (1985) L789.

[136] S.Y. Tong, H. Huang and C.M. Wei, in: Chemistry & Physics of Solid Surfaces VIII Eds. R. Vanselow and R. Howe (Springer, Berlin, 1990) p. 395.

Surface Science 299/300 (1994) 375–390
North-Holland

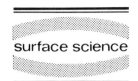

surface science

Multiple scattering theory of electron diffraction

J.B. Pendry

The Blackett Laboratory, Imperial College, London SW7 2BZ, UK

Received 8 April 1993; accepted for publication 12 May 1993

In the early 1960's surface science set itself some fundamental goals: to make a quantitative science out of surface crystallography; to understand the nature of electronic structure and bonding at surfaces; and to enhance the tools available for study of surfaces. The effort has very much been a collective one, reflected in the wide authorship of the present volume. Here I contribute to the picture my personal perspective on developments in the past 30 years of surface science, and describe some of the highlights in my own research and that of my close colleagues.

1. The challenge facing electron diffraction in 1964

Surfaces have a central role in science and technology and as such have always been objects of intense study. From time to time the focus of this activity changes, and 1964 was one of those pivotal points in the subject. Hitherto interest in surfaces had largely been to do, not with their detailed atomic properties which were inaccessible, but with the consequences of those properties and with mesoscale phenomena such as friction. At the same time there were pioneering spirits interested in atomic scale phenomena who worked at the techniques necessary to study them. It was from their work that the general interest in atomic and electronic phenomena at surfaces took off. Did technology call forth the science, or was science the mother of technology? Whatever the cause and effect symbiosis between the two resulted in an explosion of interest, and in the generous funding needed to carry out the ambitious programs put in place. At that time there was a belief that science and technology needed one another, a faith that is not so evident thirty years on.

The new journal, Surface Science, firmly set its cap at atomic and electronic interests and was certainly a guiding light to me. At the beginning of my PhD in October 1966 Volker Heine sent me to the Cavendish library to glean what I could from the new journal to help with my project on *theory of low energy electron diffraction*. The task was relatively simple as few volumes adorned the shelves and the avalanche of papers was still a twinkle in the editor's eye. What struck a theorist was the clearly stated objective of experimentalists like Farnsworth and Germer to find out where the atoms were at surfaces, and the near total failure they encountered with their X-ray diffraction based interpretation of electron diffraction data. Electrons just did not behave like X-rays because they were scattered too strongly by the surface and account had to be taken of multiple scattering effects. If ever there was a princess and a dragon surely surface crystallography was the princess and multiple scattering was the dragon. Theorists rode to the rescue in droves.

At centre stage was the foundation of surface crystallography as a subject which awaited a decent theory to interpret LEED data but flowing from this central objective were other questions to do with how atoms and molecules vibrated and diffused at surfaces, and how they reacted with one another. How did surface geometry control the electronic properties of a surface as in semiconductor devices, and how in turn did valence electronic structure control the atomic arrangement? These questions called forth a major program of theoretical development ranging from solving the multiple scattering problem, to understanding the electron surface interaction through local density methods, to calculation of the total energy of surfaces.

One thing I have become conscious of is how much surface science is a collective effort, and how many have contributed to its present success. Here I have been asked to write about my own contribution, but I look forward with pleasure to reading how others saw it.

2. Pre-history and the state of the theoretical art in 1964

At that time much of what we knew about surfaces was based on speculation. The best guess for atomic arrangement was based on truncation of the bulk structure and although the field ion microscope could certainly "see" atoms at the surface it could do so only in plan view and only for surfaces of materials that could withstand the powerful electric fields necessary to form the image. Models of adsorbates were equally crude: most people thought that alkali metals stuck to transition metals as ionised species with radii to match; bonding of covalent species to surfaces was via surface states. Bond lengths at surfaces were to be inferred by analogy with bulk crystallography.

Study of surface states would captivate large number of theorists in the 1970's partly because of their celebrated role in the invention of the transistor, partly because of the appealing picture they made of bond breaking and formation at the surface. Although well founded for semiconductors this picture was extended onto thin ice by some who applied it to adsorption on transition metals, where the existence of surface states was only put on a proper theoretical basis in 1971, but sadly without confirmation of any central role in bond formation for the surface state. Experimental evidence to support the theory came later with the introduction of angle resolved valence band photoemission experiments by Neville Smith, Per Olof Nilsson, and others.

The movement of atoms at surfaces had been discussed, but was the subject of debate. The "ball and springs" movement thought that there were less bonds at a surface therefore vibrational frequencies should be lower, and the top layer of atoms should relax outward. Little help was forthcoming from experiment where measurement of relaxation at a surface awaited solution of the multiple scattering problem, and interpretation of temperature dependence of diffraction data was confused. Diffusion and reaction at surfaces was the subject to be in for those with febrile imagination, not wishing to be fettered by experiment: no one new where the reaction started from, how it progressed, or in what configuration it finished. This vacuum of understanding was to produce hotly argued debates in the 1970's about whether molecules like carbon monoxide were "standing up" or "lying down" at surfaces, or just wagging around from one position to the other.

The routine achievement of ultra high vacuum and very high surface cleanliness was something of a novelty so that observation of complex diffraction patterns from "clean" surfaces was generally taken as a signal that they were not clean. Thus surface reconstruction was not recognised as the near universal phenomenon that it is today.

The problems seemed so hard and intractable that it was easy to be pessimistic about the possibility of ever making a useful statement about surface mechanisms based on fundamental investigations. Yet many of the goals set in 1964 have been achieved, and in ways that could not even have been dreamt of at that time. Experience has taught me to be an optimist, at least in surface science.

3. Foundations of low energy electron diffraction theory

In the summer of 1970 Thor Rhodin invited me to Cornell where a strong team was working on LEED which included Joe Demuth, Alex Ignatiev, Arthur Jones, and Dave Tong. In other parts of the laboratory Lester Germer and Dave

Adams were also working on LEED experiments. Bob Jones and John Strozier were investigating a pseudopotential approach to the theory. Luckily for me Alex Ignatiev had succeeded in taking some diffraction data for solid xenon which I fitted with a weak-scattering pseudopotential theory. The data [1] are shown below in Fig. 1. In solid xenon the inelastic mean free path is uniquely short relative to the inter atomic spacing limiting the amount of multiple scattering. In this material the simple "kinematic theory" with which experimentalists had been struggling for years correctly predicts peaks close to the Bragg conditions. Few other materials are so obliging: aluminium is almost as good as Tong was to show later, but the great majority of surfaces refused to give agreement with kinematic predictions. Copper became a touchstone for multiple scattering theory partly because Stig Andersson had taken some excellent data [2]. Kinematic theory did not even get close for copper, and I left Cornell encouraged by our success with xenon, largely an experimental triumph, and determined to do something about the copper problem. Others were in the same mood.

Electronic theory of surfaces started life with a history of bulk calculations. There was a more relevant experience available in high energy electron diffraction theory already well developed [3] but, probably because of the background of the main players, it was the calculation of valence band electronic structure in the bulk that set the scene for most of us. From this standpoint surface electronic structure appears rather difficult.

There are several sources of difficulty:

(i) In a surface experiment electrons of fixed energy hit a surface at a given angle, fixing the momentum parallel to the surface. This beam excites all possible Bloch waves in the surface with fixed $(E, k_\parallel)$. In contrast the methods available for calculating electronic structure would fix $(k_\parallel, k_z)$ and calculate E. To find k_z given $(E, k_\parallel)$ involved a trial and error process of guessing k_z and testing to see if it corresponded to the correct $(E, k_\parallel)$: a very messy and time consuming process when k_z may be complex. I ought to know, I made some of those calculations!

(ii) The electron–surface interaction was not well known at this time. At low energies there was the Slater prescription for calculating a po-

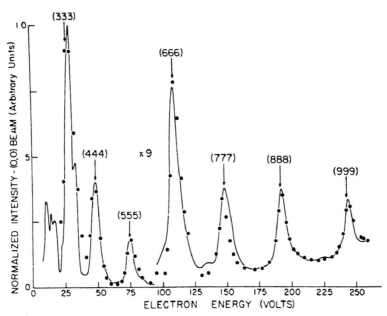

Fig. 1. Intensity versus energy curve measured for the (111) surface of xenon. The dots show the result of a weak scattering theory, and the vertical bars indicate where Bragg diffraction peaks can be expected.

tential, firmly rooted in the assumption that one was calculating in the valence band and including corrections for exchange and correlation known to be approximately energy independent in the valence band. The high energy electron diffraction people worked at energies where it was safe to assume that these corrections were small. We were caught somewhere in between. Other approaches to bulk valence band calculations were based on the concept of a *pseudopotential*. Again this concept made assumptions about the energy of states being calculated.

(iii) Inelastic scattering was known to be important. (Our Editor played a key role in bringing this to the attention of the community. I well remember him rebuking others for leaving out this term saying that he himself had learnt about it "in high school". A fine tribute to his high school!). Inelastic scattering is indeed vital to a surface calculation because it limits penetration of electrons into the solid and helps make the diffraction process surface sensitive. No guidance

was available from valence band calculations, and in high energy electron diffraction inelastic scattering is mainly due to diffuse thermal effects which are treated empirically.

Difficulty (iii) was solved by Lundqvist [4] who calculated the electron self energy in a uniform electron gas of various densities. It happens that the main contribution comes from plasmon loss terms which are not much affected by details of valence electronic structure: most materials give an imaginary part to the LEED electron of between 3 and 5 eV. We are fortunate that this term is so insensitive to the surface, otherwise we would have found the theory much harder than we did. Lundqvist's calculations have also played a central role in estimating the exchange and correlation corrections to the electron self energy which are vital to accurate valence band calculations, and are still important even at LEED energies.

I had spent a lot of time developing pseudopotential methods in collaboration with Gilles Ca-

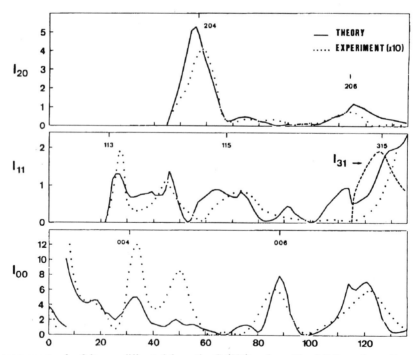

Fig. 2. Intensity energy spectra for 3 beams diffracted from the Cu(001) surface. The full lines shows Capart's pseudopotential calculation, and the dotted line Andersson's experimental data. Positions of Bragg peaks are indicated at the top of each panel.
This is the first time that theory had calculated anything that looked like experiment for a strongly multiple scattering material.

part and it was a version of these methodologies that first successfully interpreted the copper data. Gilles' calculations [5] are shown below in Fig. 2. Note that this theory correctly predicts most of the non-Bragg peaks and is therefore addressing the multiple scattering problem. The calculation was a tour de force because pseudopotential calculations require a guessing process to calculate Bloch wavevectors as discussed in difficulty (i) above. In the event this was the first and almost the last triumph for the pseudopotential theory (Strozier and Jones were to make some interesting developments which they applied to beryllium [6,7]). The method was already stretched to its limit for a clean copper surface.

Eion McRae had developed an alternative theory [8] borrowing from methods introduced by Lax for solving multiple scattering problems in optics, but it was so complex that it was, I believe, only implemented for s-wave scattering. It rapidly became clear that many phase shifts were required: Capart had used five phase shifts to construct his pseudopotential. Nevertheless Eion's theory was a pointer to the future and subsequent development would involve multiple scattering approaches. Eion was generous with his time in helping younger scientists, including myself.

A great step forward came with the layer methodologies in which the surface is decomposed into a series of layers. Each layer is formally separated by a region of constant potential, which can in principle be shrunk to zero width and so constitutes no restriction on the generality of the formulation. Two sets of plane waves defined in the region of constant potential constitute the basis set, one set travelling forwards, the other backwards. Fig. 3 makes the point. In the region of constant potential between the nth and $(n + 1)$th layers the wavefield takes the form,

$$\sum_g a_{ng}^+ \exp\left[i K_g^+ \cdot (r - nc)\right]$$
$$+ a_{ng}^- \exp\left[i K_g^- \cdot (r - nc)\right], \qquad (1)$$

where,

$$K_g^\pm = \left[k_\parallel + g, \pm \sqrt{2E - 2V_0 - |k_\parallel + g|^2} \right], \qquad (2)$$

and c is the displacement of one layer from the next (assumed constant for simplicity). In this picture we need only know the transmission $t_{gg'}^{++}$, $t_{gg'}^{--}$, and reflection $t_{gg'}^{+-}$, $t_{gg'}^{-+}$, coefficients of the layers. The superscripts denote the direction of incident and transmitted or reflected beams. These layer scattering amplitudes can be calculated either by real space methods as suggested by Beeby [9], or by a layer equivalent of the KKR method developed by Kambe [10–12]. I soon decided that these layer methods were the right way forward and adapted them to the calculation of Bloch waves. My method [13,14] had the advantage over pseudopotential methods that it automatically gave $k_z(E, k_\parallel)$ without guesswork. This gave rise to a compact and efficient computer code which could easily cope with the many phase shifts required, and the many beams between which the electron multiply scattered. My calculations for copper [14] are shown below in Fig. 4 and compared to Andersson's experiments [2]. Marcus and Jepsen at IBM Yorktown Heights were also convinced that a layer approach combined with Bloch wave methods would give good results, and soon after my paper they published calculations with the added sophistication of a Debye–Waller correction for temperature, which compared well with experiments for copper, silver, and aluminium [15]. Laramore and Duke made similar calculations for aluminium [16].

By this stage it had become apparent that a

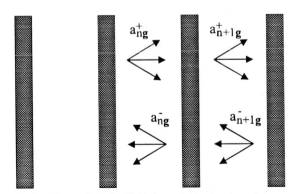

Fig. 3. The surface is divided up into a series of layers separated by regions of constant potential, and between each pair of layers the wavefield is expanded in forward and backward travelling waves.

layer multiple scattering approach which included an accurate description of elastic scattering using many phase shifts, and took account of the short inelastic path length, could accurately reproduce experiment. On the computers of the day the computational effort was substantial (even the largest machine was not as powerful as today's personal computers) and it was obvious that if multiple scattering theory was to interpret diffraction data from complex adsorbate systems, we had better get on and improve the speed of the method. The task before us seemed immense because few could guess just how powerful com-

puters would become, and how successful we should all be at improving the theory.

There are several ways of calculating the total reflection coefficient of a surface, given those of the individual layers. Originally the Bloch waves of the system were calculated, but a much simpler way is to use the "layer doubling method" first introduced in 1974 [17]. Here two layers can be combined into a single scattering entity by the multiple scattering formula,

$$t^{++}(12) = t^{++}(1)[1 - t^{+-}(2)t^{-+}(1)]^{-1}t^{++}(2), \tag{3}$$

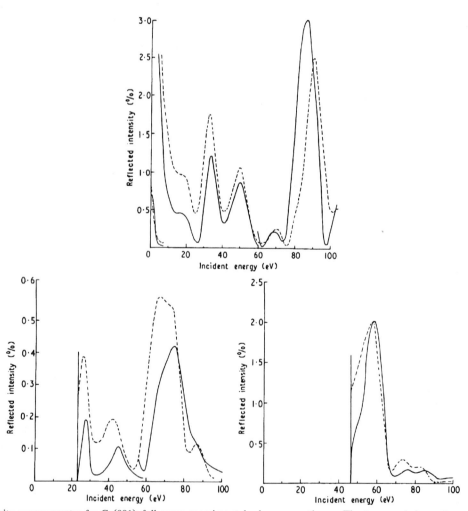

Fig. 4. Intensity energy spectra for Cu(001): full curve experiment, broken curve, theory. The top panel shows the specular beam, the next the (11) beam, and finally the (20) beam.

with similar expressions for $t^{+-}(12)$, $t^{-+}(12)$, $t^{--}(12)$. Because of the inelastic scattering we need consider only a finite number if layers and we treat the identical layers in the bulk region by first combining two bulk layers into a pair of scatterers, then combine two pairs into a quartet and so on, rapidly building up to the required thickness. In general the surface region contains several distinct layers which present no problem and can be treated by the same set of formulae.

Another approach to saving computer time was to exploit the speed of perturbation theory methods. Unfortunately the conventional Born series approach does not work very well: either it converges slowly or sometimes may actually diverge. Fig. 5 illustrates the problem. However the problem with conventional perturbation theory can be put right by recognising that all the strong scattering is in the forward direction and can easily be accounted for exactly by a trick, leaving weak back scattering as the remaining perturbation. This is the renormalised forward scattering (RFS) perturbation scheme [18,19] and is shown in action on the right hand side of Fig. 5.

The layer doubling and RFS schemes are perhaps the two most common methods used to analyse LEED data. With the introduction of these methodologies LEED has found its own place in scattering theory, and related versions of this theory have been applied to numerous spectroscopies involving low energy electrons, such as angle resolved photoemission. The same methodologies have also had considerable impact on calculation of valence electronic structure of surfaces, and associated total energies, because of the ideal adaptation to surface geometry. Surface calculations began as the ugly duckling of scattering theory, but soon matured into an elegant methodology of their own.

The goal in everyone's minds was to find atomic arrangements at surfaces using LEED theory to interpret diffraction data. In this respect LEED theory was quite special. There were several other difficult problems around at that time, such as the Kondo problem, but these differed from LEED theory in that when they were solved someone would write a review article and everyone would move on to the next problem. LEED theory was customer driven and as soon as it arrived there was a job of work to be done investigating the crystallography of surfaces which had for so long been a field for speculation. In 1972 several groups were in possession of computer codes capable of interpreting diffraction data. Stig Andersson and I made a first attempt with the c(2 × 2)Na/Ni(001) system. Stig had a clever technique whereby he could measure the specular beam at normal incidence and this enabled me to use highly symmetrised LEED codes which were more than an order of magnitude

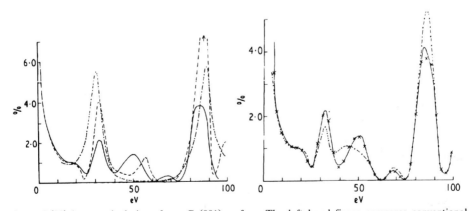

Fig. 5. Comparison of (00) beam calculations for a Cu(001) surface. The left hand figure compares conventional perturbation theory: full line, accurate Bloch wave calculation; chain curve, first order perturbation calculation; dashed curve, second order perturbation calculation. The right hand figure compares the renormalised forward scattering (RFS) perturbation theory: full line, accurate Bloch wave calculation; chain curve, one pass perturbation; crosses, two pass perturbation.

faster than the unsymmetrised variety. This considerably reduced pressure on the Cambridge computing service and early in 1972 we made the first attempt at structural analysis using the full scale multiple scattering theory. Our results [20] are shown in Fig. 6. This first exercise in surface structure determination was useful in more than one respect. It certainly showed that theory and experiment could be brought together to do the job that surface scientists had been trying to do for many years, but at the same time it illustrated the problems of trying to pin down something as mathematically precise as a surface structure with a single experiment. In fact the data set we had used was much too small for the task and although we correctly predicted that the sodium lies in the hollow site, the height was not quite right: as Demuth et al. [21] were to show using a much larger data set, it should have been $d = 2.23$ Å, not $d = 2.87$ Å.

The era of quantitative surface crystallography had begun. The first determination using a larger data set was by Forstmann et al. [22] for the $(\sqrt{3} \times \sqrt{3})30°I/Ag(111)$ structure, and throughout the 1970's structure after structure was determined by comparing LEED experiments to multiple scattering LEED theory until by the mid 1980's several hundred structures were available. Many of these are catalogued in the SCIS database [23].

It was not a decade without its problems: typically we were all struggling with inadequate computing power, and also experiments were extremely time consuming. Advances in electronics eventually were to solve both these problems. One consequence was that a lot of mistakes were

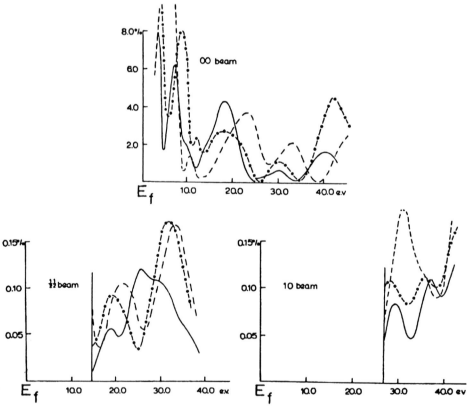

Fig. 6. LEED spectra taken at normal incidence from a c(2 × 2)Na/Ni(001) surface. Full curve, experiment; broken curve, theory, $d = 2.55$ Å, chain curve, theory, $d = 2.87$ Å.

made which attracted criticism from those outside the field. Internal competition and dissension did not help the case of LEED. In retrospect it can be seen that the problems were caused by the scale of our ambition. We now know that a typical adsorbate structure will involve the determination of perhaps a dozen atomic coordinates including the reconstruction induced by the adsorbate. Few other surface techniques come close to offering the information content necessary to do the job. I shall argue in Section 5 that recent advances have enabled us to fix these teething problems.

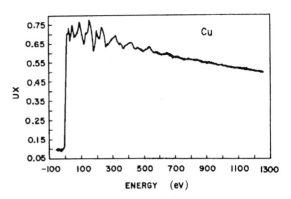

Fig. 7. K-shell absorption coefficient of copper versus the Xray photon energy after Eisenberger. The zero of energy is chosen at the K edge. Note the oscillatory structure on the edge which is caused by electron diffraction from neighbouring copper atoms.

4. Electron diffraction and the missionary spirit

Having solved the LEED problem, theorists now prepared to take on the rest of the problems of surface science. In fact the low energy electron diffraction process is central to many spectroscopies, and the layer methodology can be employed for the calculation of valence electronic structure, such as that of surface states. My personal opportunity to spread the gospel into new areas came in 1973 during a year that I spent a Bell Laboratories. I had just completed my book and was looking around for new problems when Peter Eisenberger came into my office waving some curious data taken on his powerful X-ray setup. They are shown in Fig. 7.

Sayers et al. [24] had earlier made the suggestion that the extended X-ray absorption fine structure (EXAFS) could be used as a probe of local atomic arrangements around the emitting atom: a sort of internal LEED experiment. The analogy did not escape me and in collaboration with Patrick Lee we set about putting together a soundly based theory of these oscillations. This theory did not get published for some time because Patrick decamped from Bell to Seattle for a year, but eventually appeared in 1975 [25]. That was the beginning of my long association with synchrotron radiation and was the basis for many fruitful collaborations with experimentalists at the Daresbury Laboratory, where I moved in 1975, and elsewhere. Like LEED, EXAFS and its surface partner SEXAFS proposed by Patrick in

1976 [26], are now firmly established tools for structural studies and theory plays a central role in their interpretation.

It is interesting to make a comparison of EXAFS with LEED. Many early theories of LEED tried to make use of Fourier transformation with respect to incident wave vector. They all came unstuck because of the strength of multiple scattering. In EXAFS, despite the formal analogy with LEED, multiple scattering is much less important. The emitting atom has a clear view of its main scattering partners, the nearest neighbours, and strong forward scattering never gets a look into this process. Scattering from more distant neighbours can give multiple scattering problems as we showed for copper [25], but the wavefield is already very weak far from the emitter and the process is not a strong one. As a result Fourier transformation gives sensible results in EXAFS. The spectrum can be regarded as a sort of one-dimensional hologram of the radial distribution function. Fig. 8 shows some of Stephen Gurman's results [27] in which he Fourier transformed the copper EXAFS data from Fig. 7 with various corrections. It is evident that the Fourier transform is an excellent guide to the shell structure in EXAFS. Even so for the most precise interpretation one must turn to direct fitting of the data as in LEED. The value of being able to "take a peek" at the information content of EXAFS data by Fourier transformation cannot be overestimated and must be responsible for much of the current popularity of the technique. In contrast LEED data offer us an all or nothing situation.

Despite its ease of interpretation EXAFS does have some limitations: it can only present us with radial coordinates. In some materials such as glasses, the structure within the coordinating shell is important and contains information about bond angles and other chemically sensitive quantities. A multiple scattering electron at least has chance to get a good look at the environment and generally speaking the multiple scattering component of an EXAFS spectrum contains information beyond the radial distribution function concerning the shell structure. The problem is that multiple scattering in EXAFS is almost always very weak, except for the region within 20 to 30 eV of threshold. It was for this reason that Paul Durham, Chris Hodges, and I proposed that X-ray absorption near edge structure (XANES, or NEXAFS as it is sometimes known) could be used to investigate more subtle structure in the environment of an emitting atom [28]. We ap-

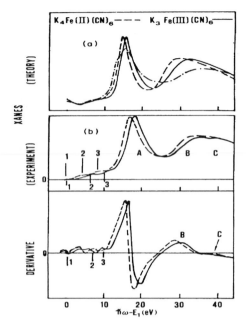

Fig. 9. (a) Calculated XANES of $[Fe(CN)_6]^{2+}$ and $[Fe(CN)_6]^{3+}$ for various trial coordinates. (solid and dashed lines) (b) the experimental XANES spectrum (upper panel) and its derivative (lower panel) for the Fe II and Fe III complexes. The zero of the energy scale is taken at the absorption maximum of the first feature of the Fe II complex. This feature shifts by 0.5 eV in the Fe II complex.

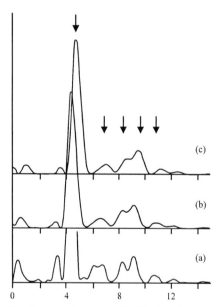

Fig. 8. Fourier transform of EXAFS data for copper: (a) without a Gaussian window (b) with a Gaussian window (c) as in (b), but with phase shift correction. The arrows mark the true shell radii.

plied the full multiple scattering theory to the problem and showed that it is possible to get a good description of the data. Fig. 9 below shows some data taken by Bianconi et al. [29] compared to theory.

It gave me great satisfaction when in a later paper Paul Durham and Samar Hasnain [30] were able to interpret XANES data from haemoglobin in various states of oxidation using a multiple scattering theory: perhaps the ultimate in complexity of structure for multiple scattering.

Dimitri Vvedensky was to apply perturbation theory to XANES and transform the theory into an even more flexible tool. His codes have been widely used to interpret a wide range of spectra [31].

At first LEED theory needed all the help it could get from other areas of electronic structure calculations, but with photoemission experiments coming on stream, particularly angle resolved photoemission using synchrotron radiation sources, LEED was able to repay its debt. In the 1970's valence band electronic structure calculations were largely about calculating eigenvalues and squaring them to find the charge density. In LEED eigenvalues are never calculated because all the time we work in a regime of strong absorption and there are no stationary states. Rather the mathematical entity with which we work is the Green's function. In that sense we are closer to the Feynman picture of quantum mechanics in which physics emerges by particles exploring all possible trajectories. This picture had a radical effect on the theory of photoemission and its implementation was largely due to the ideas developed by LEED theorists. This way of looking at the problem has the advantage that, for the first time, hole lifetimes can be properly built into the photoemission process giving a true one-step interpretation of valence band photoemission [32–36]. Some of our results from [34,36] are shown in Fig. 10.

These developments took place just as the new synchrotron sources came available and Paul Durham was able to exploit the KKR-CPA theory he learnt from Balazs Gyorffy in Bristol to give a quantitative account of photoemission from alloys [37]. Ian Moore was to apply these theories to spin polarised experiments and explain the spin reversal at the Fermi surface of nickel [38].

Perhaps the most pleasing application of LEED theory in other areas concerns the new "photonic materials", where periodic dielectric structures are designed to give a material specific electromagnetic properties. Using the right micron scale structure, materials can be turned into photonic insulators [39]. Inside these materials we find the darkest places in the universe, at least in the range of frequencies within the band gap. Even zero point fluctuations are forbidden, and the decay of excited atomic states with transitions in this range is forbidden. LEED theory suitably adapted to the electromagnetic case is ideally suited to calculating the band structure, and the reflection and transmission coefficients of these new materials, repaying the original debt LEED theory has to optical multiple scattering theory. Stefanou et al. [40] have made perhaps the most direct application of LEED theory, and MacKinnon and myself have made a real space adaptation of LEED theory [41] to calculate the reflection coefficient of a structure consisting of an array of 0.74 mm diameter cylinders, $\epsilon = 8.9$, arranged in a square lattice with spacing 1.87 mm. Our results are compared to experiments by Robertson et al. [42] in Fig. 11. The resemblance to a LEED experiment will be noted!

These structures, suitably scaled to optical wavelengths, may find applications in optoelectronics, modifying the properties of semiconductors lasers. There are already applications of periodic structures in optics where they are used as delay lines.

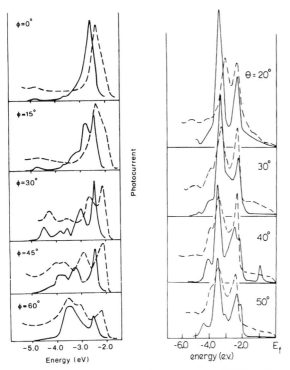

Fig. 10. Photoemission spectra (a) for a copper (1 11) surface. 16.8 eV unpolarised radiation is incident normally on the surface and the electrons are emitted at $\theta = 45°$ to the normal. The angle ϕ is measured relative to the ($\bar{2}11$) azimuth. (b) for a copper (001) surface at various polar angles in the G–K azimuth (the (110) direction). 21.2 eV unpolarised He I radiation is used. Full line: theory, dashed line: experiment. The energy is measured relative to E_F, and the imaginary part of the hole self energy is -0.054 eV.

5. The new theory of electron diffraction – 1984 onwards

As a probe of surface crystallography LEED was already very successful by the end of the 1970's. In a sense it was a victim of its success because the need to know ever more complex and interesting surface structures was placing severe demands on the methodology. In the summer of 1980 Paul Marcus and Franco Jona organised an influential conference at IBM Yorktown Heights [43]. This was a chance to review progress so far, and two things were very clear to me at the end of the conference:

(i) LEED experiments were far too time consuming for the study of really complex and difficult systems.

(ii) Even if the experimental problem could be solved, the theory was not up to interpreting the large volumes of data such systems would deliver.

In the event problem (i) was solved by applications of electronics and devices like the Erlangen DATALEED system solved the problem of fast data acquisition for complex systems. These advances will be related elsewhere in this volume.

The problem with theory is as follows: suppose that we wish to determine the coordinates of 3 atoms, each of which can occupy 10 different positions, giving a total of 10^3 trials to be made. The analogy is with trying to find the right key to open a lock. Obviously the number of trials grows rapidly with complexity: given a system containing 6 atoms, 10 trials per atom would involve 10^6 calculations.In fact the problem is worse than that because each individual calculation itself scales with the complexity of the system: something like N^3 where N is the number of atoms in the system.

Many theorists have responded to this problem with imaginative schemes with the results that we are now in a position to solve very complex surface problems containing many non-equivalent atoms per unit cell. In the early 1980's Dimitri Vvedensky, Dilano Saldin and Philip Rous were with me in London and we had established collaborations with Klaus Müller and Klaus Heinz in Erlangen, and with Gabor Somorjai and Michel Van Hove in Berkeley. During this period we developed the surface structure database the Surface Crystallographic Information Service (SCIS) [23] and worked on new methodologies for applying LEED to surface structure analysis. It was a very fruitful and enjoyable period for me personally.

The first problem we tackled was to remove the restriction that LEED can only look at ordered surfaces. Obviously if the surface is not ordered, the diffraction pattern will be diffuse, hence our new theory of diffuse LEED (DLEED) [44]. The idea is that a single atom on a surface

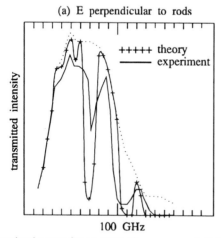
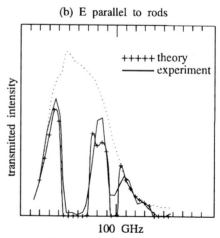

Fig. 11. Transmitted power for an array of seven rows of dielectric cylinders. The dashed curve shows the instrument response in the absence of the cylinders. Left curve (a): E perpendicular to the cylinders, right curve (b): E parallel to the cylinders.

interacts with a LEED beam in much the same way as it would emit a photoelectron in a SEXAFS experiment. There are three basic processes which we identified in Fig. 12.

The analogy with SEXAFS and conventional LEED enabled the necessary computer codes to be assembled very quickly and we soon had a working theory ready to interpret diffuse LEED data. An alternative version of the theory, the beam set neglect method, was developed with Michel Van Hove [45]. Meanwhile the Erlangen team had measured the first diffuse LEED intensities and Klaus Heinz spent the summer of 1985 with us in London applying the theory to analysis of data for O/W(001) which we did successfully to find the first disordered structure by DLEED [46].

The expanded experimental capabilities of LEED made it even more imperative that a better theory be found. Traditionally theorists have turned to perturbation theory when they want to speed up calculations and that is exactly what Philip Rous and I did when we introduced the concept of Tensor LEED [46–48]. The idea is a simple one, equivalent to the distorted-wave Born

approximation in nuclear physics. We start with a reference structure and do a full LEED calculation for that structure. This is then used as a starting point for a perturbation theory in the deviation of the structure from the reference. The theory has worked well, and is especially well suited for complex reconstructions of the substrate where many atoms may move, but by only small amounts. In collaboration with the Erlangen team Philip, Dilano and I combined Tensor LEED with DLEED and reanalysed the O/W(001) data for substrate reconstruction [46]. Our results are shown in Figs. 13. Philip developed the codes until they could speed up calculations by a factor of 10^3 or more [49,50]. Later, in collaboration with Michel Van Hove and Adrian Wander he built the Tensor LEED method into the Van Hove–Tong codes and using these codes it has become relatively routine to analyse complex substrate reconstruction especially when combined with their strategy for reaching the optimum structure using a minimum number of trials.

An example of the sort of complexity that can be achieved with tensor LEED analysis is the

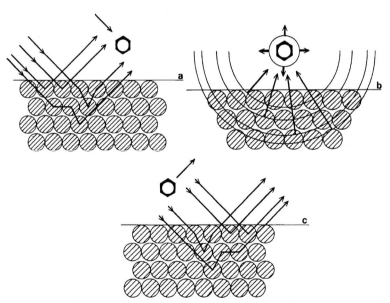

Fig. 12. Diffuse LEED as a three step process: (a) All scattering events prior to the electron meeting the adsorbed atom replace the K-shell excitation process in a conventional SEXAFS experiment. (b) All scatterings between the first and last encounter of the electron with the atom (c) All scatterings after the last encounter.

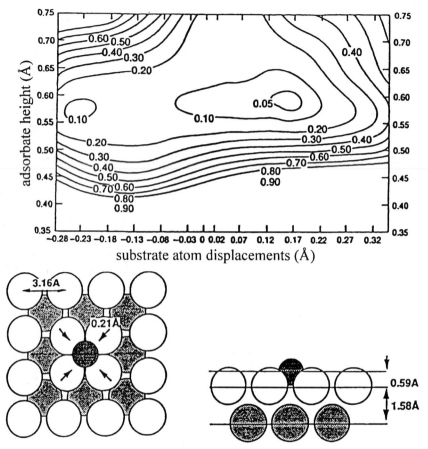

Fig. 13. Top: theory–experiment comparison of the diffuse intensity patterns from a disordered O/W(001) system displayed as an *R*-factor plot of adsorption height (d_{WO}) against displacements of the four tungsten atoms surrounding the oxygen atom (d_W). The minimum is at $d_{WO} = 0.59$ Å and $d_W = \sqrt{2} \times 0.15 = 0.21$ Å. Bottom: the structure surrounding a single oxygen atom.

p(2 × 2)C$_2$H$_2$/Ni(111) structure shown below in Fig. 14 determined by Wolfgang Oed et al. using data from Erlangen and Wolfgang's version of the tensor LEED code. Tensor LEED gives enormous gains in speed for structural analysis. It also demonstrates that LEED amplitudes are intrinsically linear in some parameter. Adrian Wander, Michel Van Hove and I were to exploit this with the linear LEED theory [51], but also there is the possibility that something which is a linear function of the structure can be inverted to give the structure directly. This is the idea behind direct methods in LEED [52,53]. The idea has been demonstrated in principle for simple systems. We are currently working on the extension to more complex systems where it offers the possibility of extending structural analysis to systems massively more complex than those we have examined so far.

6. Epilogue

How did we do? In some ways the goals of surface science have exceeded expectations. Surface crystallography is a firmly established discipline largely due in the first instance to the impact of multiple scattering theory, but increasingly drawing strength from other techniques such as X-ray diffraction. The electronic properties of surfaces can be measured with great precision through angle resolved photoemission experi-

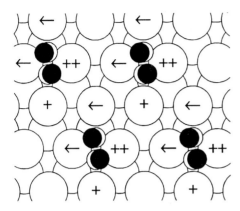

Fig. 14. The structure of p(2×2) C_2H_2 adsorbed on an Ni(111) surface as determined by Oed et al. using the Tensor LEED method. Note the complexity of the structure which includes substrate reconstruction of the order of 0.1 Å, and an asymmetrical position for the molecule, rotated away from the mirror plane by 10°. Only after all these refinements were included could an acceptable R-factor of 0.24 be obtained. The hydrogens were not considered in this analysis.

ments and theorists can justly claim to have a good understanding of bonding at surfaces. Surface scientists have been very creative in devising new probes of the surface. Scanning tunnelling microscopy is but one of a host of new surface probes that have burst upon the scene since 1964.

So there is much to be satisfied with but not much cause for complacency, because in many ways the old questions remain. David Tabor would draw the attention of young surface scientists in the Cavendish to adhesion, friction, reaction and diffusion of atoms at surfaces and encourage them to relate their studies of electronic structure, and atomic arrangements to these questions. I guess it is time to get on with our homework assignment: there are plenty of problems waiting to be solved in surface science.

References

[1] A. Ignatievs, J.B. Pendry and T.N. Rhodin, Phys. Rev. Lett. 26 (1971) 189.
[2] S. Andersson, Surf. Sci. 18 (1969) 325.
[3] P.B. Hirsch et al., Electron Microscopy of Thin Crystals (Butterworths, London, 1965).
[4] B.I. Lundqvist, Phys. Status Solidi 32 (1969) 273.
[5] G. Capart, Surf. Sci. 26 (1971) 429.
[6] J.A. Strozier, Jr. and R.O. Jones, Phys. Rev. Lett. 25 (1970) 516.
[7] J.A. Strozier, Jr. and R.O. Jones, Phys. Rev. B 3 (1971) 3228.
[8] E.G. McRae, J. Chem. Phys. 45 (1966) 3258.
[9] J.L. Beeby, J. Phys. C: Solid State Phys. 1 (1968) 82.
[10] K. Kambe, Z. Naturf. 22a (1967) 322.
[11] K. Kambe, Z. Naturf. 22a (1967) 422.
[12] K. Kambe, Z. Naturf. 23a (1968) 1280.
[13] J.B. Pendry, J. Phys. C: Solid State Phys. 4 (1971) 2501.
[14] J.B. Pendry, J. Phys. C: Solid State Phys. 4 (1971) 2514.
[15] D.W. Jepsen, P.M. Marcus and F. Jona, Phys. Rev. B 5 (1972) 933.
[16] G.E. Laramore and C.B. Duke, Phys. Rev. B 5 (1972) 267.
[17] J.B. Pendry, Low Energy Electron Diffraction (Academic Press, London, 1974).
[18] J.B. Pendry, Phys. Rev. Lett. 27 (1971) 856.
[19] J.B. Pendry, J. Phys. C: Solid State Phys. 4 (1971) 3095.
[20] S. Andersson and J.B. Pendry, J. Phys. C: Solid State Phys. 5 (1972) 241.
[21] J.E. Demuth, P.M. Marcus and D.W. Jepsen, J. Phys. C: Solid State Phys. 8 (1975) L25.
[22] F. Forstmann, W. Berndt and P. Büttner, Phys. Rev. Lett. 30 (1973) 17.
[23] J.M. MacLaren, J.B. Pendry, P.J. Rous, D.K. Saldin, G.A. Somorjai, M.A. Van Hove and D.D. Vvedensky, Surface Crystallographic Information Service (Riedel, Dordrecht, 1987).
[24] D.E. Sayers, E.A. Stern and F.W. Lytle, Phys. Rev. Lett. 27 (1971) 1204.
[25] P.A. Lee and J.B. Pendry, Phys. Rev. B 11 (1975) 2795.
[26] P.A. Lee, Phys. Rev. B 13 (1976) 5261.
[27] S.J. Gurman and J.B. Pendry, Solid State Commun. 20 (1976) 287.
[28] P.J. Durham, J.B. Pendry and C.H. Hodges, Solid State Commun. 38 (1981) 159.
[29] A. Bianconi, M. Dell'Ariccia, P.J. Durham and J.B. Pendry, Phys. Rev. B 26 (1982) 6502.
[30] P.J. Durham, A. Bianconi, A. Congiu-Castellano, A. Giovannelli, S.S. Hasnain, L. Incoccia, S. Morante, J.B. Pendry and M. Perutz, The EMBO Journal 2 (1983) 1441.
[31] D.D. Vvedensky, D.K. Saldin and J.B. Pendry, Comput. Phys. Commun. 40 (1986) 421.
[32] J.B. Pendry, J. Phys. C: Solid State Phys. 8 (1975) 2413.
[33] J.B. Pendry, Surf. Sci. 57 (1976) 679.
[34] J.B. Pendry and D.J. Titterington, Commun. Phys. 2 (1977) 31.
[35] J.B. Pendry and J.F.L. Hopkinson, J. Phys. F. 8 (1978) 1009.
[36] J.B. Pendry and J.F.L. Hopkinson, J. Phys. (France) C4 (1978) 142.
[37] P.J. Durham, J. Phys. F 11 (1981) 2475.
[38] I.D. Moore and J.B. Pendry, J. Phys. C: Solid State Phys. 11 (1978) 4615.
[39] E. Yablonovitch, T.J. Gmitter and K.M. Leung, Phys. Rev. Lett. 67 (1991) 2295.

[40] N. Stefanou, V. Karathanos and A. Modinos, J. Phys. Condensed Matter 4 (1992) 7389.

[41] J.B. Pendry and A. MacKinnon, Phys. Rev. Lett. 69 (1992) 2772.

[42] W.M. Robertson, G. Arjavalingam, R.D. Meade, K.D. Brommer, A.M. Rappe and J.D. Joannopoulos, Phys. Rev. Lett. 68 (1992) 2023.

[43] P.M. Marcus and F. Jona, Determination of Surface Structure by LEED (Plenum, New York, 1984).

[44] J.B. Pendry and D.K. Saldin, Surf. Sci. 145 (1984) 33.

[45] D.K. Saldin, J.B. Pendry, M.A. Van Hove and G.A. Somorjai, Phys. Rev. B 31 (1985) 1216.

[46] K. Heinz, D.K.Saldin and J.B. Pendry, Phys. Rev. Lett. 55 (1985) 2312.

[47] P.J. Rous and J.B. Pendry, Surf. Sci. 219 (1989) 355.

[48] P.J. Rous and J.B. Pendry, Surf. Sci. 219 (1989) 373.

[49] P.J. Rous and J.B. Pendry, Comput. Phys. Commun. 54 (1989) 137.

[50] P.J. Rous and J.B. Pendry, Comput. Phys. Commun. 54 (1989) 157.

[51] A. Wander, J.B. Pendry and M.A. Van Hove, Phys. Rev. Rapid Commun. B 46 (1992) 9897.

[52] J.B. Pendry and K. Heinz, Surf. Sci. 230 (1990) 137.

[53] J.B. Pendry, K. Heinz and W. Oed, Phys. Rev. Lett. 61 (1989) 2953.

Surface Science 299/300 (1994) 391–404
North-Holland

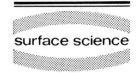

surface science

The equilibrium crystal shape and the roughening transition on metal surfaces

Edward H. Conrad

School of Physics, Georgia Institute of Technology, Atlanta, GA 30332-0430, USA

and

Thomas Engel *

Department of Chemistry, BG-10, University of Washington, Seattle, WA 98195, USA

Received 2 April 1993; accepted for publication 17 July 1993

The surfaces of metals at thermal equilibrium are characterized by a variety of defects which are thermally activated and entropically favored. In the last 10 years surface roughening transitions have been shown to exist on a large number of low-symmetry surfaces. For a given surface the transition temperature is known to scale with bulk properties. While a good deal of theoretical work has paralleled the experimental effort, direct comparisons between model potentials and experimental data are still lacking. The roughening transition on these surfaces is discussed in terms of competition with other processes such as facetting. Our work in this area is discussed and more recent developments in the field are reviewed. Unresolved problems in applying the theoretical framework to specific systems are also discussed.

1. Introduction

The earliest humans on our planet must have been fascinated by natural crystals found in sheltered geological repositories. Even after scientists were able to show that these macroscopic crystals are a direct image of the building blocks which make up the solid at the atomic level, it is a marvel that replication of $\sim 10^8$ unit cells proceeds sufficiently free of error that symmetry at the atomic level is preserved on the centimeter scale. The equilibrium shape of crystals has continued to interest generations of investigators since the advent of modern science. Many contributions in this volume concern structural aspects of matter and in particular the structure of solid surfaces. Our contribution focuses on one particular aspect of the structure of solid surfaces, the dependence of the equilibrium structure of solid surfaces on temperature due to the formation of topological defects.

The restriction to equilibrium structures is an important one so that we will exclude kinetic influences in this review. We will begin with a brief outline of crystal structure and the roughening transition that are important to our discussion. Theoretical models that can be used to describe the evolution of surface structure with temperature will be introduced at this point. We continue with a short historical summary of the concept of equilibrium crystal shapes and the work up to 1980 that lead to the current work on the surface roughening transition. This is followed by a discussion of our experiments in this area. Since 1980, there has been a rapid growth of work in this field and we highlight the major research directions and results which have

* Corresponding author.

emerged from this work. In this section we place work on the roughening transition in the broader framework of other phenomena which determine the equilibrium crystal shape such as faceting, reconstruction and surface melting. Finally we examine possible future directions for this work and its impact on "real world" phenomena.

2. Crystal structure and the roughening transition

No crystal is perfect over truly macroscopic-length scales. Steps provide a convenient way for the surfaces of crystals to adjust to departures from perfect alignment of the surface with respect to major symmetry directions. Fig. 1 shows an image of a Si(100) surface obtained in ultra-high vacuum using a scanning tunneling microscope (STM) [1]. The surface slopes downward from the lower-left to the upper-right corner. Each abrupt jump corresponds to a change in height by one atomic layer which is 1.36 Å. Over a distance of ~ 6000 Å, this surface changes in height by 8.2 Å, corresponding to a misorientation from the (100) direction by 0.01°. The considerable variation in the degree to which these step edges are straight is seen and has been discussed

Fig. 1. STM image of Si(001)(2×1) reconstruction. Image size 4000×4000 Å. $V_{tip} = +2$ V and $I_{tunnel} = 1 \times 10^{-9}$ Å [1].

Fig. 2. As for fig. 1, showing a 400×400 Å area of the surface [1].

elsewhere [2]. Examination of such a surface in the vicinity of a step edge with higher resolution will typically yield an image such as that in Fig. 2. It is seen that step edges can also accommodate to changes from major symmetry directions by incorporating kinks. One also observes the appearance of isolated defects within a terrace such as vacancies, and occasional adatoms which are located at binding sites on top of the surface plane. These elements, terraces, steps, kinks, vacancies and adatoms are essential in discussing the changing morphology of surfaces with temperature.

At any finite temperature a small concentration of defect steps, kinks, etc., will always be thermodynamically stable because of the increased configurational entropy associated with their formation. At sufficiently high temperatures the entropy term in the surface free energy overcomes the enthalpy term to create a step and the free energy to make this defect goes to zero. At this temperature (the roughening temperature, T_R), steps can spontaneously form with no energy cost and the surface is said to be thermodynamically rough.

A convenient model describing roughening on low-index metal surfaces with cubic symmetry is

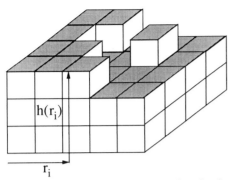

Fig. 3. Schematic representation of a surface in the SOS model. Atoms (represented by cubes) lie in vertical columns of height $h(r_i)$ that are confined to 2D lattice sites r_i [4].

the solid-on-solid (SOS) model [3]. In this description the surface is described by vertical columns of atoms of integer height $h(r_i)$ confined to 2D lattice sites at a distance $r_i = (n_i a, m_i b)$ (n and m are integers) from an arbitrary origin as shown in Fig. 3. The excess energy associated with deviations from an otherwise flat surface is due to the sum over all interactions between neighboring columns. In the SOS model the interaction only depends on the height difference between columns:

$$H = \frac{J}{2} \sum_{i,j} U(h_i - h_{i+j}). \tag{1}$$

The sum is over nearest-neighbor columns, thus the SOS Hamiltonian is inherently a short-range interaction. J is the interaction energy between two adjacent atoms when $\Delta h = \pm 1c$.

As will be shown below, surface roughening implies a specific type of long-range disorder on the surface. Since roughening does not refer to short-range atomic roughness due to isolated adatoms or kinks at step edges, the local form of the potential should not be too important as long as it reproduces the long-range distortions of the surface. An appropriate choice for U near and above the roughening transition is given by the discrete Gaussian SOS (DGSOS) model

$$U(h_i - h_{i+j}) = (h_i - h_{i+j})^2. \tag{2}$$

This potential has the desired property that the leading order term in its Fourier transform (corresponding to long wavelength distortions) is exactly that expected for an elastic continuum model [3].

Chu and Weeks [3] have shown that this model undergoes a continuous Kosterlitz–Thouless (KT) roughening transition [5]. The KT transition is an infinite order transition, meaning that all temperature derivatives of the free energy are continuous. The important prediction of the KT transition is that fluctuations of the vertical position of the surface diverge in a specific way. Above T_R, the height–height correlation function is given by [3,6]

$$\langle (h(r) - h(0))^2 \rangle = 2X(T) a_z^2 \ln(\rho), \tag{3a}$$

$$\rho^2 = \frac{(c_x x)^2}{a_x^2} + \frac{(c_y y)^2}{a_y^2}, \tag{3b}$$

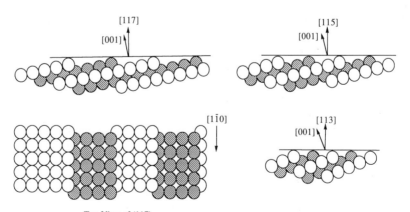

Fig. 4. Examples of three (11m) surfaces of a fcc crystal. The (117), (115) and (113) surfaces make an angle of 11.42°, 15.79°, and 25.24° with the (001) surface, respectively [4].

where x and y are now the unit cell vectors in the surface plane. c_x and c_y are general anisotropy parameters related to the surface tension differences in the x and y directions [6]. $X(T)$ is the roughness exponent, which is a function of temperature. The exponent, however, takes on a universal value at T_R [3,6,7].

$$X(T) = \frac{1}{\pi^2}. \tag{4}$$

Eq. (3a) shows that long-range correlations are divergent. This is in contrast to atomically rough surfaces consisting of vacancies, steps, or adatoms that still have a finite correlation length. The roughening temperature for the SOS model is found from simulations to be $kT_R/J = 1.48$ [8]. Historically, surface roughening of metals was first observed on stepped $(11m)$ fcc surfaces (Fig. 4). This is simply because the roughening temperature is much lower on these surfaces than on higher symmetry surfaces like the (001) facet. On the $(11m)$ surfaces the important energy is a kink-pair formation energy that costs only two broken bonds compared to four broken bonds for vacancy formation on the (001) surface. The model most suitable to describe roughening on these surfaces is the terrace–step–kink model (TSK) [6,9]. The TSK model has the advantage that step–step interactions can be explicitly included. These interactions play a more important role on (113) surfaces of Cu and Ni [10,11].

The roughening transition on $(11m)$ surfaces is illustrated schematically in Fig. 5. The low-temperature surface $(T \ll T_R)$ consists of (001) terraces of width $ma_0/2$, separated by monatomic steps (a_0 is the nearest-neighbor spacing). These surfaces are stabilized by a sufficiently long-range repulsion between terrace edges preventing the steps from collapsing [6,12,13].

As the temperature is increased, isolated kinks form along the terrace edges because of the entropy gain in the free energy. At T_R the free energy to make a step goes to zero and the surface takes on the structure seen in Fig. 5b. Above T_R, kinks spontaneously form and the terrace edges meander with a fractal structure. At still higher temperatures, the meander entropy overcomes the energy to make an adatom–

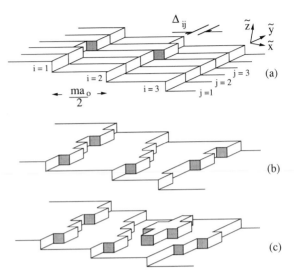

Fig. 5. Schematic view of the roughening transition on a $(11m)$ surface. (a) $T > T_R$, step edges are straight with a low density of isolated kinks. (b) $T > T_R(11m)$, the step edges meander. (c) $T < T_R(001)$, nested islands with fractal boundaries form [4].

vacancy pair and nested islands form (Fig. 5c). Like the meandering terraces, the islands have fractal boundaries. This corresponds to the roughening of the (001) terraces. In the TSK model, the $(11m)$ surface is described by a series of columns *parallel* to the surface and the energy for a configuration of steps and kinks is [9]

$$H = \sum_{i,j} V\left(|\Delta_{i,j} - \Delta_{i,j+1}|\right) + \sum_{i,j} \sum_{k,l} U(\Delta_{i,j}, \Delta_{k,l}), \tag{5}$$

where $\Delta_{i,j}$ is the deviation of the jth element on the ith step from its $T = 0$ K configuration ($\Delta = x/a_0$).

The first term in Eq. (5) is the energy to create a kink of length $|\Delta_{i,j} - \Delta_{i,j+1}|$. The second term is the interaction of the jth column on the ith terrace with all other elements on all other terraces. Note that the displacement of a column by 1 D changes the local surface height by $\pm h$, where h is the height of the surface measured normal to the $(11m)$ plane ($h = a_0 \sin \theta$, where $\cos \theta = m/(m^2 + 2)^{1/2}$). Therefore, calculating $\langle \Delta^2 \rangle$ is equivalent to calculating $\langle h^2 \rangle$. No analytic expression exists for $\langle \Delta^2 \rangle$ starting from the Hamiltonian in Eq. (5). Several simplifications, however, can be made to obtain a tractable solution. Villain et al. [6], have assumed that the step–step interactions (the second term in, Eq. (6)) only involve atoms in the same column ($j = l$) and between neighboring steps ($k = i \pm 1$). They further assumed, as in the DGSOS model, only long wavelength fluctuations of the terrace edges are important. This allows Eq. (5) to be written as

$$H = W_0 \sum_{i,j} |\Delta_{i,j} - \Delta_{i,j+1}|^2 + \frac{\omega}{2} \sum_{i,j} |\Delta_{i,j} - \Delta_{i+1,j}|^2,$$

(6)

where W_0 is the energy to make a kink and ω is the energy per length of a terrace to displace a step. Thus, the energy to make a kink n atoms wide is $E_n = 2W_0 + n\omega$. This Hamiltonian also cannot be solved analytically but a closely related one of the Sine–Gordon type has been shown to be similar in the limit that $W_0 \gg kT \gg \omega$. In this limit, Villain et al. [6], have calculated the $\langle h^2 \rangle$ to be

$$\langle (h(r) - h(0))^2 \rangle = \frac{2kTa_z^2}{\pi\sqrt{\beta_x\beta_y}} \ln(\rho),$$

(7a)

$$\rho^2 = \left(\frac{y}{a_y}\right)^2 \left(\frac{\beta_x}{\beta_y}\right)^{1/2} + \left(\frac{x}{a_x}\right)^2 \left(\frac{\beta_y}{\beta_x}\right)^{1/2}.$$

(7b)

From Fig. 5, $x = ma_0$ and $y = a_0$. β_x and β_y are the normalized surface stiffness in the x and y direction, respectively ($\beta_\tau = [\gamma + \tau(\partial^2\gamma/\partial\tau^2)] \times a_x a_y$; τ refers to the x or y coordinate). $X(T)$ in Eq. (3a) can be identified from Eq. (7a):

$$X(T) = \frac{kT}{\pi\sqrt{\beta_x\beta_y}}.$$

(8)

The surface stiffness also has a universal behavior

$$\sqrt{\beta_x\beta_y}\,|_{T=T_R} = \frac{\pi kT}{2},$$

(9)

so that $X(T_R)$ again takes the universal value $1/\pi^2$ at the roughening transition. Villain et al. [6], showed that in the limit $W_0 \gg kT \gg \omega$, the TSK model within the Gaussian approximation further predicts that the surface stiffness is related to W_0 and ω

$$\frac{\beta_y}{\beta_x} = \frac{kT}{2\omega} e^{W_0/kT}, \text{ and } \beta_x = \omega.$$

(10)

3. Facets, equilibrium crystal shape, and roughening

The phenomenon of surface roughening is closely related to the equilibrium shape that a crystal assumes. The equilibrium crystal shape (ECS) for a solid of volume V is the three-dimensional shape that minimizes the surface free energy. The minimization procedure leads to the well known Wulff construction that relates the angular dependence of the surface tension $\gamma(\hat{n})$ (energy/area) to the ECS [14].

The crystal shape is found by drawing planes perpendicular to $\hat{n}$ at a distance $|\gamma(\hat{n})|$ from the origin. The ECS is the inner envelope of these planes as illustrated in Fig. 6. Whenever a sharp cusp appears in the ECS (i.e., where the slope of $\gamma(\hat{n})$ changes discontinuously), a stable facet forms (Fig. 6b). The existence of a facet implies that isolated steps are not stable on these planes. The thermodynamic description of faceting was developed originally by Herring [15]. If a surface is miscut by an angle α_0 from a low-index direction whose surface normal is $\hat{n}_0$, the miscut facet will be stable only if the free energy cannot be

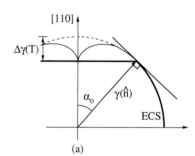

(a)

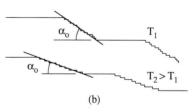

(b)

Fig. 6. (a) Wulff construction for an ECS using the polar plot of the surface tension. (b) An ECS with square facets produced by cusps in $\gamma(\hat{n})$ [4].

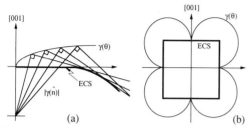

Fig. 7. (a) Wulff construction with a cusp, $\Delta\gamma$, showing a portion of the ECS. (b) The corresponding surface of a crystal miscut by an angle $0 < \alpha < \alpha_0$. The surface phase separates into a flat low-index face separated by a rough surface – "hill and valley" structure. The rough surface makes an angle α_0 with respect to the flat phase. Because the depth of the cusp is temperature dependent, α_0 depends on temperature [40].

lowered by phase separating into a "hill and valley" structure consisting of faceted planes with surface normals $\hat{n}_i$. Quantitatively, the stability condition is

$$A_0\gamma(\hat{n}_0) < \sum_i \gamma(\hat{n}_0), \tag{11}$$

$$A_0\hat{n}_0 = \sum_i A_i\hat{n}_i, \tag{12}$$

where A_i is the area of the ith facet. This is illustrated in Fig. 7. At high temperatures the ECS is nearly spherical because of entropy. At intermediate temperatures a cusp forms in the low-index direction. Surfaces with a miscut angle between 0 and α_0 will phase separate into a flat low-index face and a rough surface having steps (Fig. 7). The rough surface will make an angle α_0 with the low-index face and the relative amount of the two phases will be given by the lever rule. The angle α_0 is determined by the tangent of $\gamma(\hat{n})$ at the point where the stable facet intersects the $\gamma(\hat{n})$ plot (i.e., by the depth of the cusp $\Delta\gamma$). Near a cusp, the free energy is given by [16,17]

$$\gamma(\theta, T) = \gamma_0(T) + \frac{\mu(T)}{a_z}|\tan\theta|$$

$$+ \frac{B(T)}{a_y a_z^3}|\tan\theta|^3, \tag{13}$$

where γ_0 is the surface tension of the low-index face, a_z is the height of a step on the low-index face, and a_y is the unit vector perpendicular to

the miscut direction (i.e., along the step edge). The second term in Eq. (13) is the increase in surface energy due to step formation. The density of steps is proportional to $|\tan\theta|$; so that $\mu(T)$ is the energy per unit length to form an isolated single atomic height step. The third term in Eq. (13) takes into account interactions between steps. The connection between the ECS and the roughening transition is found in the second term of Eq. (13). Note that as long as $\mu(T) \neq 0$ the slope of $\gamma(\theta)$ is discontinuous at $\theta = 0$ and a cusp develops. This is equivalent to saying that isolated steps are not stable on the facet. The roughening transition, which involves the formation of stable isolated steps must therefore correspond to the temperature where the sharp cusp in the surface tension disappears [16]. Near the roughening transition $\mu(T)$ has been shown to go to zero exponentially [14]

$$\mu(T) \approx \exp\left(-\frac{C}{\sqrt{T_R - T}}\right), \tag{14}$$

where C is a constant.

These general considerations of the existence of well defined equilibrium crystal facets and of their thermal excitations have been a major focus of recent research in surface science. We will restrict our consideration to studies carried out in ultrahigh vacuum on surfaces which have been shown to be well ordered on an atomic length scale and free of impurities. Both direct imaging techniques like STM and LEEM and diffraction techniques including LEED, RHEED, X-ray and atom diffraction have been used to investigate the formation of defect structures including random step arrays, faceting and roughening. It would seem at first glance that direct imaging techniques would be far more appropriate to investigate aperiodic phenomena. While this is true in some cases, diffraction techniques have the advantage that ensemble averages are obtained directly. This is particularly true in cases such as the roughening transition on metals where $\langle(h(r) - h(0))^2\rangle$ diverges weakly with distance (see Eq. (7a)).

The roughening transition was first proposed by Burton et al. [18]. In their model a 1D solid surface is considered to be in equilibrium with its gas phase. As in the SOS model the solid can be described by vertical columns of atoms. At the surface a column position is either occupied by an atom or not. In equilibrium, the boundary between the gas and the solid is therefore described by a two-state model. Stimulated by Onsager's solution of the Ising model Burton and Cabrera derived an expression for the thickness of partially occupied surface layers as a function of temperature. In this model the transition temperature between a flat surface with no atoms missing and a surface composed of vacancies and adatoms is related to the roughening temperature.

It was immediately realized that the existence of a roughening transition would have a profound effect on phenomena such as crystal growth. However, a direct comparison between the model of Burton and Cabrera and experimental results was not available until the classical work carried out on solid He crystals between 1979 and 1982 was published [19,20]. These experiments were performed on macroscopic surfaces. Similar experiments on metal and semiconductor surfaces were not feasible because purity and kinetic effects could not be sufficiently controlled. Clearly, it would be necessary to use the diffraction methods, which surface scientists had painstakingly developed over the years, to examine whether roughening could be observed at the atomic level.

4. Work at the University of Washington on the Roughening Transition, 1980–1985

Looking back, it is interesting to note that we did not set out to perform experiments on surface roughening. This is an illustration of how one can be led into a new field by unexpected experimental results growing out from the original research direction. It is a useful lesson for those who plan and fund research to have another illustration that directed research has its limits and that careful work can yield dividends through unexpected paths. One of us (T.E.) had become interested in probing the extent to which quantitative surface structural information could be extracted

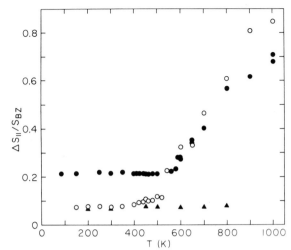

Fig. 8. Peak width (FWHM) in momentum space normalized to the first Brillouin zone as a function of temperature. (○): [$1\bar{1}0$] azimuth, $S_\perp = 21.7$ Å^{-1}, out-of-phase condition, (●): [$5\bar{5}2$] azimuth, $S_\perp = 14.1$ Å^{-1}, out-of-phase condition, (▲): [$5\bar{5}2$] azimuth, $S_\perp = 9.3$ Å^{-1}, in-phase condition [26a].

from helium diffraction experiments. The strength of the technique is its spectacular surface sensitivity. Only the topmost layer of atoms is probed on reasonably close-packed surfaces [21]. This strength is also one of its weaknesses. The registry information with the underlying substrate is not readily accessible as in LEED because of the lack of surface penetration.

In order to see whether we could find evidence for inward relaxation at step edges, we decided to investigate the Ni(115) surface. The He diffraction spectra which we observed were unexpected in several ways. The fact that the specular beam was broader than the ($\bar{1}0$) beam was clear evidence for steps on this (already stepped) surface (see Fig. 4). When we looked at the temperature dependence of the specular peak half width, we obtained the results shown in Fig. 8. Under in-phase scattering conditions, the peak half-width is independent of temperature. At the out-of-phase conditions, the half-width rises dramatically with temperature beginning near 500 K. Having studied the LEED literature [22,23] we knew that this meant that the atoms continue to occupy 2D lattice positions at high temperatures but that there is a substantial increase in the step density. The association of these observations with a roughening transition occurred through two separate inputs. One was a speculation by J. Lapujoulade [24] that a similar effect observed on Cu surfaces might be due to the onset of roughening. The second was the publication of a paper by G. Blatter [25] in which the effect on helium diffraction lineshapes of a surface undergoing a roughening transition was calculated. These two stimuli prompted us to engage in discussions with Marcel den Nijs and Eberhard Riedel, who were located 100 yards away in the Physics Department. This collaboration resulted in two papers on the surface roughening of $(11m)$ fcc metal surfaces [7a,26a]. Regrettably, there was an error in the analysis, communicated to us by J. Villain, which led us to quote a value for the transition which was too low. The correct value turned out to be 450 K [7b,26b]. These experiments provided the first conclusive evidence of a roughening transition on a well characterized surface. The experiments were very difficult and we found that we had to take extraordinary care to ensure that the surfaces were free of impurities. We found that the roughening temperature was sensitive to impurities whose coverage was not detectable with Auger electron spectroscopy. Why this is so can be seen in Fig. 9. A very small concentration of impurities can pin steps so that they cannot undergo the large scale motion required to achieve equilibrium above the roughening temperature. This is one of a number of experimental difficulties which must be overcome to obtain an accurate estimate of roughening temperature and its dependence on orientation.

5. Recent work on surface roughening

Initial evidence for roughening was found on $(11m)$ surfaces of fcc metals as described above. Because the potentials for metals are much less corrugated than on surfaces which have directed covalent bonds, roughening transitions will be more likely to occur below the melting temperature on metals than semiconductors. Conclusive evidence for roughening has also been obtained on fcc (110) planes as described below. Anomalous diffusion behavior and the observation of

Fig. 9. STM image from Ag(113). Image size 4000×4000 Å, $V_{tip} = 25$ mV, $I_{tunnel} = 2 \times 10^{-9}$ Å. Note localized contamination which appears to pin steps [27].

large anharmonicity in surface vibration amplitudes on more closely packed surfaces such as the (100) and (111) planes of the fcc lattice suggest that roughening may also be occurring here. However, the evidence is not as clear cut as for the $(11m)$ and (110) surfaces.

Direct evidence for surface roughening on $(11m)$ metal surfaces was first presented for Ni(115) by helium atom scattering lineshape studies [7,26]. At about the same time helium scattering studies of Cu (113), (115), and (117) surfaces observed an anomalous decay in the diffraction intensities that was postulated to be due to step formation [24]. The temperature behavior of the diffraction intensity was analyzed by Villain et al. [6], according to the behavior predicted by the TSK model below the roughening temperature.

However, the nature of the disorder remained in question. This is because any analysis based on intensity measurements alone is subject to a number of unreliable interpretations. Anomalous temperature behaviors due partly from anharmonic potentials leading to large non-linear terms in the vibrational amplitudes of surface atoms [28–30], structural changes resulting from thermal expansion [31], point defect formation, and a large contribution from thermal diffuse scattering, all influence the diffracted intensities [32,33]. Later lineshape studies on Cu $(11m)$ surfaces, however, strengthened the claim that these surfaces do have a roughening transition. Helium atom scattering [11,26,34–38] and surface X-ray studies [39,10] on both Ni $(11m)$ and Cu $(11m)$ surfaces observed a change in the diffraction lineshapes consistent with a logarithmic divergence in the height–height correlation function as given in Eq. (7a).

An important parameter one wishes to extract from these systems is the energy to create a kink W_0. At the moment, ab initio calculations for defect formation energies are sorely lacking. It seems reasonable that if these energies can be measured, they would provide an important experimental benchmark to compare with emerging defect energy calculations. Unfortunately, experimental values reported for the kink formation energy have varied considerably in the literature. This is due primarily to the different ways in which these parameters are derived from the experimental data. For instance, although the TSK model of Villain et al. [6] reasonably describes the (113) surfaces, values ranging from $3400\ \mathrm{K} > W_0 > 800\ \mathrm{K}$ and $560\ \mathrm{K} > \omega > 3.5\ \mathrm{K}$ have been reported by the same helium scattering group using different data analysis [6,11]. An X-ray scattering experiment, which is much less influenced by inelastic effects analyzing intensity data alone has reported $W_0 = 2100\ \mathrm{K}$ and $\omega = 86$ K [39]. These large ranges reflect both the insensitivity of the fitting procedure. to a lesser extent the error bars in the data, and more important these results point out the inconsistencies within the TSK model. With these problems of the TSK model in mind, a list of measured roughening temperatures and the estimates for W_0 and ω are summarized in Table 1. The data is an average over several authors' values and the error bars reflect both the spread in the reported values as well as the accuracy of the individual reported results.

In addition to $(11m)$ surfaces of fcc metals the (110) surfaces have also shown evidence of surface roughening. Cu(110) was the first transition

Table 1
Estimated values of the kink formation energy and step–step interaction for various $(11m)$ metal surfaces

Surface	T_R (K)	W_0 (K)	ω (K)	Ref.
Cu(113)	620 ± 10	2100 ± 75	86 ± 10	[39]
	720 ± 50	800 ± 50	560 ± 50	[12]
Cu(115)	380 ± 50	1500 ± 250	120 ± 20	[34]
	380 ± 50	850 ± 200	120 ± 20	[35]
Cu(117)	< 315	[a]	[a]	[6]
Ni(113)	750 ± 50	1700 ± 250	200 ± 100	[37]
	755 ± 30	2650 ± 800	110 ± 50	[10]
Ni(115)	450 ± 50	1700 ± 250	35 ± 25	[37]
Ni(117)	400 ± 50	[a]	[a]	[38]

[a] Not estimated.

metal to be studied near its melting temperature. Helium atom scattering experiments observed an anomaly in the temperature behavior of the diffraction peak intensities above 570 K [28]. The investigators suggested that the cause was the formation of a substantial fraction of thermally activated point defects.

X-ray scattering studies by Mochrie [40] also showed an anomaly that was originally attributed to surface roughening. Follow up X-ray Fresnel reflectivity experiments by Ocko and Mochrie [41] showed that the earlier measurements were due to a faceting transition caused by a 0.8° misorientation of the surface normal away from the [110] direction. Ocko and Mochrie concluded that the low-temperature surface consists of large (110) terraces separated by small faceted high step density regions ("hill and valley structure") that preserve the average miscut orientation [41]. At higher temperatures entropy drives the breakup of the (110) terraces and collapses the hill and valley structure. A faceting transition has also been observed on Ag(110) [42]. The existence of reversible faceting on crystals with miscuts less than 0.2° is important to note. It underlines the point that studies on "supposedly" atomically flat samples are more the exception rather than the rule.

More recent atom scattering studies of Cu(110) have shown that Cu(110) roughens near 950 K [45]. Similarly, definite evidence of a high-temperature step-disordered phase has been seen on

Ni(110). High q-resolution low-energy electron diffraction (LEED) experiments show that this surface loses vertical order above 1300 K [30].

It is interesting to note that the formation of steps on both Ni(110) and Cu(110) is preceded by large anharmonicity in the surface atomic vibrations. For Ni(110) the anharmonicity is observed to occur near 900 K well before the disordering transition [30]. Anharmonic anomalies at high temperatures have also been observed in diffraction experiments from (11m) [24,26] and (110) [28,30,51] fcc metal surfaces prior to their roughening transitions. Simulation studies confirm many of the qualitative features of the enhanced vibrations. Jayanthi et al. [52] calculated the quasi-harmonic free energy for Cu(001), (110), and (111) surfaces using Morse-type potentials. They showed that an instability in the free energy, as a function of inter-atomic spacing, occurred below T_m. At the instability temperature the local minimum in the free energy versus atomic separation disappears and the vibrations become unbounded. While the potentials used are oversimplified for metals, they do show that the instability exists and that the instability temperature, T_A/T_m, is not too different for (110) and (001) surfaces. Experimentally the Cu(110) [28], Ni(110) [30] and Ni(001) [31] vibrations become soft at nearly the same relative temperature, consistent with the calculations (see Table 2). The relationship of the anharmonicity to the disordering transitions is not yet known, but it seems that its appearance in all of the above systems should be more than a coincidence.

Besides the diffraction evidence for a roughening transition on Cu(110) and Ni(110), diffusion studies indicate an increased adatom mobility on these surfaces occurring at the same temperature as the roughening temperature. Bonzel et al. [53,54] have measured the surface diffusion of an adatom on argon RF-sputtered surfaces of Cu and Ni(110) using optical microscopy. They find that the surface diffusion coefficient becomes isotropic and increases dramatically at the same temperature. The observed temperatures, 1060 K for Cu and 1270 K for Ni, correspond to the roughening temperatures measured in the diffraction studies.

Table 2
Critical temperatures of (110) metals

Surface	Δ/kT[43]	T_c/T_m	T_a/T_m	T_m (K)
1 × 1 surfaces				
Ni(110)	0.47	0.75[30]	0.5[30]	1726.1
Pd(110)	−0.25	0.75[44]	b	1825.1
Cu(110)	0.43	0.74[45]	0.4[28]	1356.1
Ag(110)	−0.23	~0.80[42]	b	1233.9
Pb(110)	NC	0.75[46]	b	600.7
		0.65[45]	b	
Al(110)	NC	~0.70[48]a	0.66[48]a	933.5
In(110)	NC	0.69[47]	b	430.0
2 × 1 surfaces				
Pt(110)	−1.22	0.53[49]	b	2042.1
Au(110)	−1.05	0.49[50]	b	1336.1

T_c is the critical temperature for disordering as described in the text. T_a is the temperature where an anharmonicity is observed in the vibrational amplitude. NC: Not calculated. References are given in brackets.
[a] Estimated from peak intensity decay.
[b] Not measured.

Grazing angle X-ray scattering experiments have also suggested that the Ag(110) surface disorders near 720 N [55]. However more recent X-ray work by Robinson et al. [42], using a larger finite domain sample, indicates that a logarithmic height-height correlation function never fits the diffracted lineshapes up to at least 800 K. Robinson et al. showed that much of the temperature dependence of the diffraction lineshapes can be explained by a continuous faceting of the (110) surface. They estimated the roughening temperature by extrapolating their data to the point where the cusp in the ECS disappears: $T_R \approx 992$ K ($T_R/T_m = 0.8$). Although it is only approximate, this value of T_R is consistent with other (110) metal surfaces that show a reduced roughening temperature T_R/T_m near 0.75 (see Table 2).

One of the most extensively investigated low-index surfaces has been Pb(110). It has been studied by ion scattering, LEED, RHEED, emissivity, helium scattering, and X-ray reflectivity. Clear evidence of both a surface roughening transition and the formation of a quasi-liquid layer have been obtained by several investigators. Frenken and van der Veen were the first to show the formation of a molten layer on Pb(110) within a few degrees of its bulk melting temperature ($T_m = 600.7$ K) [56].

Inelastic helium scattering measurements also showed two other important results [51]. First, the elastic intensity was observed to decay rapidly above 500 K, suggesting a strong anharmonicity in the Pb surface potential, similar to that observed on Ni and Cu (110) and (11m) surfaces. Second the diffusion coefficient was found to be isotropic in spite of the structural asymmetry of the (110) surface, consistent with the optical measurement on Cu(110) and Ni(110) [53,54]. While the surface melting transition has been clearly demonstrated by both ion scattering and inelastic helium scattering, the onset of disorder at 500 K was not adequately described by either technique.

High q-resolution LEED studies by Yang et al. [57] suggest that the Pb(110) surface undergoes a roughening transition near the onset of disorder observed in the ion scattering experiments. Their work clearly shows a change in the diffraction lineshapes at the out-of-phase diffraction condition without any measurable change in the in-phase peak widths. The line shapes were fit assuming a height–height correlation function similar to Eq. (7a). The roughening temperature was determined from the temperature at which $X(T) = 1/\pi^2$. This analysis found $T_R = 415$ K. The LEED measurements also showed that the diffraction intensities began to decay above 520 K which is very close the premelting temperature observed in the ion scattering experiments. For Pb(110) it is clear that a roughening transition precedes the formation of the surface melt.

Preliminary LEED studies of Al(110) show an anomalous decay of the diffraction peak intensities near 620 K. Von Blanckenhagen et al. [48], measured the temperature dependence of an off specular diffraction beam near the in-phase direction. Since they measured near an in-phase diffraction condition, it is unlikely that the decay can be due to step formation. The decay is more likely due to a large anharmonicity in the surface atom potential although vacancy formation cannot be ruled out. Ion scattering studies from Al(110) show a behavior similar to that observed on Pb(110) [58]. An initial disordering of the top layer begins near 800 K. By 900 K the number of disordered layers diverges suggesting the formation of a surface melt. Denier et al. [58] found that the liquid layer thickness diverges logarithmically with temperature, consistent with theoretical predictions [59]. For now it is certainly safe to say that if step formation does occur on Al(110) it must begin above 700 K.

Another interesting area in the work on roughening is the effect of a surface reconstruction on the roughening transition. Au(110), which has a 2×1 missing row reconstruction, was the first such system to be investigated. LEED measurements indicated that the (2×1) low-temperature ordered phase transformed to a (1×1) disordered phase by a 2D Ising transition [50]. More recent work on Pt(110), however cautions against that interpretation and suggests that further measurements on the Au(110) system will be required.

Studies by Robinson and co-workers [49] indicate that Pt(110) does not have a simple Ising transition even though the measured critical exponents agree with the theoretical predictions for an Ising model. Their surface X-ray scattering experiments showed that the half-order diffraction peaks have an oscillatory shift in $q_{\parallel}$ as a function of $q_{\perp}$. This behavior is indicative of surface steps [60]. More importantly, they showed that the 2×1 reconstruction disappeared at the same temperature the peak shift began to increase. Robinson et al. found that the (1.5, 0.06, 0.06) rod $a^* = 1.60$ Å^{-1}, had a temperature dependence consistent with an Ising transition [49]. While the temperature dependence of the peak intensities and FWHM can be reconciled with an Ising transition Robinson et al. [49] pointed out that the peak shift cannot. The shift in the half-order peaks signals a change in the step density. A similar peak shift was not observed on Au(110) but may in fact not have been resolvable with the poor sample quality. Whether or not steps form on Au(110) remains an open question.

The reason that a simple Ising model is inadequate is that the formation of steps introduces four additional degrees of freedom making a total of six possible topological defect steps (1a, 1b, 2a, and 2b in Fig. 10) and two antiphase boundaries (3a and 3b), shown in Fig. 10. The existence of six

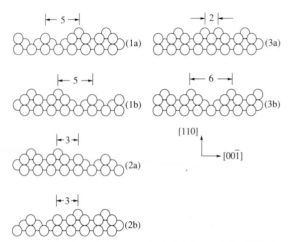

Fig. 10. Side view of the 6 topological defects of the (110) fcc missing row structure. (1a) and (1b) are the low-energy (111) steps. (2a) and (2b) are a pair of high-energy steps. (3a) and (3b) are two Ising antiphase walls. The parallel phase shifts between flat 2×1 regions is given for each defect in units of $a_0/2$ (a_0 is the NN distance) [48].

possible ground state configurations rules out a 2-state Ising model.

The simultaneous loss of the reconstruction, signaled by the decay of the half-order peak intensity, and the increasing step density above T_c, is inconsistent with both a pre-roughening transition [62] and a reconstructed rough phase. Villain and Vilfan [61] suggest that steps form in bound pairs and unbind at a critical temperature above the Ising T_c. Den Nijs has recently included the step energy differences in a 4-state chiral clock step model [63]. The high-temperature phase in this model (with non-zero chirality) is shown to be an incommensurate fluid phase. That is, the surface loses the reconstruction order and roughens at the same temperature.

6. Outlook for future work

As outlined above, there is substantial evidence for the occurrence of the roughening transition on "real world" metal surfaces. The time from the theoretical formulation in 1952 to the first experimental observations in 1975–1980 seem in retrospect quite long. This period also coincides with the maturing of surface science as a discipline in which atomically clean surfaces of well defined orientation can be generated reproducibly. There has been a rapid convergence of theory and experiment towards an understanding of the equilibrium crystal shape as well as the competing phenomenon of reconstruction, faceting, surface melting, and roughening.

The next step in this process calls for an acknowledgement of the real world aspects of crystalline perfection and the refinement of models which describe step–step interactions on surfaces. For instance, metal surfaces are seldom single crystalline over distances greater than a few millimeters and individual crystallites are joined by low-angle grain boundaries. How does the presence of these defects, or the distortions in the local crystal structure around a defect of any kind, influence the phenomena we have discussed? Another failing of the theory is that long-range interactions have been excluded in both the SOS and TSK models. This is an important omission that, while making simulation calculations easier, results in tenuous comparisons to experimental data that attempt to extract defect formation energies.

The effects of longer-range interactions on surface morphology can be illustrated by the work of den Nijs [62]. By including next nearest neighbor interactions in the restricted solid-on-solid (RSOS) model den Nijs has predicted a new topological phase known as the "pre-rough" phase [62]. The inclusion of longer-range elastic interactions is certain to lead to new disordered surface phases.

In a similar vein, recent experiments by Frohn et al. [64] have found a temperature regime in which surface steps have an attractive interaction. Incorporating this observation of step–step interactions in a realistic potential will give us the ability to predict equilibrium surface configurations as a function of temperature in a way that the models outlined in Section 2 cannot. The synthesis of all of this information will allow us to go beyond the global prediction of whether a phase transition exists at all to a prediction of the temperature at which it occurs and how it competes with a range of other topological morphologies, all of which affect the equilibrium crystal shape. This is a challenge to theorists and experimentalists alike.

Acknowledgement

Our joint and separate research on these issues has been supported by the Divisions of Chemistry and Materials Research of the National Science Foundation: (for E.H.C.) No. DMR-8703750 and No. DMR-9211249.

References

[1] K.E. Johnson, P.K. Wu, M. Sander and T. Engel, Surf. Sci. 290 (1993) 213.

[2] B.S. Swartzentruber, Y.-W. Mo, R. Kariotis, M.G. Lagally and M.B. Webb, Phys. Rev. Lett. 65 (1990) 1913.

[3] See J.D. Weeks, in: Ordering in Strongly Fluctuating Condensed Matter Systems, Ed. T. Riste (Plenum, New York, 1980).

[4] E.H. Conrad, Prog. Surf. Sci. 39 (1992) 65.

[5] J.M. Kosterlitz and D.J. Thouless, J. Phys. C 6 (1973) 1181;
J.M. Kosterlitz, J. Phys. C 7 (1974) 1064.

[6] J. Villain, D.R. Grempel and J. Lapujoulade, J. Phys. F 15 (1985) 809.

[7] (a) M. den Nijs, E.K. Riedel, E.H. Conrad and T. Engel, Phys. Rev. Lett. 55 (1985) 1689;
(b) M. den Nijs, E.K. Riedel, E.H. Conrad and T. Engel, Phys. Rev. Lett. 57 (1986) 1279.

[8] W.J. Shugard, J.D. Weeks and G.H. Gilmer, Phys. Rev. Lett. 41 (1978) 1399.

[9] H. van Beijeren and I. Nolden, in: Structured and Dynamics of Surfaces, Eds. W. Schommers and P. von Blanckenhagen (Springer, Heidelberg, 1987).

[10] I.K. Robinson, E.H. Conrad and D.S. Reed, J. Phys. (Paris) 51 (1990) 103.

[11] B. Salanon, F. Fabre, J. Lapujoulade and W. Selke, Phys. Rev. B 38 (1988) 7385.

[12] K.H. Lau and W. Kohn, Surf. Sci. 65 (1977) 607.

[13] M.B. Gordon and J. Villain, J. Phys. C 12 (1979) L151.

[14] For a review, see M. Wortis, in: Chemistry and Physics of Solid Surfaces VIII, Eds. R. Vanselow and R.F. Howe (Springer, Berlin, 1988) p. 367.

[15] C. Herring, Phys. Rev. 82 (1951) 87.

[16] C. Jayaprakash, W.F. Saam and S. Teitel, Phys. Rev. Lett. 50 (1983) 2017.

[17] C. Jayaprakash C. Rottman and W.F. Saam, Phys. Rev. B 30 (1984) 6549.

[18] W.K. Burton N. Cabrera and F.C. Frank, Philos. Trans. R. Soc. London A 243 (1951) 299.

[19] J.E. Avron L.S. Balfour, C.G. Kuper, J. Landau, S.G. Lipson and L.S. Schulman, Phys. Rev. Lett. 45 (1980) 814;
S. Balibar, D.O. Edwards and C. Laroche, Phys. Rev. Lett. 42 (1979) 782;
P.E. Wolf, S. Balibar and F. Gallet, Phys. Rev. Lett. 51 (1983) 1366.

[20] S. Balibar and B. Castings, Surf. Sci. Rep. 5 (1985) 87.

[21] T. Engel and K.H. Rieder, in: Stuctural Studies of Surfaces, Vol. 91, Ed. G. Höhler (Springer, Berlin, 1982).

[22] M. Henzler, in: Electron Spectroscopy for Surface Analysis, Ed. H. Ibach (Springer, Berlin, 1979).

[23] C.S. Lent and P.I. Cohen, Surf. Sci. 139 (1984) 121.

[24] J. Lapujoulade, J. Perreau and A. Kara, Surf. Sci. 129 (1983) 59.

[25] G. Blatter, Surf. Sci. 145 (1984) 419.

[26] (a) E.H. Conrad, R.M. Aten D.S. Kaufman, L.R. Allen, T. Engel, M. den Nijs and E.K. Riedel, J. Chem. Phys. 84 (1986) 1015;
(b) E.H. Conrad, R.M. Aten, D.S. Kaufman, L.R. Allen, T. Engel, M. den Nijs and E.K. Riedel, J. Chem. Phys. 85 (1986) 4756.

[27] K.E. Johnson and T. Engel, unpublished.

[28] P. Zeppenfeld, K. Kern, R. David and G. Comsa, Phys. Rev. Lett. 62 (1989) 63.

[29] E.H. Conrad, D.S. Kaufman, L.R. Allen, R.M. Aten and T. Engel, J. Chem. Phys.83 (1985) 5286.

[30] Y. Cao and E.H. Conrad, Phys. Rev. Lett. 64 (1990) 447.

[31] Y. Cao and E.H. Conrad, Phys. Rev. Lett. 65 (1990) 2808.

[32] J.T. McKinney, E.R. Jones and M.B. Webb, Phys. Rev. 160, suppl. 3 (1967) 523.

[33] E.H. Conrad. L.R. Allen D.L. Blanchard and T. Engel, Surf. Sci. 184 (1987) 227.

[34] F. Fabre, D. Gorse B. Salanon and J. Lapujoulade, J. Phys. (Paris) 48 (1987) 1017.

[35] F. Fabre, D. Gorse J. Lapujoulade and B. Salanon, Europhys. Lett. 3 (1987) 737;
F. Fabre, B. Salanon and J. Lapujoulade, in: The Structure of Surface II, Eds. J.F. van der Veen and M.A. Van Hove (Springer, Berlin, 1988) p. 520;
J. Lapujoulade and B. Salanon, in: Phase Transitions in Surface Films, Eds. H. Taub, G. Torzo, H. Lauter and S.C. Fain, Jr. (Plenum New York), to be published.

[36] F. Fabre B. Salanon and J. Lapujoulade, Solid State Commun. 64 (1987) 1125.

[37] E.H. Conrad, L.R. Allen, D.L. Blanchard and T. Engel, Surf. Sci. 187 (1987) 265.

[38] D.L. Blanchard, D.F. Thomas, H. Xu and T. Engel, Surf. Sci. 222 (1989) 477.

[39] K.S. Liang, E.B. Sirota, K.L. D'Amico, G.J. Hughess and S.K. Sinha, Phys. Rev. Lett. 59 (1987) 2447.

[40] S.G.J. Mochrie, Phys. Rev. Lett. 59 (1987) 304.

[41] B.M. Ocko and S.G.J. Mochrie, Phys. Rev. B 38 (1988) 7378.

[42] I.K. Robinson, E. Vlieg, H. Hörnis and E.H. Conrad, Phys. Rev. Lett. 67 (1991) 1890.

[43] S.M. Foiles, Surf. Sci. 191 (1987) L779.

[44] H. Hörnis and E.H. Conrad, to be published.

[45] K. Kern, in: Phase Transitions in Surface Films, Eds. H. Taub, G. Torzo, H. Lauter and S.C. Fain, Jr. (Plenum, New York), to be published.

[46] H.-N. Yang, T.-M. Lu and G.-C. Wang, Phys. Rev. Lett. 63 (1989) 1621.

[47] J.C. Heyraud and J.J. Mètois, J. Cryst. Growth 82 (1987) 269.

[48] P. von Blanckenhagen, W. Schommers and V. Voegel, J. Vac. Sci. Technol. A 5 (1987) 649.

[49] I.K. Robinson, E. Vlieg and K. Kern, Phys. Rev. Lett. 63 (1989) 2578.

[50] J.C. Campuzano M.S. Foster, G. Jennings, R.F. Willis and W. Unertl, Phys. Rev. Lett. 54 (1985) 2684.

[51] J.W.M. Frenken, J.P. Toennies and Ch. Wöll, Phys. Rev. Lett. 60 (1988) 1727.

[52] C.S. Jayanthi, E. Tosatti and L. Pietronero, Phys. Rev. B 31 (1985) 3456.

[53] H.P. Bonzel N. Freyer and E. Preuss, Phys. Rev. Lett. 57 (1986) 1024;
H.P. Bonzel, in: Surface Mobilities on Solid Materials, Ed. Vu Thien Binh (Plenum, New York, 1988).

[54] H.P. Bonzel and E. Latta, Surf. Sci. 76 (1978) 275.

[55] G.A. Held, J.L. Jordan-Sweet, P.M. Horn, A. Mak and R.J. Birgeneau, Phys. Rev. Lett. 59 (1987) 2075.

[56] J.W.M. Frenken and J.F. van der Veen, Phys. Rev. Lett. 54 (1985) 134.

[57] H.-N. Yang, T.-M. Lu and G.-C. Wang, Phys. Rev. Lett. 62 (1989) 2148.

[58] A.W. Denier van der Gon, R.J. Smith, J.M. Gay, D.J. O'Connor and J.F. van der Veen, Surf. Sci. 227 (1990) 143.

[59] R. Lipowsky, Phys. Rev. Lett. 49 (1982) 1575;
R. Lipowsky and W. Speth, Phys. Rev. B 28 (1983) 3983.

[60] P. Fenter and T.M. Lu, Surf. Sci. 154 (1985) 15.

[61] J. Villain and I. Vilfan, Surf. Sci. 199 (1988) 165.

[62] M. den Nijs, Phys. Rev. Lett. 64 (1990) 435.

[63] M. den Nijs, in: Phase Transitions in Surface Films, Eds. H. Taub, G. Torzo, H. Lauter and S.C. Fain, Jr. (Plenum, New York), to be published.

[64] J. Frohn, M. Giesen, M. Poensgen, J.F. Wolf and H. Ibach, Phys. Rev. Lett. 67 (1991) 3543.

Surface Science 299/300 (1994) 405–414
North-Holland

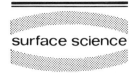

surface science

Phases of helium monolayers: search and discovery

J.G. Dash *, M. Schick and O.E. Vilches

Department of Physics FM-15, University of Washington, Seattle, WA 98195, USA

Received 8 April 1993; accepted for publication 21 June 1993

In this article we describe from a historical perspective the experimental and theoretical interplay amongst the three authors, their students, and many colleagues which led to the initial discoveries of several of the phases of ^{4}He and ^{3}He adsorbed on graphite.

This article is a *memoir á trois,* an account of an unusually fertile period in the professional lives of the authors and their many students and colleagues. Whereas the scientific results have been published and reviewed elsewhere [1], we take this opportunity to describe the interplay of people and ideas; how the discoveries were made as well as their content.

1. J.G.D.: The early days; first signs of several two-dimensional phases

An experimental program began at The University of Washington during the early 60's, as a search for evidence of quantum effects in helium monolayers. There was a long history of experimental and theoretical interest in ultrathin adsorbed films in general and thin helium films in particular. Several studies had explored the variations of thermal and transport properties with thickness, but had focused mainly on thicknesses well above the monolayer range [2]. Many years earlier, Felix Bloch [3] and Rudolph Peierls [4] had predicted the impossibility of long range order in 2D ferromagnets and crystals at $T = 0$. Osborne [5] and Ziman [6] examined an ideal Bose gas in a thin slab geometry, showing that

momentum condensation was suppressed as the thickness became much smaller than the lateral dimensions.

I might have known of some of these papers in 1960, when I arrived at the University of Washington to start a low temperature laboratory, but they were not uppermost in my mind then, since the lab's first project was an extension of work I had been doing at Los Alamos, on Mössbauer spectroscopy. The monolayer interest began a little later, stimulated by a talk I heard about what was then a hot topic in nuclear physics: the theory of nuclear matter [7]. The theory had recently developed considerably, so that more quantitative predictions were possible. The trouble was, the theory was of *infinite nuclei,* without finite size effects. Problems connected with size and surface effects belong to a quite different domain of physics, and blending the two presented a considerable problem. So as a test of the theory without surface effects, Brueckner and Gammel [8] applied it to liquid ^{3}He, also a fermion fluid, but one that could be prepared and studied in macroscopic amounts. It occurred to me that, since it was not possible to make infinite nuclear matter, it might be possible to make finite ^{3}He in droplets so small that they could in some ways act like nuclei. Such a simulation might be feasible if one works very hard, but all it served for then was as a stimulus to begin think-

* Corresponding author.

SSDI 0039-6028(93)E0326-P

ing more about surfaces, quantum effects, and dimensionality. An experimental test of the Bose gas slab model seemed reasonable as a thesis topic (it was really pretty *un*reasonable!), and David Goodstein was eager to try. The technique we chose was calorimetry, similar to an earlier study by Frederikse [9], which had shown that the lambda point specific heat anomaly in multilayer ^{4}He films adsorbed on a high area powder became progressively lowered and rounded with decreasing thickness. Our goal was to extend the range to monolayer thicknesses and lower temperatures, where Frederikse's signals had vanished into the noise. We were fortunate to have George Halsey in Chemistry as a neighbor, to help us learn the business of adsorption, but even so, progress was slow. Eventually, we learned how to make calorimeter cells of sintered copper, and how to achieve detectable specific heat signals with samples of only a few cc's STP of ^{3}He or ^{4}He. The principal results were measurable heat capacity signals in monolayer films at temperatures extending down below 0.25 K [10]. The net film contributions to the total calorimeter signals were only a few percent, but they showed a consistent trend, varying as T^2 over a wide temperature range. Moreover, the ratio of characteristic temperatures was approximately equal to the square root of the reciprocal isotopic mass ratio. We were delighted to get these signatures of 2D behavior, but it was not the phase we had set out to find. The film was acting as a 2D harmonic solid rather than a gas; furthermore, it contradicted predictions of atomic localization in surface sites at low temperature; that would give specific heat signatures $\exp(-\text{const.}/T)$. The discrepancy brought the realization that these theories were classical, which motivated an examination of quantum hopping between surface sites. This was a tight binding model of non-interacting atoms in simple square well potentials, with well depth and barrier parameters based on typical He adsorption data [11]. The result predicted considerable tunneling mobility from site to site, comparable with a classical 2D gas at low T, and convinced us that the single particle ground state is a 2D gas. (We did not know then that many years earlier, Lennard-Jones and Devonshire [12]

had predicted considerable helium mobility on a realistically corrugated substrate. Several detailed calculations of mobility and band structure of He on graphite have been carried out by now. See the review by Cole et al. [13]; a graph from this review is reprinted in Fig. 1.) We concluded that the gas phase ought to be detectable at lower coverages and higher temperatures than the range of the low temperature solid, i.e. above its 2D melting point. Alec Stewart and I later understood that the non-localization indicated by the T^2 heat capacities of the dense adsorbed films does not necessarily indicate mobility of non-interacting atoms, but might be due instead to misregistry, where the intrinsic structure of the monolayer solid is incommensurate with that of the substrate [14]. At any rate, this imperfect understanding motivated a search for the missing phases. We extended the range of exploration to higher temperatures and lower coverages, looking for melting anomalies. But the solid-like behavior was surprisingly persistent: Stewart found it continuing even at coverages down to 10% of a layer. However, these heat capacities were not quite ideal T^2; the effective Debye temperature de-

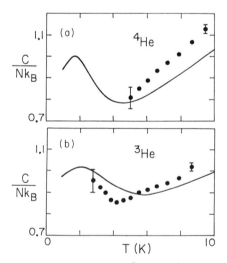

Fig. 1. Band structure effects of ^{3}He and ^{4}He in graphite. Specific heat measurements extrapolated to the limit of zero coverage, compared to the theory of non-interacting ^{4}He and ^{3}He on graphite. Solid points are intercepts from experimental data, and curves are predictions of band-structure calculations [13].

creased appreciably with T. The work reawakened Rudi Peierls' interest in two-dimensional physics, and that began our long association, for us an enormously stimulating tutelage: *he taught us how to think about films*. To explain the temperature-dependent Debye temperatures, Peierls suggested a "two-patch model", representing the film as a mixture of two 2D solids with different characteristic temperatures [15]. This theory, which gave better fits to the data, became the first of a series of models of heterogeneity in condensed phases of adsorbed films. (The theory of heterogeneity has a long history in the literature of physisorption, but it has been mainly concerned with the description of vapor pressure isotherms of non-interacting localized atoms, with little attention to effects on condensed phases of films and their phase transitions.) With the idea that we could make the 2D solid melt if we just went to a high enough temperature and a low enough coverage, David Princehouse [16] looked for a melting peak in submonolayer ^{4}He in a special 1 kg (!) calorimeter of sintered copper substrate. But even at 5 K and 1/20 layer there were no anomalies: indeed, at low coverages, the heat capacity was linear in T. At that, George Halsey and his student Narendra Roy, proposed a neat explanation [17]: the adsorbed atoms occupied strong-binding sites preferentially, in a manner analogous to the progressive filling of single particle states in a degenerate Fermi gas. The linear-in-T heat capacity was then proportional to the density of states with respect to binding energy. The explanation convinced us to search for more uniform substrates. We tried other preparation recipes for copper sponges, then a silver sponge, with no evident improvement in uniformity. We then remembered that several studies, particularly those of George Halsey and his students, had obtained remarkable adsorption isotherms of Kr on graphitized carbon black powder [18]. In contrast to the smooth sigmoid shapes of typical "BET" vapor pressure isotherms, these showed distinct steps, each step corresponding to the adsorption of a single monolayer. The steps are indications of exceptional substrate homogeneity, and their absence in typical isotherms is due to the smearing by variations in adsorption

energy. At the time, the conventional wisdom in the adsorption community was that heterogeneity was limited to the first layer, so that preplating with another or the same substance could make a substrate effectively uniform. But we had had no such indications in films on Ne- and Ar-preplated Cu. Graphite seemed worth a try, but we worried that a powder would be unsuitable for calorimetry, due to poor heat transfer. We then consulted David Fischbach [19], an authority on carbons, and fortunately for us a colleague in the Department of Ceramics. Our problem was to prepare a high area graphite powder with good thermal conductivity. Fischbach brought out a thin flexible sheet of material with an unusual, shiny black appearance. A friend in Union Carbide had just sent the sample of this experimental product, manufactured from chemically exfoliated crystals of graphite, then rolled into lightly compressed sheets. It was called Grafoil [20]; we thought it promising, and gratefully accepted the gift. Mike Bretz, who had just finished apprenticing with Stewart, and ready for his own thesis topic, constructed a combination adsorption cell calorimeter. After a series of adsorption isotherms with Ar and He to measure the adsorption area, the heat capacity measurements began, with a coverage of about one-third of a monolayer of ^{4}He. The first point, at 4.2 K, was a surprisingly large signal compared to all of the previous, solid-like data. We estimated the specific heat on the spot: the value was very nearly $C/Nk = 1$, our first sign of the 2D gas [21] (see Fig. 2). Actually, we just hoped it was the gas phase: conviction came only after Bob Siddon and Michael Schick [22] calculated the heat capacity of 2D interacting quantum gases, which showed remarkable agreement with the ^{3}He and ^{4}He results. The comparison is shown in Fig. 3.

Grafoil and other exfoliated graphite substrates opened the way to a succession of discoveries. A few weeks after the first appearance of the 2D gas, Bretz found a spectacular specific heat peak at higher coverage [23]. With Chuck Campbell's insight, we suggested that it was an order–disorder transition to a registered phase, the same structure $(\sqrt{3} \times \sqrt{3})R30°$ that Lander and Morrison [24] had seen in the LEED pattern

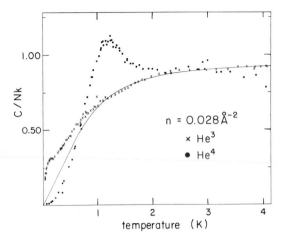

Fig. 2. Specific heat of low coverage ^{3}He and ^{4}He in Grafoil exfoliated graphite. In ^{4}He the peak is associated with 2D vapor–liquid or vapor–commensurate condensation, but in ^{3}He, there is no indication of condensation above 50 mK. The solid line shows the specific heat of an ideal 2D Fermi gas fitted to the ^{3}He data. For details, see Ref. [25]. The most recent measurements of these specific heats have been reported by Greywall and Busch [34,35].

of Br on graphite. We became convinced when Hickernell et al. [25] found identical anomalies in ^{3}He monolayers at nearly the same temperature

and coverage, and Rollefson [26] detected changes in NMR signals corresponding to mobility changes in ^{3}He around the transition. And then, with still further increases in coverage, the ordering peak disappeared, to be followed by a series of new anomalies, increasing dramatically in magnitude and rising in temperature to more than 9 K [27] (see Fig. 4). At about the same time, David Goodstein and his student Robert Elgin discovered the same strong high density peaks in their Grafoil calorimeter at Caltech [28]. The agreement between the data sets from the two labs was remarkable. And we agreed on the interpretation: the peaks were the long searched for melting anomalies!

2. O.E.V.: What happens when you land in the right place at the right time!

The beautiful experiments on the helium isotopes adsorbed on graphite done in 1971 across the room by the Bretz, Dash and Huff [1] collaboration had a profound impact on the future of our lab. They had found most of the phases we know of today for ^{4}He and ^{3}He above 0.8 K. It

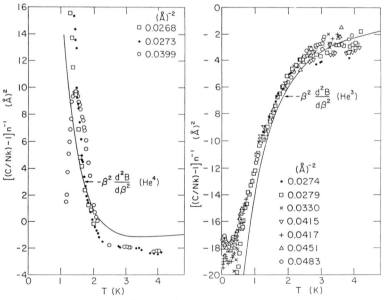

Fig. 3. Second virial coefficients of two-dimensional ^{3}He and ^{4}He, calculated by Siddon and Schick [22], compared to low coverage heat capacity data measured by Bretz, Dash, Hickernell, McLean and Vilches [1].

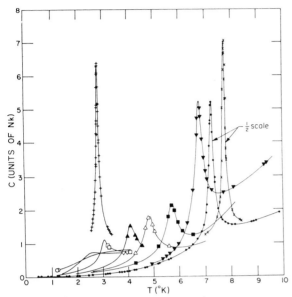

Fig. 4. Melting and ordering anomalies of ^{4}He–graphite [27]. Areal densities in units of (Å^{-2}) are: (+) 0.0637; (○) 0.075; (▽) 0.0873; (□) 0.0942; (▲) 0.0967; (△) 0.0991; (■) 0.104; (▼) 0.108; (●) 0.115; (×) 0.133. The peak at 0.0637 (Å^{-2}) corresponds to the ordering transition to the $\sqrt{3}$ phase in a Grafoil cell (compare to data in Fig. 5, using a much more homogeneous expanded graphite [40]). The rest of the figure shows melting peaks of incommensurate 2D solids. Elgin and Goodstein [28] have explained that the dramatic increase in peak strength at the two highest coverages comes from second-layer promotion due to expansion in melting.

seemed to all of us that what was needed was to extend the measurements to lower temperature. I was (and am!) lucky to have been a student and post-doctoral fellow of John Wheatley at Illinois and San Diego during the years of the development of the dilution refrigerator, and had just finished building one at Washington with my two students Gene McLean and Don Hickernell and a retired faculty member, Paul Higgs. We were testing it by measuring the specific heat of ^{4}He and ^{4}He–^{3}He mixtures adsorbed on a copper sponge [29]. In a short time we made "cell B" patterned after the Bretz and Dash cell, added a CMN thermometer and the first DC-SQUID magnetometer made by SHE Inc., and on we went [25]. By 1974 McLean and Hickernell had departed, Susanne Hering and Steven Van Sciver were on board, and we had mapped the mono-

layer heat capacity of the two isotopes and four coverages of their mixtures between 0.05 and 4.2 K, as well as the second layer and several multi-layer coverages of ^{3}He [30–33]. The monolayer ^{4}He and ^{3}He data had been extended to higher temperatures at Caltech by Goodstein and his student Robert Elgin, who added vapor pressure measurements and made the first comprehensive thermodynamic analysis of the monolayer properties [28]. In recent times, Dennis Greywall, having done beautiful measurements that extended the temperature range to the mK range, had discovered new phases in both ^{4}He and ^{3}He [34,35] and has mapped the thermodynamic properties of these two monolayers over almost four decades in temperature.

Three important things occurred in the first six months of 1975, while I was on leave in Brazil! Van Sciver, trying to interpret his multilayer ^{3}He data, worked with Greg Dash and came up with the seeds for the complete and incomplete wetting conditions, to be pursued in the 80's by Greg and Jacqueline Krim. Selden Crary, a new student in my group, in search of a substrate "other than graphite" decided to follow the earlier idea of Stewart and Dash, already tried in isotherms by Lerner and Daunt [36], of plating the graphite substrate with a monolayer of Ar. Lastly, in Brazil, I met an eager young student, Olegario Ferreira, who was to come back to the USA for one year to do his dissertation in my group.

In the next year, the plating studies turned out to be very fruitful. Selden measured not only the heat capacity of ^{4}He/Ar/gr [37], but also ^{4}He/Ne/gr and a few coverages of ^{3}He on the plated substrates [38]. Olegario then did a comprehensive study of ^{4}He/Kr/gr for two different incommensurate Kr platings, and a substantial number of ^{3}He/Kr/gr measurements [39].

During a short visit and seminar, Michael Bretz reported on his new data for the specific heat of ^{4}He at the order–disorder transition on *ZYX* exfoliated graphite [40]. The peak was double the height of what had been observed previously, and had been carefully measured to where rounding by finite size effects could be clearly observed. The data was finally good enough to obtain a value for the critical exponent α, which Michael reported to be close to $1/3$, not the Ising model's $\alpha = 0$. His measurements, shown in Fig. 5, also emphasized the importance of the substrate quality.

A new student with an experimental–theoretical bent, Manu Tejwani, was anxious to try the three-state Potts–Ising experiment being discussed by Domany, Schick and Walker (as reviewed below). A quick look at Olegario's data showed us that on incommensurate Kr the order–disorder transition occurred at 4.5 to 5 K, too high a temperature to measure accurately due to desorption. We then decide to "lower the peak temperature" by expanding the spacing of the Kr monolayer. A commensurate monolayer of Kr/gr, well characterized using LEED by Samuel Fain and his student Martin Chinn in our department [41], was chosen as the expanded lattice. A large test piece of very expanded graphite (now called graphite foam) had been given to us by Mike Dowell of Union Carbide and was used for a new cell.

As told below, the results proved very rewarding. After some trial and error, we were able to measure the two different critical exponents for the same ^{4}He coverage, without and with the Kr plating [42], see Fig. 6. In addition, the transition occurred at essentially the same temperature for the plated and unplated substrates, as expected from the lattice gas models.

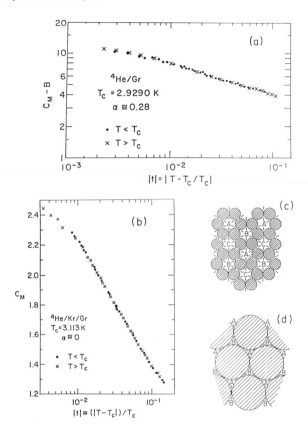

Fig. 6. Critical behavior of the heat capacities of (a) ^{4}He on bare graphite and (b) ^{4}He on Kr-plated graphite, showing the effect of substrate symmetry at the order–disorder transition. On bare graphite the substrate sites form a triangular array with three distinct sublattices (c). On Kr-plated graphite the film can order in two distinct sublattices (d). The critical heat capacity, in (a) is a power-law, with an exponent approximately equal to that of the theoretical three-state Potts model ($\alpha = 1/3$); in (b) the heat capacity is logarithmic, consistent with the two-state (Ising) model [42,48].

The fruitful collaboration, and being in the right place at the right time, had produced a beautiful piece of work. Perhaps more lasting, it had tied a large group of friends for life.

3. M.S.: the apotheosis of Renfrew B. and his eponymous model

The first measurements of the specific heats of ^{4}He and ^{3}He adsorbed on graphite were exciting

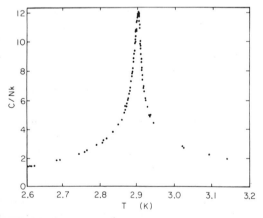

Fig. 5. Specific heat of ^{4}He on highly ordered synthetic graphite at the transition to the commensurate $(\sqrt{3} \times \sqrt{3})$R30° phase, measured by Bretz [40].

because they saturated at a value near unity which clearly showed that the adsorbed system was behaving like a two-dimensional, nearly ideal, gas on a rather weak substrate. Almost as exciting was the fact that the specific heats, shown in Fig. 2, approached unity from *below*. Because the densities of the systems were low, measured either with respect to the inverse square of the hard-core radius or of the thermal wavelength, a virial expansion of thermodynamic properties was appropriate. The contribution to the specific heat from the second virial coefficient of a classical gas is positive, so that the specific heat approaches unity from *above*. Thus, it was clear that the quantum nature of the gas was crucial to the experimental results. Robert Siddon, then a graduate student, calculated the second virial coefficients of two-dimensional bosons and fermions interacting via different realistic potentials. This was an interesting exercise in quantum mechanics, and it was fortunate that Rudi Peierls was visiting at the time to remind us of Levinson's theorem and the like, and what was known from analogous calculations of bulk systems. In addition to providing rather good fits to the data (see Fig. 3) the calculation [22] yielded some surprises and some insights. Among the former was the fact that the virial coefficients differed considerably from their classical values even at temperatures as high as 60 K. Among the latter was that it was the attraction which brought the specific heat below unity. Further, spin and statistics became important below 3 K, and this explained why the specific heat of ^{3}He continued to fall with decreasing temperature whereas that of ^{4}He began to rise. Due to the different statistics, fermions and bosons interact with different relative angular momenta; the zero momentum state favored by the bosons is sensitive to the repulsive core which leads to the increase in specific heat, while the $l = 1$ state favored by the fermions samples the attractive tail and leads to the decrease. The virial analysis also indicated that the ^{4}He gas was unstable to condensation at about $T = 1.4$ K. Finally, the fact that the high temperature specific heats do not asymptote to unity precisely is due to the band structure of the helium energies brought about by the periodic

potential provided by the graphite [43] (see Fig. 1).

The next phase to be detected after the gas was the $(\sqrt{3} \times \sqrt{3})$R30° registered phase. The specific heat signal (shown in Fig. 4 by the (+) symbol), which appeared to diverge logarithmically with temperature, reminded us all of the order–disorder transition in the two-dimensional Ising model. We knew, of course, that the graphite provided a triangular array of adsorption sites, and that the hard-core repulsion forbade simultaneous occupation of nearest-neighbor adsorption sites by helium atoms. This would translate to an Ising model with some sort of antiferromagnetic interaction, and we were not quite sure what to expect of such a system. Chuck Campbell was then a post-doctoral fellow at the University of Washington, and he and I investigated a lattice model of this system within a Bethe–Peierls approximation. We found [44] a phase diagram which was reasonable, but disappointing. It concluded that the transition to the ordered phase was always first order, whereas the experiment clearly indicated a continuous transition. Three years later, Schlomo Alexander [45] stated that the transition of this system was the same as that of the three-state Potts model, a remark which was totally opaque to me, never having heard of the three-state Potts model. Furthermore, I knew very well that the system could be treated as a lattice gas which was fully equivalent to an Ising model. I therefore ignored Alexander's remarks because I did not understand them. They proved, however, to be the harbingers of a revolution in statistical mechanics which was about to sweep over us: the introduction by Ken Wilson of the Renormalization Group. A year later, Michael Wortis took a sabbatical with us, and lectured on the new techniques as expounded by Wilson and others. My reaction to these new methods which promised to supplant all older ones with which I was familiar was probably typical: I wished that they would go away! What eventually changed my mind was the introduction by Dorus Niemeijer and Hans van Leeuwen of real space renormalization group methods, and a few enlightening conversations about finite cluster methods with Michael Nauenberg and K. Subbaro. These methods seemed ideally suited to the study of two-dimensional systems, and they promised all thermodynamic information about them, not just the critical exponents of the phase transition. Michael Wortis, my graduate student James Walker, and I applied them to the study of the order–disorder transition of helium on graphite. It was great fun, and the calculations were really quite simple. They were done on desk-top calculators readily available then and long since supplanted by programmable pocket versions. We not only obtained a reasonable phase diagram, but this time we found the transition to be continuous, in agreement with experiment. We also calculated specific heats along the transition curve and obtained reasonable, if not spectacular, agreement with experiment. It was the first application of renormalization group techniques to a real experimental system [46].

During the same period, our colleague Eberhard Riedel had a remarkable post-doctoral fellow working with him, Eytan Domany. It was he who finally explained to me what Schlomo Alexander had said several years before. The Ising model we were studying was an *antiferromagnetic* one, and its transition to the $(\sqrt{3} \times \sqrt{3})R30°$ phase would not be in the universality class of the *ferromagnetic* Ising model, but of the ferromagnetic three-state Potts model as Schlomo had predicted. (N.B. The adjective *ferromagnetic* refers to interactions in the Potts model, and not those in the underlying helium/graphite system.) With that, it was clear why the earlier Bethe–Peierls calculation had found a first-order transition; the free energy of the three-state Potts model has a cubic term in its order parameter expansion, and Landau had long ago shown that this inevitably led to a first-order transition if fluctuations were neglected. That the transition was actually continuous, as was shown rigorously by Rodney Baxter [47], showed how important fluctuations were in two dimensions. Once Eytan had clarified the symmetry argument of Alexander's, we thought it an excellent idea to see what other universality classes could be represented by transitions occurring in adsorbed systems. In a series of papers [48–51], authored by various subsets of Domany, Riedel, Walker, Bob Griffiths

and me, we determined the ordered structures produced by transitions corresponding to various universality classes. The list was not long, but included several very interesting ones, such as those of the four-state Potts model, the *xy* model with a cubic anisotropy (for which the exponents were non-universal), and the Heisenberg model with a cubic anisotropy pointing to either the corners or faces of the cube. The latter was of interest to Sam Fain in his experiments of N_2 adsorbed on graphite [52]. Bernard Nienhuis, another unusually gifted post-doctoral fellow, Riedel and I determined [53] that this transition was first order, a result in accord with experiment. Certainly, the experiment that showed most convincingly the power of the symmetry ideas of Landau was one that Domany, Walker and I proposed [48]. We suggested comparing the results of two adsorption experiments. In one, helium is adsorbed on graphite to yield the $(\sqrt{3} \times \sqrt{3})R30°$ phase. The lattice of graphite adsorption sites can be decomposed into three sublattices, and the helium chooses one of them. As there are three ground states, the transition is in the universality class of the three-state Potts model. That this class was measurably different from that of the familiar Ising model had been shown by Michael Bretz [40] who found the value of the specific heat exponent $\alpha = 0.36$. We suggested that the results of this experiment be compared to one in which krypton is first adsorbed on the graphite epitaxially in the same $(\sqrt{3} \times \sqrt{3})R30°$ pattern. Then the helium is introduced. Of the three sublattices on the graphite, one of them is occupied by the krypton. The latter presents to the helium a honeycomb array of adsorption sites which decompose into two sublattices, in exactly the same positions as the two graphite sublattices which the krypton did not occupy. The helium can then order into *precisely* the same pattern as it does on the bare graphite, at *precisely* the same nearest-neighbor distance. The only difference is that the helium now chooses between only two possible sublattices of the pre-plated graphite, as opposed to the three choices on the bare graphite. That is, only the symmetry has changed. The result is that the transition should now be in the class of the ferromagnetic two-state Potts model, i.e. the Ising model, so that the specific heat should diverge logarithmically. This experiment was carried out by Oscar and two of his students, Manu Tejwani and Olegario Ferreira [42] and they observed the desired change in transition (see Fig. 6).

The constant interplay between experiment and theory among us at that time was extremely rewarding, and no result was more personally satisfying to me than that one. It gives me some insight into the feeling experienced years later when, although working in a very different discipline, Renfrew B. Potts was persuaded to attend a physics colloquium, and heard the news.

Acknowledgement

Research by the authors was and is being supported by the NSF, currently grants DMR 9220729 (O.E.V. and J.G.D.), and DMR 9220733 (M.S.).

References

[1] J.G. Dash, Phys. Rep. 38c (1978) 178;
J.G. Dash and M. Schick, in: Liquid and Solid Helium, Vol. II, Eds. K.H. Benneman and J.H. Ketterson (Wiley/Interscience, New York, 1978);
O.E. Vilches, Ann. Rev. Phys. Chem. 31 (1980) 463;
J.G. Dash, Films on Solid Surfaces (Academic Press, New York, 1975);
M. Bretz, J.G. Dash, E.C. Hickernell, E.O. McLean and O.E. Vilches, Phys. Rev. A 8 (1973) 1589; A 9 (1974) 2814.

[2] F.D. Manchester, Rev. Mod. Phys. 39 (1967) 383.

[3] F. Bloch, Z. Phys. 61 (1930) 206.

[4] R.E. Peierls, Ann. Inst. Henri Poincaré 5 (1935) 177.

[5] M.F.M. Osborne, Phys. Rev. 76 (1949) 396.

[6] J.M. Ziman, Philos. Mag. 44 (1953) 548.

[7] L. Wilets, Bull. Am. Phys. Soc., Ser. 11, 6 (1961) 154.

[8] K.A. Brueckner and J.L. Gammel, Phys. Rev. 109 (1958) 1040.

[9] H.P.R. Frederikse, Physica 15 (1949) 860.

[10] D.L. Goodstein, W.D. McCormick and J.G. Dash, Phys. Rev. Lett. 15 (1965) 447.

[11] J.G. Dash, J. Chem Phys. 48 (1968) 2820; J. Low-Temp. Phys. 1 (1969) 173.

[12] J.E. Lennard-Jones and A.F. Devonshire, Proc. R. Soc. London A 158 (1937) 242.

[13] M.W. Cole, D.R. Frankl and D.L. Goodstein, Rev. Mod. Phys. 53 (1981) 199.

[14] G.A. Stewart and J.G. Dash, Phys. Rev. A 2 (1970) 918.

[15] R.E. Peierls, private communication;
G.A. Stewart and J.G. Dash, J. Low-Temp. Phys. 5 (1971) 1.

[16] D.W. Princehouse, J. Low-Temp. Phys. 8 (1972) 287.

[17] N.N. Roy and G.D. Halsey, J. Low-Temp. Phys. 4 (1971) 231.

[18] J.H. Singleton and G.D. Halsey, J. Phys. Chem. 58 (1954) 330, 1011.

[19] D.B. Fischbach, private communication.

[20] Grafoil is the tradename of a product marketed by Union Carbide Carbon Products Division, 270 Park Ave., New York, NY.

[21] M. Bretz and J.G. Dash, Phys. Rev. Lett. 26 (1971) 963.

[22] R.L. Siddon and M. Schick, Phys. Rev. A 9 (1974) 907.

[23] M. Bretz and J.G. Dash, Phys. Rev. Lett. 27 (1971) 674.

[24] J.J. Lander and J. Morrison, Surf. Sci. 6 (1967) 1.

[25] D.C. Hickernell, E.O. McLean and O.E. Vilches, Phys. Rev. Lett. 28 (1972) 789.

[26] R.J. Rollefson, Phys. Rev. Lett. 29 (1972) 410.

[27] M. Bretz, G.B. Huff and J.G. Dash, Phys. Rev. Lett. 28 (1972) 729.

[28] R.L. Elgin and D.L. Goodstein, Phys. Rev. A 9 (1974) 2657.

[29] D.C. Hickernell, E.O. McLean and O.E. Vilches, J. Low-Temp. Phys. 23 (1976) 143.

[30] S.W. Van Sciver and O.E. Vilches, Phys. Lett. A55 (1975) 191.

[31] S.V. Hering, S.W. Van Sciver and O.E. Vilches, J. Low-Temp. Phys. 25 (1976) 793.

[32] S.W. Van Sciver and O.E. Vilches, Phys. Rev. 18 (1978) 285.

[33] S.W. Van Sciver, Phys. Rev. 18 (1978) 277.

[34] D.S. Greywall, Phys. Rev. B 41 (1990) 1842.

[35] D.S. Greywall and P.A. Busch, Phys. Rev. Lett. 67 (1991) 3535.

[36] G.J. Goellner, J.G. Daunt and E. Lerner, J. Low-Temp. Phys. 21 (1975) 347.

[37] S.B. Crary and O.E. Vilches, Phys. Rev. Lett. 38 (1977) 973.

[38] S.B. Crary, PhD Dissertation, University of Washington, 1978.

[39] O. Ferreira, Doctoral Dissertation, Universidade Estadual de Campinas, Campinas, SP, Brazil, 1978.

[40] M. Bretz, Phys. Rev. Lett. 38 (1977) 501.

[41] M.D. Chinn and S.C. Fain, Jr., J. Vac. Sci. Technol. 14 (1977) 314.

[42] M.J. Tejwani, O. Ferreira and O.E. Vilches, Phys. Rev. Lett. 21 (1980) 152.

[43] G. Vidali, G. Ihm, H.-Y. Kim and M.W. Cole, Surf. Sci. Rep. 12 (1991) 135.

[44] C.E. Campbell and M. Schick, Phys. Rev. A 5 (1972) 1919.

[45] S. Alexander, Phys. Lett. A 54 (1975) 353.

[46] M. Schick, J.S. Walker and M. Wortis, Phys. Rev. B 16 (1977) 2205.

[47] R.J. Baxter, J. Phys. C: Solid State Phys. 6 (1973) L445.

[48] E. Domany, M. Schick and J.S. Walker, Phys. Rev. Lett.
     38 (1977) 1148.
[49] E. Domany, M. Schick, J.S. Walker and R.B. Griffiths,
     Phys. Rev. B 18 (1978) 2209.
[50] E. Domany and E.K. Riedel, Phys. Rev. Lett. 40 (1978)
     561.

[51] M. Schick, Prog. Surf. Sci. 11 (1981) 245.
[52] R.D. Diehl, M.F. Toney and S.C. Fain, Jr., Phys. Rev.
     Lett. 48 (1982) 177.
[53] B. Nienhuis, E.K. Riedel and M. Schick, Phys. Rev. B 27
     (1983) 5625.

Surface Science 299/300 (1994) 415–425
North-Holland

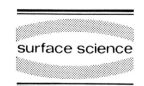

surface science

Stepwise isotherms and phase transitions in physisorbed films

André Thomy [a],* and Xavier Duval [b]

[a] *Laboratoire de Métallurgie Physique et Science des Matériaux, URA-CNRS 155, Université de Nancy 1, BP 239, F-54506 Vandoeuvre, France*
[b] *Laboratoire Maurice Letort, CNRS, 405, rue de Vandoeuvre, F-54600 Villers-lès-Nancy, France*

Received 30 April 1993; accepted for publication 21 July 1993

The present paper outlines the evolution during the last thirty years of research in the field of physisorption of gases on solids with a homogeneous surface. It relates this evolution as lived by some researchers of a physical chemistry laboratory whose objective was not originally to study physisorption as such, but to use it as a method to determine the specific surface area of solids participating in gas reactions. Consequently, the aim is not to give a complete review even of only those results obtained from adsorption isotherms, but simply to recall the way which led to the discovery of several of the most typical adsorption phenomena, i.e. : "gas–liquid–solid" and "commensurate–incommensurate" 2D transitions, 2D polymorphism, wetting transitions and specific behaviour of mixed films.

1. Beginning 1963...

It is just thirty years ago that results obtained with exfoliated graphite as an adsorbent started to give a new impetus to the study of two-dimensional (2D) phases adsorbed on solids. This occurred at the "Laboratoire de Cinétique Hétérogène" of the "Ecole Nationale Supérieure des Industries Chimiques" (ENSIC) in Nancy, France. Further research after 1965 was carried out at the "Centre de Cinétique Physique et Chimique" of the CNRS in Villers-lès-Nancy, which in 1977 was given the name "Laboratoire Maurice Letort" [#1]. These investigations were developed in relation with some ten groups and more especially those of Y. Larher, R. Kern, M. Bienfait, J. Suzanne, B. Mutaftschiev and C. Marti.

By 1963 it had indeed been found that the exfoliation of graphite by thermal dissociation of its intercalation compound with ferric chloride ($FeCl_3$) results in a considerable increase in surface area and also in a much more homogeneous surface than in the original sample [1]. For instance the specific surface area of a natural graphite could be increased by a factor of more than 200 and the heterogeneity originating essentially from the edges of the graphite layers was considerably reduced. Such results were obtained with all graphites whose layers were accessible for intercalation by molecules such as $FeCl_3$ (or $SbCl_5$ which we used later). At that time and since then, we benefited from the experience of the "Laboratoire de Chimie Minérale Appliquée" (ENSIC and Université de Nancy) then directed by Professor A. Hérold in the synthesis of intercalation compounds.

2. From specific surface area determination to stepwise isotherms

In the early sixties there was practically only one technique for studying the adsorption of gases on solids. This was "adsorption volumetry" with which were determined adsorption isotherms N_a

---

* Corresponding author.
[#1] Professor Letort, member of the Academie des Sciences had been director of ENSIC where he developed research in chemical kinetics and chemical engineering.

SSDI 0039-6028(93)E0435-W

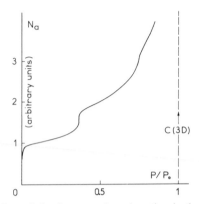

Fig. 1. One of the first stepwise adsorption isotherms published in 1952 by Bonnetain et al. showing the 2nd and 3rd layering transitions of CH_4 on graphite at 77 K (from Ref. [5]). N_a: amount adsorbed; P: pressure of the gas in equilibrium with the film; P_0: saturated vapor pressure in equilibrium with the 3D phase. $P/P_0 = 1$ corresponds to 3D condensation $= C(3D)$.

$= f_T(P)$, N_a being the amount adsorbed on a given mass of adsorbent, P the pressure of the three-dimensional (3D) gas in equilibrium with the film at temperature T. It is only since the work of Morrison and Lander (low energy electron diffraction on films of rare gases adsorbed on graphite – 1966) [2] that local probes became of general use in the study of adsorbed films on a given face of a single crystal of which some have been used for the determination of adsorption isotherms [3,4].

Adsorption volumetry, giving access to properties of the film and the adsorbing surface as a whole, was commonly used for the determination of the specific surface area of powders by the so-called BET method. This was used in Professor Letort's laboratory already in 1950 in a research program on the chemical reactivity of graphites. In order to increase the sensitivity of the method, methane was chosen as adsorbate instead of nitrogen, because of its much lower vapor pressure at 77 K (about 10 Torr as compared to 760). Attention was soon attracted by several steps present on the isotherms at the same pressures with all graphites and graphitized carbons (Fig. 1). It was conjectured that these steps indicated 2D condensation within the film. Such phase changes had been found long before in monolayers on water, but sought on solids in

vain or with controversial results. A prevailing opinion was that expressed a few years later by Landau and Lifschitz : "The adsorption on a solid surface is of no interest owing to the inhomogeneity which it almost always possesses" (Statistical Mechanics, 1959).

The flaky crystals of graphite not only expose predominantly the basal planes, but this surface was also known from reactivity studies to be remarkably unreactive and therefore not liable to contamination. Other examples of stepwise isotherms obtained on a series of adsorbents with a layer structure (MoS_2, $CdCl_2$, ...) showed conclusively that they were not artefacts [5].

These results lead the laboratory to study phase transitions in adsorbed films as such. In addition, they incited the use of physical adsorption, not only as a means of measuring specific surface area, but also to estimate the surface homogeneity of solids of interest (especially graphite). Indeed, one could hope to estimate the fraction of the total surface area constituted by basal planes of the layered solid and responsible for the isotherms steps. However standard solids of homogeneous surface had to be available.

3. The advantages of exfoliated graphite and of the (0001) graphite face

In 1954, Polley et al. [6] had shown that carbon blacks graphitized by heat treatment provided the best homogeneous surface at that time. After this work and for more than ten years, graphitized carbon blacks were considered as standard adsorbents and used as such in many studies, specially by G.D. Halsey, A.V. Kiselev, J.H. de Boer, S. Ross and their collaborators.

Graphite exfoliated from intercalation compounds proved itself superior to graphitized carbon blacks. It allowed the observation of an increasing number of phase transitions (more and sharper steps (Fig. 2)) and with appreciably shorter times to reach equilibrium. The advantages of exfoliated graphite are due to the larger sizes of uniform areas (parts of (0001) faces) than in graphitized carbon blacks, and to less propensity to intergranular condensation (cf. Ref. [7]).

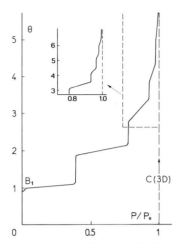

Fig. 2. Krypton adsorption isotherm at 77.3 K on exfoliated graphites (specific surface area: 20 to 85 m^2/g instead of 0.4 m^2/g for an untreated natural graphite). (From Ref. [1].) Detail of the upper steps above the third is due to Goffinet (cf. Ref. [7]); concerning the first step, see Fig. 5.

But the exfoliation of graphite afforded an additional important advantage: it was a simple means to make changes of the ratio between the part of the surface made of the edges of graphitic layers (non uniform part of the surface) and the part constituted by the basal planes (uniform part of the surface made of (0001) faces). It was then possible, more easily and rigorously than with graphitized carbon blacks, to approach by limiting curves the case of an ideal uniform graphitic surface. The availability of standard surfaces allowed the estimation of the degree of heterogeneity of any graphite (cf. Ref. [7]).

By means of exfoliation, it also became easier to show whether a singularity in adsorption isotherms indicated a phenomenon occurring on the uniform part of the graphite surface or not. It was thus proved that the "sub-step" A_1' D_1' (cf. Fig. 5) discovered in 1959 in the krypton isotherm at 90 K [8] corresponded to a transition effectively occurring on the (0001) face of graphite during the build-up of the first krypton monolayer. This "sub-step" led us to the idea that the same physisorbed monolayer could undergo several successive 2D transitions, which was effectively observed later on to be the case in many systems.

In addition, it should be emphasized that re-

sults obtained with exfoliated graphite dispelled doubts that one might still have about the cleanliness of the uniform part of the graphite surface resulting from treatment prior to an adsorption experiment (degassing at 800°C under vacuum down to a residual pressure of the order of 10^{-6} Torr). These doubts were to be ultimately eliminated after confirmation of results pertaining to krypton and xenon films by experiments on single crystals (electron spectroscopy [3], ellipsometry [4], electron microscopy [9], electron diffraction [10]).

The availability of very homogeneous surfaces of exfoliated graphite, allowed us, for all adsorbates used and for a given surface area, to compare amounts adsorbed at different points of isotherms as in elementary surface stoichiometry. We were then able to draw conclusions about film densities and the possible nature of the transitions. In this procedure, the krypton isotherm at 77 K (Fig. 2) was systematically used as a standard and more especially the amount of krypton adsorbed at point B_1.

Graphite remained our preferred adsorbent. Firstly, for the reasons related to exfoliation, but also on account of the new phenomena which it revealed (successive 2D transitions of the type "gas–liquid–solid" and more particularly "commensurate–incommensurate" transitions). Graphite soon also appeared as an adsorbent lending itself to various studies by means of miscellaneous techniques due to its intrinsic properties and the different forms it may take. Especially interesting is exfoliated graphite compressed into sheets of type "Grafoil" (Union Carbide) or "Papyex" (le Carbone Lorraine) profitably used by Dash and collaborators [11] in heat capacity measurements of helium films in the late sixties and currently used in neutron and X-ray diffraction experiments.

Finally, we realized that the (0001) face of graphite was an exceptional "vessel" for the study of the 2D phase transitions owing to its great energetic uniformity (shallow potential wells able to hinder only slightly the mobility of adsorbed molecules and their interactions).

Independently, Y. Larher and J.P. Coulomb with their collaborators carried out exemplary

investigations, either thermodynamical (from adsorption isotherms) or structural (by neutron
diffraction) on lamellar dihalides and magnesium
oxide respectively.

4. Lamellar di-halides: an adjustable adsorbent

Lamellar halides of type MX_2 (M = metal ion,
X = halogen ion) constitute a whole series of
solids whose basal plane surface presents a structure and a relief resembling a (111) face of a rare
gas or methane crystal. Moreover, the distances
between nearest potential wells change progressively within values which include the distances
between nearest molecules of argon, krypton,
xenon, and methane in the (111) plane of their
3D crystal. Larher and collaborators [12] have
made use of these adsorbents to elucidate the
conditions under which the substrate relief may
affect the film properties and if so to specify the
nature of this effect.

This systematic and precise investigation was
based on adsorption isotherms on powder adsorbents, some of which had a surface homogeneity
comparable to that of exfoliated graphite (cf. Fig.
3 and Ref. [12]).

5. A square lattice substrate: MgO(100)

J.P. Coulomb and collaborators have succeeded in synthetizing MgO powders of very ho-

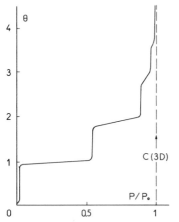

Fig. 3. Krypton adsorption isotherm at 73 K on $CdBr_2$ (from
Larher [13]).

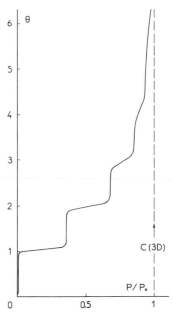

Fig. 4. Adsorption isotherm of CH_4 at 77.3 K on MgO. An
example of layer-by-layer growing mode from a square (100)
interface. (from Madih et al. [16]). In this case, the first
monomolecular layer undergoes only one first order phase
transition between a 2D lattice gas and a "commensurate" 2D
solid, as generally do the 2nd, 3rd ... layers.

mogeneous surface (cf. Fig. 4). Precise sets of
isotherms of rare gases and methane allowed a
thermodynamical characterization of the films,
completed by a structural study [14–16]. One of
the interests of this work is that it is the only
complete investigation carried out on a crystal
face with a square lattice as the surface of MgO
particles growing in the gas phase is composed of
(100) faces. Contrary to the cleavage face of
graphite (or boron nitride which has also been
used as adsorbent) and dihalides, such a surface
tends therefore to impede the formation of a 2D
structure like that of a (111) plane of the 3D
crystal of a rare gas or methane. Furthermore the
diameter of CH_4, about 4.20 Å, is very close to
the parameter of a square sublattice of sites, viz.
4.21 Å, whereas the diameter of krypton is 3.4%
below and that of xenon is 4.5% above this value.

For these reasons and owing to the great surface homogeneity of the MgO powders used in
the experiments, results have been obtained which
also bring more light on the conditions under
which surface corrugation affects film properties.

6. Successive phase transitions within a single monolayer

For a long time adsorption isotherms had been generally determined at a few convenient and conventional temperatures (especially 77 K). It is only from the late fifties that sets of isotherms began to be determined systematically at different temperatures with the use of cryostats (see, for instance Ref. [17]).

In our laboratory, the first and promising results obtained with exfoliated graphite prompted us as early as 1963 to build several cryostats enabling accurate measurements in a temperature range extending from 63 to above 200 K.

Until the mid seventies our attention was focused on successive phase transitions within the first monolayer of more and more complex adsorbates: rare gases (except helium), N_2, O_2, NO, SF_6, hydrocarbons (CH_4, C_2H_6, C_2H_4, C_2H_2 C_6H_{12}) and some of their halogenated derivatives...

As mentioned above, graphite was our preferred adsorbent, but we also used some dihalides (in NO adsorption studies) and more systematically boron nitride, a substrate differing very little from graphite.

6.1. "Gas–liquid–solid" and "commensurate–incommensurate" 2D transitions

Two diagrams are at the origin of many of our exchanges and collaborations with researchers of different laboratories in France and abroad: these are the 2D diagrams of krypton and xenon shown Figs. 5 and 6. The diagrams, the resulting thermodynamical data and their interpretation as to the nature of phase transitions, are the outcome of a number of studies not only thermodynamical [3,4,18–22] but also structural [2,9,10,23–27] and have been the subject of as many theoretical works [28–40] (many other papers would be worth quoting).

We had thought for some time (1964–1974) that krypton had provided us with the first example of a diagram including a 2D triple point and a 2D critical point. It turned out later that the domain of the supposed 2D liquid krypton was

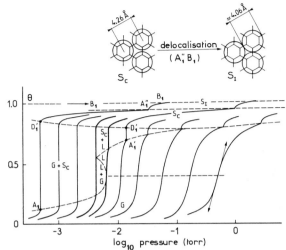

Fig. 5. Redrawn diagram showing the evolution of krypton isotherms in the first layer domain between 77 and 110 K on a homogeneous graphite surface (from Refs. [18–20], see also Ref. [7]). θ: surface coverage taken equal to 1 when the first monomolecular layer is complete (at point B_1). Isotherm temperatures: 77.3; 79.8; 82.3; 84.8; 86.0; 88.0; 91.8; 96.6; 102.6; 109.5 K G = 2D gas; L = 2D liquid; S_c = 2D commensurate solid; S_I = 2D incommensurate solid. According to Larher [20] the krypton layer undergoes two first-order phase transitions (G--→L and L--→S_c), but in a very narrow temperature range (1 K ~ 86–84.8 K). According to Butler et al. [21] the liquid phase would not even exist (notion of "incipient" triple point). The "sub-step" $A_1''B_1$ corresponds to a transition between a commensurate solid (S_c) and an incommensurate solid (S_I).

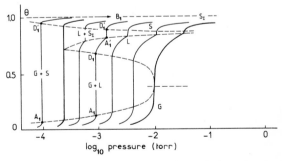

Fig. 6. Redrawn diagram showing the evolution of xenon isotherms in the first layer domain between 97 and 117 K on a homogeneous graphite surface (from Refs. [3,18], see also Ref. [7]). Isotherm temperatures: 97.4; 100.1; 102.4; 105.4; 108.3; 112.6; 117.0. Between 100.1 and 117 K, the layer undergoes two first order phase transitions where it passes successively with increasing θ through three distinct 2D states denoted as "gaseous" (G), "liquid" (L) and "incommensurate solid" (S_I) (cf. Ref. [7]).

very narrow [20] or even non-existent [21] (cf. Fig. 5). Anyway, we have thus to deal with a very peculiar diagram with no equivalent in 3D, considering in addition the "commensurate–incommensurate" transition A''_1B_1.

The xenon diagram of Fig. 6 must consequently be considered as really the first 2D diagram with clear evidence of a "normal" triple point and a critical point. About ten other diagrams of this type were later obtained with rare gases and methane on graphite [18], boron nitride [41,42], lamellar dihalides (cf. Ref. [12]), magnesium oxide [14–16,43].

Generally the following relations hold between the temperatures of 2D and 3D triple points and critical points:

$$T_t(2D)/T_t(3D) \sim 0.6,$$

$$T_c(2D)/T_c(3D) \sim 0.4.$$

T_t and T_c being respectively the triple point and critical point temperatures 2D and 3D for the adsorbate–adsorbent pair considered. These ratios are in agreement with theories and simulations with the hypothesis of an energetically uniform adsorbing surface, i.e. with no potential wells (cf. Ref. [12]). In other words, the 2D "gas–liquid–solid" transitions observed with rare gases and methane, for which the ratios considered are valid, would correspond to the ideal case of a plane substrate on the atomic scale, chemically inert and in no way impeding interactions between adsorbed molecules.

The results obtained on lamellar dihalides and MgO(100) show the conditions where the film exhibits a 2D ideal behaviour with triple point and critical point, as for the xenon/graphite pair, and those where the underlying structure appreciably or even drastically affects its properties (cf. Ref. [12]).

6.2. 2D Polymorphism

Adsorbate molecules which are neither spherical nor globular may form several 2D solids on a given adsorbent surface, owing to the possibility for admolecules to take different positions on the surface, even if this is without relief (e.g. "lying

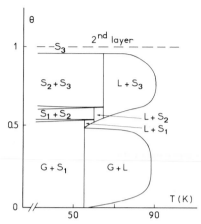

Fig. 7. Tentative schematic phase diagram corresponding to the NO/graphite couple in the first monomolecular layer (from adsorption isotherms and neutron diffraction patterns, cf. Refs. [47,51]). Three different 2D solids have been observed (S_1, S_2, S_3) in addition to a 2D liquid (L).

flat" or "standing up" molecules with an adsorbate like ethane for example). Evidence of 2D polymorphism was first obtained on graphite with oxygen and nitrogen by structure analysis [44–46].

The adsorption of nitrogen monoxide (NO) has given rise to an extensive thermodynamical investigation using isotherms on graphite, boron nitride and lamellar dihalides by Matecki et al. [47,48] and Enault et al. [49]. Films on graphite have been more completely analysed by neutron diffraction at ILL (Grenoble) by Suzanne and coworkers [50,51]. The outcome has been a peculiarly complex diagram of 2D phases owing to the discovery of three different 2D solids in addition to a 2D liquid (Fig. 7).

Other similar complex diagrams were subsequently found; that of the C_2H_6/graphite pair being among the first (Regnier et al. began its study as early as 1966) [52–56].

7. From the 2D to the 3D state: continuity or discontinuity

It has been customary to use the term "wetting" to describe the evolution from the 2D to the 3D state, whatever the state of the adsorbate under the conditions used, 3D liquid ($T > T_t(3D)$) or 3D solid ($T < T_t(3D)$).

7.1. Complete wetting – incomplete wetting

An isotherm such as on Fig. 2 includes five sharp steps showing that at least five monomolecular layers form successively when the 3D gas pressure in equilibrium with the film is increased. It is noted that the steps become less and less sharp as the saturated vapor pressure P_0 of the adsorbate is approached. There are two reasons for this: on the one hand the steps become closer and closer to P_0 (with rare gases or methane $P_n/P_0 \sim$ constant$/n^3$, P_n being the pressure of the "vertical" part of the nth step); on the other hand in the case of powders, an intergranular condensation may occur as P_0 is approached, increasing the slope of the isotherms between the steps and so blurring them.

As a consequence an isotherm such as on Fig. 2 is consistent with a complete wetting allowing for the film to grow continuously from the 2D to the 3D state (Frank–Van der Merwe type of growth). This is the more likely as the structure and density of the monolayer at completion are close to those of a plane of the 3D phase which forms at P_0.

In order to observe the formation of as many monomolecular layers as possible, some researchers have determined isotherms on single crystals, so as to eliminate interparticle condensation encountered with powders. The method used in this connection is ellipsometry which allows to explore a much larger pressure domain than methods using electron spectroscopies. Measure-

ments near P_0 are nevertheless delicate since the adsorbing surface studied is not necessarily the coldest place in the system. Hess, Knorr and their collaborators [57–62], following Quentel et al. [4] have obtained the most remarkable results with this method (Fig. 8).

As a whole the thermodynamical results from adsorption isotherms tend to show that one generally has to deal rather with "incomplete wetting" in the domain of crystal growth of the adsorbate $(T < T_t(3D))$; the number of layers would then be limited (Stranski–Krastanov type of growth). In this case the 3D phase does not form in the continuity of the thickening film, but grows probably near surface defects. The explanation is that the surface generally imposes a 2D state different, at least in density, from any plane in the 3D crystal which forms under pressure P_0 (cf. Ref. [63]). This occurs more particularly with non spherical or non globular adsorbates as well on surfaces with strong relief such as the cleavage faces of lamellar dihalides as on more energetically uniform surfaces like the (0001) face of graphite.

7.2. Increase with temperature of the number of adsorbed layers-wetting transition

Adsorption isotherms have shown that for a given adsorbate–adsorbent pair an increase in temperature may shift incomplete wetting to complete wetting. The first typical example is with the ethylene–graphite (0001) pair where the

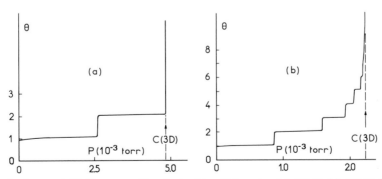

Fig. 8. Two adsorption isotherms of CF_2Cl_2 on graphite obtained by ellipsometry. (a): 100 K; (b): 115 K; ($T_t(3D)$ of $CF_2Cl_2 = 116$ K). The system shows a similar wetting behaviour as the C_2H_4/graphite couple (cf. Fig. 9, Section 7.2) (from Volkmann and Knorr (1991), cf. Ref. [62]).

maximum number n of adsorbed monolayers increases with temperature, and this, more and more steeply and tending to infinity at 104 K (Fig. 9, Ref. [64]). This temperature, at which incomplete wetting shifts to complete wetting, is termed the *wetting transition temperature* T_w In the example considered, T_w turns out to be equal, or very close to the 3D triple point temperature: $T_w \sim T_t$ (3D).

Furthermore, the maximum thickness of the film below T_t varies approximately as $(T_t - T)^{-1/3}$ as expected for liquid like van der Waals multilayers [69].

This property is not peculiar to the ethylene/graphite pair; it may thus indicate a behaviour having some generality (cf. Refs. [60,62,63,70]).

7.3. No wetting

In the above example of the ethylene/graphite pair, lowering the temperature from 104 K decreases the number of adsorbed layers which falls to 1 below 80 K.

Examples have been found where not even a single condensed monolayer is formed below some definite temperature. Such a case is the CO_2/graphite pair investigated in detail by Terlain and Larher [71]: below about 100 K not even one dense layer of CO_2 forms on graphite (0001).

This example suffices to point out that when physisorption is used to characterize a solid surface, the different types of surface constituting the whole surface may be a priori quite unequally covered, some being covered with a dense monolayer, whereas others may remain practically bare.

8. Stepwise isotherms on metal powders

From the late fifties, our laboratory undertook the determination of adsorption isotherms of some gases on metal powders at low temperatures. As a consequence of their reactivity, especially towards oxygen, this investigation turned out to be more difficult than with graphite and other layered solids. Very few investigators have undertaken this type of work during the last thirty years (notably van Dongen et al. [72]). Nevertheless, Génot succeeded in obtaining interesting and original results with powders prepared by ingenious methods and exposing preferentially the densest crystal face (adsorption of Kr, CF_4, CH_3F, CH_3Cl, CH_3Br, C_2H_6, SF_6 on powders of Ag, Fe, U, Mn, Cd, Zn, Cr, Co, Cu) [73–77]. The features of some isotherms (e.g. Fig. 10) lead him to conceive different types of diagrams, including the diagram (b) of Fig. 9 which corresponds to an increasing number of layers

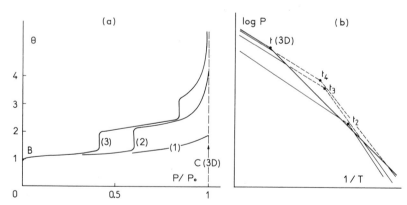

Fig. 9. Adsorption of ethylene on graphite. (a) Adsorption isotherms of C_2H_4 on exfoliated graphite (1): 77.3 K; (2): 90.0 K; (3): 105.8 K (from Ref. [64]). (b) Corresponding schematic diagram log P versus $1/T$. In thick continuous lines: log P_0 versus $1/T$ curves. In thin continuous lines: log P versus $1/T$ curves corresponding to the 2nd, 3rd, 4th adsorbed layers; in dashed lines: virtual parts of these curves (i.e. situated in domain of supersaturation). (From Ref. [64].) The C_2H_4/graphite system has also been studied by X-ray diffraction [65,66], neutron diffraction [67], and NMR [68].

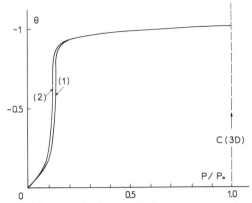

Fig. 10. Adsorption isotherms of SF_6 on copper powder (1): 127.6 K; (2): 143.4 K. The adsorption enthalpy in the quasi-vertical part of the step is smaller than the 3D condensation enthalpy (as for the 2nd, 3rd, ... steps of the isotherms corresponding to C_2H_4/graphite couple). (From B. Génot et al., Ref. [75].)

with temperature, and which includes virtual 2D triple points [75].

9. From pure films to mixed films

In 1967 our laboratory began experiments similar to those carried out previously namely by Halsey and collaborators [78–80]. In these experiments, adsorption isotherms of a gas X were determined on a surface preplated to some extent with an other adsorbate Y, the equilibrium partial pressure of which, P_Y, being negligible as compared to P_X at the isotherm temperature.

The Kr/Xe/graphite system was the first that we investigated. It was shown that dissolved xenon results in a stabilisation of the "commensurate–incommensurate" transition characterizing the krypton film [81]. This system has given rise to a theoretical work by Villain and Moreira [82].

An interesting behaviour is presented by the Kr/SF_6/graphite system. Like the metals investigated by Génot [74,75] graphite is very imperfectly wetted by SF_6 since it adsorbs no more than a single monolayer (cf. Refs. [83–85]). This layer is then displaced via a first order phase transition by krypton which admittedly wets completely the graphite (0001) face [86]. This is shown by isotherms of Fig. 11 and a structure analysis

(X-ray diffraction by Croset, Marti et al.; unpublished results). More or less similar results have been obtained by Dupont-Pavlovsky and Menaucourt et al. [87–90] with other systems the behaviour of which can be explained by thermodynamical considerations [91].

Coadsorption of two gases on uniform surfaces with rather similar thermodynamical properties has until now been barely investigated experimentally. The difficulty lies in the determination of the adsorbed amounts of each gas as well as the partial pressures in equilibrium with the film. The system $(Kr + CH_4)$/graphite has recently been theoretically and experimentally investigated by Hommeril and Mutaftschiev [92]. The ideality of 2D solutions has been studied, and multiple phase transitions have been observed in the mixed monolayer.

10. Conclusion

The last thirty years have seen considerable progress with our understanding of physisorption of gases on solids, with the simultaneous enrichment of the field of surface and interface phenomena, as well as the more general field of phase transitions.

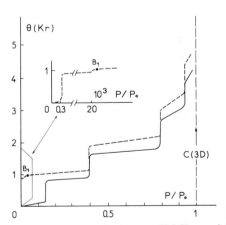

Fig. 11. Krypton adsorption isotherm at 77.3 K on exfoliated graphite previously covered with one SF_6 monolayer. In dashed line: krypton adsorption isotherm obtained on bare graphite (from Ref. [86]). At the end of the first transition, SF_6 has been displaced but still occupies some 15% of the uniform part of the graphite surface.

This progress is due to the success in preparing crystal powders with very homogeneous surfaces which allowed rigorous thermodynamical studies from sets of adsorption isotherms determined by "volumetric" techniques. It is also due to the development, during the same period, of techniques for surface analysis which have enabled to check the cleanliness of the surfaces investigated, to determine adsorption isotherms on single crystals and finally to reveal film structures.

Solids with lamellar structure have played a prominent role in these studies, and above all graphite because of its outstanding surface properties coupled with its also outstanding anisotropy.

Exfoliated graphite itself deserves special distinction, since it revealed several of the most typical adsorption phenomena ("gas–liquid–solid" and "commensurate–incommensurate" 2D transitions, wetting transitions, displacement of a 2D films by an other 2D film, ...).

With regard to lamellar dihalides and magnesium oxide MgO(100), a systematic and precise utilisation of their isotherms has allowed for a better understanding of the consequences of substrate corrugation on film properties. A comparison of results obtained with the above adsorbents allowed to determine the conditions where a film behaves as if the substrate were plane on the atomic scale.

At the present time, research is developing towards more and more complex systems namely as to the nature of adsorbates, which leads to 2D states also more and more complex and difficult to identify. As a result, thermodynamical studies based on adsorption isotherms are giving way more and more to structural investigations. This evolution is very similar to that followed formerly by research on 3D phases.

References

[1] X. Duval and A. Thomy, C.R. Acad. Sci. Paris 259 (1964) 4007;
A. Thomy, Thesis, Nancy (1968).

[2] J. Lander and J.J. Morrison, Surf. Sci. 5 (1966) 1963.

[3] J. Suzanne, J.P. Coulomb and M. Bienfait, Surf. Sci. 40 (1973) 414; 44 (1974) 141.

[4] G. Quentel, J.M. Rickard and R. Kern, Surf. Sci. 50 (1975) 343.

[5] L. Bonnetain, X. Duval and M. Letort, C.R. Acad. Sci. Paris 234 (1952) 1363.

[6] M.H. Polley, W.D. Schaeffer and W.R. Smith, J. Phys. Chem. 57 (1953) 469.

[7] A. Thomy, X. Duval and J. Regnier, Surf. Sci. Rep. 1 (1981) 1.

[8] A. Thomy, Thèse 3ème cycle, Nancy (1959).

[9] J.A. Venables, H.M. Kramer and G.L. Price, Surf. Sci. 55 (1976) 373.

[10] M.D. Chinn and S.C. Fain, Phys. Rev. Lett. 39 (1977) 146.

[11] J.G. Dash, Films on Solid Surfaces, The physics and chemistry of physical adsorption (Academic Press, New York, 1975).

[12] Y. Larher, Surface Properties of Layered Structures, Ed. G. Benedek (1992) (Kluwer, Dordrecht, 1992), p. 261.

[13] Y. Larher, J. Coll. Interface Sci. 37 (1971) 836.

[14] J.P. Coulomb, T.S. Sullivan and O.E. Vilches, Phys. Rev. B 30 (1984) 4753.

[15] J.P. Coulomb, K. Madih, B. Croset and H.J. Lauter, Phys. Rev. Lett. 54 (1985) 1536.

[16] K. Madih, B. Croset, J.P. Coulomb and H.J. Lauter, Europhys. Lett. 8 (1989) 459.

[17] D.M. Young and A.D. Crowell, Physical Adsorption of Gases (Butterworths, London, 1962).

[18] A. Thomy and X. Duval, J. Chim. Phys. 67 (1970) 1101.

[19] A. Thomy, J. Regnier and X. Duval, in: Thermochimie, Colloq. Intern. CNRS, Marseille, 201 (1972) 511.

[20] Y. Larher, J. Chem. Soc. Faraday Trans. I, 70 (1974) 320.

[21] D.M. Butler, J.A. Litzinger, G.A. Stewart and R.B. Griffiths, Phys. Rev. Lett. 42 (1979) 1289.

[22] J. Regnier, J. Rouquerol and A. Thomy, J. Chim. Phys. 72 (1975) 327.

[23] C. Marti, B. Croset, P. Thorel and J.P. Coulomb, Surf. Sci. 65 (1977) 532.

[24] M.D. Chinn and S.C. Fain, J. Vac. Sci. Technol. 14 (1977) 314.

[25] S.C. Fain and M.D. Chinn, J. Phys. (Paris) 10 (1977) C4-99.

[26] T. Ceva and C. Marti, J. Phys. (Paris) 39 (1978) L221.

[27] E.M. Hammonds, P. Heiney, P.W. Stephens, R.J. Birgeneau and P. Horn, J. Phys. C 13 (1980) L 301.

[28] J.P. Stebbins and G.D. Halsey, J. Phys. Chem. 68 (1964) 3863.

[29] F. Tsien and G.D. Halsey, J. Phys. Chem. 71 (1967) 4012.

[30] S. Toxvaerd, Mol. Phys. 29 (1975) 373.

[31] G.L. Price and J.A. Venables, Surf. Sci. 49 (1975) 264; 59 (1976) 509.

[32] J.A. Venables and P.S. Schabes-Retchkiman, J. Phys. (Paris) 10 (1977) C4-105.

[33] G.D. Halsey, J. Phys. Chem. 81 (1977) 2076.

[34] B. Mutaftschiev and A. Bonissent, J. Phys. (Paris) 10 (1977) C4-82.

[35] F.A. Putnam, J. Phys. (Paris) 10 (1977) C4-115.
[36] A.N. Berker, S. Ostlund and F.A. Putnam, Phys. Rev. B 17 (1978) 3650.
[37] S. Toxvaert, J. Chem. Phys. 69 (1978) 4750.
[38] S. Ostlund and A.N. Berker, Phys. Rev. Lett. 42 (1979) 843.
[39] J. Villain, in: Ordering in Two Dimensions, Ed. S.K. Sinha (North-Holland, Amsterdam, 1980) p. 123.
[40] J.A. Barker, D. Henderson and F.F. Abraham, Physica A 106 (1981) 226.
[41] J. Regnier, A. Thomy and X. Duval, J. Coll. Interface Sci. 70 (1979) 106.
[42] C. Bockel, A. Thomy and X. Duval, Surf. Sci. 90 (1979) 109.
[43] J.P. Coulomb, in: Phase Transitions in Surface Films 2, Eds. H. Taub et al. (Plenum, New-York, 1991) p. 113.
[44] J.P. Mc Tague and M. Nielsen, Phys. Rev. Lett 37 (1976) 596.
[45] T.T. Chung and J.G. Dash, Surf. Sci. 66 (1977) 559.
[46] J. Eckert, W.D. Ellenson, J.B. Hastings and L. Passel, Phys. Rev. Lett. 43 (1979) 1329.
[47] M. Matecki, A. Thomy and X. Duval, J. Chim. Phys. 71 (1974) 1484.
[48] M. Matecki, A. Thomy and X. Duval, Surf. Sci. 69 (1977) 596.
[49] A. Enault and Y. Larher, Surf. Sci. 62 (1977) 233.
[50] J. Suzanne, J.P. Coulomb, M. Bienfait, M. Matecki, A. Thomy, B. Croset and C. Marti, Phys. Rev. Lett. 41 (1978) 760.
[51] J.P. Coulomb, J. Suzanne, M. Bienfait, M. Matecki, A. Thomy, B. Croset and C. Marti, J. Phys. (Paris) 41 (1980) 1164.
[52] J. Regnier, Thesis, Nancy (1976).
[53] J.P. Coulomb, J.P. Biberian, J. Suzanne, A. Thomy, G.J. Trott, H. Taub, H.R. Danner and F.Y. Hansen, Phys. Rev. Lett. 43 (1979) 1878.
[54] H. Taub, G.J. Trott, F.Y. Hansen, H.R. Danner, J.P. Couloumb, J.P. Biberian, J. Suzanne and A. Thomy, in: Ordering in Two Dimensions, Ed. S.K. Sinha (North-Holland, Amsterdam, 1980) p. 91.
[55] J. Regnier, J. Menaucourt, A. Thomy and X. Duval, J. Chim. Phys. 78 (1981) 629.
[56] J.M. Gay, J. Suzanne and R. Wang, J. Chem. Soc. Faraday Trans. 82 (1986) 1669.
[57] H.S. Nham, M. Drir and G.B. Hess, Phys. Rev. B. 35 (1987) 3675.
[58] H.S. Nham and G.B. Hess, Phys. Rev. B 38 (1988) 5166.
[59] H.S. Youn and G.B. Hess, Phys. Rev. Lett. 64 (1990) 443.
[60] G.B. Hess, in: Phase Transitions in Surface Films 2, Ed. H. Taub et al., (Plenum, New York, 1991) p. 357.
[61] J.W.O. Faul, U.G. Folkmann and K. Knorr, Surf. Sci. 227 (1990) 390.
[62] K. Knorr, Phys. Rep. 214 (1992) 113.
[63] M. Bienfait, Surf. Sci. 162 (1985) 411.

[64] J. Menaucourt, A. Thomy and X. Duval, in: Phases Bidimensionnelles Adsorbées, J. Phys. (Paris) 10 (1977) C4-195.
[65] M. Sutton, S.G.J. Mochrie and R.J. Birgeneau, Phys. Rev. Lett. 51 (1983) 407.
[66] S.G.J. Mochrie, M. Sutton, R.J. Birgeneau, D.E. Moncton and J.M. Horn, Phys. Rev. B 30 (1984) 363.
[67] S.K. Satija, L. Passel, J. Eckart, W. Ellenson and H. Patterson, Phys. Rev. Lett. 51 (1983) 539.
[68] J.Z. Larese and R.J. Rollefson, Surf. Sci. 127 (1983) L 172.
[69] J. Krim, J.G. Dash and J. Suzanne, Phys. Rev. Lett. 52 (1984) 640.
[70] I.C. Bassignana and Y. Larher, Surf. Sci. 147 (1984) 48.
[71] A. Terlain and Y. Larher, Surf. Sci. 125 (1983) 304.
[72] R.H. van Dongen, J.H. Kasperma and J.H. de Boer, Surf. Sci. 28 (1971) 237.
[73] B. Génot and X. Duval, C.R. Acad. Sci. Paris 265 (1967) 285.
[74] B. Génot and X. Duval, J. Chim. Phys. 69 (1972) 1238.
[75] B. Génot and X. Duval, J. Chim. Phys. 70 (1973) 134.
[76] B. Génot, J. Chim. Phys. 70 (1973) 1565.
[77] B. Génot, Thesis, Nancy (1974).
[78] J.H. Singleton and G.D Halsey, J. Phys. Chem 58 (1954) 1011; 58 (1954) 330.
[79] T. Kwan, M.P. Freemann and G.D. Halsey, J. Phys. Chem. 59 (1955) 600.
[80] C.F. Prenzlow and G.D. Halsey, J. Phys. Chem. 61 (1957) 1158.
[81] J. Regnier, C. Bockel and N. Dupont-Pavlovsky, Surf. Sci. 112 (1981) 770.
[82] J. Villain and J.G. Moreira, J. Phys.: Condensed Matter 3 (1991) 4587.
[83] M. Bouchdoug, J. Menaucourt and A. Thomy, J. Chim. Phys. 91 (1984) 381.
[84] M. Bouchdoug, T. Ceva, C. Marti, J. Menaucourt and A. Thomy, Surf. Sci. 162 (1985) 426.
[85] C. Marti, T. Ceva, B. Croset, C. de Beauvais and A. Thomy, J. Phys. (Paris) 47 (1986) 1517.
[86] M. Bouchdoug, J. Menaucourt and A. Thomy, J. Phys. (Paris) 47 (1986) 1797.
[87] A. Razafitianamaharavo, N. Dupont-Pavlovsky and A. Thomy, J. Phys. (Paris) 51 (1990) 91.
[88] A. Razafitianamaharavo, P. Convert, J.P. Coulomb, B. Croset and N. Dupont-Pavlovsky, J. Phys. (Paris) 51 (1990) 1961.
[89] J. Menaucourt and C. Bockel, J. Phys. (Paris) 51 (1990) 1987.
[90] M. Abdelmoula, T. Ceva, B. Croset and N. Dupont-Pavlovsky, Surf. Sci. 272 (1992) 167.
[91] B. Mutaftschiev, Phys. Rev. B. 10 (1989) 849.
[92] F. Hommeril and B. Mutaftschiev, Croatica Chem. Acta 63 (1990) 489.

Surface Science 299/300 (1994) 426–432
North-Holland

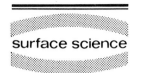

surface science

Making sense of surface structure

Peter J. Feibelman

Sandia National Laboratories, Albuquerque, NM 87185, USA

Received 8 March 1993; accepted for publication 30 April 1993

Electronic-structure calculations based on the local-density-functional approximation generally agree with measured surface atomic geometries and electron energy-level dispersions. Such calculations provide insight, not easily accessible from experiment alone, into the nature of bonding at surfaces. The result is that we can often interpret and sometimes even predict both static surface geometry and the mechanisms of surface dynamical processes.

Surface scientists aim to develop a predictive understanding of surface structure, electronic properties and elementary dynamical processes, laying the groundwork for improved control of numerous, complex, technologically important phenomena. The enormous economic significance of surface-related technologies, from catalytic surface chemistry to semiconductor devices, from materials growth and stability to friction, lubrication and wear, explains the continued vitality of and financial support for surface science research.

Despite its impressive objectives, the first two decades of surface science research were largely devoted to developing tools for preparing and characterizing samples, and for converting acquired data into useful information. One could hardly hope to analyze trends or verify theories, to say nothing of contributing to improved technology, before there was a reliable surface science information base. Thus, from the late 1960's through the early 1980's, the focus of the field was largely on solving "the surface structure problem", i.e., on developing a crystallography of single-crystal surfaces, and in parallel, on learning the elements of surface electronic and vibrational behavior.

By the mid-1980's, the surface structure problem had largely been solved through the development of a host of experimental techniques, based on electron diffraction, X-ray diffraction, ion scattering or one of various microscopies. The consequence is that many surface structures are now known. Via ARUPS (angle-resolved ultraviolet photoemission spectroscopy), HREELS (high resolution electron energy loss spectroscopy), IRAS (infrared absorption spectroscopy), He atom scattering, etc., electronic surface states and surface phonon bands are known as well. Thus, the character of the field in the early 1990's is dramatically different from what it was even ten years ago.

For theorists, the availability of a meaningful information base has meant that computational methods for predicting surface behavior can be tested and improved. Thus, perhaps most importantly, the local density functional (LDF) approximation [1] has been shown to be a reliable tool for predicting surface structures, and also for interpreting electronic surface state dispersions. This success has spawned two important efforts. One is aimed at systematizing our understanding of surface behavior. The other is devoted to enhancing the numerical efficiency of LDF calculations. Now that massively parallel computers have become available, the latter effort promises to make it possible to study systems and/or processes of considerable complexity and interest.

The first efforts to "make sense" of surface structure were responses to the surprising discovery of superlattice periodicity in low energy electron diffraction (LEED) from various faces of Si and Ge [2–5]. The observation of extra diffraction beams is definitive evidence that surface atoms are arranged in a superlattice – you do not need a theorist to tell you that – and begs for some kind of rationalization. Schlier and Farnsworth explained the 2×1 superstructure on Si(001) as a dimerization that would reduce the number of dangling bonds from two per surface Si atom on the 1×1 surface to one on the reconstructed face [2]. Seiwatz applied similar reasoning in proposing a π-bonded chain model for cleaved Si(111), which also manifests a 2×1 LEED pattern [5]. Again the reduction of the number of dangling bonds, in this case by the formation of π-bonds, was assumed to be the driving force behind the unexpected surface reconstruction. In detail, neither of these two models was quite right, as shown by far more extensive calculations and experiments than were possible in the 1960's. Nevertheless, the basic nature of the reconstructions was guessed correctly by applying chemical common sense.

The idea that tying off dangling bonds is an important principle in determining surface atomic arrangement is a manifestation of the intimate relation between surface electronic and geometric structure. At a time when the surface structure problem was still not under control, and because adsorbed H atoms are often not easy to "see" in surface sensitive experiments [6], D.R. Hamann and I proposed using the regularities of this relation to determine a H/metal surface structure by comparing theoretical and ARUPS surface state band positions and dispersions [7]. (A similar, though somewhat less extensive study of H/Pd(111), had previously been done by S.G. Louie [8].) Assuming that the surface electronic properties correspond to the geometry in a simple way, one might thereby gain important structural information. The basic ideas in the analysis were once again drawn from elementary theoretical chemistry. The study was made possible by the collaboration of F.J. Himpsel, who obtained ARUPS data for both clean and H-covered Ti(0001).

Since Ti has a strong affinity for H, a first interesting question was whether Himpsel's data, which were taken for a H(1×1) adlayer, correspond to 1 or to 2 H adatoms per surface Ti. Comparison of LDF energy bands with the data resolve that issue easily. If there are 2 H's per surface unit cell, there should be *two* H(1s)-derived surface state bands, corresponding to 1s bonding and antibonding molecular orbitals and these bands should be broad because of the close proximity of the H's. Since Himpsel observed nothing like this, one can conclude that there was only one adsorbed H atom per surface Ti.

In seeking to determine the position of the single ad-H, we discovered an interesting feature of the H-induced levels. Near $\bar{\Gamma}$, for H in a 3-fold hollow, the numerical calculations imply that there should be two H-induced states if the H lies high above the surface, while only one state should be present if the H sits closer to the outer Ti plane. As the H is moved along the surface normal closer to the surface plane, the splitting of the two H-induced states increases, and eventually the upper state disappears, blending into a band of bulk Ti states. The reason for this behavior is that the two surface states correspond to bonding and anti-bonding orbitals formed from the H(1s) and Ti($3d_{3z^2-r^2}$) orbitals, and the band splitting is therefore related to the degree to which these two orbitals overlap. When the H sits relatively high above the surface, its 1s orbital overlaps the nodal plane of the Ti($3d_{3z^2-r^2}$) state. Thus, the bonding–anti-bonding splitting is small and both levels can be seen. If the H sits lower, then the overlap between the H and Ti orbitals increases. The surface state splitting therefore increases, pushing the anti-bonding state up into the Ti bulk bands. By fitting to the positions of the *two* states observed by Himpsel, we find, satisfyingly, that the height of the H above the surface corresponds to the bondlength of TiH_2.

Our study of H adsorption on Ti(0001) included one of the first uses of total energy calculations to determine a surface structure. One problem with the determination of the H-position by comparing to ARUPS levels is that several adsorption geometries give level positions and dispersions that correspond reasonably well to those observed in Himpsel's experiment. Some have the H's above the surface, one has them below. (That this is possible is suggested by the reflection symmetry of the $3z^2 - r^2$. orbital in the $x–y$ plane.) The binding energies for the overlayer geometries, however, are generally about an eV greater than those for underlayer sites. This makes it easy to exclude the possibility that underlayer sites are occupied at low temperatures, and to choose the best from among the overlayer geometries.

In recent years, the prediction of surface geometry via LDF total energy calculations has become commonplace. This fact, together with numerous measurements of surface structure, invites attempts to discover and interpret "trends" and also provides some excitement when an "anomaly" is found. Many metal surfaces are 1×1. This reduces the determination of surface structure to learning how the interlayer spacings near a surface are different from the corresponding inter-planar spacing in the bulk metal. Two trends are common: (1) outer layer separations alternate – the first interlayer separation is generally contracted relative to the bulk, the second expanded, and so on, and (2) the more open crystal surfaces have larger relaxations. The first of these trends can be explained by a bond-order bond-length argument [9]. The first interlayer separation contracts because the valence electrons of the outer atomic layer, in the absence of a "next" layer to which to bond, rehybridize into stronger "back-bonds," to the layer below. Since the first-to-second layer bonds become stronger, they also become shorter, giving rise to the observed contraction. Continuing, if the second layer atoms gain energy from forming stronger bonds to the outermost layer, they do so by shifting some of their valence electrons from bonds to the third layer. Thus, the second-to-third layer bonds are weakened and the second to third layer separation expands. This explains the alternation that is observed.

A more "physics-" than "chemistry-oriented" argument for outer layer contraction, originally suggested by Finnis and Heine [10], explains the second trend as well. The idea is that at a surface, the electron gas smooths its corrugation along the surface in order to reduce its kinetic

energy. This "Smoluchowski effect" [11] corresponds to moving electronic charge from the regions above surface atoms to the hollows between them. Since this transfer of charge moves it closer to the rest of the metal, the Smoluchowski smoothing attracts positive ion cores closer to the metal as well, resulting in an outer layer contraction. The contraction is larger for open surfaces, because the smoothing of charge corrugations is larger for them.

Although this discussion appears to provide a nice example of a "trend" and its interpretation, recent studies of Be(0001) remind us that we are only just beginning to systematize our understanding of surface structure. Davis et al. [12] find that in contrast to all previous cases, the outer Be interlayer spacing is 6% *expanded* relative to bulk Be. More recently, Sprunger et al. [13] have found a similar result for Mg(0001), though the expansion in this case is only 1.9%. Other close-packed metal surfaces have been found to be expanded, but only by a fraction of a percent. Such a small expansion is arguably a "detail" whose interpretation is and will remain difficult for some time to come. The large expansions of the Be(0001) and Mg(0001) surfaces are really exceptional and beg for immediate consideration. I have therefore applied the linearized augmented plane wave (LAPW) method [14] to the case of Be [15]. More recently, Wright et al. [16] have studied Mg, using a parallel implementation of a plane-wave pseudopotential (PWPP) method.

These recent calculations show that the reason for the unusual surface relaxations of the Group-IIA metals is that Group-II atoms have a filled valence shell. To bond strongly, one of the metal atom's valence s-electrons must be promoted to a p-state, at an energetic cost of about 2.7 eV [17]. This means that in a close-packed surface layer of a Group-II metal, wherein the atoms are missing three neighbors, there is less s- to p-promotion. Thus, because the surface atoms profit less from the promotion, the degree of promotion is reduced. This means that the surface atoms have less power to form bonds, and move away from the remainder of the metal. That this should occur is already suggested by the often-used piece of surface science wisdom: *Surface atom behavior*

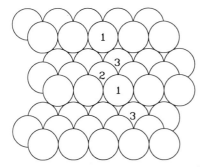

Fig. 1. Schematic of the outer layers of an ideal fcc (331) surface. Labels 1, 2, and 3 indicate the first, second and third layer atoms. Note that atom 2 is at the center of a narrow close-packed terrace, while the step-edge atom, 1, and the step-bottom atom, 3, have nearest neighbors located just where they would be on a fcc (110) surface.

is intermediate between that of the same atoms in bulk and in the gas phase. Be and Mg obey an inverse bond-order bond-length correlation. That is, Be and Mg dimer bond lengths are longer than the nearest-neighbor separations in the corresponding bulk metals [18]. This is just the opposite of what is observed for all non-Group-II elements, for which electronic promotion is not a prerequisite for bonding, and should have told us to expect the observed outer layer expansions.

Another interesting anomaly relative to the usual trends in outer layer relaxations is exhibited by the Al(331) surface. Instead of an alternation in the relaxations, on this surface both the first and second interlayer separations contract while the third expands [19]. J.S. Nelson and I found a simple explanation for this result in terms of the phenomenon of screening [20].

The (331) surface of a fcc crystal can be thought of as an ordered, "stepped" surface, comprised of narrow fcc (111) terraces separated by steps whose local coordination is like a fcc (110) surface (see Fig. 1). In fact this description of a fcc (331) surface, as a superposition of (111)- and (110)-like regions, suggests the explanation of the observed, unusual relaxation pattern. Because of screening, each atom "feels" the presence of its nearest neighbors most strongly. Forces exerted by second and more remote neighbors are much weaker. Accordingly, to a first approximation, the first layer atoms of an ideal (331) surface do not

"know" that they are not the first layer atoms of a (110) surface – their immediate neighbors are located just where they would be if the surface really were the (110) and exert a strong inward force in the [110] direction. Similarly, the third layer atoms of the (331) surface are subject to a force very close to what they would experience if they were second layer atoms on Al(110), namely a strong outward force along the [110] direction. (These inward and outward forces give rise to the large, alternating relaxations of Al(110)'s outer layer spacings [20].) Finally, the second layer atoms on Al(331), which lie in the middle of this surface's (111) terraces, are subject to the same weak force that accounts for the very slight relaxation observed on Al(111). Thus on Al(331), the atoms of layer 1 move inward, those of layer 2 move hardly at all, and the 3rd layer atoms relax outward. This explains why the first and second layer separations contract while the third one expands.

Once we believe in the reliability of our predictive methods, and in our understanding of the physics that underlies surface structure, we may even be able to discover systematic problems in experimental analyses. LEED studies of Rh(001), for example, by Oed et al. [21] and by Hengrasmee et al. [22]. report that the outer layer separation of this surface is close to ideal, but perhaps slightly expanded relative to the bulk. This result contradicts the usual trend, and unlike the cases of Be(0001) and Mg(0001), there seems to be no reason for it. Moreover, two independent LDF calculations [23,24] agree that the outer layer spacing of the Rh(001) surface should be several percent contracted. Such calculations are generally very reliable. Is there then some special reason that they are inapplicable to Rh(001), or is there a systematic problem in the experimental data?

Unwilling to accept the former possibility lightly, Hamann and I surmised that the experimental surface was contaminated with H [23]. H_2 is generally a contaminant in UHV systems. It easily cracks and sticks on Rh, and H is hard to see in LEED when it does not reconstruct the underlying metal [6]. In addition, by tying off unsaturated metal valence, H tends to "heal" surface relaxation. Thus it was not surprising to find, via LAPW calculations, that a H monolayer would reduce the inward force on the outer Rh's of Rh(001) by a factor of roughly 4. The final experimental word on the correctness of our guess is not yet available. (The analysis of LEED data taken at temperatures ≥ 400 K, where H is guaranteed not to be adsorbed on a transition metal surface, is not straightforward, because phonon effects are substantial, anisotropic and atomic layer dependent. On the other hand, to obtain H-free data at low temperatures one must flash the sample, quench and take data quickly.) But the message is clear in any case – we now have enough confidence in our theoretical understanding of surface structure that we can demand that the experimental results make sense, or at the very least that all the experimental loose ends are tied.

Without doubt, the most satisfying occupation for a theorist is to predict a previously unknown phenomenon. Interpretation of experiments is generally desirable and often useful, a worthy undertaking, but prediction suggests real success in understanding nature. A particularly important, though "elementary" process in the growth of materials, as well as in surface chemical phenomena, is surface atom diffusion. To simulate a dynamical phenomenon such as diffusion, one needs to know: (1) what are the forces on the moving atoms as a function of their position, and (2) given the force law, how will the atoms move? The second question can now be addressed by molecular dynamics methods that, at least when classical equations of motion are appropriate, are

highly developed. The first question requires knowing the total energy function for the system under study. Since this implies the same kind of work that goes into predicting surface structure, methods that reliably predict surface structure can also be expected to give a reasonable account of diffusion mechanisms and barrier energies.

On the assumption that the "tried and true" LDF method would correctly predict a diffusion barrier for an Al adatom hopping on an Al(001) surface, I calculated the binding energy of the adatom in its optimal 4-fold hollow site, and again at its optimal position above a two-fold bridge separating two hollows [25]. The energy difference should represent the minimum barrier to the diffusion process. The result, however, is 0.65 eV, an unsatisfyingly large energy. Barriers of this magnitude have been measured using field ion microscopy (FIM), for *transition metal atom* self-diffusion [26]. Al is a soft metal, and thus should have a lower barrier. As a matter of fact, Tung's FIM studies imply barriers of roughly 0.45 eV for single atom self-diffusion on Al(110), (331) and (311) [27].

As usual, when a result makes no sense, if the theoretical method is reliable, then there must be something missing in the physics. The physics that gives rise to the large barrier energy is clear. In a two-fold bridge configuration, the three valence electrons of the ad-Al cannot all form bonds, or at least they can only do so by packing into a relatively small spatial region, at a substantial cost in electrostatic repulsion energy. Thus the cost of diffusion is the cost of passing through a configuration where the adatom's coordination

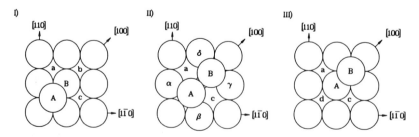

Fig. 2. Schematic snapshots of a concerted substitutional diffusion event on a fcc (001) surface – adatom A replaces the substrate atom B as one moves from panel I to panel III. At worst (panel II) A and B both have three near neighbors. For example, A has neighbors, α, β and B. As a result of the concerted process, the adatom appears to have made a diagonal or "checkerboard hop", to site b, along the [100] direction.

is reduced from four to only two nearest neighbors. As a way of avoiding the corresponding energetic bottleneck, I considered the possibility of a "concerted substitutional diffusion" mechanism (Fig. 2) in which the ad-Al substitutes for a surface atom at the same time that the latter emerges onto the surface. In such a process, all atoms always have at least three nearest neighbors, which is just what trivalent Al atoms need to saturate their valence. The transition state for the concerted process should therefore correspond to a much lower barrier energy than that for ordinary hopping over a bridge.

The LDF calculation I performed verified this idea. The barrier to concerted diffusion is only about 0.2 eV! In addition, the concerted process has a "signature" that should make it particularly easy to observe in a FIM. The geometry of the transition state is only favorable for the formation of three bonds per moving atom if the atoms move along a [100] or [010] direction on the surface. Concerted motion along the usual [1$\bar{1}$0] and [110] directions is unfavorable, because, as in the case of ordinary hopping, it requires passing through atomic geometries in which electrons have to squeeze into a relatively small spatial region to bind to positive ion cores. As a consequence, if concerted substitution is occurring, the diffusing atom will appear to move at low temperatures as though on a checkerboard. If the adatom starts on a "white square" it will never appear on a "black" one. Unfortunately, because Al is a soft metal, surface diffusion on Al is very difficult to observe in a FIM. But checkerboard hopping has been seen both for Pt diffusion on Pt(001) [28] and for Ir on Ir(001) [29], proving that concerted substitutional diffusion really happens, and confirming once again the usefulness of simple bond-counting arguments in making educated guesses about the energy of a surface as a function of its atomic arrangement.

Nowadays, we are beginning to see simulations *based on first-principles forces* [30], that involve substantial numbers of particles. With massively parallel computers on the scene, we can expect to see simulations attempted that are of increasing complexity and interest [31]. At the same time, advances in surface microscopies now provide us with "movies" of the nucleation of phase transformations and of surface chemical phenomena [32]. Using modern algorithmic and computational power to analyze and make sense of the wealth of experimental information that is becoming available represents a challenging and important task for the next decade of surface science.

Acknowledgement

This work, performed at Sandia National Laboratories, was supported by the US Department of Energy under contract DE-AC04-76DP00789.

References

[1] For a review, see: The Theory of the Inhomogeneous Electron Gas, Eds. S. Lundqvist and N.H. March (Plenum, New York, 1983).

[2] R.E. Schlier and H.E. Farnsworth, J. Chem. Phys. 30 (1959) 917.

[3] D. Haneman, Phys. Rev. 121 (1961) 1093.

[4] J.J. Lander, G.W. Gobeli and J. Morrison, J. Appl. Phys. 34 (1963) 2298.

[5] R. Seiwatz, Surf. Sci. 2 (1964) 473.

[6] If it adsorbs in a superlattice structure without strongly perturbing the underlying metal, surface H produces fractional order LEED beams that can be used to determine its location. Analysis of such a case yielded an early H-adsorption geometry by K. Christmann, R.J. Behm, G. Ertl, M.A. Van Hove and W.H. Weinberg, J. Chem. Phys. 70 (1979) 4168, who investigated p(2×1)H/Ni(111).

[7] P.J. Feibelman, D.R. Hamann and F.J. Himpsel, Phys. Rev. B 22 (1980) 1734.

[8] S.G. Louie, Phys. Rev. Lett. 42 (1979) 476.

[9] L. Pauling, The Nature of the Chemical Bond (Cornell University Press, Ithaca, NY, 1960) 3rd ed.

[10] M.W. Finnis and V. Heine, J. Phys. F 4 (1974) L37.

[11] R. Smoluchowski, Phys. Rev. 60 (1941) 661.

[12] H.L. Davis, J.B. Hannon, K.B. Ray and E.W. Plummer, Phys. Rev. Lett. 68 (1992) 2632.

[13] P.T. Sprunger, K. Pohl, H.L. Davis and E.W. Plummer, to be published.

[14] O.K. Andersen, Phys. Rev. B 12 (1975) 3060;
D.R. Hamann, Phys. Rev. Lett. 42 (1979) 662;
L.F. Mattheiss and D.R. Hamann, Phys. Rev. B 33 (1986) 823.

[15] P.J. Feibelman, Phys. Rev. B 46 (1992) 2532.

[16] A.F. Wright, P.J. Feibelman and S.R. Atlas, Surf. Sci., in press.

[17] C.E. Moore, Atomic Energy Levels, Circ. Natl. Bur. Standards 467, Vol. I (US Gov. Printing Office, Washington, DC, 1949) p. 12.

[18] Ref. [15] contains a table of dimer bond lengths and nearest-neighbor distances, as well as references wherein the former were published.

[19] D.L. Adams and C.S. Sørenson, Surf. Sci. 166 (1986) 495.

[20] J.S. Nelson and P.J. Feibelman, Phys. Rev. Lett. 69 (1992) 1568.

[21] W. Oed, B. Dötsch, L. Hammer, K. Heinz and K. Müller, Surf. Sci. 207 (1988) 55.

[22] S. Hengrasmee, K.A.R. Mitchell, P.R. Watson and S.J. White, Can. J. Phys. 58 (1980) 200.

[23] P.J. Feibelman and D.R. Hamann, Surf. Sci. 234 (1990) 377.

[24] M. Methfessel, D. Hennig and M. Scheffler, Phys. Rev. B 46 (1992) 4816.

[25] P.J. Feibelman, Phys. Rev. Lett. 65 (1990) 729.

[26] D.W. Bassett and P.R. Webber, Surf. Sci. 70 (1978) 520.

[27] R. Tung, Thesis, University of Pennsylvania, 1980.

[28] G.L. Kellogg and P.J. Feibelman, Phys. Rev. Lett. 64 (1990) 3143.

[29] C. Chen and T.T. Tsong, Phys. Rev. Lett. 64 (1990) 3147.

[30] For a recent example, see: A. Pasquarello, K. Laasonen, R. Car, C. Lee and D. Vanderbilt, Phys. Rev. Lett. 69 (1992) 1982.

[31] A parallel implementation of a plane wave LDF calculation was given by J.S. Nelson, S.J. Plimpton and M.P. Sears, Phys. Rev. B 47 (1993) 1765. Demonstration projects involving structural calculations, on parallel computers, for the Si(111)7×7 structure have been reported by K. Brommer, M. Needels, B. Larson and J.D. Joannopoulos, Phys. Rev. Lett. 68 (1992) 1355 and by I. Stich, M.C. Payne, R.D. King-Smith, J.S. Lin and L.J. Clarke, Phys. Rev. Lett. 68 (1992) 1351.

[32] See, e.g., L. Ruan, F. Besenbacher, I. Stensgaard and E. Laegsgaard, Phys. Rev. Lett. 69 (1992) 3523;
L.P. Nielsen, F. Besenbacher, E. Laegsgaard, I. Stensgaard, Phys. Rev. B 44 (1991) 13156;
A.W. Denier van der Gon and R.M. Tromp, Phys. Rev. Lett. 69 (1992) 3519.

Surface Science 299/300 (1994) 433–446
North-Holland

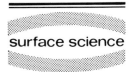

Geometrical and chemical restructuring of clean metal surfaces as retrieved by LEED

K. Heinz

Lehrstuhl für Festkörperphysik, Universität Erlangen-Nürnberg, Staudt-Strasse 7, D-91058 Erlangen, Germany

Received 25 March 1993; accepted for publication 6 May 1993

Quantitative structure determination of clean metal surfaces by LEED during the last three decades is reviewed. First, the development of experimental techniques necessary for the fast and reliable measurement of electron diffraction intensities is described. Emphasis is on the structure of geometrically reconstructed clean metals as well as on the monitoring of the corresponding structural phase transitions and their kinetics. The appearance of chemical reconstruction (segregation) in binary alloy surfaces is described as well.

1. The need and realization of fast and easy LEED intensity measurements

The development of effective computer codes describing the multiple scattering of electrons in low energy electron diffraction (LEED) marks the start of quantitative surface crystallography in the early 1970's. Up to this time the LEED experiment was ahead of theory, i.e. well-resolved spectra of intensity versus energy $I(E)$ or intensity versus angle of incidence $I(\theta, \phi)$ could be measured but lacked reliable reproduction by model calculations. However, soon the increasing accuracy of calculated spectra required equal accuracy of the experiment. It was found that a number of experimental conditions and quantities influence the spectra in a critical way. The most important among these are sample misalignment, residual gas adsorption, incomplete subtraction of background intensities and – for surfaces and adsorbates sensitive to electron radiation – structural damage by electron stimulated cracking of surface chemical bonds. Misalignments by errors in the polar angle as low as 1° can cause large intensity changes, too large even by the standards of the 1970's [1]. Electron stimulated cracking of intramolecular bonds happens, e.g., in the

adsorbate system $c(2 \times 2)CO/Ni(100)$. An electron dose of the order of 100 $\mu A\,s/mm^2$ already causes considerable changes which become dramatic with larger doses [1,2].

These error sources called for the development of fast and easy to handle techniques of measurement which would reduce the electron dose and allow for routine tests of the accuracy of measurement [1,3,4]. So far, intensities had been recorded by the classical Faraday cup [5,6] or the spot photometer [7]. In both cases measuring times of the order of hours and corresponding electron doses were implied. This is because detectors had to be moved mechanically within ultrahigh vacuum or outside the chamber, respectively, in order to track a diffraction spot moving with sweeping energy or varying angles of incidence. There are a number of modifications of these classical methods such as the replacement of the Faraday cup by a channeltron [8], the intermediate storage of the LEED pattern energy by energy on a photographic film with subsequent intensity evaluation by a computer controlled microdensitometer [9–11] or automated but still mechanical tracking of the diffraction spots [12–16]. However, though they helped to facilitate and speed up the measurement they marked no real breakthrough. This eventually came by the use of video cameras as optical detectors in the mid 1970's and by the use of position sensitive devices about a decade later. Both techniques avoid the mechanical tracking of moving spots. They are shortly reviewed in the following.

1.1. Video based methods

The main idea of the video based methods is to view the diffraction pattern on the display type LEED screen by a video camera whose electronic signal is evaluated under on-line computer con-trol. There is automated electronic tracking of the moving spot under computer control. The first realization of the technique introduced by the Erlangen group [17] needed some twenty video half frames, i.e. about 0.4 s to yield the intensity of a single spot at a single energy. A video recorder can be used to store all spots of a single frame for subsequent off-line evaluation [18]. Faster on-line access to intensity data was achieved through the availability of fast digital electronics in the 1980's, which allowed one to take data with a rate of 20 ms per beam and

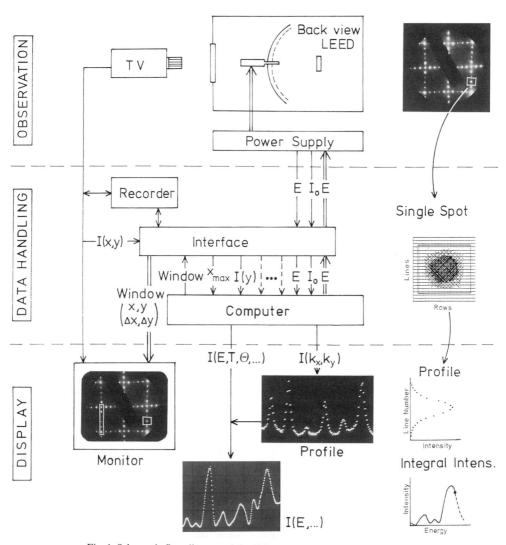

Fig. 1. Schematic flow diagram of the Erlangen video LEED method [18–21].

energy [19,20]. The latest development [21] achieves this rate also for digitization and storage of full video frames allowing fast on-line access to, e.g., full screen diffuse intensity distributions. It is emphasized that during the years different groups have developed and used similar video based systems [20–27].

The setup of the Erlangen system [19–21] is schematically displayed in Fig. 1. From the back of a glass screen the video camera views the diffraction pattern and passes the corresponding electronic signal to a control monitor as well as to an interface/computer system. The latter digitizes the signal within a window preselected in size and position by the operator and displayed on the monitor for optical control. The window can degenerate to a slit of arbitrary orientation in order to yield spatial spot profiles or blow up to the full video frame size in order to store the full information of the pattern. For measurements of integrated spot intensities the window is made to enframe the spot wanted. The signal is integrated within the window whereby the average intensity determined at the window's edges is used for background correction. The window's position is readjusted at each energy so that the position of the maximum intensity of the spot centers the window. This allows automated tracking of the spot when it moves with the energy sweeping under computer control. The whole procedure takes the time of a video half frame, i.e. 20 ms per spot and energy, so that the measurement of a spectrum of, e.g., 300 intensity-energy points can be completed within 6 s. However, in order to improve the signal/noise ratio in case of weak superstructure spots it is sometimes necessary to add several subsequent video frames. Then the measuring time increases accordingly. At a primary beam current of the order of typically 1 $\mu A/mm^2$ and a total of 10 spots the minimum electron dose implied is of the order of 100 μA s/mm², if one adds some time for initial adjustment of the electron beam. Yet, the use of an image intensifier camera allows one to reduce the primary beam and so the dose by up to two orders of magnitude. However, this is still too large for sensitive adsorbates as, e.g., C_2H_4 [21,30] and called for lower dose methods.

1.2. Methods based on channel plates and position sensitive devices

In the 1980's channel plates and position sensitive devices were introduced for low dose LEED intensity measurements. A channel plate positioned in front of the screen was used even earlier [31] and full screen intensity maps were obtained by the use of an optical multichannel analyzer [32,33]. These maps could be stored on disk for subsequent off-line evaluation to yield integrated beam intensities. However, the time and storage capacity needed for the data transfer were rather large. With the development going on, dual channel plates (chevrons) [34] provided even further increased sensitivity. The display type screen was replaced by position sensitive devices such as the resistive anode or the wedge-and-strip anode [35] which allowed one to obtain digitized intensity maps by computer evaluation of their signals [36,37].

The most sensitive system currently in use is that developed by the Berkeley group [38,39]. Primary currents lower than 1 pA can be allowed by a gain of the chevron array larger than 10^7 and by the use of digital electronics to count the electrons arriving at the wedge-and-strip anode. For a reasonable value of the signal/noise ratio the measurement of a LEED pattern at a single energy typically takes 10–300 s [39–41] dependent on the intensity level. Though this is by about three orders of magnitude higher than needed by the TV method, the total electron dose to which the surface is exposed is considerably less even when the TV method uses an intensifier camera. Off-line evaluation is applied to produce $I(E)$ spectra though on-line evaluation is possible in principle. This lack of automated on-line tracking of moving spots prevents immediate feedback to the experiment in order to allow, e.g., for correction and adjustment of parameters of the measurement. This is provided by the TV method, which because of its speed allows also for time resolved intensity measurements. However, it cannot be applied to surfaces which tolerate only extremely low electron doses. On the whole, the last decades have produced very fast and sensitive techniques which complement each other. If one includes techniques providing high spatial resolution for spot profile analysis [3,42] (not reviewed in this contribution) they can deal with almost every situation where LEED intensity measurements are wanted in surface structure investigations. They allow for high accuracy and precision and opened new fields of applications.

2. Clean metal surfaces: a test case for experimental and theoretical LEED

Low index surfaces of clean elemental metals were the first candidates for quantitative surface structure determination by LEED in the early 1970's. These surfaces seemed to be rather inert to the impinging electron beam and were believed to be more or less bulk terminated. Theory had developed to much increased accuracy, particularly by the improvement of atomic scattering potentials, sophistication of computational schemes for multiple scattering, consideration of non-structural parameters and introduction of layer relaxation. This improvement as well as the development of quantitative measures for the comparison of theory and experiment (R-factors) caused the question about the experimental accuracy to come up.

As mentioned in the introductory part, sample misalignment can have a dramatic influence on the reliability of experimental spectra. Yet, it is rather difficult to adjust the angle of incidence to better than ±0.5°. A way out of this dilemma came by the procedure to adjust the direction of incidence as precisely as possible and then measure and average symmetrically equivalent beams [1,43]. Of course, this is only possible when the incident beam direction lies within a symmetry plane or coincides with a symmetry axis of the surface. This together with the saving of computer time for full dynamical calculations made symmetric primary beam incidence, particularly normal incidence, the most frequently applied experimental arrangement up-to-date. As an example Fig. 2 demonstrates the result of symmetry averaging for a selected beam diffracted from clean Ni(100) which shows fourfold symmetry for

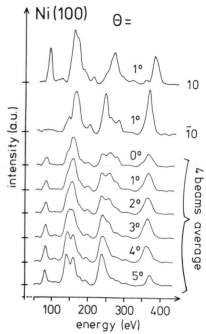

Fig. 2. Result of averaging of symmetrically equivalent beams from Ni(100) for different polar angles θ [1,19].

normal incidence. Though there are considerable differences between spectra of ±1° off normal incidence the average of the four symmetrically equivalent beams reproduces the "exact" spectrum very well. This holds up to misalignments of even 3°, i.e. it is only for larger off normal angles that intolerable changes appear. Of course this critical angle may depend on the particular surface under investigation.

Comparison of spectra of symmetrically equivalent beams serves also to adjust the angle of incidence as precisely as possible prior to measuring and averaging. Repeated and stepwise adjustment followed by R-factor comparison of beams eventually leads to an optimal orientation of the sample. This procedure can be automated using step motors or performed by hand. Of course, for practical reasons a fast technique of intensity measurement is required. Routine measurement of all symmetrically equivalent beams and immediate calculation of the mutual Pendry R-factors [44] by the computer controlling the measurement is recommended. Sample adjustment is not stopped until R-factors are below 0.1, a value which has never been achieved in a theory–experiment fit. Also, this R-factor value should not be exceeded when different measurements, i.e. spectra remeasured after "identical" sample preparations at different days, are compared. For clean metal surfaces this accuracy and reproducibility is today's standard. Of course, it is only achieved if the same care is also applied to minimize residual gas adsorption and optimize background subtraction. Also, the correct measurement of the electron energy (voltage) seems to be important. This has been demonstrated by a joint program of five different laboratories triggered and directed by F. Jona and P.M. Marcus in 1980: Spectra measured from clean Cu(100) by the different laboratories differed by rigid shifts on the energy scale by as much as 10 eV though otherwise there was good agreement on the above-mentioned level [45,46].

Today, experimental and computational accuracy have reached a level where for low index metal surfaces the agreement between theory and experiment is nearly perfect. As an example, Fig. 3 displays experimental and calculated best fit spectra for Ni(111) for an energy range up to 512 eV [21]. The Pendry R-factor is $R_P = 0.16$ and

can be reduced to $R_P = 0.13$ by using an energy dependent real part of the inner potential. In turn this accuracy can also be used to determine the angles of incidence if it cannot be adjusted experimentally because of lacking symmetry, i.e. for high index surfaces or oblique incidence.

Vertical multilayer relaxation with frequently (but not necessarily) alternating sign is the prominent structural feature of clean surfaces though in some rare cases also parallel registry shifts of layers were determined [4,47]. With the momentum transfer being predominantly normal to the surface, LEED is most sensitive to vertical layer relaxations. Frequently, the accuracy claimed for the determination of interlayer distances is – dependent on the layer depth – in the percentage region or even less, i.e. 0.01–0.05 Å. Whilst close packed surfaces tend to be nearly unrelaxed (in the above example Ni(111) is bulk terminated within the limits of error, i.e. ± 0.02 Å), multilayer relaxation develops and increases in relative amplitude and layer depth with increasing roughness (openness) of the surface [4,46]. So, for fcc(210) which is the most open type of surfaces investigated by LEED, relaxations of the first layer distance of -23% for Pt(210) [48] and -16% for Al(210) [48] are reported. For Pt(210) the small interlayer distance (bulk value 0.88 Å) together with the strong multiple scattering makes routine computational schemes such as layer doubling break down. Other approaches, e.g. real space multiple scattering theories, become necessary [48–51]. The precise determination of the multilayer relaxation of clean metal surfaces and its dependence on the surface roughness is an important input to solid state theory in order to yield a quantitative understanding of the physics of metal surfaces. However, surfaces become even more exciting when they reconstruct.

3. Geometrical restructuring of clean metals

Creation of a surface means to truncate chemical bonds. For the covalent, i.e. spatially aligned bonds of semiconductors this is a dramatic change. It normally leads to a drastic rearrangement of atoms whereby the original translational symme-

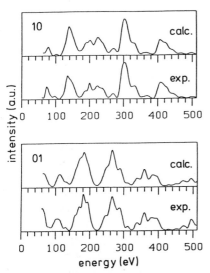

Fig. 3. Best fit agreement between theory and experiment for two selected beams from the almost unrelaxed surface Ni(111) [21].

try parallel to the surface is broken. This surface reconstruction is characterized by additional spots in the LEED pattern. Metals with their undirected metallic bonds, however, have a high chance to find their new structural minimum of free energy without restructuring, i.e. by vertical multilayer relaxation only. Nevertheless, about a dozen clean and low index metal surfaces are known to reconstruct [52–54]. Most reconstructions were first observed by LEED, but in the meantime many other methods such as, e.g. ion and atom scattering, high resolution electron and field ion microscopy, X-ray diffraction and scanning tunneling microscopy have considerably contributed to solve the real space structures. The reconstructions not doubted today are those of the (100) and (110) surfaces of iridium, platinum and gold, the Au(111) surface and the (100) surfaces of chromium, molybdenum and tungsten. The different reconstructions of the fcc(100), fcc(110) and bcc(100) surfaces are very much similar for the various materials, though different in detail. For fcc(100) and fcc(110) the average density of surface atoms changes from reconstruction, whilst for bcc(100) only local displacements of atoms take place. Though for a long time it was not clear whether impurities cause the superstructures observed, it is commonly accepted today that the reconstruction is an intrinsic property of the clean surface. Nevertheless, impurities as well as defects can play an important role in stabilizing certain phases of reconstruction (e.g. Ref. [55]).

3.1. Reconstruction of fcc metals

The reconstructions of Pt(100) and Au(100) were observed already in the mid 1960's [56–60] and only a few years later Ir(100) was found to reconstruct as well [61,62]. The reconstructions have the same strange feature in common: The top layer atoms, which have a quadratic unit cell when the surface is bulk terminated, rearrange towards a new equilibrium state by forming a quasi-hexagonal close packed structure. The corresponding gain in energy seems to be larger than the strain energy caused by the mismatch to the quadratic substrate. The increased density of

atoms is met by the formation of steps. The simplest type of reconstruction is that of Ir(100) which in thermal equilibrium shows a 5×1 superstructure as displayed in Fig. 4 [63]. Obviously, the interaction between first and second layer atoms is strong enough to cause sufficient distortion of the hexagonal first layer in order to make it form a coincidence structure on the quadratic substrate. Fig. 4 also displays the corresponding real space model for one of two orthogonal domains. The quasi-hexagonal layer is considerably buckled. The angle between the hexagonal unit mesh vectors is only about 59° rather than the ideal value of 60°. The model is confirmed by several LEED structure determinations carried out in the early 1980's [64–67]. At that time the solution of a structure with 5 atoms in the surface unit cell was a demanding task. A best fit spectrum for a selected beam also given in Fig. 4 displays the level of agreement reached [67].

The LEED patterns of Pt(100) and Au(100) are similar to each other but different to that of Ir(100) [68]. There are a number of different reconstruction phases whereby 5th order spots observed for Ir(100) are split in complicated ways. The reason for that is that the top layer is incommensurate with respect to the second layer obviously because of decreased interaction between the two layers. The angle between hexagonal unit mesh vectors is now nearly ideal but the hexagon is slightly and nearly isotropically contracted as compared to the hexagon of the (111) face [63,68]. The hexagonal unit mesh vectors can be aligned to one of the quadratic substrate vectors or rotated by about 0.7°, which is kind of rotational self-epitaxy and makes the pattern even more complex (see, e.g. Ref. [63]). Both for platinum and gold a number of different phases are reported depending on sample preparation and the presence of defects, particularly steps (for a survey see Ref. [68]). Frequently it is tried to describe the structures in terms of large coincidence super cells, e.g. (5×20), (5×26), $c(26 \times 68)$ or by equivalent matrix notation. Very detailed and high resolution investigations on Au(100) were recently performed by X-ray scattering [69,70]. Interestingly and surprisingly, also the Au(111) surface was found to reconstruct [71] even though it

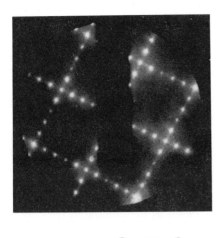

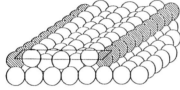

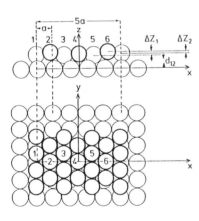

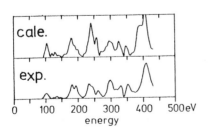

Fig. 4. LEED pattern and surface reconstruction model of Ir(100)5×1. For the $(0\frac{6}{5})$ beam the experimental and best fit spectra are compared. The best is for $\Delta z_1 = 0.5$ Å, $\Delta z_2 = 0.28$ Å and $d_{12} = 1.975$ Å [63,67].

is already hexagonally close packed when bulk terminated. There is a $(23 \times \sqrt{3})$ superstructure with atomic distances contracted by about 4.5% with respect to the bulk atomic distance. Recent high resolution X-ray diffraction experiments show that kinks play an important role and that at 880 K there is a phase transformation to an isotropically compressed surface layer [72].

The (110) surfaces of iridium, platinum and gold reconstruct as well showing a (1×2) superstructure in the LEED pattern. Though observed as early as 1967 for gold [73] it was only by the end of the 1970's that the real space structure was unambiguously resolved. Models in which atomic rows along the $[1\bar{1}0]$ direction were arranged to pair, to buckle or to miss completely were extensively discussed until quantitative LEED structure determination made a decision in favour of the missing row model with every second atomic row in the $[1\bar{1}0]$ direction missing [74–76]. Meanwhile this model was confirmed by application of a number of other structure sensitive methods. Different from the reconstructions of the (100) surfaces it is observed that the reconstruction is not limited to the top layer but extends into the surface. The second and third layers exhibit some pairing and mutual buckling of rows, respectively, and simultaneously there is strong vertical multilayer relaxation. Impressive and quantitative reproduction of these results could be achieved by first principle calculations [77]. Glue model calculations [78] indicate that (1×3) or (1×4) missing row reconstructions are not much higher in surface energy and in fact such superstructures have been observed by, e.g. LEED and ion scattering (e.g. Refs. [74,55]). Their appearance seems to depend on sample preparation as well as on the presence of impurities [55].

3.2. Reconstruction of bcc metals

Indications that the W(100) surface is reconstructed came as early as 1971 [79] but it took a couple of years to become clear that the reconstruction of W(100) as well as of Mo(100) were properties of the clean surfaces [80,81]. While W(100) shows a c(2×2) superstructure near and below room temperature (the same structure was reported also for Cr(100) [82]), the reconstruction of Mo is more complex but related [83]. The large majority of structural LEED investigations is on W(100).

Detailed structural information about W(100) came first by a LEED investigation of a stepped surface [84]. It was concluded that different domains with p2mg symmetry cause the observed c(2 × 2) pattern implying a structural model in which surface atoms are laterally displaced to form zig-zag chains (Fig. 5). In spite of this strong symmetry argument also other models were tested as vertical displacements or such to form atomic dimers. An early LEED intensity analysis ruled out vertical displacements and determined the intralayer displacement for the zig-zag model to

be about 0.2 Å [85]. More recent LEED and X-ray diffraction analyses [86,87] confirmed this value (0.21 Å) and found that the reconstruction extends to the second layer with, of course, a weaker zig-zag displacement (0.04 Å). However, it was found that LEED performed at normal incidence could not distinguish between the zig-zag and the dimer model [86]. This is because of the reduced sensitivity of LEED to in-plane displacements *and* the averaging over different domains of reconstruction orthogonal to each other. However, the zig-zag model is clearly favoured by intensity analysis for oblique incidence of the primary beam as demonstrated recently [88]. Yet hydrogen adsorption makes the atoms switch to the dimer arrangement [84,88–91].

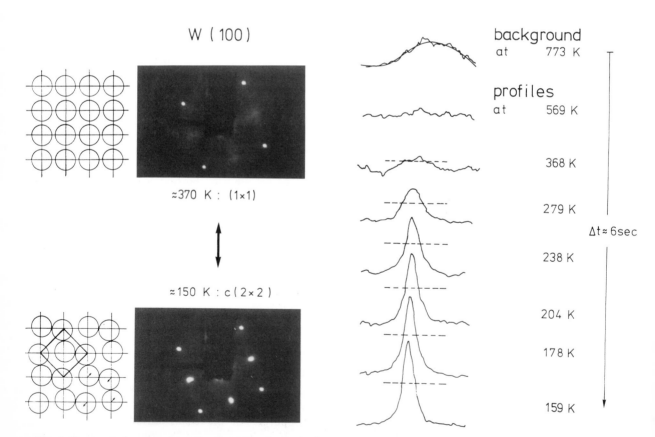

Fig. 5. Structural phase transition $(1 \times 1) \to c(2 \times 2)$ of W(100) by a rapid temperature quench monitored by profiles of the $\frac{1}{2}\frac{1}{2}$ spot. The background signal determined at 773 K is subtracted from profiles taken at lower temperatures [1].

4. Structural phase transitions of clean metals monitored by LEED

4.1. Equilibrium transitions

With the static structures quantitatively determined, surface scientists became interested in the transitions themselves, i.e. the transitions from the unreconstructed (1×1) to the reconstructed phase. Practically all reconstructed metal surfaces display (1×1) LEED patterns at sufficiently high temperatures, i.e. there are temperature driven phase transitions. In the case of W(100) this was easy to observe as in some cases the (1×1) phase is observed already slightly above room temperature. The temperature dependent equilibrium transition $(1 \times 1) \leftrightarrow c(2 \times 2)$ was already monitored in the early LEED investigations of reconstructed W(100) [84]. Temperature dependent spot intensities and half widths were recorded showing that spots sharpen and increase in intensity with decreasing temperature reflecting the growth of reconstructed domains. Similar data from a similar investigation [92,1] are displayed in Fig. 5. Rapid cooling was applied in order to minimize hydrogen adsorption from the residual gas. The transition region is rather broad and

superstructure spot intensities can be observed well above room temperature possibly due to short range order [93]. Investigations on stepped surfaces show that the transition temperature is about 211 K for large terraces but increases with decreasing terrace width. The transition itself is supposed to be second order with a critical exponent $\beta = 0.144$ [93]. The high temperature (1×1) phase seems to be disordered [94–96] with similar but statistic displacements of top layer atoms as in the reconstructed phase [97].

The high temperature structural phase transitions $(2 \times 1) \leftrightarrow (1 \times 1)$ of Au(110) and Pt(110) are of the order–disorder type as well [98–102]. The top layer is only half a monolayer of atoms so that there is no trouble with mass transport when the (2×1) missing row structure forms. For gold, detailed LEED investigations with independent measurement of 3 critical exponents have shown that the transition belongs to the 2D Ising universality class with a transition temperature of about 650 K [99,100]. However, for Pt(110) non-Ising behaviour was observed and a roughening transition favoured instead [101]. However, for both cases the transition seems to be strongly influenced by surface impurities, particularly tin, which can change the transition temperature by as much as 200 K [103].

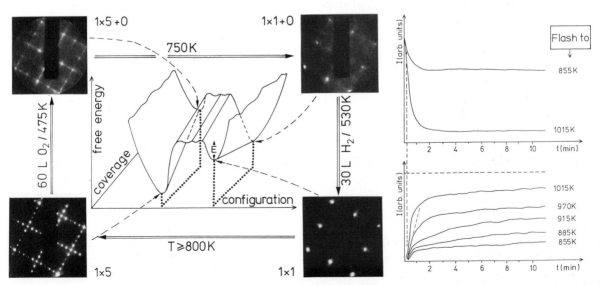

Fig. 6. Reconstruction and deconstruction of Ir(100) (left) and kinetics of the phase transition $(1 \times 1) \rightarrow (5 \times 1)$ monitored by LEED intensities (right) [21,106].

4.2. *Transition kinetics*

Simultaneous with the observation of structural phase transition scientists became interested in the time scale on which they develop [104,105]. For clean metal surfaces again the (100) and (110) surfaces of platinum, iridium and gold were excellent candidates as they can be prepared in their (1×1) phases also at room temperature. These phases are metastable and result by adsorption of certain gases on the reconstructed surface, i.e. restoration of surface chemical bonds restores the bulk-like surface structure. Careful annealing makes the surface deconstruct. The adsorbed gases can be made to desorb subsequently or can be chemically removed and the clean but unreconstructed surface results. Annealing makes the surface reconstruct again. If the temperature is raised rapidly the *kinetics* of the transition can be monitored by measurement of the development of LEED intensities with time [105]. As an example Fig. 6 shows the deconstruction and reconstruction loop for Ir(100) $(1 \times 1) \leftrightarrow (5 \times 1)$ [21,105,106]. A schematic free energy surface is used to indicate the stability or metastability of the different phases. The time dependent development of LEED intensities for the (1×1) surface flashed to different temperatures is displayed as well. From the initial slope of the curves the activation energy for the transition can be determined (0.88 eV for Ir(100)). Though not falling into the scope of the present article it should be mentioned that by careful adjustment of residual gas pressure and temperature structural oscillations can be triggered, as reported, e.g., for Pt(100) [107,108], Pt(110) [109] and other surfaces. Also, wave-like spatial propagation of structural oscillations were reported (for a recent review see Ertl [110] and [135]).

5. Chemical restructuring of alloys

With the knowledge that the structure of elemental metal surfaces – even when reconstructed – can be reliably determined, surface scientists started to investigate the structure of alloys in the mid 1980's. Additional to geometrical restructuring they can reconstruct also chemically, i.e. there can be segregation or depletion of one or the other of the constituents and this can be layer dependent. Also, surfaces can exhibit domains with different elemental terminations. Altogether, this opens a large variety of possible structures. Yet, up to now only bimetallic alloys are investigated with respect to both their geometrical and chemical structure (for a review see Ref. [111]).

In case of geometrically *and* chemically ordered samples no segregation is observed. Surface reconstruction with simultaneous appearance of LEED extra spots is rare. Because of the larger unit cell compared to elemental low index metals there can be strong vertical buckling without causing extra spots (the same may happen with high index elemental surfaces). NiAl is a good example of what can happen [112]: For the (110) face with its mixed layers a rippling of the first two layers is reported, i.e. Al atoms are moved out of the surface with respect to Ni atoms by 0.20 and 0.02 Å for the first and second layer, respectively. For NiAl(100) with a sequence of elemental layers only Al terminated samples exist whilst for NiAl(111) domains of both terminations exist. For samples of the $AuCu_3$ structure such as Ni_3Al [113] or Pt_3Ti [114], whose (100) and (110) surfaces consist of alternating mixed and elemental layers, both mixed layer terminations with buckling in the outermost layer (Ni_3Al) and elemental layer terminations with buckling only in the second layer (Pt_3Ti) can exist. In all cases there is additional multilayer relaxation of layer distances similar to elemental metals.

The treatment of chemically, i.e. substitutionally disordered alloys seemed to be a real challenge for LEED intensity analysis because the disorder destroys the periodicity parallel to the surface. Fortunately, however, the average T-matrix approximation (ATA) [115], which uses the average of the scattering matrices of the different atoms and takes advantage of the positional periodicity of atoms, turned out to be a very good approximation [116,117]. When applied to calculate the diffraction matrices of the different surface layers, the ATA method allows one to determine the statistical composition of each layer separately making LEED also sensitive to segregation, i.e. chemical restructuring of a surface. So, LEED provides layer by layer resolution of both the geometry and composition of a surface.

In nearly all cases investigated so far the top layer is enriched in one of the constituents, the second layer is depleted. Damped oscillations around the bulk concentration are the rule and can extend over several layers as found for varying bulk composition for, e.g. $Pt_x Ni_{1-x}$ [111,116, 118,119] and $Mo_x Re_{1-x}$ [120,121]. It seems that the larger of the atoms is driven to the top layer and that the compositional oscillation tries to compensate for the demixing caused by the segregation [122] in agreement with theoretical predictions [123]. However, also monotonic concentration profiles have been reported, e.g. for $Pt_{80}Fe_{20}(111)$ [124]. Also, segregation can be influenced by the crystallographic face. Whilst for $Pt_{50}Ni_{50}(100)$ and (111) the first layer is almost pure Pt and the second layer almost pure Ni, the segregation is reversed for the (110) surface (for a detailed comparison see Ref. [111]). Also, whilst the reconstruction observed on clean Mo(100) is inhibited on $Mo_{95}Re_{50}$ [120,121], the top layer of $Pt_{50}Ni_{50}(100)$ shows a reconstruction similar to that of clean Pt(100) [111]. A reconstruction similar to that of Pt(110) was also found for $Pt_{80}Fe_{20}(110)$ [125]. In general it seems that substitutionally disordered alloys exhibit a huge variety of different structures.

6. Outlook

In view of the large amount of data available at present with respect to the crystallographic and chemical structure of clean metal surfaces it seems that not much is left to do. However, one should remember that there are only few results for very open, i.e., stepped surfaces (even for elemental metals) which leave a large field open for quantitative structure determination. Also, surface structures of metals so far were mostly investigated in a restricted temperature range, i.e. between room and liquid air temperature. Other surface reconstructions might still be undetected. Also, the large variety of structures found

for binary alloys is likely to explode when going to multicomponent metallic systems. This includes also nonmetal–metal compounds as, e.g., metallic silicides with their technical importance in surface and interface systems. For the determination of the surface segregation of disordered ternary or higher componental compounds, the application of the ATA method within the LEED intensity analysis may be still too time consuming. It is therefore worth noting that recently the perturbation method Tensor LEED, which allows the fast perturbative calculation of intensities for structures differing by small atomic displacements [126–128] (see also Pendry [136]), could be extended to "chemical displacements", i.e. chemical substitution of atoms [129]. The method should easily fit into the structural search procedures based on Tensor LEED introduced for the LEED analysis of complex structures [128,130] (see also Van Hove and Somorjai [137]). Last but not least there is the important field of ultrathin epitaxial metallic films, whose structural research has started in recent years but which is still in its infancy. Additionally, little is known about the structure of defects on clean surfaces. In principle this could be retrieved by the diffuse LEED technique [131–133], which has proved to be successful for the retrieval of the local structure of disordered adsorbates (for a review see Ref. [134]). The method requires the dominance of a single type of defects on the surface, so much progress could come if experimentalists could prepare samples accordingly. On the whole, the structure of clean metal surfaces seems to remain an interesting field at least for another decade.

References

[1] K. Heinz and K. Müller, in: Structural Studies of Surfaces, Springer Tracts of Modern Physics, Vol. 91 (Springer, Berlin, 1982) p. 1.

[2] K. Heinz, E. Lang and K. Müller, Surf. Sci. 87 (1979) 595.

[3] M.G. Lagally and J.A. Martin, Rev. Sci. Instrum. 54 (1983) 1273.

[4] M.A. Van Hove, W.H. Weinberg and C.-M. Chan, Low-Energy Electron Diffraction (Springer, Berlin, 1986).

[5] C.J. Davison and L.H. Germer, Phys. Rev. 30 (1927) 305.

[6] H.E. Farnsworth, Phys. Rev. 34 (1929) 679.

[7] F. Jona, Discuss. Faraday Soc. 60 (1975) 210.

[8] S.P. Weeks, C.D. Ehrlich and E.W. Plummer, Rev. Sci. Instrum. 48 (1977) 190.

[9] P.C. Stair, T.J. Kaminska, L.L. Kesmodel and G.A. Somorjai, Phys. Rev. B 11 (1975) 623.

[10] D.C. Frost, K.A.R. Mitchell, F.R. Shepherd and P.R. Watson, J. Vac. Sci. Technol. 13 (1976) 1196.

[11] K. Ueda, G. Lempfuhl and E. Reuber, Z. Naturforsch. 34a (1979) 648.

[12] R.L. Park and H.E. Farnsworth, Rev. Sci. Instrum. 35 (1964) 1592.

[13] H.J. Bechthold, E. Kreutz, E. Rickus and N. Sotnik, Verh. Dtsch. Phys. Ges. (VI) 13 (1978) 585.

[14] W. Moritz and D. Wolf, Surf. Sci. 88 (1979) L29.

[15] T. Kanaji, H. Nakatsuka, T. Urano and Y. Taki, Surf. Sci. 86 (1979) 587.

[16] W. Berndt, Rev. Sci. Instrum. 53 (1982) 221.

[17] P. Heilmann, E. Lang, K. Heinz and K. Müller, Appl. Phys. 9 (1976) 247.

[18] E. Lang, P. Heilmann, G. Hanke, K. Heinz and K. Müller, Appl. Phys. 19 (1979) 287.

[19] P. Heilmann, E. Lang, K. Heinz and K. Müller, in: Determination of Surface Structure by LEED, Eds. P.M. Marcus and F. Jona (Plenum, New York, 1984) p. 463.

[20] K. Müller and K. Heinz, in: The Structure of Surfaces, Eds. M.A. Van Hove and S.Y. Tong (Springer, Berlin, 1985) p. 105.

[21] K. Heinz, Prog. Surf. Sci. 27 (1988) 239.

[22] D.G. Welkie and M.G. Lagally, Appl. Surf. Sci. 3 (1979) 272.

[23] H. Leonhard, A. Gutman and K. Hayek, J. Phys. E 13 (1980) 297.

[24] V.E. Carvalho, M.W. Cook, P.G. Cowell, O.S. Heavens, M. Prutton and S.P. Tear, Vacuum 34 (1984) 893.

[25] F. Jona, J.A. Strozier and P.M. Marcus, in: The Structure of Surfaces, Eds. M.A. Van Hove and S.Y. Tong (Springer, Berlin, 1985) p. 92.

[26] J.W. Anderegg and P.A. Thiel, J. Vac. Sci. Technol. A 4 (1986) 1367.

[27] T. Guo, R.E. Atkinson and W.K. Ford, Rev. Sci. Instrum. 61 (1990) 968.

[28] D.F. Ogletree, G.A. Somorjai and J.E. Katz, Rev. Sci. Instrum. 57 (1986) 3012.

[29] D.I. Adams, S.P. Andersen and J. Buchhardt, in: The Structure of Surfaces III, Eds. S.Y. Tong, M.A. Van Hove, K. Takayanagi and X.D. Xie (Springer, Berlin) p. 156.

[30] K. Hammer, Ph.D. Thesis, Erlangen (1985).

[31] M.D. Chinn and S. Fain, Jr., J. Vac. Sci. Technol. 14 (1976) 314.

[32] D.G. Welkie and M.G. Lagally, Appl. Surf. Sci. 3 (1979) 272.

[33] S.P. Weeks, J.E. Rowe, S.B. Christmann and E.E. Chaban, Rev. Sci. Instrum. 50 (1979) 1249.

[34] P.C. Stair, Rev. Sci. Instrum. 51 (1980) 132.

[35] C. Martin, P. Jelinsky, M. Lampton, R.F. Mailina and H.O. Anger, Rev. Sci. Instrum. 52 (1981) 1068.

[36] E.G. McRae, R.A. Malic and D.A. Kapilow, Rev. Sci. Instrum. 56 (1985) 2077.

[37] E.G. McRae and R.A. Malic, Surf. Sci. 177 (1986) 74.

[38] G.S. Blackman M.-L. Xu, D.F. Ogletree, M.A. Van Hove and G.A. Somorjai, Phys. Rev. Lett. 61 (1988) 2352.

[39] D.F. Ogletree, G.S. Blackman, R.Q. Hwang, U. Starke, G.A. Somorjai and J.E. Katz, Rev. Sci. Instrum. 63 (1991) 104.

[40] U. Starke, K. Heinz, N. Materer, A. Wander, M. Michl, R. Döll, M.A. Van Hove and G.A. Somorjai, J. Vac. Sci. Technol. A 10 (1992) 2521.

[41] U. Starke, private communication.

[42] U. Scheithauer, G. Meyer and M. Henzler, Surf. Sci. 178 (1986) 441.

[43] H.L. Davis and J.R. Noonan, J. Vac. Sci. Technol. 20 (1982) 842.

[44] J.B. Pendry, J. Phys. C 13 (1980) 937.

[45] F. Jona, Surf. Sci. 192 (1987) 414.

[46] F. Jona, P. Jiang and P.M. Marcus, Surf. Sci. 192 (1987) 414.

[47] F. Jona and P.M. Marcus, in: The Structure of Surfaces II, Eds. J.F. van der Veen and M.A. Van Hove (Springer, Berlin, 1988) p. 90.

[48] X.-G. Zhang, M.A. Van Hove, G.A. Somorjai, P.J. Rous, D. Tobin, A. Gonis, J.M. MacLaren, K. Heinz, M. Michl, H. Lindner, K. Müller, M. Ehsasi and J.H. Block, Phys. Rev. Lett. 67 (1991) 1298.

[49] D.L. Adams, V. Jensen, X.F. Sun and J.H. Vollesen, Phys. Rev. B 38 (1988) 7913.

[50] X.-G. Zhang, P.J. Rous, J.M. MacLaren, A. Gonis, M.A. Van Hove and G.A. Somorjai, Surf. Sci. 239 (1990) 103.

[51] P. Pinkava and S. Crampin, Surf. Sci. 233 (1990) 27.

[52] P.J. Estrup, Springer Ser. Chem. Phys. 35 (1984) 205.

[53] J.E. Inglesfield, Prog. Surf. Sci. 20 (1985) 105.

[54] H. Ohtani, C.-T. Kao, M.A. Van Hove and G.A. Somorjai, Prog. Surf. Sci. 23 (1986) 155.

[55] W. Hetterich, U. Korte, G. Meyer-Ehmsen and W. Heiland, Surf. Sci. Lett. 254 (1991) L487.

[56] S.B. Hagström, H.B. Lyon and G.A. Somorjai, Phys. Rev. Lett. 15 (1965) 491.

[57] H.B. Lyon and G.A. Somorjai, J. Chem. Phys. 46 (1967) 2539.

[58] D.G. Fedak and N.A. Gjostein, Phys. Rev. Lett. 16 (1966) 171.

[59] D.G. Fedak and N.A. Gjostein, Surf. Sci. 8 (1967) 77.

[60] P.W. Palmberg, in: The Structure and Chemistry of Surfcaes, Ed. G.A. Somorjai (Wiley, New York, 1969) p. 291.

[61] J.T. Grant, Surf. Sci. 18 (1969) 228.

[62] A. Ignatiev, A.V. Jones and T.N. Rhodin, Surf. Sci. 30 (1972) 573.

[63] P. Heilmann, K. Heinz and K. Müller, Surf. Sci. 83 (1979) 487.

[64] M.A. Van Hove, R.J. Koestner, P.C. Stair, J.P. Bibérian, L.L. Kesmodel, I. Bartos and G.A. Somorjai, Surf. Sci. 103 (1981) 218.

[65] E. Lang, K. Müller, K. Heinz, M.A. Van Hove, R.J. Koestner and G.A. Somorjai, Surf. Sci. 127 (1983) 347.

[66] W. Moritz, F. Müller, D. Wolf and H. Jagodzinski, in Proc. 9th Int. Vacuum Congr. and 5th Intern. Conf on Solid Surfaces, Madrid (1983) p. 70.

[67] N. Bickel and K. Heinz, Surf. Sci. 163 (1985) 435.

[68] M.A. Van Hove, R.J. Koestner, P.C. Stair, J.P. Bibérian, L.L. Kesmodel, I. Bartos and G.A. Somorjai, Surf. Sci. 103 (1981) 189.

[69] S.G.J. Mochrie, D.M. Zehner, B.M. Ocko and D. Gibbs, Phys. Rev. Lett. 64 (1990) 2925.

[70] D. Gibbs, B.M. Ocko, D.M. Zehner and S.G.J. Mochrie, Phys. Rev. B 42 (1990) 7330.

[71] D.M. Zehner and J.F. Wendelken, in: Proceedings of the VII International Vacuum Congress and the III International on Solid Surfaces, Eds. R. Dobrozemsky et al. (Berger und Söhne, Vienna, 1977) p. 517.

[72] K.G. Huang, D. Gibbs, D.M. Zehner, A.R. Sandy and S.G.J. Mochrie, Phys. Rev. Lett. 65 (1990) 3313.

[73] D.G. Fedak and N.A. Gjostein, Acta Met. 15 (1967) 827.

[74] W. Moritz and D. Wolf, Surf. Sci. 88 (1979) L29; 163 (1985) L655.

[75] C.-M. Chan, M.A. Van Hove, W.H. Weinberg and E.D. Williams, Solid State Commun. 30 (1979) 47; Surf. Sci. 91 (1980) 440.

[76] D.L. Adams, H.B. Nielson, M.A. Van Hove and A. Ignatiev, Surf. Sci. 104 (1981) 47.

[77] K.-M. Ho and K.P. Bohnen, Phys. Rev. Lett. 59 (1987) 1833.

[78] M. Garofalo, E. Tosatti and F. Ercolessi, Surf. Sci. 188 (1987) 321.

[79] K. Yonehara and L.D. Schmidt, Surf. Sci. 25 (1971) 238.

[80] T.E. Felter, R.A. Barker and P.J. Estrup, Phys. Rev. Lett. 38 (1977) 1138.

[81] M.K. Debe and D.A. King, J. Phys. C 10 (1977) L303.

[82] G. Gewinner, J.C. Peruchetti, A. Jaegle and R. Riedinger, Phys. Rev. Lett. 43 (1979) 935.

[83] R.A. Barker, S. Semancik and P.J. Estrup, Surf. Sci. 94 (1980) L162.

[84] M.K. Debe and D.A. King, Phys. Rev. Lett. 39 (1977) 708; Surf. Sci. 81 (1979) 193.

[85] R.A. Barker, P.J. Estrup, F. Jona and P.M. Marcus, Solid State Commun. 25 (1978) 375.

[86] H. Landskron, N. Bickel, K. Heinz, G. Schmidtlein and K. Müller, J. Phys. (Condensed Matter) 1 (1989) 1.

[87] M.S. Altmann, P.J. Estrup and I.K. Robinson, J. Vac. Sci. Technol. A 6 (1988) 630; Phys. Rev. B 38 (1988) 5211.

[88] G. Schmidt, H. Zagel, H. Landskron, K. Heinz, K. Müller and J.B. Pendry, Surf. Sci. 271 (1992) 416.

[89] R.A. Barker and P.J. Estrup, J. Chem. Phys. 74 (1981) 1442.

[90] D.A. King and G. Thomas, Surf. Sci. 92 (1980) 201.

[91] D.A. King, Phys. Scr. T 4 (1982) 34.

[92] P. Heilmann, K. Heinz and K. Müller, Surf. Sci. 89 (1979) 84.

[93] J.F. Wendelken and G.-C. Wang, J. Vac. Sci. Technol. A 2 (1984) 888; Phys. Rev. B 32 (1985) 7542.

[94] I. Stensgaard, L.C. Feldman and P.J. Silverman, Phys. Rev. Lett. 42 (1979) 247.

[95] A.J. Melmed, R.T. Tung, W.R. Graham and G.D.W. Smith, Phys. Rev. Lett. 43 (1979) 1521.

[96] T.T. Tsong, and J. Sweeney, Solid State Commun. 30 (1979) 767.

[97] J.B. Pendry, K. Heinz, W. Oed, H. Landskron, K. Müller and G. Schmidtlein, Surf. Sci. 193 (1988) L1.

[98] J.C. Campuzano, A.M. Lahee and G. Jennings, Surf. Sci. 152/153 (1985) 68.

[99] J.C. Campuzano, M.S. Foster, G. Jennings, R.F. Willis and W. Unertl, Phys. Rev. Lett. 54 (1985) 2684.

[100] J.C. Campuzano, G. Jennings and R.F. Willis, Surf. Sci. 162 (1985) 484.

[101] I.K. Robinson, E. Vlieg and K. Kern, Phys. Rev. Lett. 63 (1989) 2578.

[102] M.S. Daw and S.M. Foiles, Phys. Rev. Lett. 59 (1987) 2756.

[103] E.G. McRae, T.M. Buck, R.A. Malic and G.H. Wheatley, Phys. Rev. B 36 (1987) 2341.

[104] M. Grunze and H.J. Kreuzer, Kinetics of Interface Reactions, Springer Ser. Surf. Sci. 8 (Springer, Berlin, 1987).

[105] K. Heinz, in Ref. [104] p. 202.

[106] K. Heinz, G. Schmidt, L. Hammer and K. Müller, Phys. Rev. B 32 (1985) 6214.

[107] G. Ertl, P.R. Norton and J. Rüstig, Phys. Rev. Lett. 49 (1982) 177.

[108] M.P. Cox, G. Ertl, R. Imbihl and J. Rüstig, Surf. Sci. 134 (1983) L517.

[109] M. Eiswirth and G. Ertl, Surf. Sci. 177 (1986) 90.

[110] G. Ertl, Surf. Sci. 287/288 (1993) 1.

[111] Y. Gauthier and R. Baudoing, in: Surface Segregation and Related Phenomena, Eds. P. Dowben and A. Miller (CRC Press, Boca Raton, FL, 1990) p. 169.

[112] H.L. Davis and J.R. Noonan, Phys. Rev. Lett. 54 (1985) 566; Proc. Int. Conf. Structure of Surfaces, Amsterdam (Springer, Berlin, 1987) p. 152.

[113] D. Sondericker, F. Jona and P.M. Marcus, Phys. Rev. B 33 (1986) 900; 34 (1986) 6770, 6775.

[114] A. Atrei, L. Pedocchi, U. Bardi, G. Rovida, M. Torrini, E. Zanazzi, M.A. Van Hove and P.N. Ross, Surf. Sci. 261 (1992) 64.

[115] B.L. Györffy and G.M. Stocks, in: Electrons in Disordered Metals and Metallic Surfaces, Eds. P. Phariseau,

B.L. Györffy and L. Scheire (Plenum, New York, 1979) p. 89.

[116] Y. Gauthier, Y. Joly, R. Baudoing and J. Rundgren, Phys. Rev. B 31 (1985) 6216.

[117] S. Crampin and P.J. Rous, Surf. Sci. Lett. 244 (1991) L137.

[118] R. Baudoing, Y. Gauthier, M. Lundberg and J. Rundgren, J. Phys. C 19 (1986) 2825; Phys. Rev. B 35 (1987) 7867.

[119] Y. Gauthier, R. Baudoing and J. Jupille, Phys. Rev. B 40 (1989) 1500.

[120] H.L. Davis, D.M. Zehner, B. Dötsch, A. Wimmer and K. Müller, Bull. Am. Phys. Soc. 36 (1991) 705.

[121] B. Dötsch, A. Wimmer, L. Hammer, K. Müller, H.L. Davis and D.M. Zehner, Verh. Dtsch. Phys. Ges. (VI) 26 (1991) 1335.

[122] Y. Gauthier, R. Baudoing-Savois, J.M. Bugnard, U. Bardi and A. Atrei, Surf. Sci. 276 (1992) 1.

[123] B. Legrand, G. Tréglia and F. Ducastelle, Phys. Rev. B 41 (1990) 4423.

[124] P. Beccat, Y. Gauthier, R. Baudoing-Savois and J.C. Bertolini, Surf. Sci. 238 (1990) 105.

[125] R. Baudoing-Savois, Y. Gauthier and W. Moritz, Phys. Rev. B 44 (1991) 12977.

[126] P.J. Rous, J.B. Pendry, D.K. Saldin, K. Heinz, K. Müller and N. Bickel, Phys. Rev. Lett. 57 (1986) 2951.

[127] P.J. Rous and J.B. Pendry, Surf. Sci. 219 (1989) 355; 373.

[128] P.J. Rous, Prog. Surf. Sci. 39 (1992) 3.

[129] R. Döll, M. Kottcke and K. Heinz, Phys. Rev. B, in press.

[130] P.J. Rous, M.A. Van Hove and G.A. Somorjai, Surf. Sci. 226 (1990) 15.

[131] J.B. Pendry and D.K. Saldin, Surf. Sci. 145 (1984) 33.

[132] K. Heinz, D.K. Saldin and J.B. Pendry, Phys. Rev. Lett. 55 (1985) 2312.

[133] D.K. Saldin, J.B. Pendry, M.A. Van Hove and G.A. Somorjai, Phys. Rev. B 31 (1985) 1216.

[134] K. Heinz, Vacuum 41 (1990) 328.

[135] G. Ertl, Surf. Sci. 299/300 (1994) 742.

[136] J.B. Pendry, Surf. Sci. 299/300 (1994) 375.

[137] M.A. Van Hove and G.A. Somorjai, Surf. Sci. 299/300 (1994) 487.

Surface Science 299/300 (1994) 447–453
North-Holland

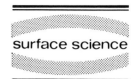

LEED and clean metal surfaces: personal reminiscences and opinions

P.M. Marcus

IBM Research Center, Yorktown Heights, NY 10598, USA

Received 2 June 1993; accepted for publication 12 July 1993

The early history of surface structure determination of clean metals from LEED intensities at IBM Research and in collaboration with SUNY at Stony Brook (1964–1975) is sketched. The reasons for the success of LEED intensity analysis in establishing the structure of clean metal surfaces at that time are discussed. A selective overview of the rich phenomenology of metal surface structure as we know it today is given. Personal comments are made on important current and future developments in LEED theory.

1. Introduction

In the spirit of the celebration of the 30 year anniversary of Surface Science, this paper combines a sketch of some early history of the development of surface structure determination by analysis of LEED intensity measurements (LEED analysis for short) with a personal view of current knowledge of the structure of clean metal surfaces and of LEED theory. The history focuses on the work at IBM Research that the writer participated in, which later became a collaboration with Franco Jona at SUNY at Stony Brook; it is not a general history of studies of metal surfaces by LEED.

Now that LEED analysis has become a widely-recognized source of reliable surface structures, it is instructive to recall the beginnings of the modern era of LEED analysis nearly 30 years ago, which was a time of great uncertainty about the future of LEED analysis, but also a time of great excitement. Much skepticism was expressed initially about the validity of a theory of LEED intensities adequately quantitative for structure determination and many objections were raised. Over the years of successfully application all objections have been disposed of, notably by

the remarkably close fit to experiment on clean metal surfaces. Accordingly the conditions for a reliable structure determination are now well understood. However, we should keep in mind that surface structure problems of increasing complication lie ahead, and that not all LEED structure analyses succeed. There are still important theoretical questions to answer, some of which will be noted in the last section. Between the historical sketch and the future problems of LEED theory are sections which survey clean metal surface structure and which offer some comments on the current state of LEED theory.

2. Historical sketch of LEED analysis of clean metals at IBM Research

Work on LEED at IBM Research began in 1964 when Franco Jona acquired one of the first commercial LEED machines made by Varian [1]. Jona quickly produced sharp LEED patterns and reproducible intensity–energy curves on various crystal surfaces, including clean metal surfaces such as Al and Ag. Jona then came to the theoretical physics group at IBM Research for help in interpreting the LEED intensity curves. Don

Jepsen and the writer started to develop a theory of the intensity. It is worth mentioning that this early experimental work preceded the use of Auger measurements to check surface impurity content, and it is a remarkable accomplishment that Jona developed a technique that gave consistent results suitable for intensity analysis without the use of Auger electron spectroscopy [2].

Looking back at the state of theoretical understanding of LEED intensities at that time is to recall that there was no quantitative theory of the intensities, even though LEED had been known and used for 40 years to characterize surfaces and to study surface symmetry. We did not understand the origin of secondary peaks, i.e., peaks other than the Bragg peaks, the importance of attenuation of electrons at LEED energies in crystals, the roles of the inner potential, the ion-core potential, the surface potential and the vibrations of the atoms.

However, we did have the advantage of understanding the current band structure theory, which made effective use of multiple scattering theory and we recognized that LEED theory could be regarded as an extension of band theory to higher-energy electronic states, which were scattering states rather than bound states. Hence the methods of band theory could be applied. We benefited in our application of band theory from the presence of A.R. Williams at IBM Research, who instructed us in the procedures for dealing with spherical waves and the calculation of Gaunt coefficients. We were also aided by attendance at the LEED Seminars, which went on for many years and were forums for lively discussion of the rapidly evolving field of LEED. These seminars introduced us to the work of other theorists, including E.G. McRae, J.B. Pendry, C.B. Duke, S.Y. Tong, and many others, which helped clarify our theoretical understanding, e.g., McRae told us about the new formulation of the LEED problem by K. Kambe, Duke made us aware of the importance of electron attenuation, so that we made sure that our programs had introduced sufficient attenuation. After our first attempt to use just beam expansions in the LEED problem was found inadequate to handle the strong scattering of ion cores [3a], we turned to the exten-

sion of the KKR method of band theory to periodic layers (nets) developed in Kambe's papers [4], which showed us how to evaluate structure sums by Ewald methods for a Bravais net (a net with just one atom type).

A formulation of the LEED problem and a computer program for calculating LEED intensities evolved from the influences noted above, and was applied first to clean metal surfaces. The clean metal surfaces were truly the touchstone of our success in showing that LEED intensities could quantitatively fit observed intensities [3b–3e]. These first successes in the quantitative fitting of experiment were made possible by a fortunate choice of a number of approximations, which are instructive to examine with hindsight. These approximations are:

(1) the formulation of the LEED problem as the multiple scattering of an incident plane-wave electron from spherical ion cores embedded in a uniform attenuating medium at a certain inner potential (i.e., the muffin-tin model of band theory extended to higher electron energies, which require many more phase shifts to compute scattering than are needed in band theory);

(2) a no-reflection boundary condition at the vacuum interface, i.e., beams with enough perpendicular energy to propagate in vacuum passed the interface without reflection, but with refraction (hence the unknown details of the surface potential between the crystal and vacuum were not needed, assuming the transition was smooth);

(3) treatment of the real and imaginary parts of the inner potential as adjustable fitting parameters (thereby avoiding the difficult problems of electron interactions);

(4) taking over the ion-core potential from muffin-tin band theory to use as a transferable form factor (thereby making the strong assumption that the potential does not have significant energy dependence in the LEED range; the core potential provides an explicit spherically-symmetric scattering potential);

(5) using only relative intensities within each LEED beam (i.e., discarding the overall level of intensity relative to the incident beam as being controlled by unknown factors);

(6) ignoring energies near the vacuum level

(they are too dependent on the valence electron contribution to the potential; this contribution is uncertain, varies as the structure changes, and is nonspherical);

(7) treating the lattice motion as a Gaussian distribution of displacements from mean positions depending on a single adjustable Debye theta (a "blurred atom" model, which made the atom effectively larger and the phase shifts complex).

The assumptions have been refined in later work, but proved to be adequate at the time to make quantitative analysis of LEED intensities possible. The early studies not only gave a good quantitative fit, but led quickly to the realization that significant relaxations of surface layers were present, i.e., changes in layer spacings which preserved the translational symmetry of each layer. Such changes did not affect the 1×1 spot pattern, hence intensity analysis was required to determine the relaxations [5].

Further advances in the analysis of clean metal surface structure at IBM Research came soon after the work on Al and Ag with the arrival of J. Demuth, who brought to IBM Research his LEED data on Ni, both clean and with chalcogenide overlayers, that he had measured at Cornell University with T. Rhodin. The work on clean Ni not only gave a very good fit of theory to experiment [6], it showed conclusively that Ni(110) was relaxed and it refined the analysis by determining the energy dependence of the real and imaginary parts of the inner potential. However, only the first layer spacing was relaxed in this early work on Ni (and Al). The success with clean Ni permitted the analysis of overlayers on Ni to proceed with confidence.

There are two additional significant chapters in the study of clean metal surfaces by LEED at IBM Research, both carried out in a collaboration with F. Jona after his move to the State University of New York at Stony Brook in 1969. One chapter is the remarkable sequence of structure determinations of six surfaces of Fe, carried out with J. Sokolov [7], which extended the LEED analysis of surfaces to high-index surfaces. This work showed the development of deep multilayer relaxations on more open surfaces and the occurrence of parallel relaxation, i.e., parallel components of the translation of a surface layer with respect to bulk, whenever the symmetry of the layer permitted such movement.

A second chapter, still ongoing, is the extension of the study of clean metal surfaces to coherent epitaxial films of metals. If the films are sufficiently thick that the behavior of the surface layers is not affected by the interface with the substrate, then the surface of the film may be considered to occur on a bulk specimen of the film material. Such a film may be called a bulk film, even if it has only a few atomic layers, usually six or more. Two new features of such bulk epitaxial films are that they may be in metastable phases, not available macroscopically, e.g., fcc Fe, bcc Co, and that the film can be under large epitaxial strain (strain in the plane of the film, but with the thickness allowed to equilibrate freely). Such films have been produced with as much as 9% strain in the film plane [8]. LEED analysis finds the bulk spacing, as well as the relaxation of the surface layers. Then the bulk spacing can be compared with the elastic strain calculated from the elastic constants (more precisely, the strain ratio, defined as the ratio of the perpendicular strain to the parallel strain, can be compared with the value expected from Poisson's ratio).

3. Significance and status of clean metal surface studies

The clean metal surface structure studies at IBM Research and elsewhere by LEED and other surface techniques have now produced a substantial body of quantitative structure results on perhaps a hundred distinct surfaces. The number is enhanced by the possibility of studying an indefinite number of surfaces of a given bulk crystal, by the existence of many surface phases whose formation depends on temperature and the conditions of growth, and by the use of coherent epitaxy to stabilize new phases. Truly impressive correspondence between theory and experiment on clean metal surfaces has been achieved in the work of J. Noonan and H. Davis [9], who paid particular attention to the accuracy of normal incidence data, and to refining the fitting parameters. In some cases the theoretical intensity curves reproduce all details of the experimental curves – peak positions, shapes and amplitudes. These close fits to experiment provide strong evidence for the accuracy not only of the theory, but also of the experimental data.

A particularly important recent development (last 10 years) in the writer's opinion is the conjunction of experimental surface structure determination with theoretical surface structure determination by first-principles total-energy calculations. Such calculations are self-consistent in both the electronic charge distribution and in the nuclear positions in the surface layers. This powerful combination of experiment and theory serves to verify and stimulate both activities.

LEED analysis has proved very well suited to the study of relaxation on clean metal surfaces in which layers translate with respect to bulk, but do not change translational symmetry, and also to the study of reconstruction in which layers change their translational symmetry. The LEED analysis is sensitive enough to find structure four or five layers deep, accurate enough to find the atomic positions within a few hundredths of an ångström, and complete enough to find the positions of all atoms in surface layers with minimal deviation from bulk positions (say 1% of the near-neighbor distance). The result has been to produce a rich body of structural phenomenology, quite astonishing in view of the initial skepticism which greeted the first indications of relaxation on metal surfaces. A brief survey will exhibit some of this richness.

We start by observing that almost all metal surfaces show significant relaxation, and more open surfaces can show large multilayer relaxations [7]. The changes in interlayer spacings show various patterns of compression and expansion from bulk spacing; the first spacing is usually, but not always, compressed. In fact a number of cases of substantial expansion of the first spacing have been found recently, e.g., fcc Fe(001) [10], fcc Pd(001) [11], hcp Be(0001) [12] for which various explanations have been suggested: an increased magnetic moment of the first layer compared to

bulk [10], subsurface H or a magnetic first layer [11], unusual bonding involving the promotion of an electron [12].

Reconstruction is a more drastic change from bulk structure than relaxation (essentially a new surface phase) and is less common on clean metal surfaces. Among the observed cases are a number of surprises. Not only can reconstruction be achieved by relative motion of sublayers of the bulk structure, as for W(001)c(2 × 2) [13], but reconstructions exist with surface layers that are less dense than bulk layers, e.g., missing-row structures like Au(110)2 × 1 [14] (every alternate row along [$\bar{1}$10] is missing), but also with surface layers that are more dense than bulk layers, like Au(001)5 × 28 (among others) [15] which has a quasihexagonal first layer. The different periodicity of the first layer requires large cells in order to lock coherently to the second layer and produces significant buckling.

LEED analyses of the surface layers of epitaxial metal films can be used to prove that the film has a certain structure, and to find the strain in the film. From the measured strain some elastic properties can be deduced (such as Poisson's ratio) [16]. But the LEED analysis can also be used in a negative way to show that films of a few layers have not grown layer-by-layer, since no such structure fits the LEED data ("does not pass the LEED test") [17].

The structure of clean metal surfaces is frequently complicated by the presence of domains, i.e., regions of the same simple order, but that differ from each other in some respect, such as rotation, inversion, registration with the bulk structure, height above the bulk structure, etc. The presence of such domains can give rise to pseudosymmetry in which the surface appears to be more symmetric than the atomic-scale structure. Domains impose an additional step on the theoretical calculation, namely averaging over the domains; the possibility of unequal weights for the domains introduces additional fitting parameters. A striking example of domain formation occurs in the growth of Fe films on Ru(0001). The Fe film grows (after a few layers) as bcc(110) in six domains, three differ by 120° rotations and three more by inversion of the first three, giving rise to a characteristic split pattern of five spots [18].

The surfaces of metal compounds have additional degrees of freedom and additional structural features. Thus a compound surface like NiAl(110) (CsCl structure) shows a rippled surface with the plane of the Al atoms displaced outward 10% of the bulk spacing from the plane of the Ni atoms [19]. The surface of a compound like $Ni_3Al(001)$ (Cu_3Au structure), which is made up of alternate layers of pure Ni and mixed atoms (half Ni, half Al), can be shown always to have a mixed layer as its outside layer [20].

LEED does not give a detailed view of surface disorder, but can provide a measure of surface vibration amplitudes through evaluation of the effective Debye theta required to fit the LEED data. A qualitative idea of the amount and type of structural disorder is obtained from the diffuseness and angular profile of the LEED spots.

4. Some developments in LEED theory

Among the improvements of the layer-KKR procedure for calculating LEED intensities made since the early papers, the introduction of composite layers was essential. This extension of the layer-KKR made it possible to treat close-lying layers, which required too many beams to be described with beam expansions (i.e., expansions in plane waves of the same energy and the same reduced wave number in the surface plane). The composite-layer procedure handles a set of closely-spaced layers as a unit with spherical-wave scattering theory. The paper by Tong and Van Hove on the combined-space method [21] introduced such a procedure and applied it to closely-spaced adsorbate–substrate layers. The substrate, however, had a large enough spacing between layers for the beam expansion to be used. In the case of higher-index surfaces of clean metals all the layers are closely spaced and the composite-layer treatment is needed for the entire lattice. In the paper of Jepsen et al. [22] the units are thick composite layers used throughout the lattice, thick enough so that the centers of the units are far enough apart for a beam expansion to hold. This

procedure was incorporated in a program called CHANGE, which was satisfactorily applied to the relaxation of surfaces with layer spacings as small as 0.64 Å (bcc Fe(210) [23]).

Various approximate procedures have proved useful in providing more efficient calculation of LEED intensities at the expense of accuracy. Prominent among these procedures are the quasidynamic method and tensor LEED. The quasidynamic method uses the kinematic (single scattering) approximation to describe scattering by individual layers, but solves the interlayer scattering accurately, i.e., the method neglects multiple scattering in the layer planes, but not between layer planes; it does only the easier part of the multiple scattering problem accurately, hence is much faster than the full dynamical calculation. A compact formulation of the quasidynamic method and illustration of its accuracy for Cu(001) is given by Lin et al. [24].

Tensor LEED does an accurate full-dynamical calculation at one reference structure, but then explores intensities for structures around the reference structure by perturbation formulas based on the calculation of the reference structure. Simple forms, linear in the changes in structural parameters, or linear in simple functions of the changes (spherical harmonics and spherical Bessel functions), permit very rapid evaluation of intensities in a useful region around the reference structure [25]. A directed search at electronic speeds can then be used to minimize the R-factor, which measures the goodness of fit to experiment [26]. This procedure has made it possible to handle many more structural parameters than the older methods based on examination of the fit to experiment at each stage of the intensity calculation. This very useful development then makes possible structure determination of large cells with many structural parameters, such as occur in high-index surfaces of clean metals.

5. Suggested future developments of LEED theory

Despite the impressive advances in intensity calculations in recent years, which have greatly extended the capabilities of LEED analysis, there still is substantial room for improvement of the theory. First we note that no demonstrated improvement in the basic procedure for accurate full-dynamical calculation of LEED intensities has been found since the introduction of composite-layer ideas into the layer-KKR procedure some 15 years ago. Large gains in speed of calculation have come from faster computers with larger memories, which can store more quantities that are used repeatedly. New approaches to the efficient solution of the multiple-scattering problem in layered structures would be valuable.

There is still no quantitative theory of LEED intensities down to the vacuum level. Fortunately such a theory is not needed for structure determination. Such a theory would involve details of the surface potential, i.e., the potential in the transition region from crystal to vacuum. After the structure has been found by the LEED analysis described above, based on energies well above the vacuum level, the surface potential itself could be determined by fitting the calculated intensities to experiment in the energy range close to the vacuum level with the potential included among the fitting parameters. Note that the intensity calculations in this very low energy range would require calculation of scattering from non-spherical scatterers, and the phase-shift matrices and T-matrices would no longer be diagonal. A truly quantitative theory of LEED intensities, which included determination of the surface potential, should also improve the precision of the structure parameters, since the precision is now limited by the uncertainties in the scattering potential.

Finally, we note the desirability of coordinating experimental structure determination with first-principles total-energy calculations that find the surface structure by minimizing the total energy (spin-polarized for magnetic materials). Examples of metal surfaces studied both experimentally and by first principles theory are the relaxed surfaces Al(110) [27], Al(331) [28], and the reconstructed surfaces Au(110)2 × 1 [29], W(001)c(2 × 2) [30]. Such coordination checks the accuracy of both the LEED analysis and of the basic assumptions of the total-energy calculations of the ground-state structure, namely the local-density approximation. This continued interaction of structure determination by experiment and by basic theory, especially for the simple cases of relaxation of clean metal surfaces, offers the prospect of critical tests of the local-density approximation and of any improved approximation. Such tests are of fundamental importance to condensed matter physics.

References

[1] Some details of the early history of electronic structure theory at IBM Research, including a section on LEED studies, appeared in the 25th anniversary issue of the IBM Journal of Research and Development by P.S. Bagus and A.R. Williams, IBM J. Res. Dev. 25 (1981) 793.

[2] The paper on the Al intensity measurements, F. Jona, IBM J. Res. Develop. 14 (1970) 444, has become a "Citation Classic".

[3] (a) P.M. Marcus and D.W. Jepsen, Phys. Rev. Lett. 20 (1971) 925;
(b) D.W. Jepsen, P.M. Marcus and F. Jona, Phys. Rev. Lett. 26 (1971) 1365;
(c) D.W. Jepsen, P.M. Marcus and F. Jona, Phys. Rev. B 5 (1972) 3933;
(d) D.W. Jepsen, P.M. Marcus and F. Jona, Phys. Rev. B 6 (1972) 3684;
(e) D.W. Jepsen, P.M. Marcus and F. Jona, Phys. Rev. B 8 (1973) 5523.

[4] K. Kambe, Z. Naturforsch. 22 a (1967) 322, 422; 23 a (1968) 1280.

[5] Relaxation of the first layer spacing was already detected in Al(110) (Ref. [3d]). The admirable early book Low Energy Electron Diffraction by J.B. Pendry (Academic Press, New York, 1974) leaves the impression that layer spacings of clean crystals expand (page 220).

[6] J.E. Demuth, P.M. Marcus and D.W. Jepsen, Phys. Rev. B 11 (1975) 1460.
Publication of the work on clean Ni was delayed by the striking success of a series of papers on chalcogenide overlayers on Ni surfaces that started with J.E. Demuth, D.W. Jepsen and P.M. Marcus, Phys. Rev. Lett. 31 (1973) 540. Of course the clean Ni analysis was done before the adsorbed structures.

[7] An overview paper on the work on Fe surfaces is that by J. Sokolov, F. Jona and P.M. Marcus, Solid State Commun. 49 (1984) 307.
The last paper in the series is by J. Sokolov, F. Jona and P.M. Marcus, Phys. Rev. B 31 (1985) 1929.

[8] An example is the growth of Cu on Pt(001), Y.S. Li, Quinn, H. Li, D. Tian, F. Jona and P.M. Marcus, Phys. Rev. B 44 (1991) 8261.

[9] See, for example, the work on Al(110), J.R. Noonan and H. Davis, Phys. Rev. B 29 (1984) 4349.

[10] Surface relaxation of fcc Fe is determined on 10-layer films grown on Cu(001), S.H. Lu, J. Quinn, D. Tian, F. Jona and P.M. Marcus, Surf. Sci. 209 (1989) 364.

[11] J. Quinn, Y.S. Li, D. Tian, H. Li, F. Jona and P.M. Marcus, Phys. Rev. B 42 (1990) 11348.

[12] H.L. Davis, J.B. Hannon, K.B. Ray and E.W. Plummer, Phys. Rev. Lett. 68 (1992) 2692.

[13] J.A. Walker, M.K. Debe and D.A. King, Surf. Sci. 104 (1981) 405.

[14] W. Moritz and D. Wolf, Surf. Sci. 163 (1985) L655.

[15] Y.-F. Liew and G.-C. Wang, Surf. Sic. 227 (1990) 190.

[16] Strain analysis in the cases of Cu films of 10 or more layers epitaxially grown on Pd(001) and Pt(001) shows that the film cannot be bcc Cu, even though the bcc Cu(001) structure is a closer match to the substrate; both cases are discussed in ref. [8].

[17] The conclusion that layer-by-layer growth did not occur in the early growth of Fe on a single crystal of Ag(001) was shown both by failure to find a fit to the data and by deterioration of the LEED pattern by deposit of up to five layer-equivalents of Fe; however, a 12-layer film could be fitted quantitatively in the paper by H. Li, Y.S. Li, J. Quinn, D. Tian, J. Sokolov, F. Jona and P.M. Marcus, Phys. Rev. B 42 (1990) 9195.

[18] D. Tian, H. Li, F. Jona and P.M. Marcus, Solid State Commun. 80 (1991) 783.

[19] H.L. Davis and J.F. Noonan, in: The Structure of Surfaces II, Eds. J.F. van der Veen and M.A. Van Hove (Springer, Berlin, 1988) p. 152.

[20] D. Sondericker, F. Jona and P.M. Marcus, Phys. Rev. B 33 (1986) 900.

[21] S.Y. Tong and M.A. Van Hove, Phys. Rev. 16 (1977) 1459.

[22] D.W. Jepsen, H.D. Shih, F. Jona and P.M. Marcus, Phys. Rev. 22 (1980) 814.

[23] J. Sokolov, F. Jona and P.M. Marcus, Phys. Rev. B 31 (1985) 1929.

[24] A compact formulation of the quasidynamic method which uses the exact interlayer matrix reflection equation, and solves that equation for Cu(001) by a simple iteration technique is given by R.F. Lin, P.M. Marcus and F. Jona, in: The Structure of Surfaces III, Eds. S.Y. Tong, M.A. Van Hove, K. Takayanagi and X.D. Xi (Springer, Berlin, 1991) p. 150.

[25] P.J. Rous, J.B. Pendry, D.K. Saldin, K. Heinz, K. Müller and N. Bickel, Phys. Rev. Lett. 57 (1986) 2951.

[26] P.J. Rous, M.A. Van Hove and G.A. Somorjai, Surf. Sci. 226 (1990) 15.

[27] K.M. Ho and K.P. Bohnen, Phys. Rev. B 32 (1985) 3446.

[28] J.S. Nelson and Peter J. Feibelman, Phys. Rev. Lett. 68 (1992) 2188. This paper shows good agreement with LEED results and contradicts its own assertion "that (LEED) $I-V$ analysis produces controversy more reliably than believable results". This statements is, in the writer's opinion, certainly wrong applied to modern work, which now has well-developed standards for reliable structure determination.

[29] K.M. Ho and K.P. Bohnen, Europhys. Lett. 4 (1987) 345.

[30] Rici Yu, H. Krakauer and D. Singh, Phys. Rev. B 45 (1992) 8671.

Surface Science 299/300 (1994) 454–468
North-Holland

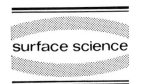

surface science

Strain effects on Si(001)

M.B. Webb

Department of Physics, University of Wisconsin–Madison, 1150 University Avenue, Madison, WI 53706, USA

Received 4 May 1993; accepted for publication 24 July 1993

There has been considerable progress in our understanding of the steps and their configuration on the Si(001) surface. Much of this has come from the realization of the role of the intrinsic surface stress of the reconstructed surface. LEED and STM experiments have given us information about the microscopic and statistical properties of the steps on this surface. First-principles theory has been able to calculate surface stresses and treat the elasticity problem. From these, statistical mechanical calculations can describe the surface step morphology. This paper is an overview of some of these developments.

1. Introduction

Surface stress of crystals has been discussed since the time of Gibbs [1]. There has since been consideration of the effects of mechanical stresses and strains on a large number of diverse surface phenomena. Some among these are: the accommodation of the natural misfit between a substrate and an adsorbed overlayer by introduction of heavy or light walls considered by Frank and van der Merwe [2]; the energy reduction by Novako–McTaque rotation of the overlayer [3], the influence of misfit in determining the overlayer growth mode and the resulting critical thickness of pseudomorphic layers and mechanisms of strain relief [4], the interaction between adsorbed atoms due to elastic deformations of the substrate [5], the interaction between steps on the surface due to the interaction of the strain fields they induce in the substrate [6]; and the reconstruction of surface atomic structures of both clean and over-layered surfaces [7].

It is not my present purpose to, nor could I, review this whole body of work. Rather, I hope to give a retrospective overview, much narrower in scope, of recent work which has led to an understanding of the effect of the intrinsic surface stresses on surface morphology. On surfaces with different reconstructed domains which have different intrinsic surface stresses, discontinuities in the stress at domain boundaries produce long range strain fields extending into the bulk. The strain energy in these fields can be minimized by an optimum configuration of the boundaries thus giving an effective interaction between the boundaries. The reconstructed Si(001) surface is such an example where the different intrinsic surface stress tensors of the 2×1 and 1×2 domains give stress discontinuities at the domain boundaries which in this case are monatomic height steps. The resulting step–step interaction is then an important ingredient in the statistical mechanics of the surface morphology.

In this paper we want to trace the development of this understanding. We begin by discussing a few examples of work on strain as an important ingredient of the physics of surface reconstruction since this work motivated what follows. Then we discuss experiments using externally applied strain which showed unexpected results for Si(001). These experiments led to theoretical work recognizing the role of the long range strains extending into the substrate. We outline this theory and additional experiments which verify its predictions. We then discuss two more contributions to the step energies, their formation energies and the short-range direct interaction between them. Using these ingredients, we then discuss the statistical mechanics of the step configurations on samples of intermediate vicinality. We end with a brief discussion of the evolution from single to double height steps with increasing miscut and of the very large scale step meandering on very flat surfaces. The overview will be from the perspective of our own group with the, at least implied, apologies to those whose perspectives may be quite different.

2. Strain and reconstruction

My earliest recollection of hearing about strain at surfaces is from some meeting in the mid-60's when Conyers Herring asked a speaker if he had considered strain in his treatment of a surface reconstruction. The question drew a blank from both the speaker and the audience. However, by about the mid 70's a number of people were considering surface stress as a driving force for surface reconstruction. Though the reconstruction is not explicitly a part of the main thrust of this paper, we mention a few of these early developments because they motivated both the theoretical efforts to calculate intrinsic surface stresses and the related experiments.

The (001) surface of Si had been extensively studied since Schlier and Farnsworth [8] had first observed the 2×1 reconstruction of this surface and a number of structural models had been proposed. While electronic structure calculations for a simple dimer model were consistent with the photoemission data, analyses of low-energy electron diffraction (LEED) data were not. In fact, very general properties of the diffraction, like relative intensities and energy widths of features in IV profiles, indicated that the reconstruction extended a number of atomic planes below the surface [9]. In 1978, Appelbaum and Hamann addressed this inconsistency [10]. They used an adaptation of the Keating model for bulk Si in which there are bond stretching and bond bending forces whose coefficients are fit using bulk properties. For the simple dimerized surface, they minimized the elastic energy and predicted appreciable displacements from bulk positions for atoms within the first five atomic layers and then showed that this model was reasonably consistent with the LEED data. While previously the substrate had been treated as rigid, this was perhaps the earliest realization of strain extending into the substrate.

Between 1984 and 1986, during the flurry of activity in the search for the structure of the Si(111)7×7 surface following the STM images by Binnig and Rohrer [11], Yamaguchi [12], Tromp [13], and others [14] followed Appelbaum and Hamann using empirical interatomic potentials to minimize the strain energy for various proposed models of the reconstruction.

The effects of strain on the surface reconstruction were also shown in a number of experiments using MBE grown thin films and alloys [15]. Thin semiconductor films are highly strained when grown epitaxially on lattice mismatched substrates and the strain depends on the film thickness and composition. While the reconstruction of the Si(111) surface is 7×7, the Ge(111) surface has a c(2×8) structure. The Ge lattice parameter is 4% larger than that of Si. Grossmann et al. used LEED, Rutherford backscattering, and glancing angle X-ray diffraction to study thin, MBE grown, layers of Ge on Si(111). For thicker films the strain is relieved by dislocations, but there remains a residual strain which decreases with film thickness. These authors observed that the surface of a pure Ge thin film had a 7×7 instead of its usual c(2×8) reconstruction for thicknesses up to those corresponding to a measured lateral compressive strain of somewhere between 0.22% and 0.36%. Thick Ge films reverted to the natural Ge c(2×8) surface. They showed that the 7×7 structure was very similar to the Si structure, that it was not due to islanding nor surface segregation of Si, and that it depended explicitly on the strain and not the thickness.

There were several other experiments. It was observed that Si–Ge alloys could exhibit a 5×5 reconstruction analogous to the Si 7×7 structure [16] and that Si films grown on a Si–Ge alloy substrate, so under tension, showed a 5×5 reconstruction [17].

While until 1987 the theoretical treatments of the surface stress had used empirical potentials, in that year both Needs [18], for metals, and Vanderbilt [19], for semiconductors, calculated intrinsic stress tensors from first principles.

The surface stress tensor is defined as

$$\sigma_{ij}^{surf} = \frac{1}{A_0} \frac{\partial E}{\partial \eta_{ij}}, \qquad (1)$$

where E is the surface energy per surface cell, η_{ij} is the surface strain tensor, i and j label directions in the surface, and A_0 is the surface cell area. σ can be anisotropic. A positive value of σ is a tensile stress indicating the surface would prefer a smaller lattice constant than the substrate; negative σ is compressive indicating the surface prefers a larger lattice constant.

Vanderbilt calculated total energies, forces, and stresses within the local density approximation and using norm-conserving pseudopotentials and Wigner exchange–correlation. He calculated stress tensors for various simple model structures of Si(111) which contained elements of the actual 7×7 dimer–adatom–staking fault (DAS) reconstruction. For a 1×1 surface he found a weak compressive stress much smaller than had been previously expected [20]. Also, faulting the surface reduced this compressive stress. For a 2×2 adatom structure he found a large tensile stress – the three surface atoms bound to the adatom move inward resulting in the tensile stress. A dimer–stacking fault (DS) cell was too large for the first principle calculations at that time, but Vanderbilt used a Keating model to show that dimers also contributed a tensile stress. From these results, he concluded that a compressive intrinsic stress was not the driving force for the reconstruction. He then went on to include the effect of an external strain such as those in the thin film and alloy experiments mentioned above. He considered 1×1 and $(2n + 1) \times (2n + 1)$ DS adatom-free surfaces and added the energy due to the external strain. His plot of the energy versus externally-applied isotropic strain is shown in Fig. 1. "Externally-applied compression reduces the (unfavorable strain interaction), in-

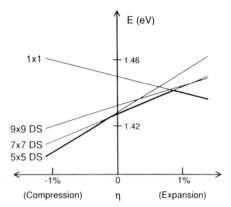

Fig. 1. Linearized surface formation energy per 1×1 cell as a function of isotropic external strain in the surface plane for adatom-free surfaces of Si(111).

creasing the relative stability of the 7×7 and 5×5 DAS structures."

3. External strain experiments

Motivated by the work summarized above, our group began to think about applying stress mechanically rather than utilizing alloys or thin films. In the earlier experiments each strain point was a new sample preparation and the strain was an average of a locally varying strain. It was clearly desirable to be able to apply an external, uniform, and continuously variable strain (more correctly you apply a stress, but we continue with the common usage) so that strain was an independent and continuous variable. With such a capability one would hope to be able to do a large number of interesting experiments. These might include: determining the temperature–strain phase diagram for reconstructed surfaces; lifting the degeneracy of orientationally equivalent domains and perhaps preparing single domain surfaces; observing the effect of strain on surface imperfections and steps and their distributions; observing the effect of strain on surface segregation in which differences in atomic sizes are important; and perhaps altering the modulated incommensurate phases of adsorbed overlayers by changing the natural misfit. Our first goal was to see if we could find the structural phase transformations that had been predicted by Vanderbilt.

There are several mechanical schemes one might consider to strain a sample in ultrahigh vacuum. The simplest, and the one we ended up using, is just to bend a cantilevered bar by loading the free end. This produces a uniaxial surface strain varying linearly along the length of the bar and, so there is a constant strain gradient. The concave surface is under compression. One could measure the surface properties as a function of strain either by observing at different places along the bar with a fixed deflection of the end or by looking at a fixed place as a function of loading. This allows separating effects of strain from those of a strain gradient. To look at the effect of strain without a gradient, one could load the center of a bar supported at both ends. For a surface like Si(111) one would rather have an isotropic strain, in which case one might squeeze a round wafer between two concentric rings of different diameter. This latter is more difficult to implement.

At room temperature Si samples fail in fracture at several tenths of a percent strain. At elevated temperatures there is a yield point beyond which the sample plastically deforms or creeps. Thus the accessible strains are small. While these available strains are in the range where changes in the reconstruction had been predicted and had been observed, the experiment seemed risky. One would like to be able to magnify the effect of the strain. For surfaces which undergo structural phase transitions with temperature, one might hope to measure the changes of the transition temperature since the free energy versus temperature curves of the different phases will move differently with strain and the temperature of their intersection may change rapidly. This turned out to be unnecessary.

Men et al. [21] implemented the simplest scheme. It consisted of a carousel mounted on a rotary feedthrough into which one end of each of twenty samples was rigidly clamped. The carousel is rotated so that the free end of a sample is brought into the gap between two wedge shaped Ta anvils. The anvil assembly is moved up or down with a precision linear feedthrough to bend the sample. The value of the strain is taken from the elasticity theory for the loaded beam, $\epsilon_{surf}(x) = 3zt(L-x)/2L^3$ where $\epsilon(x)$ is the surface strain at a distance x from the clamped end, L and t are the sample length and thickness, and z is the displacement of the loaded end. For our samples $z = 1.0$ mm corresponds to $\epsilon(0) = 0.1\%$. Considerable care is required both in the alignment and shaping of the anvils to make uniform contact and avoid twist and in the sample preparation to avoid any stress concentrations. With these precautions the experiment is surprisingly easy. Different samples break at different deflections, but usually a maximum strain of about 0.3% is achieved.

The first experiments were on Si(111) looking for the 7×7 to 5×5 transition predicted by Vanderbilt [19]. No such transition was observed for the available strain. This may have been because the isotropic component is only a fraction of the uniaxial strain.

Next, we looked at Si(001) with the hope of lifting the degeneracy in the surface energy of the 2×1 and 1×2 orthogonal domains. These experiments showed dramatic and unexpected changes in the relative populations of the two domains with very small strain. These observations led to many of the developments we want to describe in this paper.

Fig. 2 is a schematic of the Si(001) surface. Neighboring surface atoms dimerize giving a doubled periodicity in one direction, i.e. a 2×1 surface. On terraces separated by odd numbers of monatomic height steps the dimerization directions are orthogonal. Sufficiently flat Si(001) shows both 2×1 and 1×2 domains which give distinct half-order superlattice LEED reflections. At normal incidence, the relative integrated intensities of these reflections is a measure of the relative population of the two domains.

There are two types of monatomic height steps, named by Chadi [22] as S_A and S_B, which have different atomic configurations. S_A steps are parallel to the dimer rows on their upper terrace and have the lower step formation energy; S_B steps are perpendicular to their upper terrace dimer rows. There are two S_B step structures, rebonded and non-rebonded, with the rebonded one energetically favored. S_A and S_B steps alternate down the staircase on a vicinal sample. Nominally S_A steps are generally smooth while nominally S_B steps are rough. Kinks in a step are segments of the other type of step as we see in the STM image of Fig. 3. We will identify whole steps as "rough" or "smooth" and reserve S_A and S_B to denote the individual segments.

There are also double height steps, D_A and D_B, so their upper and lower terraces have the

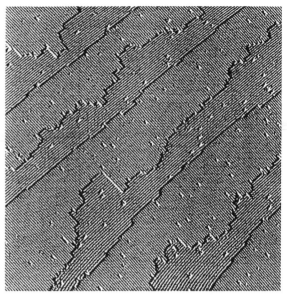

Fig. 3. A derivative mode image of Si(001). The surface is miscut by 0.5° toward the [110] direction. From left to right, areas that slope up are displayed as bright and areas that slope down are displayed as dark. The staircase goes down from the upper left to the lower right. The image is ~ 1000 Å across. The "smooth" steps are predominantly S_A; the average direction of the "rough" step is that of an S_B step. Kinks in a step are segments of the opposite type.

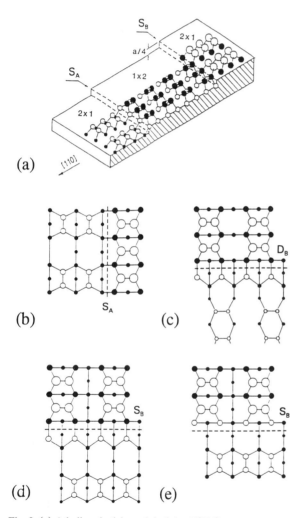

(a)

(b)   (c)

(d)   (e)

Fig. 2. (a) A ball and stick model of the Si(001) reconstructed surface. The orientation of the dimer bonds alternate on terraces separated by monatomic height steps. (b), (c), (d), (e) Plan views of the S_A, D_B, S_B rebonded and S_B non-rebonded steps on Si(001) respectively. The dashed lines indicate the step positions The level of the atoms is indicated by the size of the circles. The open circles are atoms with dangling bonds.

same symmetry. For vicinal surfaces miscut toward [110] by more than a few degrees, the surface is nearly a single domain surface with D_B steps. (D_A steps have a much higher, energy and do not appear on the equilibrium surfaces.) This change with miscut is called the "single to double step transition". The D_B steps have a lower energy than an S_A–S_B pair which raises the question of why there should be monatomic height steps at all for small miscut. The resolution of this question appears later.

Straining the sample at room temperature produced no change in the relative populations, which, as we shall see, is because of the slow kinetics. However, external strain at elevated temperatures produced the dramatic results shown in Fig. 4. This is a plot of the integrated intensities of the two half-order LEED reflections as functions of the applied strain for a nominally flat sample (terrace widths > 500 Å). The sum of the two intensities is unchanged as it

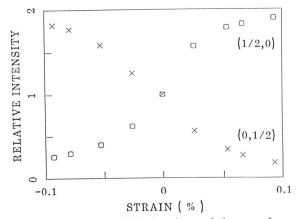

Fig. 4. The asymmetry of the domain populations as a function of strain. The relative intensity gives the relative domain population. The domain compressed along the dimer bond is favored.

should be since the intensities reflect the area occupied by each domain. In this case 90% of the surface is occupied by the majority domain for 0.1% strain. Reversing the strain reverses the sense of the population asymmetry. The sense of the effect is that the domain compressed along the dimer bond is favored. The population asymmetry is stable at room temperature if the sample is cooled under strain.

These results presented a considerable puzzle at the time and the underlying physics was unknown so a number of experiments were done to characterize the phenomenon [23,24]. One could speculate that the effect was some direct interaction between the strain gradient in the bent bar and the local dipole strain field associated with the steps. (A step would be expected to translate only in a strain gradient since in a uniform strain its energy would be independent of position.) By measuring the population asymmetry as a function of position along the bar for different deflection of the end, it was demonstrated that the effect was due to the strain and not the strain gradient. Also it was demonstrated that the effect depended on the direction of the strain relative to the dimer bond direction but was independent of the direction relative to the steps.

While the kinetics of developing the population asymmetry are temperature dependent, the measured steady state asymmetry as a function of strain was found to be temperature independent. This suggested that it is the mechanical rather than the entropic term in the free energy which is important.

The question arose of what sort of rearrangement of the steps was responsible for the population asymmetry. The particular LEED apparatus had insufficient resolution to characterize the step configurations when the terrace widths were the order of 500 Å so a device was built to strain samples and observe them with a scanning tunneling microscope [25]. We discuss extensive STM measurements later, but in these early experiments it was established that the striped phase and the number of steps were preserved upon straining. The existing steps just move with respect to each other without crossing or entangling. This is shown in Fig. 5.

The final of these exploratory experiments was to measure the kinetics of establishing the steady state domain populations after applying or removing the applied strain. The time dependence of the superlattice intensity follows a simple exponential relaxation to the new steady state value. It was very surprising at the time that the measured time constants were identical for both the development of the population asymmetry after applying strain and for the relaxation of the asymmetry after removing the applied strain. This was unexpected because, at that time, one thought the effect might be due to the external strain lifting the degeneracy of the surface energy of the two domains. If that were true, upon removing the strain there would no longer be a driving force for equalizing the populations except an entropic one, and the two time constants would be different.

At that time some aspects of the phenomenology seemed clear. Somehow the domain boundaries, in this case the steps, arranged themselves to minimize a mechanical energy. This optimum arrangement gave equal domain populations in the absence of external strain but, under external strain, it gave unequal populations which depended on the orientation of the domains relative to the strain. However, the underlying physics was not known. In a hallway conversation with David Vanderbilt at the 1988 March meeting of

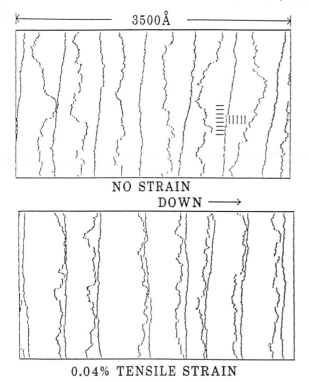

3500Å

NO STRAIN

DOWN ⟶

0.04% TENSILE STRAIN

⟨l⟩ = 270Å

80% in majority domain

Fig. 5. STM images of a Si(001) surface before and after application of strain. The images were processed so that only the steps are visible. The sample is miscut by 0.3°, and the average terrace width is 270 Å. In the upper panel the domain populations are equal to within 1%. The bottom panel is the same surface after a tensile strain of 0.04% and 80% of the surface is in the favored domain.

the APS we puzzled about these results, and he wondered if it could involve some distortions of the substrate. Later that year, Alerhand, Vanderbilt, Meade and Joannopolous, (AVMJ), published a letter "Spontaneous Formation of Stress Domains on Crystal Surfaces" [26]. Unfortunately, we had all been unaware of an earlier publication of similar results by Marchenko [6].

4. Alerhand, Vanderbilt, Meade, Joannopolous theory

AVMJ considered surfaces with orientationally inequivalent domains with anisotropic intrinsic surface stress tensors. Si(001) is an example. They consider a striped phase with alternating 2×1 and 1×2 domains of widths $(1 + p)l$ and $(1 - p)l$ so l is the average terrace width and p is the fractional displacement of a step from the midpoint between its neighbors. Then at the domain boundaries, steps in this case, there is a discontinuity in the surface stress and thus a delta function of force density, $f_i(\rho) = (d/d\rho_j)\sigma_{ij}$. The strain field in response to this force density is written in terms of the elastic Green's function. The elastic relaxation energy per unit area is then the integral of the force density times the displacement,

$$E_{el} = -\frac{1}{2L^2} \int\int d^2\rho \; d^2\rho' \chi_{ij} f_i(\rho) f_j(\rho'). \quad (2)$$

Here χ is the Green's function and the ρ's are vectors in the surface. In this case the force is $F_0 = \pm(\sigma_\parallel - \sigma_\perp)$ where the σ's are the components of the intrinsic stress tensor parallel and perpendicular to the dimer bond direction. One can think of alternating force monopoles at the steps.

Including the energy associated with an applied external strain, ϵ^{ext}, the energy per unit length of step is evaluated to be

$$E(p) = \lambda_{S_A} + \lambda_{S_B} + 2aplF_0\epsilon^{ext}$$
$$- 2\lambda_\sigma \ln\left[\frac{l}{\pi a}\cos\left(\frac{\pi p}{2}\right)\right], \quad (3)$$

where $\lambda_\sigma = F_0^2(1-\nu)/2\pi\mu$, μ and ν are the bulk modulus and Poisson ratio, and a is microscopic cut-off taken to be a lattice constant. λ_{S_A} and λ_{S_B} are the creation energies per unit length due to the local atomic and electronic structure but not including the long range strain field interaction between steps.

If the energy is minimized with respect to p holding l fixed, as suggested by the STM observations, the optimum p is given by

$$p_0 = \frac{2}{\pi}\tan^{-1}\left(\frac{2alF_0\epsilon^{ext}}{\pi\lambda_0}\right). \quad (4)$$

AVMJ also pointed out that, for zero applied strain and thus zero p, the energy becomes negative for a large enough terrace width, l_0, and so a flat surface should be unstable against the spontaneous formation of a striped phase with alternating 2×1 and 1×2 domains separated by monatomic height steps. The alternating domain walls reduce the elastic energy, and for a large enough l this more than offsets the step creation energy. This then is the answer to the question posed earlier of why surfaces with low miscut angles prefer to have alternating 2×1 and 1×2 domains with pairs of monatomic height steps rather than a single domain with only double height D_B steps of lower energy. This is in a way analogous to the formation of magnetic domains in a ferromagnet to reduce the energy in the external field. We return to this predicted instability later. It is this instability that had been noticed by Marchenko earlier [6].

AVMJ then calculated the intrinsic surface stress tensor for Si(001) using a semi-empirical tight-binding theory. Their results were $\sigma_{\parallel} = 0.035$ eV/Å^2, $\sigma_{\perp} = -0.035$ eV/Å^2 so that $F_0 = 0.07$ eV/Å^2. (Others have calculated somewhat different values [27,28].)

AVMJ predicted that, with an applied strain, there would be unequal domain populations as seen in the LEED experiments. Of the quantities appearing in Eq. (4), p_0 and ϵ^{ext} are measured in the LEED experiments and l is measured from the STM images. The only remaining quantity is F_0. Subsequent measurements of p_0 done on a series of vicinal samples yielded the correct functional dependence of p on l and on ϵ^{ext} and a value of F_0 of 0.07 ± 0.01 eV/Å^2. This is certainly fortuitous agreement with the AVMJ prediction considering the approximations in the theory and experiments. Actually the theory was done for an isotropic elastic continuum, and AVMJ chose a value of $(1 - \mu)/\pi\nu$ of 0.6 Å^3/eV to approximate Si as an isotropic solid. Later, from fitting data for a number of properties, we revised the values of F_0 and this elastic constant ratio. However, all the qualitative behavior and this semi-quantitative agreement leave little doubt that the explanation involving the long range strain fields extending into the substrate has captured the primary physics of the phenomenon. We see that a step can be thought of as moving in

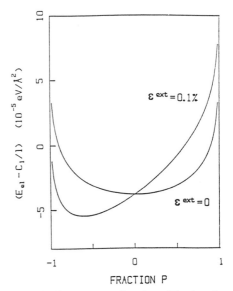

Fig. 6. The elastic energy as a function of the domain population asymmetry parameter, p. The symmetric curve is for zero applied strain and the asymmetric curve is for a strain of 0.1%.

a potential well between its neighbors an example of which is shown in Fig. 6.

5. Direct step interactions

There are at least two more ingredients we need to be able to understand the surface morphology. One is the direct and rather short-range step–step interaction and a second is the step and kink formation energies. We begin with the former.

Earlier LEED and STM observations showed that for vicinal surfaces with miscuts between 0.5° and 1°, i.e. 75 Å $< l <$ 150 Å, there are unequal domain populations even in the absence of external strain [29]. These are terraces generally narrower than those discussed above but still wide enough that there is nearly negligible occurrence of double steps. Marchenko [6] had earlier pointed out that force dipoles would occur at steps, and Yip et al. [28], using Stillinger–Weber interatomic potentials, had demonstrated that there were appreciable dipoles for the rebonded S_B and D_B steps on Si(001). Chadi had discussed the local strain around the rebonded steps [14]. Pelke and Tersoff [30] gave an approximate analytical expression for the energy per unit step length for straight steps as a function of the displacement of an S_B step from the midpoint between its S_A neighbors. This adds a term $-\sqrt{3\lambda_\sigma\lambda_d}\,(a/l)\tan(\pi p/2)$ to the energy of Eq. (3). The energy is asymmetric and the sense of the asymmetry is that the S_B step is attracted to down-hill neighbor. The energy minimum now occurs for an optimum p_0 given by

$$p_0 = \frac{1}{\pi}\sin^{-1}\left(\frac{a}{l}\sqrt{\frac{3\lambda_d}{\lambda_\sigma}}\right). \qquad (5)$$

An example of the energy as a function of p including the direct step–step interaction is shown in Fig. 7. We will see below that the coefficient of this dipole force term can be estimated from STM data taken for these intermediate terrace widths.

6. Step and kink creation energies

In 1987, Chadi [14] discussed the structure and energetics of the steps on the Si(001) surface using semi-empirical, tight-binding total energy calculations. A number of his predictions have since been confirmed by STM experiments: the D_A step has a very high energy and indeed all observed double steps are S_B; of the two atomic

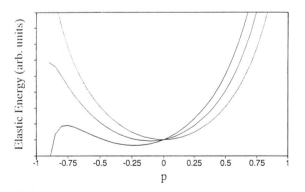

Fig. 7. The elastic energy, including the direct step–step interaction versus p for different values of this interaction. As the direct interaction increases, the potential becomes more asymmetric and its minimum moves to the left.

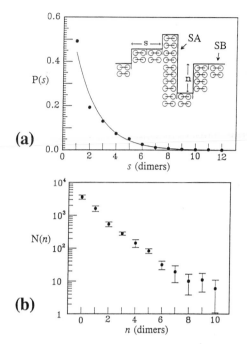

Fig. 8. Measured kink separation and kink length distributions, $P(s)$ and $P(n)$. The inset defines the kink separation, s and the kink length, n. The solid line is the $P(s)$ expected for independent kinks.

configurations for the S_B step, rebonded and nonrebonded, the latter has the lowest energy despite having considerable strain associated with it; and the energy of the D_B step formation energy is less than the sum of the S_A and S_B energies. His predicted step creation energies were $\lambda_{S_A} = 0.02$ eV/$2a$, $\lambda_{S_B} = 0.3$ eV/$2a$, and $\lambda_{D_B} = 0.05$ eV/$2a$, or a ratio of $\lambda_{S_B}/\lambda_{S_A}$ of about 15. This then qualitatively explains the observation mentioned in the introduction that nominally S_A steps are smooth and S_B are rough. The kinks in the rough step are S_A segments which are energetically inexpensive.

An experimental indication of the relative S_A and S_B energies came from the observation of the equilibrium shape of one atom high islands [31]. The observed island aspect ratio is about 3 and so smaller than that expected from Chadi's values. More detailed experimental information comes from the STM measurements [32] of the statistical distributions of kink separations and kink lengths as shown for the rough step in Fig. 8.

These data are for a surface miscut by 0.3° and so for a rather wide average terrace width of 250 Å. Because of the dimer structure, both the separations and lengths come in units of $2a$. The simplest interpretation of the kink separation distribution, $P(s)$, is that the kinks are statistically independent. This would give $P(s) = P_k(1 - P_k)^{s-1}$ where P_k is the probability of a kink at a potential kink site. This assumption is supported by the plot of this function shown as the solid line in Fig. 8a, where P_k is taken as the measured fraction of kinked sites and there is no adjustable parameter. It is also true that there are no observed correlations between the kink lengths or directions. Under this assumption, the number of kinks of length n is given by $N(n) \propto \exp(-E(n)/kt)$ where $E(n)$ is the energy of a kink of length n. $N(n)$ is well fit by this function with $E(n) = n\lambda_{S_A} + \lambda_c$. Thus λ_{S_A} is the S_A step formation energy per length $2a$ and λ_c is an additional energy, common to kinks of all lengths, most simply interpreted as a corner energy. From the fits to the data, $2a\lambda_{S_A}/kT = 0.74 \pm 0.06$ and $\lambda_c = 1.1 \pm 0.3$. Similar analysis for the smooth step gives $2a\lambda_{S_B}/kT = 2.0 \pm 0.2$. The ratio of the S_B to S_A step energies is again about 3 consistent with the island shapes.

In these experiments the samples were annealed and quenched and so the step configuration should approximate the equilibrium step at some "freeze out" temperature which we need to know in order to extract the λ values. From additional experiments described below, we estimate the "freeze out" temperature appropriate for the kink distributions is about 625 K, which gives $\lambda_{S_A} = 0.04$ eV$/2a$, $\lambda_{S_B} = 0.11$ eV$/2a$, and $\lambda_c = 0.06$ eV$/2a$.

For these wide terraces it is the Boltzmann factor which limits the occurrence of long kinks rather than the confinement of a step between its neighbors or the long-range strain fields. Indeed the same experiments on samples with narrower terraces show the suppression of long kinks due to this confinement.

It is striking that the step creation energies are only of the order of tenths of an eV compared to bond energies of several eV. This is, of course, because of the atomic configuration of the steps

on the reconstructed surface which introduces no additional broken bonds. This is very different from simple kink–ledge–terrace models where energies are just due to the coordination number, and here, for Si(001), one can expect that subtle effects can influence the surface morphology.

It is also noteworthy that the corner energy is comparable to the step creation energy per dimer length so that the energy of short excitations is increased appreciably. This makes the free energy of the steps fall much more slowly with temperature than a simple Ising-like model would predict.

7. Surface morphology

With the above ingredients, one would hope to describe the surface morphology at least over ranges where the meandering of the smooth steps is small and for average terrace widths large enough that the double step contribution is negligible. Combining the above contributions, the Hamiltonian describing a rough step meandering between two straight neighbor is:

$$
\begin{aligned}
H = \sum_i \Big\{ & \lambda_{S_A} + \lambda_{S_B} + \lambda_{S_A} |h_i - h_{i+1}| \\
& + \lambda_c(1 - \delta_{h_i, h_{i+1}}) - 2\lambda_\sigma \ln\left[\frac{l}{\pi a} \cos\left(\frac{\pi h_i}{2l} \right) \right] \\
& + \lambda_d \left(\frac{a}{2l} \right)^2 - \sqrt{3\lambda_\sigma \lambda_d}\, \frac{a}{l} \tan\left(\frac{\pi h_i}{2l} \right) \Big\},
\end{aligned} \tag{6}
$$

where h_i is the displacement of the end of the ith dimer row from the midpoint between neighboring steps. The first and second terms are the step creation energies, the third and fourth terms are the kink energy including the corner energy, the fifth term is the AVMJ energy, and the last two terms are the direct step–step interaction energy. This Hamiltonian neglects fluctuations in the smooth steps and any changes in the AVMJ elastic energy for non-straight walls.

With this model Hamiltonian one can use the transfer matrix formalism to calculate the statistical mechanical properties of the equilibrium step configurations and by fitting extract the parame-

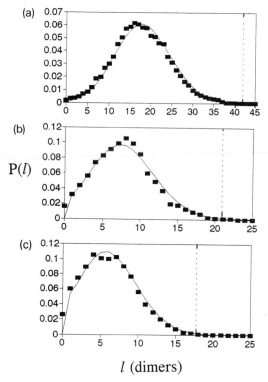

Fig. 9. Terrace width distributions for intermediate miscut angles: (a) 0.5°, (b) 0.95°, (c) 1.15°. The solid lines are fits to the data calculated with the transfer matrix using the Hamiltonian of Eq. (6). All three distributions have been fit with a single set of parameters. The vertical dashed line is the measured average S_A–S_A separation.

distribution of segments of the step of length ξ whose average positions populate the AVMJ potential. Then the distribution is

$$P(p) \propto \exp\left(-\frac{\xi E(p)}{kT}\right) \propto \left[\cos\left(\frac{\pi p}{2}\right)\right]^{K}, \qquad (7)$$

where $K = 2\lambda_\sigma \xi / kT$. ξ is found to scale with l and is about $l/2$. A plot of Eq. (7) is essentially indistinguishable from the solid line in Fig. 9a.

This simple result is to be compared to

$$P(p) \propto \left[\cos\left(\frac{\pi p}{2}\right)\right]^{2}, \qquad (8)$$

which would be expected for steps meandering freely between their neighbors. In this case the fitted K is 6.1. Thus the terrace width distribution is considerably narrower than it would for steps moving freely between neighbors and this is the consequence of the AVMJ potential due to the long range strain fields. This simple interpretation can be verified by course graining the transfer matrix [33,34].

For the narrower terrace widths, the distribution is asymmetric due to the direct step–step interactions. This allows a measure of λ_d. The best fits give $\lambda_d / \lambda_\sigma = 125$.

8. Single to double step evolution

There are at least two aspects of the step configurations which have not been included in the above discussion. One of these is the evolution from single- to double-stepped surfaces as the miscut angle increases. As we have seen for wide terraces, the surface is a striped phase of 2×1 and 1×2 domains separated by alternating S_A and S_B steps, however, for miscuts larger than several degrees the surface is a single domain containing double steps. There have been a number of diffraction experiments measuring the relative domain populations which show this [35]. This is also illustrated in STM images for various miscut angles in Fig. 10. Fig. 11 shows minority domain terrace width distributions measured from such images. The value of $P(l)$ at $l = 0$ indicates the fraction of double steps, S_B.

ters. All told, the data consists of the domain population asymmetries with external strain, the kink separation and length distributions, and the terrace width distributions, where each of these is measured for a range of vicinalities or average terrace widths. One example of the fitting is shown in Fig. 9, where terrace width distributions are shown for different miscuts. The solid lines are the fits from the transfer matrix calculations all using a single set of parameters. Similarly good fits are obtained for other properties [33]. The best overall fits give $\lambda_\sigma = 5 \times 10^{-3}$ eV/2a corresponding to $F_0 = 0.035$ eV/Å^2.

For the wide terraces, Fig. 9a, the terrace width distribution is essentially symmetric between the neighboring steps, i.e. the short-range direct interactions are still unimportant. This distribution can be though of as a simple Boltzmann

As discussed above, AVMJ showed that a flat single domain surface is unstable against alternating domain formation for large enough l or small enough mistcut. The same would be true for a vicinal, single domain surface containing D_B steps only the energy opposing the strain relief is the dissociation of D_B steps into an S_A–S_B pair. This mechanical, and so $T = 0$ K, calculation predicted that the single domain surface would be stable to much smaller miscut angles than observed experimentally [36]. It was then realized that, because of the small excitation energy for kinks in the rough step, the S_B *free* energy falls rapidly with temperature due to the entropy term. Taking this into account, the single to double step transition would move to higher miscut angle with increasing temperature [37,28]. Bartelt et al. [38] suggested from similar considerations that there should be a region of coexistence of spatially separated single and double stepped phases. However, all these treatments lead to first order transitions with either an abrupt transition between or coexistence of single and double stepped

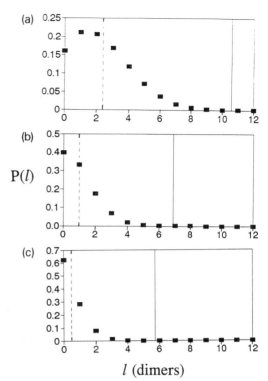

Fig. 11. Measured 1×2 terrace width distributions for large miscut angles. (a) 2.0°, (b) 3.0°, (c) 3.5°. The value of the average S_A–S_A separation is shown as the vertical line on the right, and the average 1×2 terrace width is displayed as the vertical line on the left.

phases contrary to the experimental observations of a smooth evolution.

Pehlke and Tersoff [39] added two elements to these considerations. First, rather than restricting the structure to purely single and double step phases as done previously, they show that, because of the direct step–step interaction, phases consisting of sequences of single and double height steps have lower energy than the separated phases. The particular sequences which give minimum energy depend on the miscut and these form a "devil's staircase". Secondly they point out that a D_B step can be treated as a bound pair of single steps with the addition of a contact interaction to give the correct step energy. At finite temperature the double step may dissociate locally with the S_B member then free to meander.

With these additional ingredients, these authors treat a model consisting of structures of up

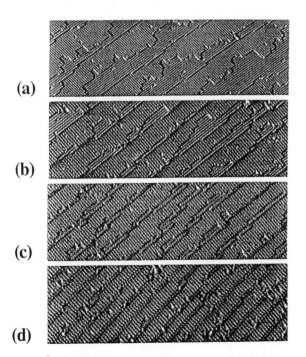

Fig. 10. STM images of the evolution from single- to double-height steps with increasing miscut angle. (a) 1.0°, (b) 2.0°, (c) 3.0°, (d) 5.25°.

to two pairs of steps and again use the transfer matrix to determine the free energy and construct a temperature-miscut phase diagram. They suggest a second-order transition with a critical temperature of about 500 K. The experiments observe a surface characteristic of some "freeze-out" temperature which is likely higher than this T_c so only a continuous change with miscut would be expected as is observed. Whether the details of this theory and model are correct must be determined by further experiments. However, and perhaps more important, the new ingredients of intermixed single and double step structures and of the dissociation and recombination of single step pairs are certainly consistent with observations. This, at least qualitatively, seems to resolve the earlier questions about the single- to double-step evolution.

9. Strain for meandering steps

While the Hamiltonian in Eq. (6) gives a reasonably consistent description of the step morphology for the range of vicinalities studied, there are worrisome aspects. In particular, the strain energy terms are derived considering only straight parallel steps and so are treated as a one-dimensional problem while the problem is basically two-dimensional. Aumann et al. [40] using LEED to study the domain populations found that the above Hamiltonian was not adequate for large miscut samples at high temperatures which they attributed to the failure to account for the meandering of the S_A steps.

Men et al. [41] did very preliminary numerical evaluations of energy for a single sine wave modulated step and saw that the elastic energy actually fell with increasing modulation amplitude. Clearly at large amplitudes the modulation results in alternating 2×1 and 1×2 terraces in the direction perpendicular to the original step.

Related to this question of meandering steps, but on a much larger length scale, beautiful experiments were done by Tromp and Reuter [42] who used low-energy electron microscopy (LEEM) to observe samples cut very close to the (001) direction. The average terrace widths were

1000 Å or more. Their images showed that the surface had large regions of narrowly spaced (~ 300 Å) and rather straight single-height steps coexisting with regions widely spaced (1000–2000 Å) wavy steps.

Tersoff and Pehlke [43] showed that, for large l, straight steps are unstable against long wave length modulations and this was due to the long-range strain fields. They considered the elastic energy of equally spaced sinusoidal steps. They numerically calculated the elastic relaxation energy as a function of the amplitude, A, and wavelength, λ, of the modulation. For $l = 1000$ Å a minimum energy occurs for $\lambda \approx 0.3A$, and the energy per surface area is about 3 or 4 meV/Å^2 below that for straight steps. As a function of l, the modulated steps have energy lower than the straight steps for l greater than about 200 Å. The optimum amplitude scales as about $0.8l$. For large amplitude modulations, the steeply ascending segments of the adjacent steps are closer together than l and can be like the critical step separation, l_0 for the instability of a flat surface discussed by AVMJ. The strain energy is thus reduced by the modulation rather than by the introduction of extra steps alternating up and down as envisioned by AVMJ.

The Tersoff and Pehlke theory then explains the major features of the Tromp and Reuter observations and illustrates the importance of accounting for the step meandering. It seems that similar consideration of the elastic energies for crooked walls on a much shorter length scale is important. This problem is difficult however. The kinks in steps occur on the scale of a few dimer spacing and have appreciable corner energies likely associated with local strains. One probably needs to go beyond elastic continuum theory and use either first principles total energy calculations or empirical potentials for more complex step configurations.

10. Discussion

In the last ten years there has been a major advance in our understanding of surface morphology in general. Here we have traced some of these advances for the Si(001) surface with emphasis on the role of surface strain. The advances have come experimentally from the ability of the STM to image surface steps surface with atomic resolution and measure their statistical properties and from the use of externally applied strain as an independent variable. Theoretically, there is now the ability to calculate intrinsic stresses from first-principles total energy calculations and from empirical potentials. These then have allowed solutions of the elasticity problem and of the statistical mechanics of the surface steps. From this there is now a considerably more complete understanding of the surface morphology. A number of questions, like the single- to double-step transition with increasing miscut, have been answered, at least qualitatively. This work also led to the surprising realization that a completely flat surface can be unstable against spontaneous domain and step formation.

Earlier discussions of the interactions between steps on surfaces were limited to consideration of the direct short-range interactions due to their detailed structure and local strains, to their effective entropic repulsion which increases their configurational entropy, and, in some cases, effective kinetic interactions due, for example, to unequal growth rates at inequivalent steps. Now, the work described above has identified an additional contribution to the interaction in cases where different surface domains have different intrinsic surface stress tensors. Discontinuities in the surface stress at domain boundaries produce strain fields extending into the bulk and the elastic energy can be relaxed by an optimum boundary configuration. This interaction is then an important ingredient in the statistical mechanics of the surface morphology for Si(001) and is the dominant factor for low vicinality surfaces.

While our discussion has been about Si(001), the same phenomena are certainly important for other systems. Narasimhan and Vanderbilt [44] have suggested that the herringbone reconstruction of the Au(111) surface results from the spontaneous formation of stress domains. Vanderbilt and Wickham [45] and Tersoff and Tromp [46] have discussed the stability and shape of coherent islands growing on a lattice mismatched substrate

due to elastic relaxation. Hornis et al. [47] have observed that the lowest energy configuration of the Pd(110) surface consists of semi-ordered up and down steps presumably due to the elastic step–step energy. Interestingly, Vanderbilt [48] has suggested that there is an electrostatic analog to the spontaneous domain formation by surface stress driven by difference in the work functions of the domains which may explain the stripped phase of alternating Cu 1×1 and C-2×1-O domains [49]. I expect that we will see many more examples.

Finally, the experiments with externally applied strain turned out to be surprisingly simple though the achievable strain was limited. With fine probes like the STM or the very narrow incident beams used in high resolution LEED and photoemission microscopy, one might do similar experiments on whisker surfaces where much larger surface strains can be had. Then a number of interesting experiments, like those mentioned in the introduction, could be done. There are also other closely related experiments: Martinez et al. [50], Tromp [51], and Sander and Ibach [52] have all measured changes is surface stress upon adsorption by observing the deflection of thin samples. I also expect that we will see many more examples where experimentalists will have strain as another knob to turn.

Acknowledgements

I would like to acknowledge many contributions by J. Men, B.S. Swartzentruber, N. Kitamur, R. Kariotis, and M.G. Lagally. This work has been supported under grant number NSF No. DMR-9104437.

References

[1] J. W. Gibbs, Collected Works, Vol. 1 (Longmans, London, 1928).
[2] E.C. Franck and J.H. van der Merwe, Proc. Roy. Soc. 198 (1949) 205.
[3] A.D. Novaco and J.P. McTague, Phys. Rev. B 20 (1979) 2469.
[4] J.H. van der Merwe, Crit. Rev. Solid State Mater. Sci. 17

(1991) 187;
R. Hull and J.C. Bean, Crit. Rev. Solid State Mater. Sci. 17 (1992) 507.

[5] K.H. Lau and W. Kohn, Surf. Sci. 65 (1977) 607.

[6] I. Marchenko and A. Ya. Parshin, Sov. Phys. JETP 521 (1980) 129;
I. Marchenko, Sov. Phys. JETP 33 (1981) 381.

[7] For early examples concerning semiconductors see: J.C. Phillips, Phys. Rev. Lett. 45 (1980) 905;
E.G. McRae, Surf. Sci. 163 (1985) L766.

[8] R.E. Schlier and H.E. Farnsworth, J. Chem. Phys. 30 (1959) 917.

[9] T.D. Poppendieck, T.C. Nogc and M.B. Webb, Surf. Sci. 75 (1978) 287.

[10] J.A. Appelbaum and D.R. Hamann, Surf. Sci. 74 (1978) 21.

[11] G. Binnig, H. Roher, Ch. Gerber and E. Weibel, Phys. Rev. Lett. 50 (1983) 120.

[12] T. Yamaguchi, Phys. Rev. B 30 (1984) 1992.

[13] R.M. Tromp, Surf. Sci. 155 (1985) 432.

[14] R.M. Himpsel, Phys. Rev. B 27 (1983) 7782;
Guo-Xin Qian and D.J. Chadi, J. Vac. Sci. Technol. B 4 (1986) 1079.

[15] T. Ichikawa and S. Ino, Surf. Sci. 136 (1984) 267;
J.C. Bean, L.C. Feldman, A.T. Fiory, S. Nalahara and I.K. Robinson, J. Vac. Sci. Technol. A 2 (1984) 426;
H.J. Grossmann, J.C. Bean, L.C. Feldman, E.G. McRae and I.K. Robinson, Phys. Rev. Lett. 55 (1985) 1106; J Vac. Sci. Technol. A 3 (1985) 1633.

[16] H.J. Grossmann, J.C. Bean, L.C. Feldman and W.M. Gibson, Surf. Sci. 55 (1984) L175.

[17] A. Ourmazd et al., Phys. Rev. Lett. 57 (1986) 1332.

[18] R.J. Needs, Phys. Rev. Lett., 58 (1987) 53.

[19] D. Vanderbilt, Phys. Rev. Lett. 59 (1987) 1456.

[20] E. Pearson, T. Takai, T. Halicioglu and W.A. Tiller, J. Cryst. Growth 70 (1984) 33.

[21] F.-K. Men, W.E. Packard and M.B. Webb, Phys. Rev. Lett. 61 (1988) 1469.

[22] D.J. Chadi, Phys. Rev. Lett. 59 (1987) 1691.

[23] M.B. Webb, F.-K. Men, B.S. Swartzentruber, R. Kariotis and M.G. Lagally, Surf. Sci. 242 (1991) 23.

[24] M.B. Webb, F.K. Men, B.S. Swartzentruber, R. Kariotis and M.G. Lagally, in: Kinetics of Ordering and Growth at Surfaces edited by M.G. Lagally (Plenum, New York, 1990).

[25] B.S. Swartzentruber, Y.W. Mo, M.B. Webb and M.G. Lagally, J. Vac. Sci. Technol. A 7 (1989) 2901.

[26] O.L. Alerhand, D. Vanderbilt, R.D. Meade and J.D. Joannopolous, Phys. Rev. Lett. 61 (1988) 1973.

[27] M.C. Payne, N. Roberts, R.J. Needs, M. Needels and J.D. Joannopoulos, Surf. Sci. 211/212 (1989) 1.

[28] T.W. Poon, S. Yip, P.S. Ho and F.F. Abraham, Phys. Rev. Lett. 65 (1990) 2161.

[29] X. Tong and P.A. Bennett, Phys. Rev. Lett. 67 (1991) 259;
E. Schroeder-Bergen and W. Ranke, Surf. Sci. 259 (1991) 323.

[30] E. Pehlke and J. Tersoff, Phys. Rev. Lett. 67 (1991) 465.

[31] Y.-W. Mo, B.S. Swartzentruber, R. Kariotis, M.B. Webb and M.G. Lagally, Phys. Rev. Lett. 63 (1989) 2393.

[32] B.S. Swartzentruber, Y.-W. Mo, R. Kariotis, M.G. Lagally and M.B. Webb, Phys. Rev. Lett. 65 (1990) 1913.

[33] B.S. Swartzentruber, N. Kitamura, M.G. Lagally and M.B. Webb, Phys. Rev. B 47 (1993) 13432.

[34] R. Kariotis and M.G. Lagally, Surf. Sci. 248 (1991) 295.

[35] Ref. [29]. Also for STM observations see: P.E. Wierenga, J.A. Kubby and J.E. Griffith, Phys. Rev. Lett. 59 (1987) 2169;
J.E. Griffith and G.P. Kochanski, Solid State Mater. Sci. 16 (1990) 255.

[36] D. Vanderbilt, O.L. Alerhand, R.D. Meade and J.D. Joannopolous, J. Vac. Sci. Technol. B 7 (1989) 1013.

[37] O.L. Alerhand, A.N. Berker, J.D. Joannopolous, D. Vanderbilt, R.J. Hamers and J.E. Demuth, Phys. Rev. Lett. 64 (1990) 962.

[38] N.C. Bartelt, T.L. Einstein and C. Rottman, Phys. Rev. Lett. 66 (1991) 961.

[39] E. Pehlke and J. Tersoff, Phys. Rev. Lett. 67 (1991) 1290; and Ref. [30].

[40] C.E. Aumann, J. De Miguel, R. Kariotis and M.G. Lagally, Phys. Rev. B 46 (1992) 10257.

[41] M.B. Webb, F.-K. Men, B.S. Swartzentruber and R. Kariotis, Bull. Am. Phys. Soc. 36 (1991) 910.

[42] R.M. Tromp and M.C. Reuter, Phys. Rev. Lett. 68 (1992) 820.

[43] J. Tersoff and E. Pehlke, Phys. Rev. Lett. 68 (1992) 816.

[44] S. Narasimhan and D. Vanderbilt, Phys. Rev. Lett. 69 (1992) 1564.

[45] D. Vanderbilt and L.K. Wickham, Mater. Res. Soc. Symp. Proc. 202 (1991) 555.

[46] J. Tersoff and R.M. Tromp, Phys. Rev. Lett. 70 (1993) 2782.

[47] H. Hornis, J.R. West, E.H. Conrad and R. Ellialtioglu, Phys. Rev. B, to be published.

[48] D. Vanderbilt, Surf. Sci. Lett. 268 (1992) L300.

[49] K. Kern et al., Phys. Rev. Lett. 67 (1991) 855.

[50] R.E. Martinez, W.M. Augustyniak and J.A. Golovchenko, Phys. Rev. Lett. 64 (1990) 1035.

[51] A.J. Schell-Sorokin and R.M. Tromp, Phys. Rev. Lett. 64 (1990) 1039.

[52] D. Sander and H. Ibach, Phys. Rev. B 43 (1991) 4263.

Surface Science 299/300 (1994) 469–486
North-Holland

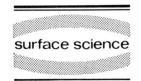

surface science

Thirty years of atomic and electronic structure determination of surfaces of tetrahedrally coordinated compound semiconductors

A. Kahn

Department of Electrical Engineering, Princeton University, Princeton, NJ 08540, USA

Received 20 April 1993; accepted for publication 13 May 1993

Surface Science has evolved considerably since the early 1960's. Experimental and theoretical techniques pioneered twenty to thirty years ago have been developed into sophisticated tools which today allow systematic and almost routine determinations of atomic geometries, electronic structures, chemical compositions and surface energies of semiconductor surfaces. In this article, I review the part of the history of this work devoted to tetrahedrally coordinated compound semiconductor surfaces.

1. Introduction

In this article, I examine the evolution of our understanding of atomic geometries and electronic structures of tetrahedrally coordinated compound (TCC) semiconductor surfaces over the past thirty years. The sum of all the work devoted to this subject is considerable and a detailed account of this period would exceed by far the boundaries set by the length of this article. It is therefore with great respect for all experimental and theoretical studies not mentioned herein that I select some examples associated with the most significant advances in the field.

The theme of the paper is that remarkable progress has been made in thirty years. The technologies for experimental and theoretical surface investigations have evolved at a pace which was difficult to foresee in the early 1960's. While early experiments were limited to observations of new surface symmetries and qualitative indications of atomic displacements, today the availability of sophisticated electron and ion diffraction techniques, of tools such as the scanning tunneling microscope, and of advanced energy minimization and electronic structure calculations allows a

systematic, almost routine, approach to the quantitative determination of complex surface atomic displacements and associated local and extended electronic structures. This progression and the ever increasing pace of gathering quantitative information on surfaces has led to the recognition and formulation of very general principles which govern TCC surface atomic geometries.

The organization of this article is essentially chronological. I identify three decisive periods of the history of surface structure determination and, within each period, illustrate the progress made with ground-breaking studies. In Section 2, I review the pioneering work done on polar and non-polar surfaces with electron diffraction and electronic structure measurements in the 1960's and early 1970's. This work laid the experimental foundations on which quantitative methods could be developed in the 1970's. Section 3 is devoted to the second decade (early 1970's to mid 1980's), during which quantitative structure determination techniques were fully developed and applied mostly to non-polar semiconductor surfaces, leading to the initial formulation of structural scaling laws and of principles governing the occurrence of surface relaxations. In the final section, I review the period extending from the mid 1980's to

SSDI 0039-6028(93)E0327-Q

the present. This period is marked by further developments in experimental techniques, in particular scanning tunneling microscopy, and broadening activity on the structure of polar surfaces of III–V compounds and cleavage surfaces of II–VI and I–VII compounds which led to the confirmation and extension of the principles governing semiconductor surface atomic geometries and electronic structures.

2. The early years (1960–1975)

2.1. Polar surfaces

2.1.1. III–V compounds

Investigations of surface atomic geometries of TCC started around 1960 with examinations of (111) surfaces of III–V zincblende semiconductors. Studies of cleavage (110) III–V surfaces followed rapidly. One may consider the reasons why the (111) surfaces, and not the lower index (100) surfaces, were chosen for these initial studies. The first was probably one of immediate interest. The (111) plane is the cleavage plane for diamond-structure elemental semiconductors, and the comparison between the structure of (111) III–V surfaces and that of Si(111) and Ge(111) surfaces investigated in 1959 by Schlier and Farnsworth [1] was a natural path at the onset of this research. The second reason might have been more practical. The topology of the (111) surface appeared at the time to favor the formation of one dangling bond (DB) per surface atom and the preservation of the tightly bound and closely spaced double layer (Fig. 1a). Thus, the surface composition, anion or cation, was thought to depend essentially on the crystal orientation. The atomic connectivity on the (100) surface, on the other hand, is more symmetric (Fig. 1b). Each surface atom is connected to two second layer atoms and exhibits two DBs. As a consequence, the surface composition can evolve continuously from full anion to full cation, a situation which was considerably more difficult to control with the surface preparation techniques available at the time. Only with the development of molecular beam epitaxy (MBE) pioneered by Cho [2,3] and

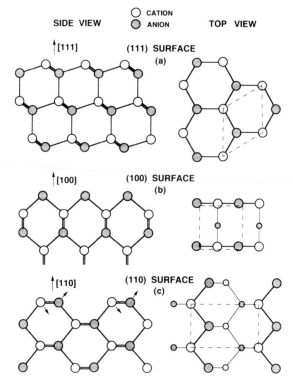

Fig. 1. Schematic illustrations of the zincblende lattice showing side and top views of the truncated (a) (111) surface; (b) (100) surface; (c) (110) surface. The (1×1) unit cell is indicated in each case by dashed lines. Arrows in panel (c) indicate atomic movements compatible with a bond-length-conserving relaxation. The double lines illustrate the parallel projections on the plane of the figure of two bonds from the same atom.

Arthur [4,5] in the early 1970's was the composition of the (100) surface brought under some control, leading to the observation of complex sequences of reconstructions spanning the full range of relative anion-to-cation surface stoichiometry.

Haneman performed the first low-energy electron diffraction (LEED) studies of the (111)-cation and ($\overline{111}$)-anion surfaces of InSb [6] and GaSb [7] in 1960. These surfaces were prepared by the argon ion bombardment and annealing (IBA) technique developed by Farnsworth et al. [8], in an experimental apparatus consisting of a vacuum vessel evacuated with a diffusion pump and equipped with molybdenum getters, an off-axis electron gun, a Faraday cup for LEED intensity measurements and an ion bombardment system. The ultimate pressure achievable at the time was of the order of 10^{-9} Torr, roughly 20 times higher than pressures routinely obtained in modern ultra-high vacuum (UHV) systems. Haneman established that these polar surfaces, like Si and Ge [1], were reconstructed and identified (2×2) reconstructions on both cation and anion surfaces. (Throughout this article, we will use a terminology consistent with the following definition: *reconstruction* is associated with a space group symmetry different at the surface than in the bulk, leading to the appearance of non-integral beams in diffraction experiments; *relaxation* designates surface atomic displacements which conserve the symmetry, shape and size of the bulk unit cell, giving rise to a (1×1) diffraction pattern.) This work was later expanded by MacRae [9] who benefited from improved vacuum technology (pressure $\sim 5 \times 10^{-10}$ Torr) and showed that the (111)-cation and $(\overline{1}\overline{1}\overline{1})$-anion surfaces of GaSb and GaAs exhibited (2×2) and (3×3) structures, respectively. The $(\overline{1}\overline{1}\overline{1})$ surface was found to be considerably less stable under annealing than its (111) counterpart and to decompose easily into (110) facets, a trend confirmed a few years later by Grant and Haas [10] for InAs.

None of the reconstructions observed by Haneman and MacRae were analyzed quantitatively before 1984 [11]. Yet, very insightful considerations about the atomic geometries already were given in the 1966 paper. MacRae suggested that the reconstructions resulted from the requirement to lower the high energy of these polar surfaces, and proposed a model for the (2×2) structure consisting of an ordered array of cation vacancies formed to decrease the polar character of the surface (Fig. 2a). He correctly pointed out the need to saturate the surface DBs and suggested that the remaining surface cations underwent a displacement toward the anion layer accompanied by a rehybridization of surface bonds from tetrahedral sp^3 to planar sp^2. This prediction was remarkably close to the (111)-Ga GaAs structure quantitatively determined by LEED two decades later [11] (Fig. 2b and Section 3.3).

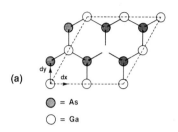

GaAs(111) - (2x2)

(a)

$\bullet$ = As
$\circ$ = Ga

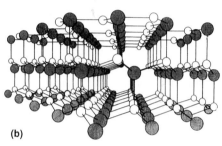

(b)

Fig. 2. (a) Schematic of the Ga vacancy model proposed by MacRae [9] and confirmed by Tong et al. [11]. The vacancy creates 12-membered GaAs rings (Section 3.3). (b) Perspective representation of the (2×2) reconstructed Ga(111) GaAs surface. The cation vacancies appear in one out of two cation–anion rows in the top atomic plane (after Tong et al. [11]).

2.1.2. II–VI compounds

Polar surfaces of II–VI TCC were investigated in the second half of the 1960's. Surfaces of wurtzite CdS, CdSe and ZnO, and zincblende ZnS were studied first because of their impact on the physics and chemistry of gas adsorption on insulators. Considering long range electrostatic interactions in the crystal, Mark and coworkers [12,13] predicted that these surfaces could be stabilized only through the introduction of a $\frac{1}{4}$ monolayer of electronic surface charge, obtained via vacancy reconstruction, charged surface states, or large scale reconstruction such as facetting of the surface. The idea of energy-lowering reconstruction ensuring surface charge neutrality was, therefore, already at the forefront for II–VI and III–V polar surfaces alike. It was to be formalized later on by Harrison with the concept of surface auto-compensation [14]. The predictions on the wurtzite surfaces were qualitatively confirmed by most LEED observations. Campbell et al. [15] observed (2×2) reconstructions on (0001)-Cd and (000$\bar{1}$)-S CdS surfaces prepared by IBA, and (2×2) and (4×4) reconstructions on the (0001)-Cd CdSe surface, compatible with the $\frac{1}{4}$ monolayer requirement. Chang and Mark [16,17] reported integral diffracted spots for the (0001)-cation surfaces of ZnO and CdS in a pattern exhibiting a six-fold symmetry indicative of a non-ideal termination, a poorly developed $(\sqrt{3} \times \sqrt{3})R30°$ reconstruction for the (000$\bar{1}$)-S CdS surface and the formations of facets and steps under various annealing conditions.

The list of examples which could be cited is long. They would underscore the fact that, at the end of this first decade, the occurrence of reconstruction on polar surfaces of tetrahedrally coordinated III–V and II–VI compounds was recognized as a general phenomenon. Although specific atomic geometries were far beyond the capabilities of experimental and theoretical techniques available at the time, the models proposed to explain the occurrence of these reconstructions already embodied the same concept as the models used today to rationalize reconstructions, namely surface auto-compensation by saturation of surface dangling bonds and minimization of surface energy by the resulting surface relaxations.

2.2. Non-polar cleavage surfaces

2.2.1. III–V compounds

The importance of cleavage surfaces in the history of semiconductor surface physics cannot be overstated. Non-polar surfaces such as the (110) zincblende surface are naturally charge neutral. They offer the combination of a high degree of structural and stoichiometric perfection and the absence of intrinsic surface states in the fundamental energy gap. This latter property, although accepted only in the mid 1970's, was established by Van Laar and Scheer [18] as early as 1967 and contributed to making the (110) surface an essential testing ground for experimental and theoretical investigations of surface electronic

properties during the next twenty years. The two ground-breaking studies outlined below are therefore among the most important contributions to the physics of TCC surfaces of this first decade.

The first was devoted to structure. The (110) surface consists of equal numbers of anions and cations arranged in zig-zag chains in which each atom is bound to two surface and one second layer atom (Fig. 1c). As would be discovered later, this topology allows activationless displacements of the surface atoms through bond-length-conserving rotations of the anion–cation surface bonds (indicated by arrows in Fig. 1c). MacRae and Gobeli [19] used (110) surfaces prepared by cleaving in UHV $(5 \times 10^{-10}$ Torr) following a technique developed by Lander et al. [20]. Second generation LEED optics with post-acceleration, spherical grids and fluorescent screen, and tracking spot photometer for intensity versus energy measurements were already available for this work. Pointing out the absence of non-integral order beams, MacRae and Gobeli measured the unit cell dimensions to show that these non-polar (110) surfaces were unreconstructed. They performed the first structure factor analysis of LEED intensities on a specially designed analog computer. They demonstrated that the intensity of the $(h0)$ beams, for h odd, should be very weak for unrelaxed surfaces of GaAs and InSb for which anion and cation scattering factors are almost identical (quasi-forbidden beams). The observation of the contrary was the first evidence of atomic relaxation on (110) surfaces. The first quantitative LEED structure determination on GaAs(110), performed a decade later [21,22], confirmed these conclusions.

The second set of experiments was devoted to electronic properties. Gobeli and Allen [23] and Van Laar and Scheer [18] performed contact potential difference (CPD) measurements using a vibrating Kelvin probe, and photothreshold measurements to investigate work function and band bending on cleaved GaAs(110) as a function of doping density, doping type and temperature. In a remarkably insightful paper, Van Laar and Scheer [18] demonstrated that the Fermi level (E_F) could be shifted by 1 eV or more from the upper to the lower part of the gap at the (110) surface of GaAs by going from n-type to p-type doping. They were first to conclude that, unlike Si(111) and Ge(111) surfaces [24,25], the (110) GaAs surface could be obtained without significant band bending and was devoid of *intrinsic* states in the fundamental gap.

The two key ingredients of the physics of the (110) zincblende surface, i.e. the occurrence of a surface relaxation and the absence of intrinsic states in the gap, were therefore recognized by 1967, although the structure was still unknown and the link between them had not been established. The exceedingly important result on the absence of gap states unfortunately was challenged incorrectly in the early 1970's by virtue of experiments flawed by the presence of cleavage-induced surface defects. The controversy which ensued led to several years of uncertainty concerning the electronic properties of GaAs(110). This subject is reexamined in Section 3.2.

Scheer and Van Laar [26] tackled the problem of adsorbate-induced changes in band bending on GaAs(110). By depositing Cs on n- and p-GaAs, they showed that E_F stabilization resulted from, but did not precede as had been suggested by Spitzer and Mead [27], the adsorption of foreign species on the surface. They correctly postulated that GaAs belonged to a group of semiconductors for which the adsorption of small amounts of any species on the (110) surface created gap states which stabilized E_F near midgap, irrespective of doping type. This report was among the first to address the problem of the formation of Schottky barriers, a subject which was to monopolize tremendous research energy during the following twenty years [28].

In retrospect, it is clear that the studies of MacRae and Gobeli and of Van Laar and Scheer had a considerable impact on the fundamental understanding of the structure and electronic properties of cleavage surfaces, as well as on the development of LEED, CPD and photoemission techniques. They represent the opening chapter of what was to become the single most successful effort to determine the atomic geometries and electronic structures of compound semiconductor surfaces and interfaces.

2.2.2. II–VI compounds

The early 1970's also were marked by the first structural studies of cleavage $(10\bar{1}0)$ and $(11\bar{2}0)$ surfaces of II–VI wurtzite compounds. The wurtzite $(10\bar{1}0)$ surface consists of anion–cation dimers with each constituent bound to one atom in the surface and two atoms in the second layer (Fig. 3a). An activationless bond-length-conserving relaxation is obtained by rocking the dimer along its axis (indicated by arrows in panel (a)). The $(11\bar{2}0)$ surface embodies a more complex chain structure with four atoms per unit cell (Fig. 3b). Each surface atom is bound to two atoms in the chain and one atom in the second layer. The rectangular cell has a glide-plane symmetry resulting in missing diffracted beams in LEED [29]. This seemingly more rigid structure can relax

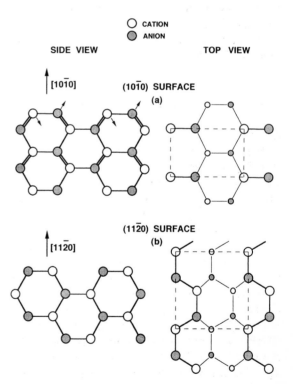

Fig. 3. Schematic illustrations of the wurtzite lattice showing side and top views of the truncated (a) $(10\bar{1}0)$ surface; (b) $(11\bar{2}0)$ surface. The (1×1) unit cells are indicated by dashed lines. Arrows in panel (a) indicate atomic movements compatible with bond-length-conserving bond rotation. The double lines indicate the parallel projections on the plane of the figure of two bonds from the same atoms.

while conserving bond lengths with a puckering of the unit cell (as determined in the late 1980's).

Cleavage initially was reported for the $(10\bar{1}0)$ surface of ZnO and CdS [30]. Shortly thereafter, Mark and coworkers used LEED to determine the unit cell symmetries and to measure diffracted intensities for the two non-polar $(10\bar{1}0)$ and $(11\bar{2}0)$ surfaces of ZnO and CdS [16,29] prepared by IBA. The mirror and glide-plane symmetries of the $(10\bar{1}0)$ and $(11\bar{2}0)$ unit cells were shown to be preserved at the surface, confirming the absence of reconstruction noted for the (110) surfaces. Diffracted intensities computed from a single-scattering ("kinematical") structure factor analysis were compared to experimental data in an effort to assess the existence of surface relaxation. Unlike for the III–V (110) surfaces, however, no compelling evidence for relaxation was found, leaving the relaxation of wurtzite-structure cleavage faces a subject which was to remain without a clear understanding until the late 1980's (Section 4.2).

3. The development years (1975–1985)

3.1. The techniques

Experimental and theoretical techniques for studying surface atomic and electronic structures were undergoing very rapid progress throughout the 1970's. Maturing UHV technologies were driving operating pressures down to levels where surface cleanliness could be maintained for days (high 10^{-11} Torr range). Ultraviolet photoemission spectroscopy (UPS) for valence band and shallow core level spectroscopy was becoming increasingly widespread with the availability of gas discharge sources coupled to UHV systems [31–38]. Angle resolved photoemission spectroscopy (ARPES) for band structure mapping followed in the mid 1970's [39–41]. Finally, synchrotron storage rings providing photon wave-length tunability in ranges extending from the UV to the X-ray became accessible to surface scientists. Several techniques utilizing photon energy scanning were developed: partial yield spectroscopy (PYS) [42] and constant initial state (CIS) spectroscopy [43], which probe empty states below and above the

vacuum level, respectively; constant final state (CFS) spectroscopy [43], which probes the initial density of filled states. Tunability also increased the power of discrimination between bulk and surface effects by allowing continuous variation of the escape depth of photoemitted electrons [28].

The structure determination techniques also were improved during this period. Most of the LEED computational formalisms were in place in the mid 1970's [44–47]. Yet, structure determinations were hindered by the prohibitive computing time and cost associated with multiple scattering analyses. Techniques to reduce the effect of multiple scattering by averaging methods [48–51] were developed to perform LEED intensity analyses with considerably less computer intensive single scattering computations. The best known of these averaging schemes, i.e. the constant momentum transfer averaging (CMTA) technique [48,49], was successfully applied to the (110) surfaces of GaAs [52] and InSb [53] and was instrumental in revealing the multi-layer relaxations present at these surfaces. The averaging techniques became obsolete at the end of the 1970's when faster computers and streamlined LEED codes decreased the computing cost. Efforts were then devoted to improving the reliability of the technique by using improved scattering potentials [54,55] and reliability factors to systemize comparisons between experimental and calculated intensities [56–58]. By the early 1980's LEED had become a powerful technique for the determination of surface atomic geometries, and was to remain without contest the most prolific tool for semiconductor surface structure determination during the next decade.

Techniques involving atom [59,60] and ion [61,62] scattering were developed in the 1970's and played a significant role in the quest for semiconductor surface structures. The former were instrumental in providing data charge densities associated with atomic positions. The latter were applied more directly to structure determination because of the simple modeling of the classical high energy ion–atom scattering process. Medium energy (~ 100 keV) and high energy (~ 1 MeV) channeling and blocking configurations were used very successfully to obtain atomic registry and displacements on a number of III–V surfaces [63–66].

Finally, theoretical models for minimization of ground-state energy were also developed in the 1970's and applied with great success to the determination of atomic geometries and electronic structures of TCC. In all these models, the total energy is expressed and minimized as a function of structural variables and the surface electronic structure is a by-product extracted from the charge density and surface state energies corresponding to the total minimum energy. Calculations involving quantum chemistry of clusters [67,68], tight binding energy models (TBE) [69–72]

and pseudo-potential methods [73,74] were applied to a number of III–V and II–VI surfaces and provided valuable insight into the relaxation mechanisms operating at these surfaces.

3.2. The zincblende (110) surface

3.2.1. GaAs(110)

Progress in the determination of surface atomic structures and electronic properties during the second decade can best be reviewed against this background of emerging techniques by examining the case of (110) surfaces. Reviews of this subject have been given by Kahn [75] and Duke [76]. The history of GaAs(110) has demonstrated better

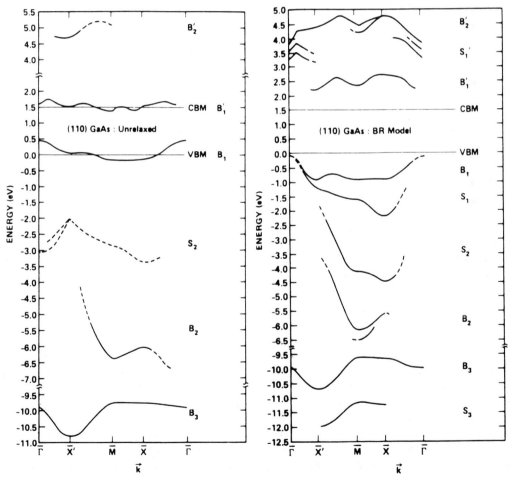

Fig. 4. Surface bands of GaAs(110) calculated for the unrelaxed surface and for bond relaxation (BR) model embodying a 0.65 Å vertical shear of the top layer (after Chadi [91]).

than any other during these thirty years the arduous process of ensuring convergence between all aspects of a same physical system and, as such, is worth a concise review.

In spite of the 1967 CPD and photothreshold work by Van Laar and Scheer [18], the absence of intrinsic states in the gap of GaAs(110) was still being debated in the early 1970's. Using similar techniques, Dinan et al. [77] concluded that a band of acceptor states overlapping with the conduction band was extending ~ 0.8 eV down into the gap. This result was confirmed at the time by several PES [78–80] and PYS [81,82] measurements on GaAs. Acceptor states were also found in the upper part of the gap on InP, but not on GaSb [83]. The level of confusion was heightened when surface state calculations performed for the unrelaxed geometry confirmed these results [84–86]. The fundamental problem with these calculations was, of course, the utilization of an inappropriate atomic structure.

The situation was resolved by four key experiments presented in 1976. First, Van Laar and Huijser confirmed their claims for GaAs and InAs, but indicated the presence of acceptor states in the upper part of the GaP gap [87]. Second, Spicer et al. [88], using synchrotron radiation PES, showed that the states previously reported in the GaAs gap were extrinsic and related to cleavage defects. Third, Duke et al. [21,22] reported the first multiple scattering analysis of GaAs(110) LEED data, which identified the surface atomic geometry as consisting of a bond-length-conserving rotation relaxation ("rotational relaxation") of the surface with the As (Ga) moving up (down) with respect to the truncated surface plane. The introduction of this geometry, and refined versions thereof, in theoretical calculations of the surface electronic structure [89–93] immediately produced shifts of the anion- and cation-dangling bond states back into the valence and conduction bands, respectively, in agreement with the new experimental results (Fig. 4). Finally, Gudat and Eastman [94] reconciled the PYS [81,82] and electron energy loss spectroscopy (EELS) [95] data, which appeared to support the presence of empty states in the GaAs(110) gap, with the new CPD and PES results. They applied

Zincblende (110)

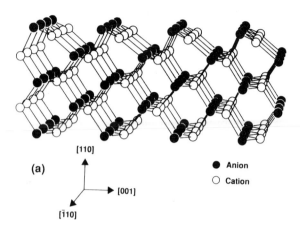

(a)

● Anion
○ Cation

[110]
[001]
[1̄10]

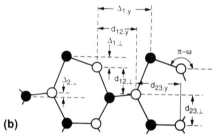

(b)

Fig. 5. (a) Perspective representation of the relaxed (110) surface of zincblende TCC (after Duke [76]); (b) Schematic of the structure showing first and second layer relaxations and the bond rotation angle ω.

an exciton binding energy correction to the Ga 3d-to-empty-state transition measured in PYS and EELS, and showed that it corresponded to a final (empty) state energy in the conduction band, in agreement with the hypothesis of a gap free of intrinsic states.

Subsequent LEED analyses of the GaAs(110) structure introduced bond length relaxations [96,97] and second layer counter-rotation (As down, Ga up) [52,98–100] (Fig. 5). This structure was repeatedly challenged in the late 1970's and early 1980's on the basis of comparisons between calculated and measured UPS spectra [101,102], inverse photoemission spectroscopy measurements [103,104], and high energy ion channeling experiments [105,106]. To resolve these issues, refinement analyses of the LEED data were performed [107] with improved scattering potentials

[54,55] and the sensitivity and accuracy of LEED for determining atomic displacements perpendicular or parallel to the surface were re-evaluated [108]. The GaAs structure determined by Meyer et al. [100] was confirmed with only minor modifications by each of these re-evaluations.

3.2.2. Principles governing surface atomic relaxations

In the years following this GaAs period, the atomic geometries of the (110) surfaces of most III–V and II–VI compounds were determined via LEED and ion scattering. They included AlP [109], AlAs [110], GaP [108,111], GaSb [112,113], InP [114,115,116], InAs [117,118], InSb [119], ZnS [120,121], ZnSe [122], ZnTe [123] and CdTe [124]. All these surfaces were found to exhibit a large top layer distortion corresponding approximately to a bond-length-conserving activationless rotation relaxation with an angle $\omega = 29° \pm 3°$ and varying top contraction and second layer relaxation. For bond-length-conserving relaxations, ω and $\Delta_{1\perp}$, the top layer vertical shear defined in Fig. 5, are simply related by $\Delta_{1\perp} = (a_0/4) \sin \omega$. These structures were in good agreement with predictions based on quantum chemical calculations [125,126] or TBE minimization calculations [69,70,127]. Further, they produced equally good agreement between calculations of anion-derived filled surface state dispersion and ARPES measurements on GaAs [128–130], InP [131], ZnSe [132] or CdTe [133], and calculations of cation-derived empty surfaces states and inverse photoemission measurements on GaAs [134] and GaP [135] (Ga-derived empty state overlapping with the gap for the latter). By the mid 1980's, the cleavage (110) surfaces of zincblende compounds constituted a thoroughly investigated set of surfaces on which an unprecedented level of convergence had been obtained between theoretical, experimental, structural and electronic data.

The availability of this extensive set of structural data and the realization that the gap of most III–V and II–VI (110) surfaces was free of intrinsic states led to the recognition of a number of general rules. First, as a lesson from the GaAs story, the direct link between the elimination of DB states from the band gap (except for GaP)

and the surface relaxation was established. The charge redistribution resulting from the relaxation was producing filled anion-derived surface and back bonding states and empty cation-derived surface states, making the surface naturally auto-compensated. In the process, the surface state energies were shifted away from the gap into the valence band and conduction band (Fig. 4). The energy gained in the process was sufficient to drive the relaxation. Second, it was realized that, unlike originally proposed [22,136], these atomic geometries did not scale with the bulk ionicity nor depend specifically on the reduced-coordination (or small molecule) chemistry of the surface anion and cation species. An early interpretation of the relaxation mechanism had attributed the GaAs bond rotation to the requirement of moving the three-fold coordinated surface As and Ga atoms into configurations more compatible with their small molecule configurations, namely pyramidal p bonding for As and planar sp^2 bonding for Ga. On the count of this picture, high ionicity II–VI compounds with different small molecule cation and anion chemistries should have exhibited different surface relaxations. In fact, isoelectronic compounds with very different ionicities [137], such as ZnSe and GaAs, or CdTe and InSb, were found to exhibit identical relaxations within the accuracy of the structure analyses. Finally, it was established that the principal parameter of these atomic structures, i.e. the top layer vertical shear $\Delta_{1\perp}$ scaled linearly with the bulk structural parameters (Fig. 6a), leaving the bond rotation angle ω approximately constant for all compounds investigated [138].

This combination of rules led to the conclusion that the atomic relaxation of these surfaces was driven primarily by the electronic energy gained upon bond rotation, the very nature of the relaxation being dictated by the specific surface topology. The concept of a universal relaxation mechanism for TCC surfaces was emerging. The extension of this concept to other types of surfaces, however, required the recognition that non-polar surfaces, on which the redistribution of charges between equal numbers of anion and cation DBs occurs naturally, are ideal candidates for activa-

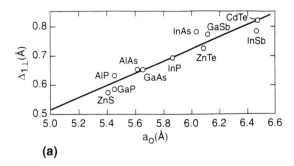

(a)

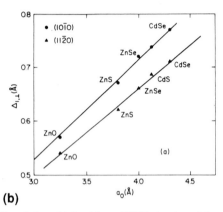

(b)

Fig. 6. Correlation obtained from LEED structure determinations between $\Delta_{1\perp}$ and lattice constant a_0 for (a) the (110) surfaces of zincblende compounds (Section 3.2; Duke [138]); (b) the $(10\bar{1}0)$ and $(11\bar{2}0)$ surfaces of wurtzite compounds (Section 4.2; after Duke [193]).

tionless energy-lowering relaxations. Polar surfaces, on the other hand, require reconstructions to create a *non-polar environment* upon which charge exchange between anion and cation DBs can take place and an energy lowering relaxation can be applied. Such transformations were indeed observed on (111) surfaces (next section) and, to the best of our present understanding, on the (100) surfaces (Section 4.1).

3.3. Polar surfaces

The atomic geometry of the (2×2) (111)-cation surfaces of GaAs [11,139], GaP [140] and InSb [141] was investigated at the end of this second decade. LEED and X-ray structure determinations produced structures consisting of a $\frac{1}{4}$

monolayer of cation vacancies with relaxation of the remaining top layer atoms toward the second layer and lateral displacement of the second layer As to form quasi-planar twelve-membered As–Ga rings, in good agreement with the motif suggested by MacRae [9] (Fig. 2). Interestingly, the local As–Ga configuration and electronic structure in the ring were found to be similar to those of on the (110) surface chain, as confirmed by TBE minimization calculations [142] and ARPES data [143,144]. The reconstruction was here again driven by the lowering of the electronic state energy. The resulting surface was auto-compensated, in agreement with Harrison's criterion of stability [14].

Finally, the end of this second decade was also marked by attempts to determine atomic structures of the (100) surface. Considerable work had been devoted since the early 1970's to establishing the relationship between atomic reconstruction and surface stoichiometry on this surface [4,5,145–148]. The (100) phase diagram, however, is still under debate in 1993. Photoemission experiments were performed on the (2×4) [149] and $c(4 \times 4)$ [150,151] reconstructions. Larsen et al. suggested that the latter was due to symmetric dimers of As adsorbed on top of an As-terminated surface, in contrast with the total energy minimization calculation by Chadi et al. [152] which suggested a combination of symmetric and asymmetric dimers in the As-layer (no chemisorption). The $c(4 \times 4)$ and other reconstructions of the (100) surfaces were to be extensively studied and their structure partially resolved with the help of scanning tunneling microscopy (STM) during the next decade (Section 4.1).

4. The maturity years (1985–1993)

The last eight years of work on TCC semiconductor surfaces were marked by extraordinary new possibilities offered by the development of scanning tunneling microscopy (STM) and related techniques. Through real space visualization of intricate atomic arrangements in large unit cells and simultaneous measurements of electronic properties of these surfaces with atomic resolution, STM opened entirely new avenues for semi-conductor surface studies. The most important and spectacular applications to compound semi-conductors have been to the determination of atomic arrangements on the (100) III–V surfaces. Structure determinations by LEED and total energy minimization calculations also were extended to non-polar surfaces of zincblende I–VII and wurtzite II–VI compounds. The confirmation of the structural mechanisms and scaling laws uncovered with previous investigations (Section 3) led in the early 1990's to extensions of the principles on atomic and electronic structures of TCC surfaces.

4.1. Polar surfaces

The technologically important (100) surfaces of III–V and II–VI compounds have been studied since the early 1970's. Several surface phases obtained during MBE growth [2–5,153] as a function of As versus Ga flux or after growth [146,148,154] had been identified prior to 1980. Yet, their precise As versus Ga compositions were still under debate in the early 1980's. In order of decreasing As/Ga surface concentration, the most commonly observed reconstructions were the c(4 × 4), (2 × 4)-c(2 × 8), (1 × 6), (4 × 6) and (4 × 2)-c(8 × 2). The shear size of the unit cells and the uncertainty of the corresponding surface compositions were impeding quantitative structure determinations, although the proliferation of MBE growth systems linked to PES and electron diffraction techniques in the 1980's was giving impetus to the study of these complex reconstructions.

Chadi used TBE minimization [155] to derive the As-stabilized (2 × 4) structure based on two different arrangements of symmetric As dimers and missing dimer rows in the top layer. Good agreement was obtained between the surface states calculated with this structure and photoemission results [149]. No surface states were found in the energy gap for this structure. A similar As dimer arrangement was also proposed by Farrell et al. [156] based on electron diffraction studies. The calculation also predicted the c(2 × 8) structure, which is a small perturbation

of the (2 × 4) structure, to be slightly more stable than the (2 × 4).

The unambiguous confirmation of the As dimer structure of the (2 × 4)-c(2 × 8) surface was first obtained by Pashley et al. [157] using STM. These GaAs surfaces were prepared by sublimation of an encapsulating As cap pre-deposited in the MBE growth chamber [158]. Rows of As-dimer arranged in (2 × 4) or c(2 × 8) unit cells with one missing dimer per cell were identified at the time. Biegelsen et al. [159] later expanded on these results by examining the c(4 × 4), c(2 × 8), (2 × 6) and c(8 × 2) MBE surfaces grown in-situ. The As-saturated c(4 × 4) structure was found to correspond to sets of three As dimers adsorbed on a monolayer of As, in agreement with the suggestion of Larsen et al. [150]. The (2 × 4) unit cell was found to consist of a mix of two and three As dimers per cell with one or two missing rows. Finally, the Ga-rich c(8 × 2) structure was attributed to groups of two Ga-dimers in the top layer (Fig. 7).

The details of some of these atomic arrangements have remained controversial. Distribution of areas with one versus two missing As-dimers in the (2 × 4)-c(2 × 8) structure have been observed [157,160]. Cation–anion mixing in the top layer has been suggested [161] and the relationship between structure and stoichiometry has not been completely resolved. Kinetics is thought to play a crucial role in the formation of these surfaces. Finally, the detailed geometry of the dimers, recently investigated [162], must be addressed with experimental techniques. It remains, however, that the basic anion or cation dimer building block of these reconstructions, predicted by theory in 1987 [155], was unambiguously confirmed by STM. The formation of dimers is consistent with one of the guiding principles governing these semiconductor surfaces, namely the requirement to saturate DBs and to lower the energy of the surface by transforming them into bonding states. Furthermore, the STM data on the distribution of dimers and missing dimers inside the unit cell have led to the demonstration that most of these surfaces are in compliance with the principle of dangling bond saturation and surface auto-compensation [163]. The application of this principle

GaAs (100) – c (4x4)

Top View

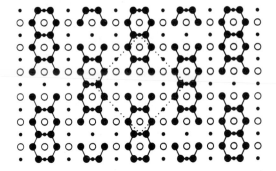

GaAs (100) – c (2x8)

Top View

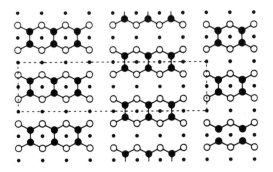

GaAs (100) – c (8x2)

Top View

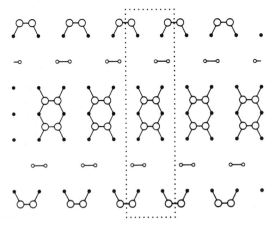

using electron counting techniques has since been recognized as a simple but powerful way to catalogue possible surface atomic arrangements and compositions for TCC semiconductors. Work function and surface dipole measurement results compatible with charge transfers imposed by auto-compensation have been recently obtained on the c(4 × 4), c(2 × 8) and c(8 × 2) GaAs(100) surfaces [164,165].

Scanning tunneling spectroscopy (STS) provided key measurements for unraveling the electronic properties of the (100) surface. Pinning of E_F near midgap on n-type (100) GaAs had been observed by PES for a range of reconstructions [166–168], although theory predicted no intrinsic gap states [155]. Pashley and Haberern using STM and STS found the charged acceptor-like pinning centers to be due to kink-defects in the rows of As-dimers on Si-doped GaAs [169,170], each carrying a single negative electron charge [171].

Studies on the (100) surface of GaAs and other compounds are expected to continue in the 1990's. The atomic geometries of most of the observed phases remain to be determined quantitatively. Band bending on clean p-type GaAs surfaces is not understood, and recent PES investigations suggest that highly doped surfaces might exhibit very small band bending [172]. The newly developed capability to perform high resolution STM at high temperature [173,174] should also provide in the years to come information on ad-atom and vacancy diffusion crucial for epitaxial growth.

STM imaging was also essential in developing the complex structure models for the (2 × 2) and $\sqrt{19} \times \sqrt{19}$ As(111) surfaces of GaAs [175]. A motif consisting of adatom As trimers bound to the complete As layer was inferred from images of the (2 × 2) ($\overline{111}$) surface. A similar (2 × 2) trimer arrangement was proposed for both (2 × 2) (111)-Ga and (2 × 2) ($\overline{111}$)-As on the basis of total energy minimization calculation [176,177], but the cation vacancy model [11] remains at

Fig. 7. Models of the c(4×4), (2×4)-c(2×8) and (4×2)-c(8×2) structures derived from STM studies (after Biegelsen et al. [159]).

present the accepted one for the (111) surface. A complex distribution of large Ga–As rings resulting from rearrangements of the top bilayer was deduced from the STM observations of the $\sqrt{19} \times \sqrt{19}$ structure.

4.2. Non-polar surfaces

STM and STS were extensively applied to (110) zincblende surfaces in the late 1980's [178,179]. Although these studies confirmed the spatial relationship between structure and electronic surface states, their impact on the field was smaller than in the case of polar surfaces given the high level of understanding of the (110) surface prior to these experiments. A more important role was, and still is, played in the area of adsorption on these surfaces, in particular with respect to problems related to the formation of Schottky barriers [180,181].

Quantitative structure determinations of TCC surfaces were extended after 1987 to the cleavage surfaces of wurtzite II–VI compounds and, more recently, to the (110) surface of highly ionic zincblende I–VII compounds. LEED and low energy positron diffraction (LEPD) structure determinations were performed on the $(10\bar{1}0)$ surface of CdSe [182–184] and to the $(11\bar{2}0)$ surfaces of CdS [185] and CdSe [183,184,186,187], and TBE minimization calculations were applied to the two cleavage faces of CdS, CdSe, ZnO, ZnS and ZnSe [188–192].

These structure determinations extended to the II–VI wurtzite surfaces the relaxation principles developed for the zincblende surfaces. First, all surfaces were found to relax with large surface distortions. On the $(10\bar{1}0)$ surfaces, dimer tilt angles between 17° and 23° were identified, depending on the structure determination technique (Fig. 8). Bond length contractions in the top layer also were suggested by the energy minimization calculations [188–192]. On the $(11\bar{2}0)$ surfaces, extensive puckering of the chain with rotation angles of the cation–anion–cation planes of the order of 30° were determined (Fig. 9). A two-angle rotation relaxation has been described by Kahn et al. [185] for this surface. On both

Wurtzite $(10\bar{1}0)$

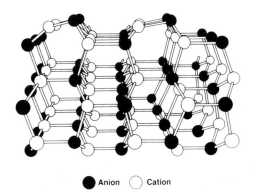

● Anion ○ Cation

Fig. 8. Perspective representation of the relaxed $(10\bar{1}0)$ surface of wurtzite TCC (after Duke [193])

surfaces, the relaxations were found to bring anions and cations in local configurations similar to those found on the (110) surface. A detailed review of these structures was given by Duke [193]. Second, like on (110) surfaces, these activationless relaxations were found to be driven by a lowering of the surface state energy. The TBE calculations [188–192], in accord with ARPES data [194], predicted a significant lowering of the anion-derived filled surface state energy below the valence band maximum on both surfaces upon relaxation (Fig. 10). Finally, the top layer vertical shear, like $\Delta_{1\perp}$ on the (110) surface, was shown to depend linearly on bulk parameters, thus con-

Wurtzite $(11\bar{2}0)$

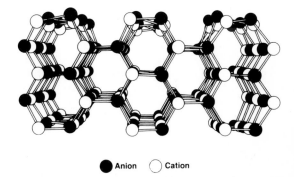

● Anion ○ Cation

Fig. 9. Perspective representation of the relaxed $(11\bar{2}0)$ surface of wurtzite TCC (after Kahn et al. [185]).

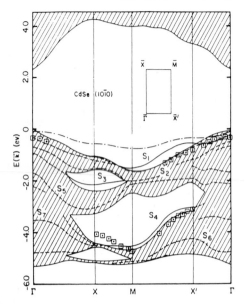

Fig. 10. Surface electronic band structure of CdSe(10$\bar{1}$0). Solid lines and dashed lines are surface state and surface resonance dispersion curves, respectively, calculated for the relaxed geometry. The dot-dashed curve is the calculated S_1 state dispersion for the unrelaxed geometry. The squares represent ARPES data [194] (after Wang et al. [189]).

pounds CuCl and CuBr. The spectroscopic ionicities of these compounds are among the highest in the zincblende group (second only to AgI and CuI), and their structure determination was to provide a stringent test on the recently renewed claim that surface relaxations should depend on compound ionicity. Contrary to the theoretical predictions of a strong dependence on ionicity by Tsai et al. [195], however, LEED studies of CuCl [196] and CuBr [197] produced (110) relaxations with top layer vertical shears compatible with covalent compounds (Fig. 11). Although an increase in surface bond-length contractions was detected in these ionic materials [196–198], the results confirmed the lack of primary role of ionicity in determining the main relaxation parameters of cleavage surfaces. By demonstrating an invariance of the (110) surface structure for compounds with drastically different bulk electronic properties and chemistry, these results emphasized the universal aspect of TCC surface relaxations driven by electronic energy and guided by surface topology.

firming the scaling laws formulated for the zincblende surfaces (Fig. 6b).

The latest chapter in this history of TCC surfaces was recently written with the study of the (110) surfaces of the zincblende I–VII com-

5. Synopsis

The body of work reviewed in this article is a strong, although incomplete, testimony to the

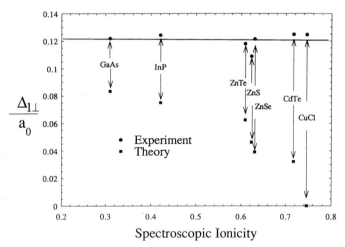

Fig. 11. Normalized zincblende (110) first-layer vertical shear, $\Delta_{1\perp}/a_0$, as a function of compound spectroscopic ionicity [137]; (●) correspond to structures experimentally determined by LEED; (■) correspond to the theoretical prediction of ionicity dependence by Tsai et al. [195]. The solid line is the best least-squares fit to the experimental values (after Kahn et al. [196]).

tremendous growth and progress made over the past thirty years in the field of Surface Science in general, and in determination of atomic geometry and electronic properties in particular. The development of sophisticated experimental and theoretical techniques has produced a detailed understanding of many of these surfaces at the microscopic level. More important, the accumulation of knowledge over these thirty years has led to the recognition that deep structural and electronic commonalities exist between seemingly very different surfaces, whether polar or non-polar, III–V, II–VI, or I–VII, zincblende or wurtzite. These commonalities on surface auto-compensation, energy lowering relaxation or reconstruction mechanisms, and universal scaling laws have now been formally expressed in terms of principles which are believed to govern most semiconductor surfaces. As Surface Science embarks on its fourth decade, these principles will guide future studies of elemental as well as compound semiconductor surfaces, and will be tested critically by the ever increasing atomic-level understanding of these surfaces.

Acknowledgements

The author is greatly indebted to Dr. C.B. Duke for many years of a wonderful collaboration on the determination of many of the surface structures mentioned in this article. Support from the National Science Foundation (DMR-90-18521) is also gratefully acknowledged.

References

[1] R.E. Schlier and H.E. Farnsworth, J. Chem. Phys. 30 (1959) 917.
[2] A.Y. Cho, J. Appl. Phys. 41 (1970) 2780.
[3] A.Y. Cho, J. Appl. Phys. 42 (1971) 1074.
[4] J.R. Arthur, Surf. Sci. 43 (1974) 449.
[5] A.Y. Cho and J.R. Arthur, Progress in Solid State Chemistry, 10 (1975) 157.
[6] D. Haneman, J. Phys. Chem. Solids 14 (1960) 162.
[7] D. Haneman, Proceedings of the International Conference on Semiconductor Physics, Prague, 1960 (Publishing House of the Czechoslovak Academy of Sciences, Prague 1961)
[8] H.E. Farnsworth, R.E. Schlier, T.H. George and R.M. Burger, J. Appl. Phys. 29 (1958) 1150.
[9] A.U. MacRae, Surf. Sci. 4 (1966) 247.
[10] J.T. Grant and T.W. Haas, Surf. Sci. 28 (1971) 669; J. Vac. Sci. Technol. 8 (1971) 94.
[11] S.Y. Tong, G. Xu and W.N. Mei, Phys. Rev. Lett. 52 (1984) 1693.
[12] J.D. Levine and P. Mark, Phys. Rev. 144 (1966) 751.
[13] R.W. Nosker, P. Mark and J.D. Levine, Surf. Sci. 19 (1970) 291.
[14] W.A. Harrison, J. Vac. Sci. Technol. 18 (1979) 1492.
[15] B.D. Campbell and H.E. Farnsworth, Surf. Sci. 10 (1968) 197;
B.D. Campbell, C.A. Hague and H.E. Farnsworth, in: Proceedings Fourth Int. Materials Symposium, Ed. G.A. Somorjai (Wiley, New York, 1969) pp. 33–1.
[16] S.C. Chang and P. Mark, Surf. Sci. 48 (1974) 293.
[17] S.C. Chang and P. Mark, J. Vac. Sci. Technol. 12 (1975) 629.
[18] J. Van Laar and J.J. Scheer, Surf. Sci. 8 (1967) 342.
[19] A.U. MacRae and G.W. Gobeli, J. Appl. Phys. 35 (1964) 1629.
[20] J.J. Lander, G.W. Gobeli and J. Morrison, J. Appl. Phys. 34 (1963) 2298.
[21] A.R. Lubinsky, C.B. Duke, B.W. Lee and P. Mark, Phys. Rev. Lett. 38 (1976) 1058.
[22] C.B. Duke, A.R. Lubinsky, B.W. Lee and P. Mark, J. Vac. Sci. Technol. 13 (1976) 761.
[23] G.W. Gobeli and F.G. Allen, Phys. Rev. 137 (1965) A245; and in: Semiconductors and Semimetals, Vol. 2, Eds. R.K. Willardson and A.C. Beer (Physics of III–V Compounds) (Academic Press, New York, 1966).
[24] J. Van Laar and J.J. Scheer, in: Proc. Int. Conf. on Physics of Semiconductors, Exeter, 1962 (The Institute of Physics and the Physical Society of London, London, 1962) p. 827.
[25] G.W. Gobeli and F.G. Allen, Phys. Rev. 127 (1962) 141.
[26] J.J. Scheer and J. Van Laar, Solid State Commun. 5 (1967) 303.
[27] W.G. Spitzer and C.A. Mead, J. Appl. Phys. 34 (1963) 3061.
[28] L.J. Brillson, Surf. Sci. Rep. 2 (1982) 123.
[29] P. Mark, S.C. Chang, W.F. Creighton and B.W. Lee, Crit. Rev. Solid State Sci. 5 (1975) 189.
[30] R.A. Powell, W.E. Spicer and J.C. McMinamin, Phys. Rev. B 6 (1972) 3056.
[31] N.V. Smith, CRC Crit. Rev. Solid State Sci. 2 (1971) 45.
[32] D.E. Eastman, in: Techniques in Metal Research VI, Ed. E. Passaglia (Interscience, New York, 1972) p. 413.
[33] W.E. Spicer, Comments Solid State Phys. 5 (1973) 105.
[34] J.E. Rowe, S.B. Christman and E.E. Chaban, Rev. Sci. Instrum. 44 (1973) 1675.
[35] D.E. Eastman and J.E. Demuth, Jpn. J. Appl. Phys. Suppl. 2, Pt. 2 (1974) 827.
[36] D.E. Eastman and M.I. Nathan, Phys. Today 28 (1975) 44.

[37] J.L. Freeouf and D.E. Eastman, CRC Crit. Rev. Solid State Sci. 5 (1975) 245.

[38] B. Feuerbacher and R.F. Willis, J. Phys. C (Solid State Phys.) 9 (1976) 169.

[39] N.V. Smith and M.M. Traum, Phys. Rev. B11 (1975) 2087.

[40] T. Grandke, L. Ley and M. Cardova, Phys. Rev. Lett. 33 (1977) 1033.

[41] E. Dietz and D.E. Eastman, Phys. Rev. Lett. 41 (1978) 1674.

[42] W. Gudat and C. Kunz, Phys. Rev. Lett. 29 (1972) 169.

[43] G.J. Lapeyre, A.C. Baer, J. Hermanson, J. Anderson, J.A. Knapp and P.L. Gobby, Solid State Commun. 15 (1974) 1601.

[44] C.B. Duke, Adv. Chem. Phys. 27 (1974) 1.

[45] J.E. Demuth, P.M. Marcus and D.W. Jepsen, Phys. Rev. B 11 (1975) 1460.

[46] J.B. Pendry, in: Low Energy Electron Diffraction (Academic Press, London, 1974).

[47] S.Y. Tong, Prog. Surf. Sci. 7 (1975) 1.

[48] M.G. Lagally, T.C. Ngoc and M.B. Webb, Phys. Rev. Lett. 26 (1971) 1557.

[49] T.C. Ngoc, M.G. Lagally and M.B. Webb, Surf. Sci. 35 (1973) 117.

[50] D. Aberdam, R. Baudoing and C. Gaubert, Surf. Sci. 57 (1976) 715.

[51] M.B. Webb and M.G. Lagally, Solid State Phys. 128 (1973) 301.

[52] A. Kahn, E. So, P. Mark and C.B. Duke, J. Vac. Sci. Technol. 15 (1978) 580.

[53] J.L. Yeh, PhD Thesis, Princeton University (1979), unpublished.

[54] R.J. Meyer, C.B. Duke and A. Paton, Surf. Sci. 87 (1980).

[55] W.K. Ford, C.B. Duke and A. Paton, Surf. Sci. 115 (1982) 195.

[56] M.A. Van Hove, S.Y. Tong and M.H. Elconin, Surf. Sci. 64 (19877) 85.

[57] E. Zanazzi and F. Jona, Surf. Sci. 62 (1977) 61.

[58] R.J. Meyer, C.B. Duke, A. Paton, J.C. Tsang, J.L. Yeh, A. Kahn and P. Mark, Phys. Rev. B 22 (1980) 6171.

[59] M.J. Cardillo, C.S.Y. Ching, E.F. Greene and G.E. Becker, J. Vac. Sci. Technol. 15 (1978) 423.

[60] T. Engel and K.H. Rieder, Proc. 4th Int. Conf. Solid State Surf., Cannes, 1980, p. 801.

[61] T.M. Buck, in: Methods of Surface Analysis, Ed. A.W. Czanderna (Elsevier, Amsterdam, 1975) p. 75.

[62] W.D. Mackintosh, in: Characterization of Solid Surfaces, Eds. P.F. Kane and G.B. Larrabee (Plenum, New York, 1976) p. 403.

[63] R.S. Williams, J. Vac. Sci. Technol. 20 (1982) 770.

[64] R.S. Williams, B.M. Paine, W.J. Schaffer and S.P. Kowalczyk, J. Vac. Sci. Technol. 21 (1982) 386.

[65] H.J. Gossmann, W.M. Gibson and L.C. Feldman, J. Vac. Sci. Technol. A 1 (1983) 1059.

[66] R.M. Tromp, J. Vac. Sci. Technol. A 1 (1983) 1047.

[67] W.A. Goddard III, J.J. Barton, A. Redondo and T.C. McGill, J. Vac. Sci. Technol. 15 (1978) 1274.

[68] J.J. Barton, W.A. Goddard III and T.C. McGill, J. Vac. Sci. Technol. 16 (1979) 1178.

[69] D.J. Chadi, Phys. Rev. Lett. 41 (1978) 1062.

[70] D.J. Chadi, Phys. Rev. B 19 (1979) 2074.

[71] K.C. Pandey and J.C. Phillips, Phys. Rev. 13 (1976) 750.

[72] K.C. Pandey, Phys. Rev. Lett. 47 (1981) 1913.

[73] J. Ihm and M.L. Cohen, Solid State Commun. 29 (1979) 2074.

[74] J. Ihm and J.D. Joannopoulos, Phys. Rev. Lett. 47 (1981) 679; J. Vac. Sci. Technol. 21 (1982) 340.

[75] A. Kahn, Surf. Sci. Rep. 3 (1983) 193.

[76] C.B. Duke, in: Surface Properties of Electronic Materials, Eds. D.A. King and D.P. Woodruff (Elsevier, Amsterdam, 1988) p. 69.

[77] J.H. Dinan, L.K. Galbraith and T.E. Fischer, Surf. Sci. 26 (1971) 587.

[78] D.E. Eastman and W.D. Grobman, Phys. Rev. Lett. 28 (1972) 1378.

[79] E.P. Gregory, W.E. Spicer, S. Ciraci and W.A. Harrison, Appl. Phys. Lett. 25 (1974) 511.

[80] W.E. Spicer and P.E. Gregory, CRC Crit. Rev. Solid State Sci. 5 (1975) 231.

[81] D.E. Eastman and J.L. Freeouf, Phys. Rev. Lett. 33 (1974) 1601; 34 (1975) 1624.

[82] J.F. Freeouf and D.E. Eastman, CRC Crit. Rev. Solid State Sci. 5 (1975) 245.

[83] W.E. Spicer, P.W. Chye, P.E. Gregory, T. Sukegawa and I.A. Babalola, J. Vac. Sci. Technol. 13 (1976) 233.

[84] J.D. Joannopoulos and M.L. Cohen, Phys. Rev. B 10 (1974) 5075.

[85] J.R. Chelikowsky and M.L. Cohen, Phys. Rev. B 13 (1976) 826.

[86] C. Calandra and G. Santoro, J. Phys. C (Solid State Phys.) 8 (1975) L86.

[87] J. Van Laar and A. Huijser, J. Vac. Sci. Technol. 13 (1976) 769.

[88] W.E. Spicer, I. Lindau, P.E. Gregory, C.M. Garner, P. Pianetta and P.W. Chye, J. Vac. Sci. Technol. 13 (1976) 780.

[89] J.R. Chelikowsky, S.G. Louie and M.L. Cohen, Phys. Rev. B 14 (1976) 4724.

[90] D.J. Chadi, Phys. Rev. B 18 (1978) 1244.

[91] D.J. Chadi, Phys. Rev. B 18 (1978) 1800; J. Vac. Sci. Technol. 15 (1978) 1244.

[92] C. Calandra, F. Manghi and C.M. Bertoni, J. Phys. C (Solid State Phys.) 10 (1977) 1911.

[93] J.R. Chelikowsky and M.L. Cohen, Solid State Commun. 29 (1979) 267.

[94] W. Gudat and D.E. Eastman, J. Vac. Sci. Technol. 13 (1976) 831.

[95] R. Ludeke and L. Esaki, Phys. Rev. Lett. 33 (1974) 653.

[96] S.Y. Tong, A.R. Lubinsky, B.J. Mrstik and M.A. Van Hove, Phys. Rev. B 17 (1978) 5194.

[97] B.J. Mrstik, S.Y. Tong and M.A. Van Hove, J. Vac. Sci. Technol. 16 (1979) 1258.

[98] A. Kahn, G. Cisneros, M. Bonn, P. Mark and C.B. Duke, Surf. Sci. 71 (1978) 387.

[99] P. Mark, A. Kahn, G. Cisneros and M. Bonn, CRC Crit. Rev. Solid State Mater. Sci. 8 (1979) 317.

[100] R.J. Meyer, C.B. Duke, A. Paton, A. Kahn, E. So, J.L. Yeh and P. Mark, Phys. Rev. B 19 (1979) 5194.

[101] K.C. Pandey, J.L. Freeouf and D.E. Eastman, J. Vac. Sci. Technol. 14 (1977) 904.

[102] J.A. Knapp, D.E. Eastman, K.C. Pandey and F. Patella, J. Vac. Sci. Technol. 15 (1978) 1252.

[103] V. Dose, H.J. Gossmann and D. Straub, Phys. Rev. Lett. 47 (1981) 608.

[104] V. Dose, H.J. Gossmann and D. Straub, Surf. Sci. 117 (1982) 387.

[105] H.J. Gossmann and W.M. Gibson, Surf. Sci. 139 (1984) 239.

[106] H.J. Gossmann and W.M. Gibson, J. Vac. Sci. Technol. B 2 (1984) 343.

[107] C.B. Duke, S.L. Richardson, A. Paton and A. Kahn, Surf. Sci. 127 (1983) L135.

[108] C.B. Duke, A. Paton, W.K. Ford, A. Kahn and J. Carelli, Phys. Rev. B 24 (1981) 562.

[109] C.B. Duke, A. Paton, A. Kahn and C.R. Bonapace, Phys. Rev. B 28 (1983) 852.

[110] A. Kahn, J. Carelli, D. Kanani, C.B. Duke, A. Paton and L. Brillson, J. Vac. Sci. Technol. 19 (1981) 331.

[111] B.W. Lee, R.K. Ni, N. Masud, X.R. Wang and M. Rowe, J. Vac. Sci. Technol. 19 (1981) 294.

[112] C.B. Duke, A. Paton and A. Kahn, Phys. Rev. B 27 (1983) 3436.

[113] L. Smit, R.M. Tromp and J.F. van der Veen, Phys. Rev. B 29 (1984) 4814.

[114] J.C. Tsang, A. Kahn and P. Mark, Surf. Sci. 97 (1980) 119.

[115] R.J. Meyer, C.B. Duke, A. Paton, J.C. Tsang, J.L. Yeh, A. Kahn and P. Mark, Phys. Rev. B 22 (1980) 6171.

[116] S.P. Tear, M.R. Welton-Cook, M. Prutton and J.A. Walker, Surf. Sci. 99 (1980) 598.

[117] C.B. Duke, A. Paton, A. Kahn and C.R. Bonapace, Phys. Rev. B 27 (1983) 6189.

[118] L. Smit and J.F. van der Veen, Surf. Sci. 100 (1986) 183.

[119] R.J. Meyer, C.B. Duke, A. Paton, J.L. Yeh, J.C. Tsang, A. Kahn and P. Mark, Phys. Rev. B 21 (1980) 4740.

[120] C.B. Duke, R.J. Meyer, A. Paton, A. Kahn, J. Carelli and J.L. Yeh, J. Vac. Sci. Technol. 18 (1981) 866.

[121] C.B. Duke, A. Paton and A. Kahn, J. Vac. Sci. Technol. A 2 (1984) 515.

[122] C.B. Duke, A. Paton, A. Kahn and D.W. Tu, J. Vac. Sci. Technol. B 2 (1984) 366.

[123] C.B. Duke, R.J. Meyer, A. Paton, P. Mark, E. So and J.L. Yeh, J. Vac. Sci. Technol. 16 (1979) 647;
R.J. Meyer, C.B. Duke, A. Paton, E. So, J.L. Yeh, A. Kahn and P. Mark, Phys. Rev. B 22 (1980) 2875.

[124] C.B. Duke, A. Paton, W.K. Ford, A. Kahn and G. Scott, Phys. Rev. B 24 (1981) 3310.

[125] C.A. Swarts, W.A. Goddard III and T.C. McGill, J. Vac. Sci. Technol. 17 (1980) 982.

[126] C.A. Swarts, T.C. McGill and W.A. Goddard III, Surf. Sci. 110 (1981) 400.

[127] C. Mailhiot, C.B. Duke and D.J. Chadi, Surf. Sci. 149 (1985) 366.

[128] R.P. Beres, R.E. Allen and J.D. Dow, Solid State Commun. 45 (1983) 13.

[129] G.P. Williams, R.J. Smith and G.L. Lapeyre, J. Vac. Sci. Technol. 15 (1978) 1252.

[130] A. Huijser, J. Van Laar and T.L. Van Rooy, Phys. Lett. A 65 (1978) 337.

[131] G.P. Srivastava, I. Singh, V. Montgomery and R.H. Williams, J. Phys. C 16 (1983) 3627.

[132] A. Ebina, T. Unno, Y. Suda, H. Koinuma and T. Takahashi, J. Vac. Sci. Technol. 19 (1981) 301.

[133] Y.R. Wang, C.B. Duke, K.O. Magnusson and S.A. Flodström, Surf. Sci. 205 (1988) L760.

[134] D. Straub, M. Skibowski and F.J. Himpsel, Phys. Rev. B 32 (1985) 5237.

[135] D. Straub, M. Skibowski and F.J. Himpsel, J. Vac. Sci. Technol. A 3 (1985) 1484.

[136] C.B. Duke, R.J. Meyer and P. Mark, J. Vac. Sci. Technol. 17 (1980) 971.

[137] J.C. Phillips, in: Bonds and Bands in Semiconductors (Academic Press, New York, 1973) p. 42.

[138] C.B. Duke, J. Vac. Sci. Technol. B 1 (1983) 732.

[139] S.Y. Tong, W.N. Mei and G. Wu, J. Vac. Sci. Technol. B 2 (1984) 393.

[140] S.Y. Tong, G. Xu, W.Y. Hu and M.W. Puga, J. Vac. Sci. Technol. B 3 (1985) 1076.

[141] J. Bohr, I. Feidenhaus, M. Nielson, M. Toney, R.L. Johnson and I.I. Robinson, Phys. Rev. Lett. 54 (1985) 1275.

[142] D.J. Chadi, Phys. Rev. Lett. 52 (1984) 1911.

[143] E. Tekman, O. Gulseren, A. Örmeci and S. Ciraci, Solid State Commun. 56 (1985) 501.

[144] R.D. Bringans and R.Z. Bachrach, Phys. Rev. Lett. 53 (1984) 1954.

[145] R. Ludeke and A. Koma, CRC Crit. Rev. Solid State Sci. 5 (1975) 259.

[146] K. Jacobi, G. Steinert and W. Ranke, Surf. Sci. 57 (1976) 571.

[147] J. Massies, P. Devoldere and N.T. Linh, J. Vac. Sci. Technol. 15 (1978) 1353.

[148] R.Z. Bachrach, in: Crystal Growth, Ed. B. Ramplin (Pergamon, Oxford, 1980) p. 221;
R.Z. Bachrach, R.S. Bauer, P. Chiaradia and G.V. Hansson, J. Vac. Sci. Technol. 18 (1981) 797.

[149] P.K. Larsen, J.F. van der Veen, A. Mazur, J. Pollman, J.H. Neave and B.A. Joyce, Phys. Rev. B 26 (1982) 3222.

[150] P.K. Larsen, J.H. Neave, J.F. van der Veen, P.J. Dobson and B.A. Joyce, Phys. Rev. B 27 (1983) 4966.

[151] J.F. van der Veen, P.K. Larsen and J.H. Neave and B.A. Joyce, Solid State Commun. 49 (1984) 659.

[152] D.J. Chadi, C. Tanner and J. Ihm, Surf. Sci. 120 (1982) L425.

[153] L. Däweritz and R. Hey, Surf. Sci. 236 (1990) 15.
[154] P. Drathen, W. Ranke and K. Jacobi, Surf. Sci. 77 (1978) L162.
[155] D.J. Chadi, J. Vac. Sci. Technol. A 5 (1987) 834.
[156] H.H. Farrell, J.P. Harbison and L.D. Peterson, J. Vac. Sci. Technol. B 5 (1987) 1482.
[157] M.D. Pashley, K.W. Haberern, W. Friday, J.M. Woodall and P.D. Kirchner, Phys. Rev. Lett. 60 (1988) 2176.
[158] S.P. Kowalczyk, D.L. Miller, J.R. Waldrop, P.G. Newman and R.W. Grant, J. Vac. Sci. Technol. 19 (1981) 255.
[159] D.K. Biegelsen, R.D. Bringans, J.E. Northrup and L.E. Swartz, Phys. Rev. B. 41 (1990) 5701.
[160] V. Bresser-Hill, M. Wassermeier, K. Pond, R. Maboudian, G.A.D. Briggs, P.M. Petroff and W.H. Weinberg, J. Vac. Sci. Technol. B 10 (1992) 1881.
[161] J. Falta, R.M. Tromp. M. Copel, G.D. Pettit and P.D. Kirchner, Phys. Rev. Lett., in press.
[162] T. Ohno, Phys. Rev. Lett. 70 (1993) 631.
[163] M.D. Pashley, Phys. Rev. B 40 (1989) 10481.
[164] W. Chen, M. Dumas, D. Mao and A. Kahn, J. Vac. Sci. Technol. B 10 (1992) 1886.
[165] R. Duszak, C.J. Palmstrom, L.T. Florez, Y.N. Yang and J.H. Weaver, J. Vac. Sci. Technol. B 10 (1992) 1891.
[166] D. Mao, A. Kahn, G. LeLay, M. Marsi, Y. Hwu, G. Margaritondo, M. Santos, M. Shayegan, L.T. Florez and J.P. Harbison, J.. Vac. Sci. Technol. B 9 (1991) 2083.
[167] G. LeLay, D. Mao, A. Kahn, Y. Hwu and G. Margaritondo, Phys. Rev. B 43 (1991) 14301.
[168] C.J. Spindt, M. Yamada, P.L. Meissner, K.E. Miyano, A. Herrera, W.E. Spicer and A.J. Arko, J. Vac. Sci. Technol. B 9 (1991) 2090.
[169] M.D. Pashley and K.W. Haberern, Phys. Rev. Lett. 67 (1991) 2697.
[170] M.D. Pashley and K.W. Haberern, Ultramicroscopy 42–44 (1992) 1281.
[171] M.D. Pashley, K.W. Haberern and R.M. Feenstra, J. Vac. Sci. Technol. B 10 (1992) 1874.
[172] X. Yang, R. Cao, J. Terry and P. Pianetta, unpublished.
[173] R.M. Feenstra, A.J. Slavin, G.A. Held and M.A. Lutz, Phys. Rev. Lett. 66 (1991) 3257.
[174] I. Ohdomari, Appl. Surf. Sci. 56–58 (1992) 20.
[175] R.D. Bringans, D.K. Biegelsen, L.-E. Swartz and J.E. Northrup, in: The Structure of Surfaces III, Eds. S.Y. Tong, M.A. Van Hove, K. Takayanagi and X.D. Xie (Springer, Berlin, 1991) p. 555.
[176] E. Kaxiras, Y. Bar-Yam, J.D. Joannopoulos and K.C. Pandey, Phys. Rev. B 35 (1987) 9625.

[177] E. Kaxiras, Y. Bar-Yam, J.D. Joannopoulos and K.C. Pandey, Phys. Rev. B 35 (1987) 9636.
[178] R.M. Feenstra and A.P. Fein, Phys. Rev. B 32 (1985) 1394.
[179] R.M. Feenstra and J.A. Stroscio, J. Vac. Sci. Technol. B 5 (1987) 923.
[180] R.M. Feenstra, J. Vac. Sci. Technol. B 7 (1989) 925.
[181] R.M. Feenstra and P. Martensson, Phys. Rev. Lett. 61 (1988) 447.
[182] C.B. Duke, A. Paton, Y.R. Wang, K. Stiles and A. Kahn, Surf. Sci. 197 (1988) 11.
[183] T.N. Horsky, G.R. Brandes, K.F. Canter, C.B. Duke, S.F. Horng, A. Kalin, D.L. Lessor, A.P. Mills, A. Paton, K. Stevens and K. Stiles, Phys. Rev. Lett. 62 (1989) 1876.
[184] T.N. Horsky, G.R. Brandes, K.F. Canter, C.B. Duke, A. Paton, D.L. Lessor, A. Kahn, S.F. Horng, K. Stevens, K. Stiles and A.P. Mills, Phys. Rev. B 46 (1992) 7011.
[185] A. Kahn, C.B. Duke and Y.R. Wang, Phys. Rev. B 44 (1991) 5606.
[186] A. Kahn and C.B. Duke, in: The Structure of Surfaces III, Eds. S.Y. Tong, M.A. Van Hove, K. Takayanagi and X.D. Xie (Springer, Berlin, 1991) p. 566.
[187] C.B. Duke, D.L. Lessor, T.N. Horsky, G. Brandes, K.F. Canter, P.H. Lippel, A.P. Mills, A. Paton and Y.R. Wang, J. Vac. Sci. Technol. A 7 (1989) 2030.
[188] Y.R. Wang, C.B. Duke and C. Mailhiot, Surf. Sci. 188 (1987) L708.
[189] Y.R. Wang, C.B. Duke, K. Stevens, A. Kahn, K.O. Magnusson and S.A. Flodström, Surf. Sci. 206 (1988) L817.
[190] Y.R. Wang and C.B. Duke, Surf. Sci. 192 (1987) 309.
[191] Y.R. Wang and C.B. Duke, Phys. Rev. B 36 (1987) 2763.
[192] Y.R. Wang and C.B. Duke, Phys. Rev. B 37 (1988) 6417.
[193] C.B. Duke, J. Vac. Sci. Technol. A 10 (1992) 2032.
[194] K.O. Magnusson and S.A. Flodström, Phys. Rev. B 38 (1988) 6137.
[195] M.H. Tsai, J.D. Dow, R.P. Wang and R.V. Kasowski, Phys. Rev. B 40 (1989) 9818.
[196] A. Kahn, S. Ahsan, W. Chen, M. Dumas, C.B. Duke and A. Paton, Phys. Rev. Lett. 62 (1992) 3200.
[197] A. Kahn, U. Dassanayake, C.B. Duke and D.L. Lessor, to be published.
[198] D.L. Lessor, C.B. Duke, A. Kahn and W.K. Ford, J. Vac. Sci. Technol. A 11 (1993) 2205.

Surface Science 299/300 (1994) 487–501
North-Holland

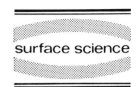

surface science

Adsorption and adsorbate-induced restructuring: a LEED perspective

M.A. Van Hove * and G.A. Somorjai

Materials Sciences Division, Lawrence Berkeley Laboratory, and Department of Chemistry, University of California, Berkeley CA 94720, USA

Received 22 March 1993; accepted for publication 9 June 1993

The history of the surface crystallography of adsorbates on single crystals is reviewed, with emphasis on the development and contributions of low-energy electron diffraction (LEED). Highlighted aspects include: progress in LEED instrumentation and theory; atomic and molecular adsorption on metals and semiconductors; coadsorption; disordered overlayers; adsorbate-induced relaxations and reconstructions; and cluster-like bonding.

1. Introduction

Surface crystallography by low-energy electron diffraction (LEED) has advanced our understanding of surfaces on two main fronts: the determination of *accurate atomic locations* in the clean substrate, in adsorbed atomic and molecular layers, and at interfaces; and the uncovering of *new surface phenomena* like multilayer relaxations and reconstructions in clean surfaces, segregation at surfaces, adsorbate-induced relaxations and reconstructions, cluster-like bonding of adsorbates, coadsorption, as well as coverage- and temperature-dependent alterations of surface chemical bonds. It is fair to say that combined studies of surface crystallography and surface phenomena provide the synergism for a much broader understanding of surface structure and bonding. This was our approach in Berkeley over the past decades.

The dramatic developments in surface structure determination over the last 30 years have led to a large body of information that was totally unavailable previously [1–7]. This was enabled by many advances in experimental and theoretical techniques, and by the emergence of much more powerful computers. Consequently, approxi-

mately 400 surface structures are now known in detail, over half of which are adsorbate structures [7].

We review here the achievements in the area of adsorbate structure determination and related surface phenomena, with a historical perspective that emphasizes the growing capabilities of surface crystallography. A major conceptual evolution has resulted and will be highlighted: the abandonment of the image of a *rigid substrate* in favor of that of a *flexible substrate* that reacts to the presence of any adsorbate, in the form of relaxations or reconstructions.

We shall focus on, but not limit ourselves to, the developments accomplished with LEED. This technique has been applied to a vast majority of the solved adsorbate structures, and has achieved the greatest level of completeness and detail in surface crystallography. Great progress continues to be achieved in developing LEED, with which dozens of unknown structural parameters can now be fit simultaneously and automatically, allowing complex relaxations to be determined overnight on today's workstations [8,9].

Many other techniques besides LEED have also contributed to the present structural knowledge of adsorbates, including primarily: scattering, channeling and blocking of ions at low, medium and high energies (LEIS, MEIS, HEIS),

* Corresponding author

surface extended X-ray absorption fine structure (SEXAFS), photoelectron diffraction (PD) and X-ray diffraction (XRD), all of which are described elsewhere in this volume. The competitive use of different techniques has been extremely beneficial to ensure reliability and accuracy and should continue.

In Section 2, we shall give a broad-brush historical overview of the evolution of LEED. We shall highlight more specific developments of the last 30 years in Section 3. We sketch the major early findings relating to the adsorption structures of atoms on metal and semiconductor surfaces in Sections 4 and 5, while addressing the adsorption structure of molecules in Section 6. Disordered adsorbates and coadsorption are treated in Sections 7 and 8. Relaxations and reconstructions induced by adsorbates are then discussed in Sections 9 and 10. We conclude in Section 11 with an outlook into the future.

2. A historical perspective

Following the discovery of low-energy electron diffraction by Davisson and Germer in 1927, Farnsworth and his co-workers labored painstakingly to monitor the directions and intensities of diffracted beams. Their studies were very difficult due to poor vacuum and slow detection systems, like the Faraday cup. But they should be credited for the invention of the technique of ion-sputter cleaning of surfaces, which is still in widespread use today, for the measurement of the inelastic electronic mean free path, and for the discovery of the reconstruction of clean semiconductor surfaces, Ge and Si, among other significant contributions.

LEED became a popular technique in the early 1960s thanks to two main developments: the rediscovery by Germer of the post-acceleration technique to detect diffracted electrons, and the widespread availability of ultra-high vacuum. The reconstruction of clean transition metal surfaces was discovered in our laboratory in Berkeley, and ordered structures of atomic adsorbates could be routinely detected on both semiconductor and metal surfaces. Careful measurement and analy-

sis of diffracted beam intensities revealed that kinematic (single-scattering) theory was inadequate for LEED. McRae in the mid 1960s demonstrated the need for a dynamical (multiple scattering) theory of LEED. By the late 1960s, thanks to the efforts of many solid state theorists, the dynamical theory of LEED was firmly established and able to reproduce experiment with high precision.

As a result, a strong collaboration between theorists and experimentalists brought about the era of surface structure determination which continues to flourish and expand to this day. This capability benefited greatly from the emerging techniques for surface composition analysis like Auger electron spectroscopy and X-ray photoelectron spectroscopy. The surface crystallography of clean surfaces was followed by the crystallography of adsorbed atoms in 1972 and of adsorbed molecular species in 1976. The surface structures of compound semiconductors and of alloys were investigated, as was the periodic step-terrace arrangement of high-Miller-index surfaces.

The significant computational demands of surface crystallography by LEED encouraged the introduction of a variety of powerful methods, like beam-set-neglect in Berkeley and Tensor LEED in London. These, coupled with the simultaneous phenomenal increase of computing power, enabled us to tackle surface structures of ever increasing complexity. Thereby, the structure of coadsorbed species could be determined to yield information on adsorbate–adsorbate interactions, while distortions within adsorbed benzene illustrated the catalytic role of the metal substrate. And the observation of adsorbate-induced relaxations and reconstructions has brought to light the active structural participation of the substrate in surface processes. The approximately 400 surface structures known to date have provided detailed insights into such diverse and important surface processes as catalysis, corrosion, crystal growth, tribology and interface states, among others. LEED has contributed a lion's share of such structural information, which forms the foundation on which modern surface science develops its concepts.

3. Evolution of LEED instrumentation, theory and computation

The early LEED measurements of the intensity of diffracted electron beams were mostly conducted with the cumbersome Faraday cup. In the late 1960s progress came in the form of post-acceleration of diffracted electrons to produce visible diffraction patterns on a fluorescent display. This provided a powerful visual characterization of the ordering and disordering properties of the surface under examination. Such LEED displays are now among the most universal tools of the surface scientist. Spot photometers were often used to measure spot intensities directly from the display.

Our group in Berkeley has contributed significantly to the introduction of more efficient techniques of data acquisition in the mid 1970s: the photographic method [10] and the video camera [11], both of which record data displayed on the fluorescent screen. The video camera is currently the standard for LEED intensity measurements, thanks to its ease of installation and operation outside the vacuum chamber. It is also sensitive enough to measure diffuse LEED intensity distributions due to disordered surfaces. The rear-view LEED instrumentation [12] displayed a much larger part of the diffraction pattern than the conventional setup, and yet remains relatively little used.

The position-sensitive detector, proposed for LEED in 1980 and developed at Berkeley, among other laboratories [13], replaces the fluorescent screen, counts individual electrons, and loads the experimental data in digital form directly into computer memory; hence the name "digital LEED". It also operates with extremely low incident electron beam currents, reducing beam-induced damage by many orders of magnitude and allowing experiments directly on insulating samples. It is also ideal for measuring diffuse intensities.

More recently, the low-energy electron microscope (LEEM) and related techniques were introduced [14]. They allow imaging of surface features on the nanometer scale, by exploiting differences in diffraction due to different local struc-

tures. The coexistence of different phases and the presence of linear defects like steps are particularly visible in LEEM.

The interpretation of measured LEED intensities was found to require a relatively sophisticated theory, which was more akin to the theory of the electronic band structure than to the theory of X-ray diffraction. The applicable principles were finally established in the late 1960s [15–18], and include a muffin-tin atomic scattering potential, a self-consistent (i.e. converged) multiple scattering formalism, a damping correction for inelastic scattering and a Debye–Waller factor for vibrational effects.

New theoretical methods and great strides in computing power are two main factors which have contributed to the impressive increases in the capability of LEED in the ensuing two decades. Rapid perturbative schemes, in particular renormalized forward scattering (RFS) [15], set the stage for later developments. Simple use of symmetry allowed somewhat larger unit cells to be analyzed efficiently [15,16,19], such as with (2×2) overlayers on simple metal surfaces or the (1×5) reconstruction of Ir(100) [20]. Systematic use of symmetry at all levels of computation [21] would permit even larger unit cells to be tackled, including, for example, the famous (7×7) reconstruction of Si(111) with its 200 non-bulk-like atoms in the surface unit cell [21c]. The combined-space method [22] was developed to deal with the frequent case where some interlayer spacings are small, including the situation where atomic layers which are planar in the bulk become non-planar (buckled) at the surface. This formulation also was very useful for dealing with molecular adsorbates, a topic of particular interest to us in Berkeley.

We made larger unit cell sizes, like (3×3) or larger, accessible in a relatively simple manner by approximations such as the beam-set-neglect (BSN) method [23,24]. BSN became very powerful for the study of coadsorption of different adsorbates, such as molecules. BSN also allowed us to study incommensurate superlattices more readily than before [25]. Furthermore, BSN provided an effective approach to the problem of disordered overlayers and the interpretation of their diffuse LEED intensity distribution [24,26]. This approach has proven very useful in our own work on disordered molecular overlayers, in particular when using diffuse LEED intensities as a function of energy at fixed parallel momentum transfer, the currently favored approach. A more exact, but computationally more demanding, 3-step method was also developed for diffuse LEED [26]. We also introduced a real-space multiple scattering theory of LEED to deal with stepped surfaces and adsorption thereon, situations which proved troublesome for earlier methods [27].

The small but universal and often complex relaxations that occur at surfaces could be determined with the Tensor LEED (TLEED) method [28] and with related methods used in bulk X-ray diffraction, especially after we coupled TLEED with automated search schemes [8,29]. This removed the prior severe limitations of "manual" trial-and-error searches, making it possible to fit dozens of parameters simultaneously. It also allowed the automatic exploration of any breaking of the substrate-like symmetry, as can occur with adsorption in low-symmetry sites, molecular tilting, and layer buckling or lateral relaxations in the substrate. TLEED in addition provided access to atomic diffusion and (harmonic or anharmonic) vibrations at surfaces [30]. TLEED is also at the root of a "direct method" [30] that attempts to solve a structure in one closed step rather than in an iterated fashion, but in practice unfortunately still needs an iterative approach.

We were closely involved with the introduction of reliability factors (R-factors) in the mid 1970s to deliver a convenient single figure of merit of the fit to experiment for any given structure [16b,31]. At present, practically all theory–experiment comparisons are quantified in this manner, and automated search schemes are based on minimizing an R-factor.

The very large improvements in computer power have benefited LEED immensely. In the early 1970s, LEED calculations typically optimized only one or perhaps two structural parameters in a matter of days. The advent of supercomputers around 1980 made the fitting of 5 to 6 parameters manageable. In the mid to late 1980s, mini- and microcomputers became powerful

enough to run LEED calculations: thus, an IBM PC-AT equipped with a coprocessor board was used in Berkeley to analyze the multilayer-relaxed missing-row reconstruction of Pt(110) in 1988 in a few days of dedicated time, compared to a comparable amount of wall-clock time on a shared supercomputer [32]. Now, in the early 1990s, we run complex automated-search LEED computations overnight on a workstation, fitting 30 or more structural parameters in the process [8]. Soon, massively parallel computers will accelerate the analyses even further; on the simplest level, the assignment of LEED calculations at single energies to different computer nodes would already reap order-of-magnitude decreases in computer time, with only a few days of reprogramming.

4. The structure of atoms adsorbed on metal surfaces

The first adsorbate structures to be analyzed by LEED were for Na on Ni(100) in 1972 [33] and I on Ag(111) in 1973 [34]. These were rapidly followed by many studies of O [35a–35c] and S [35a,c,d] on Ni(100), which became "workhorses" for many years to come, used repeatedly to explore the capabilities of new techniques and to refine existing techniques. In the early 1970s typically only one parameter was fit: the adatom–substrate bond length. One tested only a few high-symmetry sites: the hollow site with four- or three-fold coordination, the bridge site with two-fold coordination, and the top site with one-fold coordination.

Soon several trends emerged for the structure of atomic adsorption on the low-Miller-index faces of fcc metals [5]:

(1) Well-ordered overlayer superlattices were common for simple fractional coverages that correspond to periodic occupation of equivalent adsorption sites. Typical superlattices were c(2×2); p(2×2), (2×1) and ($\sqrt{3} \times \sqrt{3}$)R30° on low-index surfaces. In the early 1980s, LEED crystallography was extended to disordered overlayers as well, cf. Section 7.

(2) Assuming an unreconstructed and unrelaxed substrate often gave good results. Later, adsorbate-induced relaxations and reconstructions were to be discovered, cf. Sections 9 and 10.

(3) Adsorption mostly occurred in the highest-coordination site available, maximizing the number of adsorbate–substrate bonds, regardless of the adatom valency. Some exceptions to the rule of maximum coordination emerged with small atoms like O adsorbed on fcc(110) surfaces. Later work would uncover much greater complications in such structures: substrate reconstructions due to the adsorbate (cf. Section 10). There are rare reports of asymmetrical adsorption sites in simple structures, but such results are usually ephemeral or unconfirmed, and are usually due to fitting an insufficient number of structural parameters.

(4) Bond lengths fell within about 0.1 Å of the sum of standard covalent radii. In fact, bond lengths were soon found to fit simple rules based on the Pauling bonding principles, whereby the bond length depends on the coordination. Such rules [36] (developed primarily for atomic adsorption) even permit one to spot inconsistent, and therefore probably incorrect, results and to predict better adsorption models. A coverage dependence of adsorbate–substrate bond lengths was found later, e.g. for Cs on Ag(111) by SEXAFS and K on Ni(100) by LEED [37]: in the former case, an impressive change from 3.20 to 3.50 Å was observed upon changing the coverage of Cs from 0.15 to 0.3 monolayers [37a].

(5) A few cases of adatom penetration below the top metal layer were found. Thus, for N on Ti(0001) [38], the adatoms occupy octahedral sites between first and second metal layer, forming an atomically thin surface film with the bulk TiN structure: such results illustrated the early stages of compound formation.

(6) An even earlier stage of compound formation occurred with small atoms like C on Ni(100) [39]: they penetrate between atoms of the topmost metal layer, becoming nearly coplanar with the metal atoms in this layer. This forms a compound monolayer, which however may be accompanied by large relaxations in the metal atom positions, cf. Section 9.

5. The structure of atoms adsorbed on semiconductor surfaces

For a long time, LEED was largely absent from the structure determination of adsorbates on semiconductor surfaces, apart from the identification of diffraction patterns. Only recently has LEED been successfully applied to these cases. The problem was that the substrate undergoes substantial relaxations and often reconstructs upon adsorption, and that it is therefore difficult to identify the correct qualitative structural model. Other techniques which do not depend so much on knowledge of the substrate structure were more successful, especially SEXAFS [40], which however gave little information beyond adsorption site and adatom–substrate bond length. But recently, thanks to automated searches or patient non-automated searches, such structures have become easily accessible to LEED.

The number of solved adsorbate structures on semiconductors (or other non-metallic surfaces) is small compared to that on metals. Two primary groups of structures have been investigated: non-metal (mainly halogen) adsorption, and metal adsorption.

Halogen adatoms (and hydrogen as well) frequently appear to occupy dangling-bond sites on Si and Ge(111), as detected primarily by SEXAFS. Thus, H, Cl, Br and I choose the capping sites on Si(111) [40], terminating the bulk semiconductor lattice with the least structural disturbance. Hydrogen in particular is well known to cap Si dangling bonds and to thereby remove the clean-surface reconstruction at a sufficiently high coverage.

Among the solved structures of metal adsorbates (which are primarily chosen to study semiconductor/metal interfaces) are Ga [41] and Al [42] on Si(111), at 1/3-monolayer coverage. These are characterized by T_4-site adsorption, cf. Fig. 1. This behavior seems characteristic of large-radius

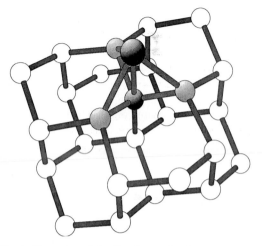

Fig. 1. Geometry of the T_4 adatom site, in perspective view: the adatom is drawn dark, sitting on top of a central atom (dark gray), around which four substrate atoms are positioned approximately tetrahedrally: three in the surface plane (light gray) and one in the second bilayer (white).

metal adatoms on Si and Ge(111) (this is the same site where the Si and Ge adatoms reside in the reconstructions of clean Si(111)-(7 × 7) and

Ge(111)-c(2 × 8) [43]). By contrast, while the smaller B on Si(111) also creates essentially the same T_4-site geometry, the boron atoms exchange places with the Si atoms directly below them, becoming subsurface atoms buried under the Si adatoms [44]. This allows more favorable strain fields in the bonds and bond angles near the adsorbate.

6. The structure of adsorbed molecules

The first structural analyses by LEED of molecules adsorbed on a metal surface were conducted in Berkeley around 1976, for the adsorption of acetylene (C_2H_2) and ethylene (C_2H_4) on Pt(111) [45]. A definitive structure for this system was published in 1979 [46], proposing ethylidyne (CCH_3) as the stable adsorbed species, cf. Fig. 2: the surface thus induced a molecular rearrangement from the gas-phase species. Since ethylidyne is particularly stable, it is no surprise that it can be formed on a number of metals, although

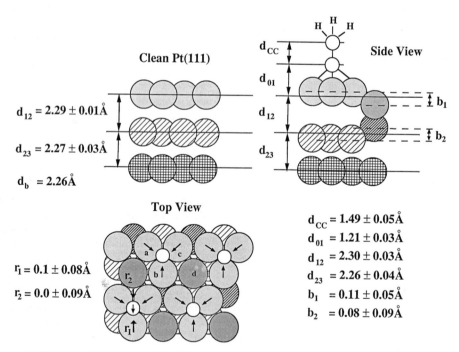

Fig. 2. Structure of Pt(111)-(2 × 2)-C_2H_3, as obtained from automated Tensor LEED, in side and top views, compared with the clean-surface structure.

not on all (e.g., a Ni surface readily breaks the C–C bond).

Starting in 1978, a simpler molecular adsorption structure was intensely studied: Ni(100)-c(2 × 2)-CO [47]. It would become another workhorse system for many structural studies. In fact, over the years CO would be studied by many groups on many metal surfaces, on several crystal faces and at several coverages, yielding a relatively complete picture of the remarkably sensitive bonding of carbon monoxide to metals [3,5,48,49]. Dissociation of CO was verified by LEED in a few cases, as on Fe(100) [50a], although a tilted intact molecule could also be prepared, as seen by photoelectron diffraction [50b]. Most structural work with CO addressed the molecularly bonded state. The intact molecule was found to mostly stand perpendicularly to the surface, with the carbon end bonding to the metal. Recent refinements using automated searches on Ru(0001) [51] and Rh(111) [52] favor a tilting of the molecule by about 9° from the surface normal, probably due to thermal wagging. Static tilting is found for tightly packed CO molecules, as in Ni(110)-pmg(2 × 1)-2CO [53], cf. Fig. 3.

Striking is the variety of adsorption sites of CO on different metal surfaces: top, bridge and hollow sites, corresponding to one-, two- and three-fold coordination (four-fold coordination seems very rare). The adsorption site varies not only from metal to metal, but also from one face to another face of the same metal, and also from one coverage to another coverage on the same face of the same metal, and furthermore from one coadsorbate to another [48]. An excellent correlation with C–O stretch vibration frequencies has helped considerably in understanding CO adsorption.

More recently, by contrast, it has become apparent that NO (nitric oxide), does not show a corresponding simple correlation between adsorption site and vibration frequencies. NO on Ni(111) [54] and Pt(111) [54b] has been found to occupy three-fold hollow sites, oriented on average perpendicular to the surface, with again (presumably thermal) tilting by about 9°.

In Berkeley we found that benzene adsorbs parallel to close-packed metal surfaces, like Pt(111) [55], Rh(111) [56] and Pd(111) [57]. The metal can break the six-fold rotational symmetry of adsorbed benzene, for example in favor of a marked Kekulé distortion when the molecule adsorbs over a three-fold symmetrical site [56], cf. Fig. 4, or more generally with a distortion that matches the symmetry of the adsorption site.

A general trend with hydrocarbon adsorption is the lengthening of C–C bonds [40b,48,55,56], especially those parallel to the surface, as in π-bonding, representing a weakening of those bonds, at the expense of strong metal–C bonds. Another description of this behavior is the changing of sp and sp^2 hybridization toward sp^2 and sp^3 hybridization of the carbons, respectively, i.e. a lowering of bond order between the carbons. These observations also agree with similar results in organometallic complexes.

A particularly important factor with molecular adsorption is temperature. Molecules often react to form different adsorbed species by internal bond breaking (dissociation) and formation (recombination) as the temperature is varied. Two-dimensional phase transformations, i.e. changes of ordered or disordered arrangements as is frequently observed with atomic adsorbates, are relatively unimportant for adsorbed molecules compared to chemical reactions.

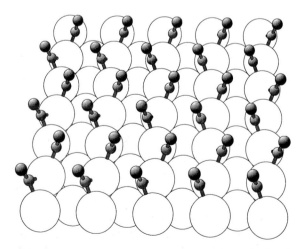

Fig. 3. Perspective view of carbon monoxide on Ni(110). CO molecules are tilted by 17° from the surface normal in two opposite directions due to steric crowding.

7. The structure of disordered adsorbates

It was realized in the early 1980s in our group and in Pendry's group in London that LEED could be applied to the problem of disordered overlayers [24,26], as long as it is of the lattice-gas type, i.e. if it has identical adsorption sites for all adsorbates (or only a few distinct sites). The local adsorption structure could be analyzed with the help of the diffuse intensities caused by the disorder; hence the name diffuse LEED or DLEED. Disordered overlayers are very common, probably more common than ordered overlayers. Some adsorbates, particularly molecules, cannot be

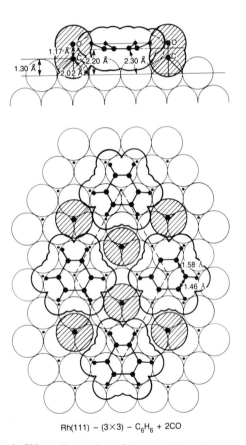

Rh(111) – (3×3) – C_6H_6 + 2CO

Fig. 4. Side and top views (above and below) of benzene coadsorbed with CO on Rh(111) in a 1:2 ratio, forming a unit cell which is outlined. Van der Waals radii are assumed for overlayer atoms. CO molecules are shaded. Large dots indicate C and O positions, medium dots indicate guessed hydrogen positions and small dots represent metal atoms in the second metal layer.

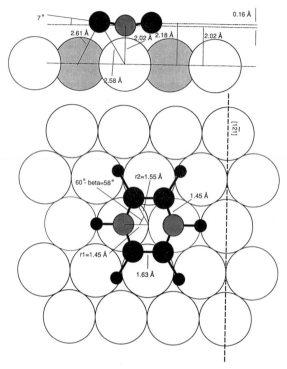

Fig. 5. Top and side views of disordered benzene on Pt(111), including distortions away from the gas-phase benzene geometry (H positions are guessed). The side view (at top) shows the out-of-plane buckling of the carbon ring, as well as selected C–metal bond lengths. The top view (at bottom) shows C–C bond lengths and ring radii.

made to order at any temperature because of competing reaction pathways. This includes many important reaction intermediates.

The first such structure determination was performed for O adsorbed on W(100) [58]. The O atoms were found to occupy four-fold hollow sites, which themselves were later discovered to be laterally contracted toward the adsorbate, through a Tensor LEED analysis [59]. DLEED was next applied to CO adsorption on Pt(111) [60], at a coverage such that most CO molecules occupy top sites and the rest bridge sites, in accordance with vibrational measurements. Similar analyses were subsequently performed for CO_2 on Ni(110) [61] and for benzene on Pt(111) [62], cf. Fig. 5. In the latter case, we detected distortions within the benzene molecule, including a "butterfly" or boat-shaped bending of the carbon ring.

8. Coadsorption

At least as early as 1976, it was found that two different adsorbates could mix and form the two-dimensional equivalent of ordered alloys or compounds. This was observed with Na and S coadsorbed on Ni(100) [63], and the structure was solved with LEED in three different conditions of Na and S coverage. Of particular interest was the arrangement whereby an adsorbate of one type (e.g., Na) was surrounded by adsorbates of the other type, and vice versa; this is similar to an ionic lattice, in which positive ions bond to negative ions, and vice versa.

We later discovered similar structures in larger numbers for molecular coadsorption, and several of these were also structurally analyzed by LEED. For instance, benzene and CO could be coadsorbed in very well ordered overlayers on Rh [56] (cf. Fig. 4), Pt [55] and Pd(111) [57]. And ethylidyne ordered well with CO or NO on Rh(111) [64]. Here again, adsorbates of one type surrounded themselves with adsorbates of the other type.

At the same time, we found a number of coadsorption combinations which do *not* order. The guiding principle that emerged for ordering [65] was whether the coadsorbates were of opposite charge-transfer character (this correlates well with the sign of work function changes). If one adsorbate is a charge donor (e.g., a hydrocarbon), while the other is a charge acceptor (e.g., CO), then mixed ordering is favored. But if both adsorbates are of the same type (both donors, or both acceptors), no ordering takes place; in this case, it may be that segregation into independent regions is favored.

Thus, coadsorbate mixing and ordering seems to be related to strong interadsorbate interactions, akin to the ionic interactions in 3D compounds. That these interactions are strong is evidenced by the fact that many of the mixed ordered layers are resistant to attack by air at atmospheric pressure for periods of hours.

Of interest also is the structural effect of one adsorbate on the other, as we observed, for instance, for CO coadsorption with benzene [56] (cf. Fig. 4) or ethylidyne [64] on Rh(111). The coadsorption changes the adsorption site of CO from top or bridge to hollow in both of these cases. This is presumably the effect of charge transfer from the hydrocarbon to the CO; the same effect is seen (by vibrational analysis) for coadsorption of CO with alkali metals, strong donors [49].

9. Adsorbate-induced substrate relaxations

A general feature of the foregoing discussions has been the tacit assumption that adsorbates do not affect the substrate structure. This assumption can not be justified when one realizes that the adsorbate–substrate bond is often very strong, frequently even stronger than the substrate–substrate bond. As a result, one should not only expect the substrate to modify the internal structure of adsorbates like molecules. One should also expect the converse: the phenomenon of adsorbate-induced relaxation and/or reconstruction of the substrate.

It was not until the late 1970s that the effect of adsorbates on the substrate structure started to be noticed, and then primarily in the form of relaxations of substrate interlayer spacings. It had been found some years earlier that the more "open" or "corrugated" clean surfaces, like the (110) surfaces of fcc metals, exhibit a contraction of the topmost interlayer spacing (this would be generalized to multilayer relaxations in the early 1980s) [66]. This contraction of the topmost interlayer spacing was now found to decrease considerably upon adsorption, often to near the bulk value or even to a slightly larger spacing [67]. This spacing expansion was found to increase with adsorbate coverage, as is well illustrated in the case of H adsorption on Cu(110) [68]. These atomic displacements amounted to typically 0.05 to 0.10 Å. The effect can be understood intuitively as the adsorbate relieving to some extent the strong asymmetry that the bare surface creates for the outermost substrate atoms: rather

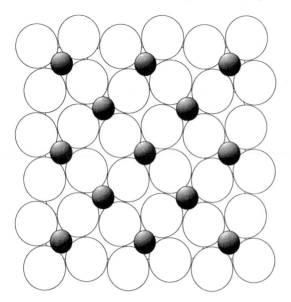

Fig. 6. Top view of Ni(100)-p4g(2×2)-2C, with C shown as small dark spheres.

than having only bonds toward the bulk, the outermost substrate atoms now also have bonds on the external side, toward the adsorbates, simulating a more bulk-like environment and reducing the asymmetry of the bare surface.

Similar effects were later also observed with some adsorbates on certain semiconductor surfaces: thus, Sb and Bi overlayers on GaAs(110) [69] remove the characteristic tilting rotation of surface bonds of the clean surface [70].

The existence of adsorbate-induced relaxations *parallel* to the surface became most obvious in 1979 with the case of carbon on Ni(100), forming a p4g(2×2) structure [39], cf. Fig. 6. In this instance, glide-plane symmetry was observed through systematic extinction of certain diffracted beams. This implied lateral displacements of Ni atoms; at that time the amount of displacement could not be determined reliably, but it was expected and later confirmed to be considerable, of the order of 0.5 Å [71]. In 1981, another case emerged with sulfur adsorbed on Fe(110) [72]: an unusually detailed and painstaking LEED analysis found that the sulfur atoms rearranged the Fe atoms in such a way as to create a nearly four-fold

symmetrical hollow site for its adsorption, rather than the more elongated rhombus-shaped site or the three-fold coordinated site available on the unrelaxed surface.

The next significant type of adsorbate-induced relaxation to be observed was buckling, as with O or S on Ni(100) [73a–73d] or with adatoms on Si(111) [41,42]. The adatom would change the position perpendicular to the surface of nearby substrate atoms more than that of distant atoms in the same layer. For Ni(100)-c(2×2)-O at 1/2 monolayer, for example, the second metal layer buckles, such that its Ni atom closest to the O adatom moves away from the adatom, i.e. deeper into the surface, relative to the other Ni atom in the c(2×2) unit cell [73a]. It was also observed in these studies that the adsorbate-induced relaxations increase markedly with increasing coverage. Similar effects have since been observed in Berkeley whenever we looked for them: Mo(100)-c(2×2)-S, Re(0001)-(2×2)-S, Re(0001)-(2$\sqrt{3}$ × 2$\sqrt{3}$)R30°-6S and Pt(111)-(2×2)-O [73e–73g].

Another type of relaxation is row pairing, which occurs, for example, with hydrogen adsorption on certain fcc(110) surfaces. On Ni and Pd(110), adjacent close-packed metallic ridges move closer together by about 0.2 Å, creating a narrower channel in which H may be able to bond to more metals atoms [74].

With automated structure search routines [8], which we now routinely use in our LEED work, it is found that adsorbate-induced relaxations are the rule rather than the exception. In fact, there is no physical reason why such relaxations should not appear, given the usually strong adsorbate–substrate interactions. These relaxations are very often of the order of 0.05 to 0.10 Å in metals and perhaps double that in semiconductors.

Recently, we have also found molecule-induced relaxations, and these are very similar to those due to adatoms. For instance, (2×2) overlayer structures of ethylidyne [75] (cf. Fig. 2) and oxygen [76], each at 1/4 monolayer, produce almost identical relaxations in Pt(111). Furthermore, molecular tilting from the surface normal of the order of 5–10° is observed [51,52,54b,75,77], which is probably of thermal character, i.e. vibrational.

The phenomenon of adsorbate-induced relaxations strongly suggests to us the picture of a cluster-like bonding of adsorbates. Namely, the bonding appears to be very local, involving primarily the nearest neighbor atoms in the substrate. Next-nearest neighbors respond already much less to the presence of an adsorbate. The cluster-like picture is also consistent with structural and vibrational comparisons of bonding at metal surfaces and in complexes containing just a few metal atoms: the local structures and properties of adsorbates on surfaces and metal clusters are remarkably similar [78].

10. Adsorbate-induced substrate reconstructions and unreconstructions

A note on terminology: We have reserved the term "relaxation" for all cases where substrate atoms are displaced by a small amount, such that no metal–metal bonds are broken or made, as discussed in Section 9. The term "reconstruction" will describe instances where the substrate must break or make bonds to change from the ideal (1×1) bulk termination to the new adsorbate-induced structure [79].

The *removal* of a clean-surface reconstruction is one well-known major effect induced by adsorbates. It takes place with many adsorbates on semiconductor surfaces, particularly when the adsorbate can cap dangling bonds, as with hydrogen and halogen adsorption, cf. Section 5. It is also observed with electronegative adsorbates, like O and to some extent CO, on the (100) and (110) faces of the metals Ir and Pt [80].

The *creation* of a substrate reconstruction by an adsorbate is even more commonplace. Often a small fraction of a monolayer is sufficient to make the entire surface reconstruct. Electron donors, like alkali metals, induce reconstructions on certain metal surfaces, like Ni, Cu, Pd and Ag(110) [81]. In this case, the (1×2) "missing-row" structure is produced, in which every other close-packed ridge of the substrate is removed. (In fact, it now appears from STM studies that a better name would have been "added-row" structure, since it seems to be formed by addition

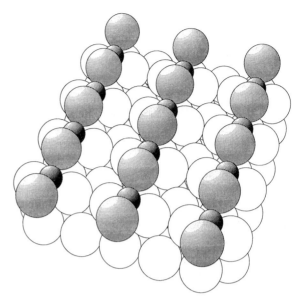

Fig. 7. Perspective view of Cu(110)-(2×1)-O, with O shown as small dark spheres.

rather than by removal of substrate atoms, by means of surface diffusion [82].) The alkali adatoms themselves seem to generally form overlayers. By contrast, surface alloying is also possible, as with Cu(100)-c(2×2)-Pd and Pd(100)-c(2×2)-Mn [83].

Small electronegative adsorbates on metal surfaces can also cause reconstructions, in which the adatoms tend to become incorporated as in a compound. This is well illustrated with the O-induced reconstructions of Ni(110) [84], Cu(110) [85] and Fe(211) [86], which consist of another type of missing-row structure. In these cases, the oxygen adatoms occupy positions midway between metal atoms in adjacent ridges, forming strings of alternating oxygen and metal atoms, while every other metal atom in each ridge is missing, cf. Fig. 7. A similar adsorption geometry has been found for O on reconstructed Cu(100) [87] and on stepped Cu(410) [88] (the latter structure, analyzed by photoelectron diffraction, is so far probably the only case of adsorption at a step solved in some detail).

On semiconductors, adsorbates not only induce or remove reconstructions, they also frequently change reconstructions. This is seen, for example, with Al [42] and Ga [41] on Si(111), whose T_4-site geometry was discussed in Section 5.

11. Conclusions and future directions

The rapid evolution of surface crystallography of adsorbates, in particular with LEED, has led to detailed structural knowledge about a wide variety of adsorption processes, from ordering to adsorbate-induced relaxations and reconstructions. Such knowledge opens insight into a multitude of surface and interface processes, from oxidation and catalysis to tribology.

The recent capability of automatically fitting dozens of parameters simultaneously will accelerate the acquisition of similar information for other categories of adsorption than studied so far. Thus, atomic and molecular structures need to be solved at the catalytically important step and other defect sites. It is important to solve series of adsorption structures on oxides and other compounds, as well as the growth of compounds due to adsorption. Also lacking now is structural information on molecular adsorption on semiconductors. Scarce at present are complete physisorption structures on any substrate. Of interest for tribology in particular is the study of buried or sandwiched adsorbate layers. Furthermore, biological molecules and systems can be usefully studied as adsorbates on substrates that fix their orientation and allow other preparations and measurements with surface science techniques.

The role of a multitechnique approach will remain highly important, especially as one addresses more complex systems. A severe shortcoming at present is the lack of a true "direct method" for crystallographic characterization; STM images are often difficult to interpret crystallographically, and it is not clear whether the exciting technique of electron holography [89] will ever achieve this goal for non-trivial structures. In any event, techniques like HREELS and STM will remain extremely useful for providing qualitative structural information as a first stage towards a more detailed structural determination.

Acknowledgments

This work was supported by the Director, Office of Energy Research, Office of Basic Energy Sciences, Materials Sciences Division of the US Department of Energy under Contract No. DE-AC03-76SF00098.

References

[1] M.A. Van Hove, S.W. Wang, D.F. Ogletree and G.A. Somorjai, in Advances in Quantum Chemistry, Vol. 20, Ed. P.O. Löwdin (Academic Press, San Diego, 1989) p. 1.

[2] J.M. MacLaren, J.B. Pendry, P.J. Rous, D.K. Saldin, G.A. Somorjai, M.A. Van Hove and D.D. Vvedensky, Surface Crystallographic Information Service: A Handbook of Surface Structures (Reidel, Dordrecht, 1987).

[3] M.A. Van Hove, in: Structure of Electrified Interfaces, Eds. J. Lipkowski and P.N. Ross (VCH, New York, 1993), p. 1.

[4] G.A. Somorjai and M.A. Van Hove, in: Investigations of Metal Surfaces, Physical Methods of Chemistry, Vol. 9B, Eds. B.W. Rossiter and R.C. Baetzold (Wiley, New York, 1993) p. 1.

[5] M.A. Van Hove, in: Structure of Solids, Ed. V. Gerold, Vol. 1 of Series Materials Sciences and Technology (VCH, Weinheim, 1993) p. 483.

[6] P.R. Watson, J. Phys. Chem. Ref. Data 16 (1987) 953; 19 (1990) 85; 21 (1992) 123.

[7] (a) P.R. Watson, M.A. Van Hove and K. Hermann, NIST Surface Structure Database Ver. 1.0, NIST Standard Reference Data Program, Gaithersburg, MD, USA (1993);
(b) P.R. Watson, M.A. Van Hove and K. Hermann, book version derived therefrom, in preparation.

[8] M.A. Van Hove, W. Moritz, H. Over, P.J. Rous, A. Wander, A. Barbieri, N. Materer, U. Starke and G.A. Somorjai, Surf. Sci. Rep. 19 (1993) 191.

[9] M.A. Van Hove, W. Moritz, H. Over, P.J. Rous, A. Wander, A. Barbieri, N. Materer, U. Starke, D. Jentz, J.M. Powers, G. Held and G.A. Somorjai, Surf. Sci. 287/288 (1993) 428.

[10] P.C. Stair, T.J. Kaminska, L.L. Kesmodel and G.A. Somorjai, Phys. Rev. B 11 (1975) 623.

[11] (a) P. Heilmann, E. Lang, K. Heinz and K. Müller, in: Determination of Surface Structure by LEED, Eds. P.M. Marcus and F. Jona (Plenum, New York, 1984) p. 463;
(b) F. Jona, J.A. Strozier, Jr. and P.M. Marcus, in: The Structure of Surfaces, Eds. M.A. Van Hove and S.Y. Tong, Springer Series in Surface Science (Springer, Heidelberg, 1985) p. 92;
(c) D.F. Ogletree, G.A. Somorjai and J.E. Katz, Rev. Sci. Instrum. 57 (1986) 3012.

[12] L. de Bersuder, Rev. Sci. Instrum. 45 (1974) 1569.

[13] (a) P.C. Stair, Rev. Sci. Instrum. 51 (1980) 132;
(b) E.G. McRae, R.A. Malic and D.A. Kapilow, Rev. Sci. Instrum. 56 (1985) 2077;
(c) D.F. Ogletree, G.S. Blackman, R.Q. Hwang, U. Starke and G.A. Somorjai, Rev. Sci. Instrum. 63 (1992) 104.

[14] E. Bauer, Surf. Sci. 299/300 (1994) 102.

[15] J.B. Pendry, Low-Energy Electron Diffraction (Academic Press, London, 1974).

[16] (a) M.A. Van Hove and S.Y. Tong, Surface Crystallography by Low Energy Electron Diffraction: Theory, Computation and Structural Results (Springer, Berlin, 1979);
(b) M.A. Van Hove, W.H. Weinberg and C.-M. Chan, Low-Energy Electron Diffraction: Experiment, Theory and Structural Determination, Vol. 6 of Springer Series in Surface Sciences (Springer, Berlin, 1986).

[17] L.J. Clarke, Surface Crystallography: An Introduction to LEED (Wiley, London, 1985).

[18] (a) S.Y. Tong, Surf. Sci. 299/300 (1994) 358;
(b) J.B. Pendry, Surf. Sci. 299/300 (1994) 375.

[19] M.A. Van Hove and J.B. Pendry, J. Phys. C 8 (1975) 1362.

[20] M.A. Van Hove, R.J. Koestner, P.C. Stair, J.P. Bibérian, L.L. Kesmodel, I. Bartoš and G.A. Somorjai, Surf. Sci. 103 (1981) 189, 218.

[21] (a) J. Rundgren and A. Salwén, Comput. Phys. Commun. 9 (1975) 312;
(b) W. Moritz and D. Wolf, Surf. Sci. 163 (1985) L655;
(c) S.Y. Tong, H. Huang, C.M. Wei, W.F. Packard, F.K. Men, G. Glander and M.B. Webb, J. Vac. Sci. Technol. A 6 (1988) 615.

[22] (a) S.Y. Tong and M.A. Van Hove, Phys. Rev. B 16 (1977) 1459;
(b) M.A. Van Hove and S.Y Tong, in: Determination of Surface Structure by LEED, Eds. P.M. Marcus and F. Jona (Plenum, New York, 1984) p. 43.

[23] M.A. Van Hove, R.F. Lin and G.A. Somorjai, Phys. Rev. Lett. 51 (1983) 778.

[24] M.A. Van Hove, in: Chemistry and Physics of Solid Surfaces VII, Eds. R.F. Howe and R. Vanselow, Vol. 10 of Springer Series in Surface Sciences (Springer, Berlin, 1988) p. 513.

[25] Z.P. Hu, D.F. Ogletree, M.A. Van Hove and G.A. Somorjai, Surf. Sci. 180 (1987) 433.

[26] D.K. Saldin, J.B. Pendry, M.A. Van Hove and G.A. Somorjai, Phys. Rev. B 31 (1985) 1216.

[27] (a) X.-G. Zhang, P.J. Rous, J.M. MacLaren, A. Gonis, M.A. Van Hove and G.A. Somorjai, Surf. Sci. 239 (1990) 103;
(b) X.-G. Zhang, M.A. Van Hove, G.A. Somorjai, P.J. Rous, D. Tobin, A. Gonis, J.M. MacLaren, K. Heinz, M. Michl, H. Lindner, K. Müller, M. Ehsasi and J.H. Block, Phys. Rev. Lett. 67 (1991) 1298;
(c) P. Pinkava and S. Crampin, Surf. Sci. 223 (1990) 27.

[28] (a) P.J. Rous and J.B. Pendry, Surf. Sci. 219 (1989) 355;
(b) P.J. Rous and J.B. Pendry, Surf. Sci. 219 (1989) 373;
(c) P.J. Rous, Progr. Surf. Sci. 39 (1992) 3.

[29] P.J. Rous, M.A. Van Hove and G.A. Somorjai, Surf. Sci. 226 (1990) 15.

[30] (a) J.B. Pendry and K. Heinz, Surf. Sci. 230 (1990) 137;
(b) J.B. Pendry, K. Heinz and W. Oed, Phys. Rev. Lett. 61 (1988) 2953;
(c) K. Heinz, W. Oed and J.B. Pendry, Phys. Rev. B 41 (1990) 10179.

[31] M.A. Van Hove and R.J. Koestner, in: Determination of Surface Structure by LEED, Eds. P.M. Marcus and F. Jona (Plenum, New York, 1984) p. 357.

[32] E.C. Sowa, M.A. Van Hove and D.L. Adams, Surf. Sci. 199 (1988) 174.

[33] S. Andersson and J.B. Pendry, J. Phys. C 5 (1972) L41.

[34] F. Forstmann, W. Berndt and P. Büttner, Phys. Rev. Lett. 30 (1973) 17.

[35] (a) J.E. Demuth, D.W. Jepsen and P.M. Marcus, Phys. Rev. Lett. 31 (1973) 540;
(b) S. Andersson, B. Kasemo, J.B. Pendry and M.A. Van Hove, Phys. Rev. Lett. 31 (1973) 595;
(c) M. Van Hove and S.Y. Tong, J. Vac. Sci. Technol. 12 (1975) 230;
(d) C.B. Duke, N.O. Lipari, G.E. Laramore and J.B. Theeten, Solid State Commun. 13 (1973) 579.

[36] K.A.R. Mitchell, S.A. Schlatter and R.N.S. Sodhi, Can. J. Chem. 64 (1986) 1435.

[37] (a) G.M. Lamble, R.S. Brooks, D.A. King and D. Norman, Phys. Rev. Lett. 61 (1988) 1112;
(b) H. Wedler, M.A. Mendez, P. Bayer, U. Löffler, K. Heinz, V. Fritzsche and J.B. Pendry, Surf. Sci. 293 (1993) 47.

[38] H.D. Shih, F. Jona, D.W. Jepsen and P.M. Marcus, Phys. Rev. Lett. 36 (1976) 798.

[39] J.F. Onuferko, D.P. Woodruff and B.W. Holland, Surf. Sci. 87 (1979) 357.

[40] (a) P.H. Citrin, J. Phys. (Paris) C8 (1986) 437;
(b) P.H. Citrin, Surf. Sci. 299/300 (1994) 199.

[41] A. Kawazu and H. Sakama, Phys. Rev. B 37 (1988) 2704.

[42] H. Huang, S.Y. Tong, W.S. Yang, H.D. Shih and F. Jona, Phys. Rev. B 42 (1990) 7483.

[43] K. Takayanagi, to be published.

[44] H. Huang, S.Y. Tong, J. Quinn and F. Jona, Phys. Rev. B 41 (1990) 3276.

[45] L.L. Kesmodel, P.C. Stair, R.C. Baetzold and G.A. Somorjai, Phys. Rev. Lett. 36 (1976) 1316.

[46] L.L. Kesmodel, L.H. Dubois and G.A. Somorjai, J. Chem. Phys. 70 (1979) 2180.

[47] S. Andersson and J.B. Pendry, Surf. Sci. 71 (1978) 75;
S. Andersson and J.B. Pendry, J. Phys. C 13 (1980) 3547.

[48] H.Ohtani, M.A. Van Hove and G.A. Somorjai, in: The Structure of Surfaces II, Eds. J.F. van der Veen and M.A. Van Hove (Springer, Berlin, 1988) p. 219.

[49] See also: J.T. Yates, Jr., Surf. Sci. 299/300 (1994) 731.

[50] (a) F. Jona, K.O. Legg, H.D. Shih, D.W. Jepsen and P.M. Marcus, Phys. Rev. Lett. 40 (1978) 1466;
(b) R.S. Saiki, G.S. Herman, M. Yamada, J. Osterwalder and C.S. Fadley, Phys. Rev. Lett. 63 (1989) 283.

[51] H. Over, W. Moritz and G. Ertl, Phys. Rev. Lett. 70 (1993) 315.

[52] A. Barbieri, M.A. Van Hove and G.A. Somorjai, to be published.

[53] D.J. Hannaman and M.A. Passler, Surf. Sci. 203 (1988) 449.

[54] (a) L.D. Mapledoram, A. Wander and D.A. King, Chem. Phys. Lett. 208 (1993) 409;
(b) N. Materer, A. Barbieri, D. Gardin, J.D. Batteas, M.A. Van Hove and G.A. Somorjai, Phys. Rev. B 48 (1993) 2859.

[55] D.F. Ogletree, M.A. Van Hove and G.A. Somorjai, Surf. Sci. 183 (1987) 1.

[56] (a) M.A. Van Hove, R.F. Lin and G.A. Somorjai, J. Am. Chem. Soc. 108 (1986) 2532;
(b) R.F. Lin, G.S. Blackman, M.A. Van Hove and G.A. Somorjai, Acta Crys. B 43 (1987) 368.

[57] H. Ohtani, M.A. Van Hove and G.A. Somorjai, J. Phys. Chem. 92 (1988) 3974.

[58] K. Heinz, D.K. Saldin and J.B. Pendry, Phys. Rev. Lett. 55 (1985) 2312.

[59] P.J. Rous, J.B. Pendry, D.K. Saldin, K. Heinz, K. Müller and N. Bickel, Phys. Rev. Lett. 57 (1986) 2951.

[60] G.S. Blackman, M.-L. Xu, D.F. Ogletree, M.A. Van Hove and G.A. Somorjai, Phys. Rev. Lett. 61 (1988) 2352.

[61] G. Illing, D. Heskett, E.W. Plummer, H.-J. Freund, J. Somers, Th. Lindner, A.M. Bradshaw, U. Buskotte, M. Neumann, U. Starke, K. Heinz, P.L. de Andres, D.K. Saldin and J.B. Pendry, Surf. Sci. 206 (1988) 1.

[62] A. Wander, G. Held, R.Q. Hwang, G.S. Blackman, M.L. Xu, P. de Andres, M.A. Van Hove and G.A. Somorjai, Surf. Sci. 249 (1991) 21.

[63] S. Andersson and J.B. Pendry, J. Phys. C 9 (1976) 2721.

[64] G.S. Blackman, C.T. Kao, B.E. Bent, C.M. Mate, M.A. Van Hove and G.A. Somorjai, Surf. Sci. 207 (1988) 66.

[65] C.-T. Kao, C.M. Mate, G.S. Blackman, B.E. Bent, M.A. Van Hove and G.A. Somorjai, J. Vac. Sci. Technol. A 6 (1988) 786.

[66] P.M. Marcus, Surf. Sci. 299/300 (1994) 447.

[67] J.F. van der Veen, R.M. Tromp, R.G. Smeenk and F.W. Saris, Surf. Sci. 82 (1979) 468.

[68] A.P. Baddorf, I.-W. Lyo, E.W. Plummer and H.L. Davis, J. Vac. Sci. Technol. A 5 (1987) 782.

[69] W.K. Ford, T. Guo, D.L. Lessor and C.B. Duke, Phys. Rev. B 42 (1990) 8952.

[70] A. Kahn, Surf. Sci. 299/300 (1994) 469.

[71] (a) Y. Gauthier, R. Baudoing-Savois, K. Heinz and H. Landskron, Surf. Sci. 251/252 (1991) 493;
(b) A.L.D. Kilcoyne, D.P. Woodruff, A.W. Robinson, Th. Lindner, J.S. Somers and A.M. Bradshaw, Surf. Sci. 253 (1991) 107.

[72] H.D. Shih, F. Jona, D.W. Jepsen and P.M. Marcus, Phys. Rev. Lett. 46 (1981) 731.

[73] (a) W. Oed, H. Lindner, U. Starke, K. Heinz, K. Müller and J.B. Pendry, Surf. Sci. 224 (1989) 179;
(b) W. Oed, H. Lindner, U. Starke, K. Heinz, K. Müller,

D.K. Saldin, P. de Andres and J.B. Pendry, Surf. Sci. 225 (1990) 242;
(c) U. Starke, F. Bothe, W. Oed and K. Heinz, Surf. Sci. 232 (1990) 56;
(d) W. Oed, U. Starke, F. Bothe and K. Heinz, Surf. Sci. 234 (1990) 72;
(e) P.J. Rous, D. Jentz, D.J. Kelly, R.Q. Hwang, M.A. Van Hove and G.A. Somorjai, Vol. 24 of Springer Series in Surface Science (Springer, Berlin, 1991) p. 432;
(f) D. Jentz, A. Barbieri, M.A. Van Hove and G.A. Somorjai, in preparation;
(g) U. Starke, N. Materer, A. Barbieri, R. Döll, K. Heinz, M.A. Van Hove and G.A. Somorjai, Surf. Sci. 287 (1993) 432.

[74] G. Kleinle, M. Skottke, V. Penka, G. Ertl, R.J. Behm and W. Moritz, Surf. Sci. 189/190 (1987) 177.

[75] U. Starke, A. Barbieri, N. Materer, M.A. Van Hove and G.A. Somorjai, Surf. Sci. 286 (1993) 1.

[76] U. Starke, N. Materer, A. Barbieri, M.A. Van Hove, G.A. Somorjai, R. Döll, M. Michl and K. Heinz, to be published.

[77] A. Barbieri, M.A. Van Hove and G.A. Somorjai, in: The Structure of Surfaces IV, Eds. X.D. Xie, S.Y. Tong and M.A. Van Hove (World Scientific, Singapore, 1994), in press.

[78] M.R. Alberts and J.T. Yates, Jr., The Surface Scientist's Guide to Organometallic Chemistry (American Chemical Society, Washington, 1987).

[79] See also: (a) K. Heinz, Surf. Sci. 299/300 (1994) 433;
(b) P.J. Estrup, Surf. Sci. 299/300 (1994) 722.

[80] (a) G.A. Somorjai and M.A. Van Hove, Cat. Lett. 1 (1988) 433;
(b) G.A. Somorjai and M.A. Van Hove, Progr. Surf. Sci. 30 (1989) 201.

[81] (a) R.J. Behm, D.K. Flynn, K.D. Jamison, G. Ertl and P.A. Thiel, Phys. Rev. B 36 (1987) 9267;
(b) M. Copel, W.R. Graham and S. Yalisove, Solid State Commun. 54 (1985) 695;
(c) C.J. Barnes, M.Q. Ding, M. Lindroos, R.D. Diehl and D.A. King, Surf. Sci. 162 (1985) 59;
(d) B.E. Hayden, K.C. Prince, P.J. Davie, G. Paolucci and A.M. Bradshaw, Solid State Commun. 48 (1983) 325.

[82] D.J. Coulman, J. Wintterlin, R.J. Behm and G. Ertl, Phys. Rev. Lett. 64 (1990) 1761.

[83] (a) S.C. Wu, S.H. Lu, Z.Q. Wang, C.K.C. Lok, J. Quinn, Y.S. Li, D. Tian, F. Jona and P.M. Marcus, Phys. Rev. B 38 (1988) 5363;
(b) D. Tian, R.F. Lin, F. Jona and P.M. Marcus, Solid State Commun. 74 (1990) 1017.

[84] (a) H. Niehus and G. Comsa, Surf. Sci. 151 (1985) L171;
(b) G. Kleinle, J. Wintterlin, G. Ertl, R.J. Behm, F. Jona and W. Moritz, Surf. Sci. 225 (1990) 171.

[85] J.A. Yarmoff, D.M. Cyr, J.H. Huang, S. Kim and R.S. Williams, Phys. Rev. B 33 (1986) 3856;
R. Feidenhans'l, F. Grey, R.L. Johnson, S.G.J. Mochrie, J. Bohr and M. Nielsen, Phys. Rev. B 41 (1990) 5420;

S.R. Parkin, H.C. Zeng, M.Y. Zhou and K.A.R. Mitchell, Phys. Rev. B 41 (1990) 5432.

[86] J. Sokolov, F. Jona and P.M. Marcus, Europhys. News 1 (1986) 401.

[87] H.C. Zeng and K.A.R. Mitchell, Surf. Sci. Lett. 239 (1990) L571.

[88] K.A. Thompson and C.S. Fadley, Surf. Sci. 146 (1984) 281.

[89] (a) S.A. Chambers, Surf. Sci. Rep. 16 (1992) 261;
(b) C.S. Fadley, Surf. Sci. Rep. 19 (1993) 231;
(c) M.A. Van Hove, in: Equilibrium Structure and Properties of Surfaces and Interfaces, Eds. A. Gonis and G.M. Stocks (Plenum, New York, 1992) p. 231;
(d) M.A. Van Hove, in: Applications of Multiple Scattering Theory to Materials Science, Eds. W.H. Butler, P.H. Dederichs, A. Gonis and R.L. Weaver, Mater. Res. Soc. Symp. Proc. 253 (1992) 471;
(e) M.A. Van Hove, MSA Bulletin 23 (1993) 119.

Surface Science 299/300 (1994) 502–524
North-Holland

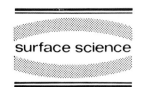

surface science

Surface steps and surface morphology: understanding macroscopic phenomena from atomic observations

Ellen D. Williams

Department of Physics, University of Maryland, College Park, MD 20742-4111, USA

Received 30 April 1993; accepted for publication 24 July 1993

The classical thermodynamics of surfaces is presented in terms of an atomic level picture in which steps serve as the fundamental unit. The important energy parameters which determine step behavior are defined, and their determination by experimental techniques is discussed. The connection between the relatively simple behavior of individual steps, and the complex macroscopic morphologies that are possible on surfaces is shown explicitly. The energy scales governing macroscopic morphologies are then evaluated in terms of the known energies relevant to step behavior. The well-known sensitivity of surface morphology to environment is shown to follow naturally from the understanding of step behavior. The mass transport required for transformations in surface morphology is similarly defined in terms of motion of steps. The experimental techniques which can be used to evaluate the kinetics of individual step motion and its relationship to driven mass transport are described.

1. Introduction

There is increasing evidence, as illustrated throughout this volume, that the surface is not simply a static substrate upon which the processes of growth and reaction occur, but often actively participates in these processes through rearrangements in structure. These rearrangements can occur as local atomic displacements [1], as large scale reconstructions [2], or as changes in surface nanostructure in which layers of atoms are displaced to new positions on the surface [3]. The latter possibility, changes in surface nanostructure, or morphology, is the subject of this contribution. It is particularly appropriate for the thirtieth anniversary of Surface Science, because the fundamental concepts needed to understand the problems of surface morphology were all put in place during the nineteen-fifties and early sixties by Herring [4], Mullins [5] and many others [6–10]. However, quantitative application of these ideas had to await the development of both experimental and theoretical methods of surface science. As I will describe below, we now have the experimental capabilities to observe surface

morphologies at an atomic level, and a theoretical understanding of the important underlying physical properties governing surface morphology. These advances allow us to use measurements or calculations of a relatively few fundamental physical parameters to understand a wide range of macroscopic behavior.

There are three general contexts in which surface morphology is important; surface/interface stability, step reactivity, and surface mass transport. While these initially appear to be completely different problems, they can in fact all be related through a description of surface thermodynamics in terms of the step structure. The first of these three canonical problems is illustrated in Fig. 1, which shows that specifying the surface orientation is not sufficient to specify the morphology. In this image, a Si surface, which has a macroscopic misorientation of 4° away from the low-index (111) orientation, exposes a "hill-and-valley" morphology, consisting of surfaces of two different surface orientations, the (111) and a surface about 20° away from the (111). The question of what determines such morphology is fully addressed by classical thermodynamics, as dis-

SSDI 0039-6028(93)E0440-6

cussed in Section 2. The formalism presented there allows one to understand why changes in morphology can be triggered easily by small changes in environment, as has long been observed during chemical adsorption [11–13], and as is of increasing concern in attempts to grow epitaxial layers with well-characterized interface structure [14–16].

The second problem of surface morphology, that of the reactivity of steps, arises because the atoms at steps sites have a lower coordination than those on terraces. Thus steps are expected to act as sites of special reactivity for adsorption [17], sites of modified activation barriers for diffusion [18], and as sources or sinks for any removal or addition of mass on the surface [6]. The same physical properties, lower coordination or higher energy, that make steps interesting physically, also makes them important in developing an atomic description of surface morphology. Thus

while a classical thermodynamic description of surface morphology requires no atomistic model of the surface behavior, a statistical mechanical description in terms of atomic-layer steps proves to be the most productive approach both in terms of developing physical insight and in terms of describing the practical processes that one most wants to understand. The use of direct experimental observations of step behavior to determine the energies governing step behavior within this formalism is described in Section 2.1.

Finally, the third problem of surface morphology, that of surface mass transport, arises during growth or other processes modifying surface structure, as well as during transformations from metastable to stable morphologies. The driving force for surface mass transport, and the kinetics depend both on the thermodynamics governing orientational stability, and the specific physical properties of the steps. The connection to surface

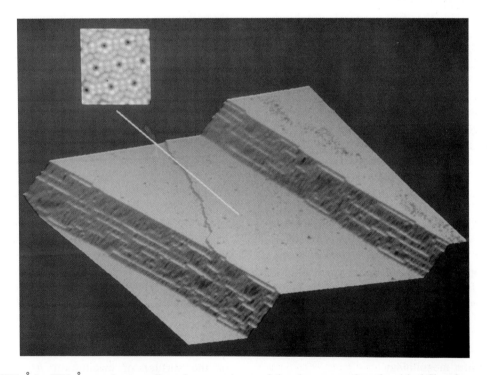

Fig. 1. A 2000 Å by 2000 Å scanning tunneling microscope image of the phase separation of a vicinal Si surface, vertical scale expanded by approximately ×10. At high temperature this surface consisted of a uniform density of steps (The net surface orientation is 4° from (111) towards [$\bar{2}$11].) The two "phases" which appear at low temperature are (7 × 7) reconstructed (111) facets and unreconstructed step bunches. There are 10 steps in each step bunch shown. The surface normal of the step bunches is temperature dependent [27,157].

thermodynamics arises because the driving force for surface diffusion is a gradient in the surface chemical potential. The connection to the physical properties of steps arises because moving material on the surface requires displacing layers of atoms, which in turn occurs through attachment and detachment of atoms at step edges. As a result, the mechanisms of surface mass transport can be related to the thermal motion of steps in equilibrium. Thus, as will be discussed in Section 3, characterization of equilibrium surface morphology and the mechanisms of equilibrium step motion provides the information needed to describe surface mass transport.

Realization of the importance of experimental studies of step structure paralleled the development of theories of surface morphology [18–21]. Early experimental confirmation of the importance of steps in surface properties was provided by electron microscopic investigations using step decoration techniques [22]. Diffraction measurements provided the first general quantitative probe of step structure under controlled UHV conditions [23–25] and remain the most useful techniques for characterizing the average surface morphology [26–35]. The development in the 1980's of UHV-compatible techniques for imaging surface structure directly [36–40] now provides the experimental complement to diffraction needed to understand surface morphology at an atomic level. In the following, I will give an overview of the theories of surface morphology [4,5,41–44], emphasizing those aspects for which direct application to experimental observations can be made. I will also give illustrative (but by no means comprehensive) examples of recent theoretical and experimental work which show our developing qualitative understanding of the important factors in surface morphology, and the potential for developing a quantitative, predictive capability.

2. Equilibrium morphology

Surface morphology arises on solids (as opposed to liquids) because the surface tension depends on the crystallographic orientation of

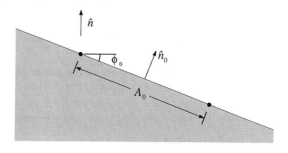

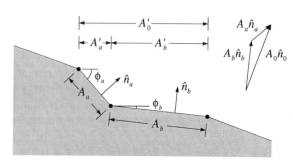

Fig. 2. A surface will be unstable with respect to faceting if the total surface tension decreases in going from the macroscopically flat surface shown in the upper panel to the hill-and-valley structure shown in the lower panel. The requirement of conservation of macroscopic orientation, given by Eq. (1), is illustrated in the insert.

the surface. As a result, the surface morphology of minimum area (i.e. a flat surface) is not necessarily the morphology which minimizes the free energy. Herring explicitly addressed the problem of the equilibrium morphology of a macroscopic surface of arbitrary orientation [45]. The important physics of his approach is illustrated in Fig. 2. The requirements for a surface of a given macroscopic orientation $\hat{n}_0$ to break up into new orientations $\hat{n}_a$ and $\hat{n}_b$ are simply that the net orientation is conserved, and the total surface tension is reduced:

$$A_0\hat{n}_0 = A_a\hat{n}_a + A_b\hat{n}_b, \tag{1}$$

$$A_0\gamma(\hat{n}_0) > A_a\gamma(\hat{n}_a) + A_b\gamma(\hat{n}_b), \tag{2}$$

where γ is the surface tension and A_i is the area of the surface of orientation $\hat{n}_i$. Several formalisms can be used to determine the morphology that minimizes the surface tension, the most common being the construction [42,46] for the equilibrium crystal shape.

It is clear, therefore, that prediction of surface morphology requires knowledge of the orientational variation of the surface tension. Experimental determinations of the anisotropy of surface tension can be made, although the number of such studies is limited due to their difficulty [20,47,48]. A rough feeling for how the surface tension (which at $T = 0$ equals the surface energy) varies with orientation can be obtained by using a rigid lattice model for the solid and counting the number of broken bonds for each surface orientation. Kern [43] gives a detailed review of the relationship between models of varying complexity and the orientational stability: generally, low-index (high symmetry) surfaces are the most stable, and the inclusion of longer range interactions results in the stabilization of lower symmetry surface orientations. Bond breaking models obviously overestimate the energy of surfaces, as they do not include the energy savings involved in surface reconstructions. Ideally, this problem would be addressed by first principles calculations, which are increasingly feasible for the large unit cells of low-index surfaces [49,50]. Practically, first principles calculations may be used to provide reference values for a few orientations. Useful semi-quantitative information about the orientation-dependence of the surface tension can be obtained by semi-empirical techniques such as the embedded atom method [51], as illustrated by the results of Wolf [52], shown in Fig. 3.

Calculations of surface energies are a necessary starting point for understanding surface morphology. However, because entropy is very important in surface morphology, energy calculations, which correspond to structures at zero temperature, do not allow an understanding of the behavior of real surfaces. There are a variety of approaches, such as molecular dynamics simulations [53,54] which one can use to address this issue. The most immediately useful approach for an experimentalist attempting to obtain physical insight is to describe the orientational variation of the surface tension in terms of the physical properties of steps [42,55]. In the following section (2.1), I will describe in some detail how one can determine the important energies governing the

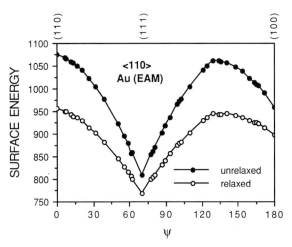

Fig. 3. Embedded-atom-method calculations of the variation of the surface energy (equal to the surface tension at $T = 0$) with surface orientation for Au. The closed circles show the values for bulk-terminated surface structures, and the open circles the values when relaxation of lattice spacing perpendicular to the surface is allowed. (Figure from the work of D. Wolf, Ref. [52]).

structure of a clean stepped surface. This detailed description will set the stage for the subsequent section (2.2), where the conditions under which an external perturbation, such as adsorption, can alter the surface morphology are described.

2.1. Simple stepped surfaces

The first step in developing a description of the surface tension based on the properties of steps is to assume that there are cusps in the surface tension at a relatively small number of low index orientations, and that between these orientations, the surface tension varies smoothly. One can then describe changes in orientation in terms of changes in step density, and introduce the temperature dependence of the surface tension by treating the entropy of the steps correctly [55,56]. The variation of the surface tension with step density, $\tan \phi$, should then have a leading term which is due to the contribution of the terraces between the steps, a second term due to the free energy cost of the steps, and a third term due to any step–step interactions [42]. The inade-

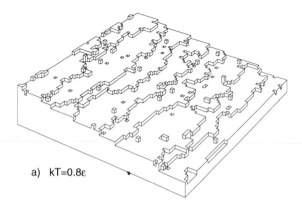

a) kT=0.8ε

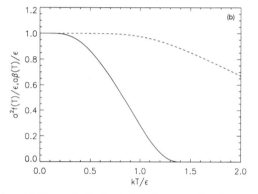

Fig. 4. (a) Monte Carlo simulation of a stepped surface in the SOS model at $kT = 0.8\epsilon$, where ϵ is the kink energy. (Figure from the work of Bartelt et al. [158].) (b) Calculation of the variation of the surface tension $f^0(T)$ (dashed line) and step energy $\beta(T)$ (solid line) for the body centered solid-on-solid model, based on the work of van Beijeren [62,159].

quacy of zero-temperature calculations of the surface tension is immediately apparent if one considers the temperature dependence of these terms separately. Because thermal excitations at the edge of a step are much less costly than those from a terrace, as illustrated in Fig. 4a, the contribution to the temperature variation of the surface tension of vicinal surfaces due to the steps is much larger than that of the terrace, as shown in Fig. 4b. Thus one cannot simply extrapolate relative zero-temperature energies for high and low-index surfaces to non-zero temperature [57].

The correct treatment of the problem of the entropy of steps as a function of step density is not obvious, and was first addressed explicitly by Gruber and Mullins [55]. They noted that at non-zero temperature, a step can wander due to the thermal excitation of kinks, as illustrated in Fig. 4a, thus generating configurational entropy which lowers the free energy of the isolated step, as illustrated in Fig. 4b. However, when there is a train of steps, the wandering is constrained because one expects that a step crossing, which would create an overhang, would be energetically unfavorable. Thus, the entropy of step wandering makes the largest favorable contribution to the surface tension when the step density is lowest, an effect which is referred to as the "entropic step repulsion". This entropic repulsion has the same effect as a true energetic repulsion which falls off in strength as the square of the distance between the steps. Subsequent work provided a quantitative description of how this entropic repulsion (as well as any energetic interaction which is not of shorter range) influences the surface tension [44,58]. The result is that the step interaction term is proportional to the third power of the step density, giving an overall expression for the reduced surface tension [58]:

$$f(\phi, T) = \gamma^0(T) + \frac{\beta(T)}{h} |\tan \phi|$$

$$+ g(T)|\tan \phi|^3, \qquad (3)$$

where $\gamma^0(T)$ is the surface tension of the terraces between the steps, $\beta(T)$ is the free energy cost per unit length of an isolated step, h is the step height, and $g(T) |\tan^3\phi|$ is the free energy cost per unit area due to step–step interactions [59,60]. Within a lattice model for the surface, as illustrated in Fig. 4, exact expressions are available for the temperature dependence of both the step formation energy [61,62] and the step interaction energy [63] in terms of the kink energy [58]. Thus given a tabulation of calculated zero-temperature energies for the facet, step and kink, as illustrated in Table 1, one can estimate both the thermal and orientational variation of the surface tension.

The advent of experimental techniques which allow direct imaging of steps allows the statistical mechanical description outline above to be tested directly. In particular, it is now possible to deter-

Table 1
The energetics of steps on Ag(100) and Ag(111) as obtained from the embedded atom method [155]

Surface orientation	Facet energy $\gamma^0(T=0)$ (meV/Å^2)	Step-staircase direction [a]	Step energy $\beta(T=0)$ (meV/Å)	Kink energy ϵ (meV)
(100)	44	{011}	36	102
		{001}	49	
(111)	39	{$\bar{2}$11}	65	102
		{2$\bar{1}\bar{1}$}	66	99
		{1$\bar{1}$0}	76	

[a] The staircase direction is the vector perpendicular to the average step edge, pointing in the downhill direction.

mine a partition function experimentally by compiling statistical distributions of the kink structure and the terrace widths. From such measurements one can test whether the step-wandering is consistent with thermal equilibrium and deduce the energetics governing surface morphology. A particularly beautiful example, in which a static kink structure was measured with atomic resolution, was performed by Swartzentruber et al. for kinks on stepped Si(100) [64]. They showed that the kink structure obeys Boltzmann statistics, and deduced the kink energy. Generally, even if steps cannot be imaged with atomic resolution, the mean-square displacements of steps can be measured microscopically, and interpreted in terms of a chosen model for kink excitations to give a kink energy [65–67]. Examples of such analyses are the STM measurements of "frizzled" step edges on Ag and Cu surfaces [68], on Au surfaces [69], and the REM measurements of step wandering on Si(111) by Alfonso et al. [70]. The possible complexity of equilibrium structure for an anisotropic system has been shown using LEEM to image steps on Si(001) [71,72].

Direct measurements of the terrace width distribution contain information about both the thermal excitations of the step edge and the nature of the step–step interactions. As mentioned above, bringing steps near to one another decreases the amount of wandering, and thus the configurational entropy, leading to an effective entropic repulsion between steps, even when there is no direct energetic interaction. This entropic repulsion also manifests itself in the spatial distribution of the steps. A step which is midway between its neighbors has more configurational entropy than one that is near a neighboring step, and thus the distribution of step-separations will be peaked near the average step-separation [73–75]. The distribution for the case where all the steps are wandering simultaneously can be solved exactly for moderate temperature, and is shown as the solid curve in Fig. 5 [74]. This scaled curve is universal: in other words it will be the same for any system regardless of the average step separation and the kink energy. Deviations from this universal form thus can be used to deduce the presence of true energetic interactions between the steps, as illustrated by the dashed curves in Fig. 5 [74]. If there are attractive interactions (of insufficient strength to overcome the entropic repulsion and cause step coalescence), the distribution becomes broader and peaked at a value smaller than the average step separation. If there are repulsive interactions, the distribution becomes narrower, and the peak moves to slightly larger value than the average step separation.

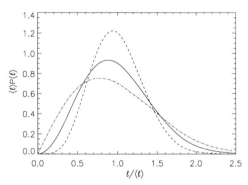

Fig. 5. Distribution of step–step separations calculated using the free-fermion approximation (Figure from B. Joos et al. [74]). $P(L)$ is the probability of observing two steps separated by a distance L. The behavior of freely wandering steps is shown by the solid curve. The distribution for steps with a repulsive energetic interaction in addition to the entropic behavior is shown by the dashed curve. The distribution for steps with an attractive energetic interaction in addition to the entropic behavior is shown by the dash-dot curve.

Quantitative measurement of the shape of the distribution function can thus provide information about the nature of the interaction. Unfortu-

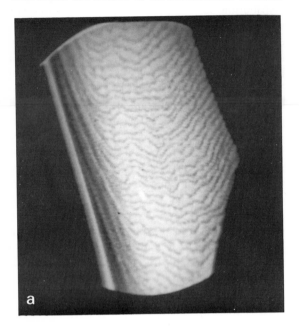

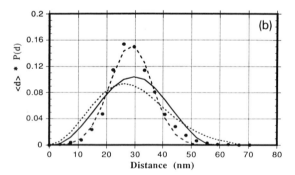

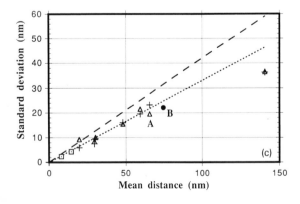

nately, except for a few special cases, analytical expressions relating the shape of the distribution to the strength of the interaction do not exist. However, for repulsive interactions, a Gaussian distribution is an excellent approximation [65] to the analytical form. The result is generally true for any repulsive interaction [76–79], but assumes an especially simple form if the interaction between steps is of the type

$$U(x) = Ax^{-n}, \tag{4}$$

where x is the distance between steps measured perpendicular to the average step edge direction. In this case, the distribution of step separations is approximately a Gaussian of width

$$w = \left(\frac{kTb^2(T)}{8n(n+1)Aa} \right)^{1/4} l^{(n+2)/4}, \tag{5}$$

where l is the average step–step separation, $b^2(T)$ is the diffusivity which is measured from observations of the step wandering, and a is the minimum kink–kink separation [65]. A measurement of the width of the distribution as a function of the average step–step separation thus in principle allows a determination of both the form (value of n) and magnitude (value of A) of the step interactions.

Terrace width distributions have been measured for several systems using STM and REM. Distributions characteristic of repulsive step–step interactions [70,73,78,80], attractive step–step interactions [81], and non-interacting steps [82] have been observed. The most extensive data set has

Fig. 6. Studies of the distribution of steps on a Si(111) surface at 900°C, measured by Alfonso et al. [70] using reflection electron microscopy. (a) A train of steps of average separation approximately 300 Å. Note the field of view is greatly compressed in the direction roughly parallel to the step edges. (b) The measured probability distribution function of the step–step separation for the train of steps of a, shown in comparison with the expected distribution for entropically interacting steps. The data are well fit by a Gaussian. (c) The width of the distributions measured for trains of steps with average separations varying from approximately 300 Å to 1400 Å. The linear behavior suggests a $1/l^2$ form for the energetic repulsions between the steps. Figures from the work of Alfonso et al. [70].

been obtained by Alfonso et al. for steps on Si(111) using REM [70], which allows a large range of step separations to be studied by virtue of its large (compared to STM) field of view. The results are illustrated in Fig. 6. The terrace separation distribution has a Gaussian shape, and is substantially narrower than the distribution for non-interacting steps, which indicates the presence of repulsive step–step interactions. The dependence of the width of the distribution on the average step separation is roughly linear, suggesting that the repulsion falls off as the square of the distance ($n = 2$ in Eqs. (4) and (5)) [70,83,84], with a magnitude of $A \approx 0.2$ eV Å. This repulsive interaction increases the magnitude of the step-repulsion coefficient in Eq. (3) by a factor of about 4 above the value of 4 meV/Å^2 that would arise for purely entropically interacting steps at 900°C, to a value of 16 meV/Å^2.

While it is difficult to prove the form of the step–step interaction using Eq. (5), because the dependence on the exponent n enters as $n/4$, physical predictions for step–step interactions through stress [85,86] or dipole form [56,87] all suggest a repulsive inverse square relationship [76]. Stress mediated interactions are expected to be comparable or larger in magnitude than dipole interactions, and this provides a useful method for estimating the magnitudes of step–step interactions. The physical basis for stress-mediated interactions is illustrated in Fig. 7. Due to the different number of neighboring atoms for surface and bulk atoms, the lattice constant which minimizes the energy will generally be different for the surface than the bulk. (This can be shown intuitively by considering a Lennard-Jones interaction [52,88]). The surface atoms can accommodate this difference by relaxation in the direction

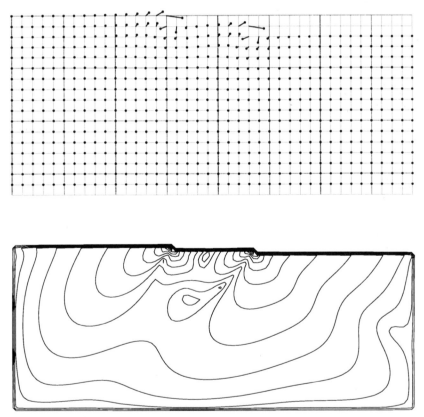

Fig. 7. The atomic displacements near two step on a surface, calculated in a simple ball and spring model of atomic interactions are shown (a) as real-space displacements magnified ×20 for clarity, and (b) contours of constant log-displacement, showing the extent of the strain field into the bulk.

perpendicular to the surface, but are forced geometrically to maintain the lateral periodicity of the bulk. As a result, the surface is under a stress, which is the effective force holding the atoms in registry with the bulk. When there is a step on the surface, the atoms near the step can relax towards the preferred lattice constant, creating a strain (or displacement) field as shown in Fig. 7. This relaxation lowers the energy of the surface. However, when a second step (in the same direction) is created on the surface, the atomic displacements which lower the energy, oppose the displacements of the first step, and the net gain of relaxation energy is reduced. Thus there is an effective step–step repulsion. While this is an atomic picture, the form and magnitude of the stress-mediated step-interactions can be calculated using bulk elasticity theory, and are accurate down to atomic length scales [89,90]. The relationship between the step interaction strength and the elastic parameters is:

$$A = \frac{2(1 - \sigma^2)}{\pi E} \tau^2 h^2, \qquad (6)$$

where σ is Poisson's ratio, E is Young's modulus, τ is the surface stress and h is the step height

[86,91]. The values of Poisson's ratio and the Young's modulus are bulk parameters which can be obtained from various tabulations [92]. The surface stress depends on the details of the surface structure, and must be measured or calculated directly. Increasingly there are good calculations [93,94] of the surface stress, and measurements of changes in the surface stress during chemical adsorption or deposition [95–98]. There is also the possibility of determining stress from direct observations of the strain field [99,100]. Some of the available values are listed in Table 2, along with the corresponding values of the step-interaction coefficient A. There is good agreement between values of the step interaction coefficient determined from measured step distributions [70,73,101–103] and values determined from the surface stress using Eq. (6). Thus preliminary estimates of the strength of step–step interactions can be obtained from the tabulations of surface stresses which are becoming increasingly available [52,90,93,94,104–106].

Thus within the formalism described above, we have a capability for understanding and qualitatively predicting the behavior of steps on vicinal surfaces. With good calculations or measure-

Table 2
Calculated values of the absolute surface stress are listed in the upper section, along with the corresponding step-interaction coefficient (as in Eqs. (4) and (6); measured changes in surface stress due to adsorption are listed in the lower section of the table

Surface	Stress (eV/$\mathring{A}^2$)	Ref.	Step interaction coefficient A (eV $\mathring{A}$) (calculated [156])
Si(111)-7 × 7	0.186	[93,95]	0.18
Si(111)-As 1 × 1	0.178	[93]	0.16
Si(111)-2 × 2	0.130	[93]	0.09
Al(111)	0.078	[94]	0.04
Ir(111)	0.331	[94]	0.08
Pt(111)	0.349	[94]	0.26
Pb(111)	0.051	[94]	0.05
Au(111)	0.173	[94]	0.13

Surface + adsorbate	Change in stress	Ref.	
Ni(100)-c(2 × 2) S	− 0.41	[96]	–
Ni(100)-c(2 × 2) O	− 0.47	[96]	–
Ni(100)-c(2 × 2) C	− 0.53	[96]	–
Si(100)-O	+ 0.02	[97]	–
Si(111)-O	− 0.45	[97]	–

ments of kink energies and step-interaction energies, this capability can become quantitative. This formalism however provides no mechanism for understanding how non-uniform step distributions, which correspond to the hill-and-valley structure illustrated in Figs. 1 and 2b, arise. For a clean, uniform surface, the smoothly varying form of Eq. (3) is consistent with stable vicinal surfaces, a stability which is in fact observed experimentally for most clean stepped surfaces [26]. As described in the following section, non-uniform distributions of steps can arise when an external perturbation changes the shape of the surface-tension curve.

2.2. Non-uniform step distributions

As suggested by Eqs. (1) and (2), the break-up of a surface into a hill-and-valley structure is

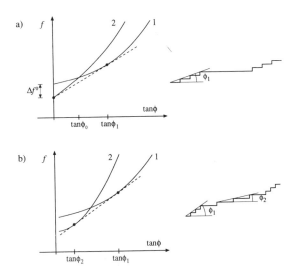

Fig. 8. Schematic illustration of intersecting curves of the reduced surface tension as a function of the step density, and the corresponding orientational phase separation. Curves 1 and 2 may represent surfaces with different composition, or different reconstruction, either of which is likely to depend on the step density in its effect on the surface tension. (a) If the surface tension at zero step density decreases due to the change in conditions from curve 1 to curve 2, while the step energy (and thus the initial slope of the curve) increases, phase separation between the flat surface and a surface of higher misorientation can occur. (b) If the surface tension at zero step density decreases and the step–step interaction energy (and thus the curvature) increases, phases separation can occur between orientations of higher and lower step density.

analogous to the phase separation of a solution. In this case the surface orientation, or the step density, plays the same role for surfaces as does the concentration in describing the phase separation of a solution. It can be shown that orientational phase separation will occur when the reduced surface tension is not a monotonically increasing, convex function of the step density [9]. Such a non-convex reduced surface tension curve could arise for a clean surface if there were attractive interactions between steps [81,87], or if reconstructions stabilized specific terrace widths [53]. However, the experimental observation that most clean stepped surfaces are stable, while faceting is frequently observed during chemical adsorption suggests another mechanism by which a non-convex surface tension can be created. As suggested by Cahn [107], if some perturbing process, such as adsorption, completely alters the shape of the surface tension curve, while not shifting the absolute energy dramatically, then the process can result in the formation of intersecting curves as shown in Fig. 8. If we consider the curve labeled 1 as the curve of the clean surface, and curve 2 as that of the surface in the presence of some specific concentration of an adsorbate, then the tie bar construction shows that phase separation will occur between not only different orientations, but different surface compositions.

It is not difficult to imagine the physical conditions which might in general drive orientational phase separation: as a simple case consider the transition, illustrated in Fig. 8a, between a uniformly stepped surface, and a surface with a hill-and-valley structure consisting of a low-index facet and bunches of steps [108]. This process can occur if a perturbation such as adsorption, lowers the surface tension of the facet but increases the step or kink energy. Conversely, a similar effect would occur if the perturbation affected the facet surface tension unfavorably, while lowering the step or kink energies. These physical ideas can be quantified by assuming that the generic form of Eq. (3) applies to both the clean and the perturbed surface. For this case it is relatively easy to set up the equations describing the tie bar [109]. The requirement that the tie bar is tangent to the curve at point 1, gives the requirement on how much the surface tension must be changed by the perturbing process in order to cause the phase separation. The result shows that the energy scale for this process is set by the necessity of overcoming the increased step–step repulsions in the step bunches:

$$\Delta\gamma^0 = \gamma_1^0 - \gamma_2^0 = 2g_1(T)|\tan \phi_1|^3. \qquad (7)$$

A second requirement is that the initial slope of curve 2 must be steeper than the slope of the tie bar (which is in turn tangent to curve 1). The initial slope is, of course, set by the energy cost of isolated steps, so this mathematical requirement tells us how much the perturbing process must

change the step energy in order to cause faceting. The resulting inequality is:

$$\beta_2(T) - \beta_1(T) > 3g_1(T)h_1 \tan^2\phi_1, \qquad (8)$$

where the step heights are the same in both phases. Thus both Eqs. (7) and (8) show that the magnitude of the step-interactions sets the energy scale for step bunching to occur. The magnitude of the change in the surface tension and the step energy required to allow orientational phase separation can be estimated using the values such as those listed in Tables 1 and 2. The result is that approximately a 0.1% change in the facet energy coupled with approximately a 10% change in the step energy is needed to cause phase separation

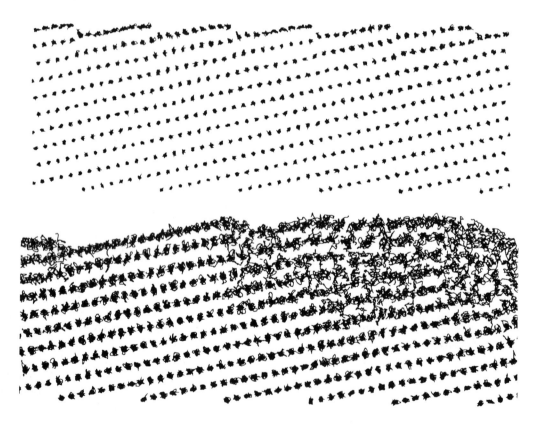

Fig. 9. Molecular dynamics simulation of a Au(534) surface [54]. At low temperature, $T = 0.37T_m$, the stepped surface is orientationally stable. At elevated temperature, $T = 0.99T_m$, a surface melting transition lowers the surface tension of the surfaces of larger misorientation relative to that of the flat surface, causing the orientational phase separation shown. The intersecting reduced surface tension curves for this phase separation are similar to those of Fig. 8a. Figure from the work of Bilalbegovic et al. [54].

between a flat surface and a step bunch. Such a mechanism has been proposed for the orientational phase separation of vicinal Si(111) [61], Pb(111) [110,111] and for Au(111) near the melting temperature [48,54,112]. For Si(111), which is illustrated in Fig. 1, and for Au(111) at moderate temperature, the perturbing process is the surface reconstruction, so that the faceted structures consist of reconstructed (111) terraces separated by unreconstructed groups of steps. For Pb(111) and Au(111) close to the bulk melting temperature, the phase separation is driven by surface melting as illustrated in Fig. 9. In this case, the faceted structure consists of unmelted (111) terraces, separated by melted regions of higher misorientation angle, for which Eq. (3) is of course no longer a valid description.

The more general case of phase separation between surfaces of two different step densities [109,113], illustrated in Fig. 8b, is governed by a similar pair of equations:

$$\Delta\gamma^0 = \gamma_1^0 - \gamma_2^0 = 2g_1(T)|\tan \phi_1|^3$$
$$- 2g_2(T)|\tan \phi_2|^3, \qquad (9)$$

$$\beta_2(T) - \beta_1(T) = 3g_1(T)h_1 \tan^2\phi_1$$
$$- 3g_2(T)h_2 \tan^2\phi_2. \qquad (10)$$

This result shows that, within the formalism of Eq. (3), step phase separation can occur if a perturbing process causes large changes in step interactions, which could arise for instance from adsorption-induces changes in surface stress, as listed in Table 2. Phase separation between two low-symmetry orientations has been observed for vicinal Pb(111) [114,115]. In this case, the perturbing process is surface melting, so that phase separation occurs between an unmelted surface of low step density, and a melted surface of higher misorientation angle, as illustrated in Fig. 10. In this case, of course, Eq. (3) is not appropriate to describe the free energy of the melted surface. However, the driving mechanisms for the phase separation can still be understood in general physical terms as due to changes in the initial slope and curvature of the surface tension curve.

Observations of faceting induced by adsorption are common, and can be understood in terms

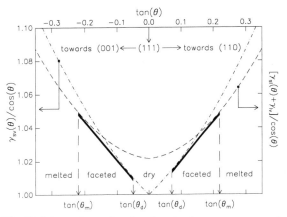

Fig. 10. Intersecting reduced surface tension curves proposed for the orientational phase separation of Pb(111) [114], observed using medium energy ion scattering. The decreased curvature of the curve for the melted surfaces leads to phase separation between two surfaces of non-zero step density, as suggested in Fig. 8b. Figure from the work of van Pinxteren et al. [114].

of the formalism describe above. It is thus important to consider how one can use the procedure described above to make predictions about the stability of stepped surfaces with respect to changes such as adsorption or deposition. In this context, it is worthwhile to emphasize that the curves illustrated in Fig. 8 correspond to constant coverages. Under an equilibrium vapor pressure (chemical potential) of an adsorbate, phase separation will occur between different orientations containing different coverages of the adsorbate. These different orientations correspond to the points of tangency illustrated in Fig. 8. To determine whether such a phase separation occurs, it is necessary to know how much adsorption shifts the surface tension curve. One can estimate the change in surface tension upon adsorption from the change in the surface energy at $T = 0$: if the bond energy of an adsorbate is E per atom independent of coverage, then the change in surface tension upon adsorption is

$$\Delta\gamma(T = 0) = \theta E / A_{uc}, \qquad (11)$$

where θ is the coverage and A_{uc} is the unit surface area. Thus for typical covalent bond energies of approximately 1 eV, the shift in the surface tension curve required to allow faceting as in

Eqs. (7) or (9), corresponds to small differences in coverage (≈ 0.1 ML) between the two orientations. In order to quantify this estimate for any real experimental system, one would measure how much the surface tension (on the flat surface, $\tan \phi = 0$) changes during adsorption using the Gibbs adsorption equation:

$$\left(\frac{\partial \gamma}{\partial \mu_i} \right)_{\mu_j, T} = -\Gamma_i, \tag{12}$$

where Γ_i is the surface excess (or coverage) of the adsorbed material, and μ_i is its chemical potential. If the adsorbate is present as an ideal gas in equilibrium with the surface, then a measurement of the coverage versus the pressure at constant temperature on a flat surface can be integrated immediately to obtain the change in the surface tension γ^0 as a function of coverage. The second requirement for phase separation to occur is that either the step energy (Eq. (8)) or the step-interaction energy (Eq. (10)) increase in the phase for which the surface tension of the flat surface has decreased. An increase in the step energy with increasing coverage will occur if adsorbates bind more strongly to the terraces than to the step edges, and vice versa. The signature of such a change should be apparent in direct images of the step wandering: for instance, if an adsorbate is strongly bound at a step edge, the kink excitation cost should increase and step wandering should decrease (or vice versa). A change in the step-interaction energy will occur if adsorption changes the amount of step wandering and thus the entropic interaction, or if there is an adsorbate-induced change in the surface stress. Such changes will be readily observable from measurements of step-separation distributions. For any system of interest, a relatively small number of measurements of step behavior during adsorption or deposition using direct imaging techniques will provide the information needed to understand the thermodynamics governing surface morphology over a broad range of conditions.

The formalism outlined above provides both a qualitative guide to the issues governing equilibrium morphology, and a prescription for the types of measurements or calculations which must be performed to quantify the issues for a given system of interest. It should be readily apparent that the equilibrium morphology of any given system is not robust: the energy balance governing whether a step structure is stable is determined by the relatively small energies of the step–step interactions. Thus the frequent reports of adsorption-induced faceting are not surprising [12,13, 116]. However, to evaluate the importance of morphological changes in surface processes fully, one also needs to consider the ease with which they occur. In other words, for surface morphology to play a dynamic role in surface processes (rather than just defining the structure of the static substrate), the kinetics of surface morphological changes must be accessible under ordinary surface processing conditions.

3. Step kinetics and surface morphology

The kinetics of changes in morphology are governed by the rates with which individual atoms move from site to site on the surface, and by the driving force for mass transport. These two factors appear in the equation describing the surface diffusive flux J through D, the self-diffusion coefficient, and the gradient in the chemical potential μ [4,5]:

$$J = -\frac{nD_c}{\Omega_0 kT} \nabla \mu, \tag{13}$$

where n is the concentration of the diffusing species, and Ω_0 is the atomic volume of the diffusing species. For a given non-equilibrium morphology, the gradient of the chemical potential can be described in terms of the orientations, or step densities, exposed on the structure as a function of the position x along the structure [5]:

$$\mu(x) - \mu_0 = \frac{\Omega_0}{R} \left(\gamma(\phi) + \frac{\partial^2 \gamma}{\partial \phi^2} \right) \Bigg|_x, \tag{14}$$

where μ_0 is the reference chemical potential, the values of the surface tension γ and its derivative are evaluated for the local step density at point x, and R is the radius of curvature at point x [117].

The relationship of these classical equations to mass transport on surfaces described in an atomic picture of steps and terraces, as shown in Fig. 4a, is an area of active investigation. The first question in this problem is the meaning of the diffusion coefficient in a process in which there are many elementary rate processes. Specifically, one needs to consider atomic motion along a step edge (kink motion), atomic attachment and detachment between the step and terrace, atomic motion across steps, and atomic motion along the terraces as all contributing to the diffusion process [18,118]. The second question is how to apply the continuum description of a chemical potential gradient to a surface which consists of discrete series of terraces and wandering steps [119,120]. In the following section (3.1) I will describe how direct observations of the kinetics of equilibrium step wandering can be used to determine the important atomic process in surface self-diffusion. The relationship of this information to experimental and theoretical studies of non-equilibrium mass transport will be discussed in Section 3.2.

3.1. Mechanisms of step motion

The same elementary processes which allow steps to fluctuate in thermal equilibrium, will also govern surface mass transport. Thus there is substantial motivation for trying to determine the nature and rates of the elementary processes from direct observations of the kinetics of step motion. True imaging techniques such as REM, LEEM, and plan-view TEM [121], because of the speed of image acquisition and large field of view, allow real time observations of step motion. Notable have been observations of the evolution of step structure during growth or etching, and of real-time fluctuations in the positions of steps under equilibrium conditions [70,122–125]. A striking example is provided in the REM [70] observations of the fluctuations of step position on Si(111) which occur over distances of hundreds of ångströms, as illustrated in Fig. 11 [126]. More recently, as STM techniques have improved in both thermal range and speed of image acquisition, direct observations of the kinetics of step

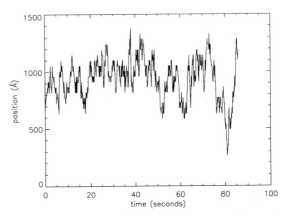

Fig. 11. Real-time observations of the step wandering on Si(111) at 900°C measured by Alfonso et al. [70] using REM can be digitized and used to extract the displacements of the step position as a function of time. The data shown here are for a step in a train of steps of average separation 1400 Å. Figure prepared from data from the work of Alfonso et al. [126].

motion with atomic resolution have also become possible [68,69,127–129].

The potential for determining atomic mechanisms for step motion was demonstrated in the stunning observations of frizzled step edges observed on Ag(111) using STM [68,130]. Ibach and coworkers showed that the rough-appearing step edges were a result of step motion on the time scale required for the scan perpendicular to the step edge. Quantitative analysis of those data, and similar later data on Cu(100) has been performed in terms of a model in which step motion occurs via atomic diffusion (kink diffusion) along the step edge only. By assuming non-interacting diffusing kinks, Poensgen et al. [68] have essentially shown that the standard deviation of the step edge position can be related to the kink concentration P_1, the kink lifetime τ, and the rate ν with which the kink moves by one elementary step:

$$\sigma_f^2 = P_1 \sqrt{\frac{2\nu\tau}{\pi}} . \tag{15}$$

The observations on Ag and Cu are in agreement with the predictions of this model, as are more recent observations of step motion on Au [69], in which it was shown that the "frizzled" appearance of the step edges can be removed by increasing the speed at which the STM data are acquired.

Such step motion via kink diffusion is one of many special cases for step motion that have recently been described [66,131,132]. Because kink excitation is physically expected to require less energy than removal of atoms from a step edge, step motion via kink diffusion alone is expected at low temperature, with processes involving atomic detachment from the step becoming important with increasing temperature. These new processes allow the fluctuations of the step edge to become more rapid, as illustrated for the high-temperature data shown in Fig. 11. As different processes become important, the time dependence of the fluctuations, as well as the length scale of the fluctuations changes, leading to methods of distinguishing the different processes experimentally. One quantitative measure of the time scale of the fluctuations is provided by the time-dependent correlation function of step edge fluctuations [67,133,134] which can be calculated from direct measurements of the edge position as a function of time:

$$G(t) = \langle x^2 \rangle - \langle x(t)x(0) \rangle, \tag{16}$$

where x is the position of the step measured perpendicular to the average direction y of the step edge. The length scale of the fluctuations parallel to the step edge can be determined experimentally by calculating the Fourier components x_q of the measured step displacements:

$$x(y, t) = \sum_q x_q \exp(iqy), \tag{17a}$$

and the related Fourier correlation function $G_q(t)$ [44,134].

$$G_q(t - t') = \left\langle \left| x_q(t) - x_q(t') \right|^2 \right\rangle. \tag{17b}$$

The Fourier correlation function will increase with time with an amplitude and time constant that will both be larger for larger wavelengths. The dependence of the function on the physical parameters of the steps will be:

$$G_q(t) = \frac{2kT}{L\tilde{\beta}q^2} \left[1 - \exp\left(\frac{-t}{\tau(q)} \right) \right], \tag{18}$$

where L is the length of the step analyzed, $\tilde{\beta}$ is the step stiffness (which is related to the step energy β, and the step diffusivity $b^2(T)$ defined in Section 2.1), and $\tau(q)$ is the time constant for fluctuations of wavelength $\lambda = 2\pi/q$. Thus from Fourier correlation function, both the step stiffness and the time constant can be determined. As discussed below the exact dependence of the time constants on wavelength provides a method of differentiating different mechanisms of step motion.

Pimpinelli et al. [132] have recently derived the wavelength dependence for a number of physical mechanisms of step motion. These mechanisms demonstrate the changes in behavior that occur as different rate limiting steps for step motion become important. A few of these mechanisms are compared below to illustrate the potential for distinguishing the important elementary processes:

3.1.1. Kink-diffusion-limited step motion

When atomic motion is limited to diffusion along the step edge, the time scale of the step fluctuations is set by the time constant (or diffusion constant) for atomic hops along the step. The resulting motion of the step will display a strong correlation between different positions along the step leading to strong fluctuations at short times and small wavelengths. The experimental signature of the mechanism is a $t^{1/4}$ variation of the correlation function, and a λ^4 variation of the time constants of the different wavelengths of the fluctuations [132–135].

3.1.2. Step / terrace-exchange-limited step motion

Once the temperature is high enough to allow atoms to detach from steps, then the competing time constants for hopping on and off the step, and for diffusion on the terrace must be considered. In this case there will be some equilibrium concentration of adatoms on the terraces. Physically, the activation energy for hopping off of the step edge is expected to be larger than for terrace diffusion [18]. Thus at intermediate temperatures, the time scale for step motion is set by the time constant for step/terrace exchange. In this case, when an atom hops off of a step onto the terrace,

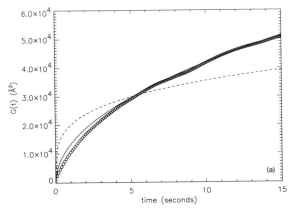
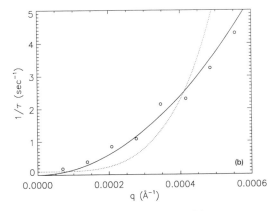

Fig. 12. Analysis [134] of real-time fluctuations of step motion on Si(111) at 900°C, such as shown in Fig. 11. (a) The correlation function $G(t)$ (Eq. (16)) determined by averaging over approximately 800 data sets such as that of Fig. 11. The solid and dashed curves show the expected $t^{1/2}$ behavior for terrace-exchange equilibration, and $t^{1/4}$ behavior for kink-diffusion equilibration respectively. (b) The inverse of the time constant for fluctuations of different wavelength λ of the step motion as a function of wave vector $q = 2\pi/\lambda$. The solid and dashed curves show the expected behavior for both terrace-exchange equilibration and long wavelength diffusion limited equilibration (q^2), and kink-diffusion equilibration (q^4) respectively.

the resulting increase in the adatom concentration decays rapidly via diffusion onto the terrace. Thus any given attachment/detachment event has little effect on subsequent events, and correlations in the step fluctuations will be relatively weak. The experimental signatures for this case have been determined for an open system in which the concentration of adatoms on the terraces is maintained constant (for instance by exchange with the bulk). The resulting variation of the correlation function with time goes as $t^{1/2}$, and the variation of the time constants with wavelength goes as λ^2 [131,133–135].

3.1.3. Terrace-diffusion-limited step motion

At very high temperatures, the difference in activation energies for various processes becomes unimportant, and one would expect that the rate of step/terrace exchange and the rate of terrace diffusion would become the same. In this case when an atom hops off of a step, the resulting local increase in the adatom concentration will not have time to decay away completely before the next step/terrace exchange event. Thus the probability of a terrace adatom hopping back onto the step is locally increased, and the probability of another atom hopping off of the step is locally decreased. As a result the fluctuations of

the step are somewhat damped. For fluctuations of wavelength much less than the distance l to a neighboring step ($\lambda \ll l$), the decay of the local concentration increase occurs primarily via adatom diffusion to another point on the step. In this case the variation of the correlation function with time goes as $t^{1/3}$, and the variation of the time constant with wavelength goes as λ^3 [131,132,135]. For fluctuations of wavelength greater than the step-separation ($\lambda \gg l$), the local concentration increase can also decay via diffusion to the other step edge. This reduces the damping of long wavelength fluctuations, leading to a time constant for the fluctuations which is proportional to both l and λ^2 [132]. The conditions for the crossover between the two regimes are still a matter of investigation.

The application of this type of analysis to the experimental data of Fig. 11 is illustrated in Fig. 12 [134]. The experimentally determined time correlation function clearly fits a $t^{1/2}$ form much better than a $t^{1/4}$ form, as shown in Fig. 12a. The wavelength dependence of the time constant for step fluctuation, shown in Fig. 12b, also fits a q^2 form much better than q^4. If one interprets the data in terms of the terrace/exchange case described above, the fit to the data gives a time constant for the rate-limiting attachment/

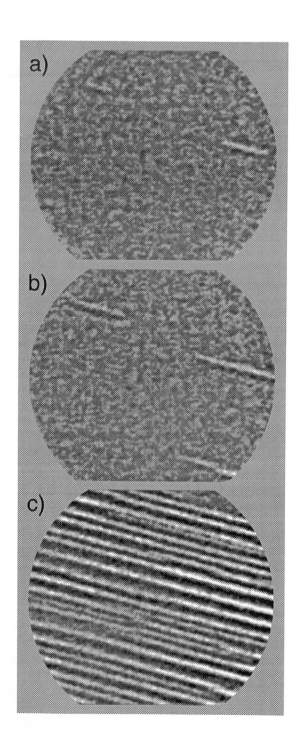

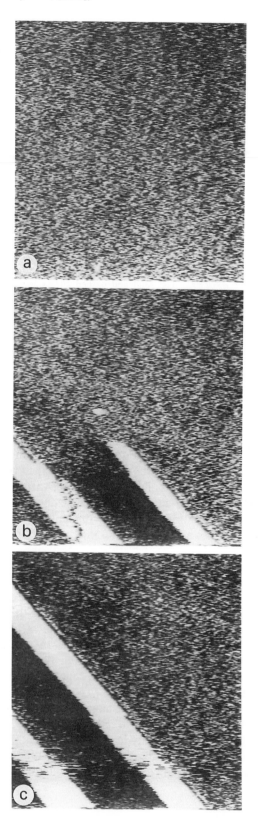

detachment events of approximately 1 μs [134]. However, the step-separation dependence of the fluctuations further suggests that there is a proportionality to the step separation, in which case terrace diffusion as a rate limiting step is indicated [132]. The fit to the data in this model yields a value of the diffusion constant times the adatom density, $D_s\rho_s$ of 10^8 s^{-1} [132]. Thus direct observations of step motion provide a means of determining both the rate limiting atomic process and the related time constant.

3.2. Mass transport

Directional surface diffusion, or mass transport, will occur, for instance, when a non-equilibrium structure is prepared by mechanical means, or when a transition occurs due to a change in temperature or ambient species. An example of the first case would be the stability of a micro-fabricated structure, and an example of the second case would be the rate of a transition from a stable to a faceted morphology. The basic understanding of the stability of artificial structures was set by Mullins' application of Eqs. (13) and (14) to the problem of the decay of an initial sinusoidal profile to a flat surface [5,135]. This problem was originally solved for the case where the variation of the surface tension is not singular (e.g. the low index surface is above its roughening temperature, so that the cusp of Eq. (3) does not occur), and the diffusion constant of Eq. (13) is defined for atomic motion on the rough surface. The result for decay via surface diffusion only is that the amplitude of the sinusoid decays exponentially with time, with a time constant that decreases as the fourth power of the wavelength of the sinusoid.

As a result of Mullins' calculation, experimen-

tal studies of artificially fabricated profiles have been performed both to confirm the predicted wavelength dependence, and to determine diffusion coefficients [136–147]. In the course of such work, it was found that as the initial sinusoid decays, it often changes shape to expose large flat regions, which correspond to facets of the equilibrium crystal shape [48,142,146,148]. This indicates that there are cusps in the variation of the surface tension with orientation, so that the application of Eqs. (13) and (14) must be handled with care. This can be done, given a specific form for the surface tension, with the result for a surface tension of the form of Eq. (3), that the variation of the chemical potential of an nonequilibrium structure is [119,120,149]:

$$\mu(x) = -6g(T)|\tan \phi(x)|\frac{d \tan \phi(x)}{dx}. \quad (19)$$

This is the appropriate form to be used for interpreting the driving force for diffusion, in Eq. (13). However, the fact that the temperature is low enough to allow facets to form, indicates that the diffusion coefficient of Eq. (13) cannot be evaluated as if diffusion were occurring on a uniform rough surface. The distinct structure of terraces and steps must be considered explicitly, as illustrated in STM studies of the atomic structure of fabricated sinusoids on Si(001) [144,145].

Similar considerations about the meaning of the diffusion coefficient in terms of the step structure also arise in evaluating the kinetics of transitions in morphology. The canonical example of such a transition, considered by Mullins [150], is the formation of a facet (for instance following a decrease in temperature, a change in reconstruction, or during adsorption, as described in Section 2.2) on a surface with an initially uniform step density. The use of direct imaging tech-

Fig. 13. A reconstruction-driven change in surface morphology on vicinal Si(111) has been observed in real time using both LEEM [151] and high-T STM [153]. The sequence of panels to the left shows LEEM images with 3 μm field of view of the facet structure at times 1 s, 3 s, and 60 s after a the temperature has been abruptly dropped through the transition temperature to 848°C. Each image required approximately 0.03 s for acquisition. The sequence of panels to the right measured by Suzuki et al. [153] shows high-T STM images with a 900 Å × 1000 Å field of view measured immediately following a temperature drop to 730°C. Each image required approximately 16 s for acquisition. In both sets of images, the bright regions are areas of 7 × 7 reconstruction, and the dark regions are areas where the surface structure is still that of the high-temperature phase. The figures in the right hand panel are taken from the work of Suzuki et al. [153].

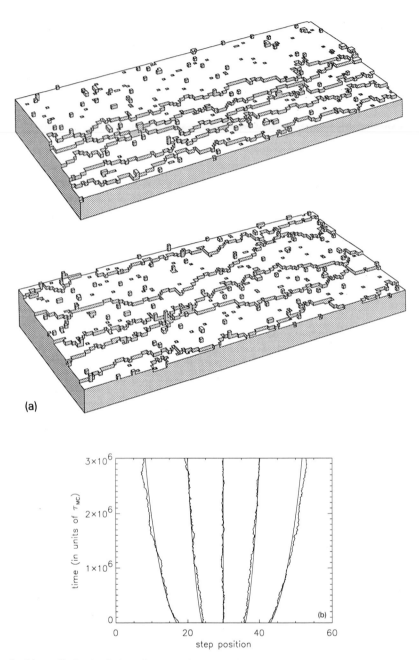

Fig. 14. The results of a Monte Carlo simulation of step motion on a 60×120 atom lattice within the SOS model at $kT = 0.9\epsilon$, where ϵ is the kink energy [154]. At this temperature the rates for step/terrace exchange and for terrace diffusion are comparable so that step motion is terrace-diffusion limited. (a) The simulated structures, top: shortly after 5 steps were placed in a uniformly spaced bunch, with average separation 6 lattice spacings, and botton: after the steps had equilibrated for 3×10^6 Monte Carlo steps. (b) The average positions of the group of five steps as a function of time following their initial placement in uniform bunch. The fluctuating lines show the positions calculated during the simulation. The solid lines show the result of the diffusion calculation described in the text. Figures taken from the work of Bartelt et al. [154].

niques now makes it possible to observe such transformations directly, as illustrated in Fig. 13 which shows LEEM [151] and high-T STM [152,153] images of the formation of a facet on vicinal Si(111), following a drop in temperature below the reconstructive phase transition. The experimental observations are exactly as predicted by Mullins [150], who recognized that the formation of a hill-and-valley structure would proceed by a nucleation event, followed by rapid growth of the new facet in the direction parallel to the steps, and slow growth in the direction perpendicular to the steps.

When interpreting the diffusion coefficient D_c of Eq. (13) for a stepped surface, the same considerations about rate limiting processes occur as in the discussion of step fluctuations in Section 3.1 [118–120,132,149]. An example of how the appropriate diffusion coefficient can be determined and used is illustrated in Fig. 14, which shows the result of a Monte Carlo simulation of step equilibration in a closed system [154]. The physical parameters of the simulation were chosen so that terrace diffusion is the rate limiting process in step motion. The equilibrium fluctuations of the steps follow the λ^3 dependence described in the previous section for terrace-limited diffusion at large step-separations. By analyzing these fluctuations, the value of the diffusion coefficient times the adatom concentration, $D_s\rho_s$ was determined, as one can do experimentally for microscopic observations of step motion. A group of steps was then placed in a non-equilibrium configuration, a closely spaced bunch of steps, as shown in the upper panel of Fig. 14a, and at $t = 0$ in Fig. 14b, and allowed to equilibrate. The final configuration of the steps is shown in the lower panel of Fig. 14a, and the evolution of the average step position with time is shown by the dashed lines in the Fig. 14b. A calculation of the time evolution, using Eqs. (13) and (19) with the "measured" value of $D_s\rho_s$ within the framework described by Rettori and Villain [119], gives an excellent fit to the simulation, as shown by the solid lines. It is interesting that while the two-dimensional fluctuations of the steps are essential to their motion, a one-dimensional calculation, in which the step entropy is appropriately described

via Eqs. (3) and (19), is sufficient to describe their motion. From a practical point of view, the stages of the simulation described above illustrate a process that can be followed experimentally to understand macroscopic changes in morphology. For transformations driven by perturbations such as adsorption or reconstruction, many additional considerations will arise, but the same physical framework will be useful for addressing the problems.

Our understanding of kinetic processes in surface morphology is less complete at this point than our understanding of equilibrium behavior. However, as in the case of equilibrium behavior, many years of effort have provided experimental confirmation of the early predictions about surface kinetic behavior. Recent work on the theory of surface kinetics, coupled with direct observations of step kinetics offer the very real expectation of a predictive capability for rates of surface morphological transformations.

4. Summary

Surfaces are far more interesting than their simplest description as infinitely periodic two-dimensional structures would suggest. The realization of this fact, and its importance in all real surface processes, dates back to the earliest days of surface science. As a result of many years of experimental and theoretical effort, it is now possible to measure and understand the complexity of surface morphology within a simple and quantitative formalism based on the physical properties of steps on the surface. Given estimates of a relatively small numbers of important parameters governing step behavior, it is possible to approach any given problem with a reasonable qualitative understanding of what role morphology and changes in morphology can be expected to play.

Acknowledgments

The author thanks the DOD, ONR, NSF-MRG and NSF-FAW for support during the preparation of this manuscript, and also very gratefully acknowledges extensive discussions with Dr. N.C. Bartelt on the topics of the manuscript, as well as his permission to use Figs. 7, 12 and 14 prior to their publication elsewhere.

References

[1] G.L. Kellogg and P.J. Feibelman, Phys. Rev. Lett. 64 (1990) 3143.

[2] J.V. Barth, H. Brune, G. Ertl and R.J. Behm, Phys. Rev. B 42 (1990) 9307.

[3] G. Ertl, Science 254 (1991) 1750.

[4] C. Herring, in: Structure and Properties of Crystal Surfaces, Eds. R. Gomer and C.S. Smith (University of Chicago Press, Chicago, 1953) p. 5.

[5] W.W. Mullins, in: Metal Surfaces, Eds. W.D. Robertson and N.A. Gjostein (American Society for Metals, Metals Park, OH, 1962) p. 17.

[6] W.K. Burton, N. Cabrera and F.C. Frank, Philos. Trans. Roy. Soc. (London) 243A (1951) 299.

[7] J.W. Cahn, Acta Met. 8 (1960) 554.

[8] A.A. Chernov, Sov. Phys.-Crystallogr. 3 (1959) 225.

[9] A.A. Chernov, Sov. Phys. Usp. 4 (1961) 1116.

[10] N. Cabrera, Surf. Sci. 2 (1964) 320.

[11] J.C. Tracy and J.M. Blakely, Surf. Sci. 13 (1968) 313.

[12] G.A. Somorjai and M.A. Van Hove, Prog. Surf. Sci. 30 (1989) 201.

[13] M. Flytzani-Stephanopoulos and L.D. Schmidt, Prog. Surf. Sci. 9 (1979) 83.

[14] P.I. Cohen, G.S. Petrich and P.R. Pukite, in: Kinetics of Ordering and Growth at Surfaces, Ed. M.G. Lagally (Plenum, New York, 1990).

[15] T.R. Ohno and E.D. Williams, J. Vac. Sci. Technol. B 8 (1990) 874.

[16] S.A. Chalmers, A.C. Gossard, P.M. Petroff and H. Kroemer, J. Vac. Sci. Technol. B 8 (1990) 431.

[17] J.E. Reutt-Robey, D.J. Doren, Y.J. Chabal and S.B. Christman, Phys. Rev. Lett. 61 (1988) 2778.

[18] R.L. Schwoebel and E.J. Shipsey, J. Appl. Phys. 37 (1966) 3682.

[19] F.W. Young and L.D. Hulett, in: Metal Surfaces, Eds. W.D. Robertson and N.A. Gjostein (American Society for Metals, Metals Park, Ohio, 1962) p. 375.

[20] J.M. Blakely, Introduction to the Properties of Crystal Surfaces (Pergamon, New York, 1973).

[21] G. Ehrlich and K. Stolt, Annu. Rev. Phys. Chem. 31 (1980) 603, and references therein.

[22] H. Bethge, in: Kinetics of Ordering and Growth at Surfaces, Ed. M.G. Lagally (Plenum, New York, 1990) p. 125, and references therein.

[23] M. Henzler, Surf. Sci. 19 (1970) 159.

[24] J.E. Houston and R.L. Park, Surf. Sci. 26 (1971) 269.

[25] W.P. Ellis, Surf. Sci. 45 (1974) 569.

[26] H. Ohtani, C.-T. Kao, M.A. Van Hove and G.A. Somorjai, Prog. Surf. Sci. 23 (1986) 155.

[27] R.J. Phaneuf and E.D. Williams, Phys. Rev. Lett. 58 (1987) 2563.

[28] T.E. Madey, K.-J. Song and C.-Z. Dong, Surf. Sci. 247 (1991) 175.

[29] J. Lapujoulade, J. Perreau and A. Kara, Surf. Sci. 129 (1983) 59.

[30] M. den Nijs, E.K. Riedel, E.H. Conrad and T. Engle, Phys. Rev. Lett. 55 (1984) 1689.

[31] D.Y. Noh, K.I. Blum, M.J. Ramstad and R.J. Birgeneau, Phys. Rev. B 44 (1991) 10969.

[32] F.K. Men, W.E. Packard and M.B. Webb, Phys. Rev. Lett. 61 (1988) 2469.

[33] C.E. Aumann, D.E. Savage and M.G. Lagally, Surf. Sci. 275 (1992) 1.

[34] J. Falta and M. Henzler, Surf. Sci. 269/270 (1992) 14.

[35] Y.L. He, H.N. Yang, T.M. Lu and G.C. Wang, Phys. Rev. Lett. 69 (1992) 3770.

[36] B. Binnig and H. Rohrer, Rev. Mod. Phys. 59 (1987) 615.

[37] E. Bauer, M. Mundschau, W. Święch and W. Telieps, Ultramicroscopy 31 (1989) 49.

[38] N. Osakabe, Y. Tanishiro and K. Yagi, Surf. Sci. 109 (1981) 353.

[39] J.J. Metois, S. Nitsche and J.C. Heyraud, Ultramicroscopy 27 (1989) 349.

[40] A.V. Latyshev, A.L. Aseev, A.B. Krasilnikov and S.I. Stenin, Surf. Sci. 227 (1990) 24.

[41] A.A. Chernov, Modern Crystallography III: Crystal Growth (Springer, Berlin, 1984).

[42] M. Wortis, in: Chemistry and Physics of Solid Surfaces VII, Eds. R. Vanselow and R.F. Howe (Springer, Berlin, 1988) p. 367.

[43] R. Kern, in: Morphology of Crystals, Ed. I. Sunagawa (Terra Scientific Publishing, Tokyo, 1987) p. 79.

[44] P. Nozieres, in: Solids far from Equilibrium, Ed. C. Godreche (Cambridge University Press, Cambridge, 1991) p. 1.

[45] C. Herring, Phys. Rev. 82 (1951) 87.

[46] C. Rottman and M. Wortis, Phys. Rep. B 124 (1984) 241.

[47] J.C. Heyraud and J.J. Métois, Acta Metall. 28 (1980) 1789.

[48] U. Breuer and H.P. Bonzel, Surf. Sci. 273 (1992) 219.

[49] K.D. Brommer, M. Needels, B.E. Larson and J.D. Joannopoulos, Phys. Rev. Lett. 68 (1992) 1355.

[50] I. Stich, M.C. Payne, R.D. King-Smith, J.-S. Lin and L.J. Clarke, Phys. Rev. Lett. 68 (1992) 1351.

[51] M.S. Daw and M.I. Baskes, Phys. Rev. B 29 (1984) 6443.

[52] D. Wolf, Surf. Sci. 226 (1990) 389.

[53] A. Bartolini, F. Ercolessi and E. Tosatti, Phys. Rev. Lett. 63 (1989) 872.

[54] G. Bilalbegovic, F. Ercolessi and E. Tossati, Surf. Sci. 280 (1993) 335.

[55] E.E. Gruber and W.W. Mullins, J. Phys. Chem. Solids 28 (1967) 875.

[56] V.V. Voronkov, Sov. Phys.-Crytallogr. 12 (1968) 728.

[57] Another way to state this is that relative zero-T energies

only remain a good approximation below the roughening temperature of the surfaces, and high-index surfaces are likely to have low or zero roughening temperatures, see: E. Conrad and T. Engel, Surf. Sci. 299/300 (1994).

[58] C. Jayaprakash, C. Rottman and W.F. Saam, Phys. Rev. B 39 (1984) 6549.

[59] In this equation it is mathematically conveniently to define a reduced surface tension, which is the projection of the true surface tension in the low-index plane $f_i = \gamma(\hat{n}_i)/(\hat{n}_i \cdot \hat{n})$.

[60] The expression in Eq. (3) explicitly refers to a variation of step density along a fixed high-symmetry direction in the reference plane. The more general case in which variations in the azimuthal angle are included can be treated explicitly as well.

[61] E.D. Williams and N.C. Bartelt, Science 251 (1991) 393.

[62] H. van Beijeren and I. Nolden, in: Structure and Dynamics of Surfaces II, Eds. W. Schommers and P. von Blanckenhagen (Springer, Berlin, 1987) p. 259.

[63] C. Jayaprakash, W.F. Saam and S. Teitel, Phys. Rev. Lett. 50 (1983) 2017.

[64] B.S. Swartzentruber, Y.-W. Mo, R. Kariotis, M.G. Lagally and M.B. Webb, Phys. Rev. Lett. 65 (1990) 1913.

[65] N.C. Bartelt, T.L. Einstein and E.D. Williams, Surf. Sci. 240 (1991) L591.

[66] N.C. Bartelt, T.L. Einstein and E.D. Williams, Surf. Sci. 276 (1992) 308.

[67] R. Kariotis and M.G. Lagally, Surf. Sci. 248 (1991) 295.

[68] M. Poensgen, J.F. Wolf, J. Frohn, M. Giesen and H. Ibach, Surf. Sci. 274 (1992) 430.

[69] L. Kuipers and F.W.M. Frenken, private communication, 1993.

[70] C. Alfonso, J.M. Bermond, J.C. Heyraud and J.J. Metois, Surf. Sci. 262 (1992) 371.

[71] R.M. Tromp and M.C. Reuter, Phys. Rev. Lett. 68 (1992) 820.

[72] R.M. Tromp and M.C. Reuter, Phys. Rev. Lett. 47 (1993) 7598.

[73] X.-S. Wang, J.L. Goldberg, N.C. Bartelt, T.L. Einstein and E.D. Williams, Phys. Rev. Lett. 65 (1990) 2430.

[74] B. Joos, T.L. Einstein and N.C. Bartelt, Phys. Rev. B 43 (1991) 8153.

[75] R. Kariotis, Surf. Sci. 248 (1991) 306.

[76] An exception is the repulsive interaction due to anisotropic stress that arises for instance on Si(001). The nature of this interaction, and its effect on step–step distributions are described in the following three references.

[77] O.L. Alerhand, D. Vanderbilt, R.D. Meade and J.D. Joannopoulos, Phys. Rev. Lett. 61 (1988) 1973.

[78] B.S. Swartzentruber, Phys. Rev. 47 (1993) 13432.

[79] J.J. de Miguel, C.E. Aumann, S.G. Jaloviar, R. Kariotis and M.G. Lagally, Phys. Rev. 46 (1992) 10257.

[80] S. Rousset, S. Gauthier, O. Siboulet, J.C. Girard, S. de Cheveigne, M. Huerta-Garnica, W. Sacks, M. Belin and J. Klein, Ultramicroscopy 42–44 (1992) 515.

[81] J. Frohn, M. Giesen, M. Poensgen, J.F. Wolf and H. Ibach, Phys. Rev. Lett. 67 (1991) 3543.

[82] Y.-N. Yang, B.M. Trafas, R.L. Seifert and J.H. Weaver, Phys. Rev. B 44 (1991) 3218.

[83] S. Balibar, C. Guthmann and E. Rolley, Surf. Sci. 283 (1993) 290.

[84] Because of the weak dependence on n of Eq. (5), we cannot definitively rule out other values of n.

[85] J.M. Blakely and R.L. Schwoebel, Surf. Sci. 26 (1971) 321.

[86] V.I. Marchenko and A.Y. Parshin, JETP Lett. 52 (1980) 129.

[87] A.C. Redfield and A. Zangwill, Phys. Rev. B 46 (1992) 4289.

[88] R. Shuttleworth, Proc. Phys. Soc. A 63 (1950) 444.

[89] T.W. Poon, S. Yip, P.S. Ho and F.F. Abraham, Phys. Rev. Lett. 65 (1990) 2161.

[90] D. Wolf and J.A. Jaszczak, Surf. Sci. 277 (1992) 301.

[91] For simplicity, we neglect the contribution to the interactions due to the step face contribution to the step elastic dipole moment.

[92] H.P.R. Frederiske, in: American Institute of Physics Handbook, Ed. D.E. Gray (McGraw-Hill, New York, 1972).

[93] R.D. Meade and D. Vanderbilt, Phys. Rev. Lett. 63 (1989) 1404.

[94] R.J. Needs, M.J. Godfrey and M. Mansfield, Surf. Sci. 242 (1991) 215.

[95] R.E. Martinez, W.M. Augustyniak and J.A. Golovchenko, Phys. Rev. Lett. 64 (1990) 1035.

[96] D. Sander, U. Linke and H. Ibach, Surf. Sci. 272 (1992) 318.

[97] D. Sander and H. Ibach, Phys. Rev. B 43 (1992) 4263.

[98] A.J. Schnell-Sorokin and R.M. Tromp, Phys. Rev. Lett. 64 (1990) 1039.

[99] H. Sato, Y. Tanishiro and K. Yagi, Appl. Surf. Sci. 60/61 (1992) 367.

[100] O. Pohland, X. Tong and J.M. Gibson, J. Vac. Sci. Technol. A 11 (1993) 1837.

[101] G.-C. Wang and M.G. Lagally, Surf. Sci. 81 (1979) 69.

[102] E.D. Williams and N.C. Bartelt, Proc. Workshop on Surface Disordering, Eds. R. Jullien, J. Kertész, P. Meakin and D.E. Wolf (Nova Science Publishers, 1992) p. 103.

[103] E.D. Williams, R.J. Phaneuf, N.C. Bartelt, W. Święch and E. Bauer, Mat. Res. Soc. Symp. Proc. 238 (1992) 219.

[104] D. Vanderbilt, Phys. Rev. B 36 (1987) 6209.

[105] P. Gumbsch and M.S. Daw, Phys. Rev. B 44 (1991) 3934.

[106] R.C. Cammarata, Surf. Sci. 279 (1992) 341.

[107] J.W. Cahn, J. Phys. (Paris) 43 (1982) C6-199.

[108] In general, on a clean surface there can be three surface phases coexisting with each other, because the surface normal has two independent components (analogous to a two component fluid). The discussion in the text is restricted to two phase coexistence.

[109] E.D. Williams, R.J. Phaneuf, J. Wei, N.C. Bartelt and T.L. Einstein, Surf. Sci. 294 (1993) 219.

[110] P. Nozieres, J. Phys. (Paris) 50 (1989) 2541.

[111] J.-J. Metois and J.C. Heyraud, Ultramicroscopy 31 (1989) 73.
[112] G. Bilalbegovic, F. Ercolessi and E. Tosatti, Europhys. Lett. 18 (1992) 163.
[113] Without loss of generality, we are considering faceting along a fixed azimuth.
[114] H.M. van Pinxteren and J.W.M. Frenken, Europhys. Lett. 21 (1992) 43.
[115] H.M. van Pinxteren and J.W.N. Frenken, Surf. Sci. 275 (1992) 383.
[116] D.G. Vlachos, L.D. Schmidt and R. Aris, Phys. Rev. B 47 (1993) 4896.
[117] Where $1/R(x) = -\ddot{z}/(1 + \dot{z}^2)^{3/2}$ and $\dot{z} = \tan \phi$ and $\ddot{z}$ denote the first and second derivatives of the surface height with respect to x.
[118] M. Ozdemir and A. Zangwill, Phys. Rev. B 45 (1992) 3718.
[119] A. Rettori and J. Villain, J. Phys. 49 (1988) 257.
[120] M. Ozdemir and A. Zangwill, Phys. Rev. B 42 (1990) 5013.
[121] G. Honjo, K. Takayanagi, K. Kobayashi and K. Yagi, J. Cryst. Growth 42 (1977) 98.
[122] M. Mundschau, E. Bauer, W. Telieps and W. Święch, Surf. Sci. 223 (1989) 413.
[123] K. Yagi, A. Yamana, H. Sato, M. Shima, H. Ohse, S. Ozawa and Y. Tanishiro, Prog. Theor. Phys. Suppl. 106 (1991) 303.
[124] F.M. Ross and J.M. Gibson, Phys. Rev. Lett. 68 (1992) 1782.
[125] A.V. Latyshev, A.L. Aseev, A.B. Krasilnikov and S.I. Stenin, Surf. Sci. 213 (1989) 157.
[126] J.-J. Metois, private communication, 1993.
[127] S.-I. Kitamura, T. Sato and M. Iwatsuki, Nature 351 (1991) 215.
[128] R. Kliese, B. Rottger, D. Badt and H. Neddermeyer, Ultramicroscopy 42–44 (1992) 824.
[129] N. Kitamura and M.B. Webb, Bull. Am. Phys. Soc. 38 (1993) 509.
[130] M. Giesen, J. Frohn, M. Poensgen, J.F. Wolf and H. Ibach, J. Vac. Sci. Technol. A 10 (1992) 2597.
[131] G.S. Bales and A. Zangwill, Phys. Rev. B 41 (1990) 5500.
[132] A. Pimpinelli, J. Villain, D.E. Wolf, J.J. Métois, J.C. Heyraud, I. Elkinani and G. Uimin, Surf. Sci. 295 (1993) 143.
[133] N.C. Bartelt, J.L. Goldberg, T.L. Einstein and E.D. Williams, Surf. Sci. 273 (1992) 252.
[134] N.C. Bartelt, J.L. Goldberg, T.L. Einstein, E.D. Williams, J.C. Heyraud and J.-J. Métois, Phys. Rev. B, in press.
[135] W.W. Mullins, J. Appl. Phys. 30 (1959) 77.
[136] J.M. Blakely and H. Mykura, Acta Method 10 (1962) 565.
[137] P.S. Maiya and J.M. Blakely, Appl. Phys. Lett. 7 (1965) 60.
[138] J.M. Blakely, Appl. Phys. Lett. 11 (1967) 335.
[139] H.P. Bonzel and N.A. Gjostein, Appl. Phys. Lett. 10 (1967) 258.
[140] H.P. Bonzel and N.A. Gjostein, J. Appl. Phys. 39 (1968) 3480.
[141] K. Yamashita, H.P. Bonzel and H. Ibach, Appl. Phys. 25 (1981) 231.
[142] H.P. Bonzel, E. Preuss and B. Steffen, Appl. Phys. A 35 (1984) 1.
[143] E. Preuss, N. Freyer and H.P. Bonzel, Appl. Phys. A 41 (1986) 137.
[144] C.C. Umbach, M.E. Keefe and J.M. Blakely, J. Vac. Sci. Technol. A 9 (1990) 1014.
[145] C.C. Umbach, M.E. Keeffe and J.M. Blakely, J. Vac. Sci. Technol. A 11 (1993) 1830.
[146] H.P. Bonzel, U. Breuer and B. Boigtlander, Surf. Sci. 272 (1992) 10.
[147] M.E. Keefe, C.C. Umbach and J.M. Blakely, Bull. Am. Phys. Soc. 38 (1993) 567.
[148] H.P. Bonzel, N. Freyer and E. Preuss, Phys. Rev. Lett. 57 (1986) 1024.
[149] P. Nozieres, J. Phys. (Paris) 48 (1987) 1605.
[150] W.W. Mullins, Philos. Mag. 6 (1961) 1313.
[151] R.J. Phaneuf, N.C. Bartelt, E.D. Williams, W. Święch and E. Bauer, Phys. Rev. Lett. 21 (1991) 2986.
[152] H. Hibino, T. Fukuda, M. Suzuki, Y. Homma, T. Sato, M. Iwatsuki, K. Miki and H. Tokumoto, Phys. Rev. B 47 (1993) 13027.
[153] M. Suzuki, Y. Homma, H. Hibino, T. Fukuda, T. Sato, M. Iwatsuki, K. Miki and H. Tokumoto, J. Vac. Sci. Technol. A 11 (1993) 1640.
[154] N.C. Bartelt, T.L. Einstein and E.D. Williams, in preparation.
[155] R.C. Nelson, T.L. Einstein, S.V. Khare and P.J. Rous, Surf. Sci. 295 (1993) 462.
[156] D. Toth, N.C. Bartelt and E.D. Williams, in preparation.
[157] R.J. Phaneuf, E.D. Williams and N.C. Bartelt, Phys. Rev. B 38 (1988) 1984.
[158] N.C. Bartelt, T.L. Einstein and E.D. Williams, Surf. Sci. 244 (1991) 149.
[159] H. van Beijeren, Phys. Rev. Lett. 38 (1977) 993.

Surface Science 299/300 (1994) 525–540
North-Holland

surface science

Electronic structure of semiconductor surfaces and interfaces

F.J. Himpsel

IBM T.J. Watson Research Center, P.O. Box 218, Route 134, Yorktown Heights, NY 10598, USA

Received 2 March 1993; accepted for publication 4 June 1993

This article discusses how our understanding of the electronic structure of semiconductor surfaces has evolved over the last decades. It traces the evolution of concepts, such as surface states, surface core level shifts, and charge transfer. Examples from silicon surfaces show how much our capabilities for detecting surface states have grown, from the first indirect evidence of surface states to the full picture in momentum space and real space. An extrapolation to the future forecasts a shift from the analysis of surfaces towards the synthesis of new structures, using advanced semiconductor processing techniques.

1. Concepts: asking the right questions

1.1. What are surface states?

The idea of surface states has been around since the Thirties. By using simple model potentials Tamm [1] and others showed that extra states could develop at a surface with energies located in a gap of bulk states. To get analytical solutions it was necessary to resort to simple, one-dimensional model potentials, which were truncated at the surface. Such a truncation is equivalent to keeping the bulk Hamiltonian, but changing the boundary conditions. Consequently, a complex wave vector comes into play at the surface and leads to evanescent waves. For small damping these surface state wave functions look just like a slightly modulated bulk solution (Fig. 1, [1,2]).

The other extreme in viewing surface states has been the chemical concept of a dangling bond, which implies a wave function localized at a single surface atom. In this case the bulk potential is strongly perturbed at the surface, creating a new potential well that supports the surface state. For semiconductors with their highly directional bonds such a picture seems more appropriate. Early calculations with realistic wave functions confirmed the expectations from the simple bro-

ken bond model, as shown in Fig. 2 [3–5]. Covalent semiconductor surfaces should be metallic in this picture since each broken bond orbital is half-filled. The first real space pictures of semiconductor surfaces with the scanning tunneling microscope (STM) by Binnig et al. [6] visualized these broken bonds stunningly (Fig. 3).

Starting from the broken bond picture it has been puzzling that many group IV semiconductor surfaces show semiconducting behaviour, instead of being metallic. We will discuss three options that have been debated over the years. First of all, it is always possible to have a band gap with an even number of electrons per surface unit cell.

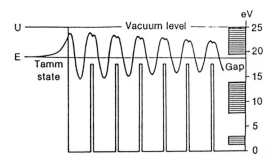

Fig. 1. Tamm's original model potential for a surface from the Thirties (bottom curve [2]), and the corresponding surface state wavefunction (top curve, from ref. [1]). It resembles an exponentially-damped bulk wavefunction.

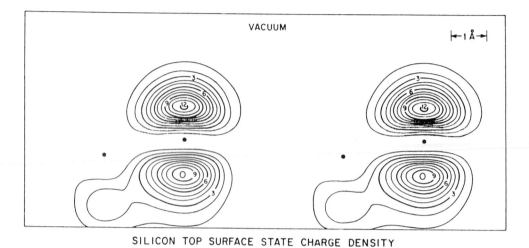

Fig. 2. Calculated charge density of broken bonds at a silicon surface showing their localized character (from Appelbaum and Hamann [4]). The dots represent surface atoms.

In order to pair up their broken bonds, the surface atoms are forced to rearrange themselves into sometimes exotic patterns, explaining the enlarged surface unit cells of semiconductor surfaces. An extreme example is the Si(111)7 × 7 surface, which has long been a playground for ideas of semiconductor reconstruction. As another option, group IV semiconductor surfaces may play the same game as ionic III–V and II–VI surfaces, which avoid half-filled dangling bond orbitals by creating a stable lone pair at the anion plus an empty orbital at the cation (Fig. 4 top, [7]). In such an "electronic" reconstruction mechanism the energy gained from opening a gap is offset by the Coulomb energy required for separating the charges, and it has been debated

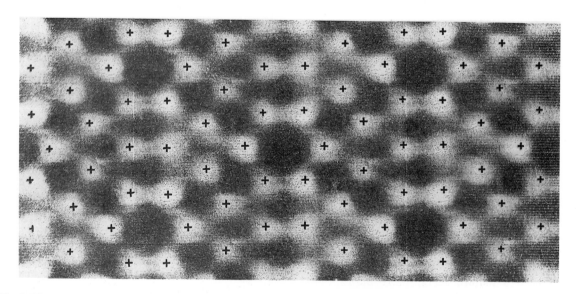

Fig. 3. First scanning tunneling microscope picture of dangling bonds at the Si(111)7 × 7 surface by Binnig et al. [6]. The bright spots correspond to enhanced tunneling through Si adatoms.

ZINCBLENDE (110)

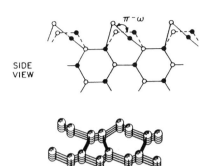

SIDE
VIEW

Si(111) - 2×1

Fig. 4. Reconstruction of semiconductor surfaces. Heteropolar surfaces (top) tend towards a disproportionation of charge, leading to a stable lone pair at the surface anions (upper row of open circles) and to a change in bond angles (from Lubinsky et al. [7]). Homopolar surfaces (bottom) need to change their bond topology to tie up broken bonds (from Ancilotto et al. [11], see also Pandey [9]). Vacuum is on top in both pictures.

for many years if such a mechanism can be viable for covalent semiconductors. Examples for reconstructions involving charge transfer are the buckling model of Si(111)2 × 1 [8], the tilted chain model for Si(111)2 × 1 (Fig. 4 bottom, [9–12]), and the asymmetric dimer model for Si(100)2 × 1 [13–19]. Our current understanding may be summarized by saying that a reconstruction driven purely by charge transfer, as in the first example, is unfavorable for group IV semiconductors. However, a charge transfer on top of a re-bonding reconstruction, as in the other two examples, is actually needed at group IV surfaces to achieve the lowest energy. The clearest case is the Si(111)2 × 1 surface [10–12], where the π-bonded chain tilts fully to achieve a geometry similar to the GaAs(110) surface [7] (see Fig. 4). This tilt favors charge transfer from the down-atom to the up-atom by providing a planar sp^2 geometry for the down-atom and a 90° s^2p^3 bond angle for the up-atom. The Si(100)2 × 1 and Ge(100)2 × 1 surfaces also appear to gain energy by having charged, asymmetric dimers, which stabilize a

c(4 × 2) structure at low temperatures [17–19]. However, the energy gain is so small in this case that the asymmetric dimers flip back and forth at room temperature. A third, still somewhat speculative mechanism for opening a surface band gap is the effect of correlation. Even for a surface with an odd number of electrons per unit cell, where band theory would predict a half-filled, metallic band, one can have a splitting into two single-electron bands via Coulomb and correlation effects. This situation is encountered in transition metal oxides, including parent compounds of high temperature superconductors. This idea keeps popping up, but it has been difficult to find hard evidence for strong Coulomb splittings at semiconductor surfaces. The best candidates right now are localized surface orbitals that become half-filled by adsorbing alkali atoms [20,21].

Calculations have come a long way since the early tight binding models. A major drawback of early work was the lack of self-consistency, both with respect to charge transfer, and with respect to the atomic positions. In the first case, a small accumulation of charge could distort the results by shifting energy levels electrostatically. In the second case, a small deviation from the ideal bond length would drastically change the splitting between the bonding and antibonding combinations of two orbitals. Although it was often tried to draw conclusions about the (then unknown) atomic geometry by comparing tight binding calculations with experimental results on surface state energies, these comparisons had a less than average hitting rate in predicting surface geometries. With the availability of ever-increasing computer power in the eighties the proper, self-consistent calculations became feasible and have enjoyed great success ever since (see Section 2.1). The most advanced method in predicting the bond geometry at semiconductor surfaces is simulated annealing with the Car–Parrinello technique [11,22], where the computer decides without human bias with structures ought to be preferred by Nature.

Another important concept that could only be explored with extensive computer power is that of quasiparticles. Until a few years ago, practically all band calculations produced energy eigenvalues for the ground state, even the self-consistent local density and Hartree–Fock calculations. Such ground state eigenvalues are fictitious quantities, since one always has to excite an electron to probe its energy, i.e., one has to create a hole to probe an occupied state and add an extra electron to probe an unoccupied state. It is surprising that the ground state eigenvalues represent experimental energy bands reasonably well. The main exception is the band gap, which comes out about a factor of two too large in Hartree–Fock and a factor of two too small in local density theory. (The discrepancy between the two calculation methods alone shows that the concept of ground state energy eigenvalues is fictitious.) Germanium even becomes a metal in local density theory. A rigid shift of the empty states cures most of the problems, but to do the job properly one has to calculate the self-energy that is induced by putting an extra electron into the solid or taking one out. Although the concept was laid down more than two decades ago by Hedin and Lundquist [23], such quasiparticle calculations have just become feasible for semiconductor surfaces. The results for silicon surfaces by Northrup, Hybertsen, and Louie [12] match the data within a typical experimental and theoretical accuracy of a tenth of an eV. The next step, i.e. the calculation of a system with two excited particles, such as an electron–hole pair in optical absorption, is just beginning to become accessible to first principles quasi-particle calculations.

The discussion of surface states using one-dimensional model potentials may give the impression that it is trivial to distinguish surface from bulk states. In reality, this distinction becomes more and more blurred when going to more realistic calculations and to the experimental results. Even simple models show that there is a continuous transition from surface to bulk states at a band edge, since the decay length of the evanescent waves diverges when approaching the bulk band. Taking into account that electrons (or holes) have a finite lifetime and, therefore, a finite mean free path, it is impossible to distinguish a surface state from a bulk state if they are less than a lifetime width apart, i.e. if the mean free path is shorter than the decay length of the evanescent state. In the dangling bond picture another ambiguity arises due to surface resonances. If a dangling bond orbital happens to lie at an energy where bulk states exist (and if the momentum parallel to the surface $k_{\parallel}$ and symmetry are identical), then the dangling bond orbital is able to hybridize with a bulk band and, strictly speaking, becomes a bulk state. This happens quite often in semiconductors, such that theorists have gone on to classifying surface resonances as surface states whenever their wave function is localized more than a certain percentage in the first layer (about 70%, depending on the theorist). Experimentalists do not have this luxury. Their localization criterium is the "crud test". By offering adsorbates to the suspected dangling bond one hopes to make it disappear or at least change its energy. However, the transmission of bulk electrons through the surface is also affected by the adsorbate, making the crud test somewhat ambiguous. Two additional criteria are available to the experimentalist which are based on the two-dimensional character of true surface states. They have to be located in a symmetry gap of bulk states at a given $k_{\parallel}$, and they cannot exhibit energy dispersion when the momentum perpendicular to the surface is changed. This requires tuning the photon energy in photoemission or inverse photoemission experiments. Therefore, a tunable light source, such as synchrotron radiation, or a tunable light detector have become very useful in identifying surface states and, at the same time, mapping the bulk band structure.

1.2. Core level shifts and local bonding

Not only valence states, but also core levels change their energy at surfaces. Core level shifts can have a variety of causes, such as band bending, charge transfer, Madelung potential, and change in relaxation. They can be viewed as the effect of various charges on the electrostatic potential at the site of the core as shown in (Fig. 5) ref. [24]. There have been various attempts to sort out the different shifts experimentally as well as theoretically. Williams and Lang [25] decomposed the core level shift between atom and solid into three mathematically well-defined components,

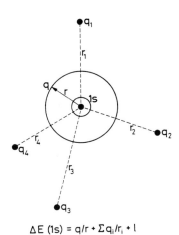

$$\Delta E \,(1s) = q/r + \Sigma q_i/r_i + l$$

Fig. 5. Model of electrostatic core level shifts induced by a charge q in a valence orbital and charges q_1 to q_4 on adjacent atoms (from Siegbahn [24]). The valence charge tends to dominate since it sits closest to the core.

i.e. chemical, configuration, and relaxation shift. For a surface as intermediate case, one can expect a similar decomposition. The chemical and configuration shifts may be lumped into the term initial state shift, since they occur already "before" ionizing the core level (the Madelung potential is also part of this initial state shift). The relaxation shift is a final state shift since it reflects the response of the solid to the core hole. This shift is much harder to calculate since the core hole breaks the translation symmetry of the solid. First principles calculations of the complete surface core level shift have, therefore, only become available for simple metals [26], where they agree well with experiment [27]. An interesting connection has been made by Rosengren and Johansson [28] between surface core level shifts and heats of surface segregation using a Born–Haber cycle and the $Z + 1$ approximation, where a core-ionized atom with atomic number Z is replaced by a valence-ionized $Z + 1$ atom.

On the experimental side there are several quantities that can be extracted from the various contributions to the core level shift. Therefore, the goal has been to separate the effects experimentally. The band bending shift has been a convenient tool for determining the Fermi level pinning and the Schottky barrier at interfaces (see the articles by Brillson, Mönch, and Spicer [29]. It only affects the overall energy calibration of the spectrum in surface sensitive experiments, where the probing depth is much smaller than the band bending region. The Fermi level position in the gap can be extracted by comparing with a reference surface. The remaining effects discriminate between surface and bulk components. The initial state shift is correlated with the local bonding of a surface atom, such as coordination and charge transfer. Pretty spectra of inequivalent carbon atoms in molecules by Siegbahn's group [24] encouraged the application to surfaces. In order to extract the initial state shift one has to eliminate the final state shift. For this purpose the concept of the Auger parameter has been tried, where the difference between the shifts of the core level and a corresponding Auger line is taken [30]. Having two experimental shifts, one hopes to determine the initial and final state contributions separately. The initial state shifts of Auger and core electrons are equal to first order while the final state shifts scale 3:1. This holds because the atom is doubly charged after Auger decay and induces twice the screening change, resulting in a relaxation energy four times as large as for the core hole. From this one has to subtract the single unit of relaxation energy that went already to the core electron. The factor of three can be seen particularly clearly for rare gas atoms adsorbed on metal surfaces, where the relaxation shift can be interpreted quite simply by the image charge [31].

In many semiconductors the dielectric constant is so high that the core hole is almost completely screened, both in the bulk and at the surface. Thus the initial state effect dominates, as in metals [26]. Only when the dielectric constant of the surface decreases, e.g. by oxidation of Si, is there a significant reduction in the relaxation at the surface, causing an extra shift to higher binding energy. With the initial state effect dominating one expects core level shifts to reflect mainly the charge transfer. This was quantified for bulk Ge compounds, where the core level shift was found to be proportional to the charge transfer after subtracting the Madelung energy of the

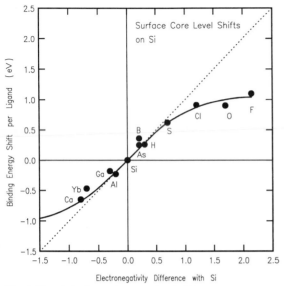

Fig. 6. Trends in the surface core level shift versus electronegativity difference for adsorbates on silicon surfaces (see ref. [33]). These shifts can be used to pinpoint surface chemical reactions.

crystal field [32]. The charge transfer was derived from the electronegativity according to Pauling. Coming to surfaces we find the same story. As shown in Fig. 6 the core level shifts at Si surfaces are related to the electronegativity difference between adsorbate and substrate [33]. At high electronegativity difference the relation becomes non-linear due to a saturation of the charge transfer, consistent with Pauling's assignment of charges.

The concept of charge transfer appears to be obvious, but it has resisted quantification upon closer inspection. A clear definition of atomic charge requires assigning a certain volume to each atom. The catch is that this volume depends again on the charge, because the negative ion is larger than the positive ion. Even cases close to the ionic or covalent limits have remained controversial, e.g. the adsorption of highly electropositive alkali atoms and the surfaces of homopolar semiconductors. In the ionic case there has been a long-standing debate between the old Gurney model of ionic adsorption [34] and proponents of a covalent picture. While common sense would predict ionic bonding of the alkalis (what else

would be ionic if not them?), the Si $2p$ core level measurements fail to show the expected large shifts. In the covalent limit it has been argued that there is nothing to be gained at a homopolar semiconductor surface by separating charges. On the other hand, core level shifts as large as -0.8 eV have been found on clean Si(111)7 × 7 [33]. Taking the shift of 1.1 eV induced by fluorine as ionic, the shift at the Si(111)7 × 7 surface falls closer to the ionic than the covalent limit. It also fits in nicely with the idea of a lone pair at the so-called rest atom sites, which avoids a half-filled dangling bond [35]. Debates on charge transfer at surfaces are still going on, and it appears that we might forever get lost in semantics when trying to quantify charge transfer.

2. Detection methods: towards the complete picture

2.1. Detecting dangling bonds

The existence of surface states on semiconductors had become clear in the 1940's from electrical measurements. They were needed to explain rectification, and were a nuisance in the development of the transistor, because it was nearly impossible at the time to control surfaces in reproducible fashion (see Bardeen [26]). With the arrival of ultrahigh vacuum technology in the 1960's it became finally feasible to study clean semiconductor surfaces. As an example for many surfaces, we follow the story of one of the favorites, i.e. the cleaved Si(111)2 × 1 surface. Fig. 7 presents a few snapshots that give a fast-forward replay, starting with early hints of surface states and ending with the state of the art.

Measurements of the Fermi level position at the surface versus types of doping by Allen and Gobeli [37] provided evidence for a high density of surface states in the gap that pins the Fermi level at the surface (Fig. 7a). Such electrical measurements were quite popular in early issues of Surface Science [38]. Filled and empty surface states were required to explain the data, with the filled states just above the valence band edge [38], and the empty states just below the center of the

a

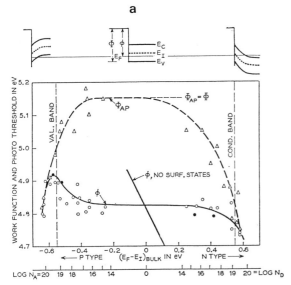

Fig. 7. Evolution of our knowledge of surface states, demonstrated for the Si(111) 2 × 1 cleavage surface. (a) Early indirect evidence of surface states in the gap from the pinning of the Fermi level (from Allen and Gobeli [37]). The difference between the photothreshold and the work function gives the position of the fermi level above the valence band maximum.
It does not change for a wide range of doping levels.

gap. Jumping all the way to Fig. 7g we see that these positions are indeed close to the edges of the surface state bands as we understand them today. Surface Fermi level measurements on cleaved GaAs(110) by van Laar and Scheer [39] showed a different behaviour for heteropolar surfaces, i.e. no Fermi level pinning by surface states. This difference fits nicely into the simplistic picture of dangling bond orbitals presented in Section 1.1. For covalent semiconductors one has half-filled dangling bond orbitals giving rise to both occupied and empty surface states, while in ionic semiconductors the dangling bonds disproportionate into a lone pair and an empty state.

Direct spectroscopic evidence for surface states on Si(111)2 × 1 came from optical results by Chiarotti et al. [40] (Fig. 7b). They observed a strong optical transition from occupied to empty surface states at about 0.5 eV photon energy, i.e. well below the bulk band gap of Si. The onset of the absorption at about 0.3 eV gave the band gap of the surface band structure. Its value was some-

what underestimated compared to current best estimates of 0.5 eV [41], possibly due to a lowering of the optical transition energy by electron hole interaction. A quantitative understanding of this effect remains one of the challenges in this field [12].

Optical experiments do not provide the absolute position of surface states since they sample a combination of occupied and empty states and are affected by electron hole interaction. Nevertheless, a variety of surface optical probes were found that narrow down the absolute position of surface states [42,43]. The direct way to probe occupied states turned out to be photoelectron spectroscopy. The first results from Si(111)2 × 1

b

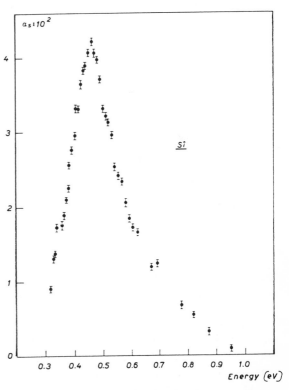

Fig. 7b. Observation of optical excitations from occupied to empty surface states, showing a surface state band gap (from Chiarotti et al. [40]). The data points represent the difference in absorption between a clean and oxidized surface.

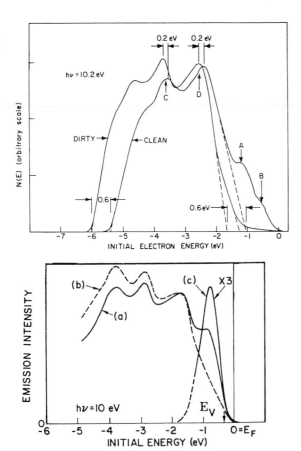

Fig. 7c. Determination of the energy distribution of surface states with photoelectron spectroscopy by Eastman and Grobman [44] (bottom) and Wagner and Spicer [45] (top). Surface states close to the top of the valence band become quenched by adsorbates, while bulk features farther down remain.

showed surface states from 0.3 to 1.2 eV below the Fermi level, which were quenched by surface contamination (Fig. 7c; Eastman and Grobman [44], Wagner and Spicer [45]). These results encouraged the widespread use of photoemission to detect surface states. The technique has been the mainstay in probing surface states ever since.

Extra information came from the angular dependence of photoemission, which provided the momentum as an extra quantum number for the electrons. Early work by Gobeli, Allen, and Kane

[46] on the angle- and polarization-dependence of the photoelectric yield from Si(111)2 × 1 inspired a series of experiments to map out the momentum-dependence of surface state emission [47–50]. After some three-fold, bulk-like symmetry patterns [46,47], the experiments eventually settled onto a two-fold pattern, representing the two-fold symmetry of a single 2 × 1 surface domain (Fig. 7d, [49]). Photoelectrons from the top of the occupied surface state band were found at emission angles corresponding to the short boundary of the rectangular surface Brillouin zone. This type of emission pattern was incompatible with previous structural models for Si(111)2 × 1. Either a new structural model was called for, or the traditional band picture had to be given up. Both avenues were pursued. The second path was tried with a highly correlated, antiferromagnetic ordering of the spins in the broken bonds, but the solution came with a new structural model by Pandey [9], i.e. the π-bonded chain (Figs. 4 and 7e). The photoemission data (full circles in Fig. 7e) could now be rationalized, except for a "controversial shoulder" (open circles in Fig. 7e). With more extensive calculations and measurements the agreement between theory and experiment kept improving, and the interest in controversial shoulder faded. The latest results are shown in Fig. 7g, where data points from angle-resolved photoemission [50], as well as inverse photoemission [51] are plotted together with band dispersions from quasiparticle calculations [12].

The π-bonded chain model for the Si(111)2 × 1 surface (Fig. 4) is a good example of the contortions that semiconductor surfaces go through in search of partners for their broken bonds. Atoms switch between the first and second layer to be able to form a π-bond, which is highly unusual in silicon chemistry. The filled and empty surface states correspond to the bonding and antibonding π and π^* band, respectively. An interesting feature of this model is that the surface states are close to one-dimensional. The chain structure resembles that of polyacetylene, except for the lack of dimerization. Its one-dimensional nature shows up directly in the momentum distribution in Fig. 7d, which changes only along the chin

d

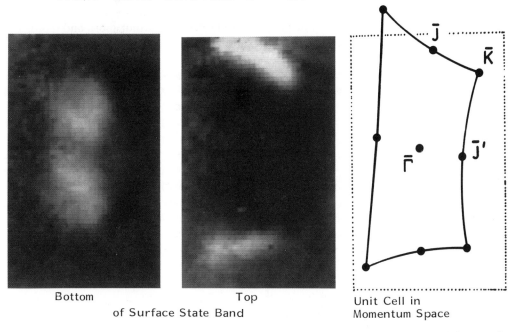

Momentum Distribution of Photoelectrons

Si (III) – (2×I) SURFACE STATES

Bottom   Top   Unit Cell in
of Surface State Band   Momentum Space

Fig. 7d. Momentum distribution of surface states at the top and bottom of the surface state band from angle-resolved photoemission (Himpsel, Heimann, and Eastman [49]). Bright areas represent high emission of photoelectrons. The top of the valence band lies along the short boundary of the surface Brillouin zone (right).

(up-down in the picture), and not perpendicular to it.

With the advent of scanning tunneling microscopy the broken bonds could be seen directly in real space [6,52,53]. The two pictures of Si(111)2 × 1 in Fig. 7f by Stroscio, Feenstra, and Fein [53] were taken by tunneling via occupied and empty surface states, respectively. As the position of the cross-hair shows, the two surface state orbitals are located on different atoms. This real space information can in principle be translated into momentum space and vice versa, e.g. by parameterization with tight binding wave functions. This closes the loop between real space and momentum space.

After understanding the energy band structure of surface states one can proceed to the dynam-

ics, which may be viewed as the imaginary part of the energy. This has become possible with pump–probe experiments in two-photon photoemission [54–56]. Electrons excited at a semiconductor surface trickle down in a complex series of jumps between different local minima in the bulk conduction bands, and eventually end up at the bottom of the surface state band as the lowest, and most stable excited state, before they recombine with a hole.

2.2. Empty states and inverse photoemission

In the early 1980's, angle-resolved photoemission started to become more common place, and was being applied to a wide variety of surfaces. As the information started to pile up it became

e

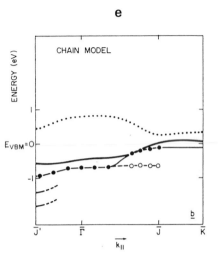

Fig. 7e. Band dispersion of surface states for the π-bonded chain structure compared to the photoemission data points (from Pandey [9]). An occupied π-band (full line) and an unoccupied π^*-band (dotted) are separated by a gap.

suddenly clear that there was a lot known about occupied states but very little about their unoccupied counterpart. Although the existence of unoccupied states was clear from optical experiments [40,57,58], their energies were shifted by electron–hole interaction and their momenta could not be determined. This uncertainty started to bother the author more and more, and eventually he quizzed all the spectroscopists he could get hold of about possible techniques to probe unoccupied states. Eventually he got a promising lead from P.O. Nilsson at a Gordon Conference on Electron Spectroscopy in Wolfeboro (1980). He alerted him to a paper on bremsstrahlung isochromat spectroscopy in the ultraviolet, buried in a conference proceedings [59]. Undaunted by the unpronouncable code word (it helped to be familiar with German) the author dug up the paper and decided this was the method of choice for solving the problem with the unoccupied

f

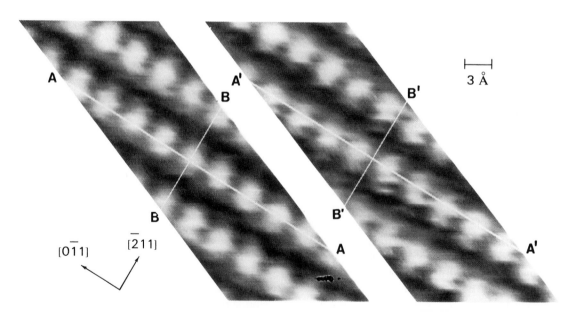

Fig. 7f. Real space image of the dangling bonds, showing different spatial locations of occupied and empty surface states (From Stroscio, Feenstra, and Fein [53]). The picture also demonstrates the chain-like, one-dimensional nature of the surface.

g

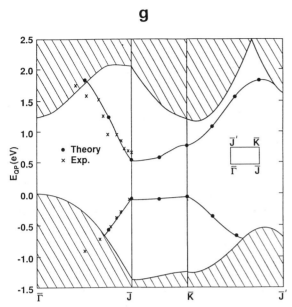

Fig. 7g. Quasiparticle band structure (lines and full circles, from Northrup, Hybertsen, and Louie [12]), together with photoemission [50] and inverse photoemission [51] data points (crosses). The hatched areas represent regions of bulk bands.

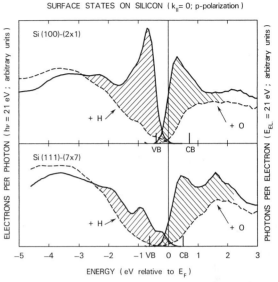

Fig. 8. Occupied and unoccupied broken bond states for two silicon surfaces, determined with photoemission and inverse photoemission (from ref. [61]). Surface states (hatched) are quenched by exposure to hydrogen (+H) and oxygen (+O).

states. Essentially, it had all the advantages of photoemission in probing a complete set of quantum numbers for the electrons, and could be made surface sensitive by using low energy electrons. After a series of trials with different detectors the first signal from an unoccupied surface state on silicon was seen two years later [60], and soon some respectable spectra were produced that showed a nice symmetry between bonding and antibonding dangling bond states (Fig. 8) [61]. Surface states on GaAs had been already detected earlier by Dose, Gossmann, and Straub [62], using a unique design for the photon detector. The new technique eventually became popular after adopting the more easily pronounced name of inverse photoemission.

Inverse photoemission has continuously tried to catch up with its cousin photoemission, but has always lagged behind by about a decade. This is illustrated in Table 1, where some comparable milestones are listed for both techniques, selected with a little bias to support the author's assertion. It begins already with the original discoveries in

the last century by Hertz [63] and Röntgen [64]. After Einstein explained the maximum kinetic energy E_{kin}^{max} of the photoelectrons as quantum effect [65], it took ten years to bring the work function ϕ to the other side of the equation and

Table 1
Photoemission versus inverse photoemission

Discovery			
1887	Hertz	1895	Röntgen
1888	Hallwachs		
Quantum effect			
1905	Einstein	1915	Duane, Hunt
	$E_{kin}^{max} = h\nu - \phi$		$h\nu^{max} = E_{kin} + \phi$
Valence band features at surfaces (UV light)			
1964	Berglund, Spicer	1977	Dose
Full theory (single step)			
1964	Adawi	1981	Pendry
Semiconductor surface states			
1972	Eastman, Grobman	1981	Dose, Gossmann, Straub (GaAs)
1972	Wagner, Spicer	1983	Fauster, Himpsel (Si)

explain the high-energy cutoff $h\nu^{\max}$ of the bremsstrahlung spectrum as quantum effect, too [66]. When the two techniques were applied to surface-sensitive studies by Berglund and Spicer [67], and Dose [68], the time lag had actually increased. It became even larger for the development of the theoretical framework [69,70]. Coming back to our topic we find that the first observations of surface states on semiconductors [44,45,60,62] occurred again with a ten year time lag. The explanation of this effect demonstrates how a physical constant influences the progress of science. The cross section for inverse photoemission is four orders of magnitude smaller than that of photoemission. While the matrix elements are the same, the phase space integration in Fermi's Golden Rule cuts down the inverse photoemission cross section by a factor α^2 ($\alpha \approx 1/237$ is the fine structure constant). Thus it appears that four orders of magnitude less intensity sets an experimentalist back by a decade. Why theorists seem to have the same problem remains an open question.

To further digress on this tangent, it is worth noticing that the same, 10 year time constant governs not only the detection, but also the production of photons. The light source of choice in photoemission experiments is synchrotron radiation (see Section 2.3). It started in the 1960's with parasitic use at electron synchrotrons using currents in the 100 μA range. A new generation of storage rings, dedicated to producing synchrotron radiation, came along in the 1970's, boosting currents up to 1 A. In the 1990's a third generation of undulator-based light sources is being built that exhibits a brilliance again 10^4 times larger than its predecessors. The fourth generation will be based on free-electron lasers, and should be so powerful that surfaces will start to evaporate.

2.3. Surface core levels and synchrotron radiation

The structure and composition of many surfaces and interfaces encountered in technology is much too complex for performing complete angle-resolved photoemission studies and for calculating the band structure from first principles. When long-range order is lost one has to resort to simpler probing techniques and must concentrate on determining the short-range order and the local properties, such as coordination, valence, and oxidation state. Core levels are well-suited for this purpose. They are not affected by the loss of long-range order because of their localized nature. As shown in Section 1.2, the core level binding energy has a simple relation with chemical properties, e.g., charge transfer.

The use of synchrotron radiation has given a great boost to the field. Monolayer surface sensitivity can be achieved by tuning the photon energy to 30–50 eV above threshold, where the photoelectrons come out with the minimum escape depth [71]. Under these conditions the escape depth can be as short as 3 Å (two layers) compared to about 20 Å (more than 10 layers) in conventional X-ray photoelectron spectroscopy (XPS). This short escape depth makes it possible to detect a monolayer of chemically altered atoms at a surface or an interface. A classic example has been the SiO_2/Si interface (Fig. 9), whose quality is one of the main reasons for the dominance of silicon technology. Intermediate states of Si at the interface were first inferred from XPS (Fig. 9 top, [72]), but it took synchrotron radiation to clearly resolve them (Fig. 9 bottom, [73]). Their distribution gives important clues about the connectivity of the interface, which cannot be obtained from traditional structural tools that require long range order. Many structural models could be rejected outright because they gave the wrong distribution oxidation states. It is interesting to note that the three intermediate oxidation states of Si are unstable in the bulk and disproportionate into Si and SiO_2. The geometrical constraints at the interface thus give rise to unusual chemistry.

Such core level work on surface and interface chemistry, together with Fermi level pinning measurements (see ref. [29]), comprise a large fraction of the current activity at soft X-ray synchrotron light sources. While current synchrotron light sources are capable of producing enough intensity to go down to the intrinsic lifetime width for the Si 2p level, they are not yet adequate for deeper core levels, e.g. the C 1s, which is crucial in organic chemistry. A new generation of undu-

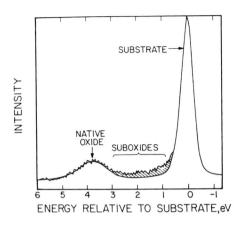

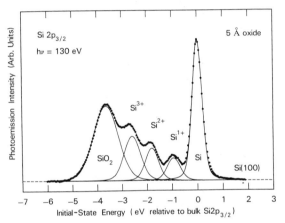

Fig. 9. Use of core level spectroscopy to identify the bond topology at the SiO_2/Si interface. X-ray photoelectron spectroscopy (top, [72]) suggested three suboxides, corresponding to 1, 2, and 3 oxygen ligands. They can be clearly resolved by tuning into the minimum of the photoelectron escape depth using synchrotron radiation (bottom, [73]).

lator-based synchrotron light sources is currently being built to access the whole Periodic Table with intrinsically-limited resolution.

2.4. So many choices and still not enough

The past thirty years of surface science have led to an enormous proliferation of surfaces techniques, leading to an alphabet soup of acronyms. And yet there exist issues where all these techniques fail. In semiconductor devices one needs to reduce the density of electrically-active defects at critical surfaces and interfaces to less than 10^{-4} of a monolayer. A typical case is the inter-

face between the channel and the gate insulator in a field effect transistor. Most surface techniques are incapable of detecting such a small amount, particularly if it is buried at an interface or localized in a micron-size device structure. Electrical measurements, on the other hand, can easily detect these changes by finding out whether a device performs properly. They do not provide many clues about the cause of the problem, though. For example, even a question as simple as the atomic number of an impurity cannot be answered easily. Here lies an open ground for combining sensitive electrical measurements with informative surface spectroscopy techniques to get the best out of both.

3. Preparation methods: from analysis to synthesis

3.1. The early days

With only a few rudimentary surface analysis methods available, the preparation of well-defined surfaces was a hunt in the dark. In the 1960's it has just become possible to keep surfaces clean, but to get good surfaces in the first place was unexplored territory. The safest way to obtain a clean semiconductor surface was cleaving. This explains the early popularity of cleaved $Si(111)2 \times 1$ and $GaAs(110)1 \times 1$ (see Section 2.1). Sputter-annealing was successful for metal surfaces. As it turned out later, the sputter damage in semiconductors cannot be annealed out as well as in metals, and today the best methods of cleaning silicon do not involve any sputtering. Another problem was posed by trace amounts of metal impurities, which could alter the surface reconstruction, e.g. the Ni-stabilized $Si(111)\sqrt{19} \times \sqrt{19}$ structure [13]. When the author started in this business, his first Si surfaces showed beautiful Cu 3d bands, despite the fact that there was no copper in the sample mount. As it turned out, the copper came from a feedthrough, diffused from there to a tungsten filament that provided electrons for electron bombardment heating, evaporated from the filament to the back of the Si wafer, and eventually diffused to the front.

3.2. Learning from device processing

Most of today's techniques for preparing well-defined semiconductor surfaces can be traced to the technology used in processing semiconductor devices. The criteria for electrical performance have been more stringent than those of the surface analysis methods, such as Auger spectroscopy and low energy electron diffraction (LEED). Only after the arrival of scanning tunneling microscopy did it become apparent how defective our semiconductor surfaces actually were. Many device processing recipes were developed by trial and error, and surface analysis methods served to shed some light onto why some recipes worked and others did not. An example in the group IV arena is passivation of Si surfaces with HF solutions [74–76]. A simple dip into a 10% HF solution in ultrapure water produced a hydrogen-passivated surface without any traces of the native SiO_2 layer [75], which was used for the growth of perfect epitaxial layers. With an ammonia buffer the solution etched away all the asperities on a Si(111) surface, and produced extremely flat surfaces [76]. In both cases it was crucial to use electronic grade reagents. Otherwise the Si wafers picked up a Langmuir–Blodgett type film of hydrocarbons, and gave more carbon contamination after heat-cleaning than untreated surfaces. Some of the new device processing techniques, such as molecular beam epitaxy (MBE) of III–V materials, allowed the preparation of completely new surface structures with variable surface stoichiometry. A more forgiving growth method than MBE is chemical vapor deposition (CVD), where the species to be deposited (e.g. Si) is bound in a relatively inert molecule (e.g. SiH_4), compared to a Si atom with four broken bonds in Si MBE (which is really atomic and not molecular beam epitaxy). The practical consequences are significant. While Si MBE requires an extremely-well outgassed growth chamber, Si CVD is much less sensitive to contamination. Not only is the SiH_4 molecule less reactive than a Si atom, it also passivates the surface at low temperatures with hydrogen [75]. Thus, surface scientists have a lot to learn from device people. They are in the process of return-

ing the favor by coming up with new deposition techniques that are based on the understanding of the surface reactions during CVD.

3.3. The future: engineering materials at an atomic scale

The future of this exchange between device technology and surface science appears to be headed for ever better control of the deposition process. Ideally one would like to control the growth of a film layer by layer. Buzzwords such as atomic layer epitaxy (ALE), digital growth and etching, and atom craft are very much in vogue these days. Scanning tunneling microscopy has even made it possible to move single atoms along the surface and create atomic patterns [77,78]. If these patterns could be connected to the macroscopic world and be mass-produced, we might have a whole new set of devices on an atomic scale. For the time being we are just beginning to control growth in the direction perpendicular to the surface, using the technique of atomic layer epitaxy (ALE). Basically, this is a chemical vapor deposition (CVD) process where two steps alternate, i.e. the adsorption of a self-saturating monolayer, and the re-activation of the surface by chemical means, irradiation, or heat. With computer-controlled gas handling systems chugging along, it is now possible to deposit compound semiconductor films with atomic precision [79]. Surface science can help a lot to understand what is going on at the surface during the two process steps and to suggest better ways of doing it.

In order to foresee what might be achieved by depositing semiconductors layer by layer it is helpful to look at the electronic structure of interfaces. After all, a solid with alternating layers of different materials consists exclusively of interfaces. The electronic structure of interfaces becomes exotic as materials with widely different bonding are forced together, particularly combinations that do not form alloys. For example, by putting the ionic insulator CaF_2 onto the covalent semiconductor Si one produces a rather strange interface bond, with Ca in the unusual $+1$ oxidation state [80]. As a result, the band gap of the interface Si–Ca layer differs grossly from that of either Si or CaF_2. It exceeds the gap of Si by more than a factor of two, and is five times smaller than that of CaF_2. Practical implications might come from adding an extra atomic layer between two semiconductors (or insulators) in order to change the band offset [81]. Another possibility is the use of superlattices to achieve specific device properties [82–85]. Close to reality are atomically-engineered structures in the area of magnetism and magnetic data storage. Multilayers of magnetic materials with nonmagnetic spacers exhibit a "giant" magnetoresistance, which can be used in magnetoresistive reading heads. The antiferromagnetic coupling required for this effect is determined by the thickness of the spacer layer, and can switch to ferromagnetic by just adding a single atomic layer to the spacer. Another interesting avenue leads into even lower dimensions, i.e. the growth of quantum wires by propagation of steps [84].

Learning from the past 30 years, one can try to predict the future development of surface science. The author sees a rapid expansion of surface synthesis techniques, such as atomic layer epitaxy. The traditional course of surface science, i.e. the addition of more and more sophisticated surface analysis techniques, appears to have reached a saturation point. Setting aside the dilute cases mentioned in Section 2.4 we currently have enough techniques to solve just about any surface structure, including the electronic states, as long as we make a concerted effort. Therefore, it makes sense to put the arsenal of techniques to work on the task of synthesizing new structures with interesting and useful properties. One would hope surface physics could get into a position similar to organic chemistry, where molecules are designed for specific purposes.

References

[1] I. Tamm, Phys. Z. Sowjetunion 1 (1932) 733.
[2] The curve shown in Fig. 1 is a plot of Tamm's surface state wavefunction, in: C.J. Chen, Introduction to Scanning Tunneling Microscopy (Oxford University Press, Oxford, 1993).
[3] R.O. Jones, Phys. Rev. Lett. 20 (1964) 992.
[4] J.A. Appelbaum and D.R. Hamann, Rev. Mod. Phys. 48 (1976) 479.
[5] M. Schlüter, J.R. Chelikowsky, S.G. Louie and M.L. Cohen, Phys. Rev. Lett. 34 (1975) 1385.
[6] G. Binnig, H. Rohrer, Ch. Gerber and E. Weibel, Phys. Rev. Lett. 50 (1983) 120.
[7] A.R. Lubinsky, C.B. Duke, B.W. Lee and P. Mark, Phys. Rev. Lett. 36 (1976) 1058;
For an overview of semiconductor surface structures see: C.B. Duke, in: Physical Structure of Solid Surfaces, Ed. W.N. Unertl, in press.
[8] D. Haneman, Phys. Rev. Lett. 121 (1961) 1093.
[9] K.C. Pandey, Phys. Rev. Lett. 47 (1981) 1913.
[10] F.J. Himpsel, P.M. Marcus, R. Tromp, I.P. Batra, M.R. Cook, F. Jona and H. Liu, Phys. Rev. B 30 (1984) 2257.
[11] F. Ancilotto, W. Andreoni, A. Selloni, R. Car and M. Parrinello, Phys. Rev. Lett. 65 (1990) 3178.
[12] J.E. Northrup, M.S. Hybertsen and S.G. Louie, Phys. Rev. Lett. 66 (1991) 500;
see also the article by S.G. Louie, Surf. Sci. 299/300 (1994) 346.
[13] R.E. Schlier and H.E. Farnsworth, J. Chem. Phys. 30 (1959) 917.
[14] D.J. Chadi, Phys. Rev. Lett. 43 (1979) 43.
[15] J. Pollmann, R. Kalla, P. Krüger, A. Mazur and G. Wolfgarten, Appl. Phys. A 41 (1986) 21.
[16] J. Dabrowski and M. Scheffler, Appl. Surf. Sci. 56 (1992) 15.
[17] Y. Enta, S. Suzuki and S. Kono, Phys. Rev. Lett. 65 (1990) 2704.
[18] R.A. Wolkow, Phys. Rev. Lett. 68 (1992) 2636.
[19] S.D. Kevan, Phys. Rev. B 32 (1985) 2344.
[20] N.J. DiNardo, T. Maeda Wong and E.W. Plummer, Phys. Rev. Lett. 65 (1990) 2177.
[21] O. Pankratov and M. Scheffler, Phys. Rev. Lett. 70 (1993) 351.
[22] R. Car and M. Parrinello, Phys. Rev. Lett. 55 (1985) 2471.
[23] L. Hedin and S. Lundquist, in: Solid State Physics, Vol.23, Eds. F. Seitz, D. Turnbull and H. Ehrenreich (Academic Press, New York, 1969) p. 1.
[24] K. Siegbahn, in: Electron Spectroscopy, Ed. D.A. Shirley (North Holland, Amsterdam, 1972) p. 15.
[25] A.R. Williams and N.D. Lang, Phys. Rev. Lett. 40 (1978) 954.
[26] P.J. Feibelman, Phys. Rev. B 39 (1989) 4866. Recently, a calculation for Si(100) has become available: E. Pehlke and M. Scheffler, to be published.
[27] R. Nyholm, J.N. Andersen, J.F. van Acker and M. Qvarford, Phys. Rev. B 44 (1991) 10987.
[28] A. Rosengren and B. Johansson, Phys. Rev. B 23 (1981) 3852.
[29] See the articles by L.J. Brillson, Surf. Sci. 299/300 (1994) 909;
W. Mönch, Surf. Sci. 299/300 (1994) 928.
[30] C.D. Wagner, in: Electron Spectroscopy of Solids and Surfaces, Faraday Discuss. Chem. Soc. 60 (1975) 291.

[31] G. Kaindl, T.-C. Chiang, D.E. Eastman and F.J. Himpsel, Phys. Rev. Lett. 45 (1980) 1808: Although there has been a fair amount of controversy over this simple image charge interpretation, recent results seem to confirm it; K. Jacobi, Surf. Sci. 192 (1987) 499.

[32] G. Hollinger, P. Kumurdjian, J.M. Mackowski, P. Pertosa, L. Porte and Tran Minh Duc, J. Electron Spectrosc. Relat. Phenom. 5 (1974) 237.

[33] F.J. Himpsel, B.S. Meyerson, F.R. McFeely, J.F. Morar, A. Taleb-Ibrahimi and J.A. Yarmoff, Proceedings of the Enrico Fermi School on Photoemission and Absorption Spectroscopy of Solids and Interfaces with Synchrotron Radiation, Eds. M. Campagna and R. Rosei (North Holland, Amsterdam, 1990) p. 203.

[34] R.W. Gurney, Phys. Rev. 47 (1935) 479.

[35] J.E. Northrup, Phys. Rev. Lett. 57 (1986) 154.

[36] J. Bardeen, Phys. Rev. 71 (1947) 717.

[37] F.G. Allen and G.W. Gobeli, Phys. Rev. 127 (1962) 150.

[38] G. Heiland and H. Lamatsch, Surf. Sci. 2 (1964) 19;
G.W. Gobeli and F.G. Allen, Surf. Sci. 2 (1964) 402;
D.E. Aspnes and P. Handler, Surf. Sci. 4 (1966) 353;
M. Henzler, Phys. Status Solidi 19 (1967) 733.

[39] J. van Laar and J.J. Scheer, Surf. Sci. 8 (1967) 342.

[40] G. Chiarotti, G. Del Signore and S. Nannarone, Phys. Rev. Lett. 21 (1968) 1170;
G. Chiarotti, S. Nannarone, R. Pastore and P. Chiaradia, Phys. Rev. B 4 (1971) 3398.

[41] R.M. Feenstra, Phys. Rev. B 44 (1991) 13791.

[42] W. Müller and W. Mönch, Phys. Rev. Lett. 27 (1971) 250.

[43] J. Clabes and M. Henzler, Phys. Rev. B 21 (1980) 625.

[44] D.E. Eastman and W.D. Grobman, Phys. Rev. Lett. 28 (1972) 1378.

[45] L.F. Wagner and W.E. Spicer, Phys. Rev. Lett. 28 (1972) 1381.

[46] G.W. Gobeli, F.G. Allen and E.O. Kane, Phys. Rev. Lett. 12 (1964) 94.

[47] J.E. Rowe, M.M. Traum and N.V. Smith, Phys. Rev. Lett. 33 (1974) 1333;
The main structure seen in this work was later identified as a bulk peak by R.I.G. Uhrberg, G.V. Hansson, U.O. Karlsson, J.M. Nicholls, P.E.S. Persson, S.A. Flodström, R. Engelhardt and E.-E. Koch, Phys. Rev. Lett. 52 (1984) 2265.

[48] F. Houzay, G.M. Guichar, R. Pinchaux and Y. Petroff, J. Vac. Sci. Technol. 18 (1981) 860.

[49] F.J. Himpsel, P. Heimann and D.E. Eastman, Phys. Rev. B 24 (1981) 2003.

[50] R.I.G. Uhrberg, G.V. Hansson, J.M. Nicholls and S.A. Flodström, Phys. Rev. Lett. 48 (1982) 1032.

[51] P. Perfetti, J.M. Nicholls and B. Reihl, Phys. Rev. B 36 (1987) 6160.

[52] R.J. Hamers, R.M. Tromp and J.E. Demuth, Phys. Rev. Lett. 56 (1986) 1972.

[53] J.A. Stroscio, R.M. Feenstra and A.P. Fein, Phys. Rev. Lett. 57 (1986) 2579.

[54] G.D. Kubiak and K.W. Kolasinski, Phys. Rev. B 39 (1989) 1381.

[55] N.J. Halas and J. Bokor, Phys. Rev. Lett. 62 (1989) 1679.

[56] M. Baeumler and R. Haight, Phys. Rev. Lett. 67 (1991) 1153.

[57] D.E. Eastman and J.L. Freeouf, Phys. Rev. Lett. 33 (1974) 1601.

[58] G.J. Lapeyre and J. Anderson, Phys. Rev. Lett. 35 (1975) 17.

[59] P.O. Nilsson and C.G. Larsson, Jpn. J. Appl. Phys. 17 (1978) 144.

[60] F.J. Himpsel, Th. Fauster and G. Hollinger, Surf. Sci. 132 (1983) 22.

[61] F.J. Himpsel and Th. Fauster, J. Vac. Sci. Technol. A 2 (1984) 815.

[62] V. Dose, H.-J. Gossmann and D. Straub, Phys. Rev. Lett. 47 (1981) 608.

[63] H. Hertz, Ann. Phys. 31 (1887) 983;
W. Hallwachs, Ann. Phys. 33 (1888) 301.

[64] W.C. Röntgen, Sitzgsber. Med.-Phys. Ges. Würzburg 1895 (1895) 137.

[65] A. Einstein, Ann. Phys. 17 (1905) 132.

[66] W. Duane and F.L. Hunt, Phys. Rev. 6 (1915) 166.

[67] C.W. Berglund and W.E. Spicer, Phys. Rev. 136 (1964) A1030, A1044.

[68] V. Dose, Appl. Phys. 14 (1977) 117.

[69] I. Adawi, Phys. Rev. A 134 (1964) 788.

[70] J.B. Pendry, Phys. Rev. Lett. 45 (1980) 1356.

[71] I. Lindau and W.E. Spicer, J. Electron Spectrosc. Relat. Phenom. 3 (1974) 409.

[72] F.J. Grunthaner and P.J. Grunthaner, Mater. Sci. Rep. 1 (1986) 65.

[73] F.J. Himpsel, F.R. McFeely, A. Taleb-Ibrahimi and J.A. Yarmoff, Phys. Rev. B 38 (1988) 6084.

[74] E. Yablonovitch, D.L. Allara, C.C. Chang, T. Gmitter and T.B. Bright, Phys. Rev. Lett. 57 (1986) 249.

[75] B.S. Meyerson, F.J. Himpsel and K.J. Uram, Appl. Phys. Lett. 57 (1990) 1034.

[76] P. Jakob and Y.J. Chabal, J. Chem. Phys. 95 (1991) 2897.

[77] D.M. Eigler and F.D. Schweizer, Nature 344 (1990) 524.

[78] I.W. Lyo and Ph. Avouris, Science 253 (1991) 173;
M. Aono, Science 258 (1992) 586.

[79] J. Nishizawa, J. Cryst. Growth 115 (1991) 12.

[80] T.F. Heinz, F.J. Himpsel, E. Palange and E. Burstein, Phys. Rev. Lett. 63 (1989) 644.

[81] G. Ceccone, G. Bratina, L. Sorba, A. Antonini and A. Franciosi, Surf. Sci. 251 (1990) 82.

[82] L. Esaki and R. Tsu, IBM J. Res. Dev. 14 (1970) 61;
L.L. Chang, L. Esaki, W.E. Howard and R. Ludeke, J. Vac. Sci. Technol. 10 (1973) 11.

[83] K. Ploog, A. Fischer and H. Künzel, J. Electrochem. Soc. 128 (1981) 400.

[84] P.M. Petroff, A.C. Gossard and W. Wiegmann, Appl. Phys. Lett. 45 (1984) 620.

[85] See the article by F. Capasso and A.Y. Cho, Surf. Sci. 299/300 (1994) 878.

Surface Science 299/300 (1994) 541–550
North-Holland

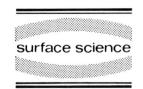

surface science

Electronic surface states investigated by optical spectroscopy

G. Chiarotti

Department of Physics, University of Rome "Tor Vergata", Rome, Italy

Received 27 April 1993; accepted for publication 6 July 1993

The contribution of optical spectroscopy to our knowledge of electronic surface states and surface reconstruction is reviewed with special emphasis to its historical development. The first detection of surface states in Ge(111)2 × 1 and Si(111)2 × 1 by our group in the years 1968–1971 is discussed in detail. Optical methods showed already in those early years that the structure of the states in the above surfaces consists of two bands, symmetrically located with respect to the Fermi level at the surface, separated by a gap of 0.35 eV (in Si). Comparison with the results obtained with other techniques is discussed.

The use of polarized light allowed the testing of the various models proposed for the 2 × 1 surface of elemental semiconductors, giving a strong evidence to the validity of the so-called chain model of the surface.

Open questions concerning the polar character of the 2 × 1 surfaces and its connection to earlier results on electric-field modulated reflectivity are also discussed.

An outline of the theory of surface differential reflectivity as well as results for other semiconductors are presented.

1. Introduction

The detection of intrinsic surface states (i.e. states caused, in otherwise perfect crystals, by the termination of the potential and/or by the re-arrangement of the atoms at the surface) was a central problem for surface physics in the 1960's.

Surface states were already predicted for a 1D potential in 1932 by Tamm [1] and, for a more realistic situation suitable for semiconductors, in 1939 by Shockley [2]. Intrinsic surface states are localized at the surface and their wave function decays exponentially on both sides of the surface plane.

In a famous paper of 1947, Bardeen [3] discussed the effect of surface states on the surface conductivity of semiconductors and explained the failure of operation of what we could now call a (rudimentary) MOS transistor [4] as due to the trapping of electrons at surface states. This work marked the discovery of the point-contact transistor [5].

Besides opening a new era in electronics and in world history, the above group of works fos-

tered a great deal of research in surface physics, especially on the so-called "real surfaces" (i.e. surfaces mechanically polished and treated in air with chemical etchants). They were primarily aimed at understanding and controlling the process of conductivity in the surface channel that severely impaired the working of the newly discovered junction transistor [6,7].

It soon became clear, however, that the states detected on real surfaces were connected with impurities or defects and had little to do with the intrinsic states predicted by Tamm and Shockley. It is enough to mention here that their number was found of the order of 10^{11}–10^{13} cm^{-2}, at least two or three orders of magnitude smaller than the number of surface atoms.

The first qualitative evidence of the existence of intrinsic surface states came in 1962 through the work of Allen and Gobeli [8] who compared the values of work function and photoemission threshold for a number of cleaved Si samples with various dopings. They found that both functions stay constant over a large range of n- and p-doping, showing that the Fermi level was pinned at the surface, a signature of the presence of a large density of surface states.

From a quantitative point of view, however, the results of Allen and Gobeli were compatible with several distributions or surface states (both with and without a surface gap) and were obtained under the hypothesis that the photoelectric threshold is a measure of the distance between the valence band maximum and the vacuum level, disregarding the fact that the threshold is entirely dominated by the surface states themselves.

2. The early experiments of the Rome group

The first direct evidence of the existence of intrinsic surface states in semiconductors came from spectroscopic measurements done in our group (G. Del Signore, S. Nannarone, P. Chiaradia and myself) in 1968 for Ge(111)2 × 1 and in 1971 for Si(111)2 × 1 [9–11]. They showed the existence of two bands of surface states (one below and one above the Fermi level at the surface) separated by a gap. Optical data alone are not enough to determine the position of the surface bands with respect to the bulk structure; nevertheless, a distribution of states consisting of two bands symmetrically located with respect to the so-called neutrality point (for Si(111)2 × 1 approximately 0.35 eV above the valence band maximum) separated by a gap of 0.3 eV is compatible with both Allen and Gobeli's and our results. A similar distribution holds for Ge(111)2 × 1 where now, however, the neutrality point is near the top of the valence band [12]. Such distributions of states have remained valid throughout the years and have been substantiated by a number of experiments and theoretical calculations.

In order to illustrate the beginning of our work in the 1960's, it must be said that optical techniques are not specially sensitive to surface properties. Light interacts only weakly with surfaces and penetrates into the solid for a length of the order of α^{-1} (α being the absorption coefficient), a distance much larger than the thickness of the layer where surface states are localized.

Trying to observe surface states with optical spectroscopy seemed hopeless at that time. Nevertheless, spectroscopy experienced in the 1960's a considerable advancement due to the development of a new technique called modulation spectroscopy [13]. This method consists of modulating periodically a parameter (electric or magnetic fields, stress, temperature, etc.) from which some specific optical property depends and detecting the change of reflectivity or transmittivity of the sample at the frequency of modulation. The use of phase-sensitive amplifiers greatly enhanced the sensitivity of this technique.

We contributed to the early development of this spectroscopy through a method that we could call "modulation of population" [14,15]. In one of its varieties suitable for surface physics, an AC electric field is applied perpendicularly to the surface of a semiconductor. Because of charge accumulation (or depletion) at the surface, the band bending and therefore the distance of the states from the Fermi level are periodically changed. As a consequence, the population of such states and the intensity of any optical transition occurring among them is modulated with the same frequency.

Experiments of modulated reflectivity were done in our and other groups in those years on "real" surfaces of semiconductors [14,16,17].

When the ultra-high-vacuum technique developed to the point of making ionic pumps and Alpert gauges commercially available, we thought of extending our experiments to "clean" surfaces in order to detect Tamm-like states.

A sample of nearly intrinsic Ge, cut in the shape shown in Fig. 1, was cleaved in UHV along the (111) plane with the double wedge technique as sketched in the figure. An AC voltage was applied between the two halves of the sample and a beam of monochromatic light of energy below the bulk gap was totally reflected several times on the freshly cleaved surface, as shown in the figure. A weak signal with a complex spectral dependence was observed, corresponding to a value of $\Delta I/I$ of the order of 10^{-4}.

However, the spectral dependence was different from what we expected and, more than that, the phase (or the sign) of ΔI was apparently

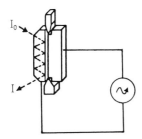

Fig. 1. Outline of the experiment of modulated reflectivity. Differential reflectivity (in the internal mode) is obtained with the same set-up, without the external electric field.

wrong. We thought we were modulating the population of a filled state in the gap and then we were expecting a negative ΔI when the voltage on the field-effect electrode (the other half of the sample) was positive. This was contradicted by the experiments. The results were not published since we could not understand them.

However, during the various normalization procedures for correcting for the spectral dependence of I (in absence of the external voltage), we realized that, in a well-defined spectral region, the intensity I increased steadily when the surface was exposed to oxygen, as if the oxidation destroyed absorbing states in the surface layer.

The results are shown in Fig. 2, where the edge of the band-to-band absorption of cleaved Ge(111) is plotted, for the clean and fully oxidized surfaces [9]. The absorption constant:

$$\alpha_s = \frac{1}{N}\left[\ln(I_0/I)_{\text{clean}} - \ln(I_0/I)_{\text{ox}}\right] \quad (1)$$

(where N is the number of multiple reflections) is shown in Fig. 3 and reveals a well-defined peak at energies below the bulk gap, which we associated to transitions among surface states.

Similar results were obtained for Si(111)2 × 1 and are shown in Fig. 4 [10].

For $N\alpha_s \ll 1$:

$$\alpha_s \simeq \frac{1}{N}\frac{I_{\text{ox}} - I_{\text{clean}}}{I_{\text{clean}}}. \quad (2)$$

It is seen from Figs. 3 and 4 that, for both Si and Ge, the relative change of intensity due to the absorption by surface states is, for a single reflection, approximately 4% at the position of the peak and then easily detectable.

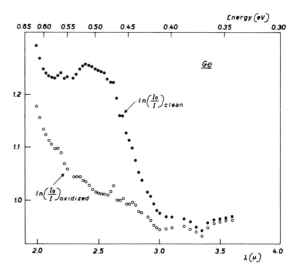

Fig. 2. Logarithm of the ratio I_0/I as a function of wavelength in Ge(111)2 × 1 for the clean surface and for the same surface after oxidation. From Ref. [10].

The value $\alpha_s = 0.04$ corresponds to an optical cross-section for the surface transition σ_s of approximately 0.5×10^{-16} cm^2 ($\sigma_s = \alpha_s/N_0$, N_0 being the density of surface atoms), a rather high value comparable, however, with the optical

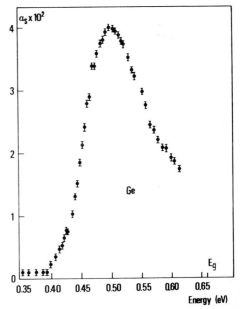

Fig. 3. Absorption constant α_s versus photon energy, obtained from the data of Fig. 2, for Ge(111)2 × 1. After Ref. [10].

cross-section found in peaks of the fundamental absorption of bulk semiconductors [18].

This simple analysis shows that surface states can indeed be detected with optical spectroscopy, though only under favorable conditions similar to those found for peaks of the bulk spectrum. These conditions occur for symmetry-allowed, direct transitions between bands running nearly parallel to each other for a consistent portion of the surface Brillouin zone, so as to give a large value of the joint density of states. The last condition has been later verified by angle resolved photoemission spectroscopy [19] and by detailed calculations [20]. In a way, Figs. 3 and 4 are the first examples of a critical point in the 2D surface joint density of states.

In the following years the above results were confirmed by various techniques. In Si(111)2 × 1 the filled band below the Fermi energy was observed in photoemission spectroscopy by Eastman and Grobman [21] and by Wagner and Spicer [22] in 1972. Photoconductivity experiments done by

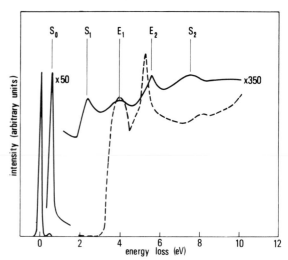

Fig. 5. High resolution electron energy loss spectrum for Si(111)2×1. Surface state transitions are labelled S_0, S_1, S_2; bulk transitions E_1, E_2. The loss function calculated from bulk data is shown as a dashed line. After Ref. [25].

Mönch and colleagues gave evidence to the existence of surface states in Si(111) in 1971 [23] and of a band identical to that of Fig. 4 in 1980 [24]. A direct confirmation came in 1975 from high-resolution electron energy loss spectroscopy through the work of Rowe, Ibach and Froitzheim [25]. Their results for Si(111)2 × 1 are shown in Fig. 5. The strongest inelastic peak S_0 corresponds to the transitions shown in Fig. 4, the small energy difference (0.52 eV against 0.45 eV) being due to the fact that energy loss spectroscopy is a measure of Im $1/(\hat{\epsilon} + 1)$, while optical spectroscopy measures $-$Im $\hat{\epsilon}$, $\hat{\epsilon}$ being the surface (complex) dielectric function. The agreement is further stressed by the same behaviour of the two peaks upon annealing that transforms the 2 × 1 structure into the 7 × 7. In both cases the peak disappears [26], presumably because the 7 × 7 reconstruction displays a metallic character. Similar results were obtained also in Ge [27].

More recently, with the development of STM techniques, it has been possible to tunnel electrons from surface states into an external metallic tip (or from the tip to the surface states) observing the surface structure at atomic level. By applying a bias to the sample it is possible to dis-

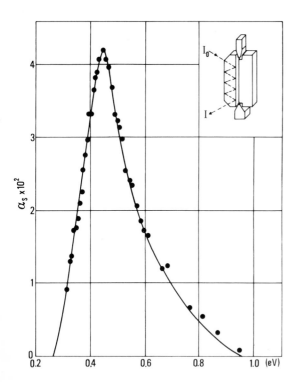

Fig. 4. Absorption constant α_s versus photon energy for Si(111)2×1. After ref. [10].

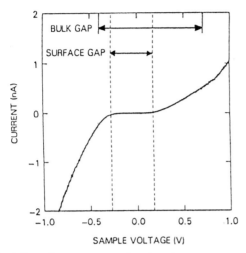

Fig. 6. Tunneling current as a function of sample voltage. The 0.45 eV wide flat spot in the current arises from a surface state band-gap, indicated together with the bulk band-gap at the top of the figure. From Ref. [28].

place the energy levels of the surface with respect to the Fermi energy of the tip. When the position of the Fermi level of the tip corresponds to a gap in the electronic structure of the surface, it is not possible to tunnel electrons from or to the tip and the current becomes zero. This is shown clearly in Fig. 6, where results of Feenstra et al. for Si(111)2 × 1 [28] are reported. It is interesting to notice that the surface gap (~ 0.5 eV) is nearly the same as that obtained in optical and energy loss experiments. This result is important since it sets an upper limit to excitonic effects in optical measurements. The extent of the excitonic contribution to the optical gap has been the subject of a long controversy in the 1970's [29]; the problem is still open since recent results of inverse photoemission seem to be at variance with those reported in Fig. 6 [30].

3. The problem of surface reflectivity

A further step in the use of optical techniques for studying the electronic structure of surfaces came when results like those of Figs. 3 and 4 were obtained with a single external reflection [31]. This allowed the extension of the range of observation above the bulk gap [32,33] and, more

important, the use of polarized light for testing the symmetry of the absorbing states [34,35].

The extension to energies above the bulk gap requires a better consideration of surface reflectivity.

In a general way, results obtained with internal or external reflections can be given in terms of the reflectivity R as:

$$\frac{\Delta R}{R} = \frac{R_{\text{clean}} - R_{\text{ox}}}{R_{\text{ox}}}. \tag{3}$$

A classical theory of differential reflectivity, especially suitable for surface physics, has been developed in 1971 by McIntyre and Aspnes [36]. They assume a three media system (1: external; 2: surface layer; 3: bulk) having each a definite dielectric function $\hat{\epsilon}_j = \epsilon_j' - i\epsilon_j''$, and work out a solution under the hypothesis $d \ll \lambda$; d being the thickness of the surface layer. For normal incidence they obtain:

$$\frac{\Delta R}{R} \equiv \frac{R(d) - R(0)}{R(0)} = -\frac{8\pi d}{\lambda} \text{Im} \frac{\hat{\epsilon}_1 - \hat{\epsilon}_2}{\hat{\epsilon}_1 - \hat{\epsilon}_3}. \tag{4}$$

$R(d)$ represents the reflectivity when a surface layer of thickness d is present on top of the bulk. $R(d)$ is then equivalent to R_{clean} in (3). For energies below the gap, both the oxide layer and the bulk do not absorb so that $\epsilon_{\text{ox}}'' = \epsilon_3'' = 0$ and $R_{\text{ox}} = R(0)$. Definition (4) is then identical to (3) that contains measurable quantities. For energies below the gap expression (4) becomes:

$$\frac{\Delta R}{R} = -\frac{4n_1 n_2}{n_1^2 - n_3^2} \frac{\omega}{c}(\epsilon_2'' d) = -\frac{4n_1 n_2}{n_1^2 - n_3^2}(\alpha_s d), \tag{5}$$

where n_j are the refractive indexes of the three media. It is seen from (5) that $\Delta R/R$ gives immediately the imaginary part of the surface dielectric function, a quantity suitable for theoretical evaluation. It is also seen that $\Delta R/R$ changes sign by interchanging 1 and 3, that is in going from internal to external reflectivity:

$$\left(\frac{\Delta R}{R}\right)_{\text{int}} = -n\left(\frac{\Delta R}{R}\right)_{\text{ext}}, \tag{6}$$

n being the refractive index of the bulk. Moreover $(\Delta R/R)_{\text{int}} < 0$ and $|(\Delta R/R)_{\text{int}}| > |(\Delta R/$

$R)_{ext}$ |. All such conditions are verified experimentally [31].

For energies above the gap the expressions are more complex, though, in a wide range of energies, $\Delta R/R$ still gives the imaginary part of the dielectric function of the surface [37].

Microscopic theories of reflectivity have also been developed [38,39] yielding essentially the same results, at least for normal incidence [40].

Surface state transitions at energies above the gap have been observed in Si(111)7 × 7 [32], Si(111)2 × 1, Ge(111)2 × 1 [33], GaAs(110), GaP(110) [41], and InP(110) [42]. Ellipsometric measurements that give evidence to surface state transitions above the gap have also been reported for various Si surfaces [43] and for GaAs(110) [44].

The results of differential reflectivity of Si and Ge above the gap have been interpreted in terms of transitions between surface states occurring in various parts of the surface Brillouin zone [45,46].

4. Models for surface reconstruction of Ge and Si(111)2 × 1

A very important problem for understanding the structure of surface states is the assumption of a model for the surface reconstruction or relaxation. LEED patterns showed in the early days that semiconductor surfaces are very often reconstructed [47,48].

Surface reconstruction is one of the most challenging problems of surface physics. Understanding the great variety of surface reconstructions and their transformation upon annealing and contamination is a formidable task. At present a general theory does not seem to be available, though attempts have been done in this direction [49]. The procedure of minimization of surface free energy, though in principle viable, is very difficult and can be done only in the frame of a given model [50]. In semiconductors, the criterion that reconstruction tends to reduce the number of free dangling bonds is generally accepted.

Cleaved surfaces of elemental semiconductors show a 2 × 1 reconstruction. Several models that combine experimental findings with theoretical

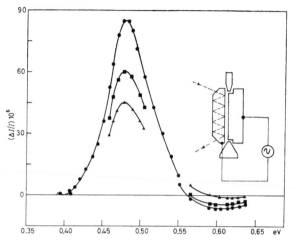

Fig. 7. $\Delta I/I$ in the electric-field modulated optical absorption as a function of the energy of the photons for Ge(111)-2×1. (●) clean surface, (■), (▲) after 900 and 2100 L of O_2. Data are from Ref. [52].

ideas have been proposed over the years. In the 70's the model most commonly accepted for the 2 × 1 surfaces was the so-called Haneman's or buckling model [51]. In such a model, rows of atoms in the [011] directions are alternatively displaced up and down. Because of the reduction of symmetry, the sp^3 bonds de-hybridize in the buckled structure, the raised atoms becoming more s-like and the lowered ones more p_z-like. An energy structure with two bands is derived that could explain the optical results. Moreover the raised s-like atoms, having lower energy, become negatively charged and vice versa, so that the surface acquires a polar character.

When the quantitative details of the model became clear, especially through the work of Pandey and Phillips [29], we realized that our earlier results on electric field modulation that we had never published because of apparent internal inconsistency, were in fact a proof of the polar character of the 2 × 1 surface! Fig. 7 shows such results, obtained in the configuration of Fig. 1. $\Delta I/I$ (the amplitude of the AC signal normalized for the light intensity) is plotted as a function of the energy of the photons for the clean Ge(111)2 × 1 surface and for various oxygen contaminations [52].

It is seen that the curves are reminiscent of the derivative of α_s. The explanation runs as follows: because of the polar character of the surface, the AC field modulates the buckling and then modifies in a periodic way the distance between the surface bands, i.e. the position of the peak of Fig. 3. In a linear approximation (which seems reasonable because of the weakness of the applied field in comparison to internal atomic fields), the signal should be proportional to the derivative of α_s.

The sign (phase) is also correct: a positive voltage applied to the field-effect electrode increases the buckling (since the raised atoms are negatively charged) giving rise to a positive ΔI for energies below the maximum of α_s and to a negative ΔI for energies above it.

The small deviations of the curves of Fig. 7 from the derivative of α_s could be due to the superposition of the effect of the modulation of population [14] and/or to a different sensitivity of the surface bands to the external field in the various regions of the surface Brillouin zone.

The experiment, of course, did not prove the validity of Haneman's model. It simply proved that the surface gap was associated (as in Haneman's model) to charge transfer among the atoms involved in the reconstruction and that the surface was polar.

Haneman's model was widely accepted in those years. At the beginning of the 80's, however, angularly resolved photoemission showed that the filled surface band displayed a considerable dispersion over most of the surface Brillouin zone [53,54]; a characteristic that was difficult to explain quantitatively on the basis of the buckling model in which the dangling bonds are in next-nearest-neighbor position and interact only weakly.

For this reason Pandey proposed in 1981 a new model consisting of surface zig-zag chains of π-bonded atoms running along [110] directions, with dangling bonds at nearest-neighbor positions [55]. Adjacent chains are well separated (~ 6 Å) giving rise to a marked anisotropy of the surface, as shown in the model of Figs. 8a and 8b.

At the beginning Pandey's model was not easily accepted since it requires the breaking of

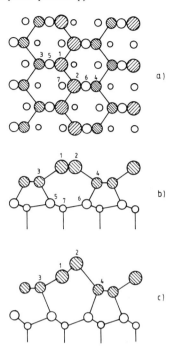

Fig. 8. Position of atoms in the reconstructed Si(111)2×1 according to the chain model: (a) top view; (b) side view; (c) side view for a buckled chain.

bonds in order to form the chains. The acceptance increased when Northrup and Cohen showed that the process of transforming the ideal surface into the π-bonded chains could be done in a quasi-adiabatic way with a very small activation energy [20].

After conflicting evidence [56,57], a very convincing proof came from differential reflectivity measurements done in our group with polarized light [34,35] and, independently, by Olmstead and Amer with a method called photothermal displacement spectroscopy [58,59]. In fact Haneman's and Pandey's models have different symmetry properties and therefore behave differently with respect to the absorption of polarized light. The former shows a maximum absorption for light polarized perpendicularly to the direction of the raised atomic rows (electric vector along [$\bar{2}11$] directions), the latter for light polarized parallel to the chains (electric vector along [$0\bar{1}1$] directions) [60,61].

In experiments with polarized light it is necessary that the reconstructed surface consists of a single domain (out of the possible three). This can be obtained by cleaving the sample under special conditions. In such a case the LEED pattern gives directly the axes of the reconstruction. Fig. 9 presents our results [34] that show a dramatic dependence of $\Delta R/R$ on the polarization, the light being reflected (absorbed) only if the electric vector is parallel to the [0$\bar{1}$1] direction, i.e. parallel to the chain axis. The results of Fig. 4 obtained with unpolarized light are close to the average value of $\Delta R/R$ for parallel and perpendicular polarizations. Pandey's model received then a strong experimental support.

A definite proof came a few years later when the chains were observed directly, with atomic resolution, by STM [28].

The experiment of electric-field-modulated absorption shown in Fig. 7 demonstrate, however, that the optical gap depends upon the transfer of charge among the atoms of the reconstructed surface. Such a transfer was absent (or present to a very limited extent) in the original model of

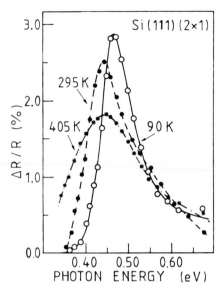

Fig. 10. Surface differential reflectivity as a function of photon energy in Si(111)2×1 at three different temperatures. After Ref. [63].

Pandey who assumed that the inequivalence of atoms labelled 1 and 2 in Fig. 8 (necessary for opening a gap in the surface band) was due to the different positions of atoms in the third and fourth layer below 1 and 2. This seemed too small a perturbation for opening a gap of approximately half an eV. Various modifications have then been introduced, the most widely accepted being a small buckling along the chain (Fig. 8c) with the raised atoms negatively charged [62], as required by the results of Fig. 7.

A further experiment that suggests a polar character for the 2 × 1 reconstructed surface is the strong temperature dependence of the width and peak position of the curves of Fig. 4, shown in Fig. 10 from Ref. [63]. The reflectivity curves of Fig. 10 closely resemble the absorption spectra of the F-center in alkali halides, a well-known example of strong electron–phonon interaction. In both cases, the width $W(T)$ depends on temperature as:

$$W(T) = W(0)\left[\coth(\hbar\omega_0/\kappa T)\right]^{1/2}, \qquad (7)$$

where ω_0 is the characteristic Einstein frequency of the phonon spectrum of the surface [64]. From the data of Fig. 10, $\hbar\omega_0$ turns out to be approxi-

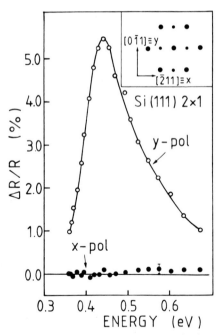

Fig. 9. Differential reflectivity spectra for Si(111)2×1 single domain surface for light polarized along [0$\bar{1}$1] and [$\bar{2}$11] directions. The inset shows the LEED pattern (with integer and half-integer spots). After Ref. [34].

mately 10 meV. An optical phonon with that energy was found, approximately at the same time, in Si(111)2 × 1 by Toennies and colleagues by inelastic He scattering [65].

On the other hand, it should be pointed out that "ab initio" calculations [66] done by the Car–Parrinello molecular dynamical method give, for the 2 × 1 structures of Si and Ge, a slightly buckled chain, the amount of buckling being of the same magnitude as that obtained from the data of Fig. 7 [52], while the energies of the surface phonons are found precisely around 10 meV.

In conclusion it can be said that optical spectroscopy marked the entire development of surface physics, contributing with many crucial experiments to our knowledge of electronic surface states.

References

[1] I. Tamm, Z. Phys. 76 (1932) 849.
[2] W. Shockley, Phys. Rev. 56 (1939) 317.
[3] J. Bardeen, Phys. Rev. 71 (1947) 717.
[4] W. Shockley and G.L. Pearson, Phys. Rev. 74 (1948) 232.
[5] J. Bardeen and W.H. Brattain, Phys. Rev. 75 (1949) 1208.
[6] W.H. Brattain and J. Bardeen, Bell Syst. Tech. J. 82 (1953) 1.
[7] J. Bardeen, R.E. Coovert, S.R. Morrison, J.R. Schrieffer and R. Sun, Phys. Rev. 104 (1956) 47.
[8] F.C. Allen and G.W. Gobeli, Phys. Rev. 127 (1962) 150.
[9] G. Chiarotti, G. Del Signore and S. Nannarone, Phys. Rev. Lett. 21 (1968) 1170.
[10] G. Chiarotti, S. Nannarone, R. Pastore and P. Chiaradia, Phys. Rev. B 4 (1971) 3398.
[11] For a historical lay-out of this early stage of research, see for example, F. Bechstedt and R. Enderlein, Semiconductor Surfaces and Interfaces (Akademie-Verlag, Berlin, 1988) p. xvi.
[12] For a discussion on relative positions of surface and bulk bands see, G. Chiarotti, in: Surface Science, Trieste Lectures 1974, Vol. 1 (International Atomic Energy Agency, Vienna 1975) p. 423; see also Ref. [52].
[13] See for example, M. Cardona, Modulation Spectroscopy, Solid State Physics, Suppl. 11 (Academic Press, New York, 1969).
[14] G. Chiarotti, G. Del Signore, A. Frova and G. Samoggia, Nuovo Cimento 26 (1962) 403.
[15] G. Chiarotti and U.M. Grassano, Phys. Rev. Lett. 16 (1966) 124.
[16] N.J. Harrick, Phys. Rev. 125 (1962) 1165.

[17] G. Samoggia, A. Nucciotti and G. Chiarotti, Phys. Rev. 144 (1966) 749.
[18] For example, in bulk Ge at 4.2 eV the absorption coefficient is $\sim 2 \times 10^6$ cm^{-1} corresponding to a cross-section of 0.45×10^{-16} cm^2.
[19] J.M. Nicholls, P. Mårtensson and G.V. Hansson, Phys. Rev. Lett. 54 (1985) 2363.
[20] J.E. Northrup and M.L. Cohen, Phys. Rev. Lett. 49 (1982) 1349.
[21] D.E. Eastman and W.D. Grobman, Phys. Rev. Lett. 28 (1972) 1378.
[22] L.F. Wagner and W.E. Spicer, Phys. Rev. Lett. 28 (1972) 1381.
[23] W. Müller and W. Mönch, Phys. Rev. Lett. 27 (1971) 250.
[24] J. Assmann and W. Mönch, Surf. Sci. 99 (1980) 34.
[25] J.E. Rowe, H. Ibach and H. Froitzheim, Surf. Sci. 48 (1975) 44.
[26] G. Chiarotti, P. Chiaradia and S. Nannarone, Surf. Sci. 49 (1975) 315.
[27] H. Froitzheim, H. Ibach and D.L. Mills, Phys. Rev. B 11 (1975) 4980.
[28] R.M. Feenstra, W.A. Thompson and A.P. Fein, Phys. Rev. Lett. 56 (1986) 608.
[29] K.C. Pandey and J.C. Phillips, Phys. Rev. Lett. 34 (1975) 1450.
[30] A. Cricenti, S. Selci, K.O. Magnusson and B. Reihl, Phys. Rev. B41 (1990) 12908.
[31] P. Chiaradia, G. Chiarotti, S. Nannarone and P. Sassaroli, Solid State Commun. 26 (1978) 813.
[32] P.E. Wierenga, A. van Silfhout and M.J. Sparnaay, Surf. Sci. 87 (1979) 43.
[33] S. Nannarone, P. Chiaradia, F. Ciccacci, R. Memeo, P. Sassaroli, S. Selci and G. Chiarotti, Solid State Commun. 33 (1980) 593.
[34] P. Chiaradia, A. Cricenti, S. Selci and G. Chiarotti, Phys. Rev. Lett. 52 (1984) 1145.
[35] S. Selci, P. Chiaradia, F. Ciccacci, A. Cricenti, N. Sparvieri and G. Chiarotti, Phys. Rev. B 31 (1985) 4096.
[36] J.D.E. McIntyre and D.E. Aspnes, Surf. Sci. 24 (1971) 417.
[37] S. Selci, F. Ciccacci, G. Chiarotti, P. Chiaradia and A. Cricenti, J. Vac. Sci. Technol. 5 (1987) 327.
[38] M. Nakayama, J. Phys. Soc. Jpn. 39 (1974) 265.
[39] A. Bagchi, R.C. Barrera and A.K. Rajagopal, Phys. Rev. B 20 (1979) 4824.
[40] R. Del Sole, Solid State Commun. 37 (1981) 537.
[41] F. Ciccacci, S. Selci, G. Chiarotti, P. Chiaradia and A. Cricenti, Surf. Sci. 168 (1986) 28.
[42] S. Selci, A. Cricenti, A.C. Felici, L. Ferrari and C. Goletti, Phys. Rev. B 43 (1991) 6757.
[43] F. Meyer, Phys. Rev. B 9 (1974) 3622.
[44] R. Dorn and H. Lüth, Phys. Rev. Lett. 33 (1974) 1024.
[45] F. Casula and A. Selloni, Solid State Commun. 37 (1981) 825.
[46] A. Selloni, P. Marsella and R. Del Sole, Phys. Rev. B 33 (1986) 8885.

[47] R.E. Schlier and H.E. Farnsworth, J. Chem. Phys. 30 (1959) 917.
[48] J.J. Lander, G.W. Gobeli and J. Morrison, J. Appl. Phys. 34 (1963) 2298.
[49] E. Tosatti, Festkörperprobleme XV (1975) 113.
[50] J.C. Chadi, Phys. Rev. Lett. 41 (1978) 1062.
[51] D. Haneman, Phys. Rev. 121 (1961) 1093.
[52] G. Chiarotti and S. Nannarone, Phys. Rev. Lett. 37 (1976) 934.
[53] F.J. Himpsel, P. Heimann and D.E. Eastman, Phys. Rev. B 24 (1981) 2003.
[54] R.I.G. Uhrberg, G.V. Hansson, J.M. Nicholls and S.A. Flödstrom, Phys. Rev. Lett. 48 (1982) 1032.
[55] K.C. Pandey, Phys. Rev. Lett. 47 (1981) 1913.
[56] R.M. Tromp, L. Smit and J.F. van der Veen, Phys. Rev. Lett. 51 (1983) 1672.
[57] R. Feder, Solid State Commun. 45 (1983) 51.
[58] M.A. Olmstead and N.M. Amer, Phys. Rev. Lett. 52 (1984) 1148.

[59] M.A. Olmstead, Surf. Sci. Rep. 6 (1986) 159.
[60] R. Del Sole and A. Selloni, Solid State Commun. 50 (1984) 825.
[61] D.J. Chadi and R. Del Sole, J. Vac. Sci. Technol. 21 (1982) 319.
[62] J.E. Northrup and M.L. Cohen, Phys. Rev. B 27 (1983) 6553.
[63] F. Ciccacci, S. Selci, G. Chiarotti and P. Chiaradia, Phys. Rev. Lett. 56 (1986) 2411.
[64] A discussion of the problem and a derivation of Eq. (7) is given by: M. Lannoo and P. Friedel, in: Atomic and Electronic Structure of Surfaces: Theoretical Foundations, Vol. 16 of Springer Series in Surface Sciences (Springer, Berlin, 1991) p. 221.
[65] U. Harten, J.P. Toennies and Ch. Wöll, Phys. Rev. Lett. 57 (1986) 2947.
[66] F. Ancillotto, W. Andreoni, A. Selloni, R. Car and M. Parrinello, Phys. Rev. Lett. 65 (1990) 3148.

Surface Science 299/300 (1994) 551–562
North-Holland

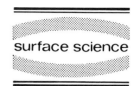

surface science

Surfaces probed by nonlinear optics

Y.R. Shen

*Department of Physics, University of California and Materials Sciences Division, Lawrence Berkeley Laboratory,
Berkeley, CA 94720, USA*

Received 30 April 1993; accepted for publication 14 June 1993

Optical second harmonic and sum-frequency generation has recently been proven to be a very versitile probe for surface and interface studies. Here a brief account is given on how the technique was developed and numerous applications found in our laboratory.

1. Introduction

The advance of surface science relies critically on our ability to characterize a surface or interface. Therefore development of effective tools for surface probing is of tentamount importance in the field. Over the past decades, many powerful techniques have been invented for surface characterization [1]. Thermal desorption spectroscopy, low energy electron diffraction, photoemission spectroscopy, electron energy loss spectroscopy, and scanning tunneling microscopy are just a few examples. Among the existing surface probes, those using optical beams are particularly attractive. They are applicable to all interfaces accessible by light and consequently not restricted to surfaces situated in vacuum. Moreover, with highly collimated light beams, they are suitable for in-situ remote sensing of a surface. Thus, it is not surprising that ellipsometry and optical reflection spectroscopy have been widely adopted for surface and interface studies. Unfortunately, these techniques are not intrinsically surface specific; inadequacy in their discrimination against the bulk signal often limits their applications.

Nonlinear optics is a field born in 1961 when Franken et al. [2] first demonstrated second har-

monic generation (SHG) in a quartz crystal. Since then, it has become increasingly more mature as optical technology progresses. In numerous instances, nonlinear optical effects have been developed as viable means for material studies. One naturally would ask if nonlinear optics could also be exploited to probe surfaces or interfaces. Indeed quite a number of surface nonlinear optical techniques have been adopted for surface studies. Those that directly interrogate a surface or interface are coherent anti-Stokes Raman scattering [3], stimulated Raman gain spectroscopy [4], second harmonic generation, and sum (or difference) frequency generation [5]. The first two arise from third-order nonlinearities. Being allowed both in bulks and at surfaces, they generally do not have the needed surface specificity. The last two are second-order nonlinear optical processes. They are forbidden in media with inversion symmetry, but are then necessarily allowed at surfaces or interfaces. This unusual surface specificity renders these processes ideal for surface probing.

There already exist in the literature many review articles on surface studies by optical second-harmonic generation (SHG) and sum-frequency generation (SFG) [5,6]. In this paper, I will give a brief account on how the techniques were developed and applications found in our laboratory.

2. Basic theory

Let us first describe briefly the theory behind surface SHG and SFG. The details can be found in Ref. [7]. Consider SFG in a medium by two laser beams at frequencies ω_1 and ω_2. Via the nonlinear response, the beams induce in the medium a nonlinear polarization having a frequency component at $\omega_1 + \omega_2$, which can be expressed as

$$P^{(2)}(\omega = \omega_1 + \omega_2)$$
$$= \chi^{(2)}(\omega_1 + \omega_2) : E(\omega_1) E(\omega_2). \tag{1}$$

Eq. (1) shows that if the medium possesses an inversion symmetry, then $\chi^{(2)}$ must vanish in the electric-dipole approximation. As a source term in the wave equation, $P^{(2)}$ is responsible for the SFG.

Let us consider, as an example, the case of SFG from reflection depicted in Fig. 1, in which medium 1 is linear but medium 2 is nonlinear.

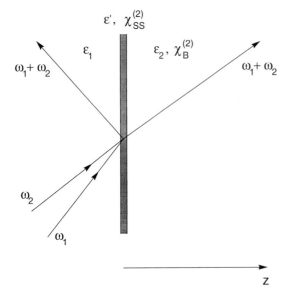

Fig. 1. Geometry of sum-frequency generation from an interface in the reflected direction. The polarization sheet $P_s^{(2)}(\omega)$ is imbedded in a thin interfacial layer of dielectric constant ϵ'.

The solution of the wave equation is found to be [8]

$$E_p(\omega)$$
$$= i\left(\frac{2\pi\omega}{c}\right)\left[L_{xx}(\omega)\chi^{(2)}_{s,xjk}L_{jj}(\omega_1)L_{kk}(\omega_2)\right.$$
$$\left. + \left(\frac{k_x(\omega)}{k_{1z}(\omega)}\right)L_{zz}(\omega)\chi^{(2)}_{s,zjk}L_{jj}(\omega_1)L_{kk}(\omega_2)\right]$$
$$\times E_{1j}(\omega_1)E_{1k}(\omega_2),$$

$$E_s(\omega) = i\left(\frac{2\pi k_1^2(\omega)}{k_{1z}(\omega)\epsilon_1(\omega)}\right)$$
$$\times L_{yy}(\omega)\chi^{(2)}_{s,yjk}L_{jj}(\omega_1)L_{kk}(\omega_2)$$
$$\times E_{1j}(\omega_1)E_{1k}(\omega_2), \tag{2}$$

where the subscripts p and s refer to beam polarizations, $k_1(\omega)$ is the wavevector at frequency ω medium 1, $E_1(\omega_i)$ is the incoming field at ω_i in medium 1, and L_{ii} is the transmission Fresnel coefficient for the field

$$L_{xx}(\omega_i) = \frac{2\epsilon_1(\omega_i)k_{2z}(\omega_i)}{\epsilon_2(\omega_i)k_{1z}(\omega_i) + \epsilon_1(\omega_i)k_{2z}(\omega_i)},$$

$$L_{yy}(\omega_i) = \frac{2k_{1z}}{k_{1z} + k_{2z}},$$

$$L_{zz}(\omega_i) = \frac{2\epsilon_1 k_{1z}(\epsilon_2/\epsilon')}{\epsilon_2 k_{1z} + \epsilon_1 k_{2z}}. \tag{3}$$

If the dielectric constants satisfy the relation $\epsilon_1 = \epsilon' = \epsilon_2$, then $L_{ii} = 1$ and Eq. (2) reduces to the well-known result for a radiating dipole sheet imbedded in a uniform dielectric medium. Thus, the physical picture underlying Eq. (2) is clear. The radiation field appears to be generated by an induced polarization sheet $P_s^{(2)}(\omega) = \chi_s^{(2)}:E_1(\omega_1)\cdot E_1(\omega_2)$ with a linear dielectic constant ϵ' sandwiched between media 1 and 2. However, $P_s^{(2)}(\omega)$ or $\chi_s^{(2)}(\omega)$ actually has contributions from both the interface and the bulk, namely,

$$\chi^{(2)}_{s,ijk} = \chi^{(2)}_{ss,ijk} + \chi^{(2)}_{B,ijk}/\left[k_{2z}(\omega) + k_{2z}(\omega_1)\right.$$
$$\left. + k_{2z}(\omega_2)\right]F_i(\omega)F_j(\omega_1)F_k(\omega_2), \tag{4}$$

where

$$F_\alpha(\Omega) = \begin{cases} \epsilon_2(\Omega)/\epsilon'(\Omega), & \text{for } \alpha = z, \\ 1, & \text{for } \alpha = x, y. \end{cases}$$

The first term on the right of Eq. (4) is the interface contribution and the second term the bulk contribution. The former comes from, at most, a few monolayers at the interface while the latter comes effectively from a bulk layer of about a reduced wavelength $(1/k)$ thick adjoining the interface. It is clear from Eq. (4) that the bulk contribution will dominate if $\chi_B^{(2)}$ is allowed (in the electric-dipole approximation), as in media without inversion symmetry. Consequently, for surface studies with SHG or SFG, we normally require bulk media with inversion symmetry. We should, however, remark that the surface and bulk of a crystal generally have different structural symmetries. This leads to $\chi_{ss}^{(2)}$ and $\chi_B^{(2)}$ with different symmetric forms, i.e., vanishing elements for one but not for the other. It is then possible to choose a proper combination of beam polarizations that can effectively suppress the bulk contribution and make $\chi_s^{(2)}$ surface specific [9].

From Eq. (2), the sum-frequency output can be readily calculated. If the input is in the form of pulses with pulse width T and the beam overlapping cross section at the interface is A, then the output in terms of photon/pulse is given by

$$S(\omega) = \frac{8\pi^3\omega \sin^2\theta_\omega}{c^3\hbar[\epsilon_1(\omega)\epsilon_1(\omega_1)\epsilon_1(\omega_2)]^{1/2}}$$
$$\times |e^\dagger(\omega)\cdot\chi_s^{(2)}:e(\omega_1)e(\omega_2)|^2$$
$$\times I_1(\omega_1)I_1(\omega_2)AT. \qquad (5)$$

Here $e(\Omega) = L\cdot\hat{e}(\Omega)$, $\hat{e}(\Omega)$ is a unit vector describing the field polarization at Ω, θ_ω is the reflection angle of SFG from the surface normal, and $I_1(\omega_i)$ is the input laser intensity at ω_i. As an example, consider the case with $I_1(\omega_1)\approx I_1(\omega_2) \approx 10^9$ W/cm^2, $T\approx 10$ ps, $A\approx 1$ mm^2 (corresponding to a pulse energy of 100 μJ and a fluence of 10 mJ/cm^2), and a typical $\chi_s^{(2)}\approx 10^{-15}$ esu for a surface monolayer. Substitution of these values in Eq. (5) yields a signal strength of $\sim 10^5$ photons/pulse. Since the noise of a detection system can be lower than 0.1 photon/pulse, the signal-to-noise ratio is better than 10^6. This example shows that SHG and SFG can easily have a submonolayer sensitivity.

The surface SHG or SFG measurement allows us to deduce $\chi_s^{(2)}$. In the case of negligible bulk

contribution, $\chi_s^{(2)}$ reflects the properties of the interface. The microscopic expression for $\chi_s^{(2)}$ in the electric-dipole approximation is (see first reference of Ref. [6], p. 16)

$$\chi_{s,ijk}^{(2)} = \int\left\{\sum_{g,n,n'}\left(-\frac{e^3}{\hbar^2}\right)\right.$$
$$\times\left[\frac{\langle g|r_i|n\rangle\langle n|r_j|n'\rangle\langle n'|r_k|g\rangle}{(\omega-\omega_{ng}+i\Gamma_{ng})(\omega_2-\omega_{n'g}+i\Gamma_{n'g})}\right.$$
$$+\frac{\langle g|r_i|n\rangle\langle n|r_k|n'\rangle\langle n'|r_j|g\rangle}{(\omega-\omega_{ng}+i\Gamma_{ng})(\omega_1-\omega_{n'g}+i\Gamma_{n'g})}$$
$$\left.\left.+6 \text{ other terms}\right]\rho_{gg}^0\right\}g(\Omega)\,\mathrm{d}\Omega, \qquad (6)$$

assuming a surface layer of localized molecules. Here, ρ_{gg}^0 is the population in the $\langle g|$ state, ω_{ng} and Γ_{ng} are the transition frequency from $\langle g|$ to $\langle n|$ and the associated damping constant, and $g(\Omega)$ is a distribution function of the parameter, or set of parameters, Ω, with $\int g(\Omega)\,\mathrm{d}\Omega = N =$ density of surface molecules. Through $\chi_s^{(2)}$, we can learn about properties of the surface molecules. For example, from the resonant enhancement of $\chi_s^{(2)}$, we can obtain spectroscopic information about the molecules, and from the relative values of different elements of $\chi_s^{(2)}$, we can deduce information about the molecular orientation and arrangement.

As surface probes, optical SHG and SFG have a number of clear advantages. Asides from being highly surface specific, sensitive to submonolayers, and applicable to all interfaces accessible by light, they can also be used for nondetrimental in-situ remote sensing of a surface in a hostile environment. They have high spatial, temporal, and spectral resolutions. With ultrashort laser pulses, they can be employed to monitor ultrafast surface dynamics. These unusual features have led to many unique applications which we will discuss later.

3. Early history

Nonlinear optical reflection from a boundary surface was first studied theoretically by Bloembergen and Pershan [10] in a classical paper in 1962. Subsequently, SHG by reflection from metal [11] and semiconductor surfaces [12] were measured and compared with the theory. Attention was then drawn to the problem of understanding the physical origin of nonlinearities responsible for the SH reflection. Bloembergen and coworkers [12,13] suggested that surface nonlinearities can arise from the response of a surface layer of atoms or molecules to a rapidly varying field across the surface. The picture perceived was, however, rather vague. Experimental results were also confusing because metal and semiconductor surfaces used were heavily contaminated and poorly defined. Insulating solid and liquid surfaces were also studied [14], but surface nonlinearities were hardly emphasized. Most of the theoretical work during this period focused on SH reflections from metals [15,16]. Among them, Rudnick and Stern [16] first recognized the importane of broken inversion symmetry at an interface in their free electron model calculation. Again, comparison between theory and experiment was rendered meaningless because of lack of exprimental data on clean metal surfaces.

Brown and Matsoka [17] actually did measure SHG from a freshly evaporated silver film in vacuum. During the course of the experiment, they noticed that as soon as the film was exposed to air, the SH signal decayed rapidly to a much lower value. This was an indication that SHG could be highly surface sensitive. Later, along the same line, Chen et al. [18] performed a much better controlled experiment studying SHG from a clean Ge(111) surface in ultrahigh vacuum. They observed a drastic increase of the signal in response to monolayer deposition of Na on the surface. This suggested that SHG could have monolayer sensitivity and might be used for surface probing. However, their papers published in Optics Communications and Japanese Journal of Applied Physics – Supplement hardly received any attention. Not until 1980 was the surface sensitivity of SHG rediscovered and the possibility of SHG for surface studies truly explored.

In the late 1970's, our laboratory at Berkeley was heavily involved in the study of surface electromagnetic waves [19], in particular, surface plasmon waves on silver films. We were interested in the enhancement of nonlinear optical effects using surface plasmons. Around the same time, surface enhanced Raman scattering (SERS) was discovered [20]. It was found that for molecules adsorbed on roughened Ag surfaces, the Raman intensity could increase by $\sim 10^6$. This surprising effect caught the fancy of many physicists and chemists. Two mechanisms were proposed to explain the observed enhancement, one due to molecule/substrate interaction and the other due to surface local field enhancement, but experimentally it was difficult to separate the two. It dawned on us that SHG could provide an answer to the problem. The reasoning was as follows. Raman scattering is a two-photon transitions process and is therefore a nonlinear optical effect. If one nonlinear optical effect is enhanced by local-field enhancement, so will the others. One should then expect SHG from a roughened Ag surface to be also strongly enhanced if the surface local-field enhancement is the dominant mechanism for SERS. Unlike Raman scattering, SHG can be detected from a bare Ag surface. Thus molecular adsorbates on Ag are not needed; any enhancement of SHG from a bare but roughened Ag surface must come from the local-field enhancement. Our SHG measurements showed that the local-field enhancement indeed played a dominant role in SERS; it yielded a $\sim 10^4$ enhancement in SHG from a roughened Ag surface [21]. With such a large enhancement, we were convinced that surface-enhanced SHG, like SERS, could be employed to monitor in-situ adsorption and desorption of molecules on electrodes in an electrochemical cell [22]. The initial trial experiment was more than successful. The signal was so strong that even without the surface enhancement, the monolayer adsorption could have been detected. This led us to the subsequent experiments studying monolayer adsorbates on smooth substrates and the development of SHG and SFG as surface probes.

4. Development of the techniques

The result of the first SHG experiment monitoring in-situ adsorption and desorption in electrolysis is presented in Fig. 2 [22]. The electrochemical cell was composed of an Ag and a Pt electrode immersed in a 0.1 M KCl solution. During the oxidation half cycle, AgCl was formed and deposited on the Ag electrode; during the reduction half cycle, AgCl was reduced back to Ag and Cl. From the amount of charges passing through the circuit, it was possible to estimate the number of AgCl molecules present on the Ag electrode. Repetitive electrochemical cycling roughened the Ag surface and effected the surface enhancement. Adsorption and desorption of AgCl monolayers on the Ag electrode could then be readily observed by SHG. As shown in Fig. 2, the SH signal increases sharply when the first one or two AgCl monolayers appear on Ag, and drops precipitously when the last one or two AgCl monolayers on Ag are reduced. Because SHG is highly surface specific, it is not sensitive to the

change in the AgCl overlayer beyond the first few surface monolayers. The observed weak variation of SHG in the middle part of the oxidation–reduction cycle is presumably due to change in surface roughness resulting from multilayer oxidation and reduction.

In the above experiment, 10 ns laser pulses at 1.06 μm focused to 0.2 mJ/0.2 cm^2 on the surface were used. The signal detected was as large as 3×10^5 photons/pulse. This suggested that a much weaker laser beam could be employed for the same experiment. Even a 20 mW CW diode laser would be strong enough. This was actually demonstrated in an on-site experiment presented as a post-deadline paper in the VIIth International Laser Spectroscopy Conference in Maui [23]. It was also clear from the experiments that even without surface enhancement, monolayer adsorption on electrodes could be detected [24]. More recently, SHG has been adopted to monitor epitaxial growth of metal overlayers on electrodes in an electrochemical cell [25]. SFG spectroscopy has been used to identify molecular species adsorbed on electrodes in an electrolytic process [26].

The realization that SHG was sensitive to surface monolayers prompted us to consider using SHG for surface spectroscopy. For demonstration, we chose rhodamine dye molecules as adsorbates in a trial experiment because they have very strong electronic transitions both in the visible and in the near UV [24]. Fig. 3 displays the observed SHG spectra from half monolayers of rhodamine 6G and rhodamine 110 adsorbed on quartz. The two peaks are present due to resonant enhancement of the $S_0 \rightarrow S_2$ transitions of the two rhodamine dyes. The signal strength at the peaks is $\sim 10^4$ photons/pulse for 10 ns input laser pulses focused to 1 mJ/10^{-3} cm^2 at the surface. It indicates that even less than 1% of a rhodamine monolayer is readily detectable.

In subsequent years, a large number of experiments were carried out to explore the potential of SHG as a surface tool. It was shown that with different input/output beam polarizations, SHG could yield information on the average orientation of molecular adsorbates [24,27]. The nonlinear susceptibility $\chi^{(2)}_{s,ijk}$ is directly related to the

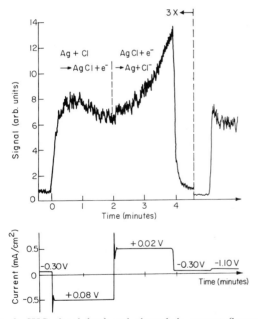

Fig. 2. SHG signal (top) and electrolytic current (bottom) from a silver electrode in a 0.5M KCl solution as functions of time during an electrolytic cycle. The voltages listed on the lower curve indicate the potential of the silver electrode with respect to a saturated calomel electrode. The SHG signal is particularly sensitive to the initial deposition and final reduction of AgCl layers on the electrode. To the left of the vertical dashed line, the SHG signal is exaggerated by a factor of three. (After Ref. [21].)

nonlinear polarizability $\alpha^{(2)}_{\xi\eta\zeta}$ of molecules by a coordinate transformation if the surface nonlinearity is dominated by contribution from the molecules and the molecule–molecule interaction is negligible

$$\chi^{(2)}_{s,ijk} = N_s \langle (\hat{i}\cdot\hat{\zeta})(\hat{j}\cdot\hat{\eta})(\hat{k}\cdot\hat{\zeta})\rangle \alpha^{(2)}_{\xi\eta\zeta}, \qquad (7)$$

where $\hat{\xi}$, $\hat{\eta}$, $\hat{\zeta}$ define the molecular coordinate axes, the angular brackets denote an average over the molecular orientational distribution, and N_s is the surface density of adsorbed molecules. Measurements of $\chi^{(2)}_{s,ijk}$ by SHG with known N_s may allow the determination of dominant elements of $\alpha^{(2)}_{\xi\eta\zeta}$ as well as the average molecular orientation if an orientational distribution is assumed. This was illustrated by a number of different examples [24,27–29]. In one case, SHG was used to probe phase transitions of a monolayer of pentadecanoic acid floating on water [28]. It was found that the so-called liquid-expanded to liquid-condensed transition of the monolayer was connected with a sudden change in the molecular tilt angle. By measuring the phase of $\chi^{(2)}_{s,ijk}$ [27,30], the polar orientation of surface molecules (heads up or down) could also be determined [31].

The possibility of using SHG to probe surface structural symmetry was also explored [32]. It was realized that the symmetry of $\chi^{(2)}_s$ should directly reflect the surface symmetry except that it cannot distinguish surfaces with 4mm, 6mm, and isotropic symmetries. As a demonstration, SHG from a Si(111) surface was measured as a function of sample rotation about the surface normal [32]. The signal exhibited a 3-fold symmetry in correspondence to the 3-fold structural symmetry as expected. This particular feature of SHG was utilized to monitor laser melting of a Si(111) surface by Shank et al. [33] and surface reconstruction of Si(111) [34] and amorphous-crystalline growth of a Si adlayer and a silicide monolayer [35] by Heinz et al.

In order to convince traditional surface scientists that SHG is a viable surface probe, it was necessary to try out the technique on well-defined surfaces in UHV. The first experiment was on CO and O adsorption on Rh(111), carried out with the help of Somorjai's group [36]. In-situ monitoring of the adsorption processes by SHG was indeed possible with a fairly high signal-to-noise ratio. The result of O/Rh(111) is reproduced in Fig. 4. The data can be well fit by the

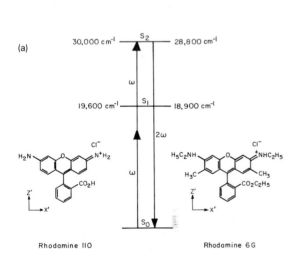
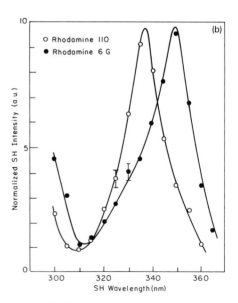

Fig. 3. Resonant second-harmonic generation in Rhodamine 110 and Rhodamine 6G: (a) the resonant process in the two dyes with energy levels corresponding to the absorption line center for the molecules dissolved in ethanol; (b) the experimental SH spectrum m the region of the $S_0 \rightarrow S_2$ transition for submonolayers of the dye molecules adsorbed on fused silica. (After Ref. [24].)

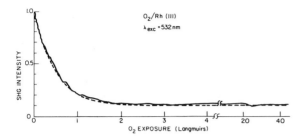

Fig. 4. SHG signal from Rh(111) during O_2 exposure. Solid curve, experimental result; dashed curve, theoretical fit. (After Ref. [36].)

simple Langmuir adsorption model, assuming $\chi_s^{(2)} = \chi_{s0}^{(2)} + \Delta\chi_s^{(2)}$, where $\Delta\chi_s^{(2)}$ is linearly proportional to the surface density of oxygen atoms. This is in agreement with the result obtained by other methods. In the case of CO/Rh(111), SHG was found to be sensitive to adsorption of CO at different sites. Later, detection of adsorption and desorption of O/Si(111) by SHG in UHV was also demonstrated [37]. Many other experiments on metal and seminconductor surfaces in UHV have been carried out subsequently by different groups interested in a variety of surface problems.

To show that SHG as a surface probe was not limited to media with inversion symmetry, an experiment was performed to see whether adsorption of Sn on GaAs could be detected [9]. Because the structural symmetries of surface and bulk are different, the bulk SH signal from GaAs can be greatly suppressed by selecting an appropriate input/output polarization combination and SHG could again appear surface-sensitive. Indeed, monolayer adsorption of Sn on GaAs was easily detectable.

A series of experiments were also conducted to explore the versatility of SHG as a surface probe. It was shown that the technique could be used to study molecular monolayers adsorbed at almost all interfaces. At the air/water interface, SHG was employed to monitor phase transitions [28] and polymerization [38] of molecular monolayers. At a liquid/solid interface, the dynamics of adsorption of molecules from the·solution to the solid substrate could be followed by SHG [27]. To demonstrate the applicability of SHG to

biology, retinel molecules adsorbed at the air/water interface and buried in membranes were both detected [39]. An experiment was also performed to show that SHG could be used for surface monolayer microscopy [23]. (See Fig. 5.)

SHG as a surface spectroscopic technique is certainly very attractive. Unfortunately, to achieve monolayer sensitivity, it is necessary to have the output in the UV or visible where sensitive photo-detectors are available. This limits the applications of SHG surface spectroscopy to electronic transitions. For studies of surface vibrational transitions, infrared-visible SFG instead of SHG must be used. With the presence of two independent input laser beams, SFG is even more versatile than SHG as a surface tool. Our first experiment on a surface vibrational spectroscopy using SFG was only partially successful [40]. A C–H stretch peak was observed in the SFG spectra of a nominally cleaned fused quartz plate. The experiment was carried out in Y.T. Lee's laboratory using an optical parametric amplifier as the tunable IR source that we had developed for another project. It was believed that alkoxide molecules were abundant in the air owing to the presence of several mechanical pumps running to maintain a molecular beam machine in the same room, and adsorption of alkoxides on quartz could have produced the C–H peak. More controlled experiments were later performed with deposition of known molecular monolayers on quartz: first with a dye monolayer using a line-tunable CO_2 laser as the IR source [41] and then with organic molecular monolayers on quartz [42] (see Fig. 6) as well as at the air/water interface [43] using an optical parametric amplifier as the tun-

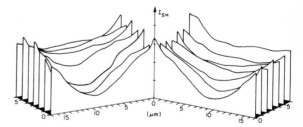

Fig. 5. Second harmonic image of a laser ablated hole in a Rhodamine 6G dye monolayer on fused quartz. (Courtesy of G.T. Boyd.)

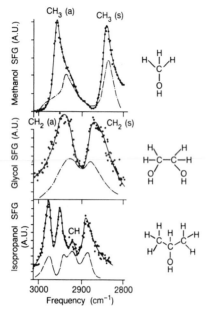

Fig. 6. Sum-frequency generation as a function of infrared input frequency for three adsorbed species, CH_3OH, $C_2H_4(OH)_2$, and C_3H_7OH, on fused silica. The spectra were obtained with a visible input at 0.532 μm and a tunable input around 3 μm. The peaks correspond to the various hydrocarbon (C–H) stretch modes, symmetric (s) and asymmetric (a); FR denotes a structure resulting from Fermi resonance between $CH_3(s)$ and the overtone on the C–H bending modes. For comparison the Raman spectra of the species in the liquid phase (dashed-dot curves) are also shown. (After Ref. [42].)

able IR source. Harris et al. [44] at Bell Labs also conducted a similar experiment on a Langmuir–Blodgett film. The success of these experiments led to much excitement. It opened the door to many hitherto unexplored areas of surface science, in particular, the possibility of studying ultrafast surface dynamics with picosecond or subpicosecond SFG [45].

5. Some unique applications

SHG and SFG as surface probes are now well established. Many laboratories have adopted the techniques for surface studies in various disciplines. New, interesting applications of the techniques have been found. For example, SHG has been used to probe surface states of metals [46] and surface magnetization [47]. With ultrashort laser pulses, SHG and SFG have both been employed to study ultrafast surface reactions and surface dynamics [45,48]. There is no room here for us to discuss all the progresses in the field. Instead, I shall describe, in the remaining part of the paper, a few unique applications studied in our lab illustrating the potential of the techniques.

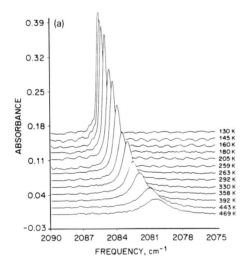

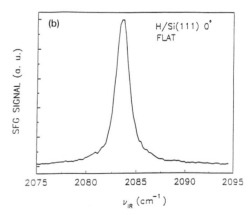

Fig. 7. (a) Infrared adsorption spectra (p-polarization) of an H-terminated Si(111) surface as a function of substrate temperature. (After Ref. [49].) (b) SFG spectrum of the Si–H stretch mode of H/Si(111) at room temperature. (After Ref. [50], Morin et al.)

We shall focus on surface vibrational spectroscopy using infrared–visible SFG. One question that would immediately come to our mind is how this technique compares with infrared reflection spectroscopy which is commonly used for IR surface spectroscopic studies. Fig. 7 shows the spectra of the H–Si stretch mode for H/Si(111) measured by these two different techniques. The one in Fig. 7a was obtained by infrared reflection spectroscopy using a multiple reflection geometry depicted in the inset [49]. The sample dimensions were $0.5 \times 19 \times 38$ mm^3 and internal reflections were used. In comparison, the one in Fig. 7b was obtained by SFG with a single reflection from the surface [50]. It is clear that per reflection, SFG had a better signal/noise ratio than IR spectroscopy. The advantage of SFG is more obvious if we are interested in vibrational spectra of adsorbates on diamond. In this case, IR spectroscopy with multiple reflections is not very

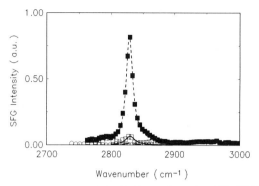

Fig. 8. SFG spectrum of the C–H stretch mode of H/C(111) at room temperature. (After Ref. [51].)

practical since large diamond crystals are difficult to find (for research), and SFG is certainly a more viable spectroscopic method. Fig. 8 is an example, showing that for H/C(111), the C–H stretch vibration can be easily detected [51]. The

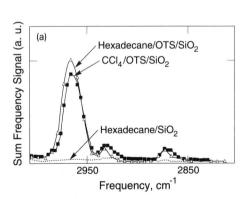

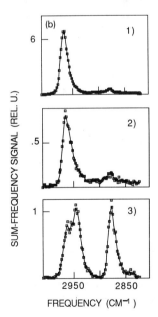

Fig. 9. (a) SFG spectra of different interfaces obtained with a p-polarized visible input at 0.532 μm and a p-polarized tunable input around 3 μm. Dashes: hexadecane/silica; solid squares: CCl$_4$/OTS/silica; triangles: hexadecane/OTS/silica; OTS denotes a monolayer of octadecyltrichlorosilane adsorbed on silica. The solid lines are guides for the eye. OTS, when present, dominates the spectra. (b) Sum-frequency spectra at the air/OTS/silica interface for various polarization combinations of the visible (0.532 μm) and infrared input beams. (1) visible: p-polarized, infrared: p-polarized; (2) visible: p, infrared: s; (3) visible: s, infrared: p. The peaks in the spectra can be assigned to C–H stretch modes of the terminal methyl group of OTS. The peak at 2878 cm^{-1} originates from a CH$_3$ s-stretch, that at 2964 cm^{-1} from a CH$_3$ a-stretch and that at 2942 cm^{-1} from the Fermi resonance between CH$_3$(s) and the overtone of the C–H bending mode. Peaks for the CH$_2$ stretches on the alkane chain of OTS are hardly visible for reasons of symmetry. (After Ref. [53].)

technique is now being used to study in-situ how fragments of methane molecules, thermally dissociated by a hot filament, adsorb and react on a diamond surface under various conditions [52]. The results hopefully would provide us with some insight on the CVD diamond growth process.

An important feature of SFG as a surface spectroscopic technique is its surface specificity. Fig. 9 gives a striking example [53]. For the hexadecane/fused quartz interface, no structure is detectable in the 2800–3000 cm^{-1} spectral range although liquid hexane is known to have strong absorption bands in this region arising from the CH_2 stretch modes. Clear spectral peaks appear if a monolayer of octadecyltrichlorosilane (OTS) molecules is adsorbed at the interface. This spectrum clearly originates from the OTS monolayer. The peaks can be identified as the stretch modes associated with the CH_3 end group of the OTS molecules. It is interesting to note that although OTS has 17 CH_2 groups on its hydrocarbon chain, they do not contribute appreciably to the SFG spectrum because of the approximate inversion symmetry in the CH_2 group arrangement about the vertical chain. If liquid hexane is replaced by CCl_4 which has no vibrational mode in the same spectral region, the observed SFG spectrum hardly changes, again indicating that the spectrum must have come from the OTS monolayer. This example illustrates most vividly the surface-specific power of SFG.

The unusually strong surface specificity of SFG provides us with a means to probe vibrational spectra of pure liquid/vapor interfaces. There exists tremendous interest in the question of how molecules are oriented at such an interface; the surface vibrational spectrum could yield some information, but it is not obtainable by IR reflection or other conventional spectroscopic techniques because the spectrum would be overwhelmed by the bulk liquid. With SFG spectroscopy, the task becomes simple. Displayed in Fig. 10 is the vibrational spectrum of a pure water/vapor interface [54]. It exhibits a sharp peak at ~ 3680 cm^{-1} that can be assigned to a free OH stretch mode and a structured broad band in the 3000–3600 cm^{-1} region due to bonded OH stretch modes. The presence of the

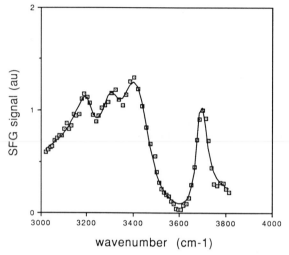

Fig. 10. SFG spectrum of pure water interface with sum-frequency, visible, and infrared radiation s-, s- and p-polarized, respectively. The theoretical fit is given by the solid line. (After Ref. [54].)

free OH peak immediately allows us to conclude that a significant fraction of surface water molecules has one of the OH bonds pointing out of the liquid. This is further confirmed by the observation that the free OH peak can be completely suppressed by floating an alcohol or fatty acid monolayer on the water surface.

6. Present status and future prospects

The past decade has witnessed the successful growth of SHG and SFG as viable surface tools. It has been well proven that they can be applied to all interfaces (including buried solid/solid and liquid/liquid interfaces) and to a wide variety of surface problems in various disciplines. Many more possible applications are yet to be explored. Surface specificity together with high spatial, temporal, and spectral resolutions has certainly made SHG and SFG most attractive as surface probes. With ultrashort laser pulses, they have opened up an exciting new area of research in surface science, e.g., ultrafast surface dynamics. With tunable lasers, they have created a great opportunity for new surface spectroscopic studies.

The techniques are presently limited by the tuning range of available lasers. This limitation will be alleviated when free electron lasers become more accessible to users. Surface spectroscopy with two tunable lasers makes doubly resonant SFG feasible and will allow more interesting variations in the spectroscopic studies, including some coherent transient surface effects. Since IR–visible SFG is capable of detecting selective surface species, it can be a powerful tool for dynamic studies with mixed adsorbed species, for example, in-situ monitoring of surface reactions. On surface microscopy, SFG provides a means to spatially resolve the distribution of different molecular species in a surface layer.

Acknowledgment

This work was supported by the Director, Office of Energy Research, Office of Basic Energy Sciences, Material Sciences Division of the US Department of Energy, under contract No. DE-AC03-76SF00098.

References

[1] See, for example, A. Zangwill, Physics at Surfaces (Cambridge University Press, Cambridge, 1988);
G. Somorjai, Chemistry in Two Dimensions: Surfaces (Cornell University Press, Ithaca, NY, 1981).

[2] P.A. Franken, A.E. Hill, C.W. Peters and G. Weinrich, Phys. Rev. Lett. 1 (1961) 118.

[3] C.K. Chen, A.R.B. de Castro, Y.R. Shen and F. De Martini, Phys. Rev. Lett. 43 (1979) 946;
W.M. Hetherington, N.E. Van Wyck, E.W. Koenig, G.I. Stegeman and R.M. Fortenberry, Opt. Lett. 9 (1984) 88;
W.M.K.P. Wizikoon, Z.Z. Ho and W.M. Hetherington, J. Chem. Phys. 86 (1984) 4384.

[4] J.P. Heritage and D.L. Allara, Chem. Phys. Lett. 74 (1980) 507.

[5] See, for example, Y.R. Shen, Nature 337 (1989) 519.

[6] Y.R. Shen, The Principles of Nonlinear Optics (Wiley, New York, 1984) ch. 25: Ann. Rev. Mat. Sci. 16 (1986) 69; Nature 337 (1989) 519; Ann. Rev. Phys. Chem. 40 (1989) 327;
G.L. Richmond, J.M. Robinson and V.L. Shannon, Prog. Surf. Sci. 28 (1988) 1;
T.F. Heinz and G.A. Reider, Trends Anal. Chem. 8 (1989) 235;
T.F. Heinz, in: Nonlinear Surface Electromagnetic Phenomena, Eds. H.E. Ponath and G.I. Stegeman (North-Holland, Amsterdam, 1991) ch. 5;
J.F. McGilp, J. Phys. Condensed Matter 1 (1989) SB85.

[7] See, Y.R. Shen, Ann. Rev. Phys. Chem. 40 (1989) 327; and references therein.

[8] C.C. Wang, Phys. Rev. 178 (1969) 1475.

[9] T. Stechlin, M. Feller, P. Guyot-Sionnest and Y.R. Shen, Opt. Lett. 13 (1989) 389.

[10] N. Bloembergen and P.S. Pershan, Phys. Rev. 128 (1962) 606.

[11] F. Brown, R.E. Parks and A.M. Sleeper, Phys. Rev. Lett. 14 (1965) 1029;
F. Brown and R.E. Parks, Phys. Rev. Lett. 16 (1966) 507;
N. Bloembergen, R.K. Chan and C.H. Lee, Phys. Rev. Lett. 16 (1966) 986;
H. Sonnerberg and H. Heffner, J. Opt. Soc. Am. 58 (1968) 209;
G.V. Krivoshchekov and V.I. Stroganov, Sov. Phys. Solid State 9 (1968) 2856; 11 (1969) 89.

[12] N. Bloembergen, R.K. Chang, S.S. Jha and C.H. Lee, Phys. Rev. 174 (1968) 813; Erratum 178 (1969) 1528.

[13] N. Bloembergen and R.K. Chang, in: Physics of Quantum Electronics, Eds. P.L. Kelley, B. Lax and P.E. Tannenwald (McGraw-Hill, New York, 1966) p. 80;
N. Bloembergen and Y.R. Shen, in: Physics of Quantum Electronics, Eds. P.L. Kelley, B. Lax and P.E. Tannenwald (McGraw-Hill, New York, 1966) p. 119.

[14] C.C. Wang and A.N. Duminski, Phys. Rev. Lett. 20 (1968) 668.

[15] S.S. Jha, Phys. Rev. Lett. 15 (1965) 412; Phys. Rev. 140 (1965) A2020.

[16] J. Rudnick and E.A. Stern, Phys. Rev. B 4 (1971) 4274.

[17] F. Brown and M. Matsuoka, Phys. Rev. 185 (1969) 985.

[18] J.M. Chen, J.R. Bower, C.S. Wang and C.H. Lee, Opt. Commun. 9 (1973) 132;
J.M. Chen, J.R. Bower and C.S. Wang, Jpn. J. Appl. Phys. Suppl. 2 (1974) 711.

[19] F. De Martini and Y.R. Shen, Phys. Rev. Lett. 36 (1976) 216;
C.K. Chen, A.R.B. de Castro and Y.R. Shen, Opt. Lett. 4 (1979) 393;
Y.R. Shen and F. De Martini, in: Surface Polaritons, Eds. V.M. Agranovich and D.L. Mills (North-Holland, Amsterdam, 1982) p. 629 and references therein.

[20] M. Fleischmann, P.J. Hendra and A J. McQuillan, Chem. Phys. Lett. 26 (1974) 163.

[21] C.K. Chen, A.R.B. de Castro and Y.R. Shen, Phys. Rev. Lett 46 (1981) 145.

[22] C.K. Chen, T.F. Heinz, D. Ricard and Y.R. Shen, Phys. Rev. Lett. 46 (1981) 1010.

[23] G.T. Boyd, Y.R. Shen and T.W. Hansch, in: Laser Spectroscopy VII, Eds. T.W. Hansch and Y.R. Shen (Springer, Berlin, 1985), p. 322;
G.T. Boyd, Y.R. Shen and T.W. Hansch, Opt. Lett. 11 (1986) 97.

[24] T.F. Heinz, C.K. Chen, D. Ricard and Y.R. Shen, Phys. Rev. Lett. 48 (1982) 478.

[25] V.L. Shannon, D.A. Koos and G.L. Richmond, J. Chem. Phys. 87 (1987) 1440; Appl. Opt. 26 (1987) 3579;

J.M. Robinson and G.L. Richmond, Electrochim. Acta 34 (1989) 1639;
J. Miragliotta and T.E. Furtak, Phys. Rev. B37 (1988) 1028;
D.A. Koos, J. Electrochem. Soc. 136 (1989) 218C.

[26] P. Guyot-Sionnest and A. Tadjeddine, Chem. Phys. Lett. 172 (1990) 341.

[27] T.F. Heinz, H.W.K. Tom and Y.R. Shen, Phys. Rev. A 28 (1983) 1883.

[28] Th. Rasing, Y.R. Shen, M.W. Kim and S. Grubb, Phys. Rev. Lett. 55 (1985) 2903.

[29] Th. Rasing, G. Berkovic, Y.R. Shen, S. Grubb and M.W. Kim, Chem. Phys. Lett. 130 (1986) 1;
S. Grubb, M.W. Kim, Th. Rasing and Y.R. Shen, Langmuir 4 (1988) 452.

[30] J.J. Wynne and N. Bloembergen, Phys. Rev. 188 (1969) 1211.

[31] K. Kemnitz, K. Bhattacharyya, J.M. Hicks, G.R. Pinto, K.B. Eisenthal and T.F. Heinz, Chem. Phys. Lett. 131 (1986) 285.

[32] H.W.K. Tom, T.F. Heinz and Y.R. Shen, Phys. Rev. Lett. 51 (1983) 1983.

[33] C.V. Shank, R. Yen and C. Hirlimann, Phys. Rev. Lett. 51 (1983) 900.

[34] T.F. Heinz, M.M.T. Loy and W.A. Thompson, Phys. Rev. Lett. 54 (1985) 63.

[35] T.F. Heinz, M.M.T. Loy and W.A. Thompson, J. Vac. Sci. Technol. B 3 (1985) 1467;
T.F. Heinz, M.M.T. Loy and S.S. Iyer, Proc. Mater. Res. Soc. Symp. 55 (1987) 697.

[36] H.W.K. Tom, C.M. Mate, X.-D. Zhu, J.E. Crowell, T.F. Heinz, G.A. Somorjai and Y.R. Shen, Phys. Rev. Lett. 52 (1984) 348.

[37] H.W.K. Tom, C.M. Mate, X.-D. Zhu, J.E. Crowell, Y.R. Shen and G.A. Somorjai, Surf. Sci. 172 (1986) 466.

[38] G. Berkovic, Th. Rasing and Y.R. Shen, J. Chem Phys. 85 (1986) 7374.

[39] T. Rasing, J. Huang, A. Lewis and Y.R. Shen, Phys. Rev. A 40 (1989), 1684;
J. Huang, A. Lewis and T. Rasing, J. Chem. Phys. 92 (1988) 1756.

[40] H.W.K. Tom, PhD Dissertation, University of California at Berkeley, 1984.

[41] X.-D. Zhu, H. Suhr and Y.R. Shen, Phys. Rev. B 35 (1987) 3047.

[42] J.H. Hunt, P. Guyot-Sionnest and Y.R. Shen, Chem. Phys. Lett. 133 (1987) 189.

[43] P. Guyot-Sionnest, J.H. Hunt and Y.R. Shen, Phys. Rev. Lett. 59 (1987) 1597.

[44] A.L. Harris, C.E.D. Chidsey, N.J. Levinos and D.N. Loiacono, Chem. Phys. Lett. 141 (1987) 350.

[45] See, for example, A.L. Harris and N.J. Levinos, J. Chem. Phys. 90 (1989) 3878;
A.L. Harris, L. Rothberg, L.H. Dubois, N.J. Levinos and L. Dahr, Phys. Rev. Lett. 64 (1990) 2086;
P. Guyot-Sionnest, P. Dumas, Y.J. Chabal and G.S. Higashi, Phys. Rev. Lett. 64 (1990) 2156;
P. Guyot-Sionnest, Phys. Rev. Lett. 67 (1991) 2323.

[46] M.Y. Jiang, G. Pajer and E. Burstein, Surf. Sci. 242 (1991) 306;
L. Urbach, K.L. Percival, J.M. Hicks, E.W. Plummer and H.L. Dai, Phys. Rev. B 45 (1992) 3769;
M.Y. Jiang, PhD Dissertation, University of Pennsylvania, 1992.

[47] R.P. Pan and Y.R. Shen, Chin. J. Phys. 25 (1987) 175;
R.P. Pan, H.D. Wei and Y.R. Shen, Phys. Rev. B 39 (1989) 1229;
W. Hubner and K.H. Benneman, Phys. Rev. B 40 (1989) 5973;
J. Reif, J.C. Zink, C.M. Schneider and J. Kirschner, Phys. Rev. Lett. 67 (1991) 2878.

[48] E.V. Sitzmann and K.B. Eisenthal, J. Chem. Phys. 92 (1988) 4579.

[49] D. Dumes, Y.J. Chabal and G.S. Higashi, Phys. Rev. Lett. 65 (1990) 1124.

[50] M. Morin, P. Jacob, N.J. Levinos, Y.J. Chabal and A. Harris, J. Chem. Phys. 96 (1992) 6203;
P. Guyot-Sionnest, P. Dumas, Y.J. Chabal and G.S. Higashi, Phys. Rev. Lett. 64 (1990) 2156.

[51] R.P. Chin, J.Y. Huang, Y.R. Shen, T.J. Chuang, H. Seki and M. Buck, Phys. Rev. (Rapid Commun.) 45 (1992) 1522.

[52] J.Y. Huang, R.P. Chin, Y.R. Shen, T.J. Chuang and H. Seki, Quantum Electron Laser Sci. Conference, Baltimore, MD, 1993, paper QWE6.

[53] P. Guyot-Sionnest, R. Superfine, J.H. Hunt and Y.R. Shen, Chem. Phys. Lett. 144 (1988) 1.

[54] R. Superfine, J.Y. Huang, Q. Du and Y.R. Shen, in: Laser Spectroscopy X, Eds. M. Duclay, E. Giacobino and G. Camy (World Scientific, Singapore, 1992) p. 117.

Surface Science 299/300 (1994) 563–586
North-Holland

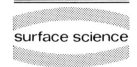

surface science

Quantum transport in semiconductor surface and interface channels

S. Kawaji

Department of Physics, Gakushuin University, Mejiro, Toshima-ku, Tokyo 171, Japan

Received 29 June 1993; accepted for publication 3 August 1993

Advanced silicon technology developed in the early sixties brought us a two-dimensional electron system in silicon inversion layers adjacent to silicon dioxides in metal–oxide–semiconductor field-effect-transistors. In the early eighties the advancement of molecular-beam-epitaxy brought us another two-dimensional electron system in gallium arsenide inversion layers adjacent to gallium aluminum arsenide. Thus semiconductor surface and interface channels have provided us outstanding stages on which two-dimensional electron systems played the leading role in the progress of quantum transport where the quantum mechanical properties of the conduction electrons influence the transport processes. This article describes the progress in the understanding of the quantum transport over the last thirty years from the birth of the two-dimensional system till the present in as far as the author was concerned with.

1. Introduction

After the discovery of the electron by J.J. Thomson in 1897, Drude derived theoretically the Wiedemann and Franz law in metals in 1900. Drude's theory assumed a free electron gas in metals. If one assumes that 2τ is the average time that an electron spends between two collisions with metal ions, the electrical conductivity is given by

$$\sigma = \frac{Ne^2\tau}{m},\tag{1}$$

where $-e$ and m are the charge and the mass of an electron and N is the number of free electrons per unit volume. The time between collisions is of the same order of magnitude as the time of relaxation τ introduced to solve the Boltzmann equation which determines the distribution of particle positions and velocities. Drude's assumption of a free electron gas in metals was placed on a sound basis by the introduction of the Bloch electron concept in periodic lattices in 1928 and the energy band concept for conduction elec-

trons by Wilson in 1931. The mainstream of understanding of electron transport in conductors until 1953 was described by Wilson in his book The Theory of Metals [1].

The theoretical treatments of the transport properties of electrons in conductors described in Wilson's book are mostly quasi-classical approaches to the problems. They do not deal with the more interesting quantum mechanical properties of conduction electrons such as the wave functions of these electrons and their interference in transport processes. The theory of irreversible processes developed by Kubo in 1957 gives an elegant quantum statistical mechanical treatment of transport properties of conduction electrons [2]. In a different approach, Landauer derived a quantum mechanical formalism of electrical conductivity [3].

Until the end of the 1950s researches into the electrical transport properties associated with surfaces were basically applications to surface phenomena of experimental and theoretical techniques developed in studies of bulk properties. However, the progress which has been made in the understanding of quantum transport in semiconductor surface and interface channels in the last thirty years was not a consequence of applications of results obtained in the researches of bulk phenomena. The semiconductor surface and interface channels themselves have enabled the progress in the understanding of quantum transport. In other words, the quantum transport researches should not have made significant progress without the development of surface and interface channels. In this article, I describe the progress in the understanding of the quantum transport in surface and interface channels over the last thirty years from the birth of the two-dimensional system to the present in as far as I was concerned with.

To begin with, the transient stage from the classical to the quantum transport in semiconductor surfaces is described. After a brief description of the two-dimensional electron gas created in semiconductor surface and interface channels in section 3, the progress in the quantum transport in these systems is indicated. The topics will be limited to those with which I personally was concerned with. The goal is to convey my personal experiences and involvement with researches in this field rather than to provide a rigorous historical survey.

2. Development of quantum effects in classical surface transport in semiconductors

It has been known since the middle of the 1930s that the average resistivity of thin wires and films of metals is larger than the resistivity of the bulk material when the physical dimensions are smaller than the mean free path l. In the age of germanium transistors in the 1950s, the mobility of carriers in semiconductor surface space charge layers was discussed in terms of surface scattering in connection with the problem of surface leakage in p–n–p junction transistors [4].

If the thickness of a surface space charge layer is small enough to be comparable with the the de Broglie wave length of electrons, effects of quantization of electron motion perpendicular to the surface are expected as shown in Fig. 1 for the case of an n-channel inversion layer on a Si(001) surface. The motion of the carriers in the direction (z-direction) perpendicular to the surface is quantized due to the small thickness of the space charge layer; each of the corresponding energy levels E_α ($\alpha = 0, 1, 2, \cdots$) is topped with the continuum of kinetic energies of the motion in the (x, y)-plane, forming a two-dimensional (2D) subband E_α.

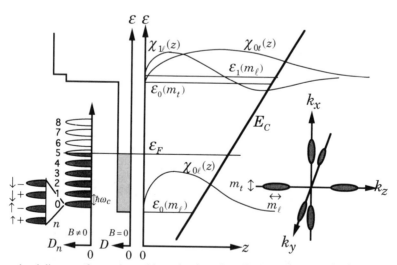

Fig. 1. Schematic energy level diagram of an n-channel inversion layer in a Si-MOSFET on a Si(001) surface. D and D_n are density of states without and with a strong magnetic field perpendicular to the interface, respectively. The Landau quantum number is denoted by n. Each Landau level with the same n splits into four Landau levels: ↑ and ↓ refer to parallel and antiparallel spin to the field, + and − refer to lower and higher valley.

It is interesting to quote here Schrieffer's statement in the abstract of his review article: "Conduction in narrow channels where quantization effects may be important is considered and it is concluded that surface scattering most likely broadens the discrete levels into a continuum and a simple theory should be adequate to describe the mobility in this limit" [4].

The role of surface scattering in the classical treatment based on a simple model is to reduce the mean free path to the width of the surface channel. In this model, the width of the surface channel is proportional to the reciprocal of the surface field [4]. The surface field is proportional to the areal density of the surface carriers. Therefore, the surface conductance cannot be controlled by the gate voltage in field effect transistors. Actually, however, the transistor performance of silicon–metal–oxide field-effect transistors (Si-MOSFETs) itself revealed that surface scattering in the classical treatment is inadequate in the age of silicon technology starting from the early 1960s [5].

Quantum effects associated with the motion of carriers perpendicular to the surface due to the narrow potential well at the semiconductor surface were discussed by several authors in the beginning of the 1960s; i.e., Kobayashi et al. [6], Greene [7], Handler and Eisenhouer [8], Murphy [9], Kawaji, Huff and Gatos [10], and Kawaji and Gatos [11,12]. In Refs. [10–12], we discussed possible effects of quantization of motion in the x–y plane due to a magnetic field in addition to electric quantization in the z-direction.

The simplest quantum effect to consider in surface transport is probably the broadening of the given space charge layer, and the consequent raising of the surface mobility [7]. In a strong inversion or accumulation layer, the charge distribution is maximum at the surface in the classical picture of the space charge layer. When the continuum of the energy levels for electrons with three-dimensional freedom is separated into a series of two-dimensional subbands whose bottoms are E_α ($\alpha = 0, 1, 2, \cdots$), the center of charge distribution moves toward the bulk because the electron wave function associated with each E_α is zero at the surface. Consequently, the

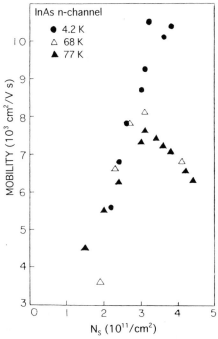

Fig. 2. Electron mobility versus surface electron concentration N_s in an n-channel inversion layer on InAs. (After Kawaji and Kawaguchi [13].)

quantum effect reduces the influence of surface scattering compared with the classical case. Such a quantum effect has been observed in the behaviour of the light hole mobility in germanium surfaces by Handler and Eisenhouer [8] and in p-channel Si-MOSFETs by Murphy [9]. In 1966 a clear increase of electron mobility with increase in surface electron concentration was observed in a naturally formed n-channel inversion layer of a p-type InAs surface at 4 K and it was explained by 2D ionized impurity scattering in a 2D electron gas by Kawaji and Kawaguchi [13] as shown in Fig. 2. In the same year clear evidence of a two-dimensional electron gas in an interface channel was demonstrated in a Shubnikov–de Haas measurement of silicon MOSFETs by Fowler, Fang, Howard and Stiles [14].

3. Two-dimensional (2D) electron gas in surface and interface channels

In Si-MOSFETs, an insulator film (SiO_2) is sandwiched between a metal film, called the gate, and a silicon substrate. In an n-channel device, whose structure is shown schematically in Fig. 3a, a current can flow between the two heavily doped n$^+$-type contacts S and D, called the source and the drain, respectively, only when an n-type conduction layer (channel) is produced on the surface of the p-type Si substrate adjacent to the SiO$_2$ film, by applying a positive gate voltage V_G against the Si substrate which is usually in equilibrium with the source. The conduction layer is called the inversion layer because the conduction type is inverted from the substrate. In a GaAs/Al$_x$Ga$_{1-x}$As $(x \approx 0.3)$ heterojunction system, called a HEMT (high electron mobility transistor) [15], an n-type conduction layer is produced on the surface of undoped GaAs adjacent to a thin undoped Al$_x$Ga$_{1-x}$As layer as shown schematically in Fig. 3b.

At low temperatures and at low electron concentration, the electrons in the inversion layer behave as a two-dimensional (2D) system because the separation between the quantized energy levels associated with their motion in the direction (z-direction) perpendicular to the conduction layer or the channel is large enough to set the

Fermi level below the first excited level; i.e., $E_F - E_0 < E_1 - E_0 \gg k_B T$ as shown in Fig. 1 [16–18].

The 2D system is characterized by a constant (energy-independent) density of states expressed as $D_2 = g_v g_s m / \pi \hbar^2$ where g_v is the valley degeneracy and g_s is the spin degeneracy. When a strong magnetic field B is applied perpendicular to the channel, the 2D motion of the electrons is quantized into Landau levels whose energies are separated by $\hbar \omega_c$, where $\omega_c = eB/m$ is the cyclotron frequency. Quantum oscillations with a constant period of conductance in the inversion layer observed by varying the gate voltage in a fixed magnetic field by Fowler et al. [14] shows clearly the 2D nature of the conduction electrons (Fig. 4).

The 2D nature of electrons in Si inversion layers can be seen even at moderate temperatures and in moderate electron concentrations. Figs. 5a and 5b show the significant effect of surface quantization calculated by Stern [17].

One of the expected quantum mechanical properties of 2D-electrons confined in an inversion layer is their phonon scattering mobility. In

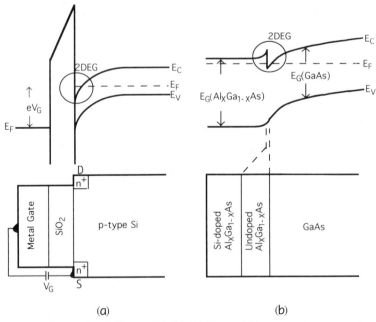

(a)   (b)

Fig. 3. Schematic structure and energy band diagram. (a) Si-MOSFET and (b) GaAs/Al$_x$Ga$_{1-x}$As $(x \approx 0.3)$ heterostructure.

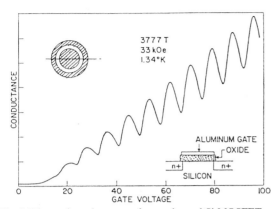

Fig. 4. Diagonal conductance of an n-channel Si-MOSFET on a Si(001) surface with a Corbino disk geometry in the presence of a magnetic field perpendicular to the interface. In this case spin and valley splittings are not resolved. (After Fowler, Fang, Howard and Stiles [14].)

1950 Bardeen and Shockley [19] studied theoretically the acoustic phonon scattering mobility of electrons in semiconductor crystals. In 1968, Fang and Fowler [20] made an extensive experimental study of electron mobility in Si-MOSFETs. I found signs of phonon scattering in their results of the temperature dependence and the electron concentration dependence of the electron mobility near room temperature, and proposed a 2D version of the Bardeen–Shockley theory where the reciprocal of the relaxation time or the scattering probability $1/\tau$ depends on the areal electron concentration N_s and the temperature T as $1/\tau \propto \langle z \rangle^{-1} T \propto N_s^{1/3} T$ [21]. Here, $\langle z \rangle$ is the thickness of the inversion layer which is proportional to $N_s^{-1/3}$ for $N_s \gg N_{dep}$ where N_{dep} is the areal concentration of ionized impurities in the depletion layer [16]. The dependences of the mobility on T and N_s were approximately in accord with experimental results by Fang and Fowler. The N_s-dependence is shown in Fig. 6. The T-dependence reflects the number of phonons excited at T in the high temperature approximation and the 2D energy-independent density of states.

The quantum mechanical nature of the 2D electron–phonon interaction appears in the term $\langle z \rangle$. As shown by the rigorous calculation by

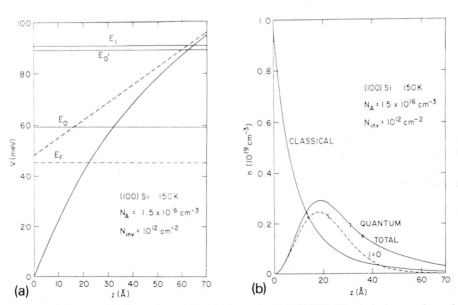

Fig. 5. (a) Electron potential energy (solid curve) near the interface of a Si-MOSFET. The dashed curve shows the contribution from the acceptor ions alone. The horizontal line indicates the position of the Fermi level E_F and the bottoms of the three lowest subbands. These results are obtained from a numerical self-consistent solution of the Schrödinger equation and Poisson's equation. (b) Classical and quantum mechanical charge densities for the case illustrated in (a). The dashed curve gives the contribution of the lowest subband to the quantum mechanical charge density. Each curve is marked by a vertical bar at the average value of z. (After Stern [17].)

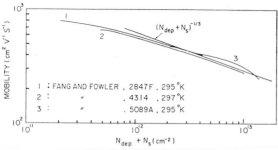

Fig. 6. Electron mobility versus $(N_{dep} + N_s)$ in an n-channel inversion layer in a Si-MOSFET on a Si(001) surface. N_{dep} is the number of ionized acceptor charges in the depletion layer and N_s the number of conduction electrons in the inversion layer, per unit area, respectively. Curves 1, 2 and 3 are after Fang and Fowler [20]. The solid line is calculated by use of $\Xi = 23$ eV for the deformation potential. (After Kawaji [21].)

Ezawa, Nakamura and Kuroda (Ref. [22], and also Refs. [23,24]), this term reflects the uncertainty principle, $\langle z \rangle \Delta k_z \geq 1$; i.e., small spreading of the wave function of the electrons in the inversion layer causes a considerable spreading of the Fourier components of the envelope function in the momentum space in the z-direction Δk_z. In this case of electron–phonon interaction, the electron system is two-dimensional, while the phonon system is essentially three-dimensional. Therefore, in the calculation of the matrix element for the interaction, momentum conservation gives a severe restriction only for the components of the phonon momentum in the direction parallel to the surface. So the matrix element contains the z-component of the phonon wave vector q_z, and we need additional integration over q_z in the calculation of the scattering probability. The result of this integration is proportional to the spreading Δk_z. This means that the number of phonons which can interact with the 2D electrons increases with decrease in the spreading of the electron wave function in the z-direction.

We can follow the progress in understanding of 2D systems in semiconductor surface and interface channels in the proceedings, published in the journal Surface Science, of a series of conferences of EP2DS (Electronic Properties of Two-Dimensional Systems) held at Brown University (1975) [25], Berchtesgaden (1977) [26], Lake Yamanaka (1979) [27], New London (1981) [28],

Oxford (1983) [29], Kyoto (1985) [30], Santa Fe (1987) [31], Grenoble (1989) [32] and Nara (1991) [33].

4. Anderson localization: strong localization in 2D systems

Since Anderson's paper entitled "Absence of Diffusion in Certain Random Lattices" published in 1956 [34], effects of disorder in the band tail and metal–insulator transitions have been studied extensively in 3D systems. The electron system in the inversion layer at a Si-MOSFET is the best material in the research of Anderson localization, because it is a 2D system and its Fermi level can be controlled by the gate voltage in a single specimen.

When random potentials due to impurities are incorporated in the crystal lattice, the electronic states near the edges of the energy band cannot extend in the crystal lattice and form localized states. In a metal, the random potentials scatter the conduction electrons and the mean free path of the electrons l becomes finite even for a non-vibrating lattice. When the random potentials are so strong that the electronic mean free path becomes smaller than the Fermi wave length $\lambda_F = 2\pi/k_F$, metallic conduction of electrons is not expected [35].

It is expected that there exists an energy E_c (the mobility edge) separating the extended and localized states. At absolute zero, the conductivity is zero when the Fermi level lies below the mobility edge and the conductivity is finite when the Fermi level lies above the mobility edge. At finite temperatures, when the Fermi level lies below the mobility edge, conduction may occur via electrons excited to the mobility edge from the Fermi level or by hopping between localized states.

In a 2D system, taking into account the fact that the areal density of electrons is expressed as $N_s = k_F^2/2\pi$, the Drude conductivity is described by

$$\sigma_s = \frac{N_s e^2 \tau}{m} = \frac{e^2}{2\pi\hbar}(k_F l), \qquad (2)$$

where the mean free path of electrons is given by $l = \hbar k_F \tau / m$. The product $k_F l$ cannot be significantly smaller than unity in metallic conduction, and if we take $k_F l \approx 1$ as the minimum value, we obtain a minimum metallic conductivity of $0.16 e^2 / \hbar$. Results of numerical studies by Licciardello and Thouless show that $\sigma_{min} \approx 0.12 e^2 / \hbar$ [36]. In 3D systems, since the electron concentration is $N_3 = k_F^2 / 3\pi^2$, and the same argument leads to an extra factor of k_F giving $\sigma_3 = N_3 e^2 \tau / m = (e^2 / 3\pi^2 \hbar)(k_F l) k_F$, we cannot discuss so clearly the minimum metallic conductivity as for 2D systems.

In 2D electron systems in MOS inversion layers, the Fermi level can be controlled with respect to the mobility edge by changing the gate voltage. The metal–insulator transition can be observed with a single specimen by changing the gate voltage. The minimum metallic conductivity can be found either from the temperature independent conductivity at certain gate voltage or from extrapolation of those curves in the activated conduction region to $1/T = 0$. Fig. 7 shows the transition between the metallic conductivity and the activated conductivity exhibited by a Si-MOSFET reported by Pollitt, Pepper and Adkins in 1975 [37].

Although the result in Fig. 7 appears to favour the mobility edge model described above, there seems to be a wide difference in experimental results depending on samples and a wide difference in the interpretations of the conduction mechanisms in the non-metallic region of the conductivity in Si-MOSFETs. For details, see the chapter on "Activated Transport" in a review by Ando, Fowler and Stern [18].

In 1979 Abrahams, Anderson, Ricciardello and Ramakrishnan [38] developed a scaling theory of Anderson localization and successfully explained experimental results in the weakly localized regime which will be described in the next section. We tried to interpret Si-MOSFETs data at low electron densities in terms of the scaling theory [39,40]. The scaling theory had predicted for 2D systems at absolute zero, that the conductance decreases first logarithmically with sample size L and falls off exponentially with L when the conductance reaches a certain critical value

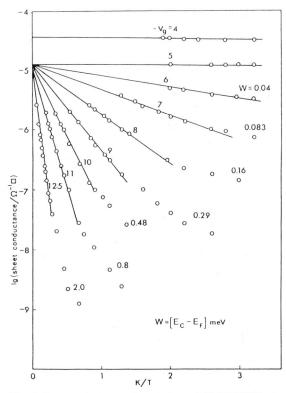

Fig. 7. Log conductance of an n-channel Si-MOSFET near the conduction threshold versus $1/T$. V_g refers to the applied gate voltage. (After Pollit, Pepper and Atkins [37].)

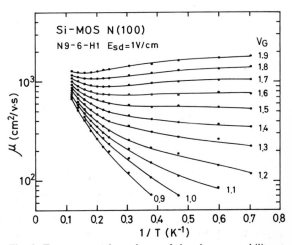

Fig. 8. Temperature dependence of the electron mobility at various gate voltages V_G. The gate voltage at the conduction threshold is 0.37 V and $dN_s/dV_G = 1.44 \times 10^{11}/\text{V} \cdot \text{cm}^2$. (After Hoshi and Kawaji [40].)

σ_c. Anderson, Abrahams and Ramakrishnan [41] extended the theory to finite temperatures in the absence of magnetic field by replacing the sample size L by the inelastic diffusion length $L_\epsilon = (\mathscr{D}\tau_\epsilon)^{1/2}$, where $\mathscr{D}$ is the diffusion coefficient and τ_ϵ is the inelastic scattering time of electrons, when $L_\epsilon < L$. In the logarithmic regime or the weakly localized regime, they made a perturbation calculation. For small conductance Abrahams et al. [38] conjectured an exponential falling

off of the conductance as $\sigma(L) = \sigma_a \exp(-\alpha L)$. However, no theory exists in this regime. Our idea was that the strongly temperature dependent conductivity at low electron densities shows exponential localization of a metallic state at finite temperatures, and this behaviour may be expressed by $\sigma(L) = \sigma_a \exp(-\alpha L)$ if we replace L by an appropriate scale length, which we took L_ϵ [40]. We tried to interpret the temperature dependence of the electron mobility shown in Fig.

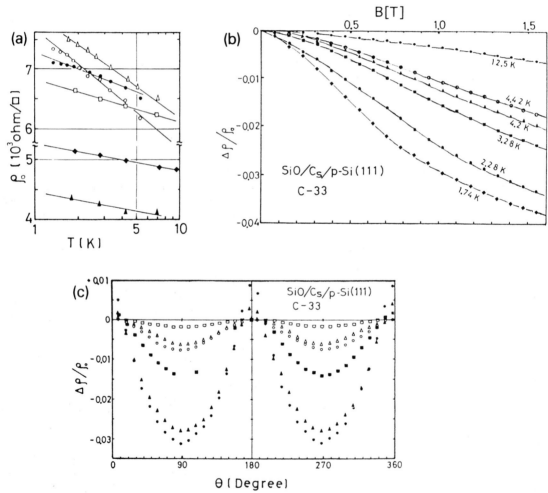

Fig. 9. Electrical properties of an n-channel inversion layer on Cs-covered Si(111) surfaces. (a) Resistivities versus log T. (b) Magnetoresistance of a sample in (a). The magnetic field is applied perpendicular to the surface and the current. (c) Angular dependence of the negative magnetoresistance at temperatures of 4.2 K (squares), 2.3 K (triangles), and 1.7 K (circles). Open symbols indicate $B = 0.31$ T and full symbols indicate $B = 1.25$ T. The magnetic field is rotated in the plane normal to the current direction and $\theta = 0$ at $B \parallel$ surface. (After Kawaguchi, Kitahara and Kawaji [42].)

8, and obtained results not inconsistent with the results in the weak localization except for the N_s-dependence of τ_ϵ; i.e., τ_ϵ in low N_s region increases with decreasing N_s while τ_ϵ is proportional to N_s in the high N_s region.

5. Anderson localization: weak localization and negative magnetoresistance

In 1977, Kawaguchi, Kitahara and Kawaji [42] reported interesting behavior of the resistivities and magnetoresistances observed in Cs-covered p-Si(111) inversion layers. As shown in Fig. 9a, the temperature dependence of the resistivity shows a log T dependence between 1 and 10 K. The sign of the magnetoresistance is negative and its field dependence is stronger at lower temperatures as shown in Fig. 9b. Moreover, the negative magnetoresistance depends on the field component normal to the surface, as shown in Fig. 9c, because $\delta\rho/\rho_0 = -sB^2 \sin^2\theta$ where B is the flux density and θ is the field direction measured from the direction in the surface. Observations in Fig. 9c clearly show that the phenomenon arises not from the electron spin but from the orbital motion of the electrons. These observations can be well understood in the light of the weak-localization theoretical framework established in 1979.

In the presence of impurities, constructive interference of scattered electron waves along the same trajectory but in the opposite directions causes localization of the electron. When we apply a magnetic field perpendicular to the trajectory, change in the phase of the electron wave function destroys the interference and gives rise to the increase in the conductivity.

Let us consider a 2D electron system at low temperatures with impurities which scatter electrons [43]. An electron propagates as a wave packet constructed from wave functions described as $\psi(r, t) \propto \exp[i(p \cdot r - \epsilon t)/\hbar]$. In quantum diffusion of an electron from a point P to another point Q, the wave packet propagates taking many different trajectories suffering from scattering by impurities. When the probability amplitude of the trajectory $i = 1, 2, 3, \cdots$ is

denoted by A_i, the total transport probability W is given by

$$W(P \neq Q) = \left| \sum_i A_i \right|^2 = \sum_i |A_i|^2 + \sum_{i \neq j} A_i A_j^*. \quad (3)$$

The term $\sum_i |A_i|^2$ is the sum of transport probabilities over all trajectories and leads to the Drude conductivity σ_0 given by Eq. (1). The second term $\sum_{i \neq j} A_i A_j^*$ is an interference term which is usually zero because the path difference between the different trajectories is larger than the Fermi wave length. An exception is the case P = Q in which the electron wave packet propagates in opposite directions along a closed path in the trajectories i and j without a change in phase. In this case the total transport probability is given by

$$W(P = Q) = \sum_{i=j} |2A_i|^2 = 4 \sum_{i=j} |A_i|^2 = 2 \sum_i |A_i|^2. \quad (4)$$

Due to the quantum mechanical interference, the probability of returning to the starting point for quantum diffusion is twice that for classical diffusion. Therefore, the electrical conductivity needs a quantum correction term σ_L' in addition to the Drude conductivity term σ_0; i.e., $\sigma = \sigma_0 + \sigma_L'$.

Abrahams, Anderson, Ricciardello and Lama-krishnan [38] found that the correction term σ_L' is given by $\sigma_L' = -a \ln(L/L_0)$ where L is sample size and L_0 is a microscopic length such as the mean free path. Perturbation calculations of the correction term σ_L' performed by several authors [41,44] led to the result $\sigma_L' = (-e^2/\pi^2\hbar) \ln(L/L_0)$.

The length of the closed trajectory which makes constructive interference is limited by a phase breaking length where the phase of the electron wave function suffered a change. As discussed by Thouless, a typical phase breaking process is inelastic scattering of the electron [45]. After the electron diffuses a length $L_\epsilon = (\mathcal{D}\tau_\epsilon)^{1/2}$, where $\mathcal{D}$ is the diffusion coefficient in a 2D system given by $\mathcal{D} = l^2/2\tau$ and τ_ϵ is the inelastic scattering time, the quantum interference is destroyed and the electron escapes from one localized state to another localized state with different energy. Thus the sample size L in the quantum correction term should be replaced by L_ϵ and the conductivity correction is given by

$$\sigma_L' = -\frac{e^2}{\pi^2\hbar}\ln\left(\frac{L_\epsilon}{L_0}\right) = -\frac{e^2}{2\pi^2\hbar}\ln\left(\frac{\tau_\epsilon}{\tau}\right). \quad (5)$$

Following Khmel'nitskii [43], we can derive the quantum correction to the Drude conductivity by using the following simple physical picture. The solution to the classical 2D diffusion equation leads to the probability of an electron returning to the initial unit area after time t of $1/(\mathscr{D}t)$. The areal element is given by $k_F^{-1}v_F\,dt$ when we take the width of the wave packet as k_F^{-1}. Then the probability of an electron returning to the starting point via the classical diffusion process leads to a correction σ_L' to the 2D Drude conductivity $\sigma_0 = (e^2/2\pi^2\hbar)(2\pi\epsilon_F\tau/\hbar)$ given by

$$\frac{\sigma_L'}{\sigma_0} = -\int_\tau^{\tau_\epsilon}\frac{1}{\mathscr{D}t}\frac{v_F}{k_F}\,dt$$

$$= -\frac{2}{k_Fl}\ln\left(\frac{\tau_\epsilon}{\tau}\right) = -\frac{\hbar}{\epsilon_F\tau}\ln\left(\frac{\tau_\epsilon}{\tau}\right). \quad (6)$$

Here we used the following relations: $\mathscr{D} = l^2/2\tau$ and $l = v_F\tau$. This simple discussion leads to Eq. (5) except for a numerical factor of 2π.

If τ_ϵ depends on T as $\tau_\epsilon \propto T^{-p}$ and τ is independent of T, then from Eq. (5) we have a $\ln T$ dependent term given by

$$\sigma(T) = \sigma_0 + \frac{pe^2}{2\pi^2\hbar}\ln T. \quad (7)$$

When an external magnetic field B is applied perpendicular to the 2D system, the electron momentum is given by the electromagnetic momentum $p + eA$ where A is a vector potential and $B = \nabla \times A$. Then an additional phase $(e/\hbar)\int A\,dr$ appears in the wave function. When we express this integral around the closed trajectory as $\oint A\,dr = BL_m^2$ and if the relation $eBL_m^2/\hbar \approx 1$ is satisfied, then the interference condition is destroyed. So we can consider that L_m is the phase breaking length. When the condition $L_m < L_\epsilon$ is fulfilled, L_ϵ in Eq. (5) should be replaced by L_m. Then we have the following approximate

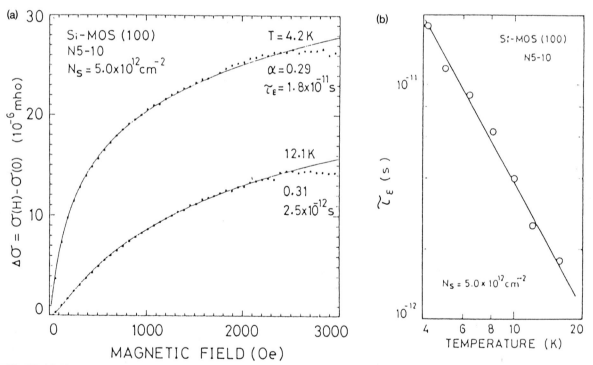

Fig. 10. (a) Field dependence of the magnetoconductivity of a Si-MOSFET. Experimental data (dots) are fitted to Hikami, Larkin and Nagaoka's theory [46] (solid curves). (b) Temperature dependence of the inelastic scattering time τ_ϵ extracted from the magnetoconductivity. (After Kawaguchi and Kawaji [48].)

formula for the change in the conductivity due to the magnetic field:

$$\Delta\sigma(B) = \sigma(B) - \sigma(0) = \frac{e^2}{2\pi^2\hbar}\ln\left(\frac{eB\mathcal{D}\tau_\epsilon}{\hbar}\right).$$
(8)

We can find excellent agreement between the experimental temperature dependences in Fig. 9a and Eq. (7) when we assume $p = 1$ in $\tau_\epsilon \propto T^{-p}$ ($p = 0.9 \pm 0.3$ in Eq. (7)). We can also find good agreement between the experimental magnetic field dependence and the temperature dependence of the field dependence in Fig. 9b and Eq. (8) when we assume also $\tau_\epsilon \propto T^{-p}$. We stress here that the angular dependence of the negative magnetoresistance in Fig. 9c is evidence that the negative magnetoresistance comes from the destruction of the phase coherence of weakly localized electrons.

Explicit expressions of the magnetoresistance in weak localization in 2D systems are given by Hikami, Larkin and Nagaoka [46] and Altshuler, Khmel'nitskii, Larkin and Lee [47]. Right after the derivation of explicit expressions of the magnetoresistance by Hikami et al., Kawaguchi and Kawaji fitted the experimental magnetoconductivity in Si(001) MOS inversion layers with $\mu_{\rm peak}(4.2$ K$) = 13\,000$ cm^2/V·s to Hikami et al.'s formula and obtained the beautiful results shown in Fig. 10 [48].

Right after the success of the weak localization theory in 2D systems, Kawabata [49] derived a theoretical expression for magnetoconductivity due to weak localization in 3D systems and successfully explained the negative magnetoresistance observed in impurity conduction of Ge by Sasaki [50] which has been left unexplained for long. Thus studies of quantum transport in 2D systems in surface and interface channels has opened a road for understanding one of the fundamental properties of conduction electrons in solids.

Inelastic scattering at liquid helium temperatures refers to electron–electron scattering [51, 52]. Therefore, we can measure the temperature of 2D electron systems in a thermally nonequilibrium state with a lattice by comparing the τ_ϵ

extracted from the negative magnetoresistance data measured under nonequilibrium conditions with that extracted from the negative magnetoresistance data measured under thermally equilibrium conditions. Kawaguchi and Kawaji [53] applied the negative magnetoresistance effect to the measurement of the electron temperature in Si MOS inversion layers at high source–drain electric fields in order to evaluate the deformation potential constants in electron–phonon coupling.

For further studies of this topic, see Refs. [51,52].

6. Quantum transport in strong magnetic field in 2D systems

6.1. Landau levels and diagonal conductivity in 2D systems

When a strong magnetic field is applied perpendicular to the 2D system, the continuum of the 2D energy levels coalesces into a series of quantized Landau levels: $\epsilon_n = \hbar\omega_c(n + 1/2)$, ω_c is the cyclotron frequency and $n = 0, 1, 2, 3, \cdots$ is the Landau quantum number (Fig. 1). At sufficiently low temperatures and in sufficiently strong magnetic fields, it is possible to realize the extreme quantum limit condition ($k_BT < \Gamma < \hbar\omega_c$), where Γ is the broadening of a Landau level. Table 1 summarizes numerical values in units of degrees Kelvin for the following energies: the

Table 1

Quantities related to transport properties of 2D systems in Si(001) and GaAs inversion layers in strong magnetic fields (m(Si) = $0.19m_0$, g(Si) = 2, m(GaAs) = $0.068m_0$, g(GaAs) = 0.52)

	Si(001)	GaAs
$\hbar\omega_c$(K)/B(T)	7.070	19.75
$\Gamma_{\rm SCBA}$(K)[μ(m^2 V^{-1} s^{-1})/B(T)]$^{1/2}$	5.642	15.76
Γ_τ(K)μ(m^2 V^{-1} s^{-1})	3.536	9.880
$g\mu_B$(K)/B(T)	1.344	0.349
ΔE_v(K)[10^{16}/(N_s(m^{-2}) $+(32/11)N_{\rm depl}$(m^{-2})]	1.7	–
l_B(A)B(T)$^{1/2}$		256.6
eB/h(m^{-2})/B(T)		2.418×10^{14}

separation between adjacent Landau level centers given by $\hbar\omega_c$, the Landau level broadening given in the self-consistent Born approximation by $\Gamma_{SCBA} = [(2/\pi)(\hbar\omega_c)(\hbar/\tau)]^{1/2}$ [54], the simple level broadening given by $\Gamma_\tau = \hbar/2\tau$, the spin Zeeman splitting given by $g\mu_B B$ and the valley splitting in the Si(001) inversion layer ΔE_v approximately given by the electric breakthrough mechanism [55]. In Table 1, numerical values of the radius of the ground Landau orbit $l_B = (\hbar/eB)^{1/2}$ and the degeneracy of a Landau level $N_L = eB/h$ are also given.

The quantities given in Table 1 show that 1 K is a low enough temperature to realize the extreme-quantum-limit condition in an n-channel MOS inversion layer on a Si(001) surface in a field of 10 T as far as the separation of Landau levels with different Landau indices is concerned. In an inversion layer in the GaAs/Al$_x$Ga$_{1-x}$As ($x \approx 0.3$) heterojunction interface, it is easier to realize the extreme-quantum-limit condition than in Si MOS inversion layers because $m = 0.068 m_0$ and $g = 0.52$ [56] and no valley degeneracy occurs in that system.

The diagonal conductivity σ_{xx} is only the conductivity tensor which can be directly measured by using the concentric source- and drain-electrode structure called a Corbino disk as shown in Fig. 11. As mentioned earlier, Fowler et al. [14] first showed clearly the existence of a 2D electron system by observing Shubnikov–de Haas oscillations in σ_{xx}. Their experiments have shown that the larger the Landau level index the larger the peak value of σ_{xx}. Theoretical study of σ_{xx} in the Landau levels in 2D systems with short range scatterers, carried out by Ando, Matsumoto and Uemura [57], showed that the peak value of σ_{xx} of the Landau level with the Landau quantum number n is given by $e^2(n + 1/2)/\pi^2\hbar$. Their result was roughly confirmed by use of a wide rectangular sample by Kobayashi and Komatsubara [57].

The classical expression for the diagonal magnetoconductivity is given by

$$\sigma_{xx} = \frac{N_s e^2 \tau/m}{1 + (\omega_c \tau)^2}. \tag{9}$$

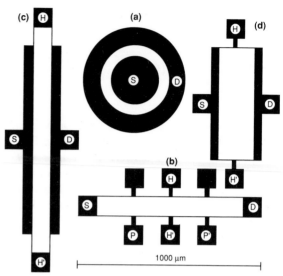

Fig. 11. Typical electrode structures for galvanomagnetic experiments for 2D systems. (a) Corbino disk, (b) Hall bar, (c) wide Hall current bar and (d) wide Hall bar.

In the strong field limit $\hbar\omega_c \gg 1$, we have $\sigma_{xx} \approx N_s m/B^2\tau$ which shows that the conductivity along the electric field is proportional to the scattering probability which is valid in the quantum mechanical derivation of the diagonal conductivity in 2D systems explained in the following [54,58].

The conductivity of a degenerate electron gas can be described by $\sigma = e^2 \mathscr{D} D(\epsilon_F)$ where $\mathscr{D}$ is the diffusion coefficient and $D(\epsilon_F)$ is the density of states at the Fermi level (see appendix) [59]. The diffusion coefficient of a 2D system is given by $\mathscr{D} = l^2/2\tau$, where l is the mean free path and τ is the mean free time. When we assume the density of states can be described by a semi-ellipse whose peak is given by $D(\epsilon_F)$ and whose half width is given by Γ, the total number of states in the Landau level is $N_L = eB/h = \pi D(\epsilon_F)\Gamma/2$. When we substitute the radius of the Landau orbit $l_n = l_B\sqrt{2n+1}$ to the mean free path along the electric field after a scattering event in the diffusion coefficient, and when we use the relation $2\Gamma\tau \sim \hbar$ based on the uncertainty principle $\Delta\epsilon \, \Delta t \sim \hbar$, then we have

$$\sigma_{xx}(n, \text{peak}) \approx \frac{2e^2}{\pi^2\hbar}\left(n + \tfrac{1}{2}\right). \tag{10}$$

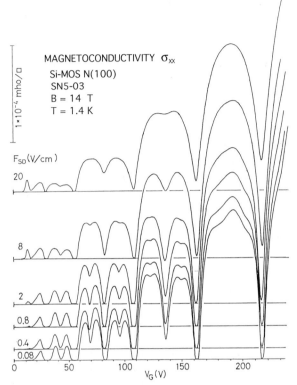

Fig. 12. Gate voltage dependence of the diagonal conductivity σ_{xx} of a Si(001) inversion layer for various source–drain fields F_{SD}. (After Kawaji and Wakabayashi [60].)

This is twice the theoretical result by Ando et al.: $\sigma_{xx}(n, \text{peak}) = e^2(n + 1/2)/\pi^2\hbar$ [54,57].

Fig. 12 shows the gate voltage dependence of σ_{xx} for various source–drain electric fields F_{SD} [60]. At low F_{SD} in Fig. 12, the peak values of σ_{xx} for the Landau levels with Landau quantum number $n = 0$ and 1 are in accord with the theoretical result, while those of higher Landau levels are larger than the theoretical values. This is because the energy levels belonging to adjacent Landau levels are mixed with each other due to the level broadening.

6.2. Valley splitting

In Fig. 12, the peaks of σ_{xx} belonging to the Landau levels with $n = 0$ and 1 are well separated. This shows that energy levels belonging to opposite valleys along the k_z-axis (Fig. 1) are

splitted. In an n-channel inversion layer on Si(001), the Landau level with the Landau quantum number n is four-fold degenerate as $(n \uparrow +)$, $(n \uparrow -)$, $(n \downarrow +)$ and $(n \downarrow -)$ levels where the $\uparrow$ and $\downarrow$ denote spin state and $+$ and $-$ are valley indices (Fig. 1).

The valley splitting in the inversion layer also comes from the quantum mechanical properties of the electrons; i.e., the uncertainty principle in position and momentum in the z-direction. The small spreading of the wave function of the electrons in the inversion layer causes a considerable spreading in the Fourier components of the envelope function in the momentum space in the z-direction around the bottoms of two valleys, $(0, 0, +k_1)$ and $(0, 0, -k_1)$. This spreading in the k_z space causes a tunneling through the Γ-point between the states in the different valleys. This tunneling causes interaction between the two bands in the bulk crystal. As in the case of a hydrogen molecule where real space tunneling of electrons which belong to different atoms produces bonding and antibonding states, the degeneracy of the energy levels belonging to different valleys is lifted in the inversion layer. This is the valley splitting due to the electric breakthrough mechanism studied by Ohkawa and Uemura [55,61]. As shown in Table 1, the valley splitting energy is small but enhancement of the valley splitting due to many body effects makes it appreciable in the Shubnikov–de Haas oscillations [55].

6.3. Localization in Landau levels in 2D systems

When random potentials are incorporated in such a system under extreme-quantum-limit conditions where gap regions exist in the density of states between the boundaries of each Landau level, localized states are expected to exist near the lower and higher edges of each Landau level. If the range of the random potential δ is much larger than the radius of the Landau orbit with Landau quantum number n, $\delta \gg l_n = l_B\sqrt{2n+1}$, it is easy to see that the centers of the Landau orbits near the bottoms and the tops of the random potentials move along closed trajectories lying on equi-potential lines. When the range of the random potentials is short ($\delta < l_n$), they also localized near the lower and higher edges of the Landau levels. These localized states exhibit exponential localization or strong localization [62,63]. The extended states which exist near the center of each Landau level are expected to show a different behavior from those in the absence of strong magnetic fields.

In Fig. 12, there are gate voltage regions where $\sigma_{xx} = 0$. This fact reveals that immobile or localized electrons exist at the edges of Landau levels. The number of localized electrons evaluated from the gate voltage region where $\sigma_{xx} = 0$ is plotted against magnetic field in Fig. 13. In Fig. 13, the number of localized or immobile electrons $N_{immobile}$ denoted by the Landau quantum number n is the sum of the localized states at the lower edge of the $(n\uparrow +)$ Landau level and the localized states at the upper edge of the $(n - 1\downarrow -)$ Landau level. For the $n = 0$ case, $N_{immobile}$ gives only the number of localized electrons at the lower edge of the $(0\uparrow +)$ level. The straight lines in Fig. 13 are plots of $N_{immobile}(B, n) = (eB/h)/(2n + 1)$. This result shows that the larger the radius of the Landau orbit the smaller the number of localized states. This is one of the fundamental properties of localization in Landau levels in 2D systems.

7. Quantum Hall effect

7.1. Quantization of the Hall conductivity

The classical expression for the Hall conductivity is given by

$$\sigma_{xy} = -\frac{N_s e}{B} + \frac{\sigma_{xx}}{\omega_c \tau}. \tag{11}$$

When i Landau levels (i is an integer) are completely filled and the next higher Landau level (($i + 1$)th Landau level) is empty; i.e., $N_s = ieB/h$, not any electron in the filled Landau levels can be scattered because the energy states in the empty Landau levels are energetically too high. Then we have $\sigma_{xx} = 0$ and the Hall conductivity is given by

$$\sigma_{xy} = -i\frac{e^2}{h}, \, (i = 1, 2, \cdots). \tag{12}$$

Let us consider the case where the system contains long range random potentials, $\delta \gg l_n$ [64]. When ΔN_s electrons ($\Delta N_s \ll B/h$) are added to the system, electrons enter in the ($i + 1$)th Landau level and are trapped in closed loops around the bottoms of the random potentials. Then, the added electrons do not contribute to

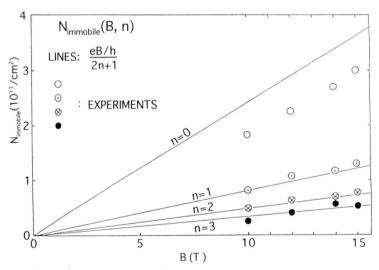

Fig. 13. Magnetic field dependence of the concentration of immobile electrons $N_{immobile}(B, n)$ associated with the lower edge of the Landau level with Landau quantum number n and the higher edge of the Landau level with the Landau quantum number $(n - 1)$. (After Kawaji and Wakabayashi [60].)

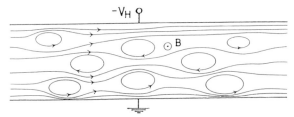

Fig. 14. When the range of random potentials is larger than the radius of the cyclotron orbit, some electrons are trapped around the tops and the bottoms of the potentials, still moving along equi-potential lines. Other electrons move freely along equi-potential lines which percolate through the sample. (After Kawaji [64].)

the conductivity. In the opposite case, when $\Delta N_s'$ electrons ($\Delta N_s' \ll eB/h$) are extracted from the ith filled Landau level, the $\Delta N_s'$ holes in the ith Landau level are also trapped in closed loops around the tops in the random potentials and they do not contribute to the conductivity. As shown in Fig. 14, localized electrons and holes are encircled by each equi-potential line and the electric field in each extended region along the y-direction, $F_j(\text{extend})$, is stronger than the average electric field F. In the extended regions, the diagonal conductivity is zero and the Hall conductivity is given by Eq. (12). Then the Hall conductivity or the Hall conductance is given by

$$G_H(i) = \frac{I}{V_H} = \frac{\sum F_j(\text{extend})(ie^2/h)}{\sum F_j(\text{extend})} = \frac{ie^2}{h},$$

(13)

or the Hall resistance is given by

$$R_H(i) = \frac{V_H}{I} = \frac{\sum F_j(\text{extend})}{\sum F_j(\text{extend})(ie^2/h)} = \frac{h}{ie^2}.$$

(14)

In resistance measurements, $\sigma_{xx} = 0$ corresponds to $\rho_{xx} = 0$ from the tensor transformation $\rho_{xx} = \sigma_{xx}/(\sigma_{xx}^2 + \sigma_{xy}^2)$. Similarly, $\rho_{xy} = h/ie^2$ from $\rho_{xy} = -\sigma_{xy}/(\sigma_{xx}^2 + \sigma_{xy}^2)$.

The qualitative discussion described above was verified theoretically for short-range scatterers in 1975, by Ando, Matsumoto and Uemura [65] who performed a theoretical study of the Hall effect in Landau levels in 2D systems with attractive

and repulsive scatterers based on the Kubo formula. They obtained the following results at absolute zero:

(1) In the self-consistent Born approximation (SCBA), the classical expression is valid when $1/\omega_c\tau$ is replaced by $\Gamma/\hbar\omega_c$.

(2) In the SCBA, when the Fermi level lies in the energy gap between i filled Landau levels and the next higher Landau level, the Hall conductivity is given by Eq. (12) and the diagonal conductivity σ_{xx} becomes zero. In other words, when each Landau level is completely filled, the second term in Eq. (11) vanishes and the Hall conductivity is not affected by the presence of impurity scattering.

(3) In the higher order approximation, the impurity bands are separated from each Landau level when the concentration of scatterers is sufficiently small. In this case, the Hall conductivity is also given by Eq. (12) and $\sigma_{xx} = 0$ when the Fermi level lies in any energy gap in the density of states between adjacent main Landau levels. This means that electrons which fully occupy impurity bands do not contribute to the Hall

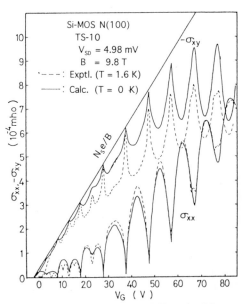

Fig. 15. Gate voltage dependences, Hall conductivity σ_{xy} and diagonal conductivity σ_{xx} of a Si-MOSFET measured by using a wide Hall bar (broken lines) and calculated results (solid lines). (After Kawaji, Igarashi and Wakabayashi [67].)

current, while those which occupy the main Landau level carry the same Hall current as that obtained when all electrons in the Landau level, $N_s = eB/h$, move freely.

We remark here that Ando et al.'s result (3) predicts the quantum Hall effect and the following two characteristic features of the σ_{xy} plateaus: (a) The Hall conductivity is given by Eq. (12) when $\sigma_{xx} = 0$. (b) Eq. (12) holds even if i Landau levels are not completely filled, contrary to the expectation that $|\sigma_{xy}| = N_s e/B < ie^2/h$ from Eq. (11). In other words, localized electrons near the lower edge of a Landau level and localized holes in the higher edge of a Landau level play the same role in the Hall conductivity. This is called electron–hole symmetry in the quantum Hall effect.

Experimental study of the Hall effect in Landau levels in Si(001) n-channel inversion layers was carried out by Igarashi, Wakabayashi and

Kawaji [66] at $T = 1.6$ K and $B = 9.8$ T and detailed results of an analysis of the experiments were published by Kawaji, Igarashi and Wakabayashi [67]. Igarashi et al. [66] used Si-MOSFETs with a pair of symmetrical Hall probes at the middle between the source and drain electrodes of a Hall bar whose width and distance apart are 600 μm and 200 μm, respectively (the wide Hall bar in Fig. 11). Their results confirmed those of Ando et al. (2) when the Fermi level lies between the Landau levels with $n = 0$ and 1, and between the Landau levels with $n = 1$ and 2. The line shape of σ_{xx} is well reproduced by the theory and the overall behavior of σ_{xy} is also satisfactorily explained by the theory as shown in Fig. 15 [66,67].

It is not easy to derive experimentally σ_{xy} values corresponding to σ_{xx} peaks because we have no technique like the Corbino disc for σ_{xx}. After several efforts [68,69], Wakabayashi and

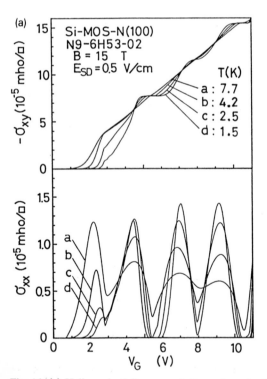

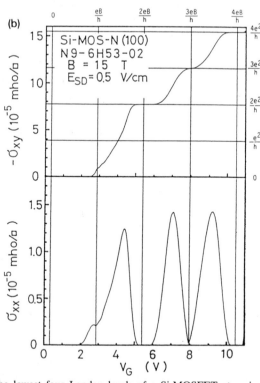

Fig. 16. (a) Hall conductivity σ_{xy} and diagonal conductivity σ_{xx} in the lowest four Landau levels of a Si-MOSFET at various temperatures measured by using the Hall current method and (b) those at $T = 1.5$ K which shows the electron-hole symmetry in the localization. (After Kawaji and Wakabayashi, [72].)

Kawaji [70,71] confirmed Eq. (11), by the Hall current method, in which $1/\omega_c\tau$ is replaced by $\Gamma_{SCBA}/\hbar\omega_c$, where Γ_{SCBA} is the Landau level broadening in the self-consistent Born approximation.

In 1980 Kawaji and Wakabayashi [72] obtained the results in Fig. 16 which show clearly that (a) localization causes quantized Hall conductivity with (b) electron–hole symmetry, as expected by Ando et al. (3).

7.2. Quantized Hall resistance and resistance standards

In 1980 von Klitzing, Dorda and Pepper [73] made a precision measurement of the Hall resistance $R_H = V_H/I$ given by Eq. (14) and showed that $R_H(4) = h/4e^2$ with an accuracy of 3 ppm to the recommended value. Von Klitzing et al. pro-

posed that this is a new method for the determination of the fine structure constant $\alpha = \mu_0 ce^2/2h$ where μ_0 is the permittivity of vacuum and c the light velocity. This proposal opened a new era in the electrical resistance standards.

After von Klitzing et al.'s proposal, standards laboratories all over the world carried out high precision measurements of the quantized Hall resistances $R_H(4)$ of Si-MOS inversion layers, and $R_H(4)$ and $R_H(2)$ of inversion layers in GaAs/Al$_x$Ga$_{1-x}$As ($x \approx 0.3$) heterostructure interfaces. Their efforts have produced two fruits: One is 1985 Nobel Prize in Physics for Klaus von Klitzing and another is the international resistance standard based on the quantized Hall resistance started from 1990.

In October 1988, the Comité International des Poids et Mesures recommended the following value denoted by R_{K-90} be adopted as a conven-

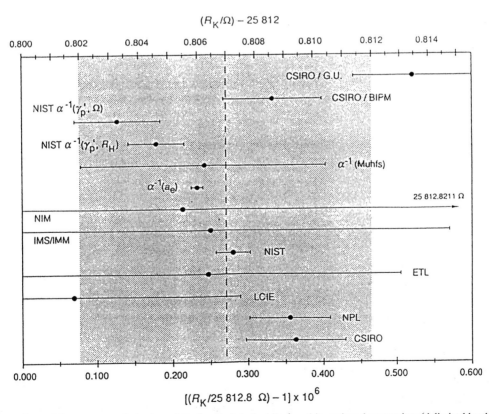

Fig. 17. Comparison of the recommended value of R_K (vertical dashed line) and its assigned uncertainty (delimited by the shading) with the values of R_K and their standard deviation uncertainties. (After Taylor and Witt [75].)

tional value for the von Klitzing constant R_K and this value be used from 1st January 1990 as the international reference standard of electrical resistance [74]:

$$R_{K\text{-}90} = 25812.807 \ \Omega. \tag{15}$$

Here, the von Klitzing constant R_K is the quotient of the Hall potential difference divided by the current corresponding to the plateau $i = 1$ in the quantum Hall effect or $R_K \equiv i \times R_H(i)$. The recommended value of $R_{K\text{-}90}$ was selected by the Working Group on the Quantum Hall Effect in the Comité Consultatif d'Électricité (Coordinator: B.N. Taylor, NBS) in June 1988 as shown in Fig. 17. [75] R_K values, based on the quantum Hall effect, were reported from laboratories all over the world to the Working Group before June 1988. The Working Group called these values the direct values. The weighted mean of seven direct values reported from laboratories where they have their own calculable capacitor [76] to realize the unit of 1 Ω_{SI} was $R_K = 25\,812.806\,97 \pm 0.000\,56 \ \Omega$. The uncertainty arises from uncertainties in the reported R_K values which depend mainly on uncertainties arising from design and construction details of the calculable capacitor and associated impedance bridges used in the Ω realization experiments. Before 1988 (and still now) the most accurate value of the inverse fine-structure constant α^{-1} is obtained from the experimental measurement of the electron-magnetic-moment anomaly a_e (relative uncertainty of 4×10^{-9}) [77] and the theoretical expression for a_e (relative uncertainty of 7×10^{-9} arising from numerical integration of the term derived by quantum electrodynamics (QED)) [78]. This gives the most precise result currently available, assuming $R_K = \mu_0 c \alpha^{-1}/2$; i.e., $R_K(a_e) = 25\,812.805\,99 \pm 0.000\,21 \ \Omega$ [75]. The Working Group called this value the indirect value. The recommended value of $R_{K\text{-}90}$ was taken as the simple mean of the weighted mean of direct values and the indirect R_K value.

It is known that the best present reference standards of electrical resistance are based on the quantized Hall resistances. Theoretical understanding of the quantum Hall effect has made progress [79–83]. However, we can find an appre-

ciable difference between the center value of $R_{K\text{-}90}$ and the indirect value of $R_K(a_e)$ including its uncertainty of 0.005 ppm. In order to eliminate the discrepancy between these values, further studies are necessary.

8. Fractional quantum Hall effect

The fractional quantum Hall effect discovered in 1982 by Tsui, Stormer and Gossard [84] in $GaAs/Al_xGa_{1-x}As$ heterostructure interfaces, as shown in Fig. 18, is not only one of the most fascinating phenomena in 2D systems but also one of the most interesting phenomena in solid state physics. It is similar to the integral quantum Hall effect except that the plateau in ρ_{xy} and the

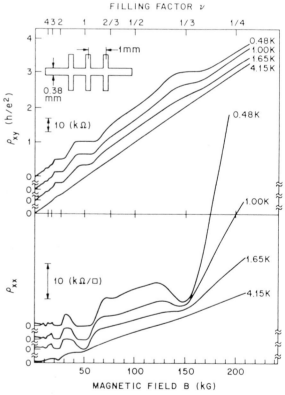

Fig. 18. Fractional quantum Hall effect appearing at the filling factors $\nu = 1/3$ and $2/3$ in the Hall resistivity ρ_{xy} and the diagonal resistivity ρ_{xx} in a $GaAs/Al_xGa_{1-x}As$ heterostructure with electron mobility 9 m^2/V·s. (After Tsui, Stormer and Gossard [84].)

minimum in ρ_{xx} appear at fractional filling factors ν of a Landau level. At an early stage, $\nu = 1/3$ and $2/3$ effects appear more prominently than other fractional effects [85]. When higher mobility samples were prepared, and measurements were extended to lower temperatures and stronger magnetic fields, an increasing number of the fractional quantum Hall effect corresponding to various filling factors ν with odd denominators were observed (Fig. 19), some of which are: $\nu = 1/3$, $2/5$, $3/7$, $4/9$, $5/11$, $\cdots$; $2/3$, $3/5$, $4/7$, $5/9$, $6/11$, $\cdots$; $1/5$, $2/9$, $3/13$, $\cdots$ [86].

It is clear that electron–electron interaction must be responsible for the fractional quantum Hall effect, creating energy gaps in the excitation spectrum at fractional filling, because for non-interacting electrons there are only gaps at integer values of ν. For fractional quantum Hall effect at $\nu = 1/(2m + 1)$, where m is an integer, Laughlin introduced a trial wave function describing the highly correlated quantum fluids and he explained the excitation gap of the quasi-particles

with fractional charges at $\nu = 1/3$, $1/5$, $\cdots$ [87]. The energy gaps at other fractions are obtained in a hierarchical scheme in which a daughter state is obtained from condensation of quasi-particles of the parent state [88,89].

Chang et al. [90] observed the energy gaps around $\nu = 2/3$ by measurement of $\rho_{xx} = \exp(-\Delta/T)$ at temperatures between 0.7 K and 66 mK in a field up to 10.6 T. The values Δ of Chang et al. obtained are too small compared with the theoretically predicted value by Laughlin [87]. We carefully measured the temperature dependence of ρ_{xx} and ρ_{xy} around $\nu = 1/3$ and $2/3$ at temperatures between 1 K and 0.1 K in a field up to 15.5 T for samples with $N_s = 1.1 \times 10^{11}$ cm^{-2} and $\mu = 1.1 \times 10^6$ cm^2/V $\cdot$ s [91]. As shown in Fig. 20, we found that the log ρ_{xx} versus $1/T$ line has a break at both the $1/3$ and the $2/3$ state. We claimed that the activation energy Δ_1 in the high temperature region is the one which can be compared with the theoretical excitation energy. Our result $\Delta_1(1/3)$ was about half Laughlin's theoretical result. However, Laughlin's calcula-

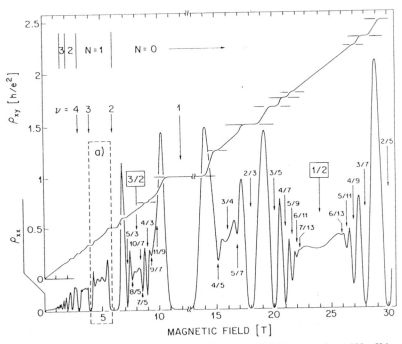

Fig. 19. Many structures appear in the Hall resistivity ρ_{xy} and the diagonal resistivity ρ_{xx} at about 100 mK in a GaAs/Al$_x$Ga$_{1-x}$As heterostructure with electron mobility 130 m^2/V $\cdot$ s. (After Willet et al. [86].)

tion was made at the high field limit. With a finite field correction, a three dimensionality correction, and the effect of Landau level mixing, the agreement between theory and experiment became good [52,92]. However, the experimental value of $\Delta_1(2/3)$ was much smaller than the theoretical result even when these corrections were taken into account. As described below, for a field $B = 7.4$ T we observed that $\Delta_1(2/3)$ was too low. We extended the experiments further by studying the sample dependence and magnetic field dependence [93–95], and also Boebinger et al. [96,97] performed extensive experiments. However, the activation energies at $\nu = 1/3$ and $2/3$ observed at fields lower than about 10 T were much smaller than the theoretical results. Recently Willet et al. [98] made similar measurements for a sample with high electron concentration and high electron mobility ($\mu = 5 \times 10^6$ cm^2/V·s) and found good result for $\Delta(2/3)$ which was observed at 11 T.

Quite recently, Jain has proposed a composite fermion approach which describes the fractional

quantum Hall effect by the integer quantum Hall effect of composite fermions consisting of electrons bound to an even number of flux quanta [99–101]. We can see the behavior of the composite fermions in ρ_{xx} at $\nu = 1/2$ and $3/2$ in Fig. 19. The composite fermion theory should explain the excitation energy in the fractional quantum Hall effect by the cyclotron energy of the composite fermion [117].

Rapid progress is being made in understanding the fractional quantum Hall effect and related phenomena [33,118].

9. Search for the Wigner crystallization in semiconductor 2D systems

In the paper entitled "On the interaction of electrons in metals" published in 1934, Wigner pointed out that the electrons would settle in lattice configurations which correspond to the absolute minima of the potential energy if the system can be expanded sufficiently so that the Coulomb energy predominates over the kinetic energy [102]. In usual 3D systems, it is hard to find a material for realization of the Wigner crystal. 2D systems whose electron concentration is controllable are potential materials to observe the Wigner crystallization. Actually, in 1979, Grimes and Adams [103] found evidence for a liquid-to-crystal phase transition in a sheet of electrons on a liquid- He surface by observations of RF electric field resonances due to excitation of standing capillary waves (ripplons).

Searches for electron crystallization or electron solids due to electron–electron interactions has been done in 2D systems in semiconductor surface and interface channels by many scientists. We first proposed pinning of the Wigner crystal as a possible explanation of the immobile electrons in the magnetoconductivity in Si-MOSFETs [60]. Lozovik and Yudson [104] made a theoretical prediction on the possibility of 2D electron crystallization of the inversion layer electrons in strong magnetic fields in a density region where this is impossible in the absence of fields. Theoretical studies were made, in connection to our experiments, on the possibility and properties of

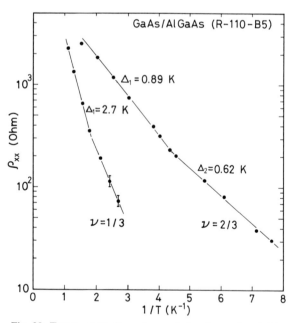

Fig. 20. Temperature dependence of the diagonal resistivity minima at $\nu = 1/3$ ($B = 14.6$ T) and $\nu = 2/3$ ($B = 7.4$ T) in a heterostructure with $\mu = 1.2 \times 10^6$ cm^2/V·s. (After Kawaji, Wakabayashi, Yoshino and Sakaki [91].)

2D electron crystallization of the inversion layer electrons in strong magnetic fields by Fukuyama [105] and Tsukada [106]. In 1977, we made an attempt to describe the activation-type temperature dependence observed in σ_{xx} in the lower edge of the lowest Landau level of Si-MOSFETs in $B = 10$ T and $T = 4.2$–1.5 K by a simple model based on the diffusion of Schottky defects in 2D Wigner crystal [107].

In the end of the 1980s, the advancement of the technology to produce 2D systems with very high electron mobilities encouraged scientists to reopen searches for Wigner crystallization or electron solids due to electron–electron interactions in $GaAs/Al_xGa_{1-x}As$ heterostructures [108–111] and Si-MOSFETs [112]. Works quite recently carried out appear in the Proceedings of EP2DS-9 published in 1992 [33].

One of the signs of the Wigner solid in high mobility heterostructures was a thermally activated insulating phase that surrounds the $\nu = 1/5$ fractional quantum Hall effect state which can be seen for temperature $T \to 0$, $\rho_{xx} \to \infty$ and intrinsic Hall resistance $\rho_{xy} = B/N_s e$, independent of T [108]. Quite recently, Kivelson, Lee and Zhang proposed a global phase diagram which covers the universal behavior of ρ_{xx} and ρ_{xy} in the integral and fractional quantum Hall effect including the effect of disorder [113]. They introduced an insulating phase which they denote a "Hall insulator" on the insulating phase in 2D systems, as a state in which, for $T \to 0$, $\rho_{xx} \to \infty$, and σ_{xx} and $\sigma_{xy} \to 0$, but ρ_{xy} tends to a constant value, which corresponds roughly to a normal Hall coefficient [113]. They proved this result for non-interacting electrons and argued it to be true for interacting systems [114]. Although their theory is still incomplete, it is possible that the insulating phase, which is considered to be a Wigner solid is a "Hall insulator". In a search for the $\nu = 1/7$ fractional quantum Hall state in a heterostructure with $\mu = 3.5 \times 10^5$ cm^2/V · s at a temperature down to 20 mK, we found that ρ_{xx} diverges with decreasing temperature but ρ_{xy} stays approximately $B/N_s e$ in the region $\nu < 1/3$ as shown in Fig. 21 [115]. This is probably an experimental support for the Hall insulator. Goldman, Shayegan and Tsui have also observed

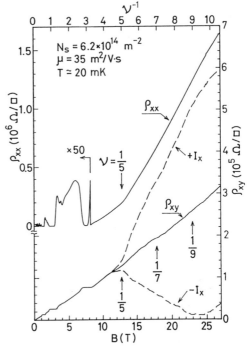

Fig. 21. Diagonal resistivity ρ_{xx} and Hall resistivity ρ_{xy} versus magnetic field B. Broken lines show the anomalous Hall voltage at positive and negative polarities of the sample current. The solid line is the average of the broken lines. Deviations of ρ_{xy} from the classical Hall resistivity $B/N_s e$ show structures at $\nu = 1/3$, 1/5 and 1/7. (After Wakabayashi et al. [115].)

similar behavior in the region $\nu < 1/5$ [116]. See also Refs. [119,120].

In connection to the researches in Si-MOSFETs carried out by D'Iorio, Pudalov and Semenchindky [112], the transport property of this system at low electron densities in the absence of magnetic field should be an interesting subject [121].

Wigner crystallization is a very interesting subject of study but further efforts are necessary to achieve conclusive results.

10. Concluding remarks

It was surprising for me to learn how much progress has been made in the quantum transport in semiconductor surface and interface channels in the last three decades. More surprising is the fact that the researches in this field are still making progress. New observations are appearing and new theories are being created. It is my pleasure to conclude this report with the observation that the study of quantum transport is scientifically sound and shall continue to be so in future as long as we have further advancement of technology.

Acknowledgements

I express my gratitude to Y. Kawaguchi, J. Wakabayashi and other members of my group in Gakushuin University. I express also my gratitude to my colleagues all over the world associated with the quantum transport in semiconductor surface and interface channels. Last but not least, I express my sincere gratitude to Professor Harry C. Gatos in MIT where I initiated my researches in this field.

Appendix

Suppose a simple metal in which both a gradient in the electron concentration N and a gradient in the electrostatic potential ϕ exist but the sum of the electrical current due to electron diffusion J_D and the electrical current due to field J_E is kept zero: $J_D + J_E = 0$. Then we have $\sigma = e^2 \mathcal{D} D(\epsilon_F)$ because J_D and J_E are given by

$$J_D = e\mathcal{D} \ \text{grad} \ N = e^2 \mathcal{D} D(\epsilon_F) \ \text{grad} \ \phi, \quad \text{(A.1)}$$

$$J_E = -\sigma \ \text{grad} \ \phi. \quad \text{(A.2)}$$

In a 3D system, $\mathcal{D} = l^2/3$ and $N = 2\epsilon_F D(\epsilon_F)/3$. In a 2D system, $\mathcal{D} = l^2/2$ and $N = \epsilon_F D(\epsilon_F)$. Therefore, we have $\mathcal{D} D(\epsilon_F) = N\tau/m$ in both systems.

References

[1] A.H. Wilson, The Theory of Metals (Cambridge University Press, Cambridge, 1953).

[2] R. Kubo, J. Phys. Soc. Jpn. 12 (1957) 570.

[3] R. Landauer, IBM J. Res. Dev. 1 (1957) 223, Z. Phys. B 68 (1987) 217.

[4] J.R. Schrieffer, Semiconductor Surface Physics, Ed. R.H. Kingston (University of Pennsylvania Press, Philadelphia, 1957) p. 55.

[5] S.R. Hoffstein and F.P. Heiman, Proc. IEEE 51 (1963) 1190.

[6] A. Kobayashi, Z. Oda, S. Kawaji, H. Arata and K. Sugiyama, J. Phys. Chem. Solids 14 (1960) 37.

[7] R.F. Greene, Surf. Sci. 2 (1964) 101.

[8] P. Handler and S. Eisenhour, Surf. Sci. 2 (1964) 64.

[9] N.St.J. Murphy, Surf. Sci. 2 (1964) 86.

[10] S. Kawaji, H. Huff and H.C. Gatos, Surf. Sci. 3 (1965) 234.

[11] S. Kawaji and H.C. Gatos, Surf. Sci. 6 (1967) 362.

[12] S. Kawaji and H.C. Gatos, Surf. Sci. 7 (1967) 215.

[13] S. Kawaji and Y. Kawaguchi, J. Phys. Soc. Jpn. 21 Suppl. (1966) 336.

[14] A.B. Fowler, F. Fang, W.E. Howard and P.J. Stiles, Phys. Rev. Lett. 16 (1966) 901; J. Phys. Soc. Jpn. 21 Suppl. (1966) 331.

[15] T. Mimura, S. Hiyamizu, T. Fujii and K. Nambu, Jpn. J. Appl. Phys. 19 (1980) L225.

[16] F. Stern and W.E. Howard, Phys. Rev. 163 (1967) 816.

[17] F. Stern, CRC Crit. Rev. Solid State Sci. 4 (1974) 499.

[18] T. Ando, A.B. Fowler and F. Stern, Rev. Mod. Phys. 54 (1982) 437.

[19] J. Bardeen and W. Shockley, Phys. Rev. 80 (1950) 72.

[20] F. Fang and A.B. Fowler, Phys. Rev. 169 (1968) 619.

[21] S. Kawaji, J. Phys. Soc. Jpn. 27 (1969) 906.

[22] H. Ezawa, S. Kawaji, T. Kuroda and K. Nakamura, Surf. Sci. 24 (1971) 659.

[23] H. Ezawa, T. Kuroda and K. Nakamura, Surf. Sci. 24 (1971) 654.

[24] H. Ezawa, S. Kawaji and K. Nakamura, Jpn. J. Appl. Phys. 13 (1974) 126; 14 (1975) 921(E).

[25] Proc. of the EP2DS, Eds. J.J. Quinn and P.J. Stiles, Surf. Sci. 58 (1976).

[26] Proc. of the 2nd EP2DS, Eds. G. Dorda and P.J. Stiles, Surf. Sci. 73 (1978).

[27] Proc. of the 3rd EP2DS, Ed. S. Kawaji, Surf. Sci. 98 (1980).

[28] Proc. of the 4th EP2DS, Ed. F. Stern, Surf. Sci. 113 (1982).

[29] Proc. of the 5th EP2DS, Ed. R. Nickolas, Surf. Sci. 142 (1984).

[30] Proc. of the 6th EP2DS, Ed. T. Ando, Surf. Sci. 170 (1986).

[31] Proc. of the 7th EP2DS, Ed. J.M. Worlock, Surf. Sci. 196 (1988).

[32] Proc. of the 8th EP2DS, Ed. J.C. Maan, Surf. Sci. 229 (1990).

[33] Proc. of the 9th EP2DS, Ed. M. Saitoh, Surf. Sci. 263 (1992).

[34] P.W. Anderson, Phys. Rev. 109 (1958) 1492.

[35] N.F. Mott and E.A. Davis, Electronic Processes in Non-Crystalline Materials (Clarendon, Oxford, 1979).

[36] D.C. Licciardello and D.J. Thouless, Surf. Sci. 58 (1976) 89; J. Phys. C 8 (1975) 4158.

[37] S. Pollit, M. Pepper and C.J. Adkins, Surf. Sci. 58 (1976) 79.

[38] E. Abrahams, P.W. Anderson, D.C. Ricciardello and T.V. Ramakrishnan, Phys. Rev. Lett. 42 (1979) 673.

[39] S. Kawaji, M. Namiki and N. Hoshi, J. Phys. Soc. Jpn. 49 (1980) 1637.

[40] N. Hoshi and S. Kawaji, Surf. Sci. 113 (1982) 189.

[41] P.W. Anderson, E. Abrahams and T.V. Ramakrishnan, Phys. Rev. Lett. 43 (1979) 718.

[42] Y. Kawaguchi, Y. Kitahara and S. Kawaji, Surf. Sci. 73 (1978) 520.

[43] D.E. Khmel'nitskii, Physica B 126 (1984) 235.

[44] L.P. Gorkov, A.I. Larkin and D.E. Khmel'nitskii, Pisma Zh. ETF 30 (1979) 248 [JETP Lett. 30 (1979) 228].

[45] D.J. Thouless, Phys. Rev. Lett. 39 (1977) 1167.

[46] S. Hikami, A.I. Larkin and Y. Nagaoka, Progr. Theor. Phys. 63 (1980) 707.

[47] B.L. Altshuler, D.E. Khmel'nitskii, A.I. Larkin and P.A. Lee, Phys. Rev. B 22 (1980) 5142.

[48] Y. Kawaguchi and S. Kawaji, J. Phys. Soc. Jpn. 48 (1980) 699.

[49] A. Kawabata, Solid State Commun. 34 (1980) 432; J. Phys. Soc. Jpn. 49 (1980) 628.

[50] W. Sasaki, J. Phys. Soc. Jpn. 20 (1965) 825.

[51] S. Kawaji, Prog. Theor. Phys. Suppl. No. 84 (1985) 178.

[52] S. Kawaji, Surf. Sci. 170 (1986) 682.

[53] Y. Kawaguchi and S. Kawaji, Jpn. J. Appl. Phys. 21 (1982) L709.

[54] T. Ando and Y. Uemura, J. Phys. Soc. Jpn. 36 (1974) 959.

[55] F.J. Ohkawa and Y. Uemura, Surf. Sci. 58 (1976) 254.

[56] W. Duncan and E.E. Schneider, Phys. Lett. 71 (1963) 23.

[57] T. Ando, Y. Matsumoto, Y. Uemura, M. Kobayashi and K.F. Komatsubara, J. Phys. Soc. Jpn. 32 (1972) 859.

[58] Y. Uemura, Jpn. J. Appl. Phys., Suppl. 2, Part 2 (1974) 17.

[59] A.A. Abrikosov, Lectures on the Theory of Metals (Hindustan Publishing Corporation, Bombay, 1968) p. 42.

[60] S. Kawaji and J. Wakabayashi, Surf. Sci. 58 (1976) 238.

[61] Y. Uemura, Surf. Sci. 58 (1976) 1.

[62] T. Ando and H. Aoki, J. Phys. Soc. Jpn. 54 (1985) 2238.

[63] T. Ando, Prog. Theor. Phys. Suppl. No. 84 (1985) 69.

[64] S. Kawaji, Proc. Int. Symp. Foundation of Quantum Mechanics, Tokyo, 1983 (Physical Society of Japan, Tokyo, 1984) p. 327.

[65] T. Ando, Y. Matsumoto and Y. Uemura, J. Phys. Soc. Jpn. 39 (1975) 279.

[66] T. Igarashi, J. Wakabayashi and S. Kawaji, J. Phys. Soc. Jpn. 38 (1975) 1549.

[67] S. Kawaji, T. Igarashi and J. Wakabayashi, Prog. Theor. Phys. Suppl. No. 57, 9 (1975) 176.

[68] S. Kawaji, Surf. Sci. 73 (1978) 46.

[69] J. Wakabayashi and S. Kawaji, J. Phys. Soc. Jpn. 44 (1978) 1839.

[70] J. Wakabayashi and S. Kawaji, J. Phys. Soc. Jpn. 48 (1980) 333.

[71] J. Wakabayashi and S. Kawaji, Surf. Sci. 98 (1980) 299.

[72] S. Kawaji and J. Wakabayashi, in: Physics in High Magnetic Fields, Eds. S. Chikazumi and N. Miura (Springer, Berlin, 1981) p. 284.

[73] K. von Klitzing, G. Dorda and M. Pepper, Phys. Rev. Lett. 45 (1980) 449.

[74] T.J. Quinn, Metrologia 26 (1989) 69.

[75] B.N. Taylor and T.J. Witt, Metrologia 26 (1989) 47.

[76] G.W. Small, IEEE Trans. Instrum. Meas. IM-36 (1987) 190.

[77] R.S. Van Dyck, Jr., P.B. Schwinberg and H.G. Dehmelt, Phys. Rev. Lett. 59 (1987) 26.

[78] T. Kinoshita, IEEE Trans. Instrum Meas. IM-38 (1989) 172.

[79] H. Aoki and T. Ando, Solid State Commun. 38 (1991) 1079.

[80] R.B. Laughlin, Phys. Rev. B 23 (1991) 5632.

[81] B.I. Halperin, Phys. Rev. B 25 (1982) 2185.

[82] E. Prange and S.M. Girvin, Eds., The Quantum Hall Effect (Springer, Berlin, 1987).

[83] M. Stone, Ed., Quantum Hall Effect (World Scientific, Singapore, 1992).

[84] D.C. Tsui, H.L. Stormer and A.C. Gossard, Phys. Rev. Lett. 48 (1982) 1559.

[85] H.L. Stormer, A. Chang, D.C. Tsui, J.C.M. Hwang, A.C. Gossard and W. Wiegmann, Phys. Rev. Lett. 50 (1983) 1393.

[86] R. Willett, J.P. Eisenstein, H.L. Stormer, D.C. Tsui, A.C. Gossard and J.H. English, Phys. Rev. Lett. 59 (1988) 1776.

[87] R.B. Laughlin, Phys. Rev. Lett. 50 (1983) 1395.

[88] F.D.M. Haldane, Phys. Rev. Lett. 51 (1983) 605.

[89] T. Chacraborty and P. Pietilinen, The Fractional Quantum Hall Effect, Vol. 85 of Springer Series in Solid State Science (Springer, Berlin, 1988).

[90] A.M. Chang, M.A. Paalanen, D.C. Tsui, H.L. Stormer and J.C.M. Hwang, Phys. Rev. B 28 (1983) 6133.

[91] S. Kawaji, J. Wakabayashi, J. Yoshino and H. Sakaki, J. Phys. Soc. Jpn. 53 (1984) 1915.

[92] D. Yoshioka, Surf. Sci. 170 (1986) 125.

[93] J. Wakabayashi, S. Kawaji, J. Yoshino and H. Kawaji, Surf. Sci. 170 (1986) 136.

[94] J. Wakabayashi, S. Kawaji, J. Yoshino and H. Sakaki, J. Phys. Soc. Jpn. 55 (1986) 1319.

[95] J. Wakabayashi, S. Kawaji, J. Yoshino and H. Sakaki, J. Phys. Soc. Jpn. 56 (1987) 3005.

[96] G.S. Boebinger, A.M. Chang, H.L. Stormer and D.C. Tsui, Phys. Rev. Lett. 55 (1985) 1606.

[97] G.S. Boebinger, A.M. Chang, H.L. Stormer, D.C. Tsui, J.C.M. Hwang, A. Cho and G. Weimann, Surf. Sci. 170 (1986) 129.

[98] R.L. Willet, H.L. Stormer, D.C. Tsui, A.C. Gossard and J.H. English, Phys. Rev. B 37 (1988) 8476.

[99] J.K. Jain, Phys. Rev. Lett. 63 (1989) 199.

[100] J.K. Jain, Surf. Sci. 263 (1992) 65.

[101] J.K. Jain, Advan. Phys. 41 (1992) 105.

[102] E. Wigner, Phys. Rev. 46 (1934) 1002.

[103] C.C. Grimes and G. Adams, Phys. Rev. Lett. 42 (1979) 795; Surf. Sci. 98 (1980) 1.

[104] Yu. E. Lozovik and V.I. Yudson, Sov. Phys. JETP Lett. 22 (1975) 11.

[105] H. Fukuyama, Solid State Commun. 19 (1976) 551.

[106] M. Tsukada, J. Phys. Soc. Jpn. 42 (1977) 391.

[107] S. Kawaji and J. Wakabayashi, Solid State Commun. 22 (1977) 87.

[108] H.W. Jiang, R.L. Willett, H.L. Stormer, D.C. Tsui, L.N. Pfeiffer and K.W. West, Phys. Rev. Lett. 65 (1990) 633.

[109] V.J. Goldman, M. Santos, M. Shayegan and J.E. Cunningham, Phys. Rev. Lett. 65 (1990) 2189.

[110] F.I.B. Williams, P.A. Wright, R.G. Clark, E.Y. Andrei, G. Deville, D.C. Glattli, O. Probst, B. Etienne, C. Dorin, C.T. Foxon and J.J. Harris, Phys. Rev. Lett. 66 (1991) 3285.

[111] H.W. Jian, H.L. Stormer, D.C. Tsui, L.N. Pfeiffer and K.W. West, Phys. Rev. B 44 (1991) 8107.

[112] M. D'Iorio, V.M. Pudalov and S.G. Semenchinsky, Phys. Rev. B 46 (1992) 15 992.

[113] S. Kivelson, D.H. Lee and S.C. Zhang, Phys. Rev. B 46 (1992) 2223.

[114] S.C. Zhang, S. Kivelson and D.H. Lee, Phys. Rev. Lett. 69 (1992) 1252.

[115] J. Wakabayashi, A. Fukano, K. Hirakawa, S. Kawaji, H. Sakaki, Y. Koike and T. Fukase, J. Phys. Soc. Jpn. 57 (1988) 3678.

[116] V.J. Goldman, M. Shayegan and D.C. Tsui, Phys. Rev. Lett. 61 (1988) 881.

[117] R.R. Du, H.L. Stormer, D.C. Tsui, L.N. Pfeiffer and K.W. West, Phys. Rev. Lett. 70 (1993) 2944.

[118] Proc. of the 10th EP2DS, Ed. B.D. McCombe, Surf. Sci. (1994), in press.

[119] V.J. Goldman, J.K. Wang, Bo Su and M. Shayegan, Phys. Rev. Lett. 70 (1993) 647.

[120] T. Sajoto, Y.P. Li, L.W. Engel, D.C. Tsui and M. Shayegan, Phys. Rev. Lett. 70 (1993) 2321.

[121] V.M. Pudalov, M. D'Iorio, S.V. Kravchenko and J.W. Cambell, Phys. Rev. Lett. 70 (1993) 1866.

Surface Science 299/300 (1994) 587–611
North-Holland

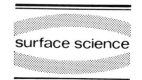

surface science

Helium atom scattering spectroscopy of surface phonons: genesis and achievements

Giorgio Benedek [a] and J. Peter Toennies [b]

[a] *Dipartimento di Fisica dell'Università, Via Celoria 16, I-20133 Milano, Italy*
[b] *Max-Planck-Institut für Strömungsforschung, Bunsenstrasse 10, D-37073 Göttingen, Germany*

Received 28 April 1993; accepted for publication 30 August 1993

New experimental techniques for high-resolution inelastic helium scattering have within the last thirteen years made possible measurements of the dispersion curves of surface phonons in insulators, semiconductors and metals. In all three types of materials a number of anomalies have been observed indicating that the interactions between atoms in the surface region differ significantly from those in the bulk. The transition metals invariably show an unexpected and anomalously soft surface phonon branch below the longitudinal acoustic band while in insulators such as LiF, semiconductors (e.g. Si, GaAs and GaSe) and charge density wave systems (e.g. $TaSe_2$ and TaS_2) other substantial deviations from the predicted ideal surface behavior are observed. These findings have stimulated new theoretical investigations based on microscopic model calculations where the role of surface electrons is explicitly taken into account.

1. Introduction

More than a century ago Lord Rayleigh demonstrated mathematically [1] that the long-wave component of earthquakes is due to sagittal elastic waves travelling along the surface. This discovery is commonly taken as the starting point of surface phonon physics. Very soon afterwards the theory of Rayleigh waves and other types of surface waves in semi-infinite elastic continua branched out beyond the realm of geophysics. Solid state physicists and materials scientists soon became interested because of possible applications of surface acoustic waves for applications in delay lines, optoacoustic systems and signal processing devices [2]. The theoretical research, initiated by Rayleigh's fundamental work was extended into two directions: In the long wavelength limit the theory was applied to different geometries, as well as to anisotropic and piezoelectric materials [3]. In the short wavelength limit the effect of the discrete lattice structure of solids was considered [4]. In 1912 Born and von Kármán [5] provided the first microscopic theory of bulk phonons in a three-dimensional crystal lattice and introduced harmonic force constant models.

Surface vibrational modes were however first calculated microscopically much later in 1948 by Lifshitz and Rosenzweig [6] using the Green's function (GF) method. Their basic idea was to consider the free surface as a perturbation of a lattice with cyclic boundary conditions and thereby derive the surface lattice vibrations from the spectrum of bulk vibrations. In the 1960s this method received various elegant formulations and extensions by Maradudin et al. [4] and with Wallis and Dobrzynski [7], by Garcia-Moliner (GF matching method [8]), Armand (GF generating coefficient method, [9]) and by Allen (continued fraction method [10]). In this period an appreciable number of applications to systems with short-range interactions were reported [11–15]. The GF method has the advantage of reducing the size of the dynamical problem from that of the whole crystal to that of the perturbation space corresponding to the surface and is therefore particularly suitable for materials with short-range forces. In 1973 one of us (GB) devised an extension of the GF method for ionic crystals with long-range Coulomb forces, such as alkali halides [16–18], which was later applied to crystals with complex many-body interactions such as the refractory superconductors [19].

Another theoretical method for treating surface lattice dynamics, conceptually similar to Lord Rayleigh's approach for continuous media is based on trial surface-localized waves with a number of adjustable parameters corresponding to the largest order of interacting neighbors. An extension to long-range forces was first given by Feuchtwang [20] and later applied to metal surfaces by Bortolani et al. [21]. Subsequently Khater and Szeftel and coworkers combined this real-space matching with a group theoretical analysis [22] with successful applications to superstructures, defects and disorder on metal and alloy surfaces [23]. Already in the late 1950s Wallis used this method to predict the existence of optical surface modes with shear vertical (SV) polarization (normal to the surface) in diatomic crystals [24]. More recently, optical surface modes polarized in the surface plane were predicted by Lucas [25] and are now called Lucas modes (LM).

Wallis and Lucas modes, which also exist in non-polar diatomic crystals like silicon, are called *microscopic* because their penetration inside the crystal remains comparable to the lattice distance even at infinite wavelength. In contrast, the penetration depth of other modes like the Rayleigh waves is proportional to the wavelength in the long-wave (continuum) limit and in this case the waves are referred to as *macroscopic*. Macroscopic optical modes may exist in polar crystals as mixed phonon–photon modes (surface polaritons). Surface phonon polaritons were discovered theoretically, in the late sixties, by Fuchs and Kliever (FK modes) [26,27].

With the advent of fast computers the calculation of surface vibrations by the direct diagonalization of the dynamical matrix for a sufficiently thick infinitely extended slab consisting of many infinitely extended two-dimensional layers of interacting atoms has become more expedient than the original GF method. It has the further advantage that it offers a more direct physical picture of the dispersion of surface phonons. This method was developed by de Wette and coworkers in the early 1970s, who then used it to perform the first realistic calculations of surface phonon dispersion curves. Their systematic studies on the low-index surfaces of Lennard-Jones fcc crystals [28–30] and of alkali halides [31–33] represent a basic reference point. With the slab method Alldredge et al. discovered another surface acoustic mode with shear-horizontal (SH) polarization (in the surface plane), orthogonal to the sagittally polarized Rayleigh waves [34].

In the early 1970s, as a result of these theoretical developments, there was considerable understanding on the nature and dispersion of long wavelength surface phonons for ideal surfaces in different classes of crystals, but not a single piece of experimental evidence for surface phonons in the dispersive region. Their elusive nature challenged spectroscopists for about fifteen years from about 1965 to 1980. Neutron scattering used since 1955 as the most efficient tool for the study of bulk phonons [35], is rather insensitive to surfaces. Thus it is not surprising that attempts to measure surface phonons with neutron scattering from powders with large specific area have met with little success [36].

Surface-sensitive probes like low-energy electrons and atoms were considered soon as candidates for surface-phonon spectroscopy. The suggestion to use helium atom scattering (HAS) for measuring surface phonons goes back to two theoretical studies by Cabrera, Celli and Manson (1969) [37] and Manson and Celli (1971) [38]. These authors gave a theoretical demonstration that helium atom scattering from single phonons is the dominant process at thermal collision energies.

This work immediately stimulated several groups to carry out experiments. Fisher and coworkers [39] in 1971 were the first to clearly resolve energy losses in the time-of-flight scattering spectra of He from LiF(001). At about the same time Miller and coworkers at La Jolla also succeeded in resolving structures associated with single surface phonons near the zone origin [40]. Through a careful analysis of out-of-plane angular distributions Williams and Mason at Ottawa were able to determine the long-wavelength part of the Rayleigh wave dispersion in LiF(001) [41] and NaF(001) [42]. Boato and Cantini [43] also devoted much effort to extracting information on surface phonons from the high-resolution angular distributions of helium and neon scattering from LiF(001). This was one of the hot subjects at the 1973 session of the Enrico Fermi International School at Varenna [44]. Benedek then proposed an interpretation of the complex inelastic features seen in Boato and Cantini's data in terms of Van Hove singularities, i.e., kinematical focusing from Rayleigh and Lucas modes at the symmetry points of the surface Brillouin zone [45–47]. Cantini et al. [48] provided an alternative, very attractive explanation of the inelastic features in terms of inelastic resonances involving Rayleigh waves out to the zone boundary.

All these early experiments were, however, hampered by inadequate velocity resolution ($\Delta v/v \geqslant 5\%$) and detector sensitivity. The first successful helium atom scattering direct measurements of surface phonon dispersion curves out to the zone boundary were carried out in Göttingen by Brusdeylins, Doak and Toennies on LiF(001) in 1980 [49–51] using a time of flight (TOF) method. High resolution dispersion curves, which first revealed an important anomalous phonon resonance, were then measured with helium scattering on Ag(111) by Doak, Harten and Toennies in 1983 [52]. In the HAS measurements energy transfers of up to about 30–50 meV with a resolution of 0.5 to 1 meV are achieved similar to the resolution in neutron studies of bulk phonons. These improved experiments were made possible by the introduction of a new mode of operation of helium atom nozzle beam sources which provided a relative velocity resolution of 1%, which is almost a factor 5 to 10 better than in previous experiments [53], and the use of a carefully designed system of differential pumping to increase the signal to noise ratio of the mass spectrometer detector. Neon atom beams have been also used for surface phonon spectroscopy, though with a

lower resolution, by Feuerbacher and Willis who measured, in 1981, the Ni(111) Rayleigh wave dispersion curve out to the zone boundary [54].

The first attempts with low-energy electrons go back even earlier. In 1965 thermal diffuse electron scattering was first attributed to inelastic scattering from Rayleigh modes [55]. Then in 1967 Probst and Piper [56] demonstrated for the first time the ability of electrons to resolve adsorbate vibrations in the energy distribution of the scattered electrons and thereby launched the field of electron energy loss spectroscopy (EELS). The advantages of going over to the impact regime at higher incident electron energies (100–200 eV) for measurements of surface phonon dispersion curves was first pointed out in 1980 by Li, Tong and Mills [57]. The first such measurements were reported in 1983 by Lehwald, Szeftel, Ibach and coworkers [58], who succeeded in measuring a full dispersion curve for one direction in Ni(100) with an electron beam with an energy halfwidth of only 7 meV.

The competition between HAS and EELS has stimulated their technical progress and the knowledge of surface dynamics has greatly benefited from their complementarity, which permits a full coverage of the vibrational spectrum and an insight into the nature of surface interactions. HAS, which can cover the range of energies of phonons of clean surfaces ($\hbar\omega \leqslant 30$–40 meV), is, thanks to its greater resolution, more suitable to the study of ordinary surface phonons and more subtle effects like Kohn anomalies and soft modes. EELS, on the other hand, is an ideal tool for higher frequency phonons like optical and gap modes and adsorbate vibrations. The second complementarity concerns the scattering mechanisms: while helium atoms are scattered by the phonon-induced electron charge oscillations, electrons (like neutrons and X-rays) are scattered by the ion cores. Thus in the wide spectral range, where the same lattice vibrations are accessible to both techniques, the amplitude of the accompanying charge-density oscillations seen by HAS can in principle be directly compared with those of the ion cores. From the direct comparison valuable information on electron–phonon interactions can be expected.

Also with respect to other probes, HAS clearly carries a richer information on the surface dynamical structure which could be fully appreciated through the theoretical analysis. The theoretical tools which we developed in recent years, such as the GF technique and more recently the multipole expansion method for surface dynamics together with a consistent calculation of the HAS amplitudes were essential for the HAS technique to deploy its full potential for surface phonon spectroscopy.

In this account we summarize the theoretical background (Sections 2–4) and the experimental breakthroughs (Sections 5 and 6) which led us to elucidate certain basic aspects of surface dynamics. The last Section offers a brief outlook on the present status and future of HAS studies on surface dynamical processes. The present report updates a previous review article [59] devoted to the study of surface interatomic forces through surface phonon spectroscopy, and complements our extended survey which we have recently completed for Surface Science Reports [60].

2. A few concepts on surface phonons

The existence and nature of surface phonons can be demonstrated by treating the surface as a perturbation of the periodic lattice and solving the resulting problem with the GF or the matching methods (see Section 1). A more common method is to calculate the lattice dynamics for a crystal slab formed by an increasing number N_z of parallel infinitely extended periodic atomic planes. In the latter case, for a monatomic system, one has to solve the following eigenvalue problem [29],

$$M\omega_{Qv}^2 u_\alpha(l_3, Qv) = \sum_{\beta l_3'} R_{\alpha\beta}(l_3 l_3', Q) u_\beta(l_3', Qv),$$

(1)

where M is the atomic mass, Q the two-dimensional phonon wavevector parallel to the atomic planes and $R_{\alpha\beta}(l_3 l_3', Q)$ is the two-dimensional Fourier transform of the force constant matrix. The index $l_3 (= 1 \cdots N_z)$ labels the planes in the

slab (total number $= N_z$) and v labels the $3N_z$ modes. From the solution of Eq. (1) the eigenfrequencies ω_{Qv} and eigenvectors $u_\alpha(l_3, Qv)$ are obtained. The eigenvectors $u_\alpha(l_3, Qv)$ are normalized to unity and determine the polarization of the displacement field.

The dispersion curves (ω_{Qv} versus Q) as obtained by solving Eq. (1) for a monatomic fcc (111) slab of increasing thickness ($N_z = 1$–30) are displayed in Fig. 1 [61]. They were calculated for a nearest neighbor (nn) pairwise interaction characterized by a radial force constant β_1 and a nn tangential force constant $\alpha_1 = 0$. For one single layer we have three modes: one with transverse polarization out of the plane (shear vertical (SV)) and two modes with an in-plane longitudinal (L) polarization and an in-plane transverse (T) polarization. The latter mode is commonly referred to as a shear horizontal (SH) mode.

As one adds more and more layers these modes increase in number. Beyond about 10 layers a definite structure in the dispersion curves becomes apparent with single modes well separated from dense bands. In the limit of $N_z = \infty$ the former are the surface modes, while the bands are attributed to bulk modes projected onto the surface.

If the two surfaces of the slab are identical the localized frequencies occur as pairs of nearly degenerate branches. The spectral localization of the eigenfrequencies is related to the spatial localization of the eigenvectors in the surface region: the larger the separation of the new frequencies away from the bulk band frequencies, the greater the localization of the corresponding displacement field at the surface of the crystal. When the penetration depth is much less than the slab thickness the mode pair is practically degenerate. However, the degeneracy only becomes exact for all surface localized modes for $N_z \to \infty$; in this case we may speak of single surface modes for a semi-infinite lattice.

Generally speaking one can assign each surface branch to the bulk band from which it originates, and also its polarization reflects approximately that of the parent band. Along the symmetry directions of the surface Brillouin zone the sagittal plane, defined by the surface phonon wavevector direction and the normal to the surface, is often a mirror-symmetry plane as is the case for the fcc (001) surface along the $\langle 100 \rangle$ and $\langle 110 \rangle$ directions and for the fcc (111) surface, along the $\langle 11\bar{2} \rangle$, but *not* along the $\langle 1\bar{1}0 \rangle$ direction. For a mirror-symmetry plane the displacement vectors of the surface modes have either even or odd symmetry with respect to the sagittal plane. We speak in this case of sagittal ($\perp$) and shear-horizontal polarization, respectively.

Since the bulk bands partially overlap each other, a surface mode originating from a given band may overlap with another band and as a result of hybridization with the bulk band the surface mode may not be localized. In this case we speak of *surface resonances*.

In di- or polyatomic crystals the optical phonon bands generate additional surface phonon branches of optical character. As a textbook example we display in Fig. 2 the full set of phonon

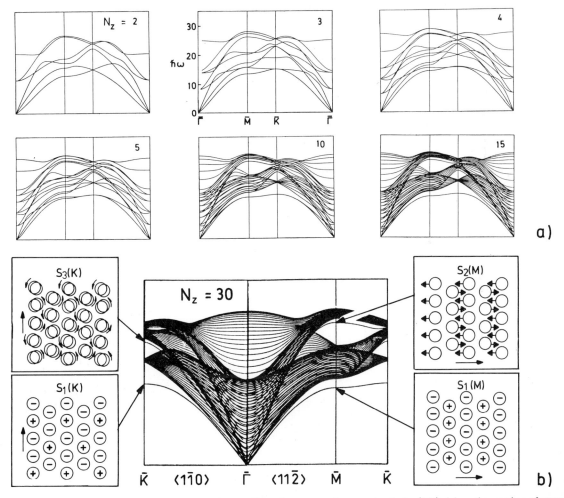

Fig. 1. Part (a) shows the evolution of the surface phonon dispersion curves of a monatomic fcc (111) slab as the number of atomic is increased from 1 to 15 [61]. Part (b) shows the dispersion curves for $N_z = 30$ where the behaviour for an infinitely thick slab is approached and the distinction between the isolated surface modes and the closely spaced bulk bands becomes apparent. For this case the displacement patterns of the surface atoms (top view) for the surface modes at the zone-boundary symmetry points are shown in the insets to the left and right. Small arrows indicate displacements parallel to the surface which can be either linearly ($S_2(\bar{M})$) or circularly ($S_3(\bar{K})$) polarized; $\pm$ indicate vertical displacements. The long arrow in each set denotes the wavevector direction.

dispersion curves for the surface of a diatomic crystal, NaF(001), as calculated by the Green's function method [62,63]. Along the symmetry di-

rections $\overline{\Gamma M}$ ($\langle 100 \rangle$) and $\overline{\Gamma X}$ ($\langle 110 \rangle$) the modes are either sagittal or shear horizontal and are plotted separately in Figs. 2a, 2b and 2d, 2e, re-

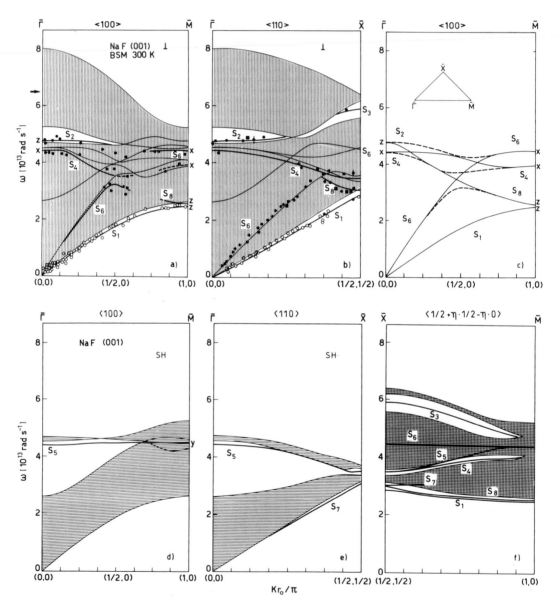

Fig. 2. Parts (a) and (b) show the calculated and HAS experimental surface phonon dispersion curves of NaF(001) along the two symmetry directions $\overline{\Gamma M}$ ($\langle 100 \rangle$) and $\overline{\Gamma M}$ ($\langle 110 \rangle$) for sagittal polarization ($\perp$). The open circles show HAS measurements from ref. [51], and the black dots are measurements from Ref. [64]. The heavy lines are surface modes and the shaded areas correspond to bulk bands projected onto the surface. Part (c) illustrates the complex hybridization scheme of the sagittal plane resonances along $\overline{\Gamma M}$. Parts (d) and (e) show the calculated dispersion curves for the shear horizontal (SH) polarized modes. Part (f) shows the mode frequencies along the zone-edge $\overline{XM}$. Calculations were made with the GF method and using the breathing shell model [62].

spectively, whereas along the zone boundary $\overline{MX}$ the sagittal and the SH component are mixed together and are shown superimposed in Fig. 2f. Heavy lines are surface modes; thin lines are the edges of the bulk bands (shadowed areas). When a dispersion curve enters a band the surface localized mode transforms into a resonance. Altogether six different bulk bands for the acoustical and optical modes and the three polarizations are present. From each of the six bulk bands we actually find at least one surface mode, either localized or resonant. Most of the sagittal surface modes have been observed experimentally by HAS (open circles for the Rayleigh waves, black circles for the other surface modes) [64].

According to a standard labelling [63,60] the sagittal mode S_1 corresponds to a Rayleigh wave in the continuum limit ($Q \rightarrow 0$). The SH mode S_7 (Fig. 2e) is an acoustic surface mode with parallel (in-plane) polarization occurring along the $\overline{\Gamma X}$ directions in cubic crystals. The quasi-longitudinal acoustic mode S_6 peeled off from the LA lower edge is a resonance due to the mixing with the TA bulk band modes. The sagittal resonance S_8, crossing the LA bulk band in both directions, may appear in some crystals as a folded prolongation of the Rayleigh wave. Its intensity is appreciable in crystals with nearly equal ionic masses (NaF, KCl, RbBr) [65]. Other mechanisms connected with the hybridization of the bulk phonon branches have been shown to give surface crossing modes in crystals with unequal masses such as NaCl and KBr [66]. S_4 and S_5 form the pair of Lucas modes (LM) [25] and are degenerate at the zone center ($\overline{\Gamma}$ point). For $Q \rightarrow 0$ along any direction, S_4 and S_5 become linearly polarized and exactly longitudinal or transverse in the surface plane, respectively. The optical mode S_2 is quasi-SV everywhere and exactly z-polarized (SV, perpendicular to the surface) at $\overline{\Gamma}$ and $\overline{M}$ [24]. At the zone boundary, S_2 turns out to be associated with the bulk TO_1 mode, whereas another mode, S_3, comes from the edge of the bulk LO mode. Surprisingly in RbI(001) the S_3 mode appears *above* the LO band [67] which can be attributed to surface elastic relaxation [66]. Elastic relaxation may induce additional modes involving second or deeper layers below the surface. Besides

the SV optical microscopic modes there is also the macroscopic FK mode, whose expected position is shown by an arrow in Fig. 2a.

3. Fundamental interactions from surface phonon dispersion curves

The theoretical analysis of measured surface phonon dispersion curves provides both information on the interatomic forces between atoms in the surface region as well as a stringent test of current dynamical models and of approximations made in ab-initio calculations of interatomic forces. The surface dynamics is more sensitive to many features of models and approximations than bulk dynamics because interactions which are suppressed in the bulk owing to the high symmetry become activated by the lower symmetry at the surface.

Moreover surface dynamics studies are useful for structural assessment due to an important link to the equilibrium structure which provides severe constraints on the force constants [4,17,68]. For example, one of the force constants appearing in the force constant matrix Eq. (1), called the surface tangential force constant, can be shown to be related to surface relaxation [69] and to the macroscopic surface stress [70]. In fact surface stress provides another way of describing the changes in the tangential force constants within the surface layer.

In covalent semiconductors the equal share of the bond charge between the two newly created surfaces of the cleaved crystal leaves half-filled bands consisting of dangling bonds at the surfaces. This situation is unstable against a symmetry-breaking reconstruction provided the surface temperature is much smaller than the splitting between bonding and antibonding surface states induced by reconstruction. As a result tetrahedrally coordinated semiconductors such as Si and Ge exhibit a variety of reconstructed surface phases. Here the analysis of the zone boundary surface phonons enabled us to discriminate between subtle structural differences. For example the buckling angles of π-bonded chains in Si(111)2 × 1 [71] or GaAs(110) [72] could be determined in this way. The structural effects on surface vibrations for the III–V family have been extensively investigated by Wang and Duke [73].

In metals the partitioning of the free electrons between the two newly created surfaces also generally results in an unstable situation leading to a rearrangement of electrons in real space and on the Fermi surface. Again the appearance of surface bands crossing the Fermi level may lead to a large electron susceptibility. At the corresponding Q-vectors there may be a strong electron–phonon coupling which results in kinks or sharp depressions in the surface phonon dispersion curves which are generally referred to as Kohn anomalies [74]. In extreme cases the phonon frequencies can go to zero resulting in structural instabilities leading to a reconstruction. This intimate connection between structure and electronic distortion is not surprising since in metals the interactions between ion cores is mostly mediated by the interposed electrons. Even for unreconstructed surfaces the actual changes in surface force constants required to explain experimental surface phonon dispersion curve frequencies are so significant that they can only be described in terms of many-body effects due to the redistribution of the electronic charge. Finnis and Heine [75] first called attention to the connection between the electronic charge redistribution at the surface and the inward relaxation of the outermost layer. The change of surface radial force constants is then directly related to the relaxation.

Along this line extensive analysis of metal surface interactions, including the effect of changing force constants at the surface on the frequencies of surface modes, has been carried out by many authors [76,77], and particularly by the Modena group [78–80], on the basis of Born–Von Kármán models in which several two-body and angle-bending force constants are used as fitting parameters. However attempts to simultaneously fit its frequency and the corresponding inelastic HAS peak intensities in many cases gave rather unphysical surface force constant changes. This ultimately led to a new dynamical scheme based on the multipole expansion of the electron density in order to account for the electronic degrees of freedom [81].

4. The coupling of He atoms to surface vibrations of metals via electrons

While in closed-shell systems such as ionic and van der Waals solids the main terms are the direct two-body interatomic force constants, in metals and covalent semiconductors the dominant interaction is mediated by the interposed electrons and takes on a many-body character. Valence electrons which (almost) rigidly follow the atomic motion, as in closed-shell systems, contribute to a strong repulsion between neighboring atoms. However repulsive force constants get softened if a deformation of the electron clouds accompanies the atomic displacements. Thus force constants in metals and semiconductors crucially depend on the electronic susceptibility (polarizability) with respect to the atomic motions. The knowledge of the electron density oscillations induced by the phonon displacement of a given wavevector and frequency is of primary importance in condensed matter physics [82] and is the key point in first-principle treatments of surface lattice dynamics based on the density response method [83,84].

HAS appeared to be an ideal probe of these electron density oscillations at the surface. Unlike other probes such as electrons in the impact regime or neutrons which interact with the ion cores and nuclei, respectively, helium atoms are inelastically scattered by the electron density at the Fermi level as modulated by the atomic motion of surface phonons. The observation by means of HAS of the anomalous longitudinal surface resonance first in noble metals [52], then in transition metals [85], brought attention to this phenomenon. This led us to suggest [81,86,87] that the large intensity of the inelastic HAS time-of-flight peak associated with this mode provides direct evidence for a large modulation of the surface electron density by the longitudinal motion of surface atoms. The softening of the longitudinal surface modes and the large HAS inelastic intensities are both accounted for quantitatively by the pseudocharge model [86,87].

In the multipole expansion method the electron density $n(r)$ is partitioned into cellular components $n_{lj}(r)$ (*pseudocharges*) where l labels the lattice unit cells (of volume v_c) and j a set of suitably chosen special points r_{lj} within each unit cell. The dynamical oscillation $\delta n_{lj}(r)$ of each component is then expanded in a multipole series about its special point

$$\delta n_{lj}(r) = \sum_\Gamma c_\Gamma(lj) Y_\Gamma(r - r_{lj}),\qquad (2)$$

where each basis function Y_Γ is the product of an appropriate radial function and an angular harmonic function which transforms according to the Γth irreducible representation of the point group at r_{lj}.

The coefficients $c_\Gamma(lj)$ are the coordinates of the electronic motion and act in the lattice dynamics as time-dependent variables dynamically coupled to the ionic displacements $u(l)$. In the adiabatic and harmonic approximations such a coupling is given, in matrix notation, by the linear equation

$$c = -H^{-1}T^+ u,\qquad (3)$$

where the force constant matrices T and H describe the multipole–ion and multipole–multipole interactions. They are defined as integrals over the functional derivatives of the total energy E with respect to $n(r)$

$$T_\alpha(\Gamma; l'lj) = \frac{1}{v_c} \int d^3 r\, \frac{\delta}{\delta n(r)} \frac{\partial E}{\partial u_\alpha(l')}$$
$$\times Y_\Gamma(r - r_{lj}),\qquad (4)$$

$$H(\Gamma\Gamma'; ljl'j') = \frac{1}{v_c^2} \iint d^3 r\, d^3 r'\, \frac{\delta^2 E}{\delta n(r)\delta n(r')}$$
$$\times Y_\Gamma(r - r_{lj}) Y_{\Gamma'}^*(r' - r_{l'j'}).$$
$$(5)$$

The Fourier-transformed ion displacement vectors $u(l_3, Q\nu)$ obey the eigenvalue equation (1) with the ion–ion force constant matrix R replaced by the effective force constant matrix

$$R^* \equiv R - TH^{-1}T^+,\qquad (6)$$

The additional term $-TH^{-1}T^+$ accounts for the response of the electrons to the ionic motion and is responsible for all effective long range ion–ion interactions, although the multipole interaction matrices T and H have short range character.

The integrals in Eqs. (3) and (4) containing the functional derivatives of the total energy with respect to $n(r)$ can in principle be obtained from ab-initio density functional calculations. Since such calculations have not been performed so far we have treated the integrals as fitting parameters in a phenomenological approach. The success of the pseudocharge model suggests that with an appropriate choice of the special points an accurate mapping of the electron density oscillations can be achieved with a comparatively small number of independent multipole components.

At the same time the multipole expansion method also provides us with a convenient scheme for accounting for the inelastic coupling of the He atoms to the pseudocharge oscillations. The He atom–surface potential $v(r)$ and its modulation $\delta v(r)$ induced by phonon displacements must take account of the strong repulsive exchange interaction of He atoms with the comparatively weak electron density of about 10^{-4} e$^-$/a.u.^{3} which prevails at distances of about 3–4 Å above the top plane of atoms [88,89]. Thus the helium atoms are in fact inelastically scattered by the time-dependent oscillations of the surface electron density in response to the vibrations of the underlying nuclei. The electron density $n(r)$, and hence the scattering potential, is of course a complicated function of nuclear positions. To date most treatments [90] have approximated the scattering potential by a sum of effective two-body potentials between the He atom and each surface atom.

The pseudocharge model provides a many-body inelastic interaction potential between the He atom and the surface by assuming a direct proportionality of the potential to the modulation of the electron density

$$\delta v(r) = A\, \delta n(r) = A \sum_{lj\Gamma} c_\Gamma(lj) Y_\Gamma(r - r_{lj}),\qquad (7)$$

where A is a proportionality factor, e.g. the Esbjerg and Nørskov constant [89]. The inelastic potential $\delta v(r)$ can be made proportional to the ionic displacements through Eq. (2):

$$\delta v(r) = -\sum_{l'\alpha} \left[A \sum_{\Gamma lj} Y_\Gamma(r - r_{lj}) \right.$$
$$\left. \times (\Gamma lj | H^{-1}T^+ | l'\alpha) \right] u_\alpha(l').\qquad (8)$$

The term in square brackets defines the effective coupling force between the He atom and the l'th surface ion. Thus the effective forces are mediated by the electronic response function H^{-1}, which is readily identified with the electron susceptibility in the multipole representation, and by the electron–phonon interaction matrix T^{+}. This approach thus indicates that inelastic HAS phonon peak intensities contain important information on the Q-dependent surface electron susceptibility. In the future it should be possible to extract this information from a careful analysis of the experimental spectra.

We have recently used the phenomenological parameterized version of the pseudocharge model to re-calculate the inelastic He scattering coupling matrices for copper surfaces [86,87,91]. In addition to the He interaction with the localized electron charge, due largely to d-electrons rigidly bound to the ion cores which gives de facto a direct He-atom–surface ion interaction, we included the He interaction with 13 equivalent pseudocharges per unit cell located at the midpoints between nearest neighbor atoms. The latter interaction receives contributions from the dipolar and quadrupolar deformations of the pseudocharges. The complex interplay of the three contributions to the scattering amplitude, i.e., the ion (rigid d-electrons), the dipolar and quadrupolar (deformable sp–d electrons) contributions, nicely accounts for the observed anomalous intensities observed in copper surfaces (see Section 6). This new formalism fully replacing the previous concepts of spring constants at metal surfaces and two-body collisions of He atoms with surface atoms. At the same time we hope that it will foster a consistent ab-initio treatment of surface dynamics and HAS intensities.

5. Experimental methods

At the present time both HAS and EELS are being used to measure surface phonon dispersion curves and therefore a brief comparison between the two techniques is in order. The kinematics of He atoms and electrons used in these experiments are in principle very similar [92]. For phonon studies electrons are accelerated to energies of 100–300 eV at which the electron wavevector is about equal to that of a reciprocal lattice vector. This mode of operation is called impact scattering in contrast to the more usual dipole scattering with energies of less than about 10 eV [93]. By virtue of their factor 10^4 greater mass, He atoms have the same wave vectors as electrons at energies which are 10^{-4} smaller or about 20 meV. It is a fortunate circumstance that at energies of 200–300 eV the electron penetration is only two or three layers. At the much lower energies of He atom scattering there is of course no penetration at all. Thus He atom scattering is one of the few surface techniques and the only diffractive probe which is sensitive only to the outermost layer and is absolutely non-destructive.

A schematic diagram of an apparatus used in Göttingen for He atom inelastic surface scattering [92,94] is given in Fig. 3. In the He atom machine a nearly monoenergetic beam of atoms is formed by expanding the gas from a high pressure (~ 100 atm) through a small orifice (~ 10 μm) into a vacuum. In the expansion the gas acquires a sharply peaked velocity distribution with a most probable velocity about 30% greater than in the source chamber [53]. The most probable velocity can be varied anywhere from 2000 m/s (~ 60 meV, $k_i = 11$ Å, $\lambda = 0.5$ Å) down to ~ 700 m/s (~ 8 meV, $k_i = 4$ Å^{-1}, $\lambda = 1.5$ Å) by simply changing the temperature of the source from 300 to 40 K, respectively. For this range of incident beam energies the energy half width lies between 1 and 0.2 meV. Lower beam energies down to about 1 meV have been reported [95] but because of cluster formation the relative energy width increases and the intensities are lowered. The energy distribution of the scattered atoms is usually measured by the time-of-flight technique. For this purpose the incident beam is mechanically chopped by a rotating disk chopper and the time of arrival of the scattered atoms at the mass spectrometer detector is analyzed by a multichannel scaler. Extensive differential pumping (as shown in Fig. 3) is required to prevent helium gas from diffusing into the detector chamber, and in practice it has been possible to keep the diffuse

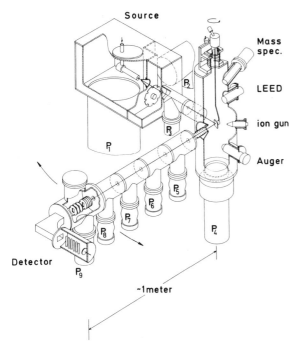

Source

Mass spec.

LEED

ion gun

Auger

P_1

P_2

P_3

P_4

P_5

P_6

P_7

P_8

P_9

Detector

~1meter

Fig. 3. Schematic view of the helium scattering apparatus used in Göttingen [92]. P_1-P_9 are differential pumping stages. The monoenergetic helium beam is emitted from the nozzle and after scattering from the sample a part of the intensity enters the detector arm. Rotation of the sample and/or the detector arm allows the angular distribution of scattered He atoms to be mapped out. For time-of-flight studies the beam is pulsed by passing it through a rapidly rotating chopper disc.

helium background partial pressure down to $<$ 10^{-15} Torr (< 10 helium atoms/cm^3). This background places a lower limit on the smallest detectable beam of about 5 counts/s. The straight-through incident beam intensity is estimated to be about 10^8-10^9 counts/s illustrating the large range in sensitivity.

For standard electron energy loss spectrometers used in phonon studies [93,96,97] the energy resolution typically is about 5–7 meV at $E_i = 100$ eV. The best resolution achieved recently in phonon measurements is of the order of 2–3 meV [98]. The range of angles which have been probed are typically $\theta_f = 55-65°$ and θ_i between $135 - \theta_f$ and $90 - \theta_f$. The count rates in electron scattering are much the same for zone boundary

phonons as in the helium atom experiments [92]. The available electron currents, which are limited by space charge, are smaller than the He atom flux by about a factor 10^5-10^6. This is compensated for, however, by the 100% detection efficiency for electrons compared to the $10^{-5}-10^{-6}$ efficiency of the mass spectrometer used to detect He atoms. Thus compared to EELS, HAS has a much greater energy resolution. The advantage of EELS compared to HAS is that EELS can access modes up to several 100 meV, which is of great interest for characterizing adsorbates. HAS is able to detect phonons up to about 30 meV which covers the maximum frequencies of the surface phonons of most clean solid surfaces.

Considerable care must be exercised in evaluating the phonon energy transfer spectra from both types of experiments. In some cases it is possible to determine the phonon dispersion curves with the aid of the equations for the conservation of energy and parallel momentum directly from the experimental spectra [92]. This procedure can, however, lead to erroneous results because of unresolved nearby lying peaks, the multiphonon background and apparatus smearing effects, etc. For this reason it is customary to carry out direct comparisons of the measurements with either calculated densities of states or preferably with a complete simulation of the measured spectra. The procedures are basically the same in the case of He atom scattering and electron scattering, despite the fact that the theories used for calculating the inelastic intensities are quite different.

These procedures are schematically illustrated in the case of experimental He atom time-of-flight spectra by the flow chart in Fig. 4. Starting from the top, a set of coupling constants which provides a best fit of the bulk dispersion curves and possibly the bulk elastic constants is used in a first trial calculation. From the solution of the dynamical matrix the surface phonon dispersion curves and polarization vectors for this set of force constants are determined. Then the distorted wave Born approximation is used to obtain the inelastic reflection coefficient for one phonon events with an effective interaction potential [90]. As discussed in the previous section to explain

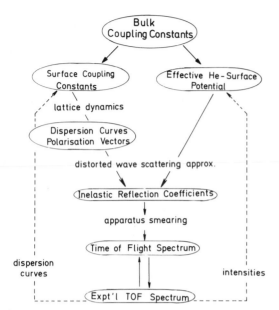

Fig. 4. Schematic diagram showing the theoretical steps and computational pathways in the best fit simulation of experimental He atom time-of-flight spectra.

the anomalous longitudinal mode observed on transition metals it is necessary to include the coupling to the electronic pseudocharges at the surface. In the final step in Fig. 4 the inelastic reflection coefficients are averaged over the apparatus resolution. These calculations are then repeated with modified coupling constants for the surface layers and for modified potential parameters until a best fit of the time-of-flight distributions is achieved.

6. Experimental results

In this section we discuss some exemplary high-resolution HAS experiments on phonons on single crystal surfaces based on the time-of-flight (TOF) technique, with special consideration to the questions solved and to the new theoretical issues raised. For a more detailed and comprehensive description of surface phonon dispersion curve measurements the reader is referred to the reviews in Refs. [59,60] (all systems), Ref. [92] (metals), and Ref. [99] (layered compounds).

6.1. Alkali halides: the assessment phase

The first measurement of the Rayleigh wave dispersion curve in LiF(001) by Brusdeylins, Doak and Toennies [49–51], stimulated new interest in the theory of surface phonons. Previous calculations for alkali halide surfaces, based on either the shell model or the breathing shell model (BSM) in the framework of the GF method [16–18] or the slab method [31–33], could now be tested and were found to be in overall good agreement with the HAS data [100,101]. Fig. 2 shows a comparison between experiment and theory for NaF, which is similar to this important prototype system. The experimental observation of a 10% lower Rayleigh mode frequency near the $\overline{M}$-point zone boundary in LiF led to various theoretical improvements including elastic relaxation, changes in the surface electronic structure and many-body effects [100,62,66]. Unlike neutron scattering, where the wave vector is conserved in three dimensions and only a finite number of discrete bulk phonons are probed, in helium atom scattering only the parallel wave vector is conserved and surface-projected bulk-phonon bands are observed in addition to surface phonon peaks. In order to extract the rich information contained in helium TOF spectra and to convert energy loss spectra into phonon densities we undertook the calculation of the inelastic reflection coefficient [100]. As shown in Fig. 5 Green's function surface phonon densities combined with the reflection coefficient for a hard potential predict TOF spectra in quite good agreement with the experimental spectra. Besides the sharp peaks corresponding to Rayleigh waves, the TOF spectra show continuous regions attributed to bulk phonons, mostly at small parallel wavevectors.

The good agreement between theory and experiment made it possible to assign the observed modes. Moreover it provided a confirmation of the assumptions on which the theory is based and indicated that under these experimental conditions multiphonon scattering processes play only a minor role. Important deviations, of which an example is shown in Fig. 5, consisting in a strong enhancement of some phonon peaks for certain kinematic conditions, could then be attributed to

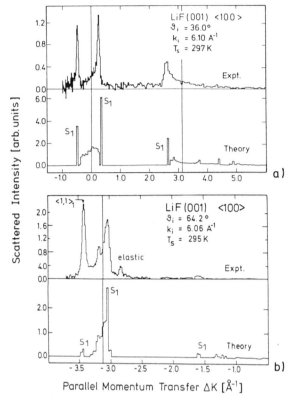

Fig. 5. Comparison of experimental and calculated inelastic HAS spectra projected onto the parallel momentum transfer ΔK axis for LiF(001) along the $\langle 100 \rangle$ direction for two different experiments indicated by the incident angles θ_i, the incident wavevectors k_i and surface temperatures T_s. The theoretical Rayleigh wave peaks are labelled by S_1. In (b) the large discrepancy at $\Delta K = -3.45$ Å^{-1} is due to a $\langle 1, 1 \rangle_1$ selective adsorption resonance in the entrance channel which strongly enhances the Rayleigh peak [100].

an increase in inelastic intensity due to bound state resonances in the initial state [102–105]. The weaker optical modes were often identified by exploiting the same enhancement phenomenon, for example in NaF(001) [64] and LiF(001) [101].

6.2. Semiconductors: surface phonons and structure

The first helium atom TOF investigations of the semiconductor surfaces such as Si(111)2 × 1 [106] and GaAs(110) [107,72] by Harten et al. in

1986 led to the discovery of a flat nearly dispersionless surface phonon branch with frequencies of about 10 meV. This branch roughly corresponds to a folded Rayleigh branch which is due to the doubling of the surface unit cell with respect to the bulk. Despite this seemingly simple origin its quantitative explanation has challenged theoreticians no less than the interpretation of the optical mode at 55 meV discovered earlier by EELS in Si(111) [108]. The two-fold problem of the silicon surface has been tackled with different techniques such as semi-empirical tight-binding [109,110], bond-charge model [71] and the powerful Car–Parrinello quantum molecular dynamics [111].

Actually Si(111)2 × 1 exhibited a surface phonon spectrum deviating significantly from the predictions for the ideal surface [14,112]. In view of the alternating sequence of five- and seven-atom rings occurring in the π-bonded 2 × 1 reconstructed surface [113] we expected stiffer modes associated with the more compact five-atom rings, and softer modes with the expanded seven-fold rings.

This was confirmed by a bond charge model (BCM) calculation for a 22 layer slab [71] which explained both the 10 meV branch and the optical mode at 55 meV. The almost isotropic dipolar activity of the latter was found to arise from the *vertical* motion of atoms in the interface region. The BCM analysis also made it possible to acquire structural information on the surface. The small surface phonon gap at the zone boundary along the $\overline{\Gamma S}$ direction where folding occurs was found to be a sensitive measure of the surface π-bonded chain buckling. The measurements of Harten et al. [106] gave a buckling amplitude of 0.20 Å in agreement with density functional calculations [114], but significantly smaller than the value of 0.30 Å previously deduced from fits to LEED data [115]. In GaAs(110) the calculated surface phonon frequencies were also found to depend strongly on the geometrical arrangement of surface atoms [72]. This situation allowed a determination of some surface crystallographic parameters from an accurate fit of the HAS dispersion curves [107,116] in good agreement with previous LEED data [117].

6.3. Metals: insight into electron–phonon interactions at surfaces

The first helium time-of-flight (TOF) measurements for a metal surface were carried out for Ag(111) [52]. This surface was chosen because of the relative inertness of silver and the negligible relaxation ($< 1\%$) of the close packed metal surfaces so that a nearly ideal behavior was anticipated [118]. The experiments surprisingly re-

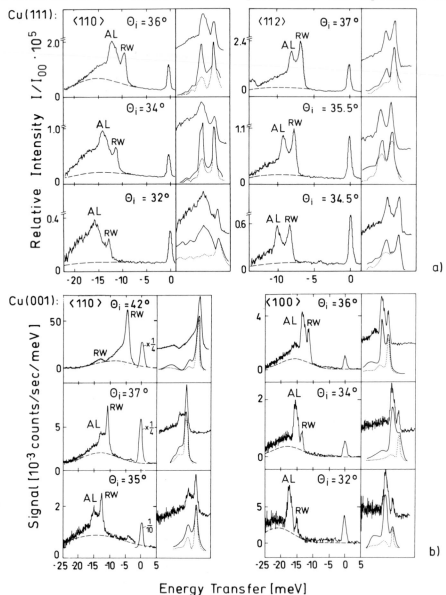

Fig. 6. Some representative HAS energy-loss spectra for Cu(111) (a) [87] and Cu(001) (b) [91] for different incidence angles θ_i along the two symmetry directions. AL: longitudinal resonance; RW: Rayleigh wave. The complete energy-loss spectra are shown on the left together with the multiphonon background (dashed line). On the right the inelastic part of the spectra showing the sharp one-phonon peaks, after subtracting off the multiphonon background, is compared to the theoretical inelastic intensity calculated with the pseudocharge model. The solid line has been calculated for scattering from both ion cores and pseudocharges whereas the broken line has been calculated for scattering from the ion cores only.

vealed two sharp inelastic peaks of comparable intensity of which the one with lower energy transfer agreed nicely with the predicted Rayleigh mode. The other mode lying at about half way between the Rayleigh mode and the longitudinally polarized bulk band edge was the anomalous longitudinal (AL) resonance we mentioned in Section 4. The subsequent observation of this mode on Cu(111) [119] and an even softer mode on Au(111) [119–121] led to considerable theoretical activity in search of a simple explanation. Careful Born–von Kármán fits of the intensities in the experimental TOF spectra carried out by the group of Bortolani and Santoro, starting in 1984, gave first layer intraplanar force constants $\beta_{\parallel}$ which were reduced by 69% for Cu(111) [122,120], 52% for Ag(111) [123,120] and 70% for Au(111) [120,121] as compared to the bulk value.

The interpretation of the HAS data in terms of extreme force constant changes was questioned in 1988 by Hall et al. [124] since they were able to explain the EELS data for the Cu(111) anomalous longitudinal mode in the $\overline{\Gamma M}$ direction by only a moderate 15% reduction of $\beta_{\parallel}$. This apparent contradiction was called the Bortolani–Mills paradox [125]. More recent HAS experiments on Cu(001) [91] and Ag(001) [126] reveal an AL mode which is five times more intense than the Rayleigh mode in the $\overline{\Gamma M}$ direction. The intensity of the longitudinal resonance is weak along $\langle 110 \rangle$ but much larger than the RW peak in the $\langle 100 \rangle$ direction [91] (Fig. 6). Whereas the anisotropy is more moderate in the (111) surface (Fig. 6). Surprisingly, even though EELS has been able to detect the AL mode on Cu(111), at least for the $\overline{\Gamma M}$ direction, there is no reported EELS evidence for the very strong AL mode on Cu(001) and Ag(001). Since no set of surface force constants no matter how strongly modified could be found to explain these results, these differences appear to be related to the more complicated multiple scattering interaction of electrons. Electrons we recall also probe the second and third layer vibrations, whereas helium atom are only sensitive to charge deformations produced by vibrations at the surface.

Calculations based on dynamical models in which electron many-body effects were included

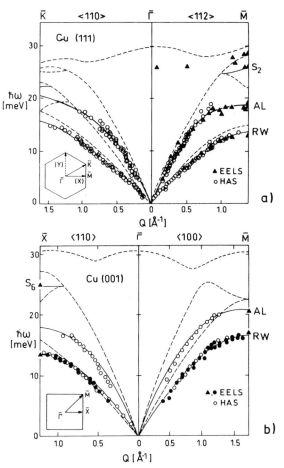

Fig. 7. Comparison of surface phonon dispersion curves of Cu(111) [87] (a) and Cu(001) [91] (b) calculated by the pseudocharge multipole model along each of the two symmetry directions (solid lines) with the experimental He scattering data (circles) and EELS data (triangles). Broken lines represent the edges of the bulk bands.

either in the form of multipoles [81,86,87], such as shown in Fig. 7, or in the embedded-atom scheme [127], were able to approximately explain the new mode without invoking anomalously large changes in the surface force constants. These and other theories indicated that the AL surface resonance appears to be a direct consequence of surface charge redistribution. This point raised an interesting discussion at a recent Solvay Symposium after the review presented by Toennies [59]: the calculations of Euceda et al. for Cu(111) surface electronic structure [128] and Kleinman's remarks pinpointed the larger sd–p hybridization as the leading mechanism. Thus it is the large response of the conduction electrons to the atomic motion, rather than a large oscillation of the surface atoms themselves which causes the AL

resonance. The pseudocharge model accounts very well for the large AL-mode intensities and their anisotropy (Fig. 6, right columns). According to this model both the frequency (Fig. 7) and the amplitude of the AL resonance can be simul-

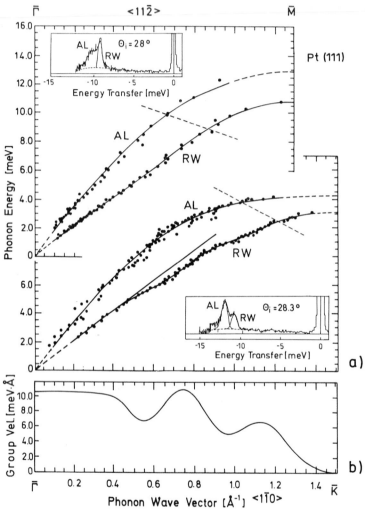

Fig. 8. (a) Helium atom scattering measurements of surface-phonon dispersion curves for Pt(111) [85,129]. The upper two curves show the longitudinal resonance (AL) and Rayleigh (RW) mode for the $\overline{\Gamma}\overline{M}$ ($\langle 112 \rangle$) direction and the bottom two curves for the $\overline{\Gamma}\overline{K}$ ($\langle 110 \rangle$) direction. The beam energies used in both experiments varied between 10 and 32 meV. The target temperature in the $\langle 112 \rangle$ measurements was 400 K, and in the $\langle 110 \rangle$ measurements 160 K. The curves show the best-fit with a nine term Fourier expansion. The solid line in the lowest set of RW data corresponds to a group velocity of 10.5 meV Å which is slightly less than the bulk transverse velocity for the equivalent direction and polarization. The insets show two HAS time-of-flight spectra (converted into an energy-loss scale) for the two scan curves represented by the broken lines. (b) The Rayleigh-mode group velocity as a function of phonon wave vector from the best-fit Fourier expansion of the measured Rayleigh (RW) dispersion curve along the $\overline{\Gamma}\overline{K}$ direction shown directly above in (a).

taneously explained for both surfaces by an enhanced dipolar and quadrupolar susceptibility of surface electrons [86,87] and a coupling of the He atoms to the corresponding deformations of the

pseudocharges. This explanation is conceptually consistent with an enhanced sd–p hybridization.

When large electron oscillations involving electron–hole pair excitations are in resonance with

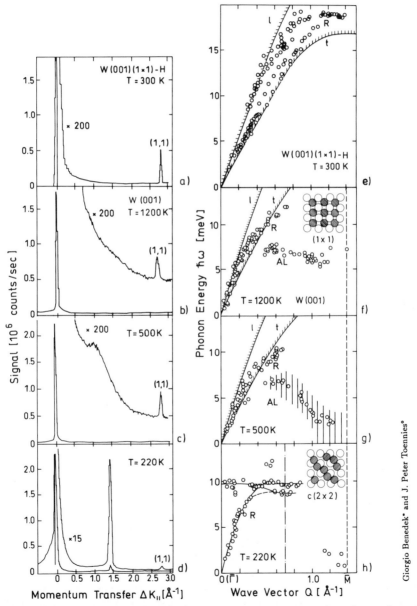

Fig. 9. The changes in the HAS angular distributions (left) and in the corresponding surface phonon dispersion curves (right) at three different temperatures for the W(001) surface (b–d, f–h) [138] and for a (1 × 1) H atom covered surface at T = 300 K (a, e) [141].

the atomic motion, Kohn anomalies may occur in the dispersion curves of surface phonons. In particular we addressed the question whether Kohn anomalies may affect the dispersion curve of Rayleigh waves, possibly also in crystals which do not show anomalies in the bulk. The search was naturally addressed to transition metals with unfilled d-bonds at the surface.

In the transition metals discussed so far, besides the surprising observation of an anomalous longitudinal resonant branch, no Kohn anomaly was found in the measured dispersion curves, consistent with the regularity of the bulk dispersion curves. Thus it came somewhat as a surprise when the first clear evidence for Kohn-type anomalies was found in the RW branches of Pt(111), which was extensively studied by He atom scattering by three groups [85,129–131] at about the same time. The experimental results (Fig. 8) revealed a similar two-peak energy loss structure as found on the other (111) surfaces. Again the AL resonance along $\langle 1\bar{1}0 \rangle$ appeared to be much more intense than the Rayleigh mode, just as in Cu(111). Through several improvements in technique over the early work the group in Göttingen in 1985 was able to greatly increase the precision of the Pt(111) measurements, which made it possible to determine the shape of the Rayleigh dispersion curve with root-mean-square deviations of about 0.09 meV (90 μeV!) [85]. The group velocity as obtained from a nine-term Fourier expansion fit of the Rayleigh mode dispersion curve (Fig. 8b) showed two anomalies at about 0.55 and 1.0 Å^{-1}. They are much stronger and sharper than in the bulk transverse branch of equal direction and polarization, where two weak anomalies occur at 0.7 and 1.8 Å^{-1} [132]. From Born–von Kármán model calculations in which a sufficient number of force constants is included to fit the bulk Kohn anomaly the surface anomaly has been identified as a projection of the bulk anomaly [129]. The observed shift to lower wavevectors of surface anomalies with respect to the bulk is consistent with theoretical studies of surface anomalies in other systems such as the refractory superconductor TiN(001) [133].

The (001) surface of another refractory material, the bcc transition metal W, which is probably

the best known system showing what appears to be a nearly perfect two-dimensional mode induced phase transition [134,135], has recently been studied in Göttingen by helium scattering [136–138]. At temperatures above about 500 K the clean surface has an ideal (1×1) structure as evidenced by sharp first-order HAS diffraction peaks up to 1200 K and by sharp TOF inelastic single phonon features up to 1900 K [136,138]. At temperatures between 450 and 280 K a certain amount of disorder may be present as suggested by a broadening of the features in the HAS TOF spectra. As the temperature drops below $T_c = 280$ K, a $(\sqrt{2} \times \sqrt{2})R45°$ structure attributed to the formation of zig-zag chains normal to the $\bar{\Gamma}\bar{M}$ direction is observed in both LEED [139] and HAS [140,136–138].

Fig. 9 summarizes the HAS results on the angular distributions and the dramatic changes in the surface phonon dispersion curves with temperature above and below the reconstruction transition temperature. Also shown at the top are the corresponding distributions for the H-atom (1×1) saturated surface [141]. On the clean surface already at the highest temperature studied $(T = 1200$ K) there is clear evidence for an anomalous mode which on the basis of theoretical simulations and experimental analysis appears to be longitudinally polarized [142]. Its shape is reminiscent of the observations on the fcc Cu(111) and Cu(001) surfaces discussed above but, as opposed to these systems, the longitudinal mode lies below the Rayleigh mode. On lowering the temperature this mode decreases significantly in frequency and already at $T = 500$ K approaches zero within the half-width of the energetically broad peaks at $Q \approx 1.2$ Å^{-1}. Then at 220 K after the transition has occurred, as indicated by the appearance of a sharp $1/2$ order diffraction peak, a new system of dispersion curves is found which is symmetric with respect to a new Brillouin zone boundary at $Q = 0.7$ Å^{-1}. Clearly the reconstruction is driven by a soft mode which develops in the form of a large, temperature-dependent incommensurate anomaly when T approaches 280 K from above. Below the transition temperature the surface is stabilized in the reconstructed commensurate phase characterized by two regular phonon branches originating from the branch folding [143]. The upper branch is an optical mode with the same polarization as the original RW branch. When the surface is saturated with a hydrogen monolayer the reconstruction disappears and at the same time, as shown in Fig. 9e, the surface modes especially the longitudinal mode stiffen. Calculations reveal that the phonons of the H(1 × 1) saturated surface have dispersion curves very similar to those expected for an ideal bulk termination [141]. Apparently the H-atoms have the effect of stabilizing the surface electrons to energies well below the Fermi level.

All the evidence now appears to indicate that the reconstruction is driven by a charge density wave (CDW) mechanism very similar to that used to explain the reconstruction in the quasi-two-dimensional layered compounds such as $TaSe_2$ and TaS_2 [99,144]. According to this model the electron–phonon coupling is strong enough to induce a full softening of the phonon and a Peierls instability in the electronic bands. For such a strong coupling to occur large nearly parallel "nesting" regions of the electronic states at the Fermi surface must be present in an extended zone scheme. In this picture electronic states sharply localized in K-space are involved implying a delocalization in real space. A recent high-resolution photoelectron experiment [145] clearly indicates that the nesting should be strong enough. This interpretation is also supported by the HAS observation of a weak diffraction peak

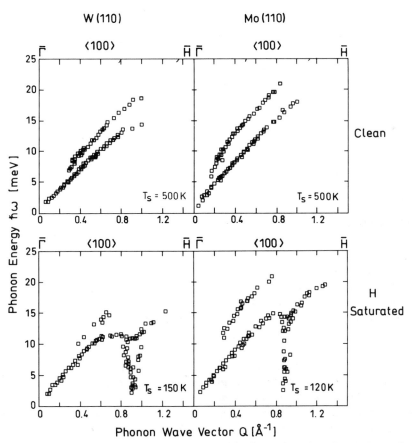

Fig. 10. Giant anomalies in H-saturated W(110) and Mo(110) surfaces along the ⟨100⟩ direction observed at low temperature (below) as compared with the regular surface phonon branches of the clean surfaces at 500 K (above) [148,149].

at $\Delta K \approx 1$ Å^{-1} already at temperatures as high as 500 K (see Fig. 9c) which was found by three different groups [140,136,138,146]. With lowering temperature this peak shifts to larger ΔK and goes over to the half-order diffraction peak at $\Delta K = 1.4$ Å^{-1}. Since this shifted peak could not even be found in especially dedicated LEED experiments [147] it has been attributed to diffraction of the He atoms from the surface charge density wave, made possible by the extreme sensitivity of HAS to the weak electronic corrugations produced by the CDW at the surface [136].

Recently very sharp giant Kohn anomalies have been found for the first time along the $\langle 100 \rangle$ direction on the hydrogen saturated W(110) and Mo(110) surfaces [148,149] and the results are shown in Fig. 10. Such anomalies show all the characteristics expected for a strong electron–phonon coupling in one dimension [150]. An interpretation of this anomaly in terms of a CDW as in the case of W(001) is not obvious since the necessary nesting of the Fermi-surfaces does not appear to be present [151]. We are intrigued by the fact that on the W(110) surface hydrogen has just the opposite effect of destabilizing the structure than on the W(001) surface where the surface phonons are hardened.

Several transition metal chalcogenides with a layered structure are known to support charge density waves (CDW) below a certain transition temperature T_c and display strong Kohn-type anomalies in the bulk. Therefore it was of some interest to look for similar anomalies in the surface phonon branches and whether they occur at the same wave vector. This question has triggered a series of interesting HAS studies on transition metal dichalcogenides and layered high-T_c superconductors, recently reviewed by Skofronick and Toennies [99]. A distinct shift of the anomaly in the RW branch with respect to the corresponding bulk anomaly has been clearly observed by HAS and carefully studied in 2H-TaSe$_2$(001) [152]. However the anomalous softening of the Rayleigh branch shows up only in the fairly narrow temperature range between 70 and 140 K and reaches its deepest point, at one-half of the reduced Brillouin zone, at about 110 K, whereas the bulk anomaly, which occurs at 2/3 of the zone in the

longitudinal branch reaches its deepest value at the bulk CDW transition temperature (~ 120 K). Thus we found a clear distinction between the bulk and the surface phonon anomalies both in position and minimum temperature and no direct correlation between the two effects. According to the Kohn-type anomaly mechanism, we considered the positional shift of the surface anomaly with respect to the bulk anomaly as indicative of a surface electronic state cutting the Fermi surface at a wavevector different from that of the bulk Fermi wavevector.

Below the CDW transition temperature phonons may be diffracted by the CDW superlattice [46] and show replicas of the ordinary phonon branches at wave vectors displaced by a CDW reciprocal vector g. This effect has been observed by He scattering in systems with large CDW amplitudes such as 1T-TaS$_2$(001) in its low temperature (120 K), commensurate $\sqrt{13} \times \sqrt{13}$ CDW phase [153,99]. The phonon dispersion curves of this system ($T_c = 543$ K) have been measured along both the crystallographic and CDW symmetry directions and could be easily fitted by a single sinusoidal dispersion curve centered at the different superstructure g-vectors contained in the irreducible part of the lattice Brillouin zone [153]. These data represent, to our knowledge, the first observation of phonons coupled to a charge density wave.

7. Present and future of HAS

Our effort for more than one decade was devoted to the development of experimental and theoretical tools for HAS phonon studies of clean surfaces of pure single crystals of relatively simple insulators, semiconductors and metals. These phonons characterize the half-space termination of the bulk. Because of its unique surface sensitivity helium scattering can also be applied to the study of the phonons of more complex systems, the vibrations of adsorbate layers and those of thin films during epitaxial growth.

In systems with large unit cells the diffraction patterns become more closely spaced and the detection and resolution of phonon dispersion curves place severe demands on the experiment. The first HAS study of surface phonons on a stepped surface has been reported in 1990 by Lock, Toennies and Witte for Al(221) [154]. The measurements of Lock et al. with HAS revealed a rich phonon structure and, particularly, a new surface mode for wavevector propagation along the step edges, in qualitative agreement with theoretical predictions. Complex reconstructions with large periods as well as the surface of intrinsically complex systems such as the high-T_c superconductors [99] have become accessible to HAS studies thanks to the recent progress in the HAS experimental technique.

Helium atom scattering is ideally suited to monitor layer by layer thin film growth [155] and also the structure and quality of the film surfaces. The mechanical properties of ultrathin films are technologically important but not amenable to study by conventional techniques (Brillouin scattering, neutron scattering, acoustical attenuation etc.). Recently we were able to measure with HAS the phonon branches of Na epitaxial films on Cu(001) of thickness varying from 2 to 30 monolayers [156]. These studies, besides providing the first evidence for the Rayleigh wave of this important free electron metal, showed that the Rayleigh velocity was 21–50% larger than predicted for the bulk as a result of the stress induced by the substrate. This was confirmed by the additional observation of confined acoustical resonances corresponding to longitudinal standing waves normal to the surface (organ pipe modes) with frequencies 20% larger than in the bulk [156].

The vibrations of physisorbed and chemisorbed atoms and molecules with low frequency modes not accessible by other techniques represent another broad field of investigation for inelastic HAS studies. The first experiments were reported by Sibener et al. for rare gases on Ag(111) [157]. For monolayer coverages a dispersionless mode was observed which corresponds to a simple Einstein oscillator vibration normal to the surface. Similar results have been obtained for Kr and Xe on graphite [158]. The frequency is of great value for determining rare gas atom–metal

surface potentials [159]. The external vibrations (e.g., frustrated translations) of molecules such as CO on Pt(111) [160], Ni(001) [161] and Fe(110) [162] and NH_3 on Fe(110) [162] as well as C_6H_6 on Rh(111) [163] have been investigated and first identified with HAS. Even systems consisting of very large molecules such as Langmuir–Blodgett layers [164,165] and C_{60} monolayers [166] have also been the objects of preliminary experiments. With further improvements in resolution and sensitivity a growing number of more complex systems will become accessible.

One attractive feature of surface studies is the reduced two dimensions of the surface which exposes new aspects suppressed in the three-dimensional bulk. One-dimensional systems can also be prepared on surfaces and a number of examples usually involving step edges of vicinal surfaces as in the fabrication of quantum wires have already been explored. The interpretation of vibrational modes in one dimension will be greatly simplified, while on the other hand, electron–phonon coupling effects are expected to be much more pronounced.

The inherent high resolution of helium atom scattering can also be used to advantage to study dynamical processes other than vibrations at surfaces. We recall that the elastic incoherent peak seen in all TOF spectra contains valuable information on defects [167]. Recently it was demonstrated that the energetic broadening of this peak and the wave vector dependence of this broadening provides detailed information on the atomic scale details of diffusing atoms [168]. For these studies special techniques were used to increase the apparatus through-put for the elastic peak. By cooling the source to 30 K an energetic half-width of about 80×10^{-6} eV (0.6 cm^{-1}) could be achieved [168]. A more striking demonstration of the power of this technique was provided by recent experiments involving adsorbates on metal surfaces [169,170]. The most detailed results were obtained for Na on Cu(001) [170] where the wave vector dependence indicated a predominantly single jump mechanism. Moreover, from Monte Carlo analysis the contribution of two- and three-jump processes could be identified. In addition the temperature dependence provided information on the activation energy (which in this case was only 51 meV). Moreover the diffusion attempt frequency could also be obtained from the TOF data. This experiment provided for the first time a complete microscopic description of the diffusion process on surfaces.

Below the diffusive regime the larger thermal oscillations of surface atoms as compared to bulk atoms, are responsible for enhanced surface anharmonicity and lead to a measurable decrease of the surface phonon frequencies with increasing temperature. This is another important aspect of surface dynamics which could be explored recently by HAS [171]. In magnetic systems the anharmonic effects are entangled with the effects of magnetization through magnetostriction. This leads to a peculiar temperature dependence of surface phonon frequencies, as recently found for Fe(110) by means of HAS spectroscopy [172]. In view of the great interest in magnetic thin films and superstructures, HAS studies of surface magnetic effects on (and possibly by means of) surface phonons look very promising. This is the present status of HAS studies. Work in these and other areas although still in the early stages hold great promise for the future.

Acknowledgments

We thank Drs. N.S. Luo and Paolo Ruggerone for a very productive and fruitful collaboration on the theoretical aspects. We are grateful to them and Profs. V. Celli, A. Khater, J.R. Manson and J.G. Skofronick for reading and commenting on the manuscript. Our collaboration was made possible by a joint award of the Alexander-von Humboldt Stiftung and the Max-Planck Gesellschaft.

References

[1] Lord Rayleigh, Proc. London Math. Soc. 17 (1887) 4.
[2] A.A. Maradudin, in: Nonequilibrium Phonon Dynamics, Ed. W.E. Bron (Academic Press, New York, 1985) pp. 395–600.
[3] G.W. Farnell, in: Physical Acoustics, Vol. 6, Eds. W.P. Mason and R.N. Thurston (Academic Press, New York, 1970) p. 109; in: Acoustic Surface Waves, Ed. A.A. Oliner (Springer, Berlin, 1978) p. 13.
[4] A.A. Maradudin, E.W. Montroll, G.H. Weiss and I.P. Ipatova, Theory of Lattice Dynamics in the Harmonic Approximation, Solid State Physics, Suppl. 3 (Academic Press, New York, 1971).
[5] M. Born and Th. von Kármán, Phys. Z. 13 (1912) 297.
[6] I.M. Lifshitz and L.M. Rozenzweig, Zh. Eksp. Teor. Fiz. 18 (1948) 1012, for an English version see I.M. Lifshitz, Nuovo Cim. Suppl. 3 (1956) 732.
[7] A.A. Maradudin, R.F. Wallis and L. Dobrzynski, Handbook of Surfaces and Interfaces, Vol. 3 (Garland, New York, 1980).
[8] F. Garcia-Moliner, Ann. Phys. (Paris) 2 (1977) 179.
[9] G. Armand, Phys. Rev. B 14 (1976) 2218;
    see also G. Armand and P. Masri, Surf. Sci. 130 (1983) 89.
[10] R.A. Allen, Surf. Sci. 76 (1978) 91.
[11] A.A. Maradudin and J. Melngailis, Phys. Rev. 133 (1964) A1188.
[12] L. Dobrzynski and G. Leman, J. Phys. (Paris) 30 (1970) 116.
[13] S.W. Musser and K.H. Rieder, Phys. Rev. B 2 (1970) 3034.
[14] W. Goldhammer, W. Ludwig, W. Zierau and C. Falter, Surf. Sci. 141 (1984) 139.
[15] J.E. Black, B. Lacks and D.L. Mills, Phys. Rev. B 22 (1980) 1818.
[16] G. Benedek, Phys. Status Solidi (b) 58 (1973) 661.
[17] G. Benedek, Surf. Sci. 61 (1976) 603.
[18] G. Benedek and L. Miglio, in: Surface Phonons, Eds. F.W. de Wette and W. Kress, Vol. 27 of Springer Series in Surface Sciences (Springer, Berlin, 1991).
[19] G. Benedek, M. Miura, W. Kress and H. Bilz, Phys. Rev. Lett. 52 (1984) 1907.
[20] T.E. Feuchtwang, Phys. Rev. 155 (1967) 731.
[21] V. Bortolani, F. Nizzoli and G. Santoro, in: Lattice Dynamics, Ed. M. Balkanski (Flammarion, Paris, 1978).
[22] J. Szeftel and A. Khater, J. Phys. C 20 (1987) 4725;
    J. Szetel, A. Khater, F. Mila, S. d'Addato and N. Auby, J. Phys. C 21 (1988) 2113.
[23] A. Khater, N. Auby and R.F. Wallis, Physica B 167 (1991) 273;
    A. Khater, H. Griemech, J. Lapujoulade and F. Fabre, Surf. Sci. 251/252 (1991) 381.
[24] The method is reviewed by: R.F. Wallis, in: Atomic Structure and Properties of Solids, Ed. E. Burstein (Academic Press, New York, 1972).
[25] A.A. Lucas, J. Chem. Phys. 48 (1968) 3156.
[26] R. Fuchs and K.L. Kliever, Phys. Rev. 140 (1965) A2076;
    K.L. Kliever and R. Fuchs, Phys. Rev. 144 (1966) 495;
    150 (1966) 573.
[27] A.D. Boardman, Ed., Electromagnetic Surface Modes (Wiley, New York, 1982).
[28] R.E. Allen, G.P. Alldredge and F.W. de Wette, Phys. Rev. B 4 (1971) 1648.

[29] R.E. Allen, G.P. Alldredge and F.W. de Wette, Phys. Rev. B 4 (1971) 1661.

[30] R.E. Allen, G.P. Alldredge and F.W. de Wette, Phys. Rev. B 4 (1971) 1682.

[31] T.S. Chen, G.P. Alldredge and F.W. de Wette, Solid State Commun. 10 (1972) 941.

[32] T.S. Chen, F.W. de Wette and G.P. Alldredge, Phys. Rev. B 15 (1977) 1167.

[33] F.W. de Wette and W. Kress, Eds., Surface Phonons, Vol. 21 of Springer Series in Surface Sciences (Springer, Berlin, 1991).

[34] G.P. Alldredge, Phys. Lett. A 41 (1972) 291.

[35] B.N. Brockhouse and A.T. Stewart, Phys. Rev. 100 (1955) 756.

[36] K.H. Rieder and E.M. Hörl, Phys. Rev. Lett. 20 (1968) 209;
K.H. Rieder and W. Drexel, Phys. Rev. Lett. 34 (1975) 148.

[37] N. Cabrera, V. Celli and R. Manson, Phys. Rev. Lett. 22 (1969) 346.

[38] J.R. Manson and V. Celli, Surf. Sci. 24 (1971) 495.

[39] S.S. Fisher and J.R. Bledsoe, J. Vac. Sci. Technol. 9 (1971) 814; Surf. Sci. 46 (1974) 129.

[40] R.E. Subbarao and D.R. Miller, J. Vac. Sci. Technol. 9 (1972) 808;
J.M. Horne and D.R. Miller, Phys. Rev. Lett. 41 (1978) 511.

[41] B.R. Williams, J. Chem. Phys. 55 (1971) 3220; 56 (1972) 1895.

[42] B.E. Mason and B.R. Williams, J. Chem. Phys. 55 (1971) 3220; 61 (1974) 2765; Jpn. J. Appl. Phys., Suppl. 2, Pt. 2 (1974) 557.

[43] G. Boato and P. Cantini, in: Dynamic Aspects of Surface Physics, Ed. F.O. Goodman (Compositori, Bologna, 1974) p. 707.

[44] See the book referred to in Ref. [43].

[45] G. Benedek, Phys. Rev. Lett. 35 (1975) 234.

[46] G. Benedek, L. Miglio and G. Seriani, in: Helium Atom Scattering from Surfaces, Ed. E. Hulpke (Springer, Berlin, 1992) p. 207.

[47] G. Benedek and G. Boato, Europhys. News 8 (1977) 5.

[48] P. Cantini, G.P. Felcher and R. Tatarek, Phys. Rev. Lett. 37 (1976) 606.

[49] G. Brusdeylins, R.B. Doak and J.P. Toennies, Phys. Rev. Lett. 44 (1980) 1417.

[50] G. Brusdeylins, R.B. Doak and J.P. Toennies, Phys. Rev. Lett. 46 (1981) 437.

[51] G. Brusdeylins, R.B. Doak and J.P. Toennies, Phys. Rev. B 27 (1983) 3662.

[52] R.B. Doak, U. Harten and J.P. Toennies, Phys. Rev. Lett. 51 (1983) 578.

[53] J.P. Toennies and K. Winkelmann, J. Chem. Phys. 66 (1977) 3965;
G. Brusdeylins, H.-D. Meyer, J.P. Toennies and K. Winkelmann, in: Progress in Astronautics and Aeronautics, Vol. 51, Ed. J.L. Poetter (AIAA, New York, 1977) p. 1047.

[54] B. Feuerbacher and R.F. Willis, Phys. Rev. Lett. 47 (1981) 526.

[55] J. Aldag and R.M. Stern, Phys. Rev. Lett. 14 (1965) 857;
R.F. Wallis and A.A. Maradudin, Phys. Rev. 148 (1966) 962;
D.L. Huber, Phys. Rev. 153 (1967) 772.

[56] F.M. Probst and T.C. Piper, J. Vac. Sci. Technol. 4 (1967) 53.

[57] C.H. Li, S.Y. Tong and D.L. Mills, Phys. Rev. B 21 (1980) 3057.

[58] S. Lehwald, J.W. Szeftel, H. Ibach, T.S. Rahman and D.L. Mills, Phys. Rev. Lett. 50 (1983) 518;
J.M. Szeftel, S. Lehwald, H. Ibach, T.S. Rahman, J.E. Black and D.L. Mills, Phys. Rev. Lett. 51 (1983) 268.

[59] J.P. Toennies, in: Solvay Conference on Surface Science, Ed. F.W. de Wette, Vol. 14 of Springer Series in Surface Sciences (Springer, Berlin, 1989).

[60] J.P. Toennies and G. Benedek, Surf. Sci. Rep., to be published.

[61] Ch. Wöll, Dissertation, University of Göttingen, Max-Planck-Institut für Strömungsforschung, Bericht 18 (1987).

[62] G. Benedek, G.P. Brivio, L. Miglio and V.R. Velasco, Phys. Rev. B 26 (1982) 497.

[63] G. Benedek and L. Miglio, in: Ab-Initio Calculation of Phonon Spectra, Eds. J.T. Devreese, V.E. van Doren and P.E. Van Camp (Plenum, New York, 1983);
L. Miglio and G. Benedek, in: Structure and Dynamics of Surfaces II, Eds. W. Schommers and P. von Blanckenhagen (Springer, Berlin, 1987) p. 35.

[64] G. Brusdeylins, R. Rechsteiner, J.G. Skofronick, J.P. Toennies, G. Benedek and L. Miglio, Phys. Rev. Lett. 54 (1985) 466.

[65] G. Benedek, L. Miglio, G. Brusdeylins, J.G. Skofronick and J.P. Toennies, Phys. Rev. B 35 (1987) 6593.

[66] F.W. de Wette, in: Surface Phonons, Eds. F.W. de Wette and W. Kress, Vol. 27 of Springer Series in Surface Sciences (Springer, Berlin, 1991) p. 67.

[67] S.A. Safron, W.P. Brug, G.G. Bishop, J. Duan, G. Chern and J.G. Skofronick, J. Electron Spectrosc. Relat. Phenom. 54/55 (1990) 343.

[68] B. Lengeler and W. Ludwig, Solid State Commun. 2 (1964) 83.

[69] A. Franchini, Dissertation, University of Modena, 1975/86.

[70] C. Herring, in: The Physics of Powder Metallurgy, Ed. W.E. Kingston (McGraw-Hill, New York, 1951) p. 162 ff;
For a recent treatment of surface stress see: A. Zangwill, Physics at Surfaces (Cambridge Univ. Press, Cambridge, 1989) p. 10.

[71] L. Miglio, P. Santini, P. Ruggerone and G. Benedek, Phys. Rev. Lett. 62 (1989) 3070.

[72] P. Santini, L. Miglio, G. Benedek, U. Harten, P. Ruggerone and J.P. Toennies, Phys. Rev. B 42 (1990) 11942.

[73] Y.R. Wang and C.B. Duke, Surf. Sci. 205 (1988) L755.

[74] E.J. Woll, Jr. and W. Kohn, Phys. Rev. 126 (1962) 1693.

[75] M.W. Finnis and V. Heine, J. Phys. F (Met. Phys.) 4 (1974) L37.

[76] J.E. Black, in: Dynamical Properties of Solids, Vol. 4, Eds. G.K. Horton and A.A. Maradudin (Elsevier, Amsterdam, 1990) pp. 179–279.

[77] M.H. Mohamed, L.L. Kesmodel, B.M. Hall and D.L. Mills, Phys. Rev. B 37 (1988) 2763; B.M. Hall, D.L. Mills, M.H. Mohamed and L.L. Kesmodel, Phys. Rev. B 38 (1988) 5856.

[78] V. Bortolani, A. Franchini, F. Nizzoli and G. Santoro, Phys. Rev. Lett. 52 (1984) 429.

[79] V. Bortolani, A. Franchini and G. Santoro, Surf. Sci. 189/190 (1987) 675.

[80] V. Bortolani, G. Santoro, U. Harten and J.P. Toennies, Surf. Sci. 148 (1984) 82; G. Santoro, A. Franchini, V. Bortolani, U. Harten, J.P. Toennies and Ch. Wöll, Surf. Sci. 183 (1987) 180; V. Bortolani, A. Franchini and G. Santoro, private communication.

[81] C.S. Jayanthi, H. Bilz, W. Kress and G. Benedek, Phys. Rev. Lett. 59 (1987) 795.

[82] L.J. Sham, in: Dynamical Properties of Solids, Vol. 1, Eds. G.K. Horton and A.A. Maradudin (North-Holland, Amsterdam, 1974) p. 301.

[83] C. Beatrice and C. Calandra, Phys. Rev. B 10 (1983) 6130.

[84] A.G. Eguiluz, A.A. Maradudin and R.F. Wallis, Phys. Rev. Lett. 60 (1988) 309.

[85] U. Harten, J.P. Toennies, Ch. Wöll and G. Zhang, Phys. Rev. Lett. 55 (1985) 2308.

[86] C. Kaden, P. Ruggerone, J.P. Toennies and G. Benedek, Nuovo Cim. 14D (1992) 627.

[87] C. Kaden, P. Ruggerone, J.P. Toennies, G. Zhang and G. Benedek, Phys. Rev. B 46 (1992) 13509.

[88] E. Zaremba and W. Kohn, Phys. Rev. B 13 (1976) 2270; Phys. Rev. B 15 (1977) 1769.

[89] N. Esbjerg and J.K. Nørskov, Phys. Rev. Lett. 45 (1980) 807.

[90] A. Levi and V. Bortolani, Riv. Nuovo Cim. 9 (1986) 1.

[91] G. Benedek, J. Ellis, N.S. Luo, A. Reichmuth, P. Ruggerone and J.P. Toennies, Phys. Rev., in press.

[92] J.P. Toennies, in: Surface Phonons, Eds. F.W. de Wette and W. Kress, Vol. 21 of Springer Series in Surface Sciences (Springer, Berlin, 1991).

[93] H. Ibach and D.L. Mills, Electron Energy Loss Spectroscopy and Surface Vibrations (Academic Press, New York, 1982)

[94] G. Lilienkamp and J.P. Toennies, J. Chem. Phys. 78 (1983) 5210.

[95] H. Buchenau, J. Northby, E.L. Knuth, J.P. Toennies and C. Winkler, J. Chem. Phys. 92 (1990) 6875.

[96] P.A. Thiry, M. Liehr, J.J. Pireaux and R. Caudano, J. Electron Spectrosc. Relat. Phenom. 39 (1986) 69.

[97] H. Ibach, Electron Energy Loss Spectrometers – "The Technology of High Performance" (Springer, Berlin, 1991).

[98] H. Ibach, M. Balden, D. Bruchmann and S. Lehwald, Surf. Sci. 269/270 (1992) 94.

[99] J.G. Skofronick and J.P. Toennies, in: Surface Properties of Layered Structures, Ed. G. Benedek (Kluwer, Amsterdam, 1992) p. 151.

[100] G. Benedek, J.P. Toennies and R.B. Doak, Phys. Rev. 28 (1983) 7276.

[101] G. Bracco, M. D'Avanzo, C. Salvo, R. Tatarek and F. Tommasini, Surf. Sci. 189/190 (1987) 684.

[102] P. Cantini and R. Tatarek, Phys. Rev. B 23 (1981) 3030.

[103] D. Evans, V. Celli, G. Benedek, J.P. Toennies and R.B. Doak, Phys. Rev. Lett. 50 (1983) 1854.

[104] G. Brusdeylins, R.B. Doak and J.P. Toennies, J. Chem. Phys. 75 (1981) 1784.

[105] G. Lilienkamp and J.P. Toennies, J. Chem. Phys. 78 (1983) 5210.

[106] U. Harten, J.P. Toennies and Ch. Wöll, Phys. Rev. Lett. 57 (1986) 2947.

[107] U. Harten and J.P. Toennies, Europhys. Lett. 4 (1987) 833.

[108] H. Ibach, Phys. Rev. Lett. 27 (1971) 253.

[109] O.L. Alerhand and E.J. Mele, Phys. Rev. Lett. 59 (1987) 657; O.L. Alerhand and E.J. Mele, Phys. Rev. B 37 (1988) 2536.

[110] A. Mazur and J. Pollmann, in: Phonons 89, Eds. S. Hunklinger, W. Ludwig and G. Weiss (World Scientific, Singapore, 1990) p. 943.

[111] F. Ancillotto, W. Andreoni, A. Selloni, R. Car and M. Parrinello, Phys. Rev. Lett. 65 (1990) 3148.

[112] W. Ludwig, Jpn. J. Appl. Phys., Suppl. 2, Pt. 2 (1974) 879.

[113] K.C. Pandey, Phys. Rev. Lett. 47 (1981) 1913; 49 (1982) 223.

[114] J.E. Northrup and M.L. Cohen, Phys. Rev. Lett. 49 (1982) 1349; J. Vac. Sci. Technol. 21 (1982) 333.

[115] F.J. Himpsel, P.M. Marcus, R. Tromp, I.P. Batra, M.R. Cook, F. Jona and H. Liu, Phys. Rev. B 30 (1984) 2257.

[116] R.B. Doak and D.B. Nguyen, J. Electron Spectrosc. Relat. Phenom. 44 (1987) 205.

[117] C.B. Duke, S.L. Richardson, A. Paton and A. Kahn, Surf. Sci. 127 (1983) L135.

[118] G. Armand, Solid State Commun. 48 (1983) 261.

[119] U. Harten, J.P. Toennies and Ch. Wöll, Faraday Discuss. Chem. Soc. 80 (1985) 137.

[120] V. Bortolani, G. Santoro, U. Harten and J.P. Toennies, Surf. Sci. 146 (1984) 82.

[121] G. Santoro, A. Franchini, V. Bortolani, U. Harten, J.P. Toennies and Ch. Wöll, Surf. Sci. 183 (1987) 180.

[122] V. Bortolani, A. Franchini and G. Santoro, unpublished.

[123] V. Bortolani, A. Franchini, F. Nizzoli and G. Santoro, Phys. Rev. Lett. 52 (1984) 429.

[124] M.H. Mohamed, L.L. Kesmodel, B.M. Hall and D.L. Mills, Phys. Rev. B 37 (1988) 2763; B.M. Hall, D.L. Mills, M.H. Mohamed and L.L. Kesmodel, Phys. Rev. B 38 (1988) 5856.

[125] J.P. Toennies, Superlattices Microstruct. 7 (1990) 193.

[126] N. Bunjes, J.P. Toennies and G. Witte, unpublished.

[127] M.S. Daw and M.I. Baskes, Phys. Rev. B 29 (1984) 6443;

J.S. Nelson, M.S. Daw and E.C. Sowa, Phys. Rev. B 40 (1989) 1465.

[128] A. Euceda, D.M. Bylander and L. Kleinman, Phys. Rev. B 28 (1983) 528.

[129] V. Bortolani, A. Franchini, G. Santoro, J.P. Toennies, Ch. Wöll and G. Zhang, Phys. Rev. B 40 (1989) 3524.

[130] K. Kern, R. David, R.L. Palmer, G. Comsa and T.S. Rahman, Phys. Rev. Lett. B 33 (1986) 4334.

[131] D. Neuhaus, F. Joo and B. Feuerbacher, Surf. Sci. Lett. 165 (1986) L90.

[132] D.H. Dutton, B.N. Brockhouse and A.P. Miller, Can. J. Phys. 50 (1972) 2915.

[133] G. Benedek, M. Miura, W. Kress and H. Bilz, Phys. Rev. Lett. 52 (1984) 1907.

[134] A. Fasolino and E. Tosatti, Phys. Rev. B 35 (1987) 4264; C.Z. Wang, A. Fasolino and E. Tosatti, Phys. Rev. B 37 (1988) 2116.

[135] D.A. King, Phys. Scr. 74 (1983) 34.

[136] H.-J. Ernst, E. Hulpke and J.P. Toennies, Phys. Rev. Lett. 58 (1987) 1941.

[137] H.-J. Ernst, E. Hulpke and J.P. Toennies, Europhys. Lett. 10 (1989) 747.

[138] H.-J. Ernst, E. Hulpke and J.P. Toennies, Phys. Rev. B 46 (1992) 16081.

[139] T.E. Felter, R.A. Barker and P.J. Estrup, Phys. Rev. Lett. 38 (1977) 1138.

[140] B. Salanon and J. Lapujoulade, Surf. Sci. Lett. 173 (1986) L613.

[141] H.-J. Ernst, E. Hulpke, J.P. Toennies and Ch. Wöll, Surf. Sci. 262 (1992) 159.

[142] E. Hulpke and D. Smilgies, Phys. Rev. B 42 (1990) 9203.

[143] H.J. Schulz, Phys. Rev. B 18 (1978) 5756.

[144] R.V. Coleman, Z. Dai, W.W. McNairy, C.G. Slough and C.W. Wang, in: Surface Properties of Layered Structures, Ed. G. Benedek (Kluwer, Amsterdam, 1992) p. 27.

[145] K. Smith, G. Elliott and S. Kevan, Phys. Rev. B 42 (1990) 5385.

[146] E.K. Schweizer and C.T. Rettner, Surf. Sci. Lett. 208 (1989) L29.

[147] E. Bauer, private communication.

[148] E. Hulpke and J. Lüdecke, Phys. Rev. Lett. 68 (1992) 2846.

[149] E. Hulpke and J. Lüdecke, Surf. Sci. 287/288 (1993) 837.

[150] H. Böttger, Principles of the Theory of Lattice Dynamics (Physik-Verlag, Weinheim, 1983) p. 72.

[151] R.H. Gaylord, K.H. Jeon and S.D. Kevan, Phys. Rev. Lett. 62 (1989) 2036.

[152] G. Benedek, G. Brusdeylins, C. Heimlich, L. Miglio, J.G. Skofronick, J.P. Toennies and R. Vollmer, Phys. Rev. Lett. 60 (1988) 1037.

[153] G. Brusdeylins, F. Hofmann, R. Ruggerone, J.P. Toennies, R. Vollmer, G. Benedek and J.G. Skofronick, in: Phonons 89, Eds. S. Hunklinger, W. Ludwig and G. Weiss (World Scientific Singapore, 1990) p. 892.

[154] A. Lock, J.P. Toennies and G. Witte, J. Electron Spectrosc. Relat. Phenom. 54/55 (1990) 309.

[155] L.J. Gómez, S. Bourgeal, J. Ibáz and M. Salmeron, Phys. Rev. B 31 (1985) 2551.

[156] G. Benedek, J. Ellis, A. Reichmuth, P. Ruggerone, H. Schief and J.P. Toennies, Phys. Rev. Lett. 69 (1992) 2951.

[157] K.D. Gibson, S.J. Sibener, B.M. Hall, D.L. Mills and J.E. Black, J. Chem. Phys. 83 (1985) 4256.

[158] J.P. Toennies and R. Vollmer, Phys. Rev. B 40 (1989) 3495.

[159] J.A. Barker and C.T. Rettner, J. Chem. Phys. 97 (1992) 5844.

[160] A.M. Lahee, J.P. Toennies and Ch. Wöll, Surf. Sci. 177 (1986) 371.

[161] R. Berndt, J.P. Toennies and Ch. Wöll, J. Electron Spectrosc. Relat. Phenom. 44 (1987) 183.

[162] J.P. Toennies, Ch. Wöll and G. Zhang, J. Chem. Phys. 96 (1992) 4023.

[163] J.P. Toennies, G. Witte and Ch. Wöll, Phys. Rev. Lett., in press.

[164] V. Vogel and Ch. Wöll, J. Chem. Phys. 84 (1986) 5200; Thin Solid Films 159 (1988) 429.

[165] J.R. Manson and J.G. Skofronick, Phys. Rev. B 47 (1993) 12890.

[166] D. Schmicker, S. Schmidt, J.G. Skofronick, J.P. Toennies and R. Vollmer, Phys. Rev. B 44 (1991) 10995.

[167] B.J. Hinch, A. Lock and J.P. Toennies, in: Kinetics and Ordering and Growth at Surfaces, Ed. M.G. Lagally (Plenum, New York, 1990) p. 77.

[168] J.W.M. Frenken, B.J. Hinch, J.P. Toennies and Ch. Wöll, Phys. Rev. B 41 (1990) 938.

[169] B.J. Hinch, J.W.M. Frenken, G. Zhang and J.P. Toennies, Surf. Sci. 259 (1991) 288.

[170] J. Ellis and J.P. Toennies, Phys. Rev. Lett. 70 (1993) 2118.

[171] G. Benedek and J.P. Toennies, Phys. Rev. (Rapid Commun.) B 46 (1992) 13643.

[172] G. Benedek, J.P. Toennies and G. Zhang, Phys. Rev. Lett. 68 (1992) 2644.

Surface Science 299/300 (1994) 612–627
North-Holland

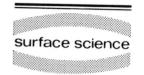

surface science

Surface phonons: theoretical developments

R.F. Wallis

Institute for Surface and Interface Science and Department of Physics, University of California, Irvine, CA 92717, USA

Received 30 March 1993; accepted for publication 20 April 1993

A review is presented of the theoretical aspects of surface phonons. The increase in sophistication of the methods employed is traced from early work based on simple force constant models through Green's function methods coupled with shell, bond-charge, and pseudoparticle models to the present-day emphasis on first-principles calculations. Numerous examples of the interplay between theory and experiment are pointed out.

1. Introduction

During the last fifteen years there has been an enormous increase in interest in the vibrational properties of crystals with free surfaces. There are several reasons for this surge in activity. From the experimental point of view, the development of experimental techniques such as inelastic electron scattering and inelastic helium atom scattering has made available surface phonon dispersion curves for a great many crystal surfaces. From the theoretical point of view, the investigation of surface effects on the vibrations of crystals has been greatly facilitated by the availability of supercomputers and the development of many-body theoretic techniques. It is now becoming possible to make first-principles calculations of the force constants for the interactions of a surface atom with neighboring atoms and from them obtain surface phonon dispersion curves spanning the entire surface Brillouin zone.

The experimental side of surface phonons is being ably handled in this volume by Toennies and Ibach. In the present article the theory of surface phonons will be developed from both the continuum and discrete lattice points of view. The various techniques that have been applied in the theoretical calculations for insulators, semiconductors and metals will be presented. Comparisons of theoretical and experimental results will be given.

2. General remarks on surface modes of vibration

Generally speaking, the creation of a free surface tends to lower the normal-mode frequencies and produce a class of modes called *surface modes*, in which the displacements of the atoms are relatively large at the surface and decrease in essentially exponential fashion going away from the surface. These effects may be viewed as arising in the following manner. Starting from a periodic crystal, a free surface can be created by setting to zero the interactions coupling atoms on opposite sides of a plane lying between two adjacent lattice planes. This decrease in coupling constants produces a lowering of the normal mode frequencies in accordance with Rayleigh's theorem [1]. Furthermore, surface-mode frequencies are typically split off from the bottom of a band of bulk-mode frequencies having the same values of the wave vector components parallel to the surface. Thus, surface modes may appear below the bulk continuum or in gaps in the bulk continuum. Under certain circumstances surface modes may lie within the bulk continuum if their polar-

SSDI 0039-6028(93)E0307-G

ization vectors are orthogonal to those of nearby bulk modes or even above the bulk continuum.

3. Elastic continuum theory of surface vibrational modes

The birth of the subject we now call surface phonons may be said to have occurred in 1887 when Lord Rayleigh [2] published his pioneering paper on waves propagating along a stress-free planar surface of an isotropic elastic continuum. The displacement vector for a Rayleigh surface wave lies in the *sagittal plane* defined by the normal to the surface and the direction of propagation. The endpoint of the displacement vector executes an ellipse in the sagittal plane as time evolves. The Rayleigh wave speed c_R is determined by the equation

$$p^6 - 8p^4 + 24p^2 - 16 - 16r^2(p^2 - 1) = 0, \quad (1)$$

where $p = c_R/c_t$, $r = c_t/c_l$, and c_R, c_t, and c_l are the speeds of the surface, bulk transverse, and bulk longitudinal waves, respectively. One finds that c_R is smaller than c_l and c_t. The dispersion relation is given by

$$\omega = c_R |k_\parallel|, \quad (2)$$

where ω is the angular frequency of the wave, and $k_\parallel$ is the wave vector parallel to the surface.

Very few real crystals are isotropic (tungsten is nearly so). Since the elastic modulus tensor is a fourth-rank tensor, even cubic crystals are in general elastically anisotropic. The theory for cubic crystals was developed by Stoneley [3]. The general procedure is similar to that for the isotropic case, but new features arise in the results as a consequence of the anisotropy. Three decay constants are typically required to characterize a surface wave, rather than two as in the isotropic case. When the elastic moduli satisfy certain relationships, complex decay constants may occur giving rise to so-called generalized surface waves [4]. Also, it may happen that one of the three decay constants is pure imaginary rather than real or complex and the corresponding contribution to the displacement does not decay exponentially into the interior of the crystal, giving rise to a

pseudosurface wave or leaky surface wave [5]. The nondecaying contribution to the displacement field leads to a flow of energy away from the surface into the interior of the solid and to an attenuation of the wave as it propagates along the surface. The wave vector $k_\parallel$ therefore acquires an imaginary part. In many cases the attenuation is so small that the leaky surface wave behaves as a true surface wave insofar as experimental observation is concerned.

4. General remarks about the microscopic theory of surface vibrations

The foundations of the microscopic or lattice dynamical theory of the atomic vibrations of bulk crystals were laid in 1912 by Born and von Karman [6]. Two world wars intervened before the first publications appeared on the lattice dynamics of surface vibrations by Lifshitz and Rosenzweig [7] and independently by Wallis [8].

The discussion in Section 3 is applicable when the wavelength of the surface wave is large compared to the lattice constant of the crystal. When this condition is not satisfied, account must be taken of the discrete atomic character of the crystal. In many calculations of surface vibrational properties, it proves to be convenient to use a crystal slab having a specific number of atomic layers rather than a semi-infinite crystal.

If we restrict ourselves to the harmonic approximation, the equation of motion for the displacement $u(l\kappa)$ of the κth atom in the lth unit cell can be written as

$$M_\kappa \frac{\partial^2 u_\alpha(l\kappa)}{\partial t^2} = -\frac{\partial \Phi}{\partial u_\alpha(l\kappa)}$$
$$= -\sum_{l'\kappa'\beta} \Phi_{\alpha\beta}(l\kappa, l'\kappa') u_\beta(l'\kappa'),$$
$$(3)$$

where M_κ is the mass of the κth atom and the $\Phi_{\alpha\beta}(l\kappa, l'\kappa')$ are force constants. The solution of Eq. (3) is complicated by two factors. First, there is no periodicity of the atomic sites in the direction normal to the surface. Second, the force constants $\Phi_{\alpha\beta}(l\kappa, l'\kappa')$ are not necessarily the same as the corresponding coefficients for the infinite crystal if atoms $l\kappa$ and/or $l'\kappa'$ are located at or near the surface. Nevertheless, we can exploit the translational periodicity parallel to the surface by seeking a solution of the form

$$u_\alpha(l\kappa) = M_\kappa^{-1/2} v_\alpha(\kappa, \boldsymbol{k}_\parallel) \, e^{i\boldsymbol{k}_\parallel \cdot \boldsymbol{R}_\parallel(l)} \, e^{-i\omega t}, \qquad (4)$$

where l stands for the pair of integers l_1, l_2 and $\boldsymbol{R}_\parallel(l)$ is the position vector of the lth lattice site in the two-dimensional lattice parallel to the surface. Substitution of Eq. (4) into Eq. (3) yields a set of equations in the amplitudes $v_\alpha(\kappa, \boldsymbol{k}_\parallel)$,

$$\omega^2 v_\alpha(\kappa, \boldsymbol{k}_\parallel) = \sum_{\kappa'\beta} D_{\alpha\beta}(\kappa, \kappa'; \boldsymbol{k}_\parallel) v_\beta(\kappa', \boldsymbol{k}_\parallel), \quad (5)$$

where

$$D_{\alpha\beta}(\kappa, \kappa'; \boldsymbol{k}_\parallel)$$
$$= (M_\kappa M_{\kappa'})^{-1/2}$$
$$\times \sum_{l'} \Phi_{\alpha\beta}(l\kappa, l'\kappa') \, e^{i\boldsymbol{k}_\parallel \cdot [\boldsymbol{R}_\parallel(l') - \boldsymbol{R}_\parallel(l)]} \qquad (6)$$

are the elements of the dynamical matrix of the crystal slab.

Eqs. (5) are linear and homogeneous; consequently, the condition for a nontrivial solution is that the determinant of the coefficients of the amplitudes $v_\alpha(\kappa, \boldsymbol{k}_\parallel)$ be zero:

$$\left| D_{\alpha\beta}(\kappa, \kappa'; \boldsymbol{k}_\parallel) - \omega^2 \delta_{\alpha\beta} \right| = 0. \qquad (7)$$

Eq. (7) is the dispersion relation for the normal modes of vibration of the crystal with surfaces. It specifies the frequency ω as a function of the wave vector $\boldsymbol{k}_\parallel$ for each branch j. Once the normal mode frequencies $\omega(\boldsymbol{k}_\parallel j)$ are calculated, they can be substituted back into eq. (5) and the amplitudes $v_\alpha(\kappa, \boldsymbol{k}_\parallel j)$, or eigenvector components, for each normal mode determined. One finds that the normal modes can be classified as surface modes, bulk modes or mixed modes.

For $\boldsymbol{k}_\parallel$ oriented along a high symmetry direction in the surface Brillouin zone (SBZ), a further classification of the normal modes can be made. Modes in which the atomic displacements lie in the sagittal plane defined by the normal to the surface and the direction of $\boldsymbol{k}_\parallel$ are referred to as sagittal modes. Modes in which the displacements are normal to the sagittal plane are referred to as shear horizontal modes.

5. Simple model calculations

The simplest system to analyze is the linear monatomic chain with nearest neighbor central interactions and free ends. If all force constants in the chain have the same value α, no surface mode exists [9]. This is a direct consequence of Rayleigh's theorem. If the force constant coupling an end atom to its interior neighbor is increased to α', a surface mode appears for $\alpha' > 4\alpha/3$ in the region above the maximum frequency of the infinitely long periodic chain [10].

For the linear diatomic chain with free ends the situation is quite different. There is a gap between the acoustic and optical branches. If there is a light atom at a free end, the lowest frequency mode of the optical branch of the periodic chain descends into the gap when the surface is created and becomes a surface mode. Its frequency ω_s is specified by [8]

$$\omega_s^2 = \frac{\alpha(m + M)}{mM}, \qquad (8)$$

where m and M are the masses of the lighter and heavier atoms, respectively, in a chain of $2N$ atoms. The displacement amplitudes decrease essentially exponentially from the end with the lighter atom to the end with the heavier atom. The relationship of the surface mode frequency to that of bulk modes is shown in Fig. 1. We see that ω_s^2 lies exactly in the middle of the gap. By reducing the mass of the end atom with the lighter mass one can shift the surface mode from

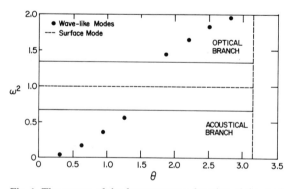

Fig. 1. The square of the frequency as a function of the mode index θ for a diatomic chain with end atoms different (after ref. [8]).

the gap to the region above the top of the optical branch [11].

For two- and three-dimensional monatomic lattices a simple model is the Montroll–Potts–Rosenstock–Newell (MPRN) model which consists of nearest neighbor central and noncentral interactions. Simple square and simple cubic lattices based on the MPRN model do not exhibit surface modes on (01) and (001) surfaces, not even Rayleigh waves [12]. The reason is that the equations of motion for the x-, y-, and z-components of displacement are uncoupled. To get a Rayleigh wave one must have coupling between the displacement components in the sagittal plane. Ludwig and Lengeler [13] have shown how Rayleigh waves can be introduced into the MPRN model by a suitable modification of the surface force constants that renders the model rotationally invariant and introduces the necessary coupling of the displacement component. However, it is not necessary to change the surface force constants to get a Rayleigh wave if the surface makes an oblique angle with the bonds between nearest neighbor atoms [14].

An alternative to the Ludwig–Lengeler procedure is to build rotational invariance into the model automatically by associating the forces acting on the displaced atoms with changes in the distance between atoms and the angle between atomic bonds. A simple application of this idea was made by Gazis et al. [4], who treated the monatomic simple cubic lattice with nearest and next-nearest neighbor central forces plus angle-bending interactions involving pairs of nearest neighbors. The three force constants involved can be determined in terms of the elastic moduli by passing to the long wavelength limit. Specific calculations were carried out for surface waves propagating in the [100] direction on a (001) surface of KCl (considered monatomic). The results exhibit the dispersion of the Rayleigh wave expected in a lattice dynamical theory. Another interesting feature is the transition of the wave from an ordinary Rayleigh wave to a generalized Rayleigh wave at a critical wave number. The model just described has been applied to (001) surfaces of body-centered cubic crystals such as vanadium and iron [15].

Ordinarily, diatomic crystals have ionic character, and the long-range Coulomb interaction between the ions must be taken into account. Some qualitative features, however, may be studied using models with only short-range interactions. A general formulation of the problem based on Green's function methods was presented in their 1948 paper by Lifshitz and Rosenzweig [7], who noted the existence of both acoustical and optical surface modes. Wallis [12] applied the MPRN model to finite two- and three-dimensional diatomic crystals of the rock-salt structure using both a perturbative and a Green's function procedure. It was found that a band of surface-mode frequencies lies in the gap between the acoustical and optical branches. These modes are closely related to the surface mode of the linear diatomic chain with nearest-neighbor interactions. Furthermore, for the finite crystals considered, edge and corner modes were found in which the atomic displacements are localized near edges and corners, respectively. Typically, the frequencies of edge modes lie below those of the surface modes mentioned above, and the frequencies of corner modes lie below those of edge modes. This is physically reasonable, in view of the decreasing number of interactions with neighbors possessed by an atom as it changes from a surface atom to an edge atom to a corner atom.

A surface-mode calculation using a rotationally invariant model for a rock-salt crystal was carried out by Wallis et al. [16], who found a surface branch in the gap between the acoustical and optical branches. It was noted by both Wallis et al. and by Lifshitz and Rosenzweig that the surface modes with frequencies in the gap can give rise to a peak in the infrared absorption spectrum of the crystal.

6. Rare-gas solids

Calculations based on simple models with short range interactions typically do not give a good representations of the vibrational modes of real crystals. However, a case where realistic calculations can be made consists of rare-gas solids which crystallize in the face-centered cubic struc-

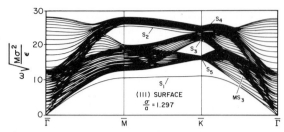

Fig. 2. Frequency versus wave vector curves for an 11-layer slab of a monatomic fcc crystal with two (111) surfaces. The surface modes are denoted by S_i (after ref. [17]). The Lennard-Jones potential parameters are σ and ϵ.

ture. The interatomic interactions are well described by Lennard-Jones potentials of the 6–12 type, from which the dynamical matrix can be constructed. With the computing power that has been available in recent years, calculations of surface and bulk mode frequencies and eigenvectors are quite feasible.

Such calculations have been carried out for slabs of rare-gas solids by Allen et al. [17]. Typical results for the dispersion curves of a slab bounded by (111) surfaces are shown in Fig. 2. The Rayleigh surface mode (mode S_1) lying below the bulk continuum is clearly evident. Allen et al. also found additional surface modes of non-Rayleigh type whose frequencies lie in gaps in the spectrum of the bulk modes. Examples are the S_2 and S_3 modes shown in Fig. 2. The atomic displacements of the non-Rayleigh surface modes are primarily parallel to the surface, whereas those of Rayleigh modes are primarily perpendicular to the surface.

Hexagonal close-packed crystals have also been treated by Allen et al. [18] with similar results. These same authors have given [19] a general formulation of the surface-vibrational-mode problem. They discuss the circumstances under which the displacement ellipse of the particles lies in the sagittal plane and when surface modes can have frequencies lying within bulk-mode bands. A nomenclature for surface modes was presented that has become the standard.

Detailed calculations have been reported by Allen et al. [20] for monatomic fcc crystals with (111), (100) and (110) surfaces. They find a number of surface-mode branches for each surface:

five for the (111) surface, at least nineteen for the (100) surface, and ten for the (110) surface. These results are probably dependent, to some extent, on the model employed which includes interactions between all pairs of atoms through a Lennard-Jones potential. Some of the surface modes found by Allen et al. [20] are primarily localized in the second layer from the surface, or even a deeper layer, rather than in the surface layer itself. These various surface modes can be regarded as "peeling off" in succession from the bulk branches. The effects of surface force-constant changes were also considered.

7. Ionic crystals

Since diatomic crystals that are insulators or semiconductors typically have ionic character, a realistic lattice-dynamical model must include the Coulomb interactions between the ions, as well as interactions associated with the polarizability of the ions. The long-range character of the Coulomb interactions makes the calculation of the surface-mode frequencies somewhat difficult, because the effects of the surface are not localized within a few atomic layers of the surface. Furthermore, the possibility of retardation effects associated with the electromagnetic field introduces an aspect to the problem not encountered when only short-range interactions are considered.

We can formulate the surface mode problem with Coulomb interactions by simply noting that the coupling constant matrix $\Phi_{\alpha\beta}(l\kappa, l'\kappa')$ can now be written as the sum of Coulomb and short-range parts:

$$\Phi_{\alpha\beta}(l\kappa, l'\kappa') = \Phi_{\alpha\beta}^{c}(l\kappa, l'\kappa') + \Phi_{\alpha\beta}^{s}(l\kappa, l'\kappa').$$

$$(9)$$

The Coulomb part has the form

$$\Phi_{\alpha\beta}^{c}(l\kappa, l'\kappa') = \frac{\partial^2}{\partial R_{\alpha}(l\kappa)\partial R_{\beta}(l'\kappa')}$$

$$\times \left(\frac{q_{\kappa}q_{\kappa'}}{|\boldsymbol{R}(l\kappa) - \boldsymbol{R}(l'\kappa')|} \right)\Bigg|_{\boldsymbol{u}(l\kappa)=0},$$

$$(10)$$

where q_κ, $q_{\kappa'}$ are the electrical charges on ions of type κ, κ', respectively. The short-range part can be treated in terms of central forces and angle-bending forces, as was done earlier. The calculation of the surface mode frequencies then requires the solution of the secular Eq. (7) using the coupling constants specified by Eqs. (9) and (10).

Fuchs and Kliewer (FK) [21] carried out a calculation for a slab of a NaCl-type crystal with a (001) surface for both rigid and polarizable ions. The short-range interactions were taken to be NN central forces. FK neglected the changes in the short-range coupling constants and the relaxation of the ions near the surfaces, but took full account of the Coulomb interactions. However, a simpler procedure, as pointed out by Kliewer and Fuchs [22], is to treat the problem macroscopically in terms of Maxwell's equations and the appropriate boundary conditions. The effects of retardation can be included in a straightforward fashion.

Consider a slab of thickness $L = 2d$ with surfaces normal to the z-direction at $z = \pm d$ and immersed in vacuum. The material is assumed to be optically isotropic with a dielectric constant $\epsilon(\omega)$ taken to be that of the bulk medium. Kliewer and Fuchs found a macroscopic surface optical mode whose dispersion relation is

$$\epsilon(\omega) = \frac{\alpha}{\alpha_0}\left(\frac{e^{-\alpha d} \mp e^{\alpha d}}{e^{-\alpha d} \pm e^{\alpha d}}\right),\qquad(11)$$

where

$$\alpha^2 = k_\parallel^2 - (\omega^2/c^2)\epsilon(\omega),\qquad(12a)$$

$$\alpha_0^2 = k_\parallel^2 - (\omega^2/c^2).\qquad(12b)$$

This mode is hereafter called the *Fuchs–Kliewer mode*.

To complete the calculation, $\epsilon(\omega)$ must be specified. For cubic crystals of the rocksalt or zincblende structure, we have

$$\epsilon(\omega) = \epsilon_\infty(\omega_L^2 - \omega^2)/(\omega_T^2 - \omega^2),\qquad(13)$$

where ω_T and ω_L are the limiting long-wavelength frequencies of transverse and longitudinal optical phonons, respectively, and ϵ_∞ is the dielectric constant at frequencies large compared

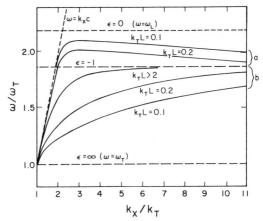

Fig. 3. Frequencies of transverse optical surface modes versus wave vector for a slab of LiF bounded by (001) surfaces. The bulk transverse optical phonon frequency is ω_T, the dielectric function of LiF is ϵ, and the thickness of the slab is L (after ref. [22]).

to ω_L, but small compared to interband electronic transition frequencies. The dispersion relations obtained by solving Eqs. (11) and (13) for several thicknesses of LiF are shown in Fig. 3. For a given thickness, there are two surface modes corresponding to the two surfaces. As $L \to \infty$, the two modes coalesce, because the coupling between the front and back surfaces disappears.

An independent treatment of the optical modes of vibration of ionic crystals, finite in one or more dimensions, has been given in a series of papers by Englman and Ruppin [23]. They generalize the results of FK to crystals of diverse shapes using the same assumptions as FK.

It will be recalled that FK made two important approximations in their treatment, namely, they neglected the effect of the surface on the short-range iterations and they assumed the wavelength to be large compared to the lattice spacing. Lucas [24] avoided the first of these approximations in an investigation of optical surface modes in thin slabs of an NaCl-type crystal with (001) surfaces. Nearest-neighbor central forces were assumed for the short-range interactions, and the absence of neighbors beyond the boundaries was taken into account in the equations of motion for the surface atoms. However, relaxation of the ions near the surfaces and retardation were neglected. The ions were assumed to be rigid, and the appropriate Coulomb sums were evaluated in plane-wise fashion. Since each lattice plane contains equal numbers of positive and negative ions and is overall neutral, the interactions between planes decrease exponentially with separation, and the convergence of the interplanar interactions is therefore very rapid.

Lucas carried out specific calculations only for $k_\parallel = 0$, where $k_\parallel$ is the two-dimensional wave vector parallel to the surface. He found a transverse optical (TO) surface mode (displacements parallel to the surface), known as the *Lucas mode*, with a frequency very slightly below that of the transverse optical bulk mode for $k = 0$. This result is in contrast to that of FK, who found that the transverse optical surface mode is not localized at the surface for $k_\parallel = 0$. This difference is attributable to the inclusion by Lucas of corrections to the short-range forces acting on surface ions due to their smaller number of neighbors. However, Lucas found no longitudinal optical (LO) surface modes (displacements perpendicular to the surface) for $k_\parallel = 0$, in agreement with FK.

A more refined calculation for the NaCl slab with (001) surfaces has been carried out by Tong and Maradudin (TM) [25] who assumed NN central forces and Coulomb forces between rigid ions. Proper account was taken of the number of short-range interactions of surface atoms, and the possibility of changes in the interplanar spacing near the surface was included in the formulation, but retardation was neglected. Calculations were made for wave vectors $k_\parallel$ covering the entire two-dimensional first Brillouin zone.

If we neglect the doubling of the surface modes, due to the presence of two surfaces, then we can characterize the results of TM as follows. For propagation in the [100] direction, there are four surface branches. One branch is acoustical in character and corresponds to Rayleigh waves. There are three optical surface branches. A high-frequency branch lies between ω_L and ω_T and approaches a frequency somewhat below ω_L as $k_\parallel \to 0$. This branch would be significantly affected by retardation. A low-frequency branch lies below the bulk optical mode frequencies for

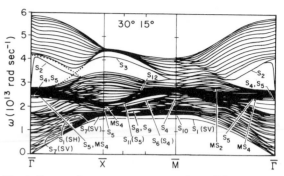

Fig. 4. Frequencies versus wave vector for a 15-layer slab of NaCl bounded by (001) surfaces. The surface modes are labeled by S_i and mixed modes by MS_i (after ref. [28]).

small $k_\parallel$, approaches a frequency just below ω_T as $k \to 0$, and is polarized in the sagittal plane. An intermediate-frequency branch has the same frequency at $k_\parallel = 0$ as the low-frequency branch, but rises into the region between ω_L and ω_T, and is polarized parallel to the surface. TM calculated the deviation of the slab frequency distribution from the bulk distribution and found pronounced peaks corresponding to the surface branches.

The investigations of FK and TM do not lead to an entirely consistent picture of the surface optical modes of an ionic crystal slab. The problem of reconciling these results has been undertaken by Jones and Fuchs [26] and Chen and coworkers [27,28]. The following picture emerges as shown in Fig. 4. The high-frequency branch of TM splits as $k_\parallel \to 0$ into two branches, one approaching ω_L and the other ω_T. These are the FK modes. (TM had not used a sufficiently small $k_\parallel$ to reveal the splitting.) Near $k_\parallel = 0$, the FK modes enter the bulk LO and TO continua and become "mixed" or pseudosurface waves. The low-frequency branch enters the *acoustical* bulk continuum about midway to the Brillouin zone boundary and becomes a mixed-mode branch. The intermediate branch does not couple to bulk modes, and remains a pure surface branch out to the zone boundary.

Chen and coworkers [27,28] have investigated several directions of propagation and have found a variety of surface modes in gaps in the bulk frequency spectrum. They have also utilized the shell model for an NaCl slab, and find that the FK modes are pseudosurface waves for all wave vectors. The other surface optical modes are still true surface modes near $k_\parallel = 0$. The shell-model calculations have been extended [29] to RbF and RbCl, the rocksalt structure materials having the largest gaps between acoustical and optical branches. Well defined *microscopic* surface-mode branches [7,8], with frequencies lying within the gap, were found for the (001) surfaces. Surface modes associated with the (001) surface of LiF have also been studied [30] using the shell model. A collection of results for the (001) surface of LiF, MgO, NaCl, NaF, NaI, RbCl, and RbI have been published by Chen et al. [31] for the unrelaxed surface and for these materials plus KBr, KCl, KF, KI, NaBr, RbBr, and RbF by Kress et al. [32] for the relaxed surface.

An important stimulus to the study of surface phonons in ionic crystals was provided by Toennies and collaborators who used inelastic helium atom scattering to determine experimental surface phonon dispersion curves for LiF(001) [33], KCl(001) [33], NaCl(001) [34], NaF(001) [35,36], and MgO(001) [37]. In the case of NaCl(001), the experimental points lie between the theoretical results of Chen et al. [31] mentioned above and those of Benedek and Miglio [38] based on a Green's function approach using the breathing-shell model. For LiF(001) and NaF(001) the experimental results agree well with the results calculated by Benedek et al. [39] using the Green's function method. In the case of NaF(001) (and other nearly equimass alkali halides), the data can be interpreted in terms of a folding back of the SBZ of a simple cubic lattice at the midpoint of the $\overline{\Gamma M}'$ direction [36]. The experimental data for KCl(001) are in reasonably good agreement with the Green's function results of Benedek and Galimberti [40]. The data reveal the "crossing mode" predicted by the latter authors. In the case of MgO the experimental results agree well with the theoretical calculations.

The shell model employed by de Wette and coworkers and by Benedek and coworkers approximates the electron distribution of an ion by an infinitesimally thin shell surrounding the ion core. The shell can be displaced relative to the core of the ion and is coupled to the core by Hooke's law forces. A given shell is coupled to

the cores of neighboring ions and to neighboring shells by additional Hooke's law forces.

The dynamical equations of the perfect crystal can be written in the form [39]

$$M\omega^2 u = \phi_{zz}u + \phi_{zy}d, \qquad (14a)$$

$$m\omega^2 d = \phi_{yz}u + \phi_{yy}d, \qquad (14b)$$

where u is the displacement vector of the cores, d is a vector describing the shell displacements with respect to the cores, M and m are core and shell effective masses, respectively, ϕ_{zz} is the rigid ion force-constant matrix, ϕ_{zy} and ϕ_{yz} are the coupling matrices between core displacements and shell displacements and deformations, and ϕ_{yy} is the shell-shell coupling matrix. Within the framework of the adiabatic approximation, one can set the shell mass m equal to zero in Eq. (14b) and then eliminate d from Eqs. (14a) and (14b) to give

$$M\omega^2 u = \left(\phi_{zz} - \phi_{zy}\phi_{yy}^{-1}\phi_{yz} \right)u. \qquad (15)$$

From the eigenvalues and eigenvectors of Eq. (15) one can construct the Green's function and from it obtain the eigenquantities for the crystal with a surface. In carrying out this procedure it is important to impose the conditions of infinitesimal translational and rotational invariance [39].

Electron energy loss spectroscopy has been applied to the determination of surface phonon spectra of the (001) surface of rock-salt structure crystals by Oshima and coworkers. Their investigation of MgO [41] revealed the S_1 (Rayleigh) mode and the S_2 (Wallis) mode. The latter mode was observed at the $\overline{\Gamma}$ point at the bottom of the optical bands. Other (001) surfaces of rock-salt structure crystals investigated by Oshima and coworkers are TiC, ZrC, HfC, NbC, and TaC. In the case of TiC [42] the Wallis mode appears near the middle of the gap between the acoustical and optical branches, consistent with early calculations [8].

A more complicated ionic crystal structure that has received recent attention is the perovskite structure. Schröder, de Wette and collaborators have calculated surface phonon dispersion curves for the (001) surface of several examples of this structure using a shell model with Born–Mayer repulsive interactions. In the cases of KMnF$_3$ and

KZnF$_3$ [43] the surface modes exhibit a very rich spectrum. A soft surface mode was found at the $\bar{M}$ point for KMnF$_3$ that can give rise to surface reconstruction. This reconstruction has been observed by Toennies and Volmer [44] using inelastic helium atom scattering. It has a transition temperature of 191 K. A similar investigation has been carried out for SrTiO$_3$ [45].

8. Semiconductors

Semiconductors are much more difficult to treat theoretically than ionic insulators. The interatomic interactions in semiconductors are associated with covalent bonds and cannot be represented by simple two-body interactions such as Coulomb interactions. Three-body interactions associated with the deformation of the angle between two bonds are important, as indicated by the large deviations from Cauchy's relation found in the elastic moduli.

Early calculations of surface phonon dispersion curves for Si(111) were made by Ludwig [46] who used an empirical model containing central forces out to fourth neighbors, angle-bending forces involving changes in covalent bond angles, and a mixed interaction involving angle bonding and bond stretching. The force constants for the interior of the crystal were determined by fitting them to the phonon dispersion curves of bulk silicon. All force constants involving a surface atom were obtained by multiplying the corresponding bulk value by the same factor Q, which was determined by fitting the frequency (55 meV) of the Ibach surface mode [47]. Ludwig used a variety of methods to calculate the surface mode frequencies and eigenvectors. Both the Rayleigh mode and a second surface mode below the bulk continuum were found. In addition, a number of surface modes in gaps in the bulk continuum occur.

Ludwig also considered the reconstructed Si(111)2 × 1 surface. He suggests various possibilities for the origin of the Ibach mode. His work has been extended by Goldammer et al. [48] to Si(111)7 × 7 using a Green's function procedure.

A different theoretical approach to the vibrational properties of reconstructed silicon surfaces has been provided by Allan and Mele [49] who considered the (001)2 × 1 surface. They utilized the tight binding method of Chadi [50] to calculate the total electronic energy and added to this a nearest-neighbor elastic interaction to model the ion–ion repulsion. By assuming that the tight binding parameters vary inversely as the square of the interatomic separations, Allan and Mele were able to calculate the dynamical matrix and thence the normal mode frequencies for the slab considered. Allan and Mele found the Rayleigh mode and several other surface modes below the bulk continuum. A variety of surface modes appear in gaps in the bulk continuum (see also Mazur and Pollman [51]). Significant changes of ~ 20–30% were found in the surface force constants compared to the bulk values. Alerhand and Mele [52] have extended the work of Allan and Mele by adding a Hubbard-type on-site electron–electron interaction to the Hamiltonian. Switching on this interaction causes the lowest gap surface mode to disappear. The Si(001)2 × 1 surface has also been studied by Tiersten et al. [53] who calculated spectral densities for surface and bulk atoms using the Keating model [54].

The tight-binding approach was applied to Si(111)2 × 1 by Alerhand et al. [55] who interpreted the surface mode observed experimentally by Ibach as a longitudinal optical mode of π-bonded chains on the surface. This work was extended by Alerhand and Mele [56] who added a Hubbard-type on-site electron–electron interaction to the Hamiltonian and were able to account for a nearly dispersionless surface mode at 10.5 meV observed by Harten et al. [57] using inelastic helium atom scattering. Alerhand and Mele interpret this mode as a transverse "bouncing" vibration of the π-bonded chain. The dispersionless character is attributed to mode softening near the zone boundary arising from polarization effects.

An alternative interpretation of the Ibach and Harten et al. surface modes has been presented by Miglio et al. [58] who employed the bond-charge model of Weber [59] to calculate the surface and bulk modes for a slab. Miglio et al. state that the folding back to the $\bar{\Gamma}$ point of the $\bar{M}$-point

Rayleigh and Lucas modes of the ideal 1×1 surface produces the Harten et al. and Ibach modes, respectively. This interpretation is consistent with recent experimental data of del Pennino et al. [60].

The Car–Parrinello ab initio molecular dynamics method has been applied to Si(111)2×1 by Anciletto et al. [61]. Their results yield both the Ibach mode at 55 meV and the Harten et al. mode at 10 meV.

The bond-charge model of Weber [59] has been used by Santini et al. [62] to investigate surface phonons on the Ge(111)2×1 reconstructed surface. The surface phonon spectrum resembles quite closely that of the corresponding Si system. The Ge(111)8×8 surface has been discussed by Zierau et al. [63].

The (110) surface of GaAs has been studied theoretically by Goldammer and Ludwig [64] who use a model involving short-range central and angle-bending forces plus forces arising from the interaction of dynamically induced electric dipoles. Similar results have been obtained by Santini et al. [65] with their bond-charge model. Of particular interest are relatively dispersionless surface modes or resonances at approximately 10 and 13 meV that correlate rather well with modes observed experimentally by Harten and Toennies [57] and by Doak and Nguyen [66]. The similarities between GaAs(110) and Si(111)2×1 can be attributed to the similar surface geometries of the two systems.

Calculations of surface phonon dispersion curves for the layered semiconductor GaSe have been carried out by Brusdeylins et al. [67] using a shell model. Significant changes in the surface force constants from their bulk values are required in order to fit the large separation of the surface mode frequencies from those of the associated bulk band observed experimentally by inelastic helium atom scattering. This large separation is at first glance unexpected for a layered compound.

9. Metals

As in the case of semiconductors, phonons in metals for the most part cannot be described by two-body interactions with a simple physical significance. The free electron gas plays an important role in the violation of Cauchy's relation by the elastic moduli of a typical metal [68]. A full understanding of phonons in metals requires a first-principles calculation of the electronic ground state energy from which the force constant matrix can be obtained.

In the early days of surface phonons, first principles calculations for metals could not be made because the necessary computing power did not exist. Recourse was therefore made to empirical models. In particular, the bulk phonons of nickel are remarkably well described by a nearest-neighbor central force model [69,70]. This model was used by Clark et al. [69] and by Wallis et al. [70] to calculate the mean-square displacements of surface atoms for comparison with experimental data from low-energy electron diffraction measurements on nickel [71].

Other metals than nickel typically require more complicated models. The Rayleigh mode for iron (001) and vanadium (001) was investigated by Gazis and Wallis [15] using a model, the CGW model [72], which contains nearest and next-nearest neighbor central interactions plus angle-bending interactions involving two pairs of nearest neighbors sharing a common atom. An extension of this model to include third and fourth neighbor central interactions was made by Cheng et al. [73] in connection with an analysis of surface-atom mean-square displacements determined by LEED for vanadium (001). The analogue of the CGW model for face-centered cubic metals was employed by Castner et al. [74] to investigate surface-atom mean-square-displacements of rhodium (001) and (111) surfaces.

The advent of inelastic electron and helium-atom scattering in the early 1980's as a means of determining surface phonon dispersion curves stimulated increased activity in the use of empirical models to calculate these curves for metals. Armand [75] obtained both dispersion curves and mean-square-displacements for the fcc (111) surface with a nearest-neighbor central-force model and a Green's function procedure. The bcc (001) surface has been studied by Castiel et al. [76], Bortolani et al. [77], and Black et al. [78].

A specific case of interest is W(001) which exhibits surface reconstruction to W(001)c(2 × 2). Fasolino and Tosatti [79] used a model containing first- and second-neighbor central potentials in the bulk and a nearest-neighbor central potential in the surface to calculate the phonon dispersion curves of the unreconstructed surface. By suitably choosing the surface force constants, they obtained unstable surface modes at the $\overline{M}$ point corresponding to the observed reconstruction. A similar model has been employed by Reinecke and Ying [80] to determine the surface phonon dispersion curves for the reconstructed W(001) surface. The tight-binding approach has been applied by Wang and Weber [81] to the study of surface phonons on W(001). They find that the maximum surface phonon instability occurs at $\overline{M}$ as expected for the c(2 × 2) reconstruction and consistent with the surface phonon mode observed experimentally by Ernst et al. [82]. The bcc (110) and (111) surfaces have been investigated by Black et al. [83] and by Black [84], respectively.

Mention has already been made of the important studies of fcc surfaces by de Wette and coworkers. Work more specifically directed toward metals includes that of Castiel et al. [76], Lehwald et al. [85], Black et al. [78], and Moretto et al. [86] for the (001) surface. The (110) surface has been considered by Black et al. [87], Bracco et al. [88], Tatarek et al. [89], Lahee et al. [90], and Balden et al. [91]. The fcc (111) surface has received a great deal of attention. We mention, among others, the work of Armand [75], Black et al. [92], Bortolani et al. [93], and Mohamed et al. [94]. We comment specifically on the latter two works. Bortolani et al. treated the (111) surfaces of Ag and Au for which surface phonon dispersion curves had been determined experimentally by Harten et al. [95]. The model employed by Bortolani et al. consists of central interactions out to second neighbors plus angle-bending interactions. In order to fit all the experimental features, particularly the longitudinal resonance, they found it necessary to reduce the lateral force constants in the surface layer for Ag and Au by 52 and 70%, respectively. In the case of Cu(111), a softening of 50% has been suggested by Harten et al. [95] and of 70% in a private communication

by Santoro and Borolani [94]. A quite different conclusion was reached by Mohamed et al. for Cu who were able to fit EELS data with only a 15% reduction in the surface force constant. This fit included the S_2 gap surface mode at $\overline{M}$. Further discussion of this discrepancy will be presented later. Other (111) surfaces that have been analyzed are Al by Lock et al. [96], Ni by Menezes et al. [97], and Pt by Bortolani et al. [98].

The controversy concerning force-constant changes on Cu(111) illustrates the desirability of calculating these quantities from first principles. A step in this direction has been taken by Jayanthi et al. [99] who introduced the pseudoparticle model which provides a localized description of many-body interactions. The pseudoparticle is a coreless particle whose deformation corresponds to the multipole expansion of the electron density about the site between adjacent nuclei where the electron density is a minimum. In a fashion similar to that found with the shell and bond-charge models, the pseudoparticles are coupled to neighboring nuclei by harmonic forces. A change in the pseudoparticle-ion force constant at the surface was determined from the work function. Invoking the adiabatic approximation, one can eliminate the pseudoparticle coordinates from the equations of motion and solve the resulting set of equations for the surface and bulk phonon frequencies and eigenvectors of a slab bounded by the surfaces of interest. The results for the (111) surfaces of Cu, Ag, and Au are in good agreement with the experimental data of Harten et al. [95].

In a recent paper, Kaden et al. [100] present a refined calculation for Cu(111) using the pseudoparticle model. They obtain very good agreement with the experimental data for the Rayleigh mode, the longitudinal resonance, and the S_2 gap mode at $\overline{M}$. A reduction of $\sim 30\%$ in the lateral surface force constant was found. The intensity of the longitudinal resonance is explained in terms of enhanced electron deformabilities in the surface region.

Another approach to the inclusion of many-body interactions in the theory is the embedded atom method (EAM) introduced by Daw and Baskes [101]. This method draws upon density functional theory and is based on the idea that the energy of an impurity in a host crystal is a functional of the electron density of the unperturbed, pure host. One can treat pure crystals by considering each atom as an impurity in the host consisting of all other atoms.

In the EAM the total energy of the crystal is written as

$$E_{\text{tot}} = \sum_i F_i(\rho_{\text{h},i}) + \tfrac{1}{2} \sum_{ij}' \varphi_{ij}(R_{ij}), \qquad (16)$$

where F_i is the embedding energy of atom i, $\rho_{\text{h},i}$ is the electron density of the host at the position R_i but without atom i, φ_{ij} is a short-range pair potential, and the prime on the second sum in Eq. (16) signifies that the term $i = j$ is omitted. The embedding energy F_i is expressed as a parameterized function of the density $\rho_{\text{h},i}$ with the parameters determined by fitting quantities such as the lattice constant and elastic moduli of the pure bulk crystal. If the crystal has a surface, the embedding energy of a surface atom differs from that for a bulk atom because the electron density is different for the two types of atoms.

The EAM has been applied to the calculation of surface phonon dispersion curves of Cu(111) and Ag(111) without fitting any experimental surface data [102]. The agreement with the experimental results of Mohamed et al. [94] is quite good. Nelson et al. [102] obtained a softening of the lateral surface force constant of 11%, considerably less than that suggested by Harten et al. [95], and more in line with the 15% softening of Mohamed et al. Nelson et al. note the existence of an avoided crossing of sagittal plane modes that forces down the longitudinal resonances and removes the necessity for a large softening of the lateral surface force constant.

10. First-principles calculations for metals

The advent of enhanced computing power during the 1980's in the form of supercomputers made possible for the first time the calculation of surface force constants and surface phonon dispersion curves of metals from first principles. In 1986 Ho and Bohnen published results of a calcu-

Table 1
Comparison of the values of the energies of the surface phonons measured by Toennies and Wöll for Al(110) at the $\overline{X}$ and $\overline{Y}$ points in the SBZ with the theoretical values obtained by Equiluz et al. and by Ho and Bohnen (energies are in millielectronvolts)

	$\overline{X}$		$\overline{Y}$	
	Unrelaxed	Relaxed	Unrelaxed	Relaxed
Equiluz et al. [a]	15.8	19.0	10.0	13.0
			12.4	13.8
Ho and Bohnen [b]	15.1	17.2	8.7	10.4
			13.2	13.9
Toennies and Wöll [c]		14.8		9.3
				13.5

[a] Ref. [110].
[b] Ref. [103].
[c] Ref. [104].

lation of surface mode frequencies of Al(110) [103]. Their procedure consisted of calculating the total electronic ground state energy using the density functional method in the local density approximation (LDA). The electron–ion interaction was represented by ab initio pseudopotentials. By displacing the surface layer of atoms in directions parallel and perpendicular to the surface, interlayer force constants involving the surface layer and interior layers were calculated using the Hellman–Feynman theorem. Intralayer surface force constants were obtained by displacing the surface atoms in a manner corresponding to the zone-boundary wave vectors $\overline{X}$ and $\overline{Y}$ in the SBZ. The couplings between inner layers were assumed to have the bulk values.

From the force constants thus determined, the surface mode frequencies at the $\overline{X}$ and $\overline{Y}$ points were calculated. The results together with the experimental values of Toennies and Wöll [104] are presented in Table 1. The agreement is reasonably good. By fitting the calculated force constants to a Born–von Karman model, Ho and Bohnen were able to calculate the surface phonon dispersion curves in the principle directions in the SBZ.

The work just described has been extended by Bohnen and Ho to the Al(001) surface [105]. From the calculated force constants, surface phonon frequencies were obtained at the $\overline{X}$ and

$\overline{M}$ points. The value predicted at $\overline{X}$ is in very good agreement with the subsequent experimental data of Mohamed and Kesmodel [106]. A further extension has been carried out by Rodach et al. for the Na(110) surface [107]. Unfortunately, no experimental data exist for surface phonon frequencies of sodium. Additional calculations have been made for Cu(001) and Ag(001) [108] and compared with experimental data.

Very recently, an interesting calculation has been reported by Chen et al. [109] for Cu(111) and Ag(111) based on the procedure of Ho and Bohnen. Chen et al. find a shear vertical resonance (SVR) lying in the bulk continuum along the $\overline{\Gamma M}$ and $\overline{\Gamma K}$ directions that they identify with the upper surface mode observed by Harten et al. [95] using inelastic helium atom scattering. These first-principles calculations give a lateral surface force constant that is softened with respect to the bulk by $\sim 13\%$ for Cu(111) and by $\sim 8\%$ for Ag(111). The value for Cu(111) is in line with those of Nelson et al. [102] and Mohamed et al. [94], but much smaller than the much quoted value of 50–70% [94].

An alternative approach to the calculation of surface phonon dispersion curves is provided by the self-consistent screening theory based on linear response theory combined with density functional theory. In a perturbative version of this approach, Eguiluz et al. [110] considered an aluminum slab bounded by (110) surfaces and treated the electron–ion interaction, represented by a local pseudopotential, as a perturbation. The zero-order problem is a jellium slab, whose density response function for noninteracting electrons $\chi^{(0)}(r, r')$ was calculated in the LDA from the eigenvalues and eigenfunctions of the Kohn–Sham equations of density functional theory. The density response function for interacting electrons $\chi(r, r')$ was obtained from $\chi^{(0)}(r, r')$ by solving an integral equation whose kernel consists of the bare Coulomb interaction plus the exchange and correlation potential. If $V_1(r)$ is the difference between the potential due to the discrete array of pseudo-ions and that due to the jellium, the total ground state energy E_{gs} of the conduction electrons to second order in $V_1(r)$ can be written as

$$E_{gs} = E_{gs}^{(0)} + \int d^3 r \, n_0(r) V_1(r) \qquad (17)$$

$$+ \tfrac{1}{2} \int d^3 r \int d^3 r' V_1(r) \chi(r, r') V_1(r'), \quad (18)$$

where $n_0(r)$ is the jellium electron number density and $E_{gs}^{(0)}$ is the electronic ground state energy for $V_1 = 0$. Differentiating Eq. (17) twice with respect to the components of displacement of the ions from their equilibrium positions yields the electronic contribution to the atomic force constants. On adding to this the contribution from the direct ion–ion interaction, one obtains the dynamical matrix of the slab.

The surface mode frequencies obtained for Al(110) at the $\overline{X}$ and $\overline{Y}$ points are presented in Table 1 for both relaxed and unrelaxed surfaces. The values for the unrelaxed surface are in rather good agreement with the theoretical results of Ho and Bohnen and the experimental results of Toennies and Wöll, but the values of the lowest frequency surface mode for the relaxed surface are appreciably higher than those of the other workers. The surface phonon dispersion curves of Eguiluz et al. are shown in Fig. 5.

The self-consistent screening procedure of Eguiluz et al. has been applied to the Al(001) surface by Gaspar and Eguiluz [111]. They find that surface stress and surface force-constant stiffening are not required to produce agreement of the calculated Rayleigh wave frequency at $\overline{X}$ with experiment.

Since the perturbative pseudopotential approach discussed above contains no terms beyond

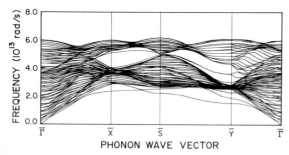

Fig. 5. Calculated surface-phonon dispersion curves along high-symmetry directions in the SBZ for an unrelaxed 27-layer aluminum slab bounded by (110) surfaces (after ref. [110]).

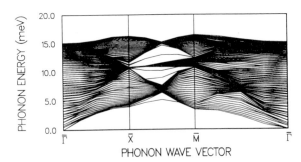

Fig. 6. Calculated surface-phonon dispersion curves along high-symmetry directions in the SBZ for a 51-layer sodium slab bounded by (001) surfaces (after ref. [112]).

second order in the electron–ion perturbation $V_1(r)$, the results involve two-body interactions between atoms, but no three-body and higher-body interactions. This limitation has been removed by Quong et al. [112] who included the electron–ion interaction in the Kohn–Sham equations whose solutions enter into the electron density response function. The electron–ion perturbation is therefore taken into account to all orders. The dynamical matrix is calculated in a fashion analogous to that employed in the perturbative procedure. Numerical calculations have been made by Quong et al. for sodium slabs bounded by (001) surfaces. The results for the surface phonon dispersion curves are given in Fig. 6. Surface modes occur both below the bulk continuum and in the gaps within the bulk continuum. As noted earlier no experimental surface phonon dispersion curves exist for sodium surfaces.

Acknowledgment

The author wishes to acknowledge support by National Science Foundation Grant DMR-89181 84.

References

[1] A.A. Maradudin, E.W. Montroll, G.H. Weiss and I.P. Ipatova, Theory of Lattice Dynamics in the Harmonic Approximation, 2nd ed. (Academic Press, New York, 1971).

[2] Lord Rayleigh, Proc. London Math. Soc. 17 (1887) 4.

[3] R. Stoneley, Proc. Roy. Soc. (London) A 232 (1955) 447.

[4] D.C. Gazis, R. Herman and R.F. Wallis, Phys. Rev. 119 (1960) 533.

[5] T.C. Lim and G.W. Farnell, J. Appl. Phys. 39 (1968) 4319;
G.W. Farnell, in: Physical Acoustics, Vol. 6, Eds. W.R. Mason and R.N. Thurston (Academic Press, New York, 1970) p. 109.

[6] M. Born and Th. von Karman, Phys. Z. 13 (1912) 297.

[7] I.M. Lifshitz and L.N. Rosenzweig, Zh. Eksp. Teor. Fiz. 18 (1948) 1012;
I.M. Lifshitz, Nuovo Cimento Suppl. 3 (1956) 732.

[8] R.F. Wallis, Phys. Rev. 105 (1957) 540.

[9] M. Born, Proc. Phys. Soc. (London) 54 (1942) 362.

[10] R.F. Wallis, Prog. Surf. Sci. 4 (1973) 233.

[11] J. Hori and T. Asaki, Prog. Theor. Phys. 31 (1964) 49.

[12] R.F. Wallis, Phys. Rev. 116 (1959) 302.

[13] W. Ludwig and B. Lengeler, Solid State Commun. 2 (1965) 83.

[14] S.L. Cunningham, Surf. Sci. 33 (1972) 139.

[15] D.C. Gazis and R.F. Wallis, Surf. Sci. 5 (1966) 482.

[16] R.F. Wallis, D.L. Mills and A.A. Maradudin, in: Localized Excitations in Solids, Ed. R.F. Wallis (Plenum, New York, 1968) p. 403.

[17] R.E. Allen, G.P. Alldredge and F.W. de Wette, Phys. Rev. Lett. 23 (1969) 1285.

[18] R.E. Allen, G.P. Alldredge and F.W. de Wette, Phys. Rev. B 6 (1972) 632.

[19] R.E. Allen, G.P. Alldredge and F.W. de Wette, Phys. Rev. B 4 (1971) 1648.

[20] R.E. Allen, G.P. Alldredge and F.W. de Wette, Phys. Rev. B 4 (1971) 1661.

[21] R. Fuchs and K.L. Kliewer, Phys. Rev. 140 (1965) A 2076.

[22] K.L. Kliewer and R. Fuchs, Phys. Rev. 144 (1966) 495; 150 (1966) 573.

[23] R. Englman and R. Ruppin, Phys. Rev. Lett. 16 (1966) 898; J. Phys. C 1 (1968) 614, 1515;
R. Ruppin and R. Englman, J. Phys. C 1 (1968) 630.

[24] A.A. Lucas, J. Chem. Phys. 48 (1968) 3156.

[25] S.Y. Tong and A.A. Maradudin, Phys. Rev. 181 (1969) 1318.

[26] W.E. Jones and R. Fuchs, Phys. Rev. B 4 (1971) 3581.

[27] T.S. Chen, G.P. Alldredge and F.W. de Wette, Solid State Commun. 8 (1970) 2105.

[28] T.S. Chen, G.P. Alldredge, F.W. de Wette and R.E. Allen, Phys. Rev. Lett. 26 (1971) 1543; Phys. Rev. B 6 (1972) 627.

[29] T.S. Chen, G.P. Alldredge and F.W. de Wette, Solid State Commun. 10 (1972) 941.

[30] T.S. Chen, G.P. Alldredge and F.W. de Wette, Phys. Lett. A 40 (1972) 401.

[31] T.S. Chen, F.W. de Wette and G.P. Alldredge, Phys. Rev. B 15 (1977) 1167.

[32] W. Kress, F.W. de Wette, A.D. Kulkarni and U. Schröder, Phys. Rev. B 35 (1987) 5783.

[33] G. Brusdeylins, R.B. Doak and J.P. Toennies, Phys. Rev. B 27 (1983) 3662.

[34] G. Benedek, G. Brusdeylins, R.B. Doak, J.G. Skofronick and J.P. Toennies, Phys. Rev. B 28 (1983) 2104.

[35] G. Brusdeylins, R. Rechsteiner, J.G. Skofronick, J.P. Toennies, G. Benedek and L. Miglio, Phys. Rev. Lett. 54 (1985) 466.

[36] G. Benedek, L. Miglio, G. Brusdeylins, J.G. Skofronick and J.P. Toennies, Phys. Rev. B 35 (1987) 6593.

[37] G. Brusdeylines, R.B. Doak, J.G. Skofronick and J.P. Toennies, Surf. Sci. 128 (1983) 191.

[38] G. Benedek and L. Miglio, in: Ab Initio Calculation of Phonon Spectra, Eds. J.T. Devreese, V.E. van Doren and P.E. van Camp (Plenum, New York, 1982) p. 215.

[39] G. Benedek, G.P. Brivio, L. Miglio and V.R. Velasco, Phys. Rev. B 26 (1982) 497.

[40] G. Benedek and F. Galimberti, Surf. Sci. 71 (1978) 87.

[41] C. Oshima, T. Aizawa, R. Souda and Y. Ishizawa, Solid State Commun. 73 (1990) 731.

[42] C. Oshima, R. Souda, M. Aona, S. Otani and Y. Ishizawa, Surf. Sci. 178 (1986) 519.

[43] R. Reiger, J. Prade, U. Schröder, F.W. de Wette, A.D. Kulkarni and W. Kress, Phys. Rev. 39 (1989) 7938.

[44] J.P. Toennies and R. Vollmer, Phys. Rev. 44 (1991) 9833.

[45] U. Schröder, J. Prade, F.W. de Wette, A.D. Kulkarni and W. Kress, Superlatt. Microstruct. 7 (1990) 247.

[46] W.E.W. Ludwig, Jpn. J. Appl. Phys. Suppl. 2 (1974) 879.

[47] H. Ibach, Phys. Rev. Lett. 27 (1971) 253.

[48] W. Goldammer, W. Ludwig, W. Zierau and C. Falter, Surf. Sci. 141 (1984) 139.

[49] D.C. Allan and E.J. Mele, Phys. Rev. Lett. 53 (1984) 826.

[50] D.J. Chadi, Phys. Rev. Lett. 41 (1978) 1062.

[51] A. Mazur and J. Pollman, Surf. Sci. 225 (1990) 72.

[52] O.L. Alerhand and E.J. Mele, Phys. Rev. 39 (1987) 5533.

[53] S. Tiersten, S.C. Ying and T.L. Reinecke, Phys. Rev. B 33 (1986) 4062.

[54] P.N. Keating, Phys. Rev. 145 (1966) 637.

[55] O.L. Alerhand, D.C. Allan and E.J. Mele, Phys. Rev. Lett. 55 (1985) 2700.

[56] O.L. Alerhand and E.J. Mele, Phys. Rev. Lett. 59 (1987) 657.

[57] U. Harten, J.P. Toennies and Ch. Wöll, Phys. Rev. Lett. 57 (1986) 2947.

[58] L. Miglio, P. Santini, P. Ruggerone and G. Benedek, Phys. Rev. Lett. 62 (1989) 3070.

[59] W. Weber, Phys. Rev. B 15 (1977) 4789.

[60] U. del Pennino, M.G. Betti, C. Mariani, S. Nannarone, C.M. Bertoni, I. Abbati and A. Rizzi, Phys. Rev. B 39 (1989) 10380.

[61] F. Ancilitto, A. Selloni, W. Andreoni, S. Baroni, R. Car and M. Parrinello, Phys. Rev. B 43 (1991) 8930.

[62] P. Santini, L. Miglio, G. Benedek, P. Ruggerone and J.P. Toennies, Surf. Sci. 241 (1991) 346.

[63] W. Zierau, W. Goldammer, C. Falter and W. Ludwig, Superlatt. Microstruct. 1 (1985) 55.

[64] W. Goldammer and W. Ludwig, Surf. Sci. 211/212 (1989) 368.

[65] P. Santini, L. Miglio, G. Benedek, U. Harten, P. Ruggerone and J.P. Toennies, Phys. Rev. B 42 (1990) 11942.

[66] R.B. Doak and D.B. Nguyen, J. Electron Spectrosc. 44 (1987) 205.

[67] G. Brusdeylins, R. Rechsteiner, J.G. Skofronick, J.P. Toennies, G. Benedek and L. Miglio, Phys. Rev. B 34 (1986) 902.

[68] E.G. Brovman and Yu. M. Kagan, in: Dynamical Properties of Solids, Eds. G.K. Horton and A.A. Maradudin, Vol. 1 (North-Holland, Amsterdam, 1974) p. 191.

[69] B.C. Clark, R. Herman and R.F. Wallis, Phys. Rev. 139 (1965) A860.

[70] R.F. Wallis, B.C. Clark and R. Herman, Phys. Rev. 167 (1968) 652.

[71] A.U. MacRae, Surf. Sci. 2 (1964) 522.

[72] B.C. Clark, D.C. Gazis and R.F. Wallis, Phys. Rev. 134 (1964) A1486.

[73] D.J. Cheng, R.F. Wallis, C. Megerle and G.A. Somorjai, Phys. Rev. B 12 (1975) 5599.

[74] D.G. Castner, G.A. Somorjai, J.E. Black, D. Castiel and R.F. Wallis, Phys. Rev. B 24 (1981) 1616.

[75] G. Armand, Solid State Commun. 48 (1983) 261.

[76] D. Castiel, L. Dobrzynski and D. Spanjaard, Surf. Sci. 59 (1976) 252.

[77] V. Bortolani, F. Nizzoli and G. Santoro, in: Lattice Dynamics, Ed. M. Balkanski (Flammarion, Paris, 1978) p. 302.

[78] J.E. Black, D.A. Campbell and R.F. Wallis, Surf. Sci. 115 (1982) 161.

[79] A. Fasolino and E. Tosatti, Phys. Rev. B 35 (1987) 4264.

[80] T.L. Reinecke and S.C. Ying, Phys. Rev. B 43 (1991) 12234.

[81] X.W. Wang and W. Weber, Phys. Rev. Lett. 58 (1987) 1452.

[82] H.-J. Ernst, E. Hulpke and J.P. Toennies, Phys. Rev. Lett. 58 (1987) 1941; Phys. Rev. B 46 (1992) 16081.

[83] J.E. Black, V.T. Huynh, D.J. Cheng and R.F. Wallis, Surf. Sci. 192 (1987) 541.

[84] J.E. Black, in: Dynamical Properties of Solids, Eds. G.K. Horton and A.A. Maradudin, Vol. 6 (North-Holland, Amsterdam, 1990) p. 179.

[85] S. Lehwald, J.M. Szeftel, H. Ibach, T.S. Rahman and D.L. Mills, Phys. Rev. Lett. 50 (1983) 518.

[86] P. Moretto, M. Rocca, U. Balvusa and J. Black, Phys. Rev. B 41 (1990) 12905.

[87] J.E. Black, A. Franchini, V. Bortolani, G. Santoro and R.F. Wallis, Phys. Rev. B 36 (1987) 2996.

[88] G. Bracco, R. Tatarek, F. Tommasini, U. Linke and M. Persson, Phys. Rev. B 36 (1987) 2928.

[89] R. Tatarek, G. Bracco, F. Tommasini, A. Franchini, V. Bortolani, G. Santoro and R.F. Wallis, Surf. Sci. 211/212 (1989) 314.

[90] A.M. Lahee, J.P. Toennies and Ch. Wöll, Surf. Sci. 191 (1987) 529.

[91] M. Balden, S. Lehwald, H. Ibach, A. Ormeci and D.L. Mills, Phys. Rev. B 46 (1992) 4172.

[92] J.E. Black, F.C. Shanes and R.F. Wallis, Surf. Sci. 133 (1983) 199.

[93] V. Bortolani, A. Franchini, F. Nizzoli and G. Santoro, Phys. Rev. Lett 52 (1984) 429; V. Bortolani, G. Santoro, U. Harten and J.P. Toennies, Surf. Sci. 148 (1984) 82.

[94] M.H. Mohamed, L.L. Kesmodel, B.M. Hall and D.L. Mills, Phys. Rev. B 37 (1988) 2763.

[95] U. Harten, J.P. Toennies and Ch. Wöll, Faraday Discuss. Chem. Soc. 80 (1985) 137.

[96] A. Lock, J.P. Toennies, Ch. Wöll, V. Bortolani, A. Franchini and G. Santoro, Phys. Rev. B 37 (1988) 7087.

[97] W. Menezes, P. Knipp, G. Tisdale and S.J. Sibener, Phys. Rev. B 41 (1990) 5648.

[98] V. Bortolani, A. Franchini, G. Santoro, J.P. Toennies, Ch. Wöll and G. Zhang, Phys. Rev. B 40 (1989) 3524.

[99] C.S. Jayanthi, H. Bilz, W. Kress and G. Benedek, Phys. Rev. Lett. 59 (1987) 795.

[100] C. Kaden, P. Ruggerone, J.P. Toennies, G. Zhang and G. Benedek, Phys. Rev. B 46 (1992) 13509.

[101] M.S. Daw and M.I. Baskes, Phys. Rev. B 29 (1984) 6443.

[102] J.S. Nelson, M.S. Daw and E.C. Sowa, Phys. Rev. B 40 (1989) 1465.

[103] K.-M. Ho and K.-P. Bohnen, Phys. Rev. Lett. 56 (1986) 934; Phys. Rev. B 38 (1988) 12897.

[104] J.P. Toennies and Ch. Wöll, Phys. Rev. B 36 (1987) 4475.

[105] K.-P. Bohnen and K.-M. Ho, Surf. Sci. 207 (1988) 105.

[106] M.M. Mohamed and L.L. Kesmodel, Phys. Rev. B 37 (1988) 6519.

[107] T. Rodach, K.-P. Bohnen and K.-M. Ho, Surf. Sci. 209 (1989) 481.

[108] Y. Chen, S.Y. Tong, J.S. Kim, L.L. Kesmodel, T. Rodach, K.-P. Bohnen and K.-M. Ho, Phys. Rev. B 44 (1991) 11394.

[109] Y. Chen, S.Y. Tong, K.-P. Bohnen, T. Rodach and K.-M. Ho, Phys. Rev. Lett 70 (1993) 603.

[110] A.G. Eguiluz, A.A. Maradudin and R.F. Wallis, Phys. Rev. Lett. 60 (1988) 309.

[111] J.A. Gaspar and A.G. Eguiluz, Phys. Rev. B 40 (1989) 11976.

[112] A.A. Quong, A.A. Maradudin, R.F. Wallis, J.A. Gaspar, A.G. Eguiluz and G.P. Alldredge, Phys. Rev. Lett. 66 (1991) 743.

Surface Science 299/300 (1994) 628–642
North-Holland

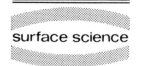

surface science

Diffusion of individual adatoms

Gert Ehrlich

Coordinated Science Laboratory and Department of Materials Science and Engineering, University of Illinois at Urbana Champaign, Urbana, IL 61801, USA

Received 29 April 1993; accepted for publication 18 May 1993

E.W. Müller's invention of the field ion microscope in the 1950s for the first time made it possible to see single metal atoms adsorbed on crystals, and in the 60s this lead to diffusion studies on individual adatoms. The background and beginning of experiments probing the motion of single adatoms is sketched briefly, as are some of the significant developments in this field that have taken place over the last three decades and have provided quite a detailed picture of atomic behavior.

1. Introduction

In the standard view of growth phenomena from the vapor, which dates back some 60 years to Stranski and Volmer [1–3], the transport of adatoms toward lattice steps is one of the primary events leading to growth. In view of the intense interest during the last decade in the growth of crystals and overlayers, it is natural that atomic diffusion on surfaces has become a topic of interest and activity in its own right: there is now available a significant body of information dealing with the diffusion of single metal atoms on crystal planes.

The situation 30 years ago was remarkably different. At that time, techniques for examining surface properties on the atomic level did not exist, and no experimental data on the behavior of single adatoms could be found in the literature. The migration of metal atoms, especially of the alkali and alkaline earths, had been examined by various macroscopic methods starting in the 1930s, primarily in connection with thermionic emission phenomena [4]. Although valuable in establishing the importance of diffusion over surfaces, these early studies generally were not done on well-defined crystal planes, and long-range interactions played an important part in these measurements, so that the overall understanding of the individual atomic events was limited. The development of the field emission microscope in the late 1930s began to change this situation [5,6], but World War II intervened.

In the immediate post-war years, observations of diffusion in metal deposits were started again. In Berlin, Stranski and other notable scientists had found a home in what is now the Fritz Haber Institute [3,7]. In Stranski's group, Erwin Müller, who had been closely involved in the development of the field emission microscope, undertook extensive observations on the transport of tungsten deposited on tungsten surfaces in such a microscope [8]. Some years later, Drechsler and Vanselow [9,10] at the same institute extended observations to tantalum, molybdenum, and nickel, and further refined the work on tungsten. These studies already provided interesting insights into the importance of surface structure in affecting migration, and gave some ideas about the energetics of atom transfer from lattice steps. Quantitative data on atomic events in diffusion was beyond the power of the techniques available. The picture began to change in the 1950s, with Erwin Müller's invention [11] of the field ion microscope (FIM). This was the first instrument to depict surfaces with atomic resolution, and it was immediately clear that the FIM would change the scientific landscape.

SSDI 0039-6028(93)E0373-3

The ability to visualize individual atoms for the first time opened the possibility of obtaining direct information about the migration of atoms on crystals, and this hope has, in fact, been realized. Essentially all the quantitative information about the behavior of individual metal atoms now available has been derived from observations in the FIM. How studies on the motion of individual atoms began and evolved will be the main theme of this personal account. The overall state of various diffusion studies has been reviewed only recently [12–14], and will therefore be of only incidental concern.

2. Background to single–atom studies

In a paper [15] entitled "Experiments with atomic crystal building blocks in the FIM", published in 1957, not too long after Müller moved to Penn State and perfected the field ion microscope, he not only outlined the principles of the technique, but also sketched areas of research that could be fruitfully explored. He stressed field evaporation and the condensation of individual atoms on surfaces, and emphasized the possibility of measuring the binding energy of atoms by determining the field strength at which they were removed from the surface. Müller was also interested in ion bombardment of crystals, and the feasibility of examining the damage caused by single-ion impacts. In this connection, he pointed out that by "stepwise heating of the tungsten tip at 700 to 800 K, the changes brought about by the surface diffusion of individual atoms can be studied after cooling". This technique, he felt, would be especially useful for following the annealing of defects induced by ion bombardment.

In his subsequent efforts, Müller [16] concentrated upon perfecting the operation of the microscope, gaining a better understanding of the physics of the imaging process and of field evaporation, and developing more powerful instruments, such as the atom probe, the combination of the FIM with a time-of-flight mass spectrometer. Among the many subjects examined in his laboratory, the study of defects of one kind or another was emphasized. That nothing further

was done to examine the migration of atoms is not too surprising. Much of his early effort was devoted to surveying the operation and capabilities of the microscope, and a premium was therefore placed on easy replacement of samples. The earliest microscopes, which had been of all-glass construction and capable of achieving ultrahigh vacuum conditions, were soon replaced by demountable models, often relying on greased joints to make access simple. In subsequent studies in Müller's laboratories, high fields were maintained on the surface to field ionize impinging gases that might otherwise contaminate the sample. These are not conditions appropriate for examining the ordinary thermal properties of atoms at the surface, and diffusion studies developed elsewhere.

In the early postwar years, J.H. Hollomon had created at the GE Research Laboratory in Schenectady a department devoted to modern materials studies. In 1953 I joined David Turnbull's group there and immersed myself in studies of chemisorption phenomena on metal surfaces. After developing thermal desorption methods and quantitatively characterizing adsorption processes on macroscopic surfaces, it became clear that more microscopic techniques would be desirable [17]. This led to adsorption studies using the field emission microscope [18], and, less successfully, the field ion microscope [19]. Although my own efforts were focussed upon chemisorption, David Turnbull was deeply involved with diffusion phenomena, both in the bulk and at surfaces, because of their importance in the kinetics of growth and transformations. Occasionally Charles Frank or Nicolas Cabrera would visit the laboratory, and I had the opportunity of talking with them about their work on crystal growth [20]. In this marvelously stimulating environment, it was natural to take at least a brief look at metal atoms in our FIM.

3. Early work on single metal atoms

At the beginning of the 60s, my assistant Frank Hudda and I began to examine the behavior of single tungsten atoms deposited on a tungsten surface from a nearby evaporator. The first studies, in 1963, were exploratory and devoted to transient diffusion during deposition, a topic that has again become of interest recently [21–24]. Through our involvement in the kinetics of chemisorption, Bruce McCarroll and I [25] had become concerned about how atoms from the vapor give up their energy on colliding with a surface to eventually equilibrate with it. Calculations with one-dimensional models suggested that for atoms striking their own lattice, energy transfer would be quite an efficient process. Metal atoms deposited on their own lattice should therefore come to rest close to their original point of impact. If this view was indeed correct, then on an emitter tip illuminated by atoms from one side, the deposit should follow the shadow line. Experiments with tungsten atoms from an evaporator at ∼ 3000 K falling on a tungsten surface at ∼ 20 K suggested that, within the limits of our ability to define the geometry, the tungsten atoms did not transgress significantly beyond the shadow line, and that just a few collisions with the lattice served to immobilize the atoms on the surface [26,27].

At the same time these experiments were underway, other laboratories were also beginning to use the FIM for exploring surface diffusion. S.S. Brenner [28], at US Steel, surveyed changes in the shapes of field-evaporated tungsten tips on heating, in an endeavor to establish how transport over the surface depended upon surface structure. More detailed experiments of this sort were subsequently undertaken at the National Chemical Laboratories in England, where David Bassett observed the atomic rearrangement of field-evaporated tungsten surfaces [29]. He was able to establish that changes occurred in different temperature regimes for different crystallographic regions, and to relate these to the diffusion of atoms bound at different types of lattice steps. Studies of transient diffusion were also started up in a collaboration between groups at Yale and the National Bureau of Standards [30,31]. Their work, published in 1965, again suggested that tungsten atoms condensing on atomically rough surfaces of tungsten came to rest in the immediate vicinity of the point of impact.

At the GE Research Labs, observations on

chemisorption phenomena were continuing. Trevor Delchar had arrived at the laboratory from England in 1963 and had initiated some of the first single-crystal studies [32]. Frank Hudda and I were therefore able to concentrate on new experiments, probing the mobility of W atoms on different planes of tungsten. Atoms were deposited on a cold emitter and their location was established by imaging with helium ions. Thereafter, the surfaces were heated, in the absence of any applied fields, and displacements were noted after allowing the surface to cool again to cryogenic temperatures. The results were surprising [27]. Atoms became mobile around room temperature, that is at less than 1/10 the melting point of tungsten. On the rough (211) surface, mobility of tungsten atoms already was pronounced at temperatures at which atoms deposited on the smoothest plane of tungsten, the (110), did not undergo any displacements. More quantitative measurements were obviously required for further explorations, and above all a technique had to be devised to deduce diffusivities from observations on individual atoms. In the past, regardless of observational technique, surface diffusion had always been examined in a chemical potential gradient. These were the first observations of surface diffusion in an equilibrium system. How should the diffusion of a few atoms on a small crystal plane, less than 100 Å across, be analyzed?

As a graduate student, I had run across Chandrasekhar's article "Stochastic problems in physics and astronomy" [33]; based on what I had learned there I considered two possibilities. One was to describe the motion of the atoms as a random walk, and to extract the diffusivity from the measured mean-square displacement $\langle \Delta x^2 \rangle$, relying on the Einstein relation [34]

$$\langle \Delta x^2 \rangle = 2Dt. \tag{1}$$

Quite another method of analysis was also possible. That was to measure the mean-square successive difference $\langle [n(\tau) - n(0)]^2 \rangle$ in the number of atoms n present in a small area element A a time interval τ after the start of observations, and to extract the diffusivity D much as Smoluchowski had analyzed the diffusivity of colloidal particles. At the beginning of the century, Smoluchowski [35] showed that for an area element surrounded by an infinite reservoir of atoms, values of the mean-square successive difference, and of $\langle (\Delta n^2) \rangle$, the mean amplitude of fluctuations in the number of particles, are related to the diffusivity D through the after-effect factor P by

$$\langle [n(\tau) - n(0)]^2 \rangle / [2\langle (\Delta n^2) \rangle] = P. \tag{2}$$

The after-effect factor is in turn tied to the diffusivity D by

$$P = 1 - 1/(4\pi D\tau A)$$
$$\times \iint_A \exp\left[-(R_1 - R_2)^2/(4Dt)\right] \, dR_1 \, dR_2. \tag{3}$$

The diffusivity on the plane of interest can therefore be derived from observations of the number of atoms present in the area element A. This second method was rejected, however, as the small planes on which observations are made in the FIM seemed to violate Smoluchowski's assumptions.

The Einstein relation between the mean-square displacement $\langle \Delta x^2 \rangle$ of an adatom and the diffusivity D in eq. (1) also has its limitations. It is valid only for a random walk on a large plane on which atoms are not constrained by the presence of boundaries. To make this approach applicable to studies on the small planes accessible in the FIM, corrections to the simple random walk relation therefore had to be worked out [36]. For mean-square displacements small compared to the plane diameter a, the Einstein relation was replaced by the approximation

$$\langle \Delta x^2 \rangle \approx 2Dt\left[1 - (4/3a)(4Dt/\pi)^{1/2}\right]. \tag{4}$$

This made it possible to extract the diffusivity D from the mean-square displacement measured on a plane of finite size.

Diffusion experiments on tungsten atoms deposited on different planes of tungsten were carried out over the course of a few years, and were reported in 1966 [37]. It was quite a different era from now. Quantitative observations were made on three differently structured planes, (110), (211),

and (321), the only ones for which quantitative information is available to this day; qualitative observations, about the much lower mobility on (310) and (111) were reported as well. It was demonstrated that atomic motion was diffusive – the mean square displacement increased linearly with time t, as expected, and interactions with plane boundaries were briefly examined. The temperature dependence of the diffusivity was analyzed according to the usual Arrhenius relation [38]

$$D = D_0 \exp(-E_A/kT). \tag{5}$$

Here E_A is the barrier opposing jumps, and D_0 is a prefactor, related to the entropy of activation ΔS_A and the jump length ℓ by

$$D_0 = \nu_0 \ell^2 \exp(\Delta S_A/k), \tag{6}$$

where ν_0 is a vibrational frequency of the adatom.

The results were surprising. On (321) and (110) planes, behavior conformed to that for a random walk between nearest-neighbor sites, for which D_0 is expected to be $\sim 10^{-3}$ cm^2/s. The activation energy for diffusion on the (110) plane, the most densely packed plane of the bcc lattice, amounted to only 22 kcal/mol, compared with the heat of vaporization of > 200 kcal/mol, or with an activation energy of > 40 kcal/mol reported for surface rearrangement. Diffusion of atoms over a flat plane obviously required little energy. Motion on the much rougher (321) occurred over a barrier of only 20 kcal/mol. On the (211) plane, whose structure is close to that of (321), diffusion was definitely anomalous, with a prefactor $D_0 \approx 2 \times 10^{-7}$ cm^2/s and a barrier of only 13 kcal/mol. The activation energies differed from anything expected for models based on pairwise interactions between atoms. However, one observation was entirely in agreement with expectations [9] – the directional dependence of atom motion. On the channelled planes (211) and (321), diffusion was always observed along the direction of the dose-packed [$\bar{1}11$] rows of the substrate, just as for a marble running along a trough. In contrast, on the (110) plane, motion appeared two-dimensional.

Details of the atomic jump processes involved in diffusion were also probed. In experiments on small planes, atoms execute only a few jumps during any given diffusion interval. The usual Gaussian expression [33] for the probability of finding an atom at a distance Δx from the origin would therefore not be valid. With the help of Bruce McCarroll, a close colleague at GE, the distribution of displacements expected in a random walk with a small number of jumps, itself subject to fluctuations, was derived as [36]

$$p_x = \exp\left[-\langle N \rangle I_x(\langle N \rangle)\right]; \tag{7}$$

here $\langle N \rangle$ is the average number of jumps during the diffusion interval, and $I_z(u)$ the modified Bessel function of order z. Unbeknownst to us, Feller at Princeton had just arrived at the same result [39]. The very sparse data that had been gathered about the distribution of displacements appeared to be in agreement with this relation, suggesting that diffusion conformed to the traditional picture of a random walk between nearest-neighbor sites.

Although the analysis of these experiments was quite sophisticated, the experiments themselves were carried out with rudimentary instrumentation. That the experiments were done at all owed everything to the skill and persistence of Frank Hudda. In the earliest work, images of the surface under study were obtained by allowing the helium ions created at the surface to fall directly on a Willemite screen. These images were recorded on a Polaroid camera extracted from an old oscilloscope and required exposures on the order of ten minutes in absolute darkness. The small darkroom, where the ultrahigh vacuum system was housed, initially did not have air-conditioning. During the summer, temperatures would rise well above 100 F, yet it was crucial to remain alert, as we were operating at voltages up to 30 kV, using glass equipment cooled with liquid hydrogen. It was therefore natural to maximize the amount of information obtained by doing experiments with several atoms deposited on one plane. The attendant problems of assigning displacements were recognized at the time; the more subtle effects of interactions upon diffusion were not.

With information about the energetics of atomic diffusion in hand, it was obviously of interest to compare this with the binding energy of adatoms on the same planes. Müller [15,16] had proposed that the removal of metal atoms from a surface by high fields was limited by evaporation of metal ions over a so-called Schottky saddle. If this is true, the desorption energy can be simply deduced from measured values of the field required for evaporation at low temperatures. Experiments to derive binding energies of tungsten atoms on various planes of tungsten were undertaken by Kirk [40], and evaporation fields were determined on (110), (211), (310), (321), and (411) planes. It is now recognized that the basic assumption of desorption over a Schottky barrier is not valid [41] and that binding energies cannot be obtained this simply, but neither the measured desorption fields, nor the apparent binding energies, showed any unusual differences between the (211) and (321) planes. To account for the surprisingly low activation energy for diffusion that had been observed on W(211), it was proposed that adatoms diffuse along a [$\bar{1}11$] channel during a fluctuation: the channel atoms move outward so that the adatom can then jump from one binding site to another over a much reduced energy barrier. This model did not, however, account for the unusually low prefactor D_0 found for the diffusion of tungsten atoms on this plane.

These early results were soon related to mass transfer processes on macroscopic surfaces [42], but at GE examination of individual atoms came to an end in 1968, as the author moved to the University of Illinois. Interesting studies on single metal atoms had in the meantime begun at other laboratories. Late in 1968, Plummer and Rhodin [43] reported on field evaporation studies of individual atoms, done at Cornell. This work greatly expanded on the GE studies, in dealing with a variety of different metals held on the more prominent planes of tungsten, but still was subject to the same fundamental limitations. In the next few years, detailed measurements of atomic diffusion using the FIM began to appear from Imperial College, where David Bassett had moved from the National Chemical Lab.

In an impressive study, Bassett and Parsley [44] deposited Ta, Mo, W, Re, Ir and Pt atoms on a tungsten emitter. For Ta, W, and Re, they were able to deduce activation energies and prefactors on W(110), (211), and (321) planes; for Ir atoms, only the behavior on the first two planes was characterized. The overall techniques were quite similar to the earlier studies at GE, and their results for tungsten atoms were in very good agreement with the previous data. Again the activation energy on (211) was much lower than on (321), while the behavior on that plane was quite close to that found on (110). This pattern was observed not only for W atoms, but also for Ta and Ir, but not Re. All but the latter again showed unusually low prefactors. The directional dependence of diffusion for all the atoms studied conformed to what had been found earlier for tungsten – one-dimensional on the channeled (211) and (321) planes, two-dimensional motion on (110). Bassett and Parsley were able to correlate the activation energy for the chemically different adatoms with the binding energies reported for these atoms by Plummer and Rhodin [43]. We now recognize that this correlation is not soundly based. It is of interest to note that neither do the diffusion barriers for the different adatoms follow the trends in the heat of vaporization of the respective elements – clearly there is room for further study here.

All of these early observations were done on tungsten, a material relatively simple to clean and also quite robust, for which surface studies date back to the 1920s. To establish a reasonable understanding of surface diffusion, it obviously was important to extend measurements to other substrates, and this effort was pioneered by Guy Ayrault at Illinois. On transferring activities from GE to Urbana, the plan had been to concentrate entirely on the chemisorption of gases on crystals. In keeping with this focus, three different projects were started. Probe hole field emission microscopy was put into place to examine chemisorption on smooth, low-index planes. Molecular beam lines were built to do studies of diffraction and also of chemical reactions on large-scale surfaces, and an atom probe system was set up to provide chemical information about chemisorption on the atomic level. Field ion microscopic examination of atomic behavior was not in the original plan; that it was continued at

Illinois was due entirely to Guy Ayrault. Because of a deep interest in radiation damage, he had started work with one of my colleagues and had begun to build a field ion microscope system. The project was entirely in his own hands, with occasional help from Mike Wald (who had done FIM studies at Cambridge University, but was working on different projects nearby). This, however, proved too great an undertaking for a single student and in the spring of 1969 he joined the surface group. Here he initiated diffusion studies on rhodium, a material of considerable catalytic interest, whose behavior in chemisorption was being examined by others in the group.

Rhodium is an fcc metal, and the intent of the diffusion studies was to determine if the pattern of behavior found on tungsten would extend to different materials. Instrumentation for these experiments was much improved. The introduction of channel plate image intensifiers built into the microscope [45–46] made it much quicker to collect data. Studies on rhodium would otherwise have been extremely hard. Neon had to be used as imaging gas, so as to operate at lower fields which would not perturb the adsorbed atoms. Direct imaging on an unintensified phosphor screen would have made it difficult to gather adequate statistics. Another improvement had been the change to liquid helium as a coolant for the FIM. This had become inevitable, as at Illinois liquid hydrogen was not readily available in quantities sufficient for our needs. With these upgrades, Ayrault [47] was able to obtain quantitative data for the self-diffusion of rhodium atoms on many different planes of rhodium – (111), (311), (110), (331) and (100). Quite early in this work it became evident that diffusion parameters differed when only a single Rh atom was present on a Rh(110) plane, and when observations were made on a plane with several atoms on it. All further experiments were therefore done with only a single atom per plane, an improvement possible because of the rapid imaging achieved with channel plates.

Ayrault's studies, which established the first diffusion parameters for isolated single atoms, yielded surprisingly simple results. The highly directional nature of diffusion, first found on tungsten, was preserved on this fcc metal: on the channeled (110), (331), and (311) planes, with structures reminiscent of W(211) and (321), atom motion was observed to occur along the channels in the substrate. Quite different from tungsten was the relation between surface roughness and the activation energy of diffusion – the barrier to the diffusion of rhodium atoms increased as the structure of the surface became rougher on the atomic scale. Behavior on the atomically smooth (111) plane was especially surprising. On the smoothest plane of tungsten, that is on W(110), diffusion occurred in the same general temperature range as for W(321), which is quite a rough surface. On Rh(111), however, atom motion already set in at cryogenic temperatures, in the vicinity of ~ 50 K. On rougher surfaces, Rh(100) for example, diffusion only began at room temperature. In fact, the progression of activation barriers from planes of one structure to another was in good agreement with calculations based on pairwise interactions represented by a Morse potential. Prefactors were generally ~ 10^{-3} cm²/s, as expected for atomic jumps between nearest neighbors.

At the end of roughly the first decade of diffusion studies on single atoms, considerable quantitative information had been obtained. Diffusion parameters had been determined on different planes of both a bcc and an fcc metal, and it was clear that atom motion already set in at rather low temperatures. Movement of atoms seemed quite simple, except on the (211) plane of tungsten: diffusion occurred by random transitions between nearest-neighbor sites. The atomic geometry of the substrate provided an excellent guide to directional preferences. On rhodium, an fcc metal, simple bonding schemes even afforded semi-quantitative predictions for the energetics of diffusion. Activation energies on tungsten, a bcc metal, were not in accord with this simple scheme, but that was not surprising; the pair bonding approximation should not work for metals in any event. The correlation between the barrier to diffusion and the apparent binding energy of atoms (derived from field evaporation studies) gave hope that a better understanding of atomic behavior would emerge for bcc surfaces as well.

4. Later developments – the second decade

Actual developments in surface diffusion did not follow the trends that had emerged from the early diffusion studies. For one thing, the focus of a considerable part of the overall effort shifted away from single atoms to examining interactions between atoms and to the behavior of clusters formed by association of several atoms. Stimulated by conversations with A.J.W. Moore from CSIRO, we had at GE briefly looked for, but not detected, the nucleation of clusters when there were several atoms present on a surface [17]. David Bassett at Imperial College, however, was able to observe the formation of small clusters of different adatoms on several planes of tungsten [48]. His work opened up a new field of activity, devoted to understanding the diffusion of clusters, as well as to characterizing the interactions between adatoms on a surface. Much effort has gone into such studies [49]. Pursuing these developments here would take us too far afield, especially since the effort devoted to single-atom motion had expanded as well, as groups at Penn State, under T.T. Tsong, at the University of Pennsylvania, under W.R. Graham, and at Osaka University, under S. Nakamura, undertook such work.

Initially, these studies served to amplify the existing picture. Considerable improvements were made in the techniques for controlling and calibrating the temperature; at Illinois, this happened largely through the efforts of David Reed [50]. With the improved equipment, W.R. Graham while at Illinois [51] made observations on a single W atom diffusing on W(211). In contrast to the initial studies, where there had been many atoms present on a plane, these observations revealed quite ordinary behavior – a prefactor D_0 of ~ 10^{-3} cm²/s, and an activation energy of ~ 18 kcal/mol, rather similar to what had been found on W(321). The mystery of anomalous diffusion on W(211) was solved.

In collaboration with Kaj Stolt [52], measurements were also done for a single Re adatom on W(211). The diffusion characteristics did not differ significantly from the earlier values obtained at Imperial College [44]. However, the distribution of distances covered in diffusion on W(211) was also measured and compared with values predicted for a model in which jumps always occurred between nearest-neighbor sites, for which eq. (7) should be appropriate. Within the large statistical scatter inevitable for a small data base, agreement was found to be reasonable, and it was concluded that on this plane, diffusion of Re involved primarily single jumps. Similar comparisons were also done by Coulman at Illinois for W atoms diffusing on W(211) [53]. The distribution of displacements expected if adatoms can jump between nearest neighbors at the rate α, and between second-nearest neighbors at the rate β, had been derived by Mark Twigg [54] as

$$p_x = \exp\left[-2(\alpha+\beta)t\right] \sum_{j=-\infty}^{\infty} I_j(2\beta t) I_{x-2j}(2\alpha t).$$

(8)

Comparison of eq. (8) with the admittedly limited data again suggested that the predominant mechanism of diffusion involved transitions between adjacent binding sites. Tsong and Casanova [55] at Penn State later carried out this type of analysis for W on W(110), and for this plane as well came to the same conclusion: atomic motion occurred predominantly between adjacent sites. It should be noted that primarily because of the tediousness of recording data and then properly analyzing it, the number of observations in all these studies was quite small. These experiments were useful, however, in demonstrating that diffusion did not occur by a series of long jumps over the surface. For deducing reliable jump rates from the displacement distribution, an order of magnitude more data would have been required.

During the 1970s there also accumulated quite a number of incidental measurements of atom diffusivities on planes of tungsten previously studied. These have been reviewed by Bassett [56], and did not significantly change the overall picture apparent from earlier measurements. Diffusion studies were brought to a new level of refinement through the work of Flahive and Graham at the University of Pennsylvania [57,58]. In their studies on W(111), (211) and (321), they identified the geometry of the sites at which atoms

were bound. Having done this, they also succeeded in deriving diffusion parameters for Ni on W(111), as well as for W on W(211); the extensive measurements for the latter, for a well-defined geometry, were in reasonable agreement with earlier (and less detailed) results. At roughly the same time, Crewe and his group at Chicago [59] developed an entirely new way of looking at the motion of atoms on surfaces. They were able to detect single atoms of heavy metals such as silver, gold, or uranium on graphite by using a high resolution scanning transmission electron microscope; they also observed the occasional movement of these atoms over the surface [60]. These were heroic experiments, under very difficult conditions, and yielded some quantitative estimates for atomic hopping rates [61]. Although a tremendous technical triumph, these observations were not continued and have not provided significant quantitative information about diffusion on surfaces.

By far the most interesting and important results came about from what seemed, at the start, a routine extension of diffusion studies to another fcc metal. Rhodium had been examined early in the 70s, Bassett and Coulson [62] briefly looked at diffusion on iridium, and later in the 70s Bassett and Webber [63] decided to examine platinum, another catalytically important metal amenable to study in the FIM. Although their work was somewhat restricted by the high temperature (77 K) at which the surface was imaged, Bassett and Webber carried out an extensive examination in which they observed diffusion of platinum atoms on the (113), (110), and (133) planes of platinum; in addition, more limited experiments were done with Ir and Au atoms. The pattern of behavior observed for the platinum atoms was similar to what had previously been noted for rhodium atoms on rhodium, but with one truly startling exception. On Pt(110), diffusion of platinum as well as iridium atoms, but not gold atoms, was two-dimensional – that is, atoms were *not* confined to moving along the surface channels formed by lattice atoms close-packed along [1$\bar{1}$0].

Bassett and Webber [63] offered two possible explanations for this surprising effect. At the diffusion temperature, chance fluctuations occur in the lateral positions of the atoms in the channel walls. During an especially large fluctuation, a gap could open in the wall, allowing the adatom to slip through. In another, likelier scenario, a lattice atom was assumed to jump out of the wall and into the adjacent channel; the vacancy so created in the wall can then be filled by the adatom. Bassett and Webber also speculated that the cross-channel diffusion observed could in some way be related to the reconstruction of the Pt(110) plane. Although the mechanism of diffusion on Pt(110) was not revealed by this study, Bassett and Webber had discovered a new, unexpected, and exciting phenomenon.

While these studies were under way, John Wrigley had started on his PhD research at Illinois, in which he was going to examine cluster motion on an fcc metal [64]. For ease of observation he picked iridium in preference to rhodium, for which diffusion of single atoms had already been studied. In exploratory experiments with iridium atoms deposited on Ir(110), he found that the atoms did not move along the [1$\bar{1}$0] channels – instead, transitions always appeared to take place into the adjacent channel. This, of course, was quite contrary to expectation. I for one suspected some artifact, so John Wrigley refocussed his work to understanding how cross-channel diffusion occurred. Soon thereafter, Bassett and Webber's exciting results appeared [63], and John Wrigley devised a clever way of testing the mechanism at work, by observing cross-channel motion for adatoms chemically different from the substrate. If cross-channel diffusion takes place by the adatom jumping, either over the channel walls, or through a gap opened up by a large fluctuation, it is the adatom that appears in the adjacent channel. On the other hand, if in diffusion the adatom changes place with a lattice atom that has moved into the adjacent channel, then the chemical identity of the adatom appearing in the neighboring channel will have undergone a change. Wrigley [65] deposited tungsten atoms on Ir(110), and in a series of experiments in the atom probe was able to establish that after a cross-channel event it was an iridium atom that was sitting on the surface, not a tungsten atom.

That, of course, demonstrates that diffusion occurred by an exchange mechanism, and this conclusion was buttressed by a further observation. After diffusion, the adatom can be removed by field evaporation, and the composition of the surface layer can then be tested atom-by-atom in a time-of-flight analysis. When this was done following a cross-channel event, tungsten was usually found present in the first surface layer, just as expected if the adatom had changed place with a lattice atom. Tungsten was never found in the surface if cross-channel motion had not taken place.

In experiments on exchange diffusion using a foreign adatom, the adatom moves into the lattice, and it is a lattice atom that continues the diffusion. That implies that the conditions necessary to initiate the diffusion of tungsten on Ir(110) should be different from those required to continue diffusion, which occurs via iridium atoms. Even prior to establishing the exchange mechanism by atom probe analysis, Wrigley [64] found that self-diffusion of iridium atoms occurred at temperatures $T > 330$ K. However, when a tungsten atom was deposited on Ir(110), cross-channel motion was already observed at $T > 270$ K. To continue atom motion thereafter required temperatures $T > 350$ K. These later transitions were always between the channel into which tungsten had originally been deposited, and its neighbor, into which the first transition had occurred. Only at much higher temperatures, ~ 400 K, was diffusion to other parts of the plane observed. After the atom probe experiments, the significance of the observations was clear: the iridium atom continuing the motion obviously was in some way bound to the buried tungsten atom.

The generality of cross-channel diffusion on fcc(110) planes, presumably by an exchange mechanism, was soon demonstrated in very impressive experiments by Tung and Graham at the University of Pennsylvania [66]. Up to that time, diffusion had been explored on reasonably refractory metals, for which experiments were easy to do. Tung and Graham set out to characterize atom motion on nickel and also aluminum; they selected these materials, despite the fact they are hard to work with, in the hope that theoretical calculations would in the near future be feasible. Despite considerable difficulties, they were able to determine diffusion parameters on Ni(311), (331), and (110), and to estimate the diffusion barrier on Ni(100) and (111). The measured values were found to be in rather poor agreement with calculations based on Morse potentials. What was most interesting, however, was that on Ni(110), and apparently also on Al(110), diffusion was two-dimensional. This was surprising, as in previous studies of diffusion on fcc (110) planes the correlation between cross-channel diffusive motion and the restructuring of the (110) plane had been emphasized. Neither Ni(110) nor Al(110) are known to reconstruct [67]. It also appeared from these measurements that it is one-dimensional diffusion on Rh(110) [47] that is unusual – cross-channel motion is rather a more widespread phenomenon.

Observation of cross-channel diffusion created considerable interest, and for the first time attracted significant theoretical effort. Halicioglu [68] at Stanford did statics calculations for atom motion on the (110) plane of a Lennard-Jones crystal. He concluded that the transition state for diffusion differed from that envisioned by Bassett and Webber [63]. The adatom, together with a lattice atom slightly displaced into the adjacent channel, formed a symmetrical, dumbbell-shaped structure. That such entities actually played a role in self-diffusion was almost immediately demonstrated in molecular dynamics simulations, by DeLorenzi, Jacucci, and Pontikis at Trento [69,70] and by Mruzik and Pound at Stanford [71]. The former studies also extensively compared vacancy and adatom movement on different fcc planes, and suggested that especially at higher temperatures, adatom hopping between nearest-neighbor sites would give way to longer jumps over the surface.

In this second decade of studies on the migration of individual adatoms, it became clear that atomic diffusion was a much more complicated phenomenon than thought earlier: it involved a variety of different mechanisms, about which even qualitatively reliable predictions were difficult to make. One hopeful development took place almost unnoticed at the end of this period: more

modern equipment, allowing automatic control of diffusion experiments, became accessible [72], raising hopes for an expansion of the quantitative data available on diffusion phenomena at surfaces.

5. The third decade

The first experiments taking advantage of improved instrumentation to accumulate statistically significant data on diffusion of individual atoms were aimed at an old subject – diffusion on W(211). Previous experiments on chemically different atoms on this plane suggested that there were large variations in the prefactor D_0 as well as in the barrier to diffusion from one system to the next. This could possibly be indicative of different diffusion mechanisms, and so S.C. Wang at Illinois [73] undertook a reexamination of the diffusion of Re, W, Mo, Ir and Rh on W(211). A rather surprising result emerged from what is at present the most extensive study of the diffusion of single atoms: there were no significant differences in the dynamics of atom motion, that is in the prefactor, for chemically different atoms on W(211). To a rather good approximation, the prefactor for all the atoms tested was close to 10^{-3} cm^2/s. Furthermore, the barrier to surface diffusion could be related quite simply to the heat of vaporization of the atoms from their parent metal. The startling differences in D_0 in the literature had to be attributed to problems in the early experiments, not to any real physical effects.

With improved capabilities for gathering data, Wang et al. [74] also carried out analyses of the distance distribution of diffusing atoms, the first in which statistically significant amounts of data were accumulated. Experiments were done with most of the adatoms for which diffusion had been studied on W(211). These demonstrated what had earlier only been surmised, namely that jumps to nearest-neighbor sites completely dominated diffusion on that plane.

It appears from these studies that at least on W(211), diffusion occurs in a very simple way, by the random motion of adatoms between nearest-neighbor sites, over a barrier related to the energy of vaporization of the parent metal. This simple picture is deceptive, however. That became evident from the work of DeLorenzi at Illinois, who in 1989 created a video from molecular dynamics simulations of atom motion on a bcc (211) plane. This revealed very clearly how complicated the diffusion process really is even when atoms jump just between nearest-neighbor sites. The early idea, that the surface channels on this plane might widen out in a fluctuation to permit easy passage of an adatom [40], appears dose to the mark: in the molecular dynamics simulations, a successful diffusion event involves the concerted movement of several lattice atoms together with the adatom. It must also be noted parenthetically that detailed diffusion experiments, on W(211) as well as on other surfaces, have been done over only a narrow temperature interval; such experiments do not preclude more unusual behavior at higher temperatures, for example [75].

That the results available on W(211) cannot serve as a guide even to behavior on other planes of tungsten has recently been shown by M.F. Lovisa at Illinois [76], who carefully characterized the distribution of displacements observed in the diffusion of iridium and also tungsten atoms on W(110). These measurements demonstrate that at lower temperatures, diffusion occurs by single jumps along the close-packed directions. At higher temperatures, $T \geq 340$ K for tungsten atoms, double jumps along the close-packed directions make quite a significant contribution on W(110). For diffusion on macroscopic crystals, where transport over distances larger than those studied in the FIM are important, a much more varied behavior can be expected.

Another interesting development has been the renewed interest in diffusion by an exchange mechanism. The stage for this was set in 1985, when DeLorenzi and Jacucci [77] reported new molecular dynamics simulations of surface diffusion; they had extended their work to bcc surfaces, using a metallic potential. The simulations revealed an unexpected effect – on the bcc (100) plane, DeLorenzi and Jacucci found that "in addition to conventional nearest-neighbor jumps

between surface sites, the adatom undergoes migration events reminiscent of exchange processes...". Jumps over the surface, along $\langle 001 \rangle$, occasionally alternated with transitions in which the adatom displaced a lattice atom from its site, and the former then continued the diffusion process. In these transitions, the net displacement was along $\langle 011 \rangle$. Diffusion by an atom exchange mechanism was *not* limited to a channelled plane like the fcc (110) surface; it was rather a general phenomenon, also "contributing to atomic diffusion on isotropic crystal surfaces". Five years later, the importance of atomic exchange in diffusion was confirmed by Feibelman at Sandia [78]. He concluded from ab initio calculations on Al(100) that a transition state would be favored in which the adatom and a slightly displaced lattice atom form a dumbbell structure. When this decomposes, an atom may appear in a unit cell diagonal to the original position of the adatom. A plot of the sites occupied by atoms during diffusion on fcc (100) should, according to this mechanism, give a c(2 × 2) mesh. Such a c(2 × 2) mesh has indeed been observed in very nice studies on Ir(100) by Chen and Tsong at Penn State [79], and also in experiments on Pt(100) by Kellogg at Sandia [80].

Friedl et al at Erlangen [81] have pointed out that a c(2 × 2) mesh for the sites observed in diffusion on an fcc (100) plane does not by itself establish that diffusion occurs by an exchange mechanism. Much the same mesh could be obtained as a consequence of a c(2 × 2) rearrangement of the (100) surface in the vicinity of the adatom, without invoking atomic exchange for diffusion. Recently, however, experiments have been done with chemically different adatoms on Ni(100) at Sandia [82], and on Ir(100) at Penn State [83]; these do demonstrate exchange between the adatom and a lattice atom. It is a reasonable assumption that in self-diffusion also this is the likely mechanism.

The interest in diffusion on fcc (100) has generated new studies on fcc (110) planes as well. In work with Pt atoms on Ni(110), Kellogg [84] verified what had previously only been an assumption, namely that cross-channel motion on this plane occurred by an exchange mechanism. Most interesting, however, is a study of the distribution of displacements for Ir on Ir(110), by Tsong and Chen [85]. The results are clearly not in agreement with the idea of a dumbbell intermediate made up of adatom and lattice atom, as proposed in various simulations of diffusion on fcc (110). How exchange diffusion actually occurs is still not clear. Energy estimates [86] using the embedded atom method indicate, however, that the balance between ordinary hopping diffusion and exchange processes can be quite close, making it difficult to predict in advance how an unexplored material will behave. The old suggestion that reconstruction and exchange diffusion are connected has been revived repeatedly, but is not in agreement with the fact that exchange occurs on Ni and also Al(110), which do not reconstruct. The best guide at the moment appears to be that if cross-channel diffusion is observed on the (110) plane of an fcc metal, diffusion will also occur by an exchange mechanism on the (100) plane.

A new area of investigation to open in the third decade of single-atom studies is the detailed exploration of atomic behavior on fcc (111) planes. Diffusion on such surfaces is of special interest, as on the atomic level this is the smoothest plane known for a metal. It is complicated by the fact that on (111) planes at least two very similar sites, labelled fcc and hcp, respectively, are available for occupation. A prerequisite to detailed studies on fcc (111) planes is the ability to ascertain the sites at which metal atoms are actually held on the surface. By mapping the positions occupied after diffusion of iridium atoms on Ir(111), Wang at Illinois [87] was able to establish that self-adsorption of single iridium atoms occurs preferentially at hcp sites, but that occasionally fcc sites were filled as well. In the course of these studies it also became clear that atoms held at different sites could be readily distinguished just by observing the image spot produced in the FIM. An Ir adatom at an hcp site has a triangular image with its apex pointed along $[2\bar{1}\bar{1}]$; at an fcc site, the orientation is reversed [88,89].

With these capabilities it has become possible to quantitatively examine the detailed jump processes contributing to diffusion on Ir(111). Even though the (111) plane is very smooth, Wang [88] showed that atoms still migrate in a sequence of jumps from hcp to fcc site and then again on to hcp. At low temperatures at least these are the only events important, and quantitative values for the barrier height, and also frequency factors, have been measured for Ir, as well as W and Re adatoms [90]. The activation energy for motion of Ir, for example, is unusually low, amounting to less than $1/20$ the binding energy of an atom on Ir(111). Perhaps the most interesting result concerns the magnitude of the prefactors in the diffusivity. There appear to be significant differences in D_0 for different metal atoms on the same plane, and the prefactors consistently are below the magic value of 10^{-3} cm^2/s.

To finish this survey of developments in the third decade of atomic diffusion studies, it may be appropriate to note that the work on Ir(111) provides an excellent example of the importance of luck in research. The detailed studies at Illinois were made possible because Ir adatoms prefer hcp sites, but occasionally do occupy fcc positions. This led to an assignment of the favored sites, and everything else followed quite simply. Had the first studies been done on other adatoms, such as W or Re, which now are known to occupy hcp sites on Ir(111) under almost all conditions, the work would have come to an end. The group at Erlangen [91] had the bad luck to start their explorations of the Ir(111) face with tungsten atoms. Although they did excellent work, their attempts at site identification were frustrated.

6. Reprise

This survey of some highlights in 30 years research on the diffusion of individual atoms on metals should make it clear that, even though the field has developed fairly randomly, we know infinitely more now than at the beginning of the 1960s. After three decades of work, the atomic events in the migration of atoms over surfaces are probably better understood than in any other process important on surfaces. One central theme to emerge from this effort is that atomic motion in surface diffusion is complex, and can occur in different ways, which were not anticipated at the beginning of this era. Another obvious conclusion, however, is that the ability to see individual atoms confers tremendous powers for characterizing atomic events, the full potential of which has *not* yet been explored.

One of the tangible consequences of past work is that we now have available quantitative tabulations of diffusion parameters for different systems [92]. Values of prefactors and activation energies are very important, both for quantitative predictions of atom transport on crystals and to reveal qualitative trends that may allow extrapolation to materials not yet studied. In this regard, the state of the literature is *not* satisfactory, however. Only a very limited number of materials has been examined. Worse than that, even though instrumentation has constantly improved, and in some instances excellent agreement has been achieved between studies in different laboratories by different investigators, such agreement is far from general. In fact, rather little effort has been devoted during the last decade to building up a reliable data base for diffusion phenomena by taking advantage of modern techniques.

Much remains to be done to characterize the diffusion of atoms on crystals, and the hope is that in the future this will be done not just on metals, but also on other materials, using a larger variety of techniques. Almost all the quantitative data presently available has been gathered by field ion microscopy. Some information about atom motion, on semiconductors [93,94], has been obtained indirectly from studies with the STM and, as already noted, observations have been made of diffusion on graphite, using high resolution scanning electron microscopy [59]. Further development of methods to provide reliable direct data on atom motion on solids would certainly be desirable, and is certain to yield important information.

Finally it should be noted that early studies of single adatoms were strongly influenced by a desire to better understand the growth of crystals. During the last decades, work on atomic diffusion has progressed to such an extent that this initial goal is gradually being realized – observations on single atoms are beginning to provide new insights into how atoms actually incorporate into crystals [95].

Acknowledgements

This account was made possible through support from the National Science Foundation under Grant DMR 91 01429. It is a pleasure to acknowledge the contributions, both in the past and in the preparation of this manuscript, from the many people that have been involved in diffusion studies on adatoms.

References

[1] O. Knacke and I.N. Stranski, Prog. Met. Phys. 6 (1956) 181.
[2] M. Volmer, Kinetik der Phasenbildung (Steinkopff, Dresden, 1939).
[3] R. Lacmann, in: Berliner Lebensbilder Naturwissenschafftler, Eds. W. Treue and G. Hildebrandt (Colloquium Verlag, Berlin, 1987) p. 329.
[4] Early studies have been reviewed by G. Ehrlich and K. Stolt, Annu. Rev. Phys. Chem. 31 (1980) 603.
[5] F. Ashworth, Adv. Electron. 3 (1951) 1.
[6] E.W. Müller, Ergeb. Exakten Naturwiss. 27 (1953) 290.
[7] Ber. Mitt. Max-Planck-Ges. Heft 7 (1986) 9.
[8] E.W. Müller, Z. Phys. 126 (1949) 642.
[9] M. Drechsler, Z. Elektrochem. 58 (1954) 334.
[10] M. Drechsler and R. Vanselow, Z. Kristallogr. 107 (1956) 161.
[11] R.H. Good and E.W. Müller, in: Handbuch der Physik, Vol. XXI-1, Ed. S. Flügge (Springer, Berlin, 1956) p. 176.
[12] H.P. Bonzel, in: Diffusion in Solid Metals and Alloys, Ed. H. Mehrer, Landolt–Börnstein NS., Vol. 26 (Springer, Berlin, 1990) p. 717.
[13] R. Gomer, Rep. Prog. Phys. 53 (1990) 917.
[14] G. Ehrlich, Surf. Sci. 246 (1991) 1.
[15] E.W. Müller, Z. Elektrochem. 61 (1957) 43.
[16] E.W. Müller and T.T. Tsong, Field Ion Microscopy Principles and Applications (Elsevier, New York, 1969).
[17] G. Ehrlich, Adv. Catal. 14 (1963) 255.
[18] G. Ehrlich and F.G. Hudda, J. Chem. Phys. 35 (1961) 1421.
[19] G. Ehrlich and F.G. Hudda, J. Chem. Phys. 36 (1962) 3233.
[20] W.K. Burton, N. Cabrera and F.C. Frank, Philos. Trans. R. Soc. London A 243 (1951) 299.
[21] W.F. Egelhoff and I. Jacob, Phys. Rev. Lett. 62 (1989) 921.
[22] S.C. Wang and G. Ehrlich, J. Chem. Phys. 94 (1991) 4071.

[23] D.E. Sanders, D.M. Halstead and A.E. DePristo, J. Vac. Sci. Technol. A 10 (1992) 1986.

[24] P. Blandin and C. Massobrio, Surf. Sci. Lett. 279 (1992) L219.

[25] B. McCarroll and G. Ehrlich, J. Chem. Phys. 38 (1963) 523.

[26] G. Ehrlich, in: Metal Surfaces: Structure, Energetics and Kinetics (American Society of Metals, Metals Park, OH, 1963) p. 221.

[27] G. Ehrlich, Brit. J. Appl. Phys. 15 (1964) 349.

[28] S.S. Brenner, in: Metal Surfaces: Structure, Energetics and Kinetics (American Society of Metals, Metals Park, OH, 1963) p. 305.

[29] D.W. Bassett, Proc. R. Soc. London A 286 (1965) 191.

[30] T. Gurney, F. Hutchinson and R.D. Young, J. Chem. Phys. 42 (1965) 3939.

[31] R.D. Young and D.C. Schubert, J. Chem. Phys. 42 (1965) 3943.

[32] T.A. Delchar and G. Ehrlich, J. Chem. Phys. 42 (1965) 2686.

[33] S. Chandrasekhar, Rev. Mod. Phys. 15 (1943) 1.

[34] J.R. Manning, Diffusion Kinetics for Atoms in Crystals (Van Nostrand, Princeton, 1968).

[35] M. von Smoluchowski, Sitzungsber. Oesterr. Akad. Wiss. Math.-Naturwiss. Kl. 123 (1914) 2381.

[36] G. Ehrlich, J. Chem. Phys. 44 (1966) 1050.

[37] G. Ehrlich and F.G. Hudda, J. Chem. Phys. 44 (1966) 1039.

[38] C.P. Flynn, Point Defects and Diffusion (Clarendon, Oxford, 1972).

[39] W. Feller, SIAM Soc. Ind. Appl. Math. J. Appl. Math. 14 (1966) 864.

[40] G. Ehrlich and C.F. Kirk, J. Chem. Phys. 48 (1968) 1465.

[41] H.J. Kreuzer, in: Chemistry and Physics of Solid Surfaces VIII, Vol. 22 of Springer Series in Surface Science, Eds. R. Vanselow and R. Howe (Springer, Berlin, 1990) p. 133.

[42] G. Ehrlich, Soc. Chem. Ind. Monograph 28 (1968) 13.

[43] E.W. Plummer and T.N. Rhodin, J. Chem. Phys. 49 (1968) 3479.

[44] D.W. Bassett and M.J. Parsley, J. Phys. D 3 (1970) 707.

[45] P.J. Turner, P. Cartwright, M.J. Southon, A. van Oostrom and B.W. Manley, J. Phys. E 2 (1969) 731.

[46] A. van Oostrom, Philips Res. Rep. 25 (1970) 87.

[47] G. Ayrault and G. Ehrlich, J. Chem. Phys. 60 (1974) 281.

[48] D.W. Bassett, Surf. Sci. 23 (1970) 240.

[49] G. Ehrlich and F. Watanabe, Langmuir 7 (1991) 2555.

[50] D.A. Reed and G. Ehrlich, Surf. Sci. 151 (1985) 143.

[51] W.R. Graham and G. Ehrlich, Phys. Rev. Lett. 31 (1973) 1407.

[52] K. Stolt, W.R. Graham and G. Ehrlich, J. Chem. Phys. 65 (1976) 3206.

[53] Cited by G. Ehrlich, J. Vac. Sci. Technol. 17 (1980) 9.

[54] M.E. Twigg, Statistics of 1-Dimensional Atom Motion with Next-Nearest Neighbor Transitions, Masters Thesis, Department of Metallurgy, University of Illinois-Urbana, 1978.

[55] T.T. Tsong and R. Casanova, Phys. Rev. B 22 (1980) 4632.

[56] D.W. Bassett, in: Surface Mobilities on Solid Materials, Ed. Vu Thien Binh (Plenum, New York, 1983) p. 63.

[57] P.G. Flahive and W.R. Graham, Thin Solid Films 51 (1978) 175.

[58] P.G. Flahive and W.R. Graham, Surf. Sci. 91 (1980) 463.

[59] A.V. Crewe, Science 221 (1983) 325.

[60] M. Isaacson, D. Kopf, M. Utlaut, N.W. Parker and A.V. Crewe, Proc. Natl. Acad. Sci. USA 74 (1977) 1802.

[61] M. Utlaut, Phys. Rev. B 22 (1980) 4650.

[62] D.W. Bassett, in: Surface and Defect Properties of Solids, Vol. 2, Eds. M.W. Roberts and J.M. Thomas (The Chemical Society, London, 1973) p. 34.

[63] D.W. Bassett and P.R. Webber, Surf. Sci. 70 (1978) 520.

[64] J.D. Wrigley, Surface Diffusion by an Atomic Exchange Mechanism, PhD Thesis, Department of Physics, University of Illinois at Urbana-Champaign, 1982.

[65] J.D. Wrigley and G. Ehrlich, Phys. Rev. Lett. 44 (1980) 661.

[66] R.T. Tung and W.R. Graham, Surf. Sci. 97 (1980) 73.

[67] M.A. Van Hove, W.H. Weinberg and C.-M. Chan, Low-Energy Electron Diffraction – Experiment, Theory and Surface Structure Determination (Springer, Berlin, 1986).

[68] T. Halicioglu, Surf. Sci. 79 (1979) L346.

[69] G. DeLorenzi, G. Jacucci and V. Pontikis, in: Proc. ICSS-4 and ECOSS-3, Cannes, 1980, Vol. I, Eds. D.A. Degras and M. Costa, p. 54.

[70] G. DeLorenzi, G. Jacucci and V. Pontikis, Surf. Sci. 116 (1982) 391.

[71] M.R. Mruzik and G.M. Pound, J. Phys. F 11 (1981) 1403.

[72] H.-W. Fink and G. Ehrlich, Surf. Sci. 143 (1984) 125.

[73] S.C. Wang and G. Ehrlich, Surf. Sci. 206 (1988) 451.

[74] S.C. Wang, J.D. Wrigley and G. Ehrlich, J. Chem. Phys. 91 (1989) 5087.

[75] See, for example, D.-S. Choi and R. Gomer, Surf. Sci. 230 (1990) 277.

[76] M.F. Lovisa and G. Ehrlich, Bull. Am. Phys. Soc. 37 (1992) 190, abstract C82.

[77] G. DeLorenzi and G. Jacucci, Surf. Sci. 164 (1985) 526.

[78] P.J. Feibelman, Phys. Rev. Lett. 65 (1990) 729.

[79] C. Chen and T.T. Tsong, Phys. Rev. Lett. 64 (1990) 3147.

[80] G.L. Kellogg and P.J. Feibelman, Phys. Rev. Lett. 64 (1990) 3143.

[81] A. Friedl, O. Schütz and K. Müller, Surf. Sci. 266 (1992) 24.

[82] G.L. Kellogg, Surf. Sci. 266 (1992) 18.

[83] C.L. Chen, T.T. Tsong, L.H. Zhang and Z.W. Yu, Phys. Rev. B 46 (1992) 7803.

[84] G.L. Kellogg, Phys. Rev. Lett. 67 (1991) 216.

[85] T.T. Tsong and C.L. Chen, Phys. Rev. B 43 (1991) 2007.

[86] C.L. Liu, J.M. Cohen, J.B. Adams and A.F. Voter, Surf. Sci. 253 (1991) 334.

[87] S.C. Wang and G. Ehrlich, Phys. Rev. Lett. 62 (1989) 2297.

[88] S.C. Wang and G. Ehrlich, Surf. Sci. 224 (1989) L997.

[89] S.C. Wang and G. Ehrlich, Surf. Sci. 246 (1991) 37.

[90] S.C. Wang and G. Ehrlich, Phys. Rev. Lett. 68 (1992) 1160.
[91] O. Schütz and K. Müller, J. Phys. (Paris) 50 (1989) C8.
[92] T.T. Tsong, Atom-probe Field Ion Microscopy (Cambridge University Press, Cambridge, 1990).

[93] H.B. Elswijk, A.J. Hoeven, E.J. van Loenen and D. Dijkkamp, J. Vac. Sci. Technol. B 9 (1991) 451.
[94] Y.W. Mo, J. Kleiner, M.B. Webb and M.G. Lagally, Phys. Rev. Lett. 66 (1991) 1998.
[95] G. Ehrlich, Appl. Phys. A 55 (1992) 403.

Surface Science 299/300 (1994) 643–655
North-Holland

surface science

Time-resolved measurements of energy transfer at surfaces

R.R. Cavanagh *, E.J. Heilweil and J.C. Stephenson

NIST, Gaithersburg, MD 20899, USA

Received 31 March 1993; accepted for publication 9 June 1993

Developments in time-resolved measurements of energy transfer at surfaces are reviewed. Picosecond and femtosecond measurements of vibrational and electronic relaxation at surfaces are highlighted. Experimental results for vibrational relaxation of simple adsorbates on metals, semiconductors, and insulators are reviewed, and relaxation mechanisms such as electron–hole pair formation, multiphonon relaxation and image dipole damping are considered. Energy transfer involving excited electronic states of molecules on liquid and solid dielectric surfaces and the relaxation of surface electronic states on semiconductors and of image states on metal surfaces are discussed.

1. Background / introduction

As illustrated in numerous other articles in this volume, the last thirty years have witnessed tremendous advances in spectroscopic probes of surfaces. The emergence of these tools, particularly with their newfound capacity to identify the atomic and molecular species present at a surface, has played a significant role in a broad range of scientific and technological frontiers. The subdiscipline of surface chemistry alone can supply a long list of accomplishments that have exploited these probes. The strength of surface bonding has frequently been revealed by the positions and widths of spectral features. Similarly, the disappearance of specific molecular spectral transitions has been taken as evidence for chemical transformation associated with specific bonds. These same tools have played a major role in establishing the conceptual foundation for surface chemistries based on non-thermal activation of surface bonds [1–3].

By the early 1980's it was widely recognized that the lifetimes of adsorbate modes could play a central role in selected spectroscopies. Most models treated isolated oscillators or assumed a perfectly uniform surface layer. Even within these simplifying assumptions, it was clear that explicit account would have to be taken of substrate mediated damping effects [4–6]. As the theoretical and experimental methods improved, questions continued to arise as to details of the damping mechanisms, and the utility of the information carried by the associated spectral band widths and band shapes. If the newfound spectral resolution was to be fully utilized, the factors that ultimately determine the spectral response would need to be established. For excited electronic states, the lifetimes could range from a few milliseconds on semiconductors and insulators, to a few femtoseconds on metallic substrates. For adsorbate vibrations on metal substrates, it was also widely recognized that electron–hole pair damping (i.e. the lost oscillator quantum couples to metal electrons by promoting an electron from a filled orbital below E_F to a vacant orbital above E_F) could lead to lifetimes on the order of picoseconds [7]. However, as molecular vibrational transitions typically lie below 0.5 eV, this electronic damping channel will not contribute to vibrational relaxation on insulators or most semiconductors. For higher lying excitations, such as the extensively studied optical transitions in free molecules, electron–hole pair damping could also come into play. It was in the early 1980's that serious experimental efforts began to concentrate

* Corresponding author.

not only on spectral positions, but also on characterizing spectral widths and shapes [8–13]. While careful band shape analyses of the newly emerged static spectroscopies provided some insight into the operative energy transfer processes, their application did not provide a method for following the mechanistic details of the steps involved [14–18].

As the ability of different spectroscopies to provide a static identification of the molecular composition and adsorbate interactions present in adlayers became more common, a parallel activity to exploit dynamical aspects of the measurements emerged. It was widely recognized that individual interfacial bonds were formed and broken on the timescale of a molecular vibration, i.e., several femtoseconds [19]. In the specific case of adsorbate vibrational features, it was evident that with sufficient spectral resolution and the presence of suitably narrow absorption features, it would be possible to distinguish the energy content of an adsorbate through the observation of subtle spectral changes related to bond anharmonicities and excited state population density. Based on the vast gas phase and solution literature, it was envisaged that different surface species, including reaction intermediates, could be identified and monitored during the course of surface reactions. Unfortunately, none of the widely accepted surface sensitive probes had the temporal resolution to follow the evolution of chemical bonds on the short timescales of individual bonding events.

In the sections that follow, the application of ultrafast laser techniques to investigate energy transfer processes at interfaces is considered. For an assessment of the impact of short-pulse lasers on surface heating [20] and nonlinear optical spectroscopy [21], the reader is directed to alternate sources.

2. Historical and theoretical approaches

2.1. Classical electromagnetic damping by surfaces

Starting in 1966, Drexhage reported studies of the effects of metal surfaces on excited electronic state damping [22,23]. He prepared samples with a fluorescent Eu^{3+} organometallic compound separated from a metal surface by a Langmuir–Blodgett organic film of controllable thickness. Following excitation with a UV flashlamp, a photomultiplier time-resolved the Eu^{3+} emission at 615 nm. By monitoring the fluorescence lifetime (τ) as a function of spacer-layer thickness, he demonstrated a strong change in τ with distance (d) from the surface. For large d ($d \geq 1000$ nm), the natural lifetime $\tau_0 \approx 800$ μs was observed; as d decreased in the range 1000–20 nm, τ oscillated about τ_0, and for $d \leq 20$ nm τ decreased rapidly as energy transferred to the metal substrate. The quality of the data was sufficiently high that it became the focus of extensive theoretical modeling [24]. These combined studies established that for layers thicker than 10 nm, and in the absence of adsorbate–adsorbate interactions, classical electrodynamics were capable of accounting for the observed excited state quenching. Since this formalism uses only the molecular transition dipole moment and the substrate-specific complex dielectric constant $\epsilon(\omega)$, it applies to quenching of both electronic and vibrational states on insulator, semiconductor, and metal surfaces.

As the spacer layer approached tens of monolayers, both the experimental approach and the assumptions in the theoretical model were no longer capable of giving a realistic account of the actual damping. The need to account for the local nature of the field and the inability to accurately locate the image plane became limiting factors for the model. From an experimental perspective, these fluorescence measurements were usually done under less than ideal conditions with regard to surface characterization, such that coverage uniformity, adlayer structure, and sample homogeneity were frequently unverified [25]. However, the work provided valuable stimulation in a number of areas that sought to exploit and control energy transfer at monolayer coverages. These areas ranged from semiconductor electrode surfaces and solar energy conversion [26], photochemistry in zeolites [27], and desorption induced by electronic transitions (DIET) [28].

Parallel efforts to explore electronic excited

states were underway in two distinctly different realms of surface science: the domain of electron scattering [29–31]; and the field of photoelectrochemistry [32–34]. In the case of metallic substrates, excited state relaxation can be mediated by excitation involving the continuous density of states on either side of the Fermi level with characteristic relaxation times anticipated to lie in the femtosecond domain. For semiconductor electrodes, complex transport properties [35] and the timescales for carrier redistribution have severely limited the application of conventional band shape studies to these problems. In the DIET field the lifetimes of the excited electronic states involved were usually sufficiently short (< 1 fs) that time-resolved methods would not have been tractable [28]. However, theoretical efforts in the DIET area are extensive, building on the large base of non-time-resolved data that reflect the short timescale of the system response [36].

2.2. Vibrations

Detailed theoretical treatments of excited state damping have largely focused on vibrational relaxation processes. The overwhelming attention to vibrations is likely a reflection of the extensive body of experimental data available [37], and the superior resolution afforded by instrumental developments in associated infrared and electron or atom scattering probes. As the static factors that contribute to the band shape became better understood, it was evident that the band shape of a homogeneous adlayer remained to be firmly established. The extent to which excited state populations decay through anharmonic coupling to lower frequency modes or through resonant electronic coupling was not obvious. Previous energy transfer studies in condensed phases pointed to the importance of multiphonon relaxation [38] with numerous examples now known in the surface science community [39]. For metallic substrates, the potential for electron–hole pair generation to provide a new relaxation channel demanded more advanced theoretical models than those previously developed for phonon damping.

Historically, a fascination with metallic sub-

strates had arisen due to their relative experimental simplicity [40], their connection with catalytic reactions, and their apparent theoretical tractability within a jellium model [41]. Consequently, there have been an abundance of studies of adsorbate vibrations on these substrates. The apparent simplicity of these systems was quickly dispelled as the complexities that contribute to spectral band shapes of coupled adsorbates at interfaces were evaluated. Models ranging from dipole–dipole coupling [42] to local representations of the polarizability [43] have been explored in terms of their impact on adsorbate vibrational band shapes. While these theories can account for symmetric or asymmetric band shapes, there remain numerous approximations in the theoretical models that limit their comparison with experimental observations. Nonetheless, band shape analysis has certainly contributed to an increased understanding of the complex factors that underlie observed band widths and shapes. Of greatest concern, however, has been the limited ability of theory to account for sample dependent inhomogeneous effects manifest by the overlayer [18].

For symmetric Lorentzian absorption bands of adsorbates, the natural widths $\Delta\nu$ are given by $2\pi c\Delta\nu$ (FWHM, cm^{-1}) $= 1/T_1 + 2/T_2^* \equiv 2/T_2$, where T_2 is the total dephasing time, T_1 is the energy decay time and T_2^* is the pure dephasing time ($T_2^* \rightarrow \infty$ as the temperature $T \rightarrow 0$). Early high resolution electron energy loss spectroscopy (HREELS) and infrared reflection absorption spectroscopy (IRAS) measurements of the widths of vibrational bands implied rather short relaxation times ($\Delta\nu = 1$ cm^{-1} corresponds to $1/T_1 + 2/T_2^* = 1/5.2$ ps). For instance, values of $\Delta\nu$ in the range 10–30 cm^{-1} for ordered adlayers of diatomics in atop sites on metals (e.g. CO/Pt(111)) were reported in both HREELS and IRAS experiments. Although these $\Delta\nu$ values are now recognized to be too large by factors of 3 or so, these pioneering measurements formed the basis for our present understanding of energy relaxation (T_1) and dephasing (T_2) processes. Since these early experiments, instrumentation has greatly improved so that spectral resolution and sensitivity are much better. Also, there is now a better appreciation of the large influence that crystal defects and adlayer disorder may have on $\Delta\nu$.

The values of $\Delta\nu$ measured in these early studies were frequently attributed to vibrational energy damping of the excited state. It was only in the early 1980's that the community became widely aware of other dephasing effects [17,18]. Detailed analysis of the temperature dependence of the position and width of adsorbate modes on a variety of surfaces has now demonstrated the presence of the exchange dephasing broadening mechanism [44] in which population in low frequency modes that are anharmonically coupled to higher energy modes causes shifting and broadening of the high energy absorption spectrum. In favorable cases it has been possible to identify the low frequency mode responsible for the dephasing [37]. Of great significance was the clear demonstration that static spectral measurements did not necessarily address the population lifetime of the excited state. Alternative techniques would be required if population lifetimes were to be established.

Simultaneous with the theoretical and experimental advances that had been going on in the surface science discipline, a parallel revolution was underway in the ultrafast laser field [45]. Major advances in pulse duration and tunability showed the potential for making time-domain measurements of excited state relaxation phenomena a viable alternative to frequency domain studies. In the sections that follow, the contributions from time-resolved studies of energy transfer at surfaces are highlighted.

3. Vibrational energy transfer

From the observed spectral band widths of adsorbates it seemed certain that laser pulses of very short duration would be necessary to determine T_1 or T_2 from a time-resolved measurement. The first application of picosecond lasers to study monolayers on surfaces was reported by Heritage and colleagues who studied CN on Ag and *p*-nitrophenol on Al_2O_3 [46]. Their picosecond Raman gain spectra exploited the intense electric field of the ultrashort pulse, but did not

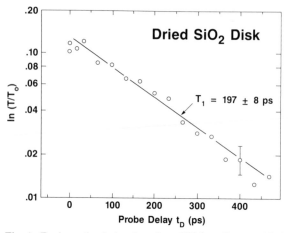

Fig. 1. T_1 decay for isolated surface OH ($v = 1$) on a dried evacuated SiO_2 disk at 293 K. The exponential transmission recovery as a function of probe delay (τ_d) yields a lifetime of $T_1 = 197 \pm 8$ ps (1σ) from a least-squares fit (after Ref. [47]).

characterize the time-resolved response of the system. Although different ultrafast optical spectroscopies [second harmonic generation (SHG) and IRAS] have subsequently proven more useful in probing surface energy transfer, the type of picosecond laser used by Heritage (synchronously pumped tunable dye laser) has been improved and used in all studies to date of vibrational energy transfer on single crystal surfaces.

The first time-resolved study of the vibrational energy relaxation time T_1 of a surface mode was for the damping of OH($v = 1$) bound to high surface area SiO_2 [47]. All experiments were performed in the transmission configuration that was so successfully applied to supported catalysts since the pioneering work of Eischens [48]. Tunable infrared pulses of approximately 20 ps in duration were used to excite the OH($v = 1$) band and to monitor the return of the absorption to its initial unexcited value. A typical example of the time-resolved response is shown in Fig. 1. It was found that at room temperature, the $v = 1$ lifetime of isolated surface OH groups was 220 ± 20 ps, and that this lifetime varied by only a factor of two depending on whether the SiO_2 was in a vacuum or suspended in a variety of solvents. Examination of the temperature dependence [49] of T_1 revealed a strong decrease in T_1 as the sample temperature rose above 500 K. The relatively long lifetime (10^4 oscillations of the OH bond) and its temperature dependence both lent support to a multiphonon decay process (the temperature dependence for a decay of the type $\nu_0 \to p\nu_1$ increases with thermal occupation $n(T)$ in ν_1 as $T_1^{-1} \alpha \langle n \rangle^p$). Given the electronic structure of SiO_2 (an insulator with a 5 eV gap), and the large mismatch between the OH stretch mode (3750 cm^{-1}) and the next highest vibrational mode in the system (Si–O stretch 1100 cm^{-1}), a relaxation pathway was proposed, involving one quantum of OH stretch excitation decaying into four lower frequency vibrational modes. Comparable results have been observed for OH groups in zeolites [50]. These first time-resolved experiments demonstrated the perils of band shape analysis. Even though inhomogeneous broadening was expected to contribute to the width ($\Delta\nu \approx 8$ cm^{-1} FWHM) in these experiments, the contribution from lifetime broadening is found to be negligible, only 0.03 cm^{-1}. Clearly, inhomogeneous broadening is capable of dominating spectral band shapes.

A second series of time-resolved measurements on high surface area samples was initiated in 1986 with the application of laser sources in the 2000 cm^{-1} region. A series of measurements was undertaken that characterized the vibrational relaxation of CO stretch modes in metal carbonyl compounds, examining the molecules when they were dilute in a variety of solvents [51] and when they were supported on SiO_2 [52]. These systems marked two significant advances over previous work. It became possible to explore the coupling between different CO modes in the same molecule, suggestive of the coupling that would be anticipated to occur between adjacent adsorbate sites on planar surfaces. In addition, by varying the number of metal atoms in the compound, it became possible to test for any role played by metal particle size in the vibrational relaxation process. The available data on the metal carbonyls in solution suggest that the different high frequency CO stretch modes in a molecule couple together rapidly [e.g., 3 ps for $Rh(CO)_2(C_5H_7O_2)$ in CCl_3H solution], and the coupled CO modes then relax on a much longer timescale by multiphonon transfer to low fre-

quency vibrations [53]. For six metal carbonyls studied [$Cr(CO)_6$ simplest; $Rh_6(CO)_{16}$ largest], the slow relaxation times in solution were very similar (e.g., 600 ± 300 ps in CCl_3H). The excited state decay rate of the metal carbonyl when attached to SiO_2 at very low surface coverage was found to increase by a factor of four over the rate found for the same compound in solution [53]. The mechanism by which adsorption on SiO_2 accelerates the rate of vibrational damping has not been proven; however the classical theories mentioned above (Förster and image dipole damping) [24,54] predict energy transfer to SiO accepting modes close to the observed rate.

In 1990, pump–probe measurements employing surface sum-frequency generation as the probe were used to characterize the vibrational relaxation of the high frequency (~ 2084 cm^{-1}) hydrogen stretch of H/Si(111) (atomically perfect hydrogen terminated silicon) [55]. The long lifetime (950 ps at room temperature), and extensive studies of the temperature dependence of the band shape and position [56], led to the identification of multiphonon relaxation as the mechanism responsible for the decay of excited state population. This work was extended to probe the relaxation of modes associated with H atoms bound to steps of vicinal Si(111) surfaces in the dihydride and monohydride configurations [57]. After picosecond IR excitation, the dihydride modes (C_2 and C_3) relax rapidly (~ 100 ps), as seen in Fig. 2. The terrace sites (mode A) exhibit a reduced T_1 compared to that found for the atomically perfect surface. Because of dipole–dipole (Förster) energy transfer between terrace and step modes, the relaxation rate of the terrace modes increases as the step density increases (e.g. T_1 decreases to 420 ps when the step density is 1 per 2.0 nm). By contrast, T_1 for monohydride step sites is long (~ 1100 ps) and so a high density of such sites does not significantly influence T_1 for the terrace sites, despite similar terrace–step coupling strengths. Photon echo techniques were also used to determine the pure dephasing time of the H/Si(111) system [58]. Although the absorption is very narrow ($\Delta\nu \approx 0.26$ cm^{-1} at $T = 120$ K), the observed width is dominated by inhomogeneous broadening, with energy decay making a negligi-

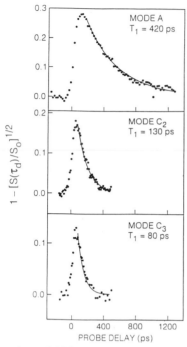

Fig. 2. Experimental SFG normalized transient response of the Si–H stretching modes of the dihydride-stepped H/Si(111)9° surface at room temperature. Mode A (2083 cm^{-1}) = terrace top sites; modes C_2 (2103 cm^{-1}) and C_3 (2136 cm^{-1}) are perpendicular and parallel dihydride modes at the steps. Solid points: experimental data; solid lines: exponential fits (after Ref. [57]).

ble (0.004 cm^{-1}) contribution to the (0.12 cm^{-1}) homogeneous width ($T = 120$ K). This series of beautiful studies makes the vibrational dynamics of H/Si(111) the best characterized single crystal system. The extremely slow damping times facilitated the observation of effects such as site-to-site energy transfer. Although similar processes must occur for molecules chemisorbed on metals (e.g., top site–bridge site transfer for CO/Pt(111)), the extremely fast electron–hole pair damping rates have to date precluded similar observations on metals.

Semiclassical models for the vibrational relaxation rate indicated that lifetimes greater that 10^{-9} s were to be anticipated for adsorbed hydrogen on semiconductor surfaces [59]. Using both the static and time-resolved properties of this system as benchmarks, the theorists have been able to test their models of excited state proper-

ties. A recent first principles calculation of the spectroscopy of the excited adlayer using a realistic vibrational potential showed a two phonon bound state [60], in agreement with the original experimental interpretation [61]. A boson Hubbard model [62] was used to calculate the spectral signatures of the excited states as they depopulate and perhaps form biexcitons. Molecular dynamics models have also been used to follow the multiphonon pathway for damping of the excited state population [63].

What may prove to be the long lifetime limit in the field of time-resolved characterization of surface vibrational relaxation is found in the damping of CO adlayers on NaCl [64]. Because of the absence of near resonant electronic excitations, the large frequency mismatch between the CO stretch and the NaCl phonons (2134 versus 164 cm^{-1}), and formation of a weak NaCl–CO bond (and corresponding absence of high frequency local M–CO bending and stretching modes, such as occur for CO chemisorbed on metals), it was found that relaxation through infrequent multiphonon events can be remarkably slow. In fact, energy pooling was a major effect in experiments where the $v = 0 \rightarrow 1$ mode was optically pumped. In those experiments, emission from higher lying vibrational levels was readily detected. Taking the observed time constant of 4.3 ms for emission at the $v = 2 \rightarrow v = 0$ frequency and correcting for emission from higher excited states, a phonon relaxation rate of 2×10^2 s^{-1} was established ($T = 22$ K). This result is only a factor of six faster than the radiative rate of CO in the gas phase, indicating the minimal influence played by the NaCl.

The preceding studies were for systems where multiphonon decay is the only plausible relaxation mechanism. For molecules on metals, it was generally believed that damping due to electron–hole pair formation would dominate the relaxation rate for high frequency ($\nu > 2\Theta_D$) modes. However, for two of the first adlayers to be studied on metal surfaces, vibrational relaxation was dominated by intramolecular coupling, rather than direct electron–hole pair damping. The terminal methyl group CH stretch in a Cd stearate (i.e., –(CH$_2$)$_{16}$CH$_3$) Langmuir–Blodgett monolayer on evaporated silver was studied [65]. Fast (2.5 ps) and slow (165 ps) biexponential decays were observed and attributed to complex intramolecular decay processes (presumably involving Fermi resonance mixing of the CH stretch and CH_3 bending modes also responsible for fast relaxation of similar molecules in solution). The large CH_3–surface distance (20 Å) also decreased the likelihood of electron–hole pair coupling. However, similar behavior was found for an ordered monolayer of CH_3S on Ag(111) [66] (fast decay ~ 3 ps; T-dependent slow decay of 55–90 ps). A consistent interpretation of the data was rapid relaxation of the CH stretch mode (2968 cm^{-1}) to the overtone ($v = 2$) of the CH_3 bend $v = 1452$ cm^{-1}), followed by multiphonon relaxation of the bend to lower frequency molecular modes (creation of 3 phonons fits the T-dependence).

The first strong time-resolved evidence for the influence of electron–hole pair damping on the vibrational relaxation of adsorbates came in measurements that contrasted the CO ($v = 1$) lifetimes of supported metal carbonyls with model catalyst systems. For the model catalysts, samples were prepared where the size of the metal aggregates was estimated to be about 3 nm. In such Rh/SiO_2 samples, where both isolated $Rh(CO)_2$ sites and CO/Rh_{island} sites are present, the associated lifetimes were found to be 120 and < 20 ps, respectively [67]. Analogous measurements on $CO/Pt/SiO_2$ indicated a T_1 of 7 ± 2 ps for CO bound to platinum aggregates of nominally 3 nm diameter [68].

While the 120 ps value of T_1 for the $Rh(CO)_2$ sites was consistent with the timescale for multiphonon relaxation processes found in the supported carbonyl cluster compounds, the short timescale observed for the larger particles was not. However, the measured short relaxation times were in reasonable agreement with the theoretical predictions for the relevant timescale for electron–hole pair damping. Such observations for the model catalyst systems were highly suggestive of the existence of a critical aggregate size, below which the electron–hole pair damping mechanism was inoperative [69]. More significantly, these measurements indicated that temporal resolution on the order of 1 ps was necessary to quantitatively address this damping mechanism. It was also clear that for time-domain techniques to contribute on an equal footing with frequency domain studies it would be essential to measure single crystal surfaces where site homogeneity could be evaluated.

Two separate pump–probe approaches based on single reflection methods were published in 1990 where resonant IR pulses were used to generate vibrationally excited CO adsorbates on metal surfaces [70,71]. Aside from the choice of Pt(111) versus Cu(100) substrates, the primary difference between the two measurements lay in the method used to probe the excited state population. One approach relied on sum frequency generation (SFG) to sample the CO ($v = 1$) population [71] while the other utilized linear infrared absorption spectroscopy [70]. Both methods reported picosecond transients: T_1 values for CO/Pt(111) and CO/Cu(100) were initially reported as 4.9 ± 1.6 [70] and 3 ps [71], respectively; later studies revised the values to $2.2 \pm .3$ [72] and 2.0 ± 1 ps [73], respectively. A time-resolved measurement using surface SFG of the total dephasing time T_2 of CO/Cu(111) at $T = 93$ K gave $T_2 = 2.0 \pm 0.3$ ps and an exponential coherence decay [74]. The IRAS full width of 4.5 cm^{-1} (Ref. [75]) implies $T_2 = 2.3$ ps so this result was not surprising. In light of the earlier T_1 measurements for CO/Cu(100) and CO/Pt(111) [70–73], it was assumed for CO/Cu(111) that the dephasing was dominated by T_1.

For the CO/Pt(111) system [72] a detailed study was made of the infrared spectral shape as a function of pump–probe time delay and pump power. To ascertain the transient response on the subpicosecond timescale while maintaining sufficient spectral resolution to establish changes in the 3.5 cm^{-1} (FWHM) spectral feature, it was necessary to introduce the ability to spectrally disperse the transform-limited probe pulse (~ 15 cm^{-1} FWHM) after it had reflected from the sample. In the cited work, a resolution of 1 cm^{-1} was utilized. These measurements established the complex spectral characteristics associated with excited vibrational modes of highly coupled oscillators. Fig. 3 depicts the transient absorption

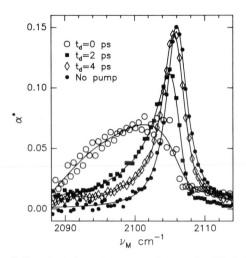

Fig. 3. Transient absorption spectra for top site CO/Pt(111) as a function of time delay following excitation of the adlayer by an 0.8 ps IR pulse (after Ref. [72]).

spectrum measured at various pump–probe time delays t_d. A distinct shift to lower wavenumber and clear broadening and asymmetry are apparent, as is the rapid return of the spectrum to its initial character ($T_1 = 2.2$ ps). As a consequence of strong adsorbate–adsorbate coupling in the excited layer, the spectrum of the excited state is best described as a one-phonon band that shifts with increasing excited state population as opposed to contributions from a spectrally fixed one-phonon band and a distinct two-phonon bound state [76]. These results should be compared with the observations for H/Si(111) where the appearance of a two-phonon bound state was noted [77]. The difference occurs because for H/Si(111), the anharmonicity of the Si–H bond ($-2\Gamma \approx 77$ cm^{-1}) is much greater than the lateral interaction energy (band dispersion width $\Delta\Omega \approx 7.2$ cm^{-1}), while for CO/Pt(111) the reverse is true ($\Delta\Omega \approx 35$ cm$^{-1} > -2\Gamma \approx 24$ cm^{-1}). Both systems represent different extremes of excited-state spectral response. These measurements elegantly demonstrated the influence that the complex excited-state spectral behavior can have on conventional measurements of excited state recovery [78]. In addition, they point to the need to account for the excited state spectral

response when designing measurements of the temporal evolution of excited state population.

As a result of time-resolved measurements, there has been a new impetus for theoretical assessment of the details of the underlying relaxation mechanism. Emergence of the experimental capability to test for predicted changes in the excited state lifetimes should provide a valuable method for assessing the applicability of competing theoretical models. Calculations aimed at evaluating the importance of nonadiabatic coupling in the quenching of these vibrationally excited adlayers are emerging [79,80]. The application of nonempirical molecular-orbital theory to the vibrational damping problem at metal surfaces is one example. Specifically the relaxation rates and mechanisms of all four local modes of CO/Cu(100) were investigated by ab initio calculations [79] on Cu$_n$CO cluster models. Nonadiabatic rates are approximately proportional to matrix elements of the type $|\langle \text{LUMO} | d / dQ_i | \text{HOMO} \rangle|^2$ which are localized near the CO molecule and hence can be calculated for small clusters and used (along with a suitably smoothed local density of states at the Fermi level) to simulate the extended surface. In the low temperature limit, slow ($T_1 \approx 100$ ps) nonadiabatic ("electron–hole pair") damping was predicted for the frustrated translation (40 cm^{-1}) and M–C stretch (345 cm^{-1}) modes, with rapid ($T_1 = 2$–3 ps) damping of the frustrated rotation (285 cm^{-1}) and C–O stretch (2085 cm^{-1}) modes. Subsequent stochastic dynamics simulations used a "molecular dynamics with electronic friction" approach (that self-consistently treated both phonon and electron–hole damping mechanisms) to examine the temperature dependence of the relaxation for these modes [80]. The theoretical results are that at elevated temperature, the relaxation of the two "slow" modes increased rapidly; the total relaxation rate could not be expressed as a sum of phonon and electron/hole pair rates; and temperature dependent measurements alone are insufficient to distinguish between phonon and electron damping mechanisms.

The relaxation times and coupling mechanisms of these low frequency modes have recently been explored in time-resolved studies of the CO/

Pt(111) system [81] and in state-resolved desorption studies of NO/Pd(111) [82]. In the CO/Pt(111) experiments, visible ps laser pulses incident on the surface (not resonant with any surface vibrational states) create a transient increase in the near-surface electron temperature ($\Delta T \approx 80$ K); the hot electrons may then equilibrate with (excite) both bulk phonon modes and the local CO/Pt modes, while the population of these modes is probed by picosecond IRAS. Since ΔT is small, of the four local modes only the low frequency frustrated translation (60 cm^{-1}) is significantly excited. If coupling existed solely between the translational mode and hot electrons, a coupling time of $\tau_e = 2 \pm 1$ ps was deduced, while if this mode coupled only to lattice phonons, fitting the data required $\tau_{lat} < 1$ ps. Because of the rapid electron–phonon coupling in Pt these two mechanisms could not be clearly separated. Similar experiments for CO/Cu, for which the electron–phonon coupling strength is twenty times less, should furnish valuable tests of the preceding calculations.

4. Electronic relaxation

In terms of surface science efforts, time-resolved measurements of excited electronic state dynamics have met with varying degrees of success when addressing the energy transfer properties of both clean and adsorbate covered solids. Frequently, the characterization is complicated by the transient response of bulk carriers. Numerous time-resolved studies have now appeared that establish the carrier lifetimes in bulk semiconductor and quantum well systems [83] and electron phonon coupling times in metals [84]. Most commonly used optical techniques, such as transient reflectivity changes, average the sample response over the penetration depth of the pump–probe laser pulses, typically 10–100 nm for visible light. Hence, the separation into "surface" and "substrate" which is clear for the vibrations of CO/metal or H/Si, is generally less distinct in the ultrafast studies of carrier relaxation. Through the application of probes with enhanced surface sensitivity, it is becoming possible to distinguish the bulk response from that characteristic of the surface.

Since the experiments of Drexhage [22] and the development by Silbey [24] and colleagues of the classical electromagnetic theory of non-radiative decay, there have been numerous short pulse laser studies [85–100] of the decay rates of electronically excited states of fluorescent dye molecules on dielectric, semiconductor, and organic crystal surfaces. These studies were motivated by the interest in finding dye sensitizers (with absorption matching the solar spectrum) to put on semiconductor surfaces to enhance conversion of solar photons into carriers [98], and by applications in color photography. For dye molecules on insulators (i.e., bandgap $\gg S_1$, the energy of the dye excited electronic state), the dynamics of S_1 could be understood in terms of molecule to molecule Förster energy transfer and quenching by dye aggregates (dye molecules do not generally form ordered adlayers, but monomers, dimers and higher aggregates are present on the surface). For dyes separated by spacer layers from semiconductor surfaces, the dependence of decay rate on distance was consistent with classical theory. Excited molecules bound directly to the surface of a semiconductor (e.g., ZnO) or organic crystalline substrate (e.g., anthracene), or to the surface of colloidal semiconductors (e.g., TiO_2), decay rapidly due to energy transfer to the substrate with creation of bulk carriers.

Rates of energy transfer involving electronically excited dye molecules have also been measured at the vacuum–liquid interface using the nonlinear optical technique of surface second harmonic generation [21] to probe the S_1 population relaxation following excitation by visible picosecond laser pulses. Rates of electronic excitation transfer from donor molecules (Rhodamine 6G) to acceptor molecules (DODCI) were measured for a density of acceptors of about 10^{12} cm^{-2} on the surface of liquid water [99], and were interpreted in terms of Förster energy transfer. Using the same SHG technique, rotational reorientation [100] of R6G on the water surface was measured and found to be slower (~ 1 ns) than the rotational relaxation time of R6G in bulk water (~ 200 ps). This is in contrast to the S_1 lifetime of DODCI at the water surface, which was found to be less (220 ps) than in bulk water (520 ps) [101]. Although time-resolved studies mentioned in this and the preceding paragraph involve molecules at surfaces, most of the results can be understood by energy transfer mechanisms well studied in solution and in bulk solids, and hence do not particularly highlight surface or interface-specific phenomena.

Angle and time-resolved measurements of the transient response of excited carriers in surface-localized states were first reported in 1985 by a group at AT&T Bell Laboratories [102]. They were able to directly observe the transient population of a normally unoccupied surface resonance in InP(110) using bichromic two-photon photoemission. With the k-space mapping afforded by this approach, the distinction between bulk and surface effects was greatly simplified. Insight into carrier recombination times at semiconductor surfaces has increased through new from time-resolved probes [103,104]. Characteristic of these measurements is the work on GaAs(110) that has followed surface states and their related electron scattering characteristics [104]. Through time-resolved two-photon photoemission measurements, it has been possible to confirm the existence of theoretically predicted bands and to establish the timescale for surface intervalley scattering ($\overline{\Gamma}-\overline{X} \approx 0.4 \pm 0.1$ ps). The same technique has been used to observe dynamics of surface states on various Si(111) [105], Si(100) [105], and Ge(111) [106] reconstructions.

A notable example of the utility of the bichromic two-photon photoemission technique is the work of Schoenlein et al. [107]. They employed 20 fs duration pulses to study surface effects in the spectral and temporal properties of image states on Ag(100) and Ag(111). On Ag(100) these experiments demonstrated that the lifetime of the $n = 1$ state lies in the range 15–35 fs. The lifetime increases to 180 fs upon going from the $n = 1$ to the $n = 2$ state, in agreement the n^3-scaling predicted theoretically. By contrast, damping of $n = 1$ and 2 image states above the Ag(111) surface occurs in < 20 fs. The faster relaxation is probably due to greater penetration of the image

state wavefunction into the crystal for Ag(111) compared to Ag(100).

From the perspective of new opportunities in surface chemistry, possibly the most intriguing advances have come from optical pumping of adsorbates bound to metal surfaces. A variety of experiments [2,108] have shown that carrier mediated photoprocesses dominate the non-thermal channels of laser initiated reactions on metal and semiconductor surfaces. For systems with novel photoreaction channels, or for mechanisms that demand electronic temperatures that deviate strongly from the corresponding bulk temperature, it has been possible to establish the involvement of optically generated carriers [3]. Although the excited state lifetimes are expected to fall in the femto- to picosecond regime, several different laboratories have been able to characterize the strength of the coupling of the adsorbate to the excited substrate [80–82].

In two examples, the timescales for laser induced non-thermal desorption events were explored. The approach used by Heinz et al. at IBM utilized state-resolved gas phase detection techniques and the nonlinear fluence dependence of the laser-induced desorption yield to follow NO desorption from Pd(111) [109]. In these measurements, two visible 400 fs pulses were overlapped on the sample and the desorption signal was monitored as function of their relative arrival time. The resultant 600 fs width of the correlation feature indicated that desorption was mediated by the transient electronic temperature and was not driven by the lattice bath. These measurements provided evidence for reaction channels that required multiple electronic excitations, a class of reactions that had not previously be proposed. In analogous experiments by Prybyla et al. at AT&T Bell Labs, second harmonic generation was used to monitor the adsorbate coverage following laser induced desorption of CO from Cu(111) [110]. In these experiments, 100 fs dye laser pulses were used to desorb CO, and the resulting disappearance of CO was followed through the magnitude of the second harmonic signal. They concluded that CO leaves the surface within a period of approximately 300 fs, providing independent evidence for novel surface chemistry that is accessible through excited electronic transients. In spite of questions regarding the applicability of thermal models for the transient carrier energy distributions [111], chemical opportunities are clear.

5. Conclusion

We have briefly reviewed time-resolved studies of the rates and mechanisms of vibrational and electronic energy transfer at surfaces from the earliest studies (about three decades ago) to the present. Although new, the application of very high time resolution techniques to surface dynamics is a growing field. The use of short-pulse lasers in surface science has been limited in the past because of their substantial difficulty of operation. Ultrafast lasers are becoming commercially available, flexible, and "user friendly". For instance, within a few years, it may be feasible to obtain IRAS or SFG spectra with subpicosecond time resolution as easily as one may now obtain static IRAS spectra with a commercial FTIR instrument.

It has recently been possible in gas and liquid phases to probe reactants, products, and reaction intermediates spectroscopically on the timescale that chemical bonds are formed or broken [19,112–114]. Although the issues of sensitivity and ease of sample preparation are much more severe, similar experiments for monolayers on metal and semiconductor surfaces are now quite feasible. We expect it will soon be possible to combine the spatial resolution of STM or other scanning probe microscopies with the high "atomic and molecular timescale" resolution of ultrafast lasers to study the dynamics of highly localized structures (perhaps individual molecules) on surfaces. The only long range (30 year timescale) prediction which seems certain is that ultrafast time-resolved studies will become routine in the surface science community.

References

[1] T.J. Chuang, Surf. Sci. Rep. 3 (1983) 1;
    R.M. Osgood, Jr., Ann. Rev. Phys. Chem. 34 (1983) 77.

[2] W. Ho, Surf. Sci. 299/300 (1994) 996.

[3] R.R. Cavanagh, D.S. King, J.C. Stephenson and T.J. Heinz, J. Phys. Chem. 97 (1993) 786.

[4] G.P. Brivio and T.B. Grimley, J. Phys C 10 (1977) 2351;
    Surf. Sci. 89 (1979) 226;
    M.A. Kozhushner, V.G. Kustarev and B.R. Shub, Surf. Sci. 81 (1979) 261.

[5] B. Hellsing and M. Persson, Phys. Scr. 29 (1984) 360;
    B.N.J. Persson, J. Phys. C 11 (1978) 4251;
    B.N.J. Persson and M. Persson, Surf. Sci. 97 (1980) 609.

[6] A.G. Eguiluz, Phys. Rev. B 30 (1984) 4366.

[7] B.N.J. Persson and M. Persson, Solid State Commun. 36 (1980) 175.

[8] B.N.J. Persson and R. Ryberg, Phys. Rev. B 24 (1981) 6954.

[9] B.E. Hayden and A.M. Bradshaw, Surf. Sci. 125 (1983) 767.

[10] M. Trenary, K.J. Uram, F. Bozso and J.T. Yates, Jr., Surf. Sci. 146 (1984) 269.

[11] B.N.J. Persson and R. Ryberg, Phys. Rev. Lett. 54 (1985) 2119.

[12] B.N.J. Persson, F.M. Hoffmann and R. Ryberg, Phys. Rev. B 34 (1986) 2266.

[13] D.C. Langreth and M. Persson, Phys. Rev. B 43 (1991) 1353.

[14] Y.J. Chabal, Phys. Rev. Lett. 55 (1985) 845.

[15] D.C. Langreth, Phys. Rev. Lett. 54 (1985) 126.

[16] Ž. Crljen and D.C. Langreth, Phys. Rev. B 35 (1987) 4224.

[17] R.G. Tobin, Surf. Sci. 183 (1987) 226.

[18] J.W. Gadzuk and A.C. Luntz, Surf. Sci. 144 (1984) 429.

[19] A.H. Zewail, Science 242 (1988) 1645.

[20] N. Bloembergen, Mat. Res. Soc. Symp. Proc. 51 (1986) 3;
    J. Hicks, in: Laser Spectroscopy and Photochemistry on Metal Surfaces, Advances in Physical Chemistry (World Scientific, Singapore, 1993).

[21] Y.R. Shen, Surf. Sci. 299/300 (1994) 551.

[22] K.H. Drexhage, M. Fleck, H. Kuhn, F.P. Schäfer and W. Sperling, Ber. Bunsenges. Phys. Chem. 70 (1966) 1179;
    H. Bucher, K.H. Drexhage, M. Fleck, H. Kuhn, D. Möbius, F.P. Schäfer, J. Sonderman, W. Sperling, P. Tillman and J. Wiegard, Mol. Cryst. 2 (1967) 199;
    K.H. Drexhage, H. Kuhn and F.P. Schäfer, Ber. Bunsenges. Phys. Chem. 72 (1968) 329;
    K.H. Drexhage, J. Lumin. 1/2 (1970) 693.

[23] D.H. Waldeck, A.P. Alivisatos and C.B. Harris, Surf. Sci. 158 (1985) 103.

[24] R.R. Chance, A. Prock and R. Silbey, in: Advances in Chemical Physics, Vol. 37, Eds. S.A. Rice and I. Prigogine (Wiley–Interscience, New York, 1978) p. 1.

[25] R. Rossetti and L.E. Brus, J. Chem. Phys. 73 (1980) 572.

[26] M. Grätzel and K. Kalyanasundaram, Eds., Kinetics and Catalysis in Microheterogeneous Systems (Decker, New York, 1991);
    V. Ramamurthy, D.F. Eaton and J.V. Caspar, Acc. Chem. Res. 25 (1992) 299, and references therein.

[27] N.J. Turro, Pure Appl. Chem. 58 (1986) 1219; Tetrahedron 43 (1987) 1589.

[28] T.E. Madey, Surf. Sci. 299/300 (1994) 824.

[29] Ph. Avouris, N.J. DiNardo and J.E. Demuth, J. Chem. Phys. 80 (1984) 491.

[30] Ph. Avouris and R.E. Walkup, Ann. Rev. Phys. Chem. 40 (1989) 173.

[31] R.E. Palmer, Prog. Surf. Sci. 41 (1992) 51.

[32] H. Gerischer, in: Photovoltaic and Photoelectrochemical Solar Energy Conversion, Vol. 69 of NATO Advanced Study Institute Series B: Physics, Eds. F. Cardon, W.P. Gomes and W. Dekeyser (Plenum, New York, 1981) p. 199.

[33] D.S. Boudreaux, F. Williams and A.J. Nozik, J. Appl. Phys. 51 (1980) 2158.

[34] A.J. Nozik, Ed., Photoeffects at Semiconductor–Electrolyte Interfaces, ACS Symp. Ser. 146 (Am. Chem. Soc., Washington, DC, 1981).

[35] L.J. Richter and R.R. Cavanagh, Prog. Surf. Sci. 39 (1992) 155.

[36] G. Beta and P. Varga, Eds., Desorption Induced by Electronic Transitions: DIET IV (Springer, New York, 1990) and references therein.

[37] F. Hoffmann, Surf. Sci. Rep. 3 (1983) 107;
Y.J. Chabal, Surf. Sci. Rep. 8 (1988) 211.

[38] L. Risenberg and H.W. Moos, Phys. Rev. 174 (1968) 429;
R. Englman, Nonradiative Decay of Ions and Molecules in Solids (North-Holland, Amsterdam, 1979);
A. Nitzan, S. Mukamel and J. Jortner, J. Chem. Phys. 60 (1974) 3929;
C.B. Harris, R.M. Shelby and P.A. Cornelius, Phys. Rev. Lett. 38 (1977) 1415.

[39] E.J. Heilweil, M.P. Casassa, R.R. Cavanagh and J.C. Stephenson, Ann. Rev. Phys. Chem. 40 (1989) 143.

[40] R.G. Greenler, J. Chem. Phys. 44 (1966) 3120.

[41] M. Persson and B. Hellsing, Phys. Rev. Lett. 49 (1982) 662;
B. Hellsing and M. Persson, Phys. Scr. 29 (1984) 360.

[42] P. Hollins and J. Pritchard, Prog. Surf. Sci. 19 (1985) 275;
B.N.J. Persson and R. Ryberg, Chem. Phys. Lett. 174 (1990) 443;
P. Hollins, Surf. Sci. Rep. 16 (1992) 51.

[43] W. Chen and W.L. Schaich, Surf. Sci. 218 (1989) 580;
R. Tobin, Phys. Rev. B 45 (1992) 12110.

[44] R.M Shelby, C.B. Harris and P.A. Cornelius, J. Chem. Phys. 70 (1979) 34.

[45] For examples of laser advances, see, Ultrafast Phenomena I–VII, Springer Series in Chemical Physics (Springer, Berlin).

[46] J.P. Heritage, J.G. Bergman, A. Pinczuk and J.M. Worlock, Chem. Phys. Lett. 67 (1979) 229;
B.F. Levine and C.G. Bethea, IEEE J. Quantum. Electron. QE-16 (1980) 85;
J.P. Heritage and D.L. Allara, Chem. Phys. Letts. 74 (1980) 507;

B.F. Levine, C.V. Shank and J.P. Heritage, IEEE J. Quantum. Electron. QE-15 (1979) 1418.

[47] E.J. Heilweil, M.P. Casassa, R.R. Cavanagh and J.C. Stephenson, J. Chem. Phys. 81 (1984) 2856; 82 (1985) 5216.

[48] W.A. Pliskin and R.P. Eischens, J. Phys. Chem. 59 (1955) 1156.

[49] M.P. Casassa, E.J. Heilweil, J.C. Stephenson and R.R. Cavanagh, J. Chem. Phys. 84 (1986) 2361.

[50] C. Hirose, Y. Goto, N. Akamatsu, J. Kondo and K. Domen, Surf. Sci. 283 (1993) 244.

[51] E.J. Heilweil, R.R. Cavanagh and J.C. Stephenson, J. Chem. Phys. 89 (1988) 230.

[52] E.J. Heilweil, J.C. Stephenson and R.R. Cavanagh, J. Phys. Chem. 92 (1988) 6099.

[53] J.D. Beckerle, M.P. Casassa, R.R. Cavanagh, E.J. Heilweil and J.C. Stephenson, Chem. Phys. 160 (1992) 487.

[54] Th. Förster, Ann. Phys. 2 (1948) 55.

[55] P. Guyot-Sionnest, P. Dumas, Y.J. Chabal and G.S. Higashi, Phys. Rev. Lett. 64 (1990) 2156.

[56] P. Dumas, Y.J. Chabal and G.S. Higashi, Phys. Rev. Lett. 65 (1990) 1124.

[57] M. Morin, P. Jakob, N.J. Levinos, Y.J. Chabal and A.L. Harris, J. Chem. Phys. 96 (1992) 6203.

[58] P. Guyot-Sionnest, Phys. Rev. Lett. 66 (1991) 1489.

[59] J.C. Tully, Y.J. Chabal, K. Raghavachari, J.M. Bowman and R.R. Lucchese, Phys. Rev. B 31 (1985) 118.

[60] X.-P. Li and D. Vanderbuilt, Phys. Rev. Lett. 69 (1992) 2543.

[61] P. Guyot-Sionnest, Phys. Rev. Lett. 67 (1991) 2323.

[62] B.N.J. Persson, Phys. Rev. B 46 (1992) 12701.

[63] H.D. Gai and G.A. Voth, J. Chem. Phys. 99 (1993) 740.

[64] G.E. Ewing, Acc. Chem. Res. 25 (1992) 292.

[65] A.L. Harris and N.J. Levinos, J. Chem. Phys. 90 (1989) 3878.

[66] A.L. Harris, L. Rothberg, L.H. Dubois and L. Dhar, Phys. Rev. Lett. 64 (1990) 2086;
A.L. Harris, L. Rothberg, L. Dhar, N.J. Levinos and L.H. Dubois, J. Chem. Phys. 94 (1991) 2438.

[67] E.J. Heilweil, R.R. Cavanagh and J.C. Stephenson, J. Chem. Phys. 89 (1988) 5342.

[68] J.D. Beckerle, M.P. Casassa, R.R. Cavanagh, E.J. Heilweil and J.C. Stephenson, J. Chem. Phys. 90 (1989) 4619.

[69] J.W. Gadzuk, Appl. Phys. A 51 (1990) 108;
E. Blaisten-Barojas and J.W. Gadzuk, J. Chem. Phys. 97 (1992) 862.

[70] J.D. Beckerle, M.P. Casassa, R.R. Cavanagh, E.J. Heilweil and J.C. Stephenson, Phys. Rev. Lett. 64 (1990) 2090.

[71] S.F. Shane, L. Rothberg, L.H. Dubois, N.J. Levinos, M. Morin and A.L. Harris, Ultrafast Phenomena VII Vol. 53 of Springer, Series in Chemical Physics, Eds. C.B. Harris, E.P. Ippen, G.A. Mourou and A.H. Zewail (Springer, Berlin, 1990) p. 362.

[72] J.D. Beckerle, R.R. Cavanagh, M.P. Casassa, E.J. Heil-

weil and J.C. Stephenson, J. Chem. Phys. 95 (1991) 5403;
R.R. Cavanagh, J.D. Beckerle, E.J. Heilweil and J.C. Stephenson, Surf. Sci. 269/270 (1992) 113.

[73] M. Morin, N.J. Levinos and A.L. Harris, J. Chem. Phys. 96 (1992) 3950.

[74] J.C. Owrutsky, J.P. Culver, M. Li, Y.R. Kim, M.J. Sarisky, M.S. Yeganeh, A.G. Yodh and R.M. Hochstrasser, J. Chem. Phys. 97 (1992) 4421.

[75] R. Raval, S.F. Parker, M.E. Pemble, P. Hollins, J. Pritchard and M.A. Chesters, Surf. Sci. 203 (1988) 353.

[76] J.C. Kimbal, C.Y. Fong and Y.R. Shen, Phys. Rev. B 23 (1981) 4946.

[77] P. Guyot-Sionnest, Phys. Rev. Lett. 67 (1991) 2323.

[78] H. Kasai and A. Okiji, Surf. Sci. 283 (1993) 233.

[79] M. Head-Gordon and J.C. Tully, J. Chem. Phys. 96 (1992) 3939;
M. Head-Gordon and J.C. Tully, Phys. Rev. B 46 (1992) 1853.

[80] J.C. Tully, M. Gomez and M. Head-Gordon, J. Vac. Sci. Technol. A 11 (1993) 1914.

[81] T.A. Germer, J.C. Stephenson, E.J. Heilweil and R.R. Cavanagh, J. Chem. Phys. 98 (1993) 9986.

[82] J.A. Misewich and T.J. Heinz, in press 1993.

[83] R.R. Alfano, Ed., Semiconductors Probed by Ultrafast Spectroscopy, Vols. 1, 2 (Academic Press, New York, 1984).

[84] S.D. Brorson, A. Kazeroonian, J.S. Moodera, D.W. Face, T.K. Cheng, E.P. Ippen, M.S. Dresselhaus and G. Dresselhaus, Phys. Rev. Lett. 64 (1990) 2172;
H.E. Elsayed-Ali, T. Juhasz, G.O. Smith and W.E. Bron, Phys. Rev. B 43 (1991) 4488;
P.B. Allen, Phys. Rev. Lett. 59 (1987) 1460;
K.M. Yoo, X.M. Zhao, M. Siddique, R.R. Alfano, D.P. Osterman, M. Radparvar and J. Cunniff, Appl. Phys. Lett. 56 (1990) 1908;
J.G. Fujimoto, J.M. Lin, E.P. Ippen and N. Bloembergen, Phys. Rev. Lett. 53 (1984) 1837;
H.E. Elsayed-Ali, M.A. Pessot, T.B. Norris and G.A. Mourou, in: Ultrafast Phenomena V, Vol. 46 of Springer Series in Chemical Physics (Springer, Berlin, 1986) p. 264.

[85] R.L. Crackel and W.S. Struve, Chem. Phys. Lett. 120 (1985) 473.

[86] P.A. Anfinrud, T.P. Causgrove and W.S. Struve, J. Phys. Chem. 90 (1986) 5887.

[87] P.A. Anfinrud, R.L. Crackel and W.S. Struve, J. Phys. Chem. 88 (1984) 5873.

[88] N. Tamai, T. Yamazaki and I. Tamazaki, Chem. Phys. Lett. 147 (1988) 25.

[89] S.R. Meech and K. Yoshihara, Chem. Phys. Lett. 174 (1990) 423; J. Phys. Chem. 94 (1990) 4913.

[90] K. Femnitz, T. Murao, I. Yamazaki, N. Nakashima and K. Yoshihara, Chem. Phys. Lett. 101 (1983) 337.

[91] D.R. Haynes, K.R. Helwig, N.J Tro and S.M. George, J. Chem. Phys. 93 (1990) 2836.

[92] D.R. Haynes, A. Tokmakoff and S.M. George, J. Chem. Phys. in press.

[93] N. Nakashima, K. Yoshihara and F. Willig, J. Chem. Phys. 73 (1980) 3553.

[94] P.V. Kamat, Langmuir 6 (1990) 512.

[95] Y. Nosaka, H. Miyama, M. Terauchi and T. Kobayashi, J. Phys. Chem. 92 (1988) 255.

[96] C. Arbour, D.K. Sharma and C.H. Langford, J. Phys. Chem. 94 (1990) 331.

[97] E.L. Quitevis, M.-L. Horng and S.-Y. Chen, J. Phys. Chem. 92 (1988) 256.

[98] M.S. Wrighton, Acc. Chem. Res. 12 (1979) 303.

[99] E.V. Sitzmann and K.B. Eisenthal, J. Chem. Phys. 90 (1989) 2831.

[100] A. Castro, E.V. Sitzmann, D. Zhang and K.B. Eisenthal, J. Phys. Chem. 95 (1991) 6752.

[101] E.V. Sitzmann and K.B. Eisenthal, J. Phys. Chem. 92 (1988) 4579.

[102] R. Haight, J. Bokor, J. Stark, R.H. Storz, R.R. Freeman and P.H. Buchsbaum, Phys. Rev. Lett. 54 (1985) 1302;
J. Bokor, R. Haight, R.H. Storz, J. Stark, R.R. Freeman and P.H. Buchsbaum, Phys. Rev. B 32 (1985) 3669.

[103] J. Bokor and N.J. Halas, IEEE J. Quantum Electron. QE-25 (1989) 2550;
J. Bokor, R. Storz, R.R. Freeman and P.H. Bucksbaum, Phys. Rev. Lett. 57 (1986) 881.

[104] R. Haight and J. Silberman, IEEE J. Quantum Electron. 25 (1989) 2556; Phys. Rev. Lett. 62 (1989) 815.

[105] M.W. Rowe, H. Liu, G.P. Williams, Jr. and R.T. Williams, Phys. Rev. B 47 (1993) 2041.

[106] M. Baeumler and R. Haight, Phys. Rev. Lett. 67 (1991) 1153.

[107] R.W. Schoenlein, J.G. Fujimoto, G.L. Eesley and T.W. Capehart, Phys. Rev. Lett. 61 (1988) 2596; Phys. Rev. B 41 5436 (1990); B 43 (1991) 4688.

[108] A. Bradshaw, Surf. Sci. 299/300 (1994) 49.

[109] J.A. Prybyla, T.F. Heinz, J.A. Misewich, M.M.T. Loy and J.H. Glownia, Phys. Rev. Lett. 64 (1990) 1537;
F. Budde, T.F. Heinz, M.M.T. Loy, J.A. Misewich, F. de Rougemont and H. Zacharias, Phys. Rev. Lett. 66 (1991) 3024.

[110] J.A. Prybyla, H.W.K. Tom and G.D. Aumiller, Phys. Rev. Lett. 68 (1992) 503.

[111] W.S. Fann, R. Storz, H.W.K. Tom and J. Bokor, Surf. Sci. 283 (1993) 221.

[112] D.F. Padowitz, W.R. Merry, R.E. Jordan and C.B. Harris, Phys. Rev. Lett. 69 (1992) 3583.

[113] Y. Chen, L. Hunziker, P. Ludowise and M. Morgen, J. Chem. Phys. 97 (1992) 2149;
S.I. Ionov, G.A. Brucker, C. Jaques, L. Valachovic and C. Wittig, J. Chem. Phys. 97 (1992) 9486;
R.E. Walkup, J.A. Misewich, J.H. Glownia and P.P. Sorokin, J. Chem. Phys. 94 (1991) 3389.

[114] A.H. Zewail, Faraday Disc. Chem. Soc. 91 (1991) 207;
J.L. Jerek, S. Pederson, L. Banares and A.H. Zewail, J. Chem. Phys. 97 (1992) 9046.

Surface Science 299/300 (1994) 656–666
North-Holland

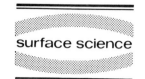

surface science

Dynamics of gas–surface reactions

Stephen Holloway

Surface Science Research Centre, The University of Liverpool, P.O. Box 147, Liverpool, L69 3BX, UK

Received 28 May 1993; accepted for publication 30 August 1993

Although there have been copious studies of gas–surface adsorption systems during the last thirty years, there is little doubt that during this period the subject has been more concerned with structural and kinetic properties than with studies of dynamical phenomena. This is now changing, both from an experimental and a theoretical point of view where there have been rapid advances particularly in the past decade. This review will concentrate primarily in the application of theoretical method to the treatment of reactive scattering of simple molecules from metal surfaces. Steric effects in the dissociation reaction, and the effects of placing energy in internal degrees of freedom for the molecular reactant will be discussed. The implications of surfaces that offer a wide range of activation barriers to adsorption will also be presented. Finally, some thoughts on future directions are given.

1. Introduction

During the lifetime of *Surface Science* there has been a dramatic change in approach for those who are interested in the steps that lead to an atom, or molecule, becoming attached to a solid surface. Looking back to Vol. 1, it is interesting to view the titles through the advantage of hindsight; "Kinetics of adsorption of oxygen on CdS single crystals" [1], "Adaptation of a lattice liquid model to gas adsorption phenomena" [2], "Desorption rates of atoms and ions" [3], "Frumkin's isotherm" [4]. These are four of the initial contributions that make a clear statement of the field in that year, 1964. It was the *kinetic* behaviour of reactants near solid surfaces that was the focus for studying the interaction of chemically active species with surfaces. Typically, experiments were made by varying the gas pressure at fixed temperature; the degree of adsorption was determined and an isotherm constructed. A modelling procedure was then initiated that has since become very familiar: A plausible scheme for the steps in the reaction was presented; rate equations were then cast and rate limiting steps (reaction bottlenecks) determined; finally an abridged set of the rate equations would be solved for several sets of

parameters. Eventually a parameter matrix would be found to reproduce the experimental observations to the required degree of precision. Now perhaps for some areas within the field this is still the *modus operandi*, but for others this is definitely *not* the case. What has happened? What has really changed?

There have been a number of diverse advances in both experimental and theoretical work that have permitted a far more *microscopic* view of the encounter of a molecule with a surface. Dynamics has taken over from kinetics. This is not to say that back in 1964 (or even further [5]) inquiring minds were not conceptualizing as to why it is that a molecule might fall apart on one surface, yet remain intact on another. It was, however, simply not possible to compare state-resolved theoretical predictions with the available kinetic data because of the vast degree of averaging required to convert cross sections into rate constants. Experiments were simply not incisive enough to favour one particular model over another; this has now changed. With the advent of sophisticated laser preparation and detection schemes, frequently borrowed from the world of gas-phase dynamics, it is now possible to design and carry out state-to-state experiments that can

SSDI 0039-6028(93)E0523-W

really discriminate between theoretical models to an extraordinary degree. Theory and experiment can now progress hand-in-hand with one another *on an equal basis* while back in 1964 they could not.

To describe in detail the advances in experimentation are far beyond the scope of this article (and author!) and readers are referred to some of the many reviews that have appeared within the past ten years [6,7]. In short what has happened is that ultrahigh-vacuum technology has been combined with laser spectroscopy to produce experiments that can measure quantum state specific information of molecules. Instead, we shall be making a trip from 1964 to the present (April 1993), examining along the way some of the key developments that have taken place on the theoretical/computational front.

There has been a major shift in the last thirty years in the entire approach employed for studying the natural sciences. While the terms "experimental" and "theoretical" have been employed previously to encompass the entire endeavour, this is actually not a true representation of the current state of play. There has appeared a new player in the game; the simulator. The simulator sits between the experimentalist and the "true" theorist investigating areas to which neither of them have access. Even though current experimental methods enables the exact motion of the nuclei in a reaction to be followed [8], there are nevertheless certain microscopic details that will never [9] be accessible to direct investigation, such as the time-dependence of the charge state of a reacting species or the correlated motion of a small cluster of atoms. At the other extreme, even though it is possible to write the "exact" Hamiltonian to describe the coupled electronic and nuclear motions as a reaction evolves, it is frequently impossible to find a solution and the heart of the theoretical method is to make a simplified calculation, amenable to solution, that will "contain the essential physics" of the problem. The simulator sits between these two camps. Using as input, data obtained from both theoretical and experimental studies, the simulator will construct computational algorithms that are capable of modelling the experimental problem, and

spend untold hours of cpu time in order to perform his or her version of reality.

Glancing back to *Surface Science* Vol. 1, there are no simulation papers but there is one paper that deals with the scattering of a He atom from a corrugated surface; "The total force of elastic scattering" [10]. This was a venerable problem first investigated by Devonshire [11] as paper number V in the series "The interaction of atoms and molecules with solid surfaces" which emerged from the Lennard-Jones group at Cambridge (UK). The structure of Beder's paper typifies the approach to dynamical problems during the first decade of *Surface Science*. A form for the interaction potential between an atom and a surface is postulated. In this instance a corrugated Morse potential which quickly metamorphoses into a square well plus corrugated hard wall in order that an approximate solution to the time-independent Schrödinger equation may be obtained. Eleven pages of algebra then follow before the required determinate solutions are obtained for the quantum mechanical forces and fluxes for the scattered particles. For good measure, these are then compared with their classical equivalents. Finally, the model is mapped onto the scattering of He/LiF with potential parameters taken from the 1936 paper of Devonshire. Numerical solutions were obtained in steps: The coefficients appearing in the wavefunction were evaluated using a Frieden desk computer. The scattering equations were then solved using an IBM 7090 main frame using a method of successive approximations. Finally, diffraction intensities were evaluated on a Honeywell 800 computer. In total, the paper comprises no less than 37 journal pages with 70 numbered equations!

The most recent *Surface Science* that I have is Vol. 283, from March 1 1993, ironically the proceedings of the Third ISSP International Symposium on Dynamical Processes at Solid Surfaces, held in Tokyo in April 1992. The fourth paper in the volume, "Classical trajectory study of fast H_2 scattering from Ag(111)" [12] is but one of the half dozen simulation papers. This paper has no equations but states "Newton's equations are solved exactly for a nonrotating, nonvibrating, randomly oriented molecule approaching a platelet of Ag atoms...". After this a brief description of the potential energy surface is given followed by "Results and discussion". What a world of difference. Interestingly, this paper is from the experimental group of Aart Kleyn at FOM Laboratories in Amsterdam who are simulating their own data by running a standard molecular dynamics code. This is an important development that could only have happened because of the major advances in computer technology.

It is interesting to compare computing resources in 1964 and now. Tom Grimley, who was one of the pioneer theorists in the field of process dynamics at surfaces, was on the Chemistry staff in at Liverpool in 1964. The University of Liverpool had no Computer Laboratory, but there was a subdepartment in Mathematics called "Computational Mathematics". They had an English Electric KDF 9 mainframe computer which was operated partly as a service to other departments. At that time its price was approximately £300 000. It stood completely alone with no campus networking let alone national or international connections. Two implementations of the programming language Algol were available to users, but no real operating system, as we know it today, was implemented. Input and output was via 1 inch paper tape, and 10 punch personnel were employed using Frieden Flexowriters for data and programme preparation. There was a primitive lineprinter for output but it was so slow that it was seldom used. There was a magnetic disc with a memory capacity of 32 kbytes which was the size of a large circular coffee table. The whole machine took up a complete floor of the Department of Mathematics. Access to the machine was by serial batch mode processing only with two shifts per day. Performance judged by todays standards was approximately 200 KIPS, ($\sim 10^{-3}$ MFLOP). Contrast this to a state-of-the-art Unix workstation ca. 1993 which would cost £25 000. This machine is located on a desktop, has 128 Mbytes of memory, a Gbyte disc for storage and performs at 30 MFLOP. It is networked to the world via Ethernet and can be accessed by multiple users in a research group. It will have a Fortran and a C compiler and additionally will have a suite of graphics programs for traditional data presentation (x versus y) and visualization software for making movies. It is impossible to compare these two facilities in any reasonable way. Simulation techniques and methods have only been made possible by the advent of such an enormous technological leap. Unfortunately, as a word of caution, computational power does not equate to good science and problem selection and methodology still are every bit as important now as in 1964.

The remainder of this article will be organised as follows; Section 2 contains a short introduction into reaction classifications followed by subsections covering the various degrees of freedom that enter into the simplest of surface reactions, dissociative adsorption. Electron–hole pair and phonon excitations in the substrate, followed by molecular translations, vibrations and rotations. Finally, Section 3 gives some brief conclusions.

2. Surface reactions

There has in fact been little theoretical work over the last thirty years on the microscopic dynamics of "real" surface reactions. What I mean by this is best illustrated by examining the classical division of reactions that has been inherited from catalytic chemistry [13]: Consider the reaction A(gas) + B(gas) → AB(gas) in the presence of a surface. If, say, species A of the reactants adsorbs on the surface and becomes thermalized (or accommodated) then a *Langmuir–Hinshelwood* process is said to be in operation if species B likewise is adsorbed and thermalizes before the two partners unite before desorbing as AB. If species B does not adsorb, but "plucks" A from its adsorption site during its brief encounter with the surface then the reaction is said to have been of the *Eley–Rideal* type. Of course this strict division is nonsense, and in reality there will be an entire spectrum of possibilities where atoms adsorb and react without falling deep into the adsorption well but perhaps undergo one or two surface bounces before the products are formed and desorption occurs [14]. Nevertheless, it is only very recently that detailed microscopic models have appeared to address such problems. This is primarily due to the fundamental limitation on describing long-time phenomena by molecular dynamics [15]. The timescales for molecular motion and thermalization are orders of magnitude apart and integrating equations of motion for times during which equilibration (*and* desorption) occur are prohibitively expensive. This puts severe limitations on the microscopic modelling of L–H reactions. For E–R reactions the situation is quite different since only one of the reactants is in contact with the surface heat bath and the motion of the other partner is not dissimilar from an unreactive scattering trajectory [16,17].

Of course to motivate such calculations, experimental evidence for a class of reactions is necessary and it is only within the last two or three years that genuine E–R reactions have been confirmed with interesting measurements of product internal state distributions [18–21]. There is little doubt that during the next decade more E–H processes will be discovered and their study will be a fertile area for research. Reactions of the L–H type, on the other hand, are relatively commonplace and here it would be of interest to investigate the internal state distribution of the products to see if they are solely determined by the surface temperature, or if they retain memory of their formation dynamics. Again, experimental motivation might be necessary to instigate theoretical modelling.

2.1. Dissociation dynamics

The "first principles" kinetic study by Stoltze and Nørskov of ammonia synthesis illustrates why reaction dynamics is of use in interpreting real processes [22]. In this work, the authors were able to deploy results from surface science experiments in a system of rate equations and produce results that were in good agreement with data obtained from a scaled down catalytic reactor. In this analysis, the rate limiting step for the reaction is the dissociative adsorption of N_2 on the metal surface. There is little doubt that the dissociation of diatomics on metal surfaces is the primary reaction to receive attention from the theoretical community over the last decades. The

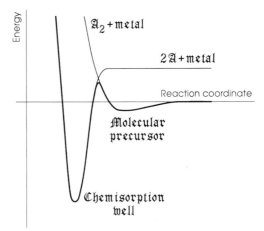

Fig. 1. "The interaction of a molecule with a metal" is how the Lennard-Jones potential energy curves was originally captioned. This venerable model has been extensively used (and abused) over the past three decades to describe simple dynamical processes.

most illustrious theoretical model in this respect is the potential energy surface (PES) presented by Lennard-Jones who was investigating the dissociation of H_2 on different metal surfaces, notably Ni [23]. Fig. 1 shows the now celebrated potential energy curves corresponding to the two diabatic states: Molecule–surface and atom–surface. Their widespread application to dissociative adsorption dynamics has been catalogued in many previous articles and it shall suffice to say that considerable insight has been gained from the model. On the down side, however, abuse of the model has in some cases given rise to extreme tunnel vision where its limited dimensionality has not been fully considered in the analysis of data. In the original article, the abscissa was the distance of the molecule from the metal surface and in the Appendix Lennard-Jones stressed that the curves would change depending on the impact parameter within the surface unit cell. Although nothing concerning the *internal* molecular degrees of freedom or those of the surface was discussed in the paper, this has not stopped countless research workers reading far more into the model than is actually there. Over the last decade, there has been an evolutionary movement within the theoretical community to address this problem and include greater dimensionality in the modelling.

As a good example of dimensionality being neglected, consider the coexistence of two processes that have been observed in the scattering of H_2 from Ni surfaces [24,25]: Diffraction and dissociation. The diffraction of He atoms from surfaces has a rich history in surface structure determination, dating back to the work of Estermann and Stern in 1928 [26]. It was found that H_2 molecules also exhibited diffraction when scattering from a variety of metal surfaces where the interaction is dominated by the physisorption potential (see Fig. 1). Now within the Lennard-Jones model such a process would be explained by the molecule scattering from the vacuum side of the activation barrier. If it had sufficient energy to traverse the barrier then it would spontaneously dissociate. The scattered intensity would appear as a step function in translational energy. How then can diffraction and dissociation occur for a beam of molecules at fixed translational energy?

2.2. Electron–hole pair creation

The semi-infinite surface of a metal contains a continuum of excitations of the conduction band electrons and it was hypothesized that these could couple to the motion of an atom outside a surface [27]. There appeared in the late seventies and early eighties a myriad of publications concerning the inelastic scattering of molecules and atoms from surfaces, where energy was dissipated by the excitation of electron–hole pairs [28]. Much of the formal treatment for this process was taken over from the closely related X-ray edge singularity problem in that a time-dependent potential (the gas atom) acts on the electron gas (surface) in a non-adiabatic fashion [29]. The main question to be answered is what the strength of this process might be. The key to answering this lies in the form of the electronic structure of the adsorbate. From chemisorption calculations, it has been shown that as atoms (or molecules) approach a surface, their sharp electronic states broaden due to interactions with the conduction band states [30]. In addition, they shift in energy

with a general rule of thumb being that ionization potentials shift upwards, while affinity levels shift down. This is simplest to see in the limit where image forces are a good description of the electrostatics. For adsorbates that have relatively sharp local densities of states when they are near the Fermi level (e.g. Cl_2), electronic non-adiabaticity is a significant process, as evidenced by the experimental observation of chemiluminescence and exoelectrons [31]. For adsorbates with deep lying states (e.g. He), the electronic mechanism on metal surfaces is of little importance [32]. For curve crossing problems such as the H_2 reaction depicted in Fig. 1, it has been shown that the electron affinity resonance typically has a width of 1 eV when crossing the Fermi level. This gives rise to an avoided crossing with electronically adiabatic progression from the molecular state to the atomic one [33]. While this subject is by no means a finished story, I think that at least for the H_2/Ni problem posed above, to claim that electronic non-adiabaticity is "the only plausible explanation for the observation" [34] is probably an overstatement and due consideration of other degrees of freedom is necessary.

2.3. Phonon creation

Needless to say, when a gas atom strikes a surface, there will be a "mechanical" dissipation of translational energy into the phonon continuum. Again this is a topic that has been extensively investigated over the past thirty years in *Surface Science* both from an an analytic as well as a molecular-dynamic point of view. What conclusions have ensued? In the classical limit the collision can only be an inelastic event; the familiar no-loss line, which results in diffraction, is a purely quantum phenomenon [35]. Again the magnitude of the energy loss is a question of the strength of the coupling. This will depend, in general, upon both the mass of the gas atom and its energy as well as properties originating from the surface such as the mass of a constituent atom, its temperature and the cohesive forces that bind the substrate together. In most instances, surface motion will need to be included in any theoretical description, particularly when pronounced surface temperature effects occur. For "heavy" molecules dissociating on a metal surface, it has been nicely illustrated that surface motion can give rise to enhanced reactivity as a consequence of the motion of the activation barrier along the reaction path [36]. This, however, was not found for the dissociation of hydrogen on surfaces of the 3d metals because of the large disparity in masses [37]. Probably in this case, because of the separation in timescales, thermal "roughness" will be the chief factor that phonon excitation exerts on the light molecule.

Dissipating energy to the phonon subsystem can give rise to the interesting dynamical possibility of trapping into a molecular state at the surface and this continues to be an interesting theme in surface dynamics [38]. For many years, kinetic data has been analysed under the assumption that molecules that dissociate can initially undergo a branching; some fraction will immediately dissociate (direct) while the rest will trap as molecules and after some residence time on the surface, dissociate [13]. Moreover, the kinetic analysis will sometimes indicate that there appears to be no barrier to separate the molecular (precursor) from the dissociated state [39]. Again recourse to Fig. 1 only confuses the issue since if there is no barrier, then what mysterious force drives the molecule into the molecular well when the atomic state is immediately accessible? The answer probably lies in the inadequacies of the one-dimensional PES. If we assume that molecules approaching the surface with different bond orientations, experience different activation barriers then it is not at all unreasonable to expect that the fraction feeling low barriers will directly dissociate, while the rest may dissipate enough energy to the surface phonons to trap intact [40]. While on the surface, the heat bath will then provide sufficient momentum and energy to allow the molecule to visit wide regions of phase space whence eventually low barriers may be found and dissociation occur. Again, by considering new degrees of freedom not present in Fig. 1, it is possible to rationalize what initially appeared to be totally irreconcilable results without having to resort to stretching the simple model beyond the limits of credulity.

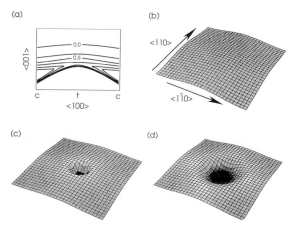

Fig. 2. (a) Contour map of the model potential for the dissociative adsorption of H_2 on a metal surface used in the study of Karikorpi et al. [41]. The cut shown links the centre and top sites where an activation barrier is located. Contour values are expressed as fractions of the minimum barrier height. (b)–(d) Surfaces of constant total energy plotted across the unit cell. The surfaces represent the loci of classical turning points for normally incident molecules with fixed energy. The corners show centre sites and the middle is an atop site. In (b) the beam energy is 20% below the barrier height while in (c) and (d) the beam energy is 5% and 25% over it. Since portions of the PES are open and closed for a particular energy, this implies that diffraction and dissociation can occur simultaneously.

2.4. Molecular translations

The Lennard-Jones model includes the molecular translational degree of freedom explicitly in the abscissa. As mentioned previously, in the Appendix of the paper a potential surface was presented that included variations in the potential within the surface unit cell. This is an interesting effect in itself and Fig. 2 shows how activation barriers might change from site to site for a molecule incident on the (100) face of an fcc metal [41]. This figure shows the "classical turning surfaces" for a range of initial energies for an H_2 molecule incident normally to the physical surface. It has here been assumed that an average over the internal degrees of freedom has been made with the consequence that they do not appear explicitly in the figure. For values of the translational energy below the activation barrier, the whole surface appears as a corrugated repulsive wall which results in the molecule being scattered back into the gas phase. For energies 5% above the barrier, a small hole has opened up which will allow some fraction of the incident flux to dissociate, while the remainder will scatter intact. As the translational energy is increased still further, even more of the surface facilitates dissociative adsorption. In this way, it is relatively straightforward to rationalize simultaneous dissociation and scattering, and it is left to the calculators of potential energy surfaces to fill in the exact details for particular systems of interest.

2.5. Molecular vibrations

Within the last decade, there has been an explosion of interest in how molecular translational motion is coupled with vibrational motion near a surface to facilitate dissociation [42]. Just over twenty years ago, there appeared the first supersonic molecular beam–surface experiment by the Stickney group at MIT [43]. Although they did not know it at the time, this group unleashed on the world the H_2/Cu system which has become a paradigm for an activated adsorption reaction. This work, which is one of the most cited dynamics papers (over 220 at the time of writing), showed that the dissociative adsorption of H_2 was translationally activated and exhibited a pronounced isotope dependence. The Balooch paper initiated great theoretical and experimental interest since it demonstrated the feasibility of measuring a cross section for a simple reaction which could be directly compared with a molecular simulation. Following these experiments, there appeared a number of theoretical papers that extended the Lennard-Jones model to explicitly include the molecular vibrational degree of freedom. I think that the first was by McCreery and Wolken [44,45] who used a parameterized potential of the LEPS form to describe $H_2/W(001)$. This form of potential had been used extensively in studying gas-phase reactions (e.g. $F + H_2$, $D + H_2$) and was employed by Polanyi for discussing how reactivity could be linked to particular topological features in a potential surface [46]. In the Wolken work [45], various cuts of the six-dimensional (static-surface) PES were calculated and a limited number of classical simulations were per-

formed to illustrate qualitative dynamical effects. The majority of calculations placed an H atom on the surface and fired a second in the head-on geometry. This is the E–R reaction that was discussed earlier. In addition, the penultimate paragraph stated, "A few trajectories were also computed for the case of a hydrogen molecule aimed at the bare tungsten surface with a collision energy of 0.87 eV. Observed outcomes included the dissociation of atoms on the tungsten surface and the rebound of the undissociated hydrogen molecule from the surface". So even in this work, almost 20 years ago, simultaneous dissociation and scattering were observed when additional molecular dimensionality was included in the calculation.

Classical trajectory studies for a rigid surface were also performed by Gelb and Cardillo, this time explicitly for H_2/Cu to compare with the experimental results [47–49]. In a trio of publications the effects of changing the LEPS potential parameters were studied. One of the factors they considered was how placing energy in the vibrational coordinate affected the dissociation probability. A 40% enhancement was found for a fixed translational energy.

Surface motion was included via the generalized Langevin method in a number of dissociation studies by the group of DePristo. This work, which has recently been reviewed [50], contains many interesting insights into dissociation dynamics, particularly for "heavy" molecules (sic) where classical mechanics is more of a valid approximation than say for hydrogen. The dissociative adsorption of $N_2/W(110)$ is an interesting case in point where experimentally it was found that unlike most other systems, the probability for dissociation did not scale with the "normal" translational energy, but rather the total [51]. It had been suggested that this could arise as a consequence of a stable molecular adsorption state [52] but Kara and DePristo showed that such a state was not necessary and it was sufficient to have an activation barrier that was constrained in the vibrational coordinate [53]. With such a topology, it was particularly difficult for the molecule to gain access to the dissociation channel and translation-to-vibrational coupling was very inefficient. In these calculations it was essential to include the high degree of inelastic scattering with the surface vibrations before a meaningful comparison with experimental results could be made.

The time-dependent quantum "age" began with rather a modest calculation by Jackson and Metiu in 1986 for the dissociation of $H_2/Ni(100)$, again with a PES of the LEPS form [54]. In this study the split operator method [55] for solving the time-dependent Schrödinger equation was employed in two dimensions; the molecular centre of mass coordinate and the molecular vibrational coordinate. It was assumed that the molecule approached the surface in the broadside configuration, which has become the surface science analogue of the collinear collision geometry favoured in gas-phase studies. The surface was assumed static. Like the Balooch paper, this work opened the door to countless studies and the method has been widely exploited to investigate the dependence of the dissociation dynamics on molecular degrees of freedom on a variety of surfaces. One such study was to address the relationship between PES topology and reactivity à la Polanyi [56]. Fig. 3 shows model "elbow" PES's for activated adsorption. In this case a highly idealized form for the potential was chosen and the effects of the barrier location on the dissociation were studied in detail. The results demonstrated that for barriers encountered "early" in the reaction path, translational energy promoted bond fission while energy placed into molecular vibrations left the dissociation probability unaffected. For "late" barriers, vibrational energy was also efficacious in promoting dissociation since near the corner of the PES translations and vibrations become strongly coupled and the resulting motion assists in surmounting the barrier.

Although limited in their application to experiment because of the low dimensionality, these calculations have helped to establish a vocabulary which enables complex motion on multi-dimensional hypersurfaces to be discussed. For example, returning to the H_2/Cu system, we now know that the barrier is "late" since seeded beam experiments have shown that molecular vibrations aid dissociation [57–59]. More recently, it

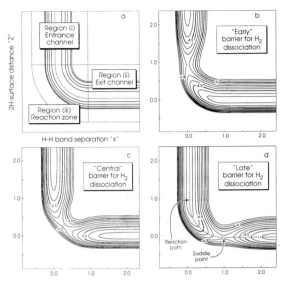

Fig. 3. (a) A contour diagram showing the essential features of the PES used by Halstead and Holloway in there study of the effects of surface topology on the dissociation of a diatomic [56]. (b)–(d) correspond to PES's with activation barriers being early, central and late in the reaction coordinate. "Elbow" potentials of this type include both translational and vibrational energy and have been popular in the last decade to extend ideas beyond the one-dimensional model shown in Fig. 1.

has been possible to place even greater restrictions on the shape of elbow PES's by the observation of vibrationally excited scattered molecules [60,61]. In addition to the height and location of the barrier along the reaction path, the degree of vibrational excitation depends crucially on the curvature of the path before the barrier is reached [62]. In this way, in the absence of first principles calculations of PES's, it should be possible to determine some of the critical features that give rise to reactivity.

2.6. Molecular rotations

It is perhaps obvious that since the coordinate that describes molecular vibrations and dissociation are the same, there should be a strong coupling near to the surface; this is not true for rotations and perhaps because of this, there have been far fewer studies of the influence of rotational motion on dissociative adsorption. Experimentally, what little information that has been reported has come from application of detailed balance to associative *desorption* experiments where it was observed that rotational energy hindered dissociation [63,64]. In their trajectory study, Gelb and Cardillo found that in general, modest rotational excitation had little effect on the dissociation probability [47]. Obviously, the results from simulations will reflect the detailed form of the PES and rotational motion is probing the dependence of the barrier on the angle at which the molecular bond makes with the surface. There appears to be a consensus that molecules will tend to have a low barrier to dissociation when in the broadside configuration, and a higher one when oriented normal to the surface [65]. One immediate consequence of this fact is that rotational motion that has angular momentum, J, normal to the surface (helicopters) will give rise to greater dissociation than motion having J parallel to it (cartwheels). The PES that will result from these elementary considerations strongly couples the translational and rotational degrees of freedom. Consequently when dissociation and scattering are occurring simultaneously, the fraction of atoms that are backscattered should show significant rotational excitation [66]. In addition to this dependence of the PES on the polar angle, variations in the azimuthal angle should also give rise to pronounced effects in scattering distributions. Detailed calculations have shown that the symmetry of the transition state for dissociation should be mirrored in the rotational selection rules of scattered molecules [67]. These can be experimentally determined by measurement of the alignment moments of the scattered flux. This is a most interesting prospect, which is only made possible because of the fixed orientation that the surface presents to the incident molecular beam; no such possibility exists in the gas phase because reacting species cannot be presented in such a well defined manner.

Recent results of the dependence of the dissociation probability for H_2/Cu on rotational energy show an interesting behaviour [68]. For fixed translational energy, rotational energy initially appears to inhibit dissociation while for higher J values it appears to promote it. Arguments have been made that placing energy in cartwheel rotations will decrease dissociation because the molecular bond will spend less time in the correct configuration for dissociation to occur [69,70]. At high J the situation may be rationalized by considering the dynamics of a helicopter rotation near a structureless surface. Far from the surface, the molecule has a well defined angular momentum which is conserved up to the point where the molecule is scattered. In the vibrational coordinate it follows the reaction path around the elbow. This, in turn, will result in an increased moment of inertia for the molecule which means that the rotational energy is decreased. The energy difference will flow into the translational–vibrational coordinate and assist in surmounting the activation barrier. These qualitative ideas await detailed high-dimensional calculations. It might be hoped that experimental results of this sort will place even greater restrictions on the form that the PES might take.

2.7. Potential energy surfaces

Up until this point, little has been said as to how potential energy surfaces have been obtained. This has been deliberate, and to end this section a few words need to be said concerning this omission. Dynamics calculations can be of the highest quality, but in order to compare with experimental data, a good understanding of the relationship between PES topology and surface electronic structure is essential. We have not yet arrived at this point and the overwhelming majority of published potential energy surfaces are of too low a quality to be gainfully employed in dynamical simulations. Up until now, dynamicists have shown skill in justifying choices for potential surfaces based almost solely on intuition! Casting a glance at the gas-phase community (something that has been of enormous help over the last decades) we see that multi-dimensional dynamical simulations using ab initio potentials from high quality electronic structure calculations have produced excellent agreement with state-resolved reaction measurements; the system? $H + H_2 \rightarrow H_2 + H$ [71]. Since even replacing H by F gives rise to severe problems, one might ask what hope is there for replacing H by W(100)? I think that because of the great strides that have been made in the development of algorithms for calculating total energies [72,73], the future looks bright. For example, I know that currently there are two independent calculations being performed for the H_2/Cu system and probably the flood gates are about to open for adsorption problems in general. The efficiency of the Car–Parinello method when combined with the widespread availability of CPU time on standard or parallel processing computers makes the future appear interesting. The real task will then be the procedural ones of creating parametric forms for such PES's (no trivial exercise) and then performing the dynamical calculations.

3. Conclusions

There is little doubt that we have come a long way in our understanding of elementary surface reactions since *Surface Science* first appeared. Naturally in this whirlwind look at the subject I have not been able to give credit to the many workers who have made valuable contributions to the field: To these I apologise. I have attempted to focus on issues that have been, and continue to be subjects of debate, perhaps at the exclusion of others that readers will believe are more important. As a graduate student between 1973–1976, performing dynamical calculations, I felt far away from the mainstream of activities in surface science which then was in its electron spectroscopy mode. I now feel that even though the subject has mushroomed, dynamical studies are somewhere near to the centre of interest. I am moreover convinced that within the next decades, our understanding of process dynamics will make great leaps forward, particularly in the area of surface reactions. There are still important bottlenecks to our understanding, for example, how to describe the influence of quantized lattice vibrations and electron–hole pair excitations on the motion of an incident molecule, but these, and other challenges are probably what keeps our interest in the subject.

Acknowledgements

I thank all of my many colleagues and coworkers over the past two decades who have shared (and shaped) my interests. Particular thanks go to Bill Gadzuk who on countless occasions has given me new insights into problems and patiently shown me the unity that exists in dynamics. Additionally, thanks also go to Laurie Schonfelder, Kieth Sharman and Tom Grimley who provided me with information related to computing in Liverpool in 1964. Finally, my thanks go to George Darling for reading the manuscript.

References

[1] C. Sébenne and M. Balkanski, Surf. Sci. 1 (1964) 22.
[2] J.M. Honig and W.H. Kleiner, Surf. Sci. 1 (1964) 71.
[3] J.D. Levine and E.P. Gyftopoulos, Surf. Sci. 1 (1964) 225.
[4] B.B. Damaskin, Surf. Sci. 1 (1964) 318.
[5] H. Eyring, J. Chem. Phys. 3 (1935) 107.
[6] J.A. Barker and D.J. Auerbach, Surf. Sci. Rep. 4 (1979) 1.
[7] H. Zacharias, Int. J. Mod. Phys. B 4 (1990) 45.
[8] M. Grueble and A.H. Zewail, Physics Today 43 (1990) 24.
[9] Probably never is a word which shouldn't be used in this sense.
[10] E. Beder, Surf. Sci. 1 (1964) 242.
[11] A.F. Devonshire, Proc. Roy. Soc. A 156 (1936) 37.
[12] U. van Slooten, E.J.J. Kirchner and A.W. Kleyn, Surf. Sci. 283 (1993) 27.
[13] D.O. Hayward and B.M. Trapnell, Chemisorption (Butterworths, London, 1964).
[14] J. Harris and B. Kasemo, Surf. Sci. 105 (1981) L281.
[15] J.C. Keck, Adv. Chem. Phys. 13 (1967) 85.
[16] P. Kratzer and W. Brenig, Surf. Sci. 254 (1991) 275.
[17] B. Jackson and M. Persson, J. Chem. Phys. 96 (1992) 2378.
[18] R.I. Hall, I. Cadez, M. Landau, F. Pichou and C. Schermann, Phys. Rev. Lett. 60 (1988) 337.
[19] P.J. Eenshuistra, J.H.M. Bonnie, J. Los and H.J. Hopman, Phys. Rev. Lett. 60 (1988) 341.
[20] E.W. Kuipers, A. Vardi, A. Danon and A. Amirav, Phys. Rev. Lett. 66 (1991) 116.
[21] C.T. Rettner, Phys. Rev. Lett. 69 (1992) 383.
[22] P. Stoltze and J.K. Nørskov, Phys. Rev. Lett. 55 (1985) 2502.
[23] J.E. Lennard-Jones, Trans. Faraday Soc. 28 (1932) 333.
[24] H.J. Robota, W. Vielhaber, M.C. Lin, J. Segner and G. Ertl, Surf. Sci. 155 (1985) 101.
[25] D.O. Hayward and A.O. Taylor, J. Phys. C 19 (1986) L309.
[26] I. Estermann and O. Stern, Z. Phys. 61 (1930) 95.

[27] E. Muller-Hartmann, T.V. Ramakrishnan and G. Toulouse, Solid State Commun. 9 (1971) 99.

[28] O. Gunnarsson and K. Schönhammer, in: Chemistry and Physics of Solid Surfaces IV, Eds. R. Vanselow and R. Howe (Springer, Berlin, 1982) p. 363.

[29] G.D. Mahan, Solid State Phys. 29 (1974) 75.

[30] J.W. Gadzuk, Surf. Sci. 6 (1967) 133.

[31] B. Kasemo, E. Törnqvist, J.K. Nørskov and B.I. Lundqvist, Surf. Sci. 89 (1979) 554.

[32] O. Gunnarsson and K. Schönhammer, Phys. Rev. B 25 (1982) 2514.

[33] J.K. Nørskov, J. Vac. Sci. Technol. 18 (1981) 420.

[34] J. Harris and S. Andersson, Comments At. Mol. Phys. 29 (1993) 83.

[35] J. Harris, Phys. Scr. 36 (1987) 156.

[36] J. Harris, J. Simon, A.C. Luntz, C.B. Mullins and C.T. Rettner, Phys. Rev. Lett. 67 (1991) 652.

[37] H.A. Michelsen, C.T. Rettner and D.J. Auerbach, Surf. Sci. 272 (1992) 65.

[38] D.J. Auerbach and C.T. Rettner, in: Kinetics of Interface Reactions, Eds. M. Grunze and H.J. Kreuzer (Springer, Berlin, 1987) p. 125.

[39] C.B. Mullins and W.H. Weinberg, J. Chem. Phys. 92 (1990) 4508.

[40] This is sometimes emotively referred to as massive steric hinderance.

[41] M. Karikorpi, S. Holloway, N. Henriksen and J.K. Nørskov, Surf. Sci. 179 (1987) L41.

[42] S. Holloway and G.R. Darling, Comments At. Mol. Phys. 27 (1992) 335.

[43] M. Balooch, M.J. Cardillo, D.R. Miller and R.E. Stickney, Surf. Sci. 46 (1974) 358.

[44] J.H. McCreery and J.G. Wolken, J. Chem. Phys. 63 (1975) 4072.

[45] J.H. McCreery and J.G. Wolken, J. Chem. Phys. 64 (1976) 2845.

[46] J.C. Polanyi, Acc. Chem. Res. 5 (1972) 161.

[47] A. Gelb and M.J. Cardillo, Surf. Sci. 59 (1976) 128.

[48] A. Gelb and M.J. Cardillo, Surf. Sci. 64 (1977) 197.

[49] A. Gelb and M.J. Cardillo, Surf. Sci. 75 (1978) 199.

[50] A.E. DePristo and A.Kara, Adv. Chem. Phys. 77 (1990) 163.

[51] D.J. Auerbach, H.E. Pfnür, C.T. Rettner, J.E. Schlaegel, J. Lee and R.J. Madix, J. Chem. Phys. 81 (1984) 2515.

[52] J.W. Gadzuk and S. Holloway, Chem. Phys. Lett. 114 (1985) 314.

[53] A. Kara and A.E. DePristo, Surf. Sci. 193 (1988) 437.

[54] B. Jackson and H. Metiu, J. Chem. Phys. 86 (1986) 1026.

[55] M.D. Feit, J.J.A. Fleck and A. Steiger, J. Comp. Phys. 47 (1982) 412.

[56] D. Halstead and S. Holloway, J. Chem. Phys. 93 (1990) 2859.

[57] B.E. Hayden and C.L.A. Lamont, Phys. Rev. Lett. 63 (1989) 1823.

[58] G. Anger, A. Winkler and K.D. Rendulic, Surf. Sci. 220 (1989) 1.

[59] C.T. Rettner, D.J. Auerbach and H. Michelsen, Phys. Rev. Lett. 68 (1992) 1164.

[60] A. Hodgson, J. Moryl, P. Traversaro and H. Zhao, Nature 356 (1992) 501.

[61] C.T. Rettner, D.J. Auerbach and H. Michelsen, Phys. Rev. Lett 68 (1992) 2547.

[62] G.R. Darling and S. Holloway, J. Chem. Phys. 97 (1992) 734.

[63] G.D. Kubiak, G.O. Sitz and R.N. Zare, J. Chem. Phys. 83 (1985) 2538.

[64] L. Schröter, H. Zacharias and R. David, Phys. Rev. Lett. 62 (1989) 571.

[65] P.K. Johansson, Surf. Sci. 104 (1981) 510.

[66] S. Holloway and B. Jackson, Chem. Phys. Lett. 172 (1990) 40.

[67] S. Holloway and X.Y. Chang, Faraday Disc. Chem. Soc. 91 (1991) 425.

[68] H.A. Michelsen, C.T. Rettner and D.J. Auerbach, Phys. Rev. Lett. 69 (1992) 2678.

[69] S. Holloway, J. Phys. (Condens. Matter) 3 (1991) S43.

[70] J.N. Beauregard and H.R. Mayne, Chem. Phys. Lett. 205 (1993) 515.

[71] A. Kuppermann and Y.M. Wu, Chem. Phys. Left. 205 (1993) 577.

[72] R. Car and M. Parrinello, Phys. Rev. Left. 55 (1985) 2471.

[73] P. Feibelman, Phys. Rev. Lett. 63 (1989) 2488.

Surface Science 299/300 (1994) 667–677
North-Holland

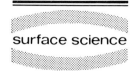

surface science

The dynamics of adsorption and desorption

John C. Tully

AT&T Bell Laboratories, Murray Hill, NJ 07974, USA

Received 30 April 1993; accepted for publication 2 June 1993

Molecular dynamics simulations, in concert with experiments, have greatly enhanced our understanding of the elementary dynamical processes of adsorption and desorption at solid surfaces. The observations of anomalously large desorption prefactors, of deviations from integer-order kinetics, of non-thermal energies of desorbing molecules, of deviations from "normal energy scaling", and of the contributions of internal energy to adsorption all are consequences of the multidimensional nature of gas–surface interactions. These behaviors are revealed clearly through the multidimensional picture afforded by molecular dynamics simulations. This paper highlights some of the issues and their resolution, with some personal perspectives.

1. Introduction

The development of an atomic level understanding of elementary dynamical processes at surfaces has been one of the central themes of the journal *Surface Science* throughout its 30 years. In 1964 at the time of the first issue of *Surface Science*, our understanding was based heavily on a few seminal concepts; concepts which still strongly influence our thinking. In this paper, I will present a personal view of how these concepts have evolved into our current state of understanding of the dynamics of adsorption and desorption.

My own participation in this evolution was spurred by two Bell Laboratories "In-Hours Courses". The first, in 1975, on "Surface Physics", was taught by John Arthur and Truman Brown, who introduced me to the recent experimental developments that were beginning to permit quantitative investigation of the interactions of adsorbates with well-characterized single crystal surfaces. Notwithstanding the prevailing prejudice that single-crystals in UHV were not "real surfaces", it was immediately obvious that these experimental capabilities would be the key to understanding chemical mechanisms at surfaces. I was pushed to action in 1977 by Ned Greene

(then on sabbatical at Bell Labs) and Mark Cardillo, who convinced me to join them in teaching an In-Hours course entitled "Kinetics and Dynamics of Gas-Phase and Gas–Surface Reactions". This was presumptuous on our part since, as I recall, attendees were to include Walter Brown, Len Feldman, Homer Hagstrum, Don Hamann, George Gilmer, Jack Rowe, Michael Schlüter, Jim Patel, Mort Traum, Norman Tolk, and a number of others who knew more than we about surface science. In the process of preparing material for this course, it became apparent that gas-phase and surface chemical reactions were controlled by many of the same factors, the locations and energies of transition states, the rates and pathways of energy flow, the lifetimes of intermediates, etc. In the hope of focusing attention on these issues, we convinced the Gordon Research Conferences to start a new conference called "Dynamics of Gas–Surface Interactions", which first met in 1979. This conference, early participants of which are shown in Fig. 1, is still continuing successfully. Since that first conference, the dynamics of gas–surface interactions has been the main focus of my research.

Much of my effort in this area has been in developing and applying molecular dynamics type computer simulations. Molecular dynamics allows one to effectively "watch" as a molecule approaches the surface, searches for a binding site, and summons the energy to break apart or escape from the surface (see Fig. 2). Furthermore, the method can treat dynamical processes in full dimensionality, thereby avoiding dependence on restrictive or questionable simple models. Our ability to simulate, on the computer, the detailed motions of individual atoms through the course of a surface chemical reaction has progressed substantially in recent years. Advances include development of "generalized Langevin" methods for reducing the numbers of degrees of freedom [1], introduction of techniques for efficiently treating "infrequent events" such as passage over energy barriers [2], and extension to processes involving electronic transitions such as charge exchange and electron–hole pair excitations [3]. Most importantly, progress in ab initio and semiempirical techniques for computing force fields have greatly enhanced the predictive power of molecular dynamics [4].

The primary contributions towards understanding elementary dynamical processes at sur-

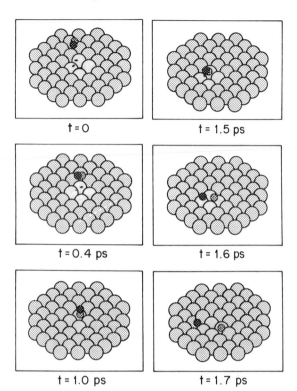

Fig. 2. Selected frames of a computer generated movie "Dynamics of Gas–Surface Interactions", J.C. Tully and K. Knowlton (Bell Laboratories, Murray Hill, NJ, 1978). The frames illustrate the dissociative adsorption of NO on a Pt(111) surface. The darker atom is the oxygen, and the lighter one the nitrogen.

Fig. 1. The "Beams Brothers", Mark Cardillo and Giacinto Scoles, or vice-versa. Photograph taken by Ulrich Harten.

faces, however, have been experimental. Prior to the first issue of *Surface Science*, chemical reactions at surfaces were usually studied under conditions that were so complex that there was little hope of making direct contact with molecular level theory. Any theory, with an adjustable parameter or two, could fit the experiments. Since 1964, more and more studies have been carried out under controlled conditions using nearly perfect single crystal surfaces and quantitative probes not only of thermal reaction rates, but of molecular level details; energy and angular dependences, lifetimes, identity of intermediates, and even quantum state specificity [5]. These experiments provide a crucial testing ground for theory and/or simulation. Computer simulations, on the other hand, when employed with judgment and properly tested against detailed experimental data, can provide an accurate and unbiased multidimensional view of adsorbate and surface atom motions. The results of simulations can be directly compared with experimental results; i.e., exactly the same quantity that is measured experimentally can be computed in the simulation. The insights extracted from such calculations can be essential input to the development of simple models that correctly incorporate the underlying physical behavior. The models and principles underlying the dynamics of adsorption and desorption that have emerged during the first 30 years of the journal *Surface Science*, in part through the application of molecular dynamics simulations, are the subject of this paper.

2. Adsorption and desorption in one dimension

In a 1932 paper Lennard-Jones [6] presented the picture that still underlies most molecular level discussions of dissociative adsorption. Consider the schematic potential energy diagrams of Fig. 3. The curves labeled A + B represent the sum of the interaction energies of separated atoms A and B with the surface. The curves labeled AB represent the interaction energy of the molecule AB with the surface.

The crossing of the two curves is designated the "transition state", the place where the energies of the dissociated state and molecular state are equal. In Fig. 3a, the crossing point lies below the asymptotic energy of molecule AB, so this case depicts non-activated adsorption. In Fig. 3b, the crossing lies above the asymptote, so additional energy is required for the gas-phase molecule to reach the crossing point and dissociate; i.e., the figure depicts activated adsorption. The AB curves in both figures exhibit a shallow binding potential corresponding to a weakly-bound molecular state, a so-called "intrinsic precursor" [7].

The one-dimensional picture of molecule–surface interactions was developed by Logan and Stickney into a very useful quantitative model of energy exchange and sticking, the "cube model" [8]. The main assumption of the cube model is

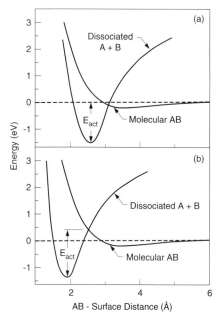

Fig. 3. Conventional Lennard-Jones picture of dissociative adsorption of a diatomic molecule AB on a surface: (a) non-activated; (b) activated.

that the molecule experiences no forces in the directions parallel to the surface; i.e, the surface is absolutely flat. This assumption leads to simple expressions for the energy and angular distributions of scattered molecules that frequently reproduce experimental trends. An expression for the sticking (adsorption) probability can be obtained under the assumption that a molecule sticks if the sum of its kinetic energy associated with normal motion and potential energy is reduced below zero. This one-dimensional definition of sticking is inadequate for quantitative purposes, but has proved useful as a qualitative guide.

The aforementioned sticking probability, which we will denote P_s, is the key quantity that characterizes adsorption. The sticking probability can be defined uniquely only if the surface temperature is 0 K. At non-zero surface temperatures all molecules which strike the surface will eventually escape, so the difference between directly scattered and stuck molecules becomes a question of timescale. A crucial contribution of molecular dynamics simulations to understanding sticking has been the ability to compute exactly the same quantity that is measured, whether it be angular distributions, velocity distributions, time-delays, etc. [9], independent of any arbitrary definition of sticking probability.

A second quantity, the accommodation coefficient, α, is a measure of the energy exchange upon impact of a molecule with the surface:

$$\alpha = (E_i - E_f)/(E_i - 2k_B T_s), \tag{1}$$

where E_i and E_f are the mean kinetic energy of the incident and scattered molecules, respectively, T_s is the surface temperature, and k_B is Boltzmann's constant. The factor of 2 in the denominator arises because the mean kinetic energy of Maxwellian gas molecules crossing a dividing plane is $2k_B T$. From Eq. (1), α is 0 if no energy is exchanged ($E_f = E_i$), and α is unity if molecules are completely thermalized during a single encounter with the surface ($E_f = 2k_B T_s$). There has been an unfortunate tendency to equate α with P_s. These two quantities are equal if, first, those molecules which do not stick exchange no energy at all, and second, those molecules which do stick subsequently desorb from the surface with kinetic energy $2k_B T_s$. Neither of these conditions is satisfied in general.

If there is a significant coverage of adsorbates on the surface, adsorption sites for incoming molecules may be blocked. Langmuir [10] proposed that the sticking probability, P_s, should be proportional to the fraction of unfilled sites on the surface:

$$P_s = (1 - \theta). \tag{2}$$

Under the assumption that the desorption rate is proportional to the fractional coverage of adsorbates on the surface, θ, this leads immediately to the Langmuir adsorption isotherm (the kinetic derivation) [11], relating coverage to equilibrium pressure, P:

$$P(\theta, T) = b(T)[\theta/(1 - \theta)], \tag{3}$$

where b is a temperature dependent constant. An isotherm is a relationship that gives the coverage as a function of pressure at constant temperature. For convenience, we have inverted this expression to obtain pressure in terms of coverage and temperature. Even prior to the first

volume of Surface Science, a number of experimental measurements of sticking probabilities were found to deviate from Eq. (2). Kisliuk [12] proposed the existence of an "extrinsic precursor" state; i.e., a weakly-bound mobile state on top of filled sites, and showed how such a precursor could increase the sticking probability over the simple Langmuir prediction. While this extrinsic precursor picture continues to be invoked today, its quantitative consequences are still under examination by experiment and simulation.

Desorption is obviously the reverse of adsorption, and this was recognized at the outset [10]. Nevertheless, there was little effort prior to 1964 to develop a unified quantitative description of the two processes. Instead, desorption was generally viewed as a simple activated process, with the thermal desorption rate R_D given by

$$R_D = \nu \theta^n \exp(-E_a/k_B T_s). \tag{4}$$

Frenkel is sometimes credited with being the first to apply this expression to thermal desorption, but Eq. (4) is, of course, simply the classic Arrhenius expression [13]. In Eq. (4), ν was originally viewed as an "attempt frequency", typically assumed to be of the order of a vibrational frequency, $\sim 10^{12}$ s^{-1}. E_a is the activation energy for desorption, as shown schematically in Fig. 3 for both activated and non-activated adsorption. The "desorption order" n was generally assumed to be unity for desorption of intact molecules and two if desorption required recombination of two adsorbed species.

The ideas described above were the foundations of the molecular level understanding of adsorption and desorption in 1964, the time of the first issue of Surface Science. But even in that very first issue, Honig and Kleiner [14] proposed extensions of Eqs. (3) and (4) through application of a lattice-gas model, and Surface Science has chronicled the further developments of these ideas to the present time.

3. Adsorption and desorption in many dimensions

From 1964 to the present there have been numerous experimental results published in Sur-

face Science and elsewhere that demonstrate inadequacies of the classic one-dimensional pictures of adsorption and desorption. I list a few of them here. Molecular beam experiments have shown dramatic breakdowns of "normal energy scaling" for both intact and dissociative sticking probabilities [15]. Thus the velocity parallel to the surface plane has an effect on the outcome, in contrast to the one-dimensional models. The advent of "state-selective" measurements of the rotational and vibrational energies of molecules escaping from the surface, and the ability to state-select molecules prior to collision with the surface, have demonstrated conclusively the critical role of internal degrees of freedom in promoting sticking [16]. A striking illustration [17] of the effect of surface corrugation is the observation of preferential "clockwise" or "counterclockwise" molecular rotation after scattering from the surface. There should be no preference for sense of rotation if the surface were flat.

Breakdowns of the simple picture of thermal desorption are at least as numerous. One frequently encountered result [18] is the observation of desorption prefactors of order 10^{16} s^{-1}, far larger than any vibrational attempt frequency and, as discussed below, a direct consequence of the multiple dimensionality of the desorption process. Similarly, violations of the expectation of integer reaction order are the rule, rather than the exception. Dramatic examples are the work of Opila and Gomer [19] demonstrating zero-order desorption for Xe on W(110) due to islanding, and of Pfnür et al. [20], showing how adsorbate–adsorbate interactions profoundly influence the coverage dependence of the desorption rate, and the apparent desorption activation energy and prefactor. The observation of non-thermal (non-Boltzmann) energy distributions of thermally desorbed molecules further emphasizes the need to extend our thinking beyond the standard models [21].

All of these observations can be reconciled by consideration of the multiple dimensionality of the dynamics of adsorption and desorption. This would surely have become apparent from experiment alone, but the point was brought home emphatically by molecular dynamics simulations.

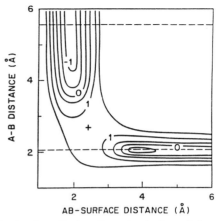

Fig. 4. A schematic two-dimensional view of the process illustrated in Fig. 3b. The solid curves are equal potential energy contours plotted as a function of the center-of-mass distance of the molecule AB from the surface plane, and the A–B bond distance. The dashed lines denote the cuts across the potential energy hypersurface that approximately correspond to the curves shown in Fig. 3b.

As a first step towards extending our thinking to more than one dimension, consider a typical multi-dimensional gas–surface potential energy hypersurface (PEHS) of the type employed in molecular dynamics simulations. A schematic illustration of the PEHS governing dissociative adsorption is shown in Fig. 4. Although only a single additional degree of freedom of the many involved, the molecular vibrational coordinate, has been considered in this illustration, the picture that emerges is fundamentally different from that implied in the Lennard-Jones picture of Fig. 3. It is clear from Fig. 4 that the dissociative and molecular curves of Fig. 3b do not intersect each other. In fact, they are not distinct curves at all, but rather are cuts across different regions of the same multidimensional potential surface. The dissociation of a molecule is better described by a concerted motion along this hypersurface than by an artificial transition at a hypothetical localized crossing point. The two-dimensional view of Fig. 4 begins to make clear that the reaction probability can be effected by the curvature of the reaction path, by the location as well as the height of the reaction barrier, and by the distribution of energy among molecular internal and transla-

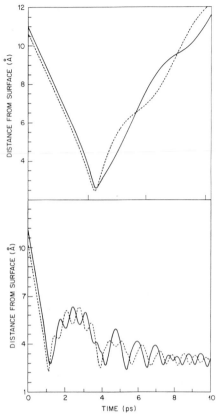

Fig. 5. Typical trajectories for N_2 colliding with Ag(111). Surface temperature and initial N_2 internal energy are both 0 K. Initial N_2 translational enery is 6.3 kJ/mol, and the incident direction is normal. The dashed and solid curves show the distances of the two N atoms from the surface plane as a function of time. The upper frame represents a direct scattering event. The lower frame represents an event in which so much energy has been transferred from translation into rotational modes that the molecule traps on the surface.

tional modes. Even this view is over-simplified, of course. The roles of initial orientation of the molecule, of molecular rotational energy, of surface atom vibrations, of surface corrugation, etc., require a full multidimensional description of the reaction dynamics. It is for this reason that molecular dynamics, the simulation of atomic motions in full dimensionality, has played such a central role in understanding adsorption and desorption. As an illustration, Fig. 5 shows two computed trajectories for N_2 incident on the Ag(111) surface [22]. The two trajectories illustrated have identical initial conditions except for the initial orientation of the N_2 bond axis. The fact that one of these impacts results in scattering and the other in trapping underscores the role of rotational degrees of freedom in adsorption and desorption.

As another illustration of the way underlying mechanisms can be identified and elucidated through molecular dynamics simulations, consider a recent observation of the breakdown of normal energy scaling in the trapping of Ar on Pt(111) [23]. Within the one-dimensional picture of adsorption, the component of adsorbate momentum in the surface plane is assumed to be conserved during a collision; only the component of momentum normal to the surface is altered. If this is true, then trapping or sticking probabilities can depend only on the component of momentum normal to the surface, not on the total momentum. This behavior is termed "normal energy scaling", and has been observed to hold approximately in many experiments and simulations. Recent experiments confirm this behavior for Ar on Pt(111) at low surface temperatures, but at higher surface temperatures they observe a significant breakdown of normal energy scaling [23]. While at higher temperatures the surface would be expected to become dynamically "roughened", the magnitude of the deviation from normal energy scaling appeared much too large to be accounted for by this effect. Molecular dynamics simulations on this system [24] confirm the experimental finding of approximate normal energy scaling of the trapping probability at low temperature. At the higher temperatures, trapping is not uniquely defined. The residence time of trapped atoms on the surface is so short at these temperatures (e.g., 45 ps for $T_s = 273$ K) that the distinction between directly scattered and trapped atoms becomes blurred. For the present discussion, let us arbitrarily designate as "direct" those trajectories for which the atom is scattered after only one encounter with the surface, and as "trapped" those trajectories which bounce more than once on the surface before escaping. The usual assumption is that trapped molecules ultimately desorb with an angular distribution that is symmetric with respect to the surface normal; i.e., they lose memory of the angle and velocity with which they initially impinged on the surface. This, in fact, was the assumption on which the method for extracting trapping probabilities from the molecular beam experiments was based. The experimental trapping–desorption flux was determined from the intensity of scattering near the surface normal. The computer simulations show that the angular distribution for the "trapped" (multiple bounce) trajectories for Ar incident at 60 degrees on a 273 K platinum surface is far from symmetric about the surface normal. Thus the trapped atoms desorb without complete loss of memory of their initial conditions. Complete loss of memory of their normal momentum occurs very quickly; the relaxation time for normal energy is calculated to be about 5 ps for this case. But the relaxation time for in-plane energy is calculated to be about 100 ps, longer than the mean residence time of 45 ps. Thus we are observing a kind of "quasi-trapping" in which the normal energy of the atoms thermalizes quickly and the atoms subsequently desorb in an apparent ordinary thermal first-order process. But they continue to glide laterally across the surface throughout their entire residence time. The angular distribution of desorbing quasi-trapped atoms thus is not symmetric about the surface normal, but peaks somewhere between the specular reflection angle and the surface normal. As a result the initial experimental procedure of deducing trapping from the intensity near the surface normal underestimates trapping for atoms incident at large angles compared to those incident nearer the normal angle. This gives the appearance of a deviation from normal energy scaling toward total energy scaling. Thus the reported apparent deviation from normal energy scaling was not a result of the roughening of the surface at higher temperatures, but rather is due to the flatness that results in the retention of memory of the in-plane momentum of trapped atoms.

4. Desorption and detailed balancing

For an equilibrium system of a gas in contact with a solid surface at temperature T and pressure P, the rate of arrival of molecules on the surface must be equal to the rate of departure. Let us construct a "dividing plane" parallel to the surface at a distance of 10 or 20 Å, i.e. a large enough distance that the interaction energy of the molecule with the surface is negligible at the dividing plane. The equilibrium one-way flux of molecules through the dividing plane must be an upper limit to the thermal desorption rate; any adsorbate which escapes from the surface into the gas phase must pass through the dividing plane. This equilibrium one-way flux is denoted the equilibrium rate, R_{eq}, frequently identified as the "transition-state theory rate" [25]. R_{eq} is in general larger than the true desorption rate at equilibrium because some molecules from the gas-phase incident on the surface may simply bounce off and pass back through the dividing plane without ever having been adsorbed. Thus non-adsorbed molecules can contribute to R_{eq}. The true equilibrium thermal desorption rate, R_D, is given by:

$$R_D(\theta, T) = P_s(\theta, T) R_{eq}(\theta, T), \qquad (5)$$

where $P_s(\theta, T)$ is the thermally averaged sticking probability. This is an exact expression. It is usually couched in the language of classical mechanics, but for this case where the dividing plane is located in the non-interacting region, Eq. (5) is an exact quantum mechanical result, provided P_s and the equilibrium flux, R_{eq}, are calculated properly by quantum mechanics. Since this paper is intended to highlight contributions of classical mechanical molecular dynamics, we will consider only the classical mechanical limit of Eq. (5). The equilibrium one-way flux R_{eq} in Eq. (5) is an equilibrium thermodynamic quantity; all dynamical information about desorption is contained in the sticking probability P_s. As discussed above, the definition of sticking probability is not unique. Thus the desorption rate is not unique to the same extent. This is unavoidable, of course; whatever rule one wishes to choose to classify a molecule as adsorbed must be employed consistently to both the acts of adsorption and desorption. For this discussion, we will retain our usual rule: a molecule is adsorbed on the surface if it loses all memory of how it got there [26]. Alternative definitions may be equally convenient.

Eq. (5) applies rigorously to the situation where gas and solid are at equilibrium at temperature T and pressure P. In the laboratory, desorption is commonly studied in a vacuum. The validity of the equilibrium expression for this non-equilibrium situation has been frequently questioned, and in a few cases experiments have been attempted to test its validity. For the most part, such studies have been a waste of effort. The experimenters are not questioning the validity of time-reversal symmetry in nature. Rather, they are questioning whether a desorbing molecule is effected by the presence or absence of gas molecules in the neighboring vapor. But consider the case where desorption occurs on a typical experimental timescale, say 10^{-2} s. For reasonably large sticking probability, the pressure of the gas required to maintain constant coverage would be of order 10^{-4} Torr. At this pressure a gas-phase molecule will strike the surface in the vicinity of the adsorbate about once every 10^{-3} s, or a few times during the residence time of the adsorbate on the surface. The only question being asked implicitly by these experimenters is, therefore, are these few gas phase impacts required to provide the energy to maintain the adsorbate in local thermal equilibration? But energy relaxation times on the surface are invariably much faster than 10^{-8} s, so the gas-phase collisions are completely insignificant. Thus, what these "test of detailed balancing" experiments actually test are the accuracy of the experiments. It is only when desorption occurs on a timescale faster than energy equilibration times that the presence or absence of the gas phase could be of any appreciable consequence.

For an ideal gas, the one-way equilibrium flux, R_{eq} through a dividing plane is given by

$$R_{eq}(\theta, T) = (k_B T/h)(2\pi m k_B T/h^2)Q_{gas}^{in}$$
$$\times \exp(\mu_{ads}(\theta, T)/k_B T), \qquad (6)$$

where m is the mass of the molecule, Q_{gas}^{in} is the partition function for the internal degrees of freedom of the molecule, and μ_{ads} is the chemical potential of the molecule in the gas phase. Eq. (6) is a completely classical mechanical expression,

so Planck's constant, h, that seems to appear in Eq. (6) is actually not there. That is a different story [25]. At equilibrium, the chemical potential must be equal in all coexisting phases, so we can just as well take μ_{ads} to be the chemical potential of the adsorbate, and we have labeled it accordingly. Eqs. (5) and (6) are the basic expressions relating adsorption and desorption.

The observed anomalous prefactors and strange coverage dependences of desorption can be easily understood within the framework of Eqs. (5) and (6). For non-interacting adsorbates with no site blocking (any number of adsorbates allowed per surface site),

$$\mu_{ads} = -k_B T \ln Q_{ads} - E_{ads} - k_B T \ln \theta, \qquad (7)$$

where Q_{ads} is the partition function for the adsorbate confined to its binding site and E_{ads} is its binding energy. Substituting Eq. (7) into (6) gives

$$R_{eq}(\theta, T) = (k_B T/h)(2\pi m k_B T/h^2)(Q_{gas}^{in}/Q_{ads})$$
$$\times \exp(-E_{ads}/k_B)\theta. \qquad (8)$$

In most situations, the sticking probability does not vary wildly with temperature or coverage, so the dominant contribution to the desorption rate, Eq. (5), is given by Eq. (8). This is similar to Eq. (4), with the same exponential term and a first-order dependence on coverage, but there is no explicit attempt frequency. Instead, the prefactor is implicitly contained in the partition function ratio. For example, for a simple molecular desorption, say CO from a smooth metal surface, we see immediately from Eq. (8) why a large prefactor should be expected. At room temperature, the factor kT/h is 5×10^{12} s^{-1}. The vibrational partition function for the constrained "frustrated rotational mode" of the adsorbate, assuming a typical frequency of 400 cm^{-1}, is a factor of about 20 greater than that of the free rotor appropriate for the transition state. Since there are two such modes, the prefactor will be increased by a factor of about 400. Similarly, the frustrated translational modes contribute roughly another factor of ten over the two-dimensional translational partition function that appears explicitly in Eq. (8), making the expected prefactor 2×10^{16} s^{-1}. This result is completely obvious, resulting from the increase in entropy of the gas molecule relative to the constrained adsorbate; i.e., it is a direct consequence of the role of additional degrees of freedom.

The observed dramatic deviations from simple first- or second-order behavior can be also understood within this framework. The effects of adsorbate–adsorbate interactions appear directly in the adsorbate chemical potential, μ_{ads}, of Eq. (6). For example, a simple Bragg–Williams lattice-gas model to represent attractive adsorbate–adsorbate interactions has been employed to calculate μ_{ads} [28]. This model exhibits a first-order 2D-gas to 2D-islanding phase transition as coverage is increased. Within this model, the desorption rate deviates strongly from first-order, and in fact, shows qualitative similarity to the experimental behavior of Pfnür et al. [20]. In fact, this provides an explanation for a type of "compensation effect"; i.e., the observation that desorption rates are frequently found to be approximately equal under conditions where the individual Arrhenius parameters vary considerably. At a first order phase transition, the free energies, and thus desorption rates, of the two phases are equal. However, the entropies and enthalpies that control the prefactors and activation energies can be quite different in the two phases.

The observed nonthermal energy distributions of desorbing molecules can be easily understood in terms of Eq. (5), as well [26]. At equilibrium the internal and translational energy distributions of all molecules crossing the dividing plane are Boltzmann at the surface = gas temperature T. If the sticking probability P_s were unity, then the only molecules crossing the dividing plane in the outward direction would be desorbing molecules, and thus they would be thermal at the surface temperature. If, however, the sticking probability is non-zero and energy dependent, then this is no longer true. The total flux of molecules through the dividing plane must still be Boltzmann, but some of these molecules are ones which have directly scattered from the surface, and some have adsorbed and subsequently desorbed. The sum of all molecules passing through the dividing plane must be Boltzmann, but the individual frac-

tions corresponding to scattered and desorbed molecules, alone, need not be. For example, for ordinary intact molecular adsorption it is usually found that molecules incident with high kinetic energy tend to scatter, while those with low kinetic energy preferentially stick. Thus thermally desorbed molecules must represent primarily the low-energy part of the Maxwellian distribution; i.e., they must be "cold" compared to $k_B T_s$, even though they have completely equilibrated to the surface temperature prior to desorption. This behavior has been thoroughly examined by molecular dynamics simulations [26]. Similarly, if incoming molecules must surmount an energy barrier in order to adsorb, typical of dissociative adsorption, then high energy molecules (perhaps including vibrational and rotational excitation as well as translation), will have higher sticking probabilities than low energy ones. Thus, desorbing molecules will be "hotter" than the surface temperature. Both of these cases have been observed experimentally. Similarly, if the transfer of energy from translation to molecular rotation makes a significant contribution to sticking, as illustrated in Fig. 5, then desorption must involve transfer of rotational energy to translation, likely resulting in rotational "cooling" of desorbed molecules [29].

5. Langmuir paradox

The Langmuir adsorption isotherm, Eq. (3), is frequently obtained via the kinetic derivation from Eq. (2). However, an isotherm is a thermodynamic relationship, and should properly be derived from thermodynamics, not kinetics. The statistical thermodynamic derivation is straightforward [11,30]. The derivation is based on the assumption that adsorbate molecules are non-interacting, and each is confined to a localized binding site, with no more than one molecule per site. The chemical potential differs from that of Eq. (7) only in the third term, resulting from site exclusion:

$$\mu_{ads} = -k_B T \ln Q_{ads} - E_{ads}$$
$$- k_B T \ln[\theta/(1-\theta)]. \qquad (9)$$

Substituting into Eq. (6) gives

$$R_{eq}(\theta, T) = (k_B T/h)(2\pi m k_B T/h^2)(Q_{gas}^{in}/Q_{ads})$$
$$\times \exp(-E_{ads}/k_B)[\theta/(1-\theta)]$$
$$= b[\theta/(1-\theta)]. \tag{10}$$

This results in the Langmuir adsorption isotherm, Eq. (3), with an explicit expression for the proportionality constant, b. The existence of two derivations of the Langmuir isotherm, the thermodynamic one from Eq. (9) and the kinetic one using Eq. (2), has led to considerable confusion. One way to express this confusion is in the form of a paradox. The way I usually state the paradox is, *how can the Langmuir term appear twice in the desorption rate, in both the equilibrium and the dynamic factors*. An equivalent but perhaps more revealing way to state the paradox is the following: The mean residence time, τ, of a molecule on the surface is

$$\tau = \theta N_s/R_D. \tag{11}$$

Assuming unit sticking probability and Eq. (10), τ is proportional to $(1 - \theta)$; i.e., *the residence time of an adsorbate molecule is affected by the presence of other adsorbates, even though the adsorbates have been assumed to be non-interacting*. The resolution of this paradox rests in properly combining the sticking probability with the equilibrium flux through the dividing plane, Eq. (5). If adsorbates are truly non-interacting except for site-blocking, then the sticking probability cannot be unity. For the sticking probability to be unity, molecules which come down on filled sites must still stick, implying some sort of precursor picture such as the Kisliuk model mentioned above. But detailed balancing then requires that desorbing molecules must visit filled sites on their way out, and the fact that these sites are filled must decrease the chance that the activated molecule will remain adsorbed; i.e., the adsorbates do interact, at least during their escape. The alternative limit is that adsorbates truly do not interact at all, even during the acts of adsorption and desorption. We can enforce this by constructing little cell walls around each binding site that extend tens of ångströms into the vacuum. Once in a cell, the molecule will not interact with molecules in other cells. But now, if a gas molecule enters a filled cell, it cannot stick. Thus the sticking probabilty must be proportional to $(1 - \theta)$, Eq. (2), and the desorption rate given by Eq. (5) is simply proportional to θ. While this discussion might seem obvious, confusion about these simple concepts has repeatedly appeared in the pages of *Surface Science*, even in recent years [31].

6. Closing remarks

Throughout this paper, I have implied that molecular dynamics simulations have been crucial in elucidating the underlying principles of adsorption and desorption. Certainly molecular dynamics has made a contribution, but in no way do I wish to slight the experimenters who, after all, have carried out the true simulations. I therefore wish to acknowledge my experimental co-authors who not only have kept my calculations honest, but who have made the work much more enjoyable and rewarding. Experimental co-authors whom I would like to particularly mention include D.J. Auerbach, S.L. Bernasek, M.J. Cardillo, Y. Chabal, P.J. Estrup, E.F. Greene, W. Heiland, J.S. Kraus, A.C. Kummel, L.C. Feldman, R.J. Madix, D. Menzel, C.B. Mullins, C.T. Rettner, H. Schlichting, J.A. Serri, G.O. Sitz, N.H. Tolk, J.T. Yates and R.N. Zare. My former postdoctoral associates deserve final mention: D.J. Doren, E.K. Grimmelmann, M. Head-Gordon, C. Lim, R.R. Lucchese, C.W. Muhlhausen, P. Nordlander and M. Shugard.

References

[1] S.A. Adelman and B.J. Garrison, J. Chem. Phys. 65 (1976) 3751;
J.D. Doll and D.R. Dion, J. Chem. Phys. 65 (1976) 3762;
J.C. Tully, J. Chem. Phys. 73 (1980) 1975.
[2] J.E. Adams and J.D. Doll, J. Chem. Phys. 74 (1981) 1467;
E.K. Grimmelmann, J.C. Tully and E. Helfand, J. Chem. Phys. 74 (1981) 5300.
[3] J.C. Tully and M. Head-Gordon, in: Desorption Induced by Electronic Transitions, Eds. A. Burns, E. Stechel and D. Jennison (Springer, Berlin, 1992) p. 150.
[4] For a recent review, see S. Holloway and J. Nørskov, Bonding at Surfaces (Liverpool University Press, Liverpool, 1991).

[5] J.A. Barker and D.J. Auerbach, Surf. Sci. Rep. 4 (1985) 1.

[6] J.E. Lennard-Jones, Trans. Faraday Soc. 28 (1932) 333.

[7] A. Cassuto and D.A. King, Surf. Sci. 102 (1981) 388.

[8] R.M. Logan and R.E. Stickney, J. Chem. Phys. 44 (1966) 195.

[9] M. Head-Gordon and J.C. Tully, Surf. Sci. 268 (1992) 113.

[10] I. Langmuir, J. Am. Chem. Soc. 40 (1918) 1361.

[11] A.W. Adamson, Physical Chemistry of Surfaces, 4th ed. (Wiley, New York, 1982) p. 521 ff.

[12] P. Kisliuk, J. Phys. Chem. Solids 3 (1957) 95; 5 (1958) 78.

[13] S. Arrhenius, Z. Phys. Chem. 4 (1889) 226.

[14] J.M. Honig and W.H. Kleiner, Surf. Sci. 1 (1964) 71.

[15] M.P. D'Evelyn, H.-P. Steinruck and R.J. Madix, Surf. Sci. 180 (1987) 47;
J. Lee, R.J. Madix, J.E. Schlaegel and D.J. Auerbach, Surf. Sci. 143 (1984) 626.

[16] E.W. Kuipers, M.G. Tenner, A.W. Kleyn and S. Stolte, Surf. Sci. 211/212 (1989) 819.

[17] G.O. Sitz, A.C. Kummel, R.N. Zare and J.C. Tully, J. Chem. Phys. 89 (1988) 2572.

[18] R.J. Gorte, L.D. Schmidt and J.L. Gland, Surf. Sci. 109 (1981) 367.

[19] R. Opila and R. Gomer, Surf. Sci. 112 (1981) 1.

[20] H. Pfnür, P. Feulner, H.A. Engelhardt and D. Menzel, Chem. Phys. Lett. 59 (1978) 481.

[21] K.C. Janda, J.E. Hurst, J.P. Cowin, L. Wharton and D.J. Auerbach, Surf. Sci. 130 (1983) 395;
J.A. Prybyla, T.F. Heinz, J.A. Misewich and M.M.T. Loy, Surf. Sci. Lett. 230 (1990) L173.

[22] J.C. Tully, C.W. Muhlhausen and L.R. Ruby, Ber. Bunsenges. Phys. Chem. 86 (1982) 433.

[23] C.B. Mullins, C.T. Rettner, D.J. Auerbach and W.H. Weinberg, Chem. Phys. Lett. 163 (1989) 111.

[24] M. Head-Gordon, J.C. Tully, C.T. Rettner, C.B. Mullins and D.J. Auerbach, J. Chem. Phys. 94 (1991) 1516.

[25] J.C. Keck, Adv. Chem. Phys. 13 (1967) 85;
P. Pechukas, in: Dynamics of Molecular Collisions, Part B, Ed. W.H. Miller (Plenum, New York, 1976) p. 269.

[26] J.C. Tully, Surf. Sci. 111 (1981) 461.

[27] D.J. Doren and J.C. Tully, Langmuir 4 (1988) 256.

[28] P.J. Estrup, E.F. Greene, M.J. Cardillo and J.C. Tully, J. Phys. Chem. 90 (1986) 4099.

[29] C.W. Muhlhausen, L.R. Williams and J.C. Tully, J. Chem. Phys. 83 (1985) 2594.

[30] R.H. Fowler and E.A. Guggenheim, Statistical Thermodynamics (Cambridge University Press, Cambridge, 1952).

[31] A. Cassuto, Surf. Sci. 203 (1988) L656;
K. Nagai, Surf. Sci. 203 (1988) L658.

Surface Science 299/300 (1994) 678–689
North-Holland

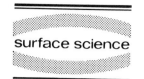
surface science

Chemisorption on metals: a personal review

David A. King

Department of Chemistry, University of Cambridge, Lensfield Road, Cambridge CB2 1EW, UK

Received 6 July 1993; accepted for publication 28 July 1993

This review is an unashamedly personal account, in outline, of a thirty year period in surface science which has seen a dramatic transformation in the understanding of solid surfaces and the process of chemisorption.

1. Introduction

In 1960, the year I began work as a research student, the world of chemisorption was very simple, and catalysis an almost unfathomable black art. Based principally on reading the work of Langmuir in the USA and J.K. Roberts in the UK, I understood that the solid surface should be treated as a checkerboard. Much discussion ensued about adsorbate–substrate bond energies; charge transfer; lateral interactions between adsorbed species; and sticking probabilities. But the surface itself was regarded as presenting a static, ideal bulk termination structure to the adsorbate. Now (of course) we know that solid surfaces are structurally interactive during chemisorption and catalysis. Even clean single crystal surfaces often themselves exhibit structures quite distinct from the ideal bulk termination structure, a restructuring which is brought about to reduce the strain generated when two surfaces are formed by splitting a solid along a particular {hkl} plane. Chemisorption invariably alters this restructuring process.

Back in 1960, we also had a rather primitive view of kinetic processes at solid surfaces. Now we understand such processes in terms of a sometimes rather daunting interplay of intrinsic and extrinsic precursors; virgin (metastable) intermediate states and stable chemisorbed states; and adsorption and desorption induced restructuring. Even the processes involved in the non-reactive scattering of a state-controlled molecule, such as NO, from a single crystal surface have come under detailed scrutiny.

What has happened over the past 30 years can be summarised very simply. In 1960 no spectroscopic or crystallographic data was available for adsorbates on well-defined surfaces: we were free to play microscopic guessing games with our macroscopic data on polycrystalline surfaces. With the development of a vast and powerful array of surface sensitive techniques and theoretical/computational methodologies, problems have been subjected to detailed, systematic scrutiny. Much of the guess-work of yesteryear has, thankfully been removed. The harvest of this work has already been gathered by industrialists, principally in the manufacture of microchips. Now even the wilderness of heterogeneous catalysis, the cornerstone of the chemical industry and one of the remaining bastions of empiricism, is beginning, reluctantly, one senses, to give way to the encroachment of high technology.

Playing some role in these developments and, more particularly, establishing close friendships with many gifted people in the process, has given me no small amount of pleasure. It is impossible to convey the shared sense of excitement that accompanied each breakthrough, however minor it may have turned out to be in retrospect; and it is probably inappropriate in this journal to describe some of the celebrations that accompanied these events.

SSDI 0039-6028(93)E0445-Z

It was thus my good fortune to begin my Ph.D. before surface science had emerged from its infancy. The most fruitful section of my thesis, the literature survey, had led me into the early papers of Gert Ehrlich on the flash filament technique – the precursor to temperature-programmed desorption (TPD) – and on field emission microscopy. My review also took me through all of the extensive work on the mechanism of the ammonia synthesis reaction, especially the work of Emmett and his coworkers establishing the nitrogen chemisorption step as rate-determining and the subsequent controversy between Horiuti and a Dutch group. Although the problems tackled on high area catalysts were intriguing, I had no doubt that my interests lay along the path laid out by J.K. Roberts [1], Gert Ehrlich [2] and Bob Gomer [3] leading towards fundamental studies on metal surfaces. The opportunity to pursue this came in the form of a Shell Scholarship which took me from South Africa to Imperial College, London. This was doubly fortunate, as I had become *persona non grata* with the apartheid authorities in the country of my birth.

2. Measuring and modelling sticking probabilities: 1963–74

At Imperial College, thanks largely to a combination of Gert Ehrlich's descriptions of his glass ultrahigh vacuum system [2], the talents of a remarkable glassblower, John Preston, and collaboration with David Hayward [4] and F.C. Tompkins, we developed the first accurate technique for measuring sticking probabilities (s) of gases on vacuum-deposited metal films [5]. Although the technique is incomparably accurate over the wide range $1 \leq s \leq 10^{-4}$, and gives accurate coverages as well, this was, in one sense, not a particularly timely piece of work. Polycrystalline metal films were being rapidly overtaken as substrates by work on polished single crystal surfaces. But our technique was the forerunner to our subsequent development, in Norwich, of a molecular beam reflection detector technique for measuring sticking probabilities as a function of surface coverage on single crystal surfaces [6].

This technique is still unsurpassed for accuracy and, dubbed the King and Wells technique, is now widely used in many laboratories. I was also pleased that the observation by Tony Horgan, my first Ph.D. student, of a sharp minimum in the sticking probability for the O_2/Ni system, at an adatom coverage of about $\frac{1}{4}$ of a monolayer [7], was the precursor to a number of detailed studies performed on single crystal nickel surfaces, later reviewed by Dick Brundle [8].

With the new molecular beam technique our first target in 1971 was the $N_2/W\{100\}$ system [6]. This was to be the start of a long, fertile affair with W{100}, which continues to the present day. Armed with a mass of sticking probability versus coverage data, obtained by independently varying the substrate temperature, the beam temperature, and the beam incidence angle, we developed a kinetic model which included mobile precursor states, as in the earlier work of Kisliuk, but in which we also included, for the first time, the influence of short range order in the overlayer [9]. A single kinetic expression was thus derived which not only provided an excellent fit to all of our kinetic data (as shown in Fig. 1) but also encompassed a description of existing structural studies, showing the formation of a c(2×2) structure at a fractional coverage $\theta = 0.5$, and our own desorption spectra. The model was based on the introduction of a pairwise lateral repulsive interaction energy term between adatoms in n.n. positions on an ideal (checkerboard) surface with fourfold symmetry. This work led directly to my first full structural study by LEED, on W{100} (described below), in an attempt to provide a quantitative structural basis for the kinetic model. Although the surface turned out to be quite unlike a checkerboard, the kinetic model has stood the test of time, despite a period of dispute with my (now!) close friend Dan Auerbach [10,11]. Later work by Charlie Rettner [12], using a state-of-the-art molecular beam scattering system, bore out our model in almost every detail: even our most precocious assumption, that the trapping probability into the weakly held precursor state was independent of substrate temperature, proved to be correct to within 10% over a range of ~ 1000 K! In later work, based on LEED studies by David Adams and Lester Germer [13] and our molecular beam studies, we extended the model to cover all crystal planes of tungsten [14]. This work may still stand as the most comprehensive description of dissociative adsorption kinetics.

3. Collaborations with Yates and Madey: 1969–1971. RAIRS, ESD and β-CO on W

At Gabor Somorjai's major surface science conference in Berkeley in 1968 [15], on my first trip to the USA, I was initiated into the massive strides being made with LEED and Auger electron spectroscopy; and I met John Yates and Ted Madey, who were doing some beautiful TPD work on the CO and N_2/W systems. The upshot of this meeting was six months sabbatical leave in 1969 from my post at the University of East Anglia with the surface science group at the National Bureau of Standards, just outside Washington DC. This proved to be a remarkably formative and productive period for me. We conducted a study of oxygen adsorption on polycrystalline W using TDS with line-of-sight mass spectrometric analysis, picking up the full range of W_xO_y and O desorption products [16]; and we also measured a giant isotope effect in electron stimulated desorption [17], confirming the Menzel–Gomer–Red-

head model for this process. In addition, I joined the weekly meetings of the Surface-Science-for-Lunch-Bunch, orchestrated by Russ Young (then working on the profilometer, the direct precursor to the scanning tunnelling microscope), and attended by Bill Gadzuk, Ward Plummer, Al Melmed, Ced Powell and John and Ted. I had never been exposed to anything like this before. Every week we ruthlessly dissected the major papers being published in the field. Those considered to be making the most important advances (outside our group, that is) were invited to give talks to the Lunch Bunch, where they were generally not just grilled, but skewered. Once I had got over the initial shock – their behaviour would have been considered rather rude at that time in England – I was exhilarated. I realised that the progress of the scientific revolution that we were participating in required that much of the shoddy thinking that had been masquerading as science in the field of surface chemistry should be directly, even rudely, confronted. (Less forgivable, though, the arrogance that some of us were learning to display derived also from the realisation that we had our hands on a significantly superior approach, and felt the need to rub in the salt.)

I attended the Gordon Conference organised by Bob Gomer at Meriden in August 1969. Even at that time it was clear to those of us attending

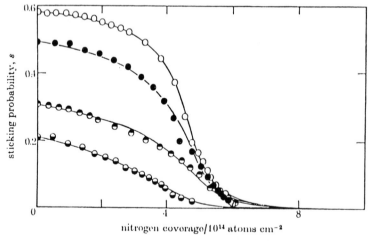

Fig. 1. Dependence of dissociative sticking probability for N_2 on W{100} on coverage at substrate temperatures between 300 K (O) and 773 K (◓). Points: experimental data; lines: best fit curve from a precursor model including lateral repulsive interactions between n.n. adatoms [9].

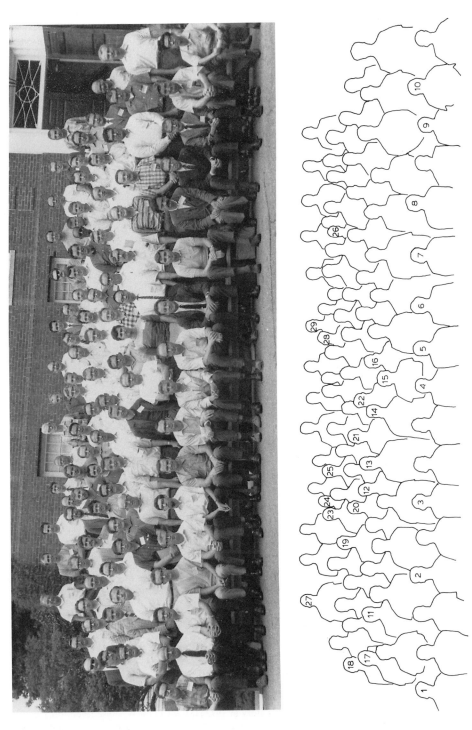

Fig. 2. Attendants at a Gordon Conference in Meriden, New Hampshire, USA August 1969. Some people identified: 1. Charlie Duke (now Editor, Surface Science); 2. Ward Plummer; 3. Denis Newns; 4. Bob Gomer, Chairman; 5. Leo Falicov; 6. Ian McRae; 7. Peder Estrup; 8. G.D. Halsey; 9. Gert Ehrlich; 10. Thor Rhodin; 11. Paul Palmberg (in the tie); 12. Russ Young; 13. Ted Madey; 14. Dietrich Menzel; 15. Lanny Schmidt; 16. Ernst Bauer; 17. Bill Gadzuk; 18. Frank Propst; 19. John Gregory; 20. The author; 21. Homer Hagstrum; 22. Charlie Tracy; 26. Denis Degras; 27. Milt Scheer; 28. John Yates; 29. Frank Goodman.

that we were experiencing the beginning of a new phase in surface science. A superficial glance at the photograph of the attendees (Fig. 2) correctly conveys the impression of a youthful field. The old guard was represented by such as Homer Hagstrum and G.D. Halsey, but the dominant characters there are still active contributors today. (Disturbingly, not one woman's face on that photograph...)

Subsequent to this period at the NBS, John Yates spent one year on sabbatical leave with me in Norwich, in 1970–1. We planned to make an attempt at obtaining IR spectra from adsorbates on a metal foil, and succeeded in this attempt,

observing (with a trained eye) the C–O stretch for α-CO on W: our first reflection–absorption infrared (RAIR) spectra [18] (Fig. 3). In fact, however, John Pritchard, with Mike Chesters and Malcolm Sims [19], at Queen Mary College, London, were the first to publish data with this technique, which has subsequently proved to be perhaps the most powerful tool for characterising molecular adsorbates on single crystal surfaces.

While John was with me, one of my Ph.D. students, Gus Goymour, and I were working on the consequences of lateral pairwise interactions between adatoms for desorption kinetics. Up to this time multiple peaks in desorption spectra

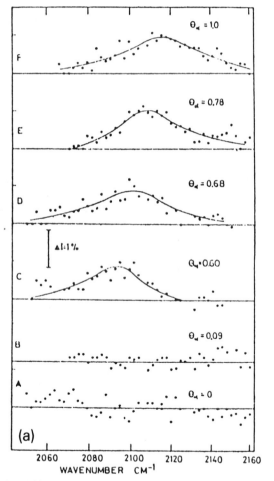

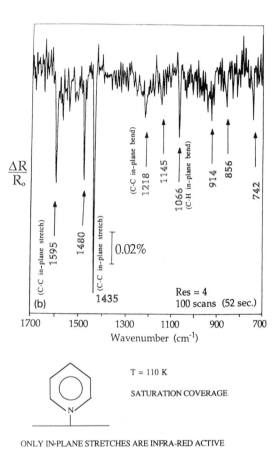

Fig. 3. (a) The first published RAIRS data from the author's laboratory in 1972 [18], α-CO on W; noise level $\sim 0.5\%$. (b) Recent data obtained by Sam Haq in the author's laboratory in 1991, for pyridine on Ni{100} at monolayer coverage, 110 K; noise level $\sim 0.01\%$.

had been attributed to the presence of distinct adsorption states in the adlayer, but now we showed [20] that interactions between identical adatoms could also give rise to multiple peaks. Carefully analysing desorption spectra and available LEED data for the β states of CO on W surfaces, we also concluded that these states were fully dissociated into a mixed adlayer of C and O adatoms [20]. At the time, this was a heresy; Gomer, Ehrlich, Redhead and others had always maintained that chemisorbed CO was not dissociated. Madey and Yates [21] had concurred with this, attributing their demonstration of complete isotopic mixing in the β-CO states to a 4-centre complex, and John was not easily persuaded when we came to write a joint paper on the subject for a Royal Society Discussion [22]. Over the following years, as spectroscopic data mounted up, all the major players did eventually line up behind the Goymour–King conclusion, despite some resistance [23]. This was not an insignificant battle: for example, in discussions of the mechanism of the Fischer–Tropsch process, it was only after this work that it became recognised that the mechanism involved dissociated CO.

4. Liverpool, 1974–1988. LEED, RAIRS; HREELS; kinetics; ARUPS; SEXAFS; and W{100}

Early in 1974 I was funded by the UK SRC to build a LEED chamber to conduct quantitative surface structural studies. With a Ph.D. student, Frank Marsh, I had already begun an investigation into the use of Fourier transform techniques to provide a direct inversion from LEED *I–V* spectra to surface structures, following the work of David Adams and Uzi Landman [24]. Although we eventually demonstrated that this was a fruitless exercise [25], this theoretical work, combined with the arrival of the LEED chamber, placed us in a strong position to exploit the advances in the full dynamical analysis of LEED intensities which had culminated in the publication of John Pendry's book in 1974 [26].

Before the new LEED chamber was delivered, however, I had been surprised and flattered by the offer of the oldest chair of physical chemistry in the UK, at Liverpool. This was, I believe,

largely the work of Tom Grimley, the founding father and guru of chemisorption theory. I accepted, and moved with my family and research group in September 1974. Mark Debe, a postdoctoral fellow sent to me from Milwaukee by Bob Greenler (who had taught me all about selection rules in RAIRS while he was on sabbatical leave in Norwich in 1973 – and about sky phenomena) immediately began work on W{100}. That crystal stayed in our LEED chamber for six years. The unexpected story that unfolded has been reviewed by me in 1983 [27], and again very recently, with Jacques Jupille [28]. Essentially, it marked the end of the checkerboard model for adsorption. The ideal, bulk termination structure of W{100} is less stable than either a zig-zag displacive structure, the Debe–King clean surface structure, which we reported in 1977 [29], or a $c(2 \times 2)$ W dimer structure, which is stabilised by a small coverage of H adatoms, which Graham Thomas and I reported in 1980 [30]. W{100} became the object of a great deal of theoretical and experimental attention, which continues to the present day [28]: the archetypal restructuring surface.

Armed with an angle-resolved photoelectron spectrometer, we first tackled the relatively stable W{110} surface, gathering data over the entire surface Brillouin zone and, with theoretical help from John Inglesfield at Daresbury, successfully assigning a prominent surface state to the $m = 1$ d-orbital of atomic W, essentially the same origin as the so-called Swanson hump surface state first identified on W{100} [31]. We also demonstrated that these tight binding electronic surface states in relative band gaps possessed directional character, and should be described as *dangling bond states*, by analogy with the band gap surface states found on semiconductors. Subsequently, our interest turned to the electronic driving force for the surface phase transition on W{100} [32], before using ARUPS [33] to make a determination of adsorbate molecular tilt angles. In 1982 we published the first accurate determination of a tilt angle for adsorbed species, for CO on Pt{110} at monolayer coverage, using orthogonal plane ARUPS. Since then, numerous structural determinations have shown that CO adsorption at monolayer coverage on the fcc{110} transition metal surfaces invariably yields a (2×1) glide line structure, with every CO molecule tilted away from the surface normal to minimise inter-adsorbate Pauli repulsion.

We derived a great deal of pleasure from our RAIRS system – albeit a string and sealing wax job by comparison with current FTIR instruments – in those early days in Liverpool. Ron Shigeishi obtained data from CO on Pt{111}, demonstrating a coverage-induced frequency shift of ~ 60 cm^{-1} in the C–O stretch [34]. Building on the earlier IR work of Eischens and coworkers on high area catalysts, using isotopic mixtures Alison Crossley and I outlined and used a simple experimental method – the infinite isotopic dilution method – to determine the extent of the shift attributable to *dipole coupling* [35]. This method has been subsequently used as the standard procedure by many others [36]. Our work also produced a flurry of activity from theoreticians; our modelling was essentially OK with one major omission: we had not accounted for coupling to the *image* dipoles [37]!

At this time RAIRS was rather overshadowed as a technique for obtaining vibrational spectra of adsorbates by high resolution electron energy loss spectroscopy (HREELS). In 1967 Frank Propst had published a remarkable, one-off paper establishing HREELS as a viable technique for obtaining vibrational spectra; but I was told by his student Pyper that the instrument was very temperamental. The technique was very successfully revived by Harald Ibach some years later, and in 1980, with Peter Hofmann joining me as a postdoc from Alex Bradshaw's group in Berlin, we also began working with it, collecting data both in and out of the incidence plane [39]. One highlight from this work was the application of the newly determined electron impact (and dipole) selection rules to establish to orientation of the formate species on Pt{110} [40].

Work on *kinetic processes* at surfaces also continued well at this time. Having worked on the implications of lateral interactions for adsorption and desorption kinetics I became interested in modelling the influence of these interactions on diffusion kinetics. This work was carried through

successfully with a Ph.D. student, Michael Bowker, who also obtained experimental data using our molecular beam deposition technique to generate sharp coverage gradients [41]. On study leave in Milwaukee in 1977 for three months, I produced a review [42] essentially pulling together the work on lateral interactions in adsorption, desorption and diffusion kinetics, and including the influence of precursor states on both adsorption and desorption [43] kinetics. (A few years later, Mick Morris, Michael Bowker and I attempted a fully comprehensive review of kinetic processes at surfaces [44].) This work was brought to a head through my close friendship with Bob Cassuto, at the CNRS in Nancy. Together we produced detailed rate expressions for adsorption and desorption processes which encompassed precursor states, lateral interactions and direct (i.e. non-precursor mediated) adsorption and desorption [45].

One further kinetic study from this period deserves particular mention: working with our molecular beam system, using mixed beams of NO and CO and a Pt{100} target, Sunder Singh Boparai and I made the first observations of an *oscillatory reaction* on a single crystal surface [46]. This particular baton was quickly taken over, in succession, by Lanny Schmidt, Peter Norton and Gerhard Ertl, and continues to be a source of fascination to many, including those whose interests lie outside surface science [47].

In the 1980s a range of quantitative structural studies of clean surface and adsorbate-covered systems were conducted at Liverpool using both LEED intensity analyses and the newly available technique of SEXAFS. The latter was the beginning of my love–hate relationship with synchrotron radiation sources which continues to the present day. With Geraldine Lamble as a long-suffering but tenacious research student, we obtained accurate adsorbate–substrate bond lengths for Cl on Ag surfaces [48] and reported a unique result for Cs on Ag{111} [49], showing a large variation in bond length with coverage. In this work we stressed the importance of anharmonic effects in accurate bond length determinations. To avoid these, we pointed out, data should be taken at low temperatures.

The period 1974 to 1988 saw a big change in surface science activity. The beginning of the period was still dominated by technique development, but soon these new quantitative techniques were being directly applied to the task of forming a data base from which a deeper understanding of surface adsorption and reaction processes could be developed. Noting this change, Phil Woodruff and I began editing our series of volumes on "The Chemical Physics of Solid Surfaces and Heterogeneous Catalysis", which is now into its seventh volume [50].

Before leaving Liverpool I had established the Leverhulme Centre for Innovative Catalysis there, which has developed into a viable, productive research centre in heterogeneous catalysis. There was by this time such a range of activity in surface science in Liverpool that the local team was subsequently successful in a bid for status as an SERC-supported Interdisciplinary Research Centre.

5. The Cambridge years, 1988–present. Microcalorimetry; supersonic beams; RAIRS; tensor and holographic LEED

A phone call from the Vice Chancellor at Cambridge in the summer of 1987 conveyed the offer (without interview, as is the custom at this ancient place of learning) of their Chair of Physical Chemistry. This plunged my wife, Jane Lichtenstein, and me into a most difficult decision. We both felt a strong commitment to Liverpool; but in the end the potential of the move to Cambridge proved overwhelming. A Fellowship at St. John's College followed rather quickly on the acceptance of the Chair, and we moved to Cambridge in May 1988. The research group and equipment followed in September.

The move to Cambridge coincided with a series of new developments in my research. In part, these were catalysed by the move itself. In retrospect, I heartily recommend the traumatic experience of a major move at mid-life to anyone considering it.

The first of these new departures was the development of what is still the world's only *single crystal adsorption microcalorimeter*. On a visit to the outstanding surface physics groups of Flemming Besenbacher, Ivan Stensgaard and David Adams at Aarhus in Denmark I learnt about their ability to prepare free-standing, ultra-thin (~ 2000 Å) single crystals from a range of metals with unrestricted choice of crystal face. For many years I had toyed with the idea of building a single crystal calorimeter, and after the Aarhus visit I realised that these crystals could form the basis of a new instrument. Chris Borroni-Bird, a brave and very able Ph.D. student, took the project to a successful conclusion [51]. With the calorimeter adsorption heats, and heats of surface reactions, can be accurately measured as a function of coverage, with a sensitivity to 0.01 monolayer doses. We have already applied it to adsorption, coadsorption and oxidation on metal single crystal surfaces [52]. The potential applications appear to be without bound: even differential scanning calorimetry is looming as a possibility.

Our work on dynamics and kinetics has also made giant strides, with the design and construction of a dual *supersonic beam surface scattering* system. The object was to provide the data required to establish the microscopic mechanism of adsorption coupled to adsorbate-induced restructuring processes, and to model catalytic processes including possible restructuring. Our work on the CO-induced restructuring of Pt{100}, and modelling of the $CO + O_2$, $CO + NO$ and $NH_3 + O_2$ reactions on this surface, including oscillations, has already proved handsome reward for the early efforts of Andrew Hopkinson on this system [53].

Before leaving Liverpool, Rasmita Raval had joined my research group and, not before time, we built up an entirely new *RAIRS* system, based on a Mattson interferometer. Now at last we had caught up with the advances made elsewhere, but particularly by Mike Chesters at Norwich [54], since our early development of this technique. The immediate reward was the discovery of surface restructuring induced by a non-dissociating adsorbed molecule, CO: over the coverage range from 0.3 to 0.7 monolayers the Pd{110} surface was restructured to the missing row structure characteristic of the {110} clean surfaces of Pt and Au [55]. Now, too, the work could be extended to molecules other than CO! For example, the spectrum from pyridine on Ni{100}, obtained in only 52 s, is shown in Fig. 3. Looking to the immediate future, the system is now armed with a new high pressure transfer cell, capable of operating at up to 25 atm and thus realising the full potential of this technique in its ability to remove the so-called pressure gap between surface science and catalysis: in situ analysis of adsorbates on single crystals at high pressures.

We also began a programme of work to investigate the use of pulsed lasers in the analysis of surfaces. Using two-photon photoemission, we were the first to obtain an experimental measurement of significant surface temperature rises induced by a laser pulse, on a nanosecond time scale [56]. Essentially, we fitted a Fermi–Dirac distribution function to the high energy edge of the photoemission spectrum, yielding the electronic temperature of the system.

Work at the Daresbury synchrotron radiation source continued, with our attention now drawn to the use of NEXAFS to determine the orientation of molecules at surfaces, the technique pioneered by Joe Stöhr [57]. Up to this time, selection rules had been applied on the assumption that the surface molecule could be treated as frozen in space. In our analysis we now included the large amplitude, low frequency vibrational modes of the adsorbed species [58]. This refinement brings into question some earlier work from which static tilt angles were derived, and, as in our earlier SEXAFS work, we concluded that data should be collected at low temperatures to reduce vibrational amplitudes.

Perhaps surprisingly, some of the most important advances in recent years have occurred in surface crystallographic analysis based on LEED. Klaus Müller and Klaus Heinz, at Erlangen, produced a very refined data collection system (Auto LEED), based on a sensitive CCD camera, which we acquired. With this, data collection times were reduced by a very useful margin. Using a Faraday cup in 1975, a useful set of beams took us months to collect; now, the same data could be collected in minutes. On the theoretical front, the most important development has been tensor LEED,

by Philip Rous and John Pendry. The automated version of the tensor LEED program, written by Adrian Wander [59], who joined my group as a postdoctoral fellow from Gabor Somorjai and Michel Van Hove early in 1992, has reduced the amount of computing time required for a structural analysis by a factor of more than 1,000 [60]. Armed with these advances, LEED has finally emerged as a routine surface structural technique without rival, with the capability for tackling quite formidable structural problems. And it is not limited to well-ordered systems. As shown by Klaus Heinz [61] and by ourselves [62], LEED intensity spectra can be obtained, and analysed, from completely disordered overlayers. LEED intensity spectra are dominated by short range order. Of course, as in X-ray diffraction analyses of solid structures, structural input from other techniques, such as RAIRS and STM, will always be important input for LEED analyses.

Over the years many attempts have been made to obtain a direct transform from experimental LEED intensity spectra to real space structures, and a Ph.D. student with me, Peijun Hu, has recently returned to this problem. Following John Barton [63], we began working on the application of holographic transforms to Auger and photoemission intensity maps [64], and this work culminated in the first successful direct inversion of LEED I–V spectra to real space 3D structures [65]. At the moment this is not a routine procedure: prior to application of the holographic transform, the data are treated by a plane wave atomic scattering filter function, which includes a phase correction and removes unwanted multiple scattering.

Based on these technique developments, the efforts of my group are currently concentrated on the dynamics and kinetics of bond making and breaking processes at surfaces, backed up by full structural analyses. We are also involved in two rather speculative technique developments. The first is based around a liquid-helium-cooled Fourier transform infrared emission spectrometer, coupled to a supersonic pulsed beam source, for dynamic studies. The second is a surface acoustic wave resonance spectrometer, built at Cambridge in collaboration with Andrei Boronin and Viktor Ostanin from Novosobirsk and a Ph.D. student, Thanos Mitrelias. The latter may provide us with a simple, efficient route to vibrationally exciting selected adsorbate modes to influence catalytic processes, hopefully heralding an important new stage in the control of chemistry at well-defined surfaces.

6. Conclusions

While the developments in surface science over the past 30 years have been quite staggering, the future holds even more promise. This is particularly true for the dynamics of chemical processes at surfaces, where I believe we are currently on the brink of powerful new developments. There are some clear signs that this will happen. On the theoretical front I would cite the molecular dynamics simulation work of Uzi Landman on the one hand [66], in which complex surface processes have been simulated, and on the other hand the application of a fully ab initio molecular dynamics approach, based on density functional theory, to simulate a surface chemical reaction – Cl_2 dissociation on Si{111}(2 × 1) – by Mike Payne at Cambridge and Mike Gillan at Keele [67]. In the latter work, a large surface response to the molecular dissociation process is found, including local rehybridization leading to restructuring. Experimentally, the work of Flemming Besenbacher [68] and others showing real-time snapshots of surface restructuring processes in atomic detail is also a clear pointer to the future.

All of these techniques together now provide us with a reliable, reductionist approach to surface problems. For a given reactive system of interest, the following questions can be tackled in turn.

(1) Where are the atoms comprising a surface? And why are they there? What structural changes accompany surface processes?

(2) What are the bond energies involved, and what is the nature of the surface chemical bonds?

(3) What are the factors controlling the bond making and breaking processes at surfaces?

(4) How can we *control* these bond making and breaking processes?

The next thirty years should prove at least as interesting as the last as, furnished with the answers to the first three of these questions, we (and our successors) come to grips with the fourth.

Acknowledgements and apologies

The encouragement, and the distractions, from my family have kept me going at all times. Of course, none of the work described here would have been done without all those who have informed and inspired me, helped me in collaborations, and, especially, worked as members of my research group. I have, throughout my career, been particularly fortunate in having an extraordinary range of lively, talented colleagues to draw on. This brief review is necessarily somewhat randomly selective, and I apologise to those whose contributions are not explicitly mentioned. Rather a long discourse would be needed to set the record straight.

References

[1] J.K. Roberts, Some Problems in Adsorption (Cambridge University Press, Cambridge 1939).

[2] G. Ehrlich, Adv. Catal. 14 (1963) 255.

[3] R. Gomer, Field Emission and Field Ionization (Harvard University Press, Harvard, MA, 1961).

[4] The outstanding textbook covering early work in the field is D.O. Hayward and B.M.W. Trapnell, Chemisorption (Butterworths, London, 1964).

[5] D.O. Hayward, D.A. King and F.C. Tompkins, Chem. Commun. (1965) 178; Proc. R. Soc. (London) A297 (1967) 305;
D.A. King, Surf. Sci. 9 (1968) 375.

[6] D.A. King and M.G. Wells, Surf. Sci. 29 (1972) 454.

[7] A.M. Horgan and D.A. King, Surf. Sci. 23 (1970) 259.

[8] C.R. Brundle and J.Q. Broughton, in: The Chemical Physics of Solid Surfaces and Heterogeneous Catalysis, Vol. 3A, Eds. D.A. King and D.P. Woodruff (Elsevier, Amsterdam, 1990) p. 131.

[9] D.A. King and M.G. Wells, Proc. R. Soc. (London) A339 (1974) 245.

[10] K.C. Janda, J.E. Hurst, C.A. Becker, J.P. Cowin, L. Wharton and D.J. Auerbach, Surf. Sci. 93 (1980) 270.

[11] P. Alnot and D.A. King, Surf. Sci. 126 (1983) 359.

[12] C.T. Rettner, H. Stein and E.K. Schweizer, J. Chem. Phys. 89 (1988) 3337.

[13] D.L. Adams and L.H. Germer, Surf. Sci. 27 (1971) 21.

[14] S.P. Singh-Boparai, M. Bowker and D.A. King, Surf. Sci. 53 (1975) 55.

[15] G.A. Somorjai, Ed., The Structure and Chemistry of Solid Surfaces (Wiley, New York, 1969).

[16] D.A. King, T.E. Madey and J.T. Yates, Jr., J. Chem. Phys. 55 (1971) 3274.

[17] T.E. Madey, J.T. Yates, Jr., D.A. King and C.J. Uhlaner, J. Chem. Phys. 52 (1970) 5215.

[18] J.T. Yates, Jr. and D.A. King, Surf. Sci. 30 (1972) 601.

[19] M.A. Chesters, J. Pritchard and M.L. Sims, Chem. Commun. (1970) 1454.

[20] C.G. Goymour and D.A. King, Faraday J. Chem. Soc. Trans. I, 69 (1973) 749.

[21] T.E. Madey, J.T. Yates, Jr. and R.C. Stem, J. Chem. Phys. 42 (1965) 1372.

[22] D.A. King, C.G. Goymour and J.T. Yates, Jr., Proc. R. Soc. (London) A 331 (1972) 361.

[23] P.L. Young and R. Gomer, Surf. Sci. 44 (1974) 277.

[24] D.L. Adams and U. Landman, Phys. Rev. Lett. 33 (1974) 585.

[25] F.S. Marsh and D.A. King, Surf. Sci. 79 (1979) 445.

[26] J.B. Pendry, Low Energy Electron Diffraction (Academic Press, New York, 1974).

[27] D.A. King, Phys. Scr. T 4 (1983) 34.

[28] J. Jupille and D.A. King, in: The Chemical Physics of Solid Surfaces, Vol. 7, Eds. D.A. King and D.P. Woodruff (Elsevier, Amsterdam, 1993) ch. 3.

[29] M.K. Debe and D.A. King, Phys. Rev. Lett. 39 (1977) 708.

[30] D.A. King and G. Thomas, Surf. Sci. 92 (1980) 201.

[31] M.W. Holmes, D.A. King and J.E. Inglesfield, Phys. Rev. Lett. 42 (1979) 394;
M.W. Holmes and D.A. King, Proc. R. Soc. (London) A376 (1981) 565.

[32] J.C. Campuzano, D.A. King, C. Somerton and J.E. Inglesfield, Phys. Rev. Lett. 45 (1980) 1649;
J.C. Campuzano, J.E. Inglesfield, D.A. King and C. Somerton, J. Phys. C (Solid State Phys.) 14 (1981) 3099.

[33] P. Hofmann, S.R. Bare, N.V. Richardson and D.A. King, Solid State Commun. 42 (1982) 645.

[34] R.A. Shigeishi and D.A. King, Surf. Sci. 58 (1976) 379.

[35] A. Crossley and D.A. King, Surf. Sci. 68 (1977) 528.

[36] Reviewed by P. Hollins, Surf. Sci. Rep. 16 (1992) 54.

[37] G.D. Mahan and A.A. Lucas, J. Chem Phys. 68 (1978) 1344.

[38] F.M. Propst and T.C. Piper, J. Vac. Sci. Technol. 4 (1967) 53.

[39] S.R. Bare, P. Hofmann, M. Surman and D.A. King, J. Electron Spectrosc. Relat. Phenom. 29 (1983) 265;
D.A. King, J. Electron Spectrosc. Relat. Phenom. 29 (1983) 1.

[40] P. Hofmann, S.R. Bare, N.V. Richardson and D.A. King, Surf. Sci. 133 (1983) L459.

[41] M. Bowker and D.A. King, Surf. Sci. 71 (1978) 583; 72 (1978) 208; 94 (1980) 564.

[42] D.A. King, CRC Crit. Rev. Solid State Mater. Sci. 7 (1978) 167.

[43] D.A. King, Surf. Sci. 64 (1977) 43.

[44] M.A. Morris, M. Bowker and D.A. King, in: Comprehensive Chemical Kinetics, Vol. 19, Eds. C.H. Bamford, C.F.H. Tippert and R.G. Compton (Elsevier, Amsterdam, 1984) pp. 1–181.

[45] A. Cassuto and D.A. King, Surf. Sci. 102 (1981) 388.

[46] S.P. Singh-Boparai and D.A. King, Proc. 4th Int. Conf. Solid Surf.; Suppl. Rev. Le Vide No. 201 (1980) 403.

[47] R. Imbihl, in: Optimal Structures in Heterogeneous Reaction Systems, Vol. 44, Ed. P.J. Plath (Springer, Berlin, 1989).

[48] G.M. Lamble and D.A. King, Philos. Trans. R. Soc. (London) 318 (1986) 203.

[49] G.M. Lamble, R.S. Brooks, D.A. King and D. Norman, Phys. Rev. Lett. 61 (1988) 1112.

[50] D.A. King and D.P. Woodruff, Eds., The Chemical Physics of Solid Surfaces and Heterogeneous Catalysis, Vols. 1–7, (Elsevier, Amsterdam, 1980–1993).

[51] C.E. Borroni-Bird, N. Al-Sarraf, S. Andersson and D.A. King, Chem. Phys. Lett. 183 (1991) 516;
C.E. Borroni-Bird and D.A. King, Rev. Sci. Instrum. 62 (1991) 243.

[52] N. Al-Sarraf, J.T. Stuckless, C.E. Wartnaby and D.A. King, Surf. Sci. 283 (1993) 427;
N. Al-Sarraf, J.T. Stuckless and D.A. King, Nature 360 (1992) 243.

[53] X.-C. Guo, J.M. Bradley, A. Hopkinson and D.A. King, Surf. Sci. Lett. 292 (1993) L786;
A. Hopkinson, J.M. Bradley, X.-C. Guo and D.A. King, Phys. Rev. Lett. 71 (1993) 1597;
A. Hopkinson and D.A. King, J. Chem. Soc., Faraday Trans. No. 96, in press.

[54] M.A. Chesters, J. Electron Spectrosc. Relat. Phenom. 38 (1986) 123.

[55] R. Raval, S. Haq, M.A. Harrison, G. Blyholder and D.A. King, Chem. Phys. Lett. 167 (1990) 391;
P. Hu, L. Morales de la Garza, R. Raval and D.A. King, Surf. Sci. 249 (1991) 1;
R. Raval, S. Haq, G. Blyholder and D.A. King (Proceedings, Vibrations VI), J. Electron Spectrosc. Relat. Phenom. 54/55 (1990) 629.

[56] S.S. Mann, B.D. Todd, J.T. Stuckless, T. Seto and D.A. King, Chem. Phys. Lett. 183 (1991) 529.

[57] J. Stöhr, NEXAFS Spectroscopy (Springer, Berlin, 1992).

[58] J. Singh, W. Walter, A. Atrei and D.A. King, Chem. Phys. Lett. 185 (1991) 426.

[59] A. Wander, M.A. Van Hove and G.A. Somorjai, Phys. Rev. Lett. 67 (1991) 626.

[60] For example: A. Wander, P. Hu and D.A. King, Chem. Phys. Lett. 201 (1993) 393.

[61] K. Heinz, U. Starke and F. Bothe, Surf. Sci 243 (1991) L70.

[62] P. Hu, C.J. Barnes and D.A. King, Phys. Rev. B 45 (1992) 13595.

[63] J. Barton, Phys. Rev. Lett. 61 (1988) 1356.

[64] P. Hu and D.A. King, Nature 353 (1991) 831;

P. Hu, C.J. Barnes and D.A. King, Chem. Phys. Lett. 183 (1991) 521;
P. Hu and D.A. King, Phys. Rev. B 46 (1992) 13615.

[65] P. Hu and D.A. King, Nature 360 (1992) 655.

[66] J.P. Cheng and U. Landman, Science 260 (1993) 1304.

[67] A. de Vita, I. Sitch, M.J. Gillan, M.C. Payne and L.J. Clarke, Phys. Rev. Lett. 71 (1993) 1276.

[68] F. Bessenbacher, in: The Chemical Physics of Solid Surfaces and Heterogenous Catalysis, Vol. 7, Eds. D.A. King and D.P. Woodruff (Elsevier, Amsterdam, 1993) ch. 14.

Surface Science 299/300 (1994) 690–705
North-Holland

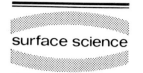

surface science

Theory of adsorption and adsorbate-induced reconstruction

J.K. Nørskov

Physics Department, Technical University of Denmark, DK 2800 Lyngby, Denmark

Received 5 April 1993; accepted for publication 30 April 1993

A brief personal account is given of the development of the theory of chemisorption based on a combination of local density functional calculations and modelling within the effective medium theory. Starting from the calculations of atomic and molecular adsorption on jellium surfaces in the late seventies, a description is given of the effective medium theory and its application to an understanding of trends in atomic chemisorption, adsorbate–adsorbate interactions, adsorbate-induced surface reconstructions, and molecular adsorption. The main emphasis will be on the physical picture of adsorption that has been developed in this way. The results of the effective medium theory are contrasted with results of the full local density functional calculations for atomic and molecular adsorption and dissociation that have appeared in the last few years. Finally, I discuss how the present understanding of adsorption phenomena can be directly applied to an understanding of trends in heterogeneous catalysis.

1. Introduction

The theory of chemisorption on metal surfaces has developed enormously over the last thirty years. It was only in the late sixties that the first simple models were formulated and some of the basic concepts used to describe the chemisorption electronic structure and bond were introduced. In the seventies the first self-consistent calculations of adsorption on model surfaces – jellium or small metal clusters – appeared, and it was only in the eighties that local density functional calculations of binding energies and vibrational properties using a realistic description of the metal began to appear. Even though such calculations are still rare and absolutely non-trivial we have seen in the beginning of the nineties the appearance of the first realistic local density functional calculations of activation barriers for diffusion and reactions on surfaces.

Along with this tendency towards better and better ab initio calculations of chemisorption properties there has been a further development of models of chemisorption and approximate theories. Such a development has been important for at least two reasons. First, there are a large number of situations, where full self-consistent calculations are not feasible. This is the case for systems that are very complex or large (with steps or islands, for instance) or in situations where time dependent or finite temperature properties are simulated. The second reason is that it is through models and approximate methods that our *understanding* of surface phenomena is developed. The large scale ab initio calculations are extremely important in providing *bench marks* for comparison to experiment and for simpler theoretical approaches. Even though the ab initio calculations involve an approximate treatment of electron exchange and correlation effects, the quality of the results are often so that they provide *quantitative* predictions. Such calculations do, however, not a priori provide much insight. The result of one calculation cannot be brought to say anything about another system without some model or approximate scheme which can generalize the results and extract the most important physical parameters in the problem.

The present paper will present a personal account of how a physical picture of adsorption, the effective medium theory, has developed through an interplay between self-consistent calculations

SSDI 0039-6028(93)E0272-V

and more approximate modeling. It will briefly be described how the method can describe and systematize a large number of adsorption phenomena, and how it allows going one step further in complexity to treat dynamics, finite temperature problems, and even trends in heterogeneous catalysis.

2. From jellium to real metals

In the seventies the first self-consistent calculations of chemisorption on a semi-infinite metal substrate were performed. They all used a theorist dream of a metal, jellium, where the positive ion cores are smeared out as a positive background. The jellium model can be considered a generalization of the Sommerfeld model which is an extremely useful model of a bulk metal, in particular for the simple, or free-electron metals. A number of clean surface properties turned out to be well described by the model [1], and it was therefore natural to use it as a model for a prototype simple metal in studies of chemisorption. A number of studies by the IBM group (Lang and Williams) [2,3] and the Göteborg/Århus group (Lundqvist, Gunnarsson, Hjelmberg, Johansson, and Nørskov) [4,5] mapped out the electronic structure for a number of atomic and molecular chemisorption systems. Concepts like resonance broadening of adsorbate levels that had been introduced in the previous studies using for instance the Newns–Anderson model Hamiltonian [6] were given a more quantitative content, and in particular the trends in resonance width and position with varying adsorbate, substrate density, and adsorbate–substrate distance were mapped out.

One particularly important effect, that these jellium calculations pinpointed, was the downshift of adsorbate valence levels when the adsorbate approaches the surface. Fig. 1 shows the variation of adsorbate valence levels as a function of distance outside a jellium surface. As indicated in the figure, the levels follow the effective one-electron potential of the clean surface quite closely.

In principle, the one-electron energy eigenvalues that come out of a density functional calcula-

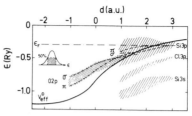

Fig. 1. The variation of the resonance position and width for O, Cl and Si outside a jellium surface ($r_s = 2$) as a function of the distance of the adsorbates outside the surface. The full curve shows the variation of the effective one-electron potential. From Ref. [7].

tion has no direct physical interpretation. Nevertheless, they are often found to represent the excitation energies in metallic systems quite closely. The trend seen in Fig. 1, which is found for a large number of adsorbates, has been used to represent the variation of the affinity levels of adsorbates with distance to the surface. These variations are all-important in determining the charge transfer during adsorption or desorption, and has been used as a basis for modeling of ionization probabilities in sputtering and ion-surface scattering and of the probability of electron–hole excitations during adsorption [8–10].

The down-shift of the valence levels also shows up in calculations for an H_2 molecule approaching a jellium surface and completely controls the dissociation process [5]. Consider first the potential energy contour plot of the potential felt by an H_2 molecule approaching a jellium surface. In the model calculation the effects of the lattice are included in first-order perturbation theory. The contour plot shows three minima. One far from the surface, where the molecule is weakly bound to the surface (physisorbed) and retains the bond strength of the free molecule. Closer to the surface there is a second molecular minimum. The molecular bond has been stretched a little and the molecular vibration is softer than for the free molecule. The last minimum has the two atoms far from each other and represent the atomic chemisorption minima. The minima are separated by barriers representing activation energies for the adsorption and dissociation process.

In Fig. 2 the changes in the adsorbate-induced density of states during the dissociation are included. It can be seen that the dissociation is driven by the gradual filling of the anti-bonding H_2 state which is shifted down through the Fermi level as the molecule approaches the surface. The initial repulsion that the molecule feels upon approaching the surface is due the Pauli repulsion between a closed-shell adsorbate (H_2) and the surface. This interaction is quite similar to the repulsion other closed shell adsorbates like He feels at a metal surface. But for H_2, the anti-bonding state starts filling, and this changes the repulsion. Notice that the molecularly chemisorbed state is characterized by the molecule having a partly filled anti-bonding state, and during the dissociation the bonding and anti-bonding states move together and eventually coalesces into the atomic adsorbate-induced resonance.

On this basis it was suggested to make a refinement of the traditional Lennard-Jones picture [11] of molecular dissociation as shown in the last part of Fig. 2. As the molecule approaches the surface it is a neutral object behaving like other closed shell molecules or atoms. The interaction with the surface is repulsive apart from the physisorption minimum. The second molecular minimum corresponds to a partly negative molecule

and the barrier separating depends on the crossing of the curves for the neutral and the negatively charged curves. Finally, if the molecule had been dissociated outside the surface a new potential corresponding to two atomic adsorbates would appear. The activation barrier for dissociation depends on the crossing of the negative molecule and atom curves.

Whereas the actual potential energy surface in Fig. 3 has turned out not to be correct in detail (the molecularly adsorbed state does not exist for H_2 adsorption on simple metal surfaces, see later) the general picture of the adsorption process and the role of the down-shift of the anti-bonding state has turned out to be valid both for H_2 dissociation, but also more generally for molecular dissociation reactions. For other molecules like O_2, CO or N_2 a molecularly adsorbed state with a partly filled ($2\pi^*$) anti-bonding state is well described theoretically and experimentally [12]. In particular, it is observed that the adsorbed molecule has a lower stretch frequency than in the gas phase as expected when the anti-bonding levels are partly filled [13].

In spite of the influence these calculations had on the development of concepts in the theory of chemisorption, the substrate jellium has one

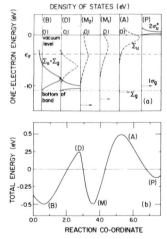
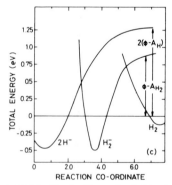

Fig. 2. A one-dimensional representation of the two-dimensional potential in Fig. 2. The potential is shown along a coordinate following the lowest energy path toward the surface. Above the one-dimensional potential the development of the adsorbate-induced density of states along the reaction path is shown. To the right, it is shown how the one-dimensional potential can be constructed from a series of diabatic potential energy curves. ϕ and A refer to the metal work function and the adsorbate affinity, respectively. $\phi - A$ is then the energy required to form the negative ion by taking the electron from the Fermi level of the metal and placing it on the adsorbate far from the surface. From Ref. [16].

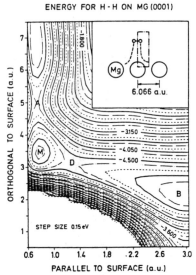

ENERGY FOR H-H ON MG (0001)

STEP SIZE 0.15 eV

Fig. 3. A contour plot from a model calculation of the potential energy of an H_2 molecule adsorbing and dissociating on a Mg(100) surface. From Ref. [5].

drawback. There are no adsorption experiments on jellium in the literature. Not even simple metals, which are the closest real life can get to jellium, are used very much in experimental chemisorption studies. The favorites are the noble and transition metals, which are most interesting from a technical point of view as catalysts. In order to make contact to experiment it was therefore necessary to either do much more elaborate calculations for real metal substrates or to see if the jellium results could be generalized in some other way. The effective medium theory is an attempt to do the latter.

3. The effective medium theory

Along with the adsorption calculations on jellium surfaces, the same theoretical approach was used to study the interaction of impurities with bulk jellium, and it was noticed that the interaction energy of an impurity as a function of jellium density had a characteristic form (see Fig. 4) which allowed an estimate of the binding energy of the atom in question in other less homogeneous surroundings [14]. The idea is that the

binding energy is a function $E_c(n)$ mainly of the local electron density n of the substrate at the position of the atom in question. Soon this was formulated as a formal perturbation theory, and a set of comparisons with full adsorbate on jellium calculations was made [15].

Independent of the development of the effective medium theory a similar approach, the quasi-atom approach, was introduced by Stott and Zaremba [21]. Later various similar approaches have been introduced under various names, the corrected effective medium theory of DePristo and co-workers [22], the embedded atom method by Daw, Baskes, and Foiles [23], the Finnis–Sinclair method [24], and the glue model of Tossati and co-workers [25]. Common to all of these is the idea of a density-dependent embedding function determining part of the binding energy. Since this function is non-linear the interaction energy is not a linear function of the number of neighbors of a certain kind. This means that the strength of any bond is dependent on the position of the other atoms – the bonding thus cannot be desribed through pair-wise interactions. The main advantage of these methods over previous pair interaction models is that they are not much more computationally demanding, but include the many-atom interactions that are essential for an understanding of adsorption and surface reconstructions as I will illustrate in the following Section.

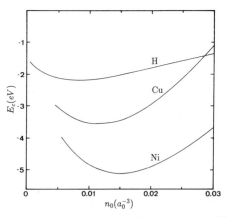

Fig. 4. The cohesive function $E_c(n)$ for hydrogen, nickel and copper calculated within the local density approximation. The curves are similar for other reactive atoms. From Ref. [18].

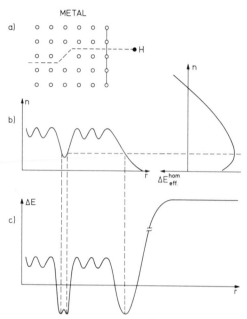

Fig. 5. A schematic illustration of the potential energy variation of a hydrogen atom moving from far outside the surface, through the surface and into the bulk. From Ref. [16].

In Fig. 5 I show qualitatively how the concept of a binding energy $E_c(n(r))$ which depends primarily on the electron density $n(r)$ from the surroundings at the position r of the atom in question leads to a very simple and powerful picture of the hydrogen–metal interaction. A similar picture can be used for other adsorbates as well.

By combining the $E_c(n)$ function of hydrogen with a schematic representation of the electron density of a metal as a function of distance from the surface, one can construct a potential energy diagram. It is seen that the chemisorption minimum comes about as a direct reflection of the minimum of the $E_c(n)$ curve. We can use Fig. 5 to follow the hydrogen atom further into the metal. Since most metals have interstitial electron densities that are higher than the optimum density in $E_c(n)$, the variation in energy inside the metal is a direct reflection of the variation in the metal electron density. This explains immediately why interstitial hydrogen is usually found to occupy the most open sites in the lattice, why the barrier for diffusion is smaller in the open bcc

metals than in the close packed fcc metals, why interstitial hydrogen usually induces a local expansion of the metal lattice, and why all defects that have lower than average electron densities like vacancies, grain boundaries and voids are found to act as traps for hydrogen.

Closed shell atoms like He or Ar has a cohesive function that grows almost linearly with the electron density. This means that e.g. the He–surface interaction potential is approximately proportional to the metal electron density, a feature that has been extensively used in modeling He scattering experiments [17].

Based on density functional theory the effective medium idea has been developed into an approximate theory of bonding [19,20]. It has been shown that the total energy of a system of atoms can be written approximately [20]

$$E_{tot} = \sum_i E_{c,i}(\tilde{n}_i) + \Delta E_{AS} + \Delta E_{1el}. \tag{1}$$

Here the first term on the right-hand side is a sum over all the atoms in the system of the cohesive function discussed above. The two other terms give the difference between the real system under study and the reference system where $E_c(n)$ has been calculated. The so-called atomic sphere correction ΔE_{AS} measures primarily the electrostatic energy difference while ΔE_{1el} measures the difference in the one-electron energy sum for the two systems.

In the following I will concentrate on the physical picture of adsorption that has emerged from these considerations. I will include results of full calculations (taking into account all three terms in Eq. (1)) to illustrate specific points, but otherwise refer to the original papers for details.

4. Trends in atomic chemisorption

For a start, let us concentrate on the first term in Eq. (1). It is found generally, that this terms determines the bond lengths in the system, indicating that a chemisorbed atom should not have a fixed chemisorption bond length to a given metal, but rather should seek out the position above the surface where the electron density from the sur-

Table 1
Bond lengths (d) and vibrational frequencies perpendicular to the surface ($\omega_\perp$) for hydrogen and oxygen outside the (111) and (100) surfaces of Ni (from ref. [26])

	d (Å)		$\omega_\perp$ (meV)	
	H	O	H	O
Ni(111)	1.65	1.88	139	72
Ni(100)	1.83	1.96	74	46

roundings is exactly the optimum one where the $E_c(n)$ function has its minimum. This means that bond lengths for adsorbates in three-fold coordinates sites should be smaller than in four-fold coordinates sites because three atoms contribute less electron density at the adsorbate site than four atoms at a given adsorbate–metal distance. This is borne out by experimental observations. Table 1 shows this for both hydrogen and oxygen adsorbed on various Ni surfaces.

Full local density functional calculations have also shown that the concept of an optimum metal density for adsorption is indeed useful [27]. Table 2 shows that for a number of systems the metal electron density at the calculated equilibrium adsorption site is essentially constant.

The first term in Eq. (1) also provides a usefull relationship between the frequency for vibrations perpendicular to the surface $\omega_\perp$ and the gradients of the electron density. Using that the density derivative of the cohesive function is zero at equilibrium, we get that [30]

$$\omega_\perp \sim \sqrt{\frac{\mathrm{d}^2 E_c}{\mathrm{d}z^2}} = \sqrt{\frac{\mathrm{d}^2 E_c}{\mathrm{d}n^2}} \frac{\mathrm{d}n}{\mathrm{d}z}. \qquad (2)$$

Table 2
Calculated clean surface charge densities at the calculated H equilibrium site, calculated vibrational frequencies and the ratio of the frequency to the density gradient (from Ref. [27])

	$n_{\text{metal}}(r_H)$	$\omega_\perp$	$\dfrac{\omega_\perp}{\mathrm{d}n_{\text{metal}}(r_H)/\mathrm{d}z}$
	(a.u.$^{-3}$)	(meV)	
H/Cu/Ru-hcp	0.0134	120	12
H/Cu/Ru-fcc	0.0134	130	13
H/Cu(111)-unstretched	0.0152	160	12
H/Cu(111)-stretched	0.0147	120	12

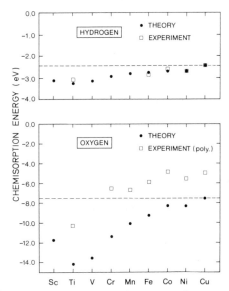

Fig. 6. Experimentally deduced chemisorption energies for hydrogen and oxygen on the 3d transition metals. The trends are the same in the 4d and 5d series. The theoretical results shown are from a model calculation within the effective medium theory. From Ref. [26].

This dependence is again confirmed by the local density functional calculations in Table 2. Another consequence of Eq. (2) is that $\omega_\perp$ should depend on the coordination number through the density derivative $\mathrm{d}n/\mathrm{d}z$. One can show that $\mathrm{d}n/\mathrm{d}z \sim z/d$ where z is the distance of the adsorbate above the surface and d is the bond length [30]. From the experimental bond lengths of Table 1 we deduce a ratio of the frequencies on the (111) (three-fold coordinated) and the (100) (four-fold coordinated) of 1.8 for H and 1.6 for O. The corresponding ratios for the experimentally determined frequencies in Table 1 are 1.9 and 1.6. for the two cases.

The first term in Eq. (1) thus gives a qualitative understanding of atomic chemisorption including a detailed picture of the metal and adsorbate properties determining the chemisorption bond length and the vibrational frequencies. The depth of the minimum in the cohesive function in Fig. 4 also gives the overall size of the adsorption energy, but it does not account for the variations in the binding energies from one metal to another. This is illustrated in Fig. 6. To understand these variations one must include the one-electron energy correction in Eq. (1). Simple estimates are, however enough to give a very nice account of the experimental observations. It turns out that the decisive factor is the degree of filling of the anti-bonding adsorbate – metal-d states, which is given by the number of d-electrons of the metal.

5. Adsorption-induced reconstructions

During the late eighties and the early nineties it was realized that surface reconstructions are of enormous importance for an understanding of many surface problems. Many clean metal surfaces reconstruct, but in particular scanning tunneling microscopy (STM) has opened our eyes to the fact that a large number of simple gas adsorbates also restructure the surface and often very strongly by breaking a number of metal–metal bonds during adsorption [29].

The effective medium theory has offered a possibility to understand both the clean surface and the adsorption-induced reconstructions. In the following I will briefly review the general picture that has evolved from this approach.

First, consider the clean surface reconstructions. One of the most studied reconstructions is the (1×2) missing row reconstruction of the fcc(110) surfaces [31]. The reconstruction is observed on Au and Pt while the other metals in the Ni and Cu groups do not reconstruct in this way. In the reconstruction every second of the close-packed rows on the (110) surface are removed as illustrated in Fig. 7.

To understand the origin of the (1×2) missing row reconstructions consider again the plot in Fig. 4 of the energy versus electron density for an

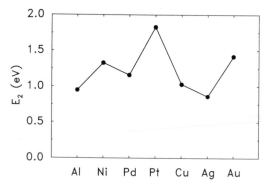

Fig. 8. The value of the curvature E_2 of the E_c function for a number of metals. E_2 is proportional to the bulk modulus.

atom. An fcc metal atom in the bulk will have 12 nearest neighbors contributing a total electron density around any atom of exactly the optimum value n_0 at the equilibrium lattice constant. Atoms at the surface will have fewer nearest neighbors and thus feel a lower than optimum electron density. This means that surface atoms have a higher energy than bulk atoms (part of the surface energy), and it explains why there is a general tendency for contraction of the first interlayer spacing at metal surfaces. The curvature of the cohesive function is given primarily by the kinetic energy of the conduction electrons – the smaller the volume available, the higher the kinetic energy, and the higher the curvature (measuring the energy change with volume). The curvature determines the bulk modulus of the metal in question. The curvatures calculated for various metals are plotted in Fig. 8. It is seen how the curvature is smaller for the 4d metals Pd and Ag than for the 3d metals Ni and Cu, because the former have considerably smaller lattice constants. The 5d metals Pt and Au have essentially the same lattice constants as the 4d's, but the 5d shells are bigger, and the volume available to the valence (6s) electrons in the 5d is therefore smaller.

The fact that the cohesive function has a curvature means that the bond strength depends on the number of other neighbors as discussed above. One consequence of this is that the system can gain energy without changing the number of nearest neighbor bonds by creating a situation where some atoms have more than the average

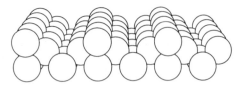

Fig. 7. The (1×2) missing row reconstruction of an fcc(110) surface.

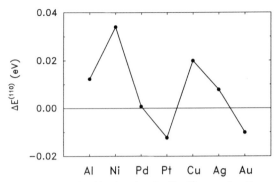

Fig. 9. Energy gained by forming the (1×2) reconstruction on various fcc(110) metal surfaces. From Ref. [32].

number of bonds and some less. This is what happens on the (110) surfaces. On the unreconstructed surface the atoms in the first layer have seven neighbors and those in the second layer have eleven. The rest have twelve neighbors as in the bulk. On the reconstructed surface the first layer atoms have seven neighbors, the second layer atoms have nine and the third layer atoms have eleven. If we denote the energy of an atom with N neighbors by $E(N)$ and use $E(12)$ as the energy zero, then the reconstruction energy per (1×2) unit cell is

$$\delta E^{1 \times 2} = 2(E(7) + E(11)) - (E(7) + 2E(9)$$
$$+ E(7)) \simeq -4 \frac{d^2 E}{dN^2}. \qquad (3)$$

The reconstruction energy is therefore a direct measure of the curvature of the cohesive function, and the tendency for reconstruction should be closely related to the trends in the curvatures shown in Fig. 8 [31].

Fig. 9 shows that a full calculation within the effective medium theory including relaxations and interactions beyond nearest neighbors confirms this picture and also give the result that only Au and Pt should reconstruct as observed experimentally. The tendency of Pt and Au to reconstruct is therfore related to the large 5d shell, and in this way the present picture of the driving force behind the missing row reconstructions make contact to the picture derived by Ho and Bohnen [28].

Turning now to adsorbate-induced reconstructions, the reconstruction energies can conveniently be separated into the energy of reconstructing the *clean* metal surface and the difference in chemisorption energy for the adsorbate on the reconstructed and unreconstructed surface:

$$\Delta E_{rec} = \Delta E_{rec}^{metal} + \Delta E_{chem}. \qquad (4)$$

Adsorbate-induced reconstructions can now be divided into two classes [29]. The first consist of the cases where the energy required to reconstruct the clean surface is small. This includes situations where the adsorbates just induce a small local rearrangement of the surface atoms. It is for instance clear from the discussion above, that the fact that the adsorbates will contribute to the electron density around the surface metal atoms will lift or invert the contraction of the first interplanar distance of the clean surface. Other local rearrangements may also occur. Also a number of reconstructions involving long range mass transport belong to this category. Fig. 9, for instance, shows that even the (110) surfaces that do not reconstruct spontaneously are only a few hundredths of an eV from doing it. Clearly, the adsorption energy only has to be marginally larger on the reconstructed Al, Cu, Ni, Ag, or Pd surfaces to reconstruct them. This is the case with the hydrogen or alkali-induced (1×2) reconstruction of several of these metals [34,35]. The larger chemisorption energy is mainly due to the

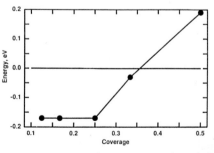

Fig. 10. The K chemisorption energy on the missing row reconstructed Cu(110) surface minus the chemisorption energy on the unreconstructed surface as a function of the K coverage. The adsorption of K is seen to favor reconstruction of the surface up to a coverage of about one half. From Ref. [34].

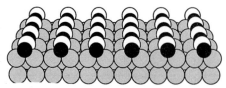

Fig. 11. The equilibrium structure for the missing row reconstruction of Cu(110) with half a monolayer of oxygen adsorbed. The O atoms are shown smaller than the Cu atoms.

possibility of a larger coordination number of the adsorbate on the more open reconstructed surface. Fig. 10 shows how the chemisorption energy of K on the missing row reconstructed Cu(110) surface is much larger at low coverages. At higher coverage, the K–K repulsion makes the reconstruction less favorable.

The second class consists of reconstructions where the reconstruction energy of the clean surface is substantial. One very extensively studied case is the (2×1) missing row reconstructions of

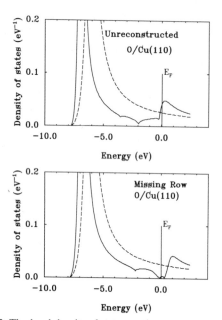

Fig. 12. The local density of states on the oxygen atoms at the equilibrium position at the unreconstructed and the reconstructed Cu(110) surface. The local density of states in the homogeneous electron gas is shown dashed for comparison. On the reconstructed surface the anti-bonding states formed by hybridization between the oxygen p-state and the metal d-states are less occupied and therefore this structure is more stable than the unreconstructed one. From Ref. [36].

the Cu(110) surface. The reconstruction is illustrated in Fig. 11. Here the energy cost of reconstructing the clean surface is large (0.3 eV) because nearest neighbor Cu–Cu bonds are broken in the process [36]. The chemisorption energy of oxygen on the reconstructed surface therefore has to be much larger than on the unreconstructed one. In the effective medium theory the reason for this effect can be traced to the one-electron energy contribution to the chemisorption energy in Eq. (1). The one-electron energy contribution is not important for the clean Cu surfaces because the d-bands of Cu are completely filled. On the unreconstructed surface the the oxygen levels will interact with the d-bands, but because the d-bands are well below the Fermi level, both the bonding and the anti-bonding levels will be filled and there is no net contribution to the bond energy. On the reconstructed surface, on the other hand, the Cu d-bands are higher in energy and that means that the anti-bonding O–Cu-d states are above the Fermi level as seen in Fig. 12. There is thus an extra contribution to the bond energy on the reconstructed surface, which more than compensates for the energy cost if breaking the Cu–Cu bonds [36].

6. Molecular adsorption and dissociation

The molecular adsorption and dissociation process is more complex to handle theoretically than the atomic adsorption process mainly because there are more ionic coordinates to vary before a complete picture of adsorption is reached. Progress in the theoretical treatment of molecular adsorption and dissociation has therefore been slower. Since the first calculations of the electronic structure of H_2 dissociating over jellium surfaces in the late seventies and early eighties [5], molecular adsorption has mainly been dealt with using small metal clusters to model the surface [37]. It is only in the early nineties that calculations of molecular dissociation over an infinite substrate in the form of a slab has been possible [38–40].

Fig. 13 shows a two-dimensional representation of the potential energy surface for H_2 disso-

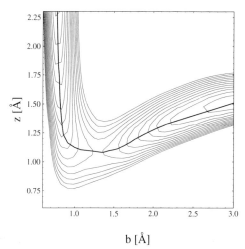

Fig. 13. Contour plot of the potential energy surface for an H_2 molecule dissociation over an Al(110) surface. From Ref. [39].

ciating over Al(110). The process is found to be activated in good agreement with molecular beam scattering experiments [41]. It is worth pointing out that the potential only shows one barrier for dissociation and not two as found for the jellium model. This may be due to shortcomings of the jellium model as pointed out by Harris [42], or it may simply be due to shortcomings in the numerical accuracy of the jellium calculation far from the surface.

Irrespective of these considerations, the calculations for H_2 dissociating over Mg and Al [39,40] have confirmed the general picture of the adsorp-

tion and dissociation process developed on the basis of the jellium calculations. It was suggested from the jellium calculations (Fig. 2) that the driving force in the dissociation process is dominated by the gradual filling of the anti-bonding H_2 states. Fig. 14 shows the electron density rearrangements for an H_2 molecule far outside an Al(110) surface and at the top of the barrier for dissociation. It is seen how, at large distance from the surface, the metal electrons are expelled from the region of the bonding H_2 state due to the Pauli repulsion. The interaction is purely repulsive at these distances. Closer to the surface, on the other hand, it can be seen how extra charge is transferred to the anti-bonding state of the adsorbing molecule.

The effective medium theory has also been used as an approximate method that allows comparisons of different systems quite easily [18,43]. Fig. 15 shows the calculated potentials for H_2 dissociating on different facets of Ni and Cu. It can be observed that there are large differences between Cu and Ni and (smaller) between the open and close-packed facets. Ni shows nonactivated adsorption on the open (110) surface and a small barrier in the entrance channel for the (111) surface. Cu on the other hand shows activated adsorption on both facets. All these findings are in complete agreement with experiment. The calculation clearly shows the smaller barrier on Ni to be related to the interaction between the partly filled Ni d-bands and the

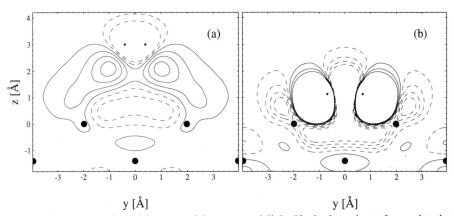

Fig. 14. Charge density difference plot $(n_{total}(r) - n_{surface}(r) + n_{molecular}(r))$ for H_2 far from the surface and at the transition state outside Al(110). From Ref. [39].

anti-bonding H_2 state which is around the Fermi level during the dissociation process (i.e. the one-electron correction is again important in determining *differences* between metals). On Cu the d-bands are well below the Fermi level and this interaction is much weaker. The difference between the different facets are found to be a reflection of the different work functions and in the position and width of the d-bands at the surface.

The lower barrier for dissociation on Ni than on Cu is observed quite generally for a number of adsorbates. The explanation given above is also part of a more general picture of the trend in molecular adsorption energies and dissociation barriers along the transition metal series. Based on the same kind of effective medium theory ideas as for the atomic adsorbates one can argue that the trends are governed to a large extent by the coupling between the adsorbate valence electrons and the metal d-bands [44]. The main difference between the adsorbed atoms and the adsorbed molecules lies in the position of the valence levels relative to the Fermi level. For the simple gas atoms the valence levels are always well below the Fermi level. This gives rise to the linear trend seen in Fig. 6. For the molecules the valence levels that interacts most strongly with the metal d-bands will often be the anti-bonding molecular levels because they are partly filled and therefore lie around the Fermi level where the d-states are also situated.

Fig. 16 shows the result of a calculation of the adsorbate–metal-d interaction within the Newns–Anderson model [44]. Both the atomic chemisorption case (the adsorbate level well below the Fermi level) and the molecular adsorption case (the adsorbate level at the Fermi level) are included. In the molecular case the interaction is seen to increase to the left of the noble metals, but it bends over and the interaction becomes weaker further to the left.

Most potentials for dissociation reactions have only been determined in two-dimensions. For ab initio calculations this is due to the limited computer power and for approximate model potentials due to the limited possibilities of imagining how to construct reasonable potentials in all de-

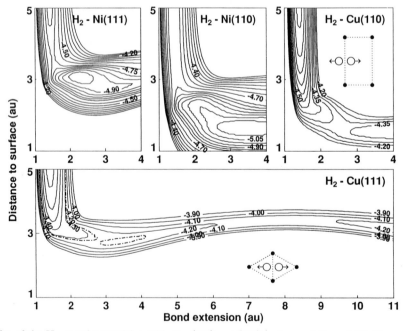

Fig. 15. Contour plot of the H_2–metal potential outside the (111) and (110) surfaces of Cu and Ni. The potential is shown in a plane perpendicular to the surface in the geometries indicated by the inserts. From Ref. [43].

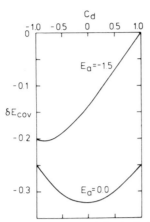

Fig. 16. A Newns–Anderson model calculation of the one-electron energy $E_{1\,el}$ for an adsorbate state interacting with a d-band as a function of the d-band filling. The hopping matrix element is kept fixed and so is the d-band width. Two cases are shown. One is for an adsorbate level well below the d-band ("atomic chemisorption") and one where it is at the Fermi level ("molecular adsorption"). From Ref. [44].

grees of freedom. The simplicity of the effective medium energy calculations has helped in this respect. The main point that these calculations have raised is the importance of including *all* the degrees of freedom for a detailed description of the dynamics. There is not a single activation barrier for an adsorption reaction, but a distribution of barriers as shown in Fig. 17. In a thermal adsorption experiment the Boltzmann distribution of incoming energies will single out the low-

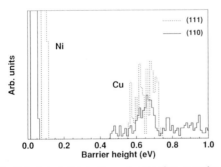

Fig. 17. Calculated distribution of activation energies for H_2 dissociation over the (111) and (110) faces of Cu and Ni. The distribution $f(E)\,dE$ is calculated as the fraction of the unit cell area where a molecule impacting would experience a barrier between E and $E + dE$. Only impacts with the molecule lying flat on the surface are included and the distribution is averaged with respect to the orientation of the molecular axis parallel to the surface. From Ref. [43].

est barrier, but in a beam experiment one can actually sample a large part of the distribution [43].

7. Adsorbate–adsorbate interactions and the poisoning and promotion of the adsorption process

The electro-positive alkali adsorbates and the very electro-negative adsorbates like oxygen, sulfur, or chlorine are known to change adsorption energies and adsorption rates of simple gas molecules like CO, O_2, H_2 or N_2 significantly [45]. Typically, the electro-positive adsorbates increase the molecular adsorption energies and increase the dissociation rates while the electro-negative adsorbates have the opposite effect. For polar molecules like H_2O or NH_3 the picture is more complex, but still very well studied experimentally [45]. These effects are closely related to the promotion and poisoning effects observed in catalytic reactions, where additives to the catalyst or to the gas feed can increase or decrease the turn over rates or change the selectivity [45].

The theoretical approaches have been divided in two camps. One suggestion is that the indirect interaction through the metal surface is most important. It has been argued that the changes in the surface density of states by the pre-adsorbed atoms should induce changes in the interaction of the surface with the adsorbing molecule [46,47].

The other suggestion is that the interaction is dominated by electro-statics [48]. This is a tempting idea since the adsorbates that are known to induce the large changes in the binding and dissociation of small molecules are either very electro-positive or electro-negative. They induce large electro-static potentials when adsorbed due to electron transfer to or from the surface. This is quantified by the calculated electrostatic potentials for adsorbed atoms on jellium surfaces shown in N. Lang's article in this volume [3]. An adsorbing molecule with a dipole moment μ will experience an energy change due to a pre-adsorbed atom with an induced electrostatic potential $\phi(r)$ of [48]

$$\delta E \simeq \mu \, d\phi(r)/dr. \tag{5}$$

First, consider the usual case where the induced dipole moment of the adsorbing molecule is due to charge transfer to or from the surface. In this case μ is perpendicular to the surface and it is only the perpendicular derivative $d\phi(r)/dz$ at the position of the molecule that enters. Fig. 18 shows the electrostatic potentials due to an electro-positive (K) and an electro-negative (S) adsorbate as a function of distance from a jellium surface. For simple molecules like CO, O_2, N_2, or H_2 the *chemisorbed* (as opposed to physisorbed) molecule usually induces a charge transfer from the surface to the molecule. This is evidenced by measured work function shifts and calculations. The charge transfers are dominated by the filling of the anti-bonding levels discussed above. Such molecules are therefore *stabilized* by pre-adsorbed alkalies and *destabilized* by pre-adsorbed electronegative atoms, and the barrier for dissociation is expected to be affected in the same way. This is in general agreement with experimental observations [48,51]. For adsorbates with an internal dipole moment like H_2O and NH_3 the situation is more complex. They can rearrange on the surface in response to the electrostatic potential of the preadsorbed atoms. This has been described in great detail in a number of reviews [51].

It has long been realized that the electrostatic interactions alone gives a good account of both the trends and the absolute order of magnitude

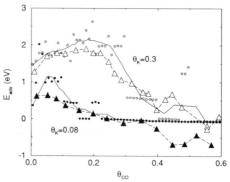

Fig. 19. Heat of adsorption with the optimal distribution of K at any instant (the best of about 5 possibilities investigated). $\theta_K = 0.3$ (open circles/full line), $\theta_K = 0.08$ (filled circles/dashed line) and $\theta_K = 0$ (thin line) monolayers. The calculated points show some scatter due to finite size effects and because only a limited number of configurations have been tested. The curve through the calculated points have been determined as a box average over ten points. The only input into the calculation are the independently measured dipole moments and polarizabilities of the two adsorbates. The experimental results from Ref. [49] are included as open ($\theta_K = 0.3$) and solid ($\theta_K = 0.08$) triangles.

of interactions between pre-adsorbed atoms and adsorbing simple molecules [48,51]. It is, however, only recently that experimental data from single crystal micro-calorimetric measurements have appeared that are detailed enough for a quantitative comparison between theory and experiment [49]. Fig. 19 shows the experimental results together with a simulation of the interactions based solely on the electrostatics. Apart from the direct electrostatic interaction between the rigid adsorbates, the mutual polarization of the two kinds of adsorbates have been included. The electrostatic potential of a preadsorbed atom can also change the transfer of the molecule. This effect is *second order* in the electrostatic potential [51]. It turns out not to be important for a description of the overall trends and order of magnitude but to be essential for a correct description of the details [50].

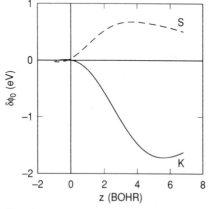

Fig. 18. The electrostatic potential due to K and S shown as a function of distance outside the jellium ($r_s = 2$) surface a lateral distance of 5 bohr from the atoms. From Ref. [48].

8. Implications for heterogeneous catalysis

One of the main motivations for the enormous efforts in surface science over the last thirty years has been the hope that a more detailed under-standing of adsorption on metal surfaces would lead to new possibilities in understanding hetero-geneous catalysis. One of the main questions has been whether it is possible at all to extrapolate between the ultra-high vacuum conditions and single crystal substrates of surface science to high pressures, high temperatures and often quite complex structures of industrial catalysts at reac-tion conditions.

There is now a rapidly growing body of evi-dence that this can be done. Let me discuss briefly one of the earlier attempts to go all the way from surface science to heterogeneous catal-ysis that I was involved in [54]. Consider the ammonia synthesis reaction

$$N_2 + 3 H_2 \rightleftharpoons 2 NH_3. \tag{6}$$

Industrially the process takes place at 100–300 atm, and around 700 K over an Fe based catalyst with Al_2O_3 added as a structural promoter in addition to a number of other promoters. The most important of these is potassium which in-creases the activity significantly (an electronic promoter) [52].

The basic steps in the process are very simple. They include the adsorption and dissociation of H_2 and N_2, the sequential addition of hydrogen to the adsorbed nitrogen and the desorption of ammonia. All of the steps and the intermediates, chemisorbed hydrogen and nitrogen has been studied very thoroughly under surface science conditions [53]. Based solely on this information an attempt has been made to predict the activity of the industrial catalyst under industrial condi-tions. The comparison of the model predictions to experiments is shown in Fig. 20. Over the full range of conditions obtainable in the chemical reactor there is a surprisingly good agreement indicating that the essential part of the reaction is well described by the model.

It turns out that there are two factors that are important in determining the reaction rate. One is the N_2 sticking probability. The N_2 dissociation turns out to be rate limiting, i.e. it is the slowest step, and the total rate of the reaction is there-fore equal to the dissociation rate. Since the sticking probability is essentially the same as the

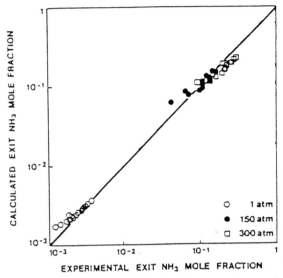

Fig. 20. Comparison of calculated and measured ammonia production over a commercial iron-based catalyst for a broad range of temperatures, pressures, N/H ratios and gas flows. From Ref. [54].

dissociation rate it is not strange that it enters in an essential way in the overall reaction rate. The other factor is the N chemisorption energy. Chemisorbed atomic nitrogen is by far the most stable reaction intermediate. The surface is therefore mainly covered by chemisorbed N (up to 90%) and the number of free sites on the surface where new nitrogen can adsorb is given by the N coverage. The N coverage is determined by the N chemisorption energy.

The question now arises, why the preferred catalyst is based on Fe rather than another metal. This can be checked by considering the variation in the N_2 sticking probability and N binding en-ergy from one metal to the next. The trends in the former over the transition metals will be given mainly by the trends in the activation en-ergy for dissociation. This is shown in Fig. 16 to have a parabolic dependence on the number of d-electrons. The N chemisorption energy, on the other hand, depends roughly linearly on the num-ber of d-electrons. The curvature of the activa-tion energy dependence and the slope of the chemisorption energy dependence can be esti-mated from independent measurements of the

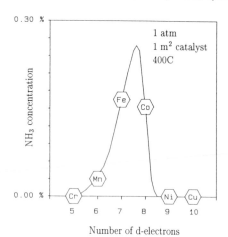

Fig. 21. The calculated ammonia production for a fixed set of reaction conditions as a function of the number of d-electrons. From Ref. [55].

sticking probability and chemisorption energy on other metals than Fe [55].

Given the kinetic model behind the results in Fig. 20 the ammonia production rate can then be calculated as a function of the number of d-electrons. The result in Fig. 21 shows that this produces a *volcano curve* with a maximum around seven d electrons, that is close to Fe. Fig. 21 quantifies the conventional picture of the origin of the volcano curves [56]. On the right-hand side of the maximum the ammonia production decreases because the rate of N_2 dissociation goes down as a consequence of the increase in the activation energy for dissociation in Fig. 16. To the left, on the other hand, the dissociation rate increases, but since the N chemisorption bond also increases in strength the number of free sites where N_2 can dissociate decreases so fast that the overall rate decreases. We are thus at a point where we can start saying something about the factors that determine the reactivity of a given surface.

9. Concluding remarks

In the present paper, I have tried to give a very brief review of the research relating to adsorption on metal surfaces that I have been involved in over my approximately fifteen years in research.

This work has to a large extend been done in collaboration with a number of extremely inspirering people. Let me mention, in cronological order, Bengt Lundqvist, Stig Andersson, Bengt Kasemo, Hans Hjelmberg, Olle Gunnarsson, Dennis Newns, Norton Lang, Art Williams, Matti Manninen, Martti Puska, Stephen Holloway, Bulbul Chakraborty, Peter Nordlander, Henrik Topsøe, Bjerne Clausen, Jens Rostrup-Nielsen, Eric Törnqvist, Per Stoltze, Karsten Jacobsen, Flemming Besenbacher, Ole H. Nielsen, Hans Skriver, Peter Ditlevsen, Ole Bøssing, Charlotte Ovesen, Bjørk Hammer, Lars Hansen, N. Chetty, Kurt Stokbro, J. Sethna, and Kent Gundersen.

References

[1] N.D. Lang and W. Kohn, Phys. Rev. B 1 (1970) 4555.
[2] N.D. Lang and A.R. Williams, Phys. Rev. B 18 (1978) 616.
[3] N.D. Lang, Surf. Sci. 299/300 (1994) 284.
[4] H. Hjelmberg, O. Gunnarsson and B.I. Lundqvist, Surf. Sci. 68 (1977) 158.
[5] H. Hjelmberg, B.I. Lundqvist and J.K. Nørskov, Phys. Scr. 20 (1979) 192;
J.K. Nørskov, A. Houmøller, P. Johansson and B.I. Lundqvist, Phys. Rev. Lett. 46 (1981) 257;
P. Johansson, Surf. Sci. 104 (1981) 510.
[6] D.M. Newns, Phys. Rev. 178 (1969) 1123.
[7] B.I. Lundqvist, Vacuum 33 (1983) 639.
[8] J.K. Nørskov and B.I. Lundqvist, Phys. Rev. B 19, (1979) 5561.
[9] J.K. Nørskov, J. Vac. Sci. Technol. 18 (1981) 420;
N.D. Lang and J.K. Nørskov, Phys. Scr. T6 (1983) 15.
[10] M.L. Yu and N.D. Lang, Phys. Rev. Lett. 50 (1983) 127.
[11] J.E. Lennard-Jones, Trans. Faraday Soc. 28 (1932) 333.
[12] G. Blyholder, J. Phys. Chem. 68 (1964) 2772;
W. Sesselmann, B. Woratschek, G. Ertl, J. Küppers and H. Haberland, Surf. Sci. 146 (1984) 17.
[13] H. Ibach and D.L. Mills, Electron Energy Loss Spectroscopy and Surface Vibrations (Academic Press, New York, 1982).
[14] J.K. Nørskov, Solid State Commun. 24 (1977) 691.
[15] J.K. Nørskov and N.D. Lang, Phys. Rev. B 21 (1980) 2131.
[16] J.K. Nørskov and F. Besenbacher, J. Less-Common Met. 130 (1987) 475.
[17] N. Esbjerg and J.K. Nørskov, Phys. Rev. Lett. 45 (1980) 807.
[18] J.K. Nørskov, J. Chem. Phys. 90 (1989) 7461.
[19] J.K. Nørskov, Phys. Rev. B 26 (1982) 2875.
[20] K.W. Jacobsen, J.K. Nørskov and M.J. Puska, Phys. Rev. B 35 (1987) 7423;

K.W. Jacobsen, Comments Condensed Matter Phys. 14 (1988) 129.

[21] M. Stott and E. Zaremba, Phys. Rev. B 22 (1980) 1564.

[22] S.B. Sinnot, M.S. Stave, T.J. Raeker and A.E. DePristo, Phys. Rev. B 44 (1991) 8927.

[23] M.S. Daw and M.I. Baskes, Phys. Rev. Lett. 50 (1983) 1285.

[24] M.W. Finnis and J.E. Sinclair, Phil. Mag. A 50 (1984) 45.

[25] F. Ercolessi, E. Tosatti and M. Parinello, Phys. Rev. Lett. 57 (1986) 719.

[26] J.K. Nørskov, Rep. Prog. Phys. 53 (1990) 1253.

[27] P.J. Feibelman and D.R. Haman, Surf. Sci. 182 (1987) 2291;
D.R. Hamann, J. Electron Spectrosc. Relat. Phenom. 44 (1987) 1.

[28] K.M. Ho and K.P. Bohnen, Phys. Rev. Lett. 59 (1987) 1833.

[29] F. Besenbacher and J.K. Nørskov, Prog. Surf. Sci., in print.

[30] P. Nordlander, S. Holloway and J.K. Nørskov, Surf. Sci. 136 (1984) 59;
B. Chakraborty, S. Holloway and J.K. Nørskov, Surf. Sci. 152/153 (1985) 660.

[31] J.K. Nørskov and K.W. Jacobsen, The Structure of Surfaces II, Eds. J.F. van der Veen and M.A. Van Hove, Vol. II of Springer Series in Surface Sciences (Springer, Berlin, 1987) p. 118.

[32] J.K. Nørskov, K.W. Jacobsen, P. Stoltze and L.B. Hansen, Surf. Sci. 283 (1993) 227.

[33] O.B. Christensen and K.W. Jacobsen, Phys. Rev. B 45 (1992) 6893.

[34] K.W. Jacobsen and J.K. Nørskov, Phys. Rev. Lett. 60 (1988) 2496.

[35] K.W. Jacobsen and J.K. Nørskov, Phys. Rev. Lett. 59 (1987) 2764.

[36] K.W. Jacobsen and J.K. Nørskov, Phys. Rev. Lett. 65 (1990) 1788.

[37] T.H. Upton and W.A. Goddard, Phys. Rev. Lett. 46 (1982) 1635;
P.S. Bagus, K. Herman and C.W. Bauschlicher, Jr., J. Chem. Phys. 80 (1984) 4378;
P.E.M. Siegbahn, M.R.A. Blomberg and C.W. Bauschlicher, J. Chem. Phys. 81 (1984) 1373;
J. Harris and S. Andersson, Phys. Rev. Lett. 55 (1985) 1583;

P.A. Schultz, C.H. Patterson and R.P. Messmer, J. Vac. Sci. Technol. A 5 (1987) 1061;
J. Muller, Phys. Rev. Lett. 59 (1987) 2943.

[38] P.J. Feibelman, Phys. Rev. Lett. 67 (1991) 461.

[39] B. Hammer, K.W. Jacobsen and J.K. Nørskov, Phys. Rev. Lett. 69 (1992) 1971;
B. Hammer, K.W. Jacobsen and J.K. Nørskov, to be published.

[40] D. Bird, to be published.

[41] H.F. Berger and K.D. Rendulic, Surf. Sci. 253 (1991) 325.

[42] J. Harris, to be published.

[43] C. Engdahl, B.I. Lundqvist, U. Nielsen and J.K. Nørskov, Phys. Rev. B 45 (1992) 11362.

[44] S. Holloway, B.I. Lundqvist and J.K. Nørskov, Proc. Int. Congr. on Catalysis, Berlin, 1984, Vol. 4, p. 85.

[45] For a review, see e.g.
(a) H.P. Bonzel, A.M. Bradshaw and O. Ertl, Eds., Physics and Chemistry of Alkali Adsorption (Elsevier, Amsterdam, 1989);
(b) D.A. King and P. Woodruff, Eds., The Chemical Physics of Solid Surfaces and Heterogeneous Catalysis, Vol. 6 (Elsevier, Amsterdam, 1993).

[46] P.J. Feibelman and D. Hamann, Phys. Rev. Lett. 52 (1984) 61.

[47] J.M. MacLaren, D.D. Vvedensky, J.B. Pendry and R.W. Joyner, J. Chem. Soc., Faraday Trans. I, 83 (1987) 1945.

[48] J.K. Nørskov, S. Holloway and N.D. Lang, Surf. Sci. 137 (1984) 65; 150 (1985) 24.

[49] N. Al-Sarraf, J.T. Stuckless and D.A. King, Nature, in press.

[50] O.B. Christensen and J.K. Nørskov, to be published.

[51] J.K. Nørskov in Ref. [45].

[52] A. Nielsen, An investigation of Promoted Iron Catalysts for the Synthesis of Ammonia (Jul. Gjellerup, Copenhagen, 1950);
B.S. Clausen, S. Mørup, H. Topsøe, R. Candia, E.E. Jensen and A. Baranski, J. Phys. (Paris) Colloq. 37 (1976) C6-245.

[53] G. Ertl, in: Crit. Rev. Solid State Mater. Sci. (CRC Press, Boca Raton, 1982) p. 349.

[54] P. Stoltze and J.K. Nørskov, Phys. Rev. Lett. 55 (1985) 2502.

[55] J.K. Nørskov and P. Stoltze, Surf. Sci. 189/190 (1987) 91.

[56] See, e.g., G.C. Bond, Catalysis by Metals (Academic Press, London, 1962).

Surface Science 299/300 (1994) 706–721
North-Holland

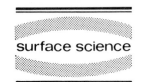

surface science

Adsorption on metals: a look from the not-too-far East

A.G. Naumovets

Institute of Physics, Academy of Sciences of Ukraine, Prospect Nauki 46, UA-252650 Kiev 28, Ukraine

Received 24 March 1993; accepted for publication 29 June 1993

This paper contains several essays on the issues which attracted much attention in studying adsorption on metals since the mid-1960's. Among them are the nature of adsorption interactions, adsorption/desorption kinetics, charge transfer, structural specificity in adsorption, 2D crystallography and phase transitions in overlayers, search for manifestations of these transitions in various surface properties of technological importance. Some emphasis is placed on the works in the former Soviet Union where lack of modern equipment was to be compensated by enthusiastic work on home-built apparatus (sometimes unique in construction and performance).

1. Prologue

Prior to praising the progress achieved in the last three decades in studies of adsorption on metals, it will be in order to render homage, even if briefly, to our outstanding predecessors whose works created prerequisites for the advances recorded in *Surface Science*.

I. Langmuir's pioneering investigations in the 1920–1930's discovered spectacular phenomena of charge transfer, work function changes and strong lateral interaction in adlayers on metals. S. Davison and L. Germer provided surface scientists with LEED, a powerful tool for explorations into surface structure. R. Gurney put forward a very productive model which gave a quantum-mechanical key to understanding the nature of chemisorptive bonding on metals. E. Müller invented field emission and field ion microscopes which allowed one to show very obviously the strong crystallographic specificity of adsorption, to study adsorption/desorption phenomena in high electric fields and to directly track individual adatoms. In the Soviet Union, rich experience in surface studies was accumulated in laboratories engaged in the research of electron and ion emission and development of electron tubes (S.A. Vekshinskii, P.I. Lukirskii, N.D. Morgulis, P.G.

Borziak, L.N. Dobretsov, N.I. Ionov, A.P. Komar, G.N. Schuppe). A far from complete list of the methods devised in these laboratories includes: a technique which allows one to simultaneously examine the properties of thin films of various thickness and/or composition ("wedge technique") [1,2]; photoelectron spectroscopy [3]; time-of-flight mass-spectrometry [4]; a multitude of surface ionization techniques [5].

In spite of these memorable efforts, the progress in adsorption studies prior to the 1960's was dramatically hampered by inaccessibility of ultra-high vacuum technology and of appropriate methods for surface characterization (actually only mass-spectrometry was available to check the chemical composition of surfaces). Starting conditions on the eve of the appearance of *Surface Science* were as follows: (1) most of the earlier data were suspected of being influenced by uncontrollable impurities; (2) quantitative works made with metal single-crystal substrates could be counted on one's fingers; (3) we were almost in the dark about the structure of adlayers and the phase transitions in them (altogether some 30 LEED works were made with adlayers on metals, the majority of them with oxygen on nickel [6]).

Along with a basic interest in adsorption phe-

SSDI 0039-6028(93)E0383-6

nomena (the nature of adsorption bonds, interactions between adsorbed species, phase transitions in two dimensions), intensive studies in this field were prompted by many practical reasons. It will be recalled that adsorption on metals represents an initial stage in processes which are widely used in various technologies (catalysis, production of ultra-high vacuum, activation of electron and ion emission, deposition of protective coatings etc.) or, in some other cases, are highly undesirable (corrosion).

Below I shall remind readers of some of the most notable advances made in studying adsorption on metals during the past 30 years. I apologize to those who may find my choice of these advances too subjective. However, *audi alteram partem* – hear the other side, the side which until recently was semi-isolated from the rest of the world and for this reason may have perceived some events in its own way.

2. Structural specificity in adsorption: going from polycrystalline to single-crystal substrates

Before the 1960's, substantial difference in adsorption properties of different crystal planes was demonstrated (predominantly by field emission microscopy) quite convincingly, but only on a qualitative level. To quantify information on structural specificity of adsorption, both microcrystals (tips) and macrocrystals were utilized. In the former case, field emission microscopes with probe holes were employed to measure the work function versus concentration of adatoms (degree of coverage). This technique proved to be especially effective in application to metal overlayers. The sticking probability of such adsorbates on metal substrates is equal to unity [7,8], which facilitates the determination of the absolute adsorbate concentration [9]. In particular, very graphic data were obtained by this method for lithium on low-index crystal planes of tungsten [10]. It was found that work function versus overlayer density curves are very diversified in shape: monotonically descending on W(111), showing a deep minimum on W(110) and quite complicated (including first a falling branch, then a wide ter-

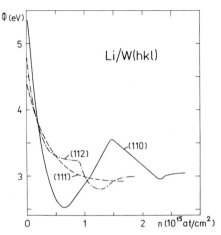

Fig. 1. Work function versus lithium concentration on tungsten crystal planes [10,11].

race and finally a nonmonotonic segment with a shallow minimum) on W(112). Subsequently these results were confirmed in the experiments carried out by a contact potential technique with tungsten macrocrystals (Fig. 1) [11].

Since macroscopic single crystals of refractory metals, which were favorite substrates by that time, were not readily available until the 1960's, some investigators made such crystals for themselves. For instance the method of recrystallization in an advancing temperature gradient was employed to convert polycrystalline tungsten ribbons to single crystal ones with a (113) surface orientation [12–14]. It should be noted that the first extended surface studies with macrocrystals were started in Tashkent where such crystals were produced by a metallurgical plant [15] (see also the later works of this group [16,17]). Once a broad assortment of metal single crystals became commercially available, intensive investigations of adsorption on such substrates started in many laboratories. It has since been demonstrated many times that the substrate atomic structure is of primary importance in adsorption phenomena. For example in the case of metals on metals, the coverage dependencies of the work function and heat of adsorption for different planes of a metal crystal are often much more distinct than the properties of structurally similar planes of different crystals [18,19] (Fig. 2). Structural sensitivity

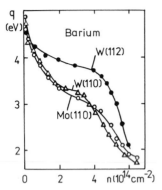

Fig. 2. Heat of adsorption of Ba on W and Mo crystal planes versus concentration of Ba adatoms [18].

was found to be even more pronounced in the case of gas adsorption on metals (see Section 5).

3. Adsorption of metals on metals: ionic? metallic?

The question about the nature of adsorption bonds is obviously of prime importance for understanding the whole variety of adsorption phenomena. On studying cesium on tungsten, Langmuir suggested that major characteristics of this and similar systems (a strong reduction of the work function with increasing coverage, the existence of a deep work function minimum, predominantly ionic evaporation of cesium adatoms at small coverages) are due to ionic bonding. R. Gurney refined this model quantum-mechanically by introducing the concepts of the virtual level and non-integer ion charge. Until the 1960's, only one experimental way was known to determine the dipole moment of an adsorption bond: the application of the Helmholtz formula to the data on the work function change. However, it was not clear in advance up to which coverage this procedure may be used, i.e. the work function change may be fully ascribed to the potential drop in the electrical double layer created by adatoms. Moreover, it was found that passing to ultra-high vacuum conditions resulted generally in a less deep work function minimum which sometimes nearly disappeared [8]. This gave grounds to some investigators to surmise that this

minimum would completely disappear under truly clean conditions and that, consequently, the polarity of the adsorption bond might be a much less important factor than it had been believed, even for electropositive adsorbates. In other words, it was proposed to interpret the coverage dependence of the work function as a trivial result of a gradual substitution of one (substrate) surface by another (adsorbate) surface having a different work function, without appreciable effects of charge transfer. This point was hotly debated, at least in the Soviet Union. It was therefore very desirable to find an independent method for the experimental evaluation of the dipole moment of adatoms.

Such a method was discovered almost simultaneously in the USSR and USA [20,21]. It was found that an inhomogeneous electric field created at a covered surface causes a drift of adatoms, the direction of which is reversed if the direction of the field is reversed. For instance electropositive species drift at small coverages to the top of a positively charged tip, but are expelled from the top if the tip is charged negatively (Fig. 3). The directions of the drift of electronegative adatoms, under otherwise indentical conditions, are opposite. Some time later, it was also found that the drift directions are opposite at the adatom concentrations below and above the value corresponding to the work function minimum (Fig. 3) [22]. This behavior of adatoms was attributed to the effect of the electric field on the energy of adsorption. If the adsorption bond is characterized by a dipole moment p and polarizability α, then the change of the adsorption energy is equal to $(pF) + \alpha F^2/2$, with F being the field strength [23]. The linear term is dominant in the effect considered here. A detailed evaluation of the experimental data in the framework of this model allowed one to extract information on the dipole moment and its change with increasing coverage [24]. The obtained results agreed satisfactorily with the dipole moments determined from the work function changes. This has lent a strong support to the viewpoint that the polarity of the adsorption bond does play an important role, at least when the electronegativities of the adsorbate and substrate differ suffi-

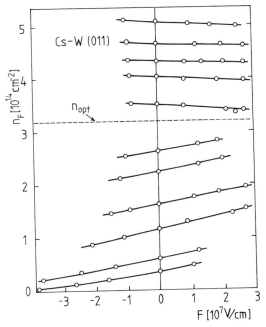

Fig. 3. Effect of electric field on equilibrium concentration of Cs adatoms on the (001) plane of a tungsten tip. n_{opt} is Cs concentration at the work function minimum [24].

ciently. It may be added that a clear-cut asymmetrical field effect has been found also in the activation energy of surface diffusion of adatoms [25].

These experimental works were being performed in parallel with theoretical studies on the nature of adsorption bonds among which the investigations of N.D. Lang deserve special mention (see also his article in this volume [114]). He made use of a jellium model for the description of both the substrate and an electropositive ad layer and, in spite of its apparent simplicity, was able to show very graphically strong charge transfer effects at small coverages and nearly complete "metallization" of the adlayer at coverages approaching a monolayer [26]. The work function minimum is tied to the transition region between these two regimes [27]. It is interesting to note that the mechanisms and conditions of metallization of ad layers have been continuing to attract considerable attention up to the present days, and the discussion on the point whether the adsorption of electropositive species is ionic or co-

valent has recently flared up with a new strength [28]. Luckily, the adherents of the covalent bond admit it to be "strongly polarized" so the controversy is rather reduced to the question where the border-line should be traced between the substrate and adatom. What is really objective, is the charge density spatial distribution which is characterized by the existence of some dipole moment per adatom influencing the work function and manifesting itself in the lateral interaction of adatoms. The last three decades was the time of intensive research into the mechanisms of this interaction, too.

4. Indirect lateral interactions, or substrate as a communicating facility for adsorbed species

The lateral interaction of adsorbed particles is one of the most important factors which makes adsorption phenomena so varied and exciting. Indeed, it determines the formation of two-dimensional adsorbate structures, phase transitions in them, the coverage dependencies of the work function, heat of adsorption, kinetics of adsorption and surface diffusion, catalytic activity, etc. Information on lateral interactions was quite scarce at the outset of *Surface Science*.

From the theoretical standpoint, three kinds of lateral interaction were considered: (i) the van der Waals interaction (universal in its manifestations, but really important predominantly in physisorbed layers); (ii) electrostatic interaction (dipole–dipole at large distances) essential in the case of strongly polar adsorption bond, and (iii) short-range exchange interaction operating between nearest-neighbor adatoms. In 1958, Koutecky [29] predicted the existence of the lateral interaction via the electrons of the substrate (indirect interaction). This idea, however, had not attracted much attention until the appearance of the works of Grimley and his coworkers [30,31]. Since then a lot of theoretical studies have been done which have revealed an important role of the indirect electron interaction in adsorption on metals, and investigated its dependence on distance and electronic structure of the substrate (see, e.g., the reviews in Refs. [32,33]).

Experimental data seem to have confirmed such predicted peculiarities of this interaction as strong anisotropy correlating with the shape of the Fermi surface of the substrate, oscillatory behavior and long-range character. The bulk of quantitative information on these points has so far been extracted from experiments performed with the aid of the field ion microscope (see G. Ehrlich's review in this volume [115] and Refs. [34–37]). With this device one is able to visually follow the movements of individual adatoms, to determine the probability $P(\mathbf{R})$ of finding two atoms separated by a specified vector $\mathbf{R}$ and hence to derive the interaction energy of the pair of adatoms. Such observations may well be placed among the top-level achievements of modern experimental physics.

Another approach to gaining information on lateral interactions relied on analysis of the symmetry of two-dimensional adatom lattices and evaluation of phase transitions in them. For instance, it was found that the same adsorbate produces distinct sets of two-dimensional lattices on substrates which have different chemical nature, but nearly identical atomic structure. Such comparison was made, e.g., for alkaline-earth atoms on W and Mo whose lattice constants differ by no more than 0.6% [38–40]. These results manifested the existence of lateral interaction which is anisotropic and dependent on the substrate. The indirect electron interaction seems to be the most probable mechanism which could be responsible for such features. Quantitative data on lateral energies for a number of systems have been obtained by fitting simulated data on phase transitions to experimental ones (see, e.g. Ref. [41]).

One of the interesting observations was that strong anisotropy of the substrate atomic corrugation results in pronounced orientational dependence of lateral interaction. For example, highly anisotropic "chain" structures are formed by many adsorbates on surfaces with furrowed potential relief like bcc(112), fcc(110) and hcp(10$\bar{1}$0) surfaces [33,40]. Fig. 4 shows one of such structures, Mo(112)-(1 × 9)Sr, in which the ratio of the periods along and across the furrows amounts to ~ 5.5 [42]. The wide spacing between the chains

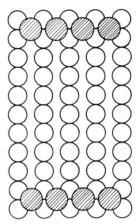

Fig. 4. A p(1 × 9) structure of a Sr adatoms on Mo (112) [42].

(24.6 Å in Fig. 4) is attributed to superposition of the dipole–dipole (repulsive) and indirect (oscillatory) interaction [33]. The latter is supposed to decay as r^{-1} since the Fermi surface of molybdenum has flattened segments perpendicular to the [111] axis directed along the furrows.

The electron indirect interaction is not the only possible kind of substrate-mediated interaction. It has been realized that all sorts of quasiparticles excited in a solid can effect exchange interactions between adatoms, but of course of differing strengths. For example, recently there has been a growing interest in the processes of mesoscopic self-organization in surface structures, with a scale ranging from tens of ångströms to approximately micrometers. In this phenomenon, a continuous (uniform) surface layer is separated into periodically repeating domains of different density (and structure) or orientation. The resulting energy gain is determined by reduction of the elastic energy which is associated with the strain induced in the substrate by an adsorbed layer. This reduction more than compensates the energy expenditure necessary to form the domain walls [43].

Another kind of the driving force of mesoscopic self-organization was predicted to operate in adlayers which cause strong change in the work function. In this case the formation of domains with differing adsorbate density (and work function) allows one to reduce the electrostatic energy associated with contact potential fields [44,45].

The effects discussed above are assumed to explain the development of striped domain structures with a periodicity of about 100–150 Å detected quite recently in the oxygen adlayers on Cu(110) [46]. One might anticipate that mesoscopic self-organization processes on surfaces will be a promising line of investigation in the immediate future.

5. Down the temperature scale. Hunting for long-period structures and precursor states

As stated in the previous section, interactions of adparticles on metals have been a major focus of interest in the last three decades. In particular, theoretical predictions of long-range lateral forces stimulated attempts to experimentally verify their existence. For instance it was interesting to reveal manifestations of the dipole–dipole interaction which was foretold to be dominant in the case of strongly polar bonding. It is clear that lateral repulsion must result in the formation, at low coverages, of two-dimensional lattices with large interatomic spacings. However, the interaction energy at large spacings should be anyway rather weak so the disordering temperatures of such structures were expected to be low. This reasoning posed quite naturally the problem of LEED investigations at temperatures well below room temperature, but it was necessary to overcome the erroneous belief that nothing interesting would happen on cooled surfaces because of low mobility of adsorbed species.

The fact that, for many years, commercial LEED equipment was not fitted with cooling facilities was a sort of reflection of this opinion. Our laboratory had no such equipment at its disposal, but this cloud had a silver lining, and we have built an all-glass LEED apparatus for ourselves which first provided cooling of the samples with liquid nitrogen [47] and, a few years later, with liquid helium as well [48]. In so doing we relied on our long experience with observations of various overlayers in the field emission microscope which showed that at least alkali and alkaline-earth adatoms have quite a noticeable mobility on close-packed surfaces even at liquid nitro-

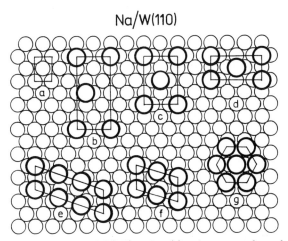

Fig. 5. A unit cell of W(110) surface (a) and structures formed by Na adatoms on W(110) (b–g) [47,49].

gen temperature. This gave good reason to expect that rarefied two-dimensional structures could be observable in an equilibrium ordered state. Indeed, as the coldfinger holding a crystal was filled with liquid nitrogen, one got the impression that the scales fell from one's eyes: quite a number of structures disordered at room temperature produced sharp diffraction spots upon cooling. The first experiments of this type were made in 1969–1970 with Na and Cs on W(110) [47,49,50] (Fig. 5). Since then much LEED data has been obtained for various systems at low temperatures which revealed clear manifestations of long-range repulsive forces in two-dimensional lattices (see, e.g., the reviews in Refs. [39,40,51–53]). For example, in electropositive adlayers on close-packed metal surfaces a number of structures were found with the nearest neighbor spacing of ∼ 10 Å [47–50,52–54]. It has been claimed recently that potassium adatoms on graphite can arrange a lattice with the shortest interatomic distance of ∼ 60 Å [55].

An unexpected result was the observation of the first-order phase transitions in the adlayers with a pronounced repulsive interaction [49]. An explanation of this finding was suggested in the framework of a depolarization model [56]. Since that time many such transitions have been observed in strongly polar adlayers [33,39,40,52,53]. Recently, attention to this issue has been attracted again in connection with the prediction of mesoscopic self-organization in these systems (see Section 4).

Another important goal pursued in the low-temperature studies of adsorption on metals was the elucidation of the mechanisms of adsorption of gases. A deep insight into kinetics of gas adsorption is of the utmost significance for catalysis, vacuum technologies, corrosion protection, materials science (e.g. hydrogen embrittlement), power generation (storage of hydrogen in metals) etc.

To explain the constancy of the sticking probability of gas molecules observed in a wide range of coverages, Langmuir introduced the concept of the weakly bound precursor state in which the molecule is temporarily trapped (over a filled site) prior to coming to the chemisorption state [57]. In 1957, Kisliuk [58] extended this concept assuming the precursor states to be possible over empty adsorption sites as well, but these important predictions had to wait for solid experimental substantiation until the 70's–80's. Since that time, many data on coverage and temperature dependence of the sticking probability have been obtained for various gas/crystal pairs. Such results are exemplified in Fig. 6a for oxygen on

W(110) in the temperature range 5–300 K [59,60]. The data have been taken with a molecular beam technique and give evidence for oxygen adsorption via an intrinsic (existing over empty sites) precursor state which, judging from thermal desorption spectra, may correspond to physisorption or weak molecular chemisorption of oxygen. The behavior of the sticking probability of O_2 on W(100) is qualitatively different (Fig. 6b) and favors a "direct" chemisorption of oxygen into the first monolayer without intermediate trapping into precursor states.

The most informative data on the mechanisms of gas adsorption have been obtained with the aid of molecular beam techniques in which a surface is exposed to a monoenergetic beam and one is able to measure both the angular and energy distribution of reflected and desorbed molecules. The orientation of the beam relative to the surface can be changed, the incident molecules can be excited to various states with a laser, and the results of the measurements can be correlated with electron-spectroscopic and structural information. We will not enter into further particulars and refer the readers to the reviews prepared for this volume by the investigators in this area. The results obtained to date have uncovered great diversity and exciting profundity of dynamical processes in adsorption on metals.

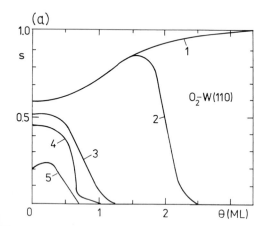
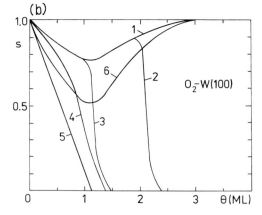

Fig. 6. Sticking probability of oxygen versus coverage (in monolayers). (a) W(110), crystal temperatures T_{cr} (in K) are 5(1), 25(2), 60(3), 78(4) and 300(5), O_2 temperature $T_g = 300$ K; (b) W(100), T_{cr}(K) are 5(1), 35(2)), 78(3), 200(4), 350(5), 5(6). T_g is 200 K for (1–5) and 550 K for (6) [59,60].

6. "Black chambers" for adsorption / desorption studies

In the elucidation of gas–solid interaction as well as adsorption and desorption kinetics, a great body of information has been gained by thermal desorption spectroscopy (TDS). TDS has been evolved from the so-called flash-filament method [61] proposed to determine the amount of the adsorbed gas from the pressure rise caused by a sharp heating of the sample. Now TDS is one of the well-founded methods to study adsorption/desorption kinetics and chemical transformations experienced by particles interacting with a surface [62]. However, in the mid-1960's very controversial results were obtained in TDS investigations of the interaction of oxygen with refractory metals. For example, some workers recorded mass-spectrometrically that oxygen was desorbed from tungsten in the form of O_2 molecules while others observed desorption of atomic oxygen, various tungsten oxides and (only in a minor quantity) of molecular oxygen [63,64]. Similar results were obtained with oxygen on molybdenum.

To unravel this puzzle and to separate desorption effects from artifacts due to reactions on the walls of the vacuum chamber, Ptushinskii and Chuikov designed an apparatus which had some special features (Fig. 7): (1) The sample under study was placed within a vacuum chamber whose walls were covered with an active sorbent (freshly evaporated titanium film) and cooled with liquid nitrogen. Such a "black chamber" irreversibly bound all molecules incident on its walls. (2) Desorption products were analyzed with a high-speed time-of-flight mass-spectrometer. The sample was installed immediately in front of the ion source of the mass-spectrometer (line-of-sight configuration), so the products were detected during their single flight through the ionization space.

Experiments carried out within this chamber showed that in the case when oxygen is present on the tungsten (110) surface at a concentration not exceeding 7×10^{14} atoms per cm^2, it is completely evaporated in the atomic form. Oxygen in excess of this concentration is incorporated into tungsten oxides which are desorbed at signifi-

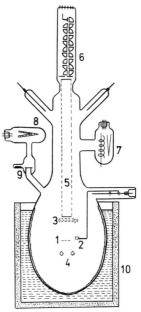

Fig. 7. "Black chamber" for desorption studies [64]: (1) substrates; (2) movable electron gun; (3) ionizator; (4) Ti evaporators; (5) ion flight space; (6) electron multiplier; (7) vacuum gauge; (8) O source; (9) vacuum valve; (10) vessel with liquid N_2.

cantly lower temperatures thus causing a corrosive ablation of the substrate material. As to O_2 molecules, their fraction in the desorption products was found to be below 1%. It was therefore concluded that O_2 molecules detected in abundance in earlier TDS experiments within "non-black" chambers were due to a secondary chemical reaction: recombination of desorbed oxygen atoms on the chamber walls and subsequent release of O_2 into vacuum [64–66].

This episode from the history of surface science showed very clearly how many reefs threaten experimentalists in their attempts to gain undistorted information about surface processes. "Black" vacuum chambers simulating free space in the laboratory are of crucial importance in experiments with molecular beams which are expected to reveal unknown subtleties of the mechanisms of adsorption and desorption. It is pertinent to note that essentials of the molecular beam technique as applied to determination of adsorption and desorption rates, adatom concentration and sticking probability have been elaborated by Zingerman et al. in the early 1960's [8,67].

7. Two-dimensional phases. What then?

In today's three-dimensional materials science, no material is recommended for use in any critical structure, which is calculated to be efficient and long-living, without a close examination of its phase diagram. Although surface science and surface technologies have not yet reached such a level of maturity, the last three decades have brought a great body of information on the correlation between the phase state of metal (clean or covered) surfaces and their various physico-chemical properties. Most of the structural information has been gained with the aid of LEED. This method attracted enormous interest in the early 60's, mainly thanks to the appearance of the apparatus with visual indication of LEED patterns [68,69] and under the impact of J.J. Lander's milestone review published in *Surface Science* [70].

One of the oldest points under study was the dependence of the work function upon coverage degree and atomic structure of an adsorbed layer. Taylor and Langmuir [71] investigated this dependence for cesium on tungsten and stated that the work function minimum is reached at a submonolayer coverage ($\theta \approx 0.67$), but they had no structural data. The concentration dependencies of the work function measured for single-crystal substrates in the mid-sixties provided some evidence that the work function variation does correlate with the adlayer structure. First, the dependencies obtained for different planes of the same crystal generally differ rather strongly, and in some cases are even qualitatively distinct (see, e.g., Fig. 1). Second, the peculiarities seen in the work function curves (minima, breaks, steps, etc.) in many cases were found to closely correspond to coverage degrees expressed by simple fractions like 1/4, 1/3 or 1/2 [72,73]. This suggested that they might be tied to the formation of some adatom lattices commensurate with the substrate structure. However, the first attempts to verify this hypothesis in LEED experiments were con-

fusing: for example, no ordered structures were detected at coverages corresponding to the work function minima in the case of alkali metals adsorption on the (111), (110) and (100) surfaces of nickel [74].

Two observations helped to solve the riddle. First, the early LEED observations were made only at room temperature at which most of the alkali metal adlayers (except for rather dense ones) are disordered. Experiments carried out at liquid nitrogen temperature did generally confirm the appearance of distinct adsorbate structures predicted at the work function minima on the basis of the measured stoichiometries [47,49,75]. Second, careful measurements have shown that if a homogeneous (single phase) alkali metal adlayer undergoes an order–disorder transition, the work function remains constant to within $\sim 10^{-2}$ eV (the decrease in the work function caused by the deposition of such adlayers amounts to several eV) [47,49,76]. These findings were interpreted as follows: the work function is sensitive mainly to short-range order which is retained in the (long-range) order–disorder transition and bears a pronounced imprint of the substrate structure [76]. This conclusion has been corroborated by experiments carried out with heterogeneous (two-phase) adlayers. When such an adlayer is heated and changes to a one-phase disordered state, a distinct short-range order is established and the work function shows an appreciable variation. Utilizing this effect, Kolaczkiewicz and Bauer [77] developed a sensitive method of detection of surface first-order phase transitions.

Hence it is evident that the knowledge of the phase diagrams of adlayers is essential both for the physical understanding of the behavior of the work function versus coverage and for an expedient choice of the operating conditions of a surface providing a desired work function. A similar statement holds for the energy of adsorption (see, e.g., Refs. [19,39,40,78]). It is notable that, at sufficiently low temperatures, even the electrical resistance of a metal sample is sensitive (via surface scattering of electrons) to the structure of an adsorbed layer [79].

With the advancement of photoelectron and low-energy electron loss spectroscopies, there

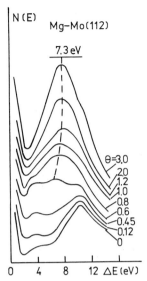

Fig. 8. Evolution of electron energy loss spectra with increasing Mg coverage on Mo(112). Mg surface plasmon peak (7.3 eV) emerges as θ surpasses the value 0.5 corresponding to a p(1×2) structure [85].

came the time when the problem of interrelation between the surface electronic structure and surface phase transitions was attacked. Here we shall only note that, in practical terms, the problem "surface electronic structure versus surface phases" is approached from two directions.

First, one is often interested in a tunable modification of the electron properties of a surface. This aim can be attained by deposition of a suitable adsorbate, usually in submonolayer amounts (see, e.g., Fig. 8). An important case is promotion of catalysts. A great deal of effort has been going into the investigation of its detailed mechanisms ([80,81], see also the reviews of D. Goodman and G. Somorjai in this volume [116]). Further examples are: activation of the thermionic emission from film-coated cathodes, comprising metal substrates covered with electropositive ad layers, through the addition of controllable amounts of electronegative species [82,83]; "tuning" of the electronic structure of electropositive ad layers to maximize the photoemission yield [27,84,85]. In all the above-mentioned cases, the most beneficial surface properties (with respect

to a specified goal) correspond to some optimum phase state of an adsorbed layer.

Another possible approach to attaining desired surface electronic properties is to use an ultrathin (one–two monolayers) film deposited on an appropriately chosen substrate. The properties of such films are strongly influenced by the substrate, both due to electronic effects and through modification of their atomic structure, and may differ very substantially from the surface properties of corresponding bulk samples (see, e.g., Refs. [86,87]). This line of research, called wittily "modern alchemy" [88], is closely related to the studies of epitaxy and growth mechanisms of thin films. In the case of metals on metals, these problems were investigated most comprehensively by E. Bauer and his coworkers (see, e.g., Refs. [19,89,90]). A review of advances in understanding crystal growth mechanisms is presented in this volume by J. Venables [117]. A striking result obtained in the studies of lateral interactions, growth modes and epitaxy for metals on metals is the strong diversity of the observed behavior even as one compares very similar adsorbates (e.g. gold and silver) or substrates (like planes of tungsten and molybdenum), not to mention different planes of the same crystal. This emphasizes considerable potentialities in tailoring surface electron properties within the approach considered.

A rapidly expanding research field is investigation of magnetic properties of ultra-thin films grown on metal substrates [91]. The stage for it has been set by the development of powerful techniques enabling one both to measure the magnetization and to visualize magnetic domain structure of such films [92]. The magnetization of the ultra-thin films prepared under specified conditions has been found to lie perpendicular to the film plane. Thus the hopes are pinned on such films in efforts to implement a high-density magnetic storage of information.

Let us now touch briefly on a further phenomenon whose intimate connection with surface phase transitions has been observed and investigated in the last three decades: surface diffusion. We shall not consider the diffusion of individual adatoms which is discussed in detail by G. Ehrlich in this volume [115]. Instead, we concentrate on

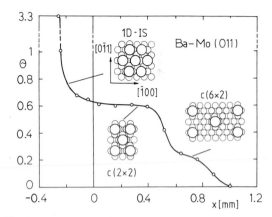

Fig. 9. A coverage profile formed in diffusion of Ba out of initial step-like profile on Mo(011). c(6×2), c(2×2) and one-dimensional incommensurate Ba structures are shown [93,94].

surface diffusion at medium and large coverages when rather dense two-dimensional phases of adatoms appear on a surface. If one deals with diffusion proceeding in the presence of an adsorbate concentration gradient (surface heterodiffusion), the phases emerge immediately in the course of the diffusion. Such effects (multiphase diffusion) have been known and examined over many years in the interior of crystals, but it was only comparatively recently that their surface counterparts were revealed (for a review, see Ref. [93]). It has been realized that under the conditions of local equilibrium, the concentration profile (coverage versus distance) corresponding to a stage of surface diffusion represents a display of all two-dimensional adsorbate phases which arise at various coverages and a given diffusion temperature.

Due to strong lateral interactions the diffusivity (and even the diffusion mechanism) vary with coverage, and this variation is reflected in the shape of the coverage profiles (Fig. 9). In particular, the flattened sections ("terraces") correspond to coverage intervals with the highest diffusivity. Of special interest are diffusion mechanisms which are expected to operate at sufficiently high coverages and involve concerted movements of a number of adjacent adatoms. One of such mechanisms is predicted for the transition region between commensurate and incommensurate structures. It is suggested that incommensurate do-

main walls (solitons) existing in this region may act as mass carriers [93–96]. Evidently, the phase sensitivity of surface diffusion must affect the kinetics of all surface processes which comprise a diffusion stage.

The aim of this section was to exemplify the relationship "surface properties versus surface structure" utilizing the data obtained with metal-based adsorption systems. It is safe to forecast that future high technologies relying of surface phenomena will not be able to dispense with careful consideration of surface phase diagrams.

8. Surface ionization on metals. From Langmuir–Saha to Ionov–Zandberg

In this section we shall dwell briefly on the studies of surface ionization, or ion desorption, an area of research which has been not so popular in the West, but intensively explored in the former Soviet Union. They have been concentrated mainly in the A.F. Ioffe Physico-Technical Institute in Leningrad (now again Saint Petersburg). This city has long-standing traditions in emission electronics built up not only in the A.F. Ioffe Institute, but also at the University, Polytechnical Institute, "Svetlana" works and other laboratories. Initiated by A.A. Lebedev, P.I. Lukirskii and S.V. Vekshinskii, investigations of emission phenomena were continued by L.N. Dobretsov, V.M. Dukel'skii, N.I. Ionov, A.P. Komar, A.R. Shul'man, I.L. Sokol'skaya, E.Ya. Zandberg (this list of the names is not complete).

The works on surface ionization (SI) in the 1960's focused on its characteristics in the case of inhomogeneous surfaces and in the presence of electric fields (up to the field strengths at which "conventional" field desorption sets in). Thermal equilibrium between the surface and evaporated ions was examined (this point attracted special interest in the case of many-atom molecules) [97]. Threshold and hysteresis effects occurring in SI at some critical coverages were also studied in detail (these works have been made in Kharkov by Chaikovskii et al. [98]). This basic research has created a solid foundation for developing many SI techniques meant for surface characterization.

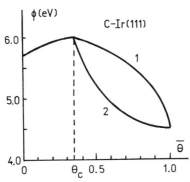

Fig. 10. Work function of Ir(111) versus average carbon coverage. (1) and (2) are effective ϕ values determined by surface ionization and thermal electron emission methods [104].

Ionov, Zandberg and their coworkers have proposed altogether about thirty SI-based methods for examining various surface properties and processes [5,97,99–103]. Some of them have important advantages, and below a few examples will be considered.

First of all, it has been realized that the data on SI must be evaluated with proper allowance for inhomogeneity of the surface. The work function nonuniformity intervenes in SI immediately, in accordance with the Langmuir–Saha formula, and via contact electric fields which (i) retard the ions emitted from the patches with the highest emission and (ii) cause the Schottky effect at the low-emission patches. To judge the work function nonuniformity (ϕ − contrast) of a surface, it will suffice to compare the thermal electron current stemming predominantly from the low-ϕ patches and the current of positive ions which, in the case of atoms that are hard to ionize (with the ionization energy > ϕ), are generated mainly at the high-ϕ patches. Of course it is important to mass-analyze the ions in order to exclude spurious effects which may be caused by easily ionizable impurities.

The most interesting results are obtained by this method in studies of the first-order phase transitions in overlayers, although the possibility of characterization of clean surfaces by it is obviously also important. The most salient feature of this method is its applicability at high temperatures when both coexisting two-dimensional

phases may be disordered and therefore cannot virtually be examined by such techniques as LEED, STM, etc. Fig. 10 depicts an example of the first-order phase transition tracked by measuring the work function contrast in a carbon adlayer on Ir(111) [104]. The transition sets in at $\theta_c = 0.35$, and a two-dimensional carbon gas ($\phi = 6.0$ eV) coexists with a condensed (graphite-like) phase ($\phi = 4.5$). This method can readily be applied to investigate the kinetics of growth or 2D sublimation of the densely packed islands and thus to determine the corresponding 2D binding energy in the islands and the activation energy of surface diffusion (such data were also obtained earlier for some systems by Shrednik and his collaborators at the A.F. Ioffe Institute who utilized field emission microscopy [105]).

In the event that the ionized particles arise from molecules due to a catalytic reaction of dissociation on a surface, the ion current can be used to estimate the catalytic activity of the surface with respect to this reaction. One can also investigate the effect of various factors (e.g. surface additives or impurities) upon the activity.

In particular, this technique has been employed to elucidate the role of adsorbed carbon in the activity of metal catalysts used in heterogeneous dissociation reactions of molecules [106]. CsCl molecules were chosen as the probe and Ir(111) as the catalyst. On the clean Ir(111) surface, CsCl molecules were found to dissociate with probability close to unity and produce Cs^+ ions (Fig. 11). The 2D carbon gas ($\theta < 0.35$) does not virtually affect the dissociation since the ion

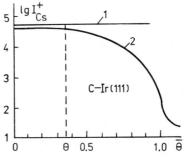

Fig. 11. Cs ion current versus average C coverage on Ir(111) in the case of surface ionization of Cs atom beam (1) and of Cs atoms originating from dissociated CsCl molecules (2) [106].

current is found to remain nearly the same. However, the 2D condensed carbon phase having a graphite-like structure acts as a catalyst poison, as signaled by a drastic decrease of I_{Cs^+}.

SI measurements combined with other techniques (thermal electron emission, TDS, AES, etc.) allowed one to detect at least four carbon forms on metal surfaces (single carbon adatoms, clusters, carbides and graphite) which reveal strongly differing properties. Especially 2D graphite islands on Ir, Pt, Ni and other metals manifest interesting characteristics. In particular it has been suggested [101,103] that each island resembles an overturned saucer in shape: while the central part of the island is physisorbed and somewhat elevated, its edge atoms are chemisorbed and lowered down to have a closer contact with the substrate. Foreign adatoms can be trapped within the cavity under the island. Reactive properties of the edge and central carbon adatoms in such islands are inferred to be quite distinct. A comprehensive review on this issue has been published recently by Tontegode [103].

A highly fruitful research field was discovered in SI studies when it was found that many organic molecules are readily ionized on solid surfaces [107]. In fact, in most cases not the primary molecules incident on a surface are ionized with the highest probability, but rather the products of their reactions on the surface (fragments and some associates, predominantly with hydrogen) [97]. On the whole, however, the SI spectra are much simpler than those obtained with electron impact ionization, which offers an important advantage in analysis of organic compounds. Especially low ionization energies (comparable with such for alkali metals) are characteristic of amines, hydrazines and other organics containing heteroatoms from the Va group. Efficient SI sensors have been developed to detect these adverse compounds in air for ecological and industrial safety purposes. For example, ternary alkilamines and their derivatives can be "smelt" in air in concentrations of $\sim 10^{-14}$–10^{-15} g/l which is close to a dog's olfactory sensitivity [97].

Examination of SI of organic compounds provides ample information on surface chemical reactions. In the last thirty years it has grown into a rather broad cross-disciplinary area at the junction of surface physics, emission electronics and organic chemistry. Works on it are now advancing not only in Saint Petersburg, but also in U. Arifov Institute of Electronics in Tashkent [108] and in some other laboratories (see, e.g., [109,110]).

Quite recently, SI has found a very promising application in studies of photodissociation of molecules on solid surfaces [111]. In cases when one of the fragments of the dissociated molecule is readily ionizable on the surface (in contrast to the original molecule), the acts of photodissociation can be recorded with a unique sensitivity. For example, in examining a model photodissociation reaction of cesium halides on a graphite monolayer adsorbed on Ir, Zandberg et al. [111] have attained a SI detection sensitivity about 10^6 times higher than that achieved by other methods. This enabled them, utilizing a low-power laser, to investigate the photodissociation (PD) of single admolecules on an opaque (conductive) substrate, to measure the spectral and temperature dependence of PD, to find the change in the PD energy caused by the substrate, to determine the orientation of the adsorbed molecules from polarization measurements etc. It is evident that this SI-based technique has large potentialities in application to PD of many organic admolecules, too.

Finally, worthy of mention is the possibility of SI detection of neutral atoms arising in desorption induced by electron transitions [112,113]. For easily ionizable particles, this method provides a very high sensitivity and allows one to investigate in detail the neutral component of desorbed species whose behavior, until recently, has remained poorly understood (see also the paper of T.E. Madey in this volume [118]).

9. Concluding remarks

Within the last thirty years our knowledge of adsorption on metals has been largely enriched. It has been fully realized on an enormous experimental basis that, coming from the gas phase to a surface, an atom or molecule enters a peculiar two-dimensional world with its own gases, liquids and crystals (probably liquid crystals, too) and laws governing transitions between them. The variety of these states is striking and exciting, and has revealed a surprisingly intricate character of surface interactions. In the spear-head were the investigations on the nature of the interactions, on the overlayer structure and phase transitions, on structural manifestations in adsorption/desorption kinetics and all possible properties of adparticles as well as surfaces covered by them. The results obtained have found numerous applications, from cathodes, ion sources and getter pumps to catalysts, gas detectors and microelectronic devices. Furthermore, they provide the basis necessary for the development of new technologies in the era of low-dimensional systems and operations with individual atoms and molecules.

Prophesying in science is a risky business. However, it is probably rather safe to predict that investigations of dynamical processes in adsorption, desorption, surface diffusion, phase transitions and all kinds of surface excitations will continue to be in the mainstream of surface studies.

Happy birthday and good dynamics to you, *Surface Science*!

Acknowledgements

I am grateful to C.B. Duke, the Editor of Surface Science, for honoring me with the invitation to contribute to this commemorative volume. I also thank Yu. Ptushinskii for useful comments as well as V.N. Bykov, V.V. Cherepanov and A.A. Marchenko for their help in the preparation of the typescript.

References

[1] S.V. Vekshinskii, A New Method for Metallographic Investigation of Alloys (OGIZ Editors, Moscow, 1944), in Russian.

[2] N.D. Morgulis, P.G. Borziak and B.I. Dyatlovitskaya, Izv. Akad. Nauk SSSR, Ser. Fiz. 12 (1948) 126.

[3] P.G. Borziak, Tr. Inst. Fiz. Akad. Nauk Ukr. SSR 2 (1952) 3.

[4] Ye.I. Agishev and N.I. Ionov, Zh. Tekh. Fiz. 28 (1958) 1775.

[5] E.Ya. Zandberg and N.I. Ionov, Poverkhnostnaya Ionizatsiya (Nauka, Moscow, 1969), in Russian. English translation: E.Ya. Zandberg and N.I. Ionov, Surface Ionization (Israel Program of Sci. Translations, Jerusalem, 1971).

[6] T.W. Haas, G.J. Dooley, III, J.T. Grant, A.G. Jackson and M.P. Hooker, Prog. Surf. Sci. 1, Pt. 2 (1971) 155.

[7] Yu.G. Ptushinskii, Zh. Tekn. Fiz. 28 (1958) 1402.

[8] Ya.P. Zingerman, V.A. Ishchuk and V.A. Morozovskii, Fiz. Tverd. Tela 3 (1961) 1044.

[9] V.N. Shrednik and E.V. Snezhko, Fiz. Tverd. Tela 6 (1964) 1501, 3410.

[10] V.M. Gavrilyuk and V.K. Medvedev, Sov. Phys. Solid State 8 (1966) 1439.

[11] V.K. Medvedev and T.P. Smereka, Sov. Phys. Solid State 16 (1974) 1046.

[12] J. Eisinger, J. Chem. Phys. 29 (1958) 1154.

[13] V.M. Gavrilyuk and V.K. Medvedev, Fiz. Tverd. Tela 4 (1962) 2372.

[14] V.S. Ageikin, Yu.G. Ptushinskii and B.P. Polozov, Fiz. Tverd. Tela 12 (1970) 221.

[15] G.N. Schuppe, Electron Emission from Metal Single Crystals (SAGU Editors, Tashkent, 1959), in Russian.

[16] N.A. Gorbatyi, L.V. Reshetnikova and V.M. Sultanov, Fiz. Tverd. Tela 10 (1968) 1185.

[17] A. Khakimov and N.A. Gorbatyi, Zh. Tekh. Fiz. 48 (1978) 621.

[18] L.A. Bolshov, A.P. Napartovich, A.G. Naumovets and A.G. Fedorus, Sov. Phys. Usp. 20 (1977) 432.

[19] E. Bauer, in: The Chemical Physics of Solid Surfaces and Heterogeneous Catalysis, Vol. 3B, Eds. D.A. King and D.P. Woodruff (Elsevier, Amsterdam, 1984) p. 1.

[20] V.M. Gavrilyuk and A.G. Naumovets, Sov. Phys. Solid State 5 (1963) 2043.

[21] L.W. Swanson, R.W. Strayer and F.M. Charbonnier, Surf. Sci. 2 (1964) 177.

[22] E.V. Klimenko and A.G. Naumovets, Surf. Sci. 14 (1969) 141.

[23] M. Drechsler, Z. Elektrochem. 61 (1957) 48.

[24] E.V. Klimenko and A.G. Naumovets, Sov. Phys. Solid State 13 (1971) 25.

[25] G.G. Vladimirov, B.K. Medvedev and I.L. Sokolskaya, Fiz. Tverd. Tela 12 (1970) 539.

[26] N.D. Lang, Phys. Rev. B 4 (1971) 4234.

[27] E.W. Plummer and P.A. Dowben, Prog. Surf. Sci. 42 (1993) 201.

[28] H. Ishida, Surf. Sci. 242 (1991) 341.

[29] J. Koutecky, Trans. Faraday Soc. 54 (1958) 1038.

[30] T.B. Grimley, Proc. Phys. Soc. 90 (1967) 751.

[31] T.B. Grimley and S.M. Walker, Surf. Sci. 14 (1969) 395.

[32] T.L. Einstein, Crit. Rev. Solid State Mater. Sci. 7 (1978) 261.

[33] O.M. Braun and V.K. Medvedev, Sov. Phys. Usp. 32 (1989) 328.

[34] T.T. Tsong, Surf. Sci. Rep. 8 (1988) 127.

[35] D.W. Bassett, Surf. Sci. 53 (1975) 74.

[36] H.-W. Fink, K. Faulian and E. Bauer, Phys. Rev. Lett. 44 (1980) 1008.

[37] F. Watanabe and G. Ehrlich, J. Chem. Phys. 95 (1991) 6075.

[38] Yu.S. Vedula, V.V. Gonchar, A.G. Naumovets and A.G. Fedorus, Sov. Phys. Solid State 19 (1977) 1505.

[39] A.G. Naumovets, Sov. Sci. Rev. A Phys. 5 (1984) 443.

[40] A.G. Naumovets, in: The Chemical Physics of Solid Surfaces and Heterogeneous Catalysis, Vol. 7, Eds. D.A. King and D.P. Woodruff (Elsevier, Amsterdam, 1993).

[41] L.D. Roelofs and D.L. Kriebel, J. Phys. C 20 (1987) 2937.

[42] V.K. Medvedev and I.N. Yakovkin, Sov. Phys. Solid State 20 (1978) 537.

[43] V.I. Marchenko, JETP Lett. 33 (1981) 381.

[44] D. Andelman, F. Brochard, P.-G. de Gennes and J.-F. Joanny, C.R. Acad. Sci. Paris 301 (1985) 675.

[45] D. Vanderbilt, Surf. Sci. 268 (1992) L300.

[46] K. Kern, H. Niehus, A. Schatz, P. Zeppenfeld, J. George and G. Comsa, Phys. Rev. Lett. 67 (1991) 885.

[47] A.G. Naumovets and A.G. Fedorus, Sov. Phys. JETP Lett. 10 (1969) 6.

[48] O.V. Kanash, A.G. Naumovets and A.G. Fedorus, Sov. Phys. JETP 40 (1975) 903.

[49] V.K. Medvedev, A.G. Naumovets and A.G. Fedorus, Sov. Phys. Solid State 12 (1970) 375.

[50] A.G. Fedorus and A.G. Naumovets, Surf. Sci. 21 (1970) 426.

[51] K. Christmann, Surf. Sci. Rep. 9 (1988) 1.

[52] K. Müller, G. Besold and K. Heinz, in: Physics and Chemistry of Alkali Metal Adsorption, Eds. H.P. Bonzel, A.M. Bradshaw and G. Ertl (Elsevier, Amsterdam, 1989) p. 65.

[53] R.D. Diehl, in: Phase Transitions in Surface Films, Eds. H. Taub, G. Torzo, H. Lauter and S.C. Fain, Jr. (Plenum, New York, 1991) p. 97.

[54] D.A. Gorodetsky and Yu.P. Melnik, Surf. Sci. 62 (1977) 647.

[55] Z.Y. Li, K.M. Hock and R.E. Palmer, Phys. Rev. Lett. 67 (1991) 1562.

[56] L.A. Bolshov, Sov. Phys. Solid State 13 (1971) 1404.

[57] I. Langmuir and J.B. Taylor, Phys. Rev. 40 (1932) 463.

[58] P. Kisliuk, J. Phys. Chem. Solids, 3 (1957) 95.

[59] B.A. Chuikov, V.D. Osovskii, Yu.G. Ptushinskii and V.G. Sukretnyi, Surf. Sci. 213 (1989) 359.

[60] Yu.G. Ptushinskii and B.A. Chuikov, Poverkhnost 9 (1992) 5.

[61] J.A. Becker and C.D. Hartman, J. Phys. Chem. 57 (1953) 157.

[62] V.N. Ageev and N.I. Ionov, Prog. Surf. Sci. 5, Pt. 1 (1975) 1.

[63] Yu.G. Ptushinskii and B.A. Chuikov, Ukr. Fiz. Zh. 9 (1964) 1035.

[64] Yu.G. Ptushinskii and B.A. Chuikov, Surf. Sci. 6 (1967) 42; 7 (1967) 90.

[65] N.P. Vas'ko, Yu.G. Ptushinskii and B.A. Chuikov, Surf. Sci. 14 (1969) 448.

[66] V.N. Ageev, Poverkhnost 3 (1984) 5.

[67] Ya.P. Zingerman and V.A. Morozovskii, Fiz. Tverd. Tela 3 (1961) 123.

[68] E.L. Scheibner, L.H. Germer and C.D. Hartman, Rev. Sci. Instrum. 31 (1960) 112.

[69] D.A. Gorodetskii and A.M. Kornev, Ukr. Fiz. Zh. 6 (1961) 422.

[70] J.J. Lander, Surf. Sci. 1 (1964) 125.

[71] J.B. Taylor and I. Langmuir, Phys. Rev. 44 (1933) 423.

[72] V.N. Shrednik, Radiotekh. Elektron. 5 (1960) 1203.

[73] V.M. Gavrilyuk, A.G. Naumovets and A.G. Fedorus, Sov. Phys. JETP 24 (1967) 899.

[74] R.L. Gerlach and T.N. Rhodin, Surf. Sci. 17 (1969) 32.

[75] V.K. Medvedev, A.G. Naumovets and T.P. Smereka, Surf. Sci. 34 (1973) 368.

[76] A.G. Fedorus and A.G. Naumovets, Surf. Sci. 93 (1980) L98.

[77] J. Kolaczkiewicz and E. Bauer, Phys. Rev. Lett. 53 (1984) 485.

[78] M.V. Loginov and M.A. Mitsev, Poverkhnost 5 (1987) 37.

[79] P.P. Lutsishin, T.N. Nakhodkin, O.A. Panchenko and Yu.G. Ptushinskii, Zh. Eksp. Teor. Fiz. 82 (1982) 1306.

[80] H.P. Bonzel, Surf. Sci. Rep. 8 (1988) 43.

[81] M.P. Kiskinova, Surf. Sci. Rep. 8 (1988) 359.

[82] E.V. Klimenko and A.G. Naumovets, Sov. Phys. Techn. Phys. 24 (1979) 710.

[83] J.-L. Desplat and C.A. Papageorgopoulos, Surf. Sci. 92 (1980) 97, 119; 104 (1981) 643.

[84] P. Feibelman, Prog. Surf. Sci. 12 (1982) 287.

[85] G.A. Katrich, V.V. Klimov and I.N. Yakovkin, Ukr. Fiz. Zh. 37 (1992) 429.

[86] M. El-Batanouny, D. Hammann, S. Chubb and J. Davenport, Phys. Rev. B 27 (1983) 2575.

[87] P.J. Berlowitz and D.W. Goodman, J. Vac. Sci. Technol. A 6 (1988) 634.

[88] G.E. Rhead, Contemp. Phys. 24 (1983) 535.

[89] E. Bauer, Appl. Surf. Sci. 11/12 (1982) 479.

[90] J. Kolaczkiewicz and E. Bauer, Surf. Sci. 175 (1986) 487.

[91] U. Gradmann, J. Magn. Magn. Mater. 100 (1991) 481.

[92] H. Pinkvos, H. Poppa, E. Bauer and J. Hurst, Ultramicroscopy (1992).

[93] A.G. Naumovets and Yu.S. Vedula, Surf. Sci. Rep. 4 (1985) 365.

[94] Yu.S. Vedula, A.T. Loburets and A.G. Naumovets, Sov. Phys. JETP 50 (1980) 391.

[95] I. Lyuksyutov, A.G. Naumovets and V. Pokrovsky, Two-Dimensional Crystals (Academic Press, Boston, MA, 1992).

[96] J. Beben, Ch. Kleint, R. Meclewski and A. Pawelek, Surf. Sci. 213 (1989) 224, 451.

[97] U.Kh. Rasulev and E.Ya. Zandberg, Prog. Surf. Sci. 28 (1988) 181.

[98] E.F. Chaikovskii, G.M. Pyatigorskii and G.V. Ptitsyn, Zh. Tekh. Fiz. 34 (1965) 1132.

[99] N.I. Ionov, Prog. Surf. Sci. 1, Pt. 3 (1972) 237.

[100] E. Ya. Zandberg, Zh. Tekh. Fiz. 44 (1974) 1809.

[101] N.A. Kholin, E.V. Rut'kov and A.Ya. Tontegode, Surf. Sci. 139 (1984) 155.

[102] E.V. Rut'kov and A. Ya. Tontegode, Surf. Sci. 161 (1985) 373.

[103] A.Ya. Tontegode, Prog. Surf. Sci. 38 (1991) 201.

[104] E.Ya. Zandberg, A.Ya. Tontegode and F.K. Yusifov, Zh. Tekh. Fiz. 41 (1971) 2420.

[105] V.N. Shrednik, G.A. Odisharia and O.L. Golubev, J. Cryst. Growth 11 (1971) 249.

[106] E.Ya. Zandberg, A.Ya. Tontegode and F.K. Yusifov, Zh. Tekh. Fiz. 42 (1972) 171.

[107] E.Ya. Zandberg and N.I. Ionov, Dokl. Acad. Nauk SSSR 141 (1961) 139.

[108] E.G. Nazarov and U.Kh. Rasulev, Nonstationary Processes in Surface Ionization (Fan, Tashkent, 1991), in Russian.

[109] U. Gemmeren and F.W. Roellgen, Org. Mass Spectrom. 22 (1987) 468.

[110] T. Fujii, H. Ishii and H. Jimba, Int. J. Mass Spectrom. Ion Processes 93 (1989) 73.

[111] E.Ya. Zandberg, M.V. Knat'ko, V.I. Paleev and M.M. Sushchikh, in: Optical Radiation Interaction with Matter, Eds. A.M. Bonch-Bruevich, Vi.I. Konov and M.N. Libenson, Proc. SPIE, 1440 (1990) 292.

[112] V.N. Ageev, Yu.A. Kuznetsov and B.V. Yakshinskii, Fiz. Tverd. Tela 24 (1982) 349.

[113] V.N. Ageev, O.P. Burmistrova and Yu.A. Kuznetsov, Usp. Fiz. Nauk 158 (1989) 389.

[114] N.D. Lang, Surf. Sci. 299/300 (1994) 284.

[115] G. Ehrlich, Surf. Sci. 299 /300 (1994) 628.

[116] D. Goodman, Surf. Sci. 299/300 (1994) 837; O. Somorjai, Surf. Sci. 299/300 (1994) 849.

[117] J. Venables, Surf. Sci. 299/300 (1994) 798.

[118] T.E. Madey, Surf. Sci. 299/300 91994) 824.

Surface Science 299/300 (1994) 722–730
North-Holland

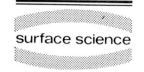

surface science

Surface phases of reconstructed tungsten (100) and molybdenum (100)

Peder J. Estrup

Department of Chemistry and Department of Physics, Brown University, Providence, RI 02912, USA

Received 29 June 1993; accepted for publication 28 July 1993

After several decades of experimental and theoretical work it has become clear that on most solids, including metals, the surface is subject to reconstruction, i.e. the outermost substrate atoms rearrange in response to changes in temperature, adsorbate coverage, or other variables. Studies of W(100) and Mo(100) have been very useful in this development. Early LEED studies of adsorption on these surfaces revealed a large number of distinct phases and it proved impossible to explain their formation in terms of conventional models which assumed the substrate to be "passive". The subsequent discovery that the surface is reconstructed in these phases necessitated a reinterpretation of the results and showed that the effective adatom–substrate and adatom–adatom interactions depend on metal atom displacements. It has been found that, in general, substrate structural changes have a strong – sometimes dominant – influence on the kinetic and thermodynamic surface properties.

1. Introduction

Our knowledge of surface phenomena has advanced greatly during the last 30 years. Although many fundamental questions concerning surface structure and dynamics were formulated much earlier [1], answers had to await the development of experimental techniques capable of exploring the interface on the atomic scale. The burgeoning field of semiconductor electronics provided a powerful motivation for such developments, and studies of silicon and germanium surfaces dominated research in the early 1960's. However, it was not long before work on metal surfaces began to attract comparable attention, spurred by interest in classical problems such as thermionic emission, as well as the need to solve long-standing problems in chemisorption and heterogeneous catalysis.

In experimental surface science metals are usually easier to handle than semiconductors and insulators, and among the metals, tungsten and molybdenum are particularly convenient because a clean surface often can be obtained on these crystals merely by heating to high temperatures. Furthermore, with few exceptions, all the tech-

niques that have been developed are applicable to W and Mo and can conveniently be tested using surfaces of these metals. Thus, over the years more techniques have been applied to W(100) than to any other surface. There are more fundamental reasons, however, why the literature contains at least 500 publications on this surface orientation alone. It has turned out that W and Mo substrates provide some of the best model systems for experimental studies of phenomena such as metal surface reconstruction, two-dimensional displacive phase transitions, and the interplay between substrate rearrangement and chemisorption. The topic of this article is the understanding we have gained from studies of these surfaces, as seen from my personal perspective.

2. LEED studies of chemisorption

It is fitting that the very first article to appear in Surface Science, by Marsh and Farnsworth, described results obtained by low energy electron diffraction (LEED) [2]. Especially after the introduction of display screens [3] to allow instanta-

SSDI 0039-6028(93)E0439-Z

neous observation of a diffraction pattern, LEED became the single most important surface diagnostic technique. The new structural information, provided in a visual form, had an impact on the field comparable to that of STM in recent years [4]. As is documented in Lander's influential review article, published in 1965 [5], in a very short time it became a realistic possibility to propose and test atomic models of the surfaces of solids.

Among the earliest applications of display-LEED to metals were the studies by Germer et al. [6] and May [7] of the W(110) surface, the densest and presumably most stable orientation for a bcc crystal. Experiments were done not only

on the clean surface but also on the effects of adsorbed oxygen and carbon monoxide, each of which produced LEED patterns with a periodicity different from that of the substrate.

An even richer behavior was observed in the LEED studies of W(100) which Anderson and I began around 1964 [8]. The great appeal to the imagination of these LEED results is perhaps evident from the examples shown in Fig. 1. The occurrence in each experiment of fractional-order extra spots in the LEED pattern suggested that, in addition to the adsorbate–substrate (AS) interactions holding the adsorbed species to the tungsten surface, adatom–adatom (AA) forces existed

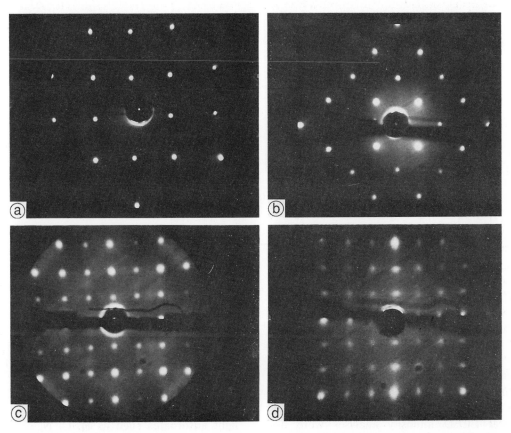

Fig. 1. LEED patterns from a tungsten (100) surface under various conditions. (a) (1×1) pattern of clean W(100) above room temperature. (b) $c(2 \times 2)$ or $(\sqrt{2} \times \sqrt{2})R45°$ pattern produced by adsorption of nitrogen. The same pattern can be induced by many other adsorbates, including hydrogen, carbon monoxide, carbon, alkali or thorium atoms. (c) (2×2) pattern due to exposure to oxygen, followed by heating of the sample. (d) (3×3) pattern which may be formed by partial cracking of hydrocarbons on the surface. (e) (4×1) pattern due to oxygen adsorption at room temperature. (f) Complex pattern produced by acetylene at elevated temperature. (g) Complex pattern formed by adsorption of chlorine at room temperature. (h) Pattern formed by oxygen adsorption on a W(100) surface with pre-adsorbed Th. (Figure continued on next page.)

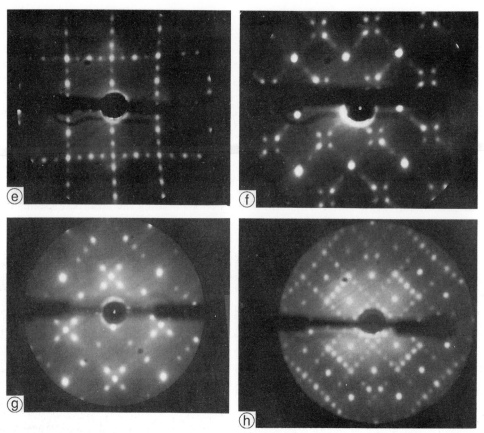

Fig. 1. (continued).

which were strong enough to cause the adsorbate to organize into a well-ordered two-dimensional (2D) arrangement. Apparently these AA forces are different for each system and they could not be predicted based on the three-dimensional properties of the species involved. A study of the geometry and stability of each surface phase would seem to be a promising approach to an understanding of these interactions.

A complete determination of a surface structure is a difficult task even today. Methods for dynamical LEED intensity analyses were not available when the data in Fig. 1 were obtained. Furthermore, the early studies of chemisorption were done almost entirely without the benefit of complementary probes, such as Auger electron spectroscopy (AES) [9], which are now considered indispensable. HREELS (high resolution electron energy loss spectroscopy) which is one of the most useful means of characterizing the bonding and site geometry of an adsorbate, did not become available until much later [10]. As a consequence, interpretations of a LEED pattern were speculative and sometimes controversial.

It was known that in the case of high energy electrons the backscattered intensity falls off rapidly with decreasing atomic number. Germer et al. [6] and May [7] assumed that this effect was important also at low energies, so much so that light adsorbates such as H, C, N, or O, would contribute negligibly to the observed LEED intensity. It was therefore postulated that whenever gas adsorption produced new spots in a LEED pattern, a place-exchange had occurred, i.e., an adsorbate atom had been substituted for a metal substrate atom. Thus, the new 2D periodicity was

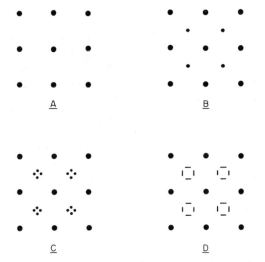

Fig. 2. Diagram of LEED patterns produced by hydrogen adsorption on W(100) at room temperature. (A) (1×1), clean W(100). (B) c(2×2)-H, also referred to as ($\sqrt{2} \times \sqrt{2}$)-H. (C) Incommensurate structure. (D) Pattern with streaks in 1/3-order positions.

due to missing atoms in the outermost layer of the solid.

This type of reconstruction is now thought to occur in some metal-on-metal systems but it appears to be rare. In the case of gas adsorption on W(100) we dismissed this mechanism primarily on the basis of our LEED results for hydrogen adsorption [11]. Fig. 2 shows schematically the changes in the diffraction pattern of the clean surface as the hydrogen coverage increases. Initially a c(2×2) structure forms (Fig. 2B); with additional adsorption each of the extra (1/2, 1/2) features splits into a quartet of spots (Fig. 2C) which move apart continuously and finally turn into streaks (Fig. 2D). These observations imply rather remarkable changes in the 2D organization of the scattering matter. An explanation in terms of a substitutional reconstruction would require that tungsten atoms were continuously removed and, furthermore, that at the same time all the remaining atoms in the top layer rearranged to produce the new periodicity, all the while maintaining long-range order. Conceivably this could be achieved with proper annealing of the sample, but the structural changes occur rapidly at room temperature; in fact, for practical purposes, the rate is limited only by the arrival of hydrogen from the ambient gas to the W(100) surface.

In our search for an alternative explanation, we briefly considered the possibility of substrate distortions [11] but – in the absence of specific data to the contrary – we adopted the working hypothesis that adsorbed hydrogen *can* produce detectable LEED intensities [12].

3. Phase diagram of H/W(100)

The description of an adsorbate is greatly simplified if it can be treated as an overlayer on top of a passive substrate. One can then invoke the existence of an AS interaction varying along the surface with the period of the substrate lattice, giving rise to adsorption sites with a specific and fixed geometry and co-ordination. This allows the adsorbate to be considered as a lattice gas which may condense into ordered arrangements depending on the AA interaction, the density, and the temperature. Thus, a contact is provided between surface diffraction data and the statistical mechanical results obtained for 2D systems.

Our observations on H/W(100) seemed to offer a very good test of these ideas. The picture of the c(2 × 2) structure formed at low coverage was equivalent to that of a 2D zero-field antiferromagnet. (If the W(100) square array of sites is thought of as a chess-board, all the black squares are occupied (spin up) and all the white squares are empty (spin down).) When only nearest-neighbor interactions occur, the thermodynamic properties of this Ising system are known: The c(2 × 2) structure will have a critical temperature T_c which depends on the coverage θ, and at $\theta = 0.5$ the pair interaction energy is given by $J = 1.76kT_c$.

The first measurements of the temperature dependence of the diffracted intensity were consistent with an order–disorder phase transition, as predicted [13]. Some years later, when precise methods for determination of the hydrogen coverage had become available [14,15] we decided to explore the phases over a wider temperature and coverage range. Fig. 3 shows the resulting phase

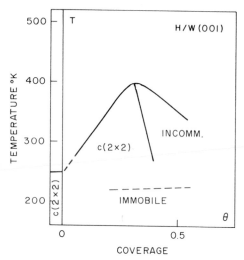

Fig. 3. Phase diagram for the H/W(100) surface showing the regions in the $T-\theta$ plane where different structures are stable. The incommensurate phase is characterized by split 1/2 1/2 spots in the LEED pattern. No ordered phases are formed at $T \leq 200$ K where the H adatoms appear to be immobile. The coverage is relative to the number of W atoms in the surface, i.e. $\theta = 0.5$ means 5.0×10^{14} H atoms/cm^2.

diagram [16] which demonstrates that the simple Ising model fails for this system: The maximum stability occurs at $\theta \approx 0.3$ and not at $\theta = 0.5$ as required; in fact the diagram is highly asymmetric about this value, and the incommensurate phases are not accounted for. Clearly, the interactions at the surface are considerably more complex than allowed for in a conventional lattice gas model. We have found this to be true for all adsorbates not only on W(100) but also on Mo(100); we have also discovered the reason: the underlying assumption that the substrate can be treated as a passive matrix of sites is false.

4. Displacive reconstruction

When Hagstrum was developing the technique of ion neutralization spectroscopy he planned to use W(100) as the test sample. However, he complained to me that he had difficulties getting a

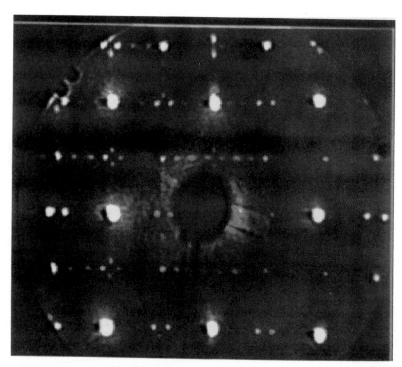

Fig. 4. LEED pattern from a clean Mo(100) surface at low temperature. The extra spots are due to a single domain of the $c(7\sqrt{2} \times \sqrt{2})$ reconstruction.

clean surface; the samples persisted in giving a faint c(2 × 2) LEED pattern. He found our standard prescription of heating the crystal in oxygen to be unsatisfactory and eventually he switched to nickel [17].

The same problem had been encountered by us, as well as by other researchers. Yonehara and Schmidt, in particular, observed the c(2 × 2) pattern in the course of their studies of hydrogen adsorption [18]; they attributed it to hydrogen that dissolved in the bulk at high temperature and segregated at the surface when the sample cooled, but they also considered the possibility of structural changes in the substrate surface.

Investigations by Felter of the closely related Mo(100) surface provided us with the solution to the puzzle. When this sample was cooled to $T \leq$ 250 K a new structure forms; and since effects of common impurities, including hydrogen, could be ruled out in this case, we concluded that a temperature-induced reconstruction of the clean surface must occur. The transformation is very rapid and completely reversible so that transport of Mo atoms over long distances cannot be involved. We therefore proposed that the surface undergoes a type of reconstruction in which the lattice is distorted by small displacements of each surface atom from its nominal lattice position [19]. Re-examination of the W(100) then led to the conclusion that a similar displacive reconstruction occurs also on that surface [19]. The faint but bothersome c(2 × 2) pattern observed in earlier work is thus a property of the clean surface; the extra LEED spots disappear at higher temperature when $T \gg T_c$ or in the presence of oxygen which quenches the reconstruction.

On the basis of careful LEED intensity measurements Debe and King also came to the conclusion that W(100) undergoes reconstruction. Furthermore, from observations of the symmetry of the diffraction pattern, they determined the direction of the atomic displacements [20]. A model of the surface is shown in Fig. 5A.

The low-temperature phase of Mo(100) was at first thought to be much more complex, perhaps even incommensurate [19]. Its LEED pattern, which we have studied again in the course of very recent experiments [21], is shown in Fig. 4. Actu-

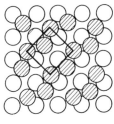

A.

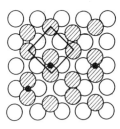

B.

Fig. 5. (A) Model of the clean reconstructed W(100) ($\sqrt{2} \times \sqrt{2}$) surface. The W atoms are displaced by 0.23 Å (exaggerated in drawing) along [11] directions to form zig-zag chains. (B) Model of the hydrogen induced ($\sqrt{2} \times \sqrt{2}$) structure. The magnitude of the W atom displacements is the same as for the clean surface but the direction is switched to [10] so that dimers are formed, providing sites for H atoms (small filled circles).

ally the structure is quite similar to that of Fig. 5A, but with an additional modulation which gives it a c($7\sqrt{2} \times \sqrt{2}$) unit cell.

These remarkable transitions are restricted to the outermost surface layers. Thus, for W(100) the displacements in first layer are ~ 0.25 Å but are reduced to 0.05 Å already in the second layer [22]. Calculations show that this reconstruction lowers the energy by about 0.1 eV per surface atom, and that the main contribution to the stabilization comes from a decrease in the number of "dangling bonds", i.e. the reconstruction allows for a higher co-ordination of the surface atoms [23].

When the temperature is raised an order–disorder transition occurs in which the long-range c(2 × 2) periodicity is gradually lost; however, the atomic displacements persist locally well above room temperature. It is interesting, that as a result, even in the absence of defects or impurities, the (100) surface never has the atomic arrangement of an ideal square net [24].

Since the discovery of the W(100) and Mo(100) reconstructions, an enormous amount of work, by many different groups, has been directed at determining the nature, origin, and consequences of the transitions. The work includes detailed investigations of the surface crystallography by many techniques, calculation and measurement of the electronic surface states responsible for the reconstruction, searches for soft surface phonons, and measurements of the critical phenomena accompanying the phase transition. It is beyond the scope of the present article to attempt a review of the literature; suffice it to state that especially W(100) has played a role comparable to that of Si(111) as a testing ground for both experimental and theoretical surface science.

5. Adsorbate effects

The W(100) and Mo(100) surface phase transitions not only provided us with a wealth of interesting problems in surface physics, they also forced us to change our view of chemisorption. In particular we observed large effects of adsorbates on the substrate structural changes; the converse must then also be true: the substrate changes will alter the chemisorption behavior. This is true for all types of reconstruction but the effects for W(100) and Mo(100) are particularly striking. Unlike the transformations observed on many fcc substrates [25], on the bcc crystals the structural changes are displacive and conserve substrate atom density. The activation energy for the reconstruction is negligible and the substrate responds readily to changes in temperature or adsorbate coverage.

An example is provided by the phase diagram in Fig. 3 which shows that the clean $c(2 \times 2)$ structure is stabilized by the adsorption of hydrogen. We found that the adsorbate causes a modification of the substrate geometry consisting of a switching of the W atom displacements into a new direction, as shown in Fig. 5B, thereby providing more stable bridge-sites for bonding of the

H atoms [27]. It can be seen that this arrangement immediately implies a non-traditional type of adatom–adatom interaction: if a H atom were to occupy a bridge site other than that offered by a dimer, a different distortion would be induced which would tend to destabilize the reconstructed surface. Simultaneous occupation of nearest-neighbor sites is therefore not favored; i.e. the substrate reconstruction creates an effective short-range AA repulsion.

The reconstruction also changes the effective adatom–substrate interaction in the H/W(100) system. At a coverage beyond half a monolayer nearest-neighbor pairs can no longer be avoided so that annihilation of the dimers must begin, with a consequent loss of the reconstruction stabilization energy. It follows that the adatom binding will be weaker at high coverage. The effect is very large; we found the desorption energy to decrease by a factor of 2 as the coverage was increased [26].

The effects of substrate distortion on the AA interactions are even more dramatic in the case of H/Mo(100) [27]. The diagram shown in Fig. 6 contains a myriad of phases as the temperature or hydrogen coverage is varied. For each phase

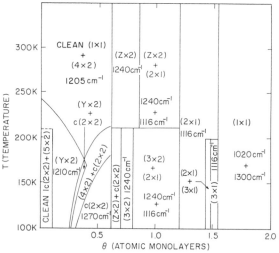

Fig. 6. Phase diagram in the coverage–temperature plane for the H/M(100) system. For the phases labeled $(y \times 2)$ and $(z \times 2)$ y and z vary continuously. The wavenumbers in the diagram denote the perpendicular vibration frequency of the hydrogen adatoms.

the hydrogen vibration frequency, measured by IR spectroscopy, is indicated. The value is a measure of the substrate distortion amplitude; the higher the frequency, the shorter is the distance between the two Mo atoms in the dimer to which the H atom is bound. The data show that – as for W(100) – the dimers open up at high coverage. They also show that even at a fixed coverage, $\theta = 0.5$ for example, the dimer configuration depends on temperature. Thus, an adsorption site with fixed properties does not exist, and we therefore should not expect conventional lattice gas models to be successful in describing the surface.

Finally, Fig. 7 summarizes some additional features of chemisorbed hydrogen on Mo(100). A priori, it would seem difficult to come up with a more straight-forward system: the simplest possible atom is placed on a single-crystal surface of high symmetry. Nonetheless, the behavior is astonishingly complex. The electron-stimulated desorption cross section goes through a maximum which depends on temperature; the work function is not a unique function of coverage; the sticking probability is non-linear; and the shape of the desorption trace would ordinarily be interpreted to indicate three different binding states. These results would indeed be hard to explain if

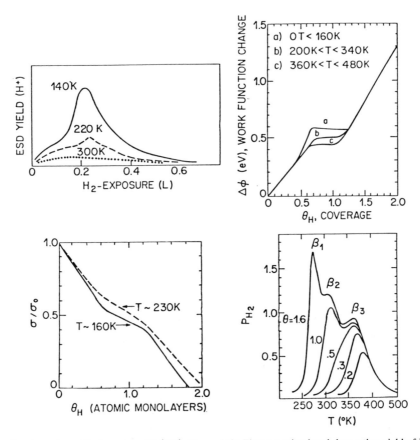

Fig. 7. Results for chemisorption of hydrogen on Mo(100). Upper left: Electron-stimulated desorption yield of H^+ as a function of initial exposure and of temperature. Upper right: Work function change as a function of coverage, in three different temperature regions. Lower left: Relative sticking coefficient of hydrogen as a function of hydrogen coverage at two different temperatures. Lower right: Temperature programmed desorption (TPD) traces showing the development of three desorption peaks at high coverage.

the only degrees of freedom were those corresponding to vibration or translation of H atoms on a passive substrate. However, as we have learned – sometimes painfully slowly – through our structural studies, substrate degrees of freedom can be sufficiently active to dominate the behavior of an adsorbate, and may therefore profoundly influence processes such as adsorption, desorption, diffusion and film growth.

6. Concluding remarks

Surfaces are highly interesting objects of enormous technological importance. The combination has motivated a very large body of work during the past three decades, and the field has advanced greatly.

In any period where knowledge is expanding rapidly, many existing models and beliefs must become outdated. In this article I have considered examples of this, dealing with surface reconstruction, phase transitions and chemisorption.

It is not true any more that the only surfaces we can claim to understand are those we have not studied carefully. Nonetheless, almost every new experiment gives surprises. I do not think that this indicates lack of success; rather it is due to our ability to consistently probe at a deeper level, and it is a sign of the continuing vitality of the field.

Acknowledgements

Many students and collaborators have made invaluable contributions to the research mentioned in this article. Space limitations prevent me from giving proper credit to each one, but I am deeply grateful to them.

References

[1] See, for example: I. Langmuir, J. Am. Chem. Soc. 38 (1916) 2221.

[2] J.B. Marsh and H.E. Farnsworth, Surf. Sci. 1 (1964) 3.

[3] L.H. Germer and C.D. Hartman, Rev. Sci. Instrum. 31 (1960) 784.

[4] In 1963 J.J. Lander (Bell Telephone Laboratories) produced a movie of changes, in real time, of the Si(111) LEED pattern as function of temperature and various adsorbates. It was shown to great effect at many national meetings, symposia, and seminars.

[5] J.J. Lander, Prog. Solid State Chem. 2 (1965) 26.

[6] L.H. Germer, R.M. Stern and A.U. MacRae, in: Metal Surfaces (American Society of Metals, Metals Park, OH, 1963).

[7] J.W. May, Ind. Eng. Chem. 57 (1965) 18.

[8] P.J. Estrup and J. Anderson, Surf. Sci. 8 (1967) 101, and references therein.

[9] P.W. Palmberg and T.N. Rhodin, J. Appl. Phys. 19 (1968) 2425.

[10] See H. Ibach, Surf. Sci. 299/300 (1994) 116.

[11] P.J. Estrup and J. Anderson, J. Chem. Phys. 45 (1966) 2254.

[12] This is indeed the case as has been shown for H/Ni(111) and H/Pd(111). However, the LEED intensity of the extra feature is an order of magnitude smaller than that of the "normal" spots.

[13] P.J. Estrup, in: The Structure and Chemistry of Solid Surfaces, Ed. G.A. Somorjai (Wiley, New York, 1969) Ch. 19.

[14] T.E. Madey, Surf. Sci. 36 (1973) 281.

[15] I. Stensgaard, L.C. Feldman and P.J. Silverman, Phys. Rev. Lett. 42 (1979) 247.

[16] R.A. Barker and P.J. Estrup, J. Chem. Phys. 74 (1981) 1442.

[17] H.D. Hagstrum, Science 178 (1972) 275.

[18] K. Yonehara and L.D. Schmidt, Surf. Sci. 25 (1971) 238.

[19] T.E. Felter, R.A. Barker and P.J. Estrup, Phys. Rev. Lett. 38 (1977) 1138.

[20] M.K. Debe and D.A. King, J. Phys. C 10 (1977) L303; Phys. Rev. Lett. 39 (1977) 708.

[21] R.S. Daley, T.E. Felter, M.L. Hildner and P.J. Estrup, Phys. Rev. Lett. 70 (1993) 1295.

[22] M.S. Altman, P.J. Estrup and I.K. Robinson, Phys. Rev. B 38 (1988) 5211; I.K. Robinson, A.A. MacDowell, M.S. Altman, P.J. Estrup, K. Evans-Lutterodt, J.D. Brock and R.J. Birgeneau, Phys. Rev. Lett. 62 (1989) 1294.

[23] L.D. Roelofs, T. Ramseyer, L.I. Taylor, D. Singh and H. Krakauer, Phys. Rev. B 40 (1989) 9147.

[24] C.Z. Wang, E. Tosatti and A. Fasolino, Phys. Rev. Lett. 60 (1988) 2661.

[25] P.A. Thiel and P.J. Estrup, in: CRC Handbook of Surface Imaging and Visualization (CRC, Boca Raton, FL, 1994), in press.

[26] A. Horlacher Smith, R.A. Barker and P.J. Estrup, Surf. Sci. 136 (1984) 327.

[27] J.A. Prybyla, Y.J. Chabal and P.J. Estrup, J. Chem. Phys. 94 (1991) 6274; Surf. Sci. 290 (1993) 413.

Surface Science 299/300 (1994) 731–741
North-Holland

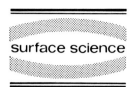

surface science

Chemisorption on surfaces – an historical look at a representative adsorbate: carbon monoxide

John T. Yates, Jr.

Surface Science Center, Department of Chemistry, University of Pittsburgh, Pittsburgh, PA 15260, USA

Received 3 May 1993; accepted for publication 19 July 1993

The study of the interaction of molecules with clean surfaces extends back to the work of Irving Langmuir. In this historical account, the development of selected experimental methods for the study of molecular adsorption will be discussed. This will be done by historically reviewing research on one of the most well-studied adsorbate molecules, carbon monoxide. Many of the modern surface science techniques have first been used to study chemisorbed carbon monoxide, and the CO molecule is employed even today as a test molecule for currently developing surface measurement instruments such as the low temperature STM. In addition to being a good test molecule for new surface measurement techniques, adsorbed carbon monoxide is one of the centrally important molecules in the field of heterogeneous catalysis where the production of synthetic fuels and useful organic molecules often depends on the catalytic behavior of the adsorbed CO molecule. Interestingly, the carbon monoxide molecule also serves as a bridge between surface chemistry on the transition metals and the field of organometallic chemistry. Concepts about the chemical bonding and the reactive behavior of CO chemisorbed on transition metal surfaces and CO bound in transition metal carbonyls link these two fields together in a significant manner. The carbon monoxide molecule has been the historical focal point of many endeavors in surface chemistry and surface physics, and research on adsorbed carbon monoxide well represents many of the key advances which characterize the first thirty years of the development of surface science.

1. Introduction

The carbon monoxide molecule has been an historical focal point in the development of the field of surface science for three basic reasons: (1) it is a simple molecule which has the highest dissociation energy of any molecule; (2) it is a technologically significant molecule in the field of heterogeneous catalysis and has a long history in this area; (3) CO forms a major historical linkage to the field of organometallic chemistry through the metal carbonyls, where the carbonyl ligands often closely resemble chemisorbed CO species on transition metal surfaces.

This review is not a full review of the surface chemical properties of CO; the reader is referred to full reviews concerned with CO exclusively [1,2]. Instead, I employ the CO molecule as a focal point for the development of a number of key experimental methods which characterize the development of the field of surface science itself.

These experimental methods have given us a good view of the behavior of chemisorbed CO on the metals; in addition, because of the relative simplicity of CO, these studies have provided needed understanding and calibration information for the experimental surface measurement methods themselves.

2. Early history of the observation of chemisorbed CO

Looking back in time, a surface chemist invariably confronts the monumental contributions of Irving Langmuir, who is certainly the founder of the field of surface science [3]. In 1922, Langmuir studied the catalytic reaction of CO and O_2 over a Pt wire, finding the inhibitory effect of CO on the reaction rate [4,5]. He correctly conceived that the CO molecule interacted with the Pt surface as a chemically-bound species, possibly

SSDI 0039-6028(93)E0436-X

due to unsaturated valencies of the carbon atom. In the late 1930's and 1940's, Brunauer and Emmett [6,7] studied CO adsorption on iron catalysts employed for nitrogen chemisorption. The isoelectronic character of CO and N_2 may have been the driving force for these studies, since understanding the adsorption behavior of nitrogen was a major goal of their work. Evaporated metal films were also employed in the early days for adsorption studies, and Rideal and Trapnell [8] observed the irreversible adsorption of CO on tungsten films in 1951. It is very likely that these films were contaminated because of the poor vacuum conditions of the day. Studies of the heat of adsorption of CO were also carried out calorimetrically using eggshell glass calorimeters with deposited (and likely contaminated) metal thin films on their inner surface [9]. Kummer and Emmett [10] employed radioactive ^{14}CO to investigate the possibility of multiple chemisorbed species and multiple types of chemisorption sites on iron catalysts. They added isotopic CO doses sequentially, and then analyzed the gas as it desorbed to see if complete randomization occurred. They found that the catalysts had a memory for the order of addition of the CO isotopes. In this work, they first suggested that chemisorbed bridging-CO and terminal-CO might be present together, drawing the analogy to the bonding of CO ligands in the higher iron carbonyls.

The first measurements of the quantum behavior of a chemisorbed species on a metal surface were performed by Eischens and coworkers [11,12]. Using transmission IR spectroscopy through supported metal catalysts containing chemisorbed CO, they measured the vibrational frequency of the C–O bond in these surface species, and also drew the analogy between the vibrational frequencies observed and the terminal-CO and bridging-CO ligands known to be present in metal carbonyls. These experiments were considered by others to be impossible before they were performed, since it was well known that the finely divided metals would absorb strongly in the IR making them black for transmission experiments [13]. The method works because of the transmission of scattered IR radiation past many adsorbed species. In my opinion,

the work of Eischens represents the point of departure from macroscopic measurements of surface phenomena to the molecular domain of experimental investigation of adsorbed molecules. The immense power of the vibrational spectroscopies for understanding surface bonding has its origin in these early investigations. I began my studies in surface chemistry using transmission IR spectroscopy to investigate the adsorption of CO on Ni/Al_2O_3 catalysts [14]. Work along these lines was also carried out at the same time by Sheppard and coworkers at Cambridge [15]. The development of various types of surface vibrational spectroscopy is treated in Ref. [16].

An interesting and important model for CO bonding to transition metal surfaces was first suggested by Blyholder in 1964 [17]. He perceived that donor–acceptor bonding between the CO and the metal surface would take place and that this could offer an explanation for the various C–O frequencies observed. His model has been the basis for our thinking about CO bonding for the last 30 years. He postulated that it was not necessary to rehybridize the CO to a $C(sp^2)$ electronic configuration to produce low frequency C–O modes; instead various sites could donate more or less electrons to the antibonding $CO(2\pi^*)$ levels to achieve the desired C–O bond strength. Blyholder's basic ideas are still widely employed, although we now know that rehybridization of CO also occurs when CO chemisorptive bonding involves multiple metal atom sites, compared to single site bonding for terminally bonded CO.

3. Thermal desorption of CO from polycrystalline tungsten

The thermal desorption of adsorbates from polycrystalline tungsten surfaces was first employed by Taylor and Langmuir in 1933 as a method for the study of chemisorbed Cs on W [18]. Twenty years later, Ehrlich and coworkers pioneered the use of the "flash filament" technique for the study of N_2, CO and Xe adsorption on polycrystalline W filaments [19]. Fast rates of temperature programming ($\sim 10^3$ K s^{-1}) were used to reduce the possibility of surface diffusion occurring from one crystal plane to another during the time of desorption, confusing the assignment of different "desorption states" to different exposed crystal planes. Redhead [20] employed much slower rates of temperature programming (~ 35 K s^{-1}) and in so doing, achieved additional resolution of the binding states which had been less well resolved by the faster desorption rates of Ehrlich. Fig. 1 shows the desorption spectra achieved [20], where a low temperature CO state (designated α) and at least three β states are resolved. The understanding of the origin of these states was a driving force in surface science research for the next 15 years.

One critical early experiment investigated the extent of "isotopic mixing" in chemisorbed carbon monoxide on polycrystalline W using both temperature-programmed desorption and flow techniques, in which the mixed isotopes, $^{13}C^{16}O$ and $^{12}C^{18}O$, were adsorbed and a mass spectrometer was used to observe scrambling of the isotopic atoms within the desorbing CO molecules. This type of experiment with CO was not a new idea, having been employed on Fe [21] and Ni [22] catalysts previously, but without the rapid temperature programming afforded by the use of a filament substrate. The first experiment, done

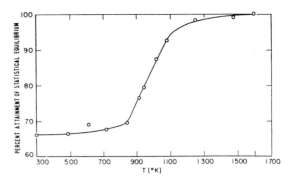

Fig. 2. Isotopic mixing between $^{12}C^{18}O$ and $^{13}C^{16}O$ as a function of the tungsten filament temperature in flowing gas [23].

by Ted Madey and myself at the beginning of our 19 years of continuous collaboration, is shown in Fig. 2, and demonstrates that only CO desorbing above about 850 K is isotopically mixed [23]. At the time of this experiment there was overwhelming belief that CO in the β states did not dissociate on W. This belief was based on field emission (no "O" patterns) and thermal desorption (no second-order kinetics). Since CO possesses the highest chemical bond strength of any molecule, we rejected the CO dissociation hypothesis and employed a bimolecular isotopic exchange model involving "inclined" CO molecules bound via both C and O atoms to surface W atoms. While it has now been proven that β-CO desorption is in fact due to the recombination of C(a) and O(a), produced from dissociated CO, it is of interest to note that based on more modern methods of investigation, "inclined" CO species are common on the early transition metals and exhibit highly weakened C–O bonds due to bonding of both C and O moieties to multiple substrate metal atoms [24]. Indeed, early transition metal carbonyls which bond both C and O moieties are now well recognized [24,25].

4. Field emission studies of CO chemisorbed on W – early work

The field emission microscope, with its use of ultrahigh vacuum technology, as well as its sensitivity to adsorption, has provided an historical pathway leading to the opening of the surface

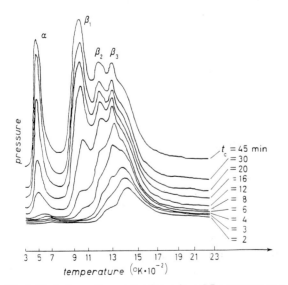

Fig. 1. Flash desorption spectra for various CO coverages on a tungsten filament [20].

science field [26]. After the invention of the field electron microscope in 1937 by Müller [27], the adsorption of carbon monoxide on tungsten was first studied by Müller using this method [28]. Gomer [29], and Ehrlich et al. [30], and Klein [31], quickly followed with studies of average work function changes and adsorbate surface migration behavior on the tungsten tip. These studies showed in various ways that multiple binding states of CO existed on the tungsten surface, but did not reveal the inherent dissociation of chemisorbed CO. These studies, and many more like them, were carried out in glass ultrahigh vacuum systems with low pumping speeds compared to the large, rapidly pumped metal vacuum systems of today. The use of the field emission microscope as a detector of surface contamination from poor vacuum, when combined with both the inverted Bayard–Alpert ionization gauge and modern high sensitivity mass spectrometers [32], firmly established the study of chemisorption on surfaces uncontaminated from the vacuum-a condition which did not exist in the earlier work on evaporated thin films and catalyst surfaces. One example of a glass vacuum apparatus designed to combine molecular beam dosing, temperature-programmed desorption and field emission to measure pulses of desorbing gas from a single crystal [33] comes from Gomer's laboratory and is shown in Fig. 3. This apparatus operated under liquid hydrogen. Professor Gomer once remarked to me (jokingly?) that the reason he liked surface science so much was the glass blowing challenge which apparatus like this presented to him personally.

5. Electron-stimulated desorption studies – early work

The interaction of electrons with surfaces provides a basis for a number of surface science methods which have developed historically. Among these, LEED, Auger spectroscopy, work function measurements, field electron emission, and a variety of electron scattering techniques all involve the passage of electrons through the surface region. It is therefore not surprising that

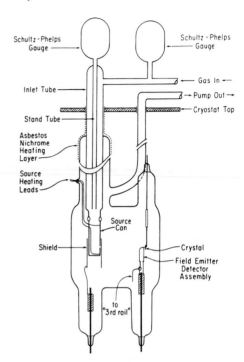

Fig. 3. Pyrex thermal desorption apparatus. A field emitter detector is used to intercept and measure pulses of desorbing gas from the crystal under study. The source can be used to deposit the gas initially on the single crystal [33].

early work in surface science examined the interaction of electrons with adsorbed molecules. Adsorbed carbon monoxide was, and is, one of the favorite target molecules for these studies. About 1000 papers have been written on electron-stimulated desorption (ESD), and a recent review covering the literature up to 1990 has been written [34]. More than 60 ESD papers deal with chemisorbed CO on 31 different types of surfaces.

The first interest in ESD phenomena was derived from two sources. Paul Redhead acquired an interest in the subject because of his work on ionization gauges, and the extension of their sensitivity to lower and lower pressure limits. It was found in careful studies that residual positive ion currents, originating from electron bombardment of the grid containing adsorbates, was responsible for a background effect in these gauges [35]. Robert Gomer and Dietrich Menzel came at the problem from a different perspective, using field

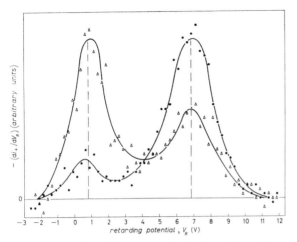

Fig. 4. Comparison of ion energy distributions at two adsorption temperatures for CO on polycrystalline W. $I_e = 20 \ \mu A$; $V_e = 100$ V; (Δ): full coverage CO at 195K; ($\bullet$): full coverage CO at 300 K [44].

electron emission to detect changes in adsorbed layers when a field emission tip was bombarded with electrons [36–38]. While the studies by Redhead, and Menzel and Gomer were key studies for the development of great interest in ESD, there were also earlier studies of this phenomenon [39–42]. Moore, for example, employed a mass spectrometer to detect O^+ ions originating from polycrystalline W or Mo targets containing chemisorbed CO and subjected to electron bombardment [42]. An early review of the beginnings of ESD research may be found in Ref. [43].

The existence of several adsorption states for CO on polycrystalline W, as seen originally in temperature-programmed desorption (Fig. 1), was also observed in ESD. This is graphically seen in a measurement of the ion energy distribution from CO layers prepared at 195 and at 300 K, and shown in Fig. 4. The low energy ions, now known to be CO^+, are desorbed along with higher energy ions (O^+). At 195 K adsorption temperature, CO^+ dominates; the opposite is true for 300 K adsorption [44]. Similar results were reported by Redhead [45], using a retarding potential analyzer which formed the basis for the design of the analyzer employed by ourselves [44]. A review of the historical development of electron-stimulated desorption is given by Ted Madey in this journal.

6. LEED and HREELS studies – CO on W – early work

The earliest example of CO chemisorption studies on a single crystal comes from the work of Davisson and Germer [46–49] who showed that the electron diffraction pattern from a Ni(111) crystal was altered upon the chemisorption of CO. For a number of years, LEED studies were carried out exclusively by Farnsworth at Brown University, until the advent of the post acceleration LEED imaging method [50,51]. Since this development, CO has been a favorite molecule for study by LEED, and many hundreds of metal crystals have been investigated using CO as the adsorbate [2].

The earliest example of the use of HREELS for the study of CO chemisorption comes, in fact, from the first demonstration of the HREELS principle by Propst and Piper [52]. They observed for CO adsorption on W(100) that the C–O stretching vibration could not be seen when the W(100) surface was in vacuum at about 450 K, suggesting that CO dissociation occurs upon chemisorption on W. A weakly bond CO state giving a weak C–O stretching vibration was found only under equilibrium gas pressure conditions. Fig. 5 shows their first spectrum of CO on W(100). This work was done with recognition of the earlier infrared work of Eischens on CO vibrational spectroscopy on catalytic surfaces.

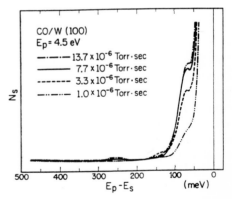

Fig. 5. Energy distribution of electrons scattered from W(100) after exposure to CO [52].

7. The discovery that CO dissociates on W

The recognition that CO dissociates on W came slowly at first from a number of different directions. I have already mentioned the pioneering work of Propst and Piper in 1968 which did not observe the expected 250 meV energy loss peak for strongly chemisorbed CO on W(100) [52]. This observation could be explained away, however, by considering the CO to be inclined on W(100) and, hence, not contributing a normal intensity C–O stretching mode at the normal loss energy, in agreement with the "inclined" structures proposed by Madey, Yates and Stern [23].

King, Goymour and Yates [53] reasoned that the monolayer of β-CO occupied 2 W atoms per CO molecule. This argument was based on integration of the evolution of CO in the β-CO state (a dangerous procedure for coverage estimates, even today), and on quantitative measurements of the CO coverage on W(100) by Anderson and Estrup [54], Clavenna and Schmidt [55], and on O_2 coverage measurements reported by Madey [56]. In the King et al. paper [53], a notable quote is: "The absolute coverage measurements can thus only be reconciled with the LEED data, accepting an unreconstructed model, if it is assumed that β-CO is dissociatively adsorbed, the diffraction centers being C and O atoms". I can remember the intense discussion of this sentence in the paper, and my insistence that all other explanations for non-dissociated CO be included in the final section of the paper also, since this conclusion about CO dissociation was heresy at the time and would put us into conflict with Gomer and Ehrlich who had advocated non-dissociative CO adsorption on W.

The first reflection IR work directed at CO adsorption on tungsten was carried out at the University of East Anglia while I was a visiting scholar in Dave King's laboratory [57]. The CO stretching modes for two different α-CO states were detected using a prewar Grubb Parsons prism-grating IR spectrometer and a W ribbon surface [58]. It was found, by comparison of the IR intensity and the thermal desorption yield, that partial conversion of the α_2-CO state (2090 cm^{-1}) to the β-CO state occurred irreversibly on

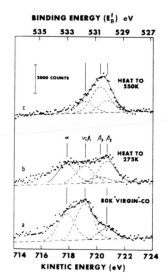

Fig. 6. (a) O(1s) spectra for monolayer CO chemisorption on W(100) at ~ 80 K; (b) virgin-CO to β-CO conversion by heating to 275 K; (c) desorption of α-CO by heating to 550 K [59].

warming the fully covered surface, and we now understand that this was due to CO dissociation. A similar conversion process was also observed later by XPS [59] as shown in Fig. 6, and a schematic drawing of the postulated CO dissociation process on W(100) is given in Fig. 7. These XPS results first showed that the O(1s) binding energy of the oxygen species derived from CO dissociation was identical to that derived from the dissociative adsorption of O_2 [59].

A definitive study of the character of β_3-CO on W(100) was carried out by Ward Plummer and his associates in the mid-1970's using ultraviolet

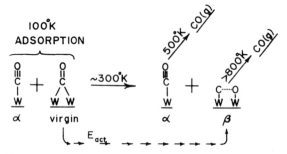

Fig. 7. Simplified picture of possible bonding structures for chemisorbed CO on tungsten. The structure of virgin α-CO is speculative, and a number of β-CO species exist together [59].

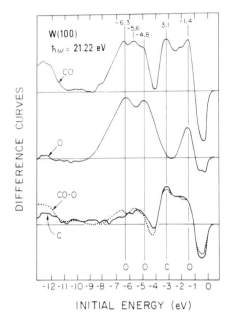

Fig. 8. UPS difference spectra for β_3-CO, O and C adsorbed in the c(2×2) structure on W(100). The oxygen curve follows O_2 adsorption and heating to 1500 K. The carbon curve follows C_2H_4 adsorption and heating to 1500 K [60].

photoelectron spectroscopy and a hemispherical retarding field electron energy analyzer [60]. This careful study, built on previous work reported here, compared the photoelectron spectrum from β_3-CO with a composite spectrum produced by adding the spectrum of adsorbed carbon and adsorbed oxygen as shown in Fig. 8. The key comparison is obtained in the bottom difference spectrum, where the difference between CO and O spectra is compared to a carbon spectrum produced from the adsorption of C_2H_4 on W(100) at 1500 K. The agreement conclusively shows that β_3-CO dissociates into C(a) and O(a) after heating to 1100 K. The strong evidence for CO dissociation is temperered somewhat by a cautionary statement referring to indirect bonding effects which may be present, leading to the c(2 × 2) structure. Similar measurements made for CO on W(110) were not so successful in demonstrating dissociative adsorption of CO in the β_2 state.

Thus, we can see that over more than a ten year period, the question of CO dissociation on tungsten surfaces occupied the attention of many

workers in the field of surface science and served as the test case for thermal, vibrational and electronic surface spectroscopies which were just coming into use.

8. Vignettes of more recent studies of CO surface behavior

8.1. ESDIAD studies of CO on Pt

The ESDIAD method for observing chemical bond orientations in chemisorbed surface species has been highly successful, and has been reviewed elsewhere in this issue by Ted Madey. We have recently used ESDIAD to monitor the growth of one-dimensional islands of CO on the steps of the Pt(112) surface [61–64]. As the coverage of CO is increased on the steps, tilt angles parallel to the steps are observed first as the molecules sterically interact; when step sites become filled, the tilt angles of the CO molecules revert to the orthogonal direction, up and down the steps of the crystal, as seen in Fig. 9. This observation is consistent with the idea that lateral tilting will occur when empty sites exist on the steps; when the steps are filled, the molecules can avoid each other only by tilting forward and backward. In another study, by examining the anisotropy of the librational amplitude of the isolated CO molecules on the steps, it was found that the largest freedom for libration occurs in the direction parallel to the steps. These measurements were made using both CO^+ and CO^*, a metastable CO species which is ejected with high relative yield during ESD of CO on Pt [65]. The CO^* species is ideal for ESDIAD studies because of its uncharged state, leading to only small dipole image interaction effects with the substrate.

8.2. IRAS studies of electron-stimulated migration (ESM) of CO on Pt

About 1000 papers have been written in which electron-stimulated desorption effects have been studied. Recently, using IRAS methods, we have observed and characterized the ESM process, in

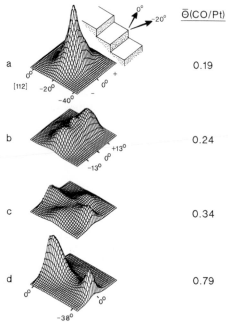

$\overline{\Theta}$(CO/Pt)

a   0.19

b   0.24

c   0.34

d   0.79

Fig. 9. ESDIAD studies of CO chemisorption on the steps of Pt(112). The metastable CO* species is being imaged. The coverages are average CO coverages on the stepped crystal surface. CO tilt angles to left and right ($\sim 13°$) occur at partial step filling, followed by CO tilt angles up and down stairs (0°; $-38°$) as the steps fill. Steric effects cause the orthogonal tilt directions to be occupied for different CO coverages. ESDIAD patterns for CO adsorption on the terrace sites are not shown [61].

which electronic excitation of a CO molecule causes it to move across the surface to be trapped at a vacant step site [66]. Fig. 10 shows the experimental results, where two isotopically different CO molecules have been sequentially adsorbed on the step and terrace sites of a Pt(335) surface. Electron bombardment leads to the desorption of CO from both types of sites, and to the transfer of the CO isotope on the terrace sites to the step sites. It is likely that the ESD process feeds a mobile precursor to adsorption and that this mobile species is preferentially trapped at the steps. This idea accounts for the observed one way transfer between terrace and step sites in the ESM process. The overall cross section of the ESM process is similar to that of the ESD process for CO on this surface.

8.3. Anisotropic surface diffusion of CO on Ni(110) using optical second harmonic diffraction

A new technique for studying surface diffusion has been used with CO to demonstrate the differing diffusion kinetics of CO on the highly corrugated Ni(110) surface [67]. Two interfering laser beams are used to etch a thermal gradient onto the Ni(110) crystal originally containing a monolayer of chemisorbed CO. A surface concentration grating with a period of about 20 μm may be etched in any desired azimuth on the crystal. Upon raising the crystal temperature, surface diffusion tends to level the concentration gradients, and the progress of this kinetic process can be

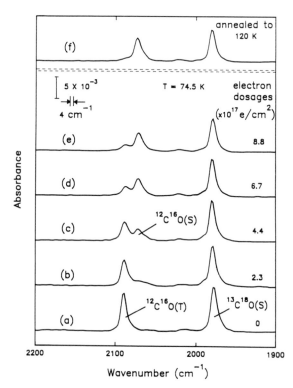

Fig. 10. Electron-stimulated migration (ESM) of CO from terrace sites to step sites. The experiment begins by filling the step sites on Pt(335) with $^{13}C^{18}O$ and then filling the terrace sites with $^{12}C^{16}O$. Electron bombardment desorbs CO from both types of sites and causes $^{12}C^{16}O$ (T) to convert to $^{12}C^{16}O$ (S). Annealing to 120 K causes the process to culminate. The ESM process is observed at a crystal temperature of 74.5 K, below the temperature needed for migration of CO by a thermal process on the time scale of this experiment [66].

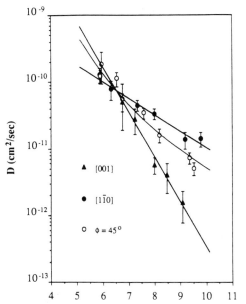

Fig. 11. Diffusion coefficient, D, versus reciprocal temperature in an Arrhenius plot for CO diffusion on Ni(110) along two principal crystal directions and in a direction bisecting the two ($\phi = 45°$) [67].

monitored by second harmonic diffraction. An Arrhenius plot of the results in three azimuthal directions is shown in Fig. 11. Diffusion along the ridges (the $\langle 1\bar{1}0 \rangle$ direction) is governed by an activation energy of 1.1 kcal/mol. Along the

$\langle 001 \rangle$ direction, diffusion occurs with an activation energy of 3.1 kcal/mol. There is a large difference in the preexponential factors, giving a preexponential factor in the $\langle 001 \rangle$ direction about 1000 times greater than in the $\langle 110 \rangle$ direction. This difference is tentatively attributed to a more shallow potential well at the saddle point for diffusion across the channels (long bridge sites) of the Ni(110) surface; diffusion along the ridge Ni sites occurs from atop sites to short bridge sites with a lower preexponential factor and a lower activation energy than in the orthogonal direction.

8.4. STM observation and manipulation of single CO molecules on the Pt(111) surface

The observation and manipulation of single adsorbate molecules on single crystal surfaces in the low temperature STM forms a pinnacle of achievement recognized worldwide for its beauty and its directness. One example [68] of this work is delightful to contemplate, as shown in Fig. 12. Here a single CO molecule is moved across a Pt(111) surface and is then imaged by the STM tip which was used for manipulation of the CO. It was found that the CO could be imaged in two forms – a short molecule and a tall molecule, as shown in the figure. These two forms of

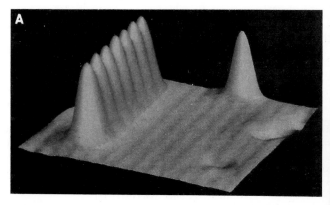

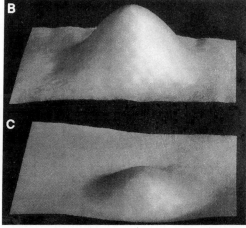

Fig. 12. (A) The first hand-built adsorbate structures-Xe on Ni(110) at 4 K; (B) CO on Pt(111), one site; (C) CO on Pt (111), another site. The adsorbate molecules can be moved from site to site using the STM tip. Two binding configurations are detected for CO on Pt(111) [68].

chemisorbed CO are assigned to the bridged form and the terminal form, present on multiple or single Pt sites, respectively. Irving Langmuir would truly be proud of this singular accomplishment, but I am sure he would also be asking about the next step in our modern inquiry into the nature of chemisorbed species.

9. Concluding remarks

Only a small portion of the work in surface science which is focussed on the chemisorbed CO molecule has been treated in this limited review. Studies of the surface properties of the CO molecule have been carried out over the past 70 years with two main goals. These are: (1) to understand the chemistry of the chemisorbed molecule itself, and (2) to use the CO molecule to devise and test new surface measurement methods which form a central focus in the field of surface science. The continued focus on chemisorbed CO suggests that still more beautiful and interesting surprises may be in store for us in the future.

Acknowledgements

I thank the Office of Basic Energy Sciences of The Department of Energy and the Air Force Office of Scientific Research for the support of the work described from my laboratory in this article.

References

[1] R.R. Ford, Carbon Monoxide Adsorption on Transition Metals, in: Adv. Catal. Relat. Subj., Vol. 21 (Academic Press, New York, 1970) p. 51.
[2] J.C. Campuzano, The Adsorption of Carbon Monoxide by the Transition Metals, in: The Chemical Physics of Solid Surfaces and Heterogeneous Catalysis Vol. 3, D.A. King and D.P. Woodruff, Eds. Part A, (Elsevier, Amsterdam, 1990) p. 389.
[3] The reader is referred to a complete collection of Irving Langmuir's work in: C.G. Suits and H.E. Way, Eds., The Collected Works of Irving Langmuir (Pergamon, Oxford 1960).
[4] A discussion of this is given elsewhere in this volume by G. Ertl.
[5] I. Langmuir, Trans. Faraday Soc., 17 (1922) 621.
[6] S. Brunauer and P.H. Emmett, J. Am. Chem. Soc. 57 (1935) 1754.
[7] S. Brunauer and P.H. Emmett, J. Am. Chem. Soc. 62 (1940) 1732.
[8] E.W. Rideal and B.M.W. Trapnell, Proc. R. Soc. London, Ser. A205 (1951) 409.
[9] O. Beeck, W.A. Cole and A. Wheeler, Disc. Faraday Soc. 8 (1950) 314.
[10] J.T. Kummer and P.H. Emmett, J. Am. Chem. Soc. 73 (1951) 2886.
[11] R.P. Eischens, S.A. Francis and W.A. Pliskin, J. Phys. Chem. 60 (1956) 194.
[12] R.P. Eischens and W.A. Pliskin, in: Adv. Catal. Relat. Subj., Vol. 10 (Academic Press, New York, 1958) p.1.
[13] R.P. Eischens, Private communication.
[14] J.T. Yates, Jr. and C.W. Garland, J. Phys. Chem. 65 (1961) 617.
[15] For an excellent review of the development of vibrational spectroscopy for CO chemisorption studies, see, N. Sheppard and T.T. Nguyen, in: Advances in Infrared and Raman Spectroscopy, Vol. 5, Eds. R.J. Clarke and R.E. Hester (Heyden, London, 1978).
[16] J.T. Yates, Jr. and T.E. Madey, Eds., Vibrational Spectroscopy of Molecules on Surfaces, in: Methods of Surface Characterization, Vol. 1 (Plenum, New York, 1987).
[17] G. Blyholder, J. Phys. Chem. 68 (1964) 2722.
[18] J.B. Taylor and I. Langmuir, Phys. Rev. 44 (1933) 423.
[19] A comprehensive review of Ehrlich's work may be found in: Adv. Catal. Relat. Subj. Vol. 14 (Academic Press, New York, 1963) p. 255.
[20] P.A. Redhead, Trans. Faraday Soc. 57 (1961) 641.
[21] A.N. Webb and R.P. Eischens, J. Am. Chem. Soc. 77 (1955) 4710; J. Chem. Phys. 20 (1952) 1048.
[22] J.T. Yates, Jr., J. Phys. Chem. 68 (1964) 1245.
[23] T.E. Madey, J.T. Yates, Jr. and R.C. Stern, J. Chem. Phys. 42 (1965) 1372.
[24] This type of bonding has been observed on Cr(110), Fe(100), Mo(100), and Mo(110). For a recent paper detecting inclined CO species on Mo(110) as intermediates before CO dissociation, see, M.L. Colaianni, J.G. Chen, W.H. Weinberg and J.T. Yates, Jr., J. Am. Chem. Soc. 114 (1992) 3735.
[25] W.A. Herrmann, H. Biersack, M.L. Ziegler, K. Weidenhammer, R. Siegel and D. Rehder, J. Am. Chem. Soc. 103 (1981) 1692.
[26] R. Gomer, Field Emission and Field Ionization (Harvard, University Press, Cambridge, MA, 1961).
[27] E.W. Müller Z. Phys. 106 (1937) 541.
[28] E.W. Müller, S. Flugge and F. Trendelenburg, Exakt. Naturwiss. 27 (1953) 330.
[29] R. Gomer, J. Chem. Phys. 28 (1958) 168.
[30] G. Ehrlich, T.W. Hickmott and F.G. Hudda, J. Chem. Phys. 28 (1958) 506.
[31] R. Klein, J. Chem. Phys. 31 (1959) 1306.

[32] P.A. Redhead, J.P. Hobson and E.V. Kornelsen, The Physical Basis of Ultrahigh Vacuum (Chapman and Hall, London, 1968).

[33] C. Kohrt and R. Gomer, J. Chem. Phys. 52 (1970) 3283; Surf. Sci. 24 (1971) 77; 40 (1973) 71.

[34] R.D. Ramsier and J.T. Yates, Jr., Surf. Sci. Rep. 12 (1991) 243.

[35] P.A. Redhead J. Vac. Sci. Technol. 7 (1970) 182.

[36] D. Menzel and R. Gomer, J. Chem. Phys. 40 (1964) 1164.

[37] D. Menzel and R. Gomer, J. Chem. Phys. 41 (1964) 3311.

[38] D. Menzel and R. Gomer, J. Chem. Phys. 41 (1964) 3329.

[39] Y. Ishikawa, Rev. Phys. Chem. Jpn. 16 (1942) 83, 117.

[40] Y. Ishikawa, Proc. Imp. Acad. (Tokyo) 19 (1943) 380, 385.

[41] Y. Ishikawa, Proc. Imp. Acad. (Tokyo) 18 (1942) 246, 390.

[42] G.E. Moore, J. Appl. Phys. 32 (1961) 1241.

[43] T.E. Madey and J.T. Yates, Jr., J. Vac. Sci. Technol. 8 (1971) 525.

[44] J.T. Yates, Jr., T.E. Madey and J.K. Payn, Nuovo Cimento Suppl. 5, series 1 (1967) 558.

[45] P.A. Redhead, Nuovo Cimento Suppl. 5, series (1967) 1, 586.

[46] C.J. Davisson and L.H. Germer, Phys. Rev. 30 (1927) 705.

[47] C.J. Davisson and L.H. Germer, Phys. Rev. 30 (1927) 634.

[48] C.J. Davisson and L.H. Germer, Proc. Natl. Acad. Sci. 14 (1928) 317.

[49] C.J. Davisson and L.H. Germer, Proc. Natl. Acad. Sci. 14 (1928) 619.

[50] E.J. Scheibner, L.H. Germer and C.D. Hartman, Rev. Sci. Instrum. 31 (1960) 112.

[51] L.H. Germer and C.D. Hartman, Rev. Sci. Instrum. 31 (1960) 784.

[52] F.M. Propst and T.C. Piper, J. Vac. Sci. Technol. 4 (1967) 53.

[53] D.A. King, C.G. Goymour and J.T. Yates, Jr., Proc. R. Soc. (London) A 331 (1972) 361.

[54] J. Anderson and P.J. Estrup, J. Chem. Phys. 46 (1967) 563.

[55] L.R. Clavenna and L.D. Schmidt, Surf. Sci. 33 (1972) 11.

[56] T.E. Madey, Surf. Sci. 36 (1973) 281.

[57] J.T. Yates, Jr., R.G. Greenler, I. Ratajezykowa and D.A. King, Surf. Sci. 36 (1973) 739.

[58] J.T. Yates, Jr. and D.A. King, Surf. Sci. 30 (1972) 601.

[59] J.T. Yates, Jr., N.E. Erickson, S.D. Worley and T.E. Madey, in: The Physical Basis of Heterogeneous Catalysis, Eds. E. Drauglis and R.I. Jaffee (Plenum, New York, 1975) p. 75.

[60] E.W. Plummer, B.J. Waclawski, T.V. Vorburger and C.E. Kuyatt, Prog. Surf. Sci. 7 (1976) 149.

[61] M.A. Henderson, A. Szabó and J.T. Yates, Jr., Chem. Phys. Lett. 168 (1990) 51.

[62] M.A. Henderson, A. Szabó and J.T. Yates, Jr., J. Chem. Phys. 91 (1989) 7245.

[63] M.A. Henderson, A. Szabó and J.T. Yates, Jr., J. Chem. Phys. 91 (1989) 7255.

[64] J.T. Yates, Jr., M.D. Alvey, M.J. Dresser, M.A. Henderson, M. Kiskinova, R.D. Ramsier and A. Szabó, Science 255 (1992) 1397.

[65] M. Kiskinova, A. Szabó, A.M. Lanzillotto and J.T. Yates, Jr., Surf. Sci. Lett. 202 (1988) L559.

[66] H.J. Jänsch, J. Xu and J.T. Yates, Jr., J. Chem. Phys. 99 (1993) 721.

[67] Xu-Dong Xiao, X.D. Zhu, W. Daum and Y.R. Shen, Phys. Rev. Lett. 66 (1991) 2352.

[68] J.A. Stroscio and D.M. Eigler, Science 254 (1991) 1391.

Surface Science 299/300 (1994) 742–754
North-Holland

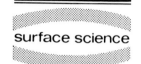

surface science

Reactions at well-defined surfaces

G. Ertl

Fritz-Haber-Institut der Max-Planck-Gesellschaft, Faradayweg 4–6, D-1000 Berlin 33, Germany

Received 22 March 1993; accepted for publication 14 May 1993

Early work (mainly by Langmuir) has erected the conceptual framework for chemical reactions occurring at well-defined solid surfaces and forming the basis for heterogeneous catalysis but experimental verification has been enabled only during the past three decades. My first contribution to Surface Science appeared in 1967 and concerned the interactions of various molecules with Cu single-crystal surfaces, mainly by using LEED. This technique revealed not only structural information, but also was applicable for studying kinetic phenomena as exemplified somewhat later with the catalytic oxidation of carbon monoxide on Pd(110). Some of the questions raised in this early work could be answered only quite recently or were revealed to be even more complex than anticipated. This holds in particular for catalytic reactions under steady-state conditions for which a wealth of phenomena of nonlinear dynamics, ranging from oscillatory or chaotic kinetics to spatio-temporal pattern formation, may occur.

1. Introduction: the Langmuir period

Well characterized (usually single crystalline) surfaces under low-pressure conditions are nowadays generally accepted as suitable model systems for studying the elementary processes underlying chemical reactions at surfaces which are of considerable practical relevance for phenomena such as heterogeneous catalysis, oxidation etc. This "surface science approach" became possible through the development of UHV techniques and of the arsenal of tools from surface physics which started about thirty years ago. Some of the basic underlying principles were, however, formulated much earlier. It was in particular Langmuir who introduced the concept of chemical interactions at well-defined surfaces without being able to conduct the appropriate experimental studies [1]:

"Most finely divided catalysts must have structures of great complexity. In order to simplify our theoretical consideration of reactions at surfaces, let us confine our attention to reactions on plane surfaces. If the principles in this case are well understood, it should then be possible to extend the theory to the case of porous bodies. In general, we should look upon the surface as consisting of a checkerboard..."

The idea of the "checkerboard" onto which chemisorption of particles from the gas phase may take place can first be found in Langmuir's laboratory notebook on April 21, 1915 [2]:

"1. The surface of a metal contains spaces according to a surface lattice.

2. Adsorption films consist of atoms or molecules held to the atoms forming the surface lattice by chemical forces.".

Based on these principles the famous Langmuir adsorption isotherm was derived, relating the equilibrium concentration (= coverage) of an adsorbed species with the corresponding partial pressure in the gas phase and with temperature [3]. Although intended to be valid for energetically a priori homogeneous surfaces as only presented by single-crystal planes, the Langmuir isotherm and kinetic expressions derived therefrom is widely and successfully applied for modeling the behavior of "real" surfaces. Ironically, careful measurements with single-crystal surfaces reveal that the adsorption equilibrium practically never follows the Langmuir isotherm. This is due to the fact that interactions between the adsorbed particles come into play which cause the adsorption energy to become coverage dependent and, in particular, frequently prevent the adsorbed

SSDI 0039-6028(93)E0375-5

particles to simply occupy at random the sites offered by the "checkerboard" up to saturation of the monolayer. The latter is usually not characterized by a ratio between the density of adsorbates and that of the surface atoms (= the coverage) of 1, and the operation of interactions between adsorbed species is inter alia reflected by the wealth of phases with long-range order and associated phase transitions as studied by LEED and other techniques. Langmuir himself recognized the complication of his model [4]:

"The properties of adsorbed molecules may often determine the amount adsorbed where the forces acting between adjacent adsorbed molecules are comparable with these holding the molecules on the surfaces".

As we now know, these interactions may be caused either by the modifications of the electronic properties of the substrate in the vicinity of an adsorbed species (indirect i.) or by orbital overlap as well as electrostatic interactions between them (direct i.). The latter become particularly manifest with systems of chemisorbed alkali-metal atoms, an area of research again established by Langmuir and also recognized by him in its basic nature [5]:

"When a cesium atom comes close to a tungsten surface, the tungsten robs the cesium atom of its valence electron and leaves it in the form of a univalent positive ion. This ion, however, tends to be held by a strong force to the tungsten surface because of the negative charge induced in the metallic surface".

Structural investigations for alkali-metal overlayers on extended single-crystal surfaces, being dominated by dipole–dipole repulsions, can already be found in early issues of Surface Science [6] and the whole field is still subject of intense experimental and theoretical studies [7].

There is another aspect invalidating the lattice gas concept which not only became manifest with alkali-metal adsorption, but whose general significance was recognised early: adsorbate-induced reconstruction of the structure of the substrate. Langmuir [3] speculated already in 1916:

"The atoms in the surface of a crystal must tend to arrange themselves so that the total energy will be a minimum. In general, this will involve a shifting of the positions of the atoms with respect to one another".

Based on observations with the field ion microscope, Sachtler coined the term "'corrosive chemisorption'" which was a heavily debated issue at a Discussion Meeting of the Faraday Society in 1966 [8]. Already from the first applications of the display-type LEED method, after the renaissance of this technique, adsorption-induced surface reconstruction was invoked to account for the high intensities of the "extra" spots [9]. The author published his first paper in Surface Science 6 (1967) [10]. Among others, it concerned LEED observations on the chemisorption of oxygen on a Cu(110) surface, and from the relative intensities of the diffraction spots it was likewise concluded that these were due to displacements of Cu atoms from their original positions within the unit cells. The 2×1- and $c(6 \times 2)$-phases observed had been the subject of numerous investigations in the meantime, and only recently conclusive structure determinations could be terminated [11]. Elucidation of the mechanism for the formation of the 2×1-phase by real-space imaging by means of scanning tunneling microscopy (STM) revealed that this is of the "added row" rather than "missing-row" type as a consequence of the presence of highly mobile Cu adatoms on the terraces, even at room temperature [12]. It is now evident that a metal surface is by no means a rigid lattice but that its structure may be subject to continuous fluctuations under the influence of the chemisorption bond.

In turn also the bonds within adsorbed molecules will be affected by the strong interaction with the surface, eventually leading to bond breaking or to the formation of new ones. Usually dissociative chemisorption of a diatomic molecule is rationalized in terms of a schematic one-dimensional potential diagram of the type shown in Fig. 1b and as first proposed by Lennard-Jones [13]. This implies the possible existence of an energy barrier marked by the crossing of the two potential curves characterizing the phenomenon of "activated adsorption" [14]. More illustrative than the Lennard-Jones potential is a representation of the potential energy as a function of two coordinates, namely the distance x of the

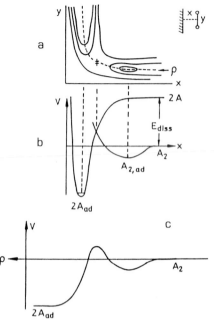

Fig. 1. Schematic potential diagram for dissociative chemisorption of a diatomic molecule with its axis parallel to the surface. (a) Two-dimensional contour plot. (b) One-dimensional Lennard-Jones diagram. (c) Variation of the potential along the reaction coordinate ρ.

molecule from the surface and the interatomic separation y within the molecule (Fig. 1a) [15]. The shallow minimum at large x then represents the molecular "precursor", the transition state ($\ddagger$) corresponds to the crossing of the two curves in Fig. 1b, and the valley at small x extending to increasing y to the dissociated state. The dashed line in Fig. 1a marks the reaction coordinate ρ, i.e. the path for the reaction with the minimum energy requirement. The variation of the potential along ρ is plotted in Fig. 1c.

This is just a special example for the description of the progress of a chemical reaction along the concepts of "reaction dynamics" as formulated first in 1931 by Eyring and Polanyi [16]. The multi-dimensional potential surface exhibits local minima for certain sets of nuclear coordinates representing the reaction intermediates and local maxima representing the transition states. Knowledge about the nature of intermediates and their transformation steps denotes the "reaction mechanism". Passage across the potential surface is

associated with continuous transformation of energy between the various degrees of freedom, as may be investigated experimentally by combination of molecular beam and laser-spectroscopic techniques [17]. Theoretical modeling, on the other hand, will provide a description of the reaction kinetics [18].

In general, kinetics describes the dependence of the rate of a chemical reaction on the externally variable parameters, viz. the partial pressures p_i and temperature T in the case of a surface reaction, $r = f(p_i, T)$. If the mechanism of a reaction is known in detail, the kinetics will be a natural consequence, but not vice versa! Conclusions about the mechanism from purely kinetic measurements may be very misleading, as demonstrated by numerous examples of surface catalyzed reactions.

"The reaction which takes place at the surface of a catalyst may occur by interaction between molecules or atoms adsorbed in adjacent spaces on the surface, (...) or it may take place directly as a result of a collision between a gas molecule and adsorbed molecule or atom on the surface".

In this way Langmuir [1] classified the two basic types of bimolecular surface reactions which are now denoted as Langmuir–Hinshelwood (LH, i.e. $A_{ad} + B_{ad} \rightarrow P$) and Eley–Rideal (ER, i.e. $A_{ad} + B_{gas} \rightarrow P$) mechanism, respectively. This conclusion was based on experimental investigations on the catalytic oxidation of carbon monoxide, $CO + \frac{1}{2}O_2 \rightarrow CO_2$, over a Pt wire [19].

He had found, like others before him [20], that in this case the steady-state reaction rate is proportional to the ratio of the partial pressures of O_2 and CO, $r \propto p_{O_2}/p_{CO}$. This puzzling rate law, whereafter an *increase* in the concentration (= partial pressure) of one of the reactants (CO) causes a *decrease* of the rate, was attributed by Langmuir to the inhibition of oxygen adsorption by the presence of adsorbed CO [1] whose cover-

---

[1] Langmuir's fascinating intuition is reflected by another remark in his laboratory notebook from 1915 [21]:
"*The CO molecules striking the clean Pt surface adhere to it (perhaps not all but a fraction a_1) probably largely because of the unsaturated character of their carbon atoms*".

age of course increases with p_{CO}. Since there is apparently no such inhibition effect for CO by the presence of adsorbed oxygen, he suggested that the reaction in fact proceeds through the ER mechanism by collision of CO from the gas phase with chemisorbed O atoms. This conclusion was wrong as will be outlined below. The reason is that with a surface reaction the rate will not primarily be governed by the partial pressures of the gaseous reactants, but rather by the concentrations and mutual configurations on the surface of the adsorbed intermediates. Implications of these effects on the overall rate expression are discussed in Weinberg's contribution [18], and it is evident that only experimental information about the state of the surface during the reaction will help to solve this problem.

2. The surface science approach: catalytic CO oxidation on platinum metals

After the possibility to monitor the occurrence of chemical reactions at well-defined single-crystal surfaces had been demonstrated in several cases [23], the first systematic study of a catalytic reaction along the "surface-science approach" was performed by the author together with his first graduate student, P. Rau [24]. It concerned the catalytic oxidation of CO at a Pd(110) surface, whereby the state of the surface was monitored by LEED, while simultaneously the stationary reaction rate (with the system operated under steady-state flow conditions) was recorded by means of a quadrupole mass spectrometer at various CO and O_2 partial pressures as well as temperatures. Conclusions about the nature of the surface under reaction conditions were drawn by comparing the LEED patterns with those observed after separate adsorption of the reactants.

Fig. 2 shows the variation of the reaction rate r with temperature T at a fixed composition of the gas-phase $p_{O_2} = 1 \times 10^{-7}$ Torr, $p_{CO} = 1.6 \times 10^{-7}$ Torr. (Note that in those days Surface Science was still a trilingual journal, hence, this paper was written in German.) Below about 100°C the LEED pattern shows spots from the CO adlayer signaling high coverages of this species. With

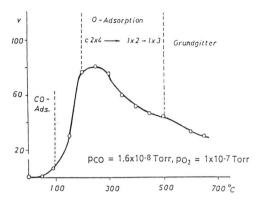

Fig. 2. Variation of the rate of CO_2 production on a Pd(110) surface with temperature under steady-state flow conditions with constant CO and O_2 partial pressures. The state of the surface was simultaneously monitored by LEED [22].

increasing temperature the rate of CO desorption increases and, hence, the stationary CO coverage drops, the inhibiting effect for oxygen adsorption is weakened. Since the reaction requires chemisorbed O atoms the rate hence increases continuously until it reaches its maximum in the temperature range between 200 and 300°C.

At even higher temperatures also depletion of the surface from oxygen comes increasingly into play, as reflected by passing through the sequence of oxygen-derived LEED patterns with decreasing coverages, until above 500°C the surface concentrations are so low that only the substrate spots are visible. Measurements at constant temperature with variation of either p_{CO} or p_{O_2}, on the other hand, confirmed qualitatively the previous findings about the rate law $r \propto p_{O_2}/p_{CO}$, with no indication for the occurrence of noticeable inhibition by adsorbed oxygen. Hence, again (erroneously!) the operation of the ER mechanism was suggested.

Subsequent detailed coadsorption experiments extending even to lower temperatures [25] revealed that the situation is definitely more complex: CO tends to form densely packed adlayers which inhibit dissociative oxygen adsorption for which a large ensemble of neighboring free surface atoms is required. The chemisorbed O atoms, on the other hand, form rather open structures which do not noticeably affect additional uptake of CO. Hence, even with the Langmuir–

Hinshelwood mechanism the reaction rate does not decrease with increasing p_{O_2}, even if the surface is saturated with adsorbed O atoms. The situation is illustrated schematically in Fig. 3. Under these circumstances a clear distinction between the two possible mechanisms on the basis of stationary measurements becomes impossible but requires time-resolved experiments: with the ER mechanism product formation takes place in a collision, i.e. on a time scale of $\sim 10^{-12}$ s, which is much shorter than the flight time in the gas phase, while with the LH mechanism the CO molecule becomes first adsorbed and might have a measurable (even if short) average lifetime on the surface before the product molecule is formed.

Since the latter is instantaneously released into the gas phase a finite delay time between impact of CO and detection of CO_2 directly proves the operation of the LH mechanism. Detailed experi-

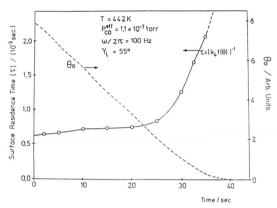

Fig. 4. Modulated molecular-beam experiment on the catalytic CO oxidation on Pt(111) [27]. For explanation see text.

ments by applying a modulated molecular beam relaxation technique yielding this kind of information were performed with Pd(111) [26] as well as with Pt(111) [27].

The result of a typical experiment is reproduced in Fig. 4. At $t = 0$, an O-covered Pt(111) surface at 442 K is exposed to a periodically chopped CO beam with an effective partial pressure of 1.1×10^{-7} Torr. From the phase lag between the incident CO and released CO_2 molecules the average surface lifetime of CO_{ad} is determined to be about 5×10^{-4} s, which definitely rules out an ER-type mechanism. Near full consumption of the amount of O_{ad} present on the surface, τ even increases continuously since CO molecules diffusing across the surface need longer and longer for a successful collision with O_{ad}. (Note that the lifetime of CO_{ad} against desorption is of the order of about 10 s at this temperature.)

The basic mechanism underlying catalytic CO oxidation on surfaces of the platinum group metals is illustrated schematically by Fig. 3 and has been verified for a large number of systems [28]. The potential diagram indicates that the enthalpy for the overall reaction is essentially released during the chemisorption steps. However, the adsorption energies and the activation energy for the LH step as well as the sticking coefficients are dependent on the coverages of the two surface species which, in addition, will not be randomly distributed. Hence, modeling of the kinet-

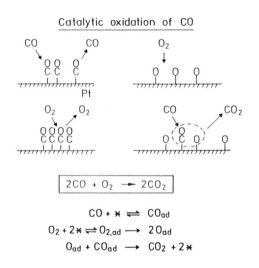

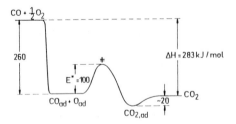

Fig. 3. Schematic reaction mechanism and potential diagram for catalytic CO oxidation on platinum.

ics in terms of simple Langmuir-type descriptions seems not to be feasible. Nevertheless, this could be achieved successfully, even for high pressures, with parameters for the elementary steps derived from surface-science studies [29]. For high CO coverages the rate can indeed be approximated by a relation $r \propto p_{O_2}/p_{CO}$ and becomes thus independent of the total pressure as has recently nicely been verified for pressure variations over 6 orders of magnitude [30].

Perhaps the most spectacular example is offered by the catalytic synthesis of ammonia for which reaction detailed information about the individual reaction steps obtained from UHV studies with Fe single-crystal surfaces [31] enabled successful modeling of the kinetics in industrial plants [32].

Although the CO oxidation reaction described in some detail represents probably the simplest surface-catalyzed reaction (except for isotope exchange reactions of the type $H_2 + D_2 \rightleftarrows 2HD$), more detailed inspection reveals a series of complications.

2.1. Influence of the surface structure

Fig. 5 shows the variation of the rate of CO_2 formation on Pt(110) as a function of p_{CO} for fixed values of the two other control parameters, p_{O_2} and T [33]. The full curve was recorded with a "flat" surface, while the broken line resulted from experiments with a sample exhibiting a significant concentration of steps on the surface. At

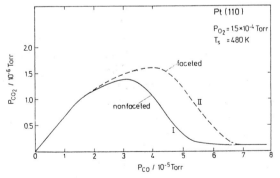

Fig. 5. Variation of the rate of CO_2 formation with CO partial pressure at fixed values for p_{O_2} and T on a Pt(110) surface which is either flat (full line) or faceted (broken line) [31].

low CO pressures the surface is mainly covered by adsorbed oxygen. The rate increases linearly with p_{CO}, since it is limited by adsorption of CO in this range. Since the sticking coefficient for CO is close to unity and is not noticeably affected by the presence of steps, the data for both types of samples coincide in this range: the reaction is structure-insensitive. With further increase of p_{CO}, however, the surface becomes increasingly covered by adsorbed CO which inhibits oxygen adsorption, so that the rate passes through a maximum and then decreases roughly $\propto 1/p_{CO}$. Now the rate is determined by dissociative chemisorption of oxygen. The sticking coefficient for this process is increased by the presence of surface defects, and, hence, for identical conditions, the faceted surface exhibits a higher reactivity: now the reaction is structure-sensitive.

To complicate the situation even further: the data of Fig. 5 were in fact obtained with the sample after different total reaction periods. For conditions in the high CO-coverage regime the rate (at fixed control parameters) continuously increases, which effect is accompanied by a continuous splitting of the LEED spots signaling the formation of facets composed of periodic step and terrace configurations [33].

With the present system this structural modification of the surface is not simply a consequence of the thermodynamic tendency to minimize the surface free energy, since heating in either one of the reacting gases alone restores the initial flat structure. Microscopically, the restructuring is based on the continuous lifting of the reconstruction of the Pt(110) surface by adsorbed CO and its reactive removal. Computer simulations with such a model revealed indeed the formation of a periodic hill-and-valley structure [34]. Direct imaging by STM of this type of reaction-induced surface morphology is shown in Fig. 6 for Pt(210) which had been subject to the same reaction [35]. The crucial role of the microscopic adsorbate-induced transformation of the surface unit cell for the occurrence of kinetic oscillations and related phenomena will be outlined in the next section.

Changes of the surface topography under the influence of an ongoing catalytic reaction have already been observed quite frequently [36], and

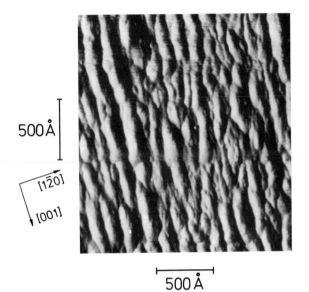

500 Å

Fig. 6. Morphology of a Pt(210) surface as probed by STM after catalytic CO oxidation for 1 h at 450 K with $p_{O_2} = 1.5 \times 10^{-4}$ Torr and $p_{CO} = 5 \times 10^{-4}$ Torr [33].

indeed even Langmuir reported about such an effect associated with CO oxidation at a Pt wire [1]:

> "*Closer examination shows that the values for k tended to increase steadily, indicating that the filament was undergoing a progressive change in the direction of becoming a better catalyst. (...) There is good evidence that the effect is caused by changes in the structure of the surface itself, brought about by the reaction. After the wire has been used, the surface becomes very rough*".

2.2. Participation of subsurface oxygen species

Contrary to the above statements valid for low-pressure conditions, an inhibitory effect of oxygen on the reaction rate has indeed been found at higher pressures [37]. This is attributed to the formation of another type of oxygen which is not adsorbed *on* the surface but rather located in sites *below* the topmost atomic layer and which reduces the sticking coefficient as well as the adsorption energy for CO [38]. Although definite structural information is still missing, there is general agreement that with O/Pd(110) the LEED patterns formed for coverages > 0.5 [22]

are associated with such "subsurface" species which, however, are distinctly different from genuine bulk oxides [39]. The participation of these species in the catalytic CO oxidation on Pd(110) is, inter alia, made responsible for the occurrence of kinetic oscillations at O_2 pressure above $\sim 10^{-3}$ Torr [38], as well as for the observed change of the hysteresis behavior of the reaction rate at elevated oxygen pressures [40,41].

2.3. Transient oxygen species: "hot" adatoms

Hints for the participation of another oxygen species with quite different nature stem from TPR experiments by Matsushima [42]. After adsorption of CO on an oxygen-precovered Pt(111) surface at low temperature the temperature of the sample was continuously raised and the evolution of CO_2 monitored by mass spectrometry. If the adsorbed oxygen species was atomic, CO_2 formation showed up as a peak around 300 K. If, however, the experiment was started with molecularly adsorbed oxygen, CO_2 evolution was observed already at considerably lower temperature, namely about 150 K. O_2 forms two types of molecular adsorbates on the Pt(111) surface with their molecular axis parallel to the surface and vibration frequencies of 87 meV (bridge species = peroxo type) and 108 meV (a top site = superoxo type) [43]. If such an overlayer is heated up, dissociation takes place at 150 K – just the temperature at which also CO_2 production is observed in the presence of coadsorbed CO. This suggests the following interpretation in terms of Fig. 1: the energy released during dissociation of the molecular precursor (in the case O_2/Pt(111) of the order 3 eV) will be preferentially channeled into motions of the O atoms parallel to the surface ("hot" adatoms). Upon collision with a co-adsorbed CO molecule, the excess energy will suffice to overcome the activation barrier for CO_2 formation (< 1 eV) so that instantaneous product formation takes place.

This picture gains additional support by experiments performed by Mieher and Ho [44]. Molecularly adsorbed oxygen on Pt(111) may also be dissociated photochemically by absorption of UV photons [45]. If a composite O_2/CO overlayer at

100 K is irradiated by UV light, again the release of CO_2 is observed, while no such reaction occurs with (ground-state) atomically adsorbed oxygen. The explanation is quite analogous as before, in that hot adatoms are now the result of photochemical dissociation as a consequence of light-induced electronic excitation.

Direct evidence for the formation of hot adatoms in the course of dissociative adsorption was recently found in an STM study on the interaction of oxygen with an Al(111) surface [46]. Thermally equilibrated O adatoms are completely immobile at room temperature on this surface. However, dissociative chemisorption was found not to produce pairs of adjacent adatoms, but these separate from each other by at least about 100 Å before the excess energy is dissipated. This corresponds to a lifetime of the "hot" adatoms of the order of more than 10^{-12} s. At higher coverages, on the other hand, equilibrated adatoms are very efficient in stopping the ballistic particles. (Because of the equal masses the conditions for momentum and energy transfer are favorable.) As a consequence, clusters of adhering adatoms are formed as can be seen from the STM image reproduced in Fig. 7, despite the fact that thermally accommodated O adatoms are immobile. (A single particle is, e.g., discernible near the lower boundary of this image; it did not change its position over prolonged periods of observation.)

3. Oscillatory kinetics and spatio-temporal self-organisation

The rate of a chemical reaction occurring under continuous-flow conditions at constant external control parameters (temperature, partial pressures) is not necessarily stationary, but may vary with time – even if there is no irreversible change of the state of the surface. First observations of oscillatory kinetics were reported in 1970 for the catalytic CO oxidation over supported platinum catalysts [47], and since then numerous studies of this type with "real" catalysts were performed [48]. Experiments with the same reaction on well-defined Pt surfaces under UHV con-

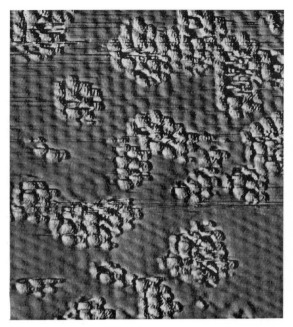

Fig. 7. STM image of an Al(111) surface with chemisorbed O atoms forming 1×1 islands [45].

ditions were first described in 1982 [49]. For special conditions the rate of CO_2 formation was found to vary periodically with time for Pt(100) and a polycrystalline Pt wire, but not with a

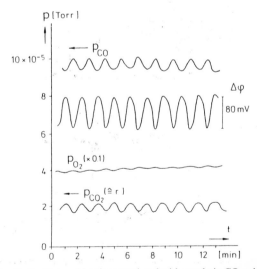

Fig. 8. Oscillatory kinetics associated with catalytic CO oxidation on a polycrystalline Pt wire under steady-state flow conditions [48]. $T = 502$ K, $p_{O_2} = 4 \times 10^{-4}$ Torr, p_{CO}(Av.) = 1×10^{-4} Torr.

defect-free Pt(111) surface. As an example, Fig. 8 shows data taken with a platinum wire for the rate of CO_2 formation r and for the simultaneously recorded CO and O_2 partial pressures, as well as of the variation of the work function, $\Delta\varphi$. The partial pressures vary out of phase with r because of the varying gas consumption. $\Delta\varphi$, on the other hand, reflects the oxygen coverage and changes parallel to r. With Pt(100) the conditions for oscillations were found to coincide with those for the CO-induced structural transformation of the surface which process was suggested to determine the mechanism of this effect. The coupling of periodic structural transformation and kinetic oscillations was directly verified by parallel LEED and work-function measurements on Pt(100) as shown in Fig. 9 [50]. The adsorbate-induced re-

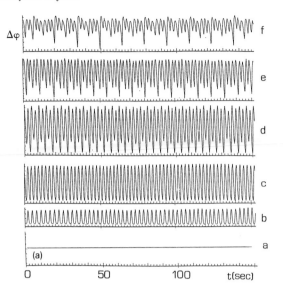

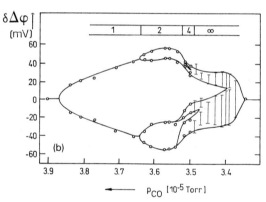

Fig. 10. (a) Temporal variation of the work function during CO oxidation on Pt(110) at $T = 540$ K, $p_{O_2} = 7.5 \times 10^{-6}$ Torr. The CO partial pressure was varied between 3.90×10^{-5} Torr ((a): constant rate = fixed point) and 3.42×10^{-5} Torr ((f): chaotic behavior). (b) Bifurcation diagram derived from the experimental data [54].

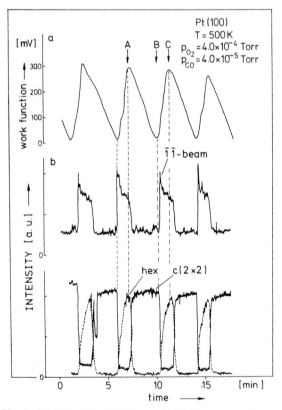

Fig. 9. Catalytic CO oxidation on Pt(100) under oscillatory conditions [49]. (a) Variation of the work function (= reaction rate) with time. (b) Simultaneously recorded intensities of characteristic LEED spots, signaling the periodic hex $\rightleftarrows$ c(2×2)-1×1 transformation of the surface structure.

construction mechanism does not require continuous transport of surface atoms over large distances, as might be suggested by the varying density of surface atoms, but it involves indeed only displacements over only a few lattice constants as demonstrated by STM observations [51]. In the meantime, numerous reports on oscillatory kinetics with various reactions occurring on single-crystal surfaces (in part also governed by other mechanisms) appeared in the literature [52]. A particularly rich scenario is offered by the CO oxidation on Pt(110) [53] which has recently been studied in great detail, both experimentally and theoretically [54]. An interesting aspect concerns the transition from regular periodic oscillations to chaos via a sequence of period doublings (Feigenbaum transition) by variation of one of the control parameters as reproduced in Fig. 10 [55]. Still other effects are observed if one of the control parameters is periodically modulated with small ($\sim 1\%$) amplitudes and varying frequency, causing forced oscillations with either fixed (= entrainment) or continuously varying (= quasi-periodicity) phase relation [56] which are well reproduced by a theoretical model based on the kinetics of the individual reaction steps [57].

Description of merely the integral temporal

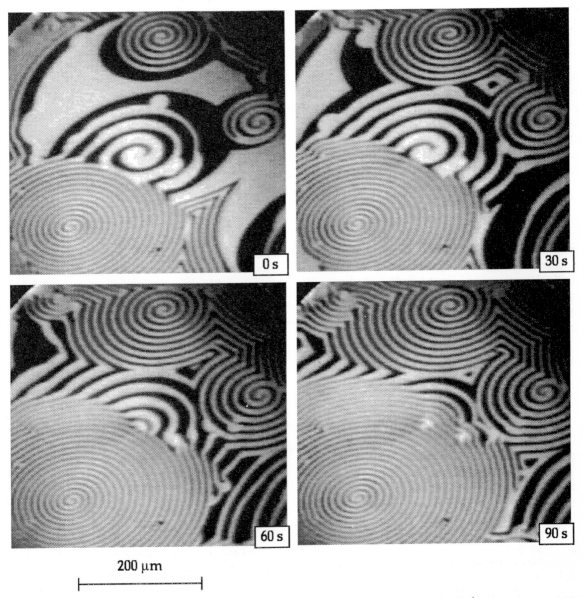

Fig. 11. PEEM images from a Pt(110) surface during catalytic CO oxidation at $T = 448$ K, $p_{O_2} = 4 \times 10^{-4}$ mbar and $p_{CO} = 4.32 \times 10^{-5}$ mbar, exhibiting the evolution of oxygen-concentration spirals with varying wavelengths and rotation periods [63].

behavior in terms of coupled, nonlinear ordinary differential equations is, however, incomplete since in general the surface concentrations will not only be functions of time but also of the spatial coordinates on the surface. Experimental verification of phenomena of spatio-temporal pattern formation was first achieved by a scanning LEED technique [58], and later with considerably improved resolution with scanning photoemission microscopy [59], as well as photoemission electron microscopy (PEEM) [60] and low-energy electron microscopy (LEEM) [61]. Most recently even field electron and field ion microscopy were demonstrated to produce time-resolved information nearly down to the atomic scale [62]. The interesting features of the concentration pattern occur, however, on the *mezoscopic* length scale of μm, where they can be followed in real time most conveniently by PEEM [54]. The contrast of this technique is based on the difference in (local) work function and, hence, areas in different stages of adsorption can readily be distinguished. With Pt(110), for example, oxygen-covered regions emit much fewer electrons (and, hence, appear dark on the fluorescence screen) than those which are predominantly covered by adsorbed CO. From the wealth of patterns observed, Fig. 11 shows, as an example, the development of spirals [63]. Analysis and theoretical modelling of these phenomena within the conceptual framework of nonlinear dynamics is a topic of strong current activity.

4. Conclusions

This article was not intended as a review, but rather as a subjective view on the development of a field illustrated by selected examples. Although detailed information is so far still restricted to fairly simple systems, closer inspection shows that even with those the situation may become far more complex than originally believed: the structure (and, hence, reactivity) of the surface may be modified by the strong bonds formed with the adsorbates. Adsorbate–adsorbate interactions, on the other hand, will modify the energetics and kinetics beyond the simple concepts of the Lang-

muir approach. As a consequence, even mean-field type approximations, according to which rates are just formulated in terms of relations of the concentrations (= coverages), may become questionable, so that computer simulations employing Monte Carlo techniques are more appropriate [18]. The results described in the previous section illustrate how completely new phenomena of spatio-temporal self-organization on a mezoscopic scale can come into play. Finally, there is still lack of truly time-resolved experimental observation of surface reaction dynamics, although there have been promising applications of fs laser techniques to elucidate the energy flow between surface and adsorbate [64]. With homogeneous reactions in the gas phase this goal has nearly been reached [65].

Nevertheless, studies of reactions on well-defined surfaces are nowadays considered as appropriate models for "real" catalysis. To quote just two representatives from the latter area:

"Surface science has shown that catalytic molecular design is possible. (...) Increasingly single-crystal studies define the active site structure desired in the industrial catalyst" [66].

And

"Probably the most important conceptual benefit which studies of this sort have to offer to catalytic science as a whole is a framework of mechanistic concepts, principles, and insights, which should be part of the intellectual equipment of anyone working in the field of heterogeneous catalysis" [67].

References

[1] I. Langmuir, Trans. Faraday Soc. 17 (1922) 607.
[2] I. Langmuir, Laboratory Notebook, April 21, 1915. Quoted from G.L. Gaines and G. Wise, in: Heterogeneous Catalysis – Selected American Histories, Eds. B.H. Davis and W.P. Hettinger (American Chemical Society, Washington, DC, 1983) p. 13.
[3] I. Langmuir, J. Am. Chem. Soc. 38 (1916) 2221.
[4] I. Langmuir, J. Am. Chem. Soc. 40 (1918) 1361.
[5] I. Langmuir and K.H. Kingdon, Science 37 (1923) 58.
[6] (a) R.L. Gerlach and T.N. Rhodin, Surf. Sci. 10 (1968) 446; 17 (1969) 32; 19 (1970) 403;
   (b) A.U. MacRae, K. Müller, J.J. Lander and J. Morrison, Surf. Sci. 15 (1969) 483.

[7] H.P. Bonzel, A.M. Bradshaw and G. Ertl, Eds., Physics and Chemistry of Alkali Metal Adsorption (Elsevier, Amsterdam, 1989).

[8] A.A. Holscher and W.M.H. Sachtler, Disc. Faraday Soc. 41 (1966) 29, and subsequent discussion remarks by G. Ehrlich and others.

[9] L.H. Germer, A.U. MacRae and C.D. Hartman, J. Appl. Phys. 32 (1961) 2432.

[10] G. Ertl, Surf. Sci. 6 (1967) 208.

[11] (a) S.R. Parkin, H.C. Zheng, M.Y. Zhou and K.A.R. Mitchell, Phys. Rev. B 41 (1990) 5432;
(b) R. Feidenhans'l, F. Grey, M. Nielsen, F. Besenbacher, F. Jensen, E. Læsgaard, I. Stensgaard, K.W. Jacobsen, J.K. Nørskov and R.L. Johnson, Phys. Rev. Lett. 65 (1990) 2027.

[12] (a) D.J. Coulman, J. Wintterlin, R.J. Behm and G. Ertl, Phys. Rev. Lett. 64 (1990) 1761;
(b) F. Jensen, F. Besenbacher, E. Læsgaard and I. Steensgaard, Phys. Rev. B 41 (1990) 10233.

[13] J.E. Lennard-Jones, Trans. Faraday Soc. 28 (1932) 333.

[14] (a)H.S. Taylor, J. Am. Chem. Soc. 53 (1931) 578;
(b) G. Ehrlich, in: Chemistry and Physics of Solid Surfaces, Vol. VII, Eds. R. Vanselow and R.F. Howe, (Springer, Berlin, 1988) p. 1 – review.

[15] See, e.g., G. Ertl, Ber. Bunsenges. Phys. Chem. 86 (1982) 425.

[16] H. Eyring and M. Polanyi, Z. Phys. Chem. 12 B (1931) 279.

[17] See, e.g., R.R. Cavanagh, E.J. Heilweil and J.C. Stephenson, Surf. Sci. 299/300 (1994) 643;
S. Holloway, Surf. Sci. 299/300 (1994) 656;
J. Tully, Surf. Sci. 299/300 (1994) 667.

[18] See, e.g., C.H. Kang and W.H. Weinberg, Surf. Sci. 299/300 (1994) 755.

[19] I. Langmuir, Trans. Faraday Soc. 17 (1922) 621.

[20] (a) M. Bodenstein and F. Ohlmer, Z. Phys. Chem. 53 (1905) 166;
(b) M. Bodenstein and C.G. Fink, Z. Phys. Chem. 60 (1907) 1.

[21] I. Langmuir, Laboratory Notebook, Feb. 2, 1915, loc. cit.

[22] G. Ertl and P. Rau, Surf. Sci. 15 (1969) 443.

[23] (a) G. Ertl, Surf. Sci. 7 (1967) 309;
(b) P.J. Estrup and J. Anderson, J. Chem. Phys. 49 (1968) 523;
(c) M. Onchi and H.E. Farnsworth, Surf. Sci. 11 (1968) 203.

[24] G. Ertl and P. Rau, Surf. Sci. 15 (1969) 443.

[25] H. Conrad, G. Ertl and J. Küppers, Surf. Sci. 76 (1978) 323.

[26] T. Engel and G. Ertl, J. Chem. Phys. 69 (1978) 1267.

[27] C.T. Campbell, G. Ertl, H. Kuipers and J. Segner, J. Chem. Phys. 73 (1980) 5862.

[28] T. Engel and G. Ertl, Adv. Catal. 28 (1979) 1.

[29] (a) S.H. Oh, G.B. Fisher, J.E. Carpenter and D.W. Goodman, J. Catal. 100 (1986) 360;
(b) A.G. Sault and D.W. Goodman, Adv. Chem. Phys. 76 (1989) 153.

[30] L. Kieken and M. Boudart, Catal. Lett. 17 (1993) 1.

[31] G. Ertl, in: Catalytic Ammonia Synthesis, Ed. J.R. Jennings (Plenum, New York, 1991) p. 109.

[32] (a) P. Stoltze and J.K. Nørskov, Phys. Rev. Lett. 55 (1985) 2502; Surf. Sci. 197 (1988) 230;
(b) M. Bowker, I. Parker and K. Waugh, Appl. Catal. 14 (1985) 101; Surf. Sci. 197 (1988) L233;
(c) J.A. Dumesic and A.A. Trevino, J. Catal. 116 (1989) 119.

[33] S. Ladas, R. Imbihl and G. Ertl, Surf. Sci. 197 (1988) 153.

[34] R. Imbihl, A.E. Reynolds and D. Kaletta, Phys. Rev. Lett. 67 (1991) 275.

[35] M. Sander, R. Imbihl, R. Schuster, J.V. Barth and G. Ertl, Surf. Sci. 271 (1992) 159.

[36] M. Flytzani-Stephanopoulos and L.D. Schmidt, Prog. Surf. Sci. 9 (1979) 83.

[37] P.J. Berlowitz, C.H.F. Peden and D.W. Goodman, J. Chem. Phys. 92 (1988) 5213.

[38] (a) J. Goschnik, M. Grunze, J. Loboda-Cackovic and J.H. Block, Surf. Sci. 189/190 (1987) 137;
(b) S. Ladas, R. Imbihl and G. Ertl, Surf. Sci. 280 (1993) 14.

[39] (a) J.W. He, U. Memmert, K. Griffiths and P.R. Norton, J. Chem. Phys. 90 (1989) 5082;
(b) M. Jo, Y. Kuwahara, M. Onchi and M. Nishijima, Chem. Phys. Lett. 131 (1986) 106.

[40] (a) M. Ehsasi, C. Seidel, H. Ruppender, W. Drachsel, J.H. Block and K. Christmann, Surf. Sci. 210 (1989) L198;
(b) S. Ladas, R. Imbihl and G. Ertl, Surf. Sci. 219 (1989) 88.

[41] (a) M. Ehsasi, M. Berdau, T. Rebitzki, K.-P. Charlé, K. Christmann and J.H. Block, to be published;
(b) N. Hartmann and R. Imbihl, to be published.

[42] T. Matsuhima, Surf. Sci. 127 (1983) 403.

[43] (a) J.L. Gland, B.A. Sexton and G.B. Fisher, Surf. Sci. 95 (1980) 587;
(b) H. Steininger, S. Lehwald and H. Ibach, Surf. Sci. 123 (1982) 1;
(c) D.A. Outka, J. Stöhr, W. Jarke, P. Stevens, J. Solomun and R.J. Madix, Phys. Rev. B 35 (1987) 4119;
(d) I. Panas and P. Siegbahn, Chem. Phys. Lett. 153 (1988) 458;
(e) W. Wurth, J. Stöhr, P. Feulner, X. Pan, K.R. Bauchspiess, Y. Baba, E. Hudel, G. Rocker and D. Menzel, Phys. Rev. Lett. 65 (1990) 2426.

[44] W.D. Mieher and W. Ho, J. Chem. Phys. 91 (1989) 2755.

[45] (a) X. Guo, L. Hanley and J.T. Yates, Jr., J. Chem. Phys. 90 (1989) 5200;
(b) L. Hanley, X. Guo and J.T. Yates, Jr., J. Chem. Phys. 91 (1989) 7220;
(c) X.Y. Zhu. R. Hatch, A. Campion and J.M. White, J. Chem. Phys. 91 (1989) 5011;
(d) A. Hoffmann, X. Guo, J.T. Yates, Jr., J.W. Gadzuk and C.W. Clark, J. Chem. Phys. 90 (1989) 5793.

[46] (a) H. Brune, J. Wintterlin, R.J. Behm and G. Ertl, Phys. Rev. Lett. 68 (1992) 624;
(b) H. Brune, J. Wintterlin, J. Trost, G. Ertl, J. Wiechers and R.J. Behm, J. Chem. Phys., submitted.

[47] (a) P. Hugo, Ber. Bunsenges. Phys. Chem. 84 (1970) 121;
(b) H. Beusch, P. Fieguth and E. Wicke, Chem. Ing.-Techn. 44 (1972) 445.

[48] (a) M. Sheintuch and R.A. Schmitz, Catal. Rev. 15 (1977) 107;
(b) M.G. Slinko and M.M. Slinko, Catal. Rev. 17 (1978) 119;
(c) F. Razón and R.A. Schmitz, Catal. Rev. 28 (1986) 89;
(d) F. Schüth, B.E. Henry and L.D. Schmidt, Adv. Catal. 39 (1993) 51.

[49] G. Ertl, P.R. Norton and J. Rüstig, Phys. Rev. Lett. 49 (1982) 177.

[50] M.P. Cox, G. Ertl, R. Imbihl and J. Rüstig, Surf. Sci. 134 (1983) L517.

[51] (a) T. Gritsch, D. Coulman, R.J. Behm and G. Ertl, Phys. Rev. Lett. 63 (1989) 1086;
(b) A.E. Reynolds, D. Kaletta, G. Ertl and R.J. Behm, Surf. Sci. 218 (1989) 452.

[52] (a) G. Ertl, Adv. Catal. 37 (1990) 213;
(b) R. Imbihl, in: Optimal Structures in Heterogeneous Reaction Systems, Ed. P.J. Plath (Springer, Heidelberg, 1989) p. 26.

[53] M. Eiswirth and G. Ertl, Surf. Sci. 177 (1986) 90.

[54] (a) G. Ertl, Science 254 (1991) 1750;
(b) G. Ertl, Surf. Sci. 287/288 (1993) 1.

[55] M. Eiswirth, K. Krischer and G. Ertl, Surf. Sci. 202 (1988) 565.

[56] (a) M. Eiswirth and G. Ertl, Phys. Rev. Lett. 60 (1988) 1526;
(b) M. Eiswirth, P. Möller and G. Ertl, Surf. Sci. 208 (1989) 13.

[57] K. Krischer, M. Eiswirth and G. Ertl, J. Chem. Phys. 96 (1992) 9161; 97 (1992) 302.

[58] M.P. Cox, G. Ertl and R. Imbihl, Phys. Rev. Lett. 54 (1985) 1725.

[59] H.H. Rotermund, S. Jakubith, A. von Oertzen and G. Ertl, J. Chem. Phys. 91 (1989) 4942.

[60] (a) M. Mundschau, M.E. Kordesch, B. Rausenberger, W. Engel, A.M. Bradshaw and E. Zeitler, Surf. Sci. 227 (1990) 246;
(b) H.H. Rotermund, W. Engel, M.E. Kordesch and G. Ertl, Nature 343 (1990) 355.

[61] W. Święch, W. Engel, C.S. Rastonjee, B. Rausenberger, A.M. Bradshaw and E. Zeitler, to be published.

[62] V. Gorodetskii, W. Drachsel and J.H. Block, Catal. Lett. 19 (1993) 223.

[63] S. Nettesheim, A. von Oertzen, H.H. Rotermund and G. Ertl, J. Chem. Phys. 98 (1993) 9977.

[64] See, e.g., R.R. Cavanagh, E.J. Heilweil and J.C. Stephenson, Surf. Sci. 299/300 (1994) 643;
W. Ho, Surf. Sci. 299/300 (1994) 996.

[65] J.C. Williamson and A.H. Zewail, Proc. Natl. Acad. Sci. USA 88 (1991) 5021.

[66] J.A. Cusumano, in: Perspectives in Catalysis, Eds. J.M. Thomas and K.I. Zamaraev (Oxford-Blackwell, London, 1992) p. 1.

[67] M. Boudart, preface to: Catalysis – Science and Technology, Vol. 4, Eds. J.R. Anderson and M. Boudart (Springer, Heidelberg, 1983).

Surface Science 299/300 (1994) 755–768
North-Holland

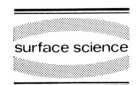

surface science

Kinetic modeling of surface rate processes

H. Chuan Kang

Department of Chemistry, National University of Singapore, 10 Kent Ridge Crescent, Singapore 0511, Singapore

and

W. Henry Weinberg *

Department of Chemical Engineering, University of California, Santa Barbara, CA 93106, USA

Received 5 April 1993; accepted for publication 2 June 1993

A hierarchical review is given of the modeling of surface rate processes. The following levels of sophistication of kinetic modeling are discussed (in the following order): (1) Langmuirian adsorption and desorption with random adsorbate distributions; (2) precursor-mediated adsorption and desorption in the absence of lateral interactions employing both kinetic and statistical approaches; (3) Langmuirian and precursor-mediated adsorption and desorption, with lateral interactions, treated using mean-field approximations; and (4) Monte Carlo simulations. Although this discussion encompasses the past sixty years of research in the field, emphasis is placed on the considerable advances in our understanding that have been made in the past thirty years.

1. Introduction

The kinetics of surface rate processes can be rather complex due to the effects of precursor states and lateral interactions. However, for "direct" Langmuirian adsorption and desorption in a system in which the adsorbed molecules are randomly distributed on the surface, rate processes at the surface are straightforward to analyze. For instance, the rate of adsorption is given by

$$r_a = k_a F (1 - \theta)^n, \tag{1}$$

and the rate of desorption is given by

$$r_d = k_d \theta^n. \tag{2}$$

In these expressions θ is the fractional coverage of the adsorbed species, k_a is the rate coefficient of adsorption (a cross section in this case),

F is the flux, given by $p/(2\pi m k_B T)^{1/2}$, of the adsorbing species from the gas phase, and k_d is the rate coefficient of desorption. For molecular adsorption and desorption, the exponent n is equal to unity, whereas for dissociative adsorption and recombinative desorption, n is equal to two. The rates given above are the rate per site for the first-order case and the rate per nearest-neighbor pair of sites for the second-order case. The form of the dependence of the rate of adsorption and desorption on the fractional coverage should be noted. Specifically, for first-order adsorption and desorption, r_a is proportional to $(1 - \theta)$ and r_d is proportional to θ, while for second-order adsorption and desorption, r_a is proportional to $(1 - \theta)^2$ and r_d is proportional to θ^2. These fractional coverage dependences reflect the Langmuirian notion of requiring a single vacant (occupied) site for a first-order adsorption (desorption) event, and a pair of vacant (occupied) sites for a dissociative adsorption (recombinative desorption) event. The adsorption iso-

---

* Corresponding author.

therm can be obtained by equating r_a and r_d. When n is equal to one, we obtain the Langmuir adsorption isotherm,

$$\theta/(1-\theta) = k_a F/k_d. \tag{3}$$

Langmuirian adsorption has been briefly delineated here to provide a better perspective of the influence, discussed below, of precursor states and lateral interactions on surface rate processes. We shall first review some of the basic ramifications that the existence of a precursor state has upon chemisorption, desorption, and other rate processes at surfaces. Following this, we shall discuss the developments, since 1960, in modeling the kinetics of precursor-mediated rate processes. Then we shall review the application of the quasi-chemical approximation to treat the effects of lateral interactions between adsorbed molecules. Finally, the use of Monte Carlo simulations in modeling surface rate processes will be reviewed.

2. Precursor-mediated adsorption and desorption

When a molecule is incident upon a solid surface, inelastic collisions (resulting in phonon excitation or electron–hole pair creation, for instance) can lower its energy until it is negative relative to the energy zero of a molecule infinitely far from the surface and at rest. The molecule can then be trapped in a shallow, weakly bound potential well of a precursor state. The potentials involved in precursor-mediated molecular and dissociative chemisorption are shown schematically in Fig. 1. In general, one should expect such precursor states to mediate chemisorption and desorption. An important exception to this scenario is the activated dissociative chemisorption of molecules with sufficiently high incident translational energy. In this case the molecule cannot be expected to become trapped in a shallow well. The trajectory followed by such a molecule on the multi-dimensional potential energy surface will be different from that followed by a molecule with low-incident translational energy. Particularly, the apparent activation energy barrier en-

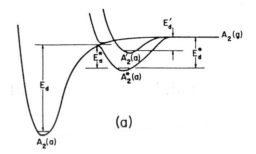

(a)

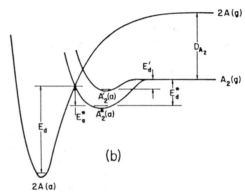

(b)

Fig. 1. Schematic one-dimensional potential energy diagrams depicting the general case of precursor-mediated chemisorption and desorption. (a) Molecular chemisorption and (b) dissociative chemisorption. The activation energies for adsorption are denoted by E_a^*. Activation energies for desorption are denoted by E_d' and E_d^* for desorption from the extrinsic and intrinsic precursor states, respectively [20].

countered in such a case is expected to be higher than that for precursor-mediated chemisorption.

Once trapped in the precursor state, a molecule will begin to accommodate thermally to the surface. The time required for this thermal accommodation is short, typically allowing the molecule to make fewer than ten hops on the surface [1]. Tully has demonstrated that for xenon atoms with a translational energy corresponding to 1960 K and incident upon a Pt(111) surface at 773 K, the motion normal to the surface is thermally accommodated to the surface temperature after ~ 20 ps [2]. The motion parallel to the surface is thermally accommodated within 100 ps. Thus, thermal accommodation of molecules in the precursor state occurs rather rapidly. Since this is typically the case, the degrees of freedom of the molecule in the precursor state rapidly attain the temperature of the surface. Hence, the probability of chemisorption would depend to a large extent upon the surface temperature. On the other hand, if chemisorption occurs "directly", i.e., without the molecule first trapping in a precursor state, the gas temperature rather than the surface temperature would exert a greater influence upon the probability of chemisorption.

Since the precursor well is typically shallow, we expect the lifetime of a molecule in a precursor state be very short at sufficiently high temperatures. If this is the case, the concentration of precursor molecules will be very low. However, even in such a case, the existence of precursor states changes the very nature of the adsorption event. With a precursor state present, the adsorption kinetics, for instance, are not just quantitatively but qualitatively different from the case when there is no precursor state. As an example, the probability of chemisorption of cesium on tungsten shows a very weak dependence on fractional surface coverage for coverages up to almost saturation, whereupon the probability of chemisorption decreases rapidly [3,4]. This is in sharp contrast to the linear decrease with fractional coverage that occurs for Langmuirian adsorption of a monatomic species (and other instances of first-order adsorption). Indeed, it is this deviation of the probability of chemisorption from the expected dependence on the fractional coverage in Langmuirian adsorption that prompted Langmuir, himself, to visualize a precursor state intermediate to chemisorption [3,4]. Thus, the importance of accounting for these ephemeral states is indisputable.

The idea of precursor states originated from classic papers by Langmuir when he attempted to understand the probability of chemisorption of alkali metals on tungsten [3,4]. Almost concurrently, Lennard-Jones argued for the existence of a physically adsorbed precursor state intermediate to chemisorption on a surface [5]. This was in the late 1920's and early 1930's, and this concept of precursor-mediated chemisorption was revived by Becker twenty years later to analyze the chemisorption rate of nitrogen on tungsten [6,7]. Following this, both Ehrlich and Kisliuk, in the late 1950's, presented two important mathematical descriptions of chemisorption rates using a kinetic and a statistical approach, respectively [8,9].

Erhlich developed a model in which a molecule trapped in an extrinsic precursor state, i.e., a precursor state that exists above an occupied chemisorption site, can hop across the surface until it encounters a vacant chemisorption site at which it becomes chemisorbed with probability of unity. The kinetic description is based on a pseudo-steady-state approximation for the concentration of molecules in the precursor state. Kisliuk extended the model by including the existence of an intrinsic precursor state, i.e., a precursor state that exists above an unoccupied surface site. Assuming that the chemisorbed molecules are randomly distributed on the surface, the rate of chemisorption can be expressed in closed form by considering the adsorption, desorption, and migration probabilities for a precursor molecule at each site.

Since 1960, this seminal work of Becker, Ehrlich, and Kisliuk has been refined by several other groups culminating in a unified treatment of both adsorption and desorption for both the kinetic and the statistical approaches [10–17]. We shall review these unified descriptions by applying them to the case of molecular chemisorption and desorption mediated by two kinds of energetically equivalent precursors, one intrinsic and the other extrinsic. In the case of adsorption, the model can be described by

$$A(g) \underset{k_d^*}{\overset{\xi F_A}{\rightleftarrows}} A^*(a) \overset{k_a^*}{\rightarrow} A(a),$$

where ξ is the probability of trapping of the incident molecule $A(g)$ into either of the precursor states $A^*(a)$, F_A is the incident flux of A molecules (proportional to the pressure of $A(g)$), k_d^* is the rate coefficient of desorption from either of the precursor states, and k_a^* is the rate coefficient of chemisorption from the intrinsic precursor state. For desorption the model may be generalized by including one additional step, viz.,

$$A(g) \underset{k_d^*}{\overset{\xi F_A}{\rightleftarrows}} A^*(a) \underset{k_a^*}{\overset{k_d}{\leftrightarrows}} A(a),$$

where k_d is the rate coefficient of desorption from the chemisorbed state into the intrinsic precursor state. It will be seen that the kinetic and the statistical approaches yield identical rate expressions when applied to the same model, a point which has been discussed previously [18].

In the kinetic approach applied to adsorption [16], we construct rate equations describing the time dependence of the fractional coverages of the various species on the surface. Then, using the pseudo-steady-state approximation for the fractional coverage θ^* of the precursor states, we can write

$$\mathrm{d}\theta^*/\mathrm{d}t = -k_d^*\theta^* - k_a^*\theta^*(1 - \theta_A) + \xi F_A = 0, \tag{4}$$

where θ_A is the fractional coverage of the chemisorbed A molecules. Hence, the rate of chemisorption is given by

$$r_a = k_a^*\theta^*(1 - \theta_A)$$
$$= k_a^* \, \xi F_A(1 - \theta_A)/[k_d^* + k_a^*(1 - \theta_A)]. \tag{5}$$

Similarly, for desorption the pseudo-steady-state approximation for the concentration of the precursor yields

$$\mathrm{d}\theta^*/\mathrm{d}t = -k_d^*\theta^* - k_a^*\theta^*(1 - \theta_A) + k_d\theta_A = 0. \tag{6}$$

The desorption rate is, therefore, given by

$$r_d = k_d^*\theta^* = k_d^* k_d\theta_A/[k_d^* + k_a^*(1 - \theta_A)]. \tag{7}$$

If the precursor is chemisorbed much more rapidly than it is desorbed into the gas phase, then $k_a^*(1 - \theta_A) \gg k_d^*$. The condition $k_a^*(1 - \theta) \gg k_d^*$ implies that even in the case of a high fractional coverage of the chemisorbed species, the trapped precursor has a sufficiently low desorption rate that it can eventually find a vacant chemisorption site on the surface. Thus, this condition implies a situation in which the influence of a precursor state is strongest. For this situation, the rates of chemisorption and desorption are given by

$$r_a = \xi F_A, \tag{8}$$

and

$$r_d = (k_d^* k_d/k_a^*)[\theta_A/(1 - \theta_A)]. \tag{9}$$

Note that the adsorption rate is independent of the fractional coverage of the chemisorbed species, and is determined entirely by the probability of trapping into the precursor state.

On the other hand, if a precursor molecule is more likely to desorb into the gas phase than to chemisorb, then $k_a^*(1 - \theta_A) \ll k_d^*$. In this case the rates of chemisorption and desorption are given by

$$r_a = \xi F_A(1 - \theta_A)(k_a^*/k_d^*), \tag{10}$$

and

$$r_d = k_d\theta_A. \tag{11}$$

These rate expressions represent the "pseudo-Langmuirian limit". In the case of activated chemisorption, where the energy barrier to chemisorption is larger than the energy barrier to desorption into the gas phase, a precursor molecule might not have a lifetime sufficiently long to find a vacant chemisorption site. In this case it will be more likely to desorb than to chemisorb. Therefore, Eqs. (10) and (11) are expected to be valid in the case of activated adsorption.

Note that the same adsorption isotherm applies to both of the cases discussed above. This can be seen by equating the rates of chemisorption and desorption for each case to give the same Langmuirian form

$$\theta_A/(1 - \theta_A) = \xi k_a^* F_A/k_d k_d^*. \tag{12}$$

Detailed balance necessitates that the rates of chemisorption and desorption that are equated to obtain the adsorption isotherm are the rates for a model with the same precursor states mediating chemisorption and desorption. As required thermodynamically, the isotherm can also be obtained by considering the equilibrium between the gas phase and the chemisorbed state, rather than equating rates that apply far from equilibrium, as we have done here. If the equilibrium situation is considered, then it is clear that, regardless of whether chemisorption and desorption are mediated by a precursor state or not, the adsorption isotherm will have the Langmuirian form.

In the statistical approach [17], the fate (chemisorption, desorption, or migration) of a molecule at each hop after it has impinged upon the surface is considered. The probability for it to be trapped in a precursor state is equal to ξ. From the precursor state it has a probability of chemisorption equal to $k_a^*(1 - \theta_A)/K$, a probability of desorption equal to k_d^*/K, and a probability of migration to another precursor state equal to k_m^*/K where K is given by $[k_a^*(1 - \theta_A) + k_d^* + k_m^*]$. A molecule in the precursor state can become chemisorbed where it was originally trapped from the gas phase, or it can become chemisorbed after executing any number of hops. Thus, the adsorption rate is given by

$$r_a = \xi F_A k_a^*(1 - \theta_A)\left[1 + (k_m^*/K)\right.$$
$$\left. + (k_m^*/K)^2 + \ldots\right]/K, \tag{13}$$

where the first term corresponds to chemisorption of a precursor into the site at which it is trapped, the second term corresponds to chemisorption after one hop, the third term corresponds to chemisorption after two hops, and so on. Summing the geometric series gives

$$r_a = \xi F_A k_a^*(1 - \theta_A)/K[1 - k_m^*/K]$$
$$= k_a^* \xi F_A(1 - \theta_A)/[k_d^* + k_a^*(1 - \theta_A)], \tag{14}$$

which is the same result obtained above by the kinetic approach.

In the case of desorption, similar considerations allow us to write the desorption rate as the geometric series

$$r_d = k_d \theta_A k_d^*\left[1 + (k_m^*/K) + (k_m^*/K)^2 + \ldots\right]/K$$
$$= k_d k_d^* \theta_A/[k_a^* + k_d^*(1 - \theta_A)]. \tag{15}$$

In both the kinetic and the statistical approach, it is assumed that the fractional coverage of the precursor molecules is negligibly low. This is eminently reasonable for systems to which this modeling has been applied, but it is not necessarily true in general. The fractional coverage of the precursor depends upon the temperature (via k_a^* and k_d^*) and the flux from the gas phase. At sufficiently low temperatures the lifetime and the fractional coverage of precursors may become observably high. For example, for carbon monoxide adsorbed on Ni(111), the (physically adsorbed) precursor molecule has been isolated at a surface temperature of 6 K [19].

Clearly, the above procedures can also be extended to more complicated cases of precursor-mediated chemisorption and desorption. Generally, we may consider the precursor states as either intrinsic precursors that exist above vacant surface sites or extrinsic precursors that exist above occupied chemisorption sites. Extrinsic precursors should be regarded as physically adsorbed molecules, while intrinsic precursors can be either physically adsorbed or molecularly chemisorbed. This latter possibility of a molecularly chemisorbed precursor arises in dissociative chemisorption, and this precursor state, itself, may have yet a different physically adsorbed precursor state. Such a situation can also be treated using the methods we have described above. The chemisorption and desorption rates for a rather extensive list of physically interesting models, including dissociative chemisorption and recombinative desorption with various types of precursor states, have been given by Weinberg [20]. There is also a discussion in Ref. [20] of the dissociative chemisorption of methane on Ni(111) and the dissociative chemisorption of carbon monoxide on nickel in terms of precursor-mediated chemisorption. Particularly in the latter case, an apparent conflict among experimental data [21–24] was clarified using the ideas we have discussed here.

3. Effects of lateral interactions

We have thus far discussed only situations where the distribution of chemisorbed and precursor molecules on the surface is random. Generally, however, this is not the case because of lateral interactions between the adsorbed molecules. For low fractional coverages this is not an important issue because the adsorbed molecules, generally well separated from one another, are expected not to interact [*1]. However, for suffi-

---

[*1] This simple argument breaks down in the presence of strong *attractive* interactions which can lead to island formation and high local coverages, even when the macroscopic surface coverage is moderately low.

ciently high coverages it becomes important to consider lateral interactions. Thus, in general, there are correlations between the positions of adsorbed molecules due to lateral interactions. For instance, in the case of two kinds of adsorbed molecules with fractional coverages of θ_A and θ_B, the probability P_{AB} of finding a nearest-neighbor AB pair is not simply equal to $2\theta_A\theta_B$, but is rather dependent on the nature of the lateral interactions and the temperature of the surface. Indeed, at sufficiently low temperatures ordering of the adsorbed molecules may occur. In order to model the kinetics of rate processes, it is, therefore, necessary to account for the correlations that result from the lateral interactions.

Various types of mean-field approximations have been widely used to treat the spatial correlations that arise in systems of interacting adsorbed molecules. Generally, the idea in these approximations may be expressed as follows. Rather than accounting exactly for the lateral interactions between all the adsorbed molecules, the lateral interactions are treated exactly only for each selected cluster of sites. The interactions with the molecules outside each cluster are approximated by an average field acting on the molecules in the cluster. Typically, this field is determined self-consistently. In the modeling of adsorption, mean-field approximations using clusters of various sizes have been utilized.

The simplest approximation, which accounts for the correlation between nearest-neighbor pairs of sites, is to consider a nearest-neighbor pair as a cluster. The quasi-chemical approximation [25], which has been widely applied to various models of adsorption and desorption, is a mean-field approximation which accounts exactly for the interactions within a nearest-neighbor pair of sites on the lattice. Here, we will briefly review the application of the quasi-chemical approximation to a model with dissociative adsorption and recombinative desorption.

There are three types of nearest-neighbor pairs of sites on the lattice: occupied–occupied, occupied–vacant, and vacant–vacant. If we consider the adsorption of A_2 to occur dissociatively when a gas-phase molecule impinges upon a nearest-neighbor pair of vacant sites, then the rate of

adsorption per surface site is $zk_aF_{A_2}P_{OO}/2$, where P_{OO} is the probability of finding a vacant–vacant nearest-neighbor pair, and z is the coordination number of the lattice.

In the absence of lateral interactions, we expect P_{OO} to be equal to $(1 - \theta_A)^2$, and, hence, the adsorption rate per site to be equal to $2k_aF_{A_2}(1 - \theta_A)^2$ for a square lattice $(z = 4)$. Clearly, in the presence of lateral interactions, which imply that the adsorbate is not distributed randomly on the surface, P_{OO} is not equal to $(1 - \theta_A)^2$. The problem of determining the rate of adsorption then involves understanding the statistics of the adsorbed overlayer and calculating P_{OO}.

In the quasi-chemical approximation the probabilities for the three types of nearest-neighbor pairs are related by

$$P_{AA}P_{OO}/P_{AO}^2 = \left[\exp(-E_{AA}/k_BT)\right]/4, \qquad (16)$$

where A denotes an occupied site and O denotes a vacant site, and E_{AA} is the nearest-neighbor lateral interaction energy between two nearest-neighbor adatoms. This equation is readily obtained by considering the equilibrium between AA, AO, and OO nearest-neighbor pairs. These pair probabilities also satisfy the following conservation equations,

$$P_{AA} + P_{AO} + P_{OO} = 1, \qquad (17)$$

and

$$2P_{AA} + P_{AO} = 2\theta_A. \qquad (18)$$

These equations yield expressions for the pair probabilities P_{AA}, P_{AO}, and P_{OO} in terms of the temperature T, the fractional coverage θ_A, and the lateral interaction energy E_{AA}. For the probability P_{OO} of a vacant–vacant nearest-neighbor pair, we obtain the following expression

$$P_{OO} = 1 - \theta_A - \alpha, \qquad (19)$$

where

$$\alpha = \left\{1 - \left[1 - 2\beta\theta_A(1 - \theta_A)\right]^{1/2}\right\}/\beta, \qquad (20)$$

and

$$\beta = 2\left[1 - \exp(-E_{AA}/k_BT)\right]. \qquad (21)$$

Using the quasi-chemical approximation for P_{OO}, it is possible to write an expression for the adsorption rate which takes into account, at least approximately, the correlation between the adsorbate particles. This gives, for a square lattice, an adsorption rate of

$$r_a = 2k_a F_{A_2}(1 - \theta_A - \alpha).$$ (22)

It is instructive to consider three cases: zero lateral interactions, strong repulsions, and strong attractions. In the limit of zero lateral interactions ($E_{AA} = 0$), we obtain an adsorption rate proportional to $(1 - \theta_A)^2$, as expected for second-order Langmuirian adsorption.

In the limit of strong repulsive interactions (E_{AA} large and positive) two coverage regimes are observed. When $E_{AA} \rightarrow \infty$, P_{AA} is equal to zero, cf. Eq. (16). Although adsorption occurs at vacant nearest-neighbor sites, once the A_2 molecule has been dissociatively adsorbed, the A atoms migrate to occupy sites which have no occupied nearest-neighbor sites. Then, using the conservation equations, Eqs. (17) and (18), we obtain $P_{OO} = (1 - 2\theta_A)$, which applies for $\theta_A < 0.5$. When the fractional coverage is equal to or exceeds 0.5, P_{OO} is equal to zero, since there are no remaining vacant-nearest-neighbor pairs of sites. The surface exhibits perfect $c(2 \times 2)$ ordering at a fractional coverage of 0.5. Hence, the adsorption rate is given by

$$r_a = 2k_a F_{A_2}(1 - 2\theta_A), \quad \text{for } \theta_A < 0.5,$$ (23)

and

$$r_a = 0, \quad \text{for } \theta_A \geq 0.5.$$ (24)

In the limit of strong attractive interactions, P_{AO} tends to zero in the thermodynamic limit. Hence, $P_{OO} = (1 - \theta_A)$ from the conservation equations, Eqs. (17) and (18). Thus, the rate of adsorption is equal to

$$r_a = 2k_a F_{A_2}(1 - \theta_A),$$ (25)

which is a pseudo-first-order rate. This occurs because the lattice gas undergoes phase separation, and the lattice is divided into a domain of occupied sites and a domain of vacant sites.

We have considered the dependence of the adsorption rate on the fractional coverage through P_{OO}. It is obvious that the desorption rate, in the presence of lateral interactions, will be a function of the fractional coverage through P_{AA}, i.e.,

$$r_d = 2k_d P_{AA}.$$ (26)

Therefore, using the Polanyi–Wigner equation [26] to describe the desorption rate will result in an activation energy which is dependent on the fractional coverage. Indeed, the concept of lateral interactions was first invoked by Roberts [27] over fifty years ago to explain just this dependence. The variation in the rate of desorption as a function of fractional coverage has been modeled rather extensively by several authors [13,28–35] using the quasi-chemical approximation or other similar mean-field type approximations. Thermal desorption spectra have been simulated for many systems, including systems with only one species, systems with two coadsorbed but non-reactive species, and systems with coadsorbed and reactive species. An example is the modeling of nitric oxide adsorbed on platinum and nickel by Bridge and Lambert [33], where the quasi-chemical approximation was used to compute the statistics of the adsorbed layer, i.e., the probabilities of pair occupancies. These were then used as parameters in a Runge–Kutta scheme to propagate the rate equations.

4. Precursor states and lateral interactions

We have reviewed at some length the general approaches taken in dealing with precursor states and in dealing with lateral interactions between adsorbate molecules. Thus, it should now be clear how a system, in which the effects of *both* lateral interactions *and* precursor states are important, can be treated. We briefly consider the case of dissociative adsorption and recombinative desorption of A_2 mediated by a precursor state. The intrinsic and extrinsic precursors are assumed to be energetically equivalent, as before. In this case the rate equation for the fractional coverage of the precursor is given by

$$d\theta^*/dt = \xi F_{A_2} - k_d^* \theta^* - k_a^* \theta^* P_{OO}.$$ (27)

With the pseudo-steady-state approximation, we obtain an adsorption rate of

$$r_a = 2\xi F_{A_2} k_a^* P_{OO}/(k_d^* + k_a^* P_{OO}), \qquad (28)$$

where the dependence of r_a on the adsorbate concentration can be obtained easily using the quasi-chemical approximation for P_{OO}. Similarly, the rate of desorption is given by

$$r_d = 2k_d k_d^* P_{AA}/(k_d^* + k_a^* P_{OO}). \qquad (29)$$

To provide a sense of the effects of both lateral interactions (via β) and the presence of precursor states (via k_d^*/k_a^*), we present in Fig. 2 the dependence of the relative probability of chemisorption (frequently referred to, lamentably, as the relative "sticking" probability), given by $r_a/r_a(\theta_A = 0)$, upon the coverage for this system. Equivalent results have been obtained by King [13], using Kisliuk's statistical approach, for nitrogen adsorption on tungsten.

Clearly, more complicated models can be considered within the same framework reviewed above. For example, the intrinsic and extrinsic precursor states can have different desorption and chemisorption rates, or the molecules in the precursor states can be allowed to interact with chemisorbed molecules or with each other. Typically, however, precursor–precursor lateral inter-

actions can be neglected to a very good approximation because of their normally low fractional coverages.

5. Monte Carlo simulations

Although the methods reviewed above for treating surface rate processes are extremely useful, it is apparent that the complexity of the rate equations quickly increases with the sophistication of the model that is used. For instance, it may be necessary to extend the range of the lateral interactions, or to make the interactions non-isotropic, or to model systems with more than one interacting species. Moreover, the approaches which we discussed above have a *fundamental* limitation in the sense that the average environment of a molecule or site is first calculated, and then the rate of the surface process under consideration is evaluated on the basis of this average environment. The effects of variation in the local environment of an adsorbed molecule are taken into account, approximately, in this manner. In addition, these approaches implicitly assume that the configurations of the adsorbed molecules relax infinitely quickly. Thus, regardless of what the microscopic rates of diffusion,

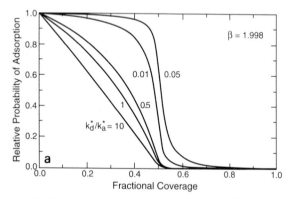
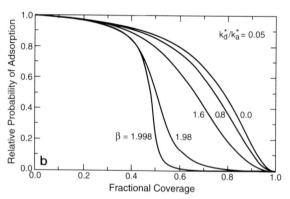

Fig. 2. The dependence of the relative probability of adsorption, $r_a/r_a(\theta = 0)$, on the fractional coverage is shown here for precursor-mediated dissociative chemisorption in the presence of lateral interactions between nearest-neighbor chemisorbed particles. In (a) the interaction energy parameter β (see text) is held fixed, while in (b) the ratio of the desorption rate coefficient to the adsorption rate coefficient for the precursor state, k_d^*/k_a^*, is held fixed. In (a) each curve is labeled with its corresponding value of k_d^*/k_a^*, while in (b) each curve is labeled with its corresponding value of β. Note that $\beta = 1.998$ corresponds to $E_{AA}/k_B T = 6.908$. Similar results have been obtained by King [13].

adsorption, desorption, and reaction may be, when modeling thermal desorption spectra, for instance, the configuration of the adsorbed overlayer is assumed to be relaxed completely at all times. These limitations can be overcome by using Monte Carlo simulation methods.

There are powerful motivations for using Monte Carlo methods in treating surface rate processes. As we have seen above, one of the principal issues which must be considered in order to have an accurate description of surface rate processes is the statistics of the adsorbed overlayer. Mean-field type approximations, such as the quasi-chemical approximation, may be useful but do not give exact results. Monte Carlo simulations, on the other hand, allow the statistical part of the problem to be solved exactly, at least in principle. By using Monte Carlo methods, the local environment of each adsorbed molecule or vacant site is considered independently, in contrast to the surface-averaged environment that is obtained in the quasi-chemical approximation. It is also relatively easy to gain microscopic insight from Monte Carlo simulations of a surface process. Furthermore, these simulations are straightforward to implement, even for complicated lattice-gas models. Finally, in Monte Carlo simulations, relaxation of the configurations of the adsorbed overlayer is allowed to proceed on their natural time scales once the microscopic rates of adsorption, desorption, reaction, and surface diffusion are defined. In a Monte Carlo simulation, the adsorbed overlayer configuration is not assumed a priori to be relaxed completely. We shall illustrate these points in detail below, following a review of how Monte Carlo simulations may be performed. An overview of Monte Carlo simulations, in general, may be found in Ref. [36].

We denote the configuration of a lattice gas by C and the probability distribution of configurations by $P(C)$, and we allow transitions between an initial configuration C_i and a final configuration C_f to occur with probability $\omega(C_i, C_f)$. Then, at thermal equilibrium, detailed balance requires that the transition probabilities satisfy

$$\omega(C_i, C_f) P_{eq}(C_i) = \omega(C_f, C_i) P_{eq}(C_f), \quad (30)$$

where the equilibrium probability distribution of configurations is given by the Boltzmann distribution

$$P_{eq}(C) = Z^{-1} \exp[-H(C)/k_B T], \quad (31)$$

where Z is the partition function and H is the Hamiltonian. An easy and effective way to generate equilibrium configurations is the Metropolis algorithm [37] in which the transition probability between two configurations, from C_1 to C_2, is given by

$$\omega(C_1, C_2) = \exp[-\Delta E_{12}/k_B T], \quad \text{for } \Delta E_{12} > 0,$$
$$= 1, \quad \text{for } \Delta E_{12} \leq 0,$$

where ΔE_{12} is the change in the energy of the lattice gas when the configuration changes from C_1 to C_2. Alternatively, the transition probabilities may be given by the Kawasaki algorithm [38], namely,

$$\omega(C_1, C_2) = \exp(-\Delta E_{12}/2k_B T)$$
$$/[\exp(-\Delta E_{12}/2k_B T)$$
$$+ \exp(\Delta E_{12}/2k_B T)]. \quad (32)$$

It is also possible to consider an activation energy formulation of the transition probability $\omega(C_1, C_2)$. Let the activation energy be E_{12} for the microscopic event which changes the configuration of the lattice gas from C_1 to C_2. Then the transition probability is given by

$$\omega(C_1, C_2) = \exp(-E_{12}/k_B T). \quad (33)$$

The reverse transition, from C_2 to C_1, will have its own activation energy E_{21}, which is related to the forward activation energy E_{12} and the energy change ΔE_{12} by

$$\Delta E_{12} = E_{12} - E_{21}. \quad (34)$$

It can be seen that all three prescriptions for $\omega(C_1, C_2)$ generate configurations according to the equilibrium probability distribution $P_{eq}(C)$, i.e., detailed balance, as articulated by Eq. (30), is satisfied.

Although the Metropolis and the Kawasaki algorithms should strictly be used only to generate configurations for computing equilibrium properties, they have also been widely used to simulate time-dependent processes. In this connection it is important to clarify the relation between the number of iterations in the simulation and the real time that is being simulated. In the limit of a large number of simulation runs, the number of iterations may be taken to be linearly proportional to the simulated time, although the constant of proportionality depends on the nature of the probabilities used in updating the configurations. In many investigations of time-dependent processes, the probabilities have been computed using the prescriptions suggested by Metropolis [37] or Kawasaki [38]. These, however, are generally not suitable for such studies because, although detailed balance is satisfied, these algorithms have a different time scale for each pair of transitions ($C_1 \leftrightarrow C_2$). This issue and the general issue of relating the simulated time to the number of Monte Carlo iterations have been discussed in detail by Kang and Weinberg [39], and by Fichthorn and Weinberg [40].

An algorithm for a lattice gas approaching equilibrium through adsorption, desorption, and surface diffusion has been discussed in detail by Fichthorn and Weinberg [40]. This algorithm, based on the activation energy prescription for the transition probabilities $\omega(C_1, C_2)$, will be described briefly here. First, a configuration is generated that is defined by some macroscopic parameters, for example, a specified fractional surface coverage and a spatial distribution corresponding to some specified temperature. For this configuration a list of all possible transitions, i.e., all possible desorption, adsorption, and migration events, and their rates k_i, is constructed. Here, we use a single index i to denote a transition between two configurations. A transition is then picked according to its probability $\omega_i = k_i / \Sigma_i k_i$, and the lattice-gas configuration is updated. In this way a sequence of configurations can be generated. By repeating this procedure many times, each time starting with a configuration which satisfies the same set of macroscopic parameters (surface temperature and fractional coverage, for example), an ensemble of trajectories in configuration space is generated. Each step in these simulations corresponds to a time increment given by the reciprocal of the average, over the ensemble, of the transition rates k_i. Averages

over this ensemble then provide the time evolution of macroscopic properties. We have described a procedure which uses strictly Monte Carlo methods. Combinations of Monte Carlo simulations with rate equations [41] or with the quasi-chemical approximation [42] are also possible and have been employed effectively.

Various different initial conditions can be used to simulate the initial states in different physical situations. For example, if it is desired to model molecular beam reflectivity experiments, which measure the adsorption probability, the simulations start with an initially vacant lattice [43]. If it is desired to simulate thermal desorption spectra with a particular initial fractional coverage, the lattice is first populated at the specified fractional coverage and then allowed to equilibrate at the initial temperature of the adsorbed overlayer [41,42]. Laser-induced thermal desorption experiments to measure surface diffusion coefficients can be simulated with a lattice configuration containing an initially vacant circular region [44]. Generally, periodic boundary conditions are used for the lattices that are simulated. There is also a need to account for finite-size effects when correlation lengths in the lattice gas become comparable to the sizes of the lattices used in the simulations [36].

As a result of the versatility of Monte Carlo simulations, the method has been widely applied to many problems in surface science. Monte Carlo simulations were used in the early 1970's by Ertl and Küppers [45] and by Adams and Germer [28] to study the formation of ordered adsorbed overlayers with lateral interactions in order to compare with experimental results from LEED. Subsequently, they have also been used extensively to study thermal desorption [33,41,42,46–49], adsorbate islanding effects on reactivity [46,50–52], effects of surface heterogeneity on reactivity [48], the influence of defects on surface processes [53], surface diffusion in laser-induced thermal desorption experiments [44], precursor-mediated adsorption [43], and, most recently, kinetic segregation of reactants in surface reactions [52,54,55]. We shall discuss one of these applications in detail in order to illustrate the utility of Monte Carlo simulations.

An interesting application is the study of pre-cursor-mediated chemisorption and desorption of molecular nitrogen on Ru(001), as reported in Ref. [41]. The influence of lateral interactions on the kinetics, energetics, and configurations of the adsorbed overlayer was investigated. In this work the formation of the adsorbed overlayer was sim-ulated by sequentially populating randomly cho-sen lattice sites. This simulates the trapping of gas-phase nitrogen into a physically adsorbed precursor state. The molecules in the precursor state are allowed to desorb, diffuse, or chemisorb. Diffusion occurs by allowing the precursor molecules to migrate on the surface by hopping to nearest-neighbor sites. Chemisorption is al-lowed to occur only if the site occupied by the precursor molecule is not already occupied by another chemisorbed molecule. The relative probabilities for each of these three events are dependent on the local configuration of the ad-sorbed overlayer, and they vary from one ad-sorbed molecule to another. For each type of local configuration, these relative probabilities may be calculated from the prescribed lateral interaction strengths, binding energy, and energy barrier for hopping.

Clearly, Monte Carlo simulations can provide a complete microscopic account of the adsorbed overlayer configurations. In particular, the neigh-borhood of each adsorbed molecule is known completely. This is in contrast to the approach discussed earlier, where information concerning the adsorbed overlayer configuration is obtained using the quasi-chemical approximation. The lat-ter yields only P_{AA}, P_{OO}, and P_{AO}, i.e., three numbers representing only the average character of each nearest-neighbor pair.

Using the Monte Carlo procedure described above, Hood et al. [41] were able to generate configurations corresponding to the state of ad-sorbed N_2 overlayers on Ru(001) at the beginning of a TDS experiment. The thermal desorption spectrum was computed using a combination of Monte Carlo simulations and deterministic rate equations. In particular, molecules were selec-tively removed from the lattice according to their probabilities of desorption. Quantitative agree-ment between the computed thermal desorption spectra and experimental data was obtained for all adsorbate coverages, as may be seen in Fig. 3. More importantly, the Monte Carlo simulations reveal that, at the same coverage, smaller adsor-bate islands are present during chemisorption than during thermal desorption. This reflects the dominance of trapping during chemisorption which results in the formation of numerous small islands, as opposed to the dominance of relax-ation (annealing) during thermal desorption which results in the formation of fewer large islands. This is illustrated in Fig. 4 where, particularly, panels (c) and (e), and (b) and (f) should be compared. Even when supplemented by the quasi-chemical approximation, an approach which uses only rate equations to study this system would not have so readily provided such micro-scopic insights.

This advantage of Monte Carlo simulations has been extensively exploited in many investiga-tions. The effects of islanding on the reactivity of the adsorbed overlayer have been studied by Stiles and Metiu [52], and by Silverberg et al. [50], among others. Ensemble and electronic effects on

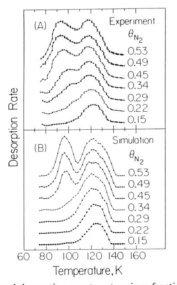

Fig. 3. Thermal desorption spectra at various fractional cover-ages of N_2 on Ru(001), previously reported in Ref. [41]. (a) Experimental results, and (b) simulation results using a com-bination of Monte Carlo simulations and deterministic rate equations.

catalytic activity have been investigated by Reynolds et al. [51]. Sales and Zgrablich [48], Gupta and Hirtzel [47], and Rivera and Hirtzel [56] have considered the influence that heterogeneity of adsorption site energy and structure has on rate processes. Bowler and Hood [53] have investigated the effects of surface steps and defects on surface diffusion. Mak et al. [44] have also used Monte Carlo simulations to model surface diffusion in connection with LITD experiments. In all these investigations the essential microscopic picture of the surface configuration is provided by the Monte Carlo simulations.

We mentioned earlier another important advantage of Monte Carlo simulations, namely, the adsorbed overlayer configuration is allowed to relax on its natural time scale. In the simulations of nitrogen on Ru(001), this relaxation was easily implemented by allowing the molecules to migrate, having specified the lateral interactions and the energy barriers [41]. If a rate equation approach is taken, a diffusion-reaction equation with a coverage-dependent diffusion coefficient would have to be dealt with. Even then, effects of the local environment of each adsorbed molecule would be considered only in an average manner. The effect of diffusion on the reactivity of an adsorbed overlayer has also been considered by Silverberg and Ben-Shaul [46]. Here, again, the simulations automatically take into account relaxation of the adsorbed overlayer configuration.

If there are some degrees of freedom of the adsorbed overlayer which relax much more rapidly than others, then Monte Carlo simulations may be combined with a quasi-chemical approximation. The fast degrees of freedom are accounted for using the quasi-chemical approximation, and the slow degrees of freedom are simulated using Monte Carlo methods. An example, studied by Silverberg and Ben-Shaul [46], is a coadsorbed system in which one species has a diffusion rate that is much higher than the diffusion rate of the other species and also much higher than the reaction rate. Thus, the configuration of the "fast" diffusing species relaxes very quickly, and Monte Carlo simulations are not required for that part of the system.

Finally, we review the use of Monte Carlo simulations in currently active investigations of reactive lattice-gas systems in which reactant segregation occurs even though lateral interactions are absent [54,55,57,58]. Interest in this area was sparked by the Ziff, Gulari, and Barshad model of a monomer–dimer Langmuir–Hinshelwood re-

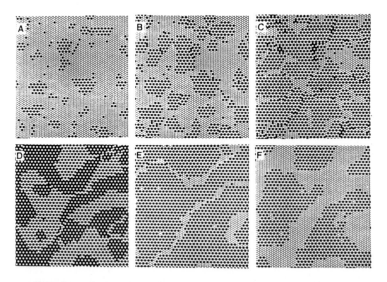

Fig. 4. Typical lattice gas configurations obtained from simulation studies of N_2 on Ru(001) [41]. The N_2 fractional coverages are (a) 0.06, (b) 0.15, (c) 0.29, (d) 0.53, (e) 0.29, and (f) 0.15. The configurations shown in (a) to (d) were obtained during adsorption, while the configurations in (e) and (f) were obtained during thermal desorption.

action which exhibited two poisoned phases and a reactive phase [59]. Since then, similar models of monomer–dimer, dimer–dimer, and monomer–monomer systems have been investigated. In the monomer–monomer systems with no desorption, segregation of the reactants occurs because of the reaction between nearest-neighbor pairs of unlike species. In this case the reactants form islands which grow indefinitely. The structure of these fractal islands has been investigated by several workers [54,55,57]. Since reaction may occur only at the edges of these islands, the reactivity of the lattice decreases with time. Recently, this decrease in reactivity has been related to the roughening of an initially flat interface between two semi-infinite domains of reactants [60].

6. Summary

We have provided a hierarchical review of the modeling of surface rate processes proceeding from the simplest non-interacting Langmuirian systems, through precursor-mediated adsorption–desorption systems with and without lateral interactions, to complicated laterally interacting systems which are conveniently handled only by Monte Carlo simulations. This hierarchy is also approximately chronological. Much of the effort in this area has focused on the treatment of precursor states and the statistics of the adsorbed overlayer. Although the basic framework for describing precursor-mediated adsorption and desorption, without the complication of lateral interactions, was established rather early by Ehrlich [8] and Kisliuk [9], refinements of this framework continued until the late 1970's, resulting in both a kinetic and a statistical approach which are equivalent. We have discussed these approaches in some detail.

We have also discussed the treatment of the consequences of lateral interactions in adsorbed overlayers using mean-field approximations such as the quasi-chemical approximation. This yields, for instance, the probabilities of occurrence of each possible type of nearest-neighbor pair on the surface. Such statistical information can then be used in combination with rate equations to provide a quantification of the kinetics of surface processes. We have illustrated this by reviewing the description of precursor-mediated adsorption and desorption in the presence of lateral interactions.

Monte Carlo simulations of surface rate processes have also been reviewed. We discussed the technical details of implementation and the ability of such simulations to provide microscopic insight into the kinetics of surface processes. In order to illustrate this, the modeling of nitrogen adsorption and desorption on Ru(001) using Monte Carlo simulations was specifically reviewed. We also briefly discussed the more recent application of Monte Carlo simulations to study reactant segregation in reactive lattice gases. Looking into the future, we expect that the rather recent formulation of dynamic Monte Carlo simulations [39–41] will result in their widespread adoption, perhaps eventually superseding their analytic, but approximate, precursors.

Acknowledgements

H.C. Kang acknowledges the support of grant RP3920608 from the National University of Singapore. W.H. Weinberg acknowledges the support of the National Science Foundation (grant CHE-9003553) and the Department of Energy (grant DE-FG03-89ER14048). We also acknowledge helpful discussions with Jim Evans, Kristen Fichthorn, and Baoqi Meng.

References

[1] J.A. Barker and D.J. Auerbach, Faraday Disc. Chem. Soc. 80 (1985) 277.
[2] J.C. Tully, Faraday Disc. Chem. Soc. 80 (1985) 291.
[3] I. Langmuir, Chem. Rev. 6 (1929) 451.
[4] J.B. Taylor and I. Langmuir, Phys. Rev. 44 (1933) 423.
[5] J.E. Lennard-Jones, Trans. Faraday Soc. 28 (1932) 333.
[6] J.A. Becker, in: Structure and Properties of Solid Surfaces, Eds. R. Gomer and C.S. Smith (University of Chicago Press, Chicago, 1953) p. 459.
[7] J.A. Becker and C.D. Hartman, J. Phys. Chem. 57 (1953) 153.
[8] G. Ehrlich, J. Phys. Chem. 59 (1955) 473.

[9] P. Kisliuk, J. Phys. Chem. Solids 3 (1957) 95; 5 (1958) 78.

[10] P.W. Tamm and L.D. Schmidt, J. Chem. Phys. 52 (1970) 1150; 55 (1971) 4253.

[11] L.R. Clavenna and L.D. Schmidt, Surf. Sci. 22 (1970) 365.

[12] C. Kohrt and R. Gomer, J. Chem. Phys. 52 (1970) 3283.

[13] D.A. King and M.G. Wells, Surf. Sci. 23 (1971) 120; Proc. R. Soc. (London) A 339 (1974) 245.

[14] M.R. Shannabarger, Solid State Commun. 14 (1974) 1015; Surf. Sci. 44 (1974) 297; 52 (1975) 689.

[15] D.A. King, Surf. Sci. 64 (1977) 43.

[16] R. Gorte and L.D. Schmidt, Surf. Sci. 76 (1978) 559.

[17] A. Cassuto and D.A. King, Surf. Sci. 102 (1981) 388.

[18] K. Schönhammer, Surf. Sci. 83 (1979) L633.

[19] M. Shayegan, E.D. Williams, R.E. Glover and R.L. Park, Surf. Sci. 154 (1985) L239.

[20] W.H. Weinberg, in: Kinetics of Interface Reactions, Eds. H.J. Kreuzer and M. Grunze (Springer, Berlin, 1987)·p. 94.

[21] H.P. Steinrück, M.P. d'Evelyn and R.J. Madix, Surf. Sci. 172 (1986) L561.

[22] R. Rosei, F. Ciccacci, R. Memeo, C. Mariani, L.S. Caputi and L. Papagno, J. Catal. 83 (1983) 19.

[23] D.W. Goodman, J. Vac. Sci. Technol. 20 (1982) 522.

[24] D.W. Goodman, R.D. Kelley, T.E. Madey and J.M. White, J. Catal. 64 (1980) 479.

[25] R. Fowler and E.A. Guggenheim, Statistical Thermodynamics (Cambridge University Press, Cambridge, 1952).

[26] P. Redhead, Vacuum 12 (1962) 203; D.L. Adams, Surf. Sci. 42 (1974) 12.

[27] J.K. Roberts, Some Problems in Adsorption (Cambridge University Press, Cambridge, 1939).

[28] D.L. Adams and L.H. Germer, Surf. Sci. 27 (1971) 21.

[29] J. Küppers, Vacuum 21 (1971) 393.

[30] C.G. Goymour and D.A. King, J. Chem. Soc. Faraday Trans. I, 69 (1973) 736, 749.

[31] K. Christmann, G. Ertl and T. Pignet, Surf. Sci. 54 (1976) 365.

[32] M.E. Bridge and R.M. Lambert, J. Catal. 46 (1977) 143; Surf. Sci. 63 (1977) 315.

[33] M.E. Bridge and R.M. Lambert, Proc. R. Soc. (London) A 370 (1980) 545.

[34] B. Hellsing and A. Mällo, Surf. Sci. 144 (1984) 336.

[35] V.P. Zhdanov, Surf. Sci. 111 (1981) L662, 63; 148 (1984) L691; 165 (1986) L31; 171 (1986) L461.

[36] Monte Carlo Methods in Statistical Physics, Ed. K. Binder, Topics in Current Physics, Vol. 7 (Springer, Berlin, 1986); Applications of Monte Carlo Methods in Statistical Physics, Ed., K. Binder, Topics in Current Physics, Vol. 36 (Springer, Berlin, 1987).

[37] N. Metropolis, A.W. Rosenbluth, M.N. Rosenbluth, A.H. Teller and E. Teller, J. Chem. Phys. 21 (1953) 1087.

[38] K. Kawasaki, Phys. Rev. 145 (1966) 224.

[39] H.C. Kang and W.H. Weinberg, J. Chem. Phys. 90 (1989) 2824; Acc. Chem. Res. 25 (1992) 253.

[40] K.A. Fichthorn and W.H. Weinberg, J. Chem. Phys. 25 (1991) 1090.

[41] E.S. Hood, B.H. Toby and W.H. Weinberg, Phys. Rev. Lett. 55 (1985) 2437.

[42] M. Silverberg and A. Ben-Shaul, J. Chem. Phys. 87 (1987) 3178.

[43] H.C. Kang and W.H. Weinberg, J. Chem. Phys. 92 (1990) 1397.

[44] C.H. Mak, S.M. George and H.C. Andersen, J. Chem. Phys. 88 (1988) 4052.

[45] G. Ertl and J. Küppers, Surf. Sci. 21 (1970) 61.

[46] M. Silverberg and A. Ben-Shaul, Chem. Phys. Lett. 134 (1987) 491; J. Stat. Phys. 52 (1988) 1179; Surf. Sci. 214 (1989) 17.

[47] D. Gupta and C.S. Hirtzel, Mol. Phys. 68 (1989) 583; Chem. Phys. Lett. 149 (1988) 527; Surf. Sci. 210 (1989) 322.

[48] J.L. Sales and G. Zgrablich, Surf Sci. 187 (1987) 1; Phys. Rev. B 35 (1987) 9520.

[49] S.J. Lombardo and A.T. Bell, Surf. Sci. 206 (1988) 101; 224 (1989) 451; Surf. Sci. Rep. 13 (1991) 1; B. Meng and W.H. Weinberg, in preparation.

[50] M. Silverberg, A. Ben-Shaul and F. Robentrost, J. Chem. Phys. 83 (1985) 6501.

[51] A.E. Reynolds, J.S. Ford and D.J. Tildesley, Surf. Sci. 166 (1986) 19; 191 (1987) 239.

[52] M. Stiles and H. Metiu, Chem. Phys. Lett. 128 (1986) 337.

[53] A.M. Bowler and E.S. Hood, J. Chem. Phys. 97 (1992) 1250; 1257.

[54] O.M. Becker, M. Bennun and A. Ben-Shaul, J. Phys. Chem. 95 (1991) 4803.

[55] H.C. Kang, M.W. Deem and W.H. Weinberg, J. Chem. Phys. 93 (1990) 6841; H.C. Kang and W.H. Weinberg, Phys. Rev. E 47 (1993) 1604; B. Meng, W.H. Weinberg, and J.W. Evans, Phys. Rev. E, in press.

[56] P.J. Rivera and C.S. Hirtzel, Chem. Eng. Commun. 108 (1991) 333.

[57] R.M. Ziff and K. Fichthorn, Phys. Rev. B 34 (1986) 2038.

[58] P. Meakin and D.J. Scalapino, J. Chem. Phys. 87 (1987) 731.

[59] R.M. Ziff, E. Gulari and Y. Barshad, Phys. Rev. Lett. 56 (1986) 2553.

[60] H.C. Kang and W.H. Weinberg, Phys. Rev. E, in press.

Surface Science 299/300 (1994) 769–784
North-Holland

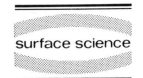

surface science

Chemisorption and reactions at metal surfaces

M.W. Roberts

School of Chemistry and Applied Chemistry, University of Wales, Cardiff CF1 3TB, UK

Received 12 April 1993; accepted for publication 30 April 1993

The late 1950s saw the emergence of two schools of thought, one based on the clean surface concept and the other on the more traditional bulk catalyst approach to surface chemistry. It is the synergistic influence of these two approaches that guided much of the author's research over the last 30 years – *an attempt to understand molecular behaviour at well defined metal surfaces but with a view to appreciate the mechanistic subtleties of heterogeneous catalysis.*

Direct evidence for the atomic nature of solid surfaces and their adsorption behaviour was extremely limited in 1960 and this article attempts to trace how improved experimental methods became available to study relatively simple metal–gas interactions. The limitations of kinetic studies based on indirect measurements (of the gas phase) were recognised and the possible advantages of monitoring the surface work function or photoemission characteristics seen as a possible way forward. The latter resulted in the development of a multi-photon source UHV compatible photoelectron spectrometer and much of this article describes how a detailed understanding was obtained from electron spectroscopy of molecular and dissociative chemisorption, the nature of chemisorbed oxide overlayers at metal surfaces, adsorbate activation and the role of precursor or transient states in the mechanisms of surface reactions.

1. Setting the scene

The topic of my PhD thesis was to explain-he role of chemisorbed sulphur in the catalytic formation of nickel carbonyl – was it an electronic mechanism or was the intermediate compound theory more appropriate? Since there was little difference between Beeck's reported heat of adsorption (35 kcal mol^{-1}) for carbon monoxide at an evaporated nickel film surface and the heat of formation of each Ni–CO bond in $Ni(CO)_4$ the energetics were finely balanced. The possibility of surface sulphur withdrawing electrons from the nickel was considered to be one way – the electronic mechanism – by which the energetics might be tipped in favour of carbonyl formation and involving a step-wise process: $Ni(CO)_2 \rightarrow Ni(CO)_3 \rightarrow Ni(CO)_4$. The alternative intermediate compound theory invoked the participation of $Ni(CO)_2S$ and competition between its decomposition and further reaction with $(CO)_g$ to give the volatile tetracarbonyl. This was the challenge put to me by my supervisor Keble Sykes. It was 1952,

and the concept of atomically clean surfaces was just beginning to emerge as being central to our understanding of heterogeneous catalysis and the chemistry and physics of metals. The proceedings of a conference held at Lake Geneva, Wisconcin, appeared in 1953 – "Structure and Properties of Solid Surfaces", edited by Gomer and Smith [1] – provided a clear pointer to the developments that might occur over the next decade. The early calorimetric/accommodation coefficient work of J.K. Roberts in Cambridge with tungsten, the studies of Beeck in the Shell Laboratories at Emmeryville with evaporated films and the catalyst work of Emmett were debated at the conference with two schools of thought – those in the clean surface camp and those with interests in bulk catalyst behaviour. It was fortunate that my postgraduate research put me at the very heart of these central issues in surface chemistry before they had become fashionable – ultrahigh vacuum techniques, clean surfaces, the role of surface structure, energetics, surface dipoles, impurities and chemical reactivity. But a serious drawback was the *absence of direct experimental methods* available for investigating the nature of chemisorbed species, the first glimmer of hope coming in 1956 from the Texaco Research Laboratories with the publications of Eischens and Pliskin [2] reporting infrared spectra for chemisorbed carbon monoxide on iron surfaces. According to Eischens [3] much of the credit for this development goes, however, to a paper by Buswell, Krebs and Rodebush [4], some 20 years earlier, which showed how infrared spectroscopy could distinguish between adsorbed water and hydroxyl species at montmorillinite.

The Beeck evaporated metal film approach [5] was extended by Trapnell [6] in England, and the work of J.K. Roberts and Langmuir developed by Ehrlich [7] at G.E. Schenectady. Precursor states in adsorption were discussed by Kisliuk [8] while Beeck's calorimetric studies, using the advantages of metal films, were extended by Brennan, Hayward and Trapnell [9]. Two books were published with the title "Chemisorption" – one the proceedings of a conference edited by Garner [10] and the other by Trapnell [11]. The d-band theory of chemisorption had been proposed by Dowden [12] and patterns of catalytic reactivity developed; while Mignolet [13] and Tompkins [14] reported work function (surface potential) changes for the chemisorption of CO and H_2 and also for the physical adsorption of xenon. Surface characterisation at the atomic level was not possible and it was left to individuals to design their own criteria for surface definition – in the case of the preparation of atomically clean nickel powder [15] necessary for a fundamental study of carbonyl formation it was a comparison between the monolayer estimated by the physical adsorption of krypton and the "fast" – non-activated – dissociative chemisorption of H_2 at $-183°C$ – each H adatom being assumed to be chemisorbed by a surface nickel atom.

Interest in the mechanism of metal oxidation had been stimulated for three reasons: the oxidation theory of Cabrera and Mott [16], the preparation of clean surfaces which necessitated the removal of oxide overlayers and the need to understand the physics and chemistry of the silicon and germanium oxide interfaces. In 1956 Schlier

and Farnsworth [17] in a conference devoted to the physics of semiconductor surfaces reported a LEED investigation of Ge(100) where the use of argon-ion bombardment for surface-cleaning was (first?) reported. This paper and also the subsequent one by Handler [18] was followed by an extensive discussion of whether LEED would be useful in studies of chemisorption, the effectiveness of argon-ion bombardment for clean surface preparation, the phenomenon of thermally induced faceting of single crystals and oxide film formation through the place-exchange of oxygen at metal surfaces.

These were my perceptions, as I recall them, of the more significant aspects of the scientific scene in surface chemistry as the 1960s approached with the Faraday Discussion meeting (8, 1950) highlighting many of the issues relevant to heterogeneous catalysis through contributions by E.K. Rideal, D.D. Eley and K.J. Laidler. I was also fortunate to have listened to Sir Hugh Taylor's lecture "Scientific problems of surface catalysis" given in Burlington House, London, to the Chemical Society in April 1953. His theme was clean metal surfaces and technical catalysts – a not inappropriate topic for discussion 40 years later. What then has changed?

2. The emergence of photoelectron spectroscopy in surface chemistry

Siegbahn [19] published his first evidence for chemical shifts in the binding energies of metals and oxides during the period 1957–59. He was however "bothered by uncertainties due to the chemical state" and for some years concentrated more on problems in atomic physics. It was a study of $Na_2S_2O_3$ which gave two well resolved K photoelectron lines from sulphur, and attributed to S^{2-} and S^{6+} species, that he regarded as the first conclusive evidence for the separation of two electronically distinct atoms through shifts in their core-electron binding energies. Subsequently the C(1s) spectrum of ethyltrifluoroacetate became a landmark for the future potential of X-ray photoelectron spectroscopy and the acronym ESCA – electron spectroscopy for chemical analysis (but later changed to – for chemical applications) was coined. In 1967 the book "ESCA – Atomic, molecular and solid state structure studied by means of electron spectroscopy" was published [20] followed in 1969 by "ESCA applied to free molecules" [21]. At that time several firms started to develop commercial instruments and at the University of Bradford I initiated discussions in 1969 with Vacuum Generators at East Grinstead. The experimental challenge was clear and taking a lead from our earlier photoemission studies with nickel [22], our interest in clean metal surfaces, the X-ray data of Siegbahn and the ultraviolet photoelectron spectroscopic studies of the gas phase by Turner [23] and Price [24], we set out to design with VG a multi-photon (UV and X-ray) ultrahigh vacuum spectrometer suitable for the preparation and study of the chemistry of well defined atomically clean metal surfaces. At about this time Bordass and Linnett [25] reported the UV-induced photoelectron spectrum of methanol adsorbed on a (contaminated) tungsten surface, using a Perkin–Elmer analyser, while Thomas et al. reported [26] an O(1s) spectrum for chemisorbed oxygen on graphite with an AEI spectrometer and Delgass et al. [27] reviewed data relevant to catalysts. In 1972 at a Royal Society meeting we described the first multiphoton source UHV spectrometer [28] and reported O(1s) and Hg(4f) spectra as a function of exposure of CO_2 and mercury to gold surfaces at 80

K. This provided evidence (Fig. 1) for the sensitivity of XPS for adsorbed molecules as a function of their surface coverage, while the density of electron states close to the Fermi level was shown to be altered by surface impurities (fig. 1). Both had important implications for surface chemistry, the latter being relevant to theories of heterogeneous catalysis prevalent at that time.

Estimating the concentrations of surface species σ from the intensities of core-level spectra [29,30] paved the way for mechanistic studies of chemisorption and surface reactions (Eq. (1)). For the chemisorption of oxygen at a nickel surface

$$\sigma = \frac{I_m N \cos(\phi) \mu_s \rho \lambda}{I_s \mu_m M_s}, \qquad (1)$$

where I_m, I_s are the intensities of the Ni(2p) and O(1s) peaks; $\mu_{m,s}$ the Ni(2p) and O(1s) photoionization cross-sections; M_s the molecular weight of the substrate; λ the electron inelastic mean free path; ϕ the angle of collection (with respect to the surface normal) of the photoelectrons; N Avogadro's number and ρ the density of the nickel substrate. No other comparable experimental method – the only serious restriction being its insensitivity to H(a) – was available and provided the chemist with his traditional armoury to unravel reaction mechanisms – concentration data as a function of time, pressure and tempera-

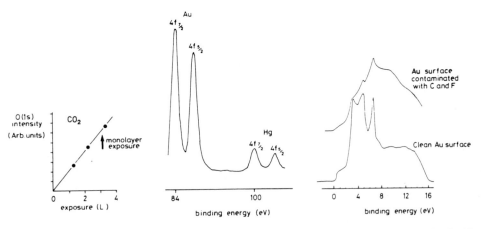

Fig. 1. Establishing the surface sensitivity of XPS for adsorption: CO_2 and Hg adsorption on Au; electron density of states of Au and contaminated Au surfaces by He(I) spectroscopy [28a].

ture. The surface chemistry scene had been changed!

It was (and still is) frequently argued that in the context of heterogeneous catalysis or "real chemistry" a serious limitation of photoelectron spectroscopy was its inability to follow a surface reaction at high pressures. It, however, emerged that there was a rich chemistry that needed to be explored at low temperatures (80–200 K) and most of our effort went in studying molecular behaviour – bond breaking and bond making – under these conditions. Nevertheless, we designed with VG Scientific an in situ photoelectron spectrometer capable of taking spectra at pressures some 10^5 greater than in the more conventional spectrometer [31]. Relatively little work has been reported at high pressures, other than that from the Catalysis group at Novosibirsk, although recently Grunze, Dwyer, Nassir and Tsai [32] have described a high pressure photoelectron spectrometer with an alternative design for the high pressure cell to that in Ref. [31].

3. Facile bond cleavage and the chemisorption of diatomic molecules

Prior to 1970 our understanding of the bonding of diatomic molecules to metal surfaces was almost entirely dependent on kinetic studies of desorption – the one exception being carbon

monoxide where infrared spectroscopy had played an important part following Eischen's pioneering studies in the 1950s. Nevertheless, infrared spectra provided only positive information for the molecularly adsorbed state – it was not diagnostic of dissociative chemisorption. A combination of in situ X-ray and UV-photoelectron spectroscopies was a way forward – He(I & II) spectroscopies providing information on the molecular orbitals of the adsorbed state and core-level spectra through shifts in binding energies distinguishing between C(a) and CO(a). Eastman and Cashion [33] were the first to report UV-induced (He radiation) spectra for carbon monoxide chemisorbed on nickel but the assignments of just two peaks at 7 and 11 eV to the three molecular orbitals 5σ, 1π and 4σ was not straightforward. Suggestions by Lloyd [34] and Mason and his colleagues [35] were confirmed by the synchrotron radiation studies of Gustafsson et al. [36] – the degenerate 5σ and 1π orbitals being assigned to the 7 eV peak and the 4σ orbital to that at 11 eV. Evidence was obtained for the interplay between molecular and dissociative states of CO chemisorption based on the presence or absence of orbital structure in the valence spectra and chemical shifts in the core-level C(1s) and O(1s) binding energies. A spectra data base evolved and patterns of reactivity across the Periodic Table emerged – a new experimental strategy for exploring two-dimensional chemistry using a combination of in situ XPS and UPS.

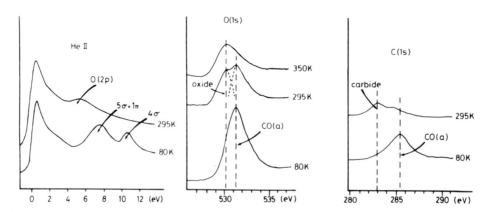

Fig. 2. Molecular and dissociative chemisorption of carbon monoxide at an iron surface at low temperatures: in situ XPS and UPS studies [37].

The case of carbon-monoxide chemisorption at iron surfaces illustrated (Fig. 2) all the advantages of this approach [37], the molecule being adsorbed in the molecular state at 80 K but dissociating at or close to 295 K. Dissociation was accompanied by binding energy shifts in both the C(1s) – 285.3 to 282.9 eV – and O(1s) – 531.3 to 530.1 eV – spectra and also by loss of the 7 and 11 eV peaks in the He(II) spectra characteristic of the $(5\sigma + 1\pi)$ and 4σ orbitals of molecularly adsorbed CO. Furthermore, the role of preadsorbed sulphur on inhibiting bond cleavage was reported for the first time and interpreted in terms of the Dewar–Chatt or Blyholder models for metal–CO bonding. These observations were of direct relevance to mechanistic aspects of Fischer–Tropsch catalysis and the resurgence of interest in the carbide theory. Correlations were established between the onset of dissociative chemisorption and the heat of CO adsorption [38] while Jones and McNicol [39] recognised the implications of the photoelectron results to Fischer–Tropsch catalysis. The latter was subsequently reviewed by Röper [40] in the context of the information gleaned from XPS and UPS. Photoelectron spectroscopy was clearly emerging from being just of academic interest to becoming a significant part of the industrialists' armoury – although it is as a means of surface analysis at the atomic level that it has been used most extensively by the chemical and semiconductor-physics based industries.

Although infrared spectroscopy had established (see the work of Pritchard discussed in ch. 9 of Ref. [71]) that CO was molecularly adsorbed at single crystal copper surfaces, nitric oxide – with one extra electron – was unexpectedly found by XPS and UPS to exhibit a rich surface chemistry [41]. Through quantitative analysis of core and valence level spectroscopies, two molecular states of nitric oxide – bent and linear – were delineated, one of which (the bent) was a precursor to a dissociated state. Difference N(1s) and O(1s) spectra also provided definitive evidence for the formation of $N_2O(a)$ at low temperatures. Following cleavage of the nitrogen–oxygen bond there was, in contrast to that observed at iron [42] and aluminium surfaces [43], no evidence for

N(a), the preferred pathway being an addition reaction to form $N_2O(a)$. Whether the formation of nitrous oxide in the adsorbed state at 80 K involved two discrete steps – the formation of a transient N(s) atom followed by the addition reaction $N(s) + NO(l) \rightarrow N_2O(a)$ – or the concerted reaction $NO(l) + NO(l) \rightarrow N_2O(a) + O(a)$ where NO(l) is a linearly bonded molecule, has as yet not been established. That surface structure played a role in the chemistry was evident from Demuth and Eastman's observation [44] that only when the Ni(111) surface was roughened did nitrogen-bond cleavage occur – a concept that had already been recognised as of general significance for the dissociative chemisorption of dihydrogen by Somorjai [45] through studies of well defined stepped platinum single crystal surfaces.

Stimulated by such observations – and particularly of CO and NO dissociation at iron surfaces – we investigated the chemisorption of dinitrogen by polycrystalline iron. The XPS results were surprising in that bond cleavage was established at low temperatures – adding weight to the idea that dissociation of dinitrogen might not be the rate-determining step in ammonia synthesis [46]. Of more general significance was the manner in which XPS revealed the sharpness of the transition which occurred just above 80 K and the need for an atomic level explanation of thermal activation of molecular adsorbates. Somorjai [47] has recently drawn attention to our inadequate understanding of this aspect of surface chemistry in the context of the dissociative chemisorption of CO, NO, N_2 and C–H bond cleavage in hydrocarbons. There is little doubt that much of the thermodynamic driving force must come from surface reconstruction – as indeed has been well recognised in dioxygen dissociation leading to the formation of an oxide overlayer.

4. Oxygen chemisorption and metal oxidation

Kinetic studies of oxygen chemisorption at iron surfaces [48] enabled parameters in the Cabrera–Mott theory to be extracted which led me to look for more direct experimental evidence from work function measurements for the central feature in the theory, namely the field F given by V/X, where V is the oxygen surface potential and X the overlayer oxide thickness. It turned out that work function data provided evidence for facile low temperature place-exchange [49] – surface reconstruction – and in (for example) the case of

oxygen chemisorption at nickel in the formation of a "semiconductor-like oxide overlayer" at low oxygen pressures (10^{-6} Torr) at 295 K [50–54]. Although "oxidation" had been clearly identified, a significant feature was that the initial interaction ($\theta < 0.2$) led to just oxygen chemisorption

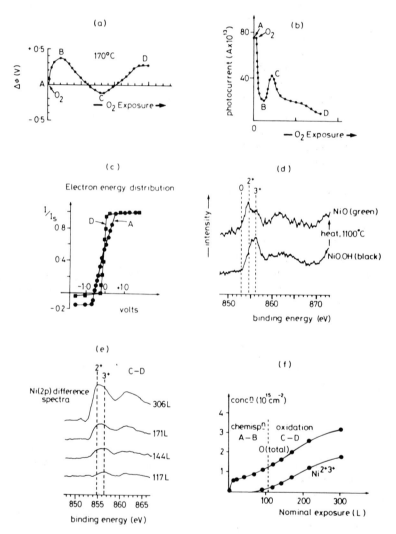

Fig. 3. Studies of the nickel–oxygen system [22,50,51,54] during the period 1963 to 1985: (a) work function change during exposure to oxygen at 343 K; (b) photoemission ($h\nu = 6.2$ eV) as a function of exposure at 295 K; (c) stopping potentials and "oxide" formation at 295 K; (d) Ni(2p) spectra for thermal conversion of NiO · OH to NiO and the assignment of Ni^{2+} and Ni^{3+} states; (e) Ni(2p) difference spectra as a function of exposure of Ni(100) to oxygen at 295 K; (f) formation of Ni$^{2+,3+}$ states at Ni(100) as a function of surface oxygen concentration at 295 K. Regions A–B correspond to "chemisorption", B–C to oxygen "place-exchange" and C–D to "oxidation".

(Fig. 3a) and only when the coverage increased above this value, even at 343 K, was there evidence for the onset of oxidation [51].

The work function change at 295 K was, somewhat surprisingly, shown to be sensitive to the oxygen pressure, suggesting a close relationship between the stoichiometry (defective nature) of the chemisorbed (oxide?) overlayer and the surface potential [50]. No longer could we think of oxygen chemisorption as a uniform layer of chemisorbed adatoms with a characteristic surface potential (Fig. 3a) and we turned to photoemission as a possible way of discriminating between chemisorption and oxidation. We adopted two different approaches – measurements of (a) the photoelectric yield and (b) of the energy distribution of the photoelectrons – the results of which [22,51,52] had a significant influence on the direction of our research during the period 1965–1975 when we also benefited from keeping abreast of the rapid developments occurring in the area of the surface physics of semiconductors and particularly of silicon and germanium where somewhat analogous problems were being addressed but in the context of understanding the effect of bulk doping and surface states.

Photoelectric yield data as a function of oxygen exposure (Fig. 3b) confirmed the view that place-exchange of chemisorbed oxygen present at 80 K could be thermally activated on raising the temperature to only ~ 120 K, while electron energy distribution data indicated that the electronic structure of the surface was "semiconductor-like" – after an atomically clean polycrystalline nickel surface was exposed to oxygen (10^{-6} Torr for 60 s) at 295 K [22,51,52]. The latter was based on the observation [22] of a shift in the "stopping potential" V_0 given by $eV_0 = \phi_c - h\nu$ where ϕ_c is the work function of the spherical gold collector and $h\nu$ the photon energy (Fig. 3c). It was considered unlikely that degradation of the energy of the photoelectrons was by inelastic scattering and that δ – the difference between the two stopping potentials – is a consequence of a change in the electronic structure arising from surface "oxidation". We estimated the oxygen uptake to be equivalent to two unit cells of NiO or about 8 Å at 295 K and an oxygen pressure of

$\sim 10^{-6}$ Torr. We concluded [22,51,52] that the photoelectron escape depth from the nickel-chemisorbed oxygen overlayer at 295 K was, for $h\nu = 6.2$ eV, about 8 Å and certainly no greater than 20 Å if it was an oxygen deficient oxide. It was against this background that we obtained funding in 1970 from the Science Research Council for the development of a UHV compatible photoelectron spectrometer for surface studies although it is worth recalling that our first application in 1969 had been turned down "on the grounds that the technique was unlikely to be surface sensitive"!

As soon as it was established that core-level spectroscopy provided a means, through shifts in electron binding energies, of recognising the formation of surface "oxide" – for example through shifts in Al(2p) binding energies in oxygen chemisorption at aluminium surfaces [43] – we addressed the questions we had raised earlier regarding oxygen chemisorption at nickel and whether or not we could obtain further evidence for the development of a defective oxide overlayer. Would the Ni(2p) spectra provide information pertinent to the question of whether or not Ni^{2+} and Ni^{3+} states were present in the "oxide" overlayer? The first attempt by Brundle and Carley [53] at the University of Bradford was limited by the spectral accumulation method available to us at that time – a single scan and a pen-recorder! With the availability of microprocessors and the development of suitable software we returned to this problem using nickel single crystals [54], having first gained experience of the photoelectron spectroscopy of defective bulk nickel oxides [55,56] – including both "green" NiO and the non-stoichiometric "black" oxide. Thermally induced conversion of an oxygen-excess black oxide, and characterised by a Ni(2p) peak at 856.2 eV, to the near-stoichiometric green oxide, was accompanied by the emergence of a second Ni(2p) peak at a binding energy of 854.7 eV (Fig. 3d). Ni^0 has a Ni(2p) binding energy of 853 eV and so we assigned [54] the 856.2 eV peak to Ni^{3+} and that at 854.7 eV to Ni^{2+}. Subsequently, Cimino et al. [57], using a combination of XPS and EPR, confirmed the presence of O^- species in bulk nickel oxides and correlated their presence with

XPS evidence for Ni^{3+} while very recently Kotsev and Ilieva [58] determined non-stoichiometric oxygen in nickel oxide by a titrimetric method involving the reduction of Ni^{3+} to Ni^{2+}. A photoelectron data base was therefore available in the early 1980s to explore the interaction of oxygen with Ni(100) and Ni(210) surfaces [54] – the latter being atomically rough and considered to be more susceptible to reconstruction than the Ni(100) surface.

The intensity of O(1s) relative to Ni(2p) spectra provided a direct measure of the concentration of surface oxygen present as a function of exposure – and therefore the sticking probability – while Ni(2p) difference spectra (Fig. 3e) provided evidence for the development of Ni^{2+} and Ni^{3+} states. Three distinct regimes were observed [54] with Ni(100): a fast initial oxygen uptake, followed by a slower process leading to the emergence of intensity in the Ni(2p) difference spectra for exposures greater than about 100 L (Fig. 3f). The latter was attributed to "oxidation" through the development of Ni^{2+} and Ni^{3+} states with angular dependent studies suggesting that Ni^{3+} species were concentrated more "at the surface" than the Ni^{2+}. In the case of the more reactive Ni(210) surface it was possible to estimate the separate contributions of Ni^{3+} and Ni^{2+} species within the "oxide" overlayer. Furthermore, it was suggested that the Ni^{3+} states are formed by a redox type reaction (Eq. (2)) and compatible with the observation that stoichiometric high surface area "green nickel oxide" is converted to

$$Ni^{2+}(s) + \tfrac{1}{2}O_2(g) \rightarrow Ni^{3+}(s) + O^-(s) \qquad (2)$$

"black nickel oxide" on exposure to low pressures of oxygen at room temperature a process that is known to be highly exothermic (250 kJ mol^{-1} of oxygen). In our single crystal experiments the O(1s) spectrum also developed asymmetry on the high binding energy side (531 eV) which is compatible with the above redox reaction and symptomatic of excess surface oxygen with a different charge. The three stages recognised by XPS are therefore: oxygen chemisorption ($\theta < 0.6$) and suggested to involve the Ni(4s4p) band and occurring with a high sticking

probability; a much slower oxidation process leading to Ni^{2+} states and lastly a redox reaction resulting in Ni^{3+} states being formed with a total oxygen uptake of $(3–4) \times 10^{15}$ atoms cm^{-2}. Al-Sarraf et al. [59], measuring for the first time heats of adsorption and sticking probabilities, delineated four regions for the Ni(100)–oxygen interaction: first a high heat (500 kJ mol^{-1}) chemisorption region (up to $\theta \approx 0.35$), followed by a slow oxidation process where the heat falls slowly to ~ 100 kJ mol^{-1}, but then by an increase in the heat of adsorption to 400 kJ mol^{-1}, and finally a low heat (100 kJ mol^{-1}) region where the sticking probability is ~ 0.03. Al-Sarraf et al. [59] attribute the latter region to molecular adsorption on the surface of the oxide. Other than the molecular adsorption stage – for which we have no evidence – there are close similarities between the concepts that have emerged for nickel oxidation over the last thirty years from photoemission, work function and photoelectron spectroscopy (Fig. 3).

A similar strategy was adopted to search for redox states at oxide overlayers at titanium surfaces [60] with Ti(2p) spectra providing evidence for discrete localized Ti^{2+}, Ti^{3+} and Ti^{4+} present in overlayers no more than three unit cells thick.

5. Activation of adsorbates by chemisorbed oxygen and the mechanism of surface reactions

The debate between the "clean surface" and "bulk catalyst" approaches to surface chemistry, described by Wheeler [61] some 40 years ago, led naturally to the proposition that chemisorption at oxygen-contaminated metal surfaces was slow and activated. On the other hand clean metal surfaces were characterised by "fast" non-activated adsorption. Oxygen was therefore perceived to be a surface "poison" and the general view was that every effort should be made to study metal surfaces that were completely free of it – hence, the use of "flash" filaments (Ehrlich) and evaporated metal films (Beeck) and as a prerequisite for such work the development of ultrahigh vacuum techniques. However, this did not prevent us exploring how adsorbed oxygen influenced the chemistry of an initially clean metal surface and following a kinetic study [62] of the interaction of H_2S with pre-oxidized lead surfaces, which had resulted in some unexpected results, we returned to this system 10 years later with the advantage of single crystals, low energy electron diffraction and X-ray photoelectron spectroscopy [63]. Not only did chemisorbed oxygen activate an other-

wise unreactive H_2S molecule but it participated in facile H-abstraction – acting as a base – resulting in the chemisorptive replacement of chemisorbed oxygen and the creation of a sulphide overlayer at low temperatures. The O(1s) and S(2p) spectra taken as a function of time and temperature provided the first example (as far as we are aware) of this type of surface reaction [63]

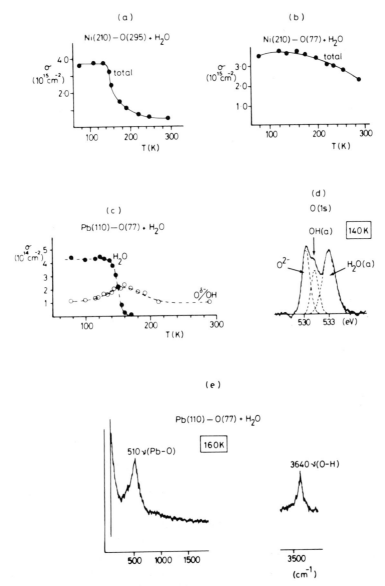

Fig. 4. The role of surface oxygen in the activation of $H_2O(a)$ and the inactivity of the "perfect oxide" overlayer [74]: (a) the desorption of H_2O from Ni(210)–O(295); (b) the high reactivity of H_2O at Ni(210)–O(77); (c)–(e) XPS and HREEL evidence for hydroxyl formation at Pb(110)–O(77).

and with LEED to follow structure a rather satisfying model for the role of oxygen in adsorbate activation emerged [64]. It is worth noting that a sulphide overlayer generated by the high-temperature induced dissociative chemisorption of H_2S – i.e., in the absence of surface oxygen – exhibited a very different LEED structure [64] from that formed through oxygen-activation. It is an area of surface science that still remains to be explored – the generation of novel interfacial structures through activation of adsorbates by oxygen. Not surprisingly oxygen activation of H_2S led to the study of other adsorbates – the next being water – and considerable activity worldwide of this general phenomenon. Of particular note have been the contributions of Madix [65] to this subject, although initially restricted to XPS and UPS, the availability of HREELS made it possible to distinguish between H-bonded water and hydroxyl formation [66a]. This was a problem recognised in the Cu(111)–O–water system where shifts in the O(1s) binding energies [66b] were not sufficiently definitive to make a distinction between the various possible states: $H_2O(a)$, OH(a) and H-bonded H_2O. The role of chemisorbed oxygen in being able to induce order in molecularly adsorbed water was also recognised at about this time through the development of ESDIAD by Doering and Madey [67] with a review being published by Thiel and Madey [68]. Activation of adsorbates by preadsorbed oxygen is now recognised to be a wide spread and subtle phenomenon, the surface oxygen being able to play a dual role – a promoter and a poison (Fig. 4). It was this that prompted us to explore whether the chemical reactivity of oxygen was coverage dependent, if it was, then whether this was related to surface oxidation or reconstruction, and, furthermore, whether a distinction should be made between the chemistry of *preadsorbed* and *coadsorbed oxygen*.

6. Precursor states, surface transients and reaction mechanisms

A central theme in our studies of the chemistry of metal surfaces has been the use of low temperatures. This largely stemmed from an early appreciation of the relative ease (negligible activation energy) with which at least some metals reconstructed [49–51] to form a chemisorbed oxygen overlayer but also from investigations of the sticking probabilities of N_2 and CO with molybdenum [69,70] in the form of evaporated films. An inverse dependence of sticking probability on temperature was found [71], in keeping with Ehrlich's classical studies with tungsten filaments and Kisliuk's statistical model [8] based on the assumption that two kinds of precursor states can exist: the so-called intrinsic state existing above an empty site and the extrinsic state existing above an occupied surface site. Weinberg [72] has provided a thorough review of the concepts and realities of precursor states in chemisorption – relying largely on kinetic studies. We did not pursue the kinetic approach preferring to use more direct experimental methods to search for evidence, first for the existence of precursor states and secondly whether such states participated in surface reactions. Work function studies [73] had indicated that at low temperatures a weakly adsorbed molecular state of dinitrogen was present at 80 K at a molybdenum surface which had associated with it a positive dipole pointing away from the surface thus lowering the work function. This provided experimental evidence for a molecular state which conceivably might participate in the kinetics of dinitrogen dissociation and inherent in the Lennard-Jones picture of chemisorption. Kinetic models based on precursor states were developed for a whole range of chemisorption systems [71,72] and, as emphasised by Weinberg [72], although useful models emerged they are limited by the large number of parameters that can appear in their formulation. It was partly for this reason that we took a more direct approach to search for evidence for precursor states or surface transients choosing the dissociative chemisorption of dioxygen as a model system. This was stimulated by our observations of the activation of adsorbates by chemisorbed oxygen and in particular the high reactivity of chemisorbed oxygen present at (for example) a nickel surface at 80 K compared with the reactivity of the chemisorbed oxygen layer formed at 295

K (Fig. 4). Could the oxygen species present at 80 K be considered to be precursors to the formation of the chemisorbed oxygen overlayer? The same had been observed (Fig. 4) for oxygen chemisorption at a lead surface [74]. It was to explore this that we used the probe-molecule or chemical trapping approach to provide possible evidence for oxygen transients [75]. Ammonia, and to a lesser extent pyridine [76], carbon monoxide [77] and water, have been used as probe molecules – the prerequisite being that the probe-molecule is unreactive to both the atomically clean metal and oxide overlayer surfaces [78]. If we consider [75] the dynamics of the dissociative chemisorption of dioxygen as a sequence of steps (Eq. (3)):

$$O_2(g) \to O_2(s) \to O_2^{\delta-}(s) \to 2O^{\delta-}(s)$$

$$\to 2O^{2-}(a), \tag{3}$$

where following accommodation of dioxygen at the surface, $O_2(s)$, electron transfer from the metal occurs leading to transient molecular and atomic states of oxygen, $O_2^{\delta-}(s)$ and $O^{\delta-}(s)$ prior to the formation of the final chemisorbed state designated as $O^{2-}(a)$ and possibly involving surface reconstruction. It should be recalled that $O^- + e \to O^{2-}$ is a highly endothermic process (820 kJ) in the gas phase and its formation in the solid state involves a substantial Madelung term. Our assignment of the final chemisorbed state as $O^{2-}(a)$ does, therefore, not necessarily imply the development of the full 2e charge although surface reconstruction would favour it.

The chemical trapping or molecular probe approach is best appreciated in terms of what is an intuitively appealing picture of the adsorption process [78–81] based on the surface residence time, $\tau = \tau_0 \exp(E_{des}/RT)$, where $\tau_0 \approx 10^{-13}$ s and E_{des} is the activation energy for desorption, and the site residence time given by $\tau_s = \tau_0 \exp(E_{diff}/RT)$ where E_{diff} is the activation energy for surface diffusion or hopping. For a typical probe molecule with $E_{des} = 40$ kJ mol^{-1} and $E_{diff} \approx 20$ kJ mol^{-1} the values of τ and τ_s are roughly 10^{-6} and 10^{-10} s, respectively at room temperature. Furthermore the surface coverage

will be given by $\theta = N\tau$, where N is the molecular impact rate, so that at a pressure (e.g., of NH$_3$) of 10^{-6} Torr, $\theta = 10^9$ molecules cm^{-2} at room temperature for a E_{des} value of ~ 40 kJ mol^{-1}. A molecule will, therefore, during its sojourn at the surface visit some 10^4 sites. If under dynamic conditions the steady state surface concentration of NH$_3$(s) is maintained at (say) 10^9 molecules cm^{-2} then trapping potentially reactive oxygen transients is highly favourable remembering that (for example) in the gas phase the reaction NH$_3$ + O$^- \to$ NH$_2$ + OH$^-$ occurs with almost unit collisional efficiency. This was the basis of the methodology adopted using both photoelectron and vibrational spectroscopies to characterise the trapped stable chemisorbed species generated. Model systems were chosen ensuring in particular that the probe molecule (ammonia) was chemically specific in its reactivity – unreactive to both the atomically clean and metal oxide overlayers. In the case of Mg(0001) evidence for O$^{\delta-}$(s) transients generated in the dissociative chemisorption of dioxygen was obtained with N(1s), O(1s) and electron energy loss spectroscopies providing evidence for the formation of amide, hydroxyl and chemisorbed oxygen species [75]. Analogous chemistry was observed in the study of the dissociative chemisorption of nitric and nitrous oxides [75] confirming the participation of oxygen atoms in H-abstraction. In the case of Zn(0001) ammonia–dioxygen mixtures (Fig. 5) kinetics were observed typical of precursor mediated reactions – reaction rates increasing with decreasing temperature – to generate amide and hydroxyl species [79]. Furthermore, the rate of cleavage of the dioxygen bond was nearly 10^3 times faster in the presence of ammonia than with pure O$_2$(g) consistent with the rate determining step involving O$_2^{\delta-}$(s) and the surface complex [O$_2^{\delta-}$- - - - - -NH$_3$](s). These coadsorption studies [75–81] have drawn attention to the need to consider the chemistry of the total system, with predictions based on the known reactivities of the individual molecular components being misleading. It is through such experiments that a better understanding of heterogeneous catalysis could be achieved, with low energy pathways available through precursor states present at

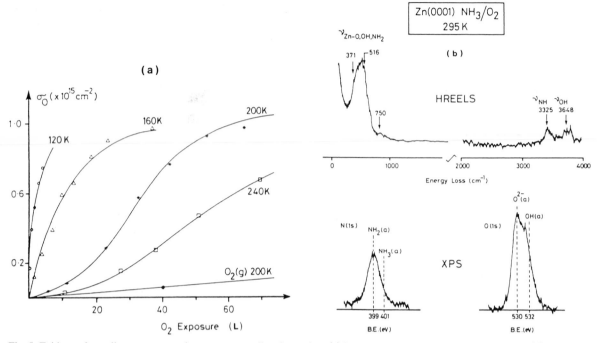

Fig. 5. Evidence for a dioxygen–ammonia precursor mediated reaction: (a) inverse temperature dependence of rate; (b) typical XP and HREEL spectra during coadsorption.

very low surface concentrations dominating the chemistry (Fig. 6).

Two general aspects of dioxygen surface chemistry have therefore emerged. The first with Mg(0001) involves hot oxygen transients $O^{\delta-}(s)$ with significant surface life-times and undergoing rapid surface diffusion so that they exhibit special chemistry prior to being chemisorbed [75]. Indirect evidence that this was a possible realistic model was already available [82] from optical simulation studies of the surface structures observed in the dissociative chemisorption of dioxygen at Cu(210) – but had not been considered by us as having any consequences for chemical reactivity. Following dissociation of the dioxygen molecule optical simulation of the LEED patterns suggest that (a) there must be considerable diffusion of a molecular precursor state and (b) more significantly that subsequent to dissociation a correlated or semi-correlated diffusion of oxygen adatoms by a hopping mechanism occurring over quite long distances (~ 10 nm) is necessary to generate the defective structures necessary to

account for the observed streaked LEED patterns.

The second aspect involves molecular oxygen, in a pre-bond cleavage state, forming complexes with the ammonia probe-molecule and providing, as with Zn(0001), a low energy pathway to the chemisorbed state [78,79]. All the experimental

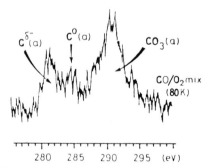

Fig. 6. Carbon–oxygen bond cleavage for carbon monoxide exposed to aluminium at 80 K effected by surface oxygen transients: C(1s) evidence for carbonate, carbon and carbide formation [77].

evidence points to the precursor complex being present "at immeasurably low" concentrations which Weinberg [72] has emphasised is a feature of the kinetic modelling of precursor-mediated chemisorption processes. In the case of the Zn–O_2–NH_3 system [79,80] the rate of formation of the dioxygen–ammonia complex reflects the number of visits the ammonia probe molecule makes to all surface sites and the fraction of these sites occupied by the $O_2^{\delta-}(s)$ transient. This leads to the following expression:

$$\frac{d\left(O_2^{\delta-}\text{- - -}NH_3\right)}{dt}$$

$$= \frac{10^{-15}P_{NH_3}}{\left(2\pi mkT\right)^{1/2}}$$

$$\times \exp\left(\frac{\Delta H_{NH_3}}{RT}\right)\frac{\nu_{diff}}{\nu_{des}}\exp\left(-\frac{E_{diff}}{RT}\right)\sigma_{O_2^{\delta-}} \quad (4)$$

for the rate of complex formation where ΔH and E_{diff} are the enthalpies and activation energies for adsorption and surface diffusion of ammonia, ν_{diff} and ν_{des} are pre-exponential factors (sometimes assumed to be of comparable magnitude to the vibrational frequencies) for diffusion and desorption of ammonia and $\sigma_{O_2^{\delta-}}$ the steady state surface concentration of the $O_2^{\delta-}$ transient. Mathematical modelling of these reactions [83] illustrates how experimentally immeasurably small concentrations of precursor or transient states, undergoing rapid surface diffusion, can lead to very efficient low energy pathways to chemisorbed products. It is an aspect of experimental surface chemistry that requires a new approach; STM is a possible way forward.

Although the hot oxygen-transient mechanism has received further support from studies of the coadsorption of dioxygen and ammonia at Cu(111) and Cu(110) surfaces [84,85], they raised the issue as to whether isolated oxygen adatoms exhibited special chemistry compared with oxygen-adatom clusters or islands. Is it the case that under coadsorption conditions ammonia molecules undergoing rapid surface diffusion intercept highly reactive oxygen adatoms $O^{\delta-}(s)$ present at very low surface concentrations before they aggregate to form relatively inactive clusters?

$O_2(g) \rightarrow O^{\delta-}(s)$	RAPIDLY DIFFUSING HOT OXYGEN ADATOMS
$O^{\delta-}(s) \rightarrow O^{\delta-}(a)$	ISOLATED OXYGEN ADATOM IN FINAL CHEMISORBED STATE
$O^{\delta-}(a) \rightarrow O^{\delta-}(a)$	GROWTH OF OXYGEN CLUSTERS
$O^{\delta-}(a) \rightarrow O^{2-}(a)$	OXIDATION and SURFACE RECONSTRUCTION
$O^{\delta-}(s)$ $+NH_3(s)$ $\rightarrow$ $NH(a)$ $+H_2O(g)$	MECHANISM OF IMIDE FORMATION

To maintain high reactivity in (for example) the oxydehydrogenation catalytic reaction $NH_3(g) + O_2(g) \rightarrow NH(a) + H_2O(g)$ at a Cu(111) surface, it is clearly advantageous to ensure that oxygen-cluster formation is limited [85], since this favours the development of unreactive O^{2-}-like species (isoelectronic with Ne) while isolated oxygen adatoms $O^{\delta-}(a)$ might be expected to be more O^--like (isoelectronic with F). When pread-

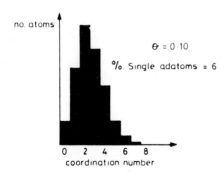
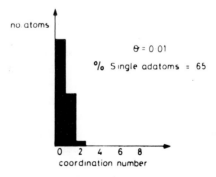

Fig. 7. Modelling the structure of the chemisorbed oxygen overlayer: with increasing oxygen coordination or cluster size, the effectiveness of the oxygen-induced formation of chemisorbed imide species decreases [86].

sorbed oxygen ($\theta = 0.1$) present at Cu(110) is exposed to ammonia at 300 K only a fraction of the oxygen participates in the formation of NH(a) while when θ approaches unity there is little or no evidence for a fast oxydehydrogenation process [86]. What was surprising is the implication that an appreciable fraction of the chemisorbed oxygen aggregates to form inactive oxygen-clusters even at low oxygen coverages ($\theta = 0.1$). This stimulated us to model, using a Monte Carlo approach, the dissociative chemisorption of dioxygen with relatively simple instructions (cf. Ref. [82]) for the fate of the oxygen atom $O^{\delta-}(s)$ after bond cleavage. Only for very low coverages ($\theta = 0.01$) was the oxygen predominantly present as isolated atoms (fig. 7) and for $\theta = 0.1$ the majority were present as clusters of 3 or 4 adatoms [87]. These observations are supported by the STM studies of Brune et al. [88] for the chemisorption of oxygen at Al(111); at $\theta = 0.0014$ virtually all the oxygen is present as isolated adatoms while for $\theta = 0.18$ only a small minority are present as single oxygen atoms with clusters (or islands) involving up to 27 oxygen atoms being observed.

The object of the coadsorption studies [75–81] was to establish whether in the chemisorption of oxygen, species were generated which had sufficiently long surface life-times for them to be considered as distinct chemical identities and more significantly exhibited unique reactivities compared with the final chemisorbed state. The experimental evidence for the model systems suggest that under coadsorption conditions the reactions are more akin to those that might occur in a two-dimensional gas rather than by either Langmuir–Hinshelwood or Eley–Rideal mechanisms [75]. The facile reactions clearly involve low energy pathways which are available, provided the oxygen reacts before it attains the "final chemisorbed state". It follows that predictions based on the known reactivities of the individual components of a gaseous mixture will be at best restrictive and at the worse misleading, emphasising the need to bridge the gap between the static surface science and the dynamic molecular beam approaches to the study of chemistry at metal surfaces. The latter is an aspect recently touched

on by Mullins, Rettner and Auerbach [89] in studies of the dynamics of the oxidation of CO at Pt(111) by an atomic oxygen beam.

7. Conclusions

The study of chemisorption and chemical reactivity has been transformed over the last 30 years into a more precise science through the development of a plethora of surface sensitive spectroscopies and from which – for personal reasons – photoelectron spectroscopy has been largely singled out in this article. Although developments of experimental techniques arose partly from the need of the semiconductor industry the influence of chemists wishing to search for possible correlations between structure and reactivity has also been important. In this context developments in the field of infrared spectroscopy – an aspect not touched upon here in any detail – have exceeded all expectations over the last fifteen years, with phenomenal improvements in both resolution and sensitivity.

Although it was inevitable that attention was given initially to the chemisorbed state per se the bigger challenge is to understand chemical reactivity using in situ dynamic experimental methods. A start on this has already been made and must be the way forward to a better understanding of heterogeneous catalysis.

References

[1] R. Gomer and C.S. Smith, Structure and Properties of Solid Surfaces (Chicago University Press, Chicago, 1953).
[2] R.P. Eischens and W.A. Pliskin, Adv. Catal. 10 (1958) 1.
[3] R.P. Eischens, in: Heterogeneous Catalysis, Eds. B.H. Davis and W.P. Hettinger, Jr., ACS Symp. Ser. 222 (1983).
[4] A.B. Buswell, K. Krebs and W.H. Rodebush, J. Am. Chem. Soc. 59 (1937) 2603.
[5] O. Beeck, A.E. Smith and A. Wheeler, Proc. R. Soc. London A 177 (1940) 62.
[6] B.M.W. Trapnell, Proc. R. Soc. London A 218 (1953) 566.
[7] G. Ehrlich, J. Chem. Phys. 34 (1961) 29.
[8] P. Kisliuk, J. Phys. Chem. Solids 3 (1957) 95; 5 (1958) 78.

[9] D. Brennan, D.O. Hayward and B.M.W. Trapnell, Proc. R. Soc. London A 256 (1960) 81.

[10] W.E. Garner, Chemisorption (Butterworths, London, 1957).

[11] B.M.W. Trapnell, Chemisorption (Butterworths, London, 1955).

[12] D.A. Dowden, in: Chemisorption, Ed. W.E. Garner (Butterworths, London, 1957).

[13] J.C.P. Mignolet, Disc. Faraday Soc. 8 (1950) 326.

[14] See: R.V. Culver and F.C. Tompkins, Adv. Catal. 11 (1959) 104.

[15] M.W. Roberts and K.W. Sykes, Proc. R. Soc. London A 242 (1957) 534; Trans. Faraday Soc. 54 (1958) 548.

[16] N. Cabrera and N.F. Mott, Rep. Prog. Phys. 12 (1948) 163.

[17] R.E. Schlier and H.E. Farnsworth, in: Semiconductor Surface Physics, Ed. R.H. Kingston (University of Pennsylvania Press 1957).

[18] P. Handler in Semiconductor Surface Physics Ed. R.H. Kingston (University of Pennsylvania Press, 1957).

[19] K. Siegbahn, Nobel lecture 1981 (Nobel Foundation, 1982) p. 113.

[20] K. Siegbahn, C. Nordling, A. Fahlman, R. Nordberg, K. Hamrin, J. Hedman, G. Johanson, T. Boegmark, S.E. Karlsson, I. Lindgren and B. Lindberg, ESCA – Atomic, Molecular and Solid State Structure Studied by Means of Electron Spectroscopy, Nova Alta Regiae Soc. Sci. Ups. Ser. 20 (1967).

[21] K. Siegbahn, C. Nordling, G. Johansen, J. Hedman, P.F. Heden, K. Hamrin, U. Gelius, T. Bergmark, L.O. Werme, R. Manne and Y. Baer, ESCA Applied to Free Molecules (North-Holland, Amsterdam, 1969).

[22] C.M. Quinn and M.W. Roberts, Trans. Faraday Soc. 61 (1965) 1775.

[23] D.W. Turner, Philos. Trans. R. Soc. A 268 (1970) 7.

[24] W.C. Price, in: Chemical Spectroscopy and Photochemistry in the Vacuum Ultraviolet, Eds. C. Sandorfy, P.J. Ansloos and M.B. Robin (Reidel, New York, 1974).

[25] W.T. Bordass and J.W. Linnett, Nature London 222 (1969) 660.

[26] J.M. Thomas, E.L. Evans, M. Barber and P. Swift, Trans. Faraday Soc. 67 (1971) 1875.

[27] W.N. Delgass, T.R. Hughes and C.S. Fadley, Catal. Rev. 4 (1970) 179.

[28] (a) C.R. Brundle and M.W. Roberts, Proc. R. Soc. London. A 331 (1972) 383;
(b) C.R. Brundle, M.W. Roberts, D. Latham and K. Yates, J. Electron Spectrosc. 3 (1974) 241.

[29] T.E. Madey, J.T. Yates Jr. and N.E. Erickson, Chem. Phys. Lett. 19 (1973) 487.

[30] A.F. Carley and M.W. Roberts, Proc. R. Soc. London A 363 (1978) 403;
M.W. Roberts, Adv. Catal. 29 (1980) 55.

[31] R.W. Joyner, M.W. Roberts and K. Yates, Surf. Sci. 87 (1979) 501.

[32] M. Grunze, D.J. Dwyer, M. Nassir and Y. Tsai, in: Surface Science of Catalysis, Eds. D.J. Dwyer and F.M. Hoffmann, ACS Symp. Ser. 482 (1992).

[33] D.E. Eastman and J.K. Cashion, Phys. Rev. Lett. 27 (1971) 1520.

[34] D.R. Lloyd, Faraday Disc. Chem. Soc. 58 (1974) 136.

[35] T.A. Clark, I.D. Gay, B. Law and R. Mason, Chem. Phys. Lett. 31 (1975) 29.

[36] T. Gustafsson, E.W. Plummer, D.E. Eastman and J.L. Freehouf, Solid State Commun. 17 (1975) 391.

[37] K. Kishi and M.W. Roberts, J. Chem. Soc. Faraday Trans. I, 71 (1975) 1721.

[38] R.W. Joyner and M.W. Roberts, Chem. Phys. Lett. 29 (1974) 447.

[39] A. Jones and B.D. McNicol, J. Catal. 47 (1977) 384.

[40] M. Röper, in: Catalysis in C_1 Chemistry, Ed. W. Keim (Reidel, Dordrecht, 1983);
see also: M. Smutek and S. Cerny, Int. Rev. Phys. Chem. 3 (1983) 263.

[41] D.W. Johnson, M.H. Matloob and M.W. Roberts, J. Chem. Soc. Chem. Commun. (1978) 40; J. Chem. Soc. Faraday Trans. I, 75 (1979) 2143.

[42] K. Kishi and M.W. Roberts, Proc. R. Soc. London A 352 (1976) 289.

[43] A.F. Carley and M.W. Roberts, Proc. R. Soc. London A 363 (1978) 403.

[44] J.E. Demuth and D.E. Eastman, IBM Res. RC 4736 (1974).

[45] G.A. Somorjai, in: Chemistry in Two Dimensions: Surfaces (Cornell University Press, Ithaca, NY, 1981).

[46] D.W. Johnson and M.W. Roberts, Surf. Sci. 87 (1979) L255;
G. Ertl, M. Weiss and S.B. Lee, Chem. Phys. Lett. 60 (1979) 391.

[47] G.A. Somorjai, Catal. Today 12 (1992) 343.

[48] M.W. Roberts, Trans. Faraday Soc. 57 (1961) 99.

[49] M.W. Roberts, J. Chem. Soc. Quart. Rev. 16 (1962) 1;
see also: M.W. Roberts and C.S. McKee, Chemistry of the Metal–Gas Interface (Oxford University Press, Oxford, 1978).

[50] C.M. Quinn and M.W. Roberts, Nature 200 (1963) 648;
Trans. Faraday Soc. 60 (1964) 899.

[51] M.W. Roberts and B.R. Wells, Trans. Faraday Soc. 62 (1966) 1608; Disc. Faraday Soc. 41 (1966) 162.

[52] M.W. Roberts, Surf. Defect Prop. Solids 1 (1972) 144.

[53] C.R. Brundle and A.F. Carley, Chem. Phys. Lett. 31 (1975) 423.

[54] A.F. Carley, P.R. Chalker and M.W. Roberts, Proc. R. Soc. London A 399 (1985) 167;
C.T. Au, A.F. Carley and M.W. Roberts, Philos. Trans. R. Soc. London A 318 (1986) 61.

[55] M.W. Roberts and R.St.C. Smart, J. Chem. Soc. Faraday Trans. 80 (1984) 2957.

[56] L.M. Moroney, M.W. Roberts and R.St.C. Smart, J. Chem. Soc. Faraday Trans. I, 79 (1983) 1769.

[57] A. Cimino, D. Gazzoli, V. Indovina, M. Inversi, G. Moretti and M. Occhiuzzi, in: Structure and Reactivity of

Surfaces, Eds. C. Morterra and A. Zechina (Elsevier, Amsterdam, 1988) p. 48.

[58] N.K. Kotsev and L.I. Ilieva, Catal. Lett. 18 (1993) 173.

[59] N. Al-Sarraf, J.T. Stuckless, C.E. Wartnaby and D.A. King, Surf. Sci. 283 (1993) 427.

[60] A.F. Carley, J.C. Roberts and M.W. Roberts, Surf. Sci. Lett. 225 (1990) L39.

[61] A. Wheeler, in: Structure and Properties of Solid Surfaces, Eds. R. Gomer and C.S. Smith (Chicago University Press, Chicago, IL, 1953).

[62] J.M. Saleh, B.R. Wells and M.W. Roberts, Trans. Faraday Soc. 60 (1964) 1865.

[63] K. Kishi and M.W. Roberts, J. Chem. Soc. Faraday Trans. I, 71 (1975) 1721.

[64] R.W. Joyner, K. Kishi and M.W. Roberts, Proc. R. Soc. London A 358 (1977) 223.

[65] E.M. Stuve, R.J. Madix and B.A. Sexton, Surf. Sci. 111 (1981) 11.

[66] (a) K. Prabhakaran, P. Sen and C.N.R. Rao, Surf. Sci. 169 (1986) L301;
(b) C.T. Au, J. Breza and M.W. Roberts, Chem. Phys. Lett. 66 (1979) 340.

[67] D. Doering and T.E. Madey, Surf. Sci. 123 (1982) 305.

[68] P.A. Thiel and T.E. Madey, Surf. Sci. Rep. 7 (1987) 211.

[69] M.W. Roberts, Trans. Faraday Soc. 59 (1963) 698.

[70] C.S. McKee and M.W. Roberts, Trans. Faraday Soc. 63 (1967) 1418.

[71] See: M.W. Roberts and C.S. McKee, Chemistry of the Metal–Gas Interface (Oxford University Press, Oxford, 1978) Ch. 8, for detailed discussion of the work of King and others.

[72] W.H. Weinberg, in: Kinetics of Interface Reactions, Eds. M. Grunze and H.J. Kreuzer (Springer, Berlin, 1987) 94.

[73] C.M. Quinn and M.W. Roberts, J. Chem. Phys. 40 (1964) 237.

[74] A.F. Carley, S. Rassias and M.W. Roberts, Surf. Sci. 135 (1983) 35.

[75] C.T. Au and M.W. Roberts, Nature 319 (1986) 206;
J. Chem. Soc. Faraday Trans. I, 83 (1987) 2047, and discussion p. 2085.

[76] A.F. Carley, M.W. Roberts and S. Yan, Catal. Lett. 1 (1988) 265.

[77] A.F. Carley and M.W. Roberts, J. Chem. Soc. Chem. Commun. 355 (1987).

[78] M.W. Roberts, J. Chem. Soc. Rev. 15 (1989) 451; J. Mol. Catal. 74 (1992) 11.

[79] A.F. Carley, M.W. Roberts and S. Yan, J. Chem. Soc. Faraday Trans. 86 (1990) 2701.

[80] M.W. Roberts, Catal. Today 12 (1992) 501.

[81] M.W. Roberts, Appl. Surf. Sci. 52 (1991) 133.

[82] C.S. McKee, L.V. Renny and M.W. Roberts, Surf. Sci. 75 (1978) 92.

[83] P.G. Blake and M.W. Roberts, Catal. Lett. 3 (1989) 379;
A.F. Carley, M.W. Roberts and M. Tomellini, J. Chem. Soc. Faraday Trans. I, 87 (1991) 3563.

[84] B.A. Afsin, P.R. Davies, A. Pashuski and M.W. Roberts, Surf. Sci. Lett. 259 (1991) L724.

[85] A. Boronin, A. Pashuski and M.W. Roberts, Catal. Lett. 16 (1992) 345.

[86] B.A. Afsin, P.R. Davies, M.W. Roberts and D. Vincent, Surf. Sci. 284 (1993) 109.

[87] A.F. Carley, P.R. Davies, M.W. Roberts and D. Vincent, to be published.

[88] H. Brune, J. Wintterlin, R.J. Behm and G. Ertl, Phys. Rev. Lett. 68 (1992) 624.

[89] C.B. Mullins, C.T. Rettner and D. Auerbach, J. Chem. Phys. 95 (1991) 8649.

Surface Science 299/300 (1994) 785–797
North-Holland

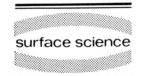

surface science

Through the labyrinth of surface reaction mechanism: a personal account, 1964–1992

Robert J. Madix

Departments of Chemical Engineering and Chemistry, Stanford University, Stanford, CA 94305, USA

Received 1 March 1993; accepted for publication 19 April 1993

The events leading to the current study of the mechanism of surface reactions on metals are discussed. This article is written with graduate students in mind, with particular attention on the aspects of the development of surface science that lead to significant advances in the capacity to understand complex reactions of organic substrates adsorbed on solid surfaces. The field is traced from the advent of the postacceleration low energy electron diffraction through the methods used to unravel the mechanism of complex surface reactions used today. It is a personal account of the course of surface science and the hallmark developments in the field that affected my research most significantly. Examples are given that trace the development of studies of the mechanism of surface reactions over the last three decades.

1. Early developments leading to activity in surface science

In the spring of 1964 my PhD work on molecular beam studies of the dynamics of oxygen reactive etching reactions of single crystal germanium with Professor Michel Boudart at the University of California was coming to a close. It is clear that my graduate work with single crystals shaped the course of my later interests. During the course of that work I had become very interested in the expanding work in the surface science of germanium and silicon surfaces, and it was timely that an international meeting on the topic of surface science was to be held at Brown University that summer. Since I was to begin a postdoctoral appointment with Professor Carl Wagner at the Max Planck Institute for Physical Chemistry in Göttingen, Germany, the next fall, it was opportune for me to attend the meeting, and Professor Boudart was kind enough to send me. For me this meeting was the beginning of the field of surface science. Indeed the proceedings of this meeting comprise one of the first volumes of the journal to which this volume of *Surface Science* pays tribute [1].

At that time one of the primary topics of interest was low energy electron diffraction and the structure of silicon surfaces. The school of Professor H.H. Farnsworth was prominent, and there was a great deal of discussion of the Faraday cup method for collecting diffraction intensities at this meeting [2]. Though this method was ultimately to give essential information for structural determination, at the time it was cumbersome to employ and not of general utility to those interested in more qualitative aspects of structure – such as overlayer unit cell sizes and symmetries. Soon thereafter the *postacceleration display method of LEED* was developed; this advancement in low energy electron diffraction was one of three that propelled activity in the study of single crystal surfaces – the field that we now identify as surface science.

At about this time Varian Associates began to sell ion-pumped ultrahigh vacuum chambers equipped with postacceleration LEED display systems, simple crystal manipulators and gas handling systems. The *commercial availability of UHV* was the second factor spurring the advancement of surface science. A bakeable quadrupole mass spectrometer could be easily added to this sys-

SSDI 0039-6028(93)E0446-2

tem, and thus the rudimentary equipment for studies of surface reactions, which necessitated measurement and control of the gas phase composition, was available. Within three years *retarding field Auger spectroscopic analysis* appeared [3], and this method for surface analysis, the third crucial development, combined with LEED, afforded the control of surface structure, surface composition and gas composition. The only remaining element for studying *reactions* on surfaces was a method for measuring their rates.

Our first attempt at studying reaction rates at Stanford concerned the etching rates of germanium by oxygen using a Varian system purchased in 1965. We were engaged in effusive molecular beam studies of the reaction probabilities of silicon and germanium at high temperatures with oxygen molecules, atoms, ozone and halogen molecules [4]. Postreaction electron micrograph replication images revealed that the surfaces were severely etched by these reactions to expose (111) surfaces, independent of the initial orientation of the surface (Fig. 1) [5]. We were able to follow the initial stages of these etching reactions by observing the evolution of the (111) microfacets with LEED. In order to determine the rates of these reactions we monitored the intensities of the diffraction features due to the microfacets photographically. However, rate constants obtained in this fashion were inaccurate, so we sought better methods.

2. The advent of relaxation methods for the study of surface reactions on single crystal surfaces

During my postdoctoral year with Professor Karl Wagner in Germany, in 1964, I was closely associated with the group of Professor Eigen. I became very interested in his studies of fast reactions using perturbation displacements from equilibrium. It occurred to me that similar methods could be used to study reactions with molecular beams on single crystal surfaces by using pressure modulation to perturb the system about the *steady state*. I had an ardent interest in molecular beam studies of reaction dynamics during my graduate work fueled both by my dissertation work and by

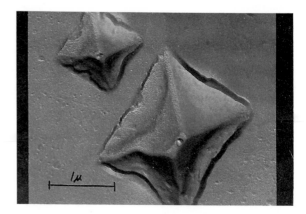

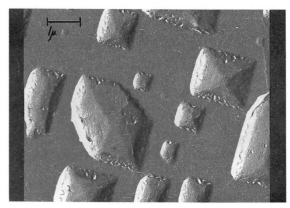

Fig. 1. Etch pits in Ge(100) formed by high temperature reaction of dioxygen.

my proximity to and friendships with members of the group of Professor Dudley Herschbach at Berkeley. Furthermore, one of the successful methods used at the time by Professor Wagner at Göttingen to study catalytic reactions on metal foils was a step change in reactant partial pressure to produce a measurable response in electrical conductivity of the foil, which could then be related to the thermodynamic activity of the reactants at the surface and the reaction kinetics. I was convinced that understanding surface reactivity on a fundamental level required studies of reaction kinetics with well defined surfaces using surface science techniques.

As a step in this direction we initiated molecular beam relaxation spectroscopy (MBRS) in 1970 with studies of reactions of oxygen atoms with germanium to form volatile germanium oxide

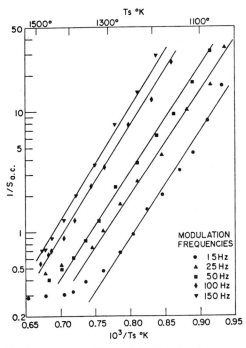

Fig. 2. The inverse amplitude of the first Fourier component of the product waveform of SiCl due to the imposition of a squarewave molecular beam flux of Cl_2 at the surface.

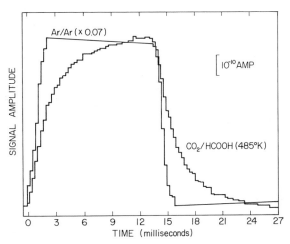

Fig. 3. Waveforms obtained by a boxcar integrator for argon reflected from Ni(110) and CO_2 due to formic acid decomposition on Ni(110).

(GeO) [6]. This was accomplished by retrofitting the effusive molecular beam system we had used in the reactive etching studies with a squarewave chopper to provide the pressure modulation of the gas striking the surface and using a quadrupole mass spectrometer and phase sensitive detection to monitor the first Fourier component of the time-dependent flux of product emanating from the surface in response to the input time-varying reactant flux. By conducting such experiments at different surface temperatures, we were able to accurately determine the rate constant for the rate-limiting step in the reaction, which was desorption of the product molecule (GeO). In quick succession we studied the high temperature reactions between the halogens and silicon, and soon we applied this method to high temperature reactions of refractory metals (Fig. 2) [7]. There was a flurry of activity from several laboratories in this area of study, and MBRS has been used to great advantage since these times [8]. We obtained wave forms for products species

in 1972 (Fig. 3) with the clear indication that waveform analysis methods, commonly used in both chemical and electrical engineering at the time, could be applied to studies of surface reactions on single crystal surfaces by MBRS [9]. Independently, such analysis was proposed by Foxon, Boudry and Joyce [10]. Fourier wave form analysis to very high order has been applied in several cases, beginning with the work of Merrill and Sawin [11] and most recently by Sibener and Padowitz (Fig. 4) [12]. Among its many applications, this method has been used to (1) reveal that preexponential factors for desorption of sim-

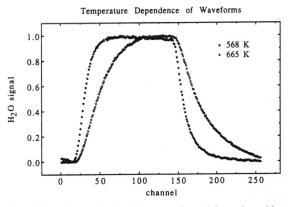

Fig. 4. Waveforms obtained for water formed from the oxidation of hydrogen on Ru(100).

ple molecules such as GeO from Ge and CO from Ni(110) can greatly exceed 10^{13} s^{-1} in accord with the expectations of transition state theory [13], (2) establish that CO oxidation on palladium surfaces (and others) occurs by reaction of *coadsorbed* CO and oxygen (the so-called Langmuir–Hinshelwood mechanism) [14], and (3) give substantial insight into high temperature etching reactions of elemental semiconductors [15], for example. During the course of this work it became apparent to us, however, that studies of reactions in the frequency domain using Fourier analysis of the response to an input forcing function must be limited to rather simple reactions, since it was not possible to obtain *unique* mechanistic detail by back transformation. We were thus led to try other methods to understand more complex surface reaction mechanisms. The foundation for our next advance rested on work that I became interested in as a graduate student.

3. The study of complex surface reactions

In 1963 Dr. Gert Ehrlich, then at the research laboratories of General Electric, published an article in Advances in Catalysis that was of great significance to me [16]. I was particularly interested in this work as a graduate student, because it clearly pointed the direction for measuring the rates of reactive processes on surfaces with low surface areas which might provide information based on studies with single crystals relevant to heterogeneous catalysis. This so-called "flash desorption" technique was performed by first adsorbing a gas on a metal wire in an otherwise evacuated glass bulb, evacuating the bulb once again, and then resistively heating the filament rapidly to desorb the gas into a closed volume. In a classic experiment Dr. Ehrlich and coworkers measured the evolution of the pressure of N_2 evolved from a tungsten filament with time measured by an ionization gauge, revealing the presence of several kinetic routes for desorption with different energetics (Fig. 5). Equations were published which related the shapes of these curves to the kinetic parameters for desorption. Shortly later Redhead extended this method to a pumped

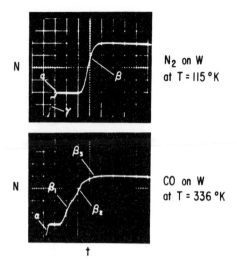

Fig. 5. Oscilloscope traces for the flash desorption of N_2 and CO from tungsten filaments.

volume, yielding flash desorption peaks which were the derivatives of the pressure build-up curves and from which activation energies and preexponential factors could also be derived [17]. Work began in earnest in the late 1960's studying the desorption kinetics of CO and H_2 from metallic single crystal surfaces. During this time period substantial advances were made in understanding the kinetics of desorption of these gases from metal surfaces [18], and the first reliable values for metal–gas bond energies became available (Fig. 6) [19]. It soon became apparent that there were significant coverage effects on desorption rates [20], and that only in the limit of low coverage were simple desorption kinetics to be expected. These developments are discussed in greater detail in other articles in this volume.

However, none of this work addressed the more complex reactions related to heterogeneous catalysis. In fact there was little known about the mechanism of reactions on metal surfaces. There was even less known about rate constants for elementary steps on surfaces. There were serious problems with the determination of the kinetics and mechanism on a molecular level because reactions were normally studied under steady state conditions on metallic catalysts supported on high surface area materials. Rate expressions were always the result of the combination of

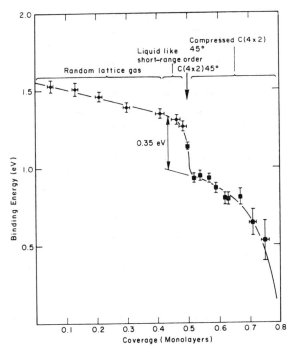

Fig. 6. The binding energy of CO as a function of coverage on Pd(100).

many steps, often including transport effects, and isolation and identification of *the* rate-determining step was extremely difficult [21]. Furthermore, surface concentrations of reactive intermediates were not measurable. Without these concentrations it was impossible to determine preexponential factors – critical quantities for understanding the mechanism of elementary reaction steps. Only in cases where the surface was "judged" by the kinetic order to be saturated with an intermediate (a zero-order reaction) could the concentration be *guessed* and the preexponential factor estimated. Arguments concerning catalytic mechanism and rates often degenerated into cross accusations about surface cleanliness.

In this backdrop in 1970 two of my students and I began a series of studies that provided the foundation for the study of the kinetics and mechanism of complex reactions on metal surfaces. The method we developed we eventually called temperature-programmed reaction spectroscopy (TPRS). In favorable cases it enables one to determine the identity of rate-determining

reactive intermediates and the rate constant parameters for the slowest (rate-determining) step in complex reaction sequences on surfaces. It was an outgrowth of the "flash desorption" methods, but it embodied additional logic necessary to identify the rate-determining reactive intermediates and, thus, to understand complex surface chemical reactions.

The logic of temperature-programmed reaction spectroscopy is simple, but powerful. Mass spectrometry is used to differentiate the rate of evolution of different products from the surface. This method provides a complete map of the reaction network. Initially, we performed experiments by tuning the quadrupole mass spectrometer to a *specific*, single, mass-to-charge ratio and following the evolution of each product from the surface in separate, sequential, experiments with an x–y recorder, plotting the ion fragment intensity from the mass spectrometer against the surface temperature (or time) during the programmed temperature ramp. During the linear temperature program reactions reach their peak rate when the temperature is approximately equal to $16E$, where E is the activation energy (in kcal/mol) for the reaction in question. Thus, for complex reactions in which a multiplicity of reactions may be involved, the different reaction channels are separated in temperature (time) by approximately $16\Delta E$, where ΔE (in kcal/mol) is the difference in activation energies of any two reaction channels. Since temperature differences of a few degrees Kelvin are easily measured in such an experiment, activation energy differences of a fraction of a kcal/mol can be measured (assuming, for the sake of example, that the preexponential factors are equal). Consequently, not only were the kinetic parameters for reaction(s) of the rate-limiting reaction intermediate(s) involved readily determined [22], but also the method allowed accurate determinations of the kinetic isotope effect for the study of the surface reaction mechanism [23]. Furthermore, an important corollary follows; namely, two products showing the same peak shape and peak temperature in a temperature-programmed reaction spectrum are the products of the same rate-limiting reaction step, and, therefore, result from the re-

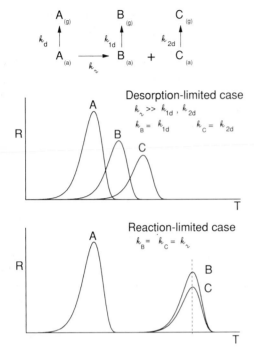

Fig. 7. A schematic representation of temperature-programmed reaction spectroscopy, showing rate-limiting surface reactions.

grammed desorption (TPD) and similar terms were inappropriate for such studies, unless desorption was indeed the only rate-limiting step measured for product evolution into the gas phase.

Our first experiments were primitive. We converted a UHV molecular beam apparatus we were using for studies of angular distributions of O_2 scattering from Si(111) measurements and inserted a Ni(110) crystal to begin our TPRS studies. The temperature "programmer" was the hand of the experimenter on the dial of a powerstat; the powerstat was quickly turned to a preset point, the current through a filament behind the crystal heated instantly to incandescence and radiatively heated the crystal, suspended effectively in thermal isolation a mm away, at a nearly linear rate (this trick was taught to us by Professor L.D. Schmidt who was on sabbatical at Stanford at that time). After some initial trials, we placed the mass spectrometer close to the crystal in order to increase the flux of product into the ionizer of the mass spectrometer. The products were detected in flow-through mode in the ionizer in order to monitor the product flux, and thereby the rate of product formation.

We began our work with the study of the decomposition of formic acid. This reaction had long been a test reaction for activity of supported metal catalyst, and there were primitive theories of catalysis based on studies of this reaction. Furthermore, we expected this reaction to exhibit simple kinetics and to provide a basis for defining the method. In addition we anticipated study of the effects of surface modification by adsorbed species, such as carbon, oxygen and sulfur, on the kinetics and mechanism of the reaction in order to provide a better understanding of catalyst poisoning.

Even at the outset, the immense level of effort we put into these experiments was exciting, fun and very rewarding. We were in unexplored territory, a field so fertile that even mistakes or mishaps turned into adventures that lead us to one discovery after another. John Falconer (now Professor of Chemical Engineering at the University of Colorado) and Jon McCarty (now Director of Catalysis Research at SRI International)

action of (a) common intermediate(s). Consequently, if the relative number of molecules of each of the products evolved in this rate-limiting step can be determined by quantitative mass spectrometry, the composition of the rate-limiting reaction intermediate can be determined. In many cases this composition leads straightforwardly to the identification of the rate-determining reactive intermediate. Due to the high resolving power of the experiment for each reaction channel and the fact that we were able to probe complex reactions occurring on the surface which were not in themselves *desorption*, we dubbed the method *temperature-programmed reaction spectroscopy*. The measurement of rates of *surface reactions* could be straightforwardly distinguished from *desorption* events by separately determining the peak shapes and temperatures for desorption of the products species in separate experiments (Fig. 7). We specifically intended this terminology (TPRS) to indicate that *reactions* occurring *on the surface* were being observed. The term temperature-pro-

worked as a team on our only UHV system, working extremely long hours, pausing only to catch up on a night's sleep before plunging headlong into another siege of experiments that would last days at a time. After an initial round of experiments we met for several hours and laid out a plan for studies that encompassed both clean Ni(110) and different ordered overlayers of carbon and oxygen on Ni(110). We relied heavily on the scant AES and LEED work that had been published on overlayer structures on this surface to define the systems we were to study. Jon and John split the 24 hours of the day, overlapping at some times, working independently at others. The equipment was relatively simple, and the most problematic instrument was the x–y recorder, which failed repeatedly. They became expert with this device, and repaired it on the fly time after time, often in the wee hours of the morning, so that the experiments would not be interrupted. They understood the importance of getting results expediently while the apparatus cooperated. There was real beauty to the simplicity of these experiments. We had a great time, and we marveled at and pondered the results that we achieved day after day. In all, we worked a year without a significant pause, in order to perform a fraction of the experiments we had mapped out in a few hours. That's approximately 0.3% inspiration, 99.7% perspiration. Adding the inspirational efforts in interpretation of the results brings the tally to about 1.0% inspiration. But what fun.

Much came of this work, which I shall not reiterate in any detail here, since it has all been published. We discovered surface kinetic explosions [24], a phenomenon in which the rate of a surface reaction accelerates in time isothermally, and which has been observed many times subsequently by others. We pioneered the use of isotopic labelling and the kinetic isotope effect for reactions on single crystal surfaces to dissect reaction mechanisms – in their case, the formic acid decomposition [25]. We showed that adsorbed carbon completely altered the kinetics and mechanism of the reaction, illustrating that the carbidic surface retained a high reactivity, but a thin graphitic layer completely eliminated reactivity, leading only to reversible adsorption of the

acid [26]. These studies illustrated that poisoning of nickel by carbon required graphitization for this reaction and signaled loud and clear the general potential of such studies on metallic single crystals and their relevance to catalyst poisoning. All the while we relentlessly improved the method, until it was possible to make accurate estimates of the relative product yields and to nail down the nature of the intermediates involved. We were thus able to absolutely identify the formate intermediate on the Ni(110)-(2 × 1)-C, produced by simple dehydrogenation and to show that the decomposition proceeded by dehydration on the clean Ni(110) surface. In these initial studies the foundation was laid for subsequent studies of the kinetics and mechanism of surface reactions on well defined metal surfaces. In succeeding studies we showed that the reactivity of nickel carbide (Ni(110)-(2 × 1)-C) and Cu(110) were very similar [27], that adsorbed oxygen modified the surface so as to gradually suppress the dehydration reaction in favor of dehydrogenation [28], and that similar patterns of reactivity occurred for acetic acid [29], and, by inference, for other carboxylic acids. Except for the special circumstances that produced the anhydride intermediate, reaction of the carboxylic acid with the surface produced an adsorbed carboxylate and hydrogen. With heating the carboxylate decomposed to yield CO_2 and dehydrogenation products of the remaining fragment. In many respects these reactions paralleled decomposition reactions of the corresponding bulk salts.

It was soon thereafter that we discovered the phenomenon of "oxygen activation" in selective oxidation reactions. Israel Wachs (now Professor of Chemical Engineering at Lehigh University) began experiments with methanol on copper. He had difficulty reacting methanol with clean Cu(110). However, he soon discovered that methanol reacted readily with a surface partially covered with oxygen, producing several products, including formaldehyde and hydrogen. In a few weeks he mapped out the reaction mechanism for the oxidative dehydrogenation of methanol on copper (Fig. 8), making extensive use of isotopic labelling [30]. In his reading of the literature, he found that decomposition of methanol over silver

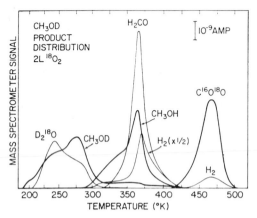

Fig. 8. The temperature-programmed reaction spectrum of methanol oxidation to formaldehyde and hydrogen by reaction of methanol with preadsorbed oxygen on Cu(110). The isotope distributions clearly reveal the mechanistic pathway.

catalysts required oxygen, which lead to the formation of formaldehyde. Soon we had shown that methanol oxidation on silver proceeded via a series of steps initiated by oxygen-induced removal of the hydroxyl hydrogen from methanol and that the rate-limiting step was C–H bond rupture in the resulting methoxy species to liberate formaldehyde. In a series of experiments we mapped out the reactions the products as well, giving the mechanism for the selective oxidation of methanol and ethanol over metallic silver catalysts [31]. These experiments led to the extensive investigation into the physical organic chemistry of selective oxidation on silver [32].

As one of our interests was the chemical modification of metal surfaces by adsorbed species, we began studies of the decomposition of alcohols and related molecules on clean and adsorbate-modified Ni(100). One of our first interesting observations was that adsorbed sulfur led to a nearly linear decline in the reactivity of Ni(100) for methanol decomposition to CO and H_2 [33]. Furthermore, above 0.25 monolayer of adsorbed S, significant quantities of formaldehyde formed, and at 0.38 monolayer sulfur *only* formaldehyde and hydrogen formed as products. From the product stoichiometry we deduced the reactive intermediate to be a methoxy (CH_3O), which was stabilized by the presence of sulfur. As the sulfur

coverage built up, cleavage of C–H bonds became more difficult, and the methoxy species survived intact to higher temperatures. When the C–H bond was finally thermally activated, the binding energy of the formaldehyde formed was much less than the activation energy for cleavage of a second C–H bond, and formaldehyde desorbed immediately; on the clean surface further reaction to CO and H_2 occurred, however, because the activation energy for C–H bond cleavage was low. These experiments were exceedingly tedious, because the TPRS peaks contained mass fragments that were cracking fractions of larger species as well as products themselves, and the mass spectra had to be decomposed to give the relative product yields in order to determine the rate-limiting reactive intermediate. Experiments had to be repeated many times, since we were detecting only a single mass/charge in each temperature-programmed sequence.

4. Experimental impact of minicomputers on studies of reactions on surfaces

At this time it became obvious that we could improve our technique immensely by multiplexing our mass spectrometer to detect several mass/charge ratios during the single temperature sweep, taking advantage of the small laboratory computers (the DEC PDP 11 series) that were available. So, when in the mid-1970's the Department of Chemical Engineering at Stanford purchased several minicomputers to be shared among the faculty, we built an analogue switching device that allowed us to monitor four (and later eight) masses repeatedly throughout the temperature sweep. This development enabled us to quantify the product yields in TPRS for more complex reactions. With the use of these small computers experiments became possible that we could not have performed previously, because all reaction channels could be monitored in a single sweep, eliminating the need (nearly impossible) to repeat experiments at identical heating rates, pumping speeds, etc. We were only limited now by our ability to anticipate the identity of the products, which was still necessary, since we were using a manually set, analogue device to choose the settings on the quadrupole filter.

5. The merger of TPRS and EELS

At this time it is important to add another historical perspective. In the mid-1960s I was fortunate to attend the Physical Electronics Society Meeting in Minneapolis, Minnesota. I nearly jumped out of my seat when Professor Frank Probst of the physics department of the University of Illinois presented the vibrational spectra for hydrogen atoms, nitrogen atoms, CO and H_2O adsorbed on tungsten, which he had obtained by scattering low energy electrons from the surface and measuring the energy distribution of the scattered electrons [34]. As a graduate student I had investigated the possibility of applying infrared absorption methods to the identification of species adsorbed on single crystals, but the progress reported in the literature for reflection methods was very discouraging. The magnitude of the accomplishment of Professor Probst was immediately apparent. I was so excited about this work that I arranged to visit his lab immediately after the meeting (one could change air travel without monetary penalty then), hoping that the visit would enable me to apply this method to problems of my immediate interest. I was shown the apparatus, I subtracted my estimate of the years it would take me to construct the identical apparatus and get it to work (let alone a significant improvement), from the years left ticking on my tenure clock, noted the negative number, and decided it was inadvisable to pursue it. Without substantial technical help with the electron optics, it was an impossible task for me. It was indeed several years before the first higher resolution EELS spectrometers useful for studying adsorbed reaction intermediates emerged from the laboratory of Professor Harold Ibach at the Kernforschungsanlage in Jülich, Germany [35]. It was apparent from the first appearance of these instruments that they would add a significant component to the study of surface reactions and the identification of surface reactive intermediates, though it was very difficult to construct a working apparatus without the help of someone familiar with the pitfalls.

So I was delighted when, a few years later, Dr. Brett Sexton expressed interest in collaborating on EELS studies of reactive intermediates on silver. Brett had been a student with Professor Gabor Somorjai at the University of California in Berkeley, a postdoctoral student with Professor Ibach and was a research scientist in the Division of Physical Chemistry department of General Motors in Warren, Michigan. He had begun a series of studies, some of which were based on our previous suggestions for intermediates in oxidation reactions on copper and silver. We first set out together to examine molecular oxygen on Ag(110) [36], which we had identified by isotope cross-mixing TPD experiments, and formate on Ag(110) [37], which had been identified by TPRS. The EELS results showed that the O–O bond order in adsorbed dioxygen was reduced to that of a single bond by adsorption; EELS enabled us to see reactions at temperatures below which any species were evolved into the gas phase, which made these reactions inaccessible to TPRS. It was then abundantly clear that EELS and TPRS would become inextricably entwined in our future research. As techniques they would grow to compliment one another in nearly every study of reactions of gases on metal surfaces.

As the result of this initial collaboration, Eric Stuve (then a graduate student in my laboratory and now Professor of Chemical Engineering at the University of Washington) went to the General Motors Research Labs in Warren. I am deeply grateful to the management of General Motors for fostering this interaction, which lasted several years. It had a profound effect on the course of my research. In a short time Eric studied the reaction of H_2O, acetylene, CO_2 and formaldehyde with oxygen dissociatively adsorbed on Ag(110) [38]. This research was very illuminating; overall it agreed with our previous TPRS results on the surface chemistry (there were no surprises here), but the EELS provided information on the bonding and structure of the adsorbed reactive intermediates. The nature of the information available in the combined TPRS-EELS experiment lead to an advanced generation of problems in my laboratory that required *both* methods.

6. Mechanistic discovery with full mass spectrometer multiplexing

At about the same time another event occurred which propelled surface mechanistic studies to their current level. Due to the experiments of Mark Barteau (now Professor of Chemical Engineering at the University of Delaware) and Dr. Mike Bowker (now Assistant Director of the Leverhulme Center for Innovative Catalysis) in my laboratory, we had shown that the reactions of simple molecules with oxygen dissociatively adsorbed on Ag(110) were the result of an acid-base reaction – a proton transfer from the acidic molecule to the oxygen, which served as a strong base [39]. Mark was in the process of establishing a scale which we could use to predict the reactivity of select organic substrates with this "active" oxygen [40]. One interesting result of these studies was that acetylene reacted with active oxygen, whereas ethane did not, even though the C–H bond energy in ethane exceeded that in acetylene. To see if this was a general property of oxygen adsorbed on Group II-B metals, we studied the reaction of acetylene with adsorbed oxygen on Cu(110) [41]. The results were confusing. First, unlike Ag(110), reaction occurred with the clean surface, and products were evolved. Our interpretation of the mass spectra indicated that we formed ethylene (mass/charge 28) and hydrogen. Though oxygen appeared to enhance reactivity, the stoichiometry of the products did not add up. In particular, there was excess hydrogen evolved. At the time my group met periodically with the group of Professor Earl Muetterties, then at the chemistry department at UC Berkeley. During the time of these meetings they began the study of acetylene reactions on palladium, knowing that tricyclization of acetylene to benzene occurred in palladium complexes in solution. Soon he and his students had shown that this cyclization reaction occurred on palladium single crystal surfaces. Some time later Dr. Neil Avery demonstrated that the same tricyclization

reaction occurred on Cu(110) and that in our experiments we had observed the cracking fractions of benzene, not ethylene and hydrogen. Our inability to identify this product pointed to a serious deficiency in the experimental method of TPRS.

This misinterpretation occurred since we were literally blind to the appearance of unexpected products because of the manual multiplexing method we used to drive the mass spectrometer. If we did not anticipate the product, we did not preset the necessary parameters for switching of the quadrupole, and we did not detect it. Clearly, a change needed to be made to eliminate such oversights. Immediately we set out to remedy this problem. I discussed this problem with Professor Cynthia Friend at Harvard, and within a few weeks a graduate student in her laboratory, Albert Liu, and one in mine, M.L. Burke had produced computer-driven repetitive scans of the quadrupole over the entire mass range as the

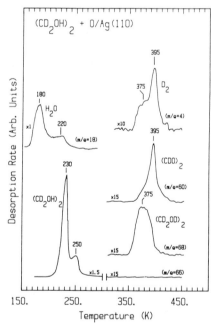

Fig. 9. The temperature programmed reaction spectrum for ethylene glycol oxidation to the dialdehyde on Ag(110). The selective reaction to form the dialkoxide is revealed by the selective formation of H_2 at the lower temperature, and the two-step dehydrogenation of the dialkoxide near 400 K is evident.

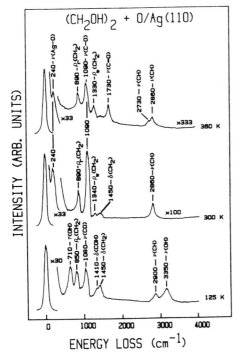

Fig. 10. EELS spectra showing the formation of the dialkoxide and the partial dehydrogenation to form the surface-bound monoaldehyde at 360 K.

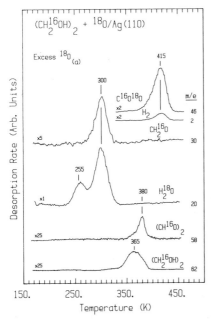

Fig. 11. TPRS for more extensive oxidation of the dialkoxide of ethylene glycol in excess adsorbed oxygen on Ag(110). Attack of the carbon skeleton to form formaldehyde and surface-bound formate is evident.

temperature was ramped. The intensity-time information over the mass range from 1–200 amu was output from the quadrupole to a computer for subsequent display and analysis. The Harvard version was written for the (then) new IBM PCXT; ours was written for the DEC PDP1103. This approach to TPRS immediately provided the technique for the *discovery* of surface reactions that could not be anticipated in advance and, of course, greatly reduced the work necessary for quantitative TPRS by accumulating all the rate data for all ion fragments for all products simultaneously. Overall, an improvement was made that revolutionized the technique – a change in response to an error resulting from a prior limitation in the technique.

The combined use of TPRS and EELS is nicely illustrated by the study of the oxidation of ethylene glycol (HOCH$_2$CH$_2$OH) on Ag(110) [42]. The oxidation proceeds via a dialkoxide intermediate, ethylenedioxy silver, formed by the selective removal of the –OH protons by adsorbed oxygen atoms below 170 K. Excess diol and molecularly held water desorb between 170 and 300 K, so that the dialkoxide can be isolated (Fig. 9). Further reaction of the dialkoxide begins above 350 K. The EELS indeed reveal loss of the O–H stretch upon annealing the diol to 300 K with prominent vibrational modes assignable to the symmetric CO stretch (1090 cm^{-1}), the CH$_2$ wag (1340 cm^{-1}) and the C–H symmetric stretch (2860 cm^{-1}). At 375 K there is indication of cleavage of the first C–D bond in the dialkoxide. Indeed, the

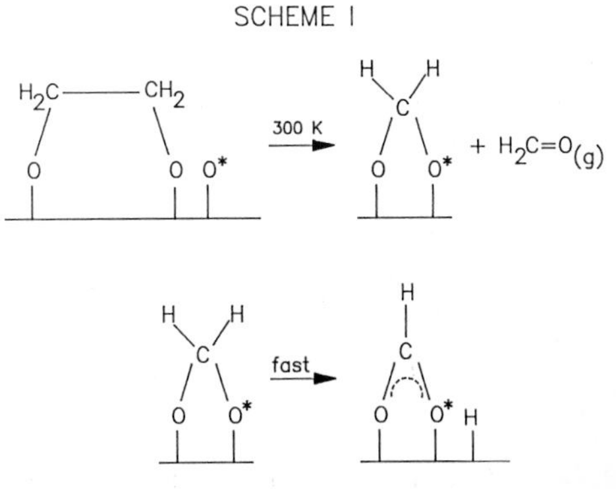

SCHEME I

Fig. 12. The mechanism of secondary oxidation of the dialkoxide.

SCHEME II

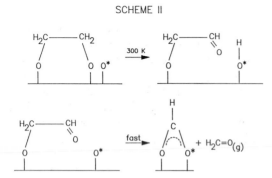

Fig. 13. Illustration of the rate-determining step for secondary oxidation of the dialkoxide.

loss of a single D in the dialkoxide is evident in EELS following annealing the surface to 360 K; both the C=O stretch (1730 cm^{-1}) and the aldehydic C–H stretch (2730 cm^{-1}) of adsorbed O=CHCH$_2$O– are evident (Fig. 10). The rate constant for cleavage of the second C–D bond to eliminate the dialdehyde (O=CHCH=O) from the surface is $10^{12.8}$ exp[$-23\,100(\text{cal/mol})/RT$] s^{-1}.

In the presence of excess oxygen the C–C bond is attacked and cleaved. TPRS reveals the

formation of both formaldehyde (CD$_2$O) and CO$_2$ (Fig. 11). The temperature at which CO$_2$ is liberated suggests it arises from decomposition of surface formate. TPRS alone then suggests that the activation of the C–C bond in the dialkoxide results from the attack of surface oxygen upon one of the carbons, leading to surface-bound formate and expulsion of formaldehyde (Fig. 12). However, the kinetic isotope effect, k_H/k_D, for this latter step was 8, indicative of the direct involvement of H(D) in the rate-limiting step. Thus, it appears that either transfer of hydrogen from carbon to surface oxygen or transfer of hydrogen from carbon to the metal induced by surface oxygen is the rate-limiting step (Fig. 13). The EELS clearly shows the formation of formate subsequent to the liberation of formaldehyde (Fig. 14).

7. Summary

The intent of this article is to trace the development of the study of complex surface reactions since the first volumes of *Surface Science* appeared. The tools to understand the mechanism of surface reactions are now at hand. With the precision of modern mass spectrometry and the capabilities of small computers to process large amounts of information, even very complex reaction processes can be understood. By using judiciously isotopically labelled species, the reaction paths of functional groups within molecules can be identified and the rate constants for rate-limiting steps in complex reaction cycles quantified. Reaction intermediates can be isolated and their structure and bonding studied with a variety of spectroscopies. We are poised to understand surface reactivity in unprecedented detail and to generate a general understanding of the reactivity of surfaces.

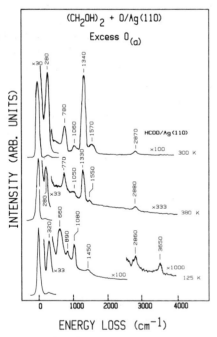

Fig. 14. EELS showing the formation of the surface formate in the secondary oxidation of ethylene dialkoxide.

References

[1] Proc. Conf. on the Physics and Chemistry of Solid Surfaces, Brown University, June 21–26, 1964, in: Surf. Sci. 2 (1964).

[2] It is most appropriate that the first article to appear in Vol. 1 of Surface Science is authored by Professor Farnsworth (J.B. Marsh and H.E. Farnsworth, Surf. Sci. 1 (1964) 3).

[3] R.E. Weber and W.T. Peria, J. Appl. Phys. 39 (1967) 4355.

[4] R.J. Madix and R. Korus, Trans. Faraday Soc. 64 (1968) 2514;
R.J. Madix and A. Susu, J. Catal. 28 (1973) 316.

[5] N.T. Batkin and R.J. Madix, Surf. Sci. 7 (1967) 109.

[6] R.J. Madix, R. Parks, A.A. Susu and J.A. Schwarz, Surf. Sci. 24 (1971) 288.

[7] R.J. Madix and J.A. Schwarz, Surf. Sci. (1971) 264.

[8] M.P. D'Evelyn and R.J. Madix, Surf. Sci. Rep. 3 (1984) 1.

[9] R.J. Madix, Surf. Sci. 45 (1974) 696.

[10] C.T. Foxon, M.R. Boudry and B.A. Joyce, Surf. Sci. 46 (1974) 317.

[11] H. Sawin and R.P. Merrill, J. Vac. Sci. Technol. 19 (1981) 40.

[12] D.F. Padowitz and S.J. Sibener, J. Vac. Sci. Technol. 9 (1991) 2289.

[13] C.R. Helms and R.J. Madix, Surf. Sci. 52 (1976) 677.

[14] T. Engel and G. Ertl, Adv. Catal. 28 (1979) 1.

[15] M.P. D'Evelyn, M.N. Nelson and T. Engel, Surf. Sci. 186 (1987) 75;
J.R. Engstrom and T. Engel, Phys. Rev. B 41 (1990) 1038;
K. Ohkubo, X. Igari, S. Tomodu and I. Kusunoki, Surf. Sci. 260 (1992) 44.

[16] G. Ehrlich, Adv. Catal. 14 (1963) 255;
See also G. Ehrlich and T.W. Hickmont, Nature 177 (1956) 1045;
T.W. Hickmont and G. Ehrlich, J. Chem. Phys. 24 (1956) 1263.

[17] P.A. Redhead, Vacuum 12 (1962) 203; Trans. Faraday Soc. 57 (1961) 641.

[18] See, for example, P. Tamm and L.D. Schmidt, J. Chem. Phys. 51 (1969) 5352.

[19] J.C. Tracy and P.W. Palmberg, J. Chem. Phys. 51 (1969) 4852.

[20] D.A. King, Surf. Sci. 47 (1975) 384.

[21] See, for example, the review of the formic acid decomposition: P. Mars, J.J.F. Scholten and P. Zwietering, Adv. Catal. 14 (1963) 35.

[22] More detailed analysis gives accurate values for both activation energies and preexponential factors.

[23] R.J. Madix and S.G. Telford, Surf. Sci. 227 (1992) 246.

[24] J. Falconer, J. McCarty and R.J. Madix, Surf. Sci. 42 (1974) 329.

[25] J. Falconer and R.J. Madix, Surf. Sci. (1974) 473.

[26] J. McCarty and R.J. Madix, J. Catal. 38 (1975) 402.

[27] D.H.S. Ying and R.J. Madix, J. Catal. 61 (1980) 48.

[28] S.W. Johnson and R.J. Madix, Surf. Sci. 66 (1977) 189.

[29] R.J. Madix, J. Falconer and A. Susco, Surf. Sci. 54 (1976) 6.

[30] I.E. Wachs and R.J. Madix, J. Catal. 53 (1978) 208.

[31] I.E. Wachs and R.J. Madix, Surf. Sci. 76 (1978) 531;
Appl. Surf. Sci. 1 (1978) 1978.

[32] R.J. Madix, Science 233 (1986) 1159.

[33] R.J. Madix, S.B. Lee and M.J. Thornburg, J. Vac. Sci. Technol. A 1 (1983) 1254.

[34] F.M. Probst and T.C. Piper, J. Vac. Sci. Technol. 4 (1967) 53.

[35] See, for example, H. Froitzheim, H. Ibach and S. Lehwald, Phys. Rev. B 14 (1976) 1362, and references therein.

[36] B.A. Sexton and R.J. Madix, Chem. Phys. Lett. 76 (1980) 294.

[37] B.A. Sexton and R.J. Madix, Surf. Sci. 105 (1981) 177.

[38] E.M. Stuve, B.A. Sexton and R.J. Madix, Surf. Sci. 111 (1981) 11; 123 (1982) 491; 119 (1982) 279.

[39] M. Barteau, M. Bowker and R.J. Madix, Surf. Sci. 94 (1980) 303.

[40] M.A. Barteau and R.J. Madix, Surf. Sci. (1982) 262.

[41] D. Outka, C.M. Friend, S. Jorgensen and R.J. Madix, J. Am. Chem. Soc. 105 (1983) 3468.

[42] A.J. Capote and R.J. Madix, J. Am. Chem. Soc. 111 (1989) 276; Surf. Sci. 214 (1989) 276.

Surface Science 299/300 (1994) 798–817
North-Holland

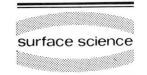

surface science

Atomic processes in crystal growth

John A. Venables

School of Mathematical and Physical Sciences, University of Sussex, Brighton BN1 9QH, UK
and
Department of Physics and Astronomy, Arizona State University, Tempe, AZ 85287, USA

Received 5 April 1993; accepted for publication 30 July 1993

The thermodynamic and kinetic processes which are involved in the early stages of crystal growth are discussed, with especial reference to vapor deposition of thin films. The atomic processes taking place during deposition are described in terms of rate and diffusion equations; the concept of "competitive capture" is outlined, where adatoms are forced to choose between competing sinks. The use of microscopy and surface physics techniques to study nucleation in films is emphasised. Examples of island (Volmer–Weber), layer (Frank–van der Merwe) and layer plus island (Stranski–Krastanov) growth in metal/insulator, metal/semiconductor and semiconductor/semiconductor deposition systems are given.

1. Introduction

1.1. Crystal growth in "Surface Science"

This article provides a short review of progress in understanding the atomic processes which occur during crystal growth, with particular reference to the production of thin (epitaxial) films. In the space available, the article can only be illustrative; it draws heavily on a longer book chapter [1], which can be used as supplementary reading, or as an alternative reference. In this article there is a bias towards articles in "Surface Science", thus charting the role of this particular journal. I do not, of course, claim to be unbiased; I use references where I have been involved, or which I have found of special interest.

In many senses vapor deposition (of thin films) is the simplest case of crystal growth, since it involves the formation of a dense, solid phase from a dilute, gaseous phase. The solid–vapor interface can be observed by a whole array of surface science and microscope-based experimental techniques. The theoretical formulations are also moderately simple: thermodynamic and statistical mechanical models can be formulated and tested by experiment. But "real life" is typically considerably more complicated than these simple

models; in particular, the relationship between "understanding" crystal or film growth in atomistic terms, and producing "better" films or crystals is as complex as the interplay between science and technology generally. I am not discussing the technology here, but rather aiming to discuss the scientific language in which such efforts can be discussed.

Crystal growth is a broad field, which merits several journals of its own [2]. Much of this work concerns the growth of large scale crystals, from the melt, or by regrowing or recrystallizing solids. Many phenomena, such as impurity segregation at the crystal–melt interface, dendrite and cell formation, are fascinating, whose description contains the properties of the solid–melt or grain boundary interface. But this type of crystal growth is also governed by bulk concentration and temperature gradients; with some notable exceptions, this is not the place to start if one searches simplicity at the atomic level.

The distinction of the first paper to be listed in the index of Surface Science under "Crystallization" goes to K. Lehovec [3]. This paper (which also contains a reference to one H. Gatos) descibes some calculations on the role of band-bending at a semiconductor interface in promoting selective adsorption of charged impurities at

SSDI 0039-6028(93)E0453-2

the growing interface. In the context of "surfactant growth" or "delta doping", this is a completely "modern" problem. As "Surface Science" came of age, circa 1985, the approach to a similar problem in molecular beam epitaxy (MBE), itself a subset of the original problem, but now with major economic activity behind it, was very different [4]. The questions needing answers had been posed more precisely, there were many new results, and the expected roles of surface and interface atomic and electronic structure were modelled, and better understood. But it is still a complex problem, and general solutions are elusive.

Crystallization and Crystal Growth are minor entries in the index to Surface Science throughout the thirty year period. Early entries under Epitaxy and Nucleation are more numerous; more recent searches need to include MBE, Clusters, Thin Films and Surface thermodynamics, Diffusion, Melting and Roughness. These categories will reveal the main references to "atomic processes in crystal growth".

1.2. Concerns and language

Within these categories, there are a range of concerns and of language used to describe the phenomena. Surface thermodynamics has been applied, both to growth mode criteria, and as an element of classical nucleation theory. The first use of surface energies to describe the three modes of crystal growth (layer, island and layer plus island) of material A on a substrate B was made by Bauer in 1958, and has been reviewed many times [1,5,6]. These distinctions can be understood qualitatively in terms of relative surface energies, γ_A and γ_B, as illustrated in Fig. 1. If $\gamma_A + \gamma^* < \gamma_B$ we get island growth (Fig. 1a); the growth of B on A would then proceed directly as islands (Fig. 1b). However, if we have layer growth initially, then it is often the case that the effective interfacial energy γ^* increases with thickness of the layer, for example due to strain in this layer. In such a case, the thermodynamic conditions for layer growth are terminated after a certain layer thickness. Further growth of the layers is then in

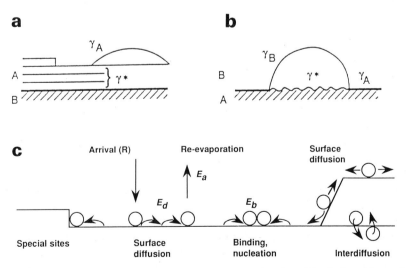

Fig. 1. (a) Growth of A on B, where $\gamma_A < \gamma_B$; misfit dislocations are introduced, or islands form after a few layers have been deposited. (b) Growth of B directly onto A as islands. The interfacial energy γ^* repesents the excess energy over bulk A and B integrated through the interfacial regions. (c) Schematic diagram of processes and characteristic energies in nucleation and growth on surfaces.

competition with growth of the more stable islands. This situation is very common, so that
many crystal growth systems can be classified as
layer plus island, or Stranski–Krastanov growth.

The above classification assumes no interdiffusion of, or chemical reactions between, A and B,
but in many cases of technological interest, for
example the production of A/B/A superlattices,
this may not be so. In these cases, the reason why
the end-product is not the equilibrium compound
of A and B is due to kinetic limitations during
and after deposition. One of the difficulties of
describing the whole deposition process, and the
final structures of the film in general, is that
there are many possible types of kinetic effects,
each with associated length and time scales which
can be very different.

1.3. Atomic processes

The atomic processes responsible for nucleation and growth of thin films on a substrate are
indicated in Fig. 1c. Atoms arrive from the vapor
at a rate R, or at an equivalent gas pressure p,
such that $R = p/(2\pi mkT)^{1/2}$, where m is the
atomic mass, k is Boltzmann's constant and T
the absolute temperature of the vapor source.
This creates single adatoms (or admolecules),
whose areal density $n_1(t)$ increases initially as
$n_1 = Rt$.

At the highest temperatures, these adatoms
will only stay on the surface for a short time, the
adsorption residence time τ_a, during which time
they migrate over the surface with diffusion coefficient D. This time is determined by the adsorption energy, E_a, and is conventionally written as

$$\tau_a^{-1} = \nu_a \exp(-E_a/kT), \tag{1}$$

where ν_a is an atomic vibration frequency, of
order 1–10 THz. A simple expression for the
diffusion constant appropriate to two-dimensional surface diffusion, in terms of the diffusion
energy E_d and frequency ν_d (typically somewhat
less than ν_a), is

$$D = (\nu_d a^2/4) \exp(-E_d/kT), \tag{2}$$

where a is the jump distance, of the order of the
surface repeat distance, say 0.2–0.5 nm.

The number of substrate sites visited by an
adatom in time τ_a is $D\tau_a/N_0$, where N_0 is the
areal density of such sites, of the same order as
a^{-2}. The rms displacement of the adatom from
the arrival site before evaporation is

$$x = (D\tau_a)^{1/2}$$
$$\simeq a(\nu_d/\nu_a)^{1/2} \exp[(E_a - E_d)/2kT]. \tag{3}$$

Since typically E_a is several times E_d, (x/a)
can be large at suitably low temperatures. Then,
in their migration over the surface, the adatoms
will encounter other atoms. Depending on the
size of the binding energy between these atoms,
and on their areal density n_1, they will form small
clusters, which may then grow to form large clusters of atoms on the surface, in the form of two-
or three-dimensional islands. This binding energy
between a pair of atoms, E_b, and the energy of
the critical cluster, E_i, are centrally important to
the understanding of nucleation and growth processes on surfaces.

Subsequently, surface- and inter-diffusion processes may occur, as illustrated on the right-hand
side of Fig. 1c. It may also be that the migrating
adatoms encounter special sites, such as surface
vacancies or the steps indicated at the left-hand
side of the figure. These processes are more
difficult to model in detail, because they are
highly specific to the systems studied, such as the
nature of any chemical reactions at the interface,
or the exact orientation of the substrate surface.
Interchange reactions at the surface are likely
where the deposit is more strongly bound than
the substrate, and/or where the substrate plane
is more open.

1.4. Thermodynamics and kinetics

Despite the use of thermodynamic classification and arguments, crystal growth is a non-equilibrium kinetic process in which one or more
steps are rate-limiting. The thermodynamic limit
is illustrated by the equilibrium vapor pressure,
p_e, of bulk material (A). By equating the chemical
potential, μ, of the low pressure vapor, which is
exact, and the solid in some approximation, we
can find an expression for p_e. Using the Einstein

approximation, in the high temperature limit, for the lattice vibrations of the solid of frequency ν, we obtain the standard result

$$p_e = (2\pi m)^{3/2} \nu^3 (kT)^{-1/2} \exp(-L_0/kT). \quad (4)$$

The vapor pressure is dominated by the sublimation energy L_0, which is much larger than kT at all temperatures below the melting point. However, it is also influenced strongly (ν^3) by the lattice vibrations, so that models which ignore such effects cannot be correct in detail. This equilibrium result is independent of the state of the surface, which acts only as an intermediary between the vapor and the solid.

When the vapor and solid are not in equilibrium, the nature of the surface does play a significant role, as first shown by Burton, Cabrera and Frank (BCF) in 1951 [7]. They showed that the condensation coefficient, α_c, in the Hertz–Knudsen equation

$$dn/dt = \alpha_c(R - R_e), \quad (5)$$

where n is the total areal density of atoms condensed, is a function of the step structure of the surface in the absence of island nucleation. At low supersaturation S, defined by p/p_e or R/R_e, such that $\Delta\mu = kT \ln S$,

$$\alpha_c = (2x/d) \tanh(d/2x). \quad (6)$$

where d is the step separation and $x = (D\tau_a)^{1/2}$ as in Eq. (3). This is equivalent, in the small-x limit, to adatom capture in a zone of width x on either side of a step, with re-evaporation of the adatoms which are further away from the step. When $x \gg d$, all atoms are captured, and the condensation coefficient tends to unity. BCF went on to show that at low $\Delta\mu$, growth did not occur at a measurable rate on a flat surface, but was mediated by (screw) dislocations which produce spiral arrays of steps on a surface [8].

At higher supersaturations, adatoms will come together before they reach the steps, and if they are bound strongly enough, will form nuclei which can develop into islands. The rate limiting step is the formation of "critical nuclei" of size i, which is defined as the size which is more likely to grow than decay [9]. As explained in more detail elsewhere [1,5], we can apply statistical mechanics, in either a "classical" or an "atomistic" version to

calculate the density of critical nuclei, n_i. The atomistic expression, in terms of the single adatom density n_1, and the (free) energy of the critical cluster, E_i, is

$$n_i/N_0 = C_i(n_1/N_0)^i \exp(E_i/kT), \quad (7)$$

where C_i is a statistical weight of order 1–10.

The use of atomistic expressions was prompted by the realisation that, in many deposition experiments the driving force $\Delta\mu$ can be so large that the critical nucleus is only a single atom [5,10]. Thus $i = 1$ represents the extreme kinetic limit on a perfect surface. The case $i = 0$ can arise on a defective surface; this means that adatoms diffuse to defect sites, and that the clusters nucleate from such filled sites. Only for the lowest substrate temperatures is it possible to suppress surface diffusion of adatoms, and in this limit the film would grow simply by accreting atoms which stick where they fall! This does not happen at temperatures involved in the growth of less reactive thin films for practical purposes. But the observation of stationary single adatoms of refractory metals has been performed by field ion microscopy (FIM) for many years [11]. More recently, scanning tunneling microscopy (STM) has been used to observe xenon and other atoms, which, despite their low diffusion energy, do not move on liquid-helium-cooled substrates, unless they are "pushed" by the STM tip [12].

Equivalent "classical" expressions to Eq. (7) have been formulated in terms of surface energies for three-dimensional (3D) or "edge" energies for 2D nuclei [1,5]. These become more realistic as the nuclei become large (many hundreds of atoms), close to equilibrium conditions. In this limit a reliable model has to be consistent with the equilibrium vapor pressure. This is not entirely straightforward, and we need definitions of quantities such as C_i and E_i in (7), and others such as the adsorption residence time τ_a in (1), which do not introduce internal contradictions [13].

2. Nucleation and growth theory

Nucleation on a surface has been discussed in both classical thermodynamic and in atomistic terms, and both have a long history. In the island growth mode, the critical nucleus size is typically only a few atoms at most, and larger clusters have a 3D form. In this case all the rate-limiting processes occur on the original substrate surface, and an atomistic treatment is appropriate. This approach has been developed to study the (2D) nucleation processes occurring in layer growth, and the (mostly 2D) processes which occur in Stranski–Krastanov growth. Here, the outline arguments are given in a uniform notation [13] so that subsequent examples can be understood. Over the period discussed, there have been many simulations of crystal growth, each with their own assumptions and terminology; we make no attempt to do justice to this large body of work.

2.1. Rate equations for cluster densities

We follow the density of clusters containing j atoms, n_j by writing coupled differential equations describing the rate at which n_j changes due to the various processes taking place on the surface. The simplest case is n_1 in the high temperature limit when only re-evaporation is considered. In this case, we can write

$$dn_1/dt = R - n_1/\tau_a, \tag{8}$$

and the solution, for times $t \gg \tau_a$, is simply $n_1 = R\tau_a$. If this is all that happens, nothing nucleates

or grows on the surface, so we must consider additional processes. We just consider single adatoms to move, and all other clusters to be stationary, when we find

$$dn_j/dt = U_{j-1} - U_j, \tag{9}$$

where U_j is the net rate of capture of adatoms by j-sized clusters. In atomistic terms we may write U_j as the difference between a "diffusion capture" and a "decay" term, as

$$U_j = \sigma_j D n_1 n_{j-1} - n_j\{\nu_d \exp[-(\Delta E_j + E_d)/kT]\}. \tag{10}$$

Here, σ_j is a "capture number" and $\Delta E_j = E_j - E_{j-1}$. If we substitute for the diffusion coefficient D from Eq. (2), and put $U_j = 0$ for all $j \leq i$, then by repeated application of Eq. (10) we obtain Eq. (7) for the density of i-sized clusters. This corresponds to the condition of "local equilibrium" between n_i and n_1. Eq. (10) is a reminder that we need to be careful about including "back reactions" in our model, since failure to do so can result in inconsistencies with the thermodynamic limits.

It is inconvenient to consider each individual atomic size for large clusters. We can therefore group clusters together, within a given size range, or even more drastically, lump all clusters $j > i$ together as "stable" clusters of density n_x, result-

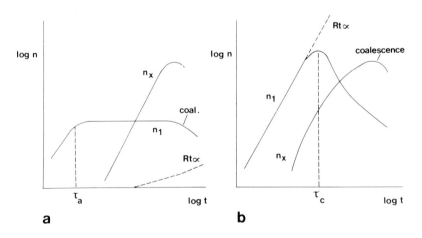

Fig. 2. Evolution of the single-atom density (n_1), the stable cluster density (n_x) and the total number of atoms condensed ($Rt\alpha$) as a function of deposition time for (a) high temperatures and (b) low temperatures. The re-evaporation time τ_a, capture time τ_c and the limitation of n_x by coalescence are indicated [5].

ing in

$$dn_x/dt = U_i = \sigma_i D n_1 n_i. \qquad (11)$$

The initial evolution of the "nucleation density", $n_x(t)$, can therefore be obtained from the coupled Eqs. (8) and (11) in the high temperature limit. A more detailed calculation of the cluster size distribution could be obtained from a more extensive set of coupled equations of the form (9).

To be realistic, we must include other processes. These include, at least, the loss of adatoms to growing clusters in (8), by diffusion capture and by "direct impingement". We also need to limit the cluster density in (11) by clusters coalescing. We then obtain an equation for the maxi-

mum, or saturation, cluster density n_x as a function of the experimental variables R and T, and the energies introduced in the above equations, which can be compared with experimental observations [5,13].

With these processes included the evolution of n_1 and n_x with time is as shown schematically in Fig. 2. At high temperatures (Fig. 2a), n_1 is constant and n_x increases linearly for $t > \tau_a$ (as in Eqs. (8) and (11)) until coalescence sets in. The condensation coefficient, i.e. the proportion of the impinging atomic dose which ends up in the deposit, $\alpha(t)$, is initially very small. At low temperatures (Fig. 2b) when there is no re-evaporation so that $\alpha = 1$, n_1 increases linearly until it is

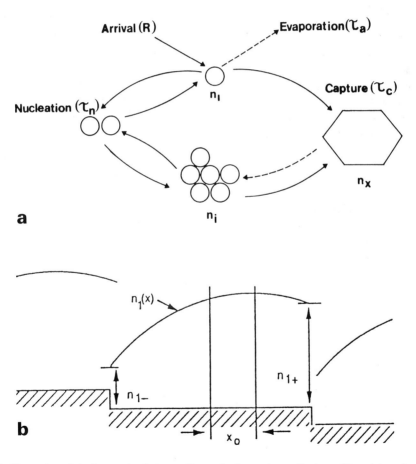

Fig. 3. (a) Schematic illustration of the interaction between the nucleation and growth stages. The adatom density n_1 determines the critical cluster density n_i; however, n_1 is itself determined by the arrival rate R in conjunction with the various loss processes described in the text [13]. (b) Capture of adatoms by steps, and the resulting adatom profile $n_1(x)$ [14].

limited by diffusion capture by the previously nucleated clusters, after a time τ_c. This is an example of "competitive capture" of adatoms, which is described in the next section.

2.2. Adatom capture and regimes of condensation

The processes considered in Fig. 1c can be visualised atomistically as in Fig. 3a. Atoms arrive from the vapor. They may evaporate, but they may alternatively start the nucleation chain of small clusters. Once some stable clusters have been nucleated, this opens up another channel for loss of adatoms, namely diffusive capture by stable clusters. When these clusters cover a fraction, Z, of the substrate, then yet another channel "direct impingement" from the vapor is also possible. These processes can be incorporated into the rate equations by writing

$$dn_1/dt = R(1 - Z) - n_1/\tau_a - n_1/\tau_n - n_1/\tau_c. \tag{12}$$

The loss of adatoms during nucleation itself (τ_n) is numerically negligible, so we may write the "steady state" solution for $n_1(t)$ as

$$n_1 = R\tau(1 - Z), \tag{13}$$

with $\tau^{-1} = \tau_a^{-1} + \tau_c^{-1} + \dots$, where the continuation ... means that we can consider adding other competing mechanisms, which will add like resistances in parallel.

The loss of adatoms to stable clusters is a diffusion problem which can be solved in various approximations [5,10] resulting in

$$\tau_c^{-1} = \sigma_x D n_x, \tag{14}$$

where the capture number σ_x is typically of order 5–10. To complete the circle, we need a further equation to subtract the coalescence rate, U_c, from Eq. (11), and to express it in terms of Z. Using $U_c = 2n_x \, dZ/dt$ with dZ/dt given by all the cluster growth terms in (13), and the shape of the clusters, then we can derive general expressions for the maximum cluster density [5,13].

These equations for the maximum cluster density are of the form

$$n_x \sim (R/\nu)^p \exp(E/kT), \tag{15}$$

where p and E are given in Table 1 for 2D and 3D clusters. In the "extreme incomplete" condensation regime, growth by direct impingement is most important; in "complete" condensation re-evaporation is negligible. In between we have the "initially incomplete" regime, where most cluster growth occurs by diffusive capture, at least initially.

2.3. Competing sinks: step capture and nucleation

The formulation of Eq. (12) can be extended to consider other competing processes. For example, real surfaces have both steps and point defects, which can act as traps for diffusing adatoms. It is instructive to formulate the step capture problem in this language, and to consider the competition between nucleation on the terraces and incorporation at steps. This problem, which is closely related to the original BCF formulation, is exploited in growth of semiconductor films by MBE, where RHEED oscillations occur if we have nucleation on the terraces, and do not occur if we have growth by step flow. The arguments

Table 1
Parameter dependencies of the maximum cluster density in various regimes of condensation (from Ref. [5])

Regime	3D islands	2D islands
Extreme incomplete	$p = 2i/3$ $E = (2/3)[E_i + (i + 1)E_a - E_d]$	$p = i$ $E = [E_i + (i + 1)E_a - E_d]$
Initially incomplete	$p = 2i/5$ $E = (2/5)(E_i + iE_a)$	$p = i/2$ $E = (E_i + iE_a)$
Complete	$p = i/(i + 2.5)$ $E = (E_i + iE_d)/(i + 2.5)$	$p = i/(i + 2)$ $E = (E_i + iE_d)/(i + 2)$

presented here are given in more detail in Ref. [14]. Fuller rate and diffusion equation treatments, which use slightly different non-linear terms and boundary conditions are given in Ref. [15]. In particular, Fuenzalida shows that one can solve numerically for the size and spatial distribution of clusters on the terraces, taking step motion into account consistently, for the case when both up- and down-steps are perfect sinks; Kajikawa et al. explicitly consider anisotropic diffusion.

The diffusion problem involved in the flow of adatoms into a step is illustrated schematically in Fig. 3b, for the important case of complete condensation, when all the arriving atoms end up in the deposit. In this case the diffusion zones, given by Eq. (3), overlap, and the adatom concentration profile during deposition is an inverted parabola. The capture probability is typically not equal at an up-step and a down-step, and this means that the center of the diffusion profile is displaced from the middle of the terrace by a distance x_0, given by

$$x_0 = (D/Rd)(n_{1+} - n_{1-}),\qquad(16)$$

where n_{1+} and n_{1-} are the adatom concentrations in equilibrium with the steps. We expect the up-step to be a better sink than the down-step, so $n_{1+} > n_{1-}$ as drawn.

With the profile as shown, we can readily work out the mean adatom density in steady state conditions, n_1, which is given by

$$n_1 = n_{1e} + Rd^2/(12D),\qquad(17)$$

where n_{1e}, the steady state concentration in equilibrium with the steps, is $(n_{1+} + n_{1-})/2$. If the diffusion problem is formulated for strong sinks ($n_{1e} = 0$), but including the possibility of nucleation on the terraces, we can see that this is the same problem as solved by BCF, but with τ_a (Eq. (1)) replaced by τ_c (Eq. (14)). This means that there is a zone, on either side of the step, of width $(D\tau_c)^{1/2}$, within which nucleation is depressed due to capture by steps. This denuded zone width, l, is just $(\sigma_x n_x)^{-1/2}$.

An equivalent way to view adatom capture by steps is to add an extra term τ_s^{-1} to Eq. (12). Then the last term in Eq. (17) corresponds to $R\tau_s$. Moreover, we can see that step-flow growth will be more important than nucleation on the terraces if $\tau_s < \tau_c$, i.e. if

$$(d^2/12D)(\sigma_x D n_x) < 1,$$

$$\text{or}\quad (d/l) < \sqrt{12} \approx 3.5.\qquad(18)$$

Thus the transition from nucleation, at lower temperatures, to step flow, at higher temperatures, happens when the denuded zone width, which increases with increasing substrate temperature, becomes a certain fraction (about a third according to Eq. (18)) of the step spacing. More elaborate treatments might include the moving boundary condition at the steps, which produces skewed cluster size and position distributions, and will modify numerically, but not qualitatively, the conclusions of Eq. (18) [15].

2.4. Pattern formation: ripening and other effects

In the quantitative parts of the above discussion, we have assumed that clusters with size $j > i$ are "stable", and that the initial nucleation events, at least on large terraces, occurred at random positions. These assumptions are good for relatively short times, but as the deposition proceeds, and even more as the deposited film is left at elevated temperature for long times, a degree of "self-organisation" asserts itself. Small differences in free energy can make themselves felt via subtle effects.

The film as deposited is rarely close to equilibrium. In the island growth mode, the equilibrium state is when all the deposited material is in one large island. The driving force to approach this equilibrium is the reduction in surface energy. Thus, for example, coalesence of two islands typically results in one island which reorganises its shape in an attempt to minimize its surface energy. Depending on the extent of (surface) diffusion, around the island, the substrate will become re-exposed. If this happens during deposition, secondary nucleation may occur in the spaces so created.

A second example related to island growth can be seen in relation to Fig. 3a. After the capture of adatoms by "stable" clusters, the adatom concentration n_1 decreases with time as the rate of capture goes up, even though the typical effect via Eq. (14) implies a fairly slow variation. Thus, in this period, the critical nucleus size, i, increases, again slowly. This in turn means that clusters which were stable no longer are.

The classical description of this effect is "Ostwald Ripening" in which large clusters grow, and small clusters disappear, via exchange of, in this case adatoms. This is the Gibbs–Thomson effect on the (2D) vapor pressure of the island, and depends on the island radius r, in contact with the substrate as

$$n_{1e}(r) = n_{1e}(1 + 2\gamma\Omega/kTr), \qquad (19)$$

where Ω is the atomic volume of the deposit, γ is a suitable surface energy, which depends on cluster shape, and the density n_{1e} is as in Eq. (17). This adatom density is governed by a (2D) evaporation energy, L, which is the difference between the sublimation energy L_0 of the deposit and the adatom adsorption energy E_a. When the transfer of atoms between the islands is diffusion, rather than interface limited, the rate of atom transfer between small and large islands is proportional to $(n_{1e}D)$, and the activation energy for coarsening is simply $(L + E_d)$ [14]. Incorporation of such effects into the description of nucleation processes is discussed in Ref. [16]. Other high temperature effects, including surface roughening and surface melting, are discussed in Ref. [17].

3. Island growth: metals on insulators and metals

Many experiments have been done to test ideas about island growth, and to abstract energy values from a comparison of these experiments with the formulation presented in Section 2. Typically, they have used ex-situ transmission electron microscopy (TEM) techniques to examine the deposit after preparing the deposit in UHV [18]. In the cases of the noble metal on alkali halides, which have been extensively studied [5,10], the island distribution is "fixed" in-situ by evaporation of an amorphous carbon film. After removing the sample from the vacuum system, the substrate is dissolved in water, leaving the metal

islands attached to the thin carbon film, which is then examined by TEM.

3.1. Ag, Au and Pd on alkali halides

A full review of early work on these noble metal systems has been given in Ref. [5], where there is also an extensive tabulation of energy values deduced from experiment. For silver and gold, the adsorption energy, E_a, of the atoms is in the range 0.5–0.9 eV, with the Au values somewhat higher than the Ag values, and errors for particular deposit–substrate combinations < 0.1 eV. These values are much lower than the binding energy of pairs of Ag or Au atoms in free space, which are accurately known, having values 1.65 ± 0.06 and 2.29 ± 0.02 eV respectively [19]. We can easily see from these values why we are dealing with island growth, and why the critical nucleus size is nearly always a single atom. The Ag or Au adatoms re-evaporate readily above room temperature, but if they meet another adatom they form a stable nucleus which grows by adatom capture.

Further experiments have been undertaken more recently to study the details of particular combinations. In addition to more accurate energy values, these experiments showed that several other surface processes can occur in addition to those considered explicitly in Section 2. It is very difficult to tell, simply from looking at the TEM pictures, whether the nuclei form at random on the terraces, or whether they are nucleated at defect sites. The classic way to distinguish true random nucleation, with $i = 1$, is to check that the nucleation rate is proportional to R^2 at high temperatures when $n_1 = R\tau_a$, as in Eqs. (8) and (11). But there are several other possibilities, including the creation of surface defects during deposition, which might mimic this effect [10].

As substrate preparation techniques have improved, lower nucleation densities which saturate earlier in time have been observed. This has been associated with the absence of defects, and the mobility of small clusters. From detailed observations as a function of R, T and t, some energies for the motion of these clusters has been extracted. Qualitatively, it is easy to see that if all

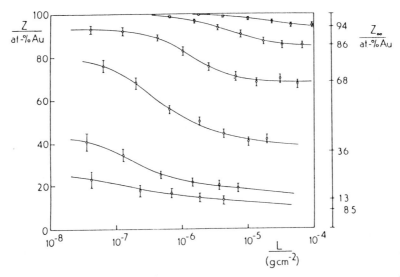

Fig. 4. Fractional Au content in Au–Ag alloys on NaCl(100), as a function of the total condensed mass per cm². The corresponding compositions of the incoming atomic beams are shown on the right-hand axis. $R_{Ag} = 5 \times 10^{13}$ atoms cm⁻² s⁻¹; R_{Au} was varied; $T_{NaCl} = 573$ K. From Ref. [20].

the stable adatom pairs move quickly to join pre-existing larger clusters, then there will be a major suppression of the nucleation rate.

Two further types of experiment are of interest. The first is the study of alloy deposits, which has now been performed for three binary alloy pairs, formed from Ag, Au and Pd on NaCl(100) [20]. In such experiments the atoms with the higher value of E_a, namely Au in Ag–Au, or Pd in Pd–Ag and Pd–Au, form nuclei preferentially, and the composition of the growing film is initially enriched in the element which is most

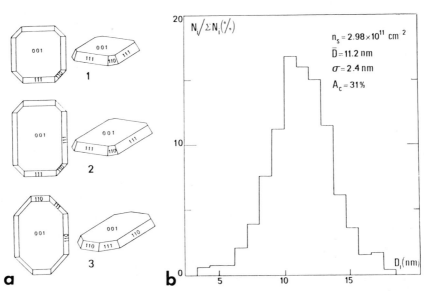

Fig. 5. Epitaxial Pd islands on MgO, after some coalescence, as studied by TEM: (a) shapes and orientation of islands; (b) size distribution and nucleation density. From Ref. [23].

Table 2

Best values (in eV) of adsorption and diffusion energies of Ag, Au and Pd adatoms on NaCl(100) from nucleation experiments (abstracted from Refs. [10,20]; values without error bars are derived from combinations of data; see text for definition of δE, and discussion)

	δE [20]	$(E_a - E_d)$	E_a	E_d
Ag		0.22	0.41	0.19
	0.11 ± 0.03			
Au		0.33 ± 0.02 [10]	0.49 ± 0.03 [10]	0.16 ± 0.02 [10]
	0.12 ± 0.03			
Pd		0.45	0.78	0.33
	0.25 ± 0.05			
Ag				

strongly bound to the substrate. The composition of the films was measured by X-ray fluorescence and energy dispersive X-ray analysis, as shown for the pair Ag–Au deposited at 300°C in Fig. 4. The composition only approaches that of the deposition sources at long times, or under complete condensation conditions.

These experiments can be analysed to yield energy differences δE, where $\delta E = \delta E_x - \delta E_y$, and $\delta E_x, \delta E_y = (E_a - E_d)_{x,y}$ for the two components. Values of δE have been obtained for the pairs, namely Au–Ag: 0.11 ± 0.03; Pd–Au: 0.12 ± 0.03; Pd–Ag: 0.25 ± 0.05 eV [20]. It is clear that these experiments measure, very accurately, differences in the integrated condensation coefficients, $\alpha_{x,y}(t)$, and that these quantities are determined by the diffusion distances (cf. Eq. (3)) of the corresponding adatoms. Coupled with nucleation density measurements, the data give particular values for E_a and E_d for these three elements on NaCl(100), as given in Table 2.

A second interesting type of experiment concerns the interaction between nucleation on the terraces and on cleavage steps. A detailed analysis gave a value 0.23 ± 0.03 eV for the binding energy of Au atoms to steps on NaCl(100) surfaces [21]. Many studies of forces between clusters have been made, both on terraces and at steps, which lead to non-uniform distributions of clusters [22].

3.2. Metals on oxide surfaces

Dispersed islands of specific transition metals on various oxide surfaces are used as catalysts for a wide range of chemical processes. Studies of "model catalysts" are therefore of great impor-

tance, and especially if microscopic observation is combined with gas reaction studies. A good example is the recent TEM study of the morphology of Pd particles grown on MgO, and the molecular beam mass spectrometry study of the interaction of CO with these particles [23].

To make such a study quantitative, the shapes and size distributions were determined as shown in Fig. 5. The Pd particles have (100) top faces, with different amounts of {111} and {110} inclined faces in contact with the substrate. The density, n_x around 3×10^{11} cm^{-2} is typical for deposition at $T = 150$–200°C, and the size distribution is characteristic of complete condensation, plus a small amount of coalesence. Surface diffusion around the islands is sufficient to form a polyhedral shape, but is low enough that coalesced islands remain elongated. The residence times of CO molecules on the Pd particles were determined as a function of temperature, to deduce their adsorption energy. This value, 30.8 kcal/mol or 1.34 eV, was independent of particle size down to diameters around 5 nm, but rose sharply to around 1.6 eV for 2 nm particles. Typically useful catalysts have particles in this smaller size range, and the adsorbing molecules can modify their geometry quite drastically during the life of the catalyst. In particular, the small metal particles can move and coalesce under the influence of the reacting gases, and the catalyst then needs to be regenerated, i.e. the particles needs to be redispersed to be effective again.

3.3. Metallic systems involving interdiffusion

Many metal–metal deposition systems should follow the island growth mode, if the surface energy of the deposit is greater than that of the substrate. Since surface energies are highly correlated with cohesive energies, the islands of the strongly bound material, once formed, could lower their energy by allowing themselves to be coated with a thin skin of substrate material! This corresponds to a curious form of interdiffusion, in which islands or layers, rather than single atoms, bury themselves in the substrate.

At low temperatures this will not happen, because the substrate atoms will not diffuse. However, recent STM studies of surface steps on noble metals have shown that steps can move quite rapidly, even around room temperature. It is tempting to speculate that the difficulties various groups have experienced in producing well-defined thin films of magnetic metals (Fe, Co, Cr, etc.) on noble metal surfaces (Cu, Ag, etc.) is related in some way to effects of this nature. Not only do such magnetic metals have higher surface energies, in general, than the substrates, but they also undergo structural phase changes with increasing thickness. Classical surface science techniques have been extensively applied to these systems, and microscopic studies are beginning to unravel the competing effects [24].

4. Layer growth: metal and semiconductor homoepitaxy

The opposite case to island growth is layer growth. If there is a nucleation barrier, as there certainly is on a perfect terrace [7], then we have nucleation of 2D islands, followed by the completion of successive layers. We might suppose that each layer is completed before the next one begins, but this only happens if nucleation is extremely slow, and growth very rapid. In practice, there will be a distribution of different heights over the surface, which corresponds to a surface which is more or less "rough", typically for kinetic rather than thermodynamic [25] reasons. This roughness depends not only on diffusion over the terraces, but also on the mechanism by which the atoms can surmount energy barriers, and become incorporated at the edges of the ML-thick islands.

4.1. Metal–metal systems

Until the advent of the STM, it was very difficult to observe monolayer thick nuclei, except in special cases by reflection electron microscopy (REM) and TEM, where high atomic number deposits were used. An early example obtained by TEM is shown in Fig. 6 [8], where the Au ML islands have decorated the steps, and have also

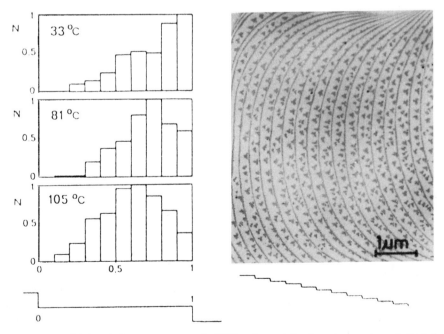

Fig. 6. Distribution of Au nuclei, in relation to steps on a Ag(111) surface, at the three temperatures indicated. The sense of the step train corresponds to the TEM image of the Au islands. Note that the up-steps are decorated with a continuous thin strip of Au, and what is plotted is the island position histogram on the terraces [8].

nucleated on the terraces. The accompanying diagrams show that the positions of the nuclei with respect to the steps have been measured, and that the island density peaks near the down-steps. This effect is temperature dependent in the range 33–105°C; at the higher temperature, the down-step becomes almost as good a sink as the up-step, and the position distribution becomes more symmetric. This Au/Ag(111) system should approximate to Ag/Ag(111) within the first monolayer, but will of course show differences at a later stage.

This step effect has been implicated in Pt–Pt(111) [26], where instead of the expected two regimes as a function of deposition temperature, originally seen by RHEED, there are actually four seen by helium beam diffraction. At the highest temperatures growth proceeds by step flow, with no diffraction oscillations. Next, there is the expected transition to nucleation of monolayer islands on the terraces, in line with Eq. (18), which gives well defined oscillations. At lower temperatures, the oscillations disappear again. The authors associate such behaviour with a transition to a new kind of island growth, or a pronounced kinetic roughening, due to the retention of adatoms on the upper terraces, because of the barriers at the edge of the monolayer islands. At even lower temperatures, the oscillations characteristic of "re-entrant layer" growth reappear.

This series of effects is extremely interesting, but the explanations so far are quite qualitative [26]. However, at the lowest temperatures, the edges of the monolayer islands become irregular, so that specific diffusion paths (i.e. gaps in the reflecting barrier) become available for the adatoms. This, coupled with the small size and close proximity of the islands, results in the reappearance of a type of "kinetically-limited" layer growth at the lowest temperatures. The results serve as a reminder that the diffusion energy, E_d, is often very low, in the 0.1 eV range or less, so that it is impossible to suppress adatom diffusion at normal growth temperatures. The promotion of smoothness in epitaxial films via localized diffusion, "downward funneling" and specific bonding geometries on fcc (100) surfaces is discussed in Ref. [27].

4.2. Semiconductor homo-epitaxy

The classic substrate for semiconductor growth is Si(100), since this has the simplest structure, and is the substrate used for growth of most practical devices. Typically device growers use a surface which is tilted off-axis by about 2–4°, to form a vicinal surface which contains a regular step array. The reason for this is precisely to promote step-flow growth, and to suppress random nucleation on terraces: nucleation is not wanted, because it increases the possibility of incorporating defects such as dislocations threading the films which have bad electrical properties!

Si(100) has a basic (2×1) reconstruction, which arises from dimerization to reduce the density of dangling bonds. This reconstruction reduces the symmetry of the surface, and results in diffusion and growth properties which are very anisotropic, and alternate across single height steps. Thus in general there are two orthogonal surface domains, and only surfaces with double height steps can give single domain surfaces. The steps are also rebonded in various ways, with two different single height steps, denoted S_A and S_B, which have very different step energies. In addition, for larger miscut angles, double-height steps are preferentially formed.

Thus nucleation and growth on this surface is intrinsically quite complicated. In principle, the dimer reconstruction has to be broken and re-formed and redimerized as each layer is grown, so there could be nucleation barriers at many stages of growth. However, at normal growth temperatures (400–650°C), the dimerization is not the rate-limiting step. There have been several studies of Si/Si(100) growth, primarily using STM [28], in addition to spot profile analysis using LEED and RHEED [29].

In one STM study, the nucleation density was observed as a function of deposition rate and substrate temperature, and an analysis similar to that of Section 2 performed, but taking into account the diffusion anisotropy, and the anisotropy in binding at the edges of the monolayer islands [30]. This $N(T)$ data is shown in Fig. 7a. The low temperature region, with a slope of 0.165 eV, is consistent with a critical nucleus size $i = 1$, and a

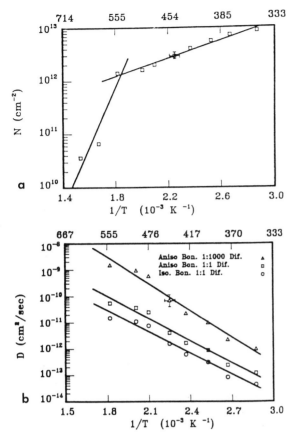

Fig. 7. (a) Nucleation density $N(T)$ abstracted from STM images of submonolayer Si/Si(100) growth. (b) Diffusion coefficients abstracted from the low temperature data in (a) according to different models: triangles, preferred model, with anisotropic bonding and diffusion; squares, anisotropic bonding, isotropic diffusion; circles, both bonding and diffusion isotropic [30].

diffusion energy, in the "easy" direction parallel to the dimer rows, $E_d = 0.67 \pm 0.08$ eV. This diffusion coefficient data is shown in Fig. 7b, where several different models are evaluated. There seems to be a transition in the nucleation density, at $T > 550$ K, to $i = 3$, and the corresponding value E_3 around 1.4 eV. A higher crical nucleus size (perhaps $i = 5$) is also seen on Si(111) at $T > 650$ K [30].

In addition to these observations of nucleation on the (100) terraces, the expected nucleus-free, or denuded, zones next to steps have been observed and quantified. The first terraces to show such effects are the (2×1) teraces, where the fast

diffusion direction is towards the steps. This is expected when nucleation and step capture are in competition, in the case of anisotropic diffusion; however, many details remain to be explored.

4.3. Compound semiconductors

The growth of compound semiconductors such as GaAs by MBE and other techniques, has been reviewed many times [31]. Recently, questions of layer growth versus nucleation on terraces have been addressed [15], as well as alloy segregation and pattern formation at steps [32]. It is clear that the detailed atomic mechanisms involved are complicated, but atomic calculations have developed to the point where the energies associated with the elementary atomic and "molecular" kinetic processes can be studied in detail. There is no space here to elaborate on such experiments and calculations, and the reader is referred to the references quoted as a starting point.

5. Layer plus island growth examples

The Stranski–Krastanov, or layer plus island, growth mode occurs for many deposit–substrate combinations, including metals, semiconductors, gases condensed on layer compounds, and others. Here, three examples taken from metal, metal–semiconductor and semiconductor hetero-epitaxy are given. The competing processes involved are discussed, and some characteristic energy values are deduced.

5.1. Metal heteroepitaxy: Ag/W(110)

This system has been studied by several surface techniques, and the nucleation and growth characteristics have been examined in detail by UHV-scanning electron microscopy (SEM) [33]. In this system, 2 ML of Ag form first, and then flat Ag islands grow in (111) orientation. The results of several nucleation density experiments are shown in Fig. 8, in comparison with a calculation of the type outlined in Section 2.2.

Condensation is complete in this system, except at the highest temperatures studied, and the

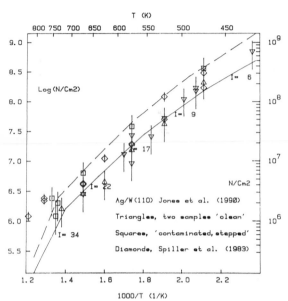

Fig. 8. Nucleation density $N(T)$ abstracted from UHV-SEM images of Ag/W(110), with superimposed calculation (full line) for $E_a = 2.1$, $E_b = 0.25$ and $E_d = 0.135$ eV. The dashed line has $E_d = 0.185$ eV [33].

Table 3
Adsorption (E_a) and diffusion (E_d) energies (in eV) for adatoms, and pair binding (E_b) energies for dimers and trimers on Ag/Ag(111), from experiment and calculated using "Effective Medium" theory (from Ref. [33])

	E_a	E_d	Dimer E_b (2)	Trimer E_b (3)
Theory	2.23	0.12	0.29	0.27
Experiment	2.2 ± 0.1	0.15 ± 0.10	0.25 ± 0.05	

critical nucleus size is in the range 6–34, increasing with substrate temperature. Energy values were deduced $E_a = 2.2 \pm 0.1$, $E_d = 0.15 \pm 0.1$ and

$E_b = 0.25 \pm 0.05$ eV, from a goodness of fit to the data. What, however, makes these values interesting is that they can be compared with the best available calculations of metallic binding. In Table 3, comparison is made with an Effective Medium theory calculation for Ag/Ag(111), to which this system approximates, and the agreement is striking.

In particular, the results demonstrate the nonlinearity of metallic binding with increasing coordination number. In the simplest nearest neighbor "bond" model, the adsorption energy on (111) corresponds to 3 bonds, or half the sublimation energy for a fcc crystal. So for Ag, with $L_0 = 2.95$ eV, such a model would give $E_a = 1.47$ eV, wheras the actual value is much larger. The same effect

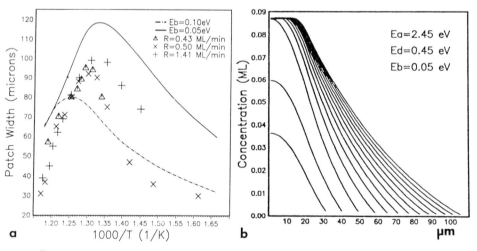

Fig. 9. Width of the $\sqrt{3}$ "patch" on Ag/Si(111) after a 5 ML dose through a 20 μm mask at various deposition temperatures, T_0 (a) Experimental data at $T = 700$ K, showing a peak at around 770 K, with the patch width around 100 μm. Superimposed calculations for $R = 1.41$ ML/min, $E_a = 2.45$, $E_d = 0.45$ eV and E_b as indicated. (b) Calculated single adatom profiles during deposition, against patch width for dose increments $\Delta\theta = 1/3$ ML, starting at $\theta = 2/3$ ML, for $R = 1.41$ ML/min, $T = 700$ K, and the energy values shown [38].

is at work in the high binding energy of Ag_2 molecules, quoted in Section 3.1 in connection with island growth experiments. However, the last bonds to form are much weaker, so that in this case E_b is much less than $L_0/6 = 0.49$ eV. This is a general feature of "Effective Medium" or "Embedded Atom" calculations on metals, which are finding wide application [34].

There are still some curious features of this system, such as how the transition from 2D nucleation to 3D growth occurs, and the precise role of surface defects, including steps on such surfaces. The evidence to date is that the 2D–3D transition occurs after nucleation is essentially complete, and that steps have a big effect on atomic motion within the first 2 ML of silver, but much less on top of this intermediate layer. The more perfect the substrate, the flatter the islands are. This is probably related to the difficulty of islands growing in height, without threading dislocations which may be generated at steps. Some effects in the first monolayers of the related, but seemingly more reactive systems Cu and Au deposited onto W and Mo(110) have been studied by LEEM [35].

5.2. Metal / semiconductor systems: Ag / Si(111)

This system has been studied by many surface techniques, and by microscopy, including STM [36]. At substrate temperatures above 460 K, an intermediate layer with the $(\sqrt{3} \times \sqrt{3})R30°$ reconstruction is formed during deposition, and then Ag islands grow. The coverage of the layer (called the $\sqrt{3}$ structure for convenience) has proved to be extremely controversial; the nucleation and growth of the islands has been extensively investigated by UHV-SEM. The reason for the interest is that the interface between the metal and the semiconductor is abrupt. Unlike other noble and near-noble metals (Cu, Au, Ni, Pd and Pt) on silicon, there is no tendency for Ag and Si to intermix or form silicides.

In addition to nucleation and growth studies of the type described in Section 5.1 [37], UHV-SEM and AES techniques have been used to study the extent of surface diffusion, and the coverage of the intermediate layer [14,38]. Diffusion can be studied by depositing the silver through a mask of holes, and observing the width of the patch as a function of deposition and annealing treatments. This data set is shown in Fig. 9a.

The data indicate that, using a 20 μm wide mask, there is a deposition temperature region around 750 K where the observed width of a "patch" with the $\sqrt{3}$ structure is around 100 μm after a silver dose of 5 ML. Diffusion distances of Ag on Si(111)7 × 7 are very short, but diffusion distances on top of the $\sqrt{3}$ layer can be very long, greater than 50 μm in this case. This is an almost macroscopic distance for an atomic process! Quantitative analysis of these results shows that at temperatures higher than this peak, the Ag adatoms diffuse rapidly, and also desorb, so that the measured diffusion length is just $(D\tau_a)^{1/2}$ in Eq. (3), and the temperature dependence gives $(E_a - E_d)$ around 2.0 eV. However, at lower temperatures, the diffusion distance is limited by nucleation of islands and subsequent capture of the adatoms, as in Eq. (14).

In this system the critical nucleus size is of order 100 atoms in the temperature range considered, and this is associated with very low values of $E_b < 0.1$ eV. A calculation of the adatom concentration across such a patch as a function of dose is shown in Fig. 9b, for the values $E_a = 2.45$, $E_d = 0.45$ and $E_b = 0.05$ eV [38]. It can be seen that, for these values, the adatom concentration approaches 0.1 ML; the values are just sufficient to observe nucleation, in the center of the patch only, at $T = 700$ K. Shifts in the parameters by ± 0.05 eV will change this situation markedly; note that the upward curvature of the latter curves indicate the re-evaporation is significant at this temperature.

The binding energy of the $\sqrt{3}$ layer is also known from isothermal desorption experiments to be around 3.05 eV, so that the $\sqrt{3}$ layer is more strongly bound than a silver island by about 0.1 eV, and than an adatom by around 0.6 eV, if $E_a = 2.45$ eV. So this is a system where there are many aspects to competitive capture of adatoms: they can diffuse, evaporate, nucleate crystals, and also exchange with the intermediate $\sqrt{3}$ layer. A quantitative analysis of this latter effect, in conjunction with AES layer coverage data taken during deposition and annealing, shows that the various effects observed can be understood if there are substantial activation barriers on the way to forming a perfect ($\theta = 1$ ML) $\sqrt{3}$ layer. The STM micrographs [36] show that at RT, Ag islands form without disrupting the 7×7 structure; at $T > 460$ K, the $\sqrt{3}$ layer forms, but it has a

coverage < 1 ML, and many determinations have given θ close to $2/3$. At higher deposition temperatures $T > 650$ K, θ may approach 1 ML, but at higher temperatures still, above 800 K, this layer starts to desorb. There are questions about the role of lower coverage phases (3×1, and

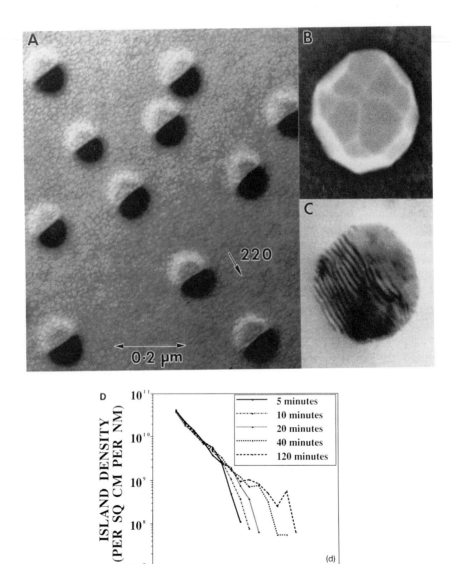

Fig. 10. Island formation in vicinal Ge/Si(100). (a) Ex-situ bright field TEM image, showing coherent islands. The strong black-white contrast parallel to the operating refection $g = 220$ indicates a radial dilatational strain field. (b) UHV-SEM and (c) UHV-STEM images of a single dislocated island, diameter $\simeq 100$ nm, showing (b) facets and (c) moiré fringes indicative of misfit dislocations. (d) Size distributions of islands on annealing a 5 ML room temperature deposit to 375°C for the times indicated [41].

maybe 5×2) and their microstructure. It is a major challenge to encompass the wealth of detail shown by recent STM and LEEM pictures [36,39] within testable kinetic models [38].

Since we are dealing with kinetic processes, the history of the sample is important, and the balance between competing effects can be shifted. For example, if the sample surface is defective, and nucleation is promoted, then we may expect that the $\sqrt{3}$ layer will not be completed when nucleation starts. The nucleation lowers the adatom concentration, which means that the driving force to complete the layer will be much reduced. An assymetry of this form between the adatom concentration during nucleation and annealing is most marked. With the values given here, the adatom density during annealing (Eq. (19)) is around 100 times less than that present during deposition (Fig. 9b) at $T = 700$ K.

5.3. Semiconductor hetero-epitaxy: Ge / Si(100)

In the above Ag/Si(111) example, the intermediate layer was a maximum of 1 ML thick, has a quite different structure than either the substrate or the deposit, and should perhaps best be regarded as a monolayer compound. At the other extreme, Ge on Si has the same structure, differing only in the lattice parameter by 4%, with the Ge slightly less strongly bound. This can be seen in the measured sublimation energies, $L_0 = 3.85 \pm 0.02$ and 4.63 ± 0.04 eV/atom respectively, and in the binding energies of the diatomic molecules, 2.63 ± 0.1 eV for Ge_2 and 3.21 ± 0.1 eV for Si_2 [19]. The surface energy of Ge is lower than that of Si, and deposition of Ge initially occurs as layers.

However, growth beyond the first few ML will build up substantial strain due to the 4% mismatch, and after a certain thickness, the Ge would prefer to grow as islands in which the strain has been relieved by misfit dislocations [6]. The route to this state is quite complicated, but can be understood qualitatively by reference to Fig. 1a. The equilibrium Ge layer thickness has been measured, after annealing, to be 3 ML [40–42]. But it is possible to grow the layer much thicker than this, and the first islands to form are not

dislocated, but are coherent with the underlying layers. This can be seen in TEM pictures, taken ex-situ after UHV preparation, as shown in Fig. 10a. The strong black-white contrast is due to the bending of the substrate (Si) lattice caused by the Ge island, and indicates a radial strain, which also has a component normal to the substrate [40]. At deposition temperatures above 500°C, where surface diffusion is rapid, the size to which these coherent islands grow is markedly dependent on the presence of other sinks within the diffusion distance [14,41].

Dislocated islands can be nucleated, prefentially from the larger coherent islands, or at impurity particles; once nucleated these islands form the strongest sinks, they grow rapidly and the supersaturation in the (> 3 ML) Ge layer reduces. At a temperature of 500°C, diffusion distances are of order 5 μm, whereas below 400°C this figure drops below 0.5 μm. An example of a facetted, dislocated island is shown in the UHV-STEM images in Figs. 10b and 10c. Similar effects are seen when Ge films, grown at RT to thicknesses above 3 ML are annealed at comparable temperatures, although the detailed mechanisms and diffusion coefficients will be different. Fig. 10d shows a series of size distributions taken at different annealing times at $T = 375$°C [41]. Initially, there are no large (> 10 nm radius) islands, but as annealing proceeds the bigger islands grow rapidly, while the size distribution of the smaller islands (< 10 nm radius) stays constant. This evidence suggests that the material for the rapid growth of the dislocated islands occurs primarily from the supersaturated layer rather than from the coherent islands; in particular, the strain fields are effective in keeping out migrating adatoms.

These effects in Ge/Si(100) have also been studied by STM [42], and compared with the layer growth system Si/Si(100) [28,30]. These two systems start out by "looking" the same, but as growth proceeds, they have to diverge. The basic driving force is the need to expand the surface to accommodate the larger lattice parameter of bulk Ge; the initial result may be a fine scale "buckled" surface, and/or missing dimers, and/or an intermixed Ge/Si surface layer. Various "kinetic pathways" to the evolution of the final film have been identified, such as "hut clusters", but the exact correspondence between the structures observed by the various microscopic techniques [40–42] is not yet entirely clear.

6. Discussion and conclusions

My aim in this article has been to discuss the "language" in which the atomic processes in crystal growth, and in particular the nucleation and growth of vapor-deposited films, can be understood. I have concentrated on illustrative surface physics and microscope based experiments, and their interpretation with the help of rate equation modelling. From my particular viewpoint, the subject is in good shape if we are in a position to use such models to explain the key experiments, and to abstract atomic parameters. The output is then, essentially, a series of energies, for example adsorption (E_a), diffusion (E_d) and pair-binding (E_b) energies in the first instance. To complete the circle, these values can be compared with detailed atomic/cluster quantum mechanical calculations, as has been done in a few cases.

Although the description of such theoretical treatments and experiments has been brief, I hope the reader can see that substantial progress has been made, for some selected deposition systems, over the last decade. Most notably, the microscopic techniques have advanced over this period, especially with the application of UHV microscopy including STM to the "simpler" growth systems. In terms of technique, we now have several ways of examining all three growth modes on a microscopic scale. We also have MBE and related fabrication techniques as a major "technology push". As we look into these growth modes and individual surface systems in more detail, we find that many are not at all simple in the expected sense, but contain many subtleties due to surface reconstructions, diffusion anisotropy, steps, etc. Several of these effects can be described by focusing on the fate of adatoms in terms of "competitive capture"; this is a useful unifying idea, which has been quantitatively developed.

Acknowledgements

I am grateful to many colleagues and co-workers whose work is cited for permission to use figures, and for comments on the manuscript. The published experimental work cited from my group in Arizona has been supported by NSF, and in Sussex by SERC.

References

[1] J.A. Venables, in: Microstructural Evolution of Thin Films, Eds. H.A. Atwater and C.V. Thompson (Academic Press, New York, 1993) ch. 1.
[2] For example: Crystal Research and Technology (Kristall und Technik), Akademie Verlag, since 1966; Journal of Crystal Growth, North-Holland, since 1967; Progress in Crystal Growth and Characterisation of Materials, Pergamon, since 1977.
[3] K. Lehovec, Surf. Sci. 1 (1964) 165.
[4] S.A. Barnett and J.E. Greene, Surf. Sci. 151 (1985) 67.
[5] J.A. Venables, G.D.T. Spiller and M. Hanbucken, Rep. Prog. Phys. 47 (1984) 399;
For previous reviews, see R. Kern, G. LeLay and J.J. Metois, Current Topics in Materials Science, Vol. 3, Ed. E. Kaldis (North-Holland, Amsterdam, 1979) 139;
E. Bauer and H. Poppa, Thin Solid Films, 12 (1972) 167;
E. Bauer, Z. Kristalogr. 110 (1958) 3720.
[6] J.W. Matthews, Ed., Epitaxial Growth, Parts A and B (Academic Press, New York, 1975).
[7] W.K. Burton, N. Cabrera and F.C. Frank, Philos. Trans. Roy. Soc. A 243 (1951) 293.
[8] H. Bethge, in: Kinetics of Ordering and Growth at Surfaces, (Ed. M.G. Lagally, NATO ASI Series B 239 (Plenum, New York, 1990) p. 125;
M. Klaua, in: "Electron Microscopy in Solid State Physics" Eds. H. Bethge and J. Heydenreich (Elsevier, Amsterdam, 1987) p. 454.
[9] R.D. Gretz, Surf. Sci. 6 (1967) 468;
C. van Leeuwen and P. Bennema, Surf. Sci. 51 (1975) 109.
[10] J.L. Robins and T.N. Rhodin, Surf. Sci. 2 (1964) 346;
B. Lewis, Surf. Sci. 21 (1970) 273, 289;
J.L. Robins, Appl. Surf. Sci. 33/34 (1988) 379;
R. Conrad and M. Harsdorff, Int. J. Electron. 69 (1990) 153.
[11] G. Ehrlich, Surf. Sci. 246 (1990) 1; Scanning Microsc. 4 (1990) 829;
T.T. Tsong, Surf. Sci. 299/300 (1994) 153;
G. Ehrlich, Surf. Sci. 299/300 (1994) 628.
[12] D.M. Eigler and E.K. Schweizer, Nature 344 (1990) 524;
H. Rohrer, Surf. Sci. 299/300 (1994) 956;
R. Feenstra, Surf. Sci. 299/300 (1994) 965;
C. Quate, Surf. Sci. 299/300 (1994) 980.

[13] J.A. Venables, J. Vac. Sci. Technol. B 4 (1986) 870; Phys. Rev. B 36 (1987) 4153.

[14] J.A. Venables, J.S. Drucker, M. Krishnamurthy, G. Raynerd and T. Doust, Mater. Res. Soc. Symp. 198 (1990) 93.

[15] A.K. Myers-Beaghton and D.D. Vvedensky, Surf. Sci. Lett. 240 (1990) L599; Phys. Rev. B 42 (1990) 5544;
V. Fuenzalida, Phys. Rev. B 44 (1991) 10835;
Y. Kajikawa, M. Hata, T. Isu and Y. Katayama, Surf. Sci. 265 (1992) 241;
C.N. Luse, A. Zangwill, D.D. Vvedensky and M.R. Wilby, Surf. Sci. Lett. 274 (1992) L535, L529;
S. Harris, Phys. Rev. B 47 (1993) 10738.

[16] M. Zinke-Allmang, L.C. Feldman and M.H. Grabow, Surf. Sci. Rep. 16 (1992) 377.

[17] M. Bienfait, Surf. Sci. 272 (1992) 1, and references therein.

[18] J.A. Venables, D.J. Smith and J.M. Cowley, Surf. Sci. 181 (1987) 235;
E. Bauer, Surf. Sci. 299/300 (1994) 102.

[19] K.A. Gringerich, I. Shim, S.K. Gupta and J.E. Kingcade, Surf. Sci. 156 (1985) 495.

[20] A. Schmidt, V. Schunemann and R. Anton, Phys. Rev. B 41 (1990) 11875;
R. Anton, A. Schmidt and V. Schunemann, Vacuum 41 (1990) 1099

[21] A.D. Gates and J.L. Robins, Surf. Sci. 116 (1982) 188; 191 (1987) 492, 499.

[22] J.C. Zanghi, J.J. Metois and R. Kern, Surf. Sci. 52 (1975) 556;
J.C. Zanghi, Surf. Sci. 60 (1976) 425;
A.D. Gates and J.L. Robins, Surf. Sci. 194 (1988) 13.

[23] C.R. Henry, C. Chapon, C. Duriez and S. Gorgio, Surf. Sci. 253 (1991) 177, 190;
C.R. Henry, C. Chapon, C. Goyhenex and R. Monot, Surf. Sci. 272 (1992) 283.

[24] C.H. Chen and F.J. Sansalone, Surf. Sci. 163 (1985) L688;
M. Poensgen, J.F. Wolf, J. Frohn, B. Vicenzi, M. Geisen and H. Ibach, Surf. Sci. 249 (1991) 233; 274 (1992) 430;
B. Voigtlander, G. Meyer and N.M. Amer, Surf. Sci. Lett. 255 (1991) L529;
M.T. Kief and W.F. Egelhoff, Jr., Phys. Rev. B 47 (1993) 10785.

[25] R. Kariotis and M.G. Lagally, Surf. Sci. 216 (1989) 557;
R. Kariotis, J. Phys. A 22 (1989) 2781;
R. Kariotis and G. Rowlands in Ref. [8], p. 313.

[26] R. Kunkel, B. Poelsema, L.K. Verheij and G. Comsa, Phys. Rev. Lett. 65 (1990) 733;
G. Comsa, Surf. Sci. 299/300 (1994) 77.

[27] J.W. Evans, Phys. Rev. B 43 (1990) 3897;
D.E. Sanders and A.E. DePristo, Surf. Sci. 254 (1991) 341;
D.M. Halstead and A.E. DePristo, Surf. Sci. 286 (1993) 275.

[28] M.G. Lagally, Y.W. Mo, R. Kariotis, B.S. Swartzentruber and M.B. Webb, in Ref. [8], p. 145;
Y.-W. Mo and M.G. Lagally, Surf. Sci. 248 (1991) 313.

[29] R. Altsinger, H. Busch, M. Horn and M. Henzler, Surf. Sci. 200 (1988) 235.

[30] Y.W. Mo, J. Kleiner, M.B. Webb and M.G. Lagally, Phys. Rev. Lett. 66 (1991) 1998; Surf. Sci. 268 (1992) 275;
for Si(111), see: B. Voigtlander and A. Zinner, Surf. Sci. Lett. 292 (1993) L775.

[31] See e.g., B.A. Joyce, D.D. Vvedensky and C.T. Foxon, in: Handbook of Semiconductors (1993), in press;
J.R. Arthur, Surf. Sci. 299/300 (1994) 818.

[32] Y.T. Lu and H. Metiu, Surf. Sci. 254 (1991) 290.

[33] G.W. Jones, J.M. Marcano, J.K. Nørskov and J.A. Venables, Phys. Rev. lett. 65 (1990) 3317;
G.D.T Spiller, P. Akhter and J.A. Venables, Surf. Sci. 131 (1983) 517.

[34] K.W. Jacobsen, J.K. Nørskov and M.J. Puska, Phys. Rev. B 35 (1987) 7423;
D.E. Sanders and A.E. DePristo, Surf. Sci. 260, (1992) 116;
J.K. Nørskov, K.W. Jacobsen, P. Stolze and L.B. Hansen, Surf. Sci. 283 (1993) 277;
C. Massobrio and P. Blandin, Phys. Rev. B 47 (1993) 13687.

[35] M. Mundschau, E. Bauer, W. Telieps and W. Śchwięch, Surf. Sci. 213 (1989) 381.

[36] H. Neddermeyer, Crit. Rev. Solid State Mater. Sci. 16 (1990) 309;
K.J. Wan, X.F. Lin and J. Nogami, Phys. Rev. B 47 (1993) 13700.

[37] J.A. Venables, J. Derrien and A.P. Janssen, Surf. Sci. 95 (1980) 411;
M. Hanbucken, M. Futamoto and J.A. Venables, Surf. Sci. 147 (1984) 433;
see also, S. Ino, T. Yamanaka and S. Ito, Surf. Sci. 283 (1993) 319.

[38] G. Raynerd, T.N. Doust and J.A. Venables, Surf. Sci. 261 (1992) 251;
G. Raynerd, M. Hardiman and J.A. Venables, Phys. Rev. B 44 (1991) 13803.

[39] A. Denier van der Gon and R.M. Tromp, Phys. Rev. Lett. 69 (1992) 3519.

[40] D.J. Eaglesham and M. Cerullo, Phys. Rev. Lett. 64 (1990) 1943.

[41] M. Krishnamurthy, J.S. Drucker and J.A. Venables, J. Appl. Phys. 69 (1991) 6461; Mater. Res. Soc. Symp. 202 (1991) 77; 263 (1992) 3.

[42] Y.W. Mo, D.E. Savage, B.S. Swartzentruber and M.G. Lagally, Phys. Rev. Lett. 65 (1990) 1020;
J. Knall and J.B. Pethica, Surf. Sci. 265 (1992) 156;
F. Iwawaki, M. Tomitori and O. Nishikawa, Surf. Sci. 266 (1992) 285.

Surface Science 299/300 (1994) 818–823
North-Holland

Molecular beam epitaxy of compound semiconductors

John R. Arthur

Department of Electrical and Computer Engineering, Oregon State University, Corvallis, OR 97331, USA

Received 29 June 1993; accepted for publication 24 July 1993

Molecular beam epitaxy (MBE) is a textbook application of surface science which makes direct use of many of the tools for characterizing clean surfaces. The development of MBE coincided with the development of these tools and thus the history of MBE is an important part of the history of modern surface science. A subjective recounting of the early history of MBE is presented, along with some biased conjectures about the future.

Molecular beam epitaxy (MBE) is a textbook application of surface science which makes direct use of many of the tools for characterizing clean surfaces. The development of MBE coincided with the development of these tools and, thus, the history of MBE is an important part of the history of modern surface science. In this treatise, I will describe the unique surface science community which existed at the Bell Telephone Laboratories in the 1960's and how this environment was essential to the expansion of MBE from a laboratory experiment to an industrial process. I will also argue that while MBE has opened the door to new fields of low-dimensional physics and electronics, the process is fundamentally a powerful means of examining vapor–surface reactions under highly controlled conditions.

As we approach the thirtieth birthday of *Surface Science*, it is astonishing to me to realize that it was just about 32 years ago that I began my professional career at what was then the Mecca of surface physics, Bell Labs, with a newly printed PhD degree in physical chemistry. My graduate studies on catalytic reactions on clean surfaces had been carried out at Iowa State University under the direction of Robert Hansen. There I learned the essentials of clean surface studies by mastering the art of glass blowing and the fabrication of leak proof uranium glass-to-tungsten

seals. As I walked with great awe to my new office at Bell (which had previously been occupied by J.A. Becker, one of the pioneers of field emission microscopy – Joe had just died, leaving me with an amazing collection of memorabilia and an ancient glass vacuum system), I was accosted by an elderly gentleman who rushed out of a darkened room to grab my sleeve and pull me into the darkness to show me some dimly glowing spots on a phosphor screen. While I watched, the existing spots faded and new ones appeared which, I was informed, represented nickel atoms reconstructing the surface of a nickel crystal due to the reaction with oxygen [1]. Such was my introduction to low energy electron diffraction, modern surface science, and just as importantly, to Lester Germer. I was later to spend many exciting days with Lester climbing some of the classic mountain routes in the northeast, and listening to his tales of the exploits of such other climbing legends as Bill Shockley. Lester was well known for his tendency to talk incessantly about physics while climbing and about climbing while in the laboratory.

The group I joined at Bell was headed by Homer Hagstrum and included Germer, Walter Brattain, Al MacRae, Fred Allen and Garth Gobeli, and had been formed to use various physical probes (photons, electrons and ions) to study bulk

SSDI 0039-6028(93)E0441-V

band structure. They had just come to realize that the phenomena they were observing were in fact characteristic of surface electronic structure and, furthermore, were reproducible and interesting even though only indirectly related to bulk effects predictable by the theories then being generated by Evan Kane and Jim Phillips. Hagstrum recognized that one of the obstacles to studies of surfaces was the limited amount of information provided by a single experiment. Surfaces were non-reproducible and difficult to clean, and it was necessary to use several tools to process and characterize the crystal. Thus, over a period of several years, due to the limited budget and the need to fabricate in house much of the hardware, Hagstrum and Gordon Becker built a system for his ion neutralization experiments that included not only a sputtering gun for surface preparation but also a LEED system for surface characterization [2,3]. Each of the various experiments was accessible by rotating the sample from port to port. The other innovative feature of this system was that it was built of stainless steel rather than glass. Until then, all the UHV systems at Bell had been mercury-pumped pyrex systems, baked at 400°C for several days. Allen and Gobeli had raised some concern over this procedure when they observed that transfer of boron from the pyrex to their Si surfaces during the bake produced a p-type skin, but this seemed to be an unavoidable consequence of the use of the otherwise ideal container material. Of course there were problems with overbaked glass systems that broke or melted occasionally [4].

My first experiments at Bell involved a continuation of my thesis work on field emission microscopy but with the use of semiconductor emitters rather than metals. This switch was made because at that time there was little interest in the chemistry of metal surfaces at Bell, because it appeared as though all problems with long-lived vacuum tubes had been solved. Semiconductors, particularly the III–V materials, were in vogue. It turned out to be a very painful and lengthy process to produce semiconductor field emission and field ion tips that survived the fabrication process, particularly when the tips were fabricated from GaAs and GaP, but I was fortunate to have

the services of a very patient technician, John LePore, who spent weeks in carefully grinding the blanks. To study some simple surface processes I began to use unfocused atomic beams to coat one side of the emitter tip with atoms to measure the subsequent surface migration at elevated temperatures after the manner of Gomer and coworkers [5]. I became aware of a new vacuum tool, the quadrupole mass spectrometer (QMS), that had just become commercially available, and experiments utilizing the QMS seemed much easier than what LePore and I were attempting. Because of the cost ($14 000), my management decided we could only afford to lease the equipment rather than buy it outright. Nevertheless, we obtained a QMS and incorporated it into a system for measuring the transient behavior of atomic beams reflected off surfaces of the fashionable compound semiconductors, GaAs and GaP. The beams were chopped with a mechanical shutter interposed between the source and the sample surface in order to define precisely the period during which atoms were incident on the surface. A particularly novel feature of this experiment was that the QMS was mounted so that if could be rotated around the sample and could thus sample either the incident beam of atoms or the beam reflected from the surface. For the purpose of cleaning the sample surface, an ion gun was provided as well. (At the time, an open-ended Bayard–Alpert gauge with the axis directed toward the sample was considered quite adequate for this purpose.) The standard semiconductor cleaning recipe had been developed by Farnsworth et al. [6] and consisted of several hours of sputtering with Ar interspersed with prolonged annealing and outgassing, the whole procedure lasting for several days. As a result of the fact that the QMS in my experiment could be moved between the sample and the sputtering gun, I learned to my amazement that during the standard static sputtering process, the Ar ambient quickly became contaminated with methane and CO so that in just a few minutes the ion beam consisted predominantly of C containing ions. The repeated sputter and anneal used by Farnsworth et al. essentially acted to purge the system gradually of outgassing contaminants.

However, when the Ar was supplied dynamically, with continual pumping, the ion beam was entirely Ar^+ and the surface could be cleaned in a minute or less; however, the real understanding of surface cleanliness came later.

The definition of a "clean" surface changed dramatically during the 1960's. At first, it meant "clean by the standards of your experiment", i.e., it was strictly an operational definition. My early beam experiments defined the surface cleanliness by the observations of desorbing species in the QMS. When no more oxide desorption was observed and when the Group V elements As or P began to evaporate, it was assumed that the surface was clean. However, in 1970 it became possible to define cleanliness more objectively with the availability of Auger electron spectroscopy (AES) [7]. Those of us using LEED equipment could with the addition of some simple electronics obtain reasonably good AES spectra to establish the real cleanliness of our surfaces as well as the crystallinity. A later, improved version of the beam transient experiment incorporated the QMS and new AES/LEED equipment in a metal vacuum chamber that provided higher pumping speeds and faster sample insertion and vacuum processing.

These experiments were probably the most directly successful and rewarding that I have ever carried out. The equipment worked right from the start and a large amount of data was generated on adsorption and desorption from GaAs substrates. The key observation was that unlike the Group III metals that adsorbed with unity sticking coefficients, the Group V elements showed complete reflection at elevated substrate temperatures; however, if submonolayer quantities of Group III metals were present first, the Group V elements showed initially high sticking. Obviously a rapid surface reaction was occurring between the metallic and non-metallic adsorbates. Since there was considerable interest at that time in epitaxial growth methods, it was obvious to attempt to grow GaAs from beams of Ga and As, relying on the reflection of excess As to insure that the growth was stoichiometric [8,9]. Initially the measure of success was the achievement of single crystal growth as determined by post-growth X-ray and electron diffraction. These results showed that frequently the films were single crystal but that sometimes they were twinned or polycrystalline. It was clear that in situ techniques were desirable in order to determine quickly the optimum values of the significant growth parameters.

Great progress was made when it became possible to examine the surface contamination problem quantitatively using AES. It was evident immediately that surface polishing and etching were critical to the achievement of good growth and that what worked for one growth process was not necessarily optimum for another. For example, the surface preparation for liquid phase epitaxy (LPE) was not desirable for MBE because the final HCl rinse left surface carbon. At about that time Richard Henderson had just completed some very farsighted work on the epitaxy of Si grown from silane beams in which he used an electron diffraction or RHEED (for reflection high energy electron diffraction) system in the growth chamber [10]. After some extensive negotiations, Al Cho and I were able to obtain Henderson's system when he left Bell to go to Hughes Research. Al then proceded to use the system to demonstrate the value of in situ electron diffraction [11]. This now obvious development in MBE technology was probably the single most important step in making it a viable growth process because with the use of RHEED the crystal grower closed the feedback loop between the effect of growth parameters (substrate temperature, beam flux and composition) and the crystalline nature of the film. RHEED was more useful than LEED because the geometry did not interfere with the incident atomic beams; thus, RHEED could be used continuously while LEED could only be used by stopping the growth and cooling the sample. However, an unexpected feature came from the fact that the forward scattering of electrons in RHEED is far more efficient than the back-scattering in LEED. Thus, the RHEED diffraction pattern is much brighter and is not obscured by the light from the source ovens. It also became apparent that because of the low angle of incidence used in RHEED, the diffraction patterns are very sensitive to the surface morphology and roughening of the surface is immediately evident.

The point I am attempting to make is that the climate at Bell Labs in the 1960's was ideal for realizing this application of surface science. There were many colleagues involved in surface studies with a large collective experience in what worked and what did not and how to tell the difference. The atmosphere was competitive but not destructive and there was much sharing of ideas. The lunch hour discussions alternated between bitter debates over the Vietnam War and intense expositions of new results. While the original Surface Physics group gradually dispersed to other activities, there were continual enthusiastic replacements including Dave Aspnes and Mark Cardillo who brought new insights to experimental approaches. As the quality of the epitaxial material improved, a demand for semiconductor films for other experiments developed. Most important, the necessary tools for surface characterization had appeared at the right time.

It is also important to recognize that there was considerable prior science on which to draw. Several authors had demonstrated that homoepitaxy of Si on Si could be accomplished in a UHV environment [12,13]. There had also been attempts to grow GaAs layers which had been only partially successful due to the problems with substrate cleanliness [14–16].

Our original simplistic view of the value of MBE for the growth of semiconductor devices was that since impurities could be added during film growth, doping profiles could be achieved which were not possible by standard diffusion or ion implantation techniques. It became clear fairly soon that the low temperatures used in MBE reduced the interdiffusion of impurities and, thus, made possible very abrupt impurity profiles [17]. Cho showed that even more significant was the ease with which high quality hetero-interfaces between materials with dissimilar bandgaps could be grown [18]. With the work of Esaki and coworkers [19] and of Dingle et al. [20] it became evident that the low growth temperatures and clean surface conditions used in MBE made possible entirely new structures involving a sequence of heterointerfaces, i.e. superlattices, in which composition and doping could be varied at will. The relative ease with which it was possible to do "bandgap engineering" using MBE quickly produced an explosion of interest in heterostructure devices and low-dimensional physics at laboratories throughout the world.

At the same time that the heterostructures were being studied, a series of elegant molecular beam experiments was carried out by Foxon and Joyce [21,22] to unravel the complex kinetics of the reaction of As molecules with GaAs surfaces. Studies of the adsorption and surface segregation of doping impurities have been a continuing area of interest since it was demonstrated that strain effects could cause large impurities such as Sn to redistribute at the growth interface [23–27].

However, in my opinion, the emphasis in MBE during the last decade has shifted away from surface studies to device fabrication. The ease with which complex heterostructure devices can be produced and the general trend in research toward applications has resulted in less attention directed toward understanding all of the details of surface processes. There have certainly been some notable exceptions to this statement. The availability of UHV scanning tunneling microscopes has made it possible to examine the atomic structure of growth interfaces [28–30] in remarkable detail. However, these studies are in a relatively early stage compared to similar work on Si [31]. Tsao has given a good review of the present understanding of the MBE process [32].

The intense interest in decreasing the size and/or number of device dimensions has led to improvements in a number of growth techniques all of which now provide good dimensional control. It is no longer possible to claim that MBE is the only method available for fabricating quantum structures. For the large scale manufacture of heterostructure devices, it is clear that techniques such as MOCVD are more appropriate than MBE since the throughput of wafers is much higher.

Nevertheless, there remain some significant materials challenges for which MBE is uniquely suited. In particular the fundamental surface reaction processes of adsorption, desorption, and reaction have not been studied for many of the molecular species used in the various vapor phase growth processes. MBE, with the combination of an UHV environment, a large selection of surface analytic tools, and the ability to produce and maintain a highly perfect and reproducible substrate surface, is the obvious method for examining these kinetic processes. The ability to determine the spatial distribution of very small quantities of electrically active impurity is also important.

As an example of this type of approach, we are studying the reaction of CBr_4 beams with GaAs and AlGaAs surfaces. Chin et al. [33] have shown that CBr_4 provides a useful source of carbon acceptors to a growing GaAs film. It has been found in a number of laboratories that the usual MBE acceptor, Be, has a sufficiently high diffusion coefficient in GaAs and AlGaAs that the lifetime of high power junction devices is limited by diffusion of Be. Thus, there has been much interest in carbon as an alternative acceptor. A suitable source for carbon has been somewhat difficult to find, since it is important for the source material to have a high incorporation ratio into the film and it is also important for the source to provide atoms rather than molecules to avoid the incorporation of C clusters. CBr_4 seems to be a nearly ideal source for this purpose. We have found that at substrate temperatures above 500°C the dissociation of CBr_4 is 100% with a C incorporation ratio of unity into the GaAs and that doping levels in excess of 10^{20} cm^{-3} are readily obtained in either GaAs or AlGaAs [34]. Most surprising, however, is the fact that at room temperature, there is total reflection of CBr_4 from either GaAs or AlAs, so that no adsorption can be detected. Thus, there is very little problem with "memory effects" for CBr_4 vapor doping. We are presently in the process of examining the detailed decomposition kinetics.

My attempt in this discourse has been to show that a specific application of surface science, MBE, originated in parallel with modern vacuum and surface analysis technologies, and matured as these technologies matured and as important applications for the products were discovered. As a result of the interest in low-dimensional physics and electronics, other growth techniques were

improved to provide similar final results. I have only considered the application to III–V compound semiconductors, yet the MBE approach has proven to have wide applicability to a range of materials from metals to high-T_c superconductors. It is remarkable how universally the Frank–van der Merwe layer-by-layer growth mechanism seems to hold for vapor growth of low-index crystal planes. The MBE approach simply provides a suitable environment for studying this process. It is my contention that the state-of-the-art vacuum–surface experiment is one which includes a means for regenerating or repairing the crystal surface by vapor growth, and, thus, whether the process is called MBE or surface science has only to do with the desired output.

There have been many supporters over the years who have encouraged or contributed to the research field we describe as MBE. However, the initial experiments might never have progressed further if it were not for the vision and enthusiastic support of Dr. John Galt who at that time was the Director of the Semiconductor Electronics Research Laboratory at Bell Laboratories. I owe a special debt to John. ·

References

[1] L.H. Germer, E.J. Schneiber and C.D. Hartman, Philos. Mag. 5 (1960) 222.

[2] H.D. Hagstrum, Phys. Rev. 122 (1961) 83.

[3] H.D. Hagstrum, D.D. Pretzer and Y. Takeishi, Rev. Sci. Instrum. 36 (1965) 1183.

[4] C. Davisson and L.H. Germer, Phys. Rev. 30 (1927) 705. It should be noted that in the original paper by Davisson and Germer which reported the diffraction of an electron beam from a Ni crystal, the success of the experiment was described as due to recrystallization of the Ni substrate as a result of heating to remove the oxide caused by failure of the pyrex container during the original bakeout!

[5] R. Gomer, R. Wortman and R. Lundy, J. Chem Phys. 26 (1957) 1147.

[6] H.E. Farnsworth, R.E. Schlier, M. George and R.M. Burger, J. Appl. Phys. 26 (1955) 252.

[7] R.E. Weber and W.T. Peria, J. Appl. Phys. 38 (1967) 4355.

[8] J.R. Arthur, J. Appl. Phys. 39 (1968) 4032.

[9] J.R. Arthur and J.J. LePore, J. Vac. Sci. Technol. 6 (1969) 545.

[10] R.C. Henderson, W.J. Polito and J. Simpson, Appl. Phys. Lett. 16 (1970) 15.

[11] A.Y. Cho, J. Appl. Phys. 42 (1971) 2074.

[12] H. Widmer, Appl. Phys. Lett. 5 (1964) 108.

[13] H.C. Abbink, R.M. Broudy and G.P. McCarthy, J. Appl. Phys. 39 (1968) 4673.

[14] K.G. Gunther, Z. Naturforsch. 13 (1958) 1081.

[15] J.E. Davey and T. Pankey, J. Appl. Phys. 35 (1964) 2033.

[16] J.E. Davey and T. Pankey, J. Appl. Phys. 37 (1966) 1507.

[17] A.Y. Cho and F.K. Reinhart, J. Appl. Phys. 45 (1974) 1812.

[18] A.Y. Cho, Appl. Phys. Lett. 19 (1971) 467.

[19] L.L Chang, L. Esaki, W.E. Howard and R. Ludeke, J. Vac. Sci. Technol. 10 (1973) 11.

[20] R. Dingle, H.L. Stormer, A.C. Gossard and W. Wiegmann, Appl. Phys. Lett. 33 (1978) 665.

[21] C.T. Foxon and B.A. Joyce, Surf. Sci. 50 (1975) 434.

[22] C.T. Foxon and B.A. Joyce, Surf. Sci. 64 (1977) 293.

[23] C.E.C. Wood and B.A. Joyce, J. Appl. Phys. 49 (1978) 4853.

[24] S.A. Barnett and J.E. Greene, Surf. Sci. 151 (1985) 67.

[25] E.F. Schubert, J.M. Kuo, R.F. Kopf, A.S. Jordan, H.S. Luftman and L.C. Hopkins, Phys. Rev. B 42 (1990) 1364.

[26] J.C. Bean, Appl. Phys. Lett. 33 (1978) 654.

[27] H. Jorke, Surf. Sci. 193 (1988) 569.

[28] M.D. Pashley, K.W. Haberern, W. Friday, J.M. Woodall and P.D. Kirchner, Phys. Rev. Lett. 60 (1988) 2176.

[29] D.K. Biegelsen, R.D. Bringans, J.E. Northrup and L.-E. Swartz, Phys. Rev. B 41 (1990) 5701.

[30] E.J. Heller and M.G. Lagally, Appl. Phys. Lett. 60 (1992) 2675.

[31] B.S. Swartzentruber, R. Kariotis, M.B. Webb and M.G. Lagally, Phys. Rev. Lett. 63 (1989) 2393.

[32] J.Y. Tsao, Materials Fundamentals of Molecular Beam Epitaxy (Academic Press Boston, MA, 1993).

[33] T.P. Chin, P.D. Kirchner, J.M. Woodall and C.W. Tu, Appl. Phys. Lett. 59 2865 (1991) 2865.

[34] L. Ungier, Y. Dang, D. Schulte, J.R. Arthur, J. Ebner and G. Pubanz, to be published.

Surface Science 299/300 (1994) 824–836
North-Holland

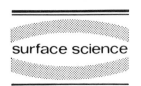

History of desorption induced by electronic transitions

Theodore E. Madey

Department of Physics and Astronomy and Laboratory for Surface Modification,
Rutgers, The State University of New Jersey, Piscataway, NJ 08855-0849, USA

Received 4 May 1993; accepted for publication 8 June 1993

Desorption induced by electronic transitions (DIET) encompasses electron- and photon-stimulated desorption (ESD and PSD) of atoms, molecules and ions from surfaces. In this paper, we focus on the key experimental and theoretical developments that have led to a fundamental understanding of DIET processes. We emphasize the effects of ionizing radiation, i.e., electrons and photons with energies ≥ 10 eV. The first DIET studies were occasioned mainly by the observation of anomalous peaks in mass spectrometers and spurious signals in ionization gauges. These observations were followed in the early 1960's by systematic studies of Redhead, and Menzel and Gomer, who independently proposed a Franck–Condon excitation model for electron-stimulated desorption of ions and neutrals from surfaces. In the years after this seminal work, ESD and PSD developed as fields of active interest to surface scientists. In addition to providing insights into the fundamental mechanisms linking atomic motion and electronic energy dissipation at surfaces, DIET investigations are continuing to impact upon radiation damage processes in areas as diverse as X-ray optics, semiconductor electronics, surface analysis and synthesis of molecules in interplanetary space.

1. Introduction

Desorption induced by electronic transitions (DIET) encompasses electron- and photon-stimulated desorption (ESD and PSD) of atoms, molecules and ions from surfaces. In ESD and PSD, energetic electrons or photons (with energies ranging from ~ 10 eV to more than 1000 eV) bombard surfaces containing a terminal layer of bulk atoms or adsorbed monolayers of atoms or molecules, and cause electronic excitations in the surface species. These excitations can result in the desorption of ions, ground state neutrals, or electronically and vibrationally excited species from the surface. Under some circumstances, even ion bombardment (which generally causes desorption by momentum-transfer processes) can cause DIET.

Why is DIET of importance in surface science? There are a number of reasons [1]. In the chemical and structural analysis of surfaces using electron or photon beams, the radiation-induced rupture of surface bonds is a nuisance to be avoided or at least, minimized. Damage-producing processes can compete with information-producing events during measurements, and often perturb accurate analysis. There are, moreover, substantial benefits to surface science arising from DIET processes. For example, direct information about the geometric structure of surface molecules can be obtained from measurements of the angular distributions of ions released during ESD or PSD: the directions of ion desorption are determined by the orientation of the surface bond ruptured by the excitation. PSD studies using tunable synchrotron radiation reveal the fundamental electronic excitations that cause desorption. Moreover, studies of DIET provide new insights into radiation damage processes that impact on areas as diverse as quantitative surface analysis, partial pressure measurements, molecular formation in interstellar space, UV and X-ray lithography.

In this paper, we trace the roots of ESD and PSD, from the early days of the 20th century to the early 1980s, and emphasize the key ideas and critical experiments that have led to advances in understanding the physics and chemistry of DIET.

The details of theories and experimental methods are discussed elsewhere; the interested reader is guided to a number of comprehensive reviews and conference proceedings, a sampling of which is given in Refs. [1–13]. We focus here on DIET processes stimulated by ionizing radiation ($E > 10$ eV). In another chapter in this volume, historical developments in surface photochemistry induced by ultraviolet and visible radiation are surveyed [14].

This paper is a look backward in time and we focus on past achievements and conceptual developments rather than on exciting contemporary work. It is not intended to be encyclopedic; rather, it is a personal account, and in some instances, personal insights are provided.

2. DIET – The first observations

In the years before reliable UHV surface science measurements on well-characterized surfaces, there were a number of isolated reports of DIET processes. (Parenthetically we note the term "DIET" is a relatively new one, having been introduced by Tolk et al. in 1982 [9]. Even the terms ESD and PSD did not appear in the literature until the late 1960's to early 1970's.) The release of gases and ions from surface layers during electron bombardment has been known and studied for decades. In most cases, experimenters were motivated by a desire to understand practical aspects of electron–surface interactions in vacuum tubes, mass spectrometers and vacuum gauges. However, with few exceptions, the surfaces were poorly characterized and there was little understanding of the physics of the excitation and desorption processes.

Most early ESD studies concerned the desorption of ions from surfaces. The first ESD report appears to have been in 1918, when Dempster [15] observed that electron bombardment of salts (aluminum phosphate on a Pt foil) released ions that he identified in a magnetic sector mass spectrometer. Plumlee and Smith [16a] in 1950 and Young in 1960 [16b] observed efficient desorption of O^+ from oxidized metal surfaces with desorption yields as high as 10^{-5} ions/e^-. In the early

1960s, Marmet and Morrison [17], Robins [18] and Alpert and deSegovia [19] discovered almost simultaneously in mass spectrometric studies that ions may be formed on the walls of the ionization source, at energies different from those corresponding to the ionized states of the free molecules. In experiments that were to have important implications for subsequent studies of surface science, Redhead [20] recognized that positive ions were formed by electron bombardment of chemisorbed oxygen in Bayard–Alpert ionization gauges, leading to serious errors in UHV pressure measurements. An extensive study of ions desorbed during electron bombardment of CO on Mo and W ribbons was carried out by Moore [21], who found that the cracking pattern of ions from surface CO (mainly O^+) was different from that of gas phase CO (mainly CO^+).

The first systematic study of desorption of neutrals was published in 1942 by Ishikawa [22], who reported the desorption of hydrogen from a Pt foil. He interpreted his measurements of desorption yield versus energy in terms of the electronic excitations in the hydrogen molecule – the first suggestion that electronic excitations play a role in desorption! In 1957, Wargo and Shepherd [23] bombarded alkaline earth oxides (MgO, BaO and SrO) with low energy electrons and detected the evolution of molecular oxygen using an omegatron mass spectrometer; they measured threshold energies for the process and postulated that electronic excitations were involved in desorption of oxygen. Degras [24] and Petermann [25] reported small increases in CO partial pressure upon bombardment of CO-covered surfaces.

Other indirect evidence for ESD effects came from field ion microscopy. Mulson and Müller [26] and Ehrlich and Hudda [27] reasoned that reductions in surface coverage and changes in adsorbed layers on field ionization emitters were due to bombardment of the emitter by electrons. These electrons were released by gas atoms ionized by high field (~ 4 V/Å) in the vicinity of the emitter surface.

By the early 1960's, the stage was set for a synthesis of a number of these diverse results and for the construction of a conceptual framework for understanding a wide range of observations.

This was provided almost simultaneously in 1964 by two research groups working completely independently and without prior knowledge of each other's work: Robert Gomer and his postdoctoral colleague Dietrich Menzel at the University of Chicago, and Paul Redhead at the National Research Council in Ottawa.

3. The dawn of a new age in 1964: Redhead, Menzel and Gomer

The papers of Redhead [28,29] and Menzel and Gomer [30,31] are landmarks in the history of DIET for several reasons. First, they each articulated the Franck–Condon excitation model for electron-stimulated desorption that has provided the jumping off point for all subsequent theories of stimulated desorption. Second, these investigators placed the indelible stamp of surface scientists on a new field. Their works represent the first comprehensive and systematic ESD studies of atomic and molecular adsorbates on clean metal surfaces under ultrahigh vacuum conditions. Third, their studies were complementary rather than redundant. Redhead focused his attention on the characterization of desorption products, i.e., positive ions from adsorbate-covered metal surfaces, whereas Menzel and Gomer studied the changes in the surface layer that were caused by electron bombardment.

Menzel and Gomer's specific motivation for studying ESD of adsorbed gases originated in previous field desorption observations in which it had been concluded that electron tunneling from adsorbate to substrate is a rapid process. They reasoned that the inverse process, electron tunneling from the metal into the adsorbate, should also be fast. If this conclusion were true, it would be confirmed (at least qualitatively) by small overall cross sections for electron-stimulated desorption resulting from "bond healing" of broken or excited bonds by electrons tunneling from the metal. They were remarkably prescient in their speculations.

Menzel and Gomer used a field electron emission technique to study the effect of low energy (15–200 eV) electrons on hydrogen, oxygen, car-

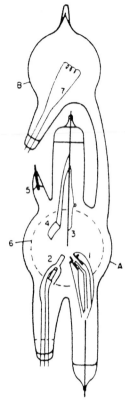

Fig. 1. Schematic diagram of the field emission tube used by Menzel and Gomer [31] in their first studies of ESD. (A) field-emission tube, looking from rear toward screen; (B) getter bulb; (1) gun assembly; (2) chemical gas source with conducting shield; (3) tip loop with potential leads; (4) collector electrode; (5) anode connection; (6) screen; (7) Ta getter wire.

bon monoxide and barium adsorbed on a tungsten tip [31]. (Note that field emission microscopy was the only surface science method in wide use at the time in which clean metal surfaces could be generated reproducibly and reliably.) Their apparatus is shown in Fig. 1. It consisted essentially of a low temperature field emission tube with a pyrex envelope that had been evacuated and sealed-off. It was equipped with a simple electron gun and chemical and/or gas sublimation sources. In order to insure ultrahigh vacuum, the sealed-off tube was immersed in liquid hydrogen or helium during operation. The entire tube had a total volume of a few hundred cm^3; the ratio of scientific results to experimental volume in this device far exceeds that of many measurements made today using sophisticated stainless steel vacuum chambers!

Menzel and Gomer determined desorption cross sections from the time dependence of field electron emission characteristics (work function and Fowler–Nordheim pre-exponential factor) during electron bombardment. They could not directly detect desorption products, but inferred electron-stimulated effects from changes in the surface layer. They found marked variations in cross section for different binding states within a given system. In the case of adsorbed CO (a system that has been well studied for many years since) three binding modes could be confirmed and differentiated by their different cross sections: $\sigma(\text{virgin-CO}) = 3 \times 10^{-19}$ cm^2, $\sigma(\beta\text{-CO}) = 5.8 \times 10^{-21}$ cm^2, and $\sigma(\alpha\text{-CO}) = 3 \times 10^{-18}$ cm^2. They also found evidence for electron-induced reactions. The conversion of molecular virgin-CO to dissociatively adsorbed β-CO occurred with cross section $\sigma(\text{virgin to } \beta) \geq 10^{-19}$ cm^2. The splitting of molecularly adsorbed hydrogen had a cross section of $\sigma(\text{H}_2) = 3.5 \times 10^{-20}$ cm^2. Adsorbed Ba was found to be very stable, with a desorption cross section $\sigma(\text{Ba}) < 2 \times 10^{-22}$ cm^2 under all conditions.

The results were interpreted in terms of electronic transitions from the adsorbed ground state to repulsive portions of excited states, followed by de-exciting transitions that can prevent desorption. (Schematic potential energy diagrams consistent with this model are shown in Fig. 2.) Menzel and Gomer recognized that all of their measured cross sections were significantly smaller than the cross sections for similar gas phase electron impact dissociation, $\sim 10^{-16}$ to 10^{-17} cm^2. They argued that the cross sections for *excitation* are of order 10^{-16} to 10^{-17} cm^2, and that the much smaller overall cross sections observed for desorption and dissociation were due to high transition probabilities to the ground state, estimated to have rates of 10^{14} s^{-1} to 10^{15} s^{-1}. This model for ESD is now known as the Menzel–Gomer–Redhead (MGR) model.

In the spring of 1963, the author (then a physics graduate student) recalls meeting Menzel and Gomer at a meeting of the American Physical

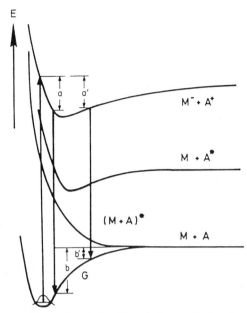

Fig. 2. Schematic of the Menzel–Gomer–Redhead model showing the potential energy curves for adsorbate levels relevant for ESD. Abscissa: distance between adsorbate and substrate; G: adsorbate ground state; (M+A)*: antibonding adsorbate state; M+A*: excited adsorbate state; M$^-$+A$^+$: ionized adsorbate state. Transitions are shown for ionic desorption only. Depending on the distance at which retunneling occurs, it leads to recapture in the ground state ($b > a$) or to desorption as a neutral particle ($b' < a'$).

Society in Buffalo, NY, where results of their first experiments were presented. The essence of their interpretation had been formulated, and they had recently become aware of the (as yet unpublished) work of Redhead across the border in Ottawa.

Redhead had a different motivation for studying ESD. His major research interests were the generation and characterization of ultrahigh vacuum (UHV), and the physical basis of processes affecting UHV. He knew that electron bombardment of chemisorbed oxygen on the grid of a Bayard–Alpert ionization gauge released positive ions that gave spurious signals in UHV pressure measurements [20]; he wanted to know more about this process. He designed a glass apparatus in which all of the positive ions liberated from an electron-bombarded sample were collected. His design also permitted determination of the distribution of ion kinetic energies, a measurement that has contributed significantly to the understanding of ESD mechanisms. A similar apparatus used later by the NBS group for ESD studies is shown in Fig. 3.

In his first experiments, Redhead studied the chemisorption of oxygen on polycrystalline Mo [28,29]. In measurements that provided a guide for many future studies, he detected positive ions (later shown to be O^+), measured their energy distribution, determined cross sections for desorption of ions and neutral species, measured the threshold energy for ion desorption, and recognized that much of the chemisorbed oxygen was ESD "inactive"; the ion signal originated mainly from species adsorbed at high exposure. He was also the first to point out that ions constitute a minority of desorbed species; the dominant des-

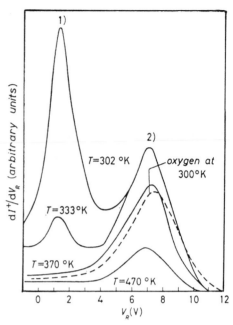

Fig. 4. Redhead's measurements of ion energy distributions for CO on W at 300 K and above [33]. The ion energy distribution for adsorbed oxygen at 300 K is also shown for comparison. Electron energy $= 100$ eV, $p = 8 \times 10^{-8}$ Torr.

orption product consists of neutrals. Typical ion energy distributions [33] are given in Fig. 4. He interpreted his results in terms of a Franck–Condon mechanism similar to that of Menzel and Gomer [31]: excitation from the ground state to a repulsive ionic state, then motion away from the surface followed by a Hagstrum type reneutralization probability of the form $R(x) = A e^{-ax}$, where R is the electron hopping rate (s^{-1}), x is the distance from ion to surface, and A and a are constants.

The MGR model provided the only conceptual framework for understanding ESD processes for nearly 14 years after its publication in 1964. As originally formulated, the model was based on a one electron valence excitation to a repulsive state followed by a reneutralization or recapture step. Since the detailed nature of the electronic excitations is not specified, the model does not have predictive powers concerning the initial excited state or about the surface atomic or electronic arrangements required for desorption to occur. On the other hand, the general definition

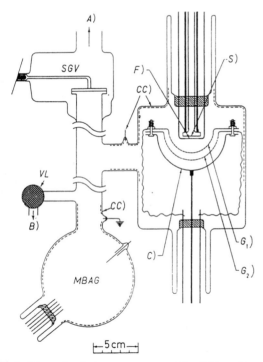

Fig. 3. Schematic of glass apparatus used for ESD studies by Madey and Yates [32,34]; it is similar to the design of Redhead [28]. (S) Tungsten surface; (F) emitter filament; (G$_1$) grid at V_e −10.5 volt; (G$_2$) grid at −285 volt; (C) collector; (CC) conductive coating; (SGV) sliding-glass valve; (MBAG) modulated Bayard–Alpert gauge; (VL) variable leak; (A) to ion pump or cryopump; (B) to auxiliary ion-pumped gas-handling system.

of the MGR model based on a Franck–Condon excitation to a repulsive final state is sufficiently broad as to encompass essentially all of the more specific excitation models that were to come later (cf. Section 4, below).

4. ESD develops as a research field

In the years immediately following the reports of Menzel, Gomer and Redhead, laboratories on several continents initiated systematic studies of ESD phenomena. The motivations were varied, from exploration of ESD mechanisms to application of the new methods. At the National Bureau of Standards (NBS; now NIST) in Washington, Madey and Yates had been studying kinetics of thermal desorption of molecules from W and Re surfaces. They were inspired by the potential of ESD as a sensitive in situ probe of kinetic processes in the adsorbed layer; it was not necessary to "destroy" the adsorbate by heating to high temperatures to detect what was on the surface. In their first studies of oxygen-containing diatomic molecules (CO, NO and O_2) on a polycrystalline W surface, they used methods similar to those of Redhead to show that ions desorbing with different kinetic energies originated from different building states (e.g., α-CO and β_1-CO) [32]. Simultaneously, Redhead [33] reported ion yields and energy distributions for CO/W. The NBS team was able to follow thermally and electron-activated dissociation processes by measurements of ESD ion yields [32]; they also used ESD to study the adsorption of oxygen on W at temperatures as low as 20 K, and characterized physisorbed and chemisorbed oxygen by their different ion yields and ion energy distributions [34]. They found that most of the chemisorbed oxygen species (β_2-state) gave little or no ESD ions, and that the ions from the ESD-active β_1-state originated from a minority of surface species, only about 3% of the adsorbed monolayer (later, it was postulated that the ESD-inactive β_2-state is adsorbed atomic O, and the ESD-active β_1-state is due to adsorption on steps and defects [35]).

The first report of ESD studies of adsorbates on macroscopic single crystals was by Zingerman

and Ishchuk in Kiev [36]. They observed periodic maxima and minima in curves of ESD total desorption cross sections as a function of electron energy, for oxygen on single crystal W. Their results were controversial at the time, but they were the first to suggest that threshold energies for ESD processes are correlated with characteristic inelastic electron energy losses due to interaction with the substrate.

Threshold measurements were recognized as being of importance to clarifying the mechanisms of ESD. Redhead [28] had reported for O/Mo that the threshold energies for desorption of ionic and neutral oxygen were identical; this suggested that near the threshold, the dominant transition is to an ionic state, and that neutrals originate from charge transfer to ions. In contrast, Menzel's measurements [37] of the thresholds for ESD of CO on W demonstrated that the situation can be much more complex. The thresholds for CO^+ and O^+ desorption were 14.6 and 20 eV, respectively, and ~ 5 eV for neutral CO desorption. This showed that desorption of CO neutrals does not proceed via Auger neutralization of desorbing CO^+ ions, but by excitation to a lower-lying neutral state.

Reports of desorption of excited neutral species were made by Baker and Petermann [38] and Redhead [33] in studies of CO on metals. Although Redhead could not distinguish unambiguously between desorption of a metastable neutral and the emission of photons from the surface, conclusive evidence for desorption of metastables was later provided by Newsham, Hogue and Sandstrom [39]. These authors used a time-of-flight technique to measure energy distributions of excited neutrals desorbed from CO/W; they identified the excited molecules as being the $a^3\Pi$-state of CO which has an internal energy of 6.01 eV.

Temperature-dependent variations in ESD ion yields [40–42] and energy distributions [42] were reported by several investigators. The effects were small ($\sim 10\%$) but reproducible, and in all cases, the results were believed to be consistent with the MGR mechanism.

An important test of the MGR model is the prediction of an isotope effect in ESD, i.e., the probabilities of ionic and neutral desorption are exponentially dependent on the $(mass)^{1/2}$ of the desorbing species [30]. Slow, more massive ions have a lower probability of escape from the surface without neutralization than faster, less massive ions. Madey, Yates, King and Uhlaner [43] examined the effect of mass on ESD for the systems ^{16}O and ^{18}O on polycrystalline and (110) tungsten samples. They measured the ratio of ionic cross sections to be

$$Q_{16}^+/Q_{18}^+ \approx 1.5 \pm 0.15.$$

This number is in quantitative agreement with the predictions of the MGR theory. On the other hand, a much smaller isotope effect for total (neutral) desorption implied that neutrals do not arise solely from reneutralization of ions, but that other options needed to be considered.

Whereas much of the research probing ESD mechanisms during the late 1960's was inspired by the work of Menzel, Gomer and Redhead, there was another parallel development that was to have important later influence on the DIET field, i.e., the effect of ionizing radiation on damage to alkali halides. Pooley [44] and Hersch [45] independently published similar explanations of how a photon (or electron) can excite an electron–hole pair in a halogen atom X beneath the surface of an alkali halide. In this mechanism, the initial excitation is followed by a focussed collision sequence that may result in neutral halogen atoms X^0 being energetically ejected from the surface. Unlike ESD of chemisorbed species in which the excitation occurs in the outermost surface layer, damage to alkali halides was believed to be initiated by an excitation below the surface, with propagation of energy to the surface causing desorption. Although there is clear relevance of this mechanism to stimulated desorption processes, there was only little contact [46,47] between the surface scientists studying ESD and the alkali halide community until the early 1980's [9].

Because of their ease of detection, most DIET studies involving ions focused on positive ions. ESD of negative ions was first reported by Moore [48] and Ayukhanov and Turnashev [49], but was not studied in a systematic way until the late 1970's by David Lichtman, J.L. Hock, James Craig

and co-workers in Milwaukee [50] and Ming Yu in Yorktown Heights [51]. In the early 1980s, Leon Sanche and co-workers [52] demonstrated for ESD of physisorbed molecules the presence of a dissociative attachment mechanism for the production of negative ions at energies of a few electron volts, well below the threshold for dipolar dissociation.

In a technically challenging experiment, Feulner, Treichler and Menzel reported in 1981 the threshold energies for ESD of neutral molecules from adsorbed monolayers of CO and N_2 on Ru [53]. Shortly after, they showed that angular distributions of neutrals are peaked about the surface normal, and that kinetic energy distributions for ground state neutrals have most probable energies below 1 eV [54].

In another area that was to blossom in later years, David Lichtman, Yoram Shapira and their colleagues performed extensive studies of photodesorption of CO_2 from oxides and other semiconducting substrates [55]. They found that bandgap radiation was necessary to induce desorption, and that the signal was linear with photon flux. This work predated the photon-stimulated desorption of ions described in Section 6.

By the late 1970's, it was also realized that DIET processes are of importance in understanding the origin of radiation damage in surface analysis using electron and photon beams (Auger electron spectroscopy, low energy electron diffraction, X-ray photoelectron spectroscopy, scanning electron microscopy, etc.). Critical doses for damage in Auger measurements were correlated with predictions based on ESD measurements, demonstrating that primary electronic excitations dominate the beam damage mechanism [56].

5. The early 1970s: ESDIAD

During the course of a study of the chemisorption of oxygen on W(100) in the early 1970s, the author [57] found, using a segmented detector, that the O^+ ions liberated in ESD had an unusual angular distribution. Although angular resolution was poor, it was clear that the desorbing ions had a sharp maximum in the direction normal to the surface. The data were consistent with a $\cos^n \theta$ distribution, with $n > 14$! Focussing of ions by the repulsive potential at the surface and directional bonding in the surface layer were mentioned as possible causes of this phenomenon [57].

To examine the ESD ion angular distributions in detail, a display-type analyzer was designed at NBS. The heart of the instrument was a sensitive two-dimensional detector based on two 25 mm microchannel plates donated to us by an applications engineer from Varian, Dom Ruggieri. A visiting scholar from Poland, J.J. Czyzewski, joined the team of Madey and Yates. The first experimental tube, a hybrid of glass and metal, was a disaster – the tube fractured into shards of glass within minutes of the final seal to the vac-

uum system. The microchannel plates were salvaged and a second tube was built (Fig. 5). Serendipitously, the first system studied was the best possible choice: oxygen on W(100). A remarkable sequence of patterns displaying sharp, rich structure was observed (Fig. 6).

The observation of sharp, distinct beams led the authors to postulate that desorption was dominated by initial state effects, viz. the direction of the ground state bond ruptured by the excitation, rather than by final state effects. The acronym ESDIAD was conceived (electron stimulated desorption ion angular distribution) and the first report was sent off to Physical Review Letters. The paper was accepted but the acronym was rejected – too many letters [58]. The term ESDIAD slipped into the literature in later papers

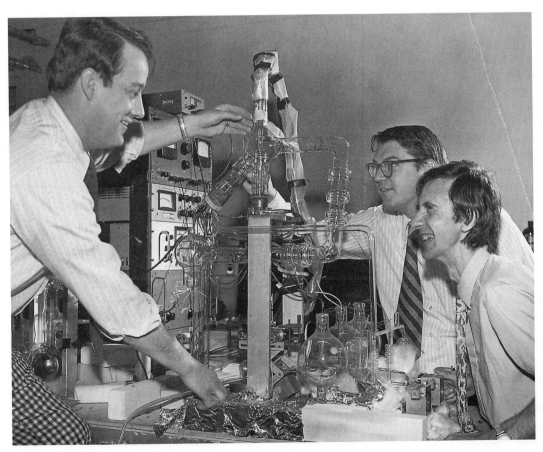

Fig. 5. 1974 photograph of the first NBS ESDIAD system, a hybrid of glass and metal. From left to right, T.E. Madey, J.T. Yates, Jr. and J.J. Czyzewski. Madey's left hand is pointing to the ESDIAD tube; the ESDIAD patterns were photographed from below.

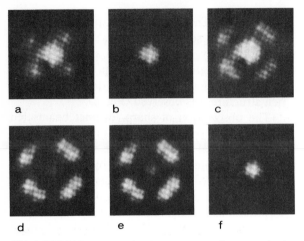

Fig. 6. ESDIAD patterns for desorption of O^+ from W(100) obtained using the apparatus of Fig. 5 [58]. This sequence shows the effect of heat treatment on ESD patterns for β_1-oxygen on W(100). Temperatures corresponding to each pattern are (a) < 400, (b) 630, (c) 705, (d) 795, (e) 865 and (f) 930 K.

[59]. Shortly after the first NBS studies, Niehus in Jülich developed a sensitive angle-resolved scanning method for quantitative ESDIAD studies [60].

The NBS group initiated a long series of experiments to test the relation between the angular distributions of ESD ions and ground state bond directions. Several visiting scientists to NBS collaborated in these studies through the late 1970s and early 1980s, including Susan Dahlberg, Jack Houston, Falko Netzer and Carsten Benndorf. The approach was to study adsorbed molecular species (CO, NO, NH_3, H_2O) whose structures had been determined independently using other techniques (angle-resolved ultraviolet photoemission spectroscopy, vibrational spectroscopy, etc.). In all cases where the structure was established ("standing up" and "tilted" CO, upright NH_3, etc.) the ESDIAD data revealed ion desorption directions dominated by the ground state bond directions (for summaries of this work, see Refs. [61,62]). Of particular importance was the discovery that coadsorption of low oxygen coverages with H_2O and NH_3 on certain surfaces (Ni(111), Ru(0001)) led to stabilization of molecular structures (azimuthal ordering) that did not occur in the absence of the oxygen [63].

The reports of angular effects in ESD stimulated the interest of theorists. Gersten, Janow and Tzoar [64] used a molecular dynamics approach to simulate ESDIAD data for O^+ desorption from W(100) and W(111), and found that the details of the angular distribution depend in a sensitive way on surface bonding sites. Clinton [65] formulated a quantum scattering theory of ESD and concluded that ESDIAD processes are dominated by initial state (ground state) structures of atoms and molecules on surfaces. Specific final state perturbations of ion trajectories due to image force and reneutralization effects were quantified in later papers [61,66].

Thus, by the early 1980s, ESDIAD was established as a useful tool for studies of molecular structure at surfaces. Even so, the number of laboratories with functioning ESDIAD apparatus up to the present time is small, due in part to the lack of availability of commercial instrumentation. Everyone who wants to do ESDIAD must build his/her own apparatus, or buy components from one of the practitioners.

6. The next steps forward:
The Knotek–Feibelman mechanism;
photon stimulated desorption

One of the most important conceptual developments in DIET was the recognition by Mike Knotek and Peter Feibelman in 1978 of an atom-specific ESD excitation mechanism with genuine predictive powers [67]. Knotek had been studying ESD of positive ions from oxide surfaces at Sandia Labs in Albuquerque. For several highly ionic materials (TiO_2, WO_3, V_2O_5), he found that the dominant desorption product was O^+, even though the formal charge state of the lattice oxygen anion is O^{2-}. The inescapable conclusion was the presence of charge transfer in the desorption process – up to three electrons in the O^{2-} to O^+ transformation. Moreover, the major thresholds for ion desorption coincided with core level excitation energies (Ti 3p, W 5p). Based on these observations, Knotek and Feibelman [67] proposed a desorption model that is particularly applicable to highly ionic systems (i.e., maximal

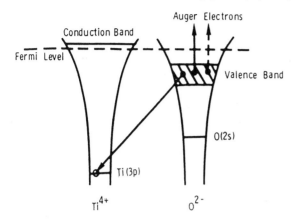

Fig. 7. Schematic of Knotek–Feibelman Auger decay model for stimulated desorption of ions. Formation of a core hole in the 3p shell of maximal valency TiO_2 is followed by interatomic Auger decay leading to formation of an O^+ ion [67].

valency systems). The primary process is the ionization of a core level on the metal cation. The interatomic Auger decay of the core hole may create a positive oxygen ion at an initially negative ion site (Fig. 7). The desorption of the positive oxygen ion results from the reversal of the Madelung potential. This Auger decay model is now known as the Knotek–Feibelman (KF) model.

In a follow up paper to their original Physical Review Letter, Feibelman and Knotek [68] showed that many features of published ESD data could be rationalized in terms of the KF Auger decay model.

There were several other reports at about the same time implicating core level excitations in electron-stimulated surface processes. Naumovets and Fedorus in Kiev [69] found that the onset for electron-stimulated disordering of Li on W(110) coincided with the Li 1s excitation energy. Franchy and Menzel [70] reported that deep core excitations (O 1s in chemisorbed CO) produced desorption of O^+, and postulated that Auger decay caused a "Coulomb explosion" in the surface molecule [71].

The formulation of the KF model generated an explosion of interest in stimulated desorption. Since the desorption process was predicted to be independent of the way the core hole is produced, several groups initiated photon-stimulated

desorption (PSD) studies using tuneable synchrotron radiation. The first successful report of O^+ desorption from an oxide surface above a core edge was by Knotek, Jones and Rehn [72] at the Stanford storage ring. Because the cross-section behavior is different, they found that thresholds in PSD are much sharper than in ESD. Madey and Roger Stockbauer convinced Dean Eastman and Friso van der Veen to use Eastman's ellipsoidal mirror analyzer at the synchrotron radiation source in Wisconsin to search for angular anisotropy in PSD. This work demonstrated that PSD and ESD ion angular distributions for O^+ desorption from W(111) are identical, and that the PSD threshold behavior is consistent with the KF model [73]. At about the same time, a collaborative group mostly from Bell Labs [74] (Woodruff, Johnson, Traum, Farrell, Smith, Benbow and Hurych) showed at Wisconsin that one can do PSD measurements using a cylindrical mirror analyzer operated in time-of-flight mode. Stockbauer, Madey and Hanson [75] used the SURF-II facility (Synchrotron Ultraviolet Radiation Facility) at NBS to compare ESD and PSD of O^+ from a Ti single crystal, and found results similar to PSD of TiO_2. In an experiment at Stanford (SSRL) Rolf Jaeger and his colleagues demonstrated that PSD ion yields can be used to monitor SEXAFS (surface-extended X-ray absorption fine structure) in an appropriate photoionization cross section. They showed this effect in the O^+ yield from Mo(100) above the Mo L_I edge [76].

Theoretical developments flourished also during this same period. Much of the interest focussed on the role of multiple hole final states (produced by Auger decay of a core hole) on the desorption process. The presence of multiple valence holes in a covalent system can result in a strongly repulsive interaction between unscreened nuclei and cause the subsequent ejection of ionic fragments. Moreover, hole–hole correlation effects can lead to intrinsically longer lifetimes for multiple hole final states than for single hole final states, as first pointed out by Cini and Sawatsky [77] in analysis of atomic-like features in Auger spectra of narrow d-band metals. This many body effect was recognized by a number of investigators (Ramaker [78], Feibelman [79], Knotek [7], Jennison [80] and others) as making the Auger-induced desorption process effective for covalent as well as ionic materials.

An extension of the MGR desorption model with application to several specific systems, including ESD of rare gases, was suggested in 1980 by Peter Antoniewicz [81]. In this model, the ion formed by an electronic excitation begins its trajectory by moving toward the surface where it is then neutralized. After neutralization, the atom exists on a strongly repulsive potential curve leading to probable desorption as a neutral (or in some cases, as an ion). In another important development, Bill Gadzuk [82] introduced a dynamics perspective into the DIET discussion, by considering the dynamics of time-dependent localized potentials and interactions in solids, and how these phenomena relate to the fundamental processes in DIET.

7. The field matures and acquires a name: DIET

By the early 1980s, the number of scientists interested in ESD and PSD processes had grown significantly and there was enough interest to warrant organizing an ad hoc conference. A Bell Laboratories/NBS team (Norman Tolk, John Tully, Morton Traum and Ted Madey) organized the first "Symposium on Desorption Induced by Electronic Transitions" in historic Williamsburg, VA. The response was excellent, and about 75 individuals convened for the 3-day meeting, May 12–14, 1982. The response was so enthusiastic that the symposium has evolved into a continuing series, on a 2 1/2 year cycle [9–13].

The introduction to the DIET-I proceedings [9] states that by 1983 "the DIET field has progressed in an exciting manner to a state of dynamic adolescence. It remains a challenge to both experimentalists and theoreticians to boldly shape this subject to an even more exciting and dynamic maturity". In the ten years since those lines were written there have been enormous strides in experiment and theory. Considering advances in quantum state-resolved desorption [83], dynamics of charge transfer [84], nuclear dynamics of de-

sorption [85], DIET of alkali halides [86], digital ESDIAD [87,88,1], ESDIAD of negative ions [89], polarization-dependent photon-stimulated desorption of ions and neutrals [90], and laser-induced photodesorption and desorption induced by multiple electron transitions (DIMET) [91], it is clear that the field is well beyond "dynamic adolescence" and has reached the young adult phase. DIET is a mature but very energetic field with great challenges ahead. A history of the next 30 years should be barely recognizable as related to the present account.

References

[1] T.E. Madey, Science 234 (1986) 316.
[2] D. Lichtman and R.B. McQuistan, Prog. Nucl. Energy, Ser. 9, 4 (1965) 95. This paper reviews ESD through 1964, with emphasis on instrumental methods.
[3] T.E. Madey and J.T. Yates, Jr., J. Vac. Sci. Technol. 8 (1971) 525.
[4] P.A. Redhead, J. Vac. Sci. Technol. 7 (1970) 182.
[5] V.N. Ageev and N.I. Ionov, Prog. Surf. Sci. 5 (part 1) (1974) 1.
[6] D. Menzel, Chem. Int. Edit. Engl. 9 (1970) 255; Nucl. Instrum. Methods B 13 (1986) 507.
[7] M.L. Knotek, Rep. Prog. Phys. 47 (1984) 1499; Phys. Today 37 (September 1984) 24.
[8] R.D. Ramsier and J.T. Yates, Jr., Surf. Sci. Rep. 12 (1991) 243. This recent comprehensive review contains over 1000 research citations, and a complete list of review and survey papers in the field of DIET, up to 1991.
[9] Desorption Induced by Electronic Transitions, DIET-I, Vol. 24 of Springer Series in Chemical Physics, Eds. N.H. Tolk, M.M. Traum, J.C. Tully and T.E. Madey (Springer, Berlin, 1983).
[10] DIET-II, Vol. 4 of Springer Series in Surface Sciences, Eds. W. Brenig and D. Menzel (Springer, Berlin, 1985).
[11] DIET-III, Vol. 13 of Springer Series in Surface Sciences, Eds. R.H. Stulen and M.L. Knotek (Springer, Berlin, 1988).
[12] DIET-IV, Vol. 19 of Springer Series in Surface Sciences, Eds. G. Betz and P. Varga (Springer, Berlin, 1990).
[13] DIET-V, Vol. 31 of Springer Series in Surface Sciences, Eds. A. Burns, D. Jennison and E.B. Stechel (Springer, Berlin, 1993).
[14] W. Ho, Surf. Sci. 299/300 (1994) 996.
[15] A.J. Dempster, Phys. Rev. 11 (1918) 316.
[16] (a) J.R. Young, J. Appl. Phys 31 (1960) 921.
(b) R.H. Plumlee and L.P. Smith, J. Appl. Phys. 21 (1950) 811.
[17] P. Marmet and J.D. Morrison, J. Chem. Phys. 36 (1962) 1238.

[18] J.L. Robins, Can. J. Phys 41 (1963) 1385.

[19] D. Alpert, Phys. Today (1962).

[20] P.A. Redhead, Vacuum 13 (1963) 253.

[21] G.E. Moore, J. Appl. Phys. 32 (1961) 1241.

[22] Y. Ishikawa, Rev. Phys. Chem. Jpn. 16 (1942) 117.

[23] P. Wargo and W.G. Shepherd, Phys. Rev 106 (1957) 694.

[24] D.A. Degras, L.A. Petermann and A. Schram, Proc. Am. Vac. Soc. 9th Nat. Symp. (1962) 497.

[25] L.A. Petermann, Nuovo Cimento Suppl. 1 (1963) 601.

[26] J.F. Mulson and E.W. Müller, J. Chem. Phys. 38 (1963) 2615.

[27] G. Ehrlich and F.G. Hudda, Philos. Mag. 8 (1963) 1587.

[28] P.A. Redhead, Can. J. Phys 42 (1964) 886.

[29] P.A. Redhead, Appl. Phys. Lett. 4 (1964) 166.

[30] D. Menzel and R. Gomer, J. Chem. Phys. 40 (1964) 1164.

[31] D. Menzel and R. Gomer, J. Chem. Phys. 41, (1964) 3311; 3329.

[32] J.T. Yates Jr., T.E. Madey and J.K. Payn, Nuovo Cimento, Suppl. Ser. 1, 5 (1967) 558.

[33] P.A. Redhead, Nuovo Cimento Suppl., Ser. 1, Vol. 5 (1967) 586.

[34] T.E. Madey and J.T. Yates, Jr., Surf. Sci. 11 (1968) 327.

[35] T.E. Madey, Surf. Sci. 94 (1980) 483.

[36] Y.P. Zingerman and V.A. Ishchuk, Fiz. Tverd. Tela 9 (1967) 3347 [Sov. Phys. Solid State 9 (1968) 2638]; 10 (1968) 3720 [10 (1969) 2960].

[37] D. Menzel, Ber. Bunsenges. Phys. Chem. 72 (1968) 591.

[38] F.A. Baker and L.A. Petermann, J. Vac. Sci. Technol. 3 (1966) 285.

[39] I.G. Newsham, J.V. Hogue and D.R. Sandstrom, J. Vac. Sci. Technol. 9 (1972) 596;
I.G. Newsham and D.R. Sandstrom, J. Vac. Sci. Technol. 10 (1973) 39.

[40] D. Menzel, Surf. Sci. 14 (1969) 340.

[41] E.N. Kutsenko, Zh. Tekhn. Fiz. 39 (1969) 942. [Sov. Phys. Tech. Phys. 14 (1969) 706].

[42] T.E. Madey and J.T. Yates, Jr., J. Chem. Phys. (1969) 1264.

[43] T.E. Madey, J.T. Yates, Jr., D.A. King and C.J. Uhlaner, J. Chem. Phys. 52 (1970) 5215.

[44] D. Pooley, Solid State Commun. London 3 (1965) 241; Proc. Phys. Soc. 87 (1966) 257.

[45] H.N. Hersch, Phys. Rev 148 (1966) 928.

[46] P. Palmberg, C.J. Todd and T. Rhodin, J. Appl. Phys. 39 (1968) 4650.

[47] P.D. Townsend and J.C. Kelley, Phys. Lett. 26A (1968) 138.

[48] G.E. Moore, J. Appl. Phys. 30 (1959) 1086.

[49] A.K. Ayukhanov and E. Turnashev, Sov. Phys. Tech. Phys. 22, (1977) 1289.

[50] J.L. Hock and D. Lichtman, Surf. Sci. 77 (1978) L184.

[51] M.L. Yu, Phys. Rev. B 19 (1979) 5995.

[52] L. Sanche, Phys. Rev. Lett. 53 (1984) 1638.

[53] P. Feulner, R. Treichler and D. Menzel, Phys. Rev. B 24 (1981) 355.

[54] P. Feulner, W. Riedel and D. Menzel, Phys. Rev. Lett. 50 (1983) 355;

P. Feulner, D. Menzel, H.J. Kreuzer and Z.W. Gortel, Phys. Rev. Lett. 53 (1984) 671.

[55] D. Lichtman and Y. Shapira, CRC Chemistry and Physics of Solid Surfaces, Vol. II (1979) p. 397.

[56] C.G. Pantano and T.E. Madey, Appl. Surf. Sci. 7 (1981) 115.

[57] T.E. Madey, Surf. Sci. 33 (1972) 355.

[58] J.J. Czyzewski, T.E. Madey and J.T. Yates, Jr., Phys. Rev. Lett. 32 (1974) 777.

[59] T.E. Madey, J.J. Czyzewski and J.T. Yates, Jr., Surf. Sci. 57 (1976)580; 49 (1975) 465.

[60] H. Niehus, Surf. Sci. 78 (1978) 667; 80 (1979) 245.

[61] T.E. Madey, Vol. 17 of Springer Series in Chemical Physics (Springer, Berlin, 1981) p. 80.

[62] T.E. Madey, F.P. Netzer, J.E. Houston, D.M. Hanson and R.L. Stockbauer, T.E. Madey et al., in: DIET-I (Ref. [9]) p. 120.

[63] F.P. Netzer and T.E. Madey, Phys. Rev. Lett. 47 (1981) 928.

[64] J. Gersten, R. Janow and N. Tzoar, Phys. Rev. Lett. 36 (1976) 610;
R. Janow and N. Tzoar 69 (1977) 253.

[65] W.L. Clinton, Phys. Rev. Lett. 39 (1977) 965.

[66] Z. Miskovic, J. Vukanic and T.E. Madey, Surf. Sci. 141 (1984) 285; 169 (1986) 405.

[67] M.L. Knotek and P.J. Feibelman, Phys. Rev. Lett. 40 (1978) 964; Surf. Sci. 90 (1979) 78.

[68] P.J. Feibelman and M.L. Knotek, Phys. Rev. B 18 (1978) 6531.

[69] A.G. Naumovets and A.G. Fedorus, Zh. Eksp. Teor. Fiz. 68 (1975) 1183; [Sov. Phys.-JETP 41 (1976) 587];
V.V. Gonchar, O.V. Kanash, A.G. Naumovets and A.G. Fedorus, Sov. Phys.-JETP Lett. 28 (1978) 330.

[70] R. Franchy and D. Menzel, Phys. Rev. Lett. 43 (1979) 865.

[71] T.A. Carlson, in: DIET-I (Ref. [9]) p. 169.

[72] M.L. Knotek, V.O. Jones and V. Rehn, Phys. Rev. Lett. 43 (1979) 300.

[73] T.E. Madey, R.L. Stockbauer, D.E. Eastman and J.F. van der Veen, Phys. Rev. Lett. 45 (1980) 187.

[74] D.P. Woodruff, P.D. Johnson, M.M. Traum, H.H. Farrell, N.V. Smith, R.L. Benbow and Z. Hurych, Surf. Sci. 104 (1981) 282.

[75] D.M. Hanson, R.L. Stockbauer and T.E. Madey, Phys. Rev. B 24 (1981) 5513.

[76] R. Jaeger, J. Feldhaus, J. Haase, J. Stohr, Z. Hussain, D. Menzel and D. Norman, Phys. Rev. Lett. 45 (1980) 1870.

[77] M. Cini, Solid State Commun. 20 (1976) 605;
G.A. Sawatzky, Phys. Rev. Lett. 39 (1977) 504.

[78] D.E. Ramaker, J.M. White and J.S. Murday, J. Vac. Sci. Technol. 18 (1981) 748; Phys. Lett. A 89 (1982) 211.

[79] P.J. Feibelman, Surf. Sci. 102 (1981) L51.

[80] D.R. Jennison, J. Vac. Sci. Technol. 20 (1982) 548.

[81] P.R. Antoniewicz, Phys. Rev. B 21 (1980) 3811.

[82] J.W. Gadzuk, in: DIET-I (Ref. [9]) p. 4.

[83] A.R. Burns, Phys. Rev. Lett. 55 (1985) 525; J. Vac. Sci. Technol. A 4 (1986) 1499.

[84] P Nordlander, in: DIET-IV (Ref. [12]) p. 12.

[85] Ph. Avouris, R. Walkup, R. Kawai, D.M. News and N.D. Lang, in: DIET-IV (Ref. [12]) p. 144.

[86] M. Szymonski, in: DIET IV (Ref. [12]) p. 270.

[87] M.J. Dresser, M.D. Alvey and J.T. Yates, Jr., Surf. Sci. 169 (1986) 91.

[88] A.L. Johnson, R. Stockbauer, D. Barak and T.E. Madey, in: DIET-III (Ref. [11]) p. 130.

[89] A.L. Johnson, S.A. Joyce and T.E. Madey, Phys. Rev. Lett. 61 (1988) 2578.

[90] S. Auer, P. Feulner and D. Menzel, Phys. Rev. B 35 (1987) 7752.

[91] T.F. Heinz, M.M.T. Loy, J.A. Misewich, D.M. Newns and H. Zacharias, in: DIET-V (Ref. [13]).

Surface Science 299/300 (1994) 837–848
North-Holland

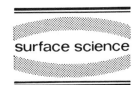

surface science

Catalysis: from single crystals to the "real world"

D. Wayne Goodman

Chemistry Department, Texas A&M University, College Station, TX 77843, USA

Received 30 April 1993; accepted for publication 29 June 1993

The "pressure and material gaps" separating UHV and technical catalytic investigations have been bridged in recent years by combining in a single apparatus the ability to measure kinetics at elevated pressures on single-crystal catalysts with the capabilities to carry out surface analytical measurements. In these high-pressure/surface analytical studies a well-defined, single-crystal plane is used to model a site or set of sites expected to exist on practical high-surface-area catalysts. This "surface science" approach has been used to study structure sensitivity, the effects of promoters/inhibitors on catalytic activity, and mixed-metal catalysis. In this article, some of the progress made during the last twenty years to integrate modern surface science into fundamental catalytic research will be reviewed, including recent efforts to simulate oxide-supported metal systems using thin oxide films.

1. Introduction

The ultrahigh vacuum (UHV) approach to the study of catalysis offers many advantages over traditional methods in that a variety of modern surface spectroscopies are now available to study in great detail the structure and composition of surfaces and to identify relevant surface species. However, a common criticism of the UHV approach has been that it is too far removed from reality, since catalytic reactions typically are carried out under quite different conditions, namely, at atmospheric (or higher) pressures and with far more complex surfaces.

These so-called "pressure and material gaps" separating UHV and technical catalytic investigations have been bridged in recent years by combining in a single apparatus the ability to measure kinetics at elevated pressures on single-crystal catalysts with the capabilities to carry out surface analytical measurements [1–6]. In these high-pressure/surface analytical studies a well-defined, single-crystal plane is used to model a site or set of sites expected to exist on practical high-surface-area catalysts. This "surface science" approach [6] has allowed a direct comparison of reaction rates measured on single-crystal surfaces with those measured on more realistic catalysts. These combined methods have also provided detailed studies of structure sensitivity, the effects of promoters/inhibitors on catalytic activity, and, in certain cases, the identification of reaction intermediates by post-reaction surface analysis. These kinds of studies have established the validity of using single-crystal surfaces as models for the more complex technical catalysts for a variety of reactions [6].

In this article, some of the progress made during the last twenty years to integrate modern surface science into fundamental catalytic research is reviewed. Specific examples from our work are used to demonstrate the advantages and relevance of single-crystal studies for modeling the behavior of high-surface-area-supported catalysts. Recent efforts to simulate oxide-supported metal systems using thin oxide films are also described. These thin oxide films offer a wide range of opportunities for investigating the more complex issues of catalysis such as metal–support interactions and quantum size effects in small metal particles using model catalysts that are realistic enough to simulate the nuances of the

SSDI 0039-6028(93)E0376-6

more complex "real world" system, yet tractable for typical surface science studies.

2. Clean metal single crystals: structure insensitivity

2.1. CO methanation

The reaction of CO with H_2 is an essential reaction in the production of synthetic natural gas from hydrogen-deficient carbonaceous materials. In addition, this reaction is an obvious starting point in studies of fuel and chemical synthesis from carbon sources. Because of the extensive investigations of the adsorption behavior of H_2 and CO on single-crystal surfaces in the surface science literature, the methanation reaction offers an attractive starting point for elevated pressure investigations using single-crystal catalysts and surface science techniques. Accordingly, this reaction has been the subject of many investigations beginning with monometallic surface studies and, more recently, with studies addressing the effects of surface additives such as poisons, promoters, and dissimilar metals.

For example, in the late seventies the steady-state specific methanation rates (methane molecules/site s) were measured on the Ni(111) and Ni(100) surfaces [7,8]. (This specific reaction rate is often referred to as a turnover frequency, TOF.) At a given temperature the rate of production of methane over an initially clean single-crystal catalyst was constant, with no apparent induction period. The kinetic data for the close-packed (111) and for the more open (100) crystal plane of Ni were strikingly similar with respect to the specific rates and activation energies. Furthermore, the single-crystal results were virtually identical to data acquired from alumina-supported nickel catalysts [7]. These extraordinary similarities in kinetic data taken under nearly identical conditions demonstrated unequivocally that there is no significant variation in the specific reaction rate or the activation energy in changing from a catalyst consisting of small metal particles to one composed of a bulk single-crystal. These studies provided convincing evidence that

the methanation reaction was indeed quite insensitive to the surface structure of nickel catalysts, and established the appropriateness of using single-crystals to model this structure insensitive reaction.

Having validated the model, post-reaction surface analysis provided unique insights into the details of the reaction mechanism. For example, Auger electron spectroscopy (AES) following the reaction showed a low level of a carbonaceous species and the absence of oxygen [7]. The Auger lineshape for the carbonaceous residue was similar to that of nickel carbide, indicating that the surface carbon was of a "carbidic" type [7]. Further experiments were carried out to study the interaction of CO with Ni(100) at different temperatures [9]. AES data showed the deposition of carbon on the surface and the absence of oxygen. Two kinds of carbon were formed on the surface from CO: a carbidic type identical to that found following a H_2/CO reaction, which formed at temperatures < 650 K, and a graphitic type, which formed at temperatures > 650 K. The carbidic type saturated at 0.5 monolayers and could be readily removed from the Ni(100) surface by heating the crystal to 600 K in 1 atm of H_2; methane was formed as product. In contrast, the graphitic type was completely unreactive toward hydrogen and led to deactivation of the catalyst. The deposition of an active carbon residue and the absence of oxygen on the nickel surface following heating in pure CO is consistent with the well-known disproportionation reaction, the Boudouard reaction,

$$2\ CO_{ads} \rightarrow C_{ads} + CO_2.$$

On Ni(100), the carbon formation data from CO disproportionation indicates a rate equivalent to that observed for methane formation in a H_2/CO mixture. Therefore, the surface carbon route to product was shown to be sufficiently rapid to account for all methane production [9].

Thus, the proposed reaction mechanism involving the dissociation of CO and the subsequent hydrogenation of the resulting carbon species (C_{ads}) accounts quite satisfactorily for the effect of pressure on the methanation rate, for the variation in the measured surface carbon level as reaction parameters are changed, and for the formation at characteristic temperature–pressure conditions of a catalyst-deactivating graphitic carbon [10,11].

Studies similar to the methanation studies outlined above for nickel have also been carried out on two faces of ruthenium: the zig-zag, open (110) and the close-packed (001) planes [12–14]. These studies confirm that CO methanation on ruthenium surfaces is structure insensitive as well, and occurs via the same mechanism as observed for nickel.

2.2. CO oxidation

The relative simplicity of CO oxidation makes this reaction an ideal model for a heterogeneous catalytic reaction. Each of the mechanistic steps (adsorption and desorption of the reactants, surface reaction, and desorption of products) has been studied extensively using surface science techniques [15]. Because many of the reaction parameters determined in UHV can be applied directly to the kinetics at higher pressures [16], CO oxidation to date best illustrates the continuity between UHV and elevated pressure kinetic studies.

The rates of CO oxidation measured on single-crystals of Rh have been measured and compared with those observed on supported Rh/Al$_2$O$_3$ catalysts [16,17]. The data show remarkable agreement between the model and supported systems with respect to the specific reaction rates and apparent activation energies. These results indicate that the kinetics of CO oxidation on Rh is insensitive to changes in catalyst surface morphology. Under the conditions of the reaction near stoichiometry, the surface of the catalyst is predominantly covered with CO so that the reaction is limited by the adsorption of oxygen [16,17]. As the temperature is increased, the reaction rate increases as more vacant sites become available for oxygen adsorption due to the higher CO desorption rate. Accordingly, the CO oxidation rate typically increases with temperature and exhibits an apparent activation energy very similar to that for CO desorption.

The behavior of the reaction has been ana-

lyzed by using a kinetic model established from UHV surface science studies of the interactions of CO and O$_2$ with Rh [16]. By using the rate constants for adsorption and desorption of CO and O$_2$ measured at UHV conditions, the kinetics of CO oxidation over Rh can be very accurately predicted at elevated pressures. The results of this model are virtually indistinguishable from the measured data, and thus represent an exceptional example of the continuity between UHV surface science and "real world" catalysis.

Further examples of the correspondence between model single crystals and supported metal catalysts have been found for Pt, Pd, Ir, and Ru [18,19]. In the case of Ru [19], post-reaction surface analysis indicates that the optimum rate of CO oxidation on Ru(001) is observed when the surface is covered by almost a monolayer of oxygen. This surface condition contrasts with that of Rh, where the optimum activity is obtained for surfaces essentially free of oxygen adatoms [17].

Although many questions remain unanswered regarding the origin of the so-called "structure insensitivity" of methanation and CO oxidation, the single-crystal results are impressive with respect to their ability to reproduce the detailed kinetic behaviors of the corresponding supported "real world" catalysts. Furthermore, the spectroscopic tools of surface science coupled with reaction kinetics on single-crystals have provided valuable new insights into the mechanisms of these reactions.

3. Clean metal single-crystals: structure sensitivity

3.1. Alkane reactive sticking on single-crystal surfaces

In contrast to methanation and CO oxidation, there are reactions whose reactivity and selectivity depend markedly on the surface geometry or the metallic particle size of the catalyst. The reactive sticking of alkanes as well as the hydrogenolysis or "cracking" of alkanes are important examples of "structure sensitive" reactions [20].

The dissociative sticking of methane has been seen to increase in the order Ni(111) < Ni(100) < Ni(110) [21]. Initial reaction rates for the Ni(110) and Ni(100) surfaces are very similar, and are ~ 7 to 10 times greater than the initial rate for the Ni(111) surface at 450 K. A comparison of these results with data of molecular beam studies suggests that the sticking probabilities of molecules with very low normal kinetic energies must be accurately known when attempting to model high-pressure processes using molecular beam techniques [21–23]. Furthermore, while dissociation of methane on Ni(111) and other close-packed transition-metal surfaces likely proceeds via the "direct" channel to dissociation, results for the reactive sticking of methane [24], ethane, propane, and butane on Ni(100) [22] show that dissociation of these molecules on this more "open" Ni surface proceeds primarily via a trapped molecular "precursor".

These so-called "bulb" experiments on alkane reactive sticking on nickel single-crystal surfaces were carried out at high incident flux conditions. Elevated pressures, in general, are required to produce measurable products, not because of the greater abundance of molecules with higher kinetic energies, but rather because of the competition that is inevitably present between desorption from the precursor, or accommodated adsorbed molecular state, and dissociation. Since activation energies for desorption of many reactants of interest (particularly saturated hydrocarbons) are usually considerably smaller than the activation energies for reaction, desorption dominates and reaction probabilities are quite small, often too small to measure at UHV conditions. For these reactants, the greater number of adsorption/desorption events at higher pressures simply serves too overcome this limitation. This behavior is common to many surface-catalyzed reactions and is the essence of the so-called "pressure gap" [25].

By extrapolating the rates of dissociative adsorption of alkanes measured in "bulb" experiments to the pressure and temperature conditions used in hydrogenolysis and steam reforming studies, it has been shown that alkane dissociation rates on the clean surface are one to two orders of magnitude larger than the rates of alkane hydrogenolysis and reforming under comparable conditions [21]. Since the alkane dissociation rates determined in the "bulb" experiments are initial rates at the limit of zero carbon coverage, these rates represent a theoretical upper limit to the rates of hydrogenolysis and steam reforming of these alkanes on unpromoted nickel catalysts.

The above studies have shown a direct relationship between the atomic corrugation of a surface and its activity toward the dissociative sticking of alkanes. Since dissociative sticking is the key initial step in the hydrogenolysis of alkanes, then a similar correlation between the atomic "roughness" and activity for hydrogenolysis reactions might be anticipated. This indeed is the case as discussed in the following section.

3.2. Hydrogenolysis of alkanes on single-crystal surfaces

The reactivity for the conversion of ethane to methane over nickel catalysts has been shown to depend critically on the particular geometry of the surface. The specific reaction rates for methane formation from ethane on the Ni(100) and Ni(111) surfaces clearly show that the more open (100) surface is far more active than the close-packed (111) surface [20]. For the Ni(100) surface, the data yield an activation energy of 100 kJ/mol. In contrast, the kinetic data for the Ni(111) surface correspond to an activation energy of 192 kJ/mol, implying that a different reaction mechanism is operative for Ni(111). It should be noted that surface analysis following reaction showed the (111) and (100) surfaces with comparable submonolayer quantities of carbon. Therefore, preferential surface carbon formation, or self-poisoning, on the (111) surface does not contribute to its lower activity [20].

The (111) surfaces are encountered more prevalently in fcc materials as the particle size is increased via successively higher annealing temperatures [20]. The results of this study then are consistent with rate measurements on supported nickel catalysts [20] which show hydrogenolysis activity to be a strong function of particle size, the larger particles exhibiting the lower rates.

The selectivity of ethane production from the hydrogenolysis of *n*-butane over iridium single-crystals has been demonstrated to scale with the concentration of low-coordination-number metal surface atoms [26,27]. The Ir(110)-(1 × 2) surface, which has a stable "missing-row" structure [26,27], has been found to produce ethane very selectively. This contrasts with the results for the close-packed Ir(111) surface, where only the statistical scission of C–C bonds has been observed. The results of this study correlate qualitatively with the observations made previously for selective hydrogenolysis of *n*-butane to ethane on supported iridium catalysts as a function of iridium particle size [28]. The results for Ir(110)-(1 × 2) model very well the small-particle limit, whereas the results for Ir(111) relate more closely to the data for the corresponding large particles (> 10 nm). By assuming particle shapes the general behavior of declining selectivity with larger particle size could be modeled accurately.

The stoichiometry of the surface intermediate leading to high ethane selectivity, based on kinetics and surface carbon coverages subsequent to reaction, is suggested to be a metallocyclopentane [26,27]. The Ir(110)-(1 × 2) or "missing-row" structure results in rows of highly coordinatively unsaturated "C_7" sites. These sterically unhindered C_7 sites can form a metallocyclopentane species (e.g., a 1,4-diadsorbed hydrocarbon species) which has been proposed as an intermediate in the central scission of butane to ethane [29,30]. Based on analogous chemistry reported in the organometallic literature [29,30], the mechanism responsible for the hydrogenolysis of *n*-butane on the Ir(110)-(1 × 2) surface is postulated to be the reversible cleavage of the central C–C bond in this metallocyclopentane intermediate. On the other hand, butane hydrogenolysis on the Ir(111) surface appears to operate via a different mechanism. First, dissociative chemisorption of butane and hydrogen occurs followed by irreversible cleavage of the terminal carbon–carbon bond of the adsorbed hydrocarbon. Further C–C bond cleavage prior to product desorp-

tion leads to the methane and ethane observed as initial products.

Although many of the atomic-level details with respect to the origin of the "structure sensitivity" of alkane hydrogenolysis still remain unresolved, this work shows clearly that a large component of this important catalytic effect is a structural one and relates to the partitioning of particular reaction sites or to facet re-distribution as a function of particle size. Most importantly, the results summarized above correlate very well with measurements on supported "real-world" catalysts and emphasize the important role that single-crystal studies can play in defining relationships between surface structure and catalytic activity/selectivity.

4. Chemically modified surfaces: poisons and promoters

The addition of impurities to a metal catalyst can produce dramatic changes in the activity, selectivity and resistance to poisoning of the catalyst. For example, the selectivity of some transition metals can be altered greatly by the addition of additives such as potassium, and the activity can be reduced substantially by the addition of electronegative species such as sulfur. Although these effects are well recognized in the catalytic industry, the mechanisms responsible for chemical changes induced by surface additives are poorly understood. An important question concerns the underlying relative importance of ensemble (steric or local) versus electronic (non-local or extended) effects. A general answer to these questions will enhance our ability to improve existing catalysts and perhaps lead to the design of more efficient catalytic materials.

4.1. Electronegative impurities

Kinetic studies [31–33] have been carried out for CO methanation over Ni(100) surfaces covered with sulfur and phosphorus impurities. The rates of CO methanation as a function of sulfur and phosphorus coverage over a Ni(100) catalyst show a non-linear relationship between the sulfur coverage and the methanation rate. A sharp decrease in catalytic activity is observed at low sulfur coverages, and the poisoning effects of the sulfur are rapidly maximized. Such is the case for the attenuation of the methanation activity by sulfur for alumina-supported nickel catalysts [34]. The initial attenuation of the catalytic activity by sulfur suggests that ten or more equivalent nickel sites are deactivated by a sulfur atom. These results can be interpreted in either of two ways: (1) an electronic effect that extends to the next-nearest-neighbor sites; or (2) an ensemble effect that requires a certain number of surface atoms to facilitate the reaction sequence. If extended electron effects are significant, then the reaction rate is expected to be a function of the relative electronegativity of the poison. On the other hand, if an ensemble of ten nickel atoms is required for the critical steps of methanation, then altering the electronegative character of the poison should produce little change in the attenuation of activity by the additive.

Substituting phosphorus for sulfur results in a marked change in the magnitude of the poisoning effect at low coverages. Phosphorus, because of its reduced electronegative character, is much less effective as a poison and influences only the four nearest-neighbor nickel atom sites. These results then support the argument that extended electronic effects, rather than ensemble or site-blocking effects, are dominant in catalytic deactivation by sulfur. Similar poisoning effects by sulfur have been observed for methanation over ruthenium and rhodium single-crystal catalysts [35,36].

4.2. Electropositive impurities

The role of electronegative impurities in poisoning Ni(100), Ru(001), and Rh(111) toward methanation activity has been discussed above. These results have been ascribed as arising, to a large extent, from an electronic effect. In the context of this interpretation it is expected that an electropositive impurity might have the opposite effect, i.e. to increase the methanation activity of a metal surface. A study of CO hydrogenation over potassium-covered Ni(100) [33,37] has

shown that this is not the case, although certain steps in the reaction mechanism are strongly accelerated by the presence of the electropositive impurity.

Kinetic measurements of CO methanation over a Ni(100) catalyst containing submonolayer quantities of potassium adatoms [37] indicate a decrease in the steady-state rate of methanation with an increase in the potassium coverage. The presence of potassium does not alter the apparent activation energy associated with the kinetics; however, potassium does change the steady-state coverage of active carbon on the catalyst. This carbon level changed from 10% of a monolayer on the clean catalyst to 30% of a monolayer for a catalyst covered with 0.1 monolayers of potassium [37].

Adsorbed potassium caused a marked increase in the steady-state rate and selectivity of Ni(100) for higher hydrocarbon synthesis [37]. At all the temperatures studied, the overall rate of higher hydrocarbon production was faster on the potassium-dosed surfaces, so that potassium may be considered a true promoter with respect to this reaction, Fischer–Tropsch synthesis. The effects of potassium upon the kinetics of CO hydrogenation over Ni(100) (i.e. a decrease in the rate of methane formation and an increase in the rate of higher hydrocarbon production) are similar to those reported for high-surface-area supported nickel catalysts [38,39]. This agreement between single-crystal nickel and supported nickel indicates that the major mechanism by which potassium additives alter the activity and selectivity of industrial catalysts is not related to the support material, but that it is rather a consequence of direct potassium–nickel interactions.

Adsorbed potassium causes a marked increase in the rate of CO dissociation on a Ni(100) catalyst [37]. There is a dramatic increase in the initial formation rate of "active" carbon or carbidic carbon via CO disproportionation as a function of potassium coverage. The relative rates of CO dissociation were determined for the clean and potassium-covered surfaces by observing the growth in the carbon Auger signal with time in a CO reaction mixture, starting from a carbon-free surface. The presence of potassium adatoms leads to a reduction of the activation energy of reactive carbon formation from 96 kJ/mol on clean Ni(100) to 42 kJ/mol on a 10% potassium-covered surface [37].

These studies of promoters and poisons on surfaces illustrate the suitability of single-crystal model catalysts for studying the effects of surface modifiers in catalysis. Further work addressing these important issues will most surely aid in the understanding of this particular important aspect of practical catalysis.

5. Alloy surfaces: bimetallic catalysts

Considerable effort has been expended in recent years to address the chemical and physical properties of mixed-metal solids. This interest, to a large extent, has been motivated by the extensive technological applications that mixed-metal systems have in catalysis. The two basic questions in these studies are: (i) what is the nature of the heteronuclear metal–metal bond?; (ii) how does the formation of this bond affect the physical and chemical properties of metals? The answers to these questions are a challenge to modern science and a prerequisite for a non-empirical design of multi-metallic catalysts for industrial applications.

For solid metals, a reduction in the atomic coordination number produces a narrowing of the valence band at the surface. As a consequence, charge must flow between the surface atoms and the bulk so that the composite system maintains a common Fermi level [40–42]. This phenomenon suggests that the properties of a bond in a bimetallic surface can be very different from those of the corresponding bond in a 3D alloy, stressing the need to investigate the nature of the surface metal–metal bond.

In the study of surface metal–metal bonds, it has been advantageous to use model bimetallic systems generated by vapor depositing one metal onto a crystal face of a second metal [43–45]. These well-defined bimetallic surfaces offer the possibility of correlating electronic and chemical properties of a system with atomic-level surface structure. The results obtained by using these well-defined models have significantly altered the way in which the metal–metal bond in bimetallic surfaces is viewed.

For example, X-ray and ultraviolet photoelectron spectroscopies (XPS and UPS) have been extensively used to investigate the core and valence levels of transition metal films supported on dissimilar transition-metal substrates [46–56]. The results of these studies show quite clearly that formation of a heteronuclear metal–metal bond can induce large changes in the electron density about a metal. These modifications in the electronic structure, in turn, affect the cohesive energy of the bimetallic bond.

Furthermore, the results of numerous studies dealing with the chemisorption of CO on well-defined bimetallic surfaces indicate that the electronic perturbations described above also modify the chemical properties of the metal overlayers. A striking correlation between changes in the CO desorption temperatures and the relative shifts in the surface core-level binding energies for supported monolayers of Pd, Ni, and Cu has been observed recently [43]. Strong electron donor–electron acceptor interactions in bimetallic bonding in these systems deactivate Pd and Ni adatoms toward CO chemisorption, whereas the same type of phenomena activate Cu adatoms. Bimetallic surfaces with the strongest Pd–substrate bonds have the weakest Pd–CO bonds. In contrast, surfaces with the strongest Cu–substrate bonds show also the strongest Cu–CO bonds [43].

In many respects, the behavior seen for the 2D metal overlayers is different from that expected for bulk metals. Results of X-ray [43] and ultraviolet [57] photoemission, work function measurements [58,59], and CO chemisorption [43,60], for pseudomorphic monolayers of Ni and Pd on W(110) show that the surface electronegativity of Pd is much lower than that of Ni. This trend is contrary to that found in several scales of bulk electronegativities [61], where Pd is more electronegative than Ni. Data reported for monolayers of Ni, Cu, and Pd on Ru(0001) [48,49,53], also indicate a sequence of surface electronegativities (Pd < Ni < Cu) opposite to that found for bulk electronegativities [61].

The experimental evidence mentioned above indicates that the nature of a metal–metal bond in a bimetallic surface is very different from that of the corresponding bond in a bulk 3D alloy. Formation of a surface metal–metal bond produces a flow of electron density toward the element with the larger fraction of empty states in its valence band. The resulting electronic modification of the interacting metals can dramatically alter the chemical (catalytic) properties of the bimetallic system [43].

The effects on catalytic activity of adding one metal to a second one has been illustrated recently via the addition of Cu to Rh(100) with respect to CO oxidation [62]. The overall rate of the reaction was observed to increase by an order of magnitude above the rate for either rhodium or copper at a Cu coverage of 1.3 monolayers. Above this coverage the rate decreases to an activity approximately equal to that of Cu-free Rh(100), a result caused by the 3D clustering of the Cu multilayers. The rate enhancement observed for submonolayer Cu deposits likely is related, to some extent, to the electronic modification of the Cu overlayer due to its interaction with the Rh(100) substrate. The interaction of overlayer Cu with Rh(100) is known to significantly increase the bonding of CO with the overlayer Cu [63], a result consistent with the enhanced catalytic properties of the overlayer for CO oxidation.

These results for mixed-metal systems have provided new information regarding the nature of the metal–metal bonding in mixed-metal interfaces (alloy catalysts). Although many subtleties of the surface chemistry of alloys remain to be addressed, model studies using surface science methods offer a new approach for elucidating many of these issues.

6. Oxide surfaces

Insulating surfaces such as those frequently encountered as catalysts or catalyst supports present problems of varying degrees to many charged-particle surface probes. For example, high-resolution electron-energy-loss spectroscopy (HREELS) is a principle technique for surface vibrational studies. However, this technique has been restricted to conducting substrates (metals and semiconductors) because of surface-charging problems encountered with insulators. Only limited studies have been carried out on highly insulating surfaces, e.g. MgO(100) [64] and Al_2O_3 (0001) [65]. In some cases, the effects of surface charging during charged-particle measurements have been compensated or stabilized with the aid of a neutralization electron gun [64,66].

Recently, the difficulty associated with surface charging has been eliminated by preparing an ultrathin, highly ordered, oxide film on the surface of a metal substrate [67]. Any charging induced in the thin film during charged-particle measurements is dissipated via the conducting substrate.

In the application of HREELS to adsorbates on ionic substrates, a second difficulty encountered is that the accompanying vibrational spectra are dominated by losses due to excitation of surface optical phonons. Since the intense multiple phonon losses generally extend over a wide vibrations frequency range of the HREELS spectra, it is not practical to observe directly adsorbate losses (which are several orders of magnitude smaller in intensity than the phonon losses) in the 0–4000 cm^{-1} spectral range.

In recent studies [68–70], a new approach has been developed to acquiring HREELS data in order to circumvent the difficulties associated with these phonon losses. By utilizing a high-energy incident electron beam in combination with an off-specular scattering geometry, this new approach enables the direct observation of weak loss features due to the excitation of adsorbates without serious interference from intense multiple surface optical phonon losses. HREELS data indicate that carboxylic acids, methanol, and water undergo heterolytic dissociation, whereas ethylene and ethane are found to adsorb associatively on MgO. However, an increase in the surface basicity of MgO, achieved by thermal treatment, results in the dissociation of ethane. Further studies [71–73] on NiO show this surface to be far less basic with respect to the dissociation of various probe molecules with acid strengths ranging from those of carboxylic acids and alcohols to alkenes and alkanes.

Very recent work [74,75] has addressed the preparation and characterization of ultrathin silicon dioxide films on a Mo(110) surface. The SiO$_2$ films following an anneal to 1400 K have properties very similar to vitreous silica as shown by reflection–absorption infrared spectroscopy (RAIRS) [74]. Because silica supports are so widely utilized in practical catalytic applications, these SiO$_2$ films offer an interesting starting material for constructing a more realistic supported-metal-particle catalyst, intermediate in complexity between metal single-crystals and the supported "real world" catalysts. Examples of model silica-supported metal catalysts are described in the following section.

These thin film oxide preparations can serve as convenient model catalysts as well. For example, recent work [76–79] has demonstrated their utility in studying the oxidative methane coupling reaction:

$$4 \, CH_4 + O_2 \xrightarrow[\sim 1000 \, K]{Li/MgO} 2 \, C_2H_6 + H_2O.$$

The kinetics of this reaction have been measured on thin MgO(100) films doped with Li. These studies show an excellent correlation between the absolute rates and activation energies of the model thin films and the corresponding rates and activation energies observed for the "real world" Li/MgO powdered catalysts [77]. Furthermore, electron energy loss studies have demonstrated a correlation between the density of F-centers at the Li/MgO surface (oxygen defect with a trapped electron pair) and the reactivity of that surface toward the above methane coupling reaction. This correlation suggests a new mechanism by which this class of materials can activate methane, the critical step in this potentially important methane coupling reaction.

These initial studies for model MgO, NiO, and SiO$_2$ catalysts are very promising with respect to the future possibilities of modeling with thin oxide films a variety of adsorption and catalytic processes that take place over oxide surfaces.

7. Metal / oxide surfaces

Recently, model silica-supported copper [80,81] and palladium [82] catalysts have been prepared by evaporating the appropriate metal onto a silica thin film. The preparation conditions define the corresponding metal-particle dispersions or average size [81]. As in the studies described in the previous section, the silica films are supported on a Mo(110) substrate [74].

The structure of the model silica-supported copper catalysts has been investigated with RAIRS and scanning tunneling microscopy (STM). The RAIRS studies of adsorbed CO indicate that there are several types of copper clusters with surface structures similar to (111), (110), and other high-index planes of single-crystal copper [80]. The STM studies show several types of copper clusters on silica and reveal images of metal clusters on the amorphous support with atomic resolution [80].

The adsorption and reaction of CO on model silica-supported palladium catalysts over a wide range of temperatures and pressures have demonstrated a continuity between the catalytic chemistry of single-crystals and small particles, as well as between the kinetics of CO oxidation at low and high pressures [82].

These kinds of model-supported catalytic studies allow the so-called "pressure and material gaps" to be bridged simultaneously. Further studies of this nature offer unprecedented opportunities to connect in a direct way studies on single-crystals at UHV conditions with analogous investigations of catalytic processes on "real world" catalysts at elevated pressures.

8. Conclusions and future prospects

An approach that combines ultrahigh vacuum surface analytical methods with an elevated pressure reactor can provide new information about the molecular details that define and control the mechanism of reactions at the gas/solid interface. Using these techniques, basic concepts of heterogeneous catalysis such as structure insensitivity and structure sensitivity can be directly addressed. For structure insensitive reactions excellent agreement can be obtained between studies on single-crystal surfaces and studies on the corresponding high-surface-area-supported catalysts, demonstrating the relevance of kinetics measured on well-ordered single-crystal surfaces for modeling the behavior of practical catalysts. For structure sensitive reactions, the activity of a particular site or set of sites can be examined and the effects of surface structure explored in atomic detail.

Model studies on single-crystal surfaces are also helpful in developing a better understanding of the effects of surface impurities, e.g. poisons and promoters, on the catalytic activity and selectivity of metals. The influence of modifiers on the surface chemistry of adsorbed reactants, products, and intermediates can be studied using UHV techniques. Such investigations have shown that this information can be related to the effects of the impurities on the catalytic behavior. Of particular interest is the possibility that these kinds of studies will help to clarify the relative importance of electronic and geometric contributions in determining the role of surface modifiers.

Combined surface science and kinetic studies have probed the nature of the metal–metal bond in mixed-metal catalysts. These investigations indicate that charge transfer is an important component in surface metal–metal bonds that involve dissimilar elements. The larger the charge transfer, the stronger the cohesive energy of the bimetallic bond and the greater the perturbation of the chemical properties of the interacting metals.

Recent studies have shown that model oxide surfaces can be prepared in thin-film form, a preparation that readily enables their exploration with a wide array of charged-particle surface techniques. Techniques have now been developed to facilitate the application of high-resolution electron-energy-loss spectroscopy to adsorbates on oxides directly. These kinds of investigations offer unprecedented opportunities to address the molecular details of the chemistry at oxide surfaces.

Finally, the addition of metals to the above oxides as supports provides a convenient method

to model important aspects of supported-metal catalysts such as support–particle interactions and quantum size effects of supported-metal particles. Using such models a host of surface science techniques can be utilized to study catalysis by metals in systems with well-defined particle sizes and morphology.

Clearly great progress has been made during the last twenty years with respect to establishing continuity between the world of surface science and the "real world" of catalysis. These two disciplines are complementary, each playing an essential role in guiding the inquiry of the other. Cooperation and synergism between these two disciplines most surely will continue to lead to new insights into the amazingly complex world of heterogeneous catalysis.

Acknowledgements

We acknowledge with pleasure the support of this work by: the Department of Energy, Office of Basic Sciences, Division of Chemical Sciences; the Gas Research Institute; and the Robert A. Welch Foundation.

References

[1] D.R. Kahn, E.E. Petersen and G.A. Somorjai, J. Catal. 34 (1974) 294.
[2] B.A. Sexton and G.A. Somorjai, J. Catal. 46 (1977) 167.
[3] D.W. Goodman, R.D. Kelley, T.E. Madey and J.T. Yates, Jr., Proc. of Symposium on Advances in Fischer–Tropsch Chemistry (American Chemical Society, Anaheim, CA, 1978).
[4] H.P. Bonzel and H.J. Krebs, Surf. Sci. 91 (1980) 499.
[5] C.T. Campbell and M.T. Paffett, Surf. Sci. 139 (1984) 396.
[6] J.A. Rodriguez and D.W. Goodman, Surf. Sci. Rep. 14 (1991) 1.
[7] D.W. Goodman, R.D. Kelley, T.E. Madey and J.T. Yates, Jr., J. Catal. 63 (1980) 226.
[8] D.W. Goodman, Annu. Rev. Phys. Chem. 37 (1986) 425.
[9] D.W. Goodman, R.D. Kelley, T.E. Madey and J.T. Yates, Jr., J. Catal. 64 (1980) 479.
[10] R.D. Kelley and D.W. Goodman, in: The Chemical Physics of Solid Surfaces and Heterogeneous Catalysis, Vol. IV, Eds. D.A. King and D.P. Woodruff (Elsevier, Amsterdam, 1980).

[11] D.W. Goodman, Acc. Chem. Res. 12 (1984) 194.

[12] D.W. Goodman and J.M. White, Surf. Sci. 90 (1979) 201.

[13] R.D. Kelley and D.W. Goodman, Surf. Sci. 123 (1982) L743.

[14] D.W. Goodman, J. Vac. Sci. Technol. 20 (1982) 522.

[15] T. Engel and G. Ertl, The Chemical Physics of Solid Surfaces and Heterogeneous Catalysis, Vol. 4, Eds. D.A. King and D.P. Woodruff (Elsevier Amsterdam, 1982).

[16] S.H. Oh, G.B. Fischer, J.E. Carpenter and D.W. Goodman, J. Catal. 100 (1986) 360.

[17] D.W. Goodman and C.H.F. Peden, J. Phys. Chem. 90 (1986) 4839.

[18] P.J. Berlowitz, C.H.F. Peden and D.W. Goodman, J. Phys. Chem. 92 (1988) 5213.

[19] C.H.F. Peden and D.W. Goodman, J. Phys. Chem. 90 (1986) 1360.

[20] D.W. Goodman, Surf. Sci. 123 (1982) L679.

[21] T.P. Beebe, Jr., D.W. Goodman, B.D. Kay and J.T. Yates, Jr., J. Chem. Phys. 87 (1987) 2305.

[22] A.G. Sault and D.W. Goodman, J. Chem. Phys. 88 (1988) 7232.

[23] X. Jiang and D.W. Goodman, Catal. Lett. 4 (1990) 173.

[24] R.A. Campbell, J. Szanyi, P. Lenz and D.W. Goodman, Catal. Lett. 17 (1993) 39.

[25] D.W. Goodman, Catal. Today 12 (1992) 189.

[26] J.R. Engstrom, D.W. Goodman and W.H. Weinberg, J. Am. Chem. Soc. 108 (1986) 4653.

[27] J.R. Engstrom, D.W. Goodman and W.H. Weinberg, J. Am. Chem. Soc. 110 (1988) 8305.

[28] K. Foger and J.R. Anderson, J. Catal. 59 (1979) 325.

[29] R.H. Grubbs and A. Miyashita, J. Am. Chem. Soc. 100 (1978) 1300.

[30] R.H. Grubbs, A. Miyashita, M. Liu and P. Burk, J. Am. Chem. Soc. 100 (1978) 2418.

[31] D.W. Goodman and M. Kiskinova, Surf. Sci. 105 (1981) L265.

[32] M. Kiskinova and D.W. Goodman, Surf. Sci. 108 (1981) 64.

[33] D.W. Goodman, Appl. Surf. Sci. 19 (1984) 1.

[34] J.R. Rostrup-Nielsen and K. Pedersen, J. Catal. 59 (1979) 395.

[35] D.W. Goodman, in: Heterogeneous Catalysis, IUCCP Conf. (Texas A&M University, College Station, TX, 1984).

[36] C.H.F. Peden and D.W. Goodman, Proceedings of the Symposium on the Surface Science of Catalysis, Eds. M.L. Devinny and J.L. Gland, ACS Symp. Ser. 288 (American Chemical Society, Washington, DC, 1984) p. 185.

[37] C.T. Campbell and D.W. Goodman, Surf. Sci. 123 (1982) 413.

[38] G.A. Mills and F.W. Steffgen, Catal. Rev. 8 (1973) 159.

[39] P. Schoubye, J. Catal. 14 (1969) 238.

[40] W.F. Egelhoff, Surf. Sci. Rep. 6 (1987) 253.

[41] D.E. Eastman, F.J. Himpsel and J.F. van der Veen, J. Vac. Sci. Technol. 20 (1982) 609.

[42] A. Zangwill, Physics at Surfaces (Cambridge University Press, New York, 1988) ch. 4.

[43] J.A. Rodriguez and D.W. Goodman, Science 257 (1992) 897.

[44] E. Bauer, in: The Chemical Physics of Solid Surfaces and Heterogeneous Catalysis, Vol. 3, Eds. D.A. King and D.P. Woodruff (Elsevier, Amsterdam, 1984).

[45] C.T. Campbell, Annu. Rev. Phys. Chem. 41 (1990) 775.

[46] M.W. Ruckman, V. Murgai and M. Strongin, Phys. Rev. B 34 (1986) 6759.

[47] W.K. Kuhn, R.A. Campbell and D.W. Goodman, J. Phys. Chem. 97 (1993) 446.

[48] J.A. Rodriguez, R.A. Campbell and D.W. Goodman, J. Phys. Chem. 95 (1991) 5716.

[49] R.A. Campbell, J.A. Rodriguez and D.W. Goodman, Surf. Sci. 256 (1991) 272.

[50] R.A. Campbell, J.A. Rodriguez and D.W. Goodman, Surf. Sci. 240 (1990) 71.

[51] J.A. Rodriguez, R.A. Campbell and D.W. Goodman, J. Vac. Sci. Technol. A 10 (1992) 2540.

[52] R.A. Campbell, J.A. Rodriguez and D.W. Goodman, Phys. Rev. B 46 (1992) 7077.

[53] J.A. Rodriguez, R.A. Campbell and D.W. Goodman, J. Phys. Chem. 95 (1991) 2477.

[54] J.A. Rodriguez, R.A. Campbell, J.S. Corneille and D.W. Goodman, Chem. Phys. Lett. 180 (1991) 139.

[55] M.L. Shek, P.M. Stefan, I. Lindau and W.E. Spicer, Phys. Rev. B 27 (1983) 7277; 7288.

[56] J.E. Houston, C.H.F. Peden, P.J. Feibelman and D.R. Hamann, Surf. Sci. 192 (1987) 457.

[57] G.W. Graham, J. Vac. Sci. Technol. A 4 (1986) 760.

[58] W. Schlenk and E. Bauer, Surf. Sci. 93 (1980) 9.

[59] J. Kolaczkiewicz and E. Bauer, Surf. Sci. 144 (1984) 495.

[60] P.J. Berlowitz and D.W. Goodman, Langmuir 4 (1988) 1091.

[61] R.E. Watson, L.J. Swartzentruber and L.H. Bennett, Phys. Rev. B 24 (1981) 6211.

[62] J. Szanyi and D.W. Goodman, Catal. Lett. 14 (1992) 27.

[63] X. Jiang and D.W. Goodman, Surf. Sci. 255 (1981) 1.

[64] P.A. Thiry, M. Liehr, J.J. Pireaux and R. Caudano, Phys. Rev. B 29 (1984) 4824.

[65] M. Liehr, P.A. Thiry, J.J. Pireaux and R. Caudano, J. Vac. Sci. Technol. A 2 (1986) 1079.

[66] M. Liehr, P.A. Thiry, J.J. Pireaux and R. Caudano, Phys. Rev. B 33 (1986) 5682.

[67] M.-C. Wu, J.S. Corneille, C.A. Estrada, J.-W. He and D.W. Goodman, Chem. Phys. Lett. 182 (1991) 472.

[68] M.-C. Wu, C.A. Estrada, D.W. Goodman, Phys. Rev. Lett. 67 (1991) 2910.

[69] M.-C. Wu and D.W. Goodman, Catal. Lett. 15 (1992) 1.

[70] M.-C. Wu, C.A. Estrada, J.S. Corneille and D.W. Goodman, J. Chem. Phys. 96 (1992) 3892.

[71] C.M. Truong, M.-C. Wu and D.W. Goodman, J. Chem. Phys. 97 (1993) 9447.

[72] M.-C. Wu, C.M. Truong and D.W. Goodman, J. Phys. Chem. 97 (1993) 4182.

[73] C.M. Truong, M.-C. Wu and D.W. Goodman, J. Am. Chem. Soc. 115 (1993) 3647.

[74] X. Xu and D.W. Goodman, Appl. Phys. Lett. 61 (1992) 774.

[75] X. Xu and D.W. Goodman, Surf. Sci. 282 (1993) 323.

[76] M.-C. Wu, C.M. Truong, K. Coulter and D.W. Goodman, J. Am. Chem. Soc. 114 (1992) 7565.

[77] K. Coulter and D.W. Goodman, Catal. Lett. 16 (1992) 191.

[78] M.-C. Wu, C.M. Truong and D.W. Goodman, Phys. Rev. B 46 (1992) 12688.

[79] M.-C. Wu, C.M. Truong, K. Coulter and D.W. Goodman, J. Catal. 140 (1993) 344.

[80] X. Xu, S.M. Vesecky and D.W. Goodman, Science 258 (1992) 788.

[81] X. Xu, J.-W. He and D.W. Goodman, Surf. Sci. 284 (1993) 103.

[82] X. Xu and D.W. Goodman, J. Phys. Chem. 97 (1993) 7711.

Surface Science 299/300 (1994) 849–866
North-Holland

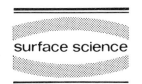

surface science

The surface science of heterogeneous catalysis

G.A. Somorjai

Department of Chemistry and Materials Sciences Division, Lawrence Berkeley Laboratory, University of California, Berkeley, CA 94720, USA

Received 12 April 1993; accepted for publication 9 June 1993

The development of a high pressure cell that can be incorporated into ultrahigh vacuum chambers permitted surface science studies of active catalysts. Model catalysts, usually single-crystal surfaces of ~ 1 cm^2 surface area, can be used to study important catalytic reactions ranging from hydrocarbon conversion over platinum to ammonia synthesis over iron. The new concepts, or phenomena, uncovered by the surface science investigations of heterogeneous catalysis of these model systems include: (1) high reaction rates and strong bonding at defects and on rough surfaces; (2) adsorbate-induced restructuring of metal surfaces that control the reaction rates; (3) the presence of a strongly chemisorbed overlayer which covers much of the active metal surface; (4) the bonding and reaction rate modifying influence of coadsorbates; and (5) the unique activity of certain oxide–metal interfaces.

1. Introduction

Heterogeneous catalysis serves as the basis for most chemical and petroleum technologies, and its importance to the life sciences and environmental protection is immense. The development of modern surface science provided the opportunity to investigate catalysts on the atomic scale and to understand the molecular ingredients that make them function. In the early 1960's, before the application of surface science to the study of heterogeneous catalysis, catalysts were viewed as a "black box" that mysteriously converted entering reactant molecules to desirable products. In the 1990's, catalyst design based on the understanding obtained by the previous 30 years of surface science studies is the norm rather than the exception.

How did this development come about? I dreamed of understanding surface catalysis on the molecular level from my earliest days as a graduate student in chemistry. In the early 1960's, the technique of low energy electron diffraction (LEED) was revolutionized by the introduction of the post-acceleration method. I saw this as an opportunity to study chemical processes, and eventually catalysis, on well-characterized single-crystal surfaces. When I arrived in Berkeley in the summer of 1964 to start my academic career as an Assistant Professor, I decided to focus my surface science research on platinum, the "father" of all catalysts (its catalytic action for the reaction $H_2 + \frac{1}{2}O_2 \rightarrow H_2O$ was discovered by Dobereiner in 1823, and was further explored by Faraday). Although my first paper in *Surface Science* [1] was published in the very first issue of the journal on a different topic of research, by 1965 I could report on the reconstruction of the (100) crystal face of platinum [2,3] and, thus, on the discovery of metal reconstruction (the reconstruction of semiconductor (Si, Ge) surfaces was known by that time).

Early chemisorption studies in my laboratory, using flat (111) crystal faces of platinum, showed very little chemical reactivity at low reactant pressures ($\sim 10^{-7}$ Torr) [4,5]. In order to study surface reactions at low pressures we developed a reactive molecular beam–surface scattering apparatus [6,7] with vacuum conditions that allowed clean single-crystal surfaces to be used as scattering targets (Fig. 1). Our early studies of reactive scattering from platinum crystal surfaces were unsuccessful [8]. Using mixed molecular beams of H_2 and D_2 we could not detect the exchange

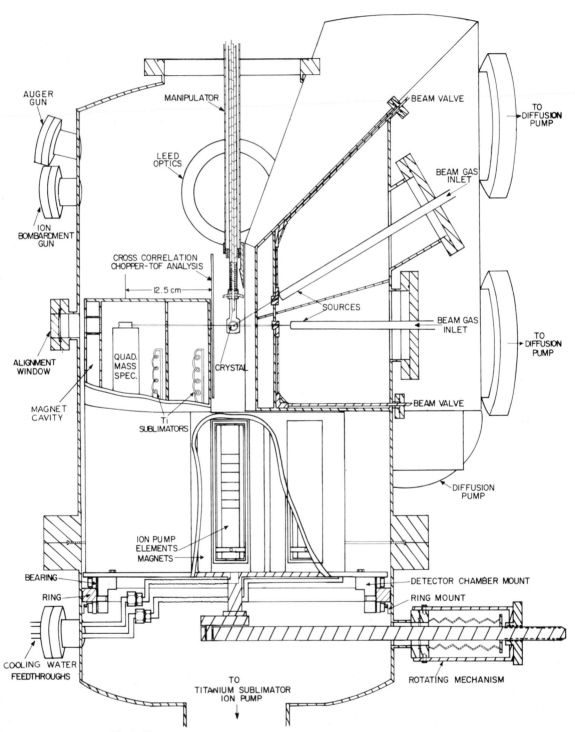

Fig. 1. The reactive molecular beam–single-crystal surface scattering apparatus.

reaction that produced HD, even though platinum catalysts were known to carry out this reaction to equilibrium below room temperature [9]. The lack of reactivity shown by chemisorption and molecular beam–surface scattering studies had to be the result of either the surface structure of the platinum crystals we used or the low pressures employed. Therefore, around 1971 we started to use high Miller index stepped (or kinked) surfaces of platinum in our surface studies (Fig. 2). Immediately, the dissociation or decomposition of chemisorbed molecules [10–12], along with the H_2–D_2 exchange, were readily observable through molecular beam–surface scattering studies [13]. Thus, the first key finding of low pressure surface science studies was the importance of surface irregularities (steps and kinks) in breaking and rearranging the chemical bonds of adsorbates.

In the mid 1970's, the major advance in instrumentation which brought surface science into the mainstream of catalysis research was our design of a high pressure cell that could be incorporated into UHV chambers utilized for surface science studies [14]. This high pressure environmental cell (Fig. 3) allowed us to perform catalytic reaction kinetics studies on small area (~ 1 cm^2) single-crystal surfaces under conditions identical to those in catalytic reactors used in the chemical technologies. (These conditions are 1–100 atm of reactant gases, 300–1000 K temperature range, using variable flow rates or batch reactor conditions.) The model catalyst was a single crystal that was properly cleaned and characterized in UHV by surface science techniques (ion sputtering, LEED, AES, XPS, thermal desorption, etc.) before and after the reactions. The appearance of the products and the disappearance of the reactants could be readily monitored by gas chromatography or a quadrupole mass spectrometer connected to the reaction chamber through a small leak. With proper design of the reaction cell, reaction rates in the range of 10^{-4}–10^2 molecules cm^{-2} s^{-1} could be monitored readily on such model single-crystal catalysts. Thus, the fields of model catalyst studies and the surface science approach to the investigation of heterogeneous catalysis were born. At the International

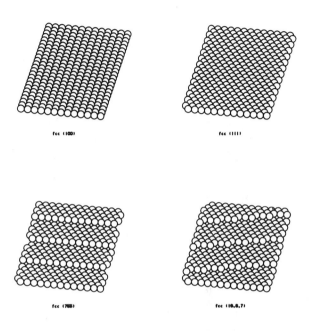

Preparation of Ordered Metal Oxide Thin Films

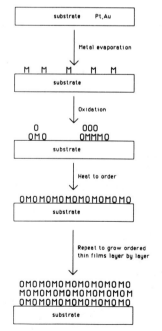

Fig. 2. Model catalyst surfaces; single crystals and thin films grown by atomic layer epitaxy.

Congress of Catalysis, held during the summer of 1992, or at the North American Catalysis Society meetings of recent years, roughly one-quarter of the scientific papers were dedicated to the surface science of heterogeneous catalysis.

Many designs of the high pressure–low pressure apparatus for combined surface science and catalytic studies became available [15,16]. The two major design types were: (a) those with a separate reaction cell, where the model catalyst could be shuttled between the reactor and the UHV system using a differentially pumped transfer rod; and, (b) those with a reaction cell built into the UHV chamber. In this latter design the model catalyst sample remained stationary while a stainless steel tubular reactor was pulled over the sample using a hydraulic press to engage a

copper gasket for sealing the space around the sample. The tube, so closed, could then be pressurized for the catalytic study to commence. In recent designs, the reactor, once closed, can also be filled with liquids so one can carry out catalytic reaction rate studies in the liquid phase [17].

There are many advantages to carrying out reaction rate studies on low surface area model catalysts. The kinetic parameters can be readily correlated with changes of surface structure and surface composition: reaction rates, deactivation times and activation energies can be directly compared with the behavior of high surface area microporous catalysts at the same time. Often, the model catalyst was more active or showed activity similar to the real catalyst [18,19]. It is

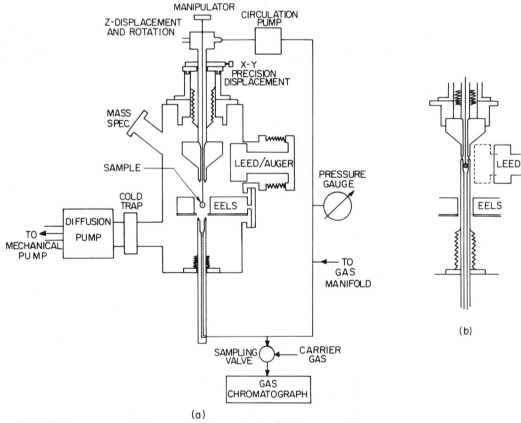

(a)

Fig. 3. The low pressure–high pressure apparatus that has a high pressure cell in the middle of a UHV chamber.

now well accepted that model catalyst behavior can be used as a reference for the evaluation of industrial catalyst systems.

Model catalysts can be prepared in thin film form, in addition to single-crystal form [20–22]. More complex oxide catalysts such as Fe_3O_4 or TiO_2 can be grown by atomic layer epitaxy (Fig. 2) on metal single-crystal surfaces [23–25]. Four to eight monolayers of the oxides exhibit catalytic behavior associated with high surface area oxide catalysts [26].

Catalyst additives, or so called promoters, have frequently been deposited from the vapor phase on metal model single-crystal catalysts to study their effects as structure or bonding modifiers. Alkali metals [27] or halogens, and oxide [28] or sulfide islands [29,30] are deposited on transition metal surfaces with variable surface concentrations, and their effects on the catalytic behavior monitored. In this manner, complex catalyst sys-

tems can be built on single-crystal surfaces or on thin films of the catalyst.

Over the past 20 years many catalyst systems and reactions have been studied using the surface science approach and understanding. A partial list of the catalyst systems studied in our laboratory is given in Table 1.

2. Concepts of the surface science of heterogeneous catalysis

New features of surface structure, composition, and bonding have been discovered as a result of these combined surface science and catalytic reaction studies. These new features either correlate with catalytic reactivity (reaction rate, selectivity, and other rate parameters) or are present on the working catalyst. This paper will present these as phenomena or new concepts that

Table 1
Model catalyst systems

Substrate	Promoters
Hydrocarbon conversion over platinum catalyst systems (n-heptane, n-hexane, methyl-cyclopentane, cyclohexane, ethylene, benzene)	
Platinum crystal faces [15,31–36]	Rhenium [78]
	Rhenium and sulfur [79]
	Potassium [80,81]
	Gold [82–85]
	Copper [86]
Rhenium crystal face	Platinum [78]
	Platinum and sulfur [79]
Ammonia synthesis over iron and rhenium catalyst systems	
Iron crystal faces [37,38]	Alumina [54,55]
	Potassium [69,70]
	Alumina and Potassium [69,70]
Rhenium crystal faces [87]	
Hydrodesulfurization of thiophene over molybdenum and rhenium catalyst systems	
Molybdenum crystal faces [39,40]	Sulfur [29], Cobalt [88–90], Carbon [91,92]
Rhenium crystal faces [41,42]	
CO, CO_2 hydrogenation over iron, rhodium, and cobalt catalyst systems	
Rhodium foil [77,93,94]	Titanium oxide [75,76,95]
	Vanadium oxide [96,97]
	Iron oxide [26]
Iron foil, thin film [71,98]	
Cobalt foil, thin film [99]	Gold [99]
Molybdenum foil [100]	

can be added to the classical chemical concepts of heterogeneous catalysis developed before the widespread use of modern surface science techniques. The roles these phenomena play during catalysis can often be explained or rationalized based on our knowledge of atomic surface structure, the dynamics of surface atoms, and of adsorbed molecules and the nature of the surface chemical bonds. Other phenomena, while they exhibit excellent correlation between certain surface properties and catalytic reactivity, await explanation through experimental and theoretical studies.

2.1. High reaction rates and strong bonding at defects and on rough surfaces

Stepped surfaces of platinum are more active for H_2–D_2 exchange and for most hydrocarbon conversion reactions than flat, low Miller index close-packed surfaces [15,31–36]. The structure sensitivity of catalytic reactions has clearly been demonstrated by studies of ammonia synthesis from nitrogen and hydrogen [37,38]. This reaction has been studied over various single-crystal surfaces of iron. This is a particularly surface structure sensitive reaction; the (111) and (211) surface orientations are approximately one order of magnitude more active than the (100) and (210) faces, and two orders of magnitude more active than the close-packed (110) face (this surface being the least active of all those studied) (fig. 4).

A more complicated example of surface structure sensitivity and insensitivity is the hydrodesulfurization reaction [39,40], a very important process used to remove sulfur from an oil feed. This reaction may be modeled by the hydrodesulfurization of thiophene to butane, butenes, and butadiene. When carried out on molybdenum [39,40] and rhenium [41,42] single-crystal surfaces, this reaction exhibits structure insensitivity over molybdenum but significant structure sensitivity over rhenium. For both of these reaction studies, more open crystal faces that have lower atomic density and, therefore, greater surface roughness are more active in carrying out the catalytic reaction (Fig. 5).

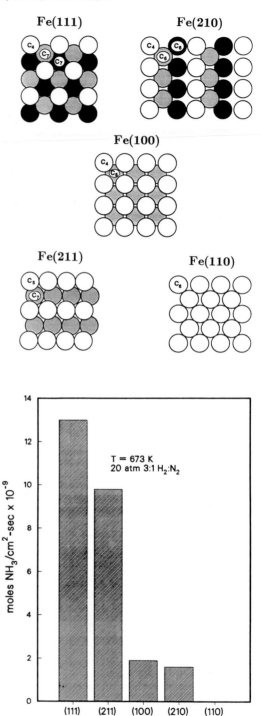

Fig. 4. Rates of ammonia synthesis for various single-crystal surfaces of iron.

Rough surfaces are not only catalytically more active, they are also more active in breaking chemical bonds. For example, a stepped surface of Ni decomposes ethylene to smaller fragments [43] at much lower temperatures (below 150 K) while the decomposition of ethylene on the Ni(111) face occurs at approximately 250 K [43]. During the temperature-programmed desorption (TPD) of H_2 from flat, stepped, and kinked surfaces of Pt, H_2 desorbs at maximum rates at the highest temperature from kink sites, then at somewhat lower temperatures from step sites, and at even lower temperatures from the flat (111) terraces [44]. This indicates higher heats of adsorption of the H atom at these defect sites. Thus, the thermodynamic driving force for dissociation is certainly greater at these sites, which can explain their enhanced bond-breaking activity.

It is difficult to understand, however, why these same strongly adsorbing sites are also very active sites for catalysis. The reaction probability of H_2–D_2 exchange on stepped surfaces is near unity at low pressures on a single scattering, while the reaction probability is below the detection limit ($< 10^{-3}$) on the flat (111) crystal face as shown by molecular beam–surface scattering studies [45,46]. How is it possible that the strongly adsorbing step sites, where H has a long residence time because of its high binding energy, are also the sites of rapid reaction turnover?

One possible explanation is that the strongly adsorbed hydrogen restructures the surface near the step, thereby creating the active site for the catalytic exchange process. At the low hydrogen pressures of these molecular beam scattering experiments, the low coverage keeps the structure of the flat part of the surface unaltered.

2.2. Metal restructuring controlled reaction rates: adsorbate–induced restructuring of metals

Chemisorption-induced restructuring can be seen very well using a small metal tip and field ion microscopy [47]. In Fig. 6 the field ion microscope picture of a rhodium tip is shown when clean and after exposure to carbon monoxide at 420 K and low pressures ($\sim 10^{-4}$ Pa). The metal

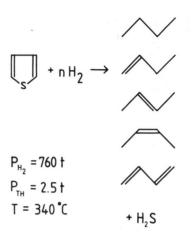

THE HYDRODESULFURIZATION
OF THIOPHENE

$P_{H_2} = 760$ t
$P_{TH} = 2.5$ t
$T = 340\,°C$

Thiophene HDS over Molybdenum and Rhenium Single Crystal Surfaces

$P_{Th} = 3.0$ Torr, $P_{H_2} = 780$ Torr, $T = 613$ K

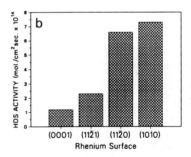

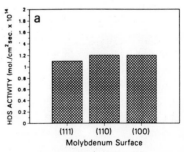

Fig. 5. (a) The structure insensitivity of the hydrodesulfurization of thiophene over molybdenum crystal surfaces. (b) The structure sensitivity of thiophene hydrodesulfurization to butene on rhenium single-crystal surfaces.

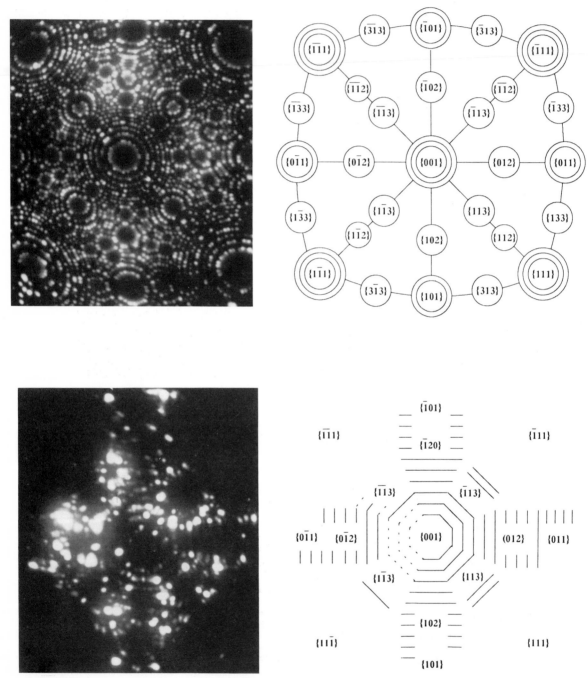

Fig. 6. Field ion micrographs (image gas: Ne, $T = 85$ K) of a (001)-oriented Rh tip before (top left) and after reaction with 10^{-4} Pa CO during 30 min at 420 K (bottom left) [47].

tip has been completely reshaped as a result of CO chemisorption. The tip becomes faceted and rougher, the step density is reduced and extended low Miller index terraces are formed.

Scanning tunneling microscopy (STM), which can be operated at high pressures (~ 2 atm) and temperatures ($\leq 240°C$), can also be used to detect surface-reaction-induced restructuring of metal surfaces [48,49]. Fig. 7 shows the structure of a Pt(110) crystal face when heated in hydrogen then heated in oxygen around one atmosphere. The platinum surface exhibits atomic details of its surface structure and ordered domains when exposed to H_2. In the presence of oxygen, the platinum surface becomes faceted and exhibits large areas composed of different crystal faces. These changes of surface structure are completely reversible; as shown when platinum is heated alternately in H_2 or in O_2.

Adsorption-induced restructuring can occur on the chemisorption time scale ($\sim 10^{-15}$ s for charge transfer or $\sim 10^{-12}$ s for vibrational times). There is evidence, however, that adsorbate-induced restructuring can occur on the time scale of catalytic reactions (seconds). CO oxidation to CO_2 [50,51] or ammonia reacting with NO to produce N_2 and H_2O [52] shows oscillatory behavior under certain circumstances of temperature and reactant partial pressures. The reaction rate alternates periodically between two values. One reason for the oscillation is the periodic restructuring of the surface. In this situation the sticking probability of one of the reactants is greater on one type of surface structure, while the sticking probability of the other reactant is greater on the surface structure of the other type. Thus, the reaction rate alternates between the two branches of the reaction; one taking place on the CO- or

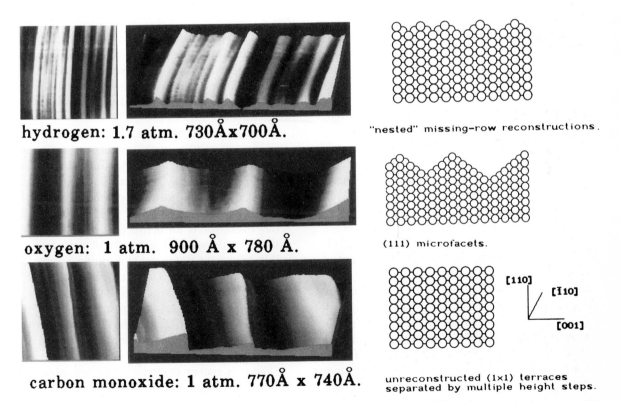

hydrogen: 1.7 atm. 730Åx700Å.

"nested" missing-row reconstructions.

oxygen: 1 atm. 900 Å x 780 Å.

(111) microfacets.

carbon monoxide: 1 atm. 770Å x 740Å.

unreconstructed (1×1) terraces separated by multiple height steps.

Fig. 7. In situ high pressure STM picture of adsorbate-induced surface reconstruction of Pt(110) under atmospheric pressures: (a) hydrogen, (b) oxygen, and (c) carbon monoxide.

NO-covered metal surface, and the other taking place on the oxygen- or ammonia-covered metal surface.

Adsorbate-induced restructuring can occur on even longer time scales (hours), involving massive restructuring of the surface by atom transport. For example, sulfur restructures the (111) crystal face of nickel until the metal surface assumes the (100) orientation [53]. Alumina restructures iron through the formation of an iron–aluminate phase to produce (111) crystal faces during ammonia synthesis, regardless of the original crystallite orientation [54,55]. Under these circumstances the chemisorption-induced restructuring can be viewed as the initial phase of a solid-state reaction whose kinetics are controlled by atom transport (diffusion). When the adsorbate-induced restructuring creates crystal faces that are beneficial to catalytic activity these adsorbates are called "structure modifiers" or structural promoters. If the reaction rate is decreasing, due to adsorbate-induced restructuring, the adsorbate is called a "catalyst poison".

Restructuring occurs in order to maximize the bonding and stability of the adsorbate–substrate complex. It is driven by thermodynamic forces and is most likely to occur when the stronger adsorbate–substrate bonds compensate for the weakening of bonds between the substrate atoms (which is the result of the chemisorption-induced restructuring process).

Substrate restructuring occurs during the chemisorption of molecules as well. The metal surface atoms that are "relaxed" by moving inward when the surface is clean, move outward during the formation of the chemisorption bond. When ethylidyne forms on the platinum (111) surface, in addition to the outward movement of surface atoms [56], the nearest-neighbor metal atoms move closer to the adsorption site, and the next-nearest-neighbor Pt atom moves inward, causing a slight corrugation of the surface, while the Pt atom underneath the adsorption site moves down and away from the carbon atom. This is shown in this volume by Fig. 2 in the paper "Adsorption and adsorbate-induced restructuring: a LEED perspective", by M.A. Van Hove and G.A. Somorjai.

Structure modifiers are often introduced as important additives when formulating the complex catalyst systems. Structural promoters can change the surface structure; which is often the key to catalyst selectivity. Alloy components may not participate in the reaction chemistry but modify structure and site distribution on the catalyst surface. Site blocking could improve selectivity as has been proven for many working catalyst systems. Sulfur and silicon, or other strongly adsorbed atoms that seek out certain active sites, can block undesirable side reactions.

2.3. The catalytically active metal surface covered with a strongly chemisorbed overlayer: coverage dependence of bonding

During most reactions the surface of the metal catalyst is covered with a strongly chemisorbed overlayer that remains tenaciously bound to the surface for 10^2–10^6 turnovers [57,58]. During hydrocarbon reactions, this carbonaceous overlayer has a composition of $H/C \approx 1$. During ammonia synthesis the overlayer is chemisorbed nitrogen. During the hydrodesulfurization of thiophene to butenes the overlayer is a mixture of sulfur and carbon. Isotope labeling of the adsorbed species (using ^{14}C, ^{35}S) has been used to determine their long residence times.

During hydrocarbon conversion reactions over platinum, over 80% of the metal surface is covered with the carbonaceous deposit. While there are suggestions that the remaining uncovered metal sites are the only active sites for catalysis, there is increasing evidence that the carbonaceous overlayer is an active part of the working catalyst. Only when it is totally dehydrogenated will it deactivate the metal by forming a cross-linked graphite coating. There are suggestions that structure insensitive reactions (hydrogenation of ethylene, for example) take place on top of the strongly bound deposit, and the metal only participates in the reaction indirectly by aiding the dissociation of molecular hydrogen.

There is recent evidence, from scanning tunneling microscopy studies in our laboratory, that the strongly bound adsorbed atoms and molecules are mobile on metal surfaces [59]. The activation energy for surface diffusion of chemisorbed species is low compared to their heats of desorption; therefore, these atoms and molecules migrate along the surface visiting various adsorption sites during their surface residence time (two-dimensional phase approximation). The STM studies provide the first direct evidence of the rearrangement of the chemisorbed layer. In Fig. 8a we show an STM picture of an ordered sulfur overlayer on rhenium (0001) single-crystal surface. In Fig. 8b the same surface is shown in the presence of CO. As the molecule coadsorbs on the metal surface, it does so by compressing the sulfur adsorbate into a higher coverage structure around the adsorption site. Thus, CO molecules adsorb by laterally displacing the sulfur atoms. Similar behavior was found on platinum crystal surfaces in the presence of coadsorbed sulfur and CO.

It is likely that the adsorption of reactants on a strongly chemisorbed overlayer-covered surface occurs by laterally displacing the overlayer, thereby creating a reaction site. As long as the activation energy for the surface diffusion (or lateral displacement) of the deposit is small, this process can always create new sites for adsorption and reaction. Only when the activation energy for surface diffusion of the overlayer becomes large will the catalyst deactivate. This happens when the overlayer polymerizes or forms a graphitic network, which occurs during hydrocarbon reactions at high temperatures.

The heat of chemisorption per atom or per molecule declines with increasing coverage for most chemisorption systems. This has been shown for potassium on a rhodium (111) crystal face [60] and for CO on several transition metal surfaces, respectively. At low coverages, potassium is strongly bound to the transition metal as it transfers electrons to it to become positively charged. With increasing coverage, adsorbate–adsorbate interaction causes repulsion among the charged species, leading to depolarization and much weakened adsorption bonds until the heat of adsorption becomes equal to the heat of sublimation of metallic potassium [61,62]. Carbon monoxide chemisorbs with its C–O bond perpendicular to the metal surface, occupying on-top and bridge sites until about one-half monolayer of coverage is reached [63]. The heat of adsorption stays relatively constant with coverage in this coverage range, indicating that very little adsorbate–adsorbate interaction is influencing the bonding of the molecule to the metal. At higher coverages, however, the molecules strongly repel each other, forcing the on-top site CO molecules to relocate [64] to maximize adsorbate–adsorbate distances, and ΔH_{ads} declines rapidly until it reaches about one-third of its value at low CO coverages.

Thus, increasing coverage of chemisorbed species not only leads to sequential filling of binding sites (the stronger binding sites filling first), as shown in the literature, but can also weaken the adsorbate–substrate bonds markedly.

Before CO Exposure

40 x 40 Angstroms

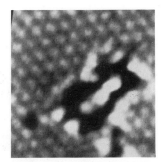

After CO Exposure

55 x 55 Angstroms

Fig. 8. (a) Scanning tunneling microscopy picture of the high coverage sulfur structure on the rhenium (0001) crystal face. (b) The same structure after coadsorption of carbon monoxide.

This effect of coverage influences the surface residence times of adsorbates and subsequently their behavior during chemisorption and surface chemical reactions.

Most catalytic reactions are carried out at high pressures under conditions of high surface coverages of adsorbed reactants and reaction intermediates. Under these circumstances the heat of desorption is low as compared to the situation at low coverages. In most surface science studies adsorbates are scrutinized at low coverages because their stability makes their study easier, and because of the low pressure conditions necessary to utilize many techniques of surface science. Thus, correlations in chemical behavior are hindered because of the different experimental conditions used.

2.4. Coadsorption

The coadsorption of two different species can lead to either attractive or repulsive adsorbate–

adsorbate interaction. The coadsorption of ethylene and carbon monoxide demonstrates the attractive interaction that can occur in the adsorbed layer. CO and C_2H_4 chemisorbed together [65] on the Rh(111) crystal face form the structure shown in Fig. 9. There are two different molecules per unit cell, indicating attraction among the molecular species of different types. Ethylene adsorption decreases the work function of rhodium while CO increases the work function of rhodium upon chemisorption [66]. Thus, C_2H_4 is an electron donor while CO is an electron acceptor on the transition metal, resulting in an attractive donor–acceptor interaction among the two types of adsorbates.

The ordering of one adsorbate by the coadsorption of another through donor–acceptor interaction is commonly observed, as shown for several coadsorbed systems. For Rh(111) the magnitude of the adsorbate–adsorbate attractive interaction is about one order of magnitude

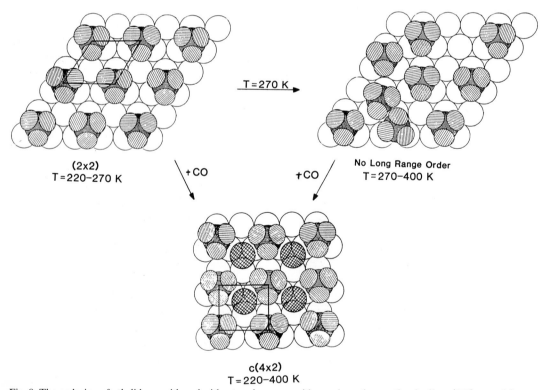

(2x2)
T = 220–270 K

T = 270 K →

No Long Range Order
T = 270–400 K

+ CO   + CO

c(4x2)
T = 220–400 K

Fig. 9. The ordering of ethylidyne, with and without carbon monoxide coadsorption on the rhodium (111) crystal face.

weaker (about 4–6 kcal/mol (17–25 kJ/mol)) than most adsorbate–substrate chemisorption bonds (about 30–60 kcal/mol (125–2500 kJ/mol)). Repulsive interaction between two donor- or two acceptor-coadsorbed molecules leads to separation of the adsorbates by island formation or disorder in the adsorbed layer.

Strong attractive interaction among adsorbates can lead to dissociation of the molecular species. This is observed during the coadsorption of potassium (donor) and CO (acceptor) on several transition metal surfaces [60,67]. Thermal desorption data indicate CO desorbing at much higher temperatures than normal in the presence of the adsorbed alkali metal, often showing a 17 kcal/mol (71 kJ/mol) increase in its heat of adsorption. The CO stretching frequency decreases with increasing dipole moment of coadsorbed donors. Isotope-labeling studies (using $^{12}C^{18}O$ and $^{13}C^{16}O$) indicate scrambling of the two isotopic species in the presence of potassium, signaling molecular dissociation, while no dissociation is apparent in the absence of potassium on rhodium at low pressures. Up to three CO molecules dissociate per potassium atom at an alkali metal coverage of 20% of a monolayer [68].

Repulsive interaction is also observed with the coadsorption of potassium and ammonia [69,70]. Both species are electron donors to transition metals. On iron, a 4 kcal/mol (17 kJ/mol) decrease in the heat of chemisorption of NH_3 is observed due to coadsorbed potassium.

Alkali metals are often used as additives during catalytic reactions. They are "bonding modifiers"; that is, they influence the bonding and, thus, the reactivity of the coadsorbed molecules. Potassium is an important promoter in CO hydrogenation reactions [71], where CO dissociation is desired and is one of the elementary reaction steps. The alkali metal also reduces the hydrogen chemisorption capacity of the transition metal. Potassium is an important promoter in ammonia synthesis for the opposite reason [69,70], because it weakens the NH_3 product molecule bonding to the metal, thereby reducing its surface concentration, which would block important reaction sites. It also aids in the dissociation of dinitrogen.

Halogen species can also be important bonding modifiers, as they are powerful electron acceptors. Indeed, they are used as promoters in several catalytic processes (for example, ethylene oxidation to ethylene oxide over silver, or during partial oxidation of methane). Nevertheless, their molecular and atomic chemisorption behavior has been studied less and, therefore, is not as well-understood as the role of coadsorbed alkali metal ions.

We would like to present one more new phenomenon of catalysis and surface science that is more difficult to correlate. Nevertheless, the structure and bonding at surfaces and the dynamics of surface restructuring are the likely reasons for its existence.

2.5. High reaction rates at oxide-metal interfaces

Oxide–metal interfaces can exhibit very high activity as compared to the metal alone whether the metal is placed on the oxide or the oxide on the metal as shown by Schwab et al. in the 1930's [72]. More recently, Tauster has reinvestigated this phenomenon using titanium oxide surfaces and found unique high catalytic activity due to oxide–metal interfaces. This phenomenon is called "strong metal support interaction" [73]. The typical oxide-induced changes in the metal activity can be readily shown for nickel if one compares the reaction rates for forming methane on this metal, from carbon monoxide and hydrogen, on various nickel catalysts that are deposited on different oxide supports. The oxide supports alone are inert for this reaction.

There are orders of magnitude changes in the reaction rate when the nickel particles are deposited on silica (low rate) as compared to nickel on titanium oxide [74] (high rate). The typical activation treatment for catalysts of this type involves reduction in hydrogen, during which the oxide encapsulates the metal, resulting in much-reduced carbon monoxide or hydrogen chemisorption activity followed by oxygen heat treatment that activates the catalysts during which the oxide forms small islands in such a way as to maximize the oxide–metal periphery area.

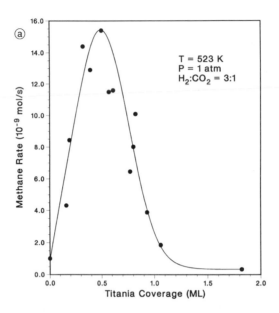

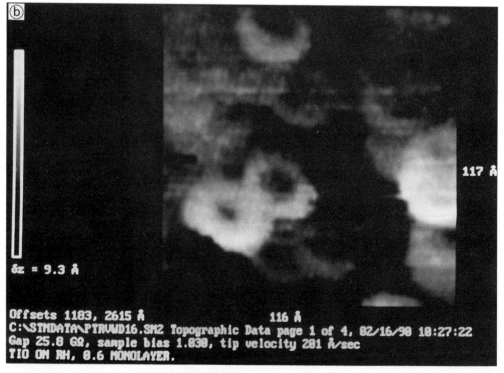

Fig. 10. (a) CO_2 hydrogenation rate over the rhodium–titanium oxide catalysts as a function of oxide coverage of the metal. (b) Scanning tunneling microscopy picture of the titanium oxide islands on rhodium. The donut-like shape is due to the increased charge density at the periphery of the oxide islands, because of the concentration of Ti^{3+} ions.

A typical reaction pattern reflecting the behavior of an active oxide–metal interface system has been reported for the CO_2/H_2 reaction on Rh onto which a TiO_x layer is deposited, and the catalytic activity is measured as a function of the oxide coverage [28,75,76]. The oxide is deposited by condensing the metal from the vapor phase and then oxidized to form TiO_x. The reaction rate is at a maximum at ~ 0.5 monolayer oxide coverage (Fig. 10a). The reaction rate increases by 14-fold with respect to the clean metal surface. If we were to give the specific turnover rates by dividing it with the surface area we should use the oxide–metal interface area instead of the total area, which is the metal plus the oxide surface area, since the oxide is inactive. Since the oxide–metal periphery area is about 10% of the total area, at 0.5 monolayer oxide coverage the specific turnover rate is two orders of magnitude greater as compared to the rate on the clean Rh metal.

Similar behavior is found for CO/H_2 and acetone/H_2 [77], while the hydrogenation of ethylene and the hydrogenolysis of ethane [76] remain unaffected by the presence of TiO_x, indicating that the carbon–oxygen bond is being activated at the oxide–metal interface. Thus any molecules that contain CO bonds are benefiting from the presence of the oxide–metal interface catalytic sites. Other oxides deposited on Rh such as vanadium oxide show similar effects; however, the increase in reaction rate is not as large in the presence of vanadium oxide as it is with the TiO_x/Rh interface.

From recent studies in my laboratory using several transition metal oxides (TiO_x, VO_x, FeO_x, NbO_x, TaO_x), the oxide-induced enhancement of the reaction rates appears to correlate with the electron accepting ability (Lewis acidity) of the oxide in its high oxidation state. Thus, the oxide may be viewed as an electron acceptor bonding modifier that alters the reactivity of polar molecules including those with C–O bonds. We carried out scanning tunneling microscopy studies to image the TiO_x islands and the oxide–metal interface. Fig. 10b shows the STM picture. The bright rings which appear at the oxide island periphery indicate large charge density associated

with the presence of Ti^{3+} ions in the partially reduced oxide at the metal interface.

3. Frontiers of the surface science of heterogeneous catalysis

3.1. Studies of surface structures and reaction dynamics

The chemically active surface undergoes adsorbate-induced restructuring that is often local (short range) and can also be diffusion controlled (long range). Structural rearrangements occur on both sides of the surface chemical bond; on the adsorbate as well as on the substrate sides. In order to obtain a complete picture of the chemical and structural changes that occur, we need techniques that can simultaneously monitor both sides of the surface chemical bonds in a time-resolved manner, with high spatial resolution. *Time resolved studies of the dynamics of substrate structure and adsorbate structure* could explore the role of restructuring in heterogeneous catalysis. *Identification of short-lived adsorbates by time-resolved spectroscopic studies* could help us determine the nature of reaction intermediates and the concentration of active sites. Photon spectroscopy techniques have this time resolution, and can be operated at high pressures to help carry out studies of this type (infrared spectroscopy and sum frequency generation, for example). Studies of *structure and bonding of adsorbed monolayers at high coverages* should be carried out, as most surface science studies up to now, focused on adsorbates at low coverages. The scanning tunneling microscope and structural studies using synchrotron radiation (EXAFS, for example) could provide this information.

3.2. Fabrication of cluster arrays as model catalysts

The demands for high selectivity and better control over the rate of catalyzed reactions forces us to rethink the process of catalyst fabrication. The use of chemical vapor deposition, ion-sputter deposition, and electron beam or X-ray beam lithographies permit the fabrication of uniform size metal and semiconductor clusters in the 30–300 Å range. This type of fabrication assures control over both particle size and spacing between catalyst particles. Compared to single-crystal surfaces, these cluster arrays provide new model systems for the study of heterogeneous catalysis which are closer in structure to working catalyst systems.

3.3. Studies of reactions and catalysts not explored by surface science

Surfaces that carry out acid–base reactions with adsorbates have yet to be explored by surface science. These are usually oxide surfaces that donate or accept protons (Bronsted acid or base) or electrons (Lewis base or acid).

The ability of the human body to differentiate between left- and right-handed isomers of molecules makes it imperative to synthesize them stereospecifically instead of producing racemic mixtures. One of the challenges for the surface science of catalysis is to develop selective catalysts for the stereospecific synthesis of drugs and fine chemicals entering the human body.

Biological catalysis provides us with many examples of processes that occur with 100% selectivity toward making a product. It is natural that we should aim for this by extending our understanding of the causes of selective catalytic surface chemistry in the production of small molecules as well. The increased selectivity of new generations of catalysts, to produce ethylene oxide from ethylene and oxygen or formaldehyde from methanol, clearly shows this trend. As surface science exerts a growing influence on the development of a new generation of partial oxidation, hydrogenation, isomerization and polymerization catalysts, increased selectivity in the production of most molecules will be the result.

Combustion at high temperatures produces nitrogen oxides. If we can carry out combustion reactions at 1500°C (at present it is being performed at 1800°C during energy production in stationary power stations) NO_x emissions can be virtually eliminated. Of course, to achieve the required high combustion rates at lower temperatures requires catalytic combustion. The catalyst not only must carry out complete combustion, but must also be stable at these high temperatures and resistant to poisoning by impurities in the coal or oil slurries used as feedstocks (sulfur, arsenic, silica, etc.). Catalytic combustion is one of the frontier areas of the surface science of catalysis. During the next decade major advances are expected in this field.

Acknowledgement

This work was supported by the Director, Office of Energy Research, Office of Basic Energy Science, Materials Science Division, of the US Department of Energy, under contract No. DE-AC03-76SF00098.

References

[1] G.A. Somorjai, Surf. Sci. 2 (1964) 298.
[2] S. Hagstrom, H.B. Lyon and G.A. Somorjai, Phys. Rev. Lett. 15 (1965) 491.
[3] M.A. Van Hove, R.J. Koestner, P.C. Stair, J.P. Biberian, L.L. Kesmodel and G.A. Somorjai, Surf. Sci. 103 (1981) 189.
[4] A.E. Morgan and G.A. Somorjai, Surf. Sci. 12 (1968) 405.
[5] A.E. Morgan and G.A. Somorjai, J. Chem. Phys. 51 (1969) 3309.
[6] L.A. West, E.I. Kozak and G.A. Somorjai, J. Vac. Sci. Technol. 8 (1971) 430.
[7] S.T. Ceyer, W.J. Siekhaus and G.A. Somorjai, J. Vac. Sci. Technol. 19 (1981) 726.
[8] L.A. West and G.A. Somorjai, J. Vac. Sci. Technol. 9 (1970) 668.
[9] L.A. West and G.A. Somorjai, J. Chem. Phys. 57 (1972) 5143.
[10] B. Lang, R.W. Joyner and G.A. Somorjai, Surf. Sci. 30 (1972) 440.
[11] B. Lang, R.W. Joyner and G.A. Somorjai, Surf. Sci. 30 (1972) 454.
[12] B. Lang, R.W. Joyner and G.A. Somorjai, J. Catal. 27 (1972) 405.
[13] S.L. Bernasek, W.J. Siekhaus and G.A. Somorjai, Phys. Rev. Lett. 30 (1973) 1202.
[14] D.W. Blakely, E. Kozak, B.A. Sexton and G.A. Somorjai, J. Vac. Sci. Technol. 13 (1976) 1091.
[15] A.L. Cabrera, N.D. Spencer, E. Kozak, P.W. Davies and G.A. Somorjai, Rev. Sci. Instrum. 53 (1982) 1893.
[16] T.G. Rucker, K. Franck, D. Colomb, M.A. Logan and G.A. Somorjai, Rev. Sci. Instrum. 58 (1987) 2292.

[17] D.E. Gardin and G.A. Somorjai, Rev. Sci. Instrum., in press.

[18] O.R. Kahn, E.E. Petersen and G.A. Somorjai, J. Catal. 34 (1974) 294.

[19] S.H. Davis and G.A. Somorjai, J. Catal. 65 (1980) 78.

[20] S.T. Oyama, A.M. Carr and G.A. Somorjai, J. Phys. Chem. 94 (1990) 5029.

[21] S. Fu and G.A. Somorjai, Surf. Sci. 237 (1990) 87.

[22] V. Maurice, M.B. Salmeron and G.A. Somorjai, Surf. Sci. 237 (1990) 116.

[23] G.H. Vurens, M.B. Salmeron and G.A. Somorjai, Surf. Sci. 32 (1989) 211.

[24] K.J. Williams, A.B. Boffa, M.B. Salmeron, A.T. Bell and G.A. Somorjai, Catal. Lett. 11 (1991) 77.

[25] W. Weiss and G.A. Somorjai, J. Vac. Sci. Technol. A, in press.

[26] W. Weiss, A.B. Boffa, J.C. Dunphy, H.C. Galloway and G.A. Somorjai, Dr. Wilhelm Heinrich Heraeus and Else Heraeus Stiftung Seminar, Bad Honnef, Germany (Springer, Berlin), in press.

[27] E.L. Garfunkel, J.E. Crowell and G.A. Somorjai, J. Phys. Chem. 86 (1982) 310.

[28] M.E. Levin, M.B. Salmeron, A.T. Bell and G.A. Somorjai, J. Chem. Soc. Faraday Trans. I, 83 (1987) 2061.

[29] A.J. Gellman, M.E. Bussell and G.A. Somorjai, J. Catal. 107 (1987) 103.

[30] R.Q. Hwang, D.M. Zeglinski, A. Lopez Vazquez-de-Parga, C. Ocal, D.F. Ogletree, M.B. Salmeron and G.A. Somorjai, Phys. Rev. B 44 (1991) 1914.

[31] D.W. Blakely and G.A. Somorjai, J. Catal. 42 (1976) 181.

[32] M.B. Salmeron and G.A. Somorjai, Surf. Sci. 90 (1980) 373.

[33] R.K. Herz, W.D. Gillespie, E.E. Petersen and G.A. Somorjai, J. Catal. 67 (1981) 371.

[34] W.O. Gillespie, R.K. Herz, E.E. Petersen and G.A. Somorjai, J. Catal. 70 (1981) 1.

[35] S.M. Davis, F. Zaera and G.A. Somorjai, J. Catal. 85 (1984) 206.

[36] F. Zaera and G.A. Somorjai, J. Am. Chem. Soc. 106 (1984) 2288.

[37] N.D. Spencer, R.C. Schoonmaker and G.A. Somorjai, J. Catal. 74 (1982) 129.

[38] D.R. Strongin, J. Carrazza, S.R. Bare and G.A. Somorjai, J. Catal. 103 (1987) 213.

[39] A.J. Gellman, M.H. Farias and G.A. Somorjai, J. Catal. 88 (1984) 546.

[40] A.J. Gellman, D. Neiman and G.A. Somorjai, J. Catal. 107 (1987) 92.

[41] M.E. Bussell, A.J. Gellman and G.A. Somorjai, J. Catal. 110 (1988) 423.

[42] M.E. Bussell and G.A. Somorjai, J. Phys. Chem. 93 (1989) 2009.

[43] M.A. Van Hove, B. Bent and G.A. Somorjai, J. Phys. Chem. 92 (1988) 973.

[44] G.A. Somorjai, Catal. Today 12 (1992) 343.

[45] R.J. Gale, M.B. Salmeron and G.A. Somorjai, Phys. Rev. Lett. 38 (1977) 1027.

[46] M.B. Salmeron, R.J. Gale and G.A. Somorjai, J. Chem. Phys. 70 (1979) 2807.

[47] A. Gaussmann and N. Kruse, Catal. Lett. 10 (1991) 305.

[48] B.J. McIntyre, M.B. Salmeron and G.A. Somorjai, Catal. Lett. 14 (1992) 263.

[49] B.J. McIntyre, M.B. Salmeron and G.A. Somorjai, Rev. Sci. Intrum., in press.

[50] R.C. Yeates, J.E. Turner, A.J. Gellman and G.A. Somorjai, Surf. Sci. 149 (1985) 175.

[51] G. Ertl, Catal. Lett. 9 (1991) 219.

[52] T. Katona and G.A. Somorjai, J. Phys. Chem. 96 (1992) 5465.

[53] J.J. McCarroll, Nature 223 (1969) 1260.

[54] S.R. Bare, D.R. Strongin and G.A. Somorjai, J. Phys. Chem. 90 (1986) 4726.

[55] D.R. Strongin, S.R. Bare and G.A. Somorjai, J. Catal. 103 (1987) 289.

[56] U. Starke, A. Barbieri, N. Materer, M.A. Van Hove and G.A. Somorjai, Surf. Sci. 286 (1993) 1.

[57] S.H. Davis, F. Zaera and G.A. Somorjai, J. Catal. 77 (1982) 439.

[58] S.M. Davis, F. Zaera, B.E. Gordon and G.A. Somorjai, J. Catal. 92 (1985) 240.

[59] G.A. Somorjai, NATO Advanced Study Workshop, Chateau de Florans, Vaucluse, France, 1992.

[60] E.L. Garfunkel and G.A. Somorjai, Alkali Adsorption on Metals and Semiconductors (Elsevier, Amsterdam, 1989) p. 319.

[61] C.-T. Kao, C.M. Mate, G.S. Blackman, B.E. Bent, M.A. Van Hove and G.A. Somorjai, J. Vac. Sci. Technol. A 6 (1988) 786.

[62] R.M. Nix and G.A. Somorjai, in: Concepts in Surface Science and Heterogeneous Catalysis (Kluwer, Dordrecht, 1989) p. 97.

[63] H. Ohtani, M.A. Van Hove and G.A. Somorjai, Appl. Surf. Sci. 33 (1987) 254.

[64] M.A. Van Hove, R.J. Koestner and G.A. Somorjai, Phys. Rev. Lett. 50 (1983) 903.

[65] G.S. Blackman, C.-T. Kao, B.E. Bent, C.M. Mate, M.A. Van Hove and G.A. Somorjai, Surf. Sci. 207 (1988) 66.

[66] C.M. Mate, C.-T. Kao and G.A. Somorjai, Surf. Sci. 206 (1988) 145.

[67] J.E. Crowell and G.A. Somorjai, Appl. Surf. Sci. 19 (1984) 73.

[68] J.E. Crowell, W.T. Tysoe and G.A. Somorjai, J. Phys. Chem. 89 (1986) 1598.

[69] D.R. Strongin and G.A. Somorjai, J. Catal. 109 (1988) 51.

[70] D.R. Strongin and G.A. Somorjai, Catal. Lett. 1 (1988) 61.

[71] D. Dwyer and G.A. Somorjai, J. Catal. 56 (1979) 249.

[72] G.-M. Schwab, in: Catalysis, Science and Technology, Vol. 2, Eds. J.R. Anderson and M. Boudart (Springer, Berlin, 1981) p. 1.

[73] S.J. Touster and S.C. Fung, J. Catal. 55 (1978) 29.

[74] C.H. Bartholomew, R.B. Powell and J.L. Butler, J. Catal. 65 (1977) 461.

[75] M.E. Levin, K.J. Williams, M.B. Salmeron, A.T. Bell and G.A. Somorjai, Surf. Sci. 195 (1988) 341.

[76] K.J. Williams, M.E. Levin, M.B. Salmeron, A.T. Bell and G.A. Somorjai, Catal. Lett. 1 (1988) 331.

[77] K.J. Williams, A.B. Boffa, J. Lahtinen, M.B. Salmeron, A.T. Bell and G.A. Somorjai, Catal. Lett. 5 (1990) 385.

[78] D.J. Godbey, F. Garin and G.A. Somorjai, J. Catal. 117 (1989) 144.

[79] C. Kim and G.A. Somorjai, J. Catal. 134 (1992) 179.

[80] E.L. Garfunkel and G.A. Somorjai, Surf. Sci. 115 (1982) 441.

[81] F. Zaera and G.A. Somorjai, J. Catal. 84 (1983) 375.

[82] J.W.A. Sachtler, M.A. Van Hove, J.P. Biberian and G.A. Somorjai, Phys. Rev. Lett. 45 (1980) 1601.

[83] J.W.A. Sachtler, J.P. Biberian and G.A. Somorjai, Surf. Sci. 110 (1981) 43.

[84] J.W.A. Sachtler and G.A. Somorjai, J. Catal. 81 (1983) 77.

[85] R. Yeates and G.A. Somorjai, J. Catal. 103 (1987) 208.

[86] R.C. Yeates and G.A. Somorjai, Surf. Sci. 134 (1983) 729.

[87] M. Asscher, J. Carrazza, M. Khan, K. Lewis and G.A. Somorjai, J. Catal. 98 (1986) 277.

[88] M.E. Bussell and G.A. Somorjai, Catal. Lett. 3 (1989) 1.

[89] C.C. Knight and G.A. Somorjai, Surf. Sci. 240 (1990) 101.

[90] G. Vurens, V. Maurice, M.B. Salmeron and G.A. Somorjai, Surf. Sci. 268 (1992) 170.

[91] M.E. Bussell and G.A. Somorjai, J. Catal. 106 (1987) 93.

[92] M.E. Bussell, A.J. Gellman and G.A. Somorjai, Catal. Lett. 1 (1988) 195.

[93] D.G. Castner, R.L. Blackadar and G.A. Somorjai, J. Catal. 66 (1980) 257.

[94] M.A. Logan and G.A. Somorjai, J. Catal. 95 (1985) 317.

[95] M.E. Levin, M.B. Salmeron, A.T. Bell and G.A. Somorjai, J. Catal. 106 (1987) 401.

[96] K.B. Lewis, S.T. Oyama and G.A. Somorjai, Surf. Sci. 233 (1990) 75.

[97] A. Boffa, A.T. Bell and G.A. Somorjai, J. Catal. 139 (1992) 602.

[98] D.J. Dwyer and G.A. Somorjai, J. Catal. 52 (1978) 291.

[99] J. Lahtinen, T. Anraku and G.A. Somorjai, J. Catal., in press.

[100] M. Logan, A.J. Gellman and G.A. Somorjai, J. Catal. 94 (1985) 60.

Surface Science 299/300 (1994) 867–877
North-Holland

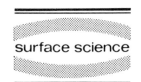

surface science

Chemical beam epitaxy – a child of surface science

Hans Lüth

Institut für Schicht- und Ionentechnik (ISI), Research Center Jülich (KFA), P.O. Box 1913, D-52425 Jülich, Germany

Received 12 April 1993; accepted for publication 28 June 1993

Chemical Beam Epitaxy (CBE) or MOMBE is currently one of the major deposition techniques in semiconductor technology. The growth process is performed in a UHV chamber under low pressure conditions and the source material is supplied by molecular beams, such that only surface kinetics are determining the chemical reactions leading to growth of the epilayer. This paper intends to give a review on the development of this deposition technique. After considering the early period, where this epitaxy method started to develop, partially from ideas being born in surface science, some milestones in the further development and basic understanding are presented. The mutual interaction between CBE/MOMBE as a deposition technique and other fields of surface science is described as well as the impact on the deposition technology of other semiconductors (e.g. for Si-based material systems). Future prospects of CBE are finally discussed, particularly in comparison with the competing techniques MBE and MOCVD (metal-organic chemical vapor deposition).

1. How it began

The term "Chemical Beam Epitaxy" (CBE) describes an important class of deposition techniques for semiconductor layer systems – in particular III–V semiconductor systems – which are performed in an ultrahigh vacuum (UHV) system using molecular beams of reactive gases as sources. III–V epitaxial source gases are chemical compounds containing the elements As, Ga, In etc. In CBE they react in complex reactions on the hot growing surface and deliver the material needed for layer growth. The present contribution is a personal view of someone who participated in the very first work which lead to the birth of this powerful technique in semiconductor technology. It thus emphasises impressions of a surface scientist about the origin and future prospects of CBE.

Already before 1980 metal-organic chemical vapor deposition (MOCVD) or MOVPE [1–3] and molecular beam epitaxy (MBE) [4–6] had become quite mature techniques for the preparation of III–V semiconductor layer systems with a growth thickness control down to atomic accu-

racy. Nevertheless the understanding of both techniques was not well developed in those days. In particular, in MOCVD – a technique performed in a cold wall reactor – the complex interplay between gas phase reactions in the hot ambient above the wafer and the surface reactions directly on the growing surface could not be studied with the experimental techniques available. On the other hand, in surface science quite sophisticated techniques such as photoemission spectroscopy (UPS/XPS), electron energy loss spectroscopy (EELS/HREELS), Auger electron spectroscopy (AES) and low energy electron diffraction (LEED) had been established to study surface reactions being related to catalysis on an atomic scale in more detail. Just at that time I was working as a guest scientist exactly in this field in the exciting atmosphere of D.E. Eastman's group at the IBM Thomas J. Watson Research Center, Yorktown Heights [7]. Another visiting scientist at that time at IBM, K.H. Bachem, was working in the field of chemical vapor deposition. He had been exposed to gas phase epitaxy techniques for III–V semiconductor layers in P. Balk's and H. Beneking's group at the Aachen Techni-

SSDI 0039-6028(93)E0384-7

cal University, where important steps to their development and understanding had been done [8].

For me the involvement in CBE or MOMBE (metal-organic MBE) as we called it later started during frequent discussions with K.H. Bachem, in which he insisted on applying all the beautiful surface science techniques to understand the chemical surface reactions in MOCVD on an atomic level, and where I thought about the realisation of convenient experimental set-ups for this purpose. At this time it was already clear to us, that the epitaxy had to be performed in an UHV environment, in order to be compatible to RHEED, UPS, AES etc., and that gases being used in MOCVD like AsH_3, trimethyl gallium (TMGa, $(CH_3)_3Ga$), triethyl gallium (TEGa) etc. had to be supplied by jet-like molecular beam sources. After coming back to the Aachen Technical University a fruitful cooperation between P. Balk's and my group started, where we realised the first UHV reactor in which GaAs layers were grown using AsH_3 and TMGa [9]. The main idea behind these experiments was to separate surface reactions, which are dominant in the UHV environment from gas phase reactions being important in the real MOCVD process and to study them. The same deposition method for GaAs in a UHV system was then confirmed by Vodjdani et al. [10]. The molecular beams used consisted of AsH_3 and the metal-organic compound TMGa for the group III component.

Nearly at the same time similar work by some American groups was initiated for more practical purposes. In order to get rid of the frequent loading procedures of arsenic crucibles in MBE and the resulting irreproducibilities in the growth process Panish used as early as 1980 gaseous sources of hydrides (AsH_3, PH_3) in an MBE set-up together with conventional solid sources for the group III component [11]. Calawa was successful in growing GaAs layers of high electrical quality ($n_{77} = 2.4 \times 10^{14}$ cm^{-3}, $\mu_{77} = 110.000$ $cm^2/V \cdot s$) by this procedure using solid Ga and precracked AsH_3 in a MBE system [12]. By this technique, MBE with conventional solid sources for the group III component and a gas line source for the hydrides with precracking facilities, some-

what later heterostructures of high quality [13,14] and quantum well structures [15] could be grown. Panish called the method gas source MBE (GSMBE). Meanwhile there is a nomenclature in use, which calls the technique with both components being gaseous sources chemical beam epitaxy (CBE), the technique with only the metal-organic group III component being the gaseous source and a solid group V source metal-organic MBE (MOMBE) and the corresponding method with solid group III component and gaseous group V hydride sources gas source MBE (GSMBE). Nevertheless many groups including our own use the term MOMBE, maybe for historical reasons, instead of CBE for a III–V layer epitaxy technique in a UHV system which is fully equiped with gas line sources.

2. Some experimental details

Some short remarks should be given about the present standard equipment for CBE or MOMBE growth of III–V semiconductor layer systems, in order to demonstrate the development of this technique from the first surface scientists little UHV chambers with leak valves as gas inlets and ion getter pumps for establishing the vacuum. Standard UHV growth chambers are now a days pumped by a combination of turbomolecular and cryo pumps, the turbo pump being protected by a liquid-nitrogen cooled trap. The UHV chamber is equipped with a LN_2 cryoshield. A rotatable crystal holder carries 2 or 3 inch wafers clamped or glued with In and heated from the backside up to temperatures between 500 and 700°C. RHEED equipment is used for in-situ monitoring of surface superstructures on the growing surface and for measuring growth rates by RHEED oscillations, i.e. intensity variations of particular Bragg spots. Mass spectrometers allow the analysis of the molecular species in the beams and the analysis of the residual gas before and during the growth process.

An important part of the MOMBE or CBE set-up is the gas inlet system. Biased by the experience with MOCVD systems which need high gas fluxes because of the relatively low

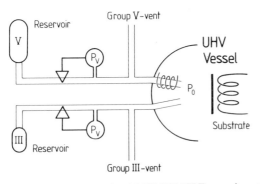

Fig. 1. Simplified schematic of MOMBE/CBE experimental set up. The gas inlet system is pressure controlled. Input pressure p_v; beam pressure $p_0 \ll p_v$; vent lines for group V and group III source gases are also indicated. The substrate is heated from the rear during growth.

vapour pressure of the metal-organics employed, bubblers and mass flow controlled systems are frequently in use. Better results however in controlling the beam flux with high accuracy are obtained by pressure controlled systems (Fig. 1). Here, material flux control is provided by adjusting the input pressure p_v of the gas injection capillary. The flux is determined by the pressure drop $p_v - p_0$ along the capillary (p_0 is beam pressure). Since $p_0 \ll p_v$ for proper designed injection geometries the material flux is only a function of the injection pressure. This pressure can be measured and controlled easily by a high accuracy capacitance manometer (baratron).

Within the UHV growth chamber the molecular beams of the gaseous source materials are formed by so-called injectors, in principle effusion jets, in which nozzle plates ensure an as homogeneous as possible beam profile. While the inlet facility for the metal-organic compound has to be heated only slightly above room temperature in order to avoid condensation, the gaseous group V starting material, the hydrides have to be precracked in the injector. Most groups including our own use so-called low pressure cracking capillaries, where the AsH_3, PH_3, etc. are thermally decomposed by means of a heated metal (Ta etc.) filament or foil [16]. A different principle is applied in the high pressure effusion source usually applied in connection with flow controlled inlet systems [17]. The hydrides are injected at a pressure between 0.2 and 2 atm through alumina tubes with fixed small leaks into the UHV chamber. The tubes are heated, such that the hydrides are effectively decomposed by gas phase collisions within the tube. On their path through the leak there is a transition from hydrodynamic to molecular flow in the UHV environment.

3. Some milestones in MOMBE

Some important findings should be mentioned which made MOMBE or CBE a so attractive epitaxy technique for III–V heterostructures that it can compete with MBE or MOCVD. In MOCVD usually GaAs layers have been grown using AsH_3 and TMGa. Used in MOMBE this material combination apparently caused extremely high p-type background doping due to incorporated carbon (C). What first seemed to be a severe drawback of MOMBE then revealed to be an extremely interesting aspect of this technique. Pütz et al. [18,19] found out, that by using TEGa instead of TMGa extremely clean GaAs material with room temperature hole concentrations between 10^{14} and 10^{16} cm^{-3} could be deposited, whereas hole concentrations around 10^{20} cm^{-3} were obtained using TMGa. Consequently TMGa could be used as a p-type carbon dopant to obtain acceptor (C) doping levels near 10^{21} cm^{-3}, a value which was never achieved by other techniques at that time (Fig. 2). This new technique of p-type doping in MOMBE [20] uses TEGa as the basic Ga source and controlled amounts of TMGa are added to achieve the desired p-type doping level (Fig. 3). By variation of the alkyl beam pressure and the TMGa/TEGa ratio the hole concentrations can be adjusted between 10^{14} and 10^{21} cm^{-3} at 300 K. The obtained hole mobilities at 300 K are comparable with the best literature values (Fig. 3). In following studies carbon furthermore appeared to be a very convenient dopant in device applications because of its low diffusion constant in GaAs. Thus MOMBE studies opened completely new pathways for extremely high p-type doping by means of carbon incorporation. Meanwhile this C-doping technique has had some impact both on MBE

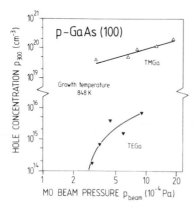

Fig. 2. Effect of metal-organic pressure p_{beam} on the background carbon doping in GaAs MOMBE/CBE. The 300 K hole concentration of GaAs epilayers grown with precracked AsH$_3$ (beam pressure 3×10^{-4} Pa) and TMGa and TEGa, respectively, is shown [18].

and MOCVD growth of GaAs. Using solid graphite sources [21] or TMAs instead of AsH$_3$ as a gaseous source [22] high carbon doping levels have also been reached in these techniques.

A further breakthrough of MOMBE or CBE in III–V technology concerns selective growth. Selective growth in the epitaxy of III–V compounds [23] means that on particular "open" areas of the substrate epitaxial growth is induced whereas other areas remain free of deposited material. It is obvious that selective growth in this sense is of virtual interest for monolithic integration of electronic and optoelectronic devices.

In epitaxy processes with gaseous sources and surface decomposition processes as in MOCVD and CBE (MOMBE) selectivity can be achieved by glassy overlayers (SiO$_2$, nitrides etc.) which are deposited on the wafer and structurized by lithographic processes. Open areas in the mask material expose the free GaAs wafer surface and enable locally restricted epitaxial growth. The underlying reason for selective growth in the openings, and no growth on the masking overlayer, is the different adsorption and/or nucleation-behaviour on the exposed III–V surface and the mask surface [24,25]. In standard MBE Ga excess is necessary for epitaxial growth; but the sticking coefficient of elemental Ga on SiO$_2$ is so high at growth temperatures, that nucleation occurs and as a consequence polycrystalline GaAs is deposited. Selective growth of GaAs in the strict sense is not possible in MBE. MOMBE is particularly suited for selective growth in comparison with MOCVD, since in contrast to MOCVD only chemical surface reactions are involved in epitaxial growth. Growth parameters such as substrate temperature and reactor pressure, of course, play a major role in enabling selective growth also in the case of MOCVD. Furthermore the kind of gaseous source material is important. The tendency of a special metal-organic to decompose within a certain temperature window selectively on the GaAs is important. The experimental results in Fig. 4, e.g., show that selective growth, i.e. good crystalline GaAs growth in the open GaAs mask window, is possible with TMGa and arsine in MOMBE at such low substrate temperatures as 770 K (up to 870 K). For the less stable TEGa compound, however, selective growth starts at higher temperatures, i.e. at around 870 K. For pressure conditions as in MOCVD with typical values between 500 and 10^5 Pa the tendency for selective growth is clearly seen to be restricted to higher temperature compared to MOMBE: with increasing pressure in the MOCVD system selectivity becomes worse and the temperature window is larger for the more stable metal-organic TMGa than for TEGa. It was furthermore found, that in MOCVD due to the importance of gas phase reactions and material transport within the gas stream above

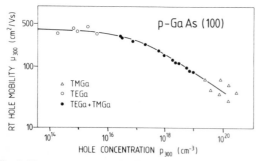

Fig. 3. Room temperature Hall mobilities μ_{300} of intentionally p-doped GaAs layers grown from precracked AsH$_3$ (beam pressure 3×10^{-4} Pa) and from TMGa only ($\triangle$), TEGa only ($\circ$), and mixtures of both alkyls ($\bullet$). Solid line: fit through best mobilities from literature [20].

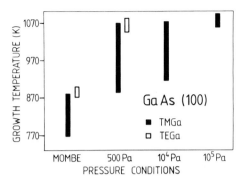

Fig. 4. Overview over temperature and pressure conditions, where selective area growth is possible in MOMBE (typical beam pressure 10^{-3} Pa) and in low pressure MOCVD (pressures between 500 and 10^5 Pa). The masking material, SiO_2, was structurized by means of optical lithography [16].

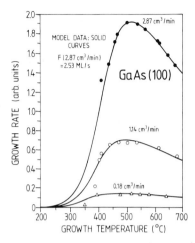

Fig. 5. Comparison of measured growth rates in GaAs MOMBE (points) versus growth temperature with results from model calculations (solid curves). Growth has been performed from precracked AsH_3 and TEGa with flow rates of 2.87, 1.14 and 0.18 cm^3/min. The corresponding coverage rates used for the model are 2.53, 1.00 and 0.16 monolayers/s [27].

the growing surface, the geometrical dimensions of the openings play a major role in selective growth, in particular when ternary compounds of a well determined stoichiometry are to be deposited [26]. The composition of the selectively deposited material depends on dimensions and directions of the mask openings. This is not the case for selective MOMBE growth, where only surface reactions are important. MOMBE or CBE is thus the ideal technique for selective growth of ternary III–V layers.

A milestone in the understanding of CBE of GaAs is provided by Robertson et al. [27], who studied surface chemical kinetics during growth. Using RHEED intensity oscillations during CBE of GaAs with TEGa and As_2 molecular beams originating from an arsine cracker these authors observed a significant variation of the GaAs growth rate with substrate temperature at a given TEGa flux (Fig. 5). For each of the incident fluxes the growth rate increases between approximately 375 and 500°C, but the rate is different for each flux. Above 500°C the growth rate decreases again with temperature. In the whole a direct proportionality between growth rate and incident flux is observed. The essential properties of this behaviour are explained by the authors as due to the surface pyrolysis of TEGa on the growing GaAs surface. In this model the existence of three adsorbed alkyl group III species is assumed:

triethyl (TEGa), diethyl (DEGa) and monoethyl (MEGa) gallium (Fig. 6). In this decomposition model initially a gas-phase TEGa molecule is trapped with a probability S_0 into a weakly bound physisorption state. After being trapped the ad-

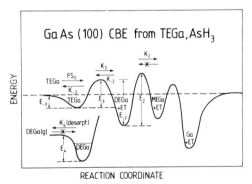

Fig. 6. Postulated reaction mechanism and reaction coordinate diagram for the TEGa surface chemical kinetics during CBE GaAs growth from precracked AsH_3 and TEGa. The different reaction constants K_i, K_{-j} and the trapping probability S_0 being used in the model are indicated as well as the activation barriers E_i and adsorption energies E_{-j} for the different species; ET means ethyl radical. The crossed reaction paths are not included in the model [27].

sorbed TEGa can either overcome the activation barrier E_3 and break the first ethyl–gallium bond, producing adsorbed diethyl gallium and ethyl radicals or desorb from the surface. The cleavage of each ethyl–gallium bond adds to the surface population of adsorbed ethyl radicals, which can eliminate atomic hydrogen (e.g. by surface β-elimination) to form ethylene. The desorption of ethylene is assumed to be very fast in comparison with the removal of atomic H. Thus the elimination of hydrogen from ethyl radicals is the rate limiting step to removal of ethyl radicals from the growth front. Within the model the different rate constants K_2, K_3, K_{-3} etc. are specified numerically and the generally observed growth rate behaviour can be described well (Fig. 5). Essential conclusions concerning the stepwise decomposition of TEGa into DEGa down to MEGa therefore seem to be correct. On the other hand the reactions of atomic hydrogen are not considered in the present formulation of the model but are expected to have a larger effect on carbon incorporation than the TEGa pyrolysis. The growth model presented by Robertson et al. [27] is a first important step to the understanding of MOMBE growth of GaAs even though the questions of carbon incorporation remain unanswered.

For heterostructure device applications it is important, that beside doping also ternary and quaternary III–V compounds can be grown in high quality. An important milestone in this respect was exhibited by Houng [28], who was the first to demonstrate the high quality CBE growth of AlGaAs/GaAs heterostructures. In a detailed study of the carbon incorporation during the AlGaAs CBE growth using TEGa, TEAl, TIBAl and AsH$_3$ Houng and his group could demonstrate that the alkyl–Al compounds appear to be the source for enhanced carbon incorporation in AlGaAs layers [29] and that an alkyl exchange reaction between TEGa and TEAl plays a major role in this incorporation mechanism. Based on this study the authors were able to grow high quality AlGaAs epilayers with reduced carbon contamination by replacing TEAl by TIBAl at growth temperatures around 560°C. With a V/III ratio of 20 modulation-doped AlGaAs/GaAs heterostructures could be grown which exhibit

electron mobilities in the 2DEG at 77 K of 88.600 cm^2/V · s at a sheet carrier concentration $n \simeq 6 \times 10^{11}$ cm^{-2}.

Gas source epitaxial techniques such as CBE (MOMBE) are particularly useful in the growth of phosphorus containing compounds, e.g. InP, GaInAsP etc., where MBE is absolutely unsuitable because of the phosphorus handling problems. In this field of P compounds, which is particularly interesting for optoelectronic applications Heinecke et al. [30,31] were able to achieve extraordinarily good results in MOMBE growth. Beside extremely high Hall mobilities also excellent optical properties (photoluminescence spectra) of the grown layers are demonstrated [30]. A detailed study of the effect of the different growth parameters, in particular the substrate temperature, lead to the preparation of high quality InP/GaInAs/ GaInAsP heterostructures, in which also a sharp profile of the critical element As at the interfaces could be demonstrated [31].

A detailed discussion of the device aspects of MOMBE and the achievements in this field is beyond the scope of this paper. Nevertheless some major breakthroughs must be mentioned. As early as 1986 Tsang demonstrated the high quality of AlGaAs/GaAs double heterostructure layers grown by CBE [32]. Threshold current densities comparable to MBE or MOCVD grown structures (< 0.5 kA/cm^2) could be achieved. Also high quality InGaAs/GaInAsP multiquantum well layers exhibiting threshold current densities as low as 860 A/cm^2 could be realized [33]. Beside layers and photodiodes [33] also GaInAs/ GaInAsP/InP heterostructure bipolar transistors (HBT) [34] as well as quantum-switched HBTs with negative differential resistance were prepared by CBE [35].

A special advantage of CBE/MOMBE with respect to device fabrication is she possibility of reaching high and sharp p-type doping profiles using carbon. This advantage of the technique was utilized for the fabrication of high quality HBTs [36] in which heavily carbon doped GaAs is used for the base. The quasi-metallic conductance behaviour of highly C-doped GaAs layers grown in MOMBE/CBE together with the possibility of good selective growth was also used recently to fabricate a new type of homoepitaxially grown permeable junction base transistor (PJBT), in which a vertical current is controlled by space charge regions around finger-like permeable base contacts formed by highly p-doped GaAs regions embedded in an n-type GaAs matrix [37].

In conclusion to this listing of device applications one might say, that MOMBE/CBE has reached a level of maturity, so that both conventional devices as well as new device concepts tailored to the MOMBE can be realized with high quality.

4. Interaction with other branches of surface science

Since MOMBE is based on complex surface reactions on the growing semiconductor surface, a better understanding of these reactions was a primary goal in all attempts to optimize the process and to use other, maybe more adapt source materials. Beside reaction studies during the real epitaxy process as in the case of Robertson et al. [27], standard surface science techniques such as reflection high energy electron diffraction (RHEED), high resolution electron energy loss spectroscopy (HREELS), and mass spectroscopy were soon applied to get a deeper insight into the surface decomposition reactions and the growth mechanism. A strong mutual interaction between classical epitaxy studies and more advanced surface characterisation techniques started to yield new impacts on both fields.

RHEED oscillation measurements, which can directly be performed on the growing surface in the epitaxial growth chamber were applied by Okuno et al. [38] to study the decomposition of TEGa in the MOMBE growth of GaAs and to compare the surface diffusion of Ga in MOMBE with that in MBE. From a detailed analysis of the damping speed of the RHEED oscillations as a function of the flux ratio of TEGa to Ga, when both beams are incident onto the surface, they conclude that part of the incident TEGa molecules pyrolyze into Ga atoms and migrate on the surface. Other molecules migrate as Ga

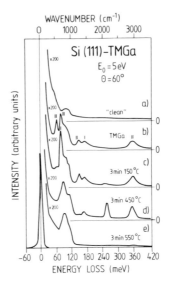

Fig. 7. High resolution electron energy loss (HREEL) spectra measured at room temperature on a Si(111) surface exposed to TMGa: (a) nearly clean surface; (b) after adsorption of 20 L TMGa at 300 K; the bars indicate IR absorption bands of solid TMGa; (c–d) after different annealing steps [39].

alkyles and pyrolyze at steps or kinks. The diffusion lengths of Ga alkyl molecules is longer than those of Ga atoms. This effect might explain some of the advantages of MOMBE over MBE, as far as layer by layer growth is concerned.

Spectroscopic techniques such as HREELS for the study of adsorbate vibrations can be used only in seperate analysis chambers, up to now rather than in-situ during the growth process itself. The first studies on surface decomposition reactions of metal-organic molecules were performed on silicon surfaces [39], since dipole induced (Fuchs–Kliewer) surface phonons as observed on GaAs surfaces do not appear on Si. Consequently the whole spectral range of energetically low lying molecule–substrate vibrations up to the CH vibrations of adsorbed alkyl radicals can be observed and detailed conclusions about split-off reactions of H etc. can be drawn. The investigations on Si surfaces are, of course, very useful for heteroepitaxial CBE growth of GaAs on Si. In the example of Fig. 7 the decomposition of TMGa is studied as a function of substrate temperature on Si(111). Characteristic features are the growth of the 84 meV loss feature due to Si–C (in

Si–CH$_x$ complexes) vibrations with increasing temperature and the simultaneous appearance of the Si–H vibrational loss at 258 meV. This characteristic pattern indicates split-off of CH$_3$ groups from the TMGa molecules, bonding of CH$_x$ to surface Si atoms and further fragmentation of the CH$_x$ groups at higher temperatures with bonding of H atoms to surface Si atoms. From such HREELS studies interesting conclusions can be drawn concerning the decomposition mechanism of metal-organics such as TMGa, TMAs, TEGa, TMAl etc. The decomposition products can be put into relation to their molecular geometry and the particular bonding strength between the carbon and the (semi) metal central atom in the molecule [39–41].

The characteristic difference between using TMGa and TEGa in MOMBE growth of GaAs, consists of the lower carbon incorporation by means of TEGa. This effect was first approached in a spectroscopic way by means of HREELS, XPS and TDS (thermal desorption spectroscopy) in an investigation of TEGa [42] and TMGa [43] decomposition on Si(100) by Lin et al. The following model for the decomposition process on Si(100) was developed: below 300 K molecular adsorption, at lower temperatures in multilayers, is revealed. For temperatures above 500 K decomposition of the adsorbed TEGa via surface β-elimination of ethylene and abstraction of the H atoms occurs, where step by step via the formation of DEGa and MEGa H atoms from the C$_2$H$_5$ groups are deposited on the Si surface and C$_2$H$_4$ molecules form which desorb. Finally the chemisorbed H also desorbs at higher temperatures to leave behind Ga atoms on the surface. In contrast to TEG, smaller amounts of hydrogen are available, when TMGa decomposes, to completely remove the carbon as methane from the surface.

It is interesting that Maeda et al. [44] come to similar conclusions concerning the TEGa decomposition on GaAs(100) surfaces from in-situ XPS studies during MOMBE growth. They also find molecular adsorption of TEGa below 450°C substrate temperature. Above 450°C TEGa decomposes on GaAs(100), possibly by a series of surface decomposition processes, where ethyl radi-

cals are split off. Further decomposition leads to desorption of ethylene and atomic H, which is adsorbed before it finally also somewhat more slowly desorbs. The remaining Ga atoms are incorporated into the new GaAs layer.

These surface science studies on the decomposition of TEGa and TMGa obviously support the essential conclusions from Robertson's et al. [27] growth studies on the GaAs MOMBE process, where a step by step decomposition of TEGa into DEGa and MEGa is concluded. The surface science investigations add the specific aspect to the reaction picture that β-elimination of ethylene and abstraction of H is important in TEGa decomposition where the C$_2$H$_5$ radicals split off an H atom to form desorbing C$_2$H$_4$ and adsorbed H. For TMGa this reaction is not possible, since too small a number of H atoms is available during the decomposition process. Surface spectroscopy studies therefore were able to make an interesting contribution to the understanding of the different behaviour of TMGa and TEGa concerning the incorporation of carbon during MOMBE growth.

5. Novel developments

The success of CBE techniques for the deposition of III–V semiconductor layer systems also lead to the development of similar experimental approaches for other semiconductor systems.

Si technology has benefitted a lot from the development of Si$_x$Ge$_{1-x}$ alloys since complex SiGe/Si heterostructures, as in the field of III–V technology, became possible, with all the new possibilities for advanced high speed devices. Although Si-MBE has been remarkably successful for this goal, also with respect to a reduction of growth temperature [45–47], it still involves some practical problems. For application purposes MBE lacks overall throughput since it utilizes a solid Si source. This makes the growth process discontinuous.

In accordance with the ideas put forward in III–V technology therefore the so-called gas source MBE (GSMBE) technique for deposition of Si and Si$_x$Ge$_{1-x}$ alloys was developed as a counterpart to Si-MBE. A UHV vessel is used as growth chamber and silane (SiH$_4$), disilane (Si$_2$H$_6$), germane (GeH$_4$) etc. are the conventional source materials, which are directed as molecular beams onto the growing surface [48–50]. By this technique epitaxial Si and SiGe films of excellent bulk crystal quality could be grown at temperatures as low as 500°C. This growth temperature is low compared to the 700°C, which are standardly used in low pressure vapor phase epitaxy (LPVPE). Even at such low growth temperatures surface defect densities below 5×10^3 cm^{-2} could be reached, while at temperatures around 600°C and above, the surfaces of the grown layers are essentially free of defects [48]. The attainment of such low temperature epitaxial growth is ascribed to the cleanliness of the UHV growth environment.

As with all gas beam techniques Si-GSMBE offers the possibility of excellent selective growth.

At a substrate temperature of 600°C, e.g., excellent Si growth was obtained from SiH$_4$ within SiO$_2$ mask openings, whereas the mask surface stayed completely free of any deposition [48].

Just recently the technique of GSMBE has also been applied to the epitaxial growth of semiconducting β-FeSi$_2$ (iron disilicide) on Si(111) [51]. β-FeSi$_2$ has gained considerable interest due to potential applications for Si based optoelectronic devices. The semiconducting β-FeSi$_2$ phase has a direct electronic gap near 0.8 eV and is therefore ideally adapted to the absorption minimum of glass fibers and the light emission of InP based III–V laser structures [52]. Beside the orthorhombic β-FeSi$_2$ phase also a metastable metallic γ-FeSi$_2$ phase with CaF$_2$ structure exists, which was observed to be stabilised as an epitaxial layer on Si [52].

GSMBE growth of β-FeSi$_2$ layers with good crystal quality was achieved by using molecular

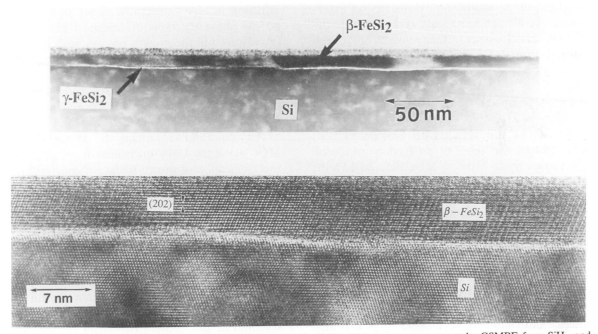

Fig. 8. Cross-section transmission electron microscopic images of a FeSi$_2$/Si heterostructure grown by GSMBE from SiH$_4$ and Fe(CO)$_5$ molecular beams. Upper part: overview image of the semiconducting β-FeSi$_2$ layer showing differently oriented domains. The bright line at the interface results from metallic γ-FeSi$_2$, which is substrate stabilised. Lower part: high resolution image with a Si $\langle 112 \rangle$ zone axis parallel to the electron beam. The thin metallic γ-FeSi$_2$ layer (1–6 ML thick) is visible between Si substrate and β-FeSi$_2$ [53].

beams of SiH_4 and $Fe(CO)_5$ (iron pentacarbonyl). They were directed onto the Si(111) or Si(100) surface at growth temperatures between 450°C and 550°C. Fig. 8 shows transmission electron micrographs of such a $FeSi_2$/Si(111) heterostructure grown by GSMBE [51,53]. In the high resolution image the excellent crystal quality, at least locally, can be seen. At the interface between the Si substrate and the β-$FeSi_2$ layer the loss of resolution is related to a metallic γ-$FeSi_2$ interlayer which grows under certain growth conditions as a substrate stabilized interlayer before the semiconducting β-$FeSi_2$ forms on top. This combination of β-$FeSi_2$ and metallic γ-$FeSi_2$ interlayer is also seen in the top overview image. From this picture the existence of several differently oriented β-$FeSi_2$ domains is revealed, whose growth is caused by the three-fold symmetry of the Si(111) substrate surface.

The presented examples show that GSMBE is developing into a very promising epitaxy technique for a variety of semiconductor and maybe also metallic layer systems. Its application, also in the field of Si technology, seems to be established, eventhough large scale application for device fabrication needs further research and development efforts.

6. Future prospects

For device production CBE has, up to now, not reached the level of perfection and wide application as do MBE or MOCVD and LPVPE in the field of SiGe/Si technology. Nevertheless the UHV based technique CBE promises a number of advantages over the competing techniques MBE and MOCVD, or LPVPE. This becomes clear, even when we restrict our consideration to III–V technology, i.e. on a comparison of MOMBE (CBE) with MBE and MOCVD.

As compared to MBE, MOMBE allows a continuous growth process without change of crucibles and subsequent bake-out procedures. Phophorus containing compounds can easily be grown in MOMBE without the problems with this element in MBE, which are related to the high vapour pressure of P. The quite accurate flux control in MOMBE by means of pressure control and leak valves facilitates the fabrication of compositionally graded layer structures in contrast to MBE, where the beam flux is determined by the crucible temperature. Lower surface defect densities without oval defects are more easily obtained in MOMBE than in MBE and, in particular, selective area growth is possible in MOMBE.

Also in comparison with MOCVD, MOMBE appears to be a quite attractive deposition technique with some remarkable advantages. In MOMBE, unlike in MOCVD, no large amounts of explosive carrier gas are needed. In addition, a higher precursor growth efficiency is given in MOMBE than in MOCVD (1%), i.e. less chemical waste needs to be disposed of. By thermally precracking the hydrides AsH_3, PH_3 etc. in MOMBE, lower V/III ratios need to be employed, i.e., less hydride quantities need to be stored. Because the hydrides are precracked the deposition process in MOMBE can run at lower temperatures than in MOCVD, which finally results in less diffusion of dopant atoms and interdiffusion in heterostructures. Since in MOMBE no gas phase reactions or transport phenomena are involved in the growth process, selective area growth does not suffer from problems associated with compositional and thickness inhomogeneity when ternary and quaternary As- and P-containing compound semiconductors are selectively deposited – a problem for MOCVD [26].

An important advantage of MOMBE over MOCVD is the UHV environment, which allows the application of in-situ surface characterisation techniques. These techniques, e.g. RHEED can be used for in-situ control of the growth process. But they can also be applied to study on structural, morphological and electronic properties of freshly grown surfaces, interfaces and thin layers, within the same vacuum conditions. This makes the UHV based CBE techniques particularly useful for fundamental studies on the growth and characterisation of novel, not well known layer systems. In this sense CBE might be the technique, where a surface scientist enters the most exciting field of layer growth with all its fascination but also bearing in mind the applications in semiconductor technology.

Acknowledgement

For helpful and critical remarks I would like to thank Hilde Hardtdegen, Karl Heinz Bachem and Markus Kamp.

References

[1] H.M. Manasevit, Appl. Phys. Lett. 12 (1968) 156.
[2] G.B. Stringfellow, in: Semiconductors and Semimetals, Vol. 22A (Academic Press, New York, 1985) p. 209.
[3] D.P. Dapkus, Ann. Rev. Mater. Sci. 38 (1978) 73.
[4] J.R. Arthur, J. Appl. Phys. 39 (1968) 4032.
[5] A.Y. Cho and J.R. Arthur, Prog. Solid State Chem. 10 (1975) 157.
[6] W.T. Tsang, in: Semiconductors and Semimetals, Vol. 22A (Academic Press, New York, 1985) p. 96.
[7] H. Lüth, G.W. Rubloff and W.D. Grobman, Solid State Commun. 18 (1976) 1427.
[8] K.H. Bachem and M. Heyen, Inst. Phys. Conf. Ser. 56 (1981) 1.
[9] E. Veuhoff, W. Pletschen, P. Balk and H. Lüth, J. Cryst. Growth 55 (1981) 30.
[10] N. Vodjdani, A. Lemarchand and H. Paradan, J. Phys. (Paris) Colloq. 43 (1982) C5-339.
[11] M.B. Panish, J. Electrochem. Soc. 127 (1980) 2730.
[12] A.R. Calawa, Appl. Phys. Lett. 38 (1981) 701.
[13] M.B. Panish and H. Temkin, Appl. Phys. Lett. 44 (1983) 785.
[14] M.B. Panish, H. Temkin and S. Sumski, J. Vac. Sci. Technol. B 3 (1985) 657.
[15] H. Temkin, M.B. Panish, P.M. Petroff, R.A. Hamm, J.M. Vandenberg and S. Sumski, Appl. Phys. Lett. 47 (1985) 394.
[16] H. Lüth, Proc. ESSDERC 1986, Cambridge (UK), Inst. Phys. Conf. Ser. 82 (1982) 135.
[17] M.B. Panish and S. Sumski, J. Appl. Phys. 55 (1984) 3571.
[18] N. Pütz, E. Veuhoff, H. Heinecke, M. Heyen, H. Lüth and P. Balk, J. Vac. Sci. Technol. B 3 (1985) 671.
[19] N. Pütz, H. Heinecke, M. Heyen, P. Balk, M. Weyers and H. Lüth, J. Cryst. Growth 74 (1986) 292.
[20] M. Weyers, N. Pütz, H. Heinecke, M. Heyen, H. Lüth and P. Balk, J. Electron. Mater. 15 (1986) 57.
[21] J. Nagle, R.J. Malik and D. Gershoni, J. Cryst. Growth 111 (1991) 264.
[22] T.F. Kuech, M.A. Tischler, P.-J. Wang, G. Scilla, R. Potemski and F. Cardone, Appl. Phys. Lett. 53 (1988) 1317.
[23] P. Balk and H. Heinecke, Proc. 4th Int. School on Physical Problems in Microelectronics, in: Physical Problems in Microelectronics, Ed. J. Kassabov (World Scientific, Singapore, 1985), p. 190.
[24] H. Lüth, Proc. European MRS Conf., Strasbourg, 1986, p. 413.
[25] H. Lüth, Inst. Phys. Conf. Ser. 82 (1986) 135.
[26] O. Kayser, J. Cryst. Growth 107 (1991) 989.
[27] A. Robertson, T.H. Chin, W.T. Tsang and J.E. Cunningham, J. Appl. Phys. 64 (1988) 877.
[28] Y.M. Houng, J. Cryst. Growth 105 (1990) 124.
[29] B.J. Lee, Y.M. Houng, J.N. Miller and J.E. Turner, J. Cryst. Growth 105 (1990) 168.
[30] H. Heinecke, B. Baur, R. Höger and A. Miklis, J. Cryst. Growth 105 (1990) 143.
[31] H. Heinecke, B. Baur, R. Höger, A. Miklis and R. Teichler, J. Cryst. Growth 111 (1991) 599.
[32] W.T. Tsang, Appl. Phys. Lett. 48 (1986) 511.
[33] W.T. Tsang, M.C. Wu, T. Tanbun-Ek, R.A. Logan, S.N.G. Chu and A.M. Sergent, Appl. Phys. Lett. 57 (1990) 2065; W.T. Tsang, J. Cryst. Growth 105 (1990) 1.
[34] W.T. Tsang, A.F.J. Levi and E.G. Burkhardt, Appl. Phys. Lett. 53 (1988) 983.
[35] M.C. Wu and W.T. Tsang, Appl. Phys. Lett. 55 (1989) 1771.
[36] C.R. Abernathy, S.J. Pearton, F. Ren, W.S. Hobson, T.R. Fullowan, A. Katz, A.S. Jordan and J. Kovalchick, J. Cryst. Growth 105 (1990) 375.
[37] J. Gräber, M. Kamp, G. Mörsch, R. Meyer, H. Hardtdegen and H. Lüth, Microelectronic Engineering 19 (1992) 131.
[38] Y. Okuno, H. Asahi, T. Kaneko, T.W. Kang and S. Gonda, J. Cryst. Growth 105 (1990) 185.
[39] A. Förster and H. Lüth, J. Vac. Sci. Technol. B 7 (1989) 720.
[40] H. Lüth, J. Vac. Sci. Technol. A 7 (1989) 696.
[41] H. Lüth and A. Förster, Proc. Int. Workshop on Science and Technology for Surface Reaction Processes, Tokyo 1992 (Phys. Soc. Japan, Tokyo, 1992) p. 63.
[42] R. Lin, T.R. Gow, A.L. Backman, L.A. Caldwell, F. Lee and R.I. Masel, J. Vac. Sci. Technol. B 7 (1989) 725.
[43] F. Lee, A.L. Backman, R. Lin, T.R. Gow and R.I. Masel, Surf. Sci. 216 (1989) 173.
[44] T. Maeda, J. Saito and K. Kondo, J. Cryst. Growth 105, (1990) 191.
[45] Y. Ohta, Thin Solid Films 106 (1983) 3.
[46] E. Kaspar, Silicon–Germanium Heterostructures on Silicon Substrates, Festkörperprobleme, Vol. 27 of Advances in Solid State Physics, Ed. P. Grosse (Viehweg, Braunschweig, 1987) p. 265.
[47] J.C. Bean, J. Cryst. Growth 81 (1987) 411.
[48] F. Hirose, M. Suemitsu and N. Miyamoto, Jpn. J. Appl. Phys. 28 (1989) L2003.
[49] H. Hirayama, T. Tatsumi, A. Ogura and N. Aizaki, Appl. Phys. Lett. 52 (1988) 1484.
[50] K. Werner, S. Butzke, J.W. Maes, O.F.Z. Schannen, J. Trommel, S. Radelaar and P. Balk, MRS Symposium Proc. 263 (1992) 261.
[51] Ch. Schäfer, B. Rösen, H. Moritz, A. Rizzi, B. Lengeler and H. Lüth, Appl. Phys. Lett. 62 (1993) 2522.
[52] A. Rizzi, H. Moritz and H. Lüth, J. Vac. Technol. A 9 (1991) 912.
[53] D. Gerthsen, Ch. Schäfer, B. Rösen, A. Rizzi, H. Moritz and H. Lüth, in: Proc. Int. Conf. on "Microscopy of semiconducting Materials", Oxford, 1993, in press.

Surface Science 299/300 (1994) 878–891
North-Holland

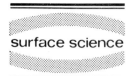

surface science

Bandgap engineering of semiconductor heterostructures by molecular beam epitaxy: physics and applications

Federico Capasso and Alfred Y. Cho

AT&T Bell Laboratories, 600 Mountain Avenue, Murray Hill, NJ 07974, USA

Received 11 May 1993; accepted for publication 19 July 1993

Bandgap engineering is a powerful technique for the design of new semiconductor materials and devices. Heterojunctions and molecular beam epitaxy allow band diagrams with nearly arbitrary and continuous bandgap variations to be made. In this way the transport and optical properties of these artificially structured semiconductors can be tailored at will. Interesting new phenomena have been discovered in these materials and a new generation of devices with unique capabilities is emerging from this approach. Finally, the tunability of band discontinuities via doping interface dipoles is discussed.

1. Introduction

During the last decade a powerful new approach for designing semiconductor structures with tailored electronic and optical properties, bandgap engineering, has spawned a new generation of semiconductor materials and of electronic and photonic devices [1]. Central to bandgap engineering is the notion that by spatially varying the composition and the doping of a semiconductor over distances ranging from a few microns down to ~ 2.5 Å (~ 1 monolayer), one can tailor the band structure of a material in a nearly arbitrary and continuous way [1]. Thus semiconductor structures with new electronic and optical properties can be custom-designed for specific applications.

In this section we summarize the milestones of research in semiconductor heterostructures from the early fifties to the late seventies. In the rest of the article we try to give the reader an intuitive appreciation of bandgap engineering and its applications through the systematic use of energy diagrams. Most of the specific examples come from our work and that of our collaborators at Bell Laboratories in the areas of electronics, photonics, optical and transport properties of artificially structured semiconductors. In the spirit of

this special issue of Surface Science, we would like to share with the readers our excitement for this interdisciplinary field of research where physics, materials science and technology complement each other.

Essential to the emergence and development of bandgap engineering have been semiconductor heterojunctions and molecular beam epitaxy. Early suggestions for using semiconductor heterojunctions to improve device performance are found in the celebrated transistor patent of Shockley [2]. A few years later, Kroemer [3] proposed the concept of a compositionally graded semiconductor. By spatially varying the stoichiometry of a semiconductor, an energy bandgap is produced that varies with position (graded bandgap). Thus "quasi-electric forces", equal to the spatial gradient of the conduction and valence band edge, respectively, are exerted on the electrons and holes. This concept is one of the earliest and simplest examples of bandgap engineering.

After the initial demonstrations of the homojunction semiconductor injection laser in the early 1960's, Kroemer suggested that carrier confinement in a low-gap region clad by wide-gap heterojunction barriers would make population inversion and laser action possible at much lower current densities [4]. The demonstration of a continuous wave (CW) heterojunction laser at 300 K [5] was made possible by the growth of high-quality AlGaAs/GaAs heterojunctions by liquid phase epitaxy and paved the way to high-performance lasers for lightwave communications.

The next breakthrough was the invention of molecular beam epitaxy (MBE) at Bell Laboratories by Cho and Arthur, an extraordinarily fruitful spin-off of Surface Science [6]. This epitaxial growth technique allows multilayer heterojunction structures to be grown with atomically abrupt interfaces and precisely controlled compositional and doping profiles over distances as short as a few tens of ångströms. Such structures include quantum wells, which are a key building block of bandgap engineering. These potential energy wells are formed by sandwiching an ultrathin lower gap layer (of thickness comparable or smaller than the carrier thermal de Broglie wavelength, which is ~ 250 Å for electrons in GaAs) between two wide-gap semiconductors (for example, AlGaAs). The spacing and position of the discrete energy levels in the well depend on the well thickness and depth. These quantum confined states were first observed in pioneering optical and transport experiments [7,8] performed at Bell Laboratories and IBM, respectively, in 1974.

If many quantum wells are grown on top of one another and the barriers are made so thin (typically < 50 Å) that tunneling between the coupled wells becomes important, a superlattice is formed [9–11]. Superlattices are new materials with novel optical and transport properties introduced by the artificial periodicity. Other superlattices with intriguing properties can be obtained by periodically alternating ultrathin n- and p-type layers (nipi (n-type intrinsic p-type intrinsic) superlattices) [11,12], or by alternating undoped layers with doped layers that have a wider bandgap (modulation-doped superlattices) [13]. In the latter structures electrons transfer from the parent donors into the lower gap layers. The spatial separation from the donors provides a high electron mobility parallel to the layers at low temperatures. Modulation doping in a single heterojunction led to a new high electron mobility field

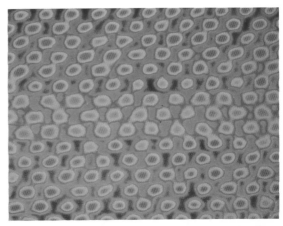

Fig. 1. High resolution (011) TEM cross section showing a lattice image of an InGaAs quantum well three molecular layers thick sandwiched between two InP lattice matched barriers, grown by Morton B. Panish at AT&T Bell Laboratories, by gas source MBE. The normal black and white contrast has been coded according to the color spectrum, the red end corresponding to higher intensities. The red spots represent tunnels between pairs of atoms. The minimum separation between the tunnels in InP is 3.4 Å. These abrupt heterointerface can be used as electron launchers in several transistor and detector applications (see Fig. 2). (Courtesy of J.M. Gibson and S.N.G. Chu.)

effect transistor [14]. Strained layer superlattices have also shown interesting properties and have potential as components of new devices [15].

It was not until the 1980's that MBE became a technology capable of producing "device quality materials" ready for commercial exploitation and semiconductor structures with arbitrary spatial control of doping and composition [16,17]. These new material structures have allowed an enormous breadth of new devices. Near infrared GaAs lasers for compact disc players, IMPATT diodes, and modulation doped field-effect transistors are among the many devices manufactured today by MBE.

2. Ballistic launching ramps

Heterojunction interfaces grown by MBE are atomically abrupt, as shown by the electron micrograph of Fig. 1. This represents a TEM cross-section of two InP/GaInAs lattice matched heterojunctions forming a quantum well, grown by gas source MBE. This material combination is currently used in lasers and detectors for light-

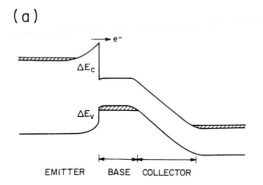

(a)

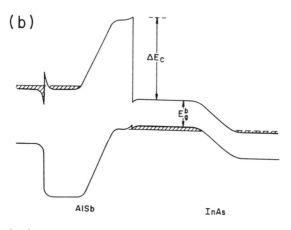

(b)

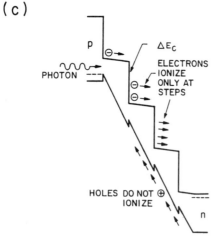

(c)

wave communications systems operating in the 1.3–1.55 μm wavelength region and is also important, as we shall see, for ultrahigh speed bipolar transistors. The abrupt heterointerfaces grown by MBE give rise to potential steps (band-discontinuities) in the conduction and valence bands, denoted by ΔE_c and ΔE_v, respectively, as illustrated in Fig. 2. These energy band diagrams show the conduction and valence band edges as a function of distance inside the materials. They are a powerful aid for the device physicist and engineer to visualize the motion of electrons and holes in these structures and the device operation.

The structures of Fig. 2 utilize abrupt heterojunctions as ballistic launching ramps for electrons to achieve improved device performance. An electron crossing these potential steps gains a kinetic energy equal to the conduction band discontinuity. This process is a ballistic acceleration over a distance equal to the interface width (~ 1 monolayer). The use of a band discontinuity as an electron launcher was first experimentally demonstrated by one of us and his collaborators in the context of enhancing the ionization probability after the step (see later in text) [18]. Shortly after, the use of these potential steps to achieve high nonequilibrium velocities was proposed [19]. Such a launching ramp has also been used to improve the high speed performance of heterojunction bipolar transistors (HBTs), as suggested by Kroemer and illustrated in Fig. 2a [20]. This was demonstrated by Chen and coworkers in an InP/GaInAs HBT grown by gas source molecular beam epitaxy [21]. They achieved a cutoff frequency f_T of 165 GHz. Electrons enter the

Fig. 2. Energy band diagrams of devices utilizing band discontinuities as launching ramps for electrons. (a) Heterojunction bipolar transistor (HBT) with abrupt emitter. The base thickness is typically in the 300–500 Å range and the collector is a few thousand Å thick. (b) Transistor structure used to demonstrate impact ionization by band discontinuities. Electrons entering the InAs base layer with a kinetic energy of 1.3 eV, greatly exceeding the InAs bandgap (0.43 eV at 77 K), can create an electron–hole pair by impact ionization. (c) Staircase solid state photomultiplier utilizing the abrupt heterointerfaces as dynodes.

GaInAs base from the InP emitter with a kinetic energy $\Delta E_c = 230$ meV, corresponding to a forward velocity approaching 10^8 cm/s. The enhanced hot electron transport in the base results in a shorter base transit time.

If the value of ΔE_c exceeds the ionization threshold energy (typically greater than the bandgap due to reasons of momentum and energy conservation) in the base material, electrons entering the base will create an electron/hole pair by impact ionization. This was originally proposed by Kroemer [22] and recently demonstrated using the structure of Fig. 2b [23]. In this device $\Delta E_c = 1.3$ eV, which greatly exceeds the electron ionization energy in InAs. In the common base configuration the observed incremental base-current at zero collector-base bias becomes negative as the emitter current is increased, providing direct evidence of impact ionization by a band-edge discontinuity in the base.

Impact ionization by band edge discontinuities can be used to implement the solid-state analog of a photomultiplier (PMT) [24]. In this structure (Fig. 2c), a compositionally graded material is used to create a sawtooth band diagram which under appropriate bias polarity acquires a staircase shape. Electrons impact ionize since ΔE_c exceeds the ionization energy in the region after the step. Holes created by ionization or photoinjected cannot ionize since the valence band steps are of the wrong sign to assist ionization and the applied bias is kept small enough that carrier cannot ionize by the action of the electric field alone. Since only one type of carrier ionizes in this structure and the avalanche is spatially localized, the avalanche noise is minimized and the structure mimics a PMT. Although the staircase PMT has not yet been implemented, the enhancement of the electron impact ionization rate α relative to the hole ionization rate β, when ΔE_c exceeds ΔE_v, was demonstrated in AlGaAs/GaAs multiquantum well (MQW) avalanche photodiodes (APD) [18]. This leads to a reduction of the avalanche noise, since the latter is known to be maximum when $\alpha = \beta$, as demonstrated recently in AlGaAs/GaAs and in AlInAs/InGaAs MQW structures [25,26]. Low noise MQW APDs for high bit rate long wave-

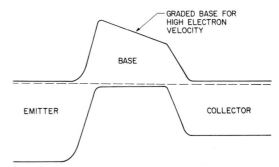

Fig. 3. Heterojunction bipolar transistor with reduced base transit time due to the bandgap grading in the base. The base thickness is typically in the 500–1500 Å range.

length fiber communications are in advanced development in Japanese companies such as Hitachi and NEC.

3. Bandgap grading

Compositional grading is another powerful device building block in bandgap engineered structures since it allows one to independently tailor the transport properties of electrons and holes [27]. One such application originally proposed by Kroemer [3] is illustrated in Fig. 3. Bandgap grading produces a force which acts on the electrons traversing the p-type base on their way to the collector. The net effect is that electrons move by drift rather than diffusion, with significantly higher velocity and therefore shorter base transit time. This was demonstrated by pulse response measurements in an AlGaAs/GaAs graded base phototransistor [28]. All optical measurements, by "pump and probe" techniques, of the velocity of minority electrons in heavily doped p-type graded gap AlGaAs with a quasi-electric field value of 8×10^3 V/cm yielded electron velocities as high as 2×10^7 cm/s [27]. The graded base transistor was first demonstrated by groups at Bell Labs [29] and Rockwell [30] in AlGaAs/GaAs and has since been used successfully in AlGaAs/GaAs [31] and in Si–Ge HBTs [32] to achieve a cutoff frequency f_T of 171 and 75 GHz, respectively.

Note that the wide gap emitter in the HBTs of Figs. 2a and 3 creates a barrier for holes in the

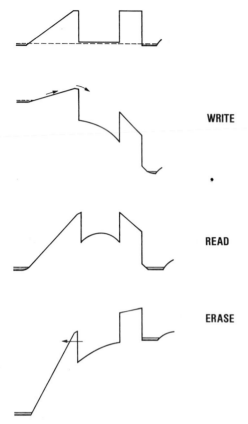

WRITE

READ

ERASE

Fig. 4. Floating gate memory utilizing a graded gap electron injector, under different operating conditions.

base. This key feature of HBTs allows one to achieve simultaneously high gain, low base resistance, and base–emitter capacitance for high-speed operation [20].

Fig. 4 illustrates the application of bandgap grading to memory devices [33]. The structure is of the floating gate type. Floating gate devices are commonly used commercially as permanent memories in silicon based electronics. In the structure of Fig. 4, a graded layer is used as an electron injector into a potential well (floating gate) where carriers are stored The stored negative charge depletes the channel underneath the floating gate, thus producing a drop in the drain current flowing through the channel. The device is maintained in this state as long as electrons remain in the well, which is determined by the thermionic emission rate over the AlAs barriers. Our mea-

surements indicate that this time constant varies from a few seconds at 300 K to an extrapolated 700 years at 77 K. The direct (GaAs)/indirect (AlAs) heterojunction strongly suppresses thermionic emission compared to the use of a direct/direct interface due to lack of lateral momentum conservation. A major difference in the operation of these devices compared to Si floating gate memories is that electrons are injected from the gate rather than the channel, adding functionality and flexibility to circuit design.

An application of graded materials to optical processing is illustrated in Fig. 5. Here a pump beam optically generates electrons and holes which are spatially separated in a time of a few picoseconds by the different quasi-electric fields

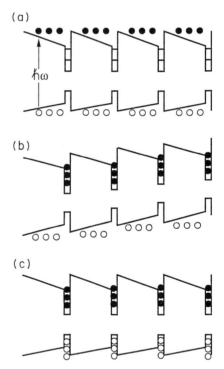

Fig. 5. New photorefractive materials based on semiconductor heterostructures. The spatial separation of photogenerated electrons and holes creates an electric field which modulates the refractive index via the electrooptic effect. (a) During photoexcitation; (b) immediately after photoexcitation, the electrons have moved to the quantum well under the influence of the conduction-band quasi-field; (c) at a later time, the holes have moved to the quantum well eliminating the space-charge field.

in the conduction and valence band [34]. The space charge electric field created by this separation persists until holes reunite with the electrons in the well in ~ 50 ps, the drift time in the graded region. If a second linearly polarized pulse is shined on the structure within such a time interval, the photoinduced space charge field will rotate the polarization of the probe beam via the refractive index modulation caused by the electrooptic effect. This structure behaves, therefore, as a new kind of photorefractive material which can be used as an optical switch when inserted between two crossed polarizers. The pump beam will activate the switch and allow the transmission of the weaker probe pulse. This effect was recently demonstrated in an AlGaAs/GaAs 50 period structure with 50 Å well and 500 Å graded regions [34]. Note that unlike conventional photorefractive materials, only two optical beams are needed and the presence of defects is not required.

4. Resonant tunneling and superlattice devices

4.1. Resonant tunneling diodes and transistors

Other fascinating phenomena and applications arise from the abruptness of heterojunction interfaces. Electrons can form standing wave patterns in the direction perpendicular to the layer so that the corresponding kinetic energy is quantized similar to the particle in a box problem in elementary quantum mechanics (quantum size effect). Of course, the motion along the plane is free so that the net effect of size quantization is that the conduction band is divided up in many subbands whose bottom is represented by the energy levels in Fig. 6a. The quantum size effect is at the basis of quantum well lasers where it can be used to shift the emission wavelength into the visible or to reduce the threshold current [35,36]. One can tunnel into the quantized states of the well if the perpendicular kinetic energy of electrons matches that of the energy levels of the well and the barriers are made sufficiently thin. In the limit of negligible scattering in the double barrier region this resonant tunneling process is concep-

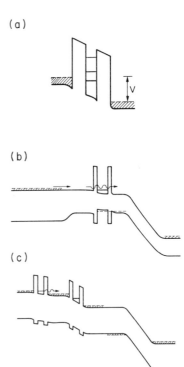

Fig. 6. Band diagrams of various resonant tunneling devices. The quantum wells barrier thicknesses are typically in the range 50–100 Å and 20–50 Å respectively. (a) Biased double barrier diode showing the states of quantum well, each corresponding to a transmission resonance. (b) Resonant tunneling bipolar transistor under operating conditions. (c) Multi-state resonant tunneling transistor.

tually similar to that giving rise to transmission resonances in an optical cavity at well defined wavelengths. In practice scattering processes tend to destroy the coherent interference so that the process in many practical situations is better described in terms of sequential tunneling into and out of the well.

It is clear from the above discussion that double barriers with heavily doped regions as contact layers behave as negative differential resistance diodes (Fig. 6a), as first shown by Chang et al. [7]. The current exhibits a sharp drop as the bias is increased corresponding to the state of the well being lowered below the conduction band edge in the emitter layer. In this situation resonant tunneling of electrons from the emitter into the well is inhibited for lack of lateral momentum conservation. In a series of pioneering experiments Sollner, Brown and coworkers at Lincoln Laboratory operated AlGaAs/GaAs resonant tunneling diodes as mixers up to 1.8 terahertz and as oscillators up to 420 gigahertz [37,38]. More recently using AlSb for the barriers and InAs for the well they have obtained oscillation frequencies of 675 GHz.

Although resonant-tunneling diodes may find application in oscillators, logic circuits require devices with input–output isolation in order to achieve gain and avoid loading of the input stages. This isolation is best accomplished with a transistor.

In general terms, a resonant-tunneling transistor is a device that uses an applied control voltage to modulate the difference between the energy levels of the quantum well and the energy of the incident electrons. Thus the resonant-tunneling current through the double barrier can be made to peak at one or more values of the control voltage, corresponding to the different energy levels [39]. Note that a transistor with such a current–voltage characteristic has multiple on and off states, corresponding respectively to the peaks and valleys of the $I-V$ curve. Although multiple-valued logic has been the subject of considerable theoretical investigation, all proposed and demonstrated circuit architectures have until now relied on conventional two-state devices or tunnel diodes. Multiple-valued logic may one day reduce the interconnection complexity of integrated circuits.

Conceptually, the simplest way to build a resonant-tunneling transistor is to form a contact with the heavily doped quantum well of a double barrier. This low-resistance contact would serve as the control terminal. This approach is, however, fraught with major technical difficulties, and attempts in this direction have had marginal success. A conceptually similar but technologically much easier approach, originally suggested by Capasso and Kiehl [39], is to incorporate a double barrier into the base of a bipolar transistor.

Fig. 6b shows the energy-band diagram for a resonant-tunneling bipolar transistor [40]. As the base–emitter voltage is increased, resonant tunneling through each subband first reaches a maximum and is then suppressed as the bottom the subband (the energy level shown in the well in Fig. 6b) is lowered below the conduction-band edge in the emitter. This should produce multiple peaks in the collector current, i.e. multiple negative transconductance.

The first operation of a resonant-tunneling bipolar transistor was reported in 1986 [40]. Room temperature operation with a single peak in the $I-V$ curve and a $3:1$ peak-to-valley ratio was demonstrated. With the band structure of Fig. 6b it is difficult, however, to achieve multiple peaks of comparable height – a useful feature for circuit applications.

To solve this problem the new structure shown in Fig. 6c was introduced [40]. In this new device, two or more double barriers are placed in the emitter rather than in the base. To achieve multiple peaks in the collector current versus base-emitter voltage space charge buildup in the quantum wells is exploited. The attendant electrostatic screening effect makes the field across the two double barriers spatially nonuniform, and higher in the well closer to the base. Thus as the bias is increased, resonant tunneling through the two wells is suppressed at two different voltages, yielding two peaks in· the $I-V$ characteristics. (When resonant tunneling is suppressed at only one well, current continuity is insured by inelastic tunneling.)

Capasso, Cho and Sivco [40] at Bell Labs have fabricated such a structure with $Al_{0.48}In_{0.52}As$ and $Ga_{0.47}In_{0.53}As$ by MBE, putting two double barriers in the emitter. At room temperature the device has demonstrated two current peaks with excellent peak-to-valley ratios, current gains in excess of 60 and a cutoff frequency $f_T \geq 24$ GHz.

The kind of transfer characteristic discussed above makes possible a class of circuits with greatly reduced complexity – requiring far fewer transistors per function than do conventional circuits. For example, a parity-bit checker for four-bit words has been built with a *single* multistate transistor [40]. Such circuits are normally used in digital communication systems to detect errors by checking if there is an odd or even number of ones (or zeros) in the binary word. Conventional 4-bit checkers require 24 transistors. Multistate transistors are also attractive for frequency multipliers. With the device shown in Fig. 6c, an input frequency of 350 MHz has been multiplied by five [40]. Other possible applications include fast analog-to-digital converters and memory cells [40].

4.2. Superlattice devices

Consider now a periodic structure with many quantum wells. The barrier layers should be thin enough to allow significant tunneling. The perpendicular kinetic energy of electrons in such a structure breaks up into allowed bands separated by forbidden gaps, just as we would expect in any periodic structure. This is of course analogous to the formation of energy bands in solids in consequence of their periodic crystal lattices.

There are, however, important differences between the natural crystal lattice and the man-made superlattice – as these structures are called [9]. First, the period of the superlattice is typically 30–100 Å, whereas the period of the crystal lattice is only a few ångströms. Therefore the energy bands of the superlattice, which arise from the coupling between the energy levels of the potential wells, are typically much narrower – on the order of 10 meV; they are called minibands. Second, the superlattice is generally one-dimensional, unlike the three-dimensional crystal lattice. (Two-dimensional superlattices with lateral structures and patterned metallic electrodes, have also been studied.)

The relation between energy and wavenumber in a miniband of width Δ can be written [9]

$$E = \tfrac{1}{2}\Delta[1 - \cos(ka)], \tag{1}$$

reflecting the periodicity of the superlattice with period a. Consequently, the electron group velocity $v(k)$ oscillates with the wavenumber k according to

$$v(k) = \frac{1}{\hbar}\frac{\mathrm{d}E}{\mathrm{d}k} = \frac{a\Delta}{2\hbar}\sin(ka). \tag{2}$$

This is in sharp contrast to the velocity of a free electron, which increases linearly with k. This band picture of transport in a superlattice is valid only if Δ is significantly larger than the collisional broadening $\hbar/\tau$ (where τ is the scattering time)

or, equivalently, if the mean free path is substantially greater than the superlattice period.

Applying an electric field F causes the wavevector k to increase linearly with time according to the quasi-Newtonian expression $\hbar\,\mathrm{d}k/\mathrm{d}t = eF$. For free electrons this would lead to a continuous increase in the velocity. But in a superlattice the velocity decreases once the wavevector k crosses $\pi/2a$, as we see from Eq. (2). Esaki and Tsu, in their seminal 1970 paper [9], showed that as a result the steady-state drift velocity also decreases, once the field F exceeds the threshold value $\hbar/ea\tau$. For such fields the electron distribution transfers to regions of the band structure where the group velocity decreases with increasing k according to Eq. (2), because of Bragg reflections from the miniband boundary. For typical superlattice periods, on the order of 100 Å and collision times τ of about 3×10^{-12} s, one gets a threshold field of 3×10^3 V/cm for the onset of negative differential resistance. This effect has proved elusive until recently. One difficulty in the observation of this phenomenon arises from the space charge injection from the contacts which gives rise to a nonuniform electric field and high field domains in the superlattice. Another problem is the presence of negative differential resistance due to the electron transfer effect into the satellite valleys, which in most III–V semiconductors occurs for values of the electric field comparable to those predicted by Esaki and Tsu in a superlattice.

The above difficulties can be overcome using the structure of Fig. 7a [41]. Here electrons can be injected at arbitrarily low density into the superlattice by controlling the base emitter bias. The superlattice is placed in the collector of a bipolar transistor, which is very low doped. Under these conditions the electric field across the superlattice remains uniform. In addition, in our bipolar transistor structure one cannot observe NDR by intervalley transfer under conditions of constant emitter current injection. Under these conditions, in fact, the decrease in velocity caused by the higher effective mass of the satellite valleys is compensated by an increase in the carrier density; the collector current is therefore not altered. On the other hand, the Esaki and Tsu

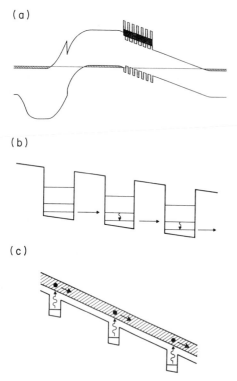

(a)

(b)

(c)

Fig. 7. Superlattice devices. (a) Heterojunction transistor used
to inject electrons into the miniband of the superlattice placed
in the collector layer. The shaded region in the superlattice
indicates the first miniband. (b) Sequential resonant tunneling
between coupled wells. (c) Quantum well infrared photodetector based on intersubband absorption between bound and
continuum states.

mechanism can be observed since the Bragg-re-
flected electrons in the negative-mass region of
the miniband give rise to an opposite flux so that
the collector current decreases while the base
current increases to maintain a constant emitter
current. We recently implemented this structure.
It had an AlInAs emitter, a quaternary AlInGaAs
base and an AlInAs(17 Å)/GaInAs(37 Å) 14
periods undoped superlattice followed by an In-
GaAs undoped layer in the collector. Negative
differential conductance was observed in the collector current measured as a function of collector
base bias, at a constant injected emitter current
[41]. This represents the first clear evidence of
the negative differential conductance by Bragg
reflection predicted by Esaki and Tsu [9].

If the field is much greater than $\hbar/ea\tau$, one
might expect the electron's wavenumber to increase to a value several times $\pi/2a$ before it is
scattered. Consequently the velocity should display oscillatory behavior in real space as well as
velocity space. This suggests the possibility for
extracting coherent radiation with an appropriately designed device (Bloch oscillator) [9].

In a recent landmark experiment [42], coherent submillimeter-wave emission from Bloch oscillations has been observed in an AlGaAs/GaAs
superlattice. The oscillation frequency was tuned
with the applied electric field from 0.5 THz to 2
THz, in accordance with the expression for the
Bloch frequency $\omega_B = eFa/\hbar$.

In superlattices with weak coupling between
wells (typically ≥ 100 Å), transport is best described in terms of tunneling between the localized states of the individual wells (Fig. 7c). This
sequential resonant tunneling process has been
clearly observed via photocurrent measurements
in a 35 period AlInAs/GaInAs superlattice with
wells and barriers about 140 Å thick [43].

Quantum states are also formed in the classical continuum above the barriers, due to quantum reflections. Barry Levine and coworkers at
Bell Labs have recently used these states to
demonstrate a new infrared detector (Fig. 7c)
sensitive in the 8–12 μm wavelength region [44].
High quality thermal images, with a noise equivalent temperature difference, have been obtained
using 128×128 arrays of these detectors multi-
plexed to Si C-MOS circuitry [45]. The potential
advantage of such detectors over conventional
HgCdTe devices lies in the virtues of GaAs/Al-
GaAs technology. This material system, which
includes the substrate, provides superior overall
uniformity, yield, and reliability. Furthermore, the
readout electronics might in future be monolithi-
cally integrated on the same substrate.

5. Coupled quantum well molecules with giant nonlinear optical properties

The structures discussed in this section can be
viewed as "quasi-molecules" with giant dipole
matrix elements and nearly equally spaced energy

(a)

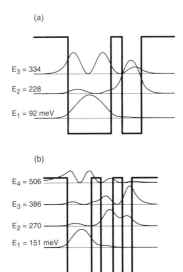

$E_3 = 334$

$E_2 = 228$

$E_1 = 92$ meV

(b)

$E_4 = 506$

$E_3 = 386$

$E_2 = 270$

$E_1 = 151$ meV

Fig. 8. Conduction band energy diagrams of a single period of the AlInAs/GaInAs coupled quantum well nonlinear optical structures. Shown are the positions of the calculated energy levels and the corresponding modulus squared of the wave functions. (a) Structure used for resonant second harmonic generation. The GaInAs wells have thicknesses of 64 and 28 Å and are separated by a 16 Å AlInAs barrier; (b) structure used for triply resonant third harmonic generation. The GaInAs wells have thicknesses of 42, 20 and 18 Å, respectively and are separated by 16 Å AlInAs barriers.

levels (Fig. 8). These characteristics are responsible for their very large nonlinear optical susceptibilities [46,47]. Both were grown in the AlInAs/GaInAs system lattice matched to InP, and only the thickest well was doped n-type. The choice of this material system facilitates the tunnel coupling between the layers due to the low effective mass of the barrier region compared to AlGaAs and provides a large band discontinuity essential for confining four states in the three-well structure separated by an energy corresponding to that of the photon from a CO_2 laser (~ 130 meV). To optically excite the quantized motion normal to the layer interfaces, one must use light with a component of the polarization normal to the layers. In our experiments this was done using linearly polarized light in a multipass waveguide structure wedged at 45°. In the second harmonic generation (SHG) experiment a coherent polarization is created at double the frequency ω of the pump wave (a CO_2 laser beam) due to the

lack of reflection symmetry of the two-well structure (Fig. 8a). This coherent polarization radiates a wave of frequency 2ω colinear with the pump. The vicinity of the pump photon energy to $E_2 - E_1$ and of the second harmonic photon to $E_3 - E_1$ produces a strong resonant enhancement of the nonlinear susceptibility $\chi^{(2)}_{2\omega}$ associated with SHG. The maximum susceptibility ($\chi^{(2)}_{2\omega}$) corresponds to exact matching, i.e. $\hbar\omega = E_2 - E_1 = E_3 - E_2$. This can be achieved using the large linear Stark effect typical of this structure [46] by applying an electric field of suitable polarity. Under these conditions a $\left| \chi^{(2)}_{2\omega} \right| = 10^{-7}$ m/V was measured, approximately three hundred times the value of $\left| \chi^{(2)}_{2\omega} \right|$ in bulk GaAs at $\lambda = 10$ μm [46].

The three-well structure (Fig. 8b) with the near equal separation of its four energy levels is suitable for triply resonant third harmonic generation (THG). In this process a pump wave at frequency ω sets up a nonlinear polarization at 3ω which coherently radiates a wave at this frequency. The nonlinear susceptibility $\chi^{(3)}_{3\omega}$ which enters the expression for the polarization is strongly enhanced when the condition $\hbar\omega = E_2 - E_1 = E_3 - E_2 = E_4 - E_3$ is met. It is important to note that a parabolic well (i.e. an harmonic oscillator potential) is unsuitable for this purpose since in such a system the electron oscillations are linear. THG experiments in the structure of Fig. 8b have found a $\left| \chi^{(3)}_{3\omega} \right| = 10^{-14}$ (m/V)^2 at 300 K [47]. At cryogenic temperatures $\left| \chi^{(3)}_{3\omega} \right|$ is four times larger. These are the highest third-order susceptibilities of any known material system. The three-coupled well structure was also used to study multiphoton electron escape from a well, the analog of multiphoton ionization of an atom [48]. In this process electrons in the biased well are photoexcited into the continuum via a CO_2 laser using a three photon transition giving rise to a photocurrent. The cross section for this process is found to be many orders of magnitude larger than in atomic and molecules.

6. Fabry–Perot electron filters and bound states in the continuum

In the previous section we have seen how confined states of quantum wells are central in

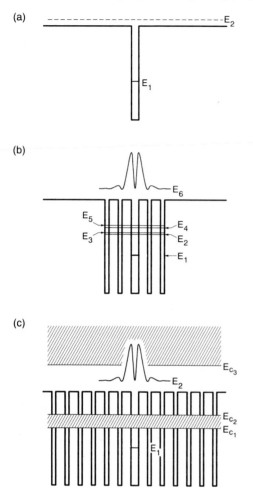

Fig. 9. Conduction band diagrams of AlInAs/GaInAs heterostructures used in the study of continuum states: (a) reference sample. Shown are the ground state of the well ($E_1 = 204$ meV) and the position (dashed line) of the first transmission resonance in the continuum ($E_2 = 560$ meV); (b) quantum well cladded by two-period quarter-wave stacks (Fabry–Perot electronic filter). Shown is $|\psi|^2$ of the localized quasi-bound state ($E_6 = 560$ meV) formed in correspondence to the transmission resonance and the positions of new states created at lower energies; (c) in the superlattice limit the $\lambda/4$ stacks behave as Bragg reflectors. The state above the well (E_2) now becomes a bound state localized by the superlattice minigap.

number of tunneling and optical phenomena. Highly localized states can also be created at energies above the barrier height in a potential well using constructive interference phenomena [49,50].

Consider first a conventional rectangular well (Fig. 9a). At energies greater than the barrier

height one has a continuum of scattering states. For discrete energies corresponding to a semi-integer number of electron wavelengths across the well one finds transmission resonances. Although at these energies the electron amplitude in the well layer is enhanced, the wave functions do not decay exponentially in the barrier, unlike the confined states of the well, but are plane-wave-like. These states can be localized in the well using as barriers stacks of layers of thickness $\lambda/4$ each where λ is the de Broglie wavelength in the layer (at the energy of the selected transmission resonance) (Figs. 9b and 9c). Constructive interference between the waves partially reflected by the heterointerfaces of the $\lambda/4$ stacks leads to the formation of a quasi-bound state above the center well (Fig. 9b). This strongly narrows the transmission resonance in analogy with a Fabry–Perot optical filter where sharp optical resonances are produced using as high reflectivity mirrors dielectric quarter-wave stacks. The degree of localization increases with the number of periods; in the structure with just two-period stacks, the wave function is already highly confined (Fig. 9b). In the superlattice limit and at low temperatures the stacks become Bragg reflectors; a minigap opens up (Fig. 9c) and the localized state becomes a bound state at energies greater than the barrier height. The prediction that certain oscillatory potentials support bound states in the continuum, due to quantum interference, was first put forth by von Neumann and Wigner in 1929 [51]. The formation of such states using superlattices was proposed in 1977 [52].

The reference sample (Fig. 9a) had twenty 32 Å InGaAs quantum wells n-type doped separated by 150 Å undoped AlInAs barriers. In the other three structures the 32 Å wells, doped to the same level, were cladded, respectively, by one-period, two-period (Fig. 9b) and six-period (Fig. 9c) $\lambda/4$ stacks consisting of 39 Å AlInAs barriers and 16 Å GaInAs wells, designed as discussed above. The phase coherence length in the superlattice structure of Fig. 9c is estimated to be ~ 300 Å at 10 K [41].

The absorption spectra of the reference sample is broad with a long wavelength cutoff determined by the height of the barrier and the energy of the ground state. In the structure with one $\lambda/4$ period the peak is considerably narrower and centered at an energy corresponding to the transition between the ground state of the well and the localized resonant state at the energy E_6. As the number of quarter-wave stacks is doubled the absorption peak does not shift and considerably narrows [50], precisely the behavior expected for a Fabry–Perot. In fact the observed narrowing (16 meV) can be quantitatively explained in terms of the reflectivity increase of the $\lambda/4$ stacks. In the structure with six periods at cryogenic temperatures, the highly localized state becomes effectively a bound state confined by Bragg reflectors from the superlattice. The absorption spectrum shows an isolated peak at 360 meV of width ~ 10 meV corresponding to the transition from the state E_1 to the state E_2 in Fig. 9c [49]. It is worth noting that the width of the transition to the confined state above the well in the two- and six-period structure is identical to that of the bound-to-bound state transition measured in a conventional 55 Å thick GaInAs well with 300 Å thick barriers, thus demonstrating the bound nature of the state above the well.

7. Interface engineering: tunable discontinuities

The ability to control band discontinuities would add a powerful degree of freedom to bandgap engineering. A number of years ago we proposed and demonstrated a method to tune interface barrier heights and band discontinuities [53]. It consists of the incorporation by MBE of ultrathin (a few monolayers) ionized donor and acceptor sheets within a few tens of ångströms from the heterojunction interface (planar doping) (Fig. 10). The electrostatic potential of this "doping interface dipole" is added to or subtracted from the dipole potential of the discontinuity. In the limit of a few atomic layer separations between the charge sheets, electrons crossing the interface "see" a new band discontinuity $\Delta E_c \pm e\Delta\phi$, where $\Delta\phi$ is the potential of the double layer (Fig. 10). Using this technique, Capasso et al. demonstrated an artificial reduction of the conduction band barrier height of the order of

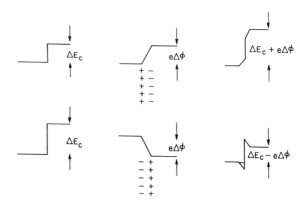

Fig. 10. Tunable band discontinuities formed from doping-interface dipoles introduced in situ by delta-doping during MBE growth. (Top) The conduction band discontinuity is increased. (Bottom) Interchange of the acceptor and donor sheets reduces the band discontinuity. Tunneling through the spike and size quantization in the triangular well plays a key role in this reduction.

0.1 eV in an $Al_{0.25}Ga_{0.75}As/GaAs$ heterojunction [53]. This lowering of the barrier produced an enhancement (by one order of magnitude) of the collection efficiency of photocarriers across the heterojunction as compared to identical heterojunctions without doping interface dipoles.

Recently, very interesting experimental and theoretical developments have confirmed the validity of the doping interface dipole approach to tuning band offsets. XPS and RHEED studies of thin (0–2 monolayers) Si interlayers fabricated in the interface region of AlAs/GaAs(001) heterostructures by MBE indicate that most of the Si atoms remain localized at the interface, exhibit long range order, and generate a local dipole of up to 0.38 eV [54]. Such a dipole can be subtracted from or added to the intrinsic AlAs/GaAs valence band offset depending on the growth sequence (AlAs on GaAs versus GaAs on AlAs) [54]. The most likely key to understanding these results is provided by two recent first-principles calculations [55,56], which show that in such heterostructures interlayer atoms occupying lattice sites on consecutive planes would exhibit a charge transfer of -1 and $+1$ electron per atom prior to dielectric screening, reflecting their valence difference relative to the matrix. The resulting predicted dipole is consistent in direction and order of magnitude with the experimental dipole.

These techniques of modifying band offsets offer tremendous potential for heterojunction devices. The performance of some of the devices described in this article (see Figs. 2a and 3) and others depends exponentially on band discontinuities. Thus artificial variations of these quantities by $1 kT$ or $2 kT$ (thermal energy where k is Boltzmann's constant and T is temperature) can significantly alter their performance.

8. Conclusions

In summary, we have illustrated the scope, power and use of bandgap engineering. This technique, in our opinion, will have far reaching implications and impact on electronics and photonics. Commercial devices which use it include multiquantum well lasers, high electron mobility transistors, avalanche photodiodes with separate absorption and multiplication and planar doped barrier diodes. Devices under advanced development and likely to be commercialized include heterojunction bipolar transistors, multiquantum well avalanche photodiodes and quantum well infrared detectors.

From a physics point of view bandgap engineered structures have displayed a richness of new electrical and optical phenomena associated with band structure tailoring and reduced dimensionality. Research in this interdisciplinary environment, where material science, physics and technology meet and complement each other, is likely to lead to more exciting developments in the near future.

Acknowledgments

It is a pleasure to acknowledge the many collaborators and colleagues who have significantly contributed to this work: F. Beltram, C.G. Bethea, T.H. Chiu, A.Y. Cho, S.N.G. Chu, J. Faist, A. Franciosi, J.M. Gibson, A.C. Gossard, I.R. Hayes, R.A. Kiehl, L.M. Lunardi, B.F. Levine, R.J. Malik, K. Mohammed, M.B. Panish, S.E. Ralph, S. Sen, D.L. Sivco, C. Sirtori, P.R. Smith, W.T. Tsang, A.S. Vengurlekar, G.F. Williams and S. Datta.

References

[1] F. Capasso, J. Vac. Sci. Technol. B 1 (1983) 457;
For a review see F. Capasso, in: Semiconductors and Semimetals, Vol. 24, Ed. R. Dingle (Academic Press, Orlando, 1987).
[2] W. Shockley, U.S. Patent 2 569 347 (1951).
[3] H. Kroemer, RCA Rev. 18 (1957) 332.
[4] H. Kroemer, Proc. IEEE 51 (1963) 1782.
[5] Zh.I. Alferov et al., Sov. Phys. Semicond. 4 (1971) 1573 [translated from Fiz. Tekh. Pouprovodn. 4 (1970) 1826];
I. Hayashi, M.B. Panish, P.W. Foy and S. Sumski, Appl. Phys. Lett. 17 (1970) 109.
[6] A.Y. Cho and J.R. Arthur, Progress in Solid-State Chemistry, Vol. 10, Eds. J.O. McCaldin and G. Somorjai (Pergamon, New York, 1975) p. 157;
M.B. Panish, Science 208 (1980) 916.
[7] L.L. Chang, L. Esaki and R. Tsu, Appl. Phys. Lett. 24 (1974) 593.
[8] R. Dingle, W. Wiegmann and C.H. Henry, Phys. Rev. Lett. 33 (1974) 827.
[9] L. Esaki and R. Tsu, IBM J. Res. Dev. 14 (1970) 61.
[10] L. Esaki and L.L. Chang, Phys. Rev. Lett. 33 (1974) 495.
[11] R. Dingle, A.C. Gossard and W. Wiegmann, Phys. Rev. Lett. 34 (1975) 1327.
[12] G. Döhler, J. Phys. Status Solidi (B) 52 (1972) 79; (b) 52 (1972) 533.
[13] R. Dingle, H.L. Stormer, A.C. Gossard and W. Wiegmann, Appl. Phys. Lett. 33 (1978) 665.
[14] H.L. Stormer, R. Dingle, A.C. Gossard, W. Wiegmann and M.D. Sturge, Solid State Commun. 29 (1979) 705;
R. Dingle, A.C. Gossard and H.L. Stormer, U.S. Patent 4 194 935 (1980);
T. Mimura, S. Hiyamizu, T. Fuji and K. Nanbu, Jpn. J. Appl. Phys. 19 (1980) L125.
[15] J.W. Matthews and A.E. Blakeslee, J. Cryst. Growth 27 (1974) 118; 29 (1975) 273; 32 (1976) 265;
For a recent review see: G. Osbourn, IEEE J. Quantum Electron. QE-22 (1986) 1677.
[16] A.Y. Cho, Thin Solid Films 100 (1983) 91.
[17] A.Y. Cho, J. Cryst. Growth 111 (1991) 1.
[18] F. Capasso, W.T. Tsang, A.L. Hutchinson and G.F. Williams, Appl. Phys. Lett. 40 (1982) 30.
[19] J.A. Cooper, F. Capasso and K.V. Thornber, IEEE Electron Device Lett. ED-83 (1982) 497.
[20] H. Kroemer, Proc. IEEE 70 (1983) 13.
[21] Y.K. Chen, R.N. Nottenburg, M.B. Panish, R.A. Hamm and D.A. Humphrey, IEEE Electron Device Lett. 10 (1989) 267.
[22] See, A.G. Milnes and D.L. Feucht, Heterojunctions and Metal–Semiconductor Junctions (Academic Press, New York, 1972) p. 29.

[23] A. Vengurlekar, F. Capasso and T. Heng Chiu, Appl. Phys. Lett. 57 (1990) 1772.

[24] F. Capasso, W.T. Tsang and G.F. Williams, IEEE Trans. Electron Devices ED-30 (1983) 381.

[25] T. Kagawa, H. Iwamura and O. Mikami, Appl. Phys. Lett. 54 (1989) 33.

[26] T. Kagawa, A. Kawamura, A. Asai, M. Naganuma and O. Mikami, Appl. Phys. Lett. 55 (1989) 993.

[27] F. Capasso, Ann. Rev. Mater. Sci. 16 (1986) 263.

[28] F. Capasso, W.T. Tsang, C.G. Bethea, A.L. Hutchinson and B.F. Levine, Appl. Phys. Lett. 42 (1983) 93.

[29] J.R. Hayes, F. Capasso, A. Gossard, R.J. Malik and W. Wiegmann, Electron. Lett. 19 (1983) 410.

[30] D.L. Miller, P.M. Asbeck, R.J. Anderson and F.H. Eisen, Electron. Lett. 19 (1983) 367.

[31] T. Ishibashi, H. Nakajima, H. Ito, S. Yamahata and Y. Matsuoka, Technical Digest of the Device Research Conf., 1990, Santa Barbara, paper VIIB-3.

[32] S.S. Iyer, G.L. Patton, J.M.C. Stork, B.J. Meyerson and D.L. Harame, IEEE Trans. Electron Devices ED-36 (1989) 2043.

[33] F. Capasso, F. Beltram, R.J. Malik and J.F. Walker, IEEE Electron Device Lett. 9 (1988) 377.

[34] S.E. Ralph, F. Capasso and R.J. Malik, Phys. Rev. Lett. 63 (1989) 2272.

[35] J.P. van der Ziel, R. Dingle, R.C. Miller, W. Wiegmann and W.A. Nordland, Jr., Appl. Phys. Lett. 26 (1975) 463.

[36] G.P. Agrawal and N.K. Dutta, Long Wavelength Semiconductor Lasers (Van Nostrand-Reinhold, New York, 1986) p. 372.

[37] T.C.L.G. Sollner, W.D. Goodhue, P.E. Tannenwald, C.D. Parker and D.D. Peck, Appl. Phys. Lett. 43 (1983) 588.

[38] E.R. Brown, T.C.L.G. Sollner, C.D. Parker, W.D. Goodhue and C.L. Chen, Appl. Phys. Lett. 55 (1989) 1777.

[39] F. Capasso and R.A. Kiehl, J. Appl. Phys. 58 (1985) 1366.

[40] For a review on quantum transistors, see, F. Capasso et al., IEEE Trans. Electron Devices ED-36 (1989) 2065; For quantum devices in general, see, F. Capasso and S. Datta, Phys. Today 43 (1990) 74.

[41] F. Beltram, F. Capasso, D.L. Sivco, A.L. Hutchinson, S.N.G. Chu and A.Y. Cho, Phys. Rev. Lett. 64 (1990) 3167.

[42] C. Waschke, H.G. Roskos, R. Schwedler, K. Leo, H. Kurz and K. Köhler, Phys. Rev. Lett. 70 (1993) 3319.

[43] F. Capasso, K. Mohammed and A.Y. Cho, IEEE J. Quantum Electron. QE-22 (1986) 1853.

[44] B.F. Levine, C.G. Bethea, G. Hasnain, J. Walker and R.J. Malik, Electron. Lett. 24 (1988) 747.

[45] F. Capasso, Microelectron. Eng. 19 (1992) 909.

[46] C. Sirtori, F. Capasso, D.L. Sivco, A.L. Hutchinson and A.Y. Cho, Appl. Phys. Lett. 60 (1992) 151.

[47] C. Sirtori, F. Capasso, D.L. Sivco and A.Y. Cho, Phys. Rev. Lett. 68 (1992) 1010.

[48] C. Sirtori, F. Capasso, D.L. Sivco and A.Y. Cho, Appl. Phys. Lett. 60 (1992) 2678.

[49] F. Capasso, C. Sirtori, J. Faist, D.L. Sivco, S.N.G. Chu and A.Y. Cho, Nature 358 (1992) 565.

[50] C. Sirtori, F. Capasso, J. Faist, D.L. Sivco, S.N.G. Chu and A.Y. Cho, Appl. Phys. Lett. 61 (1992) 888.

[51] J. von Neumann and E. Wigner, Phys. Z. 30 (1929) 465.

[52] F. Stillinger, Physica B 85 (1977) 270.

[53] F. Capasso, A.Y. Cho, K. Mohammed and P.W. Foy, Appl. Phys. Lett. 46 (1985) 664.

[54] L. Sorba, G. Bratina, G. Ceccone, A. Antonini, J.F. Walker, M. Micovic and A. Franciosi, Phys. Rev. B 43 (1991) 2450.

[55] A. Munoz, N. Chetty and R.M. Martin, Phys. Rev. B 41 (1990) 2976.

[56] M. Peressi, S. Baroni, R. Resta and A. Baldereschi, Phys. Rev. B 43 (1991) 7347.

Surface Science 299/300 (1994) 892–908
North-Holland

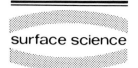

surface science

Mechanisms of GaAs atomic layer epitaxy: a review of progress

John M. Heitzinger [a], J.M. White [*,a] and J.G. Ekerdt [b]

[a] *Department of Chemistry and Biochemistry and the Center for the Synthesis, Growth, and Analysis of Electronic Materials,*
University of Texas, Austin, TX 78712, USA
[b] *Department of Chemical Engineering and the Center for the Synthesis, Growth, and Analysis of Electronic Materials,*
University of Texas, Austin, TX 78712, USA

Received 30 April 1993; accepted for publication 9 July 1993

In this article we review the surface chemistry of several precursors for the growth of GaAs. Specifically, we discuss trimethylgallium, triethylgallium, and arsine. Our discussion focuses on the importance of surface chemistry in achieving atomic layer epitaxy (ALE) of GaAs. While ALE has been the subject of numerous publications, a thorough understanding of the mechanism(s) of GaAs ALE has not been attained. We wish to illustrate how surface science experiments have helped in the search for a complete understanding. More generally, the results discussed in this paper are relevant to gas source molecular beam epitaxy, metal organic molecular beam epitaxy, organometallic vapor phase epitaxy, and chemical beam epitaxy.

1. Introduction

1.1. General background information

The continued trend towards the reduction in the feature size of electrical devices and the increasing complexity of these devices has placed stringent demands on the techniques used to fabricate them. We are now at the point where an atomic level understanding is crucial for further development. Within the field of device fabrication, the growth of GaAs has received a great deal of attention due to its superior high frequency characteristics and unique optical properties when compared to Si [1]. Much progress in the growth of semiconductor materials has been made since the early days of molecular beam epitaxy (MBE), and many new techniques have evolved from the field of MBE [2].

The development of MBE was begun in 1958 by Günther, who devised the means to deposit stoichiometric films of III–V semiconductors from molecular beams [3]. These films were deposited on glass substrates and were therefore polycrystalline. A decade later, Davey and Pankey achieved growth of single crystal GaAs using Günther's technique under better vacuum conditions and on single crystal substrates [4]. This was followed by important studies of the surface processes which occurred during growth by Arthur [5–7] and Foxon [8–11].

One of the most significant developments was the introduction of gas phase sources for crystal growth [12–19]. This led to gas source MBE (GSMBE), in which the group III component is supplied from a solid elemental source and the group V component is supplied in the gas phase. Metal-organic MBE (MOMBE), in which the group III component is supplied in the gas phase and the group V component is supplied from a solid elemental source was also developed. In addition, techniques in which both the group III and the group V components are supplied in the gas phase have been developed. These include metal-organic chemical vapor deposition (MOCVD), which is also known as organometallic vapor phase epitaxy (OMVPE), and chemical beam epitaxy (CBE), which is a low pressure

* Corresponding author.

variation of MOCVD. A subfield of these techniques (including MBE) is atomic layer epitaxy (ALE) which was first developed by Suntola and Antson for the deposition of ZnSe on glass substrates [20]. ALE is best described as a process in which the surface controls the growth, not the supply of the source. The reactants are supplied as alternating pulses of molecules or atoms and the film growth proceeds in a stepwise fashion. Ideally, a single monolayer is deposited per pulse. Using this technique, extremely precise control of the film thickness is possible.

The use of gas phase sources for the growth of III–V materials has definite advantages [21]. Precise and reproducible control of the delivery of the group III and group V element can be achieved by using mass flow controllers. In addition, more uniform films can be obtained. Dopant gases may be premixed with the group III or group V gas, allowing the stoichiometry of the film to be carefully controlled. However, the deposition process becomes much more complex when gas phase precursors are used, owing to gas phase reactions (at high pressure) and more complex growth kinetics on the surface when compared to growth using solid elemental sources.

In order to improve current deposition techniques, and thus device quality, a detailed understanding of the reaction mechanisms must be achieved. Important species must be identified and the energetics and kinetics of various reactions must be determined. Additionally, the adsorption site(s), surface structure, and stoichiometry must be investigated. Also, the sticking coefficient is an important parameter and its temperature and coverage dependence should be examined.

In this paper, we review the results of surface science experiments involving three gas phase precursors for the growth of GaAs: trimethylgallium (TMGa), triethylgallium (TEGa), and arsine. These three molecules have been the most commonly used gas phase precursors for GaAs growth, and, as such, have been studied most extensively. Since much of the research has been aimed at an understanding of ALE, we review it in this light. However, the results discussed herein are also relevant to GSMBE, MOMBE, MOCVD,

and CBE. We further limit our discussion to the GaAs(100) surface. This is because most film growth, and thus most research, is carried out on (100) substrates. Film growth on this surface allows the fabrication of rectangular chips whose edges are {110} planes, which allow for easy cleavage [1]. While there is an enormous amount of literature describing techniques and conditions for GaAs growth, the focus here is to review work that is aimed at understanding the fundamental chemical processes occurring on the surface, which are important for film growth.

The outline of the paper is as follows. We first describe three proposed mechanisms for ALE of GaAs. Conditions under which ALE of GaAs have been achieved are then summarized. Next, we briefly discuss surface reconstructions of GaAs(100) and present results of surface science studies involving TMGa, TEGa, and AsH$_3$, commenting on the importance of these results and their implications for understanding the mechanisms of GaAs ALE. Finally, we summarize and discuss future directions.

1.2. Models of GaAs atomic layer epitaxy (ALE)

The proposed mechanisms of GaAs ALE are illustrated in Fig. 1.

1.2.1. The selective adsorption model

The basis for the selective adsorption model [22–26] is the expected preference of a molecule

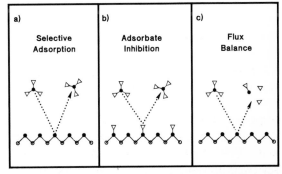

Fig. 1. Schematic representation of three proposed mechanisms for GaAs ALE. Solid circles represent Ga atoms, open circles represent As atoms, and triangles represent alkyl groups (CH$_3$ or C$_2$H$_5$). From Ref. [56].

containing a group III element (e.g., TMGa) to chemisorb on a group V element adsorption site (e.g., an As site), and vice versa. Thus, adsorption of a Ga alkyl species converts an As-terminated surface to a Ga-terminated surface, at which point adsorption of the Ga containing species stops. On an ideally terminated Ga rich surface (1 monolayer), no As sites would be available for chemisorption of Ga species. A complementary process is expected if one begins with a Ga rich surface and exposes it to an As containing molecule.

1.2.2. The adsorbate inhibition (site blocking) model

The basic feature of the adsorbate inhibition mechanism [27–30] of ALE is that the conversion of an As rich surface to a Ga rich surface, with Ga alkyl species, leaves the surface covered with adsorbates (alkyl groups, i.e., CH_3, C_2H_5, etc.). The adsorbates then inhibit any further decomposition of Ga containing species and thus, no further deposition of Ga occurs.

1.2.3. The flux balance model

In the flux balance model [31–34], Ga alkyl species convert an As rich surface to a Ga rich surface. Ga containing molecules may still adsorb and decompose on the Ga rich surface, but Ga alkyl species desorb quantitatively from the surface, i.e., the incoming flux of Ga alkyl species is "balanced" by desorption of Ga alkyl species from the surface. The desorbing Ga alkyl species need not be the same molecule which is dosed. This balance of adsorbing and desorbing Ga alkyl species prevents deposition of more than one monolayer of Ga.

1.3. Conditions used to achieve GaAs atomic layer epitaxy (ALE)

In this section we briefly describe the conditions used to achieve ALE using $TMGa/AsH_3$ and $TEGa/AsH_3$. Because this topic has been reviewed previously [35], we will not report all of the available results. Rather we wish to give the reader a feel for typical conditions used to achieve ALE.

1.3.1. The trimethylgallium and arsine system

Nishizawa and coworkers [27] were able to achieve ALE at low pressures. For a substrate temperature of 773 K and an AsH_3 injection pressure of 6×10^{-5} Torr, GaAs ALE was observed for TMGa injection pressures between 3×10^{-6} and 1×10^{-4} Torr. The operating mode was 20 s of AsH_3 exposure, 3 s evacuation, 4 s of TMGa exposure, and 3 s evacuation. The effect of AsH_3 pressure was also investigated. For a substrate temperature of 773 K and a TMGa injection pressure of 6×10^{-5} Torr, ALE growth was observed for AsH_3 injection pressures between 5×10^{-5} and 6×10^{-4} Torr.

Ozeki et al. have achieved excellent ALE growth of GaAs using "pulsed jet epitaxy" at a total pressure of 20 Torr [22,36]. ALE growth was achieved over the temperature range of 713 to 833 K. H_2 was used as a carrier gas. The partial pressure of TMGa was varied between 4×10^{-5} and 1×10^{-1} Torr. The AsH_3 partial pressure was varied between 4×10^{-4} and 2×10^{-1} Torr. Pulse durations between 1 and 25 s were used for TMGa. A typical AsH_3 pulse duration was 10 s. Hydrogen purges of 3 s were used between TMGa and AsH_3 pulses.

1.3.2. The triethylgallium and arsine system

Ohno and coworkers have achieved ALE growth of GaAs using TEGa and AsH_3 in an atmospheric pressure MOVPE chamber [37]. ALE was observed only from 613 to 623 K when 0.61 μmol/cycle TEGa and 38 μmol/cycle AsH_3 were used. An operation mode of 5 s AsH_3, 5 s H_2 purge, 5 s TEGa, and 5 s H_2 purge was used. For a substrate temperature of 623 K and an AsH_3 supply of 38 μmol/cycle, the amount of TEGa injected per cycle could be varied between 0.6 to 0.8 μmol/cycle. The growth rate was independent of AsH_3 supply over the range of 100 to 300 SCCM for a TEGa supply of 0.61 μmol/cycle and substrate temperature of 623 K.

1.4. Surface reconstructions of GaAs(100)

In this section, we briefly describe the surface reconstructions of GaAs(100) important for the work presented in this paper. This is not intended to be a comprehensive review of this topic, but rather, a simple introduction to relevant material.

The most studied and best understood surface of GaAs(100) is the (2×4) (or $c(2 \times 8)$) As rich surface. A scanning tunneling microscopy (STM) study carried out by Pashley and coworkers [38] has verified previous predictions of the structure of this surface [39]. The unit cell of the As rich (2×4) surface consists of rows of three As dimers and one dimer vacancy. The $4 \times$ periodicity in the (110) direction is due to the regular array of dimers, while the $2 \times$ periodicity in the $(\overline{1}10)$ direction is caused by the dimers themselves. The $c(2 \times 8)$ structure is due to a special arrangement of 2×4 units. We will, therefore, refer to this surface as the $c(2 \times 8)/(2 \times 4)$ As rich surface for the remainder of this paper. The As coverage of this surface is thought to be 0.75 monolayers (ML).

Another As rich surface of interest for this article is the $c(4 \times 4)$ surface. Biegelsen et al. [40] have studied this surface with STM and observe three As dimer pairs running in the (011) direction. These dimers are aligned perpendicular to the As dimers observed on the $c(2 \times 8)/(2 \times 4)$ surface. The As coverage of this surface has been estimated to be between 0.9 and 1.75 ML of As. Structures with one, two, and three As dimers in $c(4 \times 4)$ arrays have been proposed for this surface to try to explain the variation in As coverage [41,42]. Biegelsen and coworkers only observed the structure with three As dimers [40].

The Ga rich $c(8 \times 2)$ surface has also been studied by STM [40]. The 4×2 subunit consists of two dimers adjacent to one another and two missing dimers. These 4×2 subunits are arranged in such a way as to form a $c(8 \times 2)$ structure. The proposed structure would have 0.5 ML of Ga in the topmost layer. However, third layer Ga atoms will also be exposed. Auger electron spectroscopy (AES) measurements have been used to estimate the amount of As in the first atomic layer. Values between zero and 0.25 have been reported [43–45].

The (4×6) Ga rich surface has also been studied by AES. It is estimated that 0.27 ML of As reside in the surface layer. The observed (4×6) low energy electron diffraction (LEED)

pattern could be due to a superposition of a 4×1 (or $c(8 \times 2)$) and a 1×6 pattern [46]. STM images show no regions of (4×6) symmetry, but do show 4×1 and 2×6 domains [40].

For the (1×6) surface, LEED patterns typically show sharp 1×6 spots plus streaks corresponding to $2 \times$ and $3 \times$ symmetry [40]. STM images also show considerable ordering in the $2 \times$ direction, and the unit cell appears to be 2×6. Based on the STM images, Biegelsen and coworkers [40] propose a model in which the As coverage in the surface layer is 0.33. AES measurements estimate the As coverage in the first layer to be between 0.42 and 0.52 [45,47].

Finally, the (4×1) surface reconstruction has been estimated to have an As coverage in the surface layer of 0.25 ML. Frankel et al. [48] propose that the unit cell of this surface consists of three Ga dimers and one dimer vacancy. Imperfect long range order between individual unit cells along the (110) direction may be responsible for the degradation of the twofold symmetry.

2. Results

2.1. The interaction of trimethylgallium with GaAs(100) surfaces

We now describe the interaction of trimethylgallium (TMGa) with GaAs(100) surfaces. We begin by discussing experiments carried out on the Ga rich (4×6) reconstructed surface, discuss work on the Ga rich (1×6) reconstructed surface, and finally the As rich $c(2 \times 8)/(2 \times 4)$ surface.

2.1.1. The Ga rich (4×6) surface

Exposure of TMGa to a (4×6) Ga rich GaAs(100) surface at 108 K leads to molecularly adsorbed TMGa. This was demonstrated by high resolution electron energy loss spectroscopy (HREELS) and X-ray photoelectron spectroscopy (XPS) [49]. With large enough exposures, multilayers of TMGa are formed at 108 K. Heating the (4×6) Ga rich GaAs(100) surface to 200 K desorbed the TMGa multilayers, leaving a monolayer of TMGa adsorbed on the surface.

Above 200 K, part of the molecularly adsorbed TMGa dissociated, forming monomethylgallium (MMGa) and possibly some dimethylgallium (DMGa). Fig. 2 shows the temperature programmed desorption (TPD) spectrum following saturation TMGa exposure at 200 K. Some of the species adsorbed on the surface desorbed, resulting in peaks in the temperature programmed desorption (TPD) spectrum at 350 K for 99 amu (DMGa), 84 amu (MMGa), and 69 amu (Ga) [28,50–53]. Some disagreement exists about the identity of the desorbing species due to the efficient fragmentation of TMGa during electron impact ionization. Since all three signals are correlated and have a constant intensity ratio, the peaks have been attributed to desorption of a single species, TMGa [50]. Others have attributed the 350 K desorption peak to a mixture of DMGa and TMGa [28,51–53]. It is not yet clear if the desorption of TMGa at this temperature was due to associative desorption or non-dissociated TMGa desorption.

Increasing the temperature above 350 K caused additional desorption, resulting in a peak at approximately 500 K. Creighton has attributed desorption at this temperature to recombinative TMGa desorption from dissociated TMGa [50]. However, as with the desorption state at 350 K, there is some disagreement as to whether this

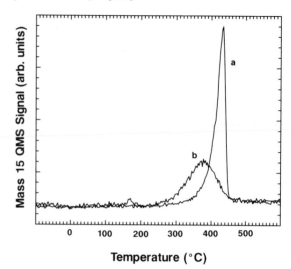

Fig. 3. Methyl radical desorption from the Ga rich (4×6) surface (curve a) and from the As rich c(2×8)/(2×4) GaAs(100) surface (curve b). Note that the temperature is reported in degrees Celsius. From Ref. [56].

peak is due solely to TMGa or a mixture of TMGa and DMGa [28,51–53].

MMGa appears to be the dominant species on the surface above 400 K, and it remains stable until 650 K [49,54]. The presence of MMGa is supported by static secondary ion mass spectroscopy (SSIMS), which showed a strong MMGa signal that cannot be accounted for by fragmentation of TMGa or DMGa [54]. Surface MMGa would result from a stepwise transfer of methyl groups from adsorbed TMGa to nearby Ga atoms. Above 650 K methyl radicals desorbed from the surface, resulting in a peak in the TPD spectrum at 710 K (440°C), as shown in Fig. 3 (curve a) [28,49–54]. Using multiphoton ionization techniques, it has been shown that the desorption species were indeed methyl radicals and not methane [29]. The narrow FWHM of the methyl radical desorption peak suggests that there is only one type of binding site for methyl groups on the (4 × 6) Ga rich surface at temperatures above 650 K [50]. A kinetic analysis of the methyl radical desorption shows the preexponential factor, ν, and activation energy for desorption, E_a, to be coverage dependent [50]. In the limit of zero coverage, E_a is found to be 46.1 ± 0.05 kcal/mol with ν equal to $1.6 \times 10^{14 \pm 0.3}$ s^{-1}. As an alterna-

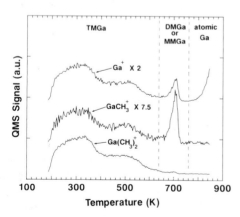

Fig. 2. TPD results for Ga(CH$_3$)$_x$ fragments (scaled appropriately) following a saturation dose of TMGa on the GaAs(100) (4 × 6) surface. The lower curve is the $m/e = 99$ (^{69}Ga(CH$_3$)$_2^+$) signal, the middle curve is the $m/e = 85$ (^{69}GaCH$_3^+$) signal × 7.5, upper curve is the $m/e = 69$ (^{69}Ga$^+$) signal × 2. From Ref. [50].

tive, Redhead analysis [55] has been used to get an approximate value for E_a, assuming a preexponential factor of 1×10^{13} s^{-1}. McCaulley et al. [52] report a value of E_a of 43 kcal/mol while Creighton and Banse [56] report 44.1 kcal/mol.

There was also a small amount of MMGa, or possibly DMGa, which desorbed at 710 K, as well as "excess" atomic Ga, which desorbed between 800 and 900 K. This Ga desorption has been termed "excess" since, following TMGa exposure, the Ga desorption signal exceeds the Ga desorption signal from the undosed surface. Interestingly, this excess Ga can be detected before methyl radical desorption by Auger electron spectroscopy (AES), but not after methyl radical desorption. This suggests that methyl radicals stabilize a gallium coverage that is higher than that obtained on the adsorbate free (4×6) Ga rich surface. Following methyl radical desorption, the "excess" Ga agglomerates into droplets [56]. No carbon was detected using AES following TPD experiments [50].

Using a molecular beam technique developed by King and Wells [57], the total amount of TMGa that reversibly and irreversibly chemisorbs on the (4×6) Ga rich GaAs(100) surface was estimated to be 1.7×10^{14} molecules cm^{-2} [50,54]. This is close to the expected Ga vacancy density of 1.56×10^{14} cm^{-2}. It was further estimated that 8×10^{13} TMGa molecules cm^{-2} are required to saturate the methyl radical desorption state at 710 K. If each TMGa molecule were to dissociate and lead to the desorption of three methyl radicals, 2.4×10^{14} methyl radicals cm^{-2} would be desorbed. This is 3 to 4 times higher than the absolute methyl radical yields reported by Donnelly and McCaulley (5.5×10^{13} to 8.1×10^{13} cm^{-2}) [51]. The values reported by Donnelly were obtained by comparison of the methyl radical desorption signal to the 25 through 27 AMU signals for ethylene desorption from triethylgallium decomposition. The discrepancy between the values reported by Creighton and Donnelly may be due to the different types of analysis.

Molecular beam techniques were also used at temperatures where GaAs growth is carried out to measure the TMGa + GaAs(100) reaction rate directly, i.e., the reactive sticking coefficient

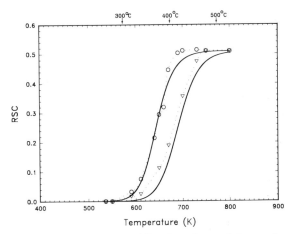

Fig. 4. The temperature and flux dependence of the reactive sticking coefficient (RSC) for TMGa impinging on a Ga rich (4×6) surface. Open circles are for a TMGa flux of 10^{12} molecules cm^{-2} s^{-1} and open triangles are for a TMGa flux of 10^{13} molecules cm^{-2} s^{-1}. The dashed lines are guides for the eye. The solid lines are calculated from a model discussed in Ref. [56].

(RSC) [56]. Fig. 4 shows the results for the (4×6) Ga rich GaAs(100) surface using two fluxes of TMGa. The RSC increases with temperature and reaches a value of about 0.5 above 700 K for both TMGa fluxes. Since the TMGa flux and the methyl radical desorption rate (from TPD results) are known, the coverage of methyl groups on the surface can be determined. At temperatures where the RSC is about 0.5, the methyl coverage is essentially zero. This means that when TMGa impinges on an adsorbate free Ga rich GaAs(100) surface at high temperature (about 700 K) it has a 50% chance of irreversibly decomposing.

Maa and Dapkus have carried out reflectance-difference spectroscopy (RDS) experiments during TMGa exposures to a Ga rich (4×6) GaAs(100) surface at 703, 723, and 773 K [25,26]. The experimental apparatus used in the work of Maa and Dapkus was sensitive to the existence of Ga dimers. Upon exposure of the surface to TMGa at a pressure of 1.2×10^{-5} Torr and an exposure time of six seconds, the RDS signal decreased rapidly at all three substrate temperatures. The RDS signal reached a lower limit, and then returned to the initial level once the TMGa dose was stopped. The time required to return to the initial level depended strongly on the substrate temperature, increasing with decreasing substrate temperature. The decrease in the RDS signal was therefore attributed to adsorption of TMGa on Ga dimer vacancies followed by decomposition of TMGa to MMGa and two methyl groups. These methyl groups bond to nearby Ga atoms causing Ga–Ga dimer bonds to break. The cleavage of the Ga–Ga dimer bonds causes a decrease in the RDS signal. Once the TMGa exposure was stopped, methyl radicals desorbed from the surface and Ga–Ga dimer bonds reformed. The rate at which this occurs depends on the temperature of the substrate. Lower temperatures resulted in a smaller desorption rate for methyl radicals and led to a longer RDS signal recovery time. Maa and Dapkus estimate the methyl group residence time to be 10, 4.5, and 1 s at substrate temperatures of 703, 723, and 773 K, respectively [26].

2.1.2. The Ga rich (1 × 6) surface

The interaction of TMGa with the (1 × 6) Ga rich GaAs(100) surface has been studied much less than the (4 × 6) Ga rich surface. Creighton and coworkers report that TPD of TMGa from the (1 × 6) Ga rich reconstructed surface is identical to that seen for the (4 × 6) Ga rich reconstructed surface [29,50]. Yu et al. [31] observed no differences in the methyl radical desorption kinetics for the (1 × 6) Ga rich surface when compared to an As rich, c(2 × 8)/(2 × 4) GaAs(100) surface using pulsed molecular beam techniques (see the discussion of the As rich c(2 × 8)/(2 × 4) GaAs(100) surface below for details). HREELS results for adsorption of TMGa on the (1 × 6) Ga rich GaAs(100) surface at 200 K show four loss peaks related to TMGa [58]. They are the CH_3 rocking mode at 707 cm^{-1}, the CH_3 symmetric deformation at 1172 cm^{-1}, the CH_3 antisymmetric deformation at 1427 cm^{-1}, and the C–H stretch at 2932 cm^{-1}. These energies and mode assignments agree with those of White and coworkers [49] for low temperature (108 K) adsorption of TMGa on the (4 × 6) Ga rich GaAs(100) surface.

Little change occurred in the vibrational spectrum of adsorbed TMGa upon annealing to 573 K. However, annealing to 623 K caused the disappearance of all loss peaks due to adsorbed TMGa. No vibrational modes attributable to adsorbed H atoms were observed under the conditions used, indicating that the methyl groups did not decompose, but presumably desorbed.

2.1.3. The As rich c(2 × 8) / (2 × 4) surface

In this section, we review work on As rich GaAs(100) surfaces, and, where possible, compare these results to Ga rich GaAs(100) surfaces.

Adsorption of TMGa on an As rich c(2 × 8)/(2 × 4) GaAs(100) surface at low temperature (approximately 200 K) leads to molecular adsorption. This is demonstrated by HREELS experiments carried out by Yu and coworkers [58]. Similar to observations on the (4 × 6) Ga rich surface, four loss peaks attributable to adsorbed TMGa were observed. Both the energies and mode assignments are the same for the As rich c(2 × 8)/(2 × 4) and the (4 × 6) Ga rich GaAs(100) surface. Annealing the adsorbed TMGa layer caused a reduction in the intensity of the loss peaks. However, loss peaks were still observed following an anneal at 723 K. No vibrational modes attributable to adsorbed H atoms were observed under the conditions used, indicating that the methyl groups did not decompose. For comparison, all loss peaks disappeared following an anneal to 623 K following TMGa adsorption on the (1 × 6) Ga rich surface [58]. Based on this, the authors claim that the bonding of TMGa and its decomposition products on an As rich surface is stronger than on a Ga rich surface.

However, TPD experiments show methyl radical desorption occurring from the As rich c(2 × 8)/(2 × 4) surface at 653 K (380°C) [52,56]. Fig. 3 shows this result (curve b), as well as methyl radical desorption from the Ga rich (4 × 6) GaAs(100) surface (curve a). The peak temperature for methyl radical desorption from the As rich c(2 × 8)/(2 × 4) GaAs(100) surface is significantly lower than observed for methyl radical desorption from the (4 × 6) Ga rich surface (710 K). This conflicts with the conclusion drawn from the HREELS annealing experiment of Yu and coworkers that the bonding of TMGa and its decomposition products on an As rich surface is stronger than on a Ga rich surface [58], and indicates that methyl groups are more strongly bound on Ga rich surfaces. The methyl radical desorption peak is also much broader for desorption from the As rich surface, compared to the Ga rich surface. The broadness of this peak implies that there is more than one binding site for methyl groups on the As rich $c(2 \times 8)/(2 \times 4)$ GaAs(100) surface. Ultraviolet photoelectron spectroscopy (UPS) [32] and molecular beam results (discussed more fully below) are consistent with this result, indicating two binding sites for methyl groups. Using Redhead analysis [55] and assuming a preexponential factor of 1×10^{13} s^{-1}, Creighton and Banse [56] estimate an activation energy for methyl radical desorption from the As rich $c(2 \times 8)/(2 \times 4)$ GaAs(100) surface of 40.4 kcal/mol. Comparison of the rate constants for methyl radical desorption from the As rich $c(2 \times 8)/(2 \times 4)$ GaAs(100) surface, $K_{As}(CH_3)$, and the Ga rich (4×6) GaAs(100) surface, $K_{Ga}(CH_3)$, shows that $K_{As}(CH_3) \approx 10\, K_{Ga}(CH_3)$. To our knowledge, no TPD results of the desorption of other species from the As rich $c(2 \times 8)/(2 \times 4)$ GaAs(100) surface have been reported.

Several studies of TMGa adsorption have been carried out using XPS in an attempt to monitor the amount of Ga deposition [25,32,59–61]. The Ga 2p/As 2p peak area ratio or the Ga 2p/As 3d peak area ratio were used to measure the relative Ga surface coverage. TMGa was exposed to the As rich $c(2 \times 8)/(2 \times 4)$ GaAs(100) surface at a variety of temperatures. For temperatures between 643 and 803 K, the Ga/As peak area ratio increased rapidly as TMGa was exposed to the surface and saturated at some level. This has been interpreted as showing self-limiting deposition of Ga due to the lack of further TMGa adsorption once a complete Ga layer is formed. Some disagreement as to the conclusiveness of the XPS results exists. It has been pointed out that if Ga droplets form they can be virtually undetectable by XPS [56].

Pulsed molecular beam studies carried out between 700 and 783 K indicate two binding sites for methyl groups on the As rich $c(2 \times 8)/(2 \times 4)$ GaAs(100) surface [31–34], in agreement with TPD [56] and UPS [32] results. The temperature

dependence of the desorption rates from these two sites were determined using Arrhenius plots. The fast desorption channel had an activation energy of 37.9 kcal/mol and a preexponential factor of 2.57×10^{13} s^{-1}. Desorption from the second channel was slower and had an activation energy of 45.0 kcal/mol and a preexponential factor of 3.02×10^{14} s^{-1}. The relative yield of the two channels was also determined. The yield of the slow channel always exceeded that of the fast channel. Desorption of methyl radicals through the fast channel has been attributed to desorption from the As sites, while desorption of methyl radicals through the slow channel has been associated with Ga [34]. MMGa was also observed to desorb from the surface. The MMGa yield increased with temperature and TMGa exposure.

The RSC of TMGa on the As rich $c(2 \times 8)/(2 \times 4)$ GaAs(100) surface was also determined as a function of TMGa exposure and substrate temperature [33,34]. For substrate temperatures of 323 K, the RSC was initially close to unity and remained high up to a critical coverage. At this point, the RSC rapidly decreased to zero. This behavior is typical of molecular precursor mediated adsorption [62]. At a substrate temperature of 623 K, the initial value of the RSC was 0.9, but dropped rapidly, reaching a limiting value of about 0.2. At a substrate temperature of 698 K, the RSC was nearly independent of the TMGa exposure and was near 0.6. Yu and coworkers [33] suggest that this high value of the RSC (0.6) illustrates the unimportance of a site blocking mechanism in ALE using TMGa. However, we point out that the flux of TMGa molecules used under typical ALE growth conditions is at least a factor of 10 to 100 larger than used in the molecular beam experiments. Under these conditions site blocking could still play a role.

RDS experiments have been performed to investigate the adsorption of TMGa on the $c(2 \times 8)/(2 \times 4)$ As rich surface at 703, 723, and 773 K [25,26]. Adsorption of TMGa on this surface led to an increase in the RDS signal, indicating the formation of Ga–Ga dimers. When the TMGa exposure was stopped, the RDS signal increased further. This is due to the desorption of methyl radicals and the subsequent formation of more Ga–Ga dimer bonds. If the TMGa exposure is stopped near the time required to deposit one monolayer of atomic Ga, a monotonic increase in the RDS signal is observed. However, if the TMGa exposure time is longer than the time required to deposit one monolayer of atomic Ga, the RDS signal increases, then a decrease in the RDS signal is observed. When the TMGa exposure is stopped, the RDS signal increases to the same level observed when the TMGa exposure time is near the time required to deposit one monolayer of atomic Ga. The decrease in the RDS signal has been attributed to the decomposition of TMGa on As sites and migration of methyl groups to nearby Ga atoms. The bonding of methyl groups to Ga atoms breaks existing Ga–Ga dimer bonds and is the cause of the decrease of the RDS signal. Once the TMGa exposure is stopped, methyl groups desorb, allowing Ga–Ga dimer bonds to reform. This causes the observed increase in the RDS signal.

2.1.4. Implications for GaAs ALE using trimethylgallium

A fairly clear picture of the processes that occur when TMGa is incident on a GaAs surface at film growth temperatures (700–800 K) emerges from the results presented above. However, there is still disagreement on some issues. Based on sticking coefficient measurements, a TMGa molecule incident on an adsorbate free GaAs(100) surface at temperatures between 700 and 800 K will have about a 50% chance of adsorbing. As stated in Section 1.2, the adsorption of the Ga containing species should stop once the As rich surface is converted to a Ga rich surface if the selective adsorption model is correct and an ideally terminated (one monolayer Ga) surface is obtained. The TMGa sticking probability data of Yu and coworkers shows that there is virtually no change in the sticking probability of TMGa as the As rich surface is converted to the Ga rich surface [33,34]. Additionally, Creighton has also shown that the sticking probability of TMGa on a Ga rich surface is high at typical GaAs growth temperatures [56]. It is argued that these results indicate that TMGa adsorbs on Ga and As sites, indicating that the selective adsorption model

cannot explain GaAs ALE when TMGa is used as the Ga source. However, no ideally terminated Ga rich surface is known to exist. Therefore, during growth, Ga dimer vacancies will be present which expose As sites in the second layer. This leaves open the possibility that selective adsorption could occur. Clearly, no agreement as to the importance of selective adsorption has been reached thus far.

Those TMGa molecules that do adsorb on the surface rapidly dissociate, forming MMGa and two methyl groups. On Ga rich surfaces, these methyl groups migrate to nearby Ga atoms, break existing Ga–Ga dimer bonds, and form MMGa. Only one type of methyl species exists on Ga rich surfaces. Two types of methyl species are observed on As rich surfaces. These are likely a MMGa species and methyl groups bonded to As atoms. Methyl radical desorption then occurs, with the desorption rate being approximately ten times faster on the As rich surface when compared to the Ga rich surface for any given temperature. This difference in the desorption rates is necessary for ALE behavior to be observed. For the adsorbate inhibition mechanism of ALE to be operable, the desorption rate of the methyl groups must be sufficiently slow (or the residence time sufficiently long) so that an appreciable amount of methyl groups can accumulate on the Ga rich surface and inhibit further adsorption or reaction of TMGa.

Fig. 5 shows Arhennius plots of methyl radical desorption data as measured by a number of groups [22,33,50,53,56,63,64]. Also included in this figure are several measurements of surface conversion and growth. While the activation energies measured by the different groups are quite similar, the absolute rates vary by nearly four orders of magnitude. Some of the variation is likely due to the measurements being done on different surfaces, but much of the variation may be due to temperature measurement errors [56]. In spite of the variation, most of the groups report rate constants between 0.1 and 10 s^{-1} at typical ALE temperatures (700–800 K). Since the flux of TMGa molecules under growth conditions is typically several orders of magnitude larger than this, and the initial sticking coefficient of TMGa is

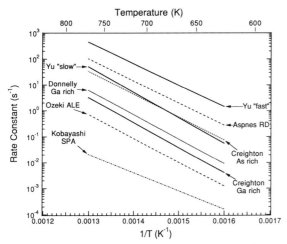

Fig. 5. Summary of literature data for methyl radical desorption kinetics and ALE growth/reaction rates. Following Creighton and Banse [56]. See text for references.

high, the surface will be saturated with methyl groups. This shows that the site blocking mechanism will play a role in GaAs ALE, especially at lower growth temperatures and higher TMGa fluxes. However, TMGa still reacts with the Ga rich surface at a finite rate. This indicates that the adsorbate inhibition mechanism cannot completely explain the ALE behavior [65].

In addition to desorption of methyl radicals, MMGa has been observed to desorb. This observation led to the proposed flux balance mechanism [31]. Yu and coworkers [31,33,34] write the TMGa reaction as:

$$(CH_3)_3 Ga \rightarrow x\, Ga + (1-x)\, CH_3 Ga_{(g)}$$
$$+ (2+x)\, CH_{3(g)}. \tag{1}$$

Here, x represents the degree of gallium deposition. Most of the Ga remains on the surface if x is close to one, while if x is near zero, Ga leaves the surface as MMGa. Thus, on going from an As rich surface to a Ga rich surface, x should change from a value near one to zero, if the flux balance mechanism is correct. Unfortunately, it has not yet been possible to quantitatively measure the MMGa and methyl radical evolution. Yu et al. [33,34] do observe a decrease in the methyl radical desorption yield and an increase in the MMGa desorption yield as the surface is converted from

As rich to Ga rich. However, the decrease in the methyl radical desorption yield may be due to a lower rate of TMGa decomposition caused by a lower reactive sticking coefficient [65]. Thus the molecular beam results are not conclusive.

2.2. The interaction of triethylgallium with GaAs(100) surfaces

Using triethylgallium (TEGa) as the group III metalorganic has been shown to reduce the amount of carbon incorporation during GaAs growth when compared to growth using TMGa [66]. This has resulted in a great deal of interest in the surface chemistry of TEGa. We discuss the surface science results of the interaction of TEGa with GaAs surfaces with emphasis on work aimed at understanding ALE.

2.2.1. The (4 × 6) Ga rich surface

Exposure of a (4 × 6) Ga rich GaAs(100) surface to TEGa at 123 K resulted in the formation of multilayers of TEGa [67]. Heating the surface resulted in the desorption of the multilayers of TEGa and produced a peak in the TPD spectrum at about 180 K. This leaves a chemisorbed monolayer of TEGa. Further heating of the surface caused part of the adsorbed TEGa to decompose and part to desorb. Broad peaks were observed in the TPD spectrum at 350 and 520 K for Ga, diethylgallium (DEGa), and TEGa. Based on the mass spectrometer fragmentation patterns of TEGa, Banse and Creighton attribute desorption from these states as due only to TEGa desorption [67]. No other Ga alkyl species are thought to desorb.

In addition to TEGa desorption, peaks in the TPD spectrum were observed for hydrogen, ethylene, ethyl radical, and ethane. These species are a result of the decomposition of TEGa. The decomposition products desorbed in the temperature range of 500 to 680 K. By integrating the areas under the TPD peaks, it was estimated that about 1.2×10^{14} C_2H_4 molecules cm^{-2}, 1.2×10^{14} H atoms cm^{-2}, and 7×10^{11} ethyl radicals cm^{-2} desorbed from the surface. Thus, approximately 4×10^{13} TEGa molecules cm^{-2} decomposed on the (4 × 6) Ga rich GaAs(100) surface.

The observation of ethane was attributed to reaction of ethyl radicals with the chamber walls.

The observation of C_2H_4 and H_2 desorption is evidence that a β-hydride elimination reaction occurs. Because desorption of both of these products is reaction rate limited, analysis of the TPD spectra yields an estimate of the activation energy of the β-hydride elimination reaction. Banse and Creighton report an activation energy for this reaction of 38.5 kcal/mol [67].

2.2.2. The (4 × 1) Ga rich surface

The adsorption and reaction of TEGa on the (4 × 1) Ga rich GaAs(100) surface is quite similar to that seen on the (4 × 6) Ga rich GaAs(100) surface. Because of these similarities, we will only briefly discuss the results here.

The exposure of TEGa to a (4 × 1) Ga rich GaAs(100) surface at temperatures below 150 K results in the formation of multilayers of TEGa [68]. The multilayers desorbed upon heating, creating a peak in the TPD spectrum at 170 K. A molecularly adsorbed monolayer of TEGa remained on the surface. Molecularly adsorbed TEGa is supported by HREELS results, which show loss peaks consistent with ethyl groups, and XPS results, which show shifts in the binding energy of the Ga 2p peak, also consistent with molecularly adsorbed TEGa [68]. Further heating of the surface caused part of the adsorbed TEGa to decompose and part to desorb. Desorption peaks were observed at 220, 350, and 500 K. Based on mass spectrometer fragmentation patterns for TEGa, the peak at 220 K was attributed to desorption of TEGa. The higher temperature desorption peaks are attributed to a mixture of TEGa and DEGa. There is also the possibility that some MEGa desorbs. This is a different conclusion than Banse and Creighton [67] reach for the (4 × 6) Ga rich surface, where all Ga alkyl desorption was attributed to TEGa.

In addition to Ga alkyl species, H_2, C_2H_4, and C_2H_6 were observed to desorb. Ethylene was found to be the predominant organic desorption product. Murrell et al. [68] estimate the activation energy for the β-hydride elimination reaction to be 37.5 kcal/mol on the (4 × 1) Ga rich GaAs(100) surface.

2.2.3. The c(8 × 2) Ga rich surface

As with the (4 × 1) Ga rich surface, the chemistry observed for TEGa adsorbed on the c(8 × 2) Ga rich GaAs(100) surface is similar to that observed on the (4 × 6) Ga rich GaAs(100) surface. Therefore, we only briefly summarize the results.

Broad desorption peaks for Ga-alkyl species (69, 98 and 127 amu) in the temperature range 273 to 570 K have been observed in the TPD spectra following exposure of TEGa to a c(8 × 2) Ga rich GaAs(100) surface near room temperature [51–53]. Based on the measured mass spectrometer fragmentation patterns, Donnelly et al. ascribe the desorption signals to being mostly due to DEGa. It is estimated that TEGa is only a minor desorption product responsible for less than 30% of the observed signal. Hydrocarbon species are also observed to desorb during TPD experiments following TEGa exposures. The major hydrocarbon product was C_2H_4, which accounted for 75% to 82% of the total hydrocarbon species desorbing. Desorption of ethyl radicals was also observed. The absence of a detectable desorption peak for mass 30 was taken as evidence that no C_2H_6 was formed. The activation energy of the β-hydride elimination reaction was estimated to be 32 ± 4 kcal/mol.

TPD and XPS results were used to determine the absolute yields of desorbing species. It was estimated that $7.6 \pm 1.6 \times 10^{13}$ Ga alkyl molecules cm^{-2} desorbed and $1.5 \pm 0.3 \times 10^{14}$ hydrocarbon species cm^{-2} desorbed (C_2H_4 plus C_2H_5). Only 25% of the Ga adsorbed on the surface from TEGa remained on the surface.

2.2.4. The c(2 × 8) / (2 × 4) As rich surface

TPD experiments following exposure of TEGa to the c(2 × 8)/(2 × 4) As rich GaAs(100) surface produce similar results when compared to the Ga rich surfaces [67]. More TEGa desorbed from the As rich surface in the temperature range of 200 to 500 K when compared to the (4 × 6) Ga rich surface. This indicates that either more TEGa can adsorb or less irreversible decomposition of TEGa occurs on the As rich surface. As with the Ga rich surfaces, ethylene, ethyl radicals, ethane, and H_2 are observed as decomposition products. The desorption temperature for these products is very similar on Ga rich and As rich surfaces. However, much less ethylene and hydrogen desorb from the As rich surface when compared to the (4×6) Ga rich surface. This suggests that Ga atoms may be important in the β-hydride elimination reaction. Approximately four times more ethyl radicals and ethane have been observed to desorb from the As rich surface when compared to the (4×6) Ga rich surface.

The reaction of TEGa with the $c(2 \times 8)/(2 \times 4)$ As rich surface has also been studied with the pulsed molecular beam technique [31,33,69]. Yu et al. observed that ethylene and ethyl radicals desorbed with the same kinetics [31,33]. The removal of ethyl ligands from the surface was modeled with two first order desorption rate expressions. The activation energy for the first desorption channel was 17.4 kcal/mol with a preexponential factor of 1.35×10^8 s^{-1} and the activation energy for the second desorption channel was 23.9 kcal/mol with a preexponential factor of 3.39×10^9 s^{-1}. The resulting rates are about an order of magnitude faster than the corresponding rates for methyl radical desorption.

The sticking coefficient of TEGa on the As rich $c(2 \times 8)/(2 \times 4)$ surface was also measured using the pulsed molecular beam technique. Above 623 K, the sticking coefficient was about 0.8 and was nearly independent of Ga coverage [33]. Monoethylgallium (MEGa) was not observed to desorb.

2.2.5. Implications for GaAs ALE using triethylgallium

Due to significantly less work being done with TEGa, the processes that occur during GaAs ALE using this precursor are not as well understood compared to our understanding of the TMGa reaction mechanisms. However, some general features can be discerned.

A gas phase TEGa molecule has a high probability of dissociatively chemisorbing on the GaAs surface at typical growth temperatures (613–623 K). Above 623 K, the sticking probability is virtually independent of Ga coverage. Based on the RSC results, it is argued that TEGa has no selectivity towards As or Ga sites on the GaAs surface and that the selective adsorption mechanism cannot explain GaAs ALE using TEGa. As in the case of TMGa, there is disagreement on this point since As sites in the second layer are available for chemisorption. Following adsorption, a TEGa molecule breaks an ethyl-Ga bond, forming DEGa and an adsorbed ethyl group. DEGa has three reaction pathways it can follow: it can desorb from the surface (although there is some disagreement about the identity of the desorbing species), react with an adsorbed ethyl group to form TEGa, or react to form MEGa and an ethyl group. The adsorbed ethyl groups can react to form ethylene and H atoms via the β-hydride elimination reaction. The H atoms react to form H_2, and both C_2H_4 and H_2 rapidly desorb from the surface. There appears to be little difference in the kinetics of the β-hydride elimination reaction on Ga and As rich surfaces. In comparison, when TMGa is used as the group III source, removal of the alkyl group (CH_3) occurs much faster on As rich surfaces when compared to Ga surfaces. An additional pathway for the removal of ethyl groups is the desorption of ethyl radicals. This appears to be a more important pathway on the As rich surface than the Ga rich surface. Based on the desorption kinetics reported by Yu et al. [31,33], the desorption rate of ethylene and ethyl radicals will be on the order of 0.1 s^{-1} at typical ALE temperatures (623 K). Since the TEGa flux used is much higher than this, site blocking will be important. Thus, as with TMGa, the adsorbate inhibition mechanism can be expected to play a role in ALE using TEGa. Currently, there is no evidence that the flux balance mechanism is operable for ALE using TEGa.

2.3. The interaction of AsH$_3$ with GaAs(100) surfaces

2.3.1. The (4×6) Ga rich surface

AsH$_3$ adsorbs molecularly on the Ga rich (4×6) surface at 115 K. This is demonstrated by HREELS results that show only one loss peak attributable to adsorbed AsH$_3$ at 2100 cm^{-1} (see Fig. 8) [70]. This has been assigned to the As–H stretching mode of AsH$_3$. XPS results, also consistent with molecularly adsorbed AsH$_3$ at 115 K, show a well-resolved As$2p_{3/2}$ peak at 1325.4 eV,

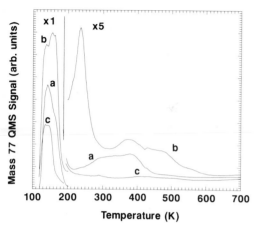

Fig. 6. Mass 77 due to AsD_3 desorption after a saturation arsine exposure to the (a) (4×6) Ga rich GaAs(100) surface, (b) $c(2 \times 8)/(2 \times 4)$ As rich GaAs(100) surface, and (c) the $c(4 \times 4)$ As rich GaAs(100) surface. From Ref. [71].

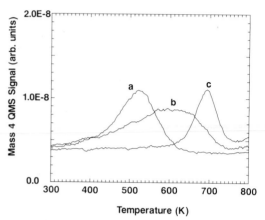

Fig. 7. D_2 desorption signals observed following a saturation arsine exposure to the (a) (4×6) Ga rich GaAs(100) surface, (b) $c(2 \times 8)/(2 \times 4)$ As rich GaAs(100) surface, and (c) the $c(4 \times 4)$ As rich GaAs(100) surface. From Ref. [71].

a 2.7 eV higher binding energy than the $As\,2p_{3/2}$ peak, due to substrate As atoms [70]. Heating the surface to 140 K caused some of the molecularly adsorbed AsH_3 to desorb, while some dissociated to AsH_2 and possibly AsH. Fig. 6 shows TPD results for mass 77 due to AsD_3 desorption following saturation AsD_3 exposures at 115 K (curve a) [70,71]. The peak at 140 K is caused by non-associative desorption of molecularly bound AsD_3. Increasing the temperature resulted in desorption peaks at 290 and 380 K [71]. Both of these peaks are attributed to recombinative desorption of AsD_3. Using Redhead analysis [55], the activation energies for desorption for these three states are found to be 12, 18, and 23 kcal/mol, respectively [71].

In addition to AsD_3 desorption, D_2 desorption is observed following AsD_3 exposures. Fig. 7 shows these results for the Ga rich (4×6) GaAs(100) surface (curve a). The D_2 desorption peak at 520 K is due to recombinative desorption following AsD_3 decomposition, and is therefore a measure of the amount of irreversibly adsorbed AsD_3. It is estimated that only 0.01 ML of AsD_3 irreversibly adsorbs [71].

The dissociation of AsH_3 at low temperature is supported by HREELS and XPS results [70]. The HREELS results for the Ga rich (4×6) GaAs(100) surface are shown in Fig. 8, left panel. As discussed above, at 115 K, a single loss peak

assigned to the As–H stretch is observed (ignoring the losses due to phonon modes). Annealing the surface to 200 K caused desorption of AsH_3,

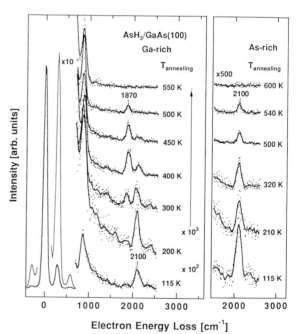

Fig. 8. High resolution electron energy loss spectra (HREELS) taken at 115 K after adsorption of AsH_3 at 115 K and annealing to the temperature indicated. Hydrogen transfer from As (2100 cm^{-1}) to Ga (1870 cm^{-1}) is observed above 250 K on the Ga rich (4×6) GaAs(100) surface. On the As rich $c(2 \times 8)/(2 \times 4)$ GaAs(100) surface, only the As–H stretch is observed. From Ref. [70].

resulting in a large decrease in the intensity of the As–H loss peak. Annealing to 300 K results in a further reduction in the size of the loss peak at 2100 cm^{-1}; a new loss peak is also observed at 1870 cm^{-1}. This has been assigned to the Ga–H stretch. The appearance of this peak shows that hydrogen transfers from As to Ga sites on the Ga rich (4 × 6) GaAs(100) surface. The XPS results point to a lower temperature for the onset of AsH$_3$ dissociation compared to HREELS experiments. Based on a deconvolution of the As 2p$_{3/2}$ peaks, Wolf et al. [70] conclude that AsH$_3$ decomposition begins as low as 140 K.

Both XPS and TPD results have been used to estimate the coverages for adsorbed AsH$_3$. The XPS results give a saturation AsH$_3$ coverage at 115 K of 0.164 ± 0.04 ML, with 0.025 ± 0.008 ML (15 ± 5%) being irreversibly adsorbed and 0.05 ML (35%) reversibly decomposed [70]. The TPD results are in general agreement, yielding a saturation AsH$_3$ coverage at 120 K of 0.10 ML, 0.01 ML (10%) irreversibly adsorbed, and 0.02 ML (20%) reversibly decomposed [71].

Banse and Creighton have shown that large (2.6 × 10^6 L) AsH$_3$ exposures to the Ga rich surface between 373 and 623 K will produce an As rich c(4 × 4) surface [72]. Subsequent TPD showed three As desorption peaks at 713, 753, and 843 K. As$_4$ desorption was detected from the low temperature (713 K) peak, while As desorbed mainly as As$_2$ from the two higher temperature states. It is estimated that up to 1.75 ML of As can be deposited.

2.3.2. The c(2 × 8) / (2 × 4) As rich GaAs(100) surface

Experiments involving AsH$_3$ have been performed on the c(2 × 8) / (2 × 4) As rich GaAs(100) surface, similar to those discussed for the (4 × 6) Ga rich surface. AsH$_3$ adsorbs molecularly on the c(2 × 8) / (2 × 4) As rich GaAs(100) surface at 115 K [70]. Heating the substrate caused desorption and decomposition of AsH$_3$ to occur. Fig. 6 shows TPD results for mass 77 due to AsD$_3$ desorption following saturation AsD$_3$ exposures at 115 K on the As rich c(2 × 8) / (2 × 4) GaAs(100) surface (curve b) [71]. In comparison to the Ga rich (4 × 6) reconstruction, more AsD$_3$

desorbed from the low temperature peak at approximately 165 K, the desorption state at 323 K, and the desorption state at 400 K. Additionally, new desorption peaks are observed at 233 and 460 K.

D$_2$ desorption following saturation AsD$_3$ exposures at 120 K on the As rich c(2 × 8)/(2 × 4) GaAs(100) surface are shown in Fig. 7 (curve b). A single desorption peak is observed at 600 K, a somewhat higher temperature than observed for D$_2$ desorption from the (4 × 6) Ga rich surface. This desorption peak is also much broader than the D$_2$ desorption peak from the Ga rich (4 × 6) GaAs(100) surface (curve a). However, the amount of D$_2$ desorption is similar.

Fig. 8 shows the HREELS data for AsH$_3$ adsorbed on the As rich c(2 × 8)/(2 × 4) GaAs(100) surface. At 115 K, a single loss peak is observed at 2100 cm^{-1} and is assigned to the As–H stretch [70]. Annealing the adsorbed layer causes a decrease in the intensity of this loss peak and it disappears by 600 K. This correlates well with the desorption of H$_2$ (D$_2$) from the surface as shown in Fig. 7. In contrast to the Ga rich (4 × 6) surface, no Ga–H stretching mode was ever observed. This is likely due to the Ga sites being blocked by As. Based on the HREELS results, it appears that the hydrogen desorption following AsH$_3$ exposures on the As rich c(2 × 8)/(2 × 4) surface is due entirely to hydrogen adsorbed on As sites.

XPS and TPD results were used to estimate AsH$_3$ coverages. The XPS results give a saturation AsH$_3$ coverage at 115 K of 0.154 ± 0.04 ML, with approximately 30% decomposing upon heating [70]. The TPD results give a saturation AsH$_3$ coverage at 120 K of 0.20 ML, with 0.01 ML being irreversibly adsorbed [71]. Thus, while the XPS results predict a similar saturation AsH$_3$ coverage for the Ga rich (4 × 6) and the As rich c(2 × 8)/(2 × 4) surface, TPD results indicate twice as much AsH$_3$ adsorbs on the As rich c(2 × 8)/(2 × 4) surface relative to the Ga rich (4 × 6) surface.

2.3.3. The c(4 × 4) As rich GaAs(100) surface

To our knowledge, TPD is the only experimental technique used to study the adsorption of

AsH$_3$ on the c(4 × 4) As rich GaAs(100) surface. Fig. 6 shows the AsD$_3$ TPD spectrum following saturation AsD$_3$ exposure to the c(4 × 4) surface at 120 K (curve c) [71]. In comparison with the Ga rich (4 × 6) (curve a) and As rich c(2 × 8)/(2 × 4) surfaces (curve b), much less AsD$_3$ is observed to desorb in the 120 to 200 K temperature range. Additionally, no appreciable high temperature desorption of AsD$_3$ is observed. A D$_2$ TPD spectrum taken concurrently with the AsD$_3$ TPD spectrum of Fig. 6 is shown in Fig. 7 (curve c). The desorption peak observed at 700 K is not due to the AsD$_3$ dosed at 120 K, but is an artifact of the AsD$_3$ dose required to create the c(4 × 4) As rich surface. This indicates that arsine is only weakly adsorbed on the As rich c(4 × 4) GaAs(100) surface.

2.3.4. Implications for GaAs ALE using arsine

Because of the low decomposition temperature, we can expect AsH$_3$ to rapidly dissociate on a Ga rich GaAs(100) surface at typical growth temperatures (700–800 K). Hydrogen atoms will transfer from As atoms to Ga atoms and rapidly recombine and desorb as H$_2$. High AsH$_3$ exposures typically used during the AsH$_3$ cycle will form a c(4 × 4) As rich surface. Once this surface is formed, the interaction of AsH$_3$ with the growing surface is very weak and may contribute to limited As deposition. This may have important implications for ALE using AsH$_3$. Previously, only the c(2 × 8)/(2 × 4) As rich surface was considered to be important during ALE [32,60]. This surface has only 0.75 ML of As, making it difficult to explain how one ML/cycle growth can be achieved. The stoichiometry issue might be resolved by considering that saturation of the desorption state at 753 K corresponds to 1 ML of As. Based on estimates of the residence time for As in the 713 K state, it was concluded that As in excess of 1 ML will desorb during the typical purge cycle during ALE, leaving a surface terminated with 1 ML of As.

3. Discussion

The work presented above indicates that we have advanced our understanding of GaAs ALE

tremendously. Many of the important reaction pathways and their energetics have been determined. An issue not fully addressed above is the stoichiometry of the surface. All evidence suggests that there are no Ga rich surface reconstructions that contain a complete monolayer of Ga or As rich surfaces that contain a complete monolayer of As (see Section 1.4). This leaves us with the question of how ideal ALE (one monolayer growth per cycle) is achieved. A possible solution for the AsH$_3$ cycle was discussed in Section 2.3. For Ga rich surfaces, Creighton suggests that alkyl groups stabilize a gallium coverage that is higher than that obtained on the adsorbate free Ga rich surface, allowing one ML/cycle growth to be obtained [56].

An alternative mechanism to achieve one monolayer growth per cycle, suggested by Dapkus, involves second layer vacancies (both Ga and As) [73]. If growth is initiated on an As rich c(2 × 8)/(2 × 4) surface, As dimer vacancies exist, which expose Ga sites. We will refer to these Ga sites as "layer 1" Ga sites and surface As as "layer 1" As sites. Exposure of a Ga alkyl species to the As rich c(2 × 8)/(2 × 4) surface results in a Ga rich surface with 0.75 ML of Ga. We refer to this Ga as "layer 2" Ga sites. Now AsH$_3$ is exposed to the surface. Three fourths of a monolayer of As will be deposited on layer 2 Ga sites, forming surface As, which we refer to as "layer 2" As sites. Additionally, 0.25 ML of As will be deposited at layer 1 Ga sites, completing the layer 1 As monolayer. The next exposure of Ga alkyl species results in 0.75 ML of Ga being deposited at layer 2 As sites and 0.25 ML being deposited at layer 1 As sites (completing the layer 2 Ga monolayer). This cycle continues and allows for one monolayer growth per cycle. This implies that the most abrupt heterojunction which can be achieved with this type of growth is two layers thick. Neither the adsorbate stabilization nor the second layer vacancy model have been verified experimentally.

At low growth temperatures, the growth per cycle falls below one monolayer. This is due to effective site blocking by alkyl groups. ALE fails at high growth temperatures for a number of reasons. First, desorption of As and Ga atoms can occur and lead to less than one monolayer growth per cycle. Second, site blocking by alkyl groups will be less efficient since the residence time of these groups will be short. Finally, for ALE using TMGa, the Ga deposition rate will exceed the MMGa desorption rate and lead to excess Ga deposition [34].

We have learned a great deal about the mechanisms of ALE of GaAs through the efforts of many researchers. Rather than any one model being exclusively responsible for ALE growth of GaAs, it appears that perhaps all three mechanisms are involved. Site blocking by adsorbed alkyl groups most certainly plays an important role in the kinetics of ALE. However, it alone cannot explain the observed ALE behavior. Flux balance and/or selective adsorption must be involved. Further work is needed to address the relative importance of these two mechanisms. Clearly, chemical selectivity between the As and Ga rich surfaces is important.

In summary, we have reviewed the proposed mechanisms of GaAs ALE and presented results that show the significance of each. Surface science techniques have played an enormously important role in developing our current level of understanding of GaAs ALE, and will undoubtedly impact future developments.

Acknowledgements

The authors would like to gratefully acknowledge support for this work by the Science and Technology Program of the National Science Foundation, grant number CHE8920120. J.M.H. wishes to thank J.R. Creighton and P.D. Dapkus for stimulating and helpful discussions.

References

[1] S.K. Ghandhi, in: VLSI Fabrication Principles (Wiley, New York, 1983).
[2] M.A. Herman and H. Sitter, in: Molecular Beam Epitaxy (Springer, Berlin, 1989).
[3] K.G. Günther, Z. Naturforsch. 13A (1958) 1081.
[4] J.E. Davey and T. Pankey, J. Appl. Phys. 39 (1968) 1941.
[5] J.R. Arthur, J. Appl. Phys. 39 (1968) 4032.
[6] J.A. Arthur, Surf. Sci. 43 (1974) 449.
[7] A.Y. Cho and J.R. Arthur, Prog. Solid State Chem. 10 (1975) 157.
[8] C.T. Foxon, M.R. Boudry and B.A. Joyce, Surf. Sci. 44 (1974) 69.
[9] C.T. Foxon and B.A. Joyce, Surf. Sci. 50 (1975) 434.
[10] C.T. Foxon and B.A. Joyce, Surf. Sci. 64 (1977) 293.
[11] C.T. Foxon, J.A. Harvey and B.A. Joyce, J. Phys. Chem. Solids 34 (1973) 1693.
[12] M.B. Panish, J. Electrochem. Soc. 127 (1980) 2729.
[13] M.B. Panish, H. Temkin and S. Sumski, J. Vac. Technol. B 3 (1985) 687.
[14] M.B. Panish, H. Temkin, R.A. Hamm and S.N.G. Chu, Appl. Phys. Lett. 49 (1986) 164.
[15] W.T. Tsang, Appl. Phys. Lett. 45 (1984) 1234.
[16] W.T. Tsang, J. Cryst. Growth 81 (1987) 261.
[17] E. Tokumitsu, Y. Kudou, M. Konagai and K. Takahashi, J. Appl. Phys. 55 (1984) 3163.
[18] E. Tokumitsu, Y. Kudou, M. Konagai and K. Takahashi, Jpn. J. Appl. Phys. 24 (1985) 1189.
[19] E. Tokumitsu, T. Katoh, R. Kimura, M. Konagai and K. Takahashi, Jpn. J. Appl. Phys. 25 (1986) 1211.
[20] T. Suntola and J. Antson, US Patent 4058430, November 15 (1977).
[21] W.T. Tsang, J. Cryst. Growth 95 (1989) 121.
[22] M. Ozeki, K. Mochizuki, N. Ohtsuka and K. Kodama, Appl. Phys. Lett. 53 (1988) 1509.
[23] Y. Sakuma, M. Ozeki, N. Ohtsuka and K. Kodama, J. Appl. Phys. 68 (1990) 5660.
[24] K. Kodama, M. Ozeki, K. Mochizuki and N. Ohtsuka, Appl. Phys. Lett. 54 (1989) 656.
[25] B.Y. Maa and P.D. Dapkus, Mater. Res. Soc. Symp. Proc. 222 (1991) 25.
[26] B.Y. Maa and P.D. Dapkus, Appl. Phys. Lett. 58 (1991) 2261.
[27] J. Nishizawa, T. Kurabayashi, H. Abe and A. Nozoe, Surf. Sci. 185 (1987) 249.
[28] J. Nishizawa and T. Kurabayashi, J. Cryst. Growth 93 (1988) 98.
[29] J.R. Creighton, K.R. Lykke, V.A. Shamamian and B.D. Kay, Appl. Phys. Lett. 57 (1990) 279.
[30] A. Watanabe, T. Kamijoh, M. Hata, T. Isu and Y. Katayama, Vacuum 41 (1990) 965.
[31] M.L. Yu, U. Memmert, N.I. Buchan and T.F. Kuech, Mater. Res. Soc. Symp. Proc. 204 (1991) 37.
[32] M.L. Yu, U. Memmert and T.F. Kuech, Appl. Phys. Lett. 55 (1989) 1011.
[33] M.L. Yu, N.I. Buchan, R. Souda and T.F. Kuech, Mater. Res. Soc. Symp. Proc. 222 (1991) 3.
[34] M.L. Yu, J. Appl. Phys. 73 (1993) 716.
[35] A. Usui and H. Watanabe, Annu. Rev. Mater. Sci. 21 (1991) 185.
[36] M. Ozeki, N. Ohtsuka, Y. Sakuma and K. Kodama, J. Cryst. Growth 107 (1991) 102.
[37] H. Ohno, S. Ohtsuka, H. Ishii, Y. Matsubara and H. Hasegawa, Appl. Phys. Lett. 54 (1989) 2000.

[38] M.D. Pashley, K.W. Haberern, W. Friday, J.M. Woodall and P.D. Kirchner, Phys. Rev. Lett. 60 (1987) 2176.

[39] P.K. Larsen and D.J. Chadi, Phys. Rev. B 37 (1987) 8282.

[40] D.K. Biegelsen, R.D. Bringans, J.E. Northrup and L.-E. Swartz, Phys. Rev. B 41 (1990) 5701.

[41] P.K. Larsen, J.H. Neave, J.F. van der Veen, P.J. Dobson and B.A. Joyce, Phys. Rev. B 27 (1983) 4966.

[42] M. Sauvage-Simkin, R. Pinchaux, J. Massies, P. Calverie, N. Jedrecy, J. Bonnet and I.K. Robinson, Phys. Rev. Lett. 62 (1989) 563.

[43] J.R. Arthur, Surf. Sci. 43 (1974) 449.

[44] R. Ludeke and A. Koma, J. Vac. Sci. Technol. 13 (1976) 241.

[45] P. Drathen, W. Ranke and K. Jacobi, Surf. Sci. 77 (1978) L162.

[46] J. Massies, P. Etienne, F. Dezaly and N.T. Linh, Surf. Sci. 99 (1980) 121.

[47] R.Z. Bachrach, R.S. Bauer, P. Chiaradia and G.V. Hansson, J. Vac. Sci. Technol. 18 (1981) 797.

[48] D.J. Frankel, C. Yu, J.P. Harbison and H.H. Farrell, J. Vac. Sci. Technol. B 5 (1987) 1113.

[49] X.-Y. Zhu, J.M. White and J.R. Creighton, J. Vac. Sci. Technol. A 10 (1992) 316.

[50] J.R. Creighton, Surf. Sci. 234 (1990) 287.

[51] V.M. Donnelly and J. McCaulley, Surf. Sci. 238 (1990) 34.

[52] J.A. McCaulley, R.J. Shul and V.M. Donnelly, J. Vac. Sci. Technol. A 9 (1991) 2872.

[53] V.M. Donnelly, J.A. McCaulley and R.J. Shul, Mater. Res. Soc. Proc. 204 (1991) 15.

[54] J.R. Creighton, J. Vac. Sci. Technol. A 9 (1991) 2895.

[55] P.A. Redhead, Vacuum 12 (1962) 203.

[56] J.R. Creighton and B.A. Banse, Mater. Res. Soc. Symp. Proc. 222 (1991) 15.

[57] D.A. King and M.G. Wells, Surf. Sci. 29 (1972) 454.

[58] A. Narmann, R.J. Purtell and M.L. Yu, Mater. Res. Soc. Symp. Proc. 222 (1991) 41.

[59] B.Y. Maa and P.D. Dapkus, J. Electron. Mater. 19 (1990) 289.

[60] B.Y. Maa and P.D. Dapkus, J. Cryst. Growth 105 (1990) 213.

[61] P.D. Dapkus, B.Y. Maa, Q. Chen, W.G. Jeong and S.P. DenBaars, J. Cryst. Growth 107 (1991) 73.

[62] P. Kisliuk, J. Phys. Chem. Solids 3 (1957) 95.

[63] D.E. Aspnes, E. Colas, A.A. Studna, R. Bhat, M.A. Koza and V.G. Keramidas, Phys. Rev. Lett. 61 (1988) 2782.

[64] N. Kobayashi and Y. Horikoshi. Jpn. J. Appl. Phys. 30 (1991) L319.

[65] J.R. Creighton and B.A. Bansenauer, Thin Solid Films 225 (1993) 17.

[66] T.F. Kuech and R. Potemski, Appl. Phys. Lett. 47 (1985) 821.

[67] B.A. Banse and J.R. Creighton, Surf. Sci. 257 (1991) 221.

[68] A.J. Murrell, A.T.S. Wee, D.H. Fairbrother, N.K. Singh and J.S. Foord, J. Appl. Phys. 68 (1990) 4053.

[69] T. Martin and C.R. Whitehouse, J. Cryst. Growth 105 (1990) 57.

[70] M. Wolf, X.-Y. Zhu, T. Huett and J.M. White, Surf. Sci. 275 (1992) 41.

[71] B.A. Bansenauer and J.R. Creighton, Surf. Sci. 278 (1992) 317.

[72] B.A. Banse and J.R. Creighton, Appl. Phys. Lett. 60 (1992) 856.

[73] P.D. Dapkus, personal communication.

Surface Science 299/300 (1994) 909–927
North-Holland

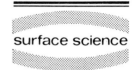

surface science

Metal–semiconductor interfaces

L.J. Brillson

Xerox Webster Research Center, 800 Phillips Road 114-41D, Webster, NY 14580, USA

Received 28 April 1993; accepted for publication 6 August 1993

Progress in understanding the chemical and electronic properties of metal–semiconductor interfaces has depended heavily on the use of surface science techniques. This article provides an overview of the wide range of atomic-scale chemical phenomena observed at metal–semiconductor interfaces and their relation to contact charge transfer and macroscopic Schottky barrier formation. Particular emphasis is given to this author's research contributions over the past two decades.

1. Historical background: the role of surface science in understanding metal–semiconductor interfaces

Metal–semiconductor interfaces have chemical, geometrical, and electrical properties different from those of their bulk constituents. It has been a goal of researchers over the last few decades to understand how these properties derive from the individual metal and semiconductor features and the processes by which they are joined. A central issue has been the intrinsic versus extrinsic nature of interface electronic phenomena, namely, whether or not the Schottky barrier formation which occurs upon contact charge exchange can be derived directly from electronic features intrinsic to the constituent semiconductor or metal media or whether extrinsic chemical phenomena play a dominant role in forming the electronic junction.

Understanding this Schottky barrier formation has been a major theme of surface science for the past three decades. Numerous reviews of the progress made using surface science techniques are available which examine this topic in much greater detail than is possible here [1–9]. Nevertheless, this progress has been limited due to (a) a strong dependence on chemical, geometrical,

and electrical structure on the processes used in preparing the interfaces, (b) complications associated with the measurement techniques used, (c) the difficulties associated with preparing well-defined (e.g., epitaxial) interface atomic structures, and (d) the multiplicity of physical models capable of accounting for the narrow range of Schottky barriers commonly reported.

The study of the Schottky barrier formation is of more than academic interest since the semiconductor rectification is a major feature of semiconductor devices. Indeed, contacts to semiconductors must satisfy a variety of requirements – (i) electronic, in terms of low or high barriers for high power or rectifying applications, respectively, low trap densities, carrier recombination, and high mobility, as well as uniformity and reproducibility; (ii) chemical, in terms of shallow depths of reaction and diffusion, lateral uniformity, low degradation with thermal stress or ambient exposure, and environmental compatibility, as well as (iii) structural, for example, high adhesive strength, extended dimensionality (e.g., multiple interface structures), and resistance to mechanical stress [7]. Thus an understanding and control of interface electrical barrier properties can greatly improve the technologist's ability to satisfy these diverse requirements. Indeed, such understanding and control has become increasingly urgent as device sizes have shrunk to ever smaller dimensions.

This review will emphasize the author's personal research initiatives and contributions during this period, within the context of major developments and conceptual advances which have taken place. These contributions and the surface science techniques which enabled them include: (a) the observation of interface states at semiconductor–adsorbate surfaces via surface photovoltage spectroscopy (SPS), their extrinsic, chemical nature, and their distribution across the semiconductor band gap [10], (b) the chemical dependence of Schottky barriers as first identified from low energy electron energy loss (LEELS) and soft X-ray photoelectron spectroscopy (SXPS) measurements – especially the interface parametrization via interface heat of reaction [11,12] and the use of reactive interlayers to control interface chemistry [13,14] and macroscopic Schottky barriers [12,15,16], (c) the correlation between interface chemistry and deep levels via cathodoluminescence spectroscopy (CLS) and SXPS [17,18] and the evolution of interface electronic and chemical features with successive nanometer-scale deposits of metal on semiconductors (e.g., Refs. [19,20]), (d) the identification of bulk semiconductor features which can affect interface states and Schottky barrier formation (e.g., Refs. [19,21]), and (e) the demonstration of wide Schottky barrier ranges for all compound semiconductor contacts via SXPS [22–24] and macroscopic internal photoemission (IP) measurements [25,26]. For more extended overviews, see Refs. [1,2,4,7–9] as well as Professor W. Mönch's companion review [40].

Much of the early work reported here was influenced by the progress made by Professor Harry Gatos and his group at M.I.T. For example, his studies on semiconductor etching clearly emphasized the influence of geometric effects associated with atomic bonding [27,28]. He and his group were responsible for major developments in the technique and theory of SPV as applied to semiconductors [29]. Furthermore, his proposals on the role of native defects on the Schottky barrier formation [30] presaged later

developments ([5,7] and Refs. therein). Certainly his professional leadership as Editor-in-Chief of Surface Science has helped motivate the chemical perpectives now dominant in interface studies and his zest for life and excitement with the world of surface science has been an inspiration to us all.

Surface science has played a central role in understanding metal–semiconductor interfaces and Schottky barrier formation. At the most basic level, ultrahigh vacuum (UHV) techniques are required in order to prepare clean surfaces and junctions, free of any contamination which could introduce new chemical, electronic, geometrical, and even morphological features. The electronic and optical techniques used to characterize interface properties also demand an environment which will not interfere with the excitation or detection of the surface-specific information. Surface science techniques have revealed strong correlations between chemical and electronic structure which become apparent only at the atomic or nanometer scale. Furthermore, the localization of these features need not be confined solely in the dimension perpendicular to the interface. Variations in interface properties are now commonly observed across the plane of the interface as well – also on a microscopic scale. Perhaps most importantly, surface science has continued to uncover and identify new phenomena on a micro-scale which alter physical properties readily measured in the macroscopic world. These new phenomena have involved unique chemical phases, altered semiconductor crystal morphology, bulk defects and impurities, variations in interface bonding epitaxy and misorientation, and the heterogeneity of chemical composition, bonding, and bond ordering. New phenomena continue to emerge as the control of surface and interface structure improves.

2. Interface charge transfer

When a metal and a semiconductor come into contact, charge transfers across their interface, resulting in band bending within the semiconductor surface space charge region. See Fig. 1, left

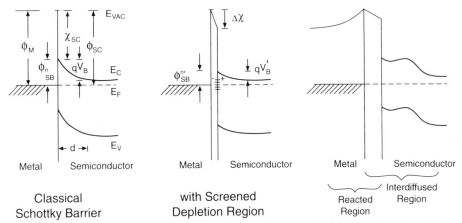

Fig. 1. Schematic energy band diagram of n-type semiconductor–metal contact for (a) abrupt junction with no trapped charge states, (b) abrupt junction with trapped charge states, and (c) extended junction with trapped charge states and intermediate dielectric phases.

panel. Here ϕ_M is the metal work function, X_{SC} is the semiconductor electron affinity, ϕ_{SC} is the semiconductor work function as defined by the potential difference between the vacuum reference energy level E_{VAC} and the Fermi level E_F, qV_B is the semiconductor band bending of thickness d and E_C and E_V are the semiconductor conduction and valence bands, respectively. For the Schottky barrier pictured here, the band bending is given by the classical expression

$$qV_B = \phi_M - \phi_{SC}. \qquad (1)$$

In general, however, band bending for a given metal–semiconductor system does not follow this expression. Early systematic measurements carried out under low vacuum conditions [31] showed a much lower dependence of semiconductor band bending on metal work function. The sensitivity of band bending to different metals was shown to vary with the semiconductor in a systematic way: high for ionically-bonded solids and low for more covalently-bonded semiconductors [32]. See Fig. 2.

In order to understand the relative insensitivity of band bending, Bardeen proposed that electronic states localized at the semiconductor surface play a role in the contact rectification. Such states could accomodate part of the voltage difference between metal and semiconductor and thereby account for the relative insensitivity of

the semiconductor band bending to different metals or to changes in applied bias [33]. Such interface dipoles ΔX are represented in Fig. 1, center panel, which illustrates the reduced, or screened, potential qV_B' within the surface space

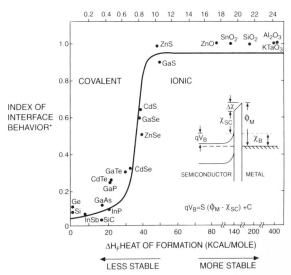

Fig. 2. Transition of interface behavior S as described in the inset as a function of electronegativity difference (upper scale) (after Kurtin et al. [32]) or, equivalently, as a function of thermodynamic heat of formation (lower scale) (after Brillson [57]) for a wide variety of semiconductors.

charge region. The interface states themselves reside within the top layer or layers of the junction and are indicated by the charges in this panel. The field effect measurements (surface conductivity versus applied gate bias in a three-terminal device structure) of Schockley and Pearson confirmed the existence of such states for the first time [34].

The phenomenon of strong Fermi level stabilization in a narrow range of energies, regardless of external electrostatic forces, has been termed Fermi level "pinning", since E_F appears to be pinned within a narrow range of energies within the semiconductor band gap. This behavior is usually associated with high densities of states (i.e., $10^{14}–10^{15}$ cm^{-2}) within the band gap whose changes in population constrain the Fermi level movement. Fermi level "pinning" is to be distinguished from Fermi level stabilization at a given energy, which is simply due to a balance of electrostatic forces. Here, a lower density of interface states permits E_F to move over a wider energy range with applied electrostatic forces.

The nature of the charge states pictured here have been a primary concern of researchers for the past forty-five years. With improvements in vacuum hardware, experimental and theoretical methods, it has been possible to prepare and analyze numerous types of metal–semiconductor interfaces. Their properties suggest more than one possible physical origin of localized states.

3. Interface charge states

Initially, the charge states at metal–semiconductor interfaces were thought to be surface states intrinsic to the semiconductor. These states were associated with the discontinuity of the lattice potential at the semiconductor–vacuum interface [31]. This idea was consistent with the trend of more E_F "pinning" for the more covalent com-

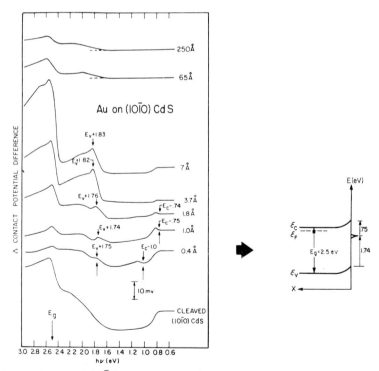

Fig. 3. Surface photovoltage spectra of CdS($10\bar{1}0$) cleaved in UHV and with increasing thickness of deposited Au. Arrows indicate changes in slope corresponding to new optical transitions involving gap states induced by the metal overlayers. The inset shows schematically the complementary transitions into and out of the gap states [10,43].

pound semiconductors, where the discontinuity of the lattice potential produced a larger disruption of the lattice wavefunctions for covalent than for ionic materials. However, these states do not appear to play a major role, since they are either not present on the clean surface or they are removed by metal deposition [1].

Another form of charge localization at the interface is associated with wave function tunneling from the metal into the semiconductor. Here the wave function localization is due to the attenuation of the wave function into the semiconductor at energies within the forbidden gap. This idea was originally proposed in the mid-1960's by Heine [35] and further developed by Louie and Cohen [36] and Flores and coworkers [37,38]. Such metal-induced gap state (MIGS) models were tied to the semiconductor band structure by Tersoff [39] and shown to account for some Au–semiconductor barrier trends semiquantitatively. MIGS plus extrinsic phenomena such as interface defects can also account for trends involving adsorbates and metal overlayers on GaAs. See the companion review on metal–semiconductor interfaces by Professor Mönch [40]. In general, it is difficult to account for the wide variation in Schottky barrier heights using a MIGS model without additional, extrinsic perturbations.

The strong influence of extrinsic phenomena on contact formation was recognized early in interface studies, starting with efforts to produce high purity semiconductors to eliminate bulk contributions to contact measurements, then in the preparation of clean semiconductor surfaces to avoid impurity effects. Surface science has played a key role in isolating and controlling such effects so that the role of extrinsic phenomena could be understood quantitatively.

Early results on metal–semiconductor interfaces were complicated by the presence of contacts prepared under low vacuum conditions. One of the pioneers in characterizing such impurities quantitatively was Peter Mark, who analyzed the electrical changes at CdS and other wide-gap semiconductor surfaces with adsorption of various gases [41,42]. Harry Gatos was another such pioneer, using the techniques of surface photovoltage spectroscopy (SPS) to identify gap states

specific to particular adsorbates on CdS and GaAs [29]. Brillson extended the SPS technique to the study of metal–semiconductor interfaces, showing the formation of discrete states deep within the semiconductor band gap with only submonolayers of deposited metal [10,43]. Fig. 3 shows that changes in contact potential difference with incident photon energy occur at well-defined energies and correspond to transitions either into or out of deep gap states [10]. The CdS band gap energy is indicated by the arrow labeled E_g. The schematic energy band diagram shown at right illustrates the corresponding transitions for overlayers of Au on CdS($10\bar{1}0$). Furthermore, the deep level features can be seen to vary with metal coverage on a monolayer scale. The presence of islanding at this interface shows that attenuation of the optical response, which disappears as the surface becomes completely metallic. Finally, these deep level features vary with the particular metal, suggesting that interface electronic properties may also depend on the specific metal.

Imperfections within the bulk or at the surface of semiconductors can be electrically active. An extensive literature exists on the properties of bulk native defects [44]. Besides vacancies, interstitials, and their complexes, deep levels can be associated with impurities and even impurity-defect complexes. In the late 1970's, Spicer et al. [45,46] Wieder [47], and Williams et al. [48] presented considerable evidence for such defects as the dominant influence on Schottky barrier formation. Gatos speculated on the role of EL2, a common mid-gap level, in causing E_F pinning at GaAs interfaces [30]. Initially, the defect model of Spicer and Lindau [45,46] was based on adsorbate-induced defect production and "pinning" in some narrow energy range at submonolayer coverage, regardless of adsorbate on a given III–V compound semiconductor. However, considerable evidence now exists for substantial variation of E_F position with different adsorbates and different preparation conditions. Currently, such interface defects for GaAs are modeled as similar to EL2, since the stabilization energy range observed for metals on GaAs(110) corresponds to transition levels of EL2 extracted from photoinduced electron spin resonance measurements [5].

The morphology of semiconductor surfaces may also contribute electrically-active interface states. Such morphological features include the steps associated with off-axis crystal surfaces [24], kinks in atomic bonding which help balance the charge neutrality with contributions from bulk doping [49], and dislocations associated with lattice relaxation of pseudomorphic epitaxial overlayers [50]. In each case, changes in band bending can be modeled via charge states localized near altered surface bond sites.

Perhaps studies of interface chemical reaction and diffusion have provided the largest body of evidence for extrinsic interface states. Surface science techniques provided the first demonstration that Schottky barrier formation could be associated with the degree of chemical reaction [11]. Fig. 4 illustrates the transition in Schottky barrier formation for a diverse set of semiconductors as a function of chemical reactivity. Here, a qualitative change occurs for different metals on the same semiconductor as a function of heat of chemical reaction, defined as

$$\Delta H_F = (1/x) H_F(M_x A) - H_F(CA) \qquad (2)$$

according to the chemical equation

$$M + (1/x)CA \rightarrow (1/x)M_x A + (1/x)C. \qquad (3)$$

Taking into account metal–cation interactions provides a further refinement [51]. This transition between two band bending regimes occurs independent of semiconductor ionicity and has been confirmed for numerous other III–V compound semiconductors such as InP [52], as well II–VI semiconductors such as CdTe [53], II–V's such as Zn_3P_2 [54], and IV–VI's such as PbTe [55]. SXPS measurements of metal–semiconductor systems under UHV conditions reveal that this transition can be associated with the presence of strong metal–anion bonding and a change in outdiffusion stoichiometry [12]. SXPS also reveals that a rate-limiting step in such outdiffusion is semiconductor dissociation, which scales with the semiconductor heat of formation [15,56]. The role of semiconductor stability in Schottky barrier formation is also suggested by the interface behavior illustrated in Fig. 2, which exhibits the same variation with semiconductor heat of formation as

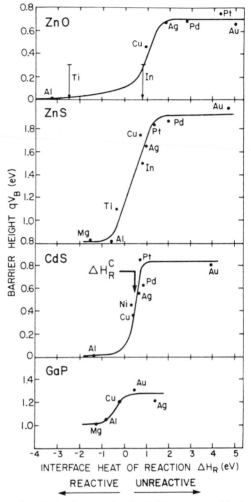

Fig. 4. Correlation between barrier heights measured by internal photoemission (after Mead [31]) as a function of thermodynamic heat of reaction ΔH_R (after Brillson [11]). A transition occurs at ΔH_R^C, observed spectroscopically. Qualitatively similar behavior occurs for a wide variety of compound semiconductors, regardless of ionicity.

with semiconductor ionicity [43,57]. Fig. 5 illustrates the reversal of stoichiometry for metals of different chemical reactivity on InP(110) and the correlation of this reversal with macroscopic Schottky barrier measurements [53]. The vertical scale denotes P-to-In concentration near the semiconductor surface as measured from SXPS core level intensities. Metals such as Ag, Pd, Au, and Cu produce anion-rich outdiffusion, whereas

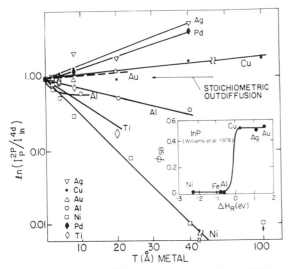

Fig. 5. Reversal of outdiffusing anion versus cation stoichiometry (extracted from SXPS core level intensities) for metals of different reactivity with InP(110). The inset illustrates a Φ_B verus ΔH_R plot for InP (after Williams and Patterson [53]) and emphasizes the correlation between anion versus cation-rich outdiffusion with Schottky barrier heights measured by conventional electrical techniques (Brillson et al. [12].

However, such defect creation depends on both the methods by which the semiconductor–metal interface and even the semiconductor itself are prepared, as will be shown in subsequent sections. Effects of surface preparation on interface chemical and geometrical structure are also pronounced, especially at low temperatures [9]. Thus, interface band bending is more controllable than such defect or defect plus MIGS models permit.

Above the monolayer scale, chemical interactions between metal and semiconductor can give rise to interface regions with localized bonding, dipole formation, defect formation, and new chemical phases unlike those of either constituent. The electrostatic effects of such physical phenomena are represented schematically in the right panel of Fig. 1.

Besides defects at a given energy, interfacial phases with similar work functions can also account for the narrow range of E_F in the GaAs band gap. Freeouf and Woodall [59] demonstrated a match between the E_F barrier positions of a wide variety of III–V compound semiconductors and the classical E_F positions expected on the basis of the corresponding anion work functions. Here, the presence of an interfacial anion phase between the metal and the semiconductor leads to similar barriers since the work function difference between media is relatively constant. Transmission electron microscopy (TEM) and light scattering measurements reveal evidence for such excess anion phases (e.g., clusters and crystallites) at GaAs interfaces [7].

A classical work function dependence is also evident from SXPS studies of metal-induced band bending for GaAs [23] and GaP [60], especially for high work function metals. Low work function behavior is complicated by a strong dependence on interface preparation and measurement artifacts [26].

Overall, both intrinsic metal-induced gap states and extrinsic, chemically-induced interface states appear to be present at metal–semiconductor interfaces. While MIGS are expected to contribute charge to the interface dipole, the pronounced chemical interactions observed via surface science techniques and the variations in band bending with different interface preparation indi-

more reactive metals such as Ni, Ti, and Al lead to cation-rich outdiffusion. Current–voltage measurements on diodes with these and other metals show a corresponding and qualitative difference in n-type barrier heights. Such results can be interpreted either in terms of dipoles associated with localized interface bonding or stoichiometry-dependent deep levels created by the reaction. In either case, Fig. 5 demonstrates the chemical basis for the contact rectification.

The new states associated with new chemical bonding at interfaces can be atomic, nanometer, or macroscopic in scale. On an atomic scale, submonolayer deposits of metal produce band bending which varies with adsorbate [40,58]. Indeed the gap state energies linked to this band bending appears to correlate with adsorbate ionicity. Only at higher metal coverage does E_F converge to a narrow range of values. For GaAs(110), such behavior may be indicative of charge states associated with charge transfer between adsorbate and semiconductor, followed by defect creation at more metallic coverages [40].

cate that extrinsic states dominate the contact rectification.

4. Evolution of interface characterization techniques

Over the past three decades, there has been a continuing refinement of the techniques used to probe Schottky barrier mechanisms. Such refinement has been rewarded by the discovery of a continuing series of new physical phenomena. Early experimental work relied entirely on macroscopic measurements such as current–voltage (J–V), capacitance–voltage (C–V), and internal photoemission. While directly related to device performance, such techniques provided little information on charge transfer on a microscopic scale. Also such techniques did not lend themselves readily to in-situ analysis of charge transfer or to process-dependent studies. With the advent of surface science techniques, it became possible to probe chemical, electronic, and geometric properties of semiconductor surfaces on an atomic scale. Work in the early 1970's focused primarily on the search for intrinsic surface states, as predicted by theory. By the mid-1970's, careful experimental work had showed that the appearance of such states in the semiconductor band gap could be associated with artifacts such as cleavage steps or excitonic corrections [1]. Reexamination of the surface energetics revealed that such states should be swept out the band gap for reconstructed surfaces. Angle-resolved XPS of clean, reconstructed surfaces confirmed the presence of such states at the band edges rather than with the gap, in agreement with theory [61].

Rutherford backscattering spectrometry (RBS) work on Si–metal interfaces in the 1970's revealed the presence of substantial interfacial phase formation on a micron scale at elevated temperatures [62]. With the increased sensitivity of XPS, researchers found significant chemical reaction and diffusion at metal–semiconductor interfaces, even near room temperature [1] or below [9]. Indeed, such chemical activity was found to exhibit systematic trends as a function of

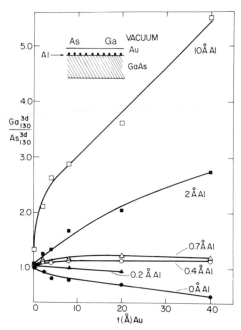

Fig. 6. Atomic-scale control of metal–semiconductor interdiffusion, as illustrated by SXPS core level ratios of Ga to As for the Au/Al/GaAs interface versus Al interlayer thickness. The inset shows the interlayer configuration schematically. The thickness of the interlayer controls the outdiffusion of Ga and As into Au and their concentration at the intimate junction [13,14].

thermodynamic heats of formation and reaction for the constituents [1,12–14]. Fig. 6 illustrates the dependence of outdiffusing constituents from the semiconductor, in particular their anion/cation stoichiometry, on the thickness of a reactive interfacial layer between the metal and the semiconductor. The effect of such a reactive metal layer, in this case, Al at the Au/GaAs(110) interface, at a III–V compound semiconductor junction, is to retard anion outdiffusion from the semiconductor bulk – a "chemical trapping" effect [13]. As already shown in Fig. 5, such microscopic changes in stoichiometry have pronounced effects on macroscopic Schottky barrier heights. XPS has also provided a measure of band bending during the inital stages of Schottky barrier formation. See, for example, early work by Spicer [63], Eastman [64], and Rowe et al. [65]. Over the years, such measurements have come to represent the primary data base for evaluating models of contact rectification.

Within the past few years, scanning tunneling microscopy (STM) has proved effective in probing gap states as a function of surface morphology and composition [66,67]. While confined to prob-

ing the near-surface region as well as XPS, STM possesses the advantage of lateral scanning on an atomic scale, thereby revealing variations in electronic properties related to atomic bond order, overlayer, and surface morphology features. STM studies have shown the presence of deep levels in

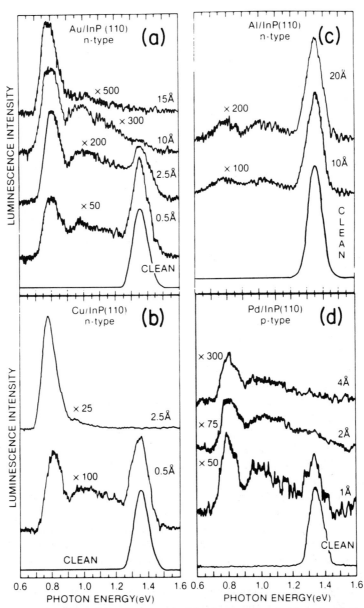

Fig. 7. Direct observation of interface states via low-energy cathodoluminescence spectroscopy for (a) Au, (b) Cu, (c) Al and (d) Pd on UHV-cleaved InP(110), illustrating the detection of optical transitions involving discrete states within the semiconductor band gap. These states depend on the particular metal and continue to change with multilayer coverages [19].

the semiconductor band gap for the clean surface and new features induced by metallic overlayers [66]. Variations of these features across the interface plane highlight the importance of local probes in understanding electronic properties sampled on a macroscopic scale.

In the mid-1980's, new techniques revealed that Schottky barrier formation was not complete with submonolayers of metal deposition but continued to evolve at nanometer overlayer thicknesses [19, 68]. While the evolution was suggested by early SXPS measurements of band bending versus thickness for 10–15 Å [68], it was confirmed by the changes in interface deep level emission observed via low energy cathodoluminescence spectroscopy (CLS) for semiconductors with overlayers several nanometers thick [19]. Fig. 7 illustrates the variation in deep level formation with increasing coverages for different metals on clean, ordered InP(110). Major changes occur at multilayer metal coverages – beyond the detection capability of more surface-sensitive techniques.

Recent measurements have sought to compare band bending/Fermi level position obtained by both microscopic and macroscopic techniques. Thus a comparison of E_F positions at 10–15 Å obtained via SXPS versus internal photoemission spectroscopy at 50–100 Å of Al on GaAs demonstrates that major changes in band bending can indeed occur at intermediate coverages [20]. Such changes appear linked to changes in near-surface stoichiometry with continuing outdiffusion. Likewise, time-dependent, internal photoemission measurements of nanometer thick metal contacts confirm the extended nature of reactive metal–semiconductor contacts [25]. Such measurements reveal the presence of multiple barriers for Al on GaAs(100), whose absolute magnitudes and relative contributions to the macroscopic barrier values varied over the course of hours at room temperature.

Lateral inhomogeneity is also evident from ballistic energy electron microscopy (BEEM) measurements of metal–semiconductor interfaces. This powerful new technique senses the reflection of ballistic electrons below the transmission threshold defined by the Schottky barrier. BEEM measurements taken through multilayer metal films reveal variations in barrier height across the lateral interface which depend on the nature of local chemical reaction [69]. Lateral barrier inhomogeneities confirm earlier suggestions of Freeouf et al. [70] and the observed variations in diode behavior for the same metal at different points on the same semiconductor surface [71]. Detailed comparisons of barriers measured by various macroscopic techniques also indicates the homogeneous nature of many metal–semiconductor contacts [72]. Correlations between different barriers and changes in local bonding as observed via transmission electron microscopy (TEM) confirm that such inhomogeneities alter the macroscopic electrical properties [73]. Besides variations in chemical bonding, elemental precipitates such as As in GaAs have been shown via internal photoemission and STM to introduce characteristic internal band bending regions within the semiconductor [74] and new deep levels localized within the precipitates themselves [75]. Overall these many experimental techniques have demonstrated the need for an even more refined view of metal–semiconductor junctions.

Theoretical work has focused primarily on the energy band structure and energetics of atomic bonding near the semiconductor surface. Such calculations have demonstrated the sensitivity of intrinsic surface states to the detailed atomic bonding at and near the vacuum–semiconductor interface. Indeed, theoretical work of the mid-1970's confirmed the need to take surface reconstruction into account explicitly in order to accurately determine the energy position of such states [1,4]. Such reconstructions vary for elemental and compound semiconductors and vary from one crystal orientation to another. Likewise, surface electronic structure can vary dramatically between reconstructions (e.g., metallic versus semiconducting for Si(100), depending on the dimerization). Chadi has shown [76] that dimerization geometries at GaAs(100) surfaces serve to minimize surface free energies. For clean GaAs(110) and other III–V surfaces, no intrinsic surface states are found within the semiconductor band gap, consistent with a surface relaxation of the topmost atomic layer of Ga and As atoms [77,78].

Hence, surface electronic structure pertinent to Schottky barrier formation depends on the atomic rearrangements needed to minimize surface free energy.

Deposition of metal on the clean semiconductor surface can change this electronic structure substantially. With the exception of epitaxical junctions, such local bonding is neglected in accounting for E_F pinning within a narrow range of band gap energies. In the case of MIGS, one extracts an interface dipole from either wave function tailing or a charge neutrality level [37,38] within the semiconductor. Here the bulk semiconductor band structure rather any interface-specific electronic structure plays a dominant role. Of course, for non-abrupt junctions, such wave function tailing may be screened by the presence of an interfacial layer [79].

Defect models of E_F "pinning" invoke the disruption of surface bonding by adsorbed species in order to introduce new localized states [45–47]. These new states may be independent of the adsorbate [45] or strongly-dependent on the near-surface change in chemical composition [5,12]. Likewise, segregation of native bulk defects to the surface as a result of changes in chemical potential may also occur. The bulk concentration of such native defects can also vary with the semiconductor growth method and subsequent processing [7,80]. Other theoretical models of Schottky barrier formation incorporate combinations of MIGS and deep level defects, as well as bond orbital shifts [81] and local disorder [82].

One can obtain important information about the densities, energies, and the electrically-active nature of interface states without assuming a particular physical nature for the states. Zur et al. [83] used a simple electrostatic model to show that the E_F stabilization was not complete until multiple metallic layers are present on the semiconductor surface, that is, until metallic screening of the interface states was complete. Pinning required densities of 10^{14}–10^{15} cm^{-2}, depending upon the distance of the charge sites from the metallic interface. Duke and Mailhiot [84,85] addressed the surface and interface dipole contributions to the Schottky barrier explicitly and showed that their overall contribution to the classical expression given as Eq. (1) was of the order 0.1–0.15 eV. They developed a self-consistent formalism which allowed a self-consistent fit of Schottky barrier versus work function to multiple donors and/or acceptors and their densities. Such a formalism is especially useful when applied to interfaces with significant variations in E_F position. Metals on GaAs(100) have provided such a range of barrier heights [23], leading to the first self-consistent electrostatic analysis of deep level energies and densities for a metal–semiconductor system [86]. It has also been applied to variable densities of interface states for the same metal on vicinal GaAs(100) [24,87], as discussed in the next section.

5. Atomic-scale control of Schottky barriers

Perhaps the best indication of progress in metal–semiconductor interfaces is the ability to predict and control Schottky barriers behavior. Substantial progress has been made in the last decade and especially in the last few years in controlling Schottky barriers on a microscopic scale. Several approaches have proven useful, namely, chemical interlayers, semiconductor crystal growth, epitaxical interface growth, surface morphology, and interface processing.

Chemical trapping by reactive layers at metal–semiconductor interfaces alters atomic stoichiometry, resulting in dramatic changes in band bending. See Figs. 5 and 6. Such interlayers act differently in III–V versus II–VI compound semiconductor systems. At III–V compound semiconductor interfaces, the barrier changes produced by reactive versus unreactive layers can be attributed to new electrically-active sites due to the reversal in stoichiometry [12] or local dipole changes with local chemical bonding [88]. At II–VI compound semiconductor junctions, reactive interlayers act to increase the local doping concentration by increasing the cation concentration near the interface. The increased doping density leads to a narrowing of and tunneling through the barrier [16,89]. As shown by the J–V characteristic in Fig. 8, a rectifying barrier of 0.8 eV for Au

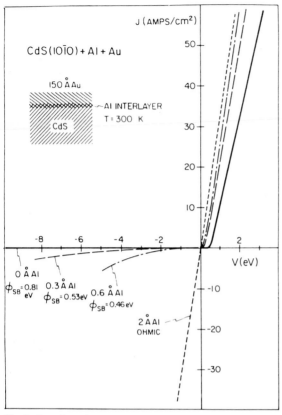

Fig. 8. Atomic-scale control of macroscopic device features, as illustrated by the J–V characteristics of Au–CdS(10$\bar{1}$0) Schottky diodes versus Al interlayer thickness. The inset shows a cross-sectional schematic diagram of the interlayer structure [16,89].

on CdS(10$\bar{1}$0) can be converted to an "ohmic" contact by the introduction of only one or two monolayers of Al. This behavior clearly demonstrates the effect of atomic scale chemistry on macroscopic electronics.

Other interlayers at metal–semiconductor junctions have proven effective as well. These include semiconductors such as Si and Ge at GaAs contacts, which provide E_F stabilization over a wide range of the GaAs band gap, depending on the interlayer doping [90,91], Inert gases such as Xe, which inhibit the chemical interactions between the deposited metal and the underlying substrate [92,93], as well as reactive interlayers at other metal–semiconductor interfaces, e.g. Ref. [53]. In all cases the variations in barrier

height have been substantial, rather than minor perturbations around some characteristic value.

Semiconductor crystal growth has proved to be a significant factor in Schottky barrier formation for compound semiconductors. Depending on the growth method and procedures used, the resultant crystal may possess different levels of native defects or defect complexes due to nonstoichiometry, strain, or impurities. Such defects can be electrically-charged and migrate toward the free surface or interface. Evidence for such surface segregation is available from deep level transient spectroscopy (DLTS) and suggests that deep level densities near the surface can reach levels sufficient to account for E_F pinning [94]. In the case of GaAs, crystals grown from the melt versus by MBE show major differences in deep level luminescent features [86]. Photoluminescence of the melt-grown GaAs (typically used for cleaved (110) studies in UHV) reveals orders-of-magnitude higher emission from states near mid-gap than the MBE-grown crystals. Likewise, such melt-grown crystals also exhibit much higher intensities of deep level features induced near the interface by metallization [95].

Associated with these different materials are major differences in Schottky barrier formation for metals on their clean surfaces. Fig. 9 illustrates the wide range of E_F stabilization obtained for the MBE-grown GaAs(100) surface [23] versus the much narrower range for the melt-grown, GaAs(110) surface [45]. Another difference between the two panels is the absence of a gap between n- versus p-type material in the (100) case versus the 0.2 eV gap in the (110) panel. Furthermore, the E_F movement in the (100) case occurs over several monolayers of coverage, consistent with interface chemical changes, versus the submonolayer stabilization evident for the (110) case. Fig. 9 illustrates not only that the differences in Schottky barrier formation can be substantial between different crystals of the same semiconductor, but also that band bending at GaAs is much more controllable than hitherto believed.

The Schottky barrier dependence on bulk states is also evident for a II–VI compound semiconductor, CdTe. In this case, the barrier height

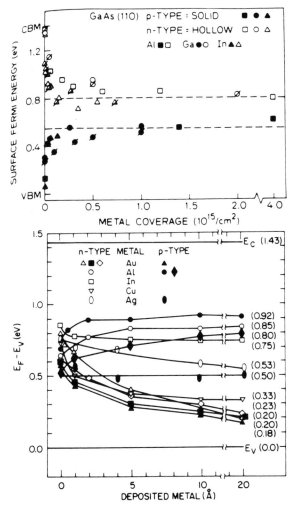

Fig. 9. Qualitative difference in metal-induced E_F movements within the GaAs band gap, measured via SXPS rigid core level shifts. Metals on UHV-cleaved melt-grown GaAs(110) (upper) exhibit a submonolayer E_F movement to a narrow mid-gap range of energies with different n- and p-type values (after Spicer et al. [45]). Metals on decapped, MBE-grown GaAs(100) (lower) show E_F movement over several monolayers to a wide range of energies with matching n- and p-type values [23,86].

can switch from 0.6 to 0.9 eV depending on the presence of a deep level evident from photoluminescence [96]. When present, E_F stabilizes at the position of this deep level in the gap; when absent, E_F stabilizes where expected on the basis of a classical work function difference. Analogous barrier differences have been reported by

Williams and coworkers [71] for different diodes on the same CdTe surface, suggesting that the concentration of such deep levels may vary across a semiconductor surface. Laser-annealed CdTe exhibits E_F energies which move between discrete positions in the band gap as the chemical composition and bonding change. These positions correspond to deep levels observed by photoluminescence spectroscopy and indicate links between E_F and the deep levels as well as deep levels and the interface chemistry [97]. Indeed, altering the interface chemistry via a reactive interlayer (e.g., Yb between Au and CdTe) alters the relative deep level concentrations [98b]. For epitaxial CdTe, bulk deep level concentrations are much lower and do not dominate surface deep level features. In this case, one can measure dramatic changes in deep levels with surface preparation [21]. Evidence now exists to suggest that similar deep level differences can account for diverse measurements of Schottky barriers on GaP [98a].

Epitaxial growth offers another avenue for Schottky barrier control as well as providing well-defined interface atomic structures for theoretical analysis. One notable example is the difference between the two epitaxical variants of the $NiSi_2$ on Si(111) interface [99], obtained by modifying the initial growth template. The atomic bonding between these two variants differs only by a rotation about an axis normal to the interface plane, yet they introduce a 0.14 eV perturbation in Schottky barrier. This structural effect is distinct from electronic perturbations which may be introduced by impurities, structural imperfections, and other artifacts. Structural effects on electronic properties are also apparent at epitaxial interfaces between Pb and Si(111) [100]. Such phenomena underscore the importance of local atomic bonding in the charge transfer at these ordered metal–semiconductor interfaces.

Barrier heights can vary over a wide range for the same metal on the same semiconductor as a function of orientation and annealing [101]. Using epitaxial overlayers of $Sc_xEr_{1-x}As$ metal alloys on GaAs(100), Palmstrøm et al. showed that it is possible to separate out the effects of strain from those of misorientation and lattice mismatch. Diodes formed from $Sc_xEr_{1-x}As$ $(0 < x < 1)$

grown with different composition and lattice mismatch show only a weak barrier dependence (less than 0.1 eV). In contrast, barriers vary by over 0.4 eV as a function of misorientation away from the GaAs [100] plane and annealing. Similarly, Φ_B variations of 0.2–0.3 eV are found for epitaxical variants of Ni/Al on GaAs [102]. Both changes in interface dipole as well as local chemical reaction may contribute to the altered barriers in these GaAs cases. Overall, even well-ordered junctions provide evidence for significant Schottky barrier variation.

Surface morphology provides yet another tool to understand and control Schottky barrier formation. Steps present at vicinal semiconductor surfaces can produce electrically-active sites. This is evident for GaAs(100) grown with different orientation directions and angles [25,87]. The dif-

ferent barriers for different vicinal surfaces of GaAs(100) span a wide energy range, as shown by the data points extracted from SXPS in Fig. 10a. A self-consistent analysis of Φ_B versus metal work function provides a family of curves corresponding to different interface state densities for levels at $E_v + 0.2$ eV and $E_v + 0.6$ eV (see inset). Electronic transitions involving such states are evident in CLS spectra for vicinal surface with Al overlayers [87]. Fig. 10b illustrates the densities of electronic states extracted from Fig. 10a plotted against the density of chemically-active (exposed As) sites. There is a one-to-one correspondence between structural and electronic features, inclusive of both misorientation angle and direction. The slope indicates a charge of two-thirds of an electron per chemically-active site, extrapolating down to a low (10^{13} cm^{-2}) density for the well-

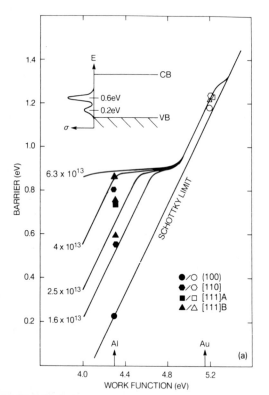
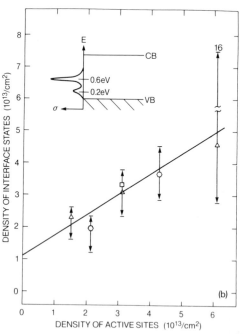

Fig. 10. Self-consistent analysis of barrier heights versus metal work function (a) and interface densities versus active structural site densities (b) for Al and Au on vicinal GaAs(100) surfaces. Barrier heights increase monotonically with misorientation angle and direction. The family of density curves for the 0.6 eV and (constant density) 0.2 eV states shown in the insets yield the interface state densities in (b), which display a near one-to-one dependence on the chemically-active structural site density [24,87].

oriented surface. STM measurements of these vicinal surfaces by Pashley and Haberern reveals densities of kinks sufficiently high to account for these electrical effects [103]. Thus, for such vicinal surfaces, electronic states calculated self-consistently and observed spectroscopically can be correlated with structural imperfections on an atomic scale.

Other morphological features have been shown to be electrically-active as well. Dislocations produced at relaxed epitaxial InGaAs overlayers on GaAs produce band bending extending away from the dislocations cores, which alter the electrical barriers measured macroscopically [50]. STM measurements of metal clusters on GaAs reveal the presence of new electronic states within the band gap, localized at the cluster edges [66,104]. STM studies have also revealed the presence of point defects associated with metals on semiconductors [105]. Even elemental precipitates such as As clusters in low-temperature GaAs have been isolated by STM and shown to possess distinct, deep levels within the GaAs band gap [75]. Hence, structural imperfections across the interface plane or even within the semiconductor bulk can be related explicitly to charge sites.

Last of the atomic-scale methods for Schottky barrier control is interface processing. Such processing includes surface passivation, surface preparation, and crystal growth. Passivation techniques have included photoelectrochemical washing of GaAs surfaces in water to remove near-surface As [106], Na_2S and NaOH treatments to produce sulfide and other intermediate chemical layers on GaAs and InGaAs [107,108], as well as HF etching to produce atomically-smooth and monohydride-terminated Si surfaces [109]. Each such wet-chemical treatment produces orders-of-magnitude reduction in surface recombination velocity (SRV) and interface state density [110].

The preparation of the semiconductor surface itself can yield quite large variations in Schottky barrier, even for the same metal on the same semiconductor. Fig. 11 illustrates such a wide Φ_B variation for Al on GaAs [20,26]. GaAs surfaces prepared at different annealing temperatures in order to obtain different LEED reconstructions provide one method of varying barriers and can

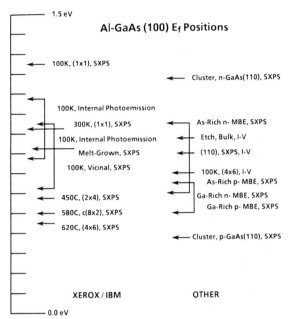

Fig. 11. Dependence of E_F stabilization energy on surface preparation for Al interfaces with GaAs. Results obtained by SXPS and internal photoemission from thermally-decapped (100) surfaces are shown in the left column for different surface reconstructions, morphologies and Ga/As stoichiometries (including results from Figs. 9 and 10). Results for Al on GaAs(100) and (110) surfaces prepared and processed by other techniques appear at right [20,26]. E_F stabilization energies vary over more than half the GaAs band gap for the same metal overlayer.

be associated with differences in interface state energies and densities observed via CLS [110]. Such differences are not observed for a less-reactive metal such as Au on GaAs and therefore suggest a chemical basis for such electronic effects. Also shown are the different E_F positions for vicinal surfaces described in Fig. 10 and the inert gas interfaces [92]. The wide range of Φ_B values obtained for Al–GaAs junctions by different measurement techniques, surface orientations, and temperatures underscores the importance of interface preparation on the resultant electronic structure.

A final and very promising area of interface processing is in the interface growth itself. Thus for the ZnSe-on-GaAs heterojunction, deep level densities of states near the junction can range over more than two orders-of-magnitude, depending on the RHEED reconstruction pattern of the GaAs prior to ZnSe overlayer growth [111]. The reduction of deep levels can be related to the promotion of Ga–Se compound formation at Ga-rich interfaces. Another growth technique involves the incorporation of near-monolayer sheets of dopants within a few monolayers of the interface (and on both sides of the interface for heterojunctions), so-called "delta doping" (see, for example, Ref. [112]). Such layers introduce strong electric fields near the junction and alter the effective barriers to charge transport. Assuming abrupt surface morphologies, the thickness of such delta-doped regions will alter the barrier widths for tunneling and thereby the effective barrier height. Such an atomic-scale technique is particularly useful for devices with shallow depth requirements. Strained layer superlattices of different semiconductors represent the latest approach to barrier control (see, for example, Ref. [113]). Charge transfer from the wider to the narrower semiconductor and tunneling through the wider gap material affords high conductivity to the contacted semiconductor and the modified band offset at the semiconductor–multilayer junction dominates the macroscopic electronic properties. The multiple interface structure also serves to inhibit the propagation of dislocations. A major strength of such growth techniques is their straightforward implementation using currently available MBE techniques.

Overall, there are now many new avenues available for controlling band bending at metal–semiconductor interfaces. The wide range of barriers achieved by many surface science and device groups underscore this controllability. The links now established between Schottky barriers, interface states, and atomic-scale chemical structure demonstrate the progress that has been made in understanding contact rectification.

6. New directions

Opportunities for improved understanding of interface states and Schottky barriers exist in several areas. The observations of interface heterogeneity described above indicate the importance of spatially-resolved interface analysis. First, this involves the identification of domains with different barrier heights. Microspectroscopy tools are now becoming available which are sensitive to band bending and chemical composition with resolution below 0.1 μm [114]. BEEM measurements can provide barrier heights on an even smaller scale, albeit without chemical sensitivity. In-situ studies could provide information on the evolution of domains of different barrier height as a function of processing. Second, these microscopic measurements must be correlated to macroscopic electrical properties. Such studies would provide experimental tests for theoretical model predictions. For instance, calculations of "pinch-off" of low barrier regions at heterogeneous surfaces could be tested against barriers as a function of observed domain size and depletion width (controlled via background doping). Third, the size and barrier heights of the domains could be related to the heterogeneous overlayer chemistry, the surface morphology, and any surface segregation of bulk species. Again, one can use in-situ analysis to identify physical correlations as the surface features change with processing. Finally, one can thereby identify processing techniques to assure uniformity (for microelectronic devices) or heterogeneity (for catalytic surfaces).

A second new direction for surface science in probing metal–semiconductor contacts involves time-resolved excitation analysis. Here, picosecond laser techniques are now available to excite deep level luminescence from interface states and thereby extract their densities, cross sections, and energies. Such work could provide a quantitative basis for the steady state luminescence observations already reported and thereby establish figures of merit for interface preparation. One could then use in-situ spectral features to evaluate starting interfaces prior to growing full device structures.

The correlation of electronic, chemical, and geometric structure on an atomic scale represents yet another exciting opportunity for surface scientists. Here, spatially-localized probes such as STM, BEEM, and microspectroscopy may provide additional structural information associated with specific interface charge states. Such analyses may also prove useful in understanding the actual operation of ever smaller and more complex device structures.

7. Perspectives

The author would like to conclude this review with several perspectives on the evolution of metal–semiconductor interfaces, the importance of surface science and the role of Professor Harry Gatos. Surface science has played an increasingly important role in exploring the properties of metal–semiconductor interfaces. Over time, the field has evolved from (a) purifying single crystals in order to separate bulk from surface effects, (b) developing UHV techniques in order to achieve clean interfaces, (c) developing electronic, optical, and other techniques to probe the interface structure, (d) developing theoretical tools to relate these observations to physical properties, (e) extending characterization techniques to atomic-scale lateral dimensions, (f) subsequently discovering the influence of surface processing on interface properties, and (g) discovering even more refined influences of bulk crystal growth on interface properties.

Currently, the simple picture of abrupt metal–semiconductor junctions pictured in solid-state textbooks has been replaced by a complex, three-dimensional system with interrelated chemical, structural, and electronic properties. Understanding such systems represents a challenge for both theorist and experimentalist. Yet it represents an opportunity since new ways have been found to control macroscopic physical properties. Still required is a predictive model for Schottky barrier formation. Perhaps the buried interface and spatially-localized probes now available will play a key role in developing such a predictive capability.

Harry Gatos can take much satisfaction from the importance which atomic-scale chemical effects have assumed in the formation of macroscopic electronic interfaces and the pioneering role he took in identifying them. Certainly, he can be proud of the role which surface science, fostered by its journal namesake, has played in achieving such progress.

The outlook for continued advances in interface research will continue to depend on the productivity and insights of surface scientists. However, the support for their work will depend on the economics of large-scale electronics manufacturing for communication and processing of information. As the value of easily communicated and processed information continues to increase, so should the need for more powerful micro- and optoelectronic devices. Therefore the identification of optimal materials for such applications, their processing, and the control of their quality from fundamental, atomic-scale principles offers a potentially exciting future for interface research.

References

[1] L.J. Brillson, Surf. Sci. Rep. 2 (1982) 123.
[2] E.H. Rhoderick and R.H. Williams, in: Metal–Semiconductor Contacts, 2nd ed., Monographs in Electrical and Electronic Engineering, Eds. P. Hammond and R.L. Grimsdale (Clarendon, Oxford, 1988).
[3] J.M. Woodall, P.D. Kirchner, J.L. Freeouf, D.T. McInturff, M.R. Melloch and F.H. Pollak, Inst. Phys. Conf. Ser., in press.
[4] L.J. Brillson, in: Basic Properties of Semiconductors, Vol. 1, Handbook on Semiconductors, Ed. P.T. Landsberg (North-Holland, Amsterdam, 1992) p. 281.
[5] W.E. Spicer, Z. Lilienthal-Weber, E. Weber, N. Newman, T. Kendelewicz, R. Cao, C. McCants, P. Mahowald, K. Miyano and I. Lindau, J. Vac. Sci. Technol. B 6 (1988) 1245.
[6] R.H. Williams, Surf. Sci. 251/252 (1991) 12.
[7] L.J. Brillson, Ed., Contacts to Semiconductor Devices (Noyes, Park Ridge, NJ, 1993); Encyclopedia of Advanced Materials, Ed. S. Mahajan (Pergamon, Oxford, 1993), in press.
[8] W. Mönch, Rep. Prog. Phys. 53 (1990) 221.
[9] V.A. Grazhulis, Prog. Surf. Sci. 36 (1991) 89.
[10] L.J. Brillson, Surf. Sci. 51 (1975) 45.
[11] L.J. Brillson, Phys. Rev. Lett. 40 (1978) 260.
[12] L.J. Brillson, C.F. Brucker, A.D. Katnani, N.G. Stoffel and G. Margaritondo, Appl. Phys. Lett. 38 (1981) 784.
[13] L.J. Brillson, G. Margaritondo and N.G. Stoffel, Phys. Rev. Lett. 44 (1980) 667.
[14] L.J. Brillson, C.F. Brucker, A.D. Katnani, N.G. Stoffel and G. Margaritondo, Phys. Rev. Lett. 46 (1981) 838.
[15] L.J. Brillson, Thin Solid Films 89 (1982) 461.

[16] C.F. Brucker and L.J. Brillson, Appl. Phys. Lett. 39 (1981) 67.

[17] L.J. Brillson, H.W. Richter, M.L. Slade, B.A. Weinstein and Y. Shapira, J. Vac. Sci. Technol. A 3 (1985) 1011.

[18] L.J. Brillson, R.E. Viturro, J.L. Shaw and H.W. Richter, J. Vac. Sci. Technol. A 6 (1988) 1437.

[19] R.E. Viturro, M.L. Slade and L.J. Brillson, Phys. Rev. Lett. 57 (1986) 487.

[20] A.D. Raisanen, I.M. Vitomirov, S. Chang, L.J. Brillson, P.K. Kirchner, G.D. Pettit and J.M. Woodall, J. Vac. Sci. Technol., in press.

[21] J.L. Shaw, L.J. Brillson, S. Sivananthan and J.-P. Faurie, Appl. Phys. Lett. 56 (1990) 1266.

[22] L.J. Brillson, M.L. Slade, R.E. Viturro, M. Kelly, N. Tache, G. Margaritondo, J.M. Woodall, G.D. Pettit, P.D. Kirchner and S.L. Wright, Appl. Phys. Lett. 48 (1986) 1458.

[23] R.E. Viturro, J.L. Shaw, C. Mailhiot, L.J. Brillson, N. Tache, J. McKinley, G. Margaritondo, J.M. Woodall, P.D. Kirchner, G.D. Pettit and S.L. Wright, Appl. Phys. Lett. 52 (1988) 5052.

[24] S. Chang, L.J. Brillson, Y.J. Kime, D.S. Rioux, D. Pettit and J.M. Woodall, Phys. Rev. Lett. 64 (1990) 2551.

[25] S. Chang, L.J. Brillson, Y.J. Kime, D.S. Rioux, G.D. Pettit and J.M. Woodall, J. Vac. Sci. Technol. A 9 (1991) 902.

[26] L.J. Brillson, I.M. Vitomirov, A. Raisanen, S. Chang, R.E. Viturro, P.D. Kirchner, G.D. Pettit and J.M. Woodall, Appl. Surf. Sci. 65/66 (1993) 667.

[27] H.C. Gatos and M.C. Lavine, J. Appl. Phys. 31 (1960) 743.

[28] H.C. Gatos and M.C. Lavine, in: Progress in Semiconductors, Vol. 9, Eds. A.F. Gibson and R.E. Burgess (Temple, London, 1965) pp. 1–45.

[29] H.C. Gatos and J. Lagowski, J. Vac. Sci. Technol. 10 (1973) 130.

[30] T.M. Valahas, J.S. Sochanski and H.C. Gatos, Surf. Sci. 26 (1971) 41.

[31] C.A. Mead, Solid-State Electron. 9 (1966) 1023, and references therein.

[32] S. Kurtin, T.C. McGill and C.A. Mead, Phys. Rev. Lett. 22 (1970) 1433.

[33] J. Bardeen, Phys. Rev. 71 (1947) 717.

[34] W. Schockley and G.L. Pearson, Phys. Rev. 74 (1948) 232.

[35] V. Heine, Phys. Rev. 138 (1965) A1689.

[36] S.G. Louie and M.L. Cohen, Phys. Rev. Lett. 35 (1975) 866.

[37] F. Flores and C. Tejedor, J. Phys. C (Solid State Phys.) 20 (1987) 145.

[38] C. Tejedor, F. Flores and E. Louis, J. Phys. C (Solid State Phys.) 10 (1977) 2163.

[39] J. Tersoff, Phys. Rev. B 32 (1985) 6968.

[40] W. Mönch, Surf. Sci. 299/300 (1994) 928.

[41] T.A. Goodwin and P. Mark, in: Progress in Surface Science, Vol. 1 (Pergamon, New York, 1972) p. 1.

[42] S. Baidyaroy and P. Mark, Surf. Sci. 30 (1972) 53.

[43] L.J. Brillson, Phys. Rev. 18 (1978) 2431.

[44] A.G. Milnes, Deep Impurities in Semiconductors (Wiley-Interscience, New York, 1973).

[45] W.E. Spicer, I. Lindau, P. Skeath, C.Y. Su and P. Chye, Phys. Rev. Lett. 44 (1980) 420.

[46] I. Lindau, P.W. Chye, C.M. Garner, P. Pianetta, C.Y. Su and W.E. Spicer, J. Vac. Sci. Technol. 15 (1978) 1337.

[47] H.H. Wieder, J. Vac. Sci. Technol. 15 (1978) 1498.

[48] R.H. Williams, R.R. Varma and V. Montgomery, J. Vac. Sci. Technol. 16 (1979) 1418.

[49] M.D. Pashley, K.W. Haberern and J.W. Gaines, Appl. Phys. Lett. 58 (1991) 406.

[50] J.M. Woodall, G.D. Pettit, T.N. Jackson, C. Lanza, K.L. Kavanaugh and J.M. Mayer, Phys. Rev. Lett. 51 (1983) 1783.

[51] J.F. McGilp, J. Phys. C 17 (1984) 2249.

[52] S. Makram-Ebeid, D. Gautard, P. Devillard and G.M. Martin, Appl. Phys. Lett. 40 (1982) 161.

[53] R.H. Williams and M.H. Patterson, Appl. Phys. Lett. 40 (1982) 484.

[54] N.C. Wyeth and A. Catalano, J. Appl. Phys. 51 (1980) 2286.

[55] J. Baars, D. Bassett and M. Schulz, Phys. Status Solidi (a) 49 (1978) 483.

[56] D.D. Wagman, W.H. Evans, V.B. Parker, I. Halow, S.M. Bailey and R.H. Schumm, Natl. Bur. Std: Technical Notes, 270-3-270-7 (US Government Printing Office, Washington, DC, 1968–1971).

[57] L.J. Brillson, J. Vac. Sci. Technol. 15 (1978) 1378.

[58] A. Kahn, K. Stiles, D. Mao, S.F. Horng, K. Young, J. McKinley, D.G. Kilday and G. Margaritondo, in: Metallization and Metal–Semiconductor Interfaces, Vol. 195, NATO Advanced Study Institute, Series B Physics, Ed. I.P. Batra (Plenum, New York, 1989) p. 163.

[59] J.L. Freeouf and J.M. Woodall, Appl. Phys. Lett. 39 (1981) 727;
J.M. Woodall and J.L. Freeouf, J. Vac. Sci. Technol. 19 (1981) 794.

[60] L.J. Brillson, R.E. Viturro, M.L. Slade, P. Chiaradia, D. Kilday, M. Kelly and G. Margaritondo, Appl. Phys. Lett. 50 (1987) 1379.

[61] G.P. Williams, R.J. Smith and G.J. Lapeyre, J. Vac. Sci. Technol. 15 (1978) 249.

[62] J.M. Poate, K.N. Tu and J.M. Mayer, Thin Films – Interdiffusion and Reactions (Wiley-Interscience, New York, 1978).

[63] W.E. Spicer, Comments Solid State Phys. 5 (1973) 105.

[64] D.E. Eastman, in: Techniques of Metals Research VI, Ed. E. Passaglia (Interscience, New York, 1972) p. 413.

[65] J.E. Rowe, S.B. Christman and H. Ibach, Phys. Rev. Lett. 34 (1975) 874.

[66] R.M. Feenstra and J.A. Stroscio, Phys. Rev. Lett. 59 (1987) 2173;
R.M. Feenstra and P. Martensson, Phys. Rev. Lett. 61 (1988) 447.

[67] M.D. Pashley, K.W. Haberem, W. Friday, J.M. Woodall and P.D. Kirchner, Phys. Rev. Lett. 60 (1988) 2176.

[68] L.J. Brillson and G. Margaritondo, in: Surface Proper-

ties of Electronic Materials, Eds. D.A. King and D.P. Woodruff, The Chemical Physics and Heterogeneous Catalysis of Solid Surfaces, Vol. 5 (Elsevier, Amsterdam, 1988) pp. 119–181.

[69] W.J. Kaiser and L.D. Bell, Phys. Rev. Lett. 60 (1988) 1406.

[70] J.L. Freeouf, T.N. Jackson, S.E. Laux and J.M. Woodall, J. Vac. Sci. Technol. 21 (1982) 570.

[71] A.E. Fowell, R.H. Williams, B.E. Richardson and T.-H. Shen, Semicond. Sci. and Technol. 5 (1990) 348.

[72] R.T. Tung, Appl. Phys. Lett. 58 (1991) 2821.

[73] J.P. Sullivan, D.J. Eaglesham, F. Schrey, W.R. Graham and R.T. Tung, J. Vac. Sci. Technol., in press.

[74] D.T. McInturff, J.M. Woodall, A.C. Warren, N. Braslau, G.D. Pettit, P.D. Kirchner and M.R. Melloch, Appl. Phys. Lett. 60 (1992) 448.

[75] A. Vaterlaus, R.M. Feenstra, P.D. Kirchner, S.D. Pettit and J.M. Woodall, J. Vac. Sci. Technol., in press.

[76] D.J. Chadi, J. Vac. Sci. Technol. A 5 (1987) 1691.

[77] C.B. Duke, in: Surface Properties of Electronic Materials, Eds. D.A. King and D.P. Woodruff, The Chemical Physics of Solid Surfaces and Heterogeneous Catalysis, Vol. 5 (Elsevier, Amsterdam, 1988) pp. 69–188.

[78] C.B. Duke, J. Vac. Sci. Technol. B 1 (1983) 732.

[79] J.L. Freeouf, J.M. Woodall, L.J. Brillson and R.E. Viturro, Appl. Phys. Lett. 56 (1990) 69.

[80] L.J. Brillson, Comments Cond. Mat. Phys. 14 (1989) 311.

[81] W.A. Harrison, J. Vac. Sci. Technol. B 3 (1985) 1231.

[82] H. Hasegawa and H. Ohno, J. Vac. Sci. Technol. B 4 (1986) 1130.

[83] A. Zur, T.C. McGill and D.L. Smith, Phys. Rev. B 28 (1983) 2060.

[84] C.B. Duke and C. Mailhiot, J. Vac. Sci. Technol. B 3 (1985) 1170.

[85] C. Mailhiot and C.B. Duke, Phys. Rev. B 33 (1986) 1118.

[86] L.J. Brillson, R.E. Viturro, J.L. Shaw, C. Mailhiot, N. Tache, J. McKinley, G. Margaritondo, J.M. Woodall, P.D. Kirchner, G.D. Pettit and S.L. Wright, J. Vac. Sci. Technol. B 6 (1988) 1263.

[87] S. Chang, L.J. Brillson, D.S. Rioux, S. Kirchner, D. Pettit and J.M. Woodall, Phys Rev. B 44 (1991) 1391.

[88] L.J. Brillson, J. Vac. Sci. Technol. 16 (1979) 1137.

[89] C.F. Brucker, L.J. Brillson, A.D. Katnani, N.G. Stoffel and G. Margaritondo, J. Vac. Sci. Technol. 21 (1982) 590.

[90] R.W. Grant and J.R. Waldrop, J. Vac. Sci. Technol. B 5 (1987) 1015;
J.R. Waldrop and R.W. Grant, Appl. Phys. Lett. 52 (1988) 1794;
J.R. Waldrop, Appl. Phys. Lett. 53 (1988) 1518.

[91] J.C. Costa, F. Williamson, T.J. Miller, K. Beyzavi, M.I. Nathan, D.S.L. Mui, S. Strite and H. Morkoc, Appl. Phys. Lett. 59 (1991) 382.

[92] G.D. Waddill, I.M. Vitomirov, C.M. Aldao and J.H. Weaver, Phys. Rev. Lett. 62 (1989) 1568.

[93] M. Vos, C.M. Aldao, D.J.W. Aastuen and J.H. Weaver, Phys. Rev. B 41 (1990) 991.

[94] A. Yahata and M. Nakajima, Jpn. J. Appl. Phys. 23 (1984) L313.

[95] J.L. Shaw, R.E. Viturro, L.J. Brillson and D. LaGraffe. J. Electron. Mat. 18 (1989) 59.

[96] J.L. Shaw, R.E. Viturro, L.J. Brillson, D. Kilday, M.K. Kelly and G. Margaritondo, J. Electron. Mat. 17 (1988) 149.

[97] J.L. Shaw, R.E. Viturro, L.J. Brillson, D. Kilday, M.K. Kelly and G. Margaritondo, J. Vac. Sci. Technol. A 6 (1988) 1579, 2752.

[98] (a) L.J. Brillson, I.M. Vitomirov, A.D. Raisanen and S. Chang, unpublished;
(b) J.L. Shaw, R.E. Viturro, L.J. Brillson and D. LaGraffe, Appl. Phys. Lett. 53 (1988) 1723.

[99] R.T. Tung, K.K. Ng, J.M. Gibson and F.J. Levy, Phys. Rev. B 33 (1986) 7077.

[100] D.R. Heslinga, H.H. Weitering, D.P. van der Werf, T.M. Klapwijk and T. Hibma, Phys. Rev. Lett. 64 (1990) 1589.

[101] C.J. Palmstrøm, T.L. Cheeks, H.L. Gilchrist, T.G. Zhu, C.B. Carter and R.E. Nahory, in: Electronic Optical and Devices Properties of Layered Structures, Eds. J.R. Hayes, M.S. Hybertsen and E.R. Weber (Materials Research Society, Pittsburgh, PA) p. 63.

[102] S.A. Chambers and V.A. Loebs, J. Vac. Sci. Technol. B 8 (1990) 724.

[103] M.D. Pashley and K.W. Haberern, Phys. Rev. Lett. 67 (1991) 2697.

[104] R.M. Feenstra, Phys. Rev. Lett. 63 (1989) 1412.

[105] L.J. Whitman, J.A. Stroscio, R.A. Dragoset and R.J. Celotta, Phys. Rev. B 42 (1990) 7228.

[106] S.D. Offsey, J.M. Woodall, A.C. Warren, P.D. Kirchner, T.I. Chappell and G.D. Pettit, Appl. Phys. Lett. 48 (1986) 475.

[107] E. Yablonovitch, C.J. Sandroff, R. Bhat and T. Gmitter, Appl. Phys. Lett. 51 (1987) 439.

[108] C.J. Sandroff, R.N. Nottenburg, J.-C. Bischoff and R. Bhat, Appl. Phys. Lett. 51 (1987) 33.

[109] G.S. Higashi, Y.J. Chabal, G.W. Trucks and K. Raghavachari, Appl. Phys. Lett. 56 (1990) 656.

[110] E. Yablonovitch and T.J. Gmitter, Proc. Electrochem. Soc. 88 (1988) 207.

[111] J. Qiu, Q.-D. Qian, R.L. Gunshor, M. Kobayashi, D.R. Menke, D. Li and N. Otsuka, Appl. Phys. Lett. 56 (1990) 1272.

[112] E.F. Schubert, J.E. Cunningham, W.T. Tsang and T.H. Chiu, Appl. Phys. Lett. 49 (1986) 292.

[113] C.K. Peng, G. Ji, N.S. Kumar and H. Morkoc, Appl. Phys. Lett. 53 (1988) 900.

[114] F. Cerrina, G. Margaritondo, J.H. Underwood, M. Hettrick, M. Green, L.J. Brillson, A. Franciosi, H. Hoechst, P.M. DeLuca, Jr. and M. Gould, J. Nucl. Instrum. Methods A 266 (1988) 303.

Surface Science 299/300 (1994) 928–944
North-Holland

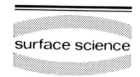

surface science

Metal–semiconductor contacts: electronic properties

Winfried Mönch

Laboratorium für Festkörperphysik, Universität Duisburg, D-47048 Duisburg, Germany

Received 18 March 1993; accepted for publication 20 August 1993

Rectification in metal–semiconductor contacts was first described by Braun in 1874. We owe the explanation of this observation to Schottky. He demonstrated that depletion layers exist on the semiconductor side of such interfaces. The current transport across such contacts is determined by their barrier heights, i.e., the respective energy difference between the Fermi level and the edge of the majority-carrier band. Since Schottky had published his pioneering work in 1938 the mechanisms, which determine the barrier heights of metal–semiconductor contacts, have remained under discussion. In 1947, Bardeen attributed the failure of the early Schottky–Mott rule to the neglect of electronic interface states. The foundations for a microscopic description of interface states in *ideal* Schottky contacts was laid by Heine in 1965. He demonstrated that a continuum of metal-induced gap states (MIGS), as they were called later, derives from the virtual gap states of the complex semiconductor band-structure. Neither this MIGS model nor any of the many other *monocausal* approaches, the most prominent is Spicer's Unified Defect Model, can explain the experimental data. In 1987, Mönch concluded that the continuum of MIG states represents the *primary* mechanism, which determines the barrier heights in ideal, i.e., intimate, abrupt, and homogeneous metal–semiconductor contacts. He attributed deviations from what is predicted by the MIGS model to other and then secondary mechanisms. In this respect, interface defects, structure-related interface dipoles, interface strain, interface compound formation, and interface intermixing, to name a few examples, were considered.

1. Introduction

The discovery of rectifying properties of metal–semiconductor contacts by Braun [1] in 1874 marks the beginning of semiconductor surface and interface physics. The technical importance of this *anomalous phenomenon* was soon realized. Large scale application of plate rectifiers based on cuprous oxide and later on selenium started as early as 1925 when a patent for such devices was issued to Grondahl [2].

A physical explanation of the *unilateral conduction* had to wait until Wilson [3] presented his quantum theory of semiconductors and the positive sign of the Hall coefficient and by this the p-type character of Cu_2O was finally established [4]. In his famous paper on the *Halbleitertheorie der Sperrschicht*, which was published by the end of 1938, Schottky [5] explained the blocking behavior of metal–semiconductor contacts by a space-charge layer on their semiconductor side which is depleted of mobile carriers. By now, this conclusion is easily derived.

In a Gedanken experiment, a metal–semiconductor contact may be created by gradually decreasing the distance between a metal and a semiconductor until eventually an intimate and abrupt interface has formed. This is illustrated in Fig. 1. The semiconductor is assumed to be non-degenerately doped n-type and to have no surface states within its band gap. The bands are thus flat up to the surface for infinite separation between metal and semiconductor.

The work functions of the metal and of the bare semiconductor generally differ so that in thermal equilibrium an electric field will exist in the vacuum gap between them. As a consequence, metal and semiconductor carry surface charges of equal density but of opposite sign. The condition of charge neutrality may be written as

$$Q_m + Q_{sc} = 0. \tag{1}$$

SSDI 0039-6028(93)E0514-U

In Fig. 1, the metal is assumed to have the larger work function. Then, the surface charge Q_m on the metal and Q_{sc} on the semiconductor have a negative and a positive sign, respectively.

The electric field enters both metal and semiconductor. However, the penetration depths are quite different. They scale with the Fermi–Dirac length of the metal and the Debye length of the semiconductor. Due to the large electron densities in metals, their screening lengths typically measure less than an ångström so that the field does not penetrate beyond the first atomic layer. For a doping level, for example, of 10^{17} cm^{-3} and room temperature, the Debye length amounts to typically 13.4 nm. Electric fields thus enter into non-degenerately doped semiconductors and, as a consequence, extended space-charge layers form. For the case assumed in Fig. 1, the space charge will be carried by positively charged donors. This is equivalent to a surface depletion of mobile electrons and an upward bending of the bands which increases the energy distance from the Fermi level to the conduction-band edge at the surface. This is the conclusion which was reached by Schottky in his famous paper mentioned above.

The current transport across such a depletion or Schottky barrier is governed by its barrier height or, as Schottky initially called it, its metal–semiconductor work function. The barrier height is defined as the energy distance between the Fermi level and the edge of the majority-carrier band, i.e.,

$$\phi_{Bn} \equiv W_{ci} - W_F \tag{2a}$$

and

$$\phi_{Bp} \equiv W_F - W_{vi}, \tag{2b}$$

where W_{ci} and W_{vi} denote the conduction-band minimum and the valence-band maximum at interfaces with n- and p-type doped semiconductors. For a specific metal–semiconductor pair, the experimental barrier heights ϕ_{Bn} and ϕ_{Bp} were always found to add up to the width of the bulk band gap of the semiconductor.

Since Schottky published his basic paper in 1938 the mechanisms determining the barrier heights in metal–semiconductor or Schottky contacts, as they are customarily named to honor Schottky's pioneering contribution to this field, have remained under discussion. In this contribution I will describe my view of the present understanding of this topic. For an extended review on metal–semiconductor contacts and a collection of most relevant papers in this field, the reader is referred to Refs. [6] and [7], respectively. More detailed presentations of specific aspects of semiconductor surface and interface physics may be found in Ref. [8].

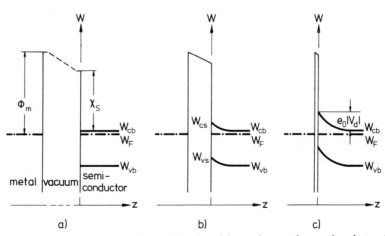

Fig. 1. Development of a Schottky barrier as a function of decreasing metal-to-semiconductor distance.

2. Determination of barrier properties

2.1. Barrier heights

Barrier heights in metal–semiconductor contacts may be evaluated from their
– current–voltage or I/V and
– capacitance–voltage or C/V characteristics as well as by
– internal photoemission and
– ballistic electron emission microscopy (BEEM). The experimental data discussed in this paper were mostly derived from I/V and C/V characteristics. Therefore, these two methods shall be discussed briefly.

Current transport in Schottky contacts is due to majority carriers and, to a first approximation, it may be described by thermionic emission over the interface barrier. Tunneling through the barrier must be considered for high doping levels of the semiconductor, i.e., for narrow space-charge layers. For moderately doped n-type semiconductor substrates, the density of the thermionic emission current may be written as

$$j = j_0 \exp(e_0 V_a / n k_B T) [1 - \exp(-e_0 V_a / k_B T)], \tag{3}$$

with the saturation current density

$$j_0 = A_R^* T^2 \exp(-\phi_{Bn}^z / k_B T). \tag{4}$$

Here, A_R^* is the effective Richardson constant, n is the so called ideality factor, and ϕ_{Bn}^z is the barrier height at zero applied bias V_a. For a derivation of these relations the reader is referred to Refs. [9] and [10].

Recently, Sullivan et al. studied the current transport across homogeneous as well as inhomogeneous metal–semiconductor interfaces by numerical simulations [11]. Inhomogeneous Schottky contacts were modeled by patches with low barrier height which are embedded in a region with larger barrier height. Here, only the ideality factor and the *effective* barrier height are of interest. The ideality factor was obtained to be close to unity, typically $n < 1.03$, for homogeneous Schottky contacts, but becomes as large as 1.25 when the diameter of the patches measures 0.06 μm with all other parameters kept un-

changed. Simultaneously, the *effective* barrier height of these patchy contacts decreased from the large value assumed for the embedding region to almost the smaller value chosen for the patches. The ideality factor and the effective barrier height were found to be almost linearly correlated. These results indicate that ideality factors n close to unity are a characteristic of homogeneous Schottky contacts. This result will be an important criterion for the selection of experimental barrier heights which will be considered in this paper.

The differential capacitance of the depletion layers in homogeneous Schottky contacts is obtained as

$$C_{dep} = \left[e_0^2 \epsilon_b \epsilon_0 N_d / 2(e_0 V_i - e_0 V_a) \right]^{1/2}, \tag{5}$$

where ϵ_b is the bulk dielectric constant. The extrapolated intercept on the abscissa of an $1/C^2$ versus V_a plot gives the interface band-bending,

$$e_0 V_i = \phi_{Bn} - (W_{cb} - W_F). \tag{6}$$

The energy distance $W_{cb} - W_F \equiv W_n$ from the Fermi level to the conduction-band minimum in the bulk is determined by the donor density N_d in the bulk. The flat-band barrier heights ϕ_{Bn} determined from the C/V characteristics of Schottky diodes are larger than their zero-voltage barrier heights ϕ_{Bn}^z evaluated from the respective I/V characteristics.

Ballistic electron emission spectroscopy (BEEM), a technique pioneered by Kaiser and Bell, utilizes the injection of electrons from a tip through a vacuum gap into the metal overlayer of Schottky contacts [12]. Provided the metal layer is sufficiently thin, a fraction of the injected electrons will reach the metal–semiconductor interface without scattering. These ballistic electrons will enter into the semiconductor only if their energy with respect to the Fermi level in the metal is larger than the barrier height of the contact or, in other words, if the voltage applied between the tip and the sample exceeds the barrier height. In patchy Schottky contacts, the length scale for the lateral resolution of BEEM is determined by the Debye length of the semiconductor and the extension of the depletion layer. However, irregularities of the metal film such as variations in thickness or chemical composition and grain boundaries will reduce the lateral resolution.

2.2. Chemical composition at metal–semiconductor interfaces

Physical models and explanations are always based on certain assumptions and idealizations. For the present case, intimate, abrupt, and homogeneous metal–semiconductor interfaces are the ideal. In general, both structural and compositional characterizations of metal–semiconductor interfaces are rather complicated and have been performed for specific cases only.

Chemical reactions were already detected in metal–selenium rectifiers by Poganski as early as 1952 [13]. Cd/Se rectifiers are the most prominent example. Just after metal evaporation their diode properties are poor but they are drastically improved by subsequent tempering at elevated temperatures. Careful studies revealed the formation of n-CdSe interlayers during such annealing treatments. In the resulting Cd/n-CdSe/p-Se sandwich structures, rectification occurs at the p-Se/n-CdSe heterojunction rather than at a metal–semiconductor interface.

During the past 40 years, quite a number of experimental techniques have been developed and improved for the determination of both the atomic arrangement and the chemical composition on surfaces which are either clean or covered with up to a few monolayers of adatoms. Typical and widely used structural probes are low-energy electron diffraction (LEED) and scanning tunneling microscopy (STM) while chemical surface compositions are routinely studied by electron-excited Auger electron spectroscopy (AES) and core-level photoemission spectroscopy (PES). These nondestructive techniques were also applied to follow the formation of metal–semiconductor interfaces. However, the applicability of electron-diffraction as well as electron-emission techniques in interface studies is principally limited by the escape depths of the electrons employed. For kinetic energies ranging between 50 and 2000 eV, the escape depths of electrons vary from 0.4 to 2 nm, respectively.

Alkali metals, which are evaporated on semiconductors and investigated at reduced temperatures, behave almost ideally in that they grow layer-by-layer. Already two layers, for example, of cesium atoms exhibit metallic behavior [14].

Most metals evaporated on semiconductor surfaces at room temperature form islands and, in some cases, it takes more than nominally 15 nm of metal deposited until the films eventually become continuous (see for example Ref. [15]). Chemical compositions at interfaces between continuous films of such metals and semiconductors can thus not be obtained by using the electron-emission spectroscopies mentioned above. Furthermore, data acquired at submonolayer coverages are not necessarily representative for real interfaces formed after evaporation of thicker metal overlayers. As an example, some results reported for GaAs and InP Schottky contacts shall be briefly mentioned. For a compilation of experimental data, the reader is referred to Ref. [16]. Many of the metals investigated were found to react with these compound semiconductors, i.e., chemical bonds at the semiconductor surface become disrupted. However, this does not necessarily imply that interfaces under thicker metal overlayers are intermixed. Most metals have larger surface free energies than gallium, indium, arsenic, and phosphorus [17,18]. Therefore, the latter atoms will segregate on surfaces of the growing metal islands and films. Furthermore, the solid solubility of the substrate atoms determines the extent to which they are dissolved in growing metal films.

Most of the 3d transition metal atoms evaporated on GaAs(110) surfaces at room temperature were found to replace surface Ga atoms [19]. Mostly, this cation exchange is limited to the top surface layer. The Ga atoms released first coalesce into islands and eventually segregate on top of the growing metal films.

Chemical compositions may also be evaluated utilizing destructive methods such as ion milling and secondary ion mass spectroscopy. The impinging ions not only remove surface atoms but also generate collision cascades in which the atoms are intermixed. These regions typically extend 2 to 5 nm below the surface. Thus, chemical compositions at interfaces are difficult to evaluate from such data.

2.3. Atomic arrangements at interfaces

A typical experimental tool for the determination of surface structures is low-energy electron diffraction (LEED). Since the escape depths of the electrons used are small this technique is not suitable for interfaces between thick and continuous metal films and semiconductors. In recent years, X-ray techniques have been developed and considerably improved for the determination of surface and even interface structures. This progress is at least partly due to new and powerful X-ray sources such as electron storage rings. The potential of grazing-incidence X-ray diffraction was demonstrated by Hong et al. who investigated buried Ag/Si(111) interface structures [20]. The clean-surface Si(111)-7 × 7 reconstruction was found to persist under 26 nm of silver deposited at room temperature. Annealing of such films at 250°C, however, transformed the 7 × 7 into a 1 × 1 interface structure. The 7 × 7 structure has the lowest surface free energy of all clean-surface Si(111) structures but is obviously metastable under thick Ag films. The Si(111):Ag($\sqrt{3} \times \sqrt{3}$)R30° structure, which is obtained, for example, by deposition of a monolayer of Ag at 500 K, did not form. Quite on the contrary, the Ag-induced ($\sqrt{3} \times \sqrt{3}$)R30° structure is destroyed by further silver deposition even at room temperature. The conversion of the 7 × 7 to the 1 × 1 interface structure increases the barrier height by 0.05 eV from 0.69 to 0.74 eV [21].

The most interesting examples are the epitaxial NiSi$_2$/Si(111) interfaces. Such interfaces can be grown with a high degree of perfection. Cross-sectional investigations with high-resolution transmission electron microscopes revealed such interfaces to be abrupt and to grow in two different orientations of the epitaxial NiSi$_2$ films with respect to the underlying Si substrate. In type-A interfaces, the lattices are identically aligned on both sides of the interface while they are rotated by 180° around the interface normal for the case of type-B contacts [22,23]. Investigations of medium-energy ion scattering [24] as well

as X-ray standing waves [25] confirmed that at both interfaces the Ni atoms are sevenfold coordinated. Rutherford backscattering provided upper limits of 1×10^{12} and 3×10^{13} Si atoms per cm^2 being displaced from lattice sites on the semiconductor side of type-A and type-B NiSi$_2$/Si(111) contacts, respectively [24]. Meanwhile it is well established that the barrier heights of the two types of NiSi$_2$/Si(111) contacts differ by 0.14 eV. Tung was the first to showthat they measure 0.65 eV for type-A and 0.79 eV for type-B interfaces prepared on samples doped n-type [26].

3. Mechanisms determining the barrier heights in Schottky contacts

3.1. No interface states: the Schottky–Mott rule

Schottky [5] proposed that rectification at metal–semiconductor contacts is due to the existence of depletion layers on the semiconductor side of such interfaces. Already a year later, Schweikert [27] reported a linear correlation between barrier heights measured with metal–selenium rectifiers and the work functions of the metals used. Such a chemical trend is obtained by quantifying the Gedanken experiment which is illustrated in Fig. 1 (see, for example, Ref. [6]).

In the vacuum gap between a metal and a semiconductor facing each other, an electric field exists due to the difference $\phi_m - \phi_{s0}$ of their work functions. The electric field penetrates into the semiconductor and its work function increases by the respective surface band-bending $e_0 V_d$. Since metals exhibit high densities of states at the Fermi level, the respective band bending at the metal surface will be extremely small and may be safely neglected. Therefore, the energy barrier across the vacuum gap, which amounts to $\phi_m - \phi_{s0}$ for infinite metal–semiconductor separation, reduces by $e_0 V_d$. Assuming the semiconductor and the metal to form a parallel-plate capacitor with plate separation d_{ms}, the surface charge densities on the semiconductor and the metal are given by

$$Q_{sc} = -Q_m = (\epsilon_0/e_0)[(\phi_m - \phi_{s0}) - e_0 V_d]/d_{ms}.$$

$$(7)$$

By solving Poisson's equation, the space-charge density in depletion layers is obtained as

$$Q_{sc} = +(2\epsilon_0\epsilon_b N_d e_0 V_d)^{1/2}, \tag{8}$$

for $e_0 V_d = W_{ci} - W_{cb} \geq 3k_B T$. By combining Eqs. (7) and (8), one obtains

$$[(\phi_m - \phi_{s0}) - e_0 V_d]^2 / e_0 V_d = 2e_0^2(\epsilon_b/\epsilon_0) N_d d_{ms}^2. \tag{9}$$

In the limit of an intimate contact, i.e., for $d_{ms} \to 0$, it follows

$$\phi_m - \phi_{s0} - e_0 V_d = 0. \tag{10}$$

The work function of a semiconductor, which has flat bands up to the surface, may be written as $\phi_{s0} = \chi_s + (W_{cb} - W_F)$, where χ_s is the surface electron affinity. Together with definition (2a) of the barrier height, Eq. (10) may be rewritten as

$$\phi_{Bn} = \phi_m - \chi_s, \tag{11}$$

which is the famous Schottky–Mott rule [28,29].

Barrier heights reported for GaAs Schottky diodes are displayed in Fig. 2. All data were evaluated from I/V curves. The ideality factors n ranged between 1.03 and 1.07. Thus, the data considered here originate either from homogeneous or from only slightly inhomogeneous Schottky diodes. To within 0.03 eV, which is the limit of experimental uncertainty, these flat-band barrier heights agree with respective values evaluated from C/V characteristics. Obviously, the experimental data are not described by the Schottky–Mott rule (11). As a trend, however, larger barrier heights seem to correlate with larger metal work-functions. There is no clustering of data points at or close to specific values. The experimental barrier heights rather scatter by 0.47 eV between 0.62 eV for Mg– and 1.09 eV for Ru–GaAs diodes. Schottky diodes were prepared on both {110}- and {100}-oriented substrates. Within the limits of experimental error, the data reveal no dependence of the barrier heights on the crystallographic orientation of the substrate surface. Details of the preparation, however, have a pronounced influence on the barrier height. In most cases, the metals were thermally evaporated. Electrochemical metal deposition, however, results in diodes with larger barrier heights compared with what is obtained when the metals are evaporated or even sputtered. Co/GaAs(001) Schottky diodes, for example, were prepared by thermal evaporation and electroless deposition of Co and were found to have barrier heights of 0.76 and 1.00 eV, respectively [30,31].

3.2. The effect of a continuum of interface states on barrier heights

The barrier heights reported by Schweikert [27] for metal–selenium rectifiers certainly showed a linear correlation with the metal work-function but the slope $S_\phi = d\phi_{Bp}/d\phi_m$ was much smaller than unity as predicted by the Schottky–Mott rule (11). Later on, similar observations were made with other semiconductors. Bardeen [32] attributed the obvious discrepancy between the experimental barrier heights of metal–semiconductor contacts and the Schottky–Mott rule (11) to electronic interface states. Electronic interface states will absorb charge which has to be added to the condition of charge neutrality, i.e., Eq. (1) may be rewritten as

$$Q_m + Q_{is} + Q_{sc} = 0. \tag{12}$$

Since Q_m and Q_{is} reside on either side of the interface an electric double layer exists at

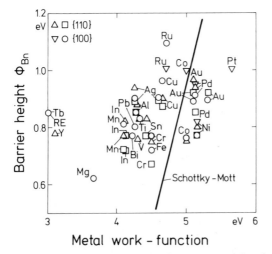

Fig. 2. Barrier heights of GaAs Schottky contacts as a function of the metal work-function. The straight line represents the Schottky–Mott rule. From Ref. [6].

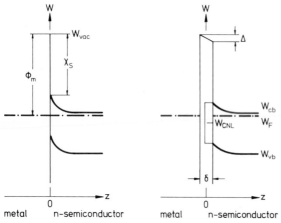

Fig. 3. Schematic band diagrams at intimate, rectifying metal–semiconductor contacts without and with electronic interface states.

metal–semiconductor interfaces. Its width δ_i is of atomic dimensions.

The effect of interface states on the barrier heights of metal–semiconductor contacts was first analyzed by Cowley and Sze [33]. They assumed a continuum of interface states with a constant density of states D_{is} across the band gap and a charge neutrality level W_{cnl}. For energies larger and smaller than W_{cnl}, the interface states have acceptor and donor character, respectively. The charge density in these interface states is then given by

$$Q_{is} = -e_0 D_{is}(W_F - W_{cnl}) = -e_0 D_{is}(\phi_{Bn}^0 - \phi_{Bn}), \tag{13}$$

where $\phi_{Bn}^0 \equiv W_{ci} - W_{cnl}$ is the barrier height for $Q_{is} = 0$, i.e., when no interface dipole exists and the Fermi level coincides with the charge neutrality level of the continuum of the interfaces states.

Fig. 3 displays band diagrams for rectifying metal–semiconductor contacts with and without interface states. The effect of interface states will be again analyzed for a parallel-plate-capacitor arrangement. The voltage drop across the interfacial double layer is related to the surface charge density as

$$\Delta = \phi_m - \chi_s - \phi_{Bn} = -(e_0/\epsilon_i\epsilon_0)Q_m\delta_i, \tag{14}$$

where ϵ_i is the dielectric constant of the interfacial layer. Inserting of Eqs. (8), (13), and (14) into

the condition of charge neutrality (12) finally gives

$$\phi_{Bn} = S_\phi(\phi_m - \chi_s) - (1 - S_\phi)\phi_{Bn}^0, \tag{15}$$

with the slope parameter

$$S_\phi = \left[1 + (e_0^2/\epsilon_i\epsilon_0)D_{is}\delta_i\right]^{-1}. \tag{16}$$

The barrier heights are again linearly related to the work functions of the metals but the slope is reduced and depends on the density of states in the continuum of interface states. For $D_{is} = 0$, the slope parameter reaches its maximum value, $S_\phi = 1$, i.e., the Schottky–Mott rule is recovered. With $D_{is} \to \infty$, the barrier height becomes independent of the metal work-function, i.e., $S_\phi = 0$. Then, the Fermi level is pinned at the charge neutrality level of the interface states.

3.3. The continuum of metal-induced gap states

At clean metal surfaces or, in other words, at metal–vacuum interfaces, the wavefunctions of the electrons are exponentially decaying into vacuum. When the vacuum is replaced by a semiconductor or, more generally speaking, a dielectric the propagation of the wavefunctions across the solid–solid interface is somewhat more complicated. At the same time when Cowley and Sze analyzed the influence of a continuum of interface states on barrier heights in Schottky contacts, Heine [34] pointed out that at metal–semiconductor contacts a continuum of metal-induced interface states will exist. He argued that these states are derived from the virtual gap states (ViGS) of the complex band structure of the semiconductor.

Schrödinger's equation may be solved not only for real but also for complex wavevectors. For the bulk band structure, only real wavevectors are relevant since otherwise the Bloch functions cannot be normalized. Complex wavevectors mean that the wavefunctions decay or grow exponentially. Such behavior becomes meaningful at interfaces since the wavefunctions of real interface states will decay exponentially to both sides of the interface and are thus normalized. Such solutions of Schrödinger's equation with complex wavevectors will have energy levels which lie within gaps of the bulk band structure. Therefore, these solutions of Schrödinger's equation are called virtual gap states (ViGS) of the complex band structure. Virtual gap states were first considered by Maue [35] for one-dimensional, linear chains of finite length by using the approximation of nearly free electrons. The respective ViGS wavefunctions may be written as

$$\psi(z) = A \exp(-qz) \cos(\pi z/a + \varphi), \qquad (17)$$

where A is a constant, φ is a phase factor which varies across the band gap, a is the lattice parameter of the chain, and q is the imaginary part of the wavevector.

With appropriate boundary conditions, real surface and interface states are derived from the continuum of virtual gap states. Real surface states at the ends of linear chains are obtained when one of the oscillatory solutions (17) can be fitted to a tail which exponentially decays into vacuum. Respective boundary conditions were derived by Maue [35]. With regard to interface states at metal–semiconductor interfaces, Heine [34] argued that
• in the energy range, where the metal conduction-band overlaps the semiconductor band-gap, the wavefunctions of the metal electrons decay into the semiconductor and
• these tails are to be described by the virtual gap states of the semiconductor band structure.
These metal-induced gap states (MIGS), as they were named later, form a continuum. They are occupied up to the Fermi level and empty above. Since the MIG states are derived from the bulk energy bands they will predominantly have donor character near to the valence-band maximum and acceptor character closer to the conduction-band minimum. The respective branch point is intuitively called the charge neutrality level of the ViGS. Charge neutrality levels of ViGS were computed first by Tejedor et al. [36] and later on by Tersoff [37] as well as Cardona and Christensen [38].

More realistic calculations of electronic properties of metal–semiconductor contacts were first performed by Louie and Cohen [40]. They considered Al–jellium/silicon contacts. Their calculations revealed four different types of electronic states to exist at such interfaces:
• In the energy region, where the conduction band of the metal overlaps the semiconductor valence-band, the states are matched and bulklike on either side of the interface.
• Below the bottom of the metal conduction-band, bulklike semiconductor states penetrate into the metal.
• Truly localized interface states, which decay to both sides of the interface, may be present in low-lying semiconductor band gaps.
• In the energy range where the metal conduction-band overlaps the energy gap of the semiconductor, the wavefunctions of the metal electrons tail into the semiconductor. This gives a continuum of metal induced-gap states (MIGS) which are occupied up to the Fermi level and empty above.
These calculations excellently confirmed Heine's conclusions which were based on simple physical concepts.

Recently, the existence of a continuum of metal-induced gap states was experimentally demonstrated by First et al. [41]. They evaporated submonolayer quantities of iron on cleaved GaAs(110) surfaces at room temperature and investigated the deposit with a scanning tunneling microscope. They found the evaporated iron atoms to coagulate in small epitaxial clusters. Islands with volumes larger than 1 nm^3 showed metallic behavior. Around such particles, metallic states detected on them were observed to overlap the semiconductor band gap in energy. These states decay exponentially as a function of distance away from the metallic iron islands. By using p- and n-type GaAs samples, occupied as well as empty gap states were probed. Experimental results published by First et al. are displayed in Fig. 4. The decay lengths of these gap states vary U-shaped across the GaAs energy gap with a minimum length of 0.34 nm near to mid-gap position. This experimental result is in excellent agreement with theoretical predictions. Already Maue [35] showed that the decay lengths $1/q$ of the ViGS vary U-shaped across the energy gap of one-dimensional chains. Louie et al. [42] studied Al–jellium/GaAs contacts by using a pseudo-

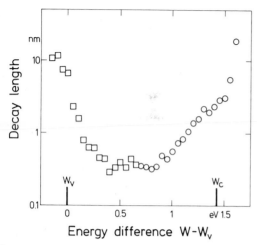

Fig. 4. Energy dependence of the decay length of gap states tailing away from metallic iron islands on cleaved GaAs(110) surfaces as determined by using a scanning tunneling microscope: (○) and (□) data recorded with samples doped p- and n-type. After First et al. [41].

potential approach. They obtained a U-shaped continuum of metal-induced gap states across the GaAs energy gap and a minimum decay length of 0.28 nm for the charge density in these MIGS.

The continuum of gap states around metallic iron islands on cleaved GaAs(110) surfaces, which was directly observed by First et al. with a scanning tunneling microscope, exhibits all the features characteristic of the continuum of metal-induced gap states.

3.4. The electronegativity concept of charge transfer at metal–semiconductor interfaces

In their study on the influence of a continuum of electronic interface states on barrier heights in metal–semiconductor contacts, Cowley and Sze [33] made no assumptions on the physical nature of these states. Already Heine [34] identified them as the MIGS continuum. The charge in the MIGS or, in other words, the charge transfer across the interface determines the interface position of the Fermi level or, in other words, the barrier height. This is explained schematically in Fig. 5.

The Q_{gs}^{mi} versus W_F diagram on the right side of Fig. 5 immediately provides a chemical trend of barrier heights in Schottky contacts provided the charge transfer at metal–semiconductor interfaces is known. No such calculations have been

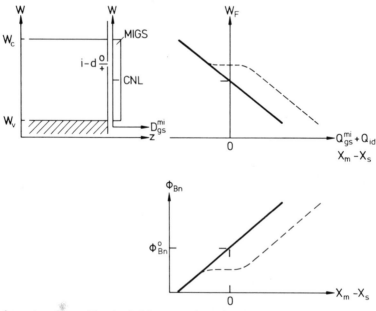

Fig. 5. Band diagram, charge transfer, and barrier height at metal–semiconductor contacts containing a continuum of interface states and interface defects of donor type in addition (schematically).

performed till now. Therefore, the present author extended Pauling's electronegativity concept of the partial ionic character of covalent bonds to semiconductor interfaces [39].

Pauling [43] correlated the ionicity Δq_1 of single bonds in diatomic molecules A–B between unlike atoms with the difference $X_A - X_B$ of the atomic electronegativities of the atoms forming the molecule. A revised version of the relation originally proposed by Pauling is that of Hanney and Smith [44]

$$\Delta q_1 = 0.16| X_A - X_B | + 0.035(X_A - X_B)^2. \quad (18)$$

In a simple point-charge model, the atoms are charged by $+\Delta q_1 e_0$ and $-\Delta q_1 e_0$, where the more electronegative atom becomes negatively charged. In a more realistic picture, the bond charge is slightly shifted towards the more electronegative atom in heteropolar diatomic molecules while it is in the middle between both atoms in homopolar diatomic molecules.

Pauling's concept proved to be useful also in solid state physics. Miedema and coworkers [45] applied it to metal alloys. The present author

used it for modeling of the charge transfer across semiconductor interfaces [39] and proposed that to first approximation the charge transfer across metal–semiconductor interfaces varies proportional to the difference $X_m - X_s$ between the electronegativities of the metal and the semiconductor. For elemental semiconductors, their atomic electronegativities may be used. In generalizing Pauling's concept, the electronegativity of compounds is taken as the geometric mean of the atomic electronegativities of their constituents. For binary compounds one then obtains

$$X_{AB} = (X_A X_B)^{1/2}. \quad (19)$$

According to relation (18), the electronegativity concept of the charge transfer at metal–semiconductor interfaces assumes that the charge Q_{gs}^{mi} in the metal-induced gap states varies proportional to the electronegativity difference $X_m - X_s$. The W_F versus Q_{gs}^{mi} or, what is the same, W_F versus $(X_m - X_s)$ diagram on the right side of Fig. 5 now provides a chemical trend of the barrier heights of metal–semiconductor contacts. For $X_m - X_s = 0$, the Fermi level will coincide with the

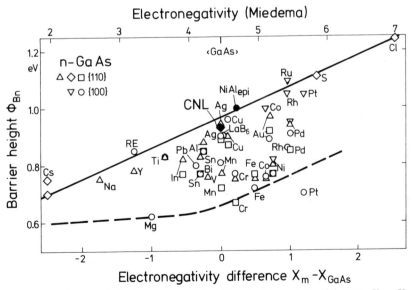

Fig. 6. Barrier heights reported for metal/GaAs contacts against the electronegativity difference $X_m - X_{GaAs}$ (Miedema electronegativities). From Refs. [6,44].

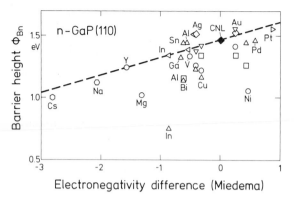

Fig. 7. Barrier heights reported for metal/GaP contacts against the electronegativity difference $X_m - X_{GaP}$ (Miedema electronegativities). From Ref. [14].

trated by the diagram in the lower right side of Fig. 5. Relation (20) explicitly states that the charge neutrality levels represent no *canonical pinning positions of the Fermi level* in metal–semiconductor contacts. According to Eq. (16), the slope S_X is determined by $D_{is}\delta_i/\epsilon_i$, i.e., by the product of the density of interface states and the width of the interface dipole layer divided by the interface dielectric constant.

There have been previous attempts to correlate barrier heights of Schottky contacts with metal electronegativities. These earlier ϕ_{Bn} versus X_m plots were still inspired by the Schottky–Mott rule in that they replaced the metal work-function by the metal electronegativity. By this, dipole contributions to the work functions, which are known to vary as a function of surface orientation, should remain disregarded. It shall be explicitly emphasized that the present correlation between barrier heights of Schottky contacts and the *difference* of the metal and semiconductor electronegativities *conceptually* differs from these

charge neutrality level of the MIG states. Thus, the barrier height will vary as

$$\phi_{Bn} = \phi_{Bn}^0 + S_X(X_m - X_s),\qquad(20)$$

for semiconductors doped n-type. Eq. (20) is illus-

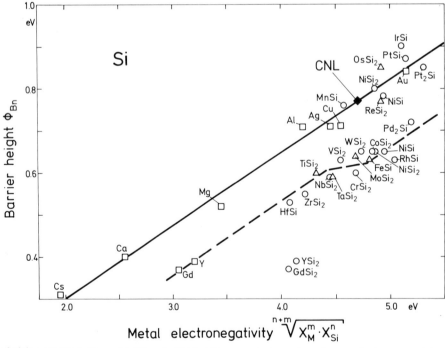

Fig. 8. Barrier heights reported for metal/Si and silicide/silicon contacts against the metal electronegativity $(X_m^m X_{Si}^n)^{1/(n+m)}$ (Miedema electronegativities). The charge-neutrality barrier height was entered at $X_{Si} = 4.7$. From Ref. [46].

earlier attempts. The present approach models the *charge transfer* between the metal and the metal-induced gap states of the semiconductor by the electronegativity difference $X_m - X_s$ and, therefore, it *predicts* the charge-neutrality barrier height ϕ_{Bn}^0 for metal–semiconductor pairs with $X_m - X_s = 0$.

In Figs. 6 to 8, barrier heights of GaAs [44], GaP [15], and Si [45] Schottky diodes are plotted as a function of the electronegativity difference $X_m - X_s$. The GaAs data are the same as in Fig. 2. All data were evaluated from I/V curves. Only the Cs/GaAs and Cs/GaP data were determined by using photoemission spectroscopy and a Kelvin probe. To obtain metallic cesium overlayers, the semiconductor substrates were held at low temperatures during Cs evaporation and subsequent measurements.

According to relation (20), the charge-neutrality barrier heights $\phi_{Bn}^0 = W_{ci} - W_{cnl}$ were entered at the electronegativity difference $X_m - X_s = 0$. Here, the charge neutrality levels of the ViGS evaluated by Tersoff [37] were used. The straight lines, which were drawn through the charge-neutrality points, represent upper limits of the barrier heights in GaAs, GaP as well as Si Schottky diodes. The arguments, which led to relation (20), suggest to assign these straight lines to the charging of the continuum of metal-induced gap states. Consequently, the present author proposed the continuum of metal-induced gap states to be the primary mechanism which determines the barrier heights in intimate, abrupt, and homogeneous Schottky contacts and deviations to lower barrier heights are attributed to other, secondary mechanisms which are effective in addition to the MIG states [46,47].

The conclusion just reached means that there is not just one physical mechanism which determines the barrier heights in metal–semiconductor contacts. In the past, always monocausal approaches were invoked to explain the nonuniform chemical trends observed with the barrier heights of Schottky contacts. Most theoreticians favored the continuum of metal-induced gap states, i.e., the MIGS model. A most prominent contender was Spicer's Unified Defect Model [48]. It postulates that the Fermi level becomes pinned at defect levels which are generated in the selvedge of the semiconductor during metal depositions. However, the experimental data obtained – and this is most important – with *Schottky diodes* do not support the Unified Defect Model. This does not mean that defects play no role in determining barrier heights in Schottky diodes.

The Effective Work Function Model of Freeouf and Woodall [49] was designed to preserve the Schottky–Mott rule. It proposes to replace the work function of the metal deposited by an effective work function. For this purpose, the formation of anion microclusters was assumed by either oxide contaminations or metal–semiconductor interactions at interfaces between gold and compound semiconductors. It was suggested to replace the metal work-function in Eq. (11) by the work function of the anions. Again, this model cannot account for the wealth of the experimental data. By the way, this model is almost equivalent to the common anion rule which is not supported by the experimental data either. This does not mean that anion aggregates play no role at all in determining barrier heights in Schottky contacts. However, there is no unambiguous experimental evidence for the existence of such microclusters at metal–semiconductor interfaces.

3.5. Mechanisms other than MIGS

The experimental data for GaAs, GaP, and Si Schottky contacts, which are displayed in Figs. 6 to 8, show deviations from the MIGS lines towards lower barrier heights. There are a number of mechanisms, which might reduce barrier heights in metal–semiconductor contacts, such as

- interface defects,
- structure-related interface dipoles,
- interface strain,
- interface compound formation, and
- interface intermixing,

to name a few examples. Some of these extrinsic mechanisms are also considered in discussions of the band line-up at semiconductor heterostructures. Here, only the influence of interface defects and of structure-related extra dipoles on the barrier height shall be discussed.

3.5.1. Interface defects

Interface defects at metal–semiconductor interfaces will become charged and have thus to be considered in the condition of charge neutrality at the interface. Relation (12) is then replaced by

$$Q_m + Q_{gs}^{mi} + Q_{id} + Q_{sc} = 0, \tag{21}$$

where Q_{id} is the charge density in interface defects. The charge density Q_m on the metal side is now balanced by $Q_s \equiv Q_{gs}^{mi} + Q_{id} + Q_{sc}$ on the semiconductor side of the interface.

The energy levels of adatoms on metal surfaces or, more generally speaking, of defects at metal–vacuum interfaces are customarily broadened into wide resonances due to the interaction of the atomic levels with the continuum of conduction-band states. The respective line widths amount to typically 1 eV. At metal–semiconductor interfaces, the interaction between the metal and defects in the semiconductor selvedge is screened by the interface dielectric function ϵ_i. Ludeke et al. [50] obtained $\epsilon_i \approx 4$. The interaction matrix element contains the Coulomb potential squared and, therefore, the broadening of defect levels at metal–semiconductor interfaces reduces to 60 meV [51]. Defects in Schottky contacts may thus be assumed to exhibit sharp levels.

The influence of donor-type defects on the barrier heights of metal–semiconductor contacts is illustrated in Fig. 5. Sharp donor levels are assumed above the charge-neutrality level of the continuum of metal-induced gap states. As long as that much negative charge is transferred to the semiconductor as to keep the Fermi level well above the defect level, all donors are neutral. With decreasing negative charge in the semiconductor, the Fermi level approaches the defect level and defects are gradually charged positively. As a result, the Fermi level becomes *intermediately pinned* at the position of the defect levels. When all defects are eventually charged the continuum of MIG states will again take up additional charge and will again determine the position of the Fermi level in the band gap as a function of charge on the semiconductor side of the interface. In the diagrams on the right side of Fig. 5, the dashed lines illustrate the influence of the donor-type defects on the Fermi-level posi-

tion within the gap or, what is the same, on the barrier height as a function of the charge density $Q_{gs}^{mi} + Q_{id}$ on the semiconductor side of the interface. Evidently, interface donors are lowering the barrier heights with respect to what is found when no interface defects are present.

Discrete interface donors contribute a net charge per unit area

$$Q_{id} = +e_0 N_{id}\left[1 - f_0(W_{id} - W_F)\right], \tag{22}$$

where N_{id} and W_{id} are the area density and the energy levels of the interface defects, respectively, and $f_0(W_{id} - W_F)$ is the Fermi-Dirac distribution function. The maximum decrease $\delta\phi_{Bn}^{max}$ of the barrier height is achieved when all interface donors are charged, i.e., for $f_0 = 0$. By inserting Eqs. (8), (13), (14), and (22) into the condition of charge neutrality (21) one obtains

$$\delta\phi_{Bn}^{max} = -\left(1 - S_\phi\right)N_{id}/D_{gs}^{mi}. \tag{23}$$

The experimental data plotted in Figs. 6 to 8 give maximum barrier-height reductions of $\delta\phi_{Bn}^{max} \simeq -0.3$ eV. With $S_\phi = 0.2$ and $D_{gs}^{mi} = 3 \times 10^{14}$ cm^{-2} eV^{-1}, which are typical parameters, one obtains a density of approximately 1×10^{14} interface donors per cm^2. This simple estimate is in excellent agreement with results from more elaborate computations by Zhang et al. [52].

The reduced barrier heights in Figs. 6 to 8 with respect to the MIGS lines may thus be explained by defects of donor type. The estimated maximum density of 1×10^{14} cm^{-2} corresponds to approximately one tenth of a monolayer. As was pointed out earlier, such defects might be fabrication-induced. As possible candidates for such donor defects, Weber et al. proposed As_{Ga} antisite defects to exist in metal–GaAs contacts [53]. Since such defects are double donors, interface acceptors have to be present in addition [54]. Such As_{Ga} defects might also be described as anion microclusters. However, no fabrication-induced defects were identified at metal–semiconductor interfaces till now.

3.5.2. Structure-related interface dipoles

The barrier heights of type-B and type-A NiSi$_2$/Si(111) contacts measure 0.79 and 0.65 eV, i.e., they differ by 0.14 eV. These data were first reported by Tung [26] and later on confirmed by many other groups. Fig. 8 reveals the data point for type-B interfaces to fit the silicon-MIGS line while type-A contacts deviate from it towards lower barrier heights. Furthermore, growth studies suggested that the interface free energy is lower for type-B than for type-A interfaces [55]. The reduced barrier height of type-A contacts cannot be attributed to interface defects. Medium-energy ion scattering studies performed by Fischer et al. gave an upper limit of 1×10^{12} displaced Si atoms per cm^2 [24]. By inserting such a low density of defects and the typical parameters used above into Eq. (23), one estimates a barrier-height decrease by 3 meV. Therefore, a mechanism other than defects has to be responsible for the reduction of the type-A barrier height in comparison with the one of type-B contacts which fits the silicon MIGS-line.

Besides defects, structure-related interface dipoles or, in other words, charge transfer across interfaces in addition to what results from the MIG states will also lead to variations in barrier heights. The present author proposed to describe such interface dipoles by an electric double layer [8]. The dipole moment $p_{i\perp}$ per interface atom may then be estimated by identifying dipole-induced variations of the barrier height by the potential drop across a dipole layer

$$\Delta\phi_{Bn}^{d} = \pm (e_0/\epsilon_i\epsilon_0)p_{i\perp}N_i, \qquad (24)$$

where ϵ_i and N_i are the dielectric constant and the number of additional dipoles per unit area at the interface, respectively. The sign of the barrier-height variation depends on the orientation of the dipoles. Their moment may be approximated by

$$p_{i\perp} = e_0\Delta q_i d_i, \qquad (25)$$

where $e_0\Delta q_i$ and d_i are the dipole charge and the dipole length, respectively. For the case of NiSi$_2$/Si(111) contacts, the dipole length is approximated by the Si bond length, the density of dipoles N_i is taken as the number of atoms per unit area in a Si(111) plane, and the interface dielectric constant is again assumed as $\epsilon_i \approx 4$ [50]. The barrier-height difference $\phi_{Bn}^{B} - \phi_{Bn}^{A} = 0.14$

eV then gives a charge

$$\Delta q_i^{A} \approx 0.017$$

for structure-related dipoles at type-A NiSi$_2$/Si(111) interfaces. The extra valence charge at interface atoms, which is attributed to structural differences between the two types of interfaces, is quite small. This estimate explains the great difficulties encountered in theoretical studies, which aim at computations of the barrier heights for type-A and type-B NiSi$_2$/Si(111) contacts, even if identical approaches such as the linear muffin-tin orbitals method in the atomic-sphere approximation or with a full-potential scheme are employed [56–58]. The barrier height of type-A interfaces turned out to be especially sensitive to variations of the interface geometry. This result is quite plausible since growth studies suggested that the interface free energy is lower for type-B than for type-A interfaces [55] and the extra charge transfer estimated above for this type of interfaces from the reduced barrier height is extremely small.

The electronegativity of NiSi$_2$ is larger than the one of silicon and, as a consequence, the charge in the MIG states is positive while a negative charge resides on the NiSi$_2$ side of the interface. Fujitani and Asano [56] confirmed this prediction by results from their computations mentioned above. The extra, structure-related dipoles are thus oriented such that they decrease the positive charge on the silicon side of type-A NiSi$_2$/Si(111) interfaces in comparison with type-B contacts.

3.6. The slope parameter

The MIGS model and the electronegativity concept of charge transfer at metal–semiconductor contacts describes the chemical trend of the respective barrier heights by the charge-neutrality barrier height $\phi_{Bn}^{0} = W_{ci} - W_{cnl}$ and the slope parameter $S_X = d\phi_{Bn}/dX_m$. This is illustrated by the diagram at the right bottom of Fig. 5. The charge neutrality levels W_{cnl} of the ViG states have been computed for many semiconductors [36–38]. This leaves the slope parameter to be discussed.

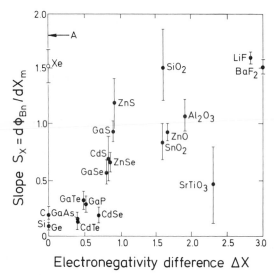

Fig. 9. Slope parameters $S_X = \mathrm{d}\phi_{Bn}/\mathrm{d}X_m$ as a function of the electronegativity differences of the constituent atoms of the semiconductors and insulators. After Kurtin et al. [59] and Schlüter [60].

In an early attempt, Kurtin et al. [59] evaluated slope parameters for some twenty different semiconductors and insulators and plotted them against the respective ionicities. Schlüter [60] reanalyzed these data and his revised set is displayed in Fig. 9. The data point for metal–xenon interfaces, which was obtained by Jacob et al. [61] some ten years later, clearly rules out the S-shaped trend which was inferred by Kurtin et al.

The present interpretation of the experimental data plotted in Figs. 6 to 8 concludes that barrier heights at abrupt metal–semiconductor interfaces are primarily determined by the charge transfer between the metal and the continuum of metal-induced gap states of the semiconductor. The ViG states of linear chains have a density of states which varies U-shaped across the band gap and is almost constant around the charge neutrality level over approximately half of the band gap. Therefore, it seems to be a reasonable assumption that the continua of metal-induced gap states also have almost constant densities of states near to their charge neutrality levels. This assumption is also justified by the experimental data of First et al. which are displayed in Fig. 4.

A model with a continuum of unspecified interface states at metal–semiconductor contacts was first considered by Cowley and Sze [33]. It is described in Section 3.2. This approach modeled the voltage drop across the interfacial dipole layer by the difference $\phi_m - \chi_s$. According to Eq. (16), the slope parameter $S_\phi = \mathrm{d}\phi_{Bn}/\mathrm{d}\phi_m$ is then determined by the product $D_{is}\delta_i$. Here, however, slope parameters $S_X = \mathrm{d}\phi_{Bn}/\mathrm{d}X_m$ are of interest. The slope parameters S_ϕ and S_X may be easily converted since the work functions and the electronegativities of metals are linearly correlated as was first pointed out by Gordy and Thomas [62]. A least-squares fit to the work functions of polycrystalline metals and the respective Pauling electronegativities gives

$$\phi_m = 1.79 X_{Paul} + 1.11 \text{ [eV]}. \tag{26}$$

Such a trend is easily explained. In a vacuum gap between two solids exhibiting different work functions, an electric field builds up. Due to this contact potential the material with the smaller and the larger work function will become charged positively and negatively, respectively. As in chemical bonds, this charge transfer again follows the sign of the electronegativity difference. By considering the empirical relation (26), Eq. (16) may be rewritten as

$$A/S_X - 1 = (e_0/\epsilon_i\epsilon_0) D_{is}(W_{cnl})\delta_i. \tag{27}$$

The coefficient A amounts to 1.79 when Pauling's and 0.93 when Miedema's electronegativities are used.

The present approach identifies the interface states with the continuum of MIGS. Then, the thickness δ_i of the interfacial dipole layer may be approximated by the decay length $1/q_{cnl}$ of the MIGS. For a one-dimensional linear chain, both $D_{gs}^{vi}(W_{cnl})$ and $1/q_{cnl}$ vary inversely proportional to the width of the respective energy gap. Therefore, it is expected that the slope parameters S_X will also be determined by the band gaps of the respective semiconductor substrates. In three-dimensional semiconductors, the width of the band gap varies across the Brillouin zone. Therefore, average band gaps have to be considered rather than the usual direct or indirect band gaps which are all referenced with respect to the valence-

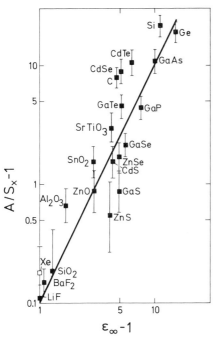

Fig. 10. Slope parameters $S_X = d\phi_{Bn}/dX_m$ (same data as in Fig. 9) as a function of the electronic susceptibilities $\epsilon_\infty - 1$ of the semiconductors and insulators. From Mönch [39].

band maximum in the middle of the Brillouin zone. The average band gap $\langle W_g \rangle$ is defined by

$$\epsilon_\infty - 1 = \left(\hbar\omega_p / \langle W_g \rangle \right)^2, \tag{28}$$

where ϵ_∞ is the electronic part of the static dielectric constant and $\hbar\omega_p$ is the energy of the plasmon of the bulk valence electrons. For the group IV and the III–V and II–VI semiconductors, the experimental plasmon energies vary by only $\pm 10\%$. Thus, the present author proposed that the slope parameter should obey a power law as a function of the susceptibility $\epsilon_\infty - 1$ [39], i.e.,

$$A/S_X - 1 \propto \left(\epsilon_\infty - 1 \right)^n. \tag{29}$$

In Fig. 10, the same slope parameters, which are displayed in Fig. 9, are plotted over the electronic susceptibilities $\epsilon_\infty - 1$. Now, a distinct trend is obtained and a least-squares fit to the experimental data points yields

$$A/S_X - 1 = 0.1 \left(\epsilon_\infty - 1 \right)^{2.0}. \tag{30}$$

The regression coefficient is 0.91. The good correlation is not obvious since the S_X data exhibit

large margins of experimental error and a large amount of scatter. Most importantly, the data point for metal–xenon interfaces now fits the trend. It shall be mentioned that the xenon data point was reported after the *semi-theoretical* relation (30) had been published. In agreement with relation (28), one finds $A/S_X - 1$ to vary proportional to $\langle W_g \rangle^4$. No such correlation is obtained when the same $A/S_X - 1$ data are plotted versus the widths of either the direct or the indirect band gaps. These findings strongly support the present approach that the charge transfer between the metal and the continuum of MIGS on the semiconductor side determines the barrier heights at metal–semiconductor interfaces and that the respective charge transfer may be modeled by the differences of the metal and semiconductor electronegativities.

References

[1] F. Braun, Pogg. Ann. 153 (1874) 556.
[2] L.D. Grondahl, U.S.Patent 160335 issued January 1, 1925.
[3] A.H. Wilson, Proc. Roy. Soc. (London) A 133 (1936) 458; 134 (1936) 277.
[4] O. Fritsch, Ann. Phys. 22 (1935) 375.
[5] W. Schottky, Naturwissenschaften 26 (1938) 843.
[6] W. Mönch, Rep. Prog. Phys. 53 (1990) 221.
[7] W. Mönch, Electronic Structure of Metal–Semiconductor Contacts (Kluwer, Dordrecht, 1990).
[8] W. Mönch, Semiconductor Surfaces and Interfaces (Springer, Berlin, 1993).
[9] S.M. Sze, Physics of Semiconductor Devices (Wiley, New York, 1981).
[10] E.H. Rhoderick and R.H. Williams, Metal–Semiconductor Contacts (Clarendon, Oxford, 1988).
[11] J.P. Sullivan, R.T. Tung, M.R. Pinto and W.R. Graham, J. Appl. Phys. 70 (1991) 7403.
[12] W.J. Kaiser and L.D. Bell, Phys. Rev. Lett. 60 (1985) 1406;
L.D. Bell and W.J. Kaiser, Phys. Rev. Lett. 61 (1988) 2368.
[13] S. Poganski, Z. Elektrochem. 56 (1952) 193; Z. Phys. 134 (1953) 469.
[14] R. Linz, H.J. Clemens and W. Mönch, J. Vac. Sci. Technol. B 11 (1993) 1591.
[15] D.E. Savage and M.G. Lagally, J. Vac. Sci. Technol. B 4 (1986) 943.
[16] Z. Lin, F. Xu and J.H. Weaver, Phys. Rev. B 36 (1987) 5777.
[17] A. Miedema, Z. Metallkd. 69 (1978) 287.
[18] L.Z. Mezey and J. Giber, Jpn. J. Appl. Phys. 21 (1982) 1569.

[19] F. Schäffler, G. Hughes, W. Drube, R. Ludeke and F.J. Himpsel, Phys. Rev. B 35 (1987) 6328.

[20] H. Hong, R.D. Aburano, D.-S. Lin, H. Chen, T.-C. Chiang, P. Zschack and E.D. Specht, Phys. Rev. Lett. 68 (1992) 507.

[21] R. Schmitsdorf, T.U. Kampen and W. Mönch, Proc. 1st Int. Conf. on Control of Semiconductor Interfaces, Karuizawa, Japan, November 8–12, 1993.

[22] D. Cherns, G.R. Anstis, J.L. Hutchinson and J.C.H. Spence, Phil. Mag. A 46 (1982) 849.

[23] J.M. Gibson, R.T. Tung and J.M. Poate, Mater. Res. Soc. Symp. Proc. 14 (1983) 395.

[24] J. Vriemoeth, J.F. van der Veen, D.R. Heslinga and T.M. Klapwijk, Phys. Rev. B 42 (1990) 9598.

[25] A.E.M.J. Fischer, E. Vlieg, J.F. van der Veen, M. Clausnitzer and G. Materlick, Phys. Rev. B 36 (1987) 4769.

[26] R.T. Tung, Phys. Rev. Lett. 52 (1984) 461.

[27] H. Schweikert, Verh. Phys. Ges. 3 (1939) 99. These results were published in Ref. [28].

[28] W. Schottky, Phys. Z. 41 (1940) 570.

[29] N.F. Mott, Proc. Cambridge Phil. Soc. 34 (1938) 568.

[30] J.R. Waldrop, J. Vac. Sci. Technol. B 2 (1984) 445.

[31] P. Allongue and E. Souteyrand, J. Vac. Sci. Technol. B 5 (1987) 1644.

[32] J. Bardeen, Phys. Rev. 71 (1947) 717.

[33] A.M Cowley and S.M. Sze, J. Appl. Phys. 36 (1965) 3212.

[34] V. Heine, Phys. Rev. 138 (1965) A1689.

[35] A.W. Maue, Z. Phys. 94 (1935) 717.

[36] C. Tejedor, F. Flores and E. Louis, J. Phys. C (Solid State Phys.) 10 (1977) 2163.

[37] J. Tersoff, Phys. Rev. Lett. 52 (1984) 465; Surf. Sci. 168 (1986) 275.

[38] M. Cardona and N.E. Christensen, Phys. Rev. B 35 (1987) 6182.

[39] W. Mönch, Festkörperprobleme (Advances in Solid State Physics) Vol. 26, Ed. P. Grosse (Vieweg, Braunschweig, 1986) p. 67.

[40] S.G. Louie and M.L. Cohen, Phys. Rev. B 13 (1976) 2461.

[41] P.N. First, J.A. Stroscio, R.A. Dragoset, D.T. Pierce and R.J. Celotta, Phys. Rev. Lett. 63 (1989) 1416.

[42] S.G. Louie, J.R. Chelikowsky and M.L. Cohen, Phys. Rev. B 15 (1977) 2154.

[43] L.N. Pauling, The Nature of the Chemical Bond (Cornell University, Ithaca, 1939/60).

[44] N.B. Hanney and C.P. Smith, J. Am. Chem. Soc. 68 (1946) 171.

[45] A.R. Miedema, F.R. de Boer and P.F. de Châtel, J. Phys. F (Metal Phys.) 3 (1973) 1558; A.R. Miedema, P.F. de Châtel and F.R. de Boer, Physica B 100 (1980) 1.

[46] W. Mönch, Phys. Rev. B 37 (1988) 7129.

[47] W. Mönch, Phys. Rev. Lett. 58 (1987) 1260.

[48] W.E. Spicer, P.W. Chye, P.R. Skeath and I. Lindau, J. Vac. Sci. Technol. 16 (1979) 1422.

[49] J.L. Freeouf and J.M. Woodall, Appl. Phys. Lett. 39 (1981) 727.

[50] R. Ludeke, G. Jezequel and A. Taleb-Ibrahimi, J. Vac. Sci. Technol. B 6 (1988) 1277; Phys. Rev. Lett. 61 (1988) 601.

[51] R. Ludeke, in: Metallization and Metal–Semiconductor Interfaces, Ed. I.P. Batra (Plenum, New York, 1989) p. 39.

[52] S.B. Zhang, S.G. Louie and M.L. Cohen, Phys. Rev. B 32 (1985) 3955.

[53] E.R. Weber, H. Ennen, U. Kaufmann, J. Windscheif, J. Schneider and T. Wosinski, J. Appl. Phys. 53 (1982) 6140.

[54] W. Mönch, Surf. Sci. 132 (1982) 92.

[55] R.T. Tung, J.M. Gibson and J.M. Poate, Phys. Rev. Lett. 50 (1983) 429; Appl. Phys. Lett. 42 (1983) 888.

[56] H. Fujitani and S. Asano, Phys. Rev. B 42 (1990) 1696.

[57] G.P. Das, P. Blöchl, O.K. Anderson, N.E. Christensen and O. Gunnarsson, Phys. Rev. Lett. 63 (1989) 1168.

[58] S. Ossicini, O. Bisi and C.M. Bertoni, Phys. Rev. B 42 (1990) 5735.

[59] S. Kurtin, T.C. McGill and C.A. Mead, Phys. Rev. Lett. 22 (1970) 1433.

[60] M. Schlüter, Phys. Rev. B 17 (1978) 5044.

[61] W. Jacob, E. Bertel and V. Dose, Europhys. Lett. 4 (1987) 1303.

[62] W. Gordy and W.J.O. Thomas, Phys. Rev. 24 (1956) 439.

Surface Science 299/300 (1994) 945–955
North-Holland

surface science
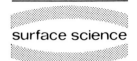

Electrochemical surface science

T.E. Furtak

Physics Department, Colorado School of Mines, Golden, CO 80401, USA

Received 12 April 1993; accepted for publication 25 May 1993

Surface science at electrified interfaces, specifically those encountered in electrochemical problems, displays a rich variety of interesting phenomena and some surprising similarities to vacuum surface science. During the last 30 years research in this area has evolved from primarily electrical measurements to encompass a broader range of more powerful techniques. It is now possible to handle a sample under water with very well-defined conditions, thus opening the way to detailed understanding through close links between experiment and theory. This article sketches the development of electrochemical surface research from a macroscopic to a microscopic science.

1. Introduction

The environment of the electrolyte–solid system is complex, combining the established difficulty of conventional surface studies with the disordered and mobile charge character of the liquid. Until recently there were no common methods which could provide the type of information equivalent to that available through LEED and Auger measurements in vacuum. Today electrochemical surface science has come to the brink of maturity. It represents one of the exciting frontiers toward which surface science will evolve in the future.

In this short article it will not be possible to cover all aspects of this how this has come to pass. Unfortunately, this means that many investigators will not be properly recognized. I apologize for this in advance. I will try to portray a personal picture of how electrochemical surface science has evolved from the mid-sixties to the present. More comprehensive and technical treatments have been published [1].

1.1. Overview

To help establish the background for understanding of electrochemical surface science I need to review some of the fundamental aspects of the situation. The most powerful feature of the electrolyte–solid interface is the control that comes through external manipulation of the interfacial potential difference, ϕ. All reactions that involve charge transfer, including many that might contaminate a surface, respond directly to ϕ. The rate (r) for a reduction process, for example, depends on ϕ, as well as on the conventional activation energy, E_a. $r = r_0 \exp\{-[E_a + |e|(\phi - \phi_0)]/k_B T\}$. Here ϕ_0 is the equilibrium potential for the reaction and r_0 contains kinetic parameters and the surface concentrations of the involved species. By convention ϕ is the potential of the sample with respect to a second, reference electrode. The latter should be chosen such that a well-defined charge transfer equilibrium can be maintained at that interface. This fixes the Fermi level of the reference electrode metal with respect to the chemical potential of an electron in the electrolyte. The accepted standard is high surface area Pt supporting $H^+ + e^- \leftrightarrow \frac{1}{2}H_2$. Although it cannot be directly measured, theoretical arguments have placed the Pt Fermi level in this system at -4.5 to -4.8 eV with respect to the vacuum [2].

By exercising control over ϕ one can significantly influence the character of the surface. A

SSDI 0039-6028(93)E0378-8

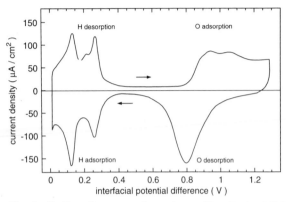

Fig. 1. Cyclic voltammetry for polycrystalline Pt in 0.5M H$_2$SO$_4$, 50 mV/s [3].

convenient diagnostic is a record of current through the sample electrode versus ϕ in an experiment where the potential is continually scanned with a linear ramp between fixed limits. This so-called cyclic voltammetry (CV) provides information that is very similar to what can be obtained in vacuum through thermal desorption spectroscopy (TDS). An example for polycrystalline Pt is shown in Fig. 1 [3]. The area under the peaks is related to the total charge involved with that process. We can see that, over a fairly large potential range, Pt can be maintained in a "clean" condition. A surprisingly large number of metals can be handled this way under water so as to avoid oxidation or direct ionic decomposition. These include Ru, Rh, Ir, Pd, Pt, Cu, Ag, Au, and Hg.

In addition to potential control, electrochemical surface science benefits from the vast difference in diffusion of material to the sample compared to the situation in vacuum. Let us see how this influences sample contamination. It is possible to produce water with impurity levels of about 5 ppb. In order to reach the surface these must cross a quiescent layer next to the sample. Without agitation the thickness of this layer is at least 500 μm. If the diffusion coefficient is 10^{-5} cm^2/s, and the sticking coefficient is unity, it would take 41 hours to create one ML of contamination on the sample under these conditions. To reduce the contamination rate to this level in a vacuum ex-

periment the base pressure would need to be less than 1.5×10^{-11} Torr!

The concept of clean surfaces under water has not been easy for vacuum surface scientists to accept. Of course, even in the ideal situation the sample is covered with water. However, it is the character of this coverage, and how things change as other species displace water, that forms one of the central issues in electrochemical surface science. Let us see how things have changed over the last 30 years.

1.2. Electrochemical surface science in 1964

A clear picture of how things stood at that time can be found in the excellent monographs by Delahay [4] and by Conway [5]. Most of the experimental data dealt with liquid Hg as the sample. This approach provided three advantages: (a) the sample surface could be kept clean by renewing the interface through an expanding metal drop, (b) the range of ϕ over which Hg is thermodynamically stable in contact with water is particularly large, and (c) the surface energy of the system could be conveniently measured through the surface tension of the liquid metal. Thermodynamic concepts were the primary theoretical tool. The measurements consisted of CV experiments, as well as other combinations of current and potential.

A working model for an electrified interface which was in common use at the time is reproduced in Fig. 2. Notice first that the metal side is structureless, with a perfect boundary. All of the concern was with the character of the liquid. The system is a "double layer" of charge, whose distributions involve electron and ion screening. Important features include the layer of partially oriented water, the "specifically adsorbed" smaller negative ions at the "inner Helmholtz plane" (IHP), and the location of the closest approaching solvated ions at the "outer Helmholtz plane" (OHP). The thickness of the "diffuse layer" beyond the OHP depends on the concentration and type of ions. This space charge region vanishes in concentrated electrolytes (of the order of one molar). In that case the entire potential difference occurs in a very localized

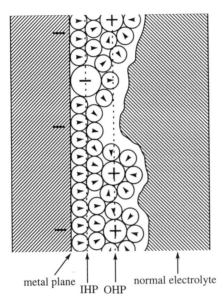

Fig. 2. Working model for the electrolyte–solid interface. Negative ions are *specifically adsorbed* beyond the requirements of electrostatics alone. Positive ions tend to remain hydrated.

metal plane IHP OHP normal electrolyte

distance. Fields of the order of 10^6 V/cm can be achieved. This is closely related to the microscopic driving force behind the impressive reaction rate control outlined earlier. It is also certain, however, that fields of this size must influence the metal. In 1964 this facet was entirely ignored.

It is clear that an important condition is that under which the metal is uncharged. For dilute solutions of F^- the so-called "potential of zero charge" (PZC) for Hg is $\phi = -0.26$ V. This coincides with the largest surface tension, or the smallest differential capacitance. It will also be the potential where the interfacial molecules have their maximum entropy. For this reason, uncharged species can more easily displace water to become adsorbed on the sample at the PZC.

The data for Hg was, and still remains, reliable. The situation at that time was quite different for other metals. Sample quality was an issue. Most studies had been done on polycrystalline surfaces. Time-dependent data were common. Little connection had been made between vacuum surface science, which was in its infancy, and

electrochemical surface science, which had not yet been born.

2. A new perspective

In the late sixties several centers emerged in which serious consideration of the solid side of the system began. These were the groups at the University of Pennsylvania (Bockris), Case-Western Reserve in Cleveland (Yeager), the Fritz-Haber-Institut in Berlin (Gerischer), and the Institute of Electrochemistry in Moscow (Frumkin). There was a recognition that new experimental information was needed. This led to some very creative approaches that can be broadly classified as either in situ or ex situ. The in situ methods were concentrated on optical techniques, primarily spectral reflectance. The ex situ methods involved analysis, using vacuum spectroscopies, of a sample that had been influenced by an electrochemical environment. This involved delicate transfer of an emersed electrode from the liquid to a UHV system. I will primarily concentrate on the in situ techniques. However, the ex situ studies have, to this day, reached a significant level of sophistication. It is now known, for example, that the double layer is preserved, as if it were a charged capacitor, as a sample is withdrawn from an electrolyte. Extensive tests have shown that strongly adsorbed material maintains its structure on an emersed electrode. A great deal is being learned through this approach. For more information on this important connection between electrochemical and vacuum surface science one can consult a number of well-written reviews [6].

2.1. Electroreflectance

One of the more direct optical methods for studying changes in an electrode with ϕ is through its differential reflectance spectrum. In 1970 this phenomenon was poorly understood. At the suggestion of Dave Lynch, at Iowa State University, I began work on performing the experiment with oriented single crystals to separate the "bound" and "free" electron contributions to the modulation. By that time a significant body of research

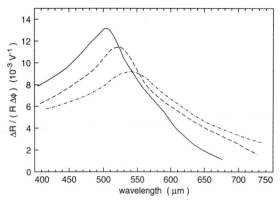

Fig. 3. Electroreflectance of polycrystalline Au in 1M HClO$_4$. The *average* potential difference was 0.2 (solid), 0.6 (dashed), or 1.0 V (dash-dotted) [8].

had already been accomplished by other groups. This was summarized at a symposium of the Faraday Society held in London, December 14 and 15, 1970 [7]. Those in attendance represented the nucleus of investigators that would carry the field for the next two decades. That meeting was the first of a variously organized series focused on new methods in electrolyte–solid science.

Fig. 3 is an example of one of the issues considered at that gathering [8]. This shows electroreflectance spectra recorded on polycrystalline Au. The peaks are concentrated in the vicinity of the interband transition threshold for Au. This was expected. The surprising feature of these data was that the peak shifted with the bias potential. Boris Cahan, and his collaborators at Case, proposed that surface states were involved, although the details were not explored. Most investigators who learned of Cahan's proposition quickly dismissed it under the assumption that surface states would be eliminated by contact with the water. Jim McIntyre and Dave Aspnes, at Bell Labs, later showed that the main features could be explained by a simple three-layer model (M-A model) in which a very thin transition zone of the metal was modulated [9]. They used the Drude approximation for the dielectric function of the metal, modulating it by the experimental surface charging. The correspondence with the data was quite good, although no clear mechanism existed for the shift of the peak.

We had chosen to work with Ag, emphasizing the (110) orientation where bound electron effects would, by symmetry considerations, be prominent. We discovered that not only did the peak shift with the average potential, but the orientation of the sample with respect to the incident optical polarization vector had a significant effect on the size of the spectrum [10]. This was even observed at normal incidence, where the free electron model predicted isotropic behavior. Our results were a clear indication that the optical measurement was probing electron effects due to the crystal potential and that the inert electrode model depicted in Fig. 2 was seriously inadequate.

These and other experiments performed in the early seventies were discussed at an ad hoc meeting sponsored by the European Physical Society in La Colle-sur-Loup, France, May 1977 [11]. The theme at the meeting was, again, optical methods. By this time the next generation of scientists, myself included, had something to add to the discussions. Notable among the attendees at that meeting was the distinguished current editor of Surface Science, Charlie Duke. This was the first time that he, and a number of other physicists, had been specifically encouraged to look at the electrochemical system as a surface science problem.

By 1977, Dieter Kolb had already performed many of his outstanding experiments at the Fritz-Haber, involving novel ways of studying the electrolyte–solid system. He showed that both bulk and surface plasmon effects could be identified in the electroreflectance of Ag. He had also learned a great deal about "underpotential deposition" (UPD) [12]. This is a phenomenon, which has its analog in UHV, that continues to fascinate people in the field today. The effect is caused by stronger bonding between dissimilar materials, compared to atom–atom interactions within the bulk deposit. It is possible to grow a stable monolayer of a foreign metal on an electrode using a growth potential that is significantly more positive than that required for multilayer deposition. At La-Colle-sur-Loup, Kolb showed differential reflectance spectra for several types of UPD layers on Ag and on Pt. It was clear from this that optical methods were very sensitive to electronic structural changes at the surface.

2.2. Photoemission

In addition to Kolb's contribution, the Berlin group was represented at La-Colle-sur-Loup by Jürgen Sass. He had been studying photoemission at electrodes, as had we in our laboratory. This phenomenon makes a particularly clear link between electrochemical and vacuum surface science. One can measure the quantum yield of photo electrons that leave the sample by collecting them in a "capture reaction", such as occurs with acid electrolytes. The operational threshold energy for emission was found to be 3.2 eV, when $\phi = 0$, independent of the type of metal. This peculiar behavior is understood as barrier penetration. The final state for the electron in the solution is defined beyond the localized influence of the interface [13]. Sass had shown photoemission to be one of the most sensitive measures of sample perfection through its sensitivity to electron scattering at the boundary.

2.3. Surface-enhanced Raman scattering

Also in attendance at the La-Colle-sur-Loup meeting was Rick Van Duyne from Northwestern University. He reported upon a curious effect, about which few of us had previously heard, involving an unusually large Raman intensity from pyridine at an electrochemically roughened Ag electrode [14]. This turned out to be the seed for an eventual torrent of interest in electrochemical phenomena, and was easily the single most important revelation of the meeting.

Van Duyne's realization was that under certain conditions the cross-section for Raman scattering from an adsorbed molecule was enhanced by more than a factor of 10^6. It did not take long for the potential impact of this phenomenon to sink in. A number of us immediately began experimenting with Raman enhancement after the La-Colle-Sur-Loup meeting. In the first experiments in our laboratory I had decided to look at CN^- on Ag. The data are shown in Fig. 4. The first line of the paper pronounced a hope which many

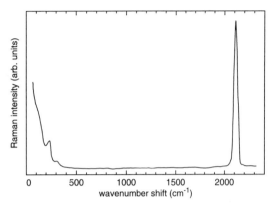

Fig. 4. Surface-enhanced Raman scattering from cyanide on polycrystalline (rough) Ag at $\phi = -0.56$ V. The electrolyte was 0.1M Na_2SO_4 + 10mM KCN [15].

of us felt: "The long sought for experimental tool for detailed chemical characterization of the solid–aqueous electrolyte interface may have at last been found" [15]. Indeed, that certainly appeared to be the case. The character of surface enhanced Raman scattering (SERS) changed dramatically as the interfacial potential was changed. This confirmed the assumption that the majority of the molecules contributing to the spectrum were in the double layer. I interpreted the vibrational features with a simple model of CN^- singly-coordinated through the carbon. That seemed to fit. However, there were many subtleties about the data, not the least of which were the extra modes at low energy.

By the time the cyanide paper was published a frenzy of attention was already under way concerning the origin of the anomalous enhancement. Beyond its importance as a valuable analytical method, SERS had an even more important impact on electrochemical science through its attraction of surface scientists and physicists. The focus for the physicists was to explain the origin of the enhancement. However, they brought with them a new way of looking at the electrochemical problem. Many of them knew little about electrolyte–solid issues prior to their concern with SERS. The result was an accelerated mixing of ideas from different disciplines. The time was right for the birth of electrochemical surface science.

3. Electrolyte–solid renaissance

At the Ames Laboratory I had been hired as an associate physicist with the charge to develop an electrochemical science group. This was promoted, primarily, by Ken Kliewer, then director of Solid State Physics at the lab. He had spent a stimulating sabbatical at the Fritz-Haber Institut, where he was exposed to the fertile atmosphere of the Gerischer influence. He was convinced that physics could do a great deal for electrolyte–solid science. The problem was to figure out how to get prominent solid state scientists to pay attention long enough to become interested. He enlisted Dave Lynch and me to help him organize a special conference where physicists and chemists would exchange ideas about electrochemistry. And thus, planning started for the International Conference on Non-Traditional Approaches to the Study of the Solid-Electrolyte Interface.

Our strategy was simple. We would choose an attractive site: Snowmass, Colorado. We would also pay a large number of prominent individuals to come. They would present tutorials from their various perspectives. We included people who had never seen an electrochemical sample chamber, as well as those who were tired of looking at them. We assembled a large amount of funding from a variety of sources, secured commitments from the invitees, and issued a broad announcement. The response was overwhelming. September 1979 was a turning point for electrochemical surface science [16]. Barriers rooted in tradition, terminology, and misunderstanding were removed. Lines of communication were established that enabled the science to grow through the next decade. During that time similar ad hoc conferences have been held in Logan, UT, USA (1982) [17], Telavi, Georgia, USSR (1984) [18], W.-Berlin, Germany (1986) [19], Bologna, Italy (1988) [20], and Asilomar, CA, USA (1990) [21]. The growth and maturation of electrochemical surface science is conveniently documented by the sequence of progress reported through the proceedings of these events.

3.1. Evolution of optical methods

The Snowmass Conference naturally contained a significant number of presentations associated with SERS. It was evident that the enhancement was associated with "active sites", not the general surface of the sample. Silver was uniquely efficient in supporting the effect, but only when it was rough. This led to the association with surface plasmon resonances. Both concepts were addressed at the meeting.

Richard Chang's group from Yale revealed the clarifying interpretation of what was really happening in the case of cyanide on Ag. Their data showed how the C–N mode evolved as a function of the applied potential [22]. They correctly associated this with surface-bonded Ag–cyanide complexes, not linearly bonded CN^-, as had I. The extra-low energy vibrations in Fig. 4 were low energy modes associated with the structure of the complex. Later we, and others, were able to show that such complexes and atomic scale defects were, in fact, the active sites. We demonstrated that electronic resonance, frequently associated with small metal clusters, provided one component of the enhancement [23].

Work on SERS at the Fritz-Haber was being conducted by Bruno Pettinger. His contribution to the Snowmass Conference demonstrated that electromagnetic resonance was also a primary factor [24]. When surface plasmons are excited they confine the electromagnetic energy to a region adjacent to the interface. Rough surfaces are required to establish the appropriate coupling condition between propagating radiation and the plasmons.

During the last fifteen years several thousand papers have been published on SERS, the majority dealing with adsorption on electrodes. However, SERS has not become the panacea many once thought it would be, primarily because of the "active site" contribution to the enhancement. Without surface enhancement Raman scattering remains an inefficient method. Alan Campion has set the standard in his laboratory at the University of Texas for non-enhanced Raman scattering from smooth samples in vacuum [25]. He is currently continuing to develop non-enhanced methods for the electrochemical environment.

The Logan, Utah meeting featured a broad collection of research projects, several having resulted from connections forged at Snowmass.

Notable at that conference was a presentation by Alan Bewick, from the University of Southampton, who had succeeded in obtaining infrared spectra from adsorbates at metal electrodes [26]. This was particularly significant in that it required no special property of the metal (unlike SERS). His approach was to perform differential reflectivity (EMIRS), or to use the optical polarization selection rule already developed for UHV (IRRAS). Both methods enabled preferential detection of the material immediately adjacent to the sample. The key feature was an ultra-thin electrolyte layer, produced by pressing the sample against an IR window. From that point on, infrared methods have grown in their sophistication and reliability [27]. Single crystal studies are now common. It is possible to perform high quality vibrational spectroscopy on adsorbates at well-defined metal electrodes.

Since the Snowmass Conference Dieter Kolb had been collaborating with Kai-Ming Ho, Bruce Harmon, and Sam Liu from the theory group at the Ames Laboratory. In Utah they reported on a body of evidence suggesting that much of the unexplained behavior in the electroreflectance of metals was caused by optical transitions involving intrinsic surface states [28]. Kolb's data contained features that shifted with ϕ in the expected way. Since water is only weakly adsorbed on Ag (about 0.4 eV/molecule) these states are preserved. However, very low concentrations of specifically adsorbed material were later shown to remove the surface state transitions [29]. Electroreflectance also helped identify surface states on Au [30], ultimately justifying Cahan's proposal 14 years earlier.

3.2. New models

Associated with the increased activity on the experimental side there was also renewed interest in developing better models for the double layer. Wolfgang Schmickler and Doug Henderson, who were then at the University of Bonn and IBM, Almaden, respectively, summarized the state of affairs by the mid-eighties [31]. The principle feature of the new efforts was a serious attempt to build a more realistic model for the metal. In the best models the metal is allowed to intermix with the first layer of molecules in the liquid. The primary point of comparison between theory and experiment has been differential capacitance as a function of temperature or potential. The new insight has shown that the metal carries a significant contribution. An interesting feature of the models is that discreteness effects become apparent. The potential does not change monotonically as a function of distance into the liquid. Rather, it exhibits oscillations about the predicted continuum result. This is a natural consequence of the finite size of the ions. In addition to this aspect, the models also show that water should tend to form clusters, similar to the hydrogen-bonded networks in ice. Recent experiments involving force measurements between parallel interfaces in the presence of water have verified this prediction [32]. It is likely that such structures are an essential part of water at any interface.

3.3. Sample preparation

Throughout the development of electrochemical surface science the field has been limited by the difficulty that one encounters preparing clean, well-ordered single crystal surfaces. This changed when, in 1980, J. Clavilier published an account of CV measurements on Pt(111) which had been prepared by a radically different method [33]. In the final step, the sample was annealed in a gas-oxygen flame, then directly quenched with pure water. Only the chosen face was allowed to touch the surface of the solution. Clavilier's CV contained features that had not been observed before, a set of reversible peaks in the middle of the double layer region of potential (Fig. 5). He proposed that the peaks involved some form of high energy adsorption of hydrogen, which required a perfect single crystal. Debate continues to this day, regarding the origin of these features. However, one thing is certain — the flame-annealing method *does* yield clean, well-ordered single crystal surfaces of Pt. This has been thoroughly verified with vacuum transfer experiments [34]. The method has now been extended to Au [35]. Prior to Clavilier's research it was very difficult for the average electrolyte–solid scientist to

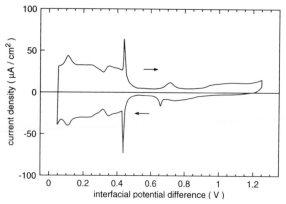

Fig. 5. Cyclic voltammetry for flame-annealed Pt(111) in 0.5M H_2SO_4, 50 mV/s [33].

produce high quality surfaces, particularly on Pt. Now the procedure, although tricky, is routine. With the flame-anneal technique it is now possible to perform reliable surface preparations that meet the standards of vacuum surface science.

4. Recent progress

Today we can say that electrochemical surface science has grown to adolescence. The second generation of electrolyte–solid scientists are now in the early part of their careers. New advances in experimental technology have brought an increased degree of confidence to the field.

In my own group we spent a great deal of time investigating the mechanisms of SERS. This theme continued throughout the beginning of my own academic career at Rensselaer Polytechnic Institute and its evolution at the Colorado School of Mines. During the last few years, however, we have initiated efforts involving other non-traditional techniques. These, and other methods under development elsewhere, represent the new frontier in electrolyte–solid science.

4.1. Optical second harmonic generation

Among the more interesting optical phenomena that occur at an interface is second harmonic generation (SHG). Geri Richmond, at Bryn Mawr and later at the University of Oregon, performed

many of the early tests, over the last ten years, which helped establish SHG as a valuable method for electrode study [36]. The technique derives its interface sensitivity from symmetry restrictions. As a second order optical effect, SHG is zero or very small in a medium possessing a center of inversion (as is the case for elemental samples and disordered liquids). Richmond demonstrated that much useful information can be extracted through straightforward analysis of the SHG intensity versus ϕ. However, there are still many unknown features about non-linear optical properties of metal surfaces. The phenomenon has become a valuable testing ground for non-local, time-dependent, density functional theory for the electrodynamics of metal surfaces [37]. In spite of this incomplete picture, one can predict the dependence of the effect on the geometry of the surface atoms. The detailed character of such a test turns out to be a very sensitive function of the sample preparation, the interfacial potential, the incident photon energy, and the overall quality of the experiment. In a recent comparison between the electrochemical and UHV environments, the Oregon group has shown that the angle-dependent character of the SHG is nearly identical (Fig. 6) [38]. This is one of the most direct pieces of evidence supporting the contention that, at the PZC, the energy structure of

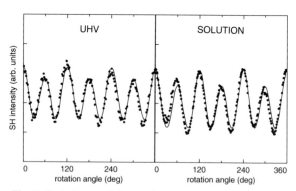

Fig. 6. Optical second harmonic generation as a function of the crystal orientation (about the surface normal) from Ag(111). Incoming light: 1.064 nm, p-polarized, 30° incidence angle. Outgoing light: 532 nm, p-polarized. UHV environment: 3×10^{-10} Torr. Solution environment: 0.25M Na_2SO_4 with ϕ at the PZC [38].

some metal electrodes are very similar to what one finds for that material in UHV.

Some time ago, Antoinette Hamelin interpreted some of the features in CV data as potential-induced reconstruction of the single crystal surface [39]. An early test of this came through the use of optical SHG. In vacuum Au(111) reconstructs from the perfectly terminated fcc structure to a $23 \times \sqrt{3}$ overlayer. Using optical SHG, Freidrich et al. observed a transformation from the normal three-fold azimuthal angle dependence to a two-fold pattern as Au(111) was made more negatively charged in an electrochemical cell [40]. This was one of the first in situ structural measurements at the electrolyte–solid interface. Some SHG results of this type have been more controversial [41]. We have studied the (111) face of Ag. Our results suggest that, under conditions of positive charging, it is possible for sulfate ions to induce a buckling of the surface so as to remove the mirror plane symmetry of the (111) face.

Today there are even more powerful structural probes. They are providing a picture of the solid side of the electrochemical boundary that is unprecedented in its resolution. For the first time it is possible to identify the atomic scale geometry of the system.

4.2. Newest structural techniques

With the advent of intense X-ray sources at synchrotron radiation facilities it is now possible to probe the structure of the electrolyte–solid boundary with the same precision that accompanied LEED in vacuum surface science many years earlier. All of the X-ray methods exploit total *external* reflection at grazing incidence. This limits the penetration of the incident X-rays into the sample, thus emphasizing interactions at the surface. A thin layer electrolyte configuration must be used, since the X-rays must go through the water diagonally.

A good demonstration of what can be accomplished was reported by Jai Wang et al., at the Brookhaven National Laboratory, who have studied potential-induced surface reconstruction of Au(111) [42]. They have identified the diffraction signature of the reversible lifting of the $23 \times \sqrt{3}$ structure. In the same experiment the density profile of the reconstructed surface was measured by the angle dependence of the X-ray reflectivity. Together these two techniques unambiguously verified that the interfacial potential plays an important role in the surface structural stability of a metal electrode.

It is also possible to perform X-ray absorption spectroscopy (XAS) on an electrode using the same grazing incidence techniques. This method has high precision and elemental specificity. It is primarily useful in the determination of short-range structure. However, these data must also be consistent with models of long-range order. In the case of Cu on Au, for example, XAS has verified the existence of several ordered superlattices which form at sub-monolayer coverage [43].

The most dramatic developments in electrochemical surface science with regard to structural measurements have come through the application of scanning tunneling microscopy (STM) and atomic force microscopy (AFM). At the time of one of the more recent "Snowmass" Conferences, the one held in Berlin, the STM had only been around for a short time. Heinrich Rohrer, one of the fathers of the technique, was an invited speaker at that meeting. The concept of the STM under water had already been demonstrated. However, those who were working with it were having trouble achieving reliable atomic scale resolution. These technical obstacles have now been overcome.

Using the AFM, Andy Gewirth and his group, at the University of Illinois, have been able to observe individual atoms on various surface orientations of Au, as well as in several UPD experiments. They were the first to identify the low coverage phases of Cu on Au(111) by actually detecting the location of the adsorbed atoms [44]. As an example of the power of these new technologies Gewirth has published data associated with the influence of Bi on the reduction of H_2O_2 [45]. It had been known that the conversion efficiency was maximized with an intermediate coverage of the foreign metal, however, the mechanism was not clear. The AFM has revealed that this is behavior is associated with a 3×3

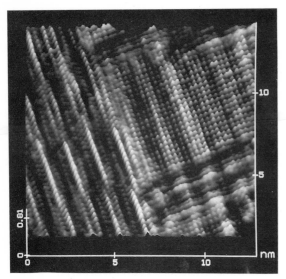

Fig. 7. Scanning tunneling micrograph of flame-annealed Au(100) in 0.1M HCLO$_4$ recorded 10 min after stepping ϕ from 0.24 to -0.06 V. Both (1×1) and (5×27) domains are evident [46].

structure, which enables a bimetallic bridge bond to form as an intermediate step.

Perhaps the most striking structural studies have been published by Xiaoping Gao, working with Mike Weaver at Purdue. They have studied the potential-induced reconstruction of Au(100) [35,46]. With stunning detail their images show a rich set of complicated reconstructions. The relationship between these structures and the underlying crystal were established in experiments where the transformation was sufficiently slow to allow observation of the development of the reconstruction in a single image (Fig. 7).

5. Future perspectives

Electrochemical surface science has reached its adolescence. The requisite structural techniques and sample preparations are now available by means of which rapid progress involving well-defined interfaces can now be made. As this takes place it is interesting to speculate about some of the major issues.

One of these points was emphasized in the closing remarks, delivered by Charlie Duke, at the original Snowmass Conference [47]. He reminded us that real understanding can be achieved only through a close link between theory and experiment. That did not exist in 1979 and, except for a few cases, has not yet been achieved. It is time for a new group of theoreticians to take on the electrochemical interface as a challenge, and to do so in concert with experimental tests.

Some of the predictions of theory remain controversial. Intrinsic surface states were originally predicted to shift with ϕ [28]. More recent calculations show that there should not be a shift – that even the surface states are screened from static fields [48]. This issue has not been resolved. Another interesting unsolved problem is the nature of the "butterfly peaks" in the CV of Pt (111) [34]. Are these associated with hydrogen adsorption or with long-range order in the lateral structure of the double layer? Other quandaries are associated with the nature of the new techniques themselves. I have already mentioned how optical SHG in metals is not well understood. It is also not totally clear how the scanning probe microscopies work under water. For example, the components of the electrolyte have not yet been observed by STM or AFM, in spite of the fact that it is known to be fairly well ordered. For reasons that are not understood, compared to vacuum it is actually easier to obtain high quality images of some materials (metals) under water. These and other puzzles are part of the picture on which a new generation of scientists is now working.

It is now possible to foresee the day when electrified interfaces are sufficiently well understood that practical technologies, involving advanced fuel cells, corrosion protection, and electrochemical synthesis, enjoy the solid base of support that now exists for UHV processing. By achieving that goal, electrochemical surface science will have reached its maturity.

References

[1] J.O'M. Bockris and A.K.N. Reddy, Modern Electrochemistry (Plenum, New York, 1970);
D.M. Kolb, J. Vac. Sci. Technol. A 4 (1986) 1294;

M.P. Soriaga, Ed., Electrochemical Surface Science (American Chemical Society, Washington, DC, 1988); H.D. Abruña, Ed., Electrochemical Interfaces (VCH, New York, 1991).

[2] R. Gomer and G. Tryson, J. Chem. Phys. 66 (1977) 4413.

[3] B.E. Conway, A. Angerstein-Kozlawska, W.B.A. Sharp and E. Criddle, Anal. Chem. 45 (1973) 1331.

[4] P. Delahay, Double Layer and Electrode Kinetics (Wiley, New York, 1965).

[5] B.E. Conway, Theory and Principles of Electrode Processes (Ronald Press, New York, 1965).

[6] A.T. Hubbard, Langmuir 6 (1990) 97; M.P. Soriaga, Prog. Surf. Sci. 39 (1992) 325.

[7] F.C. Tompkins, Ed., Proceedings of the Symposium on Optical Studies of Adsorbed Layers at Interfaces, Symp. Faraday Soc. 4 (1970).

[8] B.D. Cahan, J. Horkans and E. Yeager, Symp. Faraday Soc. 4 (1970) 36.

[9] J.D.E. McIntyre and D.E. Aspnes, Surf. Sci. 24 (1971) 417.

[10] T.E. Furtak and D.W. Lynch, Phys. Rev. Lett. 35 (1975) 960.

[11] G. Blondeau, M. Costa, A. Hugot-Le-Goff, Eds., Proceedings of the International Colloquium on Optical Properties of the Solid–Liquid Interface, J. Phys. (Paris) C-5 (1977).

[12] D.M. Kolb, Adv. Electrochem. Electrochem. Eng. 11 (1978) 125.

[13] T.E. Furtak and K.L. Kliewer, Comments Solid State Phys. 10 (1982) 103.

[14] R.P. Van Duyne, J. Phys. (Paris) C-5 (1977) 239.

[15] T.E. Furtak, Solid State Commun. 28 (1978) 903.

[16] T.E. Furtak, K.L. Kliewer and D.W. Lynch, Eds., Non-Traditional Approaches to the Study of the Solid–Electrolyte Interface (North-Holland, Amsterdam, 1980), also published as: Surf. Sci. 101 (1980).

[17] W.N. Hansen, D.M. Kolb and D.W. Lynch, Electronic and Molecular Structure of Electrode–Electrolyte Interfaces (Elsevier, Amsterdam, 1983), also published as: J. Electroanal. Chem. 150 (1983).

[18] International Conference on the Electrodynamics and Quantum Phenomena at Interfaces, Telavi, GSSR, USSR, Oct. 1–5, 1984, extended abstracts, unpublished.

[19] H. Gerischer and D.M. Kolb, Eds., Structure and Dynamics of Solid/Electrolyte Interfaces, Ber. Bunsenges. Phys. Chem. 91 (1987) 257.

[20] International Conference on Chemistry and Physics of Electrified Interfaces, Bologna, Italy, August 29 – September 2, 1988, extended abstracts, unpublished.

[21] D.J. Henderson and O.R. Melroy, Eds., The Structure of the Electrified Interface, Electrochem. Acta 36 (1991).

[22] R.E. Benner, R. Dornhaus, R.K. Chang and B.L. Laube, Surf. Sci. 101 (1980) 341.

[23] D. Roy and T.E. Furtak, Phys. Rev. B 34 (1986) 5111.

[24] B. Pettinger, U. Wenning and H. Wetzel, Surf. Sci. 101 (1980) 409.

[25] A. Campion, Annu. Rev. Phys. Chem. 36 (1985) 549.

[26] A. Bewick, J. Electroanal. Chem. 150 (1983) 481.

[27] S.M. Stole, D.D. Popenoe and M.D. Porter, in: H.D. Abruña, Ed., Electrochemical Interfaces (VCH, New York, 1991) p. 339.

[28] K.M. Ho, B.N. Harmon and S.H. Liu, Phys. Rev. Lett. 44 (1980) 591.

[29] T.E. Furtak and U. Pahk, Surf. Sci. 158 (1985) 682.

[30] S.H. Liu, C. Hinnen, C.N. Van Huong, N.R. De Tacconi and K.M. Ho, J. Electroanal. Chem. 176 (1984) 325.

[31] W. Schmickler and D. Henderson, Prog. Surf. Sci. 22 (1986) 323.

[32] J.D. Porter and A.S. Zinn, J. Phys. Chem. 97 (1993) 1190.

[33] J. Clavilier, J. Electroanal. Chem. 107 (1980) 211.

[34] F.T. Wagner and P.N. Ross, J. Electroanal. Chem. 250 (1988) 301.

[35] X. Gao, A. Hamelin and M.J. Weaver, Phys. Rev. Lett. 67 (1991) 618.

[36] G.L. Richmond, J.M. Robinson and V.L. Shannon, Prog. Surf. Sci. 28 (1988) 1.

[37] T.E. Furtak, Yeke Tang and L.J. Simpson, Phys. Rev. B 46 (1992) 1213.

[38] R.A. Bradley, S. Arekat, R. Georgiadis, J.M. Robinson, S.D. Kevan and G.L. Richmond, Chem. Phys. Lett. 168 (1990) 468.

[39] A. Hamelin, J. Electroanal. Chem. 142 (1982) 299.

[40] A. Friedrich, B. Pettinger, D.M. Kolb, G. Lüpke, R. Steinhoff and G. Marowsky, Chem. Phys. Lett. 163 (1989) 123.

[41] Y. Tang, L.J. Simpson and T.E. Furtak, Phys. Rev. Lett. 67 (1991) 2814; R. Georgiadis, R.A. Bradley and G.L. Richmond, Phys. Rev. Lett. 69 (1992) 989.

[42] Jia Wang, B.M. Ocko, A.J. Davenport and H.S. Isaacs, Phys. Rev. B 46 (1992) 10321.

[43] A. Tadjeddine, D. Guay, M. Ladouceur and G. Tourillon, Phys. Rev. Lett. 66 (1991) 2235.

[44] S. Manne, P.K. Hansma, J. Massiè, V.B. Elings and A.A. Gewirth, Science 251 (1991) 183.

[45] C-H. Chen and A.A. Gewirth, J. Am. Chem. Soc. 113 (1991) 6049.

[46] X. Gao, A. Hamelin and M.J. Weaver, Phys. Rev. B 46 (1992) 7096.

[47] C.B. Duke, Surf. Sci. 101 (1980) 624.

[48] G.C. Aers and J.E. Inglesfield, Surf. Sci. 217 (1989) 367; D.M. Kolb and R. Michaelis, J. Electroanal. Chem. 284 (1990) 507.

Surface Science 299/300 (1994) 956–964
North-Holland

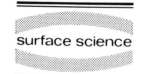

surface science

Scanning tunneling microscopy: a surface science tool and beyond

H. Rohrer

IBM Research Division, Zurich Research Laboratory, 8803 Rüschlikon, Switzerland

Received 26 May 1993; accepted for publication 4 August 1993

A short overview of the history of scanning tunneling microscopy and the principle of local probe methods is given. A selection of applications illustrates the unique and attractive features and the wide interdisciplinary nature of local probe methods. They could be the beginning of a new era in surface science.

1. Introduction

Scanning tunneling microscopy (STM) was originally intended as a way to learn about the local structural, electronic and growth properties of very thin insulating layers [1]. The term "local" meant on the scale of the inhomogeneities of these properties, believed to be no larger than a few nanometers in size – a scale that at the time was totally inaccessible with existing techniques. Electron tunneling appeared to be a promising approach, provided it could be done locally. This led in a natural, non-premeditated way to the local probe method now known as scanning tunneling microscopy. Electron tunneling already contained two of the four main technical elements of a local probe method: a strongly distance-dependent interaction and, inherently necessary, a close proximity of probe and object. A tunneling electrode in the form of a sharp conducting tip would then provide the third element, the local probe. Metal tips to achieve the desired resolution, i.e., with a radius of curvature of about 20 nm, were already in use as field emitters and in field ion microscopy. These three elements determine the resolution. The fourth and final element was the stable positioning of the probe with respect to the object and with an accuracy exceeding the desired resolution as well as within the practical interaction range of a few ångström. We expected to achieve this in a vibration-pro-

tected environment with piezo drives made from commercially available material.

Tunneling from a free probe electrode provided a new surface science tool with atomic resolution. STM is truly surface sensitive, since the tunnel current is a measure of the overlap of the electronic wave functions of probe and sample in the gap separating them. The subsurface sensitivity sometimes cited is actually more a matter of how much the subsurface manifests itself on the surface than of a true subsurface-probing capability. Atomic resolution is, so to speak, inherent in STM or, in other words, by choosing electron tunneling for the investigation of thin insulating layers, atomic resolution could not be avoided. Of course, in the beginning it was far from easy, because first of all everybody had to find a recipe for producing stable tunneling tips.

Although the first STM experiments [2] in a desiccator did not even come close to the state of the art in surface science, the role of surface science in the development of STM was instrumental. It provided the preparation methods for realizing defined and stable tip and sample conditions as well as interesting model surface structures of known periodicity. However, the surface science community by and large remained aloof towards this new surface science tool. They left it to others to pioneer the method [3], although the "splash" made by the Si(111)7 × 7 reconstruction [4] (Fig. 1) awakened curiosity and set a signal.

This reservation was quite in contrast to the pioneering spirit I had sensed a couple of years earlier when I found myself at a surface science conference. Today, however, there is a host of applications of STM and related methods in practically every domain of surface science, and we can hardly imagine surface science without them now. The many contributions in this Jubilee issue that involve STM speak for themselves.

2. Principle of local probe methods

STM and its derivates will in the following be called local probe methods in order to stress their basic element: the local experiment. Other expressions in use, such as scanning and proximal probe methods, seem somewhat narrow. Proximity is just one aspect of performing a local experiment, and scanning one-sidedly emphasizes imaging. Although local probe methods are as old a concept as that of caressing, had found useful applications such as the doctor's stethoscope and had even been worth some scientists' thoughts [5], it had not been developed systematically.

In local probe methods, the properties of objects are sensed, conditioned or changed by interactions between the probe and the object, see Fig. 2. The "localization" of the experiment is given by the active size of the probe and by the

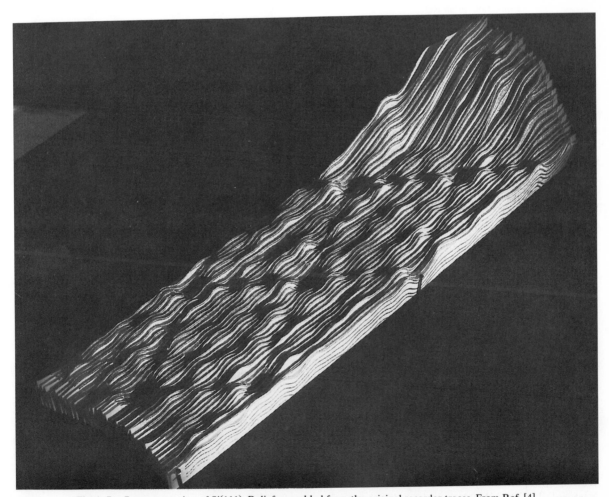

Fig. 1. 7 × 7 reconstruction of Si(111). Relief assembled from the original recorder traces. From Ref. [4].

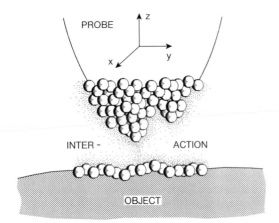

Fig. 2. Schematic of local probe methods. The circles represent atoms of the parts of probe and object, respectively, closest to each other. The probe is moved in the x, y and z-direction. From Ref. [36].

interaction distance, i.e., the distance between those parts of object and probe that interact. In the following, "object" is also used for part of an object, e.g. a surface area of atomic dimensions. For an exponential distance dependence of the interaction, the localization is of the order of $\sqrt{(D+R)/\kappa}$, where R is an effective probe size, D an effective interaction distance, and $1/\kappa$ the decay length of the interaction. To obtain atomic resolution R, D and $1/\kappa$ therefore have to be of atomic dimension. For tunneling between two bare metal surfaces, $1/\kappa$ is about 0.4 Å, thus most of the tunneling current flows between atomic-sized regions of tip and sample as indicated in Fig. 2.

The distance dependence of the interaction is the key to the sample topography. Scanning at constant interaction gives a constant-interaction contour that reflects the sample topography, provided the interaction is laterally homogeneous. However gaining access to inhomogeneities down to the atomic scale is one of the unique and attractive features of the local approach and of atomic scale imaging. The more inhomogeneous the objects of interest are, i.e., the "colorful" and interesting objects, the more important it becomes that the probe–object distance can be controlled independently of the experiment to be performed. Also, a local probe measurement usu-

ally includes different interactions, e.g. different electronic states with different wave-function overlaps contribute to the total tunnel current. The art of local probe methods is then to find an interaction suitable to control the probe–object distance and one to perform the experiment and to separate either interaction from all the others, i.e., separation into a control and a working interaction, respectively. Ideally, the control interaction should be monotonous and, for imaging, laterally homogeneous.

For most of the classical surface-science-type STM experiments, this interaction separation can be handled to a great extent by tunneling spectroscopy. The preparation methods yield compositionally well-defined surfaces of long-range homogeneity. Short-range inhomogeneities are periodic or easily recognizable, like steps and defects – yet by no means does this imply "easy" experiments. An excellent review of the history and application of scanning tunneling spectroscopy is given in the article by Feenstra in this volume [38]. Note that local tunneling spectroscopy has its own merits, besides its role in obtaining topographies from STM images. In general, separation requires simultaneous measurement of two or more quantities. In magnetic force imaging, e.g. the separation of the magnetic from the other forces can be achieved by introducing a well-defined Coulomb interaction [6], while for ambient imaging a procedure to separate the topography from electronic and elastic effects has recently been proposed that requires the simultaneous measuring of force and compliance on a constant tunnel current contour [7]. This is a beginning but we definitely have to do better. Images can be beautiful and interesting, but then so is a sphinx.

3. Experimenting with local probes

The following examples should give a taste of the richness of application possibilities and illustrate some especially attractive features. There are many more versions not mentioned and also many more still to come. The selection is somewhat arbitrary; the references reflect history, not state of the art.

A first set of applications deals with measurements, i.e., they monitor displacements, determine when contact occurs, and measure properties and perform imaging (see Figs. 3a–3c). The interactions should of course not affect the properties under consideration, although they might change others. A second set makes use of special aspects of the probe–object configuration (Figs. 3d–3f). Finally, the local probe can serve as a tool (Figs. 3g–3i). Attractive features unique to local probe methods are the following:

• The variety of interactions makes a wide variety of properties accessible.

• The environments in which local probe methods can be performed range from ultrahigh vacuum to electrolyte, including in principle any

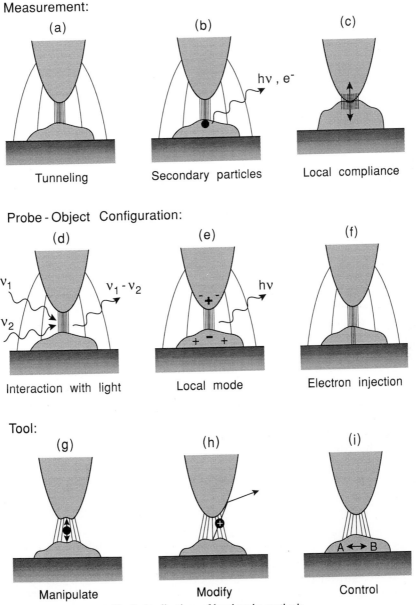

Fig. 3. Applications of local probe methods.

environment in which it is possible to move the local probe.

- Proximity and locality lead to unique conditions.
- Instrumentation can be miniaturized.
- Local probes can be made functional.

3.1. Interactions

The interactions most commonly used so far include tunneling, forces [8], and electrostatic (capacitive) and electromagnetic (e.g. near field optics [9]) interactions. Each of them comes in many variations and combinations, e.g., lateral forces for tribology [10] studies. An example for the richness of approaches is magnetic imaging, although not yet very common: applied spin-dependent tunneling due to the magnetic valve effect [11], modulation of the tunnel current at the Larmor frequency of a precessing magnetic moment [12], emission of polarized light from a tunnel junction [13], magnetic force microscopy [6], and resonant magneto-mechanic excitation [14]. The last example can be considered a beginning of "nanomechanics", as a valid complement to electronics.

Subsurface sensitivity is achieved when the interaction extends into the object, e.g. the electrostatic interaction of a conducting or polarizable probe with an electronic charge in an insulating layer [15]. This, however, results in a loss of resolution, since the probe physically cannot come closer than the object surface. Other subsurface methods include ballistic electron emission microscopy [16] (BEEM), in which ballistic electrons injected by a tunnel tip probe electronic properties at buried interfaces, or local luminescence [17] of quantum well structures, where the emitted light from the recombination of injected electrons is characteristic of both the surface band bending and the band gap in the interior.

3.2. Environment

The properties and in particular the functional states of nano-objects can depend quite sensitively on their immediate environment. The large variety of environments in which local probes can be operated is, therefore, the desired platform for a wide and interdisciplinary applicability in science and technology on the nanometer scale. For industrial applications imaging at ambient conditions is of immediate practical interest, e.g. force microscopy for the determination of surface roughness [18]. For electrochemistry, routine atomic-resolution STM and high resolution force microscopy brought a quantum leap in characterization methods [19]. For biology, it opens the possibilities of in vivo imaging and characterizing biological functions. Substantial efforts are currently being devoted to decoding DNA by imaging, where the environment is, however, not critical except maybe for preparation and immobilization. Of a different kind are the prospects of interfacing individual biological functional units, down to the molecular level, to the macroscopic world and of the ability to control their functions [20]. We have to establish communication with the biological objects and find ways to immobilize them at specific, predetermined sites in the proper environments they require for being functional (see Fig. 4 and the paragraph on "functional probes" below). Electronic communication seems quite feasible, since biological material fortunately appears much more "electron-transfer friendly" than commonly believed – scanning tunneling microscopy on alkane crystals [21] is an example as well as an interesting problem in itself. But other ways of communication also have to be explored. The road to "interfacing molecules" is long and difficult, but it is a worthwhile challenge. Just imagine, for instance, that the multibillion-dollar human genome project could essentially be miniaturized, in a first step, to a local-probe DNA imaging station and in a second step to a biological DNA reading unit with an appropriate interface to the human world [22].

3.3. Proximity and locality

One set of application deals with the absorption, emission and conversion of electromagnetic radiation in the tip–sample regime. The tip–sample configuration acts as a nonlinear element for frequency mixing [23] or higher harmonic generation [24] (Fig. 3d), or as a means to set up local modes that can be excited by the tunneling electrons and emit characteristic photons on decaying [25] (Fig. 3e). Particularly interesting is the combination of the ultrashort time domain of electromagnetic radiation with the spatial resolution capability of local probe methods.

In ballistic electron emission microscopy [16] and local luminescence [17] already mentioned above, the "tunneling interaction" merely controls the injection of a collimated electron beam into the object (Fig. 3f). The experiment then deals with processes induced by the injected electrons.

Another set of applications exploits the extreme conditions that can be achieved on a small scale. Local probes as nanotools currently receive a lot of attention (see Figs. 3g–3i). Most convenient is that the same probe can be used as sensor for measuring and imaging and as a machining and control tool. The interactions necessary for measurement and imaging, manipulation [26], modification [27], and control are tuned by adjusting the probe–object distance or changing the interaction externally, e.g. by an applied voltage. Atom-by-atom and molecule-by-molecule manipulation (Fig. 3g illustrates the atom switch [28]) require relatively weak interactions. On the other hand, electric fields can be high and local enough to tear individual atoms out of a compact surface, leading to the removal from and deposition and collection of atoms and small clusters at a predetermined location in a controlled way [29].

Finally, otherwise slow processes can become very fast, e.g. thermal relaxation times that are below nanoseconds. This is sufficient to freeze local structures in nonequilibrium configurations or to make thermal switching in combination with electronics interesting once more.

3.4. Miniaturization

The "active" part of a local probe is very small. Local probe methods, therefore, lend themselves miniaturization. The heart of force microscopy is a cantilever of submillimeter dimension; piezo-resistive sensing [30] and piezoelectric [31] or capacitive [32] actuation indeed already yield a miniaturized tool of considerable capability and autonomy.

One set of applications for such miniaturized tools is to serve as eyes, nose, ears and hands of minirobots that have to work and also have to be directed with nanometer or even subnanometer precision. Another field is parallel operation of miniaturized sensors and tools. We can see – or at least already imagine – that components, e.g. storage elements in data processing, will reach dimensions of a few nanometers. This implies large numbers, e.g. Pbit memories. To deal reasonably with such numbers, parallel operation is required, which means arrays of miniaturized sensors or actuators.

3.5. Functional probes

When studying functions of nano-objects and processes associated with them by local probe methods, it is important that the immobilization of the object and the proximity of the probe do not unduly affect function and process. When addressing the function, the object has to be adsorbed on a chemically active, often also called functionalized, substrate in the proper environment so that it retains its function, a standard procedure in biochemistry. This chemical activation of a bare substrate is usually achieved by covering it with molecules with two highly reactive chemical groups that are separated from each other by a spacer. One group has to react specifically with the surface, the other has to anchor the object. It is clear that also the probe has to be properly activated for interacting with the object. Fig. 4 sketches this "program" for interfacing nano-objects to the macroscopic world for functional biological macromolecules. In the first step shown in Fig. 4a, neither the substrate nor the probe are activated; the molecule is physisorbed directly onto the substrate [33]. This step is used for qualitative imaging and for exploring communication with the molecule. In Fig. 4b, object and probe are immersed in the proper liquid environment. Of interest here is the immobilization in a liquid environment for imaging the "true" shape of the molecule and for communication. In Fig. 4c, the molecule is immobilized on

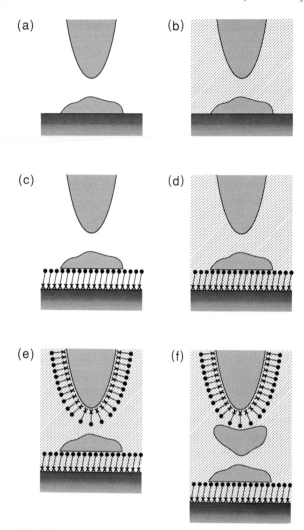

Fig. 4. Program for the chemical activation (functionalizing) of probe and substrate. (a) "Bare" configuration, (b) "functional" environment, (c) activation of substrate, here by a self-assembled monolayer, (d) configuration (c) in proper environment, (e) configuration (d) with activated probe. Steps (a) to (e) connect a functional object via the functional probe with the outside world. In (f) a functional molecule is the new probe.

a self-assembled monolayer, a problem of current interest [34]. The next steps include immobilization in the proper environment on a chemically activated substrate (Fig. 4d) and finally activation of the probe (Fig. 4e).

Another motivation for using functional probes is to defer a specific task from processing to the

experiment. In Fig. 4f, for instance, molecular recognition would be performed by a test molecule fixed to the probe. As another, experimentally maybe less involved example, consider an atom with a magnetic moment fixed to the apex of a non-magnetic tip. Deposition of a specific apex atom onto a pyramidal tip has already been achieved [35]. The modulation of the tunnel current at the precession frequency [12] would provide an atomic size probe for "effective" magnetic fields of any kind.

4. Local probe methods for surface science of the future

No doubt the proliferation of local probe methods in surface science will continue. It is moot to speculate about new applications, new approaches, and new insights – they will come in any case. The question about the future, therefore, is not so much "what type of new local probe approaches in surface science", but rather "what type of new surface science with local probe methods" will there be. The spatial resolution of local probe methods reaches the dimension of the smallest inhomogeneities of condensed matter. The statistic approach is, so to speak, replaced by an individual one. This, together with the adaptability to a large variety of environments and interactions, might well be the key to a new type of surface science for which the solid–liquid interface plays a central role. It could open the present surface science of homogenized, well-prepared, well-controlled, and reasonably well-defined surfaces to a large variety of "real" surfaces and interfaces, which can be inhomogeneous on the smallest possible scale. Local probe methods applied to electrochemistry problems is a promising beginning. It also illustrates that solid–liquid interfaces are no less interesting a variety – starting with reconstructions and depositions – than surfaces of present-day classical surface science. The liquid can serve as a new control for the state of the surface, e.g. for the protection or for the control of forces [37] and also adds a new element, so to speak a third dimension, for mass and charge transport which is crucial for self-assembly procedures and should generally lead to new growth and structuring processes.

For such a new type of surface science, however, substantial progress in many respects is required. The chemical analysis capability of local probe methods is still quite poor, and no significant improvement by a single method is in sight. Characterization of "real" surfaces and interfaces will involve different types of experiments, since to start with much less is known about the state of such surfaces and interfaces than about the well-prepared and controlled surfaces. The experiments have to be performed simultaneously for interaction separation, but especially since "real" interfaces can neither be reproduced on a local scale nor sufficiently controlled for sequential local experiments. The local approach will also produce very large data sets for representative surface samples, calling for both increased speed and parallel operation as well as for new ways of handling and analyzing data.

Science with "real" surfaces might still appear more of a desirable than a realistic goal. But who would have thought, only ten years ago, of the astonishing progress and of all the fantastic images that local probe methods made possible.

References

[1] For historical overviews, see G. Binnig and H. Rohrer, Helv. Phys. Acta 55 (1982) 726; Nobel Lecture, Rev. Mod. Phys. 59 (1987) 615;
C.F. Quate, Phys. Today 39 (1986) 26;
H. Rohrer, Ultramicroscopy 42–44 (1992) 1.

[2] G. Binnig, H. Rohrer, Ch. Gerber and E. Weibel, Physica 109&110B (1982) 2075; Appl. Phys. Lett. 40 (1982) 178.

[3] J. Moreland, S. Alexander, M. Cox, R. Sonnenfeld and P.K. Hansma, Appl. Phys. Lett. 43 (1984) 387;
S. Elrod, A.L. de Lozanne and C.F. Quate, Appl. Phys. Lett. 45 (1984) 1240;
R.M. Feenstra and A.P. Fein, Phys. Rev. B 32 (1985) 1394;
R.S. Becker, J.A. Golovchenko and B.S. Schwartzentruber, Phys. Rev. Lett. 54 (1985) 2678;
see also proceedings of the IBM Europe Institute Workshop, Oberlech (Austria), July 1985, IBM J. Res. Develop. 30 Nos. 5 and 6 (1986).

[4] G. Binnig, H. Rohrer, Ch. Gerber and E. Weibel, Phys. Rev. Lett. 50 (1983) 120.

[5] For contemplations about local probe methods, see: H.K. Wickramasinghe, Sci. Am. 261 (1989) 74.

[6] Ch. Schönenberger and S.F. Alvarado, Z. Phys. B: Condensed Matter 80 (1990) 373.

[7] D. Anselmetti, Ch. Gerber, B. Michel, H. Wolf, H.-J. Güntherodt and H. Rohrer, Europhys. Lett. 23 (1993) 421.

[8] G. Binnig, C.F. Quate and Ch. Gerber, Phys. Rev. Lett. 56 (1986) 930.

[9] D.W. Pohl, W. Denk and M. Lanz, Appl. Phys. Lett. 44 (1984) 651.

[10] R. Erlandsson, G. Hadziioannou, C.M. Mate, G.M. McClelland and S. Chiang, J. Chem. Phys. 89 (1988) 5190.

[11] R. Wiesendanger, H.J. Güntherodt, G. Güntherodt, R.J. Gambino and R. Ruf, Phys. Rev. Lett 65 (1990) 247;
J.C. Slonczewski, Phys. Rev. B 39 (1989) 6995.

[12] Y. Manassen, R.J. Hamers, J. Demuth and A.J. Castellano, Jr., Phys. Rev. Lett. 62 (1989) 2531.

[13] S. Alvarado and P. Renaud, Phys. Rev. Lett. 68 (1992) 1387.

[14] D. Rugar, C.S. Yannoni and J.A. Sidles, Nature 360 (1992) 563.

[15] B.D. Terris, J.E. Stem, D. Rugar and H.J. Mamin, Phys. Rev. Lett. 63 (1989) 2669;
Ch. Schönenberger and S.F. Alvarado, Phys. Rev. Lett., 65 (1990) 3162.

[16] W.J. Kaiser and L.D. Bell, Phys. Rev. Lett. 60 (1988) 1406.

[17] D.L. Abraham, A. Veider, Ch. Schönenberger, H.P. Meier, D.J. Arent and S.F. Alvarado, Appl. Phys. Lett. 65 (1990) 1564.

[18] N. Garcia, A.M. Baro, R. Miranda, H. Rohrer, Ch. Gerber, R. Garcia and J.L. Pena, Metrologia 21 (1985) 566.

[19] R. Sonnenfeld and P.K. Hansma, Science 223 (1987) 211;
S. Manne, P.K. Hansma, J. Massie, V.B. Elings and A.A. Gewirth, Science 251 (1991) 183.

[20] B. Michel, in: Highlights in Condensed Matter Physics and Future Prospects, Ed. L. Esaki (Plenum, New York, 1991) pp. 549–572.

[21] G. Travaglini, B. Michel, H. Rohrer, C. Joachim and M. Amrein, Z. Phys. B: Condensed Matter 76 (1989) 99.

[22] B. Michel, private communication.

[23] T.E. Sullivan, Y. Kuk and P. Cutler, IEEE Trans. Electron. Dev. 36 (1989) 2659;
W. Krieger, H. Kopperman, T. Suzuki and H. Walter, IEEE Trans. Instrum. Meas. 38 (1989) 1019.

[24] G.P. Kochanski, Phys. Rev. Lett. 62 (1989) 2285.

[25] J.H. Coombs, J.K. Gimzewski, B. Reihl, J.K. Sass and R.R. Schlittler, J. Microsc. 152 (1988) 325.

[26] D.M. Eigler and E.K. Schweizer, Nature 344 (1990) 524.

[27] D.W. Abraham, H.J. Mamin, E. Ganz and J. Clarke, IBM J. Res. Develop. 30 (1986) 492;
H. van Kempen and G.F.A. van de Walle, IBM J. Res. Develop. 30 (1986) 509.

[28] D.M. Eigler, C.P. Lutz and W.E. Rudge, Nature 352 (1991) 600.

[29] R.S. Becker, J.A. Golovchenko and B.S. Swartzentruber, Nature 325 (1987) 419;
I.-W. Lyo and Ph. Avouris, Science 253 (1991) 173.

[30] A. Tortonese and C.F. Quate, to be published.

[31] S. Akamine, T.R. Albrecht, M.J. Zdeblick and C.F. Quate, IEEE Electron. Dev. Lett. 10 (1989) 490.

[32] G. Binnig, U. Dürig, J.K. Gimzewski, D. Pohl and H. Rohrer, European Patent Application EP-A 0 290 648 'Atomic Force Sensor Head';
J. Brugger, N. Blanc, Ph. Renaud and N.F. de Rooij, Proc. Transducer '93, Yokahama, Japan, June 7–19, 1993 (in press).

[33] G. Binnig and H. Rohrer, in: Trends in Physics 1984, Vol. 1, Eds. J. Janta and J. Pantoflicek (European Physical Society, Prague, 1985) pp. 38–46.

[34] L. Häussling, B. Michel, H. Ringsdorf and H. Rohrer, Angew. Chem. Int. Ed. Engl. 30 (1991) 569.

[35] H.-W. Fink, IBM J. Res. Develop. 30 (1986) 460.

[36] H. Rohrer, Jpn. J. Appl. Phys. 32 (1993) 1335.

[37] O. Marti, B. Drake and P.K. Hansma, Appl. Phys. Lett. 51 (1987) 484;
F. Ohnesorge and G. Binnig, Science 260 (1993) 1451.

[38] R.M. Feenstra, Surf. Sci. 299/300 (1994) 965.

Surface Science 299/300 (1994) 965–979
North-Holland

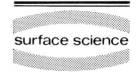

surface science

Scanning tunneling spectroscopy

Randall M. Feenstra

IBM Research Division, T.J. Watson Research Center, Yorktown Heights, NY 10598, USA

Received 4 May 1993; accepted for publication 1 June 1993

The development of the field of spectroscopic measurement with the scanning tunneling microscope (STM) is discussed. A historical review of early experimental results in this field is presented, with emphasis on the techniques for data acquisition and interpretation. The applicability of STM spectroscopic measurement to surface structural determination is addressed. The role of geometric versus electronic contributions to STM images is discussed, with reference to studies of Si(111)7 × 7, Si(111)2 × 1, and Ge(111)c(2 × 8) surfaces. It is concluded that, for semiconductor surfaces, the observed corrugations are dominated by electronic effects. Issues of dynamic range in spectroscopic measurement, and interpretation of spectroscopic images, are examined.

1. Introduction

Since its inception in 1982, the scanning tunneling microscope (STM) [1–3] has proven to be a powerful tool within the field of surface science. When used in a spectroscopic mode, the STM can probe the electronic states of a surface which are located within a few eV on either side of the Fermi level. Since this energy range is typical for surface states derived from dangling bonds on semiconductor surfaces, STM spectroscopic studies have focused primarily on those materials. Indeed, since STM images of semiconductor surface structures often depend sensitively on the voltage applied between the probe-tip and sample, it is generally necessary to perform some sort of spatially resolved spectroscopic measurement in order to deduce the geometric structure of the surface.

This article presents a historical perspective of the development of scanning tunneling spectroscopy techniques. The development of the STM and various other related methods are discussed in the preceding article by Rohrer [51]. Here, we focus on the specific application of the STM to spectroscopic studies primarily of semiconductor surfaces. There was a period of intense activity in this area in the years 1985–1987, with a number of research groups developing various techniques for acquiring and analyzing the data. We review these "early" results here in chronological order, including discussion of the applicability and relative merits of the methods. For simplicity we concentrate on the early experimental works with the STM, although, as referred to briefly throughout this paper, significant theoretical works also existed and in many cases they predated the experimental studies.

Before embarking on our historical survey of this field, it is important to put in perspective the level of knowledge which determined the course of many of the STM spectroscopic studies. In the early 1980's, a large number of reconstructions of semiconductor surfaces were known to exist, and had been characterized in many cases by diffraction and scattering methods. However, the detailed arrangement of atoms which constituted these surface structures were, in general, not definitively known. Many models had been proposed for each particular surface structure, but deciding amongst these models on the basis of available data was difficult. Into this arena came the STM [1–3] providing beautiful images of, e.g., the Si(111)7 × 7 surface with atomic resolution [4,5], and it seemed that the entire problem of surface structure determination would soon be

SSDI 0039-6028(93)E0379-9

solved. However, the situation proved to be somewhat more difficult – the STM images did not always directly lead to a unique surface structure due to limitations of resolution and/or the fact that the images themselves would depend on the applied voltage between the probe-tip and the sample. This voltage dependence is, of course, the essence of the spectroscopic capabilities of the STM, and thus can be regarded as a very useful aspect of the instrument. But, from the standpoint of determining surface structure, the voltage dependence of the STM was an unwanted complication and preferably something which could be ignored. Thus, it was all too easy to simply forget about the voltage dependence, and assume simply that the STM directly revealed the atoms on the surface. Such an assumption was at times adopted by both STM experts and nonexperts alike, and in some cases led to incorrect determinations of a surface structure. Finally, enough structures were correctly determined with the STM (and by other methods) so that some experience and intuition were gained which could be used to point the way through potential pitfalls in structural determination by STM. In short, the problem of surface structural determination, although it provides only one small application of STM spectroscopy, dominated much of the early work in the field.

It is probably worth discussing at this point one other general aspect of the STM which is important for both imaging and spectroscopic measurements, that is, the properties of the probe-tip. It is well known that the shape of the probe-tip apex can dramatically influence the appearance of STM images; on the atomic scale, individual corrugation maxima may be elongated in a particular direction by an asymmetric tip, or effects of multiple tips can distort and complicate the images. Similarly, the electronic properties of a probe-tip can influence spectroscopic measurements, leading to apparent band gaps or voltage offsets around zero volts, or in worse cases to distinct features in the spectrum. Such effects were fully appreciated in the early works. Efforts were made to clean the tips to ensure good metallic character, and enough spectroscopy data were collected using different tips and samples so

that individual spurious results could be discarded. Indeed, an inspection of the results in the early works reveals many of the highest quality spectra which have been reported to date, and most of the results have been reproduced many times in later studies. Thus, probe-tip characteristics, although they do present a difficulty in spectroscopic studies, do not represent a limitation which cannot be overcome with sufficient careful work. On the other hand, it is apparent from the number of publications that the growth of the STM spectroscopic field has been somewhat slower than other areas of STM, and one reason for this may be the stringent demands of probe-tip cleanliness required for spectroscopic work.

2. Historical survey

2.1. Conductance spectroscopy and imaging

The basic mode of operation of the STM is constant-current imaging, in which the probe-tip is raster scanned across the surface and a feedback loop adjusts the height of the tip in order to keep the tunnel current constant. The resulting tip height, s, as a function of lateral (x, y) position, constitutes a constant-current image (often called a topograph). Within this same mode of STM operation, one can envision positioning the probe-tip at some fixed lateral position, and then ramping the bias voltage between tip and sample while keeping the tunnel current constant. The resultant $s–V$ curve contains information on the spectrum of states over the applied voltage range. This type of measurement was first performed by Binnig and Rohrer, for the case of metal surfaces [3]. At high voltages > 4 V they observed a series of plateaus in the $s–V$ curves. These features were interpreted in terms field-emission resonances consisting of standing waves formed in the positive electron energy region between the top of the triangular vacuum barrier and the sample surface. This observation provided important proof of the coherent nature of the tunneling between tip and sample, which, together with the exponential decay of the tunnel current [1] demonstrated that the STM did indeed operate in a well-defined vacuum tunneling mode. Later measurements of field-emission resonances were performed by Becker, Golovchenko and Swartzentruber [6], who improved the spectral resolution by measuring conductance, dI/dV, using a modulation technique.

In addition to the field-emission resonances seen at high voltages in conductance spectra, weaker structure was observed at voltages < 4 V. This low voltage structure was identified as arising from image states in one study [7]. The first work to relate such spectral features with structurally derived electronic surface states was performed by Becker and co-workers, on the Si(111)7 × 7 surface [8]. Fig. 1a shows a topograph of the 7 × 7 structure, with the unit cell consisting of two equilateral triangles. Conductance spectra acquired over the two halves of the unit cell are shown in Fig. 2. The features labelled I and II in the spectra arise from surface states, whereas the oscillations at higher voltage arise from the above-mentioned field-emission resonances. At the time of this study, some asymmetry between the two halves of the 7 × 7 unit cell had already been reported in previous STM studies [4,5,9], although a quantitative measure of the voltage dependence of this effect was lacking. The observed asymmetry had been interpreted [9] in terms of the faulted and unfaulted halves of the 7 × 7 structure which appear in the dimer–adatom–stacking fault (DAS) model of this surface [10]. Thus, the two spectra in Fig. 2 are labelled according to which half of the unit cell they were acquired over. A clear difference between the spectra is seen, with the peak labelled II for the faulted-side spectrum being absent from the unfaulted side, and the peak labelled I being shifted between the spectra. Peak II was thus assigned to a specific electronic feature associated with the stacking fault in the DAS structure.

Imaging of these spectral features of the Si(111)7 × 7 unit cell was performed by recording the conductance signal, at a fixed voltage, simultaneously with the topography. The result is shown in Fig. 1b, for a sample bias of 2.0 V. The two halves of the unit cell, A and B, refer to the faulted and unfaulted sides of the structure re-

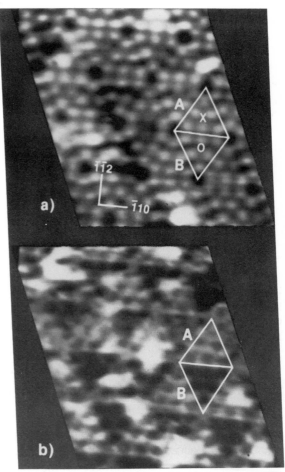

Fig. 1. (a) Grey-scale constant-current STM image of the Si(111)7 × 7 surface. Light areas are high, while dark areas are low, with a total range of 1 Å. (b) Simultaneous conductance image with light representing increased signal and dark representing decreased signal (arbitrary units). The 7 × 7 unit mesh is depicted as a rhombus built up from two equilateral triangles. The surface orientation is as shown, with the A and B designating triangular subunits with vertices pointing in the $\langle 2\bar{1}\bar{1}\rangle$ and $\langle \bar{2}11\rangle$ directions, respectively. The points x and o indicate the centers of the subunits. (From Ref. [8].)

spectively according to the DAS model. In Fig. 1b, the faulted half appears brighter in the conductance, consistent with the spectra in Fig. 2 which show higher conductance at 2.0 V for the faulted spectrum.

In the above example of conductance imaging, the voltage of 2.0 V sample bias was specifically chosen to correspond to a large spectral feature (peak I) which forms the basis for the spectral

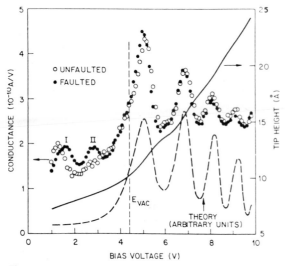

Fig. 2. Conductance spectra vs. sample bias voltage for the Si(111)7×7 surface. The filled circles and open circles denote data taken with the tip centered over faulted and unfaulted triangular subunits, respectively. The solid line shows the vacuum gap during the conductance measurements. The vacuum level is indicated, with the dashed curve showing the theoretical standing-wave conductance oscillations for the observed dependence on vacuum gap on bias voltage. (From Ref. [8].)

image. This procedure illustrates a general point regarding conductance imaging: it is always necessary to have detailed knowledge of the spectrum prior to choosing the specific imaging voltage, since otherwise the conductance image may just reflect some underlying "background" feature of the spectrum as opposed to a specific surface state feature. The same conclusion can be made for all the types of spectroscopic imaging, discussed below.

2.2. Constant-separation spectroscopy

The above examples of conductance spectroscopy and imaging, while providing clear and distinct spectroscopic results, suffer from the drawback that the measurements must be performed for only positive or negative tip–sample bias, but not both in the same scan. The reason for this limitation is that the tunnel current is kept constant during the scan, and crossing zero volts would result in tip–sample contact. The first

study to overcome this limitation was performed by Feenstra, Thompson, and Fein, in which they acquired tunneling spectra at constant tip–sample separation using an interrupted feedback method [11]. Fig. 3 shows the resulting current–voltage (I–V) curve, obtained from the Si(111)-2×1 surface. A distinct gap of width ∼ 0.5 eV is observed, and was identified as the gap between filled and empty surface state bands. The observation of this gap provided the first proof that localized surface states (i.e. with energies within the bulk gap) did indeed participate in the tunneling process. The corrugation amplitude of atomic rows seen in the STM images was found to be highly voltage dependent for bias voltages near the surface gap, thus relating the surface states with the observed corrugation.

With the ability of interrupting the feedback loop to maintain a fixed tip–sample separation, it is possible to scan the bias voltage continuously (as done in a spectrum) and also to quickly change the polarity of the voltage, e.g. between consecutive line scans in an image. This latter capability was important in early studies with the STM since residual drift in the microscope, and tip-related instabilities, made it difficult to compare images unless they were acquired consecutively or, even better, simultaneously. An experiment

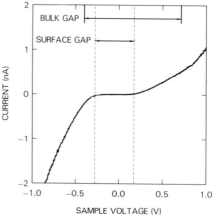

Fig. 3. Tunneling current as a function of sample voltage, measured with constant tip–sample separation on the Si(111)-2×1 surface. The 0.45 V wide flat region in the current arises from a surface state band gap, indicated together with the 1.1 V bulk band gap at the top of the figure. (From Ref. [11].)

was performed by Feenstra et al. in which the polarity of the tip–sample voltage was alternated between consecutive line scans. It was argued that the position of the observed corrugation maxima should reverse between opposite polarities for the case of the buckled model for the structure, but not for the π-bonded chain model [12]. The reversal was not seen, consistent with the latter model. Subsequent results for voltage-dependent imaging, discussed in Section 2.4 below, were more convincing since atomic features along the rows were also resolved, but the essential concept in using voltage dependence to derive structural information is provided in this early result (the same experiment had been independently suggested by Baratoff [13]).

Another method of performing tunneling spectroscopy was developed by Kaiser and Jaklevic, for their studies of Au(111) and Pd(111) surfaces [14]. In this method, the tunneling resistance is kept constant during the voltage scan, and the conductance is measured with a modulation technique. This procedure works well for metal surfaces, since a constant tunnel resistance implies a tip–sample separation which is practically constant (albeit singular at zero volts), although the method is not applicable to semiconductor surfaces since tip–sample contact would occur throughout the band gap region. Numerous features were observed in their spectra, and were interpreted in terms of the projected bulk band structure of the materials.

2.3. Current imaging tunneling spectroscopy

The method of collecting $I-V$ curves at fixed tip–sample separation was extended by Hamers, Tromp, and Demuth to provide a powerful new method of spectroscopic imaging [15]. In the method, called current imaging tunneling spectroscopy (CITS), an $I-V$ curve is acquired at every pixel within an image. The $I-V$ curves themselves can be examined to reveal the spatial localization of spectral features, or images can be formed by plotting the measured current at any voltage. Furthermore, differences between current images at neighboring voltages can be plotted to enhance the spectral resolution. An exam-

ple of this procedure, applied to the Si(111)7 × 7 surface, is shown in Figs. 4–6. Fig. 5 shows the total conductance (I/V) measured at various points over the 7 × 7 unit cell. Large features can be seen in the spectra, arising from specific spatial locations. For example, the onset observed near −0.8 V sample bias arises almost completely from the topographic minimum between the adatoms (i.e. from the rest atoms), whereas the onsets at −0.2 and +0.5 V arise from the adatoms. Some of these features can be visualized in the current images, shown in Fig. 4. In particular, the strong rest-atom feature with onset at −0.8 V gives rise to the maxima in the current image seen at −1.45 V in Fig. 4c. An asymmetry between the two halves of the unit cell is seen in the current image at 1.45 V, Fig. 4b, arising from surface states related to the stacking fault as seen in the conductivity spectra and images in Figs. 1 and 2.

Additional spectroscopic features of the 7 × 7 unit cell can be obtained by examining differential current images, as shown in Fig. 6. At −0.35 V, Fig. 6a, the differential current reveals the dangling bond state of the adatoms, with the faulted half of the unit cell appearing brighter. This is the same asymmetry which is typically seen in constant-current images (asymmetries in the unit cell were reported in all early STM studies [4,5,8,9], but the first clear voltage dependence of the asymmetry, with the faulted half appearing brighter at negative sample bias, was reported by Tromp et al. [16]) and these states near the Fermi level thus are responsible for that asymmetry. At large negative voltages, near −0.8 V, the rest atoms are seen as in Fig. 6b. At still higher negative voltages, Fig. 6c, features are observed which are interpreted as backbond states of the 7 × 7 structure.

It is important to understand that current images obtained with the CITS technique actually represent *difference* images, with the measured current being a difference between the current at the sampling voltage and that at the set-point voltage used for the constant-current topograph. In this sense, current images, as well as differential current images, can suffer from difficulties in interpretation in the same way as for conductivity

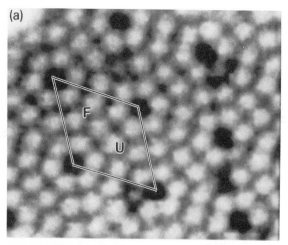

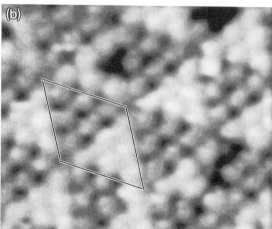

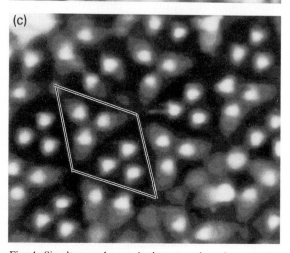

Fig. 4. Simultaneously acquired topograph and current images, for the Si(111)7×7 surface: (a) STM topograph with +2 V sample voltage, and current images with (b) +1.45 V and (c) −1.45 V applied to the sample. (From Ref. [15].)

images, discussed in Sections 2.1 and 3. Any type of spectroscopic image can be dominated by certain background features of the spectroscopy rather than genuine surface state features. This property of CITS imaging was examined by Hamers and co-workers [15,16], who argued that in cases where the topograph reflected mainly *geometric* features of the surface, then the current images would contain predominantly *electronic* information. This argument does have some qualitative basis for those systems in which it is possible to find such a topograph, but it is not possible to know a priori whether a given constant-current image does indeed contain predominantly geometric information. Subsequent work, discussed below, demonstrated that electronic ef-

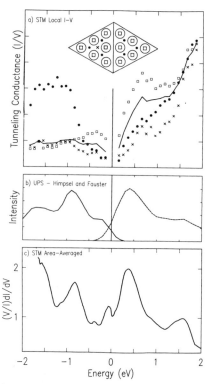

Fig. 5. (a) Constant-separation $\sigma-V$ spectra for the Si(111)-7×7 surface averaged over one unit cell (solid line) and at selected locations in the unit cell (other symbols). Crosses are from the faulted half only; others are averaged over both halves of the unit cell. (b) Surface observed using ultraviolet photoemission (solid line) and inverse photoemission (dashed line), from Refs. [47,48]. (c) Area-averaged spectroscopy results. (Adapted from Ref. [15].)

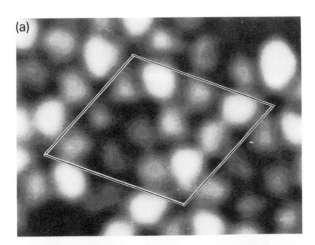

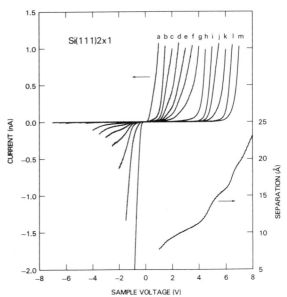

Fig. 7. Tunneling current vs. voltage for a tungsten probe-tip and Si(111)2×1 sample, at tip–sample separations of 7.8, 8.7, 9.3, 9.9, 10.3, 10.8, 11.3, 12.3 14.1, 15.1, 16.0, 17.7, and 19.5 Å for the curves labelled a–m, respectively. These separations are obtained from a measurement of separation vs. voltage, at 1 nA constant current, shown in the lower part of the figure. (From Ref. [17].)

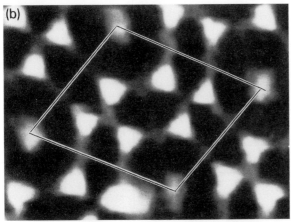

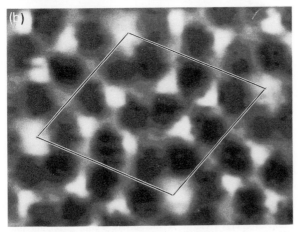

Fig. 6. CITS images of occupied Si(111)7×7 surface states. (a) Adatom state at −0.35 V, (b) dangling bond state at −0.8 V, and (c) backbond state at −1.7 V. (From Ref. [15].)

fects can and do play an important role in STM imaging at all voltages, and a simple separation between electronic and geometric effects cannot generally be obtained. On the other hand, by understanding the electronic properties of a given structure, it is often possible to use some type of voltage-dependent or spectroscopic imaging to directly confirm or deny specific structure models.

2.4. Spectroscopic normalization and polarity-dependent imaging

Subsequent work by Stroscio, Feenstra and Fein on the Si(111)2 × 1 surface provided new results in two areas of STM spectroscopic research [17]. First, detailed *I–V* measurements at a variety of different tip–sample separations were performed, as shown in Fig. 7. Surface state-density features can be seen in these *I–V* curves as the various kinks and bumps occurring between −4 and 4 V. The features are obscured by the fact that the tunneling current depends exponentially on both separation and voltage. It was found that this dependence could be effectively removed by plotting the ratio differential to total conductance, $(dI/dV)/(I/V)$, as shown in Fig. 8b where the different symbols refer to different curves from Fig. 7 (this same type of normalization was also used effectively by Lang, in his theoretical studies of metal adsorbates [18]). A

detailed spectrum of states is thus obtained, and good agreement was found between the measurements and theoretical expectations for the π-bonded chain model, shown in Fig. 8c. In addition, the inverse decay length of the tunnel current could be extracted from the data, shown in Fig. 8a, and for low voltages near the surface state band edges, this showed an enhancement over the usual values of $\sqrt{(2m\phi/\hbar^2)} \simeq 1.1$ Å^{-1}.

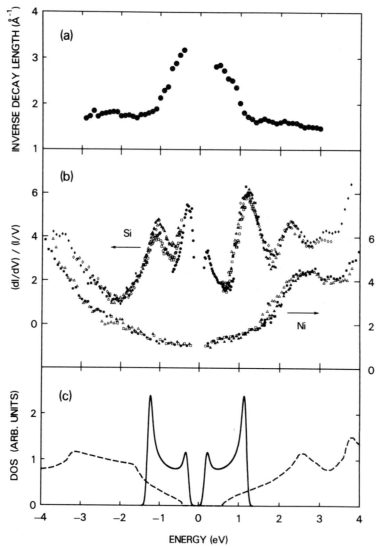

Fig. 8. (a) Inverse decay length of the tunneling current as a function of energy (relative to the surface Fermi level), for the Si(111)2 × 1 surface. (b) Ratio of differential to total conductivity for silicon and for nickel. The different symbols refer to different tip–sample separations. (c) Theoretical DOS for the bulk valence band and conduction band of silicon (dashed curve, Ref. [49]), and the DOS from a one-dimensional tight-binding model of the π-bonded chains (solid line, Ref. [50]). (From Ref. [17].)

In addition, Stroscio and co-workers presented dramatic new results for the voltage dependence of the Si(111)2 × 1 STM images. Images were acquired at voltages on either side of the surface band gap. As shown in Fig. 9, each image shows a single topographic maxima per 2 × 1 unit cell. However, the position of this maxima was found to shift by one-half a unit cell (1.92 Å) in the [0$\bar{1}$1] direction when the polarity of the voltage was changed. A smaller shift of 0.7 Å was found for the orthogonal [2$\bar{1}$$\bar{1}$] direction. This voltage dependence of the images was easily interpreted in terms of the surface state band structure for the π-bonded chain model: the surface band gap

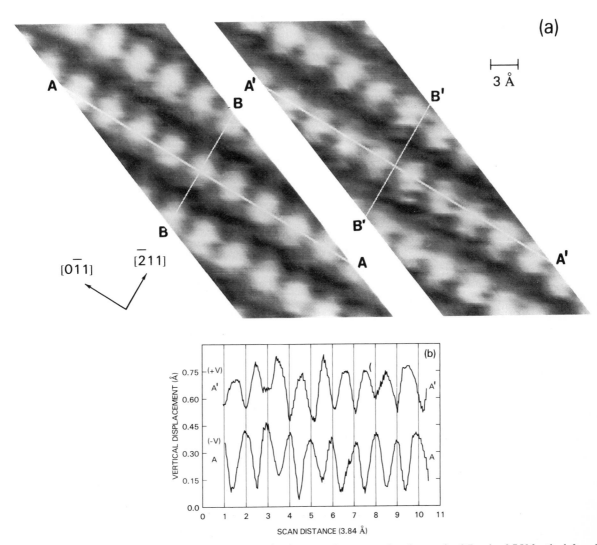

Fig. 9. Constant-current STM image acquired from the Si(111)2 × 1 surface at sample voltages of − 0.7 and + 0.7 V for the left and right images, respectively. The surface height is given by a grey-scale, ranging from 0 Å (black) to ~ 1 Å (white). (b) Surface height along the cross-sections AA and A′A′, which occur at identical lateral positions in the two images. The curve A′A′ has been shifted 0.5 Å upwards relative to AA, and the zero level on the *y* axis is arbitrary. Maxima in one cross-section correspond to minima in the other. (From Ref. [17].)

formed due to the inequivalence of the two silicon atoms in the surface unit cell, and at the band edges the wave-function was localized entirely on one or the other of the atoms. Thus, the images provided a direct connection to the electronic properties of the surface structure. Other models for the 2×1 structure were not consistent with the observed voltage dependence [11,17].

The work of Stroscio et al. provided the first clear demonstration that the STM actually imaged surface wave-functions rather than just atomic positions. At the time, it was of course known from theoretical considerations that the tunnel current consisted of the overlap of wave-functions between tip and sample, and, neglecting details of the tip, the images could be interpreted as the local state-density of the sample surface [19]. Nevertheless, it was still quite tempting to think that geometric features on the surface might dominate the images, and this viewpoint was advanced with the studies of the Si(111)7 $\times$ 7 surface discussed above. The new results for the Si(111)2 $\times$ 1 surface demonstrated conclusively that the simple interpretation in terms of geometric positions was invalid for any applied bias between tip and sample – electronic effects dominated the images. However, these electronic effects could be used to advantage in deducing the underlying structure of the surface; since the electronic properties of most simple surface can be constructed based on elementary considerations, it was still possible in many cases to directly confirm or deny specific structural models.

2.5. Other studies

From the above studies of the Si(111)7 $\times$ 7 and Si(111)2 $\times$ 1 surfaces, the major concepts of STM surface state spectroscopy were deduced. Subsequent studies of other surfaces provided additional emphasis of these concepts. For clean surfaces, the Si(001)2 $\times$ 1 surface had been studied early on in terms of its basic topography (symmetric or asymmetric dimers) [20], and voltage dependence of this surface was reported by Hamers, Tromp and Demuth [21]. A distinct minimum in the corrugation for the empty states directly on

top of the dimers was seen in both constant-current images and CITS current images, and was interpreted in terms of the node which occurs for these anti-bonding states.

The study of adsorbates on surfaces provided important new results which could be compared and contrasted with the prior studies of clean surfaces. The first adsorbate system to be studied, the $(\sqrt{3} \times \sqrt{3})$R30° structure of Ag on Si(111), turned out to be controversial since two STM groups came to different conclusions regarding the structure. Using constant-current images with positive sample bias, Wilson and Chiang observed the $\sqrt{3} \times \sqrt{3}$ structure with two topographic maxima per unit cell [22]. A simple honeycomb model, in which the observed topographic maxima corresponded to Ag atoms on the surface, was found to match the data so long as electronic effects in the images were small. The same surface was studied by van Loenen, Demuth, Tromp, and Hamers using the CITS method [23]. A similar honeycomb STM image was obtained for small negative sample biases, but for larger negative biases the corrugation was substantially reduced. CITS images at positive bias also revealed the honeycomb structure, and a surface state gap between empty and filled states was observed. It was argued that the data supported an embedded Ag-trimer, Si-honeycomb model, in which the Ag layer was embedded below a top Si layer, and the Si atoms in this top layer formed the honeycomb pattern which was seen in STM. The controversy was fueled by a subsequent STM study by Wilson and Chiang [24]. Investigations of this surface by many other techniques finally led to the proposal of a honeycomb-chained-trimer (HCT) model in which the top layer consists of Ag atoms in the HCT arrangement, below which there is a layer of Si atoms which form trimers [25]. Theoretical calculations revealed that the empty states for this structure did indeed form a honeycomb arrangement, but the maxima are not associated with individual surface atoms [26]. Rather, the corrugation maxima for empty states occurs in the middle of Ag trimers on the surface. Thus, agreement between the STM images and all other methods was finally achieved, as discussed in a reexamination of this surface by STM [27].

Although a careful examination of the voltage dependence of the STM images for Si(111)–Ag does yield results which are consistent with the known structure [27], it would be presumptuous to think that the STM could have determined this structure originally. Indeed, much of the voltage dependence was already known from the work of van Loenen et al., and furthermore, the possibility of the observed topographic maxima corresponding to Ag trimers had been considered by Wilson and Chiang. This latter possibility was rejected based on the arrangement of trimers for the simple model they considered, thus illustrating an intrinsic limitation of the STM for structural determination of complicated structures: if extensive reconstruction exists below the top surface layer, it is difficult on the basis of STM images alone to uniquely determine the geometric structure.

One clean surface structure which had been important in the development of low-energy electron diffraction techniques was the buckled geometry of GaAs(110). The basic topography of this surface had been observed in an early STM study [28], and the voltage dependence of the images was reported by Feenstra, Stroscio, Tersoff and Fein [29]. Empty states on the surface were found to be localized over the Ga atoms, and filled states over the As atoms, thus confirming an earlier theoretical prediction for this surface [19]. The spatial separation between empty and filled states provided a quantitative measure of the surface buckling. This situation for GaAs(110) is analogous to that observed earlier on Si(111)2 × 1, although the contrast between the two atoms in the unit cell arises, for Si, from a structural inequivalence, whereas for GaAs it comes, of course, from the fact that they are different atoms with As being more electronegative than Ga.

Numerous STM studies of clean and adsorbate covered surfaces followed these early works. The subsequent studies served to reinforce the concepts which had been developed in the initial investigations. We mention one study here, on the Ge(111)c(2 × 8) surface, which settled some of the controversy that had developed in earlier studies of Si(111)7 × 7 and Si(111)2 × 1. As dis-

cussed above, the STM topographs for Si(111)-7 × 7 had been interpreted primarily in terms of geometric structure, whereas the topographs for Si(111)2 × 1 were found to consist primarily of electronic structure. This difference arises, of course, in large part from the different structures of the two surfaces: the 7 × 7 structure contains adatoms (located ~ 1 Å above the underlying layer) which are significant geometric features, whereas the 2 × 1 is relatively flat (along the π-bonded chains) and thus dominated by electronic effects. The c(2 × 8) structure provides an interesting comparison, since it also contains adatoms, but the charge transfer between adatoms and rest atoms is complete so that a surface state band gap exists thereby implying possibly large electronic effects. The basic topography of the Ge(111)c(2 × 8) surface was observed in an early STM work [30], and a detailed study of the spectroscopy and voltage dependence was presented by Becker and co-workers [31]. It was found that the topographic maxima in the images shift their positions between filled and empty states. The adatoms are seen in the filled state images, but despite the geometric protuberance of the adatoms the empty state images show a minimum in topography at the adatom sites. Thus, it was concluded that electronic effects dominate in the images. The STM images could essentially be understood in a "dangling bond picture", based on the earlier studies of Si(111)2 × 1 and GaAs(110): the major source of contrast in all types of STM images of semiconductor surfaces comes from states associated with the surface dangling bonds, and such states are seen only at the specific bias voltages which access their energy levels.

Many subsequent STM works yielded spectroscopic information on new systems, and also reproduced the original early results. Examples of the latter include: for Si(111)7 × 7 and related systems, observation of the stacking fault states in conductance spectroscopy [32], and observation of the adatom and rest atom derived states by constant-separation spectroscopy [33,34]. The spectrum of states for Si(111)2 × 1 was reproduced [35,36], and similar observations were made on the Ge(111)2 × 1 surface [36]. The above-mentioned results for Ge(111)c(2 × 8) were also reproduced [37]. Thus, all of the early results survived the test of time, and they formed a reliable base on which to build an understanding of STM spectroscopy.

3. Discussion

In this section, we analyze and compare the various methods developed for acquiring spectra with the STM. For semiconductor surfaces, a very important feature of such measurements is their *dynamic range*. To fully resolve all the features in a spectrum acquired with constant tip–sample separation over the voltage range of, say, −3 to 3 V requires a dynamic range in the current or conductance of 4–6 orders-of-magnitude, with the higher values required in cases where the surface state band gap is large. Such a large variation in the current originates, at high voltages, from the voltage dependence of the tunneling transmission probability, and at low voltages from the varying state-density of the semiconductor itself. Measurement of such a wide range of current values is practically impossible in a single $I-V$ curve, since the integration time for the measurement is limited because of residual drift in the STM. Fortunately, the STM allows us to overcome this limitation simply by varying the tip–sample separation, thereby amplifying the current at low voltages and permitting measurement of the entire spectrum with the experimentally available dynamic range.

These issues regarding dynamic range can be clearly seen in the results reviewed in Section 2. For example, for the Si(111)7 × 7 surface we compare the spectrum in Fig. 2 with that in Fig. 5. The former is dI/dV acquired at constant I, thereby allowing observation at large voltages, but not permitting measurement through zero volts nor at low voltages less than about 1 V (the conductance at constant current diverges at low voltages). The latter is I/V acquired at constant tip–sample separation, so that the spectrum is limited to voltages less than about 2 V. In the region of overlap, the same features are observed (peak I in Fig. 2, observed as the onset at 1.5 V in

Fig. 5), but clearly neither spectrum displays all the spectral information. Turning now to the results of Figs. 7 and 8, it was found that multiple $I-V$ curves at various tip–sample separations could yield the entire spectrum with high spectral resolution. Note that although each measurement was performed at constant tip–sample separation, the subsequent normalization relied on using the ratio of conductance to current, thus mimicking the constant-current measurement! This interplay between constant current versus constant separation, and the issue of dynamic range which is especially vexing for wide band gap surfaces such as GaAs(110), was explored in a series of subsequent works [38]. In this author's opinion most of the issues regarding acquisition and analysis techniques are now understood, although admittedly, application of these techniques is not widespread.

Let us now consider what is arguably the most powerful aspect of the STM – spatially resolved spectroscopic measurement or spectroscopic imaging. Three methods for spectroscopic images have been reviewed in Section 2: conductance imaging, current (or differential current) imaging in the CITS mode, and voltage dependent imaging. To the casual reader it might appear that the results from any of these techniques are essentially equivalent, and it should be possible to easily transform data from one method into another. This is, however, definitely not the case! On the contrary, images obtained by these methods often yield contradictory results, and phenomenon such as contrast reversal (maxima in images appearing where the atoms or electron are *not* localized) can be easily obtained, as has been discussed in detail for the Si(111)7 × 7 and 2 × 1 surfaces [33,39,40]. The essential problem in spectroscopic imaging is illustrated in Fig. 10, where we plot a typical $I-V$ curve which might be obtained over two spatial locations on a surface [41]. In the example shown, the currents are equal at V_1, which could be the constant-current setpoint, and the $I-V$ curves are measured at the constant tip–sample separation determined by this setpoint. In curve (a), a large spectral feature is seen at V_2, which is not present in curve (b). Thus, a current image obtained in the CITS mode

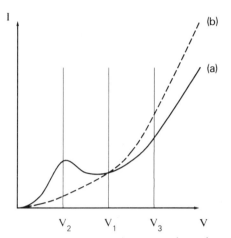

Fig. 10. Schematic illustration of current–voltage characteristics at two different spatial locations (a) and (b) on a surface. The currents are equal at the voltage V_1. At V_2, a surface-state feature appears at location (a), and it can be imaged at that voltage. An image formed at the voltage V_3 will show a maximum at spatial location (b), associated with the varying background level of the current. (From Ref. [41].)

would correctly spatially resolve this feature. However, a current image performed at some other voltage, say V_3 would produce spatial maxima over position (b). In this case, the maxima arise purely from the varying background level of the tunnel current, and have nothing to do with any real spatially resolved spectral feature. Similar comments can be made for differential current images, or for conductance images acquired at constant current. In cases where the spectral features are large, then all of the methods will produce correct spectroscopic images, but when the features are small (or nonexistent) then care must be used in the interpretation of the images.

The above problems of artifacts in spectroscopic imaging are minimized when one considers simply the voltage dependence of constant-current images. Since the tunnel current consists of contributions integrated over the entire energy range between the tip and sample Fermi levels, it is explicitly *not* a difference measurement, and thus it is less sensitive to possible background effects. On the other hand, this feature of constant-current images will, of course, produce less spectral resolution in the images. However, one particular aspect of constant-current images has

turned out to provide a powerful and general method for structural analysis, that is, comparing images obtained with positive and negative bias voltages. For surfaces which possess an energy band gap in their state-density (which is practically all semiconductor surfaces, with the exception of Si(111)7 × 7), images of filled and empty states, obtained near the band edges, often contain significant structurally dependent features. This proved to be an essential feature in STM structural determination for the Si(111)2 × 1 surface, discussed above. Even for the case of the Si(111)7 × 7 surface, in which the polarity dependence is not so striking because of the partially filled adatom states, the surface stacking fault is directly seen in the voltage dependence (the rest atoms can also be seen directly in voltage dependent imaging [31,42], although not nearly as well as can be achieved with the CITS imaging). And particularly for the case of Si(111)–Ag, it is seen above that the voltage dependence of the STM images does contain the essential structural information, which cannot be obtained by examining single topographs.

4. Summary

In this paper, we have reviewed the history of STM spectroscopy, with emphasis on the strengths and weaknesses of spectroscopic imaging as applied to surface structural determination. It is perhaps appropriate at this point to remind ourselves of the comment made in Section 1, that structural determination is only one small part of the range of application for STM spectroscopy. As illustrations of this point, we note the recent application of current imaging to the observation of band gap states induced by metal adsorbates on the GaAs(110) surface [43]. The CITS method is ideally suited to this study, since the tunnel current is zero at all locations other than around the adsorbates. Similarly, conductance imaging has proved to be ideally suited to several recent studies involving purely spectroscopic (nonstructural) features of the surface [44,45]. Finally, to generalize our scope even more, it should be remembered that STM spectroscopic measurement forms only one small area of application of the STM technique itself. Even for semiconductor surfaces, where spectroscopy finds its major application, the multitude of recent studies involving, e.g., epitaxial growth on surfaces [46], demonstrates the wide and major application of STM in areas where spectroscopic effects are small. Thus, the range of STM applications is very large, and in this work we have described one small, but powerful, application of the scanning tunneling microscope.

Acknowledgements

The author acknowledges many friends and co-workers who have provided insight into areas of STM spectroscopy. In particular, I am grateful for a collaboration with Joseph A. Stroscio during the early years of development of this field.

References

[1] G. Binnig, H. Rohrer, Ch. Gerber and E. Weibel, Appl. Phys. Lett. 40 (1982) 178.
[2] G. Binnig, H. Rohrer, Ch. Gerber and E. Weibel, Phys. Rev. Lett. 49 (1982) 57.
[3] G. Binnig and H. Rohrer, Helv. Phys. Acta 55 (1982) 726.
[4] G. Binnig, H. Rohrer, Ch. Gerber and E. Weibel, Phys. Rev. Lett. 50 (1983) 120.
[5] G. Binnig, H. Rohrer, F. Salvan, Ch. Gerber and A. Baro, Surf. Sci. 157 (1985) L373.
[6] R.S. Becker, J.A. Golovchenko and B.S. Swartzentruber, Phys. Rev. Lett. 55 (1985) 987.
[7] G. Binnig, K.H. Frank, H. Fuchs, N. Garcia, B. Reihl, F. Salvan and A. R. Williams, Phys. Rev. Lett. 55 (1985) 991.
[8] R.S. Becker, J.A. Golovchenko, D.R. Hamann and B.S. Swartzentruber, Phys. Rev. Lett. 55 (1985) 2032.
[9] R.S. Becker, J.A. Golovchenko, D.R. Hamann and B.S. Swartzentruber, Phys. Rev. Lett. 55 (1985) 2028.
[10] K. Takayanagi, Y. Tanishiro, M. Takahashi and S. Takahashi, J. Vac. Sci. Technol. A 3 (1985) 1502.
[11] R.M. Feenstra, W.A. Thompson and A.P. Fein, Phys. Rev. Lett. 56 (1986) 608.
[12] K.C. Pandey, Phys. Rev. Lett. 47 (1981) 1913; 49 (1982) 223.
[13] A. Baratoff, Physica B 127 (1984) 143.

[14] W.J. Kaiser and R.C. Jaklevic, IBM J. Res. Dev. 30 (1986) 411.
[15] R.J. Hamers, R.M. Tromp and J.E. Demuth, Phys. Rev. Lett. 56 (1986) 1972.
[16] R.M. Tromp, R.J. Hamers and J.E. Demuth, Phys. Rev. B 34 (1986) 1388.
[17] J.A. Stroscio, R.M. Feenstra and A.P. Fein, Phys. Rev. Lett. 57 (1986) 2579.
[18] N.D. Lang, Phys. Rev. B 34 (1986) 5947.
[19] J. Tersoff and D.R. Hamann, Phys. Rev. Lett. 50 (1983) 1998; Phys. Rev. B 31 (1985) 805.
[20] R.M. Tromp, R.J. Hamers and J.E. Demuth, Phys. Rev. Lett. 55 (1985) 1303.
[21] R.J. Hamers, R.M. Tromp and J.E. Demuth, Surf. Sci. 181 (1987) 346.
[22] R.J. Wilson and S. Chiang, Phys. Rev. Lett. 58 (1987) 369.
[23] E.J. van Loenen, J.E. Demuth, R.M. Tromp and R.J. Hamers, Phys. Rev. Lett. 58 (1987) 373.
[24] R.J. Wilson and S. Chiang, Phys. Rev. Lett. 59 (1987) 2329.
[25] A. Ichimiya, S. Kohmoto, T. Fujii and Y. Horio, Appl. Surf. Sci. 41/42 (1989) 82;
M. Katayama, R.S. Williams, M. Kato, E. Nomura and M. Aono, Phys. Rev. Lett. 66 (1991) 2762.
[26] Y.G. Ding, C.T. Chan and K.M. Ho, Phys. Rev. Lett. 67 (1991) 1454;
S. Watanabe, M. Aono and M. Tsukada, Phys. Rev. B 44 (1991) 8330.
[27] K.J. Wan, X.F. Lin and J. Nogami, Phys. Rev. B 45 (1992) 9509.
[28] R.M. Feenstra and A.P. Fein, Phys. Rev. B 32 (1985) 1394.
[29] R.M. Feenstra, J.A. Stroscio, J. Tersoff and A.P. Fein, Phys. Rev. Lett. 58 (1987) 1192.
[30] R.S. Becker, J.A. Golovchenko and B.S. Swartzentruber, Phys. Rev. Lett. 54 (1985) 2678.
[31] R.S. Becker, B.S. Swartzentruber, J.S. Vickers and T. Klitsner, Phys. Rev. B 39 (1989) 1633.
[32] R.S. Becker, B.S. Swartzentruber and J.S. Vickers, J. Vac. Sci. Technol. A 6 (1988) 472.
[33] Th. Berghaus, A. Brodde, H. Neddermeyer and St. Tosch, Surf. Sci. 193 (1988) 235.
[34] R. Wolkow and Ph. Avouris, Phys. Rev. Lett. 60 (1988) 1049.
[35] R.S. Becker, T. Klistner and J.S. Vickers, Phys. Rev. B 38 (1988) 3537.
[36] R.M. Feenstra, Phys. Rev. B 44 (1991) 13791.
[37] R.M. Feenstra and A.J. Slavin, Surf. Sci. 251/252 (1991) 401.
[38] For a review, see J.A. Stroscio and R.M. Feenstra, in: Scanning Tunneling Microscopy, Eds. J.A. Stroscio and W.J. Kaiser (Academic, Boston, MA, 1993) ch. 4.
[39] G. Binnig and H. Rohrer, IBM J. Res. Dev. 30 (1986) 355.
[40] J.A. Stroscio, R.M. Feenstra, D.M. Newns and A.P. Fein, J. Vac. Sci. Technol. A 6 (1988) 499.

[41] R.M. Feenstra, in: Scanning Tunneling Microscopy and Related Methods, Eds. R.J. Behm, N. Garcia and H. Rohrer (Kluwer, Dordrecht, 1990).

[42] Ph. Avouris and R. Wolkow, Phys. Rev. B 39 (1989) 5091.

[43] J.A. Stroscio, P.N. First, R.A. Dragoset, L.J. Whitman, D.T. Pierce and R.J. Celotta, J. Vac. Sci. Technol. A 8 (1990) 284.

[44] J.A. Kubby, Y.R. Wang and W.J. Greene, Phys. Rev. Lett. 65 (1990) 2165.

[45] A. Vaterlaus, R.M. Feenstra, P.D. Kirchner, J.M. Woodall and G.D. Pettit, J. Vac. Sci. Technol. B 11 (1993) 1502.

[46] For example, Y.W. Mo, J. Kleiner, M.B. Webb and M.G. Lagally, Phys. Rev. Lett. 66 (1991) 1998.

[47] F.J. Himpsel and Th. Fauster, J. Vac. Sci. Technol. A 2 (1984) 815.

[48] Th. Fauster and F.J. Himpsel, J. Vac. Sci. Technol. A 1 (1993) 1111.

[49] J.R. Chelikowsky and M.L. Cohen, Phys. Rev. B 10 (1974) 5095.

[50] R. Del Sole and A. Selloni, Phys. Rev. B 30 (1984) 883.

[51] H. Rohrer, Surf. Sci. 299/300 (1994) 956.

Surface Science 299/300 (1994) 980–995
North-Holland

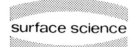

The AFM as a tool for surface imaging

C.F. Quate

Edward L. Ginzton Laboratory, Stanford University, Stanford, CA 94305-4085, USA

Received 25 June 1993; accepted for publication 5 August 1993

The atomic force microscope was introduced in 1986 as a new instrument for examining the surface of insulating crystals. There was a clear implication in the first paper that it was capable of resolving single atoms. Unambiguous evidence for atomic resolution with the AFM did not appear until 1993. In the intervening years the AFM evolved into a mature instrument that provides us with new insights in the fields of surface science, electrochemistry, biology and the technology. In this paper we will discuss the evolution of this new high resolution microscope and describe some of the events that led up to the present state-of-the-art instrument.

1. Introduction

In the Fall of 1985 Gerd Binnig and Christoph Gerber used a cantilever to examine insulating surfaces. A small hook at the end of the cantilever was pressed against the surface while the sample was scanned beneath the tip. The force between tip and sample was measured by tracking the deflection of the cantilever. This was done by monitoring the tunneling current to a second tip positioned above the cantilever. They could delineate lateral features as small as 300 Å. The force microscope [1] emerged in this way, but it evolved under the tutelage of three people: Gerd Binnig, Christoph Gerber and Tom Albrecht. Binnig and Gerber were veterans of the STM [2]. They fashioned the first instrument. Albrecht was a fresh graduate student. He fabricated the first silicon microcantilever [3].

In the Spring of 1985 we arranged for Binnig to spend a year at Stanford. In exchanges of this kind between industry and universities the central issue is always intellectual property. I resolved this in a conversation with Binnig when I said "Gerd, please don't invent anything while you are at Stanford." He replied, "Don't worry, I want to spend the year doing science. There will be no work on devices."

In the Summer of 1985 the small STM community gathered in the Austrian Alps. It was a marvelous setting with a marvelous leader, Heini Rohrer [4]. It was a learning experience for each of us. We were confronted and confounded with the images of atomic structure. The atomic arrangement on the reconstructed (111) surface of silicon appeared with great clarity when Joe Demuth presented his large area images of the 7×7 structure. John Pethica stood on the podium to describe the forces encountered when small particles were in close proximity to each other. He argued that these forces should be included in the interpretation of the images from the tunneling microscope. This was a distraction for many of us since we were struggling with tunneling current and sub-ångström motion. But not for Binnig, he was comfortable with the principles underlying tunneling current and ready for new ideas. He knew that insulators represented a huge region in the world of materials inaccessible to the STM. Within a few months of the Oberlech meeting he introduced in a system for imaging insulating surfaces with force as the main ingredient.

Picasso on teaching:

"So how do you go about teaching them something new? By mixing what they know with

SSDI 0039-6028(93)E0480-I

what they don't know. Then, when they see vaguely in their fog something they recognize, they think, "Ah, I know that." And then it's just one more step to, "Ah, I know the whole thing." And their mind thrusts forward into the unknown and they begin to recognize what they didn't know before and they increase their powers of understanding [5]."

The lectures at Oberlech are hazy in my mind, but the walk led by Heini Rohrer is still vivid. It was to begin at 9:30 on Saturday morning. Heini told us that we should be well rested, it would be a strenuous walk over at least one mountain pass. When I turned in Friday evening, Gerd and Heini were deep in conversation. On my way to breakfast the next morning I encountered Heini and Gerd in the same position discussing their philosophy of experimental science. When he saw me, Heini jumped up saying, "I must get some rest". Two hours later he showed up at the meeting place ready to go. Indeed, it was a strenuous walk but Heini insisted on maintaining a good pace at the head of the pack. Sunday I sat in front of the TV watching Boris Becker winning the Wimbledon. It was 2 o'clock in the afternoon when Heini emerged from his room muttering, "Sleep is such a wonderful thing".

The atomic force microscope emerged from its cocoon more slowly than the STM. When the Zurich group released the image of a silicon (111) 7×7 pattern, the world of surface science knew that a new tool for surface exploration was at hand. With the force microscope it was a series of incremental steps. I realized the strength of the force microscope when Albrecht measured the atomic structure of boron nitride. I began to believe when the results started to emerge from Paul Hansma's laboratory in Santa Barbara. McClelland [6] with the group at IBM/Almaden studied the force on the cantilever as the tip was brought into close contact with the sample surface. The force versus distance curve for the withdrawal cycle was distinctly different from that of the approach cycle. They concluded that there must be a thin liquid film covering the surface, and when the tip penetrated this liquid film a capillary was formed around the tip. The capillary force on the tip is large. It dominates the

other forces and produces hysteresis in the force versus distance curve.

The group with Paul Hansma in Santa Barbara reasoned that if imaging was possible with the tip penetrating the film, it would be equally possible for the system to operate with the entire cantilever immersed in fluid [7]. Their observation was profound. It carried the force microscope into electrochemistry and biology.

I was introduced to electrochemistry by the work of A. Gewirth at the University of Illinois. While he listened to Hansma describe his liquid microscope, Gewirth designed an experiment to observe the deposition of copper on gold. He traveled to Santa Barbara, and in two weeks time performed the copper on gold experiment [8] with the results shown in Fig. 1. It was a dramatic event with manifest repercussions.

We are optimistic about the prospects for molecular biology because of the work with the DNA molecule. The group with Bustamante, first in New Mexico and now in Oregon, collaborating with Helen Hansma in Santa Barbara and Keller in New Mexico have recorded meaningful and definitive images of biological molecules [9]. In a different setting Eric Henderson at Iowa State found that the liquid instrument could produce images of whole cells and delineate the internal fibril structure in the living cells [10].

Fig. 1. The AFM image illustrating deposition of copper on gold (courtesy of A. Gewirth).

In another sector of the world of science and technology, Professor T. Ohmi [11] was concerned with the roughness of silicon wafers. In modern devices the width of the lines and the thickness of the oxide beneath the gate electrodes continue to shrink. The E-field that an oxide film can sustain before breakdown is a measure of the dielectric strength of the film. If thin oxides are grown on rough silicon substrates, the gate oxide integrity is comprised by a reduction in the dielectric strength. It is, therefore, imperative to measure the roughness of silicon wafers and Professor Ohmi selected the force microscope for this task. It was a wise selection. He has determined that the RMS value of roughness must be reduced to 1–2 Å to maintain the integrity of the gate oxide.

Seven years have passed, and the atomic force microscope has emerged as a commercial instrument. The microcantilevers have been perfected. The instrument has been embraced by scientists and technologists. Electrochemists use the device to study electrolytic deposition processes, biophysicists use it to observe biological molecules in aqueous solutions. Surface scientists use it to study structures on non-conducting surfaces. Technologists use it to study fabricated structures on silicon substrates. Others believe it can be used for new high density storage devices. All of these are interesting areas but we will only examine a partial list in this paper.

2. Section of forces

The forces measured with the force microscope are discussed in the book by Israelachvili [12]. Force, as the negative gradient of the interaction potential, is easily measured with physical apparatus. For two closely spaced atoms, or small molecules, the interaction energy is described by the Lennard-Jones potential. The force between the particles as a function of their separation is illustrated in Fig. 2. There are two regions, one to the left of the potential minimum where the particles are repelled from each other and one to the right where the particles are attracted to each other.

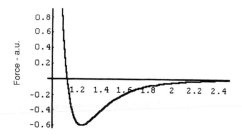

Spacing between Particles - a.u.

Fig. 2. Interatomic force between two small particles.

The geometry of the force microscope does not correspond to two closely spaced atoms, rather it resembles a small sphere above a plane. The force versus distance curve for this configuration comes from the calculation in Ref. [12], Section 10.2. The general features depicted in Fig. 2 remain intact with a repulsive force for small spacings and an attractive force for large spacings. The majority of the force microscopes operate in the contact mode where the tip moves against the "hard core" potential of the substrate. Even though the forces are small (10^{-7}–10^{-9} N), surface damage is still encountered with some samples. For those applications the attractive mode is preferred since there is no contact between tip and sample.

The attractive force, known as the van der Waals force [13], originates from the polarization of the electron cloud surrounding the atomic core. The van der Waals force for the sphere above the plane varies as the inverse square of the distance between tip and sample.

There is no known example where the van der Waals force has been measured with the force microscope. The reason is quite simple. This force diminishes as the physical volume decreases. In the force microscope with a tip of nanometer dimensions the attractive forces due to the induced polarization are small and stronger forces control the motion of the tip. The force microscope normally operates in ambient air where electrostatic charges can accumulate. Furthermore the humidity of the air leads to the formation of a thin layer of liquid over the surface of the sample and a capillary forms when the tip dips into the film. The capillary [14] and electrostatic [15] forces dominate the polarization forces of van der Waals.

In the attractive mode [16] the tip is scanned over the surface with a spacing of 50–200 Å. This spacing is controlled by monitoring the resonant frequency of the cantilever. This feature is impor-

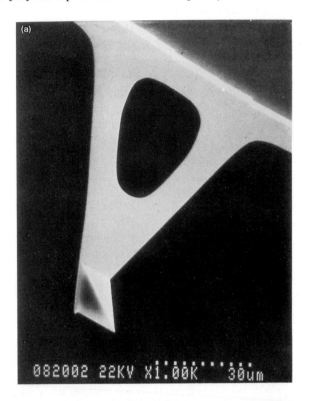

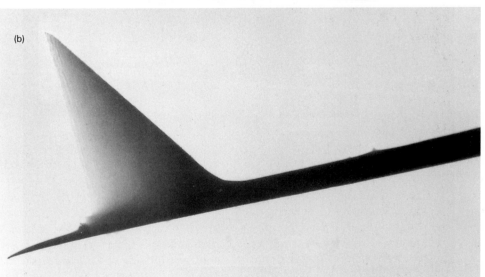

Fig. 3. (a) The heart of the AFM – the cantilever with integrated tip. (b) Enlarged view of integrated tip and cantilever (courtesy of S. Akamine).

tant, and we divert the discussion to describe the primary feature of the method.

The cantilever spring constant, k, relates the displacement, Δz, of the end of the cantilever to the force, F, applied to the end of the cantilever by the relation [17],

$$F = \Delta z. \tag{1}$$

The resonant frequency, ω_r, of a cantilever with a mass, m, is given by the relation,

$$\omega_r = \sqrt{k/m}. \tag{2}$$

This situation prevails if the force at the end of the cantilever does not vary with the z-position of the cantilever. In general, this is not the case. The force does vary with the distance between tip and sample. We can express it thus,

$$F = F_0 + (\partial F/\partial z)\Delta z = k\ \Delta z,$$
$$F_0 = (k - \partial F/\partial z)\Delta z. \tag{3}$$

We see from this that the effective spring constant changes in the presence of a gradient in the force field. In turn, resonant frequency changes to

$$\omega_r = \sqrt{(k - \partial F/\partial z)/m}. \tag{4}$$

This change provides a method for controlling the spacing between tip and sample. The cantilever, as mounted on a piezoelectric motion device, can be moved in z in such a manner as to keep the resonant frequency constant. This means that the tip moves in a region where the force gradient is constant. This implies that the tip spacing is constant since the gradient is a single-valued function of the z-spacing.

3. Cantilevers and tips

The cantilevers and tips used in the FM are fabricated with anisotropic etching of silicon. This art is part of the emerging field of micromachining [18]. When silicon is etched in a solution of KOH, the (111) face etches slowly in comparison with the other faces. This characteristic makes it

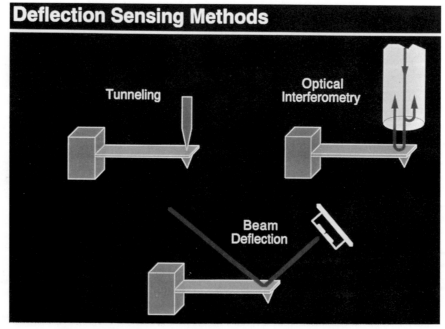

Fig. 4. Classical systems for detecting the motion of cantilevers in the AFM.

easy to etch pyramidal shapes into the (100) face of silicon. The pits are in the form of inverted pyramids. A thin film of silicon nitride deposited on this surface will conform precisely to the topography and fill in the inverted pyramid. Subsequent to the nitride deposition, the underlying silicon can be completely removed leaving only the deposited film of nitride. This forms the cantilever where the tip extends outward toward the sample.

Conversely the anisotropic etching characteristic can be used to form a tip of silicon extending outward from the surface as shown in Fig. 3. This configuration is intriguing since the tip can be sharpened with further etching. This procedure follows from the work of Marcus and Sheng [19] where they observed that the oxidation rate of silicon was dependent on the curvature of the silicon substrate. They found that a planar surface will oxidize at a faster rate than a surface with curvature. The flat sloping face of a pyramid tip oxidizes faster than the other regions. When the oxide was removed by etching, the induced curvature on the pyramidal faces results in a sharpened tip. Akamine and Quate [20] used this principle to sharpen the apex of a silicon pyramid to form a tip for the force microscope.

Beyond that a more refined technique has evolved to the point where pillar-like tips can be grown on top of the pyramidal tip. The new tips are formed when the surface is bombarded with the electron beam for an SEM. Presumably this comes from the polymerization of the residual hydrocarbons in the atmosphere of the SEM. Recently David Keller [21] in New Mexico has demonstrated the utility of this type of tip in connection with his work with biological molecules.

4. Detection of cantilever motion

The FM does not measure force; it measures the deflection of the microcantilever. The linear relation expressed in Eq. (1) between force and displacement permits us to use the displacement data to measure the force. The detecting systems for monitoring the deflection fall into several categories illustrated in Fig. 4. The first device introduced by Binnig was a tunneling tip placed above the metallized surface of the cantilever. This is a sensitive system where a change in spacing of 1 Å between tip and cantilever changes the tunneling current by an order of magnitude.

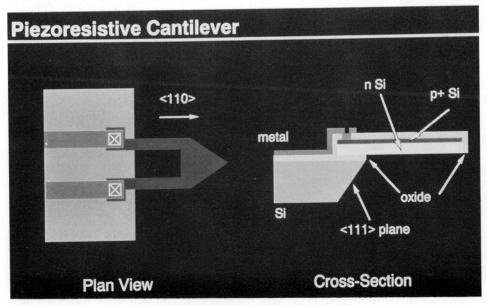

Fig. 5. The piezoresistive cantilever for the AFM.

It is straightforward to measure deflections smaller than 0.01 Å. Subsequent systems were based on optical techniques. The interferometer sketched in the upper right of Fig. 4 is the most sensitive of the optical methods, but it is somewhat more complicated than the beam-bounce method sketched in the lower figure. The beam-bounce method was introduced by Meyer and Amer [22] at Yorktown, and it is now widely used as a result of the excellent work by Alexander et al. [23] at Santa Barbara. In that system an optical beam is reflected from the mirrored surface on the back side of the cantilever onto a position-sensitive photodetector. In this arrangement a small deflection of the cantilever will tilt the reflected beam and change the position of the beam on the photodetector. A third optical system, introduced by Sarid et al. [24] in Tucson, uses the cantilever as one of the mirrors in the cavity of a diode laser. Motion of the cantilever has a strong effect on the laser output, and this is exploited as a motion detector.

The change in capacitance is a classical and sensitive method for measuring small deflections, and it is not surprising that it too has been exploited in the force microscope [25].

The most recent system is a miniature strain gauge based on the piezoresistive coefficient in crystalline silicon and fabricated as an integral part of the cantilever. This effect has been used as a strain gauge for more than 40 years [26], and it is natural to find that it is used to monitor the deflection of cantilevers in the force microscope. The cantilever is made on silicon oriented in the (100) direction. The cantilever has two legs where current flows in one leg and out the other as shown in Fig. 5. The change in current corresponding to the change in stress in the cantilever is used to monitor the deflection. In the latest device as reported by Tortonese et al. [27] the

Fig. 6. The molecular structure of didodecylbenzene (DDB) molecules on graphite imaged with the AFM in the contact mode (courtesy of H. Fuchs).

sensitivity is sufficient to detect a deflection of 0.25 Å with a bandwidth of 1000 Hz in the output circuit. This sensitivity is inferior to that of the optical systems but it is sufficient for applications in technology.

5. Contact mode

The contact mode where the tip rides on the sample in close contact with the surface is the common mode used in the force microscope. The force on the tip is repulsive with a mean value of 10^{-9} N. This force is set by pushing the cantilever against the sample surface with a piezoelectric positioning element.

A few instruments operate in UHV [28] but the majority operate in ambient atmosphere, or in liquids. It is a boon for biologists to have a high resolution microscope operating in liquid which is the natural environment for living cells [10]. It is also useful to surface scientists since liquids serve to protect surfaces from the contaminates coming from the atmosphere. In spite of these advantages most of the work has been done in air. It is a convenient environment, and the information gained from imaging in air is important. A molecular pattern of DDB molecules on a graphite substrate is shown in Fig. 6.

There is an element of controversy surrounding the high resolution images recorded in the contact mode. Many of the published images show atomic resolution, and this implies that the tip is sharp enough to interact with a single atom of the surface. The controversy is related to the force exerted on this atom by the tip. Ferrante and Smith [29] have calculated the adhesive force for several metals and show that the maximum

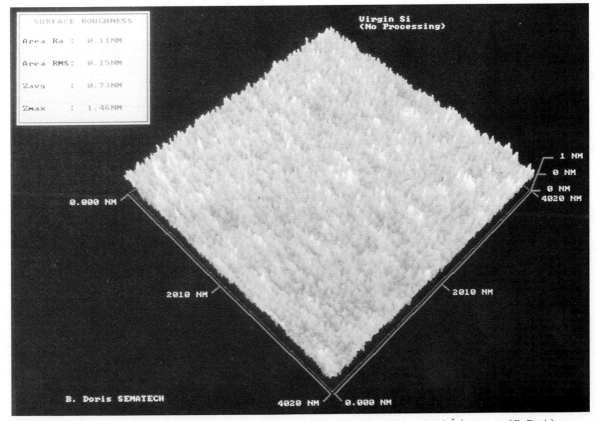

Fig. 7. The surface of an unpatterned silicon wafer in the AFM. The rms roughness is 1.5 Å (courtesy of B. Doris).

Bare Si Surface

NH₄OH:H₂O₂:H₂O = 1:1:5

NH₄OH:H₂O₂:H₂O = 0.05:1:5

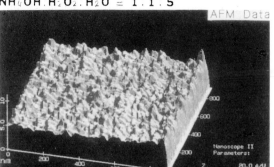

Fig. 8. The change in surface roughness when the ammonia content in the cleaning cycle is increased (courtesy of T. Ohmi).

adhesive force for Mg–Mg is 5×10^9 N/m², or about 5×10^{-10} N/atom. This force applied to a single atom in the Mg crystal would result in brittle fracture. At most temperatures Mg fails in plastic flow, and we must therefore reduce the force of a single atom tip to 10^{-10} N to avoid deformation of the surface through plastic flow. The imaging force used to record atoms with the AFM is greater than 10^{-9} N. With a limiting pressure of 5×10^9 N/m² we conclude that the

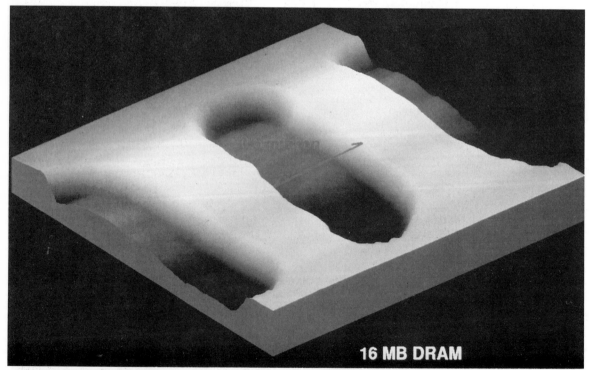

Fig. 9. The profile of a surface feature of a 16 MB DRAM. The trenches are 0.5 μm in width and 1 μm in depth (courtesy of K. Wickramasinghe).

contact area must extend over several atomic sites. How then do we rationalize the images that report atomic resolution? Heretofore, the published images have displayed perfect periodicity without defects, and the lack of defects is the key to the dilemma. We know that multiple images of periodic structure are also periodic. It is not possible to distinguish between a single image and superimposed multiple images that arise from broadened tips or from multiple tips. To resolve the ambiguity and demonstrate true atomic resolution it is necessary to reduce the imaging force to 10^{-10} N and image atomic detail on surfaces with lattice imperfections, such as step edges, or lattice vacancies. This has now been accomplished [30] in a way that we discuss in a later section.

We will bypass the marvelous work done by the electrochemists and the biophysicists and turn to the studies of surface roughness (see Fig. 7). Roughness of silicon wafers can increase during the fabrication of microchips, and it is necessary to monitor this parameter as the wafer moves through the FAB. The example from Ohmi's work [11], shown in Fig. 8, illustrates that surface roughness can increase when excess ammonia is used in one of the standard wafer cleaning steps.

Surface roughness is one facet, surface profiling is another. The illustration in Fig. 9 is typical. With the advancing state of art in microchips the line widths will shrink to 0.25 μm while the thickness of the photoresist layers will remain at 1 μm. Via's with this aspect ratio are difficult to profile with available tools. We know that the force microscope will play a role because of the work carried out by groups such as those with Wickramasinghe at IBM and Griffith at AT&T. The IBM group has used the non-contact mode

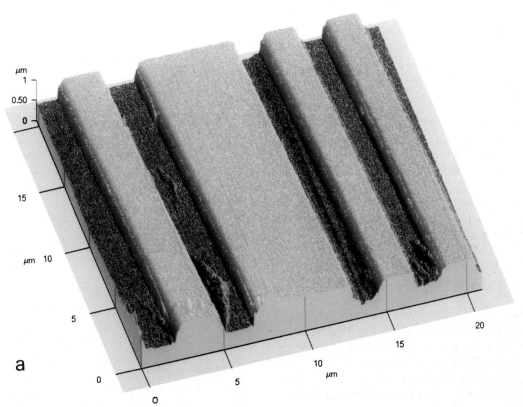

Fig. 10. The surface of soft photoresist imaged with the attractive mode. (a) The patterned unbaked resist. (b) Detail of the upper surface where the rms roughness is 17 Å (courtesy of Park Scientific Instruments). (Figure continued on next page.)

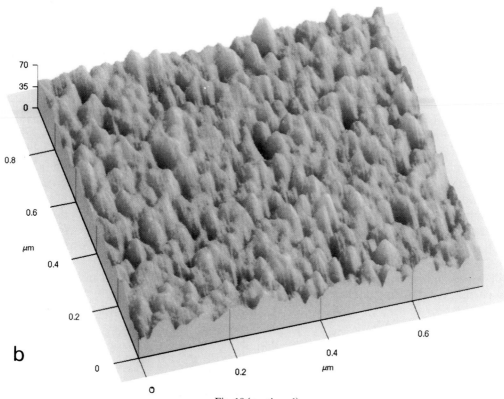

Fig. 10 (continued).

to profile trenches 0.45 μm in width and 1 μm in depth. Their goal [31] is to profile contact via's 0.25 μm in diameter (and 1 μm in depth) since these will be encountered in the microchips of the future.

The force microscope will become an indispensable tool for monitoring and profiling in the manufacture of silicon microchips [32].

6. The non-contact mode

A new era in imaging was opened when microscopists introduced a system for implementing the non-contact mode. This mode, where the tip hovers 50–150 Å above the surface, is used in situations where tip contact might alter the sample in subtle ways. The spacing between tip and sample is controlled by monitoring the resonant frequency of the cantilever as described previ-

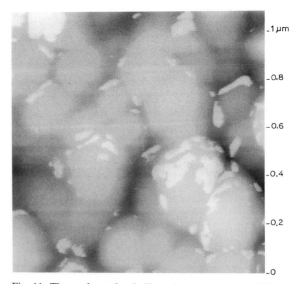

Fig. 11. The surface of polysilicon imaged with the AFM in the non-contact mode. The image size is 1 μm by 1 μm. The height in z is 150 Å from dark to light regions (courtesy of Park Scientific Instruments).

ously in Section 2. The essence of this idea is contained in the work of Israelachvili and Tabor [33]. It was introduced to the force microscope in the work of Martin, Williams and Wickramasinghe [16] and Albrecht et al. [34].

The non-contact mode is essential for examining soft samples where tip contact could change the topography. A striking example is shown in Fig. 10. This soft sample is a layer of photoresist that has not been hardened with baking. A second example of the utility of the non-contact mode is illustrated in Fig. 11. It is a polysilicon surface with the structure of the grains as the prominent feature. Grain structure of polysilicon is easily recorded in other modes but the fine

detail on the rounded hillocks is only visible with the non-contact mode.

The shift in resonant frequency of mechanically vibrating elements is a powerful method for controlling narrow spacings. It is used routinely in the magnetic force microscope where a magnetic tip is used to image magnetic field patterns. We will return with more details in the next section. The principle is also used in the near field optical microscope when it is combined with the force microscope for the purpose of controlling the spacing between the optical aperture and the sample [35].

The non-contact mode is now available in "large sample" instruments. There the probe can

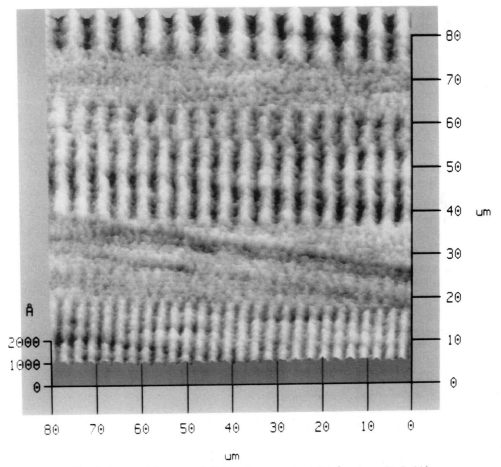

Fig. 12. Image of the magnetic field pattern on a hard disk (courtesy of I. Smith).

be positioned over any segment of an 8 inch silicon wafer. Imagine the response when this mode is introduced to the semiconductor industry, "The scanning probe, with nanometer resolution, allows the operator to examine the full area of an 8 inch wafer without touching the silicon surface." This indeed must be the threshold of the 21st century.

7. Other modes

Just as the STM spawned the AFM so it is that the AFM spawned other forms of scanning probes. The friction force microscope [36] which measures the lateral force on the scanning tip has been used to great effect by workers such as Mate at IBM and Kaneko at NTT to study the coefficient of friction for various surfaces. The capacitance microscope [37] is used to measure the variation in capacitance between the tip and sample. It has been used to profile the doping levels in silicon. The interfacial microscope [38] uses a double ended cantilever to measure the curve of force versus distance at various points in the image.

Among the members of this family the magnetic force microscope [39] (MFM) has special significance since it has spawned an offspring of its own, one that responds to signals from magnetic resonance. The magnetic force microscope, operating in air, images magnetic patterns without special preparation of the sample. The MFM is used to image the gradient of the magnetic field with a lateral resolution that is better than 0.1 μm. A spacing of 100–200 Å between tip and sample is maintained throughout the scan. In the magnetic force microscope a magnetic tip with the magnetization aligned with the axis of the tip is mounted on a flexible cantilever. The magnetic dipole of the tip interacts with the stray magnetic fields from the sample. The stray magnetic fields exert a force on the magnetized tip, and this is the information displayed in the image. In reality the gradient of stray fields alters the resonant frequency of the cantilever as described in Section 2, and it is the change in resonant frequency that is used for the information signal. The image

in Fig. 12 is typical of the work that can be done with the instrument.

The magnetic resonance force microscope [40] is the offspring of the magnetic force microscope. In this instrument a small particle of paramagnetic material is mounted on a cantilever. The unit is then immersed in a magnetic field which polarizes the spinning electrons. The electric field from a small RF coil is used to drive electron spin resonance (ESR). The magnetic field fixes the resonant frequency, and when this field is modulated, the force on the cantilever is modulated in a corresponding manner. If the modulation frequency of the magnetic field is adjusted to excite the mechanical resonance of the cantilever, the amplitude of motion in the cantilever builds to a detectable level. The modulating frequency is equal to one-half the resonant frequency since the driving force on the cantilever is proportional to the square of the gradient of the field. In the system studied by Rugar et al., the noise floor of the vibrating cantilever was 10^{-15} N. This was in vacuum at room temperature. They predict that the sensitivity can be improved by several orders of magnitude if the cantilever is redesigned and cooled to liquid helium temperatures.

The magnetic resonance force microscope has some remarkable features. Electron spin resonance at 1000 MHz can be detected with a mechanical device resonant near 10 kHz. The spatial

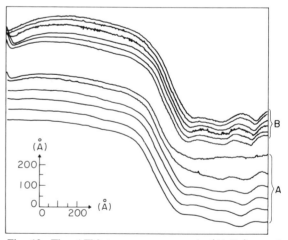

Fig. 13. The AFM traces on a ceramic (Al_2O_3) sample (courtesy of G. Binnig).

position of spins in the lateral plane can be imaged when a gradient is introduced into the magnetic biasing field. A microscope based on magnetic resonance with a sensitivity sufficient for the imaging of a single spin may be within their grasp!

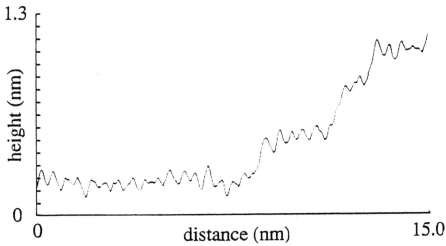

Fig. 14. True atomic structure with the AFM in the contact mode. This is the surface of calcite immersed in purified water showing the oxygen atoms as the bright spots (courtesy of F. Ohnesorge).

A second example where microwave frequencies are used to deflect a mechanical cantilever comes from the group with D. Bloom at Stanford [41]. They worked with a conventional tip mounted on a metal coated cantilever and scanned over the surface of an integrated circuit. The stray electric fields from the IC exerts a force on the cantilever. The mechanical cantilever cannot respond to microwave frequencies, nevertheless the group with Bloom devised a system to monitor the spatial variation of electric fields at microwave frequencies. They applied two RF signals, f_1 and f_2, to the circuit beneath the cantilever. The force on the cantilever is proportional to the square of the E-field. It is a "square-law" detector for force, and this is used to generate a force at $f_1 - f_2$. When the difference frequency is adjusted to the mechanical resonance, the amplitude of cantilever motion reaches a level that is detectable. The tip is scanned over the active circuit to record the spatial variations in the E-field.

8. Summary

The AFM appeared in the Winter of 1986 in a paper titled "Atomic Force Microscope" with "images" that were not images but traces from the chart of an x–y recorder as reproduced in Fig. 13. In that original paper, Binnig, with his remarkable insight, made two predictions – predictions that were heavily challenged during the process of peer review. He predicted that the instrument could be used to measure "… forces on particles as small as single atoms". He went on to predict that a cooled system could be used to measure forces as small as 10^{-18} N.

No one has yet measured forces this small but the group at IBM/Almaden is close. Their present apparatus with conventional cantilevers at room temperature has a noise floor of 10^{-15} N. It is straightforward to reduce this floor with redesigned cantilevers operating in liquid helium.

Atomic resolution in the AFM has been the goal of many studies over the years but those reports were suspect. We had to live with that suspicion until the study by Ohnesorge and Bin-

nig [30] of the IBM Laboratory in Munich appeared in June 1993. They used the AFM immersed in water to study the cleaved $(10\bar{1}4)$ face of calcite. With this arrangement they reduced the force exerted on the sample to 10^{-10} N which is a factor of 10 less than that used in conventional microscopes. Their images not only show the atomic arrangement of the uppermost oxygen atoms but they also show atomic scale defects. Atomic structure and atomic steps between terraces are common in the images from the tunneling microscope but they did not appear in the images from the AFM until Ohnesorge and Binnig published their results. Their images of calcite are shown in Fig. 14.

It has been a grand time for those associated with this era. The era is not over, there is more to come as this form of instrumentation is moved beyond imaging and employed in the fabrication of nanostructures.

References

[1] G. Binnig, C.F. Quate and Ch. Gerber, Phys. Rev. Lett. 56 (1986) 930.
[2] G. Binnig and H. Rohrer, Sci. Am. 253 (1985) 50.
[3] T.R. Albrecht and C.F. Quate, J. Appl. Phys. 62 (1987) 2599.
[4] STM Workshop, Oberlach, Austria (July 1986), Proceedings in IBM J. Res. 30 (July 1986) and 30 (Sept. 1986).
[5] F. Gilot and C. Lake, Life with Picasso (Nelson, London, 1965) p. 66.
[6] G.M. McClelland, R. Erlandsson and S. Chiang, Review of Progress in Quantitative Nondestructive Evaluation, Vol. 6B, Eds. D.O. Thompson and D.E. Chimenti (Plenum, New York, 1987) 307.
[7] O. Marti, B. Drake and P.K. Hansma, Appl. Phys. Lett. 51 (1987) 484.
[8] S. Manne, P.K. Hansma, J. Massie, V.B. Elings and A.A. Gewirth, Science 251 (1991) 133.
[9] W.A. Rees, R.W. Keller, J.P. Vesenka, G. Yang and C. Bustamante, Science 260 (1993) 1646.
[10] E. Henderson, P.G. Haydon and D.S. Sakaguchi, Science 257 (1992) 1944.
[11] T. Ohmi, M. Miyashita, M. Itano, T. Imaoka and I. Kawanabe, IEEE Trans. Electron Devices ED-39 (1992) 537.
[12] J.N. Israelachvili, Intermolecular and Surface Forces (Academic Press, New York, 1985) ch. 10.
[13] J.N. Israelachvili, Proc. R. Soc. (London) A 331 (1972) 39.

[14] R. Erlandsson, G.M. McClelland and S. Chiang, J. Vac. Sci. A 6 (1988) 266;
H.W. Hao, A.M. Baro and J.J. Saenz, Proceedings of the STM '90/NANO I Conference, Baltimore, MD, July 23–27, 1990.

[15] Y. Martin, D.W. Abraham and H.K. Wickramasinghe, Appl. Phys. Lett. 5 (1988) 1105;
J.E. Stern, B.D. Terris, H.J. Mamin and D. Rugar, Appl. Phys. Lett. 53 (1988) 2717.

[16] Y. Martin, C.C. Williams and H.K. Wickramasinghe, J. Appl. Phys. 61 (1987) 4723.

[17] D. Sarid, Scanning Force Microscopy (Oxford University Press, New York, 1991).

[18] W. Trimmer, J. of Microelectromech. Syst. 1 (1992) 1.

[19] R.B. Marcus and T.T. Sheng, J. Electrochem. Soc. 129 (1982) 1278.

[20] S. Akamine and C.F. Quate, J. Vac. Sci. Technol. B 10 (1992) 2307.

[21] D. Keller, D. Deputy, A. Alduino and K. Luo, Ultramiscroscopy 42–44 (1992) 1481.

[22] G. Meyer and N.M. Amer, Appl. Phys. Lett. 53 (1988) 2400.

[23] S. Alexander, L. Hellemans, O. Marti, J. Schneir, V. Eling, P.K. Hansma, M. Longuire and J. Gurley, J. Appl. Phys. 65 (1989) 164.

[24] D. Sarid, D. Iams, J.T. Ingle, V. Weissenberger and J. Ploetz, J. Vac. Sci. Technol. A 8 (1990) 378.

[25] R.V. Jones and J.C.S. Richards, J. Phys. E 6 (1973) 589.

[26] C.M. Harris and C.E. Crede, Shock and Vibration Handbook (McGraw-Hill, New York, 1961).

[27] M. Tortonese, R.C. Barrett and C.F. Quate, Appl. Phys. Lett. 62 (1993) 834.

[28] G. Myer and N.M. Amer, Appl. Phys. Lett. 56 (1990) 2100.

[29] J. Ferrante and J.R. Smith, Phys. Rev. B 19 (1979) 3911.

[30] F. Ohnesorge and G. Binnig, Science 260 (1993) 1451.

[31] H.K. Wickramasinghe, private communication.

[32] D. Rugar and P. Hansma, Phys. Today (Oct. 1990) 23.

[33] J.N. Israelachvili and D. Tabor, Proc. R. Soc. (London) A 331 (1972) 19.

[34] T.R. Albrecht, P. Grutter, D. Horne and D. Rugar, Proceedings of the STM'90/NANO I Conference, Baltimore, MD, July 23–27, 1990.

[35] E. Betzig, P.L. Finn and J.S. Weiner, Appl. Phys. Lett. 60 (1992) 2484;
R. Toledo-Crow, P.C. Yang, Y. Chen and M. Vaez-Iravani, Appl. Phys. Lett. 60 (1992) 2957.

[36] C.M. Mate, G.M. McClelland, R. Erlandsson and S. Chiang, Phys. Rev. Lett. 59 (1987) 1942;
T. Miyamoto and R. Kaneko, J. Vac. Sci Technol. A 9 (1991) 1336.

[37] C.C. Williams, W.P. Hough and S.A. Rishton, Appl. Phys. Lett. 55 (1989) 203;
J.R. Matey and J. Blanc, J. Appl. Phys. 57 (1985) 1437.

[38] J.A. Joyce and J.B. Houston, Rev. Sci. Instrum. 62 (1991) 710.

[39] Y. Martin and H.K. Wickramasinghe, Appl. Phys. Lett. 50 (1987) 1445;
D. Rugar, H.J. Mamin, P. Guethner, S.B. Lambert, J.E. Stern, I. McFadyen and T. Yogi, J. Appl. Phys. 68 (1990) 1169.

[40] D. Rugar, C.S. Yannoni and J.A. Sidles, Nature 360 (1992) 563.

[41] A.S. Hou, F. Ho and D.M. Bloom, Electron. Lett. 28 (1992) 2302;
G.E. Bridges and D.J. Thompson, Ultramicroscopy 42–44 (1992) 321.

Surface Science 299/300 (1994) 996–1007
North-Holland

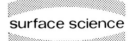

Surface photochemistry

W. Ho

Laboratory of Atomic and Solid State Physics and Materials Science Center, Cornell University, Ithaca, NY 14853-2501, USA

Received 28 April 1993; accepted for publication 19 August 1993

Surface photochemistry with photon energies less than 6.5 eV (wavelength longer than 190 nm) is a relatively new area of research with significant advances made starting about 1980. There has been a steady increase in the number of researchers and papers published in this area. This growth of activity, which combines optics and surface science, is promoted both by interest in technological applications and the desire for further fundamental scientific understanding. This contribution will trace a personal overview of the historical development of modern surface photochemistry.

1. Introduction

Photochemistry of molecules in the gas and liquid phase has been an active area of research for many years [1,2]. Interest in photo-induced processes at solid surfaces and interfaces has its origin mainly in photoelectrochemistry [3,4] and photocatalytic reactions on supported catalysts [5–7]. More recently, photons have been applied to the processing of electronic and optoelectronic materials, including photo-induced metal deposition, etching, oxidation, nitridation, and doping of semiconductors [8–10]. Along with this technological research and development, there arises the need and desire for a fundamental understanding of photo-induced processes on solid surfaces. It is in this direction that surface science, combined with optics and photon spectroscopy, has made the greatest impact. In addition to the optimization of materials growth and processing, photochemistry provides a unique window for viewing the kinetics and dynamics of chemical reactions.

2. Pre-1980 surface photochemistry

In 1974 a general discussion on *Photo-effects in Adsorbed Species* was held at the University of Cambridge and the conference was summarized in a special volume of the Faraday Discussions of the Chemical Society [11]. While one of the three sessions dealt with photoelectron spectroscopy of adsorbed molecules, the remaining two sessions were on (1) photo-adsorption, photo-desorption and photo-reactions at surfaces and (2) photochemistry of adsorbed species. The majority of the work concentrated on oxide surfaces, such as ZnO and TiO_2, and relatively large molecules were adsorbed on the surface. The photochemical processes studied included the oxidation of hydrocarbons (alkanes), the reduction and oxidation of alcohols, the oxidation of CO under ambient mixtures of O_2 and CO, the polymerization of tetrafluoroethylene, and charge transfer interaction with dye molecules. It was realized that photogenerated charge carriers were intimately involved in the photochemistry. In studies of photo-oxidation, questions concerning the charge state of oxygen and its atomic versus molecular form were specifically addressed [12].

Some of the earliest theoretical efforts in understanding photo-induced effects at solid surfaces were carried out by Many [13] and Lichtman and coworkers [14,15]. Their work pointed out the important role that charge carriers from band gap excitation play in photodesorption. The model included the effects due to band bending and impurities and provided a useful framework for discussing photo-induced processes on semiconductors.

Pre-1980 surface photochemistry is characterized predominantly by reactions involving either static or constant flow ambient reactants, and the surfaces are not well characterized. For example, photons have been used in the splitting of water over TiO_2, the reduction of nitrogen over Fe-doped TiO_2 [16], in the isotope enrichment of boron in the presence of a Ti catalyst and Pb metal powders [17], and the oxidation of CO to yield CO_2 on Pt wire [18]. However, interest in obtaining a fundamental understanding has motivated experiments to be carried out under better controlled conditions. High power pulsed lasers were used to thermally desorb molecules from surfaces. Levine et al. measured desorption of H_2O, CO, and CO_2 from W foil [19]. More recently, pulsed laser desorption of Na and Cs was observed from Ge(100) [20], H_2 from the basal planes of Ni [21], and D_2 from polycrystalline W [22].

A definitive book has been written by Ready on the effects of high power irradiation on materials [23]. The results of the calculations on the substrate temperature rise due to nanosecond pulse laser irradiation have been used extensively and verified to be consistent with experiments to date [24,25].

Experiments with lower photon intensities were also carried out. Discharge arc lamps provided an ideal photon source from which light of continuous wavelength in the range 200–700 nm was obtained. The importance of the wavelength dependence of the photoyield was realized. Photodesorption of CO from metal surfaces was observed to have very low cross sections. The substrates used were Fe, Ni, Zr, Mo [26], and W [26–28] ribbons.

Two reviews summarized the work up to 1980 [15,29]. It is possible that in some of these studies the observed photoyields were due to thermal effects or photodesorption from locally oxidized surfaces [15,29]. The presence of impurities, in particular C, is believed to give rise to photo-production of CO_2 from metal, semiconductor, and oxide surfaces [13,15,29]. Furthermore, desorp-

tion of molecules occurs when photons impinge on the walls of the vacuum chamber.

The photochemical experiments performed prior to 1980 were generally carried out under non-ideal conditions arising from insufficient vacuum and inadequate characterization of the substrate surface. Because of the low cross sections encountered in these pioneering experiments on metal surfaces, interests in surface photochemistry was low. Surface scientists, such as Menzel, White, and Ertl, did not continue their investigations. Currently, however, the research groups of White and Ertl are very active in the investigation of surface photochemistry.

3. The beginning of the surge

The electronic industry was the main driving force behind the initial surge of interests in obtaining a basic understanding of photo-induced processes on solid surfaces. In fact, the technology was far ahead of the basic understanding, and this is evident in the papers given at conferences. In particular the Materials Research Society played an important role in hosting the first symposium in November 1982 on *Laser Diagnostics and Photochemical Processing for Semiconductor Devices* [30]. The symposium consisted of eight sections: (1) photodeposition of metal structures; (2) photoetching of electronic materials; (3) diagnostics of semiconductor structures; (4) photoformation of insulators; (5) diagnostics of conventional and laser processing; (6) photodeposition of semiconductors; (7) photochemical doping; and (8) novel processes. There were 28 contributed papers and 7 invited papers, however, none of the papers were given by researchers from the surface science community, in spite of the large number of questions which could be effectively addressed through a surface science approach.

It is this lack of basic understanding of a seemingly important field and the opportunities for making significant contributions which propelled us into the study of photo-induced processes on solid surfaces. The symposium held in the following year [31] had 7 sections similar to those in the first symposium, 47 contributed pa-

pers, and 11 invited papers. Impressive achievements were presented, but again none of the
papers were given by researchers from the surface science community. Similar symposia continued to be hosted by the Materials Research Society in subsequent years. The application of lasers
in thin films and the electronic industry was documented in a recent book edited by Ehrlich and
Tsao [32].

Guided by the importance of photochemistry
in the general field of chemistry, the projection
was that surface photochemistry could become an
equally important component of surface science.
The field of surface photochemistry thus includes
the study of photodesorption, photodissociation,
and photoreaction. Photo-induced surface reactions could provide a new angle for probing the
interactions between and the evolution of
molecules on solid surfaces.

In 1983, the single person who was ahead of
everybody in using ultrahigh vacuum and surface
probes to study the basic mechanisms of photo-
induced processes on solid surfaces was Chuang,
who was an invited speaker in both the first and
second symposia hosted by the Materials Research Society [30,31]. In fact, the first authoritative review was published by Chuang in 1983,
even before other researchers started working in
this field [33]. A comprehensive review of the
field was written and included discussions on the
different mechanisms leading to photo-stimulated
surface processes. In addition, a summary of the
technical applications in catalysis, etching, chemical vapor deposition, and oxidation reactions was
given. While the use of visible and UV photons
was discussed, the major emphasis was on photodesorption and photodissociation of organic and
inorganic molecules, such as pyridine (C_5H_5N)
and SF_6, with infrared irradiation. One of the
main goals was the demonstration of photodesorption arising from resonant vibrational excitation of the adsorbed molecules, an example of
which is shown in Fig. 1 for NH_3 adsorbed on
Cu(100) at 90 K [34,35]. At the time around the
publication of the review by Chuang, the IR work
was on a much firmer foundation than visible and
UV work; there were simply more experiments
performed under controlled conditions with IR

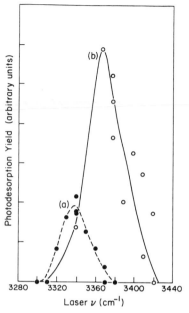

Fig. 1. Photodesorption yields of NH_3 from Cu(100) at 90 K
as a function of IR laser energy for surface coverage of about
(a) one monolayer and (b) multilayer (3.4 layers). (Reprinted
from Ref. [34].)

irradiation. Chuang was a pioneer in using surface science techniques to investigate photo-induced processes on solid surfaces, and his work
and the 1983 review have had a lasting influence
in this field.

Another pioneer and an early spokesman for
the emerging field of surface photochemistry was
Polanyi, from the University of Toronto. Being a
renowned chemist, Polanyi's involvement brought
a certain credibility to the field. His interest in
chemical dynamics led him to the key idea that
the adsorbed state offers a new environment for
studying photochemically-induced chemical reactions. Molecules are adsorbed with a restricted
set of orientations compared to those in the gas
phase and new reaction pathways can be established in the adsorbed state. The phrase "Surface
Aligned Photoreaction" or "SAP" was coined to
describe photochemistry of the adsorbed state in
the second paper published in this field from
Polanyi's group in 1986 [36], with the first in 1984
[37]. The importance of measuring the angular
distribution and the translational energy of the

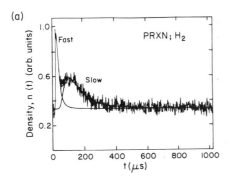

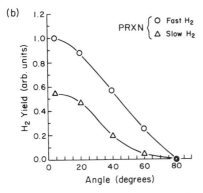

Fig. 2. Photodesorption of H_2 from pulsed laser irradiation at 222 nm of adsorbed H_2S continuously dosed on LiF(100) at 115 K. (a) Time-of-flight distribution: the smooth curves are least squares fits to the data, resolving fast and slow components. (b) Angular distribution of the fast and slow components. (Reprinted from Ref. [38].)

desorbed photoproducts by time-of-flight techniques was emphasized from the very beginning. Extremely valuable information on photochemical dynamics can be obtained from these measurements, as shown in Fig. 2 for H_2 desorption from photo-induced reaction of coadsorbed H_2S on LiF(100) [38]. In addition, Polanyi stressed the need for trajectory studies of photoreactions, in line with his prior work on chemical reaction dynamics in the gas phase [39].

The main emphasis of Polanyi's research has shifted from the dynamics of the gas phase to that of the adsorbed state, as is evident from the publications from his group. This shift reflects the interdisciplinary nature of the field of photo-induced processes on solid surfaces, encompassing researchers from chemistry, engineering, and physics.

Although some of the recent work of Polanyi's group dealt with metal surfaces, in particular Ag(111), the majority of the studies were performed on LiF(001), an insulator with weak interactions with physisorbed molecules [40]. This choice was intentional and based on the notion that such weak interactions would minimize the complications associated with a rigorous understanding of the adsorbate–substrate interaction potential in the trajectory calculations.

Looking at the problem from another angle, we concentrated on elucidating the fundamental mechanisms for photo-induced processes beginning in 1983. Although the various mechanisms had been discussed in the literature, there did not exist experiments which revealed them directly [33]. This lack of understanding of an important aspect of this emerging field was evident from the papers and discussions at the first two symposia sponsored by the Materials Research Society [30,31] and provided the original impetus for our involvement. Taking cues from laser-assisted direct writing of metal lines and thin film deposition on semiconductors using metal carbonyls [30–32] and the significant photoactivity of carbonyls in the UV [41], our first series of experiments used 514 and 257 nm radiation from an Ar ion laser for studying the photodissociation of $Mo(CO)_6$, $W(CO)_6$, and $Fe(CO)_5$ on Si(111)7 × 7 [42]. One of the key steps in this investigation was the direct measurement of the laser-induced substrate temperature rise with a thermocouple; this was made possible with the use of a CW light source. Furthermore, the temperature rise can be calculated from the thermal conduction equation and the results were in excellent agreement with the data [42,43]. These results were crucial in providing direct evidence of the nonthermal, direct adsorbate electronic excitation mechanism for the photodissociation of carbonyls with 257 nm laser irradiation. The experiments also showed how surface probes can be used for the diagnosis of surface photochemistry. Measurement of the temperature rise with pulsed laser irradiation is much more difficult and indirect [44].

In the case of metal carbonyls, direct excitation of the adsorbed molecules was determined to be the mechanism [42,43]. The photons excite the

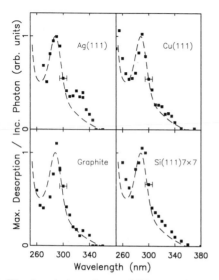

Fig. 3. Wavelength dependence of CO mass signal, normalized to the number of incident photons, from photodissociation of Mo(CO)$_6$ adsorbed on metallic Ag(111) and Cu(111), semimetallic graphite, and semiconducting Si(111)7×7. The dashed curve in each panel is the absorption spectrum of Mo(CO)$_6$ in cyclohexane solution convoluted with the monochromator transmission. (Data for Ag(111) and graphite reprinted from Ref. [45]; data for Cu(111) and Si(111)7×7 reprinted from Ref. [46].)

molecule from the ground electronic state to an excited electronic state which is unstable towards dissociation. The optically induced electronic transition leads to a charge rearrangement within the molecule which leads to nuclear motion and bond breaking. One of the consequences of the direct adsorbate excitation mechanism is that the photoreaction should not depend on the nature of the substrate, as long as the electronic structure of the adsorbate does not change appreciably. The photodissociation of Mo(CO)$_6$ was studied on four different substrates: Ag(111) [45], Cu(111) [46], graphite [45], and Si(111)7 × 7 [46], and the wavelength dependencies of the photoyields were obtained and summarized in Fig. 3. Regardless of whether the substrate is a metal, a semimetal, or a semiconductor, the wavelength dependence of the photoyields show the same distinct resonance centered at 290 nm, which corresponds to a metal-to-ligand charge transfer transition. The unique secondary peak near 330 nm for Ag(111) arises from the enhancement of

the photon field near the surface due to the unique dielectric property of silver, which leads to enhanced optical absorption by the adsorbed molecule. In these experiments, it was shown that the arc lamp has two main advantages, besides the economical costs, over pulsed lasers. First, the CW light allows direct measurement of the temperature rise of the substrate. Second, the wavelength can be chosen conveniently with filters over the range 230–900 nm.

In photocatalytic and photoelectrochemical reactions, charge carriers are produced in the substrate which subsequently interact with the adsorbed species to induce chemical reactions [3–7]. Light induced charge separation is known to be an important step in reactions within biological and chemical systems [47]. For molecules adsorbed on solid surfaces, the substrate can also absorb the photons, producing charge carriers. Those charge carriers near the surface can interact with the adsorbed molecules and cause changes in the charge distribution, which can relax through chemical reaction. The role of photogenerated carriers in photo-induced reactions had been proposed in the literature, however, experimental evidence was lacking [15,33,48].

The wavelength dependence of the photoyield was shown to be an important criterion for distinguishing between different mechanisms, as shown by experiments on metal carbonyls [45,46]. By choosing a semiconductor, charge carrier generation in the substrate should show a significant increase when the photon energy becomes larger than the bandgap. If photo-induced processes depend on charge carriers, the photoyield should show a significant increase as the threshold for bandgap excitation is crossed. However, other effects such as carrier diffusion, relaxation, and energy with respect to the energy levels of the adsorbed molecules need to be considered in addition to the concentration of carriers.

Photogenerated charge carrier induced surface reaction was shown to be the mechanism for NO photodesorption from Si(111)7 × 7 [49]. The wavelength dependence of the photoyield, reproduced in Fig. 4a, was again crucial in attributing the photodesorption and photodissociation of adsorbed NO to adsorbate interactions with photo-

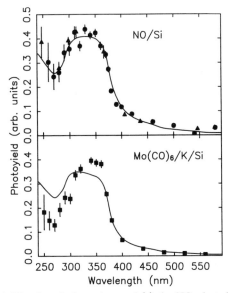

Fig. 4. Wavelength dependence of (a) the NO photodesorption yield for saturation coverage of NO on Si(111)7×7 at 85 K, and (b) CO product from photodissociation of $Mo(CO)_6$ coadsorbed with K on Si(111)7×7 at 85 K. In both cases, the solid line is the best fit using the hot-carrier model, involving the interaction of photogenerated charge carriers with the adsorbed molecules. (Data for NO/Si(111)7×7 reprinted from Ref. [49]; data for $Mo(CO)_6$/K/Si(111)7×7 reprinted from Ref. [50].)

generated hot carriers. The variations in the photoyield follow nicely the wavelength dependence of the silicon substrate optical absorption. In addition, the carriers have to be energetic and created within a distance on the order of the inelastic mean free path from the adsorbate in order to be effective in causing the reactions, i.e. they cannot relax appreciably after being generated. For semiconductors, the photoreaction can involve either electrons or holes, depending on the charge state and the occupied and empty energy levels of the adsorbate. These measurements demonstrated the decisive role surface science can play in contributing to the fundamental understanding of charge transfer induced surface photochemistry and the importance of a systematic approach.

The experiments involving NO on Si(111)7 × 7 led to the supposition that by changing the surface work function, it is possible to induce a different channel for surface reactions due to the changes in the energy levels of the adsorbate relative to those of the substrate. This scheme was realized in the photodissociation of $Mo(CO)_6$ coadsorbed with K on both Cu(111) and Si(111)-7 × 7 [50]. In the presence of K, photogenerated hot electrons attach to the adsorbed $Mo(CO)_6$ and the negatively charged molecules undergo dissociative electron attachment. This is in vivid contrast to the mechanism of direct adsorbate electronic excitation in the absence of K. The wavelength dependence of the photoyield was again instrumental in the analysis of the photochemical mechanism [50]. The similarity in the wavelength dependence of the photoyield between NO/Si(111)7 × 7 [49] and $Mo(CO)_6$/K/Si(111)7 × 7 [50] is striking, as shown by a comparison of Figs. 4a and 4b, especially in view of the results for $Mo(CO)_6$/Si(111)7 × 7 in Fig. 3 [46].

The idea of hot carriers in photo-induced surface processes gained greater acceptance as additional experiments were performed, in particular with the measurement of the translational, vibrational, rotational state distributions, and spin–orbit population of NO photodesorbed from Pt(111) [51]. In these measurements the translational and internal energy distributions of photodesorbed molecules were used to address the mechanisms of photo-induced surface processes. This work was followed by close theoretical collaboration with Gadzuk, who has become one of the main theoretical spokesmen in this field [52,53]. The theory of hot electron mediated desorption was proposed and provided a framework for the discussion and comparison of experimental results. The role of surface states within the hot electron model was demonstrated for NO on Si(111) [54]. These papers from the National Institute of Standards and Technology (NIST) have demonstrated the importance of state-resolved measurements and theoretical modeling in our understanding of the fundamental photochemical mechanisms and adsorbate dynamics.

More recently, Polanyi and coworkers have measured negative ions desorbing from the surface following UV irradiation, giving further evidence for the existence of photo-induced charge transfer photodissociation mechanism [55].

The combination of mass spectrometry, surface analysis, state-resolved measurements, and theoretical calculations, along with a judicious choice of the adsorbate–substrate systems has put the hot carrier mediated mechanism on a firm foundation. The other main mechanism involves direct excitation of the adsorbed molecule from the ground to the excited electronic states.

4. The surge and the present status

There were a few other pioneering investigations which made definitive contributions to the field. State-resolved measurements were reported for NO desorbing from Ni(100)–O [56], which constitutes the first of such measurements on photodesorbed molecules, and NO from photodissociation of N_2O on Pd(111) [57]. More recently, state-resolved measurements were extended to CO photodesorbed from Ni(111)–O [58]. By probing the translational energy distribution of the photodesorbed molecules and varying the number of layers of adsorbed molecules, Cowin and coworkers were able to gain new insights into the dynamics of carrier-induced versus direct adsorbate excitation surface photochemistry [59,60]. A great deal of work has been carried out by White and coworkers at the University of Texas in Austin, culminating in a recent comprehensive review article [61] which can be viewed as the most definitive since the review by Chuang [33]. White was involved in photodesorption experiments even before 1980 [28] but put them aside for a few years; currently White is leading one of the major efforts in the study of surface photochemistry.

The increase in research activities focusing on photo-induced surface processes is reflected in the abundance of review articles which are currently in the literature [33,40,61–72]. The majority of the studies concern photodesorption and photodissociation, which are two elementary steps in photoreaction between adsorbed molecules. One of the most exciting aspects of surface photochemistry, however, lies in the realization of photo-induced reactions between two different types of coadsorbed species on solid surfaces.

Photodesorption, photodissociation, and photoreactions involve the breaking of bonds between the adsorbate and the substrate and within the adsorbed molecule. In addition, new bonds are formed in photoreactions. Thus, an understanding of surface photochemistry provides us with new insights into the dynamics of elementary chemical reactions. Furthermore, the presence of the substrate gives rise to new channels for reactions to occur.

The production of CO_2 from UV-irradiation of O_2 coadsorbed with CO on Pt(111) at 85 K demonstrated the feasibility of such experiments on catalytically active surfaces with submonolayer coverages under ultrahigh vacuum conditions [73]. There were three events which stimulated the choice and realization of this photochemical surface reaction. In a series of papers, Polanyi's group reported results on the translational energy distributions of products from the photodissociation of CH_3Br, HBr, H_2S, and OCS physisorbed on LiF(001) [40]. In addition, photoreactions were observed, arising from the reaction of a photochemically produced H atom with a coadsorbed molecule to yield H_2 and HS or Br [36,38,74,75]. In the case of OCS, it was found that a S atom recoiling into an OCS molecule yielded S_2 and CO [76]. In all these cases, a single type of adsorbate is physisorbed on a relatively inert surface. However, these were the first experiments which revealed the potential richness of photo-induced reactions of oriented molecules on solid surfaces. Thus the conceptual framework for the study of the photochemistry of oriented molecules on solid surfaces is rooted in the work from Polanyi's group. It is not surprising that the impetus for the probing of this new area of chemical reactions comes from someone who has contributed so much to the understanding of gas phase kinetics and dynamics. This is a true reflection of the fact that surface photochemistry will be of interest to a wide audience, and in particular will bind gas phase and surface dynamics increasingly closer.

The second event involves the observation of the photodissociation of O_2 on Pt(111) by White's group [77], which I learned about from a talk given at the 1989 SPIE conference on "Photochemistry in Thin Films". Before hearing these results, we were interested in photochemistry between two coadsorbed species on a metal surface but were frustrated by the fact that the absorption spectra of interesting small molecules in the gas phase lie too far into the UV ($\lambda < 240$ nm). However, these crucial experiments demonstrated the role of the substrate in modifying the electronic properties of the adsorbate–substrate complex, leading to a red-shift such that with near-UV irradiation ($\lambda < 420$ nm) photodissociation of adsorbed O_2 occurs. The presence of the substrate could also provide a different channel of photochemistry which is not present in the gas phase.

The third event which led us to the investigation of $O_2 + CO$ photochemistry on Pt(111) was performed in our laboratory and was not related to photo-induced processes. It was found that when O_2 and CO were coadsorbed on Pt(111) at 85 K, CO_2 was produced by increasing the sample temperature to about 150 K. It is known that at 150 K, both desorption and dissociation occur for O_2. The coincidence in temperature at which these different thermally activated processes occur suggests a mechanism for the CO_2 production. At 150 K, an energetic O atom formed by the O_2 thermal dissociation reacts with an adjacent, coadsorbed CO before losing its energy down into the O chemisorption well. The CO_2 product desorbs and is detected by a mass spectrometer. Unknown to us, similar results were obtained and published by Matsushima [78]; furthermore, the angular distribution of the CO_2 desorption was measured, however, the mechanism involving hot O atoms was not proposed to explain the low temperature CO_2 production peak at 150 K, but reaction with molecular O_2 was noted.

On Pt(111) at 85 K, CO occupies the on-top site when coadsorbed with O_2. It has been found by three research groups that CO_2 is not produced when O_2 coadsorbed with CO on Pd(111) is irradiated with CW arc lamp [79], nanosecond [80] and femtosecond [81] pulsed lasers. This result is surprising in view of the many similarities in the adsorbate–substrate properties to those of Pt(111). However, a possible explanation is that

CO occupies the three-fold hollow site on Pd(111), indicating a structural sensitivity in the surface photochemistry. In the case of photodesorption of NO from Ag(111) and Cu(111), the photoactive species is NO adsorbed in the on-top site; the bridge bonded NO is not photoactive [82].

A number of other photochemical reactions between two different types of coadsorbed species on metal surfaces have since been reported. Irradiation of O_2 coadsorbed with H on Pt(111) leads to the formation of OH and H_2O which remain adsorbed on the surface at 85 K [83]. In the case of H_2S coadsorbed with CO on Cu(111) at 68 K, UV-irradiation leads to the photodesorption of H_2, CO, H_2S, HCO, H_2CO, and the formation of HCO, H_2CO, and OCS on the surface [84]. On Pt(111) at 47 K, CO_2 formation is observed with broadband irradiation of N_2O coadsorbed with CO [85]. Most recently, it has been reported that CH_4 is produced when submonolayers of CH_3Br coadsorbed with H on Pt(111) at 25 K are irradiated with 308 nm laser pulses [86]. Harrison's group at the University of Virginia has also reported the oxidation of CO coadsorbed with O_2 by measuring the translational energy distribution of the CO_2 product [87]. In addition to the photoreaction between chemisorbed O_2 and CO on Pt(111) at 85 K, photoproduction of CO_2 has been observed from irradiation of physisorbed O_2 coadsorbed with CO on Ag(110) at 30 K [88]. These experiments show that with judicious choices of the adsorbate–substrate systems, photochemical reactions can be induced and serve to provide a new perspective into elementary chemical reaction dynamics.

There are different photochemical pathways which are possible for the $O_2 + CO$ reaction on Pt(111), as shown schematically in Fig. 5. The two key issues involved are: (1) the mechanism of the photoexcitation which leads to subsequent bond changes, i.e. direct adsorbate versus carrier-induced excitations, and (2) hot oxygen atom versus molecule reaction. While it is certain that coadsorbed CO is not affected by photon irradiation, the mechanism of O_2 excitation is less clear. This uncertainty arises largely from the limited photon wavelengths ($\lambda > 190$ nm) available in the laboratories. Within the range 1000 nm $> \lambda > 230$ nm,

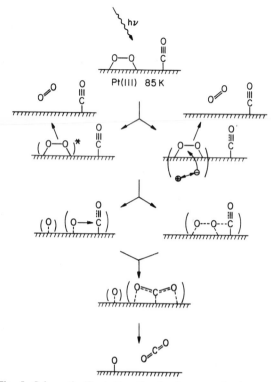

Fig. 5. Schematic illustrating the dominant mechanisms of surface photochemistry. The incident photons can initiate the photochemistry by exciting directly the adsorbed molecules or generating charge carriers, electrons and holes, which interact with the adsorbed molecules. This excitation is followed by desorption, dissociation, or reaction between coadsorbed species. Reaction can proceed via a photogenerated hot atom or an excited molecular precursor. In the case of O_2 coadsorbed with CO on Pt(111) at 85 K, the same products CO_2 and O are formed.

the dependence of the O_2 photodesorption yield from different surfaces closely resembles each other: Ag(110) [89], Pt(111) [73], O_2 coadsorbed with CO on Pt(111) [73], O_2 coadsorbed with NO on Pt(111) [90], Pd(111) [91,92], and Pd(100) [93]. Since the wavelength dependence for O_2 photodissociation and photoreaction closely follows that for O_2 photodesorption [73], the "universal" behavior of the wavelength dependence suggests that any proposed mechanism has to be able to explain this observed behavior. A similar "universal" wavelength dependence has also been observed for NO photodesorbed from Ag(111) [94], Cu(111) [94], Pt(111) [51,95], NO coadsorbed with

O_2 on Pt(111) [90], partially oxidized Ni(111) [96], fully oxidized Ni(111) [96], and Ni(111) preadsorbed with S [96]. For NO, the photodesorption results show a threshold of 1.5 eV compared to 2.5–3.0 eV for O_2. The wavelength dependence for NO, however, is very similar for different metal substrates, which is distinctly different from that for Si [49]. It remains a challenge to come up with a rigorous model for explaining these "universal" behaviors for O_2 and NO photodesorption.

5. Future prospects

The nature of the investigation of surface photochemistry can be distinguished methodologically by the type of photon sources used: continuous-wave arc lamps, nanosecond or femtosecond pulsed lasers. While photochemical mechanisms and general surface photochemistry are most conveniently and fruitfully investigated with arc lamps, nanosecond lasers have been used effectively in state-resolved measurements. It is desirable, however, to extract a basic understanding of the adsorbate–substrate dynamics from state-resolved measurements in addition to a general report of the state distributions [58,97]; theoretical modeling is crucial in reaching this goal. The use of femtosecond lasers to probe the initial stages of photochemical events is only just beginning [98–101]. However, important insights into the microscopic events leading to photodesorption, photodissociation, and photoreaction are starting to be revealed [102]. Furthermore, the branching ratios of surface photochemical reactions can be altered with femtosecond lasers [103].

One of the main issues continues to be the elucidation of the fundamental mechanisms of the initial photoexcitation and the subsequent changes in the chemical bonds. The exact mechanisms giving rise to the "universal" wavelength dependence of O_2 and NO photodesorption yields need to be pinned down. Although different models have been proposed, a rigorous treatment is still lacking. Direct measurements of the electronic transitions of the adsorbate/substrate systems will be crucial in elucidating the mechanisms. Electron energy loss spectroscopy can be applied effectively to probe electronic transitions. However, the subsequent steps following the initial excitation need to be clarified. The best hope may lie in a combined study with CW, nanosecond, and femtosecond light sources and to find trends and universal behavior in a wide range of surface photochemical systems.

The ability to vary the substrate temperature, especially reaching 10–30 K, is important in extending the nature of the adsorbate–substrate systems which can be studied. Low temperatures also allow the adsorption of weakly interacting states or precursor states which can exhibit new channels for photochemical reactions [88]. The quenching rate of the photoexcited states can be altered by the insertion of spacer layers. In addition, it is desirable to vary the substrate from a metal to a semiconductor or an insulator using the same adsorbate in order to isolate substrate effects. Femtosecond lasers, in principle, can be used to probe the dynamics of surface photochemistry in real time. The goal is to follow directly the dynamics of bond breaking and formation. Much can be learned from pioneering work done in the gas phase, however, the substrate plays an important role in surface photochemistry and there are reaction channels which are unique to the adsorbate/substrate system. It is this uniqueness which makes surface photochemistry an interesting and challenging field.

It should be stressed that one of the stumbling blocks in our understanding of surface chemistry is the difficulty in obtaining information on the geometry of the adsorbed molecules as a function of the coverage. For two different types of coadsorbed species, it is desirable to know how the two species are distributed in addition to the knowledge of their bonding geometries. Scanning tunneling microscopy operating at low temperatures is a promising technique for directly viewing the structure of the adsorbed phase.

Our understanding of surface photochemistry can also benefit from the availability of a wider range of photon energies. It is especially desirable to be able to tune the photon energy from below 1 eV up to 10 eV and simultaneously have sufficient flux to induce measurable photochemi-

cal events. While CW arc lamps will continue to serve as valuable and versatile photon sources, rapid advances made in CW and pulsed lasers offer exciting possibilities and hopes of probing new regimes of surface photochemistry: the initial and final states and the transition between them.

What can we hope to learn from the study of surface photochemistry? The emergence of this new field was stimulated initially by the potential applications of photons in the processing and growth of materials. While these technological applications remain today, our basic understanding of this field has grown substantially. This new understanding can be used to optimize technological applications. Photochemistry has been and will continue to be an important field of chemistry and biology. The adsorbed state provides us with a unique medium for probing some of the most fundamental issues concerning chemical reactions, such as charge transfer and rearrangement, and energy flow. In order to understand surface photochemistry in the most detailed way, it is necessary to connect structural properties with both thermal and photochemical reactivities. The final goal is to understand not only surface chemistry but also its relevance to reactions in the gas phase and solutions.

Acknowledgements

Our research in this exciting field of surface photochemistry has benefited from generous support by the Office of Naval Research under Grant No. N00014-90-J-1214, the National Science Foundation under Grant No. DMR-9015823, the Department of Energy under Grant No. DE-FG02-91ER14205, and the MRL Program of the National Science Foundation under Award No. DMR-9121654. In addition, critical readings of this manuscript by Amanda Killen, Robert Pelak, Rowena Young, and Frank Zimmermann is greatly appreciated.

References

[1] J.G. Calvert and J.N. Pitts, Jr., Photochemistry (Wiley, New York, 1966).

[2] H. Okabe, Photochemistry of Small Molecules (Wiley, New York, 1978).

[3] M.S. Wrighton, Chem. Eng. News Vol. 57 (1979) 29.

[4] A.J. Bard, Science 207 (1980) 139.

[5] K. Hauffe, Rev. Pure Appl. Chem. 18 (1968) 79.

[6] T. Freund and W.P. Gomes, Catal. Rev. 3 (1969) 1.

[7] Th. Wolkenstein, Adv. Catal. 23 (1973) 157.

[8] R.M. Osgood and T.F. Deutsch, Science 227 (1985) 709.

[9] I.P. Herman, Chem. Rev. 89 (1989) 1323.

[10] R.L. Woodin, D.S. Bomse and G.W. Rice, Chem. Eng. News, 68 (1990) 20.

[11] Photo-Effects in Adsorbed Species, Faraday Discussions of the Chemical Society, No. 58 (The Faraday Division Chemical Society, London, 1974).

[12] S.J. Teichner and M. Formenti, in: Photoelectrochemistry, Photocatalysis and Photoreactors, Ed. M. Schiavello (Reidel, Dordrecht, 1985) p. 457.

[13] A. Many, CRC Crit. Rev. Solid State Sci. May (1974) 515.

[14] Y. Shapira, R.B. McQuistan and D. Lichtman, Phys. Rev. B 15 (1977) 2163.

[15] D. Lichtman and Y. Shapira, CRC Crit. Rev. Solid State Mater. Sci. 8 (1978) 93.

[16] G.N. Schrauzer and T.D. Guth, J. Am. Chem. Soc. 99 (1977) 7189.

[17] C.T. Lin, T.D.Z. Atvars and F.B.T. Pessine, J. Appl. Phys. 48 (1977) 1720;
C.T. Lin and T.D.Z. Atvars, J. Chem. Phys. 68 (1978) 4233.

[18] R.F. Baddour and J. Modell, J. Phys. Chem. 74 (1970) 1392.

[19] L.P. Levine, J.F. Ready and E.G. Bernal, J. Appl. Phys. 38 (1967) 331.

[20] J.M. Chen and C.C. Chang, J. Appl. Phys. 43 (1972) 3884.

[21] K. Christmann, O. Schober, G. Ertl and M. Neumann, J. Chem. Phys. 60 (1974) 4528.

[22] J.P. Cowin, D.J. Auerbach, C. Becker and L. Wharton, Surf. Sci. 78 (1978) 545.

[23] J.F. Ready, Effects of High-Power Laser Radiation (Academic Press, New York, 1971).

[24] R.B. Hall and S.J. Bares, in: Chemistry and Structure at Interfaces: New Laser and Optical Techniques, Eds. R.B. Hall and A.B. Ellis (VCH, Deerfield Beach, 1986) p. 83.

[25] J.L. Brand and S.M. George, Surf. Sci. 341 (1986) 167.

[26] R.O. Adams and E.E. Donaldson, J. Chem. Phys. 42 (1965) 770.

[27] W.J. Lange, J. Vac. Sci. Technol. 2 (1965) 74.

[28] P. Kronauer and D. Menzel, in: Adsorption–Desorption Phenomena, Ed. F. Ricca (Academic Press, London, 1972) p. 313.

[29] B.E. Koel and J.M. White, in: Interfacial Photoprocesses: Energy Conversion and Synthesis, Ed. M.S. Wrighton, Advances in Chemistry Series, Vol. 184 (American Chemical Society, Washington, D.C., 1980) p. 27.

[30] Laser Diagnostics and Photochemical Processing for Semiconductor Devices, Materials Research Society Symposia Proceedings, Vol. 17, Eds. R.M. Osgood, S.R.J. Brueck and H.R. Schlossberg (North-Holland, New York, 1983).

[31] Laser-Controlled Chemical Processing of Surfaces, Materials Research Society Symposia Proceedings, Vol. 29, Eds. A.W. Johnson, D.J. Ehrlich and H.R. Schlossberg (North-Holland, New York, 1984).

[32] Laser Microfabrication: Thin Film Processes and Lithography, Eds. D.J. Ehrlich and J.Y. Tsao (Academic Press, New York, 1989).

[33] T.J. Chuang, Surf. Sci. Rep. 3 (1983) 1.

[34] T.J. Chuang and I. Hussla, Phys. Rev. Lett. 52 (1984) 2045.

[35] I. Hussla, H. Seki, T.J. Chuang, Z.W. Gortel, H.J. Kreuzer and P. Piercy, Phys. Rev. B 32 (1985) 3489.

[36] E.B.D. Bourdon, P.Das, I. Harrison, J.C. Polanyi, J. Segner, C.D. Stanners, R.J. Williams and P.A. Young, Faraday Disc. Chem. Soc. 82 (1986) 343.

[37] E.B.D. Bourdon, J.P. Cowin, I. Harrison, J.C. Polanyi, J. Segner, C.D. Stanners and P.A. Young, J. Phys. Chem. 88 (1984) 6100.

[38] I. Harrison, J.C. Polanyi and P.A. Young, J. Chem. Phys. 89 (1988) 1498.

[39] J.C. Polanyi, Faraday Disc. Chem. Soc. 84 (1987) 1.

[40] J.C. Polanyi and H. Rieley, in: Dynamics of Gas–Surface Interactions, Eds. C.T. Rettner and M.N.R. Ashfold (Royal Society of Chemistry, London, 1991) p. 329.

[41] G.L. Geoffrey and M.S. Wrighton, Organometallic Photochemistry (Academic Press, New York, 1979).

[42] C.E. Bartosch, N.S. Gluck, W. Ho and Z. Ying, Phys. Rev. Lett. 57 (1986) 1425;
N.S. Gluck, Z. Ying, C.E. Bartosch and W. Ho, J. Chem. Phys. 86 (1987) 4957.

[43] Z. Ying, C.E. Bartosch, N.S. Gluck and W. Ho, J. Vac. Sci. Technol. A 5 (1987) 1608.

[44] J.M. Hicks, L.E. Urbach, E.W. Plummer and H.-L. Dai, Phys. Rev. Lett. 61 (1988) 2588.

[45] S.K. So and W.Ho, J.Chem. Phys. 95 (1991) 656.

[46] Z.C. Ying and W. Ho, J. Chem. Phys. 94 (1991) 5701.

[47] Light-Induced Charge Separation in Biology and Chemistry, Eds. H. Gerischer and J.J. Katz (Verlag Chemie, Weinheim, 1979).

[48] E. Ekwelundu and A. Ignatiev, Surf. Sci. 179 (1987) 119.

[49] Z. Ying and W. Ho, Phys. Rev. Lett. 60 (1987) 57; J. Chem. Phys. 93 (1990) 9089.

[50] Z.C. Ying and W. Ho, Phys. Rev. Lett. 65 (1990) 741.

[51] S.A. Buntin, L.J. Richter, R.R. Cavanagh and D.S. King, Phys. Rev. Lett. 61 (1988) 1321.

[52] J.W. Gadzuk, L.J. Richter, S.A. Buntin, D.S. King and R.R. Cavanagh, Surf. Sci. 235 (1990) 317.

[53] J.W. Gadzuk, Phys. Rev. B 44 (1991) 13466.

[54] L.J. Richter, S.A. Buntin, D.S. King and R.R. Cavanagh, Phys. Rev. Lett. 65 (1990) 1957.

[55] St.-J. Dixon-Warren, E.T. Jensen and J.C. Polanyi, Phys. Rev. Lett. 67 (1991) 2395; J. Chem. Phys. 98 (1993) 5938.

[56] F. Budde, A.V. Hamza, P.M. Ferm, G. Ertl, D. Weide,

P. Andresen and H.-J. Freund, Phys. Rev. Lett. 60 (1988) 1518.

[57] E. Hasselbrink, S. Jakubith, S. Nettesheim, M. Wolf, A. Cassuto and G. Ertl, J. Chem. Phys. 92 (1990) 3154.

[58] M.Asscher, F.M. Zimmermann, L.L. Springsteen, P.L. Houston and W. Ho, J. Chem. Phys. 96 (1992) 4808.

[59] E.P. Marsh, M.R. Schneider, T.L. Gilton, F.L. Tabares, W. Meier and J.P. Cowin, Phys. Rev. Lett. 60 (1988) 2551.

[60] E.P. Marsh, T.L. Gilton, W. Meier, M.R. Schneider and J.P. Cowin, Phys. Rev. Lett. 61 (1988) 2725.

[61] X.-L. Zhou, X.-Y. Zhu and J.M. White, Surf. Sci. Rep. 13 (1991) 73.

[62] D.S. King and R.R. Cavanagh, in: Chemistry and Structure at Interfaces, New Laser and Optical Techniques, Eds. R.B. Hall and A.B. Ellis (VCH, Deerfield Beach, 1986) p. 25.

[63] W. Ho, Comments Cond. Mat. Phys. 13 (1988) 293.

[64] P. Avouris and R.E. Walkup, Annu. Rev. Phys. Chem. 40 (1989) 173.

[65] W. Ho, in: Photochemistry in Thin Films, SPIE Proceedings, OE/LASE '89, Vol. 1056, Eds. T.F. George, J.E. Butler, H.-L. Dai, S.M. George, J.T. Ho and T. Venkatesan (SPIE, Bellingham, 1989) p. 157.

[66] W. Ho, in: Physics and Chemistry of Alkali Metal Adsorption, Eds. H.P. Bonzel, A.M. Bradshaw and G. Ertl (Elsevier, Amsterdam, 1989) p. 159.

[67] D.S. King and R.R. Cavanagh, Adv. Chem. Phys. 76 (1989) 45.

[68] W. Ho, in: Desorption Induced by Electronic Transitions, DIET IV, Eds. G. Betz and P. Varga, Springer Series in Surf. Sci. 19 (1990) 48.

[69] H. Zacharias, Intern. J. Mod. Phys. B 4 (1990) 45.

[70] J.M. White, in Chemistry and Physics of Solid Surfaces VIII, Eds. R. Vanselow and R. Howe (Springer, Berlin, 1990).

[71] W. Ho, Res. Chem. Interm. 17 (1992) 27.

[72] L.J. Richter and R.R. Cavanagh, Prog. Surf. Sci. 39 (1992) 155.

[73] W.D. Mieher and W. Ho, J. Chem. Phys. 91 (1989) 2755; to be published.

[74] J.C. Polanyi and R.J. Williams, J. Chem. Phys. 88 (1988) 3363.

[75] C.-C. Cho, J.C. Polanyi and C.D. Stanners, J. Chem. Phys. 90 (1989) 598.

[76] St.J. Dixon-Warren, I. Harrison, K. Leggett, M.S. Matyjaszczyk, J.C. Polanyi and P.A. Young, J. Chem. Phys. 88 (1988) 4092.

[77] X.-Y. Zhu, S. Hatch, A. Campion and J.M. White, J. Chem. Phys. 91 (1989) 5011.

[78] T. Matsushima, Surf. Sci. 123 (1982) L663; 127 (1983) 403.

[79] J. Yoshinobu and J.T. Yates, Jr., private communication.

[80] E. Hasselbrink, private communication.

[81] J.A. Misewich, private communication.

[82] S.K. So, R. Franchy and W. Ho, J. Chem. Phys. 95 (1991) 1385.

[83] T.A. Germer and W. Ho, J. Chem. Phys. 93 (1990) 1474.

[84] D.V. Chakarov and W. Ho, J. Chem. Phys. 94 (1991) 4075.

[85] J. Kiss and J.M. White, J. Phys. Chem. 95 (1991) 7852.

[86] V.A. Ukraintsev and I. Harrison, J. Chem. Phys. 98 (1993) 5971.

[87] V.A. Ukraintsev and I. Harrison, J. Chem. Phys. 96 (1992) 6307.

[88] R.A. Pelak and W. Ho, to be published.

[89] S.R. Hatch and A. Campion, in: Surface Science of Catalysis: In-Situ Probes and Reaction Kinetics, Eds. D.J. Dwyer and F.M. Hoffmann (American Chemical Society, Washington, 1992) p. 316.

[90] W.D. Mieher, R.A. Pelak and W. Ho, to be published.

[91] F. Weik, A. de Meijere and E. Hasselbrink, to be published.

[92] L. Hanley, X. Guo and J.T. Yates, Jr., J. Chem. Phys. 91 (1989) 7220.

[93] N. Ohta, Y. Ohno and T. Matsushima, Surf. Sci. 276 (1992) L1.

[94] R. Franchy, S.K. So, Z.C. Ying and W. Ho, in: Desorption Induced by Electronic Transitions, DIET IV, Eds. G. Betz and P. Varga, Vol. 19 of Springer Series in Surface Science (Springer, Berlin, 1990) p. 85.

[95] L.J. Richter, S.A. Buntin, R.R. Cavanagh and D.S. King, J. Chem. Phys. 89 (1988) 5344.

[96] J. Yoshinobu, X. Guo and J.T. Yates, Jr., J. Chem. Phys. 92 (1990) 770

[97] F.M. Zimmermann and W. Ho, to be published.

[98] J.A. Prybyla, T.F. Heinz, J.A. Misewich, M.M.T. Loy and J.H. Glownia, Phys. Rev. Lett. 64 (1990) 1537.

[99] F. Budde, T.F. Heinz, M.M.T. Loy, J.A. Misewich, F. de Rougemont and H. Zacharias, Phys. Rev. Lett. 66 (1991) 3024.

[100] J.A. Prybyla, H.W.K. Tom and G.D. Aumiller, Phys. Rev. Lett. 68 (1992) 503.

[101] F.-J. Kao, D.G. Busch, D. Cohen, D. Gomes da Costa and W. Ho, Phys. Rev. Lett. 71 (1993) 2094.

[102] J.A. Misewich, T.F. Heinz and D.M. Newns, Phys. Rev. Lett. 68 (1992) 3737.

[103] F.-J. Kao, D.G. Busch, D. Gomes da Costa and W. Ho, Phys. Rev. Lett. 70 (1993) 4098.

Surface Science 299/300 (1994) 1008–1021
North-Holland

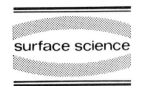

Response theory of interfaces, superlattices and composite materials

Léonard Dobrzynski

CNRS, URA 801, UFR de Physique, Université de Lille I, F-59655 Villeneuve d'Ascq Cedex, France

Received 5 December 1992; accepted for publication 23 March 1993

This short contribution to the Gatos Festschrift on "Surface Science: The First Thirty Years" describes how a simple general theory developed. It is the story of the formation of a useful mathematical language for the study of surfaces, interfaces, superlattices and composite materials. Personal accounts of the evolution of the theory are given and illustrations come mostly from the field of vibrations. This paper tries also to emphasize the benefits wrought by such a fundamental research.

1. Introduction

When looking back at the field "Surface Science", which I only know since 1966, I first remember those who opened this research area for me. I hope the readers will excuse me for expressing this in such a personal manner. But my deep conviction is that the importance of the pioneers is primordial. It is their belief, their hope and their love which opens for many others large new areas to explore. This volume dedicated to Professor Harry Gatos, the founder of the Journal Surface Science, cites many names of scientists who opened an area of the surface science field to others. This paper intends to add or to repeat a few names from my personal reminiscences which are of course subjectives and within the area of response theory.

With a background in Solid State Physics, thanks mainly to the lectures given by Professors Jacques Friedel, Pierre-Gilles de Gennes, Philippe Nozières, André Guinier,... at the Université de Paris, Orsay, I entered the field of surface science from the theoretical side. In 1966, Jacques Friedel proposed me to study the theory of surfaces with the methods used before for point defects in solids and which gave birth in particular to the well-known Friedel sum rule. This point of view is complementary to others and especially to the point of view of many experimentalists who were able to start novel investigations thanks to the progress in ultra high vacuum techniques. I learned at that time from Jacques Friedel that many new experimental results about surfaces and thin films were coming out and that the theoretical language necessary for their interpretation and their understanding remained to be developed.

As I was invited to look mostly at the vibrations of surfaces, my bibliographical researches led me to the founding papers of Lord Rayleigh [1] and Lifshitz and Rosenzweig [2], R.F. Wallis and coworkers [3]. It is remarkable that the theoretical discovery of the surface localized Rayleigh waves published in 1887 found an experimental confirmation for microscopic waves in a crystal only in 1965 [4]. Although Rayleigh's theory has been developed to explain earthquake waves and since that time surface elastic waves have been currently considered in acoustic engineering. Then numerous applications came out very fast and now many electronic devices use Rayleigh waves. This is one of the beauties of fundamental discoveries. Even if they have very little immediate impact, they remain like a precious reserve to be used for the mankind progress by future generations. Lord Rayleigh's study of the vibrations of surfaces was done within the frame of the elasticity theory. The first atomic model for surface vibrations appeared only in 1948 [2] and 1960 [3]. The theoretical developments of surface phonons studies are discussed

SSDI 0039-6028(93)E0276-2

elsewhere [5] in the same volume, as well as the corresponding experimental developments [6,7] which started around 1970.

Although at this time the activity around surface vibrations started to expand very fast and this field was offering many exciting new problems, some were turning already part of their activity towards interface and superlattice investigations. In what follows, I will try to show one way leading from surface science to interface science and composite materials science. When using the word interface one has generally in mind the interface between two different solids. But in order to define a common language valid for any composite material, it is helpful to call interface all the regions in real space bounding different homogeneous parts of a composite material. Such a definition enables one then to address with the same mathematical language the physics of point defects, surfaces, interfaces, superlattices, fiber materials, composite materials, ... This paper focuses, on the development of such a general mathematical language, illustrated by specific examples taken mostly from the field of the vibrations of solids. Section 2 is devoted to atomic models and Section 3 to continuous ones. It is conceptually easier to take the continuous limit of an atomic model than the inverse. It is clear, for example, how to find the elastic limit of a lattice dynamical model. On the other hand, the extension of an elastic model to a lattice dynamical one is not an unambiguous problem.

Having focused this contribution on a general mathematical language, I have to apologize to the authors of many important contributions to the physics of interfaces and superlattices, I do not cite here mostly because I think it is too early to summarize the developments of this field. Let me just cite a few review papers [8–15] on the interface and superlattice vibrations and recall that doping and compositional superlattices were proposed by L. Esaki and R. Tsu in 1969 [16]. Since that time many research and review papers mostly on the electronic properties of these novel materials appeared. They are not reviewed here, as the aim of this paper is only to present the development of the response theory of such materials.

2. Atomic models

The transposition to planar surfaces of the Green function formalism used before for point defects seems to have been done first by Lifshitz and Rosenzweig [2], for a simple lattice dynamical model of a crystal. Taking into account the translational invariance in directions parallel to the surface transforms the study of a planar surface to a uni-dimensional problem for a given value of the propagation vector $k_{\parallel}$ parallel to the surface. The introduction of two independent free surfaces was obtained by breaking all interactions between two adjacent nearest neighbour atomic (001) planes. This cleavage can be represented by a localized perturbation V_0 added to the matrix H_0 of the infinite crystal. Note that we define $H_0 = \omega^2 I - D_0$, where ω is the frequency, I the identity matrix and D_0 the dynamical matrix. The sum of these two matrices

$$h_0 = H_0 + V_0 \tag{1}$$

describes the two adjacent semi-infinite crystals with a free surface. The definition of the Green's or response functions G_0 and g_0 by

$$H_0 G_0 = I \tag{2}$$

and

$$h_0 g_0 = I, \tag{3}$$

where I is the unity matrix, enables to obtain the well-known Dyson equation

$$g_0(I + V_0 G_0) = G_0. \tag{4}$$

This equation permitted the calculation of the surface response functions associated to phonons, electrons, magnons and the corresponding physical properties: localized and resonant states, density of states, thermodynamical functions. Let us cite here only a few early papers [2,17–24], relying on surface response functions and some review papers and books [25–28] describing the development of this mathematical method for surfaces.

Once the surface response functions g_{s1} and g_{s2} for two different semi-infinite crystals were known, it was possible to use them for the study of planar interfaces. This was done on simple phonon [29], electron [30] and magnon [31] models and then for more realistic ones; see for example the review papers [32–34].

Let us define a block diagonal surface response function g_s such as

$$g_s = \begin{pmatrix} g_{s1} & 0 \\ 0 & g_{s2} \end{pmatrix} \tag{5}$$

and

$$h_s g_s = I, \tag{6}$$

where

$$h_s = \begin{pmatrix} h_{s1} & 0 \\ 0 & h_{s2} \end{pmatrix}. \tag{7}$$

Defining the interaction V_I coupling to the two semi-infinite crystals, the matrix describing the coupled crystals is

$$h = h_s + V_I \tag{8}$$

and the corresponding response function g defined by

$$hg = I \tag{9}$$

can be calculated with the help of Eqs. (6), (8) and (9) from

$$g(I + V_I g_s) = g_s. \tag{10}$$

This manner of studying planar interfaces was extended later to systems having two periodic planar interfaces, namely superlattices [13,35].

However it is much simpler to calculate directly the response function g for a material with an interface from the bulk response function G_{01} and G_{02} rather than from the surface response functions g_{s1} and g_{s2} [36]. In order to achieve this, one has to remark that in Eq. (4), g_0 as inverse of a block diagonal matrix h_0 (Eq. (3)) is also a block diagonal matrix. Let us call respectively A_{si} and G_{si} the truncated parts of $A_{0i} = V_{0i} G_{0i}$ and G_{0i}, $i = 1$ and 2 belonging to the same infinite crystal one wishes to keep, namely

$$A_{0i} = \begin{pmatrix} A_{si} & \\ & \end{pmatrix} \tag{11}$$

and

$$G_{0i} = \begin{pmatrix} G_{si} & \\ & \end{pmatrix}. \tag{12}$$

This enables to rewrite Eq. (4) in the following form

$$g_{si}(I + A_{si}) = G_s, \tag{13}$$

where all matrix elements belong now to the semi-infinite crystal one considers. It is possible to eliminate g_s from Eq. (10) with the help of Eq. (13), and to rewrite this equation as

$$g(I + A) = G \tag{14}$$

where

$$A = A_s + V_I G, \tag{15a}$$

$$A_s = \begin{pmatrix} A_{s1} & 0 \\ 0 & A_{s2} \end{pmatrix}, \tag{15b}$$

and

$$G = \begin{pmatrix} G_{s1} & 0 \\ 0 & G_{s2} \end{pmatrix}. \tag{15c}$$

It is finally straightforward to generalize Eq. (14) to any composite discrete material [36], constructing G and A_s of truncated parts of the G_{0i} and the $A_{0i} = V_{0i} G_{0i}$, $i = 1 \ldots N$ and by defining V_I as the interaction operator connecting the N homogeneous parts.

The interface response operator A defined by Eq. (15a) has non-zero elements only between point x of the total interface domain M and any other point x' of the space of definition of the composite material D. We then introduce the corresponding rectangular matrix $A(MD)$ which contains all the non-zero elements of A and similar notations for all the other operators.

The general Eq. (14) can then be written as

$$g(DD) = G(DD) - G(DM)\Delta^{-1}(MM)A(MD), \tag{16}$$

where

$$\Delta(MM) = I(MM) + A(MM). \tag{17}$$

This interface response theory enables also to obtain the deformations $\langle u|$ of the composite system when it is submitted to an action $\langle F|$ so that

$$\langle u|h = \langle F|,$$

where $\langle u|$ and $\langle F|$ are row vectors.

When the action on the system is nil, then $\langle u|$ is the eigenvector corresponding to the eigenvalue E. Applying the same action on both sides of the Eq. (16) provides

$$\langle u(D)| = \langle U(D)| - \langle U(M)|\Delta^{-1}(MM)A(MD), \tag{18}$$

where

$$\langle U(D)| = \langle F(D)|G(DD) \tag{19}$$

are the deformations of the reference system. Note that the action can be localized in one subsystem, and even at one single point, or be extended to the whole space of the composite.

So the knowledge of the deformation $|U(D)\rangle$ of the reference system and that of the scattering matrix $\Delta^{-1}(MM)A(MD)$ enable to obtain the deformation of the composite system.

Eq. (18) can also be used [37] for the determination of the eigenvectors corresponding to the eigenvalues E of the operator h.

Without trying to be exhaustive about the numerous applications of this general theory, let us cite two papers giving analytical examples [38,39] enabling to understand this general theory better. One application is presented in Appendix A. Its reading may help to appriciate better the general theory presented here. The importance of very simple models for the opening and development of new research areas seems to be very general in physics. This appears also very clearly when one opens classical textbooks for graduate and undergraduate students.

3. Continuous models

In order to take the continuous space limit of the above formalism established for discrete models, let us first derive within the above discrete formulation, an useful general relation. For this purpose define in the total interface domain M of the composite system the inverse $g^{-1}(MM)$ of the matrix $g(MM)$. With the help of Eqs. (15a), (16) and (17) one easily shows that for any composite system

$$g^{-1}(MM) = g_s^{-1}(MM) + V_I(MM),\tag{20}$$

where $g_s^{-1}(MM)$ is the block diagonal matrix formed out of all the $g_s^{-1}(M_iM_i)$, $1 \leq i \leq N$, the inverse in M_i of the surface response functions $g_s(M_iM_i)$.

This coupling operator V_I has no counterpart in continuum theories. Indeed, the usual continuum theories do not take into account the variations of the particle interactions on both sides of an interface between two different materials. The continuum theories just assume that the two materials in contact have certain values for their physical parameters on one side of the interface of zero thickness and other values on the other side. So it is clear from this physical discussion that the atomic parameters entering V_I have to vanish in the equations when one takes the continuum limit of an atomic theory.

Another interesting feature of Eq. (20) is that the other contribution than V_I to the inverse interface response functions $g^{-1}(MM)$ is just the block diagonal free surface $g_s^{-1}(MM)$. So for example for a discrete medium and for x and x' in the interface domain M_i of the same subsystem i

$$g^{-1}(x, x') = g_{si}^{-1}(x, x') + V_I(x, x'), \quad x, x' \in M_i \tag{21}$$

and when x and x' are in the interface domains of two different subsystems

$$g^{-1}(x, x') = V_I(x, x'), \quad x \in M_i, \ x' \in M_{i'} \quad i \neq i' \tag{22}$$

and usually V_I has non-zero matrix elements only between particles separated by a few atomic distances on each side of adjacent interface domains M_i and $M_{i'}$.

So if one wants the continuum limit of Eqs. (20), (21) and (22) of the discrete theory, it is clear that the coupling operator V_I contributions will vanish and that one has to consider different cases depending on the position of x and x' in the total interface domain M. Remember that in a complex composite system within a continuous space, one has to distinguish in general N subsystems contained in their respective domains D_i, $1 \leq i \leq N$. Each of these subsystems is bounded by an infinitely thin interface domain M_i, adjacent in general to J ($1 \leq j \leq J$) other subdomains through subinterface domains M_{ij}. Part of these subinterface domains M_{ij} may separate the D_i subsystem from the vacuum. In the discrete theory each of all the discrete interface points x_{ij} belong necessarily to only one subinterface domain M_{ij}. In continuous theories, each continuous interface point x_{ij} can be either a free surface point situated on the interface M_{ij} bounding the composite system with the vacuum, or an interface point belonging not only to the subinterface M_{ij} but also to another or several other subsystems $i' \neq i$. This continuous interface point x_{ij} can be viewed as the common continuous limit of two or several interface points of the corresponding discrete theories.

Consider now two such continuous interface points x_{ij} and x'_{ij} of the total interface domain M and their corresponding discrete points in the discrete theory. It is clear from Eq. (21), from the fact that V_I

has no counterpart in continuous theories and that the $g_{si}^{-1}(x, x')$ have non-zero elements only for x, $x' \in D_i$ that [40]

$$g^{-1}(M_{ij}, M_{i'j'}) = 0, \qquad M_{i'j'} \notin M_i, \tag{23a}$$

$$g^{-1}(M_{ij}, M_{ij'}) = g_s^{-1}(M_{ij}, M_{ij'}), \qquad j' \neq j, \tag{23b}$$

$$g^{-1}(M_{ij}, M_{ij}) = \sum_{i'} g_s^{-1}(M_{i'j'}, M_{i'j'}), \qquad M_{i'j'} \equiv M_{ij}. \tag{23c}$$

The first of the above results is a direct consequence of Eq. (22) and of the fact that $V_1(MM)$ does not exist in continuum theories. The Eq. (23b) results from Eq. (21). The last result comes also from Eq. (21), bearing in mind that in continuum theories an interface point belongs in general to several different subsystems i and that we deal with a linear response theory.

For an interface between two different media, the result (23c) can be obtained with the Green's theorem and all the ingredients for its derivation were already available in 1953 in a classical text [41]. In different forms, the equations (23) appear also in the surface green's function matching (SGFM) theory for one single interface and two single interface quantum well structures [42] and had also two other independent derivations [43,44]. The extension to multiple periodic interfaces of the SGFM theory as well as its transposition from continuous to discrete media was also achieved [45]. This theory starts from the finite medium with multiple periodic interfaces and then introduces the reflection and transmission amplitudes at the interfaces. The interface response theory presented in this paper starts from the different incomplete crystals and then couples them to construct the composite structure under study. Although all these approaches are different, if at the end they address the same physical system their results have to be equivalent.

Knowing how to obtain the inverse in the total interface space of $g(MM)$, we now have to explain how to obtain the $g_s^{-1}(M_i M_i)$ appearing in the Eqs. (23). The continuum version of Eq. (2) is in the infinite space of medium i

$$H_{0i}(x)G_{0i}(x, x') = I\delta(x - x'), \tag{24}$$

where the bulk response function $G_{0i}(x, x')$ is now a function of the two continuous variables x and x'.

The surface response operator A_{si} as defined above for discrete systems (matrix elements of Eq. (11) inside D_i) becomes in the continuum theory,

$$A_{si}(x, x') = V_{0i}(x'')G_{0i}(x'', x')|_{x''=x}, \quad x', x'' \in D_i, x \in M_i. \tag{25}$$

The expressions for the continuous versions of the cleavage operator V_{0i} can be obtained by introducing in the differential form $H_{0i}(x)$ a Heaviside function describing the discontinuity at the surface [40]. In Appendix B is given an example illustrating how to obtain the cleavage operator.

This surface is taken to be the continuous limit of a discrete one obtained by cutting all interactions between two different parts of an infinite crystal. It is therefore a perfectly reflecting surface for the elementary excitations described by $H_{0i}(x)$. Explicit expressions for $V_{0i}(x)$ were obtained in elasticity theory [40], for the Maxwell equations [46] and the Schrödinger equation for nearly free electrons [47].

Eq. (13) of the discrete interface response theory becomes in the continuous theory the following "surface" integral equation:

$$g_{si}(x, x') + \int dx'' \, g_{si}(x, x'')A_{si}(x'', x') = G_{0i}(x, x'), \quad (x, x') \in D_i \text{ and } x'' \in M_i. \tag{26}$$

The solutions of the integral Eq. (26) have in general to be obtained by numerical methods. One of these numerical methods may be to transform the integral Eq. (26) into a matrix equation by a suitable choice of a discrete number of values for x and $x' \in D_i$. The use of this method brings us to the same matrix equation for Eq. (26) as for Eq. (16) of the discrete theory.

The $g_s^{-1}(M_iM_i)$ needed in Eq. (23) can be finally obtained with the help of Eqs. (16) and (17) in the following form

$$g_s^{-1}(M_iM_i) = \Delta_s(M_iM_i)G_0^{-1}(M_iM_i), \tag{27}$$

where

$$\Delta_s(M_iM_i) = I(M_iM_i) + A_{si}(M_iM_i). \tag{28}$$

Once the $g^{-1}(MM)$ is obtained for a given composite material from Eq. (23), one can calculate by matrix inversion $g(MM)$. Any other element of $g(DD)$ can then be found as explained below. Let us introduce first an interface scattering matrix $T(MM)$ by the following relation

$$g(DD) = G(DD) + G(DM)T(MM)G(MD). \tag{29}$$

A particular value of the above equation for $g(MM)$ enables to obtain easily

$$g(DD) = G(DD) + G(DM)G^{-1}(MM)[g(MM) - G(MM)]G^{-1}(MM)G(MD). \tag{30}$$

The first demonstration of Eqs. (27) and (30) was given for a planar surface within the surface Green's function matching theory [42].

In the same manner as for the discrete systems, the deformations $|u(D)\rangle$ of a continuous system are obtained [37] from the deformation $|u(D)\rangle$ of the reference system, when multiplying by a given action $|F\rangle$ both sides of Eq. (30), namely

$$|u(D)\rangle = |U(D)\rangle - G(DM)G^{-1}(MM)|U(M)\rangle$$
$$+ G(DM)G^{-1}(MM)g(MM)G^{-1}(MM)|U(M)\rangle. \tag{31}$$

Also this expression can be used for the calculation of eigenvectors corresponding to given eigenvalues.

A few simple and analytical applications of the above theory can be found in Refs. [40,46,47].

4. Summary and prospectives

This overview of the development of a mathematical language valid for surfaces, interfaces and composite materials is of course subjective. For clarity reasons mostly, all the very helpful interactions with other formulations of surface and interface linear response theories [45,48,49] were not fully described here. These interactions were however very stimulating. It seems important to approach a given problem from different points of view. Such exchanges are often very fruitful.

The main benefit from the theoretical developments described here which led from surface science to composite material science seems to be the formulation of a mathematical tool which should be useful for many novel investigations to come. These theoretical developments were also motivated by the appearance of an increasing number of man-tailored novel composite materials. Although the first production of most of these materials was achieved within an experimental and industrial logic, their improvements will certainly benefit from a better theoretical understanding of their physical properties.

Appendix A: Response functions for an adsorbed slab in lattice dynamics

As an application of the general theory summarized in Section 3, we derive here the response function for an adsorbed slab using a simple phonon model.

A.1. Bulk phonon model

We start from an infinite simple cubic lattice of atoms of mass m_i. Let $u_\alpha(n)$ denote the α ($= 1, 2$ or 3) component of the displacement of the atom at lattice site $x(n) = a_0(n_1 \hat{x}_1 + n_2 \hat{x}_2 + n_3 \hat{x}_3)$ where a_0 is the lattice parameter and $\hat{x}_1$, $\hat{x}_2$ and $\hat{x}_3$ unit vectors. The potential energy Φ associated with the lattice vibrations of the model considered here has the form

$$\Phi_i = \tfrac{1}{2}\beta_i \sum_n \sum_p \sum_\alpha \left[u_\alpha(n) - u_\alpha(n+p) \right]^2, \tag{A.1}$$

where n ranges over all sites of the crystal, and p over the six nearest sites of the atom n.

This model is not rotationally invariant and does not give rise to Rayleigh surface waves on a (001) surface. Nevertheless, these deficiencies are unimportant for the qualitative study of many physical properties of surfaces and interfaces and in particular for the study of the transverse polarized modes we will consider here [25,27].

From the above form of the potential energy and by assuming a sinusoidal time dependence for the displacements, we obtain three uncoupled equations of motion, which we can write in the form

$$\sum_{n'} H_{0i}(nn'; \omega^2) u_\alpha(n') = 0, \quad \alpha = 1, 2 \text{ or } 3, \tag{A.2}$$

where

$$H_{0i}(n, n'; \omega^2) = \left(\omega^2 - 6\frac{\beta_i}{m_i} \right) \delta_{nn'} + \frac{\beta_i}{m_i} \sum_p \delta_{n, n'+p}. \tag{A.3}$$

The three times degenerated bulk phonon dispersion relation is

$$\omega^2 = 2\frac{\beta_i}{m_i}(3 - \cos k_1 a_0 - \cos k_2 a_0 - \cos k_3 a_0), \tag{A.4}$$

where k is the propagation vector.

A.2. Bulk response function

The bulk vibrational properties of the above crystal can be studied with the help of its bulk response function defined by

$$H_{0i} G_{0i} = I, \tag{A.5}$$

where I stands for the unit matrix.

Taking now advantage of the periodicity of the system in directions parallel to the (001) plane, we introduce the following two-dimensional vectors.

$$x_\parallel(n) = a_0(n_1 \hat{x}_1 + n_2 \hat{x}_2), \tag{A.6}$$

$$k_\parallel(n) = k_1 \hat{x}_2 + k_2 \hat{x}_2, \tag{A.7}$$

and a Fourier transformation of the response function

$$G_{0i}(n, n'; \omega^2) = \frac{1}{N^2} \sum_{k_\parallel} G_{0i}(n_3, n_3'; k_\parallel \omega^2) \exp\left[ik_\parallel(x_\parallel - x_\parallel')\right], \tag{A.8}$$

where N^2 is the number of atoms in a (001) plane.

The corresponding bulk response function may be easily obtained in closed form [27]

$$G_{0i}(n_3, n_3'; k_{\parallel}, \omega^2) = \frac{m_i}{\beta_i} \frac{t_i^{|n_3 - n_3'| + 1}}{t_i^2 - 1},$$ (A.9)

with

$$t_i = \begin{cases} \xi_i - (\xi_i^2 - 1)^{1/2}, & \xi_i > 1, \\ \xi_i + i(1 - \xi_i^2)^{1/2}, & -1 < \xi_i < 1, \\ \xi_i + (\xi_i^2 - 1)^{1/2}, & \xi_i < -1, \end{cases}$$ (A.10)

and

$$\xi_i = 3 - \cos k_1 a_0 - \cos k_2 a_0 - \frac{m_i}{2\beta_i}(\omega^2 + i\epsilon),$$ (A.11)

where ϵ is an infinitely small positive number.

A.3. The surface response operators

Let us now create a slab ($i = 2$) by removing in the infinite crystal all interactions between the atoms situated in the $n_3 = 0$ and $n_3 = 1$ planes and also between those situated in the $l_3 = L$ and $L + 1$ planes. The corresponding cleavage operator V_{02} which when added to H_{02} gives the dynamical matrix h_{02} of the slab and of the two semi-infinite crystals is

$$V_{02}(n_3 n_3') = \frac{\beta_i}{m_i}\left(\delta_{n_3 0}\delta_{n_3' 0} + \delta_{n_3 1}\delta_{n_3' 1} - \delta_{n_3 0}\delta_{n_3' 1} - \delta_{n_3 1}\delta_{n_3' 0}\right)$$

$$+ \frac{\beta_i}{m_i}\left(\delta_{n_3 L}\delta_{n_3' L} + \delta_{n_3 L+1}\delta_{n_3' L+1} - \delta_{n_3 L}\delta_{n_3' L+1} - \delta_{n_3' L+1}\delta_{n_3' L}\right).$$ (A.12)

Define now the surface response operator A_{s2} associated to this slab as formed out of the elements of

$$\boldsymbol{A}_{02} = \boldsymbol{V}_{02}\boldsymbol{G}_{02},$$ (A.13)

belonging only to the slab, namely

$$A_{s2}(n_3 n_3') = -\frac{1}{t_2 + 1}\left(\delta_{n_3 1}t_2^{n_3'} + \delta_{n_3 L}t_2^{L - n_3' + 1}\right), \quad 1 \leq n_3, n_3' \leq L.$$ (A.14)

In the same manner, one obtains the surface response operators for the semi-infinite crystal

$$A_{s1}(n_3 n_3') = -\delta_{n_3 0}\frac{t_1^{1 - n_3}}{t_1 + 1}, \quad n_3, n_3' \leq 0.$$ (A.15)

The above results (A.9)–(A.15) will enable us to derive in the next section the response function for the adsorbed slab.

A.4. The response function

The system under study here is formed out of the slab ($i = 2$, $1 \leq n_3 \leq L$) adsorbed on the semi-infinite crystal ($i = 1$, $n_3 \leq 0$).

We define the reference function $\mathbf{G}$ (Eq. (15c)) for the composite system under study here, as a block diagonal matrix whose non-zero elements are given by

$$G\left(n_3, n_3'; \mathbf{k}_{\parallel}\omega^2\right) = \begin{cases} G_{01}\left(n_3, n_3'; \mathbf{k}_{\parallel}\omega^2\right), & n_3, n_3' \leq 0, \\ G_{02}\left(n_3, n_3'; \mathbf{k}_{\parallel}\omega^2\right), & 1 \leq n_3, n_3' \leq L. \end{cases} \tag{A.16}$$

The interface coupling operator which binds the two free surface crystals together through the interaction β_1 between the interface atoms is

$$V_1(n_3, n_3') = -\frac{\beta_1}{m_1}\left(\delta_{n_30}\delta_{n_3'0} - \delta_{n_30}\delta_{n_3'1}\right) - \frac{\beta_1}{m_2}\left(\delta_{n_31}\delta_{n_3'1} - \delta_{n_31}\delta_{n_3'0}\right). \tag{A.17}$$

The interface response operator can now be calculated from Eq. (15a), where the non-zero elements of the surface response operators $\mathbf{A}_s$ are given by

$$A_s(n_3, n_3') = \begin{cases} A_{s1}(n_3 n_3'), & (n_3, n_3') \leq 0, \\ A_{s2}(n_3 n_3'), & 1 \leq (n_3, n_3') \leq L, \end{cases} \tag{A.18}$$

where the values of the A_{si} are taken from Eqs. (A.14)–(A.15).

The response function $\mathbf{g}$ of the adsorbed slab can now be calculated from the general Eqs. (14) and (16). For the evaluation of the matrix elements of $\mathbf{g}$, this equation becomes

$$g(n_3, n_3') = G(n_3, n_3') - \sum_{m,m' \in M} G(n_3, m)\Delta^{-1}(m, m')A(m', n_3'), \tag{A.19}$$

where M stands for the total interface, namely $m, m' = 0, 1, L$ and Δ is defined by

$$\Delta(m, m') = \delta_{mm'} + A(m, m'), \quad m, m' \in M. \tag{A.20}$$

In the above Eqs. (A.18)–(A.20) and in what follows, the dependence on $\mathbf{k}_{\parallel}$ and ω is not explicitly written for simplicity.

When calculating the determinant of Δ, it is convenient to define it as being proportional to

$$W = \left\{1 - \beta_1\left[\frac{t_2}{\beta_2(t_2 - 1)} + \frac{t_1}{\beta_1(t_1 - 1)}\right]\right\} - t_2^{2L}\left\{1 + \beta_1\left[\frac{1}{\beta_2(t_2 - 1)} - \frac{t_1}{\beta_1(t_1 - 1)}\right]\right\}. \tag{A.21}$$

The localized states due to an adsorbed slab 2 on the crystal 1 are obtained from $W = 0$ and are given by

$$2\left(1 - \frac{\beta_1}{\beta_1}\frac{t_1}{t_1 - 1}\right)\sinh q_2 L - \frac{\beta_1}{\beta_2}\frac{\cosh q_2\left(L - \frac{1}{2}\right)}{\sinh q_2/2} = 0, \quad \text{with } q_2 \neq i\pi. \tag{A.22}$$

Finally all the matrix elements of $\mathbf{g}$ can be expressed by the following closed form expressions

$$g(n_3, n_3') = \frac{m_1}{\beta_1}\frac{t_1^{|n_3 - n_3'| + 1}}{t_1^2 - 1} + \frac{m_1}{\beta_1}\frac{t_1^{2 - (n_3 + n_3')}}{t_1^2 - 1}\frac{1}{W}$$

$$\times \left\{\left[1 + \beta_1\left(\frac{1}{\beta_1(t_1 - 1)} - \frac{t_2}{\beta_2(t_2 - 1)}\right)\right]\right.$$

$$\left. - t_2^{2L}\left[1 + \beta_1\left(\frac{1}{\beta_1(t_1 - 1)} + \frac{1}{\beta_2(t_2 - 1)}\right)\right]\right\}, \quad \text{for } n_3, n_3' \leq 0; \tag{A.23}$$

$$g(n_3, n_3') = -\frac{\beta_1 m_2 t_1^{1 - n_3}}{\beta_1\beta_2(t_1 - 1)(t_2 - 1)W}\left(t_1^{n_3'} + t_2^{2L + 1 - n_3'}\right), \quad \text{for } n_3 \leq 0 \text{ and } 1 \leq n_3' \leq L; \tag{A.24}$$

for $1 < n_3 \leq L$ and $n_3' \leq 0$, $g(n_3, n_3')$ is obtained just by interchanging the indices n_3 and n_3' in the right hand side of Eq. (A.24);

$$g(n_3, n_3') = \frac{m_2}{\beta_2} \frac{t_2^{|n_3 - n_3'| + 1}}{t_2^2 - 1} + \frac{m_2}{\beta_2} \frac{t_2^{n_3'}}{t_2^2 - 1} \frac{1}{W} \left[1 + \beta_1 \left(\frac{1}{\beta_2(t_2 - 1)} - \frac{t_1}{\beta_1(t_1 - 1)} \right) \right] \left[t_2^{n_3} + t_2^{2L + 1 - n_3} \right]$$

$$+ \frac{m_2}{\beta_2} \frac{t_2^{(L + 1 - n_3')}}{t_2^2 - 1} \frac{1}{W} \left\{ t_2^{n_3} \left[1 + \beta_1 \left(\frac{1}{\beta_2(t_2 - 1)} - \frac{t_1}{\beta_1(t_1 - 1)} \right) \right] \right.$$

$$\left. + t_2^{L + 1 - n_3} \left[1 - \beta_1 \left(\frac{t_1}{\beta_1(t_1 - 1)} + \frac{t_2}{\beta_2(t_2 - 1)} \right) \right] \right\}, \quad \text{for } 1 \leq n_3, n_3' \leq L. \quad (A.25)$$

The above results enable to study localized phonons within absorbed slabs. An application to a more realistic model for thin epitaxial layers on (001) surfaces of bcc metals appeared recently [50], together with a fast numerical procedure for calculating the bulk response function. Resonant phonons in adsorbed slabs were also studied recently [51] with the help of the above response function. Such resonant phonons were observed recently [52] by inelastic helium atom scattering for sodium epitaxial multilayers on Cu (001) and interpreted as organ-pipe modes.

Appendix B: Any composite electronic system in the effective mass approximation

First the response function for an infinite homogeneous electronic medium will be defined, within the effective mass approximation. Then a material limited by one or several perfect reflecting surface(s) will be considered. Any composite electronic system can then be constructed out of such different pieces and its response function calculated.

B.1. One infinite electronic material

Within the free electron approximation, the Hamiltonian of a given infinite material i is

$$H_i(x) = \tfrac{1}{2} p_i v_i + E_i, \quad (B.1)$$

where $p_i = m_i v_i$ is the impulse of an electron of effective mass m_i, v its velocity, E_i a constant potential energy and x the space position. In quantum mechanics the corresponding Schrödinger equation is

$$H_i(x) = -\frac{\hbar^2}{2m_i} \nabla \nabla + E_i, \quad (B.2)$$

where $\hbar$ is the Planck constant.

The corresponding bulk response function $G_i(x, x')$ is defined by

$$[E - H_i(x)] G_i(x, x') = \delta(x - x'), \quad (B.3)$$

where $\delta(x - x')$ is the usual δ function and E the energy.

The solution of Eq. (B.3) for the bulk response function is straightforward.

B.2. An electronic material bounded by perfectly reflecting surfaces

Let us consider first an electronic material bounded by one perfectly reflecting surface M_i defined by

$$x_3 = f(x_1 x_2). \quad (B.4)$$

The material is situated in the space D_i such that $x_3 \geq f(x_1 x_2)$.

The assumption of complete reflection of the electrons at the surface M_i can be introduced in the following manner in the corresponding Hamiltonian, defined in D_i

$$h_{si}(x) = \tfrac{1}{2} p_i \left[\theta(x_3 - f(x_1 x_2)) v_i \right] + \theta(x_3 - f(x_1 x_2)) E_i, \tag{B.5}$$

where

$$\theta(x_3 - f(x_1 x_2)) = \begin{cases} 1, & \text{for } x_3 \geq f(x_1 x_2) \\ 0, & \text{for } x_3 < f(x_1 x_2). \end{cases} \tag{B.6}$$

This imposes to the electron velocity to vanish for $x_3 < f(x_1 x_2)$, in agreement with the assumption of a perfectly reflecting surface. In quantum mechanics Eq. (B.5) becomes in D_i

$$h_{si}(x) = -\frac{\hbar^2}{2m_i} \nabla \left[\theta(x_3 - f(x_1 x_2)) \nabla \right] + \theta(x_3 - f(x_1 x_2)) E_i, \tag{B.7}$$

or

$$h_{si}(x) = \theta(x_3 - f(x_1 x_2)) \left[\frac{\hbar^2}{2m_i} \nabla \nabla + E_i \right] + \delta(x_3 - f(x_1 x_2)) V_i(x), \quad \text{in } D_i, \tag{B.8}$$

where

$$V_i(x) = -\frac{\hbar^2}{2m_i} \left[-\frac{\partial f(x_1 x_2)}{\partial x_1} \frac{\partial}{\partial x_1} - \frac{\partial f(x_1 x_2)}{\partial x_2} \frac{\partial}{\partial x_2} + \frac{\partial}{\partial x_3} \right] \tag{B.9}$$

is the cleavage operator, in D_i. Define now the response function $g_{si}(x, x')$ by

$$[E - h_{si}(x)] g_{si}(x, x') = \delta(x - x'). \tag{B.10}$$

Once the bulk response function $G_i(x, x')$ is determined by Eq. (B.3), then Eqs. (25), (27) and (28) enable us to calculate the $g_{si}^{-1}(M_i M_i)$.

B.3. Any composite electronic material

When all these independent submaterials are allowed to interact, the Fermi energy has to be the same in the whole composite material. This is easily obtained by taking this Fermi energy as the energy origin in all independent pieces. Necessary for the translations of the atomic levels E_i are functions of the occupations z_i of the band and of their shapes. The calculation of these shifts requires the knowledge of the bulk density of states of each material. Charge rearrangements near the interfaces usually appear and are responsible for these energy shifts; they may create also interface self-consistency potentials. So an exact calculation requires a self-consistent procedure, which can be included in the theory outlined above.

Once the $g_{si}^{-1}(M_i M_i)$ are determined from Eq. (27) for all independent submaterials, one easily obtains the interface elements $g^{-1}(MM)$ of the composite medium from Eq. (23) and then any element of g from Eq. (30).

B.4. Layered materials

For a composite made out of different slabs, it is helpful to use a Fourier transformation in order to take advantage of the translation symmetry parallel to the slabs, Eq. (B.3) defining the bulk response function then becomes

$$\frac{F_i}{a_i \alpha_i} \left(a_i^2 \alpha_i^2 + \frac{\partial}{\partial z^2} \right) G_{0i}(k_\parallel E | z - z') = \delta(z - z'), \quad -\infty < z, z' < +\infty \tag{B.11}$$

with $x_3 = za_i$

$$\alpha_i = \left[\frac{2m_i}{\hbar^2} (E - E_i) - k_{\parallel}^2 \right]^{1/2}, \quad i = 1, 2 \tag{B.12}$$

and

$$F_i = \frac{\hbar^2}{2m_i} \alpha_i. \tag{B.13}$$

Applying the interface response theory as explained above, one obtains easily the response function for different layered materials and in particular for an infinite electronic superlattice [47] made out of two different free electrons slabs with respective potential energies E_1 and E_2 and effective masses m_1 and m_2.

References

[1] Lord Rayleigh, Proc. London Math. Soc. 17 (1887) 4.

[2] I.M. Lifshitz and L.N. Rosenzweig, Zh. Eksp. Teor. Fiz. 18 (1948) 1012;
L.N. Rosenzweig, Uch. Zap. Hark. Gos. Univ. Tr. Fiz. Mat. Otdel 2 (1950) 19.

[3] D.C. Gazis, R. Herman and R.F. Wallis, Phys. Rev. 119 (1960) 533.

[4] R.M. White and F.W. Voltmer, Appl. Phys. Lett. 7 (1965) 314.

[5] R.F. Wallis, Surf. Sci. 299/300 (1994) 612.

[6] H. Ibach, Surf. Sci. 299/300 (1994) 116.

[7] G. Benedek and J.P. Toennies, Surf. Sci. 299/300 (1994) 587.

[8] B. Djafari-Rouhani, L. Dobrzynski and P. Masri, Ann. Phys. (Paris) 6 (1981) 259.

[9] J. Sapriel and B. Djafari-Rouhani, Surf. Sci. Rep. 10 (1989) 189.

[10] M.V. Klein, IEEE J. Quant. Elec. QE - 22 (1986) 1760.

[11] B. Jusserand and M. Cardona, in: Light Scattering in Solids, V, Eds. M. Cardona and G. Güntherodt (Springer, Berlin, 1989) p. 49.

[12] M. Cardona, in: Proc. NATO ARW on Spectroscopy of Semiconductor Microstructures, Eds. G. Fasol, A. Fasolino and P. Lugli (Plenum, New York, 1990);
M. Cardona, Superlattices and Microstructures 4 (1989) 27.

[13] G. Benedek and V.R. Velasco, in: Dynamical Phenomena at Surfaces, Interfaces and Superlattices, Eds. F. Nizzoli, K.H. Rieder and R.F. Willis, Springer Series in Surface Science 3 (Springer, Berlin, 1985) p. 66.

[14] A. Huber, T. Egeler, W. Ettmüller, H. Rothfritz, G Tränkle and G. Abstreiter, Superlat. Microstruct. 9 (1991) 309.

[15] See also: 'Light Scattering in Semiconducteur Structures and Superlattices' Eds. D.J. Lockwood and J.F. Young (Plenum, New York, 1991).

[16] L. Esaki and R. Tsu, in: IBM Res. Note RC - 2418 (1969), IBM J. Res. Develop. 14 (1970) 6.

[17] J. Koutecky, Phys. Rev. 108 (1957) 13.

[18] A.A. Maradudin and J. Melngailis, Phys. Rev. 133 (1964) A 1188.

[19] R.A. Brown, Phys. Rev. 156 (1967) 889.

[20] L. Dobrzynski, Ann. Phys. (Paris) 4 (1969) 637.

[21] S.G. Davison and J.D. Levine, Solid State Phys. 25 (1970) 1.

[22] G. Allan, Ann. Phys. (Paris) 5 (1970) 169.

[23] S.W. Müsser and K.H. Rieder, Phys. Rev. B 2 (1970) 3034.

[24] D. Kalstein and P. Soven, Surf. Sci. 26 (1971) 85.

[25] A.A. Maradudin, E.W. Montroll, G.H. Weiss and I.P. Ipatova, Theory of Lattice Dynamics in the Harmonic Approximation (Academic Press, New York, 1963 and 1971).

[26] P. Lenglart, L. Dobrzynski and G. Leman, Ann. Phys. (Paris) 7 (1972) 407.

[27] A.A. Maradudin, R.F. Wallis and L. Dobrzynski, in: Surface Phonons and Polaritons, Vol. 3 of the Handbook of Surfaces and Interfaces, Ed. L. Dobrzynski (Garland, New York, 1980).

[28] G. Allan, in: Handbook of Surfaces and Interfaces, Vol. 2, Ed. L. Dobrzynski (Garland, New York, 1978) p. 299.

[29] P. Masri and L. Dobrzynski, Surf. Sci. 34 (1973) 119.

[30] E. Foo and H. Wang, Phys. Rev. B 10 (1974) 4819.

[31] B. Djafari-Rouhani and L. Dobrzynski, J. Phys. (Paris) 36 (1975) 835.

[32] B. Djafari-Rouhani, L. Dobrzynski and P. Masri, Ann. Phys. (Paris) 6 (1981) 259.

[33] J. Pollman, Adv. Solid State Phys. 20 (1980) 117.

[34] J. Pollman and A. Mazur, Thin Solid Films 104 (1983) 257.

[35] L. Dobrzynski, B. Djafari-Rouhani and O. Hardouin Duparc, J. Electron Spectrosc. Relat. Phenom. 30 (1983) 119; Phys. Rev. B 29 (1984) 3138.

[36] L. Dobrzynski, Surf. Sci. Rep. 6 (1986) 119; Prog. Surf. Sci. 26 (1987) 103.

[37] L. Dobrzynski and H. Puszkarski, J. Phys.: Condensed Matter 1 (1989) 1239.

[38] A. Akjouj, B. Sylla, P. Zielinski and L. Dobrzynski, J. Phys. C. 20 (1987) 6137.

[39] B. Sylla, L. Dobrzynski and H. Puszkarski, J. Phys.: Condensed Matter 1 (1989) 1247.

[40] L. Dobrzynski, Surf. Sci. 180 (1987) 489; Surf. Sci. Rep. 11 (1990) 139.

[41] P.M. Morse and H. Feshbach, Methods of Theoretical Physics, Vol. I (McGraw-Hill, New York, 1953).

[42] F. Garcia-Moliner and J. Rubio, J. Phys. C 2 (1969) 1789.

[43] J.E. Inglesfield, J. Phys. C 4 (1971) L 14.

[44] B. Velicky and I. Bartos, J. Phys. C 4 (1971) L104.

[45] F. Garcia Moliner and V. Velasco, Phys. Scr. 34 (1986) 257; Phys. Scr. 34 (1986) 252; Prog. Surf. Sci. 21 (1986) 93; Surf. Sci. 299/300 (1994) 332.

[46] L. Dobrzynski, Surf. Sci. 180 (1987) 505; see also for a review: M.L. Bah, A. Akjouj and L. Dobrzynski, Surf. Sci. Rep. 16 (1992) 95.

[47] L. Dobrzynski, Surf. Sci. 200 (1988) 435.

[48] A.A. Maradudin and D.L. Mills, Ann. Phys. 100 (1976) 262.

[49] M.G. Cottam and A.A. Maradudin, in: Surface Excitations, Vol. 9 of Modern Problems in Condensed Matter Sciences (North-Holland, Amsterdam, 1986).

[50] P. Zielinski and L. Dobrzynski, Phys. Rev. B 41 (1990) 10377.

[51] L. Dobrzynski, A. Akjouj, B. Sylla and B. Djafari-Rouhani, Acta Phys. Polon. A 81 (1992) 85.

[52] G. Benedek, J. Ellis, A. Reichmuth, P. Ruggerone, H. Schief and J.P. Toennies, Phys. Rev. Lett. 69 (1992) 2951.

Surface Science 299/300 (1994) 1022–1030
North-Holland

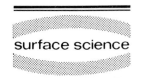

surface science

On low temperature studies of semiconductor surfaces

V.A. Grazhulis

Institute of Solid State Physics, Russian Academy of Sciences, Chernogolovka 142432, Moscow District, Russian Federation

Received 2 April 1993; accepted for publication 27 July 1993

A brief review of low-temperature investigations of semiconductor surfaces, performed in the years 1960–1990, is presented. Particular emphasis is given to the contributions of the author's group. Most attention is paid to the surface atomic structures and electronic properties of clean Ge(111) and EuX(100) surfaces (where X = O, S, Se, Te). The effects of low temperatures on the properties of metal-deposited surfaces are mentioned as well. A comparison with the room temperature results is also performed for some cases.

1. Introduction

In 1932 Tamm published the paper [1] in which he predicted the existence of localized electronic states in the vicinity of crystal surfaces due to strong perturbation of the periodical potential of the surface lattice. This paper was the "starting point" for research activities related with the electronic properties of clean surfaces of crystals. A few years later, namely, in 1939, Shockley [2] proposed another type of surface states, based on the assumption that surface atomic orbitals for some crystals can in a sense be weakly disturbed at the surface. Therefore, two alternative models have been proposed, based on strong and weak surface perturbation. Originally Shockley proposed that clean surfaces of covalent Si and Ge crystals can possess surface states characteristic for weak perturbation ("Shockley states"), rather than "Tamm states" [1]. However, the first LEED observations, already in the early 1960's (see, for instance, Lander et al. [3]), proved that clean surfaces of Si are reconstructed. The observations of the reconstruction qualitatively supposed that the atomic orbitals at the Si surface were strongly perturbed. Since that time it was clear, however, that both Tamm and Shockley approaches were too simple, in fact, to correctly describe surface electronic properties in detail. Therefore, later

on a lot of different experimental investigations were performed for clean surfaces of both semiconductors and metals to reveal the main features of the surface states and atomic structures. A kind of boom started in the early 60's when ultra-high vacuums (UHV) (10^{-10}–10^{-11} Torr) became available in many surface science laboratories. Very interesting events have been related with the investigations of clean cleaved Ge(111) surfaces (for details see below). E.g. Ge(111) surfaces have shown a lot of unique features which could be observed at low temperatures of $T = 1.2$–300 K. In the present paper we briefly discuss some of these Ge(111) results related with surface states, atomic structures, and surface electric properties at low temperatures. We shall demonstrate that low-T investigations give very often unique information which can hardly be gained at room or elevated temperatures, though some low-temperature results are even more complicated to interpret than those obtained at room or elevated temperatures. Apart from that, we briefly describe low-T investigations of clean surfaces of magnetic semiconductors EuX (here X = O, Te, Se, Te). A few remarks will also be made on the peculiarities in the metal/semiconductor interface formation at low temperatures. More detailed discussion of this large field is out of the scope of the present paper.

SSDI 0039-6028(93)E0437-Y

Appropriate information can be found in the papers by Mönch and Brillson (see this volume) devoted to the formation of metal/semiconductor interfaces. We restrict ourselves to mentioning just of the experimental results to demonstrate that low-temperature deposition of metals on semiconductor surfaces can lead to interesting effects, in addition to those which can be observed at room or elevated temperatures.

It is worth mentioning here that in the present paper particular emphasis is given to the author's group activities, related with the investigations of clean and metal-deposited Ge(111), Si(111) and $A_3B_5(110)$ surfaces.

2. Ge(111) at low temperatures

Let us consider, first of all, the Ge(111) surface structure as a function of the cleavage temperature. We shall pay most attention to the low-temperature range, $T < 100$ K. The results obtained for Ge(111) clean cleaved surfaces clearly demonstrate that the surface atomic structure and electronic properties at low temperatures can be much more complicated than at room or elevated temperatures.

Since the 60's it was generally accepted that the atomic structure of cleaved semiconductor surfaces does not essentially depend on the cleavage temperature, T_{cl}, if $T_{cl} \lesssim 300$ K (see, for instance, Refs. [4–10]). However, it turned out that at least in the case of Ge crystals this statement is not correct. Actually, it is known that just after the crystal cleavage in ultra-high vacuum at room temperature, the Ge(111) surface (and Si(111)) shows a 2×1 superstructure [4–10]. However, investigations performed in 1981–82 in Grazhulis' group [11,12] have revealed that the Ge(111) surface structure can essentially depend on the cleavage temperature. That was an unexpected result. In 1981, in Ref. [11], it was found that in UHV ($\sim 10^{-10}$ Torr), at $T_{cl} \gtrsim 40$–60 K, the cleaved Ge(111) surface possesses a 2×1 superstructure, however, at $T_{cl} < 20$–40 K the LEED picture can show no superstructural 2×1 spots. Measurements performed by Haneman and Bachrach [13,14] in 1982–83, at the fixed temperatures 300, 18 and 4 K, have also revealed that the low-T cleavages can show no superstructural spots. All these observations were interpreted as a result of a non-reconstructed 1×1 state of cleaved Ge(111)-1×1 surface or a strong disorder of the Ge(111)-2×1 structure after low-T cleavage.

Ge(111) surfaces cleaved at low T (10–300 K) in UHV were later more systematically investigated, see Refs. [15,16]. These investigations have

Fig. 1. Part of the LEED picture of low-temperature cleaved Ge(111). One can see strongly and anisotropically broadened superstructural spots.

revealed new interesting peculiarities, namely, it was found that low-T cleavages ($T_{cl} < 40$ K) can give rise to quite unusual LEED pictures with sharp main spots, but with strongly broadened superstructural ones (see Fig. 1). The broadening was not well reproducible at $T_{cl} < 40$ K so that some cleavages showed very strong broadening giving rise to even vanishing of the superstructural spots. Another interesting point was that superstructural spots broadening was essentially anisotropic and was not accompanied by the same broadening of the main spots, see Fig. 1. At present, there is no final interpretation of these interesting observations. It was suggested, however, that such unusual Ge(111) cleaved surface behaviour is due, probably, to surface overheating and quenching in the course of cleavage at low temperature [15–17]. Such interpretatoion seemed to be very interesting, however, it had to be proved.

Low-T cleavage followed by reconstruction and relaxation may actually denote strong initial "heating" of the surface due to the excitation of the non-equilibrium phonons at the crack tip and sharp surface "quenching" if the cleavage temperature T_{cl} is small. It is evident that the absence of the long-range order in the 2×1 surface system can result in strong spreading of the superstructural spots and even their "disappearance". One can expect, that such disordered surfaces can survive after the crystal cleavage if the surface cooling time is short enough compared to the annealing time needed to form and ordered 2×1 state. One can approximately estimate a cooling time $\delta\tau$ for a surface which is suddenly heated up to a high temperature.

If we assume, after Grazhulis and Kuleshov (1985, [17]), that some part of the excess energy ΔE_s (~ 1 eV/atom) is suddenly transferred to the surface due to surface phonon excitation then for the transferred energy $\mathscr{E}$ we have the ratio $\mathscr{E} \approx P\delta\tau$, where P is the heat power. The heat energy $\mathscr{E}$ can be presented in the form $\mathscr{E} = \gamma\Delta E_s N_s S$, where S is the cleaved surface area, N_s is the number of broken bond atoms per unit area, and γ is a coefficient showing what fraction of the surface excess energy is to be transferred to the bulk ($\gamma \lesssim 1$). On the other hand, we can

assume also that $P = S\kappa\Delta T/\Delta X$, where κ is the thermal conductivity and $\Delta T/\Delta X$ is the temperature gradient. Supposing, for instance, that the surface is heated up to a temperature which is of the order of the surface melting temperature, T_m, that is $\Delta T = T_s - T_{cl} \approx T_s - T_m - 2\Theta_{DS} \approx 500$–$1000$ K, where T_s is the surface temperature just after the cleavage, T_{cl} is the temperature of the bulk ($T_s \gg T_{cl}$), Θ_{DS} is the effective Debye temperature of the cleaved surface (note that the surface Debye temperature can be much lower than the bulk one (see Refs. [17,19])) and assuming also that $\Delta X \approx (1 - 10^2)\lambda_{ph}$, where λ_{ph} is the characteristic phonon length in the bulk corresponding to the cleavage temperature (at $T_{cl} \approx 10$ K, $\lambda_{ph} \approx 10^{-6}$ cm). Then, at $\gamma \approx 1$, $\Delta E \approx 1$ eV, $N_s \approx 10^{15}$ cm^{-2}, $\Delta X \approx 10^{-6}$–10^{-4} cm, $\kappa \approx 10^2$ W/mK $\approx 6 \times 10^{18}$ eV/cm·s K, and $\Delta T \approx 500$ K, one obtains $\delta\tau \approx 10^{-2}$–$10^{-10}$ s. So, the cooling of the cleaved surface may indeed be very fast and, therefore, something like an amorphous state of the broken bond system may occur if the initial (heated) state corresponds to a strong disorder.

There are other interesting effects (for details see Ref. [21]) related with low-T cleavages of Ge crystals, namely, the ones related with the electrical behaviour of Ge(111) cleaved surfaces.

Before the discussion of the low-T electrical data it is worth mentioning some of the earlier results which demonstrate important features of the Ge(111) surfaces at room temperature.

It was found (Farnsworth et al., 1955) that Ge(111) surface can show electric conductivity (σ_s) at room temperature due to spontaneous surface p-channel formation [22]. The photoemission and Kelvin probe investigations by Gobeli and Allen [28] for cleaved Ge(111) surfaces in UHV ($\sim 10^{-10}$ Torr) at room temperature appeared in qualitative agreement with this observation. It has to be pointed out, however, that the accurate position of E_F, with respect to the top of the valence band $E_v^{(s)}$, could not be established in these experiments due to the error in the measurements (± 0.05 eV). So, one could not decisively conclude wether the Ge(111) surface p-channel can survive at low temperatures or not. Actually, if $E_F - E_v^{(s)} > 0$, then at $T \to 0$ the surface hole concentration, p_s, should vanish.

However, the surface DC conductively measurements (1968, Grant and Webster [29]) performed for n-type 2 $\Omega \cdot$ cm Ge crystals, cleaved at $T = 300$ K in UHV, have shown that surface p-channels can, in fact, survive even at low temperatures ($T \approx 4.2$ K). From that result once could suppose that $E_F - E_v^{(s)} \lesssim 0$, that is the Fermi level at low temperatures is, probably, located below $E_v^{(s)}$ and the hole system is degenerated.

The Hall effect measurements in UHV have shown (see, for instance, the paper by Katrich et al. [30]) that the Ge(111) surface conductivity is due to the surface degenerated hole system and that the mobility of these holes, $\mu_p^{(s)}$, is around 500 cm^2/V·s; it was found that there are two important features in σ_s-behaviour: it slightly increases with the oxygen exposure if the latter is below disappearance of the Ge(111) surface conductivity after strong oxidation which was supposed to be due to the disappearance of the surface states giving rise to the p-channel.

On the basis of the performed investigations by different groups [31] one could state that surface p-channels can be formed for both n- and p-type crystals, which means doping does not play an important role, and "elevated" temperature cleavages ($T_{cl} = 77$–300 K) generate channels with sufficiently reproducible parameters: $\sigma_s \lesssim 10^{-4}$ Ω^{-1}/square and $\mu_p \approx 5 \times 10^2$ cm^2/V·s at 77 K.

Important additional information was supplied by the low-T studies of p-channels.

Taking into account the observations of the p-channels after elevated temperature cleavages, $T_{cl} = 77$–300 K, one could expect such channels to occur for low-T cleavages as well. Note, however, that in a UHV chamber it is rather difficult to cool a sample down to liquid He temperature. Therefore, it was suggested to perform cleavages inside the liquid He itself. The first studies of liquid-He cleaved Ge(111) surfaces were performed by Talyanskii et al. [32] and Zavaritskaya et al. [33] in 1979. In Ref. [32] the surface microwave conductivity ($\lambda \approx 3$ cm) for Ge n-type ($N_d = 10^{11}$ cm^{-3}, 5×10^{13} cm^{-3}) and p-type ($N_a = 10^{13}$ cm^{-3}) crystals cleaved in liquid He was studied. From the measured quality of the cavity, $f_0/\Delta f$, the cleaved surface conductivity could be estimated, where f_0 is the resonance frequency

of the cavity and Δf is the width of the resonance curve. It was found that Ge samples before the cleavage did not effect the cavity quality in the whole temperature range $T = 1.4$–10 K. However, cleavage in liquid He (inside the cavity inserted into liquid He) lead to a drastic increase of Δf which could be interpreted as being due to the created microwave surface conductivity ($\sigma_s^{(0)}$). It was found that at $T = 1.4$–10 K the value $\Delta f \approx \sigma_s$ is practically constant. The latter is the characteristic feature of the metallic conductivity. Thus, from the microwave measurements one could also arrive at the proposal that the degenerated conducting surface channel is formed (with the Fermi level below $E_v^{(s)}$).

The DC conductivity measurements for Ge samples cleaved in liquid He were performed in Ref. [33]. p- and n-type Ge crystals with a dopant concentration in the range from 10^{13} to 5×10^{15} cm^{-3} were studied. As a result of these investigations it was found that Ge(111) surfaces cleaved at low temperatures initially show no or very low DC conductivity, with $\sigma_s \approx 10^{-9} \Omega^{-1}$ square. This was an unexpected result. The DC conductivity starts increasing only after slight surface oxidation. From the very beginning the reason for such unusual surface behaviour was not clear. The performed Hall effect measurements for conducting surfaces have confirmed that the conductivity is due to the degenerated hole system appearing at liquid-He cleaved Ge(111) surfaces as well. The hole concentration for these surfaces varied in the range $p_s = (6$–10$) \times 10^{12}$ cm^{-2}. The maximum Hall mobility was found to be in the range $\mu_p = R\sigma_m = 250$–400 cm^2/V·s, where R_0 is the Hall coefficient and σ_m is the observed maximum surface conductivity. It is important to note that in the course of exposure to oxygen the surface conductivity for all investigated Ge cleavages was always achieving practically the same maximum value of $\sigma_s \approx \sigma_m \approx 4 \times 10^{-4} \Omega^{-1}$ square. However, after complete surface oxidation the surface conductivity was vanishing (the p-channel was disappearing).

It has to be pointed out that quantitatively the low-T results appeared to be different from those for room temperature in UHV. Actually, slight oxidation did not strongly effect room temperature cleavages, whereas liquid-He cleavages did: a few times increase of the microwave conductivity and a 4–5-order increase in magnitude of the DC conductivity was observed. The latter observations allow one to propose that just after low-T cleavage the Ge(111) surface can be strongly inhomogeneous: it seems to contain p-channel "lakes", probably, without percolation [32,33]. Let us also point out that low-T cleavages as a rule show $\sigma_s \ll \sigma_m$, whereas at $T_{cl} = 77–300$ K the observed surface conductivity $\sigma_s \approx \sigma_m$. The latter also implies that high-T cleavages produce more homogeneous surfaces. These conclusions are in agreement with those arriving from the analysis of the LEED data, see above.

At present, unfortunately, there are not quantitative studies of oxidation effects on the conductivity of Ge(111) surfaces. Therefore, the details of the oxidation effects are not clear. Note, that the oxidation at low T is accompanied by interesting physical phenomena. Actually, in Ref. [34] it was reported that one can observe two very drastic jumps on the σ_s (T_a)-curve at constant oxidation time. Here T_a is the oxidation temperature. Another important feature reported in Ref. [34] was the minimum metallic conductivity for Ge(111) surfaces, σ_{min}. The value of that conductivity was found to be $\sigma_s = \sigma_{min} \approx e^2/h \approx 4 \times 10^{-5}$ Ω^{-1}/square.

Thus, one can actually propose that in the case of Ge(111)-2 × 1 there are empty surface states overlapping with the valence band, capturing electrons and thus forming the surface p-channel, see Fig. 2. One has to note, however, that angle-resolved photoemission measurements for Ge(111)-2 × 1 surfaces [35] so far did not give any direct indications on the existence of the surface dipole layer consisting of surface captured electrons and holes of the p-channel. However, the formation of the degenerated p-channel definitely requires positioning of the Fermi level in the valence band at the surface. Therefore, additional investigations are to be performed for a better understanding of the Ge(111)-2 × 1 electronic spectrum. It is worth noting, that the position of the surface empty states with respect to the Fermi level can be sensitive to surface defects. The latter can pin the Fermi level essen-

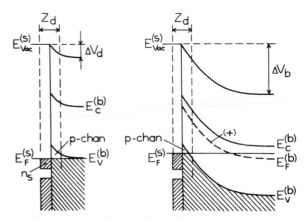

Fig. 2. Formation of the p-channel. Left: p-type Ge, right: n-type Ge (after Ref. [21]).

tially above the valence band (and, thus, prevent the formation of the p-channel [36]). This can be the reason for different results in different experiments.

In conclusion let us note, that in the case of Si(111) no p-channels could be observed due to the fact that the Fermi level is pinned deep in the gap at the surface and the surface conductivity is negligible [21].

These low-temperature Ge(111) results seem to be quite interesting and show that at low temperatures many important effects can be observed on semiconductor surfaces.

3. Low-temperature studies of clean surfaces of magnetic crystals: EuX (X = O, S, Se, Te)

There are two most interesting problems related with the investigations of surfaces on magnetic semiconductor crystals, EuX (the non-zero spin is due to Eu^{2+}, see below). The first is related with the spin polarization measurements at the surface by the use of the spin-polarized UPS technique. In these measurements it is of interest to investigate the temperature ranges above and below the Néel and Curie temperatures (T_N and T_C). Note that the crystals with low T_N and T_C are preferred due to the fact that the spin polarization of the photoelectrons can essentially be effected by the spin–phonon interactions (if the temperature is not sufficiently low). EuX crystals are good candidates for such experiments. The second problem is related with the investigations of the surface atomic structures by LEED. The atomic structure can be effected by the spin–lattice interactions as well and, therefore, one can try to use common LEED without spin polarization of the primary electron beam.

Europium compounds of the EuX (X = O, S, Se, Te) type are ionic Heisenberg magnetic semiconductors. Crystals of these compounds have a cubic lattice of the NaCl type. Their electric and magnetic properties are defined by the electrons of the half-field 4f-shell of Eu^{2+} ($^8S_{7/2}$, $L = 0$, $I = S = 7/2$). EuO and EuS crystals are ferromagnetics with the Curie temperatures $T_C \approx 69$ and 16 K, respectively. EuTe crystals are antiferromagnetics with the Néel temperature $T_N \approx 9.6$ K. EuSe crystals have a complex magnetic phase diagram with the critical temperatures located in the range of the liquid-He temperatures, $T \leq 4.2$ K. The bulk properties of these crystals have been studied very extensively in the last two decades. A lot of interesting phenomenon were disclosed.

3.1. EuO(100) clean cleaved surface by UPS technique

Experimental studies of EuO(100) cleaved surfaces at $T < T_c$ using spin-sensitive UPS technique were performed in 1972 by Sattler and Siegman [37]. Both doped and "pure" EuO crystals were investigated. Cleaved (100) surfaces were studied in UHV ($\sim 10^{-10}$ Torr) at $T \approx 10$ K. The spin polarization (P) of the photoelectrons was measured as a function of the photon energy ($h\nu$) and external magnetic field (H). The main features of the observed photoelectron polarization were the following: (i) there is no magnetic saturation; (ii) a kink occurs at 8–10 kG; (iii) doped EuO shows higher P-values than "pure" EuO. These observations lead the author to the conclusion that the EuO(100) surface is in the paramagnetic state, though the bulk is ferromagnetic at $T < T_c$. Taking into account that the atomic structure of the cleaved surfaces was not investigated [37], decisive conclusions on the nature of the

surface paramagnetic state were not possible. However, if the investigated surfaces were crystalline, then the paramagnetic state could be due to the weakening of the exchange interactions at the surface. The latter can, for instance, be due to surface relaxation and an increase of the lattice constant.

Thus, spin-sensitive UPS measurements suggest that at EuO(100) clean cleaved surfaces a paramagnetic sheet can exist even at $T \ll T_c$.

Note that the observation of the sheet was reported for evaporated polycrystalline thin films of EuS at $T < T_c \approx 16$ K as well (1974, Campagna et al. [38]). Therefore, one can probably suppose that the paramagnetic sheet is a general property of clean ferromagnetic semiconductors.

The effects of adsorption of Cs atoms on EuO(100) surface magnetization were investigated in 1975 (Meier et al. [39]).

These investigations were stimulated by the fact that in the bulk of the magnetic semiconductors short-range magnetic interactions are dominating and the surface properties (related to relaxation and reconstruction) are mainly limited to the first atomic layer. The latter suggested, see above, that the thickness of the surface paramagnetic layer is of the order of one lattice constant. However, if the layer is so thin, its properties should be sensitive to the adsorption of atoms. Actually, the surface adatoms can modify both the structure and the electron density at the surface. As a result of that the magnetization can be varied due to variation of the exchange interactions.

In Ref. [39] a freshly cleaved EuO(100) single crystal surface was covered with Cs in several steps and spin polarization of the photoelectrons was investigated. The Cs deposition and UPS measurements were performed at 4.2 K. These measurements have shown that [39]: (i) Cs deposition decreases the absolute value of the polarization, however, (ii) there is a tendency to a saturation effect at $H \approx 10$ kG. The former is due to the non-polarized contribution to the photocurrent from the Cs coverage; the latter is an indication of the EuO(100) surface magnetization (for details see Ref. [39]). Thus, the observed proximity effect suppots the suggestion that the thickness of the paramagnetic sheet at the EuO(100) surface is indeed of the order of one lattice constant, that is, the depolarization effects occur in the surface region and not in the bulk.

It has to be emphasized, that all investigated EuX crystals showed qualitatively similar LEED pictures at temperatures in the range 10–300 K. No peculiarities were observed. However, at $T \approx$ 10 K EuTe(100) surfaces have shown particular behaviour.

Note once again that EuTe is in the bulk antiferromagnetic with $T_N \approx 9.6$ K. The crystal and spin structures of EuTe in the bulk are identical to those of NiO at $T < T_N$ (in the case of NiO the Néel temperature is $T_N \approx 520$ K). One could expect the (100) surface spin structure corresponding to the (100) plane of the bulk. Due to that spin ordering, there should be doubling of the periodicity. If the primary electron beam in the LEED experiment possesses a zero spin polarization, then the two-dimensional spin lattice should give rise to the appropriate LEED superstructural picture only in the case of lattice modulation by spin–lattice interactions.

Experiments performed for cleaved EuTe(100) surfaces at temperatures in the range of 8–10 K led to very interesting results (1987–88, Grazhulis, Ionov and Kuleshov [40,41]). After cleavages at 15 K the EuTe(100) surfaces showed distinct LEED patterns which corresponded to the unreconstructed EuTe(1 × 1) surface. When the temperature was lowered (< 10 K) an interesting phenomenon was observed: extra spots were suddenly splitting off from the main ones and moving along the diagonal direction [100] towards the centre of the LEED pattern squares. Near 8 K all the extra spots were located in the centers of the 1 × 1 squares of a LEED patterns. It was also found that multiple temperature scanning caused completely reproducible changes of the LEED pattern. The formation of the final surface superstructure ($\sqrt{2} \times \sqrt{2}$)R45° was attributed, in Ref. [41], to the formation of the incommensurate helicoidal spin structure near the surface at $T < T_N$.

However, to elucidate the detailed mechanism of the surface superstructure formation, further experiments are necessary.

4. Metal–semiconductor interfaces at low temperatures

If metal atoms are deposited on clean semiconductor surfaces at room or elevated temperatures, then islands and non-sharp interfaces are very often formed due to diffusion effects. At low T, however, diffusion can be suppressed and, therefore, adatoms appear statistically distributed on the surface. Apart from that, chemical interactions of adatoms with the substrate at low T can also be suppressed if the potential barrier to form chemical bonds is high enough.

The first systematic low-T (at $T \approx 10$ K) investigations of metal/semiconductor interfaces were performed in Grazhulis' group in the 80's. More detailed description of the obtained results at $T \approx 10$ K can be found in the review [21]. In this paper, we, therefore, present just a few remarks illustrating some of the effects characteristic for "very low" temperatures. We note very briefly here only some results for Ag deposited on Si(111), Ge (111), and A_3B_5(110).

Speaking about surface atomic structures, one could probably expect that deposition of atoms on clean surfaces at $T \approx 10$ K should just give rise to continuous vanishing LEED pictures of the substrate due to the formation of statistically disordered overlayers. However, it turned out that, instead, a variety of interesting effects can be observed.

First of all, let us briefly consider Si(111)-2 × 1 and Ge(111)-2 × 1 surfaces covered by Ag at $T \approx$ 10 K. It appeared that these two crystals show essentially different behaviour at low temperatures (see Refs. [42,43]). Namely, it turned out that Ag atoms are weakly interacting with Si(111)-2 × 1 substrate at $T \approx 10$ K so that adatoms form a "suspended" overlayer. As a result of that the 2 × 1 superstructure of the substrate survives and can be observed by LEED even at an average coverage $\theta = 3$ ML. Quite opposite behaviour, however, shows Ge(111)-2 × 1–Ag. Actually, Ag atoms strongly interact with a Ge(111) surface, even at $T \approx 10$ K so that superstructural 2 × 1 spots disappear already at $\theta \approx 0.1$–0.2 ML. At $\theta \approx 1$–2 ML main 1 × 1 spots disappear as well. From these observations it was concluded that Ge(111) and Ag atoms form chemical bonds, even at $T \approx 10$ K (contrary to Si(111)–Ag) and strongly violate the 2×1 superstructure. Apart from that, at $\theta \geq 1$ ML the translational symmetry in the system of surface atoms appears essentially destroyed, also resulting in vanishing of the 1×1 main spots.

Among the A_3B_5(110)–Ag surfaces it is worth mentioning here InSb(110)–Ag, first of all. Actually, these surfaces show very interesting behaviour (1987, Aristov et al. [44–46]) at low temperatures. Investigations by LEED have disclosed that Ag deposition at $T \approx 10$ K on InSb(110) clean cleaved surfaces demonstrates behaviour similar to Ge(111), if the average coverage $\theta \leq 1.5$ ML. However, further increase of θ at $T \approx 10$ K gives rise to new effects. Actually, InSb(110)–Ag surfaces show strong disorder (no LEED pattern) up to $\theta \approx 4$ ML. However, at $\theta = \theta^* \approx 4$–4.5 ML interesting phenomena take place: the system undergoes a phase transition (at $T \approx 10$ K) to an ordered state so that suddenly new LEED reflections appear which are showing the formation of a new type crystalline Ag, namely, bcc instead of the common fcc modification (the latter is always formed if Ag deposited at room temperature on InSb(110)). Thus, the thickness of the amorphous Ag overlayer is a critical parameter for the mentioned phase transition. This was, probably, the first observation of bcc Ag formation. It is important that this new type of Ag survives after heating to 300 K; it is important also that common fcc Ag formed at 300 K does not transform to bcc after subsequent cooling to $T \approx 10$ K.

The details of many different investigations of A_3B_5(110)–Ag one can also find in the review [21].

In general it has to be noted here that investigations have revealed many interesting effects directly related with low temperatures both in the case of atomic structures and electronic properties. It is worth pointing out that band bendings (Schottky barrier) also show particular features at low temperatures, however, discussion of these effects is out of the scope of the present paper. One can find information on that matter in Ref. [21] (see also the papers by Mönch and Brillson in this volume [47]).

5. Conclusion

The performed brief analysis of low-T experimental investigations ($T \leq 20$ K) of semiconductor surfaces shows that during the last three decades great progress has been achieved in this particular field. The performed analysis shows that the low-T studies have revealed a lot of new features of clean and metal-adsorbed surfaces of semiconductors. Though in many cases "the low-temperature picture" is rather complicated, the obtained low-T results help in a better understanding of the properties of surfaces at both low and elevated temperatures. At present, low-T studies are extensively developing. Therefore, one can expect new interesting findings. Taking into account that many effects observed so far are not properly understood yet, these new findings can be important for the further development of the physics of semiconductor surfaces.

Acknowledgement

The author is grateful to Dr. V.Yu. Mukhina for her help in the preparation of the manuscript.

References

[1] I.E. Tamm, Phys. Z. Sowjetunion 1 (1932) 733.
[2] W. Shockley, Phys. Rev. 56 (1939) 317.
[3] I.I. Lander, G.W. Gobeli and J. Morrison, J. Appl. Phys. 34 (1963) 2298.
[4] W. Mönch, Surf. Sci. 86 (1979) 672.
[5] D. Haneman, Adv. Phys. 31 (1982) 166.
[6] M. Henzler, Appl. Surf. Sci. 11/12 (1982) 450.
[7] D.J. Chadi, Proc. 5th Int. Vacuum Congr., Madrid (1983) p. 80.
[8] N.P. Lieske, J. Phys. Chem. Solids 45 (1984) 821.
[9] D. Haneman, Rep. Prog. Phys. 50 (1987) 1045.
[10] B.A. Nesterenko and O.V. Snitko, Phyzicheskiye svoistva atomarno-chistoi poverkhnosty poluprovodnikov (Naukova Dumka, Kiev, 1983).
[11] V.Yu. Aristov, N.I. Golovko, V.A. Grazhulis and Yu.A. Osipyan, ECOSS-4, Final Program (1981) p. 42.
[12] V.Yu. Aristov, N.I. Golovko, V.A. Grazhulis, Yu.A. Osipyan and V.I. Talyanskii, Surf. Sci. 117 (1982) 204.
[13] D. Haneman and R.Z. Bachrach, J. Vac. Sci. Technol. 21 (1982) 337.

[14] D. Haneman and R.Z. Bachrach, Phys. Rev. B 27 (1983) 3921.

[15] N.I. Golovko, V.A. Grazhulis, V.F. Kuleshov and V.I. Talyanskii, Poverkhnost 1 (1986) 77.

[16] V.A. Grazhulis, Appl. Surf. Sci. 33/34 (1988) 1.

[17] V.A. Grazhulis and V.F. Kuleshov, Appl. Surf. Sci. 22/23 (1985) 14.

[18] B.A. Nesterenko, A.D. Borodkin, O.V. Snitko, Surf. Sci. 32 (1972) 576.

[19] B.A. Nesterenko and O.V. Snitko, Phyzicheskiye svoistva atomarno-chistoi poverkhnosty poluprovodnikov (Naukova Dumka, Kiev, 1983).

[20] Yu.S. Zharkikh and S.V. Lysochenko, Surf. Sci. 145 (1984) L513.

[21] V.A. Grazhulis, Surf. Sci. 36 (1991) 89.

[22] H.E. Farnsworth, R.E. Schiller, T.H. George and R.M. Burger, J. Appl. Phys. 26 (1955) 252.

[23] G.W. Göbeli and F.G. Allen, J. Phys. Chem. Solids 14 (1960) 23.

[24] D.R. Palmer, S.R. Marrison and C.E. Dauenbaugh, J. Phys. Chem. Solids 14 (1960) 27.

[25] D.R. Palmer, S.R. Morrison and C.E. Dauenbaugh, Phys. Rev. 129 (1963) 608.

[26] G.A. Katrich and O.G. Sarbey, Fiz. Tverd. Tela 5 (1963) 3321.

[27] P. Handler and W. Portnay, Phys. Rev. 116 (1959) 516.

[28] G.W. Gobeli and F.G. Allen, Surf. Sci. 2 (1964) 402.

[29] J.T. Grant and D.S. Webster, J. Appl. Phys. 39 (1968) 3129.

[30] G.A. Katrich, O.G. Sarbey and D.T. Taraschenko, Fiz. Tverd. Tela 7 (1965) 1352.

[31] M. Henzler, Surf. Sci. 9 (1968) 31.

[32] Yu.A. Osipyan, V.I. Talyanskii and A.A. Kharlamov, Pis'ma Zh. Eksp. Teor. Fiz. 30 (1979) 253.

[33] B.M. Vul, E.I. Zavaritskaya and E.G. Sokil, Pis'ma Zh. Eksp. Teor. Fiz. 30 (1979) 517.

[34] B.M. Bul, E.I. Zavaritskaya and V.N. Zavaritskii, Pis'ma Zh. Eksp. Teor. Fiz. 34 (1981) 371.

[35] F.J. Himpsel, Surf. Sci. Rep. 12 (1990) 1.

[36] V.A. Grazhulis, Proc. 17th Int. Conf. Phys. Semiconductors (Springer, Berlin, 1985).

[37] K. Sattler and H. Siegman, Phys. Rev. Lett. 29 (1972) 1565.

[38] M. Campagna, K. Sattler and H.C. Siegman, AIP Conf. Proc. 18 (1974) 1388; Helv. Phys. Acta 47 (1974) 27.

[39] F. Meier, D. Pierce and K. Sattler, Solid State Commun. 16 (1975) 401.

[40] V.A. Grazhulis, A.M. Ionov and V.F. Kuleshov, Appl. Surf. Sci. 33/34 (1988) 81.

[41] V.A. Grazhulis, A.M. Ionov and V.F. Kuleshov, Pis'ma Zh. Eksp. Teor. Fiz. 46 (1987) 42.

[42] V.Yu. Aristov, I.L. Bolotin, V.A. Grazhulis and V.M. Zhilin, Sov. Phys. JETP 64 (1986) 832.

[43] V.Yu. Aristov, V.A. Grazhulis and V.M. Zhilin, Poverkhnost 8 (1987) 84.

[44] V.Yu. Aristov, I.L. Bolotin and V.A. Grazhulis, Pis'ma Zh. Eksp. Teor. Fiz. 45 (1987) 49.

[45] V.Yu. Aristov, I.L. Bolotin and V.A. Grazhulis, Zh. Eksp. Teor. Fiz. 93 (1987) 1821.

[46] V.Yu. Aristov, I.L. Bolotin and V.A. Grazhulis, J. Vac. Sci. Technol. B 5 (1987) 992.

[47] W. Mönch, Surf. Sci. 299/300 (1994) 928;
L.J. Brillson, Surf. Sci. 299/300 (1994) 909.

Surface Science 299/300 (1994) 1031–1039
North-Holland

surface science

The development of surface science in China: retrospect and prospects

Xie Xide

Fudan University, Shanghai 200433, People's Republic of China

Received 4 May 1993; accepted for publication 25 June 1993

It is generally agreed that the year of 1977 marked the birth of surface science in China, therefore the length of its history of development is only half of that shown in the title of this volume.

Since 1977 laboratories with modern facilities for surface studies have been established in various universities and research institutes. Three open laboratories better equipped than others have been set up in Beijing, Xiamen and Shanghai for surface physics, surface chemistry and applied surface physics, respectively. Five National Conferences on Physics of Surfaces and Interfaces were held in 1982, 1984, 1985, 1988 and 1991. In 1993 China is going to host the Fourth International Conference on the Structure of Surfaces in Shanghai August 16–19 which will serve as a milestone in the history of development of surface science in China.

With the access to many overseas laboratories, quite a number of Chinese scientists and students have had opportunities to work and study abroad and have brought back with them experiences acquired. During the Conferences just mentioned, one could witness a number of steady progresses made over the years. In the present review, a brief description about the establishment of some major research facilities and progresses of some of the research is given with emphasis on work related to semiconductor surfaces, interfaces, superlattices, heterojunctions and quantum wells. Although the review nominally covers the development of research in surface science in China, due to the limitation of the capabilities of the author, mostly work done at Fudan University is included. For this the author would like to express her deep apology to many Chinese colleagues whose works have not been properly mentioned.

1. Introduction

Efforts for developing surface science in China started near the end of 1977. The major challenge confronted was to build up ultrahigh vacuum electron spectrometers for various kinds of surface studies. Since it was expensive to import all the facilities, the problem was solved via purchasing some facilities from abroad while designing and manufacturing some equipment in China. At present in major surface physics research laboratories, one can see domestic manufactured ultrahigh vacuum chambers, analyzers, and molecular beam epitaxy facilities together with the imported ones. The design of an 800 MeV electron storage ring for synchrotron radiation at Hefei was started in the late seventies and construction was completed in April, 1989. At present there are five beamlines available for re-

search in X-ray lithography, time-resolved spectroscopy, soft X-ray microscopy, photoelectron spectroscopy and photochemistry. Moreover, synchrotron radiation was also available at the Beijing Synchrotron Radiation Laboratory (BSRL) which is part of the Beijing Electron Positron Collider (BEPC), originally planned for high energy physics research in China. At present there are 7 photon beamlines and 10 experimental stations available for studies in topography, extended X-ray absorption fine structure (EXAFS) and fluorescence, general purpose diffractions, surface absorption X-ray spectroscopy (SAXS), photoemission, lithography, soft X-ray applications, and spectroscopies for biological studies. The other challenge confronted during the early stage of development was a shortage of qualified personnel familiar with surface physics. This problem was partially solved through various activities in international exchanges, such as sending visiting scholars to work in overseas laboratories, encouraging graduate students to work towards Ph.D. programs in overseas universities, and participating in international conferences. At the same time, graduate programs for Master's degree and Ph.D. were started in China in 1981. During the past decade, a new generation of researchers in surface physics, and surface chemistry was trained. In the present article, due to the limitation of pages available, this review is by no means complete. Only part of the research work carried out in Chinese laboratories is included. Emphasis is put on surfaces and interfaces associated with semiconductors: fields with which the present author is relatively familiar.

1.1. Research progress

With the gradual development of human resources and the building up of necessary research facilities, the scope of publications covered has been extended from depth profiling of various surfaces to studies on atomic and electronic structures of surfaces and interfaces as well as from work performed in overseas laboratories to that accomplished in Chinese laboratories. Recently both experimental and theoretical studies on man-made materials have attracted great attention. Since China is a developing country, proposals with more practically oriented motivations can get financial support with higher priority. Hence, more research projects are associated with applications of surface science.

2. Polar surfaces of III–V compound semiconductors and metal / semiconductor interfaces

2.1. Polar surfaces of III–V compound semiconductors

Since the cleaved surfaces of III–V compound semiconductors have been widely studied and the atomic and electronic structures of these surfaces are now well established, research interests have been more oriented toward the (100) and (111) surfaces of these compounds which are orienta-

tions of substrates used in device technology and are comparatively less explored up to the present. During the past decade, Wang and co-workers have studied polar surfaces of InSb, InP and GaP [1–5]. Although InP is a promising substrate material for fabricating optoelectronic and microwave devices, due to the difficulties in preparing stable clean surfaces, the polar surface of InP has attracted less attention. It was shown that the reproducibility of the surface after ion bombardment and annealing (IBA) treatment is worse than that of the cleaved surface [1]. Efforts have been devoted to detailed studies of the creation and elimination of In islands which are inevitably introduced after IBA. As a result, surfaces which are free of In islands as illustrated by electron loss spectroscopy (ELS) have been realized and detailed studies on atomic and electronic structures of InP(100) and InP($\overline{111}$) surfaces have been carried out [3,4]. The surface electronic states of the InP(100) (4 × 2) surface have been measured by ultraviolet photoelectron spectroscopy (UPS) and high resolution electron energy loss spectroscopy (HREELS) and the experimental result was qualitatively explained by a missing row-dimer atomic structure model [3]. The dispersion curve of surface dangling bond states for an InP($\overline{111}$) surface was measured by angle resolved UPS [4]. A slab model and extended Hückel theory (EHT) calculation has been used to simulate the surface and to derive the theoretical E–k relation of dangling bond state [4].

The second material chosen was GaP, an important material in the fabrication of light emitting diodes. The surface reconstruction of GaP($\overline{111}$) was studied [5]. It was found that the reconstruction depends on the annealing temperature. Below 480°C, the surface shows a 1 × 1 low energy electron diffraction (LEED) pattern with relatively strong background, whereas above 520°C, the surface becomes facetted after annealing [5]. In the temperature range of 480–520°C, a complicated LEED pattern with very sharp contrast could be observed [5]. It was found that the best fit could be reached with a $\sqrt{247} \times \sqrt{247}$ R22.7° reconstruction model [5]. This is considered to be a more reasonable one than the 17 × 17 reconstruction previously reported [6,7].

2.2. Chemisorption, metal / semiconductor and semiconductor / semiconductor interfaces

It is well known that the understanding of chemical reactions which occur at the interface between metal and semiconductors is of both practical significance and theoretical interest. Extensive theoretical studies have been carried out to understand the geometry, nature of chemisorption bonding, and the electronic structures. The results are helpful in understanding the early stage of metal deposition and to elucidate whether mixing occurs. Both cluster and slab models are used for theoretical studies. Methods include the empirical charge self-consistent EHT, the discrete variational method (DVM) for clusters, empirical tight binding method, and self-consistent linearized augmented plane wave method (LAPW) for slab models. A review of work done by the Fudan group on adsorption of various groups of metals on Si(111), Si(100) and GaAs(110) surfaces has been given by Xie and Zhang [8]. Recently, Lu et al. [9,10] performed a series of studies on adsorption of Au, Al and Fe on the wide gap β-SiC(111) and (100) surfaces. Experimental studies on transition metal / silicon (d-metal/Si) and rare earth metal/silicon (f-metal/Si) by Hsu [11] have shown that for d-metal silicides, the d-metal adatoms tend to diffuse into the silicon lattice and an intermixing overlayer is formed, the chemical reactivity of 3d metals with silicon increases with increasing number of d-electrons. Those results agreed well with those of the theoretical studies [8]. For Cr with half-filled d-shells, experimental results show that at the chromium silicide / silicon interface, the density of states near the Fermi energy is higher than that of chromium [12]. As for rare earth metals (REM), at low coverages, the chemisorption energy is small and little intermixing occurs. The effect of thermal treatment on the thin film formation of REM on silicon surface was studied [13].

Applications of HREELS to studies of reactions at the metal and semiconductor interfaces have been explored by Wang and his group. Ding et al. [14] tentatively identified the interfacial reaction products by measuring the threshold of electronic transitions in the loss spectrum. Al/GaAs and Al/GaP interface systems have been chosen as the particular systems in their study. The interactions between aluminum and Ga-contained III–V compounds have been extensively studied by means of other surface techniques such as Auger electron spectroscopy (AES), photoemission spectroscopy, and Raman spectroscopy, the results of which can be used as a comparison to that obtained by HREELS. Following the IBA treatment, GaAs(100) surface showed a sharp (1×1) LEED pattern. The primary beam used is 20 eV and the spectrum is characterized by a steep increase at the loss energy at 1.4 eV. Several different primary energies ranging from 12 to 40 eV also gave the same result. Furthermore, the value of 1.4 eV remained unchanged, even after the surface was exposed to residual gases for tens of hours. Therefore it is reasonable to consider that the threshold can be used as a measure of the band gap of bulk GaAs. For Al covered surfaces, with the increase of Al coverage, the original threshold structure at 1.4 eV becomes somewhat smoother, changing to a hollow with its rising side starting at 2.2 eV. Since the energy gap of AlAs is 2.2 eV, which coincides with the new loss peak after the Al deposition, it was concluded that the reaction product present at the interface is probably AlAs. When the thickness of deposited Al is 100 Å, the threshold moves to 1.7 eV, which is probably the band gap of a ternary alloy $Al_xGa_{1-x}As$. A similar method was used to study the reaction of Al and GaP(111) surfaces and to study the P/GaAs system [14,15], the reaction products of AlP and $GaAs_{1-x}P_x$ mixed crystal were identified respectively.

Wang et al. [16] have investigated the change of surface reconstructions during the growth of Ge/Si(111) and Si/Ge(111) heterojunctions. Surface reconstructions during the successive molecular beam epitaxial (MBE) growth of Ge/Si(111) and Si/Ge(111) systems were investigated using in-situ RHEED and AES. For Ge deposition on a Si(111) surface, it was found that a transition from a 7×7 to a 5×5 surface reconstruction occurs at a Ge thickness of 6 ML (1 nm). Beyond this thickness, at which the pseudomorphic growth is replaced by a three-dimensional island growth of Ge, the streaky pattern of RHEED is converted to a spotty (5×5) pattern. The peak intensity of Si LVV in AES remains unchanged for the overlayer thickness of Ge between 1 to 10 nm. It was suggested [16] that the Ge overlayer is now no longer commensurate with the Si substrate. Consequently the built-in strain in the epilayer is partially relieved by creating interfacial dislocations, and Si atoms from the substrate could diffuse outward into the epilayer at the growth temperature to form a GeSi alloy. The spotty (5×5) pattern could then be attributed to the surface structure of the islanding GeSi alloy. When the overlayer thickness increases from 10 to 100 nm, the surface becomes gradually flat and the streaky RHEED pattern appears again accompanied with a transition of (5×5) to (7×7) reconstruction. Since the intensities of both the Si LVV and Si KLL are very weak, the surface concentration of Si is quite small. Hence, it was suggested by Wang et al. [16] that the (7×7) structure could be a surface reconstruction of Ge stabilized by the residual strains or trace amount of Si rather than that of GeSi alloy. Above 100 nm, the Si concentration on the epilayer is no longer detectable by AES, and the surface exhibits a $c(2 \times 8)$ structure which is the most favorable reconstruction of an unstrained clean Ge surface.

The growth of Si on Ge(111) is relatively simple [16]. The AES intensities of Ge LMM and Ge MNN decrease monotonously with the increase of Si epilayer thickness. After depositing 0.2 nm Si, the $c(2 \times 8)$ pattern becomes a little diffused and a new reconstruction $(\sqrt{3} \times \sqrt{3})$R30° appears at the Si thickness of 0.5 nm. Beyond 1 nm the $(\sqrt{3} \times \sqrt{3})$R30° RHEED pattern becomes spotty, indicating the beginning of islanding growth. The surface structure changes gradually from $(\sqrt{3} \times \sqrt{3})$R30° to (1×1), weak (7×7) and finally streaky (7×7) at the overlayer thickness of 1.5, 10 and 50 nm, respectively. All these indicate that the out diffusion of Ge is less significant and plays no role on the surface reconstruction beyond the epilayer thickness of 10 nm [16].

Recently the Mn/GaAs(100) interface grown at room temperature was studied by photoemis-

sion and electron energy loss spectroscopy (EELS) [17]. From the valence band photoemission spectra at different Mn coverages, it was found that the peak of surface As dangling bond decreases very fast and disappears before the Mn coverage reaches $\theta \approx 1$ Å, therefore it could be concluded that the interface formation of Mn at an early stage is identified to be the two dimensional mode rather than the cluster mode. Chemical reaction and interface diffusion between Mn and GaAs were observed. It also was found that the interface is initially semiconducting and the transition to metallic occurs at a coverage of Mn $\theta > 2$ Å. A possible model has been proposed to account for these results [17].

3. Passivation of III–V compound semiconductor surfaces, catalysis and other developments

3.1. Passivation of GaAs by electrochemical sulfur treatment

It has been shown by Hou et al. [18] that an anodic sulfurized treatment of GaAs can be used to passivate the surface for preventing oxidation. The photoemission core level spectra show that the surface Ga and As atoms are bonded to S atoms to form a thick sulfurized layer. No oxygen uptake on the sulfurized GaAs surface is illustrated by the HREELS. The results of photoluminescence spectra show that the passivated surface has low surface recombination velocity and can be protected against photoassisted oxidation under laser illumination.

3.2. Application of surface studies to catalysis

At Fudan University studies have been focused on polycrystalline silver catalysis, which are prepared by repetitive electrolytic refining and widely used as industrial catalysts for the oxidation of methanol to formaldehyde. It was found that at room temperature, both atomic and molecular oxygen exist on electrolytic silver [19]. For dissociative adsorption electron transfer from the metal (4d,5s) valence orbits to oxygen takes place and O^- species are formed and the work function increases [19]. For molecular adsorption electron transfer from the π_u orbital to silver (spd) hybrid takes place [19]. Low temperatures and rough surfaces are favorable for the stabilization of molecular oxygen. The atomic oxygen species come from the dissociation of molecular oxygen as a precursor [19]. Poisoning of silver catalysts used for partial oxidation of methanol has been investigated by combined surface techniques, work function measurements and microreactor studies; Since iron is sometimes present in the catalyst as a ferric oxide impurity and in technical grade methanol as $Fe(CO)_3$. Results show that the oxidation path is governed mainly by surface iron loading. Once Fe_2O_3 is present at the catalyst surface, the catalytic performance degrades rapidly. The species which cause the poisoning action were identified [20]. The catalytic activity and selectivity of Ag–Pt alloys for oxidizing methanol into aldehyde were measured in a microreactor. It was shown that the Pt atoms alloyed with Ag promote the decomposition of formaldehyde. The sites related to Pt atoms on the surface of Ag–Pt alloy can promote the decomposition of formaldehyde and then decrease the yields and selectivities for formaldehyde [21].

3.3. Studies of molecular adsorption on metal surfaces using the photoacoustic spectroscopy

Using home-made facilities, laser photoacoustic spectroscopy [22,23] has been performed by Zhu and co-workers [24] to study vibrational modes of molecules adsorbed on Ag and Cu surfaces. This method has the advantage of high resolution and good sensitivity. The interaction of ethanol with clean and oxygen-preadsorbed surfaces of polycrystalline Ag was studied [24]. A significant difference in the C–O stretching mode between ethanol and ethoxy was found. An extremely narrow line feature of the ethoxy C–O mode with frequency assigned at 1047.0 cm^{-1} was observed for the first time. It also was shown that the O–H bond does not affect the adsorption state of ethanol.

3.4. Structure determination

Using a domestically manufactured LEED-AES system with a base pressure of 2×10^{-8} Pa

in its vacuum chamber, Wang et al. [25] have observed a Si(100) c(4 × 4) structure which had been previously reported by Müller et al. [26]. It was reported in Ref. [25] that the c(4 × 4) reconstruction can be obtained repeatedly for both n-type and p-type samples if the sample is maintained at a relatively narrow temperature range between 580 to 630°C for several minutes. It has also been shown by the authors that the transition between c(4 × 4) and 2 × 1 is reversible. It is also proposed that with small modification, Pandey's π bond model [27] might be used to explain all the experimental results observed.

Conventionally the LEED spectra have been widely and successfully used in solving the surface structures. By assuming the positions of surface atoms in the unit cell, the intensity of LEED spot versus the energy of incident electron (I–V) curves are calculated for certain spots taking the multiple scattering into account as expressed in the dynamical theory of LEED. In order to have the best fit to the experimental data, a trial and error approach is usually used for finding the optimal parameters which define the best fit structure. The approach becomes very cumbersome due to the time consuming computation involved. It is quite difficult to determine the optimized position of atoms for a complicated surface structure. In order to find a better way of dealing with the problem, Yang and co-workers [28], after analyzing the kinematic low energy electron diffraction (KLEED) and the method of constant momentum transfer averaging (CMTA) developed by Lagally [29], have put forward a modified method which greatly reduced the amount of computation. The method has been applied to study the surface structure of Cu(001) 1 × 1 and Si(111) $\sqrt{3} \times \sqrt{3}$ -Al [30] and the results obtained agree quite well with those obtained by the conventional methods [30].

The structure of high index Si(113) surfaces has been studied by LEED [31]. It was found that there exist three kinds of reconstruction, namely 1 × 1, 3 × 1 and 3 × 2 depending on the annealing temperature and the presence of structural defects and vacancies at the surface. Both 3 × 1 and 3 × 2 surfaces are stable at room temperature and it was suggested that high quality epitaxial layers can be obtained using Si(113) as substrates.

3.5. Interaction of ions with surfaces

Systematic detailed studies have been carried out on observing the change of surface composition after the ion bombardment of alloy surfaces, such as Au–Ni, Cu–Pt, Au–Cu, etc., by Li [32]. It was found that due to the ion surface interaction, the surface composition can be altered. The existence of a bombardment-induced Gibbsian segregation was proposed. Correlation between isotope fractionation and surface composition profile during prolonged bombardment has also been studied [33].

Interactions of ion beam with surfaces have also been used for analysis in various fields such as archaeological studies and medical diagnosis [34]. High vacuum beamline connected to 2×3 MeV tandem has also been established.

3.6. Diamond

Using various types of surface analytical tools, Lin et al. have been able to grow high quality diamond films under low pressure and relatively low temperature. It is found that through the adsorption of hydrocarbon compounds or clusters on the substrate surface and by monitoring the growth process with HREELS and other analytical tools as well as by suitable tailoring, high quality thin diamond film can be obtained and the mechanism of growth elucidated [35].

Ye Ling [36] has studied the reconstruction of diamond C(100) surface using the local density functional theory. It was found that similar to Si(100), the dimerized state is more stable with the bond length shorter than that of the bulk by 8.8%. The experimentally observed diamond C(1×1) structure is really due to the presence of hydrogen or oxygen on the surface [37]. Pan and Xia [38] have studied the electronic structure in the hydrogen-induced structural transition from diamond C(111) (2×1) to (1×1) using the ab initio DV-Xα method. Their results suggest that the role of adsorbed hydrogen in the structural transition is to initially distort the Pandey π bond

in the (2×1) reconstruction, followed by the breaking of the sp^2 hybrid $+ \pi$ bond structure and then forming the sp^3 bond between the surface carbon atoms and the adsorbed H.

3.7. Imaging the structure of surfaces with home-made scanning tunneling microscope

Yang et al. [39] have developed a scanning tunneling microscope (STM) which can be operated at atmospheric pressure. They have carried out a series of STM studies on DNA, tRNA and proteins covered with glycerol water solution. Atom-resolved images of single glycine molecules have also been observed. The microscopes thus developed are manufactured and are used by many other laboratories.

3.8. Exploring the quantum confinement through the study of light emission from porous silicon

Recently Wang and his colleagues [40,41] have carried out a series of studies on light emitting porous silicon. It was demonstrated that a porous silicon layer which has a bright light emission band in the range of 500–700 nm, exhibits a strong visible-range luminescence under the illumination of an infrared ultrashort pulse laser. A third-order nonlinear optical effect has been proposed by studying the dependence of integrated luminescence intensity on pump power [40]. Recently the shortest photoluminescence wavelength at 500 nm has been reported by a boiling water treatment of porous silicon. It was suggested that the effect of boiling water is to reduce the size of Si wires and to strengthen the skeleton of porous silicon by aqueous oxidation [41].

4. Superlattices, heterojunctions and quantum wells

The importance of developing research on superlattices was recognized by Chinese scientists in the early seventies, followed by designing and manufacturing of MBE facilities in Chinese factories. The first successful growth of GaAs/AlAs superlattices with good qualities marked the birth of research in this field [42]. Integer quantum Hall effects in $GaAs/Al_xGa_{1-x}As$ heterostructures with low electron densities were observed at 4.2, 1.3 and 0.34 K. The fractional quantum Hall effect was identified at 0.34 K [43]. The effect on transport properties of electrons at the modulation-doped $GaAs/Al_xGa_{1-x}As$ heterostructure interface of neutron irradiation was reported [44].

At present, facilities both domestic-made and imported ones for growing GaAs/AlAs, Ge/Si, Ge_xSi_{1-x}/Si, ZnSe/ZnTe, HgTe/CdTe, GaAs/GaP superlattices are available in a number of laboratories. Chinese scientists are able to monitor and control the layer by layer growth of superlattices with great accuracy by the oscillation of RHEED. Quite a number of significant results have been achieved with samples of good quality [45–51,55].

Extensive studies on confined transversal optical (TO) phonons were studied by Raman scattering. AlAs-like TO confined modes were reported for the first time [45]. It was suggested that TO phonon dispersion curves resulting from the confined TO modes are in good agreement with those of bulk GaAs and AlAs, therefore further Raman scattering measurement of superlattices is a meaningful method for determining the phonon dispersion curve of crystals. Longitudinal optical (LO) phonon in $GaAs/Al_xGa_{1-x}As$ short period superlattices also were observed by Raman scattering. In addition to the LO mode confined in the GaAs layer, the AlAs-like LO confined mode in the $Al_xGa_{1-x}As$ layer was observed [46].

The successful growth of high quality ultra-thin Ge/Si superlattices has also made the observation of some new phenomena possible. In order to grow ultra-thin multi-layer Ge/Si superlattices by the phase-locked epitaxy (PLE) method, it is necessary to understand the oscillatory behavior of RHEED intensity and its relation with the growth mechanism. Wang et al. [47] have investigated the behavior of RHEED intensity oscillations during MBE of Ge/Si superlattices on Si(100) and Si(111) substrates. Since the critical thickness of pseudomorphic growth of Ge films on Si substrate is only about six monolayers, a Ge/Si strained-layer superlattice could not be prepared satisfactorily unless the thickness of al-

ternately stacked Ge and Si layers can be precisely controlled and repeated. It is found that the RHEED intensity oscillations for continuous growth of Ge on Si(100) could last up to six periods, which correspond roughly to the critical thickness of pseudomorphic growth of Ge on Si. However, the RHEED intensity oscillations for growing Ge and Ge/Si multilayers on Si(111) seem to be more complicated than that on Si(100). Some tentative growth models have been suggested [47]. A further study on the RHEED intensity oscillations is in progress in order to exploit their full potential in controlling the growth of Ge/Si superlattices with a large number of alternating periods by PLE.

The small angle X-ray diffraction has been observed up to 18th order which is an indication of the existence of a highly smooth and flat interface [48]. The ninth folded LA phonon mode also has been observed in the Raman spectra of strained Ge_xSi_{1-x}/Si superlattices [49]. Phonon spectra of $(Si)n/(Ge)m$ were calculated and compared with the results of those of the experiments. It was found that interface alloying might account for the extra peak around the 400 cm^{-1} observed in the Raman spectra, which had not been predicted in previous theoretical calculations [50].

The band offset and interface state energy distributions are two important parameters for semiconductor heterojunction devices. Deng et al. have studied the interfacial properties of the Si/GaP(111) heterojunction grown by MBE [51]. The conduction band offset and interface charge density distribution are derived from the apparent carrier distribution obtained by the $C-V$ profiling technique [51]. The conduction band offset obtained is 0.10 eV. The result is compared with that obtained by theoretical calculation carried out by Huang et al. [52] which gave a value of $\Delta E_v = 0.97$ eV for the valence band offset. It can be concluded that the major portion of band offset for Si/GaP heterojunction occurs in the valence band.

Theoretical studies of electronic states, probabilities of optical transitions under external magnetic field and electric field have been performed by Huang [53]. By taking into account the four-fold degeneracy of the hole states, the first accurate description of excitonic states in quantum wells has been achieved [53]. After critically examining the dielectric continuum model, which was conventionally used to treat the polar interaction between electrons and optical phonons, a microscopic model taking the phonon dispersion into account was developed [54]. It was found that the interface mode and the bulk-like modes are not independent. The interface modes can be mixed with the bulk-like modes of nearby frequencies. The conventionally recognized lowest first-order bulk-like mode should be an interface mode propagating along the direction of the periodic sequence of superlattices.

(CdTe)m(ZnTe)n-ZnTe short period superlattice quantum wells (SPSQWs) were proposed and successfully grown with exceptionally good quality [55]. Modulated optical spectra showed several well-defined interband transitions indicating the quantum confinement in the SPSQW [50]. Very large optical nonlinearity was measured. As many as 13 multiphonon Raman scattering (MPRS) were observed. It was found for the first time that intersubband transitions of quantum wells interfered with the MPRS process. Excitonic luminescences with high efficiency and narrow line width in the energy range of red to blue light were observed even at room temperature [55].

5. Conclusion and prospects

It can be seen from the above discussion that research in surface science and its applications to fields such as superlattices and quantum wells have had a satisfactory beginning in China and that Chinese surface scientists have travelled a long way during the past decade. However, comparing with the world level, Chinese surface scientists still have a long way to go. More work must be performed with STM operated in ultrahigh vacuum and with facilities connected to the light source of synchrotron radiation. New facilities and original ideas have to be developed in Chinese laboratories. It is hoped that more pioneer work can be reported by the end of this century.

Acknowledgments

The author is very grateful to Professors X. Wang, K.M. Zhang, L. Ye, A.R. Zhu and Ms. J.H. Hu for their contributions and valuable support during the course of preparing this manuscript.

References

[1] X.Y. Hou, M.R. Yu and X. Wang, Chin. J. Semicond. 8 (1987) 193.
[2] X. Wang, Appl. Surf. Sci. 33/34 (1988) 88.
[3] X.Y. Hou, G.S. Dong, X.M. Ding and X. Wang, J. Phys. C: Solid State Phys. 20 (1987) L121.
[4] X.Y. Hou, G.S. Dong, X.M. Ding and X. Wang, Surf. Sci. 183 (1987) 123.
[5] X.Y. Hou, X.K. Lu, P.H. Hao, X.M. Ding, P. Chen and X. Wang, The Structure of Surfaces III, Vol. 24 of Springer Series in Surface Science, Eds. S.Y. Tong, M.A. Van Hove, K. Takayanagi and X.D. Xie (Springer, Berlin, 1991) p. 560.
[6] Yoshihiro Kumazaki, Yasuo Nakai and Noriaki Itoh, Surf. Sci. 184 (1987) L445.
[7] A.J. Van Bommel and J.E. Grombean, Surf. Sci. 93 (1980) 383.
[8] Xide Xie and Kaiming Zhang, Prog. Surf. Sci. 28 (1988) 71.
[9] Lu Wenchang, Ye Ling and Zhang Kaiming, Chin. Phys. Lett. 9 (1992) 161.
[10] Lu Wenchang, Zhang Kaiming and Xie Xide, Phys. Rev. B 45 (1992) 1048.
[11] C.C. Hsu, Surface Physics and Related Topics, Festschrift for Xie Xide, Eds. F.J. Yang, G.J. Ni, X. Wang, K.M. Zhang and D. Lu (World Scientific, Singapore, 1991) p. 194.
[12] C.C. Hsu, B.Q. Li and Susan Ding, Vacuum 41 (1990) 690.
[13] C.C. Hsu, Y.X. Wang, J. Hu, J. Ho and J.J. Qian, J. Vac. Sci. Technol. A 7 (1989) 3016.
[14] X.M. Ding, G.S. Dong, X.K. Lu, H.Y. Xiao, P. Chen and X. Wang, Appl. Surf. Sci. 41/42 (1989) 123.
[15] Xuekun Lu, Xiaoyuan Hou, Xunmin Ding, Zhongqing He and Xun Wang, Proceedings of the 21st International Conference on the Physics of Semiconductors (World Scientific, Singapore, 1993) p. 413.
[16] Xun Wang, K.M. Chen, G.L. Zhou, C. Sheng, W.D. Jiang, X.J. Zhang and M.R. Yu, Proceedings of the 20th International Conference on the Physics of Semiconductors (World Scientific, Singapore, 1990) p. 115.
[17] M. Zhang, G.S. Dong, X.G. Zhu, Z.S. Li, X.F. Jin and X. Wang, Proceedings of the 21st International Conference on the Physics of Semiconductors (World Scientific, Singapore, 1993) p. 1779.

[18] X.Y. Hou, W.Z. Cai, Z.Q. He, P.H. Hao, Z.S. Li, X.M. Ding and X. Wang, Appl. Phys. Lett. 60 (1992) 2252.

[19] X.H. Bao, S.Z. Dong and J.F. Deng, Surf. Sci. 199 (1988) 493.

[20] J.F. Deng, X.H. Bao and S.Z. Dong, J. Catal. 129 (1991) 414.

[21] S.Z. Dong, F.H. Xiao and J.F. Deng, J. Catal. 109 (1988) 170.

[22] T.J. Chuang, H. Confal and F. Trager, J. Vac. Sci. Technol. A 1 (1983) 1236.

[23] C.K.N. Patel and A.C. Tam, Rev. Mod. Phys. 53 (1983) 517.

[24] M.C. Wu, A.R. Zhu and Z.Y. Wang, Phys. Rev. B 36 (1987) 9824.

[25] Hongchuan Wang, Rongfu Lin and Xun Wang, Phys. Rev. B 36 (1987) 7712.

[26] E. Müller, E. Lang, L. Hammer, W. Grim, P. Heelman and K. Heinz, Determination of Surface Structure by LEED, Eds. P.M. Marcus and F. Jona (Plenum, New York, 1984) p. 483.

[27] K.C. Pandey, Proceedings of the 17th International Conference on the Physics of Semiconductors, San Francisco, 1984, Eds. J. Chadi and W.A. Harrison (Springer, New York, 1985) p. 55.

[28] J.F. Jia, Y.F. Li, R.G. Zhao and W.S. Yang, Acta Phys. Sin. 41 (1992) 819.

[29] M.C. Lagally, T.C. Ngoc and M.B. Webb, Phys. Rev. Lett. 26 (1971) 1557.

[30] J.F. Jia, R.G. Zhao and W.S. Yang, Acta Phys. Sin. 41 (1992) 821.

[31] Y.R. Xing, J.A. Wu, J.P. Zhang, C.Z. Liu and C.H. Wang, Acta Phys. Sin. 41 (1992) 1806.

[32] R.S. Li, Surf. Sci. 193 (1988) 373.

[33] R.S. Li, C.F. Li, W.L. Zhang, G. Liu, X.S. Zhang and D.S. Bao, Thin Films and Beam Solid Interactions, Ed. I. Huang (Elsevier, Amsterdam, 1991) p. 373.

[34] F.J. Yang and G.Q. Zhao, Ion Beam Analysis (Fudan University Press, Shanghai, 1985).

[35] Biwu Sun, Xianpin Zhang and Zhangda Lin, Phys. Rev. B 47 (1993) 9816.

[36] Ling Ye, Acta Phys. Sin. 42 (1993) 97.

[37] A.V. Hamza, G.D. Kubiak and R.H. Stulen, Surf. Sci. 237 (1990) 35.

[38] B.C. Pan and S.D. Xia, Acta Phys. Sin. 42 (1993) 320.

[39] W.S. Yang, W.J. Sun, J.X. Mou and J.J. Yan, The Structure of Surfaces III, Vol. 24 of the Springer Series in Surface Science, Eds. S.Y. Tong, M.A. Van Hove, K. Takayanagi and X.D. Xie, (Springer, Berlin, 1991) p. 237.

[40] J. Wang, H.B. Jiang, W.C. Wang, J.B. Zheng, F.L. Zhang, P.H. Hao, X.Y. Hou and X. Wang, Phys. Rev. Lett. 69 (1992) 3252.

[41] X.Y. Hou, G. Shi, F.L. Zhang, P.H. Hao, D.M. Huang and X. Wang, Appl. Phys. Lett. 63 (1993) 1.

[42] Z.G. Zhou, J.B. Liang, D.Z. Shou, Y.M. Huang and M.Y. Kong, Chin. J. Semicond. 5 (1984) 694.

[43] F.H. Yang, W.C. Cheng, H.Z. Zheng and H.P. Zhou, Chin. J. Semicond. 9 (1988) 96.

[44] Y.S. Wu, Y. Huang, J.M. Zhou and X.T. Meng, Chin. Phys. Lett. 4 (1987) 373.

[45] Z.P.Wang, H.X. Han, G.H. Li and D.S. Jiang, Chin. J. Semicond. 9 (1988) 559.

[46] Z.P.Wang, H.X. Han, G.H. Li, Z.G. Chen and Z.T. Zhong, Chin. J. Semicond. 11 (1990) 72.

[47] X. Wang, K.M. Chen, G.L. Jin, C. Sheng, G.L. Zhou, W.D. Jiang, X.J. Zhang and M.R. Yu, Surf. Sci. 228 (1990) 334.

[48] G.L. Zhou, X.L. Shen, C. Sheng, W.D. Jiang and M.R. Yu, Acta Phys. Sin. 40 (1991) 56.

[49] S.L. Zhang, Y. Jin, G.G. Qin, Z. Sheng, G.L. Zhou and T.C. Zhou, Chin. J. Semicond. 12 (1991) 448.

[50] J. Zi, K.M. Zhang and X.D. Xie, Proceedings of the 21st International Conference on the Physics of Semiconductors (World Scientific, Singapore, 1993) p. 448.

[51] R.P. Deng, W.D. Jiang and H.H. Sun, Acta Phys. Sin. 38 (1989) 1265.

[52] Chunhui Huang, Ling Ye and Xun Wang, J. Phys. (Condense Matter) 1 (1989) 907.

[53] K. Huang, Wuli (Physics) 20 (1991) 321.

[54] H. Tang and K. Huang, Chin. J. Semicond. 8 (1987) 1.

[55] K. Huang and B.F. Zhu, Phys. Rev. B 38 (1988) 13377.

[56] S.X. Yuan, J. Li, Z.L. Peng and X.Y. Chen, Proceedings of the 21st International Conference on the Physics of Semiconductors (World Scientific, Singapore, 1993) p. 819.

Surface Science 299/300 (1994) 1040–1042
North-Holland

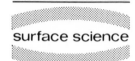

surface science

Author index

Elsevier Science B.V.

Surface Science 299/300 (1994) 1043–1054
North-Holland

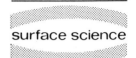

surface science

Subject index

Elsevier Science B.V.